Geotechnical Earthquake Engineering

This fully updated second edition provides an introduction to geotechnical earthquake engineering for first-year graduate students in geotechnical or earthquake engineering graduate programs with a level of detail that will also be useful for more advanced students as well as researchers and practitioners. It begins with an introduction to seismology and earthquake ground motions, then presents seismic hazard analysis and performance-based earthquake engineering (PBEE) principles. Dynamic soil properties pertinent to earthquake engineering applications are examined, both to facilitate understanding of soil response to seismic loads and to describe their practical measurement as part of site characterization. These topics are followed by site response and its analysis and soil–structure interaction. Ground failure in the form of soil liquefaction, cyclic softening, surface fault rupture, and seismically induced landslides are also addressed, and the book closes with a chapter on soil improvement and hazard mitigation. The first edition has been widely used around the world by geotechnical engineers as well as many seismologists and structural engineers.

The main text of this book and the four appendices:

- Cover fundamental concepts in applied seismology, geotechnical engineering, and structural dynamics.
- Contain numerous references for further reading, allowing for detailed exploration of background or more advanced material.
- Present worked example problems that illustrate the application of key concepts emphasized in the text.
- Include chapter summaries that emphasize the most important points.
- Present concepts of performance-based earthquake engineering with an emphasis on uncertainty and the types of probabilistic analyses needed to implement PBEE in practice.
- Present a broad, interdisciplinary narrative, drawing from the fields of seismology, geotechnical engineering, and structural engineering to facilitate holistic understanding of how geotechnical earthquake engineering is applied in seismic hazard and risk analyses and in seismic design.

Geotechnical Earthquake Engineering

Second Edition

Steven L. Kramer and Jonathan P. Stewart

CRC Press is an imprint of the
Taylor & Francis Group, an **informa** business

Designed cover image: National Information Service for Earthquake Engineering.

Second edition published 2025
by CRC Press
2385 NW Executive Center Drive, Suite 320, Boca Raton FL 33431

and by CRC Press
4 Park Square, Milton Park, Abingdon, Oxon, OX14 4RN

CRC Press is an imprint of Taylor & Francis Group, LLC

First edition published by Pearson 1996

Library of Congress Cataloging-in-Publication Data
Names: Kramer, Steven Lawrence, author. | Stewart, Jonathan P., author.
Title: Geotechnical earthquake engineering / Steven L. Kramer and Jonathan P. Stewart.
Description: Second edition. | Boca Raton, FL : CRC Press, [2025] | Includes bibliographical references and index. |
Identifiers: LCCN 2024012008 | ISBN 9781032842745 (hbk) | ISBN 9781032842776 (pbk) | ISBN 9781003512011 (ebk)
Subjects: LCSH: Earthquake engineering. | Engineering geology.
Classification: LCC TA654.6 .K72 2025 | DDC 624.1/762—dc23/eng/20240415
LC record available at https://lccn.loc.gov/2024012008

ISBN: 978-1-032-84274-5 (hbk)
ISBN: 978-1-032-84277-6 (pbk)
ISBN: 978-1-003-51201-1 (ebk)

DOI: 10.1201/9781003512011

Typeset in Times
by codeMantra

Access the Instructor Resources: www.routledge.com/9781032842745

To Diane and Alisa

Contents

Preface

The field of geotechnical earthquake engineering, like the related fields of engineering seismology and structural earthquake engineering in the broad discipline of earthquake engineering, has advanced dramatically since the first edition of this book was published. Advances have come with the occurrence and investigation of numerous earthquakes, both small and large, with the massive increases in computational power now at the hands of engineers and scientists, and with advances in the analytical tools that leverage those computational capabilities.

Advances in geotechnical earthquake engineering have been made both in research and in practice. Among the important advances since the publication of the first edition are the systematic collection and archiving of data through organized and coordinated field reconnaissance efforts following major earthquakes, the development and increasingly frequent use of advanced numerical analyses for assessment of seismic response, and the systematic extension of ground motion hazard analysis to the estimation of important measures of performance such as response, damage, and loss. This last development, which generally falls under the heading of performance-based earthquake engineering (PBEE), is most appropriately implemented in a probabilistic framework that accounts for uncertainty in the parameters and models used to predict performance.

The purpose of this book is to introduce the reader to the concepts, theories, and procedures of contemporary geotechnical earthquake engineering and its role in performance-based assessment and design. It is intended for use as a text in graduate courses on geotechnical earthquake engineering or seismic hazard and risk analysis and as a reference for practicing engineers. Recognizing that geotechnical earthquake engineering is a broad, multidisciplinary field, this book draws from seismology, geology, structural engineering, hazard and risk analysis, and other disciplines. This book places an emphasis on seismic performance and the emerging paradigm of PBEE.

This book is written at a level suitable for readers with knowledge equivalent to that of a senior (fourth-year) civil engineering student. The reader should have had basic courses in soil mechanics, structural engineering, and hydraulics; introductory courses in geology and probability/statistics would also be helpful. Many graduate students will have had courses in structural dynamics or soil dynamics by the time they begin the study of geotechnical earthquake engineering. For those without prior exposure, introductions to the nomenclature and mathematics of dynamic systems, structural dynamics, wave propagation, and probability are presented in four appendices.

ORGANIZATION

The subject matter falls into three main categories. The first five chapters present fundamental principles of seismology, ground motions, hazard analysis, and performance assessment. The next three chapters describe static and dynamic soil behavior and the characterization of that behavior to inform assessments of site effects/response and soil–structure interaction. The last three chapters focus on ground failure and mitigation of ground failure hazards.

Chapter 1 introduces the reader to the types of damage that can occur during earthquakes and to the problems they present to geotechnical earthquake engineers. Basic concepts of earthquake seismology and the terminology used to describe earthquakes and their effects are described in Chapter 2. Chapter 3 describes ground motion measurement, the parameters used to characterize strong ground motion, and methods for the prediction of those parameters. Deterministic and probabilistic seismic hazard analyses are presented in Chapter 4. Chapter 5 introduces the reader to the concepts of PBEE and the various manners in which it can be implemented.

The properties of soils that control their response to seismic loading are described in Chapter 6. Field and laboratory techniques for the measurement of these properties are also described. Chapter 7 presents methods for the estimation of site effects both through empirical and analytical methods.

Analysis of ground response during earthquakes, beginning with one-dimensional ground response analysis and moving through two- and three-dimensional dynamic response analyses are presented; both frequency- and time-domain approaches are described. Chapter 8 presents the basic concepts, effects, and procedures for analysis of soil–structure interaction including foundations, buried structures, and retaining structures.

Chapter 9 deals with liquefaction – it begins with a conceptual framework for understanding various liquefaction-related phenomena and then presents practical procedures for the evaluation of liquefaction hazards. Permanent shear-induced ground movements due to fault rupture and seismic slope instability are covered in Chapter 10; procedures for probabilistic fault rupture hazard analysis and probabilistic slope stability hazard analysis are presented. Chapter 11 introduces commonly used soil improvement techniques for the mitigation of seismic hazards.

Readers familiar with the first edition will notice several structural changes in this edition. The topic of soil–structure interaction, formerly a section within the ground response analysis chapter is now a standalone chapter with a much more detailed treatment of the subject. The seismic response of retaining structures, formerly a chapter in the first edition is now treated as what it is – an aspect of soil–structure interaction. The former Chapter 5 on wave propagation is now Appendix C with some new material on wave scattering. The new Chapter 5 deals with seismic performance and design and provides an introduction to the evolving paradigm of PBEE. Appendix D provides a more detailed treatment of probability concepts in support of the probabilistic nature of PBEE.

PEDAGOGY

The first edition of this book was the first to deal explicitly with the topic of geotechnical earthquake engineering. During its preparation, a great deal of time and effort was devoted to decisions regarding content and organization. The organization of this edition is similar to the first, but the scope has both broadened and deepened to reflect developments in the field and the requirements of practice. Because the first edition was used by a surprising number of earth scientists and "non-geotechnical" engineers, this edition has expanded its treatment of basic soil mechanics to help those whose background in soil mechanics may be limited. As with the first edition, the preparation of this text also involved a great deal of interpretation of information from a wide variety of sources. While the text reflects our own interpretation of this information, it is heavily referenced to allow readers to explore background or more detailed information on various topics.

This book contains numerous worked example problems. The example problems are intended to illustrate basic concepts and to guide readers in the solutions of common practical problems. A number of these problems involve calculations carried out to more significant figures than the accuracy of the procedures (and typical input data) would justify. Many of the important problems of geotechnical earthquake engineering, however, do not lend themselves to the types of short, well-defined problems that appear as examples in this book. More realistic situations are best handled with longer, project-oriented homework assignments based on actual case histories, which we recommend that instructors develop in consideration of the geotechnical conditions in their practice region.

Acknowledgments

As this second edition evolved from the first, it was continually improved by comments and suggestions from colleagues and students from soil dynamics and geotechnical earthquake engineering courses around the world. Their assistance is greatly appreciated. We are also grateful to colleagues who provided constructive critical reviews of full chapters in this book, including Prof. Lorne Arnold, Dr. David M. Boore, Prof. Brendon A. Bradley, Prof. Jonathan D. Bray, Dr. Zachary N. Bullock, Prof. Ashly Cabas, Dr. Antonio Araujo Correia, Prof. Brady Cox, Dr. John Douglas, Prof. Brett W. Maurer, Prof. George Mylonakis, Prof. Ellen M. Rathje, Prof. Adrian Rodriguez-Marek, Prof. Vernon R. Schaefer, Dr. David P. Teague, and Prof. Paolo Zimmaro.

Moreover, other colleagues provided information or data in response to our queries, wrote code or performed analyses to assist with generation of figures, or provided feedback on specific sections of the text, which improved the quality of this book. These include Prof. Pedro Arduino, Prof. Jack W. Baker, Prof. Ross W. Boulanger, Prof. Scott J. Brandenberg, Dr. Long Chen, Prof. Victor Contreras, Dr. Robert B. Darragh, Dr. Jennifer L. Donahue, Dr. Edward H. Field, Dr. Robert W. Graves, Dr. Nick Gregor, Prof. Michael G. Gomez, Dr. Thomas C. Hanks, Dr. Erol Kalkan, Prof. Ronnie Kamai, Prof. Tadahiro Kishida, Prof. Dongyoup Kwak, Prof. Annie O. Kwok, Dr. Andrew J. Makdisi, Dr. David Mencin, Prof. Robb E.S. Moss, Dr. Grace A. Parker, Prof. Thomas Rockwell, Prof. Lisa M. Star, Prof. Mark Talesnick, and Prof. Pengfei Wang.

Finally, we are most grateful for the support of our families whose understanding, encouragement, and support over many years of long working hours enabled the preparation of this book.

Authors

Dr. Steven L. Kramer is Professor Emeritus of Civil and Environmental Engineering at the University of Washington. He received his BS, M.Eng., and Ph.D. from U.C. Berkeley. Kramer joined the geotechnical group in the Department of Civil Engineering in 1984. His primary research interests included soil liquefaction, site response analysis, seismic slope stability, and hazard analysis. He has conducted research work in the area of performance-based earthquake engineering, specifically the integration of probabilistic response analyses with probabilistic seismic hazard analyses, and remains active in soil liquefaction research. Kramer has served as a consultant to private firms and government agencies on projects including high-rise structures, bridges, dams, seawalls, levees, underground structures, offshore structures/facilities, and nuclear facilities in the United States and abroad. He holds numerous honors and awards, including being elected into the National Academy of Engineering and being named a Distinguished Member of ASCE and an Honorary Member of the IAEE.

Dr. Jonathan P. Stewart is a Professor at the Samueli School of Engineering at UCLA. He received his BS, MS, and Ph.D. from U.C. Berkeley. He focuses his research on geotechnical earthquake engineering and engineering seismology, with emphasis on seismic soil–structure interaction, earthquake ground-motion characterization, site response, seismic ground failure, and the seismic performance of structural fills and levee embankments. Findings from his research have been widely utilized in engineering practice, including through the National Seismic Hazard Model, produced by the U.S. Geological Survey, and the American Society of Civil Engineers' guidelines for new and existing structures. He maintains an active consulting practice related to seismic hazard analysis, site response, seismic performance assessment, and geotechnical engineering for private and public agencies worldwide. His work has been recognized with best-paper awards, honorary lectures, teaching awards, and election to the National Academy of Engineering.

1 Introduction to Geotechnical Earthquake Engineering

1.1 INTRODUCTION

Earthquake engineering deals with the effects of earthquakes on people and their environment and with methods of reducing those effects. It is a relatively young discipline, many of its most important developments have occurred in the past 40–50 years. Earthquake engineering is a very broad field, drawing on aspects of geology, seismology, geotechnical engineering, structural engineering, risk analysis, and other technical fields. Its practice also requires the consideration of social, economic, and political factors. Most earthquake engineers have entered the field from structural engineering or geotechnical engineering backgrounds. This book covers geotechnical aspects of earthquake engineering. Although its primary audience is geotechnical engineering students and practitioners, it contains a great deal of information that should be of interest to the structural engineer and the engineering seismologist.

The study of earthquakes dates back many centuries. Written records of earthquakes in China date as far back as 3,000 years. Japanese records and records from the eastern Mediterranean region go back nearly 1,600 years. In the United States, the historical record of earthquakes is much shorter, about 350 years. On the seismically active west coast of the United States, earthquake records go back only about 200 years. Compared with the millions of years over which earthquakes have been occurring, humankind's experience with earthquakes is very brief.

Today, hundreds of millions of people throughout the world live with a significant risk to their lives and property from earthquakes. Billions of dollars of public infrastructure are continuously at risk of earthquake damage. The health of many local, regional, and even national economies is also at risk from earthquakes. These risks are not unique to the United States, Japan, or any other country. Earthquakes are a global phenomenon and a global problem.

1.2 BACKGROUND

Although earthquake engineers frequently find themselves poring over the details of a particular design or analysis, it is essential that they also have a "big picture" understanding of earthquakes, their effects, and the losses they produce. The earthquake-resistant design of new structures and the seismic vulnerability evaluation of existing structures typically involve professionals from many fields – seismology, geotechnical and structural engineering, architecture, construction, loss analysis, finance, and public policy. While not all of these professionals will work with all of the others, they will definitely work with some of them, and their working relationship will be improved by an understanding of how the "pieces" fit together.

An intuitive way of viewing the process by which earthquakes affect individuals and society-at-large is illustrated in Figure 1.1. When an earthquake occurs, it produces ground motion at a particular site. The motion may be weak or strong, depending on the size of the earthquake and its proximity to the site. That ground motion will cause a structure or system of interest at the site to respond dynamically. The level of that response may be low or high, depending on the motion and the characteristics of the structure. If the response is sufficiently strong (or the structure sufficiently weak), it may cause physical damage to the structure or system. That damage, in turn, will lead to losses which can be measured in terms of casualties or in economic terms. The prediction

DOI: 10.1201/9781003512011-1

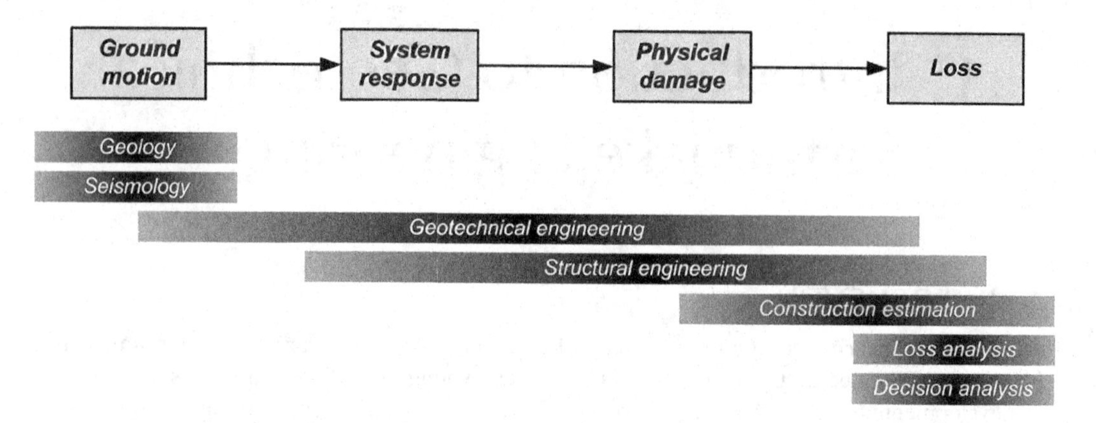

FIGURE 1.1 Schematic illustration of the components that lead to earthquake losses and the professions involved in the evaluation of each of those components.

of losses, therefore, requires the ability to predict ground motion intensity, system response, and physical damage.

Ultimately, losses are the most important measure of performance, and professionals in several fields (Figure 1.1) work toward reducing those losses as efficiently as possible. Geologists and seismologists identify and characterize seismic sources and, with geotechnical engineers, evaluate the level of ground motion expected at a particular site. Geotechnical and structural engineers evaluate the response of soil-foundation-structure systems and estimate the levels of physical damage expected to be caused by that response. Engineers also work with construction estimators and loss analysts to determine the expected losses so that decision-makers (typically, owners) can decide how to mitigate those losses. The work done by all of these groups is important and is enhanced when they understand, appreciate, and communicate about each other's activities.

A number of terms are commonly, and not always consistently, used when describing earthquakes and their effects. A *seismic hazard* is a phenomenon that has the potential to cause harm – tectonic movement, ground shaking, ground failure, and flooding are all examples of seismic hazards. In most of this book, the term "hazard" will primarily be applied to ground motions, in particular to the rates at which different levels of ground shaking can be expected to occur. Whether or not a seismic hazard results in damage depends on the *vulnerability* of structures and facilities exposed to the hazard; vulnerability can be thought of as a predisposition to suffer damage due to some hazard. Finally, *seismic risk* is the probability that damage or loss will occur and is evaluated by combining hazard and vulnerability. A framework that provides a more complete picture of hazard and risk is described in Section 1.5.

1.3 EFFECTS OF EARTHQUAKES

The most immediately visible effects of earthquakes are the images of collapsed buildings and bridges, and of injured and displaced people, i.e., of physical damage and the losses it causes. There are, however, many effects of earthquakes – some are more subtle than others but all are capable of causing significant losses.

When an earthquake occurs, seismic waves radiate away from the source and travel rapidly through the Earth's crust. When these waves reach the ground surface, they produce shaking that may last from seconds to minutes. The strength and duration of shaking at a particular site depends on the size and location of the earthquake and on the characteristics of the site. At sites near the source of a large earthquake, ground shaking can cause tremendous damage. In fact, ground shaking can be considered to be the most important of all seismic hazards because most of the other hazards are caused by ground shaking. Where ground shaking levels are low, these other seismic hazards may be low or nonexistent. Strong ground shaking, however, can produce extensive damage. Although seismic waves travel through rock over the overwhelming majority of their path from the source of an earthquake to the ground surface, the final portion of that path is often through soil, and the characteristics of the soil can greatly influence the nature of shaking at the ground surface. Soil deposits tend to act as "filters" to seismic waves by attenuating motion at certain frequencies and amplifying it at others. Since soil conditions often vary dramatically over short distances, levels of ground shaking can vary significantly within a small area. One of the most important aspects of geotechnical earthquake engineering practice involves the evaluation of the effects of local soil conditions on strong ground motion. In this book, Chapter 3 presents methods for quantifying the most important characteristics of strong ground motions, and Chapters 6 and 7 provide the background and techniques for site-specific ground motion prediction.

1.3.1 Fault Rupture

In many shallow earthquakes, fault rupture extends to the ground surface where it can produce sharp or gradual offsets in elevation and/or horizontal position. These offsets can be highly damaging to structures that lie on, across, or near the fault. Buildings have been torn apart, tilted, rotated, and racked by surface fault ruptures. Long, extended structures such as bridges, dams (Figure 1.2a), highways, and railroads (Figure 1.2b) have been pulled apart, distorted, and collapsed by fault rupture deformations. Buried pipelines have been pulled apart and sheared by fault rupture, leading to instances of fire and environmental damage. Although the effects are generally localized, many faults run through developed and heavily populated areas so the damage caused by surface fault rupture can be very serious. Fault displacement and its prediction are discussed in Chapter 10.

1.3.2 Structural Damage

Without doubt, the most dramatic and memorable images of earthquake damage are those of structural collapse. From the predictable collapse of the unreinforced masonry and adobe structures in which many residents of underdeveloped areas of the world live (Figure 1.3) to the surprising destruction of more modern construction (Figures 1.4–1.6), structural damage is the leading cause of death and economic loss in many earthquakes. However, structures need not collapse to cause death and damage. Falling objects such as brick facades and parapets on the outside of a structure or heavy objects within a structure have caused casualties in many earthquakes. Non-structural elements such as piping, lighting, and storage systems can also suffer costly damage during earthquakes.

Over the years, considerable advances have been made in the earthquake-resistant design of structures, and seismic design requirements in building codes have steadily improved. As earthquake-resistant design has moved from an emphasis on structural strength to emphases on both strength and ductility, the need for accurate predictions of ground motions has increased. In current design practice, the geotechnical earthquake engineer is often responsible for providing appropriate design ground motions to structural engineers. In this book, Chapter 7 describes the effects of local soil conditions on ground motions and provides guidance for the development of site-specific design ground motions. To allow the geotechnical engineer to better understand and appreciate the issues and options involved in structural design, a brief introduction to the seismic design of structures is presented in Chapter 5.

1.3.3 Liquefaction

Some of the most spectacular examples of earthquake damage have occurred when soil deposits have lost their strength and appeared to flow as fluids. In this phenomenon, termed liquefaction, the strength of the soil is reduced, often drastically, to the point where it is unable to support structures

FIGURE 1.2 (a) Shihkang Dam near Fengyuan City in Taiwan following the 1999 Chi-Chi earthquake, where fault rupture lifted the right abutment some 9 m higher than the left abutment (courtesy of NISEE), and (b) railroad tracks shifted horizontally by fault in Turkey following the 1999 Kocaeli earthquake. (After Ogut, 2000; used with permission of CSEG Recorder.)

FIGURE 1.3 Damage to buildings in Huaras, Peru following the 1970 Peru earthquake. The adobe structures in the foreground were destroyed, but the reinforced concrete structure in the background suffered little damage. (Photo by G. Plafker, courtesy of USGS.)

FIGURE 1.4 Collapsed portion of the reinforced concrete Hospital Juarez in Mexico City following the 1985 Mexico earthquake. (Photo by E.V. Leyendecker, courtesy of EERI.)

FIGURE 1.5 Effects of column failures at Olive View Hospital in the 1971 San Fernando earthquake. Collapse of the canopy in the foreground pinned the ambulances beneath, them, rendering them useless. (Courtesy of EERI.) Note failed first-floor columns and collapsed stairwell structure in the background.

or remain stable. Because it only occurs in saturated soils, liquefaction is most commonly observed near rivers, bays, and other bodies of water.

The term liquefaction actually encompasses several related phenomena. All involve the generation of excess porewater pressure, which can soften and weaken the soil. In some cases, the high porewater pressures lead to the development of *sand boils* (Figure 1.7), which are produced by groundwater rushing to the surface. Although not particularly damaging by themselves, sand boils indicate the presence of high porewater pressures whose eventual dissipation can produce subsidence and damaging differential settlements. When extensive liquefaction occurs, as in Christchurch, New Zealand in 2011 (Figure 1.8), the volume of ejected soil can contribute significantly to settlement damage to structures founded on shallow foundations.

Flow failures, on the other hand, can occur when the strength of the soil drops below the level needed to maintain stability under static conditions. Flow failures are therefore driven by static gravitational forces and can produce very large movements. Flow failures have caused the collapse of earth dams (Figure 1.9) and other slopes, and the failure of foundations (Figure 1.10). The 1971 San Fernando earthquake caused a flow failure in the upstream slope of the Lower San Fernando Dam (Figure 1.11) that nearly breached the dam. Thousands could have been killed in the residential area immediately below the dam if the flow failure had extended slightly deeper. Tailings dams, which often retain toxic mine waste products, have also suffered flow failures (Figure 1.12) that caused extensive physical and environmental damage.

Lateral spreading is a related phenomenon characterized by incremental displacements during earthquake shaking. Depending on the number and strength of the stress pulses that exceed the strength of the soil, lateral spreading can produce displacements that range from negligible to quite large. Lateral spreading is common around water bodies and the displacements it produces have often damaged bridge abutments, foundations, superstructures, and port facilities (Figures 1.13–1.15).

FIGURE 1.6 Reinforced concrete column at Olive View Hospital following the 1971 San Fernando earthquake. Insufficient transverse reinforcement was unable to provide adequate confinement. (Courtesy of USGS.)

Liquefaction is a complicated phenomenon, but research has progressed to the point where an integrated framework of understanding can be developed. Chapter 9 of this book presents the basic concepts with which the susceptibility, triggering conditions, and effects of all liquefaction phenomena can be understood, together with practical procedures for the evaluation of liquefaction hazards.

1.3.4 Landslides

Strong earthquakes often cause landslides. Although the majority of such landslides are small, earthquakes have also caused very large slides. In a number of unfortunate cases, earthquake-induced landslides have buried entire towns and villages (Figure 1.16). More commonly, earthquake-induced landslides cause damage by destroying buildings, disrupting bridges and other constructed facilities, or blocking rivers (Figures 1.17 and 1.18). Some earthquake-induced landslides result from liquefaction phenomena, but many others simply represent the failures of slopes that were marginally stable under static conditions. Various types of seismic slope failures, their frequency of occurrence, and procedures for their analysis are described in Chapter 10.

FIGURE 1.7 Sand boil in rice field following the 1964 Niigata earthquake. (K. Steinbrugge collection; courtesy of EERC, University of California.)

FIGURE 1.8 Extensive ejecta covering streets and yards in the Bexley area of Christchurch, NZ following the 2011 Christchurch earthquake. (Used with permission of Stuff Ltd.)

FIGURE 1.9 Liquefaction failure of Sheffield Dam following the 1925 Santa Barbara earthquake. (K. Steinbrugge collection; Courtesy of EERC, University of California.)

FIGURE 1.10 Liquefaction-induced bearing capacity failures of the Kawagishi-cho apartment buildings following the 1964 Niigata earthquake. (Courtesy of USGS.)

FIGURE 1.11 Lower San Fernando Dam following liquefaction failure of its upstream slope in the 1971 San Fernando earthquake. (K. Steinbrugge collection; courtesy of EERC, University of California.)

FIGURE 1.12 Flow failure of Tapo Canyon Tailings Dam from the 1994 Northridge earthquake. (Harder and Stewart, 1996; used with permission of ASCE.)

FIGURE 1.13 The Showa Bridge following the 1964 Niigata earthquake. Lateral spreading caused bridge pier foundations to move and rotate sufficiently for simply supported bridge spans to fall. (Courtesy of USGS.)

1.3.5 Retaining Structure Failures

Walls retaining unsaturated soils have generally performed relatively well in most earthquakes. Anchored bulkheads, quay walls, and other retaining structures, however, are often damaged in earthquakes. Damage is usually concentrated in waterfront areas such as ports and harbors. Because such facilities are often essential for the movement of goods upon which local economies rely, the business losses associated with their failure can go far beyond the costs of repair or reconstruction. The Port of Kobe, Japan suffered such extensive damage to quay walls (Figure 1.19) around the perimeters of Port Island and Rokko Island that it went from the 6th largest container port (in terms of cargo throughput) in the world to complete closure following the 1995 Hyogo-ken Nanbu earthquake. After two years, approximately US$10 billion in damage had been repaired but the Port ranked only 17th largest in the world (Chang, 2000). The seismic design of retaining structures is covered in Chapter 8.

1.3.6 Lifeline Hazards

A network of facilities that provide the services required for commerce and public health can be found in virtually any developed area. These networks, which include electrical power and telecommunications, transportation, water and sewage, oil and gas distribution, and waste storage systems, have collectively come to be known as lifelines. Lifeline systems may include power plants, transmission towers, and buried electrical cables; roads, bridges, harbors, and airports; water treatment facilities, reservoirs and elevated water tanks, and buried water distribution systems; liquid storage tanks and buried oil and gas pipelines; and municipal solid waste and hazardous waste landfills. Lifeline systems and the facilities that comprise them provide services that many take for granted but which are essential in modern industrial areas. Lifeline failures not only have severe economic consequences but can also adversely affect the environment and quality of life following an earthquake.

FIGURE 1.14 Rotation of Christchurch, NZ bridge abutment due to lateral spreading of underlying soil and restraint by bridge deck. (After Tasiopoulou et al., 2012.)

FIGURE 1.15 Lateral spreading damage to North Wharf following the 2010 Haiti earthquake. (Rathje et al., 2010; used with permission of the Geotechnical Extreme Events Reconnaissance Association.)

FIGURE 1.16 Village of Yungay, Peru, (a) before and (b) after being buried by a giant landslide in the 1970 Peruvian earthquake. The same palm tines are visible on the left side of both photographs. The landslide involved 50 million cubic meters of material that eventually covered an area of some 8,000 km^2. About 25,000 people were killed by this landslide, over 18,000 in the villages of Yungay and Ranrahirca. (K. Steinbrugge collection; courtesy of EERC, University of California.)

FIGURE 1.17 A wing of Government Hill School in Anchorage, Alaska, straddled the head scarp of the Government Hill landslide in the 1964 Good Friday earthquake. (K. Steinbrugge collection; courtesy of ERRC, University of California.)

FIGURE 1.18 Aerial view of the Chiu-fen-erh-shan landslide caused by the 1999 Chi-Chi earthquake. An estimated 50 million cubic meters of rock slid and blocked two streams. (Wang et al., 2003; used with permission of Elsevier BV.)

FIGURE 1.19 Failure of a quay wall on Rokko Island in Kobe, Japan in the 1995 Hyogo-ken Nanbu earthquake. (Photo by S. L. Kramer.)

Lifeline failures can cause disruption and economic losses that greatly exceed the cost of repairing facilities directly damaged by earthquake shaking. The 1989 Loma Prieta earthquake caused the failure of more than 750 water mains in the San Francisco Bay area (Lund and Schiff, 1991) and two 2011 earthquakes in Christchurch, New Zealand caused more than 2,500 water pipe breaks (O'Rourke, et al., 2014) and leaks leaving households without water for extended time periods. A hypothetical magnitude 7.8 earthquake on the San Andreas Fault in Southern California would be expected to lead to $1.1 billion in pipeline (water, sewer, and gas) repair costs, but a $53 billion loss due to business interruption (Jones et al., 2008).

Lifeline failures can also hamper emergency response and rescue efforts immediately following damaging earthquakes. Most of the damage in the 1906 San Francisco earthquake, for example, was caused by a fire that could not be fought properly because of broken water mains (Figure 1.20). Eighty-three years later, television allowed the world to watch another fire in San Francisco following the Loma Prieta earthquake — these fires were caused by broken natural gas pipes, and again, firefighting was hampered by broken water mains. The Loma Prieta earthquake also caused the collapse and near collapse of several elevated highways and the collapse of a portion of the San Francisco-Oakland Bay Bridge. The loss of these transportation lifelines caused gridlock throughout the area. Some of the elevated highways were still out of service five years after the earthquake. Some instances of lifeline failures are directly attributable to ground failure. Figure 1.21a shows a landslide head scarp that caused damage to electrical substation equipment (bushings). Figure 1.21b shows collapse of an electrical transmission tower from the 2023 Kahramanmaraş, Turkey earthquakes.

1.3.7 Tsunami and Seiche Hazards

Rapid vertical seafloor movements caused by fault rupture during earthquakes can produce long-period sea waves called *tsunamis*. In the open sea, tsunamis travel great distances at high

FIGURE 1.20 Fire in San Francisco exacerbated by damage to water supply pipelines: (a) 1906 San Francisco earthquake (K. Steinbrugge collection; courtesy of ERRC, University of California) and (b) 1989 Loma Prieta earthquake (Scawthorn et al., 1992).

FIGURE 1.21 Damage to: (a) electrical substation at the head of landslide scarp (photo by Jonathan Stewart, 1999) and (b) collapsed electrical transmission tower due to fault offset in Hatay, Turkey (Courtesy of B. Wham, 2023).

speeds but are difficult to detect-they usually have heights of less than 1 m and wavelengths (the distance between crests) of several hundred kilometers. As a tsunami approaches shore, however, the decreasing water depth causes its speed to decrease and the height of the wave to increase. In some coastal areas, the shape of the seafloor may amplify the wave, producing a nearly vertical wall of

FIGURE 1.22 Tsunami damage in (a) Meulaboh, Indonesia in the 2004 Sumatra earthquake (Reuters, 2004; used by permission) and (b) Kesennuma, Japan in the 2011 Tohoku earthquake (Photo by S. L. Kramer).

water that rushes far inland and causes devastating damage (Figure 1.22). The Great Hoei Tokaido Nonhaido tsunami killed 30,000 people in Japan in 1707. The 1960 Chilean earthquake produced a tsunami that not only killed 300 people in Chile, but also killed 61 people in Hawaii and, 22 hours later, 199 people in distant Japan (Lida et al., 1967). The 2004 Sumatra earthquake produced a tsunami that killed more than 200,000 people in Indonesia, Thailand, India, Sri Lanka, Myanmar, and other countries, and spurred the deployment of tsunami warning systems around the world. Earthquake-induced waves in enclosed bodies of water are called *seiches*. Typically caused by long-period seismic waves that match the natural period of oscillation of the water in a lake or reservoir, seiches may be observed at great distances from the source of an earthquake. The 1964 Good Friday earthquake in Alaska, for example, produced damaging waves up to 5 ft high in lakes in Louisiana and Arkansas (Spaeth and Berkman, 1967). Another type of seiche can be formed when faulting causes permanent vertical displacements within a lake or reservoir. In 1959, vertical fault movement within Hebgen Lake produced a seiching motion that alternately overtopped Hebgen Dam and exposed the lake bottom adjacent to the dam in 1959 (Steinbrugge and Cloud, 1962).

1.4 SIGNIFICANT HISTORICAL EARTHQUAKES

Earthquakes occur almost continuously around the world. Fortunately, most are so small that they cannot be felt. Only a very small percentage of earthquakes are large enough to cause noticeable damage, and a small percentage of those are large enough to be considered major earthquakes. Throughout recorded history, some of these major earthquakes can be regarded as being particularly significant, either because of their size and the damage they produced or because of what scientists and engineers were able to learn from them. A partial list of significant earthquakes, admittedly biased toward U.S. earthquakes and earthquakes with significant geotechnical earthquake engineering implications, is given in Table 1.1.

Much has been learned from these earthquakes, and efforts to systematically observe and document the effects of earthquakes have advanced significantly in recent years. Because geotechnical evidence, in particular, is so easily erased by rainfall or clean-up efforts, it is important that trained observers get to the site of a major earthquake as quickly as possible. Many government and professional organizations send teams of observers on reconnaissance missions following earthquakes; the Geotechnical Extreme Events Reconnaissance (GEER) Association (www.geerassociation.org) has coordinated the reconnaissance efforts of U.S. geotechnical researchers since 1999. GEER teams use GPS, Google Earth, and other technologies to rapidly post geo-located photographs, preliminary measurements, and descriptions of ground failures and other observations that, along with subsequent subsurface investigation data, form the basis for valuable case histories.

1.5 A FRAMEWORK OF UNDERSTANDING

The process illustrated in Figure 1.1 forms the backbone of a framework that can be used to develop an understanding of how earthquakes produce losses and how potential earthquake losses can be estimated. This framework can be extended, as discussed in Chapter 5, to provide a basis for performance-based evaluation and design. Development of that framework, however, requires the introduction of some terminology that allows quantification of ground motions, response, damage, and loss. The terminology used herein follows that originally proposed by the Pacific Earthquake Engineering Research (PEER) Center (Moehle and Deierlein, 2004).

1.5.1 Ground Motion

Earthquake ground motions can vary dramatically from one earthquake to another, and from one location to another in the same earthquake. The seismic responses of buildings, bridges, and other elements of infrastructure are sensitive to a number of characteristics of the ground motions they

TABLE 1.1
Significant Historical Earthquakes

Date	Location	Magnitude	Deaths	Comments
780 BC	China	-	-	One of the first reliable written accounts of a strong earthquake; produced widespread damage west of Xian in Shaanxi Province.
AD 79	Italy	-	-	Sixteen years of frequent earthquakes culminating with the eruption of Mt. Vesuvius, which buried the city of Pompeii.
893	India	-	180,000	Widespread damage; many killed in collapse of earthen homes.
1556	China	8.1 (est.)	530,000	Occurred in densely populated region near Xian; produced thousands of landslides, which killed inhabitants of soft rock caves in hillsides; death estimate of questionable accuracy.
1755	Portugal.	8.6	60,000	Lisbon earthquake; first scientific description of earthquake effects.
1783	Italy	-	50,000	Calabria earthquake; first scientific commission for earthquake investigation formed.
1811–1812	Missouri, USA	7.5, 7.3, 7.8	Several	Three large earthquakes in less than 2 months in New Madrid area, felt all across central and eastern United States.
1819	India	-	1,500	Cutch earthquake; first well-documented observations of faulting.
1857	California, USA	8.3	1	Fort Tejon earthquake; one of the largest earthquakes known to have been produced by the San Andreas Fault; fault ruptured for 250 miles (400 km) with up to 30 ft (9 m) offset.
1872	California, USA	8.5	27	Owens Valley earthquake; one of the strongest ever to have occurred in the United States.
1886	South Carolina, USA	7.0	110	Strongest earthquake to strike the east coast of the United States; produced significant liquefaction in the Charleston area.
1906	California, USA	7.9	700	First great earthquake to strike densely populated area in the United States; produced up to 21 ft (7 m) offset in 270-mile (430 km) rupture of San Andreas Fault; most damage caused by fire; extent of ground shaking damage correlated to geologic conditions in post-earthquake investigation.
1908	Italy	7.5	83,000	Messina and the surrounding area devastated; Italian government appointed engineering commission that recommended structures be designed for equivalent static lateral loads.
1923	Japan	7.9	99,000	Kanto earthquake; caused major damage in Tokyo-Yokohama area, much due to fue in Tokyo and tsunami in coastal regions; strongly influenced subsequent design in Japan.
1925	California, USA	6.3	13	Santa Barbara earthquake; caused liquefaction failure of Sheffield Dam; led to first explicit provisions for earthquake resistance in U.S. building codes.

(Continued)

TABLE 1.1 (*Continued*)
Significant Historical Earthquakes

Date	Location	Magnitude	Deaths	Comments
1933	California, USA	6.3	120	Considerable building damage; schools particularly hard-hit, with many children killed and injured; led to greater seismic design requirements in building codes, particularly for public school buildings.
1959	Montana, USA	7.1	28	Hebgen Lake earthquake; faulting within reservoir produced large seiche that overtopped earth dam
1940	California, USA	7.1	9	Large ground displacements along Imperial Fault near El Centro; the first important accelerogram for engineering purposes was recorded.
1960	Chile	9.5	2,230	Probably the largest earthquake ever recorded.
1964	Alaska, USA	9.2	131	The Good Friday earthquake; caused severe damage due to liquefaction and many earthquake-induced landslides.
1964	Japan	7.5	26	Widespread liquefaction caused extensive damage to buildings, bridges, and port facilities in Niigata; along with Good Friday earthquake in Alaska, spurred intense interest in the phenomenon of liquefaction.
1967	Venezuela	6.5	266	Caused collapse of relatively new structures in Caracas; illustrated effects of local soil conditions on ground motion and damage.
1971	California, USA	6.6	65	San Fernando earthquake; produced several examples of liquefaction, including near collapse of Lower San Fernando Dam; caused collapse of several buildings and highway bridges; many structural lessons learned, particularly regarding the need for spiral reinforcement of concrete columns; many strong motion records obtained.
1975	China	7.3	1,300	Evacuation following successful prediction saved thousands of lives in Haicheng, Liaoning Province.
1976	China	7.8	700,000	Thought to be the most deadly earthquake in history; destroyed the city of Tangshan, Hebei Province; not predicted; death estimate of questionable accuracy.
1985	Mexico	8.1	9,500	Epicenter off the Pacific Coast, but most damage occurred over 220 miles (360 km) away in Mexico City; illustrated the effect of local soil conditions on ground motion amplification and damage; subsequent studies led to better understanding of dynamic properties of fine-grained soils.
1989	California, USA	6.9	63	Loma Prieta earthquake; extensive ground motion amplification and liquefaction damage in San Francisco Bay area.
1994	California, USA	6.7	61	Northridge earthquake; occurred on previously unknown fault beneath heavily populated area; buildings, bridges, lifelines extensively damaged; produced extraordinarily strong shaking at several locations.
1995	Japan	6.9	5,300	Hyogo-ken Nanbu earthquake; caused tremendous damage to Kobe, Japan; widespread liquefaction in reclaimed lands constructed for port of Kobe; landslides and damage to retaining walls and underground subway stations also observed.

(*Continued*)

TABLE 1.1 (*Continued*)
Significant Historical Earthquakes

Date	Location	Magnitude	Deaths	Comments
1999	Turkey	7.6, 7.2	18,000	Izmit and Duzce earthquakes; strike-slip events on North Anatolian fault occurring 3 months apart; extensive damage due to surface fault rupture and liquefaction.
1999	Taiwan	7.6	2,400	Chi-Chi earthquake; reverse movement on Chelungpu Fault; extensive damage from surface offsets up to 10 m vertically; extensive landsliding in steep central mountains; liquefaction damage in coastal areas and inland areas near rivers and streams; well-instrumented event.
2004	Sumatra	9.1	228,000	Megathrust subduction event off west coast of Sumatra; caused devastating tsunami killing thousands in Indonesia, Sri Lanka, India, Thailand, and other countries.
2008	China	7.9	88,000	Wenchuan earthquake; thrust event on Longmenchuan fault in central China; extensive landsliding in steep, mountainous terrain; many collapsed structures including school buildings.
2010	Haiti	7.0	160,000	Combination of strike-slip and blind thrust event 25 km west of Port-au-Prince; collapsed 250,000 residences, 30,000 commercial buildings, and all hospitals in Port-au-Prince; liquefaction rendered port unusable for rescue and aid operations.
2010	Chile	8.8	600	Great subduction event near Concepcion; strong, long-duration shaking along coast and in Santiago; liquefaction damage to bridges in coastal areas; tsunami damage in several coastal towns. Produced approximately 30 ground motion recordings.
2010–2011	New Zealand	7.0, 6.3	400	Canterbury earthquake in September 2010, strike-slip located some 40 km west of Christchurch, followed by February 2011 Christchurch earthquake located 5 km southeast of Christchurch. Extensive liquefaction in both earthquakes, but much greater losses in second due to proximity to Christchurch.
2011	Japan	9.0	16,000	Tohoku earthquake; great subduction event off east coast of Honshu; devastating tsunami damage along northern part of east coast of Honshu, minor damage in Hawaii and California; extensive liquefaction damage in Tokyo and Chiba waterfront areas; produced many records of long-duration ground motions.
2023	Turkey	7.8, 7.7	51,000	Kahramanmaraş earthquake; left-lateral strike-slip earthquakes on East Anatolian (M7.8) fault, followed by M7.7 event on Sürgü-Çarda fault. Extensive structural collapse of URMs and reinforced concrete structures; extensive liquefaction damage in alluvial river valleys.

are subjected to. Earthquake engineers characterize the intensity of ground motions using a variety of ground motion parameters, or *intensity measures* (*IM*s). The most useful *IM*s are those to which the response of the system of interest is most closely related, and those are different for different types of systems (e.g., short buildings, tall buildings, landslides, liquefiable soil deposits, etc.). The values of those *IM*s depend on the size and location of the earthquake, the local site conditions, and a number of other factors.

1.5.2 System Response

Structures, whether comprised of steel, concrete, or soil, have mass and are compliant, and therefore respond more strongly at some frequencies than others. Their response to earthquake loading depends on the mass, geometry, stiffness, and damping characteristics of the structure and its foundation, and on the amplitude, frequency content, and duration of the ground motion. The response of a system to earthquake shaking can be expressed in terms of forces and stresses or displacements and strains, and are frequently characterized by *engineering demand parameters*, or *EDP*s. The most useful *EDP*s are those to which damage is related, so they will vary from one type of system to another.

1.5.3 Physical Damage

The response of a structure may or may not result in physical damage depending on the structure's *vulnerability*, i.e., its capacity to resist damage. The capacity may be viewed as a level of response beyond which some level of physical damage can be expected to occur. Many different types of damage can occur during an earthquake – some can be related to the structure itself, some to physical systems within the structure, and some to the contents of the structure. Consider a building, for example, subjected to earthquake shaking of various intensities. At low levels of shaking, some unrestrained objects within or attached to the structure may shift or fall – this could be as innocuous as books falling from a shelf or as serious as chemical containers falling in a laboratory. At stronger levels of shaking, sheetrock and interior partitions can crack, plumbing pipes can break, cover concrete can spall, and windows or cladding can break and fall. At even stronger levels of shaking, beams can crack, joints can fail, foundations can rock and settle, and diagonal braces can buckle. At very strong levels of shaking, ground movement can occur, foundations can fail, welded connections can fracture, columns can lose capacity and collapse can occur – such severe physical damage can lead to extremely high losses. To predict the losses associated with these and other forms of physical damage, it is necessary to identify the specific form(s) of physical damage, quantified by *damage measures*, or *DM*s, that contribute most strongly to the losses of interest, and to be able to predict the physical damage associated with the response of the system of interest.

1.5.4 Losses

Physical damage to structures and their contents results in losses. Losses can have many components – deaths and injuries, repair and replacement costs, and loss of utility for extended periods of time. Decisions regarding risk reduction through retrofitting, insurance protection, etc. are usually made on the basis of expected losses, so those losses are often characterized by decision variables or DVs.

The prevention of death and serious injury has been the fundamental basis of seismic design since it was first contemplated. While the goal of preserving life safety during earthquakes has largely been achieved in many urban areas of well-developed countries, millions of people remain at risk in the many areas around the world where earthquake-resistant design and construction are not practiced.

The economic losses associated with earthquake damage are many but can be divided into two categories – direct and indirect losses. Direct losses are those associated with the repair and/or replacement of structures and facilities damaged by earthquake shaking. Direct losses also include those associated with the contents of structures, which in some cases (e.g., museums, data centers, medical research laboratories, etc.) can be far more valuable than the structures that house them. Indirect losses include those associated with delayed or lost business, environmental damage, compromised infrastructure, etc.

Downtime, which refers to the period of time in which structures or facilities are unavailable for their intended use, is among the most important of indirect losses and can produce, for critical systems, long-term economic losses that far exceed direct losses. The loss of a major bridge or water supply line in a non-redundant system, for example, can lead to inefficiencies in moving goods, services, and people or to health problems that have very real, and very high, economic consequences.

1.6 MITIGATION OF EARTHQUAKE LOSSES

The ultimate goal of the professionals involved in earthquake engineering is to reduce the losses associated with earthquakes. Reduction of earthquake losses can be accomplished in many different ways. Individuals, private organizations, and public agencies can all plan for earthquakes, and some may establish policies that promote the reduction of losses. As such, policy and planning activities can have a large effect on earthquake loss reduction.

Such efforts can be implemented at different points in the earthquake process, as illustrated in Figure 1.23. Recognizing that earthquakes themselves and the ground shaking they produce cannot be eliminated, steps can be taken to reduce their consequences. Zoning and land use policies, for example, can be used to keep certain structures and facilities out of areas where seismic hazards are expected to be particularly severe. The Alquist-Priolo Act in California, for example, seeks to prevent the construction of buildings used for human occupancy on the surface traces of active faults by establishing Special Studies Zones; structures constructed in Special Studies Zones must have a licensed geologist's report prepared and be set back (generally at least 15 m) from known fault traces.

In many cases, it is impossible or uneconomical to move a particular structure or facility to another location. In such cases, an alternative approach is to focus on reducing the amount of damage that occurs for a given level of response. Retrofitting structures, for example, can allow them to resist greater forces and/or accommodate greater levels of deformation without incurring significant damage. Some retrofitting measures are effective and relatively inexpensive and can reduce potential losses very efficiently. Simple cable restrainers, for example, were added to bridges in California following the Loma Prieta earthquake in 1989 to prevent simply-supported spans with insufficient seat widths from falling upon unseating. Seats were also widened by welding or bolting seat extensions to existing bridges. In other cases, however, reducing damage can be prohibitively expensive.

In some cases, it may be more economical to allow damage to occur but to reduce the losses associated with that damage. Force-based design of new structures allows damage to occur (and thereby dissipate energy) in a ductile manner, i.e., in elements that are unlikely to cause collapse and that, in some cases, can be easily repaired. Decisions on whether or not to retrofit existing structures are affected by many considerations including risk tolerance, costs, potential impacts of earthquake damage, and concerns about changing the finish/architecture of historical structures.

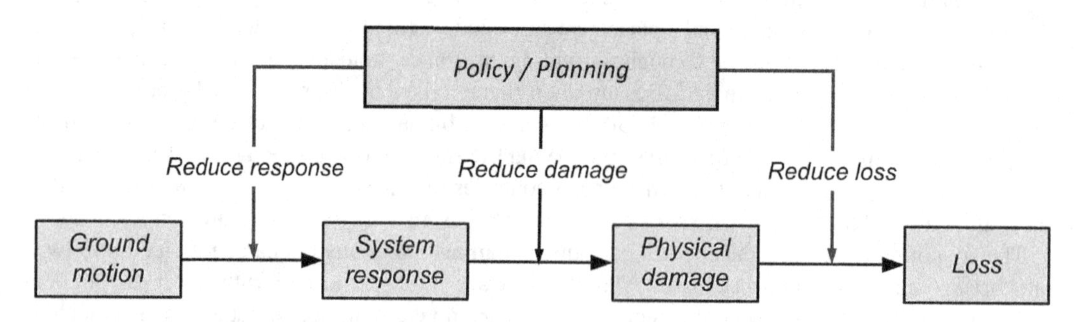

FIGURE 1.23 The relationship of policy and planning activities to the earthquake process.

When a vulnerable structure is not retrofitted, educating their occupants about earthquake safety and emergency evacuation can help reduce casualties even when physical damage does occur.

Investing in improved emergency response capabilities can also help reduce casualties and loss of life in earthquakes. From a financial standpoint, purchasing earthquake insurance can reduce the exposure to very large financial losses from earthquakes.

The development and implementation of sound policy for earthquake hazard reduction is not easy. While some successes have been made (Los Angeles Times, 2019), legislating seismic mitigation has proven to be difficult – the benefits are real, but difficult to quantify, and likely to be realized only in the long term. In a political environment where short-term demands (e.g., re-election) are paramount, legislators are often more concerned with the short-term concerns of their current constituents than investing scarce resources in long-term earthquake hazard reduction. Increased levels of seismic mitigation also place financial burdens on developers (a powerful interest group) and regulatory burdens on local governments (the agencies usually responsible for ensuring compliance), neither of whom may be enthusiastic about such changes. Earthquake engineers can and do, however, play an important role in policy implementation through their advancement of the seismic provisions in codes and regulatory guides.

1.7 PERFORMANCE

The concepts described in the previous sections all point toward a notion of *performance* during earthquakes. Performance can be described in many different ways, and the ways in which it is described can vary from stakeholder to stakeholder. To a seismologist, some measure of ground motion intensity (i.e., an *IM*) may be a good descriptor of the potential performance of a building subjected to earthquake shaking. To a structural engineer, maximum interstory drift (the relative horizontal displacement of adjacent floors, an *EDP*) would likely be a better descriptor of performance. To an estimator preparing a bid for repairs, the width and spacing of cracks in beams and columns (*DM*s) could be more useful measures of performance. Finally, to an owner, the economic loss associated with earthquake damage (a *DV*) could be the best measure of performance.

In the early days of earthquake engineering, performance was effectively measured in terms of deaths, and codes were written with the single performance objective of preserving life safety in strong earthquakes that could be anticipated. A building that teetered on the edge of collapse under very strong shaking, but did not kill anyone in or near it, would be judged to have performed successfully, even if it had to be torn down following the earthquake. Later, the concept of dual performance objectives was introduced into earthquake-resistant design – life safety was still required under very strong shaking, but an additional objective of limited damage under weaker (and more likely) shaking was added. For example, in the mid-1990s, a major study (Poland et al., 1995) described a design framework in which four discrete performance levels (fully operational, operational, life safe, and near collapse) were paired with four ground motion hazard levels (frequent, occasional, rare, and very rare) to achieve broadly acceptable performance. More recently, these concepts have evolved into what is now referred to as *performance-based design*, a paradigm in which multiple performance objectives, each corresponding to different levels of ground shaking, are to be satisfied. For example, performance objectives for building structures are often based on collapse risk (Luco et al., 2007) and, increasingly, limiting the time and expense required before a building can be re-occupied following earthquakes (EERI, 2019). These performance objectives are evaluated during design using *performance-based earthquake engineering* (Chapter 5), which considers *all* levels of ground shaking. Such procedures are attractive to a variety of stakeholders because they offer the potential for designs with consistent, predictable levels of performance.

The concepts of performance-based earthquake engineering are advancing rapidly and are increasingly underpinning earthquake engineering practice. Although the precise manner in which these concepts will be implemented into codes and regulatory guides will likely evolve over time, this book will seek to introduce and explain them in a general and useful form.

REFERENCES

Chang, S.E. (2000). "Disasters and transport systems: Loss, recovery and competition at the Port of Kobe after the 1995 earthquake," *Journal of Transport Geography*, Pergamon Press, Vol. 8, pp. 53–65. https://doi.org/10.1016/S0966-6923(99)00023-X.

Earthquake Engineering Research Institute (2019). "Functional Recovery: A Conceptual Framework with Policy Options," *EERI White Paper*, Earthquake Engineering Research Institute.

Harder, L.F. and Stewart, J.P. (1996). "Failure of Tapo Canyon tailings dam," *Journal of Performance of Constructed Facilities*, Vol. 10, No. 3, pp. 109–114. https://doi.org/10.1061/(ASCE)0887-3828(1996)10#:3(109).

Jones, L.M., Bernknopf, R., Cox, D., Goltz, J., Hudnut, K., Mileti, D., Perry, S., Ponti, D., Porter, K., Reichle, M., Seligson, H., Shoaf, K., Treiman, J., and Wein, A. (2008). "The ShakeOut scenario," *USGS Open File Report 2008-1150*, U.S. Geological Survey, U.S. Department of the Interior, 308 pp. https://doi.org/10.3133/ofr20081150.

Lida, K., Cox, D.C., and Pararas-Carayannis, G. (1967). "Preliminary catalog of Tsunamis occurring in the Pacific Ocean," *Data Report 5,* HIG-67-10, Hawai'i Institute of Geophysics and Planetology, University of Hawaii, Honolulu.

Los Angeles Times (2019). https://www.latimes.com/california/story/2019-11-22/los-angeles-soft-story-earthquake-building.

Luco, N., Ellingwood, B.R., Hamburger, R.O., Hooper, J.D., Kimball, J.K., and Kircher, C.A., 2007. "Risk-targeted versus current seismic design maps for the conterminous United States," *Proceedings, Structural Engineers Association of California 76th Annual Convention*, Lake Tahoe, CA.

Lund, L. and Schiff, A. (1991). *"TCLEE Pipeline Failure Database, Technical Council on Lifeline Earthquake Engineering,"* ASCE, Reston, VA, 36 pp.

Moehle, J.P. and Deierlein, G.G. (2004). "A framework methodology for performance-based earthquake engineering," Proceedings, World Conference on Earthquake Engineering, Vancouver, BC, Canada, Paper 679, 13 pp.

Ogut, T. (2000). "Izmit-Turkey earthquake, August 1999," *Recorder*, Canadian Society of Exploration Geophysicists, Vol. 25, No. 6. https://csegrecorder.com/editions/issue/2000-06

O'Rourke, T.D., Jeon, S.S., Toprak, S., Cubrinovski, M., Hughes, M., van Ballegooy, S., and Bouziou, D. (2014). "Earthquake response of underground pipeline networks in Christchurch, NZ," *Earthquake Spectra*, Vol. 30, No. 1, pp. 183–204. https://doi.org/10.1193/030413EQS062M.

Poland, C.D., Hill, J., Sharpe, R.L., and Soulages, J. (1995). *"Vision 2000: Performance Based Seismic Engineering of Buildings,"* Structural Engineers Association of California and California Office of Emergency Services, Sacramento, CA.

Rathje, E., Bachhuber, J., Cox, B., French, J., Green, R., Olson, S., Rix, G., Wells, D., and Suncar, O. (2010). "Geotechnical engineering reconnaissance of the 2010 Haiti earthquake, Version 1," *GEER Association Report No. GEER-021*, University of California, Berkeley, 97 pp.

Scawthorn, C.R., Porter, K.A., and Blackburn, F.T. (1992). "Performance of emergency-response services after the earthquake," *The Loma Prieta, California, Earthquake of October 17, 1989 – Marina District*, U.S. Geological Survey Professional Paper 1551-F, pp. F195–F215.

Spaeth, M.G. and Berkman, S.C. (1967). "The tsunami of March 28, 1964 as recorded at tide stations," *ESSA Technical Report C and GS 33*, U.S. Coast and Geodetic Survey, Rockville, Maryland.

Steinbrugge, K.V. and Cloud, W. (1962). "Epicentral intensities and damage in the Hebgen Lake, Montana earthquake of August 17, 1959," *Bulletin of the Seismological Society of America*, Vol. 52, No. 2, pp. 181–234. https://doi.org/10.1785/BSSA0520020181.

Tasiopoulou, P., Smyrou, E., Bal, I.E., and Gazeetas, G. (2012). "Bridge pile-abutment-deck interaction in laterally spreading ground: Lessons from Christchurch," Proceedings, 15th World Conference on Earthquake Engineering, Lisbon, 10 pp.

Wang, W.-N., Chigiro, M, and Takahiko, R. (2003). "Geological and geomorphological precursors of the Chiu-fen-erh-shan landslide triggered by the Chi-Chi earthquake in central Taiwan," *Engineering Geology*, Vol. 69, No. 1–2, pp. 1–13. https://doi.org/10.1016/S0013-7952(02)00244-2.

2 Seismology and Earthquakes

2.1 INTRODUCTION

The study of geotechnical earthquake engineering requires an understanding of various processes by which earthquakes occur and their effects on ground motion. The field of seismology (from the Greek *seismos* for earthquake and *logos* for science) developed from a need to understand the internal structure and behavior of the Earth, particularly as they relate to earthquakes. Although earthquakes are complex phenomena, advances in seismology have produced a good understanding of the mechanisms and rates of occurrence of earthquakes in most seismically active areas of the world. This chapter provides a brief introduction to the structure of the Earth, the reasons why earthquakes occur, and the terminology used to describe them. More complete descriptions of these topics may be found in a number of seismology texts, such as Gutenberg and Richter (1954), Richter (1958), Bullen (1975), Bath (1979), Aki and Richards (1980), Bullen and Bolt (1985), Gubbins (1990), Lay and Wallace (1995) and Shearer (2019). A very readable description of seismology and earthquakes is given by Bolt (2006).

2.2 INTERNAL STRUCTURE OF THE EARTH

The Earth is roughly spherical, with an equatorial diameter of 12,740 km and a polar diameter of 12,700 km, the higher equatorial diameter being caused by higher equatorial velocities due to the Earth's rotation. The Earth weighs some 5.97×10^{24} kg, which indicates an average specific gravity of about 5.5. Since the specific gravity of surficial rocks is known to be on the order of 2.7–3, higher specific gravities are implied at greater depths.

One of the first important achievements in seismology was the determination of the internal structure of the Earth. Large earthquakes produce enough energy to cause measurable shaking at points all around the world. Different types of seismic waves are produced by earthquakes, which travel through, or across the surface of, the Earth at different speeds. Waves propagating through the Earth are refracted and reflected at boundaries between different layers, reaching different points on the Earth's surface by different paths. Studies of these refractions and reflections early in the twentieth century revealed the layered structure of the Earth and provided insight into the characteristics of each layer.

2.2.1 Seismic Waves

When an earthquake occurs, different types of seismic waves are produced: *body waves* and *surface waves*. Although seismic waves are discussed in detail in Appendix C, the brief description that follows will allow the reader to continue in this chapter without referring to the more mathematical treatment of wave propagation in the appendix.

Body waves, which can travel through the interior of the Earth, are of two types: *p-waves* and *s-waves* (Figure 2.1). P-waves, also known as primary, compressional, or longitudinal waves, involve successive compression and rarefaction (dilation) of the materials through which they pass. They are analogous to sound waves; the motion of an individual particle that a p-wave travels through is parallel to the direction of travel. Like sound waves, p-waves can travel through solids and fluids. S-waves, also known as secondary, shear, or transverse waves, cause shearing deformations as they travel through a material. The motion of an individual particle is perpendicular to the direction of s-wave travel. The direction of particle movement can be used to divide s-waves into

DOI: 10.1201/9781003512011-2

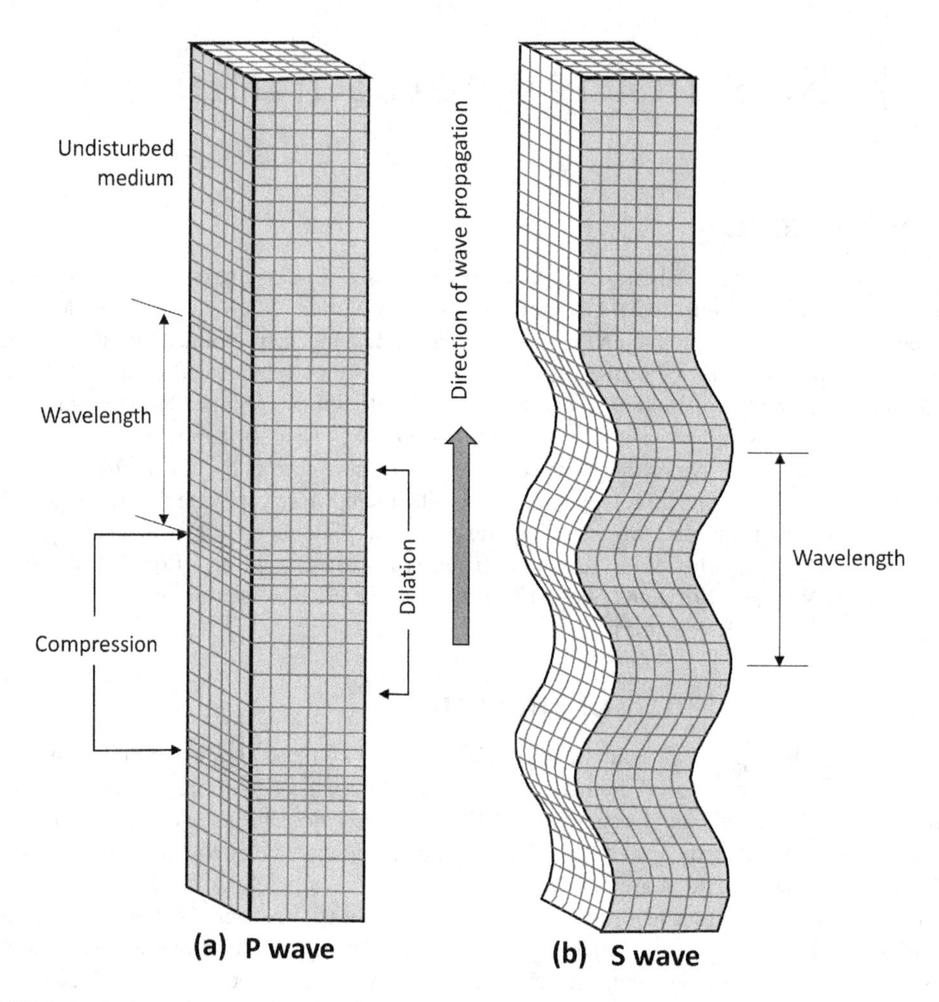

FIGURE 2.1 Deformations produced by body waves: (a) p-wave; (b) SH-wave. Shaded surfaces remain flat (no out-of-plane movement). (Modified from Bolt, 2006.)

two components, SV (particle movement in the vertical plane) and SH (particle movement in the horizontal plane). The speeds at which body waves travel vary with the stiffness of the materials they travel through. Since geologic materials are stiffest in compression, p-waves travel faster than other seismic waves and are therefore the first to arrive at a particular site. Fluids, which have no shearing stiffness, cannot sustain s-waves.

Surface waves result from the interaction between body waves and the surface and shallow layers of geologic materials. They travel along the Earth's surface with amplitudes that decrease roughly exponentially with depth (Figure 2.2). Because of the nature of the interactions required to produce them, surface waves are more prominent at distances farther from the source of the earthquake. At distances greater than about twice the thickness of the Earth's crust, surface waves, rather than body waves, will usually produce peak ground motions.

The most important surface waves, for engineering purposes, are Rayleigh waves and Love waves. Rayleigh waves, produced by the interaction of p- and SV-waves with the Earth's surface, involve both vertical and horizontal particle motion. They are similar, in some respects, to the waves produced by a rock thrown into a pond. Love waves result from the interaction of SH-waves with a soft surficial layer overlying stiffer materials and have no vertical component of particle motion.

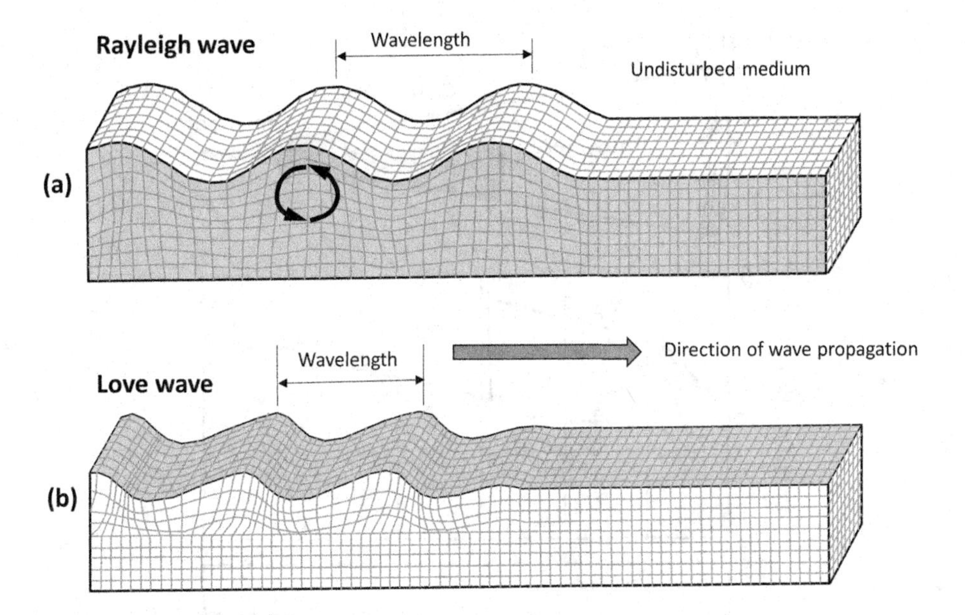

FIGURE 2.2 Deformations produced by surface waves: (a) Rayleigh wave; and (b) Love wave. Shaded surfaces remain flat (no out-of-plane movement). (Modified from Bolt, 2006.)

2.2.2 Interior of the Earth

The crust, on which human beings live, is the outermost layer of the Earth. The thickness of the crust is about 25–40 km beneath the continents (although it may be as thick as 60–70 km under young mountain ranges). In oceans, the crust has thicknesses that can be as thin as 5 km (although thicknesses of practically zero occur at spreading ridges; Section 2.3.2.1) and tends to be more uniform and denser than continental crust. The crust represents only a very small fraction of the Earth's diameter (Figure 2.3). The internal structure of the Earth is complex but can be represented by the crust, mantle, and core (Figure 2.3). The crust and a portion of the upper mantle are relatively cool and stiff and together comprise the *lithosphere*, which is cooler than the materials below it (bottom panel, Figure 2.4).

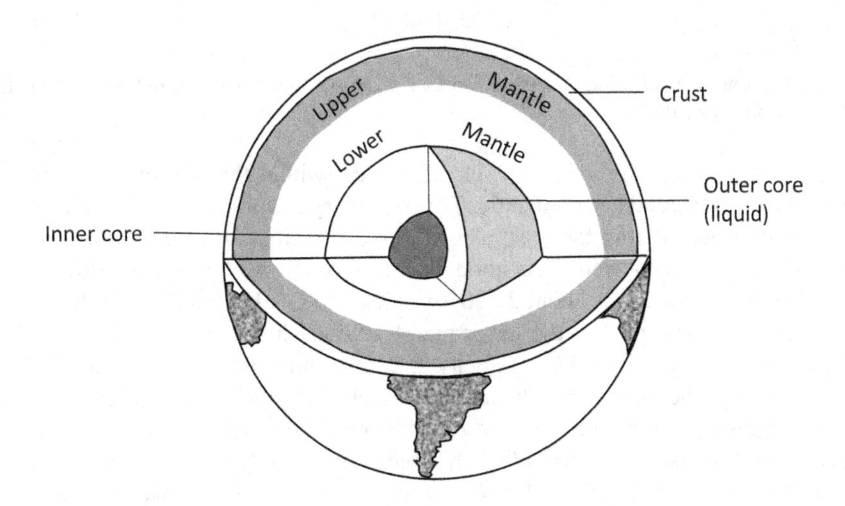

FIGURE 2.3 Internal structure of the Earth.

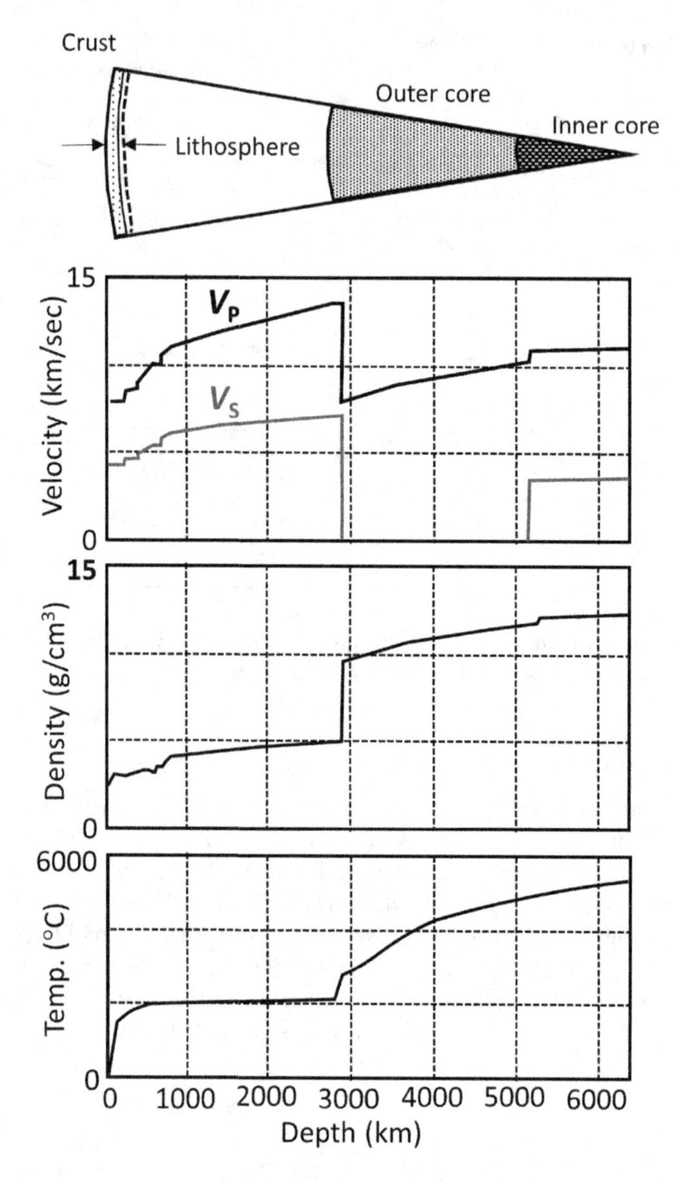

FIGURE 2.4 Approximate variation of seismic velocities, density, and temperature below the surface of the Earth. (Adapted from Eiby, 1980.)

A distinct change in wave propagation velocity occurs within the lithosphere at the boundary between the crust and the underlying mantle. This boundary is known as the Mohorovicie discontinuity, or the Moho, named after the seismologist who discovered it in 1909. Although the specific nature of the Moho itself is not well understood, its role as a reflector and refractor of seismic waves is well established. The mantle is about 2,850 km thick and can be divided into the upper mantle (shallower than about 650 km) and the lower mantle. No earthquakes have been recorded in the lower mantle, which exhibits a uniform velocity structure and appears to be chemically homogeneous, except near its lower boundary. The mantle is cooler near the crust than at greater depths but still has an average temperature of about 2,200°C. As a result, the mantle materials are in a viscous, semi-molten state. They behave as a solid when subjected to rapidly applied stresses, such as those associated with seismic waves, but can slowly flow like a fluid in response to long-term stresses. The mantle material has a specific gravity of about 4–5.

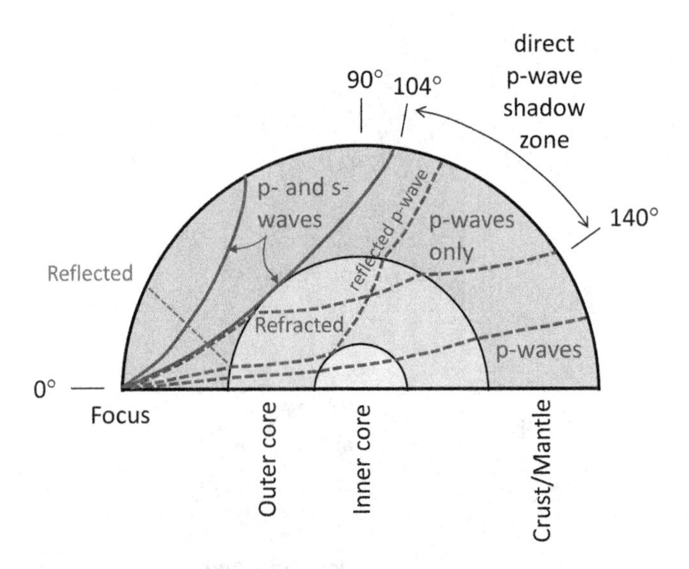

FIGURE 2.5 Seismic wave paths illustrating reflection and refraction of seismic waves from the source (focus) of the earthquake by the different layers of the Earth. Note that p- and s-waves can reach the Earth's surface between 0° and 104°, but the liquid nature of the outer core allows only p-waves to reach the surface between 140° and 180°. In the shadow zone between angles of 104° and 140°, only paths reflected from the inner core can reach the Earth's surface.

The outer core, or liquid core, is some 2,260 km thick. As a liquid, it cannot transmit shear (or s-) waves. As shown in Figure 2.4 (top panel), the s-wave velocity drops to zero at the core/mantle boundary, or Gutenberg discontinuity; note also the precipitous drop in compression (or p-) wave velocity. The outer core consists primarily of molten iron (which helps explain its high specific gravity of 9–12). The inner core, or solid core, is a very dense (specific gravity up to about 15), solid nickel-iron material compressed under tremendous pressures. The temperature of the inner core is estimated to be relatively uniform at about 5,200°C.

Figure 2.5 shows the influence of the Earth's structure on the distribution of seismic waves during earthquakes. Since wave propagation velocities generally increase with depth, wave paths are usually refracted back toward the Earth's surface. An exception is at the core-mantle boundary, where the outer core velocity is lower than the mantle velocity. Figure 2.5 shows that a wave that is just tangent to this interface arrives at an angle of 104° (angle is measured clockwise relative to a radial line through the focus), whereas a wave arriving at the core-mantle boundary at a slightly flatter angle will refract through the outer core and ultimately arrive at an angle of 140°. The intermediate zone (between 104° and 140°) is a shadow zone that cannot 'see' shaking from the earthquake, other than through reflection of p-waves off of the inner core.

2.3 CONTINENTAL DRIFT AND PLATE TECTONICS

Although observations of similarity between the coastlines and geology of eastern South America and western Africa and the southern part of India and northern part of Australia have intrigued scientists since the seventeenth century (Glen, 1975; Kearey and Vine, 1990), the theory that has come to be known as continental drift was not proposed until the early twentieth century (Taylor, 1910; Wegener, 1915). Wegener, for example, believed that the Earth had only one large continent called Pangaea 200 million years ago. He believed that Pangaea broke into pieces that slowly drifted (Figure 2.6) into the present configuration of the continents. A more detailed view of the current similarity of the African and South American continental boundaries is shown in Figure 2.7.

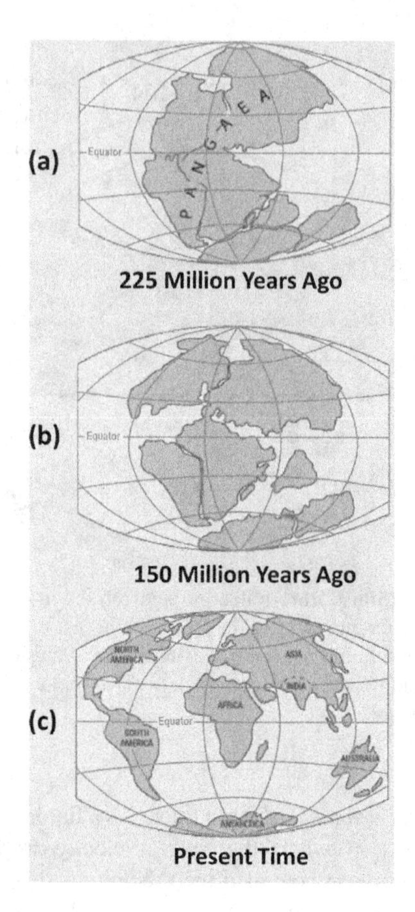

FIGURE 2.6 Wegener's theory of continental drift: (a) 225 million years ago; (b) 150 million years ago; (c) present time. (Maps from USGS, 2023.)

The theory of continental drift did not receive much attention until about 1960 when the current worldwide network of seismographs was able to define earthquake locations accurately and to confirm that long-term deformations were concentrated in narrow zones between relatively intact blocks of crust. Also, exploration of the ocean floor did not begin in earnest until after World War II, when new techniques such as deep-water echo sounding, seismic refraction, and piston coring became available. The geologic units comprising the ocean floor are young, representing only about 5% of the Earth's history (Gubbins, 1990), and relatively simple. Its detailed study provided strong supporting evidence of the historical movement of the continents as assumed in the theory of continental drift. Within 10 years, the theory of continental drift had become widely accepted and acknowledged as the greatest advance in the earth sciences in a century.

2.3.1 Plate Tectonics

The original theory of continental drift suggested images of massive continents pushing through the seas and across the ocean floor. It was well known, however, that the ocean floor was too strong to permit such motion, and the theory was originally discredited by most earth scientists. From this background, however, the modern theory of plate tectonics began to evolve. The basic hypothesis of plate tectonics is that the Earth's surface consists of a number of large, intact blocks called plates and that these plates move with respect to each other. The plates themselves are comprised of the crust and a portion of the upper mantle, being approximately 100 km in thickness. The Earth's crust is

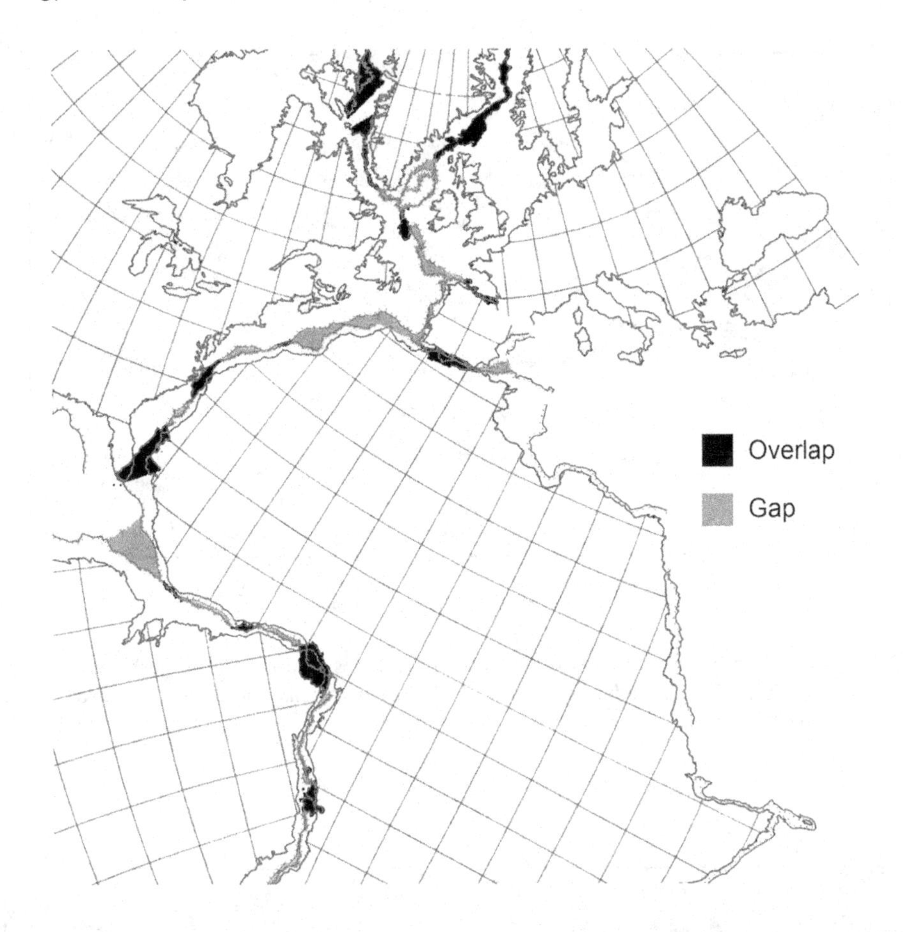

FIGURE 2.7 Statistical spherical fit proposed by Bullard et al. (1965) of several continents using the continental slopes rather than the coastlines. Areas of overlap and gap were identified. (Modified from Bullard et al., 1965.)

divided into six continental-sized plates (African, American, Antarctic, Australia-Indian, Eurasian, and Pacific) and about 14 of subcontinental size (e.g., Caribbean, Cocos, Nazca, Philippine, etc.). The major plates are shown in Figure 2.8. Smaller platelets, or microplates, have broken off from the larger plates in the vicinity of many of the major plate boundaries but are not shown here. The relative deformation between plates occurs only in narrow zones near their boundaries. This deformation between the plates can occur slowly and continuously (aseismic deformation) or can occur spasmodically in the form of earthquakes (seismic deformation). Since the deformation occurs predominantly at the boundaries between the plates, it would be expected that the locations of earthquakes would be concentrated near plate boundaries. The map of earthquake epicenters shown in Figure 2.9 confirms this expectation, thereby providing strong support for the theory of plate tectonics.

The theory of plate tectonics is kinematic (i.e., it explains the geometry of plate movement without addressing the cause of that movement). Something must drive the movement, however, and the tremendous mass of the moving plates requires that the driving forces be very large. The most widely accepted explanation of the source of plate movement relies on the requirement of thermo-mechanical equilibrium of the Earth's materials. The upper portion of the mantle is in contact with the relatively cool crust while the lower portion is in contact with the hot outer core. Obviously, a temperature gradient must exist within the mantle (see Figure 2.4). The variation of mantle density with temperature produces the unstable situation of denser (cooler) material resting on top of less

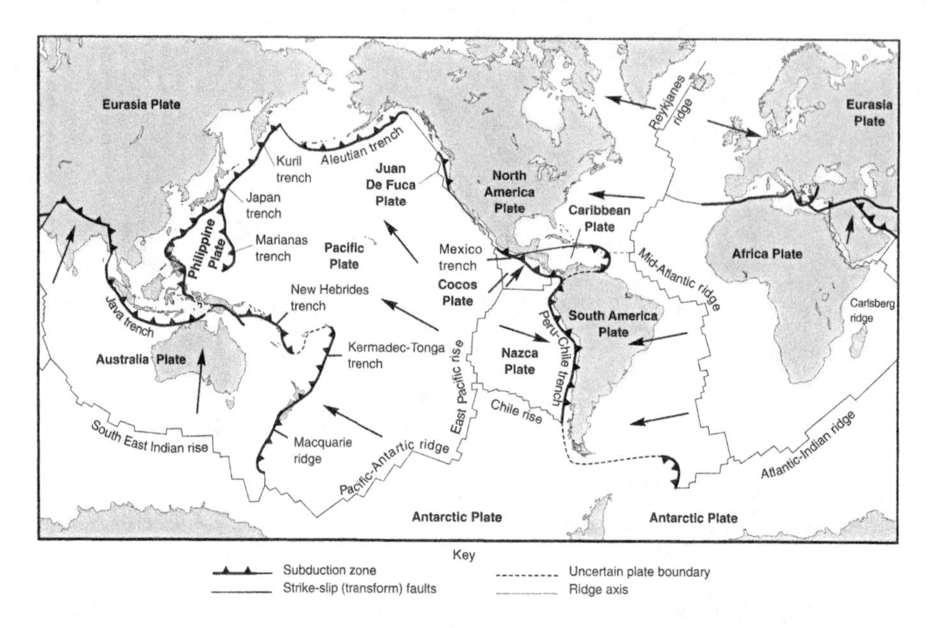

FIGURE 2.8 The major tectonic plates, divergent plate boundaries (mid-oceanic ridges), convergent boundaries (trenches), and transform boundaries of the Earth. Arrows indicate the direction of plate movement. (After Fowler, 2004.)

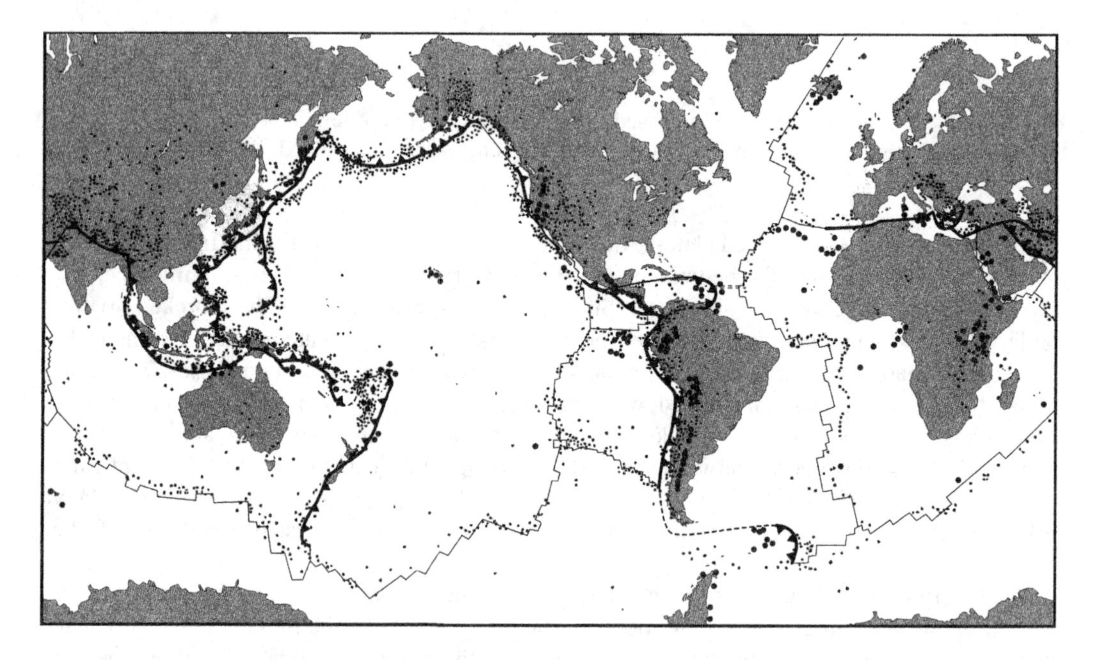

FIGURE 2.9 Worldwide seismic activity. The dots represent the epicenters of significant earthquakes colored by hypocentral depth. The great majority of earthquakes are located near plate boundaries. (After Bolt, 2006)

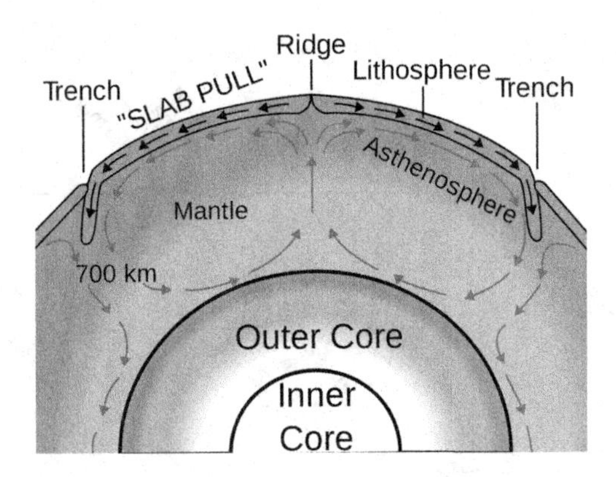

FIGURE 2.10 Convection currents in the mantle (arrows forming closed loops). Near the bottom of the lithosphere, the horizontal components of convection currents impose shear stresses, causing plate movement. Relative movements of adjacent plates can be to separate (slab pull from away from spreading ridge) or converge (subduction at trench). (USGS image from Wikimedia Commons.)

dense (warmer) material. Eventually, the cooler, denser material begins to sink under the action of gravity and the warmer, less dense material begins to rise. The sinking material gradually warms through heat conduction from the core and becomes less dense; conduction continues as the material moves laterally and eventually it will begin to rise again as subsequently cooled material begins to sink. This process is the familiar one of convection. Convection currents in the semi-molten rock of the mantle, illustrated schematically in Figure 2.10, impose shear stresses on the bottom of the plates, thus "dragging" them in various directions across the surface of the Earth. Other phenomena, such as ridge push or slab pull, may also contribute to the movement of plates (Hager, 1978).

2.3.2 Plate Boundaries

Three distinct types of plate boundaries have been identified, and understanding the movement associated with each will aid in the understanding of plate tectonics. The characteristics of the plate boundaries also influence the nature of the earthquakes that occur along them.

2.3.2.1 Spreading Ridge Boundaries

In certain areas, the plates move apart from each other (Figure 2.11) at boundaries known as spreading ridges or spreading rifts. Molten rock from the underlying mantle rises to the surface where it cools and becomes part of the spreading plates. In this way, the plates "grow" at the spreading ridge. Spreading rates range from approximately 2 to 18 cm/year; the highest rates are found in the Pacific Ocean ridges and the lowest along the Mid-Atlantic Ridge. It is estimated (Garfunkel, 1975) that new oceanic crust is currently formed at a rate of about 3.1 km^2/year worldwide. The crust, mainly young, fresh basalt, is thin in the vicinity of the spreading ridges. It may be formed by relatively slow upward movement of magma, or it may be ejected quickly during seismic activity.

Underwater photographs have shown formations of pillow lava and have even recorded lava eruptions in progress. Volcanic activity, much of which occurs beneath the ocean surface, is common in the vicinity of spreading ridge boundaries. Spreading ridges can protrude above the ocean; the island of Iceland, where volcanic activity is nearly continuous (there are 150 active volcanoes), is one example.

Basalt and other materials cool after reaching the surface in the gap between the spreading plates. As cooling occurs, iron-titanium oxide minerals within the rock preserve the direction of the Earth's magnetic field at the time of cooling (remnant magnetism). The magnetic field of the Earth is not constant on a geological time scale; it has fluctuated and reversed over geological time,

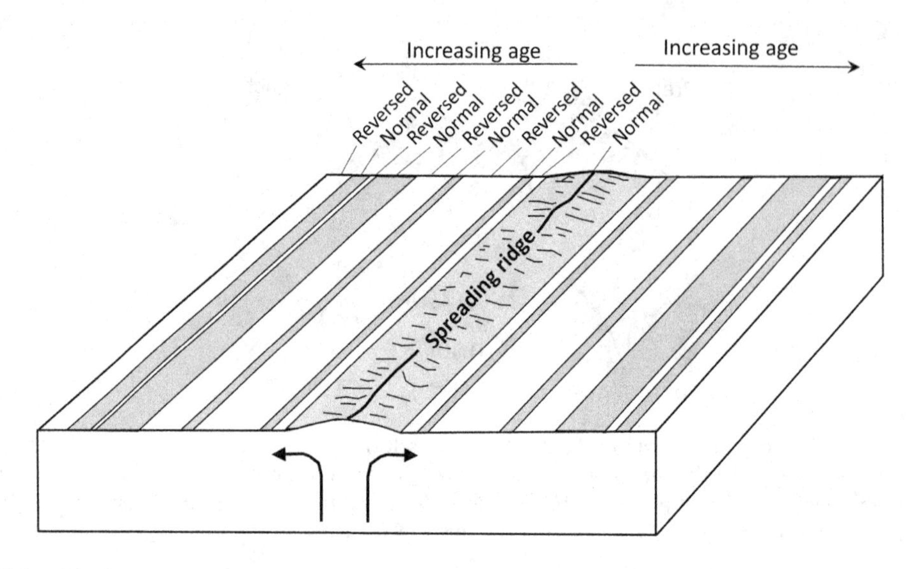

FIGURE 2.11 Spreading ridge boundary. Magma rises to surface and cools in gap formed by spreading plates. Magnetic anomalies are shown as stripes of normal and reversed magnetic polarity. (After Foster, 1971.)

thus imposing magnetic anomalies (reversals of polarity) in the rock that forms at the spreading ridge boundaries. Measurement of the magnetic field in a direction perpendicular to a spreading ridge plate boundary reveals a fluctuating pattern of magnetic intensity, as illustrated for the eastern Pacific Ocean region in Figure 2.12. These magnetic anomalies have allowed large portions of the major plates to be dated. Comparison of the ages of various materials allows the identification of the geometry and movement of various plates and has proven invaluable in the verification and acceptance of the theory of plate tectonics.

2.3.2.2 Subduction Zone Boundaries

Since the size of the Earth remains essentially constant (Bursa 1993), the creation of new plate material at spreading ridges must be balanced by the consumption of plate material at other locations. This occurs at subduction zone boundaries where the relative movement of two plates is toward each other. At the point of contact, one plate plunges, or subducts, beneath the other, as shown in Figure 2.13. Examples of major subduction zone plate boundaries include the western coasts of Central and South America, in Alaska south of the Aleutian Island chain, off the western coast of Sumatra, and off the eastern coast of Japan. The Cascadia subduction zone off the coast of Washington and British Columbia is shown in Figure 2.13.

Subduction zone boundaries are often found near the edges of continents. Because the oceanic crust is generally cold and dense, it descends beneath (i.e., *subducts*) the lighter continental crust. When the rate of plate convergence is high, a trench is formed at the boundary between plates. In fact, subduction zone boundaries are sometimes called trench boundaries. Earthquakes are generated at the sloping interface between the subducting and overriding plates (*interface* earthquakes), and can also occur within the subducting slab (*intra-slab* earthquakes), which is subject to large warping and deformation. When the rate of convergence is slow, sediments accumulate in an accretionary wedge on top of the crustal rock, thus obscuring the trench.

As a subducting plate moves toward and beneath an overriding plate, the overriding plate is deformed by the large frictional and compressive forces caused by the relative movement of the subducting plate (Figure 2.14a). The seaward portion of the overriding plate is slowly dragged downward and the landward potion slowly bulges upward. When slip occurs at the interface between plates during an earthquake, the overriding plate snaps back into its less-stressed position

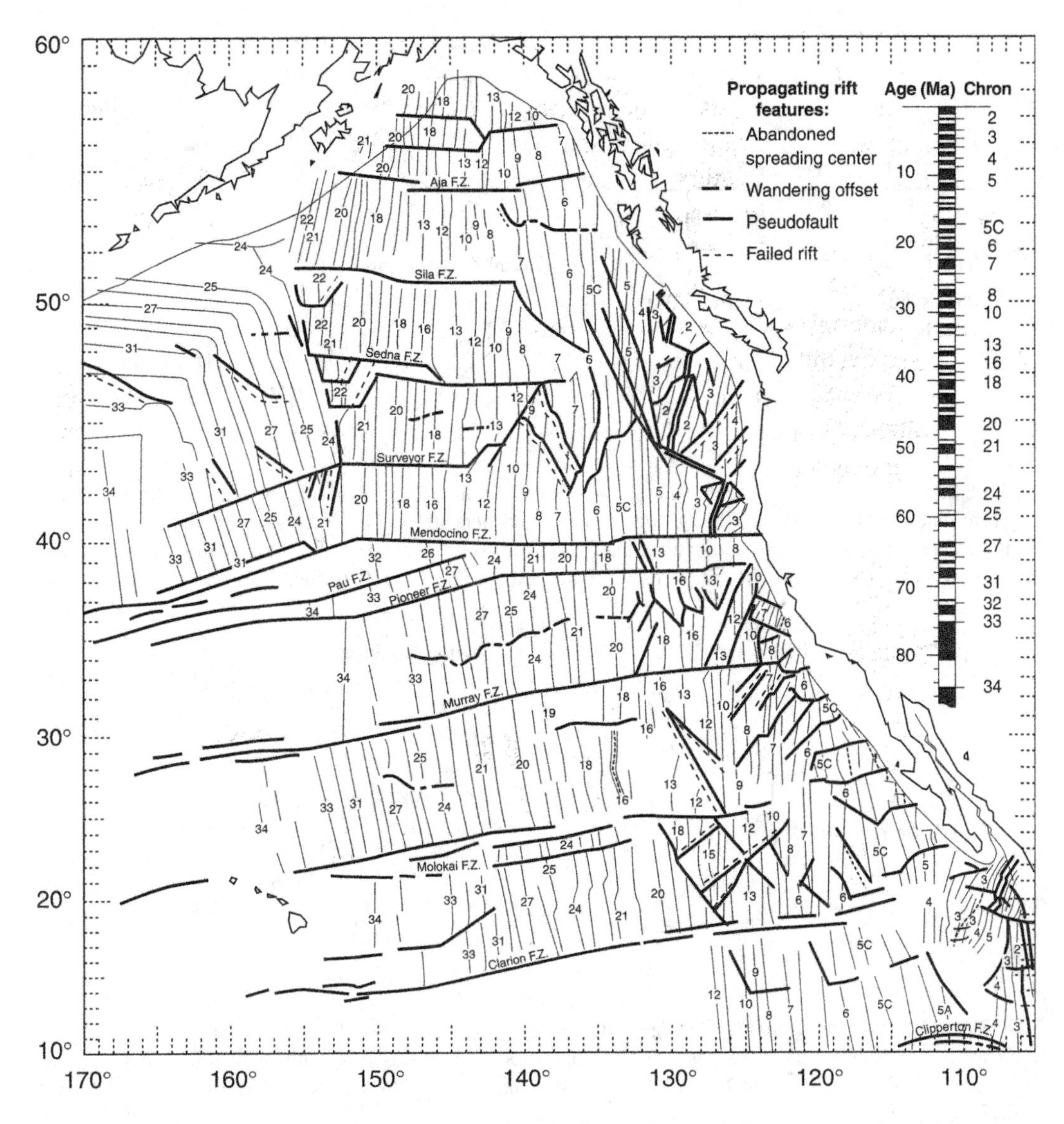

FIGURE 2.12 Magnetic anomalies in the eastern Pacific Ocean. The dark lines represent bands of common magnetic polarity. (After Atwater and Sveringhaus, 1989.)

(Figure 2.14b), lifting the water above it relative to the water above the subducting plate. This phenomenon produces tsunamis (Section 1.3.8) (the largest and most damaging tsunamis are produced by great subduction zone earthquakes) and can substantially change the elevation of coastal areas (Figures 2.14b and 2.15). Globally, the largest magnitude earthquakes are produced at subduction interfaces, including events of magnitude 9.0 and above.

As a subducting plate sinks, it warms and becomes less brittle. Eventually, the plate becomes so ductile as to be incapable of producing earthquakes; the greatest recorded earthquake depth of approximately 700 km supports this hypothesis. Portions of the subducting plate melt, producing magma that can rise to the surface to form a line of volcanoes roughly parallel to the subduction zone on the overriding plate. In recent years, sensitive instruments have detected slow movements, termed *episodic tremor and slip* (ETS), in the lower portions of subduction zone boundaries. ETS events lasting about two weeks have been regularly observed about every 14½ months below the

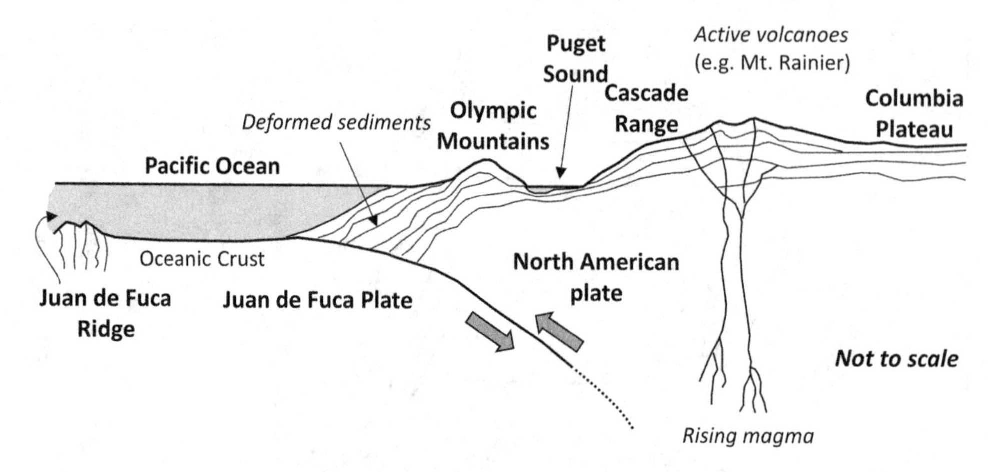

FIGURE 2.13 Cascadia subduction zone off the coasts of Washington and Oregon. The Juan de Fuca plate originates at the Juan de Fuca spreading ridge and subducts beneath the North American plate. Magma rising from the deeper part of the subduction zone has formed a series of volcanoes that run roughly parallel to the subduction zone. One of these, Mt. St. Helens, erupted explosively in 1980. (After Noson et al., 1988.)

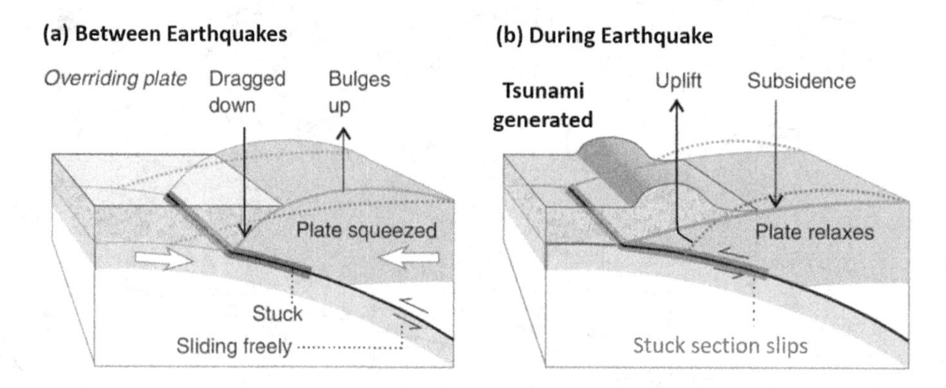

FIGURE 2.14 Plate movement (a) during subduction prior to slip, and (b) from slip in an interface subduction earthquake. Rapid change in elevation of seabed lifts localized zone of water which flows both toward and away from shore. Portions of the shore are uplifted and other portions subside. (Modified from Atwater et al., 2005.)

Cascadia subduction zone near Seattle (Rogers and Dragert, 2003); the amount of energy released in each has been the equivalent of about a magnitude 6.5 earthquake. Such events have also been observed in the subduction zone in southwestern Japan (Obara et al., 2004). Their relationship to the more sudden release of energy that produces earthquakes, i.e., whether their occurrence makes earthquakes more or less likely, is not currently understood.

When plates carrying continents move toward each other, there is no clear preference for which one subducts, and the resulting *continental collisions* can lead to the formation of mountain ranges. The Himalayas formed as the Australia-Indian plate collided with the Eurasian plate. Continental collision of the plates carrying Africa and Europe has not only produced the European Alps but also reduced the size of the Mediterranean Sea (McKenzie, 1970).

2.3.2.3 Transform Faults

Transform faults occur where plates move laterally past each other without creating new crust or consuming old crust. They are usually found offsetting spreading ridges as illustrated in Figure 2.16.

FIGURE 2.15 Example of uplifted area along the coast of Santa Maria Island following 2010 magnitude 8.8 Maule earthquake. The platform at the base of the hill had been submerged before the earthquake. The amount of uplift in this area is 2 m. (Helmholtz Association of German Research Centres, 2015; used with permission of Elsevier Science and Technology Journals.)

These transform faults are identified by offsets in magnetic anomalies and, where preserved, scarps on the surface of the crust. Magnetic anomaly offsets defining fracture zones may be observed over thousands of kilometers; however, it is only the segment of the fracture zone between active spreading ridges that is referred to as the transform fault. As illustrated in Figure 2.16b, the motion on the portions of the fracture zone that extend beyond the transform fault is in the same direction on either side of the fracture zone; hence there is generally no relative motion. These inactive portions of the fracture zone can be viewed as *fossil faults* that are not producing earthquakes.

The San Andreas fault, for example, has been characterized as a transform fault (Wilson, 1965) connecting the East Pacific ridge off the coast of Mexico with the Juan de Fuca ridge off the coast of Washington state. In reality, the geometry of transform faults is usually quite complex with many

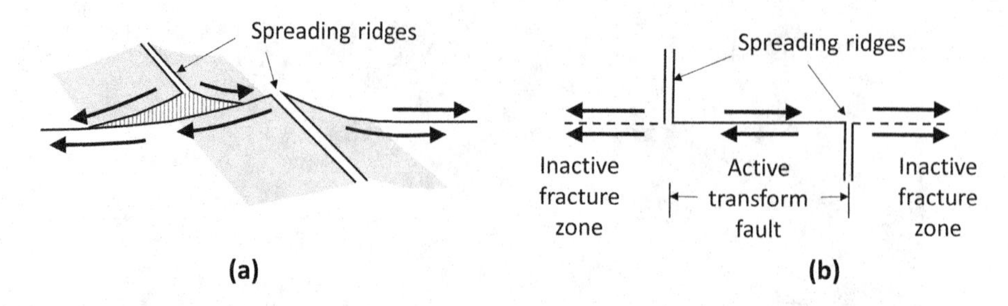

FIGURE 2.16 (a) Oblique and (b) plan views of transform fault and adjacent inactive fracture zones.

bends and kinks, and they are often divided into a number of fault segments. Their depth is typically limited but they can extend horizontally over very long distances. Long transform faults such as San Andreas are capable of producing large magnitude earthquakes up to approximately 8. Evidence of ETS has also been reported along the San Andreas fault (Nadeau and Dolenc, 2004). Other important transform faults include the Motagua fault (which separates the North American and Caribbean plates), the Alpine fault of New Zealand, the North Anatolian fault in northern Türkiye, and the Dead Sea fault system that connects the Red Sea to the Bitlis Mountains of eastern Türkiye (Kearey and Vine, 1990).

Plate tectonics provides a very useful framework for understanding and explaining movements on the Earth's surface and the locations of earthquakes and volcanoes. Plate tectonics accounts for the formation of new and consumption of old crustal materials in terms of the three types of plate movement illustrated in Figure 2.17. It does not, however; explain all observed tectonic seismicity. For example, it is known that intra-slab earthquakes (earthquakes that occur within a plate, away from its edges) have occurred on most continents. Well-known North American examples are the series of midplate earthquakes that occurred in the vicinity of New Madrid, Missouri, in 1811–1812, and the 1886 Charleston (South Carolina) earthquake. The 1976 Tangshan (China), 1993 Marathawada (India), and 2001 Bhuj (India) earthquakes are more recent examples of damaging intra-slab events.

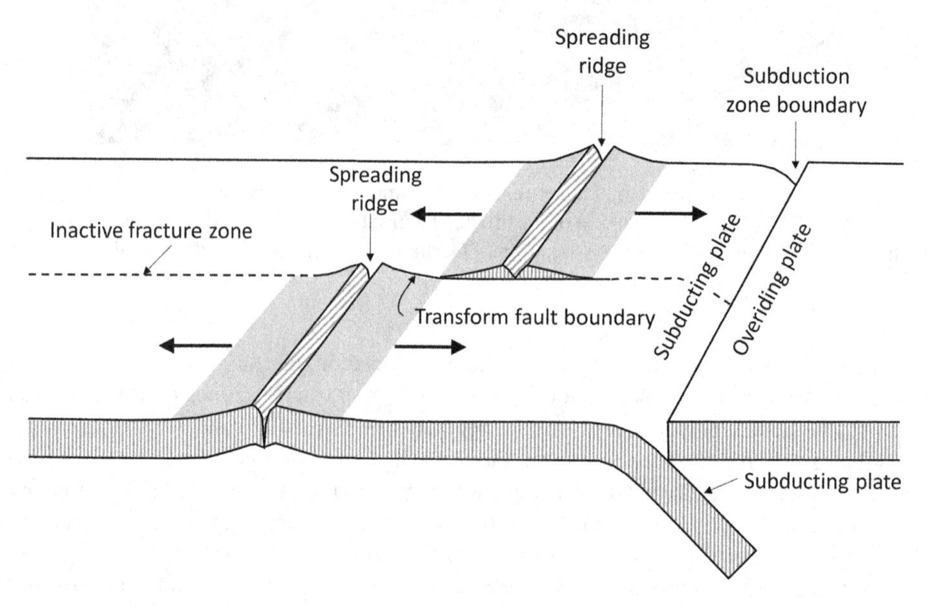

FIGURE 2.17 Interrelationships among spreading ridge, subduction zone, and transform fault plate boundaries.

2.3.3 Tectonic Regimes

For many engineering applications, it is useful to classify regions of the Earth according to the prevailing tectonics, which are referred to as *tectonic regimes*. Leaving aside mid-ocean areas and their spreading ridges, most continental, and near-continental, regions globally can be associated with one of three tectonic regimes (García et al., 2012):

Active Tectonic Regions: These regions are characterized by active seismicity associated with earthquakes at shallow depths (typically < 20 km) in the Earth's crust. Transform faults can occur within active tectonic regions, as can other fault types described in Section 2.4. Active tectonic regions are typically near plate boundaries. Examples include California, portions of Taiwan and Japan, China, New Zealand and portions of southern Europe including Italy, Greece, and Türkiye.

Subduction Zones: Similar to active tectonic regions, these regions also have active seismicity and several such regions have produced large numbers of ground motion recordings. However, subduction zones are distinguished from active tectonic regions in that the dominant mechanism is subduction, typically of an oceanic slab beneath a continental plate. Two source types are considered for subduction zones – interface events between the subducting oceanic slab and overriding continental plate (typical hypocentral depths of 15–40 km) and intra-slab events within the subducting slab (hypocentral depths ~ 40–100 km). Examples of subduction zones include the west coasts of South and Central America, the Pacific Northwest portion of the United States, Alaska south of the Aleutian island chain, eastern Japan, Taiwan, and New Zealand.

Stable Continental Regions: These regions are not near an active plate boundary and have relatively low seismicity rates. Nonetheless, the Earth's crust can be faulted in these regions, with residual stresses acting on those faults from ancient geological processes. As a result, earthquakes can and do occur in stable continental regions, although absent human-induced activity (Section 2.6), they are relatively rare. Due to the relatively low levels of crustal deformation, the shallow crust in stable continental regions can have higher stiffness and lower material damping than in areas near plate boundaries, which affects seismic wave propagation.

2.4 FAULTS

While the theory of plate tectonics generally assigns the relative movement of plates to one of the three preceding types of plate boundaries, examination on a smaller scale reveals that crustal deformations near plate boundaries can be complex. In some regions, plate boundaries are distinct and easily identified, while in others they may be spread out with the edges of the plates broken to form smaller platelets or microplates trapped between the larger plates. Locally, the movement between two portions of the crust will occur on discontinuities within the geologic structure of the crust known as *faults*. As a result, locating faults and characterizing their activity is a critical element of seismic hazard and risk analysis.

Faults may range in length from several meters to hundreds of kilometers and extend from the ground surface to depths of several tens of kilometers. Their presence may be obvious, as reflected in surficial topography, or they may be very difficult to detect. The presence of a fault does not necessarily mean that future earthquakes can be expected – movement can occur aseismically, or the fault may be inactive. Moreover, the lack of observable surficial faulting at a particular location does not ensure that earthquakes cannot originate beneath that location due to the presence of buried faults. The activity of faults is discussed in more detail in Chapter 4 (Section 4.2).

2.4.1 Geometric Notation for Faults and Earthquakes

Standard geologic notation is used to describe the orientation of a fault in space. While the surface of a large fault may be irregular, it can usually be approximated, at least over short distances, as a plane. The orientation of the fault plane is described by its strike and dip. The strike of a fault is the

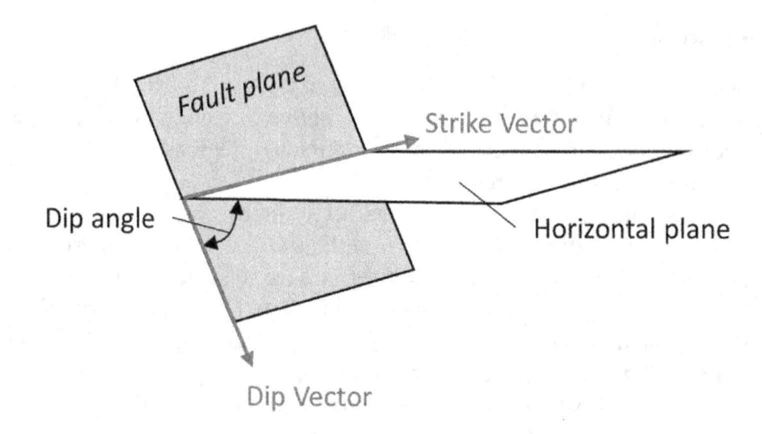

FIGURE 2.18 Geometric notation for description of fault plane orientation.

horizontal line produced by the intersection of the fault plane and a horizontal plane as shown in Figure 2.18. The azimuth of the strike (e.g., N60°E) is used to describe the orientation of the fault with respect to due north. The downward slope of the fault plane is described by the dip angle, which is the angle between the fault plane and the horizontal plane measured perpendicular to the strike. A vertical fault would have a dip angle of 90°. Since the orientation of a line has two possible azimuths (e.g., N60°E identifies the same line as N120°W), a unique strike is defined as the direction of the fault such that the dip is measured to the right (in Figure 2.18, this would be the direction pointing up and to the right) (Aki and Richards 1980).

The geometry of actual faults, however, is not as simple as implied by Figure 2.18. While they may be idealized as planes over relatively short distances, faults have kinks and bends – their strikes and dips may vary to some degree over their lengths. Changes in topography, intersections with other faults, and other features can divide long faults into a series of shorter fault segments that may rupture individually or in groups. These shorter fault segments, as discussed in Chapter 4, will tend to produce smaller earthquakes than would occur if an entire fault ruptures. Figure 2.19 illustrates the San Andreas fault in California, including segments that ruptured in 1857 and 1906. Note that the fault may be idealized as linear over some short segments, but when viewed on a broad scale it has several notable bends.

To describe the location of an earthquake, it is necessary to use accepted descriptive terminology. Earthquakes result from rupture of the rock along a fault, and even though the rupture may involve thousands of square kilometers of fault plane surface, it must begin somewhere. The point at which rupture begins, or *nucleates*, and from which the first seismic waves originate is called the *focus*, or *hypocenter*, of the earthquake (Figure 2.20). From the focus, the rupture spreads across the fault at velocities that are typically on the order of 2.4–3.0 km/s (Somerville et al., 1999). Although fault rupture can extend to the ground surface, the focus is located at some *focal depth* (or *hypocentral depth*) below the ground surface. The point on the ground surface directly above the focus is called the *epicenter*. The distance on the ground surface between an observer or site and the epicenter is known as the *epicentral distance*, and the distance between the observer and the focus is called the *focal distance* or *hypocentral distance*.

2.4.2 Fault Movement

The type of movement, or slip, occurring on a fault is usually reduced to components in the directions of the strike and dip. While some slip in both directions is inevitable, slip in one direction or the other will often predominate.

FIGURE 2.19 San Andreas fault in California showing sections that ruptured in 1857 and 1906.

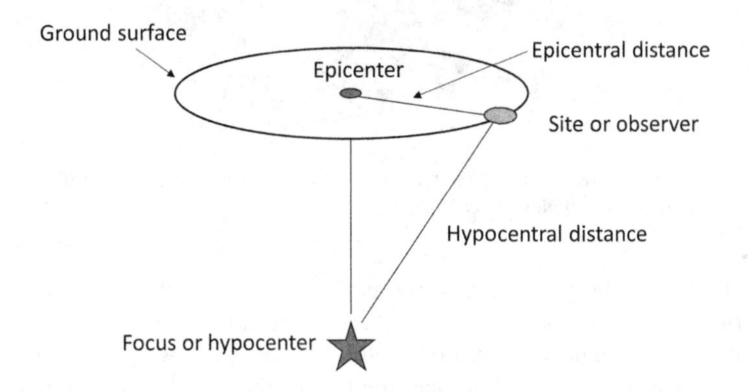

FIGURE 2.20 Notation for the description of earthquake location.

2.4.2.1 Dip-Slip Movement

Fault movement that occurs primarily in the direction of the dip (or perpendicular to the strike) is referred to as *dip-slip* movement. There are different types of dip-slip movements, classified according to the direction of movement and the dip angle of the fault. Normal faults, illustrated in Figure 2.21, occur when the horizontal component of dip-slip movement is extensional and when the material above the inclined fault (referred to as the *hanging wall*) moves downward relative to the material below the fault (the *foot wall*). *Normal faulting* is generally associated with tensile stresses in the crust and results in a horizontal lengthening of the crust. When the horizontal component of dip-slip movement is compressional and the material above the fault moves upward relative to the material below the fault, *reverse faulting* is said to have occurred. Movement on reverse faults, illustrated in Figure 2.22, results in a horizontal shortening of the crust. A special type of reverse fault is a *thrust fault*, which occurs when the fault plane has a small dip angle.

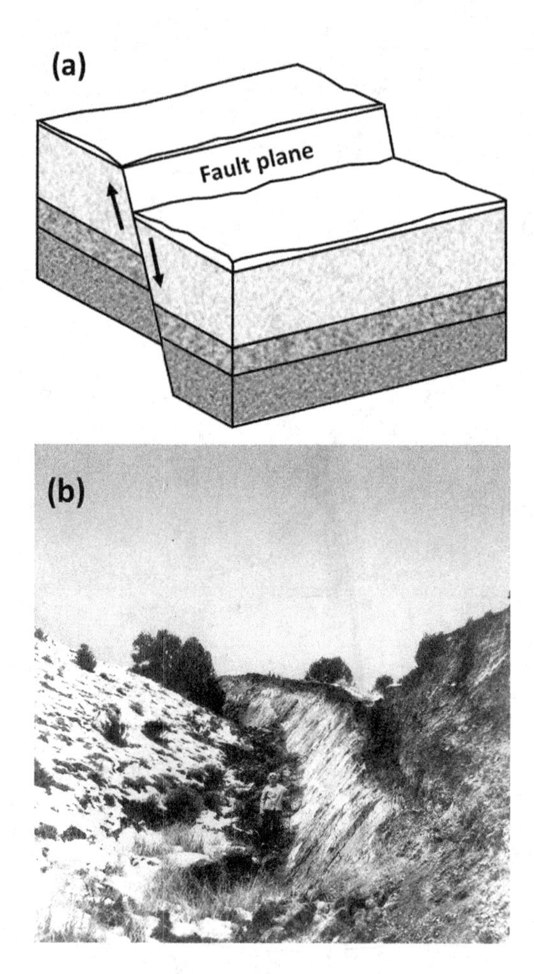

FIGURE 2.21 (a) Normal faulting (After Noson et al., 1988); (b) scarp of the normal fault that produced the 1954 Dixie-Fairview earthquake in Nevada (USGS).

Dip-slip faults do not always rupture to the ground surface as shown in Figure 2.23. When faults produce rupture that does not reach the surface, the fault is referred to as *blind*. For example, the 1994 Northridge earthquake occurred on a *blind reverse fault*, which was buried several km beneath rock or sediments ductile enough that rupture did not extend to the ground surface (Yeats and Huftile, 1995).

2.4.2.2 Strike-Slip Movement

Fault movement occurring parallel to the strike is called *strike-slip* movement. Strike-slip faults are usually nearly vertical. Strike-slip faults are further categorized by the relative direction of movement of the materials on either side of the fault. An observer standing near a right-lateral strike-slip fault would observe the ground on the other side of the fault moving to the right. Similarly, an observer adjacent to a left-lateral strike-slip fault would observe the material on the other side moving to the left. The strike-slip fault shown in Figure 2.24a would be characterized as a left-lateral strike-slip fault. The San Andreas fault in California is an excellent example of right-lateral strike-slip faulting; in the 1906 San Francisco earthquake, several roads and fences north of San Francisco were offset by nearly 6 m.

Individual strike-slip fault systems can have bends in which the fault strike changes along the fault length, or *step-overs* in which the fault is comprised of discontinuous segments. The kinematics

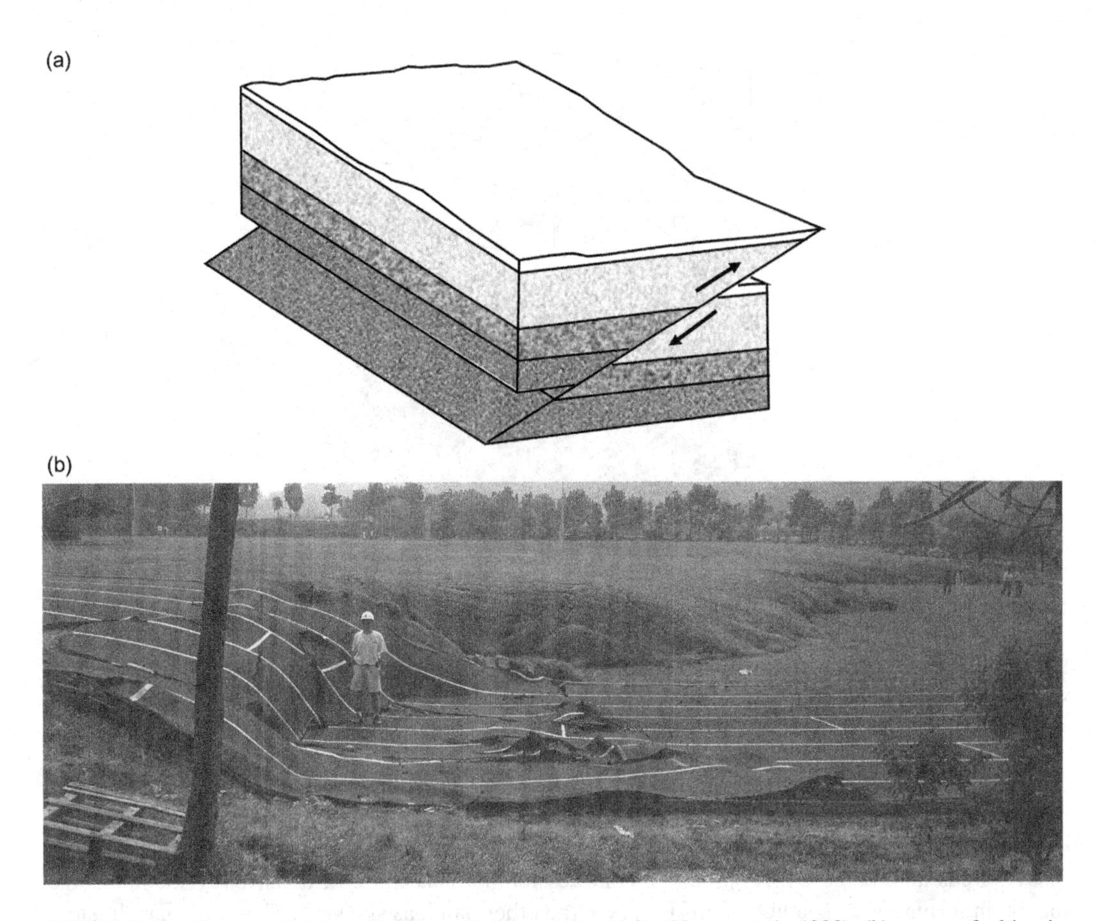

FIGURE 2.22 Reverse faulting; (a) schematic illustration (After Noson et al., 1988), (b) reverse faulting in Wufeng, Taiwan in the 1999 Chi-Chi earthquake (Photo by Steve Kramer).

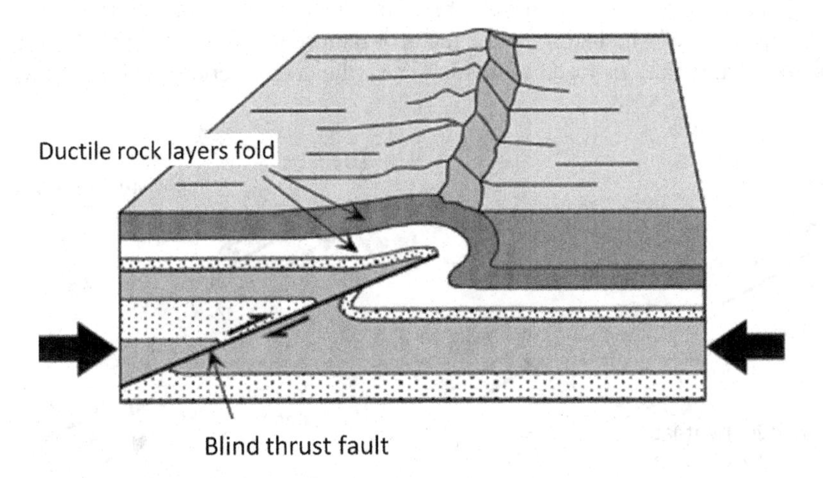

FIGURE 2.23 Blind thrust fault that produces tectonic displacement but not rupture of the ground surface. (USGS.)

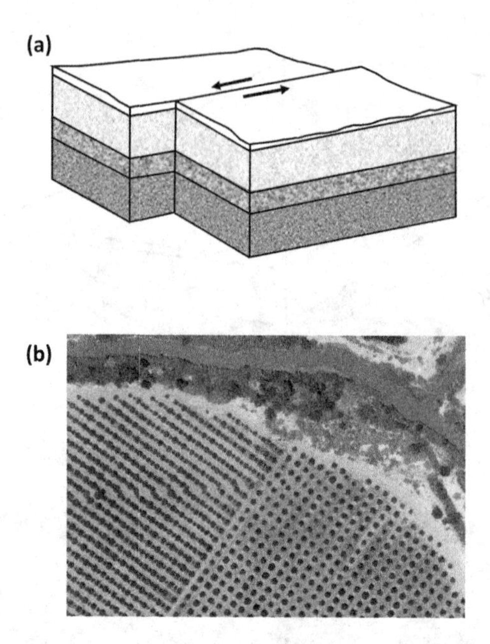

FIGURE 2.24 (a) Left-lateral strike-slip faulting (From Noson et al., 1988); (b) vertical view of trees offset by right-lateral strike-slip faulting through citrus grove in 1940 Imperial Valley earthquake (Photo from Rockwell and Klinger, 2013).

of strike-slip ruptures require that the crust experience local tension or compression in the vicinity of bends or step-overs (Figure 2.25). For example, the left-stepping "big bend" in the right-lateral San Andreas fault (Figure 2.19) produces compression that has caused extensive reverse faults and mountain formation in southern California. On the other hand, as shown in Figure 2.26, right-stepping step-overs in the right-lateral North Anatolian fault (which produced the 1999 **M**7.5 Kocaeli, Türkiye earthquake) produce extension and down-dropped grabens, which control the position of the shoreline of Ismit Bay and produce the depression occupied by Lake Sapanca (Lettis et al., 2002).

In general, fault segments may or may not be connected at depth, which introduces some uncertainty into the question of whether or not the faults or fault segments are likely to rupture individually in separate earthquakes or together in a larger earthquake (Section 4.2.2). Large earthquakes

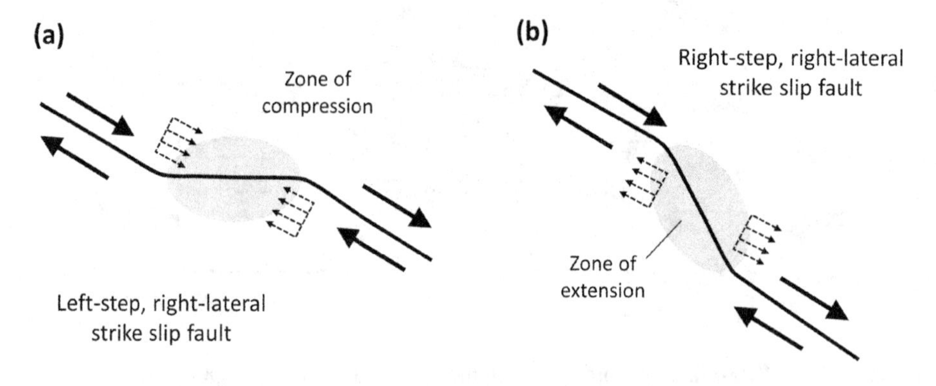

FIGURE 2.25 Crustal stress conditions caused by step-overs: (a) compression caused by left-stepping bend in right-lateral strike-slip fault, and (b) extension caused by right-stepping bend in right-lateral strike-slip fault.

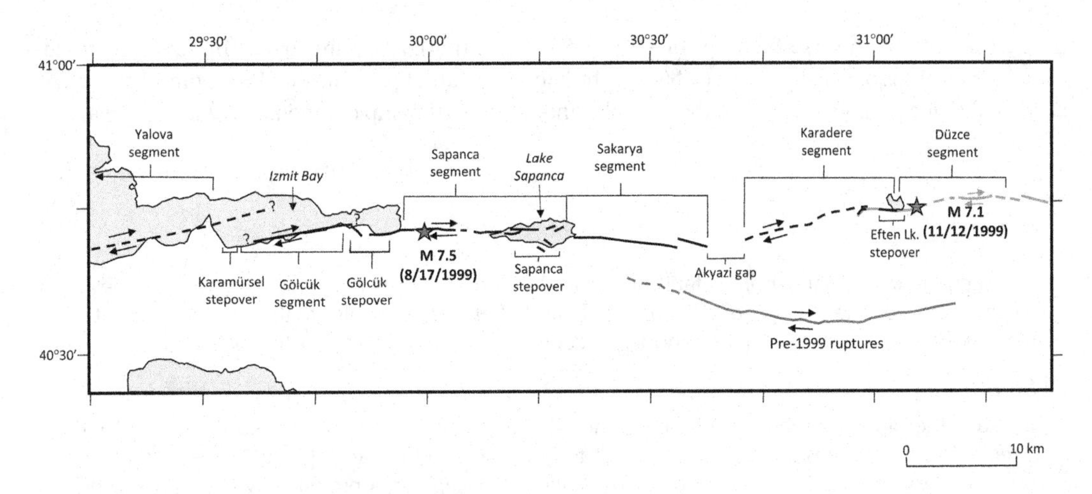

FIGURE 2.26 Portion of North Anatolian fault that ruptured in the 17 August 1999 Kocaeli earthquake. The rupture broke four distinct fault segments (from east to west – Karadere, Sakarya, Sapanca, and Golcuk) with right-stepping step-overs. (After Lettis et al., 2002.)

typically rupture multiple fault segments, but may be arrested by wide step-overs. For example, the Kocaeli earthquake rupture in Figure 2.26 occurred over four segments and through several gaps and step-overs, but was arrested by step-overs at the east and west ends. Lettis et al. (2002) found that 4–5 km wide step-overs have generally arrested strike-slip fault rupture but that smaller step-overs could be ruptured by large fault movements (approximately 1–2 km step-over width per 1 m fault displacement).

2.4.2.3 Oblique Movement

Most fault movement is neither purely strike-slip nor dip-slip, but has components of both. When both components of slip are significant, the fault movement is referred to as *oblique*. The direction of slip can be quantified using the *rake angle* illustrated in Figure 2.27. Rake is defined as the counterclockwise angle within the fault plane from a line drawn through the hypocenter in the direction of the strike to the slip direction. By convention, rake is defined as slip (movement) of the hanging wall relative to the foot wall. An observer located at the hypocenter looking in the strike direction (footwall to the left) would measure the rake angle counter clockwise relative to the strike direction (Aki and Richards, 1980), as illustrated in Figure 2.27. With this convention, pure dip-slip would

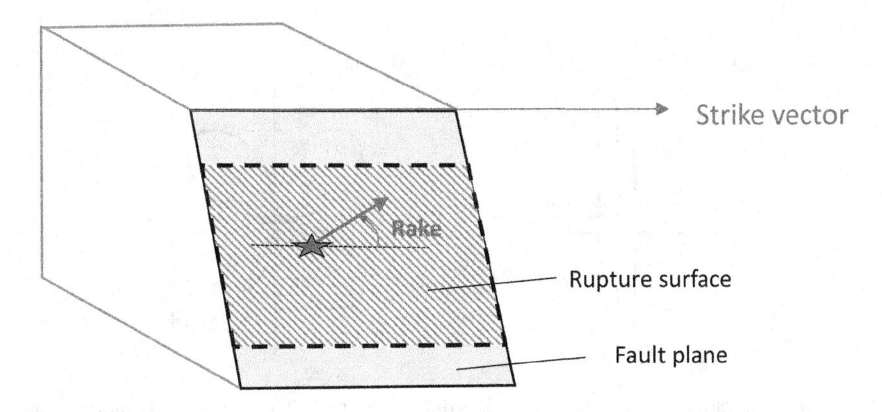

FIGURE 2.27 Planar description of earthquake fault rupture showing slip direction and rake angle. (After Aki and Richards, 1980.)

have a rake of 90° (reverse faulting) or –90° (normal faulting), and pure strike-slip faulting would have a rake of zero (left-lateral) or 180° (right-lateral). Fault slip from the 1989 Loma Prieta CA earthquake, for example, had both thrust and right-lateral strike-slip components, with a rake angle of 140°.

Example 2.1

The segment of the San Andreas fault near San Bernardino, CA is depicted by a dashed line in (Figure 2.19). For this segment of the fault, the fault dips at variable angles to the northeast. What is the strike angle? If a right-lateral earthquake occurs at this location, what is the rake angle?

Solution:

Because the dip of the fault is to the northeast, the hanging wall is to the northeast, and the direction of the strike is the angle from north to the fault with the hanging wall to the right of the observer. From Figure 2.19 at the indicated location, this angle is approximately 80 degrees west of north (written as N80°W). Because the San Andreas fault is right-lateral, the rake angle of a pure strike-slip event is 180 degrees.

2.4.2.4 Moment Tensor Solution

The nature of fault movement at the hypocenter in a specific earthquake can be determined by examining the first-arriving p-waves at a series of seismographs. Analytical solutions for the p-waves radiating away from a source idealized as a *double-couple* (Figure 2.28a) show lobes of compression and expansion oriented at 90° to each other and 45° to the axes of the double-couple (Figure 2.28b). Thus, the first arriving p-waves at some instruments will be in compression while those at other stations will be in dilatation or expansion. By comparing the locations of a three-dimensional network of stations measuring compressive and dilatational first arrivals, seismologists can determine the orientation of the planes that separate the different lobes (one of which is the fault plane) and thereby determine the nature of faulting that produced the earthquake. These planar orientations, in combination with the hypocentral location (latitude, longitude, and depth) comprise the *moment tensor solution* for an earthquake.

The three-dimensional orientation of the compressive and dilatational lobes can be plotted on a lower hemisphere stereographic projection. In such *moment tensor plots*, directions of initial compression are shown in white and initial dilatation in gray or black. The resulting diagram, sometimes referred to as a *beach ball plot*, indicates the mechanism of faulting. Figure 2.29 shows a series of beach ball plots for idealized faulting (pure strike-slip, normal, and reverse) and for one case of oblique faulting (mostly reverse with some strike-slip). Figure 2.30 shows focal mechanisms from

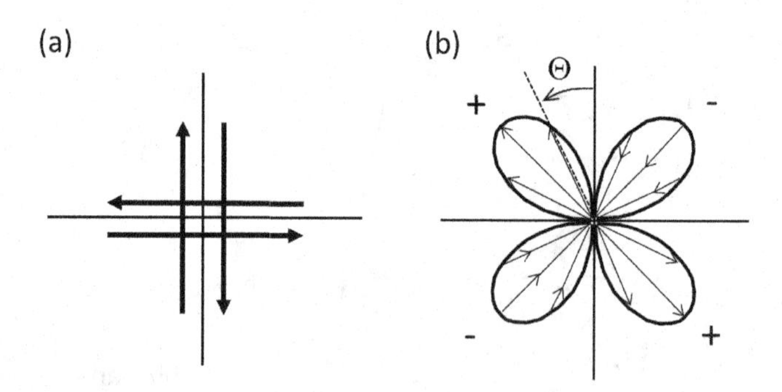

FIGURE 2.28 p-wave radiation: (a) double-couple, (b) p-wave amplitudes and particle motion (indicated by arrows within lobes). Positive lobe indicates initial tension, negative lobe initial compression. (After Stein and Wysession, 2003; used with permission of Springer Nature BV.)

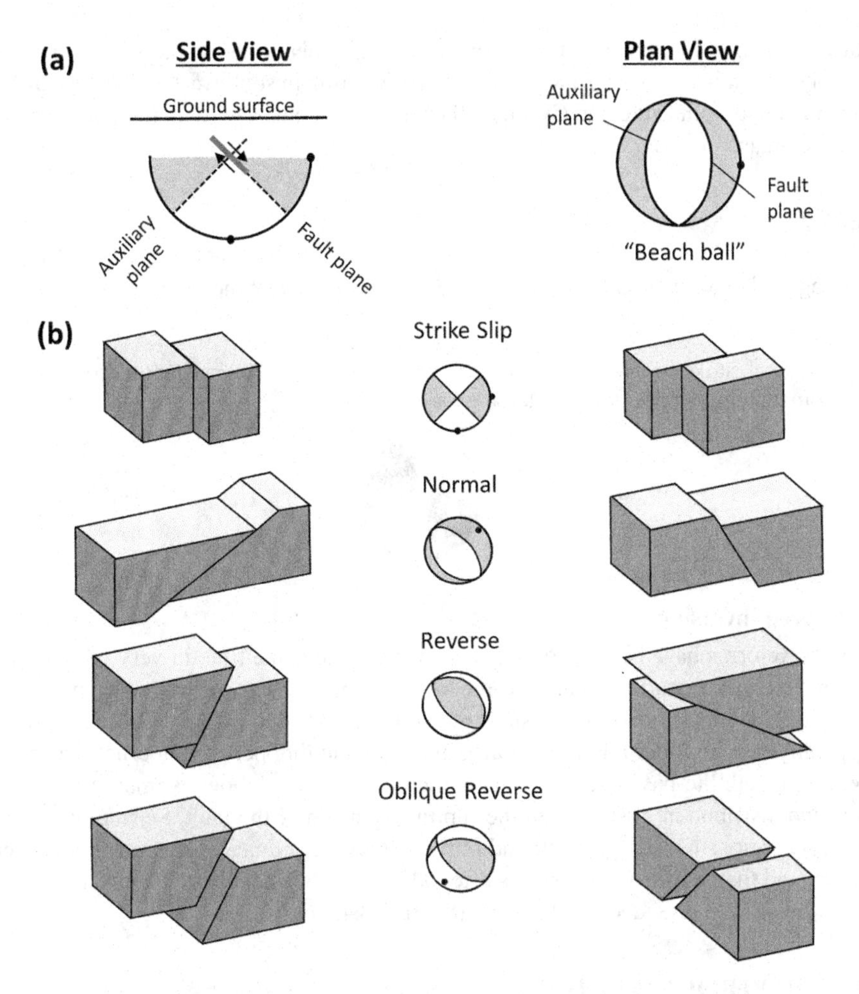

FIGURE 2.29 (a) Side and plan views of moment tensor, (b) moment tensors for different types of faulting; left and right diagrams indicate alternate fault configurations associated with the moment tensor in the center.

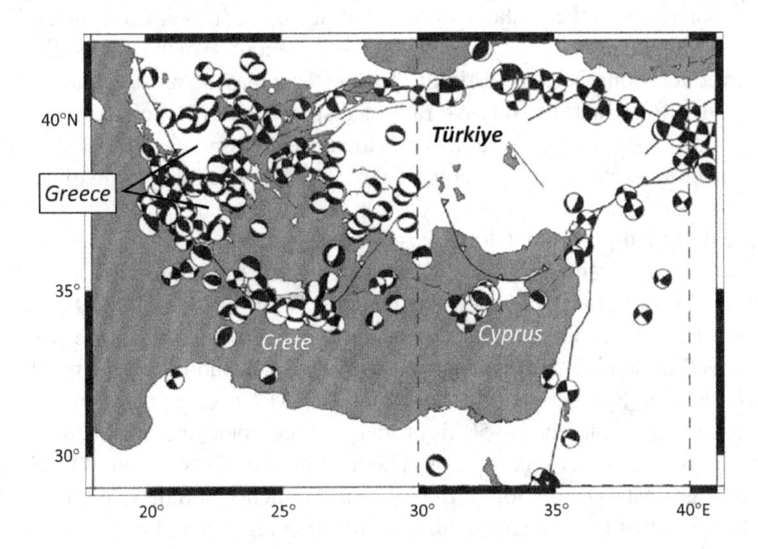

FIGURE 2.30 Moment tensors indicating focal mechanisms in eastern Mediterranean region. (Modified from Dwivedi and Hayashi 2010 and McClusky et al., 2000; used with permission of Springer Nature BV.)

a series of earthquakes in the eastern Mediterranean region. Note the strike-slip motion along the North Anatolian fault in northern Türkiye, the normal faulting in southwestern Türkiye and Greece, and the reverse faulting in Crete and Cyprus; all of these are consistent with the known plate movements in these areas.

Example 2.2

Consider again the event described in Example 2.1. Draw the event moment tensor.

Solution:

For a right-lateral, strike-slip earthquake with a strike of N80°W, the moment tensor is as shown below. Moment tensors provide two alternate fault strikes

2.4.2.5 Source Inversion

After entire waveforms have been recorded, which may take some time in very large earthquakes and at distant recording stations, multiple recordings can be used to back-calculate the rupture process over a fault surface that best fits a large suite of recorded ground motions. These source models typically also consider additional data sources, including permanent crustal displacements as inferred from satellite-based sensors. Such source inversions allow estimation of the rupture dimensions and distribution of slip along the ruptured portion of the fault. Seismological laboratories in different regions usually perform and post (online) preliminary source inversions very soon after shaking, and then refine the inversions after additional data acquisition and vetting. Examples of source inversions are provided in Chapter 3 (Section 3.4.1.1).

2.5 ELASTIC REBOUND THEORY

The plates of the Earth are in constant motion, and plate tectonics indicates that the majority of the distortion associated with their relative movement occurs near their boundaries. The long-term effects of this movement can be observed in the geologic record, which reflects displacements that have occurred over very long periods of time. These effects may be most apparent in the immediate vicinity of faults, for which the relative rate of displacement of one side relative to the other is referred to as a *slip rate* having the units of velocity (e.g., mm/year). High slip rate faults will produce more pronounced effects in the geological record, due to a higher rate of earthquakes, than low slip rate faults.

With the advent of high-precision Global Positioning Systems (GPS), displacements can also be observed over very short, even continuous, time scales with resolution down to millimeter scale. Arrays of GPS receivers (typically spaced on the order of 10 km apart) can be used to measure the direction and rate of crustal displacements during quiescent periods between earthquakes (Figure 2.31) as well as sudden displacements caused by an earthquake (Figures 2.21, 2.22, 2.24, and 4.3). Interferometric Synthetic Aperture Radar (InSAR) uses changes in radar wave phase between different times (typically 10–50 days apart) to compute the spatial distribution of displacements during the period between images. The high temporal resolution of GPS measurements complements the high spatial resolution of InSAR measurements, and they can be combined to elucidate the mechanisms that lead to earthquakes in different regions and the slip rates across faults. By combining measured surface deformations with numerical models of crustal plates, insight into subsurface deformations, even along the fault itself, can be gained.

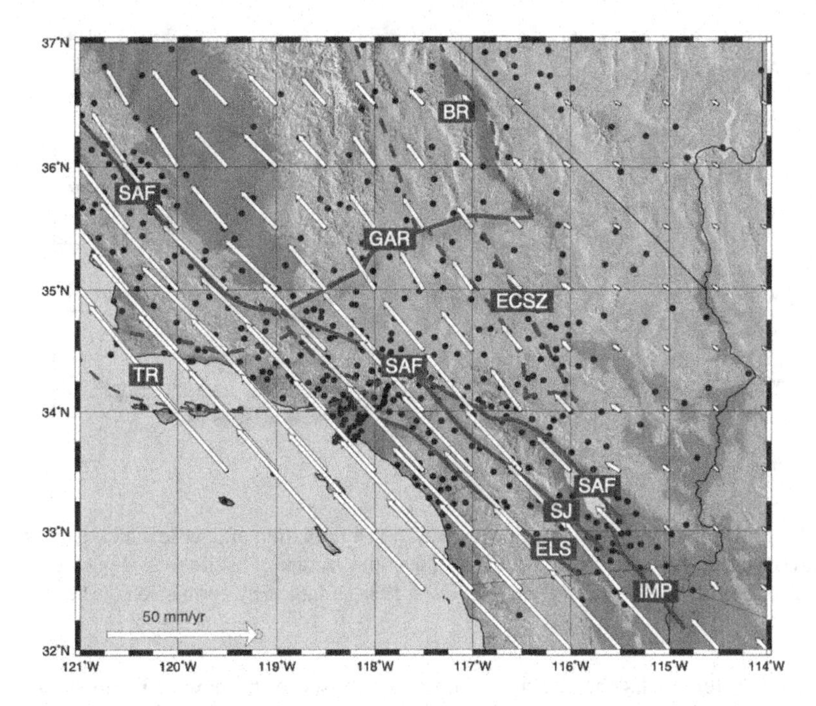

FIGURE 2.31 Crustal velocities of southern California relative to reference regions on North American plate located well east of plate boundary at the San Andreas fault (SAF). Note the decrease of velocities as the observer moves from southwest to northeast across the SAF, with maximum velocities being about 50 mm/year in coastal areas. The differences in velocities across the SAF are the approximate fault slip rates. Velocity vectors are interpolated from measured velocities at Global Navigation Satellite System stations (black dots) at 0.5° resolution using interpolation algorithm of Shen et al. (2015). TR, Transverse Range; GAR, Garlock Fault; SJ, San Jacinto Fault; ELS, Elsinore Fault; IMP, Imperial Fault; ECSZ, Eastern California Shear Zone; BR, Basin and Range. (Figure provided by UNAVCO, courtesy of David Mencin.)

As relative movement of the plates occurs, elastic strain energy is stored in the materials near the boundary as shear stresses increase on the fault surfaces that separate the plates. When the shear stress reaches the shear strength of the rock along the fault, the rock fails at a *nucleation point* and the accumulated strain energy is released. The effects of the failure depend on the nature of the rock along the fault. If it is weak and ductile, what little strain energy that could be stored will be released relatively slowly and the movement will occur aseismically. If, on the other hand, the rock is strong and brittle, the failure will be rapid. Rupture of the rock will release the stored energy explosively, partly in the form of heat and partly in the form of the stress waves that produce earthquake ground motions. The theory of *elastic rebound* (Reid, 1911) describes this process of the successive buildup and release of strain energy in the rock adjacent to faults. It is often illustrated as shown in Figure 2.32.

The nature of the buildup and release of stress is of interest. Faults are not uniform, either geometrically or in terms of material properties – both strong and weak zones can exist over the surface of a fault. The stronger zones, referred to as *asperities* by some (Kanamori and Stewart, 1978) and *barriers* by others (Aki, 1979), are particularly important. The asperity model of fault rupture assumes that the shear stresses prior to an earthquake are not uniform across the fault because of stress release in the weaker zones by creep or foreshocks. Release of the remaining stresses held by the asperities produces the main earthquake that leaves the rupture surface in a state of uniform stress. In the barrier model, the pre-earthquake stresses on the fault are assumed to be uniform. When the main earthquake occurs, stresses are released from all parts of the fault except for the

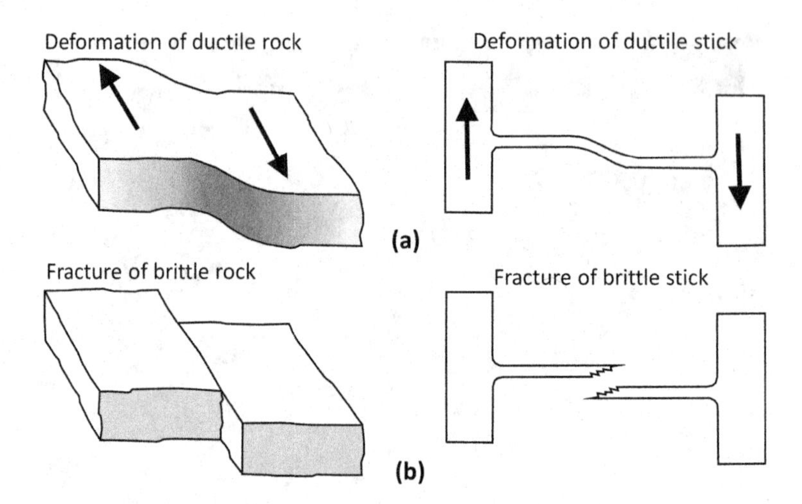

FIGURE 2.32 Elastic rebound theory of earthquakes. (a) Deformation of ductile material (rock or stick) due to relative movements of plates on the right and left sides, (b) Fracture of brittle rock along fault or brittle stick in bending. (After Foster, R.J., *General Geology*, 5/e, © 1988. Adapted by permission of Prentice Hall, Upper Saddle River, NJ.)

stronger barriers; aftershocks then occur as the rock adjusts to the new uniform stress field. Since both foreshocks and aftershocks are commonly observed, it appears that some strong zones behave as asperities and others as barriers (Aki, 1984). The engineering significance of asperities and barriers lies in their influence on ground-shaking characteristics close to the fault. A site located close to one of these strong zones may experience stronger shaking than a site equally close to the fault but further from a strong zone. At larger distances from the fault the effects of fault nonuniformity decrease. Unfortunately, methods for locating asperities or barriers prior to rupture have not yet been developed.

Example 2.3

Consider again the section of the San Andreas fault from Example 2.1. What is the approximate slip rate across the fault?

Solution:

The crustal velocities southwest and northeast of the fault are approximately 50 mm/year and 20 mm/year, respectively. The slip rate of the intermediate fault structures is the difference, or approximately 30 mm/year.

2.5.1 Relationship to Earthquake Recurrence

The theory of elastic rebound implies that the occurrence of earthquakes will relieve stresses along the portion of a fault on which rupture occurs, and that subsequent rupture will not occur on that segment until high stresses have had time to build up again. The chances of an earthquake occurring on a particular fault segment should therefore be related in some way to the time that has elapsed since the last earthquake and, perhaps, to the amount of energy that was released in that earthquake. In a probabilistic sense, then, individual earthquakes on a particular fault segment should not be considered as random, independent events in time. This characteristic is important in the prediction of future ground motions, as undertaken in seismic hazard analyses (Chapter 4).

Because earthquakes relieve the strain energy that builds up on faults, they should be more likely to occur in areas where little or no seismic activity has been observed for some time. By plotting

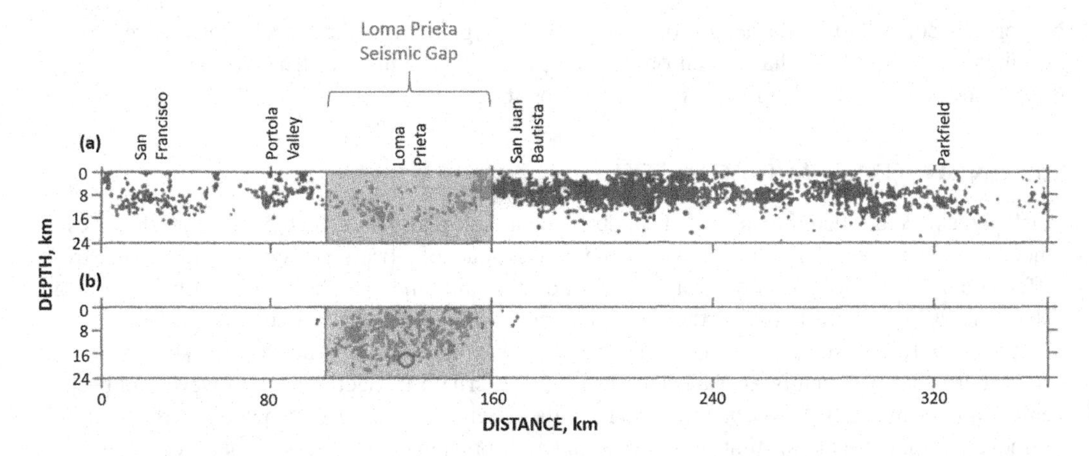

FIGURE 2.33 Cross section of the San Andreas fault from north of San Francisco to south of Parkfield: (a) seismicity in the 20 years prior to the 1989 Loma Prieta earthquake is shown with the Loma Prieta gap highlighted; (b) main shock (open circle) and aftershocks of the Loma Prieta earthquake. Note the remaining gaps between San Francisco and Portola Valley and south of Parkfield. (After Plafker and Galloway, 1989.)

fault movement and historical earthquake activity along a fault, it is possible to identify gaps in seismic activity at certain locations along faults. According to elastic rebound theory, either the movement is occurring aseismically or strain energy is building in the vicinity of these seismic gaps. In areas where the latter is known to be the case, seismic gaps should represent the most likely locations for future earthquakes (McCann et al., 1979). A number of seismic gaps have been identified around the world and large earthquakes have subsequently been observed on several of them. The 1989 Loma Prieta earthquake occurred on a fault within the San Andreas fault system (Zayante Fault) that had previously been identified as a gap, as shown in Figure 2.33. As described further in Chapter 4, knowledge of seismic gaps is used in the analysis of probabilities of future large earthquakes.

2.5.2 Relationship to Tectonic Environment

Elastic rebound also implies that tectonic environments capable of storing different amounts of energy will produce earthquakes of different sizes. Consider, for example, the tectonic environment in the vicinity of a spreading ridge plate boundary. First, the crust is thin; hence the volume of rock in which strain energy can build up is small. Second, the horizontal component of the relative plate movement is extensional; hence the normal stress on the fault plane, and with it the rupture strength, is low. Third, the rock is relatively warm and ductile, so it will not release strain energy as suddenly as in other regions. Taken together, these factors limit the total strain energy that can build up and be suddenly released at a spreading ridge boundary. These factors explain the observed absence of very large earthquakes at spreading ridge boundaries.

By the time the oceanic crust has moved from a spreading ridge to a subduction zone, it has cooled and become much thicker and stronger. Relative movement of the plates is toward each other, so high compressive normal stresses increase the rupture strength on the fault plane. Because subduction zone plate boundaries are inclined, the potential rupture area is large. All of these factors support the potential buildup of very large amounts of strain energy that, when suddenly released, can produce great earthquakes. In fact, the largest recorded earthquakes have been produced on subduction zone interfaces.

At transform faults, the rock is generally cool and brittle, but large compressive stresses do not usually develop because the faults are often nearly vertical and movement is typically of a strike-slip nature. Because the depth of transform faulting is limited, the total amount of strain energy that can be stored is controlled by the length of rupture. Very large earthquakes involving rupture lengths of hundreds of kilometers have been observed on transform faults, but truly "great" earthquakes (magnitudes of about 8.5 or greater) may not be possible.

2.6 INDUCED SEISMIC ACTIVITY

Earthquakes occur when the applied shear stress on a fault plane exceeds the available shear resistance. This shear resistance, in turn, is related to the product of a friction coefficient μ and the stress difference $\sigma_n - u$, where σ_n is the normal stress on the fault and u is the pore fluid pressure (e.g., Ellsworth, 2013). Accordingly, earthquakes can occur when applied shear stresses exceed a nominal (presumably, time-invariant) shear resistance – this is the case with the earthquakes of tectonic origin that occur principally at plate boundaries as described in Section 2.5. However, anthropogenic (human-induced) stress changes also have the potential to cause earthquakes in this manner. Moreover, a condition of constant shear stress, but anthropogenic reductions of shear resistance, can also induce shear failure and earthquakes.

2.6.1 Earthquakes Induced by Fluid Injection in Wells

A number of industrial processes, typically related to oil and gas extraction, involve fluid injection from wells into the Earth's crust (Rubinstein and Mahani, 2015):

1. Hydraulic fracturing (commonly known as *fracking*), which involves injection of fluids at high pressure so as to fracture rock, increasing fluid pathways to the well.
2. Wastewater disposal, which involves injecting waste into deep underground rock formations of high permeability. This wastewater has a number of sources, including water recovered during oil/gas extraction and spent hydraulic fracture fluid.
3. Enhanced oil recovery, which involves flooding an oil/gas-bearing reservoir with water, steam, or carbon dioxide to increase yield. This process is differentiated from fracking in that reservoir fluid pressures are not increased markedly.

Fluid injection has the potential to induce earthquakes principally by increasing fluid pressure u, which can decrease the shear resistance on a fault plane to the point that it equals crustal shear stresses, thus inducing rupture. This mechanism is not confined to areas immediately adjacent to wells, but can extend up to 10–20 km horizontally from the well and several km deeper (Rubinstein and Mahani, 2015). Most injection wells do not induce earthquakes that produce felt ground motions; the requisite conditions include the presence of a stressed fault of sufficient size in the vicinity of the well and the presence of fluid pathways from the injection point to the fault. The fluid itself need not travel from the well to the fault to induce rupture, the pressure increase from fluid injection can be sufficient.

The first earthquakes to be identified as associated with fluid injection were caused by chemical wastewater recovery at the Rocky Mountain Arsenal in the 1960s, producing events as large as **M** 4.9 (Herrmann et al., 1981). Since 2009, portions of the central and eastern US have experienced dramatically increased rates of earthquakes with **M** > 3. As shown in Figure 2.34, tectonic earthquakes, which dominate the pre-2009 data set, are broadly distributed across the region. In contrast, the strong increases in seismicity since 2009 are concentrated in central Oklahoma, southern Kansas, central Arkansas, southeastern Colorado and northeastern New Mexico, and multiple parts of Texas. Some particularly large events in this sequence include the 2011 **M**5.3 Trinidad Colorado earthquake, the 2011 **M**5.6 Prague Oklahoma earthquake, and the 2016 **M**5.8 Pawnee Oklahoma earthquake. Most of the increased seismic activity since 2009 can be traced to increased wastewater disposal since 2009 (e.g., Ellsworth, 2013; Weingarten et al., 2015), with fracking and enhanced oil recovery being relatively minor contributors.

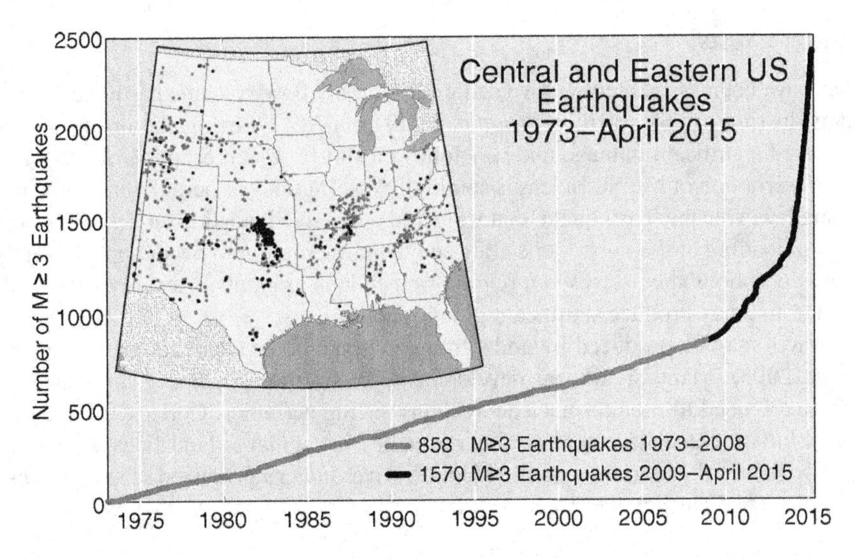

FIGURE 2.34 Cumulative number of $M > 3$ earthquakes in the central and eastern US since 1973, showing rapid increase in rate since 2009. (Rubinstein and Mahani, 2015.)

Relative to tectonic earthquakes, ground motions from induced earthquakes have been observed to have different ground motion characteristics for the same magnitude and distance in some regions (Bommer et al., 2022) and to be indistinguishable in others (Douglas et al., 2013; Atkinson, 2020). There is wide variability in results due to regional issues related to source depth and the stress change on faults from the induced earthquakes. Understanding ground motions from induced earthquakes remains an area of active scientific inquiry.

2.6.2 Reservoir-Induced Earthquakes

Reservoir-induced earthquakes have been the subject of considerable study and some controversy. Local seismicity increased significantly after the filling of Lake Mead behind Hoover Dam on the Nevada-Arizona border in 1935. When the Koyna Dam (India) reservoir was filled, local shallow earthquakes became common in an area previously thought to have been virtually aseismic. In 1967, five years after filling of the Koyna reservoir had begun, a **M**6.5 earthquake killed 177 persons and injured more than 2,000 more. Local seismicity has been observed to increase seasonally with seasonal increases in reservoir level. In 1975, seven years after the filling of Oroville Dam in an area of low historical seismicity in northern California, a swarm of earthquakes culminated in a **M**5.7 main shock. In 1981, after the construction of the High Dam, a **M**5.6 earthquake occurred in Aswan, Egypt where very little significant seismic activity had been observed in the 3,000-year recorded history of the area.

Simpson et al. (1988) found that the patterns of reservoir-induced earthquakes at seven major dams either tended to involve almost immediate increases in seismicity upon reservoir filling or delayed seismicity that appeared after several filing cycles had occurred. In the cases of nearly immediate seismicity, the earthquakes tend to occur near the reservoir and appear to be associated with the complex interaction of stress increase from the reservoir weight and increased pore pressure from elastic compression of pore space. Hence, there is potential for both increased shear demands on faults and decreased shear resistance. In cases of delayed seismicity, earthquakes can occur at greater distance and appear to be related to pore pressure increase. Hence, the principal mechanism is reduced shear resistance on faults, which require a hydraulic connection to the reservoir area (as with well injection, the water itself need not travel to the fault, only pore pressure increase is required).

2.6.3 Other Causes

Earthquakes have been associated with volcanic activity. Shallow volcanic earthquakes may result from sudden shifting or movement of magma. In 1975, a **M**7.2 earthquake on the big island of Hawaii produced significant damage and was followed shortly by an eruption of the Kilauea volcano. The 1980 eruption of Mt. St. Helens is southern Washington was actually triggered by a small ($M_s = 5.1$), shallow, volcanic earthquake that triggered a massive landslide on the north slope of the volcano. The unloading of the north slope allowed the main eruption to occur approximately 30 sec later. Volcanic eruptions themselves can release tremendous amounts of energy essentially at the Earth's surface and may produce significant ground motion.

Seismic waves may be produced by underground detonation of chemical explosives or nuclear devices (Bolt, 2006). Many significant developments in seismology during the Cold War years stemmed from the need to monitor nuclear weapons testing activities. Collapse of mine or cavern roofs, or mine bursts, can cause small local earthquakes, as can large landslides. A 1974 landslide involving 1.6×10^9 m^3 of material along the Montaro River in Peru produced seismic waves equivalent to those of a **M**4.5 earthquake (Bolt, 2006).

2.7 LOCATION OF EARTHQUAKES

As described in Section 2.4.1, an earthquake source can be partially located in space through its hypocenter, which has a particular location on the Earth's surface (the epicenter) and hypocentral depth. This section introduces the principles involved in locating seismic sources through the relatively simple case of locating the epicenter. A preliminary epicenter location can be obtained using relative arrival times of p- and s-waves at a set of at least three seismographs (Figure 2.35).

Since p-waves travel faster than s-waves, they will arrive first at a given seismograph. The difference in arrival times will depend on the difference between the p- and s-wave velocities, and on the distance between the seismograph and the focus of the earthquake (R_{hyp}), according to

$$R_{hyp} = \frac{\Delta t_{p-s}}{1/V_s - 1/V_p} \tag{2.1}$$

where Δt_{p-s} is the difference in time between the first p- and s-wave arrivals, and V_p and V_s are the p- and s-wave velocities, respectively. In bedrock, p-wave velocities are generally 3–8 km/s and s-wave velocities range from 2 to 5 km/s. At any single seismograph, it is possible to determine

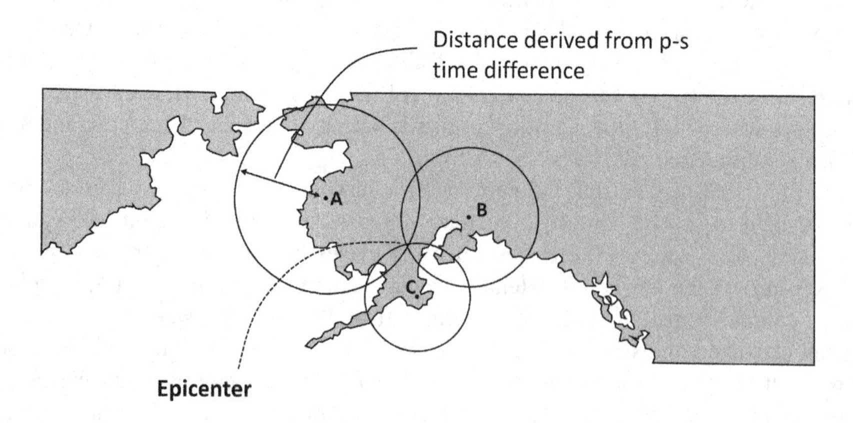

FIGURE 2.35 Preliminary location of epicenter from differential wave-arrival-time measurements at seismographs A, B, and C. Most likely epicentral location is at the intersection of the three circles. (After Foster, R.J., General Geology, 5/e, ® 1988. Adapted by permission of Prentice Hall, Upper Saddle River, NJ.)

the hypocentral distance but not the hypocentral direction. This limited knowledge is expressed graphically by plotting a circle with a radius equal to R_{hyp} as shown in Figure 2.35 (this plotting is approximate because the circle represents distance in the horizontal plane, or epicentral distance R_{epi}, whereas what is strictly computed from Equation (2.1) is hypocentral distance – the error is small when $R_{hyp} \gg$ focal depth, which is often the case). When the distance from a second seismograph is plotted as a circle about its location, the possible location of the epicenter is narrowed to the two points of intersection of the circles. A third seismograph is necessary to identify the most likely location of the epicenter, as illustrated in Figure 2.35.

More refined procedures are required to estimate hypocentral depth in addition to epicentral location. Such refinements are possible using multiple seismographs, a three-dimensional seismic velocity model of the Earth, and numerical optimization techniques. The accuracy of these techniques depends on the number, quality, and geographic distribution of the seismographs and on the accuracy of the seismic velocity model (Dewey, 1979).

2.8 SIZE OF EARTHQUAKES

The "size" of an earthquake is obviously a very important parameter, which has been described in different ways. Prior to the development of modern instrumentation, methods for characterizing the size of earthquakes were based on qualitative descriptions of the effects of earthquakes. More recently, modern seismographs have allowed the development of a number of quantitative measures of earthquake size. Since several of these measures are commonly used in both seismology and earthquake engineering, the distinguishing features of each should be understood.

2.8.1 Seismic Moment

The concept of elastic rebound theory can be used to develop a useful measure of the size of an earthquake. The *seismic moment* of an earthquake is given by (Aki, 1966):

$$M_o = \mu A \bar{D} \tag{2.2}$$

where μ is the shear modulus of the material along the fault (taken as 3×10^{11} dyne/cm^2; 1 dyne/cm^2 = 0.1 Pa), A the rupture area, and $\bar{D}$ the average amount of slip. The seismic moment is named for its units of force times length; however, it is more a measure of the work done by an earthquake at the source. As such, the seismic moment correlates well with energy release. Note that the seismic moment increases indefinitely in proportion to the area of the rupture surface – this characteristic makes it particularly useful for describing very large earthquakes, which may have rupture lengths of many hundreds of kilometers. The seismic moment can be estimated from geologic records for historical earthquakes, or obtained from the long-period components of a seismogram (e.g., Bullen and Bolt, 1985).

2.8.2 Earthquake Magnitude

The possibility of obtaining an objective, quantitative measure of the size of an earthquake came about with the development of instrumentation for measuring earthquake ground motions. In the past 80 years, the ability of seismic instruments to measure earthquake ground motions over a wide frequency range has increased dramatically. Various quantitative metrics of earthquake size, corresponding to alternate magnitude scales, can be computed using these ground motion data. Ground motion measurements are discussed in Section 3.2 – this section focuses on the use of this and other data to compute magnitude.

Many different magnitude scales have developed over the years as seismic instruments and earthquake knowledge have improved. Until the mid-1970s, magnitudes were typically derived directly from recorded ground motions. The *local* (popularly, "Richter") *magnitude*, M_L (Richter, 1935), *surface wave magnitude*, M_s (Gutenberg and Richter, 1936), *body wave magnitude*, m_b (Gutenberg, 1945), and *Japanese Meteorological Agency magnitude*, M_{JMA} (Tsuboi, 1954) are examples of such *instrumental magnitude scales*.

As the total amount of energy released during an earthquake increases, however, the ground-shaking characteristics on which instrumental magnitude scales are based do not necessarily increase at the same rate because the energy released at large distances from the recording instrument contributes little to the ground motion amplitudes that are used in those scales. As a result, for very large earthquakes that involve rupture over large fault areas, the measured level of ground-shaking at a given location does not continue to increase with increasing size of the earthquake. This phenomenon is referred to as *magnitude saturation*; the body wave and local magnitudes saturate at magnitudes of 6–7 and the surface wave magnitude saturates at about $M_s = 8$. To describe the size of very large earthquakes, a magnitude scale that does not depend on ground-shaking levels, and consequently does not saturate, is desirable. The only magnitude scale that is not subject to saturation is the moment magnitude since it is based on the seismic moment (Section 2.8.1), which is directly related to the total energy release on the fault. Moment magnitude is given by (Kanamori, 1977; Hanks and Kanamori, 1979):

$$\mathbf{M} = \frac{\log_{10} M_o}{1.5} - 10.7 \qquad (2.3)$$

where M_o is the seismic moment (Section 2.8.1) in dyne-cm (10^7 dyne-cm = 1.0 N-m). Readers should note that moment magnitude is referred to as M_w in some older literature, but the boldface **M** will be used in this text.

The relationships between the various magnitude scales can be seen in Figure 2.36. Saturation of the instrumental scales is indicated by their flattening at large magnitudes. As an example of the effects of magnitude saturation, both the 1906 San Francisco and 1960 Chile earthquakes produced ground shaking that led to surface wave magnitudes of 8.3, even though the sizes of their rupture surfaces, illustrated by the shaded areas in Figure 2.37, were vastly different. The great disparity in energy release was, however, reflected in the moment magnitudes: 7.9 for San Francisco (using the seismic moment from Ben-Menahem, 1978) and 9.5–9.6 for Chile (based on moments from Kanamori, 1977 and Cifuentes and Silver, 1989). Moment magnitude is currently the widely accepted standard for describing the size of earthquakes, such that if the term "magnitude" is used in modern earthquake engineering literature, it is assumed to be moment magnitude.

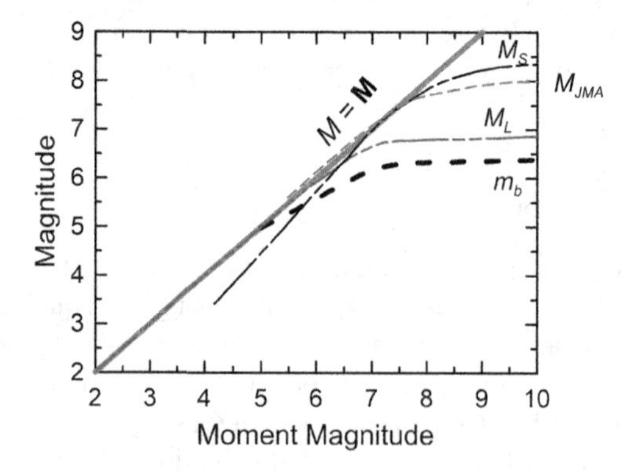

FIGURE 2.36 Saturation of various magnitude scales: **M** (moment magnitude), M_L (Richter local magnitude), M_S (surface wave magnitude), m_b (short-period body wave magnitude), and M_{JMA}, (Japanese Meteorological Agency magnitude). (After Heaton et al., 1986.)

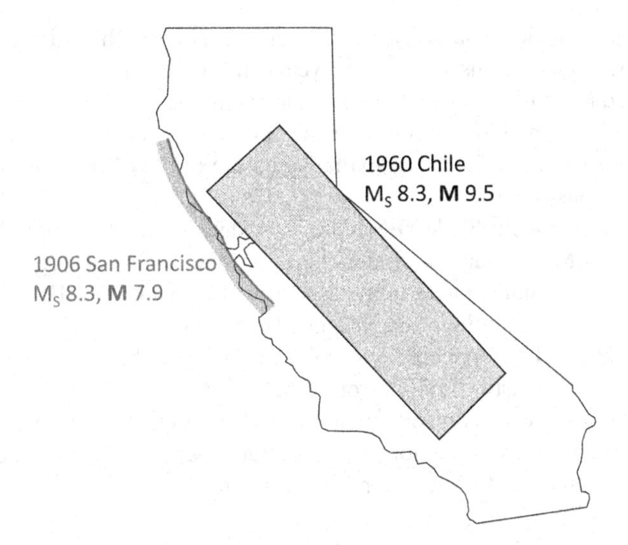

FIGURE 2.37 Comparison of relative areas of fault rupture (shaded) and magnitudes for 1906 San Francisco and 1960 Chile earthquakes. Although the shaking of both earthquakes produced surface wave magnitudes of 8.3, the amounts of energy released were very different, as reflected in their moment magnitudes. (After Boore, 1977.)

Example 2.4

An earthquake occurs across a fault surface 30 km in length (along-strike) and 10 km in width (measured down-dip) and produces an average displacement of 0.3 m. What are the seismic moment (M_o) and the moment magnitude (**M**) of this earthquake?

Solution:

The seismic moment is calculated using Equation (2.2)

$$M_o = \mu A\bar{D} = \left(3\times10^{11}\,\frac{\text{dyne}}{\text{cm}^2}\right)(30\text{ km})(10\text{ km})\left(\frac{10^5\text{ cm}}{1\text{ km}}\right)^2(30\text{ cm}) = 2.7\times10^{25}\,\text{dyne-cm}$$

The moment magnitude is then computed using Equation (2.3) as

$$\mathbf{M} = \frac{\log\left(2.7\times10^{25}\right)}{1.5} - 10.7 = 6.3$$

2.8.3 Earthquake Intensity

The oldest measure of earthquake size is the *earthquake intensity*. The intensity is a qualitative description of the effects of the earthquake at a particular location, as evidenced by observed damage and human reactions at that location. Because qualitative descriptions of the effects of earthquakes are available for much of recorded history (though for a shorter period in North America than, for example, in southern Europe or Asia), the concept of intensity can be applied to historical accounts to estimate the locations and sizes of earthquakes that occurred prior to the development of modern seismic instruments (pre-instrumental earthquakes) (e.g., Toppozada, 1975). This application has been very useful in characterizing the rates of recurrence of earthquakes of different sizes in various locations, a critical step in the evaluation of seismic hazards (Section 4.2). Intensities can also be used to estimate strong ground motion levels and for earthquake loss estimation (Wald et al., 1999b).

The Rossi-Forel (RF) scale of intensity, describing intensities with values ranging from I to X, was developed in the 1880s and used for many years. It has largely been replaced in the United States by the modified Mercalli intensity (MMI) scale originally developed by the Italian seismologist Mercalli and modified in 1931 to better represent conditions in California (Richter, 1958). The MMI scale is illustrated in Table 2.1. The qualitative nature of the MMI scale is apparent from the descriptions of each intensity level.

The European Macroseismic Scale (EMS-98) is widely used in Europe. EMS-98 has some advantages relative to MMI, in that it considers building type, vulnerability, and relative numbers of affected structures (few, many, most) in the assignment of intensities. For this reason, EMS-98 is somewhat less subjective, although assigned intensity values tend to be similar to those from MMI (Musson et al., 2010). The Japanese Meteorological Agency (JMA) also has an intensity scale, which produces numerically distinct values from MMI or EMS-98.

Earthquake intensities are usually obtained from interviews with observers after the event. Historically, the interviews were often done by mail, but in some seismically active areas, permanent observers are organized and trained to produce rational and unemotional accounts of ground

TABLE 2.1
Modified Mercalli Intensity Scale of 1931

I	Not felt except by a very few under especially favorable circumstances
II	Felt by only a few persons at rest, especially on upper floors of buildings; delicately suspended objects may swing
III	Felt quite noticeably indoors, especially on upper floors of buildings, but many people do not recognize it as an earthquake; standing motor cars may rock slightly; vibration like passing of truck; duration estimated
IV	During the day felt indoors by many, outdoors by few; at night some awakened; dishes, windows, doors disturbed; walls make cracking sound; sensation like heavy truck striking building; standing motor cars rocked noticeably
V	Felt by nearly everyone, many awakened; some dishes, windows, etc., broken; a few instances of cracked plaster; unstable objects overturned; disturbances of trees, piles, and other tall objects sometimes noticed; pendulum clocks may stop
VI	Felt by all, many frightened and run outdoors; some heavy furniture moved; a few instances of fallen plaster or damaged chimneys; damage slight
VII	Everybody runs outdoors; damage negligible in buildings of good design and construction, slight to moderate in well-built ordinary structures, considerable in poorly built or badly designed structures; some chimneys broken; noticed by persons driving motor cars
VIII	Damage slight in specially designed structures, considerable in ordinary substantial buildings, with partial collapse, great in poorly built structures; panel walls thrown out of frame structures; fall of chimneys, factory stacks, columns, monuments, walls; heavy furniture overturned; sand and mud ejected in small amounts; changes in well water; persons driving motor cars disturbed
IX	Damage considerable in specially designed structures; well-designed frame structures thrown out of plumb; great in substantial buildings, with partial collapse; buildings shifted off foundations; ground cracked conspicuously; underground pipes broken
X	Some well-built wooden structures destroyed; most masonry and frame structures destroyed with foundations; ground badly cracked; rails bent; landslides considerable from river banks and steep slopes; shifted sand and mud; water splashed over banks
XI	Few, if any (masonry) structures remain standing; bridges destroyed; broad fissures in ground; underground pipelines completely out of service; earth slumps and land slips in soft ground; rails bent greatly
XII	Damage total; practically all works of contraction are damaged greatly or destroyed; waves seen on ground surface; lines of sight and level are distorted; objects thrown into the air

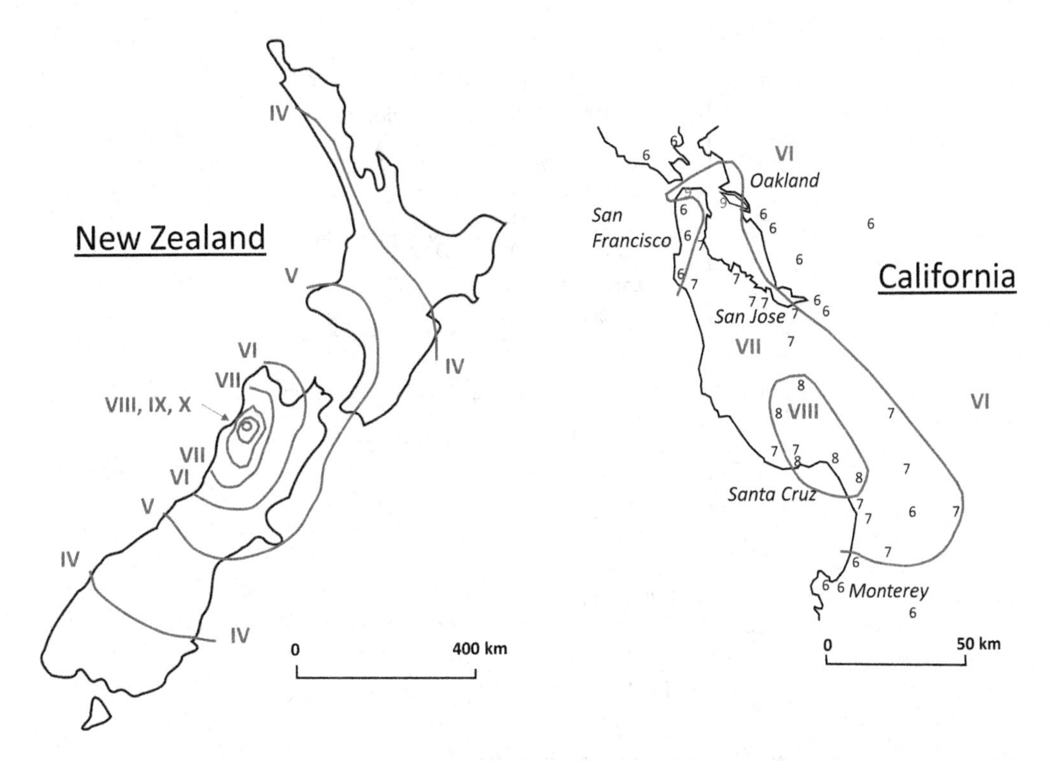

FIGURE 2.38 Isoseismal maps from (a) the 1968 Inangabua earthquake in New Zealand (After Eiby, 1980) and (b) the 1989 Loma Prieta earthquake in northern California (Modified Mercalli intensities). Local instances of intensity IX are not contoured for Loma Prieta (After Plafker and Galloway, 1989).

shaking. More recently, responses to internet-based queries such as the "Did You Feel It? website (https://earthquake.usgs.gov/data/dyfi/) have been used to estimate ground shaking intensity (Wald et al., 1999a). Since human observers and structures are scattered more widely than any seismological observatory could reasonably hope to position instruments, intensity observations provide information that helps characterize the distribution of ground shaking in a region. A plot of reported intensities at different locations on a map allows contours of equal intensity, or *isoseisms*, to be plotted. Such a plot is called an *isoseismal map* (Figure 2.38), which illustrates the variation of intensity with distance from the source region (generally located where intensities are highest). The most reliable estimates of earthquake magnitude from isoseismal maps are based on the area enclosed by isoseisms for various intensity thresholds such as V, VI, VII, and VIII (Toppozada, 1975). These areas are loosely related to source area and hence produce reasonably reliable estimates of magnitude.

2.8.4 Earthquake Energy

The total seismic energy released during an earthquake is often estimated from the relationship (Gutenberg and Richter, 1956)

$$\log_{10} E = 11.8 + 1.5M \tag{2.4}$$

where E is expressed in ergs. This relationship was later shown (Kanamori, 1983) to be applicable to moment magnitude as well. It implies that a unit change in magnitude corresponds to a $10^{1.5}$ or 32-fold increase in seismic energy. A magnitude 5 earthquake therefore would release only about 0.001 times the energy of a magnitude 7 earthquake, thereby illustrating the ineffectiveness of small

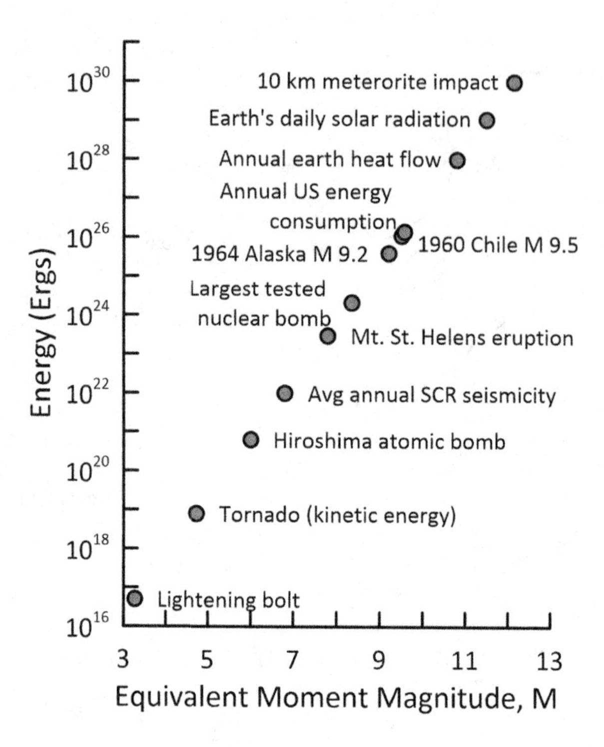

FIGURE 2.39 Relative energy of various natural and human-made phenomena.

earthquakes in relieving the buildup of strain energy that causes very large earthquakes. Combining Equations (2.3) and (2.4) (using **M**) shows that the amount of energy released during an earthquake is proportional to the seismic moment. The amount of energy released by earthquakes is often difficult to comprehend; although a single erg is small (1 erg = 1×10^{-7} J), the energy released in an atomic bomb of the size used at Hiroshima (20,000-ton TNT equivalent) would correspond to a **M**6.0 earthquake. On that basis, the 1960 Chile earthquake (**M**9.5) released as much energy as 178,000 such atomic bombs (Figure 2.39).

2.9 SUMMARY

1. The Earth has a layered structure – the surficial crust is underlain in turn by the mantle, the outer core, and the inner core. The temperature of each layer increases with depth. The temperature gradient in the mantle causes the semi-molten rock to move slowly by convection.
2. The crust consists of a number of large plates and smaller platelets. The plates move with respect to each other as a result of shear stresses that develop on the bottoms of the plates, caused by lateral movement of the convecting mantle, as well as gravitational forces.
3. Relative movement of the plates causes stresses to build up on their boundaries. As movement occurs, strain energy accumulates in the vicinity of the boundaries. This energy is eventually dissipated: generally either smoothly and continuously or in a stick-slip manner that produces earthquakes. The size of an earthquake depends on the amount of energy released.
4. There are three different types of plate boundaries and their characteristics influence the amount of strain energy that can build up near them. As a result, the different types of boundaries have different earthquake capabilities: subduction zone boundaries can produce the largest earthquakes, followed by transform fault boundaries and then spreading

ridge boundaries. Because they involve vertical plate movement and typically occur near coastlines, subduction zone earthquakes can cause tsunamis.

5. The surfaces on which relative crustal movements occur are called faults. At a particular location, a fault is assumed to be planar with an orientation described by its strike and dip. Fault movement (slip) is divided into dip-slip components (normal and reverse faulting) and strike-slip components (left-lateral and right-lateral faulting).
6. Over long distances, both the strike and dip of faults can change. Bends and step-overs in strike-slip faults create local regions of compression or extension, depending on the direction of the fault step and the nature of fault movement (right- or left-lateral).
7. Due to variations in geometry and crustal rock properties, faults may have stronger and weaker zones that release different amounts of energy during the rupture process. Source inversions can determine the sequence and spatial distribution of energy release and provide insight into spatial variations in ground-shaking intensity – but only after an earthquake has occurred.
8. The energy-releasing function of earthquakes suggests that a period of time for strain energy accumulation should be expected between large earthquakes at the same location. It also suggests that earthquakes should be most likely to occur along portions of a fault for which little seismic activity has been observed – unless the plate movement has occurred aseismically.
9. Induced earthquakes can occur when human activity increases stresses on faults or reduces the shear stress a fault can carry prior to rupture. Some common causes of induced earthquakes include fluid injection as part of industrial processes or the filling of reservoirs.
10. Earthquake magnitude is a quantitative measure of the size of an earthquake. The moment magnitude, which increases continuously with increasing source size, is the commonly accepted standard for describing earthquake size. Other magnitude scales, such as the local (Richter), body wave, and surface wave scales, are based on measured ground motion amplitudes. Because these amplitudes tend to reach limiting values, these magnitude scales may not accurately reflect the size of very large earthquakes.
11. Earthquake intensity is a qualitative measure of the effects of an earthquake at a particular location. It is related to the severity of earthquake shaking but is also influenced by other factors such as the quality of construction. Isoseismal maps can be used to describe the spatial variation of intensity for a given earthquake. Maps of this sort are useful to characterize the effects of pre-instrumental earthquakes.
12. Earthquake magnitude scales are logarithmic. A unit change in magnitude corresponds to a 32-fold change in seismic moment and energy.

REFERENCES

Aki, K. (1966). "Generation and propagation of G waves from the Niigata earthquake of June 14, 1964. Part 2. Estimation of earthquake moment, released energy and stress-strain drop from G wave spectrum," *Bulletin of the Earthquake Research Institute*, Vol. 44, pp. 73–88.

Aki, K. (1979). "Characterization of barriers on an earthquake fault," *Journal of Geophysical Research*, Vol. 86, pp. 6140–6148.

Aki, K. (1984). "Asperities, barriers, characteristic earthquakes, and strong motion prediction," *Journal of Geophysical Research*, Vol. 89, pp. 5867–5872.

Aki, K. and Richards, P.G. (1980). *Quantitative Seismology*, 2nd Edition, W.H. Freeman, San Francisco, CA.

Atkinson, G.M. (2020). "The intensity of ground motions from induced earthquakes with implications for damage potential. *Bulletin of the Seismological Society of America*, Vol. 110, No. 5, pp. 2366–2379. doi: https://doi.org/10.1785/0120190166

Atwater, B.F., Musumi-Rokkaku, S., Satake, K., Tsuji, Y., Ueda, K., and Yamaguchi, D.K. (2005). *The Orphan Tsunami of 1700. Japanese Clues to a Parent Earthquake in North America*, United States Geological Survey and University of Washington Press, 133 pp. ISBN 0-295-98535-6.

Atwater, B.F. and Sveringhaus, J. (1989). "Tectonic map of the north Pacific," in E.L. Winterer, D.M. Hussong, and R.W. Decker, eds., *DNAG: The Easter Pacific Ocean and Hawaii*, Geologic Society of America Publication, Boulder, Colorado, pp. 15–20.

Bath, M. (1979). *Introduction to Seismology*, Birkhauser, Boston, MA, 428 pp.

Ben-Menahem, A. (1978). "Source mechanism of the 1906 San Francisco earthquake," *Physics of the Earth and Planetary Interiors*, Vol. 17, pp. 163–181.

Bolt, B.A. (2006). *Earthquakes*, W.H. Freeman and Company, New York, 390 pp.

Bommer, J.J., Stafford, P.J., Ruigrok, E., Rodriguez-Marek, A., Ntinalexis, M., Kruiver, P.P., Edwards, B., Dost, B. van Elk, J. (2022). "Ground-motion prediction models for induced earthquakes in the Groningen gas field, the Netherlands," *Journal of Seismology*, Vol. 26, No. 6, 1157–1184. https://doi.org/10.1007/s10950-022-10120-w

Boore, D.M. (1977). "The motion of the ground during earthquakes," *Scientific American*, Vol. 237, No. 6, pp. 68–78.

Bullard, E.C., Everett, J.E., and Smith, A.G. (1965). "Fit of continents around Atlantic," *Philosophical Transactions of the Royal Society of London, Series A*, Vol. 258, pp. 41–75.

Bullen, K.E. (1975). *The Earth's Density*, Chapman & Hall, London.

Bullen, K.E. and Bolt, B.A. (1985). *An Introduction to the Theory of Seismology*, Cambridge University Press, Cambridge.

Bursa, M. (1993). "Global geodynamic long-term variations and expanding earth hypothesis," *Studia Geophysica et Geodaetica*, Vol. 37, No. 2, pp. 113–124.

Cifuentes, I.L. and Silver, P.G. (1989). "Low-frequency source characteristics of the great 1960 Chilean earthquake," *Journal of Geophysical Research*, Vol. 94, pp. 643–663. https://doi.org/10.1029/jb094ib01p00643.

Dewey, J.W. (1979). "A consumer's guide to instrumental methods for determination of hypocenters," *Geology in the Siting of Nuclear Power Plants*, Geologic Society of American Reviews in Engineering Geology, Vol. 4, pp. 109–117.

Douglas, J., Edwards, B., Convertito, V., Sharma, N., Tramelli, A., Kraaijpoel, D., Cabrera, B.M., Maercklin, N. and Troise, C. (2013). "Predicting ground motions from induced earthquakes in geothermal areas," *Bulletin of the Seismological Society of America*, Vol. 103, pp. 1875–1897. https://doi.org/10.1785/0120120197.

Dwivedi, S.K. and Hayashi, D. (2010). "Modeling the contemporary stress field and deformation pattern of eastern Mediterranean," *Journal of Earth Science*, Vol. 21, pp. 365–381. https://doi.org/10.1007/s12583-010-0100-6.

Eiby, G.A. (1980). *Earthquakes*, Van Nostrand Reinhold, New York.

Ellsworth, W.L. (2013). "Injection-induced earthquakes," *Science*, Vol. 341, Paper 1225942. https://doi.org/10.1126/science.1225942.

Foster, R.J. (1971). *Physical Geology*, Charles E. Merrill, Columbus, OH.

Foster, R.J. (1988). *General Geology*, 5th edition, Charles E. Merrill, Columbus, OH.

Fowler, C.M.R. (2004). *The Solid Earth: An Introduction to Global Geophysics*, 2nd Edition, Cambridge University Press, Cambridge, U.K.

García, D., Wald, D.J. and Hearne, M.G. (2012). "A global earthquake discrimination scheme to optimize ground-motion prediction equation selection," *Bulletin of the Seismological Society of America*, Vol. 102, No. 1, pp. 185–203. https://doi.org/10.1785/0120110124

Garfunkel, Z. (1975). "Growth, shrinking, and long-term evolution of plates and their implications for flow patterns in the mantle," *Journal of Geophysical Research*, Vol. 80, pp. 4425–4430. https://doi.org/10.1029/JB080i032p04425.

Glen, W. (1975). *Continental Drift and Plate Tectonics*, Charles E. Merrill, Columbus, OH.

Gubbins, D. (1990). *Seismology and Plate Tectonics*, Cambridge University Press, Cambridge, 330 pp.

Gutenberg, B. (1945). "Magnitude determination for deep-focus earthquakes," *Bulletin of the Seismological Society of America*, Vol. 35, pp. 117–130. https://doi.org/10.1785/BSSA0350030117.

Gutenberg, B. and Richter, C.F. (1936). "On Seismic Waves (third paper)," *Gerlands Bietraege zur Geophysik*, Vol. 47, pp. 73–131.

Gutenberg, B. and Richter, C.F. (1954). *Seismicity of the Earth and Related Phenomena*, Princeton University Press, Princeton, NJ, 310 pp.

Gutenberg, B. and Richter, C.F. (1956). "Earthquake magnitude: intensity, energy, and acceleration," *Bulletin of the Seismological Society of America*, Vol. 46, pp. 104–145. https://doi.org/10.1785/BSSA0320030163.

Hager, B.H. (1978). "Oceanic plate driven by lithospheric thickening and subducted slabs," *Nature*, Vol. 276, pp. 156–159. https://doi.org/10.1038/276156a0.

Hanks, T.C. and Kanamori, H. (1979). "A moment magnitude scale," *Journal of Geophysical Research*, Vol. 84, pp. 2348–2350. https://doi.org/10.1029/JB084iB05p02348.

Heaton, T.H., Tajima, F., and Mori, A.W. (1986). "Estimating ground motions using recorded accelerograms," *Surveys in Geophysics*, Vol. 8, pp. 25–83. https://doi.org/10.1007/bf01904051.

Helmholtz Association of German Research Centres (2015). "Uplifted Island," Physics.org newsletter, June 22, 2015, https://phys.org/news/2015-06-uplifted-island.html

Herrmann, R.B., Park, S., and Wang, C. (1981). "The Denver earthquakes of 1967–1968," *Bulletin of the Seismological Society of America*, Vol. 71, pp. 731–745. https://doi.org/10.1785/BSSA0710030731.

Kanamori, H. (1977). "The energy release in great earthquakes," *Journal of Geophysical Research*, Vol. 82, pp. 2981–2987. https://doi.org/10.1029/JB082i020p02981.

Kanamori, H. (1983). "Magnitude scale and quantification of earthquakes," *Tectonophysics*, Vol. 93, pp. 185–199. https://doi.org/10.1016/0040-1951(83)90273-1.

Kanamori, H. and Stewart, G.S. (1978). "Seismological aspects of the Guatemala earthquake of February 4, 1976," *Journal of Geophysical Research*, Vol. 83, pp. 3427–3434. https://doi.org/10.1029/JB083iB07p03427.

Kearey, P. and Vine, F.J. (1990). *Global Tectonics*, Blackwell, Oxford, 302 pp.

Lay, T. and Wallace, T.C. (1995). *Modern Global Seismology*, Academic Press, San Diego, CA, 521 pp.

Lettis, W., Bachhuber, J., Witter, R., Brankman, C., Randolph, C.E., Barka, A., Page, W.D., and Kaya, A. (2002). "Influence of releasing step-overs on surface fault rupture and fault segmentation: Examples from the 17 August 1999 Izmit Earthquake on the North Anatolian fault, Turkey," *Bulletin of the Seismological Society of America*, Vol. 103, pp. 19–42. https://doi.org/10.1785/0120000808.

McCann, W.R., Nishenko, S.P., Sykes, L.R., and Krause, J. (1979). "Seismic gaps and plate tectonics: seismic potential for major boundaries," *Pure and Applied Geophysics*, Vol. 117, pp. 1082–1147. https://doi.org/10.1007/bf00876211.

McClusky, S., Balassanian, S., Barka, A., Demir, C., Ergintav, S., Georgiev, I., Gurkan, O., Hamburger, M., Hurst, K., Kahle, H., Kastens, K., Kekelidze, G., King, R., Kotzev, V., Lenk, O., Mahmoud, S., Mishin, A., Nadariya, M., Ouzounis, A., Paradissis, D., Peter, Y., Prilepin, M., Reilinger, R., Sanli, I., Seeger, H., Tealeb, A., Toksöz, M.N., and Veis, G. (2000). "Global positioning system constraints on plate kinematics and dynamics in the eastern Mediterranean and Caucasus," *Journal of Geophysical Research*, Vol. 105, pp. 5695–5719. https://doi.org/10.1029/1996jb900351.

McKenzie, D.P. (1970). "Plate tectonics of the Mediterranean region," *Nature*, 226–239. https://doi.org/10.1038/226239a0.

Musson, R.M.W., Grünthal, G., and Stucchi, M. (2010). "The comparison of macroseismic intensity scales," *Journal of Seismology*, Vol. 14, pp. 413–428. https://doi.org/10.1007/s10950-009-9172-0.

Nadeau, R.M. and Dolenc, D. (2004). "Nonvolcanic tremors deep beneath the San Andreas fault," *Science*, Vol. 307, Issue 5708, p. 389. https://doi.org/10.1126/science.1107142.

Noson, L.L., Qamar, A., and Thorsen, G.W. (1988). "Washington state earthquake hazards," *Information Circular 85*, Washington Division of Geology and Earth Resources, Olympia, Washington.

Obara, K., Hirose, H., Yamamizu, F. & Kasahara, K. (2004). "Episodic slow slip events accompanied by non-volcanic tremors in southwest Japan subduction zone," *Geophysical Research Letters*, Vol. 442, No. 13, pp. 188–191. https://doi.org/10.1029/2004GL020848.

Plafker, G. and Galloway, J. P. eds. (1989). *Lessons Learned from the Loma Prieta, California, Earthquake of October 17 1989. USGS Circular 1045*, US Department of the Interior, US Geological Survey. https://doi.org/10.3133/cir1045.

Reid, H.F. (1911). "The elastic rebound theory of earthquakes," *Bulletin of the Department of Geology*, University of California, Berkeley, Vol. 6, 413–444.

Richter, C.F. (1935). "An instrumental earthquake scale," *Bulletin of the Seismological Society of America*, Vol. 25, pp. 1–32.

Richter, C.F. (1958). *Elementary Seismology*, W.H. Freeman, San Francisco.

Rockwell, T.K. and Klinger, Y. (2013). "Surface rupture and slip distribution of the 1940 Imperial Valley Earthquake, Imperial Fault, Southern California: Implications for rupture segmentation and dynamics," *Bulletin of the Seismological Society of America*, Vol. 103, No. 2A, pp. 629–640. https://doi.org/10.1785/0120120192.

Rogers, G. and Dragert, H. (2003). "Episodic tremor and slip on the Cascadia subduction zone: The chatter of silent slip," *Science*, Vol. 300, No. 5627, pp. 1942–1943. https://doi.org/10.1126/science.1084783.

Rubinstein, J.L. and Mahani, A.B. (2015). "Myths and facts on wastewater injection, hydraulic fracturing, enhanced oil recovery, and induced seismicity," *Seismological Research Letters*, Vol. 86, pp. 1060–1067. https://doi.org/10.1785/0220150067.

Shearer, P.M. (2019). *Introduction to Seismology*, 3rd Edition, Cambridge University Press, Cambridge.

Shen, Z.-K., Wang, M., Zeng, Y., and Wang, F. (2015). "Optimal interpolation of spatially discretized geodetic data," *Bulletin of the Seismological Society of America*, Vol. 105, No. 4, pp. 2117–2127. https://doi.org/10.1785/0120140247

Simpson, D.W., Leith, W.S., and Scholz, C.H. (1988). "Two types of reservoir-induced seismicity," *Bulletin of the Seismological Society of America*, Vol. 78, pp. 2025–2040. https://doi.org/10.1785/BSSA0780062025.

Somerville, P.G., Irikura, K., Graves, R.W., Sawada, S., Wald, D., Abrahamson, N.A., Iwasaki, Y., Kagawa, T., Smith, N., and Kowada, A. (1999). "Characterizing crustal earthquake slip models for the prediction of strong ground motion," *Seismological Research Letters*, Vol. 70, pp. 59–80. https://doi.org/10.1785/gssrl.70.1.59

Stein, S. and Wysession, M. (2003). *An Introduction to Seismology, Earthquakes, and Earth Structure*, John Wiley, ISBN: 978-1-118-68745-1.

Taylor, F.B. (1910). "Bearing of the tertiary mountain belt on the origin of the Earth's plan," *Bulletin of the Geological Society of America*, Vol. 21, pp. 179–226. https://doi.org/10.1130/GSAB-21-179.

Toppozada, T.R. (1975). "Earthquake magnitude as a function of intensity data in California and western Nevada," *Bulletin of the Seismological Society of America*, Vol. 65, pp. 1223–1238. https://doi.org/10.1785/BSSA0650051223.

Tsuboi, C. (1954). "Earthquake magnitude determined by ground-motion maximum amplitude," *Jishin*, 7, pp. 185–193 (in Japanese).

Wald, D. J., Quitoriano, V., Dengler, L. A., and Dewey, J. W. (1999a). "Utilization of the internet for rapid community intensity maps," *Seismological Research Letters*, Vol. 70, No. 6, pp. 680–697. https://doi.org/10.1785/gssrl.70.6.680.

Wald, D.J., Quitoriano, V., Heaton, T.H., and Kanamori, H. (1999b). "Relationships between peak ground acceleration, peak ground velocity, and Modified Mercalli Intensity in California," *Earthquake Spectra*, Vol. 15, pp. 557–564. https://doi.org/10.1193/1.1586058.

Wegener, A. (1915). *Die Entstehung der Kontinente und Ozeane*, Vieweg, Braunschweig, Germany.

Weingarten, M., Ge, S., Godt, J.W., Bekins, B.A., and Rubinstein, J.L. (2015). "High-rate injection is associated with the increase in U.S. mid-continent seismicity," *Science*, Vol. 348, pp. 1335–1340. https://doi.org/10.1126/science.aab1345.

Wilson, J.T. (1965). "A new class of faults and their bearing on continental drift," *Nature*, Vol. 207, 343–347. https://doi.org/10.1038/207343a0.

Yeats, R.S. and Huftile, G.J. (1995). "The Oak Ridge fault system and the 1994 Northridge earthquake." *Nature*, Vol. 373, No. 6513, pp. 418–420. https://doi.org/10.1038/373418a0.

3 Strong Ground Motion Characterization and Prediction

3.1 INTRODUCTION

The Earth vibrates continuously at periods ranging from milliseconds to days and amplitudes ranging from nanometers to meters. The great majority of these vibrations are so weak that they cannot be felt or even detected without specialized equipment. Such *microseismic activity* is useful for seismologists and Earth scientists but has little direct relevance to engineers. Earthquake engineers are interested primarily in *strong ground motion* (i.e., motions of sufficient strength to affect people and infrastructure). Earthquake engineering requires objective, quantitative ways of describing and predicting strong ground motions.

Earthquake ground motions are complex. At a given point, they can be completely described by three components of translation and three components of rotation. The rotational components of motions at a point are usually not measured or used in engineering analysis. Figure 3.1 shows three translational components of ground acceleration from a single site during two events having different magnitudes – a **M** 5.3 aftershock of the 1987 Whittier Narrows earthquake and the **M** 6.7 mainshock of the 1994 Northridge, California earthquake. These acceleration-time plots (referred to as *acceleration histories*) are data-intensive, containing thousands of data points at time intervals typically recorded at an increment of 0.005–0.02 sec. Earthquake engineering applications that evaluate the full time-dependent seismic response of structures use acceleration histories directly to represent the *seismic input*, a measure of the loading or deformation imposed on a system by earthquake shaking. Such *response history analyses* provide the most direct and physically meaningful insights into the dynamic response of structures, but are computationally intensive.

Fortunately, ground motions can also be described more concisely for engineering purposes using ground motion *intensity measures* (*IM*s), which describe the most important characteristics from an engineering standpoint, i.e., the *amplitude*, *frequency content*, and *duration* of the motion. A number of *IM*s are used by engineers to describe one or more of these characteristics. Many applications require multiple *IM*s because structural and geotechnical responses can be affected by multiple ground motion characteristics.

This chapter describes three principal issues related to strong ground motions – the recording instruments used for measuring ground motions and the processing that is applied to correct recorded motions for engineering application; common *IM*s used by engineers to characterize ground motion amplitude, frequency content, and duration; and models that can be used for predicting *IM*s or full acceleration histories given certain attributes of a causative earthquake event, the source-to-site distance, and site conditions. Special considerations for near-fault ground motions and procedures for describing spatial variations of ground motions at the scale of typical engineering projects, which are important for some applications, are also described. This chapter provides information essential for performing *seismic hazard analyses*, which also require source characterization and methods for combining predicted motions for potential events on one or more sources as described in Chapter 4.

DOI: 10.1201/9781003512011-3

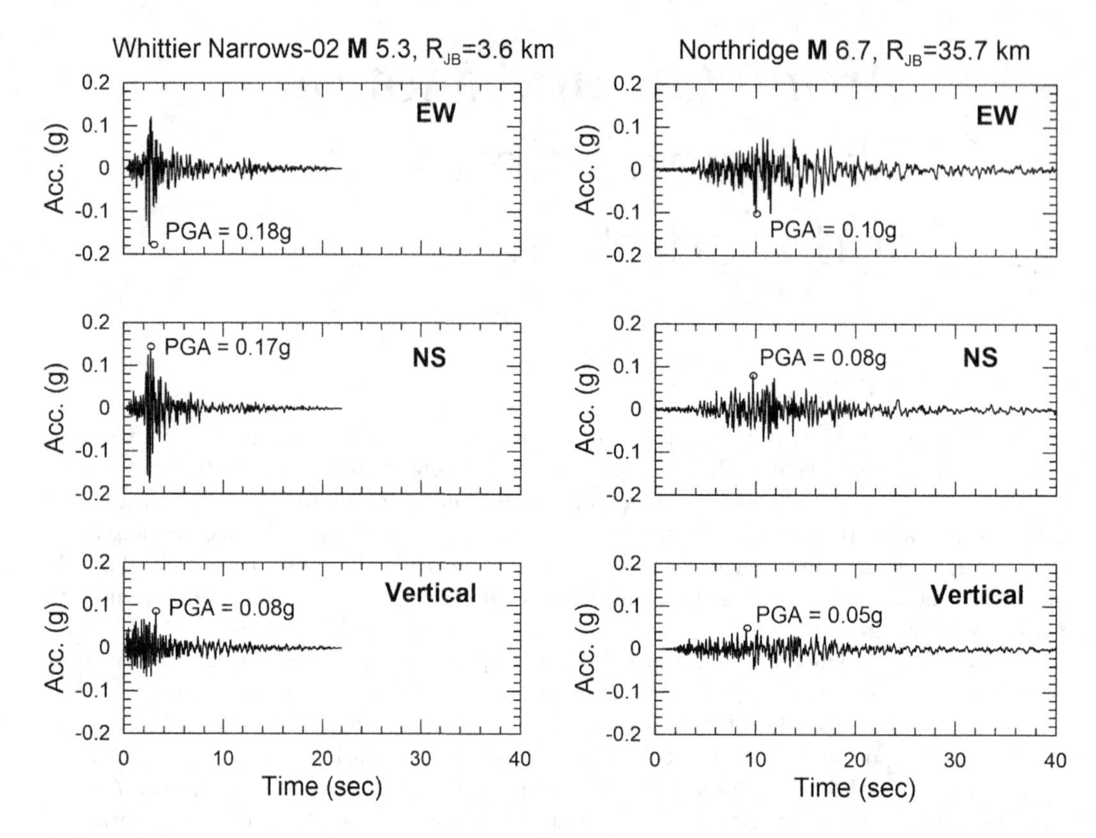

FIGURE 3.1 Acceleration histories at Alhambra-Fremont School station from nearby small magnitude (Whittier Narrows) event and more distant moderate magnitude (Northridge) event. EW and NS components refer to East-West and North-South horizontal directions, respectively.

Before proceeding further, the reader should review the topics discussed in Appendices A and B – familiarity with the concepts presented in those appendices is assumed in this chapter and the remainder of this book. Portions of Appendix C will also be referred to at certain points in this chapter.

3.2 GROUND MOTION MEASUREMENT

Seismic strong ground motion characterization requires the use of ground motion recordings from actual earthquakes. Accurate measurement of strong ground motion is critical for both seismological and earthquake engineering applications. As stated by the National Research Council Committee on Earthquake Engineering Research (Housner, 1982): "The recording of strong ground motion provides the basic data for earthquake engineering. Without a knowledge of the ground shaking generated by earthquakes, it is not possible to assess hazards rationally or to develop appropriate methods of seismic design."

3.2.1 Seismometers and Accelerometers

A number of different instruments can be used to record earthquake ground motions, and it is important to understand both the manner in which the instruments are configured and the dynamic characteristics of their recording systems.

3.2.1.1 Instrument Configurations

Seismic monitoring has taken many forms over time. Written descriptions of earthquakes and their effects date back as far as 2,000 years in Japan, southern Europe, and the Middle East (Ambraseys, 1971, 1978; Allen, 1975; Bolt, 1988; Papazachos and Papazachou, 2003) and comprise a valuable resource for analysis of earthquake size and recurrence. Instruments for measurement of ground motions can be traced as far back as 132 AD, but the nineteenth century saw substantial developments in the reliability and practicality of these instruments (Dewey and Byerly, 1969). These first instruments were *seismometers*, which record displacements of dynamic systems in response to ground motion. Early seismometers were designed to be highly sensitive and were used principally to detect and locate distant bomb detonations and earthquakes, but did not have sufficient range to record strong ground motions that would go off-scale and produce clipped (truncated) recordings. *Accelerometers* measure ground accelerations, typically with the intent of capturing strong ground motions without clipping. The first strong ground motion recordings are generally attributed to the 1933 Long Beach, California earthquake (Boore and Bommer, 2005). This section describes the operation of seismometers and accelerometers.

To better understand the operation of these instrument types, it is useful to consider two distinct designs. The first design, referred to as an inertial seismometer by Havskov and Alguacil (2010), consists of a mass-spring-damper single-degree-of-freedom (SDOF) system, as illustrated for vertical motion in Figure 3.2. Since the spring and dashpot are not rigid, the motion of the mass will not be identical to the motion of the ground during an earthquake. The relative displacement of the mass and the ground will be indicated by a recorded analog trace, usually made on a rotating drum. This same basic design was used for both seismometers and accelerometers, including the widely used SMA1 analog accelerometer.

The second, more modern design, is employed in force-balance accelerometers and *broadband* (i.e., wide frequency range) electrodynamic seismometers (Havskov and Alguacil, 2010), which have a mass attached to a magnet suspended within a coil (Figure 3.3). When the mass is accelerated, a displacement transducer instantaneously sends current to the coil surrounding the mass, generating a restoring force that offsets the inertia of the mass and pushes it back toward its initial (equilibrium) position. The mass itself undergoes very little relative displacement in this process because the restoring force keeps it essentially stationary within the reference frame of the device

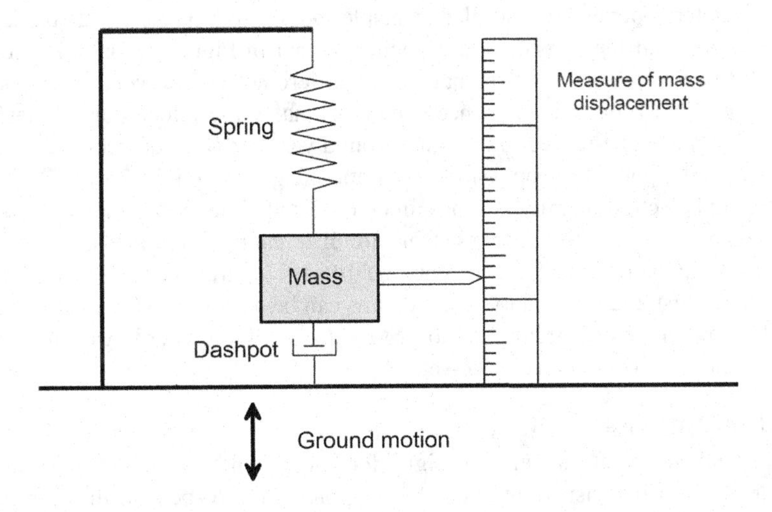

FIGURE 3.2 Simple vertical inertial seismometer (Havskov and Alguacil, 2010). The housing to which the spring and damper are attached is firmly connected the ground. When the ground shakes, the vertical displacement of the mass is recorded as a function of time. The relative displacements are related to ground displacement or acceleration, depending on instrument design. Similar systems can be constructed to measure horizontal motion.

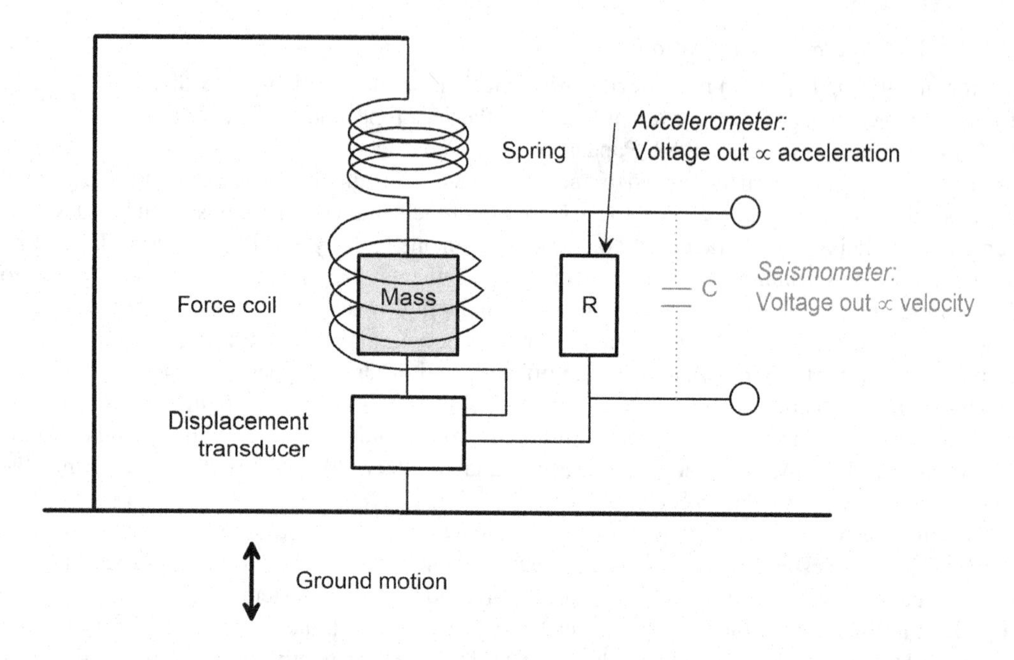

FIGURE 3.3 Schematic of electrodynamic system. (Modified from Havskov and Alguacil, 2010.) A current proportional to the displacement transducer output is produced, which forces the mass to remain stationary relative to the frame. An accelerometer operates by taking the voltage across a resistor (R) as proportional to acceleration. An electrodynamic seismometer (commonly referred to as a broadband sensor) operates by taking the voltage across a capacitor (C) as proportional to velocity.

(i.e., relative to the ground). For this reason, the spring shown in Figure 3.3 does not significantly impact the dynamics of the mass; its purpose is to maintain the static equilibrium position of the mass.

In a force-balance accelerometer, the current from the displacement transducer produces a voltage across the resistor, which is proportional to the acceleration (Figure 3.3). Broadband electrodynamic seismometers operate on a similar principle, but current in the coil is passed through a capacitor in parallel with the resistor (the capacitor shown in Figure 3.3 is only present for the seismometer and is not present in a force-balanced accelerometer). The voltage across the capacitor is a time integral of the current and hence is proportional to the velocity of the mass (Wielandt, 2002). These seismometers are used to measure ground velocity. Such devices are also referred to as *geophones* when designed for geophysical exploration (e.g., Section 6.5.2.1).

Both inertial and electrodynamic systems can be operated either as seismometers or as accelerometers. The signals produced by the two sensors are different (accelerations from accelerometers; velocity from electrodynamic seismometers), but the differences are otherwise much less significant than they have been historically because both systems can be designed to have sufficient sensitivity to record both weak and strong ground motions. Moreover, digital recordings can be readily converted from velocity to acceleration or vice-versa.

3.2.1.2 System Dynamics

Neither the inertial nor electrodynamic designs for accelerometers and seismometers measure ground motion directly but instead measure the response of a suspended mass from which the ground motion is inferred. This inference requires that the dynamic response of the oscillator system be accounted for. Consider an inertial seismometer such as the one shown in Figure 3.2. This seismometer is an SDOF oscillator whose response to shaking is given by the equation of motion (Appendix B).

$$m\ddot{u} + c\dot{u} + ku = -m\ddot{u}_g \tag{3.1}$$

where u is the recorded displacement (the relative displacement between the mass and the ground) and $\ddot{u}_g$ is the ground acceleration.

If the ground displacement is simple harmonic at a circular frequency, ω, the displacement response ratio (the ratio of trace displacement amplitude to ground displacement amplitude) will be

$$\frac{|u|}{|u_g|} = \frac{\beta^2}{\sqrt{\left(1-\beta^2\right)^2 + \left(2\xi\beta\right)^2}} \tag{3.2a}$$

where β $(= \omega/\omega_0)$ is the *tuning ratio*, ω_0 $(= \sqrt{k/m})$ is the undamped natural circular frequency, and ξ $(= c/2\sqrt{km})$ is the *damping ratio*. Figure 3.4a shows how the displacement response ratio varies with frequency and damping. For ground motion frequencies well above the natural frequency of the oscillator (i.e., large values of β), the trace amplitude is equal to the ground motion displacement amplitude. This is the approach used with traditional inertial seismometers; they have low natural frequencies so that the trace amplitude directly reflects ground displacements. The lowest frequency for which this equality holds (within a given range of accuracy) depends on the damping ratio. Because the response is insensitive to frequency and phase angles are preserved at damping ratios of 60%, SDOF displacement seismometers are usually designed with damping ratios near that value (Richart et al., 1970). Typical natural periods, $T_0 = 2\pi / \omega_0$, of seismometers used for the (now obsolete) Worldwide Standard Seismographic Network (Wielandt, 2002) were 15–90 sec.

As described in the previous section, electrodynamic broadband seismometers measure a voltage (V) that is proportional to the velocity of the oscillator. As such, the numerator in Equation (3.2a) is multiplied by $i\omega$ (or ω when considering amplitude only), which can be written as given in Equation (3.2b). To preserve the shape of the transfer function from Equation (3.2a) and Figure 3.4a, the denominator must be replaced with velocity as shown in Equation (3.2c).

$$\frac{V}{|u_g|} \propto \frac{|\dot{u}|}{|u_g|} = \frac{\omega|u|}{|u_g|} = \frac{\omega_0\beta^3}{\sqrt{\left(1-\beta^2\right)^2 + \left(2\xi\beta\right)^2}} \tag{3.2b}$$

$$\frac{V}{\omega|u_g|} = \frac{V}{|\dot{u}_g|} \propto \frac{|\dot{u}|}{|\dot{u}_g|} = \frac{\beta^2}{\sqrt{\left(1-\beta^2\right)^2 + \left(2\xi\beta\right)^2}} \tag{3.2c}$$

Accordingly, electrodynamic seismometers utilize the same response curves as traditional instruments, but the recovered ground motion is velocity and not displacement.

The acceleration response ratio (the ratio of trace displacement amplitude to ground acceleration amplitude) is given by:

$$\frac{|u|}{|\ddot{u}_g|} = \frac{1}{\omega_0^2\sqrt{\left(1-\beta^2\right)^2 + \left(2\xi\beta\right)^2}} \tag{3.3}$$

As shown in Figure 3.4b, the trace amplitude is proportional to the ground acceleration amplitude for frequencies well below the natural frequency of the instrument (i.e., low values of β). This is the approach used with accelerometers; they have high natural frequencies so that the trace amplitude directly reflects ground acceleration. An accelerometer with 60% damping will accurately measure accelerations at frequencies up to about 55% of its natural frequency. Analog accelerographs of this type typically have natural frequencies of about 25 Hz with damping ratios near 60%, with desirable

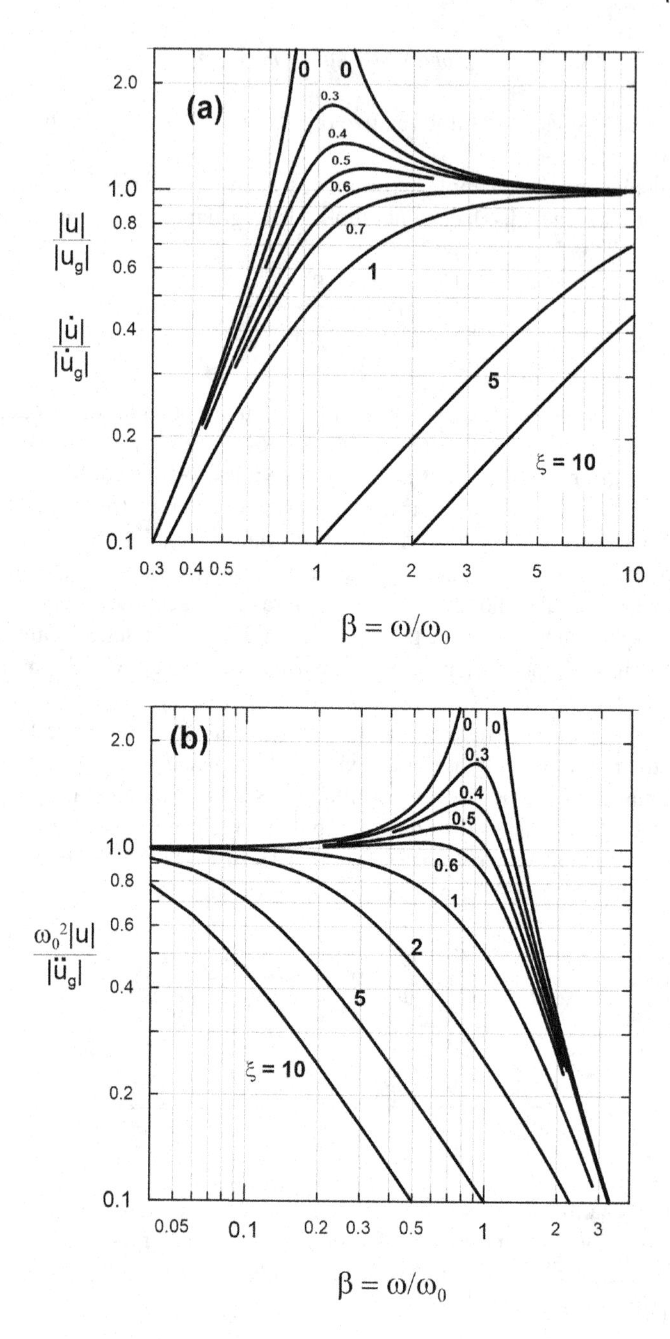

FIGURE 3.4 (a) Displacement and velocity response ratio as used for seismometers and (b) acceleration response ratio as used for accelerometers for SDOF system subjected to simple harmonic base motion.

flat response (constant acceleration response ratio) at frequencies up to about 15 Hz. Modern digital force-balance accelerometers typically have a flat frequency response between zero and 50–200 Hz (Havskov and Alguacil, 2010).

A typical *triaxial* seismometer or accelerometer station will record motion in the vertical and two perpendicular horizontal directions. Ground motions are measured with a common time basis. Time accuracy is maintained in modern instruments through global positioning system (GPS) sensors in communication with satellites using Universal Coordinated Time (the scientific equivalent of

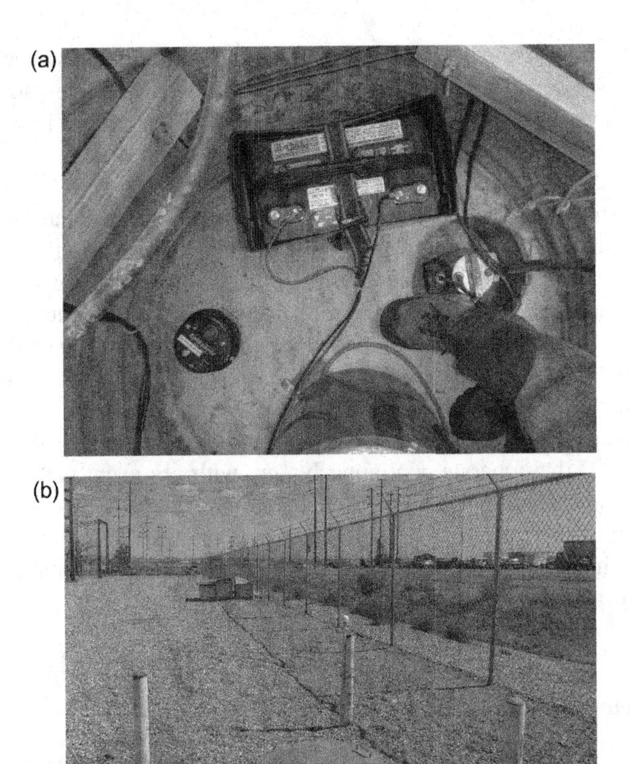

FIGURE 3.5 (a) Modern ground motion installation including (on floor, clockwise from top) battery, sump pump, broadband seismometer, digital strong motion accelerometer. The installation is approximately 0.5 m below the ground surface. Station is Antelope, Lancaster, West Ave J and 90th, Lake Hughes, CA, U.S., part of the SCSN. (b) complete installation with vault cap and protective poles. (Photos courtesy of Alan Yong, USGS.)

Greenwich Mean Time). Seismometers, accelerometers, and ancillary equipment are protected by an instrument shelter or vault (Figure 3.5). Solar panels and/or car batteries are often used to provide power for the instruments and data acquisition system.

3.2.2 Data Acquisition and Digitization

The previous section described how seismometers and accelerometers relate the motion of the ground to the relative motion of a mass. In early instruments, the relative motion of the mechanism containing the mass was recorded with a pen, stylus or reflective mirror acting on paper or photographic film attached to a rotating drum. Later-generation instruments recorded motions electronically in analog form on magnetic tape. Rather than record continuously, instruments of these types remain dormant until triggered by the exceedance of a small threshold acceleration at the beginning of an earthquake motion. As a result, wave arrivals that precede triggering are not recorded, thereby missing the P-wave arrival and introducing a *baseline error* into the acceleration record. The most well-known accelerometer of this type is the Kinemetrics SMA-1, which produced a large fraction of the strong motion recordings in international databases prior to the mid-1990s.

Digitization is required to use recorded traces from analog instruments. Originally, digitization was performed manually with paper, pencil, and an engineering scale. Semiautomatic digitizers, with which a user moved a lens with crosshairs across an accelerogram mounted on a digitizing table, were commonly used in the late 1970s. By pressing a foot-operated switch, the coordinates of the crosshairs were recorded. These forms of digitizing involved exacting and tiring work; operator accuracy and fatigue were important considerations (Hudson, 1979). The digitization process of analog records is a substantial source of noise, which is a major disadvantage of analog systems (Boore and Bommer, 2005).

Modern electrodynamic force-balance accelerometers and broadband seismometers, which began to come into operation in the 1980s, convert the analog voltage signals to a digital form internally. They record motions continuously at rates of 200–1,000 samples/sec with 10- to 24-bit resolution, saving the recorded data only if a triggering acceleration is exceeded. They have pre-event memory that preserves the initial portion of the record that is lost with triggered analog systems. As described in Section 3.2.1.2, these systems have much wider dynamic ranges (null to 50–100 Hz) than analog instruments (typical natural frequency of 25 Hz).

Digital accelerographs overcome the three major shortcomings of analog instruments (Boore and Bommer, 2005): (1) they operate continuously and, with pre-event memory, can record wave arrivals; (2) they have a wider frequency range, encompassing the full range of engineering interest; and (3) they remove the need for a manual digitization process, thus streamlining access to data and reducing noise levels in recorded ground motions. As a result of these advantages, along with their relatively low cost, digital seismometers and accelerometers have largely replaced analog instruments in most ground motion arrays worldwide.

3.2.3 Ground Motion Processing

The signals produced by accelerometers are not true ground motions, but a combination of the true ground motion, the dynamic response of the accelerometer, and *noise*. Noise consists of extraneous components of a recorded signal that reflect natural (non-seismic) ground vibrations and artifacts introduced by the recording and digitization equipment. Ground motion processing is undertaken to optimize the balance between acceptable signal-to-noise ratio and producing a corrected seismic signal having the widest possible usable frequency range (Boore and Bommer, 2005).

Ground motion processing has two broad objectives: (1) to remove high-frequency noise and correct recorded signals for the instrument response and (2) to remove low-frequency noise and baseline drift.

3.2.3.1 High-Frequency Noise and Instrument Correction

Digital ground motion time series provide acceleration or velocity on a regular time interval, referred to as the *time step*, Δt. Time series only contain information up to a limiting frequency of $1/(2\Delta t)$, which is known as the *Nyquist frequency*. For time steps of $\Delta t = 0.020$ and 0.005 sec, the corresponding Nyquist frequencies are 25 and 100 Hz, respectively. However, depending on the levels of high-frequency noise in time series, they may or may not reliably represent ground motion characteristics up to these limiting frequencies.

While all recorded earthquake ground motions contain high-frequency noise, the principal noise sources are different for older analog instruments (similar to the inertial seismometer in Figure 3.2) and relatively modern digital instruments (similar to the electrodynamic sensor depicted in Figure 3.3). In the case of an analog instrument, high-frequency noise is dominated by instrument response and digitization errors (Boore and Bommer, 2005). Recall from Section 3.2.1 that accelerometers have high natural frequencies and an *instrument response* (ratio of the relative movement of the instrument mass to ground motion) that is flat (essentially independent of frequency) for frequencies sufficiently below the natural frequency, f_0. Signal distortion related to instrument response occurs at frequencies near, and higher than, the natural frequency where the instrument response (depicted in Figure 3.4b) falls below unity. Analog accelerometers often have natural frequencies of 25 Hz. As shown in Figure 3.6, the instrument response can affect the amplitude of recorded motions for frequencies higher than approximately 2/3 of f_0 (i.e., higher than

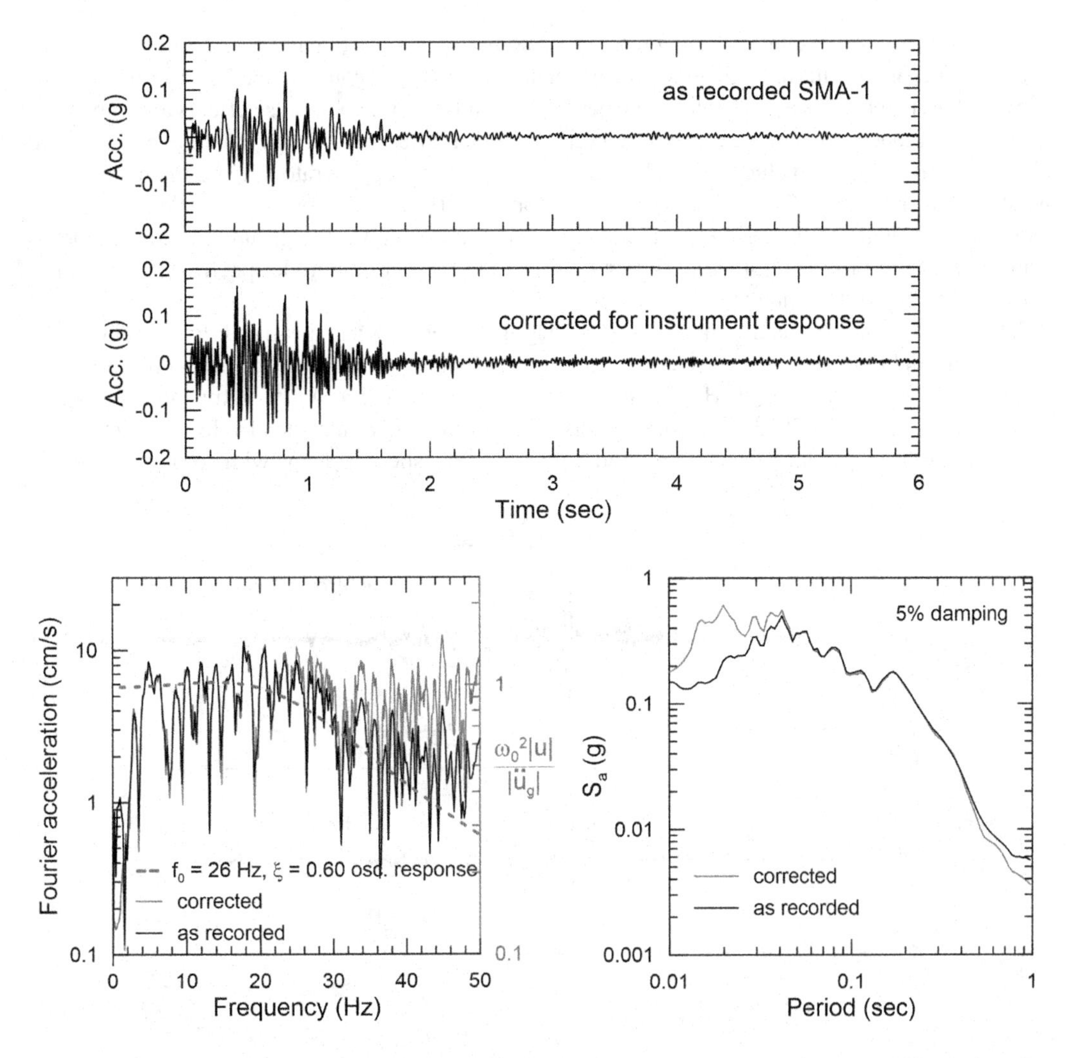

FIGURE 3.6 Acceleration histories, Fourier amplitudes, and response spectra for eastern North America recording (1982 Miramichi earthquake, **M** 4.2, Mitchell Lake Rd). This analog recording has unusually large contributions from high frequencies and hence is strongly impacted by the instrument correction in its peak accelerations and short-period response spectra. (Modified from Boore and Bommer, 2005.)

approximately 15–20 Hz). When those high frequencies represent an important contribution to the energy in a ground motion, the correction will significantly impact the appearance of the ground motion in the time domain and oscillator responses at high frequencies (used to define spectral accelerations; Section 3.3.2), as is the case for the record in Figure 3.6. Errors introduced by the digitization process include random errors in the identification of the exact mid-point of the paper or film trace (Trifunac et al., 1999) and operator error (when digitization is performed manually).

These problems are mitigated with digital instruments, for which instrument response occurs at sufficiently high frequencies (typically $f_0 = 50$–100 Hz) that instrument corrections are not needed (Douglas and Boore, 2011). Noise sources for such instruments include digitization noise and natural, non-seismic ground motions. Digitization noise is associated with the number of discrete values the digitizer uses over its (minimum to maximum) range, being quantified by the bit number (bN) of the digitizer:

$$\text{Number of discrete values} = \pm 2^{(bN-1)} \tag{3.4}$$

For example, a 12-bit resolution digitizer has ±2,048 discrete values (meaning 2,048 positive values and 2,048 negative values). Digitizers with 10- or 12-bit resolution can noticeably affect the appearance of ground motions (e.g., Figure 3.7) whereas at 24 bits, digitization noise effects are imperceptible. High-frequency sources of ground vibrations, which affect recordings from both analog and digital instruments, can include cultural sources (traffic, electrical generators, vibrating machinery), wind, and ocean waves. These vibration sources dominate the noise recorded by modern electrodynamic sensors with 24-bit digitizers. This noise can be quantified using ground motions recorded prior to the onset of seismic shaking. When converted to Fourier amplitude spectra, a flat response is typically observed as shown in Figure 3.7.

The effects of high-frequency noise are often negligible for response spectral ordinates. Figure 3.7 shows a ground motion where noise effects are evident by flattening of the Fourier amplitude spectrum beyond approximately 20 Hz. By selecting a *high-cut (low-pass) filter* (Appendix A) with a *corner frequency* (f_c) of 20 Hz, the components of the ground motion controlled by high-frequency noise are removed, producing a more natural high-frequency spectral shape with amplitude fall-off.

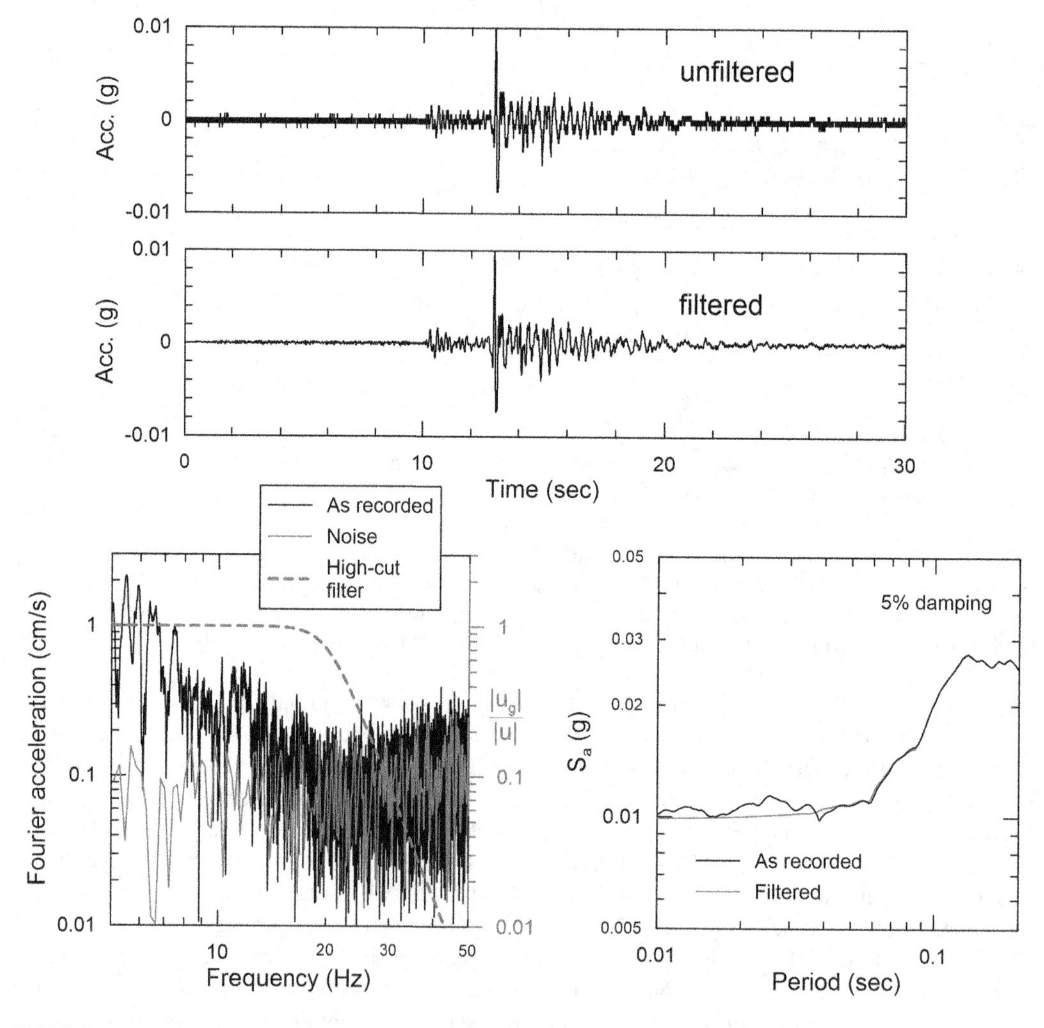

FIGURE 3.7 Acceleration history, Fourier amplitude, and response spectra for uncorrected digital record from **M** 3.9 Hengill earthquake, Oseyrarbru station (Iceland strong motion network). The Fourier spectra show that noise dominates the recording beyond approximately 20 Hz, but the application of a high-cut filter only modestly affects response spectral ordinates. (Modified from Douglas and Boore, 2011.)

Despite the substantial high-frequency noise, pseudo-spectral accelerations, S_a, for the original and filtered records are similar at short periods, even those at or below $1/f_c$. This occurs because the high-frequency oscillator response is influenced by lower-frequency, more energetic portions of the record that are unaffected by the filtering (Douglas and Boore, 2011; Akkar et al., 2011). The application of a high-cut filter is unlikely to affect high-frequency S_a ordinates (Douglas and Boore, 2011) when the peak of the Fourier amplitude spectrum of the uncorrected record is more than ten times greater than the Fourier amplitude of the noise spectrum (which applies to the record in Figure 3.7). This lack of importance of high-frequency filtering applies to approximately 95% of horizontal ground motion records and 85% of vertical ground motion records for a Euro-Mediterranean databank (Akkar et al., 2011). For this reason, high-frequency filtering is generally not applied in the widely used PEER Next Generation Attenuation (NGA) global databank for active tectonic regions (Chiou et al., 2008; Ancheta et al., 2014).

The relatively rare circumstances in which high-frequency noise effects, and hence the application of high-cut filters, are likely to affect spectral accelerations include:

- Relatively weak ground motions in which the amplitude of the noise spectrum is more than 10% of the peak of the Fourier spectrum of the uncorrected signal (Douglas and Boore, 2011).
- Motions with rich energy content at high frequencies, which may include some vertical records and near-fault records on hard rock sites. Other factors being equal, motions from stable continental regions tend to be richer in high-frequency energy than those from relatively active regions near plate boundaries.
- Motions recorded by analog sensors where the instrument correction can extend to frequencies as low as 15 Hz, especially when the conditions from the above two bullets are also met.

These high-frequency spectral accelerations are seldom of interest for ordinary structures, but can be significant for systems and components within some critical structures (e.g., nuclear power plants).

3.2.3.2 Low-Frequency Noise and Baseline Drift

As with high-frequency noise, the predominant sources of low-frequency noise are different for analog and digital instruments. Figure 3.8 shows acceleration time series from analog and digital instruments, both containing low-frequency noise. In the case of the analog recording (left side of figure), the noise is evident from non-physical features in the velocity and displacement histories, including the velocity baseline shift at about 23 sec and the single direction of ground displacement. One cause of these errors is unknown initial conditions (velocity and displacement), which are not recorded because analog instruments usually begin recording after being triggered by the exceedance of a threshold acceleration generally associated with the initial arrival of p-waves. Another possible cause is lateral movements of the recording medium (usually film) during recording and warping of the record before digitization (Boore and Bommer, 2005). In the case of the digital record, even with pre-event memory, the initial baseline is still not perfectly known due ground and instrument tilt, and possibly other sensor problems (Boore, *personal communication*, 2013). A small baseline error leads to drift as shown on the right side of Figure 3.8, which is very common, even in modern digital instruments. Low-frequency noise and baseline drift are generally not visible in acceleration histories and can only be seen after integration to velocity or displacement.

Low-frequency noise and baseline drift have practical significance for long-period ground motion *IMs* such as peak ground displacement and long period spectral ordinates (Section 3.3.2). For example, in the TCU068 record (Figure 3.8, right side), the maximum displacement in the plot is approximately a factor of 10 larger for the unfiltered as compared to the filtered record. Because long-period spectral displacements approach the peak ground displacement in the limit (Section 3.3.2.1), these

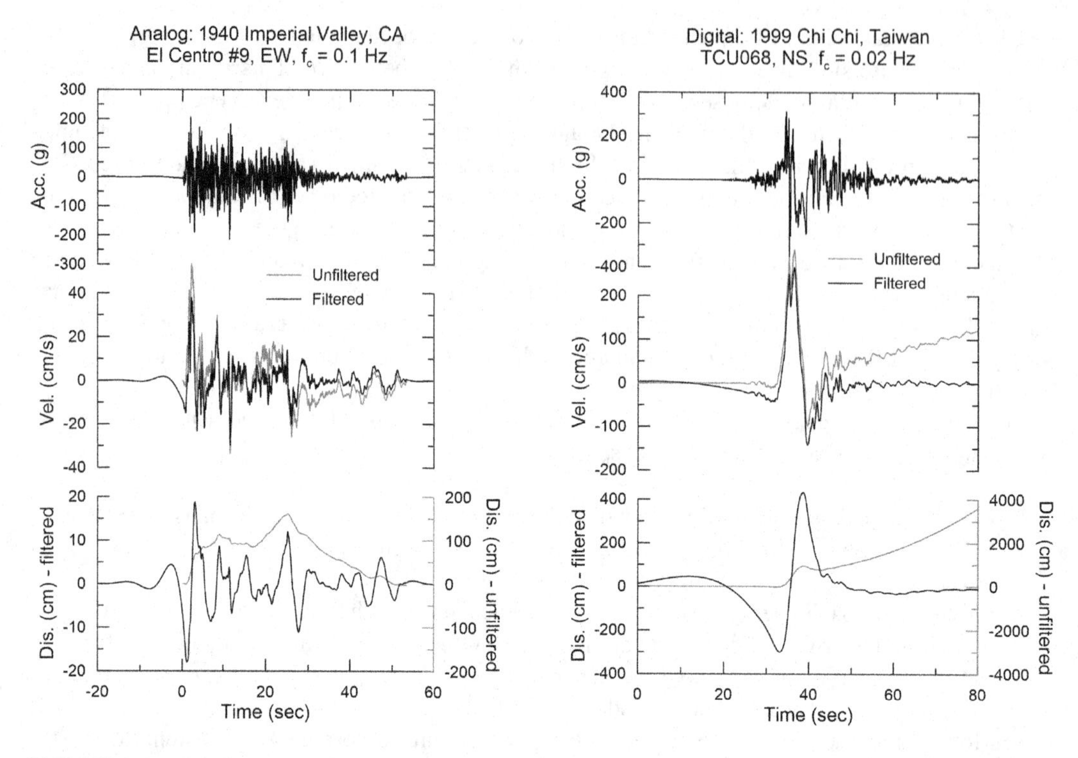

FIGURE 3.8 Acceleration, velocity, and displacement time series from analog (left) and digital (right) recordings before and after high-pass filtering (Boore, 2005). For the unfiltered records, the mean of acceleration was subtracted from the analog recording prior to integration, and the mean of pre-event memory was subtracted from the digital recording. The displacement scales for the unfiltered records are ten times larger than those of the filtered records, as indicated by the ticks to the right of the time series plots. The return of the unfiltered analog record to zero displacement is coincidental and not typical.

effects are also important for response spectra. The distinction from the high-frequency noise case is important to recognize – whereas high-frequency noise does not affect spectral ordinates for most records, the opposite is true for low-frequency noise (i.e., most records will be affected). For this reason, correction of recorded ground motions for low-frequency noise and baseline drift is nearly universal, even for data recorded by modern sensors.

Low-frequency noise and baseline drift effects are typically removed by adding zeros at the front and back ends of the record (i.e., *zero padding*) and applying *high-pass (low-cut) filters* that remove components of the ground motion at frequencies below a corner frequency, f_c. The selection of an appropriate corner frequency for a given record is an art – if set too high, desirable portions of the signal can be removed, which unnecessarily limits the applicability of the record. If set too low, displacement histories can retain non-physical low-frequency drifts. In modern ground motion databases, corner frequencies are selected on a component-by-component level by an experienced seismologist and are reported in the header information of the digital file containing the record. Engineers will seldom be required to perform this processing themselves, but it is essential that they understand the meaning of the corner frequency. Its practical significance is illustrated in Figure 3.9, where spectral displacements and pseudo-spectral accelerations of the 1940 El Centro record from Figure 3.8 (left side) are shown for the unfiltered record and versions of the filtered record with three corner frequencies. Spectral responses for the filtered records depart from that for the unfiltered record at periods $T >\sim 0.7(1/f_c)$. For this reason, in the PEER-NGA database, the lowest usable frequency is reported as $1.25f_c$. These values correspond to periods of 4 and 40 sec for the analog and digital recordings, respectively, shown in Figure 3.8.

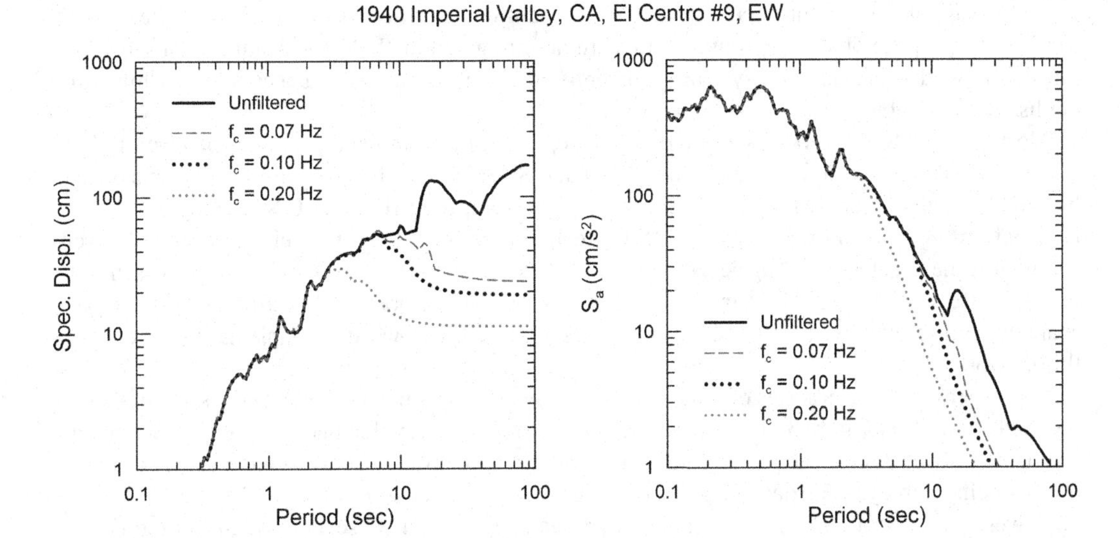

FIGURE 3.9 Displacement and pseudo-acceleration response spectra of the acceleration history recorded at El Centro Station #9 during the 1940 Imperial Valley earthquake, filtered at various corner frequencies. (Data provided by from David Boore, *personal communication*, 2013.)

As an alternative to high-pass filtering, several research groups have developed baseline correction procedures applied in the time domain. For example, the mean of a pre-event time window is typically subtracted from a digital acceleration history to produce a corrected history (Faccioli et al., 2004; Paolucci et al., 2008; Akkar and Boore, 2009). A polynomial of order n (usually $n = 5$ or 6), but without the constant and linear terms, can be fitted to a displacement time series, and its second derivative then subtracted from the corresponding acceleration time series to produce a corrected acceleration time series. Akkar and Boore (2009) described more advanced baseline removal procedures for records with a permanent static offset, caused by tectonic displacement near the causative fault or by soil ground failure. These procedures are used relatively rarely and a carefully chosen high-pass filter is generally considered the most reliable method for minimizing the effects of low-frequency noise and baseline drift (Boore, 2003a; Ancheta et al., 2014).

3.2.4 Ground Motion Instrument Networks

A ground motion instrument network is a series of accelerometers and/or seismometers distributed across a region of interest to record the spatial variations of ground shaking from earthquakes or other sources. These data are used to locate vibration sources (including earthquake hypocenters) and to produce inventories of ground motions for use by seismologists and engineers.

3.2.4.1 Surface Networks

Surface networks are maintained at global, regional, and local scales for various applications. At the global scale, the Global Seismographic Network (GSN) (https://www.iris.edu/hq/programs/gsn) is a permanent digital network of approximately 150 advanced sensors connected by a telecommunications system for earthquake monitoring and other applications. GSN is the successor to the Worldwide Standard Seismograph Network (WWSSN), which was founded in 1961 to monitor compliance with nuclear weapons test ban treaties. The GSN is especially important for earthquake detection and location, particularly in parts of the world without regional or local arrays.

Regional networks operate in many countries with significant seismic activity. Often those networks are maintained by public agencies and information on the networks and the data they produce is freely available on the internet. The most significant of those networks operating at the present time for the purpose of strong ground motion studies are given in Table 3.1. A number of other significant networks for which no system is currently in place for public data access (e.g., China) are not listed in the table.

Most of the U.S. networks listed in Table 3.1 represent the combined resources of several organizations. For example, in California, the NCSN and SCSN arrays are comprised of stations owned by the California Strong Motion Instrumentation Program (CSMIP), the U.S. Geological Survey National Strong Motion Project, and SCSN/Caltech. Figure 3.10 shows the networks' station coverage within the greater Los Angeles, CA region. These regional networks are used for estimating more precise source locations than is possible from global networks and for archiving the ground shaking from seismic events for research purposes. Figure 3.5 shows an example instrument within this network.

Local arrays provide a series of sensors at relatively close spacing, typically at the scale of tens to hundreds of meters. Figure 3.11 shows an example of such an array that had seven three-component stations within 160 m of each other and 14 more stations within 2 km. Most local arrays are temporary, being deployed for days or months (to record aftershocks) or for several years. A few are effectively permanent and are operational at present. Table 3.2 lists several long-term temporary and currently operational arrays that have produced usable data for these purposes. Data from these arrays have been used to study the spatial variations of ground motion over the length scales of interest for typical engineering projects (e.g., the size of a building or bridge).

3.2.4.2 Vertical Arrays

Arrays that include sensors installed below, as well as at, the ground surface are known as *vertical arrays*. Vertical arrays are less common than surface instruments, but are noteworthy because they provide valuable data for geotechnical earthquake engineering applications.

Subsurface sensors in vertical arrays are installed within boreholes. The deepest sensor is ideally installed within intact bedrock, but where that is not possible, such as in deep sedimentary basins, it may be installed within firm soils. Some vertical arrays include sensors at intermediate depths, which add significant value.

One motivation for the installation of vertical arrays is a relatively low-noise environment below the ground surface that is conducive to source detection. When such arrays have reasonably closely spaced sensors in the vertical direction, another major application is studies of dynamic soil behavior under in situ conditions, from which ground failure and other nonlinear phenomena can be directly observed (Zeghal and Elgamal, 1994; Elgamal et al., 1996; Holzer and Youd, 2007). Such studies require high-quality material characterization over the depth range of the array (i.e., between sensors).

Table 3.3 lists several networks that include vertical arrays. By far the largest such network is Kik-net in Japan. The Kik-net array stations are installed roughly on a grid, but with a preference for siting on rock sites because the main original purpose of the array was source detection. This array has produced thousands of recordings and has been widely used in studies of site response (e.g., Thompson et al., 2012; Ghofrani et al., 2013; Zalachoris and Rathje, 2015). Drawbacks of the Kik-net arrays are that they consist of only a single downhole sensor and the site profiles are of low resolution.

3.2.5 Ground Motion Records

Many ground motion instrument networks, including most of those listed in Tables 3.1–3.3, disseminate ground motion records online. Many records from global, regional, and local networks of seismometers and accelerometers are disseminated by the IRIS Data Management Center (https://ds.iris.edu/ds/nodes/dmc/), which is targeted to applications in seismology and geophysics as well as public education. Strong motion records, principally from accelerometers, are also disseminated by regional network operators, with varying levels of data processing and filtering.

TABLE 3.1
Selected Regional Ground Motion Networks in Operation as of April 2023 for Which the Network Information and Collected Data Are Available for Public Use

Region	Network Name or Sponsoring Organization	Sensor types (D = digital, A = analog)	# Stations	URL (4/2023)
Northern California, US	NCSN: Northern California Seismic Network	seismometers: D & A; accelerometers: D & A; downhole sensors	493	http://www.ncedc.org/ncsn/
Southern California, US	SCSN: Southern California Seismic Network	seismometers: D; accelerometers: D	565	http://www.scsn.org/index.html
Pacific Northwest, US	PNSN: Pacific Northwest Seismic Network	seismometers: D; accelerometers: D	474	https://pnsn.org/
Nevada, US	Nevada Seismological Laboratory	seismometers: D; accelerometers: D	247	http://www.seismo.unr.edu/
Arizona, US	Arizona Geological Survey	seismometers: D & A	18	https://aeic.nau.edu/index.html
New Madrid, Tennessee, South Carolina, US	CERI: Center for Earthquake Research and Information	seismometers: D & A	150	http://www.memphis.edu/ceri/seismic
Chile	Red Sismológica Nacional	seismometers: D; accelerometers: D	297	https://fdsn.org/networks/detail/C1/
Greece	ITSAK: Inst. Eng. Seism. & Eqk. Eng. Research	accelerometers: D	250	http://www.itsak.gr/en
Italy	ITACA: Multiple networks, including RSNI, RAN, and INGV	seismometers: D; accelerometers: D	1523	https://itaca.mi.ingv.it/
Japan	NIED: Nat. Research Inst. for Earth Sci. & Disaster Prevention	accelerometers: D; downhole sensors	1744	http://www.kyoshin.bosai.go.jp/
New Zealand	GeoNet: Earthquake Commission and GNS Science	seismometers: D; accelerometers: D	779	http://geonet.org.nz/
Taiwan	CWB: Central Weather Bureau	seismometers: D; accelerometers: D	870	https://scweb.cwb.gov.tw
Turkey	TNSMN: Turkish National Strong Motion Network	seismometers: D; accelerometers: D	762	https://tadas.afad.gov.tr/

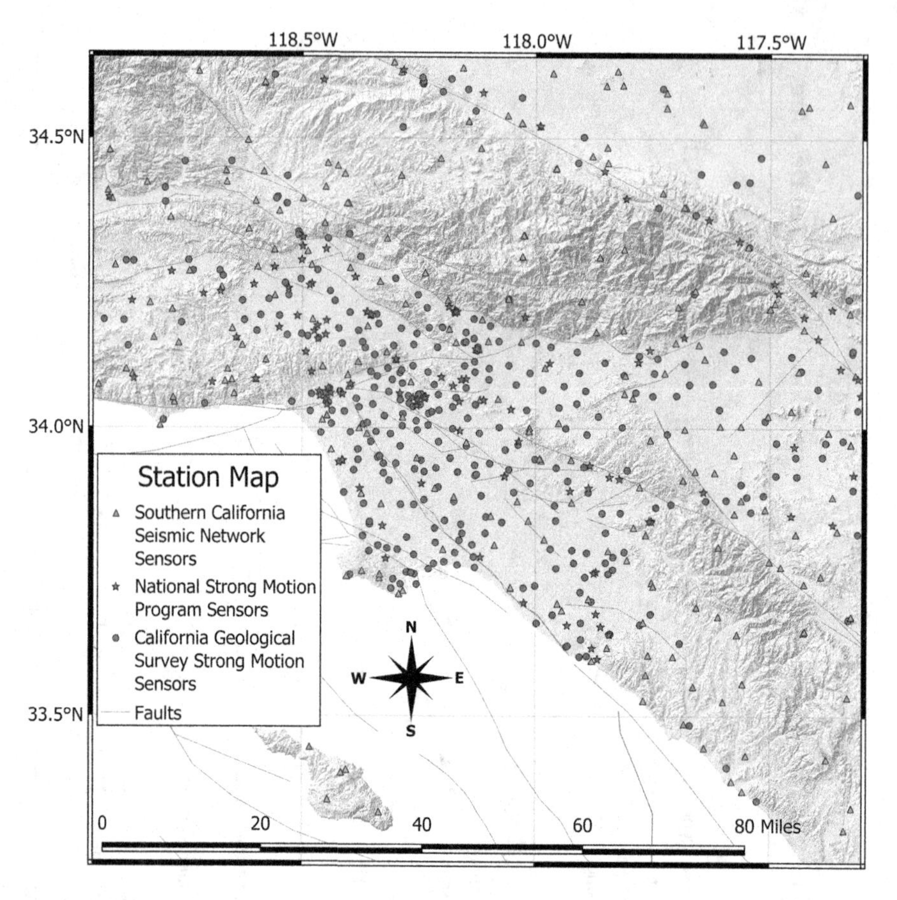

FIGURE 3.10 Map of southern California regional seismic stations in three networks.

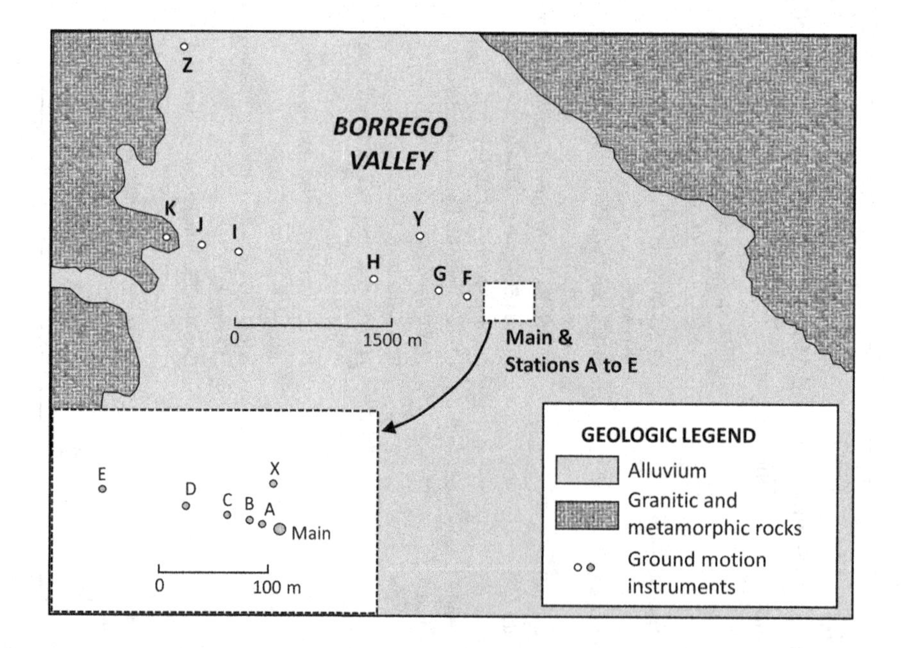

FIGURE 3.11 Configuration of the local Borrego Valley Differential Array (Kato et al., 1998).

TABLE 3.2
Examples of Local Arrays that Include Horizontal-Component Ground Motion Sensors

Region	Array	Btw.-Station Separation (m)	Seism., Accel.	Reference or URL (9/2019)
Armenia	Garni Observatory	100	S	Mori et al. (1994)
California (temp)	USGS: Imperial Valley Diff.	18–213	A	Smith et al. (1982); Abrahamson (2007)
California (temp)	USGS: Pinyon Flat	30, 100, 300	S	Owens et al. (1991)
California (temp)	UCSB: Borrego Valley Diff.	10–2,000	A	Kato et al. (1998)
California (temp)	USGS: Hollister Diff.	61–256	A	Abrahamson (2007)
California (temp)	USGS Parkfield	25–952	S, A	Fletcher et al. (1992); Abrahamson (2007)
California	EPRI Parkfield	10–191	A	Abrahamson (2007)
California	UCSB: Garner Valley	65–300	A	http://nees.ucsb.edu/facilities/GVDA
Greece	ITSAK: Euroseistest	> 220	A	Pitilakis et al. (2013)
Japan	Chiba	5–320	A	Katayama et al. (1990); Ghayamghamian & Nouri (2007)
Taiwan (temp)	SMART 1, SMART 2, LSST	3–2,000	A	Abrahamson et al. (1991)

TABLE 3.3
Major Vertical Ground Motion Arrays in Operation as of June 2024 with Publically Accessible Data

Region	Organization	# Sensor Depths	# Stations	Reference or URL
California	CSMIP: California Strong Motion Instrumentation Program; USGS	> 2	39	Afshari et al. (2019)
Greece	ITSAK	4	4	http://www.itsak.gr/en
Japan	NIED, KiK-Net	2	700	http://www.kyoshin.bosai.go.jp/

For most engineering applications, such as the selection of records for response history analysis of structures (Section 4.5), the most useful ground motion dissemination tools are those that include inventories of motions for many regions sharing a similar tectonic setting, such as active tectonic regions (e.g., California, southern Europe, parts of Japan, Taiwan, New Zealand, etc.). By including multiple regions, the size of the data inventories is increased, as are the breadth of magnitudes, site-to-source distances, and site conditions for the records in the inventory. Larger inventories covering a wider range of conditions are more likely to provide records useful for a particular engineering application. One such inventory for active tectonic regions is the PEER-NGA Ground Motion Database (https://ngawest2.berkeley.edu/site) (Ancheta et al., 2014). This inventory has the advantage of large size and uniform processing of the records using consistent protocols with respect to high-pass filtering and baseline correction. A record from the Alhambra-Fremont School site

```
PEER NGA STRONG MOTION DATABASE RECORD
Northridge-01, 1/17/1994, Alhambra - Fremont School, 90
ACCELERATION TIME SERIES IN UNITS OF G
NPTS=  3000, DT=   .0200 SEC
  -.2180011E-03   .1291326E-03   .5806444E-04  -.1951005E-02  -.1375397E-02
  -.1520930E-02  -.1118726E-02   .4013978E-03   .2410099E-02   .1457739E-02
   .8531971E-03   .2865974E-02   .2582728E-02  -.6207787E-03  -.1094474E-03
   .2047142E-02   .8061102E-03   .1112416E-02   .1541121E-02   .2519604E-02
   .2561725E-02   .2299885E-02   .1178184E-02   .1346780E-02  -.3303105E-02
  -.4010692E-02  -.2451840E-02  -.7797705E-03  -.1103842E-02  -.1753295E-02
   .2208670E-02   .1912134E-02   .1364675E-02  -.2338663E-02  -.1324505E-02
```

FIGURE 3.12 Header information and initial portion of ground motion data from Alhambra-Fremont School site as downloaded from PEER-NGA ground motion database website.

(shown previously in Figure 3.1), as downloaded from the PEER-NGA website is given in Figure 3.12. Note that the header identifies the earthquake name and date, the station name as given by the source data provider (CDMG in this case), the azimuth of the recording (090, indicating 90 degrees east from due north), the number of data points (3,000), and the time step (Δt = 0.020 sec). The data begins in the upper left and proceeds to the right across the row, then down to the next row, and so on.

Similarly compiled strong motion data sets have been assembled as part of the NGA-East project (https://ngawest2.berkeley.edu/) (Goulet et al., 2021b) and the NGA-Subduction project (https://www.risksciences.ucla.edu/nhr3/nga-subduction) (Mazzoni et al., 2022). The NGA-East data set is applicable to stable continental regions worldwide (although most of the data are from central and eastern North America), The NGA-Subduction project is applicable to subduction zones worldwide, with most data obtained from Japan, Taiwan, and South America.

3.3 GROUND MOTION INTENSITY MEASURES

Intensity measures describe amplitude, frequency content, or duration characteristics of earthquake ground motions in a concise, quantitative form. There are a number of practical uses of *IM*s in earthquake engineering. First, they provide a concise indication of important characteristics of a ground motion and are often used directly to represent seismic inputs in analysis of the seismic response of structures and geotechnical systems. Second, the most commonly used ground motion prediction methods (Section 3.5) do not provide full ground motion time series, but instead provide estimates of *IM*s given attributes of the site and its seismic environment. *IM*s have limitations in that they provide an incomplete descriptions of ground motion; a complete description requires a full time series or accelerogram, which can be used in response history analyses. Hence, the use of *IM*s represents a simplification of ground motions that is necessary for many engineering applications.

The merits of a particular *IM* relative to others lie in its relationship to the response measure(s), or *Engineering Demand Parameters* (*EDP*s), of interest (Section 5.3.2). As shown in Figure 3.13, an *efficient IM* is one that is closely related to the most useful *EDP*, i.e. one for which the uncertainty in *EDP* given *IM* is small. A *sufficient IM* is one that captures all of the useful information about the ground motion's potential to produce response, i.e., one for which additional ground motion information provides no reduction of uncertainty in *EDP*. A *predictable IM* is one that can be predicted relatively accurately from earthquake source, path, and site parameters, i.e., one for which the error term in ground motion models (Section 3.5), $\sigma_{\ln IM}$, is low. Performance predictions can be made most accurately when using *IM*s that are efficient, sufficient, and predictable.

The following sections describe typical *IM*s related to the amplitude, frequency content, and duration characteristics of ground motions. The list is not exhaustive but includes the most widely used *IM*s for seismic response characterization for geotechnical and structural systems. *IM*s are illustrated using accelerograms from the Alhambra-Fremont School site (Figure 3.1), specifically the 270 degree component from the Whittier Aftershock (**M** 5.3, distance to

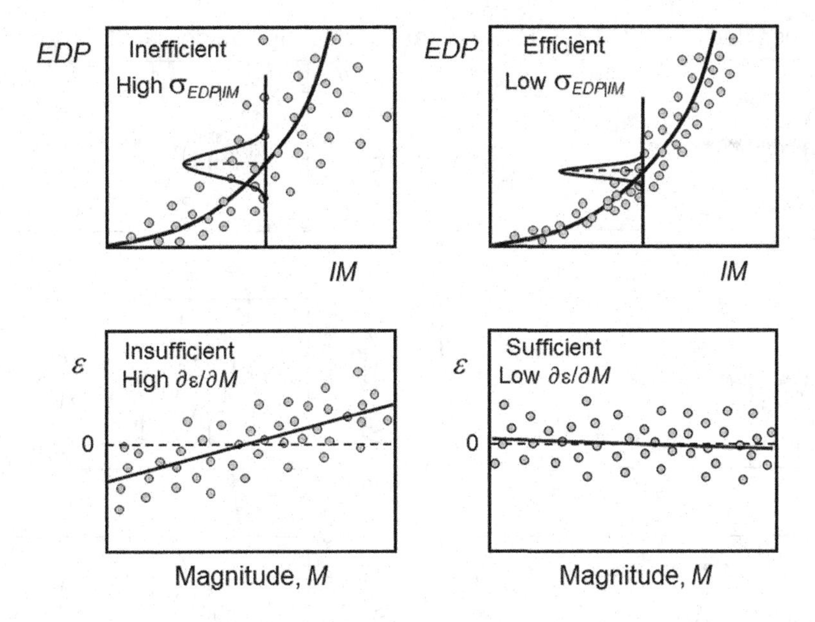

FIGURE 3.13 Illustration of inefficient, efficient, insufficient, and sufficient *IM*s. Values of ε represent vertical distances between data points and the mean *EDP* curve. Sufficiency is illustrated here with respect to magnitude.

surface projection of fault plane R_{JB} = 3.6 km) and the 90 degree component from the Northridge mainshock (**M** 6.7, R_{JB} = 36 km).

3.3.1 Amplitude Parameters

The recorded accelerograms for the east-west components of the motions at the Alhambra-Fremont School site are shown in Figure 3.14, along with velocities and displacements obtained through time integration. The most direct amplitude measures are the peak values of acceleration, velocity, and displacement, which are identified in the figure. The inherent smoothing tendency of the integration process tends to dilute high-frequency components of the motion and enhance low-frequency components of the velocity and displacement histories. For this reason, peak ground acceleration (*PGA*) is a relatively high-frequency ground motion parameter, whereas peak ground velocity (*PGV*) and peak ground displacement (*PGD*) are more sensitive to mid- and low-range frequencies, respectively. These are indicated in Figure 3.14 where the relatively low-frequency motion from the Northridge earthquake produces a larger *PGV* and *PGD* than the relatively high-frequency Whittier aftershock motion, despite the higher *PGA* for the Whittier aftershock.

3.3.1.1 Peak Acceleration

The *PGA* for a given component of ground motion is simply the largest (absolute) value of acceleration, as indicated in the accelerogram plots in Figures 3.1 and 3.14. *PGA* and other *IM*s for horizontal-component ground motions are often expressed in a manner that recognizes both horizontal directions of shaking (e.g., east-west and north-south). Figure 3.15 shows the variations in time of acceleration within the horizontal plane for the Northridge recording at the Alhambra-Fremont School site. The maximum value of acceleration does not occur in either direction of the recordings (marked 90° and 360° in the figure), but instead occurs along an axis rotated approximately 240 degrees from north. This component is referred to as the maximum component and denoted RotD100 (100 refers to the 100th percentile, or maximum value for all possible azimuths). The minimum

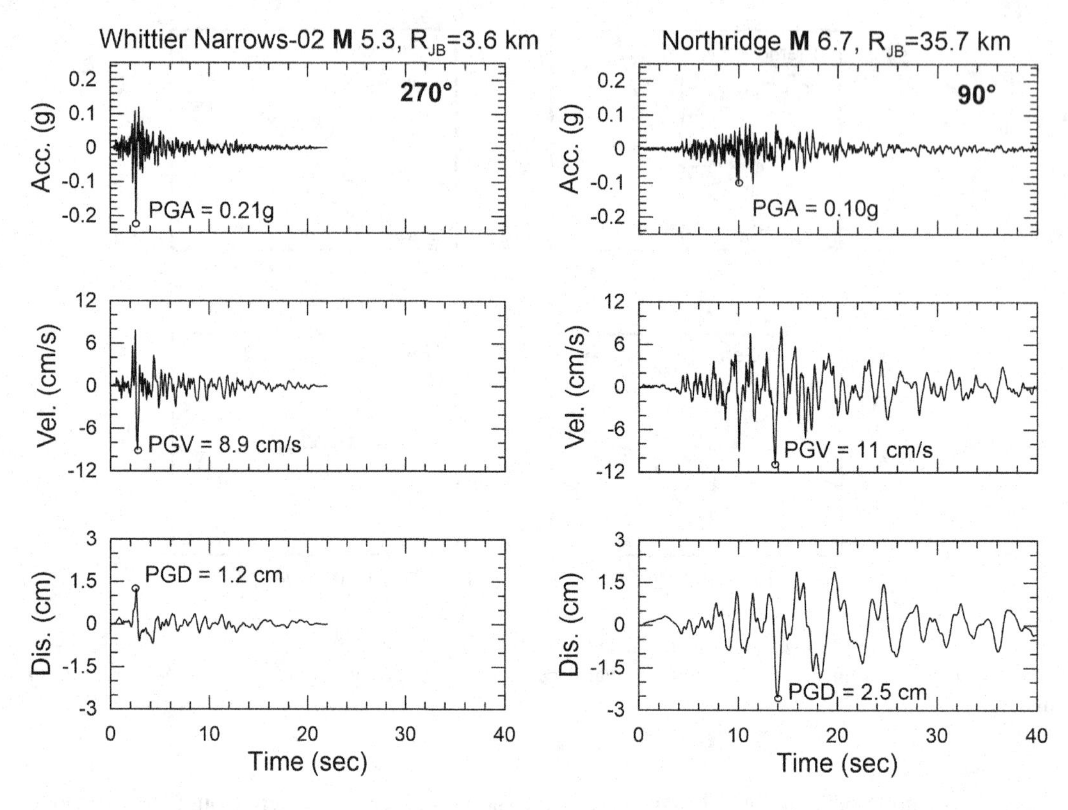

FIGURE 3.14 Acceleration, velocity, and displacement histories for east-west components of ground motion recorded at Alhambra-Fremont School during 1987 Whittier Aftershock and 1994 Northridge earthquakes. 270° and 90° refer to the horizontal angle from due north (hence, both are in the east-west direction).

component is RotD0 and the median is RotD50 (Boore, 2010). The geometric mean *PGA* (PGA_{GM}) is another parameter used to represent both horizontal directions of shaking; it is defined as

$$PGA_{GM} = \sqrt{PGA_X \cdot PGA_Y} \tag{3.5}$$

where PGA_X and PGA_Y indicate peak accelerations as recorded in perpendicular directions.

Horizontal accelerations have commonly been used to describe ground motions because of their natural relationship to the lateral inertial forces that tend to cause damage to structures; indeed the largest dynamic forces induced in certain types of structures (i.e., very stiff and elastic structures) are closely related to the horizontal *PGA*. Peak accelerations can also be related (e.g., Worden et al., 2012; Wald et al., 1999; Trifunac and Brady, 1975) to earthquake intensity (Section 2.8.3), although such relationships are more efficiently defined in terms of peak velocity, an example of which is shown in the next section. Such correlations, while far from precise, can be useful for estimating *PGA* when only intensity information is available, as in the case of earthquakes that occurred before ground motion recording instruments were available (pre-instrumental earthquakes).

Vertical accelerations have received less attention in earthquake engineering than horizontal accelerations, primarily because the margins of safety against gravity-induced static vertical forces in constructed works usually provide adequate resistance to dynamic forces induced by vertical accelerations during earthquakes. The ratio of vertical to horizontal *PGA* (i.e., the *V*/*H* ratio) has

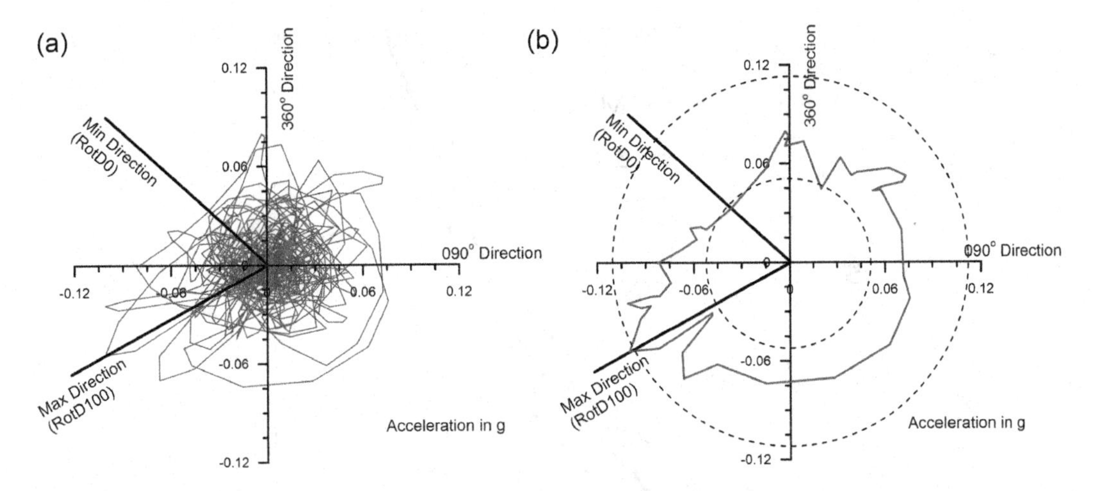

FIGURE 3.15 Variation over time of horizontal acceleration within the horizontal plane, Alhambra-Fremont School site, 1994 Northridge earthquake. The left side shows the acceleration trace, while the right side shows its envelope. RotD100 and RotD0 indicate the azimuths having the maximum (100th percentile) and minimum (0 percentile) peak accelerations, respectively. Note that the min and max directions are not necessarily orthogonal.

been found to be strongly distance-dependent, averaging approximately unity near the seismic source and dropping to values between approximately 1/2 to 2/3 for fault distances greater than approximately 50 km (Bozorgnia and Campbell, 2016). Peak vertical accelerations can be quite large; a value of 1.74*g* was measured between the Imperial and Brawley faults in the 1979 Imperial Valley earthquake. The large amplitudes of near-fault vertical motions are associated with higher frequencies than is typical for horizontal components, which results in part from richer contributions of p-waves for vertical-component motions.

Ground motions with high peak accelerations are usually, but not always, more destructive than motions with lower peak accelerations. Very high peak accelerations that last for only a very short period of time, however, often cause little damage. A number of earthquakes have produced peak accelerations in excess of 0.5*g* but caused no significant damage because the peak accelerations occurred at very high frequencies relative to the fundamental frequencies of structures and the duration of the shaking was short. *PGA* is most efficient for short, stiff structures or for problems where inertia in the ground itself is the principal consideration. For more flexible systems, *PGA* is typically not used or is supplemented by additional information to more accurately characterize the destructive potential of ground motions (i.e., *PGA* is not efficient or sufficient for predicting the response of such systems, although it is predictable).

3.3.1.2 Peak Velocity and Displacement

PGV characterizes ground motion amplitude at intermediate frequencies. Some applications of *PGV* include correlations with earthquake intensity (e.g., Worden et al., 2012 and Kaka and Atkinson 2004 for western and eastern North America, respectively), estimation of shear strains in soil through normalization by soil shear wave velocity V_s (Trifunac and Todorovska, 1996; Brandenberg et al., 2009), empirical correlations with damage in structural systems (Trifunac and Todorovska, 1997; Miyakoshi et al., 1998; Boatwright et al., 2001) and earth structures (Kwak et al., 2016), modeling of seismic lateral earth pressures on retaining structures (Section 8.6.3), and estimation of inelastic

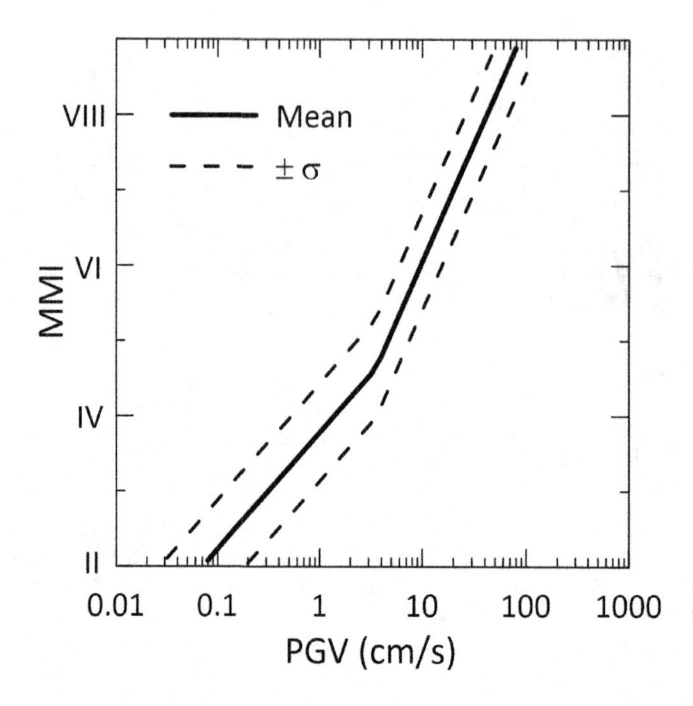

FIGURE 3.16 Relationships by Worden et al. (2012) between peak velocity and MMI derived using data from California.

responses in simple structural oscillators (Akkar and Küçükdoğan, 2008). A relationship between *PGV* and intensity is shown in Figure 3.16 (Worden et al., 2012).

PGD is generally associated with lower-frequency components of an earthquake motion. They are highly sensitive to details of the record processing, particularly the high-pass corner frequency (Section 3.2.3.2). As a result, *PGD* is not commonly used as a measure of ground motion.

3.3.2 Frequency Content Parameters

Simple analyses (Section B.5.3) illustrate that the dynamic response of compliant objects, be they buildings, bridges, slopes, or soil deposits, is very sensitive to the frequency at which they are excited. Earthquakes produce complicated loading with components of motion that span a broad range of frequencies. The frequency content describes how the amplitude of a ground motion is distributed among different frequencies. Since the frequency content of an earthquake motion will strongly influence the effects of that motion, characterization of a ground motion is incomplete without consideration of its frequency content.

3.3.2.1 Ground Motion Spectra

Any periodic function (i.e., any function that repeats itself exactly at a constant interval) can be expressed using Fourier analysis as the sum of a series of simple harmonic terms of different frequency, amplitude, and phase. Using the Fourier series (Section A.3), a periodic function, $x(t)$, can be written as

$$x(t) = c_0 + \sum_{n=1}^{\infty} c_n \sin(\omega_n t + \phi_n) \tag{3.6}$$

In this form, c_n, ω_n, and ϕ_n are the amplitude, circular frequency, and phase angle, respectively, of the nth harmonic of the Fourier series [see Equation A.12 and related text for definitions]. The Fourier series provides a complete description of the ground motion since it can be completely recovered by the inverse Fourier transform.

Fourier Spectra: A plot of Fourier amplitude versus frequency [c_n versus ω_n from Equation 3.6] is known as a *Fourier amplitude spectrum*; a plot of Fourier phase angle (ϕ_n versus ω_n) gives the *Fourier phase spectrum*. The Fourier amplitude spectrum of a strong ground motion shows how the amplitude of the motion is distributed with respect to frequency (or period). It therefore directly expresses the frequency content of the motion.

The Fourier amplitude spectrum may be narrow or broad. A narrow spectrum implies that the motion has a dominant frequency (or period) range, which can produce a smooth, almost sinusoidal time series. A broad spectrum corresponds to a motion that contains a wide range of frequencies having significant energy, which is associated with a time series with a relatively jagged, irregular shape. The Fourier amplitude spectra for the E-W components of the Alhambra-Fremont School site for the relatively low- and moderate-magnitude events (Whittier Aftershock and Northridge mainshock) are shown in Figure 3.17. The jagged shapes of the spectra are typical of those observed for individual ground motions. The shapes of the spectra for the two ground motions are similar when plotted in log-log space (as is done in Figure 3.17), but reflect different frequency contents: the Northridge (moderate magnitude and distance) spectrum is strongest at low frequencies (or long periods) while the Whittier aftershock (low magnitude and distance) spectrum is strongest at high frequencies.

As illustrated in Figure 3.17, the general shape of Fourier spectra for earthquake records plotted in log-log space tends to be flat over a central range of frequencies, with steady decay of amplitude at lower and higher frequencies. As discussed further in Section 3.6, the low-frequency limit of the flat portion occurs near a frequency that is inversely related to the cube root of the seismic moment (Brune, 1970, 1971). This result indicates that larger earthquakes produce greater low-frequency motions than do small earthquakes, which is observed from the records in Figure 3.17 where the limiting frequencies of the flat portion of the Fourier spectra are approximately 0.25 Hz for the **M** 6.7 Northridge event and 0.9 Hz for the **M** 5.3 Whittier Narrows aftershock. The rate at which the spectra fall off at high frequencies is controlled by crustal and site damping as described further in Section 3.6.

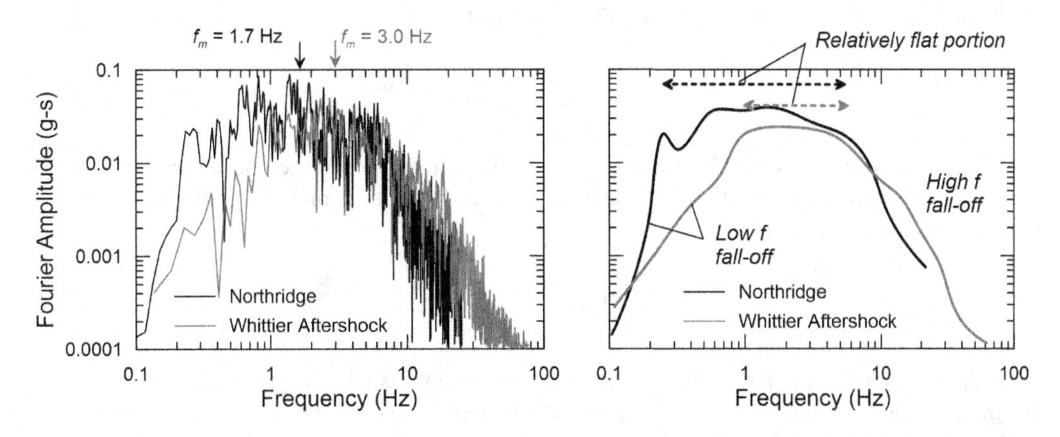

FIGURE 3.17 Fourier amplitude spectra for the E-W components of the Alhambra-Fremont School site strong motion records from the Northridge and Whittier Narrows aftershock events. Fourier spectra were obtained by discrete Fourier transform (Section A.3.3) and consequently have units of velocity. Mean frequencies (f_m) are marked for each record. Actual jagged spectra on left, smoothed spectra on right with flat portion and high- and low-frequency fall-offs indicated.

Since phase angles control the times at which the peaks of harmonic motions occur, the Fourier phase spectrum influences the variation of ground motion with time. In contrast to Fourier amplitude spectra, Fourier phase spectra from actual earthquake records do not display characteristic shapes.

Power Spectra: The frequency content of a ground motion can also be described by a power spectrum or power spectral density function. The power spectral density function can also be used to estimate the statistical properties of a ground motion and to compute stochastic response using random vibration techniques (Clough and Penzien, 1975; Vanmarcke, 1976; Yang, 1986).

The total intensity of a ground motion of duration T_d is given in the time domain by the area under the time series of squared acceleration:

$$I_0 = \int_0^{T_d} \left[\ddot{u}_g(t)\right]^2 dt \tag{3.7}$$

Using Parseval's theorem, the total intensity can also be expressed in the frequency domain, as

$$I_0 = \frac{1}{\pi}\int_0^{\omega_N} c_n^2 \, d\omega \tag{3.8}$$

where $\omega_N = \pi/\Delta t$ is the Nyquist frequency (the highest frequency in the Fourier series). The average intensity, λ_0, can be obtained by dividing Equations (3.7) and (3.8) by the duration.

$$\lambda_0 = \frac{1}{T_d}\int_0^{T_d} \left[\ddot{u}_g(t)\right]^2 dt = \frac{1}{\pi T_d}\int_0^{\omega_N} c_n^2 \, d\omega \tag{3.9}$$

Notice that the average intensity is equal to the mean-squared acceleration. The *power spectral density*, $S(\omega)$, is defined such that

$$\lambda_0 = \int_0^{\omega_N} S(\omega)\, d\omega \tag{3.10}$$

which indicates, by comparing Equations (3.9) and (3.10), that

$$S(\omega) = \frac{1}{\pi T_d} c_n^2 \tag{3.11}$$

The close relationship between the power spectral density function and the Fourier amplitude spectrum is apparent from Equation (3.11). The power spectral density is often normalized by dividing its values by the area beneath it

$$\hat{S}(\omega) = \frac{1}{\lambda_0} S(\omega) \tag{3.12}$$

where λ_0, as before, is the mean-squared acceleration. As indicated in Equations (3.9–3.11), power spectral densities are sensitive to duration, for which there is no unique definition (Section 3.3.3).

The power spectral density function is useful for characterizing earthquake ground motion as a random process. The power spectral density function by itself can describe a *stationary* random process (i.e., one whose statistical parameters do not vary with time). Actual strong motion accelerograms, however, frequently show that the intensity builds up to a maximum value in the early part of the motion, then remains approximately constant for a period of time, and finally decreases near the end of the motion. Such nonstationary random process behavior is often modeled by multiplying a stationary time series by a deterministic intensity function (e.g., Hou, 1968; Shinozuka, 1973; Saragoni and Hart, 1973). Changes in frequency content during the motion, which can be associated with arrivals of different wave types or with nonlinear soil response, have been described using an evolutionary power spectrum approach (Priestley, 1965, 1967; Liu, 1970).

Response Spectra: A third type of spectrum is used extensively in earthquake engineering practice. The *response spectrum* describes the maximum response of an SDOF system to a particular input motion as a function of the natural frequency (or natural period) and damping ratio of the SDOF system (Section B.7). Response spectra for the Alhambra-Fremont School records are illustrated in Figure 3.18, including the pseudo-spectral acceleration and velocity spectra (denoted S_a and S_v, respectively) and the displacement spectrum. The true spectral acceleration differs slightly from the pseudo-spectral acceleration for non-zero oscillator periods. While the pseudo-spectral acceleration has more direct engineering significance (it is proportional to base shear in an elastic oscillator; Clough and Penzien, 1993), because the distinction is generally small in the period range of practical interest, the subsequent text will often shorten pseudo-spectral acceleration to simply spectral acceleration or response spectral acceleration.

Response spectra may be plotted individually using arithmetic scales, or may be combined, by virtue of the relationships of Equations (B.81), in *tripartite plots*. The tripartite plot (Figure A.9) displays spectral velocity on the vertical axis, natural frequency (or period) on the horizontal axis, and acceleration and displacement on inclined axes; all four axes are logarithmic. The acceleration and displacement axes are reversed when the spectral values are plotted against oscillator natural period rather than frequency. The shapes of typical response spectra indicate that peak spectral acceleration, velocity, and displacement values are associated with different frequencies (or periods). At low frequencies (long periods) the average spectral displacement is constant and approaches *PGD*; at high frequencies, the average spectral acceleration is constant and approaches *PGA*. In between lies a range of nearly constant spectral velocity. Because of this behavior, response spectra are often divided into *acceleration-controlled* (high-frequency), *velocity-controlled* (intermediate-frequency), and *displacement-controlled* (low-frequency) portions.

Response spectra reflect strong ground motion characteristics indirectly, since they are "filtered" by the reponses of SDOF oscillators. The amplitude, frequency content, and, to a lesser extent, duration of the input motion all influence spectral values. The different frequency contents of the Alhambra-Fremont School ground motions are illustrated by the different shapes of their respective response spectra. The relatively rich energy at low frequencies from the Northridge earthquake recording (Figure 3.17) causes higher long-period response spectral ordinates (Figure 3.18), whereas the greater high-frequency energy from the Whittier Narrows aftershock recording produces larger short-period spectral ordinates. The influence of the long-period energy is most clearly seen in the spectral displacement plots, whereas the influence of the short period energy is most clear in spectral accelerations.

It is important to remember that response spectra represent only the maximum responses of oscillators at different periods. Nonetheless, for structures or systems whose responses are nearly elastic (i.e., do not involve substantial nonlinearity), spectral quantities at periods of interest (typically modal periods) generally meet all the criteria of efficiency, sufficiency, and predictability for response prediction. Even for nonlinear systems, the spectral ordinate at the elastic first-mode period is a useful *IM* and provides the principal mechanism for specifying seismic demands in codes and standards for most structures (e.g., ASCE-7, 2022).

3.3.2.2 Spectral Parameters

As an alternative to complete spectra, frequency content can be approximately characterized by individual, or scalar, *spectral parameters* representing central, or dominant, periods of the ground motion or the width of the spectrum. The most widely used of these are described below.

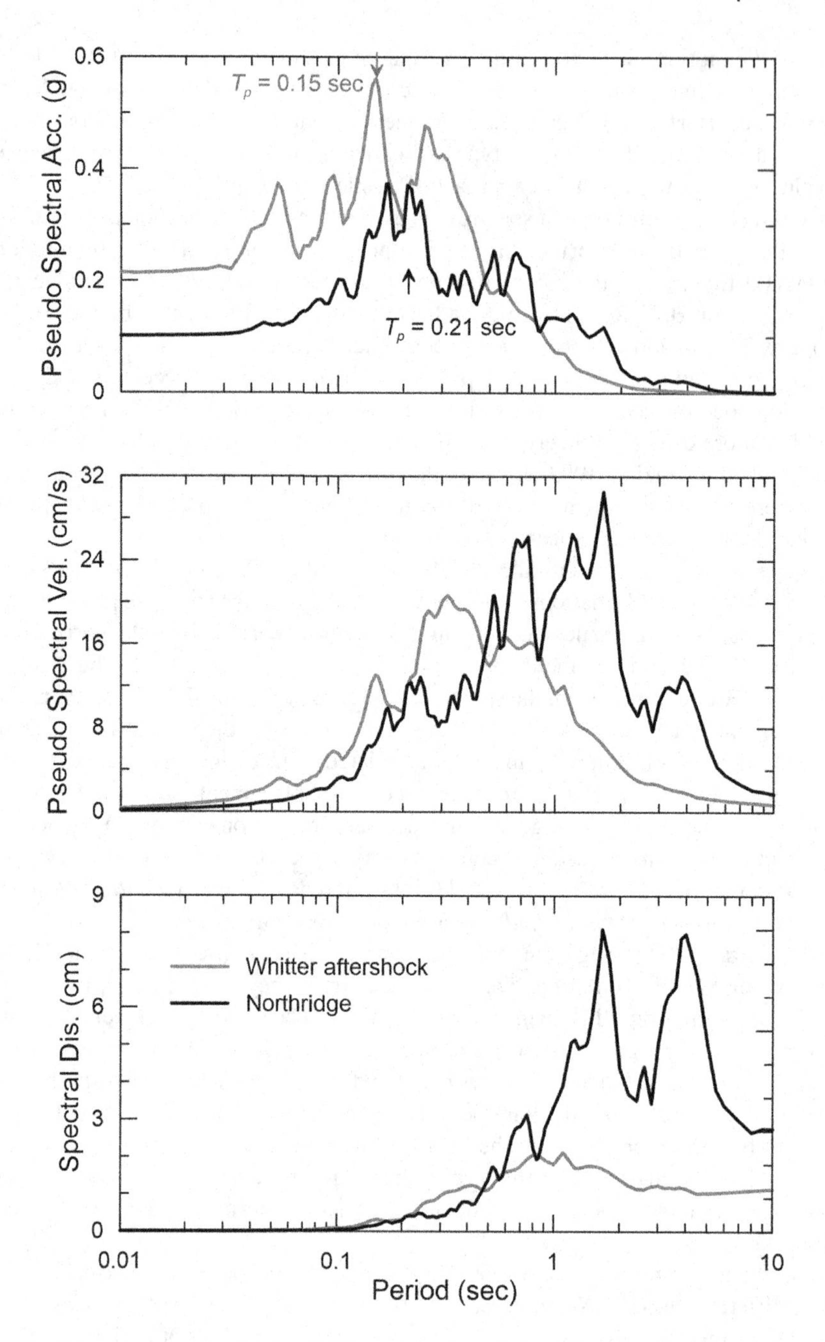

FIGURE 3.18 Response spectra for the E-W components of the Alhambra-Fremont School site strong motion records from the Northridge and Whittier Narrows aftershock events. Spectral response computed for damping ratio of 5%. Predominant periods (T_p) marked in the pseudo-spectral acceleration plots.

Predominant period: The predominant period (T_p) is the period at which the maximum 5% damped spectral acceleration occurs. Values of T_p for the Alhambra-Fremont School records are 0.15 sec for the Whittier aftershock and 0.21 sec for the Northridge event, which is characteristic of general observations showing the T_p increases with magnitude and distance (Seed et al., 1969; Rathje et al., 1998). Predominant period is seldom used at present because it is overly sensitive to potentially narrow peaks in the spectrum that may be of modest significance.

Mean period: The mean period (T_m) is defined as (Rathje et al., 1998):

$$T_m = \frac{\sum_n c_n^2 \left(\frac{2\pi}{\omega_n}\right)}{\sum_n c_n^2} \tag{3.13}$$

where c_n = Fourier amplitude coefficients (as in Section 3.3.2.1) and ω_n = discrete frequencies used in the Fast Fourier transform. The summation in Equation (3.13) occurs only between 0.25 and 20 Hz, or ω = 1.57–126 rad/s. This definition of mean period is similar to the frequency of zero crossings in random vibration theory (Boore, 1983). Mean frequencies are the inverse of mean periods, i.e., $f_m = 1 / T_m$, and are shown in Figure 3.17 for the Alhambra-Fremont School records. Note that f_m effectively represents the centroid of the Fourier amplitude spectrum on the abscissa. Mean frequencies, which are 1.69 and 3.05 Hz for Northridge and Whittier aftershock records, respectively, will in general better reflect differences in frequency content than T_p. Mean period is utilized in several applications, including for example seismic input specification for landslides (see Section 10.7.2), retaining walls (Section 8.6.3.2), and inelastic structures (Kumar et al., 2011).

Pulse Period: Ground motions recorded in close proximity to large earthquakes can exhibit pulse-like characteristics having a dominant period referred to as the *pulse period* (T_{pulse}). This parameter is described further in Section 3.7.

Bandwidth: The mean frequency can be used to locate the central value of the Fourier amplitude spectrum, which may in some cases correspond to a peak. However, it provides no information on the breadth of the spectrum or the dispersion about the mean period. The bandwidth of the Fourier amplitude spectrum is the range of frequencies over which some level of Fourier amplitude is exceeded. Bandwidth is usually measured at the level where the power of the spectrum is half its maximum value; this corresponds to a level of $1/\sqrt{2}$ times the maximum Fourier amplitude. The irregular shapes of individual Fourier amplitude spectra often renders bandwidth difficult to evaluate. It can be determined more easily for smoothed spectra.

3.3.3 Duration Parameters

For structural or geotechnical systems whose performance is measured by damage that accumulates during shaking, duration is typically a meaningful predictor of performance, along with amplitude and frequency content parameters. For example, duration has been used for response characterization in landslides (Bray and Rathje, 1998) and liquefaction-related lateral spreads (Rauch and Martin, 2000), and has been shown to affect the collapse capacity of some structural components and systems (Hancock and Bommer, 2007; Raghunandan and Liel, 2013; Chandramohan et al., 2016; Barbosa et al., 2017; Pan et al., 2018) and the nonlinear seismic response of concrete gravity dams (Wang et al., 2015). Duration also plays a role in the incremental buildup of pore pressure in liquefiable soils since longer durations lead to more cycles of loading.

The duration of a strong ground motion is related to the time required for the release of accumulated strain energy by rupture along the fault. Since fault rupture generally occurs at a characteristic velocity, the time required for rupture increases as the length, or area, of fault rupture increases. As a result, the duration of strong motion increases with increasing earthquake magnitude. In addition, because of the different speeds of seismic p-waves, s-waves, and surface waves, arrival times of energy from these wave types would be expected to diverge with distance from the fault. This causes duration to increase with distance as well as magnitude. This path effect on duration is especially important for small events, where the source duration is small. Site effects also increase duration, with soil sites typically having longer durations than rock sites at comparable distances. This is thought to result in part from complex wave propagation effects within soil deposits.

An earthquake accelerogram generally begins with p-wave arrivals and continues until the motion has returned to background noise levels. For engineering purposes, only the strong-motion portion of the accelerogram is of interest. Approaches that have been developed for measuring the duration of strong motion include the *bracketed duration* (Bolt, 1969) and *significant duration* (Trifunac and Brady, 1975). Example calculations of these parameters are shown in Figures 3.19 and 3.20. Figure 3.19 illustrates bracketed duration, which is defined as the time elapsed between the first and last (absolute) excursions beyond a specified threshold acceleration (typically $0.05g$ or $0.1g$). Bracketed duration parameters can be sensitive to the threshold accelerations and to small subevents occurring toward the end of a recording, and changes when a motion is scaled. As a result, other definitions of duration are usually preferred.

Significant duration parameters are defined as the time interval across which a specified fraction of the "energy" contained in a record is developed. In this context, energy is most often represented by the square of the ground acceleration, which is related to energy dissipation in SDOF oscillators excited by the ground motion. The integral of the square of ground acceleration is used to compute Arias Intensity (I_A) (Arias, 1970),

$$I_A = \frac{\pi}{2g} \int_0^{\infty} \left[\ddot{u}_g(t) \right]^2 dt \tag{3.14}$$

where $\ddot{u}_g$ is the acceleration history and g is the acceleration of gravity. A plot of Arias intensity, normalized by its final value, vs. time (Figure 3.20) is referred to as a Husid plot (Husid, 1969). The Husid plot, therefore tracks the build-up of energy over time in terms of normalized Arias Intensity; since it is normalized, it is not affected by scaling of ground motion amplitude. Two common measures of significant duration are time intervals between 5%–75% and 5%–95% of I_A (denoted $D_{5\text{-}75}$ and $D_{5\text{-}95}$, respectively). The corresponding points on the Husid plots are identified in Figure 3.20 along with the $D_{5\text{-}95}$ durations. Due to the modest amplitude of these motions, the significant durations are much longer than the bracketed durations. The durations show a clear increase for the larger magnitude Northridge event as compared to the Whittier aftershock.

An alternative to duration is the equivalent number of cycles of ground motion (N), which is a direct representation of demand, along with stress or strain amplitude, in laboratory testing of soils (for studies of liquefaction or seismic compression; e.g., Seed and Lee, 1966, Silver and Seed, 1971). The N parameter can also be computed from accelerograms using cycle counting algorithms (Seed et al., 1975; Liu et al., 2001; Green and Terri, 2005), which allows demands induced in laboratory tests to be related to those that would occur under field conditions. The number of cycles is correlated to duration, although the degree of correlation depends on the specific duration definition and the method for counting cycles (Hancock and Bommer, 2005; Bommer et al., 2006).

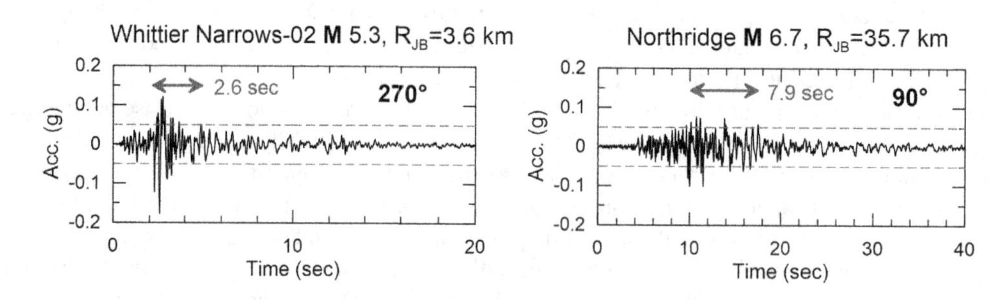

FIGURE 3.19 Evaluation of bracketed duration (acceleration threshold = 0.05g) for ground motions recorded at the Alhambra-Fremont School site during **M** 5.3 and 6.7 earthquakes. Note that the horizontal scale for the Whittier Narrows recording is modified from earlier diagrams (Figures 3.1 and 3.14) to more clearly illustrate the bracketed duration window.

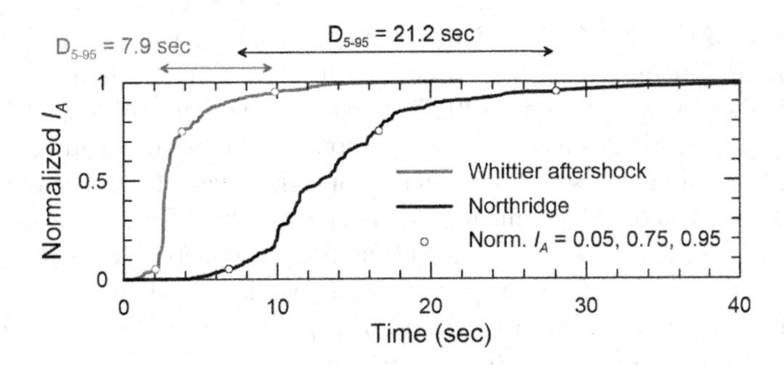

FIGURE 3.20 Husid plots showing normalized energy build up for evaluation of significant duration parameters $D_{5\text{-}75}$ and $D_{5\text{-}95}$ for ground motions recorded at the Alhambra-Fremont School site during **M** 5.3 and 6.7 earthquakes.

3.3.4 Combinations of *IM*s and Hybrid Measures of Ground Motion

*IM*s are used to quantify ground motion hazards for many types of structures. The *IM*s that are selected should ideally represent the amplitude, frequency content, and duration features of the ground motion that are critical for the given application. The individual, or scalar, *IM*s described in previous sections generally reflect only one of the three attributes of amplitude, frequency content, and duration. Hence, while these metrics are predictable, they typically are not optimized with respect to efficiency and sufficiency in prediction of the responses of structures and geotechnical systems. To overcome these limitations, some applications use combinations of *IM*s, i.e., *vector IM*s, to reflect multiple ground motion attributes. Other applications use scalar *IM*s that reflect more than one ground motion attribute, which are referred to as *hybrid scalar IM*s.

3.3.4.1 Combinations of *IM*s

When vector *IM*s are used, the accuracy and precision of response predictions should be improved, but at the expense of more complex characterization of the ground motion, (referred to as *vector* seismic hazard analysis; Section 4.4.3). The following are several example applications of *IM* combinations:

Spectral Shape: The nonlinear deformations of structural systems that respond principally in their first mode (i.e., with a large first mode participation; Section B.10.3) are well correlated to the ordinate of the response spectrum at the first-mode elastic period of the structure, T_0, and the shape of the spectrum for $T > T_0$. The form of the response spectrum that is most often used is pseudo-acceleration, so the corresponding ordinate is denoted $S_a(T_0)$. This parameter represents the amplitude of the ground motion at the period that initially is most influential on the structural response. The spectral shape is important because nonlinearities in the structure lengthen its first-mode period, so that the frequency content of the motion for $T > T_0$ affects the responses of the structure. Further explanation of spectral shape effects on the response of a yielding system is given in Section 4.5.1.2.

To represent the spectral shape for $T > T_0$, Cordova et al. (2000) combined two spectral ordinates, $S_a(T_0)$ and $S_a(2T_0)$, into a single parameter:

$$IM_{1EM} = S_a(T_0)\sqrt{\frac{S_a(2T_0)}{S_a(T_0)}} \tag{3.15}$$

where the subscript on *IM* indicates '1' for first mode, '*E*' for use of elastic period, and '*M*' for the multiplier on the elastic period. The square root term in Equation (3.15) represents the effect of spectral shape for $T = T_0$ to $2T_0$. Note that IM_{1EM} represents the geometric mean (square root of the product) of ordinates $S_a(T_0)$ and $S_a(2T_0)$. This parameter can correlate effectively with *EDPs* for first-mode dominated structures. The use of multiple spectral ordinates in a combined parameter was further explored by Luco and Cornell (2007), who used the square-root-sum-of-squares (SRSS) combination of spectral ordinates at the first and second modal periods without modification for inelastic effects (IM_{1E2E}), but with some additional complexity related to the use of modal participation factors. The application of this parameter utilizes spectral shape for $T < T_0$, and was found to be efficient and sufficient for some tall buildings with significant higher-mode responses.

The parameter combination reflecting spectral shape that has found the greatest practical application is $S_a(T_0)$ and a quantity referred to as epsilon (ε). The parameter ε is formally defined in Equation (4.26a); what is important for the present discussion is its physical significance, which is that ε represents the degree to which $S_a(T_0)$ is an outlier with respect to the mean of what would be expected for the seismic conditions under consideration (i.e., the magnitude and site-to-source distance for the controlling earthquake). The value of ε is strongly correlated to spectral shape (specifically, whether $S_a(T)$ is expected to increase or decrease) for $T > T_0$ (Baker and Cornell, 2005). This representation of spectral shape is preferable to the use of spectral ordinates at multiple periods because it avoids the use of relatively complex vector hazard analysis (Section 4.4.3.2) since ε is a standard output of single-*IM*, or *scalar*, hazard analysis. This accounts for its recent adoption in practice, as described further in Haselton et al. (2017).

Amplitude and Duration: As explained further in Section 10.9.2, seismic deformations of landslides are typically computed using the physical analogy of a block sliding on an inclined plane, with the block being rigid or compliant. In the case of rigid blocks, *IM* combinations that have been found to correlate to the sliding block displacement are *PGA* and *PGV* (Rathje and Saygili, 2009) and *PGA* and $D_{5\text{-}95}$ (Bray and Rathje, 1998). For compliant sliding blocks, the *IM* combinations are extended to consider the fundamental period of the slide mass (T_0), with proposed combinations including *PGA*, T_0/T_m, and $D_{5\text{-}95}$ (Bray and Rathje, 1998), $S_a(T_0)$ and $S_a(1.5T_0)$ (Bray and Travasarou, 2007), and *PGA* and *PGV* (Rathje and Antonakos, 2011).

Seismic inputs for various ground failure mechanisms including liquefaction, cyclic softening, and seismic compression, are represented either with a peak stress or strain (which in turn is typically evaluated from *PGA*) combined with an equivalent number of cycles *N* (Section 3.3.3). Hence, effectively the *IM* combination of *PGA* and *N* is used for such analyses.

3.3.4.2 Hybrid Scalar *IMs*

Rather than use a combination of multiple *IMs*, which generally necessitates relatively complex vector hazard analysis, the use of scalar *IMs* that capture relevant amplitude, frequency content, and duration characteristics offers obvious practical advantages. Many hybrid scalar *IMs* are proposed in the literature (Mackie and Stojadinovic, 2004; Riddell, 2007), but relatively few have been found to demonstrate adequate sufficiency and efficiency for structural or geotechnical system responses from multiple studies or have found practical applications. Some promising hybrid scalar parameters are described in this section.

Inelastic spectral ordinates: Inelastic response spectra are computed using a simplified representation of the nonlinear force-displacement relationship of a yielding structure (Section B.8). Figure 3.21a shows an elastic-perfectly plastic nonlinear relationship characterized by its elastic stiffness k and yield displacement u_y that occurs at a yield force F_y. If a particular ground motion induces a maximum displacement in the structure of u_{max}, the associated *ductility* is given as $\mu = u_{max} / u_y$ (Section B.8). A common definition of inelastic spectral ordinates is for a fixed value of μ and elastic oscillator period *T*, as shown in Figure 3.21b. Note that the vertical axis is no longer expressed as S_a, but instead is F_y normalized by structure weight *W*. For a ductility of $\mu = 1.0$, F_y/W is equivalent to S_a (explanation in Appendix B). As shown in Figure 3.21b, inelastic spectral ordinates for $T >\sim 0.05$ sec decrease rapidly with increasing μ, indicating the beneficial force-reducing effects of structural ductility, although this comes at the expense of plastic deformations (damage) in the structure.

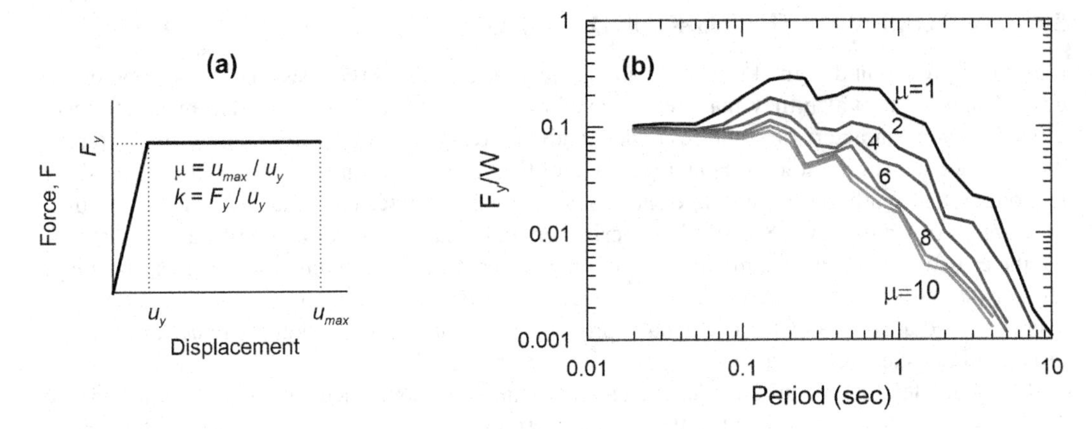

FIGURE 3.21 (a) Idealized elastic-perfectly plastic inelastic force-displacement curve for an oscillator with period $T = 2\pi\sqrt{m/k}$; (b) constant displacement ductility inelastic spectra for the Alhambra-Fremont School EW component of the 1994 Northridge earthquake.

Inelastic spectral ordinates represent the effects of both ground motion amplitude at $S_a(T_0)$ and spectral shape, which is why they are described as hybrid scalar *IM*s. This sensitivity to spectral shape occurs because ductility is affected by the frequency content of a ground motion for $T > T_0$. Inelastic spectral ordinates have been shown to be efficient and sufficient predictors of the nonlinear response of structures dominated by first-mode response (e.g., Luco and Cornell, 2007), but are not presently used to a significant degree in engineering practice.

Arias Intensity: Arias intensity, I_A, is integrated from the square of the acceleration history (Equation 3.14), and because of normalization by g, has units of velocity. The value of I_A reflects ground motion amplitude, frequency content, and duration. All other factors being equal, the I_A of different ground motions tends to increase as amplitude and duration increase and to be higher for low-frequency vs high-frequency ground motions. Arias intensity is used or has been proposed for use, in predicting landslide displacements (Howard et al., 2008), liquefaction triggering (Kayen and Mitchell, 1997), and structural damage (Cabañas et al., 1997). As noted previously, the build-up in time of I_A is used to define significant duration parameters. A drawback to the use of I_A is that it has low predictability (high $\sigma_{\ln IM}$).

Cumulative Absolute Velocity: Cumulative absolute velocity, *CAV*, is simply the area under the absolute accelerogram:

$$CAV = \int_0^{\infty} \left|\ddot{u}_g(t)\right| dt \tag{3.16}$$

As with I_A, *CAV* is sensitive to the amplitude, frequency content, and duration of the accelerogram. However, *CAV* has the advantage of improved predictability as a result of significantly lower dispersion ($\sigma_{\ln IM}$), which makes it more efficient for the prediction of response. For this reason, *CAV* will likely see increased use as an *IM* relative to I_A in the years to come. *CAV* can be computed excluding portions of the absolute accelerogram with amplitudes less than x (in units of cm/s^2) in which case it is referred to as CAV_x. Values of x that have been proposed for various applications are 25 cm/s^2 (0.025g) for post-earthquake nuclear power plant safety evaluation (USNRC, 1997; EPRI, 2006) and building settlement from liquefaction of foundations soils (Bray and Macedo, 2017) and 5 cm/s^2 (0.005g) for pore pressure generation in sands (Kramer and Mitchell, 2006). These are denoted CAV_{25} and CAV_5, respectively (the terms CAV_{STD} and CAV_{dp} are used in in lieu of CAV_{25} in the nuclear power literature). The principal application of *CAV* directly is for correlations with seismic intensity (Campbell and Bozorgnia, 2012).

3.4 FACTORS AFFECTING GROUND MOTION

Earthquake-resistant design of structural and geotechnical systems requires characterization of the ground shaking to which they may be subjected during their life. The evaluation of design-level ground motion characteristics requires understanding the behavior of seismic sources likely to produce earthquakes in the study region, prediction of the ground motion *IM*s those sources are likely to produce, and synthesis of the source and ground motion attributes in a rational framework known as *seismic hazard analysis*. Source characterization and hazard analysis are covered in Chapter 4, but the critical step of ground motion prediction is covered in the remaining sections of this chapter. For the purpose of this discussion, it will be assumed that the magnitude, location, and other source-related attributes of the earthquake producing ground motions, along with certain characteristics of the site, are known.

The characteristics of ground motion are fundamentally controlled by source, path, and site effects. *Source effects* are related to the characteristics of seismic waves emanating outward from an earthquake fault rupture (i.e., what is generated at the source), *path effects* describe the modification of seismic waves as they propagate through the Earth's crust toward the site, and *site effects* represent the influence of near-surface sediments and/or weathered rock on the ground shaking. It is essential to consider all three effects in an engineering prediction of earthquake ground shaking. This section describes the principal physical phenomena associated with source, path, and site effects.

3.4.1 Source Effects

A seismic source is a fault across which rupture, or slip, occurs in a short time interval, releasing strain energy in the Earth's crust and causing seismic waves to propagate outward. Small magnitude earthquakes (**M** < 5–5.5) occur over a sufficiently small crustal volume that the rupture location is typically idealized as a point. Larger magnitude events occur over a distributed rupture surface that is typically idealized as a plane or series of planes. As shown in Figure 3.22, the limits of the rupture on a plane are often represented with one or more rectangles and comprise a portion of a *finite fault* representation of the earthquake source. As described previously in Section 2.4.1, the intersection of a rupture plane with a horizontal plane is a horizontal line having the azimuth of the *fault strike*, whereas the vertical angle from horizontal to the fault plane, measured perpendicular to the strike, is the *fault dip angle*. While the strike and dip are useful for characterizing source geometry, other aspects of the seismic source also affect ground motions.

3.4.1.1 Slip Distribution, Rise Time, and Slip Velocity

While fault rupture is often idealized using rectangles, the details by which rupture occurs within those rectangles are both complex and significant for understanding of the ground motions that are produced. The finite fault sources shown in Figure 3.22 illustrate some features of the seismic source that affect ground motion. The contours in the figure indicates variable amounts of fault slip, ranging from null to 0.6 m for the 2004 **M** 6.0 Parkfield, CA event (Custódio et al., 2005) and null to more than 35 m for the 2011 **M** 9.0 Tohoku, Japan event (Yokota et al., 2011). Ground motion amplitudes from the Parkfield event (e.g., *PGA*, *PGV*) were highest near the portion of the fault having the largest slip, whereas Tohoku strong ground motion amplitudes were instead controlled by deeper portions of the fault (nearer the coastline) having less slip. It is illustrative to examine why the high-slip region controlled ground motion amplitudes of engineering interest in one case but not the other.

There are important elements of time in the fault slip process. The slip amounts shown in Figure 3.22 occurred over a time interval known as the *rise time*, t_r, which is specific to a particular point on the fault (Figure 3.23). For a given amount of slip, the slip velocity increases as the rise time decreases and the radiated seismic energy has higher frequencies. The Parkfield earthquake

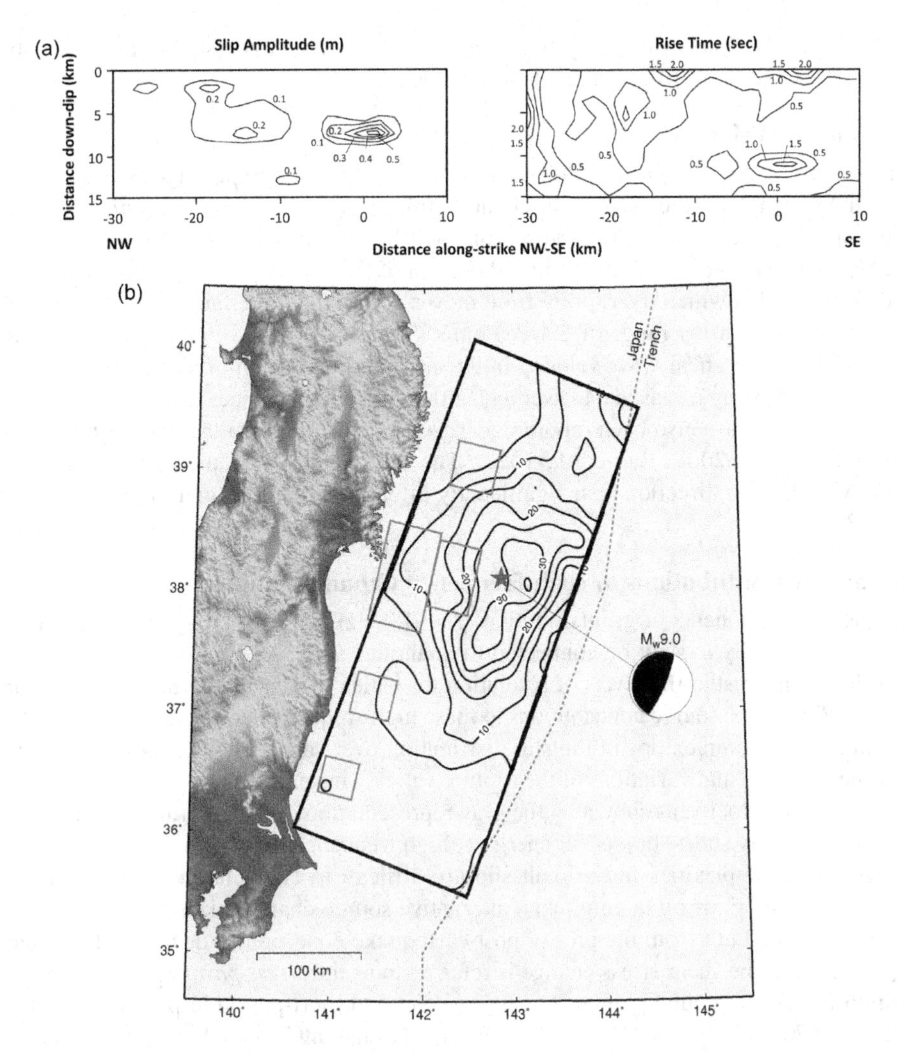

FIGURE 3.22 (a) Finite fault model for slip and rise time on rupture surface for 2004 **M** 6.0 Parkfield, CA earthquake (Modified from Custódio et al. 2005, used with permission of John Wiley & Sons); (b) finite fault model for fault slip (black contours, units are m) and strong motion generation areas having high stress drop (small rectangles) for 2011 **M** 9.0 Tohoku, Japan earthquake (Yokota et al., 2011; Kurahashi and Irikura, 2011; Stewart et al., 2013).

slip occurred over a relatively short period of time ($t_r \approx 0.5$ sec near hypocenter, which is at position [0, 7.5 km] in Figure 3.22a) with high slip velocities (maximum slip / $t_r \approx 120$ cm/s), which produced substantial energy in the frequency range of engineering interest with high peak accelerations and velocities. In the case of the Tohoku earthquake, the largest slip occurred over a relatively long time period ($t_r \approx 90$ sec near hypocenter; Yokota et al., 2011) on a shallow part of the fault. This large slip produced substantial seafloor rise and affected the subsequent tsunami, but did not produce damaging ground motion because the associated frequency content of this energy release was very low ($f < \sim 0.05$ Hz). The strong ground motions instead were principally influenced by relatively low-slip portions of the fault at depth and marked in Figure 3.22b as "strong motion generation areas" (Kurahashi and Irikura, 2011). Although the slip in these areas was smaller, the rise times were much shorter, on the order of 2–5 sec, which produced relatively high-frequency ground motions of greater engineering significance. This slip pattern for interface subduction zone earthquakes, with shallow large slip producing low-frequency ground motions and deep but lower slip producing ground motions in the frequency range of engineering interest, has also been observed for the 2010 **M** 8.8 Maule Chile earthquake (Tolga Sen et al., 2015).

3.4.1.2 Rupture Velocity

Ground motions are also affected by the manner in which the rupture propagates across a fault. As described in Section 2.4.1, the slip on a finite fault initiates at the *focus* or *hypocenter*. If t_0 indicates the rupture initiation time, for $t > t_0$ the rupture will have spread out over a portion of the fault, with the *rupture front* separating the ruptured portion of the fault from the portion that has not yet ruptured. The speed at which the rupture front moves across the finite fault is the *rupture velocity*, v_r, which is typically in the range of 2.4–3.0 km/s (Somerville et al., 1999). Those velocities are about 70%–90% of the shear wave velocity of the crust at the location of the rupture (i.e., $v_r/V_{s,crust} \approx 0.7$–$0.9$, with 0.8 being a reasonable average), although some instances of super-shear rupture in which $v_r/V_{s,crust} > 1.0$ have also been reported (e.g., Archuleta, 1984; Dunham and Archuleta, 2004; Bizzarri and Spudich, 2008; Bao et al., 2022). The direction of movement of the rupture front, combined with the slip direction, can significantly affect near-fault ground motions, as described in Section 3.7.

3.4.1.3 Source Contributions to High-Frequency Ground Motions

High-frequency components of ground motion ($f > \sim 1.0$ Hz) are affected by source processes, but in many cases can only loosely be connected to quantities such as slip amount, slip velocity, and rupture velocity (at least at the levels of resolution for which these rupture attributes can currently be inferred). Rather, the source contributions to these high-frequency components of ground motion are associated with complexities in the slip distribution over small length scales, perturbations in the slip-time function, and variations of the rupture velocity in space and time. For example, sudden changes in the slip velocity, as shown by the gray representation of the slip function in Figure 3.23, will produce relatively short "bursts" of energy at high frequencies.

Because these complexities in the fault slip are difficult to characterize, simulations of high-frequency ground motions often employ an alternative source characterization based on the shear stress change on the fault from the pre- to post-earthquake condition in lieu of fault displacement (slip). Qualitatively, the shear stress change is referred to as the *stress parameter*, $\Delta\sigma$, which correlates to high-frequency ground motion *IMs* such as *PGA* and short-period response spectra [S_a(<~1.0 sec)] (Brune, 1970, 1971; Boore, 2003b). Two earthquakes having identical magnitudes but different stress parameters will produce different ground motions, with the higher stress event producing stronger high-frequency *IMs*. A lower stress parameter event would have a larger rupture area (in order to have the same seismic moment and hence moment magnitude). The term "stress parameter"

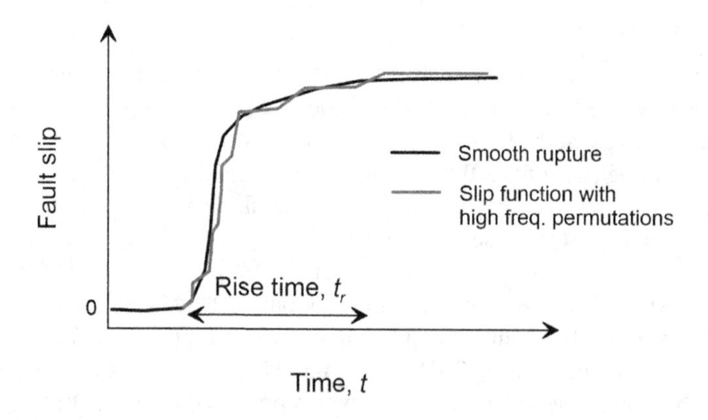

FIGURE 3.23 Slip-time function for smooth rupture and with permutations producing high-frequency ground motions showing rise time t_r.

is sometimes expressed as "stress drop," which has generated some controversy (Atkinson and Beresnev, 1997) because of the manner in which it is inferred from observations and implemented in certain simulation methods. Here the term "stress drop" qualitatively represents stress change on a fault from a seismic event whereas "stress parameter" indicates an input parameter used in simulation methods (Section 3.6.1).

3.4.1.4 Source Parameters Considered in Ground Motion Models

Ground motion models (Section 3.5) are used to predict ground motion *IM*s given certain characteristics of the source, path, and site. While all of the source attributes described above affect ground motion, few are predictable in advance of an earthquake. As described further in Section 4.2.2, the types of source parameters that may be available are those describing source geometry (length, width, depth, strike, dip), the range of magnitudes the source is capable of producing and their time-recurrence behavior, and the focal mechanisms of earthquakes on the fault (Section 2.4.2). Details of the slip distribution on a fault in time and space from future earthquakes are unknown before an earthquake occurs. In part for this reason, source attributes considered in ground motion models are generally simplified to **M**, focal mechanism, and geometric parameters.

3.4.1.5 Source Parameters Used in Ground Motion Simulations

Ground motion simulations (Section 3.6) model the rupture process (source effects), wave propagation through the Earth's crust (path effects), and wave propagation through shallow sediments (site effects) to compute ground motion time series. Simulation procedures represent the seismic source at varying levels of complexity. The simplest models utilize a point source geometry combined with a corner frequency parameter related to magnitude and stress parameter. More complex models use a finite fault with a full slip distribution, a slip velocity function, and a parameterization of rupture velocity. Many of these parameters are very difficult to estimate reliably in advance of an earthquake, which is one of the principal reasons that seismic hazard analyses have traditionally been based on ground motion estimates from ground motion models in lieu of predictions from simulations. However, some emerging approaches (currently in the research domain) use simulations to generate ground motions for seismic hazard analyses by randomizing source attributes including hypocenter location, slip distribution, and slip velocity (Graves et al., 2011; Milner et al., 2021).

3.4.2 Path Effects

In ground motion modeling, the effects of path are taken to represent the changes that occur in seismic waves as they propagate away from the source through the Earth's crust toward a site of interest. Two principal considerations influence path effects – geometric attenuation and anelastic attenuation.

3.4.2.1 Geometric Attenuation

As seismic waves propagate outward from the source, they produce ground motion over an ever increasing volume of crust. Given the finite amount of energy released at the source, this *geometric attenuation* or *geometric spreading* (Section C.5.2) of the energy necessarily reduces the amplitude of the ground shaking as the distance from the source increases. As illustrated in Figure 3.24a, for a relatively small seismic source, the rate with which the ground motion Fourier amplitude attenuates with distance R changes as the types of waves that control the motion differ at increasing distances. Body waves generally control motions at relatively short distances and surface waves are dominant at large distances. Since body waves and surface waves attenuate due to geometric spreading differently (Section C.5.2), ground motion attenuation with distance can be modeled with different rates over different distance ranges as given below (Boore, 2003b):

$$Z(R)=\begin{cases} \left(\dfrac{R_0}{R}\right)^{p1} & R \le R_1 \\ Z(R_1)\left(\dfrac{R_1}{R}\right)^{p2} & R_1 \le R < R_2 \\ Z(R_2)\left(\dfrac{R_2}{R}\right)^{p3} & R \ge R_2 \end{cases} \tag{3.17}$$

where $Z(R)$ indicates the change of amplitude relative to a reference distance, R_0 (usually taken as 1 km), and the exponents p_1 and p_3 describe geometric spreading of body waves (p_1; controls attenuation inside of R_1) and surface waves (p_3; controls attenuation beyond R_2), respectively. The exponent p_2 is for a transition zone between the two wave types that is not always used. Typical values of these exponents are p_1 = 1.0–1.3 for body waves and p_3 = 0.5 for surface waves (Boore, 2003b; Atkinson and Boore, 2014). Typical values for the distance limits are $R_1 = R_2 \approx 40$ km (no transition zone) in tectonically active regions and R_1 = 70–80 km and $R_2 \approx 130$ km in stable continental regions.

For a point source, the distance R in Equation (3.17) could in principle be taken as the closest distance to the source (R_{rup}), or alternatively as the path length of seismic waves, which may include waves reflecting off the Moho at the base of the Earth's crust. The use of R_{rup} for R is potentially problematic for finite fault representations of the seismic source, because R_{rup} becomes very small near the fault plane, suggesting effectively no along-path attenuation (i.e., $Z(R_{rup})$ approaches unity as shown in Figure 3.24b). If a large-slip patch on the fault controls the ground motion and is at a distance greater than R_{rup}, the effective path length is also greater than R_{rup}. This condition is typically accounted for by modifying the distance parameter as:

$$R=\sqrt{R_{rup}^2+h^2} \tag{3.18}$$

where h is often referred to as a *finite fault parameter* or *fictitious depth*. Values of $h > 0$ produce a saturation effect at close distances in which the attenuation flattens, as indicated in Figure 3.24a-b.

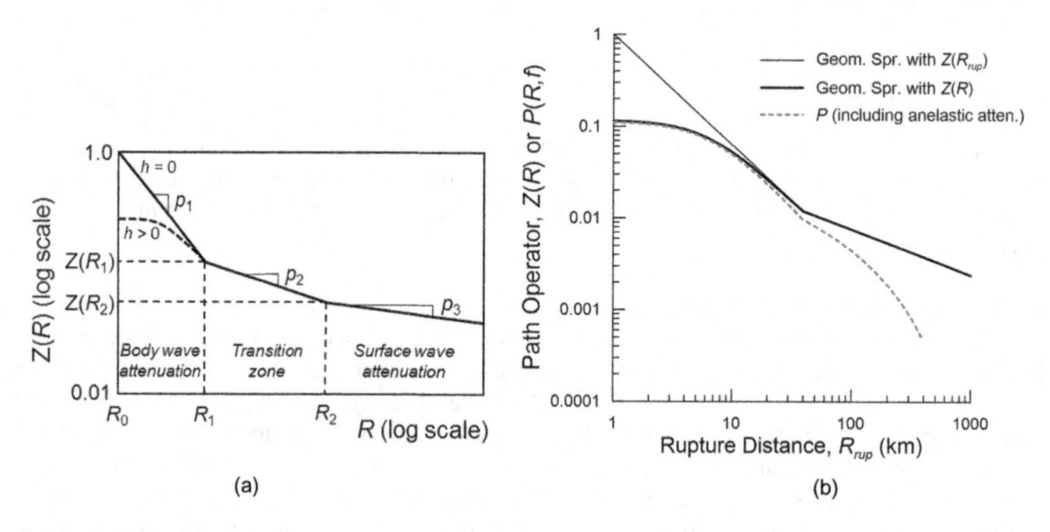

FIGURE 3.24 (a) Schematic illustration of ground motion attenuation due to geometric spreading with and without the finite fault parameter; (b) Illustration of path effects as given by idealized geometric spreading based on rupture distance [$Z(R_{rup})$], modified for near-source saturation [$Z(R)$], and modified for anelastic attenuation. Schematic computed using Equation (3.17) with p_1 = 1.2, p_3 = 0.5, R_0 = 1 km, Q = 170, f = 1 Hz, R_1 = 40 km, R_2 = 40 km, h = 6 km, and $V_{s,crust}$ = 3.5 km/s.

3.4.2.2 Anelastic Attenuation

In addition to geometric attenuation, the propagation of seismic waves is also affected by *anelastic attenuation* (or *material damping*, as discussed in Section C.5.1) in the crust. At a point in the crust, shear waves will produce a cyclic stress-strain response that may be essentially linear (modulus is independent of shear strain) but which produces hysteretic damping associated with the area inside of the stress-strain loop. This damping dissipates energy with each cycle of shear along the ray path. The wavelength λ of a shear wave of frequency f propagating at velocity $V_{s,crust}$ is given by

$$\lambda = \frac{V_{s,crust}}{f} \tag{3.19}$$

When the wavelength is short relative to the path length (i.e., large R/λ), there are many cycles of shear along the path, providing ample opportunity for even low levels of material damping to significantly attenuate seismic waves. Because high frequencies produce short wavelengths per Equation (3.19), their amplitudes attenuate more quickly over a given distance due to anelastic attenuation than low frequencies, which typically experience relatively little effects of material damping.

Wave scattering (Appendix C.5.3) also can act on waves passing through the crust as they refract around irregularities. Scattering produces an attenuation effect distinct from material damping; it is strongly frequency-dependent because the scattering produced by an irregularity is related to the degree to which its size is comparable to the wavelengths of the waves that approach it (i.e., long wavelengths do not "see" small irregularities).

Crustal damping effects arising from both material damping and wave scattering are typically characterized by the crustal quality factor, Q, which can be related to an effective damping ratio for the crustal material (D_{eff}) as

$$D_{eff}(f) = \frac{1}{2Q(f)} \tag{3.20}$$

Both the quality factor Q and D_{eff} are typically frequency-dependent, due to the effects of scattering.

The effects of crustal damping can be combined with those for geometric spreading to describe path (or attenuation) effects as (Boore, 2003b):

$$P(R,f) = Z(R)\exp\left(\frac{-\pi f R}{Q(f)V_{s,crust}}\right) \tag{3.21}$$

As shown in Figure 3.24b, the effect of crustal damping on the rate of distance decay is only significant at large distances, typically beyond about 80 km, where it causes a characteristic downward curvature in the attenuation function.

3.4.3 Site Effects

The vast majority of the path length for seismic waves traveling from source-to-site occurs through rock in the Earth's crust. As the waves approach the surface, they travel through geologic strata having progressively slower seismic velocities. This will tend to bend the wave's propagation direction upward, per Snell's Law (Section C.4.2), and change its amplitude. These and other effects of the local geology and morphology of the site on the ground motions are collectively referred to as *site effects*. Several distinct physical phenomena have the potential to contribute to site effects:

1. *Local ground response* describes the effects on the ground motion of relatively shallow sediments (typically 10–100s of m in depth) having the slowest velocities. Because the dimensions of these soft sediments are limited, the affected frequencies are relatively high

(usually above about 1 Hz). Factors contributing to local ground response will include some combination of *impedance effects*, *soil nonlinearity*, and, potentially, *resonance effects*, as described further below.

2. *Basin effects* are related to the deep structure of sediments that are present in many areas, including the Kanto Plain in Japan (which includes Tokyo), the Los Angeles and Seattle basins in the western US, and the Mississippi Embayment in the central US. Basins typically include soft sediments near the surface that transition with depth to progressively stiffer sediments, including sedimentary rock, before basement conditions (crystalline rock) are encountered. When the dimensions of basins are large (on the order of km), the affected frequencies are relatively low (less than about 1 Hz).
3. *Topographic effects* are related to irregularities in the ground surface geometry. For example, ridges and slopes can concentrate seismic energy near resonant frequencies that produce local amplification of ground motion. These effects tend to occur at high frequencies for which some fraction of the wavelength (approximately 1/4–1/6) is roughly compatible with the dimension of the morphological feature (e.g., horizontal width of slope or half width of ridge; Boore 1972, Ashford et al., 1997).

Some of the physics of local ground response and basin effects are described in Section C.6 (Appendix C), which should be reviewed before reading further. Some critical features of site response that reflect the basic problem physics are summarized below. The goal at this stage is to develop a physical understanding of these effects, which are incorporated into most of the modern ground motion models described later in this chapter. This chapter does not address site-specific analysis of site effects, which goes beyond the relatively approximate approaches adopted in ground motion models. These more complete site-specific analyses are the subject of Chapter 7.

3.4.3.1 Local Ground Response

As described in Section C.6.1, as an incident shear wave passes through a soil profile, effects related to the stiffness profile tend to increase its amplitude relative to the amplitude on a reference medium (usually rock) below the soil column. In a profile with no sharp changes in stiffness, the smooth variation of stiffness with depth would produce a relatively smooth variation of amplification with frequency as shown in Figure 3.25. This pattern of impedance-related amplification can be understood using the energy principles presented in Section C.6.1.1. In the absence of damping, the amplification of wave amplitude can be approximated by the amplification factor

$$A(f) = \sqrt{\frac{\rho_{ref} V_{ref}}{\bar{\rho}(f)\bar{V}(f)}} \tag{3.22}$$

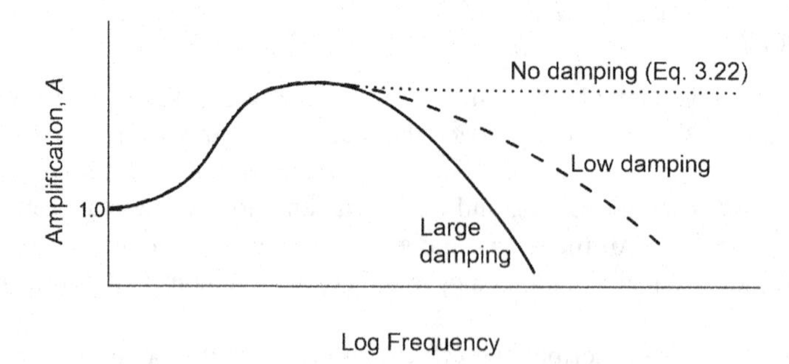

FIGURE 3.25 Schematic depiction of site amplification due to impedance effects combined with various levels of damping.

where $\rho_{ref}V_{ref}$ is the specific impedance of a reference rock layer below the soil profile and $\overline{\rho V}$ represents the average specific impedance for a depth range compatible with frequency f. That depth range can logically be considered as one-quarter of the seismic wavelength λ (Joyner and Boore, 1981; Boore, 2013). At low frequencies, the seismic wavelength may be very long relative to the soil profile thickness, in which case $\overline{\rho V}$ is dominated by the specific impedance of the reference rock and, per Equation (3.22), the amplification factor will become 1.0. As frequencies increase, wavelengths become comparable to, and eventually shorter than, the profile thickness. Assuming seismic velocities decrease toward the ground surface, $\overline{\rho V}$ becomes progressively lower with increasing frequency, increasing the site amplification (Figure 3.25). When the profile contains a sharp impedance contrast, the presence of resonance effects (C.6.1.2) complicates the amplification relative to what is shown in Figure 3.25, producing local peaks at modal frequencies as shown in Figures C.29 and C.30.

Equation (3.22) does not account for the energy dissipated by material damping and wave scattering, which always exist in actual soil and rock layers. Hence, as waves propagate upward through a soil column, the tendency for amplitude increase from impedance and resonance effects is countered by damping, as shown by the decay of amplification at higher frequencies in Figure 3.25. Like the impedance effect, damping effects are also frequency-dependent – the short wavelengths of high-frequency waves allow for multiple cycles of shearing within the soil layer, providing ample opportunity for damping to act on the wave, reducing its amplitude. Conversely, low-frequency waves have long wavelengths so that soil damping does not impact wave amplitudes as much. As described further in Section 6.6.3, the level of material damping in soil increases with strain (in combination with shear modulus reduction), which in turn is related to the amplitude (and frequency content) of the input motion for a given site profile.

Whether the net effects of impedance and damping will cause ground motion amplitudes to increase or decrease from the base to the top of a profile depends principally on the frequency of the wave and its amplitude as it enters the sediments from the crustal rock below. Figure 3.26 shows typical patterns of site amplification for *IM*s sensitive to high-frequency and low-frequency ground motions. The left side of the plots, where the input motion amplitudes for the reference site condition (PGA_X) are low, is associated with relatively small-strain ground response with low damping; in this case, the impedance effect dominates and amplification occurs for low- and high-frequency *IM*s. As the input amplitudes increase, larger-strain responses occur, producing relatively high damping. This damping can de-amplify (or at least reduce the amplification of) high-frequency *IM*s, but minimally affects low-frequency *IM*s.

In summary, ground response comprises three principal phenomena: impedance effects, resonance, and damping. These effects can be simulated through formal wave propagation analysis as described in Sections 7.5–7.6, or more commonly, through relatively simple site factors included in semi-empirical ground motion models. In the latter case, the site condition is typically represented

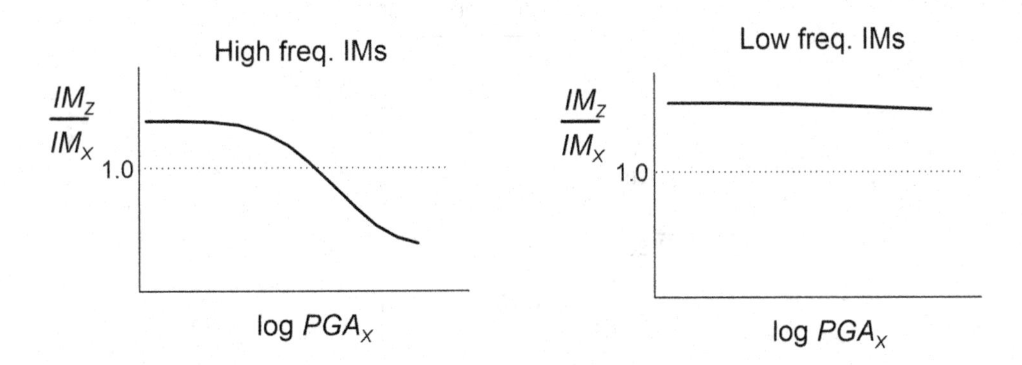

FIGURE 3.26 Schematic depiction of site factors representing the ratio of intensity measures at the ground surface (IM_Z) to those for a reference condition (typically rock) at depth (IM_X) as a function of the amplitude of shaking for the reference condition (represented by PGA_X).

by V_{S30} (Boore et al., 1994; Borcherdt, 1994; Section 7.3.3.1), which is the time-averaged shear wave velocity in the upper 30 m of the site, computed as:

$$V_{S30} = \frac{30\text{ m}}{\sum_j \Delta z_j/(V_S)_j} \tag{3.23}$$

The denominator in Equation (3.23) represents the shear wave travel time, $tt(30)$, through only the upper 30 m of the profile (i.e., $\sum_j \Delta z_j = 30$ m), with Δz_j and $(V_S)_j$ representing the thickness and velocity of layer j. V_{S30} is a relatively simple metric of site stiffness that has been found to be effective in making first-order estimates of site amplification. The value of V_{S30} has also been found to correlate well to average shear wave velocities to depths much greater than 30 m in some regions (Boore et al., 2011), which causes it to be useful for site response prediction over a wide range of periods.

Example 3.1

What is the V_{S30} of the profile in Figure E3.1?

Solution:

The table below shows depth intervals, velocities, and travel times for the upper 30 m of the profile.

Depth Interval (m)	$V_{S,j}$ (m/s)	$\Delta t = \Delta z_j / V_{S,j}$ (sec)
0–2	100	0.020
2–5	130	0.023
5–10	145	0.034
10–17	155	0.045
17–25	210	0.038
25–30	225	0.022

Using Equation (3.23), the sum of the travel times is 0.183 sec, so $V_{S30} = \dfrac{30}{0.183} = 164$ m/s

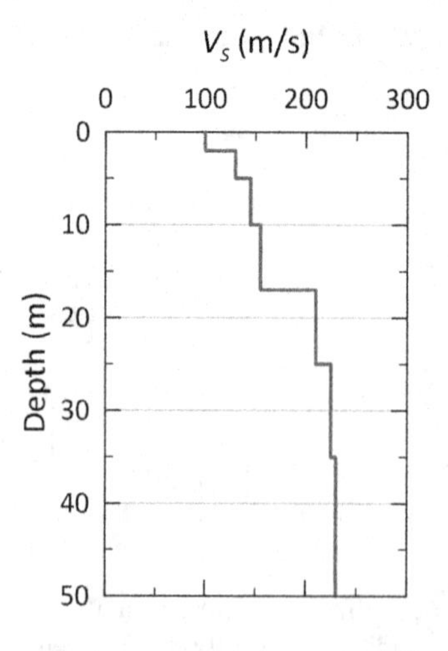

FIGURE E3.1 V_S profile for Example 3.1

3.4.3.2 Basin Effects

A basin consists of alluvial deposits and sedimentary rocks that are geologically younger and have slower seismic wave velocities than the underlying basement rock (Section C.6.2). Figure 3.27 schematically depicts two sedimentary basins – Basin 1 overlies the fault and Basin 2 is at some horizontal distance from it. Because the seismic source is located beneath Basin 1, site response in the basin is largely associated with the previously described ground response effects (impedance, damping, resonance). Wave propagation, in this case, may be reasonably well represented by a one-dimensional wave propagation analysis (Section 7.5), unless local lens-like structures that focus seismic waves (see Sections C.6.2 and 7.6.2), which is a type of basin effect, are present. Because the seismic source is located outside the perimeter of Basin 2, waves can enter that basin from the edge as well as from beneath. As described in Section 7.6.2, the waves entering from the edge can become trapped within the basin and generate surface waves that propagate across the basin, which is an effect distinct from ground response.

Basin-edge and focusing effects are site response processes caused by the three-dimensional geometry of sedimentary basins. These effects result from specific conditions of incident waves and basin geometry and thus may be different for events occurring at different locations relative to a given basin. This makes the characterization of basin effects rather complex, and simple engineering models that properly represent the physics of the problem do not presently exist. Instead, current models for basin-related amplification are formulated as simple functions of sediment depth, as discussed further in Section 3.5.2.3. Because basin effects typically involve large dimensions (on the order of km), the affected wavelengths are long, meaning low frequencies are affected. For this reason, basin effect models affect ground motion predictions typically mainly for periods longer than about 1.0 sec.

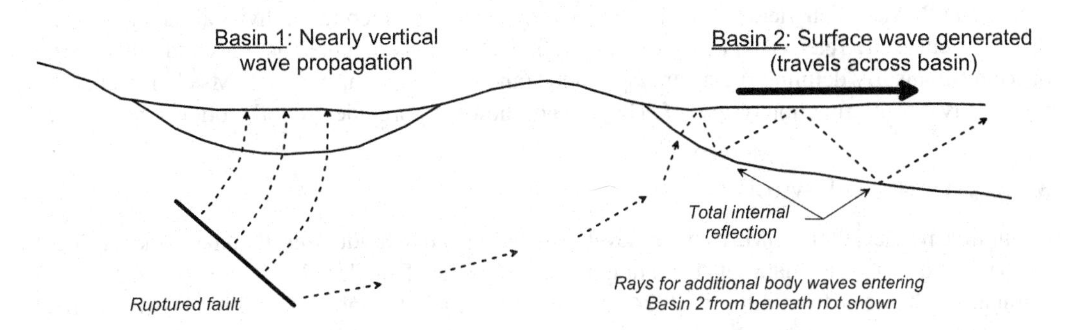

FIGURE 3.27 Schematic showing seismic body waves entering basins from beneath and from the edge. The waves entering from beneath can propagate nearly vertically as they reach the surface and are affected principally by ground response effects. The waves entering through the edge can undergo critical body-wave reflections, which generate surface waves that travel across the basin. (Modified from Choi et al., 2005.)

3.5 GROUND MOTION MODELS

Engineering seismologists and geotechnical earthquake engineers are often required to estimate ground shaking from scenario earthquake events, and do so using ground motion models, or GMMs. GMMs are mathematical expressions that predict ground motion *IM*s using various predictive variables such as source magnitude (**M**) and focal mechanism, source-site distance, and site parameters that are known to influence ground shaking. These variables are applied in equations that predict the effects of the event (or source), path, and site, leading to a generic expression for the mean *IM* as follows:

$$\mu_{\ln IM} = F_E + F_P + F_S \tag{3.24}$$

Because *IM*s tend to be lognormally distributed (Section D.7.2.3), the symbol $\mu_{\ln IM}$ is used to emphasize the fact that the expression predicts the mean value, μ, of the natural logarithm ("ln" in subscript) of the *IM*.

GMMs are developed in part using empirical data, i.e., *IM* values computed from strong motion measurements in actual earthquakes. They consist of functions selected to capture the primary physical mechanisms that control ground motions, with fitting of coefficients in those functions based on physical considerations and regressions of the data. Because they cannot capture all of the physical mechanisms, which are different for different earthquakes and different locations, and because of the variability of geologic materials, potential measurement errors, and other factors, the predictions of *IM*s are uncertain. Therefore, a GMM must predict not just the mean of an *IM*, but its distribution.

To illustrate the distribution of ground motions, Figure 3.28 shows a histogram of *PGA* data for a narrow interval of magnitude and distance (**M** 5.8–6.2, rupture distances of 30–50 km) from a large data set of ground motion recordings in active tectonic regions (i.e., NGA-West2 dataset by Ancheta et al., 2014). When plotted on a linear *PGA* axis (Figure 3.28a), the distribution is skewed to the right, whereas when plotted on a log scale (Figure 3.28b), the distribution is nearly symmetric. As a result, the distribution of ground motion data is usually taken as log-normal, with $\mu_{\ln IM}$ representing the mean in natural log units and $\exp(\mu_{\ln IM})$ in arithmetic units representing the median. The spread of the data in Figure 3.28 can be represented by a natural log standard deviation, which is denoted $\sigma_{\ln IM}$.

It is logical to expect that amplitude-related *IM*s will increase with increasing magnitude and decrease with increasing distance, and early GMMs (referred to at the time as *attenuation relationships*, and later as *ground motion prediction equations*) predicted *IM*s based only on those variables. A schematic illustration of a simple GMM is shown in Figure 3.29. The smooth curve is the regressed GMM, which yields $\mu_{\ln IM}$. The vertical distances between the individual data points and the mean curve are referred to as *residuals* ($\delta_{\ln IM}$), and $\sigma_{\ln IM}$ is computed as the standard deviation of the residuals. By defining the mean, $\mu_{\ln IM}$, and standard deviation, $\sigma_{\ln IM}$, GMMs characterize the probability density function of the *IM*. They do not, however, provide ground motion time series.

3.5.1 Historical Development

Beginning in the 1960s, GMMs were developed based on individual earthquakes and on local, regional, and worldwide data sets from multiple earthquakes. Douglas (2024) presents an extensive summary of GMMs from 1964 to 2024; the summary includes more than 500 GMMs for peak

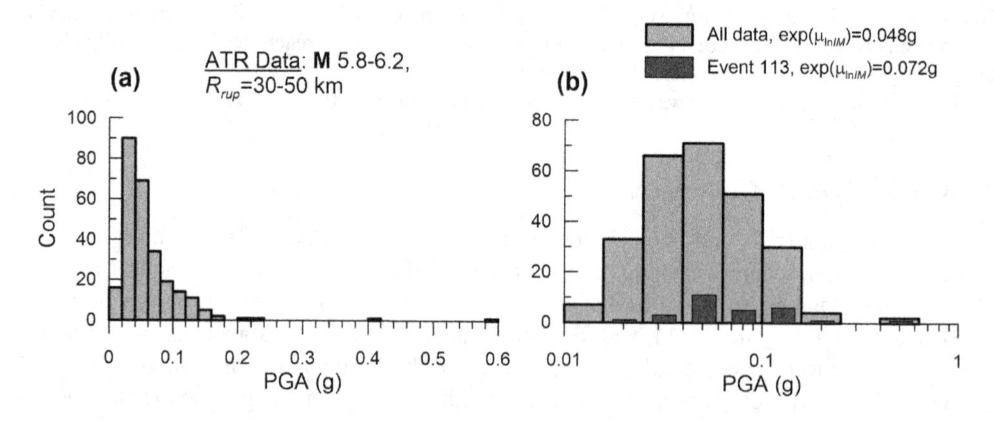

FIGURE 3.28 Histograms of *PGA* data from active tectonic region (ATR) data set plotted on (a) linear scale for *PGA* and (b) log scale.

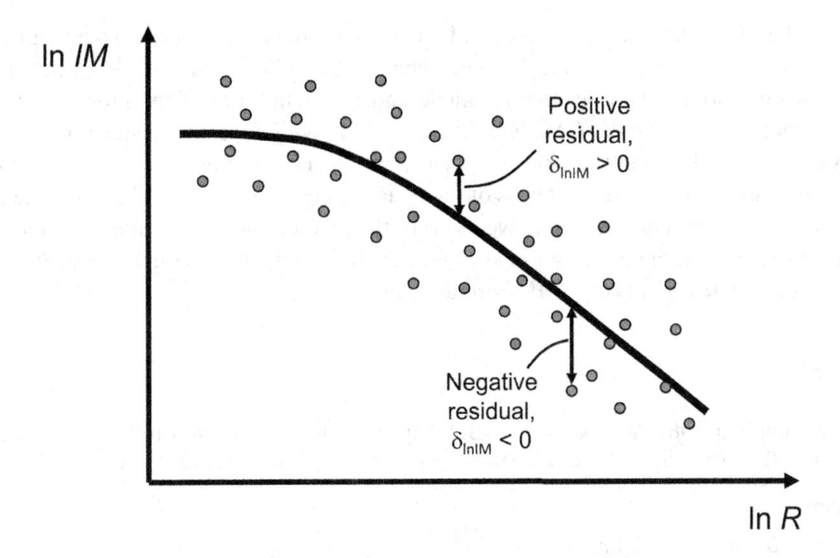

FIGURE 3.29 Schematic illustration of scattered empirical data and a regressed GMM mean curve for a particular magnitude illustrating both positive and negative residuals.

acceleration and over 320 for spectral acceleration. A brief review of the evolution of GMMs aids in the understanding of current GMMs.

Some of the earliest GMMs were developed for peak accelerations at rock sites (e.g., Schnabel and Seed, 1973; Donovan, 1973). These models were developed from small data sets, sometimes relying more on intuition and judgment than formal regression procedures. Subsequently, the derivation of GMMs from regression became routine. In 1981, Campbell used worldwide data to develop a GMM for *PGA* at sites within 50 km of fault rupture for magnitude 5.0–7.7 earthquakes. The GMM used only magnitude and distance to predict *PGA*, hence the scatter in the data included the effects of other variables (style of faulting, site conditions, etc.) now known to affect ground motions. The mean model from Campbell (1981) is:

$$\mu_{\ln PGA} = -4.141 + 0.868M - 1.09\ln\left[R_{rup} + 0.0606\exp(0.7M)\right] \tag{3.25}$$

where *PGA* is in units of *g*, *M* is the local magnitude or surface wave magnitude for magnitudes less than or greater than 6, respectively, and R_{rup} is the closest distance to the fault rupture in km. For this GMM, $\sigma_{\ln PGA} = 0.372$. Figure 3.30 shows the predicted variation of median *PGA* with magnitude

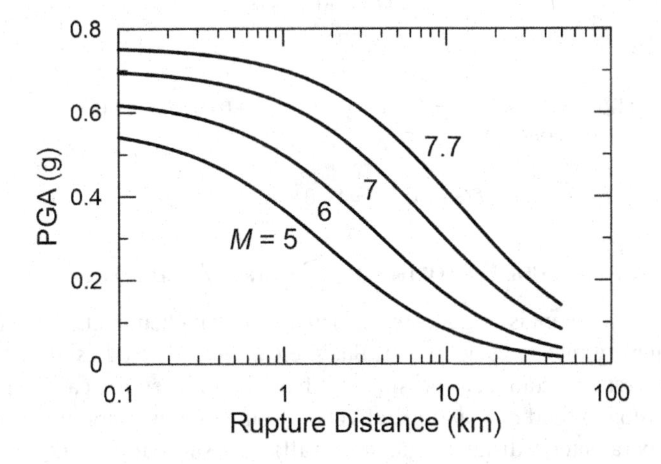

FIGURE 3.30 Variation of predicted median *PGA* with magnitude and distance from Campbell (1981) GMM.

and distance. The PGA can clearly be seen to increase with increasing earthquake magnitude and decrease with increasing source-to-site distance, which is the behavior one would intuitively expect.

As more strong motion data became available and understanding of the processes that affect ground motions increased, newer GMMs were developed. This improved understanding of earthquake physics led to the identification and incorporation of additional predictive variables, so GMMs became more complicated in terms of the equations used to describe them and the information required to use them. GMMs also moved toward the prediction of spectral accelerations, which involved computing coefficients for a range of periods of interest for civil structures (usually 0.01 sec $< T <$ 10 sec). Features of such GMMs are described next.

Example 3.2

Using the Campbell (1981) GMM (Equation 3.25), what are the median and 16th and 84th percentile values of PGA for a magnitude 6.5 earthquake and rupture distance of 15 km?

Solution:

Entering M 6.5 and R_{rup} = 15 km into Equation (3.25),

$$\mu_{\ln PGA} = -4.141 + 0.868M - 1.09\ln\left[R_{rup} + 0.0606\exp(0.7M)\right]$$

produces a median PGA of exp(−1.804) = 0.165g. Since $\sigma_{\ln PGA} = 0.372$, and recognizing that the 84th and 16th percentiles are, respectively, one (logarithmic) standard deviation above and below the (logarithmic) mean,

$$PGA_{84} = \exp\left[\mu_{\ln PGA} + \sigma_{\ln PGA}\right] = \exp[-1.804 + 0.372] = 0.24g$$

$$PGA_{16} = \exp\left[\mu_{\ln PGA} + \sigma_{\ln PGA}\right] = \exp[-1.804 - 0.372] = 0.11g$$

Example 3.3

Considering again the conditions described in Example 3.2, what is the probability that the peak acceleration would exceed 0.3g?

Solution:

The median ground motion is 0.165g and the standard deviation is 0.372. From Section D.7.2.2 in Appendix D (Equation D.52), the standard normal variate is:

$$z = \frac{\ln(0.300) - \ln(0.165)}{0.372} = 1.61$$

From Table D.1, $F_z(1.61) = P(z < 1.61) = P(PGA < 0.3g) = 1 - 0.0537 = 0.9463$.
As a result, the desired probability is

$$P(PGA > 0.3g) = 1 - 0.9463 = 0.0537$$

3.5.2 GMMs for Amplitude Parameters and Spectral Accelerations

The goal in GMM development is to model the principal physical phenomena known to affect ground motions so that intensity measures at a particular site can be predicted as accurately, i.e., without bias and with the lowest uncertainty, as possible. Such predictions require careful attention to strong motion data, regression procedures, and the effects of source parameters (magnitude, depth, focal mechanism), path parameters (distance, hanging wall), and site parameters (stiffness, thickness) on ground motion characteristics. This section examines each of these parameters, their effects on ground motions, and the *scaling relationships* that describe how ground motions vary with those parameters.

GMMs for amplitude parameters and spectral accelerations (S_a) are widely used in engineering practice. Table 3.4 presents GMMs for the three principal tectonic regimes introduced in Section 2.3.3. For active tectonic regions, Table 3.4 shows models for active regions from the NGA-West2 project (Bozorgnia et al., 2014), an example of a local model (Bindi et al., 2011), and a European model (Akkar et al., 2014). Modern GMMs will typically include the amplitude parameters of *PGA* and *PGV* along with 5%-damped S_a at periods between 0.01 and 10 sec.

3.5.2.1 Scaling with Source Parameters

As described in Section 3.4.1, the principal source parameters used in GMMs are magnitude and focal mechanism. Magnitude is generally taken as moment magnitude, **M**. Additional source parameters considered in some active tectonic region GMMs are fault dip and rupture depth.

A typical function for a relatively simple source term is as follows:

$$F_E = e_0 + e_1\mathbf{M} + e_2\mathbf{M}^2 + f(F) \tag{3.26}$$

where **M** is moment magnitude and F represents the focal mechanism (or *style of faulting*). Figure 3.31 shows the scaling (or variation) of *PGA* and $S_a(3.0)$ with magnitude at source-site distances of 1 and 50 km using a representative GMM (Boore et al., 2014).

Magnitude Scaling: Ground motion amplitudes increase with magnitude for the simple reason that magnitude is related to energy release. To develop a conceptual understanding of the trends in the plots, it is helpful initially to study results for a single regime, and active tectonic regions will be used. Ground motion data shows that *IM*s increase with increasing magnitude, but at a rate that decreases with increasing magnitude, requiring nonlinear **M**-scaling relationships typically represented in various GMMs with second-order polynomials (as given in Equation 3.26) or multiple linear (in semi-log space) segments, as illustrated in Figure 3.31. If multiple linear segments are used, $e_2 = 0$ and different values of e_0 and e_1 are given for each of several magnitude ranges. For magnitudes greater than about ~6, the variation of *IM*s with magnitude is strongly frequency- and distance-dependent. At close distances, relatively small patches of high slip (or stress drop) on a fault can control high-frequency *IM*s (such as *PGA*), causing them to be only modestly affected by the full rupture surface and to scale weakly with magnitude (e.g., see the *PGA* slope for $R_{JB} = 1$ km in Figure 3.31). This is referred to as a *magnitude saturation* effect, which diminishes with increasing distance where waves produced by larger portions of the fault surface contribute. Low-frequency *IM*s [such as $S_a(> 1.0)$] are affected by fault slip and energy release, both of which are strongly magnitude-dependent regardless of distance. For this reason, low-frequency *IM*s vary relatively strongly with magnitude.

Effect of Focal Mechanism: Figure 3.32 shows the effect of focal mechanism on the magnitude-scaling for active tectonic region earthquakes using the Boore et al. (2014) GMM. Because normal faults are associated with crustal extension, the normal stress on the fault is lower than for other fault types, leading to reduced stress drop and lower high-frequency *IM*s than for other mechanisms. In Figure 3.32, the effect of focal mechanism is shown as being constant with respect to magnitude (i.e., $f(F)$ is constant for a given mechanism), although there is some evidence that the effect disappears for magnitudes less than about 5 (Boore et al., 2014).

Effect of Source Depth: As shown in Figure 3.33, source depth is typically measured either as a depth to top-of-rupture (Z_{tor}) or hypocentral (focal) depth (Z_{hypo}). Large magnitude shallow crustal earthquakes ($\mathbf{M} > \sim 7.5$) in active tectonic regions will most often rupture to the ground surface ($Z_{tor} = 0$), but smaller events may occur anywhere within the crust. As rupture depth increases, the energy is released further from sites on the ground surface (decreasing amplitude-related *IM*s), but the stress drop may be higher (causing amplitudes to increase). As a result of these trade-offs, source

TABLE 3.4
Selected GMMs for Horizontal-Component *PGA*, *PGV*, and S_a at 5% Damping

Reference[a]	Data Source[b]	Regression Method[c]	M Range[d]	R Range (km)	R type[e]	Site Parameters[f]	Site Response Model[g]	Other Para.[h]	Comp.[i]	Period Range
				Active Tectonic Regions (ATRs)						
NGAW2: Abrahamson et al. (2014)	Global	ME	3–8.5	0–300	R_{rup}, R_{JB}	V_{S30}, $z_{1.0}$	NL	F, W, Z_{tor}, δ, *var R*, *rgn*, CR_{jb}	D50	PGA-10
NGAW2: Boore et al. (2014)	Global	2-s/ME	3–8.5 (*r*, *ss*) 3–7 (*n*)	0–400	R_{JB}	V_{S30}, $z_{1.0}$	NL	F, *rgn*	D50	PGA-10
NGAW2: Campbell & Bozorgnia (2014)	Global	ME	3.3–8.5 (*ss*), 3.3–8 (*r*), 3.3–7.5 (*n*)	0–300	R_{rup}, R_{JB}	V_{S30}, $z_{2.5}$	NL	F, W, Z_{hypo}, δ, *var R*, *rgn*	D50	PGA-10
NGAW2: Chiou & Youngs (2014)	Global	ME	3.5–8.5(*ss*), 3.5–8 (*n*, *r*)	0–300	R_{rup}, R_{JB}	V_{S30}, $z_{1.0}$	NL	F, Z_{tor}, δ, *var R*, *rgn*, *DPP*	D50	PGA-10
NGAW2: Idriss (2014)	Global	NS	5–7.9	0–150	R_{rup}	--	--	F	D50	PGA-10
Akkar et al. (2014)	Europe	ME	4–7.6	0–200	R_{JB}	V_{S30}	NL	F	gm	PGA-4
Local: Bindi et al. (2011)	Italy	ME	4–6.9	0–200	R_{JB}	*Cat.*	L	F	gm	PGA-2
				Subduction Zones (SZs)						
NGAS: Abrahamson et al. (2018)	Global	ME	5–9.5	0–1,000	R_{rup}	V_{S30}	NL	Z_{tor}, Z_t, *rgn*	D50	PGA-10
NGAS: Abrahamson & Gülerce (2022)	Global	ME	5–9.2 (*f*), 5–7.8 (*s*)	0–500	R_{rup}	V_{S30}, $z_{2.5}$	NL	Z_{tor}, Z_t, *rgn*	D50	PGA-10
NGAS: Kuehn et al. (2023)	Global	ME	5–9.5 (*f*), 5–8.5 (*s*)	10–800	R_{rup}	V_{S30}, $z_{2.5}$, $z_{1.0}$	NL	Z_{tor}, Z_t, *rgn*, *var R*, *arc*	D50	PGA-10
NGAS: Parker et al. (2022)	Global	ME	4.5–9.5 (*f*), 4.5–8.5 (*s*)	20–1,000	R_{rup}	V_{S30}, $z_{2.5}$	NL	Z_{tor}, Z_t, *rgn*	D50	PGA-10
GEM: Zhao et al. (2006)	Japan	RE	5–8.3	0–300	R_{hypo}, R_{rup}	Cat.	L	Z_{hypo}	gm	PGA-5
				Stable Continental Regions (SCRs)						
NGAE: Goulet et al. (2021a)	Sims, CENA	ME	4–8.2	0–1,500	R_{rup}	Rock	--	*rgn*, Z_{tor}, δ, R_x	D50	PGA-10

(Continued)

TABLE 3.4 (*Continued*)
Selected GMMs for Horizontal-Component *PGA*, *PGV*, and S_a at 5% Damping

Reference[a]	Data Source[b]	Regression Method[c]	M Range[d]	*R* Range (km)	*R* type[e]	Site Parameters[f]	Site Response Model	Other Para.[h]	Comp.[i]	Period Range
Hassani and Atkinson (2015)	CENA	ME	3–8.5	0–400	R_{JB}	V_{S30}	NL	*F*	D50	PGA-10
Pezeshk et al. (2018)	Sims, CENA	EW	4–8	0–1,000	R_{rup}	Rock	--	-	ns	PGA-10
Boore (2018, 2020)	Sims, CENA	ME	3–8.5	0–1,200	R_{rup}	V_{S30}	L	-	D50	PGA-10

[a] GEM = models recommended for Global Earthquake Model (Stewart et al. 2015). NGAW2, NGAE, NGAS = NGA-West2, NGA-East, NGA-Subduction GMMs.
[b] Sims = Computed ground motions from simulations; CENA = Data from Central and Eastern North America.
[c] ME = mixed effects; 2-s = two-step procedure; EW = all date points weighted equally; NS = not specified.
[d] *n* = normal-slip; *r* = reverse-slip; *ss* = strike-slip; *f* = subduction interface; *s* = subduction intra-slab.
[e] R_{rup} = rupture distance (closest distance between site and source); R_{JB} = surface projection distance; R_{hypo} = hypocenter distance. See Figure 3.36 for definitions.
[f] Cat. = site categories; Rock = reference rock (typically $V_S = 3.0$ km/s in SCRs); V_{S30} = 30 m time-averaged shear wave velocity; z_x = depth to shear wave velocity of *x* km/s.
[g] NL = site effect is nonlinear; L = site effect is linear.
[h] *F* = style of faulting factor; *W* = fault width; *rgn* = regional flag; Z_{hypo} = focal depth, Z_{tor} = depth to top of rupture; δ = fault dip; *var R* = various distance metrics, used for hanging wall terms; *DPP* = direct point parameter (used for near-fault effects); Z_t = subduction zone source type (interface, intra-slab); arc = flag for back-arc or forearc region.
[i] Component of horizontal motion (Section 3.3.1.1). gm = geometric mean; D50 = median component; ns = not stated.

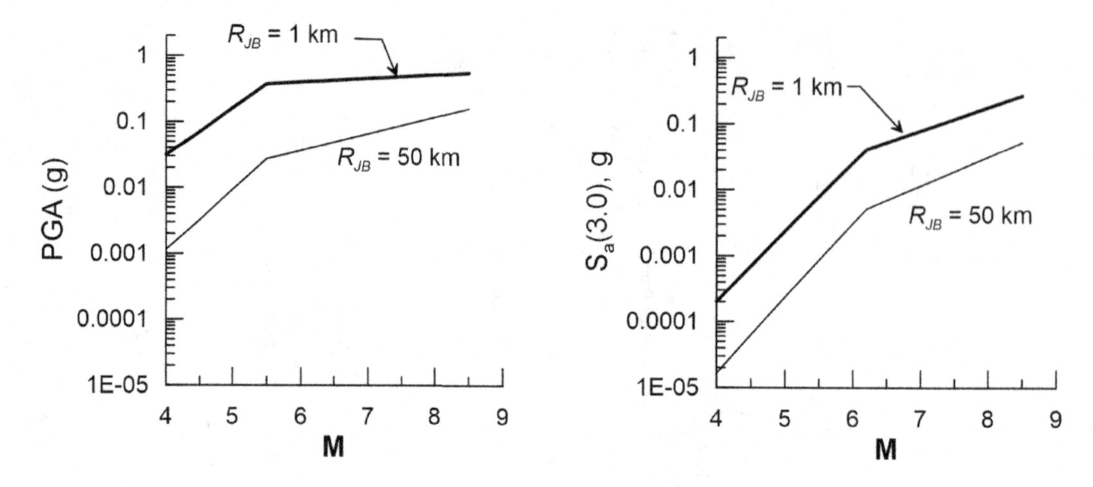

FIGURE 3.31 Effects of magnitude on median ground motions for active tectonic regions. Site condition is rock with $V_{S30} = 760$ m/s (Boore et al., 2014 GMM). Note the relatively strong sensitivity of the low-frequency *IM* ($S_a(3.0)$) to magnitude and the low sensitivity of *PGA* to **M** at short distances.

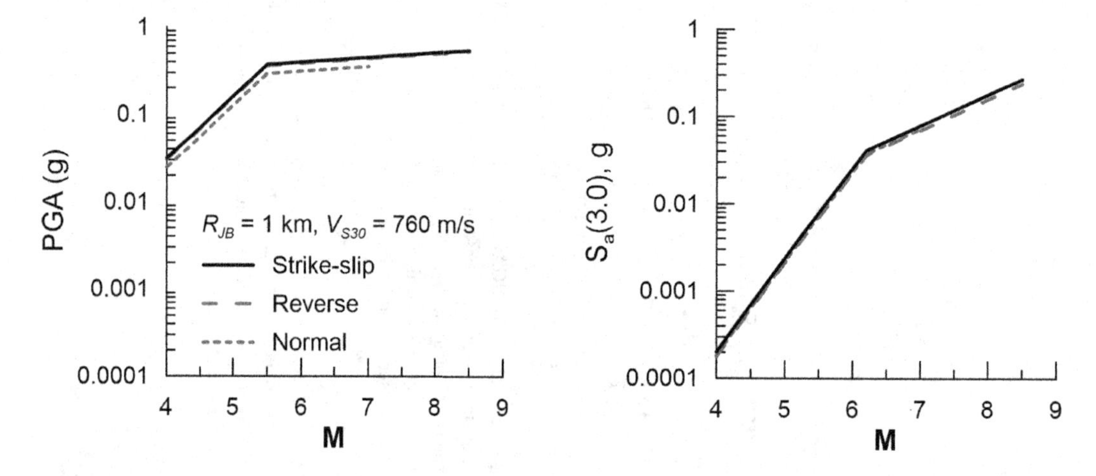

FIGURE 3.32 Effect of focal mechanism on median active tectonic region ground motions from the Boore et al. (2014) GMM.

depth effects on *IM*s are less well resolved than the other source metrics. The effects of fault dip angle are also difficult to resolve relative to focal mechanism effects and are modeled differently in current active tectonic region GMMs.

3.5.2.2 Scaling with Path Parameters

The principal path parameter used in GMMs is source-site distance. An additional path-related effect considered in some GMMs is the position of the site of interest over the hanging wall of dipping faults.

Distance Scaling: Ground motions usually decrease with increasing distance from the fault due to geometric spreading and anelastic attenuation effects (Section 3.4.2). The principal parameter used to capture these effects is source-site distance, which can be measured in a variety of ways. The most widely used distance measures in GMMs, however, are the rupture distance (R_{rup}) and the closest distance to the horizontal projection of the fault to the ground surface (R_{JB}), as shown in Figure 3.33. A typical function for a relatively simple path term is as follows:

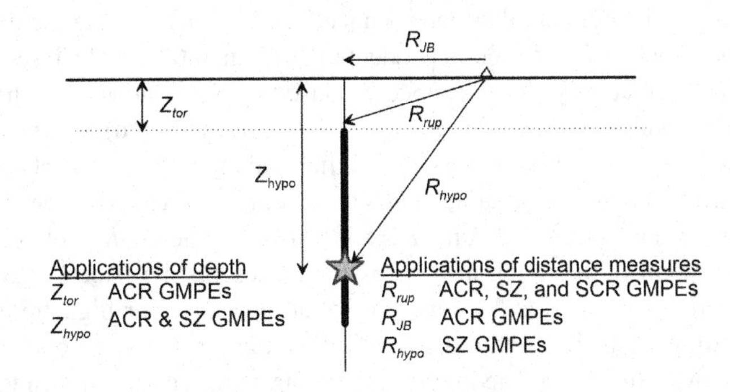

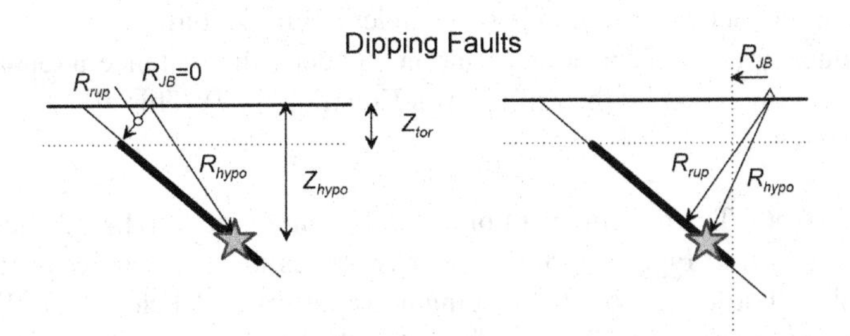

FIGURE 3.33 Depth and site-to-source distance measures used in GMMs.

$$F_P = (c_1 + c_2\mathbf{M})\ln R + c_3 R \tag{3.27}$$

where R is typically the vector sum (Equation 3.18) or sum of either R_{rup} or R_{JB} (depending on the GMM) and the finite fault term, h.

Figure 3.34 shows the variation of PGA and $S_a(3.0)$ with distance R_{JB} for **M** 5, 6, 7, and 8 strike-slip earthquakes and V_{S30} = 760 m/s. For these conditions, rupture distance (R_{rup}) matches R_{JB} when the rupture is shallow, and very similar results are obtained using alternate GMMs that use R_{rup} as the distance parameter (Gregor et al., 2014). Figure 3.34 illustrates several of the key attributes of distance attenuation introduced in Section 3.4.2:

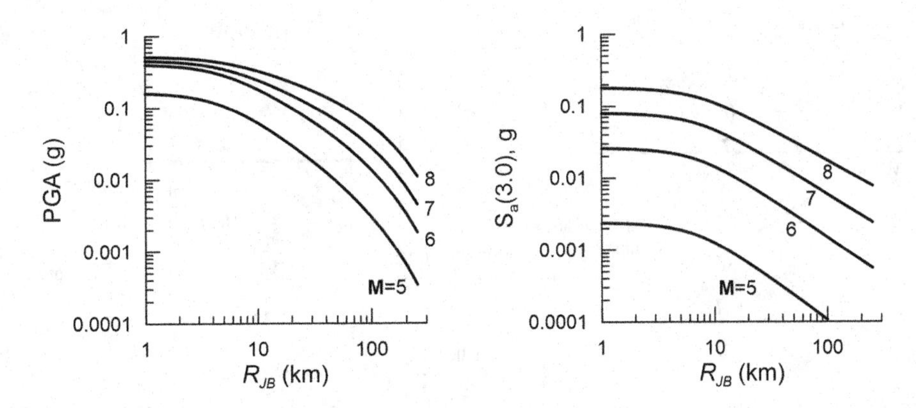

FIGURE 3.34 Effect of magnitude and distance on median ground motions using Boore et al. (2014) GMM. Vertically-dipping strike-slip earthquakes, V_{S30} = 760 m/s. Distance R_{JB} is interchangeable with R_{rup} for these conditions when the rupture depth is shallow.

- At close distances (R_{JB} < ~ 5–10 km), ground motions saturate with distance (the *IM*-R_{JB} curve is nearly horizontal) at all periods. This effect is modeled using the finite fault term, *h*, which may be period or magnitude-dependent in different GMMs. The insensitivity of ground motions to distance at very close distance is caused by the relatively low likelihood that the closest point on the fault (from which R_{rup} and R_{JB} are measured) is generating the seismic waves that control ground motion amplitudes. In effect, the site is subject to waves generated on many nearby sections of the fault, so *IM*s are not directly related to the closest distance.
- Beyond the near-fault saturation zone (R_{JB} > ~ 5–10 km), the *IM*-R_{JB} relationship transitions to a nearly constant slope (in log-log space) that represents the effects of geometric spreading [controlled by the ($c_1 + c_2\mathbf{M}$) term in Equation 3.27]. For high-frequency *IM*s, geometric spreading controls the variation with distance for distances between about 10 and 70 km. Moreover, the rate of geometric spreading is magnitude-dependent, with smaller magnitudes producing faster attenuation (i.e., steeper *IM*-R_{JB} slopes). For low-frequency *IM*s, the rate of geometric spreading is independent of magnitude.
- For high-frequency *IM*s, anelastic attenuation steepens the distance attenuation beyond about 70–80 km (modeled by the c_3R term in Equation 3.27). This effect is not observed at low frequencies.

Hanging Wall Effects: The hanging wall of a dip-slip fault (Section 2.4.2.1) lies above the dipping fault plane, as shown in Figure 3.35. Consider two sites located equally far from the closest point on the fault plane but located on the hanging and foot walls (labeled as "HW site" and "FW site" in Figure 3.35). The hanging wall site is closer to other portions of the fault and hence experiences stronger ground motions than the foot wall site. This hanging wall effect occurs principally in moderate- to high-frequency *IM*s and is generally absent for oscillator periods greater than about 3–4 sec. These effects are included in some of the active tectonic region GMMs listed in Table 3.4.

The simplest approach for modeling hanging wall effects is implicit to the use of the R_{JB} distance metric; as shown in Figure 3.33, R_{JB} is zero on the hanging wall over the surface projection of the fault. As shown in Figure 3.36, the zero R_{JB} condition gives rise to higher-amplitude ground motions for a positive R_x than for the same value of negative R_x. More complex models derived in part from finite fault simulations by Donahue and Abrahamson (2014) are used in the Abrahamson et al. (2014), Campbell and Bozorgnia (2014), and Chiou and Youngs (2014) GMMs. Those models allow for various effects including different rates of distance attenuation for positive and negative R_x (perpendicular to the fault strike) and in the strike-parallel direction (measured by R_y), as well as effects of fault dip angle (δ), Z_{tor}, focal mechanism (normal or reverse), and magnitude. As shown

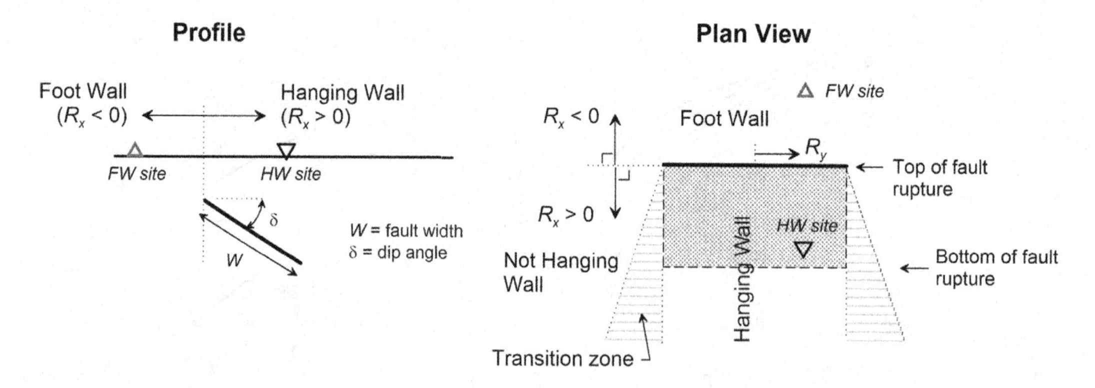

FIGURE 3.35 Schematic depiction of hanging wall and footwall of dipping fault. The HW site and FW site are drawn to have identical rupture distance (R_{rup}). The distance parameters and transition zone indicated in the figure are used in simulation-based hanging wall correction factors by Donahue and Abrahamson (2014). Distance R_y is measured from the center of the fault in the along-strike direction.

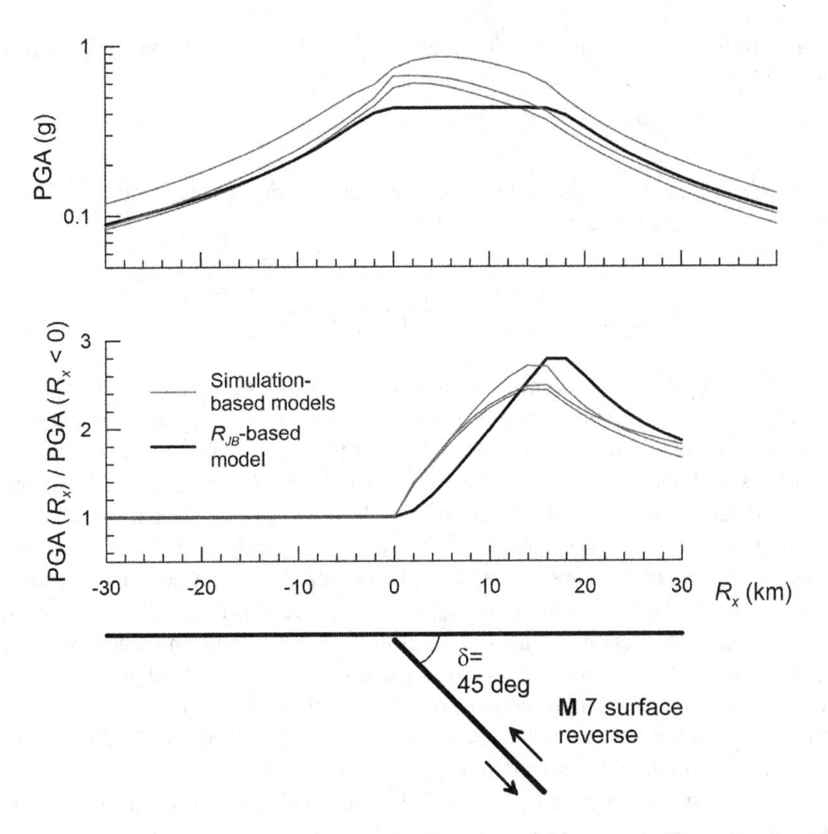

FIGURE 3.36 Peak accelerations and hanging wall effect above **M** 7 reverse-slip earthquake with $Z_{tor} = 0$. Top frame shows profiles of *PGA* at $R_y = 0$ from various models using simulation-based hanging wall term as compared to profile from R_{JB}-based hanging wall term. Middle frame shows hanging wall effect as expressed by ratio of hanging wall to footwall motions for common horizontal distance from top of rupture (R_x). The simulation-based models that are shown are Abrahamson et al. (2014), Campbell and Bozorgnia (2014), and Chiou and Youngs (2014).

in Figure 3.36, the hanging wall effect is strongest over the bottom portion of the dipping fault and extends well beyond the surface projection of the fault in the positive R_x direction (i.e., $R_x > W\cos\delta$). Although the R_{JB}-based hanging wall effect is similar to those from simulation-based models for the case shown in Figure 3.36, there are other cases (e.g., buried ruptures) where the differences are larger. Large model-to-model variability indicates substantial epistemic uncertainty, which occurs because the available data are too limited to constrain the models. This large epistemic uncertainty is an important consideration for sites near dipping faults.

3.5.2.3 Scaling with Site Parameters

The principal site parameter used in most modern GMMs is the time-averaged shear wave velocity in the upper 30 m, V_{S30}, which was defined in Equation (3.23). Exceptions include GMMs in which site conditions are represented by discrete categories (e.g., soft soil, firm soil, rock), which are correlated with V_{S30}, and GMMs that apply only for rock conditions (common for stable continental regions). Some GMMs also consider sediment depth parameters (z_x), which are defined as the depth to $V_s = x$ (typical values of x are 1.0 and 2.5 km/s) in order to better model site response for low-frequency motions, whose wavelengths are much greater than 30 m.

V_{S30}-scaling: As described in Section 3.4.3.1, V_{S30} is a site parameter that can be used with site response models in GMMs to provide first-order estimates of impedance and damping effects. Most contemporary GMMs use nonlinear site response models, meaning that the site amplification is a function of the ground motion amplitude for a reference (rock) site condition. The site response, F_S, in Equation (3.24) has, in general, two components:

$$F_S = F_{lin} + F_{nl} \tag{3.28}$$

where F_{lin} and F_{nl} represent the linear and nonlinear site responses, respectively. The linear site response term can be expressed as:

$$F_{lin} = c \ln\left(\frac{V_{S30}}{V_{ref}}\right) \tag{3.29}$$

where V_{ref} is the velocity for the reference condition (typical values are 760 m/s or ~1,000 m/s) and c is the slope of the linear site amplification term as a function of V_{S30} (in semi-log space, as depicted in Figure 3.37a). Parameter c is a negative number, and the larger its absolute value, the stronger the amplification for soft sites. Typical values of c for active tectonic regions worldwide are given in Figure 3.37b. The largest amplification (indicated by lowest, i.e., most negative value of c) occurs at relatively long periods (peak between 0.4 and 4 sec), which reflects the predominance of deep sediment sites (with long predominant periods and significant basin effects) in the databases used to regress the c terms and the correlation of V_{S30} with the average properties of deeper sediments.

The nonlinear term, F_{nl}, changes the site amplification for conditions where shaking induces strains that are sufficiently large to decrease the stiffness and increase the damping of the soil. The increased damping reduces the amplification, so F_{nl} is usually negative. Equations for F_{nl} have been derived from one-dimensional ground response simulations (Walling et al., 2008; Kamai et al., 2014b; Section 7.5), from data analyses (Choi and Stewart, 2005), and from combinations of simulations and data analysis (Seyhan and Stewart, 2014). A simple expression that captures the main features of F_{nl} is:

$$F_{nl} = f_2 \ln\left(\frac{x_{IMr} + f_3}{f_3}\right) \tag{3.30}$$

where f_2 and f_3 are illustrated in Figure 3.38a for the case where $x_{IMr} = PGA$. In Equation (3.30), x_{IMr} is the ground motion for a reference (rock) site condition (usually taken from a GMM applied for a high V_{S30} value such as 760 m/s); x_{IMr} may be parameterized with various IMs – PGA is common, as is S_a at the same spectral period for which F_{nl} is being evaluated. The parameter f_2 quantifies the degree of site nonlinearity and is typically negative. The parameter f_3 forces F_{nl} to zero as x_{IMr} goes to zero and corresponds to the approximate acceleration differentiating linear amplification at low x_{IMr} (where the F_{nl} trend is flat) to nonlinear amplification at higher x_{IMr} (where the F_{nl} trend is sloped). High absolute values of f_2 and low values of f_3 indicate high levels of nonlinearity.

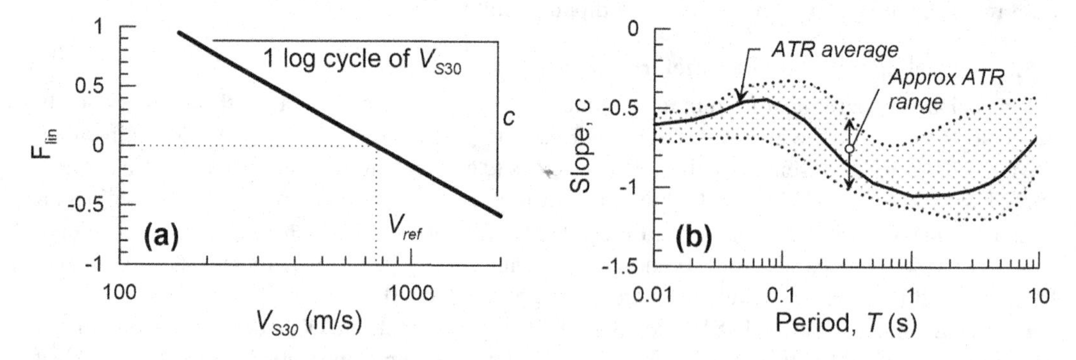

FIGURE 3.37 (a) Illustration of parameters in linear site response model, V_{ref} and c (the results shown apply to PGA); (b) trend of c vs period for active tectonic regions (ATRs).

Typical features of a complete site amplification model (including linear and nonlinear terms) are shown in Figure 3.38b. Plots of amplification [exp(F_S)] against V_{S30} for S_a(3.0) reflect linear response (i.e., the amplification is independent of x_{IMr}), whereas the corresponding plots for PGA amplification bend downward as x_{IMr} increases for low V_{S30} sites (indicating nonlinearity). Similar behavior is revealed in the plots of amplification [exp(F_S)] against x_{IMr}, which are flat for S_a(3.0) and follow the characteristic nonlinear shape in Figure 3.38a for PGA amplification.

Amplification factors for spectra acceleration of soil profiles with V_{S30} = 270 m/s and V_{S30} = 560 m/s are shown in Figure 3.39. The amplification can be seen to be period-dependent for both cases, but also to be much higher and more strongly period-dependent for the softer profile. The low-period (high frequency) amplification decreases with increasing input motion amplitude, but much more strongly for the softer profile since it exhibits much greater nonlinearity than the stiffer profile. For both cases, the curves converge and approach amplification factors of 1.0 at longer periods where the ground motion wavelengths become long relative to the soil profile thickness and the entire soil profile essentially translates back and forth with the underlying rock as a nearly rigid body.

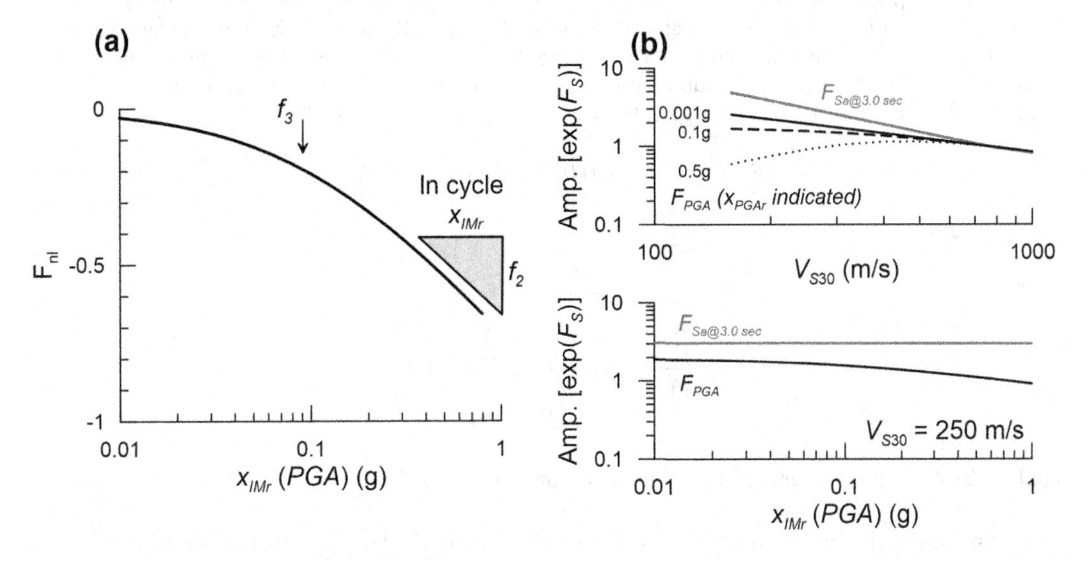

FIGURE 3.38 (a) Illustration of parameters f_2 and f_3 in nonlinear site response model (for x_{IMr} = PGA); (b) trends of amplification factor (=exp[F_s]) against V_{S30} and x_{IMr}.

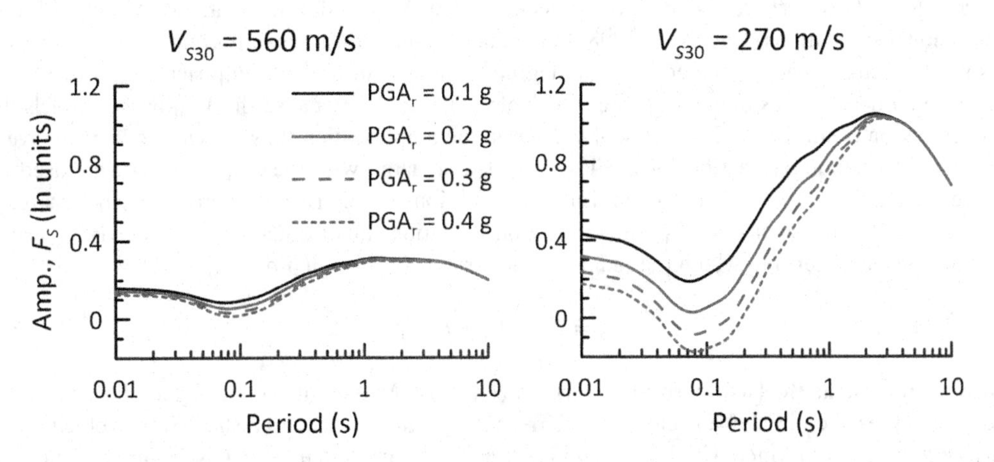

FIGURE 3.39 Variation of site amplification with period for (a) stiff soil (V_{S30} = 560 m/s) and (b) relatively soft soil (V_{S30} = 270 m/s) conditions and input peak accelerations ranging from 0.1 to 0.4g.

Example 3.4

Consider a site with $V_{S30} = 250$ m/s. What levels of site amplification of *PGA* would be expected for this site if the reference level of shaking (for rock) is 0.001*g* or 0.5*g*?

Solution:

The linear site response can be computed using the site V_{S30} and the model in Equation (3.29). Using $V_{S30} = 250$ m/s and $c = -0.6$ (from Figure 3.37b with $PGA \approx S_a(0.01)$), F_{lin} is computed as

$$F_{lin} = c\ln\left(\frac{V_{S30}}{V_{ref}}\right) = (-0.6)\ln\left(\frac{250 \text{ m/sec}}{760 \text{ m/sec}}\right) = 0.667$$

Then, the amplification in arithmetic units is exp(0.667) = 1.95.

For the weak input motion level, there is no nonlinearity so the *PGA* site amplification is 1.95. For the stronger input motion level, the total site amplification can be obtained graphically from the lower part of Figure 3.38b, which is reproduced below with a linear-axis for the ordinate and a narrower range of amplification. At $x_{IMr} = 0.5g$, the amplification is read as 1.10 in arithmetic units and $F_S = 0.095$ in ln units. Subtracting the linear amplification term from the total amplification (in ln units), the corresponding nonlinear term in ln units is (Figure E3.4)

$$F_{nl} = F_S - F_{lin} = 0.095 - (0.667) = -0.572$$

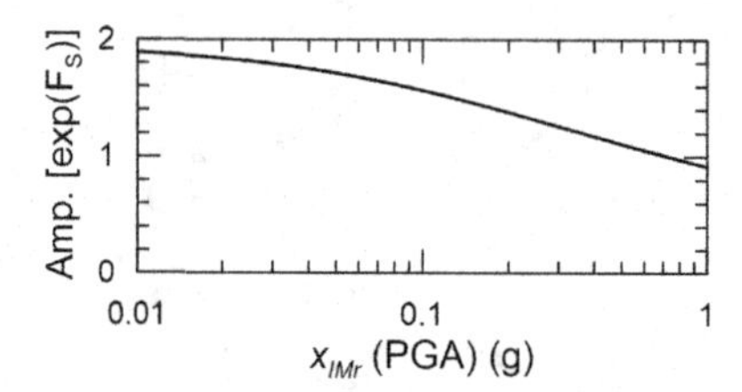

FIGURE E3.4 Expanded version of *PGA* amplification from Figure 3.38b.

Basin Depth Scaling: As discussed in Section 3.4.3.2, particular combinations of incident wave directions and sedimentary basin structure can give rise to ground motion amplification from different mechanisms than those causing ground response. This sort of amplification is commonly referred to as *basin effects*. While basin effects can be estimated with relatively complex three-dimensional simulations (Day et al., 2008; Wang and Jordan, 2014), it is important to recognize that basin effects are already included to some degree in empirical, or semi-empirical, V_{S30}-based site response models such as those illustrated in Figures 3.37–3.39. Because the V_{S30}-based models do not depend on basin-related independent variables, the amplification they provide reflects an average level of basin amplification for the V_{S30} values associated with the empirical data set used to derive the amplification model. To discriminate conditions giving rise to relatively weak or strong basin effects given a particular V_{S30}, the site response relationship of Equation (3.28) can be adjusted with an additional term in which the independent variable is basin depth:

$$F_S = F_{lin} + F_{nl} + F_z(z_x) \tag{3.31}$$

where F_z represents the basin term; this term is zero for GMMs that do not include basin effects. Depth is represented by the vertical distance from the ground surface to a shear wave velocity isosurface having $V_S = x$ km/s, which is denoted z_x (typical parameters used in GMMs are $z_{1.0}$ and $z_{2.5}$; Table 3.4).

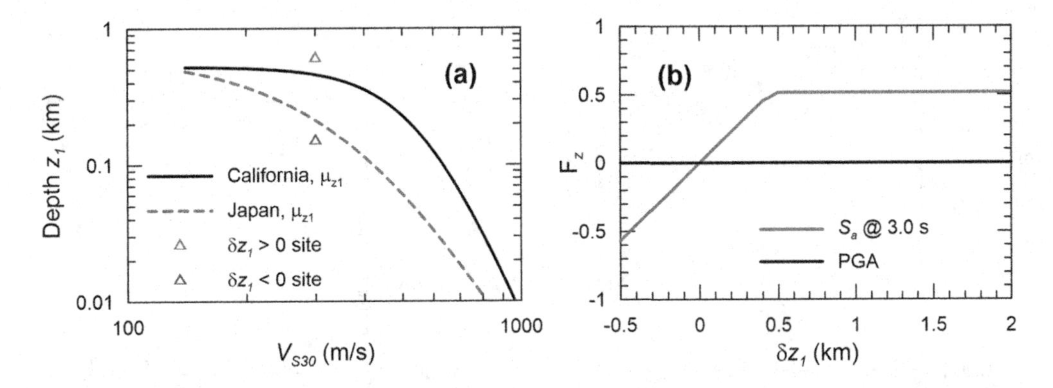

FIGURE 3.40 (a) Illustration of the correlation between V_{S30} and basin depth z_1 for Japan and California (Chiou and Youngs, 2014) and examples of sites with positive and negative δz_1; (b) example basin amplification term F_z, showing dependence on depth differential δz_1 (Boore et al., 2014).

It is important to recognize that z_x is negatively correlated with V_{S30} for a given geologic region; sites having lower V_{S30} values tend to have larger depths. Accordingly, the $F_z(z_x)$ term in Equation (3.31) should modify the average basin effect for depths z_x different from the average depth expected for the site's V_{S30}. In other words, if the site has the same depth as the average depth conditional on the site's V_{S30}, $F_z(z_x)$ should be zero. This conceptual understanding of the basin depth term is applied by replacing z_x in Equation (3.31) with the differential between the depth and the expected value given V_{S30}:

$$\delta z_x = z_x - \mu_{zx}(V_{S30}) \tag{3.32}$$

where μ_{zx} (V_{S30}) is the mean value of depth given V_{S30}. Figure 3.40a shows example relationships between z_1 and V_{S30} using data from Japan and California (Chiou and Youngs, 2014). Figure 3.40b shows an example of the F_z term when parameterized using δz_1 (Boore et al., 2014). There is no effect of basin depth at short periods in the Boore et al. (2014) model. For long-period *IM*s, unusually shallow sites ($\delta z_x < 0$) have negative F_z terms (indicating weaker than average ground motions) whereas deep sites ($\delta z_x > 0$) experience amplified shaking.

3.5.2.4 Variations of Ground Motions between Tectonic Regimes

Most continental regions globally can be associated with active tectonic regions, subduction zones, or stable continental regions (Section 2.3.3; García et al., 2012). One reason for classifying these different regimes is that characteristics of ground motions differ between regions and different GMMs are used for applications in different regimes.

The ground motion scaling relationships shown in previous sections apply for active tectonic regions. Additional GMMs applicable to subduction zones and stable continental regions are listed in Table 3.4. The same general trends hold for other regimes (ground motions increase with increasing magnitude, decreasing distance, and decreasing site stiffness), although there are some important differences in the trends that are found in subduction zones and stable continental regions. Figures 3.41 and 3.42 show the scaling of ground motions with R_{rup} for stable continental region and interface subduction zone events, respectively, with the active tectonic region results shown in the background for reference. For stable continental regions (Figure 3.41), near-fault high-frequency *IM*s are stronger than those for active tectonic regions due to higher stress drops. Distance attenuation rates are generally considered to be faster than those for active tectonic regions for the distance range controlled by shear waves (up to approximately 50 km), then a flattening occurs in the longer distance range controlled by surface waves. Anelastic attenuation effects are small for the distance range shown because of lower crustal damping (higher Q) in stable continental regions, which also contributes to the reduced distance attenuation rate for distances beyond 50 km. The differences

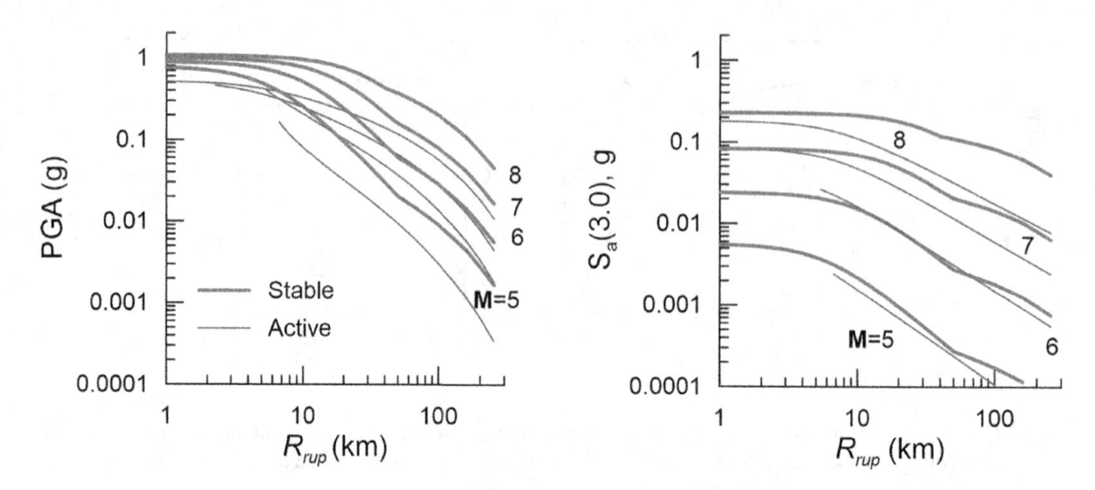

FIGURE 3.41 Effect of magnitude and distance on median ground motions for stable continental regions using Yenier and Atkinson (2015) GMM. Active tectonic region *IM*s from Boore et al. (2014) shown for comparison with R_{JB} converted to R_{rup} (Kaklamanos et al., 2011). Site condition is 760 m/s (site correction from Stewart et al., 2020 applied to adjust 3,000 m/s reference condition in Yenier and Atkinson model).

between active and stable regions are smaller at long periods for conditions often critical for hazard analyses (**M** 5–7, distance < 50 km).

For interface subduction events (Figure 3.42), ground motion amplitudes are similar to those for active tectonic regions for magnitudes of 7 or less, but are stronger for **M** 8 (the subduction models also extend to **M** 9, which is beyond the range for the active tectonic regions). This indicates stronger magnitude-scaling at large magnitudes for interface subduction events relative to active region events. Distance attenuation rates are only modestly different from those for active tectonic regions. As shown in Figure 3.43, there are notable differences between ground motions within subduction zones from interface and slab sources, with slab events having high frequency *IM*s with stronger amplitudes at the closest distances and faster attenuation rates. The faster attenuation likely results from the large depth of these events, which causes the seismic waves to travel through the upper

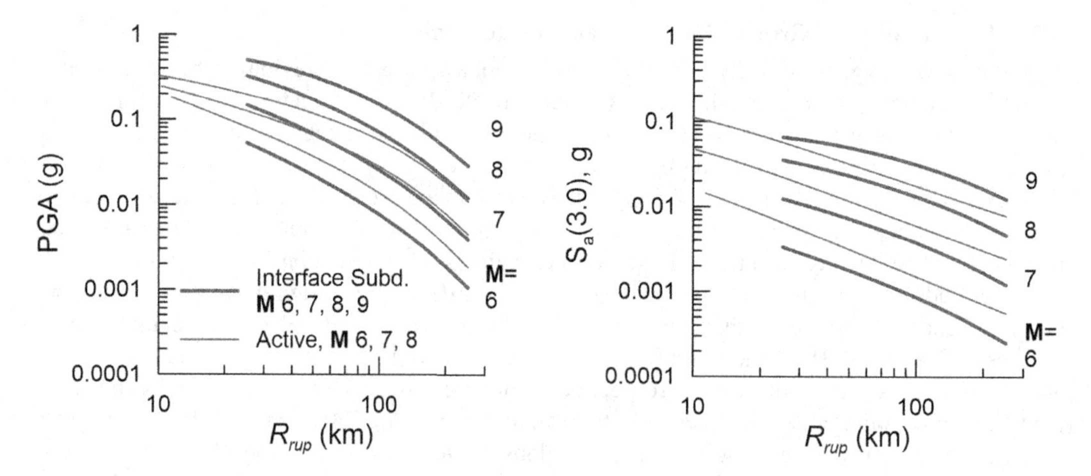

FIGURE 3.42 Effect of magnitude and distance on median ground motions for interface subduction zone earthquakes using the Parker et al. (2022) global GMM. Subduction zone results not shown for distances <25 km because interface events occur in offshore areas and near-fault data are unavailable. Active tectonic region *IM*s from Boore et al. (2014) shown for comparison with R_{JB} converted to R_{rup} (Kaklamanos et al., 2011). Rock site conditions, V_{S30} = 760 m/s.

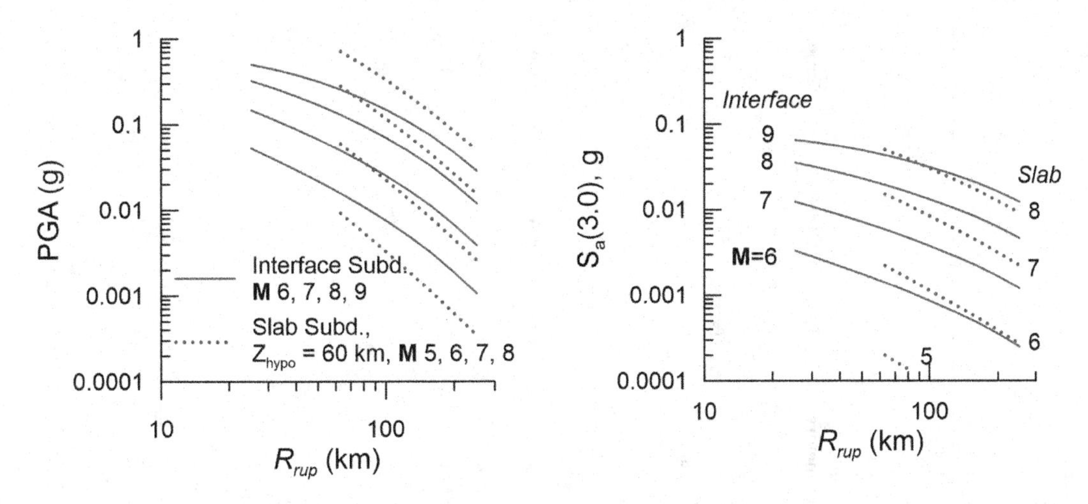

FIGURE 3.43 Effect of magnitude and distance on median ground motions for interface and slab subduction zone earthquakes using the Parker et al. (2022) global GMM. The slab model results are plotted for hypocentral depth of and minimum distance of 60 km. Rock site conditions, V_{S30} = 760 m/s.

mantle, which is softened by high heat and hence has lower Q (higher damping), before passing into the crust and toward the ground surface.

3.5.3 Other Intensity Measures

Sections 3.3.2–3.3.3 describe *IM*s for the frequency content and duration of ground motion that are used in engineering applications, along with several hybrid scalar parameters such as Arias intensity (I_A) and cumulative absolute velocity (*CAV*). As with amplitude parameters, GMMs are available that describe the effects of magnitude, distance, and site condition on these *IM*s.

3.5.3.1 Duration

Table 3.5 lists several GMMs for duration. Most models were developed for significant duration, as defined in Section 3.3.3. GMMs for duration use one of two general function types. One type takes the natural log of duration as a sum (in natural log units) of source, path, and site terms similar to Equation (3.24) (Bommer et al., 2009; Lee and Green, 2014; Bora et al., 2014). The second type is based on physical principles in which duration is the sum, in arithmetic units, of a source duration and path duration (Boore and Thompson, 2014; Afshari and Stewart, 2016). The source duration is related to the time for slip to occur across the fault surface, which in turn is related to both rise time and the ratio of fault dimension to rupture speed. Source duration is given as a function of **M**. Path duration accounts for the separation of p- and s-waves in time, which increases with distance from the fault.

Figure 3.44 shows the variation of significant duration with **M**, R_{rup}, and V_{S30} for active tectonic regions using the Afshari and Stewart (2016) model. Duration increases strongly with magnitude, which is a source duration effect. Duration increases steadily with distance, reflecting the path-spreading effect, and is also marginally higher for soil sites than for rock sites. Duration at soil sites is lower for stable continental regions than for active tectonic regions (Lee and Green, 2014), which is consistent with the higher stress drop in stable continental regions.

3.5.3.2 Mean Period

Table 3.5 lists a single GMM for mean period by Rathje et al. (2004). Figure 3.45 shows that the variation of mean period with magnitude and distance follows similar trends to those for duration – strong magnitude-dependence, steady increase with distance, and higher values for soil sites

TABLE 3.5
Selected GMMs for Duration, Frequency Content, and Hybrid Scalar *IMs*

Reference and IM	Tectonic Regime[a]	Data Source[b]	Regression Method[c]	M Range	R Range (km)	R Type[d]	Site Parameters[e]	Site Term[f]	Other Para.[g]	Comp.[h]
				Duration						
Afshari and Stewart (2016), $D_{5\text{-}95}$, $D_{5\text{-}75}$, $D_{20\text{-}80}$	ATR	Global	ME	5–8	0–200	R_{rup}	V_{S30}, z_1	L	-	gm
Bommer et al. (2009), $D_{5\text{-}95}$, $D_{5\text{-}75}$	ATR	Global	ME	4.8–7.9	0–100	R_{rup}	V_{S30}	L	Z_{tor}	arb
Lee and Green (2014), $D_{5\text{-}95}$, $D_{5\text{-}75}$	SCR	CENA	ME	4.5–7.6	0–200	R_{rup}	Cat.	L	-	arb
Boore and Thompson (2014), $D_{5\text{-}95}$	ATR	Global	subj	3–8	0–400	R_{rup}	-	-	-	gm
				Mean Period						
Rathje et al. (2004), T_m	ATR	Global	ME	4.7–7.6	5–550	R_{rup}	Cat.	L	-	vs
				Hybrid Scalar						
Campbell & Bozorgnia (2019), I_A, CAV	ATR	Global	ME	3–7.9	0–500	R_{rup}, R_{JB}	V_{S30}, $z_{2.5}$	NL	$\acute{F}$, W, Z_{hypo}, δ, $var\ R$, rgn	gm
Farhadi and Pezeskh (2020), I_A, CAV	SCR	CENA	ME	3–7.9	0–500	R_{rup}	V_{S30}	NL	-	gm
Macedo et al. (2019), I_A	SZ	Global	IM cor			Controlled by GMM for reference IM				D50

[a] ATR = active tectonic regions; SCR = stable continental regions; SZ = subduction zones.
[b] CENA = Data from Central and Eastern north America.
[c] ME = mixed effects; subj = subjective fit (by hand); IM cor = correlation with S_a and PGA.
[d] R_{rup} = rupture distance (closest distance between site and source); R_{JB} = surface projection distance.
[e] Cat. = various site categories; V_{S30} = time-averaged shear wave velocity in upper 30 m; z_x = depth to shear wave velocity of x km/s.
[f] NL = site effect is nonlinear; L = site effect is linear.
[g] F = style of faulting factor; Z_{tor} = depth to top of rupture; δ = fault dip; $var\ R$ = various distance metrics, used for hanging wall terms.
[h] Component of horizontal motion considered. gm = geometric mean; arb = arbitrary component; vs = vector sum; D50 = RotD50.

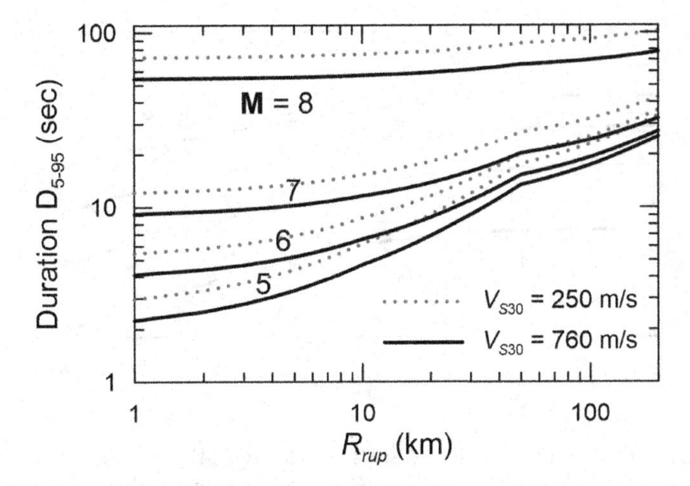

FIGURE 3.44 Variation of median significant duration (5%–95%) with magnitude, distance, and site condition as given by the Afshari and Stewart (2016) GMM.

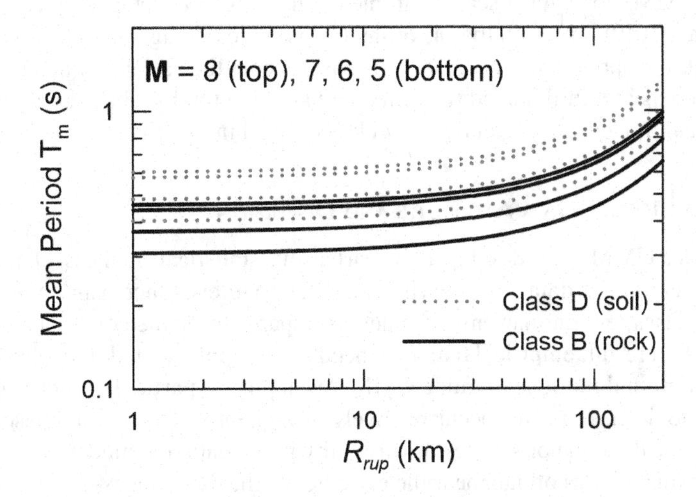

FIGURE 3.45 Variation of median values of mean period (T_m) with magnitude, distance, and site condition as given by the Rathje et al. (2004) GMM.

than for rock sites. These trends in the frequency content of ground motion are consistent with factors previously identified as affecting S_a in Section 3.5.2.

3.5.3.3 Hybrid Scalar *IM*s

As shown in Table 3.5, Campbell and Bozorgnia (2019) have presented GMMs for *CAV* and I_A using identical functional forms (but with different coefficients) to those for their 2014 model for S_a (Campbell and Bozorgnia, 2014). As shown in Figure 3.46, although these hybrid scalar parameters reflect the combined influence of amplitude, duration, and frequency content, their scaling relationships have similarities to those for short-period S_a – modest variation with magnitude for $\mathbf{M} > 6.5$ and steady decrease of the *IM*s with distance due to geometrical spreading (the GMMs do not include an anelastic attenuation term). The much faster attenuation of I_A as compared to *CAV* results from I_A being more sensitive to high-frequency components of the ground motion, which have relatively fast distance attenuation. Although not shown in Figure 3.46, the site response models

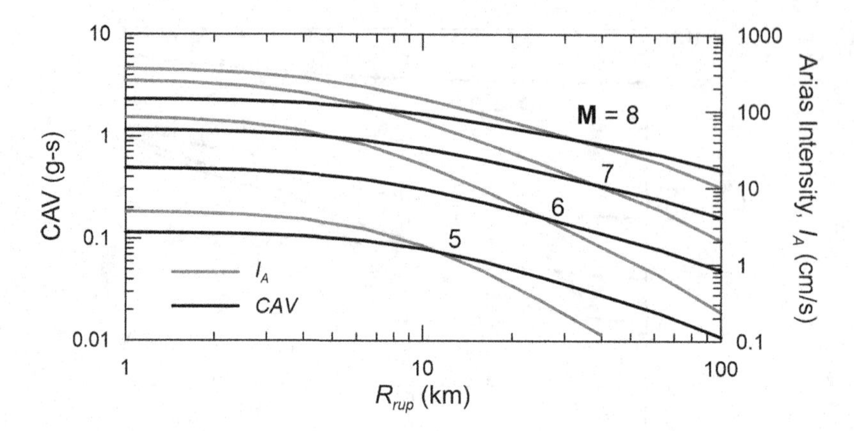

FIGURE 3.46 Variation of median cumulative absolute velocity (*CAV*) and Arias Intensity (I_A) with magnitude and distance for strike-slip faults and V_{S30} = 760 m/s as given by the Campbell and Bozorgnia (2019) GMMs.

are significantly nonlinear for soft soils, which is also similar to high-frequency *IM*s. As shown in Table 3.5, some GMMs for hybrid scalar parameters have been developed as conditional models (e.g., Macedo et al., 2019), in which the *IM* of interest is regressed against other *IM*s (typically S_a). When such models are applied, it is necessary to consider both the variability and uncertainty in the unconditioned models and in the conditional model using the total probability theorem. For these reasons, conditional models are currently not widely utilized in practice.

3.5.4 Ground Motion Data Attributes and Dispersion

Whenever possible, GMMs are developed in part using statistical analysis of recorded ground motion data (i.e., empirical data) from earthquakes. Despite ever increasing amounts of ground motion data, databases remain unevenly sampled and sparse for some conditions that are of engineering interest. As a result, empirical data may need to be supplemented with physical constraints or with simulated ground motions to guide GMM development, particularly for stable continental regions where tectonic earthquakes occur relatively infrequently. This section describes attributes of currently available data, major sources of variability in the data, the modeling of data dispersion in GMMs, and characteristics of data near the extremes of the data ranges.

3.5.4.1 Data Distributions

The availability of ground motion data over the range of magnitude and distance that are relevant for engineering applications is strongly dependent on tectonic regime. Figure 3.47 shows the availability of recordings in magnitude-distance space for the three regimes. The horizontal 'lines' of dots are in most cases single events that were recorded by many instruments at different distances. The approximate data ranges are 0–400 km and **M** 3–7.9 for active tectonic regions, 10–2,000 km and **M** 4.0–9.0 for subduction zones, and 30–1,000 km and **M** < 6 for stable continental regions.

Despite the large number of recordings in each regime, the data are poorly sampled in three respects. Table 3.6 summarizes these sampling issues and the means by which they are dealt with in GMM development. When selecting from among alternate GMMs to use for a given project, the manner by which they address these data sampling issues should be considered.

3.5.4.2 Between- and Within-Event Variability

Quantifying uncertainty in predicted ground motions is extremely important for seismic hazard analyses (Chapter 4). Some of the uncertainty in predicted motions comes from variations of ground motions among the recordings of an individual earthquake (*within-event variability*), and some

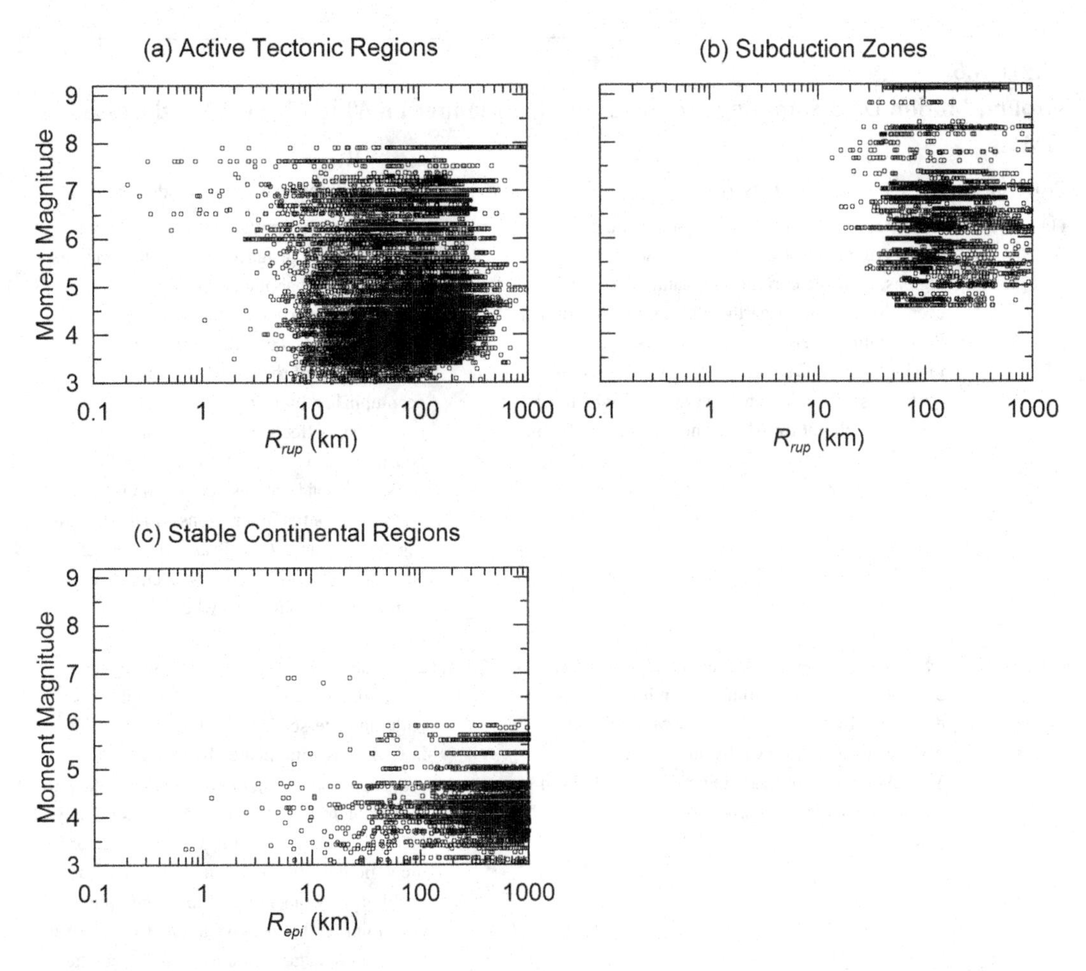

FIGURE 3.47 Distribution of ground motion recordings in (a) active tectonic regions (source: Ancheta et al., 2014); (b) subduction zones, including interface and slab events (source: Mazzoni et al., 2022); (c) stable continental regions (source: Goulet et al., 2021b). These data were gathered and synthesized for NGA projects (NGA-West2, NGA-Subduction and NGA-East).

comes from variations between different earthquakes (*between-event variability*). Ground motion variability can be partitioned into within- and between-event components, which are commonly denoted δW and η_E, respectively. Figure 3.48 shows how data for two earthquakes with the same magnitude compare to the mean prediction from a GMM. Data for Event *i* is systematically higher than the model, while data for Event *j* is lower. The between-event variability η_E is approximately the mean offset of the respective events from the model (positive for *i*, negative for *j*). Within-event variability δW comprises the remaining variability for a given data point. These two components, when taken together, sum to the total residual $\delta_{\ln IM}$,

$$\left(\delta_{\ln IM}\right)_{ik} = \eta_{E,i} + \delta W_{ik} \tag{3.33}$$

In Equation (3.33), index *i* refers to the event and index *k* to the site.

When considering a GMM for use in hazard analyses, it is important to identify whether these two sources of variability were properly considered in the model regression. Considering both sources of ground motion variability affects data weighting in regression analysis. The two extremes are (1) if between-event variability is negligible, each recording can be weighted equally and (2) if

TABLE 3.6
Ground Motion Data Sampling Problems and the Manner in Which They Are Addressed in GMMs

Problem	Description	Significance and How Addressed
(1)	Data are sparse for magnitude-distance ranges of engineering interest. In the case of active tectonic regions, there are currently no data for **M** > 7.9 events, despite **M** 8+ earthquakes often controlling the outcome of seismic hazard analyses (e.g., in much of coastal California). The problem is most acute for stable continental regions, for which data are largely absent for **M** > 6 and distance < 30 km.	GMMs are commonly regressed over the magnitude and distance range of the available data. However, applications often require a larger range, particularly for larger magnitude and closer distances, which requires extrapolation of the model. This is best accomplished by using equations for source, path, or site effects with a physical basis and constraining model coefficients based on physical considerations and/or simulations. For stable continental regions, where data are especially sparse, GMMs are based largely on simulated ground motions with empirical constraint for conditions where data are available.
(2)	The data are unevenly distributed geographically. For example, the active tectonic region inventory is dominated by California data for **M** < 5.0, yet contains no data from California for **M** > 7.3. Moreover, there are many regions, particularly in developing nations, for which data are absent altogether.	Certain attributes of GMMs exhibit regional variability, such as constant terms and path- and site-scaling. Capturing these effects in GMMs is complicated by the uneven geographic sampling of data. One approach to this problem is to develop GMMs using only local data sets. While this approach addresses the regionalization issue, it exacerbates Problem 1 by placing a strong geographic constraint on the data. An alternative is to use a broader database and have region-specific coefficients where needed.
(3)	The data sets are dominated by a few well-recorded events, particularly at the upper end of the magnitude range for each regime. For example, the active tectonic region data set has only three events with **M** > 7.5 (1999 **M** 7.6 Chi Chi Taiwan, 2002 M 7.9 Denali Alaska, 2008 **M** 7.9 Wenchuan China) and the subduction zone data set has two events with **M** > 8.5 (2010 M 8.8 Maule Chile and 2011 **M** 9.0 Tohoku Japan).	When important magnitude and/or distance ranges of data sets are dominated by a few well-recorded earthquakes, the result is sparse sampling of *between-event variability*. It is unknown whether the few available samples represent mean behavior, or if they are biased relative to the global mean that would be observed if there were many more recorded events. This is addressed by considering *within-event variability* and *between-event variability* (Section 3.5.4.2) in the regressions used to develop GMMs.

within-event variability is negligible, each event can be weighted equally. In reality, both sources of variability are significant, and regression procedures that address this issue include:

- **Two-Step Regression Procedure** (Joyner and Boore, 1993, 1994): All data are weighted equally to regress components of the model related to distance attenuation and site effects. Data for each event are then "corrected" to a common distance (typically 1 km) and site

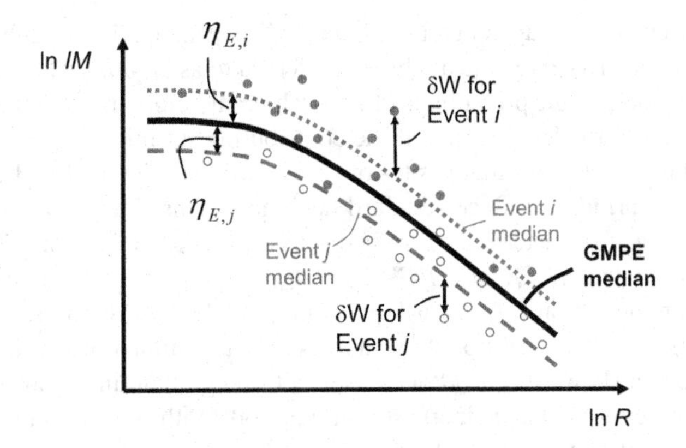

FIGURE 3.48 Schematic depiction of between- and within-event variability of two earthquakes with the same magnitude relative to the median prediction from a GMM.

condition (typically rock) and an average value of the *IM* for each event is computed. Source terms are regressed using those averaged *IM*s, which are weighted in consideration of the number of recordings (weight increases with this number) and event-to-event variability.

- **Mixed Effects Procedure** (Abrahamson and Youngs, 1992): Between-event residuals (also known as event terms), η_{Ei}, are evaluated for each earthquake, *i*. By subtracting η_{Ei} from the data, between-event variability is removed and all adjusted data points can be weighted equally. The evaluation of η_{Ei} is iterative with the regression of GMM regression coefficients, so this procedure is relatively computationally intensive.

3.5.4.3 Intensity Measure Dispersion

Both forms of ground motion variability described in the preceding section can be described by normal distributions (of ln *IM*) with means of zero (assuming the GMM is unbiased) and different standard deviations, which quantify dispersion. The dispersion (standard deviation) across all events of between-event terms η_{Ei} is denoted $\tau_{\ln IM}$ and the dispersion of within-event residuals δW_{ik} across all observations is denoted $\phi_{\ln IM}$. These two forms of variability are assumed to be statistically independent, so the total standard deviation of a GMM $\sigma_{\ln IM}$ (i.e., standard deviation of $\delta_{\ln IM}$) is given as:

$$\sigma_{\ln IM} = \sqrt{\phi_{\ln IM}^2 + \tau_{\ln IM}^2} \tag{3.34}$$

The between-event dispersion $\tau_{\ln IM}$ is principally related to the seismic source; its presence indicates that the ground motions used to develop GMMs are sensitive to source effects beyond those considered as independent variables (e.g., stress drop, slip amount, and distribution) in the source function F_E. Similarly, within-event dispersion $\phi_{\ln IM}$ represents record-to-record variability for any given earthquake that is influenced by source, path, and site effects not captured by the GMM path and site models F_P and F_S (e.g., from locally weaker or stronger distance attenuation relative to the F_P model as a result of variations in crustal structure, or site-specific variations in site response relative to F_S due to local geologic conditions).

The "ideal" GMM for a specific, individual site would be based on recordings at that site from earthquakes on all contributing seismic sources over a wide range of magnitudes and distances. Such a GMM would reflect specific attributes of those sources, features of the source-to-site wave paths, and site-specific site response. At this time, such GMMs do not exist, so in order to have data for GMM development that covers the required range of magnitudes, site-to-source distances, and site effects, data are sampled over many geographic regions. This relatively broad sampling of data comes with a price, which is that source-, path-, and site-specific effects that are present in each recording to some extent are averaged out and affect GMMs principally through their contribution to data dispersion. This approach to data analysis is referred to as *ergodic*. The GMMs presented in Tables 3.4 and 3.5 provide ergodic predictions of *IMs*, which therefore in general are neither specific to the site of interest nor to the local region of the contributing seismic sources. If additional information is introduced to the analysis that is site- or region-specific, the modified GMM is no longer ergodic, which could in principle reduce standard deviation terms. The most common example of this is the use of *non-ergodic* site terms, which is discussed further in Section 7.7.

For amplitude parameters (*PGA* and *PGV*) and S_a, the principal factor controlling ergodic between-event standard deviation $\tau_{\ln IM}$ is earthquake magnitude. GMMs for active tectonic regions are based on data sets relatively well-populated across the magnitude range where changes in $\tau_{\ln IM}$ are evident. Figure 3.49a shows that $\tau_{\ln IM}$ decreases with magnitude in the approximate range of **M** 4.5 to 5.5 and the amount of change increases with period. Within-event standard deviation $\phi_{\ln IM}$ has a more complex set of dependencies, both increasing and decreasing with magnitude depending on oscillator periods (Figure 3.49a), increasing with source-site distance beyond about 100 km (Figure 3.49b), and decreasing for low V_{S30} sites (less than about 300 m/s) relative to stiffer sites (Figure 3.49c). The increase of $\phi_{\ln IM}$ with distance is thought to be caused by regional variations in anelastic attenuation that are difficult to capture in GMMs. The decrease of $\phi_{\ln IM}$ for softer sites is caused by nonlinear site response. An unusually strong rock motion (positive outlier in data distribution) will produce stronger-than-average nonlinearity, reducing the motion at soil sites. Similarly, an unusually weak rock motion will induce weaker-than-average nonlinearity, strengthening the soil motion. These features compress the distribution of soil motions relative to rock motions, thus reducing $\phi_{\ln IM}$.

In addition to trends with predictive variables, the overall levels of data dispersion, as represented by total standard deviation $\sigma_{\ln IM}$, are of interest because a low-dispersion (i.e., predictable) *IM* is more useful as a predictor of seismic response. Figure 3.50 shows the variation of $\sigma_{\ln IM}$ with oscillator period for S_a and other *IMs*. In active tectonic regions, $\sigma_{\ln IM}$ increases with period, whereas it generally decreases beyond 0.1 sec in other regions (these differences are largely a result of regional variations in the component of variability related to ergodic site response). Standard deviations are relatively low for duration, mean period, and *CAV*, and are relatively high for Arias Intensity. Except at long periods, $\sigma_{\ln IM}$ tends to be lower in active tectonic regions than the other regions.

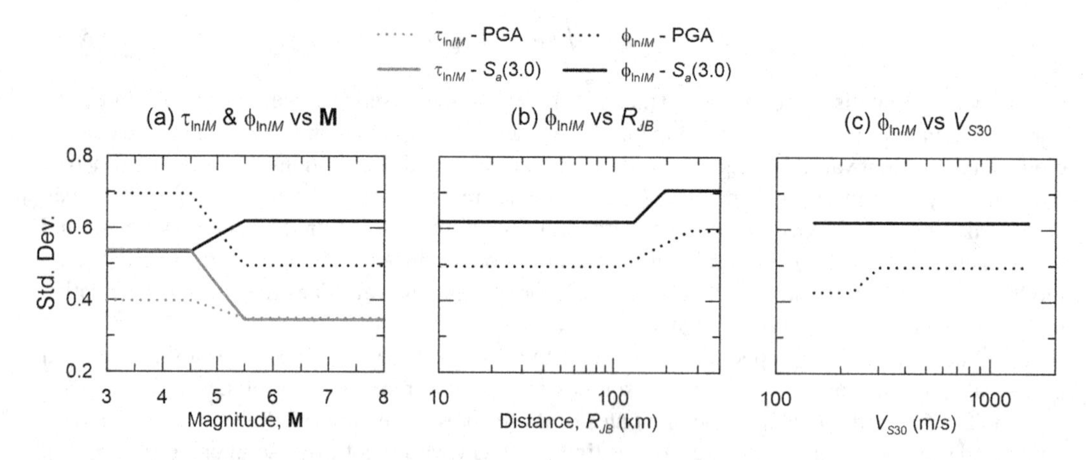

FIGURE 3.49 (a–c) Variation of between- and within-event standard deviation terms ($\tau_{\ln IM}$ and $\phi_{\ln IM}$) for *PGA* and S_a (3.0) with **M**, R_{JB}, and V_{S30} ($\tau_{\ln IM}$ depends only on **M**). Trends based on Boore et al. (2014) GMM.

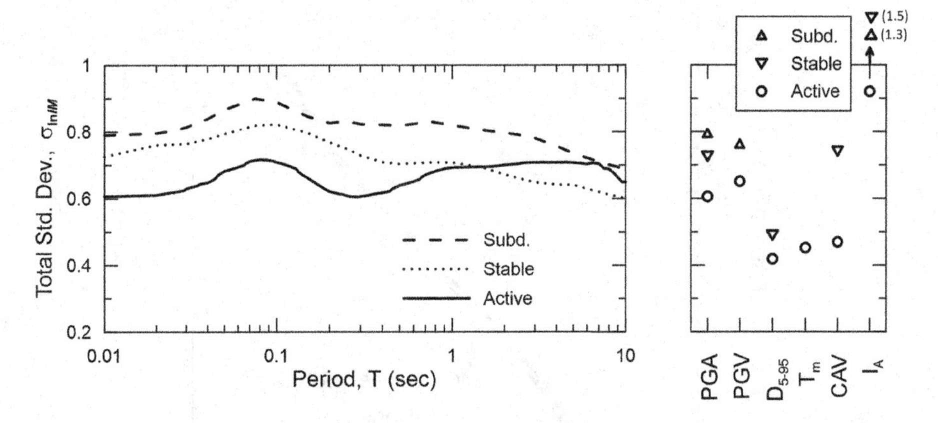

FIGURE 3.50 Variation of total standard deviation $\sigma_{\ln IM}$ across *IM*s. S_a, *PGA*, and *PGV* are based on Boore et al. (2014) for active tectonic regions (**M** > 5.5, R_{JB} < 80 km, V_{S30} > 300 m/s), Parker et al. (2022) for subduction zones (R_{rup} < 200 km, V_{S30} > 500 m/s), and Goulet et al. (2017) and Stewart et al. (2018) for stable continental regions (**M** > 7, V_{S30} < 1,200 m/s). Duration $\sigma_{\ln IM}$ from Afshari and Stewart (2016) for active tectonic regions (**M** > 6.5) and Lee and Green (2014) for stable continental regions. Mean period $\sigma_{\ln IM}$ from Rathje et al. (2004). *CAV* and I_A $\sigma_{\ln IM}$ from Campbell and Bozorgnia (2019) for active tectonic regions (**M** > 5.5) and Farhadi and Pezeshk (2020) for stable continental regions, and I_A $\sigma_{\ln IM}$ from Macedo et al. (2019) for subduction zones.

3.5.4.4 Data Distribution Near Extremes of Range

As described further in Chapter 4 (Section 4.4.3), seismic hazard analyses can be sensitive to the properties of the ground motion distribution near the extremes of its range. This tends to be the case when rare occurrences of ground motion *IM*s are used as the basis for engineering design, as is often the case for critical structures such as nuclear power plants.

As illustrated in Figure 3.28, for a particular range of controlling parameters (magnitude, distance, etc.), ground motion *IM*s from recordings tend to be lognormally distributed, with the dispersion represented by standard deviation term $\sigma_{\ln IM}$. Formal analyses of data distributions do not utilize *IM*s directly, but rather the normalized residuals of a GMM, which are computed as:

$$\varepsilon = \frac{\ln im - \mu_{\ln IM|\mathbf{M},R,V_{S30}}}{\sigma_{\ln IM|\mathbf{M},R,V_{S30}}} \tag{3.35}$$

where *im* indicates a specific value of *IM* obtained from a recording, and $\mu_{\ln IM|\mathbf{M},R,V_{S30}}$ and $\sigma_{\ln IM|\mathbf{M},R,V_{S30}}$ are the mean and standard deviation from a GMM, computed using the independent variables (**M**, R, V_{S30}) associated with the observation. If there are many observations, a percentile level can be assigned to any particular observation (i.e., value of *IM* and corresponding independent variables); the percentile is based on the percentage of the data with smaller ε values. That percentile is analogous to the cumulative distribution function (Section D.5.2) of the observed data distribution for a given value of ε (denoted $F_{obs}(\varepsilon)$). For that percentile value of the residual, there is a corresponding expected normalized residual ε' corresponding to a theoretical distribution ($F_{th}(\varepsilon') = F_{obs}(\varepsilon)$). For example, if the percentile is 95 and the theoretical distribution is normal, ε' is approximately 1.65 (meaning 1.65 standard deviations above the mean) based on standard normal distribution tables (Table D.1).

The properties of a data distribution near its tails are evaluated by plotting ε (from the data distribution) against ε' (from a corresponding theoretical distribution) for all available observations in a quantile-quantile (or Q-Q) plot. Figure 3.51 presents an example Q-Q plot for the case of within-event residuals for *PGA* using the Boore et al. (2014) GMM. If the data follow the theoretical distribution perfectly, the ε–ε' values would plot on the 45-degree line. Using this plot as a

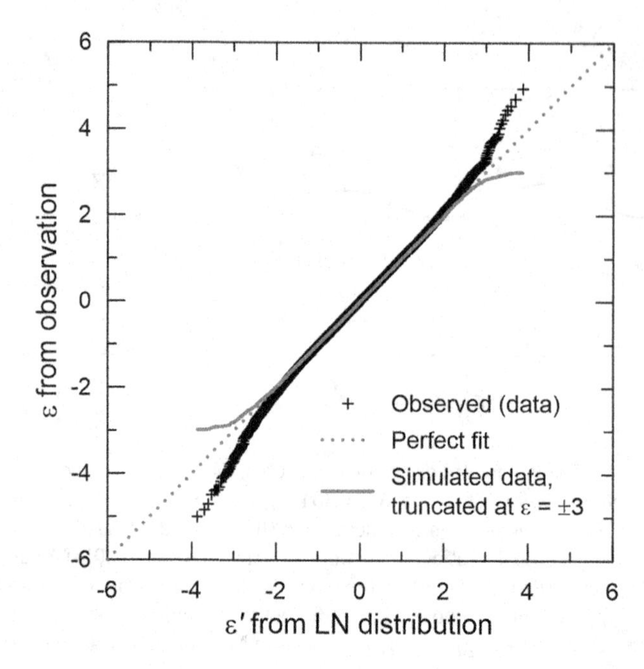

FIGURE 3.51 Quantile-quantile (Q-Q) plot of the within-event *PGA* residuals of the NGA-West2 data set as compared to Boore et al. (2014) GMM. Approximate Q-Q values that would be expected for a Gaussian data set with truncation at ±3 standard deviations are shown for comparison.

representative example of data distributions, we can now address several commonly posed questions regarding data distributions:

- *How well does the log-normal distribution fit the data?* The fit is quite good to about 2.5–3.0 standard deviations above/below the mean. This indicates that for most applications, the use of a log-normal distribution and its corresponding standard deviation provides an accurate description of the actual data distribution.
- *Can the tails of IM distributions be truncated?* The data do not support truncation. In fact, as shown in Figure 3.51 and other similar plots from the literature (e.g., Strasser et al., 2008), the tails of the distribution trend in the opposite direction of what would be expected in a truncated data set (an example of which is shown in Figure 3.51 for truncation at ±3 standard deviations from the mean). An exception is when physical constraints at a site effectively rule out the occurrence of ground motions beyond a particular level (Andrews et al., 2007; Baker et al., 2013).
- *Are log-normal representations of the distribution tails reliable?* This question remains a source of active scientific debate. As shown in Figure 3.51, the available data argue for a distribution with a "heavier tail," meaning higher concentrations of observations at high ε than implied by the log-normal distribution. However, the data are sparse and arguably inconclusive.

3.6 GROUND MOTION SIMULATIONS

As used here, the term "simulations" refers to computational methods for generating ground motion time series that are based at least in part on *physics-based modeling* of source, path, and site effects. It is important from the outset to distinguish such simulations from semi-empirical ground motion models (GMMs) – simulations provide full time series (e.g., accelerograms) while GMMs provide estimated distributions of ground motion *IM*s.

Simulation procedures can be useful tools for ground motion characterization, especially for conditions poorly represented in ground motion databases. For example, Figure 3.47 shows that databases for active tectonic regions and stable continental regions are sparsely populated for **M** > 7.5 and 6.0, respectively. In both regions, engineering applications frequently require analysis of ground shaking hazards for larger magnitude events. Hence, there is a practical need for ground motion prediction tools that can operate beyond the limits of the database. Simulations can address this need and are typically used to help solve two types of practical problems: (1) provide motions to constrain GMMs beyond empirical data limits and (2) provide acceleration time series for use in response history analyses for conditions poorly represented in empirical databases.

The following sections briefly describe two general types of simulation methods – physics-informed stochastic methods and deterministic/hybrid methods. Fully stochastic methods that omit physical considerations (apart from reference to GMMs) are not discussed because they are unlikely to extrapolate properly beyond the limits of empirical data. Vital considerations associated with the use of simulations, namely their verification, validation, and calibration are discussed. The section concludes with some example applications.

3.6.1 Physics-Based Stochastic Methods

As introduced in Section 3.4, high-frequency components of ground motion are influenced by complex (and generally unknown) features of the fault slip and wave propagation that are difficult to model explicitly but can be reasonably represented stochastically. The term 'stochastic' refers to the use of random white noise as part of the simulation process. The term 'physics-based' is used because these procedures incorporate physical principles with stochastic components. These simulations have been frequently used to produce ground motions for site-specific seismic hazard studies (e.g., Bommer et al., 2015, 2017) and to support the development of GMMs for regions with sparse recordings in the magnitude-distance range of engineering interest (e.g., Pezeshk et al., 2018; Yenier and Atkinson, 2015).

Physics-based stochastic simulation methods can operate with the seismic source represented as a point or a finite fault. Point source methods are most commonly used, and the basic elements of such methods for generating acceleration time series are illustrated in Figure 3.52. The process begins with Gaussian white noise (i.e., random numbers that are normally distributed with a mean of zero and standard deviation of one) generated for a desired record length in time (Figure 3.52a). A shaping window function is then applied to the noise (Figures 3.52a and b). Boore (2003b) suggested a window function given by

$$w(t) = a\left(t/t_\eta\right)^b \exp\left(-ct/t_\eta\right) \tag{3.36}$$

where t is time and t_η represents an approximate ground motion duration. The window function is zero at $t = 0$, rises to a peak of unity at $t = \epsilon t_\eta$, and decays exponentially for $t > \epsilon t_\eta$ to a value of η at t_η (recommended parameters are $\epsilon = 0.2$, $\eta = 0.05$, $a = 26.3$, $b = 1.25$, and $c = 6.27$). The Fourier amplitudes of the windowed noise are then computed (Figure 3.52c), and normalized to a root-mean square average of unity. Note that the Fourier amplitude spectrum of the noise is flat, which is characteristic of white noise. The Fourier amplitudes are then re-shaped by a model described further below (Figure 3.52d). The resulting Fourier amplitude spectrum can be combined with the original phase (of the windowed white noise) to produce the final accelerogram (Figure 3.52e). The accelerogram produced by this process has a realistic appearance, but should be recognized as being stationary (its frequency content does not change with time). Actual ground motions are nonstationary because p-, s-, and surface waves generally arrive at different times and with different frequency contents. Several investigators (e.g., Boore 2003c) have extended the physics-based stochastic approach to provide nonstationary time series.

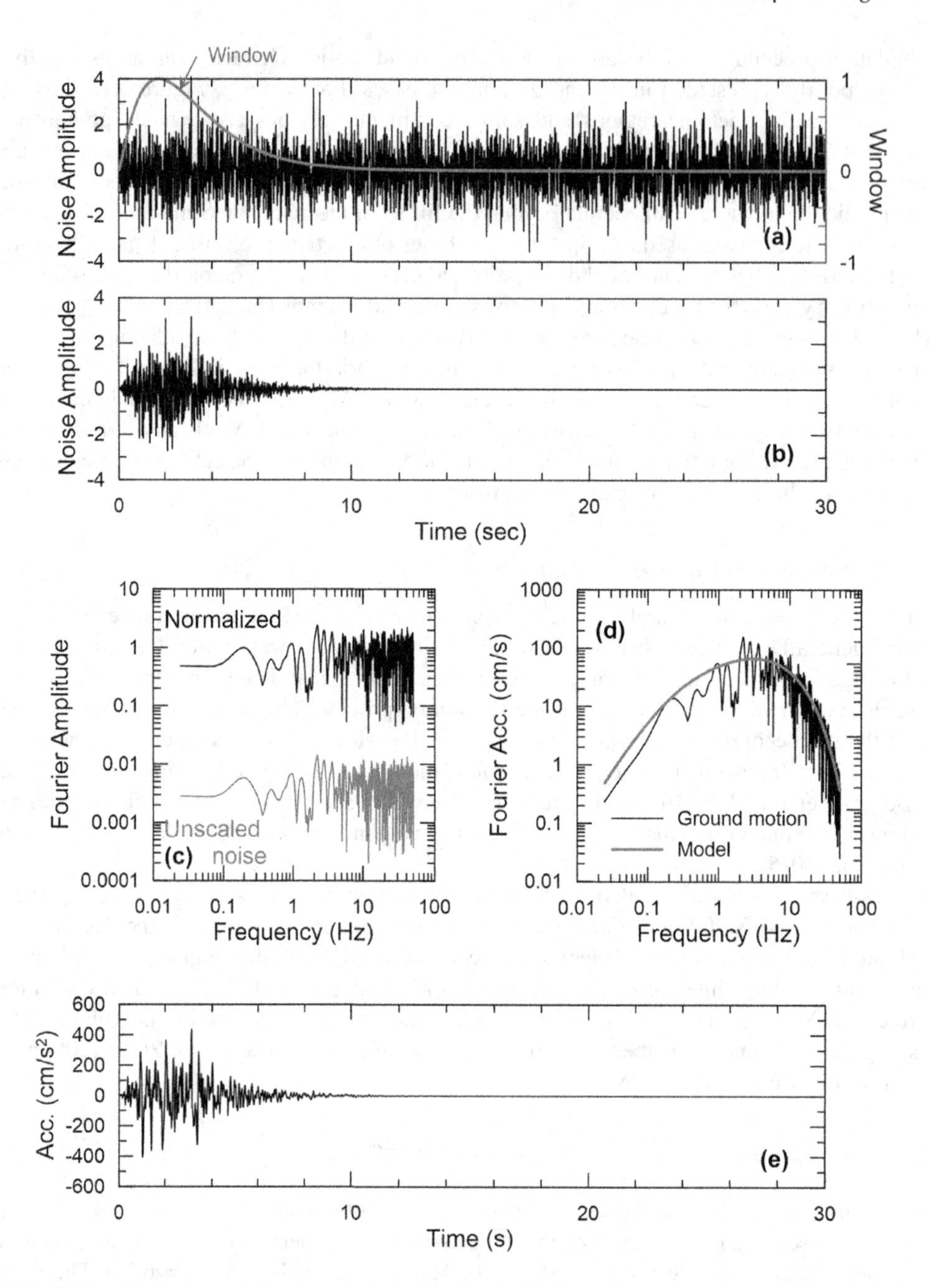

FIGURE 3.52 Illustration of physics-based stochastic model for generation of time series (adapted from Boore, 1983): (a) Gaussian noise and smooth window function; (b) Windowed noise; (c) Fourier amplitude of noise before and after normalization; (d) Fourier acceleration amplitude spectra for model and simulated ground motion; (e) Acceleration time series obtained from inverse Fourier transform. Simulations are for **M** 6.5 event, R_{rup} = 10 km, generic rock site condition, omega-square source model.

The Fourier amplitude spectrum and duration models are two critical elements of the analysis that involve earthquake physics. Point source methods of analysis for the Fourier amplitude spectrum can be traced to Aki (1967) and have taken many forms since that time (Boore, 2003b). The ground motion spectrum (denoted as Y) can be expressed as the product of source (event), path, and site terms as follows:

$$Y(\mathbf{M},R,f) = E(\mathbf{M},f)P(R,f)S(f) \tag{3.37}$$

Although different spectral quantities are represented, Equation (3.37) is similar to the formulation of the GMM mean intensity measure $\mu_{\ln IM}$ in Equation (3.24).

The source term has taken many forms, but the most widely used is the so-called "omega-squared" model for the acceleration spectrum, which is generally expressed as (Aki, 1967; Boore, 1983, 2003b):

$$E(\mathbf{M},f) = CM_0 \frac{(2\pi f)^2}{1+(f/f_a)^2} \tag{3.38}$$

where f is frequency in Hz and f_a is a corner frequency that separates the portion of the spectrum at low frequencies that ascends as f^2 from a flat portion at higher frequencies (Figure 3.52d). M_0 is seismic moment (Section 2.8.1) in units of dyne-cm, which is related to moment magnitude as (Hanks and Kanamori, 1979):

$$M_0 = 10^{(1.5\mathbf{M}+16.05)} \tag{3.39}$$

The parameter C is a constant given by (Boore, 2003b):

$$C = \frac{R_{\theta\phi}FV}{4\pi\rho_{crust}V_{S,crust}^3 R_0} \tag{3.40}$$

where $R_{\theta\phi}$ represents the radiation pattern (Boore and Boatwright, 1984 provide tables – a commonly used average value is ≈ 0.55), F (= 2) accounts for the free-surface effect, $V\,(=1/\sqrt{2})$ accounts for partitioning of the earthquake energy into two horizontal components, ρ_{crust} and $V_{S,crust}$ are the mass density and shear wave velocity of the crust (typical values are 2.8 gm/cm³ and 3.5 km/s, respectively), and R_0 (usually taken as 1 km) is a reference distance. Equation (3.40) requires consistent units – if ground motion is to be in cm/s (as a Fourier amplitude; it becomes cm/s² through the inverse Fourier transform) and ρ_{crust}, $V_{S,crust}$, and R_0 are in units of gm/cm³, km/s and km, respectively, then C should be multiplied by the factor 10^{-20}.

The corner frequency, f_a, is a critical element of the source model. A common approach is to take $f_a = f_0$, where f_0 is related to seismic moment as (Aki, 1967):

$$M_0 f_0^3 = \text{constant} \tag{3.41}$$

The corner frequency can be related to the *stress parameter*, $\Delta\sigma$ (Brune, 1970, 1971) as:

$$f_0 = 4.906 \times 10^6 V_{S,crust} \left(\frac{\Delta\sigma}{M_0} \right)^{1/3} \tag{3.42}$$

where $\Delta\sigma$ is in bars and the units of f_0, $V_{S,crust}$, and M_0 are Hz, km/s, and dyne-cm, respectively. The effective duration of the record, t_η (required for the simulation through the window function in Equation 3.36) is the sum of source and path durations. The source duration, T_d, is related to f_0 (Hanks and McGuire, 1981) as:

$$T_d = 1/f_0 \tag{3.43}$$

Recommendations on path duration for use with simulation models are given in Boore and Thompson (2014, 2015) (approximately 4 sec at 20 km and 10–25 sec, depending on tectonic regime, at 100 km). Figure 3.53a shows a strong increase of low-frequency seismic energy with

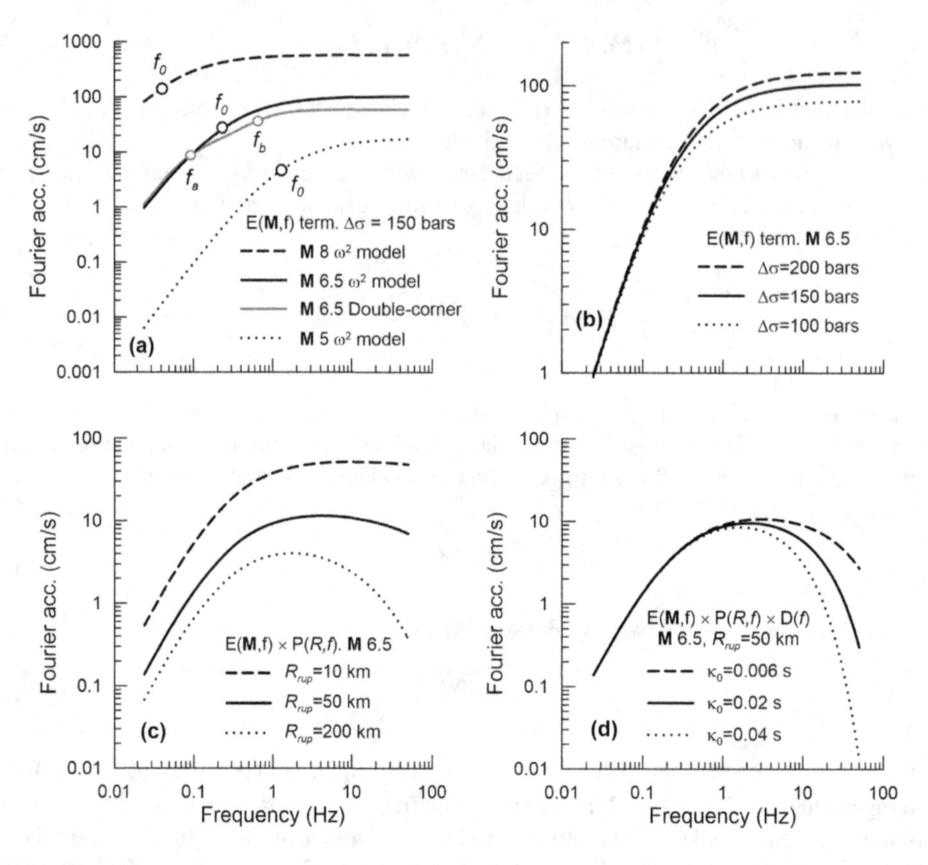

FIGURE 3.53 Effects of source, path, and site damping parameters on Fourier amplitude spectra (all computations leave out site amplification, $A(f)$): (a) effect of magnitude for omega-squared (ω^2) model and shape of double-corner model for **M** 6.5 using parameters from Atkinson and Silva (1997) (for these examples, in the ω^2 model $f_0 = f_a$); (b) effect of stress parameter on source model for **M** 6.5; (c) effect of path scaling for **M** 6.5 and $\Delta\sigma = 150$ bars; (d) effect of κ_0.

magnitude, which is generally consistent with observations as reflected by source terms in GMMs (Figure 3.31). Figure 3.53b shows the somewhat more modest effect of stress parameter $\Delta\sigma$, which increases amplitude for $f > f_0$ by increasing the corner frequency.

Also shown in Figure 3.53a is an alternative to the omega-square model, which is known as the "double-corner" model (Atkinson, 1993; Boore et al., 2014; Ji and Archuleta, 2021) in which the source term can be written as:

$$E(\mathbf{M}, f) = CM_0 (2\pi f)^2 \left(\frac{1-\chi}{1+(f/f_a)^2} + \frac{\chi}{1+(f/f_b)^2} \right) \tag{3.44}$$

where f_a is the lower corner frequency and f_b is a higher corner frequency. Both corner frequencies and χ are set empirically. Atkinson and Silva (1997) provided the following recommendations for California earthquakes:

$$\begin{aligned} \log_{10} f_a &= 2.181 - 0.496\mathbf{M} \\ \log_{10} f_b &= 1.778 - 0.302\mathbf{M} \\ \log_{10} \chi &= 2.764 - 0.623\mathbf{M} \end{aligned} \tag{3.45}$$

The double-corner model retains the main features of the omega-squared model, but the reduction at low to intermediate frequencies has provided improved matches to data in some cases.

The other elements of the physics-based stochastic model are the path and site terms, $P(R,f)$ and $S(f)$, respectively. The path term describes the effects of geometric spreading and anelastic attenuation as given by Equation (3.21). Typical features of the path term, $P(R,f)$, are given in Figure 3.24. Figure 3.53c shows the effect of path terms, which is to decrease the spectra at all periods due to geometrical attenuation, and to produce enhanced attenuation at long distance and high frequencies from anelastic attenuation.

The site term, $S(f)$, is intended to account for impedance and damping effects as seismic waves propagate toward the surface from the deep, seismogenic portions of the crust. The site term is not customarily evaluated for site-specific conditions within the simulation code, but is evaluated for relatively generic rock sites – additional adjustments for local site conditions can be made using procedures such as those given in Section 3.5.2.3 or site-specific procedures (Section 7.4). For the analysis of site term $S(f)$ it is convenient to separate amplification and damping effects (Boore, 2003b) as follows:

$$S(f) = A(f)D(f) \tag{3.46}$$

The amplification, $A(f)$, represents the impedance effect (Section 3.4.3.1). Amplification effects are often computed using the square-root impedance method as given by Equation (3.22) and shown in Figure 3.25. Boore and Joyner (1997) present generic rock and generic very hard rock profiles and computations of $A(f)$ for those profiles, with the results in Figure 3.54. The site damping function is given by:

$$D(f) = \exp(-\pi\kappa_0 f) \tag{3.47}$$

The parameter, κ_0, is set empirically to attenuate high-frequency amplitudes (Anderson and Hough, 1984) as observed in recorded ground motions as shown in Figure 3.53d. κ_0 values of 0.02 and 0.006

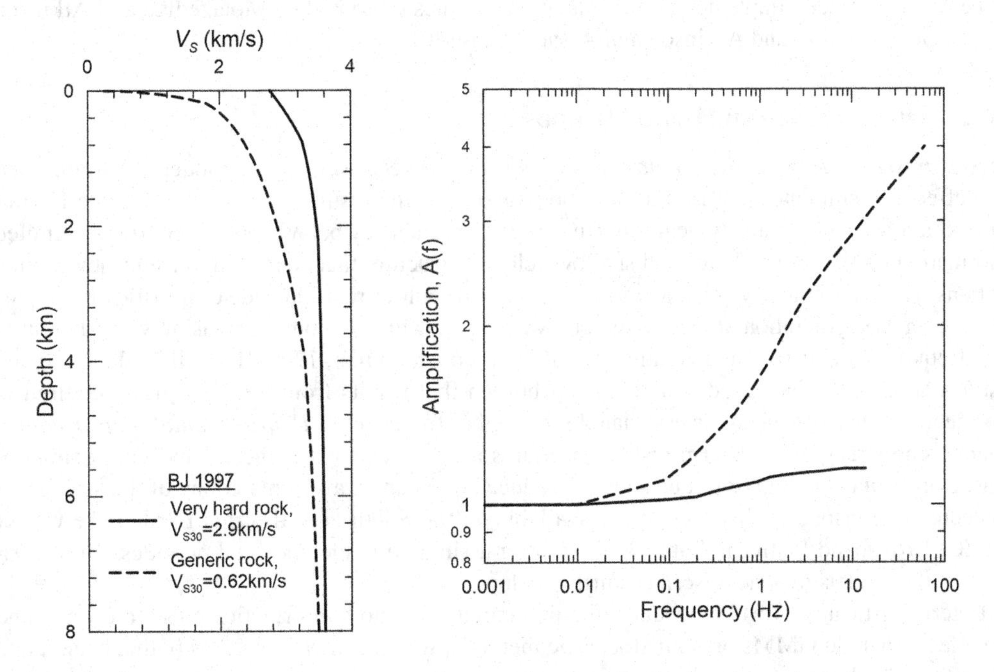

FIGURE 3.54 Generic rock and very hard rock velocity profiles and amplification functions ($A(f)$) as given by Boore and Joyner (1997).

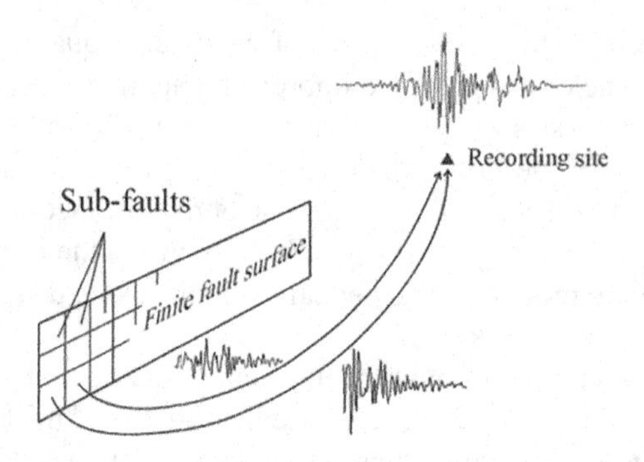

FIGURE 3.55 Use of sub-faults to characterize slip along finite fault surface in physics-based stochastic simulations (graphic from Atkinson, personal communication, 2014). Each sub-fault produces an accelerogram computed using point-source methods, which are then summed to develop the motion at the site of interest (marked as "recording site" in figure).

sec have been recommended for rock sites with V_{S30} = 760 m/s and hard rock sites applicable to stable continental regions, respectively (Campbell, 2009; Hashash et al., 2014).

Point source models of the type described above have the shortcoming of being unable to represent the geometric extent of finite fault ruptures. As shown in Figure 3.55, finite fault models overcome this limitation by discretizing the fault geometry into sub-faults, each of which is modeled as a point source (Hartzell, 1978). The rupture of different points along the fault can be delayed in time relative to the original rupture at the hypocenter using an assumed rupture velocity (v_r). The point source models used in the finite fault simulations follow the procedures just described, except that the corner frequency f_0 (= f_a in the omega-square source function, Equation 3.38) for a given sub-fault is computed using the portion of the overall seismic moment, M_0, associated with that portion of the rupture. Time series for each sub-fault are computed in the time domain, and then combined with appropriate time delays given the variable sub-fault rupture times. These finite fault procedures have been implemented in simulation routines presented by Motazedian and Atkinson (2005), Boore (2009), and Atkinson and Assatourians (2015).

3.6.2 Deterministic and Hybrid Methods

Physics-based, deterministic simulation procedures utilize seismological models of source and path effects to simulate fault rupture and propagation of the resulting waves through the Earth's crust. Such procedures are typically useful only at frequencies below about 1 Hz due to limited information on source attributes and shallow velocity structure that control high frequency ground motions. Higher frequency seismic waveforms are difficult to reproduce deterministically, in part because energy generation at the source and wave propagation become increasingly incoherent at high frequencies (e.g., Liu and Helmberger, 1985; Sato and Fehler, 1998; Hartzell et al., 1999) and also because of the increased computational burden that results from the finer grids required to represent the short wavelengths associated with higher frequencies. *Hybrid simulation procedures* leverage the strengths of deterministic procedures at low frequencies and stochastic or semi-stochastic procedures at higher frequencies to produce broadband waveforms. Many of the simulation procedures described in this section are available on the Southern California Earthquake Center (SCEC) Broadband Platform. Table 3.7 lists several simulation methods and provides a few attributes of the models that are described further below.

Deterministic and hybrid procedures require a great deal more information on source, path, and site effects than do GMMs or semi-stochastic methods, which can be viewed as both a benefit and a liability. The benefit is that the specific conditions associated with a particular source/path/site combination can be considered, presumably leading to more reliable ground motion estimates. The

TABLE 3.7
Example Simulation Methods (Implemented on SCEC Broadband Platform)

Simulation Method	Approach	Source Model	Path Model
CMU: Bielak et al. (2005); Tu et al. (2006)	Deterministic	Kinematic	1D or 3D. No scattering operator
G&P: Graves and Pitarka (2015)	Hybrid	Kinematic	1D or 3D. No scattering operator.
UCSB: Crempien and Archuleta (2015)	Hybrid	Kinematic	1D or 3D. No scattering operator.
SDSU: Olsen and Takedatsu (2015)	Hybrid	Kinematic	1D or 3D. Scattering operator.
EXSIM: Atkinson and Assatourians (2015)	Semi-stochastic	ω^2 with finite source	Section 3.4.2
CSM: Anderson (2015)	Composite source	Kinematic	1D. Scattering operator
Frankel (2009); Hartzell et al. (1999)	Hybrid	Kinematic	1D or 3D. No scattering operator

liability is that required information may be difficult to specify *a priori* (i.e., before the earthquake has occurred). Contributing to the complexity of providing simulation inputs are correlations among input parameters (e.g., longer rise time when slip increases), which are not well understood. Given these and other complexities, these analysis procedures require rigorous verification and validation to enable the computed motions to be used with confidence. The remainder of this section reviews the types of information that are required for deterministic simulations, and Section 3.6.3 addresses procedures used to check simulation methods to assess their reliability.

3.6.2.1 Source Models

In deterministic simulations, fault rupture is described kinematically or as a spontaneous dynamic process. *Kinematic source models* include spatially variable slip distributions, rise times, and rupture velocities (Section 3.4.1) determined in advance of the analysis. *Dynamic rupture models* prescribe initial fault stresses and constitutive relations for shear failure criteria, numerically trigger the initiation of slip, and then allow the rupture to propagate "naturally" across the fault surface. These procedures in principle have the ability to address the parameter correlation issue mentioned above, although their efficacy in this regard has not yet been adequately demonstrated. In some cases, the development of kinematic models is guided by the results of dynamic rupture simulations (Guatteri et al., 2004; Schmedes et al., 2010), so the outcomes of the different modeling procedures can be similar. Table 3.7 lists the source modeling approach used in the simulation methods on the SCEC broadband platform.

3.6.2.2 Path Models

Path models describe the effects of geometric spreading and anelastic attenuation, but may also include reflection and refraction at the interface of distinct rock types (large-scale heterogeneities) and scattering from small-scale heterogeneities. *Green's functions* are used in simulation procedures to collectively represent the effects of wave propagation in a crustal structure model. Green's functions are defined as the response of the Earth to a seismic point source. Green's functions use a model of the Earth's velocity structure and depend on the site and source locations. Per the *ground motion representation theorem* (Aki and Richards, 1980; Spudich and Archuleta, 1987), a fault slip model and Green's function can be used (with the inherent assumption of linearity) to compute the total ground motion at a site due to a finite earthquake source.

The geometric spreading and anelastic attenuation models described in Section 3.4.2 and used in GMMs and semi-stochastic simulations represent relatively simple closed-form Green's functions that can be calibrated (though the exponents p_1–p_3 in Equation 3.17 and the Q term in Equation 3.21) to capture average path effects for a study region. Alternatives to those equations are *analytical Green's functions* that can be computed for a particular velocity model of the Earth's crust. The simplest models assume lateral continuity of horizontal layers, and hence are referred to as 1D models (e.g., Helmberger et al., 1992; Olson et al., 1984; Luco and Apsel, 1983). 1D models can capture the contributions of multiple seismic arrivals, for example from direct and reflected ray paths, on the computed earthquake motion. More complex three-dimensional (3D) velocity models can also be used to compute Green's functions (e.g., Graves and Pitarka, 2010), which hold the potential for explicit evaluation of basin effects. Scattering functions can be included within analytical Green's functions to introduce essentially random high-frequency components of ground motion (Mai et al., 2010; Mena et al., 2010).

An alternative, referred to as the *empirical Green's function* approach (Hartzell, 1978, 1985; Irikura, 1983; Hutchings, 1994), takes a recording (at the site of interest) from a small-magnitude earthquake as the Green's function for the corresponding source/path combination. The empirical Green's function is considered more realistic than its analytical counterpart because it contains wave propagation effects of the real Earth (along the anticipated path), including scattered waves. However, the practical difficulties of this approach include high levels of noise in the weak motion recordings (particularly at low frequencies) and obtaining recordings for a sufficient number of source-path combinations to cover the conditions of interest.

Most simulation methods use analytical Green's functions, including each of those listed in Table 3.7. Table 3.7 indicates whether scattering operators and 3D velocity models are available components of the different Green's functions.

3.6.2.3 Site Models

As currently configured, ground motion simulation procedures are generally not able to capture local site effects related to ground response, including soil nonlinearity effects. Some models can include basin response within the Green's function, which is a site effect. In lieu of local, site-specific modeling, simulation codes will generally include a shallow velocity model (similar to those in Figure 3.54) with relatively firm surface velocities. Ground response through that model is generally evaluated using quarter-wavelength methods (Equation 3.22) combined with a damping function (Equation 3.47). Local ground response is then left to independent analyses of the type described in Chapter 7.

3.6.3 Verification, Validation, and Calibration

Ground motion simulations involve complex numerical models with significant potential for coding errors and for inaccurate modeling assumptions to introduce bias to computed waveforms. NIST (2014) described the general need for a tiered approach of *verification*, *validation*, and *calibration* for any complex computational procedure in earthquake engineering. Arguably, this process has been exercised more thoroughly for ground motion simulation methods than for many other applications in earthquake engineering and seismology. The process is described here for ground motion simulations, but the steps are generally applicable.

The first step is *verification*, which means comparing the predictions of analysis procedures to each other for simple and well-defined problems, or, in some cases, testing model predictions against checkable, known solutions. As illustrated in Figure 3.56, for ground motion simulations, this has typically involved comparing velocity or displacement time series (i.e., relatively low-frequency components of ground motion) from independent computational platforms for a common set of source and path conditions (Bielak et al., 2010).

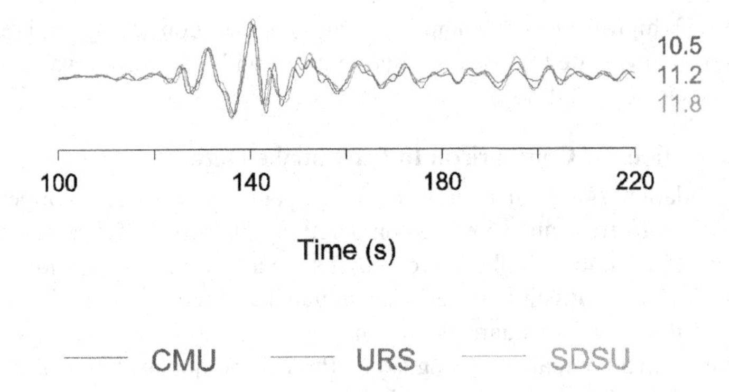

FIGURE 3.56 Comparison of velocity time series for a particular earthquake and site location using simulation procedures marked as CMU, URS, and SDSU (Table 3.7). *PGV* values are listed in the figure. This comparison comprises a portion of the verification for these simulation procedures. (From Bielak et al., 2010; used with permission of Oxford University Press.)

The second step is *validation*, which involves the comparison of analysis predictions to data. Ideally, validation should be combined with clear and transparent selection protocols for the parameters used in the analysis and should be undertaken by researchers independent of the model developers. An objective validation employing pre-defined parameter selection protocols may well reveal a certain degree of bias (i.e., misfit of the mean of model predictions to data). This then sets the stage for the third step, *calibration*. During calibration, a relatively poorly constrained parameter or parameters are adjusted in a transparent and repeatable manner to remove bias in model predictions.

This stepped, methodical approach is essential to identify bugs in the coding of software (during verification) and to guide the development of parameter selection and code usage protocols (during validation and calibration). The sections below illustrate common methods of validation and explain their usefulness for evaluating the suitability of a simulation procedure for engineering application.

3.6.3.1 Waveform Comparisons to Earthquake Data

As shown by the example in Figure 3.57, validation can be performed by comparing simulated waveforms for a particular earthquake event to recordings of actual ground motions. In most cases, these comparisons are qualitative but quantitative comparison schemes are also available (e.g., Olsen and Mayhew, 2010). Typically, velocity or displacement histories are used for these comparisons, thereby emphasizing low-frequency ground motions relatively unaffected by stochastic processes. Problems with this approach are (1) the recordings used to invert the source function are often used to demonstrate the efficacy of the simulation code, which makes good matches probable

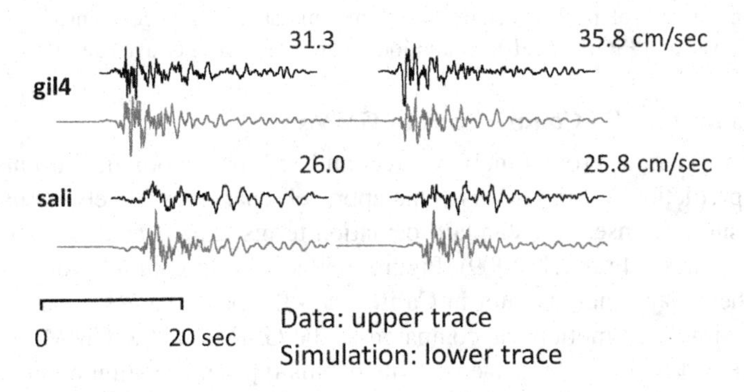

FIGURE 3.57 Comparisons of horizontal-component velocity time series (right – fault normal, left – fault parallel) for selected recordings (Gilroy No.4 and Salinas) from 1989 Loma Prieta earthquake using Graves and Pitarka (2010) simulation procedure. (Modified from Graves and Pitarka, 2010.)

but less meaningful, (2) high-frequency components are often not considered, and (3) it is often difficult with these plots to evaluate trends of ground motions with controlling variables (magnitude, source-site distance, and site condition).

3.6.3.2 Ground Motion *IM* Comparison to Earthquake Data

In this method of validation, *IM*s such as *PGA*, *PGV*, or S_a at particular oscillator periods are calculated for simulated motions from an event and compared (as a function of distance or frequency) to *IM*s from recordings. If a suitable number of recordings are available, it is possible to evaluate *bias* of the simulation based on the misfit between the simulated and recorded *IM*s in an average sense. Differences in the within-event standard deviations ($\phi_{\ln IM}$) can also be investigated. Figure 3.58 shows an example of such a comparison using Loma Prieta earthquake *PGA*s; in this case, there is no apparent bias, but the dispersion of the simulated motions is smaller than that of the data. One drawback of this approach is that recordings are generally not available for the types of earthquakes for which simulations are most valuable (i.e., large magnitudes). In addition, the circular reasoning associated with the use of an inverted source function with recordings from that same event (as described above) can diminish the value of these comparisons. Nonetheless, comparisons of this type have been used to develop recommendations for simulation procedures suitable for engineering applications (Dreger et al., 2015).

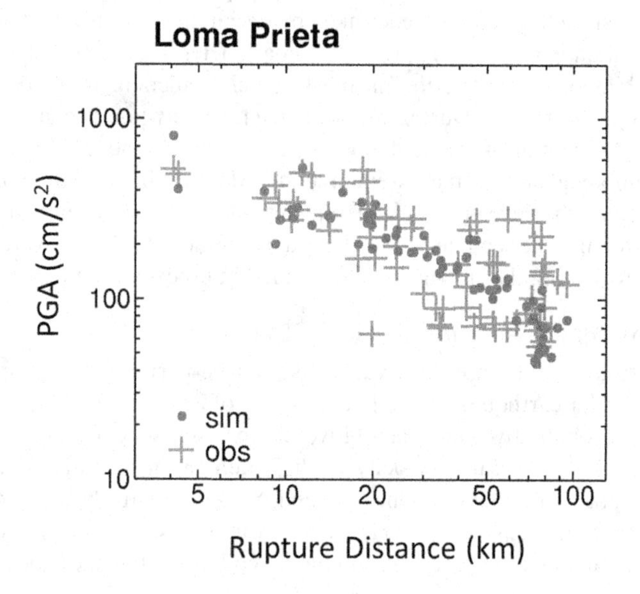

FIGURE 3.58 Comparison of peak accelerations from simulations and recordings of 1989 Loma Prieta earthquake using Graves and Pitarka (2010) simulation procedure. (From Graves and Pitarka, 2010.)

3.6.3.3 Ground Motion *IM* Comparisons to GMMs

In this method of validation, ground motions are simulated for hypothetical events, and their *IM*s are compared to predictions from GMMs. This approach enables relatively robust evaluations of distance scaling, site response, and standard deviation terms (e.g., Star et al., 2011; Nweke et al. 2022) or subsets of these (Frankel, 2009). Figure 3.59 shows an example for a simulated **M** 7.8 event on the southern San Andreas fault in California – the result shows a faster rate of distance attenuation in the simulated motions as compared to the GMMs. If the GMMs are considered to more accurately reflect real behavior, then misfits of this type in the simulations can be removed through calibration, for example by adjusting the *Q* parameter controlling crustal damping (Seyhan et al., 2013). A similar result has been obtained by calibrating the number of scatterers in analytical Green's functions to achieve desired median levels of high-frequency motions in a hybrid simulation

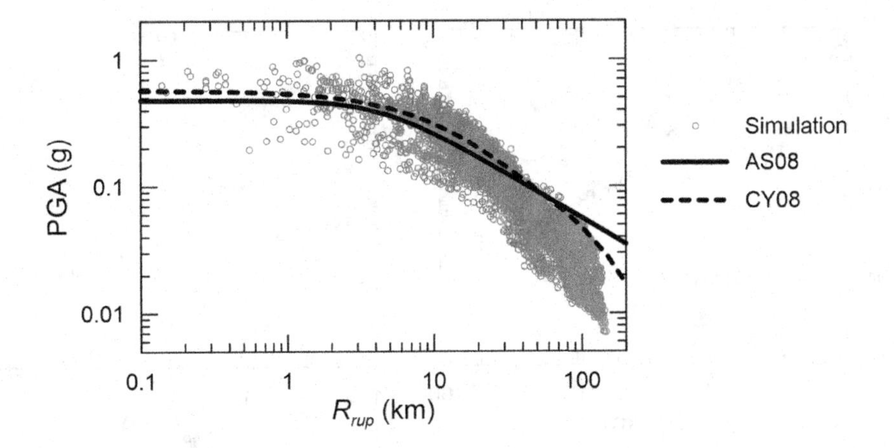

FIGURE 3.59 Variation with distance of *PGA*s at rock sites (V_{S30} = 760–1,000 m/s) from simulations of **M**7.8 earthquake on southern San Andreas fault (event used for ShakeOut scenario; Porter et al., 2011) and two GMMs. Results show faster distance attenuation of simulations as compared to GMMs. (Modified from Star et al., 2011.)

procedure (Mena et al., 2010). Comparisons of this type, along with earthquake-specific analyses, have been used (Dreger et al., 2015) to provide recommendations on simulation procedures suitable for engineering applications.

3.6.4 Example Applications

Ground motion simulation procedures of the types described in Sections 3.6.1–3.6.2 have seen two principal applications to date: (1) to guide the development of GMMs in the range of parameter space for which recorded ground motions are sparse and (2) to generate ground motion time series for conditions of engineering interest and for which recorded motions are not available. In principle, simulations can also be used to generate ground motions for hazard analysis, as demonstrated by Graves et al. (2011) and Milner et al. (2021).

As indicated in Table 3.4, ground motions from simulations have been used to support the development of GMMs for stable continental regions. The simulation procedures used for these analyses have included both point-source and finite-source physics-based stochastic methods (Section 3.6.1). The simulations have been used to generate large numbers of ground motions, considering epistemic uncertainties in key input parameters such as stress parameter and κ_0. These simulations avoid the data sampling problems otherwise inherent to GMMs derived principally from recorded data (for active tectonic regions and subduction zones) (Section 3.5.4.1). *IM*s computed from time series are used in regressions to derive GMMs. Figure 3.60 shows examples of geometric means of computed *IM*s in magnitude and distance bins and medians from GMMs for *PGA* and S_a(3.0).

Simulations are also used in the development of GMMs for relatively data-rich active tectonic regions. While the principal magnitude- and distance-dependence for these GMMs are based on data analysis, other effects that are relatively poorly resolved by the available data are often evaluated in part from simulated data. These include hanging wall effects (Donahue and Abrahamson, 2014), nonlinear site response (Walling et al., 2008; Kamai et al., 2014b), basin effects (Day et al., 2008; Frankel et al., 2009; Wang and Jordan, 2014; Moschetti et al., 2024), and magnitude-scaling beyond data limits. The simulation methods used to evaluate these effects are typically a subset of the hybrid methods listed in Table 3.7 (an exception is nonlinear site response, which has typically been evaluated using ground response analysis procedures described in Section 7.5). Figure 3.61 shows an example of how simulated motions were used to develop a model for ground motions over the hanging wall of dip-slip faults.

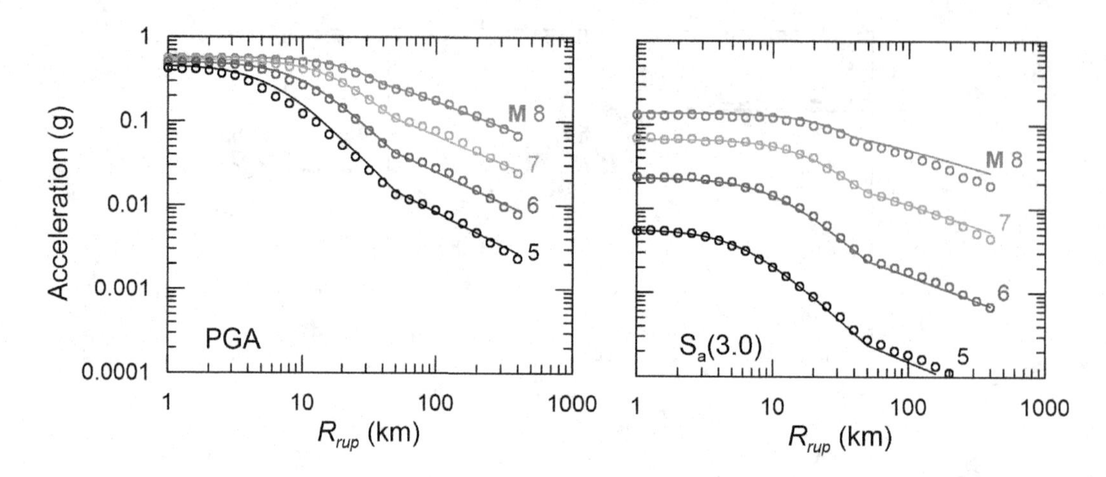

FIGURE 3.60 Binned median intensity measures from simulated ground motions for stable continental regions (symbols) and GMM medians fit to the simulated data for **M** 3-8 earthquakes (lines). Results are for stable continental region reference rock conditions having near-surface V_{S30} = 760 m/s. (Modified from Yenier and Atkinson, 2015.)

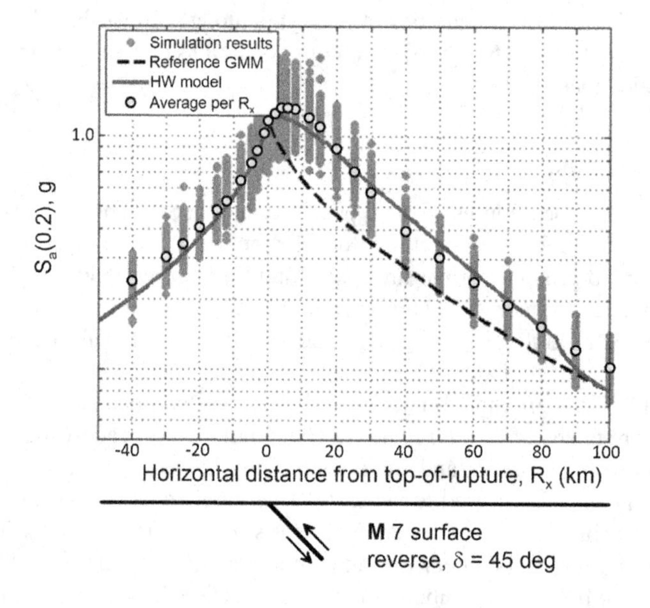

FIGURE 3.61 Simulated motions over footwall and hanging wall portions of a dipping fault, median predictions from a reference GMM that does not include hanging wall effects (model estimates are identical for positive and negative values of R_x), and model fit to the simulated data for the hanging wall portion. Bottom of figure shows dipping fault considered for the simulation. (Modified from Donahue and Abrahamson, 2014; used with permission of SAGE Publications, Ltd.)

Another application of simulations is to generate ground motion time series for conditions for which recorded ground motions may be sparse or absent (e.g., large magnitudes). For many years a synthetic seismogram generated largely through a judgment-driven process was used for large magnitude earthquakes (Seed and Idriss, 1969). This ground motion can no longer be recommended. Physics-based stochastic procedures (Section 3.6.1) are routinely used to generate time series for ground response analysis and other applications in stable continental regions such as central and eastern North America (e.g., McGuire et al., 2001; EPRI, 2013; Harmon et al., 2019). In active tectonic regions, simulated ground motions are considered an acceptable source of time series in guidelines documents (ASCE-7, 2022, Chapter 16; PEER, 2017), provided that suitable verification and validation of the simulations has been performed (Section 3.6.3; Dreger et al., 2015; Bradley et al., 2017). Simulated ground motions (from the SCEC Broadband platform), selected for compatibility with target spectra and duration, have been shown (Bijelić et al., 2018) to produce responses of tall buildings similar to those produced by natural (recorded) ground motions selected in a comparable manner. This demonstrates that, when carefully selected, simulated ground motions do not appear to produce biased structural responses.

3.7 NEAR-FAULT GROUND MOTIONS

Ground motions close to a seismic source can in principle be predicted using GMMs (Section 3.5) or simulations (3.6) by using small values of source-site distance parameters R_{JB} or R_{rup}. However, the physical processes associated with fault rupture can systematically affect ground motions in predictable ways that are not accounted for in current GMMs. The following sections describe how the fault rupture process affects ground motions and the modeling of three types of near-fault effects – (1) polarization of ground motions in particular directions, (2) spatial patterns of ground shaking in the near-fault environment that depend on the direction of fault rupture and the direction of slip on the fault, and (3) pulse-like characteristics in ground motions and their effects on S_a.

3.7.1 Physical Processes

In the near-fault environment, rupture toward (or away from) a site of interest can produce ground motion features known as *rupture directivity* effects. A conceptual appreciation of these effects can be gained by recalling the well-known Doppler effect in physics. A different feature is the sudden permanent displacement of the ground near rupturing faults, known as *fling step*.

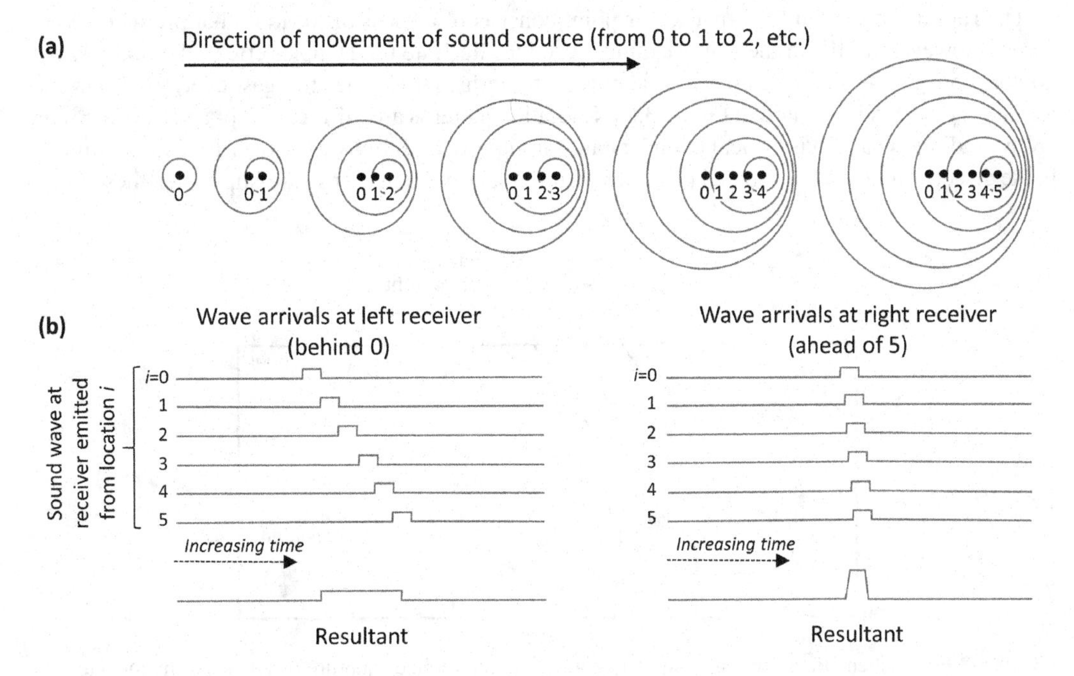

FIGURE 3.62 Illustration of Doppler effect. (a) Sound wave patterns from emissions sequenced to begin at location 0, then advance to locations 1, 2, etc.; (b) Wave arrivals at receivers behind and ahead of a moving sound source.

3.7.1.1 Doppler Effect

Consider a point source of sound that transmits at time 0. As shown on the upper left of Figure 3.62, after some small increment of time, the sound wave will have traveled out in a circular (actually spherical in three dimensions) pattern. Suppose that the sound source is moving to the right in the diagram at a constant velocity and emitting sound at successive times 1, 2, etc. Each successive emission produces its own circular pattern of sound. However, because the source is moving, the sound patterns are closely spaced (i.e., they arrive in a shorter interval of time) for points toward which the source is traveling, whereas they are widely spaced in the direction behind (away from) the direction of source travel.

The effect of these stacked, or dispersed (spread out), wave arrivals is shown at the bottom of Figure 3.62. In the "behind" (or backward) direction, the waves arrive at different times, producing a resultant that is low in amplitude and long in duration. In the forward direction, the waves arrive as a "pulse" in a short interval of time, which increases amplitude and decreases the duration of the resultant. These different attributes of waves arriving behind and ahead of a moving source comprise the *Doppler effect*, which provides a good starting point for understanding near-fault rupture directivity effects.

3.7.1.2 Rupture Directivity

Figure 3.63 shows a schematic view normal to a fault rupture plane (e.g., this would be a horizontal view for a vertically-dipping strike-slip fault). The rupture process begins with slip at the hypocenter, which produces body waves. At an arbitrary time, t, following the onset of slip, shear waves generated from the slip at the hypocenter have propagated to the *S-wave front*. As described in Section 3.4.1, the rupture velocity v_r is about 70%–90% of the crustal shear wave velocity $V_{s,crust}$, so the *rupture front* (which defines the boundary between portions of the fault that have not yet slipped and portions where slip has initiated) trails the S-wave front. The *healing front* separates portions of the fault still experiencing slip from portions where the slip process is complete. The elapsed time between passage of the rupture and healing fronts at a point is the rise time, t_r.

The rupture of the fault at a particular point consists of a shear dislocation that produces body waves. However, unlike in the sound example used to illustrate the Doppler effect, the body waves do not propagate in all directions with identical strengths; rather, the strongest energy follows the *radiation patterns* illustrated in Figure 3.64. Radiation patterns are different for p- and s-waves. The s-wave radiation pattern is critical to understanding rupture directivity and consists of lobes aligned in the direction of the fault slip and perpendicular to that direction. For strike-slip faults, these lobes

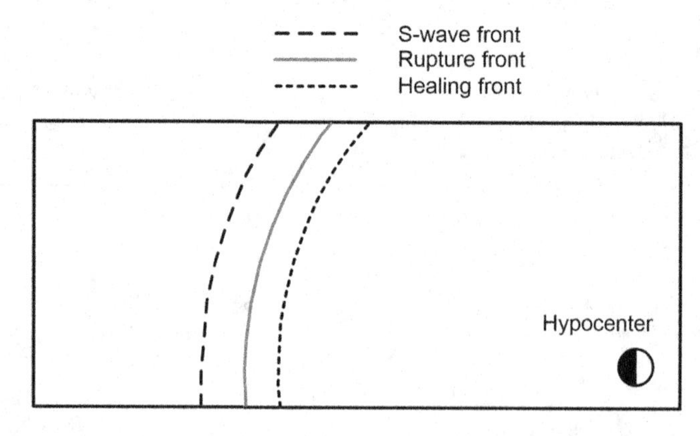

FIGURE 3.63 Schematic orthogonal view of rupturing fault surface. Rupture originates at hypocenter. At time t following the onset of slip, shear waves produced at the hypocenter have propagated to the S-wave front, the portion of the fault to the left of the rupture front has begun to slip, and the portion of the fault to the left of the healing front has concluded slip.

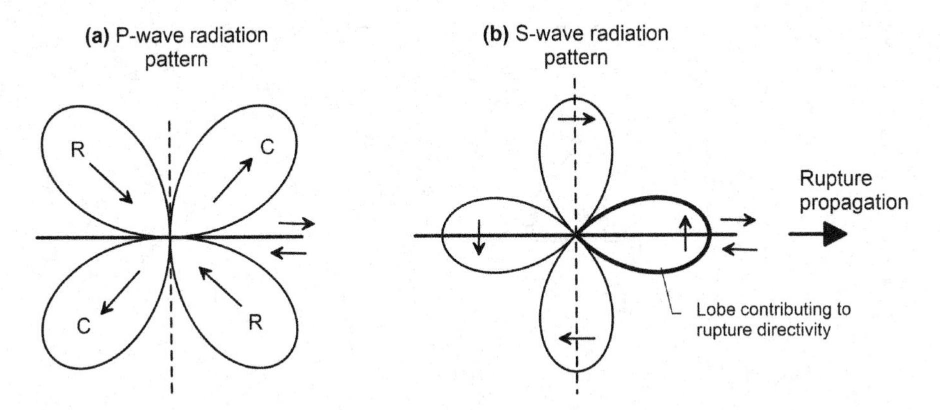

FIGURE 3.64 Plan view of radiation pattern lobes indicating direction of principal body wave propagation for (a) p-waves and (b) s-waves generated by shear dislocation on fault (horizontal line). In p-wave diagram, the compression (C) and rarefaction (R) refer to the initial strain at a point as the p-wave arrives, which is also shown by the arrows indicating the initial directions of particle motion. The same principle applies to the s-wave diagram, but the stress that is changed upon wave arrival is shear stress. In the s-wave diagram, the bulb oriented along the fault and traveling in the direction of rupture propagation will contribute to rupture directivity for rupture surfaces of extended length.

and the associated motions are polarized in the horizontal plane and are shown as SH waves. For dip-slip faults, the bulbs and motions are in the vertical plane (SV waves).

Rupture directivity can be understood as the constructive interference of shear waves generated at multiple points (and times) along an extended rupture. Only one of the four lobes shown in Figure 3.64b, the one oriented in the direction of rupture propagation, contributes to rupture directivity. The constructive interference results from the shear waves associated with the critical lobe traveling along the fault at the crustal shear wave velocity $V_{s,crust}$ and hence nearly overlapping with those generated subsequently along the rupture front (which only marginally trail the S-wave front). In other words, the critical lobes from multiple points along the rupture surface stack, one nearly atop the next, producing a strong pulse of ground motion propagating along the fault surface in the direction of the fault rupture (much like the coincident arrivals of sound waves in the forward direction in Figure 3.62). This stacking does not occur for any of the other three lobes in Figure 3.64b. Note that the direction of particle motion associated with the pulse is normal to the fault rupture, which is also normal to the fault strike for both strike-slip and dip-slip ruptures. Hence, rupture directivity effects are expected, on a conceptual level, to produce pulses of motion polarized in the strike-normal direction.

Figure 3.65 shows an example of two strike-normal motions recorded in the 1992 **M**7.3 Landers, California earthquake. The Lucerne site is located near the fault but far from the epicenter, meaning that a large portion of the fault has ruptured toward the site (marked in the figure as the *forward directivity region*). The ground motion includes a large amplitude, short duration velocity pulse. The Joshua Tree site is also located near the fault but is near the epicenter, hence the fault rupture principally moves away from this site (marked as *backward directivity region*). Shear waves arriving at this site are associated with the lobe oriented 180 degrees from the critical lobe; these do not stack but arrive in succession over a relatively long time interval, hence the relatively low amplitude and long duration motion shown in Figure 3.65. These effects are qualitatively similar to those produced by the Doppler effect described in Section 3.7.1.1.

3.7.1.3 Fling Step

Sites located within about 10 km of the surface projection of a ruptured fault, including sites on the hanging wall of dipping faults, can experience a permanent displacement, referred to as *fling step*,

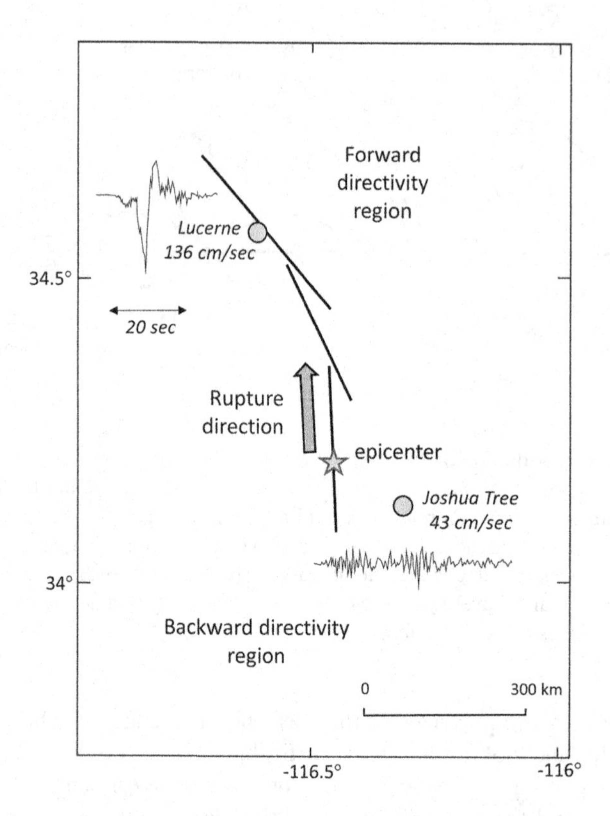

FIGURE 3.65 Map of Landers region showing location of 1992 rupture (on three fault segments), the epicenter, and the Joshua Tree and Lucerne recording stations. The velocity time series are in the strike-normal direction. Note the pulse in the forward directivity region. (Modified from Somerville et al., 1997.)

that is directly associated with the fault slip. The direction of displacement is parallel to the fault slip; hence, the displacement is principally horizontal in the fault-parallel direction for strike-slip earthquakes, upward and fault-normal on the hanging wall of thrust faults, and downward and fault-normal on the hanging wall of normal faults (Section 10.2). The displacement occurs over a time interval roughly comparable to the rise time, t_r, during which the velocity is in the direction of the fault slip. Hence, fling tends to produce single-direction velocity pulses, unlike rupture directivity where the pulses are roughly symmetric (positive and negative). Figure 3.66 shows an example fling step motion from the **M**7.5 Kocaeli, Turkey earthquake recorded in the east-west direction at the SKR station (3.1 km from the fault), which is parallel to the fault strike. The fling step in this case occurs between 8 and 12 sec in the record, producing approximately 80 cm/s of peak velocity and nearly 2 m of displacement. Relatively complex record processing techniques are required to develop fling step records as shown in Figure 3.66 from unprocessed recordings (Boore, 2000; Gregor et al., 2002). Due to large displacements occurring in short time intervals, fling has produced some of the largest recorded peak ground velocities, including values approaching 300 cm/s from the Chi-Chi (Taiwan) earthquake.

3.7.2 Ground Motion Polarization

Ground motions produce shaking in all directions within the horizontal plane. In the near-fault region, shaking can be systematically stronger in some directions (e.g., fault-normal) than others. Such polarization of ground motions can be important for structures or systems that exhibit different levels of response or resistance in different directions. Section 3.3.1.1 described how a given metric

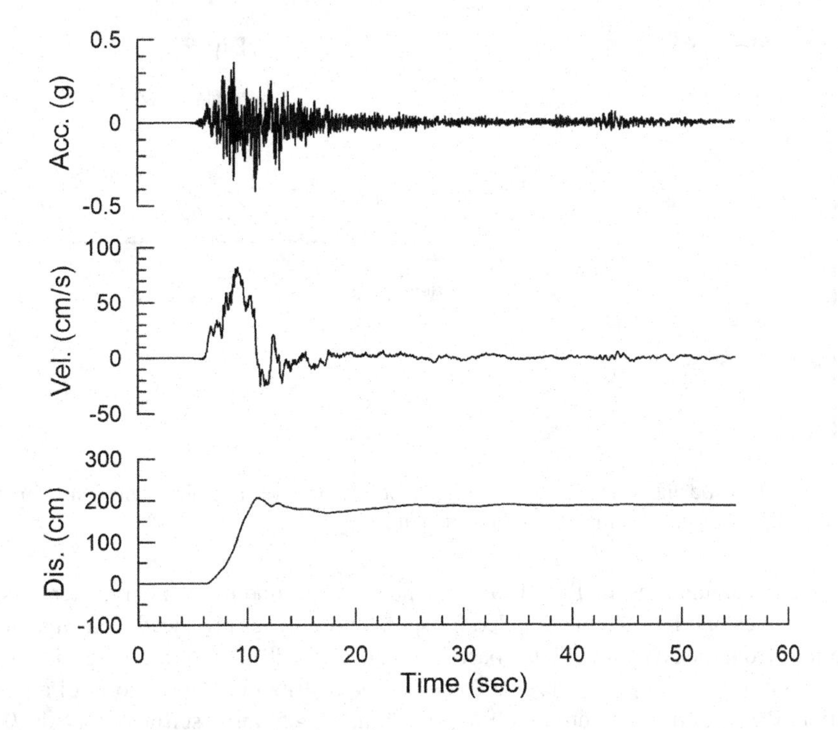

FIGURE 3.66 Example of fling step ground motion from Sakarya station (SKR) recorded during 1999 **M** 7.5 Kocaeli earthquake. Station is 3 km from the fault rupture, where fault displacements were approximately 2 m. (Data provided by E. Kalkan, personal communication, 2014.)

of ground shaking (e.g., acceleration in the case shown in Figure 3.15) will have azimuths with the largest and smallest peak amplitudes, which are referred to as the RotD100 and RotD00 motions, respectively. The median ground motion across all possible azimuths is RotD50. These metrics of shaking can also be defined for oscillator responses (i.e., response spectra) using any desired period.

Ground motion polarization refers to preferred azimuths in which peak ground motions (or oscillator responses) may occur. Polarization is quantified by identifying the angle, α, between the

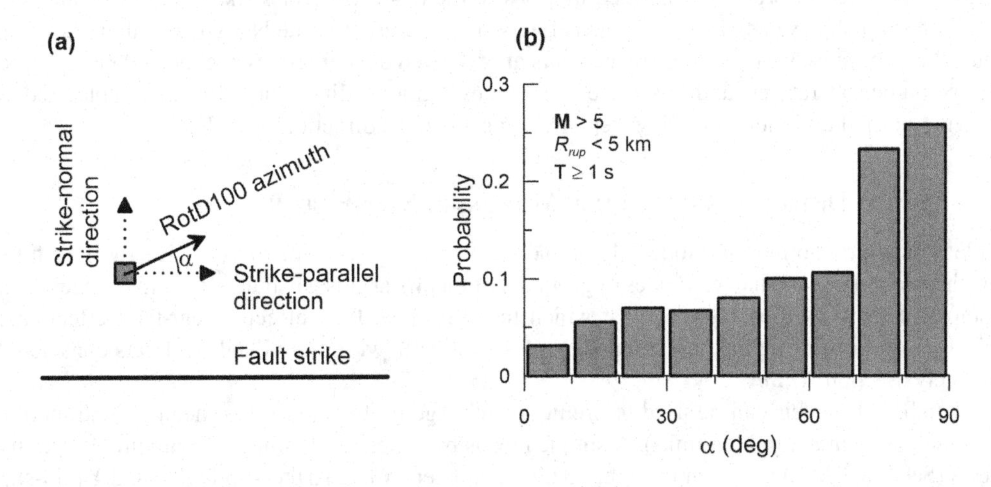

FIGURE 3.67 (a) Definition of angle, α, between fault strike and RotD100 azimuth; (b) probability mass function using data from the NGA-West2 ground motion database with **M** > 5, R_{rup} < 5 km, and $T \geq 1$ sec. (Data from Shahi and Baker, 2014b.)

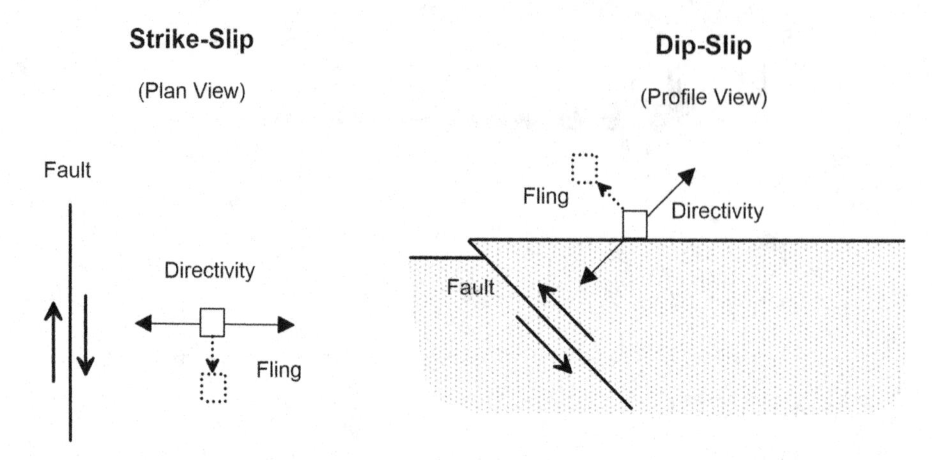

FIGURE 3.68 Schematic diagram showing the orientations of fling step and directivity pulse for strike-slip and reverse dip-slip faulting. (Modified from Stewart et al., 2001.)

fault strike and the azimuth of the RotD100 component of a ground motion parameter, as shown in Figure 3.67a. Values of α have been compiled for spectral responses produced by a large number of ground motions from the NGA-West2 database (Ancheta et al., 2014) at oscillator periods from 0 to 10 sec (Shahi and Baker, 2014b). Figure 3.67b shows the distribution (in the form of a probability mass function, PMF) of α for records with $R_{rup} < 5$ km, $\mathbf{M} > 5$, and oscillator periods $T \geq 1$ sec. The PMF shows higher probabilities for $\alpha > 70°$ than for smaller angles, indicating a tendency for ground motion polarization perpendicular to the fault strike for this range of conditions (close distances, long periods). Data show no consistent polarization, i.e., a nearly uniform distribution from 0 to 90°, for rupture distances greater than 5 km. Due to a relatively high degree of incoherence of high-frequency ground motions, RotD100 oscillator responses at short periods ($T < 0.5$ sec) tend to be arbitrarily oriented at all distances (even $R_{rup} < 5$ km).

Section 3.7.1 describes the expected polarization of ground motion from rupture directivity and fling step. As shown in Figure 3.68, the physics producing the directivity pulse suggest it should be polarized in the azimuth normal to the fault strike, which is consistent with approximately 50% of observations for the subset of data shown in Figure 3.67b. As also shown in Figure 3.68, fling step would be expected to produce polarized motions in the $\alpha \approx 0$ bin (for strike-slip events) and $\alpha \approx 90°$ bin (for dip-slip events). However, such effects are not present in the NGA-West2 data shown in Figure 3.67b because permanent displacements are removed during processing. Nonetheless, based on physical considerations, and to some extent on observations, directivity effects are polarized in the strike-normal direction, and fling step effects occur in the direction of fault slip.

3.7.3 Spatial Distribution of Ground Motions in Near-Fault Regions

While typical ground motion models do not account for the physics of rupture directivity or fling step, they do provide median estimates of ground motion *IM*s at close distances (i.e., in the near-fault region) that serve as a baseline against which to assess near-fault effects. Near-fault effects are typically examined using median component *IM*s (RotD50), with directionality effects considered separately (Section 3.7.2).

A number of models can be used to predict the change in the natural log mean ground motion (expressed as f_D in natural log units). A simple means by which to parameterize rupture directivity effects (Section 3.7.1.2) is shown in Figure 3.69. Parameters X (for strike-slip faults) and Y (dip-slip faults) represent the fraction of the fault dimension, between the hypocenter and the site, which ruptures toward the site. The applicable fault dimension is that which is aligned with the slip direction,

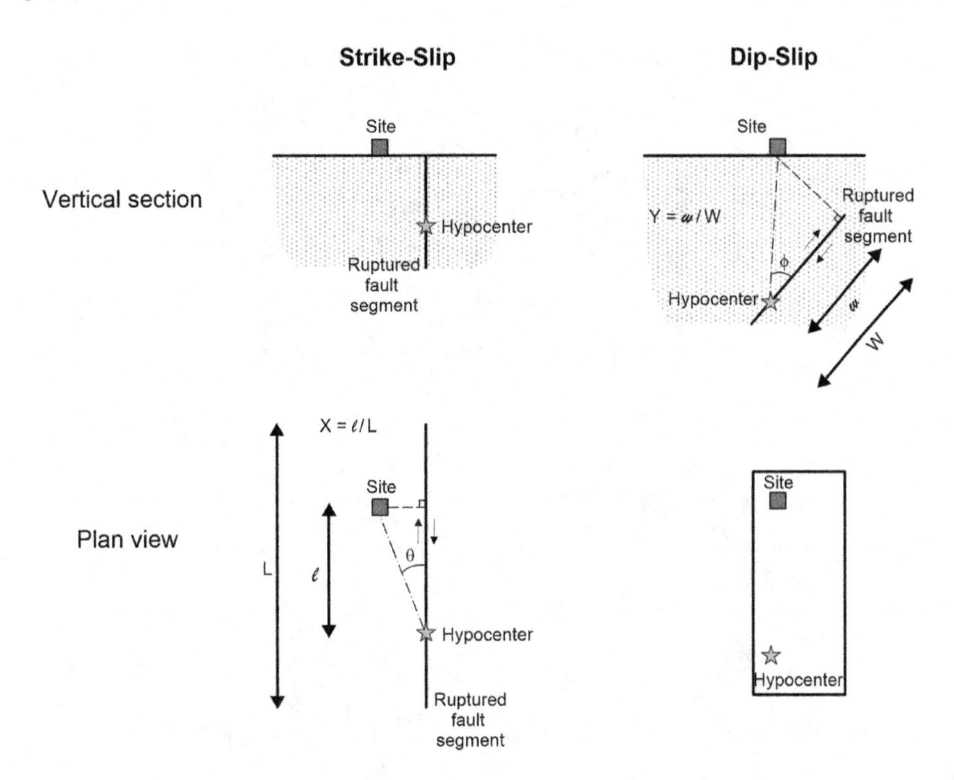

FIGURE 3.69 Definition of directivity parameters for strike-slip and reverse faults as used in Somerville et al. (1997) model.

as illustrated in Figure 3.69. This approach was used in early directivity models (Somerville et al., 1997). Small values of $X\cos\theta$ and $Y\cos\phi$ (i.e., near zero) indicate backward directivity (negative f_D, or decreased ground motions), while positive values indicate forward directivity (positive f_D, increased ground motions).

More recent directivity models use more complex parameterizations of site location relative to the fault, location of hypocenter on the fault, and fault dimensions (summarized in Spudich et al., 2014 and Donahue et al., 2019). A feature shared by these models is sensitivity of predicted near-fault effects to hypocenter location. This presents some practical difficulty, because the location of a hypocenter on the fault for future earthquakes is not predictable in advance of an earthquake. This issue can be addressed by randomizing hypocenter location (i.e., by considering all locations across a fault surface), and examining the resulting directivity effects (f_D). Figure 3.70 shows the spatial distribution (in plan) of mean directivity around a vertically-dipping strike-slip fault and a 45°-dipping dip-slip fault for magnitude 7 earthquakes. Two models are shown to provide a sense of between-model variability.

Figure 3.70 shows both higher-than-baseline ground motions ($\exp(f_D) > 1$) and lower-than-baseline ground motions ($\exp(f_D) < 1$) at different locations. For strike-slip faults, most hypocenters produce some rupture toward the ends of the faults along the strike (top and bottom of diagrams), causing forward directivity, whereas the middle of the fault, on average has backward directivity. This pattern is observed for both Models 1 and 2 in Figure 3.70, although to different degrees. For dip-slip faults, forward directivity is observed in different locations for the two models, reflecting different directivity parameterizations. Model 1 predicts forward directivity principally based on the rupture dimension toward a site, whereas Model 2 predicts directivity from the component of rupture direction that aligns with slip direction. As a result, Model 1 predicts forward directivity for locations that have, on average large rupture dimensions, which is off the ends of shallow portion of

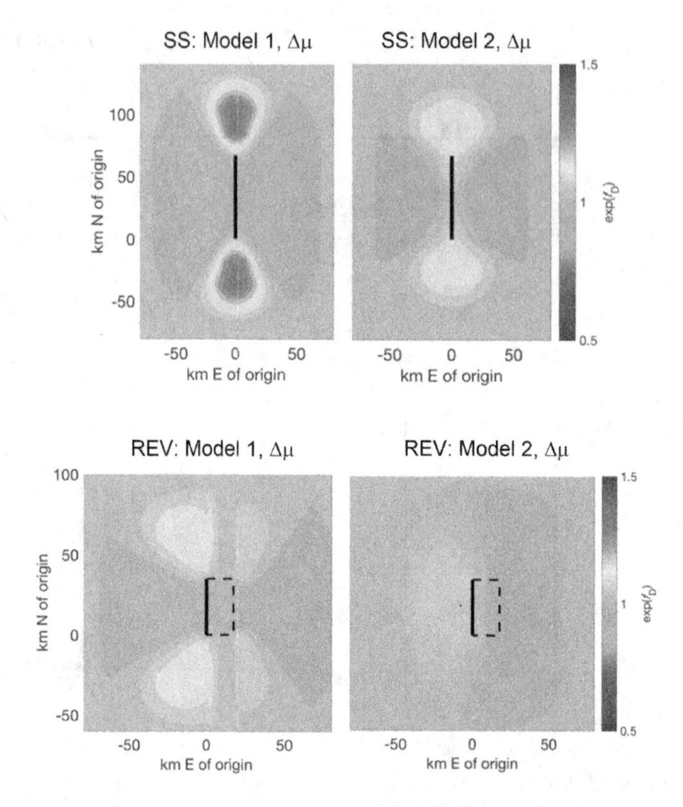

FIGURE 3.70 Changes of median response spectra by location around fault for strike-slip and reverse **M** 7 earthquakes at a period of 3.0 sec. The shallowest portion of the reverse fault is shown with a solid line. (Figure adapted from Donahue et al., 2019. Model 1 derived from Watson-Lamprey, 2018. Model 2 is described in Donahue et al., 2019.)

the fault. Since the slip direction is up-dip (toward the left), Model 2 predicts forward directivity on, and to the left of, the hanging wall. Negative directivity in this case occurs in the area to the right of the surface projection of the fault, because on average the fault ruptures away from this region. When alternate credible models produce different predictions, such as Model 1 and 2 predictions for reverse earthquakes, it is advisable to consider both types of models in seismic hazard analyses (more information in Section 4.4.3.6).

There are important effects of magnitude and period on directivity. All other factors being equal, larger magnitude earthquakes produce stronger directivity because the larger dimensions of the rupture surface for larger magnitude events provide more opportunity for the wave stacking that causes directivity. Directivity effects are negligible at short periods ($f_D \approx 0$) due to the incoherent nature of the rupture process that is most responsible for these components of motion. In many cases, directivity increases with period due to the increasing coherence of the rupture process as period increases. There can be conditions under which directivity effects increase to a peak at a particular period, and then decrease for larger periods, which are described in the next section.

3.7.4 Pulse-Like Ground Motions and Effects on S_a

While near-fault effects can influence response spectra generally at long periods, as shown in the previous section, in many cases the near-fault effect takes the form of pulses. Such pulses can result from forward rupture directivity (two-sided pulses, as at Lucerne, Figure 3.65) or fling step (single-sided pulses, as at Sakarya, Figure 3.66). In addition to their distinctive features in

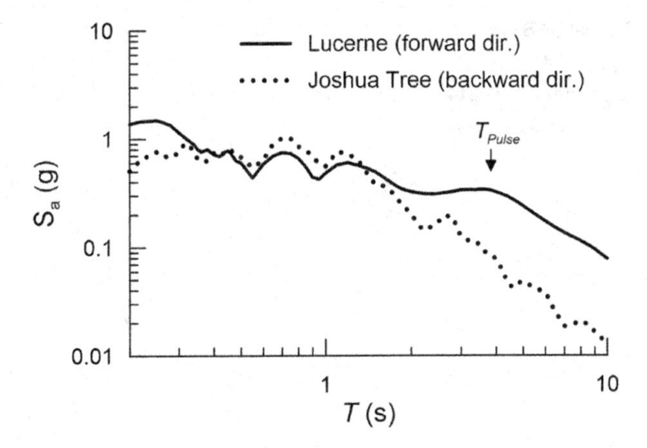

FIGURE 3.71 Response spectra of Lucerne and Joshua Tree RotD100 records from 1992 Landers earthquake; the Lucerne records exhibit spectral amplification near pulse period T_P.

velocity and displacement time series, they can also produce peaks in response spectra near a characteristic pulse period. Figure 3.71 shows a local peak in the spectra for the Lucerne record (with a forward directivity pulse at 4 second period) along with the lack of such a peak in the Joshua Tree record.

The Lucerne and Joshua Tree records illustrate a complexity of near-fault motions, which is that only some fraction exhibit pulse-like characteristics. Moreover, not all pulse-like ground motions occur in the near-field. Nonetheless, because pulse-like characteristics in ground motions can be more damaging to structures than non-pulse motions (Bertero et al., 1978; Alavi and Krawinkler, 2004; Kalkan and Kunnath, 2006), analysis of pulse effects can be an important component of ground motion characterization, especially in the near-fault environment. Engineering characterization of pulse-like ground motions has two principal components – analysis of pulse characteristics (period and amplitude) and the probability of experiencing a pulse.

3.7.4.1 Rupture Directivity Pulse Characterization

Records containing double-sided velocity pulses can be characterized by a pulse period T_{pulse} (defined from the local peak of the Fourier amplitude spectrum) and the amplification of S_a near T_{pulse}. The S_a amplification is relative to a GMM that does not consider near-fault effects (i.e., the baseline condition described in Section 3.7.3). The pulse period increases with magnitude as shown in Figure 3.72a, although there is large dispersion of data around the trend line ($\sigma_{\ln TP} = 0.61$). The S_a amplification is centered on T_{pulse} as shown in Figure 3.72b and given below (Shahi and Baker, 2011):

$$\mu_{\ln Ap} = \begin{cases} 1.131\exp\left[-3.11\left(\ln\left(T/T_{pulse}\right)+0.127\right)^2\right]+0.058 & T/T_{pulse} \le 0.88 \\ 0.896\exp\left[-2.11\left(\ln\left(T/T_{pulse}\right)+0.127\right)^2\right]+0.255 & T/T_{pulse} > 0.88 \end{cases} \quad (3.48)$$

where $\mu_{\ln Ap}$ is the mean amplification of S_a relative to a GMM, which is in natural log units and hence is additive with the GMM log mean. Note that the amplification peaks at $T/T_{pulse} = 0.88$ and attenuates away from that peak.

3.7.4.2 Rupture Directivity Pulse Probability

Pulse probabilities have been evaluated (Shahi and Baker, 2011, 2014a) by applying pulse identification algorithms to a large ground motion inventory, considering all possible horizontal orientations

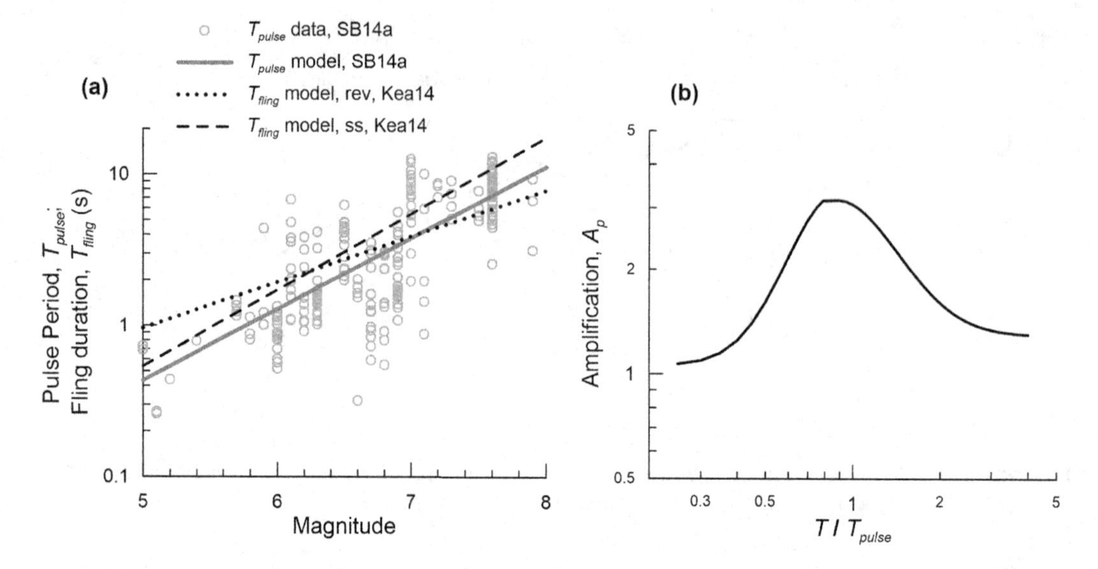

FIGURE 3.72 (a) Dependence of pulse period on magnitude, including directivity pulse data and model from Shahi and Baker (2014a, SB14a) and fling step pulse duration model (Kamai et al., 2014a: Kea14); (b) S_a amplification from directivity pulse effect (modified from Shahi and Baker, 2011). "rev" indicates reverse fault, "ss" indicates strike-slip fault.

of the ground motion. These algorithms identify double-sided pulses, as from rupture directivity, but generally cannot identify pulses from fling step (because such effects are removed from the records in the database through the use of high-pass filters). As shown in Figure 3.73a, pulse probabilities increase with the along-fault distance from the origin of rupture (hypocenter) to the point on the fault closest to the site. In the case of strike-slip faults, this distance is measured in the horizontal plane (hence the origin of the rupture is represented with the epicenter) and is denoted ℓ, as shown in Figure 3.73b. Pulse probabilities decrease with increasing source-site distance, R_{rup}. These features of the pulse probabilities are consistent with expectation for the case of forward rupture directivity.

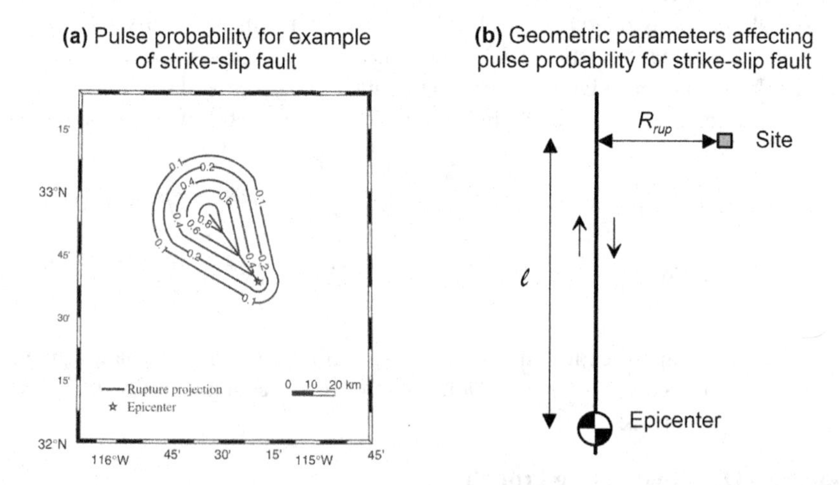

FIGURE 3.73 (a) Example of pulse probability variations around strike-slip fault and (b) geometric parameters affecting pulse probabilities for strike-slip faults. Similar parameters available for dip-slip faults. (From Shahi and Baker, 2011.)

3.7.4.3 Fling Step Pulse Effects

Ground motion records included in databases such as NGA are subjected to high-pass filtering that removes low-frequency noise but also baseline drift from fling step. As such, fling effects are absent from these data sets and from GMMs. For near-fault sites where fling effects are expected, a single-sided velocity pulse can be added to the ground motion components aligned with the slip direction. These pulses would be added in the strike-parallel direction for strike-slip earthquakes, and in the strike-normal and vertical directions (scaled according to fault dip) for dip-slip earthquakes. A pulse function (Kamai et al., 2014a) is given below and illustrated in Figure 3.74a:

$$u_f(t) = \begin{cases} 0 & t \le t_1 \\ \dfrac{u_{g,end}}{T_{fling}}(t-t_1) - \dfrac{u_{g,end}}{2\pi}\sin\left[\dfrac{2\pi}{T_{fling}}(t-t_1)\right] & t_1 < t < \left(t_1 + T_{fling}\right) \\ u_{g,end} & t > \left(t_1 + T_{fling}\right) \end{cases} \tag{3.49}$$

where t_1 is the fling step pulse start time (usually taken as shear wave arrival), $u_{g,end}$ is the static ground displacement at the site from fling step ("end" refers to displacement at the end of the ground motion), and T_{fling} is the fling step pulse duration. Figure 3.72a shows the variation of fling pulse duration with magnitude; interestingly, the available models for T_{fling} are generally similar to those for T_{pulse}.

The static displacement from fling step, $u_{g,end}$, is taken as a fraction of the average fault slip, s. Based on analysis of data from hybrid simulations, Kamai et al. (2014a) estimated the displacement ratio, $u_{g,end} / s$, as a function of rupture distance and other parameters to distinguish hanging wall from footwall sites for dipping faults. For the case of vertically-dipping strike-slip faults, this displacement ratio is given as:

$$\ln\left(u_{g,end}/s\right) = \ln(0.5) + \left(25.29\mathbf{M} - 1.577\mathbf{M}^2 - 104.22\right)\ln\left(\frac{R_{rup}+50}{50}\right) \tag{3.50}$$

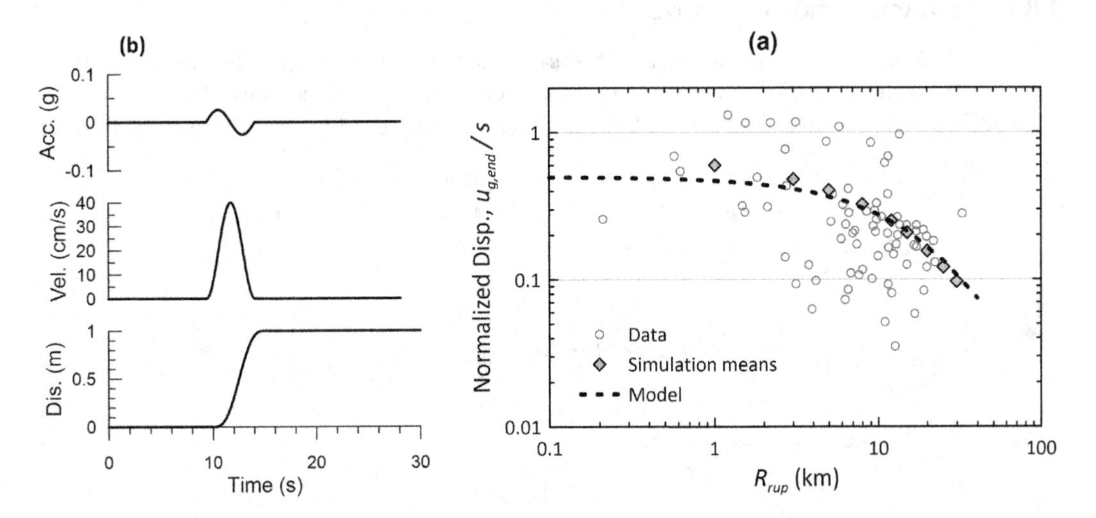

FIGURE 3.74 (a) Fling step pulse model from Kamai et al. (2014a) plotted for $u_{g,end}$ = 1.0 m, T_{fling} = 5 sec, and t_1 = 10 sec; (b) normalized fling step displacement as function of R_{rup} from data (Gregor et al., 2002), strike-slip simulations of **M** 7.5 earthquakes (From Kamai et al., 2014a), and model given in Equation (3.50).

where R_{rup} is in units of km. Medians of simulation results used for the derivation of Equation (3.50) for strike-slip faults are shown in Figure 3.74b along with available empirical data (Kamai et al., 2014a; Gregor et al., 2002). Within approximately 1 km of the fault, the displacement ratio ranges from about 1/3 to 1.0 and decays to values less than 0.1–0.2 beyond 20 km.

3.8 SPATIAL VARIABILITY OF GROUND MOTIONS

The major factors affecting ground motions are source, path and site effects (Section 3.4). When ground motions are recorded at dense arrays of seismometers in which station separations are much less than the rupture distance (R_{rup}) and the site conditions are relatively uniform, the source, path, and site effects are for all practical purposes identical. Nonetheless, variations in recorded ground motions are observed in such arrays. Figure 3.11 shows an example dense array from Borrego Valley, California, a portion of which has separation distances ranging from 10 to 160 m. Figure 3.75 shows recorded ground motions from a **M** 3.1 earthquake recorded at that array (the rupture distance of 54 km is much greater than the distances between stations). The motions do not match, having both different amplitudes (as seen by peaks of variable size) and variable phases (as seen, for example, by different arrival times of the first s-wave in the 10.9–11.0 sec time interval). The physical mechanisms that produce this variability are described in Section 3.8.1.

Spatial variations of ground motions affect seismic responses of structures in two ways: (1) differential ground motions can impose deformations upon extended structures such as pipelines, tunnels, and long-span bridges (Der Kiureghian and Neuenhofer, 1992; O'Rourke and Liu, 1999; Hashash et al., 2001); and (2) when relatively stiff structural systems are excited by spatially variable wave fields, an averaging effect occurs within the foundation that modifies the effective ground motion relative to what would be expected at a point (Section 8.4). Standard ground motion analysis procedures (Chapters 3 and 4) characterize free-field ground motions at a point, and as such do not account for the effects of spatial variability.

The following sections present a mathematical representation of spatially variable ground motions (SVGMs) and ground strains produced by SVGMs. The effects of SVGMs on ground motions experienced by structural foundations are discussed in Section 8.4 and their impact on the responses of embedded structures are discussed in Section 8.6.3.

3.8.1 CHARACTERIZATION OF SVGMs

Assuming homogenous site conditions, SVGMs are caused by *wave passage effects* and *wave scattering*. Additional variability is produced by heterogeneities in site conditions. As illustrated in Figure 3.76a, wave passage effects consist of time delays in arrivals of the wave front, which result

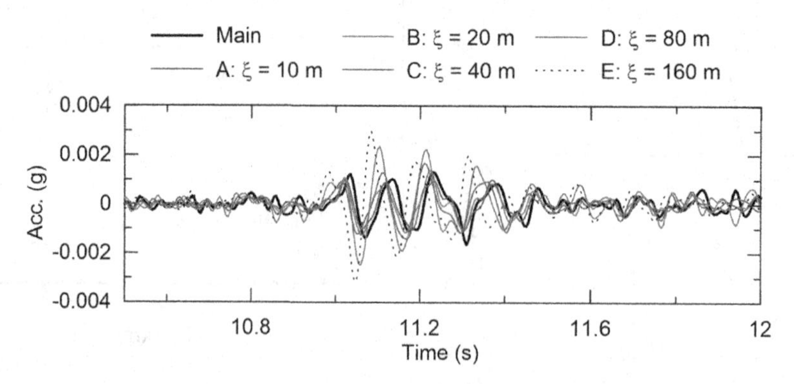

FIGURE 3.75 S-wave window of horizontal component of ground motion in direction normal to ray path for 1997 **M** 3.1 event recorded at BVDA. Stations A through E are located along a line at separation distances from the Main station as noted in the legend. (Data from Ancheta, 2010.)

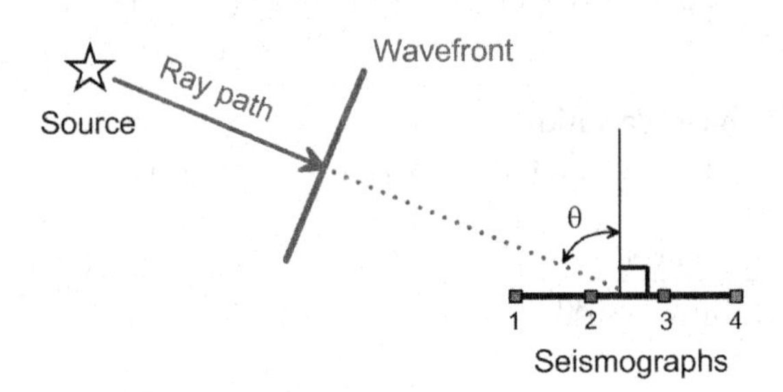

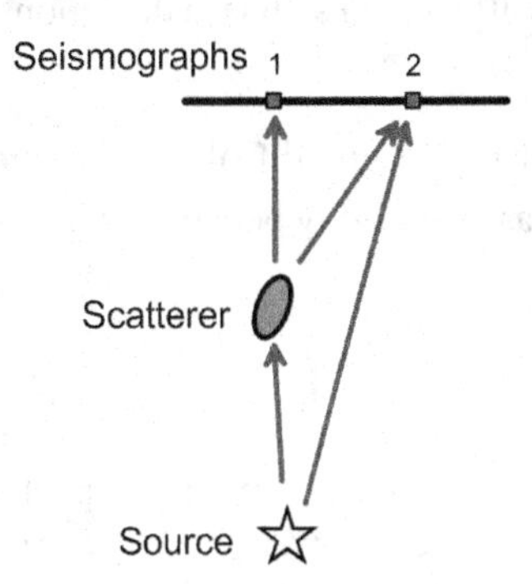

FIGURE 3.76 Schematic illustrations of (a) wave passage and (b) wave scattering. (After Zerva, 2009; used with permission of Taylor and Francis Group, LLC.)

from inclined (from vertical) body waves or horizontally propagating surface waves. Wave passage produces a time delay, Δt, between two locations that can be equivalently represented as a shift of the Fourier phase angle, ϕ_n:

$$\Delta\phi_n = \omega_n \Delta t \tag{3.51}$$

where $\Delta\phi_n$ is the phase shift, ω_n is frequency in radians/sec, and n is the frequency index for the discrete Fourier transform (notation introduced in Section 3.3.2). As illustrated in Figure 3.76b and Appendix C.5.3, wave scattering occurs as body waves are reflected and refracted by heterogeneities encountered along their travel path. Scattering causes Fourier phase and amplitude variations from station to station that are described as random (or stochastic) processes. Accordingly, SVGMs can be characterized with models for the time delay from wave passage, additional phase shift from stochastic processes (scattering and variable site conditions), and random variations in Fourier amplitude. The wave passage effect is principally sensitive to separation distance and azimuth (the effect is maximized in the direction of wave travel), whereas the stochastic effects depend only on frequency and separation distance.

3.8.1.1 Wave Passage

The critical parameter controlling wave passage is the *apparent propagation velocity*, i.e., the separation distance between two observation points divided by the difference in arrival times. The apparent velocity, $V_{app,\theta}$, is related to the apparent shear wave velocity for horizontal ray paths in the underlying rock (V_{app}) and azimuth angle (θ in Figure 3.76) as follows:

$$V_{app,\theta} = \frac{V_{app}}{\sin\theta} \tag{3.52}$$

When the observation points are aligned with the ray path ($\theta = 90°$), the apparent velocity is equal to the rock velocity; otherwise it is higher by an amount that depends on θ. Recommended values of V_{app} have been developed through analysis of data from dense arrays (LSST and BVDA; see Table 3.2) using the s-wave portions of seismograms, from which V_{app} was found to be in the range of 2.1–2.6 km/s for active tectonic regions. These velocities are comparable to crustal bedrock shear wave velocities of 2.5–3 km/s; this similarity indicates that wave passage for these arrays can be explained by inclined shear waves. Actual shear wave arrivals can have random offsets from predictions derived from $V_{app,\theta}$, which are termed arrival time perturbations (Zerva and Zervas, 2002).

These perturbations produce negligible standard deviations of arrival times for station separations $\xi < 100$ m, but standard deviations on the order of 0.05–0.1 sec for $\xi = 100$–1,000 m (Ancheta et al., 2011).

3.8.1.2 Lagged Coherency and Stochastic Phase Variations

Phase variability between two seismic signals can be expressed through a coherency function:

$$\gamma_{jk}(\omega) = \frac{S_{jk}(\omega)}{\left[S_{jj}(\omega) S_{kk}(\omega)\right]^{1/2}} \tag{3.53}$$

where S_{jj} and S_{kk} are power spectral density functions (Equation 3.11) for accelerograms at stations j and k, respectively, and S_{jk} a cross spectral density function (computed similarly to Equation 3.11 but using Fourier amplitudes from signals j and k). A coherency function describes the correlation between the signals as a function of frequency. Coherency is a dimensionless complex-valued number, the amplitude of which ($|\gamma|$) is referred to as *lagged coherency* (since the effect of time lag caused by wave passage is removed) and represents the contributions of only stochastic processes on phase. Lagged coherency values range from a lower bound representing no correlation to 1.0 representing perfect correlation. The lower bound (no correlation) value of lagged coherency depends on smoothing that is applied in the computation of the power spectral density functions, with 0.35 being a typical value. Random phase variations follow a zero-mean normal distribution with a standard deviation (σ_ϕ) that is related to lagged coherency as follows (Ancheta and Stewart, 2015):

$$E\left[\left|\gamma_{jk}(\omega)\right|\right] = e^{-0.5\sigma_\phi^2} \qquad \left|\gamma_{jk}(\omega)\right| > 0.35 \tag{3.54}$$

Array data has been used to develop lagged coherency models having two basic functional forms. One approach is to relate lagged coherency to the dimensionless ratio of station separation distance ξ to wavelength λ (e.g., Loh and Yeh, 1988; Oliveira et al., 1991; Harichandran and Vanmarcke, 1986):

$$|\gamma| = \exp(\xi/\lambda) = \exp(2\pi\omega\xi/V_{app,\theta}) \tag{3.55}$$

Subsequently, lagged coherency was related to frequency and ξ independently (Abrahamson et al., 1991; Ancheta et al., 2011) because analyses of SMART-1 and LSST data indicated that the ξ/λ normalization did not adequately capture trends in lagged coherency at short separation distances ($\xi < 200$ m) (Abrahamson, 1985, 1992). Figure 3.77 shows a lagged coherency model with these features derived from dense array data (Ancheta et al., 2011). Lagged coherency decreases strongly

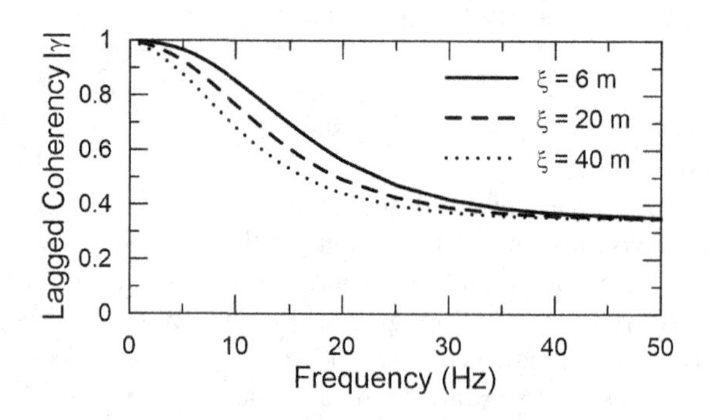

FIGURE 3.77 Model for lagged coherency (Ancheta et al., 2011).

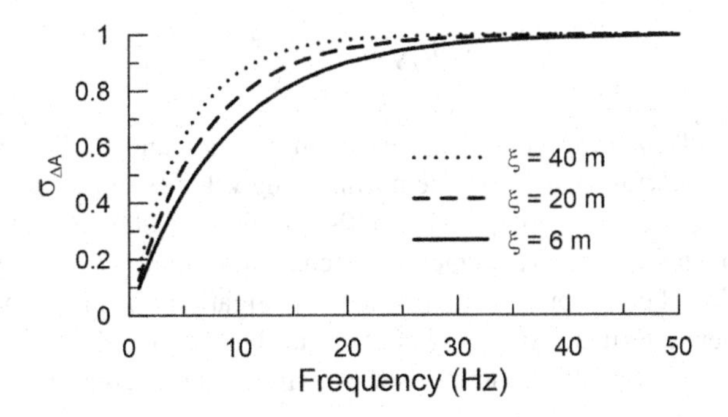

FIGURE 3.78 Model for standard deviation of natural log of Fourier amplitude derived from LSST and BVDA data (Ancheta et al., 2011).

with increasing frequency (hence, decreasing wavelength) and relatively weakly with separation distance ξ. The limiting value of lagged coherency at high frequencies represents the lagged coherency of noise using the selected smoothing function.

3.8.1.3 Amplitude Variability

Dense array recordings show that stochastic Fourier amplitude variability contributes to SVGM in addition to the phase variability (e.g., Zerva and Zhang, 1996). Amplitude variability can be quantified from differences in Fourier amplitudes between recordings at array station pairs. The difference of the natural logs of the amplitudes for a given frequency, ω_n, and station separation, ξ, is denoted $\Delta A_n(\omega_{n,}\xi)$. The quantity, ΔA_n, is normally distributed with zero mean and standard deviation $\sigma_{\Delta A}$ (ω,ξ). Figure 3.78 shows that $\sigma_{\Delta A}$ (ω,ξ) increases strongly with frequency and relatively weakly with separation distance ξ.

3.8.2 Ground Strains Produced by SVGMs

SVGMs can be represented by ground strains for relatively simplified analyses of their effects on structures such as pipelines (e.g., O'Rourke and Liu, 1999) and underground structures (Hashash et al., 2001). Ground strains result from differential displacements between two points divided by separation distance, the maximum absolute value of which is the *peak ground strain* (*PGS*).

Strain fields of various types (extensional, shear, etc.) can be derived for propagating body and surface waves using solutions of applicable wave equations (Sections C.2–C.3). Such solutions indicate that peak strain occurs at the time of peak ground velocity and is proportional to the ratio of *PGV* to the apparent wave velocity (St. John and Zahrah, 1987; Hashash et al., 2001), V_{app}. For example, if the strongest shaking (in velocity) is from shear waves, then *PGS* can be estimated as:

$$PGS = a\frac{PGV}{V_{app}} \tag{3.56}$$

where a is a constant of proportionality that depends on the alignment of the ray path relative to the alignment of the structure of interest (i.e., angle θ in Figure 3.76a) and V_{app} is the apparent shear wave propagation velocity in the ray path direction. Solutions of this type consider only the wave passage contribution to ground strains.

The combined contribution of all sources of SVGMs can be evaluated by inferring differential displacements from dense array data and normalizing by separation distance (Bodin et al., 1997; Gomberg et al., 1999; Paolucci and Smerzini, 2008). Array data from California and New Zealand (Paolucci and Smerzini, 2008) indicated that *PGS* could be estimated from the following relationship:

$$PGS = \frac{PGV}{963} \quad (3.57)$$

where PGV is in units of m/s (and has values less than 0.3 m/s) and PGS is in decimal form. The associated natural log uncertainty is 0.37. The normalizing velocity of 963 m/s is much slower than typical values of V_{app} from wave propagation (1,500–3,500 m/s; Ancheta et al., 2011), which shows that stochastic components of SVGM (lagged coherency, amplitude variability) contribute significantly to ground strains. The empirical relationship in Equation (3.57) has been found (Ancheta, 2010) to depend on separation distance and to saturate beyond PGV = 1.0 m/s. Equation (3.57) applies for relatively small separation distances (ξ = 6–10 m) and overestimates strains for larger ξ.

3.9 SUMMARY

1. The complete description of ground motion at a point requires three components of translation and three components of rotation. In practice, only translational motions are typically measured in two orthogonal horizontal directions and the vertical direction.
2. Instruments that record ground motion have two basic designs consisting of an oscillating mass (inertial seismometer) or a suspended mass within a coil that is maintained in a fixed position through electromagnetic forces generated by current in the coil. The latter design is used in modern instruments and can measure either acceleration (force-balance accelerometer) or velocity (broadband electrodynamic seismometer).
3. As-recorded signals from accelerometers or seismometers include the actual ground motion as well as errors due to noise and instrument response. These errors are reduced, but not eliminated, with modern electrodynamic instruments. Ground motion processing minimizes the effects of these errors and consists principally of applying instrument corrections and filtering signals to reduce low-frequency portions of signals dominated by noise.
4. Ground motion signals have complex waveforms. These signals can often be described adequately for engineering purposes with intensity measures that reflect the amplitude, frequency content, or duration of the ground motion. The merits of various intensity measures are judged by their effectiveness at efficiently and sufficiently predicting the seismic response of structures or other systems.
5. The most widely used intensity measures are PGA, PGV, and response spectral acceleration at various oscillator periods (S_a). PGA describes the relatively high-frequency content of the ground motion and PGV the mid-range frequency content. S_a generally reflects the frequency content of the ground motion at the oscillator period, so a response spectrum describes the ground motion amplitude over a wide period range.
6. Earthquake ground motion characteristics are controlled by source, path, and site effects. The most important source parameter is magnitude (**M**); as **M** increases, the ruptured fault area increases as well as the average slip on the fault. Larger fault sizes produce longer duration motions, while larger fault slips produce more low-frequency ground motions.
7. The most important path effects are geometric spreading and anelastic attenuation, both of which can be parameterized through various site-to-source distance metrics including Joyner-Boore distance (R_{JB}) and rupture distance (R_{rup}). As distance increases, ground motion amplitudes decrease, but durations increase. High-frequency ground motions attenuate faster with distance than low frequencies, which causes motions at large distance to have, on average, lower mean frequencies (longer mean periods) than motions at close distance.
8. Site conditions affect ground motion characteristics as a result of local ground response, basin effects, and topographic effects. Site effects are quantified by frequency-dependent

amplification functions that describe the ratio of ground shaking for the actual site condition to that for a reference (usually rock) condition.

9. Local ground response is strongly influenced by the site-specific shear wave velocity (V_S) profile, which controls impedance effects. Resonance effects can occur in profiles having a large impedance contrast, and are characterized by pronounced amplification at the fundamental mode site period. Strong shaking causes soil nonlinearity, which produces large soil damping that attenuates high-frequency ground motions but does not significantly affect low-frequency components of ground motion.
10. Basin effects are related to large-scale features of sedimentary basins, which can be many kilometers in dimension. The three-dimensional geometry of the basin controls phenomena such as basin edge effects and focusing effects. Basin edge effects occur when body waves enter basins from their edge and propagate in the direction of basin thickening. Focusing occurs when body waves propagate upwards through lens-like basin structures.
11. GMMs are semi-empirical relationships used to compute the mean and standard deviation (in natural log units) of ground motion intensity measures. All GMMs include source and path terms, and most include a site response model. GMMs have been developed for three tectonic regimes: active tectonic regions, subduction zones, and stable continental regions.
12. GMMs are based in large part on intensity measures from recordings. The databases used for GMM development include some well-recorded events and other sparsely recorded events. This uneven sampling of data necessitates statistical analyses in GMM development to address substantial data variability between events and within events. GMMs developed using simple least squares regressions that give equal weight to all data points should not be used. Data are sparse at large **M** and close distance, which often controls seismic hazard for practical applications. The manner by which GMMs extrapolate into data-poor regions can vary significantly and is one of several factors influencing epistemic uncertainty in ground motion prediction.
13. Typical source parameters used in GMMs are **M**, focal mechanism, and source depth. Amplitude-related intensity measures (*PGA*, *PGV*, S_a) increase with magnitude, but the sensitivity is much greater for low-frequency parameters than relatively high-frequency parameters. **M**-scaling is also nonlinear, being stronger at low magnitudes than for **M** > ~ 5.5–6.5. Ground motions are lower for normal fault earthquakes than for strike-slip or reverse fault earthquakes. Source depth effects are relatively poorly resolved.
14. Path effects considered in GMMs include geometric spreading, anelastic attenuation, and hanging wall effects. At close distances (under approximately 3 km), ground motions saturate and have a flat trend with distance. Geometric spreading reduces ground motions at all frequencies and dominates distance attenuation effects up to about 70 km. Anelastic attenuation affects principally high-frequency components of ground motion at distances larger than 70 km in active tectonic regions.
15. Site effects in GMMs typically use the average shear wave velocity in the upper 30 m as the principal site parameter (V_{S30}). Weak motion site amplification increases with decreasing V_{S30} (impedance effect), with the rate of increase being stronger at low frequencies than for high frequencies. Nonlinear effects, which reduce high-frequency site amplification, increase as V_{S30} decreases, potentially offsetting the amplification that would otherwise occur from impedance effects. A secondary site parameter used in some GMMs is basin depth. Ground motions tend to increase with increasing basin depth.
16. Standard deviation terms in GMMs increase with period and typically decrease with magnitude. There is no evidence that a log-normal distribution derived from these standard deviation terms should be truncated; in fact, to the extent that data depart from the theoretical log-normal distribution, it is toward "heavy tails" (larger than expected probability densities at the tails of the distribution).

17. Ground motion simulations are computational procedures that generate ground motion accelerograms based on models of source, path and site effects. Simulation procedures are especially useful to engineers for providing ground motions for conditions where recordings are sparse.
18. Physics-based stochastic simulation procedures utilize a white noise phase and prescriptive functions for the Fourier amplitude spectrum based on source, path, and site models. These functions are calibrated against data and comprise a computationally efficient means by which to generate time series and S_a.
19. Deterministic simulation procedures utilize relatively complex source models that consider the finite size of the fault, fault slip distribution, and other details of the rupture process. Path effects are described with Green's functions, which most often are computed through the simulation of wave propagation in crustal models. High-frequency components of ground motion are difficult to simulate with deterministic source and path models and are typically introduced through calibrated stochastic models.
20. Simulation procedures should not be considered a reliable source of ground motions for engineering applications without proper verification, validation, and (as needed) calibration. Verification involves testing to confirm the correct coding of simulation procedures. Validation involves testing the results of the analyses against data. Calibration involves tuning appropriate model parameters to match data when misfits are identified in validation.
21. Near-fault ground motions can have unique characteristics that depart from baseline predictions provided by GMMs (which typically do not consider the physical processes involved in near-fault ground motions). Rupture directivity occurs when a significant distance of fault rupture occurs toward a site, in which case shear waves can arrive in a double-sided velocity pulse polarized in the horizontal direction normal to the fault strike. Fling step is caused by permanent displacement of the Earth's crust near the fault in the slip-parallel direction. It takes the form of a single-sided velocity pulse that integrates (in time) to the permanent crustal displacement. These near-fault effects can be damaging to structures, and models are available for predicting their effects on response spectra.
22. Ground motions vary spatially over short length scales, even when source, path and site effects at observation points are nominally similar. These variations in ground motion are caused by wave passage effects and essentially stochastic (random) variability in Fourier amplitude and phase angle, which are most pronounced at high frequencies. These components of spatial variability can be predicted using models derived from dense array data. Peak ground strains from spatially variable motions can be estimated from the ratio of *PGV* to an empirically calibrated apparent wave velocity.

REFERENCES

Abrahamson, N.A. (1985). Estimation of seismic wave coherency and rupture velocity using the SMART 1 strong-motion array recordings, *Report No. UCB/EERC-85/02*, Earthquake Engineering Research Center, UC Berkeley.

Abrahamson, N.A. (1992). Spatial variation of earthquake ground motion for application to soil-structure interaction, *Report No. EPRI TR-100463*, Electrical Power Research Institute, Palo Alto, CA.

Abrahamson, N.A. (2007). Program on Technology Innovation: Effects of Spatial Incoherence on Seismic Ground Motions, *Report No. EPRI-1015110*, Electric Power Research Institute, Palo Alto, CA.

Abrahamson, N.A. and Gülerce, Z. (2022). "Summary of the Abrahamson and Gulerce NGA-SUB ground-motion model for subduction earthquakes," *Earthquake Spectra*, Vol. 7, No. 4, pp. 2638–2681. https://doi.org/10.1177/87552930221114374.

Abrahamson, N.A., Kuehn, N., Gülerce, Z., Gregor, N., Bozorgnia, Y., Parker, G.A., Stewart, J.P., Chiou, B., Idriss, I.M., Campbell, K.W., and Youngs, R.R. (2018). "Update of the BC Hydro subduction ground-motion model using the NGA- Subduction dataset," *Report No. 2018/02*, Pacific Earthquake Engineering Research Center, UC Berkeley. https://doi.org/10.55461/oycd7434.

Abrahamson, N.A., Schneider, J.F., and Stepp, J.C. (1991). "Empirical spatial coherency functions for applications to soil-structure interaction analyses," *Earthquake Spectra*, Vol. 7, pp. 1–27. https://doi.org/10.1193/1.1585610.

Abrahamson, N.A., Silva, W.J., and Kamai, R. (2014). "Summary of the ASK14 ground-motion relation for active crustal regions," *Earthquake Spectra*, Vol. 30, pp. 1025–1055. https://doi.org/10.1193/070913eqs198m.

Abrahamson, N.A. and Youngs, R.R. (1992). "A stable algorithm for regression analyses using the random effects model," *Bulletin of the Seismological Society of America*, Vol. 82, pp. 505–510. https://doi.org/10.1785/bssa0820010505.

Afshari, K. and Stewart, J.P. (2016). "Physically parameterized prediction equations for significant duration in active crustal regions," *Earthquake Spectra*, Vol. 32, pp. 2057–2081. https://doi.org/10.1193/063015eqs106m.

Afshari, K., Stewart, J.P., and Steidl, J.H. (2019). "California ground motion vertical array database," *Earthquake Spectra*, Vol. 35, pp. 2003–2015. https://doi.org/10.1193/070218eqs170dp.

Aki, K. (1967). "Scaling law of seismic spectrum," *Journal of Geophysical Research*, Vol. 72, pp. 1217–1231. https://doi.org/10.1029/jz072i004p01217.

Aki, K. and Richards, P.G. (1980). *Quantitative Seismology*, Vol. 1, W.H. Freeman, San Francisco, CA. https://doi.org/10.1121/1.385057.

Akkar, S. and Bommer, J.J. (2010). "Empirical equations for the prediction of PGA, PGV and spectral accelerations in Europe, the Mediterranean region and the Middle East," *Seismological Research Letters*, Vol. 81, pp. 195–206. https://doi.org/10.1785/gssrl.81.2.195.

Akkar, S. and Boore, D.M. (2009). "On baseline corrections and uncertainty in response spectra for baseline variations commonly encountered in digital accelerograph records," *Bulletin of the Seismological Society of America*, Vol. 99, pp. 1671–1690. https://doi.org/10.1785/0120080206.

Akkar, S., Kale, Ö., Yenier, E., and Bommer, J.J. (2011). "The high-frequency limit of usable response spectral ordinates from filtered analogue and digital strong-motion accelerograms," *Earthquake Engineering and Structural Dynamics*, Vol. 40, 1387–1401. https://doi.org/10.1002/eqe.1095.

Akkar, S. and Küçükdoğan, B. (2008). "Direct use of PGV for estimating peak nonlinear oscillator displacements," *Earthquake Engineering and Structural Dynamics*, Vol. 37, 1411–1433. https://doi.org/10.1002/eqe.819.

Akkar, S., Sandıkkaya, M.A., and Bommer, J.J. (2014). "Empirical ground-motion models for point- and extended-source crustal earthquake scenarios in Europe and the Middle East," *Bulletin of Earthquake Engineering*, Vol. 12, pp. 359–387. https://doi.org/10.1007/s10518-013-9461-4.

Alavi, B. and Krawinkler, H. (2004). "Behavior of moment-resisting frame structures subjected to near-fault ground motions," *Earthquake Engineering & Structural Dynamics*, Vol. 33, pp. 687–706. https://doi.org/10.1002/eqe.369.

Allen, C.R. (1975). "Geologic criteria for evaluating seismicity," *Bulletin of the Geological Society of America*, Vol. 86, No. 8, pp. 1041–1057. https://doi.org/10.1130/0016-7606(1975)86<#; 1041#: gcfes>#; 2.0.co#; 2.

Ambraseys, N.N. (1971). "Value of historical records of earthquakes," *Nature*, Vol. 232, pp. 375–379. https://doi.org/10.1038/232375a0.

Ambraseys, N.N. (1978). "Middle East - A reappraisal of the seismicity," *Quarterly Journal of Engineering Geology*, Vol. 11, pp. 19–32. https://doi.org/10.1144/gsl.qjeg.1978.011.01.03.

American Society of Civil Engineers, ASCE (2022). *Minimum Design Loads and Associated Criteria for Buildings and Other Structures*, Report ASCE/SEI 7-22, Reston, VA.

Ancheta, T.D. (2010). Engineering characterization of spatially variable earthquake ground motions, *Phd Dissertation*, University of California, Los Angeles, Dept. of Civil & Environmental Engineering.

Ancheta, T.D., Darragh, R.B., Stewart, J.P., Seyhan, E., Silva, W.J., Chiou, B.S.J., Wooddell, K.E., Graves, R.W., Kottke, A.R., Boore, D.M., Kishida, T., and Donahue, J.L. (2014). "NGA-West2 database," *Earthquake Spectra*, Vol. 30, pp. 989–1005. https://doi.org/10.1193/062913eqs180m.

Ancheta, T.D. and Stewart, J.P. (2015). "Conditional simulation of spatially variable motions on 2D grid," *Proceedings of 12th International Conference on Applications of Statistics and Probability in Civil Engineering*, Vancouver, Canada, T Haukaas (ed.), Paper No. 550 (electronic file).

Ancheta, T.D., Stewart, J.P., and Abramhamson, N.A. (2011). "Engineering characterization of earthquake ground motion coherency and amplitude variability," *Proceedings of 4th International Symposium on Effects of Surface Geology on Seismic Motion, IASPEI / IAEE*, August 23–26, 2011, University of California Santa Barbara.

Anderson, J.G. (2015). "The composite source model for broadband simulations of strong ground motions," *Seismological Research Letters*, Vol. 86, pp. 67–74. https://doi.org/10.1785/0220140098.

Anderson, J.G. and Hough, S.E. (1984). "A model for the shape of the Fourier amplitude spectrum of acceleration at high frequencies," *Bulletin of the Seismological Society of America*, Vol. 74, 1969–1993. https://doi.org/10.1785/BSSA0740051969.

Andrews, D.J., Hanks, T.C., and Whitney, J.W. (2007). "Physical limits on ground motion at Yucca Mountain," *Bulletin of the Seismological Society of America*, Vol. 97, pp. 1771–1792. https://doi.org/10.1785/0120070014.

Archuleta, R.J. (1984). "A faulting model for the 1979 Imperial Valley earthquake," *Journal of Geophysical Research*, Vol. 89, 4559–4585. https://doi.org/10.1029/jb089ib06p04559.

Arias, A. (1970). A measure of earthquake intensity, in R. J. Hansen, ed., *Seismic Design for Nuclear Power Plants*, MIT Press, Cambridge, MA, pp. 438–483.

Ashford, S.A., Sitar, N., Lysmer, J., and Deng, N. (1997). "Topographic effects on the seismic response of steep slopes," *Bulletin of the Seismological Society of America*, Vol. 87, No. 3, pp. 701–709. https://doi.org/10.1785/bssa0870030701.

Atkinson, G.M. (1993). "Earthquake source spectra in Eastern North America," *Bulletin of the Seismological Society of America*, Vol. 83, pp. 1778–1798. https://doi.org/10.1785/BSSA0830061778.

Atkinson, G.M. and Assatourians, K. (2015). "Implementation and validation of EXSIM (a stochastic finite-fault ground-motion simulation algorithm," *Seismological Research Letters*, Vol. 86, pp. 48–60. https://doi.org/10.1785/0220140097.

Atkinson, G.M. and Beresnev, I. (1997). "Don't call it stress drop," *Seismological Research Letters*, Vol. 68, pp. 3–4. https://doi.org/10.1785/gssrl.68.1.3.

Atkinson, G.M. and Boore, D.M. (2014). "The attenuation of Fourier amplitudes for rock sites in eastern North America," *Bulletin of the Seismological Society of America*, Vol. 104, pp. 513–528. https://doi.org/10.1785/0120130136.

Atkinson, G.M. and Silva, W.J. (1997). "An empirical study of earthquake source spectra for California earthquakes," *Bulletin of the Seismological Society of America*, Vol. 87, pp. 97–113. https://doi.org/10.1785/bssa0870010097.

Baker, J.W., Abrahamson, N.A., Whitney, J.W., Board, M.P., and Hanks, T.C. (2013). "Use of fragile geologic structures as indicators of unexceeded ground motions and direct constraints on probabilistic seismic hazard analysis," *Bulletin of the Seismological Society of America*, Vol. 103, pp. 1898–1911. https://doi.org/10.1785/0120120202.

Baker, J.W. and Cornell, C.A. (2005). "A vector-valued ground motion intensity measure consisting of spectral acceleration and epsilon," *Earthquake Engineering & Structural Dynamics*, Vol. 34, pp. 1193–1217. https://doi.org/10.1002/eqe.474.

Bao, H., Xu, L., Meng, L., Ampuero, J.-P., Gao, L., and Zhang, H. (2022). "Global frequency of oceanic and continental supershear earthquakes," *Nature Geosciences*, Vol. 15, pp. 942–949. https://doi.org/10.1038/s41561-022-01055-5

Barbosa, A.R., Ribeiro, F.L., and Neves, L.A. (2017). "Influence of earthquake ground-motion duration on damage estimation: Application to steel moment resisting frames," *Earthquake Engineering and Structural Dynamics*, Vol. 46, pp. 27–49. https://doi.org/10.1002/eqe.2769.

Bertero, V.V., Mahin, S.A., and Herrera, R.A. (1978). "Aseismic design implications of near-fault San Fernando earthquake records," *Earthquake Engineering & Structural Dynamics*, Vol. 6, pp. 31–42. https://doi.org/10.1002/eqe.4290060105.

Bielak, J., Ghattas, O., and Kim, E.-J. (2005). "Parallel octree-based finite element method for large-scale earthquake ground motion simulation," *Computer Modeling in Science and Engineering*, Vol. 10, pp. 99–112. https://doi.org/10.3970/cmes.2005.010.099.

Bielak, J., Graves, R.W., Olsen, K.B., Taborda, R., Ramirez-Guzman, L., Day, S.M., Ely, G.P., Roten, D., Jordan, T.H., Maechling, P.J., Urbanic, J., Cui, Y.F., and Juve, G. (2010). "The ShakeOut earthquake scenario: Verification of three simulation sets," *Geophysical Journal International*, Vol. 180, pp. 375–404. https://doi.org/10.1111/j.1365-246x.2009.04417.x.

Bijelić, N., Lin, T., and Deierlein, G.G. (2018). "Validation of the SCEC Broadband Platform simulations for tall building risk assessments considering spectral shape and duration of the ground motion," *Earthquake Engineering and Structural Dynamics*, Vol. 47, pp. 2233–2251. https://doi.org/10.1002/eqe.3066.

Bindi, D., Pacor, F., Luzi, L., Puglia, R., Massa, M., Ameri, G., and Paolucci, R. (2011). "Ground motion prediction equations derived from the Italian strong motion database," *Bulletin of Earthquake Engineering*, Vol. 9, pp. 1899–1920. https://doi.org/10.1007/s10518-011-9313-z.

Bizzarri, A. and Spudich, P. (2008). "Effects of super-shear rupture speed on the high frequency content of S-waves investigated using spontaneous dynamic rupture models and isochrone theory," *Journal of Geophysical Research*, Vol. 113. https://doi.org/10.1029/2007jb005146.

Boatwright, J., Thywissen, K., and Seekins, L. (2001). "Correlation of ground motion and intensity for the January 17, 1994 Northridge, California earthquake," *Bulletin of the Seismological Society of America*, Vol. 91, pp. 739–752. https://doi.org/10.1785/0119990049.

Bodin, P., Gomberg, J., Singh, S.K., and Santoyo, M. (1997). "Dynamic deformations of shallow sediments in the valley of Mexico, Part I: Three-dimensional strains and rotations recorded on a seismic array," *Bulletin of the Seismological Society of America*, Vol. 87, pp. 528–539. https://doi.org/10.1785/bssa0870030528.

Bolt, B.A. (1969). "Duration of strong motion," *Proceedings of the 4th World Conference on Earthquake Engineering, Santiago*, Chile, 1304–1315.

Bolt, B.A. (1988). *Earthquakes*, W.H. Freeman and Company, New York, 282 pp.

Bommer, J.J., Coppersmith, K.J., Coppersmith, R.T., Hanson, K.L., Mangongolo, A., Neveling, J., Rathje, E.M., Rodriguez-Marek, A., Scherbaum, F., Shelembe, R., Stafford, P.J., and Strasser, F.O. (2015). "A SSHAC Level 3 probabilistic seismic hazard analysis for a new-build nuclear site in South Africa," *Earthquake Spectra*, Vol. 31, pp. 661–698. https://doi.org/10.1193/060913eqs145m.

Bommer, J.J., Hancock, J., and Alarcón, J.E. (2006). "Correlations between duration and number of effective cycles of earthquake ground motion," *Soil Dynamics and Earthquake Engineering*, Vol. 26, pp. 1–13. https://doi.org/10.1016/j.soildyn.2005.10.004.

Bommer, J.J., Stafford, P.J., and Alarcón, J.E. (2009). "Empirical equations for the prediction of the significant, bracketed, and uniform duration of earthquake ground motion," *Bulletin of the Seismological Society of America*, Vol. 99, pp. 3217–3233. https://doi.org/10.1785/0120080298.

Bommer, J.J., Stafford, P.J., Edwards, B., Dost, B., van Dedem, E., Rodriguez-Marek, A., Kruiver, P., van Elk, J., Doornhof, D., and Ntinalexis, M. (2017). "Framework for a ground-motion model for induced seismic hazard and risk analysis in the Groningen Gas Field, The Netherlands," *Earthquake Spectra*, Vol. 33, pp. 481–498. https://doi.org/10.1193/082916eqs138m.

Boore, D.M. (1972). "A note on the effect of simple topography on seismic SH waves," *Bulletin of the Seismological Society of America*, Vol. 62, pp. 275–284. https://doi.org/10.1785/bssa0620010275.

Boore, D.M. (1983). "Stochastic simulation of high-frequency ground motions based on seismological models of the radiated spectra," *Bulletin of the Seismological Society of America*, Vol. 73, pp. 1865–1894. https://doi.org/10.1785/BSSA07306A1865.

Boore, D.M. (2000). "Effect of baseline corrections on displacement response spectra for several recordings of the 1999 Chi-Chi, Taiwan, earthquake," *Bulletin of the Seismological Society of America*, Vol. 91, pp. 1199–1211. https://doi.org/10.3133/ofr99545.

Boore, D.M. (2003a). "Analog-to-digital conversion as a source of drifts in displacements derived from digital recordings of ground acceleration," *Bulletin of the Seismological Society of America*, Vol. 93, pp. 2017–2024. https://doi.org/10.1785/0120020239.

Boore, D.M. (2003b). "Simulation of ground motion using the stochastic method," *Pure and Applied Geophysics*, Vol. 160, pp. 635–675. https://doi.org/10.1007/978-3-0348-8010-7_10.

Boore, D.M. (2003c). "Some notes on phase derivatives and simulating strong ground motions," *Bulletin of the Seismological Society of America*, Vol. 93, pp. 1132–1143. https://doi.org/10.1785/0120020196.

Boore, D.M. (2005). "On pads and filters: Processing strong-motion data," *Bulletin of the Seismological Society of America*, Vol. 95, pp. 745–750. https://doi.org/10.1785/0120040160.

Boore, D.M. (2009). "Comparing stochastic point-source and finite-source ground-motion simulations: SMSIM and EXSIM," *Bulletin of the Seismological Society of America*, Vol. 99, pp. 3202–3216. https://doi.org/10.1785/0120090056.

Boore, D.M. (2010). "Orientation-independent, non geometric-mean measures of seismic intensity from two horizontal components of motion," *Bulletin of the Seismological Society of America*, Vol. 100, pp. 1830–1835. https://doi.org/10.1785/0120090400.

Boore, D.M. (2013). "The uses and limitations of the square-root impedance method for computing site amplification," *Bulletin of the Seismological Society of America*, Vol. 103, pp. 2356–2368. https://doi.org/10.1785/0120120283

Boore, D.M. (2018). "Ground-motion models for very-hard-rock sites in Eastern North America: An update," *Seismological Research Letters*, Vol. 89, pp. 1172–1184. https://doi.org/10.1785/0220170218.

Boore, D.M. (2020). "Revision of Boore (2018) ground-motion predictions for central and eastern North America: Path and offset adjustments and extension to 200 m/s ≤ VS30 ≤ 3000 m/s," *Seismological Research Letters*, Vol. 91, pp. 977–991. https://doi.org/10.1785/0220190190.

Boore, D.M. and Boatwright, J. (1984). "Average body-wave radiation coefficients," *Bulletin of the Seismological Society of America*, Vol. 74, pp. 1615–1621. https://doi.org/10.1785/bssa0740051615.

Boore, D.M. and Bommer, J.J. (2005). "Processing of strong-motion accelerograms: Needs, options and consequences," *Soil Dynamics and Earthquake Engineering*, Vol. 25, pp. 93–115. https://doi.org/10.1016/j.soildyn.2004.10.007.

Boore, D.M., Di Alessandro, C., and Abrahamson, N.A. (2014). "A generalization of the double-corner-frequency source spectral model and its use in the SCEC BBP Validation Exercise," *Bulletin of the Seismological Society of America*, Vol. 104, pp. 2387–2398. https://doi.org/10.1785/0120140138.

Boore, D.M. and Joyner, W.B. (1997). "Site amplifications for generic rock sites," *Bulletin of the Seismological Society of America*, Vol. 87, pp. 327–341. https://doi.org/10.1785/bssa0870020327.

Boore, D.M., Joyner, W.B., and Fumal, T.E. (1994). "Estimation of response spectra and peak accelerations from western North American earthquakes: An interim report, Part 2," *U.S. Geological Survey Open-File Report 94-127*, 40 pp. https://doi.org/10.3133/ofr94127.

Boore, D.M., Stewart, J.P., Seyhan, E., and Atkinson, G.M. (2014). "NGA-West 2 equations for predicting PGA, PGV, and 5%-damped PSA for shallow crustal earthquakes," *Earthquake Spectra*, Vol. 30, pp. 1057–1085. https://doi.org/10.1193/070113eqs184m.

Boore, D.M. and Thompson, E.M. (2014). "Path durations for use in the stochastic-method simulation of ground motions," *Bulletin of the Seismological Society of America*, Vol. 104, pp. 2541–2552. https://doi.org/10.1785/0120140058.

Boore, D.M. and Thompson, E.M. (2015). "Revisions to some parameters used in stochastic-method simulations of ground motion," *Bulletin of the Seismological Society of America*, Vol. 105, pp. 1029–1041. https://doi.org/10.1785/0120140281.

Boore, D.M., Thompson, E.M., and Cadet, H. (2011). "Regional correlations of VS30 and velocities averaged over depths less than and greater than 30 meters," *Bulletin of the Seismological Society of America*, Vol. 101, pp. 3046–3059. https://doi.org/10.1785/0120110071.

Bora, S.S., Scherbaum, F., Kuehn, N., and Stafford, P. (2014). "Fourier spectral- and duration models for the generation of response spectra adjustable to different source-, propagation-, and site conditions," *Bulletin of Earthquake Engineering*, Vol. 12, pp. 467–493. https://doi.org/10.1007/s10518-013-9482-z.

Borcherdt, R.D. (1994). "Estimates of site-dependent response spectra for design (methodology and justification)," *Earthquake Spectra*, Vol. 10, pp. 617–653. https://doi.org/10.1193/1.1585791.

Bozorgnia, Y. and Campbell, K.W. (2004). "The vertical-to-horizontal response spectral ratio and tentative procedures for developing simplified V/H and vertical design spectra," *Earthquake Engineering*, Vol. 8, No. 2, pp. 175–207. https://doi.org/10.1080/13632460409350486.

Bozorgnia, Y. and Campbell, K.W. (2016). "Ground motion model for the vertical-to-horizontal (V/H) ratios of PGA, PGV, and response spectra," *Earthquake Spectra*, Vol. 32, pp. 951–978. https://doi.org/10.1193/100614eqs151m.

Bozorgnia, Y. and 31 other authors (2014). "NGA-West2 research project," *Earthquake Spectra*, Vol. 30, pp. 973–987. https://doi.org/10.1193/072113EQS209M.

Bradley, B.A., Pettinga, D., Baker, J.W., and Fraser, J. (2017). "Guidance on the utilization of earthquake-induced ground motion simulations in engineering practice," *Earthquake Spectra*, Vol. 33, pp. 809–835. https://doi.org/10.1193/120216eqs219ep.

Brandenberg, S.J., Coe, J., Nigbor, R.L., and Tanksley, K. (2009). "Different approaches for measuring ground strains during pile driving at a buried archeological site," *Journal of Geotechnical and Geoenvironmental Engineering*, Vol. 135, pp. 1101–1112. https://doi.org/10.1061/(asce)gt.1943-5606.0000031.

Bray, J.D. and Rathje, E.M. (1998). "Earthquake-induced displacements of solid-waste landfills," *Journal of Geotechnical and Geoenvironmental Engineering*, Vol. 124, pp. 242–253. https://doi.org/10.1061/(asce)1090-0241(1998)124#: 3(242).

Bray, J.D. and Travasarou, T. (2007). "Simplified procedure for estimating earthquake-induced deviatoric slope displacements," *Journal of Geotechnical and Geoenvironmental Engineering*, Vol. 133, pp. 381–392. https://doi.org/10.1061/(asce)1090-0241(2007)133#: 4(381).

Brune, J.N. (1970). "Tectonic stress and the spectra of seismic shear waves from earthquakes," *Journal of Geophysical Research*, Vol. 75, pp. 4997–5009. https://doi.org/10.1029/jb075i026p04997.

Brune, J.N. (1971). "Correction," *Journal of Geophysical Research*, Vol. 76, p. 5002. https://doi.org/10.1029/jb076i020p05002.

Cabañas, L., Benito, B., and Herráiz, M. (1997). "An approach to the measurement of the potential structural damage of earthquake ground motions," *Earthquake Engineering and Structural Dynamics*, Vol. 26, pp. 79–92. https://doi.org/10.1002/(sici)1096-9845(199701)26#: 1<#; 79#: #: aid-eqe624>#; 3.0.co#; 2–y.

Campbell, K.W. (1981). "Near source attenuation of peak horizontal acceleration," *Bulletin of the Seismological Society of America*, Vol. 71, pp. 2039–2070. https://doi.org/10.1785/BSSA0710062039.

Campbell, K.W. (2009). "Estimates of shear-wave Q and κ0 for unconsolidated and semiconsolidated sediments in eastern North America," *Bulletin of the Seismological Society of America*, Vol. 99, pp. 2365–2392. https://doi.org/10.1785/0120080116.

Campbell, K. W., and Bozorgnia, Y. (2011). "Predictive equations for the horizontal component of standardized cumulative absolute velocity as adapted for use in the shutdown of U.S. nuclear power plants," *Nuclear Engineering Design*, Vol. 241, pp. 2558 -2569. https://doi.org/10.1016/j.nucengdes.2011.04.020.

Campbell, K.W. and Bozorgnia, Y. (2012). "Cumulative absolute velocity (CAV) and seismic intensity based on the PEER-NGA database," *Earthquake Spectra*, Vol. 28, No. 2, pp. 457–485. https://doi.org/10.1193/1.4000012.

Campbell, K.W. and Bozorgnia, Y. (2014). "NGA-West2 ground motion model for the average horizontal components of PGA, PGV, and 5%-damped linear acceleration response spectra," *Earthquake Spectra*, Vol. 30, pp. 1087–1115. https://doi.org/10.1193/062913eqs175m.

Campbell, K.W. and Bozorgnia, Y. (2019). "Ground motion models for the horizontal components of Arias intensity (AI) and cumulative absolute velocity (CAV) using the NGA-West2 database," *Earthquake Spectra*, Vol. 35, pp. 1289–1310. https://doi.org/10.1193/090818eqs212m.

Chandramohan, R., Baker, J.W., and Deierlein, G.G. (2016). "Quantifying the influence of ground motion duration on structural collapse capacity using spectrally equivalent records," *Earthquake Spectra*, Vol. 32, pp. 927–950. https://doi.org/10.1193/122813eqs298mr2.

Chiou, B.S.-J., Darragh, R., Dreger, D., and Silva, W.J. (2008). "NGA project strong-motion database," *Earthquake Spectra*," Vol. 24, pp. 23–44. https://doi.org/10.1193/1.2894831.

Chiou, B.S.-J. and Youngs, R.R. (2014). "Update of the Chiou and Youngs NGA model for the average horizontal component of peak ground motion and response spectra," *Earthquake Spectra*, Vol. 30, pp. 1117–1153. https://doi.org/10.1193/072813eqs219m.

Choi, Y. and Stewart, J.P. (2005). "Nonlinear site amplification as function of 30 m shear wave velocity," *Earthquake Spectra*, Vol. 21, pp. 1–30. https://doi.org/10.1193/1.1856535.

Choi, Y., Stewart, J.P., and Graves, R.W. (2005). "Empirical model for basin effects that accounts for basin depth and source location," *Bulletin of the Seismological Society of America*, Vol. 95, pp. 1412–1427. https://doi.org/10.1785/0120040208.

Clough, R.W. and Penzien, J. (1975). *Dynamics of Structures*, McGraw-Hill, New York, 634 pp.

Clough, R.W. and Penzien, J. (1993). *Dynamics of Structures*, 2nd edition, McGraw-Hill, New York, 738 pp.

Cordova, P.P., Deierlein, G.G., Mehanny, S.S.F., and Cornell, C.A. (2000). "Development of a two-parameter seismic intensity measure and probabilistic assessment procedure," *2nd U.S.-Japan Workshop on Performance-Based Earthquake Engineering Methodology for Reinforced Concrete Building Structures*, Sapporo, Hokkaido, pp. 187–206.

Custódio, S., Liu, P.P., and Archuletta, R.J. (2005). "The 2004 Mw 6.0 Parkfield, California earthquake: Inversion of near-source ground motions using multiple datasets," *Geophysical Research Letters*, Vol. 32, p. L23312. https://doi.org/10.1029/2005gl024417.

Crempien, J. and Archuleta, R.J. (2015). "UCSB method for broadband ground motion from kinematic simulations of earthquakes," *Seismological Research Letters*, Vol. 86, pp. 61–67. https://doi.org/10.1785/0220140103.

Day, S.M., Graves, R.W., Bielak, J., Dreger, D.S., Larsen, S., Olsen, K.B., Pitarka, A., and Ramirez-Guzman, L. (2008). "Model for basin effects on long-period response spectra in southern California," *Earthquake Spectra*, Vol. 24, pp. 257–277. https://doi.org/10.1193/1.2857545.

Der Kiureghian, A. and Neuenhofer, A. (1992). "Response spectrum method for multiple support seismic excitation," *Earthquake Engineering and Structural Dynamics*, Vol. 21, pp. 713–740. https://doi.org/10.1002/eqe.4290210805.

Dewey, J. and Byerly, P. (1969). "Earthquake Monitoring," *Bulletin of the Seismological Society of America*, Vol. 59, No. 1, pp. 183–227. https://doi.org/10.1785/BSSA0590010183.

Donahue, J. and Abrahamson, N.A. (2014). "Simulation-based hanging-wall effects," *Earthquake Spectra*, Vol. 30, pp. 1269–1284. https://doi.org/10.1193/071113eqs200m.

Donahue, J.L., Stewart, J.P., Gregor, N., and Bozorgnia, Y. (2019). "Ground-motion directivity modeling for seismic hazard applications," *Report No. 2019/03*, Pacific Earthquake Engineering Research Center, UC Berkeley. https://doi.org/10.55461/gphh9609.

Donovan, N.C. (1973). "A statistical evaluation of strong motion data including the February 9, 1971, San Fernando Earthquake," *Proceedings of 5th World Conference on Earthquake Engineering*, Rome, Italy.

Douglas, J. (2024). GMPE compendium. https://www.gmpe.org.uk/ (last accessed 6/8/2024).

Douglas, J. and Boore, D.M. (2011). "High-frequency filtering of strong-motion records," *Bulletin of Earthquake Engineering*, Vol. 9, pp. 395–409. https://doi.org/10.1007/s10518-010-9208-4.

Dreger, D.S., Beroza, G.C., Day, S.M., Goulet, C.A., Jordan, T.H., Spudich, P.A., and Stewart, J.P. (2015). "Validation of the SCEC broadband platform V14.3 simulation methods using pseudo spectral acceleration data," *Seismological Research Letters*, Vol. 86, pp. 39–47. https://doi.org/10.1785/0220140118.

Dunham, E.M. and Archuleta, R.J. (2004). "Evidence for a supershear transient during the 2002 Denali Fault Earthquake," *Bulletin of the Seismological Society of America*, Vol. 94, pp. S256–S268. https://doi.org/10.1785/0120040616.

Electrical Power Research Institute (EPRI) (2006). "Program on technology innovation: Use of Cumulative Absolute Velocity (CAV) in determining effects of small magnitude earthquakes on seismic hazard analyses," *Report No.* 1014099, Palo Alto, CA.

Electrical Power Research Institute (EPRI) (2013). "Seismic evaluation guidance: Screening, Prioritization and Implementation Details (SPID) for the resolution of Fukushima near-term task force recommendation 2.1: Seismic," *Report No.* 1025287, Palo Alto, CA.

Elgamal, A.-W., Zeghal, M., and Parra, E. (1996). "Liquefaction of reclaimed island in Kobe, Japan," *Journal of Geotechnical Engineering*, Vol. 122, No. 1, pp. 39–49. https://doi.org/10.1061/(asce)0733-9410(1996)122#: 1(39).

Faccioli, E., Paolucci, R., and Rey, J. (2004). "Displacement spectra for long periods," *Earthquake Spectra*, Vol. 20, pp. 347–376. https://doi.org/10.1193/1.1707022.

Farhadi, A. and Pezeshk, S. (2020). "A referenced empirical ground-motion model for Arias intensity and cumulative absolute velocity based on the NGA-East database," *Bulletin of the Seismological Society of America*, Vol. 110, pp. 508–518. https://doi.org/10.1785/0120190267.

Fletcher, J.B., Baker, L.M., Spudich, P., Goldstein, P., Sims, J.D., and Hellweg, M. (1992). "The USGS Parkfield, California, dense seismograph array-UPSAR," *Bulletin of the Seismological Society of America*, Vol. 82, pp. 1041–1070. https://doi.org/10.1785/BSSA0820021041.

Frankel, A. (2009). "A constant stress-drop model for producing broadband synthetic seismograms: Comparison with the Next Generation Attenuation relations," *Bulletin of the Seismological Society of America*, Vol. 99, pp. 664–680. https://doi.org/10.1785/0120080079.

Frankel, A., Stephenson, W., and Carver, D. (2009). "Sedimentary basin effects in Seattle, Washington: Ground-motion observations and 3D simulations," *Bulletin of the Seismological Society of America*, Vol. 99, pp. 1579–1611. https://doi.org/10.1785/0120080203.

García, D., Wald, D.J., and Hearne, M.G. (2012). "A global earthquake discrimination scheme to optimize ground-motion prediction equation selection," *Bulletin of the Seismological Society of America*, Vol. 102, No. 1, pp. 185–203. https://doi.org/10.1785/0120110124.

Ghayamghamian, M.R. and Nouri, G.H. (2007). "On the characteristics of ground motion rotational components using Chiba dense array data," *Earthquake Engineering and Structural Dynamics*, Vol. 36, pp. 1407–1429. https://doi.org/10.1002/eqe.687.

Ghofrani, H., Atkinson, G.M., and Goda, K. (2013). "Implications of the 2011 M9.0 Tohoku Japan earthquake for the treatment of site effects in large earthquakes," *Bulletin of Earthquake Engineering*, Vol. 11, pp. 171–203. https://doi.org/10.1007/s10518-012-9413-4.

Gomberg, J., Pavlis, G., and Bodin, P. (1999). "The strain in the array is mainly in the plane (waves below ~ 1 Hz)," *Bulletin of the Seismological Society of America*, Vol. 89, pp. 1428–1438. https://doi.org/10.1785/bssa0890061428.

Goulet, C.A., Bozorgnia, Y., Kuehn, N., Al Atik, L., Youngs, R.R., Graves, R.W., and Atkinson, G.M. (2017). "NGA-East ground-motion models for the U.S. Geological Survey National Seismic Hazard Maps," *PEER Report 2017/03*, Pacific Earthquake Engineering Research Center, Berkeley, CA.

Goulet, C.A., Bozorgnia, Y., Kuehn, N., Al Atik, L., Youngs, R.R., Graves, R.W., and Atkinson, G.M. (2021a). "NGA-East ground-motion characterization model Part I: Summary of products and model development," *Earthquake Spectra*, Vol. 37, pp. 1231–1282. https://doi.org/10.1177/87552930211018723.

Goulet, C.A., Kishida, T., Ancheta, T.D., Cramer, C.H., Darragh, R.B., Silva, W.J., Hashash, Y.M.A., Harmon, J., Parker, G.A., Stewart, J.P., Youngs, R.R. (2021b). "PEER NGA-East database," *Earthquake Spectra*, Vol. 37, No. S1, pp. 1331–1353. https://doi.org/10.1177/87552930211015695.

Graves, R.W., Jordan, T.H., Callaghan, S., Deelman, E., Field, E.H., Juve, G., Kesselman, C., Maechling, P., Mehta, G., Okaya, D., Small, P., and Vahi, K. (2011). "CyberShake: A physics-based seismic hazard model for southern California," *Pure and Applied Geophysics*, Vol. 168, pp. 367–381. https://doi.org/10.1007/s00024-010-0161-6.

Graves, R.W. and Pitarka, A. (2010). "Broadband ground-motion simulation using a hybrid approach," *Bulletin of the Seismological Society of America*, Vol. 100, pp. 2095–2123. https://doi.org/10.1785/0120100057.

Graves, R.W. and Pitarka, A. (2015). "Refinements to the Graves and Pitarka (2010) broadband ground motion simulation method," *Seismological Research Letters*, Vol. 86, pp. 75–80. https://doi.org/10.1785/0220140101.

Green, R.A. and Terri, G.A. (2005). "Number of equivalent cycles concept for liquefaction evaluations-Revisited," *Journal of Geotechnical and Geoenvironmental Engineering,* Vol. 131, pp. 477–488. https://doi.org/10.1061/40797(172)31.

Gregor, N., Silva, W.J., and Darragh, R. (2002). Development of attenuation relations for peak particle velocity and displacement, *Report submitted to PG&E, CEC, Caltrans*. Available online at https://www.pacificengineering.org/rpts_page1.shtml (last accessed June 2012).

Gregor, N., Abrahamson, N.A., Atkinson, G.M., Boore, D.M., Bozorgnia, Y., Campbell, K.W., Chiou, B.S.-J., Idriss, I.M., Kamai, R., Seyhan, E., Silva, W., Stewart, J.P., and Youngs, R. (2014). "Comparison of NGA-West2 GMPEs," *Earthquake Spectra*, Vol. 30, pp. 1179–1197. https://doi.org/10.1193/070113eqs186m

Guatteri, M., Mai, P.M., and Beroza, G.C. (2004). "A pseudo-dynamic approximation to dynamic rupture models for strong ground motion prediction," *Bulletin of the Seismological Society of America*, Vol. 94, pp. 2051–2063. https://doi.org/10.1785/0120040037.

Hancock, J. and Bommer, J.J. (2005). "The effective number of cycles of earthquake ground motion," *Earthquake Engineering & Structural Dynamics*, Vol. 34, No. 6, pp. 637–664. https://doi.org/10.1002/eqe.437.

Hancock, J. and Bommer, J.J. (2007). "Using spectral matched records to explore the influence of strong-motion duration on inelastic structural response," *Soil Dynamics & Earthquake Engineering*, Vol. 27, No. 4, pp. 291–299. https://doi.org/10.1016/j.soildyn.2006.09.004.

Hanks, T.C. and Kanamori, H. (1979). "A moment magnitude scale," *Journal of Geophysical Research*, Vol. 84, pp. 2348–2350. https://doi.org/10.1029/jb084ib05p02348.

Hanks, T.C. and McGuire, R.K. (1981). "The character of high-frequency strong ground motion," *Bulletin of the Seismological Society of America*, Vol. 71, pp. 2071–2095. https://doi.org/10.1785/bssa0710062071.

Harichandran, R.S. and Vanmarcke, E.H. (1986). "Stochastic variation of earthquake ground motion in space and time," *Journal of the Engineering Mechanics Division*, Vol. 112, pp. 154–174. https://doi.org/10.1061/(asce)0733-9399(1986)112#: 2(154).

Harmon, J., Hashash, Y.M.A., Stewart, J.P., Rathje, E.M., Campbell, K.W., Silva, W.J., Xu, B., Musgrove, M., Ilhan, O. (2019). "Site amplification functions for central and eastern North America - Part I: Simulation dataset development," *Earthquake Spectra*, Vol. 35, pp. 787-814. https://doi.org/10.1193/091017EQS178M.

Hartzell, S. (1978). "Earthquake aftershocks as Green's functions," *Geophysical Research Letters*, Vol. 5, No. 1, pp. 1–14. https://doi.org/10.1029/gl005i001p00001.

Hartzell, S. (1985). "The use of small earthquake as Green's functions," In *Strong Motion Simulation and Earthquake Engineering Applications*, Publication 85-02, R. E. Scholl and J. L. King, ed., Earthquake Engineering Research Institute, El Centro, CA.

Hartzell, S., Harmsen, S., Frankel, A., and Larsen, S. (1999). "Calculation of broadband time histories of ground motion; Comparison of methods and validation using strong-ground motion from the 1994 Northridge earthquake," *Bulletin of the Seismological Society of America*, Vol. 89, pp. 1484–1504. https://doi.org/10.1785/bssa0890061484.

Haselton, C.B., Baker, J.W., Stewart, J.P., Whittaker, A.S., Luco, N., Fry, A., Hamburger, R.O., Zimmerman, R.B., Hooper, J.D., Charney, F.A., and Pekelnicky, R.G. (2017). "Response-history analysis for the design of new buildings in the NEHRP provisions and ASCE/SEI 7 standard: Part I - overview and specification of ground motions," *Earthquake Spectra*, Vol. 33, pp. 373–395. https://doi.org/10.1193/032114eqs039m.

Hashash, Y.M.A., Hook, J.J., Schmidt, B., and Yao, J.I-C. (2001). "Seismic design and analysis of underground structures," *Tunneling and Underground Space Technology*, Vol. 16, pp. 247–293. https://doi.org/10.1016/s0886-7798(01)00051-7.

Hashash, Y.M.A., Kottke, A.R., Stewart, J.P., Campbell, K.W., Kim, B., Moss, C., Nikolaou, S., Rathje, E.M., and Silva, W.J. (2014). "Reference rock site condition for central and eastern North America," *Bulletin of the Seismological Society of America*, Vol. 104, pp. 684–701. https://doi.org/10.1785/0120130132.

Havskov, J. and Alguacil, G. (2010). in B.L.N. Kennet, R. Magariaga, R. Marschall, and R. Wortel, eds., *Instrumentation in Earthquake Seismology*, Modern Approaches in Geophysics, Vol. 22, Springer. https://doi.org/10.1007/978-1-4020-2969-1_5.

Helmberger, D., Stead, R., Ho-Liu, P., and Dreger, D. (1992). "Broadband modeling of regional seismograms; Imperial Valley to Pasadena," *Geophysical Journal International*, Vol. 110, pp. 42–54. https://doi.org/10.1111/j.1365-246x.1992.tb00711.x.

Holzer, T.L. and Youd, T.L. (2007). "Liquefaction, ground oscillation, and soil deformation at the Wildlife Array, California," *Bulletin of the Seismological Society of America*, Vol. 97, No. 3, pp. 961–976. https://doi.org/10.1785/0120060156.

Hou, S. (1968). "Earthquake simulation models and their applications," *Report R68-17*, Department of Civil Engineering, Massachusetts Institute of Technology, Cambridge, Massachusetts.

Housner, G.W. (1982). Earthquake Engineering Research - 1982, Overview and Recommendations, National Research Council. https://nehrpsearch.nist.gov/static/files/NSF/PB83176024.pdf; https://doi.org/10.17226/19571.

Howard, J.K., Fraser, W.A., and Schultz, M.G. (2008). Probabilistic use of Arias intensity in geotechnical earthquake engineering, in D. Zeng, M.T. Manzari, and D.R. Hiltunen, eds., *Geotechnical Engineering and Soil Dynamics IV*, May 18–22, 2008, Sacramento, CA, ASCE Geotechnical Special Publication No. 181, 10 pages (electronic file). https://doi.org/10.1061/40975(318)6.

Hudson, D.E. (1979). *Reading and Interpreting Strong Motion Accelerograms*, Earthquake Engineering Research Institute, Berkeley, CA, 112 pp.

Husid, L.R. (1969). Características de terremotos, Análisis general, Revista del IDIEM 8, Santiago del Chile, pp. 21–42.

Hutchings, L. (1994). "Kinematic earthquake models and synthesized ground motion using empirical Green's functions," *Bulletin of the Seismological Society of America*, Vol. 84, pp. 1028–1050. https://doi.org/10.1785/BSSA0840041028.

Idriss, I.M. (2014). "An NGA-West2 empirical model for estimating the horizontal spectral values generated by shallow crustal earthquakes," *Earthquake Spectra*, Vol. 30, pp. 1155–1177. https://doi.org/10.1193/070613eqs195m.

Irikura, K. (1983). "Semi-empirical estimation of strong ground motions during large earthquakes," *Bulletin, Disaster Prevention Research Institute*, Kyoto University, Kyoto, Japan, Vol. 33, pp. 63–104.

Ji, C. and Archuleta, R.J. (2021). "Two empirical double-corner-frequency source spectra and their physical implications," *Bulletin of the Seismological Society of America*, Vol. 111, pp. 737–761. https://doi.org/10.5194/egusphere-egu21-13499.

Joyner, W.B. and Boore, D.M. (1981). "Peak horizontal acceleration and velocity from strong-motion records including records from the 1979 Imperial Valley, California earthquake," *Bulletin of the Seismological Society of America*, Vol. 71, pp. 2011–2038. https://doi.org/10.1785/bssa0710062011.

Joyner, W.B. and Boore, D.M. (1993). "Method for regression analysis of strong motion data," *Bulletin of the Seismological Society of America*, Vol. 83, pp. 469–487. https://doi.org/10.1785/bssa0830020469.

Joyner, W.B. and Boore, D.M. (1994). "Errata: Method for regression analysis of strong motion data," *Bulletin of the Seismological Society of America*, Vol. 84, pp. 955–956. https://doi.org/10.1785/bssa0840030955.

Kaka, S.I. and Atkinson, G.M. (2004). "Relationships between instrumental ground motion parameters and Modified Mercalli Intensity in eastern North America," *Bulletin of the Seismological Society of America*, Vol. 94, pp. 1728–1736. https://doi.org/10.1785/012003228.

Kaklamanos, J., Baise, L.G., and Boore, D.M. (2011). "Estimating unknown input parameters when implementing the NGA ground-motion prediction equations in engineering practice," *Earthquake Spectra*, Vol. 27, pp. 1219–1235. https://doi.org/10.1193/1.3650372.

Kalkan, E. and Kunnath, S.K. (2006). "Effects of fling-step and forward directivity on seismic response of buildings," *Earthquake Spectra*, Vol. 22, pp. 367–390. https://doi.org/10.1193/1.2192560.

Kamai, R., Abrahamson, N.A., and Graves, R.W. (2014a). "Adding fling effects to processed ground-motion time histories," *Bulletin of the Seismological Society of America*, Vol. 104, pp. 1914–1929. https://doi.org/10.1785/0120130272.

Kamai, R., Abrahamson, N.A., and Silva, W.J. (2014b). "Nonlinear horizontal site response for the NGA-West 2 Project," *Earthquake Spectra*, Vol. 30, pp. 1223–1240. https://doi.org/10.1193/070113eqs187m.

Katayama, T., Yamazaki, F., Nagata, S., Lu, L., and Turker, T. (1990). "A strong motion database for the Chiba seismometer array and its engineering analysis," *Structural Dynamics*, Vol. 19, No. 8, pp. 1089–1106. https://doi.org/10.1002/eqe.4290190802.

Kato, K., Takemura, T., Uchiyama, S., Iizuka, S., and Nigbor, R.L. (1998). "Borrego Valley downhole array in southern California: Instrumentation and preliminary site effect study," *Proceedings of 2nd International Symposium on the Effects of Surface Geology on Seismic Motion*, Yokohama, Japan.

Kayen, R.E. and Mitchell, J.K. (1997). "Assessment of liquefaction potential during earthquakes by Arias intensity," *Journal of Geotechnical and Geoenvironmental Engineering*, Vol. 123, pp. 1162–1174. https://doi.org/10.1061/(asce)1090-0241(1997)123#: 12(1162).

Kim, S. and Stewart, J.P. (2003). "Kinematic soil-structure interaction from strong motion recordings," *Journal of Geotechnical and Geoenvironmental Engineering*, Vol. 129, pp. 323–335. https://doi.org/10.1061/(asce)1090-0241(2003)129#: 4(323).

Kramer, S.L. and Mitchell, R.A. (2006). "Ground motion intensity measures for liquefaction hazard evaluation," *Earthquake Spectra*, Vol. 22, pp. 413–438. https://doi.org/10.1193/1.2194970.

Kuehn, N., Bozorgnia, Y., Campbell, K.W., and Gregor, N. (2023). "A regionalized partially nonergodic ground-motion model for subduction earthquakes using the NGA-Sub database," *Earthquake Spectra*, Vol. 39, No. 3, pp. 1625–1657. https://doi.org/10.1177/87552930231180906.

Kumar, M., Castro, J.M., Stafford, P.J., and Elghazouli, A.Y. (2011). "Influence of the mean period of ground motion on the inelastic dynamic response of single and multi-degree of freedom systems," *Earthquake Engineering & Structural Dynamics*, Vol. 40, pp. 237–256. https://doi.org/10.1002/eqe.1013.

Kurahashi, S. and Irikura, K. (2011). "Source model for generating strong ground motions during the 2011 off the Pacific coast of Tohoku Earthquake," *Earth Planets Space*, Vol. 63, pp. 571–576. https://doi.org/10.5047/eps.2011.06.044.

Kwak, D.Y., Stewart, J.P., Brandenberg, S.J., and Mikami, A. (2016). "Characterization of seismic levee fragility using field performance data," *Earthquake Spectra*, Vol. 32, pp. 193–215. https://doi.org/10.1193/030414eqs035m.

Lee, J. and Green, R.A. (2014). "An empirical significant duration relationship for stable continental regions," *Bulletin of Earthquake Engineering*, Vol. 12, pp. 217–235. https://doi.org/10.1007/s10518-013-9570-0.

Liu, A.H., Stewart, J.P., Abrahamson, N.A., and Moriwaki, Y. (2001). "Equivalent number of uniform stress cycles for soil liquefaction analysis," *Journal of Geotechnical and Geoenvironmental Engineering*, Vol. 127, pp. 1017–1026. https://doi.org/10.1061/(asce)1090-0241(2001)127#: 12(1017).

Liu, H. and Helmberger, D.V. (1985). "The 23#: 19 aftershock of the October 1979 Imperial Valley earthquake: More evidence for an asperity," *Bulletin of the Seismological Society of America*, Vol. 75, pp. 689–708. https://doi.org/10.1785/bssa0750030689.

Liu, S.C. (1970). "Evolutionary power spectral density of strong-motion earthquakes," *Bulletin of the Seismological Society of America*, Vol. 60, No. 3, pp. 891–900. https://doi.org/10.1785/bssa0600030891.

Loh, C.H. and Yeh, Y.T. (1988). "Spatial variation and stochastic modeling of seismic differential ground movement," *Earthquake Engineering and Structural Dynamics*, Vol. 16, pp. 583–596. https://doi.org/10.1002/eqe.4290160409.

Luco, J.E. and Apsel, R.J. (1983). "On the Green's functions for a layered half-space, I," *Bulletin of the Seismological Society of America*, Vol. 73, pp. 909–929. https://doi.org/10.1785/bssa0730040931.

Luco, N. and Cornell, C.A. (2007). "Structure-specific scalar intensity measures for near-source and ordinary earthquake ground motions," *Earthquake Spectra*, Vol. 23, pp. 357–392. https://doi.org/10.1193/1.2723158.

Macedo, J., Abrahamson, N.A., and Bray, J.D. (2019). "Arias intensity conditional scaling ground-motion models for subduction zones," *Bulletin of the Seismological Society of America*, Vol. 109, pp. 1343–1357.

Mackie, K. and Stojadinovic, B. (2004). "Improving probabilistic seismic demand models through refined intensity measures," *Proceedings of 13th World Conference on Earthquake Engineering*, Vancouver, Canada, Paper 1553.

Mai, P.M., Imperatori, W., and Olsen, K.B. (2010). "Hybrid broadband ground-motion simulations: Combining long-period deterministic synthetics with high-frequency multiple S-to-S backscattering," *Bulletin of the Seismological Society of America*, Vol. 100, pp. 2124–2142. https://doi.org/10.1785/0120080194.

Mazzoni, S., Kishida, T., Stewart, J.P., Contreras, V., Darragh, R.B., Ancheta, T.D., Chiou, B.S.J., Silva, W.J., and Bozorgnia, Y. (2022). "Relational database used for ground-motion model development in the NGA-sub project," *Earthquake Spectra*, Vol. 38, No. 2, pp. 1529–1548. https://doi.org/10.1177/87552930211055204.

McGuire, R.K., Silva, W.J., and Costantino, C.J. (2001). "Technical basis for revision of regulatory guidance on design ground motions: Hazard- and risk-consistent ground motion spectra guidelines," *Report No. NUREG/CR-6728*, US Nuclear Regulatory Commission, Washington, DC.

Mena, B., Mai, P.M., Olsen, K.B., Purvance, M.D., and Brune, J.N. (2010). "Hybrid broadband ground-motion simulation using scattering green's functions: Application to large-magnitude events," *Bulletin of the Seismological Society of America*, Vol. 100, pp. 2143–2162. https://doi.org/10.1785/0120080318.

Milner, K.R., Shaw, B.E., Goulet, C.A., Richards-Dinger, K.B., Callaghan, S., Jordan, T.H., Dieterich, J.H., and Field, E.H. (2021). "Toward physics-based nonergodic PSHA: A prototype fully deterministic seismic hazard model for southern California," *Bulletin of the Seismological Society of America*, Vol. 111, pp. 898–915. https://doi.org/10.1785/0120200216.

Miyakoshi, J., Hayashi, Y., Tamura, K., and Fukuwa, N. (1998). "Damage ratio functions for buildings using damage data of the Hyogo-ken Nanbu earthquake," *Proceedings of 7th International Conference on Structural Safety and Reliability [ICOSSARJ97]*, Kyoto, pp. 349–362.

Mori, J., Filson, J., Cranswick, E., Borcherdt, R.B., Amirbekian, R., Aharonian, V., and Hachverdian, L. (1994). "Measurements of P and S wave fronts from the dense three-dimensional array at Garni, Armenia," *Bulletin of the Seismological Society of America*, Vol. 84, pp. 1089–1096.

Moschetti, M.P., Thompson, E.M., and Withers, K. (2024). "Basin effects from 3D simulated ground motions in the Greater Los Angeles region for use in seismic hazard analyses," *Earthquake Spectra*, Vol. 40, pp 1042-1065, https://doi.org/10.1177/87552930241232372.

Motazedian, D. and Atkinson, G.M. (2005). "Stochastic finite-fault modeling based on a dynamic corner frequency," *Bulletin of the Seismological Society of America*, Vol. 95, pp. 995–1010. https://doi.org/10.1785/0120030207.

NIST (2014). "Nonlinear analysis research and development program for performance-based engineering," *Report No. NIST GCR 14-917-27*, National Institute of Standards and Technology, U.S. Department of Commerce, Washington D.C. Project Technical Committee: Deierlein, G. (Chair), P. Behnam, F. Charney, L. Lowes, J. P. Stewart, and M. Wilford.

Nweke, C.C., Stewart, J.P., Buckreis, T.E., Graves, R.W., Goulet, C.A., Brandenberg, S.J. (2022). "Validating predicted site response in sedimentary basins from 3D ground motion simulations," *Earthquake Spectra*, Vol. 38, pp. 2135-2161. https://doi.org/10.1177/87552930211073159.

Oliveira, C.S., Hao, H., and Penzien, J. (1991). "Ground motion modeling for multiple-input structural analysis," *Structural Safety*, Vol. 10, pp. 79–93. https://doi.org/10.1016/0167-4730(91)90007-v.

Olson, A.H., Orcutt, J.A., and Frazier, G.A. (1984). "The discrete wavenumber/finite element method for synthetic seismograms," *Geophysical Journal International*, Vol. 77, pp. 421–460. https://doi.org/10.1111/j.1365-246x.1984.tb01942.x.

Olsen, K.B. and Mayhew, J.E. (2010). "Goodness-of-fit criteria for broadband synthetic seismograms, with application to the 2008 Mw 5.4 Chino Hills, California, earthquake," *Seismological Research Letters*, Vol. 81, pp. 715–723. https://doi.org/10.1785/gssrl.81.5.715.

Olsen, K.B. and Takedatsu, R. (2015). "The SDSU broadband ground motion generation module BBtoolbox Version 1.5," *Seismological Research Letters*, Vol. 86, pp. 81–88. https://doi.org/10.1785/0220140102.

O'Rourke, M.J. and Liu, X. (1999). *Response of Buried Pipelines Subject to Earthquake Effects*, Multidisciplinary Center for Earthquake Engineering Research, University of Buffalo, New York. ISBN 0-9656682-3-1.

Owens, T.J., Anderson, P.N., and McNamara, D.E. (1991). "The 1990 Pinyon Flat high frequency array experiment, An IRIS Eurasian seismic studies program passive source experiment," *PASSCAL Data Report No. 91-002.*

Pan, Y., Ventura, C.E., and Liam Finn, W.D. (2018). "Effects of ground motion duration on the seismic performance and collapse rate of light-frame wood houses," *Journal of Structural Engineering*, Vol. 144, 04018112. https://doi.org/10.1061/(asce)st.1943-541x.0002104.

Paolucci, R., Rovelli, A., Faccioli, E., Cauzzi, C., Finazzi, D., Vanini, M., Di Alessandro, C., and Calderoni, G. (2008). "On the reliability of long-period response spectral ordinates from digital accelerograms," *Earthquake Engineering and Structural Dynamics*, Vol. 37, pp. 697–710. https://doi.org/10.1002/eqe.781.

Paolucci, R. and Smerzini, C. (2008). "Earthquake-induced transient ground strains from dense seismic networks," *Earthquake Spectra*, Vol. 24, pp. 453–470. https://doi.org/10.1193/1.2923923.

Papazachos, B. and Papazachou, C. (2003). *The Earthquakes of Greece*, 3rd edition, Ziti Publishing Co., Thessaloniki Greece, 286 pp. (in Greek).

Parker, G.A., Stewart, J.P., Boore, D.M., Atkinson, G.M., and Hassani, B. (2022). "NGA-subduction global ground motion models with regional adjustment factors," *Earthquake Spectra*, Vol. 38, pp. 456–493. https://doi.org/10.1177/87552930211034889.

PEER Center (2017). "Guidelines for performance-based seismic design of tall buildings, Version 2," *Report No. 2017/06*, Pacific Earthquake Engineering Research Center, UC Berkeley.

Pezeshk, S., Zandieh, A., Campbell, K.W., and Tavakoli, B. (2018). "Ground-motion prediction equations for central and eastern North America using the hybrid empirical method and NGA-West2 empirical ground-motion models," *Bulletin of the Seismological Society of America*, Vol. 108, pp. 2278–2304. https://doi.org/10.1785/0120170179.

Pitilakis, K., Roumelioti, Z., Manakou, M., Raptakis, D., Liakakis, K., Anastasiadis, A., and Pitilakis, D. (2013). "The web portal of the EUROSEISTEST strong ground motion database," *Bulletin of the Geological Society of Greece*, Vol. 47, No. 3, pp. 1221–1230. https://doi.org/10.12681/bgsg.10978.

Porter, K., Jones, L., Cox, D., Goltz, J., Hudnut, K., Mileti, D., Perry, S., Ponti, D., Reichle, M., Rose, A.Z., Scawthorn, C.R., Seligson, H.A., Shoaf, K.I., Treiman, J., and Wein, A. (2011). "The ShakeOut scenario: A hypothetical Mw7.8 earthquake on the southern San Andreas fault," *Earthquake Spectra*, Vol. 27, pp. 239–261. https://doi.org/10.1193/1.3563624.

Priestley, M.B. (1965). "Evolutionary spectra and non-stationary processes," *Journal of the Royal Statistical Society,* Series B, Vol. 27, pp. 204–237. https://doi.org/10.1111/j.2517-6161.1965.tb01488.x.

Priestley, M.B. (1967). Power spectral analysis of non-stationary random processes," *Journal of Sound and Vibration*, Vol. 6, No. 1, pp. 86–97. https://doi.org/10.1016/0022-460x(67)90160-5.

Raghunandan, M. and Liel, A.B. (2013). "Effect of ground motion duration on earthquake-induced structural collapse," *Structural Safety*, Vol. 41, pp. 119–133. https://doi.org/10.1016/j.strusafe.2012.12.002.

Rathje, E.M. and Saygili, G. (2009). "Probabilistic assessment of earthquake-induced sliding displacements of natural slopes," *Bulletin of the New Zealand Society for Earthquake Engineering,* Vol. 42, No. 1, pp. 18–27. https://doi.org/10.5459/bnzsee.42.1.18-27

Rathje, E.M. and Antonakos, G. (2011). "A unified model for predicting earthquake-induced sliding displacements of rigid and flexible slopes," *Engineering Geology*, Vol. 22, pp. 51–60. https://doi.org/10.1016/j.enggeo.2010.12.004

Rathje, E.M., Faraj, F., Russell, S., and Bray, J.D. (2004). "Empirical relationships for frequency content parameters of earthquake ground motions," *Earthquake Spectra*, Vol. 20, pp. 119–144. https://doi.org/10.1193/1.1643356.

Rauch, A.F. and Martin, J.R. (2000). "EPOLLS model for predicting average displacements on lateral spreads," *Journal of Geotechnical Engineering*, Vol. 126, pp. 360–371. https://doi.org/10.1061/(asce)1090-0241(2000)126#: 4(360).

Richart, F.E., Hall, J.R., and Woods, R.D. (1970). *Vibrations of Soils and Foundations*, Prentice Hall, Englewood Cliffs, NJ, 401 pp.

Riddell, R. (2007). "On ground motion intensity indices," *Earthquake Spectra*, Vol. 23, pp. 147–173. https://doi.org/10.1193/1.2424748.

Saragoni, G.R. and Hart, G.C. (1973). "Simulation of artificial earthquakes," *Earthquake Engineering and Structural Dynamics*, Vol. 2, No. 2, pp. 249–267. https://doi.org/10.1002/eqe.4290020305.

Sato, H. and Fehler, M.C. (1998). *Seismic Wave Propagation and Scattering in the Heterogeneous Earth*, Springer-Verlag New York, Inc., New York.

Schmedes, J., Archuleta, R.J., and Lavallée, D. (2010). "Correlation of earthquake source parameters inferred from dynamic rupture simulations," *Journal of Geophysical Research*, Vol. 115. https://doi.org/10.1029/2009jb006689.

Schnabel, P.B. and Seed, H.B. (1973). "Accelerations for rock in earthquakes in the Western United States," *Bulletin of the Seismological Society of America*, Vol. 63, No. 2, pp. 501–516. https://doi.org/10.1785/BSSA0630020501.

Seed, H.B. and Idriss, I.M. (1969). "Rock motion accelerograms for high magnitude earthquakes," *Report No. UCB/EERC 69-7*, University of California, Berkeley, 8 pages.

Seed, H. B., Idriss, I.M., and Keifer, F.W. (1969). "Characteristics of rock motions during earthquakes," *Journal of Soil Mechanics and Foundations Division*, Vol. 95, No. 5, pp. 1199–1218. https://doi.org/10.1061/jsfeaq.0001325.

Seed, H.B., Idriss, I.M., Makdisi, F., and Banerjee, N. (1975). "Representation of irregular stress time histories by equivalent uniform stress series in liquefaction analyses," *Report No. UCB/EERC 75-29*, Earthquake Engineering Research Center, UC Berkeley.

Seed, H.B. and Lee, K.L. (1966). "Liquefaction of saturated sands during cyclic loading," *Journal of the Soil Mechanics and Foundations Division*, Vol. 92, No. SM6, pp. 105–134. https://doi.org/10.1061/jsfeaq.0000913.

Seyhan, E. and Stewart, J.P. (2014). "Semi-empirical nonlinear site amplification from NGA-West 2 data and simulations," *Earthquake Spectra*, Vol. 30, pp. 1241–1256. https://doi.org/10.1193/063013eqs181m.

Seyhan, E., Stewart, J.P., and Graves, R.W. (2013). "Calibration of a semi-stochastic procedure for simulating high-frequency ground motions," *Earthquake Spectra*, Vol. 29, pp. 1495–1519. https://doi.org/10.1193/122211eqs312m.

Shahi, S.K. and Baker, J.W. (2011). "An empirically calibrated framework for including the effects of near-fault directivity in probabilistic seismic hazard analysis," *Bulletin of the Seismological Society of America*, Vol. 101, pp. 742–755. https://doi.org/10.1785/0120100090.

Shahi, S.K. and Baker, J.W. (2014a). "An efficient algorithm to identify strong velocity pulses in multi-component ground motions," *Bulletin of the Seismological Society of America*, Vol. 104, pp. 2456–2466. https://doi.org/10.1785/0120130191.

Shahi, S.K. and Baker, J.W. (2014b). "NGA-West2 models for ground-motion directionality," *Earthquake Spectra*, Vol. 30, pp. 1285–1300. https://doi.org/10.1193/040913eqs097m.

Shinozuka, M. (1973). "Digital simulation of strong ground accelerations," *Proceedings, 5th World Conference on Earthquake Engineering*, Vol. 2, pp. 2829–2838, Rome.

Silver, W.J. and Seed, H.B. (1971). "Volume changes in sands during cyclic loading," *Journal of the Soil Mechanics and Foundations Division*, Vol. 97, No. SM9, pp. 1171–1182. https://doi.org/10.1061/jsfeaq.0001658.

Smith, S.W., Ehrenberg, J.E., and E.N. Hernandez (1982). "Analysis of the El Centro differential array for the 1979 Imperial Valley Earthquake," *Bulletin of the Seismological Society of America*, Vol. 72, pp. 237–258. https://doi.org/10.1785/bssa0720010237.

Somerville, P.G., Irikura, K., Graves, R.W., Sawada, S., Wald, D., Abrahamson, N.A., Iwasaki, Y., Kagawa, T., Smith, N., and Kowada, A. (1999). "Characterizing crustal earthquake slip models for the prediction of strong ground motion," *Seismological Research Letters*, Vol. 70, pp. 59–80. https://doi.org/10.1785/gssrl.70.1.59.

Somerville, P.G., Smith, N.F., Graves, R.W., and Abrahamson, N.A. (1997). "Modification of empirical strong ground motion attenuation relations to include the amplitude and duration effects of rupture directivity," *Seismological Research Letters*, Vol. 68, pp. 199–222. https://doi.org/10.1785/gssrl.68.1.199.

Spudich, P. and Archuleta, R.J. (1987). Techniques for earthquake ground-motion calculation with applications to source parameterization of finite faults, in B.A. Bolt, ed., *Seismic Strong Motion Synthetics*, Academic Press. https://doi.org/10.1016/b978-0-12-112251-5.50009-1.

Spudich, P., Rowshandel, B., Shahi, S.K., and Baker, J.W. (2014). "Overview and comparison of the NGA-West2 directivity models," *Earthquake Spectra*, Vol. 30, pp. 1199–1221. https://doi.org/10.1193/080313eqs222m.

St. John, C.M. and Zahrah, T.F. (1987). "Aseismic design of underground structures, *Tunneling and Underground Space Technology*, Vol. 2, 165–197. https://doi.org/10.1016/0886-7798(87)90011-3.

Star, L.M., Stewart, J.P., and Graves, R.W. (2011). "Comparison of ground motions from hybrid simulations to NGA prediction equations," *Earthquake Spectra*, Vol. 27, pp. 333–350. https://doi.org/10.1193/1.3583644.

Stewart, J.P., Chiou, B.S.-J., Bray, J.D., Somerville, P.G., Graves, R.W., and Abrahamson, N.A. (2001). "Ground motion evaluation procedures for performance based design," *Report No. PEER-2001/09*, Pacific Earthquake Engineering Research Center, University of California, Berkeley, 229 pgs.

Stewart, J.P., Midorikawa, S., Graves, R.W., Khodaverdi, K., Miura, H., Bozorgnia, Y., and Campbell, K.W. (2013). Implications of Mw 9.0 Tohoku-oki Japan earthquake for ground motion scaling with source, path, and site parameters, *Earthquake Spectra*, Vol. 29, pp. S1–21. https://doi.org/10.1193/1.4000115.

Stewart, J.P., Parker, G.A., Al Atik, L., Atkinson, G.M., and Goulet, C.A. (2018). "Site-to-site standard deviation model for Central and Eastern North America," *UCLA E-Scholarship Web Report*.

Stewart, J.P., Parker, G.A., Atkinson, G.M., Boore, D.M., Hashash, Y.M.A., and Silva, W.J. (2020). "Ergodic site amplification model for central and eastern North America," *Earthquake Spectra*, Vol. 36, pp. 42–68. https://doi.org/10.1177/8755293019878185

Strasser, F.O., Bommer, J.J., and Abrahamson, N.A. (2008). "Truncation of the distribution of ground-motion residuals," *Journal of Seismol*ogy, Vol. 12, pp. 79–105. https://doi.org/10.1007/s10950-007-9073-z.

Thompson, E.M., Baise, L.M., Tanaka, Y., and Kayen, R.E. (2012). "A taxonomy of site response complexity," *Soil Dynamics and Earthquake Engineering*, Vol. 41, pp. 32–43. https://doi.org/10.1016/j.soildyn.2012.04.005.

Tolga Sen, A., Cesca, S., Lange, D., Dahm, T., Tilmann, F., and Heimann, S. (2015). "Systematic changes of earthquake rupture with depth: A case study from the 2010 Mw 8.8 Maule, Chile, earthquake aftershock sequence," *Bulletin of the Seismological Society of America*, Vol. 105, pp. 2468–2479. https://doi.org/10.1785/0120140123.

Trifunac, M.D. and Brady, A.G. (1975). "On the correlation of seismic intensity with peaks of recorded strong ground motion," *Bulletin of the Seismological Society of America*, Vol. 65, pp. 139–162. https://doi.org/10.1785/BSSA0650010139.

Trifunac, M.D., Lee, V.W., and Todorovska, M.I. (1999). "Common problems in automatic digitization of strong motion accelerograms," *Soil Dynamics and Earthquake Engineering*, Vol. 18, pp. 519–530. https://doi.org/10.1016/s0267-7261(99)00018-4.

Trifunac, M.D. and Todorovska, M.I. (1996). "Nonlinear soil response – 1994 Northridge, California earthquake," *Journal of Geotechnical and Geoenvironmental Engineering,* Vol. 122, No. 9, pp. 725–735. https://doi.org/10.1061/(asce)0733-9410(1996)122#: 9(725).

Trifunac, M.D. and Todorovska, M.I. (1997). "Northridge, California, earthquake of 1994: Density of red-tagged buildings versus peak horizontal velocity and intensity of shaking," *Soil Dynamics and Earthquake Engineering*, Vol. 16, 209–222. https://doi.org/10.1016/s0267-7261(96)00043-7.

Tu, T., Yu, H., Ramirez-Guzman, L., Bielak, J., Ghattas, O., Liu Ma, K., and O'Hallaron, D.R. (2006). From mesh generation to scientific visualization: An end-to-end approach to parallel supercomputing," *Proceedings of the 2006 ACM/IEEE conference on Supercomputing*, Article No. 91, Doi 10.1145/1188455.1188551. https://doi.org/10.1109/sc.2006.32.

U.S. Nuclear Regulatory Commission (USNRC) (1997). *Pre-Earthquake Planning and Immediate Nuclear Power Plant Operator Postearthquake Actions*, Regulatory Guide 1.166, Washington, D.C., 8 pp.

Vanmarcke, E.H. (1976). Chapter 8: Structural response to earthquakes, in C. Lomnitz and E. Rosenblueth, eds., *Seismic Risk and Engineering Decisions*, Elsevier, Amsterdam, pp. 287–338. https://doi.org/10.1016/b978-0-444-41494-6.50011-4.

Wald, D.J., Quitoriano, V., and Heaton, T.H. (1999). "Relationships between peak ground acceleration, peak ground velocity, and Modified Mercali Intensity in California," *Earthquake Spectra*, Vol. 15, pp. 557–664. https://doi.org/10.1193/1.1586058.

Walling, M., Silva, W.J., and Abrahamson, N.A. (2008). "Nonlinear site amplification factors for constraining the NGA models," *Earthquake Spectra*, Vol. 24, pp. 243–255. https://doi.org/10.1193/1.2934350.

Wang, F. and Jordan, T.H. (2014). "Comparison of probabilistic seismic-hazard models using averaging-based factorization," *Bulletin of the Seismological Society of America*, Vol. 104, pp. 1230–1257. https://doi.org/10.1785/0120130263.

Wang, G., Wang, Y., Zhou, W., and Zhou, C. (2015). "Integrated duration effects on seismic performance of concrete gravity dams using linear and nonlinear evaluation methods," *Soil Dynamics and Earthquake Engineering*, Vol. 79, pp. 223–236. https://doi.org/10.1016/j.soildyn.2015.09.020.

Watson-Lamprey, J.A. (2018). "Capturing directivity effects in the mean and aleatory variability of the NGA-West 2 ground motion prediction equations," *Report No. 2018/04*, Pacific Earthquake Engineering Research Center, Berkeley, CA. https://doi.org/10.55461/usop6050.

Wielandt, E. (2002). Chapter 5: Seismic sensors and their calibration, in P. Bormann, ed., *New Manual of Seismological Observatory Practice (NMSOP)*, GeoForschungsZentrum Potsdam, Germany, pp. 1–46.

Worden, C.B., Gerstenberger, M.C., Rhoades, D.A., and Wald, D.J. (2012). "Probabilistic relationships between ground-motion parameters and modified Mercalli intensity in California," *Bulletin of the Seismological Society of America*, Vol. 102, pp. 204–221. https://doi.org/10.1785/0120110156.

Yang, C.Y. (1986). *Random Vibration of Structures*, John Wiley and Sons, New York, 295 pp.

Yenier, E. and Atkinson, G.M. (2015). "Regionally adjustable generic ground-motion prediction equation based on equivalent point-source simulations: Application to central and eastern North America," *Bulletin of the Seismological Society of America*, Vol. 105, pp. 1989–2009. https://doi.org/10.1785/0120140332.

Yokota, Y., Koketsu, K., Fujii, Y., Satake, K., Sakai, S., Shinohara, M., and Kanazawa, T. (2011). "Joint inversion of strong motion, teleseismic, geodetic, and tsunami datasets for the rupture process of the 2011 Tohoku earthquake," *Geophysical Research Letters*, Vol. 38, p. L00G21. https://doi.org/10.1029/2011gl050098.

Zalachoris, G., and Rathje, E. M. (2015). "Evaluation of one-dimensional site response techniques using borehole arrays," *Journal of Geotechnical and Geoenvironmental Engineering*, Vol. 141, 04015053. https://doi.org/10.1061/(ASCE)GT.1943-5606.0001366

Zeghal, M. and Elgamal, A.W. (1994). "Analysis of site liquefaction using earthquake records," *Journal of Geotechnical and Geoenvironmental Engineering*, Vol. 120, No. 6, pp. 996–1017. https://doi.org/10.1061/(asce)0733-9410(1994)120#: 6(996).

Zerva, A. (2009). *Spatial Variation of Seismic Ground Motion: Modeling and Engineering Applications*. CRC Press, Taylor & Francis Group, Boca Raton, FL.

Zerva, A. and Zervas, V. (2002). "Spatial variation of seismic ground motions: An overview," *Applied Mechanics Reviews*, Vol. 55, pp. 271–297. https://doi.org/10.1115/1.1458013.

Zerva, A. and Zhang, O. (1996). "Estimation of signal characteristics in seismic ground motions," *Probabilistic Engineering Mechanics*, Vol. 11, pp. 229–242. https://doi.org/10.1016/0266-8920(96)00018-5.

Zhao, J.X., Zhang, J., Asano, A., Ohno, Y., Oouchi, T., Takahashi, T., Ogawa, H., Irikura, K., Thio, H.K., Somerville, P.G., and Fukushima, Y. (2006). "Attenuation relations of strong ground motion in Japan using site classification based on predominant period," *Bulletin of the Seismological Society of America*, Vol. 96, pp. 898–913. https://doi.org/10.1785/0120050122.

4 Seismic Hazard Analysis

4.1 INTRODUCTION

In many areas of the world, the threat to human activities from earthquakes is sufficient to require careful consideration in the design of structures and facilities. The goal of *earthquake-resistant design* is to produce a structure or facility that can withstand a certain level of shaking without excessive damage. That level of shaking is described by a *design ground motion*, which can be characterized by *design ground motion intensity measures* (*IMs*). The specification of design ground motion *IMs* is one of the most difficult and most important problems in engineering seismology and geotechnical earthquake engineering.

Much of the difficulty in specifying design ground motions comes from its unavoidable reliance on subjective decisions that must be made with incomplete or uncertain information. These decisions largely revolve around the uncertainty in the size, time, and location of future earthquakes, quantification of the important characteristics of the resulting ground motions, and definition of the boundary between acceptable and unacceptable performance of new and existing structures. If very little damage is acceptable, a relatively strong level of shaking must be designed for, and the measures required to resist that shaking can be quite expensive. If greater levels of damage are tolerable, lower design levels of shaking may be considered and the resulting design will be less expensive. Obviously, there are trade-offs between the short-term cost of providing an earthquake-resistant design and the potential long-term cost (which, for many structures, may never be realized) of earthquake-induced damage.

Seismic hazard analysis is the first step in a comprehensive seismic performance evaluation. Seismic hazard analyses involve the quantitative estimation of ground-shaking hazards at a particular site. Seismic hazards may be analyzed deterministically, as when a particular earthquake scenario is assumed, or probabilistically, in which uncertainties in earthquake size, location, and time of occurrence as well as ground motion uncertainties are explicitly considered. Other resource texts for seismic hazard analysis are McGuire (2004) and Baker et al. (2021).

4.2 EARTHQUAKE SOURCES

To evaluate seismic hazards and performance for a particular site or region, possible sources of seismic activity must be identified and their potential for generating future strong ground motion evaluated. Earthquakes occur on faults, but not all earthquakes occur on previously known or easily identified faults.

4.2.1 Identification of Earthquake Sources

Identification of seismic sources often requires detective work; nature's clues, some of which are obvious and others obscure, must be observed and interpreted. The availability of modern seismographs, GPS, and Interferometric Synthetic Aperture Radar (InSAR) networks have made observation and interpretation of current earthquakes much more convenient than in the past. The occurrence of a large earthquake is now recorded by hundreds of seismographs around the world. Within minutes, seismologists are able to estimate its magnitude, locate its rupture surface, and even evaluate fault slip distributions and other source characteristics. It is now impossible for a significant earthquake anywhere in the world to go undetected.

The current ability to identify and locate faults as earthquake sources is a relatively recent development, particularly when compared with the time scales on which large earthquakes usually occur.

DOI: 10.1201/9781003512011-4

The fact that no strong motions have been instrumentally recorded in a particular area does not mean that they have not occurred in the past or that they will not occur in the future. In the absence of an instrumental seismic record, other clues of earthquake activity must be uncovered. These may take the form of geologic and tectonic evidence, or historical (preinstrumental) seismicity.

4.2.1.1 Geologic Evidence

The theory of plate tectonics assures us that the occurrence of earthquakes is written in the geologic record, primarily in the form of offsets, or relative displacements, of various strata. The study of the geologic record of past earthquake activity is called *paleoseismology* (Wallace, 1981). In some parts of the world, this geologic record is accessible and relatively easily interpreted by trained seismic geologists. In other locations, however, the geologic record may be very complex or it may be hidden by thick layers of recent sediments that have not been displaced by seismic activity or by dense surface vegetation. The identification of seismic sources from geologic evidence is a vital, though often difficult, part of a seismic hazard analysis. The search for geologic evidence of earthquake sources centers on the identification of faults and is generally practical in active tectonic regions with high rates of seismicity and faults that extend to the surface. A variety of tools and techniques are available to the geologist, including a review of published literature; interpretation of air photos and remote sensing (e.g., infrared photograph) imagery; field investigations including logging of trenches (Figure 4.1), test pits and borings; and geophysical techniques. Criteria for the identification of faults are described in numerous reports and textbooks on structural geology, field geology, and geomorphology (Adair, 1979; Sieh, 1981; Yeats et al., 1997; CGS, 2018). Observable features that suggest faulting include (Reiter, 1990):

1. Directly observable fracture surfaces and indicators of fracturing. These include disruption of the ground surface and evidence of the movement and grinding of the two sides of the fault (slickensides, fault gouge, and fault breccia).
2. Geologically mappable indicators. These include the juxtaposition of dissimilar materials, missing or repeated strata and the truncation of strata or structures.
3. Topographic and geomorphic (surface landform) indicators (Figure 4.2). These include topographic scarps or triangular facets on ridges, offset streams or drainage, tilting or changes in elevation of terraces or shorelines, sag ponds (water ponded by depressions near strike-slip faults) and anomalous stream gradients. In heavily vegetated areas, the ability of Light Detection and Ranging (LIDAR) equipment to "see through the trees" (Figure 4.3) can reveal topographic features that would otherwise be hidden.

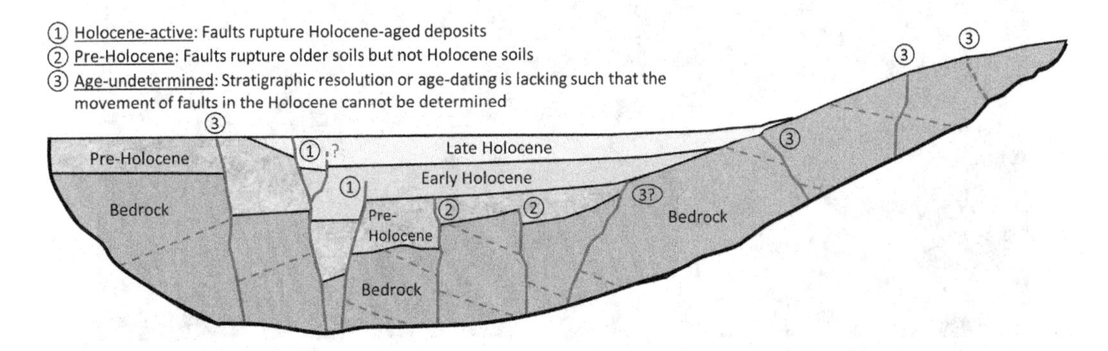

FIGURE 4.1 Fault classifications in a hypothetical trench log where Holocene-active faults break Holocene-age deposits and pre-Holocene faults break pre-Holocene age deposits, but not Holocene age deposits. The ages of movement for age-undetermined faults are unconstrained due to a lack of overlying deposits to determine the timing of the most recent fault displacement. (Modified from CGS, 2018; used by permission of California Geological Society.)

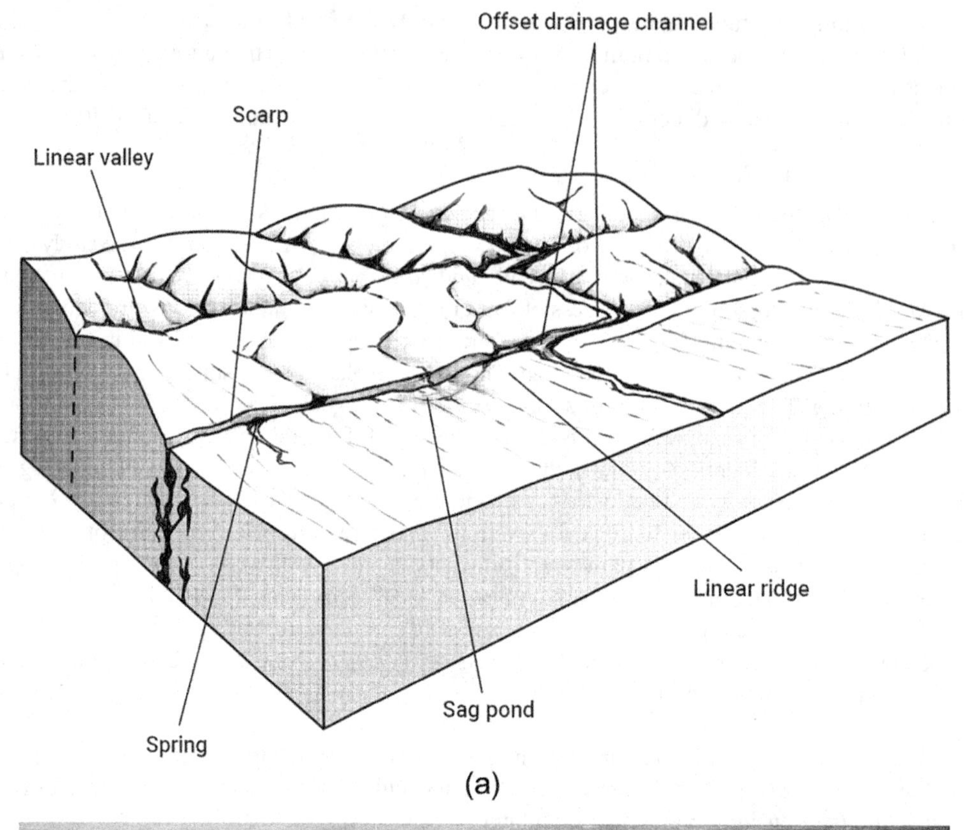

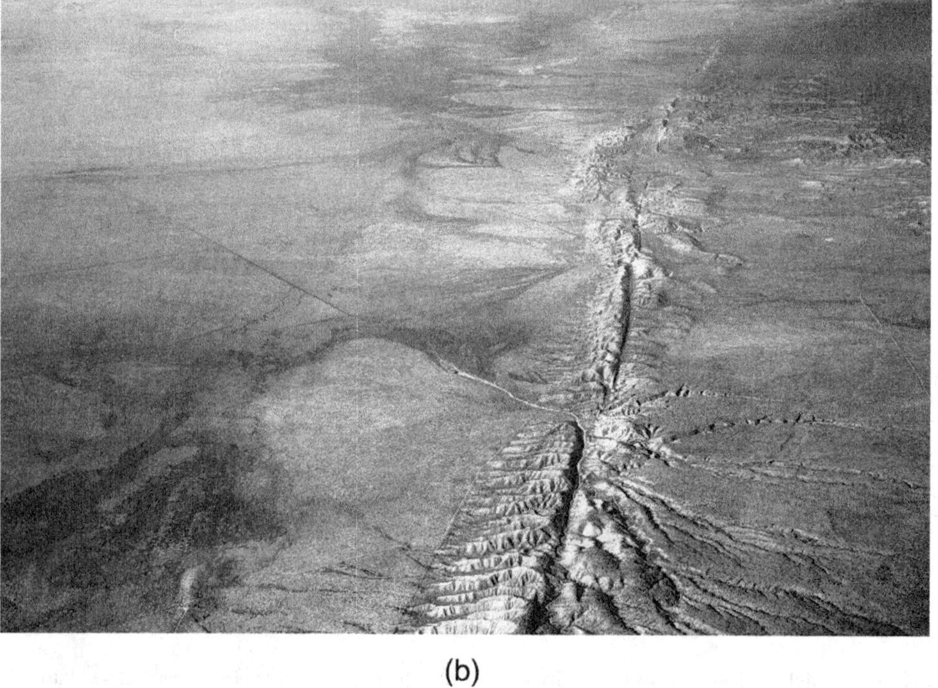

FIGURE 4.2 Typical terrain in the vicinity of a fault showing (a) topographic and geomorphic indicators of faulting (Modified from Wesson et al., 1975) and (b) an aerial view of such terrain along the San Andreas fault in the Carrizo Plain. (Photo from Wikimedia Commons, 2007.)

FIGURE 4.3 LIDAR image of surface rupture on normal fault from 2009 L'Aquila, Italy earthquake. The area shown in the image is heavily vegetated. The surface rupture was not visually apparent to reconnaissance teams. (Image produced by Ralph Haugerud, 2009. Lanzo et al., 2010; used with permission of Argo-E Group, P.C.)

4. Secondary geologic features. These include abrupt changes in groundwater levels, gradients, and chemical composition, alignment of springs or volcanic vents, and the presence of hot springs.
5. Lineaments on remote sensing imagery. These may be caused by topography, vegetation, or tonal contrasts.
6. Geophysical indicators of subsurface faulting. These include steep linear gravity or magnetic gradients, differences in seismic wave velocities, and offset of seismic reflection horizons.
7. Geodetic indicators. These include fault movement appearing in geodetic surveys as tilting and changes in the distance between fixed points.

4.2.1.2 Tectonic Evidence

Plate tectonics and elastic rebound theory (Sections 2.3 and 2.5) tell us that earthquakes occur to relieve the strain energy that accumulates in the Earth's crust as plates slowly move relative to each other. These relative plate movements are concentrated at plate boundaries, accordingly major fault sources are expected to be located near plate boundaries (Section 2.4). Moreover, the rate of relative plate movement (i.e., slip rate) should be related to the rate of strain energy accumulation and also to the rate of strain energy release (e.g., Smith, 1976; Idriss, 1985; Kreemer et al., 2014). Because strain energy is often released in the form of earthquakes, the frequency and size of earthquakes should be related to relative plate movement.

One such relationship was presented by Kanamori (1986), who related maximum magnitude at subduction plate boundaries to both the rate of convergence and the age of the subducted slab:

$$\mathbf{M} = -0.0093T + 0.143V + 8.01 \tag{4.1}$$

where T is the age in millions of years and V is the rate of convergence in cm/year. This relationship indicates that maximum magnitude increases with increasing convergence rate and decreasing age.

However, modern source characterization does not use convergence rates (or slip rates) to directly estimate earthquake magnitudes for shallow crustal earthquakes as in Equation (4.1). Instead, measured slip rates are related to moment rates, which in turn can give an indication of how often earthquakes following a certain magnitude distribution must occur to account for the measured slip (Section 4.2.2.5).

4.2.1.3 Historical Seismicity

Earthquake sources may also be identified from records of historical (preinstrumental) seismicity. The written historical record extends back only a few hundred years or less in the United States. However, in Japan, southern Europe, and the Middle East, it may extend about 2,000 years and up to 3,000 years or so in China (Ambraseys, 1971, 1978, 2009; Allen, 1975; Usami, 1975; Bolt, 1988; Papazachos and Papazachou, 2003). Historical accounts of ground-shaking effects can be used to identify past earthquakes and to estimate their geographic distributions of intensity. When sufficient data are available, the maximum intensity can be used to estimate the location of the earthquake epicenter. Similarly, the maximum intensity or the area within isoseismal contours can be used to estimate earthquake magnitude (Section 2.8.3). Although the accuracy of locations and magnitudes determined in this way depends strongly on population density and the quality of historical documentation, a geographic pattern of historic epicenters provides strong evidence for the existence of earthquake sources. Since historical records are dated, they can also be used to evaluate the rate of recurrence of mid- to large-magnitude earthquakes, or seismicity, in particular areas.

4.2.1.4 Instrumental Seismicity

On average, about ten earthquakes of magnitude greater than seven occur somewhere in the world each year (Kanamori, 1988; Ekström et al., 2012; Engdahl et al., 2020). Instrumental records from large earthquakes have been available since about 1900, although many from before 1960 are incomplete or of uneven quality. Nevertheless, instrumental recordings represent the best available information for the identification and evaluation of specific earthquake sources (such as faults) and *source zones* (i.e., areas in plan, or volumes, where earthquakes beyond a particular minimum magnitude have been observed to occur, regardless of their association with known faults). For example, Figure 4.4 shows hypocenter locations in the central and eastern US, which are heavily relied upon in the identification of source zones in this region. In relatively active regions, the alignment of instrumentally located epicenters or hypocenters tends to coincide with fault zone geometry (similar to how aftershocks can delineate source zones for large earthquakes). The most significant limitation of instrumental seismicity is the relatively short period of time over which such measurements have been made, which limits their effectiveness for the identification of faults or source zones that produce infrequent (typically large) earthquakes.

4.2.2 Characterization of Earthquake Sources

Once seismic sources have been identified, their potential for producing earthquakes with damaging ground motions must be characterized. The ground motion models (GMMs) described in Section 3.5 indicate that magnitude, distance, and style of faulting are among the factors that affect the distribution of ground motion *IM*s that can be expected during future earthquakes. Accordingly, source characterization must provide those parameters, along with the rate of earthquake occurrence, for each source.

4.2.2.1 Location, Geometry, and Activity

The location of a source is usually determined during the process of its identification. The geometries of earthquake sources depend on the tectonic processes involved in their formation. Earthquakes associated with volcanic activity, for example, generally originate in zones near volcanoes that are small enough to allow them to be characterized as *point sources*, i.e., sources whose location can be

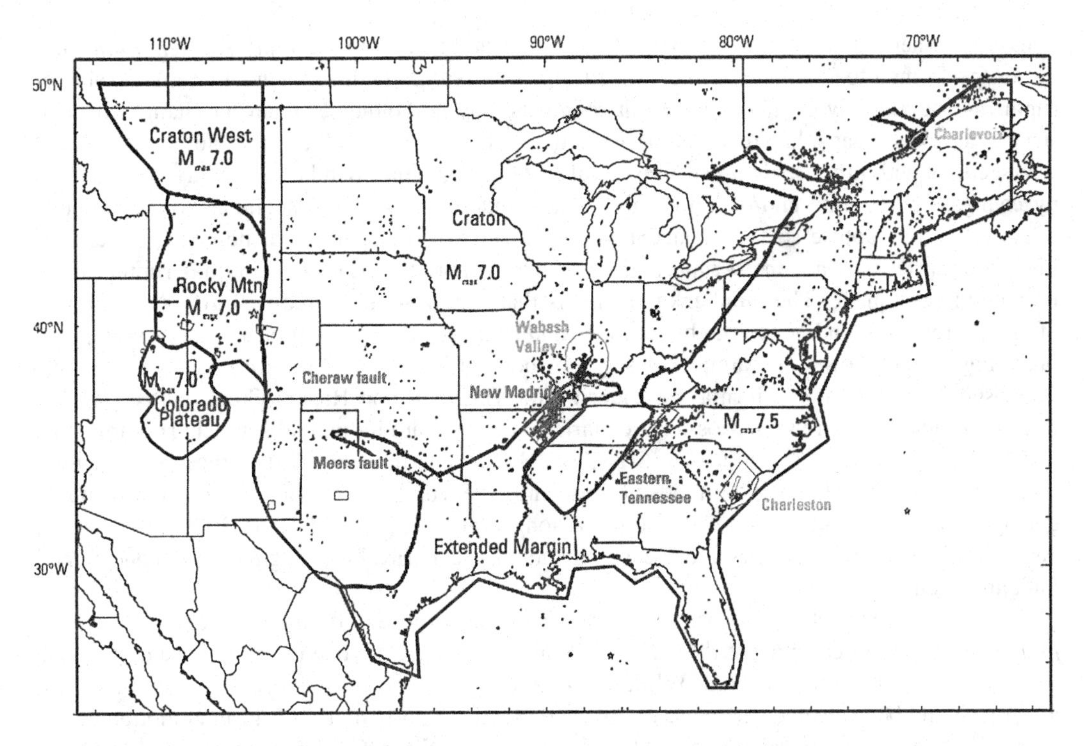

FIGURE 4.4 Hypocenter locations and aerial source map for CEUS. Black lines delineate the boundaries of source zone polygons, and colored polygons and ellipses delineate fault sources. (Petersen et al., 2008)

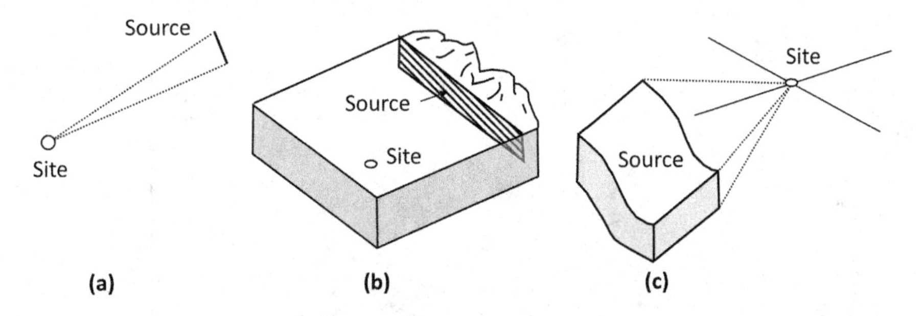

FIGURE 4.5 Examples of different source zone geometries: (a) short fault that can be modeled as a point source; (b) shallow fault that can be modeled as a rectangular source; and (c) three-dimensional source zone.

described by a latitude, longitude, and depth (e.g., Figure 4.5a). Large earthquakes on *active faults* tend to rupture over surfaces that can be described by one or more rectangles (e.g., Figure 4.5b). Fault rectangles can be considered as two-dimensional areal sources. Areas where earthquake mechanisms are poorly defined, or where faulting is so extensive as to preclude distinction between individual faults, can be treated as three-dimensional volumetric sources (e.g., Figure 4.5c). The purpose of geometrically characterizing earthquake sources as points, rectangles, or volumes is principally to enable site-source distance computation. Hence, while such characterizations necessarily represent simplifications of actual sources, they are generally adequate for the purposes of a seismic hazard analysis.

Individual faults, particularly the large active faults capable of producing large earthquakes, often have irregular geometries that can influence the manner in which they rupture in individual earthquakes. In many cases, earthquakes rupture only a portion of a given fault and therefore release less energy than they would have if the entire fault had ruptured. *Fault segmentation* studies are performed to characterize the geometries of individual fault segments that are capable of rupturing individually or in groups. Fault segments may be bounded by abrupt changes in fault strike, step-overs, gaps, bifurcations (division of single fault branch into multiple branches), and intersections with other structures (Yeats et al., 1997). Segment boundaries can also be detected by changes in slip rate and/or deformation rates and patterns. Some segment boundaries can effectively stop rupture so that adjacent segments rupture in separate earthquakes. Other segment boundaries may not arrest rupture, in which case more than one segment may rupture in an individual earthquake. The number of segments that rupture at one time will affect the amount of energy released, and hence the magnitudes and rates of recurrence of earthquakes on a given fault. Figure 4.6 shows an example of segment boundaries on several active faults in the San Francisco Bay Area in California. Considering the Hayward and Rogers Creek faults, possible large earthquake rupture scenarios for the marked segments are individual segment ruptures (SH, NH, and RC), two segments in series (SH-NH, NH-RC) and all segments together (SH-NH-RC). Identification of the relative likelihoods of those and other rupture scenarios requires considerable geological expertise and consensus building among experts (e.g., Field et al., 2009, 2014, 2024). As in the Hayward and Rodgers Creek case, earthquake scenarios involving more than one named fault are often considered.

The mere presence of a fault does not guarantee the occurrence of future earthquakes. An *active fault* poses a current earthquake threat, whereas an *inactive fault* is one on which past earthquake activity is unlikely to be repeated. While there is no formal, uniform consensus as to how fault activity should be evaluated, there are established practices, particularly for ground motion evaluations. For the purpose of seismic hazard analysis, active faults in California are defined from slip

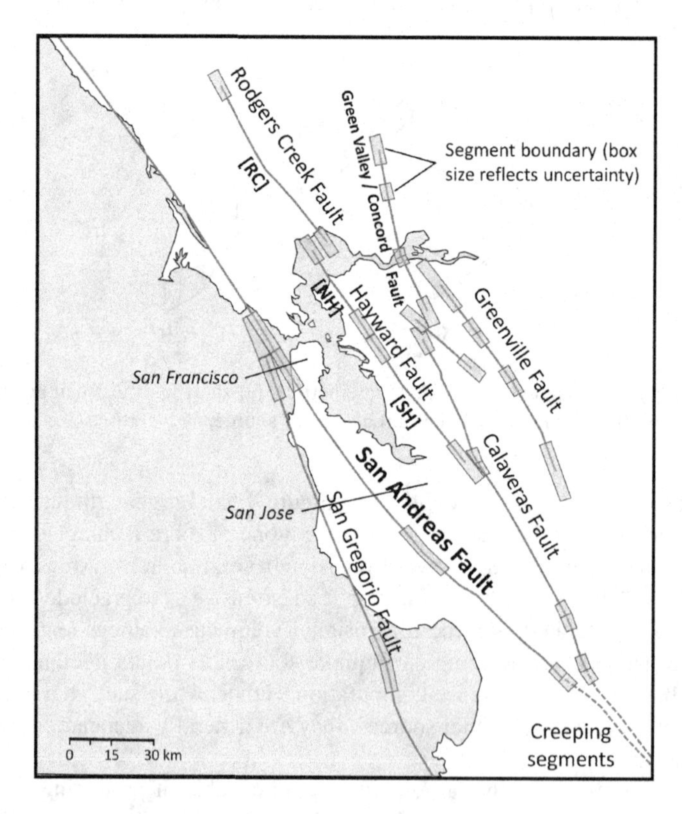

FIGURE 4.6 Major San Francisco Bay Area faults and their segment boundaries as defined by WGCEP (1999). Identified segments are SH, southern Hayward; NH, northern Hayward; RC, Rodgers Creek.

rate s, with $s > 0.1$ mm/year (Field et al., 2009). For the purpose of developing national seismic hazard maps, the US Geological Survey catalogs faults with seismic activity in Quaternary time (within the past 1.6M years; USGS, 2024). Fault activity is assessed for shorter time horizons (Holocene; within the last 11K years) for the purpose of mitigating surface fault rupture hazards (Section 10.2 and Bryant, 2010).

Volumetric source zones are used in lieu of fault sources in many regions worldwide (e.g., Woessner et al., 2015) because, whereas earthquakes may be known to occur, the geometry of the causative faults is unknown. This is often the case when the slip rates on the faults are low, causing the fault geometry to be poorly expressed in landforms and hence difficult to map (e.g., in stable continental regions). Even in seismically active regions where slip rates are relatively high, geologic conditions may cause fault mapping from surface evidence to be impractical, for example when faults are buried beneath deep sediments, such as the Puente Hills fault in Los Angeles (Shaw et al., 2002) or the Seattle fault (Johnson et al., 1994) (although, in these two cases, the faults have been located using other information sources). As a practical matter, volumetric sources are often expressed as areal source zones, especially when the distance metric to be used for hazard calculations is the closest distance to the surface projection of the fault (Joyner-Boore distance, R_{JB}). The boundaries of areal source zones are typically delineated on the basis of relative seismicity rates, as shown for example for the central and eastern US in Figure 4.4 by areas with high concentrations of events such as near New Madrid.

4.2.2.2 Earthquake Size

The sizes of earthquakes generated along a fault or within a source zone can range from imperceptible events of very small magnitudes to events rupturing one or more fault segments. While seismologists can use very small earthquakes to learn about source characteristics and crustal structure, engineers are interested in a range from some minimum magnitude ($\mathbf{M}_{min}$), below which damaging ground motions are not expected, up to the largest magnitude earthquake that a source is capable of producing. The size of larger magnitude events is often estimated using fault area (A). These $\mathbf{M}$-A relations are uncertain and are represented by a conditional probability density function, i.e., $f(\mathbf{M}|A)$, developed from empirical data. Early work often used fault length as a proxy for area, which is best suited to cases in which the rupture surface is fairly narrow, typically less than about 20km (Bonilla et al., 1984).

Minimum Magnitude: The minimum magnitude ($\mathbf{M}_{min}$) is the smallest event that is expected to produce ground motions capable of damaging structures or facilities (e.g., Bommer and Crowley, 2017). The most appropriate value of $\mathbf{M}_{min}$ for different regions and different classes of structures is generally selected subjectively. For example, $\mathbf{M}_{min}$ is generally taken as 5 in California. In much of Europe, old, brittle, unreinforced masonry structures are commonly encountered, and values of $\mathbf{M}_{min}$ in the range of 4–4.7 have been used (Beauval et al., 2008; Woessner et al., 2015). In the case of hazard analysis for anthropogenic (i.e., of human origin) events (Section 2.6), much lower magnitudes are considered to capture the sizes of earthquake typically associated with such events (e.g., Bommer et al., 2017). Green and Bommer (2019) argue that magnitudes as low as 4.5–5 should be considered in analyses of liquefaction risk.

Maximum Magnitude: The largest earthquake that a particular fault segment, or series of segments, is capable of producing ($\mathbf{M}_{max}$) occurs when the entire surface area ruptures. Empirical relationships relate the energy release from an event ("magnitude") to the size of its rupture (i.e., the "area"). As the amount of energy released also depends on the average stress drop across the rupture, magnitude-area relations incorporate assumptions regarding the variation of stress drop with magnitude.

A constant stress drop is commonly assumed within a particular tectonic region (e.g., Hanks, 1977; Baltay et al., 2011), which allows magnitude-area relations to have a log-linear form as follows

$$\mathbf{M} = a_1 \log_{10}(A) + a_2 \tag{4.2}$$

Table 4.1 shows coefficients a_1 and a_2 as estimated by several investigators based on shallow-focus (hypocentral depth<40 km) earthquakes in active tectonic regions (i.e., along plate boundaries). Other relations predict seismic moment (M_0) from area as follows:

$$\log_{10}(M_0) = a_3 \log_{10}(A) + a_4 \tag{4.3}$$

Table 4.2 shows coefficients a_3 and a_4 from Leonard (2010, 2014) both for active tectonic regions and stable continental regions. Equations (4.2) and (4.3) are conceptually alike due to the relationship between seismic moment and moment magnitude (Equation 2.3).

Fault dimensions for the relationships in Equations (4.2) and (4.3) were established from spatial patterns of aftershocks that occur within a short time interval following the mainshock and from inverted fault plane solutions. Figure 4.7 shows a relationship between rupture area and M_0 for dip-slip earthquakes in active tectonic regions along with the scatter in the data. The relationship for strike-slip earthquakes is nearly equivalent, whereas the relationship for stable continental regions predicts smaller areas for the same seismic moment due to their higher stress drops (Section 3.5.2.4). The relationships in Equations (4.2–4.3) and Tables 4.1–4.2 can be used to predict mean values of magnitude, and the standard deviations can be used can be used to compute values for various percentiles (Section D.5.2).

TABLE 4.1

Coefficients for Empirical Relationships (Equation 4.2) between Moment Magnitude, M, and Surface Rupture Area *A* (in km²)

Author (Year)	a_1	a_2	Application	σ_M
Wells & Coppersmith (1994)	0.98	4.07	All	0.24
Wells & Coppersmith (1994)	1.02	3.98	Strike-Slip	0.23
Wells & Coppersmith (1994)	0.9	4.33	Reverse	0.25
Somerville et al. (2006)	1.05	3.87	Strike-Slip	
Hanks & Bakun (2008)	1	3.98±0.03	Strike-Slip, $A \leq 537\,\text{km}^2$	
	4/3	3.08±0.04	Strike-Slip, $A > 537\,\text{km}^2$	
Thingbaijam et al. (2017)	1.05	3.47	Subduction Interface	0.16
	1.06	3.70	Strike-Slip	0.20
	0.95	4.16	Reverse	0.12
	1.23	3.16	Normal	0.22

TABLE 4.2

Coefficients for Empirical Relationships (Equation 4.3) between Seismic Moment M_0 (in N-m) and Surface Rupture Area *A* (in m²)

Author (Year)	a_3	a_4	Application	$\sim\sigma_{\log M_0}$ [a]
Leonard (2010, 2014)	1.5	6.10	Active Crustal Region, Dip-Slip	0.45
	1.5	6.09	Active Crustal Region, Strike-Slip	0.39
	1.5	6.38	Stable Continental Region	0.15

[a] Log standard deviation terms inferred from uncertainties in source documents that are reported as a range.

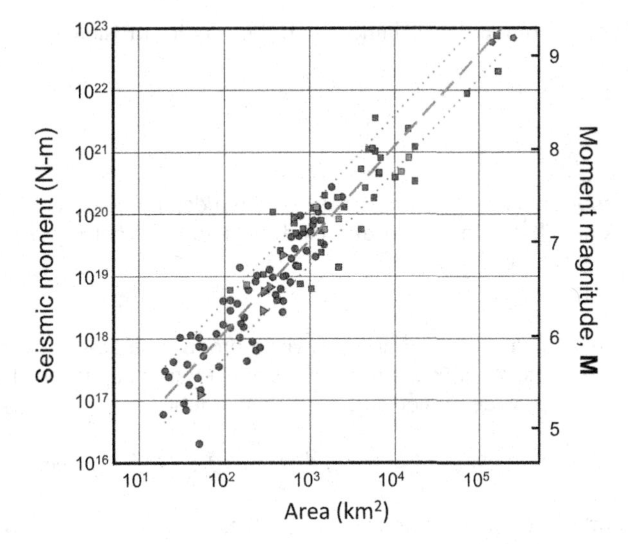

FIGURE 4.7 Relationships between rupture area and seismic moment for dip-slip earthquakes in active crustal regions (Table 4.2) showing the scatter inherent to the database. (After Leonard, 2010, used by permission of the Seismological Society of America.)

Data from several large earthquakes in active tectonic regions suggest that stress drops for low to medium magnitude strike-slip events ($\mathbf{M} \leq \sim 7$, or $A < 537\,\text{km}^2$) may be smaller than those for larger magnitude events (Hanks and Bakun, 2002, 2008). As shown in Figure 4.8, the greater energy release associated with these higher stress drops causes the **M**-A relationships to take on a steeper

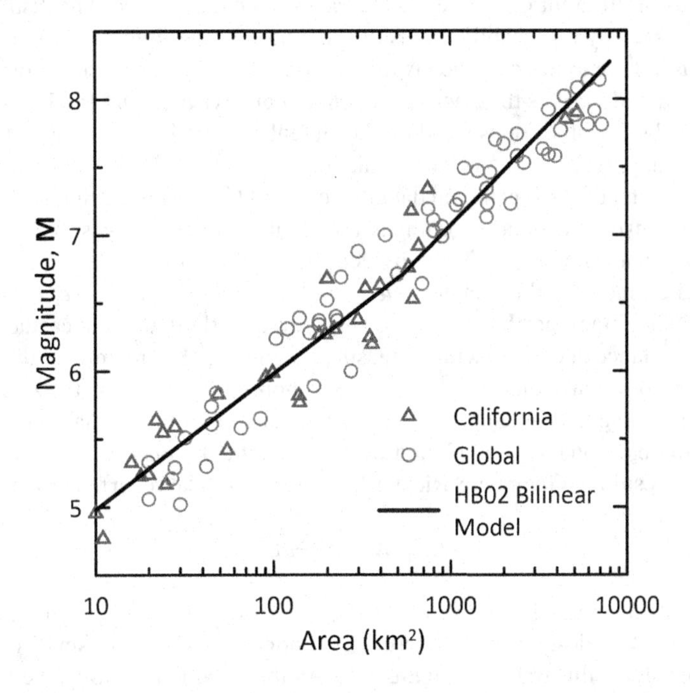

FIGURE 4.8 Relationships between moment magnitude and rupture area showing break in slope at large magnitudes indicating potentially higher stress drops in this range. (Adapted from data in Wells and Coppersmith, 1994 and Hanks and Bakun, 2002, 2008.)

slope in the large magnitude range. The "Hanks and Bakun" entry in Table 4.1 is intended to capture this effect.

Example 4.1

Consider a surface rupture on the San Andreas fault with a depth (measured down-dip) of 15 km and length of 100 km. What is the mean expected magnitude? What is the uncertainty in the magnitude estimate?

Solution:

The San Andreas fault is nearly vertical and known to produce strike-slip earthquakes (Section 2.3.2.3), so the strike-slip model of Leonards (2010) is applied using Equation (4.3) with a rupture area A of 1,500 km^2. This provides a median seismic moment of

$$M_0 = 10^{1.5\log(1{,}500)+6.09} = 7.1\times10^{19}\ \text{N-m} = 7.1\times10^{26}\ \text{dyne-cm}$$

The median magnitude is computed from Equation (2.3) as

$$\mathbf{M} = \frac{\log M_0}{1.5} - 10.7 = \frac{\log\left(7.1\times10^{26}\right)}{1.5} - 10.7 = 7.2$$

The uncertainty in the seismic moment is 0.39 (from Table 4.2). Per Equation (2.3), the corresponding error term for magnitude is 0.39/1.5 = 0.26.

4.2.2.3 Magnitude-Recurrence Relations

Once the location, geometry, and magnitude range for a seismic source have been identified, it is necessary to characterize the rate of occurrence of significant (i.e., $\mathbf{M} > \mathbf{M}_{\min}$) earthquakes and the relative likelihoods of different magnitude earthquakes from that source. The distribution of earthquake size is expressed by a probability density function for magnitude, $f_M(\mathbf{m})$, for a given source (lowercase **m** is used here to indicate the magnitude used as an argument in the probability density function). This section describes the classical Gutenberg and Richter (1944) relationship for magnitude recurrence. Subsequent sections build on the Gutenberg and Richter relationships and present procedures for evaluation of $f_M(\mathbf{m})$ and earthquake rate.

Gutenberg and Richter (1944) gathered data for southern California earthquakes over a period of many years and organized the data according to the number of earthquakes that exceeded different magnitudes during that time period. They divided the number of exceedances of each magnitude by the length of the time period to define a *mean annual rate of exceedance*, λ_m of an earthquake of magnitude **m**. Since small earthquakes occur more frequently than large earthquakes, the mean annual rate of exceedance decreases with increasing magnitude. The reciprocal of the mean annual rate of exceedance for a particular magnitude is commonly referred to as the *recurrence interval* of earthquakes exceeding that magnitude. When the logarithm of the annual rate of exceedance of southern California earthquakes was plotted against earthquake magnitude, a linear relationship was observed. The resulting Gutenberg-Richter law for earthquake recurrence was expressed as

$$\log_{10}\lambda_m = a - b\mathbf{m} \tag{4.4}$$

where 10^a is the mean annual number of earthquakes of magnitude greater than or equal to zero, and b (the b-value) describes the relative likelihoods of large and small earthquakes. The Gutenberg-Richter law is illustrated in Figure 4.9a. As the b-value increases, the number of larger magnitude earthquakes decreases compared to those of smaller magnitudes. The Gutenberg-Richter law is not restricted to the use of magnitude as a descriptor of earthquake size; epicentral intensity has also been used. Southern California recurrence data are shown in Figure 4.9b.

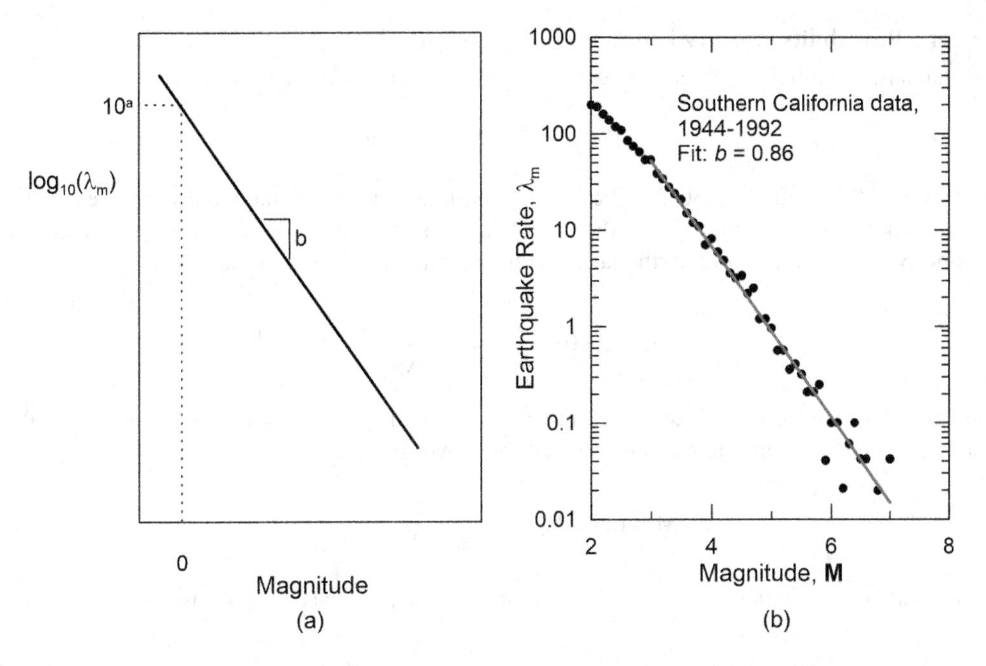

FIGURE 4.9 (a) Gutenberg-Richter recurrence law, showing the meaning of *a* and *b* parameters; and (b) application of Gutenberg-Richter law to seismicity data from southern California. (After Stirling et al., 1996.)

The *a* and *b* parameters are generally obtained by regression on a database of seismicity from the source zone of interest. Unless the source zone is extremely active or the observation period extremely long, the database is likely to be relatively sparse at large magnitudes. If both instrumental and historical events are present in the event inventory, magnitude estimates for historical events (from paleoseismic evidence or intensity-based correlations) are required. In some areas, the record of seismicity may be distorted by the presence of dependent events such as aftershocks and foreshocks (Merz and Cornell, 1973). Although such dependent events can cause significant damage, a hazard analysis is intended to evaluate the hazard from discrete, independent releases of seismic energy. Therefore, dependent events must be removed from the seismicity catalog (Gardner and Knopoff 1974) and their effects accounted for in separate analyses (Yeo and Cornell 2009). In forming an earthquake catalog for analysis, completeness of the database must also be considered (e.g., Stepp 1972) and human-induced events such as mining blasts should be removed (Wiemer and Baer 2000). Whereas the relatively short-lived instrumental database is typically complete for small events that occur frequently, it is generally incomplete for large events. Conversely, the historical record is usually incomplete for small events, which go undetected for a variety of physical and demographic reasons, but is much more complete for large events.

Example 4.2

Using Figure 4.9b, compute the recurrence interval of **M**6 earthquakes in southern California.

Solution:

At a magnitude of 6, Figure 4.9b indicates a mean annual rate of exceedance of approximately 0.15 events per year. Therefore, the corresponding return period is:

$$T_R = \frac{1}{\lambda_m} = \frac{1}{0.15\,/\,\text{year}} = 6.7\ \text{years}$$

4.2.2.4 Probability Density Functions for Magnitude

The standard Gutenberg-Richter recurrence law of Equation (4.4) can be rewritten as

$$\lambda_m = 10^{a-b\mathbf{m}} = \exp[\alpha - \beta\mathbf{m}] \tag{4.5}$$

where α=2.303a and β=2.303b. The rate of earthquakes with magnitudes greater than **m** (Equation 4.5) can be normalized by the rate of "all" earthquakes (taken as **m**>0) to calculate the probability that a given future earthquake will have a magnitude greater than **m**:

$$P(\mathbf{M} > \mathbf{m} \mid \mathbf{m} > 0) = \frac{\lambda_m}{\lambda_0} = \frac{\exp[\alpha - \beta\mathbf{m}]}{\exp[\alpha]} = e^{-\beta\mathbf{m}} \tag{4.6}$$

Equation (4.6) can be related to the cumulative distribution function for magnitude, $F_M(\mathbf{m})$, as follows (the conditionality on **m**>0 is dropped for convenience):

$$P(\mathbf{M} < \mathbf{m}) \equiv F_M(\mathbf{m}) = 1 - \frac{\lambda_m}{\lambda_0} = 1 - e^{-\beta\mathbf{m}} \tag{4.7}$$

The derivative of Equation (4.7) with respect to magnitude produces a probability density function,

$$f_M(\mathbf{m}) = \frac{d}{d\mathbf{m}} F_M(\mathbf{m}) = \beta e^{-\beta\mathbf{m}} \tag{4.8}$$

As described in Section D.7.2.4, Equation (4.8) is the exponential distribution. Hence, the Gutenberg-Richter law implies that earthquake magnitudes are exponentially distributed.

Truncated Exponential Model: The standard Gutenberg-Richter law covers an infinite range of magnitudes, from 0 to +∞. As described in Section 4.2.2.2, the effects of very small earthquakes are of little engineering interest and can be disregarded, whereas the size of the source limits the maximum magnitude that can occur. The *truncated exponential* model for $f_M(\mathbf{m})$ is based on the Gutenberg-Richter relation, but the magnitude scale is truncated at lower- and upper-bound values, $\mathbf{M}_{\text{min}}$ and $\mathbf{M}_{\text{max}}$, respectively. With these modifications, and the constraint that $f_M(\mathbf{m})$ must integrate to 1.0, the cumulative distribution function (CDF) and probability density function (PDF) for the truncated exponential model are expressed as follows:

$$F_M(\mathbf{m} \mid \mathbf{M}_{\text{min}} < \mathbf{m} < \mathbf{M}_{\text{max}}) = P(\mathbf{M} < \mathbf{m} \mid \mathbf{M}_{\text{min}} < \mathbf{m} < \mathbf{M}_{\text{max}}) = \frac{1 - \beta e^{-\beta(\mathbf{m} - \mathbf{M}_{\text{min}})}}{1 - e^{-\beta(\mathbf{M}_{\text{max}} - \mathbf{M}_{\text{min}})}}$$

$$f_M(\mathbf{m} \mid \mathbf{M}_{\text{min}} < \mathbf{m} < \mathbf{M}_{\text{max}}) = \frac{\beta e^{-\beta(\mathbf{m} - \mathbf{M}_{\text{min}})}}{1 - e^{-\beta(\mathbf{M}_{\text{max}} - \mathbf{M}_{\text{min}})}} \tag{4.9}$$

A plot of $f_M(\mathbf{m})$ for a fault with $\mathbf{M}_{\text{min}}$=5 and $\mathbf{M}_{\text{max}}$ =7.5 is shown with a dashed line in Figure 4.10a.

Maximum Magnitude Model: Some faults appear to generate earthquakes only within a relatively narrow magnitude range involving full-segment ruptures (Allen, 1968; Wesnousky et al., 1983). Such faults include certain segments of the San Andreas fault (e.g., the North Coast segment, which last ruptured in 1906, Hill et al., 1990; and the Cholame, Carrizo, and Mojave segments, which last ruptured in 1857, Hauksson et al., 2012) and portions of the Wasatch fault in Utah (Arabasz et al., 1980). Such atypical behavior on major faults may result from the last major event on the fault having ruptured to unusually large depth, into the ductile zone in the upper mantle (Jiang and Lapusta, 2016). This behavior can be described using the *maximum magnitude* model (Wesnousky et al., 1983), which consists of a truncated normal distribution for $f_M(\mathbf{m})$, with cutoff magnitudes of $\mathbf{M}_{\text{min}}$ and $\mathbf{M}_{\text{max}}$, mean magnitude of $\bar{\mathbf{M}}_{\text{max}}$, and a standard deviation of σ_m. The CDF and PDF for this model are given by:

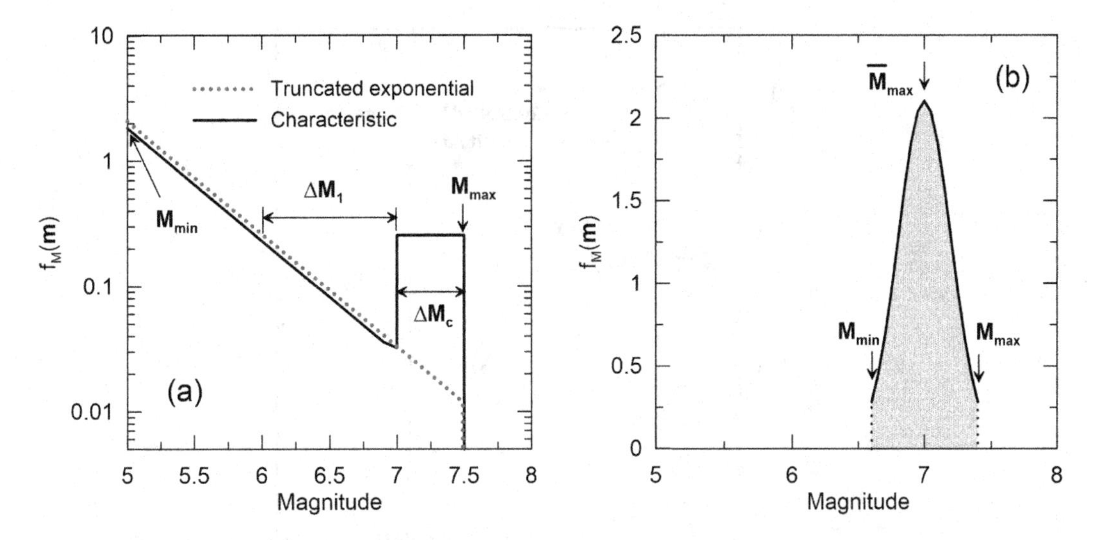

FIGURE 4.10 PDFs for magnitude, $f_M(\mathbf{m})$; (a) truncated exponential and characteristic, (b) maximum magnitude.

$$F_M(\mathbf{m} \mid \mathbf{M}_{min} < \mathbf{m} < \mathbf{M}_{max}) = P(\mathbf{M} < \mathbf{m} \mid \mathbf{M}_{min} < \mathbf{m} < \mathbf{M}_{max}) = \frac{F_Z(Z_{\mathbf{m}})}{F_Z(Z_{\mathbf{M}_{max}}) - F_Z(Z_{\mathbf{M}_{min}})}$$

$$f_M(\mathbf{m} \mid \mathbf{M}_{min} < \mathbf{m} < \mathbf{M}_{max}) = \frac{\dfrac{1}{\sqrt{2\pi}\sigma_m}\exp\left[-\dfrac{(\mathbf{m} - \bar{\mathbf{M}}_{max})^2}{2\sigma_m^2}\right]}{F_Z(Z_{\mathbf{M}_{max}}) - F_Z(Z_{\mathbf{M}_{min}})} \tag{4.10a}$$

where $F_Z(Z)$ represents the CDF of the standard normal distribution (Table D.1), with the standard normal variate as used in Equation (4.10a) defined as

$$Z_{\mathbf{m}} = \frac{\mathbf{m} - \bar{\mathbf{M}}_{max}}{\sigma_m}, \quad Z_{\mathbf{M}_{max}} = \frac{\mathbf{M}_{max} - \bar{\mathbf{M}}_{max}}{\sigma_m}, \quad Z_{\mathbf{M}_{min}} = \frac{\mathbf{M}_{min} - \bar{\mathbf{M}}_{max}}{\sigma_m} \tag{4.10b}$$

The model is depicted in Figure 4.10(b) for $\bar{\mathbf{M}}_{max} = 7.0$, $\sigma_m = 0.2$, and truncation limits set at $\pm 2\sigma_m$ from the mean, i.e., $\mathbf{M}_{min} = 6.6$ and $\mathbf{M}_{max} = 7.4$.

Hybrid Models: The Gutenberg-Richter law implies that individual faults should produce an exponential distribution of magnitudes and, hence, an exponential distribution of fault offsets. However, paleoseismic investigations using fault trenching have shown that points on individual faults and fault segments tend to displace by similar amounts in successive earthquakes. These observations suggest that individual large active faults repeatedly generate earthquakes of similar (within about one-half magnitude unit) size, known as *characteristic earthquakes* (Schwartz and Coppersmith, 1984; Schwartz, 1988).

By dating these characteristic earthquakes, their historical rate of recurrence can be estimated. Geologic evidence indicates that characteristic earthquakes occur more frequently than would be implied by extrapolation of the Gutenberg-Richter law from high exceedance rates (low magnitudes) to low exceedance rates (high magnitudes). The result is a more complex recurrence law that is governed by seismicity data at low magnitudes and geologic data at large magnitudes, as shown in Figure 4.11. These are referred to as hybrid recurrence relations because they combine different functional forms for the magnitude PDF at low and high magnitudes.

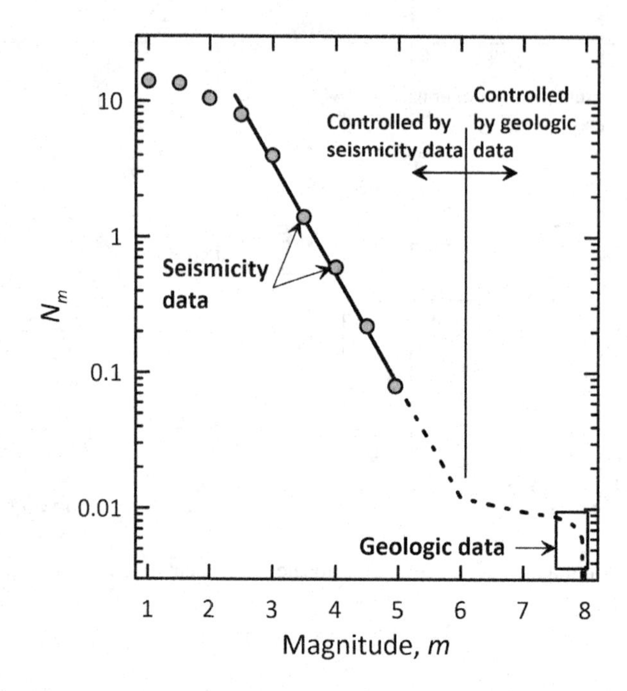

FIGURE 4.11 Inconsistency of mean annual rate of exceedance (N_m) as determined from seismicity data and geologic data. (After Youngs and Coppersmith, 1985.)

Youngs and Coppersmith (1985) developed a hybrid PDF for $f_M(\mathbf{m})$ that combines a truncated exponential magnitude distribution at low magnitudes with a uniform distribution in the vicinity of the characteristic earthquake. This distribution is shown in Figure 4.10a, where $\Delta\mathbf{M}_c$ is the increment of uniform probability density associated with the characteristic earthquake (generally taken as 0.5), and $\Delta\mathbf{M}_1$ is the magnitude offset between the lower bound of the characteristic earthquake and the portion of the truncated exponential model having the same probability density. The CDF of this distribution can be expressed as (Convertito et al., 2006):

$$F_M\left(\mathbf{m} \mid \mathbf{M}_{min} < \mathbf{m} < (\mathbf{M}_{max} - \Delta\mathbf{M}_c)\right) = \frac{1-e^{-\beta(\mathbf{m}-\mathbf{M}_{min})}}{1-e^{-\beta(\mathbf{M}_{max}-\mathbf{M}_{min}-\Delta\mathbf{M}_c)}}\frac{1}{1+C}$$

$$F_M\left(\mathbf{m} \mid (\mathbf{M}_{max} - \Delta\mathbf{M}_c) < \mathbf{m} < \mathbf{M}_{max}\right) = \frac{\beta e^{-\beta(\mathbf{M}_{max}-\mathbf{M}_{min}-\Delta\mathbf{M}_1-\Delta\mathbf{M}_c)}}{1-e^{-\beta(\mathbf{M}_{max}-\mathbf{M}_{min}-\Delta\mathbf{M}_c)}}\frac{\mathbf{m}-(\mathbf{M}_{max}-\Delta\mathbf{M}_c)}{1+C}+\frac{1}{1+C} \tag{4.11a}$$

where C is a constant defined by,

$$C = \frac{\beta e^{-\beta(\mathbf{M}_{max}-\mathbf{M}_{min}-\Delta\mathbf{M}_1-\Delta\mathbf{M}_c)}}{1-e^{-\beta(\mathbf{M}_{max}-\mathbf{M}_{min}-\Delta\mathbf{M}_c)}}\Delta\mathbf{M}_c$$

Differentiating $F_M(\mathbf{m})$ yields the PDF

$$f_M\left(\mathbf{m} \mid \mathbf{M}_{min} < \mathbf{m} < (\mathbf{M}_{max} - \Delta\mathbf{M}_c)\right) = \frac{\beta e^{-\beta(\mathbf{m}-\mathbf{M}_{min})}}{1-e^{-\beta(\mathbf{M}_{max}-\mathbf{M}_{min}-\Delta\mathbf{M}_c)}}\frac{1}{1+C}$$

$$f_M\left(\mathbf{m} \mid (\mathbf{M}_{max} - \Delta\mathbf{M}_c) < \mathbf{m} < \mathbf{M}_{max}\right) = \frac{\beta e^{-\beta(\mathbf{M}_{max}-\mathbf{M}_{min}-\Delta\mathbf{M}_1-\Delta\mathbf{M}_c)}}{1-e^{-\beta(\mathbf{M}_{max}-\mathbf{M}_{min}-\Delta\mathbf{M}_c)}}\frac{1}{1+C} \tag{4.11b}$$

Other approaches for hybrid modeling of major active faults use a maximum magnitude model (truncated normal distribution) for the characteristic earthquake coupled with a truncated exponential model for smaller magnitudes (e.g., Field et al., 2009, 2014).

4.2.2.5 Rate of Earthquake Recurrence

Two philosophies are available for modeling the rate of earthquake recurrence. The *time-independent* approach assumes that earthquakes occur as a Poisson process, which in turn assumes that the occurrence of an event (excluding foreshocks and aftershocks) has no influence on the timing of future events. Alternatively, the *time-dependent* approach holds that the occurrence of an event releases stress on the fault and thus temporarily reduces the rate of future earthquakes (excluding aftershocks). The processes by which rates of earthquake occurrence are derived using these approaches are described below.

Time-Independent Models: Earthquake recurrence has most commonly been expressed using the time-independent Poisson model. As described in Section D.7.1.2, the Poisson model provides a simple framework for evaluating probabilities of events that can be represented by a Poisson process in which the occurrence of events in any given time interval is independent of events in other time intervals. The key parameter in the Poisson model is the mean annual rate of occurrence of events. As described in Section 4.2.2.3, the mean annual rate of exceedance, λ_m is defined as the rate of occurrence of events with magnitude $>\mathbf{m}$. This rate can be defined using a fully empirical approach or a moment balance approach.

As described in Section 4.2.2.3, the empirical approach involves collecting available earthquake magnitude data and computing earthquake rates as a function of magnitude. Such data are typically observed to follow the Gutenberg-Richter law. However, it should be recognized that the observation period may be inadequate to accurately determine the rates of infrequent large magnitude events, which are often critical for engineering design. Nonetheless, this approach is widely used in seismically active areas where fault source information may be unavailable within regions having long historical records such as southern Europe (Basili et al., 2008) and Japan (Pagani et al., 2016).

In the *moment balance approach*, the Poisson rate of earthquake occurrence on a given fault can be computed based on the seismic moment (Section 2.8.1) if the slip rate, s, is known and a magnitude distribution model, $f_M(\mathbf{m})$, is selected. Differentiating Equation (2.2) with respect to time, the rate of moment build-up on the fault during periods without earthquakes is

$$\dot{M}_0 = \mu A s \tag{4.12}$$

For non-hybrid fault models, the rate of moment release is calculated as

$$\dot{M}_0 = \lambda_{\mathbf{M}_{\min}} \cdot \int_{\mathbf{M}_{\min}}^{\mathbf{M}_{\max}} f(\mathbf{m}) M_0(\mathbf{m})\, d\mathbf{m} \tag{4.13}$$

where $\lambda_{\mathbf{M}_{\min}}$ = the Poisson rate of occurrence of earthquakes with $\mathbf{m} > \mathbf{M}_{\min}$ and M_0 indicates seismic moment. The integral in Equation (4.13) represents the mean moment release for future earthquakes in the magnitude range of $\mathbf{M}_{\min}$ to $\mathbf{M}_{\max}$. The seismic moment, M_0, can be related to $\mathbf{m}$ by re-arranging Equation (2.3) as

$$M_0 = 10^{1.5\mathbf{m}+16.05} \tag{4.14}$$

Assuming that the seismic moment that builds up over time is eventually released in the form of earthquakes, the long-term rate of moment build-up must be equal to the rate of moment release. Setting Equations (4.12) and (4.13) equal, $\lambda_{\mathbf{M}_{\min}}$ can be readily calculated.

When a hybrid model for the magnitude PDF is used, the rate of the characteristic earthquake, λ_c, is typically evaluated from a paleoseismic investigation. A separate rate for the smaller magnitude events, $\lambda_{\mathbf{M}_{\min}}$, is computed to cover the magnitude range for the low-magnitude PDF (e.g., $\mathbf{M}_{min} \le \mathbf{m} < (\mathbf{M}_{max} - \Delta\mathbf{M}_c)$ for the Characteristic Earthquake model). In this case, moment build-up is still evaluated using Equation (4.12), with moment release now evaluated as

$$\dot{M}_0 = \lambda_{\mathbf{M}_{min}} \cdot \int_{\mathbf{M}_{min}}^{\mathbf{M}_{max}-\Delta\mathbf{M}_c} f(\mathbf{m}) M_0(\mathbf{m})\, d\mathbf{m} + \lambda_c \cdot \int_{\mathbf{M}_{max}-\Delta\mathbf{M}_c}^{\mathbf{M}_{max}} f(\mathbf{m}) M_0(\mathbf{m})\, d\mathbf{m} \quad (4.15)$$

The Poisson rate term $\lambda_{\mathbf{M}_{min}}$ is evaluated by equating Equation (4.12) and (4.15).

Time-Dependent Models: The assumption of independence of events inherent in the Poisson model is inconsistent with elastic rebound theory, which holds that some period of time is required for strain energy to build back up on a fault that has just produced an earthquake. A logical extension of this principle is that if a fault section has not produced a large earthquake over a long time horizon, it is more likely to produce a future earthquake. Such seismic gaps, introduced in Section 2.5.1, have produced some notable earthquakes, but whether they necessarily have a higher likelihood of future earthquakes has been a topic of debate (Kagan and Jackson, 1991).

To model the time-variable rate of earthquakes implied by elastic rebound theory, *time-dependent models* can be applied that adjust earthquake probabilities during a future time window based on the time elapsed since the previous major earthquake on the fault. These models assign a probability distribution to the time from present until the next occurrence of a major earthquake, $f_T(t)$, where t is the time to the next event. These models are used to update the rate of occurrence of large events such as characteristic earthquakes. Figure 4.12 illustrates the use of $f_T(t)$ to evaluate the rate of a characteristic event, λ_c, using a time-dependent model. In the figure, $f_T(t)$ is taken as normal with a mean of $1/\lambda_c$. If t_e years have passed since the last event, a conditional density function $f_T(t|t>t_e)$ can be constructed as:

$$f_T(t \mid t > t_e) = \frac{f_T(t)}{1 - F_T(t_e)} \quad (4.16)$$

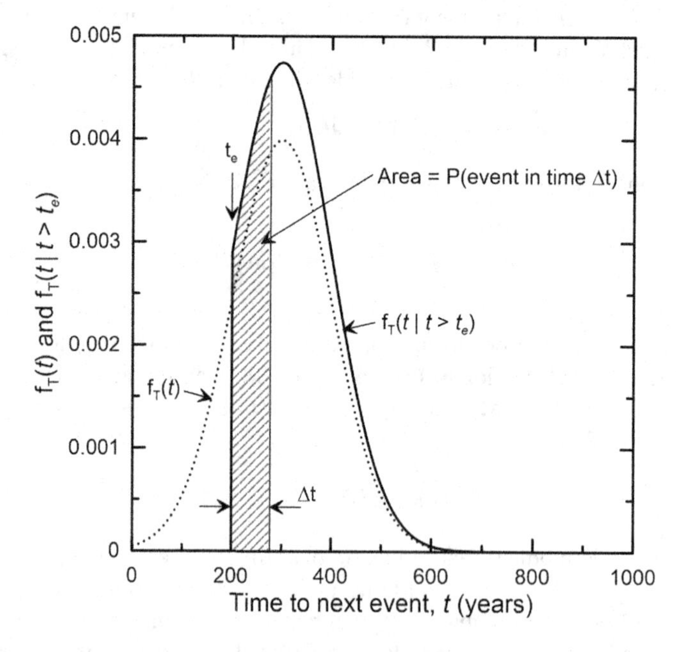

FIGURE 4.12 Illustration of PDF for time between earthquakes and its use to evaluate conditional event probability given that time t_e has elapsed since the last event with no activity. Annual rate of characteristic event, $\lambda_c = 1/300$ years.

where $F_T(t_e)$ is the CDF of time between earthquakes evaluated at time t_e. As shown in Figure 4.12, the probability of an event between time t_e and $t_e+\Delta t$, denoted $P_{E,\Delta t}$, is simply the area beneath $f_T(t|t>t_e)$ over the time interval t_e to $t_e+\Delta t$. This probability can be computed as

$$P_{E,\Delta t} = \frac{F_T(t_e + \Delta t) - F_T(t_e)}{1 - F_T(t_e)} \tag{4.17}$$

The probability from Equation (4.17) can then be converted to an updated Poisson rate of the characteristic event, $\lambda_{c|t>t_e}$, by manipulation of Equation (D.48) as:

$$\lambda_{c|t>t_e} = \frac{-\ln(1 - P_{E,\Delta t})}{\Delta t} \tag{4.18}$$

This updated rate of characteristic earthquakes can then be used in the same manner as λ_c for probabilistic seismic hazard analysis.

Field and Jordan (2015) derived earthquake probabilities using time-dependent models for the case where t_e is unknown; interestingly, they found higher probabilities than when time-independent models were used. This suggests that consideration of t_e is an important aspect of source characterization, even when specific knowledge of t_e is unavailable for a particular source.

It should be noted that moment build-up and release rates must not only balance for individual faults, but cumulative slip rates for all faults in a region should be consistent with regional geodetic and geologic data where such data exists and is reliable. This may require adjustments of slip rates for individual faults (Field et al., 2009; 2014). This topic is discussed further in Section 4.4.2.

4.3 DETERMINISTIC SEISMIC HAZARD ANALYSIS

In the early years of geotechnical earthquake engineering, the use of deterministic seismic hazard analysis (DSHA) was prevalent. A DSHA involves the development of a particular earthquake scenario upon which a ground motion hazard evaluation is based. The scenario consists of a postulated earthquake of a specified size occurring at a specified location. A typical DSHA can be described as a four-step process consisting of (modified from Reiter, 1990):

1. Identification and characterization of all earthquake sources capable of producing significant ground motions at the site. The characteristics of a potential earthquake, which take into consideration source geometry, activity, and magnitude, are developed for each source. For a particular fault, this requires selecting a single magnitude from the appropriate distribution, which will have some associated recurrence rate. Traditionally, the maximum magnitude of each source is selected (e.g.. $\mathbf{M}_{\mathbf{max}}$ in Figure 4.10), but this may be considered impractical in cases where the maximum magnitude is especially unlikely.
2. Selection of a source-to-site distance parameter for each source. The distance parameter(s) must be consistent with that in the GMM used to estimate ground motion characteristics in Step (4) of this procedure. In most DSHAs, the shortest distance between each source and the site of interest is selected.
3. Calculation of ground motions at the site of interest produced by the earthquakes identified in Step (1), which are assumed to occur at the distances identified in Step (2). Ground motions are generally expressed in terms of some *IM* such as peak acceleration, peak velocity, or spectral acceleration, the values of which are usually computed using GMMs (Section 3.5) for a selected percentile.
4. Identification of the controlling earthquake for the site, typically by comparing ground motions across all considered sources and selecting the one that produces the strongest

shaking. The controlling earthquake is described in terms of its size (usually expressed as magnitude) and distance from the site. The ground motion *IM*s produced by the controlling earthquake define the hazard from a DSHA.

The DSHA procedure is shown schematically in Figure 4.13. Expressed in these four compact steps, DSHA appears to be a very simple procedure, and in many respects, it is.

When applied to structures for which failure could have catastrophic consequences, such as nuclear power plants and large dams, DSHA is often considered to provide a straightforward framework for the evaluation of "worst-case" ground motions for the identified scenario. However, a number of subjective decisions that can strongly affect the computed *IM*s must be made in the DSHA procedure, causing the "worst-case" condition to be poorly (and often inconsistently) defined. For example, in Step (1), what combination of fault segments should be considered to rupture together for a given source? Referring to the Hayward and Rogers Creek faults shown in Figure 4.6, the worst-case would require a through-going rupture involving the SH, NH, and RC segments, even if that combination is extremely unlikely. What magnitude will then be assigned to a particular rupture scenario? As described in Section 4.2.2.2, relationships between fault area and magnitude have uncertainty. If a 95th percentile earthquake (two standard deviations above the mean) was considered, the magnitude would be approximately 0.5 (magnitude units) beyond the mean estimates.

Similarly, in Step (3), GMMs produce a distribution of ground motion *IM*s for a particular scenario earthquake. As described in Section 3.5.4, *IM*s are generally considered to be lognormally distributed, which implies that the true "worst-case" ground motion would be much larger than the median, i.e., that computed from the logarithmic mean. To account for uncertainty in ground motion predictions, DSHAs must select some percentile *IM* level (often 84th or 95th percentile, which corresponds to one or two logarithmic standard deviations above the logarithmic mean) to characterize the ground shaking from the identified scenario event.

If true "worst-case" conditions are considered, involving multi-segment through-going ruptures, the magnitude will correspond to a point on the $f_M(\mathbf{m})$ distribution with a low likelihood and high

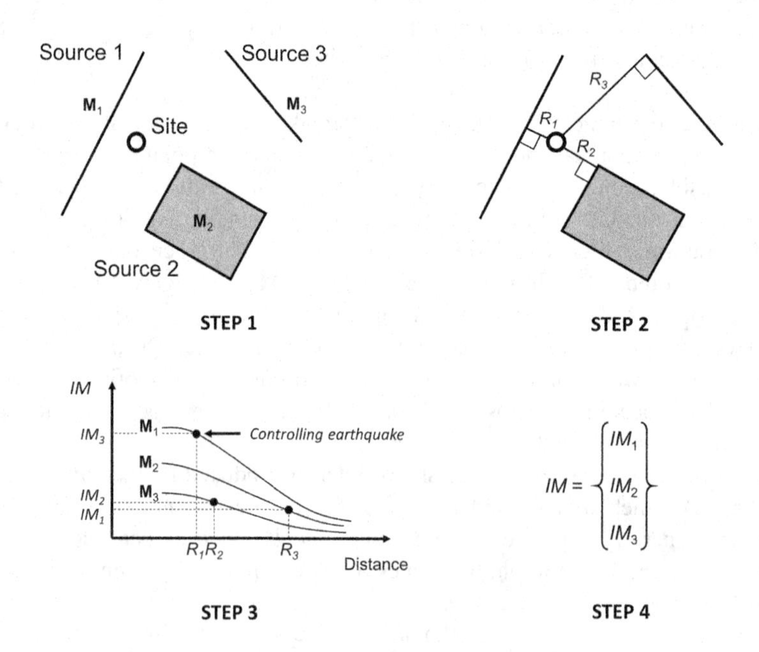

FIGURE 4.13 Four steps of a deterministic seismic hazard analysis (DSHA).

percentile *IMs* from the GMM would be used. This can cause the resulting ground motions to be so large that structures cannot be reasonably, let alone economically, designed to resist them. They can also be so strong that the likelihood of their occurrence during the useful lifetime of the structure being designed is miniscule. These factors create pressure to "back off" from the worst-case scenario, but where and how far to back off is also subjective and inherently arbitrary.

Dealing with the subjectivity inherent to DSHA requires the combined expertise and opinions of seismologists, seismic geologists, engineers, risk analysts, economists, social scientists, and government officials. The broad range of backgrounds and often divergent goals of such professionals can cause difficulty in reaching a consensus on earthquake potential. Over the years there have been many terms used to describe earthquake potential (Krinitzsky, 2002); among them the maximum credible earthquake (MCE), design basis earthquake (DBE), safe shutdown earthquake (SSE), maximum probable earthquake (MPE), operating basis earthquake (OBE), and seismic safety evaluation earthquake. The MCE, for example, is usually defined as the largest earthquake that appears capable of occurring under the known tectonic framework. The DBE and SSE are usually defined in essentially the same way as the MCE. The MPE has been defined as the maximum historical earthquake and also as the maximum earthquake likely to occur in a 100-year interval. Many DSHAs have used the two-pronged approach of evaluating hazards for both the MCE and MPE (or SSE and OBE). The Committee on Seismic Risk of the Earthquake Engineering Research Institute (EERI) has stated that terms such as MCE and MPE "are misleading ... and their use is discouraged" (Committee on Seismic Risk, 1984).

One common use of deterministic analysis is to provide a "cap" to ground motions predicted through probabilistic seismic hazard analysis for use in the USGS national seismic hazard model (Leyendecker et al., 2000). Mapped ground motions are based on probabilistic results with a risk adjustment, except in areas where DSHA provides lower values. The outcome of this process is referred to as a *risk-targeted, maximum considered earthquake ground motion*, MCE_R (Luco et al., 2015), with the objective being a 1% probability of structure collapse in 50 years. That objective is not achieved in areas where the MCE_R ground motions are set by deterministic caps, which causes confusion and controversy (Stewart et al., 2020). The DSHA magnitude for these calculations is based on disaggregations of probabilistic analysis (Section 4.4.3.5). Uncertainty in the ground motion is considered by using an 84th percentile value.

Example 4.3

The site shown in Figure E4.3 is located in the vicinity of the three independent, shallow seismic sources shown as Sources 1, 2, and 3. Using a DSHA, compute the peak acceleration considering the 50th percentile (median) ground motions.

Solution:

Source zones 1, 2, and 3 are shallow linear, areal, and point sources with the coordinates and maximum magnitudes indicated in the figure. The four-step procedure for DSHA is applied as follows:

1. The problem statement provides the location and maximum magnitude of each source. In a real DSHA, this is a difficult task.
2. Since the sources are shallow, the source-to-site distances are given below.

Source	Distance (km)
1	23.7
2	25.0
3	60.0

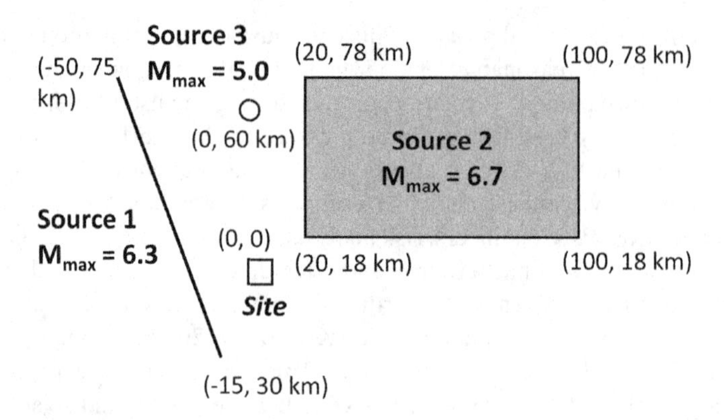

FIGURE E4.3 Positions of sources around site for Problem E4.3.

3. For this example, the peak acceleration is computed using the Campbell (1981) GMM (Equation 3.25). This leads to the ground motions below,

Source	M	Distance (km)	*PGA* (g)
1	6.3	23.7	0.09
2	6.7	25.0	0.12
3	5.0	60.0	0.01

On this basis, Source 2 produces the controlling event. Note: Though currently out of date, the Campbell (1981) relationship is used here because of its simplicity.

4. The hazard would be taken as that which results from a magnitude 6.7 earthquake occurring at a distance of 25 km, which produces $PGA = 0.12g$. Other *IM*s could be obtained from the GMMs described in Chapter 3.

4.4 PROBABILISTIC SEISMIC HAZARD ANALYSIS

Probabilistic seismic hazard analysis (PSHA) provides a framework in which seismic hazards can be computed with consideration of uncertainty in the size and location of earthquakes, their rates of recurrence, and variations of ground motion characteristics for various magnitude and distance combinations. Whereas a DSHA estimates ground motions for a controlling earthquake scenario corresponding to a single magnitude-distance pair, PSHA-based ground motions include contributions from all possible combinations of magnitude and distance, and the full range of ground motions that can be produced by each combination, all weighted by their relative likelihoods of occurrence. As such, PSHA avoids the arbitrary choices required in DSHA and provides a more complete and objective representation of seismic hazards.

Understanding the concepts and mechanics of PSHA requires familiarity with some terminology and basic concepts of probability theory. Such background information can be found in Appendix D. The PSHA methodology described in this section is similar in many respects to the well-established methods developed by Cornell (1968), Esteva (1969), Algermissen et al. (1982), and McGuire (2004).

PSHA can be described as a four-step procedure (modified from Reiter, 1990), each of which bears some similarity to the DSHA steps, as illustrated in Figure 4.14.

1. The identification and characterization of earthquake sources is identical to the first step of DSHA, except that the probability distribution and recurrence rates of the full range of earthquake magnitudes that each source is capable of producing must also be characterized. In contrast, a DSHA implicitly assumes a probability of 1.0 for the selected magnitude and 0.0 for all other magnitudes.

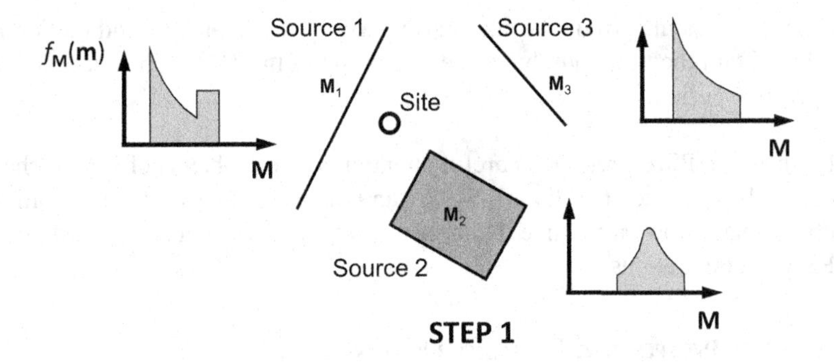

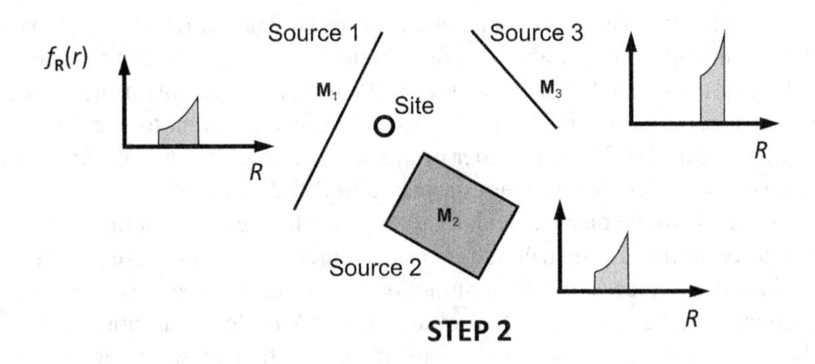

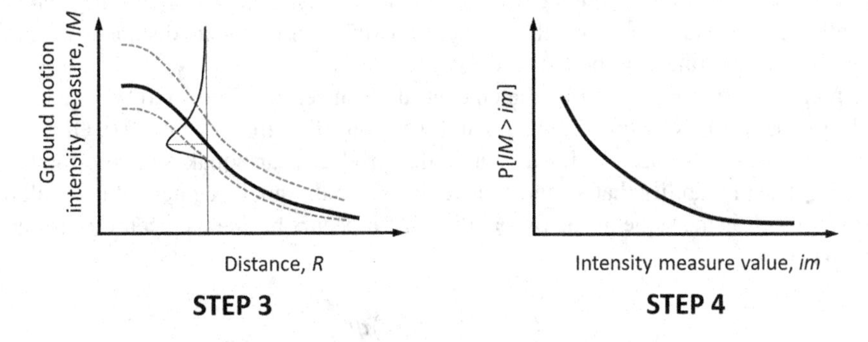

FIGURE 4.14 Four steps of a probabilistic seismic hazard analysis (PSHA).

2. The probability distribution of potential rupture locations within each source, which controls the distribution of site-to-source distances, is considered. In most cases, uniform probability distributions are assigned to hypocenter locations, implying that earthquakes are equally likely to initiate at any point within the source. These distributions are then combined with the source geometry to obtain the corresponding probability distribution of source-to-site distances, using the distance measure(s) appropriate for the GMM(s) considered in the next step. DSHA, on the other hand, typically assumes a probability of 1.0 that the portion of the source closest to the site will rupture.
3. The ground motions produced at the site by earthquakes of any size within the applicable range from Step (1), occurring at any possible hypocenter location in each source, are computed, typically with GMMs. The entire distribution of *IM* estimates from GMMs is considered in a PSHA, whereas a DSHA would typically compute an *IM* at a single, fixed percentile level.

4. Finally, the uncertainties in earthquake size, earthquake location, and ground motion *IM*s are combined to obtain the time rate that a specified level of the *IM* will be exceeded during a particular time period.

The proper performance of a PSHA requires careful attention to the problems of source characterization (Section 4.2.2), *IM* prediction (Sections 3.5–3.6), and the mechanics of the probability computations. The sections that follow emphasize the implementation of concepts described previously and describe the PSHA computations.

4.4.1 Distributions of Independent Variables Used in GMMs

A fundamental aspect of PSHA is the characterization of the range of input variables used to predict future ground motions for a particular site. In general, the three types of parameters used in ground motion prediction are related to source, path, and site conditions. The principal source parameter is magnitude. In PSHA, an appropriate magnitude distribution, generally selected from among the options in Section 4.2.2.4, is assigned to each source. Site parameters are not variable for a given location (i.e., sites have a certain condition, which does not vary from event-to-event in the same manner as source and path effects). The remainder of this section describes variability in the principal path parameter, source-to-site distance, as considered in PSHA.

Earthquakes can occur anywhere on a particular fault, so their hypocenters, or *nucleation points*, are usually assumed to be uniformly distributed across a source (i.e., fault rupture is considered equally likely to initiate at any location). The assumption of uniformity is by no means required; nonuniform distributions may be used (e.g., Mai et al., 2005). A uniform distribution within the source does not, however, often translate into a uniform distribution of source-to-site distance. Since GMMs use particular measures of source-to-site distance, the spatial uncertainty of earthquake locations must be used to compute the distribution of distance measures. The relative likelihood of realizing different values over the range of possible source-to-site distances is represented by a probability density function for distance, $f_R(r)$.

For the point source of Figure 4.15a, the epicentral distance, R, is known to be r_s; consequently, the probability that $R = r_s$ is assumed to be 1.0 and the probability that $R \neq r_s$, is 0.0. Other cases are not as simple. Consider the case of a linear source that produces earthquakes of zero rupture length (Figure 4.15b). The probability that an earthquake occurs on the small segment of the fault between $L = \ell$ and $L = \ell + d\ell$ is the same as the probability that it occurs between epicentral distances $R = r$ and $R = r + dr$; i.e.,

$$f_L(\ell)d\ell = f_R(r)dr \tag{4.19}$$

where $f_L(\ell)$ and $f_R(r)$ are the probability density functions for the variables L and R, respectively. Consequently, the distribution of epicentral distance is

$$f_R(r) = f_L(\ell)\frac{d\ell}{dr} \tag{4.20}$$

which produces a distribution of the type shown in Figure 4.15b.

Epicentral distance, however, is not a good predictor of *IM*s because it does not indicate how close the actual rupture is to a particular site of interest. Energy is not released from a single location in an actual earthquake – it is released over some finite fault rupture area and the level of shaking at a particular site depends on how close the site is to the zone of energy release. In the case of the linear source example in Figure 4.15, the applicable fault dimension is rupture length, which is commonly assumed to have a logarithmic mean that is a linear function of magnitude and to have a lognormally distributed error term. One empirical relationship for length distribution for

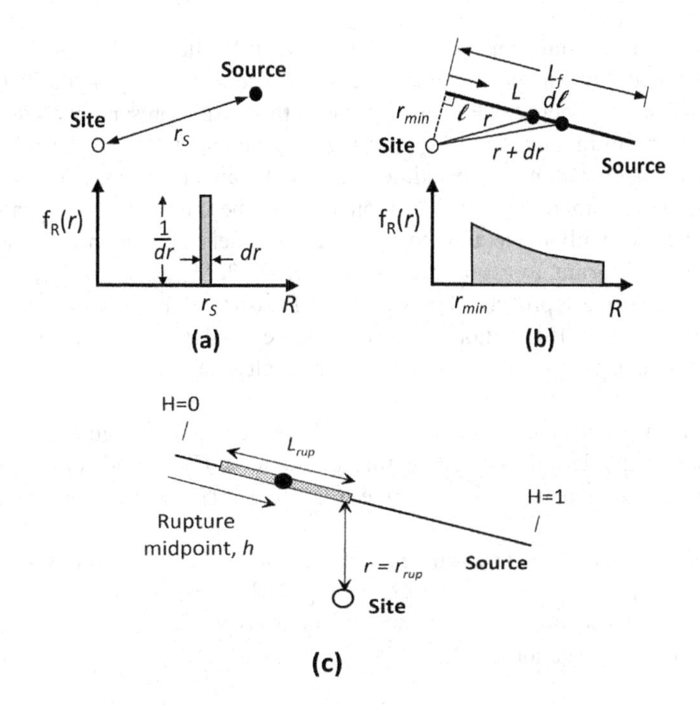

FIGURE 4.15 (a–c) Examples of variations of source-to-site distance for different source geometries

active tectonic regions has a mean of $\log_{10}(L_{rup}) = 0.74\mathbf{m} - 3.55$ (in km) and a standard deviation of $\sigma_{\log 10} = 0.23$ (Wells and Coppersmith, 1994).

When the finite size of fault ruptures is considered (e.g., Figure 4.15c), or for more complex source geometries, $f_R(r)$ is replaced with PDFs for the parameters controlling source distance. Consider, for example, the case shown in Figure 4.15c in which the distance measure of interest is the closest distance to the rupture surface, R_{rup}. That distance depends on the position of the midpoint of the ruptured portion of the fault (described by the normalized location, h) and the length of the fault rupture, L_{rup}. Taking r as the closest distance from the ruptured portion of the fault to the site, the probability of having a particular value of r is equivalent to the probability of the corresponding values of h and L_{rup}, both of which are defined by PDFs (typically uniform for h; log-normal for L_{rup}). Since the probability of r is $f_R(r)dr$, this can be expressed as:

$$P(R = r) = f_R(r)dr = f_{L_{rup}}(\ell_{rup})d\ell_{rup} f_H(h)dh \tag{4.21}$$

When using relations like Equation (4.21) in probability calculations, it is important to check that ruptures do not extend beyond the fault dimensions; for the case in Figure 4.15c in which the rupture location is specified at the midpoint, this can be achieved by taking $f_H(h)$ as zero from the ends of the fault to $L_{rup}/2$ inward from the ends. Randomization of the rupture length relationship turns out to not affect hazard results significantly (Bender, 1984), hence some hazard analyses are performed by taking fault geometric parameters as mean values. This would remove the first two terms from the analysis of $f_R(r)dr$ in Equation (4.21).

4.4.2 Earthquake Rates

To calculate the probabilities of various hazards occurring in a given time period, the distribution of earthquake occurrence with respect to time must be considered. The probability of earthquake occurrence in a specified time interval is typically modeled with a Poisson process, which assumes that earthquakes occur randomly in time. As described in Sections 4.2.2.3–4.2.2.5, the classical Gutenberg-Richter recurrence model is equivalent to an exponential magnitude PDF and a Poisson model to express time-dependence. The Poisson rate in this case represents the rate of earthquakes larger than the minimum magnitude, $\lambda_{M_{min}}$. Strategies for computation of the Poisson rate in the classical, time-independent manner, or in a time-dependent manner that considers time since the last large (characteristic) event are described in Section 4.2.2.5. The time-dependent approach allows the Poisson model to be used without violating the implications of elastic rebound theory (Section 2.5.1).

Fault models for California, as assembled by Field et al. (2009, 2014, 2015), provide instructive examples of how alternative approaches for developing Poisson rate parameters can be used in combination with the magnitude distributions described in Section 4.2.2.4. Earthquake rate parameters for piecewise rectangular faults and off-fault zones are evaluated as:

- Time-dependent models are considered for all piecewise rectangular faults, including those for which the date of the last rupture is unknown (Field and Jordan, 2015). Large ruptures on faults are also modeled as time-independent (Poisson), which is considered to be less accurate.
- For *off-fault areal sources* (i.e., sources that are not associated with any specific known fault), location-specific values of the Gutenberg-Richter *a* parameter (Equation 4.4) are used based on areal source zones with observed rates of seismicity (Section 4.2.1.4). The *b* value is taken as 1.0 (Section 4.2.2.3).

Earthquake rates on all source types are evaluated to achieve regional moment balance, in consideration of a variety of available geologic, geodetic, and seismological data.

4.4.3 Probability Computations

The results of a PSHA can be expressed in different ways. All involve some level of probabilistic computations to combine the distributions of earthquake magnitudes and site-source distances, the distributions of ground motion *IM*s for each magnitude and distance combination, and the rate of earthquake occurrence to estimate seismic hazards. A common approach involves the development of *seismic hazard curves*, which indicate the mean annual rate of exceedance of different values of a selected *IM*. The seismic hazard curves can then be used to compute the probability of exceeding the selected *IM* in a specified period of time.

PSHA calculations can be performed in a number of ways, but have historically been performed for individual (scalar) *IM*s. The response of a structure or system to earthquake shaking, however, may be more accurately predicted if more than one *IM* is known. It is possible to predict the joint occurrence of multiple *IM*s using a *vector PSHA*.

4.4.3.1 Scalar PSHA

Seismic hazard curves can be obtained for individual sources and combined to express the aggregate hazard at a particular site. The basic concept of the computations required for the development of seismic hazard curves is fairly simple. The probability of exceeding a particular value, *im*, of a single, or *scalar*, ground motion *IM* is calculated for one possible earthquake at one possible source location and then multiplied by the probability that that particular magnitude earthquake would occur at that particular location. The process is then repeated for all possible magnitudes and locations with the probabilities of each summed. The required calculations are described in the following paragraphs. For a given earthquake occurring on a given source, the probability that a single ground motion parameter, *IM*, will exceed a particular value, *im*, given that an earthquake (event *E*) has occurred, can be computed using the total probability theorem, i.e.,

$$P[IM > im \mid E] = P[IM > im \mid \mathbf{X}]P[\mathbf{X}] = \int_{\text{all } x} P[IM > im \mid \mathbf{X}] f_{\mathbf{X}}(\mathbf{x})\, d\mathbf{x} \qquad (4.22)$$

where $\mathbf{X}$ is a vector of random variables that influence IM and $f_{\mathbf{X}}(\mathbf{x})$ is the *joint pdf* (Section D.5.3) of $\mathbf{X}$. Equation (4.22) and variants thereof are often referred to as *hazard integrals*. In most cases the quantities in $\mathbf{X}$ that are varied within the integral are limited to the magnitude, $\mathbf{M}$, and distance, R, so $f_{\mathbf{X}}(\mathbf{x})$ can be written as $f_{\mathrm{M,R}}(\mathbf{m},r)$. Assuming that $\mathbf{m}$ and r are independent, then $f_{\mathrm{M,R}}(\mathbf{m},r)=f_{\mathrm{M}}(\mathbf{m})$ $f_{\mathrm{R}}(r)$, and the probability of exceedance can be written as

$$P[IM > im \mid E] = \int_{\mathbf{M}}\int_{R} P[IM > im \mid \mathbf{m},r] f_M(\mathbf{m}) f_R(r)\, d\mathbf{m}\, dr \tag{4.23}$$

where $P[IM > im \mid \mathbf{m},r]$ is obtained from the GMM and $f_M(\mathbf{m})$ and $f_R(r)$ are the probability density functions for magnitude and source-site distance, respectively. Note that if $\mathbf{m}$ and r are *not* independent, the integration would take place across the joint probability density function, $f_{\mathrm{M,R}}(\mathbf{m}, r)$.

The individual components of Equation (4.23) are, for virtually all realistic PSHAs, sufficiently complicated that the integrals cannot be evaluated analytically. Numerical integration, which can be performed by a variety of different techniques, is therefore required. One approach, used here for simplicity and clarity rather than efficiency, is to divide the possible ranges of magnitude and distance into N_M and N_R segments, respectively. The probability of exceedance can then be estimated as

$$P[IM > im \mid E] \approx \sum_{k=1}^{N_R}\sum_{j=1}^{N_M} P\left[IM > im | \mathbf{M} = \mathbf{m}_j, R = r_k\right] P\left[\mathbf{M} = \mathbf{m}_j\right] P\left[R = r_k\right] \tag{4.24}$$

where $\mathbf{m}_\mathrm{j}=\mathbf{M}_{\mathrm{min}}+(j\text{-}0.5)(\mathbf{M}_{\mathrm{max}}-\mathbf{M}_{\mathrm{min}})/N_{\mathrm{M}}$ and $r_k=r_{\mathrm{min}}+(k-0.5)(r_{\mathrm{max}}-r_{\mathrm{min}})/N_R$. This is equivalent to assuming that each source is capable of generating only N_M different earthquakes of magnitude, $\mathbf{m}_\mathrm{j}$, at only N_R different source-to-site distances, r_k. The three components on the right side of Equation 4.24 are illustrated graphically for a single $\{\mathbf{M}, R\}$ pair in Figure 4.16. The accuracy of

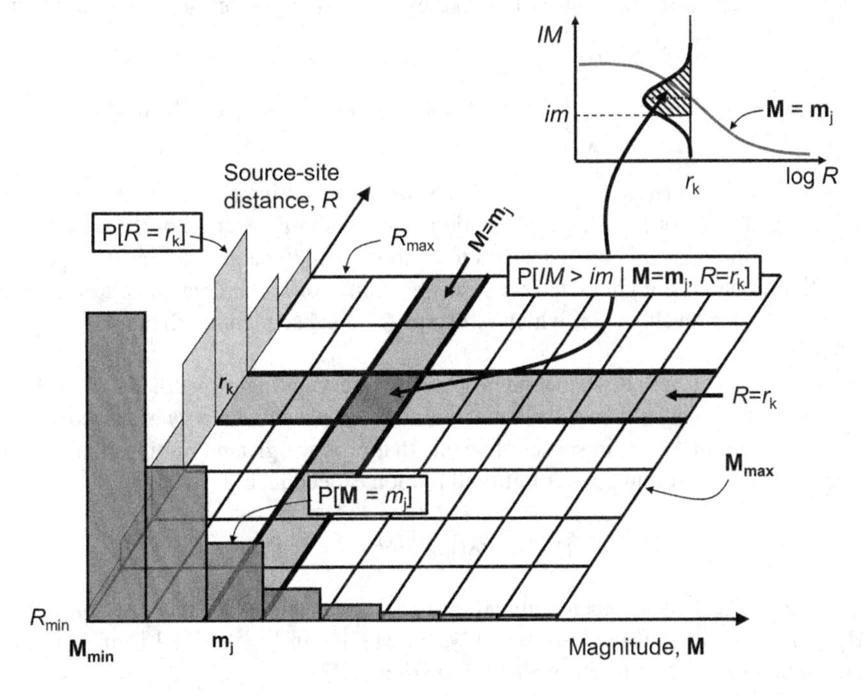

FIGURE 4.16 Schematic illustration of main components of PSHA probability calculations. Products of terms in boxes are summed over all combinations of $\mathbf{M}$ and R (in Equation 4.24) to obtain $P[IM > im \mid E]$.

this simple numerical integration procedure increases with increasing N_M and N_R. More refined methods of numerical integration (e.g., Bradley et al., 2009) will provide greater accuracy for the same values of N_M and N_R.

Equation (4.23) provides the conditional probability required to compute a hazard curve. However, a refinement of the conditional probability calculation provides useful insight into the role of ground motion variability. Since *IM*s are found to be lognormally distributed, the probability of exceeding a specific value, *im,* for a given **m**-*r* pair can be expressed as

$$P[IM > im \mid \mathbf{m}, r] = 1 - \Phi\left[\frac{\ln im - \mu_{\ln IM|\mathbf{m},r}}{\sigma_{\ln IM|\mathbf{m},r}}\right] \tag{4.25}$$

where $\mu_{\ln IM|\mathbf{m},r}$ and $\sigma_{\ln IM|\mathbf{m},r}$ are the median and standard deviation of *IM* from a GMM in natural log units conditional on magnitude **m** and distance *r.* Defining term "epsilon" as the number of logarithmic standard deviations by which the natural log of *im* exceeds the logarithmic mean for that **m**-*r* pair,

$$\varepsilon = \frac{\ln im - \mu_{\ln IM|\mathbf{m},r}}{\sigma_{\ln IM|\mathbf{m},r}} \tag{4.26a}$$

the conditional probability of exceeding *im,* which appears in Equations (4.23) and (4.24), can be written more compactly as

$$P[IM > im \mid \mathbf{m}, r] = 1 - \Phi[\varepsilon] \tag{4.26b}$$

Epsilon, ε, describes how "unusual" the value of *im* is – highly positive values represent particularly strong motions, highly negative values represent particularly weak motions, and values near zero represent typical (median-level) motions. Utilizing the parameter, ε, the probability of exceeding *im* conditional on the occurrence of earthquake event *E* in Equation (4.23) can alternatively be written as

$$P[IM > im \mid E] = \int_{\mathbf{M}}\int_{R}\int_{\varepsilon} P[IM > im \mid \mathbf{m}, r, \varepsilon] f_{M,R,\varepsilon}(\mathbf{m}, r, \varepsilon)\, d\mathbf{m}\, dr\, d\varepsilon \tag{4.27}$$

Note that the probability term in the integrand is now binary, taking on values of 1.0 if the inequality is true and 0.0 if it is not for each combination of **m**, *r,* and ε, because of the specification of ε. With an additional integral, this form appears to be more complicated than that of Equation (4.23), however, it allows convenient characterization of the values of ε that contribute to the exceedance probability at different levels of *im,* which is useful for a subsequent application (disaggregation, Section 4.4.3.5).

Equations (4.23) and (4.27) describe the probability of exceeding a specified *IM* level (*im*) given that an earthquake has occurred on a particular source. The rate at which *im* is exceeded, which can be thought of as a rate of a certain subset of events that produce ground motions that exceed *im,* is also of interest. The mean annual rate of ground motion exceedance is given by

$$\lambda_{IM}(im) = \lambda_{\mathbf{M}_{\min}} P[IM > im \mid E] \tag{4.28}$$

where $\lambda_{\mathbf{M}_{\min}} = \lambda_M(\mathbf{M}_{\min})$ is the mean annual rate of earthquakes that exceed the minimum magnitude, $\mathbf{M}_{\min}$. The conditional probability, $P[IM > im \mid E]$, can be obtained from either Equation (4.23) or, more commonly in current practice, Equation (4.27).

If the site of interest is in a region of N_S potential earthquake sources, each having its own rate of earthquakes, λ_i, with magnitudes $\mathbf{m} > \mathbf{M}_{\min}$, and the occurrences of earthquakes on the various sources are statistically independent, the total average exceedance rate for the site can be obtained by adding the rates from each of the sources

$$\lambda_{IM}(im) = \sum_{i=1}^{N_S} \lambda_i \int_{\mathbf{M}} \int_{R} \int_{\varepsilon} P[IM > im \mid \mathbf{m}, r, \varepsilon] f_{M,R,\varepsilon}(\mathbf{m}, r, \varepsilon) d\mathbf{m}\, dr\, d\varepsilon \tag{4.29}$$

Integrating numerically, the average exceedance rate can then be estimated as

$$\lambda_{IM}(im) \approx \sum_{i=1}^{N_S} \sum_{j=1}^{N_M} \sum_{k=1}^{N_R} \lambda_i P[IM > im \mid \mathbf{m}_j, r_k] f_M(\mathbf{m}_j) f_R(r_k) \Delta\mathbf{m} \Delta r \tag{4.30}$$

where $\mathbf{m}_j$, r_k, $\Delta\mathbf{m}$, and Δr are defined as before. Note that Eq. (4.30) returns to the *im* exceedance probability represented as a real number and not a 0.0 or 1.0 binary. An equivalent but more compact form of Equation (4.30) utilizing Eq. (4.26b) to replace the probability of exceedance term can be written as

$$\lambda_{IM}(im) \approx \sum_{i=1}^{N_S} \sum_{j=1}^{N_M} \sum_{k=1}^{N_R} \lambda_i \left[1 - \Phi(\varepsilon)\right] f_M(\mathbf{m}_j) f_R(r_k) \Delta\mathbf{m} \Delta r \tag{4.31a}$$

which is also equivalent to

$$\lambda_{IM}(im) \approx \sum_{i=1}^{N_S} \sum_{j=1}^{N_M} \sum_{k=1}^{N_R} \lambda_i \left[1 - \Phi(\varepsilon)\right] P(\mathbf{M} = \mathbf{m}_j) P(R = r_k) \tag{4.31b}$$

When rate calculations of the type given by Equation (4.31) are carried out for a range of *IM* levels (as given by *im*) for a given *IM*, the resulting plot of *im* against rate of exceedance is termed a *seismic hazard curve.* Hazard curves for peak ground acceleration are shown in Figure 4.17 for several regions in the US. For any particular hazard curve, relatively low values of *IM* (i.e., weak levels of shaking) occur relatively frequently and relatively high values of *IM* (strong shaking) will occur more rarely. The *return period* of a particular intensity value, *im*, is the reciprocal of its mean annual rate of exceedance, i.e.

$$T_R(im) = \frac{1}{\lambda_{IM}(im)} \tag{4.32}$$

Accordingly, areas of high hazard have short return periods for a given ground motion level, whereas areas of lower hazard have longer return periods.

As shown in Figure 4.17, areas near plate boundaries (coastal California, Seattle) are high-hazard, meaning high exceedance rates for a given *IM*. In contrast, stable continental regions (central and eastern North America) have relatively low hazard. Slopes of hazard curves are also regionally variable. As one moves to the right on a hazard curve (increasing *IM*), both the earthquake rate and conditional ground motion exceedance probability (Equation 4.23) change. In areas with low rates of seismicity and sources that are modeled with Gutenberg-Richter magnitude-frequency distributions, increasing ground motions are associated with rarer, larger magnitude earthquakes (decreasing λ_m), and higher conditional probability terms (*P*) (because larger earthquakes produce higher *IM*s). These counter-acting effects can produce relatively flat hazard curves, as in stable continental regions near faults or in seismic zones with a history of large but low-recurrence rate earthquakes (Memphis, Charleston). In contrast, if moving right on the hazard curve does not appreciably change magnitude, which can occur if hazard is dominated by relatively high-rate characteristic events, then λ_m remains nearly constant, and the *P* term controls the hazard curve slope, which is relatively steep. This occurs near plate boundaries where earthquake rates can be relatively high relative to much of the λ_{IM} range in the plot.

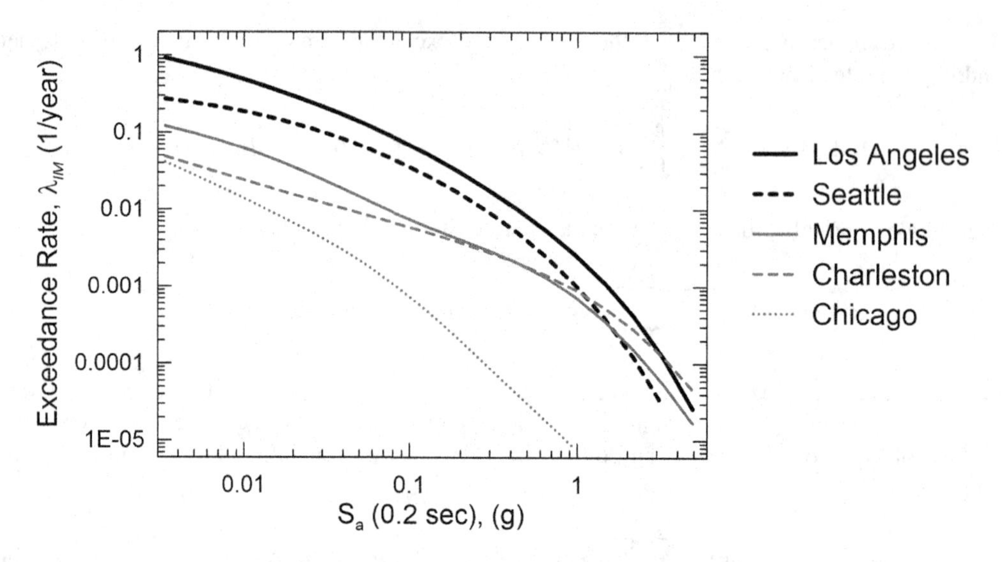

FIGURE 4.17 Hazard curves for S_a(0.2 sec) at rock site conditions (V_{S30}=760 m/s) at various locations across the US (using 2018 seismic hazard model from USGS, Petersen et al., 2020).

Example 4.4

Consider the site and fault in Figure E4.4a with the indicated location coordinates and, using the fault parameters shown in the figure, compute a seismic hazard curve for *PGA*.

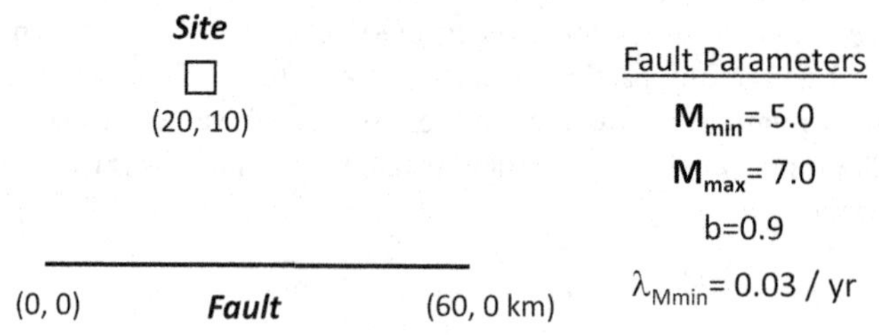

Site

(20, 10)

(0, 0) ***Fault*** (60, 0 km)

Fault Parameters

$\mathbf{M}_{\mathbf{min}}$= 5.0

$\mathbf{M}_{\mathbf{max}}$= 7.0

b=0.9

λ_{Mmin}= 0.03 / yr

FIGURE E4.4A Position of site relative to linear fault and fault parameters for Example 4.4

Solution:

The PSHA can be performed by applying the four previously described steps:

1. This step involves the identification and characterization of earthquake sources, including the characterization of the probability distribution and recurrence rates of the full range of earthquake magnitudes that each source is capable of producing. To simplify the calculation, four magnitude bins are considered as shown in the table below. Using the mid-magnitude in each bin, the probability density using a truncated exponential model is computed using Equation (4.9).

$$f_M(\mathbf{m}) = \frac{\beta e^{-\beta(\mathbf{m}-\mathbf{M}_{\mathbf{min}})}}{1-e^{-\beta(\mathbf{M}_{\mathbf{max}}-\mathbf{M}_{\mathbf{min}})}}$$

where $\beta = b \cdot \ln(10)$. The corresponding probability masses for each rupture size are computed as $P(\mathbf{m}) = f_M(\mathbf{m})d\mathbf{m}$ (where the bin width, $\mathbf{dm} = 0.5$).

Magnitude Bin	Mid-Value	$f_M(m)$	$P(m)$
5.0–5.5	5.25	1.254	0.627
5.5–6.0	5.75	0.445	0.223
6.0–6.5	6.25	0.158	0.079
6.5–7.0	6.75	0.056	0.028

These results are summarized in the portion of Table E4.4 marked as *Step 1.*

2. The second step considers alternative rupture locations within the source, which control the rupture distances that must be considered in hazard calculations. A uniform probability distribution is assigned to hypocenter locations along the fault as parameterized by the normalized position coordinate h as shown in Figure 4.15. Six bins of h are considered along with their probability masses in the table below.

h	$P(h)$
0.083	0.167
0.250	0.167
0.417	0.167
0.583	0.167
0.750	0.167
0.917	0.167

Distance calculations are affected by rupture size (length in this case) and hypocenter location h as shown in Figure 4.15. As indicated in the text below Equation (4.20), logarithmic mean rupture length is taken as $log_{10}\left(L_{rup}\right) = 0.74\mathbf{m} - 3.55$ (L_{rup} in km; from Wells and Coppersmith, 1994). Table E4.4 shows distances calculated for the six realizations of h for each magnitude, each of which has a unique L_{rup} (to simplify the calculations, uncertainty in rupture dimensions is not considered in this example). The probability mass (related to the distance PDF, $f_R(r)$) for each distance is computed using Equation (4.21) adapted to probability masses as

$$P(R=r) = f_R(r)dr = P\left(\ell_{rup}\right)P(h)$$

These results are provided in the portion of Table E4.4 marked as *Step 2.*

3. For each magnitude and distance combination, median ground motions (denoted $\mu \mid \mathbf{m}, r$) are computed using the simple Campbell (1981) GMM in Equation (3.25). These are shown in the *Part 3* portion of Table E4.4. As indicated in Example 4.3, the use of the Campbell model is for illustrative purposes only – it is a simple model that can be relatively easily applied to illustrate the steps in PSHA.
4. The probability calculation begins by computing the epsilon (ε) value corresponding to a particular target ground motion level (*im*) for each magnitude and distance combination using Equation (4.26a). A target ground motion level of 0.1g is used in Table E4.4. This value of ε is used with standard normal CDF table (Table D.1) to compute $F_{IM}(\varepsilon)$. The value $1 - F_{IM}(\varepsilon)$ is the probability that ε is exceeded, which is equivalent to the probability that the target ground motion *im* is exceeded for the particular $\mathbf{m}$ and r combination. The combined probability (integrand in Equation 4.23) is then computed. The summation provides the condition probability that $IM > im$ given that an earthquake has occurred (Equation 4.23), which is 0.281 for the target *im* value.

   The mean annual rate of ground motion exceedance is computed using Equation (4.28) as $\lambda_{IM}(im = 0.1g) = \lambda_{\mathbf{M}_{\min}} P(IM > im \mid E) = 0.03 \times 0.281 = 0.0084$/year. This result represents one point on a *PGA* seismic hazard curve as shown in Figure E4.4b. Repeating this calculation for many values of *im* provides the entire hazard curve.

TABLE E4.4
Calculations for Mean Annual Rate of Exceeding PGA Value of 0.1g

Step 1		Step 2						Step 3		Step 4			
m	P(m)	L_{rup} (km)	$P(L_{rup})$	h	$P(h)$	$r(h,L)$	$P(r)$	$\mu_{\ln PGA}\|\mathbf{m},r$	$\sigma_{\ln PGA}$	ε (im = 0.1g)	$F_{IM}(\varepsilon)$	$1 - F_{IM}(\varepsilon)$	$P(\mathbf{m})P(r)P(PGA > im\|\mathbf{m},r)$
5.25	0.627	2.16	1.0	0.083	0.167	17.14	0.1667	0.0594	0.372	1.400	0.919	0.081	0.0084
5.25	0.627	2.16	1.0	0.250	0.167	10.74	0.1667	0.0916	0.372	0.237	0.594	0.406	0.0425
5.25	0.627	2.16	1.0	0.417	0.167	10.74	0.1667	0.0916	0.372	0.237	0.594	0.406	0.0425
5.25	0.627	2.16	1.0	0.583	0.167	17.14	0.1667	0.0594	0.372	1.400	0.919	0.081	0.0084
5.25	0.627	2.16	1.0	0.750	0.167	25.92	0.1667	0.0396	0.372	2.489	0.994	0.006	0.0007
5.25	0.627	2.16	1.0	0.917	0.167	35.36	0.1667	0.0290	0.372	3.331	1.000	0.000	0.0000
5.75	0.223	5.07	1.0	0.083	0.167	15.98	0.1667	0.0925	0.372	0.210	0.583	0.417	0.0155
5.75	0.223	5.07	1.0	0.250	0.167	10.30	0.1667	0.1350	0.372	−0.807	0.210	0.790	0.0293
5.75	0.223	5.07	1.0	0.417	0.167	10.30	0.1667	0.1350	0.372	−0.807	0.210	0.790	0.0293
5.75	0.223	5.07	1.0	0.583	0.167	15.98	0.1667	0.0925	0.372	0.210	0.583	0.417	0.0155
5.75	0.223	5.07	1.0	0.750	0.167	24.59	0.1667	0.0620	0.372	1.287	0.901	0.099	0.0037
5.75	0.223	5.07	1.0	0.917	0.167	33.97	0.1667	0.0452	0.372	2.134	0.984	0.016	0.0006
6.25	0.079	11.89	1.0	0.083	0.167	13.49	0.1667	0.1518	0.372	−1.123	0.131	0.869	0.0114
6.25	0.079	11.89	1.0	0.250	0.167	10.00	0.1667	0.1912	0.372	−1.743	0.041	0.959	0.0126
6.25	0.079	11.89	1.0	0.417	0.167	10.00	0.1667	0.1912	0.372	−1.743	0.041	0.959	0.0126
6.25	0.079	11.89	1.0	0.583	0.167	13.49	0.1667	0.1518	0.372	−1.123	0.131	0.869	0.0114
6.25	0.079	11.89	1.0	0.750	0.167	21.52	0.1667	0.1021	0.372	−0.057	0.477	0.523	0.0069
6.25	0.079	11.89	1.0	0.917	0.167	30.73	0.1667	0.0737	0.372	0.821	0.794	0.206	0.0027
6.75	0.028	27.86	1.0	0.083	0.167	10.06	0.1667	0.2559	0.372	−2.526	0.006	0.994	0.0046
6.75	0.028	27.86	1.0	0.250	0.167	10.00	0.1667	0.2568	0.372	−2.536	0.006	0.994	0.0046
6.75	0.028	27.86	1.0	0.417	0.167	10.00	0.1667	0.2568	0.372	−2.536	0.006	0.994	0.0046
6.75	0.028	27.86	1.0	0.583	0.167	10.06	0.1667	0.2559	0.372	−2.526	0.006	0.994	0.0046
6.75	0.028	27.86	1.0	0.750	0.167	14.92	0.1667	0.1942	0.372	−1.785	0.037	0.963	0.0045
6.75	0.028	27.86	1.0	0.917	0.167	23.32	0.1667	0.1360	0.372	−0.827	0.204	0.796	0.0037
													0.2809 Σ
													0.0084 $\lambda_{IM} = \Sigma \times \lambda_{M_{min}}$
													0.2234 $P = 1 - \exp(-\lambda_{IM} \times \Delta t)$ where $\Delta t = 30$ years

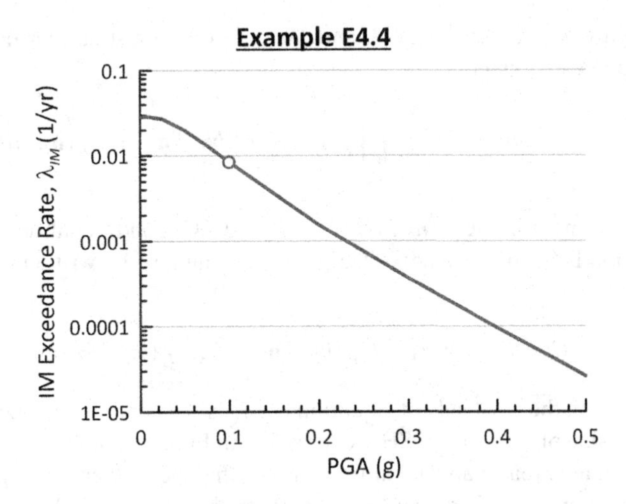

FIGURE E4.4B Hazard curve derived for Example 4.4, with the point computed in Table E4.4 marked.

4.4.3.2 Vector PSHA

As discussed in Section 1.6, earthquake engineers use *IM*s to predict the seismic response of systems of interest and characterize that response in terms of engineering demand parameters, or *EDP*s. Because no single *IM* captures all of the ground motion characteristics that influence response, there will always be some uncertainty in *EDP* for a given *IM* (i.e., in $EDP|IM$). That uncertainty can often be reduced, however, with the addition of more information about the ground motions. For example, it would seem logical (and will be discussed in Chapter 5) that the permanent displacement of an unstable slope could be predicted more accurately if both the amplitude and duration of an earthquake motion are known rather than just the amplitude.

The *joint rates of exceedance* of multiple *IM*s can be predicted using *vector PSHA*. Bazzurro and Cornell (2002) described a procedure for evaluating the mean annual joint rate of occurrence of the elements of a vector of ground motion *IM*s. The procedure is more easily understood by considering a scalar PSHA expressed in a different form. As shown in Section 4.4.3.1, the mean annual rate of exceedance of some value of a scalar $IM = im$ for a given source can be expressed (noting that, despite the boldface notation, magnitude is a scalar quantity) as

$$\lambda_{IM}(im) = \lambda_{\mathbf{M}_{\min}} \int_{\mathbf{M}} \int_{R} P[IM > im \mid \mathbf{m}, r] f_{M,R}(\mathbf{m}, r)\, d\mathbf{m}\, dr \tag{4.33}$$

The conditional probability $P[IM > im \mid \mathbf{m}, r]$ is the complementary cumulative distribution function, $G_{IM}(im \mid \mathbf{m}, r) = 1 - F_{IM}(im \mid \mathbf{m}, r)$. Differentiating the expression for the mean rate of exceedance in Equation (4.33) with respect to *IM* produces the mean rate density

$$MRD_{IM}(im) = \lambda_{\mathbf{M}_{\min}} \int_{\mathbf{M}} \int_{R} f_{IM}(im \mid \mathbf{m}, r) f_{M,R}(\mathbf{m}, r)\, d\mathbf{m}\, dr \tag{4.34}$$

where

$$f_{IM}(im \mid \mathbf{m}, r) = \frac{1}{im\,\sigma_{\ln IM|\mathbf{m},r}\sqrt{2\pi}} \exp\left[-\frac{1}{2}\left(\frac{\ln im - \mu_{\ln IM|\mathbf{m},r}}{\sigma_{\ln IM|\mathbf{m},r}}\right)^2\right] \tag{4.35}$$

for a lognormally distributed *IM*.

Considering the case where two *IM*s, IM_1 and IM_2, are of interest at a particular site, the *joint mean rate density* can be expressed as

$$MRD_{IM_1,IM_2}(im_1,im_2) = \lambda_{\mathbf{M}_{\min}} \int_{\mathbf{M}}\int_{R} f_{IM_1,IM_2}(im_1,im_2 \mid \mathbf{m},r) f_{M,R}(\mathbf{m},r)\, d\mathbf{m}\, dr \tag{4.36}$$

where $f_{IM_1,IM_2}(im_1,im_2 \mid \mathbf{m},r)$ is the *joint PDF* of IM_1 and IM_2 conditional upon $\mathbf{M}=\mathbf{m}$ and $R=r$. As indicated in Section D.5.3 of Appendix D, a joint PDF can also be written in conditional form, i.e., as

$$f_{IM_1,IM_2}(im_1,im_2 \mid \mathbf{m},r) = f_{IM_1}(im_1 \mid \mathbf{m},r) f_{IM_2|IM_1}(im_2 \mid im_1,\mathbf{m},r) \tag{4.37}$$

Note that the first term on the right side of Equation (4.37) is the conditional PDF of a single *IM* (as appeared on the right side of the scalar PSHA expression in Equation 4.33). Assuming that IM_1 and IM_2 are jointly lognormal (a reasonable assumption given that they are each marginally lognormal), the conditional distribution comprising the second term on the right side of Equation (4.37) will also be lognormal. The conditional distribution can then be expressed as

$$f_{IM_2|IM_1}(im_2 \mid im_1,\mathbf{m},r) = \frac{1}{im_2 \sigma_{\ln IM_2|im_1,\mathbf{m},r}\sqrt{2\pi}} \exp\left[-\frac{1}{2}\left(\frac{\ln im_2 - \mu_{\ln IM_2|im_1,\mathbf{m},r}}{\sigma_{\ln IM_2|im_1,\mathbf{m},r}}\right)^2\right] \tag{4.38}$$

The logarithmic mean and standard deviation of the conditional distribution are (Benjamin and Cornell, 1970)

$$\mu_{\ln IM_2|im_1,\mathbf{m},r} = \mu_{\ln IM_2|\mathbf{m},r} + \rho_{1,2}\frac{\sigma_{\ln IM_2|\mathbf{m},r}}{\sigma_{\ln IM_1|\mathbf{m},r}}\left(\ln im_1 - \mu_{\ln IM_1|\mathbf{m},r}\right) \tag{4.39a}$$

and

$$\sigma_{\ln IM_2|im_1,\mathbf{m},r} = \sigma_{\ln IM_2|\mathbf{m},r}\sqrt{1-\rho_{1,2}^2} \tag{4.39b}$$

where $\rho_{1,2}$ is the correlation coefficient for the natural logarithms of IM_1 and IM_2 and $\sigma_{\ln IM_1|\mathbf{m},r}$ and $\sigma_{\ln IM_2|\mathbf{m},r}$ are obtained from the GMM.

Substituting Equations (4.38) and (4.39) into Equation (4.37), the joint mean rate density function can be obtained. Computing this function, therefore, requires the availability of GMMs for each *IM* and the correlations between both *IM*s in the vector (e.g., Baker and Jayaram, 2008; Bradley, 2011). By integrating over the *IM* values, a corresponding joint mean rate of exceedance surface (the multi-dimensional analog to the one-dimensional curve obtained for a scalar *IM*) is obtained as:

$$\lambda_{IM_1,IM_2} = MRE_{IM_1,IM_2}(im_1,im_2) = \int_{IM_1}\int_{IM_2} MRD_{IM_1,IM_2}(im_1,im_2)\, d(im_2)\, d(im_1)$$

where the mean rate density in the integrand is given in Equation (4.36).

4.4.3.3 Finite Time Periods

The *IM* exceedance rates described in Section 4.4.3.1 can be combined with the Poisson model (Section D.7.1.2) to estimate the probability of exceeding a particular *IM* level (*im*) in a finite time interval. From Equation (D.48), the probability of exceedance of *im* in a time period, Δt, is

$$P[N \geq 1] = 1 - e^{-\lambda_{IM}(im)\Delta t} \tag{4.40}$$

Expanding $1 - e^{-\lambda_{IM}(im)\Delta t}$ in a Taylor series shows that

$$P[N \geq 1] \approx \lambda_{IM}(im)\Delta t \tag{4.41}$$

for small values of $\lambda_{IM}(im)\Delta t$ (or, equivalently, small values of $\Delta t/T_R$) that are normally of interest for design. This result is often used to approximate the annual ($\Delta t = 1$ yr) probability of exceedance as being equal to λ_{IM}.

Hazard maps show the spatial variation of a given *IM* at a particular probability level; the example in Figure 4.18 applies for the *IM* of S_a(0.2 sec), $V_{S30} = 760$ m/s site condition, and the 2% probability of exceedance in 50-year hazard level. Finally, as shown in Figure 4.19, a *uniform hazard spectrum* (UHS) can be defined for a given site by computing hazard curves for spectral accelerations at multiple periods and plotting the spectral accelerations vs. period for a given exceedance rate. All ordinates of the UHS, therefore, have the same return period, i.e., the same probability of exceedance in a particular period of time.

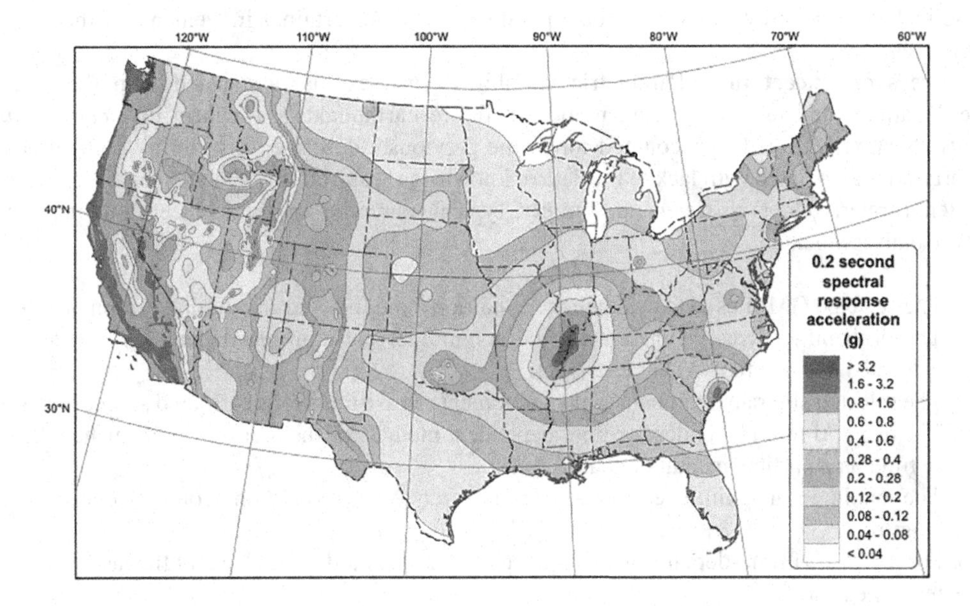

FIGURE 4.18 Shaded map of S_a(0.2 sec) spectral acceleration (expressed as a percentage of gravity) with 2% probability of exceedance in 50 years for rock sites with V_{S30}=760 m/s. (Petersen et al., 2020; used with permission of SAGE Publications, Ltd.)

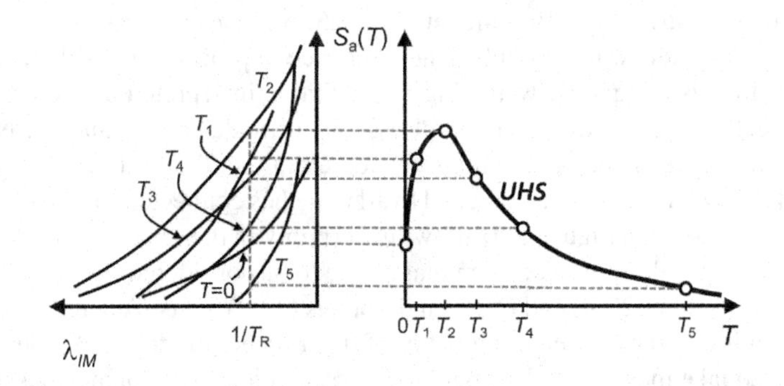

FIGURE 4.19 Schematic showing the development of uniform hazard spectrum for return period, T_R, from spectral acceleration hazard curves at multiple oscillator periods.

Example 4.5

Returning to Example 4.4, what is the probability that an acceleration of 0.1*g* would be exceeded in a 30-year period?

Solution:

This probability can be computed using the rate computed in Example 4.4 with the equation for a Poisson process (Equation 4.40):

$$P[N \geq 1] = 1 - e^{-\lambda_{IM}(im)\Delta t}$$

where N is the number of occurrences of the event and $\Delta t = 30$ years. Using $\lambda_{IM}(0.1) = 0.0084$/year and $\Delta t = 30$ years provides a probability of 0.22.

4.4.3.4 Managing Epistemic Uncertainty

PSHAs are intended to capture the effects of all uncertainties that can affect ground motions. These include variability in hazard model parameters and uncertainty in elements of the models themselves.

Sources of Uncertainty: Parametric variability, such as that associated with the magnitude, locations, and level of ground motion in future earthquakes, is a form of *aleatory variability* (Section D.8) and is accounted for in the previously described probability computations. Uncertainties associated with lack of the data or knowledge required to model the "true" behavior of a system is referred to as *epistemic uncertainty*. Typical sources of epistemic uncertainty in seismic hazard analyses include:

1. The selected GMM used to evaluate the conditional median and standard deviation of *IM*s.
2. The probability density function for the magnitudes of future earthquakes, $f_M(\mathbf{m})$ (e.g., truncated exponential vs characteristic).
3. The selected slip rate on known faults or moment release rates for distributed source zones.
4. The use of time-independent or time-dependent models for the recurrence of earthquakes in time on a particular fault or source zone.
5. The maximum magnitude earthquake for a source ($\mathbf{M}_{\mathbf{max}}$) conditioned on its size (area or length).
6. For the case of time-dependent earthquake occurrence models, the date of the last characteristic earthquake.

As discussed in Section D.8, epistemic uncertainty can be reduced by additional knowledge, but aleatory variability cannot.

Logic Trees: The use of *logic trees* (Power et al., 1981; Kulkarni et al., 1984; Coppersmith and Youngs, 1986; McGuire, 2004; Bommer et al., 2005) provides a convenient framework for the explicit treatment of epistemic uncertainty. The logic tree approach allows the use of alternative models, each of which is assigned a weighting factor that is interpreted as the probability of that model being correct. The logic tree consists of a series of nodes and branches. Nodes represent points at which models are specified and branches represent the different models specified at each node. The sum of the weighting factors assigned to all branches connected to a given node must be 1. The simple logic tree shown in Figure 4.20 allows uncertainty in the selection of GMMs, magnitude distribution models, and values of maximum magnitude to be considered. In this logic tree, GMMs by Campbell and Bozorgnia (2014) and Chiou and Youngs (2014) are considered equally likely to be correct, hence each is assigned a weighting factor of 0.5. Proceeding to the next level of nodes, the characteristic earthquake magnitude distribution is considered to be 50% more likely to be correct than the truncated exponential, so the models are assigned weighting factors of 0.6 and 0.4, respectively. At the final level of nodes, different weighting factors are assigned to different maximum

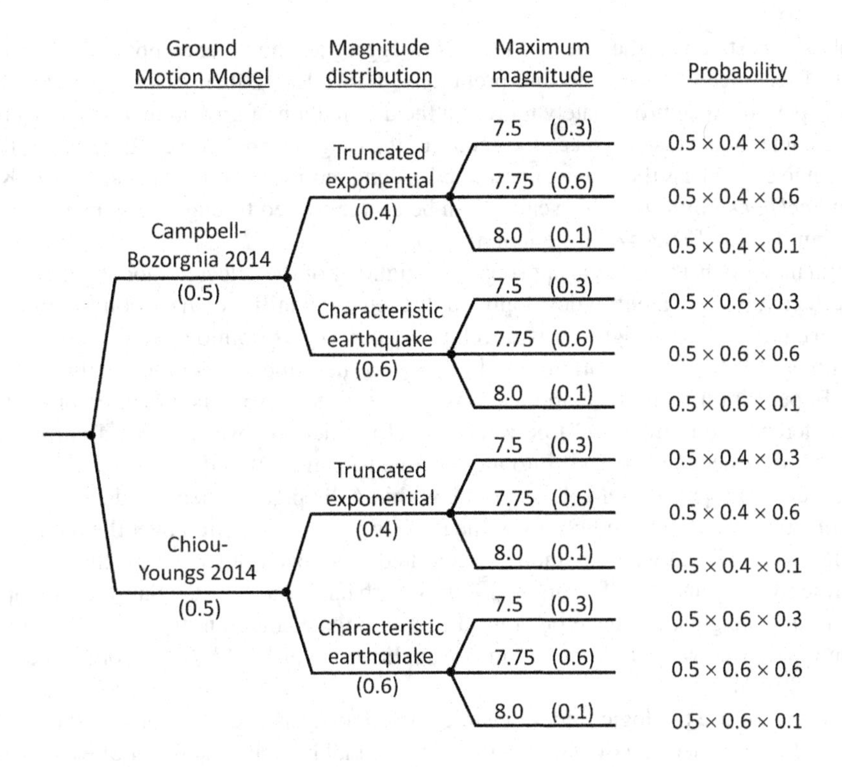

FIGURE 4.20 Simple logic tree for incorporation of model uncertainty.

magnitude values. This logic tree terminates with a total of 2×2×3=12 (no. of GMMs×no. of magnitude distributions×no. of maximum magnitudes) branches. The relative likelihood of each combination of models and/or parameters implied by each terminal branch is given by the product of the weighting factors of the terminal branch and all prior branches leading to it. Hence, the relative likelihood of the combination of the CB14 GMM, truncated exponential magnitude distribution, and maximum magnitude of 7.5 is 0.5×0.4×0.3=0.06. The sum of the relative likelihoods of the terminal branches, or of those at any prior level, is equal to 1.

The product of the weighting factors leading to each terminal branch can be viewed as probabilities or subjective degrees of belief, that all models along its path are correct. This type of treatment requires that the branches emanating from a single node of the logic tree be mutually exclusive (i.e., not a combination of models) and collectively exhaustive (the models at each node account for all possibilities). Achieving these goals is challenging, often involving modeling a process with few observations (e.g., rates and magnitude distributions of large magnitude earthquakes) exhaustively. As a result, logic trees must be designed and branches populated very carefully.

The GMM branches of the logic tree are among the most critical. There are two essential steps to the selection and weighting of GMMs for use in a logic tree. First, a GMM selection process should be undertaken whereby a reasonable number of candidate models are selected from the hundreds of models available in the literature (Douglas, 2024). This selection process will typically require adopted models to be in the appropriate tectonic regime (e.g., active tectonic region), to have been derived from well-documented databases, to be applicable for the *IM*s of interest (typically S_a for a range of oscillator periods), to be applicable for the magnitude and distance ranges of interest, to have been derived using a functional form that appropriately captures physical attributes of ground motion scaling (Sections 3.4–3.5), and to have been developed using regression procedures that correctly consider both between- and within-event sources of ground motion variability (Bommer et al., 2010; Stewart et al., 2015). For the selected models to be considered as collectively exhaustive, some applications state that the selected models in aggregate should encompass the "center, body, and range of technically defensible interpretations" (US NRC, 2012). After selecting a series of models, it is possible to approximately represent the distribution of ground motions they provide by selecting a single model (usually near the center of the range of predictions) and scaling the model up and down to encompass the range of predicted ground motions (Atkinson et al., 2014). Known as the *scaled backbone approach*, this scaling can be accomplished through adjustment of the model constant term (i.e., coefficient e_0 in Equation 3.26).

A second major consideration is appropriate weighting of the selected models in the logic tree, which is effectively equivalent to developing a discrete probability distribution to represent epistemic uncertainties. This weighting is developed from expert opinion (Scherbaum and Kuehn, 2011; Bommer, 2012), which is often based largely on the same considerations that affect model selection. Expert opinion for such applications can be informed by visualization procedures that map the differences (or "distance") between model predictions over a defined parameter space (typically, **M**, distance, and period) into a two-dimensional (2D) plane. For example, Sammon's maps (Figure 4.21) represent GMMs as points within a 2D plane, where models clustered in the same region provide similar predictions of the median of ground motion over the parameter space (e.g., models 4–6, 12–13) whereas models separated from each other within the plane are relatively distinct (e.g., 2 and 16) (Sammon, 1969; Scherbaum et al., 2010). Sammon's maps do not define logic tree weights, but have been used to guide the evaluation of GMM epistemic uncertainties and logic tree weights for some major applications of PSHA (e.g., Bommer et al., 2015; Goulet et al. 2021a).

Application: To use the logic tree, a seismic hazard analysis is carried out for the combination of models and/or parameters associated with each terminal branch. The result of each analysis is a seismic hazard curve corresponding to the "path" through the logic tree with an associated likelihood. As shown in Figure 4.22a, for a given value of *im*, there are N_{br} values of exceedance rate λ_{IM} (where N_{br} is the number of branches in the logic tree), each having a weight (or relative likelihood) as described above. To illustrate the manner in which these results are utilized to develop a composite hazard curve, consider an individual point in Figure 4.22a as having exceedance rate $\lambda_{IM,i}$ with a

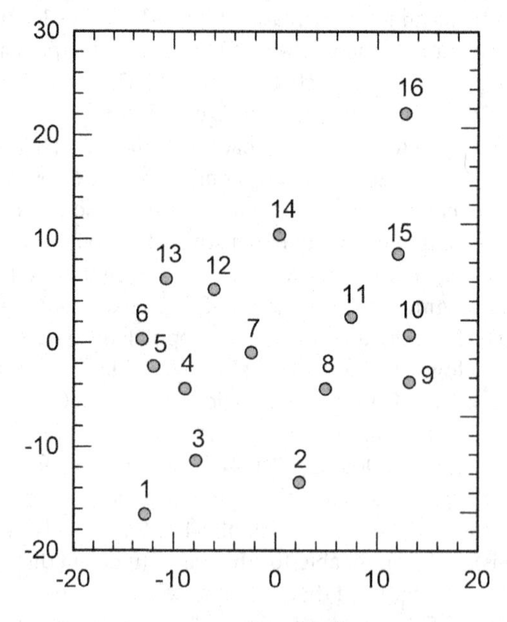

FIGURE 4.21 Sammon's map for 16 GMMs. The distances between GMMs in the map represent differences in model predictions (both log mean and standard deviation) over a prescribed range of **M**, distance, and period. (Modified from Scherbaum et al., 2010.)

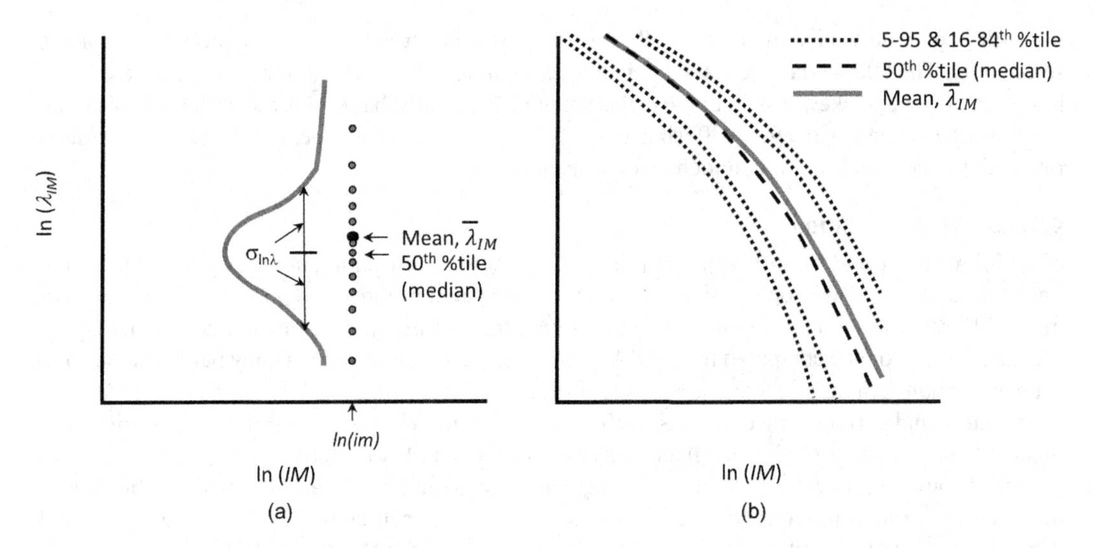

FIGURE 4.22 (a) Alternate realizations of computed annual probabilities for given *IM* level (*im*) and PDF of the distribution, reflecting epistemic uncertainty; (b) alternate representations of hazard curve in terms of a mean hazard curve and various percentile curves.

corresponding weight, w_i. The weighted average, weighted log mean (median), weighted variance, and weighted standard deviation of the probability of exceedance can then be computed as:

$$\overline{\lambda}_{IM} = \sum_{i=1}^{N_{br}} w_i \lambda_{IM,i} \tag{4.42a}$$

$$\mu_{\ln\lambda} = \sum_{i=1}^{N_{br}} w_i \ln \lambda_{IM,i} \tag{4.42b}$$

$$Var(\lambda_{IM}) = \sum_{i=1}^{N_{br}} w_i \left(\ln \lambda_{IM,i} - \mu_{\ln\lambda}\right)^2 \tag{4.42c}$$

$$\sigma_{\ln\lambda} = \sqrt{Var(\lambda_{IM})} \tag{4.42d}$$

The use of natural log mean and variance in Equation (4.42) assumes that exceedance rates are log-normally distributed, which is not always the case but usually matches results better than a normal distribution. The quantities $\overline{\lambda}_{IM}$ and $\sigma_{\ln\lambda}$ are illustrated in Figure 4.22a.

The hazard curve representing the ensemble of $\overline{\lambda}_{IM}$ values for various *im* is referred to as the *mean hazard curve*. The mean hazard curve, which is often used for design (McGuire et al., 2005), is sensitive to the dispersion from epistemic uncertainty. For example, if the distribution of exceedance rates is log-normal,

$$\overline{\lambda}_{IM} = \exp\left(\mu_{\ln\lambda} + \frac{\sigma_{\ln\lambda}^2}{2}\right) \tag{4.43}$$

(Section D.7.2.3). As such, mean hazard can be sensitive to outlier values of hazard associated with combinations of models having low weight. Another option is to base design upon a percentile value of the hazard curve (Abrahamson and Bommer, 2005). For example, as illustrated in Figure 4.22b, the 84th percentile value indicates that 84% of the hazard curves have a lower probability for the selected *im*. The 50th percentile is the median hazard curve, which is generally lower than the mean hazard curve due to skewness in the distribution. The percentile-based representation of epistemic uncertainty has the advantage of offering analysts the ability to select a percentile level commensurate with the desired level of confidence for a given project.

4.4.3.5 Disaggregation

The PSHA procedures described in the preceding sections allow computation of the mean annual rate of exceedance at a particular site based on the aggregated hazard from potential earthquakes of many different magnitudes on multiple sources occurring at many different source-site distances. The rate of exceedance computed in a PSHA, therefore, is not associated with any particular seismic source, earthquake magnitude, or source-site distance.

In many applications, however, it is useful to identify the ranges of magnitudes and distances that contribute most strongly to a given hazard level for the *IM* of interest, which can be accomplished through a process known as *disaggregation*. Disaggregation can also provide the degree to which the ground motions for a given hazard level from each combination of magnitude and distance is above the median *IM* level, expressed in terms of ε (Equation 4.26a). These quantities are used, for example, to select ground motion records for response analyses, as described in Section 4.5.

Disaggregation requires that the mean annual rate of exceedance be expressed as a function of magnitude and/or distance. Computationally, this simply involves the removal of terms from the summations of Equation (4.30). For example, the mean annual rate of exceedance for a given magnitude $\mathbf{m_j}$ and distance r_k can be expressed as:

$$\lambda_{IM}\left(im,\mathbf{m}_{\mathrm{j}},r_k\right)\approx P\left[\mathbf{M}=\mathbf{m}_{\mathrm{j}}\right]P\left[R=r_k\right]\sum_{i=1}^{N_S}\lambda_i P\left[IM>im\mid\mathbf{m}_{\mathrm{j}},r_k\right] \tag{4.44}$$

The probability terms before the summation represent the likelihood of having an earthquake with the selected magnitude and distance combination. Those terms can be computed as:

$$\begin{aligned}P[\mathbf{M}=\mathbf{m}_{\mathrm{j}}]&=f_M\left(\mathbf{m}_{\mathrm{j}}\right)\Delta\mathbf{m}\\ P[R=r_k]&=f_R\left(r_k\right)\Delta r\end{aligned} \tag{4.45}$$

The terms within the summation in Equation (4.44) represent the rates of exceedance of ground motion level *im* for magnitude $\mathbf{m_j}$ and distance r_k from each source *i*. Recall that the probability term in the summation is computed through an integration of the $f_{IM}(im|m,r)$ PDF (the integration is over ε).

To include ε in the disaggregation, Equation (4.44) is rewritten to evaluate the ground motion exceedance rate for magnitude $\mathbf{m_j}$ and distance r_k and epsilon ε_l:

$$\lambda_{IM}\left(im,\mathbf{m}_{\mathrm{j}},r_k,\varepsilon_l\right)\approx P\left[\mathbf{M}=\mathbf{m}_{\mathrm{j}}\right]P\left[R=r_k\right]P\left[\varepsilon=\varepsilon_l\right]\sum_{i=1}^{N_S}\lambda_i P\left[IM>im\mid\mathbf{m}_{\mathrm{j}},r_k,\varepsilon_l\right] \tag{4.46}$$

As in Equation (4.27), the probability in the summation is binary, being unity if $IM>im$ and zero if $IM<im$ for the given values of magnitude, distance, and ε. The probability term for ε can be computed as:

$$P\left[\varepsilon=\varepsilon_l\right]=f_{IM}\left(\mathbf{m}_{\mathrm{j}},r_k,\varepsilon_l\right)\Delta\varepsilon \tag{4.47}$$

where the conditional PDF f_{IM} is given in Equation (4.35). In disaggregations that include ε, the values of ε increase as *im* increases. For very large *im* values, only the highest **M** and lowest *R* may contribute significantly to the hazard, and may then only reach that *im* with very large ε values.

Disaggregation results are typically expressed as relative contributions (*RC*) to hazard, expressed in terms of exceedance probabilities, as a function of magnitude, distance, and sometimes, ε. Those *RC*s are computed from the rate in Equation (4.46) as follows:

$$RC = \frac{1 - e^{-\lambda_{IM}\left(im, \mathbf{m}_j, r_k, \varepsilon_l\right)\Delta t}}{1 - e^{-\lambda_{IM}(im)\Delta t}} \tag{4.48}$$

The term in the denominator utilizes the rate of special events for all magnitudes, distances, and ε values, as in Equation (4.40) to effectively normalize the individual contributions so that the sum of all *RC*s is 1.0.

Figure 4.23 shows example disaggregations for a site in downtown Los Angeles; the disaggregations are for the *IM*s of PGA and S_a (2.0 sec) at return periods of 200 years and 2,475 years. Note that these two factors must always be specified in a disaggregation (the *IM* under consideration and the hazard level, or return period). The disaggregation plots show that the ground motion hazard at the site has contributions from local, modest magnitude sources (**M**=6–7, *R*=5–20 km) that rupture infrequently and the more distant San Andreas fault (**M**=7–8, *R*=50–60 km), which ruptures relatively frequently. The San Andreas contribution is much stronger for S_a (2.0 sec) than for PGA, which is caused by the stronger scaling of long-period motions with **M** and the relatively rapid distance attenuation of high-frequency vs low-frequency components of ground motion (San Andreas events produce modest PGA at the site, but strong S_a (2.0 sec)). Comparing the 200 and 2,475 year results, the latter have larger values of ε and generally larger *RC*s from the local faults (the longer return period on ground motion allows these relatively infrequent earthquakes to contribute more strongly).

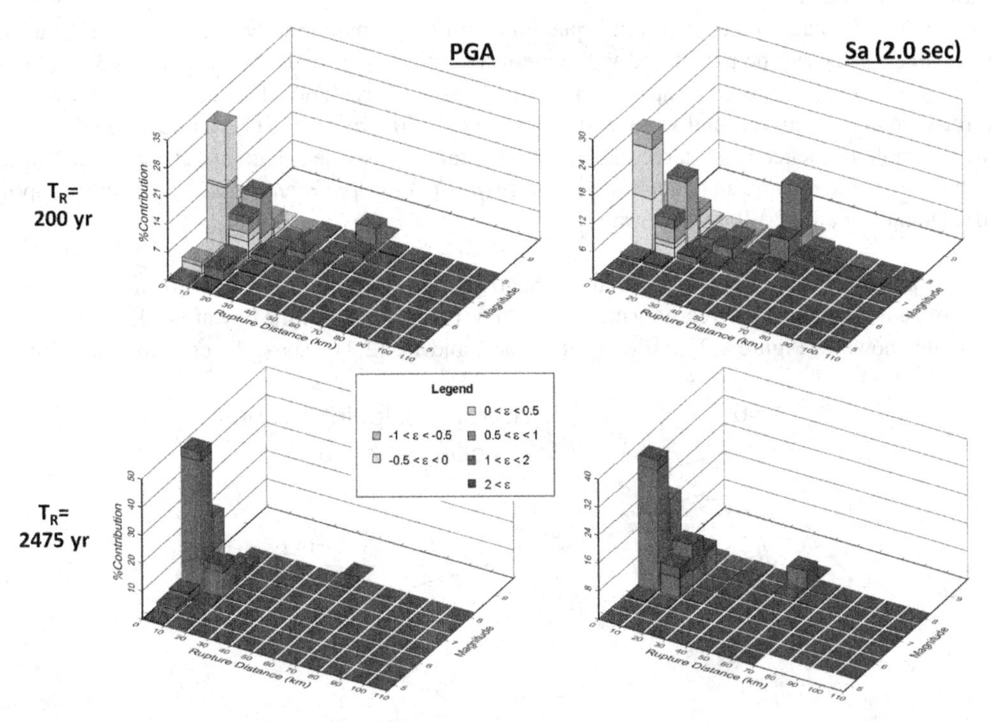

FIGURE 4.23 Example hazard disaggregations for PGA and S_a (2.0 sec) at return periods of 200 and 2,475 years for downtown Los Angeles site (V_{S30}=285 m/s).

4.4.3.6 Incorporating Near-Fault Effects

Near-fault effects have the potential to modify ground motions at close distances to large faults relative to the predictions from GMMs (Section 3.7). Customarily, scalar or vector PSHA is first performed without consideration of near-fault effects, and then disaggregation is performed to identify sources that strongly contribute to the hazard. It may be advisable to consider near-fault effects when large active faults at close distance to the site are a major contributor to the hazard. For example in Figure 4.23, the San Andreas fault contributes at distances of around 50–60 km, which is too far for near-fault effects to be significant. On the other hand, the relatively local faults at distances between 5 and 20 km could produce important near-fault effects.

Models for near-fault effects require additional parameters beyond those used in conventional GMMs. In particular, the critical parameter for rupture directivity effects is the location of the hypocenter on the ruptured fault surface, because the hypocenter location (in combination with rupture dimensions and site location) determines what fraction of the fault rupture is toward a site of interest (e.g., Figure 3.69). Figure 4.24 shows the hypocenter location using dimensionless parameter $\mathcal{H}$, which can be varied, and its effects incorporated into the hazard integral. Referring to Equation (4.22), this is accomplished by incorporating directivity into the GMM, including in $\mathbf{X}$ (the vector of parameters that affect *IM*s) the hypocenter location, and modifying the joint PDF $f_{\mathbf{X}}(\mathbf{x})$ to include a PDF on $\mathcal{H}$ as follows:

$$P[IM > im \mid E] = \int_{\mathbf{M}}\int_{R}\int_{\mathcal{H}} P[IM > im \mid \mathbf{m}, r, \mathcal{H}] f_M(\mathbf{m}) f_R(r) f_{\mathcal{H}}(\mathcal{H})\, d\mathbf{m}\, dr\, d\mathcal{H} \tag{4.49}$$

The PDF, $f_{\mathcal{H}}(\mathcal{H})$, represents the likelihood of hypocenters occurring at different locations within a finite fault. Hypocenters are more likely to occur near the center of a fault (both along-strike and down-dip) than near the edges, and their locations have been described using Weibull and Gamma distributions (Mai et al., 2005).

Extending the hazard integral as in Equation (4.49) is computationally intensive. As an alternative, scalar PSHA can be performed without consideration of directivity (Section 4.4.3.1) and the results subsequently adjusted to approximately account for directivity. This approach involves disaggregation of the conventional scalar PSHA for the *IM* of interest at the target rate to see if the hazard is strongly influenced by a large fault at close distance. If so, near-fault effects are likely, which can be incorporated using an adaptation of the *composite distribution approach* (Watson-Lamprey, 2018; Donahue et al., 2019) as follows:

1. From the disaggregation plot, find the relative contribution to the hazard from the proximate fault for which directivity effects are to be considered. For the downtown Los Angeles site shown in Figure 4.23, *IM* of S_a at 2.0 sec, and T_R=2,475 years, the contribution of the

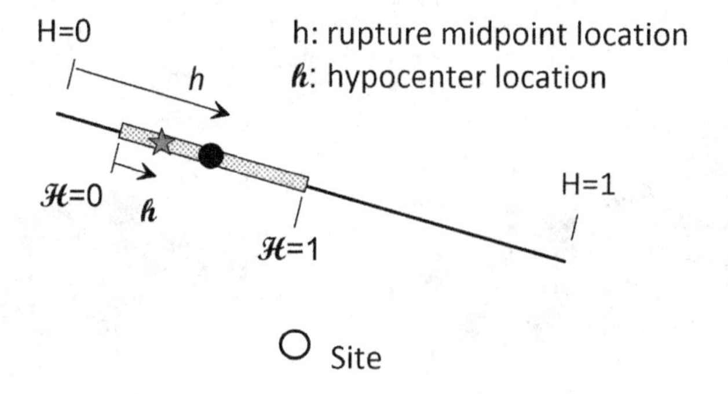

FIGURE 4.24 Schematic of fault, ruptured segment, and parameterization of the location of ruptured segment midpoint and hypocenter location.

closest fault (Puente Hills) can be taken from the bins for distances of 0–10 km, which provides $RC \approx 0.6$. Only sources that contribute significantly to the hazard ($RC > \sim 0.3$) are worth considering for possible near source effects.

2. The change of ground motion from that provided by conventional scalar PSHA can be expressed by a modified *IM* computed as:

$$f_D \approx \ln(IM_{haz}) + RC\exp(\Delta\mu_{IM}) \tag{4.50}$$

where $ln(IM_{haz})$ is the natural log *IM* from the conventional PSHA (without directivity), and f_D is the location-dependent mean change in ground motion resulting from directivity effects. Maps of the spatial variation of f_D for randomized hypocenter locations, different magnitudes, and different focal mechanisms (Figure 3.70) can be created. The Los Angeles downtown site is located on the hanging wall of the Puente Hills fault, and f_D is approximately 0.15. Given RC=0.6, the change in ground motion from Equation (4.50) is 0.1 in ln units or 10%.

4.5 GROUND MOTION SELECTION AND MODIFICATION

DSHA and PSHA (Sections 4.3–4.4) produce values of ground motion *IM*s. Many applications in earthquake engineering use those *IM*s directly; for example, spectral accelerations are used to compute base shear forces in buildings (e.g., as specified in building codes) and can be adapted to spectral displacements for pushover-type methods of analysis for structures. However, relatively sophisticated earthquake engineering applications, particularly in a performance-based design context, utilize response history analysis of geotechnical or structural systems to assess seismic demands and their distribution through the analyzed system. While in principle seismic demands could be predicted directly for a series of faults, and a range of earthquake ruptures on each fault, in combined simulations of wave propagation and structural response (Bradley, 2012), the analysis of seismic demands in practice requires the intermediate step of computing ground motion hazard in terms of *IM*s. These *IM*s are used to guide the selection (and possibly modification) of a series of acceleration histories (often colloquially referred to as 'ground motions') used in response analyses. Multiple ground motions are selected to capture a range of frequency contents, durations, amplitudes, and phases, so that the effects of those variabilities on structural response can be evaluated. The ground motions are typically taken from earthquake recordings (Section 3.2) but may include simulated motions as well (Section 3.6). The selection and modification of ground motions should consider the manner in which they will be used (e.g., for response history analysis of a structure with a certain fundamental period). For example, some common questions pertaining to the aims of response history analyses that affect the selection/modification process include:

1. Will the results of the response history analyses be used to evaluate the median response conditional on the hazard level or an alternate percentile level (typically 84th or 95th)? Accurate establishment of the median response typically requires few records, whereas response at a higher percentile level, i.e. toward the upper tail of the response distribution, requires relatively more.
2. Will the response be computed in a single horizontal direction or multi-dimensionally?
3. How many hazard levels are to be considered?
4. What modal periods (one or more) are critical for analysis of structural or system response?

The following sections describe considerations in the selection (4.5.1) and modification (4.5.2) of ground motions for engineering application.

4.5.1 Ground Motion Selection

The ground motion selection process should be preceded by a PSHA and disaggregation for the *IM*(s) of greatest interest at the appropriate hazard level. The disaggregation identifies the earthquake events that most strongly contribute to the hazard for the selected *IM* and return period. Those events are identified in magnitude/distance space. Disaggregation can also identify other attributes, such as epsilon (ε: see Equation 4.26a) that directly impact ground motions. These conditions, along with site parameters, present a large number of variables that could be considered for ground motion selection. Ground motion selection has been the subject of extensive study with many recommendations put forth. However, most of the methods fall into one of three categories: selection based principally on earthquake attributes (magnitude, distance, etc.); selection based on ground motion record attributes (amplitude, frequency content, duration); and selection based on match to a target response spectrum. These approaches are described below.

4.5.1.1 Selection Based on Earthquake Attributes

Traditional methods for ground motion selection prioritize records that are generally compatible with the disaggregation results and site condition. Numerous specific protocols for this process have been proposed. One example (Stewart et al., 2001) suggested that:

- Selected records should have magnitudes within 0.25 magnitude units of the "target" (i.e., the mean or mode from disaggregation). Style of faulting is generally not considered in record selection.
- Selected records should have generally compatible source-site distances to the target, particularly when near-fault effects are likely to be present.
- Selected records should come from sites having comparable site conditions to the target site; for example, recordings from sites in basins should not be applied to non-basin sites.

The target conditions used for selection come from the disaggregation, but may comprise multiple events. For example, the conditions shown in Figure 4.23 depict contributions from close-distance, moderate magnitude events along with more distant, large magnitude events. Both event types could be used to define target magnitudes and distances for the target site. This technique is equally applicable to single-component or multi-component ground motions.

4.5.1.2 Selection Based on Record Attributes

A better approach to ground motion selection is to identify the features of ground motions that are likely to affect their damage potential when applied to structures, evaluate parameters that represent those features from disaggregation, and then prioritize those parameters in the ground motion selection process. For most geotechnical and structural systems, the ground motion features likely to affect system performance include:

1. The *amplitude of shaking*, often characterized by the spectral acceleration at the elastic, first-mode period (T_0) of a soil-structure system.
2. *Spectral shape*, particularly the spectral amplitudes at periods longer than T_0 (to account for period lengthening due to inelastic response) and at periods shorter than T_0 (to account for higher modes that, if present, could significantly contribute to the response). These spectral shape considerations are unimportant for single-mode, elastic systems (such as the single degree-of-freedom (SDOF) oscillator used to define the response spectrum), but can be critical for yielding systems and multi-degrees-of-freedom (MDOF) systems.
3. The *duration* of shaking, which affects systems comprised of materials whose strength or stiffness degrades with the number of loading cycles (e.g., most soils and many structures).

Criterion 1 can be satisfied by choosing spectral acceleration at T_0 [$S_a(T_0)$] as the *IM* for hazard analysis and selecting records (possibly following scaling) that match the $S_a(T_0)$ value at the selected hazard level. While straightforward in principle, $S_a(T_0)$ values can be quite large for the long return periods typically used in earthquake engineering practice, and there may be relatively few recordings with comparable spectral accelerations.

To understand the effects of yielding referred to in Criterion 2, consider the simple oscillator in Figure 4.25a, with the alternate force-deflection relationships in Figure 4.25b. If an earthquake produces a structural deflection of u_0, the stiffness of the elastic structure is k, whereas the secant stiffness of the yielding structure at the same deflection level is reduced to k_{eff}. Since the vibration periods of the structures are $T_0 = 2\pi/\omega_0 = 2\pi\sqrt{m/k}$, as the stiffness is reduced by yielding of the structural system or even brittle failures of non-structural components like partition walls or brick infill walls, the period lengthens. Now, consider two earthquake events having the same spectral acceleration at T_0 (hence, the amplitudes represented in Criteria 1 are identical), but different spectral shapes, as shown in Figure 4.25c. The damage potential of Event 2 is greater than that of Event 1, because as the natural period of the structure increases as it yields, it will respond more strongly to Event 2.

These important effects of spectral shape on damage indicate that the features of earthquakes and resultant ground motions that contribute to spectral shape are important to identify from disaggregation and consider in record selection. Section 3.5 showed that magnitude, distance, and site condition affect the spectral shape in an overall sense. However, such parameters are not particularly descriptive of the attributes of individual records and have proven to be relatively ineffective in ground motion selection (Baker and Cornell, 2006). On the other hand, as shown in Figure 4.26, epsilon (ε) is strongly correlated to shape. Figure 4.26a and b shows two records with negative and positive ε at

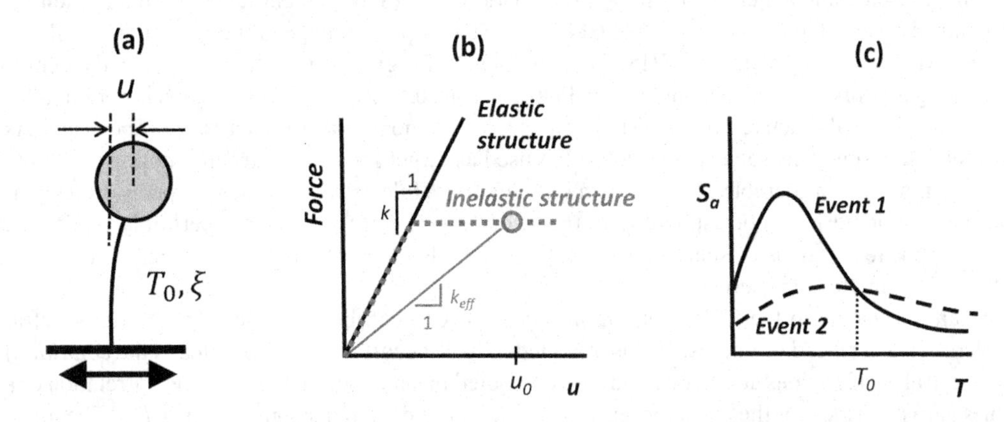

FIGURE 4.25 (a) Oscillator; (b) force-displacement relationship for elastic and inelastic oscillators; (c) ground motions from different events with different spectral shapes but common $S_a(T_0)$.

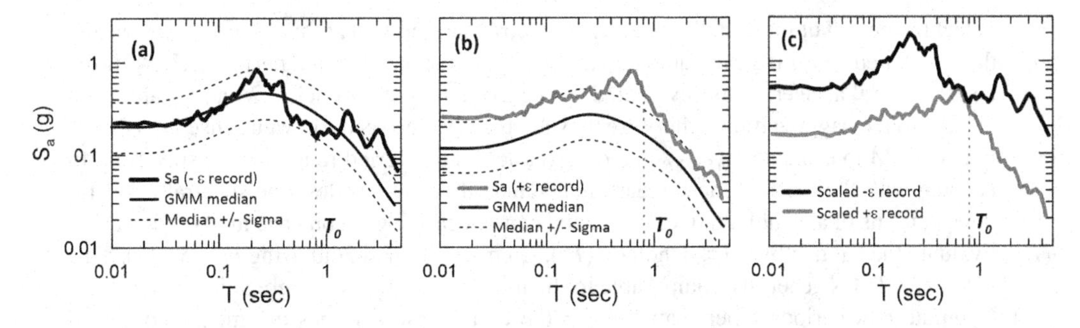

FIGURE 4.26 Illustration of the effect of epsilon (ε) on spectral shape: (a) negative ε record for T_0=0.8 sec, (b) positive ε record for T_0=0.8 sec, and (c) both records scaled to median GMM spectrum at T_0=0.8 sec.

a period of 0.8 sec. As shown in Figure 4.26c, where the two motions are scaled to the same spectral acceleration at T_0=0.8 sec, the positive ε motion has a steeper slope (i.e. 'relatively low S_a values) for $T>T_0$ than the negative ε motion. Accordingly, ε is descriptive of spectral shape and is a useful parameter to guide ground motion selection (Baker and Cornell, 2006; Goulet et al., 2007).

Based on this logic, a second approach for record selection eschews earthquake and site parameters and focuses on characteristics of the records, specifically ε and amplitude. The selection process therefore prioritizes records that require only modest scaling to match a target first-mode spectral acceleration [$S_a(T_0)$] and that have ε close to target values identified through disaggregation. The $S_a(T_0)$ and ε values can be for single-component ground motions or, if multi-component motions are desired, for a combination of horizontal components (e.g., RotD50). If the system under consideration is likely subject to cyclic degradation effects (e.g., a reinforced concrete building or a liquefiable soil deposit), duration could be considered as an additional parameter in the selection process (e.g., Chandramohan et al., 2016).

4.5.1.3 Selection to Match Target Spectrum

The logic behind the use of ε to describe spectral shape can be taken a step further by defining a full response spectrum to serve as a target, and then selecting records compatible with that target. The definition of the target and the manner by which the match of a record to that target is quantified comprise the principal aspects of this approach, which are described below.

Target Response Spectrum: Design response spectra can be produced by DSHA or PSHA (Sections 4.3 and 4.4, respectively). In either case, the spectra will most often consist of relatively extreme realizations of spectral accelerations across the full period range. Deterministic spectra are often developed using a mean + one standard deviation level (84th percentile) of ground motion for the controlling event, meaning that ε=+1.0 across the entire period range of the spectrum. Similarly, the uniform hazard spectrum, or UHS (Section 4.4.3.3), for ground motion return periods significantly longer than the recurrence intervals of hazard-controlling earthquakes (which is common for design in seismically active areas) will have positive (but nonuniform) ε across the spectrum. As a matter of convenience, such spectra are often used as target spectra for ground motion selection, but there may be considerable conservatism in doing so. While extreme values of spectral accelerations at specific periods, corresponding to the selected probability levels, are certainly realizable, simultaneous realizations of spectral accelerations at high ε levels across the full range of periods in the spectrum are extremely unlikely.

As an alternative to the UHS, *conditional spectra* (CS) can be computed. In CS, the spectral acceleration at a period of interest, T^*, is assumed to have occurred (i.e., that value is the *condition*) and spectral accelerations at other periods are computed in a manner that reflects the correlation of ε values across periods for the causative earthquake (identified from disaggregation at T^*). The mean value of the conditional spectrum is referred to as the *conditional mean spectrum* (CMS), which can be computed as follows (Baker, 2011):

1. Select the period of interest, T^*, the target spectral acceleration at that period, [$S_a(T^*)$], and the associated causative earthquake magnitude and distance for that period, [$\mathbf{M}$, R]. If the target spectral acceleration is evaluated from PSHA, then the values of $\mathbf{M}$ and R are mean values from disaggregation, otherwise they are the selected scenario values used in DSHA.
2. Use a GMM to evaluate $\mu_{\ln S_a|\mathbf{M},R}$ and $\sigma_{\ln S_a|\mathbf{M},R}$ across the period range for the spectrum. In keeping with the GMMs, these means and standard deviations describe spectral accelerations in terms of a combination of horizontal components (e.g., geometric mean or RotD50).
3. Evaluate the value of ε at T^*, denoted $\varepsilon(T^*)$, from Equation (4.26a) using $im = S_a(T^*)$. This value of $\varepsilon(T^*)$ is generally comparable to the mean value from disaggregation.
4. Compute ε at periods other than T^*. For the CMS these ε values are means computed from the correlation coefficients for spectral accelerations at different periods T and T^* [$\rho(T,T^*)$]. The mean of ε at period, T, given the value of ε at T^* is:

$$\mu_{\varepsilon(T)|\varepsilon(T^*)} = \rho(T,T^*)\varepsilon(T^*) \tag{4.51}$$

Recommended relations for ρ are given by Baker and Bradley (2017), although the following expression is reasonably accurate for periods between 0.05 and 5 sec (Baker and Cornell, 2006):

$$\rho(T,T^*) = 1 - \cos\left(\frac{\pi}{2} - \left(0.359 + 0.163 I_{T_{\min}<0.189} \ln\frac{T_{\min}}{0.189}\right)\ln\frac{T_{\max}}{T_{\min}}\right) \tag{4.52}$$

where $I_{T\max}$<0.189 is 1.0 for T_{min}<0.189 and 0 otherwise, and T_{min} and T_{max} are the smaller and larger of the two periods T and T^*, respectively.

5. Compute values of the CMS as:

$$\mu_{\ln S_a(T)|\ln S_a(T^*)} = \mu_{\ln S_a|\mathbf{M},R}(T) + \rho(T,T^*)\varepsilon(T^*)\sigma_{\ln S_a|\mathbf{M},R} \tag{4.53}$$

Values computed using Equation (4.53) are the means of CS. More generally, CS have a conditional standard deviation (Baker, 2011):

$$\sigma_{\ln S_a(T)|\ln S_a(T^*)} = \sigma_{S_a(T^*)|\mathbf{M},R}\sqrt{1-\rho^2(T,T^*)} \tag{4.54}$$

Either the CMS or the more general CS can serve as target spectra for ground motion selection, using procedures described below.

Figure 4.27a shows for an example site in Palo Alto, California, a UHS and CMS conditioned at three different periods (T^*=0.45, 0.85, and 2.6 sec). Multiple CMS conditioned at different periods may be used for structures with multiple potentially critical vibration modes. The CMS ordinates fall below those for the UHS for periods other than the matching period according to the correlation function (Equation 4.51). Figure 4.27b provides an example of CS for a matching period of T^*=2.6 sec; the CS are represented by the CMS and the CMS±one conditional standard deviation (Equation 4.54). The spectra for individual records shown in Figure 4.27b are a result of selection procedures that are described next.

Matching and Selection Procedures: Matching and selection procedures select individual ground motions whose spectra are most compatible with the target. Different procedures are used

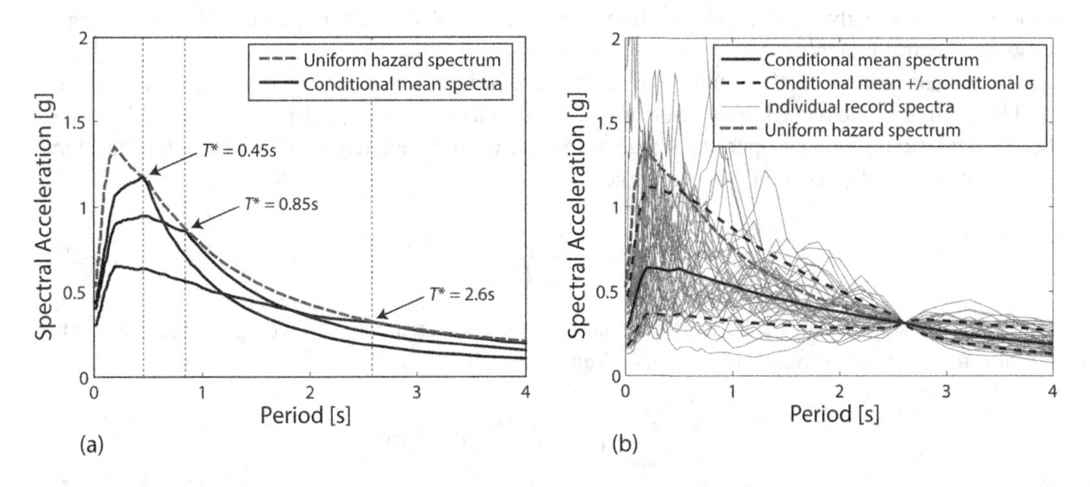

FIGURE 4.27 (a) Example Uniform Hazard Spectrum and Conditional Mean Spectra for an example site in Palo Alto, California, for a 2% in 50-year exceedance probability and with conditioning periods of T^*=0.45, 0.85, and 2.6 sec. (b) Conditional spectra for the same example with a conditioning period of T^*=2.6 sec. (Figure from Haselton et al., 2017; used with permission of SAGE Publications, Ltd.)

when the target is defined as an individual spectrum (e.g., UHS or CMS, Figure 4.27a) or as a distribution having a defined standard deviation at each period (i.e., CS, Figure 4.27b).

When the target is an individual spectrum, the goal of ground motion selection should be to match that spectrum as closely as possible. This can be done, for example, by minimizing the sum of squared differences (Wang et al., 2015), i.e., the *mean squared error*:

$$MSE = \frac{\sum_{i=1}^{n} w(T_i)\left(\ln S_{a,tar}(T_i) - SF \times \ln S_a(T_i)\right)^2}{\sum_{i=1}^{n} w(T_i)} \tag{4.55}$$

where n is the number of periods in the spectrum that are considered (e.g., often taken to encompass a period range of interest), $S_a(T)$ is the spectral acceleration of a ground motion record, $S_{a,tar}(T)$ is the individual target spectrum, SF is a scale factor that can optionally be applied to minimize MSE for a given record (details in Section 4.5.2), and $w(T)$ are period-dependent weights (the n values of weight must sum to unity). Values of MSE can be computed for all motions in a ground motion library, and the desired number of records having the lowest values of MSE can be selected. In Equation (4.55), $S_a(T)$ can be taken from individual components of ground motion or from a combination of horizontal components (e.g., geometric mean or RotD50), when multi-component motions are desired.

When the target spectrum has a distribution (CS), the selection process identifies records that collectively have the desired period-dependent means and standard deviations of the CS. Jayaram et al. (2011) described a Monte Carlo process used to generate N synthetic spectra compatible with the desired conditional means and standard deviations, and then individual records are identified that match each of the N synthetic spectra in a least-squares sense (similar to Equation 4.55, but using the synthetic spectra in lieu of a specified target). An algorithm for this procedure is presented by Jayaram et al. (2011).

4.5.2 Ground Motion Modification

Selected ground motions are often modified to improve their match to a target prior to use in response history analysis. The modification can be accomplished either through direct scaling or response spectral matching.

Direct scaling involves the multiplication of a ground motion record by a single scaling factor, *SF*. The scaling of the record in the time domain and in its response spectrum are identical because the S_a calculation is a linear operator. Values of *SF* are often selected to match the spectral acceleration at the period of interest (T^*) to the target,

$$SF = \frac{S_{a,tar}(T^*)}{S_a(T^*)} \tag{4.56}$$

Alternatively, scaling can be applied to match a target spectrum in an average sense over some period range defined by weighting factors (Wang et al., 2015):

$$\ln SF = \frac{\sum_{i=1}^{n} w(T_i)\ln\left(S_{a,tar}(T_i)/S_a(T_i)\right)}{\sum_{i=1}^{n} w(T_i)} \tag{4.57}$$

One application of such procedures is when building structures are assessed for seismic performance using response history analysis procedures, for which the mean of the record suite is required to match or exceed the target over a period range judged to significantly contribute to the structure's seismic response (ASCE-7, 2022; Haselton et al., 2017). This range is taken as extending from $0.2T_0$ (to capture potential higher mode response) to $2.0T_0$ (to capture period-lengthening effects from nonlinear response), where T_0 is the maximum first-mode elastic period of the structure (i.e., maximum of the first-mode frequencies for the two horizontal directions, or torsional response).

When ground motions are selected using earthquake attributes (Section 4.5.1.1), the use of large scale factors can lead to biased structural responses (Luco and Bazzurro, 2007). This bias occurs when the motions to which large *SF*s are applied have relatively flat spectral shapes (i.e., low ε), and hence, they produce strong nonlinear responses after scaling. On the other hand, if scaling is incorporated into a selection process that controls for spectral shape (either through the use of ε as a record attribute or a shape-appropriate target spectrum, as described in Sections 4.5.1.2 and 4.5.1.3), the resulting ground motions do not produce biased responses. Accordingly, the ground motion selection procedures described in Sections 4.5.1.2 and 4.5.1.3 can be utilized with or without the use of scaling.

Response spectral matching is the modification of a ground motion so that its response spectrum approximately matches a target

$$S_{a,\mathrm{mod}}(T) \approx S_{a,tar}(T) \tag{4.58}$$

where $S_{a,mod}(T)$ is the spectral acceleration of the modified spectrum. Two types of procedures are used for response spectral matching. Frequency-domain methods increase or decrease portions of the Fourier amplitude spectrum in an iterative fashion to change the response spectrum (e.g., Naeim and Lew, 1995); these changes affect the entire time series. Time-domain methods add or subtract wavelets with characteristic periods to change spectral accelerations (e.g., Abrahamson, 1992; Hancock et al., 2006; Al-Atik and Abrahamson, 2010). The time-domain methods are typically preferred because the wavelet modifications occur within strategically selected, limited-duration time windows, which tend to produce relatively realistic appearing modified time series.

Earthquake engineering experts are of divergent opinions about whether response spectral matching should be used to produce ground motions for response history analysis. The attractive aspect of spectrally matched motions is that the computed *EDPs* have much lower dispersion than is produced from a series of "natural" records selected according to the procedures in Section 4.5.1. This lower dispersion allows a statistically meaningful median response to be achieved with analyses using a smaller number of ground motion records than would be required with an otherwise similar suite of 'natural' (not spectrally matched) ground motions. The median response computed with spectrally matched records appears to be unbiased from that obtained from natural records, provided that the two record suites have consistent spectral shapes (Grant and Diaferia, 2013). The unattractive aspect of using spectrally matched motions is that the variability that exists in actual ground motions is lost when motions are modified to become spectrum-compatible. Response spectral matching should not be used when a desired outcome of the analysis is the dispersion of structural response. To obtain meaningful estimates of both the response median and dispersion, a record suite of direct-scaled motions compatible with the conditional spectrum should be selected.

4.6 SUMMARY

1. Earthquake-resistant design seeks to produce structures that can withstand a certain level of shaking without excessive damage. That level of shaking is described by design ground motion *IM*s, which are evaluated using seismic hazard analysis.
2. Seismic hazard analyses involve the quantitative estimation of ground motion characteristics at a particular site. They may be conducted deterministically or probabilistically.

3. Seismic hazard analyses require the identification and characterization of all potential earthquake sources that could produce significant ground motions at the site of interest. Earthquake sources may be identified on the basis of geologic, tectonic, historical, and instrumental evidence.
4. Earthquake sources can be well-defined faults (typically approximated as a series of rectangles) or relatively diffuse source zones. The future occurrence of earthquakes on a given source can be represented by a PDF for earthquake magnitude and a rate of occurrence of earthquakes larger than a minimum magnitude, $\lambda_{\mathbf{M}_{\min}}$. Earthquakes of magnitude less than $\mathbf{M}_{\min}$ are considered insignificant, i.e., incapable of producing significant damage, for engineering purposes.
5. The largest magnitude earthquakes that can occur on a given source are a function of the source dimensions, with larger faults having the potential to produce larger magnitude earthquakes. Studies of fault segmentation help to identify probable locations of large future ruptures for a given source.
6. Earthquake rate $\lambda_{\mathbf{M}_{\min}}$ can be computed by balancing the rates of moment build-up and release on the fault or, less preferably, by analysis of historical rates based on catalogs of past earthquakes.
7. DSHA involves the assumption of some scenario, which is defined by an earthquake of a particular size at a particular location for each source, for which ground motion characteristics are evaluated. In practice, DSHAs often assume that large earthquakes (rupturing full fault segments) occur at the shortest possible distance to the site within each source. DSHA also requires an assumption of the percentile level of ground motion given magnitude and distance, with the 50th percentile (median) and 84th percentile (median+one standard deviation) being common choices. The earthquake scenario that produces the most severe site motion is used to compute site-specific ground motion *IMs*.
8. DSHA provides a straightforward framework for the evaluation of strong ground motions, but the likelihood of occurrence of those ground motions is unknown. Moreover, despite popular perception to the contrary, ground motions from commonly used DSHA procedures are by no means "worst case."
9. PSHA allows the rate of earthquake occurrence on various sources to be considered along with uncertainties in the size, location, and ground motion levels. A PSHA is complete only if each of these characteristics is considered in the analysis.
10. For each source, uncertainty in earthquake location is characterized by a PDF of source-to-site distance. Evaluation of this PDF requires estimation of the geometry of the source and of the distribution of earthquakes within it.
11. The distribution of the level of shaking produced by an earthquake of a given size occurring at a given source-to-site distance is determined from GMMs. GMMs provide estimates of a natural log mean and standard deviation of ground motion *IMs* conditioned on magnitude and distance. The distribution of *IMs* is considered in PSHA.
12. PSHA involves computation of the probability of exceeding a certain *IM* level given that an earthquake has occurred, a process that samples across the range of magnitudes, distances, and conditional ground motion levels. This conditional probability of exceedance is multiplied by the earthquake rate ($\lambda_{\mathbf{M}_{\min}}$) to evaluate the rate of exceedance of ground motions (λ_{IM}) relative to a target *IM* level (denoted *im*). A plot of λ_{IM} vs *im* is referred to as a hazard curve.
13. The ground motion exceedance rate from (12) can be used with a Poisson process to compute probabilities of exceeding specified values of *IMs* for time intervals of engineering interest. Such results can be plotted in space to produce seismic hazard maps. Uniform hazard spectra represent spectral accelerations across a range of periods having a common exceedance rate.

14. PSHA should be accompanied by disaggregation to identify the locations and magnitudes of sources dominating the ground motions for a given *IM* and hazard level. The percentile level of ground motion controlling the hazard can also be evaluated, which is expressed as the number of standard deviations above (or below) the median that controls the hazard (denoted ε).
15. Limited knowledge is often available regarding key components of PSHA, such as the maximum magnitude earthquake for a given source, the rate of earthquakes, the PDF of magnitude, and the GMM that should be used for analysis. This lack of knowledge, termed epistemic uncertainty, can be captured through the use of logic trees. When logic trees are used, PSHA results can be expressed as percentile levels of hazard curves or uniform hazard spectra. The mean of the distributions is also widely used (mean hazard).
16. Ground motions can be selected and modified for use in response history analysis in a manner that is compatible with the PSHA results. In general, it is important to capture the amplitude, spectral shape, and duration of the controlling events in the ground motion selection. This can be facilitated by selecting motions for compatibility with a CS that matches the UHS at a given period and which is associated with a given magnitude, distance, and ε level.

REFERENCES

Abrahamson, N.A. (1992). "Non-stationary spectral matching," *Seismological Research Letters*, Vol. 63, No. 1, p. 30. https://doi.org/10.1785/gssrl.63.1.19.

Abrahamson, N.A. and Bommer, J.J. (2005). "Probability and uncertainty in seismic hazard analysis," *Earthquake Spectra*, Vol. 21, No. 2, pp. 603–607. https://doi.org/10.1193/1.1899158.

Adair, M.J. (1979). "Geologic evaluation of a site for a nuclear power plant," in A.W. Hatheway and C.R. McClure, eds., *Geology in the Siting of Nuclear Power Plants*, Geological Society of America Reviews in Engineering Geology, Boulder, Colorado, Vol. 4, pp. 27–39. https://doi.org/10.1130/reg4-p27.

Al-Atik, L. and Abrahamson, N.A. (2010). "An improved method for nonstationary spectral matching," *Earthquake Spectra*, Vol. 26, No. 3, pp. 601–617. https://doi.org/10.1193/1.3459159.

Algermissen, S.T., Perkins, D.M., Thenhaus, P.C., Hanson, S.L., and Bender, B.L. (1982). "Probabilistic estimates of maximum acceleration and velocity in rock in the contiguous United States," *Open-File Report 82-1033*, U.S. Geological Survey, Washington, D.C., 107 pp. https://doi.org/10.3133/ofr821033.

Allen, C.R. (1968). "The tectonic environments of seismically active and inactive areas along the San Andreas fault system," *Proceedings Conference on Geologic Problems of San Andreas Fault System*, Stanford University Press, Stanford, CA, pp. 70–82.

Allen, C.R. (1975). "Geologic criteria for evaluating seismicity," *Bulletin of the Geological Society of America*, Vol. 86, No. 8, pp. 1041–1057. https://doi.org/10.1193/070913eqs198m.

Ambraseys, N.N. (1971). "Value of historical records of earthquakes," *Nature*, Vol. 232, pp. 375–379. https://doi.org/10.1130/0016-7606(1975)86<1041:gcfes>2.0.co;2.

Ambraseys, N.N. (1978). "Middle East – A reappraisal of the seismicity," *Quarterly Journal of Engineering Geology*, Vol. 11, pp. 19–32. https://doi.org/10.1144/gsl.qjeg.1978.011.01.03.

Ambraseys, N.N. (2009). *Earthquakes in the Mediterranean and Middle East: A Multidisciplinary Study of Seismicity up to 1900*, Cambridge University Press, Cambridge, UK. https://doi.org/10.1017/s0016756810000452.

American Society of Civil Engineers, ASCE (2022). *Minimum Design Loads and Associated Criteria for Buildings and Other Structures*, Report ASCE/SEI 7-22, Reston, VA.

Arabasz, W.J., Smith, R.B., and Richins, W.D. (1980). "Earthquake studies along the Wasatch front, Utah: Network monitoring, seismicity, and seismic hazards," *Bulletin of the Seismological Society of America*, Vol. 70, pp. 1479–1499. https://doi.org/10.1785/BSSA0700051479.

Atkinson, G.M., Bommer, J.J., and Abrahamson, N.A. (2014). "Alternative approaches to modeling epistemic uncertainty in ground motions in probabilistic seismic-hazard analysis," *Seismological Research Letters*, Vol. 85, pp. 1141–1144. https://doi.org/10.1785/0220140120.

Baker, J.W. (2011). "The Conditional Mean Spectrum: A tool for ground motion selection," *Journal of Structural Engineering*, Vol. 137, No. 3, pp. 322–331. https://doi.org/10.1061/(asce)st.1943-541x.0000215.

Baker, J.W. and Bradley, B.A. (2017). "Intensity measure correlations observed in the NGA-West2 database, and dependence of correlations on rupture and site parameters," *Earthquake Spectra*, Vol. 33, pp. 145–156. https://doi.org/10.1193/060716eqs095m.

Baker, J.W., Bradley, B.A., and Stafford, P.J. (2021). *Seismic Hazard and Risk Analysis*, Cambridge University Press, Cambridge, UK. https://doi.org/10.1017/9781108425056.

Baker, J.W. and Cornell, C.A. (2006). "Spectral shape, epsilon and record selection," *Earthquake Engineering and Structural Dynamics*, Vol. 35, No. 9, 1077–1095. https://doi.org/10.1002/eqe.571.

Baker, J.W. and Jayaram, N. (2008). "Correlation of spectral acceleration values from NGA ground motion models," *Earthquake Spectra*, Vol. 24, pp 299-317. https://doi.org/10.1193/1.2857544

Baltay, A., Ide, S., Prieto, G., and Beroza, G. (2011). "Variability in earthquake stress drop and apparent stress," *Geophysical Research Letters*, Vol. 38, p. L06303. https://doi.org/10.1029/2011gl046698.

Basili, R., Valensise, G., Vannoli, P., Burrato, P., Fracassi, U., Mariano, S., Tiberti, M.M., and Boschi, E. (2008). The database of individual seismogenic sources (DISS), version 3: Summarizing 20 years of research on Italy's earthquake geology, *Tectonophysics*, Vol. 453, No. 1, pp. 20–43. https://doi.org/10.1016/j.tecto.2007.04.014.

Bazzurro, P. and Cornell, C.A. (2002). "Vector-valued probabilistic seismic hazard analysis (VPSHA)," *Proceedings, Seventh U.S. National Conference on Earthquake Engineering*, Earthquake Engineering Research Institute, Boston, Massachusetts.

Beauval, C., Bard, P.-Y., Hainzl, S., and Gueguen, P. (2008). "Can strong motion observations be used to constrain probabilistic seismic hazard estimates?" *Bulletin of the Seismological Society of America*, Vol. 98, pp. 509–520. https://doi.org/10.1785/0120070006.

Bender, B. (1984). "Seismic hazard estimation using a finite fault rupture model," *Bulletin of the Seismological Society of America*, Vol. 74, pp. 1899–1923. https://doi.org/10.1785/BSSA0740051899.

Benjamin, J.R. and Cornell, C.A. (1970). *Probability, Statistics, and Decision for Civil Engineers*, McGraw-Hill, New York.

Bolt, B.A. (1988). *Earthquakes*, W.H. Freeman and Company, New York, 282 pp.

Bommer, J.J. (2012). "Challenges of building logic trees for probabilistic seismic hazard analysis," *Earthquake Spectra*, Vol. 28, pp. 1723–1735. https://doi.org/10.1193/1.4000079.

Bommer, J.J., Coppersmith, K.J., Coppersmith, R.T., Hanson, K.L., Mangongolo, A., Neveling, J., Rathje, E.M., Rodriguez-Marek, A., Scherbaum, F., Shelembe, R., Stafford, P.J., and Strasser, F.O. (2015). "A SSHAC Level 3 probabilistic seismic hazard analysis for a new-build nuclear site in South Africa," *Earthquake Spectra*, Vol. 31, pp. 661–698. https://doi.org/10.1193/060913eqs145m.

Bommer, J.J. and Crowley, H. (2017). "The purpose and definition of the minimum magnitude limit in PSHA calculations," *Seismological Research Letters*, Vol. 88, 1097–1106. https://doi.org/10.1785/0220170015.

Bommer, J.J., Douglas, J., Scherbaum, F., Cotton, F., Bungum, H., and Fäh, D. (2010). "On the selection of ground-motion prediction equations for seismic hazard analysis," *Seismological Research Letters*, Vol. 81, pp. 783–793. https://doi.org/10.1785/gssrl.81.5.783.

Bommer, J.J., Scherbaum, F., Bungum, H., Cotton, F., Sabetta, F., and Abrahamson, N.A. (2005). "On the use of logic trees for ground-motion prediction equations in seismic hazard analysis," *Bulletin of the Seismological Society of America*, Vol. 95, No. 2, pp. 377–389. https://doi.org/10.1785/0120040073.

Bommer, J.J., Stafford, P.J., Edwards, B., Dost, B., van Dedem, E., Rodriguez-Marek, A., Kruiver, P., van Elk, J., Doornhof, D., and Ntinalexis, M. (2017). "Framework for a ground-motion model for induced seismic hazard and risk analysis in the Groningen Gas Field, The Netherlands," *Earthquake Spectra*, Vol. 33, pp. 481–498. https://doi.org/10.1193/082916EQS138M.

Bonilla, M.G., Mark, R.K., and Lienkaemper, J.J. (1984). "Statistical relations among earthquake magnitude, surface rupture length, and surface fault displacement," *Bulletin of the Seismological Society of America*, Vol. 74, pp. 2379–2411. https://doi.org/10.3133/ofr84256.

Bradley, B.A. (2011). "Correlation of significant duration with amplitude and cumulative intensity measures and its use in ground motion selection," *Journal of Earthquake Engineering*, Vol. 15, pp. 809–832. https://doi.org/10.1080/13632469.2011.557140.

Bradley, B.A. (2012). "The seismic demand hazard and importance of the conditioning intensity measure," *Earthquake Engineering and Structural Dynamics*, Vol. 41, pp. 1417–1437. https://doi.org/10.1002/eqe.2221.

Bradley, B.A., Lee, D.S., Broughton, R., and Price, C. (2009). "Efficient evaluation of performance-based earthquake engineering equations," *Structural Safety*, Vol. 31, pp. 65–74. https://doi.org/10.1016/j.strusafe.2008.03.003.

Bryant, W.A. (2010). "History of the Alquist-Priolo earthquake fault zoning act, California, USA," *Environmental and Engineering Geoscience*, Vol. 16, No. 1, pp. 7–18. https://doi.org/10.2113/gseegeosci.16.1.7.

California Geological Survey, CGS (2018). *Earthquake Fault Zones: A Guide for Government Agencies, Property Owners / Developers, and Geoscience Practitioners for Assessing Fault Rupture Hazards in California*, California Department of Conservation, Special Publication No. 42, Sacramento, CA.

Campbell, K.W. (1981). "Near source attenuation of peak horizontal acceleration," *Bulletin of the Seismological Society of America*, Vol. 71, pp. 2039–2070. https://doi.org/10.1785/BSSA0710062039.

Campbell, K.W. and Bozorgnia, Y. (2014). "NGA-West2 ground motion model for the average horizontal components of PGA, PGV, and 5% damped linear acceleration response spectra," *Earthquake Spectra*, Vol. 30, pp. 1087–1115. https://doi.org/10.1193/062913eqs175m.

Chandramohan, R., Baker, J.W., and Deierlein, G.G. (2016). "Quantifying the influence of ground motion duration on structural collapse capacity using spectrally equivalent records," *Earthquake Spectra*, Vol. 32, pp. 927–950. https://doi.org/10.1193/122813eqs298mr2.

Chiou, B.S.-J. and Youngs, R.R. (2014). "Update of the Chiou and Youngs NGA model for the average horizontal component of peak ground motion and response spectra," *Earthquake Spectra*, Vol. 30, pp. 1117–1153. https://doi.org/10.1193/072813eqs219m.

Committee on Seismic Risk (1984). "Glossary of terms for probabilistic seismic-risk and hazard analysis," *Earthquake Spectra*, Vol. 1, No. 1, pp. 33–40. https://doi.org/10.1193/1.1585255.

Convertito, V., Emolo, A., and Zollo, A. (2006). "Seismic-hazard assessment for a characteristic earthquake scenario: An integrated probabilistic-deterministic method," *Bulletin of the Seismological Society of America*, Vol. 96, pp. 377–391. https://doi.org/10.1785/0120050024.

Coppersmith, K.J. and Youngs, R.R. (1986). "Capturing uncertainty in probabilistic seismic hazard assessments with intraplate tectonic environments," *Proceedings, 3rd U.S. National Conference on Earthquake Engineering*, Charleston, South Carolina, Vol. 1, pp. 301–312.

Cornell, C.A. (1968). "Engineering seismic risk analysis," *Bulletin of the Seismological Society of America*, Vol. 58, pp. 1583–1606. https://doi.org/10.1785/bssa0580051583

Donahue, J.L., Stewart, J.P., Gregor, N., and Bozorgnia, Y. (2019). "Ground-motion directivity modeling for seismic hazard applications," *Report No. 2019/03*, Pacific Earthquake Engineering Research Center, UC Berkeley. https://doi.org/10.55461/gphh9609.

Douglas, J. (2024). GMPE compendium. https://www.gmpe.org.uk/ (last accessed June 8, 2024).

Ekström, G., Nettles, M., and Dziewonski, A.M. (2012). "The global CMT project 2004-2010: Centroid-moment tensors for 13,017 earthquakes," *Physics of the Earth Planetary Interior*, Vol. 200–201, pp. 1–9. https://doi.org/10.1016/j.pepi.2012.04.002.

Engdahl, E.R., Di Giacomo, D., Sakarya, B., Gkarlaouni, C.G., Harris, J., and Storchak, D.A. (2020). ISC-EHB 1964-2016, an improved data set for studies of earth structure and global seismicity," *Earth and Space Science*, Vol. 7, No. 1, p. e2019EA000897. https://doi.org/10.1029/2019EA000897.

Esteva, L. (1969). "Seismic risk and seismic design decisions," *Proceedings, MIT Symposium on Seismic Design of Nuclear Power Plants*, Cambridge, MA.

Field, E.H., Arrowsmith, R.J., Biasi, G.P., Bird, P., Dawson, T.E., Felzer, K.R., Jackson, D.D., Johnson, K.M., Jordan, T.H., Madden, C., Michael, A.J., Milner, K.R., Page, M.T., Parsons, T., Powers, P.M., Shaw, B.E., Thatcher, W.R., Weldon II, R.J., and Zeng, Y. (2014). "Uniform California earthquake rupture forecast, Version 3 (UCERF3)-The Time-Independent Model," *Bulletin of the Seismological Society of America*, Vol. 104, No. 3, pp. 1122–1180. https://doi.org/10.1785/0120130164.

Field, E.H., Biasi, G.P., Bird, P., Dawson, T.E., Felzer, K.R., Jackson, D.D., Johnson, K.M., Jordan, T.H., Madden, C., Michael, A.J., Milner, K.R., Page, M.T., Parsons, T., Powers, P.M., Shaw, B.E., Thatcher, W.R., Weldon II, R.J., and Zeng, Y. (2015). "Long-term time-dependent probabilities for the third uniform California earthquake rupture forecast (UCERF3)," *Bulletin of the Seismological Society of America*, Vol. 105, No. 2a, pp. 511–543. https://doi.org/10.1785/0120140093.

Field, E.H., Dawson, T.E., Felzer, K.R., Frankel, A.D., Gupta, V., Jordan, T.H., Parsons, T., Petersen, M.D., Stein, R.S., Weldon II, R.J., and Wills, C.J. (2009). "Uniform California earthquake rupture forecast, Version 2 (UCERF 2)," *Bulletin of the Seismological Society of America*, Vol. 99, No. 4, pp. 2053–2107. https://doi.org/10.1785/0120080049.

Field, E.H., Milner, K.R., Hatem, A.E., Powers, P.M., Pollitz, F.F., Llenos, A.L., Zeng, Y., Johnson, K.M., Shaw, B.E., McPhillips, D., Thompson Jobe, J., Shumway, A.M., Michael, A.J., Shen, Z.-K., Evans, E.L., Hearn, E.H., Mueller, C.S., Frankel, A.D., Petersen, M.D., DuRoss, C., Briggs, R.W., Page, M.T., Rubinstein, J.L., Herrick, J.A. (2024). "The USGS 2023 conterminous U.S. time-independent earthquake rupture forecast," *Bulletin of the Seismological Society of America*, Vol. 114, No. 1, pp. 523–571. doi: https://doi.org/10.1785/0120230120

Field, E.H. and Jordan, T.H. (2015). "Time-dependent renewal-model probabilities when date of last earthquake is unknown," *Bulletin of the Seismological Society of America*, Vol. 105, No. 1, pp. 459–463. https://doi.org/10.1785/0120140096.

Gardner, J.K. and Knopoff, L. (1974). "Is the sequence of earthquakes in Southern California, with aftershocks removed, Poissonian?" *Bulletin of the Seismological Society of America*, Vol. 64, pp. 1363–1367. https://doi.org/10.1785/bssa0640051363.

Goulet, C.A., Haselton, C.B., Mitrani-Reiser, J., Beck, J.L., Deierlein, G.G., Porter, K.A., and Stewart, J.P. (2007). "Evaluation of the seismic performance of a code-conforming reinforced-concrete frame building - from seismic hazard to collapse safety and economic losses," *Earthquake Engineering & Structural Dynamics*, Vol. 36, No. 13, 1973–1997. https://doi.org/10.1002/eqe.694.

Goulet, C.A., Bozorgnia, Y., Kuehn, N., Al Atik, L., Youngs, R.R., Graves, R.W., and Atkinson, G.M. (2021a). "NGA-East ground-motion characterization model Part I: Summary of products and model development," *Earthquake Spectra*, Vol. 37, pp. 1231–1282. https://doi.org/10.1177/87552930211018723.

Grant, D.N. and Diaferia, R. (2013). "Assessing adequacy of spectrum-matched ground motions for response history analysis," *Earthquake Engineering and Structural Dynamics*, Vol. 32, 1265–1280. https://doi.org/10.1002/eqe.2270.

Green, R.A. and Bommer, J.J. (2019). "What is the smallest earthquake magnitude that needs to be considered in assessing liquefaction hazard?" *Earthquake Spectra*, Vol. 35, No. 3, pp. 1441–1464. https://doi.org/10.1193/032218eqs064m.

Gutenberg, B. and Richter, C.F. (1944). "Frequency of earthquakes in California," *Bulletin of the Seismological Society of America*, Vol. 34, No. 4, pp. 1985–1988. https://doi.org/10.1785/bssa0340040185.

Hancock, J., Watson-Lamprey, J., Abrahamson, N.A., Bommer, J.J., Markatisa, A., McCoyh, E., and Mendis, R. (2006). "An improved method of matching response spectra of recorded earthquake ground motion using wavelets," *Journal of Earthquake Engineering*," Vol. 10, No. 1, pp. 67–89. https://doi.org/10.1080/13632460609350629.

Hanks, T.C. (1977). "Earthquake stress drops, ambient tectonic stresses, and stresses that drive plate motions," *Pure and Applied Geophysics*, Vol. 115, pp. 441–458. https://doi.org/10.1007/bf01637120.

Hanks, T.C. and Bakun, W.H. (2002). "A bilinear source-scaling model for **M**-log A observations of continental earthquakes," *Bulletin of the Seismological Society of America*, Vol. 92, No. 5, 1841–1846. https://doi.org/10.1785/0120010148.

Hanks, T.C. and Bakun, W.H. (2008). "**M**-logA observations for recent large earthquakes," *Bulletin of the Seismological Society of America*, Vol. 98, No. 1, pp. 490–494. https://doi.org/10.1785/0120070174.

Haselton, C.B., Baker, J.W., Stewart, J.P., Whittaker, A.S., Luco, N., Fry, A., Hamburger, R.O., Zimmerman, R.B., Hooper, J.D., Charney, F.A., and Pekelnicky, R.G. (2017). "Response history analysis for the design of new buildings in the NEHRP Provisions and ASCE/SEI 7 Standard: Part I - Overview and specification of ground motions," *Earthquake Spectra*, Vol. 33, pp. 373–395. https://doi.org/10.1193/032114eqs039m.

Hauksson, E., Yang, W., and Shearer, P.M. (2012). "Waveform relocated earthquake catalog for southern California (1981 to June 2011)," *Bulletin of the Seismological Society of America*, Vol. 102, pp. 2239–2244. https://doi.org/10.1785/0120120010.

Hill, D.P., Eaton, J.P., and Jones, L.M. (1990). Chapter 5: Seismicity, in R.E. Wallace, ed., *The San Andreas Fault System, California*, USGS Professional Paper 1515.

Idriss, I.M. (1985). "Evaluating seismic risk in engineering practice," *Proceedings of the 11th International Conference on Soil Mechanics and Foundation Engineering*, San Francisco, Vol. 1, pp. 255–320.

Jayaram, N., Lin, T., and Baker, J.W. (2011). "A computationally efficient ground-motion selection algorithm for matching a target response spectrum mean and variance," *Earthquake Spectra*, Vol. 27, pp. 797–815. https://doi.org/10.1193/1.3608002.

Jiang, J. and Lapusta, N. (2016). "Deeper penetration of large earthquakes on seismically quiescent faults," *Science*, Vol. 352, Issue 6291, pp. 1293–1297. https://doi.org/10.1126/science.aaf1496.

Johnson, S.Y., Potter, C.J., and Armentrout, J.M. (1994). "Origin and evolution of the Seattle Fault and Seattle Basin, Washington," *Geology*, Vol. 22, No. 1, pp. 71–74. https://doi.org/10.1130/0091-7613(1994)022<0071:oaeots>2.3.co;2.

Kagan, Y.Y., and Jackson, D.D. (1991). "Seismic gap hypothesis: Ten years after," *Journal of Geophysical Research*, Vol. 96, pp. 21419–21431. https://doi.org/10.1029/91jb02210.

Kanamori, H. (1986). "Rupture process of subduction zone earthquakes," *Annual Review of Earth and Planetary Sciences*, Vol. 14, pp. 293–322. https://doi.org/10.1146/annurev.ea.14.050186.001453.

Kanamori, H. (1988). Importance of historical seismograms for geophysical research, in W.H.K. Lee, H. Meyers, and K. Shimazaki, eds., *Historical Seismograms and Earthquakes of the World*, Academic Press, San Diego, CA, pp. 16–31.

Kreemer, C., Blewitt, G., and Klein, E.C. (2014). "A geodetic plate motion and global strain rate model," *Geochemistry, Geophysics, Geosystems*, Vol. 15, pp. 3849–3889. https://doi.org/10.1002/2014gc005407.

Krinitzsky, E.L. (2002). "How to obtain earthquake ground motions for engineering design," *Engineering Geology*, Vol. 65, pp. 1–16. https://doi.org/10.1016/s0013-7952(01)00098-9.

Kulkarni, R.B., Youngs, R.R., and Coppersmith, K.J. (1984). "Assessment of confidence intervals for results of seismic hazard analysis," *Proceedings, 8th World Conference on Earthquake Engineering*, San Francisco, CA, Vol. 1, pp. 263–270.

Lanzo, G., Di Capua, G., Kayen, R.E., Kieffer, D.S., Button, E., Biscontin, G., Scasserra, G., Tommasi, P., Pagliaroli, A., Silvestri, F., d'Onofrio, A., Violant, C., Simonelli, A.L., Puglia, R., Mylonakis, G., Athanasopoulos, G., Vlahakis, V., Stewart, J.P. (2010). "Seismological and geotechnical aspects of the Mw=6.3 l'Aquila earthquake in central Italy on 6 April 2009," *International Journal of Geoengineering Case Histories*, Vol. 1, No. 4, pp. 206–339. https://casehistories.geoengineer.org.

Leonard, M. (2010). "Earthquake fault scaling: Self-consistent relating of rupture length, width, average displacement, and moment release," *Bulletin of the Seismological Society of America*, Vol. 100, No. 5a, 1971–1988. https://doi.org/10.1785/0120090189.

Leonard, M. (2014). "Self-consistent earthquake fault-scaling relations: Update and extension to stable continental strike-slip Faults," *Bulletin of the Seismological Society of America*, Vol. 104, No. 6, pp. 2953–2965. https://doi.org/10.1785/0120140087.

Leyendecker, E.V., Hunt, R.J., Frankel, A.D., Rukstales, K.S. (2000). "Development of maximum considered earthquake ground motion maps," *Earthquake Spectra*, Vol. 16, 21–40. https://doi.org/10.1193/1.1586081

Luco, N., Bachman, R.E., Crouse, C.B., Harris, J.R., Hooper, J.D., Kircher, C.A., Caldwell, P.J., and Rukstales, K.S. (2015). "Updates to building-code maps for the 2015 NEHRP recommended seismic provisions," *Earthquake Spectra*, Vol. 31, pp. S245–S271. https://doi.org/10.1193/042015eqs058m.

Luco, N. and Bazzurro, P. (2007). "Does amplitude scaling of ground motion records result in biased nonlinear structural drift responses?" *Earthquake Engineering and Structural Dynamics*, Vol. 36, pp. 1813–1835. https://doi.org/10.1002/eqe.695.

Mai, P.M., Spudich, P., and Boatwright, J. (2005). "Hypocenter locations in finite-source rupture models," *Bulletin of the Seismological Society of America*, Vol. 95, pp. 965–980. https://doi.org/10.1785/0120040111.

McGuire, R.K. (2004). *Seismic Hazard and Risk Analysis*, EERI Monograph MNO-10, Earthquake Engineering Research Institute, Oakland, CA, 187 pp.

McGuire, R.K., Cornell, C.A., and Toro, G.R. (2005). "The case for the mean hazard curve," *Earthquake Spectra*, Vol. 21, pp. 879–886. https://doi.org/10.1193/1.1985447.

Merz, H.A. and Cornell, C.A. (1973). "Aftershocks in engineering seismic risk analysis," *Report R73-25*, Department of Civil Engineering, Massachusetts Institute of Technology, Cambridge, Massachusetts.

Naeim, F. and Lew, M. (1995). "On the use of design spectrum compatible time histories," *Earthquake Spectra*, Vol. 11, No. 1, pp. 111–127. https://doi.org/10.1193/1.1585805.

Pagani, M., Hao, K.X., Fujiwara, H., Gerstenberger, M., Ma, K.-F. (2016). "Appraising the PSHA earthquake source models of Japan, New Zealand, and Taiwan," *Seismological Research Letters*, Vol. 87, pp. 1240–1253. https://doi.org/10.1785/0220160101

Papazachos, B. and Papazachou, K. (2003). *The Earthquakes of Greece*, 3rd edition, Ziti Publishing Co., Thessaloniki Greece, 286 pp.

Petersen, M.D., Frankel, A.D., Harmsen, S.C., Mueller, C.S., Haller, K.M., Wheeler, R.L., Wesson, R.L., Zeng, Y., Boyd, O.S., Perkins, D.M., Luco, N., Field, E.H., Wills, C.J., and Ruksatles, K.S. (2008). "Documentation for the 2008 update of the United States national seismic hazard maps," *Open-File Report 2008-1128*, U.S. Geological Survey, 60 pp. https://doi.org/10.3133/ofr20081128.

Petersen, M.D., Shumway, A.M., Powers, P.M., Mueller, C.S., Moschetti, M.P., Frankel, A.D., Rezaeian, S., McNamara, D.E., Luco, N., Boyd, O.S., Rukstales, K.S., Jaiswal, K.S., Thompson, E.M., Hoover, S.M., Clayton, B.S., Field, E.H., Zeng, Y. (2020). "The 2018 update of the US National Seismic Hazard Model: Overview of model and implications," *Earthquake Spectra*, Vol. 36, pp. 5–41. https://doi.org/10.1177/8755293019878199.

Power, M.S., Coppersmith, K.J., Youngs, R.R., Schwartz, D.P., and Swan, R.H. (1981). Seismic Exposure Analysis for the WNP-2 and WNP-1/4 Site: Appendix 2.5K to Amendment No. 18 *Final Safety Analysis Report for WNP-2*, Woodward-Clyde Consultants, San Francisco, 63 pp.

Reiter, L. (1990). *Earthquake Hazard Analysis – Issues and Insights*, Columbia University Press, New York, 254 pp. https://doi.org/10.1126/science.253.5026.1429.b.

Sammon, J.W. (1969). "A nonlinear mapping for data structure analysis," *IEEE Transactions on Computers*, Vol. C-18, pp. 401–409. DOI: 10.1109/T-C.1969.222678

Scherbaum, F. and Kuehn, N.M. (2011). "Logic tree branch weights and probabilities: Summing up to one is not enough," *Earthquake Spectra*, Vol. 27, pp. 1237–1251. https://doi.org/10.1193/1.3652744.

Scherbaum, F., Kuehn, N.M., Ohrnberger, M., and Koehler, A. (2010). "Exploring the proximity of ground-motion models using high-dimensional visualization techniques," *Earthquake Spectra*, Vol. 26, pp. 1117–1138. https://doi.org/10.1193/1.3478697.

Schwartz, D.P. (1988). "Geology and seismic hazards: Moving into the 1990s," *Proceedings, Earthquake Engineering and Soil Dynamics II: Recent Advances in Ground Motion Evaluation*, Geotechnical Special Publication 20, ASCE, New York, pp. 1–42.

Schwartz, D.P. and Coppersmith, K.J. (1984). "Fault behavior and characteristic earthquakes: Examples from the Wasatch and San Andreas fault zones," *Journal of Geophysical Research*, Vol. 89, No. B7, pp. 5681–5698. https://doi.org/10.1029/jb089ib07p05681.

Shaw, J.H., Plesch, A., Dolan, J.F., Pratt, T.L., and Fiore, P. (2002). "Puente Hills blind-thrust system, Los Angeles, California," *Bulletin of the Seismological Society of America*, Vol. 92, No. 8, pp. 2946–2960. https://doi.org/10.1785/0120010291.

Sieh, K.E., (1981). A review of geological evidence for recurrence times of large earthquakes, in D. W. Simpson and T. G. Richards, eds., *Earthquake Prediction: An International Review*, Maurice Ewing Series, American Geophysical Union, Washington, D.C., Vol. 4, pp. 181–194. https://doi.org/10.1029/me004p0181.

Smith, S.W. (1976). "Determination of maximum earthquake magnitude," *Geophysical Research Letters*, Vol. 3, No. 6, pp. 351–354. https://doi.org/10.1029/gl003i006p00351.

Somerville, P., Collins, N., and Graves, R.W. (2006). "Magnitude–rupture area scaling of large strike-slip earthquakes," Final Report to U.S. Geological Survey, Grant No. 05HQGR0004.

Stepp, J.C. (1972). "Analysis of completeness of the earthquake sample in the Puget Sound area and its effect on statistical estimates of earthquake hazard," *Proceedings of the 1st International Conference on Microzonazion*, Seattle. Vol. 2.

Stewart, J.P., Chiou, B.S.-J., Bray, J.D., Somerville, P.G., Graves, R.W., and Abrahamson, N.A. (2001). "Ground motion evaluation procedures for performance based design," *Rpt. No. PEER-2001/09*, Pacific Earthquake Engineering Research Center, University of California, Berkeley, 229 pgs.

Stewart, J.P., Douglas, J., Javanbarg, M., Bozorgnia, Y., Abrahamson, N.A., Boore, D.M., Campbell, K.W., Delavaud, E., Erdik, M., and Stafford, P.J. (2015). "Selection of ground motion prediction equations for the Global Earthquake Model," *Earthquake Spectra*, Vol. 31, pp. 19–45. https://doi.org/10.1193/013013eqs017m.

Stewart, J.P., Luco, N., Hooper, J.D., and Crouse, C.B. (2020). "Risk-targeted alternatives to deterministic ground motion caps in U.S. seismic provisions," *Earthquake Spectra*, Vol. 36, pp. 904–923. https://doi.org/10.1177/8755293019892010.

Stirling, M.W., Wesnousky, S.G., and Shimazaki, K. (1996). "Fault trace complexity, cumulative slip, and the shape of the magnitude-frequency distribution for strike-slip faults: A global survey," *Geophysics Journal International*, Vol. 124, pp. 833–868. https://doi.org/10.1111/j.1365-246x.1996.tb05641.x.

Thingbaijam, K.K.S., Mai, P.M., and Goda, K. (2017). "New empirical earthquake source-scaling laws," *Bulletin of the Seismological Society of America*, Vol. 107, No. 5, pp. 2225–2246. https://doi.org/10.1785/0120170017.

U.S. Geological Survey. Quaternary fault and fold database for the United States, accessed July 1, 2024, at: https://www.usgs.gov/natural-hazards/earthquake-hazards/faults.

U.S. Nuclear Regulatory Commission (US NRC) (2012). *Practical Implementation Guidelines for SSHAC Level 3 and 4 Hazard Studies*, NUREG 2117, Washington D.C.

Usami, T. (1975) "Materials for comprehensive list of Japanese destructive earthquakes."*University of Tokyo Press*, Tokyo, p. 327 (in Japanese)

Wallace, R.E. (1981). Active faults, paleoseismology, and earthquake hazards in the western United States, in D.W. Simpson and T.G. Richards, eds., *Earthquake Prediction: An International Review*, Maurice Ewing Series, American Geophysical Union, Washington, D.C., Vol. 4, pp. 209–216. https://doi.org/10.1029/me004p0209.

Wang, G., Youngs, R.R., Power, M.S., and Li, Z. (2015). "Design ground motion library: An interactive tool for selecting earthquake ground motions," *Earthquake Spectra*, Vol. 31, pp. 617–635. https://doi.org/10.1193/090612eqs283m.

Watson-Lamprey, J.A. (2018). "Capturing directivity effects in the mean and aleatory variability of the NGA-West2 ground-motion prediction equations, *PEER Report 2018-04*. Pacific Earthquake Engineering Research Center, University of California, Berkeley, CA. https://doi.org/10.55461/USOP6050

Wells, D.L. and Coppersmith, K.J. (1994). "New empirical relationships among magnitude, rupture length, rupture width, rupture area, and surface displacement," *Bulletin of the Seismological Society of America*, Vol. 84, No. 4, pp. 974–1002. https://doi.org/10.1785/bssa0840040974.

Wesnousky, S.G., Scholz, C.H., Shimazaki, K., and Matsuda, T. (1983). "Earthquake frequency distribution and the mechanics of faulting," *Journal of Geophysical Research: Solid Earth*, Vol. 88, No. B1, pp. 9331–9340. https://doi.org/10.1029/JB088iB11p09331.

Wesson, R.L., Helley, E.J., Lajoie, K.R., and Wentworth, C.M. (1975). Faults and future earthquakes, in R.D. Borcherdt, ed., *Studies for Seismic Zonation of the San Francisco Bay Region*, Professional Paper 941-A, U.S. Geological Survey, Reston, VA, pp. A5–A30.

Wiemer, S., and Baer, M. (2000). "Mapping and removing quarry blast events from seismicity catalogs," *Bulletin of the Seismological Society of America*, Vol. 90, pp. 525–530. https://doi.org/10.1785/0119990104.

Woessner, J., Danciu, L., Giardini, D., Crowley, H., Cotton, F., Grünthal, G., Valensise, G., Arvidsson, R., Basili, R., Demircioglu, M.B., Hiemer, S., Meletti, C., Musson, R.W., Rovida, A.N., Sesetyan, K., and Stucchi, M. (2015). "The 2013 European seismic hazard model: Key components and results," *Bulletin of Earthquake Engineering*, Vol. 13, pp. 3553–3596. https://doi.org/10.1007/s10518-015-9795-1.

Working Group on California Earthquake Probabilities, WGCEP (1999). "Earthquake probabilities in the San Francisco Bay Region: 2000 to 2030-A summary of findings," *Open-File Report 99-517*, U.S. Geological Survey.

Yeats, R.S., Sieh, K.E., and Allen, C.R. (1997). *Geology of Earthquakes*, 1st edition, Oxford University Press, Oxford.

Yeo, G.L., and Cornell, C.A. (2009). "A probabilistic framework for quantification of aftershock ground-motion hazard in California: Methodology and parametric study," *Earthquake Engineering & Structural Dynamics*, Vol. 38, No. 1, pp. 45–60. https://doi.org/10.1002/eqe.840.

Youngs, R.R. and Coppersmith, K.J. (1985). "Implications of fault slip rates and earthquake recurrence models to probabilistic seismic hazard assessments," *Bulletin of the Seismological Society of America*, Vol. 75, No. 4, pp. 939–964. https://doi.org/10.1785/BSSA0750040939.

5 Seismic Performance and Design

5.1 INTRODUCTION

Geotechnical engineers are often called upon, usually as part of a team of professionals that also includes structural engineers, to both evaluate the earthquake resistance of existing structures and design earthquake-resistant new structures. The notion of earthquake resistance, however, often implies binary states of "resistant" or "non-resistant" without defining what either state really means. It is generally more helpful to think of the range of consequences of earthquake shaking or ground failure on a continuous scale of *performance*. A performance-based framework for evaluation and design can offer a clear and unambiguous path toward improved seismic designs and more useful seismic evaluations.

Geotechnical engineers have a long history of considering anticipated performance of structures in seismic evaluation and design procedures. The manner in which performance is characterized, however, has become more refined over the years. In early geotechnical practice, the notion of performance was essentially divided into poor, or unacceptable, performance considered to represent "failure" and satisfactory performance which implied a lack of failure. At that time, the evaluation or prediction of performance was generally based on forces or stresses. For example, the shear stresses required to maintain static equilibrium were compared with the available shear strength of the soil with the relationship expressed in terms of a factor of safety. Design-level factors of safety were selected with consideration of uncertainty and the consequences of failure, but were also based on experience, precedent, engineering judgment, and, in some cases, expedience. These design-level factors of safety were generally high enough that the average induced shear stresses were limited to a small enough fraction of the average shear strength that large strains, and consequently large deformations, were avoided. While geotechnical engineers recognized that serviceability was more directly related to deformations than to stresses, the fact that stresses could be predicted much more easily (and accurately) than deformations helped support the continuing use of stress-based prediction of geotechnical performance.

As the profession has developed, however, it has become possible to define, characterize, and predict performance in ways that were not previously possible. Deformation-based empirical models and numerical analyses that provide reasonable estimates of deformations are seeing increased use in practice. These developments have led to the concept of performance-based earthquake engineering (PBEE), a relatively new paradigm that is gaining widespread acceptance in the broad field of earthquake hazard mitigation. PBEE is based on the premise that structures and facilities can be designed and evaluated in such a way that their expected performance under anticipated seismic loading can be reliably predicted.

5.2 PERFORMANCE-BASED DESIGN

Early seismic design codes used preservation of life safety as the goal of earthquake-resistant design. Under that approach, a code-conforming structure that was so badly damaged by earthquake shaking that it had to be demolished was regarded to have performed satisfactorily as long as it did not kill anyone. This performance criterion, however, meant that high levels of damage, occupancy interruption, and financial loss could occur in code-compliant structures. The desire of stakeholders, such as owners, insurers, lenders, and regulators, to achieve levels of performance that

DOI: 10.1201/9781003512011-5

exceeded the life safety level specified by codes led to the development of performance-based evaluation and design. Performance-based concepts allow owners to make efficient use of their available resources for the design of new structures and retrofitting of existing structures, and to weigh the benefits of spending more money to achieve higher levels of performance than that associated with simple code conformance.

In performance-based design, earthquake engineers seek to design structures and facilities to achieve some desired and predictable level of performance during earthquakes. Different levels of performance have different costs associated with them and earthquake engineers often need to discuss the relationship between cost and performance with owners who ultimately decide what level of performance is cost-effective. This requires that earthquake professionals be able to define performance in terms that are understandable and useful to the wide range of technical and non-technical professionals who make decisions based on performance predictions. The term "performance," however, can mean different things to different people involved with seismic design. To an engineer, maximum interstory drift of that building, or permanent deformation of the ground beneath it, would likely be a better descriptor of performance. An estimator preparing a bid for repairs may consider the width and spacing of cracks in columns, beams, and slabs to be the most useful measures of performance. To an owner, the financial loss associated with earthquake damage (due to repair, replacement, and/or downtime) could be the best measure of performance. Finally, to society-at-large, deaths, injuries, and regional economic impact may be the most important indicators of performance. PBEE, as described in this chapter, offers a framework in which all of these measures of performance can be addressed so that seismic designs and evaluations can be conducted in a manner that optimizes the use of available resources.

In current practice, performance-based design is most commonly used in areas of high seismicity for structures of unconventional size, shape, structural system, or importance. While various jurisdictions require that structures be designed in compliance with the applicable code(s), code documents generally allow for deviations from the strict rules of the code if it can be shown that the alternative design would produce response that matches or exceeds that achieved by a fully code-compliant design. This often involves more sophisticated and detailed ground motion hazard analyses, geotechnical investigations and analyses, and structural analyses in order to predict performance. The process of demonstrating equivalent performance, often to peer reviewers, is frequently referred to as *performance-based design* in current U.S. practice. The focus of this chapter, however, will be on the broader interpretation of PBEE, which can be applied to all types of structures.

5.3 COMPONENTS OF PERFORMANCE

These different notions of performance lead to an intuitive way of visualizing the consequences of earthquakes. As illustrated in Figure 5.1, an earthquake produces ground motion, which leads to dynamic response of some structural or geotechnical system of interest. That response can lead to physical damage, and that damage leads to losses. The prediction of losses, therefore, requires the ability to predict ground motion intensity, system response, and physical damage. If losses are the ultimate measure of performance, seismic assessments should focus on predicting them as thoroughly, accurately, consistently, objectively, and reliably as possible. The following sections provide a general description of those components; more detailed descriptions are given in the discussion of PBEE implementation (Section 5.6).

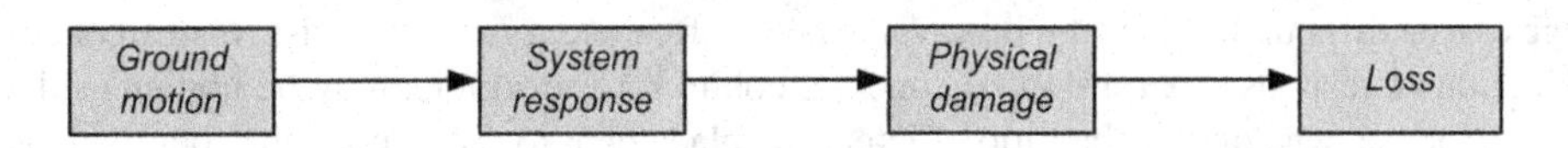

FIGURE 5.1 Schematic illustration of the progression that leads to earthquake losses.

5.3.1 Ground Motion

Ground motions excite all structures founded on or below the Earth's surface. The response of a structure to that excitation is a function of the amplitude, frequency content, and duration of the ground motion, and of the mass, stiffness, damping, and other characteristics of the structure. Chapter 3 showed how ground motion intensity measures can be used to describe ground motion characteristics, and how different *IM*s reflect different characteristics of ground motions. Structures, whether steel buildings, concrete bridges, or earth dams, are compliant and therefore respond more strongly at some frequencies than others. A stiff structure will respond more strongly to high-frequency motions, hence its response would be more accurately predicted by an *IM*, such as *PGA*, that reflects the amplitude of higher frequency components of a ground motion. The response of a more flexible structure may correlate better to lower frequency *IM*s, such as *PGV* or spectral acceleration at a longer period.

The most useful ground motion *IM*s are those to which the response of the system of interest is most closely related, i.e., those for which the variability in system response is smallest. Such *IM*s are referred to as being efficient (Figure 3.13) predictors of response. The optimum *IM*s should therefore be recognized as being different for different types of structures and different types of response.

5.3.2 System Response

Under weak shaking, structures will generally tend to respond elastically and suffer no significant damage. Under stronger shaking, inelastic response can occur and cause the structure to experience some level of physical damage. The damage to different components of a structure, however, may be related to different aspects of response. Damage to brittle elements may be controlled by peak forces, while damage to more ductile elements may be more closely related to deformations. Response can be expressed in terms of *engineering demand parameters*, or *EDP*s. To accurately predict the damage associated with system response, the response caused by earthquake ground motion should be expressed in terms of *EDP*s that are closely related to physical damage.

*EDP*s may be expressed in terms of force- or stress-related quantities such as peak bending moment (e.g., in structural beams or pile foundations) or factor of safety (for bearing capacity of a shallow foundation or stability of a slope). They may also be expressed in terms of deformation-related quantities such as permanent displacement (e.g., of a slope or foundation element) or peak curvature (of buried elements such as pipelines or piles). Force/stress-related measures of response can generally be predicted more accurately than deformation-related measures, but physical damage is usually more loosely related to deformations than forces or stresses. Multiple *EDP*s may be required to properly characterize the response of a complex soil-structure system.

5.3.3 Physical Damage

The response of a structure to a particular ground motion may or may not result in physical damage depending on the structure's capacity to resist damage. The capacity may be viewed as a level of response beyond which some form (or level) of physical damage can be expected to occur. Many different types of damage can occur during an earthquake – some can be related to the structure itself, some to physical systems within the structure, and some to the contents of the structure. Excessive levels of seismic response can cause a wide range of physical damage to structures and facilities.

Specific forms of physical damage can be expressed in terms of *damage measures*, or *DM*s. As described in Section 1.6.3, damage can take many forms ranging from minor to catastrophic. Damage can be difficult to quantify, however, since it often occurs as a stepwise, instead of continuous, function of response. A reinforced concrete column, for example, may be undamaged at low levels of response (e.g., drift – the ratio of lateral displacement to column height), but then undergo cracking, spalling, bar buckling, and collapse at successively higher levels of response. Each of these conditions can develop suddenly and can be represented as discrete *damage states.* In some cases, a degree of continuity in damage can be obtained by representing *DMs* in terms of rate, e.g., number of cracks per foot of a column or floor slab, or number of leaks per mile of a pipeline. Such *DMs* can be useful for estimating repair costs and schedules. To predict the losses associated with these and other forms of physical damage, it is necessary to identify *DMs* that are closely related to the losses of primary interest.

5.3.4 Loss

Physical damage to structures and their contents results in losses. Losses can have many components – deaths and injuries, repair and replacement costs, and loss of function for extended periods of time. Decisions regarding risk reduction through retrofitting, insurance protection, rapid repair/response provisions, etc. are usually made based on expected losses, so those losses are often characterized by *decision variables* or *DVs*. Losses can have many components, but they are divided into three categories often described informally as the "three D's" – deaths, dollars, and downtime.

Death and serious injury are the worst kinds of losses, and their prevention has been the fundamental basis of seismic design since the concept of designing for earthquakes was first contemplated. While the goal of preserving life safety during earthquakes has largely been achieved in many well-developed countries, millions of people remain at risk in many areas around the world where, for economic, political, and/or other reasons, earthquake-resistant design and construction are not routinely practiced.

Dollars refer to the economic losses associated with earthquake damage, which are often divided into two categories – *direct* and *indirect losses.* Direct losses are those associated with the repair and/or replacement of structures and facilities damaged by earthquake shaking or ground failure. Direct losses also include those associated with the contents of structures, which in some cases (e.g., museums, data centers, medical research laboratories) can be far more valuable than the structures that house them. Indirect losses include those associated with delayed or lost business, environmental damage, and compromised infrastructure. Downtime, which refers to the period of time in which structures or facilities are unavailable for their intended use, is among the most important indirect economic losses and can lead to losses that far exceed direct losses. For buildings, downtime can be characterized in days, weeks, or months, or in terms of discrete recovery states such as reoccupancy, pre-earthquake functionality, and full recovery (Bonowitz, 2011). The term *functional recovery* has been used to describe "a post-earthquake state in which capacity is sufficiently maintained or restored to support pre-earthquake functionality" (EERI, 2019; Abrahams et al., 2021). The loss of a major bridge or water supply line in a non-redundant system, for example, can lead to inefficiencies in moving goods, services, and people or to health problems that have very real, and profound, human and economic consequences. Functional recovery design objectives seek to avoid such outcomes.

5.4 PERFORMANCE CRITERIA

Whether expressed in terms of response, damage, or loss, specification of performance criteria is a critical part of performance-based design and evaluation. There are two main parts to a performance criterion, and both are equally important:

1. **A Performance Objective**: an explicit description of the desired level of performance such as a maximum displacement (for a response-level characterization of performance), a damage state such as concrete spalling (damage-level characterization), or a specified repair cost or downtime duration (loss-level characterization). Alternatively, descriptive performance objectives such as "fully operational" or "life safe" can be used.
2. **A Corresponding Occurrence Interval**: an indication of the likelihood of the performance objective being met. This can be expressed in terms of the maximum return period

at which the performance objective is met (or, more commonly, the minimum return period at which it is *not* met). Generally, higher levels of performance are expected for shorter return periods.

In many cases, multiple performance criteria are used in order to provide life safety in rare, large earthquakes, but also some measure of serviceability following more frequent smaller events. In 1959, the first edition of the Structural Engineers Association of California "Blue Book" described the intention of its lateral force requirements as ensuring that a structure would be able to resist:

- a minor level of shaking without damage (non-structural or structural),
- a moderate level of shaking without structural damage (but possibly with some non-structural damage), and
- a strong level of shaking without collapse (but possibly with both non-structural and structural damage).

In this early document, basic performance goals were specified (albeit in terms of somewhat vaguely defined levels of damage) and different performance objectives were specified for different (and equally vaguely defined) levels of ground motion.

The first document widely recognized as establishing procedures for performance-based design of new structures was the Vision 2000 report (Poland et al., 1995). Vision 2000 described procedures intended to produce structures "of predictable performance" with respect to a series of discrete ground motion hazard levels. Vision 2000 coupled four discrete performance levels (fully operational, operational, life safe, and near collapse, as described in Table 5.1) with four ground motion hazard levels (frequent, occasional, rare, and very rare). Performance criteria were established for three categories of structures – safety critical (structures with large quantities of hazardous materials, e.g., toxic, radioactive, or explosive), essential/hazardous (critical post-earthquake structures such as hospitals, emergency communications centers, police and fire stations), and basic facilities (all other structures) – as illustrated in Figure 5.2. The Vision 2000 report described the general levels of damage to various building components and provided allowable interstory drift limits associated with the four performance levels. These limits were expressed deterministically but were intended to be conservative. Thus, Vision 2000 provided for designs based on multiple levels of performance at multiple hazard levels, with performance related to deformation-related quantities (e.g., interstory drift) that are closely related to damage.

The current state of PBEE practice can be characterized by design for specific, discrete performance objectives at a small number of recurrence intervals. Unlike earlier criteria, the recurrence

TABLE 5.1
Descriptions of Vision 2000 Damage States

Damage State	Description
Fully operational	Continuous service. Negligible structural and non-structural damage.
Operational	Most operations and functions can resume immediately. Structure safe for occupancy. Essential operations protected; non-essential operations disrupted. Repair required to restore some non-essential services. Damage is light.
Life Safety	Damage is moderate, but structure remains stable. Selected building systems, features, or contents may be protected from damage. Life safety is generally protected. Building may be evacuated following earthquake. Repair possible but may be economically impractical.
Near collapse	Damage severe, but structural collapse prevented. Non-structural elements may fall. Repair generally not possible.

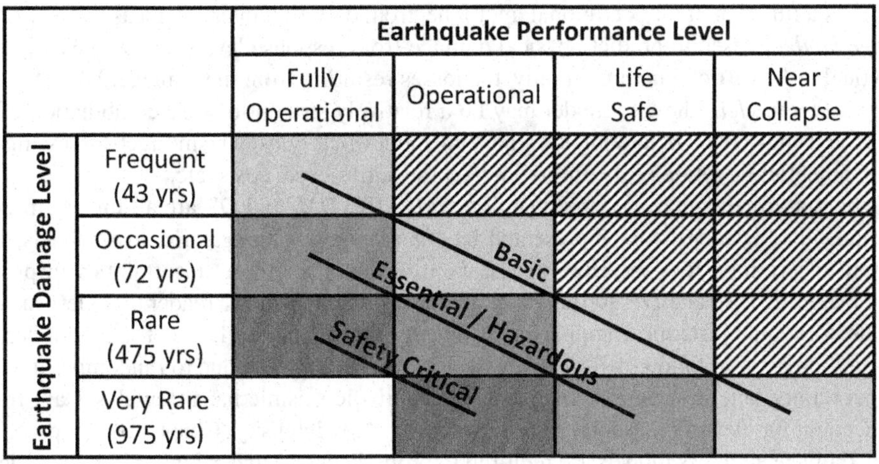

		Earthquake Performance Level			
		Fully Operational	Operational	Life Safe	Near Collapse
Earthquake Damage Level	Frequent (43 yrs)				
	Occasional (72 yrs)				
	Rare (475 yrs)				
	Very Rare (975 yrs)				

Basic
Essential / Hazardous
Safety Critical

FIGURE 5.2 Combinations of earthquake hazard and performance levels proposed by Vision 2000. (After Poland et al., 1995; used by permission of Structural Engineers Association of California.)

intervals correspond to the performance objectives themselves, not the ground motions used in the analyses. The performance objectives are generally expressed in terms of limiting values of response parameters (e.g., permanent displacements for slopes and embankments), damage levels (e.g., collapse risk for buildings), or losses (e.g., time for functional recovery of infrastructure). Some early performance criteria (e.g., 1995 NEHRP Provisions) set the ground motions associated with the design hazard levels probabilistically, and the response predictions were derived for that single level of ground motion. The implicit assumption is that limiting system response to certain "allowable" levels will limit physical damage and losses to acceptable levels, but the actual amounts of expected damage and loss are not explicitly predicted. More recent procedures, however, combine probabilistically characterized ground motions *over all return periods* with a probabilistic response or fragility model to obtain a *response hazard curve*, i.e., a relationship that shows the mean annual rate at which a slope displacement or interstory drift level would be exceeded. Such analyses consider all levels of ground motion, and account for all levels of response those motions can produce. They therefore allow estimation of the probabilities of obtaining (or not obtaining) specific levels of performance and of accounting for uncertainties by designing for an explicitly calculated probability of successful performance rather than a judgment-based minimum factor of safety. Procedures for accomplishing this are presented in the next section.

5.5 PREDICTING PERFORMANCE

A complete prediction of performance requires prediction of the response, damage, and loss associated with all anticipated levels of ground shaking. This process can be illustrated schematically as shown in Figure 5.3. A *response model*, which can range from an empirical algebraic equation to a detailed nonlinear finite element model, is used to predict the response of a soil-structure system to earthquake shaking. The response model, therefore, predicts values of the engineering demand

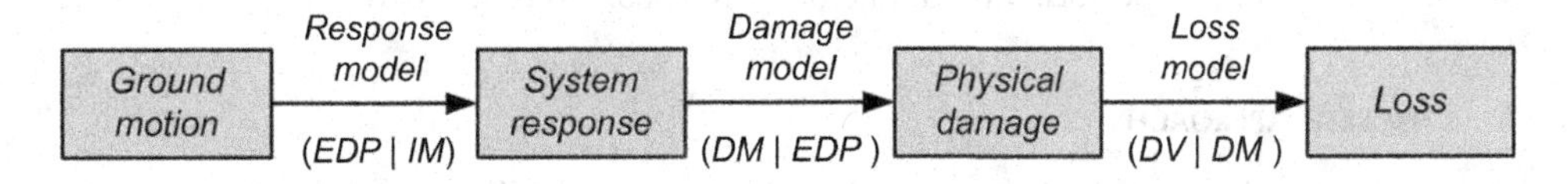

FIGURE 5.3 Schematic illustration of process by which response, damage, and loss are predicted.

parameter as a function of (or conditional upon) the ground motion intensity measure, i.e., *EDP|IM*. A *damage model* is used to predict physical damage from response levels, i.e., *DM|EDP*, and may be analytical or heuristic in nature. Finally, the losses resulting from the damage, *DV|DM*, are predicted by a *loss model*. The loss model may be a relatively straightforward combination of repair quantities and unit costs, or a complex financial model that considers indirect losses, insurance coverage, potential litigation costs, future interest rates, life-cycle costs, etc.

It is essential to recognize that the quantities *IM*, *EDP*, *DM*, and *DV* are all uncertain and that the relationships between them, represented by the response, damage, and loss models, include substantial uncertainty. It is therefore necessary to account for these uncertainties in predicting values of *IM*, *EDP*, *DM*, and *DV*. At this point, it would be useful for the reader to review the discussion of aleatory and epistemic components of uncertainty in Section D.8 as the remainder of this chapter (and others that follow) will refer to aleatory variability (i.e., due to randomness) and epistemic uncertainty (due to lack of knowledge). Probabilistic seismic hazard analyses are routinely used to account for aleatory variability in the ground motion intensity (Chapter 3), but probabilistic response, damage, and loss models are required to properly assess uncertainties in the prediction of response, damage, and loss. These probabilistic response, damage, and loss analyses can account for aleatory variability and epistemic uncertainty in a manner very similar to that used for probabilistic seismic hazard analyses.

It is critical at this point to recognize and acknowledge the long history in geotechnical practice of using "conservative" input parameters in the response analyses used to predict performance, and of conservatively interpreting the results of those predictions for design purposes. This approach has historically led to acceptably low rates of failure at least in part because it is usually applied to static problems in which performance tends to vary monotonically with the variable input parameters. It is difficult, for example, to conceive of a situation in which using a reduced soil shear strength will not lead to a more conservative assessment of the static stability of a slope. However, it leads to three important difficulties when applied to performance-based geotechnical earthquake engineering problems: (1) in the absence of standard procedures for applying conservatism, it is often applied inconsistently, (2) conservative judgments are often applied at multiple points in response analyses, leading to cascading incremental conservatism, the cumulative effects of which are seldom fully recognized, and (3) the dynamic response of soil-structure systems is much more complex than their static response and it is difficult, if not impossible in many cases, to determine how to adjust input parameters in order to produce a "conservative" result. The first two of these difficulties also apply to static problems and can lead to inconsistent designs (although a design format such as Load and Resistance Factor Design (LRFD) can minimize their consequences).

With respect to the third difficulty, and returning to the example of slope stability, decreasing the strength of the soil in a seismic slope stability analysis will also change important parameters (e.g., nonlinear stiffness and damping) that are correlated to shear strength and can strongly affect dynamic response and the mechanism(s) of instability/deformation of the slope. Depending on the relationships between the modal frequencies of the slope and the frequency content of the ground motion, the response in a particular analysis can increase or decrease in ways that cannot be predicted in advance of the analysis. As a result of these issues, the most appropriate approach is for response analyses, and the predictions of performance that derive from them, to use *accurate, unbiased parameters* as inputs to *accurate, unbiased analyses* and to account for parametric variability and model uncertainty in a manner that allows characterization of uncertainty in performance. The notion of some desired level of conservatism in the design can then be accommodated in the selection of acceptable levels of performance at one or more specified return periods.

5.5.1 Discrete Approach

A simple and straightforward way of handling aleatory variability in performance prediction is through use of the total probability theorem (Section D.4.2) employing a *discrete probability matrix*. A discrete probability matrix, also called a *stochastic matrix*, assumes that a process is *Markovian*, i.e., that the probability of being in a particular state depends only on its most recent state. In the transition from random variable $\boldsymbol{X}$ to variable $\boldsymbol{Y}=g(\boldsymbol{X})$, for example, the probability of being in a particular state, $Y=y_i$, is conditional upon being in each of the preceding states, $X=x_j$. Mathematically, this can be expressed as

$$\{P_Y\}=[P]\{P_X\} \tag{5.1}$$

where $[P]$ is the discrete probability matrix. The individual values of $\{P_Y\}$ can be computed as

$$P[y_i]=\sum_j P_{ij}P_{X,j} \tag{5.2}$$

where $P_{ij}=P[Y=y_i|X=x_j]$. In matrix form, assuming X and Y can each take on five discrete values,

$$\begin{Bmatrix} P[y_1] \\ P[y_2] \\ P[y_3] \\ P[y_4] \\ P[y_5] \end{Bmatrix} = \begin{bmatrix} P[y_1 \mid x_1] & P[y_1 \mid x_2] & P[y_1 \mid x_3] & P[y_1 \mid x_4] & P[y_1 \mid x_5] \\ P[y_2 \mid x_1] & P[y_2 \mid x_2] & P[y_2 \mid x_3] & P[y_2 \mid x_4] & P[y_2 \mid x_5] \\ P[y_3 \mid x_1] & P[y_3 \mid x_2] & P[y_3 \mid x_3] & P[y_3 \mid x_4] & P[y_3 \mid x_5] \\ P[y_4 \mid x_1] & P[y_4 \mid x_2] & P[y_4 \mid x_3] & P[y_4 \mid x_4] & P[y_4 \mid x_5] \\ P[y_5 \mid x_1] & P[y_5 \mid x_2] & P[y_5 \mid x_3] & P[y_5 \mid x_4] & P[y_5 \mid x_5] \end{bmatrix} \begin{Bmatrix} P[x_1] \\ P[x_2] \\ P[x_3] \\ P[x_4] \\ P[x_5] \end{Bmatrix} \tag{5.3}$$

The discrete probability matrix approach can be applied multiple times to go from *IM* to *EDP* to *DM* to *DV* in a complete performance evaluation. Consider a hypothetical case in which an earthquake can produce only five possible levels of shaking, five levels of response, five levels of damage, and five levels of loss. Figure 5.4 illustrates the various combinations highlighted for a case where the earthquake produces $IM=im_2$, $EDP=edp_4$, $DM=dm_3$, and $DV=dv_5$. If the highlighted path was the only possible path, i.e., if $P[IM=im_2]=P[EDP=edp_4]=P[DM=dm_3]=P[DV=dv_5]=1.0$, then the expected loss would be dv_5. However, one must consider the conditional probabilities of the four variables actually having those values. Then, the contribution of that particular path to the total expected loss would be

$$dv_{2\text{-}4\text{-}3\text{-}5} = P[IM = im_2 \mid eq]P[EDP = edp_4 \mid IM = im_2]P[DM = dm_3 \mid EDP = edp_4]\times \\ P[DV = dv_5 \mid DM = dm_3]dv_5 \tag{5.4}$$

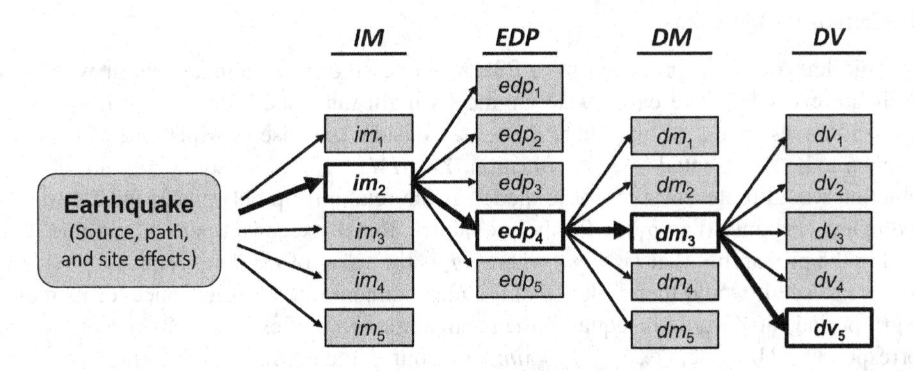

FIGURE 5.4 Illustration of discrete approach to estimation of decision variable (loss). Highlighted path is one of 625 possible paths through network.

The total expected value of *DV* would be obtained by repeating these calculations for all paths through this network – for four variables each with five possible values, there would be $5^4=625$ different paths; if each variable was assumed to take on 50 (instead of 5) values, there would be 6.25 million paths through the network. For a complicated problem, such as one involving nonlinear finite element analysis of a large soil-structure system, the required computational burden would be excessive.

Example 5.1

A hazard analysis has indicated that a hypothetical earthquake has 60% and 40% probabilities of producing peak accelerations of 0.25*g* and 0.30*g*, respectively, at a particular site. If the peak acceleration is 0.25*g*, the permanent displacement of a slope at the site has 10%, 60%, and 30% probabilities of being 10 cm, 12 cm, and 15 cm, respectively. If the peak acceleration is 0.30*g*, the permanent displacement has 20%, 70%, and 10% probabilities of being 14 cm, 18 cm, and 25 cm, respectively. Compute the expected value of permanent displacement.

Solution:

This problem is most easily solved by means of an event tree. Below, the parameter values are listed with their respective probabilities in parentheses.

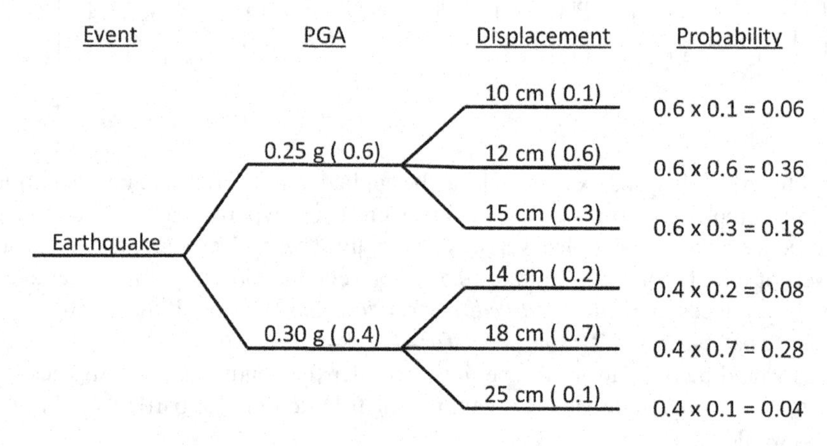

The expected value can be obtained by summing the products of the displacements and probabilities for each terminal branch of the event tree. Numerically, the expected value is given by

$$D\ (\text{cm}) = (0.06)(10)+(0.36)(12)+(0.18)(15)+(0.08)(14)+(0.28)(18)+(0.04)(25) = 14.19\ \text{cm}$$

5.5.2 Integral Approach

The seismic hazard curve produced by a PSHA defines the mean annual rate at which various *IM* levels are exceeded. The earthquake engineer's main interests, however, are in the response, damage, and losses produced by ground motions. Consider the case in which one of those quantities is taken to be represented by some function, $Y=g(IM)$ – if Y were some measure of response, this function would represent a response model. If the relationship between Y and IM was known with complete certainty (i.e., $\sigma_{Y|IM}=0$), the probability that Y exceeds a particular value, y, would be equal to the probability that $IM>im_y$, where im_y is the value of IM that produces $Y=y$ – in other words, $im_y=g^{-1}(y)$. It would then follow that the mean annual rate of exceedance (or its reciprocal, the return period) of Y would be equal to the mean annual rate of exceedance (or return period) of the corresponding IM value, i.e., $\lambda_Y(y)=\lambda_{IM}(im_y)$. Of course, the predictions resulting from such relationships always involve some level of dispersion or aleatory variability – if $\sigma_{Y|IM}>0$, there will be some non-zero probability that $Y>y$ even if $IM<im_y$ and vice versa. This variability will affect the

distribution of Y and cause $\lambda_Y(y)$ to be different than $\lambda_{IM}(im_y)$ – as a result, the return period for $Y=y$ will differ from that of $IM=im_y$ by some amount that depends on the variability in $Y|IM$.

5.5.2.1 General Approach

The integral approach to predicting performance can be developed using the total probability theorem as an extension of the PSHA procedure used to develop ground motion hazard curves. Consider a random variable, Y, that is a function of a ground motion intensity measure, IM. To account for aleatory variability in $Y|IM$, the probability that $Y>y$ if an earthquake occurs can be approximated in discrete form using the total probability theorem, i.e.,

$$P[Y > y] = \sum_{im} P[Y > y \mid IM = im]P[IM = im] \tag{5.5}$$

The term $P[IM = im]$ represents a probability mass function which, as a discrete function, has a discontinuous (stair-step) cumulative distribution function,

$$P[IM < im] = \sum_{-\infty}^{im} P[IM = im] \tag{5.6}$$

For a small number, ε, the probability mass function can be related to the cumulative distribution function by

$$P[IM = im] = P[IM < im + \varepsilon] - P[IM < im - \varepsilon] \tag{5.7}$$

or by

$$P[IM = im] = P[IM > im - \varepsilon] - P[IM > im + \varepsilon] = |\Delta P[IM > im]| \tag{5.8}$$

The differential probability on the right side of Equation (5.8) can be expressed as $\Delta P = (dP/dIM)\Delta IM$ where the quantity dP / dIM is similar to the slope of the hazard curve, as shown subsequently (the absolute value is used because the IM hazard curve has a negative slope). The probability of exceedance of y can then be written as

$$P[Y > y] = \sum_{im} P[Y > y \mid IM = im] |\Delta P[IM > im]| \tag{5.9}$$

Multiplying by the rate of minimum magnitude exceedance, $\lambda_{M_{min}}$, allows the conditional probability of exceedance to be converted to a mean annual rate of exceedance

$$\lambda_Y(y) = \lambda_{M_{min}} \sum_{im} P[Y > y \mid IM = im] |\Delta P[IM > im]| \tag{5.10}$$

or

$$\lambda_Y(y) = \sum_{im} P[Y > y \mid IM = im] |\Delta \lambda_{IM}(im)| \tag{5.11}$$

As the incremental hazard rate, $\Delta\lambda_{IM}(im)$, goes to zero, Equation (5.11) can be expressed in continuous form as

$$\lambda_Y(y) = \int_0^\infty P[Y > y | IM = im] |d\lambda_{IM}(im)| \quad (5.12)$$

This is an important result as it shows how a ground motion hazard curve can be used to compute a hazard curve for any variable for which the relationship to ground motion includes variability. By integrating over the *IM* hazard curve, the hazard curve for *Y* accounts for all levels of ground shaking – from weak levels that occur relatively frequently to strong levels that occur only rarely.

An alternative form of Equation 5.12 can be obtained through integration by parts $\left(\int u\,dv = uv - \int v\,du\right)$. Using that approach,

$$\lambda_Y(y) = P[Y > y \mid IM = im] d\lambda_{IM}(im)\Big|_0^\infty - \int_0^\infty \lambda_{IM}(im) dP[Y = y \mid IM = im] \quad (5.13)$$

The *dP* term within the integrand on the right side of Equation (5.13) is the PDF of *Y*. For the first (definite integral) term on the right side of Equation (5.13), $P[Y > y \mid IM = im] = 0$ when $im = 0$ and $d\lambda_{IM}(im) = 0$ when $im = \infty$, which means that term, as the product of those two factors, must be zero. Therefore,

$$\lambda_Y(y) = \int_0^\infty \lambda_{IM}(im) dP[Y = y \mid IM = im] \quad (5.14)$$

Equation (5.14) can be written in discretized form as

$$\lambda_Y(y) = \sum_{im} \lambda_{IM}(im) \Delta P[Y = y \mid IM = im] \quad (5.15)$$

For the purpose of numerical integration, the form of Equation (5.15) can be easier to work with than Equation (5.11) since it eliminates numerical differentiation of the *IM* hazard curve.

The function $Y = g(IM)$ can be expressed graphically by a set of *fragility curves*. The fragility curves describe the conditional probabilities, $P[Y > y | IM = im]$, of exceeding different values of *y* given some *IM*. Figure 5.5 shows a series of fragility curves for the case where *Y* is a nonlinearly increasing function of *IM* with constant aleatory variability. The positions of the fragility curves depend on the relationship between *Y* and *IM*, and their slopes depend on the level of variability in $Y|IM$ – the finer line illustrates the flatter fragility curve that would correspond to a higher value of $\sigma_{Y|IM}$.

5.5.2.2 Application to Seismic Performance

If *Y* was a measure of seismic response, i.e., an engineering demand parameter, *EDP*, the function *g*(*IM*) would represent a probabilistic response model, and Equations (5.12) and (5.14) could be written as

$$\lambda_{EDP}(edp) = \int_0^\infty P[EDP > edp \mid IM] |d\lambda_{IM}(im)| \quad (5.16a)$$

or

$$\lambda_{EDP}(edp) = \int_0^\infty \lambda_{IM}(im) dP[EDP > edp \mid IM = im] \quad (5.16b)$$

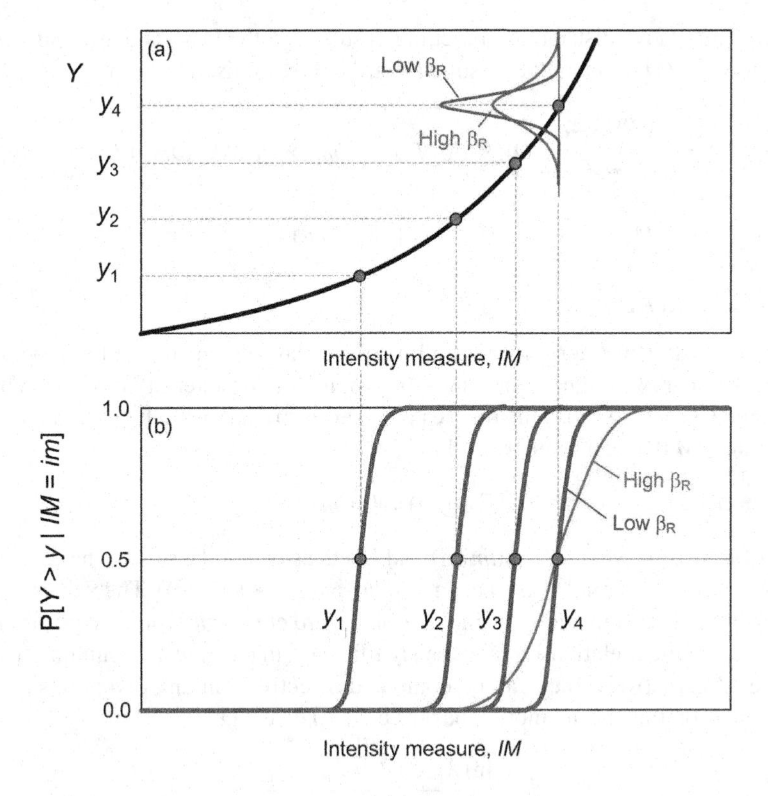

FIGURE 5.5 Schematic illustration of relationship between (a) response relationship and (b) fragility curves. The effects of aleatory variability on fragility curve shape are shown for $Y=y_4$.

to obtain a response (*EDP*) hazard curve. Note that, by integrating over the *IM* hazard curve, the *EDP* value at a particular return period considers all possible *IM* values and their individual rates of exceedance, not just the *IM* value at a single return period. It also accounts for aleatory variability in the predicted response. Characterizing response in this way allows all of the seismicity characteristics of a particular region to be accounted for in the evaluation of response – both the position and shape (e.g., slope) of the *IM* hazard curve will affect the response hazard curve.

The general process used to develop Equation (5.12) can be applied recursively to compute hazard curves for quantities that are functions of *Y*, for example,

$$\lambda_{DM}(dm) = \int P[DM > dm \mid EDP = edp]\left|d\lambda_{EDP}(edp)\right| \tag{5.17}$$

and

$$\lambda_{DV}(dv) = \int P[DV > dv \mid DM = dm]\left|d\lambda_{DM}(dm)\right| \tag{5.18}$$

Putting all of these together produces the well-known "PEER triple integral," which is often written in its most general form (Moehle and Deierlein, 2004) as

$$\lambda(\boldsymbol{DV}) = \iiint G(\boldsymbol{DV} \mid \boldsymbol{DM})\left|dG(\boldsymbol{DM} \mid \boldsymbol{EDP})\right|\left|dG(\boldsymbol{EDP} \mid \boldsymbol{IM})\right|\left|d\lambda(\boldsymbol{IM})\right| \tag{5.19}$$

In this notation, $G(a|b)$ denotes a complementary cumulative distribution function ($G(a|b)=P[A>a|B=b)$ and the bold type denotes vector quantities. From left to right, the three complementary cumulative distribution functions result from the loss, damage, and response models; the final term, $d\lambda(\boldsymbol{IM})$ is from the seismic hazard curve. In discrete form,

$$\lambda_{DV}(dv) = \sum_{k=1}^{N_{DM}} \sum_{j=1}^{N_{EDP}} \sum_{i=1}^{N_{IM}} P[DV > dv \mid DM = dm_k] \times P[DM > dm_k \mid EDP = edp_j] \times P[EDP > edp \mid IM = im_i]\Delta\lambda_{IM}(im_i) \tag{5.20}$$

5.5.3 Closed-Form Approach

Many seismic hazard curves tend to be nonlinear (concave-down) in log-log-space, but can be approximated as linear over significant ranges of ground motion intensity (Department of Energy, 1994; Shome and Cornell, 1999). This implies a power law relationship between mean annual rate of exceedance and *IM* that can be expressed as

$$\lambda_{IM}(im) = k_0 (im)^{-k} \tag{5.21}$$

In this expression, k_0 is the value of $\lambda_{IM}(im=1)$ and k is the slope of the seismic hazard curve (which, in log-log space, plots as a straight line, i.e., $\ln \lambda_{IM}(im) = \ln k_0 - k\ln(im)$). The values of k_0 and k can be obtained by fitting such a line to a ground motion hazard curve over some range of interest (while recognizing that the fitted relationship may be significantly inaccurate at ground motion levels outside that range). Alternatively, they can be computed directly from any two points $(im_1, \lambda_{IM}(im_1))$ and $(im_2, \lambda_{IM}(im_2))$ on the ground motion hazard curve, i.e., using

$$k = \frac{\ln\left[\lambda_{IM}(im_2) / \lambda_{IM}(im_1)\right]}{\ln(im_2 / im_1)} \tag{5.22a}$$

$$k_0 = \lambda_{IM}(im_1) \cdot im_1^{-k} \tag{5.22b}$$

where the values of im_1 and im_2 bracket the return period of primary interest.

Suppose the response model also takes a power law form, with a median response of

$$\widehat{EDP} = a(IM)^b \tag{5.23}$$

and a lognormal response model dispersion, $\beta_R = \sigma_{\ln EDP|IM}$ that, for the present discussion, includes both aleatory and epistemic components). Under these conditions, the resulting *EDP* hazard curve can be derived in closed form from application of Equation (5.12) as

$$\lambda_{EDP}(edp) = k_0 \left[\left(\frac{edp}{a}\right)^{1/b}\right]^{-k} \exp\left[\frac{k^2}{2b^2}\beta_R^2\right] \tag{5.24}$$

Equation (5.24) is composed of two parts, the first of which is actually the value of λ_{IM} that corresponds to the median *EDP* value (obtained by inverting Equation 5.23). The second part depends on the slopes of the *IM* hazard curve and median response model relationship (i.e., k and b) and, most significantly, on the dispersion in the response model. The second term can therefore be viewed as a "dispersion multiplier" since its value is 1.0 when $\beta_R=0$ and becomes progressively greater than 1.0 as the dispersion of response model predictions increases. This result shows that the mean annual rate of exceedance of a particular *EDP* value increases with increasing response model dispersion. Put another way, the *EDP* value corresponding to a given mean annual rate of exceedance (or return period) increases with increasing response model dispersion.

This process can be extended with similar assumptions to produce closed-form solutions for *DM* hazard and *DV* loss curves. Assuming the median damage model to be of power law form

$$\widehat{DM} = c(EDP)^d \tag{5.25}$$

with lognormal damage model dispersion ($\sigma_{\ln DM|EDP} = \beta_D$), the DM hazard curve is given by

$$\lambda_{DM}(dm) = k_0 \left[\frac{1}{a} \left(\frac{dm}{c} \right)^{1/d} \right]^{-k/b} \exp\left[\frac{k^2}{2b^2 d^2} \left(d^2 \beta_R^2 + \beta_D^2 \right) \right] \tag{5.26}$$

Similarly, assuming a power law median loss model

$$\widehat{DV} = e(DM)^f \tag{5.27}$$

with lognormal loss model dispersion ($\sigma_{\ln DV|DM} = \beta_L$), the DV hazard curve is given by

$$\lambda_{DV}(dv) = k_0 \left[\frac{1}{a} \left\{ \frac{1}{c} \left(\frac{d \cdot \lambda_{IM}}{e} \right)^{1/f} \right\}^{1/d} \right]^{-k/b} \exp\left[\frac{k^2}{2b^2 d^2 f^2} \left(d^2 f^2 \beta_R^2 + f^2 \beta_D^2 + \beta_L^2 \right) \right] \tag{5.28}$$

This equation is also composed of two parts – one that represents the value of λ_{IM} that corresponds to the median DV value (based on the ground motion hazard curve and the median values from the response, damage, and loss models), and the other that acts as a dispersion multiplier, this time accounting for dispersion in the response, damage, and loss model predictions. The dispersion multiplier depends on aleatory variabilities and epistemic uncertainties in all three (response, damage, and loss) models and on the sensitivities (through the coefficients, b, d, and f, which are the (log-log) slopes) of the respective power law models). The closed-form solution is simple and convenient but most practical problems violate its assumptions – response, damage, and loss do not increase indefinitely (the direct losses for a building, for example, are limited to its replacement value) and generally do not follow power law relationships with each other. Nevertheless, it is useful for illustrating the important role of uncertainty (and sensitivity) in the prediction of performance.

Assuming aleatory variability and epistemic uncertainty are each lognormally distributed and independent of each other, the various model dispersions can be expressed as $\beta = \sqrt{\beta_{AV}^2 + \beta_{EU}^2}$ where β_{AV} and β_{EU} represent the aleatory variability and epistemic uncertainty components of the dispersion, respectively (note that the use of the combined model for β requires that the median response, damage, and loss models represent the medians of the available relationships that might otherwise be considered in a logic tree). As a result, dispersion in the response, damage, and loss models, and hence the EDP, DM, and DV values for a given return period, can be reduced by reducing the epistemic uncertainty components of their respective dispersions. Earthquake engineers, therefore, have the opportunity to reduce uncertainty and its amplification of expected response, damage, and loss through the services they provide. For geotechnical engineers, these may include more extensive subsurface investigation, more extensive testing, and more detailed and sophisticated analyses.

Example 5.2

Using the information in Example 4.4, estimate the earthquake-induced settlement of a building that has a 5% probability of exceedance in a 50-year period. For the purposes of this example, assume that the settlement (in cm) can be represented by $s = 200\sqrt{PGA}$ with $\sigma_{\ln s|PGA} = 0.2$.

Solution:

Solving Equation (4.40) for λ, the mean annual rate of exceedance for a 5% probability of exceedance in 50 years is

$$\lambda = -\frac{\ln(1 - P[N \geq 1])}{t} = -\frac{\ln(1 - 0.05)}{50} = 0.001026$$

which corresponds to a return period of 974.8 years. Approximating the *PGA* hazard curve in the vicinity of this hazard level by a power law function, two points that bracket the mean annual rate of exceedance of interest can be selected. From Figure E4.4b, the *PGA* values for return periods of 475 years (10% probability of exceedance in 50 years) and 2,475 years (2% probability of exceedance in 50 years) are 0.18*g* and 0.29*g*, respectively. Then from Equation (5.22),

$$k = -\frac{\ln\left[\lambda_{IM}(im_2)/\lambda_{IM}(im_1)\right]}{\ln(im_2/im_1)} = -\frac{\ln\left[0.000404/0.00211\right]}{\ln(0.29/0.18)} = 3.466$$

$$k_0 = \lambda_{IM}(im_1)\cdot im_1^k = 0.00211\cdot 0.18^{3.466} = 0.00000553 = 5.53\times10^{-6}\ \text{year}^{-1}$$

Expressing the settlement model in the power law form of Equation (5.23), $a=20$ and $b=0.2$. Then, making use of Equation (5.24) and solving for $edp=s$, with $\lambda_S(s)=0.001026$,

$$s = a\left\{\left[\frac{\lambda_S(s)}{k_0\exp\left(\frac{k^2}{2b^2}\beta_S^2\right)}\right]^{-1/k}\right\}^b = 200\left\{\left[\frac{0.001026}{5.53\times10^{-6}\exp\left(\frac{(3.466)^2}{2(0.5)^2}(0.2)^2\right)}\right]^{-1/2.479}\right\}^{0.5} = 5.41\ \text{cm}$$

5.6 IMPLEMENTATION OF PERFORMANCE-BASED DESIGN AND EVALUATION

Performance-based design and evaluation can be implemented into engineering practice in a number of different ways. Different implementation approaches must address two main matters – the manner in which earthquake loading is specified and the manner in which performance is evaluated. The design process involves an iterative series of performance evaluations after each of which the predicted performance is compared with the performance objectives, and the design is modified until all performance objectives have been met. The performance evaluations may be based on measures of response, damage, or loss. The primary elements of alternative implementation procedures are illustrated in Table 5.2 and described in the following sections.

In the simplest approach, PBEE can be implemented at the response level, i.e., by specifying performance in terms of *EDP* values at different response return periods. In this approach, physical damage must be inferred from the computed response and loss must be inferred from the inferred damage. These compounded inferences may be guided by judgment and experience but must be recognized as leading to highly variable loss estimates. An intermediate approach would be to specify performance in terms of response limit states, which involves the explicit comparison of predicted and allowable response; this approach could also be used to predict *DM*s with consideration of response and damage model dispersions. Loss would still need to be inferred, but inferring loss from computed damage levels is more accurate than inferring it from computed response levels. The most complete level of implementation would involve defining performance in terms of losses (*DV*s), which would then require explicitly modeling response, damage, and loss with consideration of the dispersion inherent in each. While requiring more information and more effort, this approach would provide the most accurate (i.e., least biased) and most precise estimate of loss.

5.6.1 Scalar and Vector Approaches

The range of soil-structure systems to which performance-based principles can be applied is broad, varying from small, individual structures or soil deposits to large bridges or buildings with many structural and non-structural components. The scope and complexity of the system can influence the manner in which performance-based concepts are implemented, but the basic components are consistent.

TABLE 5.2
Main Elements of Alternative Performance Design Procedures

	Performance Evaluation Level		
Loading	**Response**	**Damage**	**Loss**
Select ground motion intensity measures (*IMs*) Select ground motion hazard levels Compute ground motion hazard curve	Select response parameters (*EDPs*) Develop performance criteria in terms of *EDPs* Select response model Develop initial design Compute response Compare computed response with performance criteria Iterate until performance criteria satisfied	Select response parameters (*EDPs*) Select damage measures (*DMs*) Characterize damage limit states Develop performance criteria in terms of *DMs* Select response model Select damage model Develop initial design Compute response Compute damage Compare computed damage with performance criteria Iterate until performance criteria satisfied	Select response parameters (*EDPs*) Select damage measures (*DMs*) Select loss measures (*DVs*) Characterize loss states Develop performance criteria in terms of *DVs* Select response model Select damage model Select loss model Develop initial design Compute response Compute damage Compute loss Compare computed loss with performance criteria Iterate until performance criteria satisfied
	Infer damage from computed response		
	Infer loss from inferred damage	*Infer loss from computed damage*	*Loss directly computed*

Losses borne by the simplest systems can often be dominated by a single quantity (such as repair cost) that results from a single damage mechanism (e.g., floor slab cracking). If that damage mechanism is associated with a single measure of response (post-earthquake settlement) that is closely correlated to a particular ground motion intensity measure (peak ground velocity), a scalar approach (one *IM*, one *EDP*, one *DM*, and one *DV*) to performance-based design/evaluation may be reasonable and appropriate. The scalar approach is relatively simple and straightforward, and will be used to illustrate implementation of performance-based concepts in the following sections.

More complex systems, however, may need multiple *IMs*, *EDPs*, *DMs*, and *DVs* in order to evaluate performance more accurately and thoroughly. Consider a large building, for example. The building has multiple structural components – beams, girders, columns, braces, shear walls, foundations, and their respective connections – each or all of which can be damaged by earthquake shaking. The building also has essential non-structural systems – exterior cladding, internal partitions, piping, electrical, fire suppression (sprinkler), stairway, elevator, and other systems – each or all of which can also be damaged by earthquake shaking. There are also the contents of the building – people (the number of which can vary dramatically over the course of a day or week, depending on the type of building) and inventory (which may be fixed or variable and may, for example in the case of a museum, have a value greater than that of the building itself). Thus, the "building" itself is a quite complicated system with a variety of important components.

The different components of a building may have different dynamic response characteristics. Depending on its stiffness and mass, the building's response may be governed by its fundamental mode of vibration or may also be influenced by higher modes, in which case spectral accelerations at multiple periods may be required as *IMs*. The residual drift of the building can be another important measure of response since excessive permanent drift can lead to significant damage resulting in high repair costs, or even require demolition of the structure. The foundation system's response may be more closely related to ground deformation/strain, so ground displacement may be required as an *IM* to accurately predict its performance. The dynamic response of piping systems, suspended ceilings, etc., and of the contents of the building are going to depend on the motions of their supports, which will vary from floor to floor; hence, individual floor spectra may be required as inputs to their analysis. While some components of the building that have similar response characteristics can be aggregated into performance groups, it is clear that multiple *IM*s and multiple *EDP*s may be required for accurate prediction of response. All of these measures of response must be recognized as having substantial aleatory variability and many of them as being correlated to each other.

Damage is even more difficult to quantify than response – it results from highly nonlinear processes that depend on complex material behavior that is often difficult to characterize, and there is often little actual damage data from which to construct empirical damage models. While response tends to occur on continuous scales, damage tends to be more discrete and is often described in a categorical manner. The different components of the building are also susceptible to different forms of damage. Damage to a reinforced concrete column is different than damage to a welded connection or to a statue on a pedestal in an art museum, so multiple *DM*s are needed for different structural and non-structural components of a building. The complexity of damage processes leads to significant uncertainty even when the response is known. Additional uncertainty comes from the fact that the future contents and/or uses of the building may not be known as tenants and technologies change.

A loss-level implementation requires a probabilistic loss model, and loss models can have different degrees of complexity. The notion of losses having direct and indirect economic, and casualty-related components (Section 5.3.4) implies that three loss models may be required, or that the three components of loss must be expressed in consistent (usually economic) terms, which involves the extremely difficult problem of defining economic losses associated with casualties. However, even direct economic losses can have multiple components – structural damage, non-structural damage, damage to contents, etc. – and these components interact with each other in complex, correlated manners that must be accounted for in a complete loss model. Similarly, indirect losses can have multiple components that are correlated with each other, and with direct losses. Losses also have the additional complexity of time-dependence as important factors such as local material and labor costs (particularly in the period shortly after a major earthquake), inflation, interest rates, and other economic factors can vary significantly over the time scales of interest in performance-based evaluation and design. These factors can introduce very high levels of variability and uncertainty into loss models.

5.6.2 Characterization of Ground Motion Intensity

Characterization of earthquake loading first requires selection of one or more ground motion intensity measures. The selection, however, must be made with consideration of the system response of greatest interest. Ground motion intensity measures should ideally be predictable, efficient, and sufficient (Section 3.3).

Intensity measure predictability refers to the aleatory variability associated with an *IM* prediction given an earthquake scenario; this variability is characterized by the standard deviation term in the ground motion model (GMM) for that *IM*. Different *IM*s have been shown to have different levels of predictability (Section 3.5.4.3). The predictabilities are highly variable – cumulative absolution velocity, *CAV*, has very good predictability while the predictability of Arias intensity is poor.

Intensity measure *efficiency* refers to the conditional aleatory variability in response given ground motion intensity (Section 3.3). Efficient intensity measures (Section 3.3) are those to which the response of interest is closely related, i.e., for which $\sigma_{\ln EDP|IM}$, is low. In a sense, efficiency can be thought of as a measure of the relevance of an *IM* to the problem of interest. Figure 5.6 shows examples of inefficient and efficient intensity measures for computed foundation settlement; here, *CAV* can be seen to be a more efficient predictor of displacement than *PGA* and $D_{5\text{-}75}$ duration by itself to be quite inefficient. *Sufficient IM*s are those for which consideration of additional parameters does

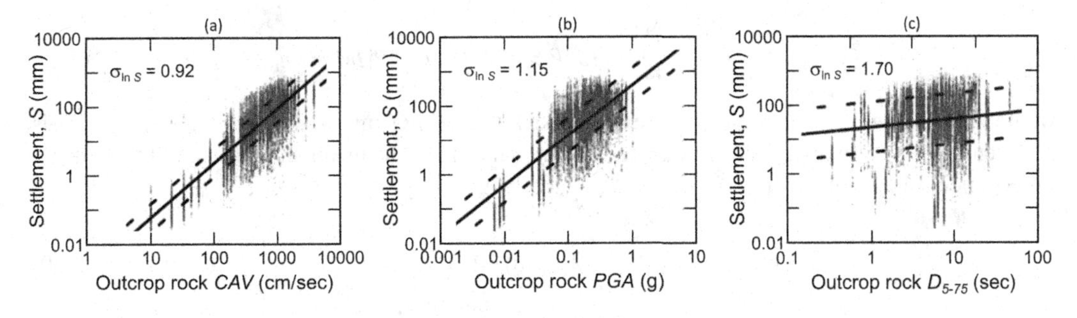

FIGURE 5.6 Correlation between computed foundation settlement and (a) *CAV*, (b) *PGA*, and (c) $D_{5\text{-}75}$. (Courtesy of Z. Bullock.) Lower dispersion of settlement corresponds to higher efficiency of its prediction.

not reduce the uncertainty in predicted response. Some measures of response are so strongly correlated to a single ground motion characteristic that their values can be accurately predicted based on a single *IM*. Other problems, however, may be influenced by multiple ground motion characteristics. Soil liquefaction (Chapter 9), for example, is affected by both the amplitude and duration of a ground motion, hence *PGA* alone is an insufficient predictor of liquefaction. In such cases, more than one parameter (e.g., *PGA* and earthquake magnitude, which serves as a proxy for duration) may be required to predict response efficiently.

The second important aspect of earthquake loading characterization concerns the number of hazard levels, or return periods, to be considered in the design. Earthquake loading can be specified in terms of a discrete number of hazard levels, such as the four levels (frequent, occasional, rare, and very rare) proposed by Vision2000 (Figure 5.2), or integrated over a continuous scale as utilized in the PEER approach (Section 5.5.2.2). Using a discrete hazard level approach requires evaluation of performance at each hazard level, hence the level of engineering effort required to develop an acceptable design increases with increasing number of discrete hazard levels. With the integral hazard approach, the variation of response over a wide range of ground shaking (i.e., that associated with *all* anticipated hazard levels), is required. This may involve many more response analyses than would be performed in a conventional design process, but it provides a more complete characterization of performance.

5.6.3 Response-Level Implementation

A response-level implementation of PBEE assumes that performance can be judged in terms of response variables such as interstory drift for buildings, peak lateral displacement for bridges, permanent ground displacement for slopes, etc. In this approach, predicted values of the response parameters, i.e., the *EDP*s, are compared with allowable values of those parameters (also known as *acceptance criteria* in structural applications) to determine whether performance objectives have been met. Exceedances of allowable values imply some level of damage, and consequently some level of loss. As illustrated in Table 5.2, however, the damage has to be inferred from the computed response, and the loss is inferred from the inferred damage.

The allowable level of response can be viewed as a *capacity*, i.e., a threshold beyond which some levels of damage and loss are anticipated. Such capacity thresholds are often referred to as *limit states* since they represent a limit beyond which a different consequence is expected. For most problems, particularly in geotechnical earthquake engineering, the capacities themselves must be considered to be variable. Representing an *EDP* level, *C*, as a random variable that describes a response capacity, the mean annual rate at which a scalar value of that capacity, $C=c$, would be exceeded can be computed using the total probability theorem,

$$\lambda_{EDP}(c) = \lambda_{\mathbf{M}_{\min}} \sum_{i=1}^{N_{IM}} P[EDP > c \mid IM = im_i] P[IM = im_i] \quad (5.29)$$

where $\lambda_{\mathbf{M}_{\min}}$ is the mean annual rate of earthquakes exceeding some minimum magnitude. If C is not accurately known, it can be treated as a random variable so the mean annual rate of the capacity exceedance limit state, i.e., $LS=EDP>C=c$, would be given by

$$\lambda_{LS} = \int_0^{\infty} \lambda_{EDP}(c) f_C(c)\, dc \quad (5.30)$$

where $f_C(c)$ is the probability density function for capacity C (equivalent to the derivative of a fragility curve). Equations (5.29) and (5.30) show that a hazard curve for limit state exceedance can be evaluated from the *IM* hazard curve, the (probabilistic) relationship between *IM* and *EDP*, and the distribution of capacity.

Design ground motions used in current U.S. building codes (e.g., NEHRP, 2020) are derived using Equation (5.30). The design objective is a specific rate of collapse (1% probability of collapse in 50 years), and calculations are performed to find a capacity that meets that objective. In these calculations, a ground motion hazard curve is combined with a collapse fragility curve that is assumed to be lognormal with a specified variability ($\lambda_{EDP|IM}$). The median of that fragility curve depends on the structural capacity or strength (which, in turn, is related to the design spectral acceleration level). For each location in a grid across the U.S., the integration is performed repeatedly for different design spectral accelerations (and hence capacities) until the target collapse risk is achieved (Luco et al. 2007). The resulting ground motion is referred to as a risk-targeted maximum considered earthquake ground motion (MCE_R).

The closed-form solution of Section 5.5.3 can be extended to account for lognormally distributed capacity (Jalayer, 2003). Using Equation (5.24), the mean annual rate at which some known load capacity, $C=c$, would be exceeded is

$$\lambda_{EDP}(c) = k_0 \left[\left(\frac{c}{a} \right)^{1/b} \right]^{-k} \exp\left[\frac{k^2}{2b^2} \beta_R^2 \right] \quad (5.31)$$

Using Equation (5.30) to account for dispersion in load capacity, the mean annual rate of limit state (capacity) exceedance is given by

$$\lambda_{LS} = \int_0^{\infty} k_0 \left[\left(\frac{c}{a} \right)^{1/b} \right]^{-k} \exp\left[\frac{k^2}{2b^2} \beta_R^2 \right] f_C(c)\, dc \quad (5.32)$$

Extracting c from the first term and moving capacity-independent terms out of the integral,

$$\lambda_{LS} = k_0 a^{k/b} \exp\left[\frac{k^2}{2b^2} \beta_R^2 \right] \left[\int_0^{\infty} c^{-k/b} f_C(c)\, dc \right] \quad (5.33)$$

The remaining term inside the integral is, by definition, the mean value of $c^{-k/b}$. Since the expected value of a lognormal random variable, Y, with median, $\hat{Y}$, and dispersion, $\beta_Y = \sigma_{\ln Y}$, raised to the power α is given by

$$E(Y^{\alpha}) = E(e^{\alpha \ln Y}) = (\hat{Y})^{\alpha} \exp\left[\frac{1}{2} \alpha^2 \beta_Y^2 \right] \quad (5.34)$$

Equation (5.33) becomes

$$\lambda_{LS} = k_0 a^{k/b} (\hat{C})^{-k/b} \exp\left[\frac{k^2}{2b^2}\beta_R^2\right] \exp\left[\frac{k^2}{2b^2}\beta_C^2\right] \quad (5.35)$$

where $\hat{C}$ is the median capacity. This relationship can be simplified to

$$\lambda_{LS} = k_0 a^{k/b} \left(\frac{\hat{C}}{a}\right)^{-k/b} \exp\left[\frac{k^2}{2b^2}\left(\beta_R^2 + \beta_C^2\right)\right] \quad (5.36)$$

Thus, the mean annual rate of limit state exceedance can be seen to increase with increasing aleatory variability in capacity as well as in response.

Example 5.3

The building for which settlement was estimated in Example 5.2 is expected to experience severe damage at a settlement that exceeds 20 cm. Assuming this descriptive measure of capacity to be lognormally distributed with $\sigma_{\ln C} = 0.3$, estimate the return period at which severe damage is likely to occur.

Solution:

Taking the 20 cm severe damage limit state boundary as the median capacity and using Equation (5.36), the mean annual rate of limit state exceedance is calculated as

$$\lambda_{LS} = k_0 \left(\frac{\hat{C}}{a}\right)^{-k/b} \exp\left[\frac{k^2}{2b^2}\left(\beta_R^2 + \beta_C^2\right)\right] = 0.000547 \left(\frac{30}{20}\right)^{-2.479/0.5} \exp\left[\frac{(2.479)^2}{2(0.5)^2}\left(0.2^2 + 0.3^2\right)\right] = 0.0027 \text{ year}^{-1}$$

which gives a limit state return period of 370.4 years.

This closed-form solution, while conceptually quite clear and simple, must be recognized as being based on a local approximation to the hazard curve because actual ground motion hazard curves tend to be nonlinear (concave-down) in log-log-space. Its accuracy, therefore, is influenced by how the assumed power law *IM* hazard curve is fit to the actual *IM* hazard curve. The original approach of fitting a line tangent to the *IM* hazard curve at the *IM* level corresponding to the median capacity (Jalayer, 2003), i.e., at $im_{\hat{C}} = (\hat{C} / a)^{1/b}$, provides a conservative result since the tangent will always be above a concave-down hazard curve. At short return periods, where actual *IM* hazard curves tend to be more linear, the approach is more accurate than at long return periods. However, with the exception of highly vulnerable structures that can fail under weak shaking, earthquake engineers are usually more interested in performance at relatively long return periods. The error associated with the tangent approach becomes excessive with increasing *IM* hazard curve curvature and dispersion of capacity (Aslani and Miranda, 2005; Bradley and Dhakal, 2008). It can be substantially reduced, however, by selecting biased (toward shorter return periods) values of the points, im_1 and im_2, used to define a secant approximation (in Equation 5.22) to the *IM* hazard curve (Vamvatsikos, 2014). Letting $edp_{\hat{C}}$ represent the *EDP* value corresponding to the median capacity, $\hat{C}$, the *IM* values

$$im_1 = \left(\frac{edp_{\hat{C}}}{a}\right)^{1/b} \exp\left[-\frac{1}{2}\frac{\sqrt{\beta_R^2 + \beta_C^2}}{b}\right] \quad (5.37a)$$

$$im_2 = \left(\frac{edp_{\hat{C}}}{a}\right)^{1/b} \exp\left[-\frac{3}{2}\frac{\sqrt{\beta_R^2 + \beta_C^2}}{b}\right] \quad (5.37b)$$

can be used with Equation (5.22) to develop an *IM* hazard curve approximation that produces a more accurate estimate of λ_{LS}.

The issue of *IM* hazard curve curvature can be more appropriately addressed, however, by relaxing the power law restriction. Bradley et al. (2007) developed a semi-analytical solution based on a hyperbolic representation of the *IM* hazard curve. Vamvatsikos (2013) derived a closed-form solution for a ground motion hazard curve that is quadratic in log-log space

$$\ln \lambda_{IM}(im) = \ln k_0 - k_1 \ln im - k_2 (\ln im)^2 \tag{5.38}$$

The mean annual rate of limit state exceedance for this case can be expressed most compactly as

$$\lambda_{LS} = k_0^{1-\phi} \sqrt{\phi} \left[\lambda_{IM}(im_{\hat{C}}) \right]^{\phi} \exp\left[\frac{k_1^2}{2b^2} q \left(\beta_R^2 + \beta_C^2 \right) \right] \tag{5.39}$$

where $\hat{C}$ is the median capacity, $im_{\hat{C}}$ is the intensity measure corresponding to the *EDP* value that produces the median capacity, i.e.,

$$im_{\hat{C}} = \left(\frac{\hat{C}}{a} \right)^{1/b}$$

$$\phi = \frac{1}{1 + 2k_2(\beta_R^2 + \beta_C^2) / b^2}$$

$$q = \frac{1}{1 + 2k_2 \beta_R^2 / b^2}$$

and

$$\lambda_{IM}(im_{\hat{C}}) = k_0 \exp\left\{ -k_1 \ln(im_{\hat{C}}) - k_2 \left[\ln(im_{\hat{C}}) \right]^2 \right\}$$

If the coefficient of the quadratic term in the hazard curve is set to zero, the values of ϕ and q go to 1.0 and the limit state exceedance rate becomes equal to the first order rate given in Equation (5.36).

5.6.4 Damage-Level Implementation

While response tends to occur on continuous, quantifiable scales, damage tends to be divided into discrete, and often descriptive, damage states. Thus damage can be described by damage probability matrices (Whitman et al., 1973) or by a finite number of fragility curves, each corresponding to a particular damage state, or category.

A damage probability matrix is a type of discrete probability matrix (Section 5.5.1) that describes the probabilities of various damage levels (or states) conditional upon various response levels. In a discrete damage-level implementation, *EDP* and *DM* take the places of *X* and *Y* in Equations (5.1)–(5.3); for an evaluation with five damage states being caused by five response levels,

$$\begin{Bmatrix} P[dm_1] \\ P[dm_2] \\ P[dm_3] \\ P[dm_4] \\ P[dm_5] \end{Bmatrix} = \begin{bmatrix} P[dm_1 | edp_1] & P[dm_1 | edp_2] & P[dm_1 | edp_3] & P[dm_1 | edp_4] & P[dm_1 | edp_5] \\ P[dm_2 | edp_1] & P[dm_2 | edp_2] & P[dm_2 | edp_3] & P[dm_2 | edp_4] & P[dm_2 | edp_5] \\ P[dm_3 | edp_1] & P[dm_3 | edp_2] & P[dm_3 | edp_3] & P[dm_3 | edp_4] & P[dm_3 | edp_5] \\ P[dm_4 | edp_1] & P[dm_4 | edp_2] & P[dm_4 | edp_3] & P[dm_4 | edp_4] & P[dm_4 | edp_5] \\ P[dm_5 | edp_1] & P[dm_5 | edp_2] & P[dm_5 | edp_3] & P[dm_5 | edp_4] & P[dm_5 | edp_5] \end{bmatrix} \times \begin{Bmatrix} P[edp_1] \\ P[edp_2] \\ P[edp_3] \\ P[edp_4] \\ P[edp_5] \end{Bmatrix} \tag{5.40}$$

TABLE 5.3
Damage Probability Matrix for Pile Foundations Supporting Bridge in Liquefiable Soils (Kramer et al., 2008)

	Pile Head Displacement Range (cm), *EDP*				
Damage State, *DM*	**< 4**	**4–10**	**10–30**	**30–100**	**> 100**
Negligible	0.95	0.05	0.00	0.00	0.00
Slight	0.05	0.80	0.20	0.05	0.00
Moderate	0.00	0.10	0.60	0.25	0.05
Severe	0.00	0.05	0.15	0.55	0.10
Catastrophic	0.00	0.00	0.05	0.15	0.85

The probabilities in each of the columns of the damage probability matrix must sum to 1.0, indicating that the damage measures are exhaustive, i.e., that a given *EDP* value must fall into one of the discrete damage levels.

As an example, damage to a five-span bridge typical of that designed by the California Department of Transportation supported by piles extending through liquefiable soils that were susceptible to lateral spreading was assessed using a sophisticated finite element analysis (Kramer et al., 2008). The pile foundations interacted with the soil through *p-y*, *t-z*, and *Q-z* springs. With an absence of empirical or analytical damage data, a group of experts was polled to aid in identifying five discrete, descriptive foundation damage states based on computed pile head displacement. The results of that poll were used in development of the discrete probability matrix shown in Table 5.3. It should be noted that the values in Table 5.3 applied to the specific soil/foundation/structure system being considered, and that a different damage probability matrix would need to be defined for a different system.

For structural elements of the bridge, experimental data from the PEER structural performance database (Berry et al., 2004) were used to establish fragility curve data for discrete damage states. For the bridge columns, four damage states were related to maximum drift ratio, as shown in Figure 5.7. Whereas the damage states and probabilities in the pile foundation damage model were both discrete, in this approach the damage states were discrete but the probabilities were continuous.

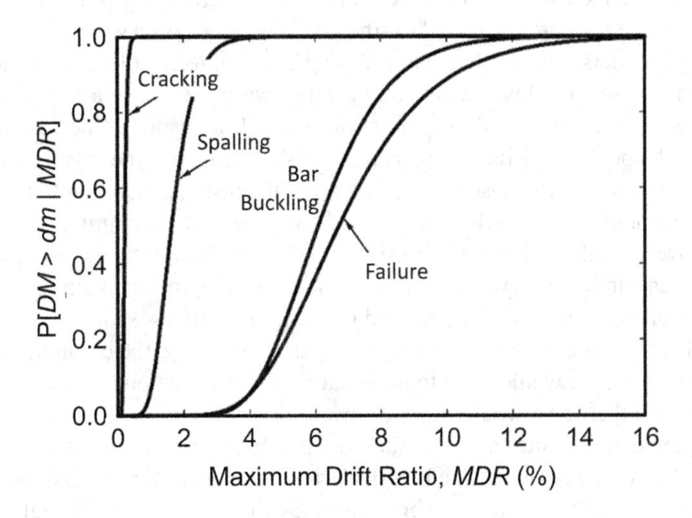

FIGURE 5.7 Fragility curves for discrete column damage states.

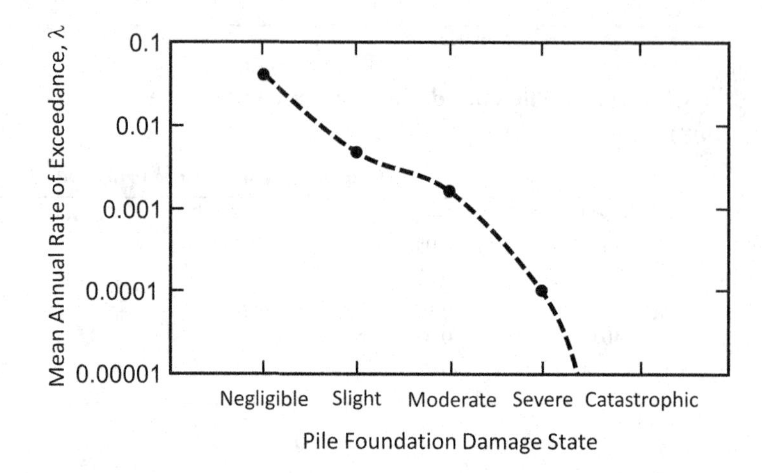

FIGURE 5.8 Categorical damage hazard curve for pile foundations supporting bridge in liquefiable soil profile.

Alternatively, damage states can be defined in terms of levels at which different expected repair actions would be triggered or at which different levels of casualties would be expected (e.g., collapse vs. non-collapse).

Both the pile foundation and bridge column damage models were convolved with the *EDP* hazard curve to produce damage hazard curves (Figure 5.8). Since the damage states are categorical, the *DM* hazard curves are also categorical in nature. By assigning a numerical damage intensity to each categorical damage state (e.g., negligible=0, slight=0.1, … catastrophic=1.0), the damage model can be formulated so that *DM* is a continuous function of *EDP*; such assignments, however, may require considerable judgment.

5.6.5 Loss-Level Implementation

The most complete evaluation of performance can be accomplished by specifying it in terms of losses. Such evaluations represent the ultimate expression of PBEE and are likely to be justified primarily for particularly large and/or important projects in the near future. Nevertheless, they provide a useful and instructive look into the future of earthquake engineering practice. They also can be used to demonstrate the economic value of earthquake engineering services.

As with damage models, loss models are often expressed in terms of discrete ranges, or categories, of loss. In such cases, the loss model can be expressed in terms of a loss probability matrix. Loss models can also be complicated by discontinuities. While response increases with increasing ground motion level, and damage increases with increasing response, the loss for some components of a soil-structure system can increase or decrease with increasing ground motion intensity and can do so in large jumps or drops. Consider a reinforced concrete bridge column, for example. The cost of repair for relatively weak shaking can consist of that associated with epoxying cracks or repairing spalls that appear under somewhat stronger shaking. Under strong shaking, rebar may buckle and repair may involve adding a steel jacket and grouting the annular space, in which case the cost of repairing spalls drops to zero. Finally, under very strong shaking, the column could be damaged beyond repair, in which case would need to be replaced and the previously described repairs would become irrelevant and their cost would drop to zero.

Loss-level implementation must also consider the specific form of the decision variable(s) used to describe the loss. Insurers and real estate investors, who may be making decisions on performance objectives for design or retrofit, have preferred metrics for describing potential losses. *Probable maximum loss* (PML) is a term that was defined differently by different entities until ASTM developed standards (E2026 for seismic risk assessments and E2557 for PML reports) in 1999. Recent updates to these standards suggest replacement of the term PML with *Scenario Expected Loss* (SEL) and *Scenario Upper Loss* (SUL), which represent losses with exceedance probabilities of 50% and 10% in either probabilistic scenarios (e.g., 475-year ground motion) or specific deterministic scenario events. The term, *probable loss* (PL) is also used to describe the loss value itself that has a 10% probability of exceedance in 50 years.

The preceding measures of loss refer primarily to individual structures or facilities. From a regional standpoint, however, the losses suffered by a community can be associated with multiple, interdependent systems such as buildings, bridges, pipelines, and electrical lines. Lifeline infrastructure systems such as transportation, water and wastewater, energy, and communications systems, are critical to the safety, security, and economic prosperity of a region and the amount of time such systems are out of service represents a loss that may dwarf the direct costs of repairing their physical damage. The concept of functional recovery (Section 5.3.4) has been developed to improve community resilience. Design for functional recovery involves consideration of both safety and recovery time and can be applied to both individual structures or regional building/infrastructure systems.

5.7 VECTOR PBEE ANALYSES

The implementation procedures discussed in the preceding section focused on scalar measures of response, damage, and loss, and such scalar measures may be reasonable and appropriate. For complex soil-structure systems, however, different components of the overall system may require individual measures of response, damage, and/or loss. Permanent displacements of unstable slopes (Chapter 10) can, for example, be predicted more accurately using a vector (*PGA* and *PGV*) intensity measure than using either (*PGA* or *PGV*) by themselves as scalar *IM*s. Flexible structures may have multiple modes of shaking excited by a given ground motion, in which case accurate prediction of their response may require determination of spectral acceleration at multiple periods. Scalar performance evaluations can be extended to consider multiple measures of ground motion, response, damage, and loss. Such analyses are often referred to as vector analyses since the *IM*s, *EDP*s, *DM*s, and/or *DV*s are represented as vector, rather than scalar, quantities.

Section 4.4.3.2 discussed vector PSHAs, which produce a vector of *IM*s with some mean annual rate (or return period) of joint exceedance. The benefit of a vector *IM* is that it can describe the characteristics of an earthquake ground motion more completely than a single, scalar *IM*. For example, a vector *IM* consisting of spectral accelerations at two periods can describe the ground motion amplitude over a broad range of frequencies that may significantly affect the response of a structure (particularly if its natural period lengthens due to damage-induced softening, or if a higher mode at a shorter period contributes significantly to its response), whereas a scalar spectral acceleration may represent the amplitude of the motion over only a narrow range of frequencies.

5.7.1 SCALAR *EDP* FROM VECTOR PSHA

The results of a vector PSHA may allow more accurate (and less variable) estimates of a single *EDP*. A response (demand) hazard curve for a scalar *EDP* can be computed for a two-component vector *IM* as

$$\lambda_{EDP}(edp) = \int_{IM_1}\int_{IM_2} P\left[EDP > edp \mid IM_1 = im_1, IM_2 = im_2\right]\left|MRD_{IM_1,IM_2}(im_1, im_2)\right| d(im_1)d(im_2) \tag{5.41}$$

where $MRD_{IM_1,IM_2}(im_1, im_2)$ is the joint *mean rate density* (Equation 4.36), or equivalently,

$$\lambda_{EDP}(edp) = \int_{IM_1}\int_{IM_2} P\left[EDP > edp \mid IM_1 = im_1, IM_2 = im_2\right] f_{IM_2|IM_1}(im_2 \mid im_1)\left|MRD_{IM_1}(im_1)\right| d(im_1)d(im_2) \tag{5.42}$$

where the conditional IM distribution can be computed as

$$f_{IM_2|IM_1}(im_2 \mid im_1) = \int_R\int_{\mathbf{M}} f_{IM_2|IM_1,\mathbf{M},R}(im_2 \mid im_1, \mathbf{m}, r) f_{\mathbf{M},R|IM_1}(\mathbf{m}, r \mid im_1)\, d\mathbf{m}\, dr \tag{5.43}$$

and the quantity, $f_{\mathbf{M},R|IM_1}(\mathbf{m}, r \mid im_1)$, represents the joint distribution, i.e., the disaggregation, of moment magnitude, **M**, and distance, R, conditional upon $IM_1 = im_1$ (Section 4.4.3.5). The quantity, $f_{IM_2|IM_1,\mathbf{M},R}(im_2 \mid im_1, \mathbf{m}, r)$, depends on the GMMs for IM_1 and IM_2 and on the correlation between the two *IMs*.

Discretizing IM_1 and IM_2 into N_{IM_1} and N_{IM_2} values, the mean annual rate of *EDP* exceedance can be approximated as

$$\lambda_{EDP}(edp) \approx \sum_{i=1}^{N_{IM_1}}\sum_{j=1}^{N_{IM_2}} P\left[EDP > edp \mid IM_1 = im_{1,i}, IM_2 = im_{2,j}\right] P\left[IM_2 = im_{2,j} \mid IM_1 = im_{1,i}\right] \cdot \left|\Delta\lambda_{IM_1}(im_{1,j})\right| \tag{5.44}$$

where $P\left[EDP > edp \mid IM_1 = im_{1,i}, IM_2 = im_{2,j}\right]$ is obtained from probabilistic response analyses (Section 5.7.3), and the conditional probability of im_2 given im_1 is computed as,

$$P\left[IM_2 = im_{2,j} \mid IM_1 = im_{1,i}\right] = \sum_{k=1}^{N_{\mathbf{M}}}\sum_{n=1}^{N_R} P\left[IM_2 = im_{2,j} \mid IM_1 = im_{1,i}, \mathbf{M} = \mathbf{m}_k, R = r_n\right] \cdot P\left[\mathbf{M} = \mathbf{m}_k, R = r_n \mid IM_1 = im_{1,i}\right] \tag{5.45}$$

and

$$\Delta\lambda_{IM_1}(im_{1,j}) = \lambda_{IM_1}\left(\frac{im_{1,j-1} + im_{1,j}}{2}\right) - \lambda_{IM_1}\left(\frac{im_{1,j} + im_{1,j+1}}{2}\right) \tag{5.46}$$

Assuming IM_1 and IM_2 are jointly lognormal (Jayaram and Baker, 2008), the first term in the summation of Equation (5.45) can be computed as

$$P\left[IM_2 = im_{2,j} \mid IM_1 = im_{1,j}, \mathbf{M} = \mathbf{m}_k, R = r_n\right]$$

$$= \Phi\left[\frac{\ln\left(\frac{IM_{2,j} + IM_{2,j+1}}{2} - \mu_{\ln IM_2|im_{1,i},\mathbf{m}_k,r_n}\right)}{\sigma_{\ln IM_2|im_{1,i},\mathbf{m}_k,r_n}}\right] - \Phi\left[\frac{\ln\left(\frac{IM_{2,j-1} + IM_{2,j}}{2} - \mu_{\ln IM_2|im_{1,i},\mathbf{m}_k,r_n}\right)}{\sigma_{\ln IM_2|im_{1,i},\mathbf{m}_k,r_n}}\right] \tag{5.47}$$

where

$$\mu_{\ln IM_2|im_{1,i},m_k,r_n} = \mu_{\ln IM_2|m_k,r_n} + \rho_{\ln IM_1,\ln IM_2} \frac{\sigma_{\ln IM_2|m_k,r_n}}{\sigma_{\ln IM_1|m_k,r_n}} \left(\ln im_1 - \mu_{\ln IM_1|m_k,r_n}\right) \tag{5.48}$$

$$\sigma_{\ln IM_2|im_1,m_k,r_n} = \sigma_{\ln IM_2|m_k,r_n} \sqrt{1-\rho^2_{\ln IM_1,\ln IM_2}} \tag{5.49}$$

and $\Phi(\bullet)$ is the cumulative distribution function of the standard Gaussian distribution.

5.7.2 Vector *EDP* from Vector PSHA

The results of a vector PSHA can also be used to compute a vector response (demand) hazard relationship, which can be useful when the relevant response of a system is described by more than one engineering demand parameter. The ability to predict multiple *EDP*s can lead to more accurate (i.e., less variable) damage estimates.

Considering a two-component vector of response, $\boldsymbol{EDP} = \{EDP_1, EDP_2\}$, the response hazard can be defined in different ways (Barbosa, 2011), for example as (i) the mean annual joint rate of events $\{(EDP_1 > edp_1)$ ***or*** $(EDP_2 > edp_2)\}$, or (ii) the mean annual rate of joint events $\{(EDP_1 > edp_1)$ ***and*** $(EDP_2 > edp_2)\}$, which obviously have different meanings and implications.

5.7.2.1 Joint Rate of Events

In some situations, a given level of damage may occur if any one of a set of multiple potential levels of response is exceeded. Considering the rate at which either $EDP_1 > edp_1$ or $EDP_2 > edp_2$ for a two-component *EDP* vector, the scalar formulation can be extended as

$$\lambda_{EDP_1 \text{ or } EDP_2}(edp_1, edp_2) = \int_{\boldsymbol{IM}} P\left[\left(EDP_1 > edp_1 \cup EDP_2 > edp_2\right) | \boldsymbol{IM}\right] \cdot \left|d\lambda_{IM}(\boldsymbol{im})\right| \tag{5.50}$$

which can be written, changing from the union to the intersection of events, as

$$\lambda_{EDP_1 \text{ or } EDP_2}(edp_1, edp_2) = \int_{\boldsymbol{IM}} \left\{1 - P\left[\left(EDP_1 \le edp_1 \cap EDP_2 \le edp_2\right) | \boldsymbol{IM}\right]\right\} \cdot \left|d\lambda_{IM}(\boldsymbol{im})\right| \tag{5.51}$$

or

$$\lambda_{EDP_1 \text{ or } EDP_2}(edp_1, edp_2) = \int_{\boldsymbol{IM}} \left\{1 - F_{EDP_1, EDP_2|IM}(edp_1, edp_2)\right\} \cdot \left|d\lambda_{IM}(\boldsymbol{im})\right| \tag{5.52}$$

where $F_{EDP_1,EDP_2|IM}(edp_1, edp_2)$ is the joint CDF of EDP_1 and EDP_2 conditional upon $\boldsymbol{IM}$, and can be computed as

$$F_{EDP_1,EDP_2|IM}(edp_1, edp_2) = \int_0^{edp_1} \int_0^{edp_2} f_{EDP_1,EDP_2|IM}(edp_1, edp_2)\, du_2\, du_1 \tag{5.53}$$

In this expression, $f_{EDP_1,EDP_2|IM}(edp_1, edp_2)$ is the joint PDF of EDP_1 and EDP_2 conditional upon $\boldsymbol{IM}$, and u_1 and u_2 represent all values of EDP_1 and EDP_2 that are less than edp_1 and edp_2, respectively. Representing joint probabilities in terms of conditional probabilities, Equation (5.53) can be written as

$$F_{EDP_1,EDP_2|IM}(edp_1, edp_2) = \int_0^{edp_1} \left[\int_0^{edp_2} f_{EDP_2|EDP_1,IM}(edp_2)\, du_2 \right] f_{EDP_1|IM}(edp_1)\, du_1 \tag{5.54}$$

where $f_{EDP_1|IM}(edp_1)$ is the marginal PDF of EDP_1 conditional upon IM, and $f_{EDP_2|EDP_1,IM}(edp_2)$ is the marginal PDF of EDP_2 conditional upon EDP_1 and $\mathbf{IM}$.

5.7.2.2 Rate of Joint Events

In other situations, a certain damage state may require the simultaneous exceedance of more than one level of response. Considering the rate at which both $EDP_1 > edp_1$ and $EDP_2 > edp_2$, the scalar formulation can be extended for a two-component EDP as

$$\lambda_{EDP_1 \text{ and } EDP_2}(edp_1, edp_2) = \int_{\mathbf{IM}} P\left[(EDP_1 > edp_1 \cap EDP_2 > edp_2) \mid \mathbf{IM} \right] \cdot \left| d\lambda_{\mathbf{IM}}(\mathbf{im}) \right| \tag{5.55}$$

where

$$P\left[EDP_1 > edp_1 \cap EDP_2 > edp_2 \mid \mathbf{IM} \right] = \int_{edp_1}^{\infty} \int_{edp_2}^{\infty} f_{EDP_1,EDP_2|\mathbf{IM}}(edp_1, edp_2)\, d(edp_1)\, d(edp_2) \tag{5.56}$$

and $f_{EDP_1,EDP_2|\mathbf{IM}}(edp_1, edp_2)$ is the joint PDF of EDP_1 and EDP_2. Switching from joint to conditional probabilities,

$$P\left[EDP_1 > edp_1 \cap EDP_2 > edp_2 \mid \mathbf{IM} \right] = \int_{edp_1}^{\infty} \left[\int_{edp_2}^{\infty} f_{EDP_2|EDP_1,\mathbf{IM}}(edp_2)\, d(edp_2) \right] f_{EDP_1|\mathbf{IM}}(edp_1)\, d(edp_1) \tag{5.57}$$

Assuming that the EDPs are marginally lognormally distributed, the EDP vector hazard can be computed knowing the marginal PDF of each scalar EDP and the correlation coefficient between the EDPs conditional upon $\mathbf{IM}$. The conditional distribution of one EDP with respect to the other (assuming lognormality) can be expressed as

$$f_{EDP_2|EDP_1,\mathbf{IM}}(edp_2 \mid edp_1, \mathbf{im}) = \frac{1}{edp_2 \sigma_{\ln EDP_2|edp_1,\mathbf{im}}} \Phi\left[\frac{\ln edp_2 - \mu_{\ln EDP_2|edp_1,\mathbf{im}}}{\sigma_{\ln EDP_2|edp_1,\mathbf{im}}} \right] \tag{5.58}$$

where

$$\mu_{\ln EDP_2|edp_1,\mathbf{im}} = \mu_{\ln EDP_2|\mathbf{im}} + \rho_{\ln EDP_1,\ln EDP_2|\mathbf{IM}} \frac{\sigma_{\ln EDP_2|\mathbf{im}}}{\sigma_{\ln EDP_1|\mathbf{im}}} \left(\ln edp_1 - \mu_{\ln EDP_1|\mathbf{im}} \right) \tag{5.59}$$

$$\sigma_{\ln EDP_2|edp_1,\mathbf{im}} = \sigma_{\ln EDP_2|\mathbf{im}} \sqrt{1 - \rho^2_{\ln EDP_1,\ln EDP_2|\mathbf{IM}}} \tag{5.60}$$

The marginal distributions, $f_{EDP_1|IM}(edp_1)$ and $f_{EDP_2|\mathbf{IM}}(edp_2)$, and correlation coefficient, $\rho_{\ln EDP_1,\ln EDP_2|IM}$, required to obtain the conditional mean and standard deviation can be obtained from probabilistic response analyses (Section 5.8).

5.8 PROBABILISTIC RESPONSE ANALYSES

The goal of probabilistic response analyses is to account for variable input parameters in prediction of the response of a soil-structure system given some level of ground shaking, i.e., to evaluate

$$P[\boldsymbol{EDP} > \boldsymbol{edp} \mid \boldsymbol{IM}] \tag{5.61}$$

where ***IM*** and ***EDP*** can be scalar or vector quantities. Probabilistic response analyses are generally performed using Monte Carlo simulation (Section D.9.3.3). The analyses are repeated using multiple realizations of the input parameters and multiple input motions with the relevant response parameters (***EDPs***) tabulated in histograms to which probability density functions can be fitted. In each realization, the input parameter values are generated to be consistent with specified distributions that reflect their respective aleatory variabilities. The number of realizations depends on the level of variability that exists and on the desired confidence level; for an *EDP* with a population mean, μ_{EDP}, and sample standard deviation, S_{EDP}, the number of simulations required to achieve a confidence level, *c*, that the error of the mean (i.e., the difference between the sample mean and the population mean, expressed as a percentage of the population mean) is less than *E* is given by (Driels and Shin, 2004)

$$n = \left[\frac{100 z_c S_{EDP}}{E \mu_{EDP}}\right]^2 \tag{5.62}$$

where z_c is the standard normal variable corresponding to *c* (for a confidence level of 95%, $z_c = 1.96$). This shows that the number of required simulations increases with increasing confidence level, increasing *EDP* uncertainty, and decreasing acceptable error.

The number of Monte Carlo analyses required to characterize uncertainties accurately can become computationally burdensome, particularly when response models are sophisticated, e.g., when they involve large, nonlinear finite element models. One approach to reducing the number of function evaluations is to fit a *response surface*, a relatively simple mathematical function, to the median response obtained from a number of analyses performed on a relatively coarse grid of influential input variable values. The analytical function, which may be as simple as a multidimensional polynomial, is then used to approximate the response at the finer grid over which numerical integration takes place. Once the response surface has been obtained, Monte Carlo methods may be used to account for the effects of uncertainty in the input variables; the response surface then takes the place of the response model so each function evaluation requires only the solution of the analytical function rather than a potentially much more time-consuming response model analysis. Obviously, the response surface is only an approximation of the response model, and the accuracy of the approximation will depend on the accuracy with which the response surface represents the output of the response model. Very simple response surface functions may not be able to accurately represent highly nonlinear response model outputs.

The primary contributors to the aleatory variability in response are variabilities in parameters used as inputs to the response model, variability in the ground motions used as inputs to the response model, and the sensitivity of the response to the input parameters. A complete performance evaluation would also include consideration of epistemic uncertainty, a major component of which would be uncertainty in the response model itself. Each of these contributors is discussed in the following sections.

5.8.1 Randomization of Input Parameters

The response of a soil-structure system predicted by some method of analysis can be sensitive to a number of its required input parameters. As a result, aleatory variability in the input parameters will lead to variability in the predicted response. Parametric variability is most commonly evaluated using Monte Carlo simulations (Section D.9.3.3) with randomized parameters; the discussion of randomization here will focus on geotechnical parameters, although actual (i.e., as-built) structural parameters also have variability.

The first step in a probabilistic response analysis is identification of which parameters should be randomized. The effect of aleatory variability in an input parameter on output parameter variability depends on (1) the dispersion of an input parameter and (2) the sensitivity of the output parameter to the input parameter. Tornado diagrams (Section D.9.1) can be used to identify influential variables for randomization; non-influential variables can generally be set at their mean or median values. The geotechnical parameters most likely to be influential are soil stiffness and damping characteristics for problems involving site response or soil-structure interaction; stiffness, damping, and shear strength for problems involving permanent deformations; and stiffness, damping, strength, and volume change (dilation-contraction) characteristics for problems involving liquefaction-susceptible soils.

Randomization of soil parameters must account for both variability and correlation. Many soil properties (e.g., density and shear wave velocity) are positively correlated, meaning that one tends to increase (or decrease) when the other increases (or decreases). Others (e.g., modulus reduction and material damping) are negatively correlated, in which case one decreases when the other increases, and vice versa. Randomization of soil properties must often also account for spatial variability, i.e., the differences in individual soil properties at different locations. In many cases, properties vary from one soil layer to another due to compositional differences, but soil properties can also vary within a given soil layer. Such variability has two components (Figure 5.9a) – a trend and a random component, i.e., $P(z)=t(z)+w(z)$. The trend may be associated with positional differences in effective stress, preconsolidation pressure, or other characteristics that generally vary in a relatively smooth manner within a soil layer. The random component represents variability about the trend and exhibits spatial correlation, i.e., a correlation between values of the same property at two locations that decreases with increasing separation distance between the locations. This behavior is not surprising – one would expect the properties at two closely spaced points to be strongly correlated (similar levels above or below the trend line) than at two points located far apart. The correlation of a variable with itself (at locations separated by a lag distance, τ), referred to as *autocorrelation*, then decreases with increasing lag distance. Vanmarcke (1977) characterized autocorrelation by means of a *scale of fluctuation,* small values of which indicate rapid fluctuations about the trend line and larger values a more slowly varying behavior. The scale of fluctuation (for a one-dimensional random field) is defined as twice the area under the correlation function, i.e.,

$$\delta = 2\int_0^\infty \rho(\tau)\,d\tau \tag{5.63}$$

where τ is the lag distance. Subtracting the trend from the parameter values in Figure 5.9a yields the random fluctuations, $w(z)$, shown in Figure 5.9b. Letting $\bar{w}$ and $\hat{\sigma}$ represent the sample mean and sample standard deviation, respectively, the sample autocorrelation function can be computed as

$$\hat{\rho}(\tau) = \frac{\sum_{i=1}^{n(\tau)}\left[w(z_i)-\bar{w}\right]\left[w(z_i+\tau)-\bar{w}\right]}{\left[n(\tau)-1\right]\hat{\sigma}^2} \tag{5.64}$$

where $n(\tau)$ is the number of pairs separated by the lag distance, τ. A number of theoretical autocorrelation models can be expressed in terms of the scale of fluctuation and it is common to determine the scale of fluctuation by fitting a theoretical autocorrelation model to the sample autocorrelation function. Some examples of theoretical autocorrelation functions are shown in Table 5.4.

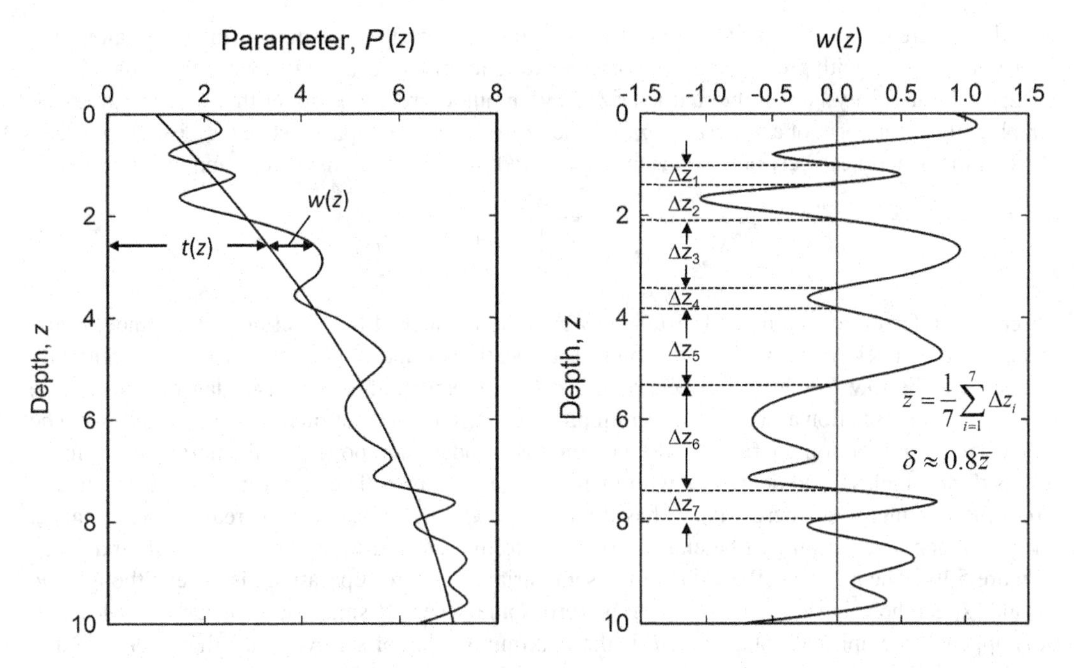

FIGURE 5.9 Variation of typical parameter with depth: (a) parameter and trend line, (b) random component of parameter after removal of trend with Vanmarcke (1977) approximation of scale of fluctuation, δ.

TABLE 5.4
Theoretical Autocorrelation Models (Vanmarcke, 1977, 2010)

Model	Autocorrelation Function
Triangular	$\rho(\tau) = \begin{cases} 1-\lvert\tau\rvert/\delta & \text{for} \quad \lvert\tau\rvert \le \delta \\ 0 & \text{for} \quad \lvert\tau\rvert > \delta \end{cases}$
Exponential	$\rho(\tau) = \exp\left[-\frac{2\lvert\tau\rvert}{\delta}\right]$
Squared exponential (Gaussian)	$\rho(\tau) = \exp\left[-\pi\left(\frac{\tau}{\delta}\right)^2\right]$

An approximate procedure for estimating the scale of fluctuation, which shows it to be related to the depth intervals between crossings of the trend line, is shown in Figure 5.9b.

The spatial variability of soils should be recognized as generally being anisotropic with less variability (i.e., longer scales of fluctuation) in the horizontal direction than in the vertical direction. Vertical variability is usually easier to characterize because subsurface explorations (e.g., soil borings, CPT soundings, etc.) are generally vertically oriented with measurements made (particularly in the CPT) at close vertical spacings. Relatively little data is available on horizontal spatial variability because few sites have been investigated with sufficiently closely spaced borings or soundings to do so. Lloret-Cabot et al. (2014) used data from 1 to 6.5 m depth in 18 CPT soundings at spacings ranging from 1.5 to 10 m on an artificial island in the Beaufort Sea to compute consistent vertical scales of fluctuation of 40–44 cm in each of three linear profiles across the island. Horizontal scales of fluctuation were on the order of 1.7 m in one section, 5.1–5.6 m in another, and 13.7–15.9 in the third; these scales of fluctuation are all significantly greater than the corresponding vertical values, and the differences between them appear to be influenced by the spacings of the CPTs along the different sections (with greater spacing corresponding to larger horizontal scales of fluctuation).

Spatial variability is usually characterized by an autocorrelation model that can be expressed graphically in the form of a *semivariogram*. The semivariance of a parameter, $P(z)$, is half the variance of the differences between values of P separated by a constant distance τ, and is defined as

$$\gamma(\tau) = \frac{1}{2|N(\tau)|} \sum_{N(\tau)} [P(z_i + \tau) - P(z_i)]^2 \tag{5.65}$$

where τ is referred to as a lagged distance, $N(\tau)$ is the number of points at lagged distance, τ, and z_i and z_j are values of the variable of interest at locations i and j. The semivariance is generally expected to be low for short lag (separation) distances and higher for larger lag distances. The autocorrelation function and semivariogram provide similar information on spatial variability. The autocorrelation function operates on the random (de-trended) component of the parameter of interest, is dimensionless, and decreases with increasing lag distance. The semivariogram is calculated from the parameter of interest itself, has dimensions, and increases with increasing lag distance. A typical semivariogram can be characterized by terms referred to as the nugget, sill, and range (Figure 5.10). The *nugget* is the value of the semivariance at zero separation distance; although one would expect the value at zero distance to be zero, some generally small value of semivariance usually appears in empirical data. The *sill* is the maximum value of semivariance that is observed at and beyond the *range* of the semivariogram. Beyond the range, the data is not spatially correlated. A number of theoretical semivariogram models are available for fitting empirical semivariance data. Since common semivariogram models can be expressed in terms of the scale of fluctuation (e.g., the exponential model can be expressed as $\rho_\tau = \exp(-2|\tau|/\delta)$ for a separation distance, τ), the fit can be used to reveal the apparent scale of fluctuation.

Spatial variability can be modeled with the concept of a random field, a set of spatially distributed numbers with prescribed probability distribution and autocorrelation behavior. Random fields (Figure 5.11) can be generated in one, two, or three dimensions and are generally "seeded" with a vector of random numbers. Either unconditional or conditioned random fields can be generated, the latter requiring that the randomized variable take on particular values (e.g., measured values) at specific locations (the measurement points). A variety of procedures, including spectral methods (Shinozuka and Deodatis, 1991, 1996; Gutjahr, 1989), turning band methods (Matheron, 1973; Mantoglou and Wilson, 1981), matrix decomposition methods (Clifton and Neuman, 1982; Davis, 1987; Baecher and Christian, 2003), and Gaussian sequential simulation (Gomez-Hernandez and Journel, 1993) are available for random field generation. Using different random seeds, a suite of random fields with the same overall probability distribution and autocorrelation behavior can be generated. A different random field can then be used in each realization of a series of Monte Carlo

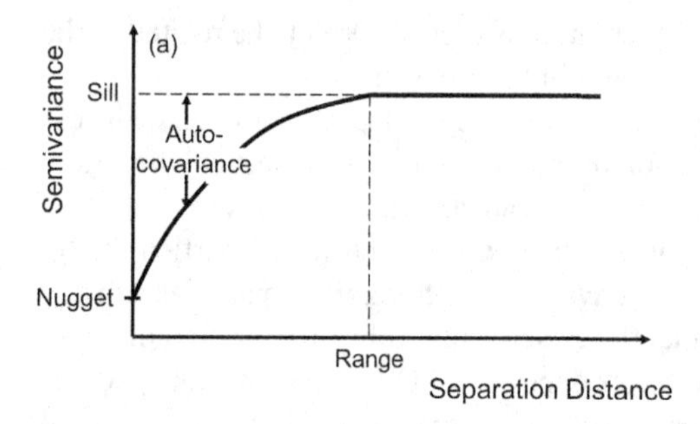

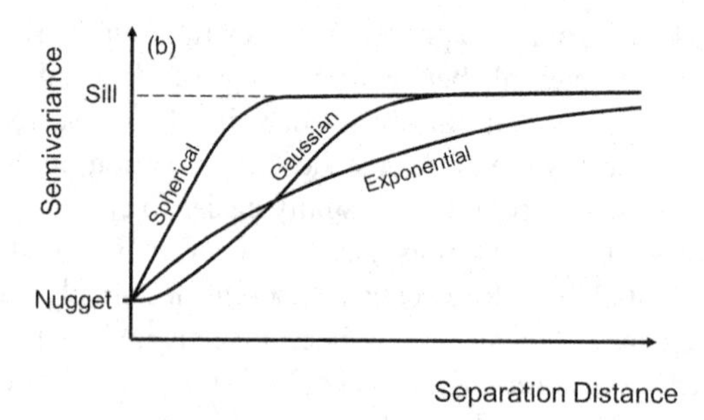

FIGURE 5.10 Schematic illustration of semivariogram: (a) illustration of nugget, sill, and range, and (b) common semivariogram models. The autocorrelation is the ratio of the autocovariance to the variance for a stationary random process.

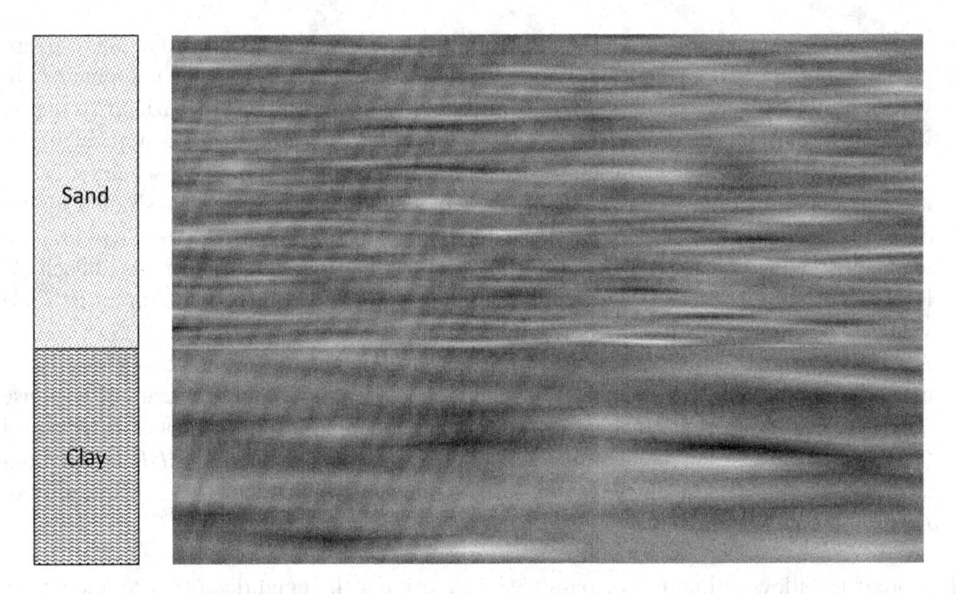

FIGURE 5.11 Illustration of two-dimensional random field for arbitrary soil parameter (e.g., penetration resistance, shear wave velocity, etc.) with layer of sand (low δ_v, low δ_h) overlying layer of clay (higher δ_v, higher δ_h).

analyses to account for the effects of spatial variability. Generation of random fields for one-dimensional site response analyses is described in detail in Section 7.5.7.1.

5.8.2 Input Motions

A large portion of the total uncertainty in seismic response is due to record-to-record variability in the ground motions, so it is necessary to apply a suite of ground motions to the response model and critical to select the motions to be used in those analyses carefully. Even spectrum-compatible motions can produce significant variability in response. Improper selection can lead to bias (high or low) and to over- or under-representation of the uncertainty in the computed response. Procedures for input motion selection and modification are described in Section 4.5; consideration of record attributes (4.5.1.2) or matching of conditional spectra (4.5.1.3) can reduce uncertainty and remove bias from the computed *EDP*(s) even when a scalar *IM* is used.

The conditional *EDP* distribution is used to calculate the mean annual rate of *EDP* exceedance. The calculation process is different, however, for scalar and vector *IM*s, and the differences can affect the manner in which ground motions are selected. Consider the case in which ground motions are characterized by a two-component vector intensity measure, $\mathbf{IM} = \{IM_1, IM_2\}$. The mean annual rate of *EDP* exceedance for this case is given by Equation (5.41). If the response of a system of interest is not influenced by IM_2, then $P[EDP > edp \mid IM_1, IM_2] = P[EDP > edp \mid IM_1]$ and there is no need for the vector PBEE analysis since the mean annual rate of exceedance would be exactly the same as given by a scalar analysis. If, however, IM_2 does affect system response, but is not used in an IM_1-based scalar analysis, the estimate of $EDP| IM_1$ will be influenced by the distribution of IM_2 values in the suite of ground motions used in the analyses. This leads to two general approaches (Baker, 2007): (1) using both IM_1 and IM_2 as a vector *IM*, or (2) selecting a suite of input motions with a sample distribution of $IM_2| IM_1$ that is equal to a target distribution of $IM_2| IM_1$ obtained from the ground motion hazard analysis.

5.8.2.1 Scalar Intensity Measure

When a scalar intensity measure is used, suites of ground motions can be developed using one of two basic procedures:

1. A suite of ground motions can be scaled to a common *IM* value and applied to the response model to predict *EDP*s conditional upon that *IM* value. The fact that the computed *EDP* values are not all the same is an indication of the record-to-record variability that is inherent in earthquake ground motions. By repeating this process for a number of *IM* values (ideally, with consideration of dominant source characteristics in selection of the motions corresponding to each *IM* value), a plot with a series of "stripes" of response data (Figure 5.12a) can be generated. Median values of *EDP* for each stripe can be used to establish a median *EDP-IM* relationship, and the distributions of residuals at each *IM* level can be used to characterize aleatory variability in *EDP*|*IM*; it is common for *EDP*|*IM* values to be characterized as lognormally distributed.
2. A series of ground motions spanning a wide range of *IM* values (again selected with consideration of source characteristics) can be identified and used as input to a series of response analyses. The resulting *EDP* values form a "cloud" of data points on a plot of *EDP* vs. *IM* (Figure 5.12b). Regression techniques can be used to establish a median *EDP-IM* relationship and the residuals of the regression can be analyzed to characterize the variability of *EDP*|*IM* around the median.

Both approaches allow estimation of parameters describing the conditional distribution of *EDP* given *IM*, i.e., the development of fragility curves. The stripes approach is generally more efficient when considering a single *IM* level, and the cloud approach when multiple *IM* levels are considered (Mackie and Stojadinovic, 2005); the use of multiple stripes, however, allows improved characterization of *IM*-dependent dispersion, which can be significant over the wide range of *IM*s considered in a PBEE analysis (Baker, 2007).

5.8.2.2 Vector Intensity Measure

A number of different approaches can be taken to the selection of ground motions that are consistent with a vector intensity measure (Baker, 2007).

A cloud method in which multiple motions representing a broad range of all *IM* components are applied to the response model can be used. The distribution of *EDP*|**IM** can be obtained by regressing *EDP* (or, usually, ln *EDP*) against all components of *IM* to find the median response, and using the median to compute the residuals from which the variability can be characterized. This approach is easily visualized for two-component *IM*s (Figure 5.13) but is easily extended to *IM*s with three

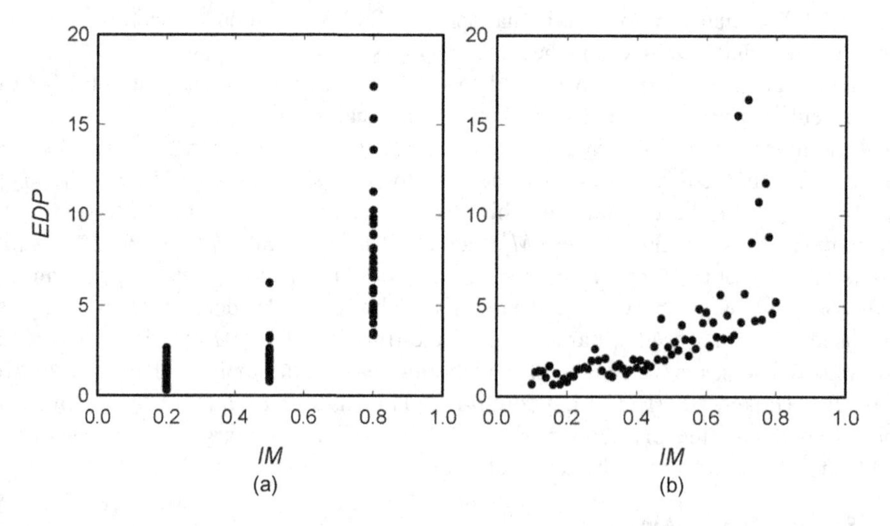

FIGURE 5.12 Schematic illustration of (a) "stripes" approach, and (b) "cloud" approach to *EDP*|*IM* characterization (Kramer, 2008).

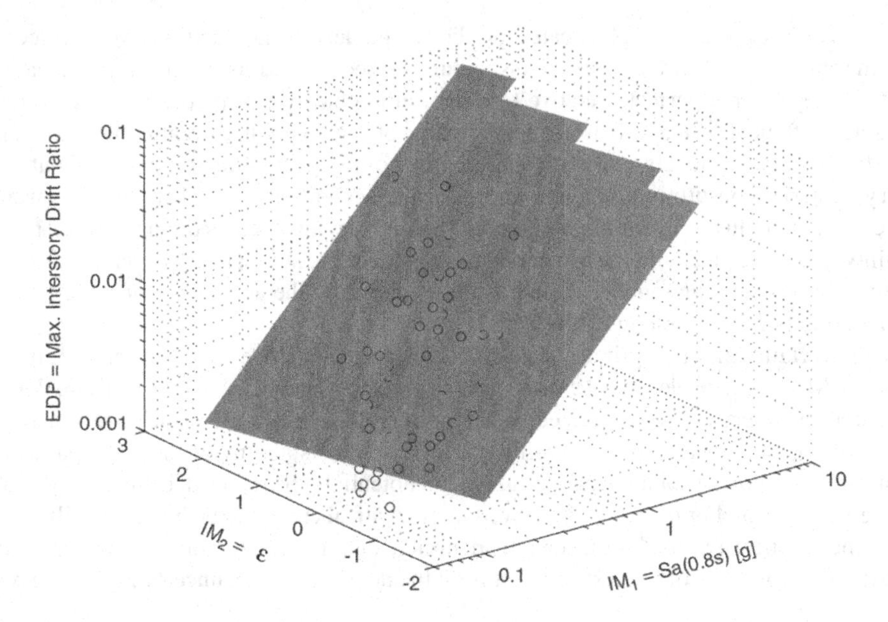

FIGURE 5.13 Estimation of median interstory drift ratio as power law (planar in log-log-log space) function of $IM = \{S_a(T = 0.8 \text{ sec}), \varepsilon\}$. (After Baker, 2007.)

or more components. To capture the interaction between *IM* components, care must be taken to ensure that all combinations of *IM* components are adequately represented in the suite of ground motions; it is possible to create a suite in which all components appear to be well distributed when viewed in a marginal (one variable at a time) sense, but lack representation of specific combinations of components. Consideration of amplitude, frequency content, and duration in the selection of input motions (Section 4.5.1.2) will help ensure reasonable correlation between *IM* components. The issue of collinearity (or multicollinearity) can be an issue when two (or more) *IM* components are closely correlated; in such cases, the correlated components can be difficult to distinguish since multiple combinations of them can produce essentially the same fit to the overall data. The overall uncertainty may not be affected, but the influence of the individual correlated components on *EDP* may be erroneous.

The response of many soil-structure systems is more closely related to one *primary IM* component than all other components. A stripe approach can be taken with a vector *IM* by scaling individual suites of records to different values of the primary *IM* component and then regressing the residuals of the response data to establish the effects of the other components for each stripe. This approach reduces collinearity problems and captures interaction between *IM* components well because a separate regression is performed for each primary component stripe. The separate regressions also allow the response at each stripe to be modeled more accurately since the burden of capturing all aspects of behavior with a single relationship, as required for the cloud method, does not exist. The stripes method, however, will generally require more response analyses than the cloud method.

A number of other approaches to ground motion selection are available. Baker (2007) provides a detailed description of many of these methods with a summary of the relative advantages and disadvantages of each.

5.8.3 Model Uncertainty

Model uncertainty results from the imperfections and idealizations of a predictive model – in the present context, a response model. The seismic response of soil-structure systems is extremely complex, and even the most sophisticated predictive models do not capture every aspect of the physics that affect it. All predictive models should be recognized as mathematical idealizations of reality – they are not perfect. Relative to the specific objective of predicting seismic response in a manner sufficient to make reasonable estimates of damage (and, eventually, loss), some are very useful. Available models make varying degrees of simplifying assumptions about problem geometry, material constitutive behavior, and boundary conditions and likely ignore variables or processes that, in reality, influence response. As a result, varying degrees of uncertainty, i.e., model uncertainty, are associated with the use of response models; because it results from a lack of knowledge about the system being modeled, and can be reduced by the acquisition of more knowledge, model uncertainty is epistemic in nature.

Model uncertainty has two primary causes: (1) missing predictive variables, and (2) inaccurate model form. Missing variables may be those not recognized as being influential or those that cannot be measured or otherwise characterized. Inaccurate model form may result from practical consideration of computational complexity/effort or lack of understanding of the basic physics of the problem. Both components of model uncertainty can potentially be reduced, by including additional predictive variables and/or the use of improved mathematical expressions, but there will usually be a limit to the number of variables that can be identified and/or measured or to the understanding of the physics of the problem of interest that will limit the degree to which uncertainty can be reduced.

5.8.4 Extension to Vector Damage and Loss Analyses

The principles used to develop the vector response procedures described in this section can be applied to perform vector damage and vector loss analyses. For a building, structural damage may be characterized individually for beams, columns, girders, and foundations, and non-structural damage may be characterized for cladding, partitions, service elements such as piping and elevator systems, and for the contents of the building. Losses also have multiple components, as previously described in the broad terms of repair costs, downtime, and casualties. Direct losses such as repair costs, however, may need to be broken down into components because the relationships between them can become complicated. Not all repair costs, for example, continuously increase with increasing response level. The cost of repairing cracking and spalling of concrete from bridge columns will increase with increasing response up to a point, but at a higher level of response the entire column may need to be replaced, in which case there is no need to repair cracks and spalls; such factors can make loss model fragility functions quite complicated.

5.8.5 Propagation of Aleatory Variability

The probabilistic response analyses described earlier in this section allow evaluation of $P[\boldsymbol{EDP}>\boldsymbol{edp}|\boldsymbol{IM}=\boldsymbol{im}]$ where ***EDP*** and/or ***IM*** can be scalar or vector quantities. *EDP* hazard curves (or hazard surfaces for vector *EDP*s) are obtained by propagating the aleatory variability in $EDP|IM$ through the integrals of Equations (5.16) (scalar *IM*) or (5.41) (vector *IM*). The simplest method of propagation (Section D.9) is direct integration (quadrature) over constant increments of *IM* (or λ_{IM}, depending on the formulation, Equation 5.12 or 5.14, that is selected). The computational efficiency, in terms of the number of function (response model) evaluations required to complete the integration, can be important when complex response models are used. Because the integrand is curved, direct integration may be efficient over some range of *IM* but inefficient in others, even for scalar *IM*s.

Computational efficiency becomes particularly important in PBEE when vectors are involved since integration is required over all uncertain variables. Given the wide ranges of the variables and the small increments required for accurate numerical integration, the number of function evaluations can become extremely large. Adaptive quadrature methods use adaptively refined *IM* increments to reduce the number of function evaluations; Bradley et al. (2009) developed an efficient magnitude-oriented adaptive quadrature method for scalar PBEE calculations. Multidimensional integration can often be accomplished more efficiently using Monte Carlo techniques, which use random sampling of a function to numerically compute an estimate of its integral. The integral is approximated by averaging samples of the function to be integrated at random points within the interval(s) of interest. The procedure converges at a rate of $N^{-1/2}$ and becomes exact in the limit as the number of random samples goes to infinity. This convergence rate is slower than that of standard numerical integration procedures in one dimension, but remains constant for multidimensional integration while the convergence rates of standard techniques decrease exponentially.

Baker and Cornell (2003) used first-order, second-moment (FOSM) procedures (Section D.9.3.2) to compute the mean and variance of $\boldsymbol{DV}|\boldsymbol{IM}$ and fit a distribution to those moments, thereby "collapsing" the interior portions of the full performance integral (Equation 5.19) to the form

$$\lambda(\boldsymbol{DV}) = \int G(\boldsymbol{DV} \mid \boldsymbol{IM}) |d\lambda(\boldsymbol{IM})| \tag{5.66}$$

which could then be integrated by one of the previously described numerical procedures. Care must be taken when applying this approach to problems such as liquefaction or structural collapse where the high level of nonlinearity is not compatible with FOSM procedures.

5.9 RELIABILITY-BASED DESIGN FORMAT

The practice of geotechnical design for static loading, particularly for foundation design, is increasingly being accomplished using reliability-based techniques in a LRFD format. LRFD recognizes that uncertainties in loads and resistances are different and characterizes them separately. Factored loads, i.e., the products of load factors and nominal loads, are required to be lower than factored resistances, which are the products of resistance factors and nominal resistances. In concept, load and resistance factors are calibrated to achieve some desired, time-invariant probability of failure (capacity exceedance). In practice, however, calibration has often been based on a target of consistency with prior design standards.

In seismic design, strong demands result from infrequent earthquake ground motions so the element of time must be brought into the process. The design level of ground motion is generally expressed in terms of a design $\boldsymbol{IM}$, i.e., one corresponding to a particular mean annual rate of exceedance or return period. That ground motion level is customarily taken as a nominal level that is used to compute dynamic response.

The basic principles of performance-based design, as implemented at the response level, can be used to develop load (or demand) and resistance (or capacity) factors. However, geotechnical damage is known to be more closely related to deformations than to forces or stresses, so a reliability-based format for deformation-based design is desirable. In such a format, the response of a system of interest could be expressed as *deformation demands* and the allowable level of deformation as *deformation capacities*. Jalayer (2003) and Cornell et al. (2002) used PBEE principles to develop a demand-and-capacity-factor (DCFD) procedure for structural design that can be adapted for geotechnical design.

5.9.1 Load and Resistance Factor Design

Load and resistance factors are intended to account for variabilities and uncertainties in loads and resistances, respectively, recognizing that they can (and usually will) be different. For seismic design, loads are induced by ground motions so the total dispersion (including both aleatory and epistemic components) in loading depends on the dispersions in ground motion and in system response given the level of ground motion. Resistances, and their dispersions, are generally independent of response.

Uncertain loads and resistances can be described by their probability density functions, as shown in Figure 5.14. A safety margin, or performance, function, g, can be defined as the difference between resistance and loading, i.e.,

$$g = R - L \tag{5.67}$$

A condition of $g<0$, therefore, indicates failure, the probability of which depends on the relative positions of the loading and resistance distributions. A limit state can be defined as the condition of incipient failure, i.e., as the point where $L=R$ and whose exceedance ($L>R$) indicates failure. Letting $LS=L-R$, the probability of failure is given by

$$P_f = P[L>R] = P[LS > 0] = 1 - P[g < 0] \tag{5.68}$$

In Figure 5.14, the nominal loads and resistances, L_n and R_n, are the expected, deterministic values. A factored load, L_f, can be defined as the product of a load factor, γ, and the nominal load. A factored resistance, R_f, is similarly defined as the product of a resistance factor, ϕ, and the nominal resistance. The load and resistance factors can be defined such that a condition of $L_f=R_f$ (which indicates a condition of incipient failure) corresponds to a target probability of failure, P_f, or target reliability index, $\beta=-\Phi-^1(P_f)$. The load factor is greater than 1.0, so the factored load is greater than the nominal load, and the resistance factor is less than one, so the factored resistance is less than the nominal resistance. In this manner, the load and resistance factors act as partial factors of safety applied individually to the nominal load and resistance.

5.9.2 Calculation of Demand and Capacity Factors

As previously discussed, deformations are generally more efficient predictors of geotechnical damage than forces, so the terms *demand* and *capacity* are used here (instead of load and resistance) to denote a more general interpretation of response. The demand is a measure of the response of a system to earthquake shaking, and the capacity is a measure of the system's ability to tolerate that demand while meeting some performance objective. In geotechnical earthquake engineering, the demand, D, would typically be some measure of permanent displacement and the capacity, C, would be a measure of allowable displacement. The notion of "failure" in this format ($g=C-D<0$) would

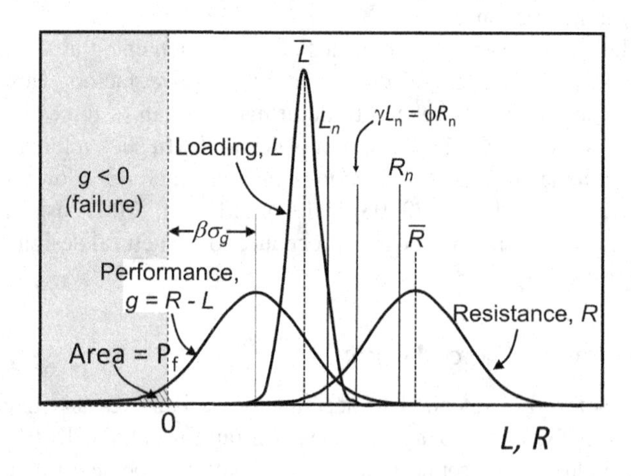

FIGURE 5.14 Illustration of relationships between loading, resistance, and performance function distributions, including nominal load, L_n, factored load, $L_f=\gamma L_n$, nominal resistance, R_n, and factored resistance, $R_f=\phi R_n$, and reliability index, β.

mean failure to achieve the performance objective, i.e., exceedance of the allowable displacement, which may or may not result in significant damage depending on the level of the performance objective. An appropriate limit state for design purposes, then, would correspond to a condition in which the demand exceeds the capacity, i.e., $LS = D > C$. A suitable design procedure would result in a user-defined, low mean annual rate or probability (or long return period) of limit state exceedance. Recalling that the mean annual rate is effectively equal to the mean annual probability for the low probabilities of failure used in seismic design, the design must satisfy $\lambda_{LS} \leq p$ for a target mean annual probability, p.

Consider a response related to the ground motion intensity by the deterministic (i.e., zero uncertainty) relationship

$$EDP = f(IM) \tag{5.69}$$

Letting $im^{\widehat{edp}} = f^{-1}(\widehat{edp})$, i.e., the IM value that would produce the median demand, $EDP = \widehat{edp}$, the median EDP hazard curve can be expressed as

$$\lambda_{EDP} = \lambda_{IM}\left(im^{\widehat{edp}}\right) \tag{5.70}$$

The median EDP hazard curve represents the "expected" demand in the absence of uncertainty. It does not, however, account for dispersion in the response, which can be quite significant for deformation demands The EDP value corresponding to the median hazard curve without response dispersion can be expressed as EDP_0. EDP_0 is not equivalent to the median EDP $\left(\text{i.e., } \widehat{EDP}\right)$ discussed previously in Section 5.5.3. The hazard curve for EDP including dispersion in response can be obtained from Equation (5.16). For a given return period, the demand including response dispersion can be expressed as EDP_R. When dispersion in response and capacity are both considered, as described in Equation (5.30), the resulting EDP at a particular return period can be expressed as EDP_{RC}. By integrating over capacity, as in Equation (5.30), the EDP_{RC} hazard curve corresponds to the median, or expected, capacity. Thus, the difference between EDP_0 and EDP_R depends on the dispersion in response and the difference between EDP_R and EDP_{RC} on the dispersion in capacity. Since dispersion increases the EDP level for a given mean annual probability (or return period), the response hazard curves for EDP, EDP_R, and EDP_{RC} would be positioned as indicated in Figure 5.15.

With this notation, *demand and capacity factors* can be defined as

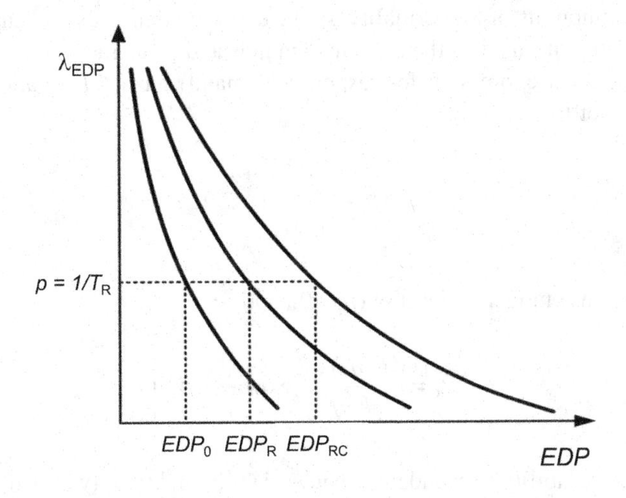

FIGURE 5.15 Schematic illustration of response (*EDP*) hazard curves for cases of median response, uncertainty in response, and uncertainties in response and capacity.

$$DF = \frac{EDP_R}{EDP_0} \tag{5.71a}$$

and

$$CF = \frac{EDP_R}{EDP_{RC}} \tag{5.71b}$$

For a given mean annual probability, p, the median response can be described as the *nominal demand*, $D_n = EDP_0$, for that return period. The nominal demand neglects response uncertainty that may cause the actual response to exceed its value. The product of the demand factor and the nominal demand is therefore equal to the *factored demand*, $D_f = DF \cdot EDP_0 = EDP_R$, which exceeds the nominal demand by an amount that depends on the uncertainty in response. For the same mean annual probability, the median capacity is the *nominal capacity*, i.e., $C_n = EDP_{RC}$. The nominal capacity, however, does not account for the fact that the actual capacity may be less than its mean value. The product of the nominal capacity and the capacity factor is the *factored capacity*, $C_f = CF \cdot C_n = EDP_R$, which is less than the nominal capacity by an amount that depends on the uncertainty in capacity. With this approach, a design that satisfies the condition

$$DF \cdot D_n \leq CF \cdot C_n \tag{5.72}$$

which indicates a factored demand less than or equal to the factored capacity ($D_f \leq C_f$), would result in the desired goal of $\lambda_{LS} \leq p$. For a typical geotechnical earthquake engineering problem, this framework would achieve the desired probability of failure by multiplying the median deformation by a factor greater than one and multiplying the median allowable deformation by a factor less than one. The factors would represent uncertainties in response and capacity separately and would be tied to a particular, user-specified mean annual probability (or return period) of limit state (demand exceeding capacity) exceedance. In this sense, the demand and capacity factors are analogous to partial factors of safety, but they are based on deformations rather than forces or stresses.

5.9.3 Closed-Form Solution

The development of closed-form expressions for the mean annual rates of response and capacity exceedance were described in Sections 5.5.3 and 5.6.3, respectively. The closed-form expressions were all based on assumptions of lognormality, so the expected demands and capacities are related to the means of their logarithms, i.e., the median demands and capacities.

Solving the closed-form expression for response (demand) hazard (Equation 5.24) at a mean annual probability, p, for the response,

$$EDP_0 = a\left(\frac{p}{k_0}\right)^{-b/k} \tag{5.73}$$

for the case of no response model variability ($\beta_R=0$), and

$$EDP_R = a\left(\frac{p}{k_0}\right)^{-b/k} \exp\left[\frac{1}{2}\frac{k}{b}\beta_R^2\right] \tag{5.74}$$

when response model variability is included. When variability in capacity is also included, Equation (5.35) can be solved to produce

$$EDP_{RC} = a\left(\frac{p}{k_0}\right)^{-b/k} \exp\left[\frac{1}{2}\frac{k}{b}\beta_R^2\right]\exp\left[\frac{1}{2}\frac{k}{b}\beta_C^2\right] \tag{5.75}$$

Then, using Equation (5.71), the demand and capacity factors can be written as

$$DF = \exp\left[\frac{1}{2}\frac{k}{b}\beta_R^2\right] \tag{5.76a}$$

and

$$CF = \exp\left[-\frac{1}{2}\frac{k}{b}\beta_C^2\right] \tag{5.76b}$$

to produce a design that will satisfy Equation (5.71). Note that, under the assumptions of the closed-form solution, the demand and capacity factors are constants, i.e., they do not vary with return period. If those assumptions are relaxed, however, the demand and capacity factors will vary with return period – with each typically deviating from 1.0 by amounts that increase with increasing return period.

Example 5.4

The building in Examples 5.2 and 5.3 is to be designed to have a 2% probability of experiencing severe (or worse) settlement-related damage over its 50-year design life. What demand and capacity factors should be used to achieve that performance level?

Solution:

From the previous examples, k=2.479, b=0.5, β_R=0.2, and β_C=0.3. Then, from Equations (5.76), multiplying the nominal demand and capacity values by the respective factors

$$DF = \exp\left[\frac{1}{2}\frac{k}{b}\beta_R^2\right] = \exp\left[\frac{1}{2}\left(\frac{2.479}{0.5}\right)(0.2)^2\right] = 1.104$$

and

$$CF = \exp\left[-\frac{1}{2}\frac{k}{b}\beta_C^2\right] = \exp\left[-\frac{1}{2}\left(\frac{2.479}{0.5}\right)(0.3)^2\right] = 0.800$$

would be expected to produce a design with the desired performance level. Note that, for the idealized assumptions of the closed-form solution, both the demand and capacity factors are independent of the hazard level (2% in 50 years, or a 2,475-year return period). This will not be the case for actual problems where the *IM* hazard curve and median response curve (*EDP* vs. *IM*) are not of power law form; in these cases, the demand and capacity factors will vary with hazard level.

5.10 SUMMARY

1. Performance of infrastructure in earthquakes can be measured in different ways that have different degrees of relevance to various stakeholders. While measures such as horizontal and vertical displacements may be of greatest interest to geotechnical engineers, measures such as cost and downtime are generally of greatest interest to owners who make decisions about seismic risk mitigation.

2. Performance objectives should be established early in a seismic design or assessment project and earthquake engineers should be prepared to discuss the relationship between cost and performance with owners and other decision-makers.
3. The fundamental components of a complete performance evaluation can be conveniently viewed as a progression from ground motion to the response of a system of interest to the physical damage caused by that response to the losses associated with that physical damage.
4. Earthquake engineers are heavily involved in predicting the response of a particular soil-structure system to earthquake ground motion. Ground motions can be characterized in many ways, i.e., by many intensity measures (*IM*s); the most useful *IM*s are those that the response is most strongly correlated to. These *IMs* can be different for different soil-structure systems.
5. Losses are often characterized in three primary categories – casualties, repair and replacement costs, and loss of functionality for extended periods of time – often referred to informally (but memorably) as deaths, dollars, and downtime. Functional recovery, a post-earthquake state in which pre-earthquake functionality has largely been restored, is an important societal objective that extends beyond the individual structures that earthquake engineering is usually involved with.
6. Performance-based design commonly considers multiple performance objectives associated with different levels of ground motion. Higher levels of performance are required for lower (and more frequent) levels of shaking and lower levels are accepted for stronger (and more rare) shaking.
7. Response models predict measures of response, commonly referred to as engineering demand parameters, or *EDP*s, conditional upon measures of ground motion (*IM*s). Damage models predict measures of damage (*DM*s) conditional upon *EDP*s. Loss models predict measures of loss, commonly referred to as decision variables (*DV*s) conditional upon *DM*s. The four variables, *IM*, *EDP*, *DM*, and *DV*, and the models that relate them are influenced by both variability and uncertainty.
8. The effects of aleatory variability and epistemic uncertainty can be propagated through a performance evaluation in a number of different ways. Aleatory variability is often accounted for by randomized simulation and epistemic uncertainty is often accounted for using a logic tree approach. Increased levels of variability and uncertainty tend to increase the probabilities of specific levels of response, damage, and loss. Put differently, they also increase the levels of response, damage, and loss for a specific probability of exceedance.
9. Expected levels of response, damage, and loss can be reduced by reducing uncertainty. In geotechnical engineering practice, this can often be accomplished by conducting more extensive subsurface investigations and performing more detailed analyses. The additional time and expense associated with these efforts is generally a very small fraction of the loss reduction they provide.
10. Performance-based design can be implemented at the response, damage, and loss levels. A complete loss-level implementation includes the explicit modeling of response, damage, and loss. A damage-level implementation explicitly models response and damage, but losses must be inferred from damage. A response-level implementation explicitly models response, but damage is inferred from response and loss is inferred from inferred damage.
11. Complex soil-structure systems may have many components whose response is sensitive to different *IM*s and is described by different *EDP*s. In such cases, a vector performance-based analysis with multiple *IM*s and multiple *EDP*s may be performed. A vector analysis requires consideration of variability and uncertainty in *IM*s and *EDP*s but also the correlation of the *IM*s and *EDP*s.
12. Parametric variability in response analyses is often accounted for by Monte Carlo simulation in which a response model is used repeatedly with different realizations of variable

input parameters. The input parameters are varied in accordance with user-specified probability distributions; joint distributions are required for parameters that are correlated. The different responses that result from each realization of input parameters are usually processed in the form of a histogram to which a probability density function can be fit.

13. The variability in response that results from Monte Carlo response analyses is a function of the uncertainty in the input parameters and the sensitivity of the response to the input parameters. For a given problem, some input parameters will influence response variability more than others, and it is useful to identify those parameters in advance to avoid simulating the variability of input parameters that do not significantly affect variability in response.
14. Performance-based design concepts can be formulated in a manner consistent with the common reliability-based design format known as load and resistance factor design. The LRFD format can be extended to a deformation-based framework through the development of demand and capacity factors that account individually for variability and uncertainty in both demands and capacities.

REFERENCES

Abrahams, L., Van Pay, L., Sattar, S., Johnson, K., McKittrick, A., Bertels, L., Butcher, L.M., Rubinyi, L., Mahoney, M., Heintz, J., Kersting, R., and McCabe, S. (2021). "NIST-FEMA post-earthquake functional recovery workshop report," *NIST Special Publication 1269*. National Institute of Standards and Technology, U.S. Department of Commerce, Gaithersburg, Maryland, 52 pp. https://doi.org/10.6028/NIST.SP.1269

Aslani, H. and Miranda, E. (2005). "Probability-based seismic response analysis," *Engineering Structures*, Vol. 27, pp. 1151–1163. https://doi.org/10.1016/j.engstruct.2005.02.015.

Baecher, G.B., and Christian, J.T. (2003). *Reliability and Statistics in Geotechnical Engineering*, Wiley, New York, 618 pp.

Baker, J.W. (2007). "Probabilistic structural response assessment using vector-valued intensity measures," *Earthquake Engineering and Structural Dynamics*, Vol. 36, pp. 1861–1883. https://doi.org/10.1002/eqe.700.

Baker, J.W. and Cornell, C.A. (2003). "Uncertainty specification and propagation for loss estimation using FOSM method," *Report No. 2003/07*, Pacific Earthquake Engineering Research Center, University of California, Berkeley, 89 pp.

Barbosa, A.R. (2011). "Simplified vector-valued probabilistic seismic hazard analysis and probabilistic seismic demand analysis: Application to the 13-story NEHRP reinforced concrete frame-wall building design example," *Ph.D. dissertation*, University of California, San Diego, 362 pp.

Berry, M., Parrish, M., and Eberhard, M. (2004). "PEER structural performance database," *User's manual* (Version 1.0), Pacific Earthquake Engineering Research Center, University of California, Berkeley, 38 pp.

Bonowitz, D. (2011). "Resilience criteria for seismic evaluation of existing buildings," *Special Projects Initiative Report*, Structural Engineers Association of Northern California (SEAONC).

Bradley, B.A. and Dhakal, R.P. (2008). "Error estimation of closed-form solution for annual rate of structural collapse," *Earthquake Engineering and Structural Dynamics*, Vol. 37, pp. 1721–1737. https://doi.org/10.1002/eqe.833.

Bradley, B.A., Dhakal, R.P., Cubrinovski, M, Mander, J.B,, and MacRae, G.A. (2007). "Improved seismic hazard model with application to probabilistic seismic demand analysis," *Earthquake Engineering and Structural Dynamics*," Vol. 36, pp. 2211–2225. https://doi.org/10.1002/eqe.727.

Bradley, B.A., Lee, D.S., Broughton, R. and Price, C. (2009). "Efficient evaluation of performance-based earthquake engineering equations," *Structural Safety*, Vol. 31, pp. 65–74. https://doi.org/10.1016/j.strusafe.2008.03.003.

Clifton, P.M., and Neuman, S.P. (1982). "Effects of kriging and inverse modeling on conditional simulation of the Avra Valley aquifer in southern Arizona," *Water Resources Research*, Vol. 18, No. 4, pp. 1215–1234. https://doi.org/10.1029/wr018i004p01215.

Cornell, C.A., Jalayer, F., Hamburger, R.O., and Foutch, D.A. (2002). "Probabilistic basis for 2000 SAC Federal Emergency Management Agency steel moment frame guidelines," *Journal of Structural Engineering*, Vol. 128, No. 4, pp. 526–533. https://doi.org/10.1061/(asce)0733-9445(2002)128#: 4(526).

Davis, M.W. (1987). "Production of conditional simulations via the LU triangular decomposition of the covariance matrix," *Mathematical Geology*, Vol. 19, No. 2, pp. 91–98. https://doi.org/10.1007/bf00898189.

Department of Energy (1994). "Natural phenomena hazards design and evaluation criteria for DOE facilities," *Report DOE-STD-1020-94*, Appendix C, U.S. Department of Energy, Washington, D.C. https://doi.org/10.2172/10158756.

Driels, M.R. and Shin, Y.S. (2004). "Determining the number of iterations for Monte Carlo simulations of weapons effectiveness," *Report NPS-MAE-04-005*, Naval Postgraduate School, Monterey, California, 16 pp. https://doi.org/10.21236/ada423541.

Earthquake Engineering Research Institute (2019). "Functional recovery: A conceptual framework with policy options," *EERI White Paper*, Earthquake Engineering Research Institute.

Gomez-Hernandez, J.J. and Journel, A.G. (1993). "Joint sequential simulation of multigaussian fields," *Geostatistics Troia'92*, Dordrecht, Kluwer Academic Publishing, Vol. 1, pp. 85–94. https://doi.org/10.1007/978-94-011-1739-5_8.

Gutjahr, A.L. (1989). "Fast Fourier Transforms for random field generation," *Technical Report No.* 4-R58-2690R, Los Alamos National Laboratory.

Jalayer, F. (2003). "Direct probabilistic seismic analysis: Implementing non-linear dynamic assessments," *Ph. D. Dissertation*, Department of Civil and Environmental Engineering, Stanford University, California.

Jayaram, N. and Baker, J.W. (2008). "Statistical tests of the joint distribution of spectral acceleration values," *Bulletin of the Seismological Society of America*, Vol. 98, No. 5, pp. 2231–2243. https://doi.org/10.1785/0120070208.

Kramer, S.L. (2008). "Performance-based earthquake engineering: Opportunities and implications for geotechnical engineering practice," *Proceedings, Geotechnical Earthquake Engineering and Soil Dynamics IV*, American Society of Civil Engineers, Sacramento, California, 32 pp. https://doi.org/10.1061/40975(318)213.

Kramer, S.L., Arduino, P., and Shin, H. (2008). "Using OpenSees for performance-based evaluation of bridges on liquefiable soils," *Report PEER 2008/07*, Pacific Earthquake Engineering Research Center, Berkeley, California, 171 pp.

Lloret-Cabot, M., Fenton, G.A., and Hicks, M.A. (2014). "On the estimation of scale of fluctuation in geostatistics," *Georisk: Assessment and Management of Risk for Engineered Systems and Geohazards*, Vol. 8, No. 2, pp. 129–140. https://doi.org/10.1080/17499518.2013.871189.

Luco, N., Ellingwood, B.R., Hamburger, R.O., et al. (2007). "Risk-targeted versus current seismic design maps for the conterminous United States," *Proceeding, Structural Engineers Association of California 76th Annual Convention*, Lake Tahoe, CA, 26–29 September.

Mackie, K.R. and Stojadinovic, B. (2005). "Comparison of incremental dynamic, cloud, and stripe methods for computing probabilistic seismic demand models," *Metropolis & Beyond: Proceedings, 2005 Structures Congress and Forensic Engineering Symposium*, American Society of Civil Engineers, New York, pp. 1–11.

Mantoglou, A., and Wilson, J.L. (1981). "Simulation of random fields with the turning bands method." *Technical Report No. 264*, Department of Civil Engineering, Massachusetts Institute of Technology, Cambridge, MA.

Matheron, G. (1973). "The intrinsic random functions and their application," *Advances in Applied Probability*, Vol. 5, pp. 439–468. https://doi.org/10.2307/1425829.

Moehle, J.P. and Deierlein, G.G. (2004). "A framework methodology for performance-based earthquake engineering," Proceedings, 13th World Conference on Earthquake Engineering, Paper No. 679, Vancouver, B.C., 13 pp.

National Earthquake Hazards Reduction Program (2020). "NEHRP recommended seismic provisions for new buildings and other structures," *FEMA P-2082-1*, 555 pp.

Poland, C.D., Hill, J., Sharpe, R.L., and Soulages, J. (1995). "Vision 2000: Performance based seismic engineering of buildings," Structural Engineers Association of California and California Office of Emergency Services, Sacramento, CA.

Shinozuka, M. and Deodatis, G. (1991). "Simulation of stochastic processes by spectral representation," *Applied Mechanics Review*, Vol. 44, No. 4, pp. 191–205. https://doi.org/10.1115/1.3119501.

Shinozuka, M. and Deodatis, G. (1996). "Simulation of multi-dimensional Gaussian Stochastic Fields by Spectral Representation," *Applied Mechanics Review*, Vol. 49, No. 1, pp. 29–53. https://doi.org/10.1115/1.3101883.

Shome, N. and Cornell, C.A. (1999). "Probabilistic seismic demand analysis of nonlinear structures," *Report No. RMS*-35, Reliability of Marine Structures Program, Department of Civil and Environmental Engineering, Stanford University, California.

Vamvatsikos, D. (2013). "Derivation of new SAC/FEMA performance evaluation solutions with second-order hazard approximation," *Earthquake Engineering and Structural Dynamics*, Vol. 42, No. 8, pp. 1171–1188. https://doi.org/10.1002/eqe.2265.

Vamvatsikos, D. (2014). "Accurate application and second-order improvement of SAC/FEMA probabilistic formats for probilistic performance assessment," *Journal of Structural Engineering*, Vol. 140, No. 2, 9 pp. https://doi.org/10.1061/(ASCE)ST.1943-541X.0000774.

Vanmarcke, E. (2010). *Random Fields: Analysis and Synthesis*, World Scientific, Singapore. https://doi.org/10.1142/5807.

Vanmarcke, E.H. (1977). "Probabilistic modeling of soil profiles," *Journal of Geotechnical Engineering*, Vol. 103, No. 11, pp. 1227–1246. https://doi.org/10.1061/ajgeb6.0000517.

Whitman, R.V., et al. (1973). "Earthquake damage probability matrices," *Proceedings, Fifth World Conference on Earthquake Engineering*, International Association for Earthquake Engineering, Rome, Italy.

6 Dynamic Soil Properties

6.1 INTRODUCTION

The nature and distribution of earthquake damage is strongly influenced by the response of soils to cyclic loading. This response depends heavily on the mechanical and, in some cases, hydraulic properties of the soil. Geotechnical earthquake engineering encompasses a wide range of problems involving many types of loading and many potential mechanisms of deformation and failure, and different soil properties influence the behavior of the soil for different problems. For many important problems, particularly those dominated by wave propagation effects, only low levels of strain are induced in the soil. For other important problems, such as those involving stability and permanent deformation, large strains are induced in the soil. The behavior of soils subjected to dynamic loading is governed by what have come to be popularly known as dynamic soil properties. While recognizing that the properties themselves are not dynamic (indeed, they apply to a host of non-dynamic problems), that term will be used in this book because of its conciseness and familiarity.

Detailed treatment of every aspect of the behavior of cyclically loaded soils is beyond the scope of a book such as this. This chapter addresses the most important aspects of their behavior in the context of the various geotechnical earthquake engineering problems addressed in the following chapters. It presents a variety of methods by which low- and high-strain soil behavior can be measured in the field and in the laboratory. The behavior of cyclically loaded soils, and different approaches to their characterization, are also described.

6.2 REPRESENTATION OF STRESS CONDITIONS BY THE MOHR CIRCLE

The cyclic response of soils to earthquake loading depends on the state of stress in the soil prior to loading and on the stresses imposed by the loading. To discuss soil properties and their relationship to the various types of cyclic loading encountered in geotechnical earthquake engineering problems, the concepts and terminology used to describe stresses must be specified. Although such concepts are ordinarily presented early in a first course on soil mechanics, their importance in the understanding and solution of geotechnical earthquake engineering problems is sufficient to warrant their presentation here.

The stress conditions at any point in a mass of soil can be described by the normal and shear stresses acting on a particular plane passing through that point. Because most normal stresses in soils are compressive (soils cannot, in general, resist tensile stresses), it is customary in geotechnical engineering to describe compressive stresses as positive. Consequently, positive shear stresses are those that tend to cause counterclockwise rotation of the body they act upon, and clockwise angles are taken as positive. Figure 6.1 illustrates the sign conventions for normal and shear stresses; σ_x and σ_y are the normal stresses acting on planes normal to the x- and y-axes, respectively, τ_{xy} (and τ_{yx},) is the shear stress in the y-direction (x-direction) on the plane normal to the x-axis (y-axis), and σ_a and τ_a are the normal and shear stresses on the plane inclined at angle α. In structural mechanics, the opposite conventions are generally used.

The notation used to describe the foregoing stresses is different than that used to develop the equations of motion for three-dimensional wave propagation in Appendix C. For that problem of solid mechanics, the standard notation of solid mechanics was used. For this chapter and the remainder of the book, the above notation, which is most commonly used in geotechnical engineering, is applied. The equivalence of the two notations was discussed in Section C.2.2.

DOI: 10.1201/9781003512011-6

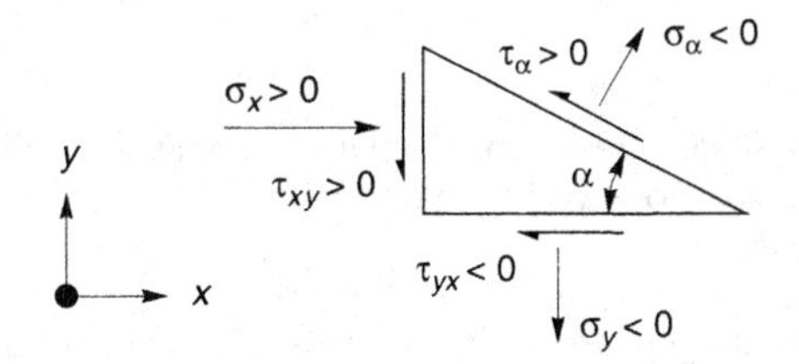

FIGURE 6.1 Sign convention for normal and shear stresses.

It is often necessary to consider the stresses on several different planes that pass through a particular point. Equilibrium requirements can be used (e.g., Holtz et al., 2011) to express the normal and. shear stresses on a plane inclined at an angle, α, to the x-axis as

$$\sigma_\alpha = \frac{\sigma_x + \sigma_y}{2} + \frac{\sigma_y - \sigma_x}{2}\cos 2\alpha - \tau_{xy}\sin 2\alpha \tag{6.1a}$$

$$\tau_\alpha = \frac{\sigma_x - \sigma_y}{2}\sin 2\alpha - \tau_{xy}\cos 2\alpha \tag{6.1b}$$

Equations (6.1) describe a circle whose center is at $\left[\sigma = \left(\sigma_x + \sigma_y\right)/2, \tau = 0\right]$ and whose radius is $\sqrt{\left[\left(\sigma_x + \sigma_y\right)/2\right]^2 + \tau_{xy}^2}$. This circle, shown in Figure 6.2, is the well-known *Mohr circle of stress.* The Mohr circle simply provides a graphical illustration of the stress conditions on all possible planes passing through an element and, as such, is very useful for understanding states of stress induced by gravity and external loading. It will be used often in the remainder of this book.

Equations (6.1) allow the stresses on planes of different inclinations to be determined analytically, but they can also be determined graphically using the pole of the Mohr circle. The pole has a useful property: *any line drawn through the pole will intersect the Mohr circle at a point that describes the shear and normal stresses on a plane parallel to that line.* Consider the element shown in Figure 6.2 subjected to a vertical normal stress, σ_y, and a horizontal normal stress, σ_x. The shear stresses on the boundaries are zero. The stress conditions on the horizontal plane are known: $\sigma = \sigma_y$ and $\tau = 0$. Since the property of the pole states that a horizontal line drawn through it must intersect the Mohr circle at a point describing those stress conditions, a horizontal line drawn through the point describing those stress conditions will intersect the Mohr circle at the pole. For the case of Figure 6.2, the point of known stress conditions is point A and the plane for which the stress conditions are known is horizontal. Consequently, a horizontal line drawn through point A must intersect the Mohr circle at the pole, labeled as point P. Once the location of the pole is known, it can be used to determine the shear and normal stresses on any plane.

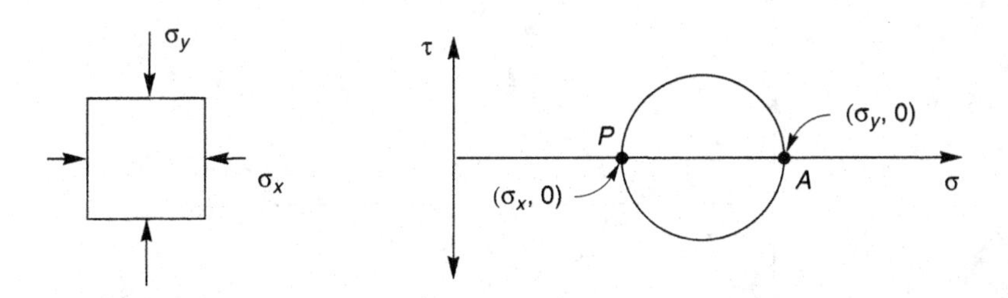

FIGURE 6.2 Mohr circle of stress for element subjected to major principal stress, σ_y, and minor principal stress, σ_x. Location of pole denoted by *P*.

Example 6.1

Compute the normal and shear stresses on a plane passing through the element shown in Figure E6.1a and inclined at 45° clockwise from horizontal.

Solution:

The stresses on the horizontal plane are $\sigma = 4$ and $\tau = +1$. Drawing a horizontal line through this point reveals the location of the pole at point P. Note that the known stresses on the vertical plane, $\sigma = 2$ and $\tau = -1$, could just as easily have been used with a vertical line to determine the location of the pole. Once the pole has been identified, the stress conditions on any plane can be determined. Drawing a line through the pole parallel to the plane of interest (Figure E6.1b) shows that the stresses on that plane are $\sigma = 4$ and $\tau = -1$.

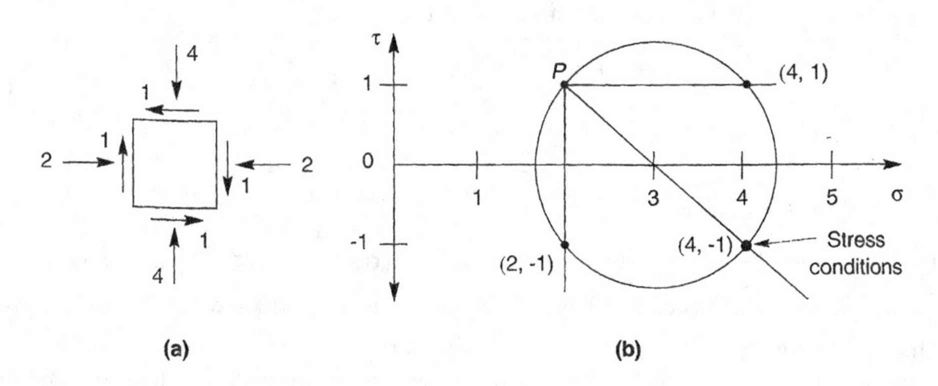

FIGURE E6.1

6.2.1 Principal Stresses

Two points on the Mohr circle are of particular interest. The points where the circle intersects the normal stress axis describe the normal stresses on planes where no shear stresses exist. Those planes are called *principal stress planes* and the normal stresses that act on them are called *principal stresses.* The *principal stress axes,* which indicate the directions in which the principal stresses act, are therefore perpendicular to the principal stress planes. The largest principal stress is the *major principal stress,* σ_1, and the smallest is the *minor principal stress,* σ_3. There is also an *intermediate principal stress,* σ_2, that can take on any value between σ_1 and σ_3; a complete Mohr diagram would include σ_2, as shown in Figure 6.3. Since the mechanical behavior of soils is much more sensitive to the relationship between σ_1 and σ_3 than to the value of σ_2, and since σ_2 and σ_3 are often nearly equal, the value of σ_2 is usually not shown.

The pole can be used to determine the orientation of the principal stress planes. The fact that the angle between two lines passing through any point on a semicircle and the "corners" of the semicircle is 90° confirms that the major and minor principal stresses act perpendicular to each other

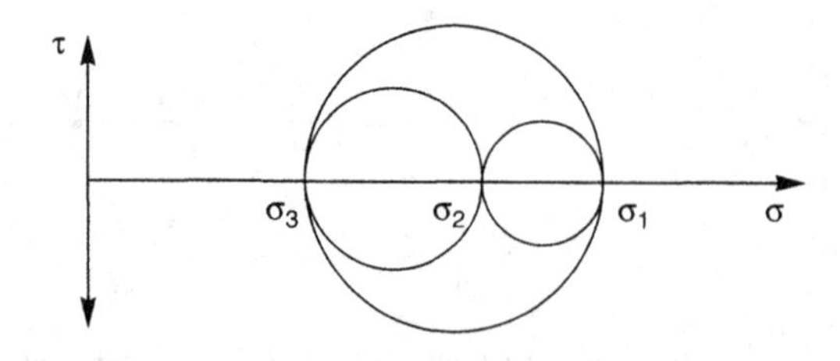

FIGURE 6.3 Mohr circles of stress including intermediate principal stress, σ_2.

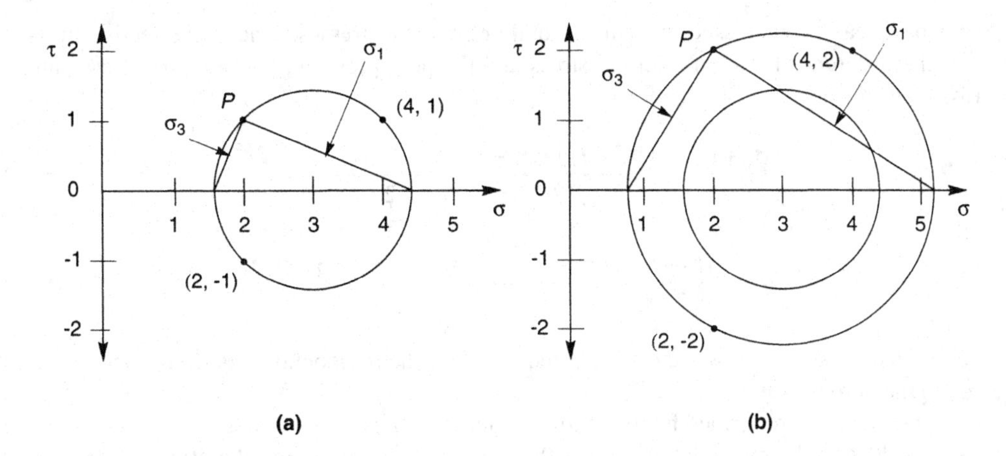

FIGURE 6.4 Orientation of principal stress axes for (a) $\tau_{xy} = 1$ and (b) $\tau_{xy} = 2$. Note the rotation of principal stress axes that accompanies the change in τ_{xy}.

(the intermediate principal stress acts mutually perpendicular to the major and minor principal stresses). Figure 6.4a shows the orientation of the principal stress axes on the element shown previously in Figure E6.1b. If the shear stress, τ_{xy}, is increased from 1 to 2 with σ_x and σ_y held constant, the Mohr circle grows to the size shown in Figure 6.4b. Note that the increase in shear stress is accompanied by rotation of the principal stress axes.

6.2.2 Stress Paths

As external loads are applied, the variation in stress conditions acting on a soil element can be tracked by plotting the Mohr circle at various stages in the loading sequence, but such a plot can quickly become cluttered for many loading sequences. It is much simpler to observe the stress conditions by plotting the variation of the position of a single point on the Mohr circle. The stress point usually selected is the top of the Mohr circle, as shown in Figure 6.5. The path taken by the stress point during loading is called the *stress path*. Since many properties of soil are dependent on the stress path induced by the applied loading, the stress path is a very useful tool in geotechnical engineering. [In a considerable body of geotechnical engineering literature, particularly that relating to the constitutive modeling of soils, the stress point is defined according to $p = (\sigma_1 + \sigma_2 + \sigma_3)/3$ and $q = \sigma_1 - \sigma_3$. Although each form has its own merits, they essentially present the same information; the form of Figure 6.5 is more commonly used in geotechnical earthquake engineering and will be used hereafter. For the case of vertically propagating shear waves, which is commonly considered in site response (Chapter 7) and liquefaction (Chapter 9) problems, plots of horizontal shear stress vs. vertical (total and effective) normal stress, which are conceptually similar to stress paths, can be quite useful.

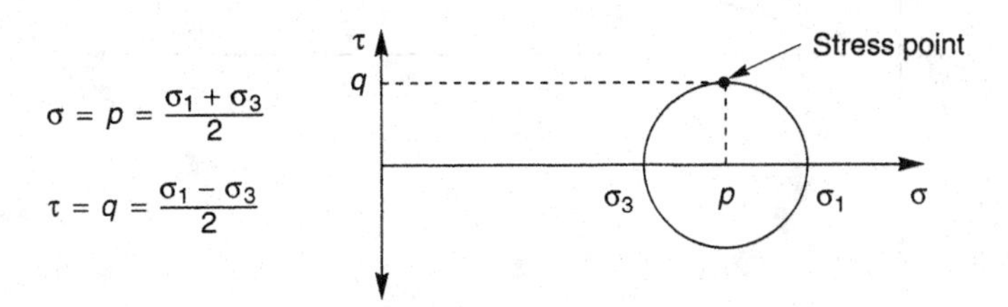

FIGURE 6.5 Location and definition of stress point on which stress path is based.

Stress paths can be expressed in terms of total or effective stresses. Since the effective stress is equal to the difference between the total stress and the pore pressure, the effective stress path is described by

$$p' = \frac{\sigma_1' + \sigma_3'}{2} = \frac{(\sigma_1 - u) + (\sigma_3 - u)}{2} = \frac{\sigma_1 + \sigma_3}{2} - \frac{2u}{2} = p - u \tag{6.2a}$$

$$q' = \frac{\sigma_1' - \sigma_3'}{2} = \frac{(\sigma_1 - u) - (\sigma_3 - u)}{2} = \frac{\sigma_1 - \sigma_3}{2} = q \tag{6.2b}$$

Total and effective stress paths are often plotted together; the horizontal distance between the two is equal to the pore pressure.

Many soil deposits are formed by the sedimentation of soil particles through water. As more and more soil is deposited, consolidation causes the volume to decrease and the effective stresses to increase. If the process is one-dimensional (i.e., if the soil particles move only in the vertical direction), the minor principal stress will be proportional to the major principal stress and the effective stress path of an element of soil below the ground surface will move from A to B in Figure 6.6. The slope of the stress path in this range is given by

$$m_0 = \frac{dq}{dp'} = \frac{\frac{\sigma_1' - \sigma_3'}{2}}{\frac{\sigma_1' + \sigma_3'}{2}} = \frac{\sigma_1'(1 - K_0)}{\sigma_1'(1 + K_0)} = \frac{1 - K_0}{1 + K_0} \tag{6.3}$$

where $K_0 = \sigma_3' / \sigma_1'$ is the *coefficient of lateral earth pressure at rest.* Equation (6.3) indicates that the effective stress path for this condition is a line that passes through the origin. If the ground surface is level, the principal stress axes will be vertical and horizontal.

After the soil has consolidated, slow (drained) external loading can cause the stress path to move in a variety of directions. If the vertical stress increases while the horizontal stress remains constant, the stress path will move in the direction labeled AC in Figure 6.6. If the horizontal stress decreases with constant vertical stress (as in the development of active earth pressure conditions), the stress path moves in the direction LE. If the vertical stress is decreased with constant horizontal stress (as beneath an excavation), the stress path moves in the AE direction, and if the horizontal stress increases with constant vertical stress (passive earth pressure conditions), the stress path moves in the direction labeled LC. For each of these idealized conditions, no shear stresses are

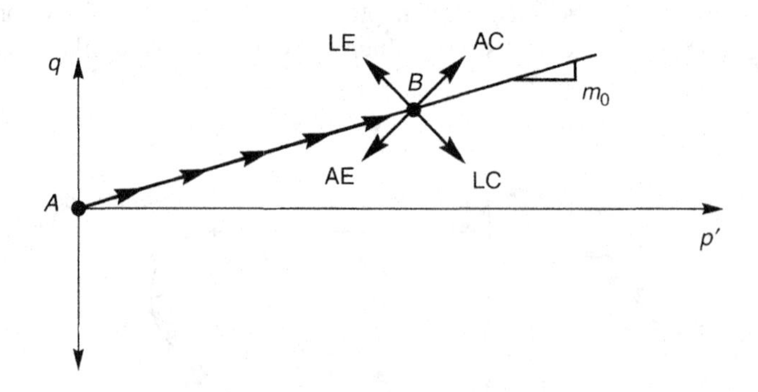

FIGURE 6.6 Effective stress paths for K_0 consolidation (*A* to *B*) and subsequent drained loading along axial compression (AC), axial extension (AE), lateral compression (LC), and lateral extension (LE) stress paths.

induced on vertical or horizontal planes; consequently, the principal stress axes remain vertical and horizontal (although they will instantaneously exchange orientations if the p'-axis is crossed).

Most realistic loading conditions involve simultaneous changes in horizontal and vertical stresses and/or the development of shear stresses on horizontal and vertical planes. Consider an element of soil beneath a level ground surface (Figure 6.7a) subjected to vertically propagating s-waves. At stage A, the element is under at-rest conditions with the Mohr circle as indicated in Figure 6.7b and the stress path at point A in Figure 6.7c. Since the major principal stress is vertical, the pole of the Mohr circle is at the point (σ'_h, 0) in Figure 6.7b. A vertically propagating shear wave will produce shear stresses on horizontal and vertical planes and distort the element as shown in stage B of Figure 6.7a. Since the shear stresses increase while the vertical and horizontal stresses remain constant, the radius of the Mohr circle increases but the center does not move (Figure 6.7b). The stress path (Figure 6.7c) moves vertically, as does the position of the pole (Figure 6.7b), which indicates that the principal stress axes are rotated from their initial vertical and horizontal positions. Since the horizontal shear stresses are cyclic in nature, their direction will reverse after going back through the $\tau_{hv} = \tau_{vh} = 0$ position in stage C. Note that the stress conditions at stage C are identical to those of stage A, and that the principal stress axes have rotated back to the vertical and horizontal positions. At stage D, the shear stresses act in the opposite direction and the principal stress axes rotate in the opposite direction as in stage B. Thus the loading induced by vertically propagating shear waves can be described by the stress path of Figure 6.7c and a principal stress axis rotation. Note that the stress path never indicates isotropic stress conditions (it never reaches the p'-axis) and that the principal stress axes rotate continuously.

The nature of principal stress axis rotation is significant. Research (e.g., Wong and Arthur, 1986; Symes et al., 1988; Sayao and Vaid, 1989) has shown that principal stress rotation can cause shear and volumetric strain by itself (i.e., even if the stress point does not move). Hence some of the strain induced by vertically propagating shear waves results from principal stress rotation; this effect is not present in many field and laboratory tests.

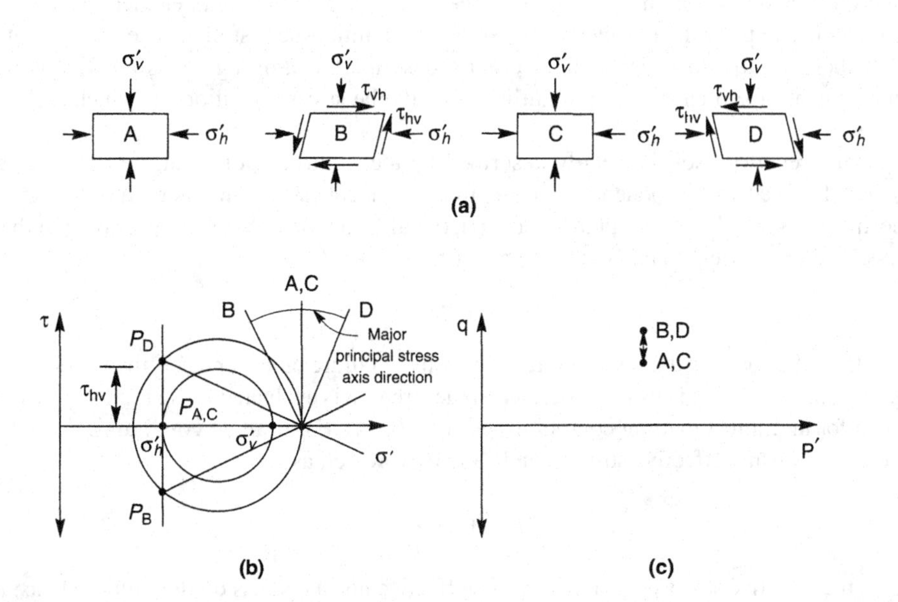

FIGURE 6.7 (a) Stress and strain conditions imposed on element of soil below level ground surface by vertically propagating shear waves at four different times; (b) Mohr circles, locations of poles, and orientations of major principal stress axis; (c) stress path.

Example 6.2

A reconstitute triaxial (Section 6.5.4.3) specimen of dry sand is consolidated isotropically to an effective confining pressure of 200 kPa, and then loaded in drained triaxial compression to a deviator stress ($\sigma_1 - \sigma_3$) of 200 kPa. At that point, the specimen is subjected to a harmonic deviator stress that oscillates between 100 and 300 kPa. Plot the total and estimated effective stress paths.

Solution:

Because the sand is dry, no pore pressures exist, so the total and effective stresses are equal. During preparation, the stresses acting on the specimen are very small, so the stress path is at point A in Figure E6.2. Isotropic consolidation takes the stress path to point B, and drained triaxial compression takes it to point C. Harmonic loading then causes the stress path to oscillate between points D and E.

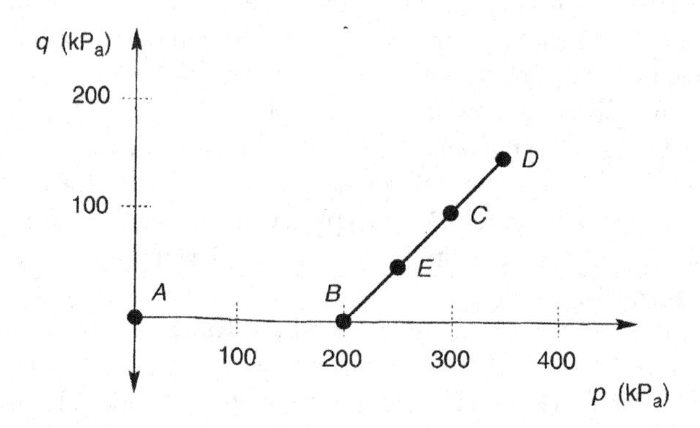

FIGURE E6.2

6.2.3 Mohr-Coulomb Failure Criterion

The Mohr circle and stress path illustrate the stress conditions acting on an element of soil. Not all stress conditions are possible, however, since soils have a finite shear strength, i.e., a limiting shear stress that they are able to resist. Failure criteria describe the limiting stress conditions beyond which an element of soil cannot be in equilibrium, i.e., the stress conditions at which the element of soil fails.

The shear strength of soil is usually described by the Mohr-Coulomb failure criterion, which states that soils have two components of shear strength – a cohesive component that is independent of the normal stress on the failure plane, and a frictional component that is proportional to that normal stress. The shear strength can therefore be written as

$$s = c + \sigma \tan\phi \tag{6.4}$$

where c is the cohesion, σ is the normal total stress on the failure plane, and ϕ is the friction angle of the soil. The above notation is usually used when the strength is evaluated using total stresses, which is common for undrained loading conditions (Section 6.2.4). For drained conditions, strengths are usually evaluated using effective stresses and can be expressed as

$$s = c' + \sigma' \tan\phi' \tag{6.5}$$

where c' is the effective stress cohesion, σ' is the effective normal stress on the failure plane, and ϕ' is the effective stress friction angle of the soil. The Mohr-Coulomb failure criterion can be expressed graphically in terms of the Mohr-Coulomb failure envelope, which plots as a straight line in τ-σ space (Figure 6.8a). An element of soil reaches the failure condition when the shear stress on some

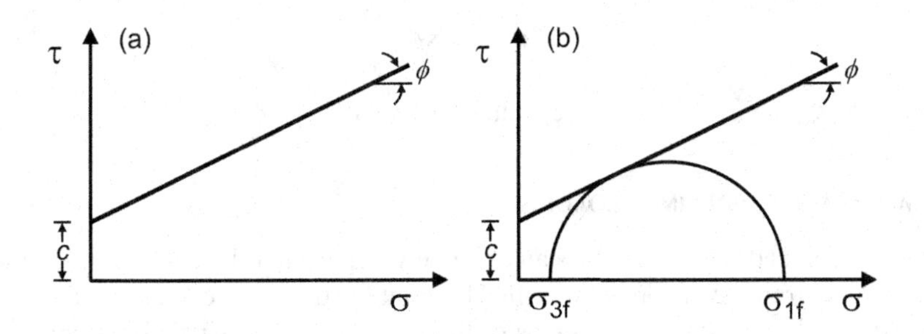

FIGURE 6.8 (a) Mohr-Coulomb failure envelope and (b) relationship between Mohr circle and Mohr-Coulomb failure envelope at failure.

plane passing through the element is equal to the shear strength of that element – geometrically, that occurs when the Mohr circle is tangent to the Mohr-Coulomb failure envelope. Under those conditions, as illustrated in Figure 6.8b, the principal stresses can be related to the Mohr-Coulomb strength parameters, c and ϕ, as

$$\frac{\sigma_{1f} - \sigma_{3f}}{2} = \frac{\sigma_{1f} + \sigma_{3f}}{2} \sin\phi + c\cos\phi \tag{6.6}$$

where the subscript "f" indicates the failure condition. Equation (6.6) can be rearranged to determine either of the principal stresses

$$\sigma_{1f} = \sigma_{3f} \tan^2(45 + \phi/2) + 2c\tan(45 + \phi/2) \tag{6.7a}$$

$$\sigma_{3f} = \sigma_{1f} \tan^2(45 - \phi/2) - 2c\tan(45 - \phi/2) \tag{6.7b}$$

Fine-grained soils are usually loaded under undrained conditions. When fully saturated, the total stress friction angle, ϕ, is zero and the strength, termed the *undrained strength*, s_u, is independent of total normal stress.

$$s = c = s_u \tag{6.8}$$

The Mohr-Coulomb failure envelope can also be plotted in stress path space, in which case it passes through the stress point (Figure 6.5) of the Mohr circle at failure. The intercept and angle of the Mohr-Coulomb failure envelope in stress path space (Figure 6.9) are given by

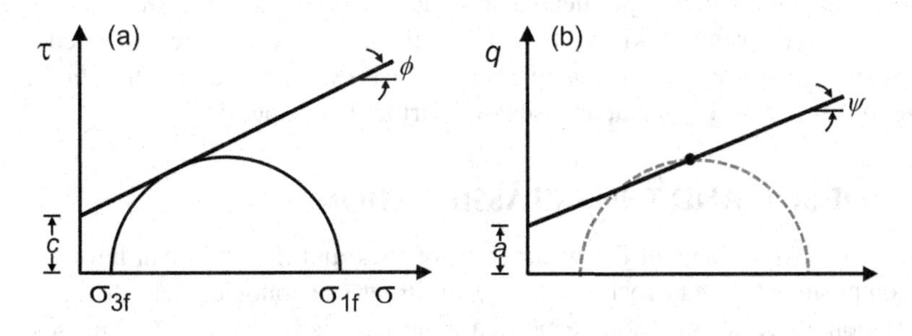

FIGURE 6.9 Illustration of cohesion intercept and friction angle expressed in (a) stress and (b) stress path space.

$$a = c\cos\phi \tag{6.9a}$$

$$\psi = \tan^{-1}(\sin\phi) \tag{6.9b}$$

6.2.4 Drained and Undrained Loading

The interaction of soil and porewater plays an extremely important role in determining the strength, stiffness, and stress-strain behavior of the soil. These interactions can occur over a broad and continuous range of conditions, but the extremes of that range are often singled out because they frequently represent critical conditions.

In a dry soil, the high compressibility of the air in the voids of the soil provides essentially no resistance to potential volume changes from either slowly or rapidly applied external loads. Even in a partially saturated soil, the high compressibility of the air bubbles that occupy part of the void space will allow volume change to occur without resistance. In a saturated soil, loading that is applied very slowly will also allow water to drain out of voids that are being compressed (or to be drawn into voids that are expanding) while it is being loaded; in this case, the porewater pressure remains constant. Each of these cases represents *drained loading*, i.e., loading that is applied so slowly that the pore (air and/or water) pressure does not change.

Loading applied so quickly that water cannot move in or out of the voids of a saturated soil during the period of load application is referred to as *undrained loading*. Because the porewater has very low compressibility (about 22,000 times lower than air), undrained loading will typically cause changes in pore pressure as the soil skeleton attempts to contract or dilate. Undrained loading occurs when the loading is applied rapidly relative to the time required for pore pressures to dissipate in a saturated soil. The time required for pore pressures to dissipate in a particular soil layer depends on (1) the permeability of the soil, (2) the thickness of the soil layer, and (3) the permeabilities of the soils that surround the layer of interest. A soil with a high permeability, like a layer of gravel, may respond under undrained conditions if it is surrounded by soils of low permeability.

The shear strength of a soil is influenced by the drainage conditions during loading. Under drained loading conditions, pore pressures do not change so it is relatively easy to determine the effective stresses in the soil. The shear strength can then be defined in terms of the effective stress Mohr-Coulomb strength parameters, c' and ϕ', as in Equation (6.5). When expressed in terms of total stresses, the Mohr-Coulomb failure criterion can be expressed as in Equation (6.4). Saturated fine-grained soils, because of their low permeability, are frequently loaded under undrained conditions. When fully saturated, changes in total normal stress are resisted by the relatively incompressible porewater, so effective stresses remain constant. Because shear strength depends on effective stresses, it is independent of changes in total stress. Therefore, for total stress analyses, $\phi = 0$ and $c = s_u$, which is known as the *undrained strength* of the soil.

Seismic loading of saturated soil is almost always reasonably assumed to occur under undrained conditions. Although partial drainage may occur during earthquake shaking in thin permeable soils or near the boundaries of thicker permeable soils, it is usually conservative to assume that undrained conditions exist during shaking. There are situations, however, where drained strengths can be lower than undrained strengths; such situations can arise when evaluating the stability of a slope underlain by soils that have liquefied, as discussed further in Chapter 9.

6.3 TYPES OF SOIL AND THEIR CLASSIFICATION

Soils are generally by-products of the weathering of rock and therefore can have many different forms and compositions. Parent rocks can have different mineralogies, and different weathering and transportation processes give soil particles a variety of sizes, shapes, roughnesses, and chemical/mechanical properties. Various depositional processes can produce soil deposits with different densities and fabrics. All of these characteristics cause different soils to exhibit different stress-strain behavior under both static and cyclic loading.

For engineering purposes, soils are usually classified, at least initially, according to grain size. From largest to smallest, terms such as boulders, cobbles, gravels, sands, silts, and clays are used to describe soils of different particle size. Boulders, cobbles, gravels, and sands are usually referred to as *coarse-grained soils*, and silts and clays as *fine-grained soils*. Boulders and cobbles are relatively rare so sands and gravels will be the coarse-grained soils of primary interest in this text. Soil grain sizes are measured by sieve analyses in which dry soil falls through a series of successively finer vibrating sieves with the amounts of soil (by weight) measured after shaking has concluded. The cumulative fractions of soil retained on sieves of a given opening size are plotted against the sieve opening size in a grain size distribution chart (Figure 6.10). The distinction between coarse-grained and fine-grained soils is usually based on a particular sieve size (a No. 200 sieve with an opening of 0.074 mm) rather than some fundamental property to which behavior is closely associated. The mean grain size, D_{50}, is the grain size for which 50% (by weight) is coarser and 50% is finer. The coefficient of uniformity, $C_u = D_{60}/D_{10}$, and coefficient of curvature, $C_c = D_{30}^2 / (D_{60} \times D_{10})$, are used to distinguish between well-graded soils ($C_u > 6$ and $1 < C_c < 3$) and poorly-graded, or uniform, soils ($C_u < 6$ or $C_c < 1$ or $C_c > 3$).

Many soils of particular interest to geotechnical earthquake engineers, e.g., *alluvial soils*, have been transported and/or deposited by water. Because the size of a particle that can be moved by water depends on the velocity of the water (according to Stokes' Law, the water velocity required to move a spherical particle varies with the square of its diameter), particles above a certain diameter will tend to drop out of suspension when the water velocity drops below a particular value. For this reason, large particles are generally found in areas where water moves, or has moved, quickly (e.g., in steep terrain or near the edges of valleys or basins) and smaller particles are found where water has moved more slowly (e.g., in lakes, marshes, or slow-moving rivers in flat terrain). Because rivers meander over time and flow quickly and slowly depending on rainfall conditions, some soil profiles can contain materials of different grain size at different depths and locations; such profiles can exhibit significant spatial variability both vertically and horizontally.

Particle shape can also be important. Coarse-grained soils, and even coarser silt particles, tend to exhibit bulky grain shapes, i.e. shapes in which the maximum and minimum dimensions do

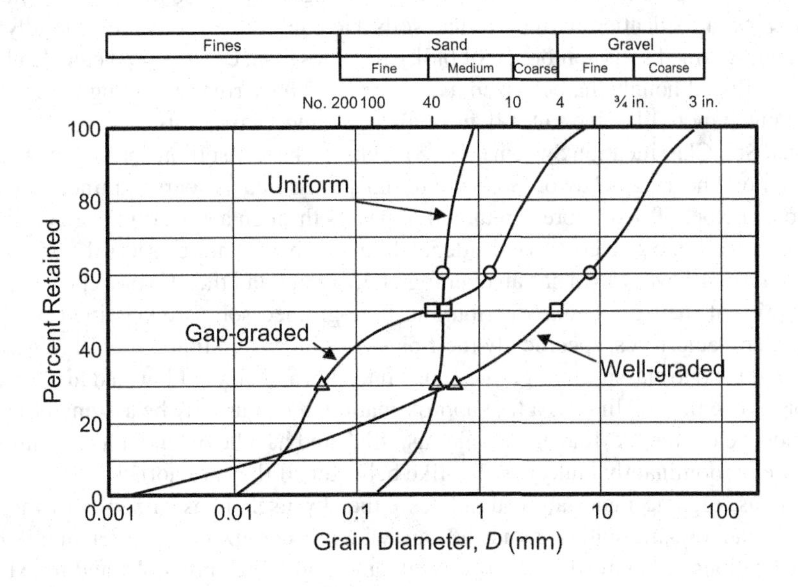

FIGURE 6.10 Grain size distribution for soils. Circles, squares, diamonds, and triangles represent D_{60}, D_{50}, D_{30}, and D_{10} values.

not differ greatly. Depending on age and geologic environment these particles may be rough and angular or they may be rounded and smooth; the angularity of natural soils tends to decrease with increasing particle size. The *specific area* (ratio of surface area to volume) of bulky-grained particles tends to be low. Fine silts and clays, on the other hand, tend to have flaky or platy shapes in which one dimension is considerably thinner than other dimensions. These particles have a much higher specific area – for example, the particles comprising 1 kg of montmorillonite (a common, and often troublesome, clay mineral) would have a surface area of about 800,000 m^2, while those comprising 1 kg of sand would have a corresponding area of about 2.3 m^2. Put differently, bulky-grained particles are "heavy" relative to their surface areas while fine-grained particles are quite "light." As a result, clays are much more sensitive to processes that affect their surfaces than are sands.

Clay minerals are comprised in part of hydrous silicate sheets with metallic cations (aluminum, magnesium, etc.). Alternate stacking configurations of these sheets with various cations give rise to different minerals and can result in positive charge deficiencies within a given particle. These deficiencies, and the manner in which the cations are spatially distributed, result in negatively charged clay particle surfaces. The negative surface charges tend to attract the positive ends of dipolar water molecules, resulting in a layer of *adsorbed water* (up to several molecules thick depending on the magnitude of the negative charge) that is closely bound to the clay particle. The forces associated with these electro-chemical interactions are much greater than the weight of a clay particle, so they have a very strong influence on the behavior of the clay. Forces caused by such interaction at the surface of a sand particle (which are much weaker due to differences in mineralogy) are miniscule in comparison with the weight of the sand particle (due to the low specific area) and therefore have virtually no effect on the behavior of the sand.

The interaction between fine-grained soil particles and porewater gives rise to a soil characteristic referred to as *plasticity*. A plastic soil has the ability to resist large deformations as a solid without cracking or crumbling. At very high water contents, i.e., above the *liquid limit*, LL, soils behave more like fluids than solids (and therefore are not plastic); at low water contents, i.e., below the *plastic limit*, PL, soils will crack and crumble rather than deform plastically. A soil will therefore have some range of water contents, numerically equal to the *plasticity index*, $PI = LL - PL$, over which it exhibits plastic behavior. This range is influenced by how strongly the clay particles attract water molecules – the stronger the attraction, the higher the plasticity index. Fine-grained soils with similar plasticity indices tend to exhibit similar engineering behavior, so the plasticity index is used in the engineering classification of fine-grained soils. Coarse-grained soils are generally nonplastic ($PI \approx 0$), and fine-grained soils comprised of bulky particles (e.g., coarser silts) can also be nonplastic. Nonplastic silts, although characterized as fine-grained by virtue of passing the No. 200 sieve, often behave much more like fine sands than smaller-grained, plastic silts.

The Unified Soil Classification System (USCS) seeks to classify soils according to characteristics that affect their engineering behavior. Soils are initially classified as coarse-grained or fine-grained (depending on whether 50% or more is retained or passes through a No. 200 sieve). Coarse-grained soils are then divided into gravels or sands (depending on whether more or less than 50% by weight is retained on a No. 4 sieve, which has an opening of 4.75 mm) and then further classified according to the uniformity of their grain size distribution. Fine-grained soils are classified on the basis of their plasticity characteristics, specifically their plasticity index and liquid limit. Figure 6.11 shows a plasticity chart used to classify fine-grained soils into silts and clays of low and high plasticity. If a material plots above the "A" line, as a first approximation, it can usually be assumed to behave like a clay. If it plots below the "A" line, it classifies as "silt," and has the potential for granular behavior (if particles are predominantly bulky) or clay-like behavior (if they are not).

In some areas, organic material accumulates either by itself or is mixed with inorganic soil particles to form an organic soil. Organic soils are widely recognized as problematic in many geotechnical applications – they tend to be soft, weak, and highly compressible, and to exhibit strong rate effects such as *creep* and *stress relaxation*. For those reasons, structures such as buildings and bridges that are underlain by organic soils are generally supported on deep foundations. Organic

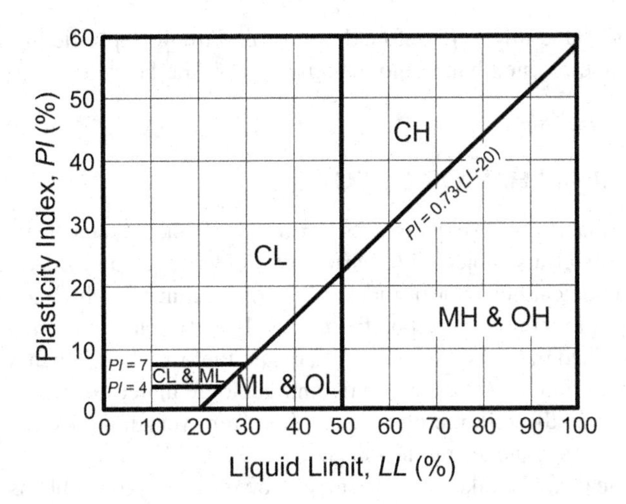

FIGURE 6.11 Plasticity chart for classification of fine-grained soils.

soils, although not common, have distinct behavioral characteristics and are therefore distinguished from inorganic soils in the USCS.

Of course, not all soils fit neatly into the specific, individual soil classes listed in the USCS. In some cases, where soil characteristics are close to the USCS boundaries, dual classifications (e.g., CL-ML, SP-ML, etc.) may be required.

The simplicity of the USCS makes it convenient and useful for many soils but it does not always identify the components of a soil that are likely to control its behavior. Common examples are mixtures of sand and gravel or sand and fines in which one size or the other may dominate the response of the mixture. As illustrated in Figure 6.12, larger and smaller particles can be distributed in different manners within a mixed soil. In some cases, the smaller particles are contained within the voids of the larger particles and have little effect on the mechanical behavior of the mixture (although they can have a dominant effect on its permeability). In other cases, the smaller particles are so plentiful that the larger particles are not in contact with each other and essentially "float" in a matrix of the smaller particles, which then control the mixture's behavior. The transition between coarser and finer particle dominance occurs in the vicinity of a limiting finer particle content, FC_L, which varies with grain size distribution and other effects but is typically in the range of 15%–40%

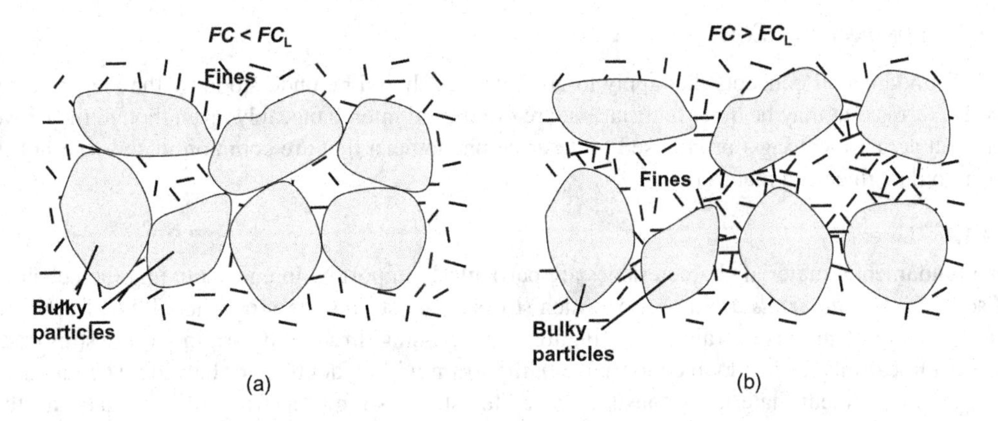

FIGURE 6.12 Schematic showing (a) soil mixture with $FC < FC_L$ in which bulky coarse-grained particles are in contact and interact directly and (b) soil mixture with $FC > FC_L$ in which fines fraction dominates mechanical behavior.

(Thevanayagam, 1998). One might, therefore, have a soil with 40% plastic fines, which the USCS would classify as coarse-grained but would actually exhibit the behavior of a plastic fine-grained soil.

6.4 STRESS-STRAIN BEHAVIOR OF SOILS

Geotechnical earthquake engineering involves a variety of phenomena that are affected by the behavior of geologic materials subjected to earthquake-induced ground shaking. That behavior is controlled by the mechanical behavior of the soil, i.e., the manner in which it deforms in response to applied stresses. The loading imposed on the materials is described in terms of stresses and the deformations are described in terms of strains. As a result, the stress-strain behavior is of great interest. Earthquake loading may be strong or weak, and geologic materials may be stiff or soft – the levels of strain induced in the soil, therefore, can range from very low to very high, and all are of interest to geotechnical earthquake engineers.

Much of geotechnical earthquake engineering can be divided into problems of site response or ground failure. Site response, i.e., the tendency for amplification or de-amplification of earthquake ground motions, is strongly affected by the stress-strain behavior at relatively low strain levels. Ground failure, which results in the permanent deformation of geological materials, is affected by stress-strain behavior at moderate to high strain levels. Because of this, it is common to divide the consideration of dynamic soil properties into low-strain and high-strain conditions.

A solid understanding of the stress-strain behavior of soils is fundamental to the understanding of soil behavior during earthquakes. While most readers of this text will have some background in soil mechanics, many will have been limited to discussions of shear strength and consolidation behavior. In conventional geotechnical engineering practice, loads are usually applied monotonically, i.e., they continuously increase (or decrease) over time until they reach some final value. In geotechnical earthquake engineering, soils are subjected to static loads, but then also to cyclic loading as the ground shakes. To predict their seismic performance, knowledge of the behavior of soils upon reversal of loading is also important.

This section presents a brief and basic description of soil stress-strain behavior and the factors that most strongly affect it. It begins at a basic level, which may be helpful for readers with limited geotechnical engineering backgrounds, and proceeds through response to monotonic and then cyclic loading. It is not an exhaustive treatment of soil mechanics but rather is intended to provide a framework in which the behavior of different types of soil, whether for site response or ground failure problems, can be understood.

6.4.1 Fundamental Concepts

A few fundamental concepts that apply to all materials should be understood by the reader. They involve terms that may be quite familiar but are often used interchangeably (even though they have different actual meanings) or are used to describe phenomena that are common in soils but not in other civil engineering materials.

6.4.1.1 Linearity

Two fundamental material characteristics are particularly important in understanding the behavior of soils. *Linear materials* are those for which stresses and strains are proportional, i.e., for which the stress-strain curve is a straight line (Figure 6.13a) passing through the origin – if the stress acting on a linear material is doubled (or halved), the strain will be doubled (or halved). The modulus (stiffness) of a linear material is constant. When the stress is removed from a linear material, the material will return to its original geometry in which the strains are zero.

Nonlinear materials exhibit variable stiffness during loading. The change in stiffness may occur suddenly, e.g., upon yielding, as illustrated in Figure 6.13b, or it may occur gradually and

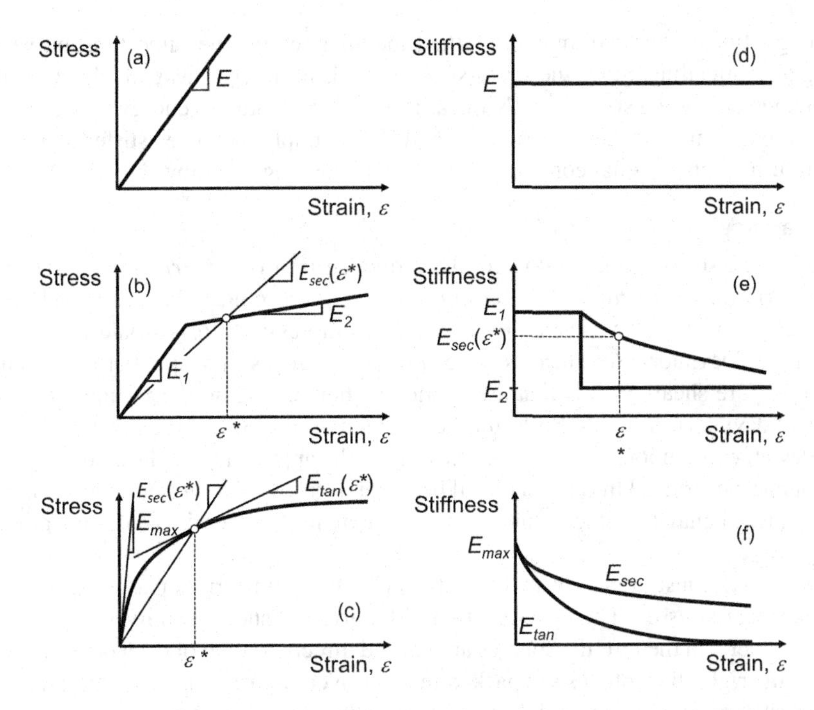

FIGURE 6.13 Schematic illustration of different types of material behavior in compression: stress-strain curves for (a) linear, (b) bilinear, and (c) continuously nonlinear behavior, and corresponding variation of tangent and secant modulus (stiffness) with strain for (d) linear, (e) bilinear, and (f) continuously nonlinear behavior

continuously as illustrated in Figure 6.13c. The stiffness of a nonlinear material varies with strain level – at some normal strain, ε^*, the stiffness can be described in terms of a secant Young's modulus, E_{sec}, or a tangent modulus, E_{tan}. The secant stiffness is simply the ratio of stress to strain at the strain of interest, while the tangent stiffness is equal to the slope of the stress-strain curve at that strain, i.e.,

$$E_{sec} = \sigma / \varepsilon \tag{6.10a}$$

$$E_{tan} = d\sigma / d\varepsilon \tag{6.10b}$$

As suggested by Figure 6.13c, the secant modulus represents an "average" stiffness of the material over the range of strain from 0 to s^*, while the tangent modulus describes the "instantaneous" stiffness at the strain level, s^*. As the strain level becomes smaller, the difference between E_{sec} and E_{tan} becomes smaller, and both converge to the same value, say E_{max}, as the strain level goes to zero. For stress-strain curves with the shapes shown in Figure 6.13b and c, both E_{sec} and E_{tan} will decrease with increasing strain level and E_{tan} will decrease more quickly than E_{sec}. For strain-softening materials, i.e., materials in which the shear stress can drop after reaching some peak value that is usually mobilized at small strains, the tangent modulus can become negative. The secant modulus, on the other hand, is always positive. Both secant and tangent moduli play important roles in geotechnical earthquake engineering analyses described later in this text.

6.4.1.2 Elasticity

Elastic materials are those that unload along the same stress-strain curve they followed upon loading. Elastic materials are usually also linear, in which case stresses and strains increase and decrease along a linear stress-strain curve – their modulus retains the same constant value during both loading and unloading. Since the stress-strain curve is perfectly linear, all elastic strain energy is recovered upon unloading so a perfectly linear material dissipates no energy – it has no damping. Not all elastic materials are linear; a rubber band, for example, becomes stiffer as it is stretched further but returns to its original configuration when the loading is removed.

6.4.1.3 Dilatancy

In solid mechanics, stresses and strains are often divided into *volumetric* and *deviatoric* components. Volumetric (or *hydrostatic* or *mean normal*) stresses produce changes in volume, but not shape, in an element of linear, elastic material. Pure compressive stress would be an example of volumetric stress. Deviatoric (or shear) stresses produce changes in shape, but not volume, of the same element – pure shear stresses that are applied without changing the volumetric stress would be examples of deviatoric stresses. Soils, unlike most materials, however, exhibit coupling of volumetric and deviatoric components of stress and strain – the application of shear stress, for example, can cause volumetric strain. This aspect of soil behavior, termed *dilatancy*, has an enormous impact on the mechanical behavior of soils and is of particular importance in geotechnical earthquake engineering.

Dilatancy is often illustrated by visualizing the behavior of uniformly packed spheres subjected to normal and shear stresses. The upper portion of Figure 6.14 shows two rows of spheres stacked on top of each other. On the left, the spheres are packed directly on top of each other in a loose configuration; on the right, the spheres are packed in a dense configuration. The void ratio, e, of a soil describes the relative volumes of voids between particles, V_v, and solid particles, V_s, i.e., $e = V_v/V_s$. As shown in the lower portion of Figure 6.14, kinematic constraints lead to vertical movement of the upper row of spheres as they move laterally in response to the applied shear stress. The initially loose assemblage becomes denser as the spheres in the upper row slide and/or roll down and to the right relative to the lower spheres; the void ratio clearly decreases. On the right side of Figure 6.14, the spheres in the upper row must move upward in order to move to the right under the applied shear stress; so the void ratio increases during shearing. Soils whose void ratio increases or decreases during shearing are described as exhibiting *dilative* or *contractive* behavior, respectively.

An actual soil consists of particles of different shapes and sizes with no regular arrangement of the type shown in Figure 6.14. Some particles may be packed tightly relative to their immediate neighbors, and others may have much larger voids adjacent to them. When sheared, a loose soil

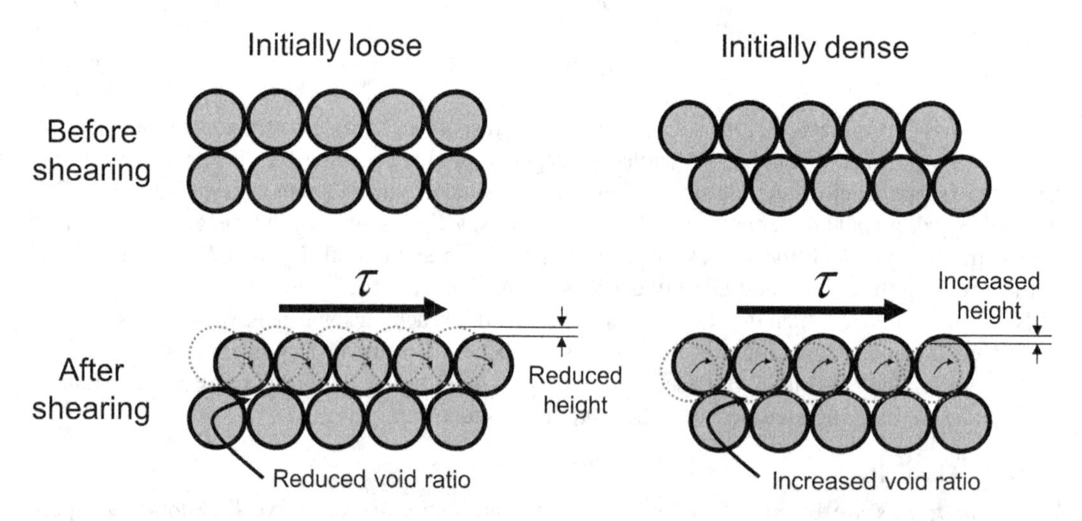

FIGURE 6.14 Schematic illustration of (left) contractive, and (right) dilative behavior upon application of shear stress.

will have more voids that contract than dilate and will therefore exhibit net contractive behavior. Similarly, a dense soil will have more dilating voids than contracting voids and will exhibit a net dilation. At some intermediate density, the expansion of dilating local voids may balance the compression of contracting voids, and the soil may shear with no net volume change, even though local dilation and contraction may be occurring.

For an element of soil in the ground, dilation is resisted by the normal stress acting on the element (Figure 6.15a), which means that external work, W, is done by dilation. The increment of external work done by deformation from the initial to final configuration is

$$dW = \tau A dx - \sigma' A dy \tag{6.11}$$

where τ and σ' are shear stress and normal effective stress, respectively, and A represents the horizontal area of the element on which the normal and shear stresses act. The corresponding incremental energy (internal work) dissipated by friction is

$$dE = \sigma' A \tan\phi_{cv} \cdot dx \tag{6.12}$$

where ϕ_{cv} is the constant-volume friction angle, i.e., the friction angle the soil would have if its volume did not change. Equating incremental work and energy and rearranging

$$\frac{\tau}{\sigma'} = \tan\phi_{cv} + \frac{dy}{dx} \tag{6.13}$$

which indicates that the shear strength of a dilative soil has a component that comes from the inherent friction of the soil, i.e., the friction mobilized during constant-volume shearing, and a component that results from dilation. This can be expressed by a friction angle ϕ:

$$\phi = \tan^{-1}\left[\tau / \sigma'\right] = \tan^{-1}\left[\tan\phi_{cv} + dy/dx\right] = \tan^{-1}\left[\tan\phi_{cv} + \tan\psi\right] \tag{6.14}$$

where $\psi = \tan^{-1}[dy/dx]$ is the *angle of dilation* of the soil. In a sense, the effects of dilation correspond to a situation in which a series of soil particles are being pushed up an inclined ramp (Figure 6.15b). In order to move the particles up the ramp, the driving force must not only overcome the frictional resistance between the particles and the ramp but also lift the particles against gravity.

6.4.2 Response to Compressive Stresses

Soils may be placed under compressive stresses in a number of circumstances. Under hydrostatic compression or one-dimensional compression, the angles between the centers of the particles do not change appreciably, but the centers of the particles may move closer together, indicating a change in volume. The following sections discuss the interaction of idealized elastic spheres, and then the actual compression behavior of clays and sands.

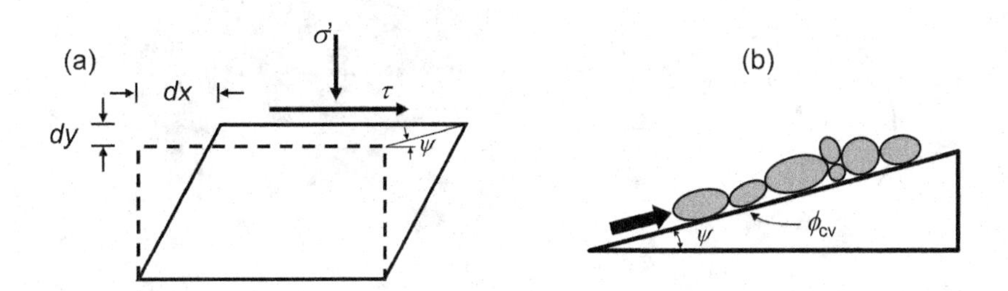

FIGURE 6.15 Shear-induced dilation: (a) behavior and notation for dilating element and (b) simplified analog illustrating role of dilation angle, ψ.

6.4.2.1 Idealized Assemblage of Spheres

The mechanical behavior of an element of soil depends on the manner in which individual soil particles interact with each other, and some insight into large-scale behavior can be gained by examining this interaction at the particle level, even under idealized conditions. Hertz (1881) studied the behavior of identical spheres of radius, R, compressed against each other (Figure 6.16a) by a normal force, N, and showed (Timoshenko and Goodier, 1951) that

$$N = \frac{2\sqrt{2}GR^{3/2}}{3(1-\nu)}\delta_N^{3/2} \tag{6.15}$$

where the shear modulus, G, and Poisson's ratio, ν, are elastic constants of the spheres themselves and δ_N is the change in distance between the centers of the spheres when the normal force is applied.

For a cubically packed array of spheres loaded along one of the packing axes (Figure 6.16b) and free to strain laterally, the average normal stress is obtained by dividing the normal force acting on a sphere by its tributary area, i.e.

$$\sigma = \frac{N}{(2R)^2} = \frac{N}{4R^2} \tag{6.16}$$

Then, the tangent Young's modulus for uniaxial loading is given by

$$E_{\tan} = \frac{d\sigma}{d\varepsilon} = \frac{dN / 4R^2}{d\delta_N / 2R} = \frac{1}{2R}\frac{dN}{d\delta_N} = \frac{3}{2}\left[\frac{2G}{3(1-\nu)}\right]^{2/3}\sigma^{1/3} \tag{6.17}$$

which suggests that the uniaxial stiffness at low strain levels should, theoretically, increase with the cube root of the axial stress – even under these idealized conditions, it is not proportional to the axial stress. Laboratory tests show that the uniaxial stiffness does increase (or, equivalently, that the uniaxial compressibility decreases) with increasing interparticle (or effective) stress. The tangent (instantaneous) bulk modulus, $K = dp' / / d\varepsilon_v$, will increase as the pressure increases – the soil becomes stiffer as the stress level increases. If the pressure reaches extremely high levels, some particle fracture may also occur and modify the soil's grain size distribution.

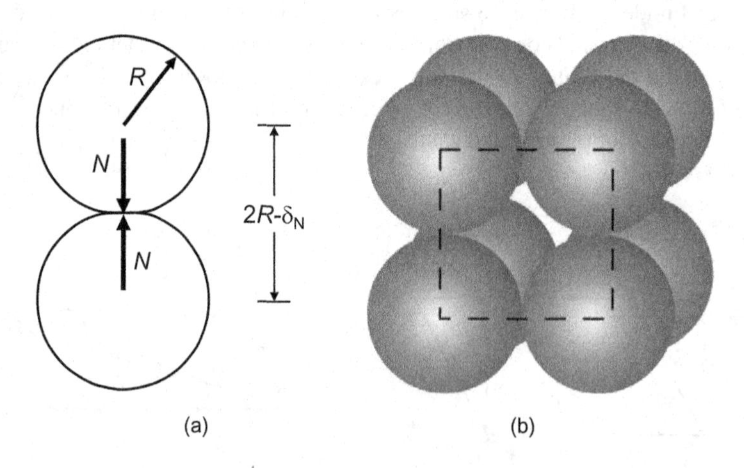

FIGURE 6.16 (a) Contact forces between spheres subjected to normal force (http://dcsymbols.com/map_cube/cube.htm) and (b) spheres in cubic packing form (Leiderman et al., 2013; used by permission of Elsevier).

6.4.2.2 Effects of Stress History

Upon deposition in a level-surfaced accreting soil deposit, an element of soil will be subjected to monotonically increasing vertical stress, which results in one-dimensional compression of the soil (Figure 6.17a). Compression behavior can be plotted in many ways – in the U.S., it is customary to plot the void ratio as a function of the logarithm of vertical effective stress for one-dimensional compression (Figure 6.17b). As the accreting soil densifies, it will follow the path ABD (i.e., the *virgin compression curve*) in Figure 6.17 upon which the vertical effective stress, σ'_{v0}, at any point will be the largest effective stress the soil has ever existed under – under these conditions, the soil is said to be *normally consolidated*. If the soil is unloaded after reaching Point B to the vertical effective stress corresponding to Point C, its effective stress is no longer the greatest effective stress it has been in equilibrium under so it is said to be *overconsolidated*. The *overconsolidation ratio*, or *OCR*, is defined as

$$OCR = \frac{\sigma'_p}{\sigma'_{v0}} \tag{6.18}$$

where σ'_p is the *preconsolidation pressure*, which is defined as the maximum vertical effective stress under which the soil has ever existed. The *OCR*, therefore, describes the extent to which a soil has been overconsolidated; a normally consolidated soil would have $OCR = 1.0$. If the vertical effective stress is then increased again, the soil will recompress to Point B (at which point it is again normally consolidated) and then continue along the virgin compression curve to Point D.

As the level of vertical effective stress increases, the soil becomes less compressible and its *constrained modulus* ($M = d\sigma/d\varepsilon$ for conditions of no lateral strain) increases. Note that the soil unloads inelastically, i.e., along a path (BC) that falls below the virgin compression curve (Figure 6.17). In the overconsolidated state (between Points B and C), the soil is much less compressible than in the normally consolidated state.

6.4.2.3 Compressibility of Clays

For clays, the virgin compression curve is unique – it is the only *state* (combination of density and effective stress) that a normally consolidated clay can exist in given its depositional structure (which is influenced principally by whether the pore fluid is freshwater or saltwater). As a result, the state of a particular normally consolidated clay can be defined by its effective stress. States above the virgin compression curve are impossible to achieve in a clay, and states below can only be reached by unloading into the overconsolidated range, such as along the path BC in Figure 6.17. Figure 6.18 shows an actual consolidation curve, including a cycle of unloading and reloading, for a typical clay.

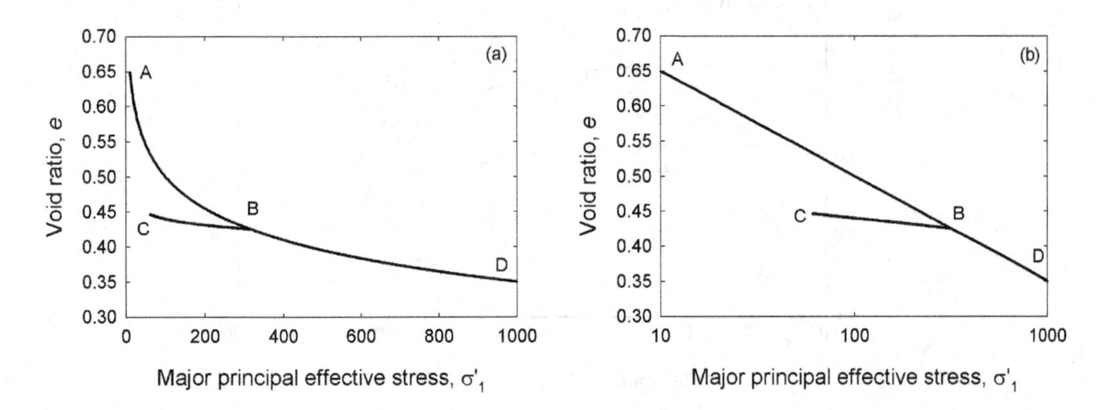

FIGURE 6.17 Idealized one-dimensional compression behavior of soils: (a) linear stress scale, and (b) log stress scale. In reality, unloading path from B to C would lie somewhat below reloading path from C to B.

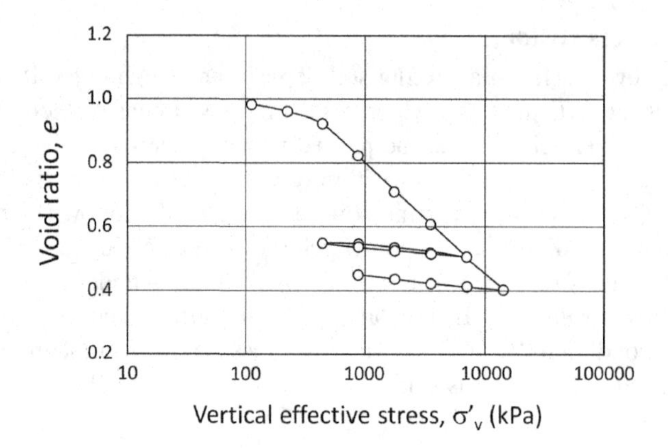

FIGURE 6.18 One-dimensional consolidation curve for clay.

Normally consolidated clays are highly compressible, so their densities are strongly affected by their effective stress level. Overconsolidated clays are much less compressible; their densities are affected by their current effective stress and their overconsolidation ratio.

6.4.2.4 Compressibility of Sands

The compression behavior of sand is more complicated. In contrast to clays, which are sedimented to initial states that lie on a unique virgin compression curve, sands deposited in different manners can have a range of different densities; hence, there are an infinite number of virgin compression curves for sand. Their subsequent behavior will depend on both the initial density and the effective confining pressure, so both are required to define the *state* of a particular sand. The looser the initial density of a sand, the greater is its compressibility (Figure 6.19). Sand compression curves tend to be flatter and more nonlinear (in e-log p space) than those of clays, at least until effective stresses

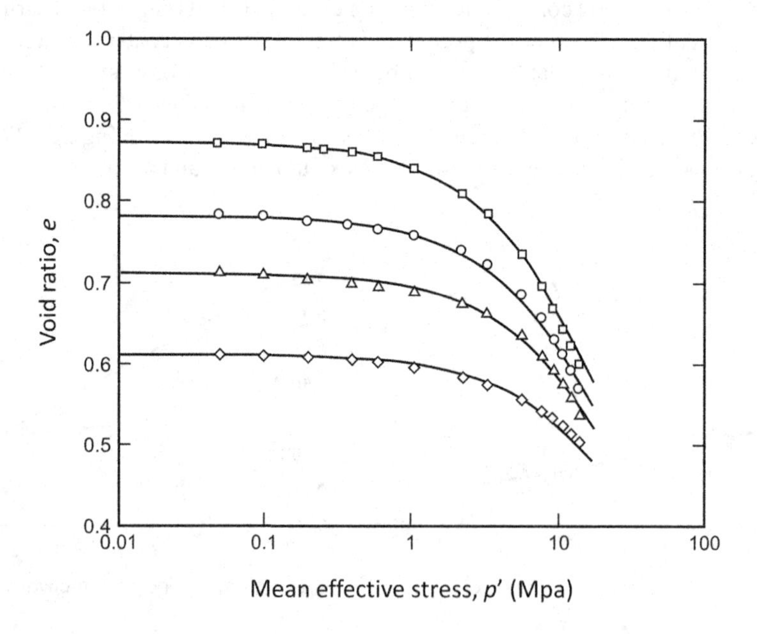

FIGURE 6.19 Isotropic compression curves for Sacramento River sand at four different initial densities. (After Sheng et al., 2008; used by permission of Canadian Science Publishing.)

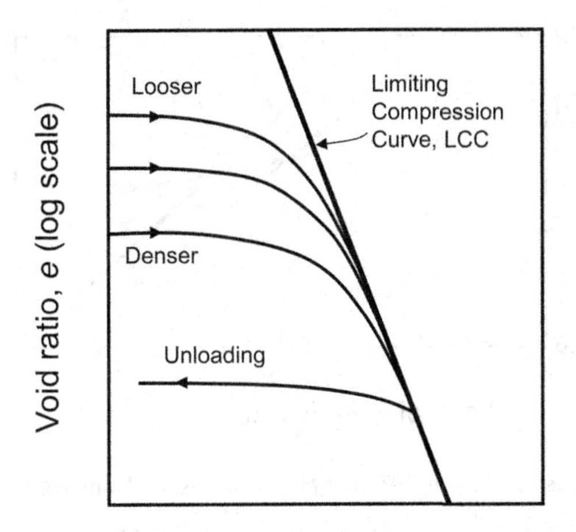

FIGURE 6.20 Limiting compression curve for sands deposited at different initial void ratios. (After Pestana and Whittle, 1995; used by permission of Thomas Telford, Ltd.)

become high at which point they appear to converge to a limiting compression curve, LCC (Pestana and Whittle, 1995) (Figure 6.20). The flatter compression curves indicate that sands are less compressible than clays – a given increase in effective stress leads to less volume change in sands than in clays. It also means, however, that a given increase in volume will lead to a much greater reduction in effective stress in a saturated sand than in a clay. Unloading-reloading behavior is also more nonlinear for sands than clays.

6.4.2.5 Compressibility of Intermediate Soils

Not all soils fall neatly into categories of sands and clays, or even coarse-grained and fine-grained. Geotechnical engineers often encounter intermediate soils (silts) that are finer than sands and coarser than clays. The distinction between coarse- and fine-grained soils is made on the basis of the fraction of the soil that passes through a No. 200 sieve. There is nothing fundamental about the opening size (0.074 mm) of a No. 200 sieve – it is essentially the finest, rugged sieve size that can be economically manufactured. Soil particles that just pass through a No. 200 sieve are bulky-grained and similar in appearance and behavior to fine sands that are retained on the No. 200 sieve. Other, finer silt particles can have high specific surface areas and exhibit relatively plastic behavior. As a result, intermediate soils have the potential to span a wide range of behaviors.

The plasticity chart in Figure 6.11 shows the wide range of plasticities that silts can exhibit, particularly at higher liquid limits. Figure 6.21 shows the results of tests on three silts from an aggregate tailings pond (Boulanger and Idriss, 2004) prepared by sedimentation from a slurry. The silts had plasticity indices of 0, 4, and 10.5 with 81%–87% passing the No. 200 sieve. Isotropic consolidation tests showed that the $PI = 0$ silt was much less compressible than the $PI = 4$ and $PI = 10.5$ silts. The compressibility of a silt, therefore, is affected by its plasticity.

6.4.2.6 Compressibility of Soil Mixtures

When soils are comprised of grains having many different sizes, for example with both coarse- and fine-grained constituents, it is important to develop an understanding of which portion of the soil's grain composition is likely to control its engineering behavior. Mixtures of different particles may be well-graded with a broad range of particle sizes or *gap-graded* as a mixture of two predominant particle sizes. Well-graded soils tend to have low compressibilities, but the compressibility of gap-graded soils depends on the nature of the particle contacts. As discussed in Section 6.3, the

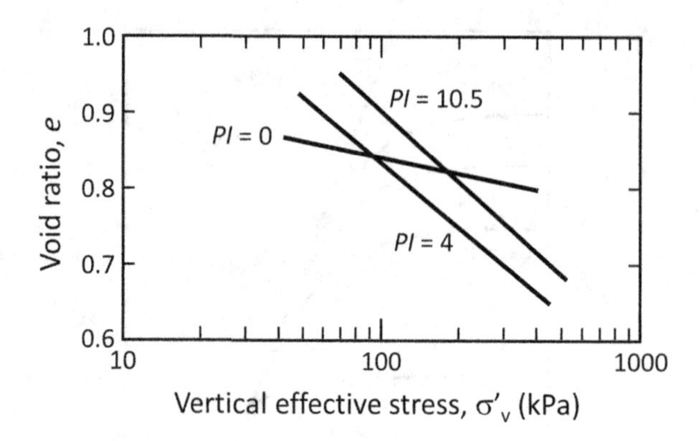

FIGURE 6.21 Compressibilities of silts of different plasticity. (After Romero, 1995.)

behavior of sand-silt or sand-clay mixtures with more than 15%–40% fines will be controlled by the fines, and hence have relatively high compressibility.

Mixtures of gravel- and sand-sized particles are also relatively common and, although their compressibilities are generally low enough not to be of primary concern, they response to static shear stresses (Section 6.4.3.5) and cyclic shear stresses (Section 6.4.4.5), particularly when saturated, can be extremely important.

6.4.3 Response to Static Shear Stresses

While compressibility plays an important role in soil mechanics, the deformations that lead to damage during earthquakes are usually dominated by shearing behavior. Although soils are generally stiff and strong in bulk or one-dimensional compression, they are much softer and weaker in shear, so their shearing behavior is of critical importance for the evaluation of performance during earthquakes. Earthquakes produce seismic waves that impose cyclic shear stresses on soils. The manner in which soils respond to cyclic shear stresses is different than, but distinctly related to, the manner in which they response to monotonically increasing stresses. To understand cyclic response, it is first necessary to develop an understanding of static response.

Geotechnical engineers are frequently required to evaluate the behavior of soils subjected to monotonic, or *static*, loading. The loading may increase monotonically, for example in the soil under a building or dam during construction, or they may decrease as they would in the soil beneath or adjacent to an excavation. In such cases, evaluation of the shear strength of the soil, which is required for capacity or stability evaluations, has historically been the most common goal of a subsurface investigation. Increasingly, however, estimation of deformations has been sought, so the characterization of the manner in which the shear strength is mobilized, i.e., the stress-strain behavior, is required. The stress-strain behavior is also important for the evaluation of the initial (static) stress conditions that exist prior to earthquake shaking since those conditions can significantly influence the response of the soil to cyclic loading.

The following sections briefly discuss the responses of different types of soil to monotonically increasing shear stresses. A number of important aspects of soil behavior are reflected in their response to static loading and also provide insight into the behavior of the soil under cyclic loading, which is discussed subsequently.

6.4.3.1 Idealized Assemblage of Spheres

The shearing resistance of a soil is mobilized through the contact forces between individual grains in a soil skeleton. Some useful insights into soil behavior can be gained from the behavior of idealized assemblages of spherical, elastic particles. In contrast to the one-dimensional compression condition shown in Figure 6.16, the application of boundary shear stresses to a series of cubically packed spheres will cause tangential forces at the particle contacts (Figure 6.22). When a tangential force, T, is applied, elastic distortion of the spheres themselves causes their centers to be displaced perpendicular to their original axes (Mindlin and Deresiewicz, 1953; Dobry et al., 1982) by an amount

$$\delta_T = \left[1-\left(1-\frac{T}{\mu N}\right)^{2/3}\right]\left[\frac{3\mu N}{4E}(2-\nu)(1+\nu)\left(\frac{3(1-\nu^2)NR}{4E}\right)^{-1/3}\right] \quad T \le \mu N \tag{6.19}$$

where μ is the coefficient of friction between the spheres. When T becomes equal to μN, gross sliding of the particle contacts occurs (though slippage of part of the contact can occur before this point). The relationship of Equation (6.19) suggests nearly linear force-displacement behavior at tangential forces lower than about 1/3 the force at which sliding occurs. Laboratory tests have shown that soils exhibit essentially linear and nearly elastic behavior at very low levels of loading. The level of shear strain at which the secant shear modulus of the soil drops to 99% of the maximum shear modulus is referred to as the *linear threshold shear strain*, γ_{tl} (Vucetic, 1994).

Gross sliding is required for permanent particle reorientation, consequently neither volume changes (drained conditions) nor pore pressures (undrained conditions) can occur when gross sliding does not occur. The strain corresponding to the initiation of gross sliding

$$\gamma_{tv} = \frac{d_T(T = fN)}{2R} = 2.08\frac{(2-\nu)(1+\nu)f}{\left(1-\nu^2\right)^{1/2} E^{2/3}}\sigma^{2/3} \tag{6.20}$$

is called the *volumetric threshold shear strain* (Vucetic, 1994). When the properties of quartz ($E = 11 \times 10^6$ psi, $\nu = 0.31, f = 0.50$) are substituted into Equation (6.19), the volumetric threshold shear strain is given by

$$\gamma_{tv}(\%) = 0.000175\sigma^{2/3} \tag{6.21}$$

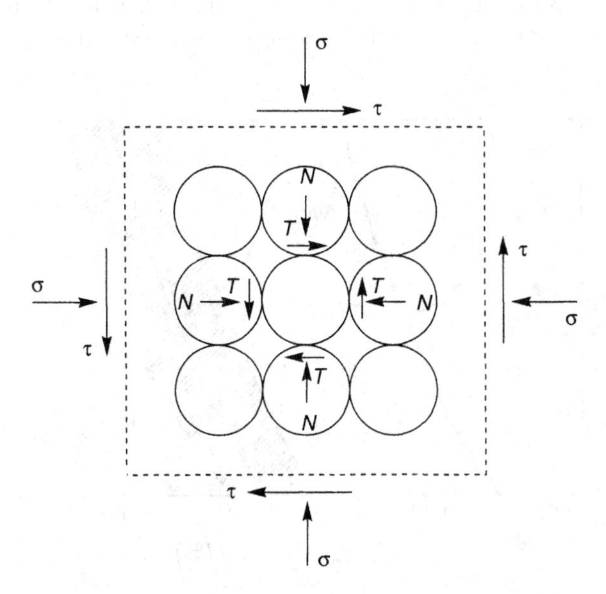

FIGURE 6.22 Cubically packed assemblage of spheres subjected to normal stress, σ, and shear stress, τ, that produce interparticle contact forces N and T. (After Dobry et al., 1982.)

where σ is in psf. For confining pressures of practical interest (500 psf to 4,000 psf), Equation (6.20) would predict a volumetric threshold shear strain between 0.01% and 0.04%.

Real soils, of course, do not consist of regular arrays of spherical particles, but the existence of a volumetric threshold shear strain very close to that predicted by Equation (6.21) has been observed experimentally for both drained (Drnevich and Richart, 1970; Youd, 1972; Pyke, 1973) and undrained (Park and Silver, 1975; Dobry and Ladd, 1980; Dobry et al., 1982) loading conditions. The simple, idealized analysis of a regular array of spheres helps illustrate the reason for its existence. Experimental evidence suggests that both the linear and volumetric threshold shear strains increase with plasticity index (Vucetic, 1994; Hsu and Vucetic, 2006); the volumetric threshold of a clay with $PI = 50$ is approximately one order of magnitude greater than that of a sand with $PI = 0$ (Vucetic, 1994). Experimental evidence also indicates (Figure 6.23) that the linear cyclic threshold shear strain, γ_{tl}, is approximately 30 times smaller than γ_{tv}.

6.4.3.2 The Critical State

Much of what is understood about the shearing behavior of soils developed from a landmark series of laboratory tests performed in the 1930s. In his pioneering work on the shear strength of soils, Casagrande (1936) performed drained, strain-controlled triaxial tests (Section 6.5.4.3) on initially loose and initially dense sand specimens. The results (Figure 6.24), which form the cornerstone of modern understanding of soil strength behavior, showed that all specimens tested at the same effective confining pressure approached the same density when sheared to very large strains. Initially loose specimens contracted, or densified, during shearing and initially dense specimens first contracted, but then quickly began to dilate. At large strains, all specimens approached the same density and continued to shear with constant shearing resistance. The void ratio corresponding to this constant density was termed the *critical void ratio*, e_c. By performing tests at different effective confining pressures, Casagrande found that the critical void ratio was uniquely related to the effective confining pressure, and called the locus of such points the *critical state line*, or CSL (Figure 6.25). By defining the state of the soil in terms of void ratio and effective confining pressure, the CSL could be used to mark the boundary between loose (contractive) and dense (dilative) states.

The equipment needed to measure pore pressure during undrained shearing was not available at the time, but Casagrande hypothesized that strain-controlled, undrained testing would produce positive excess pore pressure (due to the tendency for contraction, which could not occur under the constant-volume constraint of undrained testing) in loose specimens, and negative excess

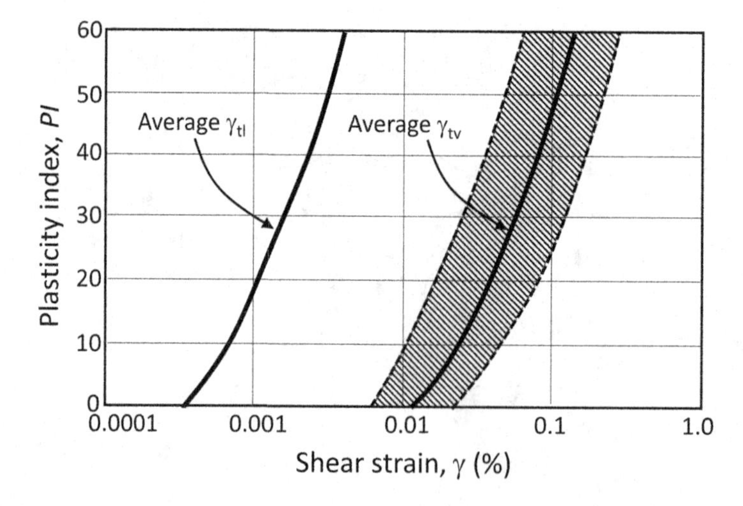

FIGURE 6.23 Variation of linear and volumetric threshold shear strains with plasticity index. (After Vucetic, 1994; used with permission of ASCE.)

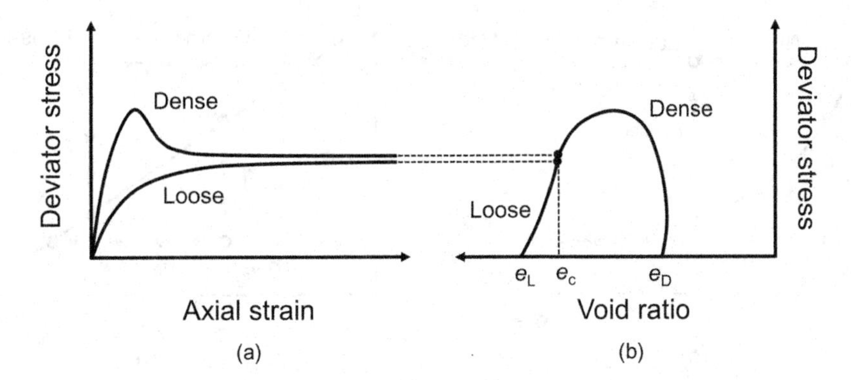

FIGURE 6.24 (a) Stress-strain and (b) stress-void ratio curves for loose and dense sands at the same effective confining pressure. Loose sand exhibits contractive behavior (decreasing void ratio) and dense sand exhibits dilative behavior (increasing void ratio) during shearing. By the time large strains have developed, both specimens have reached the critical void ratio and mobilize the same large-strain shearing resistance.

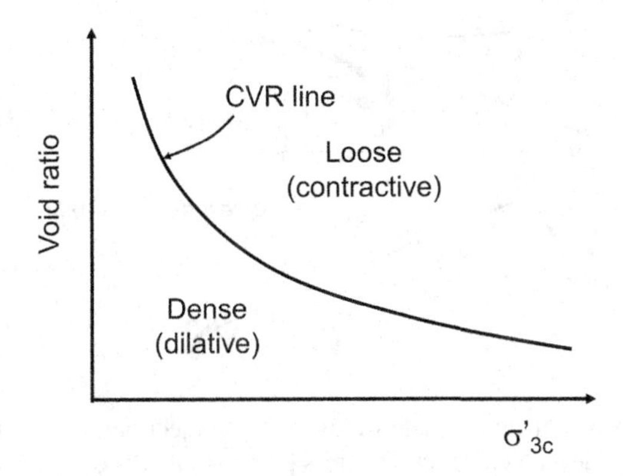

FIGURE 6.25 Use of the critical state line as a boundary between loose, contractive states and dense dilative states.

pore pressure (due to the tendency for dilation) in dense specimens (Figure 6.26) until the CSL was reached. This hypothesis was subsequently verified experimentally (e.g., Seed and Lee, 1967; Castro, 1969). The CSL therefore described the state toward which any soil specimen would migrate at large strains, whether by volume changes under drained conditions, changes in effective confining pressure under undrained conditions, or some combination under partially drained conditions.

The CSL does not control volume change, however, under low-strain vibratory loading conditions. Sands and gravels are known to densify efficiently under vibratory loading; indeed, vibratory equipment is frequently used for compaction and soil improvement (Chapter 11). In the absence of plastic fines, vibratory loading can reduce the void ratio of soils that are initially below as well as soils that are initially above the CSL.

In its most general form, the CSL can be viewed as a three-dimensional curve in e-σ'-τ (Figure 6.27) or e-p'-q space. The CSL shown in Figure 6.26a therefore represents the projection of the three-dimensional CSL onto a plane of constant τ. The CSL can also be projected onto planes of constant effective confining pressure (σ' = constant) and constant density (e = constant).

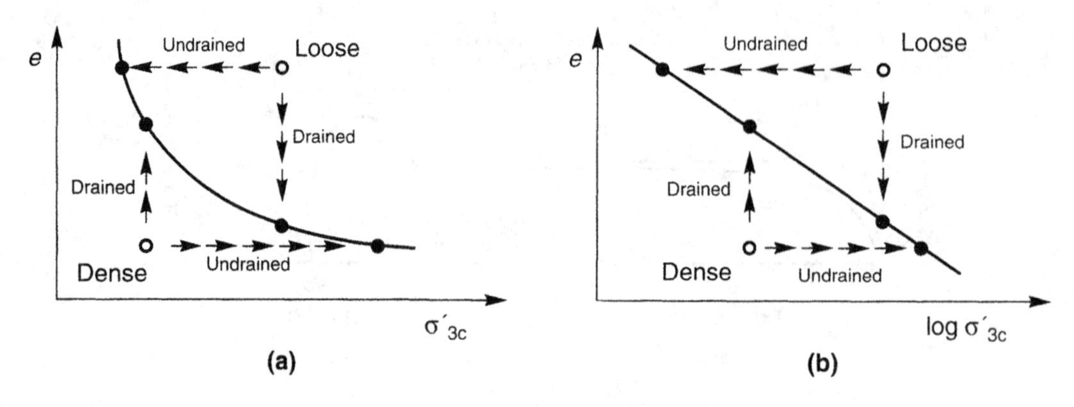

FIGURE 6.26 Behavior of initially loose and dense specimens under drained and undrained conditions for (a) arithmetic and (b) logarithmic effective confining pressure scales.

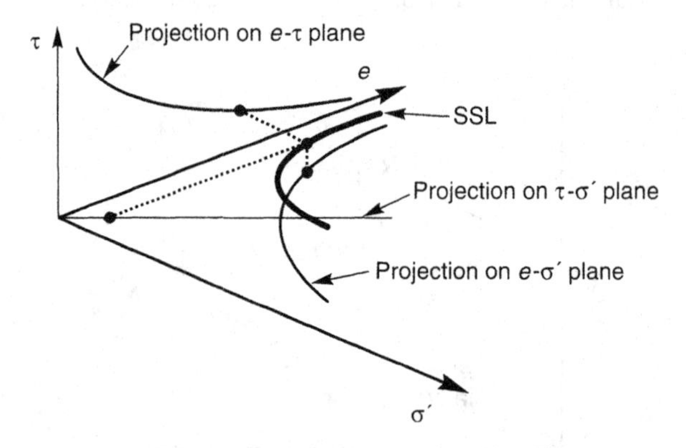

FIGURE 6.27 Three-dimensional steady state line showing projections on e-τ plane, e-σ' plane, and τ-σ' plane. A similar plot could be developed using the stress path parameters q and p' instead of τ and σ'.

6.4.3.3 Shearing Behavior of Clays

The behavior of clays is controlled by their stress history, i.e., whether they are normally consolidated or overconsolidated. Experiments indicate that the critical void ratio (or critical state) line for a typical clay lies parallel to and below the virgin compression curve (Figure 6.28). Normally consolidated clays, which lie on the virgin compression curve, are contractive under monotonic shearing and therefore tend to generate positive excess porewater pressure under undrained loading.

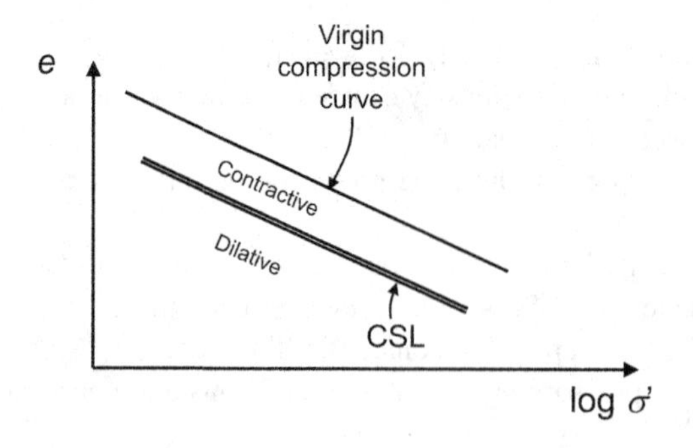

FIGURE 6.28 Relative position of virgin compression curve and critical state line for clays.

Mildly overconsolidated clays (*OCR* less than about 4), which lie between the virgin compression curve and the CSL are also contractive, but generate less porewater pressure than normally consolidated clays. Overconsolidated clays whose initial states lie below the CSL are dilative under monotonic shearing. Stress-strain curves and effective stress paths for a series of normally consolidated clay specimens consolidated to different initial effective stress levels are shown in Figure 6.29. The curves all have similar shapes, indicating similar levels of excess porewater pressure generation (relative to initial effective stress level), and show very little strain-softening behavior.

The undrained shear strength, s_u, of a normally consolidated clay is fully mobilized at the critical state (Figure 6.30), so

$$s_u = q_{cs} = p'_{cs} \tan\psi'_{cs} = p'_{cs} \sin\phi'_{cs} \tag{6.22}$$

where ϕ'_{cs} and ψ'_{cs} are the critical state effective stress friction angles in stress space and stress path space, respectively.

If the virgin compression curve is linear in e-log σ' space, its position can be described by

$$e = e_1 - C_c \log\sigma'_{vc} \tag{6.23}$$

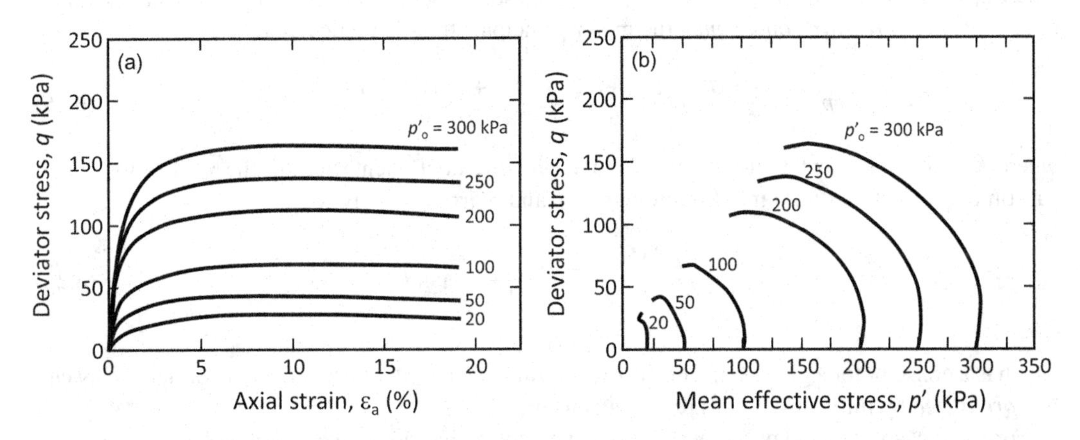

FIGURE 6.29 Stress-strain curves and effective stress paths for normally consolidated clay (PI = 36, LL = 64) at different initial effective confining pressures. (After Guo et al., 2013; used by permission of Elsevier.)

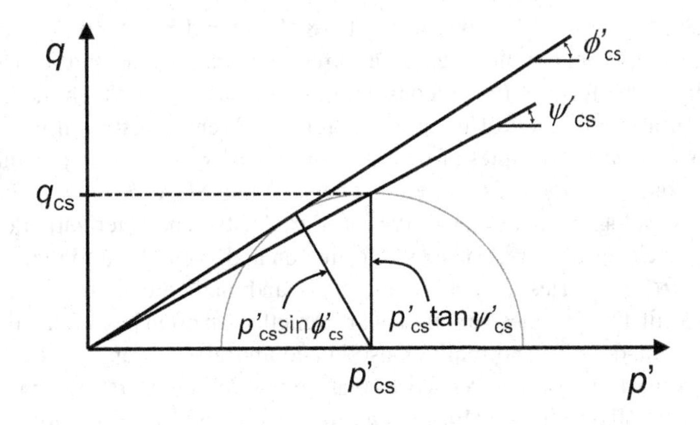

FIGURE 6.30 Friction angle expressed in terms of Mohr-Coulomb and stress path failure envelopes.

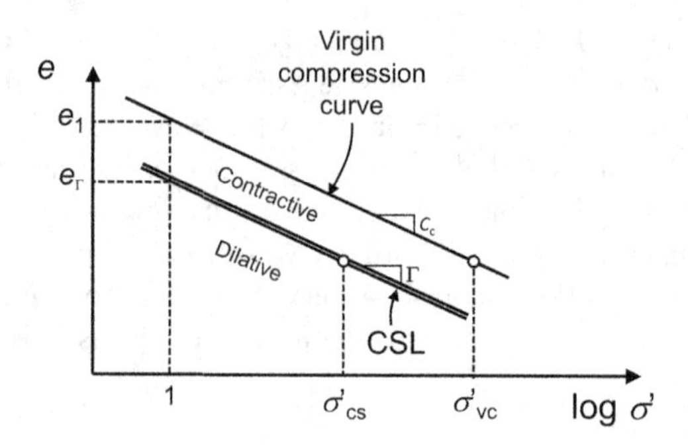

FIGURE 6.31 Relationship between the critical state line and virgin compression curve.

where e_1 is the void ratio at $\sigma'_{vc} = 1$ (Figure 6.31). If the critical state line is parallel to the virgin compression curve, it can be expressed as

$$e = e_\Gamma - C_c \log \sigma'_{cs} \tag{6.24}$$

where e_Γ is the void ratio on the CSL at $\sigma'_{cs} = 1$. For a soil under level ground (i.e., zero lateral strain, or *at-rest earth pressure conditions*), the major principal stress is vertical and

$$p'_{cs} = \frac{\sigma'_1 + \sigma'_3}{2} = \frac{\sigma'_{v0} + \sigma'_{h0}}{2} = \frac{\sigma'_{v0} + K_0 \sigma'_{v0}}{2} = \frac{1 + K_0}{2} \sigma'_{v0} \tag{6.25}$$

where $K_0 = \sigma'_{h0} / \sigma'_{v0}$ is the at-rest lateral earth pressure coefficient. Setting the void ratios equal, substituting Equation (6.25) into Equation (6.22), and rearranging gives

$$\left(\frac{s_u}{\sigma'_{v0}} \right)_{NC} = \frac{2 \cdot 10^{(e_\Gamma - e_1)/C_c}}{1 + K_0} \tan \phi'_{cs} \tag{6.26}$$

which is a constant for a given soil. Hence, the undrained strength of normally consolidated clay can be *normalized* by initial effective stress to a constant.

For overconsolidated clays, experiments have shown that normalization is still possible if the degree of overconsolidation is accounted for as

$$\left(\frac{s_u}{\sigma'_{v0}} \right)_{OC} = \left(\frac{s_u}{\sigma'_{v0}} \right)_{NC} \cdot OCR^m \tag{6.27}$$

where the exponent, m, is usually on the order of 0.8 (Ladd and Foott, 1974; Ladd, 1991). Thus, the strength of a clay is strongly influenced by its stress history, and characterization of that stress history, specifically in the form of the preconsolidation pressure upon which the *OCR* is based, is critical for the accurate estimation of undrained strength. Strength tests can be accompanied by consolidation tests on adjacent samples of the same soil in order to accurately evaluate the preconsolidation pressure on which the *OCR* is based. The so-called SHANSEP procedure of Ladd and Foott (1974) provides a logical and systematic procedure for the characterization of the undrained strength of a specific clay using stress history information and strength normalization principles.

The value of $(s_u / \sigma'_{v0})_{NC}$ varies with clay mineralogy and method of testing. If material-specific test results are not available for cohesive soils, shear strengths can often be estimated using published results for $(s_u / \sigma'_{v0})_{NC}$ and m. The normally consolidated undrained strength ratio for simple shear conditions (Section 6.5.4.3) has been related to plasticity index as $(s_u / \sigma'_{v0})_{NC} = 0.11 + 0.0037 \cdot PI$ (Skempton, 1957). Jamiolkowski et al. (1985) suggested that $(s_u / \sigma'_{v0})_{NC} = 0.23$ for clays with $PI < 60$ while Koutsoftas and Ladd (1985) and Mesri (1989) suggested values of 0.22.

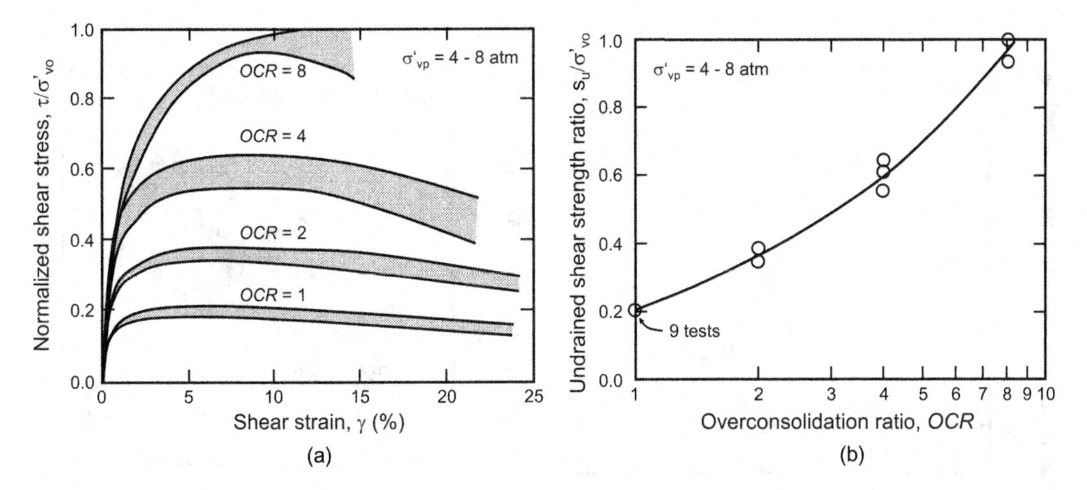

FIGURE 6.32 (a and b) Normalization of monotonic stress-strain behavior in normally consolidated and overconsolidated clays. (After Ladd and Foott, 1974; used with permission of ASCE.)

The stress-strain behavior of clays can also be normalized (Ladd and Foott, 1974), i.e., the stress-strain behavior of soils of a given *OCR* is similar after normalizing shear stress by initial vertical effective stress (Figure 6.32). The ability to normalize undrained shear strength and stress-strain behavior under monotonic loading conditions is one of the hallmarks of clay behavior.

It is not difficult to imagine that overconsolidated clays, with their very thin and flexible particles surrounded by layers of viscous adsorbed water, can move relative to each other by bending and adsorbed water layer deformation without the magnitude of dilation required by the essentially rigid particles of a dense sand. Thus, the tendencies for dilation and contraction in clays are not as strong as they are in sands. An exception to this is the case of *sensitive clays*, which have an inherently metastable structure and contract rapidly upon shearing.

It should be noted, however, that the excess pore pressures that develop under undrained conditions, whether positive (in a contractive soil) or negative (in a dilative soil) will eventually dissipate over time. When they dissipate, the strength of a contractive soil will increase as the effective stress increases, hence, the critical condition for problems such as stability evaluations will generally be the short-term (undrained) case. Heavily overconsolidated clays will dilate and develop negative excess pore pressures (and increased effective stresses) in the short term; as drainage occurs, the effective stress, and consequently the strength, decreases. As a result, the critical condition for dilative soils will generally be the long-term case. It should be noted that the criticality of short- and long-term conditions can be different for cases involving unloading (e.g., excavations) of the soil.

6.4.3.4 Shearing Behavior of Sands

The shear strength of sandy soils, whether above or below the ground water table, is typically expressed using the drained friction angle, ϕ'. Because of excessive disturbance of sands by sampling, friction angles are often evaluated using correlations with penetration resistance in lieu of sampling and strength testing in the laboratory. Two types of correlations are commonly used: those that correlate ϕ' directly with penetration resistance (from cone penetration test (CPT) or standard penetration test (SPT) – see Sections 6.5.3.1 and 6.5.3.3), and those that take friction angle as the sum of a critical state friction angle (related to mineralogy) and the difference between the peak and critical state friction angle. Figure 6.33 shows direct correlations of overburden-normalized penetration resistance to friction angle using SPT and CPT resistances.

As previously indicated, sands can be deposited in loose or dense states, so a normally consolidated sand may exist above or below the critical state line. When sheared, all sands will initially contract as the particles begin to move. As the loading continues, however, a dense sand will begin

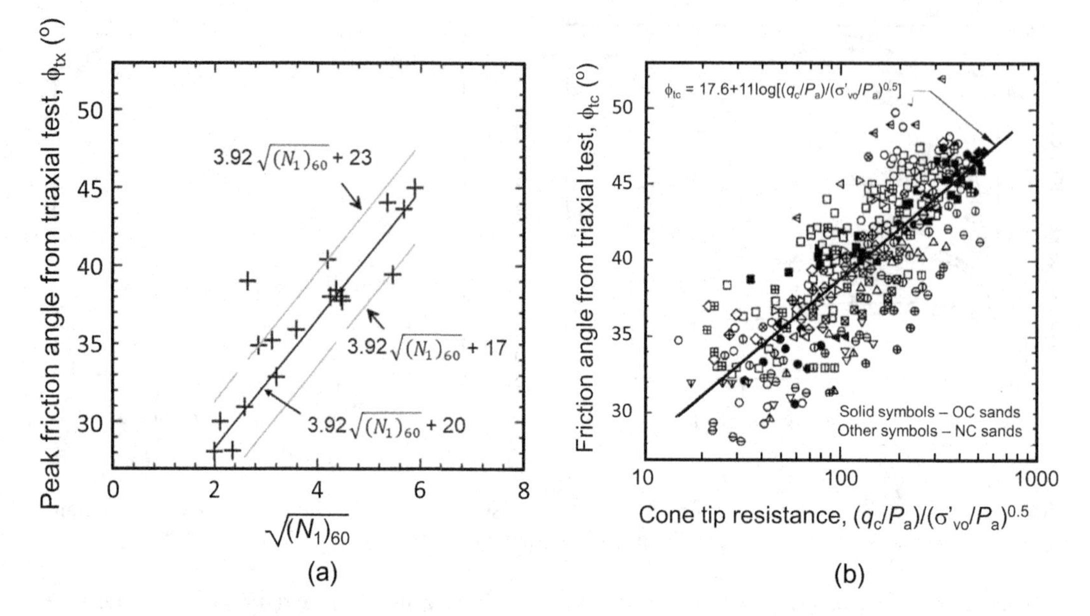

FIGURE 6.33 (a) Normalized SPT blow count versus friction angle (Adjusted from Hatanaka and Uchida, 1996 based on 78% SPT efficiency reported by Yoshimi, 1994) as measured from frozen samples; and (b) normalized CPT tip resistance versus drained triaxial friction angle (After Kulhawy and Mayne 1990; used by permission of the Electric Power Research Institute).

to dilate in order to accommodate shearing deformations. Eventually, the soil will reach the critical state, although a very high level of strain may be required to do so. Laboratory tests have shown that the critical state line for sands is generally more curved than that observed for clays, and is not as parallel to the (also curved) sand virgin compression curve. As a result, the strength and stress-strain behavior of sands cannot be normalized in the manner just described for clays.

Castro (1969) performed undrained static and cyclic tests on isotropically consolidated specimens and several static tests on anisotropically consolidated specimens. Three different types of stress-strain behavior, illustrated for isotropically consolidated specimens under stress-controlled loading in Figure 6.34a, were observed. Very loose specimens were observed to reach a peak shearing resistance at a low strain level after which the shearing resistance dropped to a constant, lower value at large strains. Dense specimens were observed to dilate and reach high shearing resistances that were still increasing when the capacity of the loading equipment was reached. Specimens of intermediate density were observed to mobilize a peak shearing resistance at low strain levels followed by a reduction in shearing resistance, followed in turn by an increase in shearing resistance. The increase in shearing resistance coincided with an increase in effective stress as the soil changed from contractive to dilative behavior. This aspect of the behavior of granular soils has received considerable attention, and different terms have been used to describe the point at which this change occurs. The transition from contractive to dilative behavior was first described by Ishihara (1993) as occurring at the *phase transformation point*, which corresponds to the instant in which the soil is neither contracting nor dilating; at this point, a tangent to the effective stress path is parallel to the total stress path (see inset in Figure 6.34b; total stress path is inclined at 45°). Further loading produced continued dilation to higher effective confining pressures and, consequently, higher large-strainstrengths. Eventually, all specimens reached a constant shearing resistance at very large strain levels; Castro and Poulos (1977) and Poulos (1981) defined the state in which a soil is shearing with constant shearing resistance, constant effective stress, constant volume (i.e., constant void ratio), and constant strain rate as the *steady state of deformation*, and postulated that the steady state is a unique function of void ratio. Therefore, any element of soil at a particular void ratio, when

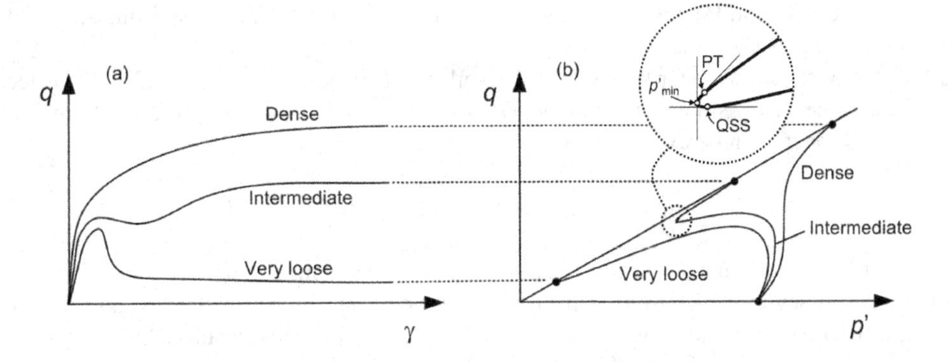

FIGURE 6.34 (a and b) Schematic illustration of undrained behavior of very loose, intermediate, and dense sands in monotonic loading tests. Inset shows differences between phase transformation, quasi-steady state, and minimum mean effective stress points.

sheared under undrained conditions, can be expected to reach the same effective confining pressure at high strain levels.

The phase transformation point has been described in some publications as corresponding to the minimum mean effective stress, p'_{min}, although the soil is still slightly contractive at that point and the instant at which it occurs depends on how p is defined. The condition in which a local minimum shearing resistance is observed at moderate strain levels in specimens of intermediate density has been referred to as the *quasi-steady state* (QSS) of deformation (Alarcon-Guzman et al., 1988). The specific locations of these points are different, as illustrated in the inset in Figure 6.34, but they are so close to each other that the various terms have come to be used interchangeably. It should be recognized, however, that phase transformation can occur without the strain-softening behavior that leads to development of the quasi-steady state; in that case, the distinction between the two terms is important. To distinguish it from the quasi-steady state, the steady state has been referred to as the *ultimate steady state* (USS) of deformation (Yoshimine and Ishihara, 1998). In this text, the term "steady state" will refer to the condition that occurs at very large strain levels, i.e., the original

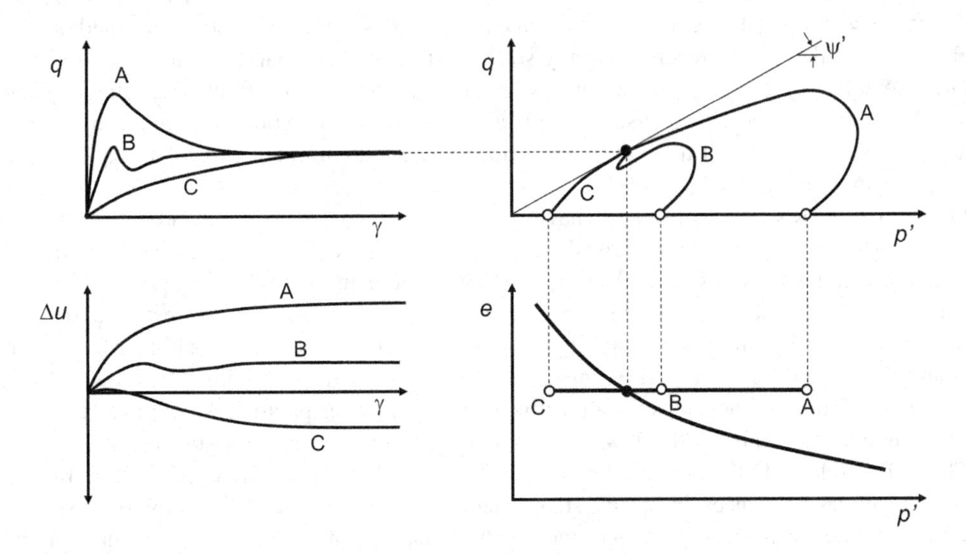

FIGURE 6.35 Illustration of the undrained behavior of three triaxial test specimens at the same void ratio but at different initial effective stress levels. All three specimens have the same void ratio, so all reach the same steady state conditions when subjected to undrained shearing.

steady state of Castro and Poulos (1977) and Poulos (1981) and the USS of Yoshimine and Ishihara (1998), not the quasi-steady state.

At the steady state, the shear strength of the soil, S_{su}, is governed by the steady state effective stress, σ'_{ss}, and the steady state friction angle, ϕ'_{ss}, (which is equivalent to ϕ'_{cv} referred to in Section 6.4.1.3). This strength can be expressed as

$$S_{su} = \sigma'_{ss} \tan \phi'_{ss} \tag{6.28}$$

The steady state friction angle depends primarily on gradation and mineralogy and ranges from about 30° to 35° (Negussey et al., 1988) for typical quartz sands. Dense sands will dilate and mobilize a higher peak friction angle (since work must be done to expand the soil against the effective stress acting on it). Following Equation (6.14), the peak friction angle of a dense sand can be expressed as

$$\tan \phi'_{peak} = \tan \phi'_{ss} + \tan \psi \tag{6.29}$$

where ψ is the dilation angle, which reflects the rate at which dilation of a soil element varies with shear strain (Figure 6.15). Bolton (1986) showed that Rowe's dilatancy relationship (Rowe, 1962) was consistent with

$$\phi'_{peak} = \phi'_{ss} + 0.8\psi \tag{6.30}$$

for $\psi < 20°$. Dilation angles are typically on the order of 10°–20° for granular soils at effective confining pressures greater than 1 atm (Bolton, 1979, 1986) but can be higher at lower effective confining pressures.

Figure 6.35 illustrates the undrained behavior of three soil specimens consolidated to the same initial void ratio under different effective confining pressures. Specimen A, which is at a high initial confining pressure, is contractive and therefore generates high positive excess pore pressure and exhibits strain-softening behavior. Specimen C, which is at a low initial effective stress, dilates with a reduction in porewater pressure and exhibits strain-hardening behavior. Specimen B at the intermediate confining pressure is contractive but then undergoes phase transformation and dilates. Specimen B develops a peak strength at a relatively low strain level, then softens until undergoing phase transformation and reaching a minimum quasi-steady state strength at an intermediate strain level, and then dilates until reaching steady state conditions. Note that, since all three specimens had the same density (or void ratio), all end up at the same steady state effective confining pressure at very large strains. The effective stress paths all move to the same point on the steady state line (SSL) and, since they are all then under the same effective confining pressure, all have the same large-strain strength.

Figure 6.36 shows the undrained behavior of three soil specimens consolidated to the same effective confining pressure, but at three different densities; Specimen B in Figure 6.35 is under the same initial conditions as Specimen B in Figure 6.36. Specimen D, which is denser than Specimen B and initially below the SSL, exhibits dilative behavior with decreasing pore pressure and increasing effective stress. Specimen D eventually reaches steady state conditions at a high effective stress level and therefore develops a high shearing resistance at very large strains. Specimen E, on the other hand, is loose and develops high positive excess porewater pressure as it moves toward the steady state where it strains with a low effective stress and low shearing resistance.

The SSL is relatively flat for most sands, which means that small differences in void ratio correspond to large differences in steady state effective stress (and, hence, steady state strength). Graphically, it means that the position of the steady state point on the failure surface moves quickly away from the origin as the void ratio of the soil decreases. Figure 6.37 illustrates the effects of soil density on the position of the steady state point and on effective stress path behavior in a three-dimensional framework.

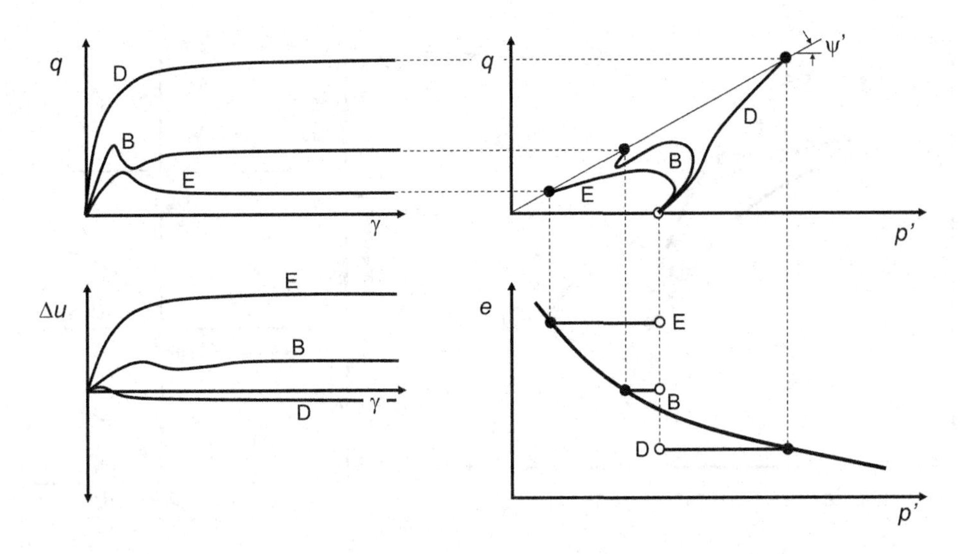

FIGURE 6.36 Illustration of the undrained behavior of three triaxial test specimens at the same initial effective stress but at different void ratios. Since all specimens have different void ratios, all reach different steady state conditions when subject to undrained shearing.

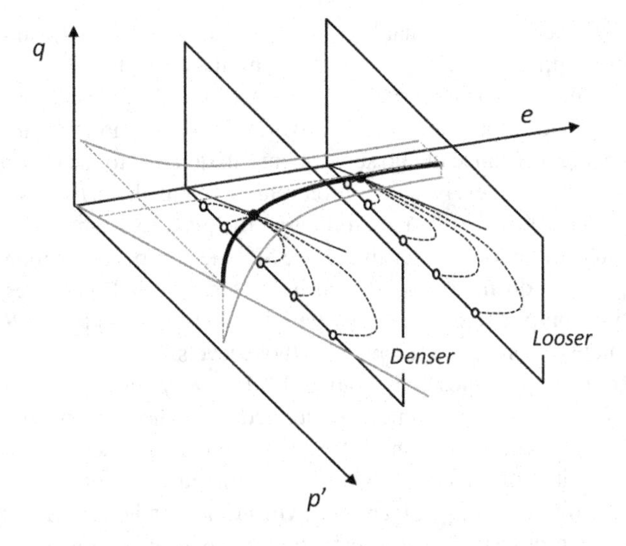

FIGURE 6.37 Three-dimensional steady-state line showing projections (in gray) on e-τ plane, e-σ' plane, and τ-σ' plane. A similar plot can be developed using stress parameters q and p' instead of τ and σ'.

Within a soil layer deposited in a consistent manner, the void ratio will decrease as the effective stress increases with depth. Yoshimine and Ishihara (1998) presented a framework that describes the relationship between the quasi-steady state and steady state at different initial effective stress levels. Figure 6.38 illustrates this framework schematically, and also shows the relative position of the initial consolidation line (ICL) as typically observed in laboratory tests on liquefiable soils. The SSL is generally somewhat steeper than the ICL and crosses the ICL at a relatively low effective stress level. If an element of soil is on the ICL at a lower initial effective stress (Case A), the sample will initially contract but then begin to dilate after reaching the quasi-steady state line (QSSL); the sample will continue to dilate until it reaches the SSL. Since there is no local minimum in the shearing resistance, phase transformation occurs without the development of a quasi-steady state. An

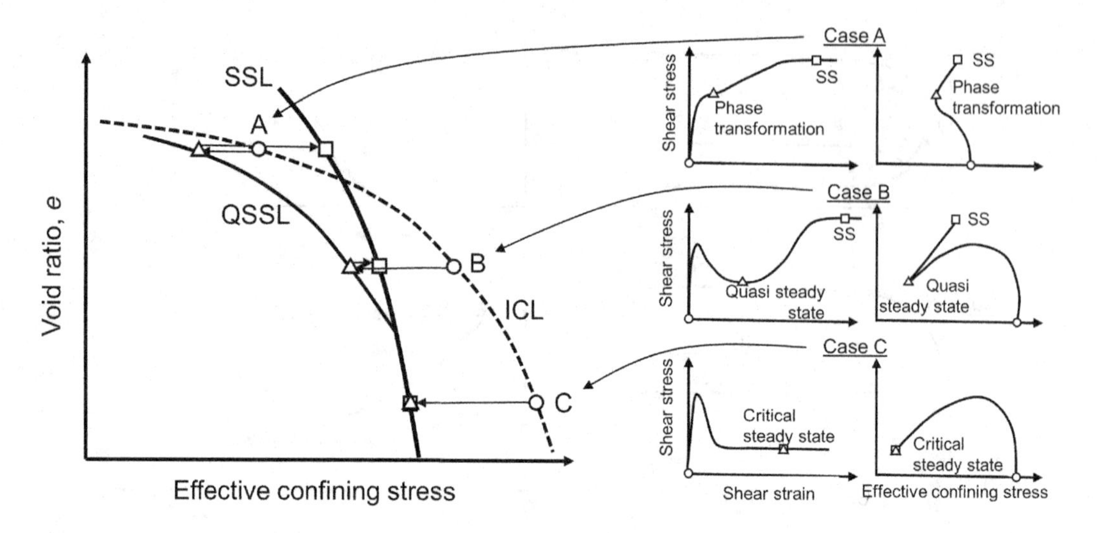

FIGURE 6.38 Stress-strain and stress path behavior for elements of soil consolidated to three different initial states and then subject to undrained monotonic shear. (After Yoshimine and Ishihara, 1998; used with permission of K. Ishihara.) Note that stress scales are not equal for the three cases – steady state strength would be highest for Case C and lowest for Case A.

element of soil consolidated to a somewhat higher initial effective stress (Case B), and hence lower void ratio, would exhibit similar behavior, but would exhibit a peak shearing resistance at low strain followed by a local minimum shearing resistance at the QSSL, followed then by dilation to the SSL. An element of soil at higher initial effective stress (Case C) would show strain-softening following a low-strain peak shearing resistance but no subsequent dilation due to the coincidence of the QSSL and SSL at high effective stress levels, a condition referred to as the critical steady state (CSS) by Yoshimine and Ishihara (1998); no dilatancy following the peak resistance is observed in the critical steady state. Because the ICL is flatter than the SSL, the soil becomes more contractive as the effective confining pressure during consolidation increases. The ICL becomes steeper as the soil approaches the limiting compression curve at very high effective stress levels (Robertson, 2017), so the contractiveness may not continue to increase at those levels.

The steady state framework proposed by Poulos (1981) is very similar to the critical state framework used for many years in soil mechanics. As defined by Poulos, the steady state is said to differ from the critical state by the formation of a "flow structure," the actual presence or absence of which is virtually impossible to verify conclusively. For the purposes of understanding soil behavior, however, the distinction is not significant and will not be made here. The term "steady state" is most commonly used in describing the liquefaction behavior of sands, and will be used in that context in this book. The term "critical state" will be used in other contexts. Both frameworks allow stress-strain and stress path behavior to be understood over a broad spectrum of soil densities and effective confining pressures.

6.4.3.5 Shearing Behavior of Soil Mixtures and Intermediate Soils

As discussed previously, when working with soils of mixed grain sizes, particularly of coarse- and fine-grained soils, it is critical to understand whether the soil behavior is governed by the coarse fraction ($FC < FC_L$) or fines fraction ($FC > FC_L$). When the coarse fraction controls, the soil behavior will follow the patterns for sands described above. When the fines fraction controls, the soil behavior will be controlled by the characteristics of the fines, which depend on their mineralogy and plasticity as described in Section 6.3.

Fines-controlled soils comprised of silts can range from being nonplastic (coarse silts) to highly plastic (finer silts with high liquid limits). Their behavior under monotonic shearing is influenced

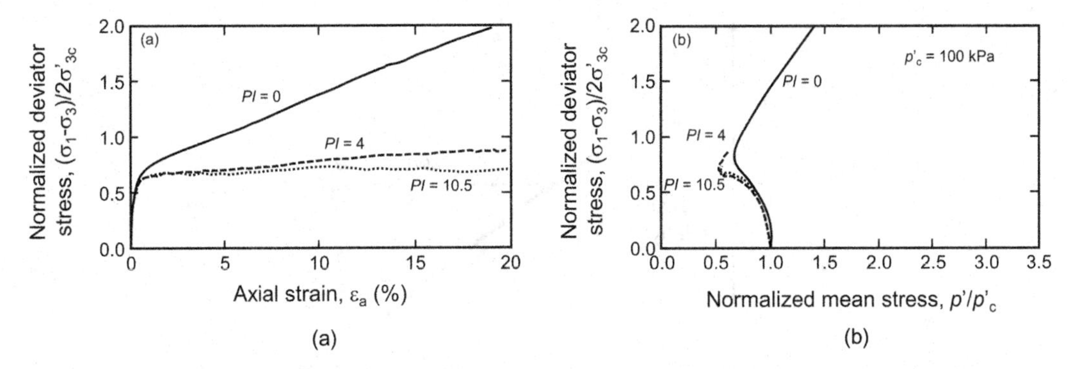

FIGURE 6.39 Normalized response of silts of different plasticity: (a) stress-strain behavior, and (b) effective stress path behavior (Boulanger and Idriss, 2004).

by their plasticity. Nonplastic silts tend to behavior much like sands, although their permeabilities are lower due to their smaller particle (hence, void) size. Nonplastic silts exhibit relatively strong volume change tendencies compared to more plastic silts – they are more contractive when loose and more dilative when dense. Normalized stress-strain curves and stress paths for triaxial compression tests on three silts from an aggregate tailings pond prepared by sedimentation from a slurry (Romero, 1995) are shown in Figure 6.39. The silts had plasticity indices of 0, 4, and 10.5 with 81%–87% passing the No. 200 sieve. The $PI = 0$ specimen behaved very much like a sand, i.e., it showed an initially contractive response followed by strongly dilative behavior throughout the latter stages of the test. The $PI = 10.5$ specimen behaved very much like a clay, showing a modest degree of contraction all the way to failure with no dilation and an essentially constant shearing resistance at large strains. The intermediate $PI = 4$ specimen showed intermediate behavior, i.e., contraction followed by weak dilation, that was closer to that of the $PI = 10.5$ specimen than the $PI = 0$ specimen. These results indicate that plastic soils exhibit weaker volume change tendencies than nonplastic soils during shearing.

6.4.3.6 Soil State

The actual behavior of soils depends on the proximity of their initial state to the critical (or steady) state line (Roscoe and Pooroshasb, 1963). For clays, that proximity is affected by stress history (e.g., *OCR*) while it is primarily affected by density (or void ratio) and effective stress level for sands. Geotechnical engineers use the term *relative density* to describe the density level of an element of granular, coarse-grained soil. The relative density of a soil with void ratio, e, is defined as

$$D_r = \frac{e_{max} - e}{e_{max} - e_{min}} \times 100\% \tag{6.31}$$

where e_{max} and e_{min} are the maximum and minimum void ratios of the soil. The relative density, therefore, provides a linear scale from the loosest ($D_r = 0$, or $e = e_{max}$) to densest ($D_r = 100\%$, or $e = e_{min}$) conditions. Soil densities are often described qualitatively with respect to relative density ranges: very loose ($D_r < 15\%$), loose ($15\% \leq D_r < 35\%$), medium dense ($35\% \leq D_r < 65\%$), dense ($65\% \leq D_r < 85\%$), and very dense ($D_r \geq 85\%$). The *void ratio range*, $e_{max} - e_{min}$, varies for soils of different gradation, typically increasing with decreasing mean grain size and increasing fines content (Cubrinovski and Ishihara, 1999, 2000). It also provides an indirect indication of volume change potential, which can affect the potential for triggering of liquefaction and the mobility of soils for which liquefaction has been triggered.

The nature of the SSL illustrates the limited applicability of absolute measures of density, such as void ratio and relative density, for characterization of sand behavior. As illustrated in Figure 6.40,

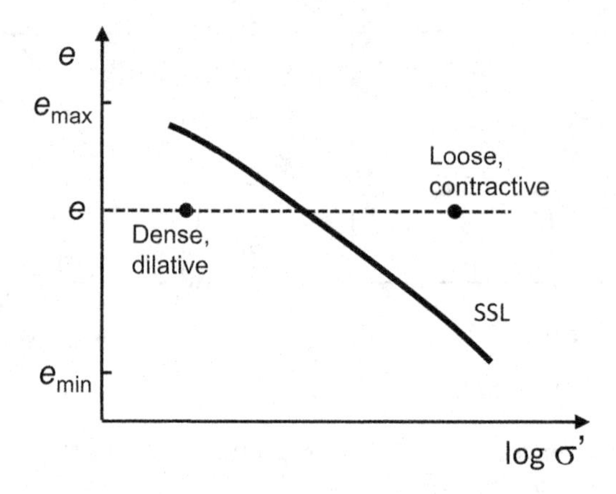

FIGURE 6.40 Illustration of two elements of soil at the same relative density, but for which very different types of behavior would be expected.

an element of soil at a specific void ratio (hence a particular density and relative density) can be contractive under a high effective confining pressure but dilative at a low effective confining pressure.

Using concepts of critical state soil mechanics, Been and Jefferies (1985) hypothesized that the behavior of a cohesionless soil was more closely related to the proximity of its initial state to the SSL than to absolute measures of density. In other words, soils in states located at the same distance from the SSL were expected to exhibit similar behavior. Been and Jefferies defined the *state parameter* as

$$\psi = e - e_{ss} \tag{6.32}$$

where e_{ss} is the void ratio of the SSL at the same effective confining pressure the soil experiences at void ratio e. When the state parameter is positive (Figure 6.41), the state plots above the SSL and the soil exhibits contractive behavior upon monotonic shearing. States plotting below the SSL have negative state parameters and exhibit dilative behavior for $\psi < -0.05$. The magnitude of the state parameter indicates the degree of contractiveness or dilatancy that the soil will exhibit – soils with high state parameter values will be highly contractive and soils with highly negative state parameters will be highly dilative.

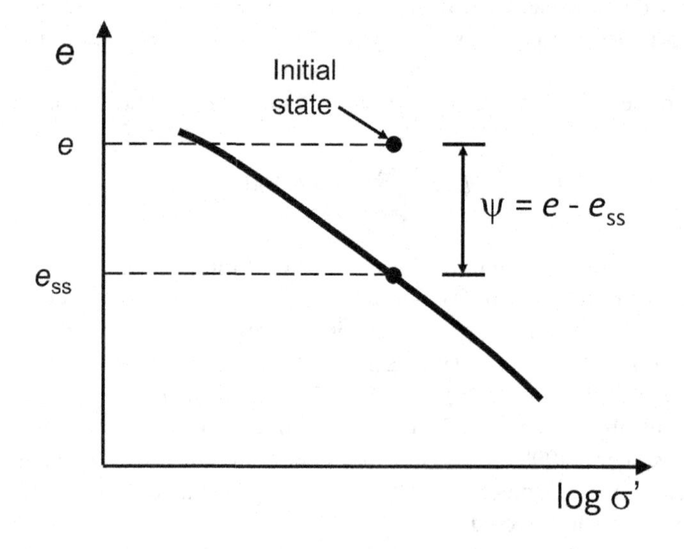

FIGURE 6.41 The state parameter: (a) basic definition and (b) illustration of suitability for characterizing volume change behavior.

Konrad (1988) examined experimental measurements of dilation rate (the ratio of incremental volumetric strain to incremental axial strain) in triaxial tests for sands with particle shapes ranging from sub-rounded to angular and found that normalizing the state parameter by $e_{max} - e_{min}$ provided improved prediction of dilatancy at $\psi < 0$ for different sands.

Bolton (1986) evaluated the dilatancy characteristics of 17 sands tested in triaxial and plane strain shear and proposed a *relative dilatancy index*

$$I_R = D_r\left(Q - \ln p'\right) - R \tag{6.33}$$

where p' = mean effective stress in kPa and Q and R are empirical constants that can be determined for individual soils from laboratory testing. The value of Q controls the slope and curvature of the implied SSL at very high effective stress levels where particle fracture may occur. Boulanger (2003b) indicated that Q ranges from 9 to 11 for silica sands with many modeled well with $Q = 10$. The value of R controls the position and slope of the SSL at lower effective stress levels; Bolton (1986) indicated that $R = 1$ provided good fits to available data. For a specific sand, accurate values of Q and R can be obtained by laboratory testing. In the absence of such testing, reasonable estimates of the average behavior of silica sands can be made using $Q = 10$ and $R = 1$.

The relative dilatancy index is positive for dense, dilative soils, negative for loose, contractive soils, and zero for soils that are neither contractive nor dilative. Using those values, and noting that $I_R = 0$ for the steady state condition of zero dilatancy, the relative density at the steady state can be seen to depend on mean effective stress, i.e.,

$$D_{r,ss} = \frac{e_{max} - e_{ss}}{e_{max} - e_{min}} = \frac{1}{10 - \ln p'} - D_r \tag{6.34}$$

Solving Equation (6.34) for e_{ss} and Equation (6.31) for e, substituting into Equation (6.32), and using Konrad's (1988) observation that $\psi/(e_{max} - e_{min})$ correlated well to shear behavior for a wide range of sands, Boulanger (2003a) developed the *relative state parameter index*

$$\xi_R = \frac{R}{Q - \ln\left(100p' / p_a\right)} - D_r \tag{6.35}$$

where p' is mean effective stress and p_a is atmospheric pressure in the same units as p'. The relative state parameter index, therefore, is consistent with steady state and state parameter concepts. Like the state parameter, the relative state parameter index is positive for contractive soils, negative for dilative soils, and zero for soils at the critical state. Because parameters R and Q generally fall within relatively narrow ranges for typical silica sands, ξ_R relates volume change characteristics to commonly measured quantities (relative density and mean effective stress). In this sense, it offers a practical and useful alternative to the state parameter when direct measurement of the state parameter is not feasible. Both the state parameter and relative state parameter index provide useful frameworks for characterizing the volume change potential of a soil loaded under drained conditions and, more importantly, the pore pressure generation potential under the undrained loading conditions that occur during earthquake shaking.

6.4.4 Response to Cyclic Shear Stresses

Prior to an earthquake, an element of soil in the field exists under some initial state of static stress that will affect its seismic response. The seismic waves produced by an earthquake superimpose dynamic stresses upon the existing static stresses, and the response of the soil will depend on both the static and dynamic stresses. The response of the soil to dynamic stresses is not unrelated to the response to static stresses, but three important factors must be considered in understanding response to dynamic loading: (1) dynamic stresses are applied more rapidly than the static stresses, (2) dynamic stresses interact with static stresses so both affect dynamic response, and (3) dynamic stresses include stress reversals, i.e., they increase and decrease and change directions repeatedly.

Since the dynamic stresses induced by earthquakes are applied quickly, earthquake loading in saturated soils is almost always applied under undrained conditions – the only exception would be for extremely permeable soils such as clean gravels that are unbounded by less permeable soils. Even clean sands can generally be considered to be loaded under undrained conditions by earthquake shaking. Dry or partially saturated soils are effectively drained, in that volume change can occur during dynamic loading.

Figure 6.42 shows two elements of soil, one beneath level ground and the other at the same depth beneath sloping ground. The element beneath level ground has no static horizontal shear stress and, therefore, no inherent tendency to deform preferentially to the left or to the right – symmetric dynamic shear stresses would not tend to cause significant accumulation of deformations in either direction. The element below sloping ground, however, has a static shear stress acting on the horizontal plane. The dynamic shear stress is then superimposed upon the static shear stress. When the dynamic shear stress reaches it maximum value, say at Point A, the soil has mobilized a greater

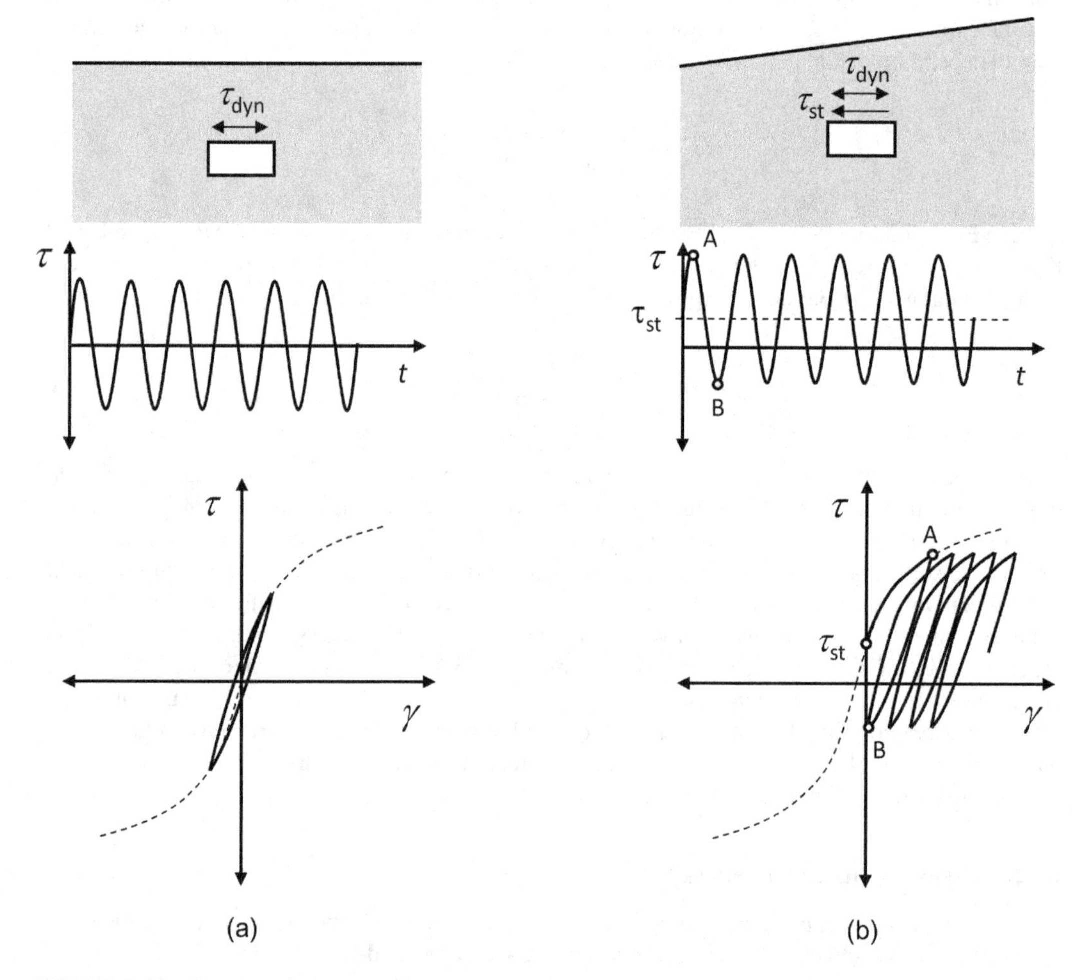

FIGURE 6.42 Stress-strain response due to cyclic loading for elements of soil under: (a) level ground with no static shear stress, and (b) sloping ground with positive static shear stress.

fraction of its shear strength than it has at its minimum value (Point B), and consequently exhibits more nonlinear response for increasing (positive) shear stresses than for decreasing (negative) shear stresses. As a result, even symmetric dynamic shear stresses tend to cause an accumulation of permanent shear strain in the element, and, hence, permanent deformation of the slope. These strains, and the deformations that result from them, can accumulate *even though the shear strength of the soil is never fully mobilized.*

Stress reversals involve changes in stiffness and the dissipation of energy, i.e., damping, through inelastic soil behavior. Stress reversals also affect dilatancy, so they will affect pore pressure generation in saturated soils.

6.4.4.1 Ground Response and Ground Failure

Most geotechnical earthquake engineering problems can be divided into *ground response* and *ground failure* categories. As described in Section 3.4.3, ground response refers to the effects of local site conditions on the characteristics (e.g., amplitude, frequency content, and/or duration) of an earthquake ground motion without the development of significant permanent deformations. Ground failure refers to situations in which significant permanent deformations – horizontal, vertical, or both – develop during and/or after strong shaking.

Ground response problems typically involve soil behavior at low to moderate strain levels, although relatively large strains can develop in soft soils subjected to strong shaking or in soils that develop significant pore pressure. Ground failure problems involve high levels of nonlinearity or even mobilization of the full shear strength of the soil, and therefore involve soil behavior at high strain levels.

The response of soils to dynamic loading is continuous over a broad range of strains and the boundary between ground response and ground failure issues is not distinct. Nevertheless, these broad classes of problems usually involve soil behavior at different levels of strain. At the same time, a number of procedures for measurement of dynamic soil properties are limited to different strain levels. As a result, it is convenient to describe the measurement of many soil properties relevant to dynamic response, and the behavior(s) those properties control, in terms of different strain ranges.

6.4.4.2 Cyclic Loading

The behavior of soils under cyclic loading conditions has been investigated through the testing of individual elements of soil under controlled boundary conditions. Such tests typically consolidate an element of soil to a known level of effective stress and then apply multiple cycles of constant amplitude, harmonic shear stress. Consider an element of dry soil subjected to harmonic loading in cyclic simple shear (Section 6.5.4.3). Upon initial loading, the shear modulus is at its highest value, G_{max}. In the first quarter-cycle (Figure 6.43a), the shear stress increases monotonically, and the stress-strain curve follows the nonlinear path it would follow in a static loading test. Since no porewater pressure is developed, the vertical effective stress remains constant. As the shear stress increases, the tangent shear modulus gradually decreases. Upon reaching the peak shear stress at Point A, the shear stress and shear strain both reverse (Figure 6.43b), and the shear modulus instantaneously increases to its maximum value, G_{max}. Since the soil is inelastic, its path upon unloading is different than its path upon original loading. As the shear stress drops and eventually changes sign, the tangent shear modulus continues to decrease as the next stress reversal, at Point B, is approached. At Point B, the direction of loading changes again and the tangent shear modulus instantaneously returns to G_{max}. Through this process, the stiffness of the soil can be seen to be affected not by the current strain level, but by the difference between the current strain and the strain at the most recent reversal. Upon reloading, the stress-strain curve begins to define a hysteresis loop that is completed when the shear stress reaches its maximum value at Point C (Figure 6.43c). For this idealized nonlinear material, Point C is at the same location as Point A.

If the soil had exhibited *cyclic degradation*, i.e., softening under constant amplitude cyclic loading, the strain at Point C in Figure 6.43c would have been slightly greater than the strain at Point

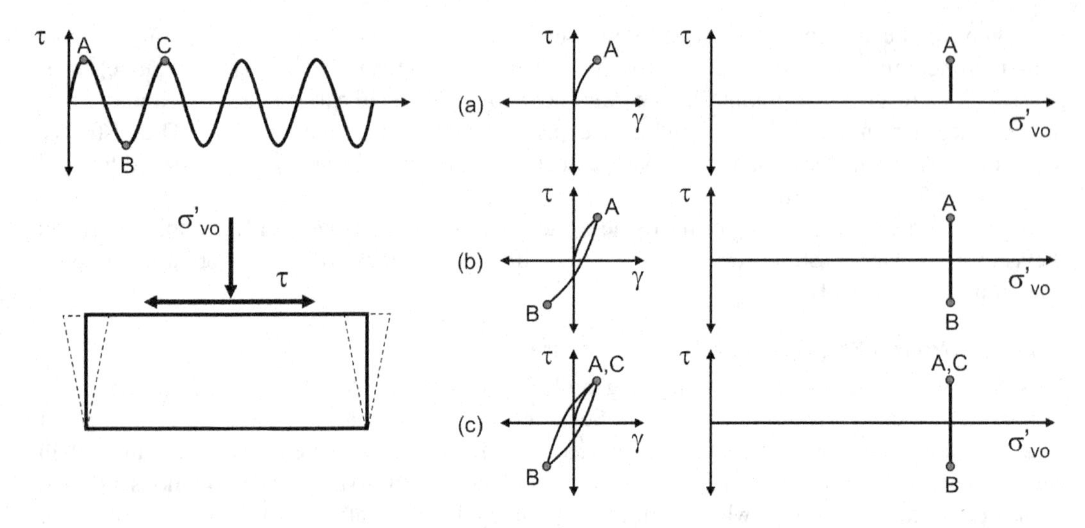

FIGURE 6.43 Response of element of non-degrading soil subjected to harmonic horizontal shear stress: (a) after one-quarter cycle, (b) after 3/4-cycle, and (c) after 5/4-cycle. Additional cycles of loading would follow the same stress-strain curve.

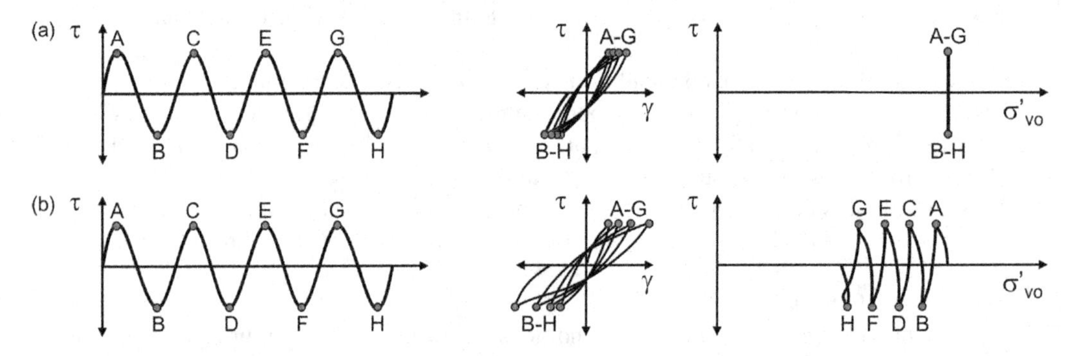

FIGURE 6.44 Response of element of degrading soil subjected to harmonic horizontal shear stress: (a) degradation with constant effective stress and (b) degradation due to pore pressure generation.

A. Successive cycles would lead to increasing strain amplitude (Figure 6.44a) even under constant effective stress conditions. If the soil exhibited *rate dependence*, the shear stress (at a given strain level) would increase with increasing strain rate (or, for harmonic loading, with increasing frequency). Finally, if the soil exhibited volumetric-deviatoric coupling, the applied shear stresses would tend to cause volumetric strain (typically contraction for low strain levels). If the soil was saturated and loaded to shear strains greater than the volumetric threshold shear strain (Section 6.4.3.1) under undrained conditions, this tendency would result in the development of excess porewater pressure and reduce the effective stress (Figure 6.44b) and stiffness of the soil.

Under transient loading conditions, the behavior of the soil is more complex. The response to small stress reversals superimposed upon larger stress cycles (Figure 6.45a) shows momentary periods of high stiffness in the small unloading-reloading loops that end when the smaller reloading loop comes back to the larger loop. Figure 6.45b illustrates the complicated response of an element to transient loading.

Figure 6.46 illustrates the behavior of a hypothetical element of saturated soil in the first part of the first cycle of undrained loading. In the first quarter-cycle of loading, the effective stress path follows the monotonic loading stress path, moving to the left as the contractive nature of the soil causes

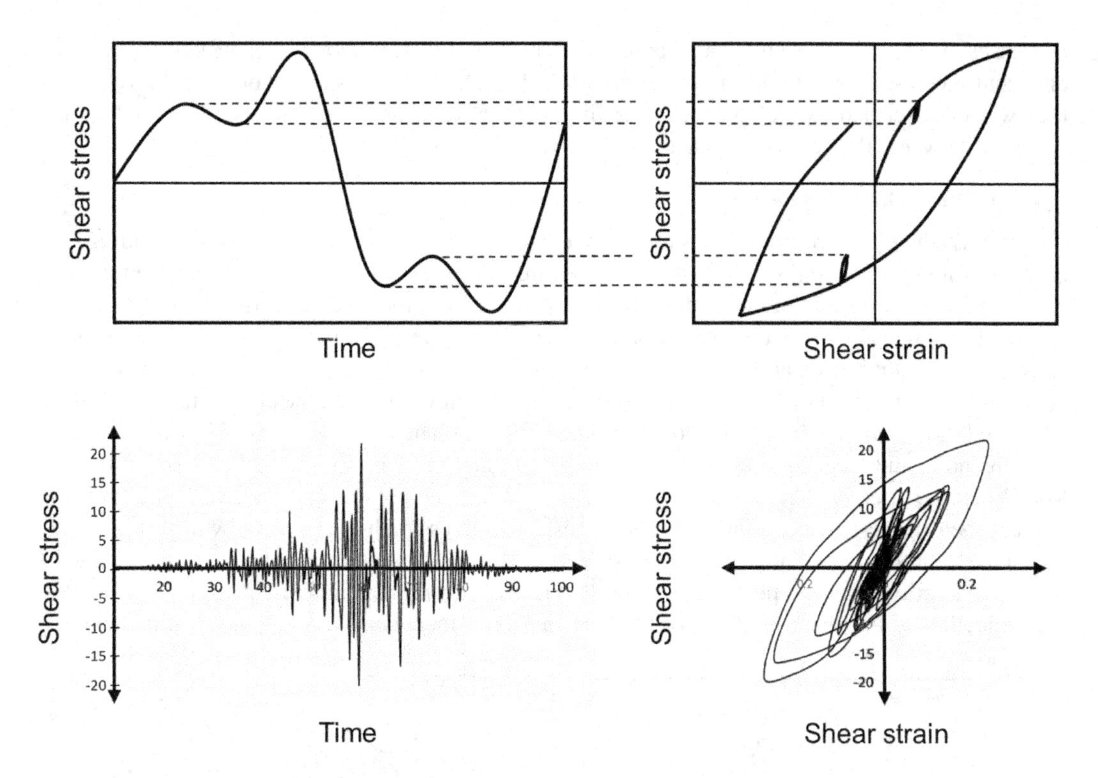

FIGURE 6.45 Response of soil to transient loading conditions: (a) small shear stress superimposed on large stress and (b) transient loading.

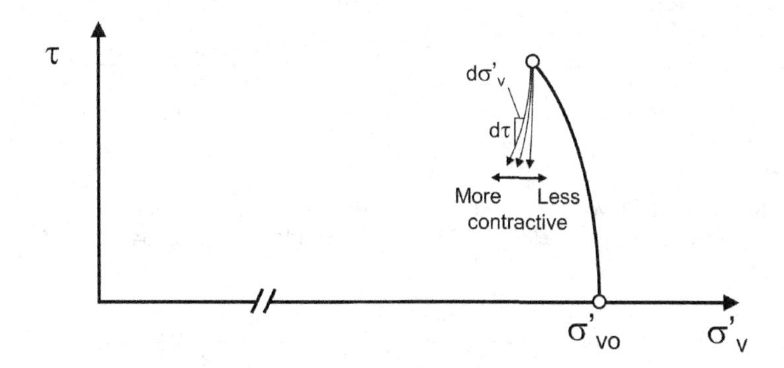

FIGURE 6.46 Influence of contractiveness on effective stress behavior following stress reversal.

the pore pressure to increase. Upon reversal of loading, the shear strain will decrease but the pore pressure will continue to increase so the stress path continues to move to the left. The slope of the stress path immediately after reversal (i.e., $d\sigma'_v/d\tau$) will depend on how contractive the soil is – if it is highly contractive (i.e., well above the CSL like a normally consolidated clay or loose sand), $d\sigma'_v/d\tau$ will be high and the stress path will move quickly to the left; if the tendency for contraction is weak, $d\sigma'_v/d\tau$ will be low and the stress path will be nearly vertical. Therefore, a more contractive soil will generate pore pressure more rapidly than a less contractive soil when subjected to the same level of cyclic loading.

The application of a series of stress cycles will therefore cause an incremental buildup of porewater pressure in each cycle, and the stress path will move to the left as the effective stress decreases. Experiments indicate that the rate of pore pressure generation (per cycle) is initially high and then decreases. As the effective stress path approaches the phase transformation line, the behavior of clay and sand specimens tend to differ and, although understanding of this behavior is still developing, they will be described separately below to help develop a framework for understanding, and distinguishing between, the behavior of each.

6.4.4.3 Behavior of Clays

Figure 6.47 shows the response of a normally consolidated clay to undrained cyclic loading. As described in the preceding section, the pore pressure increases quickly in response to the first few cycles and at a slower rate in succeeding cycles. In later cycles, the mean effective stress transitions from only decreasing to both increasing and decreasing within a given loading cycle. The level of effective stress fluctuation at the stage of the test where the stress path reaches the failure envelope is moderate – the minimum value is about half the maximum value – and the change in soil stiffness as the soil dilates up the failure envelope is relatively small. The resulting hysteresis loops are relatively broad (indicating significant energy dissipation) and the stiffness never becomes extremely low. When the effective stress reaches its minimum value, a clayey soil will tend to maintain significant shearing resistance. In many cases, the effective stress path of a clay soil will stabilize and repeat itself, producing a limiting pore pressure level and, with it, a limiting shear strain amplitude.

The generation of excess pore pressure in clays results in a reduction of effective stress that leads to a reduction of the shear modulus of the soil and to an increasing level of shear strain with an

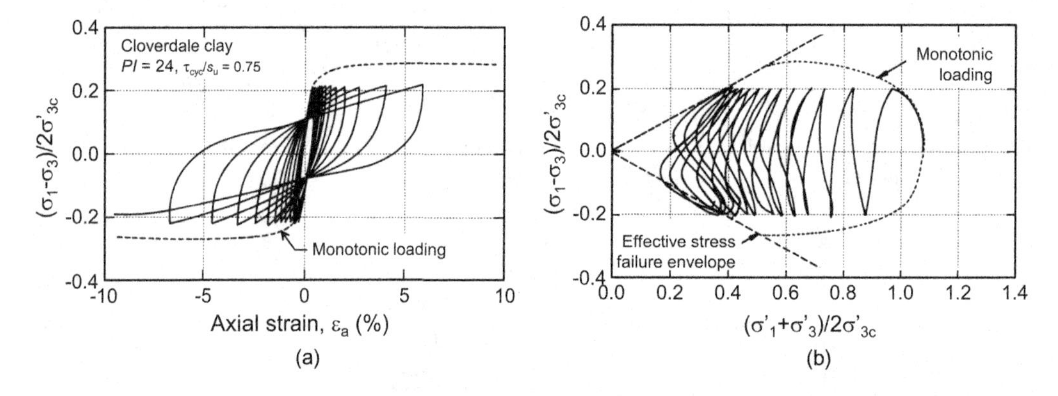

FIGURE 6.47 Response of Cloverdale clay to cyclic loading at 75% of undrained strength: (a) stress-strain behavior and (b) effective stress path behavior. (From Idriss and Boulanger, 2004 (After Zergoun and Vaid, 1994).)

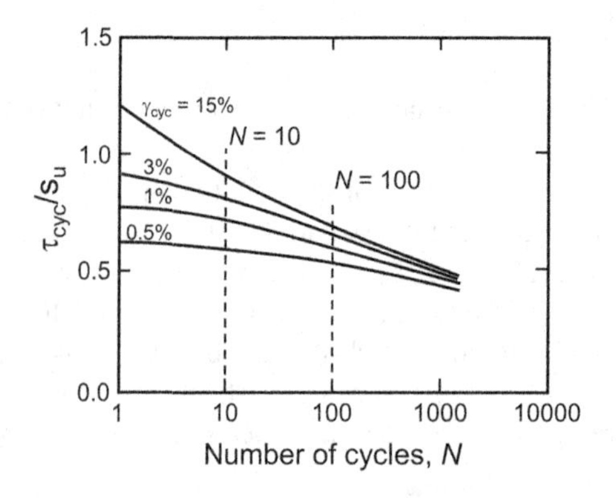

FIGURE 6.48 Variation of cyclic strain amplitude with increasing number of loading cycles for normally consolidated Drammen clay subjected to direct simple shear (DSS) testing. (After Anderson, 2015; used by permission of Taylor and Francis.)

increasing number of loading cycles. Figure 6.48 shows this behavior by plotting the ratio of applied cycle shear stress (τ_{cyc}) to undrained shear strength (S_u) from cyclic loading. The contours indicate different levels of cyclic shear strain, γ_{cyc}, and their downward slope is a manifestation of cyclic degradation. For a given value of τ_{cyc}/S_u, cyclic degradation causes the cyclic shear strain to increase with an increasing number of loading cycles.

Cyclic degradation can be simulated in a constitutive model (e.g., Pestana and Biscontin, 2000; Ni et al., 2015), but it has historically been treated in a simpler, but less direct manner. The reduction of stiffness has customarily been expressed in terms of a *degradation index* defined for soils subjected to constant-strain-amplitude cyclic loading as

$$\delta = \frac{G_N}{G_{N=1}} \tag{6.36}$$

where G_N is the secant shear modulus for the Nth cycle of loading and $G_{N=1}$ is the secant modulus for the first cycle. The value of δ has been found to decrease with increasing number of cycles in a manner that can be described by

$$\delta = N^{-t} \tag{6.37}$$

where t is a soil degradation parameter (Idriss et al., 1976; Idriss et al., 1978) that is usually modeled as a function of strain amplitude and OCR. Matasovic and Vucetic (1995) represented the degradation parameter as

$$t = s(\gamma_c - \gamma_{tv})^r \tag{6.38}$$

where γ_c is the cyclic shear strain amplitude, γ_{tv} is the volumetric threshold shear strain (Equation 6.38 only applies when $\gamma_c > \gamma_{tv}$), and s and r are coefficients that vary with OCR. Combining Equations (6.36), (6.37), and (6.38) gives

$$G_N = G_{N=1} \cdot N^{-s(\gamma_c - \gamma_{tv})^r} \tag{6.39}$$

6.4.4.4 Behavior of Sands

The behavior of sands subjected to cyclic loading is complex, but must be understood before moving on to soil liquefaction and its effects, which are discussed in detail in Chapter 9. As discussed previously, the behavior of sands is dominated by volume change tendencies, which are related to density and effective stress levels, or soil *state* (Section 6.4.3.6). These tendencies affect the behavior of both dry and saturated sands.

Dry Sand

A dry sand subjected to cyclic loading will tend to densify regardless of whether it is loose or dense – the void ratio will move to the critical void ratio if the soil develops very large, uni-directional shear strains but it will densify to void ratios lower than critical if the strains do not become very large. A loose, dry sand will densify more (and faster) than a dense, dry sand, but both will tend to densify under cyclic loading.

Figure 6.49 shows the response of a dry sand subjected to constant-amplitude, stress-controlled, cyclic simple shear loading. In the initial cycle of loading, the soil exhibits contractive behavior and is relatively soft as indicated by the large shear strains that develop in response to the applied shear stress (Figure 6.49a). Because the sand is dry, the vertical effective stress remains constant. Figure 6.49b shows how the shear and volumetric strain are related. In the early cycles of loading, the shear strains are large and the incremental volumetric strains are relatively large and purely contractive. With increasing numbers of loading cycles, however, the shear strain amplitude decreases and the incremental volumetric strain in each cycle decreases as the soil becomes denser.

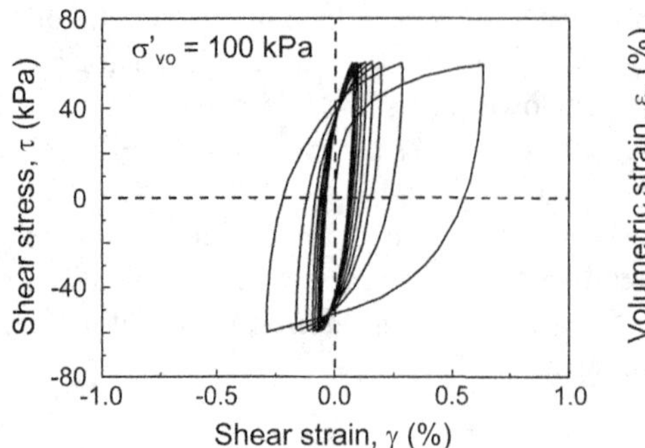

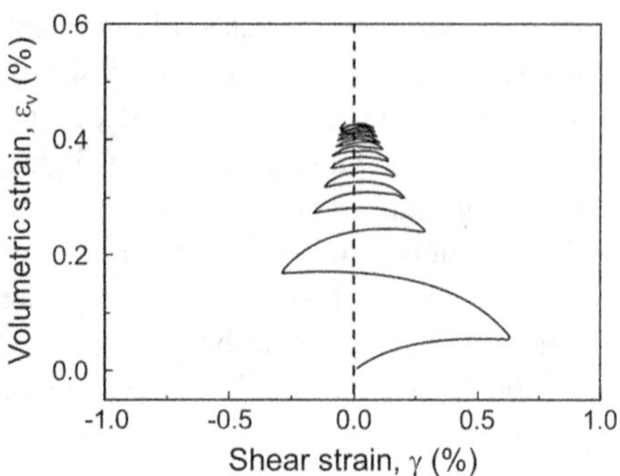

FIGURE 6.49 Stress-controlled drained cyclic torsional shear test with ten cycles of loading applied to Toyoura sand with 38% initial relative density. (After Gao and Zhao, 2015; used with permission of ASCE.)

After about 2–3 cycles, the volumetric strain can be seen to change direction at larger relative shear strain amplitudes – here the soil is dilating due to phase transformation behavior. Upon reversal after dilation, the soil contracts again until it dilates when the strain becomes large in the other direction. Eventually, the soil reaches a limiting volumetric strain and contracts and dilates repeatedly within each additional cycle of loading with no net volumetric strain at the end of the cycle.

As discussed subsequently in Sections 6.6.5.4 and 9.6.6.1, the type of volumetric strains illustrated in Figure 6.49 can occur throughout deposits of dry sand, or more generally any unsaturated soil, subjected to earthquake shaking in the field. Known as *seismic compression*, these volumetric strains, when integrated over the thickness of the unsaturated soil layer, can produce significant ground surface settlement. This type of settlement has led to damage in many past earthquakes.

Saturated Sand

The volume change tendencies that lead to densification of dry sand play a particularly critical role in the response of saturated sand to undrained cyclic loading. As discussed previously with respect to static loading, the tendency of a sand to densify, or contract, under drained loading causes the soil to generate positive excess pore pressure under undrained loading. That excess pore pressure reduces the effective stress, which softens and potentially weakens the soil.

Figure 6.50 shows a schematic illustration of the response of a saturated sand subjected to undrained, constant-amplitude, stress-controlled, cyclic simple shear loading. In the initial few cycles of loading, the soil is relatively stiff as indicated by the small shear strains that develop in response to the applied shear stress (Figure 6.50a and c). The tendency for contraction that produced rapid volumetric strain in the first few cycles of drained loading (Figure 6.49c) produces relatively rapid increases in excess pore pressure (hence, decreases in vertical effective stress), in the first few cycles of undrained loading (Figure 6.50b and d).

The conditions at the peak shear stress in different cycles are labeled in Figure 6.50. The relatively rapid increase in pore pressure can be seen in the effective stress path in Figure 6.50b – the horizontal position of the stress path indicates the value of the *pore pressure ratio*

$$r_u = \frac{\Delta u}{\sigma'_{vo}} \tag{6.40}$$

The pore pressure ratio in the 5th cycle of loading is about 0.26 but has only increased to 0.5 after 17 cycles of loading. The strain amplitude after 17 cycles, however, is still quite low, indicating that the stiffness has not decreased much even though the vertical effective stress has been cut in half. The pore pressure ratio reaches a value of 0.62 after 20 cycles; at 20 cycles, the strain amplitude is beginning to increase noticeably. In the 21st cycle, the effective stress path reaches the phase

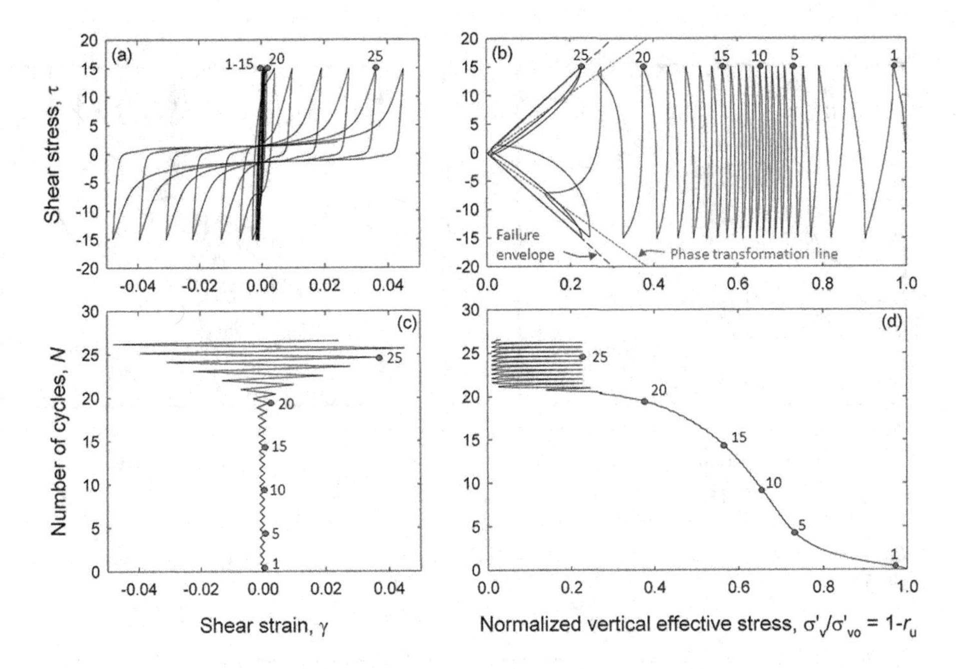

FIGURE 6.50 (a–d) Response of saturated sand subjected to constant-amplitude undrained cyclic simple shear testing.

transformation line at which point the behavior changes from contractive to dilative. It reaches this point at a high shear stress in the positive shear stress portion of the cycle and consequently doesn't dilate much before the shear stress reverses. In the negative portion of the 21st cycle, however, the soil is more contractive and thus reaches the PTL at a lower shear stress level. As the shear stress increases (becomes more negative), the soil dilates significantly, and then contracts and dilates even more significantly in the next loading cycle. In the 22nd cycle, the effective stress drops to virtually zero, a condition historically referred to as *initial liquefaction* (Seed and Lee, 1966). After about 23 cycles, the effective stress path reaches a constant state in which it remains, retracing itself in subsequent loading cycles. In each of these cycles, the soil strongly contracts and dilates, reaching a very low effective stress level twice in each loading cycle. At this stage of the test, the stress-strain curves (Figure 6.50a) reflect this behavior with very low stiffnesses being mobilized when the effective vertical stress is low, followed by a period of significant stiffening as the soil dilates and the effective stress path moves up the failure envelope. The fact that the soil momentarily has zero effective stress does not mean that it has no strength – it will develop shearing resistance as it strains and dilates and, if subjected to steadily increasing shear stress, will dilate until it reaches the critical state. Even though the effective stresses do not change from cycle to cycle after initial liquefaction has occurred, the strain amplitude continues to increase as the *fabric* of the soil is degraded.

The contractive-dilative behavior of a saturated soil subjected to undrained loading is often referred to as *cyclic mobility*. The large and rapid changes in stiffness associated with cyclic mobility can have a strong effect on the response of liquefiable soil profiles (Chapter 9).

6.4.4.5 Behavior of Intermediate Soils

Sands and clays exhibit fundamentally different behavior under cyclic loading conditions that appear to be related to their relative plasticities. Intermediate soils such as silts also exhibit behavior that is influenced by plasticity. Figure 6.51 shows stress-strain curves for silts with plasticity indices of 0 and 10.5. The nonplastic ($PI = 0$) silt showed strong volume change tendencies, reflected in effective stresses that dropped to very low values followed by strong dilation with accompanying stiffening. The stress-strain curve showed very low stiffness at low shear stress levels but rapidly

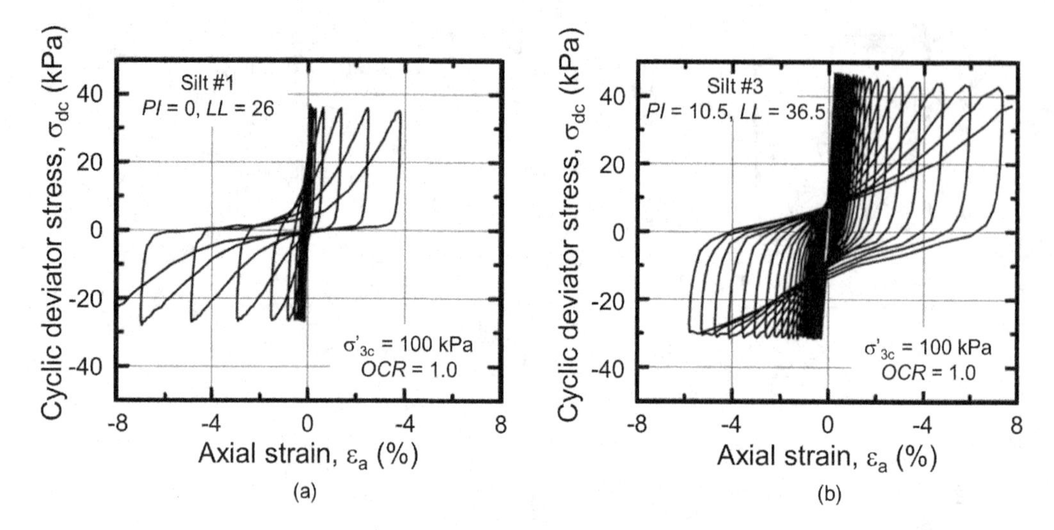

FIGURE 6.51 Stress-strain behavior in cyclic triaxial tests of (a) nonplastic silt and (b) plastic silt. (After Romero, 1995.)

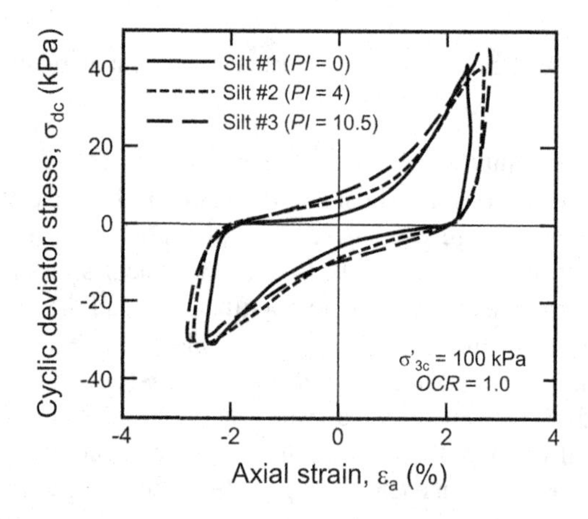

FIGURE 6.52 Hysteresis loops at approximately constant strain amplitudes from cyclic triaxial tests on silts of different plasticity. (After Boulanger and Idriss, 2004.)

increasing stiffness with increasing shear stress. The plastic silt, on the other hand, generated lower pore pressures and never became as soft as the nonplastic silt. The stress-strain loops of the plastic silt were much broader than those of the nonplastic silt, indicating a greater degree of energy dissipation, i.e., higher material damping (Section C.5.1). Figure 6.52 illustrates the effects of plasticity on stress-strain behavior; hysteresis loops extracted from tests on silts with *PI* values of 0, 4, and 10.5 at points in triaxial tests with similar strain amplitudes show differences in minimum stiffness, maximum rate of stiffening, and damping characteristics that vary with plasticity index.

6.4.5 Modeling Soil Behavior under Cyclic Loading

Prediction of the seismic response of soil deposits, and the structures located on or within them, requires the ability to predict the response of soils under cyclic loading conditions by means of dynamic response analyses. Such analyses are commonly performed for site response, soil-structure interaction, liquefaction, and seismic slope stability problems – each of which is the subject of the next four chapters of this book. A number of different approaches have been taken to modeling of the behavior of soils subjected to cyclic loading, but they can be broken down into two primary categories – equivalent linear and nonlinear models. Nonlinear models can be further subdivided into *cyclic response models* and *advanced constitutive models.* These models are described in relatively general terms here, both to introduce the different ways in which dynamic soil properties can be characterized in the remainder of this chapter and to support the description of numerical analyses oriented toward the topics of the next four chapters.

6.4.5.1 Equivalent Linear Models

Wave propagation, which is important for site response and other problems, is controlled by the stiffness and damping characteristics of a soil (Appendix C). While the nonlinear, inelastic behavior of soils is hysteretic in nature, it can be interpreted as an equivalent viscoelastic material with representative stiffness and damping characteristics. The representative stiffness is taken as the secant shear modulus, which approximates the average stiffness of the soil over a cycle of loading. The representative damping ratio is taken as the viscous damping ratio that dissipates the same amount of energy as dissipated by hysteresis in the actual soil. This characterization effectively represents the stress-strain curve of the soil by an equivalent ellipse, as shown in Figure C.22. This ellipse captures the first-order (stiffness and damping ratio) behavior well at low strain levels, but fails to capture details of soil behavior, such as the sharp changes in stiffness that occur at stress reversals. The reasonableness of the equivalent linear approximation decreases with increasing strain level.

Because stiffness decreases and damping increases with increasing strain amplitude, equivalent linear properties are strain-dependent. The highest value of shear modulus occurs at very small strains and is referred to as the maximum shear modulus, G_{max}. At these small strain levels, the secant shear modulus, G_{sec}, is equal to G_{max}, so the *modulus reduction ratio,* G_{sec}/G_{max} (more commonly written as G/G_{max}) is equal to 1.0. Nonlinearity causes that ratio to decrease with increasing strain amplitude in a manner displayed graphically by a *modulus reduction curve* (Section 6.6.1.2).

6.4.5.2 Cyclic Response Models

The nonlinear stress-strain behavior of soils can be explicitly represented by cyclic nonlinear models that seek to represent the actual stress-strain path during cyclic loading. Cyclic nonlinear models characterize the variation of the tangent modulus during loading. By forcing the tangent modulus to a value of zero as the shear stress approaches the shear strength of the soil, such models can prevent the prediction of shear stresses that exceed the shear strength of the soil.

A variety of cyclic nonlinear models have been developed; all are characterized by (1) a backbone curve and (2) a series of "rules" that govern unloading-reloading behavior, stiffness degradation, and other effects. A *backbone curve* is a monotonic stress-strain curve that the soil will follow upon initial loading; the backbone curve is defined for both positive and negative shear stresses. Some of these models have relatively simple backbone curves, such as the hyperbolic backbone curve of Equation (6.41), and only a few basic rules. More complex models may incorporate many additional rules that allow the model to better represent the effects of irregular loading, densification, pore pressure generation, or other effects. The applicability of cyclic nonlinear models, however, is generally restricted to a fairly narrow, albeit important, range of initial conditions and stress paths.

The performance of cyclic nonlinear models can be illustrated by a very simple example in which the shape of the backbone curve is described by $\tau = F_{bb}(\gamma)$. The shape of a hyperbolic backbone curve, for example, is tied to two parameters, the initial (low-strain) stiffness, G_{max}, and the (high-strain) shear strength of the soil, τ_{max}.

$$F_{bb}(\gamma) = \frac{G_{max}\gamma}{1+(G_{max}/\tau_{max})|\gamma|} \tag{6.41}$$

Other expressions [e.g., the modified hyperbolic backbone curve discussed in Section 6.6.3 or the Ramberg-Osgood model (Ramberg and Osgood, 1943)] can also be used to describe the backbone curve. Alternatively, backbone curves can be constructed from modulus reduction curves as indicated in Equation (6.75).

The quantities G_{max} and τ_{max} may be measured directly, computed, or obtained by empirical correlation. For the example model, the response of the soil to cyclic loading is governed by the following four rules:

1. For initial loading, the stress-strain curve follows the hyperbolic backbone curve given by Equation (6.41).
2. If a stress reversal occurs at a point defined by $(\gamma_{rev}, \tau_{rev})$, the stress-strain curve follows a path given by

$$\frac{\tau - \tau_{rev}}{2} = F_{bb}\left(\frac{\gamma - \gamma_{rev}}{2}\right) \tag{6.42}$$

   In other words, the unloading and reloading curves have the same shape as the backbone curve (with the origin shifted to the loading reversal point) but are enlarged by a factor of 2. These first two rules, which describe *Masing behavior* (Masing, 1926), are not sufficient to describe soil response under general cyclic loading. As a result, additional rules are needed.
3. If the unloading or reloading curve exceeds the maximum past strain and intersects the backbone curve, it follows the backbone curve until the next stress reversal.
4. If an unloading or reloading curve crosses an unloading or reloading curve from the previous cycle, the stress-strain curve follows that of the previous cycle.

Models that follow these four rules are often called *extended Masing models*. An example of the extended Masing model is shown in Figure 6.53. Cyclic loading begins at point A, and the stress-strain curve during initial loading (from A to B) follows the backbone curve as required by Rule 1. At point B, the loading is reversed and the unloading portion of the stress-strain curve moves away from B along the path required by Rule 2. Note that the initial unloading modulus is equal to G_{max}. The unloading path intersects the backbone curve at point C and, according to Rule 3, continues along the backbone curve until the next loading reversal at point D. The reloading curve then moves away from D as required by Rule 2, and the process is repeated for the remainder of the applied loading. Although this model is very simple and is expressed only in terms of effective stresses, it inherently incorporates the hysteretic nature of damping and the strain-dependence of

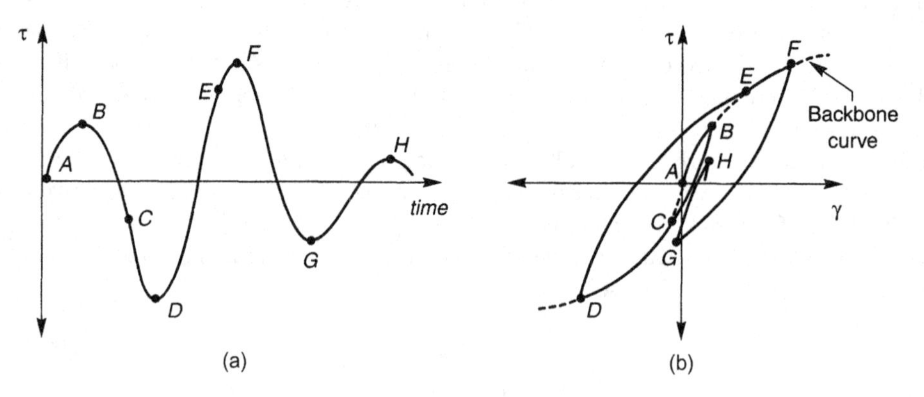

FIGURE 6.53 Extended Masing rules: (a) variation of shear stress with time; (b) resulting stress-strain behavior (backbone curve indicated by dashed line).

the shear modulus and damping ratio. Because the shape of the hysteresis loop is controlled by the shape of the backbone curve, the damping ratio is tied to the backbone curve. However, comparisons with laboratory data show that a backbone curve that matches the experimentally observed strain dependence of moduli tends to significantly overpredict damping at moderate to large strain levels and underpredict it at small strain levels; these issues are addressed further in Section 7.5.6.4. Other unloading-reloading models are also available (e.g., Iwan, 1967; Finn et al., 1977; Vucetic, 1990). To avoid spurious response at very low strain levels, some cyclic nonlinear models require the addition of a small amount of low strain damping.

Note that the cyclic nonlinear model does not require the shear strain to be zero when the shear stress is zero. As a result, cyclic nonlinear models can produce permanent shear strains, i.e., non-zero shear strains that remain after cyclic loading has ceased. Permanent strains can be caused by asymmetric cyclic loading or, more commonly, cyclic loading superimposed on static shear stresses such as those that exist below sloping ground or in the vicinity of foundations.

6.4.5.3 Advanced Constitutive Models

The most accurate and general methods for the representation of soil behavior are based on advanced constitutive models that use basic principles of mechanics to describe observed soil behavior for (1) general initial stress conditions, (2) a wide variety of stress paths, (3) rotating principal stress axes, (4) cyclic or monotonic loading, (5) high or low strain rates, and (6) drained or undrained conditions.

Such models generally require a yield surface that describes the limiting stress conditions for which elastic behavior is observed, a hardening law that describes changes in the size and shape of the yield surface as plastic deformation occurs, and a flow rule that relates increments of plastic strain to increments of stress. The Cam-Clay (Roscoe and Schofield, 1963) and modified Cam-Clay (Roscoe and Burland, 1968) models were among the first of this type. Improvements in the prediction of shear strains, particularly under cyclic loading conditions, have resulted from the use of multiple nested yield loci within the failure surface (Mroz, 1967; Prevost, 1977) and the development of bounding surface models (Dafalias and Popov, 1979; Dafalias and Herrmann, 1982) which incorporate a smooth transition from elastic to plastic behavior. Detailed treatment of such advanced constitutive models is beyond the scope of this book. The interested reader can refer to a number of sources, including Desai and Siriwardane (1984), Dafalias and Herrmann (1982), Wroth and Houlsby (1985), Wood (2004), Hicher and Shao (2008).

Discussion

Although advanced constitutive models allow considerable flexibility and generality in modeling the response of soils to cyclic loading, their description often requires many more parameters than cyclic response models. Evaluation of these parameters can be difficult, and the parameters obtained from one type of test can be different from those obtained from another. Some of the newer constitutive models, however, have been calibrated in a manner that allows them to be more readily used for site response, stability, and liquefaction problems in practice; several of these are described in later chapters dealing with those topics.

6.5 MEASUREMENT OF SOIL PROPERTIES

The measurement of dynamic soil properties is a critical task in the solution of geotechnical earthquake engineering problems. A wide variety of field and laboratory techniques are available, each with advantages and limitations with respect to different problems. Some are oriented toward the measurement of low-strain properties and others toward properties that control behavior at larger strain levels. The selection of testing techniques for the measurement of dynamic soil properties requires careful consideration and understanding of the specific problem and site conditions at hand. Efforts should always be made to use tests or test procedures that replicate the initial stress conditions and the anticipated cyclic loading conditions as closely as possible.

Soil properties that influence wave propagation and other low-strain phenomena include stiffness, damping, Poisson's ratio, and density. Of these, stiffness and damping are the most important; the others have less influence and tend to fall within relatively narrow ranges. The stiffness and damping characteristics of cyclically loaded soils are critical to the evaluation of many geotechnical earthquake engineering problems – not only at low strains but, because soils are nonlinear materials, also at intermediate and high strains. At high levels of strain, the shear strength of the soil is important, as are factors that influence it under dynamic conditions such as the rate and number of cycles of loading. Volume change (i.e., dilatancy) characteristics are also important at higher strain levels.

Following a brief discussion of the measurement of density, the following sections present the measurement of these important soil properties in field and laboratory tests. Many of the tests have been developed specifically to measure dynamic soil properties; others are modified versions of tests commonly used to measure soil behavior under monotonic loading conditions. The applicability of the various tests to dynamic soil properties is emphasized here – descriptions of their applications to static properties may be found in standard geotechnical engineering texts (e.g., Lambe and Whitman, 1969; Holtz et al., 2023; Coduto et al., 2011).

Any investigation of dynamic soil properties should be performed with due recognition of the inevitable uncertainty in measured properties. Sources of uncertainty include the inherent variability of soils (a result of the geologic environment in which they were deposited), inherent anisotropy (a function of the soil structure or "fabric"), induced anisotropy (caused by anisotropic stress conditions), drilling and sampling disturbance, limitations of field and/or laboratory testing equipment, testing errors, and interpretation errors. Some of these sources of uncertainty can be minimized by careful attention to test details, but others cannot.

6.5.1 Density

The density of a material plays a significant role in wave propagation – as indicated in Appendix C, it affects wave velocities, impedance ratios, inertial stresses, etc. While geotechnical engineers consider density very carefully in earthwork operations, relatively less attention is paid to density in wave propagation problems for two primary reasons. First, densities are relatively easy to measure with acceptable accuracy in most soils and, second, the variability of soil density is relatively low, at least in comparison with other important properties such as shear modulus and damping ratio.

There are, however, some occasions where precise knowledge of the density of the soil is important. These usually correspond to saturated cohesionless soils whose volume change behavior, and hence propensity for generating porewater pressure under undrained conditions, is extremely sensitive to their density (Section 6.4.3.4). The potential for triggering liquefaction in such soils, and the effects of liquefaction if triggered, can be strongly influenced by relatively small differences in density. Studies have shown that high-quality, fixed-piston tube sampling tends to densify loose cohesionless soils and loosen dense cohesionless soils (Broms, 1980; Seed et al., 1982; Yoshimi et al., 1989; Hofmann et al., 2000). Sample disturbance is much greater in driven samples, the retrieval of which involves substantial shear deformation and vibration.

In general, measurement of the density of cohesive soils is relatively straightforward. Cohesive soil samples can be obtained with a number of different types of samplers. If measurement of density is the sole objective, a relatively rugged and robust sampler can be used with the disturbed outer zone trimmed away prior to density measurement. If, as is usually the case, other measurements are intended, the use of thin-walled sampling tubes in fixed-piston samplers (Hvorslev, 1949; Osterberg, 1973) to minimize disturbance is advisable.

Gamma-ray geophysical logging can be used to infer the *in situ* density of a soil in a borehole. In this procedure, a probe with a radiation source (usually Cesium 137) and gamma-ray detectors is lowered into a boring. The gamma rays are attenuated by interactions with electrons in the soil's minerals, the number of which increase with increasing soil density. The level of back-scattered gamma radiation measured at the detector located at a standard distance up the probe from the source then decreases with increasing soil density. The relationship between density and detected radiation level is established by a calibration process. This process allows a virtually continuous record of the density of a soil adjacent to a borehole. The procedure does, however, require an uncased borehole, which may limit its applicability at some sites.

6.5.2 Low-Strain Field Tests

Field tests allow soil properties to be measured *in situ* (i.e., in their existing state where the complex effects of existing stress, chemical, thermal, and structural conditions are reflected in the measured properties). Thc measurement of dynamic soil properties by field tests has a number of advantages. Field tests do not require sampling, which can alter the stress, chemical, thermal, biological, and structural conditions that exist *in situ*. Many field tests measure the response of relatively large volumes of soil, thereby minimizing the potential for basing property evaluation upon small, unrepresentative specimens in spatially variable soils. Many field tests induce soil deformations that are similar to those of the problem of interest, particularly for wave propagation and foundation design problems. On the other hand, field tests do not allow the effects of conditions other than those that exist *in situ* at the time of testing to be investigated easily, nor do they allow porewater drainage to be controlled. In many field tests, the specific soil property of interest is not measured but must be determined indirectly, by theoretical analysis or empirical correlation.

Some field tests (*non-invasive* tests) can be performed from the ground surface, while others (*invasive* tests) require the drilling of boreholes or the advancement of a probe into the soil. Surface tests are often less expensive and can be performed relatively quickly in the field, although they may require significant time dedicated to complex processing in order to obtain the desired results. They are particularly useful for materials in which drilling and sampling or penetration testing is difficult. Tests associated with the drilling of a borehole, on the other hand, have the advantage of providing often vital information that is gained directly from the boring such as visual and laboratory-determined soil characteristics and water table location. Also, the interpretation of data from boreholes and their associated tests is usually more direct than that of surface tests.

Low-strain tests generally operate at strain levels that are not large enough to induce significant nonlinear stress-strain behavior in the soil, typically at shear strains below the linear threshold shear strain (about 0.001%). As such, most are interpreted using the theory of wave propagation in linear materials. Many involve the measurement of body wave velocities which can easily be related to low-strain soil moduli. Others are based on the characteristics of surface waves (Section C.3), which may be induced by mechanical impact, shakers, or explosives (*active sources*) or extracted from measured ambient vibrations (*passive sources*).

Seismic geophysical tests represent an important class of field tests for the evaluation of dynamic soil properties. Seismic tests involve the creation of transient and/or steady state stress waves and the interpretation of their behavior from measurements made at one or more different locations. In many seismic tests, a source produces a "pulse" of waves whose arrival times are measured at distant receivers. The source, which may range from a sledgehammer blow on the ground surface to a buried explosive charge, will generally produce p-waves, s-waves, and surface waves. The relative partitioning of the different wave types and their associated amplitudes depend on how and where (i.e., at or below the ground surface) the impulse is generated. Explosive sources placed within the Earth (Figure 6.54a) are rich in p-wave content, while vertical impact sources excited at the ground surface (Figure 6.54b) produce strong surface waves and minimal p-wave energy. SH-waves are produced most efficiently by striking the end of a beam pressed tightly against the ground surface (Figure 6.54c).

Since p-waves travel fastest, their arrivals at distant receivers are most easily detected and their arrival times most easily measured. S-wave resolution can be improved markedly by reversing the polarity of the impulse, as is easily accomplished for SH-waves by striking the other end of the

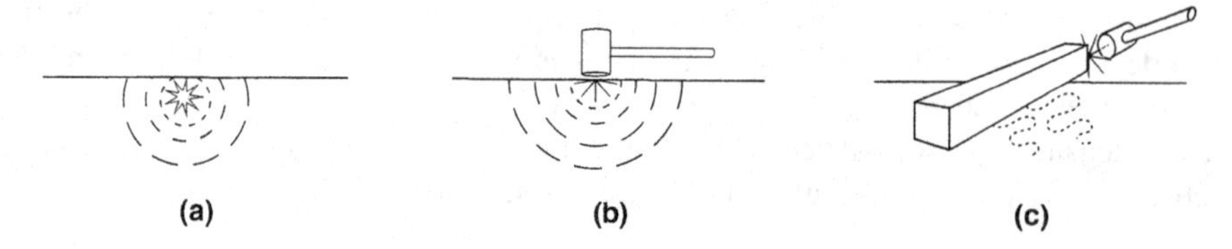

FIGURE 6.54 Different methods for the creation of impulsive disturbances for seismic geophysical tests: (a) shallow explosives; (b) vertical impact; and (c) horizontal impact.

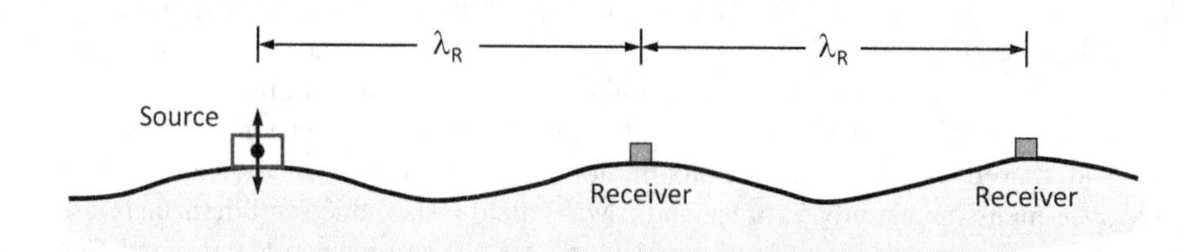

FIGURE 6.55 Rayleigh wave-induced deformation of ground surface adjacent to vertically vibrating footing. (After Richart et al., 1970.)

beam of Figure 6.54c. Since the polarity of the train of p-waves is not reversed, subtracting the reversed record from the original record will diminish the p-wave amplitudes while enhancing the s-wave amplitudes. Wave arrivals can also be enhanced by adding, or "stacking," records from multiple impulses; the random noise portions of the records tend to cancel each other while the desired signals from the actual waves are enhanced.

As indicated in Appendix C, p-wave velocity is controlled by the constrained modulus, M, which depends on the stiffness of the soil skeleton and the material in the voids. In a saturated soil, the relative incompressibility of the porewater controls the constrained modulus of the soil, so careful consideration of groundwater conditions is essential for the proper interpretation of seismic geophysical test measurements. For degrees of saturation less than about 99%, the amount of air in the voids is sufficient to render the air/water pore fluid relatively compressible so that the p-wave velocity is controlled by the constrained stiffness of the soil skeleton. P-waves travel through groundwater at about 1,500 m/sec (5000 ft/sec), depending on temperature and salinity. In soft, saturated soils p-waves will typically propagate through the soil at velocities slightly greater than the p-wave velocity of water, in which case the measured velocity is not strongly indicative of the compressive stiffness of the soil skeleton itself (Stolte and Cox, 2019). P-wave velocity measurements are sometimes used to infer the location where the soil becomes 100% saturated (which may or may not correspond to the location of the hydrostatic groundwater table), particularly in loose/soft soils, or the extent of partially saturated soils (Ishihara et al., 2001; Tsukamoto et al., 2001). Failure to consider groundwater effects can result in significant overestimation of compressional soil stiffness. The groundwater problem can largely be avoided by using s-waves which are propagated only by the soil skeleton (i.e., not by the groundwater). The effect of water on shear wave velocity is limited to its effects on total (moist) unit weight and on effective stresses (including soil suction in partially saturated soils).

6.5.2.1 Seismic Reflection and Refraction Tests

A number of important non-invasive geophysical tests based on the propagation of stress waves have been developed in the oil exploration industry. Among the most important of these are the seismic reflection and seismic refraction tests. The seismic reflection test applies an impulsive disturbance at the ground surface and measures the time required for various direct and reflected waves to reach one or more surface geophones located relatively close to the disturbance. Reflection testing is generally based on p-waves and can provide velocity and layer thickness data to substantial depths.

However, it is not an ideal method for determining near-surface soil properties. Interpretation of reflection data can be complicated by the interference of different waves arriving simultaneously having traveled at different velocities over different paths, and generally requires sophisticated filtering and data processing. Many different source and receiver locations may be required to characterize subsurface conditions at a site with some degree of subsurface complexity. Reflection methods are also best suited to detecting layers at depths greater than about 15 m (50 ft), at which reflected waves arrive well after direct waves and can be more easily detected in the recorded waveforms. The seismic reflection method is described in detail in geophysics textbooks (e.g., Dobrin and Savit, 1988; Burger, 1992).

Seismic refraction avoids the interference problem by using only the arrival time of the first arriving wave at each of a series of geophones laid out in a relatively long linear array coincident with an impulsive source. By recording arrival times from sources at both ends of the array, wave propagation velocities and layer boundary geometries can be estimated. While seismic refraction can more readily detect near-surface layering and is more commonly used for engineering purposes than seismic reflection testing, it requires that wave velocities increase with depth – a low-velocity layer between two higher velocity layers can be missed in a seismic refraction survey. Seismic refraction is described in more detail in Musgrave (1970) and U.S. Army Corps of Engineers (1995).

Seismic reflection and refraction are not commonly used to obtain the type of localized velocity data typically required for site-specific geotechnical earthquake engineering studies. This limited application results from the need for V_S measurements (which are better suited to other methods) and the need for localized, high-resolution, velocity data that is often not obtained with these tests.

6.5.2.2 Surface Wave Tests

In recent years, the use of surface wave tests has increased significantly. Early surface wave tests (Richart et al., 1970) used vertically oscillating plates to generate steady-state Rayleigh waves whose wavelength, λ_R, at a frequency, f, could be measured by moving geophones to separation distances of common phase. In a homogeneous half-space, the Rayleigh wave phase velocity, V_R, could be computed as

$$V_R = f\lambda_R \tag{6.43}$$

and the shear wave velocity computed from V_R and Poisson's ratio (Section C.3.1.1). In a homogeneous half-space, the value of V_R would be equal at all frequencies (and wavelengths). At a site at which stiffness varies with depth, however, the Rayleigh wave phase velocity will vary with frequency, i.e., it will be *dispersive* (Section C.3.4). For the common situation of stiffness increasing with depth, low-frequency Rayleigh waves will have larger wavelengths that excite deeper (and, hence, stiffer) soils and travel faster than high-frequency waves that travel through shallower and softer soils.

A plot of Rayleigh wave phase velocity versus frequency (or wavelength) is referred to as a *dispersion curve*. The shape of a dispersion curve at a particular site is related to the variation of layer thicknesses and body wave velocities with depth. The type of test illustrated in Figure 6.55, wherein a source oscillating at a fixed frequency emits steady-state vibrations and two receiver locations are moved along the ground surface until they move in phase with one another at distances of one and two Rayleigh wavelengths from the source, can be used to generate a dispersion curve by repeating the test at different loading frequencies. This process, however, tends to be quite time-consuming in the field. With the use of digital data acquisition and signal-processing equipment, a dispersion curve can be obtained from an impulsive or random noise load. Numerical modeling can then be used to identify a layered Earth model that is consistent with the measured dispersion curve. A number of techniques have been developed to measure and interpret surface wave dispersion curves. A detailed summary of best practices for the collection, processing, and interpretation of surface wave data was presented by Foti et al. (2018).

Spectral Analysis of Surface Waves (SASW) Test

The SASW test (Heisey et al., 1982; Nazarian and Stokoe, 1983; Stokoe et al., 1994) is performed by placing two vertical receivers on the ground surface in line with an impulsive or random noise source, as illustrated in Figure 6.56. The output of both receivers is recorded and transformed to the frequency domain using the fast Fourier transform. After transformation, the phase difference between receivers, $\phi(f)$, can be computed for each frequency. The corresponding travel time between receivers can be calculated for each frequency from

$$\Delta t(f) = \frac{\phi(f)}{2\pi f} \tag{6.44}$$

Since the distance between receivers, $\Delta d = d_2 - d_1$, is known, the apparent Rayleigh wave phase velocity and wavelength can be calculated as functions of frequency:

$$V_R(f) = \frac{\Delta d}{\Delta t(f)} \tag{6.45}$$

$$\lambda_R(f) = \frac{V_R(f)}{f} \tag{6.46}$$

With modern electronic instrumentation, these calculations can be performed in the field virtually in real time. The results can be used to plot the experimental dispersion curve (Figure 6.57) for a given receiver spacing. While the test should, in theory, yield good results for a single receiver spacing, practical considerations such as signal-to-noise ratios and mitigation of near-field and far-field effects dictate that several (typically six or more) different receiver spacings be used. Small receiver spacings are used to infer properties of the near-surface that are sensed by high-frequency (short wavelength) waves, while larger receiver spacings are used to infer the properties of deeper materials that are sensed by low frequency (long wavelength) waves. To balance near-field and far-field effects, the distance between the two receivers at each spacing is kept approximately equal to the distance between the first receiver and the source (i.e., $d_2 = 2d_1$). The individual dispersion curves for different receiver spacings will typically overlap each other but together form a single "field" dispersion curve that covers a wide range of wavelengths.

Identification of the thickness and shear wave velocity of subsurface layers involves the iterative matching of a theoretical dispersion curve to the experimental dispersion curve. The Haskell-Thomson transfer matrix solution (Thomson, 1950; Haskell, 1953) for a series of uniform elastic layers of infinite horizontal extent is most commonly used to predict the theoretical dispersion curve used during inversion. Calculation of the transfer matrix requires the trial subsurface model to be defined in terms of layer thicknesses, mass densities, shear wave velocities, and compression wave velocities (or Poisson's ratios) for each layer. However, the theoretical dispersion curve is most sensitive to layer thicknesses and shear wave velocities. Therefore, initial estimates

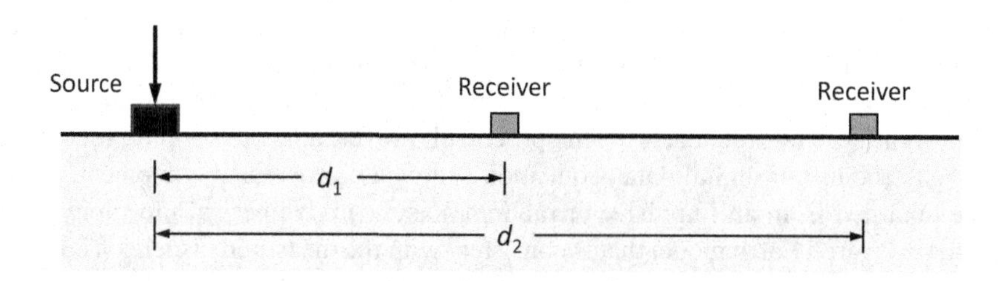

FIGURE 6.56 Typical configuration of source and receivers in the SASW test. The receiver spacing is changed in such a way that the ratio of d_2/d_1 remains near a constant value of 1.5–2 for any array length.

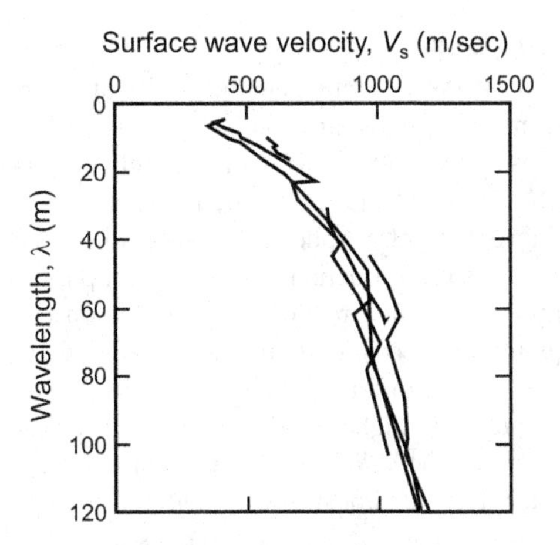

FIGURE 6.57 A group of eight dispersion curves from eight SASW receiver spacings collected using covering a wavelength range of 1–120 m (Site 267CSC, Cascia, Umbria). (After Kayen et al., 2008.)

of the thickness and shear wave velocity of each layer are adjusted until the values that produce the optimum fit to the experimental dispersion curve are identified. This identification procedure is usually referred to as *inversion* (Nazarian, 1984). As with many inversion problems, the uniqueness of the solution is a potential problem – even for a given number of assumed layers, different velocity profiles can produce nearly the same theoretical dispersion curve (Vantassel and Cox, 2021), and subsurface models with different numbers of layers can also produce equally good fits to a given experimental dispersion curve (Cox and Teague, 2016). For that reason, additional information such as depths to different layer boundaries (from boring logs or other tests), can be used to constrain the inversion and produce more accurate results. For profiles in which the shear wave velocity varies irregularly with depth, the dispersion curve may be influenced by higher-mode Rayleigh waves (Gucunski and Woods, 1991; Tokimatsu et al., 1992), which significantly complicate interpretation. Where noise levels are high relative to the impulsive disturbance, as in some urban or industrial environments, SASW measurements may be difficult to interpret. Wave stacking can help alleviate this problem.

The SASW test has a number of important advantages over other field tests. It can be performed quickly in the field, requires no borehole, can detect low-velocity layers, is especially effective at measuring seismic velocities of shallow layers, and can be used to considerable depth (>100 m) depending on the size and power of the loading source. Using very large vibroseis equipment to generate strong, low frequency (long wavelength) waves along with large receiver spacings, shear wave velocity profiles can be measured to depths of 200 m or more. To be reliably characterized, the thickness of a layer should generally be greater than about 10% of its depth. Comparison of shear wave velocity profiles obtained from SASW testing and various invasive methods such as cross-hole, down-hole and PS-logging testing have shown good agreement (Hiltunen and Woods, 1988; Lin et al., 2008, Damm et al., 2021). SASW testing is particularly useful at sites where drilling and sampling are difficult; it has been used successfully in such materials as gravels and debris flow deposits (Stokoe et al., 1988) and landfills (Kavazanjian et al., 1994). The procedure does, however, require specialized equipment and experienced operators. Its applicability is also limited to sites at which the assumptions of the Haskell-Thomson solution (e.g., horizontal layering) are at least approximately satisfied. At sites with complex stiffness profiles, e.g., with alternating layers of higher and lower stiffness, inversion based on the fundamental Rayleigh mode alone may not produce reliable results (Tokimatsu et al., 1992).

Multi-Channel Analysis of Surface Waves Test

Following the increased use of SASW, limitations in the ability of two receivers to handle the complex waveforms that existed in some profiles, particularly at noisy sites and/or those with complex subsurface conditions, led to the development of multi-channel analysis of surface wave (MASW) methods (Park et al., 1999; Miller et al., 1999). In the multi-channel approach, a linear array of receivers (typically consisting of 24–48 geophones) are placed on the ground surface in a configuration similar to that of a seismic refraction test. The multiple channels allow for rapid and semi-automated transformation of seismic records into dispersion curves (Rahimi et al., 2021), which can also be used to quantify uncertainty in the experimental data (Vantassel and Cox, 2022). Furthermore, MASW dispersion transformation methods allow for the separation of body waves and identification of fundamental- and higher-mode Rayleigh waves, which can either be removed or utilized in multi-mode inversions. MASW can be particularly useful at sites with velocity inversions (softer layers below stiffer layers) and/or sites with shallow, strong velocity contrasts, for which higher-mode surface waves can dominate the recorded response. While MASW testing is now much more commonly used in practice than SASW testing, the primary distinction from SASW is in the manner by which the experimental dispersion curve and its uncertainty are derived – the inversion procedure used to obtain a V_S profile is very similar to that used for SASW.

Refraction Microtremor Test

The Refraction Microtremor (ReMi) test (Louie, 2001; Pullammanappallil et al., 2003) is a passive-source method that uses a linear array of geophones to detect surface waves contained in an ambient noise (microtremor) wavefield. Ambient vibrations are measured with linear arrays at least 200 m long using sensors that can measure surface waves at frequencies down to 2 Hz. Like MASW, ReMi uses a two-dimensional wavefield transformation (a slowness-frequency transform in the case of ReMi) to separate fundamental- and higher-mode Rayleigh waves and thereby obtain a cleaner dispersion curve. The dispersion curve is then inverted to obtain V_s using the same approaches used in SASW and MASW inversions.

The ReMi test has the advantage of not requiring a vibration source (which reduces its cost) and being able to work in noisy environments. It can be used to profile depths of up to about 100 m (Louie, 2001) since ambient wavefields often contain low-frequency waves that cannot be easily generated using small active sources.

The wavefield under ambient vibration conditions, however, is largely unknown, which can affect the interpreted V_s profile. For example, waves from a source aligned perpendicular to the linear geophone array will reach the geophones almost simultaneously (producing almost infinite apparent phase velocities), whereas arrivals from a source aligned with the linear array will arrive at intervals associated with the true phase velocity. The ReMi method assumes that energy sources are randomly distributed so that passive noise would come equally from all directions; only the slowest phase velocities are used in the development of the site dispersion curve. This approach may be reasonable at some very noisy sites, however, microtremor measurements with two-dimensional surface arrays have shown clear, and often frequency-dependent, arrival directions in many urban and rural areas (Tokimatsu, 1995; Di Gulio et al., 2006; Rosenblad and Li, 2009; Endrun et al., 2010). Cox and Beekman (2011) found significant differences (up to about 20%) in V_s derived from ReMi due to array orientation effects, and even larger differences (up to 85%–100%) between V_s profiles obtained using ReMi with only passive sources versus those that included active sources (with random active sources always giving lower velocities than passive sources). These effects led Foti et al. (2018) to discourage the use of linear surface arrays in favor of two-dimensional surface arrays (e.g., circular, L-shaped, T-shaped, or triangular arrays) with passive sources.

6.5.2.3 Seismic Down-Hole (Up-Hole) Test

Seismic down-hole (or up-hole) tests can be performed in a single borehole. In the down-hole test, an impulse source is located on the ground surface adjacent to the borehole. A single receiver that

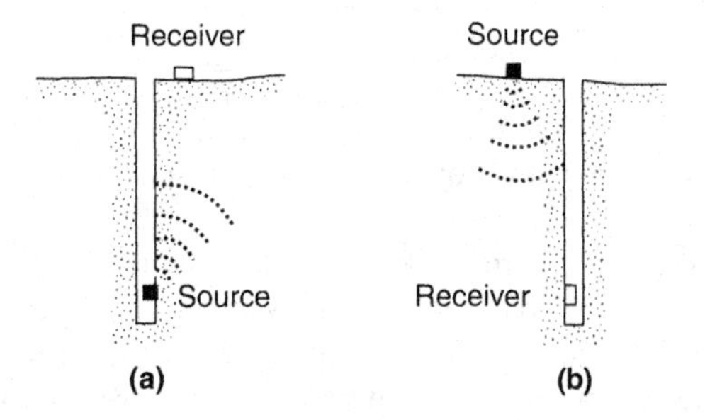

FIGURE 6.58 (a) Up-hole, and (b) down-hole test configurations.

can be moved to different depths, or a string of multiple receivers at predetermined depths, is fixed against the walls of the borehole, and a single triggering receiver is located at the energy source (Figure 6.58). The source and all receivers are connected to a high-speed recording system so that their output can be measured as a function of time. In the up-hole test, a movable energy source is located in the borehole with a single receiver on the ground surface adjacent to the borehole.

The objective of the down-hole (or up-hole) test is to measure the travel times of p- and/or s-waves from the energy source to the receiver(s). By properly locating the receiver positions, a plot of travel time versus depth can be generated (Figure 6.59). The slope of the travel-time curve at any depth represents the wave propagation velocity at that depth. In a layered profile, a significant horizontal offset between the source and receiver can result in an irregular wave travel path between the two due to refraction (Section C.4.2) caused by non-normal incidence at the layer boundaries. Interpretation of such data are usually based on approximate corrections of the measured travel times to equivalent vertical travel times,

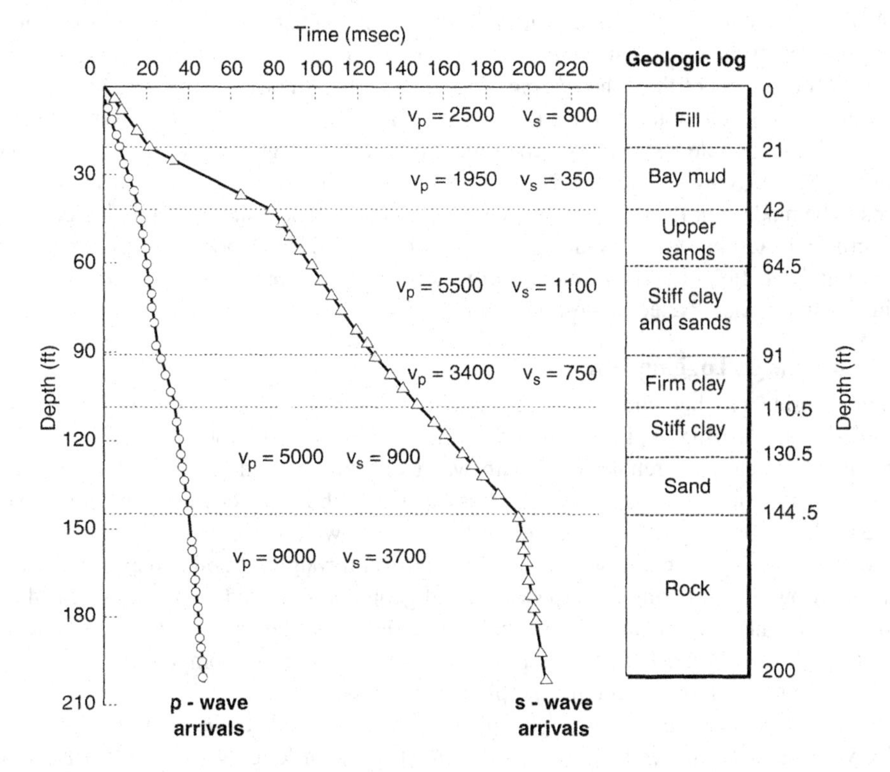

FIGURE 6.59 Travel-time curve from down-hole test in the San Francisco Bay Area (after Schwarz and Musser, 1972).

$$tt^{vert} = tt^{meas} \frac{z}{\sqrt{z^2 + h^2}} \tag{6.47}$$

where tt^{meas} = measured travel time, h = horizontal offset of the source from the borehole and z is the depth of interest.

S-waves can be generated much more easily in the down-hole test than the up-hole test; consequently, the down-hole test is more commonly used. With an SH-wave source, the down-hole test measures the velocity of waves similar to those that carry the most seismic energy to the ground surface during earthquake shaking. Because the waves must travel through all materials between the impulse source and the receivers, the down-hole test allows the detection of layers that can be hidden in seismic refraction surveys. Potential difficulties with down-hole (and up-hole) tests and their interpretation can result from disturbance of the soil during drilling of the borehole, casing and borehole fluid effects, insufficient or excessively large impulse sources, and background noise effects. The effects of material and radiation damping and wave scattering on waveforms can degrade signals, which complicates identification of s-wave arrivals at depths greater than 30–60 m (100–200 ft). It should be noted that down-hole and up-hole tests indicate velocities in a zone that is relatively local to a direct line between the source and the receiver. The measured velocities in a single hole, therefore, do not provide any indication of the potential effects of lateral variability. Multiple tests in multiple boreholes may be required to obtain a stable indication of the mean (or median) velocity profile in a spatially variable soil deposit. Uncertainty in shear wave velocities derived from down-hole tests can be estimated using procedures in Stolte and Cox (2019).

6.5.2.4 Seismic Cone Test

The seismic cone test (Robertson et al., 1985) is very similar to the down-hole test, except that no borehole is required. A seismic cone penetrometer consists of a conventional cone penetrometer (Section 6.5.3.3) outfitted with one or more geophones or accelerometers mounted above the friction sleeve. At different stages in the cone penetration sounding, penetration is stopped long enough to generate impulses at the ground surface, often by striking each end of a beam pressed against the ground by the outriggers of the cone rig (e.g., Figure 6.54c), with an instrumented hammer. With a single geophone or accelerometer, travel time-depth curves can be generated and interpreted in the same way as for down-hole tests. Using two or more receivers (Burghignoli et al., 1991; Butcher and Powell, 1996; McGillivray and Mayne, 2008), however, can allow more accurate, true interval measurements to be made over the distance between the receivers. The efficiency, and, hence, economy, of the seismic cone test has led to its increasingly common use in soils loose or soft enough to permit CPT profiling. Detection of arrival times at large depths can be complicated by damping and scattering, just as in borehole-based down-hole tests.

6.5.2.5 Suspension Logging Test

Suspension logging has become one of the preferred tests for seismic profiling for geotechnical earthquake engineering applications in the last 15–20 years. A probe 5–6 m long is lowered into a preferably uncased borehole filled with water or drilling fluid (Figure 6.60). A horizontal reversible-polarity solenoid located near the base of the probe produces a sharp, impulsive pressure wave in the drilling fluid. Upon reaching the borehole wall, the pressure wave produces both p- and s-waves in the surrounding soil. These waves travel through the soil and eventually transmit energy back through the drilling fluid to two biaxial geophones located 1.0 m apart near the top of the probe. To enhance identification of p- and s-wave arrivals, the procedure is repeated with an impulse of opposite polarity. Differences in arrival times are used to compute the average p- and s-wave velocities of the soil between the geophones.

Because the solenoid travels with the geophones down the borehole, the amplitude of the signals is relatively constant at all depths. As a result, the suspension logging test is effective at essentially any depth where a borehole is drilled, and depths of up to 2 km have been achieved (Nigbor

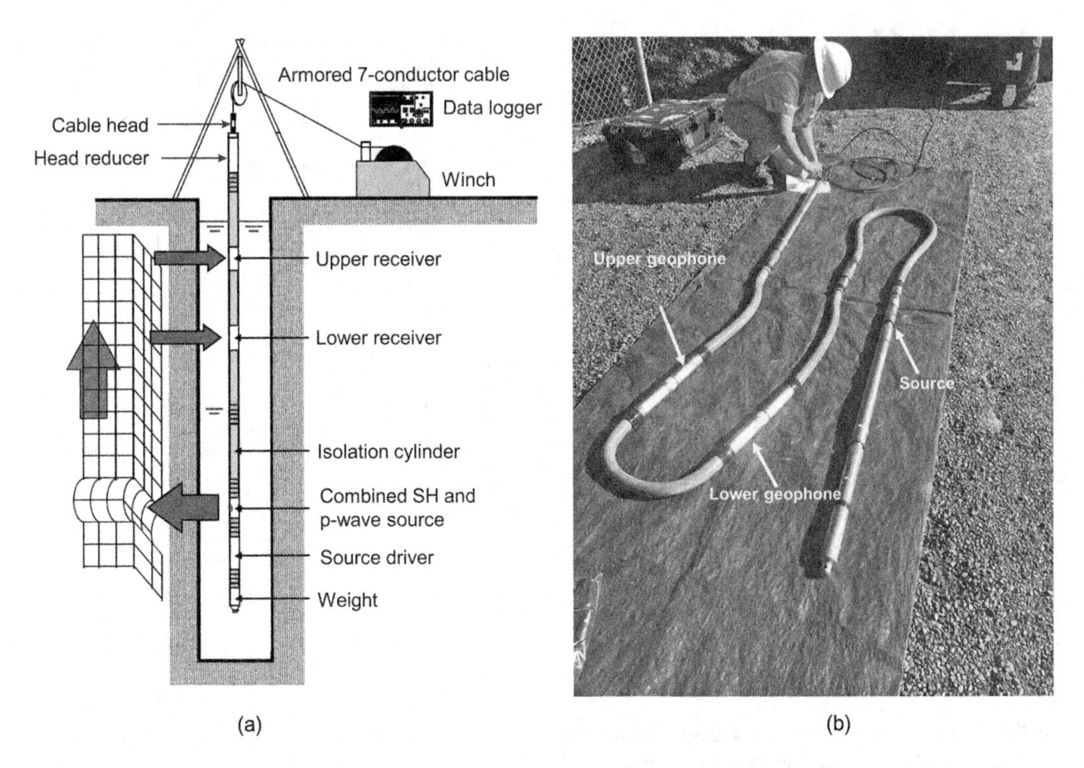

FIGURE 6.60 Suspension logging test: (a) Schematic layout (After Diehl and Steller, 2007) and (b) equipment preparation (Photo courtesy of M. Pehlivan).

and Imai, 1994). By overlapping measurement points, resolutions of less than 1 m (3.1 ft) can be obtained. This capability is particularly useful at sites that may have thin layers of soft or weak soil.

The suspension test allows measurement of wave propagation velocities, but the frequencies of the waves (500–2,000 Hz for s-waves and 1,000–3,000 Hz for p-waves) are much higher than those typically encountered in down-hole tests and of interest in geotechnical earthquake engineering. Despite these differences, seismic velocities at a common site as measured from suspension logging and down-hole methods are typically very similar, with down-hole results essentially representing a smoothed average of the high-resolution suspension logging data. Known limitations of suspension logging include the presence of tube waves that may limit its effectiveness in cased boreholes, which are often required for soft soils; and inaccurate velocity measurements at depths less than about 7 m (Stokoe et al., 2004).

The suspension logging test measures wave velocities on an extremely localized scale – the measurements indicate the stiffness of the soil in a zone very close to the edge of the boring and over only the 1 m interval between the geophones. As such, measurements from a single boring provide no indication of the potential effects of lateral variability of velocities; multiple borings would be required to characterize those effects. Because the zone in which the measurements are made is also localized in the vertical direction, interpolation across unmeasured zones is not possible (Boore, 2006) as it is (at least in an average sense) with, for example, down-hole tests.

6.5.2.6 Seismic Cross-Hole Test

Seismic cross-hole tests use two or more boreholes to measure wave propagation velocities along horizontal paths. The simplest cross-hole test configuration (Figure 6.61a) consists of two boreholes, one of which contains an impulsive energy source and the other a receiver. Cross-hole seismic tests using two seismic cones have also been performed (e.g., Baldi et al., 1988; Cox et al., 2019).

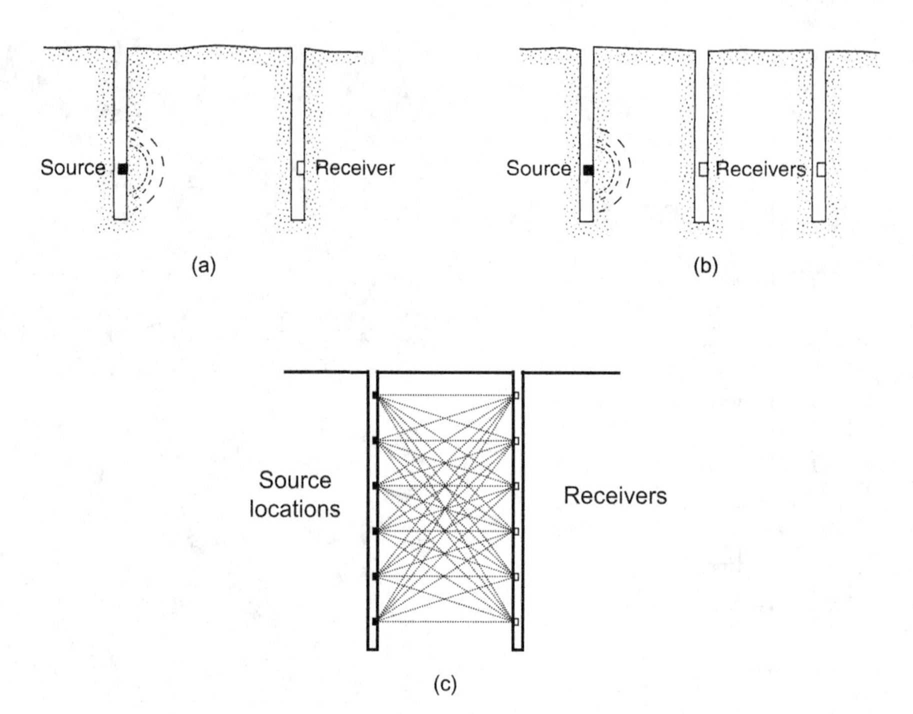

FIGURE 6.61 (a) seismic cross-hole test between two borings, (b) seismic cross-hole test with two receiver borings, and (c) seismic tomography configuration.

By fixing both the source and the receiver at the same depth in each borehole, the wave propagation velocity of the material between the boreholes at that depth is measured. By testing at various depths, a velocity profile can be obtained. When possible, the use of more than two boreholes is desirable (Figure 6.61b) to minimize possible inaccuracies resulting from trigger time measurement, casing, and backfill (material placed between the casing and the borehole wall) effects, and site anisotropy. Wave propagation velocities can then be calculated from differences in arrival times at adjacent pairs of boreholes. Arrival times can be determined by eye using points of common phase (first arrival, first peak, first trough, etc.) or by cross-correlation techniques commonly used in petroleum exploration (Roesler, 1977). Since the impulse sources must be located in the borehole, variation of the p-wave/s-wave content is more difficult than for methods in which the source is at the surface. A number of mechanical impulse sources have been used, including the driving of a SPT (Section 6.5.3.1) sampler, vertical impact loading of rods connected to borehole packers or jacks, torsional impact loading of a torque foot at the bottom of the borehole (Stokoe and Hoar, 1978), and other techniques (Applegate, 1974; Stokoe and Abdul-razzak, 1975; Auld, 1977). The best results are generally obtained when the polarity of the impulse source is reversible, hence the frequent preference for mechanical sources over explosive sources.

The cross-hole test often allows individual soil layers to be tested when layer boundaries are nearly horizontal. It can also detect hidden layers that can be missed by other techniques such as seismic refraction surveys. By placing multiple receivers in one hole and moving the source to different depths in the other hole, travel times can be measured on multiple inclined paths and interpreted using principles of *geotomography* (Fernandez and Santamarina, 2003; Santamarina and Fratta, 2005; Picozzi et al., 2009) to obtain a two-dimensional velocity field between the boreholes (Figure 6.61c).

Cross-hole tests can yield reliable velocity data to depths of 30–60 m (100–200 ft) using mechanical impulse sources, and to greater depths with explosive sources. On the other hand, cross-hole tests can be costly, due to the requirement of multiple boreholes. Also, the sensitivity of the measured velocities to source--receiver distance requires borehole deviation surveys, particularly for boreholes more than 15–20 m (50–65 ft) deep. The measured velocities may not be equal to the actual velocities when higher-velocity layers exist nearby. In such cases, more advanced methods of interpretation that can account for refraction (e.g., Butler et al., 1978) are required. Hryciw (1989) presented methods for the correction of ray-path curvature in materials of continuously varying velocity.

Amplitude attenuation measurements from cross-hole tests involving three or more boreholes have been used to compute the material damping ratio of soils (Hoar and Stokoe, 1984; Mok et al., 1988; EPRI, 1993). The procedure requires accurately calibrated and oriented receivers that are well coupled to the borehole wall. By assuming a radiation pattern, the effects of geometric attenuation (radiation damping) can be separated from the measured attenuation to leave the attenuation due to material damping. The required assumptions render such approaches best suited to sites of simple geometry and homogeneous soil conditions.

Example 6.3

Determine the SV-wave velocity from the cross-hole test trigger and geophone records shown in Figure E6.3. The trigger and geophone are located 5 m apart. The solid line represents the response from a downward impact on a mechanical source and the dotted line represents the response from an upward impact.

Solution:

There is an obvious wave arrival at the geophone at about 2 ms after impact at the source. However, close examination of the geophone record shows that the polarity of this early arrival was not influenced by the polarity of the impact; consequently, the early arrival can be identified as a p-wave. At a later point, the arrival of waves shoes polarity is reversed by the reversal of impact polarity is observed. These represent SV-waves – the arrival time of 23 ms after impact (determined graphically) indicates an SV-wave velocity of

$$V_s = \frac{x}{\Delta t} = \frac{5\text{ m}}{0.023\text{ s}} = 217\text{ m/s}$$

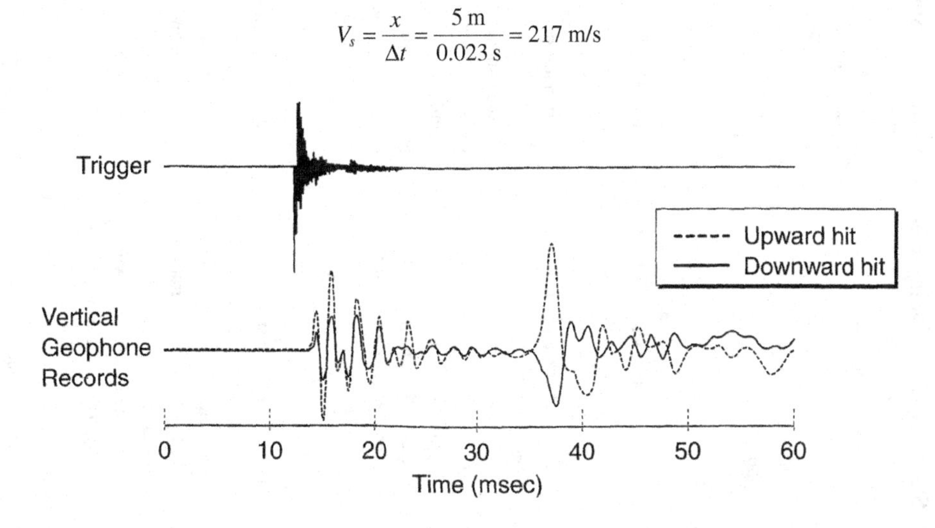

FIGURE E6.3

6.5.2.7 Discussion

As the previous sections have shown, a variety of techniques are available for measuring the shear wave velocity of soils in the field. A summary of the features of the most common measurement methods is presented in Table 6.1. Several studies (e.g., Brown et al., 2002; Xia et al., 2002; Asten and Boore, 2005; Stephenson et al., 2005; Garofalo et al., 2016a, 2016b; Damm et al., 2021) have

TABLE 6.1
Comparison of Insitu Shear Wave Velocity Measurement Methods

	Measurement Method				
Feature	**Cross-hole**	**Down-hole & Seismic CPT**	**Suspension Logger**	**SASW**	**Surface Reflection/ Refraction**
Number of holes required	2 or more	1	1	None	None
Quality control and repeatability[a]	Good	Good	Good	Good to fair; complex interpretation technique at sites with large velocity contrasts	Fair; often difficult to distinguish shear wave arrival
Resolution of variability in stiffness of soil deposits[b]	Good; constant with depth	Good to fair; decreases with depth	Good at depth; poor very close (3–6 m) to the ground surface	Good to fair; decreases with depth; provides good global average	Fair to poor; provides coarse global average
Major component of particle motion or wave propagation is in the vertical direction?	Yes, with vertically polarized shear waves	Yes, with test depth greater than distance between hole and shear-bean source	Yes, with refracted shear waves traveling parallel to vertical borehole	Yes, with vertical source	Yes, with horizontal source for reflection and vertical source for refraction
Limitations	Possible refraction problems; senses stiffer materials at test depth; most expensive test method	Possible reflection problems with shallow layers; wave travel path increases with depth	Fluid-filled hole required; may not work well near surface in cased holes and soft soils	Horizontal layering assumed; poor resolution of thin layers and soft material adjacent to stiff layers; no samples recovered	In refraction test, only works for velocity increasing with depth; no samples recovered
Other	Highly reliable test; measurements at each depth independent of other depths; well suited for tomographic imaging; independent checking of saturation with compressive waves is possible	Penetration data also obtained from seismic cone; detailed layered profile with cone	Well suited for deep borehole testing; method assumes shear waves travel in undisturbed soil	Well suited for tomographic imaging large areas and testing difficult to penetrate soils	Well suited for screening large areas; independent checking of saturation with compressive waves is possible

Source: After Andrus et al. (2004).

[a] Good quality depends on good equipment and procedural details, and good interpretation techniques for all methods.

[b] Resolution depends on test spacing for all methods.

compared the shear wave velocities measured using different procedures at the same site. The methods discussed here generally produced comparable results, with the exception of ReMi. Any can provide good results when performed carefully by experienced analysts, and variabilities between tests are generally thought to reflect natural soil heterogeneity rather than test-specific biases.

6.5.3 High-Strain Field Tests

A variety of tests and procedures are available for *in situ* testing of soils in a manner that produces significant (in some cases quite large) soil deformations. While these tests are most commonly used to measure high-strain characteristics such as shear strength, their results have also been correlated to low-strain soil properties. For geotechnical earthquake engineering problems, the SPT, CPT, vane shear test, and pressuremeter test are of particular interest.

6.5.3.1 Standard Penetration Test

The SPT is the oldest *in situ* test in geotechnical engineering and is still commonly used in a number of geotechnical earthquake engineering applications. In the SPT, a standard split-barrel sampler (Figure 6.62) is driven into the soil at the bottom of a borehole by repeated blows (30–40 blows per minute) of a 140-lb (63.6 kg) hammer released from a height of 30 in (76 cm). Several hammer types (Figure 6.63) are available. The standard SPT sampler should have a constant inside diameter; the use of samplers designed to accommodate internal sample liners can underestimate penetration resistance by 10%–20% when the liners are not in place. The sampler is usually driven 18 in. (46 cm); the number of blows required to achieve the last 12 in. (30 cm) of penetration is taken as the measured standard penetration resistance, N_m. The sample obtained in the SPT is disturbed, but can be used for soil classification purposes, including grain size, Atterberg limits, or other index tests.

The N_m value is a function of the soil type, confining pressure, and soil density, but is also influenced by the test equipment and procedures. In fact, studies have shown that different equipment

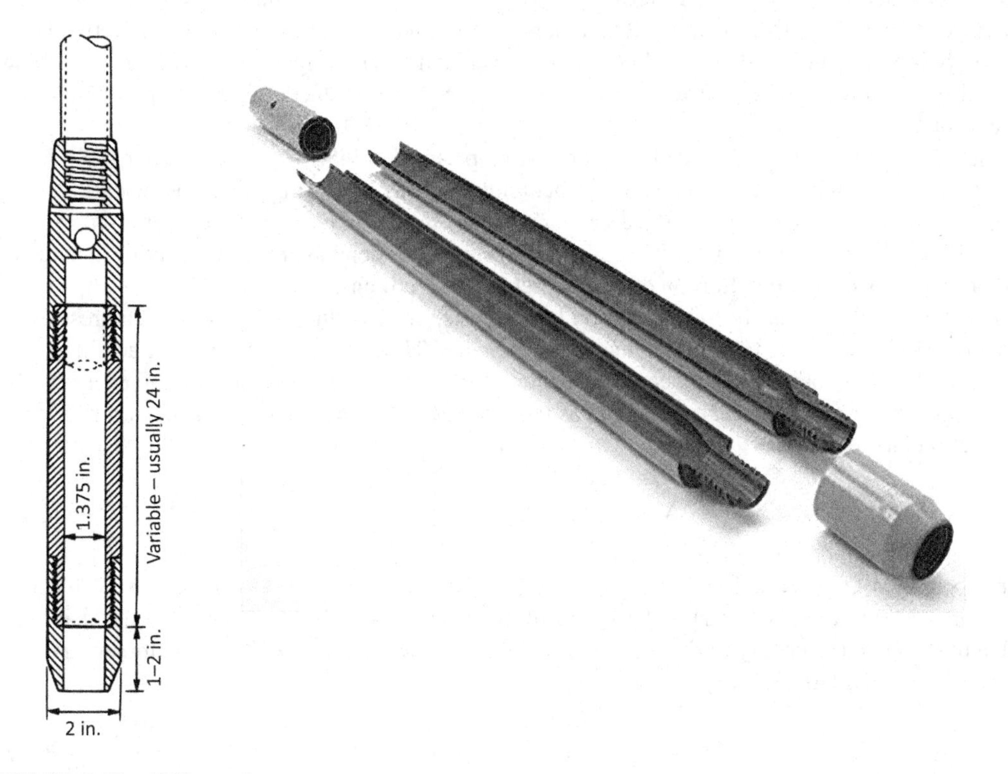

FIGURE 6.62 SPT sampler.

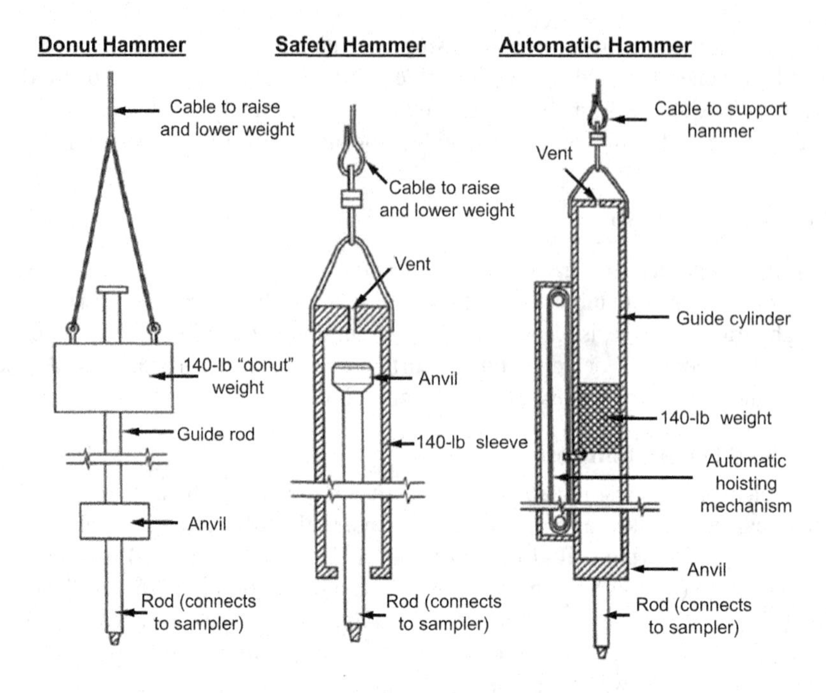

FIGURE 6.63 Common types of hammers used to drive standard penetration test samplers. (After Coduto et al., 2011; used with permission of Pearson Education, Inc.)

and procedures are quite common both within the United States and other countries, and that they strongly influence the energy delivered to the soil by each blow of the hammer (Kovacs et al., 1977; Schmertmann et al., 1978; Kovacs and Salomone, 1982), and hence the measured penetration resistance (Schmertmann and Palacios, 1979). For a given soil layer, a higher level of energy per blow would result in a lower N_m value, so efforts have been made to standardize both equipment and procedures.

Seed et al. (1985) recommended that the test be performed in 4- to 5-in.-diameter (10–13 cm) rotary boreholes with upward deflection of bentonite drilling mud using a tricone or baffled drag bit. The recommended sampler should have a constant inside diameter and be connected to A or AW [for depths less than 15 m (50 ft)] or N or NW (for greater depths) drill rods. Driving at a rate of 30–40 blows per minute with 60% of the theoretical free-fall energy delivered to the sampler was also recommended. It has become common to normalize the N value to an overburden pressure of 1 ton/ft^2 (100 kPa) and to correct it to an energy ratio of 60% (the average ratio of the actual energy delivered by safety hammers with common lift mechanisms to the theoretical free-fall energy). The corrected SPT resistance can be computed as the product of the measured resistance and a series of correction factors, i.e., as

$$(N_1)_{60} = N_m \cdot C_N \cdot C_R \cdot C_S \cdot C_B \cdot C_E \tag{6.48}$$

The correction factors are defined in Table 6.2. By instrumenting a section of rod to record acceleration (with an accelerometer) and force (with a strain gauge) as a function of time, it is possible to compute the energy of the stress wave traveling down the rod with each hammer strike as (Abou-Matar and Goble, 1997):

$$E = \int_t F(t) \cdot \dot{u}(t)dt \tag{6.49}$$

TABLE 6.2
SPT Equipment/Procedure Correction Factors

<table>
<tr><th>Factor</th><th>Correction</th><th>Comments</th></tr>
<tr><td>C_N</td><td>$C_N = \sqrt{p_a/\sigma'_{v0}}$ where σ'_{v0} is the vertical effective stress at depth of SPT test in same units as atmospheric pressure, p_a
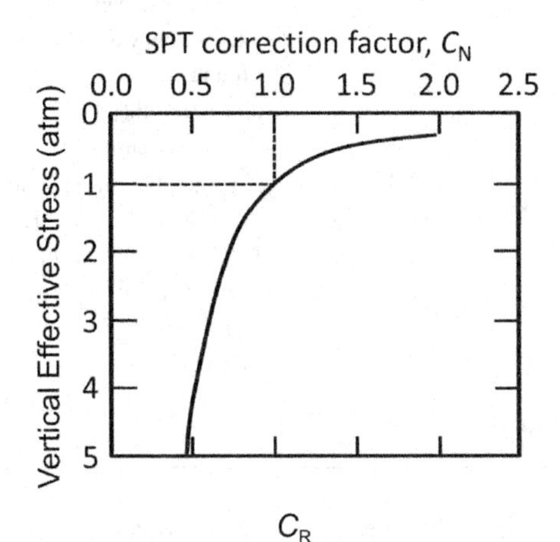
</td><td>Increased effective stress leads to increased penetration resistance for a soil of given relative density. To improve the correlation with relative density, the effect of overburden pressure can be reduced by normalizing to vertical effective stress of 1.0 atm. After Liao and Whitman (1988).</td></tr>
<tr><td>C_R</td><td>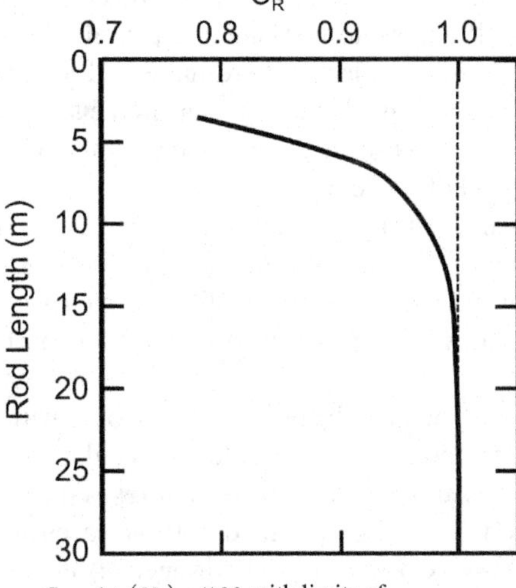
</td><td>Tension wave reflected from tip of sampler interferes with energy transfer when it reaches hammer at shallow (<10 m) depths.
Rod length is measured from point of hammer impact to tip of sampler. No rod length correction required when energy measured.
NCEER (1997); Youd et al. (2001); Seed et al. (2003)</td></tr>
<tr><td>C_S</td><td>$C_S = 1 + (N_1)_{60}/100$ with limits of $1.10 \le C_S \le 1.30$</td><td>For samplers with space for interior liners used without the liners in place, which reduces resistance to advancement of penetrometer.
Youd et al. (2001); Seed et al. (2003)</td></tr>
<tr><td>C_B</td><td>
<table>
<tr><th>Borehole diameter</th><th>C_B</th></tr>
<tr><td>65–115 mm (2.7–4.5 in.)</td><td>1.00</td></tr>
<tr><td>150 mm (6 in.)</td><td>1.05</td></tr>
<tr><td>200 mm (8 in.)</td><td>1.15</td></tr>
</table>
</td><td>Large diameter boreholes reduce confinement below sampler, thus reducing blowcounts.
Skempton (1986); Youd et al. (2001); Seed et al. (2003)</td></tr>
<tr><td>C_E</td><td>$C_E = E_R/60$
where E_R = fraction of theoretical impact energy expressed in percent</td><td>Best approach is to use direct energy measurements during sampling, otherwise use calibrated hammer (preferably with mechanical release system). If necessary, E_R can be approximately estimated from Table 6.3.</td></tr>
</table>

TABLE 6.3
Procedures for Estimating SPT Energy Correction Factor

Equipment	Approximate E_R	Approximate C_E	Comments
Safety hammer	70–120	0.7–1.2	Rope and cathead – rope not wet or excessively worn, two turns of rope around cathead, normal release
Donut hammer	50%–100%	0.5–1.0	Rope and cathead – rope not wet or excessively worn, two turns of rope around cathead, normal release
Donut hammer	70%–85%	0.5–1.0	Rope and cathead – with Japanese "throw" release
Automatic trip hammer (safety or donut)	80%–130%	0.8–1.4	$E_R > 1$ can be caused by non-zero initial hammer velocities, or drop heights larger than standard
All	–	–	For poor quality equipment and/or workmanship, further adjustments may be required

Source: After Seed et al. (1985); Skempton (1986); NCEER (1997).

where the velocity term in Equation (6.49) is obtained by integrating the recorded acceleration. This energy can be divided by the theoretical free-fall energy to compute the actual delivered energy ratio. Measurements of this sort are made for critical projects where increased accuracy in SPT blow counts is required. When these measurements are available, the measured energy ratio is substituted for the product $C_R \cdot C_E$ in Equation (6.48). In the absence of energy measurements, the energy correction factor, C_E, can be estimated with the aid of Table 6.3.

The relatively thick walls of the SPT sampler make it rugged enough to be used in stiff and dense soils. The corrected standard penetration resistance, $(N_1)_{60}$, has been correlated to many important properties of coarse-grained soils. Correlations for the properties of fine-grained soils are much less reliable, and the presence of gravel-sized particles or other obstructions can cause erroneous measurements.

Sharp increases in measured SPT resistance can indicate the presence of an obstruction such as a large particle that blocks the mouth of the sampler and effectively produces an enlarged "penetrometer." Lack of sample recovery, along with increased measured penetration resistance, can also be evidence of an obstruction. The large penetration resistances obtained in these cases are not representative of the soil behavior and should not be used. Conversely, the presence of seams or layers of fine-grained soils can lead to reductions in measured SPT resistances relative to those that would be representative of neighboring coarse-grained materials. The presence of fine-grained soils in the SPT sampler or a reduction of penetration resistance in the last portion of the test (e.g., the third of the six-inch drives) can provide a good indication of the potential effects of soft soils on measured driving resistance. For cases involving large particle obstructions or soft layer effects, correction of the measured values by interpolation/extrapolation, with careful consideration of the geologic character of the materials, may be required.

Example 6.4

A sandy site in Japan had the measured SPT resistances indicated in Table E6.4. Energy measurements indicated that the SPT equipment delivered 72% of the theoretical free-fall energy to the sampler. Assuming that the sands have an average void ratio of 0.44 and that the water table is at a depth of 1.5 m, compute the corresponding $(N_1)_{60}$ values.

Solution:

Given the void ratio of 0.44, and assuming that $G_s = 2.7$, the average dry and buoyant unit weights of the sand are 18.38 and 11.57 kN/m^3, respectively. These unit weights can be used to compute the vertical effective stress values that correspond to each depth at which the SPT resistance was measured (column 3 below). For example, the vertical effective stress at a depth of 6.2 m is given by

$$\sigma'_{v0} = (1.5\text{ m})\left(18.38\frac{\text{kN}}{\text{m}^3}\right) + (4.7\text{ m})\left(11.57\frac{\text{kN}}{\text{m}^3}\right) = 82.0\text{ kPa}$$

By converting the vertical effective stresses from units of kPa to units of tons/ft^2, the value of the correction factor at each depth can be computed by the relationship of Boulanger and Idriss (2014) in Table 9.4. Considering the measurement at a depth of 6.2 m, N_m = 9 blows/ft and $\sigma'_{v0} = 82.0$ kPa. First correcting for energy, $C_E = E_R/60 = 72/60 = 1.2$, so $N_{60} = (9)(1.2) = 10.8$ blows/ft. Anticipating that the corrected blowcount will be greater than the measured value, an initial estimate of $(N_1)_{60cs} = 14$ is made. Then,

$$m = 0.784 - 0.0768\sqrt{(N_1)_{60cs}} = 0.784 - 0.0768\sqrt{14} = 0.497$$

$$C_N = (p_a / \sigma'_{v0})^m = (101.3\text{ kPa} / 82.0\text{ kPa})^{0.497} = 1.111$$

so the computed $(N_1)_{60cs} = C_N N_{60}$ = (1.111)(10.8) = 12.0, which is lower than the initial estimate. Repeating the calculations with a revised estimate of $(N_1)_{60cs} = 12$ yields $m = 0.518$, $C_N = 1.116$, and $(N_1)_{60cs} = 12.04$, which is sufficiently close to the revised estimate. Repeating for the other depths gives the values shown in Table E6.4.

TABLE E6.4

Depth	N_m	N_{60}	σ'_{v0}	m	C_N	$(N_1)_{60}$
1.2	7	8.4	22.1	0.466	2.034	17.1
2.2	4	4.8	35.7	0.559	1.791	8.6
3.2	3	3.6	47.3	0.601	1.580	5.7
4.2	3	3.6	58.8	0.612	1.395	5.0
5.2	5	6.0	70.4	0.575	1.233	7.4
6.2	9	10.8	82.0	0.517	1.115	12.0
7.2	12	14.4	93.6	0.487	1.039	15.0
8.2	12	14.4	105.3	0.496	0.981	14.1
9.2	14	16.8	116.8	0.480	0.934	15.7
10.2	9	10.8	128.4	0.547	0.878	9.5
11.2	23	27.6	140.0	0.406	0.877	24.2
12.2	13	15.6	151.5	0.510	0.814	12.7
13.2	11	13.2	163.1	0.539	0.774	10.2
14.2	11	13.2	174.7	0.544	0.744	9.8
15.2	24	28.8	186.3	0.421	0.774	22.3
16.2	27	32.4	197.9	0.402	0.764	24.8
17.2	5	6.0	209.4	0.634	0.631	3.8
18.2	6	7.2	221.1	0.623	0.615	4.4
19.2	4	4.8	232.7	0.655	0.580	2.8
20.2	38	45.6	244.3	0.336	0.744	33.9

6.5.3.2 Becker Penetration Test

If large gravel or cobble particles in a soil layer are sufficiently prevalent that they are in contact with each other, those particles will control the behavior of the soil and the SPT should not be used. For gravelly soils, the Becker hammer penetration test (BPT) can be used in the same way that the SPT is used for sands. In a recommended BPT procedure (Harder and Seed, 1986), a closed 6.6 in. (16.8 cm)-OD (outside diameter) drill bit at the end of a 6.6 in. (16.8 cm) OD steel casing is driven into the soil by an ICE 180 diesel pile-driving hammer (with 8,100 ft-lb/blow (110 N-m/blow) rated energy). The BPT resistance is taken as the number of blows per foot of penetration, corrected for variations in the diesel hammer bounce chamber pressure (which reflects the effects of soil resistance and combustion conditions on hammer energy). By comparing the results from the BPT and SPT at the same sandy sites, Harder and Seed (1986) found that the BPT and SPT resistances were related as shown in Figure 6.64.

A significant difficulty with the standard BPT is that the diameters of the drill bit and the casing to which it is attached are the same, so the recorded penetration resistance is affected by skin friction between the casing and the surrounding soil in addition to the penetration resistance of the soil below the tip of the penetrometer. If this friction is significant and uncorrected for, it can lead to overestimation of the actual penetration resistance of the soil.

Rather than relying on bounce chamber pressures, Sy and Campanella (1994) used a pile driving analyzer (PDA) to measure the maximum transferred energy at the top of the drill string, ENTHRU. The BPT resistance can be corrected to a reference ENTHRU level of 30% of the rated energy of an ICE 180 diesel hammer

$$N_{b,30} = N_b \frac{\text{ENTHRU}}{30} \tag{6.50}$$

where ENTHRU is expressed as a percentage. To relate $N_{b,30}$ to more familiar penetration resistance measures (such as N_{60}), it is necessary to correct for the effects of shaft resistance, R_S (i.e., skin resistance along the perimeter of the casing above the tip). This is done by using PDA data in CAPWAP analyses to estimate R_S. Based on this value of R_S, $N_{b,30}$ can be correlated with SPT resistance as shown in Figure 6.65.

A modified version of the original BPT, termed the Foundex Becker penetration test (FBPT), uses an oversized, sleeved, closed-end shoe (Figure 6.66) just above which drilling mud is injected

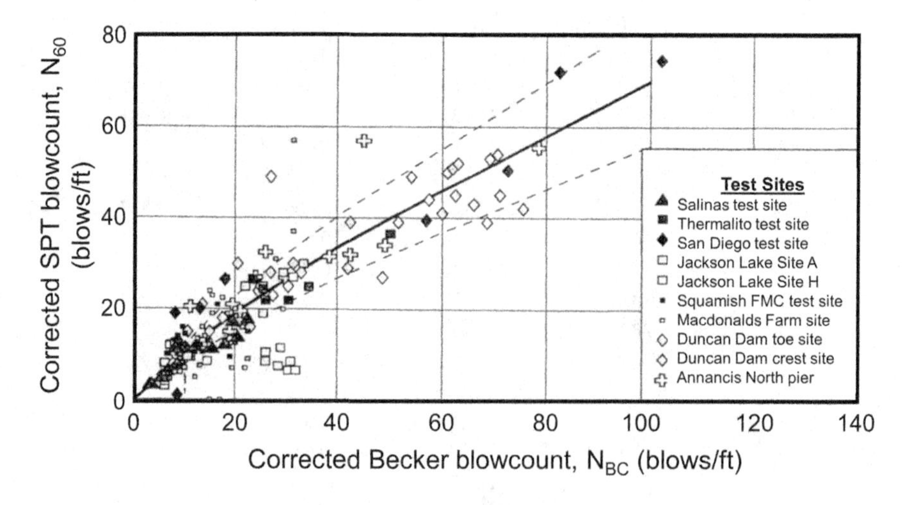

FIGURE 6.64 Relationship between corrected Becker and SPT penetration resistances (Harder and Seed, 1986).

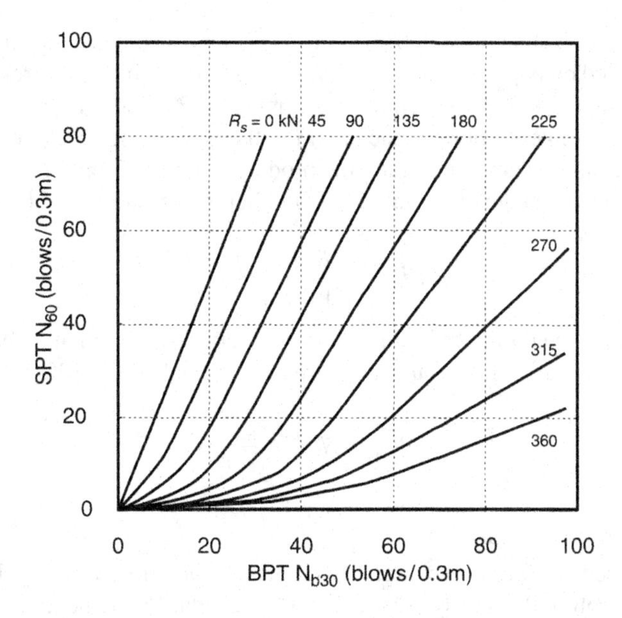

FIGURE 6.65 BPT-SPT correlations for different BPT shaft resistances. (After Sy and Campanella, 1994.) R_s represents shaft resistance.

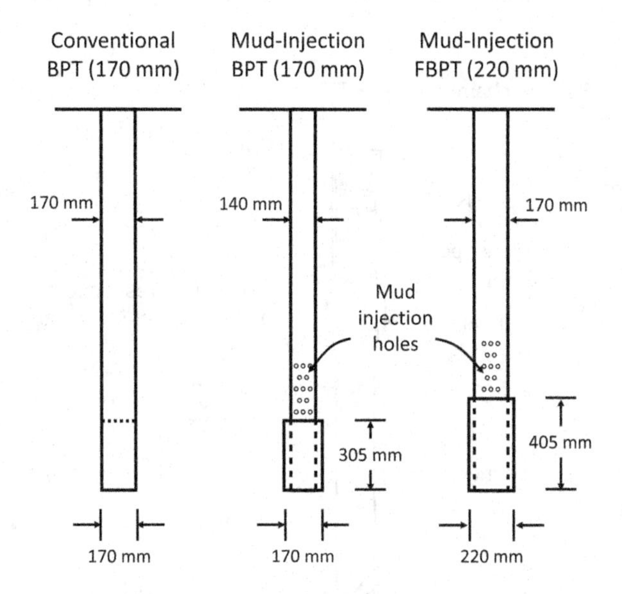

FIGURE 6.66 Comparison of configurations of conventional BPT and FBPT casings. (After Sy and Lum, 1997; used by permission of Canadian Science Publishing.)

into the soil through a series of holes in the sides of the casing. The drilling mud and oversized shoe reduce the shaft resistance along the sides of the casing so that more energy reaches the tip. Side-by-side FBPT and SPT tests (Sy and Lum, 1997) showed that the mud injection process reduced shaft resistances to about 0–60 kN such that $N_{60} \approx 1.7N_{b,30}$ for a 170 mm FBPT and $N_{60} \approx N_{b,30}$ for a 220 mm FPBT.

An instrumented Becker test (iBPT) that measures force and acceleration both at the top of the drill string and the tip of the casing has been developed to better define the energy distribution over the length of the casing (DeJong et al., 2014). Two-foot-long instrumented sections of casing, with accelerometers mounted on opposite sides and strain gauges at 90° intervals around the interior, are attached to the bottom and top of the drill string (Figure 6.67). The sensors measure the average force and acceleration from each blow of the hammer. Integrating the acceleration to obtain velocity, the energy can be obtained by integrating the product of force, F, and velocity, V. The resulting energy, expressed as a percentage of the rated ICE 180 hammer energy of 11.0 kJ, is given by

$$EFV = \frac{E}{11.0} \times 100\% \tag{6.51}$$

where the energy, E, in the numerator is conceptually identical to Equation (6.49) and expressed in units of kJ. The measured iBPT blowcounts, N_b, are then normalized to 30% reference energy

$$N_{b,30} = N_b \frac{EFV_{\max}}{30} \tag{6.52}$$

6.5.3.3 Cone Penetration Test

The CPT is widely used for characterization of subsurface conditions in geotechnical engineering practice. The CPT involves the steady penetration of a standard cone penetrometer (Figure 6.68) into the ground. The standard cone penetrometer has a conical tip with a (projected) cross-sectional area, A_c, of 10 cm² (1.55 in.²) area and 60° apex angle immediately below a cylindrical friction sleeve with a surface area, A_s, of 150 cm² (23.3 in.²) surface area. The penetrometer is pushed into

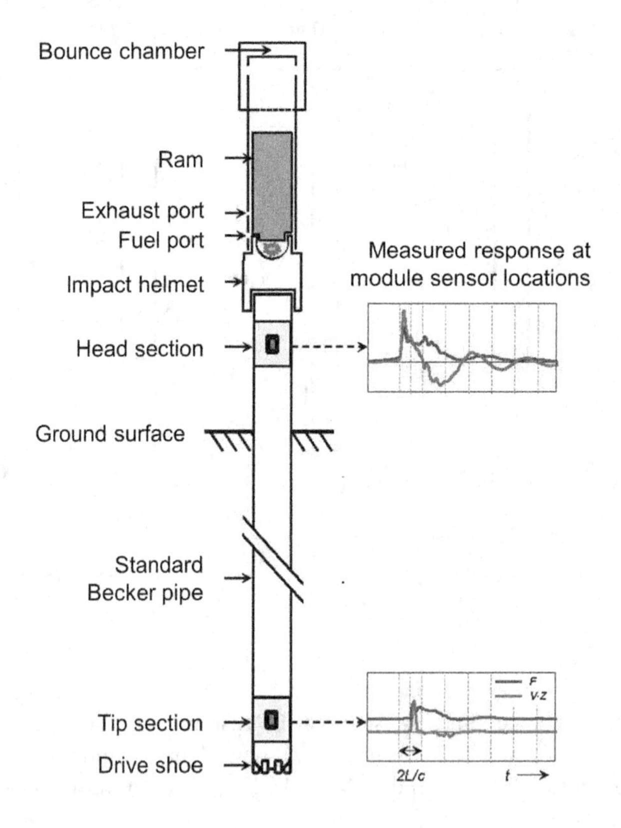

FIGURE 6.67 iBPT system: Schematic illustration of iBPT system including measurements at pile head and tip. (DeJong et al., 2017; used by permission of ASCE.)

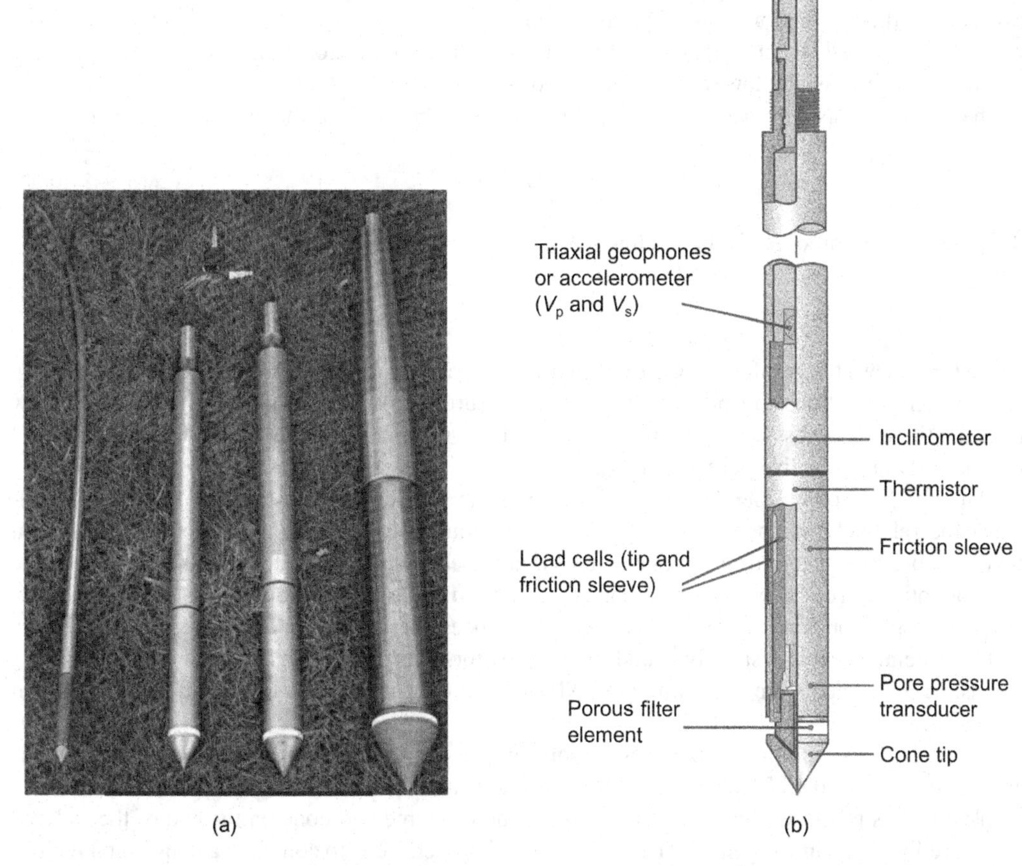

FIGURE 6.68 Typical cone penetrometers: (a) cross-sectional areas of 2 cm², 10 cm² (standard), 15 cm², and 40 cm²) (After Robertson and Cabal, 2022; used by permission of P.I. Robertson) and (b) layout with seismic cone capabilities (After Baldi et al., 1988).

the ground at a constant rate of 2 cm/s (0.8 in./s). The tip and friction sleeve are each connected to load cells that independently measure the tip load, Q_c, and sleeve load, F_s, during penetration. The cone tip resistance is expressed as $q_c = Q_c/A_c$ and the sleeve resistance as $f_s = F_s/A_s$. A *piezocone* includes one or more pore pressure transducers. A pore pressure transducer just behind the cone tip measures the pore pressure, u_2, which can be used to determine the excess pore pressure, $\Delta u = u_2 - u_0$, where u_0 is hydrostatic pore pressure. Pore pressure data can be used to correct the measured cone tip resistance for unequal area effects (Campanella et al., 1982)

$$q_t = q_c + u_2(1 - a_n) \tag{6.53}$$

where a_n is the net area ratio, which depends on the geometry of the specific CPT probe and is usually determined by calibration in the laboratory (Robertson and Cabal, 2022), but typically ranges from about 0.6 to 0.85 for soft clays and silts and at underwater sites; for sands, $a = 1.0$ so $q_t = q_c$. A similar correction can be applied to the measured sleeve friction

$$f_t = f_s - (u_2 A_{sb} - u_3 A_{st}) / A_s \tag{6.54}$$

where f_s = measured sleeve friction, u_2 = water pressure at base of friction sleeve, u_3 = water pressure at top of sleeve, A_{sb} = cross-sectional area of sleeve at base, A_{st} = cross-sectional area of sleeve at top, and A_s = surface area of sleeve. Because $A_{sb} = A_{st}$ for most cones, and both are considerably less than A_s, this correction is usually quite small and often not made. The friction ratio, $R_f = f_t/q_t$ (normally expressed in percent) is useful as an approximate indicator of soil type, typically being high in cohesive soils and low in cohesionless soils.

The net cone resistance accounts for the total stress at the depth of the tip and is defined as

$$q_n = q_t - \sigma_{v0} \tag{6.55}$$

The pore pressure ratio, B_q, is then defined as

$$B_q = (u_2 - u_0) / q_n \tag{6.56}$$

Figure 6.69 shows the results of a CPTu sounding that passes through overconsolidated silt (low tip resistance, high friction ratio, negative porewater pressure) into a clean sand (high tip resistance, low friction ratio, little excess porewater pressure) and then into a soft, silty clay (low tip resistance high friction ratio, high positive pore pressure).

CPT soundings can be performed rapidly (usually about four times faster than drilling and sampling) and relatively inexpensively. It provides a continuous profile of penetration resistance that can detect the presence of thin layers or seams that are easily missed in drilling and sampling. The slow, monotonic process by which a cone is steadily advanced into a soil deposit is simpler (and, hence, less operator-dependent) than the dynamic process of driving an SPT sampler. Thus, there are fundamental, mechanistically sound procedures for relating CPT resistance to soil properties such as the undrained strength of fine-grained soils; the CPT is a superior alternative to the SPT in such cases.

Standard CPT soundings do not provide soil samples for visual classification or laboratory testing. However, pushed CPT samplers (Robertson and Cabal, 2022) can be used to obtain disturbed samples for this purpose. These samplers replace the instrumented cone on the end of the rod and can be used with a conventional CPT rig. The typical procedure is to complete a conventional CPT, identify layers where samples are desirable, and then advance a separate sounding near the original with deployment of the sampler in the depth range of interest.

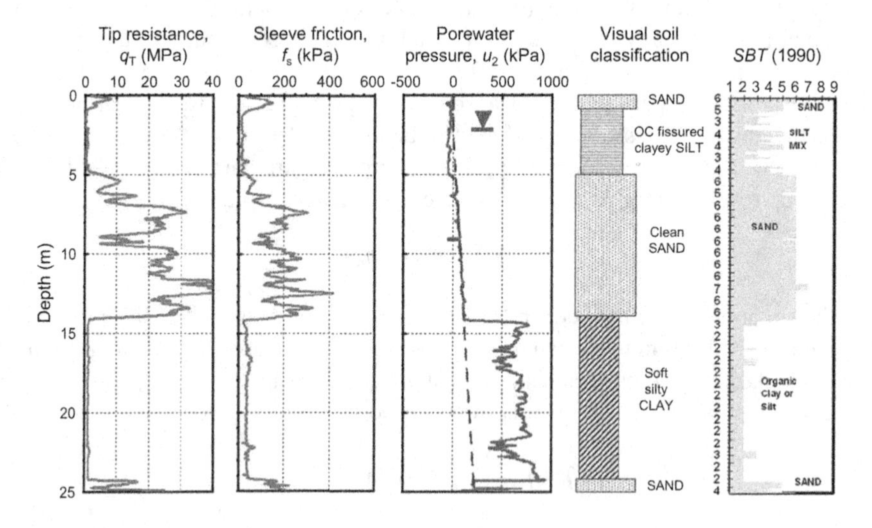

FIGURE 6.69 CPTu sounding from Steele, Missouri with soil type by visual classification and soil behavior type index (after Mayne, 2007).

The principal limitations of the CPT are that (1) it cannot be used at sites with very stiff and/or very dense soils without damaging the probe or rods; (2) the presence of over-size particles like gravels or cobbles can limit its use; and (3) it is not currently possible to obtain relatively high-quality, undisturbed samples for laboratory testing using CPT equipment.

The interpretation of CPT data typically begins by normalizing the tip resistance and sleeve friction, which makes these quantities dimensionless and allows better comparison of soils at different depths (Wroth, 1984; Houlsby, 1988; Robertson, 2012):

$$Q_{t1} = \frac{q_t - \sigma_{v0}}{\sigma'_{v0}} = \frac{q_n}{\sigma'_{v0}} \tag{6.57}$$

$$F_r = \frac{f_s}{q_t - \sigma_{v0}} = \frac{f_s}{q_n} \times 100\% \tag{6.58}$$

Note that F_r is very similar to the friction ratio R_f. Coarse-grained soils tend to produce high Q_{t1} and low F_r values relative to fine-grained soils. In order to formalize the linkages between these parameters and soil type, several soil behavior type indices have been proposed (e.g., Douglas and Olsen, 1981; Sennesett and Janbu, 1984; Robertson et al., 1986; Robertson, 1990; 2010). As shown below, these indices can be useful for estimating the behaviors associated with various soil types, but are much more approximate than the results of standard index tests (gradation and plasticity tests) performed on actual soil samples as used in conventional classification (Section 6.3).

Been and Jefferies (1992) found that boundaries between soil behavior types could be approximated by a quantitative index

$$I_c^{BJ} = \sqrt{\left(3 - \log\left(Q_{t1}\left(1 - B_q\right)\right)\right)^2 + \left(1.5 + 1.3\log F_r\right)^2} \tag{6.59}$$

with ranges for different behavior types indicated in Table 6.4.

Robertson and Wride (1998) modified the soil behavior type index for CPT data without pore pressure measurements as

$$I_c = \sqrt{\left(3.47 - \log Q_{tn}\right)^2 + \left(1.22 + \log F_r\right)^2} \tag{6.60}$$

where Q_{tn} represents a re-normalization of tip resistance by the initial vertical effective stress, this time applying a soil type-dependent exponent n as follows:

$$Q_{tn} = \left(\frac{q_t - \sigma_{v0}}{p_a}\right)\left(\frac{p_a}{\sigma'_{v0}}\right)^n \tag{6.61}$$

TABLE 6.4
Soil Behavior Types from Been and Jefferies (1992) Material Type Index, I_c^{BJ}

Soil Behavior Type	I_c^{BJ}
Gravelly sands	< 1.25
Clean to silty sand	0.24–1.80
Silty sand to sandy silt	1.80–2.40
Clayey silt to silty clay	2.40–2.76
Clays	2.76–3.22
Organic soils	> 3.22

where the exponent n varies from approximately 0.5 for clean sands to 1.0 for clays and can be estimated as

$$n = 0.381 I_c + 0.05(\sigma'_{v0} / p_a) - 0.15 \tag{6.62}$$

with a maximum value of $n = 1.0$ (at which point, Q_{tn} is equal to Q_{t1}). Note that I_c and n are co-dependent, hence they are established through an iterative procedure for a given layer in which an initial estimate of n (usually $n = 1$) is made, I_c is computed, which leads to an updated value of n. This process is repeated until n no longer changes significantly between iterations. Figure 6.70 shows how soil behavior types can be estimated from I_c. A value of $I_c = 2.60$ (Robertson and Wride, 1998) represents the approximate boundary between soil behavior type 5 (sand mixtures, i.e., silty sand to sandy silt), which are generally susceptible to liquefaction, and soil behavior type 4 (silt mixtures, i.e., clayey silt to silty clay), which are usually not susceptible.

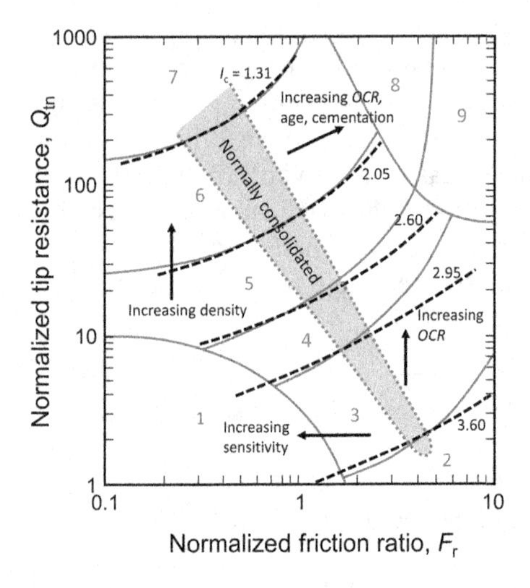

Zone	Soil Behavior Type	I_c
1	Sensitive, fine-grained	N/A
2	Organic soils – clay	> 3.6
3	Clays – silty clay to clay	2.95–3.6
4	Silt mixtures – clayey silt to silty clay	2.60–2.95
5	Sand mixtures – silty sand to sandy silt	2.05–2.6
6	Sands – clean sand to silty sand	1.31–2.05
7	Gravelly sand to dense sand	< 1.31
8	Very stiff sand to clayey sand[a]	N/A
9	Very stiff fine-grained[a]	N/A

[a] Heavily overconsolidated or cemented.

FIGURE 6.70 Normalized CPT soil behavior type (SBT_n) chart. (After Robertson, 2012; used by permission of P.K. Robertson.)

If no prior experience with CPT data exists in a particular geologic environment, soil types should be verified by drilling and sampling. If significant CPT experience has validated the soil behavior type charts against actual soil classifications, samples may not be required.

Example 6.5

A CPT was performed at the site described in Example 6.4. At a depth of 6.2 m, the measured tip and sleeve resistances were 3.2 and 0.2 MPa, respectively. Determine the Robertson and Wride (1998) soil behavior type.

Solution:

Soil behavior type is based on the soil behavior type index, I_c. Calculation of I_c requires the calculation of several quantities that contribute to it. These include the vertical stresses, which for the depth of 6.2 m are $\sigma_{v0} = 128.0$ kPa and $\sigma'_{v0} = 82.0$ kPa. Then, assuming $a_n = 1.0$ for the sandy soil, $q_t = q_c = 3.2$ MPa. The exponent, n, requires an estimate of the quantity, I_c, it is being used to compute, so an initial estimate of $I_c = 2.2$ will be used. This gives $n = 0.381 I_c + 0.05(82.0/101.3) - 0.15 = 0.729$. The dimensionless cone parameters can then be calculated as

$$Q_{tn} = \left(\frac{q_t - \sigma_{v0}}{p_a}\right)\left(\frac{p_a}{\sigma'_{v0}}\right)^n = \left(\frac{3{,}200\text{ kPa} - 128.0\text{ kPa}}{101.3\text{ kPa}}\right)\left(\frac{101.3\text{ kPa}}{82.0\text{ kPa}}\right)^{0.729} = 35.4$$

$$F_r = \frac{f_s}{q_t - \sigma_{v0}} \times 100\% = \frac{200\text{ kPa}}{3{,}200\text{ kPa} - 128\text{ kPa}} \times 100\% = 6.5$$

from which the soil behavior type index is

$$I_c = \sqrt{(3.47 - \log Q_{tn})^2 + (1.22 + \log F_r)^2} = \sqrt{(3.47 - \log 35.4)^2 + (1.22 + \log 6.5)^2} = 1.96$$

This computed value is lower than the initial estimate of 2.2, so a second iteration is required. Assuming $I_c = 2.0$ and repeating the calculations gives $n = 0.652$, $Q_{tn} = 34.8$, $F_r = 6.5$, and $I_c = 1.97$, which is sufficiently accurate for this parameter (further iterations converge to 1.972). Using the chart and table in Figure 6.70, the soil behavior type is clean to silty sand.

Measured CPT resistances can be affected by stratigraphy. As the CPT probe is pushed into the soil, it influences the stress conditions in an approximately spherical zone centered on the tip of the penetrometer. The size of the zone of influence increases with increasing soil stiffness, ranging from as low as one diameter for soft soils (e.g., soft clays) to about 15 diameters (about 0.5 m for a standard cone) for stiff/strong soils, and decreasing effective stress level (Ahmadi and Robertson, 2005). This zone of influence will affect the measured penetration resistance both ahead of and behind the tip of the cone as it passes a boundary between two different materials, for example from soft clay into a stiffer sand, or vice versa. Thus the measured CPT resistance will indicate a transition zone at what may actually be distinct boundaries between different materials. This behavior is particularly problematic for thin layers, i.e., a thin layer of soft clay within a stiffer sand or a thin layer of stiffer sand within a layer of soft clay. In such cases, the thin layer may be thinner than the transition zone, in which case the "true" penetration resistance of the thin layer is not measured. Figure 6.71 shows three soil profiles for which a cone penetrometer would pass from a relatively soft clay into an underlying layer of sand. The measured tip resistance is constant in the clay until it reaches Point A at which the penetrometer "senses" the stiffer underlying sand layer and begins to pick up resistance. As it approaches the interface between the clay and sand (Point B), the measured

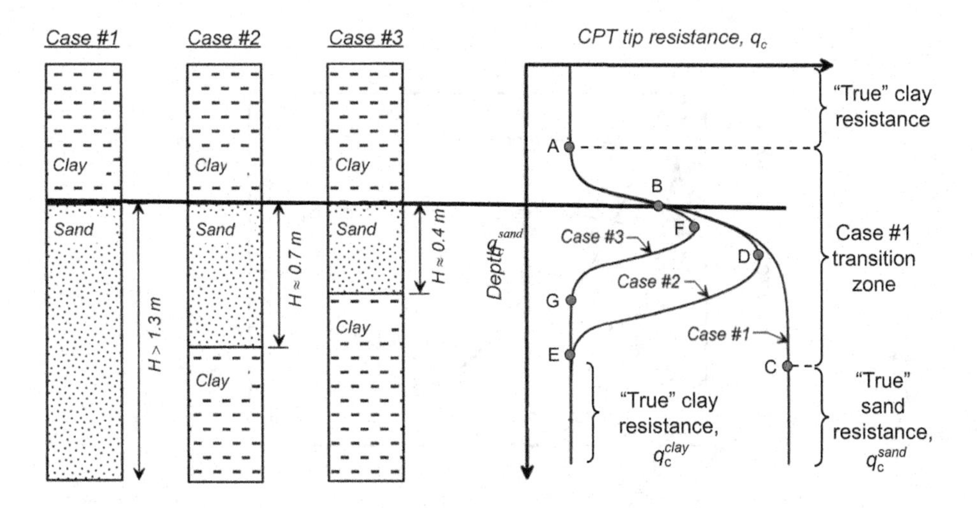

FIGURE 6.71 Schematic illustration of thin layer effects on CPT resistance. (After Idriss and Boulanger, 2008.)

tip resistance is greater than the actual "true" resistance of the clay. As the penetrometer enters the sand layer just past Point B, the tip resistance is still affected by the clay layer it has just left and is lower than the "true" resistance of the sand layer. If the sand layer is thick (Case 1), the penetration resistance will continue to increase until the zone stressed by the tip of the penetrometer has passed the interface. Below that depth (Point C), the full penetration resistance of the sand will be mobilized. If the sand layer is thin (Cases 2 and 3) relative to the transition zone, the zone stressed by the penetrometer will extend into the underlying clay layer and reduce the measured penetration resistance in the sand layer. The peak measured resistance in the sand layer (Points D and F) will not be representative of the density of the sand and can, without recognition of the thin-layer effect, cause the density of that layer to be underestimated. As the penetrometer leaves the sand layer, the measured penetration resistance will transition back to the "true" clay resistance (Points E and G). The peak measured penetration resistances can be seen to be influenced by the thickness of the sand layer; research has shown that it is also influenced by the relative strengths of the two materials. Youd et al. (2001), based on work by Robertson and Fear (1995) and Vreugdenhil et al. (1994), proposed that the "true" value of tip resistance in a thin layer of sand could be computed as

$$q_c^{sand} = K_H q_c^{peak} \tag{6.63}$$

where the correction factor, K_H (Figure 6.72), is obtained from

$$K_H = 1.0 + 0.25\left(\frac{H/d_c}{17} - 1.77\right)^2 \tag{6.64}$$

with H = thin layer thickness and d_c = cone diameter, and q_c^{peak} is the peak measured resistance in the thin layer (e.g., the values at Points D and F in Figure 6.71). Simplified elastic solutions (Vreugdenhil et al., 1994) suggest that the correction factor should increase with increasing thin-layer stiffness, but available field evidence does not support the use of correction factors greater than those given by Equation (6.67).

Boulanger and DeJong (2018) introduced an inverse filtering procedure that corrects CPT data for thin-layer and transition zone effects. The cone is assumed to act as a low-pass spatial filter that smears sharp transitions (which would be modeled by high wavenumbers, or short wavelengths) at layer boundaries. The procedure uses a filter function based on experimental studies that indicate

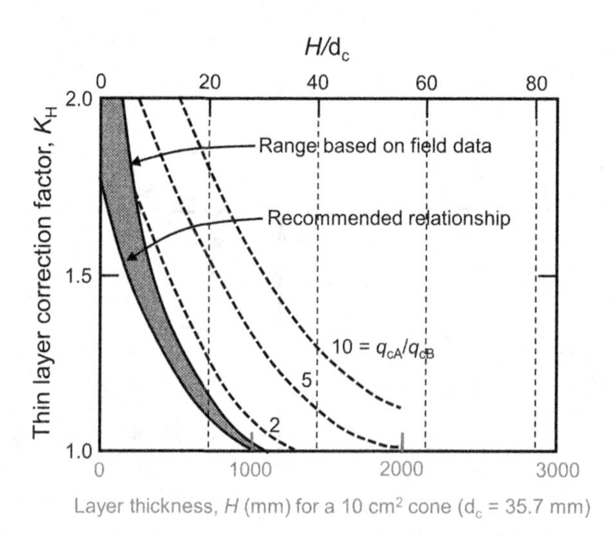

FIGURE 6.72 Correction factor for thin-layer effect. (Youd et al., 2001; used by permission of ASCE.)

normalized (by cone diameter) distances at which an advancing cone "senses" a different layer and at which the effects of a layer boundary that has been passed is no longer felt by the cone. By iteratively convolving the filter function with the measured tip resistance, an estimate of the position of the material boundary and corrected tip resistances on either side of it are obtained.

6.5.3.4 Vane Shear Test

The vane shear test is conducted *in situ* for measurement of undrained shear strength, typically in clayey soils. When a test is desired at a particular depth, drilling of a boring is halted above that depth, and a four-bladed vane (Figure 6.73) is pushed into undisturbed soils below the base of the borehole. During testing, torque applied to the rods that connect the vane to the surface shears the soil along a cylindrical surface with a radius equal to the radii of the vanes (and on planes bounded by the top and bottom of the vane). The vane can be rotated through any desired angle, even through more than 360° to produce extremely large strains on the failure surface. The torque applied to the vane and its resulting rotation are measured and recorded. The test produces a plot of torque vs rotation from which peak resistance and residual shearing resistances can be computed as shown in the inset of Figure 6.73a. Its large-straincapability makes the vane shear test very effective for measuring soil sensitivity under undrained conditions.

Measured shear strengths derived from vane shear tests are commonly corrected by multiplying the measured strength by a factor, μ (Figure 6.73b). This correction factor was developed to bring measured shear strengths into accord with undrained strengths back-computed from field case histories involving shear failure. The need for the correction is thought to result from anisotropy and rate effects that increase the strength measured in the test (high strain rates) relative to what can be mobilized in the field under static undrained conditions (low strain rates). These corrections are unlikely to remain valid for seismic conditions, as discussed further in Section 6.6.6.4.

6.5.3.5 Pressuremeter Test

The pressuremeter test (PMT) is the only *in situ* test capable of measuring stress-strain as well as shear strength behavior (Mair and Wood, 1987). The pressuremeter (Figure 6.74a) is a cylindrical device

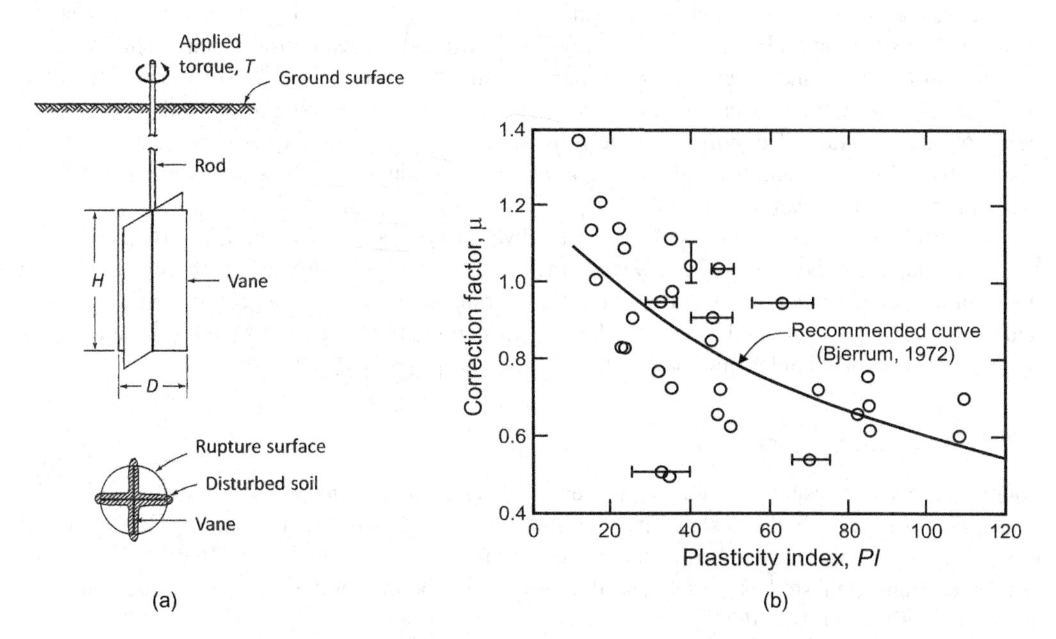

FIGURE 6.73 (a) Vane shear geometry and relations for converting torque to shear strength (After Holtz et al., 2011; reprinted by permission of Pearson Education, Inc.); (b) PI-dependent strength correction factors intended for applications involving static loading (After Ladd et al., 1977).

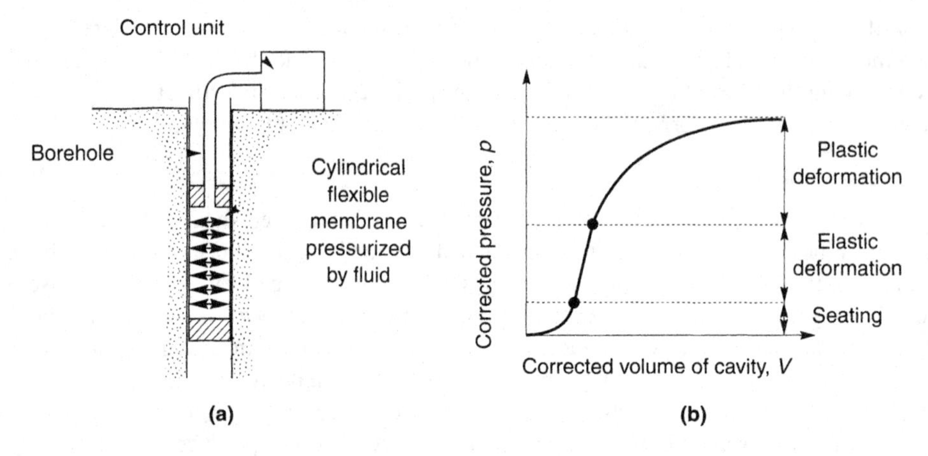

FIGURE 6.74 Pressuremeter test: (a) test setup and (b) typical pressuremeter curve. (After Mair and Wood, 1987.)

that uses a flexible membrane to apply a uniform pressure to the walls of a borehole. Deformation of the soil can be measured by the volume of fluid injected into the flexible membrane or by feeler arms for pressuremeters that use compressed gas. After correcting the measured pressures and volume changes for system compliance, elevation differences, and membrane effects, a pressure-volume curve (Figure 6.74b) can be developed. Using cavity expansion theory, the pressure-volume curve can be used to compute the stress-strain behavior. Self-boring pressuremeters, which minimize soil disturbance, and push-in pressuremeters, which can penetrate soft soils very quickly, have also been developed.

6.5.3.6 Discussion

It is important to recognize that different *in situ* tests provide information on different soil properties that relate to different aspects of soil behavior. Some provide useful information on low-strain behavior and some on high-strain behavior. Some *in situ* tests provide direct measurement of soil properties of interest and others measure characteristics to which properties of interest are indirectly related through empirical correlations. Some provide a nearly continuous indication of soil behavior as a function of depth and some provide information at discrete depth intervals. Some allow retrieval of soil samples and some do not. No individual test allows the determination of everything one might want to know about a particular soil or soil profile.

As a result, it is often necessary (and nearly always advisable) to use multiple, complementary tests to evaluate soil behavior. Shear wave velocity measurements combined with penetration test data can help characterize soil behavior over a wide range of strain amplitudes. Borings with SPT and CPT soundings performed at a particular site can capture the benefits of high resolution offered by the CPT and soil samples obtained by the SPT.

6.5.4 Laboratory Tests

Laboratory tests are usually performed on relatively small specimens that are assumed to be representative of a larger body of soil. The specimens are tested as elements (i.e., they are subjected to uniform initial stresses and uniform changes in stress or strain conditions). In other laboratory tests, specimens are tested as models and the results must be interpreted in terms of the boundary conditions acting on the model.

The ability of laboratory tests to provide accurate measurements of soil properties depends on the compatibility of the structure of the soil element in the lab with the field condition (which could differ as a result of sample disturbance) and on the ability of a laboratory test device to replicate the initial conditions and loading conditions of the problem of interest. No laboratory test can represent all possible stress and strain paths with general rotation of principal stress axes; consequently, different tests will be most suitable for different problems.

6.5.4.1 Test Specimens

Laboratory tests should ideally be performed on test specimens that are desirably uniform in composition and representative of *in situ* conditions. The extent to which these characteristics can be achieved, however, varies from soil type to soil type and from site to site depending on depositional conditions. In some cases, (relatively) undisturbed soil specimens can be obtained by drilling and sampling. In others, however, specimens (or models) must be reconstituted from disturbed or bulk samples of the *in situ* soil.

For problems involving the response of soils to be placed as fills, specimens can be constructed from bulk or disturbed samples by simulating the compaction process as closely as possible in the laboratory. When the properties of an existing soil are needed, however, the problem becomes more difficult. Tests on existing soils can be performed on undisturbed or reconstituted specimens. However, in many instances, the results of these tests will differ because of differences in soil fabric between natural and reconstituted soil specimens, even when densities and applied stresses are similar.

Soil Sampling

Dynamic soil properties are influenced by many factors, including density, stress conditions, and other factors such as soil fabric or structure, age, stress and strain history, and cementation. While the void ratio and stress conditions can be recreated in a reconstituted specimen, the effects of the other factors cannot. Since the effects of these other factors are manifested primarily at low strain levels, they are easily destroyed by sample disturbance. For the results of laboratory tests to reflect the actual behavior of the *in situ* soil as closely as possible, high-quality samples that minimize the effects of sample disturbance must be obtained.

For fine-grained soils, procedures for the preparation of test specimens by carefully trimming thin-walled tube or block samples are fairly well established (e.g., U.S. Army Corps of Engineers, 1980). Such samples are customarily described as "undisturbed samples" by geotechnical engineers, but it is understood that some degree of disturbance is inevitable. Fine-grained soils have sufficiently low permeability and/or sufficiently high plasticity that they can maintain significant stiffness and strength through the development of negative pore pressures during the sampling process. Even these soils experience some disturbance, however, as a result of shear induced by penetration of the sampler, unloading of the material during sampling and eventual extrusion from the sample tube, and reloading to the desired consolidation stresses. Each of these steps can potentially shear the specimen and therefore affect its response to subsequent undrained loading.

Sampling to mitigate disturbance of coarse-grained soils such as clean sands and gravels is more difficult. Even thin-walled sampling tubes can cause significant disturbance of clean sands, causing densification of loose sands and dilation of dense sands (Marcuson et al., 1977). The densification of a loose sand by the vibrations involved in driving the sampler and then bringing it back to the ground surface frequently causes the sample to fall out of the sampling tube. The use of block sampling (Horn, 1978) has proven effective, but the process is laborious and may not be practical below certain depths.

The most reliable way of determining the *in situ* density of the types of cohesionless soils whose behavior can be very sensitive to *in situ* density is through ground freezing and coring (Hvorslev, 1949; Singh et al., 1982; Yoshimi et al., 1978, 1984; Konno et al., 1993; Wride et al., 2000; Hofmann et al., 2000). In this approach, a freeze pipe is inserted into the ground below the bottom of a boring. The freeze pipe allows the circulation of liquid nitrogen into a reservoir within the probe (and venting of nitrogen gas). The liquid nitrogen is introduced at a rate that allows freezing of the soil to take place slowly enough that the volumetric expansion of the water upon freezing (9% increase in volume) does not fracture or otherwise disturb the soil. As the freezing front migrates away from the freeze pipe, the extra water is pushed away and a frozen zone of undisturbed soil is left behind. Conventional rock core drilling and sampling techniques can then be used to obtain samples of the frozen soil. The samples are generally maintained in a frozen condition throughout the process of transportation to a laboratory, and even insertion into a test apparatus. The samples are then slowly thawed with access to a water reservoir that allows water to be drawn into the sample as necessary due to the contraction of the water during thawing. The process, however, is quite expensive and generally used only on critical projects.

A new type of sampler, the gel-push sampler (Mori and Sakai, 2016), uses a thick, viscous polymer gel to reduce friction between the sample and the core barrel and to help preserve the soil fabric. The polymer gel acts as a drilling fluid, but is simply pushed out of the core barrel around the perimeter of the sample is it enters to remove the cuttings, cool the drill bit, and protect the sample. Several different sampler configurations, each of which is most useful for specific applications, are available. The gel then coats the sample and can protect and support it during extrusion from the sampler before being trimmed away prior to laboratory testing. Gel-push sampling has allowed retrieval of high-quality samples of sandy and gravelly soils with fines for which freezing does not work well; it has also worked well in silts and silty sands with very low CPT resistances but not in micaceous silts (Stringer et al., 2016). The procedure appears promising and is likely to see increasing use as designs and procedures are more fully developed (Taylor et al., 2012).

Specimen Reconstitution

Because undisturbed samples are so difficult to obtain, laboratory tests on coarse-grained soils are usually performed on reconstituted specimens. Reconstituted specimens are usually constructed in a rigid mold into which the soil can be placed in different ways.

In the *moist tamping* approach (Ladd, 1978; Been et al., 1991), a desired weight of dry soil is mixed with a small amount (3%–5%) of water, placed loosely in the mold, and then tamped to a desired volume. Capillary effects in the water tend to "bulk" the moist soil, allowing it to be placed at very low densities (even negative relative densities). For a triaxial test, it is common to place the soil in six layers, each tamped to a height of one inch in order to create a six-inch tall specimen. By controlling the weight and thickness of each layer, a desired (average) layer density can easily be obtained, but computed tomography (CT) scans by Thomson and Wong (2008) showed consistent trends of linear increase in void ratio within each tamped layer. Very loose moist-tamped specimens may develop a metastable, "honeycomb" structure that can produce large volumetric strains upon saturation. While relatively simple and convenient, moist tamping does not replicate the conditions under which soils are deposited in nature; whereas it can produce a desired *in situ* density, it does not produce a natural *in situ* soil fabric.

Reconstituted specimens can also be constructed by *dry pluviation*, in which the particles are rained into the mold from a constant height above the soil surface. In some cases, the soil is placed using a funnel, which must then be moved around the mold to produce a level surface; in other cases, a screened tube of diameter just smaller than the inside diameter of the mold is placed in the mold, filled with dry sand, and then slowly raised with a constant drop height. The screened tube reduces lateral movement of the soil and contributes to greater specimen uniformity. Because of potential segregation effects, dry pluviation is best suited to uniformly graded soils.

Finally, specimens can be reconstituted by *wet pluviation* (Bishop and Henkel, 1962; Vaid and Negussey, 1984). Classical wet pluviation is very similar to dry pluviation except that the mold is filled with water so that the soil rains down through the water as it is deposited in the mold. The soil may be placed continuously or in a series of layers with time for particle sedimentation between placement of each layer. The classical procedure has the potential for segregation in well-graded soils or soils containing fine particles (since the larger particles will settle faster than the smaller particles). In a variation of this procedure, sand and water are placed in a flask and the water boiled under vacuum to remove entrapped air. The flask is capped by a valved-stopper such that no air remains in the system. The mold is also filled with de-aired water. The flask is then inverted, its neck placed below the water surface in the mold and the valve opened allowing the soil to rain down through the water into the mold. Another wet pluviation technique is the slurry deposition method (Kuerbis and Vaid, 1998) in which soil and de-aired water are mixed to form a thick slurry within a cylindrical mixing tube. The tube is vigorously rotated until the slurry is thoroughly mixed and then placed into the sample mold. The mixing tube is then extracted, leaving the loose, saturated, mixed soil in the mold. The slurry deposition method can produce uniform specimens of well-graded soils and soils containing fines, which tend to segregate with the other wet pluviation techniques.

Specimen reconstitution procedures can influence test results, even when the procedures produce specimens with the same overall density. As a result, care should be taken to note the procedures used when interpreting the results of laboratory tests on reconstituted specimens. This point becomes more important with increasing coefficient of uniformity and is particularly important for materials such as silty sands whose behavior can be strongly influenced by whether the silt particles are within the voids between sand particles that are in contact with each other ($FC < FC_L$), or are between the sand grains themselves ($FC > FC_L$).

6.5.4.2 Low-Strain Element Tests

Only a limited number of laboratory tests are able to determine the properties of soils at low strain levels. These include the resonant column test, the ultrasonic pulse test, and the piezoelectric bender element test. Sample disturbance effects are known to affect the results of low-strain element tests (Chiara and Stokoe, 2006); a more detailed description of these effects is presented in Section 6.6.3.1.

Resonant Column Test

The resonant column test is the most commonly used laboratory test for measuring the low-strain-properties of soils. It subjects solid or hollow cylindrical specimens to harmonic torsional or axial loading by an electromagnetic loading system (Figure 6.75a). The loading systems usually apply harmonic loads for which the frequency and amplitude can be controlled, but random noise loading (Al-Sanad and Aggour, 1984) and impulse loading (Tawfiq et al., 1988) have also been used.

After the resonant column specimen has been prepared and consolidated, cyclic loading is begun. The loading frequency is initially set at a low value and is then gradually increased under conditions of constant torque amplitude until the response (specimen rotation, hence strain amplitude) reaches a maximum. The lowest frequency at which the response is locally maximized is the fundamental frequency of the specimen. The fundamental frequency is a function of the low-strain stiffness of

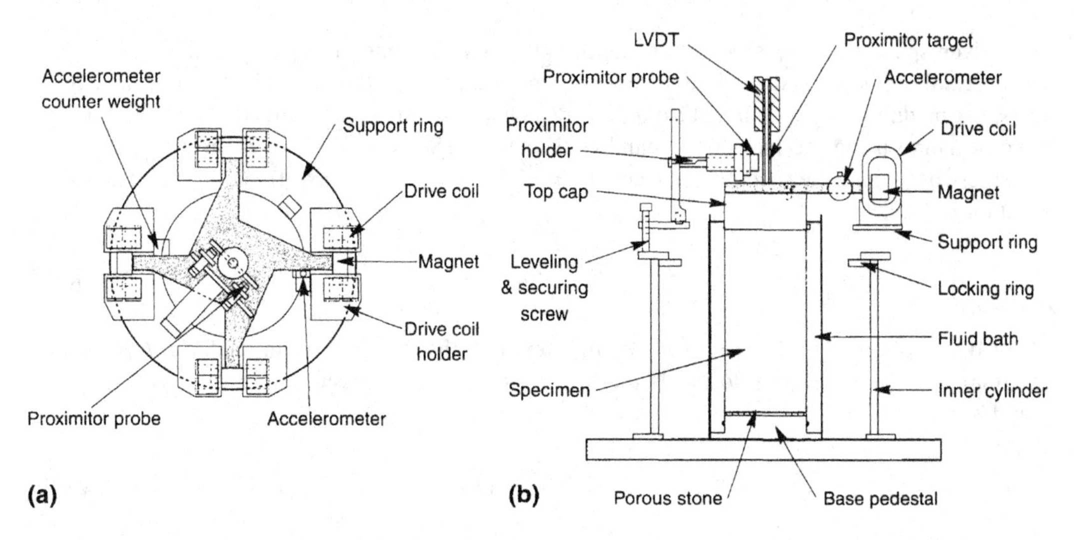

FIGURE 6.75 Typical resonant column test apparatus: (a) top view of loading system, and (b) profile view of loading system (left side magnet and coil not shown) and soil specimen. (After EPRI, 1993.)

the soil, the geometry of the specimen, and certain characteristics of the resonant column apparatus. The process begins with a low-amplitude voltage across the coils to induce small strains in the soil. The voltage is then increased after each frequency sweep to measure the response of the specimen at successively higher strain amplitudes.

To relate the shear modulus to the fundamental frequency, it is useful to consider a resonant column specimen of height, h, fixed against rotation of its base, with polar moment of inertia, J, subjected to harmonic torsional loading (Figure 6.75b). From Equation (C.8), the elastic resistance of the specimen produces a torque at its top,

$$T = GJ\frac{\partial \theta}{\partial z} = G\frac{I}{\rho}\frac{\partial \theta}{\partial z} \tag{6.65}$$

where I is the mass polar moment of inertia of the specimen. This torque must be equal to the torque applied by the loading system. If the elements of the torsional loading system connected to the top of the specimen have a mass polar moment of inertia, I_0, the applied torque is

$$T = -I_0 h\frac{\partial^2 \theta}{\partial t^2} \tag{6.66}$$

Assuming that the rotations of the specimen are also harmonic, they can be described by

$$\theta(z,t) = \Theta(z)(C_1 \cos \omega t + C_2 \sin \omega t) \tag{6.67}$$

where $\Theta(z) = C_3 \cos kz + C_4 \sin kz$ and k is the wave number (Section C.2.1.3). The zero rotation boundary condition at the base (z = 0) requires $C_3 = 0$, and the equality of Equations (6.65) and (6.66) requires, at the fundamental frequency, $\omega_n = k_n V_s$, that

$$G\frac{I}{\rho}C_4 \cos k_n h(C_1 \cos \omega_n t + C_2 \sin \omega_n t) = -I_0 h\left(-\omega_n^2 C_4 \sin k_n h\right)(C_1 \cos \omega_n t + C_2 \sin \omega_n t) \tag{6.68}$$

which can be expressed as

$$\frac{I}{I_0} = \frac{\omega_n h}{V_s}\tan\frac{\omega_n h}{V_s} \tag{6.69}$$

For a given specimen, I, I_0, and h are generally known at the time that cyclic loading begins. The fundamental frequency is then obtained experimentally, and Equation (6.69) is used to calculate V_s. The shear modulus is then obtained from $G = \rho V_s^2$. Damping can be determined from the frequency response curve using the half-power bandwidth method (Section B.6.1) or from the logarithmic decrement by placing the specimen in free vibration. For longitudinal (axial) loading, the analogous equation is

$$\frac{W}{W_0} = \frac{\omega_n h}{V_l}\tan\frac{\omega_n h}{V_l} \tag{6.70}$$

where W is the weight of the specimen, W_0 the weight of the loading system, and $V_l = E/\rho$ the longitudinal wave propagation velocity. If the loading system was massless (I_0 = 0), Equation (6.69) would become

$$V_s = \frac{2\omega_n h}{\pi} = 4 f_n h \tag{6.71}$$

where f_n is the fundamental frequency in Hertz. In this case, the rotations would follow a quarter-sine-wave pattern over the height of the specimen at the fundamental frequency. Adding the mass of the loading system results in a more linear variation of rotation and, consequently, more uniform strain conditions over the height of the specimen.

The shear strain in a solid cylindrical resonant column specimen loaded in torsion varies from zero at the centerline of the specimen to a maximum value at its outer edge. In situations in which the shear modulus varies with shear strain amplitude, the effects of nonuniform strain can be significant (Drnevich, 1967, 1972). The use of hollow specimens minimizes the variation of shear strain amplitude across the specimen.

Large-diameter resonant column devices have been used for gravelly soils (Woods, 1991) and rock (Prange, 1981). Konno et al. (1993) performed what could be described as *in situ* resonant column tests of gravelly soils at a potential nuclear power plant site in Japan. In these tests the material in two 10-m-ID (32.8 ft) circular trenches was excavated and replaced with water-filled bags to depths of 5 m (16 ft) and 9 m (30 ft). Concrete cap blocks 3 m (10 ft) and 5 m (16 ft) thick were cast on top of each of the resulting soil columns. Vibratory shakers were placed on top of the blocks near the edges; cyclic torsional loading was applied to the soil column by operating the jacks 180° out of phase. By performing frequency sweeps, the response of the soil columns could be measured at shear strain amplitudes up to 0.01%.

Example 6.6

A 6-in.-high specimen of soft silty clay with a unit weight of 105 lb/ftt[3] is tested in a resonant column device with $I/I_0 = 0.4$. From the frequency response curve shown in Figure E6.6, determine the shear modulus and damping ratio of the specimen.

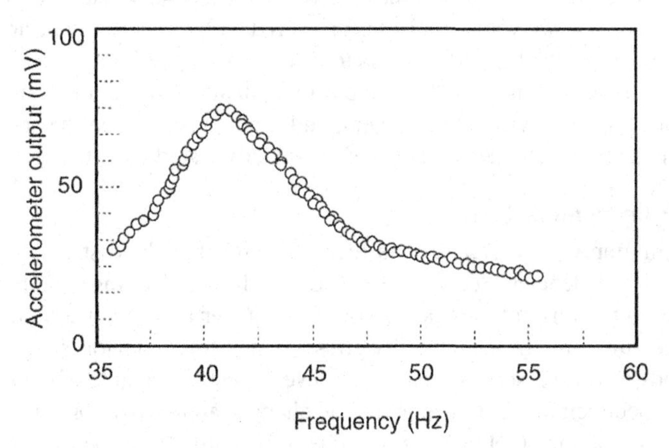

FIGURE E6.6

Solution:

The maximum amplitude of the accelerometer output occurs at the fundamental frequency of the specimen, determined graphically to be $f_n = 41$ Hz. Then Equation (6.69) can be written as

$$\frac{I}{I_0} = \frac{\omega_n h}{V_s} \tan \frac{\omega_n h}{V_s} = 0.4$$

which is satisfied when $\omega_n h/V_s = 0.593$. Then

$$V_s = \frac{\omega_n h}{0.593} = \frac{2\pi f_n h}{0.593} = \frac{2\pi(41)(0.5)}{0.593} = 212 \text{ ft / s}$$

$$G = \rho V_s^2 = \frac{105 \text{ lb/ft}^3}{32.2 \text{ ft/s}^2}(212 \text{ ft / s})^2 = 146{,}557 \text{ lb/ft}^2$$

Using the half-power bandwidth method, the half-power level is equal to $74.3\text{ mV}/\sqrt{2} = 52.6\text{ mV}$. The upper and lower frequencies at that level are 43.0 and 38.5 Hz, respectively. The damping ratio is then estimated as

$$\xi = \frac{f_{upper} - f_{lower}}{2f_n} = \frac{43.9\text{ Hz} - 38.5\text{ Hz}}{2(41\text{ Hz})} = 0.066 = 6.6\%$$

The resonant column test allows stiffness and damping characteristics to be measured under controlled conditions. The effects of effective confining pressure, strain amplitude, and time can readily be investigated. However, measurement of porewater pressure is difficult, and the material properties are usually measured at frequencies above those of most earthquake motions.

Bender Element Test

Another type of test that allows measurement of shear wave velocity on laboratory specimens makes use of piezoelectric bender elements (Shirley and Anderson, 1975; De Alba et al., 1984; Dyvik and Madshus, 1985; Da Fonseca et al., 2009). Bender elements are constructed by bonding two piezoelectric materials together in such a way that a voltage applied to their faces causes one to expand while the other contracts, causing the entire element to bend as shown in Figure 6.76. Similarly, a lateral disturbance of the bender element will produce a voltage, so the bender elements can be used as both s-wave transmitters and receivers.

In most setups, the bender elements protrude into opposite ends of a soil specimen. A voltage pulse is applied to the transmitter element, which causes it to produce an s-wave. When the s-wave reaches the other end of the specimen, distortion of the receiver element produces another voltage pulse. The time difference between the two voltage pulses is measured with an oscilloscope and divided into the distance between the tips of the bender elements to give the s-wave velocity of the specimen.

Piezoelectric bender elements have been incorporated into conventional and cubical triaxial devices, direct simple shear devices, oedometers, and model tests. Since the specimen is not disturbed during the bender element test, it can be subsequently tested for other soil characteristics.

6.5.4.3 High Strain Element Tests

At high shear strain amplitudes (greater than the volumetric threshold strain, γ_{tv}), soils generally exhibit volume change tendencies (Section 6.4.1.3). Under drained loading conditions, these tendencies are manifested in the form of volumetric strain, but under undrained conditions they result in changes in pore pressure (and, hence, effective stress). Since soil behavior is governed by effective stresses, all methods of testing soils at high strain levels must be capable of controlling porewater drainage from the specimen and measuring volume changes and/or pore pressures accurately. The problem of system compliance (volume changes due to compliance of the testing apparatus rather

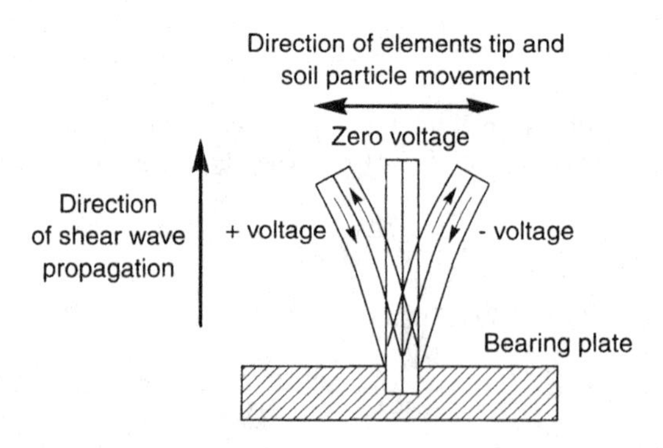

FIGURE 6.76 Piezoelectric bender element. Positive voltage causes element to bend one way, negative voltage causes it to bend the other.

than the soil), which can lead to errors in volume change/pore pressure measurement, is important in the interpretation of high-strain test results. Membrane penetration in coarse-grained soil is an important contributor to system compliance.

Cyclic Triaxial Test

Just as the triaxial compression test is the most commonly used laboratory test for the measurement of soil properties under static loading conditions, the cyclic triaxial test has historically been the most commonly used test for the measurement of dynamic soil properties of both plastic and nonplastic soils at high strain levels. In the triaxial test, a cylindrical specimen is placed between top and bottom loading platens and surrounded by a thin rubber membrane (Figure 6.77). Prior to cyclic loading, the specimen is subjected to a radial stress, usually applied pneumatically, and an axial stress. The radial stress is applied by pressurized fluid in a cell, and hence also acts vertically on the top cap of the specimen. Additional axial stress can be applied through a rod connected to a load cell, as shown in Figure 6.77. By virtue of these boundary conditions, the principal stresses in the specimen are always vertical and horizontal.

The difference between the axial stress and the radial stress is called the *deviator stress.* In the cyclic triaxial test, the deviator stress is applied cyclically, either under stress-controlled conditions (typically by pneumatic or hydraulic loaders), or under strain-controlled conditions (by servo-hydraulic or mechanical loaders). Cyclic triaxial tests are most commonly performed with the radial stress held constant and the axial stress cycled at a frequency of about 1 Hz. The amplitude of loading in a cyclic triaxial test is often expressed in the normalized form of a *cyclic stress ratio, CSR,* taken as the ratio of half the deviator stress normalized by minor principal effective stress to which the soil was consolidated prior to the onset of cyclic loading

$$CSR = \frac{\sigma_{dc}}{2\sigma'_{3c}} \tag{6.72}$$

where σ'_{3c} is the minor principal effective stress during consolidation, and σ_{dc} is the cyclic deviator stress. The quantity $\sigma_{dc}/2$ corresponds to the maximum cyclic shear stress on any plane in the specimen.

As with the static triaxial test, the cyclic triaxial test can be performed under isotropically consolidated or anisotropically consolidated conditions, thereby producing the stress paths shown

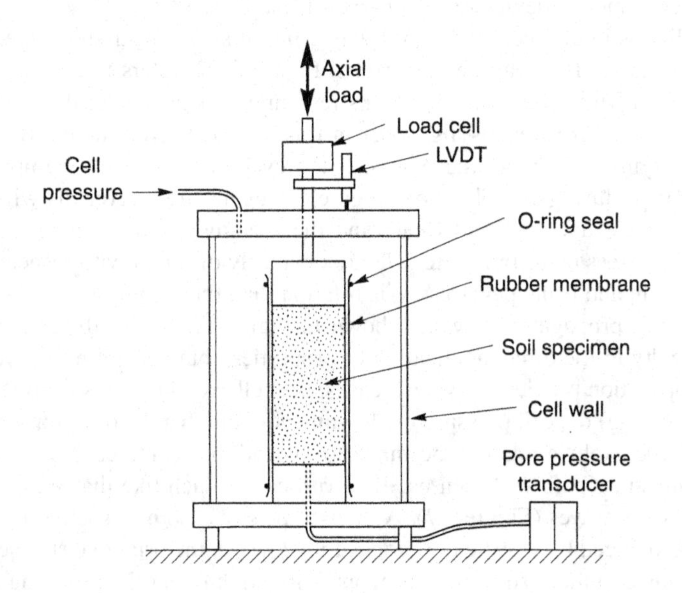

FIGURE 6.77 Typical triaxial apparatus.

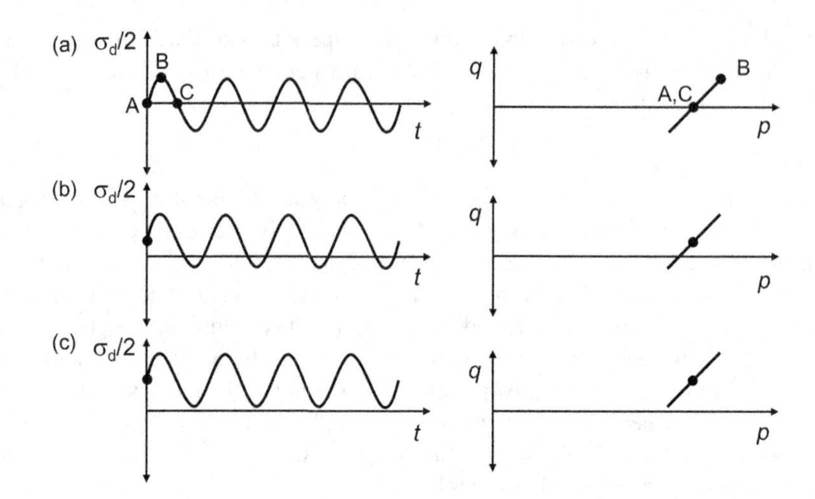

FIGURE 6.78 Time series of deviator stress and total stress paths for (a) isotropically consolidated conditions, (b) anisotropically consolidated conditions with cyclic deviator stress amplitude greater than deviator stress during consolidation (producing stress reversals), and (c) anisotropically consolidated conditions with cyclic deviator stress amplitude less than deviator stress during consolidation (no stress reversals).

in Figure 6.78. Figure 6.78a shows the cyclic deviator stress and total stress path for an isotropically consolidated specimen. Isotropically consolidated tests are commonly used to represent level-ground sites where no initial shear stresses exist on horizontal planes. The test begins with zero deviator (shear) stress (point A) and the deviator stress is initially increased. Since the axial stress is then greater than the radial stress, the major and minor principal stress axes are vertical and horizontal, respectively. After the deviator stress reaches its maximum value (point B), it decreases and approaches a value of zero (point C). Just before it reaches point C, the major principal stress axis is still vertical, but it rotates instantaneously to horizontal as point C is passed and the deviator stress becomes negative. At point C, no shear stress exists on the specimen. This process of stress reversal repeats itself throughout the test, with instantaneous 90° rotations of the principal stress axes occurring every time the deviator stress passes through zero.

To model conditions in and beneath slopes where initial static shear stresses exist, anisotropically consolidated triaxial tests can be performed. Figure 6.78b refers to an anisotropically consolidated specimen for which the cyclic deviator stress amplitude is greater than the deviator stress during consolidation. Stress reversals also exist in this situation, even though the cyclic deviator stress is no longer symmetric about the p-axis. If the cyclic deviator stress amplitude is smaller than the deviator stress during consolidation (Figure 6.78c), no stress reversals will occur. For this case, the principal stress axes will not rotate and the specimen will never reach the zero shear stress condition. The stress paths in Figure 6.78 are obviously different with respect to initial stress conditions, stress path, and principal stress axis rotation than those imposed on the element of soil subjected to vertically propagating s-waves shown in Figure 6.7. These differences illustrate the fundamental difficulty in the direct application of properties obtained from the cyclic triaxial test to actual wave propagation problems. In some cases, the cell pressure is also applied cyclically. By decreasing (or increasing) the cell pressure by the same amount that the deviator stress is increased (or decreased) by, the Mohr circle can be made to expand and contract about a constant center point. The resulting stress path will then oscillate vertically, much like that shown for the case of vertically propagating s-waves (Figure 6.7). Although the stress path of such a triaxial test can be made to match that induced by a vertically propagating s-wave, the principal stresses in the triaxial test remain constrained to the vertical and horizontal direction rather than rotating continuously as caused by the s-wave.

The stresses and strains measured in the cyclic triaxial test can be used to compute the shear modulus and damping ratio (Sections 6.6.3 and 6.6.4). The cyclic triaxial test allows stresses to be applied uniformly, although stress concentrations can exist in the specimen near the cap and base. The test also allows drainage conditions to be controlled (when the potential effects of membrane penetration are mitigated). It requires only minor modification of standard triaxial testing equipment. On the other hand, the cyclic triaxial test cannot model stress conditions that exist in most actual seismic wave propagation problems. Bedding errors and system compliance effects generally limit measurements to shear strains greater than about 0.01%, although local strain measurement (e.g., Burland and Symes, 1982; Ladd and Dutko, 1985; Goto et al., 1991; Yimsiri and Soga, 2002) can produce accurate measurements at strain levels as small as 0.0001%.

Membrane penetration effects can be important in cyclic triaxial tests of coarse sands and gravels. After consolidation, the thin triaxial membrane will penetrate the perimeter voids of coarse sand and gravel specimens. As excess pore pressures develop during cyclic loading, the net pressure on the membrane decreases and its penetration decreases. When this happens, the effective volume of the voids increases and the excess pore pressure drops below the level it would have had if true constant-volume conditions had been maintained. Because they allow the effective stresses to be higher than they would be under constant-volume conditions, membrane penetration effects can lead to inaccurate stiffness and damping measurements and unconservative estimation of liquefaction resistance (Chapter 9). Procedures have been developed for measurement (Vaid and Negussey, 1984; Kramer and Sivaneswaran, 1989a), minimization (Lade and Hernandez, 1977; Raju and Venkataramana, 1980), compensation (Seed and Anwar, 1986; Tokimatsu and Nakamura, 1986), and post-test correction (Martin et al., 1978; Kramer and Sivaneswaran, 1989b) of membrane penetration effects.

Undrained cyclic triaxial tests on loose, saturated, nonplastic soils can also be complicated by specimen nonuniformity. As high pore pressures develop in a cyclic triaxial test specimen, the soil grains tend to settle causing densification of the lower part and loosening of the upper part of the specimen. The nonuniform density leads to nonuniform strain, and eventually to thinning or necking of the upper portion of the specimen. This nonuniformity can cause considerable uncertainty in the application of cyclic triaxial test results to field conditions.

Example 6.7

A cyclic triaxial test on a saturated clay specimen produces the stress-strain loop shown in Figure E6.7. Determine the secant shear modulus and damping ratio.

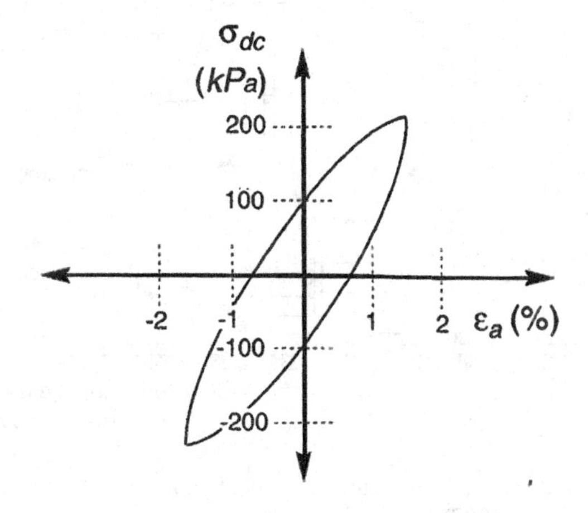

FIGURE E6.7

Solution:

Graphically, the slope of a straight line between the ends of the stress-strain curve shows that

$$E_{sec} = \frac{472 \text{ kPa}}{0.028} = 16{,}857 \text{ kPa}$$

Then, assuming that the saturated clay is loaded under undrained conditions, $\nu = 0.5$, so

$$G = \frac{E}{2(1+\nu)} = \frac{16{,}857 \text{ kPa}}{2(1+0.5)} = 5{,}619 \text{ kPa}$$

The area of the hysteresis loop is 4.52 kPa and the area of the triangle denoting the maximum strain energy is 1.65 kPa. Then

$$\xi = \frac{1}{4\pi}\frac{\text{Area of hysteresis loop}}{\text{Area of triangle}} = \frac{4.52 \text{ kPa}}{1.65 \text{ kPa}} = 0.218$$

Cyclic Direct Simple Shear Test

The cyclic direct simple shear test is capable of reproducing the common stress conditions associated with vertically propagating shear waves much more accurately than the cyclic triaxial test. It is therefore particularly useful for liquefaction, cyclic softening, and seismic compression evaluations. In the cyclic direct simple shear test, a short, cylindrical specimen is restrained against lateral expansion by rigid boundary platens (Cambridge-type device), a wire-reinforced membrane (NGI-type device), or a series of stacked rings (SGI-type device). By applying cyclic horizontal shear stresses to the top or bottom of the specimen, the test specimen is deformed (Figure 6.79) in much the same way as an element of soil subjected to vertically propagating s-waves (Figure 6.7a). The amplitude of cyclic simple shear loading can also be expressed in dimensionless form by normalizing the cyclic shear stress, τ_{cyc}, by the effective stress on the plane of maximum shear stress, i.e., σ'_{v0}. Since that plane is horizontal in the cyclic simple shear test, the cyclic stress ratio is

$$CSR = \frac{\tau_{cyc}}{\sigma'_{v0}} \quad (6.73)$$

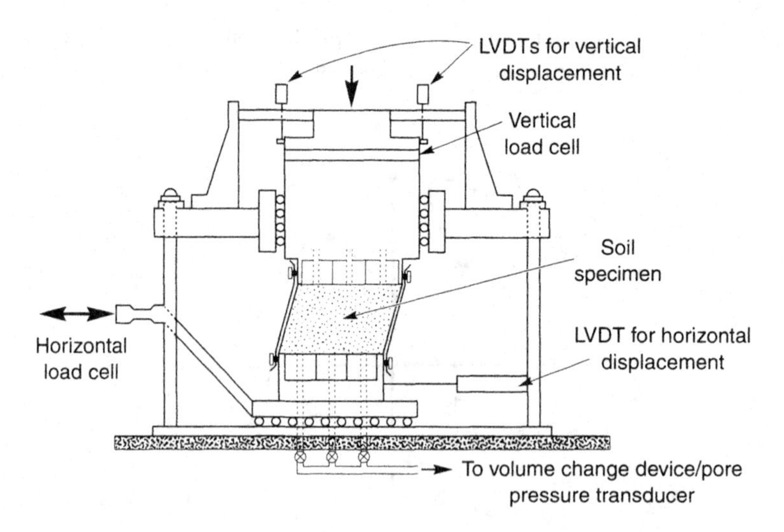

FIGURE 6.79 NGI cyclic simple shear apparatus. Soil specimen is contained within wire-reinforced rubber membrane (After Airey and Wood, 1987).

The simple shear apparatus, however, applies shear stresses only on the top and bottom surfaces of the specimen. Since no complementary shear stresses are imposed on the vertical sides, the moment caused by the horizontal shear stresses must be balanced by nonuniformly distributed shear and normal stresses. The effects of stress nonuniformity can be reduced by increasing the diameter/height ratio of the specimen; such effects are small at diameter/height ratios greater than about 8:1 (Kovacs and Leo, 1981), although ratios of 4:1 are commonly used. Most simple shear apparatuses are limited by their inability to impose initial stresses other than those corresponding to K_0 conditions and apply cyclic loading slowly and in only one horizontal direction. Modern devices can apply bi-directional loading over a wide range of frequencies under either constant height (corresponding to undrained) or constant vertical load (drained) conditions (Boulanger et al., 1993; Shafiee et al., 2017).

Because the cyclic simple shear and cyclic triaxial tests impose different loading on soil specimens, their cyclic stress ratios are not equivalent. For liquefaction testing, the two are usually related by

$$CSR_{ss} = c_r CSR_{tx} \tag{6.74}$$

where the correction factor, c_r is estimated from Table 6.5. Laboratory-measured liquefaction resistance can be expressed in terms of cyclic strength curves, i.e., plots of cyclic stress ratio versus number of cycles to liquefaction, such as those shown in Figure 9.23.

Cyclic Torsional Shear Test

Some of the difficulties associated with the cyclic triaxial and cyclic direct simple shear tests can be avoided by loading cylindrical soil specimens in torsion. Cyclic torsional shear tests allow isotropic or anisotropic initial stress conditions and can impose cyclic shear stresses on horizontal planes with continuous rotation of principal stress axes. They are most commonly used to measure stiffness and damping characteristics over a wide range of strain levels.

Ishihara and Li (1972) developed a torsional triaxial test that used solid specimens. Dobry et al. (1985) used strain-controlled cyclic torsional loading along with stress-controlled axial loading of solid specimens to develop a CyT-CAU test for the measurement of liquefaction behavior. Torsional testing of solid specimens, however, produces shear strains that range from zero along the axis of the specimen to a maximum value at the outer edge. To increase the radial uniformity of shear strains, others (e.g., Drnevich, 1967, 1972) developed hollow cylinder cyclic torsional shear apparatuses (Figure 6.80). While hollow cylinder tests offer perhaps the best uniformity and control over stresses and drainage, specimen preparation can be difficult and testing of specimens from the field is nearly impossible. For this reason, this testing is relatively uncommon and the equipment is not widely available.

6.5.4.4 Physical Model Tests

In contrast to element tests, physical model tests usually attempt to reproduce the boundary conditions of a particular problem by subjecting a small-scale physical model of a full-scale prototype

TABLE 6.5
Expressions for Correction Factor Relating Cyclic Triaxial and Simple Shear Cyclic Stress Ratios

Reference	Equation	c_r for. $K_0 = 0.4$	c_r for. $K_0 = 1.0$
Finn et al. (1971)	$c_r = (1+K_0)/2$	0.70	1.00
Seed and Peacock (1971)	Varies	0.55–0.72	1.00
Castro (1975)	$c_r = 2(1+2K_0)/3\sqrt{3}$	0.69	1.15

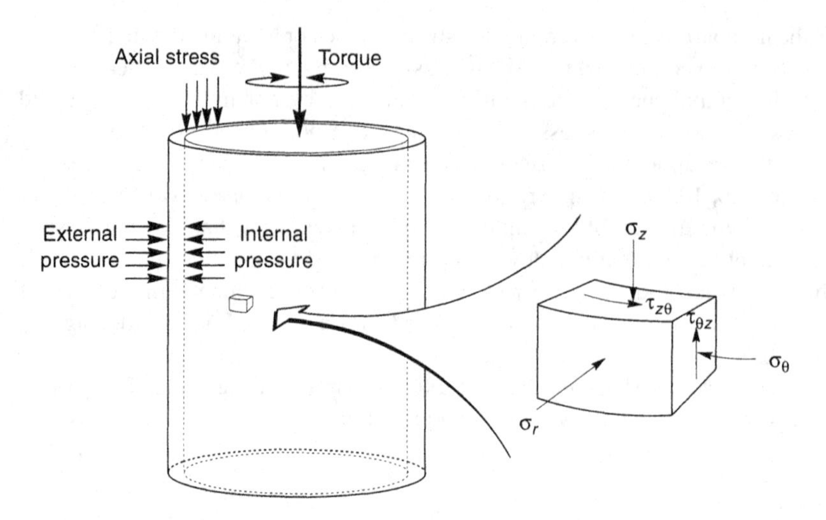

FIGURE 6.80 Hollow cylinder apparatus. The specimen is enclosed within internal and external membranes on which internal and external pressures can be applied independently. Application of cyclic torque induces cyclic shear stresses on horizontal planes.

structure to cyclic loading. Model tests may be used to evaluate the performance of a particular prototype or to study the effects of different parameters on a general problem. Model testing is generally considered to be most effective for the identification of important phenomena and deformation mechanisms and for the verification of predictive theories. It is generally not used for the design of significant structures or facilities.

The behavior of soils is sensitive to stress level; soils that exhibit contractive behavior under high normal stresses may exhibit dilative behavior at lower stress levels. One of the most significant challenges in model testing, therefore, is the problem of testing models whose stress dependency matches that of a full-scale prototype. Because this is very difficult under the gravitational field of the Earth, one common approach involves testing under increased gravitational fields. Model tests can therefore be divided into those performed under the gravitational field of the Earth (1 *g* model tests) and those performed under higher gravitational accelerations. The 1*g* tests are most commonly performed with the use of shaking tables; tests under increased gravitational fields are usually performed in a geotechnical centrifuge.

Both shaking table and centrifuge model tests have certain limitations, among the most important of which are similitude and boundary effects. The model scale factor, *N*, is taken as the ratio of prototype/model dimensions for 1*g* tests and as the acceleration level applied in centrifuge tests normalized by the Earth's gravitational acceleration (*g*). Scaling laws describe how various soil properties vary with *N* and are generally different for 1*g* vs centrifuge tests. For a given type of test, different soil properties will have different scaling laws, which means that similitude cannot be assured for all parameters simultaneously. Also, boundary effects (at the base and sides of the model) may not realistically capture prototype conditions, with artificial restraint of soil deformation or generation of reflected waves being the most common physical model boundary condition problems.

Shaking Table Tests

In the early years of geotechnical earthquake engineering, virtually all physical model testing was performed on shaking tables. Shaking table research has provided valuable insight into liquefaction, post-earthquake settlement, foundation response, and lateral earth pressure problems. Most shaking tables utilize a single horizontal translation degree of freedom but shaking tables with multiple degrees of freedom have also been developed. Shaking tables are usually driven by servo-hydraulic actuators (Figure 6.81); their dynamic loading capacities are controlled by the capacity of the hydraulic pumps that serve the actuators. Large pumps, actuators, and hydraulic accumulators are required to produce large displacements of heavy models at moderate or high frequencies. Shaking table specimens have stiff lateral restraints that are intended to mimic K_0 boundary conditions (Figure 6.81).

Shaking tables of many sizes have been used for geotechnical earthquake engineering research. Some are quite large, allowing models with dimensions of several meters to be tested. Thus shaking tables can often utilize actual, prototype soils rather than resorting to the smaller particle sizes often required for smaller scale model tests. For these large models, soils can be placed, compacted, and instrumented relatively easily. Shaking table models can be readily viewed from different perspectives during testing.

The principal limitation of shake table testing for geotechnical applications is that high gravitational stresses cannot be produced. This may cause, for example, a sand that is contractive at the high overburden stresses that exist under prototype conditions to be dilatant in a small 1 g model where stresses are much lower. Although correction procedures (e.g., Hettler and Gudehus, 1985; Iai, 1989) have been developed to aid in the interpretation of shaking table test results, in general shake-table experiments involving sands are generally not a preferred method of testing. Shake table experiments involving clays can be more meaningful, provided stress history effects between model and prototype are consistent.

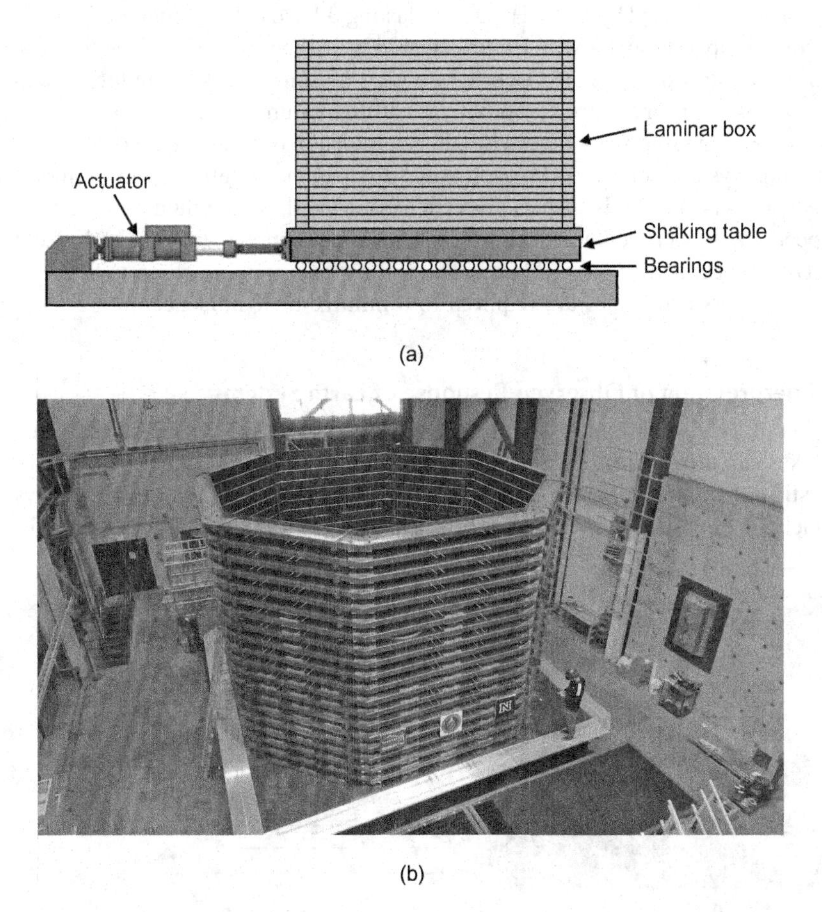

FIGURE 6.81 Typical shaking table configuration: (a) schematic illustration; used by permission of Carleton University Geoengineering Research Group) and (b) large-scale (15 ft high, 21.5 ft wide), biaxial laminar soil box at University of Nevada, Reno. (Courtesy of R. Motamed.)

Centrifuge Tests

In a centrifuge test, a $1/N$-scale model located at a distance, r, from the axis of a centrifuge (Figure 6.82) is spun at a rotational frequency, $\Omega = \sqrt{Ng / r}$, which is sufficient to raise the acceleration field at the location of the model to N times the acceleration of gravity, g. In principle, the stress conditions at any point in the model should then be identical to those at the corresponding point in the full-scale prototype.

Centrifuge tests are restricted to much smaller models than even moderate-sized shaking tables. Since the gravitational field increases with radial distance, the gravitational acceleration at the top of the model is lower than that at the bottom of the model. Since the gravitational field acts in the radial direction, the horizontal plane is curved (O'Reilly, 1991) by an amount that decreases with increasing centrifuge radius.

Similitude considerations are very important in the planning and interpretation of centrifuge tests. Scaling factors for a number of parameters are shown in Table 6.6. The scaling factors show how the speed of dynamic events is increased in the centrifuge. For example, the stresses and strains in a 30-m high prototype earth dam could be modeled with a 30-cm high centrifuge model accelerated to 100g. A common misconception of centrifuge tests is that particle sizes are also increased by N (100 in this example), although that is not the case. The particle sizes simply reflect the actual material comprising the model, there are just many fewer of them in the model than in the prototype, even if the stress conditions match. A harmonic 1-Hz base motion lasting 10 sec at the prototype scale would be modeled by a 100-Hz motion lasting 0.1 sec in the model. Because time for pore water pressure dissipation scales with length-squared, it will occur at N^2 the rate in the model as in prototype if pore fluids of the same viscosity are used (a factor of 10,000 in the present example). This faster dissipation is one of the few problems causing failure mechanisms in centrifuge models to differ from those in prototypes; otherwise, this method of model testing is considered quite effective at modeling system behaviors. Pore pressure dissipation rate effects can be more accurately modeled by using viscous fluids such as glycerin or silicon oil as pore fluids.

High-speed transducers, data acquisition systems, and cameras are required to obtain useful results in dynamic centrifuge tests. Because the scaling laws apply to all parts of the model, miniaturized transducers and cables are required to minimize their influence on the response of the model.

6.5.4.5 Interpretation of Observed Response of Earth Systems

Interpretation of the response of instrumented, full-scale structures subjected to dynamic loading or earthquakes can provide invaluable information on how soils and soil-structure systems behave under realistic dynamic loading and boundary conditions. Because a real, full-scale system rather than a model is being measured, the observed response of such systems is particularly useful for

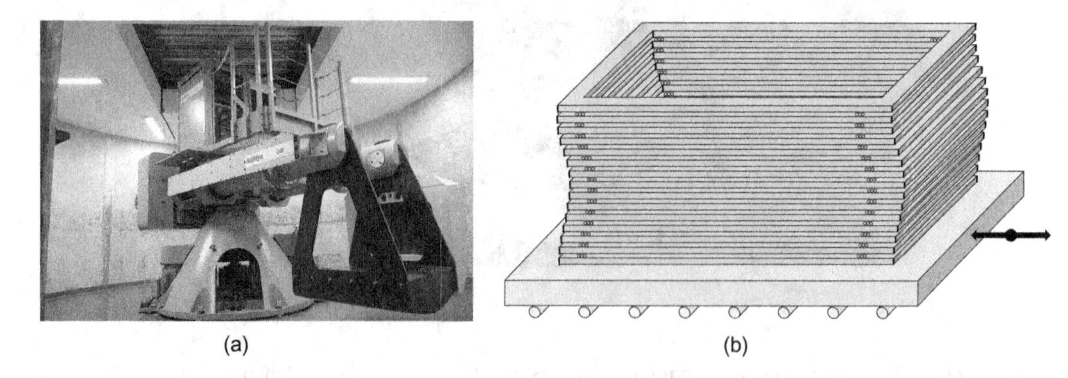

FIGURE 6.82 (a) Geotechnical centrifuge (courtesy of Network Rail) and (b) schematic illustration of laminar box.

TABLE 6.6
Scaling Factors for Centrifuge Modeling

Type of Event	Quantity	*Model Dimension* / Prototype Dimension
All events	Stress	1
	Strain	1
	Length	$1/N$
	Mass	$1/N^3$
	Density	1
	Force	$1/N^2$
	Gravity	N
Dynamic events	Time	$1/N$
	Frequency	N
	Acceleration	N
	Strain Rate	N
Diffusion events	Time	$1/N^2$
	Strain rate	N^2

Source: After Kutter and James (1989).
Values are based on the assumption that the same soils and fluid are used in the model and the prototype, and that the soil properties are not rate-dependent

understanding various phenomena in geotechnical earthquake engineering and the validation of modeling approaches.

This approach requires a good degree of instrumentation so that the levels of shaking and site response are known. Most typically this is accomplished with vertical arrays (Section 3.2.4.2), preferably with the deepest instrument located below any significant impedance contrasts in bedrock or in stiff soils with high seismic velocities. Vertical arrays may also include pore pressure transducers, inclinometer casings, and other forms of instrumentation. Recorded motions at vertical array sites can be used with a suitable ground response model (Chapter 7) to: (1) evaluate site conditions for which 1D ground response methods provide good estimates of observed site response (e.g., Thompson et al., 2012) or whether 2D or 3D analyses are required; (2) identify the dynamic soil properties (specifically, shear modulus and damping) that produce the best agreement between predicted and actual motions (e.g., Zeghal et al., 1995; Elgamal et al., 2001; Tsai and Hashash, 2009); and (3) study the relative effectiveness of alternative methods of soil behavior modeling, including equivalent linear and nonlinear methods of analysis (e.g., Kaklamanos et al., 2013, 2015; Zalachoris and Rathje, 2015; Kim et al., 2016). At sites where soils within the domain of vertical arrays are potentially liquefiable, pore pressure transducers are very useful for understanding pore pressure generation/dissipation up to, during, and following liquefaction as well as ground motions in liquefied soil (Zeghal and Elgamal, 1994; Bonilla et al., 2005; Kramer et al., 2011).

Aside from ground response, field instrumentation is also extremely useful for understanding the response of geotechnical systems including earth dams (Wieland, 2004), levees (Kishida et al., 2009c), and soil-structure systems (Stewart and Fenves, 1998; Tileylioglu et al., 2011). As vertical arrays and instrumented geotechnical structures become more common worldwide (e.g., Section 3.2.4.2), the amount of data at small and large strain conditions produced as earthquakes occur is growing rapidly. Examples of observed ground response from which dynamic soil properties can be identified are presented in Section 7.2.2.

6.6 BEHAVIOR OF CYCLICALLY LOADED SOILS

The mechanical behavior of soils can be quite complex under static, let alone seismic, loading conditions. As discussed in Section 6.4.1, soils exhibit nonlinear, inelastic stress-strain behavior. In addition, some soils exhibit rate dependence, cyclic degradation, and/or coupling between their volumetric and deviatoric responses to loading.

The propagation of waves through elastic solids has been shown (Appendix C) to be controlled by the density, stiffness, and damping characteristics of the solid. Those characteristics are also important for wave propagation in particulate materials like soils. Accounting for nonlinear, inelastic behavior of soils in the solution of response and ground failure problems, however, also requires characterization of strain-dependent stiffness and damping, volume change (dilation/contraction), and cyclic shear strength behavior of the soil.

6.6.1 Characterization of Soil Behavior

The behavior of soils in response to cyclic loading can be characterized in a number of different ways, but a standard framework has emerged over the past 50 years. The framework is rooted in the historical use of equivalent linear analyses (Section 7.5.2) but can be transformed to a framework that is convenient for nonlinear analyses. The equivalent linear framework can characterize the stiffness, strength, and damping behavior of cyclically loaded soils. That behavior, however, can also be strongly influenced by volume change (contraction/dilation) characteristics.

6.6.1.1 Density

As mentioned in Section 6.5.1, with the general exception of liquefiable soils, soil densities are usually relatively easy to measure at the levels of resolution required for Geotechnical Earthquake Engineering applications. For planning or initial calculation purposes, however, typical densities for different types of soils are presented in Table 6.7. Note that the densities of organic soils and waste materials (which often contain significant fractions of organics) are much lower than those of conventional inorganic soils. Well-graded granular soils and overconsolidated fine-grained soils tend toward the upper ends of the density ranges given.

6.6.1.2 Stiffness Behavior

Figure 6.13c showed that soil stiffness tends to decrease with increasing shear strain amplitude. Figure 6.83a shows the nonlinear stress-strain, or *backbone*, curve from a monotonic direct simple shear test on a typical soil – the shear modulus decreases from its maximum value, G_{max}, which is mobilized at very low strains, to lower values at moderate and higher strains. At very large strains,

TABLE 6.7
Typical Ranges of Density and Unit Weight for Different Soil Types

	Dry			Saturated		
Soil Type	ρ_d (Mg/m³)	γ_d (kN/m³)	γ_d (lb/ft³)	ρ_{sat} (Mg/m³)	γ_{sat} (kN/m³)	γ_{sat} (lb/ft³)
Sands and gravels	1.5–2.3	14.7–22.6	94–144	1.9–2.4	18.6–23.5	118–150
Silts and clays	0.6–1.8	5.9–17.7	37–112	1.4–2.1	13.7–20.6	87–131
Glacial tills	1.7–2.3	16.7–22.6	106–144	2.1–2.4	20.6–23.5	131–150
Crushed rock	1.5–2.0	14.7–19.6	94–125	1.9–2.2	18.1–21.6	118–137
Peats	0.1–0.3	4.9–14.7	6–19	1.0–1.1	9.8–10.8	62–69
Organics silts and clays	0.5–1.5	4.9–14.7	31–94	1.3–1.8	12.7–17.7	81–112

Source: After Holtz et al. (2011).

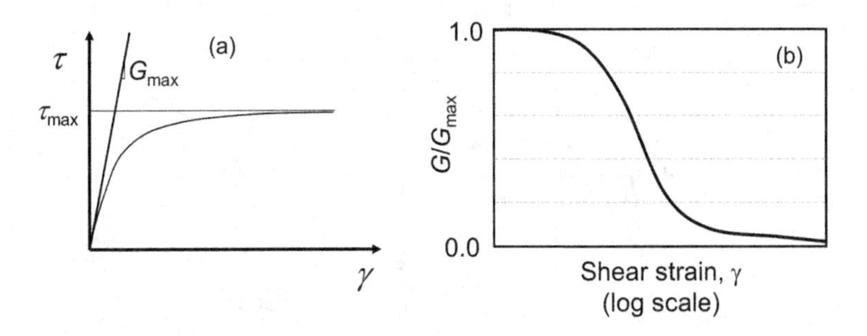

FIGURE 6.83 Illustration of soil nonlinearity: (a) hyperbolic stress-strain (backbone) curve, and (b) corresponding modulus reduction curve.

the shear strength of the soil is fully mobilized. Defining a secant shear modulus, G_{sec}, as the ratio of shear stress to shear strain, the backbone curve can be represented as

$$\tau = G_{sec}\gamma = G_{max}\left(\frac{G_{sec}}{G_{max}}\right)\gamma \tag{6.75}$$

Hence, the overall stiffness behavior of the soil can be described in terms of the maximum shear modulus that describes low-strain stiffness, and a *modulus reduction ratio* (G_{sec}/G_{max}, usually written as G/G_{max}) that describes the larger-strain stiffness behavior. A plot of modulus reduction ratio vs. shear strain, as shown in Figure 6.83b is referred to as a *modulus reduction curve.* Stiffness behavior is described in detail in Sections 6.6.2 and 6.6.3.

6.6.1.3 Strength Behavior

The degradation of soil stiffness occurs at very low shear strain levels and is usually described by modulus reduction curves typically obtained from laboratory tests that are limited to shear strain amplitudes less than about 0.5%–1.0%. The shear strength of the soil, i.e., the limiting shearing resistance, is typically mobilized at shear strains of 5%–20%. As a result, there is a large range of intermediate strain levels over which the stress-strain behavior is not well defined. In many cases, engineers have extrapolated modulus reduction curves beyond the range constrained by laboratory data without considering the effects of that extrapolation on the corresponding backbone curves. The result of this approach can be a modulus reduction curve that smoothly decreases in what appears to be a reasonable manner, but which implies shearing resistances at large strains that may grossly exceed, or fall below, the actual strength of the soil. Therefore, procedures to extrapolate the low-strain behavior obtained from dynamic laboratory tests to large-strain behavior that is consistent with measured shear strengths (Section 6.6.3.3) are required. In this way, modulus reduction curves and G_{max} values can also be used to characterize the shear strength of a soil. Strength behavior is described in detail in Section 6.6.6.

6.6.1.4 Damping Behavior

The inelastic behavior of cyclically loaded soils gives rise to energy dissipation through hysteretic behavior. Experimental evidence shows that the shapes of hysteresis loops are influenced by strain amplitude – as the shear strain amplitude increases, the loops become broader (Figure 6.84a), which indicates that the damping ratio of the soil increases with increasing strain amplitude (Figure 6.84b). Damping behavior is described in detail in Section 6.6.4.

6.6.1.5 Volume Change Behavior

As discussed in Section 6.4.3.2, monotonically loaded soils tend to be contractive (generally loose granular soils or normally consolidated fine-grained soils) or, following initial contraction at small

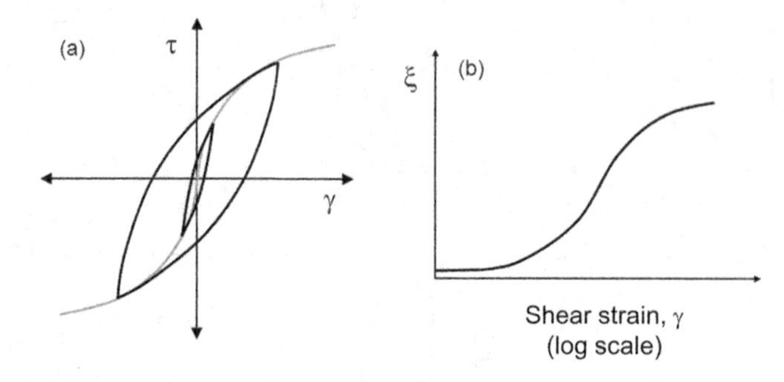

FIGURE 6.84 Illustration of soil inelasticity: (a) stress-strain loops at low and high strain levels, and (b) damping curve.

strains, dilative (dense granular soils or heavily overconsolidated fine-grained soils) at large strains. Under cyclic loading, shear strains are often so small that they produce contractive behavior in all soils, leading to densification under drained conditions and increased pore pressure under undrained conditions. Volume change behavior can be characterized in different ways for both drained and undrained conditions, as described in Section 6.6.5.

6.6.2 Low-Strain Stiffness

The shear modulus of a linear material is constant, i.e., it has the same value at high strain levels as it does at low strain levels. Soils, however, are known to be strongly nonlinear with shear moduli that decrease with increasing strain level, as shown in Figure 6.83. Laboratory tests have shown that soil stiffness can also be influenced by void ratio, mean principal effective stress, plasticity index, overconsolidation ratio, and other factors.

The low-strain shear modulus, G_{max}, is an extremely important parameter, both due to its direct influence on wave propagation at low strain levels and its role in normalizing the shear modulus at higher strain levels in modulus reduction curves. The low-strain stiffness should, whenever possible, be obtained from direct, *in situ* measurement of shear wave velocity using the types of tests described in Section 6.5.2; the equipment for performing these tests is readily available in most areas. In the event that they are not available, or for making preliminary estimates, shear wave velocities can also be obtained by empirical correlation to other commonly measured parameters. Such correlations, however, introduce significant uncertainty into the estimated low-strainstiffness.

6.6.2.1 Direct Shear Wave Velocity Measurement

The maximum shear modulus describes the stiffness of the soil at extremely low strain levels. Since most seismic geophysical tests induce shear strains lower than about $3 \times 10^{-4}\%$, the shear wave velocities measured in the field can be used to compute G_{max} as

$$G_{max} = \rho V_s^2 \tag{6.76}$$

It should be noted that Equation (6.76) was obtained from the derivation of the wave equation (Section C.2.1). In that derivation, the density affects the inertial force required to ensure dynamic equilibrium, hence, it should include the masses of all phases within a soil that accelerate as a wave passes through it. For virtually all soils, the porewater in the soil will move with the soil particles, so the total density should be used in Equation (6.76).

The use of measured shear wave velocities is the most reliable means of evaluating the *in situ* value of G_{max} for a particular soil deposit, and the seismic geophysical tests described in Section 6.5.2 are commonly used for that purpose. When measured in the field at a particular site, the stiffness of soil or rock will generally increase with depth as shown in Figure 6.85a. The increase in stiffness can be due to increased confining pressures, particle cementation, decreased weathering, increasing geologic age, or some combination of these and other factors. As a result, shear wave velocity profiles are generally observed to increase with depth. The nature of that increase, however, varies from site to site – at some sites, V_s may increase smoothly over an extended depth range, while in others it may increase in steps characterized by individual layers with relatively constant velocities. At some sites, softer soils may be covered by stiffer soils, either naturally deposited or placed as compacted fill; such conditions are referred to as *velocity inversions*. Shear wave velocities at shallow depths can vary seasonally as changes in soil moisture affect effective stresses (Roumelioti et al., 2020).

For site response problems (Section 3.4.3), ground response is influenced more strongly by low-velocity layers than high-velocity layers. In such cases, plots of *slowness* profiles, where slowness is defined as the reciprocal of velocity, can help emphasize the most important layers and aid in comparisons of different profiles measured at the same site (Brown et al., 2002). Figure 6.85 shows shear wave velocity and slowness plots for two profiles; the enhanced resolution of the softer portion of the profile in the slowness profile is apparent.

Uncertainty in Low-strain Stiffness

Because low-strain stiffness is commonly obtained from measured shear wave velocity, uncertainty in low-strain stiffness is best expressed in terms of uncertainty in shear wave velocity. This uncertainty can be evaluated when multiple V_S profiles are measured at different locations at a given site, which in general will not match. These variations in measured velocities, when based on reliable geophysical methods, represent the variability of the geologic structure of the site. Such variations are always present to varying degrees and give rise to variability in site response.

If the variations in V_S can be measured at a site using multiple profiles, a mean and standard deviation of V_S can be quantified on a site-specific basis. This is desirable because such a characterization would presumably reflect the local geologic conditions. However, in many cases only a single profile is available, and is generally assumed to represent the mean. In such situations, it is necessary to estimate the variability of V_S from relationships derived from data at sites where many V_S profiles have been measured. Because these relationships are relatively generic, they will not necessarily represent the local geologic conditions at a particular site very well.

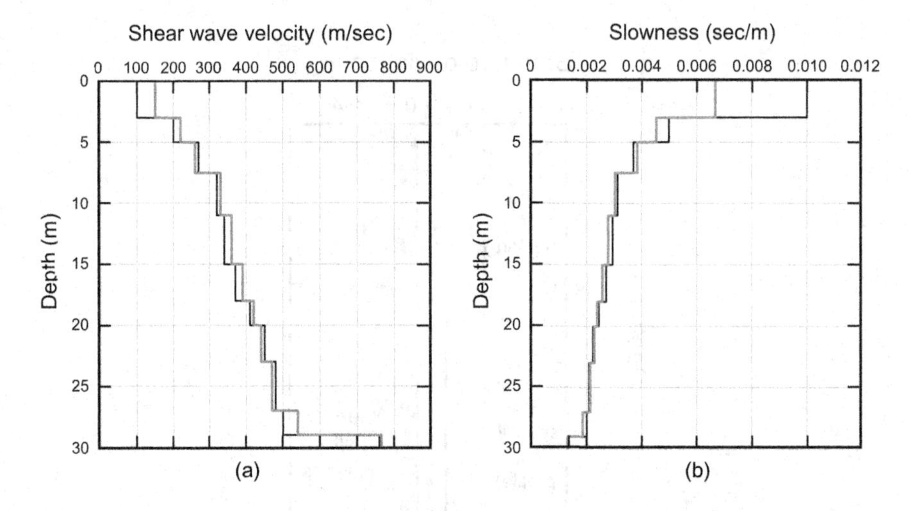

FIGURE 6.85 (a) Increase of shear wave velocity with depth at site with approximately 29 m of soil overlying weathered rock; (b) variation with depth of slowness for the same site. Slowness profile emphasizes characteristics of softer layers that often play the most important role in site response.

One such set of statistical relations based on V_S profiles in California and Georgia was given by Toro (1995). For a given depth, V_S is typically found to be lognormally distributed with a depth-dependent standard deviation ($\sigma_{\ln V_S}$). As shown in Figure 6.86, values of $\sigma_{\ln V_S}$ are provided for "generic" and "site-specific" applications. The generic results apply for horizontal separation distances between profiles on the order of several hundred meters to a few km or more, whereas the separation distances for site-specific range from 2 to 800 m. The generic $\sigma_{\ln V_S}$ is based on statistics for many sites within broadly defined NEHRP (Dobry et al., 2000) or Geomatrix (Chiou et al., 2008) site classes – they are intended for use with generic median V_S profiles for each site class. The site-specific standard deviations are intended for use with a measured V_S profile, which is generally taken as the mean for the site. The two site-specific profiles of $\sigma_{\ln V_S}$ in Figure 6.86 reflect attributes of velocity clusters (Stewart et al., 2014) and mid-layer velocities (EPRI, 2013 and Toro, 2022).

The intra- and inter-method variabilities of SASW, MASW, and ReMi techniques from five arrays have been investigated at a single site with a relatively simple velocity profile (Cox and Wood, 2011). The mean shear wave velocity profiles at the site were generally within about 12% of each other in the upper 5 m of the profile and within 20%–30% at greater depths. When the analyses were repeated using a common water table depth (obtained from p-wave velocity measurements in the MASW test), the inter-method variability dropped from a maximum of 30% to approximately 10%, which reinforces the need for accurate knowledge of groundwater conditions. More complex sites would likely show greater inter-method variability. As described previously (Section 6.5.2.2), variations of phase velocities obtained by the ReMi method, related to different array orientations and active vs. passive energy sources, can reach 100% (Cox and Beekman, 2011). The potential also exists for significant bias with that method. Investigations of intra- and inter-method variability of invasive and non-invasive methods (e.g., Garofalo et al., 2016a) have shown that V_s coefficients of variation (COV) for invasive methods were generally, but not always, lower than for non-invasive methods. COV values were typically larger near the surface and on the order of 0.2–0.35 for soil sites, and then decreased with depth to values of approximately 0.1–0.2. However, larger values of COV were calculated for a rock site, with COV values as high as 0.4–0.6 at the surface.

Uncertainty in V_{S30}

The uncertainty in V_{S30} (Section 3.4.3.1) from measured velocity profiles has been evaluated from sites with clusters of profiles, typically spaced on the order of 10 to about 100 m. This uncertainty, denoted $\sigma_{\ln V_{S30}}$, is useful for uncertainty propagation in ground motion predictions that utilize

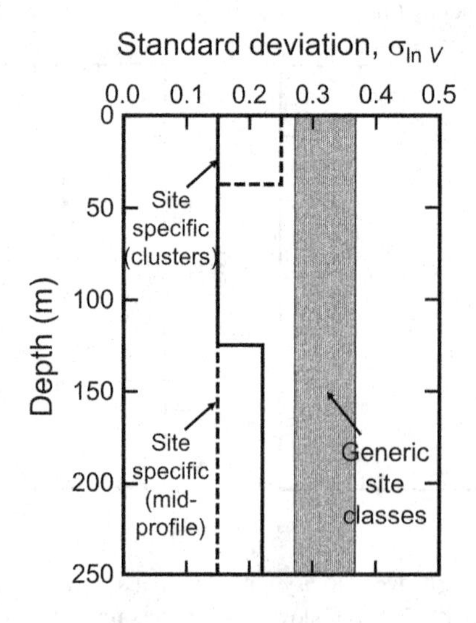

FIGURE 6.86 Variation of standard deviation of V_S with depth for generic and site-specific conditions. (Adapted from Toro, 1995, Stewart et al., 2014, and Toro, 2022.)

V_{S30}-based site response models (Section 3.5.2.3). This uncertainty can in general be influenced by variations among V_S measurement types (generally small when reliable data providers are used) and natural heterogeneity in the site conditions. Based on data compilations by Moss (2008), Seyhan et al. (2014), and Yust et al. (2018), ranges of $\sigma_{\ln V_{S30}}$ were approximately 0.02–0.12. For sites without strongly variable geological conditions, averages of about 0.06–0.07 have been found. Values of 0.1 are often applied in practice.

6.6.2.2 Empirical Correlation

The maximum shear modulus can also be estimated by empirical correlation to commonly measured *in situ* test parameters. In general, G_{max} increases with increasing soil density as do SPT and CPT penetration resistances, so it is logical to expect some correlation between G_{max} (or V_s) and penetration resistance. It should be recognized, however, that G_{max} is a low-strain parameter that can be influenced by factors such as particle cementation (generally related to the age of the deposit) and soil fabric, which are destroyed at the large strain levels (DeJong et al., 2006) induced by penetration testing. A number of empirical relationships between G_{max} and various *in situ* test parameters have been developed. Many of these correlations can be expressed as functions of penetration resistance, depth or effective stress, and material type, e.g., as

$$\ln V_s = c_0 + c_1 \ln PR + c_2 \ln f_z + c_3 \ln f_m + \varepsilon \sigma_{\ln V}, \tag{6.77}$$

where PR = penetration resistance, f_z = depth or effective stress parameter, f_m = material parameter, ε = standard normal variate, and $\sigma_{\ln V_s}$ = standard deviation of $\ln V_s$; not all relationships include all of these terms. Values of the coefficients for Equation (6.77) are shown in Table 6.8.

TABLE 6.8
Coefficients for Shear Wave Velocity (m/s) Prediction Models

Soil Type	PR	f_z	f_m	c_0	c_1	c_2	c_3	$\sigma_{\ln Vs}$
Sand[a]	N_{60}	σ'_{v0}	–	4.045	0.096	0.236	0.0	[f]
Silt[a]	N_{60}	σ'_{v0}	–	3.783	0.178	0.231	0.0	[g]
Clay[a]	N_{60}	σ'_{v0}	–	3.996	0.230	0.164	0.0	[h]
$I_c < 2.6$[b]	q_t/p_a	p_a/σ'_{v0}	$e^{1.786Ic}$	–2.488	1.0	0.25	1.0	–
$I_c > 2.6$[b]	q_t/p_a	p_a/σ'_{v0}	$e^{1.786Ic}$	–2.488	1.0	0.50	1.0	–
Holocene[c]	q_t/p_a	z	I_c	2.699	0.395	0.124	0.912	–
Pleistocene[c]	q_t/p_a	z	I_c	2.896	0.395	0.124	0.912	–
Holo., Pleist. sands[d]	$(q_t - \sigma_v)/p_a$	σ'_{v0}/p_a	e^{Ic}	1.93	0.5	0.25	0.63	–
Christchurch soils[e]	q_t/p_a	z	f_s/p_a	3.959	0.144	0.278	0.0832	[i]

Source: After BSSC (2020).
f_z column: σ'_{v0} = vertical effective stress in kPa; z = depth in meters
Notes:
[a] Brandenberg et al. (2010)
[b] Hegazy and Mayne (2006)
[c] Andrus et al. (2007)
[d] Robertson (2012)
[e] McGann et al. (2015)
[f] $\tau = 0.217, \sigma = \min\left(0.20, 0.57 - 0.07\ln\sigma'_{v0}\right), \sigma_{\ln V_s} = \sqrt{\tau^2 + \phi^2}$
[g] $\tau = 0.227, \sigma = \min\left(0.15, 0.31 - 0.03\ln\sigma'_{v0}\right), \sigma_{\ln V_s} = \sqrt{\tau^2 + \phi^2}$
[h] $\tau = 0.227, \sigma = \min\left(0.16, 0.21 - 0.01\ln\sigma'_{v0}\right), \sigma_{\ln V_s} = \sqrt{\tau^2 + \phi^2}$
[i] $\sigma_{\ln V_s} = 0.162$ for $z < 5$ m
$= 0.216–0.0108z$ for $5 \text{ m} < z < 10$ m
$= 0.108$ for $z > 10$ m

Correlations Based on Laboratory Tests

Menq (2003) tested multiple granular soils ranging from sands to gravels and found that G_{max} was influenced by grain size characteristics as well as soil density and effective confining pressure. Regression on test data from these materials suggested that G_{max} could be estimated as

$$G_{max,lab} = C_1 C_u^{-0.2} e^x \left(\frac{\sigma'_{v0}}{P_a} \right)^{n_G} \tag{6.78}$$

where C_1 = 67.1 MPa (1,400 ksf), C_u = coefficient of uniformity, e = void ratio, $x = -1 - (D_{50}/20)^{0.75}$, D_{50} = mean grain size in mm, and $n_G = 0.48C_u^{0.09}$. This relationship indicates that G_{max} increases with increasing D_{50}, C_u, and σ'_{v0}, and with decreasing void ratio.

For fine-grained soils, preliminary estimates of G_{max} can be obtained from undrained strength, plasticity index, and overconsolidation ratio (Table 6.9). Because undrained strengths are highly variable and because shear moduli and undrained strength do not vary in the same way with effective confining pressure, these results must be used carefully.

Highly organic soils and peats can behave differently than conventional inorganic soils. Their low -strain stiffnesses have been found to be affected by effective stress level, overconsolidation ratio, and organic content. Based on testing of laboratory specimens of organic soils, Kishida et al. (2009a) proposed that G_{max} could be estimated as

$$\frac{G_{max,lab}}{p_a} = A(\sigma'_{m0})^n OCR^m \tag{6.79}$$

where p_a = atmospheric pressure and A, n, and m are empirical coefficient related to organic content, OC, by

$$A \approx 121.31 - 0.9354OC + 0.0036OC^2 \tag{6.80}$$

$$n = 1 - 0.37\frac{2}{1 + \exp(OC / 23)} \tag{6.81}$$

$$m = 0.8 - 0.4\frac{2}{1 + \exp(OC / 23)} \tag{6.82}$$

where OC is in percent. Comparisons of laboratory and field measurements of V_s on the same soil indicate an under-prediction bias from the laboratory-based model, likely due primarily to disturbance effects in the laboratory samples. Kishida et al. (2009b) recommended that the bias be removed as follows:

TABLE 6.9
Values of G_{max}/s_u [a]

	Overconsolidation Ratio, *OCR*		
Plasticity Index, *PI*	**1**	**2**	**5**
15–20	1,100	900	600
20–25	700	600	500
35–45	450	380	300

Source: After Weiler (1988).
[a] Undrained strength measured in CU triaxial compression.

$$G_{max,\ field} = 1.19 G_{max,lab} \tag{6.83}$$

Correlations for V_{S30}

Measuring shear wave velocities is standard practice for contemporary site-specific subsurface investigations in seismically active areas. In areas where shear wave velocities cannot be, or have not yet been measured, several proxies for V_{S30} have been proposed. Proxies, i.e., descriptors of site condition that can be readily obtained and that have some correlation to low-strain stiffness, which have been used for this purpose include (1) mapped surface geology (e.g., Wills and Clahan, 2006); (2) geotechnical site categories (sometimes called Geomatrix site classes, Chiou et al., 2008); (3) ground slope only (Wald and Allen, 2007); (4) surface geology in combination with ground slope (Thompson et al., 2014; Wills et al., 2015; Parker et al., 2017; Ahdi et al., 2017); and (5) geomorphic terrain categories that consider ground slope, surface roughness, and convexity (Yong et al., 2012; Yong, 2016). Ground slope correlations are based on the idea that the stiffness of geologic materials is positively correlated to their strength, and that materials comprising steeper slopes are generally stronger than materials found at flat sites. As an example of proxy-based estimation of V_{S30}, Table 6.10 shows recommended values for five geotechnical site classes, based on 441 V_S profiles from California. Note that V_{S30} carries a much larger uncertainty when estimated from proxies than when measured on-site ($\sigma_{\ln VS30}$ from proxies general exceed 0.3, whereas those from measurements are generally about 0.1). Seyhan et al. (2014) found the proxies based on surface geology with slope and geotechnical conditions to provide the best performance relative to data from California, Japan, and Taiwan.

V_{S30} has also been correlated to the fundamental mode Rayleigh wave phase velocity at a wavelength of 40 m, i.e., V_{r40} (Brown et al., 2000; Martin and Diehl, 2004).

TABLE 6.10
Estimates of the Mean and Natural Log Standard Deviation of V_{S30} Using the Geotechnical Site Category Proxy of Chiou et al. (2008)

Site Category	Description	Median V_{S30} (m/s)	$\sigma_{\ln VS30}$	Mean V_{S30} (m/s)
A	Rock. Exposed at surface or < 5 m soil over rock	507	0.41	516
B	Shallow (stiff) soil. Soil profile thickness up to 20 m thick overlying rock.	425	0.43	464
C	Deep narrow soil. Soil profile at least 20 m thick overlying rock, in a narrow canyon or valley no more than several km wide.	339	0.20	345
D	Deep broad soil. Soil profile at least 20 m thick overlying rock, in a broad valley.	275	0.34	291
E	Soft deep soil. Instrument on/in deep soil profile with average V_s < 150 m/s	182	0.25	185

Numerical values are for California data as compiled by Seyhan et al. (2014).
V_{S30} has also been correlated to the fundamental mode Rayleigh wave phase velocity at a wavelength of 40 m, i.e., V_{r40} (Brown et al., 2000; Martin and Diehl, 2004).

6.6.2.3 Other Factors

A number of additional factors can affect the measurement of low-strain stiffness, or its interpretation from available measurements. Evaluation of shear modulus can be complicated by rate and time effects (Anderson and Woods, 1975, 1976; Anderson and Stokoe, 1978; Isenhower and Stokoe, 1981).

Rate Effects

Rate effects can cause G_{max} to increase with increasing strain rate; consequently, G_{max} can increase with increasing frequency under cyclic loading. The influence of strain rate on G_{max} increases with increasing soil plasticity; for San Francisco Bay mud ($PI \approx 30$–45), G_{max} increases about 4% per tenfold increase in strain rate. Rate effects can be significant when comparing G_{max} values obtained from field shear wave velocity measurements (usually made with the use of impulsive disturbances which produce relatively high frequencies) with values obtained from laboratory tests; suspension logging tests, for example, involve particularly high frequencies.

Time Effects

The shear wave velocity, and hence G_{max}, increases approximately linearly with the logarithm of time past the end of primary consolidation to an extent that cannot be attributed solely to the effects of secondary compression. The change of stiffness with time can be described by

$$\Delta G_{max} = N_G (G_{max})_{1,000} \tag{6.84}$$

where ΔG_{max} is the increase in G_{max} over one log cycle of time and $(G_{max}) = 100$ is the value of G_{max} at a time of 1,000 minutes past the end of primary consolidation. N_G increases with increasing plasticity index, *PI*, and decreases with increasing *OCR* (Kokushu et al., 1982). For normally consolidated clays, they found that N_G can be estimated as

$$N_G \approx 0.027\sqrt{PI} \tag{6.85}$$

where *PI* is in percent. Anderson and Woods (1975) showed that some of the discrepancy between G_{max} values from field and laboratory tests could be explained by time effects, and that N_G could be used to correct the G_{max} values from laboratory tests to better represent actual *in situ* conditions.

Anisotropy Effects

Although usually treated as isotropic, low-strain stiffness (and shear wave velocity) can also be affected by soil anisotropy. Soils can exhibit *inherent anisotropy*, which is due to the depositional fabric or structure of the soil skeleton and exists even under isotropic stress conditions, and/or *induced anisotropy*, which results from anisotropy of the existing (or historical) stress state. For critical problems, soils can be modeled as *cross-anisotropic*, in which their behavior is described by five parameters (White, 1965) usually determined by laboratory testing on reconstituted specimens (e.g., Stokoe et al., 1991; Kuwano and Jardine, 2002; Lee and Stokoe, 1986; Bellotti et al., 1996).

Summary

A brief summary of the effects of environmental and loading conditions on the maximum shear modulus of normally and moderately overconsolidated soils is presented in Table 6.11.

6.6.3 Higher-Strain Stiffness

The secant and tangent shear moduli (Figure 6.12c) of an element of soil both vary with cyclic shear strain amplitude. At low strain amplitudes, both shear moduli are high, but decrease as strain amplitudes increase. As described in Section 6.6.1.2, the variable stiffness of an element of soil, therefore, can be characterized by its backbone curve (Equation 6.41) or by G_{max} and a modulus reduction

TABLE 6.11
Effect of Environmental and Loading Conditions on Maximum Shear Modulus of Normally Consolidated and Moderately Overconsolidated Soils

Increasing Factor	G_{max}
Effective confining pressure, σ'_m	Increases with σ'_m
Void ratio, e	Increases with e
Geologic age, t_g	Increases with t_g
Cementation, c	Increases with c
Overconsolidation ratio, OCR	Increases with OCR
Plasticity index, PI	Increases with PI
Strain rate, $\dot{\gamma}$	No effect for nonplastic soils; increases with $\dot{\gamma}$ for plastic soils (up to ~10% increase per log cycle increase in $\dot{\gamma}$)
Number of loading cycles, N	Decreases after N cycles of large γ_c, but recovers later in time in clays; increases with N for sands.

Source: After Dobry and Vucetic (1987).

curve. Since the secant modulus, $G = \tau/\gamma$, a backbone curve can be constructed from a modulus reduction curve using Equation (6.75). Likewise, a modulus reduction curve can be obtained from a backbone curve as

$$\frac{G}{G_{\max}} = \frac{\tau}{G_{\max}\gamma} \tag{6.86}$$

While the nonlinear behavior of a soil can be characterized equally well by a backbone curve or $G_{\max}$ and a modulus reduction curve, the latter approach offers some advantages in terms of familiarity, isolation of significant factors, and direct use in certain types of response analyses (Chapter 7), and will therefore be emphasized in the remainder of this chapter.

The shear strength of a typical soil, however, is mobilized at larger strains than those typically well constrained by modulus reduction curves (Section 6.6.1.2). Using Equation (6.86), the value of $G/G_{\max}$ when the shear strength, $\tau_{\max}$, has been mobilized is

$$\frac{G}{G_{\max}} = \frac{\tau_{\max}}{G_{\max}\gamma} \tag{6.87}$$

In order to produce shear stresses consistent with the shear strength at large strain levels, the modulus reduction curve should be transitioned to the strength-based modulus reduction curve of Equation (6.87) at higher strain levels. The transition may be more easily accomplished in terms of the backbone curve, and then converted back to equivalent modulus reduction curve ordinates at large shear strains, as described further below.

Modulus reduction curves can, in principle, be measured *in situ*, measured from laboratory tests for site-specific soil materials, or derived from empirical correlations. *In situ* measurement of shear moduli at high strain levels is difficult and requires specialized equipment (Cox et al., 2009) that is not practical for routine projects. Higher strain dynamic soil properties are therefore measured in the laboratory, most commonly using resonant column testing. For critical projects, site-specific relations for modulus reduction as a function of shear strain may be obtained by a laboratory testing program (Section 6.6.3.1). In most cases, however, shear modulus behavior at higher strain levels is obtained through the use of empirical correlations to other, more easily measured soil properties that have been shown to affect modulus reduction behavior (Section 6.6.3.2).

While the shear strains mobilized during weak shaking may be low enough that $G \approx G_{\max}$ (i.e., linear response), problems of interest usually involve larger strains at which soils exhibit nonlinear behavior. Figure 6.83a shows a two-parameter backbone curve that takes the form of a simple hyperbola. As given previously in Equation (6.41), this hyperbola is completely described by the limiting shear stress τ_{max} (i.e., the nominal shear strength of the soil) and the initial shear modulus G_{max}. It is convenient to re-write Equation (6.41) for positive shear strain as:

$$\tau = \frac{G_{max}\gamma}{1+\dfrac{G_{max}}{\tau_{max}}\gamma} = \frac{G_{max}\gamma}{1+\gamma / \gamma_{ref}} \tag{6.88}$$

where $\gamma_{ref} = \tau_{max}/G_{max}$ is referred to as the *reference strain* (Hardin and Drnevich, 1972a, 1972b). In this simple model, the reference strain is the shear strain at $\tau = \tau_{max}/2$. Using Equation (6.86), the corresponding modulus reduction curve (Figure 6.83b) is given by

$$\frac{G}{G_{max}} = \frac{1}{1+\gamma / \gamma_{ref}} \tag{6.89}$$

The degree of nonlinearity is therefore controlled by the reference strain and hence is influenced by the shear strength. However, a commonly encountered difficulty is that the shear strength that is mobilized at high strain levels is poorly correlated to the small strain soil behavior that controls modulus reduction curve values at strain levels of common engineering interest (about 0.1%–1.0%).

The modeling of modulus reduction behavior can be significantly improved by the use of a three-parameter modified hyperbolic backbone curve as

$$\tau = \frac{G_{max}\gamma}{1+(\gamma / \gamma_r)^a} \tag{6.90}$$

where γ_r is a *pseudo-reference strain* defined as the strain at which $G/G_{max} = 0.5$ and a is an additional fitting parameter that influences the curvature of the backbone curve. This relationship produces modulus reduction curves of the form

$$\frac{G}{G_{max}} = \frac{1}{1+(\gamma / \gamma_r)^a} \tag{6.91}$$

Figure 6.87 shows that pseudo-reference strain is the dominant parameter controlling the modulus reduction curves; as γ_r increases, the curves shift to the right, which models the soil behavior as being more linear. Figure 6.88 shows that the exponent a controls the slope of the modulus reduction curve, and hence how rapidly the stiffness of the soil decreases once yielding begins; for example, at $\gamma = \gamma_r$, increases in a increase the slope of the modulus reduction curve.

The plots in Figures 6.87 and 6.88 also show the effect of γ_r and a on the large-strain behavior implied by the form of the modified hyperbolic function. Increases in γ_r and decreases in a lead to higher shear stresses at large strains. The shear strengths implied by modified hyperbolic model (Equation 6.90) parameters that match the low-strain (less than ~1%) behavior are typically inconsistent with the strengths measured at large strains in laboratory tests. Put another way, just as the small-strain portion of the backbone curve was poorly represented by strength-based reference strain, γ_{ref}, large strain behavior tends to be poorly represented by the modified hyperbolic model. In short, classical hyperbolic models of the sort in Equations (6.88) and (6.90), although quite simple, have a limited ability to model observed laboratory behavior of soils over wide ranges of shear strain. Procedures for adjusting modulus reduction and backbone curves for consistency with large strain (e.g., shear strength) behavior are described in Section 6.6.3.3.

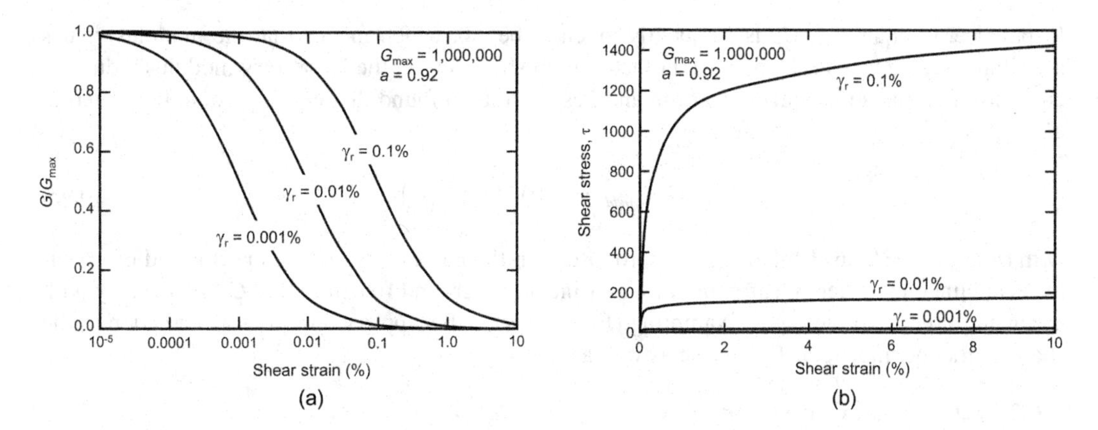

FIGURE 6.87 Influence of pseudo-reference strain, γ_r, on (a) modulus reduction curve, and (b) backbone curve.

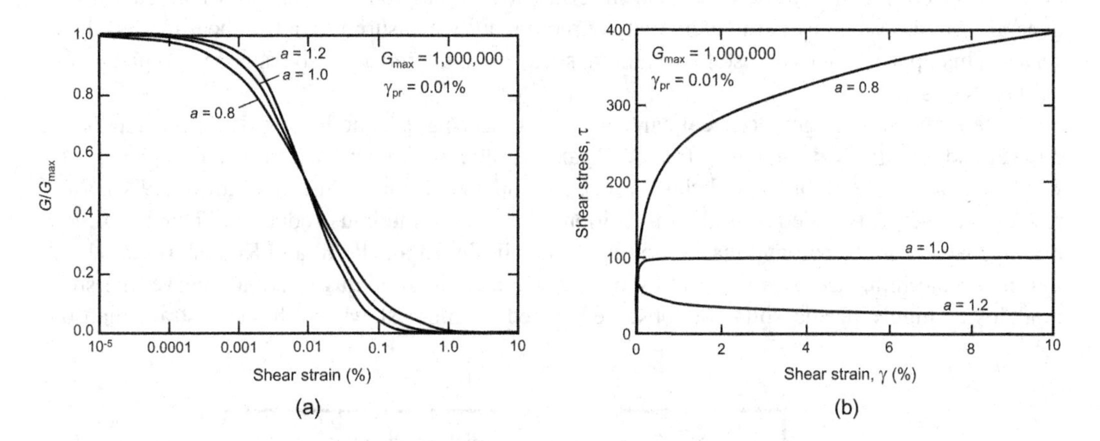

FIGURE 6.88 Influence of exponent, a, on (a) modulus reduction curve and (b) backbone curve.

6.6.3.1 Direct Laboratory Measurement

Laboratory shear moduli can be measured over a wide range of shear strain amplitudes using resonant column and torsional shear devices. The basic testing procedure, in which strain amplitudes begin at very low values and increase to high values, was described in Sections 6.5.4.2 and 6.5.4.3. The results of such a test can be described in terms of a laboratory modulus curve that shows the variation of secant shear modulus with cyclic strain amplitude. For soils, the low-strain laboratory shear modulus, $G_{max,lab}$, will generally be lower than the maximum shear modulus, $G_{max,field}$, obtained from field shear wave velocity measurements (Anderson and Woods, 1975) using Equation (6.76), with differences increasing with increasing stiffness, decreasing plasticity, and increasing depth (Chiara and Stokoe, 2006). For the case of clays sampled using relatively high-quality samplers, these differences have been postulated to result from pseudo-overconsolidation from secondary compression (Trudeau et al., 1974), which is lost in sampling.

Because field measurements of low-strainstiffness are more accurate than laboratory measurements, laboratory secant shear moduli should be corrected as

$$G_{field}(\gamma) = C_r(\gamma)\frac{G_{max,\,field}}{G_{max,\,lab}}G_{lab}(\gamma) \tag{6.92}$$

In practice, the quantity, C_r, is frequently taken to be 1.0, which means that the modulus values are simply scaled upward by a constant factor at all strain levels; the laboratory modulus reduction curve itself remains unchanged. Carlton and Pestana (2016) found that $G_{max,field}$ could be related to $G_{max,lab}$ as

$$\ln G_{max,field} = \ln\left(0.78G_{max,lab}^{1.10}\right) \tag{6.93}$$

with $\sigma_{\ln G_{max,\ field}} = 0.36$. Ishihara (1996) proposed, on the basis of cyclic tests performed on specimens obtained by different sampling methods including ground freezing, that C_r should vary with strain amplitude and the type of sampling (Figure 6.89). This approach implies a correction to the shape of the modulus reduction curve as well as to G_{max}.

6.6.3.2 Empirical Correlations

Resonant column and direct simple shear tests have been performed on many soils with different grain size characteristics, plasticity characteristics, stress histories, etc. Systematic interpretation of these tests has revealed the soil parameters that most strongly affect shear modulus reduction behavior. These results can be expressed in terms of modulus reduction or backbone curves of a general form described by parameters that are more readily measured than the shear moduli themselves. This approach can be used to estimate shear modulus reduction behavior for typical soils, or classes of soils.

In the early years of geotechnical earthquake engineering, the modulus reduction behaviors of coarse- and fine-grained soils were treated separately, and modulus reduction and damping curves were made available for broad soil classes such as sand and clay (e.g., Seed and Idriss, 1970). Later research, however, revealed a gradual transition between the modulus reduction behavior of non-plastic coarse-grained soil and plastic fine-grained soil. Zen et al. (1978) and Kokushu et al. (1982) first noted the influence of soil plasticity on the shape of the modulus reduction curve; the shear modulus of highly plastic soils was observed to reduce more slowly with shear strain than did

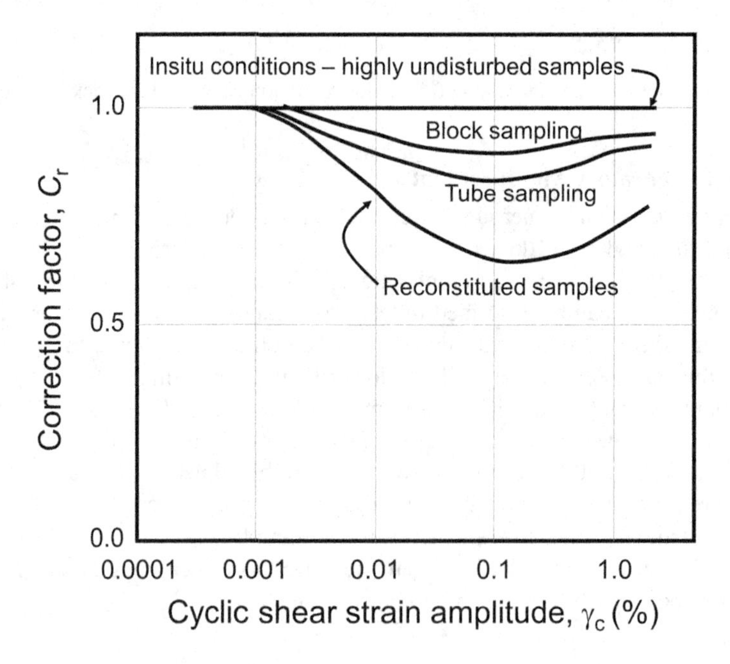

FIGURE 6.89 Strain-dependent correction factor for field shear moduli. (After Ishihara, 1996; used with permission of Oxford University Press.)

low-plasticity soils. After reviewing experimental results from a broad range of materials following the 1985 Mexico City earthquake, in which highly linear response was measured in the very plastic ($PI \approx 200$) Mexico City clay, Dobry and Vucetic (1987) and Sun et al. (1988) concluded that the shape of the modulus reduction curve was strongly influenced by the plasticity index of the soil. *PI*-dependent modulus reduction curves proposed by Vucetic and Dobry (1991) are shown in Figure 6.90.

Modulus reduction behavior has also been found to be influenced by effective confining pressure, particularly for soils of low plasticity (Iwasaki et al., 1978; Kokoshu, 1980; Ishibashi and Zhang, 1993). The Electric Power Research Institute sponsored an investigation of the dynamic properties of soils of unspecified plasticity (EPRI, 1993) that verified the effects of effective confining pressure on nonlinearity and presented modulus reduction curves in terms of different depth ranges (Figure 6.91). The EPRI curves are based on laboratory testing and comparisons of site response simulations to recorded ground motions.

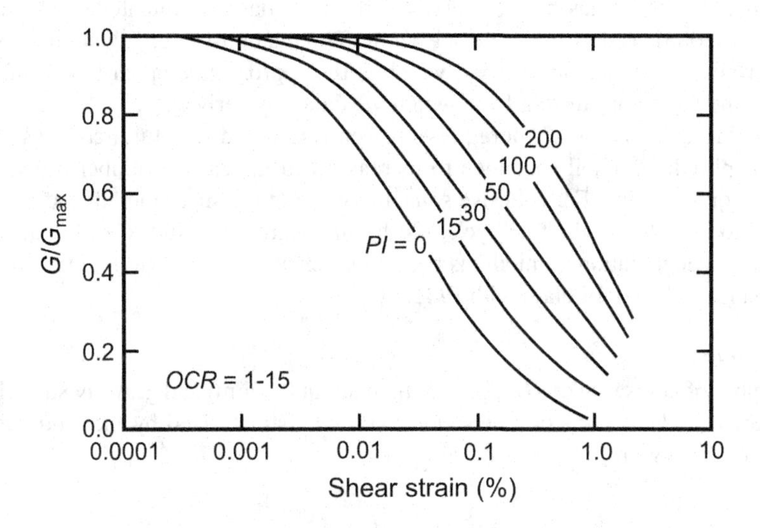

FIGURE 6.90 Modulus reduction curves of Vucetic and Dobry (1991).

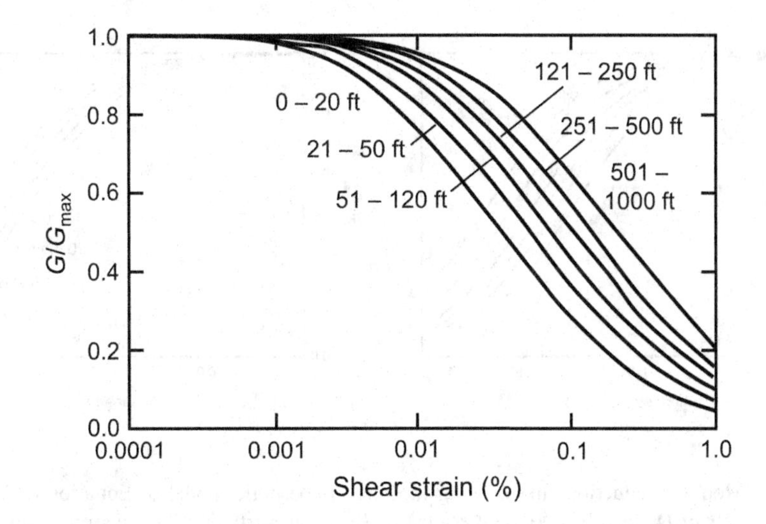

FIGURE 6.91 Depth-dependent modulus reduction curves of EPRI (1993); used with permission of EPRI.

Darendeli (2001) supplemented multiple sets of published data with additional resonant column and torsional shear tests on soil samples from multiple sites in California, South Carolina, and Taiwan to develop a database that covers a wide range of soil types, depths, densities, plasticity characteristics, and stress histories. The modified hyperbolic model (Equation 6.91) was found to fit the experimental results well with

$$\gamma_r = \left(0.0352 + 0.0010 PI \cdot OCR^{0.3246}\right)\left(\sigma'_{m0} / p_a\right)^{0.3483} \tag{6.94a}$$

$$a = 0.92 \tag{6.94b}$$

where γ is expressed in %, σ'_{m0} is the mean effective stress prior to the onset of cyclic loading in the same units as p_a, p_a is atmospheric pressure, and PI is in %. Because increasing γ_r indicates increasing soil linearity, the degree of linearity can be seen to increase with increasing PI, OCR, and σ'_{m0}. Figure 6.92 shows modulus reduction behavior for different ranges of PI and σ'_{m0}. For cohesionless soils, the Darendeli model has been superseded by a similarly formulated modified hyperbolic model by Menq (2003), as described further below. The relationship of Equation (6.94) applies to cyclic shear strain amplitudes up to 0.5%, which is the approximate upper bound of the range of strains used in the laboratory testing from which the data was derived.

As discussed in Section 6.4.4.2, pore pressure generation and structural changes can cause the shear strain amplitude of a soil specimen to increase with increasing number of cycles of stress-controlled harmonic loading. For cohesive soils, the value of the shear modulus after N cycles, G_N, can be related to its value in the first cycle, G_1, by the degradation index, δ (Equation 6.37). The effects of stiffness degradation on modulus reduction behavior are shown in Figure 6.93. Note that no degradation is predicted for clays with $PI \geq 100$.

Cohesionless Soils

Tests of a number of coarse sandy (D_{50} greater than about 0.3 mm) and gravelly soils (Menq, 2003) showed that their modulus reduction behavior could be characterized by the modified hyperbolic model (Equation 6.91) with

$$\gamma_r = 0.12 C_u^{-0.6} \left(\frac{\sigma'_{v0}}{p_a}\right)^{0.5 C_u^{-0.15}} \tag{6.95a}$$

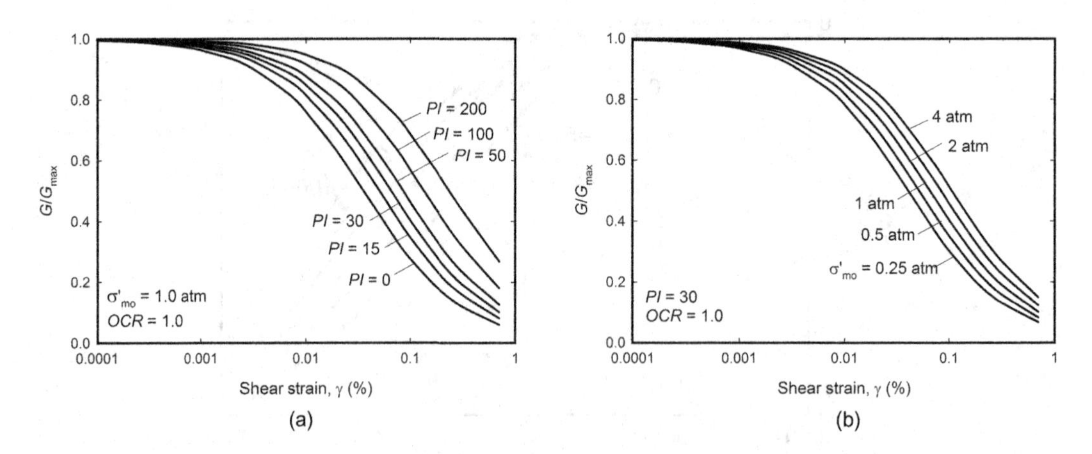

FIGURE 6.92 Modulus reduction curves using modified hyperbolic model of Equation (6.94) with coefficient set by models of Darendeli (2001): (a) effects of *PI* at low vertical effective stress, and (b) effects of vertical effective stress for nonplastic soil.

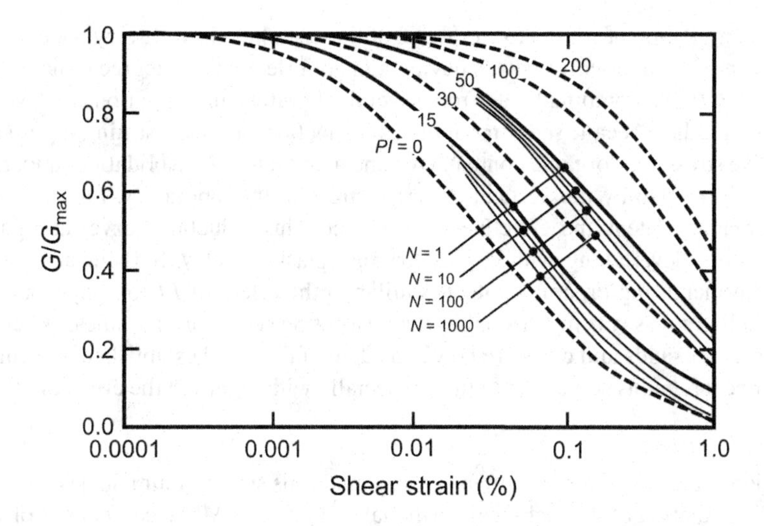

FIGURE 6.93 Effect of cyclic degradation on shear modulus. (After Vucetic and Dobry, 1989; used with permission of ASCE.)

and

$$a = 0.86 + 0.1\log\left(\sigma'_{v0} / p_a\right) \tag{6.95b}$$

where C_u = coefficient of uniformity, σ'_{v0} = initial vertical effective stress, and p_a = atmospheric pressure in the same units as σ'_{v0}. Note that soil nonlinearity in this model is unrelated to mean grain size (D_{50}), but decreases with increasing effective stress and increases with increasing C_u. Earlier work had found an effect of grain size, with coarser soils such as gravels being more linear (i.e., higher γ_r) than those for sands (Seed et al., 1986; Yasuda and Matsumoto, 1993; Rollins et al., 1998). Figure 6.94 illustrates the influence of coefficient of uniformity and initial effective stress on modulus reduction behavior.

Organic Soils

A number of investigations of modulus reduction behavior for highly organic soils and peats have been performed (Seed and Idriss, 1970; Stokoe et al., 1994; Boulanger et al., 1998; Kramer, 2000; Wehling et al., 2003; Tokimatsu and Sekiguchi, 2007; Kishida et al., 2009a). As shown in

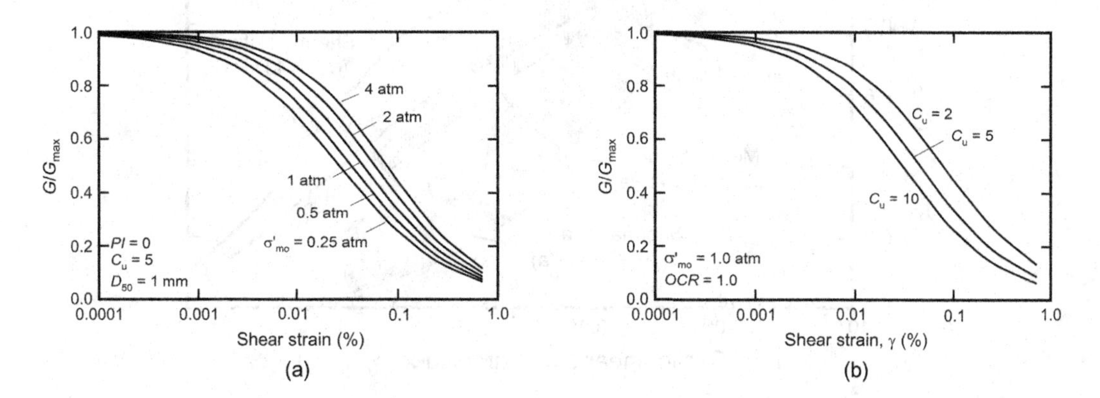

FIGURE 6.94 Modulus reduction curves of Menq (2003): (a) variation with σ'_{v0} for given $C_u = 5$ and (b) variation with C_u for $\sigma'_{v0} = 1$ atm.

Figure 6.95, a compilation of results from several of these studies shows a very wide range of modulus reduction curves, with effective stress having a major effect on the degree of nonlinearity.

Kishida et al. (2009b) combined data from several laboratory investigations to develop a regression model that predicts secant shear modulus as a function of cyclic strain amplitude, γ_c, initial vertical effective stress, σ'_{v0}, organic content, *OC*, and a laboratory consolidation ratio, *LCR*, defined as the *in situ* σ'_{v0} divided by the vertical effective stress in the laboratory. Figure 6.96 shows the influence of organic content and effective stress on modulus reduction curves at a particular vertical effective stress level along with curves for high plasticity clay. Soil linearity was observed to increase with increasing organic content, similar to the effect of *PI* for clays, but the effect of increasing effective stress was much smaller than was observed in prior studies. Whereas modulus reduction models for sands and clays are now considered fairly well established, the current state of knowledge for peat remains somewhat in flux, especially with regard to the effect of effective stress.

Municipal Solid Waste

In some locations, large deposits of human-derived materials such as municipal solid waste (MSW) may require the attention of geotechnical earthquake engineers. MSW can consist of many different types of materials, a number of which are so large or fibrous as to complicate their testing in the laboratory. Laboratory testing of MSW materials has generally involved the use of large-scale testing equipment (e.g., Matasović et al., 1998; Towhata et al., 2004; Zekkos et al., 2008a, 2008b; Towhata and Uno, 2008) and/or scalped test specimens (Athanasopoulos-Zekkos, 2008). These tests have shown that waste composition is important, and that composition can be at least approximately characterized by the amount of fibrous material (i.e., paper, plastic, and wood longer than 20 mm) they contain. Due to the difficulty of testing such materials, a number of investigators (Matasovic et al., 1995; Idriss et al., 1995; Augello et al., 1998; Matasovic and Kavazanjian, 1998; Elgamal et al., 2004) have attempted to back-calculate modulus reduction curves from the recorded response of MSW landfills. Zekkos et al. (2008a) proposed the generalized modulus reduction curves for MSW shown in Figure 6.97, which is based on limited available laboratory data. MSW can be seen to exhibit more linear behavior with increasing fractions of fibrous (> 20 mm) material and with increasing effective stress.

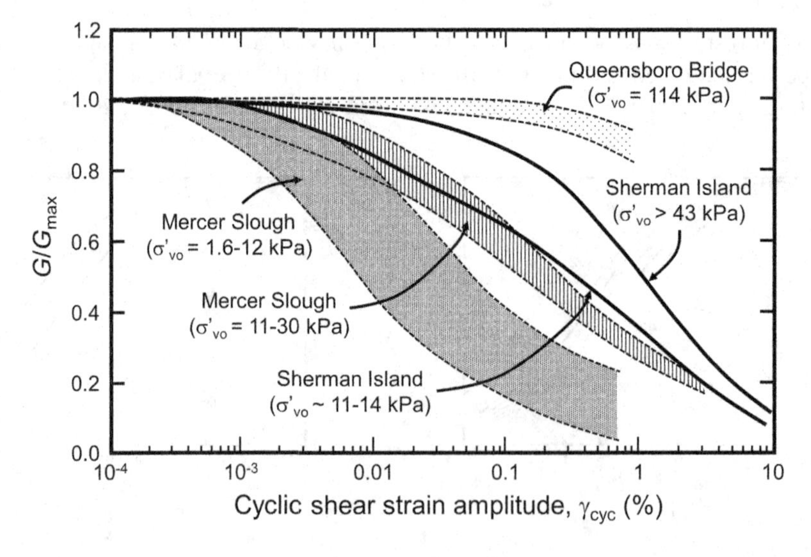

FIGURE 6.95 Laboratory test results for peats and organic soils showing strong effect of vertical effective stress prior to onset of cyclic loading, σ'_{v0}. (From Wehling et al., 2003; used with permission of ASCE.)

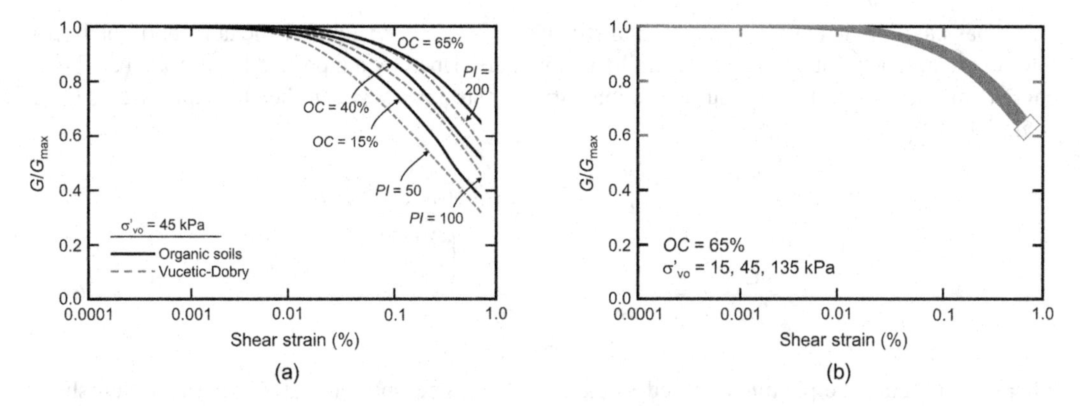

FIGURE 6.96 Modulus reduction behavior for organic soils from Kishida et al. (2009b) showing (a) effect of organic content, with clay curves of Vucetic and Dobry (1991) included for reference and (b) effect of effective stress at 65% organic content. (After Kishida et al., 2009b; used with permission of ASCE.)

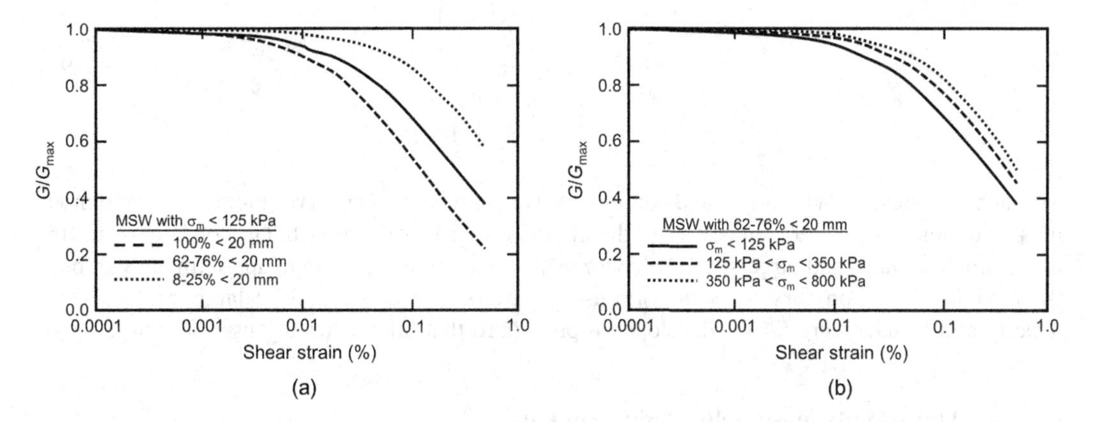

FIGURE 6.97 Generalized modulus reduction curves for MSW of different compositions (a) for mean confining stress less than 125 kPa and (b) for higher confining stresses for MSW with 62%–76% fibrous material < 20 mm. (After Zekkos et al., 2008a; used with permission of Canadian Science Publishing.)

6.6.3.3 Strength Considerations

As indicated in Section 6.6.1.2, a backbone curve can be computed from G_{max} and a modulus reduction curve. Since resonant column tests are usually limited to shear strains up to about 0.5%–1% (e.g., the Darendeli 2001 model is based on a data set with upper bound strains of about 0.5%), some degree of extrapolation is required if larger strain levels are expected to be induced in the soil. In some cases, this extrapolation is accomplished by selecting some analytical form, such as the modified hyperbola (Equation 6.91), for the backbone or modulus reduction curve. The extrapolation may imply shearing resistances that are significantly lower or significantly greater than the actual strength of the soil. While the effects of alternate representations of large-strain behavior are often nearly invisible in modulus reduction curves (e.g., see Figure 6.88a for $\gamma_c > 1\%$), they can be obvious in the corresponding backbone curves (Figure 6.88b for $\gamma_c > 1\%$). These considerations are important under strong shaking conditions because the high strain portion of the curve influences the level of shear stress induced in the soil, which in turn affects ground motion characteristics (e.g., Yee et al., 2013).

To overcome this problem, it is possible to apply a hybrid approach in which a suitable low-strain model is used up to a limiting strain level, γ_1, and larger strain behavior is then captured using a classical hyperbolic model that is asymptotic to the actual shear strength. The classical hyperbolic model has its origin at (τ_1, γ_1) with an initial stiffness (i.e., at $\gamma = \gamma_1$) equal to the tangent of the backbone curve at that point. If the low-strain behavior is described by a modified hyperbolic backbone curve (Equation 6.91), the modulus reduction curve for strains $\gamma > \gamma_1$ can then be expressed as (Yee et al., 2013):

$$\frac{G}{G_{max}} = \frac{\dfrac{\gamma_1}{1+(\gamma_1/\gamma_r)^a} + \dfrac{(G_{\gamma_1}/G_{max})(\gamma-\gamma_1)}{1+\left(\dfrac{\gamma-\gamma_1}{\gamma'_{ref}}\right)}}{\gamma} \tag{6.96}$$

where γ_r and a are the parameters used in the modified hyperbolic model, G_{γ_1} is the secant shear modulus at $\gamma = \gamma_1$, $\gamma'_{ref} = \dfrac{\tau_{ff} - \tau_1}{G_{\gamma_1}}$ and $\tau_1 = \dfrac{G_{max}\gamma_1}{1+(\gamma_1/\gamma_r)^\alpha}$. The modulus reduction ratio at $\gamma = \gamma_1$ is given by

$$\frac{G_{\gamma_1}}{G_{max}} = \frac{1+(1-a)\left(\dfrac{\gamma_1}{\gamma_r}\right)^a}{\left(1+\left(\dfrac{\gamma_1}{\gamma_r}\right)^a\right)^2} \tag{6.97}$$

This approach results in a smooth and continuous, two-part backbone curve (Figure 6.98) and modulus reduction curve. As an alternative to the modified hyperbola approach, Groholski et al. (2016) approximated small- and large-strain behavior using a quadratic equation to approximately fit both the modulus reduction curve at small strains and a target shear strength at large strains, while Yniesta and Brandenberg (2017) developed a procedure that allows for any user-backbone curve shape.

6.6.3.4 Uncertainty in Modulus Reduction Ratio

Darendeli (2001) characterized the dispersion in the data from which his modulus reduction model was developed. Assuming a normal distribution for the modulus reduction ratio at a particular strain level, the standard deviation of that ratio was expressed as

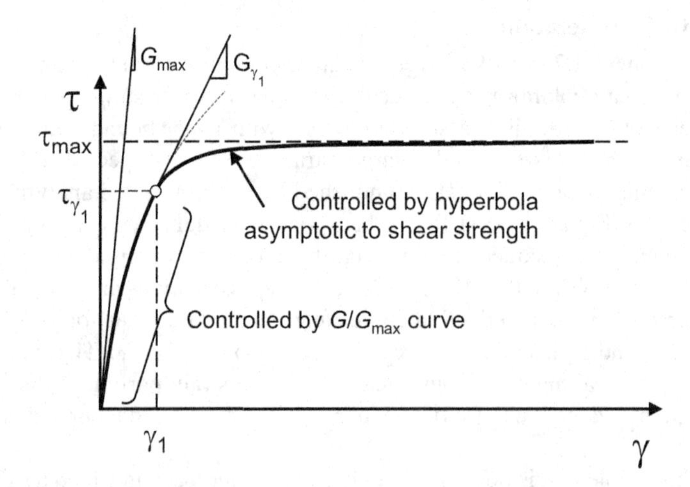

FIGURE 6.98 Backbone curve controlled by modulus reduction behavior at low strains and hyperbola asymptotic to shear strength at large strains.

$$\sigma_{NG} = 0.015 + 0.16\sqrt{0.25 - (G/G_{max} - 0.5)^2} \tag{6.98}$$

This value reaches its maximum of 0.095 at $G/G_{max} = 0.5$ and tapers down to values of 0.015 at $G/G_{max} = 1.0$ and 0.0.

Zhang et al. (2008) used a point estimation (Rosenblueth, 1975; Harr, 1987) technique to propagate uncertainties in reference strain and curvature parameter through a modified hyperbolic model (Equation 6.91) for three different geologic groups (Quaternary, Tertiary and older, and residual (saprolite) soils). The resulting uncertainties varied with soil type, shear strain amplitude, and soil plasticity and had peak values ranging from about 0.13 to 0.31.

6.6.4 Damping

A number of phenomena can cause the amplitude of a wave to decrease as it travels through a material and the term "damping" is often used to describe these phenomena. It is important to recognize, however, that the mechanisms of these phenomena are different, and that those differences may cause the effects of the phenomena to be affected by different factors. Geometric spreading (Sections 3.4.2.1 and C.5.2) produces attenuation as a result of energy spreading out in space and is not affected by material properties or frequency. Wave scattering (Section C.5.3) is affected by material properties and increases with increasing frequency. Material damping, the subject of this section, results from inelastic material behavior and, for most soils, is independent of frequency. Most practical problems will involve more than one form of damping or attenuation, hence any single damping formulation will seldom be useful in isolation.

While some attempts at field measurement of soil damping have been made (Henke and Henke, 1991; Riemer and Cobos-Roa, 2007), none have yet been proven sufficiently reliable to be used for design purposes. As a result, material damping behavior is generally obtained from laboratory tests.

6.6.4.1 Low-Strain Damping

At low strain levels ($\gamma < 0.001\%$), soils exhibit generally linear elastic behavior, which implies no energy dissipation. Experiments have shown, however, that cyclic loading does dissipate some energy, even though the mechanism(s) of dissipation is not always clear. The material damping at these low strains is referred to as a minimum damping ratio, ξ_{min}.

Like the low-strain shear modulus, G_{max}, the minimum damping ratio is affected by effective confining pressure, plasticity index, overconsolidation ratio, and frequency. In general, ξ_{min} increases with decreasing effective confining pressure, increasing plasticity index, decreasing overconsolidation ratio, and increasing loading frequency. Based on resonant column test data, Darendeli (2001) proposed that ξ_{min} be estimated as

$$\xi_{min} = \left(0.8005 + 0.0129PI \cdot OCR^{-0.1069}\right)(\sigma'_{m0})^{-0.2889}(1 + 0.2919 \ln f) \tag{6.99}$$

where f is frequency in Hz. Figure 6.99 shows the variation of ξ_{min} with effective confining pressure and plasticity index for normally consolidated soils loaded at a frequency of 1 Hz. Note that the low-strain damping ratios at a given effective confining pressure are higher for high PI than low PI values.

Menq (2003) found that the minimum damping ratio for dry coarse-grained soils was primarily affected by grain size distribution and effective stress level, and could be estimated as

$$\xi_{min} = 0.55C_u^{0.1}D_{50}^{-0.3}(\sigma'_{v0} / p_a)^{-0.08} \tag{6.100}$$

Values of ξ_{min} were noted as potentially increasing "several fold" as water is added to the soil, but quantitative values were not presented.

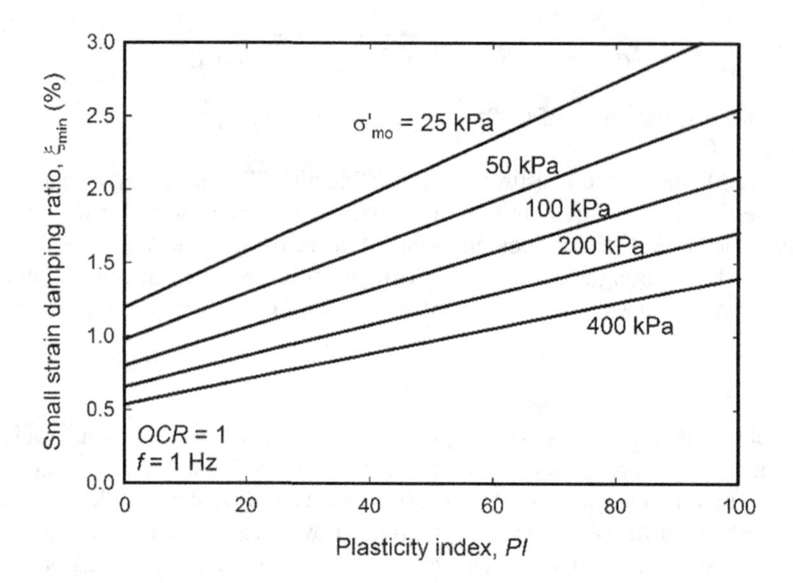

FIGURE 6.99 Variation of low-strain damping ratio with plasticity index and mean effective stress. (After Darendeli, 2001.)

Several studies of site response using weak motion data from vertical arrays (Elgamal et al., 2001; Tsai and Hashash, 2009; Yee et al., 2013; Cabas et al., 2017; Afshari and Stewart, 2019; Tao and Rathje, 2019) have found that the small-strain damping mobilized under field conditions exceeds that from laboratory-based material damping models such as those by Darendeli (2001) and Menq (2003). This additional damping is thought to arise from wave scattering phenomena, which are present under field conditions but not in small-scale laboratory tests. While small strain damping ratios of about 5% have been found to be effective in some of this prior work, it has not yet been possible to develop general recommendations for how (or if) to modify ξ_{min} predictions from models of the type discussed in this section to better match field data. Scattering effects are discussed in more detail in Section 7.5.6.2.

6.6.4.2 Higher-Strain Damping

Damping ratios at higher strain levels result from inelasticity and are also affected by soil plasticity and effective stress level. Factors that tend to increase linearity, such as increased plasticity and increased effective stress level, lead to reduced damping ratios. Mirroring the trends observed for modulus reduction, damping ratio has been observed (Figure 6.100) to decrease substantially with increasing plasticity index at a particular strain level (Sun et al., 1988; Vucetic and Dobry, 1991). Others have shown (Figure 6.101) that damping ratio also decreases with increasing initial vertical effective stress or increasing depth (Ishibashi and Zhang, 1993; EPRI, 1993).

Under harmonic loading conditions, stress-strain behavior is frequently modeled using the Masing rules (Section 6.4.5.2) in which the shapes of the unloading and reloading curves are scaled versions of the backbone curve. A scale factor of 2.0 (Equation 6.42) produces closed hysteresis loops for symmetric loading. The shape of the hysteresis loop, and consequently the damping ratio, is controlled by the shape of the backbone curve. For the simple hyperbolic model of Equation (6.88), which is equivalent to the modified hyperbolic model of Equation (6.90) with $a = 1$, the area of the hysteresis loop is given by

$$A_{loop} = 8\left[\int_0^{\gamma} \tau\, d\gamma - \frac{1}{2}\tau\gamma\right] \tag{6.101}$$

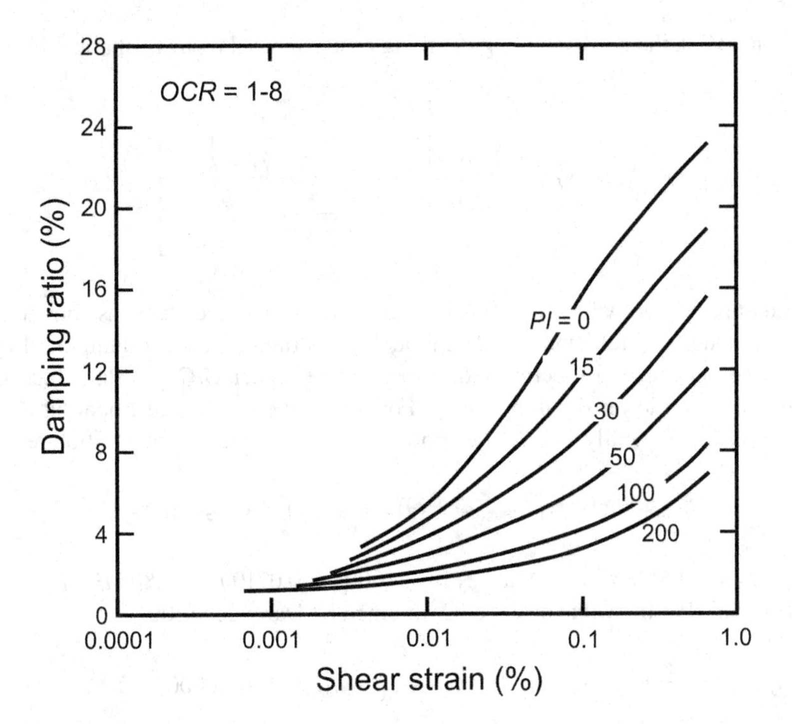

FIGURE 6.100 Variation of damping ratio with cyclic shear strain and plasticity index. (After Vucetic and Dobry, 1991; used with permission of ASCE.)

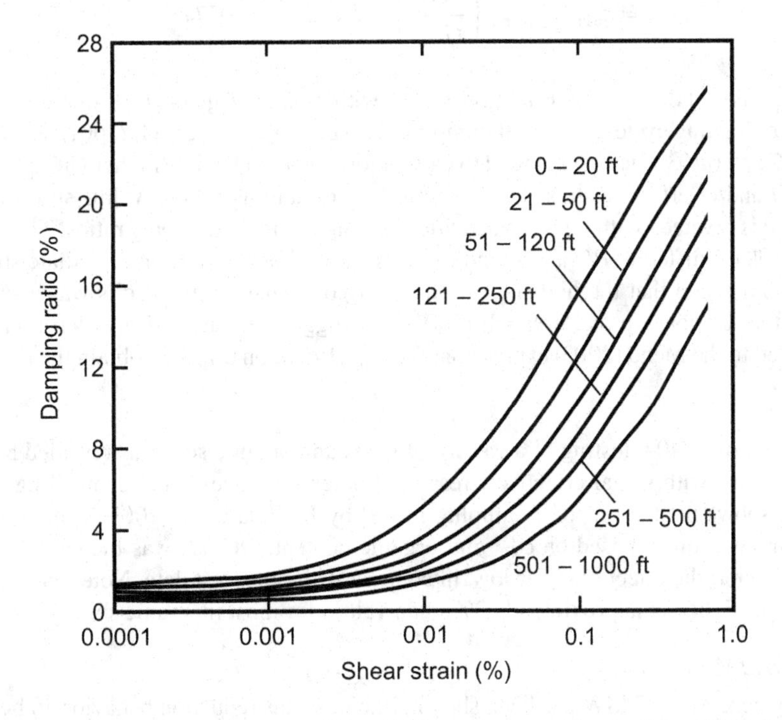

FIGURE 6.101 Variation of damping ratio with cyclic shear strain and depth. (After EPRI, 1993; used with permission of the Electric Power Research Institute.)

Using Equation (B.72), the corresponding damping ratio can be shown (Ishihara, 1996) to be

$$\xi_{M,\alpha=1}(\gamma)[\%] = \frac{100}{\pi}\left[4\frac{\gamma - \gamma_r \ln\left(\dfrac{\gamma+\gamma_r}{\gamma_r}\right)}{\dfrac{\gamma^2}{\gamma+\gamma_r}} - 2\right] \tag{6.102}$$

The Masing damping ratio, which is a function of pseudo-reference strain as shown by Equation (6.102), is often taken as a reference level in models for strain-dependent damping. Larger values of γ_r, which have been shown (Section 6.6.3) to produce higher G/G_{max} values (i.e., more linear response), also produce lower damping ratios. For $a \neq 1$, the integral in Equation (6.101) cannot generally be solved analytically, but can be approximated by numerical integration resulting in

$$\xi_{\text{Masing}}\,(\%) = c_1\xi_{\text{Masing},\,a=1} + c_2\left(\xi_{\text{Masing},\,a=1}\right)^2 + c_3\left(\xi_{\text{Masing},\,a=1}\right)^3 \tag{6.103}$$

where $c_1 = 0.2523 + 1.8618a - 1.1143a^2$, $c_2 = -0.0095 - 0.0710a + 0.0805a^2$, and $c_3 = 0.0003 + 0.0002a - 0.0005a^2$. For the value of $a = 0.92$ obtained by Darendeli (2001),

$$\xi_{\text{Masing}}\,(\%) = 1.0213\xi_{\text{Masing},\,a=1} - 0.0067\left(\xi_{\text{Masing},\,a=1}\right)^2 + 0.000061\left(\xi_{\text{Masing},\,a=1}\right)^3 \tag{6.104}$$

Darendeli (2001) proposed that the total damping ratio could be expressed as

$$\xi = \xi_{\min} + \xi_{\text{Masing}}\left(\frac{G}{G_{\max}}\right)^{0.1}(0.6329 - 0.0057\ln N) \tag{6.105}$$

where G/G_{max} is obtained from Equation (6.91) with γ_r and a given by Equations (6.94), and N = number of loading cycles. The total damping ratio curves of Darendeli (2001) are illustrated in Figures 6.102 and 6.103. Factors that tend to cause more linear behavior (higher G/G_{max} values), e.g., increasing *PI* and/or effective stress level, lead to lower damping ratios. At low strain levels, however, the trend is reversed with higher plasticities leading to higher damping ratios. Thus, the damping ratio curves for different *PI* values tend to cross each other at low to intermediate strain levels.

Menq (2003) found that the higher-strain damping of coarse-grained soils was accurately represented by the relationship of Darendeli (2001) and suggested that the Darendeli relationship for ξ-ξ_{min} be added to the Menq (2003) expressions for ξ_{min} (Equation 6.100) to obtain the total damping.

Organic Soils

As shown in Figure 6.104, testing of a variety of peats and organic soils has revealed a wide range of damping ratios, with initial effective stress (and potentially age) being controlling parameters. Figure 6.105 shows predictions of a damping model by Kishida et al. (2009b), that provides estimates of damping curves based on σ'_{v0}, and organic content, *OC*. As was the case with modulus reduction behavior, the effect of σ'_{v0} is low, in contrast with other test data. Note that the low-strain damping ξ_{min} is higher, at approximately 3%, than values for most inorganic soils.

Municipal Solid Waste

The damping behavior of MSW has been shown, like modulus reduction behavior, to be influenced by waste composition and, to a lesser degree, effective stress level. Figure 6.106 illustrates the effects of these parameters on damping ratio. Note that the low-strain damping ratios are quite high relative to those of inorganic soils.

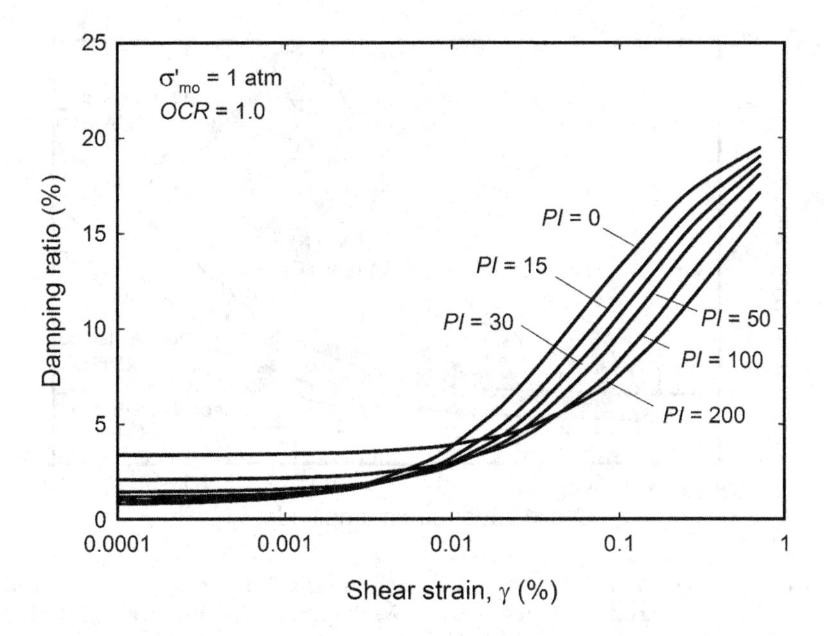

FIGURE 6.102 Variation of damping ratio with cyclic shear strain and plasticity index for soil at σ'_{m0} = 25 kPa according to Darendeli (2001).

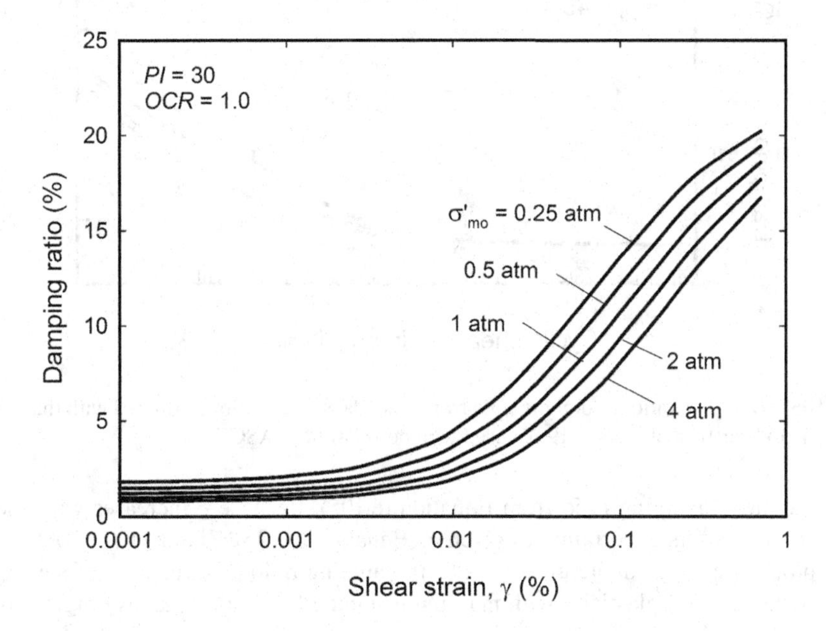

FIGURE 6.103 Variation of damping ratio of nonplastic soil with cyclic shear strain and initial vertical effective stress according to Darendeli (2001).

6.6.4.3 Uncertainty in Damping Ratio

Darendeli (2001) characterized uncertainty in damping ratio and its relationship to uncertainty in modulus reduction ratio. Assuming damping ratio to be lognormally distributed, Darendeli proposed that

$$\sigma_{\ln\xi} = 0.0067 + 0.78\sqrt{\xi} \qquad (6.106)$$

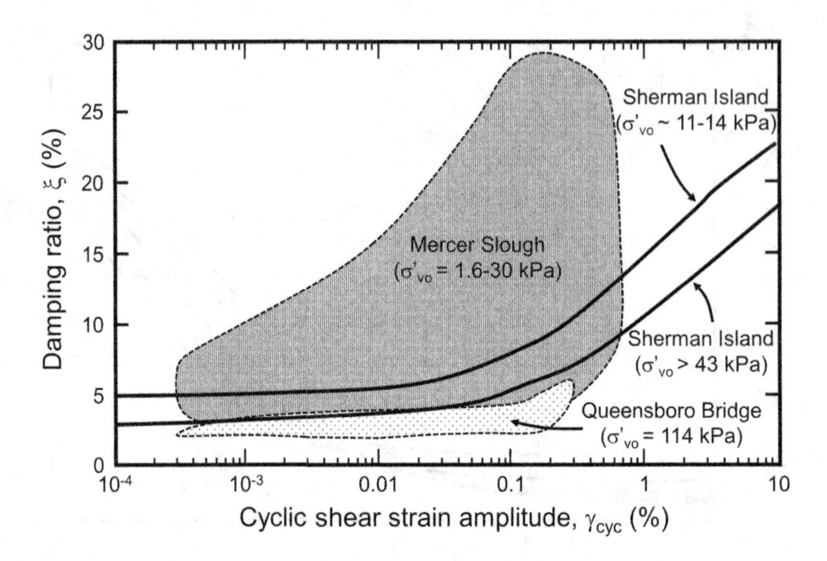

FIGURE 6.104 Damping ratio for peat and organic soils as identified from a variety of material-specific tests at various initial confining effective stresses. (After Wehling et al., 2003; used with permission of ASCE.)

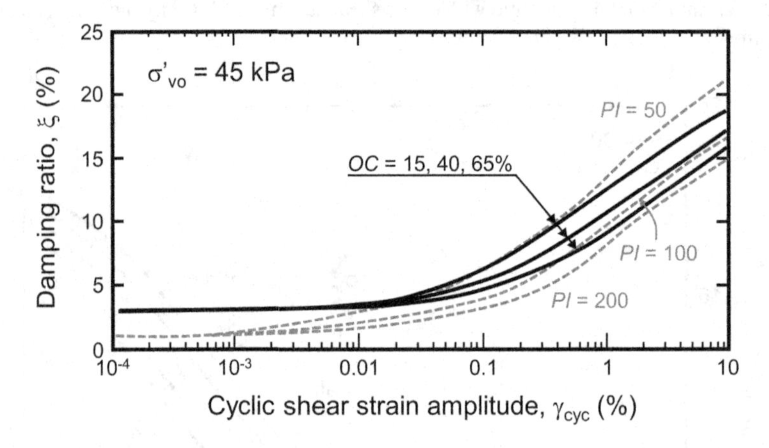

FIGURE 6.105 Damping ratio for organic soils from Kishida et al. (2009b) compared with that of inorganic soils according to Vucetic and Dobry (1991); used with permission of ASCE.

where ξ is the mean damping ratio from Equation (6.105). Because ξ increases with shear strain amplitude, so too does its uncertainty as given by Equation (6.106). Zhang et al. (2008) used point estimation procedures to estimate uncertainties in damping ratio as a function of soil type, shear strain amplitude, and soil plasticity with maximum standard deviation values ranging from about 2.5% to 7%.

Because increasing linearity corresponds to decreasing damping, the modulus reduction ratio and damping ratio should be negatively correlated to each other, i.e., an above average modulus reduction ratio should correspond to a below average damping ratio. Kottke and Rathje (2008) recommend using a correlation coefficient

$$\rho_{G/G_{\max},\xi} = -0.5 \qquad (6.107)$$

Procedures for generating pairs of correlated random variables, which can be used to generate negatively correlated modulus reduction and damping curves to be used in randomized site response

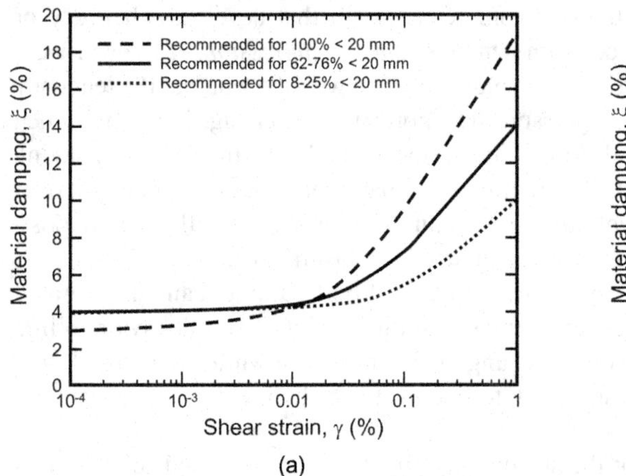

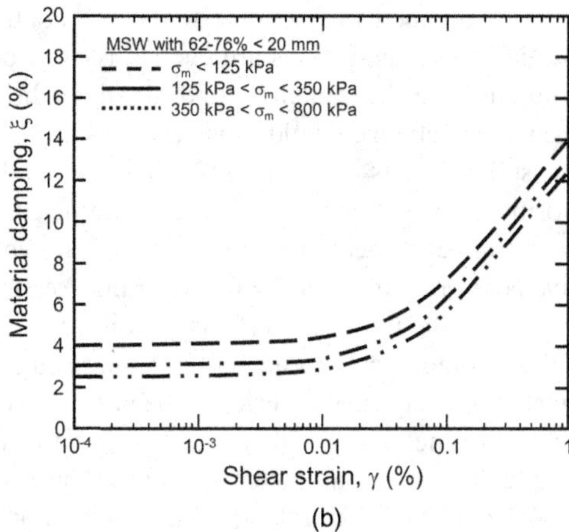

FIGURE 6.106 Generalized damping curves for MSW of different compositions (a) for mean confining stress less than 125 kPa and (b) for higher confining stresses for MSW with 62–76 fibrous material < 20 mm. (After Zekkos et al., 2008a; used with permission of Canadian Science Publishing.)

analyses of the types discussed in Section 7.5.7, are presented in Section D.9.3.3. Figure 6.107 shows examples of simulated modulus reduction and damping curves.

6.6.5 Volume Change Behavior

The degree to which a soil tends to contract or dilate when sheared is important in many geotechnical earthquake engineering applications – indeed, it controls the behavior of soils that are susceptible to liquefaction (Chapter 9). Therefore, the characterization of that tendency is extremely important. As discussed in Section 6.4.4.4, all dry or partially saturated sands, whether loose or dense, will initially contract when sheared. Loose sands will continue to contract as strain levels increase, but dense sands will begin to dilate at larger strains so the states of both will move (by contraction for the loose sand and dilation for the dense one) to the steady state under monotonically increasing shear stress. Under cyclic loading inducing low to moderate strain amplitudes, however, both loose and dense sands continue to contract and can cross or move away from the steady state line (SSL). If sheared monotonically after cyclic shearing, however, all but the very loosest will dilate back toward the steady state.

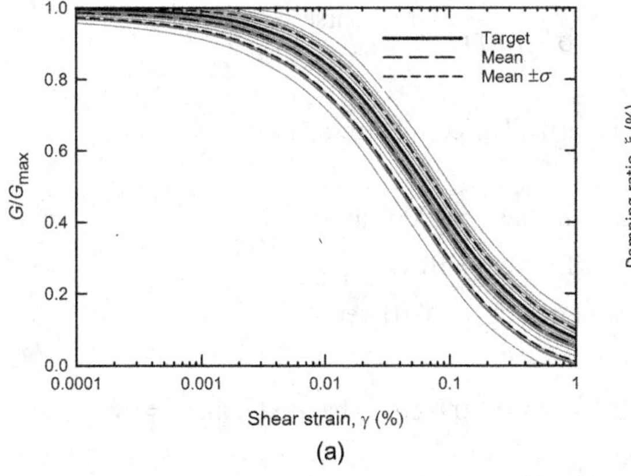

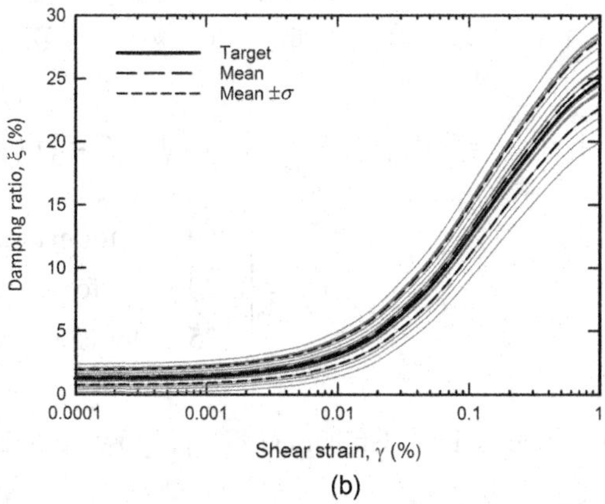

FIGURE 6.107 (a and b) 20 randomized sets of modulus reduction and damping curves for soil with $PI = 30$ and $\sigma'_{v0} = 100$ kPa using the procedure of Darendeli (2001).

As described in Section 6.4.4.4, the contractive volumetric strain that develops when dry or partially saturated sands are cyclically sheared is known as *seismic compression*. If the soil were saturated and sheared under undrained conditions, no volume change would occur, but the tendency for volume change would cause changes in pore pressure. The pore pressure change is positive when the soil responds in a contractive manner. Both soils initially above and below the SSL (i.e., having positive or negative values of state parameter, Ψ or relative state parameter index, ξ_R) contract over some range of shear strain. If saturated, no volume change can occur, and both will generate positive pore pressure over the limited range of strain often induced by seismic loading.

Due to the important effects of soil dilatancy on the response of both saturated and unsaturated soils, geotechnical engineers have long sought to characterize this aspect of soil behavior. While relative density has historically been used, volume change behavior is known to be more closely related to metrics of soil state relative to the steady state line (SSL), including the state parameter and relative state parameter index (Section 6.4.3.6).

Because the *in situ* density and position of the steady state line for coarse-grained soils are difficult to measure, procedures for estimation of state parameter and relative state parameter index have focused on their correlation to more easily measured *in situ* penetration test results. A number of procedures for estimating these metrics from penetration test results are available.

6.6.5.1 Estimation of Relative Density

Relative density, D_r (Equation 6.31), has historically been used as an indication of the state of a granular soil, but it is an incomplete indicator because it does not account for effective stress level, which affects volume change tendency. Nevertheless, a number of predictive models make use of relative density and correlations of relative density to parameters that can be measured in the field allow comparison of laboratory and field data. Correlations of relative density to SPT resistance going back to the work of Meyerhof (1957) have generally been expressed in the form

$$\frac{N}{D_r^2} = C_d \tag{6.108}$$

where N and C_d have taken various forms. Meyerhof (1957) proposed the use of Equation (6.108) with N taken as the measured SPT resistance without overburden correction, N_m, and $C_d = 17 + 24(\sigma'_{v0} / P_a)$. Skempton (1986) recommended that Equation (6.108) be used with

$$N = (N_1)_{60} = \begin{cases} N_m \cdot \dfrac{ER}{60} \cdot \dfrac{55}{60} \cdot \dfrac{2}{1+\sigma'_{v0}/100} & \text{for fine sand} \\ N_m \cdot \dfrac{ER}{60} \cdot \dfrac{65}{60} \cdot \dfrac{3}{2+\sigma'_{v0}/100} & \text{for coarse sand} \end{cases} \tag{6.109}$$

where ER = energy ratio (in percent), σ'_{v0} = initial vertical effective stress in kPa, and

$$C_d = \begin{cases} 60 & \text{for natural soils (age } > 100 \text{ years)} \\ 40 & \text{for recent fills (age } \approx 10 \text{ years)} \\ 35 & \text{for laboratory tests (age } \approx 0.01 \text{ years)} \end{cases}$$

Cubrinovski and Ishihara (1999) proposed that Equation (6.109) could be used with $N = (N_1)_{60}$ and

$$C_d = \frac{6.923}{(e_{max} - e_{min})^{1.7}} \tag{6.110}$$

For cases where minimum and maximum void ratios cannot be measured, Cubrinovski and Ishihara (1999) suggested that the void ratio range be estimated as $e_{max} - e_{min} \approx 0.23 + 0.06/D_{50}$ and listed approximate $e_{max} - e_{min}$ values of 0.625 for sand with fines ($FC \sim 20\%$), 0.41 for clean sand, and 0.30 for gravelly sand/coarse sand. Idriss and Boulanger (2007) considered the data presented by Cubrinovski and Ishihara (1999) and judged that a value of $C_d = 46$, which would correspond to $D_r = 81\%$ for $N = (N_1)_{60} \approx 30$, was reasonable for clean sand (at depths consistent with observations of liquefaction) at $\sigma'_{v0} = 100$ kPa. Idriss and Boulanger (2007) also proposed that relative density could be estimated from CPT tip resistance as

$$D_r = 0.478\left(q_{c1N}\right)^{0.264} - 1.063 \tag{6.111}$$

where q_{c1N} is the dimensionless, overburden-normalized tip resistance given in Equation (9.19).

6.6.5.2 Estimation of State Parameter

Measurement of the state parameter, defined in Equation (6.32) as $\psi = e - e_{ss}$, might seem, at first glance, to be relatively straightforward – one would obtain undisturbed samples of the soil from which (1) *in situ* void ratio would be determined and (2) shear test results would be used to determine the position of the SSL. In reality, however, direct measurement of the state parameter is extremely difficult – the successful retrieval of samples of cohesionless soils from below the water table is difficult, and the act of sampling usually changes the density of such soils. Furthermore, the position of the SSL is not easily measured in the laboratory since the steady state is reached at strain levels above that which can be reliably measured in most laboratory tests.

As a result of these factors, state parameter is usually estimated by correlation to CPT or SPT resistance. A strong advantage of the CPT for this application is that cavity expansion theory can be used to provide a mechanics-based interpretation of its measurements. CPT tip resistance can be expressed in different ways, but is usually made dimensionless by normalizing with respect to some measure of effective stress. Letting

$$Q_p = \frac{q_t - p_0}{p'_0} \tag{6.112}$$

where q_t = tip resistance after correction for unequal area effect, p_0 = mean total stress, and p'_0 = mean effective stress. Then, for level-ground conditions, Q_p can be related to the normalized tip resistance based on vertical effective stresses,

$$Q_p \approx \frac{3Q_{t1}}{1 + 2K_0} \tag{6.113}$$

where K_0 is the at-rest earth pressure coefficient under level-ground conditions (assuming $\sigma_2 = \sigma_3$ and $\sigma'_2 = \sigma'_3 = K_0\sigma'_1$) and the approximation results from lack of consideration of the difference between total mean stress and total vertical stress. Been and Jefferies (1992) found that cavity expansion theory predictions of CPT resistance could be well approximated by the simple relationship

$$Q_p = k(1 - B_q)\exp(-m\psi) \tag{6.114}$$

where k and m are sand-specific functions of the rigidity of the soil and B_q is defined in Equation (6.56). Solving for the state parameter,

$$\psi = -\frac{\ln\left(\dfrac{Q_p}{k\left(1 - B_q\right)}\right)}{m} \tag{6.115}$$

Using calibration chamber test results, Been and Jefferies (1992) related rigidity to the I_c^{BJ} soil classification index as defined in Equation (6.59). The soil classification index can be used to compute

$$k = M\left(3 + 0.85\left(34 - 10I_c^{BJ}\right)\right) \tag{6.116}$$

$$m = 11.9 - 13.3/\left(34 - 10I_c^{BJ}\right) \tag{6.117}$$

where

$$M = \frac{\bar{\sigma}_q}{\bar{\sigma}_m} = \frac{\sqrt{\frac{1}{2}\left[(\sigma_1 - \sigma_2)^2 + (\sigma_2 - \sigma_3)^2 + (\sigma_3 - \sigma_1)^2\right]}}{(\sigma_1' + \sigma_2' + \sigma_3')/3} \tag{6.118}$$

Under level ground conditions,

$$M = \frac{3(\sigma_1 - \sigma_3)}{\sigma_1'(1 + 2K_0)} \tag{6.119}$$

Robertson (2009) used data from Wride et al. (2000), Jefferies and Been (2006), and Shuttle and Cunning (2007) to estimate contours of state parameter for uncemented Holocene-age soils in Q_{tn}–F_r space. The contours are noted as being approximate since factors other than Q and F (e.g., stress state and plastic hardening for coarse-grained soils and sensitivity for fine-grained soils) can also affect state estimates. Letting

$$Q_{tn,cs} = Q_{tn} K_c \tag{6.120}$$

where Q_{tn} is as given in Equation (6.61) and

$$K_c = \begin{cases} 1.0 & \text{for} \quad I_c \leq 1.64 \\ 5.581I_c^3 - 0.403I_c^2 + 33.75I_c - 17.88 & \text{for} \quad I_c > 1.64 \end{cases} \tag{6.121}$$

Robertson (2010) proposed the simplified approximate relationship to estimate ψ from $Q_{tn,cs}$

$$\psi = 0.56 - 0.33 \log Q_{tn,cs} \tag{6.122}$$

Figure 6.108 shows contours of state parameter superimposed upon the soil behavior types of Robertson (2012).

6.6.5.3 Estimation of Relative State Parameter Index

As discussed in Section 6.4.3.6, the relative state parameter index, ξ_R, of Boulanger (2003a) uses a relative density-based relative dilatancy index (Bolton, 1986) to relate state parameter to relative density, a parameter that is both familiar and frequently measured.

Combining Equation (6.35) with the D_r relations given earlier in this section, and setting $R = 1$ and $Q = 10$, the SPT-based relative state parameter index can be described as

$$\xi_R = \frac{1}{10 - \ln(100p'/p_a)} - \sqrt{\frac{(N_1)_{60}}{46}} \tag{6.123}$$

where p' = mean effective stress in same units as p_a. A similar relationship based on CPT resistance is

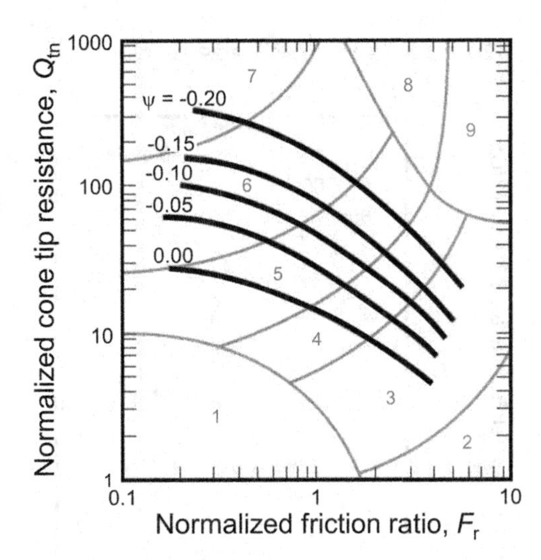

FIGURE 6.108 Contours of state parameter as functions of normalized cone tip resistance and friction ratio. (After Robertson, 2012; courtesy of P.K. Robertson.)

$$\xi_R = \frac{1}{10 - \ln(100p'/p_a)} - \left(0.086\sqrt{Q_{tn}} - 0.334\right) \tag{6.124}$$

It should be recognized that, being based on data from many different sands tested in different investigations, the correlations of Equations (6.123) and (6.124) correspond to the average behavior of sands in that database, and may not accurately represent the behavior of a specific sand whose characteristics differ significantly from the average of that population.

6.6.5.4 Element-Level Effects of Volume Change Behavior

The tendency of a soil to contract or dilate upon shearing leads to the development of volumetric strain, which can be positive (in the case of contraction) or negative (in the case of dilation). Under seismic loading, the shear strains induced in the soil at level-ground sites are usually small enough that only contraction occurs. Under drained conditions, this contraction leads to densification of the soil which is manifested at the ground surface in the form of settlement. Under the undrained conditions that exist in saturated soils subjected to earthquake shaking, the tendency for contraction leads to the development of positive excess pore pressure with resultant reduction of effective stress. After the earthquake, dissipation of the generated pore pressure will lead to contractive volume change. This section describes the effects of volume change behavior on a single element of soil under both drained and undrained conditions; the implications of those behaviors on soil deposits in the field are discussed in Section 9.6.5.

Drained Conditions

Because air is so much more compressible than water, cyclic loading generally causes dry and partially saturated soils to contract without generation of pore fluid (air or water) pressure. The densification of clean sand subjected to cyclic simple shear loading has been found to be influenced by the density of the sand, the amplitude of the cyclic shear strain induced in the sand, and the number of cycles of shear strain applied to the soil (Silver and Seed, 1971; Youd, 1972; Seed and Silver, 1972). Figure 6.109 shows how the volumetric strain after 15 cycles increases with increasing strain amplitude and decreasing relative density; most of this volumetric strain occurs in the first few cycles.

Shaking tables have been used to excite large simple shear-like test specimens with uni-directional, bi-directional (two orthogonal horizontal directions), and tri-directional (two horizontal plus vertical) shaking. Pyke et al. (1975) found that bi-directional shaking produced volumetric

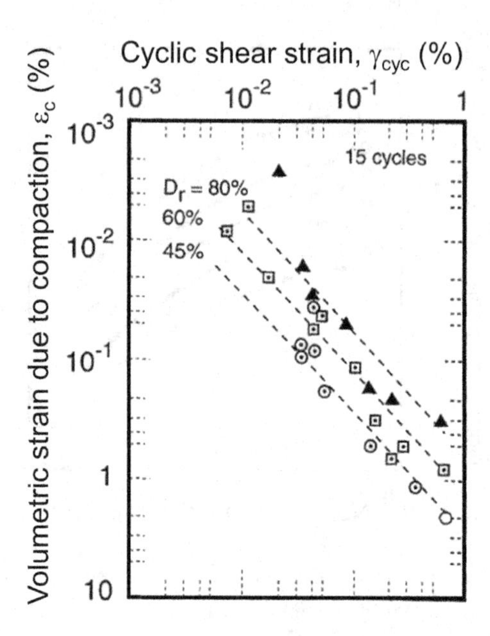

FIGURE 6.109 Relationship between volumetric strain and cyclic shear strain for a sand of different relative densities. (After Tokimatsu and Seed, 1987; used with permission of ASCE.)

strains approximately equal to the sum of the strains that would occur if each of the loading histories had been applied uni-directionally, and that adding a vertical component of acceleration resulted in an increase in volumetric strains of up to approximately 50%. Volumetric strains in 16 sands with D_{50} values ranging from 0.13 to 1.60 mm were observed to be strongly influenced by relative density and total overburden stress, but not by gradation, particle angularity, soil fabric, mineralogy, saturation, or age (Duku et al., 2008).

Fill materials are often granular with some amount of fines, and have been observed to settle in earthquakes (Seed and Lee, 1967; McClure, 1973; Slosson, 1975; Stewart et al., 2001). Whang et al. (2004) found that compacted plastic soils developed lower volumetric strain than clean sands, and that plastic soils developed smaller strains when compacted wet of optimum than when compacted dry of optimum. Tsukamoto et al. (2004) performed cyclic triaxial tests on sands containing about 20% nonplastic fines prepared at consistent void ratios but at different degrees of saturation. The specimens with lower degrees of saturation were found to develop most of their volumetric strain during cyclic loading while the more highly saturated specimens developed most of theirs after cyclic loading had ended. The total volumetric strain levels, however, were about the same, which indicated that the degree of saturation had little effect on the final volumetric strain of the soil. Whang et al. (2005) tested sand with nonplastic fines (rock flour) and found volumetric strains to be higher than those of clean sand. They also found that volumetric strain decreased with increasing degree of saturation up to $S = 30\%$, and then increased at higher degrees of saturation; the reduction in volumetric strain resulted from matric suction (Yee et al., 2014).

A simplified procedure for the estimation of volumetric strain based on tests performed on one clean sand (Silver and Seed, 1971) was developed by Tokimatsu and Seed (1987). Since that time, a number of researchers (Whang et al., 2004, 2005; Tsukamoto et al., 2004; Duku et al., 2008; Yee et al., 2014) have tested a wider variety of soils under a wider variety of conditions. These data have allowed the development of volumetric strain models that predict the volumetric strain resulting from a given number of cycles of a particular cyclic shear strain amplitude.

The mean volumetric strain resulting from 15 cycles of shear strain, which is considered to be representative of that produced by a **M** 7.5 earthquake, can be expressed as

$$(\varepsilon_{vol})_{N=15} = \begin{cases} 0 & \text{for} \quad \gamma_{cyc} \leq \gamma_{tv} \\ a(\gamma_{cyc} - \gamma_{tv})^b & \text{for} \quad \gamma_{cyc} > \gamma_{tv} \end{cases} \tag{6.125}$$

where γ_{cyc} = cyclic shear strain amplitude, γ_{tv} = volumetric threshold shear strain (Section 6.4.3.1) below which no volumetric strain occurs, a and b are coefficients that are ideally determined from high quality, material-specific laboratory tests, and all strains are in percent. An example of the variation of $\varepsilon_{v,N=15}$ with shear strain amplitude in cyclic simple shear tests is shown in Figure 6.110a.

The majority of the volumetric strain occurs in the early cycles after which the rate of volume change slows. For numbers of cycles other than 15, the volumetric strain can be computed as

$$\varepsilon_v = C_{N\varepsilon_v} \varepsilon_{v,N=15} \tag{6.126}$$

where $C_{N\varepsilon_v} = R\ln(N/15) + 1$ and R is the rate of change of ε_v with respect to N obtained from laboratory testing (e.g., Figure 6.110b).

In the absence of such testing, laboratory investigations of clean sand (Duku et al., 2008) and low-plasticity ($PI \leq 7$) silty sands (Yee et al., 2014) indicated that these coefficients can be estimated as

$$a \approx 5.38 K_{\sigma,\varepsilon} K_S K_{FC} \exp(-0.023 D_r) \tag{6.127a}$$

and

$$b \approx 1.2 \tag{6.127b}$$

where $K_{\sigma,\varepsilon}$, K_S, and K_{FC} are adjustment factors for total overburden stress, degree of saturation, and fines content, respectively, and D_r is given in percent. The values of those adjustment factors are shown in Table 6.12.

Equation (6.127a) shows that the a parameter, which controls the overall level of vertical strains from seismic compression, decreases as relative density increases and as the total vertical stress, σ_v, increases. The relative density effect results from increased dilatancy at the reference total stress of 1 atm, while the overburden effect is thought to result from the increase of bulk moduli (which control volumetric strain responses at modest shear strain levels) with increasing overburden pressure.

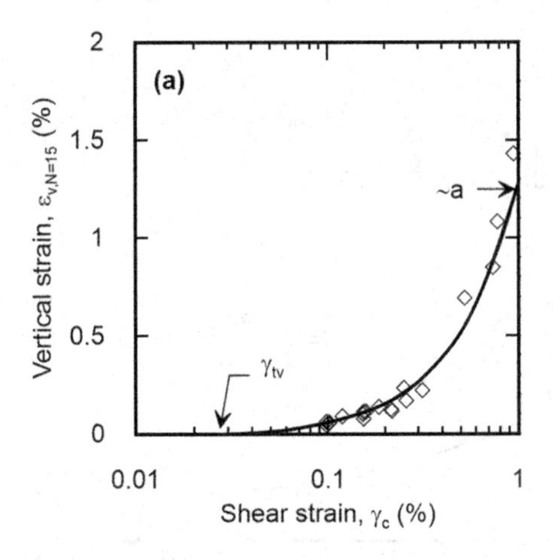

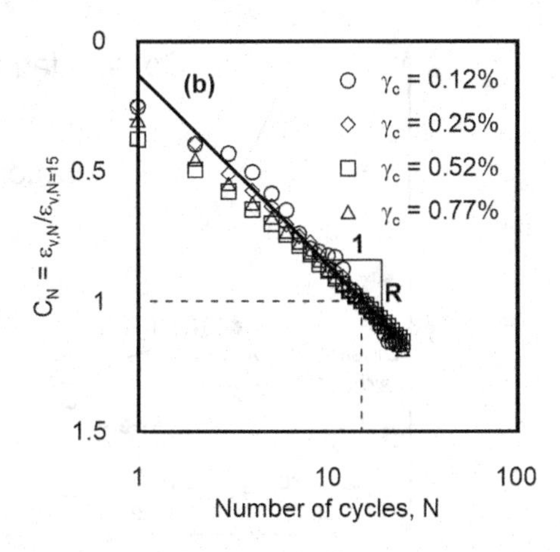

FIGURE 6.110 Volumetric strain material model fit to test results for Silica #2 sand material prepared to $D_r = 60\%$. (a) vertical strain at 15 cycles vs γ_c; (b) variation of vertical strain with the number of loading cycles N. (After Duku et al., 2008; used with permission of ASCE.)

TABLE 6.12
Adjustment Factors for Volumetric Strain Estimation for Clean Sand and Low-Plasticity Silty Sand

Adjustment Factor	Clean Sand (Duku et al., 2008)	Silty Sand (Yee et al., 2014)
Total overburden stress, $K_{\sigma,\varepsilon}$	$\left(\frac{\sigma_{v0}}{P_a}\right)^{-0.29}$	$\left(\frac{\sigma_{v0}}{P_a}\right)^{-0.29}$
Degree of saturation, K_S	1.0	$1.0 - 0.0175S$ for $S < 30\%$ 0.5 for $30 \leq S < 50\%$ $0.05S - 2.0$ for $50 \leq S < 60\%$ 1.0 for $S \geq 60\%$
Fines content, K_{FC} (%)	1.0	1.0 for $FC \leq 10\%$ $\exp[-0.041(FC - 10)]$ for $FC > 10\%$
Number of cycles, C_{Nev}	$R\ln\left(\frac{N}{15}\right)+1$ $R \approx 0.29.$	Similar to that for sand

S and FC are expressed in percent.

Undrained Conditions

When saturated soils are subject to earthquake shaking, the undrained loading conditions lead to generation of excess pore pressures, the subsequent dissipation of which can lead to compression. As illustrated schematically in Figure 6.111, generation of excess pore pressure causes the state of the soil to move from the consolidation curve (Point A) to a point at a lower effective stress (Point B) with the same void ratio. As the excess pore pressure dissipates, however, the sample will reconsolidate until the effective stress returns to its original value; the state of the soil will move from Point B to Point C, which will be at a lower void ratio than existed before earthquake shaking. The soil undergoes a volumetric strain, $\varepsilon_v = \Delta e/(1 + e_0)$. Martin et al. (1975) took the change in effective stress to be the product of the volumetric strain and the unloading-reloading modulus of the soil skeleton to develop a very early pore pressure model, which is discussed in more detail in Sections 6.6.5.4 and 9.5.7.1.

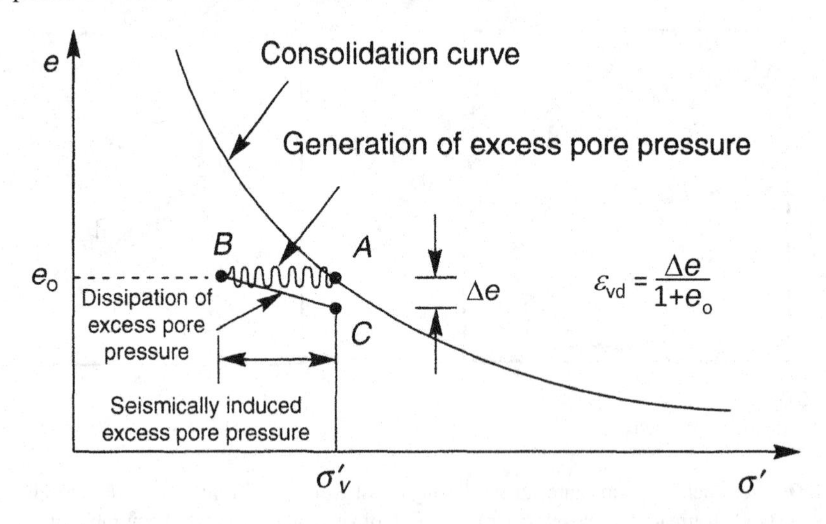

FIGURE 6.111 Process of earthquake-induced settlement from dissipation of earthquake-induced excess pore pressure.

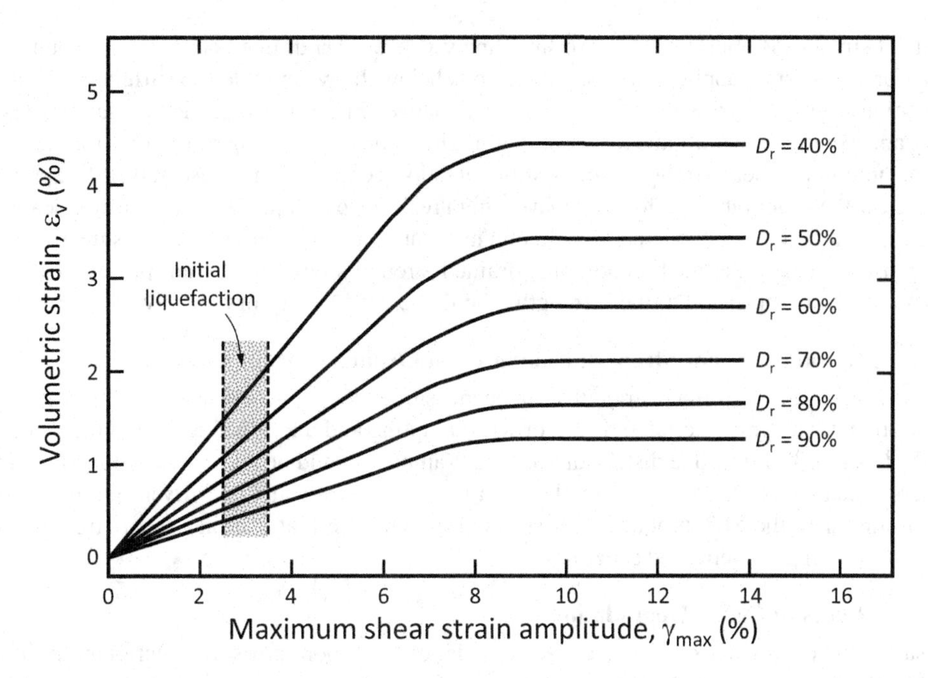

FIGURE 6.112 Development of post-cyclic volumetric strain in cyclic simple shear tests on clean sands. (After Ishihara, 1996; used with permission of Oxford University Press.)

Laboratory experiments have shown that the post-earthquake densification of saturated sand is influenced by the density of the sand, the maximum shear strain induced in it, and the amount of excess pore pressure generated by the earthquake. As would be expected, volumetric strains approach a limiting value that decreases with increasing relative density, as shown in Figure 6.112.

6.6.6 Shear Strength

The shear strength of soil is a critical parameter for many geotechnical earthquake engineering applications including stability problems (e.g., slope stability, bearing capacity of foundations) and large-strain site response. For some of these applications (e.g., site response, Newmark-type displacement analysis), the shear strength that can be mobilized during earthquake shaking is of interest. For other applications (e.g., flow sliding), the strength that is available following cyclic loading is of interest, especially when the soil is degraded as a result of changes in pore pressure or soil fabric.

The strength of nonplastic, cohesionless soils is inextricably tied to the phenomenon of liquefaction, a problem so important in geotechnical earthquake engineering that it is treated in a separate chapter (Chapter 9). Some fine-grained soils can also show significant degradation of stiffness and/or strength during earthquake shaking through cyclic softening, and their behavior under cyclic loading is also discussed in Chapter 9.

The shear strength of soils subjected to monotonic loading was described in Section 6.4.3. This section briefly reviews shear strength under typical drained and undrained conditions, and then discusses the effects of the deviations from those conditions that exist under earthquake loading. These definitions include the effects of rapid loading rates and cyclic degradation. Uncertainty in shear strength is also discussed.

6.6.6.1 Drained and Undrained Shear

During earthquakes, the rapid nature of loading will cause soils below the water table to respond under undrained conditions (Section 6.2.4) and mobilize undrained strengths. In such cases, the undrained strength is appropriate for seismic analysis even if a drained strength was found to be critical for non-seismic applications. Saturated soils below the water table can suffer cyclic strength degradation from pore pressure development and changes in fabric. In particular, sandy soils can liquefy and experience a dramatic loss of strength. This subject, including procedures for the assignment of undrained shear strengths for post-liquefaction conditions, is presented in Chapter 9. If sands below the water table are found to have a high resistance to liquefaction, then significant pore pressure generation is not expected and dilation may cause a reduction of pore pressure after cyclic loading has ended. Under such conditions, drained strengths developed from pre-event effective stresses can be appropriate. Drained strengths should be used for dry or partially saturated soils.

6.6.6.2 Monotonic Shear Strength Evaluation at Ordinary Strain Rates

For clayey materials, undrained strengths can be measured using *in situ* vane shear (Section 6.5.3.4) or pressuremeter testing (Section 6.5.3.5), or with sampling and undrained testing in the laboratory (Section 6.5.4.3). To minimize disturbance effects, samples should be taken with a thin-walled tube samplers (Shelby tube or similar). Samplers that have thick walls and are driven into the soil in a manner similar to the SPT produce excessive soil disturbance that can bias shear strengths measured in tests using the retrieved specimens.

6.6.6.3 Effects of Cyclic Degradation

Both sands and clays can experience cyclic degradation from pore pressure generation. In the case of sands, this behavior is evaluated in the context of liquefaction analysis procedures presented in Chapter 9. In the case of clays, the level of peak pore pressure ratio, $r_{u,\max}$, that can be experienced during cyclic loading is less than that for sands, generally not approaching the value of $r_u = 1.0$ associated with initial liquefaction. Nonetheless, this pore pressure generation, combined with possible soil fabric change, can significantly reduce the shearing resistance that is available during cyclic loading and post-cyclic monotonic shear. The loss of stiffness during cyclic loading due to pore pressure increase and related effects is known as *cyclic softening* (Boulanger and Idriss, 2007). Procedures for analysis of these effects are also presented in Chapter 9. This section focuses on the effects of cyclic loading on post-cyclic monotonic shear behavior.

The degradation of clay from cyclic pore pressure generation reduces the effective stress below its initial value, causing the clay to enter a state of "apparent overconsolidation" (Matsui et al., 1992). Moreover, since shear modulus is related to effective stress (Section 6.6.2), the stiffness of the soil can be expected to have decreased. These shear modulus reductions, in turn, tend to increase cyclic shear strains, which can be particularly impactful for sensitive soils that also undergo fabric change. As pointed out by Castro and Christian (1976), the ultimate (residual, high-strain) undrained shear strength of a saturated soil is controlled by its void ratio and fabric. While cyclic degradation does not affect void ratio, it can affect fabric, particularly when soils are sensitive and the cyclic loading produces large strains. In extreme cases, the shear strength can be reduced from the undrained strength to the remolded undrained shear strength.

Consider first the case of saturated clays with low sensitivity. Such a clay material at a particular void ratio will mobilize a specific undrained strength, with little influence of the history of stresses and strains by which that strength is arrived at. For such soil conditions, the undrained strength after cyclic loading would be expected to be equal to the undrained strength before undrained loading (if tested at the same strain rate). The six triaxial specimens shown in Figure 6.113 had similar void ratios (except specimen 6, which had a somewhat higher void ratio than the rest) at the end of consolidation. Specimen 1 was sheared monotonically immediately after consolidation and is shown in Figure 6.113 to have a sensitivity of about 1.0 (at least for axial strains $\leq 4\%$). Specimen 1 contracted initially but then dilated at larger strain amplitudes. Specimens 2 to 6 were first subjected to varying levels of cyclic loading, which were followed by monotonic shear without allowing drainage. Since the void ratios were nearly the same, the specimens would therefore be expected to have similar monotonic strengths. As shown by the stress-strain curves and stress paths, they

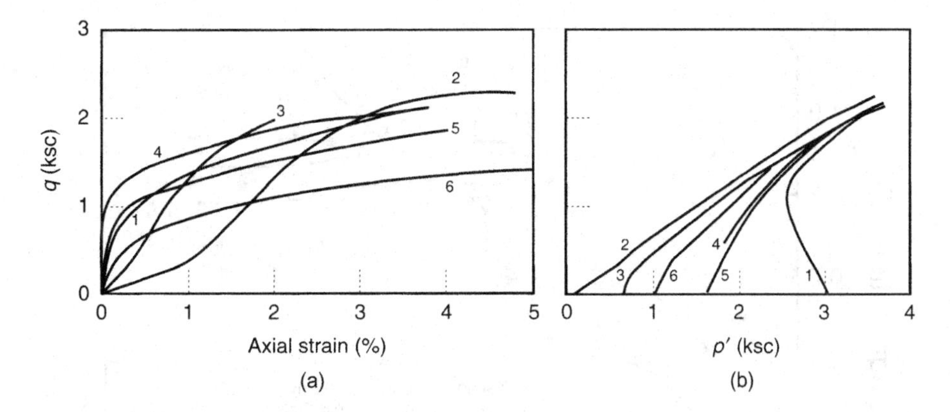

FIGURE 6.113 Effect of cyclic loading on subsequent monotonic undrained loading behavior of triaxial specimens of slightly plastic silt: (a) stress-strain behavior; (b) effective stress path behavior. Specimen 1 was tested in conventional CU test with no prior cyclic loading. Specimens 2 to 6 were subjected to different levels of cyclic loading prior to monotonic loading. Note the dilative nature of the stress paths of specimens 2–6 compared to specimen 1. (After Castro and Christian, 1976.)

behaved largely as would be expected. After being subjected to different levels of cyclic strain, all exhibited dilative behavior upon monotonic loading since their effective stresses at the end of cyclic loading were lower than their preconsolidation pressures. Their ultimate (large strain) strengths, however, were similar (except specimen 6, which was lower than the others) since they had similar effective stresses at the critical state they dilated toward. Differences in the ultimate strength can be explained by small differences in the void ratios and also by differences in the extent of structural (fabric) disturbance induced by the cyclic loading. Also evident in Figure 6.113 is reduced stiffness in the early stages of monotonic undrained loading (as compared to Specimen 1) for the elements that had previously been loaded cyclically.

Post-cyclic strength can also be influenced by mineralogy (Ajmera et al., 2019). Simple shear tests performed on mixtures of clay minerals and ground quartz and on natural soils with monotonic loading applied after cyclic loading sufficient to cause 10% double-acting shear strain showed that induced pore pressures were higher when the clay mineral was kaolinite than when it was montmorillonite. The *degradation ratio*, i.e., the ratio of post-cyclic to static undrained strength, increased with plasticity index for kaolinite mixtures, but was relatively insensitive to plasticity index for montmorillonite mixtures; the strengths of the montmorillonite mixtures, however, were sensitive to the applied cyclic stress ratio (Figure 6.114). When the tests were interpreted in terms of effective stresses after cyclic loading, the degradation ratio, was found to vary with pore pressure ratio at the end of cyclic loading as

$$\delta = \frac{s_{u,pc}}{s_u} = (1 - r_{u,cyc})^{0.247} \tag{6.128}$$

where $s_{u,pc}$ is the post-cyclic undrained strength, s_u is the static undrained strength and $r_{u,cyc}$ is the pore pressure ratio at the end of cyclic loading. The relationship of Equation (6.128) was found to be applicable to all of the tested soils (kaolinite and montmorillonite mixtures and natural soils) but with a significant degree of variability in the data.

For sensitive soils, the extent to which soil fabric is disturbed has most commonly been assumed to be influenced by the relationship between the cyclic strain amplitude and the strain at which failure occurs under monotonic loading conditions (Thiers and Seed, 1978). Substantial structural disturbance can modify the stress-strain behavior and reduce the monotonic shear strength. Thiers and Seed (1978) found that the ultimate strengths of three sensitive clays decreased by less than

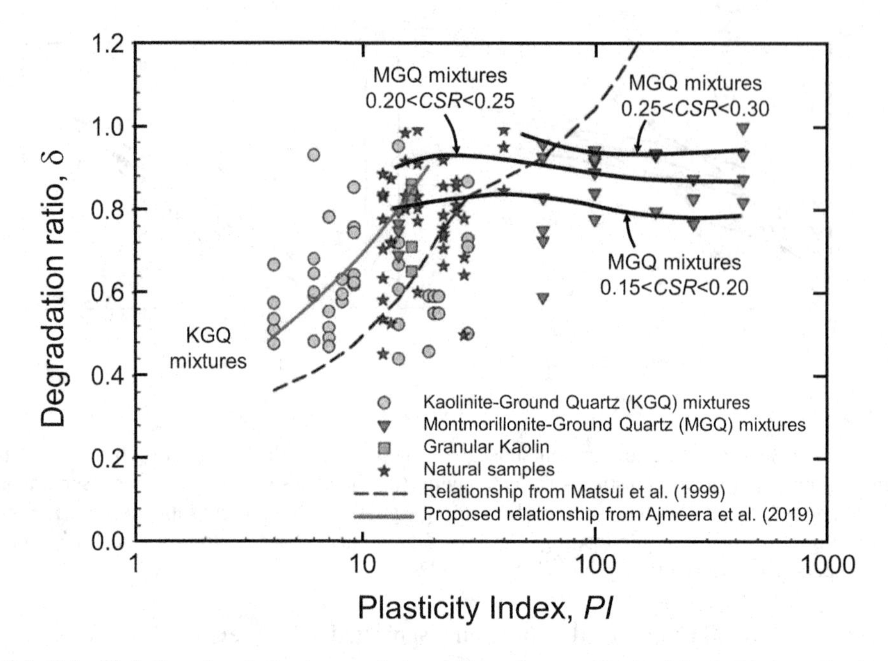

FIGURE 6.114 Variation of undrained strength degradation factor with plasticity index for soil mixtures and three natural soils. (After Ajmeera et al., 2019; used with permission of ASCE.)

10% when the cyclic strain amplitude was less than one-half of the failure strain from monotonic tests. At higher cyclic strain amplitudes, however, the reduction in strength was more dramatic, as illustrated in Figure 6.115. Similar results have been obtained by others (e.g., Koutsoftas, 1978; Ramanujam et al., 1978; Byrne et al., 1984; Erkan and Ulker, 2008a, 2008b).

Various investigators have identified factors beyond void ratio and strain amplitude that influence post-cyclic shear strength. Jitno (1990) found that the post-cyclic shear strength was also influenced

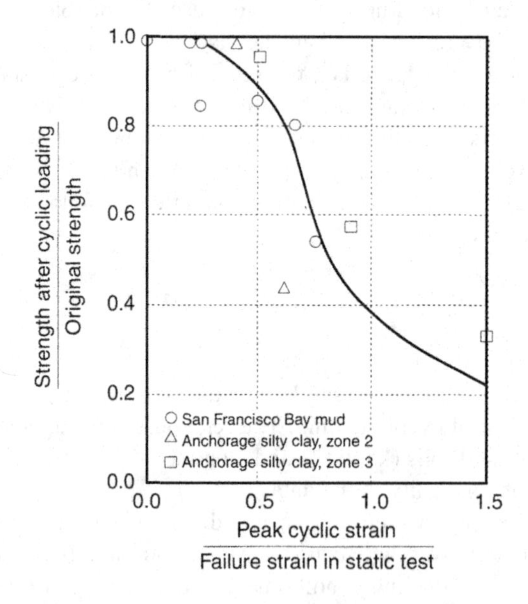

FIGURE 6.115 Effect of peak cyclic strain on monotonic strength of three clays after cyclic loading. (After Thiers and Seed, 1969.)

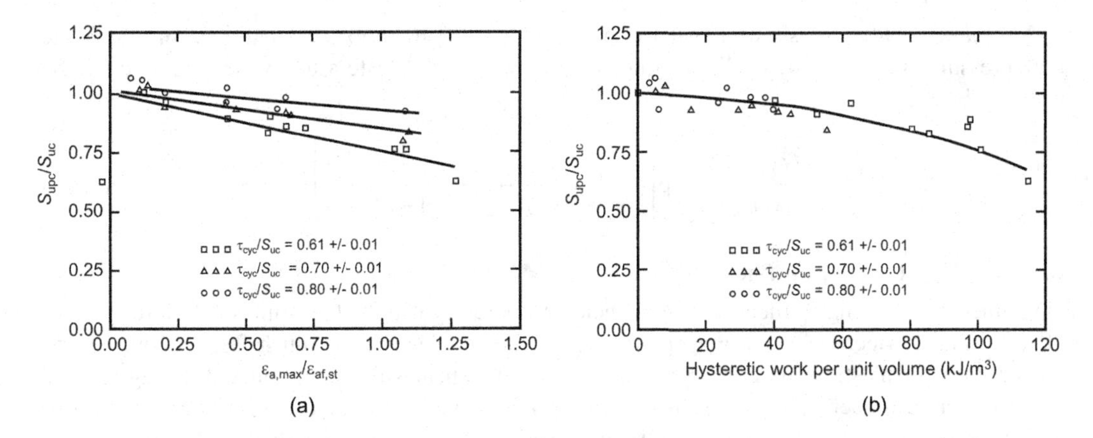

FIGURE 6.116 Reduction of undrained strength after cyclic loading as functions of (a) strain and initial stress ratio and (b) dissipated energy. (After Jitno, 1990.)

by the cyclic shear stress amplitude with lower cyclic shear stress amplitudes (hence, more cycles) for a given peak shear strain producing a greater strength loss than higher cyclic stress amplitudes (Figure 6.116a). Jitno (1990) also suggested that undrained strength loss was most closely correlated to absorbed energy, or hysteretic work (Figure 6.116b).

Hyde and Ward (1985) found the difference between original and post-cyclic monotonic strength of a silty clay (LL = 36, PL = 19) to be influenced by stress history. Specimens were consolidated one-dimensionally from a slurry, isotropically unloaded to different overconsolidation ratios, and then subjected to 10,000 cycles of loading at shear stress amplitudes ranging from 13% to 42% of the equivalent pressure (defined as the effective stress on the isotropic virgin compression curve at the void ratio of the soil following consolidation). The post-cyclic strengths of the normally and lightly overconsolidated specimens were found to be up to about 20% lower than the strengths of specimens not subjected to cyclic loading. For more heavily overconsolidated (and, hence, more dilative) specimens (typically OCR > 2.5–3.0), no significant reduction in strength due to cyclic loading was observed.

Yasuhara (1994) developed an approximate method to predict the effects of post-cyclic strength loss. Yasuhara (1994) used consolidation theory and normalized strength properties to define an apparent OCR based on the pore pressure generated by cyclic loading. That OCR was used to define a ratio of undrained strength after cyclic loading (but with no time for pore pressure dissipation) to undrained strength of a monotonically loaded normally consolidated clay

$$\frac{(s_u)_{pc}}{(s_u)_{NC}} = \exp\left[\left(\frac{\Lambda_0}{1 - C_r / C_c} - 1\right)\ln\left(\frac{1}{1 - r_{u,\max}}\right)\right] \tag{6.129}$$

where $r_{u,\max}$ is the pore pressure ratio at the end of cyclic loading, Λ_0 is a material constant and C_c and C_r are the compression and recompression indices, respectively, of the soil. Yasuhara (1994) suggested that the following ratio within the exponential term could be estimated from the plasticity index as

$$\frac{\Lambda_o}{1 - C_s / C_c} = 0.939 - 0.002PI \tag{6.130}$$

based on data by Ue et al. (1991). The Ue et al. (1991) data provides a trend whereby the ratio decreases from about 0.95 at PI = 0 to about 0.75 at PI = 80, but with a high level of scatter.

Yasuhara (1994) also showed that the post-cyclic strength ratio for conditions in which the pore pressures generated by cyclic loading were allowed to dissipate after cyclic loading could be expressed as

$$\frac{(s_u)_{pc}}{(s_u)_{NC}} = \exp\left[\left(\frac{\Lambda_0 C_s / C_c}{1 - C_s / C_c} - 1\right)\ln\left(\frac{1}{1 - r_{u,\max}}\right)\right] \tag{6.131}$$

6.6.6.4 Effects of Strain Rate

Undrained shear strengths measured in the field or laboratory usually have times to failure of about 20–30 minutes. Because the strain rate during earthquake shaking is much faster than that and the strengths of plastic soils can be rate-dependent, the shear strength mobilized during seismic shaking can be higher than that measured in the lab if cyclic degradation effects are small. For applications involving post-cyclic monotonic shear, the shear can occur relatively slowly so rate adjustments will generally not be justified. Rate adjustments to shear strength are also usually not made in sandy soils.

Rate effects on shear strength of clays have been evaluated using undrained tests at various, relatively slow, strain rates (Richardson and Whitman, 1963; Lefebvre and LeBoeuf, 1987; Sheahan et al., 1996). The fastest rates in these tests corresponded approximately to the typical 20–30-minute time to failure. These results suggested an approximately 5%–10% increase in strength for each log cycle increase in strain rate for clays with $OCR < 8$. Diaz-Rodriguez et al. (2009), using tests on Mexico City clay, found that normalized undrained strengths vary with strain rate and OCR but converged to a constant S_U / σ'_{v0} ratio of about 0.2 for strain rates less than 10^{-5}%/hr. Diaz-Rodriguez et al. (2009) proposed that normalized strength ratios at high strain rates could be estimated as

$$\frac{s_u}{\sigma'_{v0}} = \begin{cases} 0.2 + 0.1 \cdot OCR \cdot \log\left(\dfrac{\dot{\varepsilon}}{10^{-5}\% / \text{hr}}\right) & \text{for} \quad \dot{\varepsilon} > \dot{\varepsilon}_t \text{ and } 0.3 < OCR < 2.4 \\ 0.2 & \text{for} \quad \dot{\varepsilon} < \dot{\varepsilon}_t \end{cases} \tag{6.132}$$

where $\dot{\varepsilon}_t = 5 \times 10^{-2}\ \% / \text{hr}$ is a threshold strain rate. Threshold strain rates reported in the literature include $\dot{\varepsilon}_t = 5 \times 10^{-2}\ \% / \text{hr}$ for plastic Drammen clay (Berre and Bjerrum, 1973), $\dot{\varepsilon}_t = 15 \times 10^{-2}\ \% / \text{hr}$ for Haney clay (Vaid and Campanella, 1977), $\dot{\varepsilon}_t \approx 1 \times 10^{-3}\ \% / \text{hr}$ for various post-glacial clays (Graham et al., 1983), and $\dot{\varepsilon}_t = 1 \times 10^{-5}\ \% / \text{hr}$ for Mexico City clay (Diaz-Rodriguez et al., 2009). Afacan et al. (2019) found that these effects also extend to rates faster than the reference. Taken collectively, these results suggest that shear strength increases under seismic conditions are on the order of 20%–40% greater than typical undrained shear strength measurements. Such increases apply for a rapid rate of monotonic shear or for high-frequency cyclic loading in soils without significant cyclic degradation. Where cyclic degradation effects occur, they reduce the rate-adjusted shear strength.

6.6.6.5 Shear Strength Uncertainty

Ideally, the shear strength used for seismic applications would be evaluated in a series of tests in which specimens were consolidated to appropriate initial effective stresses and OCRs and then subjected to representative cyclic stresses, possibly followed by monotonic loading to failure. Armed with a sufficient number of such tests, it would be possible to characterize the shear strength uncertainty on a material-specific basis. For many projects, however, such testing programs are impractical so experience from other sites in the literature is needed to evaluate strength uncertainties.

Since the shear strength for seismic applications is related to the static undrained strength, uncertainty in post-cyclic strength is related to uncertainty in static strength. With this in mind, the effects of several sources of uncertainty should be considered for seismic shear strength uncertainty

- ***In situ* OCR**: The measurement of preconsolidation pressures from consolidation curves is subject to potentially substantial uncertainties, especially when samples are significantly disturbed (e.g., Holtz et al., 2011). The uncertainty in *OCR* directly affects uncertainty in strength ratio (via Equation 6.27).
- **Sample Disturbance**: Aside from its effect on *OCR* uncertainty, sample disturbance contributes considerable uncertainty to shear strength measurements, especially when unconsolidated-undrained testing is employed. Disturbance causes a conservative bias of unknown size to strengths evaluated using UU test methods. This source of uncertainty can be overcome by using *in situ* (e.g., vane shear) testing or CU test methods in which specimens are first consolidated to the virgin compression line before the onset of undrained shear (Ladd, 1991).
- **Rate Effects**: Rate effects increase the undrained shear strengths in clays during earthquake ground shaking (Section 6.6.6.4). However, such increases are highly uncertain (as reflected by the aforementioned range of about 1.2–1.4) because the test data upon which they are based do not resolve well the effects of compositional factors (such as plasticity). Moreover, these increases could be offset in some cases by cyclic degradation effects.

TABLE 6.13
Coefficients of Variation Reflecting Inherent Variability of Soil Shear Strength Parameters

Property (Units)	Soil Type	No. of Data Groups	No. of Tests Per Group		Property Value		Property COV (%)		Note
			Range	Mean	Range	Mean	Range	Mean	
ϕ (°)	Sand	7	29–136	62	35–41	37.6	5–11	9	a
	Clay, silt	12	5–51	16	9–33	15.3	10–56	21	
	Clay, silt	9	*	*	17–41	33.3	4–12	9	
	*	20	*	*	*	*	*	12.6	d
tan ϕ	Clay, silt	4	*	*	0.24–0.69	0.509	6–46	20	a
	Clay, silt	3	*	*	*	0.615	6–46	23	
	Sand	13	6–111	45	0.65–0.92	0.744	5–14	9	a
	*	7	*	*	*	*	*	11.3	d
ϕ (°)	Sand	*	*	*	*	*	2–5	*	b
	Gravel	*	*	*	*	*	7	*	c
	Sand	*	*	*	*	*	12	*	
S_u ⁽¹⁾ (kPa)	Fine-grained	38	2–538	101	6–412	100	6–56	33	a
S_u ⁽²⁾ (kPa)	Clay, Silt	13	14–82	33	15–363	276	11–49	22	
S_u ⁽³⁾ (kPa)	Clay	10	12–86	47	130–713	405	18–42	32	
S_u ⁽⁴⁾ (kPa)	Clay	42	24–124	48	8–638	112	6–80	32	
	*	38	*	*	*	*	*	33.8	c
S_u ⁽⁵⁾ (kPa)	Clay	*	*	*	*	*	5–20	*	b
S_u ⁽⁶⁾ (kPa)	Clayey silt	*	*	*	*	*	10–30	*	
c ⁽⁷⁾	*	*	*	*	*	*	40	*	c
S_u/σ'_{v0} ⁽⁸⁾	Clay	*	*	*	*	*	*	40	d
S_u/σ'_{v0}	Clay	*	*	*	*	*	5–15	*	b

Source: From Jones et al. (2002).

[1] Unconfined compression test; [2] Unconsolidated-undrained triaxial compression test.; [3] Consolidated isotropic undrained triaxial compression test; [4] Laboratory test not reported; [5] Triaxial test.; [6,7] No information on how the uncertainty was derived; and [8] Vane shear test.

Notes: [a] Phoon and Kulhawy (1999); [b] Lacasse and Nadim (1996); [c] Harr (1987); and [d] Kulhawy (1992).

In consideration of these sources of uncertainty, it is useful to initially characterize the uncertainty of the non-rate corrected shear strengths (Table 6.13). The COVs shown in Table 6.13 should be increased by about 20% for seismic applications to account for uncertainties in rate effects. If shear strength is not measured but is estimated using S and m values from the literature Equation (6.27), uncertainties in these parameters (as given by Ladd, 1991) should be taken into account along with the uncertainty in OCR and rate effects.

For sandy soils, the principal sources of uncertainty are the estimation of friction angle from penetration resistance (standard deviation of about 3°) and the penetration resistance itself, which will often exhibit considerable scatter from borehole-to-borehole (or sounding-to-sounding with CPT data) and with depth within a layer. The scatter of the penetration resistance reflects geological conditions related to sediment deposition and should be evaluated on a site-specific basis.

6.7 SUMMARY

1. Earthquake damage is strongly influenced by the dynamic response of soil deposits. The cyclic nonlinear, volume change, and strength characteristics of soils govern their dynamic response during earthquakes.
2. The state of stress at a point in soil can be represented by a Mohr circle, which represents the shear and normal stresses on all possible plane orientations. Principal stresses act on principal planes, which have no shear stresses. As shear is applied to the soil, the evolution of the state of stress, as represented by the top of the Mohr circle, is described by the stress path.
3. Soils tend to undergo volume change when sheared. When changes in volume are unimpeded, either by the presence of compressible air in the voids between particles or by loading applied so slowly that water can move freely out of or into the voids when they are saturated, the loading is referred to as drained loading. When shear stresses are applied to a saturated soil so quickly that water cannot move out of or into the voids, the soil volume remains constant and the loading is referred to as undrained loading.
4. Drained and undrained loading represent extremes of a spectrum of drainage conditions. In saturated soils, undrained conditions often exist in the short term and virtually always during earthquake shaking, and drained conditions exist in the long term. The adequacy of a design is generally checked for both conditions for permanent structures and facilities.
5. Coarse-grained cohesionless soils and fine-grained cohesive soils have distinct mineralogies and behavioral characteristics. It is important to be able to distinguish which portion of the soil matrix will dominate its behavior, which can be evaluated for engineering purposes on the basis of fines content relative to the limiting fines content (beyond which coarse particles become separated by fines). The mechanical behavior of soils with fines contents below the limiting fines content will be dominated by the coarse fraction and influenced by grain size characteristics and density. Soils with fines content above the limiting value will be dominated by the fine fraction and influenced by plasticity and stress history.
6. Soils exhibit nonlinear, inelastic, stress-strain behavior under cyclic loading conditions. At low strain levels, the stiffness of a soil is greatest and the damping is smallest. At higher strain levels, the effects of nonlinearity and inelasticity increase, producing lower stiffness and greater damping.
7. The state of soil is described by its position relative to the critical state line, where the critical state represents a condition where shear produces no volume change at high strain levels. When the state of stress plots above the critical state line in void ratio – effective stress space, the soil will either contract (under drained loading) or generate positive water pressure (under undrained loading) if sheared to sufficiently large strains to reach critical state. Likewise, soil states that plot below the critical state line experience dilative behavior.

Dilative soils will, after a generally brief initial contraction, expand (under drained loading) or generate reduced pore pressure (under undrained loading).

8. The behavior of clays and granular soils are fundamentally different. For clays, the shear strength and stress-deformation behavior normalizes by (i.e., are proportional to) the pre-shear consolidation stress because the virgin compression line and critical state line are parallel. The normalized strength of clays is strongly sensitive to *OCR*, which controls dilatancy. For sands, the virgin compression and critical state lines are not parallel, hence shear strength does not normalize by effective stress. Parameters that represent state (e.g., state parameter, relative state parameter index) indicate dilatancy behavior in such materials.
9. Under undrained cyclic loading, both loose and dense granular soils will generate positive pore pressures. Pore pressures will increase incrementally with increasing number of loading cycles. The rate of increase is faster for loose soils than for dense soils.
10. Initially, contractive granular soils can reach a phase transformation state at high stress ratios at which they begin to dilate. Under monotonic loading, dilation will continue until the critical state is reached at high strain levels. Under cyclic loading, a shear stress reversal in a dilating soil causes an immediate return to contractive behavior that continues until the phase transformation state is reached again (generally in the opposite direction).
11. Soils subjected to high cyclic shear stresses, or many cycles of lower cyclic shear stress, will reach the phase transformation state. Loading beyond that point will produce both increasing and decreasing pore pressures within individual cycles. The resulting variations in pore pressure will cause effective stresses, and hence soil stiffness, to both decrease and increase within individual loading cycles. With sufficient loading, effective stresses may momentarily drop to zero, at which point the stiffness of the soil may become extremely low. The application of additional shear stress, however, will cause dilation-induced stiffening and gain in shearing resistance. Individual loading cycles in this range can develop a characteristic "banana-shaped" appearance. The presence of plastic fines in a granular soil tends to suppress this aspect of behavior.
12. Characterization of stress-strain behavior for seismic applications is most often undertaken using three broad classes of stress-strain models; equivalent linear models, cyclic nonlinear models, and advanced constitutive models.
13. Equivalent linear models treat soils as linear viscoelastic materials. Nonlinear behavior is accounted for by the use of strain-dependent stiffness and damping parameters. The stiffness of the soil is usually characterized by the maximum shear modulus, which is mobilized at low strains, and a modulus reduction curve, which shows how the secant shear modulus decreases at larger strains. Damping behavior is characterized by the damping ratio, which increases with increasing strain amplitude. The shapes of the modulus reduction and damping curves are most strongly influenced by soil plasticity and effective confining pressure.
14. Cyclic nonlinear models represent the nonlinear, inelastic behavior of soils using a nonlinear backbone curve and a series of rules that govern unloading-reloading behavior. Backbone curves are usually described by simple functions that reflect the transition from the initial stiffness (at low strain levels) to the ultimate strength (at high strain levels). The unloading-reloading rules control the behavior of the model during stress reversals and ensure that it behaves in a manner similar to that exhibited by actual soils subjected to irregular cyclic loading. In contrast to equivalent linear models, cyclic nonlinear models allow permanent strains to develop. Cyclic nonlinear models can also be coupled with pore pressure generation models to predict changes in effective stress during cyclic loading. Modeling such behavior requires that the original backbone curve be degraded (softened) as pore pressures increase.

15. Advanced constitutive models use basic principles of mechanics to describe soil behavior for general initial stress conditions, a wide variety of stress paths with rotating principal stresses, cyclic or monotonic loading, high or low strain rates, and drained or undrained conditions. As such, they are much more general than equivalent linear or cyclic non-linear models. The penalty for this increased generality comes in the form of increased complexity, increased number of model parameters (some of which can be difficult to determine), and increased computational effort when incorporated into ground response or soil-structure interaction analyses. One-, two-, and three-dimensional advanced constitutive models are now available in a number of research and commercial numerical (finite element and finite difference) platforms and are seeing much greater use in practice.
16. The measurement of dynamic soil properties is an important aspect of geotechnical earthquake engineering. A variety of field and laboratory techniques are available; some are oriented toward the measurement of low-strain behavior and others toward measurement of soil behavior at high strain levels.
17. Field tests allow measurement of soil properties *in situ*; the complex effects of existing stress, chemical, thermal, biological, and structural conditions are therefore reflected in the measured soil properties. Many field tests measure the response of large volumes of soil and induce soil deformation similar to those induced by earthquakes. Because *in situ* conditions cannot be easily controlled or varied, field tests do not allow measurement of the behavior of the soil under other stress states or soil conditions.
18. A number of field tests measure low-strain soil properties, particularly wave propagation velocities. Shear wave velocities can be measured from the ground surface (non-invasive tests) and using borings or probes that extend below the ground surface (invasive tests). Surface methods are often less expensive since no borehole is required and the instrumentation and impulsive source are usually lightweight and portable (at least for relatively shallow exploration depths). Surface tests include seismic reflection, seismic refraction, and surface wave methods. Tests that take place below the surface include suspension logging, seismic cross-hole, seismic down-hole (and up-hole), and seismic cone tests.
19. Wave propagation tests vary in their spatial resolution. Because surface wave tests produce velocities measured over a distance related to the Rayleigh wavelength (short for high frequencies and long for low frequencies), the velocities represent average velocities over that distance. Borehole methods typically provide higher resolution depending on the distance between the source and the receiver. Cross-hole measurements provide an average velocity over the distance between the holes, but down-hole tests provide velocities averaged only over the distance (along a line between the source and receiver) between successive receiver positions. Hence, down-hole tests may show significant variability between different borings at laterally heterogeneous sites. Suspension logging tests can provide much higher resolution and are preferred at large depths because the source and receiver both move down the hole with constant separation distance. All borehole methods have shortcomings at shallow depths and are usually considered reliable for depths greater than about 3–5 m.
20. Other field tests, such as the standard penetration, cone penetration, vane shear, and pressuremeter tests, measure the behavior of the soil at higher strain levels. Strength and dilatancy characteristics are often inferred from these tests.
21. Laboratory tests allow the control and measurement of stresses, strains, and pore water pressures. As a result, they can often simulate anticipated initial and dynamic stress conditions better than field tests. The results of laboratory tests are influenced to varying degrees by sample disturbance. Sample disturbance has a particularly strong effect on low-strain properties.
22. Several laboratory tests measure low-strain soil properties. Among these are the resonant column test and the piezoelectric bender element test. The cyclic triaxial test, cyclic direct

simple shear test, and cyclic torsional shear test can be used to measure dynamic soil properties at higher strain levels.

23. Dynamic soil properties may also be inferred from the results of model tests or the interpretation of data from instrumented field sites. Shaking table tests can accommodate relatively large models, but their inability to produce high gravitational stresses can make extrapolation to prototype conditions difficult. Centrifuge tests can satisfy similitude requirements much better than shaking table tests but must be performed on relatively small models. Field test data from instrumented test sites represents an effective 'gold standard' for data quality, but are limited to conditions that have occurred as a result of past earthquakes.
24. For ground response applications, essential soil properties include the initial (low-strain) shear modulus, the modulus reduction ratio versus shear strain curve, and the hysteretic damping versus shear strain curve. For applications involving large strain response, dynamic shear strength is also required.
25. Initial shear modulus can be obtained from mass density and shear wave velocity. Although a number of correlations are available for initial modulus or shear wave velocity, due to its importance and ease of measurement V_S values used for analysis should be based on *in situ* measurements whenever possible.
26. Modulus reduction and damping curves are most often obtained from empirical relationships, which often take the form of hyperbolic equations. Relations are available for typical soil types ranging from gravels to clays, but also for organic soils and peats and MSW. Such relationships are generally based on experimental data extending to strains of about 0.5%; extrapolation to higher strains may produce unrealistic inferred shear stresses at larger strain levels. Procedures for modifying modulus reduction curves to be consistent with specified shear strengths (mobilized at much larger strain levels) are available and should be used when significant shear strains are expected.
27. Soil state parameters are useful to characterize dilatancy, and can be estimated for sands from *in situ* penetration resistance data combined with knowledge of the state of stress. The volume change that develops during cyclic loading under drained conditions, known as seismic compression, increases with shear strain amplitude and number of loading cycles and decreases with relative density.
28. The shear strength of soil for seismic applications will usually develop under undrained conditions for soils below the water table. A variety of field and laboratory methods can be used to evaluate strength parameters, which typically apply for relatively low strain rates and monotonic loading. The higher strain rates that exist during earthquake shaking allow clayey soils to develop higher shear resistance as a result of rate effects, although these effects can be offset by cyclic degradation. Absent cyclic degradation effects, strain rate effects can increase shear strength over typical measurements by 20%–40%. Cyclic degradation of clays may or may not significantly influence post-cyclic undrained strength depending on the soil sensitivity and the level of deformation imposed by earthquake shearing.

REFERENCES

Abou-Matar, H. and Goble, G. (1997). "SPT dynamic analysis and measurements," *Journal of Geotechnical and Geoenvironmental Engineering*, Vol., No. 10, pp. 921–928.

Afacan, K.B., Yneista, S., Shafiee, A., Stewart, J.P., and Brandenberg, S.J. (2019). "Total stress analysis of soft clay ground response in centrifuge models," *Journal of Geotechnical and Geoenvironmental Engineering*, Vol. 145, p. 04019061.

Afshari, K. and Stewart, J.P., (2019). Insights from California vertical arrays on the effectiveness of ground response analysis with alternative damping models," *Bulletin of the Seismological Society of America*, Vol. 109, pp. 1250–1264.

Ahdi, S.K., Stewart, J.P., Ancheta, T.D., Kwak, D.Y., and Mitra, D. (2017). "Development of VS profile database and proxy-based models for V_{S30} prediction in the Pacific Northwest region of North America," *Bulletin of the Seismological Society of America*, Vol. 107, No. 4, pp. 1781–1801.

Ahmadi, M.M. and Robertson, P.K., (2005). "Thin layer effects on the CPT qc measurement," *Canadian Geotechnical Journal*, Vol. 42, No. 9, pp. 1302–1317.

Airey, D.W. and Wood, D.M. (1987). "An evaluation of direct simple shear tests on clay," *Geotechnique*, Vol. 37, No. 1, pp. 25–35.

Ajmera, B., Brandon, T., and Tiwari, B. (2019). "Characterization of the reduction in undrained shear strength in fine-grained soils due to cyclic loading," *Journal of Geotechnical and Geoenvironmental Engineering*, Vol. 145, No. 5, 10 pp.

Alarcon-Guzman, A., Leonards, G.A., and Chameau, J.L. (1988). "Undrained monotonic and cyclic strength of sands," *Journal of Geotechnical Engineering*, Vol. 114, No. 10, pp. 1089–1108.

Al-Sanad, H. and Aggour, M.S. (1984). "Dynamic soil properties from sinusoidal and random vibrations," *Proceedings, 8th World Conference on Earthquake Engineering*, San Francisco, Vol. 3, pp. 15–22.

Anderson, D.G. and Stokoe, K.H. (1978). "Shear modulus: A time-dependent soil property," *Dynamic Geotechnical Testing*, ASTM STP 654, pp. 66–90. https://doi.org/10.1061/ajgeb6.0000273

Anderson, D.G. and Woods, R.D. (1975). "Comparison of field and laboratory shear modulus," *Proceedings, ASCE Conference on Insitu Measurement of Soil Properties*, Raleigh, NC, Vol. 1, pp. 69–92.

Anderson, D.G. and Woods, R.D. (1976). "Time-dependent increase in shear modulus of clay," *Journal of the Geotechnical Engineering Division*," Vol. 102, No. GT5, pp. 525–537.

Anderson, K. (2015). "Cyclic soil parameters for offshore foundation design," in V. Meyer, ed., *Frontiers in Offshore Geotechnics III*, Taylor and Francis Group, London, pp. 5–82.

Andrus, R.D., Monahan, N.P., Piratheepan, P., Ellis, B.S., and Holzer, T.L. (2007). "Predicting shear-wave velocity from cone penetration resistance," *Proceedings, 4th International Conference on Earthquake Geotechnical Engineering*, Thessaloniki, Greece, Paper No. 1454, 12 pp.

Andrus, R.D., Stokoe, K.H., and Juang, C.H. (2004). "Guide for shear wave-based liquefaction potential evaluation," *Earthquake Spectra*, Vol. 20, No. 2, pp. 285–308.

Applegate, J.K. (1974). "A torsional seismic source," *Thesis*, Colorado School of Mines.

Asten, M.W. and Boore, D.M. (2005). "Comparison of shear-velocity profiles of unconsolidated sediments near the Coyote borehole (CCOC) measured with fourteen invasive and non-invasive methods, Blind comparisons of shear-wave velocities at closely-spaced sites in San Jose, California," M. W. Asten and D. M. Boore (Editors), U.S. *Geological Survey Open-File Report 2005-1169*, part 1, 35 pp.

Athanasopoulos-Zekkos, A., Zekkos, D., and Matasovic, N. (2008). "Validation of generic municipal solid waste material properties for seismic design of landfills," *Proceedings, 4th Geotechnical Earthquake Engineering and Soil Dynamics Conference*, Sacramento, CA.

Augello, A.J., Bray, J.D., Abrahamson, N.A., and Seed, R.B. (1998), "Dynamic Properties of Solid Waste Based on Back-Analysis of OII Landfill," *Journal of Geotechnical and Geoenvironmental Engineering*, Vol. 124, No. 3, pp. 211–222.

Auld, B. (1977). "Cross-hole and down-hole vs by mechanical impulse," *Journal of the Geotechnical Engineering Division*, Vol. 103, No. GT12, pp. 1381–1398.

Baldi, G., Bruzzi, D., Superbo, S., Battaglio, M. and Jamiolkowski, M. (1988). "Seismic cone in Po River sand," *Proceedings, Penetration Testing* 1988, 1*SOPT*-1, Balkema, Rotterdam, pp. 643–650.

Been, K. and Jefferies, M.G. (1985). "A state parameter for sands," *Géotechnique*, Vol. 35, pp. 99–112.

Been, K. and Jefferies, M.G. (1992). "Towards systematic CPT interpretation," *Proceedings, Wroth Memorial Symposium*, Thomas Telford, London, pp. 121–134.

Been, K., Jefferies, M.G., and Hachey, J. (1991). "The critical state of sands," *Geotechnique*, Vol. 41, No. 3, pp. 365–381.

Bellotti, R., Jamiolkowski, M., Lo Presti, D.C.F., and O'Neill, D.A. (1996). "Anisotropy of small strain stiffness in Ticino sand," *Gèotechnique*, Vol. 46, No. 1, pp. 115–131.

Berre, T. and Bjerrum, L. (1973). "Shear strength of normally consolidated clays." *Proceedings, 8th International Conference on Soil Mechanics and Foundation Engineering, Moscow*, Vol. 1, pp. 39–49.

Bishop, A.W. and Henkel, D.J. (1962). *The Measurement of Soil Properties in the Triaxial Test*, 2nd edition, Edward Arnold Ltd., London, 228 pp.

Bolton, M.D. (1979). *A Guide to Soil Mechanics*, Macmillan, London, 439 pp.

Bolton, M. (1986). "The strength and dilatancy of sands," *Geotechnique*, Vol. 36, No. 1, pp. 65–78.

Bonilla, L.F., Archuleta, R.J., and Lavallee, D. (2005). Hysteretic and dilatant behavior of cohesionless soils and their effects on nonlinear site response: Field data observations and modeling, *Bulletin of the Seismological Society of America*, Vol. 95, No. 6, pp. 2373–2395.

Boore, D.M. (2006). "Determining subsurface shear-wave velocities: A review," Paper No. 103, *Proceedings, Third International Symposium on the Effects of Surface Geology on Seismic Motion*, Grenoble France, 21 pp.

Boulanger, R.W. (2003a). "Relating Ka to relative state parameter index," *Journal of Geotechnical and Geoenvironmental Engineering*, Vol. 129, No. 8, pp. 770–773.

Boulanger, R.W. (2003b). "High overburden stress effects in liquefaction analyses," *Journal of Geotechnical and Geoenvironmental Engineering*, Vol. 129, No. 12, pp. 1071–1082.

Boulanger, R.W., Arulnathan, R., Harder, L.F., Jr., Torres, R.A., and Driller, M.W. (1998). "Dynamic properties of Sherman Island peat," *Journal of Geotechnical and Geoenvironmental Engineering*, Vol. 124, No. 1, pp. 12–20.

Boulanger, R.W., Chan, C.K., Seed, H.B., Seed, R.B., and Sousa, J. (1993). "A low-compliance bi-directional cyclic simple shear apparatus," *Geotechnical Testing Journal*, ASTM, Vol. 16, No. 1, pp. 36–45.

Boulanger, R.W. and DeJong, J.T. (2018). "Inverse filtering procedure to correct cone penetration data for thin-layer and transition effects," *Proceedings, Cone Penetration Testing 2018*, Delft University of Technology, The Netherlands, pp. 25–44.

Boulanger, R.W. and Idriss, I.M. (2004). "Evaluating the potential for liquefaction or cyclic failure of silts and clays." *Rep. UCD/CGM-04/01*, Center for Geotech. Modeling, University of California, Davis, California.

Boulanger, R.W. and Idriss, I.M. (2007). "Evaluation of cyclic softening in silts and clays," *Journal of Geotechnical and Geoenvironmental Engineering*, Vol. 133, No. 6, pp. 641–652.

Boulanger, R.W., and Idriss, I.M. (2014). "CPT and SPT based liquefaction triggering procedures," *Report No. UCD/CGM-14/01,* Center for Geotechnical Modeling, Department of Civil and Environmental Engineering, University of California, Davis, CA, 134 pp.

Brandenberg, S.J., Ballana, N., and Shantz, T. (2010). "Shear wave velocity as a statistical function of Standard Penetration Test resistance and vertical effective stress at Caltrans bridge sites," *PEER Report 2010/03*, Pacific Earthquake Engineering Research Center, Berkeley, CA, 75 pp.

Broms, B. (1980). "Soil sampling in Europe: State-of-art," *Journal of the Geotechnical Engineering Division,* Vol. 106, No. 1, pp. 65–98. https://doi.org/10.1061/ajgeb6.0000916

Brown, L.T., Boore, D.M., and Stokoe, K.H. (2002). "Comparison of shear wave slowness profiles at 10 strong-motion sites from non-invasive SASW measurements and measurements made in boreholes," *Bulletin of the Seismological Society of America*, Vol. 92, No. 8, pp. 3116–3133.

Brown, L.T., Diehl, J.G., and Nigbor, R.L. (2000). "A simplified procedure to measure average shear-wave velocity to a depth of 30 meters (V_{S30})," *Proceedings, 12th World Conference on Earthquake Engineering, Auckland*, 8 pp.

BSSC (2020). "NEHRP recommended seismic provisions for new buildings and other structures," Volume I, Part 2 Commentary, FEMA P-2082-1, Federal Emergency Management Agency, Washington, DC, 555 pp.

Burger, H.R. (1992). *Exploration Geophysics of the Shallow Subsurface*, Prentice-Hall, Englewood Cliffs, NJ.

Burghignoli, A., Cavalera, L., Chieppa, V., and Jamiolkowski, M. (1991). "Geotechnical characterization of Fucino clay," *Proceedings, 10th European Conference on Soil Mechanics & Foundation Engineering*, Vol. 1 (Florence), Balkema, Rotterdam, pp. 27–40.

Burland, J.B. and Symes, M.J. (1982). "A simple axial displacement gauge for use in the triaxial apparatus," *Geotechnique*, Vol. 32, No. 1, pp. 62–65.

Butcher, A.P. and Powell, J.J.M. (1996). "Practical considerations for field geophysical techniques used to assess ground stiffness," *Proceedings, International Conference on Advances in Site Investigation Practice, ICE*, Thomas Telford, London, pp. 701–714.

Butler, K.K., Sloglund, G.R., and Landers, G.B. (1978). "Crosshole: An interpretive computer code for crosshole seismic test results, Documentation and examples," *Miscellaneous Paper S-78-6*, U.S. Army Engineer Waterways Experiment Station, Vicksburg, Mississippi.

Byrne, P.M., Morris, D.V., and Caldwell, J.A. (1984). "Seismic stability of a tailings impoundment on soft clayey silt deposits," *Proceedings, Eighth World Conference on Earthquake Engineering*, San Francisco, Vol. 3.

Cabas, A., Rodriguez-Marek, A., and Bonilla, L.F. (2017). "Estimation of site-specific Kappa (κ0)-consistent damping values at KiK-Net sites to assess the discrepancy between laboratory-based damping models and observed attenuation (of seismic waves) in the field," *Bulletin of the Seismological Society of America*, Vol. 107, No. 5, pp. 2258–2271.

Campanella, R.G., Gillespie, D., and Robertson, P.K. (1982). "Pore pressures during cone penetration testing," *Proceedings, 2nd European Symposium on Penetration Testing, ESPOT II*, Amsterdam, A.A. Balkema, pp. 507–512.

Carlton, B.D. and Pestana, J.M. (2016). "A unified model for estimating the in-situ small strain shear modulus of clays, silts, sands, and gravels," *Soil Dynamics and Earthquake Engineering*, Vol. 88, pp. 345–355.

Casagrande, A. (1936). "Characteristics of cohesionless soils affecting the stability of slopes and earth fills," *Journal of the Boston Society of Civil Engineers*, Vol. 23, reprinted in *Contributions to Soil Mechanics*, Boston Society of Civil Engineers, 1940, pp. 257–276.

Castro, G. (1969). "Liquefaction of sands," *Harvard Soil Mechanics Series* 87, Harvard University, Cambridge, MA.

Castro, G. (1975). "Liquefaction and cyclic mobility of sands," *Journal of the Geotechnical Engineering Division*, Vol. 101, No. GT6, pp. 551–569.

Castro, G. and Christian, J.T. (1976). "Shear strength of soils and cyclic loading," *Journal of the Geotechnical Engineering Division*, Vol. 102, No. GT9, pp. 887–894.

Castro, G. and Poulos, S.J. (1977). "Factors affecting liquefaction and cyclic mobility," *Journal of the Geotechnical Engineering Division*, Vol. 106, No. GT6, pp. 501–506.

Chiara, N. and Stokoe II, K.H. (2006). "Sample disturbance in resonant column test measurement of small-strain shearwave velocity," *Proceedings, Soil Stress-Strain Behavior: Measurement, Modeling and Analysis, Geotechnical Symposium, Roma*, pp. 605–613.

Chiou, B.S.-J., Darragh, R., Gregor, N., and Silva, W. (2008). "NGA project strong-motion database," *Earthquake Spectra*, Vol. 24, No. 1, pp. 23–44.

Coduto, D.P., Yeung, M.R., and Kitch, W.A. (2011). *Geotechnical Engineering: Principles and Practices*, 2nd edition, Prentice-Hall, Upper Saddle River, NJ, 794 pp.

Cox, B. and Wood, C. (2011) "Surface wave benchmarking exercise: Methodologies, results and uncertainties," *Proceedings, GeoRisk 2011: Geotechnical Risk Assessment & Management*, Atlanta, GA, 26–28 June, 8 pg.

Cox, B.R. and Beekman, A.N. (2011). "Intramethod variability in ReMi dispersion measurements and VS estimates at shallow bedrock sites," *Journal of Geotechnical and Geoenvironmental Engineering*, Vol. 137, pp. 354–362.

Cox, B.R., Stokoe, K.H. II, and Rathje, E.M. (2009). An in situ test method for evaluating the coupled pore pressure generation and nonlinear shear modulus behavior of liquefiable soils, *Geotechnical Testing Journal*, Vol. 32, No. 1, pp. 1–11.

Cox, B.R., Stolte, A.C., Stokoe, K.H. II, and Wotherspoon, L.M. (2019). "A direct-push crosshole test method for the in-situ evaluation of high-resolution P- and S-wave velocity," *ASTM Geotechnical Testing Journal*, Vol. 42, No. 5, pp. 1101–1132.

Cox, B.R. and Teague, D.P. (2016). "Layering ratios: A systematic approach to the inversion of surface wave data in the absence of A-priori information," *Geophysical Journal International*, Vol, 207, pp. 422–438.

Cubrinovski, M. and Ishihara, K. (1999). "Empirical correlation between SPT N-value and relative density for sandy soils," *Soils and Foundations*, Vol. 39, No. 5, pp. 61–71.

Cubrinovski, M. and Ishihara, K. (2000). "Flow potential of sandy soils with different grain compositions," *Soils and Foundations*, Vol. 40, No. 4, pp. 103–119.

Coduto, D.P., Yeung, M.R., and Kitch, W.A. (2011). *Geotechnical Engineering: Principles and Practices*, 2nd ed., Prentice-Hall, Upper Saddle River, NJ, 794 pp.

Dafalias, Y.F. and Herrmann, L.R. (1982). "Chapter 10: Bounding surface formulation of soil plasticity," in G.N. Pande and O.C. Zienkiewicz, eds., *Soil Mechanics – Transient and Cyclic Loads*, John Wiley & Sons, New York, pp. 253–282.

Dafalias, Y.F. and Popov, E.P. (1979). "A model for nonlinearly hardening materials for complex loading," *Acta Mechanica*, Vol. 21, No. 3, pp. 173–192.

Da Fonseca, A.V., Ferreira, C., and Fahey, M. (2009). "A framework interpreting bender element tests, combining time-domain and frequency-domain methods," *Geotechnical Testing Journal*, Vol. 32, No. 2, pp. 91–107.

Damm, J., Lewis, M., Stokoe, K.H., and Cox, B.R. (2021). "The use of various geophysical methods to characterize the velocity profile of a deep soil site," *Proceedings, 6th International Conference on Geotechnical and Geophysical Site Characterization*, Budapest, Hungary, 26–29 September 2021.

Darendeli, M.B. (2001). "Development of a new family of normalized modulus reduction and material damping curves," *Ph.D. dissertation*, Dept. of Civil Engineering, University of Texas.

De Alba, P., Baldwin, K., Janoo, V., Roe, G., and Celikkel, B. (1984). "Elastic-wave velocities and liquefaction potential," *Geotechnical Testing Journal, ASTM*, Vol. 7, No. 2, pp. 77–88.

DeJong, J.T., Fritzges, M.B., and Nusslein, K. (2006). "Microbially induced cementation to control sand response to undrained shear." *Journal of Geotechnical and Geoenvironmental Engineering*, Vol. 132, No. 11, pp. 1381–1392.

DeJong, J.T., Ghafghazi, M., Sturm, A., Armstrong, R., Perez, A., and Davis, C. (2014). "A new instrumented Becker Penetration Test (iBPT) for improved characterization of gravelly deposits within and underlying dams," *Journal of Dam Safety*, Vol. 12, No. 2, pp. 9–19.

DeJong, J.T., Ghafghazi, M., Sturm, A., Wilson, D.W., den Dulk, J., Armstrong, R., Perez, A., and Davis, C. (2017). "Instrumented Becker penetration test. I: Equipment, operation, and performance," *Journal of Geotechnical and Geoenvironmental Engineering*, Vol. 143, No. 9, 12 pp.

Desai, C.S. and Siriwardane, H.J. (1984). *Constitutive Laws for Engineering Materials, with Emphasis on Geologic Materials*, Prentice-Hall, Inc., Englewood Cliffs, NJ, 468 pp.

Di Giulio, G., Cornou, C., Ohrnberger, M., Wathelet, M., and Rovelli, A. (2006). "Deriving wavefield characteristics and shear-velocity profiles from two-dimensional small aperture arrays analysis of ambient vibrations in a small-size alluvial basin, Colfiorito, Italy." *Bulletin of the Seismological Society of America*, Vol. 96, No. 5, pp. 1915–1933.

Diaz-Rodriguez, J.A., Martinez-Vasquez, J.J., and Santamarina, J.C. (2009). "Strain-rate effects in Mexico City soil," *Journal of Geotechnical and Geoenvironmental Engineering*, Vol. 135, No. 2, pp. 300–305.

Diehl, J. and Steller, R. (2007). "Final data report: p- and s-wave velocity logging, Borings C4993, C4996, and C4997, Part B: Overall logs," *Report 6303-01 of WTP P-S Logging, Rev.* 1, GEOVision Geophysical Services, Vol. 2, 161 pp.

Dobrin, M.B. and Savit, C.H. (1988). *Introduction to Geophysical Prospecting*, 4th edition, McGraw-Hill Book Company, New York, 867 pp.

Dobry, R., Borcherdt, R.D., Crouse, C.B., Idriss, I.M., Joyner, W.B., Martin, G.R., Power, M.S., Rinne, E.E., and Seed, R.B. (2000). New site coefficients and site classification system used in recent building seismic code provisions, *Earthquake Spectra*, Vol. 16, No. 1, pp. 41–67.

Dobry, R. and Ladd, R.S. (1980). Discussion to "Soil liquefaction and cyclic mobility evaluation for level ground during earthquakes," by H.B. Seed, *Journal of the Geotechnical Engineering Division*, Vol. 106, No. GT6, pp. 720–724.

Dobry, R., Ladd, R.S., Yokel, F.Y., Chung, R.M., and Powell, D. (1982). "Prediction of pore water pressure buildup and liquefaction of sands during earthquakes by the cyclic strain method," *NBS Building Science Series* 138. National Bureau of Standards, 150 pp.

Dobry, R., Vasquez-Herrera, A., Mohammad, R., and Vucetic, M. (1985). "Liquefaction flow failure of silty sand by torsional cyclic tests," *Advances in the Art of Testing Soils under Cyclic Loading Conditions*, ed. V. Khosla, ASCE, New York, NY, pp. 29–50.

Dobry, R. and Vucetic, M. (1987). "Dynamic properties and seismic response of soft clay deposits," *Proceedings, International Symposium on Geotechnical Engineering of Soft Soils, Mexico City*, Vol. 2, pp. 51–87.

Douglas, B.J. and Olsen, R.S. (1981). "Soil classification using electric cone penetrometer," *Proceedings, Symposium on Cone Penetration Testing and Experience,* ASCE, St. Louis, Missouri, pp. 209–227.

Drnevich, V.P. (1967). "Effect of strain history on the dynamic properties of sand," *Ph.D. Dissertation*, University of Michigan, 151 pp.

Drnevich, V.P. (1972). "Undrained cyclic shear of saturated sand," *Journal of the Soil Mechanics and Foundations Division*, Vol. 98, No. SM8, pp. 807–825.

Drnevich, V.P. and Richart, F.E. Jr. (1970). "Dynamic prestraining of dry sand," *Journal of the Soil Mechanics and Foundations Division*, Vol. 96, No. SM2, pp. 453–469.

Duku, P.M., Stewart, J.P., Whang, D.H., and Yee, E. (2008). "Volumetric strains of clean sands subject to cyclic loads," *Journal of Geotechnical and Geoenvironmental Engineering*, Vol. 134, No. 8, pp. 1073–1085.

Dyvik, R. and Madshus, C. (1985). "Laboratory measurements of Gmax using bender elements," *Advances in the Art of Testing Soils under Cyclic Conditions*, ed. V. Khosla, ASCE, New York, NY, pp. 186–196.

Elgamal, A., Lai, T., Gunturi, R., and Zeghal M. (2004), "System Identification of Landfill Seismic Response," *Journal of Earthquake Engineering*, Vol. 8, No. 4, pp. 545–566.

Elgamal, A., Lai, T., Yang, Z., and He, L. (2001). "Dynamic soil properties, seismic downhole arrays and applications in practice," *Proceedings, 4th International Conference on Recent Advances in Geotechnical Earthquake Engineering and Soil Dynamics*, S. Prakash, ed., San Diego, CA.

Endrun, B., Ohrnberger, M., and Sawaidis, A. (2010). "On the repeatability and consistency of three-component ambient vibration array measurements." *Bulletin of Earthquake Engineering*, Vol. 8, No. 3, pp. 535–570.

EPRI (1993). "Guidelines for determining design basis ground motions," *Report TR-102293*, Electric Power Research Institute, Palo Alto, CA, Vol. 1, 488 pp.

EPRI (2013). "Seismic evaluation guidance screening, prioritization and implementation details (SPID) for the resolution of Fukushima near-term task force recommendation 2.1: Seismic," *Electrical Power Research Institute Report 1025287.*

Erkan, A. and Ulker, B.M.C (2008a). "The post cyclic shear strength of fine grained soils," *Proceedings, 14th World Conference on Earthquake Engineering*, Beijing, China, 8 pp.

Erkan, A. and Ulker, B.M.C. (2008b). "Effect of cyclic loading on monotonic shear strength of fine-grained soils," *Engineering Geology*, Vol. 87, pp. 243–257.

Fernandez, A., and Santamarina, J.C. (2003). "Design criteria for geotomographic field studies." *Geotechnical Testing Journal*, Vol. 26, No. 4, pp. 1–11.

Finn, W.D.L., Lee, K.W., and Martin, G.R. (1977). "An effective stress model for liquefaction," *Journal of the Geotechnical Engineering Division*, Vol. 103, No. GT6, pp. 517–533.

Finn, W.D.L., Pickering, D.J., and Bransby, P.L. (1971). "Sand liquefaction in triaxial and simple shear tests," *Journal of the Soil Mechanics and Foundations Division*, Vol. 97, No. SM4, pp. 639–659.

Foti, S., Hollender, F., Garofalo, F., Albarello, D., Asten, M., Bard, P.Y., Comina, C., Cornou, C., Cox, B., Di Giulio, G., Forbriger, T., Hayashi, K., Lunedei, E., Martin, A., Mercerat, D., Ohrnberger, M., Poggi, V., Renalier, F., Sicilia, D., and Socco, V. (2018). "Guidelines for the good practice of surface wave analysis: A product of the InterPACIFIC project." *Bulletin of Earthquake Engineering*, Vol. 16, No. 6, pp. 2367–2420.

Gao, Z. and Zhao, J. (2015). "Constitutive modeling of anisotropic sand behavior under monotonic and cyclic loading," *Journal of Engineering Mechanics*, Vol. 141, No. 8, 15 pp.

Garofalo, F., Foti, S., Hollender, F., Bard, P.-Y., Cornou, C., Cox, B.R., Dechamp, A., Ohrnberger, M., Sicilia, D., D. Teague, and Vergniault, C. (2016a). "InterPACIFIC project: Comparison of invasive and non-invasive methods for seismic site characterization part II: Inter-comparison between surface wave and borehole methods," *Soil Dynamics and Earthquake Engineering*, Vol. 82, No. 1, pp. 241–254.

Garofalo, F., Foti, S., Hollender, F., Bard, P.-Y., Cornou, C., Cox, B.R., Ohrnberger, M., Sicilia, D., Asten, M., Di Giulio, G., Forbriger, T., Guiller, B., Hayashi, K., Martin, A., Matsushima, S., Mercerat, D., Poggi, V., and Yamanaka, H. (2016b). "InterPACIFIC project: Comparison of invasive and non-invasive methods for seismic site characterization part I: Intra-comparison of surface wave methods," *Soil Dynamics and Earthquake Engineering*, Vol. 82, No. 1, pp. 222–240.

Goto, S., Tatsuoka, F., Shibuya, S., Kim, Y.-S., and Sato, S. (1991). "A simple gauge for local small strain measurements in the laboratory," *Soils and Foundations*, Vol. 31, No. 1, pp. 169–180.

Graham, J., Crooks, J.H.A., and Bell, A.L. (1983). "Strain rate effects in soft natural clays," *Geotechnique*, Vol. 33, pp. 327–340.

Groholski, D.R., Hashash, Y.M.A., Kim, B., Musgrove, M., Harmon, J., and Stewart, J.P. (2016). A simplified non-linear shear model to represent small-strain nonlinearity and soil strength in one-dimensional seismic site response analysis, *Journal of Geotechnical and Geoenvironmental Engineering*, Vol. 142, No. 9, 14 pp.

Gucunski, N. and Woods, R.D. (1991). "Use of Rayleigh modes in interpretation of SASW test," *Proceedings, 2nd International Conference on Recent Advances in Geotechnical Earthquake Engineering and Soil Dynamics, St. Louis*, Vol. 2, pp. 1399–1408.

Guo, J.W., Wang, J., Cai, Y., Liu, H., Gao, Y., and Sun, H. (2013). "Undrained deformation behavior of saturated soft clay under long-term cyclic loading," *Soil Dynamics and Earthquake Engineering*, Vol. 50, pp. 28–37.

Harder, L.F., Jr., and Seed, H.B. (1986). "Determination of penetration resistance for coarse-grained soils using the Becker Hammer Drill," *Rep. No.* UCB/EERC-86/06, College of Engineering, University of California, Berkeley, California, 129 pp.

Hardin, B.O. and Drnevich, V.P. (1972a). "Shear modulus and damping in soils: Measurement and parameter effects," *Journal of the Soil Mechanics and Foundations Division*, Vol. 98, No. SM6, pp. 603–624.

Hardin, B.O. and Drnevich, V.P. (1972b). "Shear modulus and damping in soils: Design equations and curves," *Journal of the Soil Mechanics and Foundations Division*, Vol. 98, No. SM7, pp. 667–692.

Harr, M.E. (1987). *Probability-Based Design in Civil Engineering*, Dover, Meniola, NY.

Haskell, N.H. (1953). "The dispersion of surface waves in multilayered media," *Bulletin of the Seismological Society of America*, Vol. 43, pp. 17–34.

Hatanaka, M. and Uchida, A. (1996). "Empirical correlation between penetration resistance and internal friction angle of sandy soils," *Soils and Foundations*, Vol. 36, No. 4, pp. 1–9.

Hegazy, Y.A. and Mayne, P.W. (2006). "A global statistical correlation between shear wave velocity and cone penetration data." Geotechnical Special Publication 149, ASCE, pp. 243–248.

Heisey, J.S., Stokoe, K.H., and Meyer, A.H. (1982). "Moduli of pavement systems from spectral analysis of surface waves," *Transportation Research Record 853*, Transportation Research Board, Washington, D.C.

Henke, W. and Henke, R. (1991). "Insitu torsional cylindrical shear test - laboratory results," *Proceedings, Second International Conference on Recent Advances in Geotechnical Earthquake Engineering and Soil Dynamics, St.* Louis, Missouri, Vol. 1, pp. 131–136.

Hertz, H. (1881). "Über die beruhrung fester elasticher köper," *J. reine u. angew Math.*, Vol. 92, pp. 156–171.

Hettler, A. and Gudehus, G. (1985). "A pressure-dependent correction for displacement results from 1 g model tests with sand," *Geotechnique*, Vol. 35, No. 4, pp. 497–510.

Hicher, P.-Y. and Shao, U. (2008). *Constitutive Modeling of Soils and Rocks*, John Wiley & Sons, Hoboken, NJ, 448 pp.

Hiltunen, D.R. and Woods, R.D. (1988). "SASW and crosshole test results compared," *Proceedings, Geotechnical Special Publication 20,* ASCE, pp. 279–289.

Hoar, R.J. and Stokoe, K.H. (1984). "Field and laboratory measurements of material damping of soil in shear" *Proceedings, 8th World Conference on Earthquake Engineering*, San Francisco, Vol. 3, pp. 47–54.

Hofmann, B.A., Sego, D.C., and Robertson, P.K. (2000). "In situ ground freezing to obtain undisturbed samples of loose sand," *Journal of Geotechnical and Geoenvironmental Engineering*, Vol. 126, No. 11, pp. 979–989.

Holtz, R.D., Kovacs, W.D., and Sheahan, T.C. (2023). *An Introduction to Geotechnical Engineering*, 3rd edition, Pearson, New York, 853 pp.

Horn, H.M. (1978). "North American experience in soil sampling and its influence on dynamic laboratory testing," *Preprint Session No. 79*, ASCE Annual Convention, Chicago, IL.

Houlsby, G (1988). *Discussion Session Contribution.* Penetration testing in the U.K., Birmingham.

Hryciw, R.D. (1989). "Ray-path curvature in shallow seismic investigations," *Journal of Geotechnical Engineering*, Vol. 115, No. 9, pp. 1268–1284.

Hsu, C.-C. and Vucetic, M. (2006). "Threshold shear strain for cyclic pore-water pressure in cohesive soils," *Journal of Geotechnical and Geoenvironmental Engineering*, Vol. 132, No. 10, pp. 1325–1335.

Hvorslev, M.J. (1949). "Subsurface exploration and sampling of soils for civil engineering purposes," *Report of Committee on Sampling and Testing, Soil Mechanics and Foundations Division*, 521 pp.

Hyde, A.F.L. and Ward, S.J. (1985). "A pore pressure and stability model for a silty clay under repeated loading," *Geotechnique*, Vol. 35, No. 2, pp. 113–125.

Iai, S. (1989). "Similitude for shaking table tests on soil-structure-fluid model in 1 g gravitational field," *Soils and Foundations*, Vol. 29, No. 1, pp. 105–118.

Idriss, I.M. and Boulanger, R.W. (2007). "SPT- and CPT-based relationships for the residual shear strength of liquefied soils," *Proceedings, 4th International Conference on Earthquake Geotechnical Engineering - Invited Lectures, K.D. Pitilakis*, ed., Thessaloniki, Springer, The Netherlands, pp. 1–22.

Idriss, I.M. and Boulanger, R.W. (2008). *Soil Liquefaction During Earthquakes*, Earthquake Engineering Research Institute, MNO-12, Oakland, California, 237 pp.

Idriss, I.M., Dobry, R., Doyle, E.H., and Singh, R.D. (1976). "Behavior of soft clays under earthquake loading conditions," *Proceedings, Offshore Technology Conference, OTC 2671*, Dallas, Texas.

Idriss, I.M., Dobry, R., and Singh, R.D. (1978). "Nonlinear behavior of soft clays during cyclic loading," *Journal of the Geotechnical Engineering Division*, Vol. 104, No. GT12, pp. 1427–1447.

Idriss, I.M., Fiegel, G., Hudson, M.B., Mundy, P.K. and Herzig, R. (1995). "Seismic Response of the Operating Industries landfill," *GSP No. 54*, ASCE, M.Y. Yegian, W.D. Liam Finn (editors), pp. 83–118.

Isenhower, W.M. and Stokoe, K.H. (1981). Strain rate dependent shear modulus of San Francisco Bay Mud, *Proceedings, International Conference on Recent Advances in Geotechnical Earthquake Engineering and Soil Dynamics, St.* Louis, Missouri, Vol. 2, pp. 597–602.

Ishibashi, I. and Zhang, X. (1993). "Unified dynamic shear moduli and damping ratios of sand and clay," *Soils and Foundations*, Vol. 33, No. 1, pp. 182–191.

Ishihara, K. (1993). Liquefaction and flow failure during earthquakes, *Geotechnique*, Vol. 43, No. 3, pp. 351–415.

Ishihara, K. (1996). *Soil Behavior in Earthquake Geotechnics*, Oxford University Press, New York, 350 pp.

Ishihara, K. and Li, S. (1972). "Liquefaction of saturated sand in triaxial torsion shear test," *Soils and Foundations*, Vol. 12, No. 2, pp. 19–39.

Ishihara, K., Tsuchiya, H., Huang, Y., and Kamada, K. (2001). "Recent studies on liquefaction resistance of sand - effect of saturation," *Proceedings, 4th International Conference on Recent Advances in Geotechnical Earthquake Engineering & Soil Dynamics*, San Diego, CA.

Iwan, W.D. (1967). "On a class of models for the yielding behavior of continuous and composite systems," *Journal of Applied Mechanics*, Vol. 34, No. E3, pp. 612–617.

Iwasaki, T., Tatsuoka, F., and Takagi, Y. (1978). "Shear modulus of sands under torsional shear loading," *Soils and Foundations*, Vol. 18, No. 1, pp. 39–56. https://doi.org/10.3208/sandf1972.18.39

Jamiolkowski, M., Ladd, C.C., Germaine, J.T., and Lancellotta, R. (1985). "New developments in field and laboratory testing of soils." *Proceedings, Eleventh International Conference on Soil Mechanics and Foundation Engineering*, San Francisco, Aug. 1985, Vol. 1, pp. 57–153.

Jefferies, M. and Been, K. (2006). *Soil Liquefaction - A Critical State Approach*, Taylor and Francis, London, UK and New York, USA.

Jitno, H. (1990). "Stress-strain and strength characteristics of clay during post-cyclic monotonic loading," *Masters Thesis*, University of British Columbia, Vancouver, BC, 116 pp.

Jones, A.L., Kramer, S.L, and Arduino, P. (2002). "Estimation of uncertainty in geotechnical properties for performance-based earthquake engineering," *Report PEER 2002/16*, Pacific Earthquake Engineering Research Center, University of California, Berkeley, 100 pp.

Kaklamanos, J., Bradley, B.A., Ehompson, E.M., and Baise, L.G. (2013). "Critical parameters affecting bias and variability in site response analyses using KiK-net downhole array data," *Bulletin of the Seismological Society of America*, Vol. 103, No. 3, pp. 1733–1749. https://doi.org/10.1785/0120120166

Kaklamanos, J., Baise, L.G., Thompson, E.M., & Dorfmann, L. (2015). "Comparison of 1D linear, equivalent-linear, and nonlinear site response models at six KiK-net validation sites," *Soil Dynamics and Earthquake Engineering*, Vol. 69, pp. 207–219. https://doi.org/10.1016/j.soildyn.2014.10.016

Kavazanjian, E., Jr., Snow, M.S., Matasovic, N., Poran, C., and Satoh, T. (1994). "Non-intrusive Rayleigh wave investigations at solid waste landfills," *Proceedings, First International Conference on Environmental Geotechnics*, Edmonton, Alberta.

Kayen, R., Scasserra, G., Stewart, J.P., and Lanzo, G. (2008). "Shear wave structure of Umbria and Marche, Italy, strong motion seismometer sites affected by the 1997 Umbria-Marche, Italy, earthquake sequence," *Open File Report 2008-1010*, Version 1.1, 51 pp.

Kim, B., Hashash, Y.M.A., Stewart, J.P., Rathje, E.M., Harmon, J.A., Musgrove, M.I. Campbell, K.W., and Silva, W.J. . (2016). "Relative differences between nonlinear and equivalent-linear 1D site response analyses," *Earthquake Spectra*, Vol. 32, 1845–1865. https://doi.org/10.1193/051215EQS068M

Kishida, T., Boulanger, R.W., Abrahamson, N.A., Driller, M.W., and Wehling, T.M. (2009c). "Seismic response of levees in the Sacramento-San Joaquin Delta," *Earthquake Spectra*, Vol. 25, No. 3, pp. 557–582.

Kishida, T., Boulanger, R.W., Abrahamson, N.A., Wehling, T.J., and Driller, M.W. (2009b). "Regression models for dynamic properties of highly organic soils," *Journal of Geotechnical and Geoenvironmental Engineering*, Vol. 135, No. 4, pp. 533–543.

Kishida, T., Wehling, T.J., Boulanger, R.W., Driller, M.W., and Stokoe, K.W. III (2009a). "Dynamic properties of highly organic soils from Montezuma Slough and Clifton Court, *Journal of Geotechnical and Geoenvironmental Engineering*, Vol. 135, No. 4, pp. 525–532.

Kokushu, T., Yoshida, Y., and Esashi, Y. (1982). "Dynamic properties of soft clay for wide strain range," *Soils and Foundations*, Vol. 22, No. 4, pp. 1–18.

Kokosho, T. (1980). "Cyclic triaxial test of dynamic soil properties for wide strain range," *Soils and Foundations*, Vol. 20, No. 2, pp. 45–60. https://doi.org/10.3208/sandf1972.20.2_45

Konno, T., Suzuki, Y., Tatoishi, A., Ishihara, K., Akino, K., and Satsuo, I. (1993). "Gravelly soil properties by field and laboratory tests," *Proceedings, Third International Conference on Case Histories in Geotechnical Engineering, St. Louis*, Vol. 1, pp. 575–594.

Konrad, J.-M. (1988). "Interpretation of flat plate dilatometer tests in sands in terms of the state parameter," *Geotechnique*, Vol. 38, No. 2, pp. 263–278.

Kottke, A. and Rathje, E.M. (2008). Technical manual for STRATA, *PEER Report No. 2008/10*, Pacific Earthquake Engineering Research Center, University of California, Berkeley, CA.

Koutsoftas, D.C. (1978). "Effect of cyclic loads on undrained strength of two marine clays," *Journal of the Geotechnical Engineering Division*, Vol. 104, No. GT5, pp. 609–620.

Koutsoftas, D.C. and Ladd, C.C. (1985). "Design strength of an offshore clay," *Journal of the Geotechnical Division*, Vol. 111, No. 3, pp. 337–355.

Kovacs, W.D., Evans, J.C., and Griffith, A.H. (1977). "Toward a more standardized SPT," *Proceedings, 9th International Conference on Soil Mechanics and Foundation Engineering, Tokyo*, Vol. 2, pp. 269–276.

Kovacs, W.D. and Leo, E. (1981). "Cyclic simple shear of large-scale sand samples: Effects of diameter to height ratio," *Proceedings, International Conference on Recent Advances in Geotechnical Earthquake Engineering and Soil Dynamics, St. Louis*, Vol. 3, pp. 897–907.

Kovacs, W.D. and Salomone, L.A. (1982). "SPT hammer energy measurement," *Journal of the Geotechnical Engineering Division*, Vol. 108, No. GT4, pp. 599–620.

Kramer, S.L. (2000). "Dynamic response of Mercer Slough peat," *Journal of Geotechnical and Geoenvioronmental Engineering*, Vol. 126, No. 6, pp. 504–510.

Kramer, S.L., Hartvigsen, A.J., Sideras, S.S., and Ozener, P.T. (2011). "Site response modeling in liquefiable soil deposits," *Proceedings, 4th IASPEI/IAEE International Symposium: Effects of Surface Geology on Seismic Motion*, University of California, Santa Barbara, 23–26 August 2011.

Kramer, S.L. and Sivaneswaran, N. (1989a). "A non-destructive, specimen-specific method for measurement of membrane penetration in the triaxial test," *Geotechnical Testing Journal, ASTM*, Vol. 12, No. 1, pp. 50–59.

Kramer, S.L. and Sivaneswaran, N. (1989b). "Stress-path-dependent correction for membrane penetration," *Journal of Geotechnical Engineering*, Vol. 115, No. 12, pp. 1787–1804. https://doi.org/10.1061/(asce)0733-9410(1989)115#: 12(1787)

Kuerbis, R. and Vaid, Y.P. (1998). "Sand sample preparation – the Slurry Deposition Method," *Soils and Foundations*, Vol. 28, No. 4, pp. 107–118.

Kulhawy, F.H. (1992). "On the evaluation of static soil properties," *Proceedings*, Stability and Performance of Slopes and Embankments – II, ASCE Geotechnical Special Publication No. 31, Vol. 1, pp. 25–55.

Kulhawy, F.H. and Mayne, P.W. (1990). Manual on estimating soil properties for engineering design, *Report EPRI EL*-6800, Electric Power Research Institute, Palo Alto, California, 308 pp.

Kutter, B.L. and James, R.G. (1989). "Dynamic centrifuge model tests on clay embankments," *Geotechnique*, Vol. 39, No. 1, pp. 91–106.

Kuwano, R. and Jardine, R. (2002) "On the applicability of cross-anisotropic elasticity to granular materials at very small strains," *Gèotechnique*, Vol. 52, No. 10, pp. 727–749.

Lacasse, S. and Nadim, F. (1996). "Uncertainties in characterising soil properties," In C.D. Shackleford, P.P. Nelson and M.J.S. Roth (eds.), *Uncertainty in the Geologic Environment: From Theory to Practice*, Geotechnical Special Publication No. 58, ASCE, New York, NY, pp. 49–75.

Ladd, C.C. (1991). "Stability evaluation during staged construction," *Journal of Geotechnical Engineering*, Vol. 117, No. 4, pp. 540–615.

Ladd, C. C., Foott, R., Ishihara, K., Schlosser, F., and Poulos, H.G. (1977). "Stress-deformation and strength characteristics," *State-of-the-Art Report, Proceedings, 9th International Conference on Soil Mechanics and Foundation Engineering, Tokyo*, pp. 421–494.

Ladd, R.J. and Dutko, P. (1985). "Small-strain measurements using triaxial apparatus," in V. Koshla, ed., *Advances in the Art of Testing Soils under Cyclic Conditions*, ASCE, pp. 148–166.

Ladd, R.S. (1978). "Preparing test specimens using undercompaction," *Geotechnical Testing Journal, ASTM*, Vol. 1, No. 1, pp. 16–23.

Ladd, R.S. and Foott, R. (1974). "New design procedure for stability of soft clays," *Journal of the Geotechnical Engineering Division*, Vol. 100, No. 7, pp. 763–786.

Lade, P.V. and Hernandez, S.B. (1977). "Membrane penetration effects in undrained tests," *Journal of the Geotechnical Engineering Division*, Vol. 103, No. GT2, pp. 109–125.

Lambe, T.W. and Whitman, R.V. (1969). *Soil Mechanics*, John Wiley & Sons, New York, 553 pp.

Lee, S.H. and Stokoe, K.H. (1986). "Investigation of low amplitude shear wave velocity in anisotropic material," *Report GR 86-6*. University of Texas, Austin.

Lefebvre, G. and LeBoeuf, D. (1987). "Rate effects and cyclic loading of sensitive clays." *Journal of Geotechnical and Geoenvironmental Engineering*, Vol. 113, No. 5, pp. 476–489.

Leiderman, K., Bouzarth, E.L., Cortez, R., and Layton, A.T. (2013). "A regularization method for the numerical solution of periodic stokes flow," *Journal of Computational Physics*, Vol. 236, pp. 187–202.

Lin, Y.C., Stokoe, K.H., and Rosenblad, B.L. (2008). "Variability in Vs profiles and consistency between seismic profiling methods: A case study in Imperial Valley, California," *Proceedings, International Conference on Site Characterization (ISC-3)*, Taipei, Taiwan, pp. 75–81.

Louie, J.N. (2001). "Faster, better: Shear-wave velocity to 100 m depth from refraction microtremor array," *Bulletin of the Seismological Society of America*, Vol. 91, No. 2, pp. 347–364.

Mair, R.J. and Wood, D.M. (1987). *Pressuremeter Testing: Methods and Interpretation*, Butterworths, London, 160 pp.

Marcuson, W.F. III, Cooper, S.S., and Bieganousky, W.A. (1977). "Laboratory sampling study conducted on fine sands," *Proceedings of Specialty Session 2, 9th International Conference on Soil Mechanics and Foundation Engineering, Tokyo*, pp. 15–22.

Martin, A.J. and Diehl, J.G. (2004). "Practical experience using a simplified procedure to measure average shear-wave velocity to a depth of 30 meters (V_{S30})," *Proceedings, 13th World Conference on Earthquake Engineering*, International Association for Earthquake Engineering.

Martin, G.R., Finn, W.D.L., and Seed, H.B. (1975). "Fundamentals of liquefaction under cyclic loading," *Journal of the Geotechnical Engineering Division*, Vol. 101, No. 5, pp. 423–438.

Martin, G.R., Finn, W.D.L., and Seed, H.B. (1978). "Effects of system compliance on liquefaction tests," *Journal of Geotechnical Engineering*, Vol. 104, No. GT4, pp. 463–479. https://doi.org/10.1061/ajgeb6.0000614

Masing, G. (1926). "Eigenspanungen und verfestigung beim Messing," *Proceedings, Second International Congress on Applied Mechanics*, Zurich, Switzerland.

Matasovic, N. and Kavazanjian, E. (1998). "Cyclic characterization of OII landfill solid waste," *Journal of Geotechnical and Geoenvironmental Engineering*, Vol. 124, No. 3, pp. 197–210.

Matasovic, N., Kavazanjian, E., Jr., and Abourjeily, F. (1995). "Dynamic properties of solid waste from field observations," *Proceedings, First International Conference on Earthquake Geotechnical Engineering*, Tokyo, Japan, Vol. 1, pp. 549–554.

Matasovic, N. and Vucetic, M. (1995). "Generalized cyclic-degradation-pore pressure generation model for clays," *Journal of Geotechnical Engineering*, Vol. 121, No. 1, pp. 33–42.

Matasović, N., Williamson, T.A., and Bachus, R.C. (1998). "Cyclic direct simple shear testing of OII landfill solid waste," *Proceedings, 11th European Conference on Soil Mechanics and Foundation Engineering*, Vol. 1, Balkema, Rotterdam, The Netherlands, pp. 441–448.

Matsui, T., Bahr, M.A., and Abe, N. (1992). "Estimation of shear characteristics degradation and stress-strain relationship of saturated clays after cyclic loading," *Soils and Foundations*, Vol. 32, No. 1, pp. 161–172.

Mayne, P.W. (2007). "Cone penetration testing state-of-practice," *Final report*, NCHRP Project 20-05, Topic 37-14, Transportation Research Board, Washington, DC, 137 pp.

McClure, F.E. (1973). *Performance of single family dwellings in the San Fernando earthquake of* February 9, *1971*, NOAA, U.S. Dept. of Commerce, May.

McGann, C.R., Bradley, B.A., Taylor, M.L., Wotherspoon, L.M., and Cubrinovski, M. (2015). "Development of an empirical correlation for predicting shear wave velocity of Christchurch soils from cone penetration test data." *Soil Dynamics and Earthquake Engineering,* Vol. 75, pp. 66–75.

McGillivray, A.V. and Mayne, P.W. (2008). "An automated seismic source for continuous-push shear wave velocity profiling with SCPT and frequent-interval SDMT," *Geotechnical & Geophysical Site Characterization*, Vol. 2, (Proc. ISC-3, Taipei), Taylor & Francis Group, London, pp. 1347–1352.

Menq, F.-Y. (2003). "Dynamic properties of sandy and gravelly soils," *Ph.D. dissertation*, Dept. of Civil Engineering, University of Texas.

Mesri, G. (1989). "Reevaluation of Su(mob) = 0.22 σ′p max using laboratory shear tests." *Canadian Geotechnical Journal*, Vol. 26, No. 1, pp. 162–164.

Meyerhof, G.G. (1957). "Discussion on soil properties and their measurement," *Session 2, Proceedings, 4th International Conference on Soil Mechanics and Foundation Engineering*, London, Vol. III, p. 110.

Miller, R.D., Xia, J., Park, C.B., and Ivanov, J. (1999). "Multichannel analysis of surface waves to map bedrock," *The Leading Edge*, Vol. 18, No. 12, pp. 1392–1396.

Mindlin, R.D. and Deresiewicz, H. (1953). "Elastic spheres in contact under varying oblique forces," *Journal of Applied Mechanics*, ASME, pp. 327–344. https://doi.org/10.1115/1.4010702

Mok, Y.-J., Sanchez-Salinero, I., Stokow, K.H., and Roesset, J.M. (1988). "In situ damping measurements by crosshole seismic method," *Proceedings, Earthquake Engineering and Soil Dynamics II – Recent Advances in Ground Motion Evaluation, Salt Lake City, UT, Geotechnical Special Publication 20, ASCE*, pp. 305–320.

Mori, K. and Sakai, K. (2016). "The GP sampler: A new innovation in core sampling," *Proceedings, Geotechnical and Geophysical Site Characterisation 5*, Sydney, Australian Geomechanics Society, pp. 99–124.

Moss, R.E.S. (2008). "Quantifying measurement uncertainty of thirty-meter shear wave velocity," *Bulletin of the Seismological Society of America*, Vol. 98, No. 3, pp. 1399–1411.

Mroz, Z. (1967). "On the description of anisotropic work hardening," *Journal of Mechanics and Physics of Solids*, Vol. 15, pp. 163–175.

Musgrave, A.W. (1970). *Seismic Refraction Prospecting*, Society of Exploration Geophysicists, Tulsa, OK, 621 pp.

Nazarian, S. (1984). "Insitu determination of elastic moduli of soil deposits and pavement systems by spectral-analysis-of-surface-waves method," *Ph.D. dissertation*, The University of Texas at Austin, 458 pp.

Nazarian, S. and Stokoe, K.H. (1983). "Use of spectral analysis of surface waves for determination of moduli and thicknesses of pavement systems, *Transportation Research Record No.* 954.

NCEER (1997). "Proceedings of the NCEER Workshop on Evaluation of liquefaction resistance of soils," T. L. Youd and I.M. Idriss, eds., *Technical Report NCEER-97-0022*, National Center for Earthquake Engineering Research, Buffalo, NY, 276 pp.

Negussey, D., Wijewickreme, W.K.D., & Vaid, Y. P. (1988). "Constant-volume friction angle of granular materials," *Canadian Geotechnical Journal*, Vol. 25, No. 1, pp. 50–55. https://doi.org/10.1139/t88-006

Ni, J., Indraratna, B., Geng, X.-Y., Carter, J.P., and Chen, Y.-L. (2015). "Model of soft soils under cyclic loading," *International Journal of Geomechanics*, Vol. 15, No. 4, 10 pp.

Nigbor, R.L. and Imai, T. (1994). "The suspension p-s velocity logging method," in R. D. Woods, ed., *Geophysical Characterization of Sites*, A. A. Balkema, Rotterdam, pp. 57–61.

O'Reilly, M.P. (1991). Cyclic load testing of soils," in M. P. O'Reilly and S. F. Brown, eds., *Cyclic Loading of Soils*, Blackie and Sons, Ltd., London, pp. 70–121.

Osterberg, J.O. (1973). "An improved hydraulic piston sampler," *Proceedings, 8th International Conference on Soil Mechanics and Foundation Engineering*, Moscow, USSR, Vol. 1, pp. 317–321.

Park, C.B., Miller, R.D., and Xia, J. (1999). "Multichannel analysis of surface waves (MASW)," *Geophysics*, Vol. 64, pp. 800–808.

Park, T. and Silver, M.L. (1975). "Dynamic soil properties required to predict the dynamic behavior of elevated transportation structures," *Report DOT-TST-75-44*, U.S. Department of Transportation.

Parker, G.A., Harmon, J.A., Stewart, J.P., Hashash, Y.M., Kottke, A.R., Rathje, E.M., Silva, W.J., and Campbell, K.W. (2017). "Proxy-based V_{S30} estimation in central and eastern North America," *Bulletin of the Seismological Society of America*, Vol. 107, No. 1, pp. 117–131.

Pestana, J.M. and Whittle, A.J. (1995). "Compression model for cohesionless soils," *Géotechnique*, Vol. 45, No. 4, pp. 611–633.

Pestana, J.M. and Biscontin, G. (2000). "A simplified model describing the cyclic behavior of lightly overconsolidated clays in simple shear," *Geotechnical Engineering Report No UCB/GT/2000-03*, University of California, Berkeley, 67 pp.

Phoon, K.-K. and Kulhawy, F.H. (1999). "Evaluation of geotechnical property variability," *Canadian Geotechnical Journal,* Vol. 36. No. 4, pp. 625–639. https://doi.org/10.1139/t99-039

Picozzi, M., Parolai, S., Bindi, D., and Strollo, A. (2009). "Characterization of shallow geology by high-frequency seismic noise tomography," *Geophysics Journal International*, Vol. 176, pp. 164–174.

Poulos, S.J. (1981). "The steady state of deformation," *Journal of Geotechnical Engineering Division*, Vol. 107, No. GT5, pp. 553–562.

Prange, B. (1981). "Stochastic excitation of rock masses," *Proceedings, 10th International Conference on Soil Mechanics and Foundation Engineering, Stockholm*, Vol. 4, pp. 89–880.

Prevost, J.H. (1977). "Mathematical modelling of monotonic and cyclic undrained clay behavior," *International Journal of Numerical and Analytical Methods in Geomechanics*, Vol. 1, No. 2, pp. 195–216.

Pullammanappallil, S., Honjas, W., and Louie, J.N. (2003). "Determination of 1-D shear wave velocities using the refraction microtremor method," *Proceedings, 3rd International Conference on the Application of Geophysical Methodologies and NDT to Transportation and Infrastructure*, Federal Highway Administration, Washington, DC.

Pyke, R., Seed, H.B., and Chan, C.K. (1975). "Settlement of sands under multi-directional loading," *Journal of the Geotechnical Engineering Division*, Vol. 101, No. GT4, pp. 379–398.

Pyke, R.M. (1973). "Settlement and liquefaction of sands under multi-directional loading," *Ph.D. dissertation*, University of California, Berkeley.

Rahimi, S., Wood, C.M., and Teague, D.P. (2021). "Performance of different transformation techniques for MASW data processing considering various site conditions, near-field effects, and modal separation," *Survey of Geophysics*, Vol. 42, pp. 1197–1225.

Raju, V.S. and Venkataramana, K. (1980). "Undrained triaxial tests to assess liquefaction potential of sands: Effect of membrane penetration," *Proceedings, International Symposium on Soils under Cyclic and Transient Loading*, Swansea, United Kingdom, pp. 483–494.

Ramanujam, N., Holish, L.L., and Chen, W.W.H. (1976). "Post-earthquake stability analysis of earth dams," *Proceedings, Earthquake Engineering and Soil Dynamics, ASCE Specialty Conference*, Pasadena, CA, Vol. 1, pp. 762–776.

Ramberg, W. and Osgood, W.R. (1943). "Description of stress-strain curves by three parameters," *Technical Note 902*, National Advisory Committee for Aeronautics, Washington, D.C.

Richardson, J.M. and Whitman, R.V. (1963). "Effect of strain-rate upon undrained shear resistance of a saturated remolded fat clay." *Geotechnique*, Vol. 13, No. 4, pp. 310–324.

Richart, F.E., Hall, J.R., and Woods, R.D. (1970). *Vibrations of Soils and Foundations*, Prentice-Hall, Inc., Englewood Cliffs, NJ, 401 pp.

Riemer, M.F. and Cobos-Roa, D. (2007). "In situ measurement of dynamic properties using the downhole freestanding shear device," *Proceedings, 4th International Conference on Earthquake Geotechnical Engineering*, Thessaloniki, Greece, Paper No. 1320, 12 pp.

Robertson, P.K. (1990). "Soil classification using the cone penetration test," *Canadian Geotechnical Journal*, Vol. 27, No. 1, pp. 151–158.

Robertson, P.K. (2009). "Interpretation of cone penetration tests - a unified approach," *Canadian Geotechnical Journal*, Vol. 46, pp. 1337–1355.

Robertson, P.K. (2010). "Soil behaviour type from CPT: An update," *Proceedings of the 2nd International Symposium on Cone Penetration Testing, CPT'10*, Huntington Beach, CA.

Robertson, P.K. (2012). "Interpretation of in-situ tests - some insights," *J.K. Mitchell Lecture, Proceedings, of ISC4*, Recife, Brazil, 22 pp.

Robertson, P.K. (2017). "Evaluation of flow liquefaction: Influence of high stresses," *Proceedings, 3rd International Conference on Performance-Based Design, ICSMGE*, Vancouver, Canada, 8 pp.

Robertson, P.K. and Cabal, K.L. (2022). *Guide to Cone Penetration Testing for Geotechnical Engineering*, 7th edition, Gregg Drilling & Testing, Signal Hill, CA, 164 pp.

Robertson, P.K. and Fear, C.E. (1995). "Liquefaction of sands and its evaluation," *Proceedings, First International Conference on Earthquake Geotechnical Engineering*, K. Ishihara, ed., A. A. Balkema, Rotterdam.

Robertson, P.K. and Wride, C.E. (1998). "Evaluating cyclic liquefaction potential using the CPT," *Canadian Geotechnical Journal*, Vol. 35, No. 3, pp. 442–459.

Robertson, P.K., Campanella, R.G., Gillespie, D., and Rice, A. (1985). "Seismic CPT to measure in-situ shear wave velocity," *Measurement and Use of Shear Wave Velocity for Evaluating Dynamic Soil Properties*, ASCE, pp. 18–34.

Robertson, P.K., Campanella, R.G., Gillespie, D., and Greig, J. (1986). "Use of piezometer cone data," *In-Situ'86 Use of Insitu testing in Geotechnical Engineering,* Specialty Publication GSP 6, ASCE, Reston, VA, pp. 1263–1280.

Roesler, S.K. (1977). "Correlation methods in soil dynamics," *Proceedings*, DMSR 77, Karlsruhe, Vol. 1, pp. 309–334.

Rollins, K.M., Evans, M., Diehl, N., and Daily, W. (1998). "Shear modulus and damping relationships for gravels." *Journal of Geotechnical and Geoenvironmental Engineering*, Vol. 124, No. 5, pp. 396–405. https://doi.org/10.1061/(ASCE)1090-0241(1998)124#: 5(396).

Romero, S. (1995). "The behavior of silt as clay content is increased," *MS Thesis*, University of California, Davis, 108 pp.

Roscoe, K.H. and Burland, J.B. (1968). "On the generalized stress-strain behavior of 'wet' clay," in J. Heyman and F. A. Leckie, eds., *Engineering Plasticity*, Cambridge University Press, pp. 535–609.

Roscoe, K.H. and Pooroshasb, H.B. (1963). "A fundamental principle of similarity in model tests for earth pressure problems," *Proceedings, Second Asian Regional Conference on Soil Mechanics, Tokyo*, Vol. 1, pp. 134–140.

Roscoe, K.H. and Schofield, A.N. (1963). "Mechanical behavior of an idealised 'wet' clay," *Proceedings, Second European Conference on Soil Mechanics, Wiesbaden*, Vol. 1, pp. 47–54.

Rosenblad, B.L. and Li, J. (2009). "Comparative study of refraction microtremor (ReMi) and active source methods for developing lowfrequency surface wave dispersion curves," *Journal of Environmental Engineering and Geophysics*, Vol. 14, No. 3, pp. 101–113.

Rosenblueth, E. (1975). "Point estimates for probability moments," *Proceedings, National Academy of Science*, Vol. 72, No. 10, pp. 3812–3814.

Roumelioti, Z., Hollender, F., and Gueguen, P. (2020). "Rainfall-induced variation of seismic waves velocity in soil and implications for soil response: What the ARGONET (Cephalonia, Greece) vertical array data reveal," *Bulletin of the Seismological Society of America*, Vol. 110, No. 2, pp. 441–451.

Rowe, P.W. (1962). "The stress–dilatancy relation for static equilibrium of an assembly of particles in contact," *Proceedings of the Royal Society of London*, Vol. A269, pp. 500–527. https://doi.org/10.1098/rspa.1962.0193

Santamarina, J.C. and Fratta, D. (2005). *Discrete Signals and Inverse Problems. An Introduction for Engineers and Scientists*, John Wiley & Sons Ltd. West Sussex, England, 350 pp.

Sayao, A.S.F. and Vaid, Y.P. (1989). "Deformations due to principal stress rotation," *Proceedings, 12th International Conference on Soil Mechanics and Foundation Engineering, Rio de Janeiro*, pp. 107–110.

Schmertmann, J.H., Smith, T.V., and Ho, R. (1978). "Example of an energy calibration report on a standard penetration test (ASTM Standard D1586-67) drill rig," *Geotechnical Testing Journal, ASTM*, Vol. 1, No. 1, pp. 57–61.

Schmertmann, J.S. and Palacios, A. (1979). "Energy dynamics of SPT," *Journal of the Soil Mechanics and Foundations Division*, Vol. 105, No. GT8, pp. 909–926.

Schwarz, S.D. and Musser, J.M., Jr. (1972). "Various techniques for making in situ shear wave velocity measurements - a description and evaluation," *Proceedings, International Conference on Microzonation, Seattle*, Vol. 2, pp. 594–608.

Seed, H.B. and Idriss, I.M. (1970). "Soil moduli and damping factors for dynamic response analyses," *Report No. EERC 70-10*, Earthquake Engineering Research Center, University of California, Berkeley.

Seed, H.B. and Lee, K.L. (1966). "Liquefaction of saturated sands during cyclic loading," *Journal of the Soil Mechanics and Foundations Division*, Vol. 92, No. SM6, pp. 105–134.

Seed, H.B. and Lee, K.L. (1967). "Undrained strength characteristics of cohesionless soils," *Journal of the Soil Mechanics and Foundations Division*, Vol. 93, No. 6, pp. 333–360.

Seed, H.B. and Peacock, W.H. (1971). "Test procedures for measuring soil liquefaction characteristics," *Journal of the Soil Mechanics and Foundations Division*, Vol. 97, No. SM8, pp. 1099–1119.

Seed, H.B. and Silver, M.L. (1972). "Settlement of dry sands during earthquakes," *Journal of the Soil Mechanics and Foundations Division*, Vol. 98, No. SM4, pp. 381–397.

Seed, H. B., Singh, S., Chan, C. K. and Vilela, T. F. (1982). "Considerations in undisturbed sampling of sands," *Journal of the Geotechnical Engineering Division*, Vol. 108, No. GT2, pp. 265–283. https://doi.org/10.1061/ajgeb6.0001243

Seed, H.B., Tokimatsu, K., Harder, L.F., and Chung, R.M. (1985). "The influence of SPT procedures in soil liquefaction resistance evaluations," *Journal of Geotechnical Engineering*, Vol. 111, No. 12, pp. 1425–1445.

Seed, R.B. and Anwar, H. (1986). "Development of a laboratory technique for correcting results of undrained traixial shear tests on soils containing coarse particles for effects of membrane penetration," *Research Report No. SR/GT/86-02*, Department of Civil Engineering, Stanford University, Stanford, California.

Seed, H.B., Wong, R.T., Idriss, I.M., and Tokimatsu, K. (1986). "Moduli and damping factors for dynamic analyses of cohesionless soils," *Journal of Geotechnical Engineering*, Vol. 112, No. 11, pp. 1016–1032. https://doi.org/10.1061/(asce)0733-9410(1986)112#: 11(1016)

Senneset, K., and Janbu, N. (1984). "Shear strength parameters obtained from static cone penetration tests," *Proceedings, Symposium on Strength Testing of Marine Sediments: Laboratory and In-Situ Measurements. ASTM 04-883000-38*, San Diego, pp. 41–54.

Seyhan, E., Stewart, J.P., Ancheta, T.D., Darragh, R.B., and Graves, R.W. (2014). "NGA-West2 site database," *Earthquake Spectra*, Vol. 30, No. 3, pp. 1007–1024.

Shafiee, A., Stewart, J.P., Venugopal, R.J. and Brandenberg, S. (2017). "Adaptation of broadband simple shear device for constant volume and stress-controlled testing," *Geotechnical Testing Journal*, Vol. 40, No. 1, pp. 15–28.

Sheahan, T.C., Ladd, C.C., and Germaine, J.T. (1996). "Rate-dependent undrained shear behavior of saturated clay." *Journal of Geotechnical Engineering*, Vol. 122, No. 2, pp. 99–108.

Sheng, D., Yao, Y., and Carter, J.P. (2008). "A volume-stress model for sands under isotropic and critical stress states," *Canadian Geotechnical Journal*, Vol. 45, pp. 1639–1645.

Shirley, D.J. and Anderson, A.L. (1975). "Acoustic and engineering properties of sediments, *Report No. ARL-TR-75-58*, Applied Research Laboratory, University of Texas at Austin.

Shuttle, D.A. and Cunning, J. (2007). "Liquefaction potential of silts from CPTu," *Canadian Geotechnical Journal*, Vol. 44, No. 1, pp. 1–19.

Silver, N.L. and Seed, H.B. (1971). "Volume changes in sands during cyclic loading," *Journal of the Soil Mechanics and Foundations Division*, Vol 97, No. SM9, pp. 1171–1181.

Singh, S., Seed, H.B., and Chan, C.K. (1982). "Undisturbed sampling of saturated sands by freezing." *Journal of the Geotechnical Engineering Division*, Vol. 108, No. 2, pp. 247–264.

Skempton, A.W. (1986). "Standard penetration test procedures and the effects in sands of overburden pressure, relative density, particle size, ageing and overconsolidation," *Geotechnique*, Vol. 36, No. 3, pp. 425–447.

Skempton, A.W. (1957). "Planning and design of new Hong Kong airport," *Proceedings, Institute of Civil Engineers*, Vol. 7, pp. 305–307.

Slosson, J.E. (1975). "Chapter 19: Effects of the earthquake on residential areas," San Fernando, California, Earthquake of 9 February 1971, Bulletin No.196, California Division of Mines and Geology, Sacramento, California.

Stephenson, W.J., Louie, J.N., Pullammanappallil, S., Williams, R.A., and Odum, J.K. (2005). "Blind shear-wave velocity comparison of ReMi and MASW results with boreholes to 200 m in Santa Clara Valley: Implications for earthquake ground-motion assessment," *Bulletin of the Seismological Society of America*, Vol. 95, pp. 2506–2516.

Stewart, J.P., Afshari, K., and Hashash, Y.M. (2014). "Guidelines for performing hazard-consistent one-dimensional ground response analysis for ground motion prediction," *PEER Report, 2014/16*. Pacific Earthquake Engineering Research Center, University of California, Berkeley.

Stewart, J.P., Bray, J.D., McMahon, D.J., Smith, P.M., and Kropp, A.L. (2001). "Seismic performance of hillside fills" *Journal of Geotechnical and Geoenvioronmental Engineering,* Vol. 127, No. 11, pp. 905–919.

Stewart, J.P. and Fenves, G.L. (1998). "System identification for evaluating soil-structure interaction effects in buildings from strong motion recordings," *Journal of Earthquake Engineering and Structural Dynamics*, Vol. 27, pp. 869–885.

Stokoe, K.H. and Abdul-razzak, K.G. (1975). "Shear moduli of two compacted fills," *Proceedings, In-Situ Measurement of Soil Properties Specialty Conference*, Raleigh, North Carolina, Vol. 1, pp. 422–449.

Stokoe, K.H., II, Bay, J.A., Rosenblad, B.L., Hwang, S.K., and Twede, M.R. (1994). "In situ seismic and dynamic laboratory measurements of geotechnical materials at Queensboro Bridge and Roosevelt Island," *Geotechnical Engineering Rep. No.* GR94-5, Civil Engineering Dept., University of Texas at Austin, Austin, Texas.

Stokoe, K.H. and Hoar, R.J. (1978). "Generation and measurement of shear waves insitu," *Dynamic Geotechnical Testing*, ASTM STP 654, pp. 3–29.

Stokoe, K.H., Joh, S.-H., and Woods, R.D. (2004). "Some contributions of insitu geophysical measurements to solving geotechnical engineering problems," *Proceedings, International Conference on site Characterization (ISC-2)*, Porto, Portugal, pp. 15–50.

Stokoe, K.H., Lee, J.N.K., and Lee, S.H.H. (1991). "Characterization of soil in calibration chambers with seismic waves," *Proceedings, First International Symposium on Calibration Chamber, ISOCCT1*, Potsdam, pp. 363–376.

Stokoe, K.H., Nazarian, S., Rix, G.J., Sanchez-Salinero, R., Sheu, J.-C., and Mok, Y.-J. (1988). "In situ seismic testing of hard-to-sample soils by surface wave method," *Proceedings, Earthquake Engineering and Soil Dynamics II - Recent Advances in Ground Motion Evaluation, Salt Lake City, UT, Geotechnical Special Publication 20, ASCE*, pp. 264–278.

Stokoe, K.H. II, Wright, S.G., Bay, J.A., and Roesset, J.M. (1994). "Characterization of geotechnical sites by SASW method," in R. D. Woods, ed., *Geophysical Characterization of Sites*, A. A. Balkema, Rotterdam, pp. 15–25.

Stolte, A.C. and Cox, B.R. (2019). "Feasibility of in-situ evaluation of soil void ratio in clean sands using high resolution measurements of Vp and Vs from DPCH testing," *AIMS Geosciences*, Vol. 5, No. 4, pp. 723–749.

Stringer, M.E., Cubrinovski, M., and Haycock, I. (2016). "Experience with gel-push sampling in New Zealand," *Proceedings, Geotechnical and Geophysical Site Characterisation 5,* Sydney, Australian Geomechanics Society, pp. 577–582.

Sun, J.I., Golesorkhi, R., and Seed, H.B. (1988). "Dynamic moduli and damping ratios for cohesive soils," *Report No. EERC-88/15*, Earthquake Engineering Research Center, University of California, Berkeley.

Sy, A. and Campanella, R.G. (1994). Becker and standard penetration tests (BPT-SPT) correlations with consideration of casing friction, *Canadian Geotechnical Journal*, Vol. 31, No. 3, pp. 343–356.

Sy, A. and Lum, K.Y.Y. (1997). "Correlations of mud-injection Becker and standard penetration tests," *Canadian Geotechnical Journal*, Vol. 34, No. 1, pp. 139–144.

Symes, M.J., Gens, A., and Hight, D.W. (1988). "Drained principal stress rotation in saturated sand," *Geotechnique*, Vol. 38, No. 1, pp. 59–81. https://doi.org/10.1680/geot.1988.38.1.59

Tao, Y. and Rathje, E.M. (2019). "Insights into modeling small-strain site response derived from downhole array data," *Journal of Geotechnical and Geoenvironmental Engineering*, Vol. 145, No. 7, 15 pp. https://doi.org/10.1061/(ASCE)GT.1943-5606.0002048

Tawfiq, K.S., Aggour, M.S., and Al-Sanad, H.A. (1988). "Dynamic properties of cohesive soils from impulse testing," *Proceedings, 9th World Conference on Earthquake Engineering, Tokyo*, Vol. 3, pp. 11–16.

Taylor, M.L., Cubrinovski, M., and Haycock, I. (2012). "Application of new 'Gel-push' sampling procedure to obtain high quality laboratory test data for advanced geotechnical analyses," *Proceedings, New Zealand Society for Earthquake Engineering Conference, Paper* No. 123, 8 pp.

Thevanayagam, S. (1998). "Effects of fines and confining stress on undrained shear strength of silty sands," *Journal of Geotechnical and Geoenvironmental Engineering*, Vol. 124, No. 6, pp. 479–491.

Thiers, G.R. and Seed, H.B. (1978). "Strength and stress-strain characteristics of clays subjected to seismic loading conditions," *Vibration Effects of Earthquakes on Soils and Foundations*, Special Technical Publication 450, ASTM, pp. 3–56.

Thompson, E.M., Baise, L.G., Tanaka, Y., and Kayen R.E. (2012). "A taxonomy of site response complexity," *Soil Dynamics and Earthquake Engineering*, Vol. 41, pp. 32–43. https://doi.org/10.1016/j.soildyn.2012.04.005

Thompson, E.M., Wald, D.J., and Worden, C.B. (2014). "A V_{S30} map for California with geologic and topographic constraints," *Bulletin of the Seismological Society of America*, Vol. 104, pp. 2313–2321.

Thomson, P.R. and Wong, R.C.K. (2008). "Specimen nonuniformities in water-pluviated and moist-tamped sands under undrained triaxial compression and extension," *Canadian Geotechnical Journal*, Vol. 45, pp. 939–956.

Thomson, W.T. (1950). "Transmission of elastic waves through a stratified solid," *Journal of Applied Physics*, Vol. 21, pp. 89–93.

Tileylioglu, S., Stewart, J.P., and Nigbor, R.L. (2011). "Dynamic stiffness and damping of a shallow foundation from forced vibration of a field test structure," *Journal of Geotechnical and Geoenvironmental Engineering*, Vol. 137, No. 4, pp. 344–353.

Timoshenko, S.P. and Goodier, J.N. (1951). *Theory of Elasticity*, McGraw-Hill Book Company, New York, 506 pp.

Tokimatsu, K. (1995). "Geotechnical site characterization using surface waves," *Proceedings, 1st International Conference on Earthquake Geotechnical Engineering*, Balkema, Brookfield, VA, pp. 1333–1368.

Tokimatsu, K. and Nakamura, K. (1986). "A liquefaction test without membrane penetration effects," *Soils and Foundations*, Vol. 26, No. 4, pp. 127–138.

Tokimatsu, K. and Seed, H.B. (1987). "Evaluation of settlements in sand due to earthquake shaking," *Journal of Geotechnical Engineering*, Vol. 113, No. 8, pp. 861–878.

Tokimatsu, K. and Sekiguchi, T. (2007). "Effects of dynamic properties of peat on strong ground motions during 2004 mid Niigata prefecture earthquake," *Proceedings, 4th International Conference on Earthquake Geotechnical Engineering*, Thessaloniki, Paper No. 1531.

Tokimatsu, K., Tamura, S., and Kojima, S. (1992). "Effects of multiple modes on Rayleigh wave dispersion," *Journal of Geotechnical Engineering*, Vol. 118, No. 10, pp. 1529–1543.

Toro, G.R. (1995). "Probabilistic models of the site velocity profiles for generic and site-specific ground-motion amplification studies," *Technical Report No. 779574*, Brookhaven National Laboratory, Upton, NY.

Toro, G.R. (2022). "Uncertainty in shear-wave velocity profiles," *Journal of Seismology*, Vol. 26, pp. 713–730.

Towhata, I., Kawano, Y., Yonai, Y., and Koelsh, F. (2004). "Laboratory tests on dynamic properties of municipal wastes," *Proceedings, 11th Conference on Soil Dynamics and Earthquake Engineering and 3rd International Conference on Earthquake Geotechnical Engineering*, Vol. 1, pp. 688–693.

Towhata, I. and Uno, M. (2008). "Cyclic shear tests of municipal waste in large triaxial device for identification of its dynamic properties." Geotechnical Special Publication 181, edited by D. Zeng, M. T. Manzari, and D. R. Hiltunen, ASCE, pp. 1–10.

Trudeau, P.J., Whitman, R.V., and Christian, J.T. (1974). Shear wave velocity and modulus of a marine clay, *Journal of the Boston Society of Civil Engineers*, Vol. 61, No. 1, pp. 12–25.

Tsai, C.C. and Hashash, Y.M.A. (2009). Learning of dynamic soil behavior from downhole arrays, ASCE, *Journal of Geotechnical Engineering*, Vol. 135, No. 6, pp. 745–757.

Tsukamoto, Y., Ishihara, K., Nakazawa, H., Kamada, K., and Huang, Y. (2001). "Resistance of partly saturated sand to liquefaction with reference to longitudinal and shear wave velocities," *Soils and Foundations*, Vol. 42, No. 6, pp. 93–104.

Tsukamoto, Y., Ishihara, K., & Sawada, S. (2004). "Settlement of silty sand deposits following liquefaction during earthquakes," *Soils and Foundations*, Vol. 44, No. 5, 135–148.

U.S. Army Corps of Engineers (1980). *Laboratory Soils Testing*, EM 1110-2-1906, Office of the Chief of Engineers, Washington, D.C., 407 pp.

U.S. Army Corps of Engineers (1995). *Geophysical Exploration for Engineering and Environmental Investigations*, EM 1110-1-1802, Office of the Chief of Engineers, Washington, D.C., 208 pp.

Ue, S., Yasuhara, K., and Fujiwara, H. (1991). "Influence of consolidation period on undrained strength of clays." *Ground and Construction*, Vol. 9, No. 1, pp. 51–62 (in Japanese).

Vaid, Y.P. and Campanella, R.G. (1977). "Time-dependent behavior of undisturbed clay." *Journal of the Geotechnical Engineering Division*, Vol. 103, No. 7, pp. 693–709.

Vaid, Y.P. and Negussey, D. (1984). "Relative density of air and water pluviated sand," *Soils and Foundations*, Vol. 24, 101–105.

Vantassel, J.P. and Cox, B.R. (2021). "SWinvert: A Workflow for Performing Rigorous Surface Wave Inversions," *Geophysical Journal International*, Vol. 224, No. 2, pp. 1141–1156.

Vantassel, J.P. and Cox, B.R. (2022). "SWprocess: A workflow for developing robust estimates of surface wave dispersion uncertainty," *Journal of Seismology*, Vol. 26, pp. 731–756.

Vreugdenhil, R., Davis, R., and Berrill, J. (1994). "Interpretation of cone penetration results in multilayered soils," *International Journal for Numerical Methods in Geomechanics*, Vol. 18, pp. 585–599.

Vucetic, M. (1990). "Normalized behavior of clay under irregular cyclic loading," *Canadian Geotechnical Journal*, Vol. 27, No. 1, pp. 29–46.

Vucetic, M. (1994). "Cyclic threshold shear strains in soils." *Journal of Geotechnical Engineering.*, Vol. 120, No. 12, pp. 2208–2228.

Vucetic, M. and Dobry, R. (1989). "Degradation of marine clays under cyclic loading," *Journal of Geotechnical Engineering*, Vol. 114, No. 2, pp. 133–149.

Vucetic, M. and Dobry, R. (1991). "Effect of soil plasticity on cyclic response," *Journal of Geotechnical Engineering*, Vol. 117, No. 1, pp. 89–107.

Wald, D.J. and Allen, T.I. (2007). "Topographic slope as a proxy for seismic site conditions and amplification," *Bulletin of the Seismological Society of America*, Vol. 97, No. 5, pp. 1379–1395.

Wehling, T.M., Boulanger, R.W., Arulnathan, R., Harder, L.F., Jr., and Driller, M.W. (2003). "Nonlinear dynamic properties of a fibrous organic soil." *Journal of Geotechnical and Geoenvironmental Engineering*, Vol. 129, No. 10, pp. 929–939.

Weiler, W.A. (1988). "Small strain shear modulus of clay," *Proceedings, ASCE Conference on Earthquake Engineering and Soil Dynamics II - Recent Advances in Ground Motion Evaluation*, Geotechnical Special Publication 20, pp. 331–335.

Whang, D.H., Moyneur, M.S., Duku, P., and Stewart, J.P. (2005). "Seismic compression behavior of non-plastic silty sands," *Proceedings, International Symposium on Advanced Experimental Unsaturated Soil Mechanics*, A. Tarantino, E. Romero, and Y. J. Cui, eds., Trento, Italy, June 27–29, A.A. Balkema Publishers, pp. 257–263.

Whang, D.H., Stewart, J.P., and Bray, J.D. (2004). "Effect of compaction conditions on the seismic compression of compacted fill soils," *Geotechnical Testing Journal*, ASTM, Vol. 27, No. 4, pp. 371–379.

White, J.E. (1965). *Seismic Waves: Radiation, Transmission, and Attenuation*, McGraw-Hill Book Company, New York.

Wieland, M. (2004). "Benefits of strong motion instrumentation of large dams," in M. Wieland, Q. Ren, and J. S.Y. Tan, eds., *New Developments in Dam Engineering*, A.A. Balkema, London, pp. 101–107.

Wills, C.G., Gutierrez, C.I., Perez, F.G., and Branum, D.M. (2015). "A next generation V_{S30} map for California based on geology and topography," *Bulletin of the Seismological Society of America.*, Vol. 105, pp. 3083–3091.

Wills, C.J. and Clahan, K.B. (2006). "Developing a map of geologically defined site-condition categories for California," *Bulletin of the Seismological Society of America*, Vol. 96, pp. 1483–1501.

Wong, R.K.S. and Arthur, J.R.F. (1986). "Sand shear by stresses with cyclic variations in direction," *Geotechnique*, Vol. 36, No. 2, pp. 215–226. https://doi.org/10.1680/geot.1986.36.2.215

Wood, D.M. (2004). *Geotechnical Modeling*, Spon Press, London, 504 pp.

Woods, R.D. (1991). "Field and laboratory determination of soil properties at low and high strains," *Proceedings, Second International Conference on Recent Advances in Geotechnical Earthquake Engineering and Soil Dynamics, St. Louis*, Vol. 3, pp. 1727–1741.

Wride (Fear), C.E., Hofmann, B.A., Sego, D.C., Plewes, H.D., Konrad, J.-M., Biggar, K.W., Robertson, P.K., and Monahan, P.A. (2000). "Ground sampling program at the CANLEX test sites," *Canadian Geotechnical Journal*, Vol. 37, No. 3, pp. 530–542.

Wroth, C.P. (1984). "Interpretation of insitu soil test," *Geotechnique*, Vol. 34, pp. 449–489.

Wroth, C.P. and Houlsby, G.T. (1985). "Soil mechanics - Property characterization and analysis procedures," *Proceedings, 11th International Conference on Soil Mechanics and Foundation Engineering*, San Francisco, Vol. 1, pp. 1–55.

Xia, J., Miller, R.D., Park, C.B., Hunter, J.A., Harris, J.B., and Ivanov, J. (2002). "Comparing shear-wave velocity profiles inverted from multichannel surface wave with borehole measurements," *Soil Dynamics and Earthquake Engineering*, Vol. 22, 181–190.

Yasuda, N. and Matsumoto, N. (1993). "Dynamic deformation characteristics of sands and rockfill materials," *Canadian Geotechnical Journal*, Vol. 30, No. 5, pp. 747–757. https://doi.org/10.1139/t93-067

Yasuhara, K. (1994). "Post-cyclic undrained strength for cohesive soils," *Journal of Geotechnical Engineering*, Vol. 120, No. 11, 1961–1979.

Yee, E., Duku, P.M., and Stewart, J.P. (2014). "Cyclic volumetric strain behavior of sands with fines of low plasticity," *Journal of Geotechnical and Geoenvironmental Engineering*, Vol. 140, No. 4, pp. 1073–1085.

Yee, E., Stewart, J.P., and Tokimatsu, K. (2013). "Elastic and large-strain nonlinear seismic site response from analysis of vertical array recordings," *Journal of Geotechnical and Geoenvironmental Engineering*, Vol. 130, No. 10, pp. 1789–1801.

Yimsiri, S. and Soga, K. (2002). "A review of local strain measurement systems for triaxial testing of soils," *Geotechnical Engineering*, Vol. 33, pp. 42–52.

Yniesta, S. and Brandenberg, S.J. (2017). "Stress-ratio-based interpretation of modulus reduction and damping curves." *Journal of Geotechnical and Geoenvironmental Engineering,* Vol 143, 06016021, 4 pp.

Yong, A. (2016). "Comparison of measured and proxy-based V_{S30} values in California," *Earthquake Spectra*, Vol. 32, No. 1, pp. 171–192.

Yong, A., Hough, S.E., Iwahashi, J., and Braverman, A. (2012). "A terrain-based site-conditions map of California with implications for the contiguous United States," *Bulletin of the Seismological Society of America*, Vol. 102, no. 1, pp. 114–128.

Yoshimi, Y. (1994). "Relationship among liquefaction resistance, SPT N-value and relative density for undisturbed samples of sands," *Tsuchi-to-Kiso*, Japan Society for Soil Mechanics and Foundation Engineering, Ser. No. 435, Vol. 42, No. 4, pp. 63–67.

Yoshimi, Y., Hatanaka, M., and Oh-Oka, H. (1978). "Undisturbed sampling of saturated sands by freezing," *Soils and Foundations*, Vol. 18, pp. 59–73. https://doi.org/10.3208/sandf1972.18.3_59

Yoshimi, Y., Tokimatsu, K., and Hasaka, Y. (1989). "Evaluation of liquefaction resistance of clean sands based on high-quality undisturbed samples," *Soils and Foundations*, Vol. 29, No. 1, pp. 93–104.

Yoshimine, M. and Ishihara, K. (1998). "Flow potential of sand during liquefaction," *Soils and Foundations*, Vol. 38, No. 3, pp. 189–198.

Youd, T.L. (1972). "Compaction of sands by repeated shear straining," *Journal of the Soil Mechanics and Foundations Division*, Vol. 98, No. SM7, pp. 709–725.

Youd, T.L., Idriss, I.M., Andrus, R.D., Arango, I., Castro, G., Christian, J.T., Dobry, R, Finn, W.D.L., Harder, L.F., Jr., Hynes, M.E., Ishihara, K., Koester, J.P., Liao, S.S.C., Marcuson, W.F., ill, Martin, G.R, Mitchell, J.K., Moriwaki, Y., Power, M.S., Robertson, P.K, Seed, R.B., and Stokoe, K.H., II. (2001). "Liquefaction resistance of soils: Summary report from the 1996 NCEER and 1998 NCEER/NSF workshops on evaluation of liquefaction resistance of soils," *Journal of Geotechnical and Geoenvironmental Engineering*, Vol. 127, No. 10, pp. 817–833.

Yust, M.B., Cox, B.R., and Cheng, T. (2018). "Epistemic uncertainty in Vs Profiles and V_{S30} values derived from joint consideration of surface wave and H/V data at the FW07 TexNet station," *Proceedings, Geotechnical Earthquake Engineering and Soil Dynamics V*, Austin, Texas, 10–13 June 2018.

Zalachoris, G. and Rathje, E.M. (2015). "Evaluation of one-dimensional site response techniques using borehole arrays," *Journal of Geotechnical and Geoenvironmental Engineering*, Vol. 141, No. 12, 15 pp. https://doi.org/10.1061/(asce)gt.1943-5606.0001366

Zeghal, M. and Elgamal, A.-W. (1994). "Analysis of site liquefaction using earthquake records," *Journal of Geotechnical Engineering*, Vol. 120, No. 6, pp. 996–1017. https://doi.org/10.1061/(asce)0733-9410(1994)120#: 6(996)

Zeghal, M., Elgamal, A.W., Tang, H.T., and Stepp, J.C. (1995). "Lotung downhole array. II: Evaluation of soil nonlinear properties," *Journal of Geotechnical Engineering*, Vol. 121, No. 4, pp. 363–378. https://doi.org/10.1061/(asce)0733-9410(1995)121#: 4(363)

Zekkos, D., Bray, J.D., and Riemer, M.F. (2008a). "Shear modulus reduction and material damping relationships for municipal solid waste based on large-scale cyclic triaxial testing," *Canadian Geotechnical Journal*, Vol. 45, No. 1, pp. 45–58.

Zekkos, D., Matasovic, N., El-Sherbiny, R., Athanasopoulos-Zekkos, A., Towhata, I., and Maugeri, M. (2008b). "Chapter 4: Dynamic properties of municipal solid waste," *Geotechnical Characterization, Field Measurement, and Laboratory Testing of Municipal Solid Waste*, ed. D. Zekkos, New Orleans, LA, ASCE, pp. 112–134.

Zen, K., Umehara, Y., and Hamada, K. (1978). "Laboratory tests and in-situ seismic survey on vibratory shear modulus of clayey soils with different plasticities," *Proceedings, Fifth Japan Earthquake Engineering Symposium, Tokyo*, pp. 721–728.

Zergoun, M. and Vaid, Y.P. (1994). "Effective stress response of clay to undrained cyclic loading," *Canadian Geotechnical Journal*, Vol. 31, pp. 714–727.

Zhang, J., Andrus, R.D., and Juang, C.H. (2008). "Model uncertainty in normalized shear modulus and damping relationships," *Journal of Geotechnical and Geoenvironmental Engineering*, Vol. 134, No. 1, pp. 24–36.

7 Site Effects and Ground Response Analysis

7.1 INTRODUCTION

As discussed in Chapter 3, earthquake ground motions are affected by source, path, and site effects. *Site effects* refer to the alteration of ground motions by geologic materials encountered by seismic waves as they approach the ground surface. There is no specific depth range for site effects, but they are commonly considered to exist within the upper several hundred meters for engineering purposes– usually a very small fraction of the total source-to-site distance. Site effects result from a number of different physical processes that can be categorized as local ground response, basin effects, and topographic effects. Local ground response refers to the effects of relatively shallow ground (upper 100m or so) on nearly vertically propagating body waves. *Basin effects* refer to the refraction of upward-traveling body waves that enter two- and three-dimensional sedimentary basins, and the generation of surface waves within those basins. *Topographic effects* are associated with the response of two- and three-dimensional topographic irregularities such as hills, valleys, ridges, bluffs, canyons, and slopes.

For many years, the importance of site effects was not widely recognized. In fact, provisions specifically accounting for site effects did not appear in building codes until the 1970s. This chapter presents cases that provide both examples of and evidence for site effects, methods for empirical and analytical prediction of site effects, and procedures for incorporating site effects into ground motion hazard analyses.

The prediction of site effects on the characteristics of earthquake ground motion is an important part of geotechnical earthquake engineering practice. Two primary approaches are available for prediction of site effects – a generic approach and a site-specific approach. The two approaches should be viewed as complementary, each having characteristics that can make it the more appropriate choice under certain conditions, but also as capable of being used beneficially together to strengthen an engineer's confidence in predictions of site effects.

7.2 EVIDENCE OF SITE EFFECTS

Site effects can influence the amplitude, frequency content, and duration of earthquake ground motions. As such, they influence the response of structures and constructed facilities, and their potential to be damaged in a particular earthquake. To the extent that site conditions change over a region affected by an earthquake, damage patterns can also change. Correlation of damage to site conditions was one of the earliest indicators of the influence of site effects – before strong motion instrumentation was available, it provided the only evidence of site effects. Even now, however, damage patterns are frequently observed to be related to site conditions. As strong motion instrumentation became available in the 1960s and 1970s, comparisons of recordings at stations located near each other but underlain by different soil/rock conditions provided further evidence of the effects of local site conditions on site response.

7.2.1 Correlation of Damage to Site Conditions

The first evidence of local site effects was based on observations of damage patterns following strong earthquakes. MacMurdo (1824) noted that "buildings situated on rock were not by any means

DOI: 10.1201/9781003512011-7

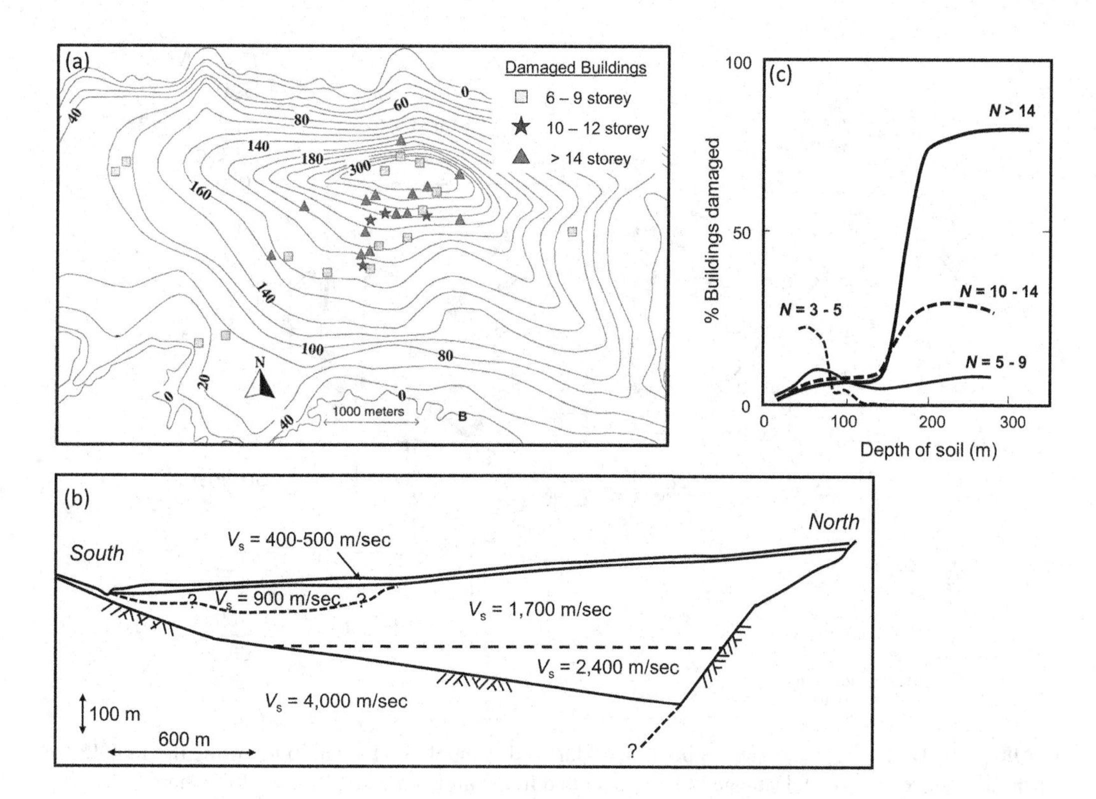

FIGURE 7.1 1967 Caracas, Venezuela earthquake: (a) Contours of depth to bedrock with locations of damaged buildings, (b) north-south cross-section through Caracas, and (c) fractions of *N*-story buildings damaged as functions of depth of underlying soil. (After Seed et al., 1972, with permission of ASCE.)

so much affected … as those whose foundations did not reach to the bottom of the soil" in the 1819 earthquake in Cutch, India. In his report on the 1857 Neapolitan earthquake, Mallet (1862) noted the effect of local geologic conditions on damage.

Much later, Seed et al. (1972) studied damage patterns in Caracas, Venezuela following the 1967 Caracas earthquake (M_w=6.4). Although the magnitude was not particularly high and the epicenter was located about 56 km from Caracas, several tall buildings partially or completely collapsed, killing more than 200 people. The city of Caracas is located in a valley filled with alluvial sediments. In the Palos Grande area, these sediments extend to depths of up to about 300 m. Examination of damage patterns following the earthquake showed that damage rates for buildings of different heights (and, therefore, different fundamental periods) were strongly related to the thickness of the alluvium beneath them (Figure 7.1). Shorter buildings (3–5 stories) had high damage rates when located in areas underlain by relatively low thicknesses of soil, and essentially no damage when located in areas underlain by greater thicknesses of soil. On the other hand, tall buildings (more than 14 stories) had low damage rates at locations underlain by thin soil layers, and extremely high damage rates in areas underlain by very thick soil deposits.

In the 1989 Loma Prieta earthquake (M_w=6.9), damage to most of San Francisco, located nearly 100 km north of the epicenter, was negligible. However, substantial damage was observed in some areas of the city, particularly areas that had been reclaimed from tideflats and marshes of San Francisco Bay. The Marina district on the north side of San Francisco, which suffered damage in the 1906 San Francisco earthquake (M_w=7.9), also suffered far more severe damage than the immediately surrounding area 83 years later in the Loma Prieta earthquake. The shoreline of the Marina district in 1857 is shown with a dashed line in Figure 7.2. To aid the construction of industrial facilities, a rock seawall was constructed in the 1890s and backfilled with dune sand, producing the

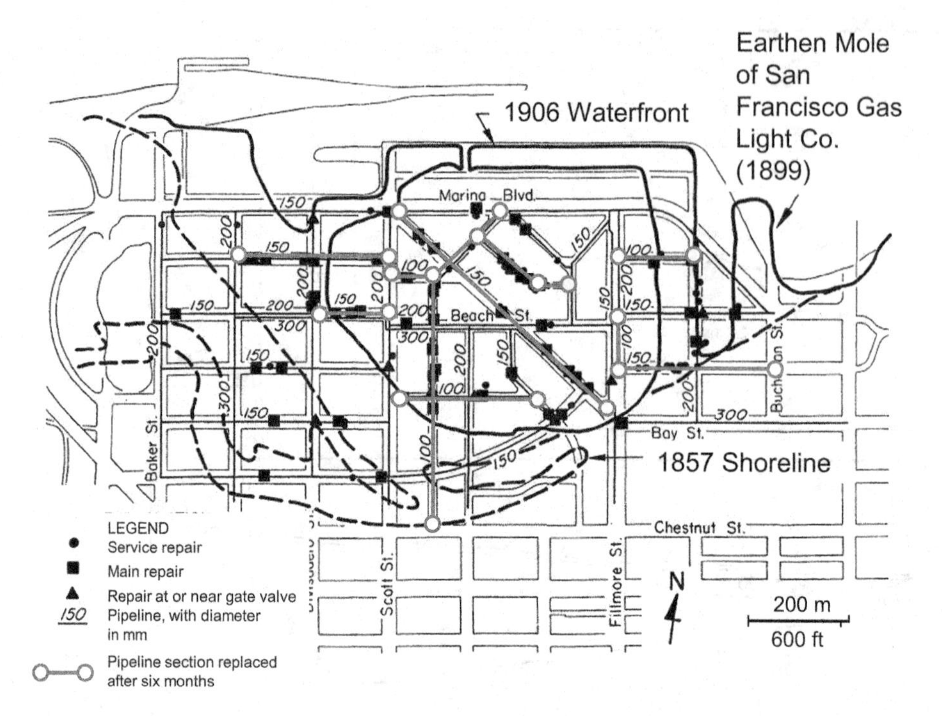

FIGURE 7.2 Locations of damaged pipelines in Marina district of San Francisco following the 1989 Loma Prieta earthquake (M_w=6.9). Damage was concentrated in reclaimed area beyond the 1857 shoreline. (After O'Rourke et al., 1991.)

shoreline marked with a solid line; this shoreline, which produced a small tidal lagoon, existed at the time of the 1906 earthquake. Although the area was not extensively developed at the time of that earthquake, timber structures were noted to have tilted and settlement to have caused a rupture of a 600-mm gas line. In 1912, the lagoon was filled with dredged sandy soils pumped from offshore deposits. The 1989 Loma Prieta earthquake also caused extensive damage in the Marina district. Site response from soft soils, including some artificial fills that liquefied, amplified ground motions and caused collapses and partial collapses to residential structures, many of which were constructed with soft first floors. Figure 7.2 shows the locations of repairs (to sheared connections, cracked mains, etc.) made to municipal water supply system facilities in the first two weeks following the earthquake, and of pipelines that were ultimately replaced about six months later. The damage can be seen to be essentially confined to the region outside the 1857 shoreline and concentrated within the former lagoon filled in 1912. Broken gas lines caused fires in the Marina district (Figure 1.20), but the damage to the water supply system left the fire trucks unable to fight them. A fire boat was called to the scene and pumped salt water from the Marina harbor area, which allowed the fire to be brought under control.

7.2.2 Comparison of Measured Surface Motions

More direct and quantitative evidence of the importance of local site conditions came with the availability of data from strong motion instruments. For example, recordings of ground motion at several locations in San Francisco were made during a nearby M=5.3 earthquake in 1957. Variations in ground motion, expressed in terms of peak horizontal acceleration and response spectra, are shown along with variations in soil conditions along a 4-mile section through the city in Figure 7.3. Ground surface motions at the rock outcrops (Market and Guerrero, Mason and Pine, Harrison, and Main) were quite similar, but the amplitude and frequency content of the motions at

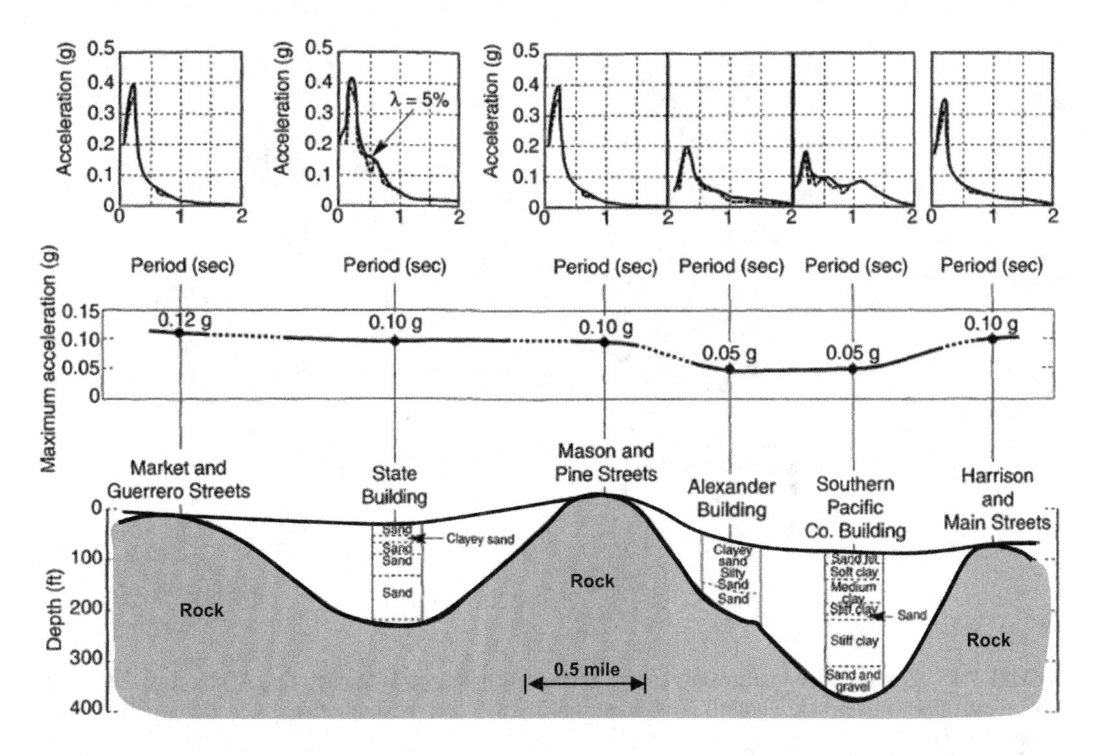

FIGURE 7.3 Variation of spectral velocity, spectral acceleration, and peak horizontal acceleration along a 4-mile section through San Francisco in the 1957 San Francisco earthquake. (After Idriss and Seed, 1968. Used by permission of the Seismological Society of America.)

sites underlain by thick soil deposits were markedly different. Examination of the peak acceleration plot shows values ranging from 0.10 to 0.12*g* at the rock sites and at the State Building, which was underlain by sandy soils. The peak accelerations at the Alexander and Southern Pacific Company buildings, which were underlain by clayey soils, were only about half as strong, at 0.05*g*. The response spectra were also significantly different. The 0.2-sec spectral accelerations at the rock sites and the State Building ranged from 0.35 to 0.40*g*, but were less than 0.2*g* at the Alexander Building and Southern Pacific Company Building sites. On the other hand, the 1.0-sec spectral acceleration at the Southern Pacific Company Building site was 3–4 times higher than at any of the rock sites, and the 0.5-sec spectral acceleration at the State Building site was twice as high as at any of the rock sites. These records show significant and consistent trends in spectral amplitude and shape for the different site conditions.

The September 19, 1985 Michoacan (M_s=8.1) earthquake caused only moderate damage in the vicinity of its epicenter (near the Pacific coast of Mexico) but caused extensive damage some 350 km away in Mexico City. Studies of ground motions recorded at different sites in Mexico City illustrated the significant relationship between local soil conditions and damaging ground motions and led to important advances in understanding the cyclic response of plastic clays (e.g., Dobry and Vucetic, 1987).

For seismic zonation purposes, Mexico City is often divided into three zones with different subsurface conditions (Figure 7.4). Shallow, compact deposits of mostly granular soil, basalt, or volcanic tuff are found in the Foothill Zone, located west of downtown. In the Lake Zone, thick deposits of very soft soils formed from the pluviation of airborne silt, clay, and ash from nearby volcanoes through the waters of ancient Lake Texcoco extend to considerable depths, as shown by the contours of Figure 7.4b. These soft soils generally consist of two soft clay (Mexico City Clay) layers separated by a 0–6-m-thick (0–20 ft) compact sandy layer called the capa dura.

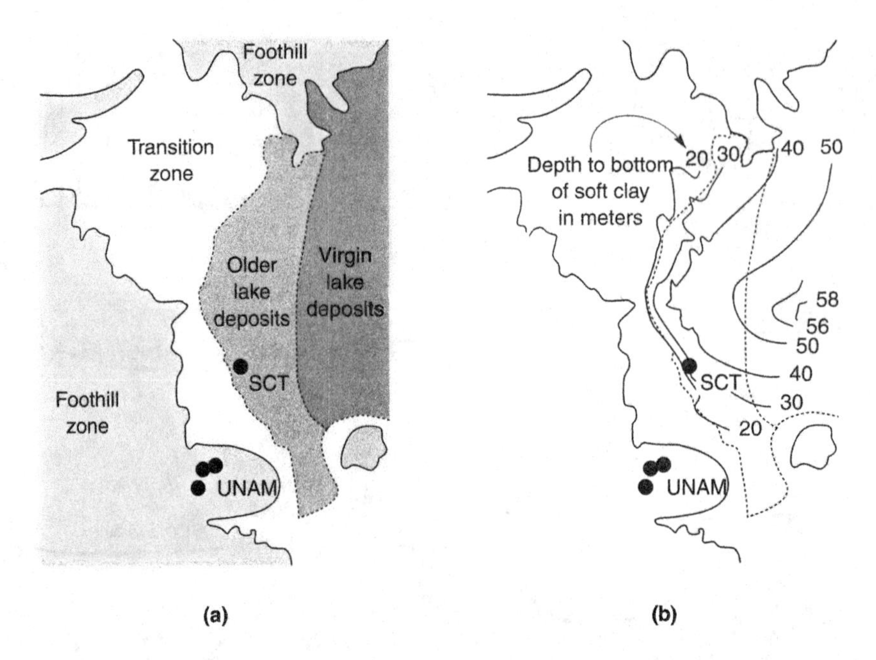

FIGURE 7.4 Strong motion instruments and geotechnical conditions in Mexico City: (a) locations of strong motion instruments relative to Foothill, Transition, and Lake Zones; (b) contours of soft soil thickness. (After Stone et al., 1987.)

Groundwater is generally found at a depth of about 2 m over most of the Lake Zone. Between the Foothill and Lake Zones lies the Transition Zone, where the soft soil deposits are thin and interspersed erratically with alluvial deposits.

Prior to 1985, a number of strong motion instruments had been deployed in Mexico City. Shown in Figure 7.4 are the locations of those at the Universidad Nacional Autonomas de Mexico (UNAM) and the Secretary of Communications and Transportation (SCT) site. The UNAM site was located in the Foothill Zone on 3–5 m (10–16 ft) of basaltic rock underlain by softer strata of unknown thickness. The SCT site was located on the soft soils of the Lake Zone.

Although the Michoacan earthquake was quite large, its great distance from Mexico City produced accelerations at the UNAM (rock) site of only 0.03g–0.04g (Figure 7.5). In the Transition Zone, peak accelerations were slightly greater than those at UNAM but still quite low. In the Lake Zone, however, the peak acceleration at the SCT site was about five times greater than at UNAM. The frequency content of the SCT motion was also much different than that of the UNAM motion; the predominant period was about 2 sec at SCT. Strong levels of shaking persisted over a very long duration at the SCT site. The response spectra shown in Figure 7.6 illustrate the pronounced effects of the Lake Zone soils: at periods of approximately 2 sec, spectral accelerations at the SCT site were about eight times greater than those at the UNAM site.

Structural damage in Mexico City was highly selective; large parts of the city experienced no damage while other areas suffered pronounced damage. Damage was negligible in the Foothill Zone and minimal in the Transition Zone. The greatest damage occurred in those portions of the Lake Zone underlain by 38–50 m (125–164 ft) of soft soil (Stone et al., 1987), where the characteristic site periods were estimated at 1.9–2.8 sec. Even within this area, damage to buildings of less than five stories and modern buildings greater than 30 stories was slight. Most buildings in the five- to 20-story range, however, either collapsed or were badly damaged. Using the crude rule of thumb that the fundamental period of an N-story building is approximately $N/10$ sec, most of the damaged buildings had fundamental periods equal to or somewhat less than the characteristic site period.

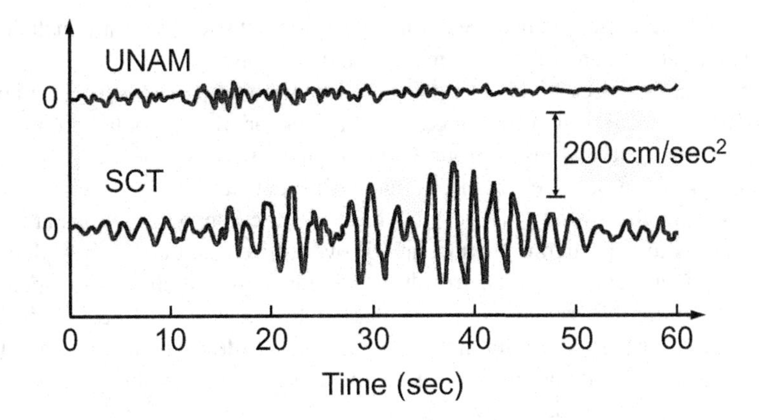

FIGURE 7.5 Time histories of acceleration recorded by strong motion instruments at UNAM and SCT sites. (After Stone et al., 1987).

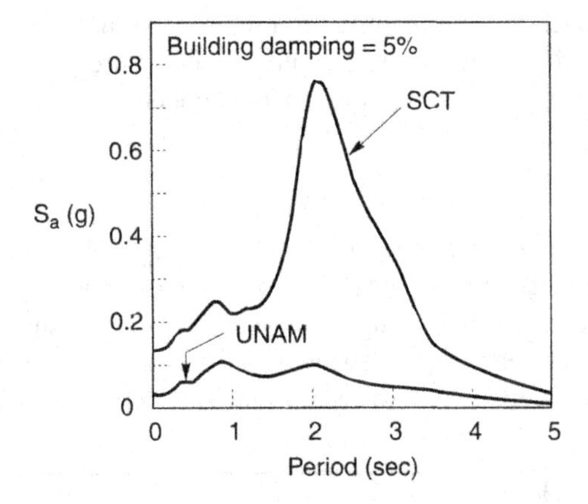

FIGURE 7.6 Response spectra computed from recorded motions at UNAM and SCT sites. (After Romo and Seed, 1986).

Accounting for the period-lengthening effect of soil-structure interaction (Section 8.3.1) and the tendency for the fundamental period of a structure to increase during a strong earthquake (due to the reduction in stiffness caused by cumulative architectural and structural damage), it seems likely that the damaged structures were subjected to many cycles of large dynamic forces at periods near their fundamental periods. This "double-resonance" condition (amplification of bedrock motion by the soil deposit and amplification of the soil motion by the structure) combined with structural design and construction deficiencies to cause locally devastating damage.

7.3 GENERIC PREDICTION OF SITE EFFECTS

Empirical procedures for predicting site effects, which are based on the analysis of recorded ground motions, did not exist until the strong motion instruments that became available in the 1960s and 1970s generated sufficient data. As more and more ground motion recordings became available, empirical procedures became more advanced. The first ground motion prediction equations (Section 3.5) predicted ground motions only for rock sites, so they required separate adjustments to account for other site conditions. Subsequent procedures predicted amplification behavior initially for a few broad categories of site conditions but later for more detailed categories or for continuous, quantitative measures of site conditions. More recently, factors accounting for amplification behavior have been folded into ground motion models (Section 3.5.2.3) so that site effects are included with magnitude, distance, and other factors that influence ground motions.

The regression analyses used to develop ground motion models (GMMs) require large amounts of ground motion data to cover the wide range of conditions for which ground motion estimates are needed. This data exists in the types of ground motion databases described in Section 3.5.4. These databases contain ground motion records from many different sites in many different earthquakes. Because very few individual strong motion sites have recorded enough motions at different magnitudes and distances to fully define their behavior over all relevant conditions, data from many sites and many events are combined. The resulting relationships, therefore, correspond best to the average conditions within the ground motion database, which are generally different than the conditions at a specific site where a ground motion estimate is needed. This approach, which is more formally known as an *ergodic* approach, predicts the response of a generic site within some site category. What is usually desired, however, is the response of a *specific* site whose characteristics (and, therefore, response) may differ systematically from the average within that site category. That objective can be achieved by means of a site-specific analysis, an approach more formally referred to as *non-ergodic*. Procedures for site-specific analyses are described in Section 7.4.

Regardless of the approach taken, it is important for geotechnical earthquake engineers to have a solid, intuitive understanding of how different sites affect different aspects of ground motions. A brief review of generic site effects helps provide such an understanding.

7.3.1 Historical

Some 40 years ago (Seed et al., 1976a), compared *PGA* values recorded at sites underlain by different types of soil profiles and found distinct trends in amplification behavior. Although data were limited and scattered, overall trends (Figure 7.7) suggested that *PGA*s at the surfaces of soil deposits were slightly greater than on rock when rock *PGA*s were small and somewhat lower at higher acceleration levels. Later, data from Mexico City and the San Francisco Bay area, and from additional ground

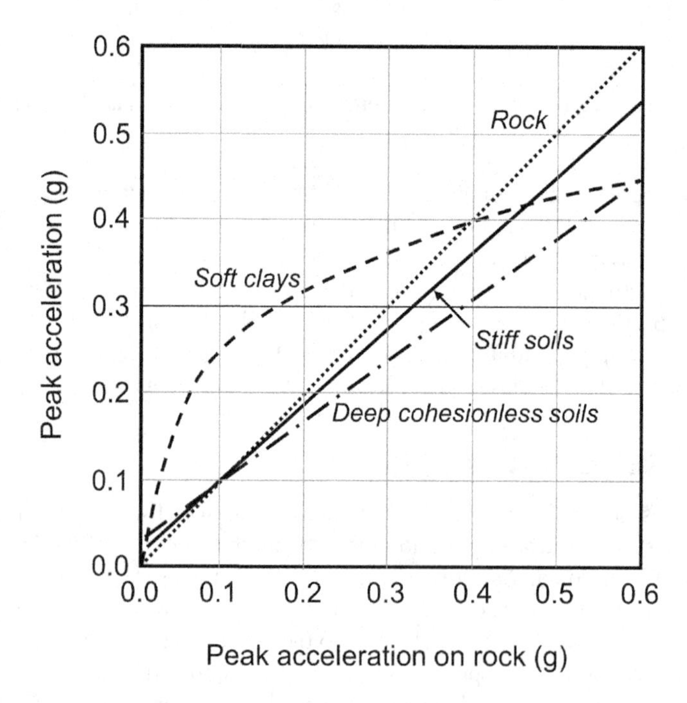

FIGURE 7.7 Early comparison of peak ground surface acceleration values for different site categories. (After Seed et al., 1976a and Idriss, 1990.)

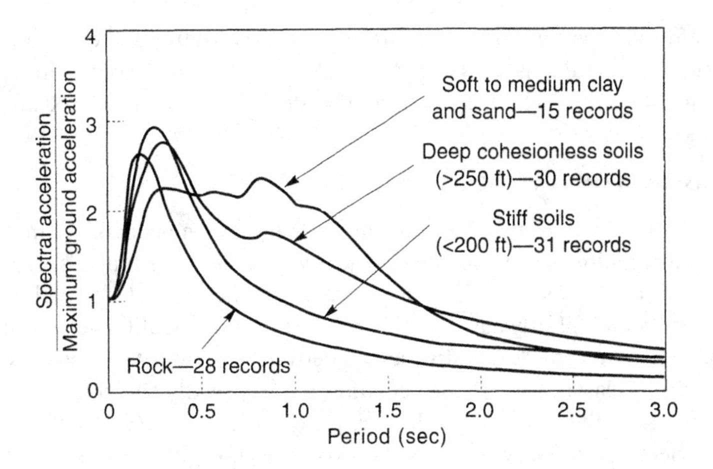

FIGURE 7.8 Normalized response spectra for four geotechnical site categories (Seed et al., 1976b).

response analyses (Idriss, 1990) led to the addition of a curve for soft clay. The curves suggested that *PGA*s on all of the soil sites were higher than on rock and stiffer soils for relatively low levels of shaking by amounts that were greater for softer profiles than for stiffer ones. They also indicated that the *PGA*s on soil dropped below that on rock at higher levels of shaking, and by amounts that were different for the different site categories. As discussed in Section 3.5.2.3, the reduction of amplification of *PGA* at stronger levels of shaking is associated with nonlinear soil behavior.

Local site conditions also influence the frequency content of surface motions and hence the nature of their response spectra. Seed et al. (1976b) also found that response spectra appeared to be different for the different site categories. After normalizing the computed spectra (i.e., dividing spectral accelerations by the *PGA*) the shapes of the normalized spectra (Figure 7.8) were found to be quite different. The effects are most apparent at periods above about 0.5 sec, where normalized spectra were much higher for soil sites than for rock sites. At longer periods, the normalized spectral accelerations increased with decreasing subsurface profile stiffness. Figure 7.8 clearly shows that deep and soft soil deposits transmit greater proportions of long-period (low-frequency) motion to the ground surface. The results showed that the use of a single response spectrum shape for all site conditions was not appropriate, a finding that strongly influenced the development of building codes and standards in the 1970s–1980s.

7.3.2 Amplification Factors

Site effects are now quantified by comparing the value of an intensity measure from a ground surface motion with the value of the same intensity measure from a motion corresponding to a reference site condition. If a particular intensity measure from the reference motion is denoted as IM_{ref} and the value from the ground surface motion is denoted IM_s, an *amplification factor* can be defined as

$$AF = \frac{IM_s}{IM_{ref}} \tag{7.1}$$

Reference motions are generally taken as recorded rock motions (Boatwright et al., 1991), most commonly "soft rock" such as that commonly associated with the western U.S., having a typical shear wave velocity of 760 m/s. Amplification factors can be computed this way when nearby instruments have recorded rock motions.

In other cases (e.g., Field and Jacob, 1995; Sokolov, 1997; Lee and Anderson, 2000; Steidl, 2000; Stewart et al., 2003), median rock spectra from GMMs are used for reference motion values; this approach, often referred to as a *non-reference site* approach, is useful for the common case in which no nearby rock motion recording is available. An amplification factor based on this approach has the advantage of being compatible with the GMM used to develop the reference rock motion.

7.3.3 Site Classification Approaches

Whether expressed directly in terms of amplification factors or embedded within GMMs, the prediction of site effects requires some consistent, objective way of describing a site. *Site classes* have frequently been used to place soil profiles in groups that behave similarly (within the group) but distinctly (from one another). Early classification procedures used qualitative descriptions based on geologic or geotechnical information to distinguish between different sites and provided amplification factors for each site class. Currently, site conditions are typically characterized using a measure of the general stiffness of the site, such as the average (harmonic mean) shear wave velocity in the upper 30m, V_{S30} (Sections 3.4.3.1 and 6.6.2). Geologic or geotechnical classification systems are now typically correlated to V_{S30}, which is then used to compute the amplification factor; the use of such classification systems is generally restricted to areas where V_S measurements are not available.

7.3.3.1 V_{S30}-Based Classification

As shown in Appendix C, wave amplitudes increase when waves travel from materials of higher specific impedance into materials of lower specific impedance. Recalling that specific impedance is the product of density and wave propagation velocity (or the square root of the product of density and modulus), and noting that the stiffness of geologic materials generally increases with depth (at a much greater rate than density), the amplification of upward-traveling waves should logically be affected by near-surface stiffness. A number of investigators (e.g., Borcherdt, 1970; Borcherdt et al., 1978; Borcherdt and Gibbs, 1976; Fumal, 1978; Borcherdt and Glassmoyer, 1994) observed that the amplification of ground motions generally increased with decreasing mean shear wave velocity of shallow (upper 30m) sediments, a quantity that became known as V_{S30} (the 30m depth over which the stiffness is characterized has no specific, fundamental significance – it is deep enough to capture wavelengths of common interest in many site response problems, and corresponds to the depth typically achieved in one day of drilling in average soil). These observations led Borcherdt (1994) to propose that V_{S30} be used to assign sites to discrete site classes for building code purposes, which are often colloquially referred to as NEHRP site classes because of their use in the NEHRP provisions (Building Seismic Safety Council, 2020). The most recent version of the site classes from the 2020 provisions is shown in Table 7.1. Over time, however, V_{S30} has become a ubiquitous parameter used on a continuous scale for representing site condition in ergodic site amplification models.

V_{S30} from Measured Shear Wave Velocities

As given previously in Equation (3.23), the value of V_{S30} is defined as the harmonic mean of the shear wave velocities in the upper 30m, or as the reciprocal of the arithmetic mean of the slowness over that depth range. Put yet another way, V_{S30} is equal to 30m divided by the shear wave travel time (the time required for a vertically propagating shear wave to travel through the upper 30m of the profile),

$$V_{S30} = \frac{30\text{ m}}{tt(30)} = \frac{30\text{ m}}{\int_0^{30} dz/V_S(z)} \approx \frac{30\text{ m}}{\sum_{i=1}^{n} \Delta z_i/V_{S,i}} \tag{7.2}$$

in m/s where $tt(30)$ is the shear wave travel time from the surface to a depth of 30m in sec, Δz_i=thickness of ith layer in m, $V_{S,i}$=shear wave velocity of ith layer, and n=number of layers comprising the upper 30m of the profile. Section 3.5.2.3 describes the use of V_{S30} in the estimation of site effects.

TABLE 7.1
NEHRP Site Classes (Building Seismic Safety Council, 2020)

Site Class	Description	V_s Range
A	Hard rock	> 1,500 m/s
B	Medium hard rock	900–1,500 m/s
BC	Soft rock	640–900 m/s
C	Very dense sand or hard clay	440–640 m/s
CD	Dense sand or very stiff clay	300–440 m/s
D	Medium dense sand or stiff clay	210–300 m/s
DE	Loose sand or medium stiff clay	150–210 m/s
E	Very loose sand or soft clay	< 150 m/s
F	Special cases requiring site-specific analyses	

Example 7.1

Downhole testing in a 100-ft-deep boring at a site in California produced the shear wave velocity profile tabulated below. Bedrock at the site, however, is known to exist at a depth of 300 ft. Estimate the shear wave velocity of the soil at the base of the deposit.

Depth (ft)	V_s (ft/s)	Depth (m)	V_s (m/s)
0–5	500	0–1.52	152.4
5–10	600	1.52–3.05	182.9
10–20	680	3.05–6.10	207.3
20–40	800	6.10–12.20	243.9
40–70	1,200	12.20–21.34	365.9
70–100	1,500	21.34–30.48	457.3

Solution:

The downhole data can be used to compute V_{S30}

$$V_{S30} = \frac{30\text{ m}}{\dfrac{1.52\text{ m}}{152.4\text{ m/s}} + \dfrac{1.53\text{ m}}{182.9\text{ m/s}} + \dfrac{3.05\text{ m}}{207.3\text{ m/s}} + \dfrac{6.10\text{ m}}{243.9\text{ m/s}} + \dfrac{9.14\text{ m}}{365.0\text{ m/s}} + \dfrac{8.66\text{ m}}{457.3\text{ m/s}}} = 294.0\text{ m/s}$$

Using Equation (7.5) and interpolating the coefficients, *c* and *d*, from Table 7.2 to the base of the soil deposit at 300 ft, or 91.5 m gives $c=0.42$ and $d=0.92$, the estimated shear wave velocity at the base of the soil deposit would be

$$V_S = 10^{c+d\log V_{S30}} = 10^{0.42+0.92\log(294.0)} = 490.7\text{ m/s}$$

The value of V_{S30} is most reliably obtained from site-specific shear wave velocity measurements using the types of geophysical tests described in Section 6.5.2. V_{S30} is correlated to similarly defined average velocities for depths other than 30 m. If the measured shear wave velocities extend to depth, z, the average shear wave velocity to that depth can be computed as

$$V_{Sz}(z) = \frac{z}{tt(z)} \tag{7.3}$$

When measured shear wave velocities are not available to the desired depth, i.e., when $z<30$ m, it is useful to estimate V_{S30} from V_{Sz} based on the shallower site characterization. Several empirical

TABLE 7.2
Coefficients for Estimation of V_{S30} for Sites at Which Velocity Data Extends to Depth, *z*

Depth, *z* (m)	*a*	*b*	$\sigma_{\log_{10} V_{S30}}$	$\sigma_{\ln V_{S30}}$
10	0.042062	1.0292	0.07126	0.164082
12	0.012571	1.0352	0.059353	0.136665
14	0.012300	1.0297	0.050086	0.115327
16	0.013893	1.0237	0.042219	0.097213
18	0.024879	1.0144	0.036365	0.083734
20	0.025439	1.0095	0.030181	0.069494
22	0.026900	1.0044	0.024087	0.055462
24	0.016891	1.0043	0.017676	0.040700
26	0.0065646	1.0045	0.011452	0.026369
28	0.00077322	1.0031	0.0055264	0.012725
30	0.0	1.0	-	-

Source: After Boore (2004).

relationships for performing this extrapolation, which reflect average velocity gradients in the region where the data was compiled, have been developed. One such relationship was developed by Boore (2004) using data from sites mostly in urban portions of California as follows:

$$\log V_{S30} = a + b \log V_{Sz}(z) \tag{7.4}$$

where a and b are empirical coefficients (Table 7.2) obtained by regression. Different models have been obtained for portions of Japan (Boore et al., 2011) and Greece (Stewart et al., 2014), which is expected due to regional variations in geological conditions.

Relationships between V_{S30} and V_{Sz} are also useful when $z > 30$ m, principally for the prediction of average velocities beyond the limits of geotechnical exploration. This can be significant for deep sites since the response to low-frequency waves (with longer wavelengths) will be affected by profile stiffness at depths greater than 30 m. Since many shear wave velocity measurements are made in borings or CPT soundings advanced for foundation design purposes, some sites may lack velocity data for depths greater than 30 m. In this case, the prediction is of V_{Sz} from V_{S30}, which can be evaluated from:

$$\log V_{Sz}(z) = c + d \log V_{S30} \tag{7.5}$$

Using data from California, the coefficients c and d and the standard deviation of the fit have been derived, as given in Table 7.3. These correlations have been found to be surprisingly robust in California (e.g., as shown by the low standard deviations in Table 7.3), which indicates that the V_{S30} parameter has strong predictive power for (i.e., is well correlated to) velocity structure at greater depth. This helps explain why V_{S30} correlates strongly to site amplification even at low frequencies, where wavelengths are often much greater than 30 m. Regional variations in geological conditions may cause correlations like Equation (7.5) to break down in some areas, weakening the predictive power of V_{S30} as a site parameter. This appears to be the case, for example, in central and eastern North America (Parker et al., 2019) where geologic conditions often cause large impedance contrasts to be present in site profiles and the dependence of site response on V_{S30} is weaker than in the western United States.

In some cases, where site-specific measurements are not available, V_{S30} can be roughly estimated using proxies such as ground slope, ground slope plus surficial geology, or more complicated terrain indicators (Section 6.6.2.2).

TABLE 7.3
Coefficients for Estimation of V_{Sz} Based on V_{S30} for $z>30$ m as Described by Boore et al. (2011)

Depth, z (m)	c	d	$\sigma_{\log V_{Sz}}$	$\sigma_{\ln V_{Sz}}$
50	0.151	0.980	0.042	0.097
75	0.327	0.942	0.071	0.163
100	0.466	0.909	0.087	0.200
150	0.713	0.831	0.112	0.258
200	0.887	0.783	0.124	0.286
300	1.188	0.661	0.114	0.262
400	1.534	0.532	0.123	0.283

Coefficients from D.M. Boore (personal communication).

Uncertainty in V_{S30}

The uncertainty in V_{S30} from measured velocity profiles has been evaluated from sites with clusters of profiles, typically spaced on the order of 10 to about 100 m. This uncertainty, denoted $\sigma_{\ln V_{S30}}$, is useful for uncertainty propagation in ground motion predictions that utilize V_{S30}-based site response models (Section 3.5.2). This uncertainty can in general be influenced by variations among V_S measurement types (generally small when reliable data providers are used) and natural heterogeneity in the site conditions. Based on data compiled by Moss (2008) and Seyhan et al. (2014), the range of $\sigma_{\ln V_{S30}}$ was approximately 0.02–0.12, with an average of about 0.06 for sites without strongly variable geological conditions. Values of 0.1 are often applied in current practice.

7.3.3.2 Geology-Based Classification

Geology-based site classification is attractive because geologic maps are readily available, at various levels of resolution, for many parts of the world. Surface geology can provide a general indication of profile stiffness, at least between the relatively broad categories shown in typical geologic maps. To better capture variations in stiffness within these broad categories, additional information can be added to supplement the basic mapped categories. Such information can include sediment thickness, texture, regional location, geologic history, and geomorphological data such as ground slope (Kottke et al., 2012). Stewart et al. (2003) used age, depositional environment, and sediment texture (Table 7.4) to classify surface geology for ground motion amplification purposes.

As part of the development of a shear wave velocity map for California, Wills et al. (2015) collected downhole, crosshole, and suspension log shear wave velocity data from multiple profiles in different geologic environments and compiled V_{S30} values for 15 geological site categories (Table 7.5). Geographic criteria were used to help define the dominant characteristics of Quaternary materials within the upper 30 m of the profile, noting that the soil maps upon which many maps of Quaternary geology are based often reflect the material only in the upper few meters. The V_{S30} values can clearly be seen to vary systematically with sediment texture (e.g., fine-grained vs. coarse-grained) and age.

7.3.3.3 Geotechnical-Based Classification

Geotechnical classification schemes are based on characteristics such as sediment type, thickness, and stiffness. Such schemes began with the four site categories of Seed et al. (1976a) shown in Figure 7.7, but were later refined by a number of researchers. Rodriguez-Marek et al. (2001)

TABLE 7.4
Geology-Based Site Classification Criteria (Stewart et al., 2003)

Age	Depositional Environment	Sediment Texture
Holocene	Fan alluvium	Coarse
Pleistocene	Valley alluvium	Fine
	Lacustrine/marine	Mixed
	Aeolian	
	Artificial Fill	
Tertiary		
Mesozoic+Igneous		

TABLE 7.5
V_{S30} Values (in m/s) for Different Site Geology Units in California from Site-Condition Map of California and from Borehole Profiles with Indicated Unit at Ground Surface

		V_{S30} for Map Units			V_{S30} from Boreholes		
Unit	Geologic Description	No. prof.	Mean V_{S30}	St. Dev.	No. prof.	Mean V_{S30}	St. Dev.
Qi	Intertidal Mud, including mud around the San Francisco Bay and similar mud in the Sacramento/San Joaquin delta and in Humboldt Bay	19	176.1	47.6	60	176.0	57.0
af/qi	Artificial fill over intertidal mud around San Francisco Bay	95	225.6	113.3	18	199.4	41.4
Qal1	Quaternary Holocene alluvium (flat)	117	228.2	48.0	273	298.5	93.9
Qal2	Quaternary Holocene alluvium (moderate)	161	293.5	73.5			
Qal3	Quaternary Holocene alluvium (steep)	114	351.9	112.2			
Qoa	Quaternary (Pleistocene) alluvium	181	386.6	145.1	113	372.8	132.1
Qs	Quaternary (Pleistocene) sand deposits	13	307.6	33.7	17	331.3	52.3
QT	Quaternary to Tertiary (Pleistocene-Pliocene) alluvial deposits such as the Saugus Formation of southern California, Paso Robles Formation of central coast ranges, and the Santa Clara Formation of the San Francisco Bay Area	17	444.0	159.7	11	456.6	193.7
Tsh	Tertiary (mostly Miocene and Eocene) shale and siltstone units such as the Repetto, Fernando, Puente, and Modelo Formations of the Los Angeles area	32	385.1	129.4	21	404.5	98.0
Tss	Tertiary (mostly Miocene, Oligocene, and Eocene) sandstone units such as the Topanga Formation in the Los Angeles area and the Butano sandstone in the San Francisco Bay Area	62	468.4	212.6	23	433.0	119.9

(Continued)

TABLE 7.5 (*Continued*)

V_{S30} Values (in m/s) for Different Site Geology Units in California from Site-Condition Map of California and from Borehole Profiles with Indicated Unit at Ground Surface

		V_{S30} for Map Units			V_{S30} from Boreholes		
Unit	**Geologic Description**	**No. prof.**	**Mean V_{S30}**	**St. Dev.**	**No. prof.**	**Mean V_{S30}**	**St. Dev.**
Tv	Tertiary volcanic units including the Conejo Volcanics in the Santa Monica Mountains and the Leona Rhyolite in the East Bay Hills	11	518.9	172.0	8	536.7	155.6
Serp.	Serpentine, generally considered part of the Franciscan complex	3	571.6	87.0	3	545.8	42.6
Kss	Cretaceous sandstone of the Great Valley Sequence in the central Coast Ranges	19	502.5	227.9	4	796.7	294.6
KJf	Franciscan complex rock, including mélange, sandstone, shale, chert, and greenstone	41	733.4	340.1	23	800.3	365.6
Xtal.	Crystalline rocks, including Cretaceous granitic rocks, Jurassic metamorphic rocks, schist, and Precambrian gneiss	35	710.1	393.8	12	614.9	228.5

Source: After Wills et al. (2015).

TABLE 7.6

Geomatrix Site Classification Based on Geotechnical Characteristics

Site Category	Description	Median V_{S30} (m/s)	$\sigma_{\ln V30}$	Mean V_{S30} (m/s)
A	Rock: Instrument on rock or <5 m soil over rock	507	0.41	516
B	Shallow (stiff) soil: Instrument on/in soil profile up to 20 m thick overlying rock	435	0.43	464
C	Deep, narrow soil: Instrument on/in soil profile at least 20 m thick overlying rock, in a narrow canyon or valley no more than several km wide	339	0.20	345
D	Deep, broad soil: Instrument on/in soil profile at least 20 m thick overlying rock, in a broad valley	275	0.34	291
E	Soft, deep soil: Instrument on/in deep soil profile with average V_{S30} < 150 m/s	182	0.25	185

Source: After Seyhan et al. (2014), Adapted from Chiou et al. (2008).

developed a system of six basic site categories based primarily on material type and stiffness with several subcategories based on sediment thickness. Geomatrix (Sadigh et al., 1993) developed a site classification system (Table 7.6) that includes five relatively broad categories – rock, shallow (stiff) soil, deep narrow soil, deep broad soil, and soft, deep soil. The distinction between narrow and broad soil deposits allows the differentiation of cases where multi-dimensional effects and basin effects may have more or less influence on ground motions.

7.4 SITE-SPECIFIC PREDICTION OF SITE EFFECTS

Generic ground motion predictions based on data from many different sites within a particular site class will necessarily represent the average response of profiles within that site class and will correspond most closely to the response of the smooth, "average profile" within it. The response of a particular site of interest, however, will be influenced in a systematic and repeatable manner by the individual characteristics of that site. Some sites may respond more strongly than the average response implied by a GMM and others may respond less strongly. Also, the dispersion of response in multiple events at a specific site will be lower than the dispersion obtained from a GMM based on data from many different sites.

More accurate and less uncertain estimates of site effects, therefore, can often be obtained by site-specific procedures. Site-specific characterization of site response can be accomplished in one of two main ways – empirically using ground motion recordings at an instrumented site, and analytically using dynamic analyses. In some situations, a third approach to characterization may be advisable – a hybrid approach in which dynamic analyses are used to define part of the amplification behavior and empirical data is used to define other parts.

7.4.1 Empirical Site-Specific Approach

A purely empirical approach to site-specific response prediction would be ideal. In that case, a site of interest would already have been instrumented with a vertical array extending to bedrock that has recorded many ground motions from a wide range of earthquakes (small to large) on all sources (nearby to distant) capable of producing strong ground motion at the site. Since source and path effects would affect both the downhole and surface instruments, ratios of surface-to-downhole spectral amplitudes (either Fourier or response spectra) could be used to characterize site-specific amplification behavior of the site (at periods less than the site period, above which the profile tends to respond as a rigid body) in each earthquake. With sufficient data, both the median amplification behavior and the dispersion in amplification could be characterized across a broad range of frequencies and input motion amplitudes. This approach would give the actual small- and large-strain amplification behavior of the site soil column, accounting for all details associated with its specific geometric and material characteristics. The approach would account for linear response under very weak shaking and nonlinear response under strong shaking and for both shallow and deep soils at the site. The effects of spatial variability, basin structure, and surface topography at that particular site would all be included. Unfortunately, instrumented sites with this amount of data do not exist.

While vertical arrays allow for a direct evaluation of site response, ground motion instruments installed only at the surface of a site can also be used to evaluate site response if the instruments have recorded a sufficient number of earthquakes (Stewart et al., 2017). In this case, the site response is evaluated using a non-reference site approach and is evaluated relative to the reference condition in the GMM (often corresponding to V_{S30}=760 m/s). Because the recordings available for a given site are likely to be low in amplitude, and thus associated with small levels of strain, they can often only be used to characterize the linear site response. The effects of nonlinearity under stronger shaking can then be assessed by numerical ground response analyses. Incorporation of this type of information into a ground motion hazard evaluation is described in Section 7.7.

7.4.2 Analytical Site-Specific Approach

When detailed information on the characteristics of a specific site is available, site-specific amplification behavior can be predicted by numerical analysis. Numerical ground response analyses provide the additional benefits of predicting not only ground surface motions but also ground motions, stresses, strains, and other important measures of response below the ground surface, which can be important for ground failure, liquefaction, and soil-structure interaction problems. Numerical analyses can account for the individual characteristics of a specific profile and the effects of those characteristics on the response of that profile to specific ground motions. For a profile with characteristics close to the average characteristics of a particular site class, a good numerical analysis will generally produce ground surface motions that are about as accurate as those produced by empirical amplification factors (Baturay and Stewart, 2003). For profiles with characteristics that deviate in some significant way from the average of a particular site class, site-specific numerical analyses can produce more accurate estimates of site response. The details of the modeling are important, though, and oversimplification of site conditions can lead to inaccurate results (e.g., Thompson et al., 2009).

Ground response analyses can be performed in a number of different ways. Some sites, with level ground surfaces and horizontal layer boundaries that extent laterally to distances significantly greater than the thickness of the profile, can be modeled as one-dimensional. Other sites may have significant two- or three-dimensional characteristics and must be modeled as such. Anticipated material behavior also plays a role in determining the type of site response analysis that should be performed. For stiff soils and/or weak motions, the stresses and strains in soil layers will be low, and linear or equivalent linear analyses can be used. For soft or loose, saturated soils and/or for strong motions, nonlinear analyses may be required. Profiles with soils capable of generating significant porewater pressure during shaking may require nonlinear effective stress analysis. All of these types of analyses are described in the remainder of this section.

7.5 ONE-DIMENSIONAL GROUND RESPONSE ANALYSIS

When a fault ruptures below the Earth's surface, body waves travel away from the source in all directions and are reflected and refracted as they reach boundaries between different geologic materials. Since the wave propagation velocities of shallower materials are generally lower than the materials beneath them, inclined rays that strike horizontal layer boundaries are usually refracted to a more vertical direction (Section C.4.2). By the time the rays reach the ground surface, multiple refractions have often bent them to a nearly vertical direction (Figure 7.9). One-dimensional ground response analyses are based on the assumption that all boundaries are horizontal and that the response of a soil deposit is predominantly caused by SH waves propagating vertically (hence with horizontal particle motion) from the underlying bedrock. For one-dimensional analyses, the soil layers and bedrock surface are assumed to extend infinitely with exactly the same properties in the horizontal direction. As a result of these assumptions, one-dimensional analyses cannot account for surface waves or for the effects of lateral inhomogeneity; the former can lead to the underprediction of low-frequency response and the latter to the overprediction of high-frequency response. These limitations in the ability of one-dimensional analyses to represent all of the physical processes that can affect the response of a particular site can lead to systematic errors, or bias, in the predicted response. The amount of bias can range from negligible to significant on a case-by-case basis, and will generally be different at different frequencies.

One-dimensional amplification arises primarily from two phenomena – conservation of energy as upward-traveling waves encounter softer materials on their way to the ground surface and *resonance*. The increase in wave amplitude (of acceleration, velocity, and displacement) that occurs when waves travel from a stiffer material into a softer material was described in Section C.4.1.

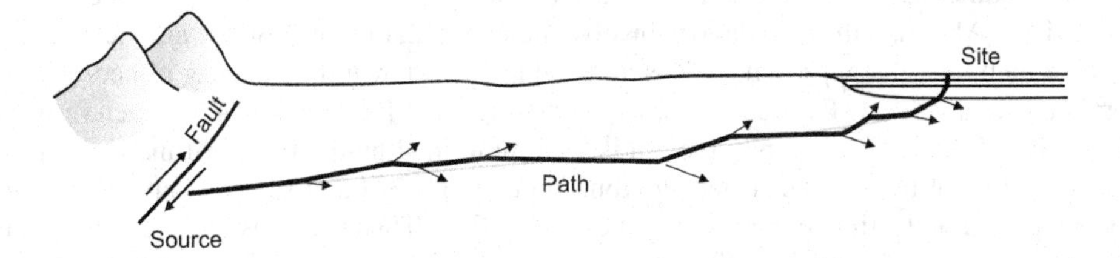

FIGURE 7.9 Refraction process that produces nearly vertical wave propagation near the ground surface.

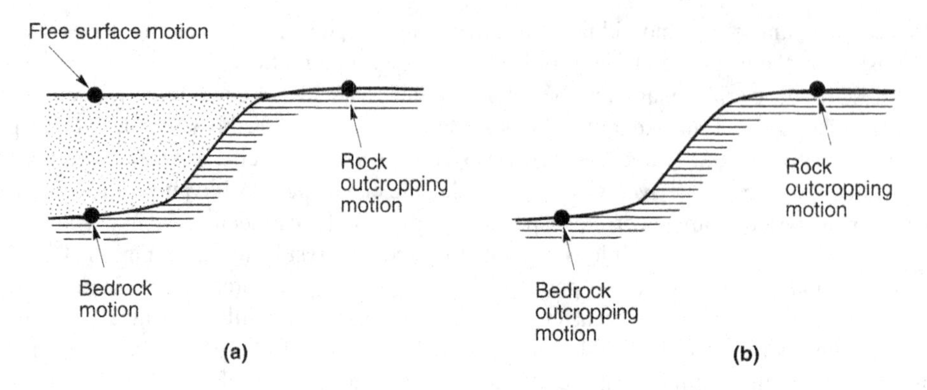

FIGURE 7.10 Ground response nomenclature: (a) soil overlying bedrock; (b) no soil overlying bedrock. Vertical scale is exaggerated.

Resonance phenomena develop in layered profiles where upward- and downward-traveling waves interfere with each other, both constructively and destructively, at different frequencies. Resonance can cause strong amplification at certain frequencies. One-dimensional ground response analyses can account for the simultaneous occurrence of both of these phenomena, along with damping effects that attenuate site amplification.

Before describing any of the ground response models, it is necessary to define several terms that are commonly used to describe ground motions. With reference to Figure 7.10a, the motion at the surface of a soil deposit is the *free surface motion*. The motion at the base of the soil deposit (also the top of bedrock) is called a *bedrock within motion*. The motion at a location where bedrock is exposed at the ground surface is called a *rock outcropping motion*. If the soil deposit was not present (Figure 7.10b), the motion at the top of the bedrock would be the bedrock outcropping motion. Consistent use of this terminology is important for understanding important principles of site response and the proper application of input motions (Section 7.5.5.1) in ground response analyses.

7.5.1 Linear Analyses

Profiles with very stiff soils subjected to relatively weak shaking may develop strains so small that the stiffness and damping of the soil material are essentially constant. Under such conditions, site response can be analyzed using linear analyses. Linear analyses of site response are most easily accomplished in the frequency domain using transfer functions. The manner in which transfer functions can be used to compute the response of single-degree-of-freedom systems is illustrated in Appendix B (Section B.5.4.2). An important class of techniques for ground response analysis is also based on the use of transfer functions. For the ground response problem, transfer functions can be used to relate various response parameters, such as displacement, velocity, acceleration, shear stress, and shear strain, to the input motion. Because it relies on the principle of superposition, this approach is limited to the analysis of linear (elastic or viscoelastic) systems. The primary effects of mild to moderate nonlinearity can be approximated using linear analyses in an equivalent linear framework (Section 7.5.2).

The basic mathematical aspects of the transfer function approach are described in Section B.5.4.2 of Appendix B. Although the calculations involve the manipulation of complex numbers in order to account for damping, the approach itself is quite simple. A known time history of bedrock (input) motion is represented as a Fourier series, usually using the FFT (Section A.3). Each term in the Fourier series of the bedrock (input) motion is then multiplied by the transfer function to produce the Fourier series of the ground surface (output) motion. The ground surface (output) motion can then be expressed in the time domain using the inverse FFT. Thus the transfer function determines how each frequency in the bedrock (input) motion is amplified, or deamplified, by the soil deposit; the imaginary part of a complex transfer function introduces the phase shift of each frequency.

The key to the linear approach is the evaluation of transfer functions. In the following sections, transfer functions are derived for a series of successively more complicated geotechnical conditions. Although the simplest of these may only rarely be applicable to actual problems, they illustrate some of the important effects of soil deposits on ground motion characteristics without undue mathematical complexity. The more complex solutions are capable of describing the most important aspects of ground response and are very commonly used in geotechnical earthquake engineering practice.

7.5.1.1 Undamped Soil on Rigid Bedrock

Section C.6.1.2 showed how a linear undamped material fixed at one end (e.g., bedrock) and free at the other (the ground surface) would respond to a harmonic motion of the fixed end. The response was expressed in terms of a transfer function whose modulus varied with frequency in the form of an amplification factor. That amplification factor was

$$|F_1(\omega)| = \frac{1}{|\cos kH|} = \frac{1}{|\cos \omega H / V_s|} \tag{7.6}$$

which is strongly frequency-dependent (Figure C.29) and implies infinite amplification at circular frequencies $\omega = \pi V_s/2H$, $3\pi V_s/2H$, $5\pi V_s/2H$, etc.

7.5.1.2 Damped Soil on Rigid Bedrock

Obviously, the type of unbounded amplification predicted by the previous analysis cannot physically occur. The previous analysis assumed no dissipation of energy, or damping, in the soil. Since damping is present in all materials, more realistic results can be obtained by repeating the analysis with damping. Assuming the soil to have the shearing characteristics of a Kelvin-Voigt solid, the wave equation can be written (Equation C.124) as

$$\rho \frac{\partial^2 u}{\partial t^2} = G \frac{\partial^2 u}{\partial z^2} + \eta \frac{\partial^3 u}{\partial z^2 \partial t} \tag{7.7}$$

where η is the viscosity of the soil. The solution to this wave equation is of the form

$$u(z,t) = Ae^{i(\omega t + k^* z)} + Be^{i(\omega t - k^* z)} \tag{7.8}$$

where k^* is a complex wave number with real part k_1 and imaginary part k_2. Repeating the previous algebraic manipulations with the complex wave number, the transfer function for the case of damped soil over rigid rock can be expressed as

$$F_2(\omega) = \frac{1}{\cos(k^* H)} = \frac{1}{\cos(\omega H / V_s^*)} \tag{7.9}$$

Unlike the undamped case described in the previous section and Section C.6.1.2, the transfer function for the damped case is complex. The imaginary part of the transfer function accounts for damping. Since the frequency-independent complex shear modulus is given by $G^* = G(1 + i2\xi)$, the complex shear wave velocity can be expressed as

$$V_s^* = \sqrt{\frac{G^*}{\rho}} = \sqrt{\frac{G(1+i2\xi)}{\rho}} = \sqrt{\frac{G}{\rho}}(1+i\xi) = V_s(1+i\xi) \tag{7.10}$$

for small ξ. Then the complex wave number can be written, again for small ξ, as

$$k^* = \frac{\omega}{V_s^*} = \frac{\omega}{V_s(1+i\xi)} = \frac{\omega}{V_s}(1-i\xi) = k(1-i\xi) \tag{7.11}$$

and finally, the transfer function, as

$$F_2(\omega) = \frac{1}{\cos k(1-i\xi)H} = \frac{1}{\cos\left[\omega H/V_s(1+i\xi)\right]} \tag{7.12}$$

Using the identity $|\cos(x+iy)| = \sqrt{\cos^2 x + \sinh^2 y}$, the amplification function (i.e., modulus of the transfer function) can be expressed as

$$|F_2(\omega)| = \frac{1}{\sqrt{\cos^2(kH) + \sinh^2(\xi kH)}} \tag{7.13}$$

Since $\sinh^2 y \approx y^2$ for small y, the amplification function can be simplified to

$$|F_2(\omega)| = \frac{1}{\sqrt{\cos^2 kH + (\xi kH)^2}} = \frac{1}{\sqrt{\cos^2(\omega H/V_s) + \left[\xi(\omega H/V_s)\right]^2}} \tag{7.14}$$

for small damping ratios. Equation (7.14) indicates that amplification by a damped soil layer also varies with frequency. The amplification will reach a local maximum whenever $kH = \omega H/V_s = \pi/2 + n\pi$ but will never reach a value of infinity since (for $\xi > 0$) the denominator will always be greater than zero. The frequencies that correspond to the local maxima are the *natural frequencies* of the soil deposit. As discussed in Section C.6.1.2, the lowest natural frequency is the *fundamental frequency* ($f_0 = V_s/4H$) and its corresponding period is the *characteristic site period* ($T_s = 4H/V_s$). The variation of amplification factor with frequency is shown for different levels of damping in Figure 7.11. This amplification factor is also equal to the ratio of the free surface motion amplitude to the bedrock (or bedrock outcropping) motion amplitude.

Comparing Figures C.28 and 7.11 shows that damping clearly reduces amplification and does so more strongly at high frequencies than at lower frequencies. High-frequency motions have shorter wavelengths than low-frequency motions, so a high-frequency component undergoes more "cycles" of stress and strain over the thickness of the soil layer than does a low-frequency component of the motion. As a result, it dissipates more energy and hence exhibits a greater reduction in amplitude, than the low-frequency component that travels the same distance.

As the frequency approaches zero, the value of $|F_1(\omega)|$ approaches 1.0, which indicates that little to no amplification of the base motion should be expected at frequencies well below the fundamental frequency (or periods well above the characteristic site period). For very low frequencies, wavelengths will be long relative to the thickness of the soil and the entire soil profile will tend to move in a nearly rigid-body fashion – the ground surface motion will be very similar to the bedrock motion. In reality, however, mechanisms not accounted for in one-dimensional analyses, subsequently discussed in Section 7.6, can cause some amplification of long-period motions.

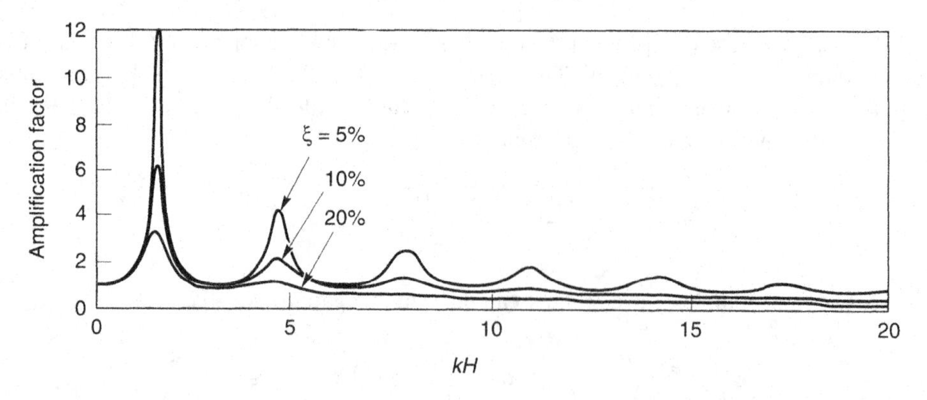

FIGURE 7.11 Influence of frequency on the steady-state response of damped, linear elastic layer.

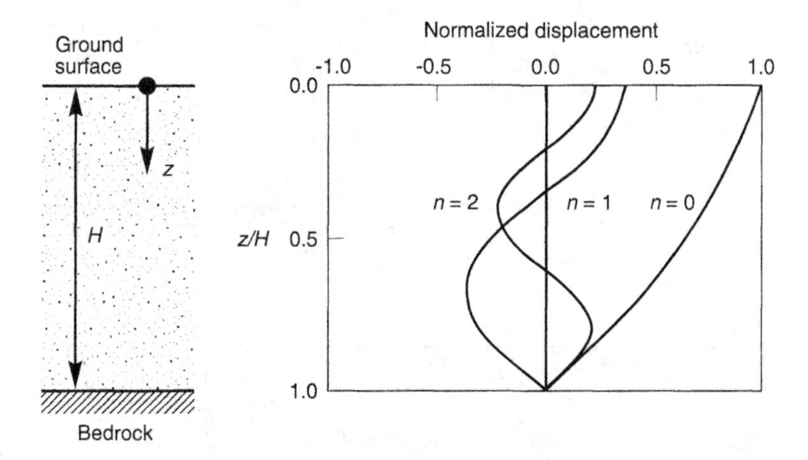

FIGURE 7.12 Displacement patterns for standing waves at fundamental ($n = 0$), second ($n = 1$), and third ($n = 2$) natural frequencies for a soil layer with $\xi = 5\%$. Displacements are normalized by maximum displacement at the fundamental frequency.

When subjected to harmonic input motion at each natural frequency, a *standing wave*, i.e., a wave in which the amplitude at each depth is constant, will develop in the soil. Normalized deformed shapes, or *mode shapes*, for the first three natural frequencies of a simple, uniform layer are shown in Figure 7.12. Note that the soil displacements are in phase at all depths in the fundamental mode, but not in the higher modes. At frequencies above the fundamental frequency, part of the soil deposit may be moving in one direction while another part is moving in the opposite direction. This phenomenon must be considered in the evaluation of inertial forces in soil masses required for seismic stability analyses (Chapter 10). The mode shapes reflect the contributions of both upward- and downward-traveling waves. At some depths, the waves interfere constructively and produce local peaks in displacement amplitude; at other depths, the waves can interfere in a destructive manner with the downward-traveling waves canceling out the upward-traveling waves. At these *nodes*, the response amplitude reaches a local minimum (zero amplitude if undamped). The amplitude of the transfer function between the depth of a node and the ground surface is infinite if damping is zero, but finite in the presence of damping due to phase lags between motions at different depths. The fact that the amplitude of motion at a node is zero, however, should not be taken to imply that waves (and energy) are not passing through those points; the nodes are specific to particular modal excitation frequencies and will not be present for other excitation frequencies. Transfer functions computed as ratios of surface-to-downhole motions recorded at vertical arrays can have strong peaks and troughs caused by constructive and destructive interference of upward- and downward-traveling waves.

Note that the displacement amplitude of the fundamental mode is considerably greater than that of any of the higher modes. The displaced shape of the fundamental mode is described by the first quarter-cycle of a cosine function, i.e., bedrock is at a depth equal to one-quarter of the wavelength of the fundamental mode. The degree of peak amplification, therefore, depends on the characteristics of the material within the upper quarter-wavelength of the motion. That depth, however, is different for different frequencies – shallower for high frequencies and deeper for low frequencies. That fact is helpful in developing an intuitive understanding of how sites are likely to respond during earthquake shaking – higher frequencies are affected by shallower material and lower frequencies are affected by deeper materials. For the purposes of the simple transfer functions described in this section, the shear modulus of the soil was assumed constant. For most actual profiles, however, stiffness tends to increase with depth.

As evidence of the importance of the characteristic site period, consider the response measured in Mexico City (Figure 7.4) in the 1985 Michoacan earthquake. As discussed in Section 7.2.2, the

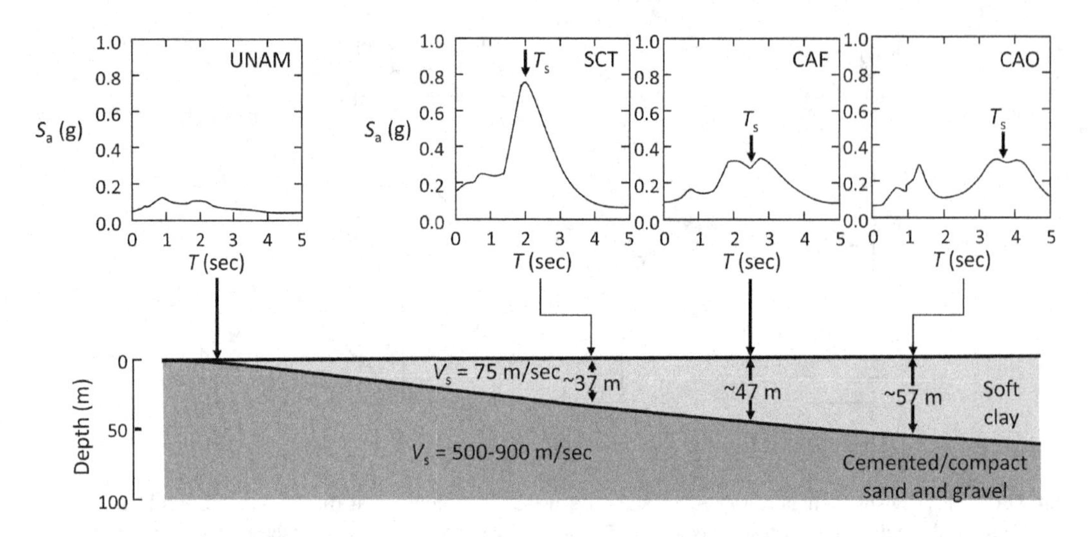

FIGURE 7.13 Illustration of effects of soft clay thickness on response spectrum amplitude and shape from Mexico City ground motion recordings. (After Dobry and Iai, 2000; used with permission of CRC Press.)

SCT site was underlain by 35–40 m (115–131 ft) of soft clay with an average shear wave velocity of about 75 m/s (250 ft/s). As a result, its characteristic site period was $T_s = 4H/V_s = 4$ (37.5 m) (75 m/s) = 2 sec, a value consistent with a strong local peak in the rock motion illustrated by the UNAM response spectrum. Figure 7.13 shows the response spectra from the UNAM and SCT sites, along with spectra from two other sites underlain by greater thicknesses of soft clay. As the thickness of the clay increased, the characteristic site periods of the various soil profiles lengthened, and the strongest levels of amplification were pushed to longer periods – 2.6 sec for the CAF site and 3.7 sec for the CAO site. The CAF and CAO motions are strongest at their respective characteristic site periods, but did not rise to the level of the SCT site because the intensity of the rock motion (see UNAM spectrum) decreased at periods beyond about 2 sec.

Figure 7.14 shows computed transfer functions and ground surface response spectra for three sites consisting of 30 m of soil with different shear wave velocities, each constant with depth and underlain by a rigid base. The velocities correspond to the centers of ASCE-7 (2022) Site Classes C, D, and E and the soils are assumed to exhibit linear behavior with 5% damping. The three profiles were subjected to the same base motion – a recorded earthquake motion from Taft Lincoln School in the 1952 Kern County earthquake (M_w 7.4). With velocities of 560, 270, and 150 m/s, the characteristic site periods can be computed ($T_s = 4H/V_s$) as 0.214, 0.444, and 0.800 sec. The three transfer functions, plotted here as functions of period rather than frequency, confirm those characteristic site periods and show that the stiffest profile amplified the lower period (higher frequency) components of the motion most strongly, and that the softest profile amplified the longer period (lower frequency) components most strongly. The nature of the amplification can be seen in the ground surface response spectra, each of which has a peak near the fundamental period of the profile. Since spectral acceleration is influenced by multiple frequencies, the amplitude and period of a particular peak in a response spectrum are not uniquely related to specific peaks in the transfer function. The acceleration histories show significant differences in amplitude and frequency content. Note that while amplification of the $V_s = 270$ m/s profile is approximately equivalent to that of the stiffer or softer profiles, the $V_s = 270$ m/s profile produces the largest response spectra. This occurs because the rock spectrum has its largest amplitudes at periods in the vicinity of the characteristic site period of the $V_s = 270$ m/s profile.

The case of a perfectly uniform layer of viscoelastic material on a rigid base is not realistic, even within the constraints of one-dimensional analyses. Rock is never perfectly rigid and soil

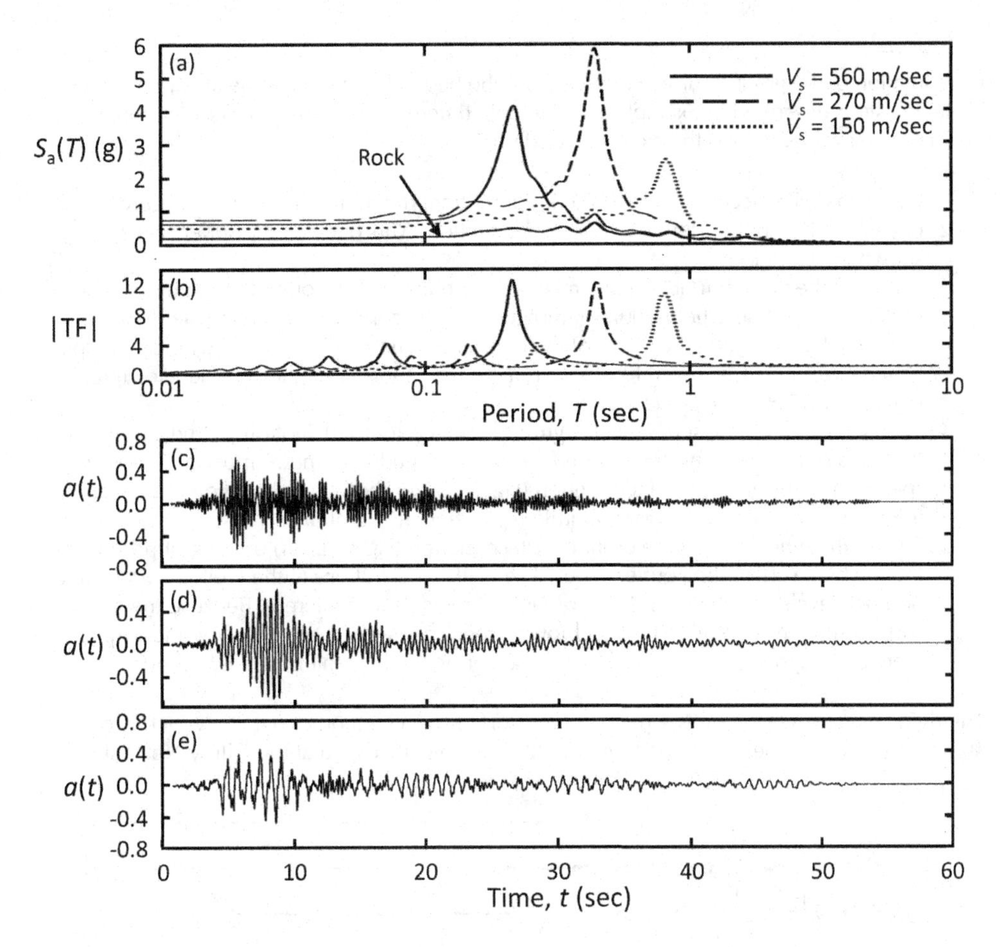

FIGURE 7.14 Effects of profile stiffness on ground surface response: (a) ground surface response spectra, (b) transfer functions, (c) ground surface acceleration for V_s=560 m/s profile, (d) ground surface acceleration for V_s=270 m/s profile, (e) ground surface acceleration for V_s=150 m/s profile.

properties, particularly soil stiffness, usually vary with depth. The basic principles of wave propagation described in Appendix C, however, can be used to account for these factors in one-dimensional ground response analyses.

Example 7.2

The site at which the Gilroy No. 2 motion from the Loma Prieta earthquake was recorded is near the location of the Gilroy No. 1 motion recording and underlain by some 540 ft of soil underlain in turn by shale and serpentinite bedrock. The shear wave velocity of the soil varies from about 300 to 600 m/s with an average velocity of 450 m/s and an average unit weight of 19 kN/m³. Assuming a damping ratio of 5% and rigid bedrock, compute the ground motion that would occur if the bedrock was subjected to the Gilroy No. 1 input motion (Figure E7.2a).

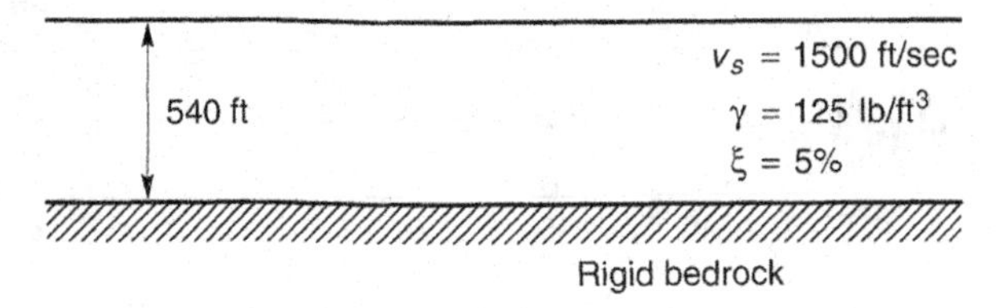

FIGURE E7.2A

Solution:

Computation of the ground surface motion from the bedrock motion can be accomplished in the same five steps described in Example 7.2. The only difference is in the transfer function, which is now complex-valued due to the existence of damping.

1. Obtain a digital accelerogram of the input motion; the motion is shown in Figure E7.2b.
2. Compute the Fourier series of the input motion. The one-sided Fourier spectrum is shown in Figure E7.2c.
3. Compute the transfer function that relates the ground surface (output) motion to the bedrock (input) motion. The modulus (amplitude) of the complex-valued transfer function is shown in Figure E7.2d. The shape of the transfer function indicates that significant amplification will occur at several natural frequencies, and that higher frequencies (greater than about 10 Hz) will be deamplified.
4. Compute the Fourier series of the ground surface (output) motion as the product of the transfer function and the Fourier series of the bedrock (input) motion. The Fourier spectrum of the ground surface motion (Figure E7.2e) shows amplification at the natural frequencies of the soil deposit and little high-frequency motion.
5. Obtain the time series of the ground surface motion (Figure E7.2f) of the ground surface motion by inverting the Fourier series. The peak accelerations at the ground surface and bedrock levels are similar, but the frequency contents are different. Because the ground surface motion is weighted toward lower frequencies, the peak velocity and displacement at the ground surface are likely to be considerably greater than at bedrock.

The rigid base analysis predicts a peak ground surface acceleration of $0.452g$, which is considerably greater than the peak acceleration of $0.322g$ actually recorded at the Gilroy No. 2 station.

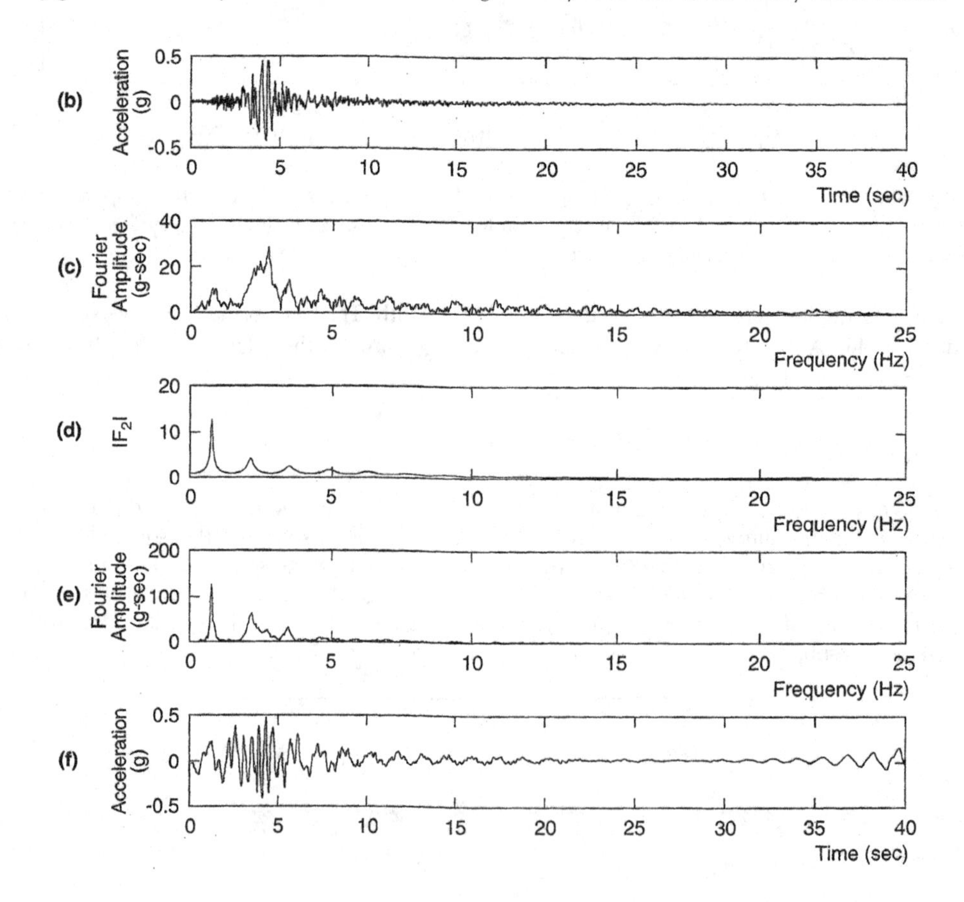

FIGURE E7.2B-F

7.5.1.3 Damped Soil on Elastic Bedrock

Section 7.5.1.2 described the response of a viscoelastic layer on rigid bedrock. If the bedrock is rigid, its motion will be unaffected by motions in (or even the presence of) the overlying soil – it acts as a fixed end (Section C.4.1) boundary. Any downward-traveling waves in the soil will be completely reflected back toward the ground surface by the rigid layer, thereby trapping all of the elastic wave energy within the soil layer.

If the rock is elastic, however, downward-traveling stress waves that reach the soil/rock boundary will be only partially reflected; part of their energy will be transmitted through the boundary to continue traveling downward through the rock. If the rock extends to great depth (large enough that waves reflected from any deeper material boundaries do not return to the soil-rock boundary soon enough, or with sufficient amplitude, to influence the response of the soil deposit), the elastic energy of these waves will effectively be removed from the soil layer. This is a form of *radiation damping*, and it causes the free surface motion amplitudes to be smaller than those for the case of rigid bedrock. Consider the case of a viscoelastic soil layer overlying a halfspace of viscoelastic rock (Figure 7.15). If the subscripts s and r refer to soil and rock, respectively, the displacements due to vertically propagating s-waves in each material can be written as

$$u_s(z_s,t) = A_s e^{i(\omega t + k_s^* z_s)} + B_s e^{i(\omega t - k_s^* z_s)} \tag{7.15a}$$

$$u_r(z_r,t) = A_r e^{i(\omega t + k_r^* z_r)} + B_r e^{i(\omega t - k_r^* z_r)} \tag{7.15b}$$

The free surface effect (Section C.6.1.2) requires that $A_s = B_s$ and compatibility of displacements and continuity of stresses at the soil-rock boundary require that

$$u_s(z_s = H) = u_r(z_r = 0) \tag{7.16}$$

$$\tau_s(z_s = H) = \tau_r(z_r = 0) \tag{7.17}$$

Substituting Equations (7.15) into Equation (7.16) yields

$$A_s\left(e^{ik_s^* H} + e^{-ik_s^* H}\right) = A_r + B_r \tag{7.18}$$

From Equation (7.17) and the definition of shear stress ($\tau = G\,\partial u/\partial z$)

$$A_s i G_s k_s^*\left(e^{ik_s^* H} - e^{-ik_s^* H}\right) = iG_r k_r^*(A_r - B_r) \tag{7.19}$$

or

$$\frac{G_s k_s^*}{G_r k_r^*} A_s\left(e^{ik_s^* H} - e^{-ik_s^* H}\right) = A_r - B_r \tag{7.20}$$

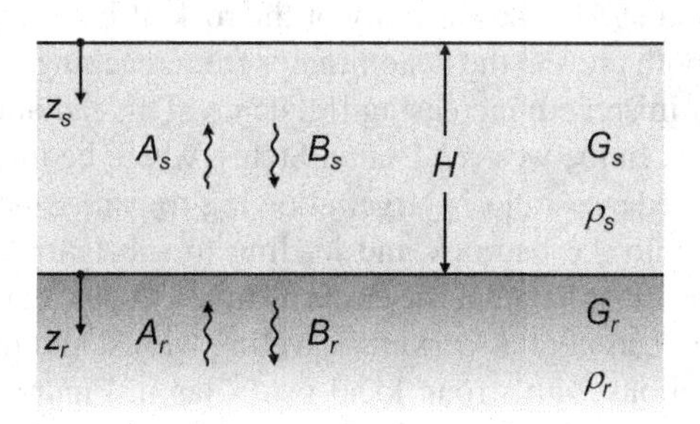

FIGURE 7.15 Nomenclature for the case of a soil layer overlying a halfspace of elastic rock.

The ratio of the products of shear modulus and wavenumber can be written as,

$$\frac{G_s k_s^*}{G_r k_r^*} = \frac{\rho_s V_{ss}^*}{\rho_r V_{sr}^*} = \alpha_z^* \tag{7.21}$$

where V_{ss}^* and V_{sr}^* are the complex shear wave velocities of the soil and rock, respectively, and α_z^* is the *complex impedance ratio* (see Section C.4.1). Solving Equations (7.18) and (7.20) simultaneously gives

$$A_r = \frac{1}{2} A_s \left[(1+\alpha_z^*) e^{ik_s^* H} + (1-\alpha_z^*) e^{-ik_s^* H} \right] \tag{7.22a}$$

$$B_r = \frac{1}{2} A_s \left[(1-\alpha_z^*) e^{ik_s^* H} + (1+\alpha_z^*) e^{-ik_s^* H} \right] \tag{7.22b}$$

Suppose that a vertically propagating shear wave of amplitude, A, traveled upward through the rock. If the soil was not present, the free surface effect at the rock outcrop would produce a bedrock outcropping motion of amplitude $2A$. With the soil present, however, the free surface motion amplitude would be

$$2A_s = \frac{4A}{(1+\alpha_z^*) e^{ik_s^* H} + (1-\alpha_z^*) e^{-ik_s^* H}} \tag{7.23}$$

Defining the transfer function, F_3, as the ratio of the soil surface amplitude to the rock outcrop amplitude.

$$F_3(\omega) = \frac{2A_s}{2A} = \frac{2}{(1+\alpha_z^*) e^{ik_s^* H} + (1-\alpha_z^*) e^{-ik_s^* H}} \tag{7.24}$$

which, using Euler's law, can be rewritten as

$$F_3(\omega) = \frac{1}{\cos\left(k_s^* H\right) + i\alpha_z^* \sin\left(k_s^* H\right)} = \frac{1}{\cos\left(\omega H / v_{ss}^*\right) + i\alpha_z^* \sin\left(\omega H / v_{ss}^*\right)} \tag{7.25}$$

The modulus of F_3 (ω) cannot be expressed in a very compact form when soil damping exists. To illustrate the important effect of bedrock elasticity, however, the amplification factor for undamped soil can be expressed as

$$F_3(\omega) = \frac{1}{\sqrt{\cos^2(k_s H) + \alpha_z^2 \sin^2(k_s H)}} \tag{7.26}$$

Note that unbounded resonance cannot occur – the denominator is always greater than zero, even when the soil is undamped. The effect of the bedrock stiffness, as characterized by the impedance ratio, on amplification behavior is illustrated in Figure 7.16. Note the similarity between the effects of soil damping and bedrock stiffness by comparing the shapes of the amplification factor curves in Figure 7.16 and those in Figure 7.11. The elasticity of the rock affects amplification similarly to the damping ratio of the soil – both prevent the denominator from reaching zero – although the bedrock stiffness effect does not diminish with increasing frequency. This radiation damping effect has significant practical importance. In the western United States, where bedrock is often not particularly stiff in many cases, the impedance ratio is large, allowing transmission of a larger proportion of downward-traveling waves into the bedrock and leading to substantial radiation damping which reduces site response. On the other hand, in the eastern United States, harder bedrock leads to lower impedance ratios, greater reflection of downward-traveling waves, and reduced radiation damping. This produces transfer functions with strong local peaks (as in Figure 7.16) that produce greater amplification at certain frequencies. As a result of these differences, design criteria established on the basis of empirical evidence from western earthquakes may be biased in the east.

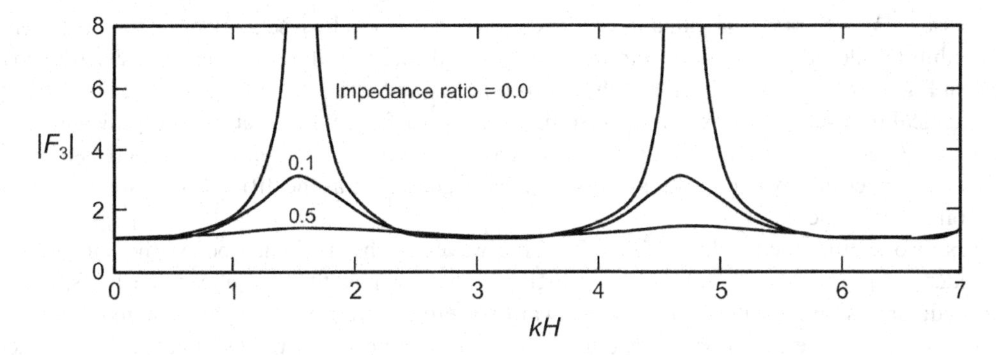

FIGURE 7.16 Effect of impedance ratio on amplification factor for case of undamped soil.

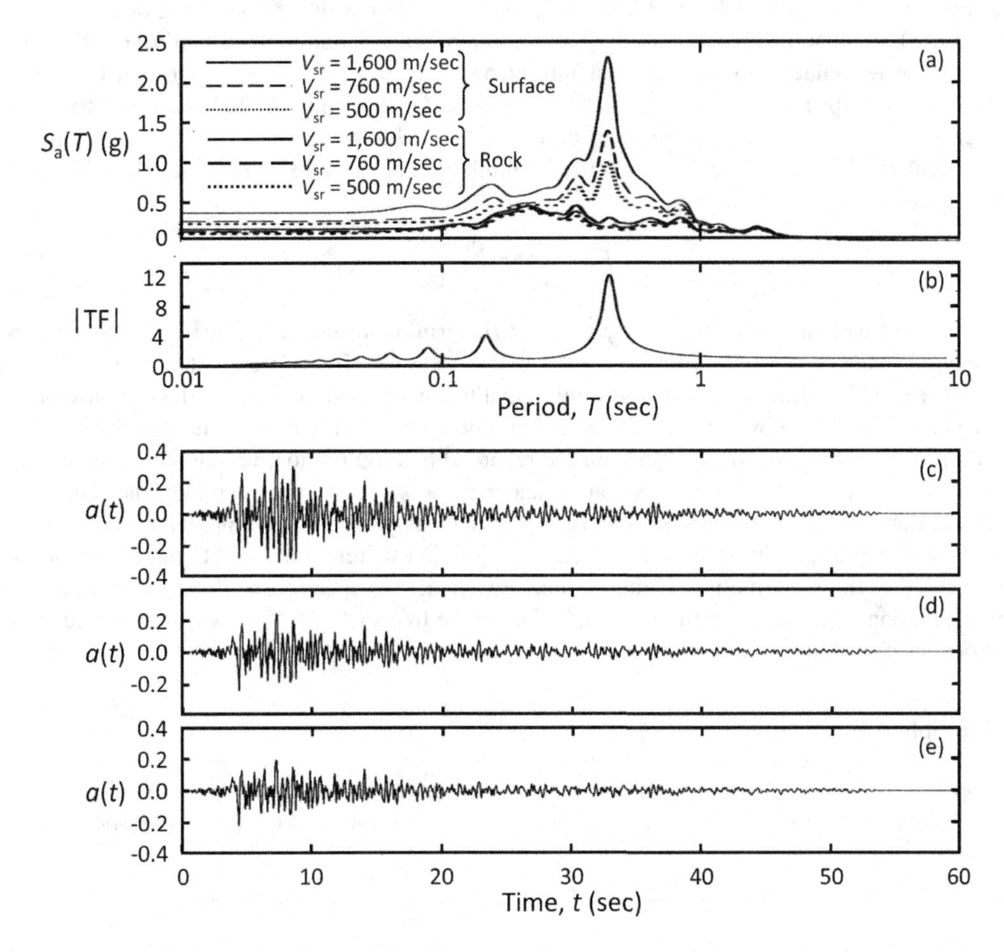

FIGURE 7.17 Effects of bedrock stiffness on ground surface response to Taft rock outcrop motion: (a) ground surface response spectra, (b) transfer functions, (c) ground surface acceleration for V_s = 1,600 m/s bedrock, (d) ground surface acceleration for V_s = 760 m/s bedrock, (c) ground surface acceleration for V_s = 500 m/s bedrock.

When performing site response analyses using recorded outcrop motions, the halfspace stiffness should be consistent with the stiffness of the material at the base of the modeled soil column. Figure 7.17 shows the results of analyses in which the same rock outcrop motion was applied to the same soil profile (constant V_s = 270 m/s) underlain by bedrock with three different stiffnesses – V_{sr}=1,600, V_{sr}=760, and V_{sr}=500 m/s. Because the bedrock stiffnesses are different, the within-profile bedrock motions are different, as indicated by the different rock spectra (bold lines in Figure 7.17a). The behavior of the soil profile was the same for each case, as evidenced by the identical transfer functions, but the resulting ground surface motions are also significantly different; as should be expected, the amplitude of the surface motion increases with increasing rock stiffness. The sensitivity to bedrock stiffness decreases, however, as the difference between the rock and soil stiffness increases.

It is also useful to consider the nature of the waves in the halfspace below the soil profile. Consider a variation of the condition in Figure 7.15 in which the soil above the halfspace is removed; for this case, the upward- and downward-traveling wave amplitudes are denoted A_r^* and B_r^*, respectively. Because the upward-traveling wave would be perfectly reflected at the free surface of the halfspace (as a downward-traveling wave of amplitude, $B_r^* = A_r^*$), the total displacement at the free surface would be $2A_r^*$. With the soil layer present (for which * notation is dropped, as in Figure 7.15), some of the upward-traveling waves will be transmitted into the overlying soil, which affects the wave reflected off the soil-rock interface so that $B_r \neq A_r$. Because the upward-traveling wave in the halfspace is the same for both cases, $A_r = A_r^*$ (assuming the halfspace has the same properties in both cases), but the motions at the top of the halfspace will be different for the two cases because of the differences in the reflected motions. The motions can be related by the transfer function

$$F_{rr^*} = \frac{u_r}{u_{r^*}} = \frac{A_r + B_r}{2A_r} \tag{7.27}$$

This transfer function is most commonly applied to determine the characteristics of the within-profile motion that is consistent with a recorded outcrop motion. If the recorded motion is from a rock outcrop with stiffness and density equal to that below a profile of interest, the response of the profile can be analyzed by applying the within-profile motion at the base of the soil column. The within-profile motion will be weaker than the recorded outcrop motion, principally at the natural frequencies of the soil profile and by amounts that increase with increasing damping and decreasing (soil/rock) impedance ratio. Most ground response analysis programs have provisions for computing the within-profile motion from an outcrop motion and therefore allow the outcrop motion to be specified as the input motion. If this is done, it is critical to specify that the input motion is an outcrop motion; otherwise, the full motion, including the free surface effect, will be imposed at the bedrock level.

Example 7.3

Repeat Example 7.2 assuming that the bedrock is not rigid. Assume a shear wave velocity of 760 m/s, a unit weight of 25 kN/m³, and 2% damping for bedrock at the site shown in Figure E7.3a.

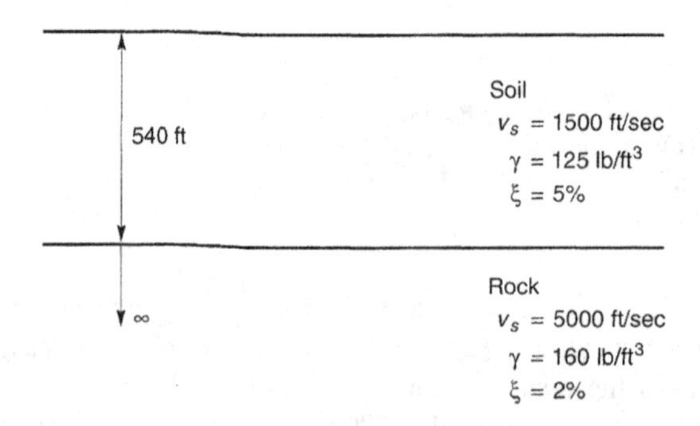

FIGURE E7.3A

Solution:

Computation of the ground surface motion from the bedrock motion can be accomplished in the same five steps described in Example 7.3. The only difference is that the transfer function in this example will include the effects of bedrock compliance.

Figure E7.3b–f shows the input motion, transfer function, and output motion in the same format as the preceding two examples. The transfer function for the compliant base case in this example is weaker than that of Example 7.2 and the resulting ground surface motion is also significantly weaker (note differences in ordinate scales used in plots).

The compliant bedrock analysis predicts a peak ground surface acceleration of 0.339*g*, which agrees well with the peak acceleration of 0.322*g* recorded at the Gilroy No. 2 station. The good agreement between peak accelerations, however, does not mean that this simple analysis has predicted all aspects of the Gilroy No. 2 motion well.

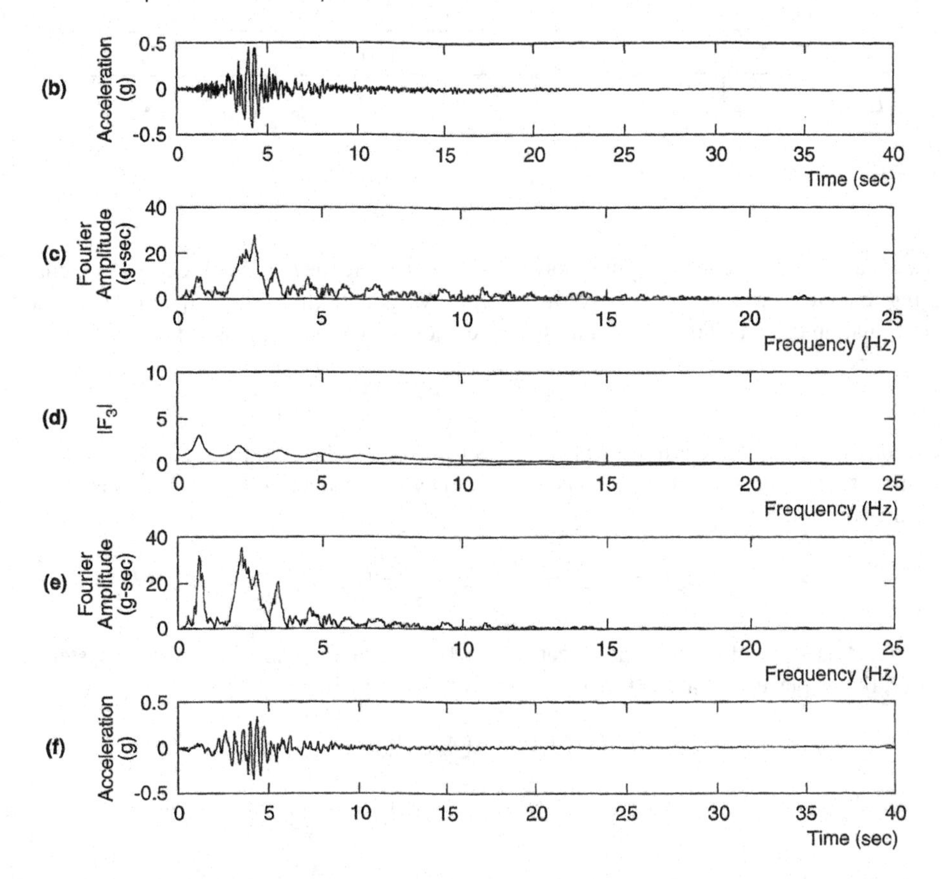

FIGURE E7.3B-F

The compliant bedrock analysis predicts a peak ground surface acceleration of 0.339*g*, which agrees well with the peak acceleration of 0.322*g* recorded at the Gilroy No. 2 station. The good agreement between peak accelerations, however, does not mean that this simple analysis has predicted all aspects of the Gilroy No. 2 motion. Comparison of the Fourier amplitude spectrum of the predicted motion (Figure E7.3e) with that of the recorded motion shows significant differences in frequency content.

7.5.1.4 Layered Soil on Elastic Rock

Real ground response problems usually involve soil deposits with stiffness and damping characteristics that vary with depth, either due to the layering of different materials or depth-dependence within individual layers. Elastic wave energy will be reflected and/or transmitted at the boundaries between these layers. Such conditions require the development of transfer functions for layered soil deposits.

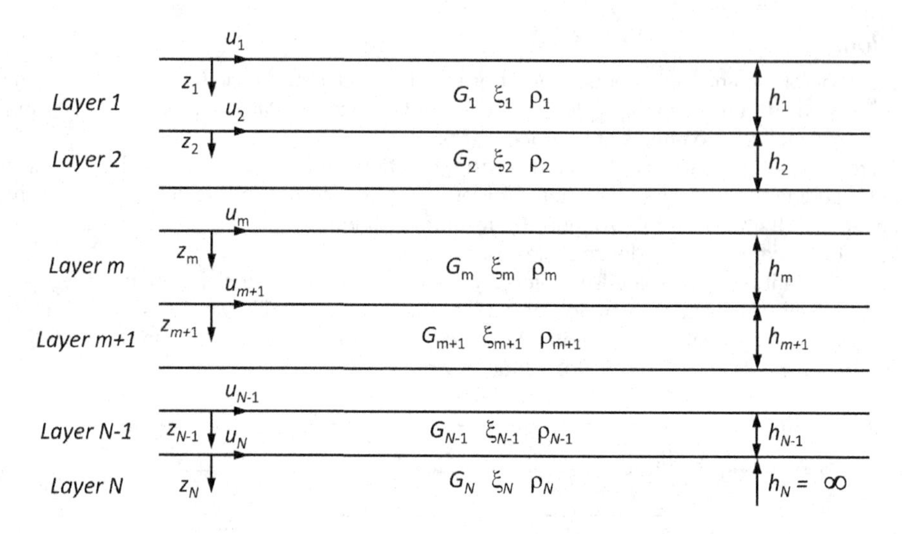

FIGURE 7.18 Nomenclature for layered soil deposit on elastic bedrock.

Consider a soil deposit consisting of N horizontal layers where the Nth layer is bedrock (Figure 7.18). Assuming that each layer of soil behaves as a Kelvin-Voigt solid, the wave equation is of the form given in Equation (C.124). The solution to the wave equation can be expressed in the form

$$u(z,t) = Ae^{i(\omega t + k^* z)} + Be^{i(\omega t - k^* z)} \tag{7.28}$$

where A and B represent the amplitudes of waves traveling in the $-z$ (upward) and $+z$ (downward) directions, respectively. The shear stress is then given by the product of the complex shear modulus, G^*, and the shear strain, so

$$\tau(z,t) = G^* \frac{\partial u}{\partial z} = (G + i\omega\eta)\frac{\partial u}{\partial z} = (1 + 2i\xi)\frac{\partial u}{\partial z} \tag{7.29}$$

where η is the viscosity of the material (Section C.5.1). Introducing a local coordinate system, Z, for each layer, the displacement at the top and bottom of layer m will be

$$u_m(Z_m = 0, t) = (A_m + B_m)e^{i\omega t} \tag{7.30a}$$

$$u_m(Z_m = h_m, t) = \left(A_m e^{ik_m^* h_m} + B_m e^{-ik_m^* h_m}\right)e^{i\omega t} \tag{7.30b}$$

Displacements at layer boundaries must be compatible (i.e., the displacement at the top of a particular layer must be equal to the displacement at the bottom of the overlying layer). Applying the compatibility requirement to the boundary between layer m and layer $m + 1$, i.e.,

$$u_m(Z_m = h_m, t) = u_{m+1}(Z_{m+1} = 0, t) \tag{7.31}$$

yields

$$A_{m+1} + B_{m+1} = A_m e^{ik_m^* h_m} + B_m e^{-ik_m^* h_m} \tag{7.32}$$

The shear stresses at the top and bottom of layer m are

$$\tau_m(Z_m = 0, t) = ik_m^* G_m^* (A_m - B_m)e^{i\omega t} \tag{7.33a}$$

$$\tau_m(Z_m = h_m, t) = ik_m^* G_m^* \left(A_m e^{ik_m^* h_m} - B_m e^{-ik_m^* h_m} \right) e^{i\omega t} \tag{7.33b}$$

Since stresses must be continuous at layer boundaries,

$$\tau_m(Z_m = h_m, t) = \tau_{m+1}(Z_{m+1} = 0, t) \tag{7.34}$$

so

$$A_{m+1} - B_{m+1} = \frac{k_m^* G_m^*}{k_{m+1}^* G_{m+1}^*} \left(A_m e^{ik_m^* h_m} - B_m e^{-ik_m^* h_m} \right) e^{i\omega t} \tag{7.35}$$

Adding (7.32) and (7.35) and subtracting (7.35) from (7.32) gives the recursion formulas

$$A_{m+1} = \frac{1}{2} A_m \left(1 + \alpha_m^*\right) e^{ik_m^* h_m} + \frac{1}{2} B_m \left(1 - \alpha_m^*\right) e^{-ik_m^* h_m} \tag{7.36a}$$

$$B_{m+1} = \frac{1}{2} A_m \left(1 - \alpha_m^*\right) e^{ik_m^* h_m} + \frac{1}{2} B_m \left(1 + \alpha_m^*\right) e^{-ik_m^* h_m} \tag{7.36b}$$

where α_m^* is the complex impedance ratio at the boundary between layers m and $m+1$

$$\alpha_m^* = \frac{k_m^* G_m^*}{k_{m+1}^* G_{m+1}^*} = \frac{\rho_m (v_s^*)_m}{\rho_{m+1} (v_s^*)_{m+1}} \tag{7.37}$$

This means that, if the motion in layer m is known and the material properties of layers m and $m+1$ are known, the motion in layer $m+1$ can be determined. The process can obviously be repeated to obtain the motions in layers $m+2$, $m+3$, etc.

At the ground surface, the shear stress must be equal to zero, which requires [from Equation 7.33a, setting $m=1$] that $A_1 = B_1$. If the recursion formulas of Equation (7.36) are applied repeatedly for all layers from 1 to m, functions relating the amplitudes in layer m to those in layer 1 can be expressed by

$$A_m = a_m(\omega) A_1 \tag{7.38a}$$

$$B_m = b_m(\omega) B_1 \tag{7.38b}$$

The transfer function relating the displacement amplitude at layer i to that at layer j is given by

$$F_{ij}(\omega) = \frac{|u_i|}{|u_j|} = \frac{a_i(\omega) + b_i(\omega)}{a_j(\omega) + b_j(\omega)} \tag{7.39}$$

Because $|\ddot{u}| = \omega|\dot{u}| = \omega^2|u|$ for harmonic motion, Equation (7.39) also describes the amplification of accelerations and velocities from layer i to layer j. Equation (7.39) indicates that the motion in any layer can be determined from the motion in any other layer. Hence if the motion at any one point in the soil profile is known, the motion at any other point can be computed.

Example 7.4

As part of a comprehensive investigation of ground motion estimation techniques, the Electric Power Research Institute performed a detailed subsurface investigation at the site of the Gilroy No. 2 recording station (EPRI, 1993). A rough approximation to the measured shear wave velocity profile is listed below.

Depth Range (ft)	Average Shear Wave Velocity (ft/s)
0–20	500
20–45	700
45–70	1,500
70–130	1,000
130–540	2,000
> 540	5,000

Assuming, as in Examples 7.2 and 7.3, an average soil unit weight of 125 lb/ft^3 and 5% soil damping, compute the expected ground surface response when the bedrock is subjected to the Gilroy No. 1 motion.

Solution:

As in the previous two examples in this chapter, this problem requires evaluation of the transfer function that relates the ground surface motion to the bedrock motion. Because of multiple reflections within the layered system, the transfer function (Equation 7.39) for this example is considerably more complicated than for the single-layered cases of the previous examples.

While the transfer function can be evaluated by hand, it has also been coded in computer programs. The program ProShake 2.0 (www.proshake.com) was used, with constant soil stiffness and damping ratio, to obtain the transfer function shown in Figure E7.4c. As in the previous examples, the Fourier series of the ground surface motion (Figure E7.4d) was computed as the product of the transfer function and the Fourier series of the input motion. Inversion of this Fourier series produces the time series of ground surface acceleration (Figure E7.4e).

Examination of Figure E7.4c shows that the transfer function for the layered system is indeed more complicated than the transfer functions for the single-layered cases of Examples 7.2 and 7.3. Resonances producing narrow, high spikes in the transfer function at frequencies of about 1.3, 3.5, and 5.5 Hz help produce a peak acceleration of 0.499*g*, which is considerably larger than the peak acceleration of 0.322*g* that was recorded at the Gilroy No. 2 station.

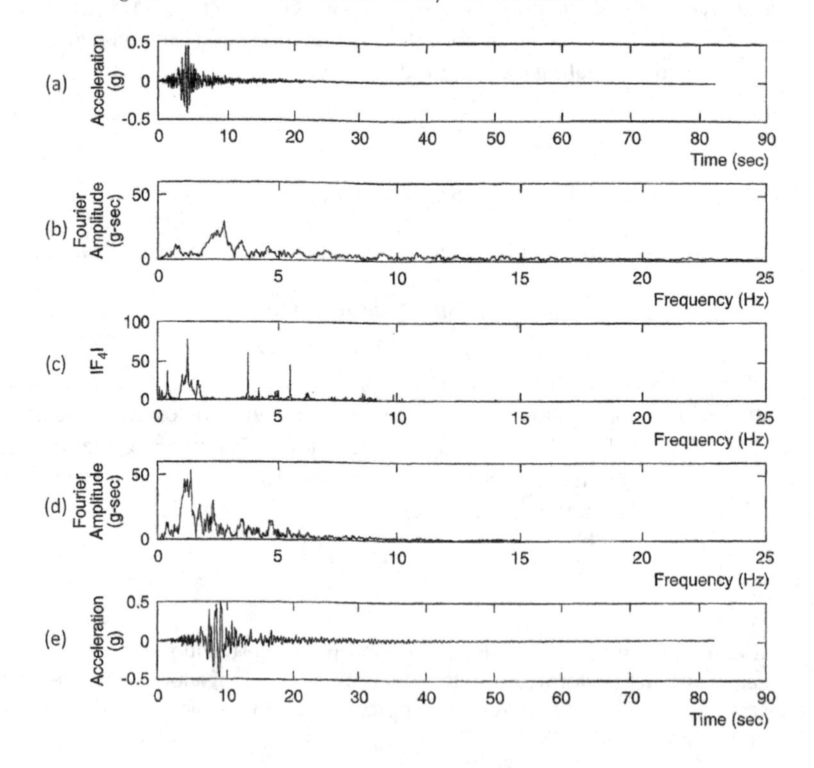

FIGURE E7.4A-E

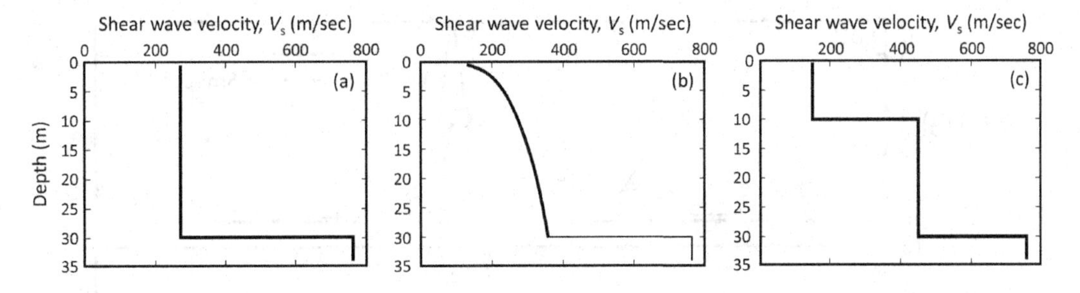

FIGURE 7.19 (a–c) Three 30-m-thick soil profiles with different velocity profiles but identical V_{S30} values (270 m/s).

The solutions for a layered profile allow detailed stratigraphy to be explicitly accounted for in a ground response analysis. Different soil profiles can have the same average stiffness, but respond differently to a particular ground motion – the distribution of that stiffness with depth, or stratigraphy, also plays a role in site response. Figure 7.19 shows three 30-m-thick soil profiles with identical V_{S30} values. One has a constant stiffness, another has a smoothly increasing stiffness, and the third consists of two soil layers of different stiffness; the shear wave travel time for all three layers is 0.111 sec, so V_{S30}=270 m/s for all three profiles. Despite having the same V_{S30} values, the transfer functions (Figure 7.20) have different shapes with actual characteristic site periods ranging from 0.34 to 0.45 sec. This period range is much smaller than that shown in Figure 7.14, but it indicates that the detailed sequence of soil layer stiffnesses, or stratigraphy, at a site can also influence site response, even for sites with the same average stiffness. Careful examination of the response spectra at periods greater than about twice the characteristic site period shows that they are virtually identical to the bedrock spectrum, indicating that the sites do not respond dynamically to periods significantly greater than the characteristic site period (or to frequencies significantly lower than the fundamental frequency) of the soil deposit. At these long periods, the wavelength of the motion is greater than the thickness of the soil profile so the layer essentially translates back and forth almost as a rigid body.

7.5.1.5 Deconvolution

Because the equivalent linear approach utilizes a linear analysis, the response at any point can be related to the response at any other point. Although the transfer functions developed in the preceding section are used to compute free surface motions from bedrock motions, transfer functions relating to motions at other depths can also be derived. This allows subsurface motions to be computed from ground surface motions using a process known as *deconvolution*. Deconvolution can be useful to derive motions for reference rock site conditions when the actual recordings are made on different (typically soil or weathered rock) conditions.

Deconvolution is typically performed using one-dimensional, equivalent linear ground response analyses that model the propagation of vertically traveling shear waves (hence, only horizontal particle motion). Mathematically, the process is relatively straightforward as the recorded surface motion is divided by the transfer function relating the surface to base (halfspace) motion. The recorded motion, however, reflects the contributions of both vertical and inclined body and surface waves and the potential nonlinearity of soils beneath the recording station. Moreover, those waves may be affected by lateral heterogeneity of ground conditions at the recording site. None of these effects are accounted for in the one-dimensional deconvolution analysis. Therefore, the peaks and troughs in the one-dimensional transfer function may not line up properly with the peaks and troughs in the Fourier spectrum of the recorded surface motion; dividing one by the other can lead to unrealistic amplitudes (either excessively high or low) in the deconvolved motion when a local peak in the surface spectrum is divided by a local trough (or vice versa) in the transfer function.

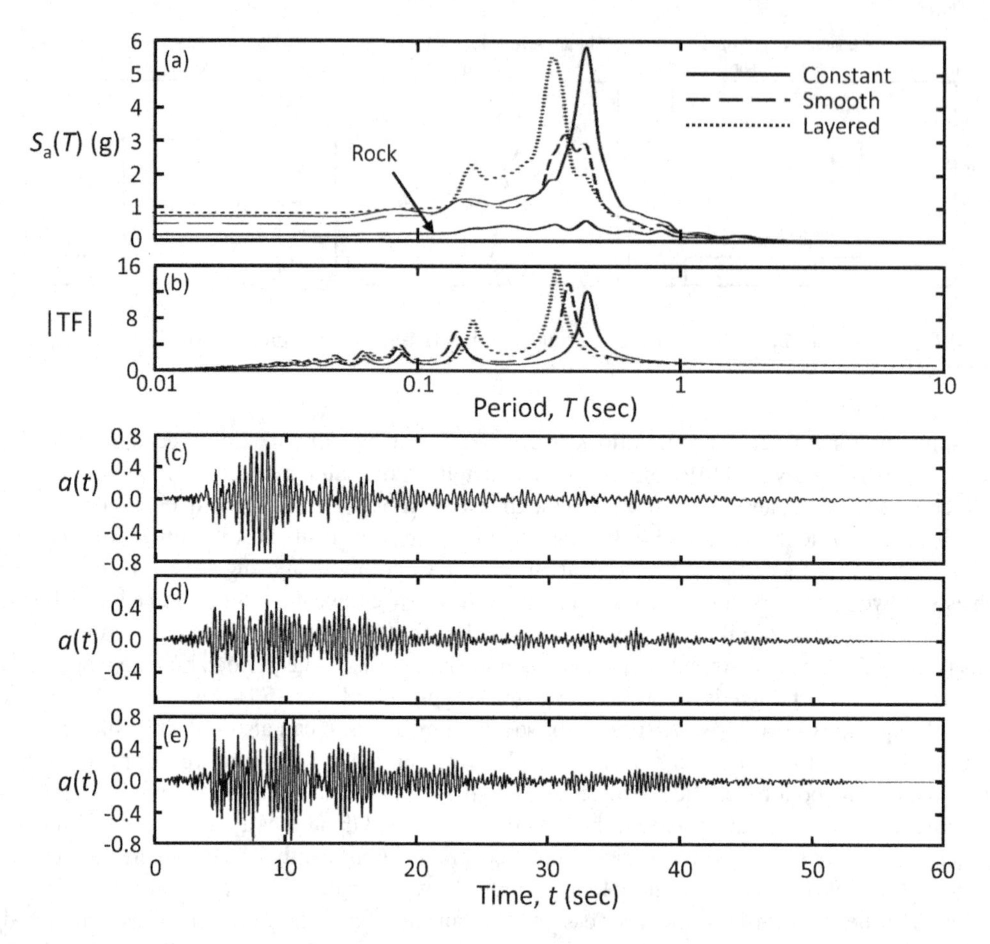

FIGURE 7.20 Effects of profile stratigraphy on ground surface response: (a) ground surface response spectra, (b) transfer functions, (c) ground surface acceleration for constant velocity profile, (d) ground surface acceleration for smoothly increasing velocity profile, (e) ground surface acceleration for layered velocity profile.

Figure 7.21 illustrates two sites in reasonable proximity (e.g., similar source-to-site distances and similar geology) in which a motion recorded at the surface of Site 1 (A_1) is to be used to predict the corresponding surface motion (A_2) at Site 2. The soil profiles at the two sites are different, however, so the recorded motion A_1 is to be deconvolved to obtain a rock motion that can be applied at the base of a model of Site 2. One option would be to deconvolve Motion A_1 down to the top of rock (at B_1) and use the resulting motion as input to a model that extended down to rock at Site 2, i.e., to apply the deconvolved Motion B_1 at B_2. If the rock at B_1 and B_2 is much stiffer than the soil above it at both sites (i.e., the impedance ratios are very large) and the shallow rock is relatively thick, the resulting input motion could produce an accurate estimate of the surface motion at Site 2 (Motion A_2). However, it is not uncommon for the upper portion of a rock layer to be weathered and hence have a stiffness that is low at its surface and increases gradually to that of the intact rock below it. Such a layer of shallow rock will contribute to the response at both sites and that contribution will be different if its thickness or stiffness is different; as a result, Motion A_1 could be deconvolved down to deeper and harder rock, thereby producing Motion C_1, which could then be applied at C_2. Motion C_1 would then account for the shallow rock at Site 1 and the ground response analysis would account for the thicker shallow rock at Site 2. The implicit assumption that Motion C_1 is the same as Motion C_2 would be better justified than the assumption that Motion B_1 is the same as Motion B_2.

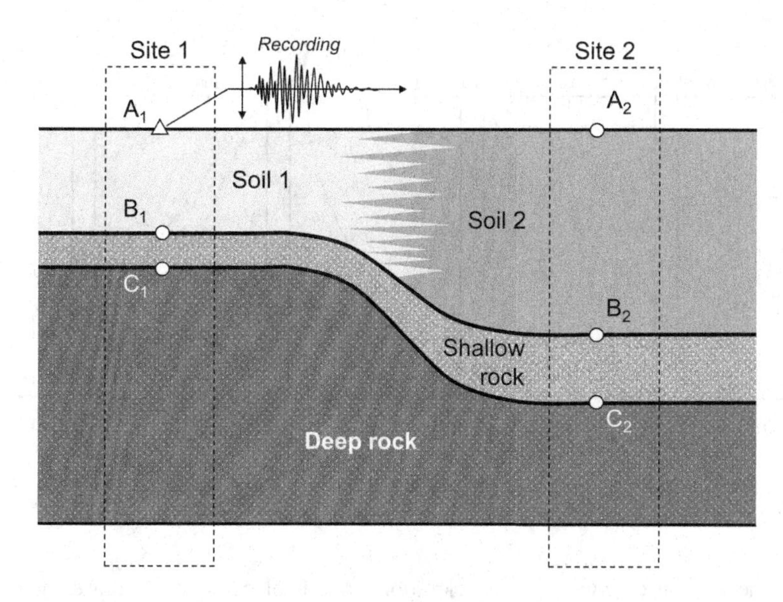

FIGURE 7.21 Schematic illustration of two sites underlain by common bedrock at depth where motion recorded at Site 1 can be used to compute corresponding motion at Site 2.

Different approaches have been used to reduce errors associated with deconvolution. Silva (1988) found that about 75% of the power (87% of the amplitude) in free surface motions at a soft soil site in California could be attributed to vertically propagating shear waves at frequencies up to 15 Hz; the remainder was attributed to scattered waves and surface waves. Silva suggested a deconvolution procedure based on: (a) initial deconvolution using a prefiltered (15 Hz low-pass) version of the surface motion scaled to 87% of its original amplitude, (b) extraction of the equivalent linear stiffness and damping values obtained in the final iteration of the initial analysis, and (c) a final deconvolution analysis using a single iteration with the stiffness and damping values from part (b) and the filtered surface motion from part (a) at 100% of its original amplitude. Markham et al. (2015) found that this procedure produced reasonable results when applied to strong motion data from the Christchurch Earthquake Sequence. Pretell et al. (2019) developed a methodology based on ratios of RVT-based (Section 7.5.2.2) transfer functions obtained by convolution through randomized profiles to produce a suite of motions that, in aggregate, reflect important effects of soil profile heterogeneity.

7.5.1.6 Soil Profile Response

One of the important benefits of site-specific analyses is that they allow evaluation of site response below, as well as at, the ground surface. Knowledge of subsurface response is important for a number of applications including soil-structure interaction (Chapter 8), liquefaction (Chapter 9), and seismic slope stability (Chapter 10). Examination of subsurface response also helps improve understanding of the manner in which sites respond to earthquake shaking, and how site characteristics influence the manner in which different ground motion characteristics vary with depth.

To demonstrate site response characteristics as a function of depth, consider the three sites with V_{S30}=270 m/s illustrated in Figure 7.19. The V_s profiles are re-plotted in Figure 7.22a. The profiles are underlain by a linear material with Vs = 760 m/s. Figures 7.22b–e show profiles of peak acceleration, peak displacement, peak shear stress, and peak shear strain for the three soil profiles. All three profiles were subjected to a ground motion that had a predominant period of about 0.4 sec, which is close to the characteristic site periods of 0.34–0.45 sec.

While the computed responses are sensitive to the details of the specific input motion used in the analyses, some features of the response are related to the characteristics of the profile itself.

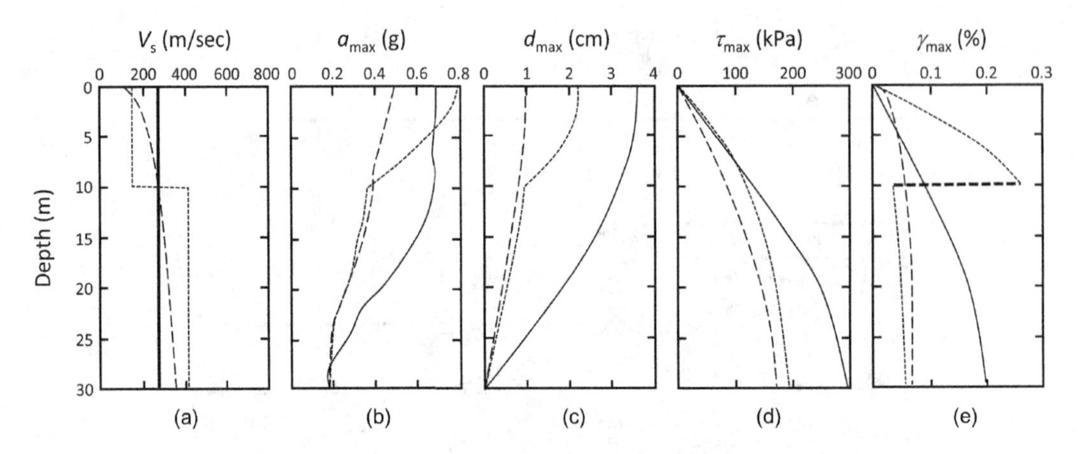

FIGURE 7.22 Profiles for three sites with V_{S30}=270 m/s: (a) shear wave velocity, (b) peak acceleration, (c) peak displacement, (d) peak shear stress, and (e) peak shear strain.

In all cases, the response (e.g., peak acceleration and displacement) decreases with depth (or, alternatively, increases as the ground surface is approached from below) due to impedance and free surface effects. The response profiles are relatively smooth and uniform for the cases with constant and smoothly increasing shear wave velocities, but show a sharp break in slope for the layered case. As indicated in Section C.4.1, deformation amplitude increases quickly when waves travel from a higher impedance (general stiffer) material into a lower impedance (generally softer) material, as occurs at a depth of 10 m for the layered profile – the peak acceleration and displacement amplitudes increase quickly at that depth. The rapid increase in displacement amplitude implies a rapid increase in shear strain amplitude, which is very clearly seen just above the impedance contrast; the shear stress profile, however, is characteristically much less sensitive to the shear wave velocity profile. An impedance contrast also exists at a depth of 30 m where the soil is underlain by stiffer bedrock. At that depth, the impedance ratio is highest for the constant velocity profile and lowest for the layered profile, and the response increases there most quickly for the profile with the highest impedance contrast.

Response spectra can also be computed for ground motions at different depths in a soil profile, and it is useful to examine how they vary with depth. Figure 7.23 shows the variation of response spectral acceleration, S_a, as a function of both period and depth over the thickness of the constant velocity profile; the darker gray zones indicate higher spectral acceleration. Figure 7.23b–d shows the corresponding response spectra at the ground surface, the middle (15 m depth) of the soil profile, and at the bottom of the profile (i.e., surface of bedrock). The grayscale map in Figure 7.23a shows a complex pattern of spectral acceleration amplitude; of particular note is the inclined band of light color that cuts diagonally across the bands of higher spectral acceleration from the upper left to lower right portions of the map. The lighter color indicates a band of lower spectral acceleration that contributes to a "hole" in the response spectrum over a relatively narrow range of frequencies. The hole occurs at low periods (high frequencies) at shallow depths and is consequently not visible in the ground surface spectrum shown in Figure 7.23b. It corresponds to longer periods (i.e., longer than the periods at which spectral acceleration is highest) at greater depth, and therefore is not prominent in the bedrock spectrum shown in Figure 7.23d. At intermediate depth, however, the hole cuts across the strongest part of the spectrum, producing a prominent depression in the 15-m-deep spectrum shown in Figure 7.23c. The source of the depression is the second-mode response node (Figure 7.15) that develops at a depth equal to $\lambda/4$ where λ is the wavelength of the second mode. For a soil profile with constant velocity, the depth of the node for harmonic loading at period, T, can be shown to be

$$z_{node} = \frac{TV_s}{4} \tag{7.40}$$

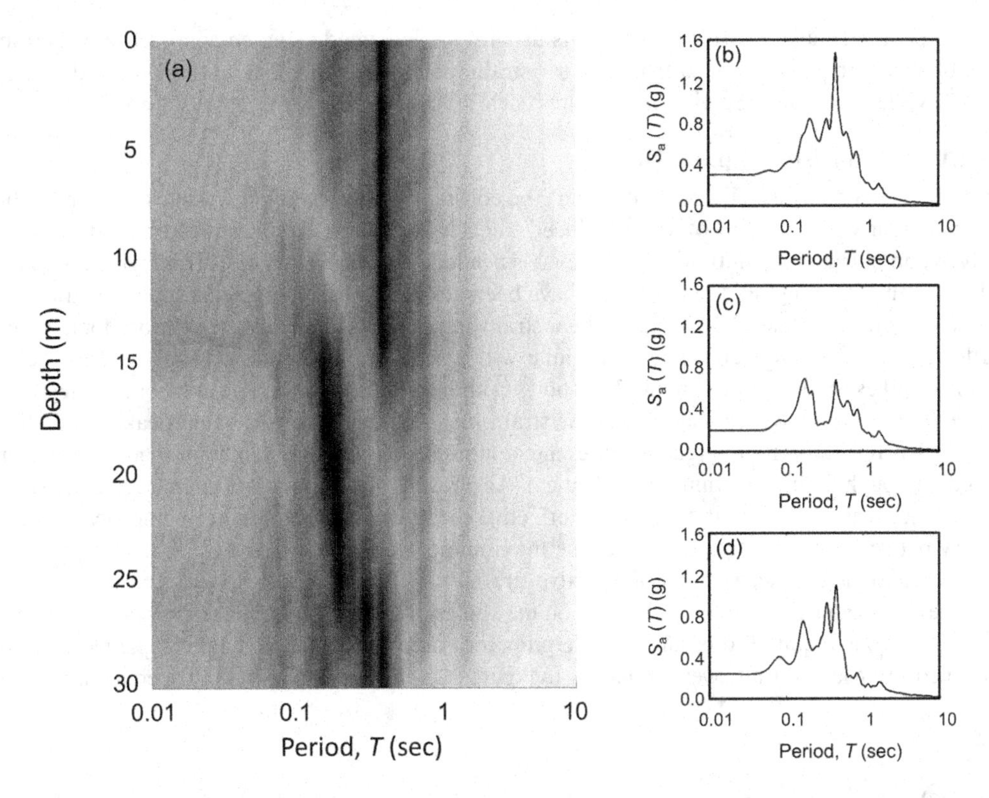

FIGURE 7.23 Depth-dependence of spectral acceleration through a $V_S=270$ m/s profile: (a) black-and-white map of spectral acceleration vs. depth and period, (b) response spectrum at ground surface, (c) response spectrum at 15 m depth, and (d) response spectrum at 30 m depth (bedrock level). Inclined line of light color represents response node, the depth of which increases with increasing wavelength (hence, period). Color scale runs from $S_a(T)=0$ (white) to $S_a(T)=1.5g$ (black).

Accounting for the fact that the motion is not simple harmonic, the depth of the light-colored band in Figure 7.23a is consistent with the depths predicted by Equation (7.40). In a softer profile, this type of band would be located at shallower depths, and additional bands due to higher-mode nodes could appear at other depths. The complexity of the patterns produced by the interaction of upward- and downward-propagating waves with the spectral characteristics of the ground motion illustrates the sensitivity of ground motions to the characteristics of the profile through which they travel.

7.5.2 Equivalent Linear Approximation of Nonlinear Response

When ground motions produce sufficiently large strains in a soil profile that nonlinear soil behavior occurs, as is typical for practical problems of interest, the linear approach must be modified to provide reasonable estimates of ground response. As discussed in Chapter 6, soils become softer and dissipate more energy as cyclic shear strain amplitudes increase. The actual nonlinear hysteretic stress-strain behavior of cyclically loaded soils can be approximated by equivalent linear soil properties. The equivalent linear shear modulus, G, is generally taken as a secant shear modulus and the equivalent linear damping ratio, ξ, as the damping ratio that produces the same energy loss in a single cycle as the actual hysteresis loop. The strain-dependent nature of these equivalent linear properties was described in Section 6.4.5.1.

In the equivalent linear approach, the shear modulus and damping ratio are adjusted in a series of iterations until their values in each soil layer are consistent with the level of strain induced in that layer. This procedure results in lower shear moduli and higher damping ratios as induced strain levels increase. Equivalent linear analysis is usually implemented using an approach in which the level of shear strain is obtained from a shear strain history, but can also be implemented using a random vibration theory (Section 7.5.2.2) approach.

7.5.2.1 Strain History Approach

Since the linear analysis requires that G and ξ be constant for each soil layer, values of G and ξ that are consistent with the level of strain induced in each layer must be determined. To address this problem, an objective definition of strain level is needed. The laboratory tests from which modulus reduction and damping ratio relationships have been developed using simple harmonic loading and characterize the strain level by the peak shear strain amplitude. The shear strain history for a typical earthquake motion, however, is highly irregular with a peak amplitude that may only be approached by a few spikes in the record. Figure 7.24 shows both harmonic (as in a typical laboratory test) and transient (as in a typical earthquake) shear strain histories that have the same peak cyclic shear strain. Clearly, the harmonic record represents a more severe loading condition than the transient record, although their peak values are identical. As a result, it is common to characterize the strain level of the transient record in terms of an effective shear strain which has been empirically found to vary between about 50% and 70% of the maximum shear strain. There are, however, different approaches to the characterization of effective strain.

Since the computed strain level depends on the values of the equivalent linear properties, an iterative procedure is required to ensure that the properties used in the analysis are compatible with the computed strain levels in all layers. Referring to Figure 7.25, the iterative procedure operates as follows:

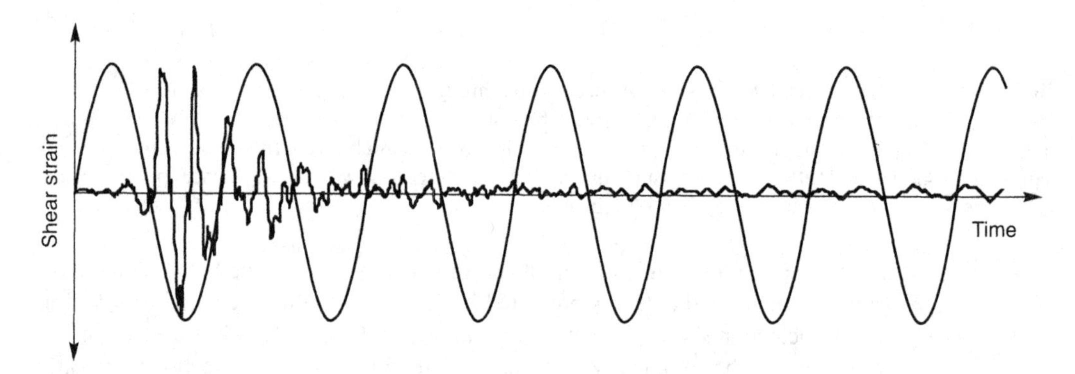

FIGURE 7.24 Two shear strain histories with identical peak shear strains. For the transient motion of an actual earthquake, the effective shear strain is usually taken as 65% of the peak strain.

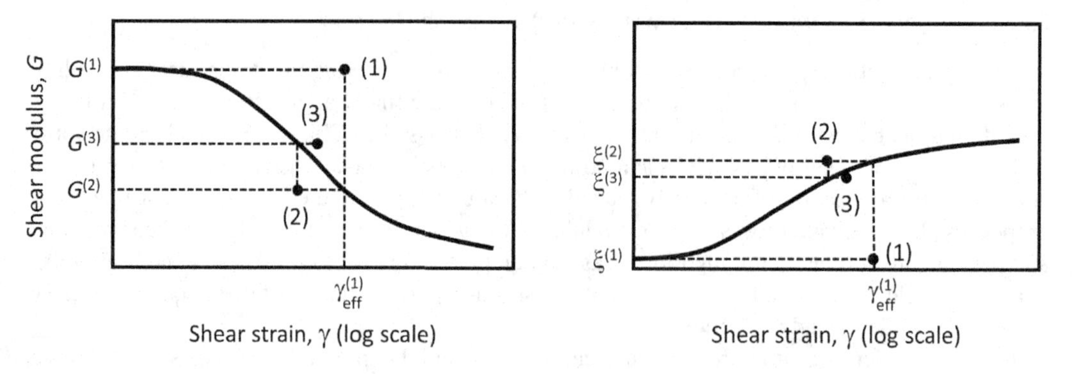

FIGURE 7.25 Illustration of modulus reduction and damping ratio changes between iterations in equivalent linear analysis.

1. Initial estimates of G and ξ, are made for each layer. The initially estimated values should correspond to the same strain level; the low-strain values are often used for the initial estimate.
2. The estimated G and ξ values are used to compute the ground response, including shear strain histories for each layer.
3. The effective shear strain in each layer is determined from the computed shear strain in each layer. The effective shear strain is most easily and commonly obtained in the time domain from the computed shear strain history. For layer j

$$\gamma_{eff\,j}^{(i)} = R_{\gamma}\gamma_{\max j}^{(i)} \tag{7.41}$$

where the superscript refers to the iteration number and R_γ is the ratio of the effective shear strain to maximum shear strain. R_γ depends on earthquake magnitude (Idriss and Sun, 1992) and should be estimated from

$$R_{\gamma} = \frac{\mathbf{M}-1}{10} \tag{7.42}$$

but is often taken as 0.65.

4. From this effective shear strain, new equivalent linear values, $G^{(i+1)}$ and $\xi^{(i+1)}$ are chosen for the next iteration.
5. Steps 2–4 are repeated (Figure 7.25) until differences between the computed shear modulus and damping ratio values in two successive iterations fall below some predetermined value in all layers. Although convergence is not absolutely guaranteed, differences of less than 5%–10% are usually achieved in three to five iterations (Schnabel et al., 1972). Some ground response analysis programs use optimization procedures, which allow more rapid and robust convergence, instead of this simple procedure.

Example 7.5

An extensive laboratory testing program conducted by EPRI (1993) produced detailed information on the modulus reduction and damping characteristics of the soils beneath the Gilroy No. 2 recording station. Although the soil conditions varied with depth, a rough approximation to the average modulus reduction and damping characteristics is given below.

Parameter	Strain (%)										
	10^{-4}	$10^{-3.5}$	10^{-3}	$10^{-2.5}$	10^{-2}	$10^{-1.5}$	10^{-1}	$10^{-0.5}$	10^{0}	$10^{0.5}$	10^{1}
G/G_{max}	1.00	1.00	1.00	0.99	0.90	0.71	0.47	0.24	0.10	0.05	0.04
ξ (%)	3.00	3.00	3.000	3.53	4.83	7.68	12.3	18.5	24.4	27.0	30.0

Repeat the analysis of Example 7.5 with the data listed above using the iterative equivalent linear approach.

Solution:

As in the case of Example 7.4, the transfer function was evaluated using the computer program ProShake (www.proshake.com). In this example, the first iteration used stiffness and damping values consistent with the modulus reduction and damping behavior listed above. After a total of eight iterations, the shear moduli and damping ratios had converged to within 1% of strain-compatible values. Because the strain-compatible shear moduli were smaller than the low-strain moduli on which the analysis of Example 7.4 was based (the iterations converged to strains at which G/G_{max} values were less than 1.0), the transfer function (Figure E7.5c) was shifted toward lower frequencies. As in the previous examples, the Fourier spectrum of the ground surface motion (Figure E7.5d) was computed as the product of the transfer function and the Fourier series

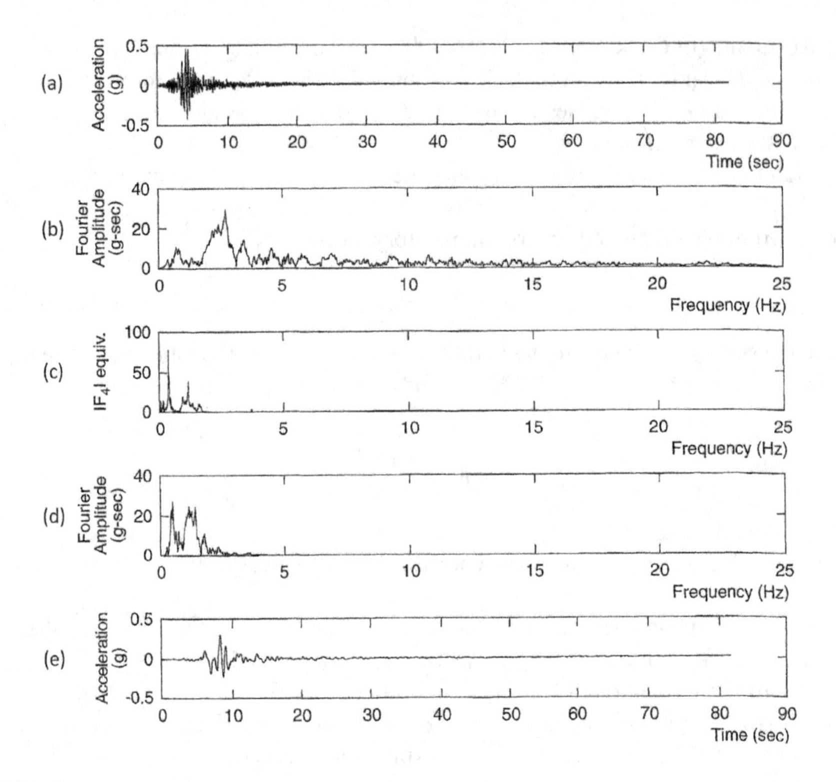

FIGURE E7.5A-E

of the bedrock motion. Inversion of this Fourier series produced the time history of ground surface acceleration shown in Figure E7.5e.

The softer soil behavior indicated by the iterative equivalent linear analysis is clearly reflected in the transfer function (Figure E7.5c), the Fourier spectrum of the ground surface motion (Figure E7.5d), and the time series of ground surface acceleration (Figure E7.5e). The predicted peak ground surface acceleration of 0.304*g* compares well with the peak acceleration of 0.322*g* recorded at the Gilroy No. 2 (soil) station. Improved predictions of the response at the Gilroy site were made with more detailed site characterization (EPRI, 1993).

Even though the process of iteration toward strain-compatible soil properties allows nonlinear soil behavior to be approximated, it is important to remember that the transfer function-based calculations still require linear soil behavior. The strain-compatible soil properties are constant throughout the duration of the earthquake, regardless of whether the strains at a particular time are small or large. The method is therefore incapable of representing the changes in soil stiffness that actually occur during the earthquake. The equivalent linear approach to one-dimensional ground response analysis of layered sites has been coded into a widely used computer program called SHAKE (Schnabel et al., 1972). Several Windows-based computer programs for equivalent linear site response analysis are now available.

As the amplitude of an input motion increases, the strain induced in a particular soil profile increases, which means that the stiffness of the soil decreases and the damping ratio increases. These changes affect the transfer function in predictable ways – the reduced stiffness leads to an increased characteristic site period (or a reduced fundamental frequency), and the increased damping ratio reduces the peak values of the transfer function as well as its sensitivity to input motion frequency. Figure 7.26 shows the results of equivalent linear analyses of a 30-m-thick sand profile with a smoothly increasing shear wave velocity profile (V_{S30}=270 m/s). The profile is subjected to the same input motion scaled to peak accelerations of 0.01*g*, 0.05*g*, and 0.20*g*.

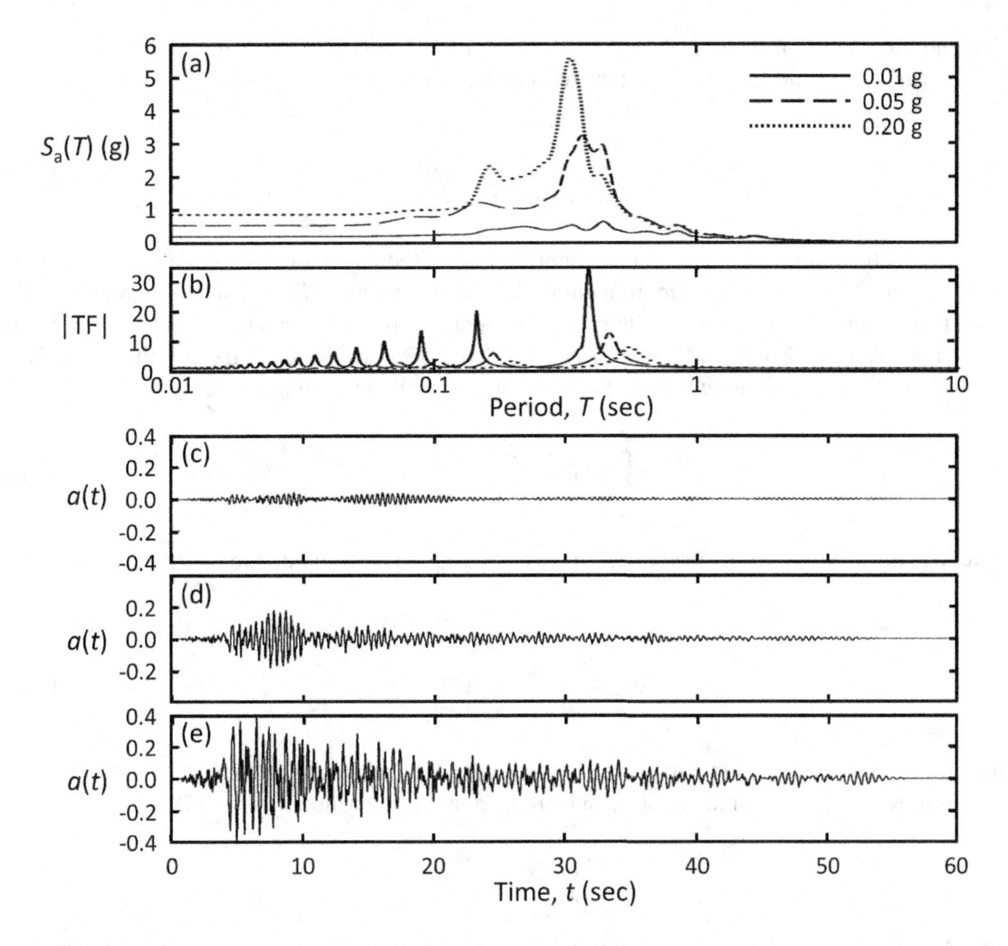

FIGURE 7.26 Effects of input motion amplitude on ground surface response: (a) ground surface response spectra, (b) transfer functions, (c) ground surface acceleration for bedrock $a_{max}=0.01g$, (d) ground surface acceleration for bedrock $a_{max}=0.05g$, (e) ground surface acceleration for bedrock $a_{max}=0.20g$.

The amplitudes of the ground surface response spectra are clearly different, as would be expected when the amplitude of one input motion is 20 times larger than that of another. The stronger input motions induce larger strains in the soil, which leads to lower strain-compatible shear moduli and higher strain-compatible damping ratios. The lower stiffness pushes the fundamental period of the profile out to longer periods; the higher damping ratio causes the peaks of the transfer function to be lower and broader, and the low-period (high frequency) values to decay rapidly. The amplitudes of the response spectra and time histories, however, do not vary by a factor of 20. The peak accelerations of the ground surface motions are $0.040g$, $0.218g$, and $0.404g$, so peak accelerations are amplified by factors of 4.0, 4.4, and 2.0. The peak shear strains for the three input motions were 0.005%, 0.045%, and 0.139%. The reduction in amplification with higher input motion amplitude results from the soil nonlinearity that becomes pronounced in the strongest of the three motions considered here, and is seen consistently in empirical data (Sections 7.3.4 and 3.5.2.3).

7.5.2.2 Random Vibration Theory Approach

In many fields of dynamics, loading is not deterministic and is therefore characterized as being random in time, or *stochastic*. Procedures for evaluation of the response of physical systems to stochastic loading have been developed (Crandall and Mark, 1963). *Random vibration theory* (RVT) provides a means for estimating peak values of a random function from the *root-mean-square* (*rms*) value and knowledge of the distribution of the function (Cartwright and Longuet-Higgins, 1956). The *rms* value of a time-varying function, $x(t)$, can be computed as

$$x_{rms} = \sqrt{\frac{1}{D_{rms}} \int_0^{D_{rms}} [x(t)]^2 \, dt} \tag{7.43}$$

where D_{rms} is the duration over which x_{rms} applies. For earthquake motions, which are non-stationary, D_{rms} is the sum of the ground motion duration (typically taken as $D_{5\text{-}75}$) and a term related to the period and damping ratio of the oscillator that accounts for the enhanced duration associated with its dynamic response (Boore and Joyner, 1984; Kottke and Rathje, 2009). Parseval's theorem relates energy in the time domain to energy in the frequency domain and can be expressed as

$$\int_{-\infty}^{\infty} |x(t)|^2 \, dt = \int_{-\infty}^{\infty} |X(f)|^2 \, df \tag{7.44}$$

where $X(f)$ is the Fourier transform of $x(t)$. This allows x_{rms} to be computed in the frequency domain as

$$x_{rms} = \sqrt{\frac{2}{D_{rms}} \int_0^{\infty} |X(f)|^2 \, df} \tag{7.45}$$

where the integer 2 appears in Equation (7.45) because the integration is over positive frequencies only. From RVT, the peak value of $x(t)$ can be related to the *rms* value (Boore, 1987) as

$$\frac{x_{\max}}{x_{rms}} = \sqrt{2\ln(N)} \tag{7.46}$$

where N is the number of extrema in the time series, which can be estimated from the characteristic frequency, f_{ch}, and duration, T, as

$$N = 2 f_{ch} T \tag{7.47}$$

The characteristic frequency can be related to the moments of the Fourier amplitude spectrum

$$f_{ch} = \frac{1}{2\pi} \sqrt{\frac{m_2}{m_0}} \tag{7.48}$$

where the kth moment can be computed, using Parseval's theorem, in the frequency domain as

$$m_k = 2(2\pi)^k \int_0^{\infty} f^k |X(f)|^2 \, df \tag{7.49}$$

Using these relationships, the peak value of a time-varying function can be estimated entirely in the frequency domain. Details of the derivations of the preceding relationships can be found in Carter and Longuet-Higgins (1956) and Boore (1983).

In an RVT-based ground response analysis, the peak shear strain, which is needed to compute the effective strain required in the process of iterating toward strain-compatible soil properties, can be estimated from the strain amplitude spectrum and duration. The effective shear strain, therefore, can be computed (in the frequency domain) as

$$\gamma_{eff} = R_\gamma \gamma_{\max} = R_\gamma \sqrt{\frac{2}{D_{rms}} \int_0^\infty |\gamma(f)|^2 \, df} \cdot \sqrt{2 \ln \frac{D_{rms}}{\pi} \sqrt{\frac{2(2\pi)^2 \int_0^\infty f^2 |\gamma(f)|^2 \, df}{2 \int_0^\infty |\gamma(f)|^2 \, df}}} \quad (7.50)$$

In concept, then, an RVT-based site response analysis does not need a time history as an input motion – rather, it simply needs the Fourier amplitude spectrum and duration of the input motion. Rathje et al. (2005) describe procedures for estimating a Fourier amplitude spectrum from a response spectrum, which allows more readily available response spectra and durations to be used as input for RVT-based analyses.

The results of an RVT analysis can be interpreted as representing the mean response of all phase spectra that can be associated with a particular Fourier amplitude spectrum, thus eliminating the need to define and analyze multiple input motions associated (e.g., by scaling or spectral matching) with some particular hazard level. The input spectra for RVT analyses are generally smooth and thus do not contain the motion-to-motion variability that is present in suites of scaled motions used as inputs to time-domain site response analyses. RVT analyses are therefore best suited for determining the median amplification behavior of a soil profile. Pehlivan et al. (2016) showed that time series and RVT equivalent linear analyses produced similar median amplification behavior, except near the characteristic site period where the RVT amplification was about 25% greater than the time series amplification. When randomized shear wave velocities were used with both, the difference was reduced to about 10%.

7.5.2.3 Frequency-Dependent Approach

Conventional equivalent linear analyses are prone to overdamping of high-frequency motions under strong shaking levels. The strain-compatible stiffness and damping characteristics of a particular layer are based on the peak strain amplitude of that layer, which occurs during the strongest part of shaking in the record. The shear modulus and damping ratio are applied over the entire duration of the motion, which misrepresents the response in the early portion of the record because they are affected by the higher strain levels that have not yet occurred. Recognizing that low-amplitude components of a motion are often associated with higher frequencies that tend to occur early in ground motions, a number of researchers (Sugito et al., 1994; Kausel and Asimaki, 2002; Yoshida et al., 2002; Meite et al., 2021) developed procedures that use a smoothed Fourier strain spectrum to assign different effective strains to different frequencies; the fact that strain spectra generally decrease at high frequencies causes higher frequency components in a motion to propagate at higher velocities and with lower damping than they would in conventional equivalent linear analyses. The frequency-dependent analyses have been observed to better match the results of nonlinear analyses than conventional equivalent linear analyses in some studies (Kausel and Asimaki, 2002) but inconsistencies between different frequency-dependent formulations and poor agreement with the results of nonlinear analyses have been found in others (Kwak et al., 2008; Zalachoris and Rathje, 2015). Procedures for correcting the results of conventional equivalent linear analyses to more accurately represent the effects of material nonlinearity on high-frequency response have been proposed (Xu and Rathje, 2021).

7.5.3 Nonlinear Approach

Although the equivalent linear approach is computationally convenient and provides reasonable results for many practical problems, it remains an approximation to the actual nonlinear, inelastic behavior of soils subjected to strong seismic ground response. An alternative approach is to analyze the nonlinear response of a soil deposit using direct numerical integration in the time domain. By integrating the equation of motion in small time steps, any linear or nonlinear stress-strain model (Section 6.4.5.1–6.4.5.2) or advanced constitutive model (Section 6.4.5.3) can be used to describe the response of the soil. At the beginning of each time step, the stress-strain relationship is referred to in order to obtain the appropriate soil properties (e.g., tangent shear moduli) to be used in that time step. By this method, a nonlinear inelastic stress-strain relationship can be followed in a set of small incrementally linear steps. By accounting for inelastic behavior following stress reversals, nonlinear methods model hysteretic damping directly.

Nonlinear analysis procedures can be distinguished by the manners in which they solve the wave propagation problem and represent nonlinear soil behavior. Two main approaches – lumped-mass and wave equation models – are used to solve the wave propagation problem, and a variety of cyclic nonlinear and advanced constitutive models are used to represent nonlinear soil behavior.

7.5.3.1 Lumped Mass Approach

Most available one-dimensional nonlinear analyses have evolved from the lumped-mass approach used in the original computer program, DESRA (Lee and Finn, 1978) developed at the University of British Columbia. In the lumped-mass approach, the soil profile is discretized into a series of lumped masses assumed to be located at the boundaries between material layers (or sublayers within a given material layer). As indicated in Figure 7.27, each lumped mass represents the lower half of the overlying sublayer and the upper half of the underlying sublayer. The stiffness and damping coefficients, however, correspond to the individual layer values. This formulation leads to the development of a set of incremental equations of motion

$$[M]\{\Delta\ddot{u}\}+[C]\{\Delta\dot{u}\}+[K]\{\Delta u\}=-[M]\{I\}\Delta\ddot{u}_b \tag{7.51}$$

where $[M]$=mass matrix, $[C]$=damping matrix, $[K]$=stiffness matrix, $\{\Delta\ddot{u}\}$=incremental (relative) acceleration vector, $\{\Delta\dot{u}\}$=incremental (relative) velocity vector, $\{\Delta u\}$=incremental (relative) displacement vector, $\{I\}_b$=unit vector, and $\Delta\ddot{u}_b$=incremental (absolute) base acceleration. The relative motions are relative to the base of the profile. Most one-dimensional lumped mass codes represent the damping matrix $[C]$ as a linear combination of the mass and stiffness matrices,

$$[C]=\alpha[M]+\beta[K] \tag{7.52}$$

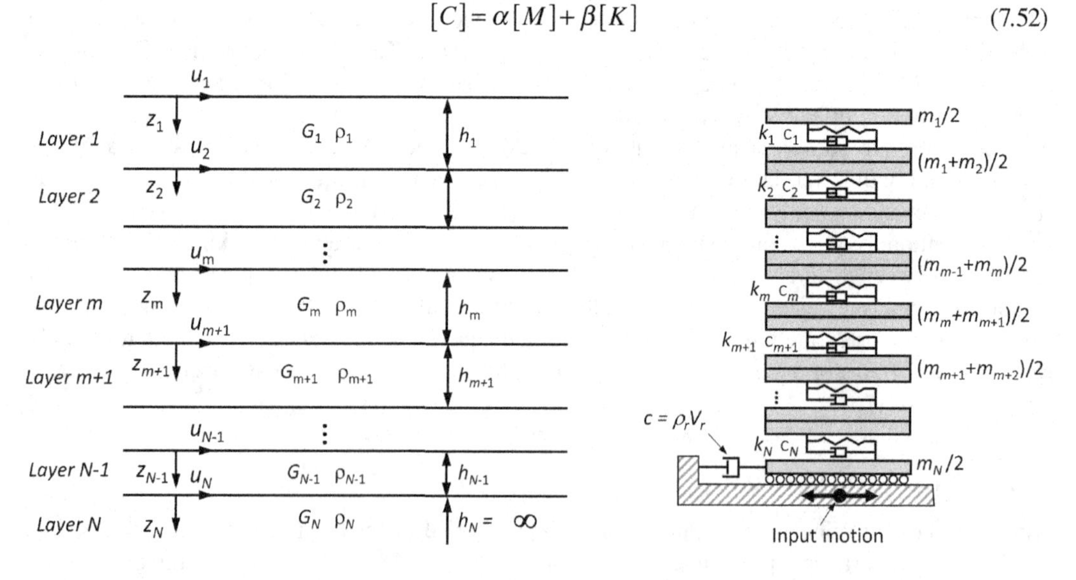

FIGURE 7.27 Illustration of the relationship between the actual profile and lumped mass model.

where α and β are coefficients that can be selected to control the variation of damping with frequency. This is known as Rayleigh damping (Section B.10.3) and is used in nonlinear analyses to represent low-strain damping (Section 6.6.4.1) and to provide numerical stability at small strain levels. The incremental equation of motion can be integrated in a series of time steps using standard methods of structural dynamics; the Newmark β method (Newmark, 1959) is commonly used in lumped-mass site response analyses.

At the end of each time step, the stiffnesses of the individual layers can be adjusted as indicated by the nonlinear stress-strain relationship for use in the next time step. By this process, the nonlinear, inelastic stress-strain behavior of each soil layer is modeled in a series of linear increments and hysteretic damping is accounted for directly.

7.5.3.2 Wave Equation Approach

The alternative approach is to treat the soil profile as a continuum and solve the wave equation rather than the equation of motion of a discrete, lumped-mass system. Consider the soil deposit of infinite lateral extent shown in Figure 7.28a. If the soil layer is subjected to horizontal motion at the bedrock level, the response will be governed by the equation of motion

$$\frac{\partial \tau}{\partial z} = \rho \frac{\partial^2 u}{\partial t^2} = \rho \frac{\partial \dot{u}}{\partial t} \tag{7.53}$$

Equation (7.53) is a partial differential equation that can be solved analytically only for very simple boundary (and initial) conditions. For virtually all practical problems, it must be solved numerically, and finite difference analysis is the most common way of doing so. A finite difference solution replaces the differential terms in a differential equation with difference terms that apply to finite increments of space and time. As the size of those increments goes to zero, the finite difference solution approaches the exact solution; for small increments, it is an approximation of the exact solution. In an explicit finite difference formulation, the conditions at a particular time step are computed directly from the conditions at the preceding time step. This makes the explicit method calculations simple and fast but, as will be shown shortly, potentially unstable when time increments (time steps) are too large. Implicit finite difference formulations evaluate the spatial derivative at both the time step of interest and the immediately preceding time step, which requires the time-consuming solution of a linear system of equations at each time step. However, implicit methods are unconditionally stable, which means that longer time steps can be used. The trade-off between fast calculations at a large number of time steps (explicit method) and slower calculations at a smaller number of time steps (implicit method) generally favors the explicit method, which is more commonly used in one-dimensional ground response analyses.

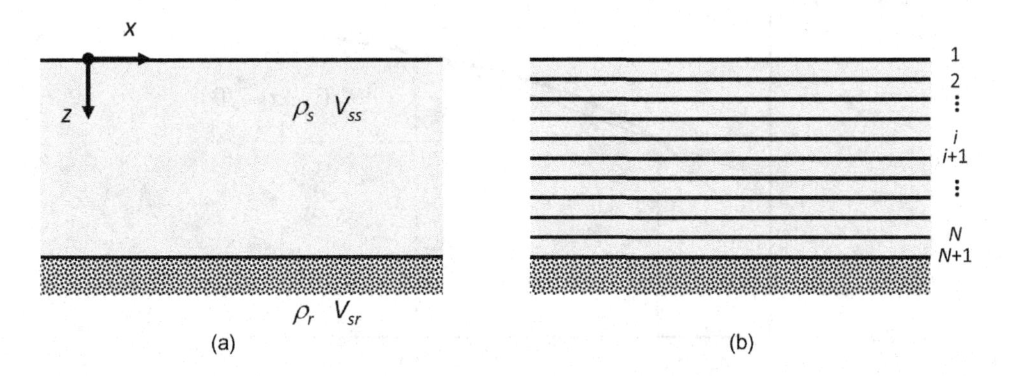

FIGURE 7.28 (a) Nomenclature for uniform soil deposit of infinite lateral extent overlying bedrock; (b) discretization of soil deposit into N sublayers – the indices are for layer boundaries.

To introduce the explicit finite difference method, consider the function, $f(x)$ shown in Figure 7.29, where x can represent space or time. The first derivative of, $f(x)$ at $x = \tilde{x}$ is given by

$$\frac{df(\tilde{x})}{dx} = \lim_{\Delta x \to 0} \frac{f(\tilde{x} + \Delta x) - f(\tilde{x})}{\Delta x} \tag{7.54}$$

A reasonable approximation to the first derivative can be made by removing the restriction of the limit and using a small but finite value of Δx. In this way, the expression of Equation (7.54) is referred to as a forward difference approximation to $df(\tilde{x})/dx$. Figure 7.29 illustrates that the forward difference actually provides a better approximation to the derivative at $x = \tilde{x} + \Delta x/2$ than at $x = \tilde{x}$ or $x = \tilde{x} + \Delta x$.

Dividing the soil layer into N sublayers of thickness, Δz (Figure 7.28b), and proceeding through time in small time increments of length, Δt, the notation $u_{i,t} = u(z = i\Delta z, t)$ can be used to write finite difference approximations to the derivatives

$$\frac{\partial \tau}{\partial z} = \frac{\tau_{i+1,t} - \tau_{i,t}}{\Delta z} \tag{7.55a}$$

$$\frac{\partial \dot{u}}{\partial t} = \frac{\dot{u}_{i,t+\Delta t} - \dot{u}_{i,t}}{\Delta t} \tag{7.55b}$$

Substituting Equations (7.55) into the equation of motion allows that differential equation to be approximated by the explicit finite difference equation

$$\frac{\tau_{i+1,t} - \tau_{i,t}}{\Delta z} = \rho \frac{\dot{u}_{i,t+\Delta t} - \dot{u}_{i,t}}{\Delta t} \tag{7.56}$$

Solving for $\dot{u}_{i,t+\Delta t}$ gives

$$\dot{u}_{i,t+\Delta t} = \dot{u}_{i,t} + \frac{\Delta t}{\rho \Delta z}(\tau_{i+1,t} - \tau_{i,t}) \tag{7.57}$$

Equation (7.57) simply shows how the conditions at time, t, can be used to determine the conditions at time, $t + \Delta t$. Using Equation (7.57) for all i, the velocity profile can be determined at time $t + \Delta t$. Using the computed velocities at the end of each time step as the initial velocities for the next time step, the repeated application of Equation (7.57) allows the equation of motion to be integrated into a series of small time steps.

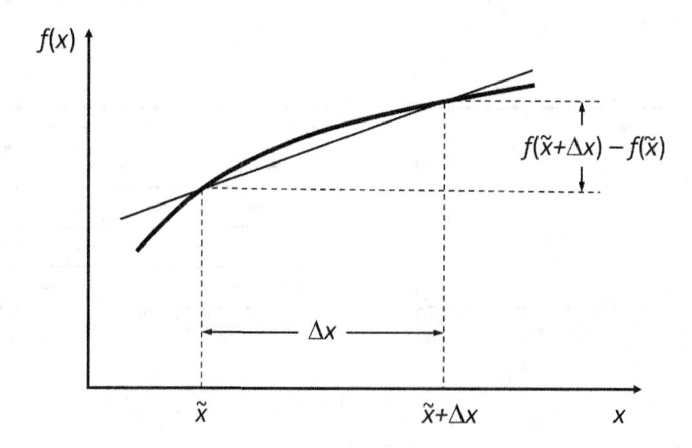

FIGURE 7.29 Forward difference approximation of $f'(\tilde{x})$ is given by slope of line passing through function at $x = \tilde{x}$ and $x = \tilde{x} + \Delta x$. Approximation becomes exact as $\Delta x \to 0$.

As with any integration problem, however, the boundary conditions must be satisfied. Since the ground surface is a free surface, $\tau_1=0$, so

$$\dot{u}_{i,t+\Delta t} = \dot{u}_{1,t} + \frac{\Delta t}{\rho \Delta z}\tau_{2,t} \tag{7.58}$$

The boundary condition at the bottom of the soil deposit depends on the nature of the underlying bedrock. If the bedrock is rigid, its particle velocity, $\dot{u}_b(t) = \dot{u}_{N+1,t}$ can be specified directly as the input motion. If the bedrock is elastic, continuity of stresses requires that the shear stress at the bottom of the soil layer, τ_{N+1}, be equal to the shear stress at the top of the rock layer, $\tau_{r,t}$. Thus

$$\dot{u}_{N+1,t+\Delta t} = \dot{u}_{N+1,t} + \frac{\Delta t}{\rho \Delta z}\left(\tau_{r,t} - \tau_{N,t}\right) \tag{7.59}$$

If an incident wave traveling upward through the rock has a particle velocity $\dot{u}_r(t)$ at the soil-rock boundary, the shear stress at the boundary is approximated (Joyner and Chen, 1975) by

$$\tau_{r,t} = \rho_r v_{sr}\left(2\dot{u}_r(t+\Delta t) - \dot{u}_{N+1,t+\Delta t}\right) \tag{7.60}$$

Substituting Equation (7.60) into Equation (7.59) gives

$$\dot{u}_{N+1,t+\Delta t} = \frac{\dot{u}_{N+1,t} + \dfrac{\Delta t}{\rho \Delta z}\left[2\rho_r v_{sr}\dot{u}_r(t+\Delta t) - \tau_{N,t}\right]}{1 + (\Delta t/\rho \Delta z)\rho_r v_{sr}}$$

Once the boundary conditions have been established, the integration calculations proceed from the bottom ($i=N+1$) to the top ($i=1$) of the soil deposit in each time step, and step by step in time. Computation of the velocity at the end of each time step, however, requires knowledge of the shear stress in that time step.

If the soil deposit is initially at rest, then $\dot{u}_{i,t=0} = 0$ and $\tau_{i,t} = 0$ for all i. When the input motion, in the form of $\dot{u}_b(t)$ (rigid bedrock) or $\dot{u}_r(t)$ (elastic bedrock), imparts some velocity to the base of the soil deposit, $\dot{u}_{N+1}$ will take on a nonzero value. In subsequent time steps, $\dot{u}_N$, $\dot{u}_{N-1}$, $\dot{u}_{N-2}$, … will all take on nonzero values as the soil deposit moves in response to the input motion. The incremental displacement in each time step is given by

$$\Delta u_{i,t} = \dot{u}_{i,t}\Delta t \tag{7.61}$$

Summing the incremental displacements allows the total displacement, $u_{i,b}$ to be determined at the beginning of each time step. The shear strain in each sublayer is given by

$$\gamma_{i,t} = \frac{\partial u_{i,t}}{\partial z} \approx \frac{u_{i+1,t} - u_{i,t}}{\Delta z} \tag{7.62}$$

If the soil is assumed to be linearly elastic, the shear stress depends only on the current shear strain, (i.e., $\tau_{i,t} = G_i\gamma_{i,t}$). If the soil is nonlinear and inelastic, however, the shear stress will depend on the current shear strain and the stress-strain history. In such cases the computed shear strain, $\gamma_{i,t}$ and the cyclic stress-strain relationship (or advanced constitutive model) are used to determine the corresponding shear stress, $\tau_{i,t}$. The integration process can then be summarized as follows:

1. At the beginning of each time step, the particle velocity, $\dot{u}_{i,t}$ and total displacement, $u_{i,t}$, are known at each layer boundary. Both are assumed to be zero at all depths at the beginning of the first time step.

2. The displacement profile is used to determine the shear strain, $\gamma_{i,t}$, within each layer. The strains will also be zero at the beginning of the first time step.
3. The stress-strain relationship is used to determine the shear stress, $\tau_{i,t}$, in each layer given the shear strain $\gamma_{i,t}$. The stress-strain curve may be linear or nonlinear. If nonlinear inelastic soil behavior is assumed, stress reversals are checked and accounted for (e.g., by application of the Masing criteria) in each layer.
4. The input motion is used to determine the motion of the base of the soil layer at time $t+\Delta t$.
5. The motion of each layer boundary at time $t+\Delta t$ is calculated, working from bottom to top. The process is then repeated from step 1 to compute the response in the next time step.

Because the particle velocities are computed at times that differ by one-half time step from those at which the shear stresses are best approximated, the explicit method can become numerically unstable if the time step is too large (i.e., if $\Delta t > \Delta z / V_{ss}$) (Davis, 1986). Most nonlinear analysis programs will compute the time step required for numerical stability from the layer thicknesses and shear wave velocities. The layer with the shortest shear wave travel time will usually control the length of the time step. To avoid very small time steps, and hence very long run times, care should be taken in discretizing the soil profile.

7.5.3.3 Effective Stress Analysis

As indicated in Section 6.6.5, some saturated soils generate significant excess porewater pressure when subjected to cyclic loading. As excess porewater pressure increases, effective stresses decrease. Since the shear modulus of the soil is a function of the effective stress (Equation 6.78), the stiffness of the soil will be reduced by the generation of excess porewater pressure. This component of stiffness reduction occurs in addition to the stiffness reduction associated with strain level due to nonlinear stress-strain behavior.

The analysis of a site containing soils expected to generate significant excess porewater pressure should account for that pore pressure and its effects on soil stiffness – such analyses are referred to as *effective stress analyses*. An effective stress analysis, therefore, must be able to represent nonlinear, inelastic soil behavior and dilatancy behavior; in addition, a good effective stress analysis should be able to compute the redistribution and dissipation of excess porewater pressure that can occur during and after shaking.

The computation of excess porewater pressure is usually accomplished in one of two ways depending on the nature of the soil model. When nonlinear, inelastic behavior is computed using a cyclic nonlinear model (Section 6.4.5.3), excess porewater pressure is usually computed using a *pore pressure model*. When an advanced constitutive model is used, excess porewater pressure can be computed by imposing a constant volume constraint within the constitutive model.

Pore Pressure Models

Cyclic nonlinear stress-strain models (Section 6.4.5.3) use an empirical backbone curve and a series of unloading-reloading rules that govern cyclic behavior. Pore pressure prediction is accomplished by pore pressure models (e.g., Martin et al., 1975; Finn and Bhatia, 1981; Matasović and Vucetic, 1992; Polito et al., 2008; Cetin and Bilge, 2011; Park et al., 2015) that can predict the generation of pore pressure under irregular cyclic loading conditions. The computed pore pressure is used to degrade, or soften, the backbone curve as the effective stress (and soil stiffness) decreases. In the Martin et al. (1975) model, for example, the pore pressure generated in an increment of undrained loading is related to the volumetric strain that would have occurred in the same loading increment under drained conditions (Figure 6.111) by

$$\Delta u = \bar{E}_r \Delta \varepsilon_{vd} \tag{7.63}$$

where $\bar{E}_r$ is the rebound modulus and $\Delta \varepsilon_{vd}$ is the incremental volumetric strain under drained conditions. The rebound modulus can be expressed as

$$\bar{E}_r = \frac{(\sigma'_v)^{1-m}}{mK_2(\sigma'_{v0})^{n-m}} \tag{7.64}$$

where σ'_v and σ'_{v0} are the current and initial vertical effective stresses and m, n, and K_2 are experimentally determined from a rebound test in a consolidometer. The incremental volumetric strain is computed as

$$\Delta\varepsilon_{vd} = C_1(\gamma - C_2\varepsilon_{vd}) + \frac{C_3\varepsilon_{vd}^2}{\gamma + C_4\varepsilon_{vd}} \tag{7.65}$$

where γ and ε_{vd} are the cyclic shear and volumetric strains, respectively, and C_1–C_4 are constants determined from the results of drained cyclic simple shear tests. Martin et al. (1981) developed a procedure for the estimation of these constants without laboratory test results.

Constitutive Models

Advanced constitutive models of soil behavior usually account for changes in effective stress by imposing a constant volume constraint for each time increment. In a simple, but typical, one-dimensional formulation, the deviatoric and volumetric strain increments, $d\varepsilon_q$ and $d\varepsilon_v$, are related to the deviatoric and mean effective stress increments, dq and dp', as

$$\begin{Bmatrix} d\varepsilon_q \\ d\varepsilon_v \end{Bmatrix} = \begin{bmatrix} \dfrac{1}{G_{\max}} + \dfrac{1}{G_t^p} & -\dfrac{1}{G_t^p}\dfrac{q}{p'} \\ \dfrac{D}{G_t^p} & \dfrac{1}{K_s} - \dfrac{D}{G_t^p}\dfrac{q}{p'} \end{bmatrix} \begin{Bmatrix} dq \\ dp' \end{Bmatrix} \tag{7.66}$$

where $G_{\max}$ is the elastic shear modulus, G_t^p is the plastic tangent shear modulus (which describes the reduction of stiffness with increasing shear strain), K_s is the bulk modulus of the soil skeleton, D is a parameter that describes the dilatancy (or contractiveness) of the soil, $p = (\sigma'_1 + \sigma'_3)/2$, and $q = (\sigma'_1 - \sigma'_3)/2$. To ensure constant volume conditions, the total incremental volumetric strain of the soil skeleton is constrained to be equal and opposite to the incremental volumetric strain of the pore fluid, i.e., any change in volume of the soil skeleton must be balanced by an equal and opposite change in the volume of the pore fluid. Changes in the volume of the pore fluid depend on changes in pore fluid pressure and the compressibility of the pore fluid, so

$$du = \frac{K_f}{n} d\varepsilon_v \tag{7.67}$$

where K_f is the bulk modulus of the pore fluid and n is the porosity of the soil. In an unsaturated soil, the pore fluid contains air; since the bulk modulus of air is essentially zero, changes in volumetric strain lead to no pore fluid pressure. Since the bulk modulus of water is very high (2.2 × 10^6 kPa) the volumetric strains associated with changes in porewater pressure in a saturated soil are virtually zero. Assuming K_0 remains constant, the incremental total volumetric stress, dp=0, so dp'=–du. With these relationships, the change in pore pressure, du, due to a change in deviatoric stress, dq, can be calculated as

$$du = \frac{K_f}{n} \frac{D/G_t^p}{1 + \left[\dfrac{1}{K_s} - \dfrac{D}{G_t^p}\dfrac{q}{p'}\right]\dfrac{K_f}{n}} dq \tag{7.68}$$

The nature of the pore pressure change depends on the dilatancy parameter, D, which is a function of the flow rule of the constitutive model. It is clear from Equation (7.68) that du=0 when D=0. A simple flow rule capable of representing the behavior of many soils could be written to ensure that the soil exhibits contractive behavior when the effective stress state is below the phase transformation line (Section 6.4.3.4), dilative when it is above it, and neither when on it. The simple expression,

$$D = \alpha(\eta_{PT} - \eta) \tag{7.69}$$

where α is a function of stress and strain history, the stress ratio, $\eta = q/p'$, and η_{PT} is stress ratio corresponding to what the phase transformation line produces. Therefore, $D > 0$ (contractive behavior) when the effective stress state is below the phase transformation line, $D = 0$ when it is on it, and $D < 0$ (dilative behavior) when the effective stress state is above the phase transformation line. This formulation allows excess porewater pressure to increase, decrease, or not change during cyclic loading depending on the relationship between the effective stress state and the phase transformation line. Such models can therefore compute the buildup of porewater pressure, and accompanying reduction in soil stiffness, that occurs in the early stages of shaking, as can the pore pressure models used in cyclic nonlinear models. However, where pore pressure models in cyclic nonlinear models indicate monotonically increasing porewater pressure and, therefore, monotonically decreasing shear moduli that degrade to nearly zero at high pore pressure ratios, advanced constitutive models can capture the dilatancy that occurs above the phase transformation line in all but extremely loose soils. As soil dilates, the effective stress, and hence the shear modulus, increases. This gives rise to the type of "banana-shaped" stress-strain loops (Figure 6.52) that are frequently seen in laboratory tests. In the field, dilatancy can result in pulses of high acceleration – in some cases, they can represent the peak acceleration and thereby control the low-period portion of a response spectrum – that are missed by simple pore pressure models used with cyclic nonlinear stress-strain models. Figure 7.30 shows an example of a recorded ground motion with strong dilation-induced acceleration spikes developing after about 18 sec.

7.5.4 Comparison of One-Dimensional Ground Response Analyses

Sections 7.5.1–7.5.3 have described one-dimensional linear, equivalent linear, nonlinear total stress, and nonlinear effective stress analyses. Under certain conditions, usually involving relatively weak ground shaking, these different analyses should and do predict consistent responses. Those conditions, however, are generally not of greatest interest for seismic design in tectonically active areas. In such areas, strong ground shaking will lead to responses that induce sufficiently large strains that equivalent linear or nonlinear analyses are required. Historically, the geotechnical earthquake engineering profession has relied upon equivalent linear site response analyses (Kramer and Paulsen, 2004; Matasovic and Hashash, 2012). However, the availability of improved nonlinear site response analyses, along with clarification of appropriate procedures

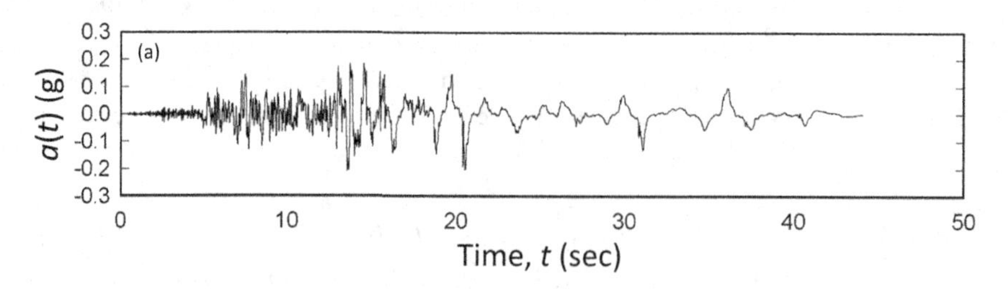

FIGURE 7.30 Time histories of recorded ground surface motion from Wildlife Liquefaction Array showing dilation-induced acceleration pulses following initiation of liquefaction.

for their use (Stewart et al., 2008), means that nonlinear analyses should be used more commonly in the future.

7.5.4.1 Total Stress Analysis

Stiff profiles and/or weak levels of shaking should produce small strains and, hence, a relatively linear response. Under such conditions, linear, equivalent linear, and nonlinear analyses should produce consistent results. As input motion amplitudes increase, strain levels increase and nonlinear behavior begins to occur – the shear modulus of the soil decreases and the damping ratio increases, and equivalent linear analyses predict site response better than linear analyses. As input motion amplitudes continue to increase, the assumptions of equivalent linear analyses become less acceptable, and the response they predict becomes less reasonable.

Figure 7.31 shows a soil profile with V_{S30} = 180 m/s and a very soft layer at 5–10 m depth. The profile was subjected to an input motion scaled to peak rock outcrop accelerations of 0.01g, 0.10g, 0.20g, and 0.40g. The weakest of these motions should be expected to produce only small strains in the soil, so the response predicted by equivalent linear and nonlinear analyses should be nearly identical. As the input motion amplitudes increase, strain levels should be expected to increase and the equivalent linear method should become less capable of representing the actual nonlinear, inelastic behavior of the soil. Figures 7.32 and 7.33 show the ground surface acceleration histories and response spectra predicted by equivalent linear and nonlinear analyses for all four input motion levels. The responses to the 0.01g input motion are virtually identical, both in terms of the acceleration histories (Figure 7.32a) and response spectra (Figure 7.33a). For higher input motion amplitudes, however, the responses predicted by the equivalent linear and nonlinear models diverge and, under higher levels of loading, become quite different. As shown in Figure 7.32d, the equivalent linear acceleration history shows evidence of strong softening and high damping within the first three seconds of the motion when the input motion is weak and the strains in the soil are still small. Because the stiffness and damping of the soil in the equivalent linear analysis are determined by the peak shear strain and then held constant throughout the entire analysis, low stiffness, and high damping ratios are assumed before the strains have reached levels that would produce such behavior. The inherent linearity of the equivalent linear analysis also causes it to respond strongly at the softened fundamental period of the soil profile, so the peak response is frequently overpredicted due to the type of *spurious resonances* seen in both the acceleration histories and response spectra. These response spectra in Figure 7.33 show increasing overprediction of response at low periods

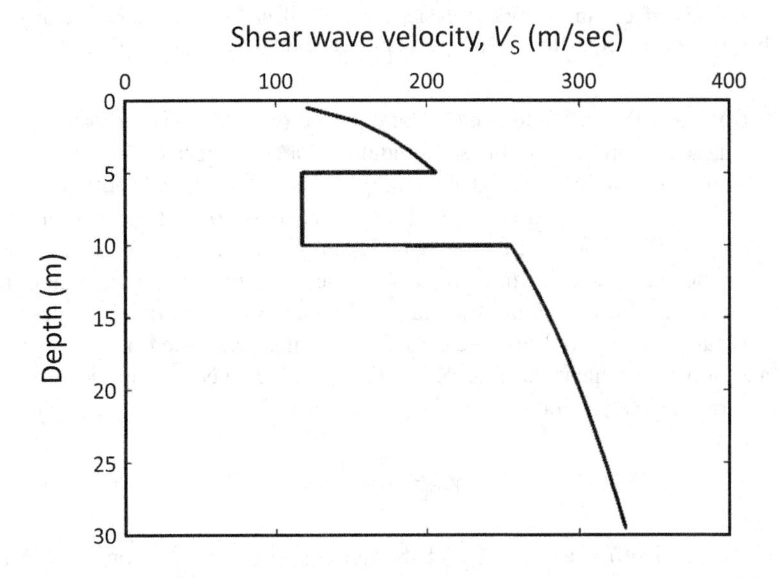

FIGURE 7.31 Shear wave velocity profile for comparison of equivalent linear and nonlinear analyses.

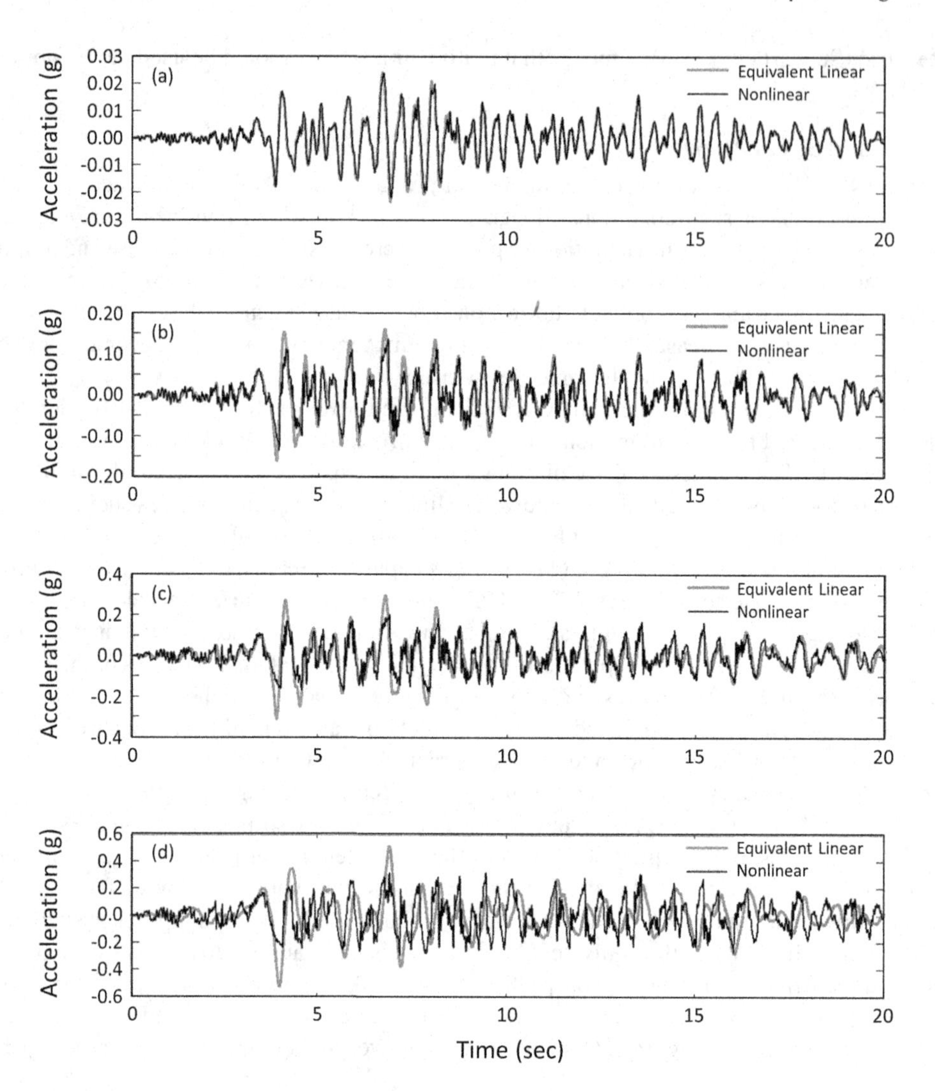

FIGURE 7.32 Ground surface time histories produced by equivalent linear and nonlinear analyses: (a) 0.01*g* input motion, (b) 0.10*g* input motion, (c) 0.20*g* input motion, and (d) 0.40*g* input motion.

(high frequencies) and at the softened fundamental period (which can be seen in Figures 7.33b–d to increase with increasing input motion amplitude), and a characteristic broad, flat shape at low periods due to overdamping of high-frequency components of the ground motion.

Linear and equivalent linear total stress analyses of response to 3,720 ground motions ranging from weak to strong at 100 vertical array sites from Japan (Kaklamanos et al., 2013) showed that the accuracy of linear and equivalent linear analyses decreased with increasing input motion amplitude, but by different amounts at different periods. Figure 7.34 shows approximate ranges of applicability of linear, equivalent linear, and nonlinear total stress analyses based on the peak shear strain computed in an equivalent linear analysis. Kim et al. (2013) related the applicability of different procedures to a strain index, γ_{ind}, defined as

$$\gamma_{ind} = \frac{PGV_r}{V_{S30}} \tag{7.70}$$

where PGV_r is the (RotD50) peak velocity of the outcropping input motion used in the analysis (Figure 7.35). Unlike the peak shear strain used by Kaklamanos et al. (2013), the strain index can be

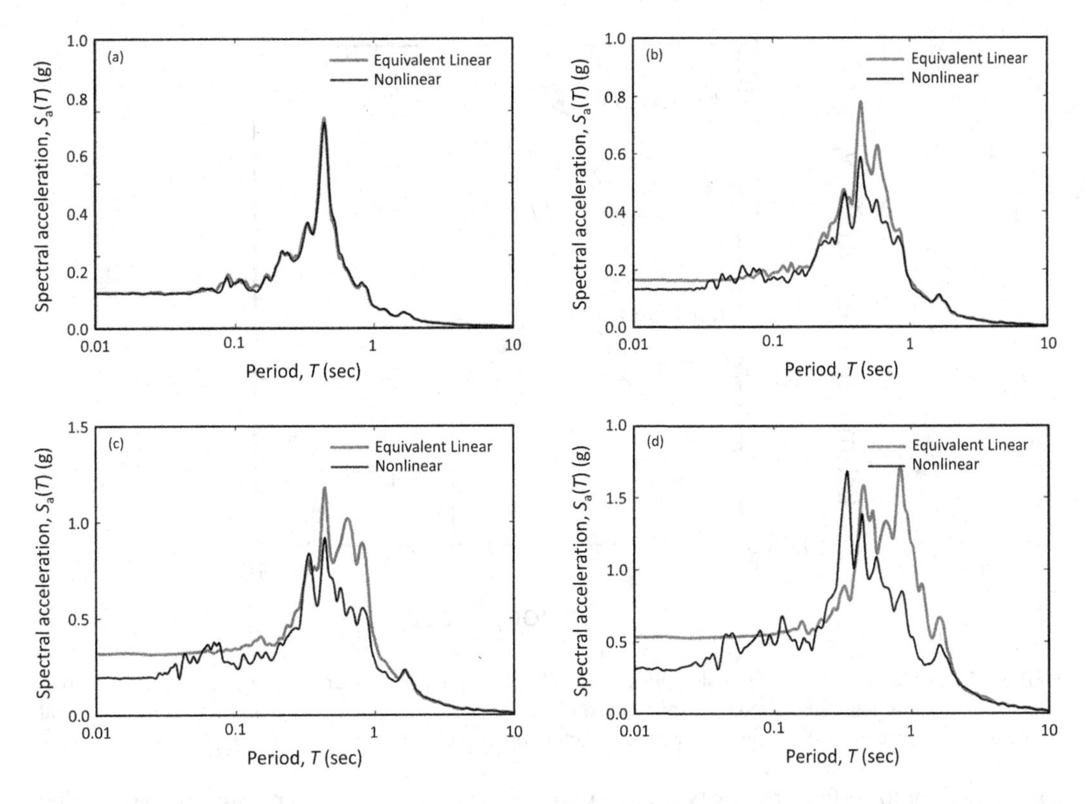

FIGURE 7.33 Ground surface response spectra produced by equivalent linear and nonlinear analyses: (a) 0.01g input motion, (b) 0.10g input motion, (c) 0.20g input motion, and (d) 0.40g input motion.

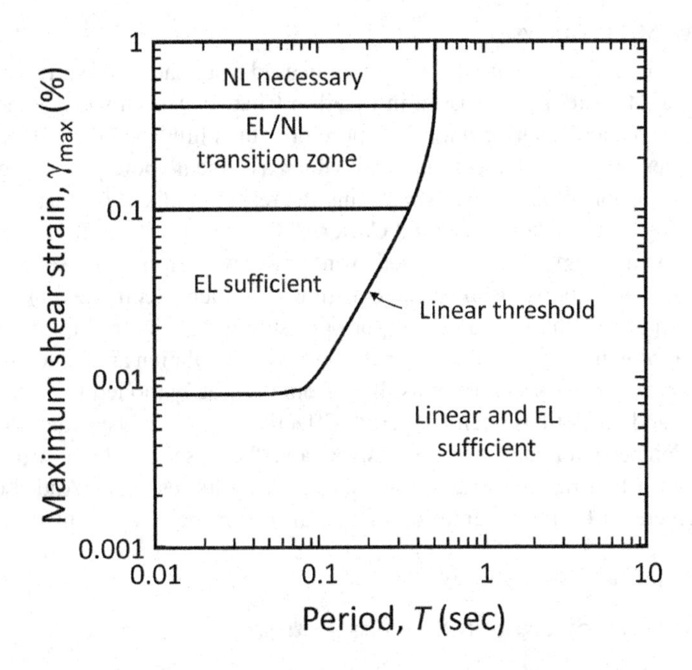

FIGURE 7.34 Approximate ranges of applicability of linear, equivalent linear (EL), and nonlinear (NL) total stress site response analyses based on maximum shear strain amplitude. (After Kaklamanos et al., 2013, with permission of Seismological Society of America.)

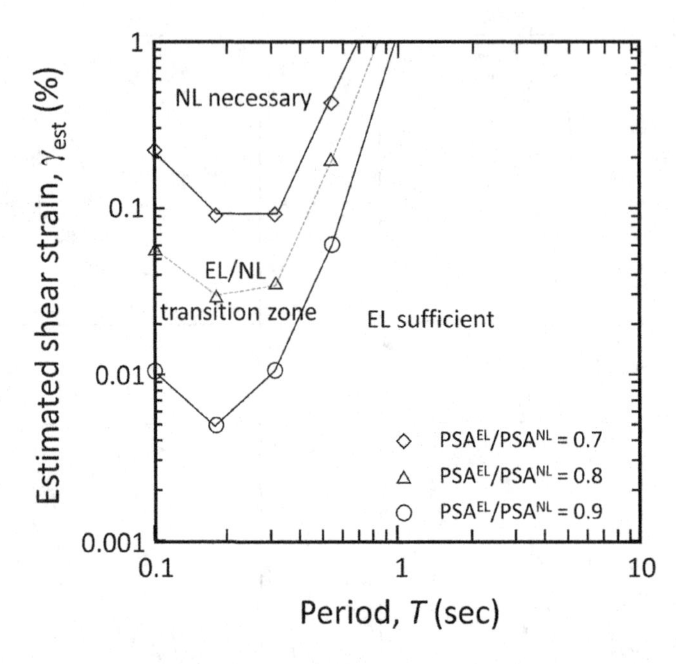

FIGURE 7.35 Approximate ranges of applicability of linear, equivalent linear (EL), and nonlinear (NL) total stress site response analyses based on strain index. (After Kim et al., 2013, with permission of International Association for Structural Mechanics in Reactor Technology.)

computed prior to performing any type of analysis. However, the peak shear strain computed using an equivalent linear analysis provides site-specific characterization of the response that provides more detail than can be inferred from a parameter based on V_{S30}.

7.5.4.2 Effective Stress Analyses

Nonlinear effective stress analyses are most commonly used when saturated soils, particularly loose, saturated soils, exist at a site. As discussed in detail in Chapter 9, such soils generate excess pore pressures during earthquake shaking and those pore pressures influence the stiffness and strength of the soil in complex ways. Total stress analyses, which assume no pore pressure generation, have been shown to be incapable of accurately simulating the response of soil profiles that generate high pore pressures during earthquakes (Youd and Carter, 2005; Kramer et al., 2011).

Pore pressure generation is typically modeled in one of two ways in effective stress analyses – using cyclic pore pressure models or using advanced constitutive models (Section 9.5.7). Cyclic pore pressure models have shown the ability to model pore pressure generation reasonably well up to pore pressure ratios of approximately 0.8 but their inability to model dilation associated with phase transformation behavior (Section 6.4.3.4) prevents them from accurately modeling response as liquefaction is approached and after it has been triggered. Effective stress analyses that do not account for phase transformation behavior can soften excessively and "base isolate" the portion of a soil profile that lies above a layer that liquefies thereby underpredicting its response (Anderson et al., 2011); this effect is exaggerated by the assumption of lateral homogeneity inherent in one-dimensional analyses.

7.5.5 Input Motions for Ground Response Analyses

Once the geotechnical model for ground response analysis is set up, it is necessary to prescribe ground motions at a boundary of the model. This aspect of the analysis is not as simple as it might appear, and careful consideration must be given to the selection and application of appropriate input motions.

7.5.5.1 Application of Input Motions

The proper application of input motions in ground response analyses is important and a frequent cause of confusion. Input motions can be developed in several different ways, and can also be applied in ground response analyses in different ways. Making sure that the method of application is appropriate for a selected motion, and for the specific computer program used to perform the analyses, is an important part of a ground response analysis. The following discussion will be based on one-dimensional analyses, but the basic concepts also apply to multi-dimensional analyses.

As discussed in Section C.6.1, the response of a one-dimensional profile to vertically propagating waves involves both upward- and downward-traveling waves. Using the notation of Section 7.5.1.4 in which the upward- and downward-traveling waves have amplitudes A_m and B_m, respectively, Figure 7.36 illustrates the contributions of the two waves at different locations in rock and soil sites. At the outcrops, the wave amplitudes are twice the amplitudes of the upward-traveling wave in order to satisfy the free-surface boundary condition. Equating A_N for the within-profile condition and outcropping condition assumes the rock profile is the same in both locations; in many cases, differences in the weathering profiles may cause the upper portion of the rock to differ at the two locations. Below the ground surface, the wave amplitudes are the sum of the upward- and downward-traveling wave amplitudes, which can interfere with each other constructively (increasing the total amplitude) or destructively (decreasing the total amplitude). As shown in Figure 7.12, destructive interference can lead to no motion at some nodes when standing waves develop under harmonic loading.

The guiding principle for the appropriate application of input motions is that the motion from the upward-traveling wave should be applied to the base of a ground response model. The boundary conditions at the base of a model, therefore, affect the manner in which input motions are applied. Most computer programs allow for the specification of rigid (fixed) and compliant bases. A rigid base model will ensure that the specified input motion is applied directly to the base of the model. A compliant base model treats the soil profile as being underlain by a semi-infinite elastic halfspace. Downward-traveling waves that are transmitted into the halfspace are "absorbed" at the compliant boundary since they continue to propagate downward (without reflection) to infinity. In a rigid base model, no downward-traveling wave can exist (any wave traveling downward through the soil would be perfectly reflected back upward by the rigid base), so the motion of the base can be interpreted as corresponding to an upward-traveling wave.

In a linear or equivalent linear analysis (e.g., as performed using the computer program SHAKE or any of its successors), the motion at any point in a soil profile can be extracted as a *within-profile*

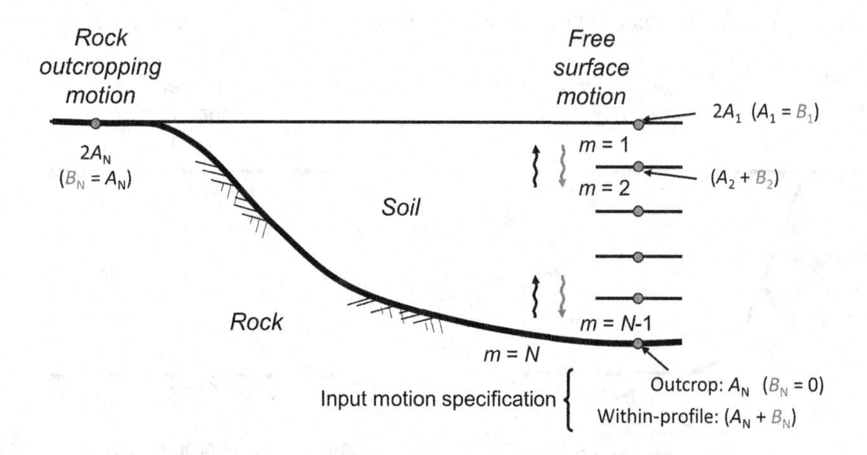

FIGURE 7.36 Illustration of contributions of upward-traveling (black) and downward-traveling (gray) waves to response at outcrop locations and base of soil profile. (After Kottke and Rathje, 2009.) Note that in most practical situations, the velocity profiles of outcropping rock and rock beneath a soil column are different. Hence, A_N values at these locations may differ.

motion or as an *outcrop motion*. The within-profile motion is the total motion at that point, i.e., the sum of the motions produced by the upward- and downward-traveling waves. The outcrop motion is the motion that would occur if all of the material above the point of interest was removed; since the point of interest would then be at a free surface (i.e., an outcrop), the outcrop motion would be twice the motion of the upward-traveling wave at that point. Note that the motion of the downward-traveling wave at that point would then be equal to the within-profile motion minus half the outcrop motion.

7.5.5.2 Sources of Input Motions

The input motions used in ground response analyses are usually defined by seismic hazard analyses. Section 4.5.1 discussed procedures for selecting ground motions for consistency with the results of a hazard analysis. Seismic hazard analyses, whether deterministic or probabilistic, make use of GMMs to predict ground motion *IM*s for different earthquake scenarios. Since GMMs are based on recorded ground surface motions, the *IM*s they predict, such as those defining a target spectrum, correspond to outcrop conditions (i.e., they include the free surface effect).

When recorded ground motions are selected (Section 4.5.1) from a ground motion database and modified (Section 4.5.2) to represent the results of a seismic hazard analysis, they are *recorded outcrop motions*. Less frequently, ground motions are recorded by instruments installed below the ground surface in borings, so the motions they record are the actual motions at that point, i.e., the sum of the motions produced by the upward- and downward-traveling waves, or *recorded downhole motions*.

In other cases, computed motions are used as inputs to ground response analyses. This is normally done in one of two ways. In the first, a recorded surface motion is deconvolved using a linear or equivalent linear analysis to some depth for use as input to another analysis, such as a nonlinear and/or multi-dimensional analysis, whose base is at that level. In the second case, a motion from a large depth may be propagated upward using a one-dimensional linear or equivalent linear analysis to some shallower depth for use as an input motion in another model (e.g., nonlinear and/or multi-dimensional) that extends only to that depth.

Application of Recorded Motions

In most cases, recorded ground motions are used as inputs to ground response analyses. Consider the site shown in Figure 7.37a at which a rock outcrop motion is recorded at Point A, a downhole bedrock within motion is recorded at Point B, and the goal is to predict the free surface motion at the top of the soil profile (i.e., at Point C). Note that even if bedrock velocities are equivalent below Points A and B, the recorded motions at Points A and B will not be equal even though both

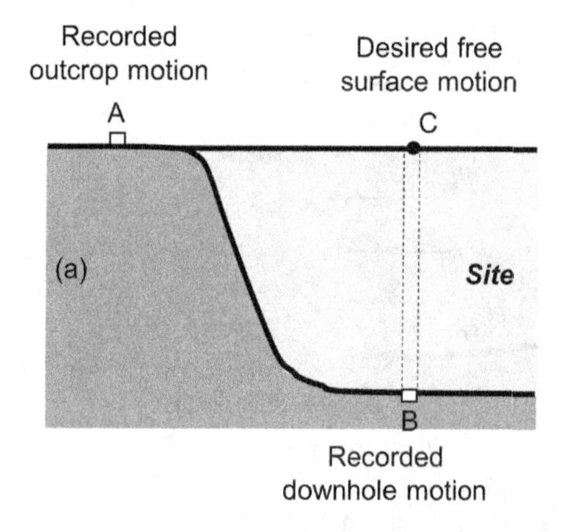

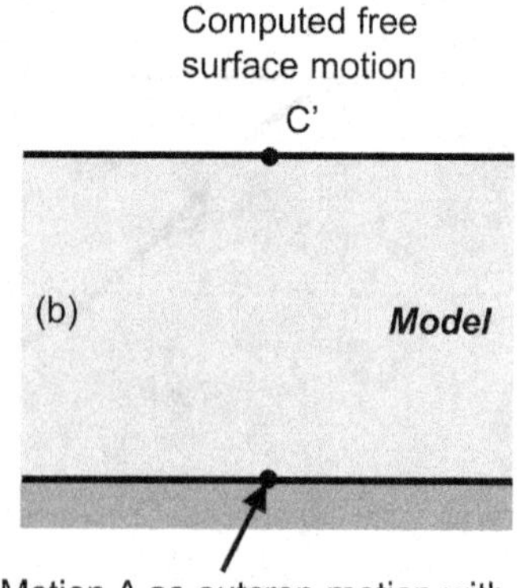

FIGURE 7.37 Specification of (a) outcrop and recorded motions as inputs to (b) one-dimensional site response analyses.

are at the top of the rock – Point A is at a free surface and Point B is not. In a ground response model (Figure 7.37b), the computed free surface motion (C′) will represent the desired free surface motion (C), to the extent that the accuracy of the response program and input parameters allow, if (a) Motion A is applied as an outcrop motion with an elastic base, or (b) Motion B is applied as a within-profile motion with a rigid base (Kwok et al., 2007).

Care must be taken to understand the manner in which input motions are handled by specific ground response analysis programs. Most programs allow an outcrop motion to be used directly, but some require the motion to be doubled to account for the energy absorbed by the halfspace. If a particular code is not familiar, it is advisable to confirm the manner in which the input motions should be applied by performing a weak-input-motion analysis (e.g., input motion scaled to 0.01g to minimize potential nonlinear effects) with the selected code and with a familiar SHAKE-type code to confirm that the computed surface motions are very close to each other.

Application of Computed Motions

Motions can also be propagated upward from a very deep level to the base of a model that extends only to a shallower depth. If bedrock is very deep, it may not be desirable (from a computational efficiency standpoint) to extend a multi-dimensional model all the way down to that level; instead, a deep one-dimensional model (Figure 7.38) can be used to propagate the bedrock motion upward to a depth corresponding to the bottom of the shallow model. In these cases, the motion at Point A can be extracted as an outcrop or a within-profile motion. If the shallow model has an elastic base, the motion at Point A should be extracted as an outcrop motion to be used as the input motion. If the shallow model has a rigid base, the motion at Point A should be extracted as a within-profile motion for use as the input motion. For this model-to-model interface to work well, at the depth of the shallow model (Point A) the seismic velocities should be high enough that material nonlinearities are small (V_S>~ 500 m/s).

As pointed out by Mejia and Dawson (2006), rigid boundaries should be used with caution. They are appropriate for use with recorded downhole motions and for profiles with strong impedance contrasts at the base of the model but can develop reflections that cause significant standing waves when the velocities of a deconvolution analysis do not match those of the second analysis exactly. If the second analysis is a two-dimensional analysis of a profile of variable soil thickness (e.g., of a

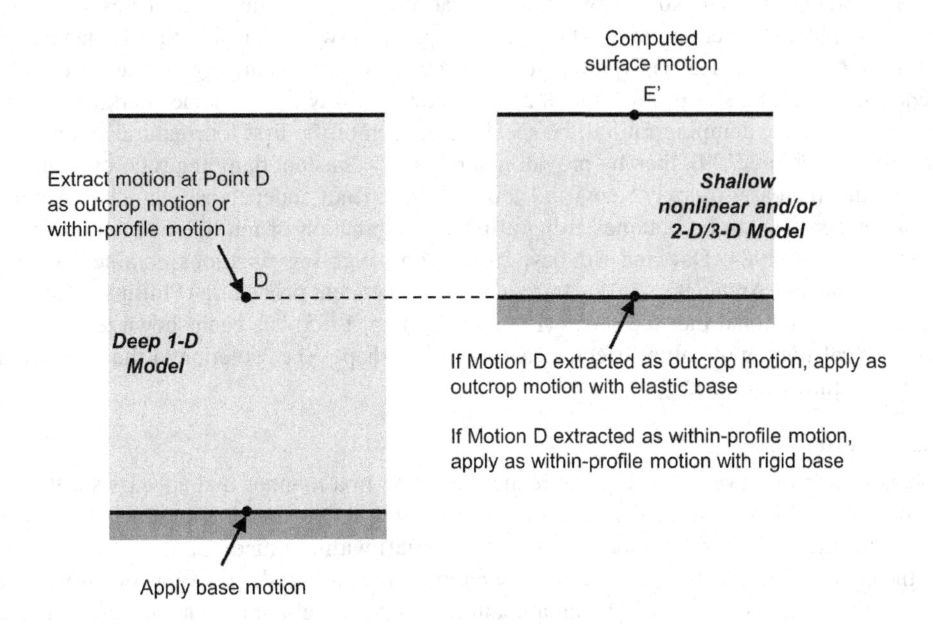

FIGURE 7.38 Propagation of motion from large depth in one-dimensional analyses for use as input to shallower one-dimensional nonlinear analysis or multi-dimensional analysis.

dam, embankment, or slope), differences in effective stress across the base of the model will lead to differences in stiffness which cannot match the single stiffness used in a one-dimensional deconvolution analysis. Thus, in most cases, the use of an elastic base is preferred. Selection of the depth of the halfspace in the shallow model, however, should be made with the recognition that the effects of geologic structure (i.e., variations in shear wave velocity such as in a zone of weathered rock or from a deep impedance contrast) below that depth will not be captured.

7.5.6 Energy Dissipation in Ground Response Analyses

At very low strain levels, laboratory element tests show that the shear modulus is essentially constant (for strains below the linear threshold strain; Section 6.4.3.1), which would imply linear behavior with no energy dissipation. The experiments also show, however, that some low-strain damping occurs. At very deep sites, seismic waves may travel for tens or even hundreds of meters through relatively stiff materials. While the strains induced in these materials may be small, their large thicknesses may allow low-strain damping to dissipate a significant amount of energy before they reach softer shallow soils – hence, the modeling of low-strain damping of these stiffer soils can be very important. For large strain problems, higher levels of hysteretic damping occur that are related to inelastic behavior (Section 6.6.4.2). Characterizing energy dissipation at both small and large strains is a critical component of ground response analysis.

7.5.6.1 Low-Strain Material Damping

Modeling low-strain material damping is straightforward in frequency domain linear and equivalent linear analyses where its effects can be represented by the imaginary part of a complex transfer function. For nonlinear analyses, however, it is more complicated. To model this low-strain damping, and to prevent a situation of undamped response, which can lead to numerical difficulties, lumped mass site response models use Rayleigh damping. Rayleigh damping assumes the damping matrix to be a linear combination of the mass and stiffness matrices and therefore is frequency-dependent. Since soil damping is not frequency-dependent, Rayleigh damping can overdamp certain frequencies and underdamp others. A number of approaches have been taken to determine Rayleigh coefficients that produce some desired target damping ratios at user-specified frequencies. For standard Rayleigh damping, a target damping ratio can be specified at two frequencies; Hudson et al. (1994) recommended setting the Rayleigh parameters to produce the target damping ratio at the fundamental frequency, f_o, of the soil profile, and at $5f_o$. An extended Rayleigh scheme (Park and Hashash, 2004) allows a target damping ratio to be specified at each of the first four natural frequencies of the soil profile (Figure 7.39), thereby providing a relatively constant damping ratio over that range (albeit one that increases quickly below and above it). Frequency-independent damping over a wide range of frequencies can be obtained using a linear combination of multiple relaxation functions (Day and Minster, 1984; Day and Bradley, 2001) with weighting functions obtained by averaging moduli (Liu and Archuleta, 2006). A frequency-independent procedure (Phillips and Hashash, 2009) based on a rational indexed series (Liu and Gorman, 1995) has been shown to be effective, but requires calculation of natural frequencies and mode shapes (i.e., solution of the eigenproblem, which can be time-consuming).

7.5.6.2 Scattering

The phenomenon of wave scattering can reduce response in a manner that appears similar to the reduction caused by low-strain damping. Wave scattering refers to the alteration of wave fields by heterogeneities (i.e., zones of softer or stiffer material) within a medium. As shown in Section C.5.3, the presence of a soft or stiff inclusion can reduce the amplitude of a transmitted wave, even under one-dimensional conditions, by an amount that depends on the dimensions of the inclusion relative to the wavelength of the wave. Waves with long wavelengths (i.e., waves at low frequency and/or traveling through very stiff materials) will not be significantly influenced by relatively small

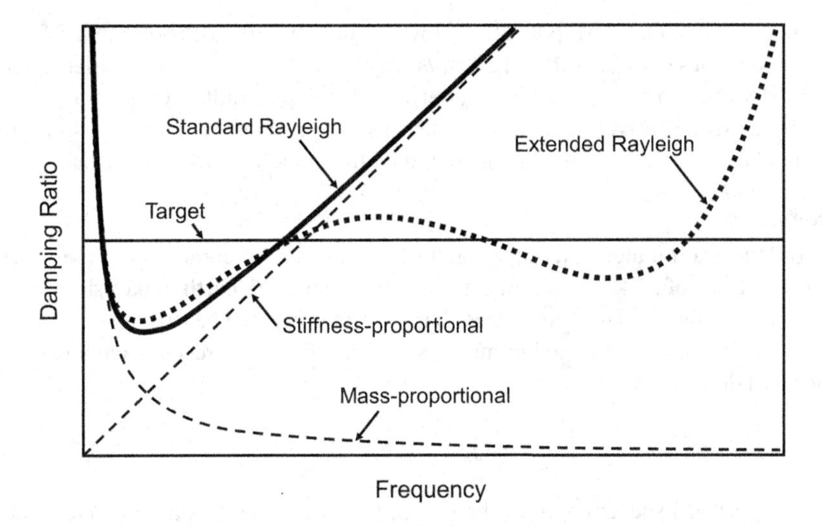

FIGURE 7.39 Variation of damping ratio with frequency for various Rayleigh damping schemes. (After Park and Hashash, 2004 with permission of Taylor and Francis.)

inclusions, but waves with wavelengths similar to the dimensions of the inclusion (i.e., waves of higher frequency and/or waves traveling through softer materials) may be strongly affected. With two- or three-dimensional heterogeneities, waves are also refracted to different paths. These phenomena occur under weak as well as strong shaking and can produce complex wave fields characterized by reduced amplitudes of high-frequency waves, particularly at shallow depths where softer soils lead to shorter wavelengths. The total energy of a wave is not reduced by scattering but it is disorganized in a manner that appears as energy loss, i.e., as apparent damping, at a given location.

Studies of recordings from vertical arrays (Elgamal et al., 2001; Tsai and Hashash, 2009; Thompson et al., 2012; Yee et al., 2013; Cabas et al., 2017) have shown that weak motions (hence, low-strain response) are systematically overpredicted by one-dimensional analyses using laboratory damping values (e.g., ξ_{min} in Equation 6.99). Asimaki and Steidl (2007) observed attenuation near the surface of five vertical arrays that were significantly higher (corresponding to minimum damping ratios of 5%–13%) than would be predicted using only laboratory-based minimum damping ratios. Accounting for scattering effects in one-dimensional analyses, therefore, leads to the need for depth-dependent minimum damping ratio profiles with higher "scattering" damping being assigned to shallower, thinner layers. Zalachoris and Rathje (2015) found that the Darendeli low-strain damping ratio (Equation 6.99) had to be multiplied by factors ranging from 1.0 to 5.0 (average of 2.98) to provide an optimal agreement for 11 sites in Japan, Taiwan, and the United States. Thompson et al. (2009) examined empirical transfer functions (ratio of surface-to-downhole Fourier amplitudes from recorded motions) from 10 earthquakes at each of 13 vertical array sites spanning a wide range of geologic conditions in Japan. The transfer functions predicted by one-dimensional ground response analyses (assuming vertically propagating SH waves) accurately represented the empirical behavior for only a few of the sites. Afshari and Stewart (2019), performing similar analyses using data from 21 vertical arrays in California, found better performance of one-dimensional analyses but similar underdamping from the Darendeli model, than had been observed by others using Japanese array data.

An alternative to the laboratory-based damping models, based on empirical studies of shallow crust energy dissipation, was provided by Campbell (2009)

$$\xi_{min}(\%) = \frac{50}{7.17 + 0.0276V_s} \tag{7.71}$$

where V_s is in m/s. This relationship produces low-strain damping ratios of 5.0%, 3.2%, and 2.4% for shear wave velocities of 100, 300, and 500 m/s, respectively. At this time, available data is insufficient to formalize recommendations for scattering-related ξ_{min} multipliers (or increments). For a given site, particularly a relatively deep one, the sensitivity of response to increased low-strain damping ratio should be investigated. This factor should be treated as a source of epistemic uncertainty.

7.5.6.3 Kappa

The effects of low-strain material damping and scattering can be combined in the spectral decay parameter, κ. Analyses of rock outcrop Fourier spectra from weak earthquakes show that acceleration spectra decay faster at high frequencies than source spectra. The difference is interpreted as resulting from crustal attenuation and damping/scattering from site response and can be modeled by an exponential decay term

$$A_o(f) = A_s(f)e^{-\pi\kappa f} \tag{7.72}$$

where A_o is the observed spectrum, A_s is the source spectrum, and κ is a parameter that describes the rate of decay of acceleration with frequency (i.e., the slope of the high-frequency portion of the acceleration Fourier spectrum). The path and site components of κ combine as

$$\kappa = \kappa_{path} + \kappa_0 \tag{7.73}$$

The path term is often modeled as being proportional to path distance R, i.e.

$$\kappa_{path} = R\kappa_r \tag{7.74}$$

The site component, also used in Equation (3.47), can be further broken down into components related to anelastic attenuation and wave scattering

$$\kappa_0 = \kappa_d + \kappa_s \tag{7.75}$$

If a wave propagates through layer i of thickness Δz_i with shear wave velocity $V_{s,i}$, the effect of material damping, ξ, in that layer can be expressed by

$$\Delta\kappa_{d,i} = \frac{2\Delta z_i \xi_i}{V_{s,i}} \tag{7.76}$$

Recalling the relationship between damping ratio and quality factor (Equation B.76), it can easily be shown that

$$\Delta\kappa_{d,i} = \frac{\Delta z_i}{V_{s,i} Q_i} \tag{7.77}$$

where Q_i is the quality factor (when derived empirically, Q_i necessarily includes effects of both material damping and scattering, in which case the appropriate kappa term on the left side of Equation 7.77 is κ_0). Since the total "kappa filter" expression above is of exponential form, the total κ_d value can be expressed as the sum of a series of $\Delta\kappa_d$ values for individual layers in a layered profile, i.e.

$$\kappa_d = \sum_{i=1}^{n} \Delta\kappa_{d,i} = \sum_{i=1}^{n} \frac{2\Delta z_i \xi_i}{V_{s,i}} = \sum_{i=1}^{n} 2t_{t,i}\xi_i \tag{7.78}$$

where $t_{t,i}$ is the layer travel time and n is the number of layers.

The value of κ_s cannot be measured independently (as κ_d can, at least conceptually, from laboratory measurements of ξ), and so is usually obtained as the difference between κ_0 and κ_d. The frequently encountered finding from vertical arrays, that laboratory-based models produce insufficient damping relative to field performance, demonstrates the importance of wave scattering under field conditions. As suggested previously, the effects of scattering can be approximated as an "equivalent material damping" effect.

7.5.6.4 Stiffness and Damping Model Consistency

The modulus reduction and damping curves used in equivalent linear analyses are well established and provide a convenient means of describing the nonlinear, inelastic stress-strain behavior observed in laboratory tests on different soils, including at large strains where consideration of shear strength is important (Section 6.6.1.3). In equivalent linear analyses that operate in the frequency domain, the stiffness and damping behaviors are independent; as a result, any damping curve can be used with any modulus reduction curve (although the arbitrary specification of the curves is not appropriate since modulus reduction and damping behavior are correlated). In nonlinear analyses, the problem is more complicated.

As discussed in Section 6.6.1, a modulus reduction curve and maximum shear modulus value can be used to construct a nonlinear backbone curve. Since most nonlinear analyses use the Masing rules (Section 6.6.4.2) or equivalent to describe unloading-reloading behavior, the shape of a hysteresis loop formed by unloading and reloading curves is controlled by the shape of the backbone curve. As a result, the damping ratio is controlled by the modulus reduction behavior – once a modulus reduction curve is selected with such an approach, the damping curve is fixed. Studies have shown (Park and Hashash, 2004; Phillips and Hashash, 2009) that a backbone curve calibrated to be consistent with an experimentally determined modulus reduction curve will, when used with the Masing rules, produce a damping curve that is significantly higher than the corresponding experimentally determined damping curve at moderate to large strain levels (Figure 7.40).

Several procedures allow for simultaneous matching of modulus reduction and damping behavior. Lo Presti et al. (2006) used a Ramberg-Osgood backbone curve with a modified Masing rule suggested by Tatsuoka et al. (1993) that does not restrict the unloading-reloading scaling factor to 2. Phillips and Hashash (2009) proposed a reduction factor that modified the unloading-reloading

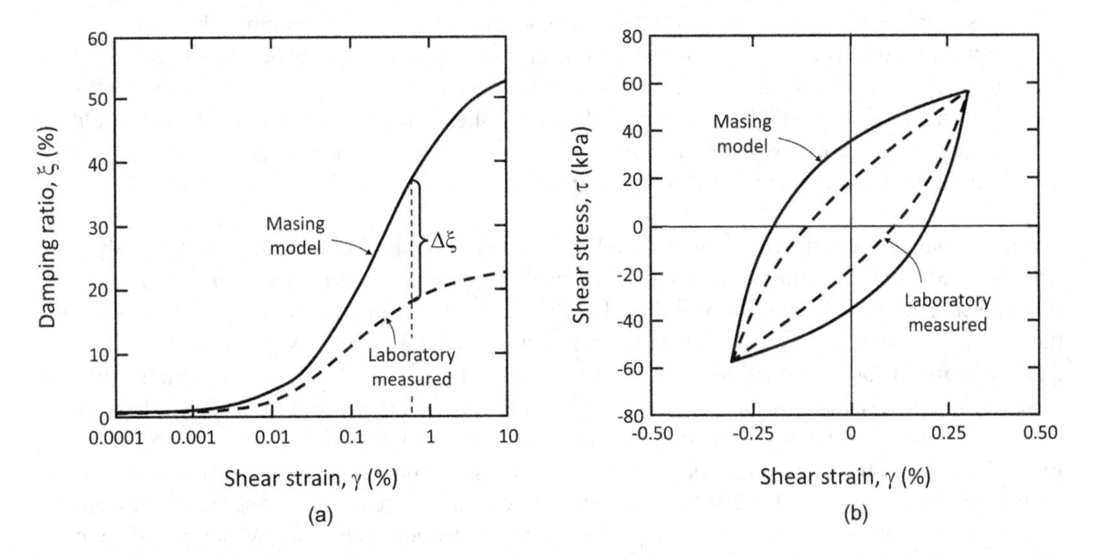

FIGURE 7.40 (a and b) Comparison of experimental and Masing-consistent damping ratios. (After Phillips and Hashash, 2009 with permission of Elsevier.) Note that Masing-consistent damping ratio is much higher than experimental values at strains greater than about 0.01 % for this case.

curves to have a shape altered from that of the backbone curve so as to produce the desired damping ratio; the alteration reduces the stiffness immediately after reversals relative to G_{max}. Arefi et al. (2013) modified the hyperbolic relationship for modulus reduction and backbone curves (Equations 6.90 and 6.91) to express a scaling factor that produces a desired damping ratio when used (even though its value may not be 2) with the Masing rules; the procedure is formulated in such a way that the shear modulus immediately after reversals is equal to G_{max}. Yniesta et al. (2017) fit a cubic spline through user-defined modulus reduction and damping data and use a coordinate transformation algorithm to match both simultaneously; the shear stress at each time step, however, needs to be computed as the root of a nonlinear equation.

7.5.7 Ground Response Variability and Uncertainty

An important aspect of ground response analysis is recognizing and considering the effects of aleatory variabilities and epistemic uncertainty. As described in Section D.8, aleatory variability represents essentially random variations in a parameter that can be measured, but not easily reduced. Epistemic uncertainty is caused by incomplete data or knowledge of a parameter or physical process.

Common sources of aleatory variability in ground response analyses include variability in modulus reduction and damping curves, random variations in V_S profiles at a site, and variations in site responses produced by alternative input motions (record-to-record variability). Common sources of epistemic uncertainty include uncertainty in the V_S profile and modeling uncertainty, which is related to lack of knowledge regarding whether one-dimensional ground response analyses provide an effective representation of site response effects that may be influenced by other factors (e.g., surface waves, basin or topographic effects, etc.).

The V_S profile is a critical element of ground response analysis and is subject to both aleatory variability and epistemic uncertainty. The distinction between these can be conceptualized as follows:

- Aleatory variability in V_S profiles represents different profile characteristics that would be measured at different locations for a particular site (Toro et al., 1997). Such variations are caused by heterogeneity in the soil profile and layer properties and are for most practical purposes not measurable given the limited extent of site exploration that is typically available.
- Epistemic uncertainty in V_S profiles is related to the potential for different measurement techniques to provide different profiles for the same location (e.g., Boore and Asten, 2008). This uncertainty would be low for a high-quality site characterization program, but would be high if more approximate procedures (such as the estimation of V_S profiles from other measured quantities, like CPT profiles) are used.

The most complete and rigorous way of handling aleatory variability in ground response analyses is to perform Monte Carlo simulations in which variable input parameters are randomized in a manner that preserves their individual probability distributions and, where appropriate, correlations with themselves between depths and with other parameters. Randomization is possible in one-, two-, and three-dimensional linear, equivalent linear, and nonlinear analyses. Its use in practice has, to date, been relatively rare except within the nuclear industry where it is common. Even there, however, its use has been largely limited to one-dimensional, equivalent linear analyses. As tools for its use become more readily available and as the benefits of properly accounting for uncertainty in ground response in probabilistic seismic hazard analyses are increasingly recognized, randomization will likely become a more common element of ground response analyses. Material and geometric characteristics are most commonly randomized, with uncertainties in ground motions accounted for by using suites of input motions that account for desired levels of ground motion uncertainty. Subsections that follow describe the randomization approach with respect to one-dimensional, equivalent linear analyses. Small-strain stiffness and site geometry are often handled together in the form of randomized shear wave velocity profiles. Soil properties at larger strain levels are handled by randomization of modulus reduction and damping curves (Sections 6.6.3.4 and 6.6.4.3).

The effects of epistemic uncertainty are considered through the use of alternative site amplification models that capture the range of possible results given limited knowledge. That can include various dynamic soil properties and variations in layer thicknesses, although a significant additional source of epistemic uncertainty is modeling uncertainty.

In many applications, the distinctions between aleatory variability and epistemic uncertainty are less important than characterizing accurately the overall dispersion of the site amplification. This can be accomplished by including both variability and uncertainty in V_S and other parameters in the Monte Carlo simulations mentioned above. The combined variations of AF can be sampled from the full suite of results (Section D.9.3.3) and then applied using a logic tree approach in PSHA (Rodriguez-Marek et al., 2021). Different sources of variability and uncertainty are described next.

7.5.7.1 Randomization of V_s Profiles

Velocity profiles are generally measured using the types of geophysical tests described in Section 6.5.2. A number of different tests, which induce different types of waves traveling along different paths and involve different interpretation procedures, are used. If used at the same site, some (test-to-test) uncertainty in measured velocities would be expected. Moreover, in complex depositional environments, the velocities at a particular depth may vary significantly (location-to-location variability) over relatively short horizontal distances. Additionally, the depths of the boundaries between different soil or rock units may vary from one location to another.

Consideration of the effects of velocity profile variability and uncertainty can involve simple sensitivity analyses or more detailed modeling of spatially variable profile characteristics. Epistemic uncertainty can be accounted for by logic tree branches for alternate V_S profile velocities that are both higher and lower (often by 10%–20%) than the best-estimate (typically taken to represent median) velocities (EPRI, 2013). This approach, which implicitly assumes that variations in velocity are perfectly correlated at all depths, underestimates site response uncertainties, which are also affected by uncertainties in additional material properties and modeling uncertainty (Rodriguez-Marek et al., 2021). Aleatory variability can be accounted for by randomizing the shear wave velocity profile. One approach is to randomize the velocities of all layers independently; this approach assumes that the variations in velocity are independent, or completely uncorrelated. A more accurate approach is to model the expected spatial variability of the velocity profile with consideration of spatial correlation.

One would expect that the chances of the velocities at two closely-spaced depths (say, 10 cm apart) both being above average would be higher than the chances that both velocities at two widely spaced depths (say 10 m apart) are both above average. In the former case, the velocities should be expected to be closely correlated to each other. In the latter case, the tendency for one or the other of the points to be above or below average would be essentially unaffected by whether the other was above or below average – they are effectively uncorrelated to (or independent of) each other. Realistic representation of the variability of shear wave velocity, therefore, requires consideration of the spatial correlation (i.e., *autocorrelation*) of velocity, which can be expected to decrease with increasing separation distance.

Toro (1995) used some 557 soil profiles to establish an ergodic layering model for layer thickness variability and a velocity model for the correlation of shear wave velocities. The velocity model accounts for uncertainty in shear wave velocity at a particular depth, and for spatial correlation of the velocities. Toro (1995) and Kottke and Rathje (2009) described the correlation coefficient for shear wave velocities of two adjacent layers as having depth- and thickness-related components

$$\rho(z,h) = \rho_z(z) + \left[1 - \rho_z(z)\right]\rho_h(h) \tag{7.79}$$

where z=depth, h=sublayer thickness, and $\rho_z(z)$ is the depth-dependent correlation coefficient defined as

$$\rho_z(z) = \begin{cases} \rho_{200}\left[\dfrac{(z+z_0)}{200+z_0}\right]^b & \text{for} \quad z \le 200\text{ m} \\ \rho_{200} & \text{for} \quad z > 200\text{ m} \end{cases} \tag{7.80}$$

with ρ_{200}=correlation coefficient at z=200 m and z_0=initial depth parameter, and $\rho_h(h)$ is the thickness-dependent correlation coefficient defined as

$$\rho_h(h) = \rho_0 \exp\left(-\frac{h}{\Delta}\right) \tag{7.81}$$

where ρ_0=correlation coefficient for zero thickness and Δ=correlation decay coefficient. Toro (1995) analyzed multiple soil profiles to identify the correlation model parameters (Table 7.4) for different generic site classes. When site-specific shear wave velocities have been measured, the values of $\beta_{V_s} = \sigma_{\ln V_s}$ in Table 7.7 should be replaced by those shown in Figure 6.86.

Toro (1995) also developed a layering model that was intended to be used for generic profiles (i.e., profiles in which specific layer boundaries have not been identified by subsurface exploration). The generic model accounts for observations that soil properties change more rapidly at shallow depths, and hence produces layers of increasing thickness with depth. The layer thicknesses are assumed to be Poisson distributed (Section D.7.1.2) with a depth-dependent rate term

$$\lambda(z) = 1.98(z+10.86)^{-0.89} \tag{7.82}$$

where $\lambda(z)$ is the rate of layer boundaries (in m^{-1}) and z is depth in meters. The mean layer thickness, $h(z)$, which is the reciprocal of the rate, is depth-dependent, but the thicknesses of adjacent layers are statistically independent of each other. Kottke and Rathje (2009) describe a procedure by which a homogeneous Poisson process (one in which $\lambda(z)$ is a constant) can be transformed into a non-homogeneous process with the desired depth-dependent characteristics. Since layer thicknesses of a homogeneous Poisson process are exponentially distributed, independent random numbers (between 0 and 1) can be used with the CDF of the exponential distribution to produce homogeneous, random layer thicknesses that can then be transformed to non-homogeneous (depth-dependent) thicknesses. By generating multiple sets of random numbers, multiple profiles with different layering patterns can be generated and used in analyses to help model uncertainty in velocity profiles.

TABLE 7.7
Coefficients for Toro (1995) Velocity Model

	V_{S30} (m/s)			
Parameter	**< 180**	**180–360**	**360–750**	**> 750**
$\beta_{V_s} = \sigma_{\ln V_s}$	0.37	0.31	0.27	0.36
ρ_0	0.00	0.99	0.97	0.95
ρ_{200}	0.50	0.98	1.00	0.42
Δ	5.0	3.9	3.8	3.4
z_0	0.0	0.0	0.0	0.0
b	0.744	0.344	0.293	0.063
Number of Profiles	27	226	169	35

Figure 7.41 shows a series of randomized soil profiles for a generic USGS Site Class C profile (with layer thickness and velocity both randomized) and a site-specific profile (in which layer boundaries were determined by subsurface investigation and considered to be known). Note that the degree of agreement between the base velocity profile and the mean (or median) of the randomized profiles will increase with increasing number of randomized profiles. Randomized profiles should be checked for reasonableness prior to their use – randomization schemes can result in physically unreasonable velocity profiles (e.g., profiles with large velocity inversions, V_{S30} values that deviate excessively from the base case value, fundamental site periods that do not agree with estimates from field testing, and unreasonable dispersion curves).

Griffiths et al. (2016a) compared dispersion curves (Section 6.5.2.2) from active and passive source surface wave testing at soft and stiff soil sites in Italy and France. Data from different linear arrays and shot locations were used to characterize median dispersion and the uncertainty about the median. Bounding velocity profiles were computed by increasing and decreasing the median velocity profile by 20%. Randomized velocity profiles were then generated using the Toro (1995) procedure. Dispersion curves were computed for the bounding profiles and each of the randomized profiles and compared with the experimentally measured dispersion curves. The dispersion curves from the bounding profiles were well outside the uncertainty range from the experimental measurements, suggesting that the ±20% bounds were too high to properly represent the subsurface stiffness characteristics of the site. Also, the dispersion curves from a significant fraction of the randomized velocity profiles fell outside the uncertainty range of the experimental data. The randomized profiles that did not match the experimental data well appeared to result from significant overestimation of the uncertainty in shear wave velocity for the shallower soil layers. Griffiths et al. (2016b) showed that randomized profiles that were outliers with respect to the experimentally measured dispersion curve also produced outliers in terms of computed response when used in ground response analyses. Dispersion curves computed for randomized velocity profiles could be used to check the reasonableness of individual profiles generated by that process. Teague et al. (2018) found that rejecting randomized profiles that do not pass certain screening criteria resulted in significantly more accurate ground response

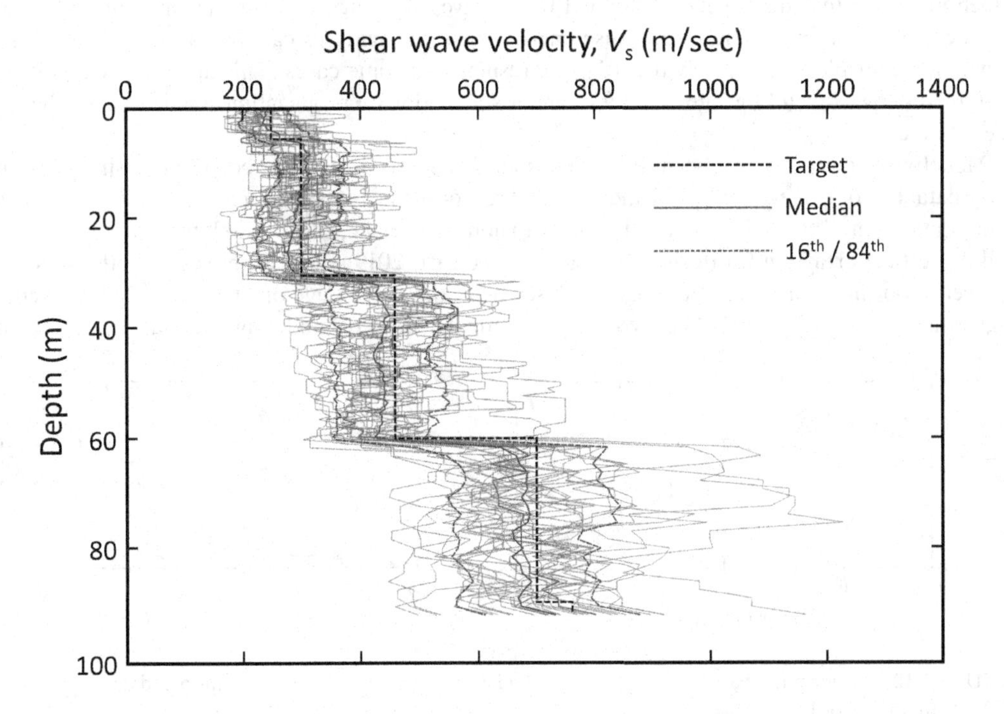

FIGURE 7.41 Illustration of randomized velocity profiles.

analyses. The screening criteria were based on the fundamental frequency of the site, which was estimated from horizontal-to-vertical spectral ratio testing, and experimental dispersion data.

Since site response is known to be sensitive to shear wave velocity, randomization of shear wave velocity in Monte Carlo simulations produces a level of uncertainty in computed ground motion *IMs* that is related to the level of uncertainty in the velocity profile. The extent to which the uncertainties produced by current randomization models match the uncertainties observed in recorded ground motions is discussed in Section 7.7.

7.5.7.2 Randomization of Modulus Reduction and Damping Behavior

While the shear wave velocity defines low-strain stiffness, and the shear wave velocity randomization described in the preceding section can allow simulation of the effects of uncertainty in low-strain stiffness, the behavior of soil at higher strain levels is also uncertain. Darendeli (2001) characterized the uncertainty in both modulus reduction and damping behavior. The uncertainties in G/G_{max} and damping ratio were given in Equations (6.98) and (6.106) and are illustrated in Figure 7.42. Note that the form of Equation (6.98) is such that the uncertainty becomes very low as G/G_{max} approaches 1.0 and 0.0, and that the uncertainty in damping ratio increases with increasing damping ratio. Uncertainty in dynamic soil properties can then be accounted for by randomizing G/G_{max} and damping ratio in the same analyses in which shear wave velocities are being randomized. As discussed in Section 6.6.4.3, the modulus reduction ratio and damping ratio are related to each other – the higher the value of G/G_{max}, the more linear the material behavior and the lower the dissipated energy, hence damping ratio. Therefore, a perturbation that produces a G/G_{max} value greater than the mean G/G_{max} value should be expected to coincide with a damping ratio that is below the mean damping ratio – in other words, G/G_{max} and damping ratio should be expected to be negatively correlated with each other. A procedure for generating correlated pairs of random variables is given in Section D.9.3.3.

7.5.7.3 Modeling Uncertainty

One-dimensional ground response analyses capture some, but not all, aspects of site response, which may also include effects of inclined body waves and surface waves from complexities in the site profile and basin effects. Because one-dimensional analyses cannot capture these effects, they may be unable to accurately describe site response in some cases. This uncertainty is related to limited knowledge (of the site conditions and associated wave propagation conditions) and hence is epistemic.

Modeling uncertainty has been quantified under small-strain (linear) conditions using vertical array data, by comparing computed and recorded responses for sites with many recordings. Studies using data from 100 vertical arrays in Japan (Thompson et al., 2012; Kaklamanos et al., 2013) and 21 vertical arrays in California (Afshari and Stewart, 2019) show examples of both excellent and very poor fit from ground response analyses, with the percentage of sites with good fits being higher in California. Figure 7.43 shows the period-dependence of modeling uncertainty, expressed

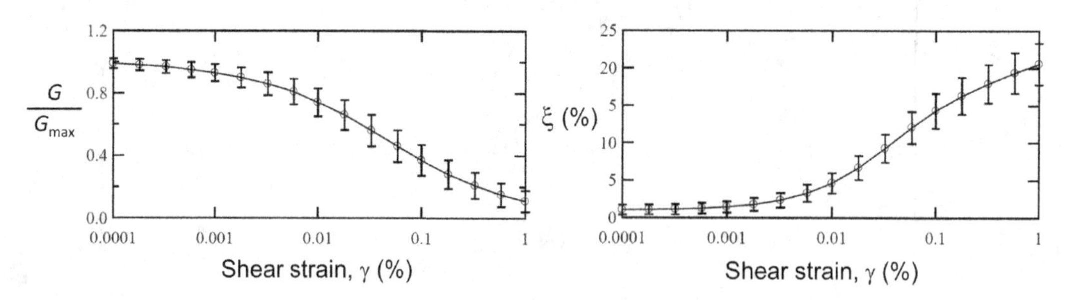

FIGURE 7.42 Uncertainty bars (+/– one standard deviation) for modulus reduction ratio and damping ratio. (After Darendeli, 2001.)

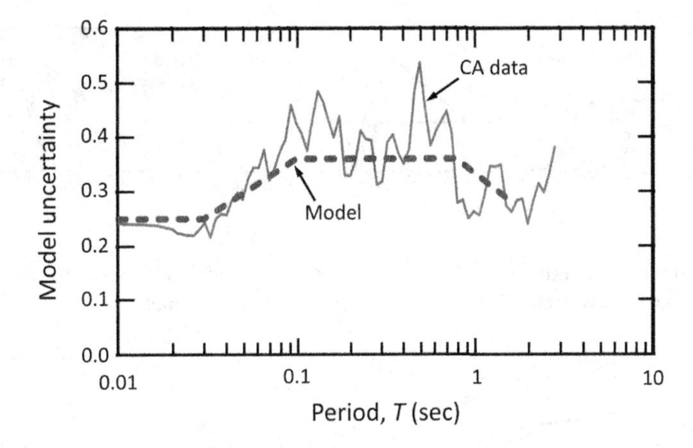

FIGURE 7.43 Model for ground response model uncertainty (ϕ_{S2S}^{m}) and site-to-site standard deviations (see Section 7.7.2.1) from ground response analysis predictions of site response from vertical arrays. (Adapted from Stewart and Afshari, 2021.)

as a site-to-site standard deviation of model residuals, for linear conditions based on data from diverse site conditions in California. Because the model residuals were affected by both modeling errors and soil property uncertainties, certain assumptions were made about the effects of soil property uncertainty and the remaining uncertainty was attributed to model errors (Stewart and Afshari, 2021). The uncertainty ranges from 0.25 to 0.35 in natural log units; subsequent work using an independent Japanese dataset generally confirmed these uncertainty levels (Bahrampouri and Rodriguez-Marek, 2023). For forward applications, additional uncertainty from material properties should be added (in variance). Lower levels of modeling uncertainty are possible in regions where ground response analyses are likely to be particularly effective, such as soft soil sites with nearly one-dimensional layering.

7.6 MULTI-DIMENSIONAL GROUND RESPONSE ANALYSIS

The methods of one-dimensional ground response analysis described in previous sections are useful for level or gently sloping sites in which waves propagate perpendicular to parallel material boundaries. Such conditions are not uncommon and one-dimensional analyses are widely used in geotechnical earthquake engineering practice. For many other problems, however, the assumptions of one-dimensional body wave propagation are not realistic. Lateral heterogeneity, sloping or irregular ground surfaces, the presence of heavy structures or stiff, embedded structures, or of basins, valleys, or canyons can all cause response to differ from those predicted by one-dimensional analyses.

7.6.1 Two- and Three-Dimensional Analysis

All site response problems are actually three-dimensional problems because earthquake ground motions are three-dimensional. However, ground response analyses rarely consider all three components of ground motion. Instead, decisions on dimensionality are typically based on geometric considerations. Level ground sites with nearly horizontal layering can be treated as one-dimensional. Sites with profiles or structures in which one dimension is significantly greater than others (Figure 7.44) can often be treated as two-dimensional plane strain problems in which no out-of-plane movement is assumed to occur. While both horizontal and vertical input motions can be applied to two-dimensional models, usually only horizontal motions are used. For other problems involving critical systems, more complex subsurface conditions, irregular topography, or potentially interacting structures (Figure 7.45), three-dimensional analyses may be required. Three perpendicular input motions can be applied to three-dimensional models.

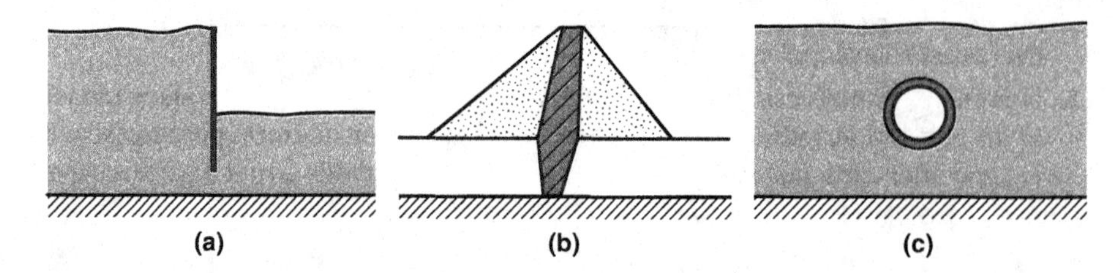

FIGURE 7.44 Examples of common problems typically analyzed by two-dimensional plane strain dynamic response analyses: (a) cantilever retaining wall; (b) earth dam; and (c) tunnel.

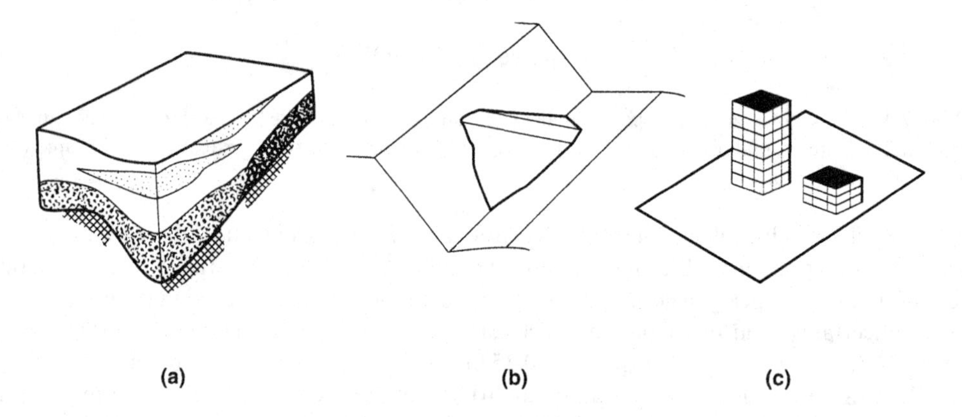

FIGURE 7.45 Three situations requiring three-dimensional dynamic response or soil-structure interaction analysis: (a) site where soil conditions vary significantly in three dimensions; (b) earth dam in narrow canyon; (c) site where the response of soil is influenced (b) response of' structures (and vice versa) and where the response of one structure may influence the response of another.

Two- and three-dimensional ground response problems can be solved by different numerical methods. Mathematically, the problem is one of solving a hyperbolic partial differential equation – the wave equation. Due to the complexity of geometries and material properties and boundary conditions, practical ground response problems must be solved numerically. Two- and three-dimensional analyses are much more complicated than one-dimensional analyses and some of the recent innovations in one-dimensional analyses (e.g., modulus reduction and damping curve consistency, frequency-independent low-strain damping, shear strength compatibility, etc.) are not yet available for multi-dimensional analyses.

The most common numerical methods for these types of problems are finite element and finite difference methods, which are simply different approaches to the numerical solution of differential equations. Broadly, finite element techniques approximate the solution of the differential equations while finite difference techniques approximate the differential equation itself. Finite element techniques can generally handle complex geometries and boundary conditions more easily than finite difference techniques and are more commonly used for two- and three-dimensional analyses. Regardless of which technique is used, several issues must receive careful consideration.

7.6.1.1 Discretization Considerations

The response of both equivalent linear and nonlinear models of all dimensions can be influenced by discretization. In particular, the use of coarse grids or meshes can result in the filtering of high-frequency components whose short wavelengths cannot be modeled adequately by widely spaced nodal points. The maximum dimension of any element should be limited to one-eighth (Kuhlemeyer and Lysmer, 1973) to one-fifth (Lysmer et al., 1975) of the shortest wavelength considered in the analysis. Figure 7.46 illustrates problems that can develop with large element sizes. To minimize such problems, the user should:

1. Determine the highest frequency of interest, f_{max}. Values of f_{max} of 25 Hz have been commonly used in engineering practice (Stewart et al., 2008).
2. For each material, compute the minimum wavelength of interest, λ_{min}, which corresponds to f_{max}

$$\lambda_{min} = \frac{V_s}{f_{max}} \tag{7.83}$$

3. Compute the maximum element dimension, d_{max}, as

$$d_{max} = \frac{\lambda_{min}}{N} \tag{7.84}$$

where N is typically on the order of 5–8; higher values of N will produce more accurate results but will also require more elements which slows down the analysis.

Note that the minimum wavelength, and hence maximum element dimension, will be smaller in softer materials. Typical meshes, therefore, will usually utilize relatively large elements in stiff materials and transition to smaller elements in softer materials. The fact that elements can become softer, thereby shortening the apparent wavelength for a particular frequency, under strong shaking should also be taken into account when determining element sizes. Of course, the geometry of the problem will dictate the final mesh geometry.

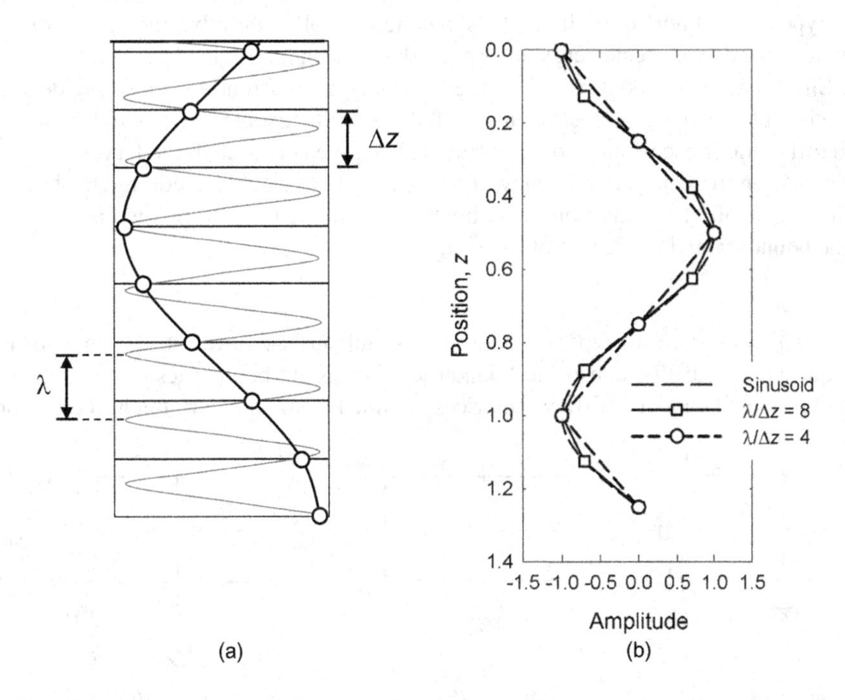

FIGURE 7.46 Discretization considerations: (a) High frequency (short wavelength) motion can be aliased to lower frequency if element thickness is not small relative to wavelength; (b) approximation of shortest wavelength (i.e., highest frequency) of interest degrades as the number of elements per wavelength decreases. A sinusoid can be represented with reasonable accuracy with $\lambda/\Delta z \geq 8$.

7.6.1.2 Boundary Conditions

For computational efficiency, it is desirable to minimize the number of elements in a finite element analysis. Since the maximum dimensions of the elements are generally controlled by the wave propagation velocity and frequency range of interest, minimizing the total number of elements often becomes a matter of minimizing the size of the discretized region. As the size of the discretized region decreases, the influence of boundary conditions becomes more significant.

For many dynamic response and soil-structure interaction problems, rigid or near-rigid boundaries such as bedrock are located at considerable distances, particularly in the horizontal direction, from the region of interest. As a result, wave energy that travels away from the region of interest may effectively be permanently removed from that region. In a dynamic finite element analysis, it is important to simulate this type of *radiation damping* behavior. The most commonly used boundaries for finite element analyses can be divided into four groups (Christian et al., 1977; Wolf, 1985; Berenger, 1994, 1996), as described in the following sections.

Elementary Boundaries

Conditions of zero displacement or zero stress are specified at elementary boundaries (Figure 7.47a). Elementary boundaries can be used to model the ground surface accurately as a free (zero stress) boundary. For lateral or lower boundaries, however, the perfect reflection characteristics of elementary boundaries can trap energy in the mesh that would, in reality, radiate past the boundaries and away from the region of interest. The resulting "box" or "bathtub" effect can produce serious errors in a ground response or soil-structure interaction analysis. If elementary boundaries are placed far enough from the region of interest, reflected waves may be damped sufficiently to negate their influence.

Local Boundaries

Section C.4.1 showed how a viscous dashpot can be used to simulate a semi-infinite region for the case of normally incident body waves. The use of viscous dashpots (Figure 7.47b) represents a common type of local boundary. It can be shown (e.g., Wolf, 1985) that the value of the dashpot coefficient necessary for perfect energy absorption depends on the angle of incidence of the impinging wave. Since waves are likely to strike the boundary at different angles of incidence, a local boundary with specific dashpot coefficients will always reflect some of the incident wave energy. Additional difficulties arise when dispersive surface waves reach a local boundary; since their phase velocity depends on frequency, a frequency-dependent dashpot would be required to absorb all their energy. The effects of reflections from local boundaries can be reduced by increasing the distance between the boundary and the region of interest.

Consistent Boundaries

Boundaries that can absorb all types of body waves and surface waves at all angles of incidence and all frequencies are called consistent boundaries. Consistent boundaries can be represented by frequency-dependent boundary stiffness matrices obtained from boundary integral equations or the

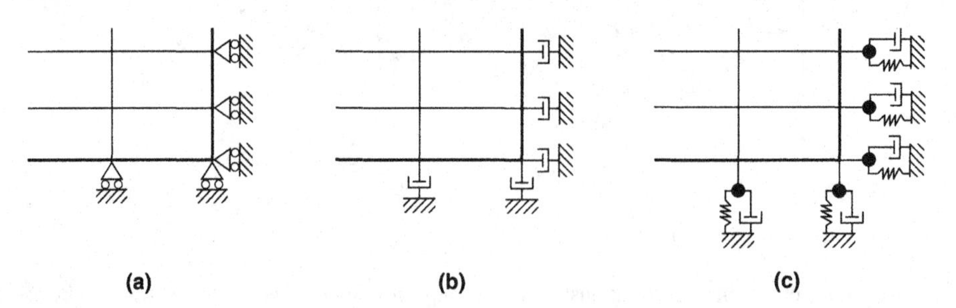

FIGURE 7.47 Three types of finite element mesh boundaries: (a) elementary boundary in which zero displacements are specified; (b) local boundary consisting of viscous dashpots; (c) lumped-parameter consistent boundary (actual lumped parameter would consist of more masses, springs, and dashpots than shown).

boundary element method. Wolf (1991), for example, developed a lumped-parameter model consisting of an assemblage of discrete springs, masses, and dashpots which can approximate the behavior of a consistent boundary. A greatly simplified example of such an assemblage is shown in Figure 7.47c.

Perfectly Matched Layer Boundaries

A perfectly matched layer, or PML (Berenger, 1994), essentially consists of a lossy (high damping) material placed around a discretized domain of interest. Originally developed for the absorption of electromagnetic waves, the PML concept was applied to stress waves for soil-structure interaction problems by Basu and Chopra (2003). A PML provides nearly perfect absorption (i.e., lack of reflection) over all angles of incidence. Early methods involved splitting the incoming wavefield and defining a non-physical anisotropic absorbing material, but more recent approaches (Roden and Gedney, 2000) do not require wave splitting and can be implemented relatively easily. Although the PML boundary can be shown to be perfect for a continuum, a small amount of reflection can occur when introduced into discrete numerical (e.g., finite element and finite difference) models.

7.6.1.3 Input Motions

Input motions for two- and three-dimensional models of structures/systems that are relatively small in plan dimension can often be analyzed with a coherent input motion applied at the base of the model. However, models of distributed infrastructure such as bridges, pipelines, dams, and levees, may extend over such longer horizontal dimensions that spatially variable input motions are required to accurately characterize response. The causes and characterization of spatially variable ground motions (SVGMs) were discussed in Section 3.8.1. Input motions that reflect wave passage and potential wave scattering effects may be required to accurately capture the response of such systems.

7.6.2 Basin Effects

Many developed areas are located on sedimentary basins – two- and three-dimensional depressions in which softer materials have been deposited above stiffer materials. In some cases, basins may consist of soil underlain by rock, and in other cases, they may consist of softer (e.g., sedimentary) rock underlain by harder (e.g., igneous) rock. Basins can exist at many scales. The Mississippi Embayment, for example, is about 900 km long and underlies portions of eight states in the southern United States. The Los Angeles basin is about 55 km long and 10 km deep, underlying several cities within Los Angeles and Orange Counties. Other basins can exist at much more local scales, such as a small river valley perhaps one km across and 100 m deep. Because they tend to be located near navigable bodies of water and their surfaces tend to be relatively flat, many basins are heavily developed and populated.

Many basins are created by surficial erosion and depositional processes, but folds and synclines can produce buried basins. Other basins are bounded by faults and are of obvious importance if the fault is seismically active. Fault-bordered basins include the Los Angeles basin, the Kanto basin, and the Seattle basin. Basins are frequently found in areas of both low and high seismicity.

Basins have been shown to affect earthquake ground motions and to be associated with locally higher levels of earthquake damage. Through different mechanisms, ground surface motions within basins can be significantly stronger and of greater duration than surface motions outside basins.

7.6.2.1 Basin Response Mechanisms

While the complexity of individual basin geometries can lead to complex interactions of different mechanisms, simplified examples can help illustrate some of the fundamental processes involved in basin response. Basin effects were briefly introduced in Section C.6.2 in order to provide some context for the GMM basin terms discussed in Section 3.5.2. A more detailed description is presented here.

Focusing

As indicated in Section C.4.2, seismic waves will be refracted upon oblique incidence to a boundary between materials of different wave propagation velocity, and an incident wave traveling through a stiffer material will be refracted at a boundary with a softer material to a direction more perpendicular to that boundary. Consider the vertically propagating SH waves encountering the base of the sedimentary basin shown in Figure 7.48. As the waves cross the boundary between the rock and the overlying sediments, they are refracted to a direction more perpendicular to the boundary. The shape of the basin causes the direction of the refracted waves to move toward the center of the basin, so the energy flux (energy per unit area) is higher near the center of the basin than it is outside the basin. As a result, the ground surface motions produced by these waves will be stronger in the middle of the basin than outside the basin.

It is important to note that the simple schematic in Figure 7.48 shows only the initial raypaths of waves that enter the basin. The rays will be reflected off the surface of the basin and will then travel back down to the bottom of the basin where they will be partially reflected and partially transmitted back into the rock. The reflected portions will travel back up and be reflected off the surface again, and so on. After a short amount of time, the wavefield within the basin, even for this very simple example, will become quite complicated.

Wave Trapping

Of course, neither basins nor incoming wave fields are as simple as the one shown in Figure 7.48. Some component incoming of wave energy may be inclined, and that will affect the directions and amplitudes of the waves that are refracted into the basin. As discussed in Section C.4.2, a critical angle of incidence exists at a particular boundary, and waves that reach the boundary at incidence angles greater than the critical angle are fully reflected at that boundary. Figure 7.49

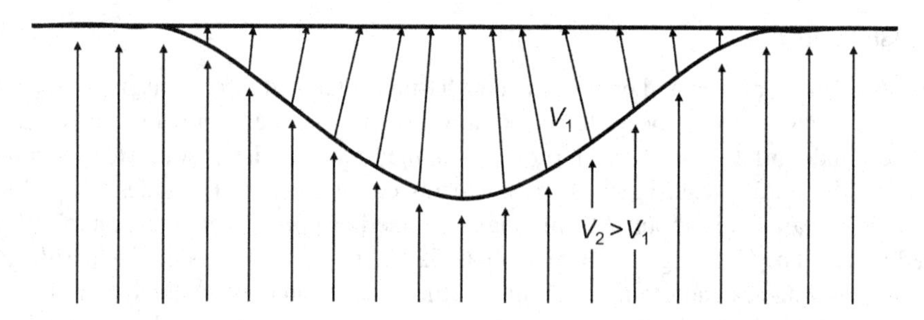

FIGURE 7.48 Basin-induced focusing due to refraction of vertically incident waves.

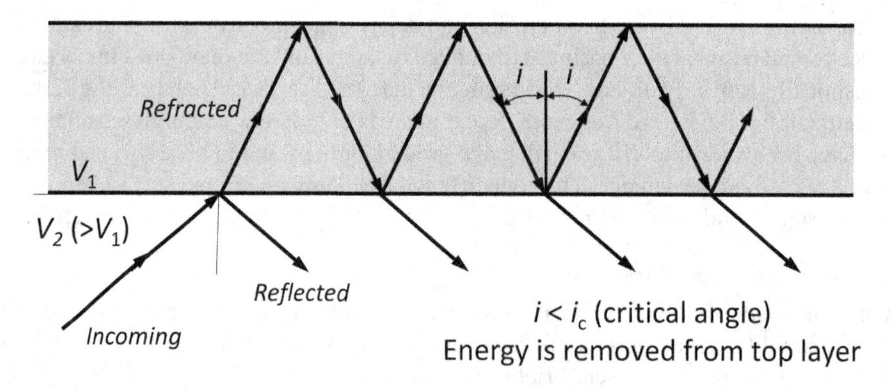

FIGURE 7.49 Reflection and refraction of wave entering soft, flat surficial layer. (After Graves, 1993 with permission of Elsevier.)

shows waves encountering a simple horizontally-layered system at a 45° angle. Part of the incident wave is reflected and part is refracted into the softer surface layer. The refracted portion travels in a more vertical direction and, after reflection off the surface, travels down to the underlying material at an angle less than the critical angle for the material boundary ($i<i_c$). As a result, part of that downward-traveling wave is reflected back upward and some is refracted back down into the underlying layer. The amplitude of the wave that remains in the surficial layer will be reduced each time it reflects off the material boundary because some of its energy "leaks" into the underlying layer.

If the same 45° incident SH-wave was approaching the same soft material at the edge of a basin, however, the nature of the waves refracted from the basin edge will be different than in the flat layer case shown in Figure 7.49. In the case of Figure 7.50, the incident wave is more normal to the basin edge so the refracted wave will be less steeply inclined than in the flat layer case. This wave will also reflect off the surface and approach the bottom of the basin at a flatter angle, i, than in the flat layer case. If that angle is greater than the critical angle ($i>i_c$), the wave will be fully reflected at the bottom of the basin. In this case, all of the energy remains in the upper layer so the wave is effectively "trapped" within the basin – none is transmitted into the underlying material. The energy in the trapped SH-wave becomes organized into surface waves called *Love waves* (Section C.3.2). If the incident wave was a p-wave or SV-wave, its interaction with the free surface would produce surface waves called *Rayleigh waves* (Section C.3.1) that would propagate across the basin. Such waves tend to travel across the basin and be reflected back at the opposite side of the basin. Surface waves that travel back and forth across a basin can significantly increase the duration of ground motions within a basin compared to those outside the basin. The incident angle at the far edge of a basin will be less than the critical angle, so energy will be transmitted into the underlying material at that location; the process of critical reflection only exists in the direction of basin thickening. Because of the large dimensions of many basins and the tendency of high-frequency (short wavelength) waves to be diminished by damping, basin effects primarily affect long-period ground motions.

7.6.2.2 Examples of Basin Effects

Significant basin effects have been observed in numerous earthquakes. Strong ground motions were generated at basin edges in both the 1994 Northridge and 1995 Kobe earthquakes. The Northridge earthquake occurred north of the Los Angeles basin, the northern boundary of which is controlled by the east-west trending Santa Monica Fault. Figure 7.51 shows a generally north-south section extending from the Santa Monica Mountains in the north down along the Pacific Ocean coastline west of downtown Los Angeles. The Santa Monica Fault can clearly be seen by the sharp drop in shear wave velocities in the upper 3 km. Recorded ground motion from locations both outside (north of) and inside the basin are also shown in Figure 7.51. While the early parts of the records are somewhat similar, the amplitudes and durations of the motions recorded within the basin can be seen to be much greater than those recorded outside the basin. Damage distributions reflected the

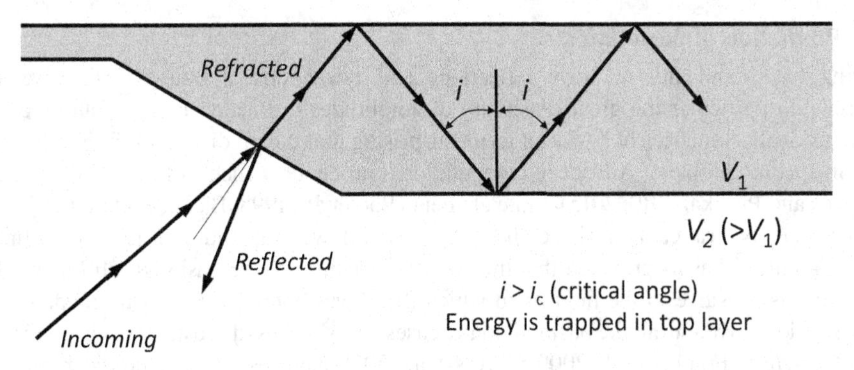

FIGURE 7.50 Reflection and refraction of wave entering soft surficial layer at basin edge. (After Graves, 1993 with permission of Elsevier.)

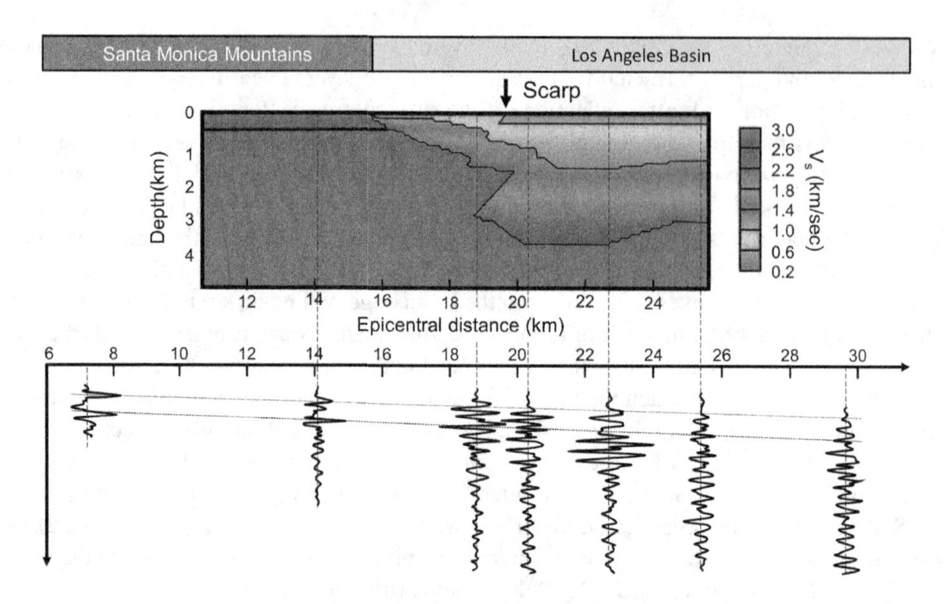

FIGURE 7.51 1 Hz low-pass filtered velocity traces from outside and inside the Los Angeles basin in the 1994 Northridge earthquake. (After Graves et al., 1998 with permission of SSA.) Note higher amplitudes and longer durations of motions recorded in the Los Angeles basin.

distribution of the recorded ground motions – the distribution of red-tagged buildings (i.e., buildings considered unsafe for occupancy) was concentrated in the area immediately south of the fault scarp in Santa Monica where basin-edge effects and focusing were significant.

The most highly developed portion of Kobe, Japan lies on a narrow alluvial plain/terrace bordered on the northwest by the Rokko mountain range and on the southeast by Osaka Bay. The Rokko fault system marks the boundary between the Rokko mountain range and the alluvial soils. The area is heavily developed with structures of similar type, age, and construction. Following the 1995 Kobe earthquake, increased rates of damage were observed in a zone about 30 km long and 1 km wide running parallel to the Rokko fault system (Figure 7.52), suggesting that ground motions were locally higher within that zone. While amplification due to the lower stiffness of the alluvial sediments undoubtedly occurred, subsequent 2-D and 3-D simulations indicated that the motions in the zone of high damage were further amplified by basin-edge effects. Constructive interference between SH waves that propagated upward from the underlying sediments and basin-edge generated surface waves, along with directivity effects (Kawase, 1996; Pitarka et al., 1998; Kawase et al., 2000), are thought to have produced strong amplification in this area.

7.6.2.3 Prediction of Basin Effects

Considering the complexity of wave reflections and refractions at boundaries between different materials, in particular the great sensitivity of amplitudes to incidence angles and the complex, three-dimensional geometries of basins, it is not surprising that the prediction of basin effects is a difficult, basin-specific problem. Advanced computational procedures such as finite difference (Graves, 1996; Graves and Pitarka, 2010, 2015), finite element (Bao et al., 1998; Bielak et al., 2010; Komatitsch et al., 2010), and spectral element (Faccioli et al., 1997) allow computation of the three-dimensional response of arbitrary basins provided that the velocity structure of the basin is well known. Because basin response is sensitive to the manner in which the waves approach the basin, models must also include the region surrounding the basin; in many cases, physics-based crustal models that include the source and path (e.g., Frankel et al., 2009; Graves et al., 2011) have been used to predict basin response.

Numerical analyses of even simple basin geometries can illustrate the complexity of basin response and the dramatic differences between response inside and outside of basins. Figure 7.53

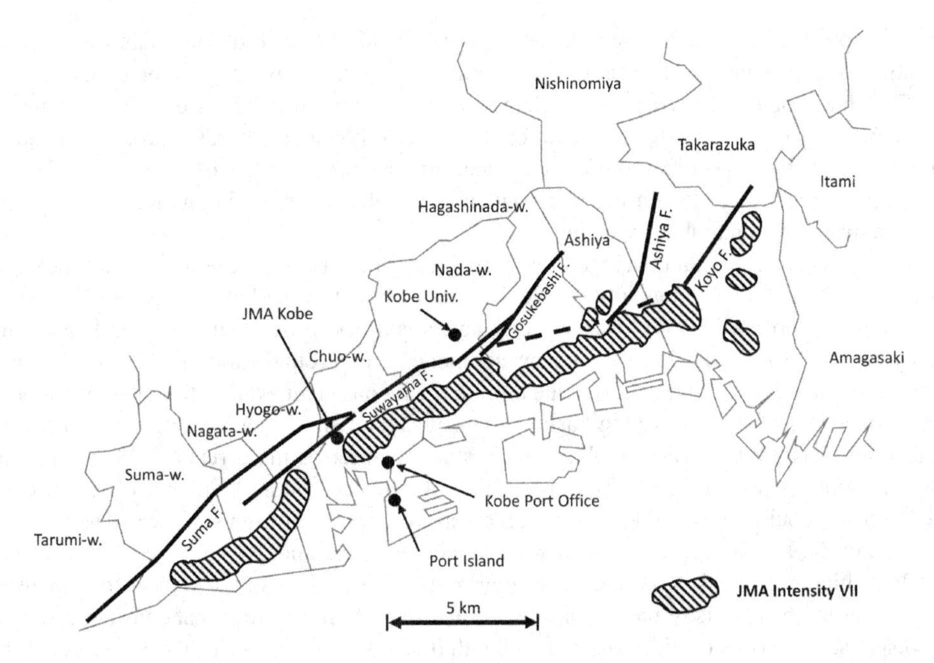

FIGURE 7.52 Distribution of damage belt, as defined by JMA intensity scale VII in which more than 30% of residential houses collapsed in the 1995 Hyogo ken Nanbu (Kobe) earthquake. (After Kawase, 1996 with permission of Seismological Society of America.)

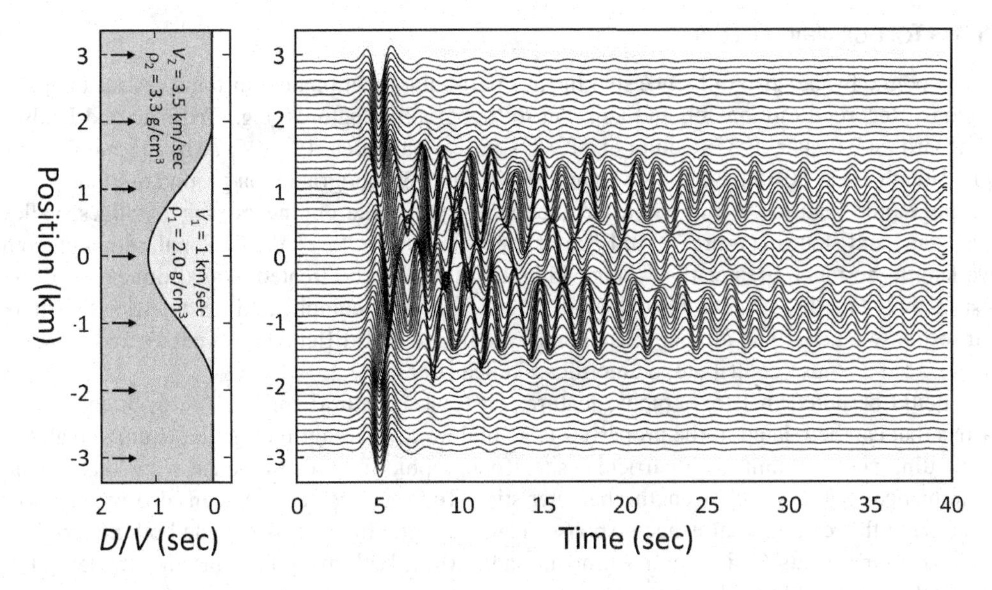

FIGURE 7.53 Response of points across surface of basin to upward-traveling pulse. (After Chen, 2007 with permission of Elsevier.)

shows the motion that results from a single Ricker pulse of motion applied at the base of a profile that included a basin with an impedance ratio of 0.12. Although the initial arrival times are nearly constant (although a little later for locations above the softer basin soils), the pulse clearly reverberates within the basin leading to large amplitudes and greatly extended durations.

Given the dimensions of most basins, many elements are required to model the domain of interest. The maximum element size depends on the minimum wavelength of interest, and the minimum wavelength depends on the wave propagation velocity and the maximum frequency of interest.

The need to predict response to high frequencies and/or the presence of materials with low wave propagation velocities can lead to short wavelengths, which can only be represented by small elements. For a basin of a given size, smaller elements obviously require the use of a greater number of elements, which leads to a greater computational burden. With current parallel processing capabilities, it is now possible to predict three-dimensional response reliably to frequencies of 1–2 Hz. With cloud computing resources becoming more widely available, the upper bound frequency for these types of analyses will undoubtedly increase.

Velocity models have been developed for a few important basins. These models are intended to capture basin geometry and elements of velocity structure that affect long-period ground motions. Available models do not include lower-velocity layers near the ground surface, and hence cannot generally be used to predict high-frequency components of ground motions. For many regions, basin velocity models are not available. Due both to the inconsistent availability of velocity models, and the complexity of the underlying basin response mechanisms, detailed empirical modeling of basin response has not yet been possible. As a result, basin terms in current GMMs are relatively simple functions conditioned on indirect indicators of bedrock depth. A number of recent GMMs (Section 3.5.2.3) contain terms that represent the depths to layers of high shear wave velocity – the terms $Z_{1.0}$ and $Z_{2.5}$ are the depths to shear wave velocities of 1.0 and 2.5 km/s, respectively. Basins tend to have higher $Z_{1.0}$ and $Z_{2.5}$ values than areas outside of basins, so those terms can be used to broadly represent the effects of basins on ground surface motions. Models that consider differences in site response for basins of different size and with different geologic histories have recently been developed (Nweke et al., 2022) and are likely to replace the simpler depth-based models in some situations.

7.6.3 Topographic Effects

Local topography can also influence shaking at a particular location and, in some cases, at adjacent locations. Recorded motions on and adjacent to topographic features (e.g., Trifunac and Hudson, 1971; Bard and Tucker, 1985; Geli et al., 1988; Chavez-Garcia et al., 1996; Asimaki and Gazetas, 2004; Hough et al., 2010) have shown significant impacts of topography on ground motion.

Topographic features of interest to geotechnical earthquake engineers include ridges, valleys or canyons, and slopes (Figure 7.54). Each of these features is three-dimensional, although some have lengths so much greater than their widths that they can be treated as two-dimensional over most of their lengths. All can have very different (and very complicated) geometries that make their response unique, and all can be mantled by soil layers of different thicknesses and velocities (Figure 7.55) that also contribute to amplification behavior.

It should be noted that topographic features, at least at natural soil/rock sites, often exist for reasons that can be related to the properties (e.g., high or low strength) of geologic units relative to surrounding geologic units at a particular site. Topographic features influence stress state, which can influence stiffness and strength characteristics. Topographic features can also include shallow layering that can also influence response. Thus, topographic effects include both material and geometric components (Asimaki and Mohammadi, 2018); both are important and neither can be completely separated from the other.

The effects of material properties on site response have been discussed at length in the earlier portions of this chapter. The following discussion focuses primarily on the geometric aspects of the topographic amplification problem.

7.6.3.1 Topographic Response Mechanisms

A topographic feature is one that deviates in some way from a level-ground condition, several examples of which are shown in Figure 7.56. In the case of the ridge and the slope, the feature extends above the level-ground plane, and in the case of the canyon, it extends below it. Alternatively, one could imagine a horizontal line tangent to the base of the canyon as a "pseudo"

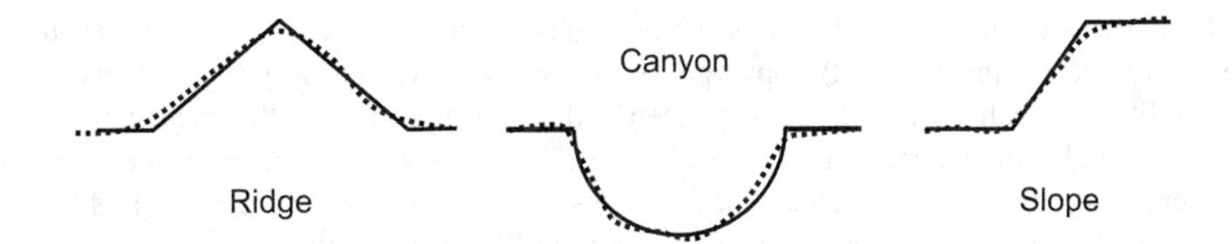

FIGURE 7.54 Actual (dashed) and idealized (solid) representations of common topographic features.

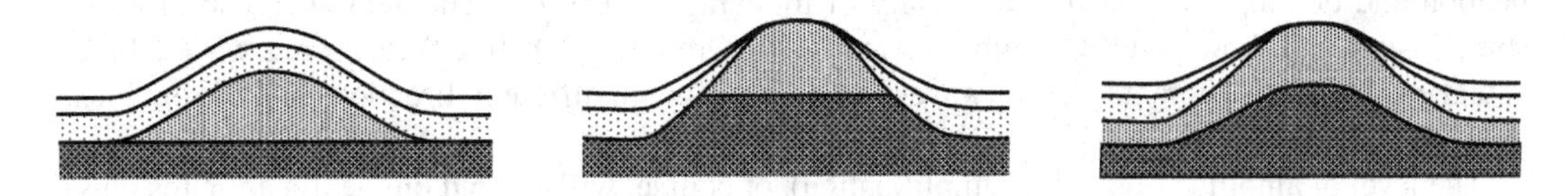

FIGURE 7.55 Topographic features with layering of uniform and variable thicknesses. Velocity contrasts and variable geometries of shallow subsurface layers can cause significant differences in response to features of similar topography. (After Geli et al., 1988 with permission of SSA.)

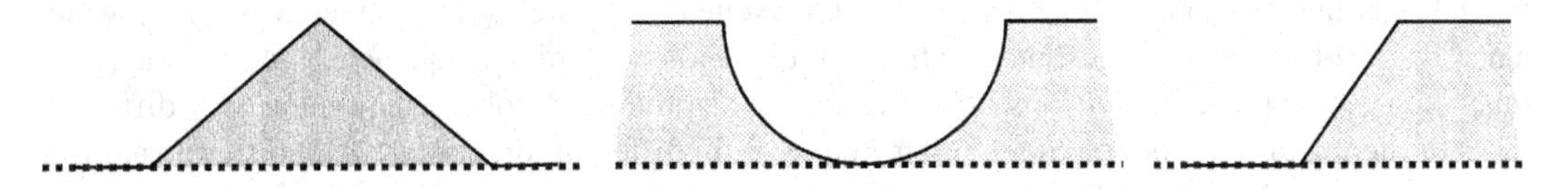

FIGURE 7.56 Idealizations of topographic features relative to actual (ridge and slope) or pseudo (canyon) level-ground surfaces (dashed lines).

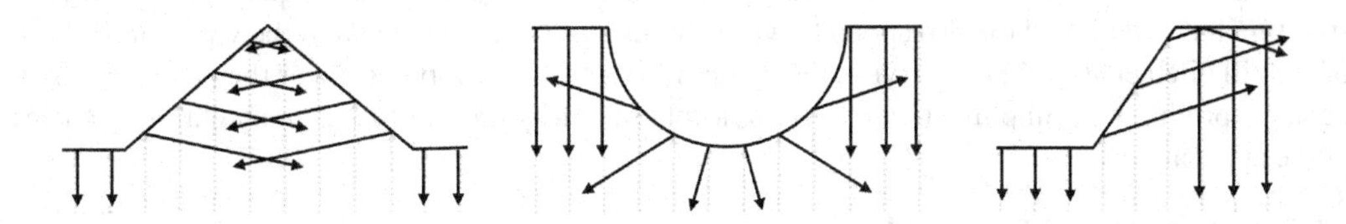

FIGURE 7.57 Reflected waves from vertically propagating incident waves for various topographic features.

level-ground surface with two opposing slopes rising above it (Figure 7.56). In each of these cases, the material above the level-ground surface has mass and stiffness and hence will respond dynamically in a frequency-dependent manner. In contrast to the common one-dimensional response case, the geometries of the zones above the level-ground surface vary with height above that surface; in each case, the width decreases with increasing elevation so that the overall stiffness of the feature will decrease with increasing elevation even if the stiffness of the material is constant.

Some of the patterns in which topographic features amplify or de-amplify ground motions can be visualized by considering the reflection of incoming waves by the irregular surfaces of the topographic features while recognizing that ground motion amplitudes are related to energy flux (i.e., energy per unit area through which waves are propagating). As energy flux increases, ground motion amplitudes increase, so zones, where wave energy is concentrated by topographic features, can be expected to experience stronger ground motions. Likewise, zones where the energy density decreases can be expected to experience weaker motions. Figure 7.57 illustrates the patterns of first reflections of upward propagating shear waves (assuming out-of-plane, i.e., perpendicular to the printed page, particle motion). To the sides of each of the features, the upward-traveling waves are reflected back downward off the horizontal ground surface; the paths remain vertical and the energy flux in these areas remains constant. Constructive interference of reflected waves can occur where wave paths (reflected and incident) converge and cross each other, leading to increased ground motion amplitudes. Such interference can be seen in the central portion of the ridge, near the edges of the canyon, and near the crest of the slope; stronger motions would be expected in these areas. Zones where waves diverge from each other can be seen near the toes of the ridge and slope, and near the base of the canyon; because the energy flux will be lower in these areas, weaker motions would be expected. This, of course, presents a very simplified representation of the response of topographic features; it is, however, one that provides some physical insight into patterns of amplification and de-amplification in the vicinity of topographic features. The approach also provides some insight into how patterns might change for different (e.g., inclined) incoming waves. In an actual earthquake, different types of waves can be traveling in different directions at different locations all at the same time.

The level of amplification (or de-amplification), of course, will depend on the dynamic response of the feature. Because the features that extend above the ground surface (actual or pseudo as shown in Figure 7.56) have finite mass and stiffness, they will have natural frequencies at which they tend to respond more strongly. Thus, the degree of amplification will be influenced by the size of the feature relative to the wavelengths of the input motions. Very low frequencies with wavelengths much larger than the topographic feature will not excite the feature itself – the feature will move in a nearly rigid-body mode in response to those frequencies. Very high frequencies with wavelengths much shorter than the dimensions of the feature will produce incoherent motions, i.e., different parts of the feature will be moving at high frequency in different directions with no significant net movement of the feature itself. At frequencies near the natural frequencies of the feature, particularly the fundamental frequency, the topographic features will response dynamically and can cause ground motions at "high points" (near the crest of a ridge or slope) to increase significantly from the level-ground response that would occur in their absence. The fundamental frequency of a hill-type structure has been found to correspond to wavelengths approximately equal to the width of the base of the hill (Geli et al., 1988). Likewise, the dynamic response at low points (near the base of a valley or slope) produces de-amplification relative to level ground, again near the fundamental frequencies of the adjacent features.

7.6.3.2 Examples of Topographic Effects

Two of the classic examples of topography come from sites relatively close to each other in southern California; both produced what were considered to be the highest acceleration levels ever recorded at their times. An accelerograph on the abutment of Pacoima Dam in southern California recorded peak horizontal accelerations of about 1.25*g* in each of two perpendicular directions in the 1971 San Fernando (M_L=6.4) earthquake, values that were considerably larger than expected for an earthquake of this magnitude. The accelerograph, however, was located at the crest of a narrow, rocky ridge (Figure 7.58) adjacent to the dam (Trifunac and Hudson, 1971). Subsequent investigations have attributed a good part of the unusually high peak accelerations to the dynamic response of the ridge itself-a topographic effect. In extreme cases, very strong topographic amplification has led to the development of *shattered ridges*, zones of rock that are shaken so hard that they fracture at the top of a ridge line (Figure 7.59).

In the 1994 Northridge earthquake (M_w=6.7), an accelerograph at the crest of a low (< 20 m high), elongated hill in Tarzana, California recorded a peak acceleration of 1.93*g*. Following the Northridge earthquake, an array of temporary seismographs was installed (Spudich et al., 1996) generally along and transverse to the axis of the hill (Figure 7.60). Figure 7.61 shows recordings made during an aftershock of the Northridge earthquake; the amplitudes of the motions can be seen to be significantly higher at and near the crest than at the edges of the hill. Thus, very strong amplification was observed for this rather unimposing topographic feature. The hill is primarily made up of a soft siltstone/claystone/shale of the Modelo formation (Spudich et al., 1996).

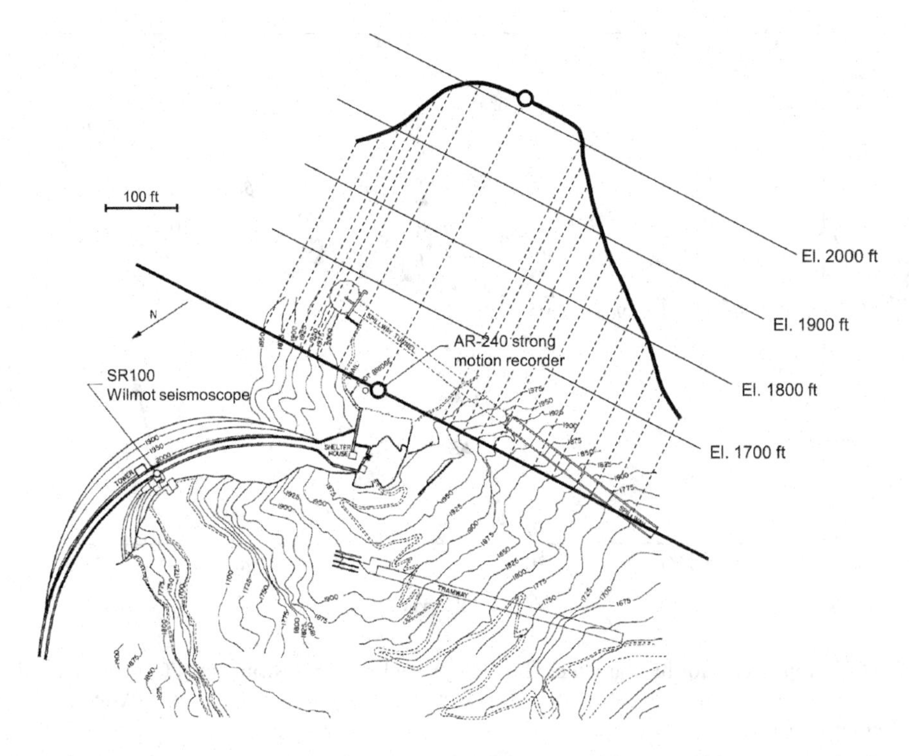

FIGURE 7.58 Pacoima Dam site illustrating geometry of abutment ridge on which strong motion instrument was located. (After Trifunac and Hudson, 1971 with permission of Seismological Society of America.)

FIGURE 7.59 (a and b) Shattered ridges in Augustana Creek area due to strong topographically-amplified motions in the 2002 Denali earthquake. (Photo courtesy of L. Cluff.)

A post-earthquake investigation of the site (Darragh et al., 1997) showed the presence of about 5 m of colluvium (V_s = 194 m/s) underlain by 12 m of decomposed shale underlain (V_s = 277 m/s), in turn, by about 40 m of less weathered shale (V_s = 433 m/s) and then slightly weathered shale (V_s = 556 m/s). Thus the hill was mantled, and underlain, by a series of successively stiffer soil and rock units which probably also contributed to the high degree of amplification observed in the Northridge earthquake.

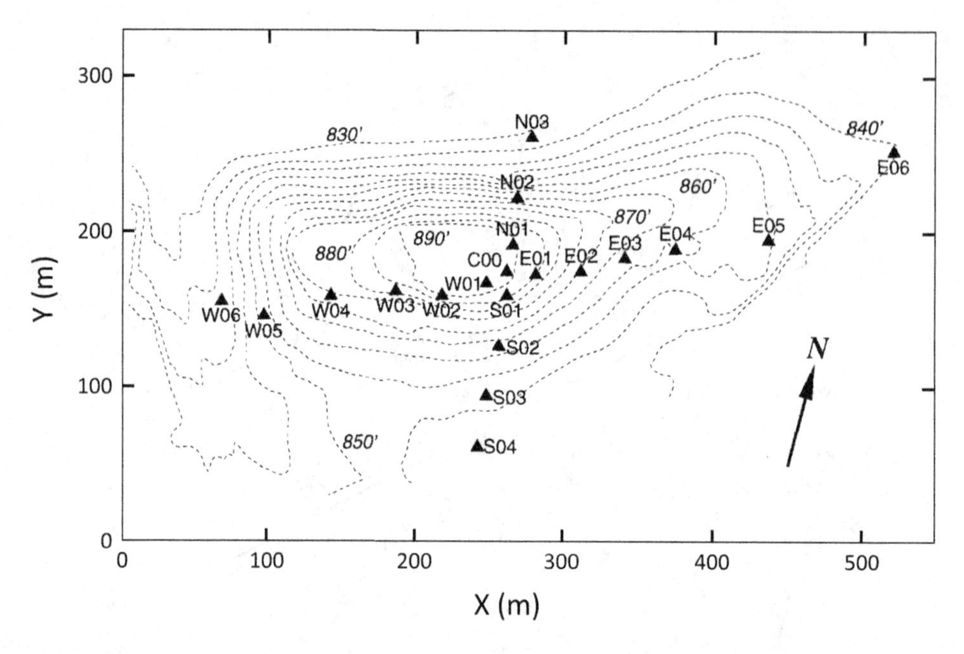

FIGURE 7.60 Topography of hill on which Tarzana ground motion instrument (C00) was located. Other instruments are temporary instruments installed to measure response in aftershocks. (After Bouchon and Barker, 1996 with permission of SSA.)

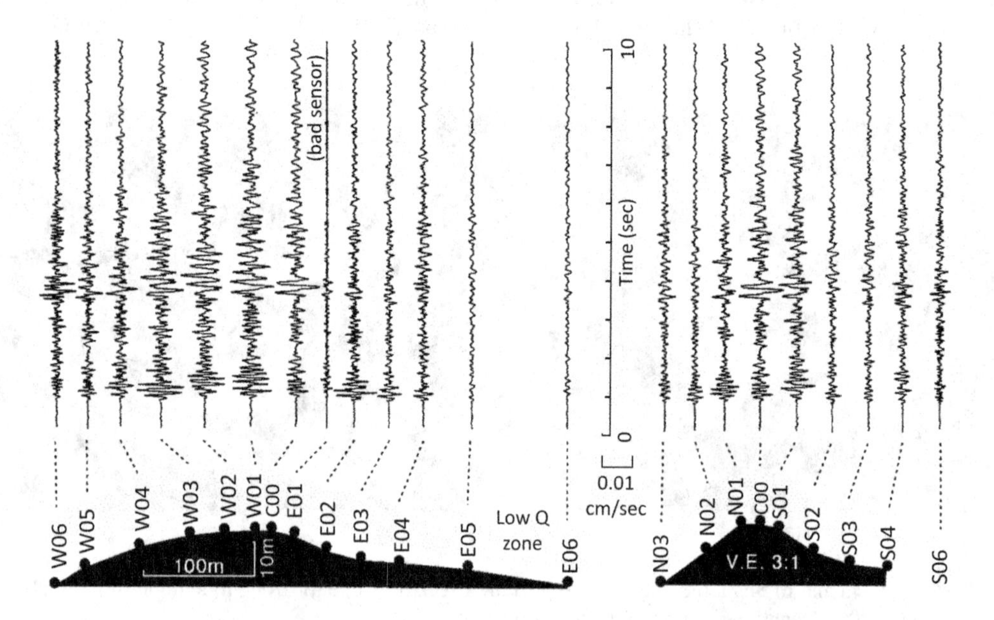

FIGURE 7.61 North-south motions recorded by temporary array instruments at Tarzana during aftershock of the 1994 Northridge earthquake. Note exaggeration of vertical scale and variation of amplitudes from edges to top of hill. (From Spudich et al., 1996 with permission of Seismological Society of America.)

7.6.3.3 Prediction of Topographic Effects

The preceding sections have indicated that topographic effects, for actual geometries and actual wavefields, are extremely complex. As with basins (Section 7.6.2), the response of a particular topographic feature is sensitive to the details of its three-dimensional geometry, the properties of the materials within, adjacent, and/or below the feature, and the nature of the wave field that excites it. These sensitivities, and the inevitable uncertainty in the factors that affect the response, make a priori prediction of topographic effects a difficult task. Nevertheless, some general insights into these effects can be gained from some relatively simple types of analyses, and tools for performing the types of detailed, site-specific analyses required to evaluate response more accurately are rapidly being improved.

Sanchez-Sesma (1990) illustrated certain aspects of amplification behavior for a simple triangular wedge with uniform shear modulus subjected to vertically propagating shear waves. For an SH-wave with out-of-plane (i.e., perpendicular to the plane of the drawing) particle motion, the amplification at the apex can be shown to be equal to $2\pi/\phi$ where ϕ is the apex angle of the wedge (Figure 7.62); for a level surface ($\phi=\pi$), therefore, the amplification factor correctly takes on the free surface value of 2. Figure 7.62 shows examples of the amplification factors for the top of a narrow ridge and the base of a broader valley; these amplification factors should be compared with the level-ground amplification factor of 2.0 to judge the degree to which the topographic feature is amplifying or de-amplifying the ground motion.

For vertically propagating shear waves with in-plane particle motion, reasonably straightforward solutions are only possible for certain apex angles and Poisson ratio values. For a 90° apex angle, Figure 7.63a shows the raypath and particle motion directions as an upward-traveling incident wave reaches the side of the wedge, is reflected to a horizontal path and then to a downward-traveling

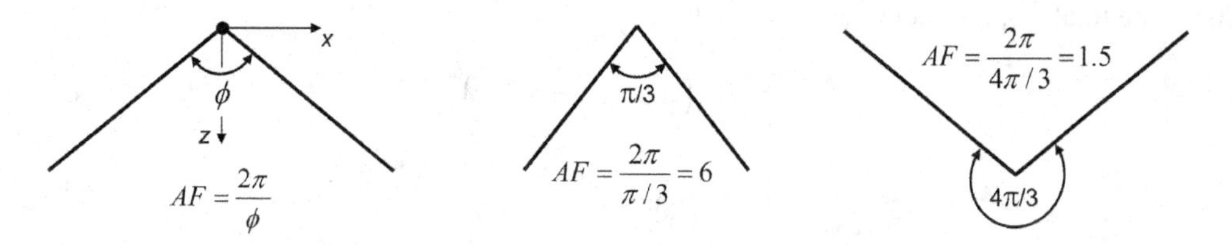

FIGURE 7.62 Amplification factors for apex of triangular wedge subjected to vertically propagating, out-of-plane shear waves. Level ground ($\phi=\pi$) amplification factor is 2.0.

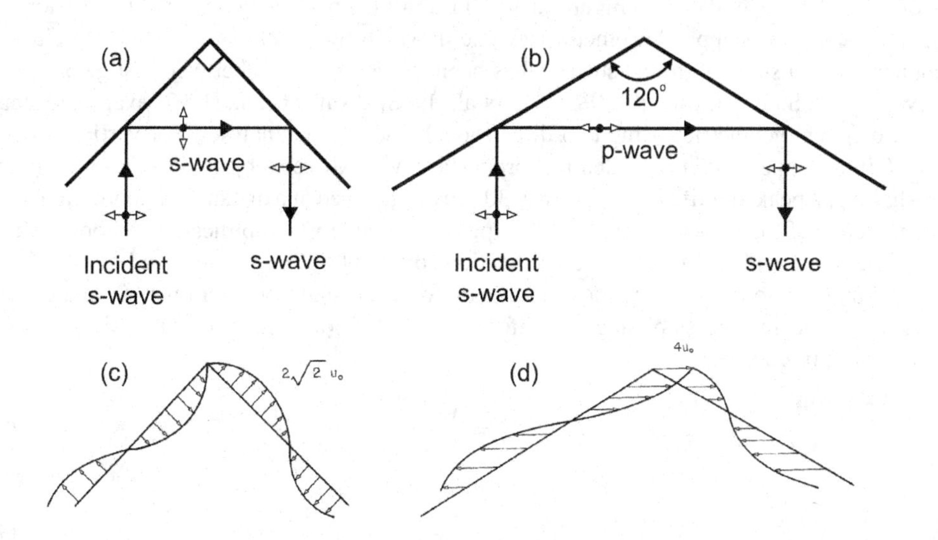

FIGURE 7.63 Illustration of response of triangular wedge to vertically incident s-waves: (a) ray paths (solid triangles) and particle motion (open triangles) for 90° apex angle, (b) ray paths and particle motion for 120° apex angle, (c) displaced shape for 90° apex angle, and (d) displaced shape for 120° apex angle. (After Sanchez-Sesma, 1990 with permission of Seismological Society of America.)

vertical path at the other side of the wedge. Letting u and w represent displacements in the x- and z-directions, respectively, particle motion of a vertically incident SH-wave given by

$$u = u_0 \exp(ikz) \tag{7.85}$$

where k is the wavenumber (ω/V_s), will be perfectly reflected at the wedge boundaries, producing a total displacement field given by

$$u = 2u_0 i \sin(kz) \tag{7.86a}$$

$$w = -2w_0 i \sin(kz) \tag{7.86b}$$

These equations indicate that the maximum surface displacement is $2\sqrt{2}u_0$ and that the apex displacement is zero. The deflected shape of the 90° wedge, shown in Figure 7.63c, shows that the surface motions are perpendicular to the wedge surface. For a 120° apex angle and assuming a Poisson's ratio of 0.25, the upward-traveling in-plane shear wave is perfectly converted from a shear wave to a p-wave that travels horizontally across to the other side of the wedge where it is converted back to a downward-traveling shear wave. The particle motion for this scenario is horizontal at all times. The total displacement field is given by

$$u = 2u_0\left(\cos k_s z + \cos k_p x\right) \tag{7.87a}$$

$$w = 0 \tag{7.87b}$$

where k_s and k_p are wavenumbers for s-waves and p-waves, respectively. These equations indicate that the maximum amplification occurs at the apex where the displacement amplitude is $4u_0$. The deflected shape of the 120° wedge, shown in Figure 7.63d, illustrates the horizontal nature of the motions.

The response of a "stepped" geometry, i.e., one in which two level zones at different elevations are connected by a sloping ground surface, has been studied by a number of investigators (Boore, 1972; May, 1980; Sitar and Clough, 1983; Geli et al., 1988). Ashford et al. (1997) examined stepped slopes of different geometries using two-dimensional viscoelastic analyses. For vertical slopes of height, H, in a stepped halfspace (identical properties everywhere) subjected to out-of-plane horizontal shaking, a peak amplification of nearly 30% was observed at a distance of approximately H/λ behind the crest of the slope with a secondary peak of about 15% amplification at about $H/\lambda = 0.7$. For in-plane horizontal shaking, peak amplifications of about 50% were observed at $H/\lambda = 0.2$ and $H/\lambda = 1.0$. Amplification levels were most significant for slope angles greater than 60° and generally decreased with decreasing slope angle. Ashford et al. (1997) also identified free-field and topographic natural frequencies,

$$f_n = \frac{V_s}{4Z} \tag{7.88}$$

$$f_t = \frac{V_s}{5H} \tag{7.89}$$

where V_s is the shear wave velocity of the material and Z and H are as indicated in Figure 7.64. The free-field natural frequency corresponds to the profile well behind the crest of the slope, and the

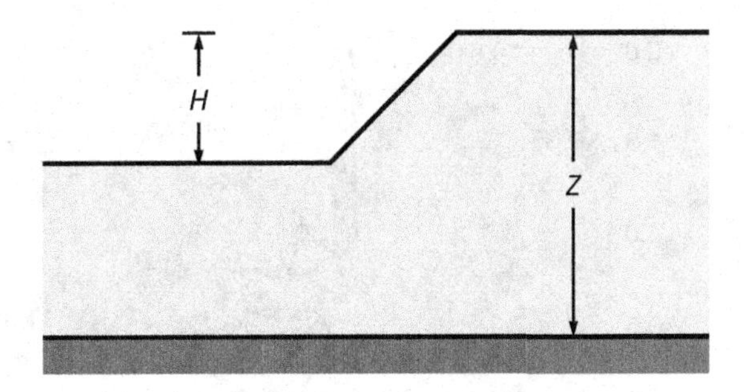

FIGURE 7.64 Notation for stepped slope. (After Ashford et al., 1997.)

topographic natural frequency corresponds to $H/\lambda=0.2$. The total ground surface amplification was found to be more strongly related to the free-field natural frequency than to the topographic natural frequency. Significant amplification can occur when the free-field and topographic natural frequencies are similar.

The symmetric triangular wedge response illustrated in Figures 7.62 and 7.63 applies only to an extremely simple idealized geometry with specific material properties and vertically incident waves. Deviations from any of these conditions, which are to be expected in any practical problem of topographic amplification, lead to multiple mode conversions, the development of surface waves, and other complexities.

The general analysis of topographic effects must be accomplished by two- or three-dimensional numerical analyses. A variety of different numerical techniques have been applied to the problem. Many are restricted to the assumption of linear elastic behavior, but many topographic features are sufficiently stiff that this is not necessarily a fatal limitation. To consider the coupled effects of topography and stratigraphy, i.e., the amplifying effects of softer materials that can cover hills and slopes, or fill the bases of valleys, finite element or finite difference techniques may be required.

Zhou and Chen (2006) used a two-dimensional boundary integral equation technique to simulate the scattering of SH waves by irregular surface topography. Figure 7.65 shows the propagation of a simple Ricker wavelet through an elastic medium with the hill-and-valley topography (rotated clockwise by 90°) shown on the left side of the figure. The wavelet propagates smoothly to the ground surface where it is reflected back downward in the level-ground areas to the sides of the irregular topography. Within the region where the ground surface is not level, however, the displacement response at the surface is considerably more complicated. The locations of the hills and valleys can be seen in the later and earlier arrivals of the incident wavelet. The hills can also be seen to have locally higher ground motion amplitudes (circled). The response of most of the valleys is obscured by the hill response, but that of the deepest valley (in box) can be seen to be weaker than the response of the level-ground portions at the sides of the profile. Maufroy et al. (2012) used a three-dimensional finite difference analysis to investigate the response of a 5 km×5 km area surrounding a low-noise underground laboratory near Apt in Provence, France. The contours of Figure 7.66a illustrate the complexity of the local topography, and a northwest-trending slice through the region shows multiple local peaks and valleys. The relative responses of Points A, B, C, and D show complex waveforms and higher response at the two points on hills (B and D) than at the nearby points (A and C) located in valleys. Note that Point A, which is at a relatively high elevation on a large ridge, but in a small swale near the edge of the ridge, had lower amplitudes than the nearby Point B, which was at very nearly the same elevation.

Eurocode 8 includes a topographic amplification factor for use in evaluating seismic slope stability. The Eurocode provisions specify that a topographic amplification factor, S_T, be applied for slopes with topographic irregularities "such as long ridges and cliffs of height greater than about 30 m." Slopes with average angles less than about 15° do not require consideration of topographic effects ($S_T=1.0$). For steeper angles, isolated cliffs and slopes should be designed with $S_T \geq 1.2$,

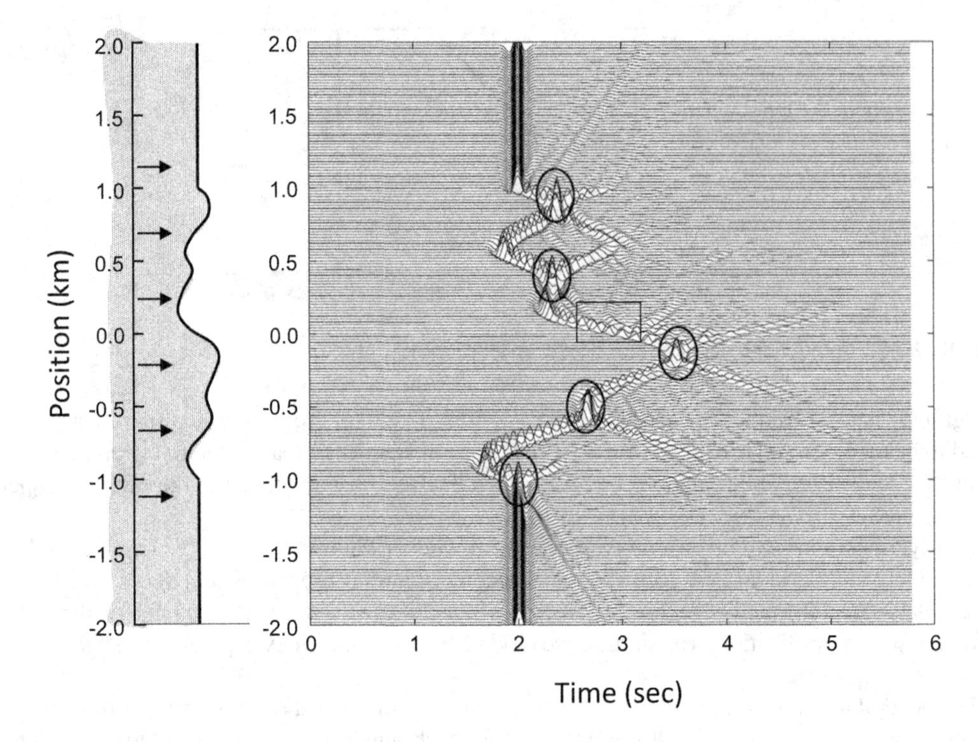

FIGURE 7.65 Ground surface response to vertically propagating SH wave in the form of Ricker wavelet through material with $V_s = 1{,}000\,\text{m/s}$ with central frequency of 5 Hz. Circled areas indicate high amplitudes associated with "hills" in topography. Boxed area indicates low amplitudes associated with the deepest "valley" in topography. (After Zhou and Chen, 2006 with permission of Oxford University Press.) Note reflections that follow the arrival of initial pulse and are diffracted into regions of level ground surface.

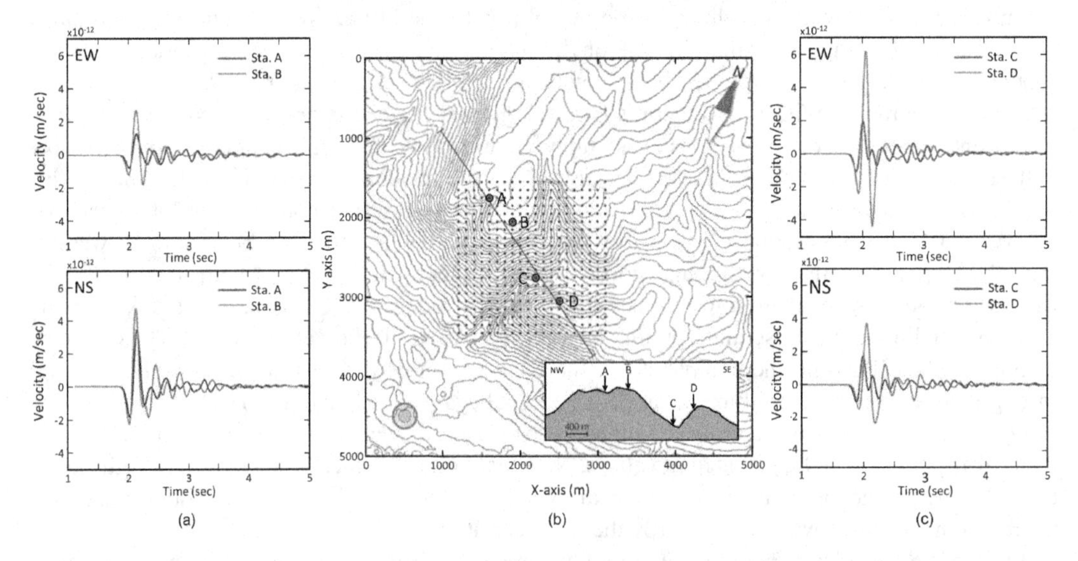

FIGURE 7.66 Three-dimensional finite difference analyses: (a) array of 400 stations (black dots) in area of variable topography, (b) response at Stations A and B, and (c) response at Stations C and D. (After Maufroy et al., 2012 with permission of Sage Publications, Ltd.)

TABLE 7.8
Coefficients for Median Topography Factor of Rai et al. (2016)

Coefficient	$T=0.0$ sec	$T=0.3$ sec	$T=1.0$ sec
c_{low}	0.0	–0.2921	–0.5911
c_{int}	0.0	–0.1346	–0.1211
c_{high}	0.0	0.0654	0.0564

ridges with crest widths significantly less than their base widths should be designed with $S_T \geq 1.4$ if the slope is greater than 30° and $S_T \geq 1.2$ if the slope is between 15° and 30°. If the slope also includes a loose surface layer, values of $S_T > 1.0$ should be further increased by at least 20%. These factors are considered to apply to the highest point of the slope and to decrease linearly to a value of 1.0 at the base.

Rai et al. (2016) used ground motions from 802 stations in a database of 188 small- to medium-magnitude ($M_w = 3$–6) earthquakes to develop an empirical model for topographic effects. Topography was characterized by a relative elevation, H_d, defined as the difference between the elevation of the recording station and the average elevation over a circular area of diameter, d. Using a value of $d = 1{,}250$ m to distinguish between sites of high, intermediate, and low topographic relief, a topography factor, f_{topo}, was defined.

$$f_{topo} = \begin{cases} c_{low} & \text{for} \quad H_d < -40 \text{ m} \\ c_{int} & \text{for} \quad -15 \text{ m} < H_d < 15 \text{ m} \\ c_{high} & \text{for} \quad H_d > 40 \text{ m} \end{cases} \tag{7.90}$$

The value of f_{topo} is in natural log units and is additive to the mean from a GMM. The median values of f_{topo} were defined at three structural periods, as indicated in Table 7.8, and linear interpolation is required for H_d values between –40 and –15 m and between 15 and 40 m. Positive values of f_{topo} indicate topographic amplification and negative values indicate topographic de-amplification. The coefficients indicate that topography has no effect on PGA ($T=0.0$ sec) and the effects at $T=0.3$ and 1.0 sec are shown in Figure 7.67. The topography factor of Rai et al. (2016) was defined with respect to the Chiou et al. (2010) GMM and its median value is shown by itself here for illustrative purposes; implementation of the Rai et al. (2016) procedure for prediction of topographic effects, including proper treatment of uncertainty, should follow the detailed description provided in that reference.

7.7 NON-ERGODIC SITE RESPONSE

For a given earthquake scenario, the ground motion at a particular site can be predicted with different levels of specificity depending on what is known about the relevant source, path, and site effects. *Ergodic models* predict ground motions based on large databases of recorded ground motions (sometimes supplemented by physics-based simulations) from many earthquakes associated with many different sources, paths, and site conditions. The sources are characterized relatively simply by quantities such as magnitude and style of faulting (Section 3.5.2.1). The source-site paths are also characterized simply by distance and potential hanging wall effects (Section 3.5.2.2). Finally, sites are characterized (Section 3.5.2.3) in a relatively crude manner by V_{S30}, basin depth, and, in some cases, topography.

The concept of ergodicity originated in statistical mechanics and holds that the average statistical properties of an ergodic system over time are equal to its average properties over some spatial

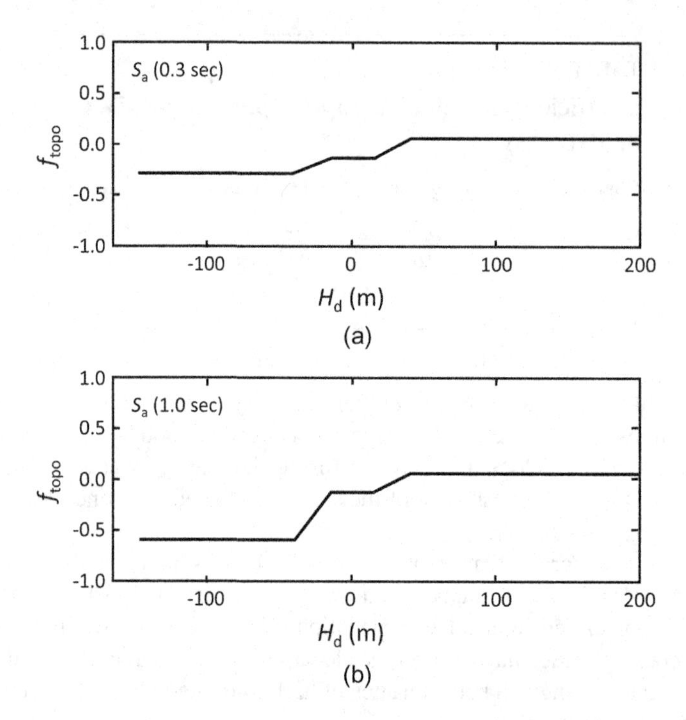

FIGURE 7.67 Variation of f_{topo} with H_d at (a) $T=0.3$ sec and (b) $T=1.0$ sec. (After Rai et al., 2016 with permission of Sage Publications, Ltd.)

domain. When applied to earthquake engineering, the short history of recorded ground motion data over time (particularly for the rare large earthquakes that are frequently of great interest for design purposes) is compensated for by collecting recorded ground motion data from earthquakes over a large (often global) spatial region. Ergodic models, such as the ground motion models described in Section 3.5, are derived in part by regressing upon all of this data to develop equations that can be used to predict ground motion intensity measures as functions of independent variables (typically magnitude, distance, style of faulting, and V_{S30}) used to characterize sources, paths, and sites. The averaging process inherent in regression means that the predicted intensity measures correspond most closely to average source, path, and site characteristics; these average, or generic, characteristics may differ significantly from the source, path, and site characteristics that influence ground motions for a specific source-path-site condition. These differences can lead to systematic, repeatable errors (or bias) in the intensity measures predicted by ergodic models. *Non-ergodic models* seek to account for specific source, path, and site conditions to produce more accurate (i.e., less biased) and less variable/uncertain predictions of ground motion intensity measures. The main focus of the following sections will be on site response, which is more amenable at present to non-ergodic treatment than source or path effects. Early efforts at examining non-ergodic path effects were made by Kuehn et al. (2019) and Abrahamson et al. (2019) and a general description of non-ergodic modeling is provided by Lavrentiadis et al. (2023).

Previous sections of this chapter have described the mechanics of site response and methods of analysis for wave propagation in the shallow crust that can be used to predict changes in ground motion at the surface of soil sites relative to reference (usually rock) sites. This site response effect is denoted AF and can, in principle, be combined in GMMs with source and path effects (Equation 3.24) to predict mean ground motions. Site response can be predicted for a given site with varying levels of specificity.

When an ergodic V_{S30}-based site amplification model is used, it is assumed that the average response of many sites with a given V_{S30} value is equivalent to the mean response of a particular site of interest. In reality, this assumption is rarely satisfied, because V_{S30} is an incomplete descriptor of site condition – as shown in Section 7.5.1.4, sites with the same V_{S30} but different stratigraphy (and also with different velocity profiles at depths greater than 30 m) can respond quite differently. In other words, ergodic site amplification models are not site-specific in the context of ground motion modeling.

Non-ergodic site response models consider site-specific features and can provide more accurate estimates of mean site response for a given location. Site-specific amplification models can be derived from analysis of recorded ground motions from a specific site of interest or from analysis of wave propagation in the shallow crust that takes that site's individual soil properties into account. This section describes procedures for developing non-ergodic site response models from either the interpretation of ground motions recorded at the site of interest (ideally, but infrequently available at present) or from site-specific ground response analyses (Sections 7.5-7.6). Also presented are the beneficial impacts of utilizing non-ergodic site response on aleatory variability and epistemic uncertainty. Models for mean site response, aleatory variability, and epistemic uncertainty are required for *partially non-ergodic seismic hazard analyses* (Section 7.8).

7.7.1 NON-ERGODIC MEAN SITE RESPONSE

Site response is commonly expressed in terms of amplification factors (Equation 7.1) that relate surface values of intensity measures to values for reference (usually rock) site conditions. The natural log mean site response for non-ergodic analysis can be expressed as (Stewart et al., 2017)

$$\mu_{\ln AF} = f_1 + f_2 \ln\left(\frac{x_{IMr} + f_3}{f_3}\right) \tag{7.91}$$

where x_{IMr} is an intensity measure (commonly *PGA*) for the reference condition that indicates the amplitude of the input motion. In Equation (7.91), the f_1 term represents the weak motion (linear) site amplification. The second term in the sum represents the effects of nonlinearity; the physical meanings of the f_2 and f_3 parameters are depicted in Figure 3.38a. Equation (7.91) is identical to the combination of Equations (3.28) and (3.30), as used in ergodic site response models; hence, the difference between ergodic and non-ergodic site response models expressed in this form is not the equation itself, but the coefficients. The following sections describe how these coefficients can be established for a particular site.

7.7.1.1 Evaluation from On-Site Recordings

When a ground motion instrument (accelerometer or seismometer) is located at or very near a site of interest, and the instrument has recorded a reasonable number of earthquakes (e.g., five or more), the data can be used to estimate portions of the site response function (typically the linear component of Equation (7.91) because weak motions are likely to be much more plentiful than motions strong enough to induce nonlinear response).

The analysis of site response from recordings requires that the data fall in the magnitude-distance range of an applicable GMM. In the case of active tectonic regions, it is generally good practice to use **M**>4 earthquakes, and site-source distances<400 km. The selected GMM should suitably capture average path effects for the application region, otherwise bias in the path term could map to erroneous assessments of site response. Intensity measures IM^{surf} are computed for all recorded events i at site k, and then residuals are computed,

$$(\delta_{\ln \mathrm{IM}})_{\mathrm{ik}} = \ln\left(IM_{ik}^{surf}\right) - \mu_{\ln IM_{ik}^{surf}} \tag{7.92}$$

where $(\mu_{\ln z})_{ik}$ is the mean prediction from a GMM (Equation 3.24) using the magnitude, distance, and site parameters associated with event i and site k. These residuals represent differences between the recorded site-specific intensity measures and the ergodic predictions of the GMM. The total residuals from Equation (7.92) are then adjusted to remove *event terms* ($\eta_{E,i}$) for all events using Equation (3.33), which provides within-event residuals, δW_{ik}. For past earthquakes that were considered in the development of GMMs, event terms may be provided as part of the GMM documentation. Figure 7.68 shows example results for 13 recordings made at the Obregon Park site in California. Residuals were computed using the Boore et al. (2014) GMM. This site has large positive residuals, indicating that the actual recorded site response is consistently and systematically stronger than anticipated from the ergodic site response model in the GMM. For example, the mean of the within-event residuals for *PGA* is about 0.7, which corresponds to underprediction by a factor of 2.0. At a different site with different characteristics, ground motion recordings may show negative residuals indicating that the actual response is consistently and systematically weaker than predicted by the ergodic site response model in the GMM.

The offset of mean within-event residuals from zero is a common feature of earthquake ground motions. This occurs because individual sites respond in particular ways that are controlled by their individual characteristics (primarily their velocity profiles and, as applicable, the properties of the sedimentary basins in which they are located). Figure 7.69 shows histograms of within-event residuals for two sites *m* and *n* that generally have negative and positive residuals, respectively. The means of the distributions represent the average misfits of the data from the ergodic model

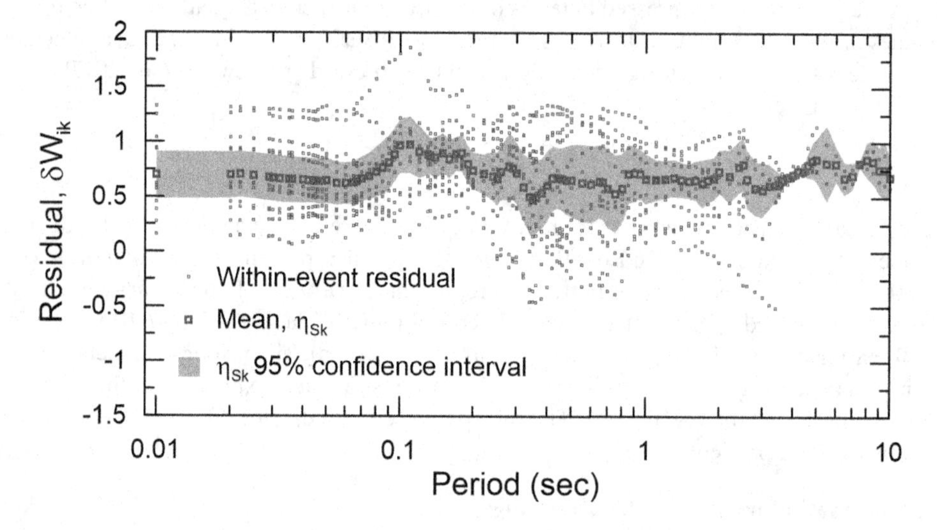

FIGURE 7.68 Within-event residuals and their mean for Obregon Park site. The large positive bias indicates under-estimation of site response from the ergodic site response model. (From Stewart et al., 2017 with permission of Sage Publications, Ltd.)

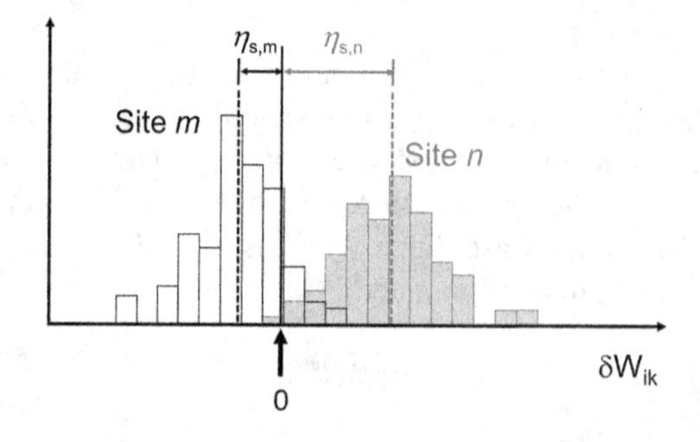

FIGURE 7.69 Histograms of within-event residuals for sites that respond more and less strongly than predicted by ergodic models. $\delta W_{ik} = 0$ corresponds to the prediction of the ergodic site model in a GMM.

used in the calculation of residuals, and are referred to as *site terms*, η_S. The site terms indicate that Site *m* systematically responds more weakly, and Site *n* more strongly, than would be predicted by ergodic models.

Assuming that these site terms are computed from relatively weak motion records that do not induce a significant nonlinear soil response, the linear component of the site amplification model (Equation 7.91) can be computed as (Stewart et al., 2017),

$$f_1 = AF_{lin}^{erg} + \eta_S \tag{7.93}$$

where AF_{lin}^{erg} is the ergodic linear site amplification (Sections 3.5.2.3 and 7.3). The linear amplification for the Obregon Park site computed using Equation (7.93) is shown in Figure 7.70 in arithmetic units (i.e., $\exp(f_1)$). The amplification is relative to the reference site condition in the ergodic model, in this case, $V_{S30} = 760\,\text{m/s}$.

7.7.1.2 Evaluation from Ground Response Analysis

When ground response analyses (GRAs) are performed for a given site, using procedures described in Sections 7.5 or 7.6, they produce a series of discrete amplification factors for a given intensity measure. When multiple input motions are used, potentially also with multiple realizations of randomized soil properties (Section 7.5.7), a distribution of x_{IMr}-AF results is obtained as shown in Figure 7.71 for two example sites with *PGA* as the intensity measure of interest. This distribution is fit with the mean amplification function in Equation (7.91); the values at low input motion amplitudes define the linear coefficient, f_1, and the values at higher amplitudes allow coefficients f_2 and f_3 to be estimated

The example ground response analysis results in Figure 7.71 show amplification of *PGA* at one stiff and one soft soil site (Obregon Park and El Centro #7, respectively). The slowly descending trend of site amplification with the input motion *PGA* demonstrates weak nonlinearity for Obregon Park and the steeper descent for El Centro #7 indicates strong nonlinearity. These are typical patterns that reflect the larger strains that develop in soils at soft sites.

The site amplification directly provided by ground response analysis reflects the surface *IM* relative to the *IM* for the site condition at the base of the modeled soil profile (i.e., the halfspace velocity), which is 540 m/s for Obregon Park and 508 m/s for El Centro #7. These conditions are different than the GMM reference condition, which is typically $V_{S30} = 760\,\text{m/s}$, hence the GRA-based

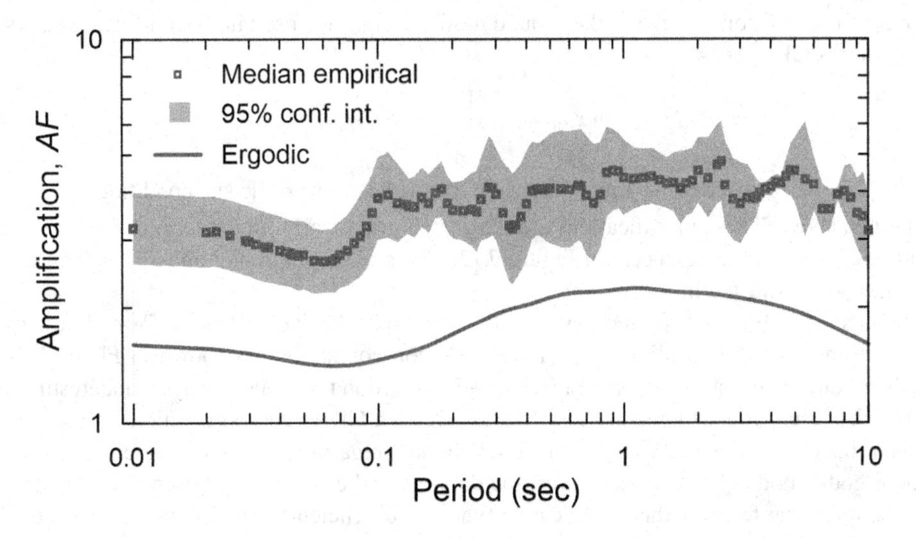

FIGURE 7.70 Ergodic and non-ergodic (site-specific) linear site amplification relative to $V_{S30} = 760$ m/s for Obregon Park site. (After Stewart et al., 2017 with permission of Sage Publications, Ltd.)

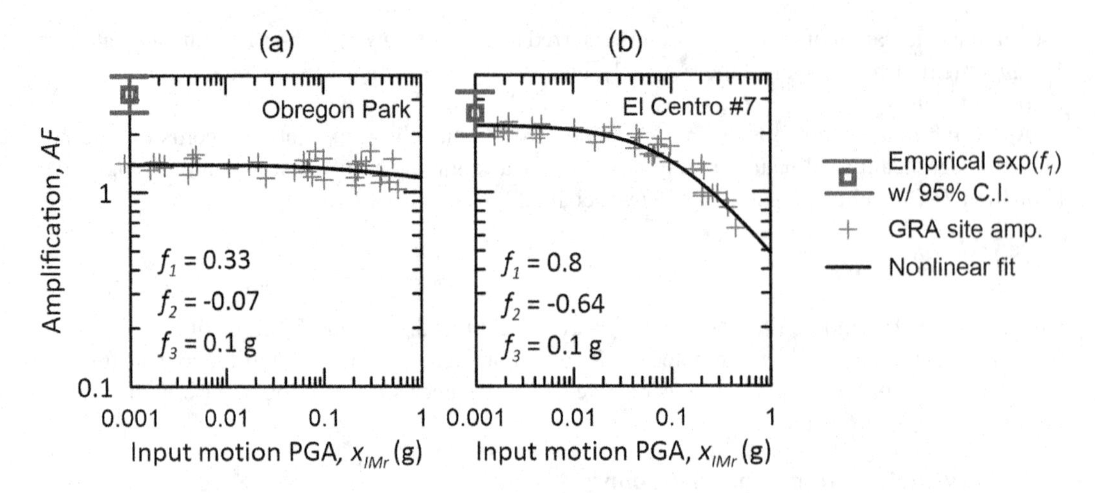

FIGURE 7.71 *PGA* amplification from ground response analyses (symbols), fit from Equation (7.91), and regression coefficients. Empirical site amplification (derived from recordings) is shown as a mean amplification with 95% confidence intervals beyond the limits of the abscissa, to indicate that this amplification is not associated with any specific value of x_{IMr}. (Adapted from Stewart et al., 2017 with permission of Sage Publications, Ltd.)

amplification requires modification. If the V_{S30} value corresponding to conditions at the base of the modeled profile is denoted V_{S30}^{base} and the ground motion *IM* for that condition (corresponding to the ground response analysis input motions) is IM^{base}, then the GRA-based site amplification is:

$$AF_{base}^{surf} = \frac{IM^{surf}}{IM^{base}} \tag{7.94}$$

where IM^{surf} is the ground surface intensity measure. The site amplification relative to the GMM reference condition can then be estimated as

$$\ln\left(AF_{ref}^{surf}\right) = \ln\left(AF_{base}^{surf}\right) + \mu_{\ln AF\left(V_{S30}^{base}\right)} \tag{7.95}$$

where $\mu_{\ln AF\left(V_{S30}^{base}\right)}$ is the mean site amplification from an ergodic model for the base-of-profile (i.e., halfspace) site condition. Likewise, the ground motion amplitude used in the nonlinear site response computation is taken as:

$$\ln x_{IMr} = \ln\left(x_{IMr}^{base}\right) - \mu_{\ln AF\left(V_{S30}^{base}\right)} \tag{7.96}$$

where x_{IMr}^{base} is the corresponding value of that *IM* for the base-of-profile site condition and $\mu_{\ln AF\left(V_{S30}^{base}\right)}$ is as defined above. These modifications are illustrated graphically (in the form of ratios rather than logarithmic sums and differences) in Figure 7.72. The site amplification and x_{IMr} values shown in Figure 7.71 were adjusted in this manner.

When a site has both ground motion recordings and has also had GRAs performed, it is possible to compare the levels of amplification for weak motion conditions. As shown in Figure 7.71, both methods produced similar results for El Centro #7, but ground response analysis underestimates the recorded site response for Obregon Park. Misfits between GRAs and observations are not unusual, and when this occurs it is necessary to make a judgment regarding which estimate of f_1 to adopt for the non-ergodic model. In most such cases, the data-derived estimates are given more weight, while basing the nonlinear terms of the amplification function (coefficients f_2 and f_3) on the results of GRAs.

As mentioned in Section 7.5, the period range of applicability of one-dimensional ground response analysis extends to a maximum of about 1.5–2 times the fundamental period of the modeled soil

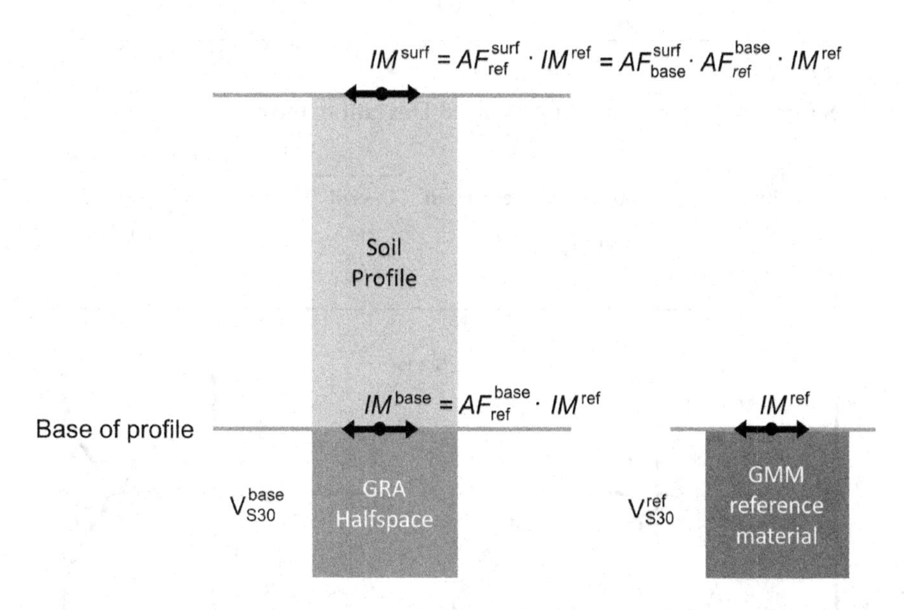

FIGURE 7.72 Schematic illustration of adjustments required to ensure consistency between halfspace velocity and reference site condition in calculation of surface intensity measure.

column. At longer periods, essentially no site response is computed because the wavelengths at these lower frequencies are so long relative to the thickness of the modeled profile that the profile responds as a nearly rigid body. Because long-period site response can be pronounced, due to site response at depths below the modeled soil column and potential basin effects, it is best to transition amplification model coefficients for these long periods to ergodic coefficients from a suitable GMM. As an alternative to transitioning to ergodic models at long periods, it is possible to extend the profile to great depth (several kilometers) using regional velocity information, in which case amplification is computed over a broad period range. An approach that has been recently applied in projects related to critical infrastructure systems uses a deep profile for the target site and another deep profile that is meant to represent the reference condition for an ergodic site amplification model; the mean site response in this case, while still using Eq. (7.91), is taken as the ratio of amplification computed for the target profile to that of the reference profile (Rodriguez-Marek et al., 2021; Williams and Abrahamson, 2021).

The assumption of constant properties in the horizontal direction that is inherent to one-dimensional GRAs should be recognized as a simplification of actual conditions. Poor one-dimensional prediction of response at a number of Kik-net sites in Japan, e.g., mismatches in the peaks of predicted and empirical (measured) amplification factors, were improved using multi-dimensional analyses with spatially randomized (in both the horizontal and vertical directions) soil properties (Thompson et al., 2009). Pehlivan et al. (2013) found that performing multiple one-dimensional analyses with randomized velocities did not accurately represent the amplification behavior of two-dimensional analyses with equivalent vertical spatial variability when velocities were also horizontally variable.

7.7.2 Aleatory Variability of Ground Motion

Standard deviation models in GMMs describe aleatory variability, or dispersion, of ground motions (Section 3.5.4). This variability is caused by unmodeled, apparently random factors related to source, path, and site effects that influence ground motions. When non-ergodic analyses are used, one or more of the previously unmodeled factors in the GMM is modeled on a source-, path-, or site-specific basis. As a result, variability is reduced. The following sections present procedures and nomenclature for characterizing aleatory variability for use with non-ergodic site response analysis.

TABLE 7.9
Summary of Residual and Standard Deviation Terms

			Within-Event		
Quantity	**Total**	**Between-Event**	**Overall**	**Site Term**	**Site Residual**
Residual term	$\delta_{\ln IM}$	η_E	δW	η_S	δS
Std. Dev. of residual	$\sigma_{\ln IM}$	$\tau_{\ln IM}$	$\phi_{\ln IM}$	ϕ_{S2S}	ϕ_{SS}

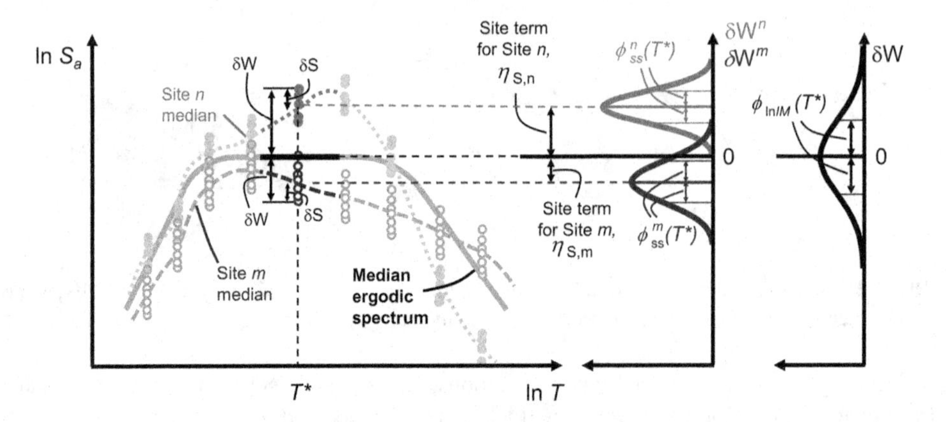

FIGURE 7.73 Illustration of notation for within-event residuals, site terms, and single-station standard deviations for two sites. Each of these quantities is period-dependent; the plot emphasizes the values at period T^* by fading portions of the spectra for periods other than T^*. See Table 7.9 for relationships between residuals and corresponding standard deviations.

7.7.2.1 Components of Aleatory Variability

The aleatory variability associated with a GMM generally has two components (Section 3.5.4.2) – the variability from event-to-event, which is represented by the standard deviation of event terms, $\tau_{\ln IM}$, and the variability within an event represented by the standard deviation of within-event residuals, $\phi_{\ln IM}$ (Figure 3.48 and Table 7.9). These two standard deviations combine to form the total ergodic standard deviation ($\sigma_{\ln IM}$) as indicated in Equation (3.34). When mean site response is evaluated using a non-ergodic model, the model's consideration of site-specific profile characteristics allows the within-event dispersion to be reduced relative to the ergodic dispersion terms provided with GMMs. Figure 7.73 illustrates response spectra for two sites, m and n, that differ systematically and consistently from a median ergodic spectrum at period T^*. Site m responds more weakly and Site n more strongly than the overall median spectrum at T^*. The scatter about the median spectra for each site is considerably less than the scatter about the median ergodic spectrum (i.e., $\phi_{SS} < \phi_{\ln IM}$) of the collective data from all sites.

To understand the rationale for reducing $\phi_{\ln IM}$ when non-ergodic site response is used, it is useful to recognize the factors contributing to within-event residuals, δW_{ik}:

1. Misfit of the site response model relative to the actual site response for some site k, given by the site term, η_{Sk}.
2. Misfit of the path model (F_P) for the source-to-site path for event i and site k.
3. Between-event variability in site responses for site k.

Each of these factors has corresponding standard deviations, but those of greatest practical interest are the standard deviation of site terms for a given site response model, i.e., the site-to-site standard deviation ϕ_{S2S}, and the remaining variability from factors 2 and 3. Because non-ergodic site response is evaluated for site-specific conditions, it is possible to remove ϕ_{S2S} from the within-event dispersion. The variability that remains, from factors 2 and 3 above, is referred to as the single-station within-event variability, ϕ_{SS}

$$\phi_{SS} = \sqrt{\phi_{\ln IM}^2 - \phi_{S2S}^2} \tag{7.97}$$

The term "single station" is used because it does not include site-to-site variability (i.e., it consists of the variability applicable to an individual site; Atkinson 2006). Figure 7.73 illustrates schematically this variability in the spread of within-event residuals for individual sites. Comparing Figures 7.73 and 3.48, it is apparent that the site-to-site and single-station variabilities are analogous to the between-event and within-event variabilities described in Section 3.5.4.2. One accounts for systematic differences in earthquakes and the other for systematic differences in site responses.

7.7.2.2 Single Station Ground Motion Aleatory Variability

When a non-ergodic site response analysis is carried out, the within-event variability is nominally the single-station variability, ϕ_{SS}. In this section, ϕ_{SS} models derived primarily from data where nonlinear site response effects are minimal are presented. For a given site, it is possible for the effective within-event variability to deviate from ϕ_{SS} due to the effects of site response nonlinearity and event-to-event variations in site response (factor 3 in Section 7.7.2.1). For this reason, the term $\phi_{\ln IM}^k$ represents the within-event variability that should be used for site k in consideration of these effects.

Figure 7.74 shows two models for period-dependent ϕ_{SS} for spectral acceleration: a model applied to several sites in the southwestern United States (SWUS) (GeoPentech, 2015), and a model for global active tectonic region earthquakes (Al Atik, 2015). These models are magnitude-dependent, and Figure 7.74 compares large-**M** values to an ergodic $\phi_{\ln IM}$ model (Boore et al., 2014). The single-station results are notably smaller, which indicates that ϕ_{S2S} is a major contributor to the overall dispersion from GMMs. This reduced variability can result in substantial reductions in ground motions at low exceedance rates (long return periods) from seismic hazard analyses.

The empirical models for ϕ_{SS} in Figure 7.74 are derived from large populations of sites, most of which are firm soils and medium stiff bedrock materials that have produced recordings of low-to-moderate amplitudes. Such models can be used for applications in which strongly nonlinear site responses are not expected, and as a result $\phi_{\ln IM}^k \approx \phi_{SS}$.

For applications involving nonlinear responses, it is important to account for the effects of nonlinearity and aleatory event-to-event variability in the site response. Using the amplification model of Equation (7.91), the variability can be computed as (Stewart et al. 2017),

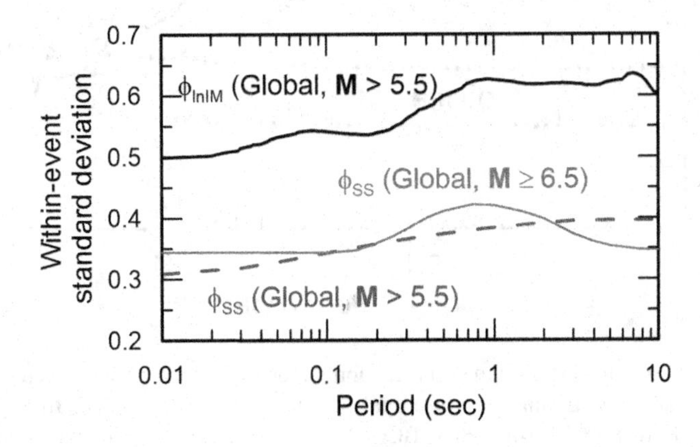

FIGURE 7.74 Models for single station and total within-event standard deviation. (From Stewart et al., 2017 with permission of Sage Publications, Ltd.)

$$\phi_{\ln IM}^{k} = \sqrt{\left(\frac{f_2 x_{IMr}}{x_{IMr} + f_3} + 1\right)^2 \phi_{SS}^2 + \phi_{\ln AF}^2} \tag{7.98}$$

In Equation (7.98), the ϕ_{SS} term represents the within-event standard deviation for the reference site condition (Figure 7.74) and $\phi_{\ln AF}$ represents aleatory event-to-event variability in site response (factor 3 above). The modifier in front of ϕ_{SS} reduces the reference site dispersion when the site response is nonlinear (i.e., when f_2 is negative). Since event-to-event variations in site response ($\phi_{\ln AF}$) affect ϕ_{SS}, in some applications, the addition of $\phi_{\ln AF}^2$ in Equation (7.98) is omitted or its values are reduced.

7.7.2.3 Site Response Aleatory Variability

Ground motion recordings from instrumented sites can be examined (Rodriguez-Marek et al., 2011; Lin et al., 2011; Kaklamanos et al., 2013) to evaluate the variability in ground motion amplification, which allows for the calculation of $\phi_{\ln AF}$ (Stewart et al., 2017). The recorded ground motion data shows that $\phi_{\ln AF}$ varies from about 0.23 to 0.30 at short periods and, as shown by the shaded band in Figure 7.75, increases only slightly with increasing period.

The variability in ground motion amplification from one-dimensional GRAs (Bazzurro and Cornell, 2004a; Kwok et al., 2008; Rathje et al., 2010; Li and Asimaki, 2011) is illustrated by the individual curves in Figure 7.75. Some of these studies included the effects of input motion variability and some did not, but all used randomized velocity profiles and randomized modulus reduction and damping curves; these studies showed that motion-to-motion variability had the greatest effect on $\phi_{\ln AF}$ but that, for a given motion, variability in shear wave velocity was most influential. The standard deviation of site amplification generally ranged from 0.35 to 0.65 at short periods, increased with increasing period to a maximum just below the characteristic site period, and then decreased at longer periods. The drop in uncertainty was quite rapid and severe when only one motion was considered (Turkey Flat and La Cienega sites) but still existed at the other sites.

These results show that one-dimensional randomizations tend to overpredict the variability of short-period amplification and underpredict the variability of amplification at long periods relative to available measured field behavior.

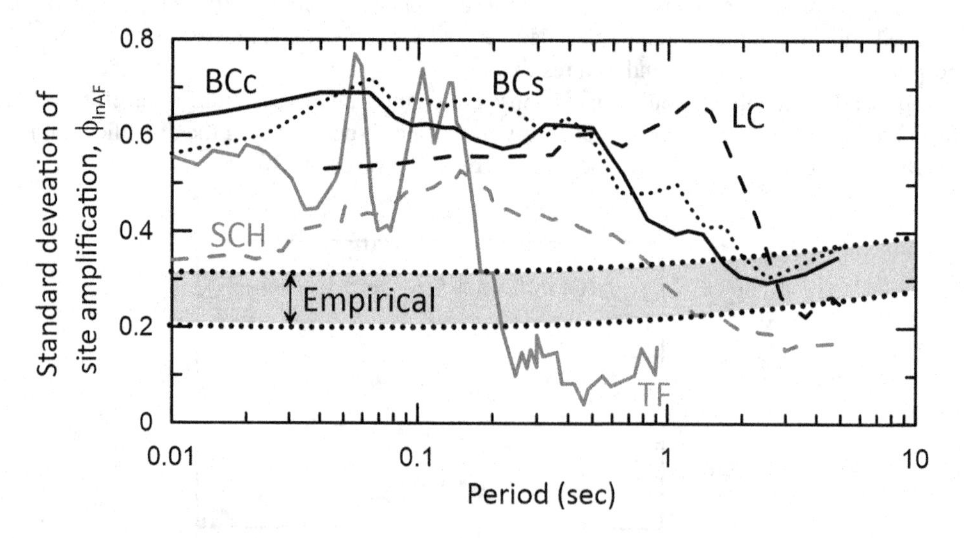

FIGURE 7.75 Variation of uncertainty in amplification factor, $\phi_{\ln AF}$, with period based on measured ground motion data (shaded band) and ground response analyses (individual curves). TF, Turkey Flat (Kwok et al., 2008); SCH, SCH10 (Rathje et al., 2010); BCs, BC04 sand, BCc, BC04 clay (Bazzurro and Cornell 2004b); LC, La Cienega (Li and Asimaki 2011). Shaded band envelopes empirical results based on recordings. (After Stewart et al., 2017 with permission of Sage Publications, Ltd.)

7.7.3 Epistemic Uncertainty

The reduction of aleatory variability applied with the use of a non-ergodic mean site response model should be accompanied by an appraisal of the degree of knowledge, or epistemic uncertainty, in the mean site response model and the within-event standard deviation model that is used. Epistemic uncertainties can be accounted for using a logic tree framework (Section 4.4.3.4).

There are different sources of uncertainty for the linear and nonlinear components of the mean model (Equation 7.91). The uncertainty in the linear component (parameter f_1 in Equation 7.91) depends upon whether the model is derived partly from recordings versus being entirely GRA-based. When f_1 is set from recordings (Section 7.7.1.1), epistemic uncertainty is represented by the standard error of the mean of the site term for site k, η_{Sk},

$$\sigma_{f_1} = \sqrt{\frac{s_k^2}{n_k}} \tag{7.99}$$

where s_k is the sample standard deviation of within-event residuals for site k (equivalent to the site-specific ϕ_{SS} values illustrated in Figure 7.73) and n_k is the number of observations. When f_1 is set by GRA, epistemic uncertainties are caused by modeling uncertainty (Section 7.5.7.3) and soil property uncertainty. Modeling uncertainty represents site-to-site variability in ground motions when a given site response model is applied and is denoted ϕ_{S2S}^m as with ergodic models (although the values tend to be lower when site-specific GRA are used). Figure 7.43 shows modeling uncertainty as derived from vertical array data (discussed in Section 7.5.7.3). Stewart and Afshari (2021) showed that in cases where V_S is measured, modeling uncertainty is often the dominant source of site response epistemic uncertainty. Thompson et al. (2009) found that poor one-dimensional predictions of response at a number of Kik-net sites in Japan were markedly improved by using three-dimensional analyses with spatially randomized (in both the horizontal and vertical directions) soil properties. As a result, the large uncertainties shown in Figure 7.43 could potentially be reduced with the use of multi-dimensional GRA, although an epistemic model for such methods of analysis is not available.

Characterization of epistemic uncertainty in the nonlinear model (F_{nl}) is directly related to uncertainty in the soil properties that produce this response, which are the modulus reduction and damping versus shear strain relations and the dynamic shear strength. If the random variability in those curves is considered in the derivation of $\phi_{\ln AF}$ (Section 7.7.2.3), then this source of parametric uncertainty should not be part of the epistemic model. Rather, alternate values of f_2 and/or f_3 derived using different models for modulus reduction and damping could be considered in the epistemic model, although this is rarely done in practice (i.e., the effects of epistemic uncertainty are usually considered mainly in the linear term).

Epistemic uncertainty of within-event standard deviation is mainly related to alternative models that could be considered for the single-site standard deviation for reference site conditions (ϕ_{SS}), as used in Equation (7.98). For example, two models for this uncertainty are shown in Figure 7.74.

7.7.4 Discussion

PSHAs are usually performed using the ergodic assumption, in which GMMs derived from global data from many events, sources, paths, and sites are used. The effects of site-to-site variability are included in the $\sigma_{\ln IM}$ values of GMMs but do not exist at actual individual sites. By means of strong motion data interpretation or GRAs, the amplification behavior of an individual site can be determined and used to both correct for bias in the median value and reduce the aleatory variability used with the GMM in a partially non-ergodic analysis. By considering additional factors (source and path) that have systematic and repeatable effects on ground motions, a fully non-ergodic analysis can be performed.

The net result of non-ergodic analyses, either partial or full, is often a reduction in the intensity measure produced by a PSHA at a given return period (although increases can also occur, particularly at the fundamental period of sites with strong impedance contrasts, where site-specific site responses are especially strong). For many structures or facilities, the reduced costs associated with constructing or retrofitting a structure for a lower ground motion will more than offset the additional effort required to perform non-ergodic analysis. Table 7.10 summarizes the essential components of each approach, and the data and models required to implement them. Table 7.10 includes the case of non-ergodic site and path combined; in this case, the non-ergodic path term is denoted η_P (analogous to η_E for events and η_S for sites).

At present, numerical GRAs appear to provide reasonable estimates of median amplification up to the fundamental period of the modeled soil column, but their uncertainty, as derived from one-dimensional GRAs is large (Figure 7.43). The large size of this uncertainty reduces to some extent the benefits of site-specific analyses for the prediction of *IM*s.

The use of non-ergodic site response procedures allows for more correct distinction between aleatory variability and epistemic uncertainty. The site-to-site variability term (ϕ_{S2S}) that is included in aleatory variability models typically applied with ergodic GMMs, is more properly taken as a source of epistemic uncertainty in a partially non-ergodic analysis. It is important to carefully consider major sources of variability and uncertainty, as described in this section, and to avoid double-counting.

7.8 INCORPORATION OF SITE EFFECTS INTO HAZARD ANALYSES

The most common application of a site-specific (or non-ergodic) site amplification model is in seismic hazard analyses to predict ground motion *IM*s. Several approaches for modifying the hazard calculation to account for these effects are available. Each of these approaches uses the same mean model (Equation 7.91), but they differ primarily in how the ground motion variability is handled. If the goal is to obtain a ground surface *IM* with a known return period, which is a primary requirement in modern earthquake engineering practice, the uncertainty must be handled very carefully.

7.8.1 Hybrid Approach

A common approach for estimating the ground surface intensity measure (IM^{surf}) at a soil site is to simply multiply the reference (usually rock) intensity measure (IM^{ref}) by the median value of the amplification factor ($\exp(\mu_{\ln AF})$ in Equation 7.91) for the level of the reference site motion. In most applications of the hybrid approach, the reference site ground motions are based on an ergodic GMM, and hence the reduction of aleatory variability is not considered. This process has been referred to as a hybrid approach (Cramer, 2003) because it combines a probabilistic rock motion with a deterministic amplification factor. It will produce an IM^{surf} value, but that value will have a smaller return period than IM^{ref} (and may have different return periods at different oscillator periods) when the site response is nonlinear. This occurs because the amplitude parameter used in the nonlinear function (x_{IMr}) is taken from the hazard curve, which for long return periods is likely to correspond to a rare realization of ground motion ($\varepsilon > 0$; where ε is defined in Equation 4.26a). When x_{IMr} is increased because of large ε, F_{nl} becomes more negative, which reduces IM^{surf}. As a result, the use of hybrid methods, which is common in some sectors of practice, has the potential to introduce underprediction bias (i.e., an unconservative error) in ground motion estimates.

Goulet and Stewart (2009) proposed a modified hybrid approach in which the reference *IM* used in the amplification factor is taken as the median value ($\varepsilon = 0$) from the GMM as computed using the mean magnitude and mean distance from a disaggregation analysis. For cases in which the controlling sources for different site conditions do not vary greatly, the modified hybrid procedure can produce ground surface spectral ordinates with return periods much closer to the rock UHS return

TABLE 7.10
Notation and Components of Variability of Empirical Ground Motion Models with and without Ergodic Assumption

Approach	Mean ln IM^{surf}	Components: Aleatory[1]	Components: Epistemic[2]	Required Data	Required Models
Ergodic (all paths, all sites)	$\mu_{\ln IM}$	$\tau_{\ln IM}$, $\phi_{\ln IM}\left(\begin{matrix}\phi_{SS},\\ \phi_{S2S}\end{matrix}\right)$	–	Global: Multiple recordings at multiple sites from motions arriving by different paths	GMM
Partially non-ergodic (all paths, single site)	$\mu_{\ln IM}+\eta_S$	$\tau_{\ln IM}$, ϕ_{SS}, $\phi_{\ln AF}$	σ_{f1} or ϕ^m_{S2S}, Soil properties	Site-specific: Multiple records of motions arriving at one site by different paths	GMM and site-specific amplification model (from recorded motions or ground response analyses with randomization to determine $\phi_{\ln AF}$)
Non-ergodic (single path, single site)	$\mu_{\ln IM}+\eta_S+\eta_P$	$\tau_{\ln IM}$, ϕ_{sp}, $\phi_{\ln AF}$	σ_{f1} or ϕ^m_{S2S}; ϕ^m_{P2P}, Soil properties	Path- and site-specific: Multiple records of motions arriving at one site from each source-to-site direction of interest	GMM, site-specific amplification model, source-to-site specific path model

Source: After Al-Atik et al. (2010) and Stewart et al. (2017).

[1] ϕ_{sp}=event-to-event variability in path effect.

[2] ϕ^m_{S2S} and ϕ^m_{P2P} are model uncertainties related to site response or source-to-site-specific path; "Soil properties" indicates effects of uncertain soil properties on $\mu_{\ln AF}$

period than those given by the hybrid procedure. For other cases, such as a soft clay site at which soil hazards may be controlled by significantly larger magnitude and/or distant events, the modified hybrid approach would not be as effective.

7.8.2 Convolution Approach

Section 5.5 discussed the extension of an *IM* hazard curve to a response quantity that is a function of the *IM* hazard curve. This approach can be applied to the problem of ground motion site amplification in which the response quantity is ground surface intensity measure, IM^{surf} (Bazzurro and Cornell, 2004b). Adapting Equation (5.12),

$$\lambda_{IM^{surf}}(im^{surf}) = \int_0^\infty P[IM^{surf} > im^{surf} \mid IM^{ref} = im^{ref}] \left| d\lambda_{IM^{ref}}(im^{ref}) \right| \tag{7.100}$$

where IM^{ref} is the quantity estimated using the reference site hazard curve. As explained in Section 5.5, the quantity $\left| d\lambda_{IM^{ref}}(im^{ref}) \right|$ is the absolute value of the slope of the reference site hazard curve at abscissa point im^{ref}.

From the definition of the amplification factor,

$$\lambda_{IM^{surf}}(im^{surf}) = \int_0^\infty P[AF \cdot im^{ref} > im^{surf} \mid im^{ref}] \left| d\lambda_{IM^{ref}}(im^{ref}) \right| \tag{7.101}$$

which is equivalent to

$$\lambda_{IM^{surf}}(im^{surf}) = \int_0^\infty P[AF > \frac{im^{surf}}{im^{ref}} \mid im^{ref}] \left| d\lambda_{IM^{ref}}(im^{ref}) \right| \tag{7.102}$$

The calculation in Equation (7.102) is referred to as *site response convolution*. If the amplification factor is lognormally distributed (as is typically assumed), the conditional probability on the right side of Equation (7.102) can be computed as

$$P[AF > \frac{im^{surf}}{im^{ref}} \mid im^{ref}] = 1 - \Phi\left[\frac{\ln(im^{surf}/im^{ref}) - \mu_{\ln AF}}{\phi_{\ln AF}}\right] \tag{7.103}$$

For applications involving non-ergodic site response, the reference hazard is computed using a single-station standard deviation and a GMM with the site condition set to the reference condition. Convolution is the most commonly used approach for the application of non-ergodic site response. The drawbacks of the approach are that nonlinearity can be overestimated (although not to the same extent as with the hybrid approach) and it does not account for differences in the contributions of different sources from soil and rock disaggregations.

7.8.3 Site-Specific PSHA

The most accurate way of accounting for site amplification and related uncertainties is to use a site-specific GMM (i.e., one modified with a non-ergodic site term and using single-station variability, i.e., $\phi_{\ln IM} = \phi_{SS}$) within the PSHA. By accounting for site effects "within the (hazard) integral," the *IM* values are integrated over all sources and all magnitudes, distances, epsilon values, etc. It is important in this approach to properly account for the correlation between x_{IMr}, as used in the nonlinear amplification function, and the *IM* of interest (Stewart et al., 2017) and to properly account for epistemic uncertainties in the site term and aleatory variability model (Section 7.7.3).

7.8.4 Example Results

To illustrate the differences between implementations performed according to the three methods described in previous subsections (hybrid, convolution, site-specific PSHA), each was implemented for the El Centro 7 site. The mean site amplification at this site for *PGA* is shown in Figure 7.71. The site has a fundamental mode response near 1.5 sec. Site response coefficients for *PGA* are $f_1=0.8$, $f_2=-0.64$, and $f_3=0.1g$ and for 1.5 sec S_a they are $f_1=1.0$, $f_2=-0.06$, and $f_3=0.1g$. The hazard calculations were performed for the reference rock site condition ($V_{S30}=760$ m/s) and using the mean site amplification functions for the two *IM*s with the three alternative approaches. The aleatory variability utilized in these analyses took the between-event standard deviation as $\tau_{\ln IM}$ from a GMM and the within-event standard deviation as ϕ_{SS}.

Figure 7.76 shows the resulting hazard curves for *PGA* and 1.5 sec S_a. These two *IM*s were selected to illustrate results for cases with nonlinearity varying from strong (*PGA*) to weak but non-zero (1.5 sec S_a). For large exceedance rate (> 0.02/year; return periods < 50 years), the soil ground motions exceed the rock motions for both *IM*s because the input rock motions are low and nonlinear effects are small. For lower rates (longer return periods), the soil *PGA* drops below that for rock due to nonlinearity, but the amount of that reduction depends on the implementation approach. The use of hybrid or convolution produces lower ground motions than site-specific PSHA. This occurs because the amplitude of shaking driving the nonlinear site response (x_{IMr}) is taken from the reference (IM^{ref}) hazard, which is "positive epsilon" (exceeds the mean), whereas the site-specific PSHA does not make this assumption. In the case of 1.5 sec S_a, for which nonlinearity effects are modest, the differences between hybrid, convolution, and site-specific PSHA results are smaller.

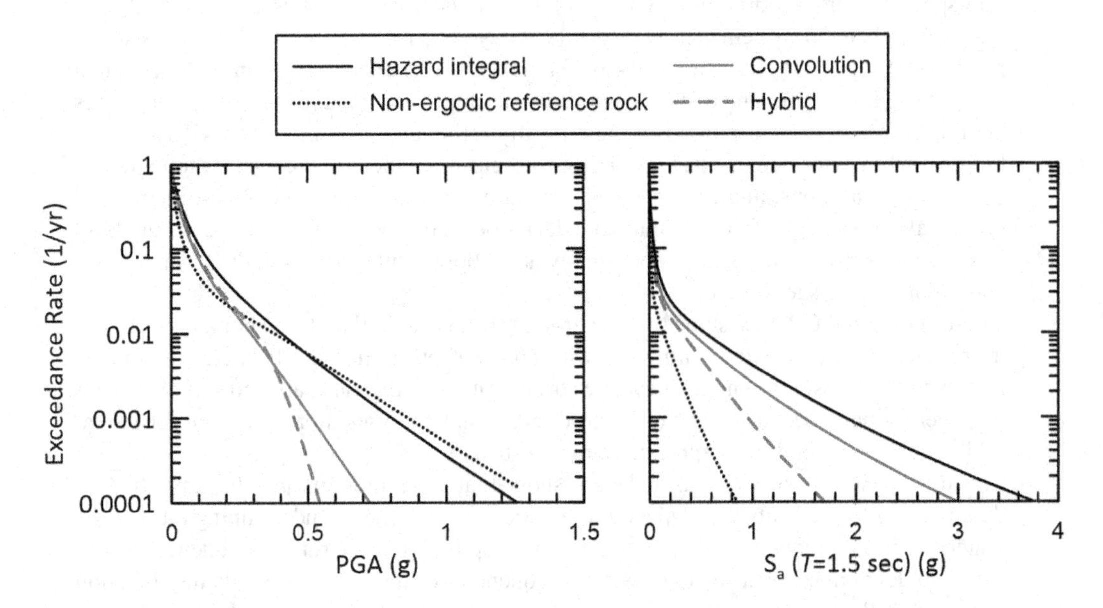

FIGURE 7.76 Hazard curves for the El Centro 7 site derived using non-ergodic site amplification models using the hybrid, convolution, and full implementation of the model within the hazard integral (site-specific PSHA).

7.9 SUMMARY

1. While site effects occur within a relatively short fraction of the total distance seismic waves travel from a source to a site, they can have a pronounced effect on the amplitude, frequency content, and duration of an earthquake ground motion.

2. Early evidence of site effects took the form of comparative damage distributions, but more direct evidence came about with the increasing availability of strong motion data in the past 50-60 years. Data from vertical and surface arrays of strong motion instruments have clearly shown how ground motions are strongly influenced by local site conditions.
3. Sites respond to earthquakes in different manners depending on their individual characteristics and the characteristics of incoming seismic waves. The relationship between the stiffnesses and thicknesses of different layers in a soil profile and the amplitudes and frequency content of the waves passing through them will control the response of the site. Even if generally similar, different sites can respond quite differently to the same input motion – and a specific site can respond differently to different input motions.
4. Site effects can be predicted generically, i.e., by characterizing the average behavior of broad categories of similar site conditions, or by site-specific techniques that account for the individual characteristics of a particular site. Generic procedures, often referred to as ergodic, can be thought of as representing the response of an average site within a particular site class. When the conditions at a particular site deviate significantly from the average conditions within its site class, non-ergodic procedures can account for the systematic and repeatable differences in site-specific response that result from those deviations.
5. Generic (ergodic) procedures classify sites into categories within which site effects are similar but between which they are different. Classification systems based on surface geology and shallow geotechnical conditions have largely been replaced by classification based on the time-averaged shear wave velocity, V_{S30}. Generic procedures predict effects associated with an "average" site within each site class.
6. Site-specific (non-ergodic) procedures account for the characteristics of individual sites that cause them to systematically and repeatably respond differently than the average response of a generic site class. Site-specific response characteristics can be determined empirically for instrumented sites subjected to multiple earthquakes but, since such sites are relatively rare, site response is more typically estimated using numerical GRAs.
7. GRAs can be performed in one, two, or three dimensions depending on surface topography, subsurface conditions, and the nature of structures of interest at a site. One-dimensional GRAs take advantage of the fact that boundaries between soil layers can be relatively level in some cases and most seismic wave energy at shallow depths arrives in the form of vertically propagating shear waves.
8. One-dimensional GRAs assume that the ground surface and all material boundaries below it are horizontal and extend infinitely with constant properties in all lateral directions. Although these assumptions can never be strictly satisfied, they are satisfied sufficiently for engineering purposes at many sites. One-dimensional analyses do not account for lateral variations in soil conditions and cannot model surface waves.
9. Linear analyses of simple elastic layers show that base motions can be amplified or de-amplified by amounts that depend on the thickness, stiffness, and damping ratio of the material and the frequency of the base motion. Amplification is strongly frequency-dependent with local maxima at various natural frequencies of the layer; the peak amplification decreases with increasing natural frequency and increasing damping ratio. Linear analyses can be performed in the frequency domain using transfer functions that relate parameters of interest (such as ground surface acceleration or velocity) to known parameters (such as base acceleration).
10. The transfer function controls the degree of amplification or de-amplification of each frequency in a ground motion. Its shape is influenced by the thickness, stiffness, and damping characteristics of each soil layer and by the properties of the underlying bedrock. Transfer functions of layered soil profiles can become complicated and are strongly influenced by the stratigraphy of the profile – soil profiles with the same average shear wave velocity but

different arrangements of the layers will have different transfer functions and therefore respond differently to the same input motion.

11. Bedrock is typically represented by an elastic halfspace in GRAs. When downward-traveling waves reflected from shallower layer boundaries reach the top of the halfspace, they can be reflected and/or transmitted at that boundary – the relative amounts are controlled by the impedance contrast at the boundary. If the halfspace is rigid, all of the energy of the downward-traveling wave will be reflected upward. If the halfspace has the same specific impedance as the material directly above it, all of the energy of the downward-traveling wave will be transmitted (with no reflection) across the boundary and into the halfspace from which it never returns. For the more common case of a halfspace that is stiffer than the material above it (but not rigid), a portion of the energy of the downward-traveling wave will be reflected upward with the remainder transmitted into the halfspace, thereby reducing the total amount of wave energy that remains in the material above the halfspace.
12. The nonlinear behavior of soil can be approximated in equivalent linear analyses in which the soil stiffness and damping characteristics are adjusted until they are compatible with the level of strain induced in the soil. Equivalent linear analyses are usually performed in the frequency domain for reasons of computational efficiency. As the power, speed, and accessibility of computers have increased in recent years, the practical significance of differences in the efficiency of one-dimensional GRAs has decreased substantially. Nonetheless, equivalent linear analyses remain popular due to their familiarity and the relative simplicity of the required soil parameters. Differences between computational speeds remain significant for multi-dimensional analyses.
13. The inherent linearity of equivalent linear analyses can lead to spurious resonances (i.e., high levels of amplification that result from coincidence of a strong component of the input motion with one of the natural frequencies of the equivalent linear soil deposit). Since the stiffness of an actual (i.e., nonlinear) soil changes over the duration of a large earthquake, such high amplification levels will not develop in the field. Under strong shaking, the dependence of the damping ratio on the peak shear strain in an equivalent linear analysis tends to overdamp high frequencies that are typically associated with lower strain amplitudes.
14. Nonlinear analyses integrate the equations of motion in small time steps. At the beginning of each time step, the stress-strain relationship is used to obtain the appropriate soil stiffness to be used in that time step. By this method, a nonlinear inelastic stress-strain relationship can be followed in a series of small incrementally linear steps. The inelastic behavior upon reversal of stresses allows hysteretic dissipation of energy to occur, thereby eliminating the need for specification of viscous damping required in equivalent linear analyses (although some non-hysteretic forms of energy dissipation maybe needed to represent low-strain damping).
15. Nonlinear models can predict the accumulation of permanent strain in individual soil layers that, when integrated over the thickness of the soil profile, produce permanent displacements. The inherent linearity of equivalent linear models prevents them from computing permanent displacements.
16. Nonlinear methods can be formulated in terms of effective stresses to allow modeling of the generation, redistribution, and eventual dissipation of excess pore pressure during and after earthquake shaking. This capability, which equivalent linear methods do not have, is critical for profiles that contain potentially liquefiable soils.
17. Nonlinear methods require a reliable stress-strain or constitutive model. Some relatively simple 1-D constitutive models can be formulated in terms of modulus reduction and damping curves, but others may require a substantial field and laboratory testing program to evaluate nonlinear model parameters.

18. Differences between the results of equivalent linear and nonlinear analyses depend on the degree of nonlinearity in the actual soil response. For problems where strain levels remain low (stiff soil profiles and/or relatively weak input motions), both analyses can produce reasonable estimates of ground response. For problems involving high strain levels, particularly problems in which the induced shear stresses approach the available shear strength of the soil, nonlinear analyses are likely to provide more reasonable results.
19. Nonlinear effective stress analyses account for pore pressure generation and its effects on soil stiffness. In liquefiable soils, dilation associated with phase transformation behavior can lead to rapid increases in stiffness that propagate sharp acceleration pulses through the liquefiable soil to the ground surface. Simple pore pressure models, which degrade the soil stiffness monotonically but do not account for dilation associated with phase transformation behavior can oversoften liquefied layers thereby unrealistically shielding overlying layers from the subsequent motion of underlying layers; such models should be used with caution for liquefiable soil profiles.
20. Prediction of site effects is complicated by aleatory variability and epistemic uncertainty. The effects of aleatory variability can be simulated using Monte Carlo analyses with randomized input parameters. Epistemic uncertainty can be reduced by conducting high-quality subsurface investigation and site characterization programs.
21. Multi-dimensional GRAs can be performed when surface topography, subsurface conditions, and structure geometry significantly violate the assumptions of one-dimensional analysis. Nonlinear multi-dimensional analyses may also be utilized when permanent deformations of a system are of interest. A number of additional factors related to numerical analysis, boundary conditions, and input motion specifications must be considered in multi-dimensional analyses.
22. Sedimentary basins can have a strong and complex effect on ground motions and their spatial distribution through focusing and wave trapping mechanisms. Body waves entering a basin can be converted to surface waves that can be trapped within and travel back and forth across the basin thereby extending ground motion durations. Ground motions within basins are frequently stronger, particularly at low frequencies than nearby motions outside the basin.
23. Ground surface topography can have a strong influence on local site effects. Hills and ridges have mass and stiffness and thus respond dynamically to earthquake shaking. Some of the highest ground motion amplitudes ever recorded have come from strong motion stations located on sharp ridges and even relatively broad hills. Sites within canyons and valleys can also respond differently than they would in a flat-ground environment.
24. Non-ergodic analyses account for site-specific conditions in order to produce a more accurate (less biased) and less variable/uncertain ground motion predictions. It is not unusual for partially non-ergodic PSHA to result in reduced ground motion values relative to ergodic estimates.

REFERENCES

Abrahamson, N.A., Kuehn, N.M., Walling, M., & Landwehr, N. (2019). "Probabilistic seismic hazard analysis in California using nonergodic ground-motion models," *Bulletin of the Seismological Society of America*, Vol. 109, No. 4, pp. 1235–1249. https://doi.org/10.1785/0120190030.

Afshari, K. and Stewart, J.P. (2019). "Insights from California vertical arrays on the effectiveness of ground response analysis with alternative damping models," *Bulletin of the Seismological Society of America*, Vol. 109, 1250–1264. https://doi.org/10.1785/0120180292.

Al Atik, L. (2015). "NGA-East: Ground motion standard deviation models for Central and Eastern North America," *Report No. PEER 2015/07*, Pacific Earthquake Engineering Research Center, University of California, Berkeley, 182 pp.

Al Atik, L., Abrahamson, N., Bommer, J.J., Scherbaum, F., Cotton, F., and Kuehn, N. (2010). "The variability of ground motion prediction models and its components," *Seismological Research Letters*, Vol. 81, No. 5, pp. 794–801. https://doi.org/10.1785/gssrl.81.5.794.

American Society of Civil Engineers, ASCE (2022). *Minimum Design Loads and Associated Criteria for Buildings and Other Structures*, Report ASCE/SEI 7-22, Reston, VA.

Anderson, D.G., Shin, S., and Kramer, S.L. (2011). "Nonlinear, effective-stress ground motion response analyses following the AASHTO guide specifications for LRFD seismic bridge design," *Transportation Research Record*, Vol. 2251, pp. 144–154. https://doi.org/10.3141/2251-15.

Arefi, M.J., Cubrinovski, M., and Bradley, B.A. (2013). "A model for nonlinear total stress analysis with consistent stiffness and damping variation," *Proceedings, 15th World Conference on Earthquake Engineering*, Lisbon, Portugal, 10 pp.

Ashford, S.A., Sitar, N., Lysmer, J., and Deng, N. (1997). "Topographic effects on the seismic response of steep slopes," *Bulletin of the Seismological Society of America*, Vol. 87, No. 3, pp. 701–709. https://doi.org/10.1785/bssa0870030701.

Asimaki, D. and Gazetas, G. (2004). Soil and topographic amplification on canyon banks and the 1999 Athens earthquake," *Journal of Earthquake Engineering*, Vol. 8, No. 1, pp. 1–43. https://doi.org/10.1080/13632460409350479.

Asimaki, D. and Mohammadi, K. (2018). "On the complexity of seismic waves trapped in irregular topographies," *Soil Dynamics and Earthquake Engineering*, Vol. 114, pp. 424–437. https://doi.org/10.1016/j.soildyn.2018.07.020.

Asimaki, D. and Steidl, J. (2007). "Inverse analysis of weak and strong motion downhole array data from the Mw7.0 Sanriku-Minami Earthquake," *Soil Dynamics and Earthquake Engineering*, Vol. 27, No. 1, pp. 73–92. https://doi.org/10.1016/j.soildyn.2005.10.010.

Atkinson, G.M. (2006). "Single-station sigma," *Bulletin of the Seismological Society of America*, Vol. 96, No. 2, pp. 446–455. https://doi.org/10.1785/0120050137.

Bahrampouri M, A Rodriguez-Marek (2023). One-dimensional site response analysis: Model error estimation, *Bulletin of the Seismological Society of America*, Vol. 113, pp. 401–416. https://doi.org/10.1785/0120210291

Bao, H., Bielak, J., Ghattas, O., Kallivokas, L.F., O'Hallaron, D.R., Shewchuk, J.R., and Xu, J. (1998). "Large-scale simulation of elastic wave propagation in heterogeneous media on parallel computers," *Computational Methods in Applied Engineering Mechanics*, Vol.152, pp. 85–102. https://doi.org/10.1016/s0045-7825(97)00183-7.

Bard, P.Y. and Tucker, B.E. (1985). "Underground and ridge site effects: a comparison of observation and theory," *Bulletin of the Seismological Society of America*, Vol. 75, pp. 905–922. https://doi.org/10.1785/bssa0750040905.

Basu, U. and Chopra, A.K. (2003). "Perfectly matched layers for time-harmonic elastodynamics of unbounded domains: Theory and finite-element implementation," *Computer Methods in Applied Mechanics and Engineering*, Vol. 192, No. 11–12, pp. 1337–1375. https://doi.org/10.1016/s0045-7825(02)00642-4.

Baturay, M.B. and Stewart, J.P. (2003). "Uncertainty and bias in ground-motion estimates from ground response analyses," *Bulletin of the Seismological Society of America*, Vol. 93, pp. 2025–2042. https://doi.org/10.1785/0120020216.

Bazzurro, P. and Cornell, C.A. (2004a). "Ground motion amplification in nonlinear soil sites with uncertain properties," *Bulletin of the Seismological Society of America*, Vol. 94, No. 6, pp. 2090–2109. https://doi.org/10.1785/0120030215.

Bazzurro, P. and Cornell, C.A. (2004b). "Nonlinear soil-site effects in probabilistic seismic hazard analysis," *Bulletin of the Seismological Society of America*, Vol. 94, No. 6, pp. 2110–2123.

Berenger, J.P. (1994). "A perfectly matched layer for the absorption of electromagnetic waves," *Journal of Computational Physics*, Vol. 114, No. 2, pp. 185–200. https://doi.org/10.1006/jcph.1994.1159.

Berenger, J.P. (1996). "Three-dimensional perfectly matched layer for the absorption of electromagnetic waves," *Journal of Computational Physics*, Vol. 127, No. 2, pp. 363–379. https://doi.org/10.1006/jcph.1996.0181.

Bielak, J., Graves, R.W, Olsen, K.B., Taborda, R., Ramırez-Guzman, L., Day, S.M., Ely, G.P., Roten, D., Jordan, T.H., Maechling, P.J., Urbanic, J., Cui, Y., and Juve, G.. (2010). "The ShakeOut earthquake scenario: Verification of three simulation sets," *Geophysical Journal International*, Vol. 180, No. 1, pp. 375–404. https://doi.org/10.1111/j.1365-246x.2009.04417.x.

Boatwright, J., Seekins, L.C., and Mueller, C.S. (1991). "Ground motion amplification in the Marina," *Bulletin of the Seismological Society of America*, Vol. 81, pp. 1980–1997.

Boore, D.M. (1972). "A note on the effect of simple topography on seismic SH waves," *Bulletin of the Seismological Society of America*, Vol. 62, pp. 275–284. https://doi.org/10.1785/bssa0620010275.

Boore, D.M. (1983). "Stochastic simulation of high-frequency ground motions based on seismological models of the radiated spectra," *Bulletin of the Seismological Society of America*, Vol. 73, pp. 1865–1894.

Boore, D.M. (1987). "The estimation of ground shaking caused by earthquakes," *Proceedings, Fifth Canadian Conference on Earthquake Engineering*, Ottawa, Canada, A.A. Balkema, Rotterdam, pp. 27–38.

Boore, D.M. (2004). "Estimating Vs30 (or NEHRP site classes) from shallow velocity models (depth < 30 m)," *Bulletin of the Seismological Society of America*, Vol. 94, pp. 591–597.

Boore, D.M. and Asten, M. W. (2008). "Comparisons of shear-wave slowness in the Santa Clara Valley, California, using blind interpretations of data from invasive and noninvasive methods." *Bulletin of the Seismological Society of America*, Vol. 98, pp. 1983–2003. https://doi.org/10.1785/0120070277.

Boore, D.M. and Joyner, W.B. (1984). "A note on the use of random vibration theory to predict peak amplitudes of transient signals," *Bulletin of the Seismological Society of America*, Vol. 74, No. 5, pp. 2035–2039.

Boore, D.M., Stewart, J.P., Seyhan, E., and Atkinson, G.M., (2014). "NGA-West 2 equations for predicting PGA, PGV, and 5%-damped PSA for shallow crustal earthquakes," *Earthquake Spectra*, Vol. 30, pp. 1057–1085. https://doi.org/10.1193/070113eqs184m.

Boore, D.M., Thompson, E.M., and Cadet, H. (2011). "Regional correlations of Vs30and velocities averaged over depths less than and greater than 30 m," *Bulletin of the Seismological Society of America*, Vol. 101, pp. 3046–3059. https://doi.org/10.1785/0120110071.

Borcherdt, R.D. (1970). "Effects of local geology on ground motion near San Francisco Bay," *Bulletin of the Seismological Society of America*, Vol. 60, No. 1, pp. 29–61.

Borcherdt, R.D. (1994). "Estimates of site-dependent response spectra for design (methodology and justification)," *Earthquake Spectra*, Vol. 10, No. 4, pp. 617–653. https://doi.org/10.1193/1.1585791.

Borcherdt, R.D. and Gibbs, J.F. (1976), "Effects of local geological conditions in the San Francisco Bay region on ground motions and the intensities of the 1906 earthquake," *Bulletin of the Seismological Society of America*, Vol. 66, No. 2, pp. 467–500. https://doi.org/10.1785/bssa0660020467.

Borcherdt, R.D., Gibbs, J.F., and Fumal, T.E. (1978). "Progress on ground motion predictions for the San Francisco Bay region, California," in Brabb, E.E., ed., *Progress in Seismic Zonation in the San Francisco Bay Region: U.S. Geological Survey Circular 807*, pp. 13–25.

Borcherdt, R.D. and Glassmoyer, G. (1994). "Influences of local geology on strong and weak ground motions in the San Francisco Bay region, California and their implications for site-specific code provisions," *USGS Professional Paper 1551-A*, The Loma Prieta earthquake of October 17, 1989 – strong ground motion, R.D. Borcherdt, ed., pp. 77–108. https://doi.org/10.3133/pp1551a.

Bouchon, M. and Barker, J.S. (1996). "Seismic response of a hill: The example of Tarzana, California," *Bulletin of the Seismological Society of America,* Vol. 86, No. 1A, pp. 66–72. https://doi.org/10.1785/bssa08601a0066.

Building Seismic Safety Council, (2020). "NEHRP Recommended Seismic Provisions for New Buildings and Other Structures, Volume 1: Part 1 Provisions, Part 2 Commentary," FEMA P-2082-1, Federal Emergency Management Agency, Washington, DC, 593 pp.

Cabas, A., Rodriguez-Marek, A., and Bonilla, L.F. (2017). "Estimation of site-specific kappa (κ 0)-consistent damping values at KiK-net sites to assess the discrepancy between laboratory-based damping models and observed attenuation (of seismic waves) in the field," *Bulletin of the Seismological Society of America*, Vol. 107, No. 5, pp. 2258–2271. https://doi.org/10.1785/0120160370.

Campbell, K.W. (2009). "Estimates of shear-wave Q and κ0 for unconsolidated and semiconsolidated sediments in Eastern North America," *Bulletin of the Seismological Society of America*, Vol. 99, pp. 2365–2392. https://doi.org/10.1785/0120080116.

Carter, D.E. and Longuet-Higgins, M.J. (1956). "The statistical distribution of the maxima of a random function," *Proceedings of the Royal Society*, Vol. 247, pp. 212–232. https://doi.org/10.1098/rspa.1956.0173.

Cartwright, D.E. and Longuet-Higgins, M.S. (1956). "The statistical distribution of the maxima of a random function," *Proceedings of the Royal Society of London, Series A, Mathematical and Physical Sciences*, Volume 237, No. 1209, pp. 212–232.

Cetin, K.O. and Bilge, H.T. (2011). "Cyclic large strain and induced pore pressure models for saturated clean sands," *Journal of Geotechnical and Geoenvironmental Engineering*, Vol. 138, No. 3, pp. 309–323. https://doi.org/10.1061/(asce)gt.1943-5606.0000631.

Chavez-Garcia, F.J., Sanchez, L.R., and Hatzfed, D. (1996). "Topographic site effects and HVSR. A comparison between observation and theory," *Bulletin of the Seismological Society of America*, Vol. 86, pp. 1559–1573. https://doi.org/10.1785/bssa0860051559.

Chen, X. (2007). "Generation and propagation of seismic SH-waves in multi-layered media with irregular interfaces," *Advances in Geophysics*, Vol. 48, pp. 191–264. https://doi.org/10.1016/s0065-2687(06)48004-3.

Chiou B.S.-J., Youngs R.R. (2008). "Chiou and Youngs PEER-NGA empirical ground motion model for the average horizontal component of peak acceleration and pseudo-spectral acceleration for spectral periods of 0.01 to 10 seconds," *Earthquake Spectra*, Vol. 24, pp. 173–215.

Chiou, B., Youngs, R., Abrahamson, N., and Addo, K. (2010). "Ground-motion attenuation model for small-to-moderate shallow crustal earthquakes in California and its implications on regionalization of ground-motion prediction models," *Earthquake Spectra*, Vol. 26. pp. 907–926. https://doi.org/10.1193/1.3479930.

Christian, J.T., Roesset, J.M., and Desai, C.S. (1977). "Chapter 20: Two- and three-dimensional dynamic analyses," in C. S. Desai and J. T. Christian, eds., *Numerical Methods in Geotechnical Engineering*, McGraw Hill Book Company, New York, pp. 683–718.

Cramer, C.H. (2003). "Site-specific seismic-hazard analysis that is completely probabilistic," *Bulletin of the Seismological Society of America*, Vol. *93*, pp. 1841–1846. https://doi.org/10.1785/0120020206.

Crandall, S.H. and Mark, W.D. (1963). *Random Vibration in Mechanical Systems*, Academic Press, New York. https://doi.org/10.1016/b978-1-4832-3259-1.50005-8.

Darendeli, M.B. (2001). "Development of a new family of normalized modulus reduction and material damping curves," Ph.D. dissertation, Dept. of Civil Engineering, University of Texas.

Darragh, R., Graizer, V., and Shakal, A. (1997). "Site characterization and site response effects at CSMIP stations: Tarzana and La Cienega near the Santa Monica Freeway (I-10)," *Report No. OSMS 96-07*, California Department of Conservation, Division of Mines and Geology, Office of Strong Motion Studies, 262 pp.

Davis, J.L. (1986). *Finite Difference Methods in Dynamics of Continuous Media*, MacMillan Publishing Company, New York, 238 pp.

Day, S.M. and Bradley, C.R. (2001). "Memory efficient simulation of anelastic wave propagation," *Bulletin of the Seismological Society of America*, Vol. 91, pp. 520–531. https://doi.org/10.1785/0120000103.

Day, S.M. and Minster, J.B. (1984). "Numerical simulation of wavefields using a Padé approximant method," *Geophysical Journal International*, Royal Astronomical Society, Vol. 78, pp. 105–118. https://doi.org/10.1111/j.1365-246x.1984.tb06474.x.

Dobry, R. and Iai, S. (2000). "Recent developments in the understanding of earthquake site response and associated seismic code implementation," *Proceedings, GeoEng2000*, Melbourne.

Dobry, R. and Vucetic, M. (1987). "Dynamic properties and seismic response of soft clay deposits," *Proceedings, International Symposium on Geotechnical Engineering of Soft Soils,* Mexico City, Vol. 2, pp. 51–87.

Elgamal, A., Lai, T., Yang, Z., and He, L. (2001). "Dynamic soil properties, seismic downhole arrays and applications in practice," *Proceedings, 4th International Conference on Recent Advances in Geotechnical Earthquake Engineering and Soil Dynamics*, S. Prakash, ed., San Diego, CA.

EPRI (1993). "Guidelines for determining design basis ground motions," *Report TR-102293*, Electric Power Research Institute (EPRI).

EPRI (2013). "Seismic evaluation guidance. Screening, prioritization and implementation details (SPID) for the resolution of Fukushima near-term task force recommendation 2.1. Seismic," EPRI report no. 1025281, December. Palo Alto, CA: Electrical Power Research Institute.

Faccioli, E., Maggio, F., Paolucci, R., and Quarteroni, A., (1997). "2D and 3D elastic wave propagation by a pseudo-spectral domain decomposition method," *Journal of Seismology*, Vol. 1, pp. 237–251. https://doi.org/10.1023/a:1009758820546.

Field, E.H. and Jacob, K.H. (1995). "A comparison and test of various site response estimation techniques, including three that are not reference site dependent," *Bulletin of the Seismological Society of America*, Vol. 85, pp. 1127–1143.

Finn, W.D.L. and Bhatia, S.K. (1981). "Prediction of seismic pore-water pressures," *Proceedings, 10th International Conference on Soil Mechanics and Foundation Engineering*, Rotterdam, The Netherlands, Vol. 3, pp. 201–206.

Frankel, A., Stephenson, W., and Carver, D. (2009). "Sedimentary basin effects in Seattle, Washington: Ground-motion observations and 3D simulations," *Bulletin of the Seismological Society of America*, Vol. 99, No. 3, pp. 1579–1611. https://doi.org/10.1785/0120080203.

Fumal, T.E. (1978). "Correlations between seismic wave velocities and physical properties of near-surface geologic materials in the southern San Francisco Bay region, California," *Open-File Report 78-1067*, U.S. Geological Survey, 114 pp. https://doi.org/10.3133/ofr781067.

Geli, L., Bard, P.-Y., and Jullien, B. (1988). "The effect of topography on earthquake ground motion: A review and new results," *Bulletin of the Seismological Society of America*, Vol. 78, No. 1, pp. 42–63. https://doi.org/10.1785/bssa0780010042.

GeoPentech (2015). "Southwestern United States ground motion characterization SSHAC level 3," *Technical Report Rev 2*, March.

Goulet, C.A. and Stewart, J.P. (2009). "Pitfalls of deterministic application of nonlinear site factors in probabilistic assessment of ground motions," *Earthquake Spectra*, Vol. 25, No. 3, pp. 541–555. https://doi.org/10.1193/1.3159006.

Graves, R.W. (1993) "Simulating realistic earthquake ground motions in regions of deep sedimentary basins," *Proceedings, 11th World Conference on Earthquake Engineering*, Paper No. 1932, Elsevier Science, Ltd., 8 pp.

Graves, R.W. (1996). "Simulating seismic wave propagation in 3D elastic media using staggered-grid finite differences," *Bulletin of the Seismological Society of America*, Vol. 86, No. 4, pp. 1091–1106. https://doi.org/10.1785/bssa0860041091.

Graves, R.W. and Pitarka, A. (2010). "Broadband ground-motion simulation using a hybrid approach," *Bulletin of the Seismological Society of America*, Vol. 99, No. 5A, pp. 2095–2123. https://doi.org/10.1785/0120100057.

Graves, R.W. and Pitarka, A. (2015). "Refinements to the Graves and Pitarka broadband ground-motion simulation method," *Seismological Research Letters*, Vol. 86, No. 1, 7 pp. https://doi.org/10.1785/0220140101.

Graves, R.W., Aagaard, B.T., and Hudnut, K.W. (2011). "The ShakeOut earthquake: Source and ground motion simulations," *Earthquake Spectra*, Vol. 27, pp. 273–291. https://doi.org/10.1193/1.3570677.

Graves, R.W., Pitarka, A., and Somerville, P.G. (1998). "Ground-motion amplification in the Santa Monica area: Effects of shallow basin-edge structure," *Bulletin of the Seismological Society of America*, Vol. 88, No. 5, pp. 1224–1242. https://doi.org/10.1785/bssa0880051224.

Griffiths, S.C., Cox, B.R., Rathje, E.M., and Teague, D.P. (2016a). "Surface-wave dispersion approach for evaluating statistical models that account for shear-wave velocity uncertainty," *Journal of Geotechnical and Geoenvironmental Engineering*, Vol. 142, No. 11, pp. 04016061. https://doi.org/10.1061/(asce)gt.1943-5606.0001552.

Griffiths, S.C., Cox, B.R., Rathje, E.M., and Teague, D.P. (2016b). "Mapping dispersion misfit and uncertainty in V s profiles to variability in site response estimates," *Journal of Geotechnical and Geoenvironmental Engineering*, Vol. 142, No. 11, https://doi.org/10.1061/(asce)gt.1943-5606.0001553.

Hough, S.E., Altidor, J.R., Anglade, D., Given, D., Janvier, M.G., Maharrey, J.Z., Meremonte, M., Mildor, B.S.-L., Prepetit, C., and Yong, A. (2010). "Localized damage caused by topographic amplification during the 2010 M7.0 Haiti earthquake," *Nature Geoscience*, Vol. 3, No. 11, pp. 778–782. https://doi.org/10.1038/ngeo988.

Hudson, M., Idriss, I.M., and Beikae, M. (1994). "QUAD4M: A computer program to evaluate the seismic response of soil structures using finite element procedures and incorporating a compliant base," User's Guide, Center for Geotechnical Modeling, Department of Civil and Environmental Engineering, University of California, Davis, CA.

Idriss, I.M. (1990). "Response of soft soil sites during earthquakes," *Proceedings, H. Bolton Seed Memorial Symposium*, J.M. Duncan, ed., BiTech Publishers, Ltd., Vancouver, British Columbia, Vol. 2, pp. 273–289.

Idriss, I.M. and Seed, H.B. (1968). "An analysis of ground motions during the 1957 San Francisco earthquake," Bulletin of the Seismological Society of America, Vol. 58, No. 6, pp. 2013–2032.

Idriss, I.M. and Sun, J.I. (1992). "SHAKE91 – A computer program for conducting equivalent linear seismic response analyses of horizontally layered soil deposits," User's Guide, University of California, Davis, 13 pp.

Joyner, W.B. and Chen, A.T.F. (1975) "Calculation of nonlinear ground response in earthquakes," *Bulletin of the Seismological Society of America*, Vol. 65, pp. 1315–1336.

Kaklamanos, J., Bradley, B.A., Ehompson, E.M., and Baise, L.G. (2013). "Critical parameters affecting bias and variability in site response analyses using KiK-net downhole array data," *Bulletin of the Seismological Society of America*, Vol. 103, No. 3, pp. 1733–1749. https://doi.org/10.1785/0120120166.

Kausel, E. and Asimaki, D. (2002). "Seismic simulation of inelastic soils via frequency-dependent moduli and damping," *Journal of Engineering Mechanics*, Vol. 128, No. 1, pp. 34–47. https://doi.org/10.1061/(asce)0733-9399(2002)128:1(34).

Kawase, H. (1996). "The cause of the damage belt in Kobe: 'The basin-edge effect,' Constructive interference of the direct s-wave with the basin-induced diffracted/Rayleigh waves," *Seismological Research Letters*, Vol. 67, No. 5, pp. 25–34. https://doi.org/10.1785/gssrl.67.5.25.

Kawase, H., Matsushima, S., Graves, R.W., and Somerville, P.G. (2000). "Strong motion simulation of Hyogo-ken Nanbu (Kobe) earthquake considering both the heterogeneous rupture process and the 3-D basin structure," *Proceedings, 12th World Conference on Earthquake Engineering*, Paper No. 990, 8 pp.

Kim, B., Hashash, Y.M.A., Kottke, A.R., Asimaki, D., Li, W., Rathje, E.M., Campbell, K.W., Silva, W.J., and Stewart, J.P., (2013). "A predictive model for the relative differences between nonlinear and equivalent-linear site response analyses," *Proceedings, 22nd Conference on Structural Mechanics in Reactor Technology (SMiRT22)*, San Francisco, CA.

Komatitsch, D., Erlebacher, G., Göddeke, D., and Michéa, D. (2010). "High-order finite-element seismic wave propagation modeling with MPI on a large GPU cluster," *Journal of Computational Physics*, Vol. 229, No. 20, pp. 7692–7714. https://doi.org/10.1016/j.jcp.2010.06.024.

Kottke, A.R., Hashash, Y.M-A., Stewart, J.P., Moss, C.J., Nikolaou, S., Rathje, E.M., Silva, W.J., and Campbell, K.W. (2012). "Development of geologic site classes for seismic site amplification for Central and Eastern North America," *Proceedings, 15th World Conference on Earthquake Engineering*, Lisbon, Portugal, 10 pp.

Kottke, A.R. and Rathje, E.M. (2009). "Technical manual for STRATA," *Report PEER 2008/10*, Pacific Earthquake Engineering Research Center, University of California, Berkeley, 84 pp.

Kramer, S.L., Hartvigsen, A.J., Sideras, S.S., and Ozener, P.T. (2011). "Site response modelling in liquefiable soil deposits," *Proceedings, 4th IASPEI/IAEE International Symposium: Effects of Surface Geology on Seismic Motion*, Santa Barbara, CA, 13 pp.

Kramer, S.L. and Paulsen, S.B. (2004). "Practical use of geotechnical site response models," Proceedings, International Workshop on Uncertainties in Nonlinear Soil Properties and their Impact on Modeling Dynamic Soil Response, PEER Center Headquarters, Richmond, CA.

Kuehn, N.M., Abrahamson, N.A., and Walling, M.A. (2019). "Incorporating nonergodic path effects into the NGA-West2 ground-motion prediction equations," *Bulletin of the Seismological Society of America*, Vol. 109, No. 2, pp. 575–585. https://doi.org/10.1785/0120180260.

Kuhlemeyer, R.L. and Lysmer, J. (1973). "Finite element method accuracy for wave propagation problems," *Journal of the Soil Mechanics and Foundations Division*, Vol. 99, No. SM5, pp. 421–427. https://doi.org/10.1061/jsfeaq.0001885.

Kwak, D.-Y., Jeong, C.-G., Park, D., and Park, S. (2008). "Comparison of frequency dependent equivalent linear analysis methods," *Proceedings, 14th World Conference on Earthquake Engineering*, Beijing, China, 9 pp.

Kwok, A.O.L., Stewart, J.P., and Hashash, Y.M.A (2008). "Nonlinear ground-response analysis of Turkey Flat shallow stiff-soil site to strong ground motion," *Bulletin of the Seismological Society of America*, Vol. 98, No. 1, pp. 331–343. https://doi.org/10.1785/0120070009.

Kwok, A.O.L., Stewart, J.P., Hashash, Y.M.A, Matasovic, N., Pyke, R., Wang, Z., and Yang, Z. (2007). "Use of exact solutions of wave propagation problems to guide implementation of nonlinear seismic ground response analysis procedures," *Journal of Geotechnical and Geoenvironmental Engineering,* Vol. 133, No. 11, pp. 1385–1398. https://doi.org/10.1061/(asce)1090-0241(2007)133:11(1385).

Lavrentiadis, G., Abrahamson, N.A. Kuehn, N.M., Bozorgnia, Y., Goulet, C.A., Babič, A., Macedo, J., Dolšek, M., Gregor, N., Kottke, A.R., Lacour, M., Liu, C., Meng, X., Phung,V.-B., Sung, C.-H., Walling, M. (2023). Overview and introduction to development of non-ergodic earthquake ground-motion models, *Bulletin of Earthquake Engineering*, Vol. 21, pp. 5121-5150, https://doi.org/10.1007/s10518-022-01485-x

Lee, M.K.W. and Finn, W.D.L. (1978). "DESRA-2, Dynamic effective stress response analysis of soil deposits with energy transmitting boundary including assessment of liquefaction potential," *Soil Mechanics Series No. 38*, Department of Civil Engineering, University of British Columbia, Vancouver, B.C.

Lee, Y. and Anderson, J.G. (2000). "A custom southern California ground motion relationship based on analysis of residuals," *Bulletin of the Seismological Society of America*, Vol. 90, pp. S170–S187.

Li, W. and Asimaki, D. (2011). "Site and motion-dependent parametric uncertainty of site-response analyses in earthquake simulations," *Bulletin of the Seismological Society of America*, Vol. 100, No. 3, pp. 954–968. https://doi.org/10.1785/0120090030.

Lin, P.-S., Chiou, B.S.-J, Abrahamson, N.A., Walling, M., Lee, C.-T., and Cheng, C.-T. (2011). "Repeatable source, site, and path effects on the standard deviation for ground-motion prediction," *Bulletin of the Seismological Society of America*, Vol. 101, pp. 2281–2295. https://doi.org/10.1785/0120090312.

Liu, M. and Gorman, D.G. (1995). "Formulation of Rayleigh damping and its extensions," *Computers and Structures,* Vol. 57, No. 2, pp. 277–285. https://doi.org/10.1016/0045-7949(94)00611-6.

Liu, P. and Archuleta, R.J. (2006). "Efficient modeling of Q for 3D numerical simulation of wave propagation," *Bulletin of the Seismological Society of America*, Vol. 96, No. 4A, pp. 1352–1358. https://doi.org/10.1785/0120050173.

Lo Presti, D.C.F., Lai, C.G., and Ignazio, P. (2006). "ONDA: Computer code for nonlinear seismic response analysis of soil deposits," *Journal of Geotechnical and Geoenvironmental Engineering*, Vol. 132, No. 2, pp. 223–236. https://doi.org/10.1061/(asce)1090-0241(2006)132:2(223).

Lysmer, J., Udaka, T., Tsai, C.F., and Seed, H.B. (1975). "FLUSH - A computer program for approximate 3-D analysis of soil-structure interaction problems," *Report No. EERC 75-30*, Earthquake Engineering Research Center, University of California, Berkeley, California, 83 pp.

MacMurdo, J. (1824). "Papers relating to the earthquake which occurred in India in 1819," *Philosophical Magazine*, Vol. 63, pp. 105–177. https://doi.org/10.1080/14786442408644477.

Mallet, R. (1862). *Great Neapolitan Earthquake of 1857*, London, 2 vols. https://doi.org/10.1093/nq/s2-v.126.437c.
Markham, C.S., Bray, J.D., and Macedo, J. (2015). "Deconvolution of surface motions from the Canterbury Earthquake Sequence for use in nonlinear effective stress site response analysis," *Proceedings, 6th International Conference on Earthquake Geotechnical Engineering*, Christchurch, NZ, 9 pp.
Martin, G.R., Finn, W.D.L., and Seed, H.B. (1975). "Fundamentals of liquefaction under cyclic loading," *Journal of the Geotechnical Engineering Division*, Vol. 101, No. GT5, pp. 423–438. https://doi.org/10.1061/ajgeb6.0000164.
Martin, G.R., Lam, I.P., McCaskie, S.L., and Tsai, C.-F. (1981). "A parametric study of an effective stress liquefaction model," *Proceedings, International Conference on Geotechnical Earthquake Engineering and Soil Dynamics*, St. Louis, Missouri, Vol. 2, pp. 699–705.
Matasovic, N. and Hashash, Y. (2012). "Practices and procedures for site-specific evaluations of earthquake ground motions: A synthesis of highway practice," *NCHRP Synthesis 428*, Transportation Research Board, Washington, D.C., 78 pp. https://doi.org/10.17226/14660.
Matasović, N. and Vucetic, M. (1992). "A pore pressure model for cyclic straining of clay," *Soils and Foundations*, Vol. 32, No. 3, pp. 156–173. https://doi.org/10.3208/sandf1972.32.3_156.
Maufroy, E., Cruz-Atienza, V.M., and Gaffet, S. (2012). "A robust method for assessing 3-D topographic site effects: A case study at the LSBB underground laboratory, France," *Earthquake Spectra*, Vol. 28, No. 3, pp. 1097–1115. https://doi.org/10.1193/1.4000050.
May, T.W. (1980). "The effectiveness of trenches and scarps in reducing seismic energy," Ph.D. Thesis, University of California at Berkeley, Berkeley, California.
Meite, R., Wotherspoon, L., McGann, C.R., Green, R.A., and Hayden, C. (2021). "An iterative linear procedure using frequency-dependent soil parameters for site response analyses," *Soil Dynamics and Earthquake Engineering*, Vol. 130, 13 pp. https://doi.org/10.1016/j.soildyn.2019.105973.
Mejia, L.H. and Dawson, E.M. (2006). "Earthquake deconvolution for FLAC," *Proceedings, 4th International FLAC Symposium on Numerical Modeling in Geomechanics*, Citeseer, Princeton, NJ, USA, pp. 4–10.
Moss, R.E.S. (2008). "Quantifying measurement uncertainty associated with thirty meter shear wave velocity (VS30)," *Bulletin of the Seismological Society of America*, Vol. 98, pp. 1399–1411. https://doi.org/10.1785/0120070101.
Newmark, N.M. (1959). "A method of computation for structural dynamics," *Journal of the Engineering Mechanics Division*, Vol. 85, No. EM3, pp. 67–94. https://doi.org/10.1061/jmcea3.0000098.
Nweke, C.C., Stewart, J.P., Wang, P., and Brandenberg, S.J. (2022). "Site response of sedimentary basins and other geomorphic provinces in southern California," *Earthquake Spectra*, Vol. 38, No. 4, pp. 2341–2370. https://doi.org/10.1177/87552930221088609
O'Rourke, T.D., Stewart, J.E., Gowdy, T.E., and Pease, J.W. (1991). "Lifeline and geotechnical aspects of the 1989 Loma Prieta earthquake," *Proceedings, Second International Conference on Recent Advances in Geotechnical Earthquake Engineering and Soil Dynamics*, St. Louis, Missouri, pp. 1601–1612.
Park, D. and Hashash, Y.M.A. (2004). "Soil damping formulation in nonlinear time domain site response analysis," *Journal of Earthquake Engineering*, Vol. 8, No. 2, pp. 249–274. https://doi.org/10.1080/13632460409350489.
Park, T., Park, D., and Ahn, J. K. (2015). "Pore pressure model based on accumulated stress," *Bulletin of Earthquake Engineering*, Vol. 13, No. 7, pp. 1913–1926. https://doi.org/10.1007/s10518-014-9702-1
Parker, G.A., Stewart, J.P., Hashash, Y.M.A., Rathje, E.M., Campbell, K.W., and Silva, W.J. (2019). "Empirical linear seismic site amplification in Central and Eastern North America," *Earthquake Spectra*, Vol. 35, pp. 849–881.
Pehlivan, M., Rathje, E.M., and Gilbert, R.B. (2013). "Influence of spatial variability on site response analysis," *Proceedings, International Conference on Earthquake Geotechnical Engineering: From Case History to Practice*, International Society of Soil Mechanics and Geotechnical Engineering, Istanbul, Turkey.
Pehlivan, M. Rathje, E.M., and Gilbert, R.B. (2016), "Factors influencing soil surface seismic hazard curves," *Soil Dynamics and Earthquake Engineering*, Vol 83, pp. 180–190.
Pitarka, A., lrikura, K., Iwata, T., and Sekiguchi, H. (1998). "Three-dimensional simulation of the near-fault ground motion for the 1995 Hyogo-ken Nanbu (Kobe), Japan, earthquake," *Bulletin of the Seismological Society of America*, Vol. 88, No. 2, pp. 428–440. https://doi.org/10.1785/bssa0880020428.
Phillips, C. and Hashash, Y.M.A. (2009). "Damping formulation for nonlinear 1D site response analyses," *Soil Dynamics and Earthquake Engineering*, Vol. 29, pp. 1143–1158. https://doi.org/10.1016/j.soildyn.2009.01.004.
Polito, C.P., Green, R.A., and Lee, J. (2008) "Pore pressure generation models for sands and silty soils subjected to cyclic loading," *Journal of Geotechnical and Geoenvironmental Engineering*, Vol. 134, No. 10, pp. 1490–1500. https://doi.org/10.1061/(asce)1090-0241(2008)134:10(1490).

Pretell, R., Ziotopoulou, K., and Abrahamson, N. (2019). "Methodology for the development of input motions for nonlinear deformation analyses," *Proceedings, 7th International Conference on Earthquake Geotechnical Engineering*, Rome, Italy.

Rai, M., Rodriguez-Marek, A., and Yong, A. (2016). "An empirical model to predict topographic effects in strong ground motion using California small to medium magnitude earthquake database," *Earthquake Spectra*, Vol. 32, No. 2, pp. 1033–1054. https://doi.org/10.1193/113014eqs202m.

Rathje, E.M., Kottke, A.R., and Ozbey, M.C. (2005). "Using inverse random vibration theory to develop input Fourier amplitude spectra for use in site response," *Proceedings, 16th International Conference on Soil Mechanics and Geotechnical Engineering: TC4 Earthquake Geotechnical Engineering Satellite Conference*, Osaka, Japan, pp. 160–166.

Rathje, E.M., Kottke, A.R., and Trent, W.L. (2010). "Influence of input motion and site property variabilities on seismic site response analysis," *Journal of Geotechnical and Geoenvironmental Engineering*, Vol. 136, No. 4, pp. 607–619. https://doi.org/10.1061/(asce)gt.1943-5606.0000255.

Roden, A. and Gedney, S.D. (2000). "Convolution PML (CPML): An efficient FDTD implementation of the CFS-PML for arbitrary media," *Microwave and Optical Technology Letters*, Vol. 27, No. 5, pp. 334–339. https://doi.org/10.1002/1098-2760(20001205)27:5<334::aid-mop14>3.0.co;2-a.

Rodriguez-Marek, A., Bommer, J.J., Youngs, R.R., Crespo, M.J., Stafford, P.J., and Bahrampouri, M. (2021). "Capturing epistemic uncertainty in site response," *Earthquake Spectra*, Vol. 37, No. 2, pp. 921–936. https://doi.org/10.1177/8755293020970975.

Rodriguez-Marek, A., Bray, J.D., and Abrahamson, N.A. (2001). "An empirical geotechnical seismic site response procedure," *Earthquake Spectra*, Vol. 17, pp. 65–87. https://doi.org/10.1193/1.1586167.

Rodriguez-Marek, A., Montalva, G.A., Cotton, F., and Bonilla, F. (2011). "Analysis of single-station standard deviation using the KiK-net data," *Bulletin of the Seismological Society of America*, Vol. 101, pp. 1242–1258. https://doi.org/10.1785/0120100252.

Rodriguez-Marek, A., Rathje, E.M., Ake J., Munson, C., Stovall, S., Weaver, T., Ulmer, K.J., and Juckett, M. (2021). "Documentation Report for SSHAC Level 2: Site Response," US Nuclear Regulatory Commission, *RIL 202114933-15*.

Romo, M.P. and Seed, H.B. (1986). "Analytical modelling of dynamic soil response in the Mexico earthquake of September 19, 1985," *Proceedings, ASCE International Conference on the Mexico Earthquakes - 1985*, Mexico City, pp. 148–162.

Sadigh, K., Chang, C.-Y., Abrahamson, N.A., Chiou, S.J., and Power, M. (1993). "Specification of long period motions: Updated attenuation relations for rock site conditions and adjustment factors for near-fault effects," *Proceedings, ATC 17-1 Seminar*, San Francisco, CA, Applied Technology Council, pp. 59–70.

Sanchez-Sesma, F.J. (1990). "Elementary solutions for response of a wedge-shaped medium to incident SH and SV waves," *Bulletin of the Seismological Society of America*, Vol. 80, No. 3, pp. 737–742.

Schnabel, P.B., Lysmer, J., and Seed, H.B. (1972). "SHAKE: A computer program for earthquake response analysis of horizontally layered sites," *Report No. EERC 72-12*, Earthquake Engineering Research Center, University of California, Berkeley, California.

Seed, H.B., Murarka, R., Lysmer, J., and Idriss, I.M. (1976a). "Relationships of maximum acceleration, maximum velocity, distance from source and local site conditions for moderately strong earthquakes," *Bulletin of the Seismological Society of America*, Vol. 66, No. 4, pp. 1323–1342.

Seed, H.B., Ugas, C., and Lysmer, J. (1976b). "Site-dependent spectra for earthquake-resistant design," *Bulletin of the Seismological Society of America*, Vol. 66, pp. 221–243. https://doi.org/10.1785/bssa0660010221.

Seed, H.B., Whitman, R.V., Dezfulian, H., Dobry, R., and Idriss, I.M. (1972). "Soil conditions and building damage in 1967 Caracas earthquake," *Journal of the Soil Mechanics and Foundations Division*, Vol. 98, No. 8, pp. 787–806. https://doi.org/10.1061/jsfeaq.0001768.

Seyhan, E., Stewart, J.P., Ancheta, T.D., Darragh, R.B., and Graves, R.W., (2014). "NGAWest2 site database," *Earthquake Spectra*, Vol. 30, pp. 1007–1024. https://doi.org/10.1193/062913eqs180m.

Seyhan, E. and Stewart, J. P. (2014). "Semi-empirical nonlinear site amplification from NGA-West 2 data and simulations," *Earthquake Spectra*, Vol. 30, No. 2, pp. 1241–1256.

Silva, W.J. (1988). "Soil response to earthquake ground motion," *EPRI Report NP-5747*, Electric Power Research Institute, Palo Alto, California.

Sitar, N. and Clough, G.W. (1983). "Seismic response of steep slopes in cemented soils," *Journal of Geotechnical Engineering*, Vol. 109, pp. 210–227. https://doi.org/10.1061/(asce)0733-9410(1983)109:2(210).

Sokolov, V.Y. (1997). "Empirical models for estimating Fourier-amplitude spectra of ground acceleration in the northern Caucasus (Racha seismogenic zone)," *Bulletin of the Seismological Society of America*, Vol. 66, No. 87, pp. 1401–1412. https://doi.org/10.1785/bssa0870061401.

Spudich, P.A., Hellwig, M., and Lee, W.H.K. (1996). "Directional topographic response at Tarzana observed in aftershocks of the 1994 Northridge, California earthquake: Implications for main shock motions," *Bulletin of the Seismological Society of America*, Vol. 86, No. 1B, pp. S193–S208. https://doi.org/10.1785/bssa08601bs193.

Steidl, J.H. (2000). "Site response in southern California for probabilistic seismic hazard analysis," *Bulletin of the Seismological Society of America*, Vol. 90, S149–S169. https://doi.org/10.1785/0120000504.

Stewart, J.P. and Afshari, K. (2021). "Epistemic uncertainty in site response as derived from one-dimensional ground response analyses," *Journal of Geotechnical and Geoenvironmental Engineering*, Vol. 147, No. 1, 13 pp. https://doi.org/10.1061/(asce)gt.1943-5606.0002402.

Stewart, J.P., Afshari, K., and Goulet, C.A. (2017). "Non-ergodic site response in seismic hazard analysis," *Earthquake Spectra*, Vol. 33, No. 4, 38 pp. https://doi.org/10.1193/081716eqs135m.

Stewart, J.P., Klimis, N., Savvaidis, A., Theodoulidis, N., Zargli, E., Athanasopoulos, G., Pelekis, P., Mylonakis, G., and Margaris, B. (2014). "Compilation of a local VS profile database and its application for inference of VS30 from geologic- and terrain-based proxies," *Bulletin of the Seismological Society of America*, Vol. 104, No. 6, pp. 2827–2841. https://doi.org/10.1785/0120130331.

Stewart, J.P., Kwok, A.O., Hashash, Y.M.A., Matasovic, N., Pyke, R., Wang, Z., and Yang, Z. (2008). "Benchmarking of nonlinear geotechnical ground response analysis procedures," *Report PEER 2008/04*, Pacific Earthquake Engineering Research Center, Berkeley, CA, 186 pp.

Stewart, J.P., Liu, A.H., and Choi, Y., (2003). "Amplification factors for spectral acceleration in tectonically active regions," *Bulletin of the Seismological Society of America*, Vol. 93, pp. 332–352. https://doi.org/10.1785/0120020049.

Stone, W.C., Yokel, F.Y., Celebi, M., Hanks, T., and Leyendecker, E.V. (1987). "Engineering aspects of the September 19, 1985 Mexico earthquake," *NBS Building Science Series 165*, National Bureau of Standards, Washington, D.C., 207 pp. https://doi.org/10.6028/nbs.bss.165.

Sugito, M., Goda, H., and Masuda, T. (1994). "Frequency dependent equi-linearized technique for seismic response analysis of multi-layered ground," *Doboku Gakkai Rombun-Hokokushu/Proceedings of the Japan Society of Civil Engineers*, Vol. 493, No. 3–2, pp. 49–58. https://doi.org/10.2208/jscej.1994.493_49.

Tatsuoka, F., Siddiquee, M.S.A., Park, C.S., Sakamoto, M., and Abe, F., (1993), "Modeling stress-strain relations of sand," *Soils and Foundations*, Vol. 33, No. 2, pp. 60–81. https://doi.org/10.3208/sandf1972.33.2_60.

Teague, D.P., Cox, B.R., and Rathje, E.M. (2018). "Measured vs. predicted site response at the Garner Valley Downhole Array considering shear wave velocity uncertainty from borehole and surface wave methods," *Soil Dynamics and Earthquake Engineering*, Vol. 113, pp. 339–355. https://doi.org/10.1016/j.soildyn.2018.05.031.

Thompson, E.M., Baise, L.G., Kayen, R.E., and Guzina, B.B. (2009). "Impediments to predicting site response: Seismic property estimation and modeling simplifications," *Bulletin of the Seismological Society of America*, Vol. 99, No. 5, pp. 2927–2949. https://doi.org/10.1785/0120080224.

Thompson, E.M., Baise, L.G., Tanaka, Y., and Kayen, R.E. (2012). "A taxonomy of site response complexity," *Soil Dynamics and Earthquake Engineering*, Vol. 41, pp. 32–43. https://doi.org/10.1016/j.soildyn.2012.04.005.

Toro, G.R. (1995). "Probabilistic models of site velocity profiles for generic and site-specific ground-motion amplification studies," *Technical Report No. 779574*, Brookhaven National Laboratory, Upton, NY.

Toro, G.R., Abrahamson, N.A., and Schneider, J.F. (1997). "Model of strong ground motions from earthquakes in Central and Eastern North America: Best estimates and uncertainties." *Seismological Research Letters*, Vol. 68, pp. 41–57. https://doi.org/10.1785/gssrl.68.1.41.

Trifunac, M.D. and Hudson, D.E. (1971). Analysis of the Pacoima Dam accelerogram: San Fernando, California, earthquake of 1971, *Bulletin of the Seismological Society of America*, Vol. 61, pp. 1393–1411.

Tsai, C.C. and Hashash, Y.M.A. (2009). "Learning of dynamic soil behavior from downhole arrays, " *Journal of Geotechnical and Geoenvironmental Engineering*, Vol. 135, No. 6, pp. 745–757. https://doi.org/10.1061/(asce)gt.1943-5606.0000050.

Williams, T. and Abrahamson, N.A. (2021). "Site-response analysis using the shear-wave velocity profile correction approach," *Bulletin of the Seismological Society of America*, Vol. 111, pp. 1989–2004. https://doi.org/10.1785/0120200345.

Wills, C.J., Gutierrez, C.I., Perez, F.G., and Branum, D.M. (2015). "A next generation VS30 map for California based on geology and topography," *Bulletin of the Seismological Society of America*, Vol. 105, No. 6, pp. 3083–3091. https://doi.org/10.1785/0120150105.

Wolf, J.P. (1985). *Dynamic Soil-Structure Interaction*, Prentice-Hall, Inc., Englewood Cliffs, NJ, 466 pp. https://onlinelibrary.wiley.com/doi/10.1002/eqe.4290140610.

Wolf, J.P. (1991). "Consistent lumped-parameter models for unbounded soil: Physical representation," *Earthquake Engineering and Structural Dynamics*, Vol. 20, No. 1, pp. 11–32. https://doi.org/10.1002/eqe.4290200103.

Xu, B. and Rathje, E. (2021). "The effect of soil nonlinearity on high frequency spectral decay and implications for site response analysis," *Earthquake Spectra*, Vol. 37, No. 2. https://doi.org/10.1177/8755293020981991.

Yee, E., Stewart, J.P., and Tokimatsu, K. (2013). "Elastic and large-strain nonlinear seismic site response from analysis of vertical array recordings," *Journal of Geotechnical and Geoenvironmental Engineering*, Vol. 139, No. 10, pp. 1789–1801. https://doi.org/10.1061/(asce)gt.1943-5606.0000900.

Yniesta, S., Brandenberg, S.J. and Shafiee, A. (2017). "ARCS: A one dimensional nonlinear soil model for ground response analysis." *Soil Dynamics and Earthquake Engineering*, Vol. 102, pp. 75–85. https://doi.org/10.1016/j.soildyn.2017.08.015.

Yoshida, N., Kobayashi, S., Suetomi, I., and Miura, K. (2002). "Equivalent linear method considering frequency dependent characteristics of stiffness and damping," *Soil Dynamics and Earthquake Engineering*, Vol. 22, No. 3, pp. 205–22. https://doi.org/10.1016/s0267-7261(02)00011-8.

Youd, T.L. and Carter, B.L. (2005). "Influence of soil softening and liquefaction on spectral acceleration," *Journal of Geotechnical and Geoenvironmental Engineering*, Vol. 131, No. 7, pp. 811–825. https://doi.org/10.1061/(asce)1090-0241(2005)131:7(811).

Zalachoris, G. and Rathje, E.M. (2015). "Evaluation of one-dimensional site response techniques using borehole arrays," *Journal of Geotechnical and Geoenvironmental Engineering*, Vol. 141, No. 12, 15 pp. https://doi.org/10.1061/(asce)gt.1943-5606.0001366.

Zhou, H. and Chen, X. (2006). "A new approach to simulate scattering of SH waves by an irregular topography," *Geophysics Journal International*, Vol. 164, pp. 449–459. https://doi.org/10.1111/j.1365-246x.2005.02670.x.

8 Soil–Structure Interaction

8.1 INTRODUCTION

It is common in structural engineering practice to assume that structures are supported on "fixed" bases, i.e., rigid foundations supported by rigid soil. Real structures, of course, are supported on bases that have some degree of flexibility, or compliance. The response of a real structure to earthquake shaking is affected by interactions between three linked systems: the structure, the foundation, and the geologic media underlying and surrounding the foundation. A seismic soil–structure interaction (SSI) analysis evaluates the collective response of these systems to free-field ground motions. The term "free-field" refers to the motions that would have occurred at a site if the structure and foundation were not present.

SSI effects reflect the differences between the actual response of the structure and the response for the theoretical, fixed-base condition. Visualized within this context, three SSI effects can be important in engineering analysis and design:

1. **Foundation Stiffness and Damping**: Inertia developed in a vibrating structure gives rise to shear, moment, and torsion at its base. These loads, in turn, generate displacements and rotations of the foundation relative to the corresponding free-field values. These relative displacements and rotations are possible because of flexibility in the soil-foundation system, which can add significantly to the overall structural flexibility in some cases. Moreover, the relative foundation-free-field motions cause stress waves to radiate away from the foundation through the supporting soil. These waves remove energy from the structure via radiation damping and hysteretic energy dissipation, which, when combined with the damping of the structure itself, may significantly increase the overall system damping. Since these effects are rooted in the structural inertia, they are referred to as *inertial interaction* effects.
2. **Variations between Free-Field and Foundation-Level Ground Motions**: The differences between motions at the base of a structure ("foundation-level") and free-field motions result from two processes: (1) *kinematic interaction*, in which the stiffness of foundation elements placed at or below the ground surface prevents the foundation from moving in the same way as the as the free-field soil even in the absence of structure and foundation inertia, and (2) relative foundation-free-field displacements and rotations associated with structure and foundation inertia.
3. **Foundation Deformations**: Foundation elements such as walls, slabs, and piles are not rigid – they can bend, twist, compress, and shear as a result of loads or displacements applied by the superstructure and/or the soil medium. Such loads or deformations represent the seismic demand for which foundation components should be designed.

The terms kinematic and inertial interaction were introduced by Professor Robert Whitman in the early 1970s (Kausel, 2010).

8.2 METHODS OF ANALYSIS

Procedures for analyzing how simple structures respond to earthquake ground motions under fixed-base conditions are presented in Appendix B. Engineers can include the effects of SSI in their response analyses by incorporating the effects of (1) foundation-soil flexibility and damping and (2) kinematic interaction on the seismic excitation. Methods of analysis that can be used to evaluate SSI effects can be categorized as either direct or substructure approaches.

DOI: 10.1201/9781003512011-8

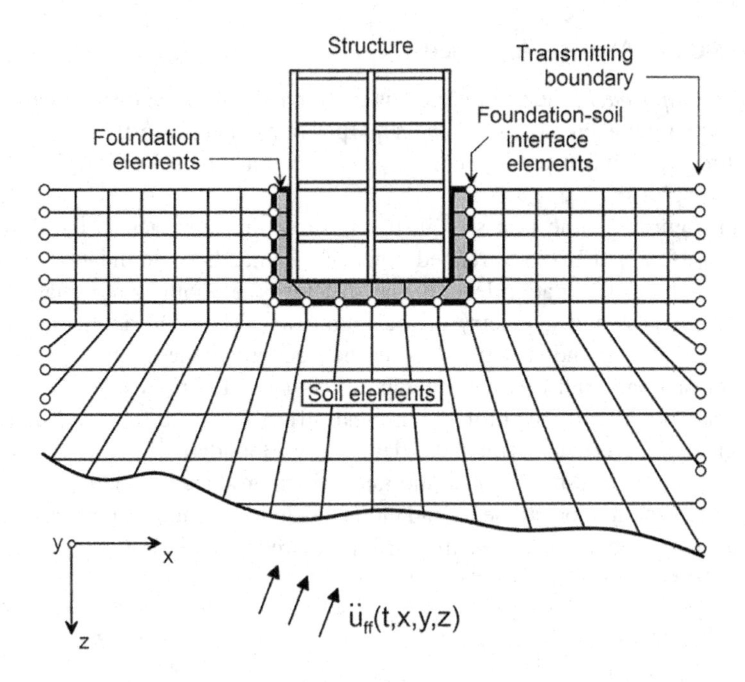

FIGURE 8.1 Schematic illustration of direct approach to the analysis of the soil–structure interaction (SSI) problem using continuum modeling by finite elements. (Figure from NIST, 2012.)

8.2.1 Direct Analysis Approach

In a *direct analysis*, the soil and structure are included within the same model and analyzed as a complete system. As schematically depicted in Figure 8.1 for the example of a building structure, the soil is often represented as a continuum (e.g., represented by finite elements) along with foundation and structural elements, transmitting boundaries at the limits of the soil mesh, and interface elements at the edges of the foundation. Ground response is evaluated using wave propagation analyses in this approach (Sections 7.5 and 7.6). Such analyses are most often performed using an equivalent-linear representation of soil properties in finite element, finite difference, or boundary element numerical formulations (Wolf, 1985; Lysmer et al., 1999). When applied in three-dimensional simulations that include fault rupture and source-to-site wave propagation in addition to SSI analysis, the domain reduction method can be used to separate the SSI analyses from the ground motion simulations (Yoshimura et al., 2002; Bielak et al., 2003).

Direct analyses can solve all three of the SSI problems (Section 8.1) in a single step. However, incorporation of the kinematic interaction component of the problem is challenging regardless of the sophistication of the computational platform, because it requires, in its most rigorous form, specification of spatially variable input motions in three dimensions (Section 3.8). Direct analyses also require software that can treat the structure, foundation, and soil with consistent levels of rigor. Historically, software for seismic analyses have been developed either in the structural domain (with great simplification of soil behavior) or in the geotechnical domain (with great simplification of structural behavior). Software that is more appropriate for direct analyses is becoming increasingly available, but it can be sufficiently complex that significant experience and expertise is required to use it effectively.

Because direct solution of the SSI problem is difficult from a computational standpoint, especially when the system is geometrically complex and/or contains significant nonlinearities in the soil and/or structural materials, its use in practice is limited to critical projects with conditions not amenable to more simplified approaches (e.g., Ellison et al., 2015).

8.2.2 Substructure Analysis Approach

In the *substructure approach*, the SSI problem is partitioned into three distinct parts that are combined to formulate a complete solution. The superposition inherent to this approach requires an assumption of linear soil and structure behavior, which in practice is provided for by using equivalent-linear soil properties.

As shown in Figure 8.2, the first step in the substructure approach is the development of a *Foundation Input Motion* (FIM) and related kinematic demands on foundation components. The FIM is the motion at the foundation level of the structure (i.e., bottom of a mat or pile cap) that accounts for the stiffness and geometry of the foundation. The FIM applies for the theoretical condition of the foundation and structure having their actual stiffness but no mass and generally involves both translational and rotational components. The FIM represents the seismic demand applied to the foundation and structural system, and differs from the free-field motion as a result of various kinematic SSI effects. Kinematic demands on foundations arise from the differences between the FIM and the free-field ground motions, which produce foundation-soil interface pressures and associated internal loads (shears and moments) in foundation components. Two important examples of kinematic demands are seismic earth pressures on retaining walls (Section 8.6.3) and bending of deep foundations (e.g., Nikolaou et al. 2001).

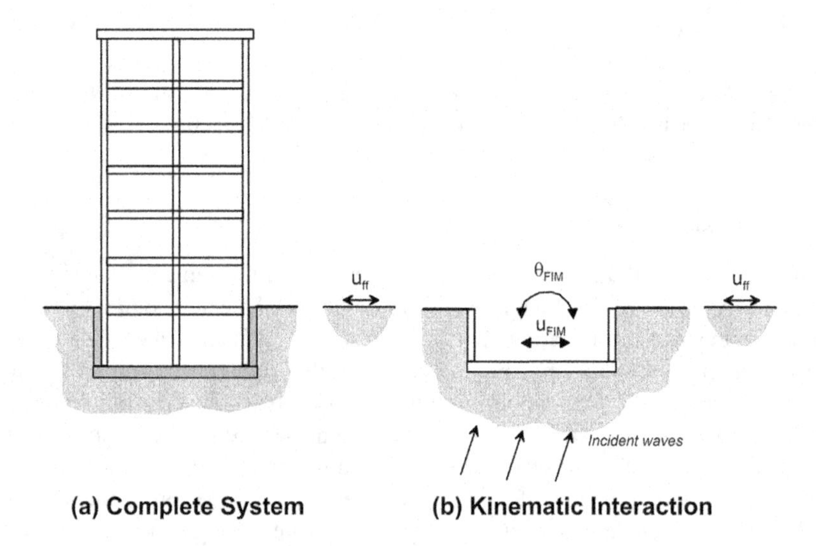

FIGURE 8.2 Schematic illustration of (a) the complete system to be analyzed using a substructure approach and (b) the kinematic interaction problem in which foundation input motions are analyzed. Shaded foundation components (Part a) have mass while open components (Part b) are massless.

Development of the FIM takes into account the free-field response of the site; in other words, the spatial and temporal variation of ground motion in the absence of the structure and foundation. This requires that the free-field earthquake ground motion is known either at a specific point (e.g., ground surface, rock-outcrop) or in the form of incident waves propagating upwards from a reference depth. Ground response analyses (Sections 7.5 and 7.6) can be performed to estimate the free-field motion along the planned soil-foundation interface. Strain-compatible soil properties (shear modulus, material damping) can be evaluated as part of these analyses.

The second step in the substructure approach is to evaluate the stiffness and damping characteristics of the foundation-soil interaction using either relatively simple impedance function models for rigid foundations (Figure 8.3a) or distributed springs and dashpots acting around the foundation (Figure 8.3b). The impedance function models are represented schematically by horizontal, vertical, and rocking springs in Figure 8.3a; in reality, these models include both stiffness and damping elements that are explained further in Section 8.3.2. The use of distributed springs and dashpots is needed for non-rigid foundations or if internal demands (e.g., moments, shears, displacements) within the foundation are required for their design. When distributed elements are used as in Figure 8.3b, there is no need for rocking springs, because the introduction of rotation to the foundation will deflect vertical and horizontal elements, thus producing resisting moment. Similarly, the presence of *coupling* between foundation degrees of freedom can be recognized at this stage – i.e., pure horizontal translation of the foundation deflects horizontal springs above the base slab level, and the associated horizontal forces generate moment relative to a reference elevation at the base slab level; similarly, if the foundation rotates without base translation, deflections of horizontal springs generates horizontal forces (shear). This coupling action only occurs for embedded foundations and is implicitly accounted for when distributed elements are used (Figure 8.3b); on the other hand, when impedance models are used (Figure 8.3a), accounting for coupling requires additional elements not depicted in the figure.

As shown in Figure 8.4, in the third and final stage of the substructure approach, the superstructure is placed atop the foundation and the system is excited through the foundation by imposing the rocking and translational components of the FIM on rigid elements (shown as stippled boxes) connected to the ends of the springs and dashpots. Note that in the case of the distributed spring/flexible foundation model (Figure 8.4b), different ground displacements over the height of the basement walls (i.e., depth of embedment) should be applied given the vertical variations of ground motion. This application of spatially variable displacements introduces a rotational component to the FIM, which is not depicted in Figure 8.4b. The superstructure and foundation masses are included in this analysis stage.

In summary, consideration of SSI effects in the assessment of the seismic response of a structure using the substructure approach begins with an evaluation of free-field soil motions and corresponding strain-compatible soil material properties and then entails: (1) evaluation of transfer functions to

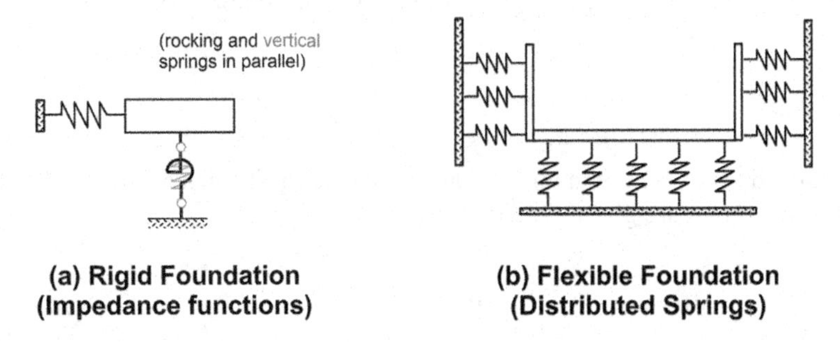

FIGURE 8.3 Schematic illustration of the foundation compliance step in the substructure approach using either (a) rigid foundation or (b) flexible foundation assumption. Note: dashpots are not shown.

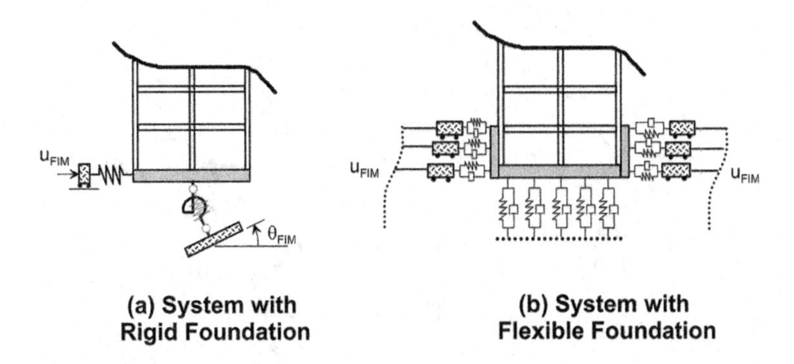

FIGURE 8.4 Schematic illustration of the final step in the substructure approach in which the superstructure is placed on the foundation on a flexible base using either (a) rigid foundation or (b) flexible foundation with distributed springs. The shading of foundation components indicates that their mass is considered at this stage of the analysis.

convert free-field motions to FIMs (and kinematic demands on embedded foundation components), (2) development of springs and dashpots (or more complex nonlinear elements) to represent the stiffness and damping at the foundation-soil interface, and (3) response analysis of the combined structure-spring/dashpot system to the FIM as input. The role of geotechnical earthquake engineers in SSI analyses is typically related to Parts (1 and 2) of this process, with Part (3) conducted jointly by structural and geotechnical earthquake engineers.

8.3 INERTIAL INTERACTION

Inertial interaction refers to displacements and rotations at the foundation level of a structure that result from inertia-driven structural forces such as base shear and moment. Inertial displacements and rotations of the foundation can be a significant source of flexibility and energy dissipation in the soil-structure system.

8.3.1 Soil-Structure System Behavior

A *rigid base* refers to a fictitious reference case of a supporting soil that cannot deform, which implies a foundation that cannot translate (horizontally or vertically) or rotate (rock or twist) relative to the free-field. Such a base would be modeled without springs, or with springs of infinite stiffness. A *rigid foundation* refers to foundation elements with infinite stiffness (i.e., not deformable).

Two structural base fixity conditions are of interest. A *fixed base* refers to rigid foundation elements on a rigid base. On the other hand, a *flexible base* condition accounts for the compliance of both the foundation elements and the soil. Consider a single degree-of-freedom structure with stiffness, k, and mass, m, resting on a fixed base, as depicted in Figure 8.5a. A static force, F, causes deflection, u:

$$u = \frac{F}{k} \tag{8.1}$$

From structural dynamics, the undamped natural vibration frequency, ω_0, and corresponding period, T_0, of the structure are given (Section B.5) by:

$$\omega_0 = \sqrt{\frac{k}{m}} \tag{8.2}$$

$$T_0 = \frac{2\pi}{\omega} = 2\pi\sqrt{\frac{m}{k}} \tag{8.3}$$

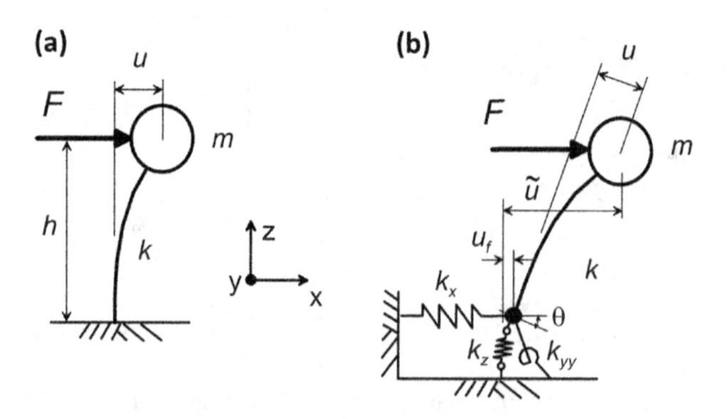

FIGURE 8.5 Schematic illustration of deflections caused by force applied to: (a) fixed-base structure; and (b) structure with vertical, horizontal, and rotational flexibility at its base. (Figure modified from NIST, 2012.)

Solving Equation (8.1) for k and substituting into Equation (8.3), an expression for the period is obtained as:

$$T_0 = 2\pi\sqrt{\frac{m}{(F/u)}} = 2\pi\sqrt{\frac{mu}{F}} \tag{8.4}$$

Now consider the same structure with vertical, horizontal, and rotational springs at its base, representing the effects of flexible (compliant) soil supporting a rigid foundation, as depicted in Figure 8.5b. The vertical spring stiffness in the z direction is denoted k_z, the horizontal spring stiffness (in the x-direction) is denoted k_x and the rotational spring (representing rotation in the x-z plane, i.e., about the yy-axis) is denoted k_{yy}. If a pseudo-static force, F, is applied to the mass in the x-direction, the structure deflects by $u=F/k$ (strictly $(F/k)\cos\theta$, but the cosine term is neglected because the angle θ is small) in the same manner as the fixed-base structure, but the base shear now also deflects the horizontal spring by $u_f=F/k_x$. Similarly, the base moment ($F \times h$) deflects the rotational spring by $\theta = Fh/k_{yy}$. Accordingly, the total deflection with respect to the free-field at the top of the structure, $\tilde{u}$ is:

$$\tilde{u} = \frac{F}{k} + u_f + \theta \cdot h \tag{8.5a}$$

$$\tilde{u} = \frac{F}{k} + \frac{F}{k_x} + \frac{Fh^2}{k_{yy}} \tag{8.5b}$$

If the total displacement in Equation (8.5b) is entered as the u term into Equation (8.4), the flexible base period, $\tilde{T}_0$ becomes

$$\tilde{T}_0 = 2\pi\sqrt{\frac{m\tilde{u}}{F}} = 2\pi\sqrt{m\left(\frac{1}{k} + \frac{1}{k_x} + \frac{h^2}{k_{yy}}\right)} \tag{8.6}$$

Combining expressions for the flexible base (Equation 8.6) and fixed-base (Equation 8.3) natural periods, the period ratio can be expressed as

$$\frac{\tilde{T}}{T} = \sqrt{\frac{k}{m}}\sqrt{m\left(\frac{1}{k} + \frac{1}{k_x} + \frac{h^2}{k_{yy}}\right)} \tag{8.7}$$

which simplifies into the classical period lengthening expression (Veletsos and Meek, 1974):

$$\frac{\tilde{T}}{T} = \sqrt{1 + \frac{k}{k_x} + \frac{kh^2}{k_{yy}}} \tag{8.8}$$

For consistency with SSI literature, the subscript "0" is dropped from the vibration periods in Equations (8.7) and (8.8). Note that $\tilde{T}/T$ goes to 1.0 as k_x and k_{yy} go to infinity (i.e., the fixed-base case), but is otherwise always greater than 1.0, indicating that the fundamental period of the soil-structure system will be greater than that of the fixed-base system. Equation (8.8) can be applied to multi-degree-of-freedom structures by taking the height, h, as the height of the center of mass for the first-mode shape (usually about two-thirds of the overall structure height). In such cases, period lengthening applies mainly to the first-mode period because higher modes produce relatively limited base shear and moment in most cases.

The dimensionless parameters controlling period lengthening of a rectangular structure with base dimensions $2B \times 2L$ are:

$$\frac{h}{V_s T_0}, \; \frac{h}{B}, \; \frac{B}{L}, \; \frac{m}{\rho_s 4BLh}, \text{ and } \nu$$

where h is the structure height (or height to the center of mass of the first-mode shape), ρ_s is the soil mass density, and ν is the Poisson's ratio of the soil. Note that it is common in SSI to denote B and L as the half-dimensions of rectangular foundations with $B \leq L$, which is different than the full dimensions commonly used in foundation engineering.

To the extent that h/T_0 quantifies the stiffness of the superstructure, the term $h/(V_s T_0)$ represents the *structure-to-soil stiffness ratio*. The term h/T_0 has units of velocity, and will be larger for stiff lateral force resisting systems, such as shear walls, and smaller for flexible structural systems, such as moment frames. The shear wave velocity, V_s, is closely related to soil shear modulus, G, being computed from $G = \rho_s V_s^2$ (Eq. 6.76). For typical building structures on soil and weathered rock sites, $h/(V_s T_0)$ is less than 0.1 for moment frame structures, and between approximately 0.1 and 0.5 for shear wall and braced frame structures (Stewart et al., 1999b). Period lengthening increases markedly with increasing structure-to-soil stiffness ratio, which is the most important parameter controlling inertial SSI effects.

The *structure aspect ratio*, h/B, and *foundation aspect ratio*, B/L describe the geometry of the soil-structure system. The *mass ratio*, $m/(\rho_s 4BLh)$, is the ratio of structure mass to the mass of soil in a volume extending to a depth equal to the structure height, h, below the foundation plan area. Equation (8.8) shows that period lengthening has no fundamental dependence on mass. The mass ratio term was introduced so that period lengthening could be related to easily recognizable characteristics such as structural first-mode period, T_0, and soil shear wave velocity, V_s, rather than structural stiffness, k, and soil shear modulus, G. The effect of mass ratio is modest, and it is commonly taken as 0.15 (Veletsos and Meek, 1974).

Using models for the stiffness of rectangular foundations (of half-width, B; half-length, L) resting on a homogeneous isotropic half-space with shear wave velocity, V_s, computed period lengthening ratios are shown for the special case of a square footing ($L = B$) in Figure 8.6a. All other factors being equal, period lengthening increases with the structure aspect ratio (i.e., increasing h/B), due to increased rocking caused by increased overturning moments. This implies that inertial SSI effects would be more significant in tall buildings, which is actually not the case. Tall buildings typically have low $h/(V_s T_0)$ ratios, which have a greater influence on inertial SSI effects. Hence, period lengthening in tall buildings is near unity. For a fixed structure aspect ratio, period lengthening decreases modestly with increasing foundation aspect ratio (i.e., increasing B/L) due to increased foundation size (and therefore stiffness) normal to the direction of loading.

In addition to period lengthening, system behavior is also affected by damping associated with soil-foundation interaction. The contribution of soil-foundation interaction to system damping is referred to as foundation damping, ξ_f. This damping has contributions from soil hysteresis (hysteretic damping) and radiation of energy, in the form of stress waves, away from the foundation (radiation damping). The flexible-base system damping, ξ_0 is related to ξ_f and fixed-base structural damping ξ_i as:

$$\xi_0 = \xi_f + \frac{1}{\left(\tilde{T}/T\right)^n} \xi_i \tag{8.9}$$

The fixed base structural damping depends on structural system type and configuration (ATC, 2010), but is often taken by default as 5%. Observations from case studies have shown that ξ_f ranges from approximately 0% to 25% (Stewart et al., 1999b). The exponent, n, on the period lengthening term in Equation (8.9) is 3 for ideally viscous material damping and is 2 otherwise (Givens et al., 2016). Use of $n = 2$ is recommended.

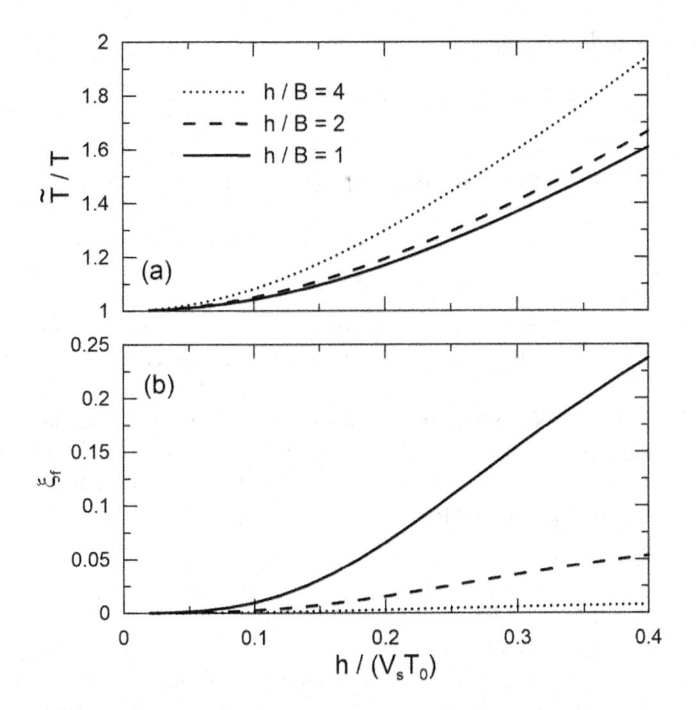

FIGURE 8.6 Plot of period lengthening ratio ($\tilde{T}/T$) and foundation damping (ξ_f) versus structure-to-soil-stiffness ratio for square foundations ($L=B$) and varying structure aspect ratios, h/B. In this plot, $\nu=0.33$, $B/L=1.0$, hysteretic soil damping $\xi_s=0$, mass ratio$=0.15$, and exponent $n=2$ (see Equations 8.9 and 8.10). (Figure modified from NIST, 2012.)

Example 8.1

Consider the simple structure in Figure E8.1 on a square foundation, $L/B = 1.0$. Find the fixed-base period (T_0), flexible-base period ($\tilde{T}$), and flexible-base damping ratio (ξ_0).

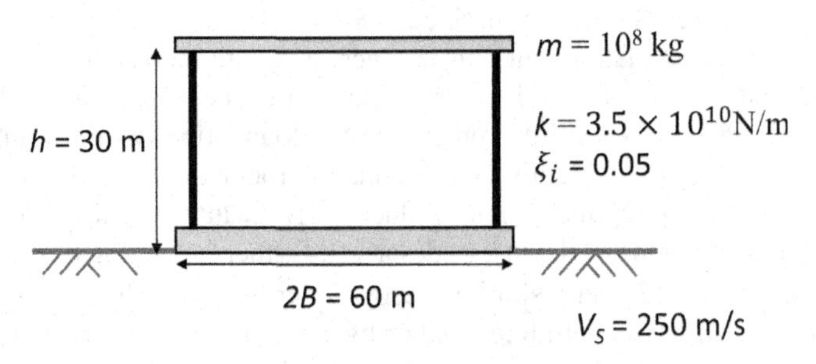

FIGURE E8.1 Soil-structure system considered in Example 8.1.

Solution:

The fixed base period is computed from Equation (8.3) as $T_0 = 2\pi\sqrt{m/k}$=0.34 sec.

Period lengthening can be evaluated from Figure 8.6 if the structure-to-soil stiffness ratio and structure aspect ratio are known. The structure-to-soil stiffness ratio is $h/(V_sT_0)$=0.35. The structure aspect ratio is $h/B = 1.0$. From Figure 8.6, period lengthening $\tilde{T}/T$ is found to be 1.5 (by convention, the "0" subscript is dropped from the periods in $\tilde{T}/T$). The foundation damping $\xi_f = 0.2$.

The flexible-base period can then be computed as

$$T_0 \times \frac{\tilde{T}}{T} = 0.51 \text{ sec.}$$

The flexible-base damping is computed using Equation (8.9) as

$$\xi_0 = \xi_f + \frac{\xi_i}{\left(\tilde{T}/T\right)^2} = 0.22.$$

In this example, the period is appreciably lengthened and the damping increases by more than a factor of four. Most building structures do not have such large effects of inertial interaction.

Givens et al. (2016) derived analytical models for foundation damping, while also explaining differences between foundation damping equations presented in prior studies (Veletsos and Nair, 1975; Bielak, 1975; Roesset, 1980; Wolf, 1985; Maravas et al., 2014). A suitable solution for the foundation damping ratio, ξ_f, for most applications is given by:

$$\xi_f = \frac{1}{\left(\tilde{T}/\left|T_x^*\right|\right)^n}\left(\xi_x + \xi_s\right) + \frac{1}{\left(\tilde{T}/\left|T_{yy}^*\right|\right)^n}\left(\xi_{yy} + \xi_s\right) \tag{8.10}$$

where ξ_s is soil hysteretic damping ratio (Section 6.6.4 and Table 8.1), ξ_x and ξ_{yy} are damping ratios related to radiation damping from translational and rotational modes (described further in Section 8.3.2), and n should generally be taken as 2. The periods T_x^* and T_{yy}^* are complex-valued fictitious vibration periods for foundation vibration (their amplitude, as used in Equation (8.10), would represent actual system period if the superstructure were rigid and the respective foundation vibrations were the only available degrees of freedom of a fictitious SDOF system):

$$\left|T_x^*\right| = 2\pi\sqrt{\frac{m}{\left|k_x^*\right|}} \qquad \left|T_{yy}^*\right| = 2\pi\sqrt{\frac{mh^2}{\left|k_{yy}^*\right|}} \tag{8.11}$$

In Equation (8.11), $\left|k_j^*\right|$ (where j refers to the x or yy indices) is the amplitude of a complex-valued impedance function, as described further in Section 8.3.2.

Figure 8.6b shows that foundation damping, ξ_f, increases with increasing structure-to-soil-stiffness ratio, $h/(V_sT_0)$ and decreases with increasing structure aspect ratio, h/B. The decrease of ξ_f with aspect ratio indicates that lateral movements of the foundation (which dominate at low h/B) dissipate energy into soil more efficiently than foundation rocking (which dominates at high h/B). The radiation damping terms, ξ_x and ξ_{yy}, are reduced significantly when a stiff bedrock layer is encountered at moderate or shallow depths, as described further in Section 8.3.2.2.

Analysis procedures for $\tilde{T}/T$ and ξ_f similar to those described above have been validated relative to observations from instrumented buildings shaken by earthquakes (Stewart et al., 1999a, b). These case studies confirm the analytical finding that the single most important parameter controlling the significance of inertial interaction is $h/(V_sT_0)$, and that inertial SSI effects are generally negligible for $h/(V_sT_0)<0.1$, which occurs in flexible structures, such as moment frame buildings, located on competent soil or rock. Conversely, inertial SSI effects tend to be significant for stiff structures, such as shear wall or braced frame buildings, located on softer soils. Many of the early advances in understanding and analysis of SSI, in fact, came from the nuclear power industry where stiff structures must be designed for extremely high performance standards.

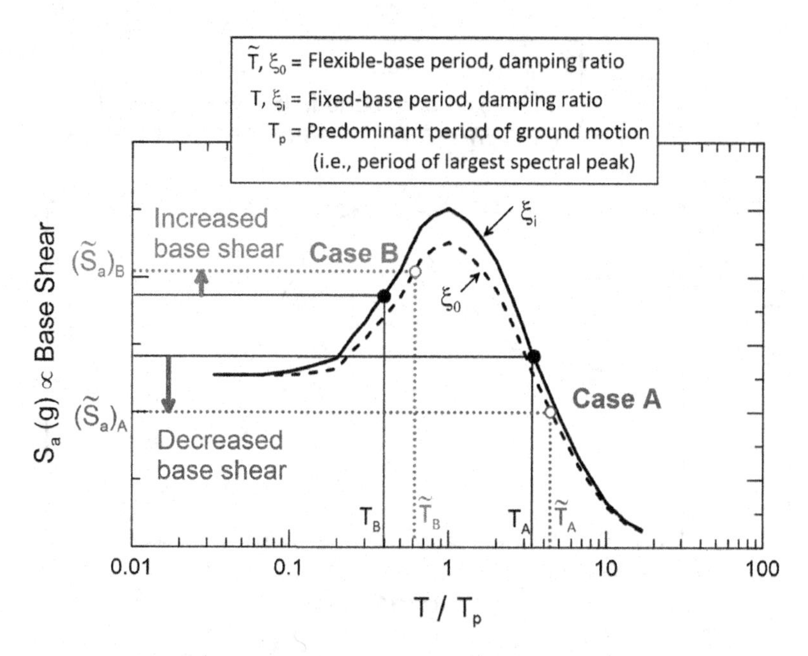

FIGURE 8.7 Illustration of effects of period lengthening and change of damping on fundamental mode spectral acceleration from which base shear is evaluated. Solid curve represents spectrum for fixed-base damping ratio. Dashed curve represents spectrum at increased damping ratio associated with a flexible-base case.

The effect of inertial SSI on the peak base shear that develops in a structure is illustrated in Figure 8.7. Because base shear for elastic response is commonly computed based on first-mode spectral acceleration, the figure depicts the variation in spectral acceleration versus normalized period on a log scale. The spectra are drawn for two effective damping ratios corresponding to flexible-base (ξ_0) and fixed-base (ξ_i) conditions. The spectral acceleration for a flexible-base structure, $\tilde{S}_a$, is obtained by entering the spectrum (dashed curve) drawn for damping ratio ξ_0 at the corresponding elongated period, $\tilde{T}$. The fixed-base counterpart is obtained using the spectral acceleration for the fixed-base period (indicated without the ~ symbol) and the fixed-base damping. The effect of SSI on base shear is influenced by the slope of the spectrum. Base shear tends to increase when the slope is positive and decrease when the slope is negative. Case *A* represents buildings with relatively long periods (e.g., $\tilde{T}_A$ in Figure 8.7) on the descending portion of the spectrum; use of $\left(\tilde{S}_a\right)_A$ (flexible base ordinate) in lieu of the fixed base ordinate typically results in reduced base shear demand. Case *B* represents a short-period structure ($\tilde{T}_B$ in Figure 8.7) on the ascending branch of the spectrum where, despite the effects of increased damping, period lengthening causes inertial SSI to increase the base shear.

Example 8.2

The structure described in Example 8.1 is subject to design ground motions with the response spectra shown in Figure E8.2. Note that the spectra are provided for two damping ratios corresponding to fixed- and flexible-base conditions. What is the change in the design spectral acceleration that would be achieved by considering inertial soil–structure interaction (SSI)?

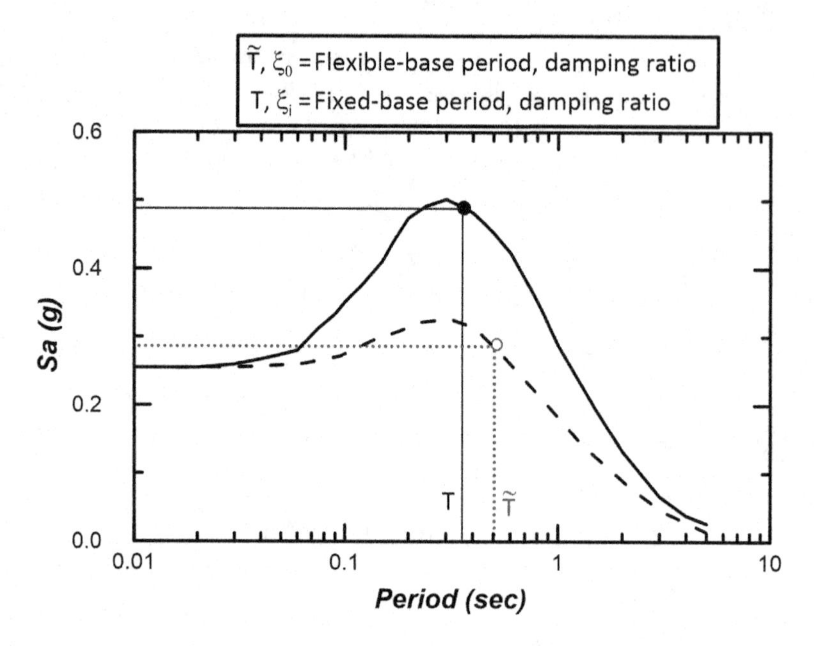

FIGURE E8.2 Effect of period lengthening and foundation damping on the first-mode spectral acceleration.

Solution:

From Example 8.1, the fixed- and flexible-base natural periods of the system were 0.34 and 0.51 sec, respectively. As shown by the solid dot in the plot, the fixed-based spectral acceleration is 0.49*g* for the fixed-base case, in which the damping ratio is 5%. For the flexible-base case, the spectrum is drawn at a higher damping ratio of 22% and the spectral acceleration at 0.51 sec (open circle) is 0.29*g*. Hence, the design ground motions are decreased by about 40%, with most of the reduction resulting from the damping increase.

8.3.2 Shallow Foundations

The stiffness and damping associated with foundation interaction with supporting soils is a fundamental consideration in SSI analysis. This sections describes the factors controlling this interaction and presents models that are used in applications.

8.3.2.1 Stiffness and Damping for Rigid Foundations and Uniform Soils

The notion of a complex stiffness that simultaneously represented the stiffness and damping characteristics of an SDOF oscillator is introduced in Section B.6.3. In analogous fashion, complex *impedance functions* can be used to represent the frequency-dependent stiffness and damping characteristics of soil-foundation interaction. Classical solutions for the complex-valued impedance function can be written as (Luco and Westmann, 1971; Veletsos and Wei, 1971):

$$k_j^* = k_j + i\omega c_j \tag{8.12}$$

where k_j^* denotes the complex-valued impedance function for a particular mode of foundation vibration indicated by index j (e.g., x or yy), and k_j and c_j denote the frequency-dependent foundation stiffness and dashpot coefficients, respectively. The dashpot with coefficient c represents the effects of damping associated with soil-foundation interaction. An alternative form for Equation (8.12) is

$$k_j^* = k_j\left(1 + 2i\xi_j\right) \tag{8.13}$$

where:

$$\xi_j = \frac{\omega c_j}{2k_j} \tag{8.14}$$

which is defined for $k_j>0$. An advantage of using ξ_j over c_j is that ξ_j can be interpreted as a fraction of critical damping (damping ratio) in the classical sense when a mass is attached to the complex spring and the system oscillates at resonance (Clough and Penzien, 1993).

The imaginary part of the complex impedance is related to phase difference between harmonic excitation and response at a given frequency, a feature that is usually associated with damping (Section B.6.3). The phase angle, ϕ_j, between force and (lagged) displacement is (Clough and Penzien, 1993; Wolf, 1985):

$$\phi_j = \tan^{-1}\left(2\xi_j\right) \tag{8.15}$$

The angle ϕ_j is also known as a *loss angle*. For example, if ξ_j is 10%, peak harmonic displacement will lag peak force by 11.3° or 0.197 radians. The corresponding time shift between force and displacement depends on oscillation frequency as $\Delta t = \phi_j \omega$.

To help conceptualize the physical meaning of the complex-valued and frequency-dependent foundation stiffness, it is useful to consider characteristics of foundation load-deformation responses as measured in field tests. Figure 8.8a shows the configuration of a simple test structure resting on soft soil, which was subjected to cyclic forces over a range of frequencies by a shaker mounted on the top slab. Figure 8.8b shows the relationship between foundation rotation and base moment, which can be evaluated for any point in time as the difference between the moment applied by the shaker force and the moments that develop from inertial forces in top and bottom slabs. The "loops" in this figure show the evolution of foundation rotation with base moment over several cycles of shaking; hence time is advancing as one moves along the loops in a clockwise fashion. Figure 8.8c of the figure shows a similar result for the base shear-displacement response. Careful interpretation of the rounded tips of the loops in both figures show that the peak load (on *y*-axis) occurs before the peak deformation (on *x*-axis) – this offset in time between load and response is a result of damping, and can be assessed either as a time- or frequency-shift per Equation (8.15). Complex numbers are used in the formulation of the stiffness to capture this damping-induced phase shift. With regard to frequency dependence, the loops in Figure 8.8 demonstrate a decrease of secant stiffness as frequency increases from 6 to 7.5 Hz as well as an increase in the relative "fatness" of the loops (which is related to damping; Section 6.6.4). Models for the k_j and ξ_j terms, described below, are intended to capture these features of foundation response.

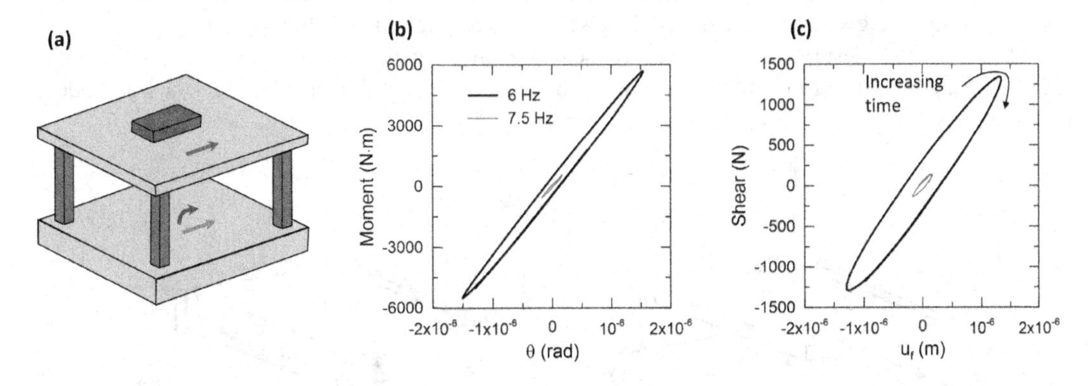

FIGURE 8.8 (a) Configuration of test structure used for cyclic forced vibration testing of SSI effects (box on roof slab represents shaker, arrows represent measured displacements and rotations); (b) base moment – foundation rotation response at two frequencies (6 Hz is near the system resonance); (c) base shear – foundation displacement response at same frequencies as in (b). (Adapted from data in Tileylioglu et al., 2011.)

Example 8.3

Derive the foundation stiffnesses in translation and rotation in response to 6 Hz harmonic loading from the recorded responses of the test structure foundation in Figure 8.8.

Solution:

The stiffnesses can be measured from the slopes of lines drawn through the tips of the force-displacement and moment-rotation loops, as shown in Figure E8.3.

For the translation case, this slope is $k_x(\omega) = \dfrac{3\text{kN}}{3.0\times10^{-6}\text{m}} = 1.0\times10^6\text{kN/m}$

For the rotation case, this slope is $k_{yy}(\omega) = \dfrac{12\text{kN-m}}{3.3\times10^{-6}\text{rad}} = 3.6\times10^6\text{kN-m/rad}$

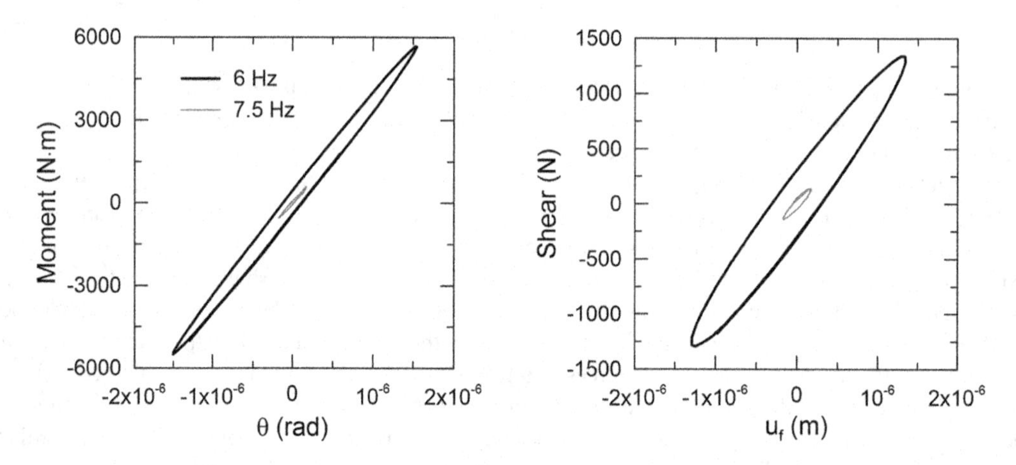

FIGURE E8.3 Secant stiffness measurement for Example 8.3.

Many impedance function solutions are available for rigid circular or rectangular foundations located on the surface of, or embedded within, a uniform, elastic, or visco-elastic half-space. In the case of a rigid rectangular foundation resting on the surface of a half-space with shear wave velocity V_s, Pais and Kausel (1988), Gazetas (1991), and Mylonakis et al. (2006) summarize impedance solutions in the literature and present equations for the stiffness and damping terms in Equation (8.12). Solutions for the 2-D case, in which excitation is in the y-direction (parallel to the "short" dimension of the footing) are given by Gazetas and Roesset (1976) and Jakub and Roesset (1977).

These solutions describe translational stiffness and damping along axes x, y, and z, and rotational stiffness and damping around those axes (denoted xx, yy, and zz) as shown in Figure 8.9a. For mode j,

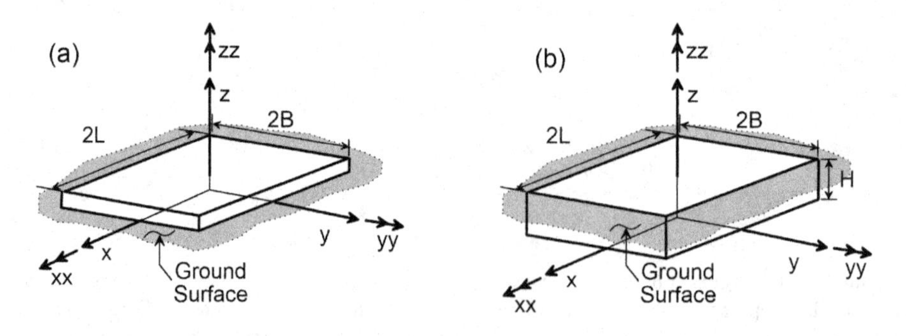

FIGURE 8.9 Geometry and axis orientations for rectangular foundations (a) at the ground surface and (b) embedded to depth H. (Adapted from NIST, 2012.)

the stiffness, k_j, is represented as the product of the static (zero frequency) stiffness of the footing resting on the surface, K_j, a dynamic stiffness modifier, α_j, and an embedment modifier, η_j:

$$k_j = K_j \alpha_j \eta_j \tag{8.16}$$

where $K_j = GB^m f(B/L, \nu)$, $\alpha_j = f(B/L, a_0)$, and $\eta_j = f(B/L, H/B)$; each of these functions are provided below in Tables 8.2 and 8.3a. In these functions, parameters G and ν are the shear modulus and Poisson's ratio of the soil, $m=1$ for translation of a rectangular foundation for the general 3D case ($m=0$ for the 2D case corresponding to excitation in y-direction for long footing, i.e., $L \to \infty$), and $m=3$ for rotation of a rectangular foundation for the general 3D case ($m=2$ for the 2D case). The shear modulus, G, is evaluated using an equivalent-linear approach and as such should reflect the effects of modulus reduction with shear strain amplitude. Approximate adjustment factors for different site classes at different *PGA* levels are shown in Table 8.1.

The maximum (or small strain) shear modulus, G_{max}, is calculated from shear wave velocity (V_s) as $G_{max} = \rho_s V_s^2$ (Eq. 6.76). An average value of V_s is generally computed across an effective profile depth, z_p, as described in Section 8.3.2.2.

Dynamic stiffness modifiers, α_j, are related to dimensionless frequency a_0:

$$a_0 = \frac{\omega B}{V_s} \tag{8.17}$$

which can be physically interpreted as the ratio of B to approximately one-sixth (actually $1/2\pi$) of the seismic wavelength for frequency ω. For time domain analysis, a single frequency must be selected to evaluate a_0-dependent foundation spring and dashpot coefficients; this can be taken

TABLE 8.1
Values of Shear Modulus Reduction and Hysteretic Soil Damping for Various NEHRP Site Classes (Section 7.3.3.1) and Shaking Amplitudes

	V_s Reduction Factor $=\sqrt{G/G_{max}}$			Value of G/G_{max}			Value of ξ_s		
	PGA (g)			PGA (g)			PGA (g)		
Site Class	**≤0.1**	**0.4**	**≥0.8**	**≤0.1**	**0.4**	**≥0.8**	**≤0.1**	**0.4**	**≥0.8**
A	1.00	1.00	1.00	1.00	1.00	1.00	na	na	na
B	1.00	0.97	0.95	1.00	0.95	0.90	na	na	na
BC	0.98	0.92	0.86	0.97	0.84	0.73	na	na	na
C	0.97	0.87	0.77	0.95	0.75	0.60	0.01	0.03	0.05
CD	0.96	0.79	0.50	0.92	0.62	0.25	0.01	0.05	0.09
D	0.95	0.71	0.32	0.90	0.50	0.10	0.02	0.07	0.15
DE	0.86	0.40	0.18	0.73	0.16	0.03	0.03	0.12	0.20
E	0.77	0.22	*	0.60	0.05	*	0.05	0.20	*
F	*	*	*	*	*	*	*	*	*

Source: Table modified from NIST (2012) and BSSC (2020a).
PGA denotes the peak horizontal acceleration for the free-field, ground surface conditions.
Note: Use straight line interpolation for intermediate values of PGA.
*, should be evaluated from site-specific analysis; na, values not specified (use of foundation damping not recommended).

as the frequency anticipated to dominate the structural response. In most cases, this will be the first-mode flexible-base frequency.

Table 8.2 provides expressions for static foundation stiffness, K_j, and embedment factors, η_j, for three translational and three rotational degrees of freedom for rigid rectangular footings (Pais and Kausel, 1988; NIST, 2012). Similar equations were provided by Gazetas (1991) and Mylonakis et al. (2006). Due to the additional normal and shear stresses that develop on the sides of the foundation, embedment increases static foundation stiffness and as a result the embedment factors in Table 8.2 only increase the static stiffness. Solutions for the 2D case are also given in Table 8.2 for one translational degree of freedom (along y-axis) and one rotational degree of freedom (xx) (Gazetas and Roesset, 1976; Jakub and Roesset, 1977).

Equations for dynamic stiffness modifiers, α_j, and radiation damping ratios, ξ_j, for rigid footings located at the ground surface are provided in Table 8.3a. The frequency dependence of these quantities is an outcome of representing the soil mass supporting the footing with discrete spring and dashpot elements. Due to its distributed mass, the medium actually has an infinite number of degrees of freedom, each with mass and associated dynamic effects. The 'condensation' of the dynamic response of that medium to discrete elements requires that their properties be frequency-dependent (i.e., to incorporate wave propagation phenomena into spring and dashpot values). This frequency dependence would disappear for massless soil, as a_0 would become zero (because $V_s = \sqrt{G_{\max}/\rho_s} \rightarrow \infty$ as $\rho_s \rightarrow 0$ in Equation 8.17), causing $\alpha_j = 1$, and $\xi_j = 0$.

Radiation damping ratios for embedded footings are provided in Table 8.3b. Dynamic stiffness modifiers are insensitive to embedment, so values for embedded foundations are the same as values given in Table 8.3a for footings located at the ground surface (i.e., $\alpha_{j,emb} = \alpha_{j,sur}$).

Figure 8.10 shows the variation of dynamic stiffness modifiers with dimensionless frequency for rigid footings located at the ground surface. In the case of translational stiffness, dynamic stiffness modifiers (α_x, α_y) are essentially unity, regardless of frequency or foundation aspect ratio. For rotational stiffness, however, dynamic stiffness modifiers for rocking (α_{xx}, α_{yy}) degrade markedly with frequency, but are relatively insensitive to aspect ratio. Figure 8.10 also shows the variation of radiation damping vs frequency for translation (ξ_x, ξ_y) and rotation (ξ_{xx}, ξ_{yy}). Translational radiation damping is only modestly affected by the direction of shaking or the aspect ratio of the foundation. The modest increase of translational damping with aspect ratio is a result of the increased foundation size (i.e., the foundation becomes larger as aspect ratio increases).

In contrast, rotational radiation damping is strongly sensitive to the direction of shaking and the aspect ratio of the foundation. Rotational damping is largely controlled by 180-degree out-of-phase vertical cyclic displacements at the edges of the foundation (assuming no separation between soil and footing). The levels of radiation damping in rotation can be influenced by destructive interference of waves emanating from the ends of the rotating foundation (in effect, the phase shift can cause the waves to "cancel out" each other, reducing radiation damping). This effect is strong for small aspect ratios, producing low levels of rotational radiation damping. As aspect ratio increases, the ends of the foundation are located further apart, and energy radiating into the soil from each end of the foundation experiences less destructive interference, thus increasing damping. At low frequencies ($a_0 < 1$ to 2), damping from rotation is generally smaller than damping from translation, although the trend reverses as frequency increases and foundations become relatively oblong. The practical significance of this effect is that translational deformation modes in the foundation, while often relatively unimportant from the perspective of overall structural system flexibility, can be the dominant source of foundation damping. When used to calculate the dashpot coefficient, c_j (Equation 8.14), the ξ_j term should be taken as the sum of radiation damping for the appropriate vibration mode (Table 8.3) and hysteretic damping (ξ_s) (from ground response analysis or Table 8.1).

TABLE 8.2
Elastic Solutions for Static Stiffness of Rigid Footings at the Ground Surface and Embedded to depth *H*

Degree of Freedom	Surface Solution		Embedment Modifier	
Translation along *z*-axis	$K_{z,\,sur} = \dfrac{GB}{1-\nu}\left[3.1\left(\dfrac{L}{B}\right)^{0.75} + 1.6\right]$	(T8.2.1-1)	$\eta_z = \left[1.0 + \left(0.25 + \dfrac{0.25}{L/B}\right)\left(\dfrac{H}{B}\right)^{0.8}\right]$	(T8.2.1-2)
Translation along *y*-axis	$K_{y,\,sur} = \dfrac{GB}{2-\nu}\left[6.8\left(\dfrac{L}{B}\right)^{0.65} + 0.8\left(\dfrac{L}{B}\right) + 1.6\right]$	(T8.2.2-1)	$\eta_y = \left[1.0 + \left(0.33 + \dfrac{1.34}{1+L/B}\right)\left(\dfrac{H}{B}\right)^{0.8}\right]$	(T8.2.2-2)
Translation along *y*-axis: 2D case	$K_{y,\,sur} = \dfrac{2.1G}{2-\nu}\left[1 + 2\dfrac{B}{z_{soil}}\right]$	(T8.2.3-1)	$\eta_y = \left[1 + \dfrac{1}{3}\dfrac{H}{B}\right]\left[1 + \dfrac{4}{3}\dfrac{H}{z_{soil}}\right]$	(T8.2.3-2)
Translation along *x*-axis	$K_{x,\,sur} = \dfrac{GB}{2-\nu}\left[6.8\left(\dfrac{L}{B}\right)^{0.65} + 2.4\right]$	(T8.2.4-1)	$\eta_x \approx \eta_y$	(T8.2.4-2)
Torsion about *z*-axis	$K_{zz,\,sur} = GB^3\left[4.25\left(\dfrac{L}{B}\right)^{2.45} + 4.06\right]$	(T8.2.5-1)	$\eta_{zz} = \left[1 + \left(1.3 + \dfrac{1.32}{L/B}\right)\left(\dfrac{H}{B}\right)^{0.9}\right]$	(T8.2.5-2)
Rocking about *y*-axis	$K_{yy,\,sur} = \dfrac{GB^3}{1-\nu}\left[3.73\left(\dfrac{L}{B}\right)^{2.4} + 0.27\right]$	(T8.2.6-1)	$\eta_{yy} = \left[1.0 + \dfrac{H}{B} + \left(\dfrac{1.6}{0.35 + (L/B)^4}\right)\left(\dfrac{H}{B}\right)^2\right]$	(T8.2.6-2)
Rocking about *x*-axis	$K_{xx,\,sur} = \dfrac{GB^3}{1-\nu}\left[3.2\left(\dfrac{L}{B}\right) + 0.8\right]$	(T8.2.7-1)	$\eta_{xx} = \left[1.0 + \dfrac{H}{B} + \left(\dfrac{1.6}{0.35 + L/B}\right)\left(\dfrac{H}{B}\right)^2\right]$	(T8.2.7-2)
Rocking about *x*-axis: 2D case	$K_{xx,\,sur} = \dfrac{\pi GB^2}{2(1-\nu)}\left[1 + \dfrac{1}{5}\dfrac{B}{z_{soil}}\right]$	(T8.2.8-1)	$\eta_{xx} = \left[1 + \dfrac{H}{B}\right]\left[1 + \dfrac{2}{3}\dfrac{H}{z_{soil}}\right]$	(T8.2.8-2)

Source: Solutions for rectangular foundations adapted from Pais and Kausel (1988), as given by NIST (2012). Solutions for 2D case apply for shaking in *y*-direction and are adapted from Gazetas and Roesset (1976) and Jakub and Roesset (1977).

Notes: Axes should be oriented such that $L \geq B$. G = shear modulus (reduced for large strain problems, e.g., Table 8.1).

TABLE 8.3A
Dynamic Stiffness Modifiers and Radiation Damping Ratios for Rigid Surface Foundations (Pais and Kausel 1988, NIST 2012)

Degree of Freedom	Surface Stiffness Modifiers		Radiation Damping	
Translation along z-axis	$\alpha_z = 1.0 - \left[\dfrac{\left(0.4 + \dfrac{0.2}{L/B}\right)a_0^2}{\left(\dfrac{10}{1+3(L/B-1)}\right) + a_0^2}\right]$	(T8.3.1-1)	$\xi_z = \left[\dfrac{4\psi(L/B)}{(K_{z,sur}/GB)}\right]\left[\dfrac{a_0}{2\alpha_z}\right]$	(T8.3.1-2)
Translation along y-axis	$\alpha_y = 1.0$	(T8.3.2-1)	$\xi_y = \left[\dfrac{4(L/B)}{(K_{y,sur}/GB)}\right]\left[\dfrac{a_0}{2\alpha_y}\right]$	(T8.3.2-2)
Translation along x-axis	$\alpha_x = 1.0$	(T8.3.3-1)	$\xi_x = \left[\dfrac{4(L/B)}{(K_{x,sur}/GB)}\right]\left[\dfrac{a_0}{2\alpha_x}\right]$	(T8.3.3-2)
Torsion about z-axis	$\alpha_{zz} = 1.0 - \left[\dfrac{\left(0.33 - 0.03\sqrt{L/B-1}\right)a_0^2}{\left(\dfrac{0.8}{1+0.33(L/B-1)}\right) + a_0^2}\right]$	(T8.3.4-1)	$\xi_{zz} = \left[\dfrac{(4/3)\left[(L/B)^3 + (L/B)\right]a_0^2}{\left(\dfrac{K_{zz,sur}}{GB^3}\right)\left[\left(\dfrac{1.4}{1+3(L/B-1)^{0.7}}\right) + a_0^2\right]}\right]\left[\dfrac{a_0}{2\alpha_{zz}}\right]$	(T8.3.4-2)
Rocking about y-axis	$\alpha_{yy} = 1.0 - \left[\dfrac{0.55a_0^2}{\left(0.6 + \dfrac{1.4}{(L/B)^3}\right) + a_0^2}\right]$	(T8.3.5-1)	$\xi_{yy} = \left[\dfrac{(4\psi/3)(L/B)^3 a_0^2}{\left(\dfrac{K_{yy,sur}}{GB^3}\right)\left[\left(\dfrac{1.8}{1+1.75(L/B-1)}\right) + a_0^2\right]}\right]\left[\dfrac{a_0}{2\alpha_{yy}}\right]$	(T8.3.5-2)
Rocking about x-axis	$\alpha_{xx} = 1.0 - \left[\dfrac{\left(0.55 + 0.01\sqrt{L/B-1}\right)a_0^2}{\left(2.4 - \dfrac{0.4}{(L/B)^3}\right) + a_0^2}\right]$	(T8.3.6-1)	$\xi_{xx} = \left[\dfrac{(4\psi/3)(L/B)a_0^2}{\left(\dfrac{K_{xx,sur}}{GB^3}\right)\left[\left(2.2 - \dfrac{0.4}{(L/B)^3}\right) + a_0^2\right]}\right]\left[\dfrac{a_0}{2\alpha_{xx}}\right]$	(T8.3.6-2)

Notes: Orient axes such that $L \geq B$. Hysteretic damping (ξ_s) is additive to radiation damping.
$a_o = \omega B/V_s$; $\psi = \sqrt{2(1-\nu)/(1-2\nu)}$, $\psi \leq 2.5$

TABLE 8.3B

Radiation Damping Ratios for Embedded Rigid Foundations (Pais and Kausel 1988, NIST 2012)

Degree of Freedom	Radiation Damping	
Translation along z-axis	$\xi_z = \left[\dfrac{4\left[\psi(L/B)+(H/B)(1+L/B)\right]}{(K_{z,emb}/GB)}\right]\left[\dfrac{a_0}{2\alpha_z}\right]$	(T8.3.1-3)
Translation along y-axis	$\xi_y = \left[\dfrac{4\left[L/B+(H/B)(1+\psi L/B)\right]}{(K_{y,emb}/GB)}\right]\left[\dfrac{a_0}{2\alpha_y}\right]$	(T8.3.2-3)
Translation along x-axis	$\xi_x = \left[\dfrac{4\left[L/B+(H/B)(\psi+L/B)\right]}{(K_{x,emb}/GB)}\right]\left[\dfrac{a_0}{2\alpha_x}\right]$	(T8.3.3-3)
Torsion about z-axis	$\xi_{zz} = \left[\dfrac{(4/3)\left[3(L/B)(H/B)+\psi(L/B)^3(H/B)+3(L/B)^2(H/B)+\psi(H/B)+(L/B)^3+(L/B)\right]a_0^2}{\left(\dfrac{K_{zz,emb}}{GB^3}\right)\left[f_{zz}+a_0^2\right]}\right]\left[\dfrac{a_0}{2\alpha_{zz}}\right]$	(T8.3.4-3)
Rocking about y-axis	$\xi_{yy} = \left[\dfrac{(4/3)\left[\left(\dfrac{L}{B}\right)^3\left(\dfrac{H}{B}\right)+\psi\left(\dfrac{H}{B}\right)^3\left(\dfrac{L}{B}\right)+\left(\dfrac{H}{B}\right)^3+3\left(\dfrac{H}{B}\right)\left(\dfrac{L}{B}\right)^2+\psi\left(\dfrac{L}{B}\right)^3\right]a_0^2}{\left(\dfrac{K_{yy,emb}}{GB^3}\right)\left[f_{yy}+a_0^2\right]}+\dfrac{\left(\dfrac{4}{3}\right)\left(\dfrac{L}{B}+\psi\right)\left(\dfrac{H}{B}\right)^3}{\left(\dfrac{K_{yy,emb}}{GB^3}\right)}\dfrac{f_{yy}}{f_{yy}+a_0^2}\right]\left[\dfrac{a_0}{2\alpha_{yy}}\right]$	(T8.3.5-3)
Rocking about x-axis	$\xi_{xx} = \left[\dfrac{(4/3)\left[\left(\dfrac{H}{B}\right)+\left(\dfrac{H}{B}\right)^3+\psi\left(\dfrac{L}{B}\right)\left(\dfrac{H}{B}\right)^3+3\left(\dfrac{H}{B}\right)\left(\dfrac{L}{B}\right)+\psi\left(\dfrac{L}{B}\right)\right]a_0^2}{\left(\dfrac{K_{xx,emb}}{GB^3}\right)\left[f_{xx}+a_0^2\right]}+\dfrac{\left(\dfrac{4}{3}\right)\left(\psi\dfrac{L}{B}+1\right)\left(\dfrac{H}{B}\right)^3}{\left(\dfrac{K_{xx,emb}}{GB^3}\right)}\dfrac{f_{xx}}{f_{xx}+a_0^2}\right]\left[\dfrac{a_0}{2\alpha_{xx}}\right]$	(T8.3.6-3)

Notes: Hysteretic damping (ξ_s) is additive to radiation damping.

$\alpha_{emb} = \alpha_{sur}$; from Table 8.3a

$a_o = \omega B/V_s$; $\psi = \sqrt{2(1-\nu)/(1-2\nu)}$, $\psi \le 2.5$; $f_{zz} = \dfrac{1.4}{1+3(L/B-1)^{0.7}}$; $f_{yy} = \dfrac{1.8}{1+1.75(L/B-1)}$; $f_{xx} = 2.2 - \dfrac{0.4}{(L/B)^3}$

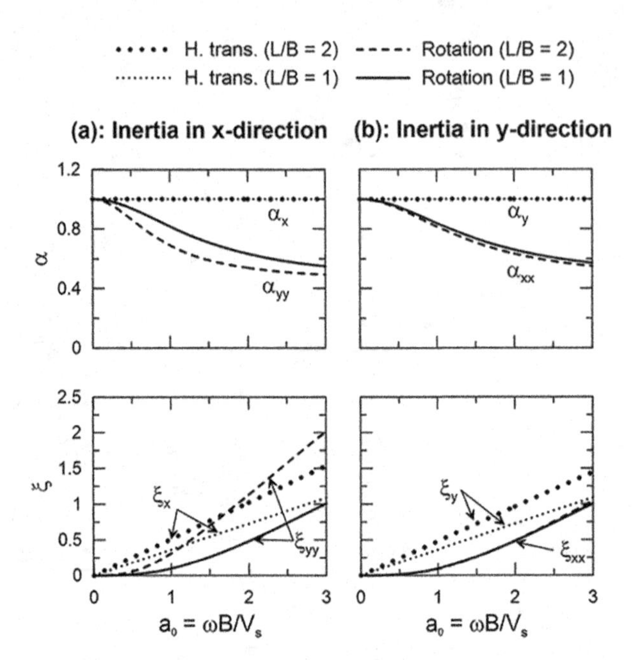

FIGURE 8.10 Plot of dynamic stiffness modifiers and damping ratios versus dimensionless frequency, for rectangular footings supported on the surface of a homogeneous half-space, with zero hysteretic damping, and ν = 0.33: (a) x-direction (long dimension of footing); and (b) y-direction (short dimension). (Figure from NIST, 2012.)

Example 8.4

The foundation for which the responses were measured in Example 8.3 has plan dimensions of 4.06 × 4.06 m. The site at which these measurements were made has an average velocity of 198 m/s and mass density of 1,800 kg/m³ over the applicable depth range for impedance calculations. Calculate the predicted foundation stiffnesses for the test structure for linear conditions using the relations in Table 8.2.

Solution:

For foundation stiffness calculations, it is best to start with the static stiffness of the foundations and then to make the dynamic adjustment per Equation (8.16).

Static stiffnesses are computed using expressions of the form $K_j = GB^m f(B/L, v)$, with equations for specific vibration modes j provided in Table 8.2.

For the case of horizontal translation, the loading direction is taken in the x-direction (because the foundation is square, it could have equivalently been taken as the y-direction). From Table 8.2, and using $G = V_s^2 \rho_s$ and taking $v = 0.3$ (surficial materials are sandy),

$$K_{x,\,sur} = \frac{GB}{2-v}\left[6.8\left(\frac{L}{B}\right)^{0.65}+2.4\right] = \frac{(1800\text{ kg/m}^3)(198\text{ m/sec})^2(2.03\text{ m})}{2-0.3}\left[6.8\left(\frac{2.03\text{ m}}{2.03\text{ m}}\right)^{0.65}+2.4\right]$$

$$= 7.8\times10^5\text{ kN/m}$$

For the case of rotation caused by excitation in the x-direction, the rotation is about the yy-axis (Figure 8.9). From Table 8.2,

$$K_{yy,\,sur} = \frac{GB^3}{1-v}\left[3.73\left(\frac{L}{B}\right)^{2.4}+0.27\right] = \frac{(1800\text{ kg/m}^3)(198\text{ m/sec})^2(2.03\text{ m})^3}{1-0.3}\left[3.73\left(\frac{2.03\text{ m}}{2.03\text{ m}}\right)^{2.4}+0.27\right]$$

$$= 3.4\times10^6\ \frac{\text{kN-m}}{\text{rad}}$$

The dynamic modifiers depend on dimensionless frequency. From Equation (8.17), $a_0 = 2\pi \times 2.03\ \text{m} \times 6\text{Hz} / 198\ \text{m/s} = 0.39$. Then, from Table 8.3,

$$\alpha_x = 1.0$$

$$\alpha_{yy} = 1.0 - \left[\frac{0.55a_0^2}{\left(0.6 + \dfrac{1.4}{(L/B)^3}\right) + a_0^2} \right] = 1.0 - \left[\frac{0.55(0.39)^2}{\left(0.6 + \dfrac{1.4}{(2.03/2.03)^3}\right) + (0.39)^2} \right] = 0.96$$

Finally, Equation (8.16) gives the dynamic stiffnesses,

$$k_x = K_{x,\,sur} \times 1.0 = 7.8 \times 10^5\ \text{kN/m}$$

and

$$k_{yy} = K_{yy,\,sur} \times 0.96 = 3.3 \times 10^6\ \text{kN-m/rad}$$

The horizontal stiffness is about 20% larger than the measured value. The rotational stiffness is about 8% lower than the measured value. Given data scatter and related experimental error, along with the assumptions implicit to the relations in Table 8.1, the experimental results are overall similar to model predictions.

Figure 8.11 shows the variation of dynamic stiffness modifiers and radiation damping ratios with dimensionless frequency for embedded foundations. In equations provided by Pais and Kausel (1988), dynamic stiffness modifiers are unaffected by embedment, as shown in Figure 8.11. Other studies have found a modest sensitivity to embedment (Apsel and Luco, 1987) for circular foundations, although the results in Figure 8.11 have generally been adopted in practice (e.g., NIST, 2012).

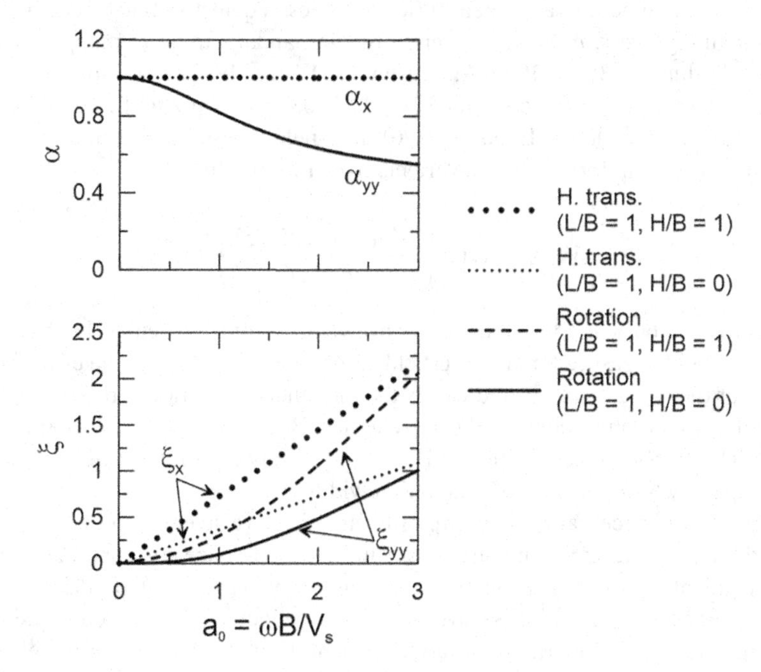

FIGURE 8.11 Variation of dynamic stiffness modifiers and damping ratios with dimensionless frequency, for square footings embedded in a homogeneous half-space, with zero hysteretic damping, and $\nu = 0.33$. (Figure adapted from NIST, 2012 using equations from Pais and Kausel, 1988.)

The elastodynamic analyses upon which Figure 8.11 is based assume perfect contact (including the development of tension, as needed) between soil and the sides (vertical faces) of the foundation, in which case waves caused by changes in normal and shear stresses on the vertical sides will radiate away from the foundation. Accordingly, the solutions indicate much higher damping levels than those for shallow foundations (Gazetas, 1991). These damping levels may not be reliable when gaps form between foundations and the adjacent soil, which reduces the potential for radiation damping from basement walls. Poor compaction adjacent to formed footings can also reduce the amount of energy lost to waves emanating from the sides of embedded foundations. In studies of field performance data (Stewart et al., 1999b), buildings shaken by earthquakes generally do not exhibit damping levels consistent with such models. As a result, the damping of embedded foundations can be conservatively estimated using models for footings located at the ground surface (Table 8.3a and Figure 8.10).

8.3.2.2 Effect of Non-Uniform Soil Profiles

Variation of soil shear modulus with depth complicates the selection of an appropriate shear wave velocity for the half-space in the calculation of static foundation stiffnesses. In most applications, V_s profiles are evaluated away from foundations (i.e., in the free-field) and show increasing stiffness with depth. To evaluate a single effective V_s value for use in computations, it is necessary to: (1) correct free-field V_s values to account for the increased overburden pressures associated with the weight of the structure; and (2) select an appropriate depth range across which to average overburden-corrected values.

Low-strain soil shear modulus, G_{max}, is known to increase with mean effective confining stress, σ'_m, as

$$G_{max} = G_0\left(\frac{\sigma'_m}{p_a}\right)^n \tag{8.18}$$

where G_0 is the shear modulus at a mean effective stress equal to atmospheric pressure, σ'_m is the effective confining stress, p_a is atmospheric pressure, and n varies from approximately 0.5 for granular soils (Hardin and Black, 1968; Marcuson and Wahls, 1972) to 1.0 for cohesive soils with plasticity index (*PI*) greater than 6.5 (Yamada et al., 2008). Recognizing that V_s is proportional to the square root of shear modulus (Equation 6.76), free-field measurements of shear wave velocity can be corrected to account for overburden effects from the structure as

$$V_{sf}(z) \approx V_s(z)\left(\frac{\sigma'_{v0}(z)+\Delta\sigma'_{v0}(z)}{\sigma'_{v0}(z)}\right)^{n/2} \tag{8.19}$$

where V_{sf} (z) is the overburden-corrected shear wave velocity for a particular depth z, V_s (z) denotes the shear wave velocity measured in the free-field at the same depth, σ'_{v0} is the effective stress from the self-weight of the soil at depth z, and $\Delta\sigma_v$ is the increment of vertical stress at depth z from the structural weight, which can be computed using classical Boussinesq stress distribution theory (e.g., Fadum, 1948). The overburden correction in Equation (8.19) is typically significant only at shallow depths (approximately 50% to 100% of foundation width 2B).

The depth interval necessary for computing an effective average profile velocity can be found by matching the static stiffnesses for a uniform half-space to those computed for various depth-dependent V_s profiles using the solutions of Wong and Luco (1985). Stewart et al. (2003) recommended that the effective average profile velocity, $V_{s,avg}$, be computed as the ratio of the depth interval (z_p) below the foundation bearing level to shear wave travel time through the depth interval (the profile is discretized into layers having thickness Δz_i and velocity $(V_s)_i$). The effective profile depth, z_p, necessary for computing $V_{s,avg}$ can be taken as the half-dimension of an equivalent square foundation matching the area (A) of the actual foundation, B_e^A, or the half-dimension, B_e^I, of an equivalent square foundation matching the moment of inertia (I_j) of the actual footing as computed below:

$$V_{s,avg} = \frac{z_p}{\sum_i \Delta z_i / \left(V_{sf}(z)\right)_i} \tag{8.20a}$$

$$\text{Horizontal}\left(x \text{ and } y\right): \quad z_p = B_e^A, \quad B_e^A = \sqrt{A/4} = \sqrt{BL} \tag{8.20b}$$

$$\text{Rocking}(xx): \quad z_p \approx B_e^I, \quad B_e^I = \sqrt[4]{3I_x/4} = \sqrt[4]{B^3 L} \tag{8.20c}$$

$$\text{Rocking}(yy): \quad z_p \approx B_e^I, \quad B_e^I = \sqrt[4]{3I_y/4} = \sqrt[4]{BL^3} \tag{8.20d}$$

This approach can, in principle, be applied to complete foundation systems by considering B and L as the half-dimensions of the entire foundation plan when the foundation system can be considered to be effectively rigid. This requires consideration of foundation flexibility effects and connectivity effects for discrete foundation elements. The following section provides additional guidance for non-rigid foundations.

When soil shear modulus increases with depth, some of the seismic energy radiating from the foundation reflects back upward toward the foundation, hence it is not "lost" from the structure as it is in the case of a uniform half-space. Impedance solutions that account for this phenomenon are available for circular foundations (Guzina and Pak, 1998 for vertical translation; Gazetas, 1991 for horizontal translation and rotation) and rectangular foundations (Vrettos, 1999 for vertical translation and rocking) for different types of heterogeneities.

Figure 8.12 shows plots of dynamic stiffness modifiers and radiation damping ratios, comparing results for uniform half-space and non-uniform profiles in which G varies with depth. The effect on radiation damping is more pronounced in rotation (Figure 8.12b) than in translation (Figure 8.12a). Also, the effect on the rotational stiffness modifier, α_{yy}, for square foundations (Figure 8.12b) is modest. Hence, the effect of variation in soil shear modulus with depth is most critical for static stiffness and radiation damping associated with foundation rocking. When rocking is an insignificant contributor to overall foundation damping, which is often the case, the practical importance of non-homogeneity in the soil profile is restricted to its effect on static stiffness.

In the extreme case of a rigid material at depth in a soil profile, radiation damping from body wave propagation disappears at frequencies lower than the fundamental frequency of the soil column (e.g., Wolf and Song, 1996; Mylonakis et al., 2006). The nonexistent wave propagation away from foundations at these low frequencies occurs because (1) no surface waves can exist in a soil stratum over bedrock at such low frequencies and (2) body waves cannot propagate downward into the rigid bedrock. While no geologic materials are actually rigid, these effects can occur when the ratio of the shear wave velocity of the firm layer to that of the soil exceeds approximately two. In such cases, the static stiffness of foundations is also increased (solutions available in Mylonakis et al., 2006).

8.3.2.3 Effect of Flexible Structural Foundation Elements

Classical impedance function solutions, such as those presented in Tables 8.2 and 8.3, strictly apply only for rigid foundations. As illustrated in Figure 8.3, soil-foundation interaction for rigid foundations can be represented by individual springs for each foundation degree of freedom. Actual foundation slabs and basement walls, however, are non-rigid structural elements. Few models for the impedance of flexible foundations are available and those that are available apply only to specific conditions. Iguchi and Luco (1982) evaluated the impedance of a circular mat of radius r_f supporting

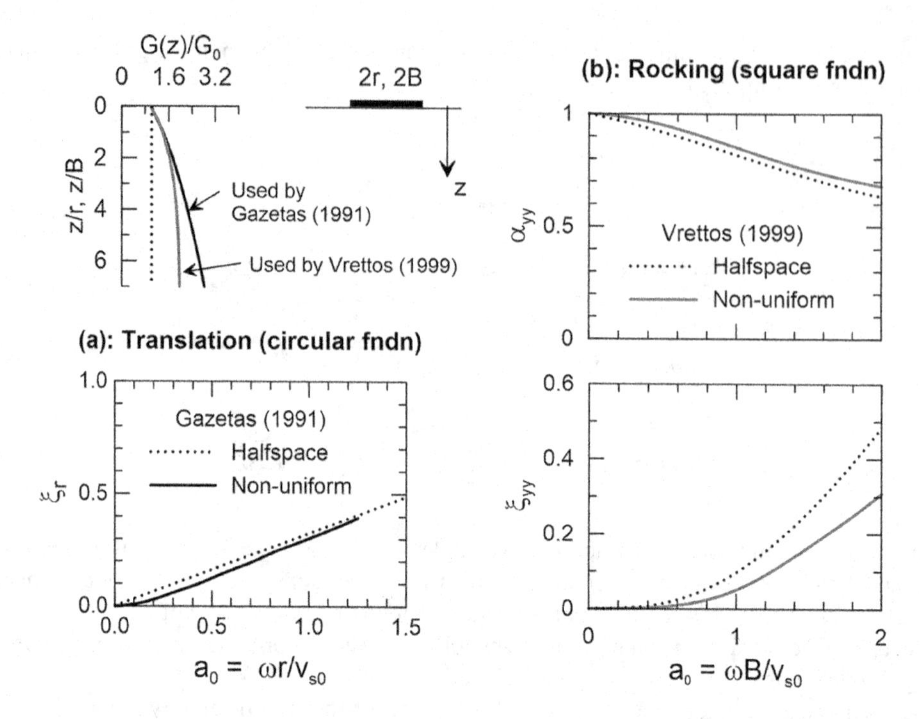

FIGURE 8.12 Plot of dynamic stiffness modifiers and radiation damping ratios versus dimensionless frequency comparing uniform half-space and non-uniform profiles in which G varies with depth: (a) translation for circular foundations (from Gazetas, 1991); and (b) rocking for square foundations (from Vrettos, 1999). (Figure from NIST, 2012.)

a rigid circular core of radius r_c (where $r_c \leq r_f$), as depicted in the top-left of Figure 8.13. Liou and Huang (1994) evaluated the impedance of a circular mat, again with a specified flexural stiffness, but this time supporting thin perimeter walls with no flexural stiffness and hinged at their connection to the mat (bottom left of Figure 8.13).

As shown in these and other studies, the effect of mat flexibility is to reduce the foundation stiffness and decrease the radiation damping. However, the amount of reduction depends on the degree of freedom being considered and the configuration of load-bearing elements on the mat. The effects of finite flexural stiffness are generally not significant for foundation vibration in translation (presumably because such movements do not involve mat flexure). Hence, the horizontal impedance can generally be evaluated using rigid foundation models. Because foundation rocking does introduce bending in foundation mats, flexible foundation effects can be important for rotational impedance. For circular mats, this rocking impedance is described by stiffness and damping ratio terms, k_{rr} and β_{rr}, respectively. Figure 8.13 compares results for rigid and flexible circular foundations supporting a rigid core or flexible perimeter wall. The flexibility of the foundation is represented by a relative soil-to-foundation stiffness ratio, ψ, taken from plate theory as:

$$\Psi = \frac{12(1-v_f^2)\,G\,r_f^3}{E_f t_f^3} \tag{8.21}$$

where G is the shear modulus of the soil (reduced as needed from G_{max} for nonlinearity), t_f is the foundation thickness, and E_f and ν_f are Young's modulus and Poisson's ratio of the foundation material (usually concrete). The case of $\Psi=0$ corresponds to a rigid foundation slab. The results in Figure 8.13 show that foundation flexibility effects are relatively modest for the case of flexible perimeter walls and most significant for the case of a rigid core.

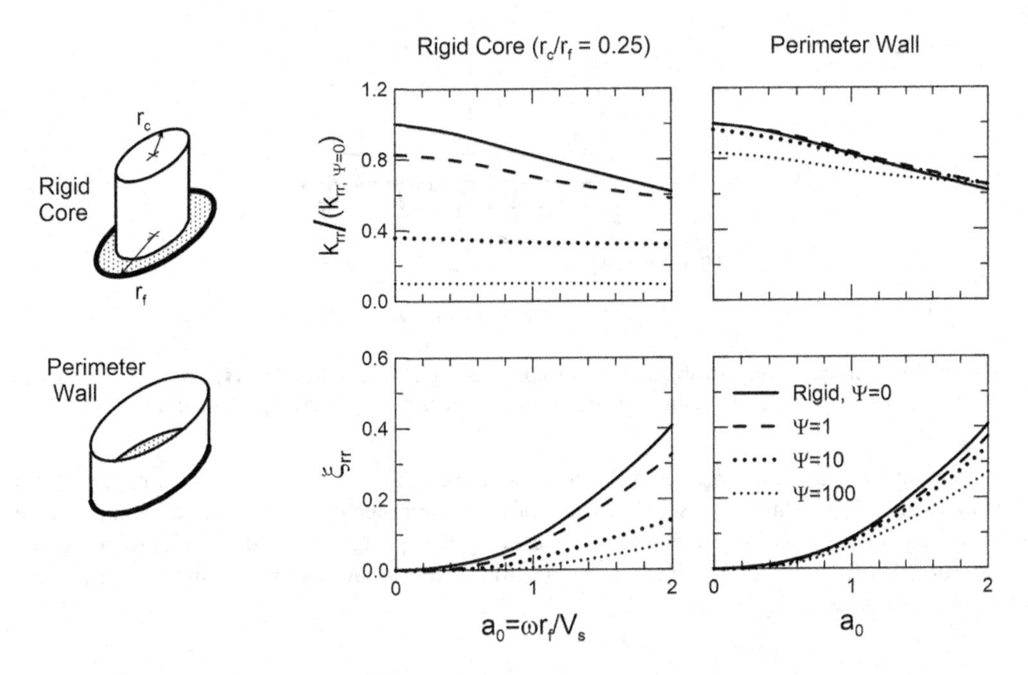

FIGURE 8.13 Effect of flexible foundation elements on rotational stiffness (k_{rr}) and rotational radiation damping ratio (ξ_{rr}) for circular foundations supporting a rigid circular core (Iguchi and Luco, 1982) and flexible perimeter walls (Liou and Huang, 1994). (Figure from NIST, 2012.)

It is common in practice to not adjust impedance functions of non-rigid foundations in the manner shown in Figure 8.13. Instead, foundations springs are distributed across the extent of the foundation, as illustrated in Figure 8.3b. Distributed springs allow the foundation to deform in a natural manner given the loads imposed by the superstructure and the spring reactions. For vertical springs, this can be accomplished by calculating the vertical translational impedance, as described above and normalizing it by the foundation area to compute stiffness intensity, k_z^i (also known as the coefficient of subgrade reaction), with dimensions of force / length^3:

$$k_z^i = \frac{k_z}{4BL} \tag{8.22}$$

Note that the term "intensity," as used here in the context of stiffness and damping coefficients, refers to area-normalized quantities that would be multiplied by tributary areas to obtain the corresponding stiffness and dashpot coefficients. A dashpot coefficient intensity can be similarly calculated as:

$$c_z^i = \frac{c_z}{4BL} \tag{8.23}$$

As illustrated in Figure 8.14, the stiffness of an individual vertical spring in the interior portion of the foundation is the product of k_z^i and the spring's tributary area dA. If this approach were used across the entire length, the vertical stiffness of the foundation would be reproduced, but the rotational stiffness would generally be underestimated. This occurs because the vertical soil reaction is not uniform, and tends to increase near the edge of the foundation for a soil with laterally uniform stiffness such as clays (e.g., Harden and Hutchinson, 2009). Using a similar process with c_z^i would overestimate radiation damping from rocking because translational vibration modes (including vertical translation) are much more effective sources of radiation damping than rocking modes.

To correct for underestimation of rotational stiffness, strips along the foundation edge (of length R_eL) can be assigned stiffer springs. When combined with springs in the interior, the total rotational stiffness of the foundation can be reproduced. There are, however, an infinite number of

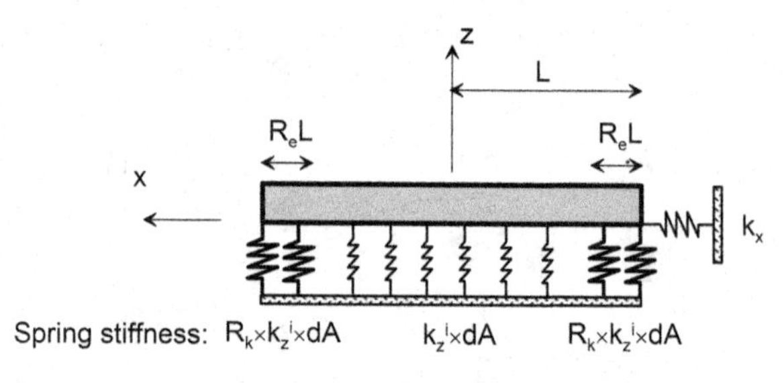

FIGURE 8.14 Vertical spring distribution used to reproduce total rotational stiffness k_{yy}. A comparable geometry can be shown in the y–z plane (using foundation dimension 2B) to reproduce k_{xx}. (Figure from NIST, 2012.)

combinations of strip width, R_eL, and stiffness multiplier, R_k, that can produce the required rotational stiffness. It is common to select a value of R_e in the range of 0.3–0.5 and compute a corresponding R_k value. Matching the moment produced by the springs for a unit foundation rotation to the rotational stiffness k_{yy} or k_{xx}, the increase in spring stiffness, R_k, can be calculated as a function of the foundation end length ratio, R_e, as:

$$\text{Rocking}(yy):\ R_{k,yy} = \frac{\left(\dfrac{3k_{yy}}{4k_z^i BL^3}\right) - (1-R_e)^3}{1-(1-R_e)^3} \tag{8.24a}$$

$$\text{Rocking}(xx):\ R_{k,xx} = \frac{\left(\dfrac{3k_{xx}}{4k_z^i B^3 L}\right) - (1-R_e)^3}{1-(1-R_e)^3} \tag{8.24b}$$

This correction for rotational stiffness, however, does not preserve the original vertical stiffness k_z. This is generally considered an acceptable trade-off because rocking is the more critical foundation vibration mode for most structures.

To correct for overestimation of rotational damping, the modified stiffness distribution can be used with the dashpot intensities over the full length and width of the foundation scaled down by a factor, R_c, computed as (NIST, 2012):

$$Rocking(yy):\ R_{c,yy} = \frac{\dfrac{3c_{yy}}{4c_z^i BL^3}}{R_{k,yy}\left(1-(1-R_e)^3\right)+(1-R_e)^3} \tag{8.25a}$$

$$Rocking(xx):\ R_{c,xx} = \frac{\dfrac{3c_{xx}}{4c_z^i B^3 L}}{R_{k,xx}\left(1-(1-R_e)^3\right)+(1-R_e)^3} \tag{8.25b}$$

The use of the above procedures for modifying vertical spring impedances will reproduce the theoretical rotational stiffness and damping through distributed vertical springs and dashpots. While this allows foundation flexibility to be accounted for, in the sense that foundation structural elements connected to springs and dashpots are non-rigid, a question that remains is whether or not the rotational impedance computed using a rigid foundation impedance function is an appropriate target in these calculations. For the case of a rigid core and flexible foundation as illustrated in Figure 8.13, it is not, whereas the rigid solution would be acceptable for the case of a flexible foundation supporting thin perimeter walls. Solutions are not currently available for many practical situations.

In the horizontal direction, the use of a vertical distribution of horizontal springs depends largely on whether the analysis is performed for shaking in one horizontal direction only (two-dimensional structural model) or two horizontal directions (three-dimensional model), and whether or not the foundation is embedded. Recommendations from NIST (2012) are as follows:

- For two-dimensional analysis of a foundation on the ground surface, the horizontal spring from the impedance function should be directly applied to the foundation, as shown in Figure 8.14 (i.e., no distributed springs).
- For two-dimensional analysis of an embedded foundation, the component of the embedded stiffness attributable to the base slab (i.e., the stiffness without the embedment modifier, k_x/η_x) should be applied to the spring at the base slab level. Distributed springs should then be positioned along the height of the basement walls with a cumulative stiffness equal to $k_x(1-1/\eta_x)$.
- For three-dimensional analysis, springs should be distributed in both horizontal directions uniformly around the perimeter of the foundation. The sum of the spring stiffnesses in a given direction should match the total stiffness from the impedance function.

More advanced guidance on the selection of horizontal and vertical springs for basement walls is provided in Section 8.6.3.2.

8.3.2.4 Foundation Capacities

Previous sections have discussed elastic spring stiffness, but have not addressed foundation capacity. The capacity assigned to a foundation spring is the limiting force that it can resist, which is related to the shear strength of the soil. In the case of vertical springs, the capacity is the unfactored (not reduced by a factor of safety) bearing capacity of the foundation distributed over the tributary area of the spring (dA). Bearing capacity is calculated considering the foundation geometry, drained or undrained shear strength parameters as appropriate, soil unit weight, and the simultaneous presence of both horizontal and vertical loads on the foundations. These concepts are discussed at length elsewhere (e.g., Soubra, 1999; Salgado, 2008; Coduto et al., 2016), and are not reviewed here. Limiting spring forces should be calculated based on soil strength, not on allowable bearing pressures that have been derived based on long-term foundation settlement limitations.

The horizontal capacity of springs located at the level of a footing or mat should reflect the unfactored sliding resistance at the slab-soil interface. The capacity of springs along basement walls should reflect the unfactored passive earth pressure (Section 8.6.2).

Shear strength parameters used for the computation of bearing capacity, sliding resistance, and passive earth pressure should be selected with due consideration of soil type, level of soil saturation, possible cyclic degradation effects, and the rapid loading rate applied during earthquakes. These issues are addressed in Section 6.6.6.

Limiting lateral and vertical capacities of foundations are usually not simultaneously realizable. This is especially important in the presence of geometric nonlinearities such as soil-foundation gapping, which is described further in Section 8.3.4.

8.3.3 Vertical Pile Foundations

Structures founded on soft soils may have pile-supported footings or mats, especially when the foundation is not embedded. The term "piles" is used here to refer both to piles driven with hammers (driven piles) and shafts that are drilled and cast-in-place (bored piles). A substantial level of engineering effort is devoted to the analysis and design of pile foundations to resist load demands associated with their installation (important for driven piles), long-term gravity loads, and potential down-drag effects. Considerations related to the selection of an appropriate type of pile for a given site condition and set of demands, how to safely install piles, and their performance under

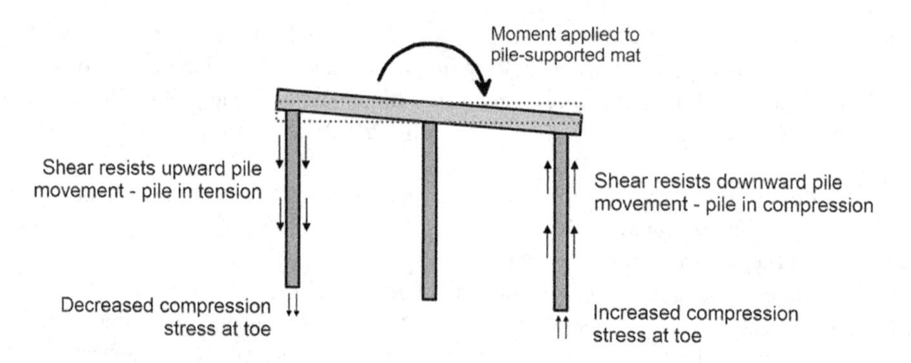

FIGURE 8.15 Resistance of pile group rocking by axial loads in pile elements offset from mat center line.

static loads are presented in foundation engineering texts such as Poulos and Davis (1980), Salgado (2008), Fleming et al. (2009), and Coduto et al. (2016). Here, the focus is on the effective stiffness and damping of vertical pile-supported foundations for SSI applications.

In substructure-type SSI analysis (Figures 8.2–8.4), pile foundations are replaced in the model of the structure by springs and dashpots (i.e., impedance functions). Impedance function models are available for single piles (horizontal and vertical modes of vibration) and pile groups (horizontal, vertical, rocking). Rocking (or coupled rocking-translation) impedance of single piles is of limited interest because rocking stiffness is typically derived from groups of piles supporting a footing or mat, which is mainly based on vertical pile responses (Figure 8.15). The pile response depicted in Figure 8.15 applies for a pinned connection in which relative mat-pile rotation occurs; if piles have a moment connection to the mat, mat rotation requires pile head rotation.

Direct methods for SSI analysis involving pile foundations typically represent the pile-soil interaction with *macro-element models* intended to capture stiffness, soil capacity, and damping characteristics. Models of this type represent extensions of so-called *p-y* and *t-z*, and *Q-z* models commonly used for the analysis of static load-deflection response of pile foundations described in the aforementioned foundation engineering textbooks. Macro-element models suitable for seismic applications are presented in Section 8.3.3.3. Models of this type are needed to evaluate demand distributions (variations of moment, shear, and axial force with depth) in the piles.

8.3.3.1 Single Piles

The impedance of single piles can be described with the notation used for shallow foundations (Equations 8.12 and 8.13). In particular, the dynamic stiffness for vibration mode j is denoted k_j^p and the corresponding dashpot, representing the effects of damping, is denoted c_j^p. The value of j can be taken as x (horizontal) or z (vertical). The springs and dashpots represented by k_j^p and c_j^p effectively replace a single pile in the numerical modeling of a foundation, as illustrated in Figure 8.16.

The dynamic stiffness of a single pile can be represented as the product of static stiffness K_j^p and a dynamic modifier α_j^p:

$$k_j^p = K_j^p \times \alpha_j^p \tag{8.26a}$$

where

$$\begin{aligned} K_j^p &= \chi_j E_s d \\ \chi_j &= \left(w_{pj} + w_{sj} + w_{bj}\right) f\left(E_p/E_s, L_p/d\right) \\ \alpha_j^p &= f\left(E_p/E_s, \rho_p/\rho_s, w_{sj}, \nu, a_0^p\right) \end{aligned} \tag{8.26b}$$

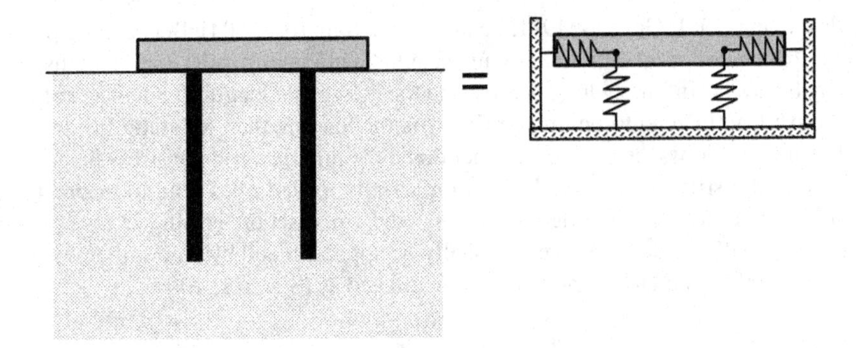

FIGURE 8.16 Schematic illustration showing the replacement of piles with equivalent springs for translational and vertical (rotational) impedance.

In Equation (8.26b), χ_j is a dimensionless constant for vibration mode j; d is the pile diameter; E_s and E_p are Young's moduli for soil and pile materials, respectively; ρ_s and ρ_p are the mass densities for soil and pile materials, respectively; ν is the Poisson's ratio of the soil; w_{pj}, w_{sj}, and w_{bj} represent weight factors that together sum to unity for pile, soil, and pile tip stiffness contributions, respectively, for vibration mode j; and a_0^p is a dimensionless frequency for piles (Kaynia and Kausel, 1982):

$$a_0^p = \frac{\omega d}{V_s} \tag{8.27}$$

Because a_0^p uses pile diameter in lieu of foundation half-width B, it is typically more than an order of magnitude smaller than a_0 (dimensionless frequency for a rigid footing at the ground surface, Equation 8.17) under the same cyclic frequency ω.

A fundamental aspect of pile response to lateral head loading is that a long pile does not deflect over its entire length, but experiences significant deflections only down to a certain depth, termed the *active pile length*, L_a (Figure 8.17a). This length is typically on the order of 10 to 20 pile diameters, depending on pile-soil stiffness contrast, soil non-homogeneity, and fixity conditions at the

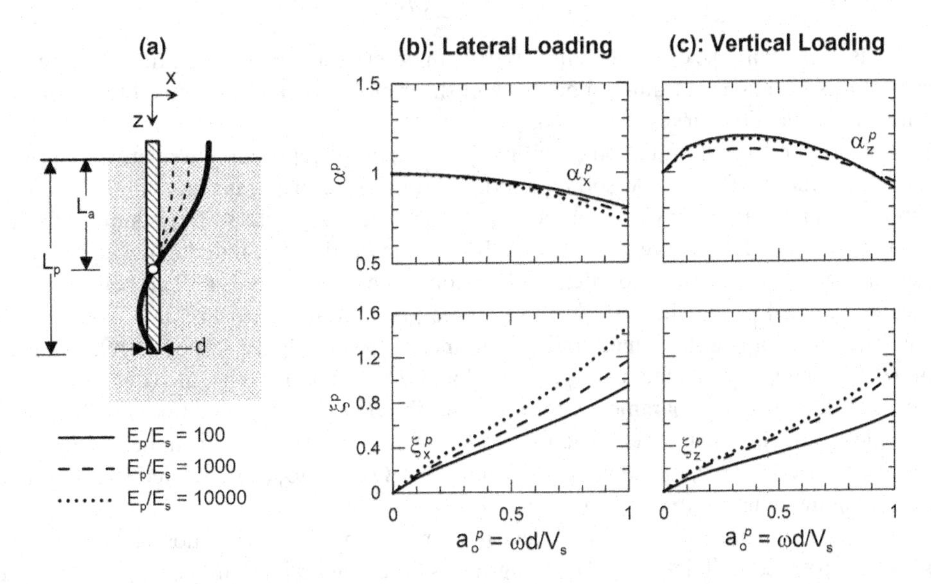

FIGURE 8.17 Plot of dynamic stiffness modifiers and damping ratios versus dimensionless frequency, for single piles in a homogeneous half-space, with ν =0.4 and ρ_p/ρ_s =1.3: (a) geometric parameters; (b) lateral loading; and (c) vertical loading. (Figure from NIST, 2012.)

pile head (Randolph, 1981; Gazetas, 1991; Pender, 1993; Syngros, 2004). Because the portion of the pile below L_a does not deflect significantly under pile head loading, piles with lengths longer than L_a essentially behave as infinitely long beams, and the actual pile length, L_p, does not affect flexural response. Active lengths tend to be greater for dynamic loading than for static loading, due to the ability of flexural elastic waves to travel further down the pile than a static stress field.

Because piles are stiff in their axial direction, axially loaded piles tend to respond over much greater depths (in excess of 50 pile diameters), and tip reaction is almost always mobilized. Accordingly, an axially loaded pile cannot usually be approximated by an infinite rod.

Active pile length in the lateral mode can be estimated as (Syngros, 2004):

$$L_a = 2.4d\left(\frac{E_p}{E_s}\right)^{0.25} \tag{8.28}$$

Similar relations have been presented by Randolph (1981), Gazetas (1991), Fleming et al. (2009), and Karatzia and Mylonakis (2017). Equation (8.28) usually results in values of active length for lateral loading of approximately 10*d* to 15*d*.

Young's modulus for soil is related to shear modulus in a 2D strain field as

$$E_s = 2G(1+\nu) \tag{8.29}$$

Small-strain shear modulus, $G_{\max}$, is evaluated from shear wave velocity, $G_{\max} = \rho_s V_s^2$ (Equation 6.76). The shear wave velocity should be taken as an average over the active length for lateral loading (Eq. 8.28) and as the actual pile length, L_p, for axial loading. These guidelines are approximate, and further research is needed to better define effective values of V_s for these conditions. Soil shear modulus, G, should be reduced relative to $G_{\max}$ for the effects of shear strain in the customary manner (e.g., Table 8.1) before evaluating E_s in Equation (8.29). As an alternative, the average shear strains in the soil at depth z, associated with the displacement of a horizontally loaded pile can be approximated as (Kagawa and Kraft, 1981; Mylonakis and Crispin, 2021):

$$\gamma(z) \approx \frac{(1+\nu)u(z)}{2.5d} \tag{8.30}$$

where $u(z)$ is horizontal pile displacement at the depth of interest (where the strain is computed). On the basis of this equation, a strain-compatible soil shear modulus can be obtained through conventional modulus reduction curves (Section 6.6.3).

The weighting factors in Equation (8.26b) (w_{pj}, w_{sj}, and w_{bj}) represent the relative contributions of the pile structural stiffness, pile-soil interaction through side-load transfer, and pile-soil interaction through tip resistance for vibration mode j. These weighting factors always sum to unity (i.e., $w_{pj}+w_{sj}+w_{bj}=1.0$), and are not required for analysis of static stiffness. Individual weighting factors are used, however, for dynamic modifiers and damping coefficients, as described below.

Equations related to the static stiffness of single piles are provided in Table 8.4a. The equations are used to compute the dimensionless parameters, χ_j, which are related to relative pile-soil moduli (E_p/E_s) and weighting factors (w_{pj}, w_{sj}, and w_{bj}). The equations for χ_j also depend on a series of additional dimensionless parameters. For lateral vibration ($j=x$), the additional variable is the dimensionless modulus of subgrade reaction, δ_x, which is related to E_p/E_s as indicated in the table. For vertical vibration ($j=z$), the additional variables are Ω and ΦL_p (both related to E_p/E_s) and the corresponding modulus parameter is δ_z.

Equations for dynamic stiffness modifiers (α_j^p terms) and damping ratios (ξ_j^p terms) for single piles are provided in Table 8.4b. Damping ratios reflect material damping in the pile and soil materials (ξ_p and ξ_s, respectively) as well as radiation damping (ξ_{rj} terms). Figure 8.17 plots the dynamic stiffness modifiers and radiation damping ratios for single piles that are obtained from these expressions.

TABLE 8.4A
Equations for Static Stiffness of Single Piles

Degree of Freedom	Surface Stiffness Modifiers		Reference
Translation along x-axis	$\chi_x = \frac{1}{2}\pi^{1/4}\delta_x^{3/4}\left(\frac{E_p}{E_s}\right)^{1/4}$	(T8.4.1-1)	Poulos and Davis (1980) Scott (1981), Mylonakis (1995)
	$\delta_x = 2\left(\frac{E_p}{E_s}\right)^{-3/40}$	(T8.4.2-1)	Dobry et al. (1982) Syngros (2004)
	$\left.\begin{array}{l} w_{px} = 1/4 \\ w_{sx} = 3/4 \\ w_{bx} = 0 \end{array}\right\}$ Long pile $\left(L_p/d > L_a\right)$	(T8.4.3-1)	Dobry et al. (1982) Mylonakis (1995) Mylonakis and Roumbas (2001)
Translation along z-axis	$\chi_z = \left(\frac{\pi\delta_z}{2}\right)^{1/2}\left(\frac{E_p}{E_s}\right)^{1/2}\frac{\Omega + \tanh\left(\Phi L_p\right)}{1 + \Omega\tanh\left(\Phi L_p\right)}$	(T8.4.4-1)	Randolph and Wroth (1978) Scott (1981)
	$\Omega = \frac{2}{\left(\sqrt{\pi\delta_z}\right)\left(1 - v^2\right)}\left(\frac{E_p}{E_s}\right)^{-1/2}$	(T8.4.5-1)	Mylonakis and Gazetas (1998), Randolph (2003), Salgado (2008)
	$\Phi L_p = \left(\frac{4\delta_z}{\pi}\right)^{\frac{1}{2}}\left(\frac{E_p}{E_s}\right)^{-1/2}\left(\frac{L_p}{d}\right)$	(T8.4.6-1)	NIST (2012)
	$\delta_z = 0.6 \;\; ; \;\; \left(L_p/d > 10,\ E_p/E_s > 100\right)$	(T8.4.7-1)	Blaney et al. (1975) Roesset (1980) Thomas (1980)
	$w_{pz} = 1 - \left(w_{sz} + w_{bz}\right)$	(T8.4.8-1)	NIST (2012)
	$w_{sz} = \frac{-2\left[\left(\Phi L_p\right)\left(\Omega^2 - 1\right) + \Omega\right] + 2\Omega\cosh\left(2\Phi L_p\right) + \left(1 + \Omega^2\right)\sinh\left(2\Phi L_p\right)}{4\cosh^2\left(\Phi L_p\right)\left[\Omega + \tanh\left(\Phi L_p\right)\right]\left[1 + \Omega\cdot\tanh\left(\Phi L_p\right)\right]}$	(T8.4.9-1)	
	$w_{bz} = \frac{2\Omega}{2\Omega\cosh\left(\Phi L_p\right) + \left(1 + \Omega^2\right)\sinh\left(\Phi L_p\right)}$	(T8.4.10-1)	

Source: From NIST (2012).

8.3.3.2 Pile Groups

When piles are used as part of a foundation system, they are usually configured in groups to support continuous mat foundations or discrete pile caps for individual load-bearing elements. The impedance of a pile group cannot be determined by the simple addition of individual pile impedances because piles within a group interact through the soil by "pushing" or "pulling" each other through waves emitted from their periphery. This is called a *group effect*, and it can significantly affect the impedance of a pile group as well as the distribution of head loads among individual piles in the group. Group effects depend primarily on pile spacing, frequency, and number of piles. Dynamic group effects are most pronounced in the elastic range and decrease in significance as the soil becomes nonlinear.

The *efficiency factor* of a pile group for vibration mode j is defined as the ratio of the impedance of the pile group to the sum of the static impedances of the individual piles within the group (Kaynia and Kausel, 1982). For the common case where all piles in the group are identical, efficiency factors can be expressed as

$$\psi_j\left(a_0^P\right) = \frac{k_j^G}{N_{piles} \times K_j^P} \quad \text{(translation modes)} \tag{8.31a}$$

TABLE 8.4B
Equations for Dynamic Stiffness Modifiers and Damping Ratios for Single Piles

Degree of Freedom	Static Stiffness Modifier		Reference
Translation along x-axis	$\alpha_x^p = 1 - \frac{3\pi}{32\delta_x}\left(\frac{\rho_p/\rho_s}{1+v}\right)\left(a_o^p\right)^2$	(T8.4.1-2)	Mylonakis and Roumbas (2001)
Translation along z-axis	$\alpha_z^p = 1 - w_{sz}\left[\left(\frac{\pi}{8\delta_x}\right)\left(\frac{\rho_p/\rho_s}{1+v}\right)\left(a_o^p\right)^2 - \frac{1}{2}\left(a_o^p\right)^{1/2}\right]$	(T8.4.2-2)	NIST (2012)

Degree of Freedom	Damping Ratio		Reference
Translation along x-axis	$\xi_x^p = \frac{1}{4}\xi_p + \frac{3}{4}\xi_s + \frac{3}{4}\xi_{rx}$	(T8.4.3-2)	Dobry et al. (1982) Mylonakis and Roumbas (2001)
	$\xi_{rx} = \left[\frac{3}{2\alpha_x^p(1+v)\delta_x}\right]\cdot\left(a_o^p\right)^{3/4}$	(T8.4.4-2)	Gazetas and Dobry (1984a,b)
Translation along z-axis	$\xi_z^p = w_{pz}\xi_p + \left(w_{sz} + w_{pz}\right)\xi_s + \xi_{rz}$	(T8.4.5-2)	NIST (2012)
	$\xi_{rz} = \frac{1}{\alpha_z^p}\left[w_{sz}\frac{1.2\pi}{4(1+v)\delta_z}\left(a_o^p\right)^{3/4} + w_{bz}\cdot 0.21\left(a_o^p\right)\right]$	(T8.4.6-2)	Gazetas and Dobry (1984b)

Source: From NIST (2012).

$$\psi_{yy}\left(a_0^P\right) = \frac{k_{yy}^G}{N_{piles}K_{yy} + K_z^P\sum x_i^2} \quad \text{(rocking about } yy\text{-axis)} \tag{8.31b}$$

$$\psi_{xx}\left(a_0^P\right) = \frac{k_{xx}^G}{N_{piles}K_{xx} + K_z^P\sum y_i^2} \quad \text{(rocking about } xx\text{-axis)} \tag{8.31c}$$

where k_j^G is the real part of the complex-valued group impedance for vibration mode j, N_{piles} is the number of piles, and K_j is the static stiffness of a single pile for mode j, x_i is the offset of pile i from the group centroid measured in the x-direction, and y_i is the offset of pile i from the group centroid measured in the y-direction. Pile group factors are complex-valued. The real part of each factor is its *stiffness efficiency*. As with other foundation impedances, there is a corresponding damping ratio defined per Equation (8.13) that is denoted ξ_j^G for pile groups. Stiffness efficiency factors are generally less than unity for low frequencies, but can increase significantly at higher frequencies under low-strain conditions. Negative efficiency factors, which suggest a phase difference of over 90° between oscillations of a single pile and oscillations of the pile group at the same frequency, are also possible. Note that these factors strictly refer to the dynamic stiffness of the group and, hence, are different from the familiar efficiency factors for static group bearing capacities in foundation engineering.

Results for horizontal and rocking oscillations are provided in Figure 8.18 for pile groups in square configurations computed using the solution by Mylonakis and Gazetas (1999). Peaks and valleys observed in the plots are due to destructive and constructive interference of the waves between piles, which tend to increase and decrease in dynamic impedance, as first identified by Wolf and Von Arx (1978), and explained by Kaynia and Kausel (1982), Nogami (1983), and Dobry and Gazetas (1988). The above effects tend to decrease with non-homogeneity and nonlinearity in the soil, as the waves emitted from the peripheries of the piles become less coherent (El-Naggar and Novak, 1994 and 1996; Michaelides et al., 1998). At the low normalized frequencies of interest in most practical problems ($a_0^P \leq 0.4$), efficiency factors are less than one and decrease as N_{piles}

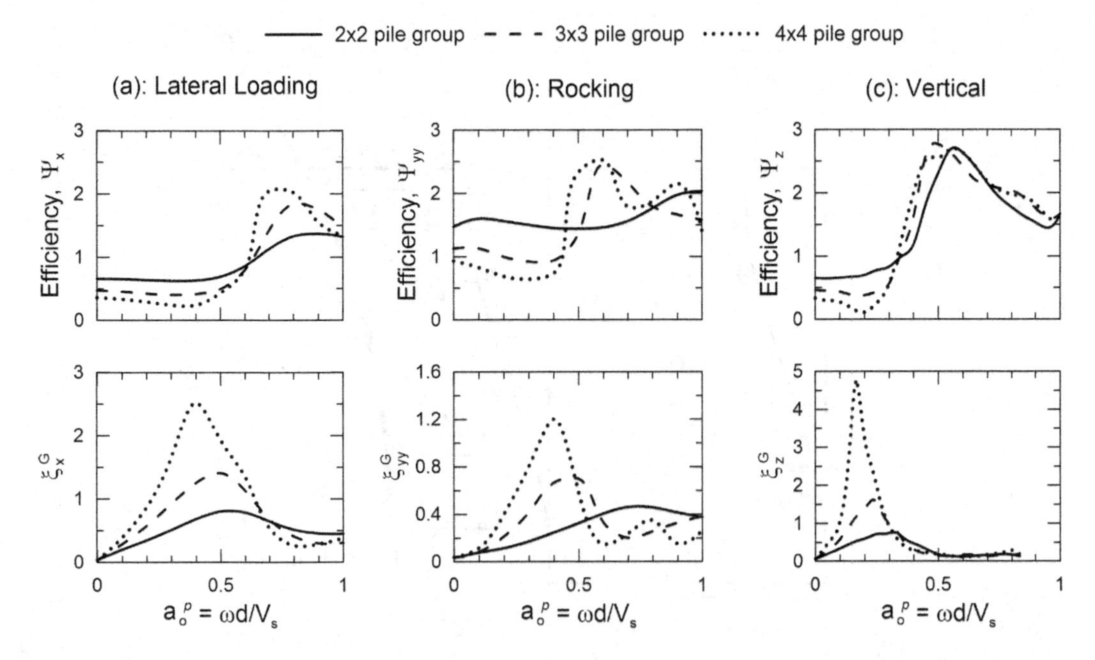

FIGURE 8.18 Plot of pile group stiffness efficiency factors (top row) and damping ratios (bottom row) versus dimensionless frequency for square pile groups for: (a) lateral loading at head of pile group under zero cap rotation; (b) moment at head of pile group, introducing rocking under zero cap translation; and (c) vertical loading at head of pile group. Lateral and rocking results are for $E_p/E_s=1{,}000$, $L_p/d=20$, $\rho_p/\rho_s=1.3$, $\nu=0.4$, (pile spacing)/$d=5$, $\xi_p=0$, and $\xi_s=0.05$. Vertical results are for $E_p/E_s=100$, $L_p/d=15$, $\rho_p/\rho_s=1.4$, $\nu=0.4$, (pile spacing)/$d=5$, $\xi_p=0$, and $\xi_s=0.05$. (Figure from NIST, 2012.)

increases as shown in Figure 8.18. Stiffness efficiencies greater than 1 for the rocking mode are due to the intrinsic out-of-phase movement of the piles located along opposite sides of the rocking axis.

An important consideration when piles are combined with shallow spread footings (i.e., pile caps) or a mat foundation is whether or not lateral resistance is provided by the shallow foundation elements in combination with the piles. Soil might be expected to settle away from the base of a shallow foundation element in cases involving clayey foundation soils and end-bearing piles, particularly when there are surface fills at the site. In such cases, lateral load resistance would derive solely from the piles. On the other hand, when soil settlement is not expected, a hybrid impedance model can be used in which lateral load resistance is provided by both shallow and deep foundation elements (e.g., in cases involving sandy soils and friction piles). Further discussion of this foundation configuration is given in NIST (2012).

8.3.3.3 Winkler-Type Macro-Element Methods

The previous subsections are applicable for use with substructure methods of SSI analysis in which pile foundations are replaced with spring (and similar) elements at the location of the pile head that are considered in the modeling of a superstructure. Such representations of pile foundations in the form of impedance functions have limitations for highly nonlinear problems and are not suitable for evaluating how demands (moment, shear, axial forces) are distributed over the length of a pile. Evaluation of these demands requires direct analysis of the supported structure and pile foundation system excited externally by free-field motions. Figure 8.19 shows such a model configuration, in which depth-dependent free-field motions are applied at the ends of elements that allow the soil to interact with the pile (Penzien et al., 1964; Kagawa and Kraft, 1980). The response of those elements are independent of each other, other than through their linkage to a common pile

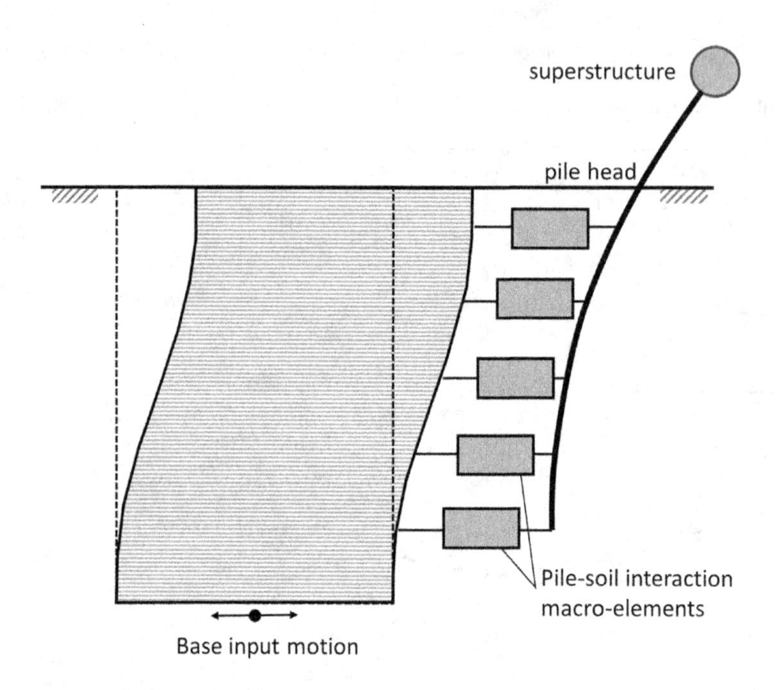

FIGURE 8.19 Schematic illustration of free-field soil column response driving the demands for a soil-pile-structure interaction analysis using macro-elements. (Modified from Boulanger et al., 1999; used with permission of ASCE.)

foundation and their excitation by a common wave field; modeling approaches of this type are known as *beam-on-nonlinear Winkler foundation* (BNWF) models (Winkler, 1867; Hayashi, 1921; Hetenyi, 1946; McClelland and Focht, 1956).

The elements representing the effects of soil-pile interaction in Figure 8.19 are referred to here as *macro-element models* because they contain one or more individual model components that capture different aspects of behavior. These macro-elements relate the force per unit length at the pile-soil interface (p in lateral direction, t in vertical) to the respective pile displacements (y for lateral, z for vertical). Macro-element models for monotonic loading of piles have been available since the 1970s for various soil types, and are colloquially known as p–y, t–z, and Q-z curves (American Petroleum Institute, 1993). These models essentially describe a nonlinear backbone curve that depends on the properties of the pile and soil; such curves can be extended to cyclic conditions through the use of extended Masing models (Section 6.4.5.2) or similar unload-reload formulations, perhaps with some reduction of the ordinate to account for cyclic degradation effects. Models of this type are not described here because they are well-covered in foundation engineering textbooks (e.g., Reese et al., 2006; Salgado, 2008) and because they omit several critical aspects of behavior for seismic applications.

Figure 8.20 shows the basic elements of a macro-element model suitable for seismic applications, in this case for horizontal excitation. Attached to the pile are three elements in series: (1) a gap element, (2) a plastic spring, and (3) a Kelvin-Voigt element containing an elastic spring and dashpot in parallel.

The *gap element* accounts for the possible formation of gaps at the soil-pile interface in the direction of loading, drag along the sides of the pile (orthogonal to the direction of loading), and interface behavior as the gap between pile and soil opens and closes. Gap behaviour is particularly relevant for shallow portions of piles in clayey soil. Taking the displacement across this element as y^g, Figure 8.21a and b shows that the closure component has no force near the origin when a gap has formed, whereas the drag component has developed resistance within this displacement range. The width of the gap in a given direction is marked in Figure 8.21b as δy^g. Once the gap is closed ($|y^g| > |\delta y^g|$), the gap/closure element exhibits rapid strain-hardening. The sum of the reactions in the drag and closure elements is p.

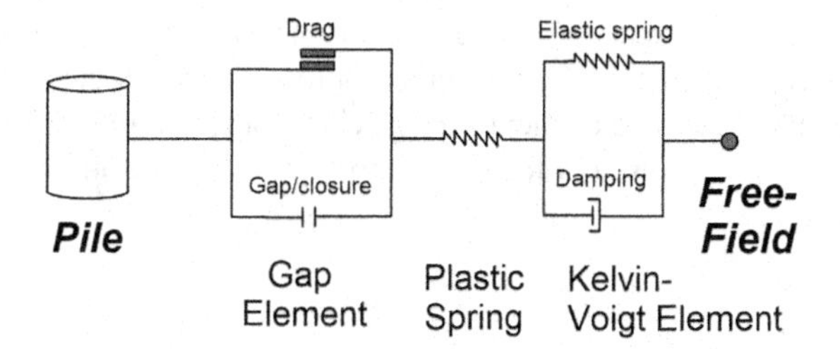

FIGURE 8.20 Basic elements of macro-element model for soil-pile interaction. In some applications, multiple Kelvin-Voigt elements are used in series.

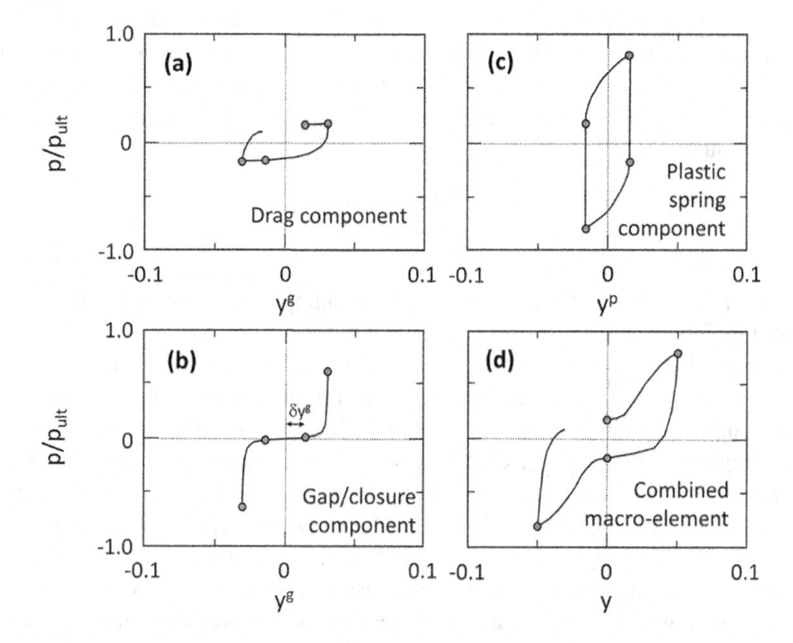

FIGURE 8.21 Load-deflection behavior of elements of macro-element model for full cycle following gap formation, (a) drag component of gap element; (b) gap/closure component of gap element; (c) plastic spring; (d) combined behavior of full macro-element. (Modified from Boulanger et al., 1999; used with permission of ASCE.)

Finite moduli and curvature in drag and gap/closure elements are applied to provide numerical stability. Alternate equations describing drag and closure behavior within the gap element are given in Nogami et al. (1992), Boulanger et al. (1999), and Taciroglu et al. (2006).

As shown in Figure 8.21c, the *plastic spring* is rigid for forces up to the elastic limit. Forces beyond that limit produce nonlinear behavior with an element displacement of y^p as the pile is advanced into the soil beyond the gap. The limiting force in the plastic spring represents the capacity of the soil to resist lateral pile movement, which is governed by the plastic flow of soil around the deflecting pile at depth and passive soil failure near the ground surface. Equations describing the nonlinear behavior of the plastic springs are often taken from classical *p–y* curves (American Petroleum Institute, 1993) (e.g., Nogami et al., 1992; Boulanger et al., 1999; Taciroglu et al., 2006), although there are several problems with those curves that may justify the use of alternative methods for critical projects (details below).

The *Kelvin-Voigt element* represents the stiffness and damping of the soil-pile interaction in the region sufficiently far from the pile that it is not affected by nonlinearities arising from relative soil-pile displacement (its response is linear visco-elastic). The displacement across the elastic spring and dashpot is denoted as y^e, and the total displacement across the macro-element is $y = y^e + y^p + y^g$. The elastic spring stiffness in the far-field element (K_e) provides for the initial stiffness of the macro-element as a whole. This stiffness is usually taken as proportional to the soil Young's modulus:

$$(K_e)_j = \delta_j E_s d \tag{8.32}$$

Values of the dimensionless factor, δ_j, depend on the relative stiffness of the pile and soil materials and are given in Table 8.4a for horizontal and vertical translation. Young's modulus of the soil, E_s, can be evaluated from shear modulus and Poisson's ratio using Equation (8.29). Note that Equation (8.32) ignores the frequency-dependence of the pile stiffness, which may be acceptable as a first-order approximation for single piles. Alternative formulations replace the single Kelvin-Voigt element shown in Figure 8.20 with three Kelvin-Voigt elements in series, which can reproduce frequency-dependent stiffness and damping behavior (Nogami and Konagai, 1988; Nogami et al., 1992).

The damper in the far-field element has a dashpot coefficient, c_j, that can be computed through manipulation of Equation (8.14) as:

$$c_j = \frac{2(K_e)_j \xi_{rj}}{\omega} \tag{8.33}$$

The damping term ξ_{rj} represents the effects of radiation damping for vibration mode j, equations for which are given in Table 8.4b.

The cumulative behavior of the macro-element is depicted in Figure 8.21d for a cycle that follows gap formation. For the initial cycle result (not shown in Figure 8.21), it is important that both the initial stiffness and the nonlinear (post-yield) features are realistic. The formulations of some traditional p–y springs (Matlock, 1970; Reese and Welch, 1975; American Petroleum Institute, 1993) have infinite slopes at zero displacement and therefore do not reliably capture pile-soil interaction stiffness characteristics (e.g., Rollins et al., 2005; Khalili-Tehrani et al., 2014). More contemporary macro-element models (Taciroglu et al., 2006; Choi et al., 2015) overcome this problem by providing initial stiffness and damping characteristics constrained by visco-elastic solutions, combined with plastic deformation and gap elements. Routines of this sort are implemented in software such as OpenSees (2011) (e.g., *pysimple3*).

Once a soil-pile interaction model like that depicted in Figures 8.19 and 8.20 has been set up for a particular pile type and soil profile, analyses can be performed for various load cases to compute head responses (nonlinear moment-rotation and shear-displacement). In some cases, it may be useful to use such results to calibrate macro-element models that reproduce the key features of pile head response. In modeling of superstructure response, such models can replace the pile and soil elements (similar to what is done with impedance functions in the substructure approach), thereby reducing the number of modeled degrees of freedom and increasing computational efficiency. Macro-element models of this sort, which are not to be confused with pile-soil interaction macro-elements at specific depths along the pile as in Figure 8.20, have been proposed by Correia and Pecker (2021) and Li et al. (2015).

8.3.4 Nonlinear Models for Shallow Foundations

Nonlinear SSI can involve geometric and material nonlinearities in the superstructure, foundation, and soil. Specific sources of nonlinearity may include: (1) yielding of seismic-force-resisting elements in the superstructure; (2) yielding of the soil, potentially exacerbated by pore-pressure induced strength loss (e.g., liquefaction or cyclic softening); (3) gapping between the foundation and the soil, such as base uplift or separation of foundation sidewalls from the surrounding material; and (4) yielding of foundation structural elements. Addressing all these issues is a formidable task, even with modern computational tools. Despite recent progress, knowledge in the area is incomplete and the subject is evolving.

Most of the research performed on nonlinear SSI has been related to structural yielding with linear, or equivalent-linear, soil (Case 1, above) or soil yielding/gapping with a linear structure (Case 2/Case 3, above). Key findings regarding the effects of nonlinearity are summarized in the following subsections.

8.3.4.1 Nonlinear Structure and Equivalent-Linear Soil

Because of difficulties associated with modeling the constitutive behavior of soil in three dimensions and wave propagation in a finite volume of geologic material under the structure without spurious wave reflections at fictitious model boundaries, most studies of nonlinear effects in SSI systems focus on superstructure nonlinearities. If structural yielding develops at relatively low-intensity input motions, or if the foundation is over-designed, significant material nonlinearities in the foundation and soil may not occur. The use of equivalent-linear representations of soil properties is justified in such analyses.

Structural response is often described in terms of *ductility* (Section B.8), which is the ratio of peak response to yield response; a structure that just reaches the point of yielding has a ductility of 1.0. Ductility can be defined from the *system ductility demand*, μ_s, or the conventional *member ductility demand*, μ, which is commonly used to define ductility for fixed-base SDOF oscillators (Section 3.3.4.2) (Priestley and Park, 1987; Paulay and Priestly, 1992). For a structure with the configuration in Figure 8.5, these ductility demands are defined as:

$$\mu_s = \frac{\tilde{u}_{\max}}{\tilde{u}_y} \tag{8.34}$$

$$\mu = \frac{u_{\max}}{u_y} \tag{8.35}$$

where $\tilde{u}_{\max}$, $\tilde{u}_y$, $u_{\max}$, u_y are the maximum earthquake-induced displacement at the top of a structure, the corresponding displacement at yield, the maximum earthquake-induced column displacement relative to a rotated foundation, and the corresponding column displacement at yield, respectively.

A key difference between these demand parameters is that member ductility demand, μ, refers exclusively to structural deformations of a member (e.g., column), whereas global or system ductility demand, μ_s, also includes movements associated with translation and rotation of the foundation, which do not cause strains (or demands) in the superstructure. Equations (8.34 and 8.35) are geometric relations, and the former always provides smaller numerical values than the latter for a given set of structural response values (Ciampoli and Pinto, 1995, Mylonakis and Gazetas, 2000). The member ductility is the preferred measure for representing nonlinear structural response.

Studies investigating the response of yielding simple oscillators, supported on foundations resting on linear or equivalent-linear soil, subjected to strong earthquake excitations (e.g., Ciampoli and Pinto, 1995; Mylonakis and Gazetas, 2000; Perez-Rocha and Aviles, 2003) have shown that nonlinear SSI generally reduces ductility demand in the superstructure relative to responses from fixed-based models. This effect can be rationalized using principles for linear SSI presented in Figure 8.7. When the period is lengthened on the descending branch of the spectrum (i.e., $T/T_p > 1$), the seismic demand is reduced, regardless of whether the structure yields or not.

However, in the case of long-period input motions that potentially place the structure on the ascending branch of the spectrum (i.e., $T/T_p < 1$), SSI-induced period lengthening may lead to an increase in ductility demand. This can be understood as a progressive resonance effect, in which the increasing effective fundamental period of the yielding structure, T, approaches the predominant period of the FIM, T_p. Evaluation of the damping effects is more complex as nonlinearity tends to reduce radiation damping and increase material damping in both the soil and the structure.

8.3.4.2 Nonlinearity in Foundation and Soil

For many seismic SSI problems, material and geometric nonlinearities in the soil are likely to occur and, perhaps paradoxically, may be beneficial to seismic structural response. This has led some researchers (e.g., Gazetas, 2006; Gajan and Kutter, 2008; Kutter et al., 2016) to propose a foundation design philosophy that allows significant yielding in the soil close to the foundation, or the foundation itself, to dissipate energy and modify the motions experienced by the superstructure (generally reduced at high frequencies). However, this approach requires control of the settlement and tilting of the structure that inevitably results from yielding of the soil. Hence, the analysis and design process considering soil nonlinearity involves optimization of the trade-offs between the potentially beneficial effects of soil yielding (especially with regard to energy dissipation) and the detrimental effects of settlement or residual tilt. In this section, several methods are described by which calculations of this type can be carried out.

SSI studies with nonlinear soil and foundation behavior can be classified into three approaches: (1) continuum models, (2) BNWF models, and (3) plasticity-based macro-models. The first approach is by far the most computationally demanding and has been employed to a limited extent (e.g., Borja and Wu, 1994; Jeremic et al., 2009).

The second and third approaches for nonlinear soil modeling are briefly described below. A basic description of the models, their input parameters, and an example comparison with experimental data are provided. A more complete description is available in Gajan et al. (2010).

Winkler-Type Nonlinear Models

BNWF models were described in Section 8.3.3.3 for application to vertical pile foundations. The same concept can be applied for soil reactions against shallow foundations such as mats and footings. As with pile foundations, the modulus of a spring (also known as modulus of subgrade reaction) is not uniquely a soil property but also depends on foundation stiffness, geometry, frequency, response mode, and level of strain.

The impedance models described in Section 8.3.2 are associated with linear springs that can be coupled with gapping and damper elements (e.g., Chopra and Yim, 1985). Nonlinear springs for shallow foundations have also been used in conjunction with gapping and damper components in combined elements similar to that shown in Figure 8.20 (Allotey and Naggar, 2003, 2007; Raychowdhury and Hutchinson, 2009).

The model by Raychowdhury and Hutchinson (2009) consists of elastic beam-column components to simulate structural footing behavior with Winkler-type macro-elements. As illustrated in Figure 8.22, a two-dimensional footing is supported by elements of this type to simulate the vertical load-displacement (q–z) behavior, horizontal passive load-displacement (p–x) behavior against the footing sides, and horizontal shear-sliding (t–x) behavior at the footing base. Moment-rotation behavior is captured by distributing vertical springs, which are stiffer at the ends for reasons discussed previously (Section 8.3.2.3), along the footing base.

Aspects of nonlinear footing-soil interaction that can be simulated with models of this type are shown in Figure 8.23, which shows simulation results and centrifuge data for a footing supporting a shear wall model building. The testing involved slow application of a horizontal force to the top of the model building, which produced a base moment (Figure 8.23a) and base shear (Figure 8.23c). Figures 8.23a and b show overturning moment and settlement as functions of footing rotation, while Figures 8.23c and d show shear force and settlement versus footing sliding displacement. The nonlinear response of the footing is shown by the limiting moment (of about 500 kN-m) and limiting shear force (about 150 kN), along with the accumulation of settlement during the tests. Impedance function models (Section 8.3.2) are unable to capture the nonlinear, inelastic load-deformation response (Figure 8.23a and c) and the accumulation of settlement observed in these tests. The traditional models can capture yielding if limiting spring capacities are applied (Section 8.3.2.4).

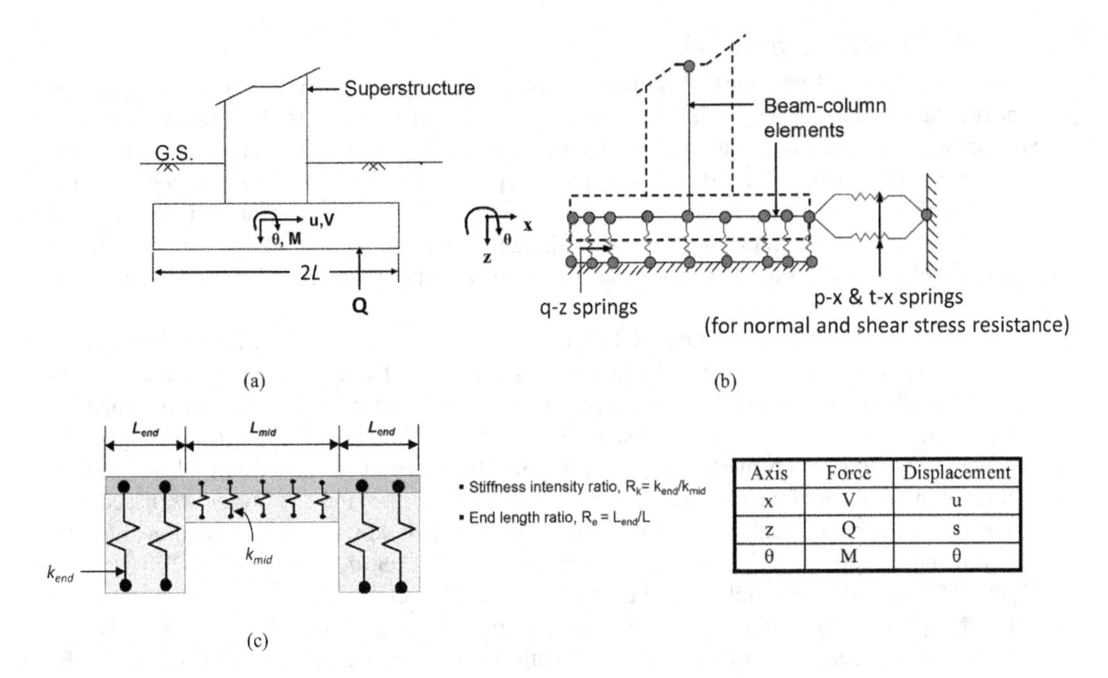

FIGURE 8.22 Schematic illustration of a BNWF model for shallow foundation: (a) foundation-superstructure system; (b) idealized model; and (c) variable vertical stiffness distribution. (Raychowdhury and Hutchinson, 2009; used with permission of John Wiley & Sons.)

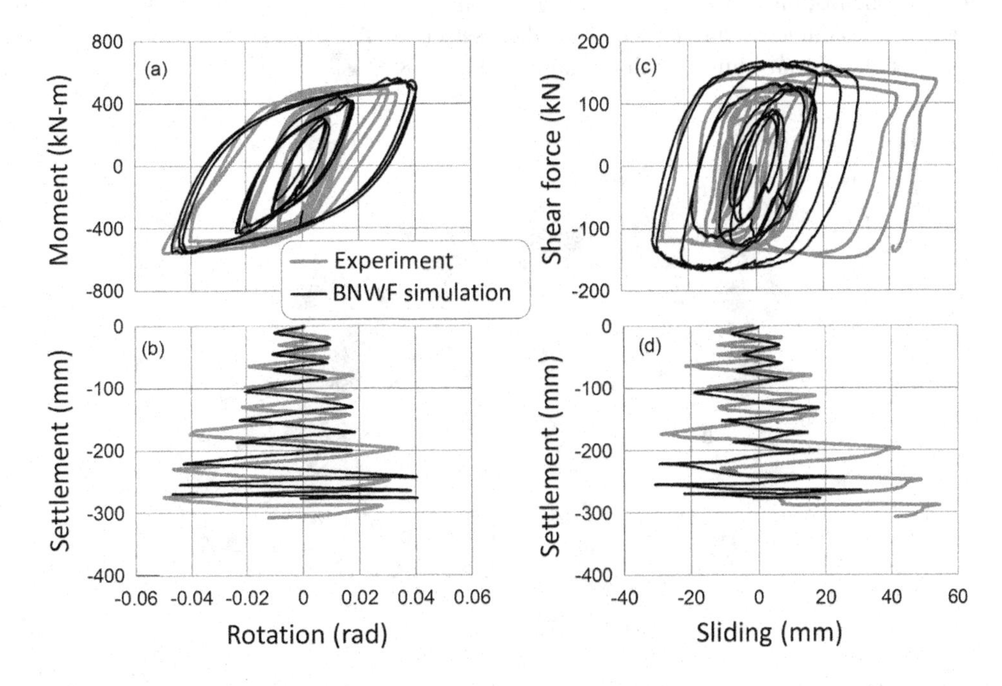

FIGURE 8.23 Comparison of BNWF model response to centrifuge tests (Thomas et al., 2005) for a shear-wall model building on dense sand. The soil relative density is D_r=80% and the vertical factor of safety of the footing against gravity loads is FS_v=2.3. (a) moment-rotation; (b) settlement-rotation; (c) shear-sliding; and (d) settlement-sliding. (Raychowdhury and Hutchinson, 2009; used with permission of John Wiley & Sons.)

Plasticity-Based Macro-Element Models

Plasticity-Based Macro-Element (PBM) theory (Nova and Montrasio, 1991) can be used to describe nonlinear behavior in multiple dimensions between two interacting systems. As applied to the nonlinear response of foundations, macro-element models comprise a single element placed at the interface between a rigid foundation and free-field soil to represent the flexibility and energy dissipation associated with SSI. From a numerical modeling standpoint, the macro-element is directly located at the footing-soil interface, replacing the rigid foundation and surrounding soil. When incremental displacements are given to the macro-element as input, it returns the corresponding incremental loads, and vice versa.

Macro-element models are based on the following main assumptions: (1) the foundation is rigid, such that its response can be described by three or six degrees of freedom in two and three dimensions, respectively; (2) a bounding yield surface exists, corresponding to general bearing capacity failure in vertical force–shear force–moment (Q-V-M) space (Figure 8.24), corresponding to the state of stress beyond which plastic flow occurs; (3) a mechanism for describing plastic flow is incorporated in the form of a hypo-plastic approach or a simple G-γ and ξ_s-γ correction based on a characteristic shear strain; and (4) uplift behavior can be modeled by means of a nonlinear elastic model allowing for separation between the footing and soil (Pecker and Chatzigogos, 2010).

Figure 8.25 shows an example of a macro-element model referred to as the contact interface model (CIM) (Gajan and Kutter, 2009). The rigid footing of length $2L$ has a gap relative to the soil on the left side of its central axis and is in contact with the soil over a length L_c. The demands on the footing in Figure 8.25 are represented by vertical force Q and the limiting moment, M_{ult}, that produces bearing capacity failure for that value of Q. The soil reaction occurs at the bearing capacity (q_{ult}), forming a resultant force of R, which is offset from the footing centroid by amount e_{max}. The model tracks the geometry of gaps and contacts along the soil-foundation interface and provides nonlinear relations between cyclic loads and displacements of the soil-foundation system during combined cyclic loading (i.e., vertical, shear, and moment). Gajan et al. (2008, 2010) presented the required model parameters and protocols for their selection. The aspects of footing-soil behavior that can be simulated are similar to those shown in Figure 8.23.

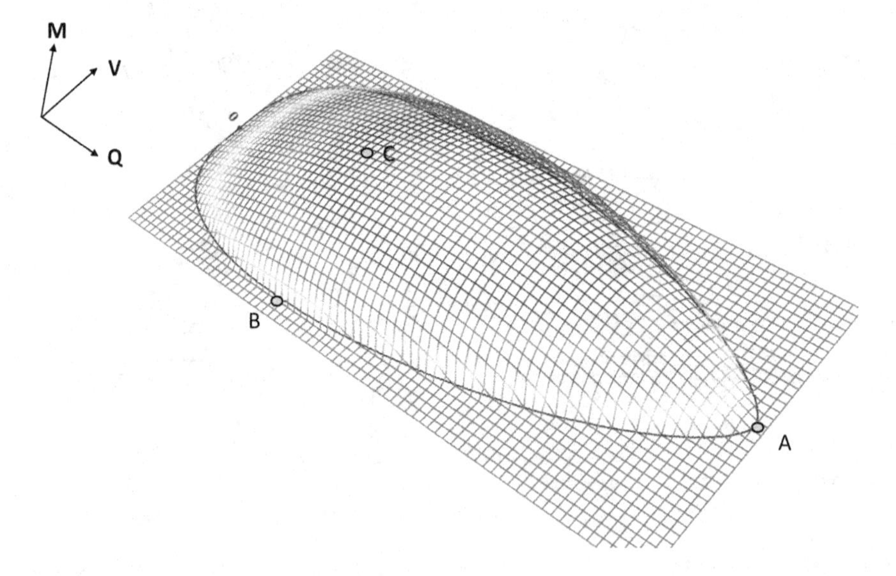

FIGURE 8.24 Bounding failure surface in vertical force (Q), shear force (V), moment (M) space. Point A shows the vertical capacity for zero shear and moment demand. Point B shows the shear force capacity for zero moment demand and a prescribed vertical force. Point C shows moment demand for zero shear force demand and a prescribed vertical force. (Modified from Cremer et al., 2001; used with permission of John Wiley & Sons.)

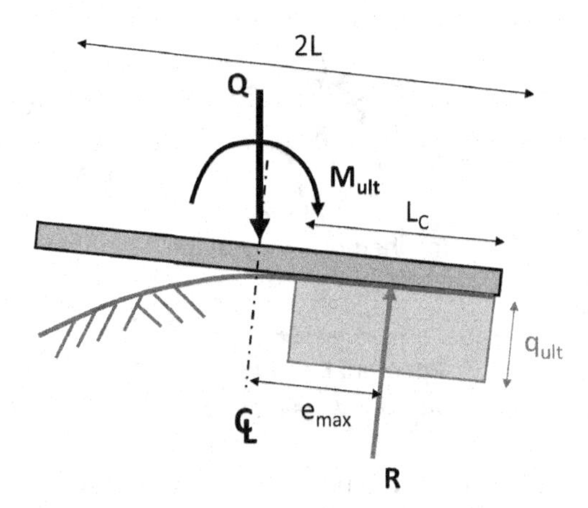

FIGURE 8.25 Schematic depiction of macro-element model at the point of a footing reaching its ultimate moment capacity, M_{ult}. (Modified from Gajan and Kutter, 2009; used with permission of ASCE).

8.4 KINEMATIC INTERACTION EFFECTS ON FOUNDATION INPUT MOTIONS

Kinematic interaction results from the presence of stiff foundation elements on or in soil, which cause foundation motions to deviate from free-field motions. One cause of these deviations is base slab averaging, which results from the restraint of spatially variable ground motions in the horizontal plane (Section 3.8) by a stiff foundation. Placement of a foundation slab across these variable motions produces an averaging effect in which the foundation motion is less than the localized maxima that would have occurred in the free-field and torsional rotations can be introduced (the latter was referred to as the "tau effect" by Newmark, 1969). Ground motion is also spatially variable in the vertical direction, as a result of wave propagation. These variabilities further modify the motions of embedded foundations (i.e., for buildings with basements) or pile-supported foundations.

The following subsections describe the phenomena involved in base slab averaging, embedment effects, and kinematic pile response and present available models for analysis of those effects. Models for kinematic interaction effects are expressed as frequency-dependent ratios of the Fourier amplitudes (i.e., amplitudes of complex transfer functions) of FIM to free-field motion. The FIM is the theoretical motion of the base slab if the near-surface foundation elements (i.e., base slabs, basement walls) and structure had no mass, and is used for seismic response analysis in a substructure approach, as shown in Figure 8.4.

8.4.1 Base Slab Averaging

Base-slab averaging results from the spatially variable nature of the free-field ground motions that would exist within the envelope of a foundation if that foundation was not present. The foundation restrains the ground motion spatial variability within its footprint due to the stiffness and strength of the foundation system. As described in Section 3.8, features of spatially variable motions include incoherence of ground motions, which is associated with differences in the Fourier phase angles, and Fourier amplitude variability. Both of these effects increase in significance with increasing frequency or decreasing wavelength. Some incoherence is deterministic (i.e., predictable) because it results from wave passage. For example, as illustrated in Figure 8.26, the non-zero vertical incidence of waves causes the waves to arrive at different points along the foundation at different times; accordingly, this is referred to as a *wave passage effect*. Investigations of wave passage at dense seismic arrays at soil sites typically indicate apparent propagation velocities, V_{app}, of approximately 2.1–2.6 km/sec (Section 3.8.1.1).

Incoherence that remains when waves are aligned to have common arrival times is stochastic, and is quantified by *lagged coherency* models (Section 3.8.1.2). As a practical matter, incoherence

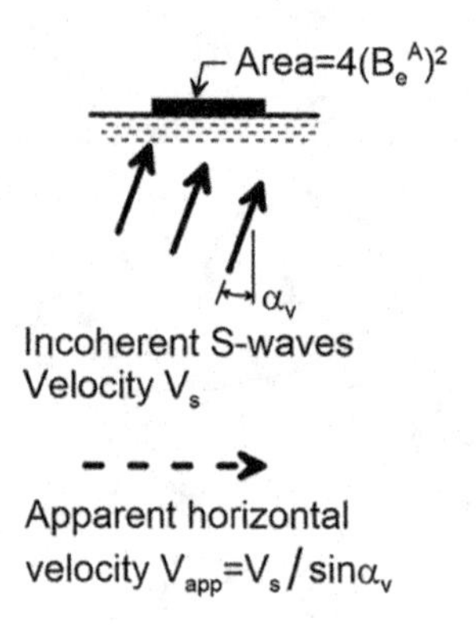

FIGURE 8.26 Schematic illustration of foundation subject to inclined shear waves producing a wave passage effect.

from wave passage and lagged coherency is always present in earthquake ground motions to some degree.

In the presence of incoherent wave fields, translational base-slab motions are reduced relative to free-field motions, and rotational motions are introduced. The reduction in translation is generally the more important outcome. The reductions of base-slab translation and the introduction of torsion and rotation in the vertical plane tend to become more significant with increasing frequency. The frequency-dependence of these effects is primarily associated with the increased effective size of the foundation relative to the seismic wavelengths at high frequencies and reductions of lagged coherency and increases of amplitude variability with increasing frequency (Figures 3.77 and 3.78).

Numerous models for predicting the relationship between foundation (u_{FIM}) and free-field ground surface motions (u_{ff0}) in the presence of inclined, but otherwise coherent, shear waves (i.e., the wave passage problem) are available. Mylonakis et al. (2006) synthesize models for the transfer function amplitude, $|H_u|$, with the following expressions:

$$u_{FIM} = |H_u| u_{ff0} \tag{8.36a}$$

where, for the purposes of Section 8.4, displacement quantities (e.g., u_{FIM}) represent their frequency-dependent Fourier amplitudes, and

$$|H_u| = \begin{cases} \dfrac{\sin\left(a_0^k\left(\dfrac{V_s}{V_{app}}\right)\right)}{a_0^k\left(\dfrac{V_s}{V_{app}}\right)} & \text{for} \quad a_0^k \le \dfrac{\pi}{2}\dfrac{V_{app}}{V_s} \\[2ex] \dfrac{2}{\pi} & \text{for} \quad a_0^k > \dfrac{\pi}{2}\dfrac{V_{app}}{V_s} \end{cases} \tag{8.36b}$$

In the above expressions, a_0^k is similar to a_0 as defined in Equation (8.17), except that the foundation dimension is related to base contact area, as

$$a_0^k = \frac{\omega B_e^A}{V_s} \quad \text{or} \quad a_0^k = \frac{2\pi B_e^A}{\lambda} \tag{8.37}$$

where B_e^A is related to foundation area, as indicated in Equation (8.20b) and λ is wavelength. The superscript k on a_0 indicates that the intended application of the dimensionless frequency is kinematic interaction. At $a_0^k = 0$, the expression in Equation (8.36b) becomes indeterminate and $|H_u|$ should be taken as unity.

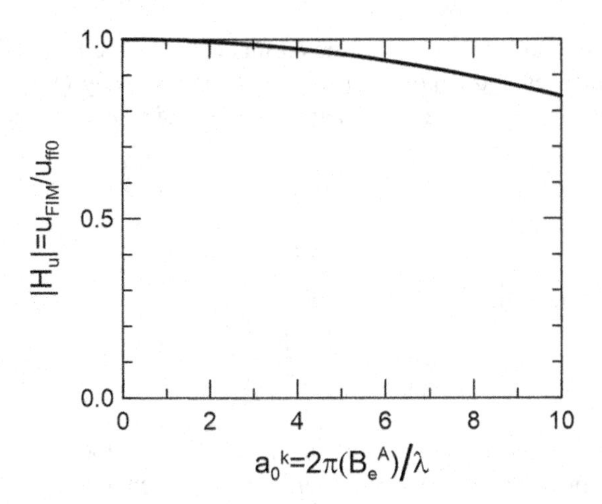

FIGURE 8.27 Transfer function amplitude between FIM and free-field motion for wave passage effect alone with $V_{app}/V_s=10$.

Because V_{app} generally falls in the range of 2.1–2.6 km/s, the velocity ratio V_{app}/V_s is approximately 10 for typical soil sites. Figure 8.27 shows the transfer function amplitude between u_{FIM} and u_g from Equation (8.36) for $V_{app}/V_s = 10$. Based on this model, wave passage alone causes relatively modest reductions of ground motions on base slabs across frequency ranges of engineering interest.

Transfer function amplitudes of recorded foundation and free-field motions are generally significantly lower at high frequencies than predicted by wave passage models. This occurs because stochastic phase variability (quantified by lagged coherency) and random variations in Fourier amplitudes often dominate the spatial variability of free-field ground motions. Two approaches for capturing these effects in the analysis of transfer functions are: (1) continuum modeling of the soil and foundation system subject to input motions with a defined coherency function as in the computer program, SASSI (Ostadan et al., 2005); and (2) application of a semi-empirical simplified model (Veletsos et al., 1997; Kim and Stewart, 2003).

In SASSI, a site- and foundation-specific model is generated in three dimensions. The foundation and soil material properties are taken as equivalent-linear. Empirical coherency models that include wave passage and lagged coherency [e.g., Abrahamson (2005) for soil sites and EPRI (2007) for hard rock sites] can be specified and their effects analyzed.

The semi-empirical model is based on a theoretical formulation of the kinematic interaction problem by Veletsos and Prasad (1989) and Veletsos et al. (1997), who applied spatially variable ground motions to rigid foundations perfectly bonded to soil. The models evaluate the response of rigid, massless circular and rectangular foundations on the surface of an elastic half-space to incoherent SH waves propagating either vertically or at an angle α_v to the vertical (Figure 8.26). The results were used to establish a relationship between transfer function amplitude and a_0^k that is essentially independent of foundation shape but varies strongly with a parameter (denoted κ_a) related to lagged coherency and wave inclination. For vertically propagating waves this transfer function amplitude can be written as (adapted from Veletsos and Prasad, 1989):

$$|H_u| = \left\{\frac{1}{b_0^2}\left[1-\exp\left(-2b_0^2\right)\left(I_0(2b_0^2)+I_1(2b_0^2)\right)\right]\right\}^{1/2} \tag{8.38}$$

where $b_0 = \left(\sqrt{4/\pi}\right)\kappa_a a_0^k$ and I_0 and I_1 are modified Bessel functions of the first kind and of zero and first order, respectively. Equation (8.38) was developed for circular foundations; the $\sqrt{4/\pi}$ term in the b_0 expression adapts a_0^k (for rectangles) to a_0 defined using the radius of a circular footing of the same area. For small and large values of the argument $\left(2b_0^2\right)$, the Bessel functions can be written in terms of power series and exponential functions, respectively (Watson, 1966); for routine application, approximations to the Bessel functions allow their contribution to Equation (8.38) to be expressed as:

$$I_0\left(2b_0^2\right)+I_1\left(2b_0^2\right)\simeq\begin{cases}1+b_0^2+b_0^4+\dfrac{b_0^6}{2}+\dfrac{b_0^8}{4}+\dfrac{b_0^{10}}{12} & for \quad b_0\le 1\\[2ex] \exp\left(2b_0^2\right)\left[\dfrac{1}{\sqrt{\pi}b_0}\left(1-\dfrac{1}{16b_0^2}\right)\right] & for \quad b_0>1\end{cases} \tag{8.39}$$

Note that the exponential terms in Equation (8.38 and 8.39) cancel for $b_0>1$.

By matching model predictions to observed variations between foundation and free-field ground motions from instrumented buildings, Kim and Stewart (2003) developed a semi-empirical model for κ_a for application with the transfer function in Equation (8.38):

$$\kappa_a = 0.00065\times V_s,\quad 200<V_s<500\text{ m/s} \tag{8.40}$$

where V_s is a representative small strain shear wave velocity for the soil beneath the foundation, which can be calculated as described in Section 8.3.2.2.

The κ_a values identified through the calibration reflect the combined effects of incoherence from wave passage and random processes. Figure 8.28 shows transfer function amplitudes, $|H_u|$, calculated using the semi-empirical approach near the upper and lower limits of κ_a along with the result obtained from wave passage alone. The figure shows that $|H_u|$ decreases significantly as κ_a increases, which is associated with increasingly spatially variable motions. Moreover, $|H_u|$ from the semi-empirical procedure is lower than from wave passage models for the range of κ_a supported by available case history data, which can be attributed to contributions of random phase and amplitude variability to base slab averaging.

The data set considered by Kim and Stewart (2003) consisted of buildings with mat, footing and grade beam, and grade beam and friction pile foundations, generally with base dimensions in the range of $B_e^A = 15$–40 m. Although the Veletsos models strictly apply to rigid foundations, the semi-empirical

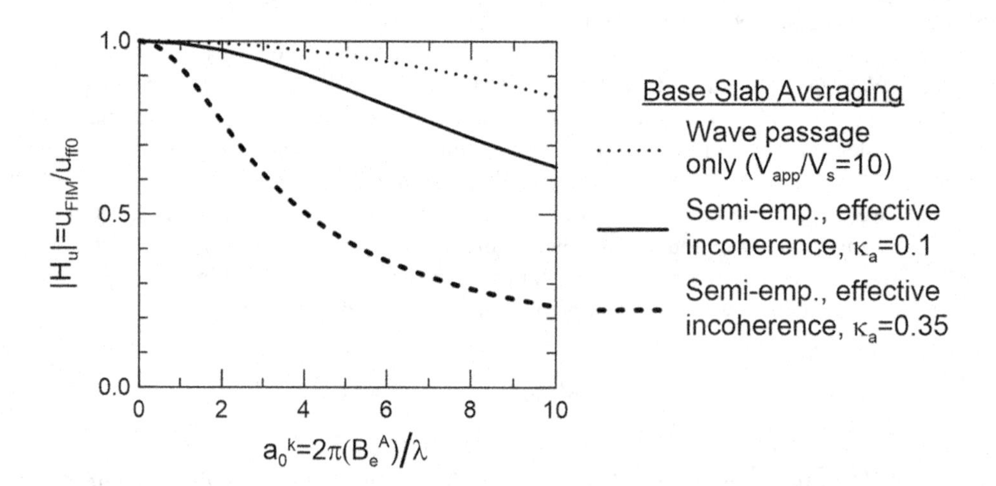

FIGURE 8.28 Transfer function amplitudes between FIM and free-field motion for wave passage only and using semi-empirical model that captures the combined impacts of all contributions to spatially variable ground motions.

model applies to the more realistic foundation conditions present in the calibration data set. Errors could occur if the semi-empirical model is applied to conditions beyond the calibration data set. In particular, the effects of incoherence in the Veletsos models are taken as proportional to wavelength, thus implying strong variation with frequency and distance. Array data indicates the variation with distance is much weaker than that associated with frequency (Abrahamson et al., 1991; Ancheta et al., 2011), so the model would be expected to overpredict the effects of incoherence (under-predict $|H_u|$) for very large foundations (opposite for small foundations). Even within the parameter space of the calibration data set, it should be recognized that the empirical model fits the data set in an average sense but would not be expected to match any particular observation. An example application showing a model comparison to a data-derived transfer function is given in the next section.

8.4.2 Embedment Effects

As described in Section C.6.1.2, free-field ground motions produced by shear waves with frequency ω propagating vertically through a uniform undamped medium of velocity V_s can be described by:

$$u_{ff}(z,t) = u_{ff0}\cos\left(\frac{\omega z}{V_s}\right)e^{i\omega t} \tag{8.41}$$

where z represents depth below the ground surface, as shown in Figure 8.29, u_{ff0} is wave amplitude at the ground surface, and t indicates time. The cosine function in Equation (8.41) describes the variation of wave amplitude with depth, while the exponential function describes the harmonic variation of displacement with time at a given depth. As shown in Figure 8.29, wave amplitude decreases with depth z according to wavelength λ, which is proportional to V_s/ω. As shown in Figure 8.29, long wavelengths (low frequencies) decay gradually with depth, such that the wave amplitude at an embedment depth H ($\ll \lambda$) will be similar to u_{ff0}. Under such conditions, the FIM and the free-field motion have similar amplitudes, meaning that the foundation moves nearly in tandem with the free-field ground motions and kinematic effects from embedment are negligible.

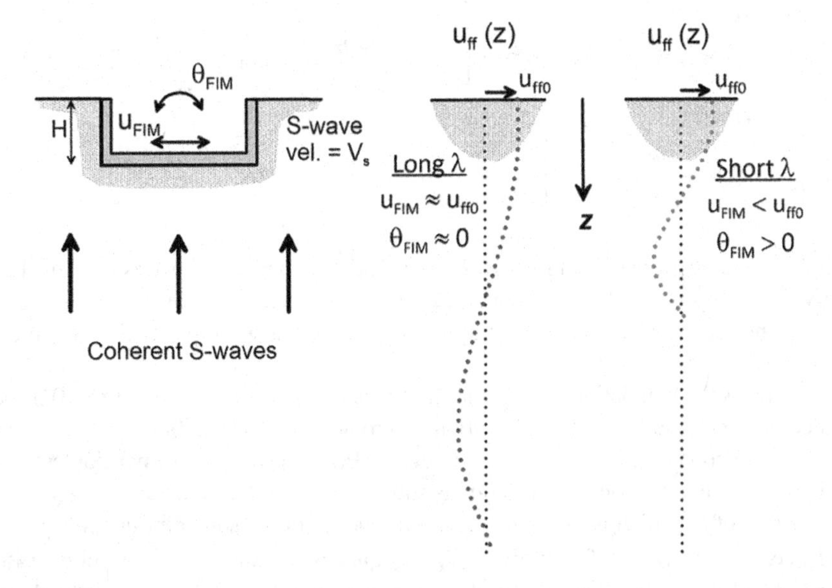

FIGURE 8.29 Embedded foundation subjected to coherent shear waves that produce free-field motion, $u_{ff}(z)$. Kinematic soil-structural interaction effects produce FIMs with translation u_{FIM} and rotation θ_{FIM}, which are sensitive to the wavelength (frequency) of the free-field motions. The depth axis shown here is used in the solution of the wave equation (depth measured from ground surface).

On the other hand, waves with short wavelengths (high frequencies) can have appreciably smaller amplitudes at foundation depth H, which reduces the FIM relative to the free-field motion. Such reductions mean the foundation and free-field motions are different, which will cause stresses to develop against the vertical faces of the foundation (such stresses are proportional to the difference in displacements), in turn producing a moment that will cause rotation in the vertical plane. The reduction of foundation translation and introduction of rotation are both kinematic SSI effects associated with foundation embedment.

Analytical solutions by Kausel et al. (1978) and Day (1978) can be used to predict kinematic SSI effects for embedded foundations. These solutions apply for rigid cylinders embedded in a uniform soil of finite or infinite thickness (half-space). When subjected to vertically propagating coherent shear waves, the embedded cylinders experience the kinematic effects of reduced base translational motion and foundation rotation. Based on those solutions, a model for transfer function amplitudes, adapted for rectangular foundation shapes, is given as (NIST, 2012):

$$|H_u|(\omega) = \frac{u_{FIM}}{u_{ff0}} = \begin{cases} \cos\left(\dfrac{\omega H}{V_s}\right) & \text{for } \dfrac{\omega H}{V_s} \le 1.1 \\ 0.45 & \text{for } \dfrac{\omega H}{V_s} > 1.1 \end{cases} \tag{8.42a}$$

$$|H_{yy}|(\omega) = \frac{\theta_{FIM} L}{u_{ff0}} = \begin{cases} 0.26\left[1-\cos\left(\dfrac{\omega H}{V_s}\right)\right] & \text{for } \dfrac{\omega H}{V_s} \le \dfrac{\pi}{2} \\ 0.26 & \text{for } \dfrac{\omega H}{V_s} > \dfrac{\pi}{2} \end{cases} \tag{8.42b}$$

$$|H_{xx}|(\omega) = \frac{\theta_{FIM} B}{u_{ff0}} = \begin{cases} 0.26\left[1-\cos\left(\dfrac{\omega H}{V_s}\right)\right] & \text{for } \dfrac{\omega H}{V_s} \le \dfrac{\pi}{2} \\ 0.26 & \text{for } \dfrac{\omega H}{V_s} > \dfrac{\pi}{2} \end{cases} \tag{8.42c}$$

where H is the embedment depth and B and L are foundation half-dimensions in plan (Figure 8.9). The velocity, V_s, in this case, should be interpreted as the average velocity across the embedment depth of the foundation (i.e., V_s taken as H divided by the shear wave travel time from depth H to the surface).

Figure 8.30 shows the foundation configuration considered in the model (Figure 8.30a) and approximate transfer function amplitudes for translation and rotation (Figure 8.30b). The model for translation reflects the reduction of ground motion with depth (Eq. 8.41) only for dimensionless frequencies up to $\omega H/V_s = 1.1$. The reduction of translation is substantial at high frequencies, reaching a constant value at about 70% of f_E, which is the fundamental frequency of the soil column between the surface and depth H, i.e., $f_E = V_s/(4H)$. The limiting high-frequency de-amplification level of 0.45 smooths through modest undulations with frequency that exist in more exact solutions, but nonetheless shows that the effects of embedment can be more important than those of base slab averaging. Foundation rotation increases with frequency and reaches a constant value at f_E. Stewart and Tileylioglu (2007) checked the predictions of the rigid cylinder models against records from nuclear reactor structures

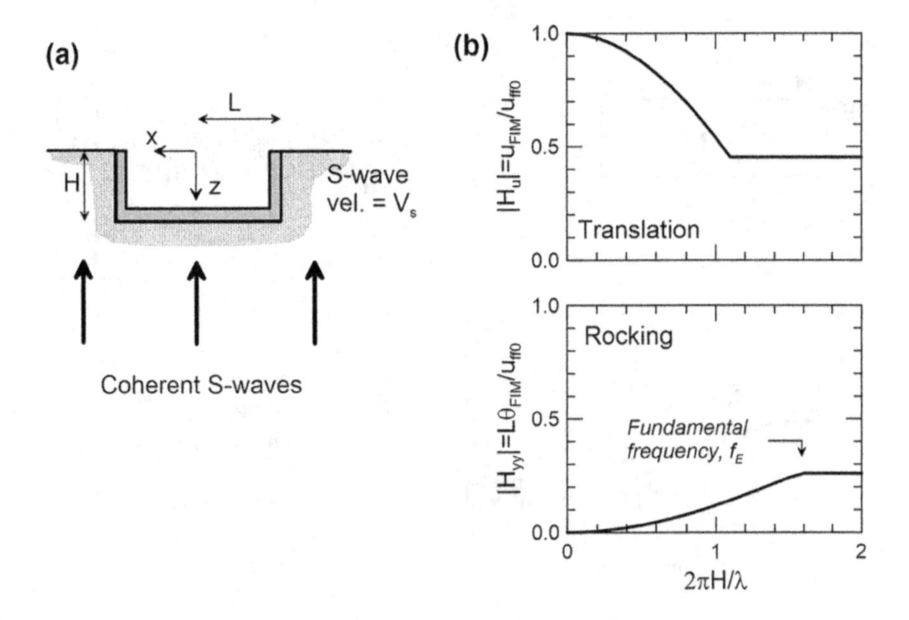

FIGURE 8.30 (a) Embedded foundation configuration; (b) transfer function amplitudes for horizontal foundation translation and rocking.

and embedded buildings. Away from frequencies strongly influenced by inertial interaction, the comparisons were favorable, indicating that these functions can be applied to realistic field conditions.

To illustrate the significance of kinematic interaction effects from base slab averaging and embedment, transfer function amplitudes were computed using recorded motions from the 1987 Whittier earthquake at the Rancho Cucamonga Law and Justice Center building (CSMIP Station Number 23497). Recordings in the reference east-west (EW) direction were used for the foundation slab (below seismic isolators near the center of the foundation at an embedment depth of 5.9 m) and at an instrument located in the free-field (approximately 101 m away from the building), as shown in Figure 8.31a. The acceleration histories shown in Figure 8.31b illustrate the general reduction of ground motions on the foundation slab in the time domain. Figure 8.31c compares the transfer function amplitudes of recordings (computed using procedures given in Mikami et al., 2008) to model predictions for base slab averaging alone and base slab averaging and embedment. Model predictions are based on Equations (8.38–8.40) and (8.42) using V_s=390 m/s, B=16.8 m, L=63.1 m, and H=5.9 m. The data have significant scatter, but trend from unity at zero frequency to about 0.2–0.4 at high frequencies. The models capture these general trends, but clearly are not a perfect fit to the data. Additional data-model validations similar to those described here are presented by Sotiriadis et al. (2020).

8.4.3 Pile Foundations

As shown in Figure 8.19, seismic waves propagating through soil deposits interact with piles through soil-pile interaction. While a perfectly flexible pile would move with the surrounding soil, the motion of a pile with non-zero flexural stiffness will differ from that of the soil column at the same depth. In the limiting case of a rigid pile and seismic waves of short wavelength (relative to pile length), the pile motion could in principle go to zero due to averaging of soil normal stresses acting in opposite directions against the pile over its length. More realistic conditions will produce reductions of pile head motions relative to free-field ground surface motions that increase with frequency, relative pile-soil stiffness, and pile head fixity against rotation, among other factors.

Solutions are available for the kinematic response of vertical piles and pile groups in elastic, and generally homogeneous, soil subjected to vertically propagating (Flores-Berrones and Whitman,

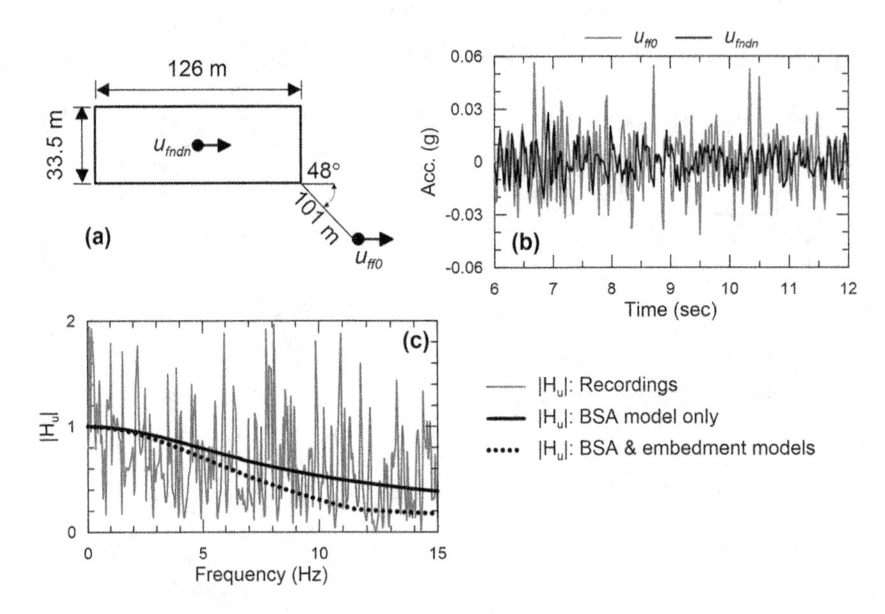

FIGURE 8.31 Recordings at foundation-level and free-field from Rancho Cucamonga Law and Justice Center during the 1987 Whittier Earthquake. (a) Foundation and sensor plan; (b) Acceleration histories; and (c) Observed and model-based transfer function amplitudes. BSA = base slab averaging (Figure from NIST, 2012.)

1982; Fan et al., 1991; Nikolaou et al., 2001; Anoyatis et al., 2013) and inclined (Barghouthi, 1984; Mamoon and Banerjee, 1990; Kaynia and Novak, 1992) coherent shear waves. These solutions can be expressed in terms of the ratios of pile head displacement to free-field displacement transfer function amplitude,

$$|I_u| = \frac{u_p(\omega)}{u_{ff0}(\omega)} \tag{8.43}$$

where $u_p(\omega)$ is the pile head lateral displacement and $u_{ff0}(\omega)$ is the free-field displacement at the ground surface. As before, displacement quantities here denote Fourier amplitudes. Separate solutions are available for free-head and fixed-head piles.

Early publications on kinematic pile response related $|I_u|$ to normalized frequency a_0^p (Equation 8.27), in which the normalizing length is the pile diameter d. With this normalization, the $|I_u| - a_0^p$ relationships depend on the ratios of pile-to-soil Young's modulus (E_p/E_s) and pile length to diameter (L_p/d) (e.g., Fan et al., 1991). Anoyatis et al. (2013) presented an alternative frequency normalization, $\omega/\left(\Lambda_j^* V_s\right)$, in which Λ_j^* has the units of 1/length and represents a complex-valued ratio of Winkler spring stiffness for mode j to pile section stiffness:

$$\Lambda_j^* = \left(\frac{(K_e)_j + i\omega c_j - \omega^2 \tilde{m}_p}{4E_p I_p}\right)^{1/4} \tag{8.44}$$

where $(K_e)_j$ and c_j are as defined in Equations (8.32 and 8.33), E_p and I_p are the pile section Young's modulus and moment of inertia, and $\tilde{m}_p$ represents the pile section mass per unit length. For most problems in earthquake engineering, the complex-valued stiffness ratio can be replaced with its static counterpart (Anoyatis et al., 2013), defined as:

$$\Lambda_j = \left(\frac{(K_e)_j}{4E_p I_p}\right)^{1/4} \tag{8.45}$$

Simple manipulation of Equation (8.45) using Equation (8.32) and assuming a circular pile section shows (Flores-Berrones and Whitman, 1982) that:

$$L_p\Lambda_j = \left(\frac{16\delta_j}{\pi}\right)^{1/4}\left(\frac{L_p}{d}\right)\left(\frac{E_p}{E_s}\right)^{-1/4} \tag{8.46}$$

where δ_j is as given in Table 8.4a (approximately two for horizontal vibration).

The physical meaning of $\omega/(\Lambda_j V_s)$ can be viewed as the ratio of a pile characteristic length $(1/\Lambda_j)$ to soil motion wavelength (proportional to ω/V_s); as this ratio increases, wavelengths become shorter relative to pile length, and the kinematic effect increases. The advantage of using dimensionless frequency $\omega/(\Lambda_j V_s)$ over a_0^p (Anoyatis et al., 2013) is that the transfer functions for fixed- and free-head piles become effectively unique (i.e., their sensitivity to E_p/E_s and L_p/d becomes small). The lack of dependence on L_p/d can be understood by recognizing active pile length as the more significant length parameter (once L_p exceeds L_a, there is little sensitivity to L_p). The lack of E_p/E_s dependence occurs because this ratio is implicit to the Λ_j definition. These transfer functions for horizontal vibration can be approximately expressed as:

Fixed-head (Flores-Berrones and Whitman, 1982; Anoyatis et al., 2013):

$$|I_u| \approx \left(1+\frac{1}{4}\left(\frac{\omega}{\Lambda_j V_S}\right)^4\right)^{-1} \tag{8.47a}$$

Free-head (adapted from Nikolaou et al., 2001 and Rovithis et al., 2009):

$$|I_u| \approx \left(1+\frac{1}{2}\left(\frac{\omega}{\Lambda_j V_S}\right)^2\right)\left(1+\frac{1}{4}\left(\frac{\omega}{\Lambda_j V_S}\right)^4\right)^{-1} \tag{8.47b}$$

The approximate equalities in Equations (8.47) occur because a real-valued shear wave velocity is taken in lieu of is complex-valued counterpart (Equation 7.10). Per Anoyatis et al. (2013), the relations in Equation (8.47) are most valid for $\Lambda_j L_p \geq 3$. These transfer functions are plotted in Figure 8.32 against $\omega/(\Lambda_j V_s)$. The reduction of $|I_u|$ with increasing frequency results from the effective averaging

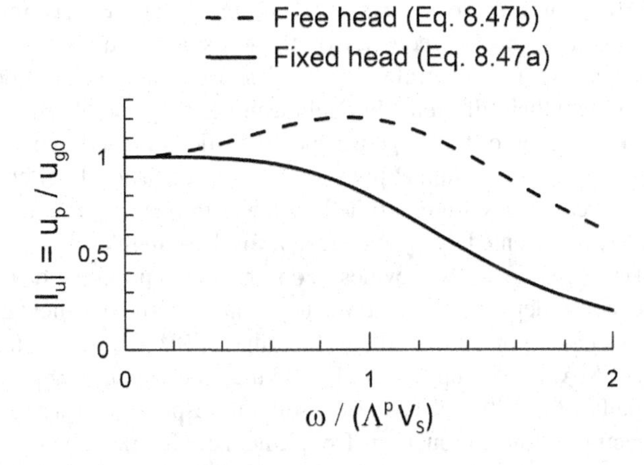

FIGURE 8.32 Pile head to free-field transfer function amplitude for horizontal ground motion as function of dimensionless frequency $\omega/(\Lambda^p V_S)$, per relations of Equation (8.47). Results applicable for long piles with $\Lambda^p L_p \geq 3$.

of vertically incoherent motions over the pile length. In the case of free-head piles, a condition of $|I_u| > 1$ (pile head displacement greater than free-field displacement) is encountered for $\omega/(\Lambda_j V_s) \approx 0.5-1.3$, which is a result of kinematic pile rotation at depth that is not restrained by soil reactions at shallower depths. For practical applications in which V_s changes with depth, a time-averaged value over the active pile length (L_a; Figure 8.17) can be used (Di Laora and Rovithis, 2014), with appropriate adjustment for nonlinear effects (Table 8.1). Corresponding values of shear modulus and Young's modulus can then be taken using Equations (6.76 and 8.29), respectively.

The transfer functions for vertical piles installed in groups and subjected to vertically propagating coherent waves are similar to those for single piles (Fan et al., 1991), hence the models in Equation (8.47) can be considered applicable for such conditions. A more difficult problem is encountered for foundation systems consisting of pile-supported base mats or grade beams of significant lateral dimensions. In such cases, the kinematic interaction problem involves base slab averaging, pile effects, and potentially embedment effects. No solutions are available for this problem, although as a first approximation the $|H_u|$ solutions in Sections 8.4.1 and 8.4.2 could be combined with the $|I_u|$ solutions in this section.

8.4.4 Application of Transfer Functions

Acceleration response spectra developed for the seismic design of structures generally apply to free-field conditions. Suites of acceleration histories also need to be developed when response history analyses are to be performed (Section 4.5). The transfer functions from kinematic interaction analyses can be used to modify free-field ground motions represented by response spectra or suites of acceleration histories to account for the interaction effects described in the preceding sections.

The response spectrum for u_{FIM} ($S_{a\text{-}FIM}$) differs from the spectrum for u_{ff0} (S_a) because of the reduction of high-frequency ground motion components from kinematic interaction. The ratio of response spectral ordinates can be estimated from the transfer function amplitude (combination of Equations 8.38, 8.42a, and 8.43, as applicable):

$$\frac{S_{a-FIM}(f)}{S_{a-ff0}(f)} \approx \begin{cases} |H_u|(f) & \text{for} \quad f \le f_L \\ |H_u|(f_L) & \text{for} \quad f > f_L \end{cases} \tag{8.48}$$

where f_L is a limiting frequency above which spectral ordinates saturate (i.e., remain approximately constant as frequency increases). The frequency on the left side of Eq. (8.48) corresponds to oscillator response (as used in response spectra) while the frequency on the right side is associated with the transfer function; despite their different physical meanings they are taken as numerically equivalent in Eq. (8.48). The saturation of the response spectral ratio is caused by response spectral ordinates converging to the peak acceleration at high oscillator frequencies. In effect, spectral ordinates at these high oscillator frequencies are controlled by lower frequency ground motion components, which are approximately represented by f_L. As a result, for those high oscillator frequencies, including *PGA*, the value of $|H_u|(f_L)$ generally provides a conservative approximation of spectral ordinates.

The limiting frequency f_L depends on the frequency content of the free-field motion. For stiff soil or rock ground motions having mean periods in the range of 0.2–0.5 sec, f_L has been found to be approximately 5 Hz (FEMA, 2005, Supplement E). On the other hand, long-period ground motions resulting from near-fault directivity pulses or soft soil site response can have much lower f_L such that no significant spectral ordinate reductions from kinematic interaction are realizable.

These differences between transfer function amplitudes and ratios of response spectra are illustrated in Figure 8.33 using the Rancho Cucamonga data presented previously in Figure 8.31. Figure 8.33a shows the EW response spectra, from which reductions of foundation motions relative to free-field motions are apparent for periods $T <\sim 0.7$ sec. ($f > 1.4$ Hz). Figure 8.33b shows the ratio

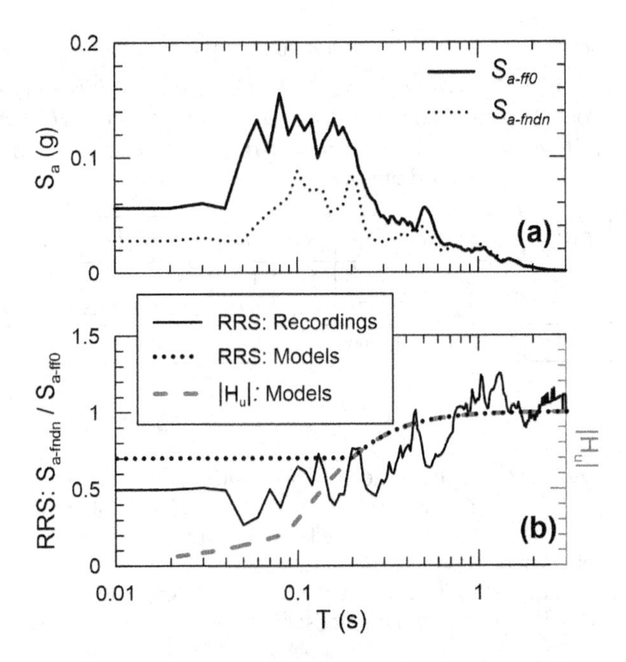

FIGURE 8.33 (a) Response spectra for EW foundation and ground recordings at site from Figure 8.31; (b) ratio of response spectra from recordings compared to model prediction for RRS and transfer function amplitude $|H_u|$. The S_a of the motion on the foundation is denoted $S_{a\text{-}fndn}$ instead of $S_{a\text{-}FIM}$ because in the real structure, the condition of zero foundation mass is not satisfied. (Figure from NIST, 2012.)

of response spectra (RRS) for the EW recordings along with: (1) the transfer function amplitude model $|H_u|$ for both base slab averaging and embedment effects; and (2) the RRS model derived from $|H_u|$ using Equation (8.48). The RRS model captures the general trends of the data, although there are significant period-to-period variations in the data and even some RRS ordinates greater than unity. The RRS data do not show the strong decrease in spectral ratios with decreasing period that is evident in $|H_u|$, which illustrates the saturation effect described above.

Example 8.5

The rigid mat foundation in Figure E8.5 has $B=15$ m and $L=30$ m and is unembedded. The average shear wave velocity in the upper layers of the site (from surface to depth B_e^A) is 250 m/sec and the strain-reduced velocity is 200 m/sec. What reduction of spectral acceleration could be expected from base slab averaging relative to the free-field ground motion at oscillator periods of 0.1, 0.2, and 1.0 sec?

Solution:

Using the semi-empirical transfer function model in Equation (8.38), the required components and solution to this problem can be compactly presented in tabular form. For each oscillator period, T, the oscillator frequency $f = 1/T$, angular frequency $\omega = 2\pi f$, dimensionless frequency

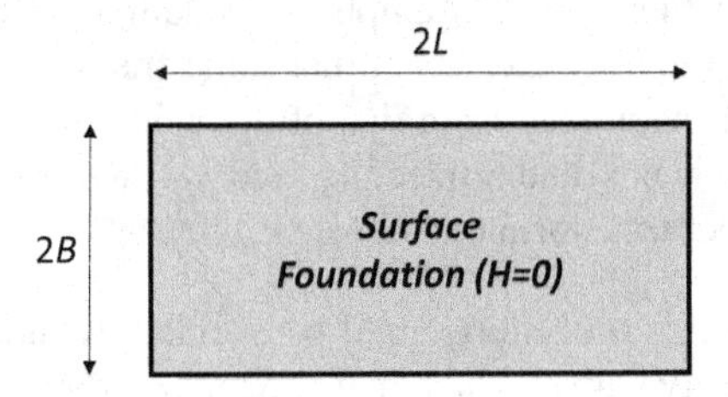

FIGURE E8.5 Plan view of foundation considered in Example 8.5.

a_0^k (Equation 8.37) and $\kappa_a = 0.00065V_s$ (Equation 8.40), are used to compute the modified dimensionless frequency $b_0 = \left(\sqrt{4/\pi}\right)\kappa_a a_0^k$. The Bessel function approximation, $I_0\left(2b_0^2\right) + I_1\left(2b_0^2\right)$, can then be computed from Equation (8.39). The transfer function ordinate H_u (Equation 8.38) and RRS S_{a-FIM}/S_{a-ff0} (Equation 8.48) can then be computed. The small-strain V_s is used with Equation (8.40) and the strain-reduced V_s is used otherwise.

T (sec)	f (Hz)	ω (rad/sec)	a_0^k	κ_a	b_0	$I_0 + I_1$	$\lvert H_u \rvert$	S_{a-FIM}/S_{a-ff0}
0.1	10	20π	6.7	0.16	1.2	8.12	0.62	0.85
0.2	5.0	10π	3.3	0.16	0.60	1.52	0.85	0.85
1.0	1.0	2π	0.66	0.16	0.12	1.01	0.99	0.99

Example 8.6

The foundation in Figure E8.6 has the same dimensions in plan as the foundation considered in Example 8.5, but now the incoming wave field can be assumed as coherent and the foundation to be embedded to a depth, $H = 6$ m. The small-strain and strain-reduced average shear wave velocities over the embedment depth are the same as given in Example 8.5 (250 and 200 m/sec, respectively). What reduction of spectral acceleration could be expected from foundation embedment relative to the free-field ground motion at oscillator periods of 0.1, 0.2, and 1.0 sec?

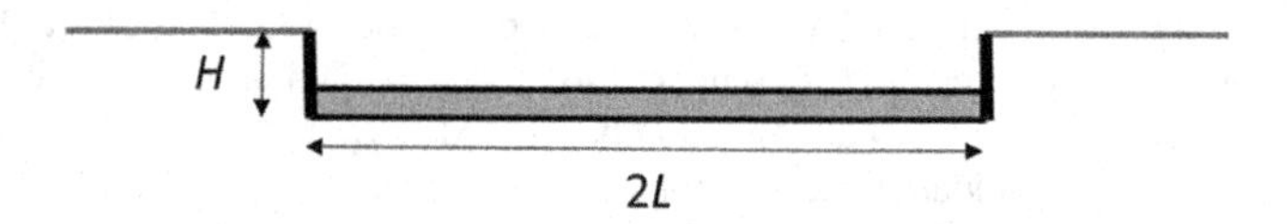

FIGURE E8.6 Side view of foundation considered in Example 8.6.

Solution:

Working in tabular form again, the first three columns of the table below repeat the values of oscillator period T, oscillator frequency f, and angular frequency ω from Example 8.5. For the embedded foundation, the dimensionless frequency $H\omega/V_s$ is used in Equation (8.42) with the strain-reduced shear wave velocity to compute the transfer function ordinate $|H_u|$. Finally, the RRS S_{a-FIM}/S_{a-ff0} can be computed from Equation (8.48).

T (sec)	f (Hz)	ω (rad/sec)	$H\omega/V_s$	$\lvert H_u \rvert$	S_{a-FIM}/S_{a-ff0}
0.1	10	20π	1.88	0.45	0.59
0.2	5.0	10π	0.94	0.59	0.59
1.0	1.0	2π	0.19	0.98	0.98

Acceleration histories representing the FIM can be modified from free-field motions by the following procedure:

1. Calculate the Fourier transforms of $\ddot{u}_{ff0}(t)$, which represents the free-field acceleration history.
2. For each frequency for which a Fourier amplitude is defined from (1), multiply the Fourier amplitude of $\ddot{u}_{ff0}$ by $|H_u|$ to estimate $\ddot{u}_{FIM}$ at that same frequency. As an optional step, the phase of $\ddot{u}_{FIM}$ could be randomized from that of $\ddot{u}_{ff0}$ using the procedures of Ancheta and Stewart (2015). If phase is not randomized, $\ddot{u}_{ff0}$ and $\ddot{u}_{FIM}$ will be coherent.
3. Perform a reverse Fourier transform to estimate $u_{FIM}(t)$.

For most practical situations, this procedure could be avoided by merely selecting and modifying ground motions for compatibility with S_{a-FIM} in lieu of S_{a-ff0}, in which case no further modifications are needed.

8.5 STRUCTURAL ANALYSIS PROCEDURES INCORPORATING SOIL–STRUCTURE INTERACTION

Consideration of SSI in the seismic evaluation and design process for structures varies from project to project. SSI effects are routinely considered in seismic analyses for bridges and nuclear facilities, but less often for buildings. The consideration, or lack thereof, reflects historical precedent to some degree, but also the manner in which SSI effects are allowed to be evaluated in seismic design documents. For example, the limited consideration of SSI for buildings in the U.S. is driven in part by the *NEHRP Provisions* (BSSC, 2020a), for which accounting for SSI is optional and can only reduce base shear demands. While incorporating these reductions can reduce construction costs and environmental impacts, many engineers justify ignoring SSI because it follows precedent and is perceived as being conservative. However, the trends in SSI practice are promising. The consideration of SSI has increased in recent years principally as a result of seismic retrofit projects in which well-informed engineers "opt in" to SSI analysis to gain better insights into the structural performance that guides retrofit decisions.

8.5.1 BUILDINGS

SSI guidelines have taken the form of force-based methods (BSSC, 2020a), displacement-based procedures (ATC, 1996; ASCE, 2023), and response history analysis methods (PEER, 2017).

8.5.1.1 Force-Based Procedures

The NEHRP Provisions describe a force-based procedure for specifying seismic demands in buildings. The procedure is used to compute a peak base shear force (V), which is the maximum cumulative seismic force at the foundation level of the structure required to place the building in dynamic equilibrium with inertial forces that develop over the height of the building. The vertical distribution of inertial loads is also estimated. These methods have two levels of sophistication. The most basic approach is known as the *equivalent lateral force procedure,* in which fundamental-mode response is assumed to be the dominant contributor to base shear. Alternatively, *response spectrum analysis procedures* (Section B.10.4) consider a number of modes sufficient to capture at least 90% of the participating mass (Section B.10.3). Inertial SSI effects can be accounted for with either force-based method as described below.

Equivalent Lateral Force Applications

Equivalent lateral force procedures in building codes evaluate seismic base shear as:

$$V = C_s\overline{W} \tag{8.49}$$

where C_s is a seismic coefficient and $\overline{W}$ is the effective seismic weight of the structure (strictly the participating mass for the first mode, but generally taken as total weight in the NEHRP Provisions for simplicity). The seismic coefficient C_s is taken as:

$$C_s = \frac{S_a(T_0)}{g}\frac{I}{R} \tag{8.50}$$

where S_a is the design spectral acceleration at the fundamental period of the building, T_0, g is the acceleration of gravity in the same units as S_a, I is an importance factor used to increase design demand levels for critical facilities, and R is a factor that accounts approximately for the effects of ductility (R depends on the type of lateral force resisting system).

The effects of inertial interaction on fundamental-mode period and damping ratio are evaluated as described in Section 8.3.1. The change in base shear is calculated as:

$$\Delta V = \left[C_s - \tilde{C}_s\left(\frac{5.6 - \ln(100\xi_0)}{4}\right)\right]\overline{W} \tag{8.51}$$

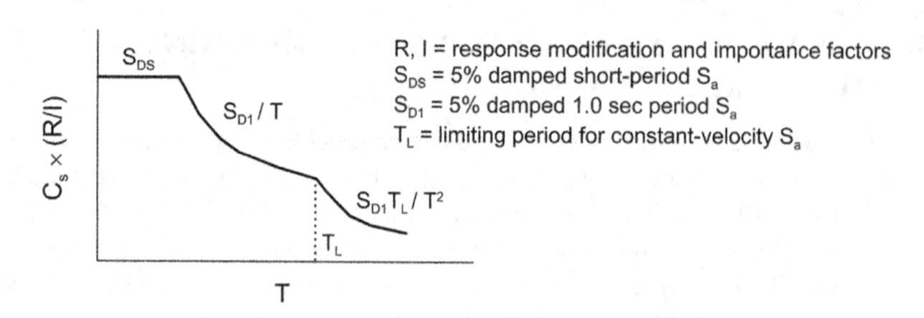

FIGURE 8.34 Schematic illustration of design response spectrum shape in NEHRP Provisions. (Figure from NIST, 2012.)

where $\tilde{C}_S = C_S(\tilde{T}_0)$ represents the seismic coefficient based on the elongated period, $\tilde{T}_0 = \left(\tilde{T}/T\right) \times T_0$ and ξ_0 is given as a decimal (not percent). The change in base shear is related to the change in seismic coefficient (or spectral acceleration, as shown in Figure 8.7). The term $\left(5.6 - ln\left(100\xi_0\right)\right)/4$ represents the reduction in spectral ordinate associated with a change of damping from the fixed base value of ξ_i=0.05 to the flexible base value, ξ_0.

The design spectrum used in conjunction with the equivalent lateral force procedure is schematically depicted in Figure 8.34. It is important to note that the spectral shape is either flat or has a negative slope with respect to period. Coupled with the requirement that ξ_0 must exceed ξ_i (discussed further below), this spectral shape ensures that utilizing the SSI provisions will never increase base shear. It is worth recalling at this point that in real structures subjected to ground motions with more realistic spectral shapes, SSI can increase or decrease base shear (Figure 8.7).

Period lengthening ($\tilde{T}/T$) in the NEHRP Provisions is calculated with Equation (8.8) using impedance expressions for lateral stiffness k_x and rotational stiffness k_{yy} similar to those given in Section 8.3.2. The shear modulus used in conjunction with equations for static foundation stiffness is reduced from the small-strain modulus G_{max} as indicated in Table 8.1. Flexible-base system damping is calculated using Equation (8.9), with the foundation damping calculated using an expression similar to Equation (8.10) and with ξ_i usually taken by default as 0.05 and $n=2$.

The 2020 version of the *NEHRP Provisions* (BSSC, 2020a) does not allow consideration of kinematic interaction effects on ground motions when seismic demands are evaluated using equivalent lateral force procedures.

Response Spectrum Analysis

The general principles related to response spectrum analyses are presented in Section B.10.4. These methods utilize a realistic spectral shape (derived from seismic hazard analysis) and require a modal analysis of the structure. The modal analysis identifies N_m vibration modes, each of which has a computed period T_i and generalized mass M_i (Section B.10.3). As such, the peak base shear for mode i can be evaluated as:

$$V_i = M_i S_a(T_i)\frac{I}{R} \tag{8.52}$$

where $S_a(T_i)$ is the spectral acceleration at the modal period, T_i, and I and R are as defined for equivalent lateral force methods. As noted previously, N_m must be selected to include all modes significantly contributing to the response – this has generally been achieved by requiring the sum of the participating masses to be at least 90% of the total weight of the structure. More recent procedures (BSSC, 2020a) require summing to 1.0W by allowing all modes with periods less than 0.05 sec to be combined at 0.05 sec.

The computation of total base shear requires the use of modal combination rules, in which the contributions of all N_m modes are considered. These rules recognize that the modal maxima do not occur at the same time. One of the most common such approaches is known as the square-root-of-sum-of-squares (SRSS) method, in which base shear is taken as:

$$V = \sqrt{\sum_{i=1}^{N_m} V_i^2} \tag{8.53}$$

More advanced procedures such as complete quadratic combination (CQC) approaches can also be used (Wilson et al., 1981). Force or moment demands on individual structural members can be computed in much the same manner as base shear – the contributions are computed for each respective mode, and then combined using SRSS or CQC approaches.

SSI affects the evaluation of base shear and other demands in two principal ways. First, the structural model used to compute modal properties (periods, mode shapes, participating masses, etc.) includes foundation spring elements. Because period lengthening is accounted for with the use of those springs, the expression used in the equivalent lateral force procedures (Equation 8.8) is not directly applied. Second, the spectrum used for evaluating mode-specific base-shears, V_i, can be modified for the effects of SSI-system damping, ξ_0, which includes foundation damping, ξ_f, as given by Equation (8.9). As with the equivalent lateral force procedure, the 2020 version of the *NEHRP Provisions* (BSSC, 2020a) does not allow consideration of kinematic interaction effects on ground motions when response spectrum analysis procedures are used.

Utilization

The methods for inertial SSI analysis in the NEHRP Provisions are optional and are infrequently used in practice. There are two primary reasons for this. First, because the equivalent lateral force procedure was written such that base shear demand can only decrease through its consideration, SSI effects have often been ignored to be conservative (and to follow precedent). Second, many engineers who have attempted to apply the method on projects have done so for major, high-rise buildings for which they felt evaluating SSI effects could provide cost savings. However, period lengthening and foundation damping effects are negligible for these tall, flexible structures, and hence the design engineers realized no benefit for their efforts. Such experiences can discourage further application on other projects. As indicated by the simple models described in Section 8.3.1, consideration of SSI effects using the guidelines actually yields the greatest benefit for short-period, stiff structures.

8.5.1.2 Displacement-Based Procedures

In displacement-based methods, the system response is represented by a lateral force-displacement relationship calculated using pushover analyses. As illustrated in Figure 8.35, pushover analyses involve the application of static lateral loads distributed over the height of the structure and the calculation of the resulting displacements from a model of the SSI system. The cumulative lateral load (the resultant, H, in Figure 8.35) is related to a reference displacement, Δ (lateral drift at the roof level, in this case) from the nonlinear pushover curve. Particular points on the pushover curve can be related to damage states in the building since the deformation of all of the structural components can be related to the reference displacement.

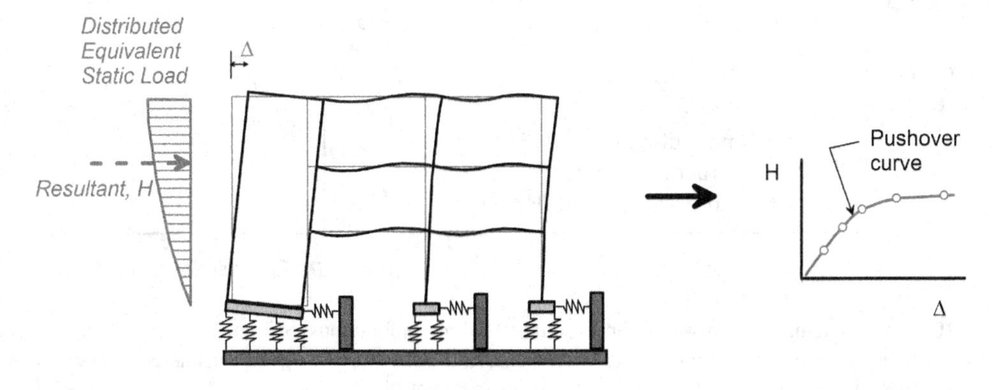

FIGURE 8.35 Schematic illustration of analysis to develop pushover curve. (Figure from NIST, 2012.)

The pushover curve can be combined with a design response spectrum in different ways (Powell, 2006) to estimate the seismic displacement of the structure. Two such methods are known as the Capacity Spectrum Method (ATC, 1996; Chopra and Goel, 1999) and the Coefficient Method (FEMA, 2000, 2005), as illustrated in Figure 8.36. FEMA (2005) uses an Equivalent Linearization Method that is conceptually similar to the capacity spectrum method (not depicted in Figure 8.36).

SSI enters displacement-based analysis procedures through three components: (1) springs used in pushover analysis; (2) reduction of free-field design spectrum for kinematic interaction (which is not allowed in the force-based methods described in Section 8.5.1.1); and (3) reduction of design spectrum for SSI system damping ratios, ξ_0, which are greater than ξ_i. The soil-foundation springs used in pushover analyses can be similar to those described in Section 8.3.2, except that dynamic modifiers are typically neglected. Distributed vertical springs are evaluated in a manner similar to that described in Section 8.3.2.3. Horizontal stiffness is not distributed but is typically represented by a single spring at the end of the foundation as shown in Figure 8.14.

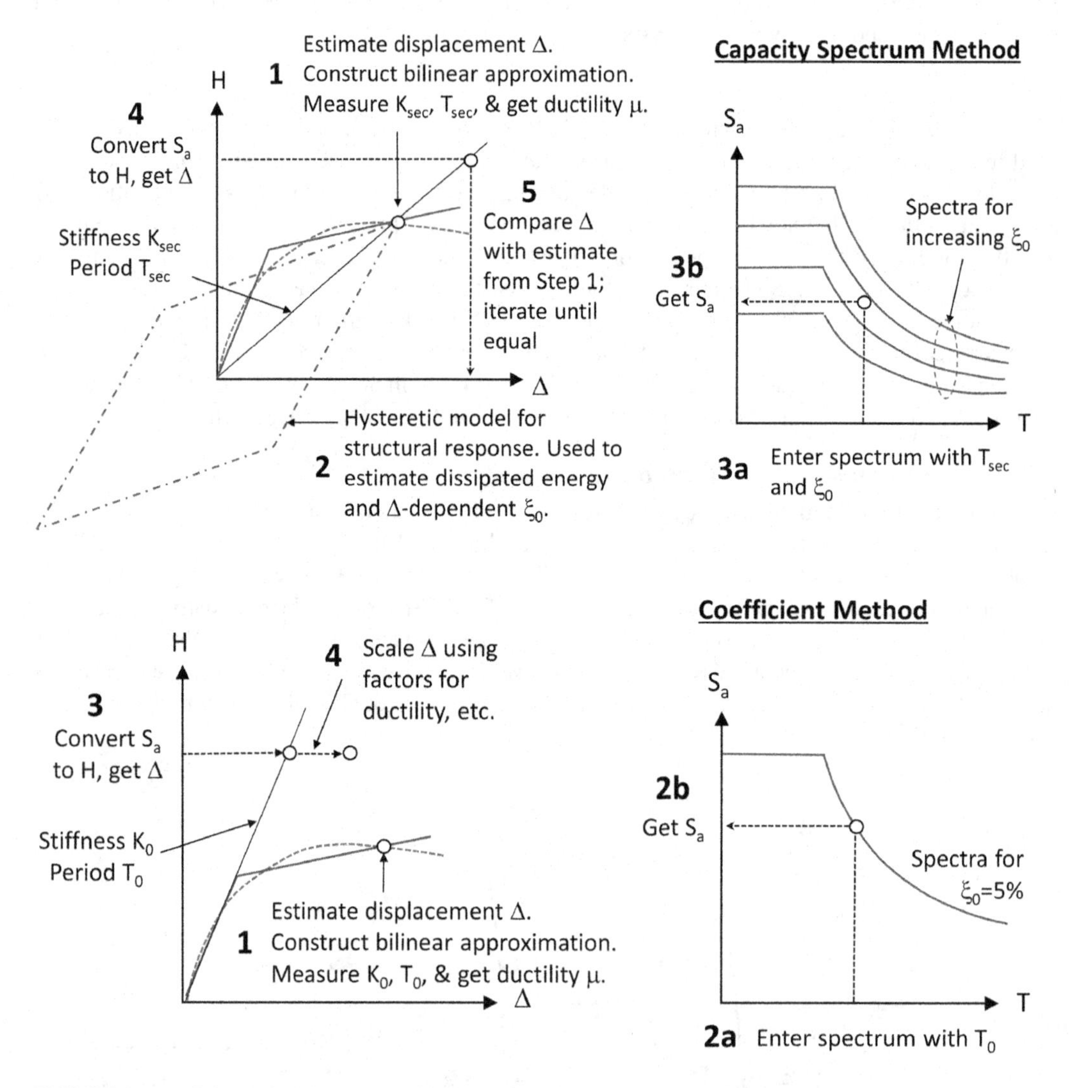

FIGURE 8.36 Schematic illustration of procedures used to combine pushover curve with design response spectrum for analysis of seismic-induced displacements in the structure. If SSI is considered, all periods are elongated by period lengthening. (Based on schematics from Powell, 2006.)

The effects of kinematic interaction on response spectral ordinates can be evaluated (ASCE-41, 2023) using transfer function amplitudes, H_u (Sections 8.4.1 and 8.4.2), as represented by spectral acceleration ratios modified from the relation in Equation (8.48). The modification allows only 75% of the predicted reductions to be taken, which is achieved by taking H_u as the sum of 0.25 and 75% of H_u values from Sections 8.4.1 and 8.4.2. This modification was a result of committee deliberations and was intended to add conservatism to the procedure.

The objective of the damping analysis is to estimate the foundation damping, ξ_f, which is then combined with fixed-base structural damping, ξ_i, to estimate the total system damping, ξ_0, using Equation (8.9). A challenge in this process is to extract ξ_0 from the results of pushover analyses in which the structural behavior is nonlinear, as shown in Figure 8.35. The evaluation of system and foundation damping follows procedures given previously, except that an effective period lengthening ratio $\left(\tilde{T}/T\right)_{eff}$ that accounts for structural system ductility is used in lieu of ordinary period lengthening (for elastic conditions) in Equation (8.9). Effective period lengthening is computed as:

$$\left(\frac{\tilde{T}}{T}\right)_{eff} = \left\{1 + \frac{1}{\mu_s}\left[\left(\frac{\tilde{T}}{T}\right)^2 - 1\right]\right\}^{0.5} \tag{8.54}$$

where μ_s is the system ductility (maximum displacement of the structural system divided by displacement at yield; Section 8.3.4.1). This ductility is also related to the R value in force-based procedures (Section 8.5.1.1).

8.5.1.3 Response History Procedures

Response history analysis procedures for assessment of structural demands subject building models to earthquake ground motion time series, requiring the computation of response at each time step. These procedures have the potential to provide the most realistic assessments of structural performance, and as such are considered permissible alternatives to the force-based and displacement-based procedures described in previous subsections (BSSC, 2020a; ASCE, 2023). However, response history procedures are computationally demanding and require considerable technical expertise to produce reliable results.

Tall Buildings Initiative (TBI) guidelines documents (PEER, 2010, 2017) provide recommendations for performing response history analyses and interpreting the results to check for acceptable performance (*acceptance criteria*). While the use of fixed-based methods is permissible in the TBI procedures, SSI can be considered following a substructure analysis procedure similar to that described in Section 8.2.2. However, the specification of input motions and the distribution of springs and dashpots depicted in Figure 8.4 is permitted to be simplified. Figure 8.37 shows two simplifications of the SSI system recommended for use in response history analysis.

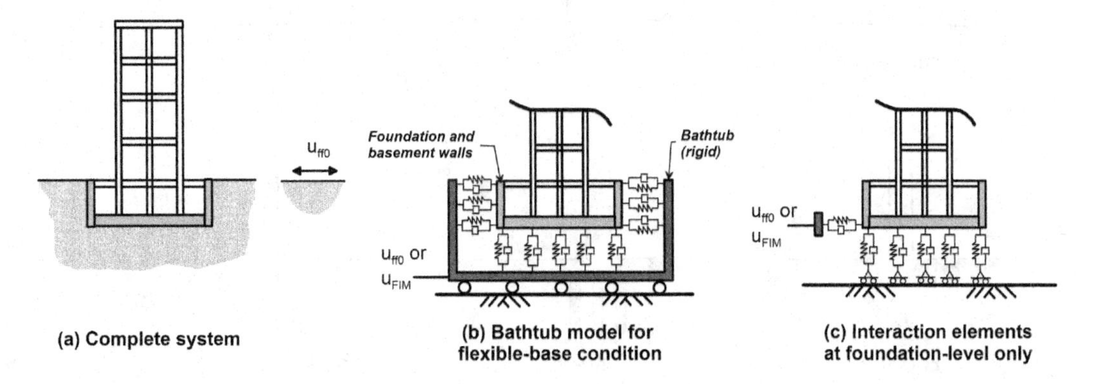

FIGURE 8.37 Schematic illustration of (a) tall building with subterranean levels, (b) substructure model in which input motions are not allowed to change over the depth of foundation embedment ("bathtub model") and (c) substructure model in which input motion is applied only at the foundation level. (Modified from PEER, 2017.)

As shown in Figure 8.37, subterranean levels are included in the structural model. Element stiffnesses and capacities for structural members such as walls, columns, and slabs are included as are the masses of subterranean elements. As shown in Figure 8.37b and c, the SSI systems also include springs and dashpots representing soil-foundation interaction. In the rigid "bathtub" model (Figure 8.37b), foundation springs are included along basement walls, whereas the alternative in Figure 8.37c localizes all lateral stiffness at the base slab. Both of the systems shown in Figure 8.37 are simplifications of the more complete substructure model (Figure 8.4); those simplifications are intended to streamline the process of developing analytical models for structures. In these procedures, the motions applied at the base of the model should be taken as the FIM (u_{FIM}), but use of the free-field ground surface motions (u_{ff0}) is also allowed.

8.5.2 Bridges

The California Department of Transportation (Caltrans) has played a key role in the development of seismic guidelines for applications to bridges, most of which are implemented in periodically updated Seismic Design Criteria (SDC; Caltrans, 2019). Seismic provisions in federal guidelines (AASHTO, 2014) are typically similar to the Caltrans SDC. Those guidelines are applied to "ordinary" bridges to account for the effects of ground shaking. Procedures for more complex bridge structures and loading applied as a result of lateral spreading of abutments are less formally codified, but can be guided by research reports (Aviram et al., 2008; Omrani et al., 2015).

8.5.2.1 Response to Ground Motions

Figure 8.38a shows a typical bridge structure configuration. Traffic flows on the bridge deck, which is supported at its ends by abutments and in intermediate areas by columns. Seismic response in the direction of traffic flow is referred to as longitudinal; the perpendicular direction is transverse. Figure 8.38b shows that foundation systems for bridge columns and abutments can consist of footings, capped pile group systems, or large-diameter shafts that serve as extensions of the bridge columns.

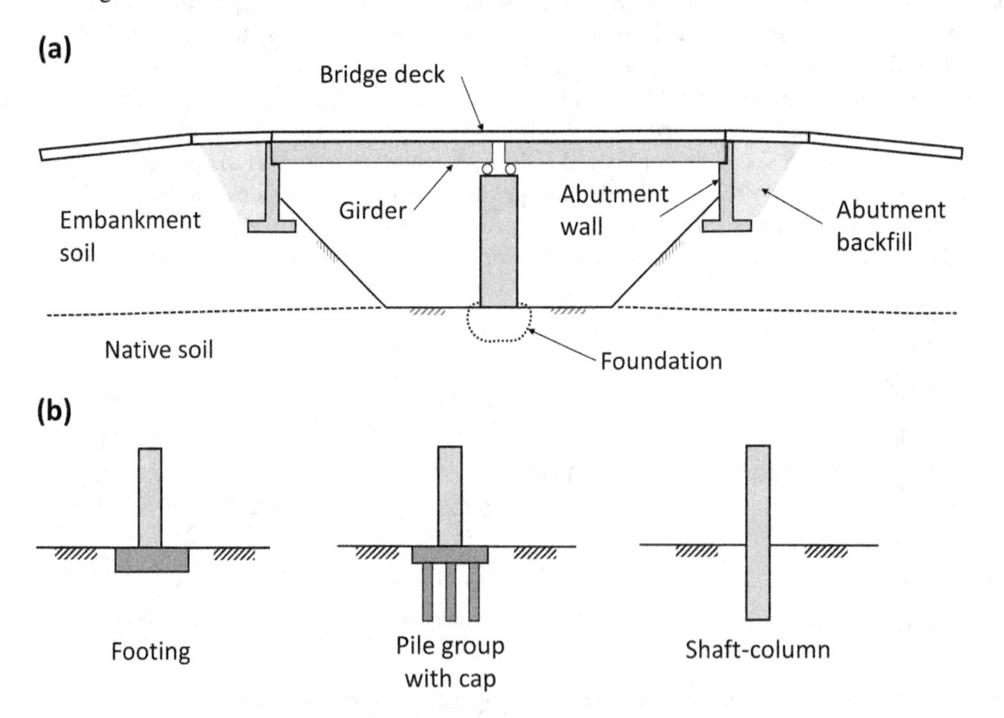

FIGURE 8.38 (a) Configuration of typical bridge structural system, including deck, girders, abutments, and support column; (b) alternate foundation support systems used for bridge columns and abutments.

SSI for bridge structures excited by earthquakes occurs at the abutments and at the base of support columns. SSI associated with abutments can have a dominant effect on bridge response (Faraji et al., 2001; El-Gamal and Siddharthan, 1998), whereas flexibility and damping for foundations at the base of support columns are generally less important. For this reason, specific guidelines for abutments are provided in the SDC whereas modeling procedures for other foundation components are not provided. The SDC applies for *ordinary standard bridges* (OSB), which have spans of limited length (<~90m) and relatively simple geometric configurations (Caltrans, 2013, 2019). These structures can be analyzed using procedures similar to those applied for buildings and described in the previous section, namely equivalent lateral force methods, response spectral methods, and displacement-based methods. Kinematic interaction effects on ground motions are not considered in these analysis methods.

In simplified analyses for OSBs, abutments are included in the bridge model due to their impact on the lateral stiffness and capacity of the bridge systems. Figure 8.39a shows typical components of seat-type bridge abutments, including the stemwall on which the bridge deck is supported, a backwall that separates the end of the bridge deck from the backfill soil, two wingwalls, and retained soil that extends well below the base of the backwall. Figure 8.39b shows a cross-section view of these components. To limit the forces transmitted to the bridge deck, abutment backwalls are often designed to break away from the stemwall when struck by the bridge deck/girders displaced in the longitudinal direction during strong earthquake shaking. Large concrete-on-concrete friction (between bridge deck end members and backwall) does not allow the backwall to uplift as it is displaced into the backfill.

When a backwall is displaced into its backfill soil, the principal source of resistance to lateral displacement is provided by shear resistance within the backfill. The ultimate capacity of the backwall is achieved when the soil is in a state of maximum passive earth pressure (Section 8.6.2.3). As such, the load-deflection behavior of the backwall-fill system is controlled by the thickness, stiffness, and shear strength of the backfill soils. Experimental evidence for cases in which the backwall displaces horizontally and is not allowed to rotate or uplift (Lemnitzer et al., 2009)

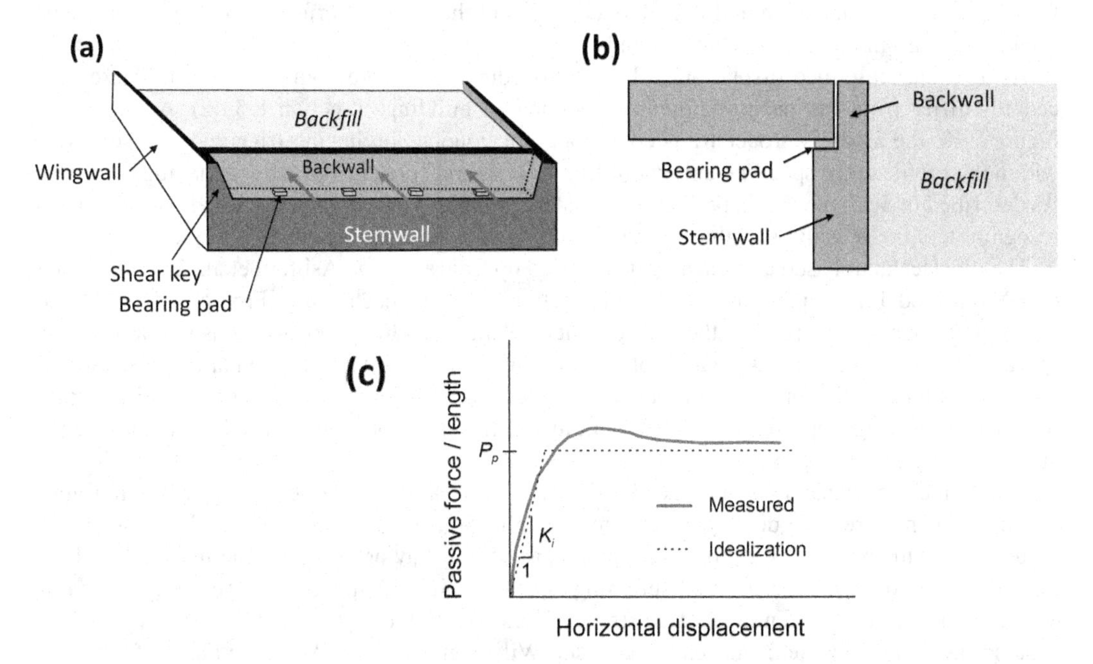

FIGURE 8.39 (a) Isometric view of typical components of seat-type bridge abutment (modified from Lemnitzer et al., 2009); (b) cross-section view of bridge abutment components; (c) backwall force-displacement backbone curve as measured and as idealized for implementation in simplified guidelines.

indicates a log-spiral failure surface geometry and nonlinear load-deflection backbone response (Figure 8.39c) characterized by an initial stiffness, K_i, and an ultimate passive resistance, P_p. In the SDC, these elastic-perfectly plastic springs (Figure 8.39c) have specified values of stiffness and capacity per unit length of wall derived from test data. For example, Caltrans (2013) provided recommended values for 1.7 m tall walls with well compacted sandy backfill (P_p=240 kN/m and K_i=29 (kN/m)/mm). The horizontal wall displacement at which full capacity develops is typically 2%–5% of the wall height.

The SDC and similar documents (e.g., AASHTO, 2014) are not applicable to complex bridge structures, such as those with long spans. For those bridges, SSI is usually considered within the framework of relatively advanced response history analyses in which ground motions are converted to displacement histories applied at the ends of BNWF macro-elements along deep foundations (Figure 8.19). These methods implicitly consider kinematic interaction effects on the input motions at the base of bridge columns and simulate moment and shear demands along pile foundations due to kinematic and inertial SSI effects. For analyses of this sort, cyclic backwall abutment response can be represented using a macro-element (similar to that shown in Figure 8.20) with the plastic spring taken from a suitable nonlinear backbone curve (Shamsabadi et al., 2010; Khalili-Tehrani et al., 2016) in combination with gap elements.

8.5.2.2 Response of Abutments to Lateral Spreading

Bridges crossing bodies of water are often located in sloping ground areas with loose, granular soils and shallow water tables. As a result, pile foundations are often required. These soils are often liquefiable, and liquefaction in the presence of static shear stresses can lead to lateral spreading (Section 9.6.5.2) that can impose displacements upon the piles (and potentially a pile cap) that may affect bridge performance. Because abutments are typically at higher elevations than the ground level at intermediate bridge bents (or columns) (Figure 8.38), the direction of the displacements is typically toward the center of the bridge. Direct dynamic SSI analyses of these problems are not commonly performed due to computational challenges associated with accurately modeling the triggering of liquefaction and lateral spreading, and the complex effects of liquefaction on the ground motions and related inertial demands.

As a result, simplified displacement-based procedures are more commonly used. These procedures differ from the pushover methods applied to buildings (Section 8.5.1.2). As shown in Figure 8.40, the analysis procedure is conditional on liquefaction having triggered in a soil layer within the profile and a specified amount of free-field lateral spread displacement having occurred. As described in Section 9.6.5.2, predictions of lateral spread displacements are highly uncertain, so recognition of the sensitivity of the analysis results to the displacement amount is essential.

The displacement-based approach, which is used by Caltrans (e.g., Ashford et al., 2011), is based on BNWF modeling of pile-soil interaction (Section 8.3.3.3). As shown in Figure 8.40, free-field soil displacements are applied at the ends of macro-elements, which typically consist solely of p–y springs due to the pseudo-static nature of the analysis. Care is required in the parameterization of p–y springs for the pile and cap/wall elements. Referring to Figure 8.40, in the firm soil at depth, ordinary p–y springs representing the strength and stiffness of soils can be used, with appropriate allowances for group effects.

Within the liquefiable soils, a reduced ordinate of p–y backbone curves is applied by multiplication of "p" by a relative density-dependent multiplier m_p, typical values of which are shown in Figure 8.41. This represents a gross simplification of the behavior of liquefiable materials, which in addition to pore-pressure induced softening near the origin in stress-strain space, also experience dilation-induced stiffening at large strains. These dilation effects have been observed in p–y behavior as well during field and centrifuge tests (Wilson et al., 2000; Weaver et al., 2005), but are not accounted for in this simplified method. Group factors are not applied to p–y backbone curves in the liquefied layer. Because the pore pressure increase in liquefied layers can affect the ultimate resistance of soil in adjacent non-liquefied layers, a smear zone is applied near the boundaries

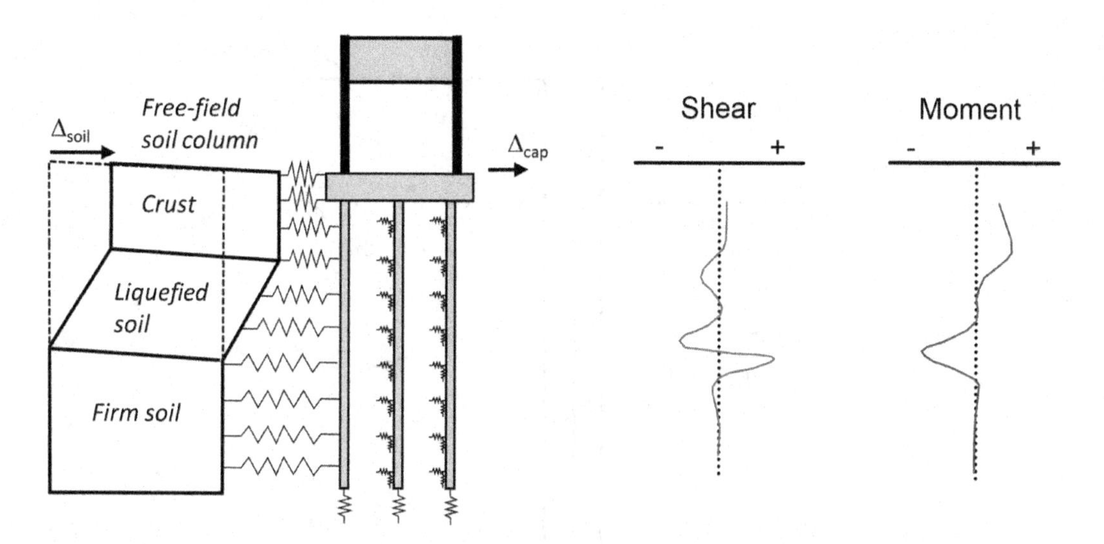

FIGURE 8.40 Schematic of displacement of free-field soil column from liquefaction-induced lateral spreading, its interaction with a pile-supported bridge bent, and representative shear and moment diagrams for the case of pile cap with rotation restrained, as simulated with a BNWF analysis. (Modified from Ashford et al., 2011.)

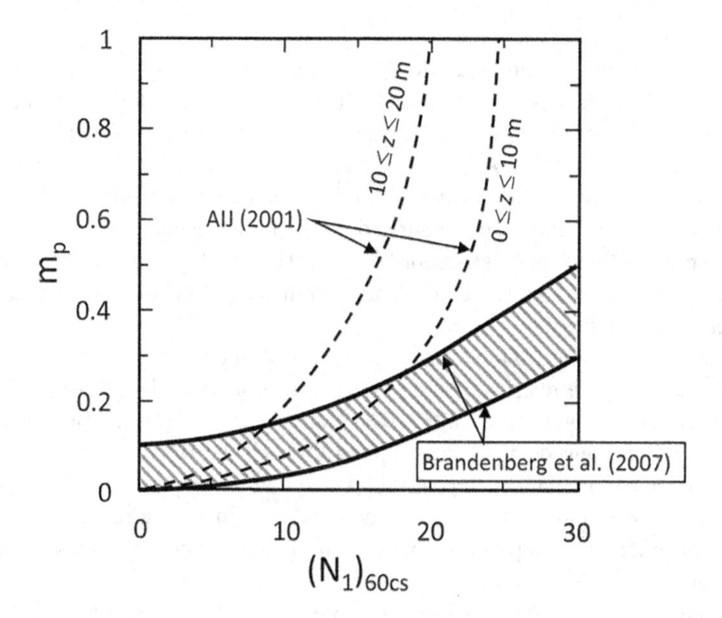

FIGURE 8.41 Dependence of p-multiplier (m_p) with relative density, as reflected by energy- and overburden-corrected SPT blow count $(N_1)_{60}$. (Modified from Boulanger et al., 2007; used with permission of Springer.)

(Figure 8.42). Three-dimensional finite element analyses (e.g., McGann et al., 2011) have shown, however, that conventional p–y curve procedures tend to overpredict pile bending moments even when a smear zone of the type indicated in Figure 8.42 is used.

In the non-liquefied crust above the liquefied layer, both pile elements and a pile cap/wall experience passive pressures (Section 8.6.2.3) that develop on the upslope face and frictional resistance develops on the sides. If the piles are spaced sufficiently closely, the spreading crust passes around the group (as opposed to flowing between the piles), causing the piles and wall/cap to effectively act as a single unit having a height equal to the crust thickness and the width of the pile group.

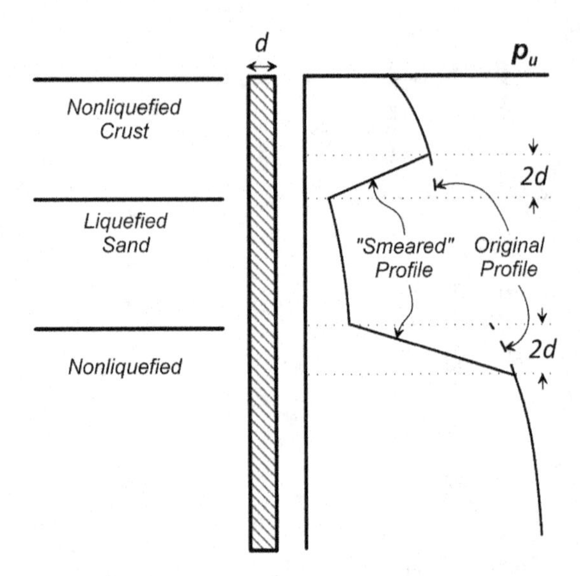

FIGURE 8.42 Modification to the profile of ultimate subgrade reaction, p_u, to account for the weakening effect the liquefied sand exerts on overlying and underlying non-liquefied layers. (Modified from Ashford et al., 2011.)

The ultimate passive earth pressures that develop against the pile/cap block are lower than those predicted by classical earth pressure theories such as Coulomb and log-spiral methods (Section 8.6.2.3) because of low frictional resistance at the base of the crust, where the soil has liquefied, and potential break-up of the crust. As such, the ultimate pressures for these conditions are represented using Rankine passive earth pressure methods, which are based on an assumed condition of zero interface friction. These ultimate pressures develop at displacements typically in the range of 25%–70% of the pile cap/block height (Ashford et al., 2011), which are much larger than the 2%–5% that is typical for passive earth pressure problems. The much softer response is a result of the soft liquefiable layer below the crust.

When the lateral dimensions (in the transverse direction) of a lateral spread feature are relatively narrow, its movement may be restricted by a pinning effect provided by pile-supported abutments. These effects are described by Martin et al. (2002), Ashford et al. (2011), and Boulanger et al. (2012).

When lateral displacement demands are applied through the BNWF model (Figure 8.40), the moment and shear demands that develop in the bridge column depend on the degree of restraint of the bridge deck. For example, if the bridge deck is restrained as a result of its axial stiffness and connection to abutments, imposed displacements in the foundation can impose large bending and shear demands. These demands from ground displacement must be combined with those from inertial loading. Restraint of deck motions from bridge system response also affects inertial demands, which are typically computed using response spectrum methods (Section B.10.4). Inertial forces are taken as the product of spectral displacement (for an oscillator period that adequately accounts for restraining effects of the bridge superstructure) and lateral stiffness of the bridge bent. Recommendations for conducting bridge bent-specific analyses that approximately account for system response are given by Ashford et al. (2011) and Turner et al. (2016).

8.5.3 Nuclear Structures

Structures associated with nuclear facilities can be extremely stiff due to shielding (which requires thick walls) and strict structural performance requirements. As a result, the structure-to-soil stiffness ratio, $h/(V_s T_0)$, is generally high, causing SSI effects to be important. Indeed, much of the current understanding of SSI effects was developed in response to the historical need for their evaluation for nuclear facilities. The American Society of Civil Engineers Standard 4 document (ASCE, 2000; 2016) provides guidance for the seismic analysis of nuclear safety-related facilities and other critical or important facilities in the United States. ASCE-4 is a companion document to ASCE Standard 43 (ASCE, 2005), which is used as a reference document for submittal of combined license applications for new reactors.

Both direct and substructure methods are allowed under ASCE-4. Referring to Figures 8.2–8.4, the analyses include consideration of kinematic interaction effects as well as foundation flexibility and damping. The guidelines emphasize response history analysis procedures. Rather than correcting free-field ground surface motions for embedment effects, input motions are provided directly at the foundation elevation using ground response analyses. The response spectrum for these motions is referred to as the *Foundation Input Response Spectrum*, which is defined for free-field conditions. The 2016 version of ASCE-4 includes provisions for considering base slab averaging effects in which input motions are specified with a defined coherency function (Section 3.8) using software such as SASSI (Lysmer et al., 1999; Ostadan et al., 2005).

Soil-foundation flexibility is represented using spring-based solutions (program CLASSI; Wong, 1979) or equivalent-linear elastic solid finite elements (SASSI). As such, the procedures for nuclear structures have generally been based on elasto-dynamic solutions to the SSI problem.

8.6 RESPONSE OF EMBEDDED STRUCTURES

The embedded foundations discussed to this point are structural elements that support loads from superstructures (i.e., the portions of a structure located above the ground surface) at some depth below the ground surface. Such foundations are generally comprised of stiff, solid, reinforced concrete elements with soil contact at their base and sides, which affects inertial and kinematic interaction effects as discussed in Sections 8.3 and 8.4. The embedded structures considered in this section include earth-retaining structures and buried hollow structures such as tunnels, culverts, tanks, and pipelines. The emphasis here is on the interaction of such structures with the surrounding ground as a result of seismic excitation.

8.6.1 Earth-Retaining Structures

Earth-retaining structures, such as retaining walls, bridge abutments, quay walls, anchored bulkheads, braced excavations, and mechanically stabilized walls, are widely used in infrastructure systems. They frequently represent key elements of ports and harbors, building structures, transportation systems, lifelines, and other constructed facilities. Earthquakes have the potential to destabilize these structures, either due to increases in earth pressures or due to large imposed displacement or force demands from ground failure or from the seismic response of attached superstructures.

8.6.1.1 Types of Earth-Retaining Structures

Many different approaches to soil retention have been developed and used successfully. Figure 8.43 shows common types of retaining walls, which are often classified in terms of their relative mass, flexibility, and anchorage conditions. *Gravity walls* are the oldest and simplest type of retaining wall. Gravity walls are thick and stiff enough that they do not bend; their movement occurs essentially by rigid-body translation and/or rotation. Figure 8.43 indicates the location of the *toe* and *heel* of the gravity wall foundation – these terms are used for all wall systems that involve a foundation system. Certain types of composite wall systems, such as crib walls and *reinforced soil walls*, are thick enough that they bend very little and consequently are often designed (with appropriate consideration of internal stability) as gravity walls. *Cantilever walls*, which bend as well as translate and rotate, require a smaller volume of structural materials and rely on their flexural and shear strength to resist lateral earth pressures. The actual distribution of lateral earth pressure on a cantilever wall is influenced by the relative stiffness and deformation of both the wall and the soil.

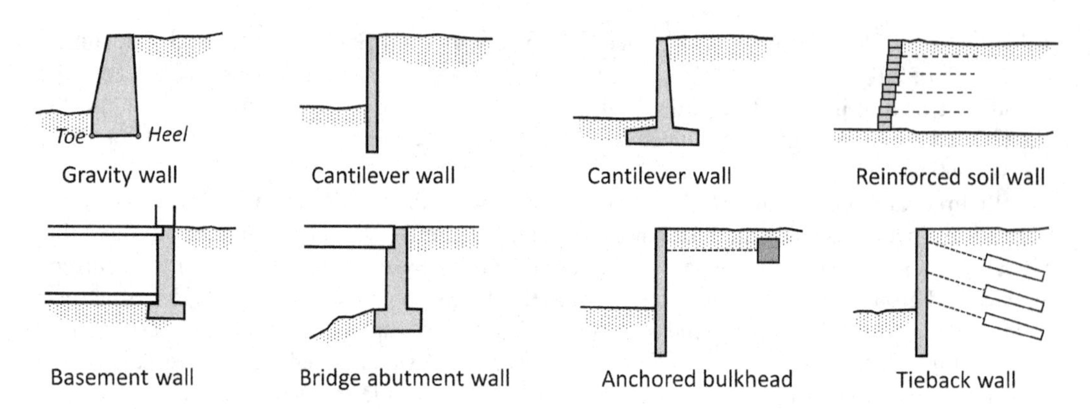

FIGURE 8.43 Common types of earth-retaining structures.

Braced walls are constrained against certain types of movement by the presence of external bracing elements. *Basement walls* and *bridge abutment walls* illustrated in Figure 8.43 may be restrained by the structures they support. *Tieback walls* and *anchored bulkheads* are restrained against lateral movement by anchors embedded in the soil behind the walls. The provision of lateral support at different locations along a braced wall may keep bending moments sufficiently low that relatively flexible structural sections can be used.

8.6.1.2 Failure Mechanisms

To design retaining walls, it is necessary to define "failure" and to know how walls can fail. Under static conditions, retaining walls are acted upon by body forces related to the mass of the wall, by soil pressures, and by external forces such as those transmitted by braces. A properly designed retaining wall will achieve equilibrium of these forces without inducing shear stresses that approach the shear strength of soils below the wall foundation. During an earthquake, however, inertial forces and changes in soil strength may violate equilibrium and cause permanent deformation of the wall. Failure, whether by sliding, tilting, bending, or some other mechanism, occurs when these permanent deformations become excessive. The question of what level of deformation is excessive depends on many factors and is best addressed on an application-specific basis.

Gravity walls usually fail by rigid-body mechanisms such as sliding and/or overturning or by gross instability (Figure 8.44). Sliding occurs when horizontal force equilibrium is not maintained (i.e., when the lateral pressures on the back of the wall produce a thrust that exceeds the available sliding resistance on the base of the wall). Overturning failures occur when moment equilibrium is not satisfied; uplift of the foundation heel and/or bearing failures at the toe are often involved.

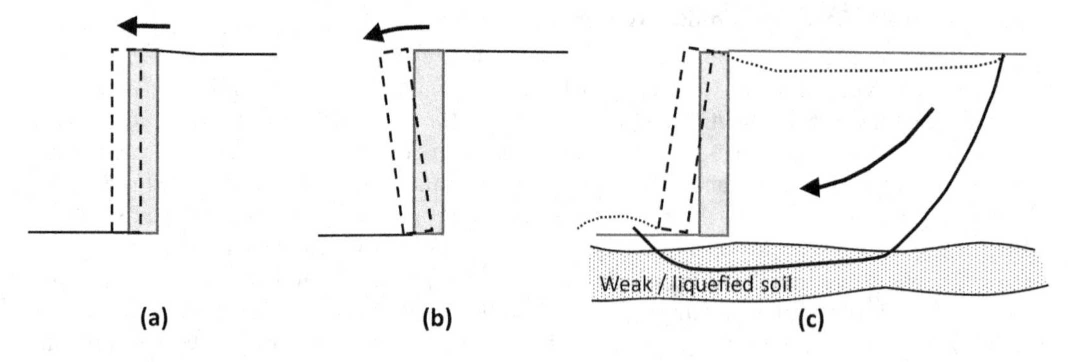

FIGURE 8.44 Typical failure mechanisms for a gravity retaining wall: (a) sliding (translational) failure; (b) overturning (rotational) failure; and (c) gross instability failure.

Gravity walls may also be damaged by gross instability of the soils behind and beneath them. Such failures may be treated as slope stability failures that encompass the wall. Composite wall systems, such as crib walls, bin walls, and mechanically stabilized walls, can fail in the same ways or by a number of internal mechanisms that may involve shearing, pullout, or tensile failure of various wall elements. Procedures for design of composite retaining wall systems against these possible failure mechanisms are given elsewhere (e.g., Salgado, 2008; Berg et al., 2009a, b; Allen and Bathurst, 2018).

Cantilever walls are subject to the same failure mechanisms as gravity walls, but can also be subject to internal (structural) failure modes, typically in flexure. Figure 8.45 depicts soil pressures and bending moments in cantilever walls. Lateral soil pressures are the subject of the following Section 8.6.2. Bending moments develop in the wall to maintain equilibrium and reach their maximum values at the base of the wall. If they exceed the flexural strength of the wall, flexural failure may occur. The structural ductility of the wall will influence the level of deformation produced by flexural failure.

Braced walls usually fail by gross instability, tilting, flexural failure, and/or failure of bracing elements. Tilting of braced walls typically involves rotation about the point at which the brace acts on the wall, often the top of the wall as in the cases of basement and bridge abutment walls (Figure 8.46a). Anchored walls can fail by "kicking out" at their toes (Figure 8.46b) due to inadequate penetration or by tilting away from the backfill due to inadequate anchor capacity (Figure 8.46c). As in the case of cantilever walls, anchored walls may fail in flexure, although the point of failure (maximum bending moment) is likely to be different. Buried anchor elements may also be susceptible to degradation from environmental factors such as corrosion. Failure of bracing elements can include anchor pullout, tierod failure, or bridge deck element buckling.

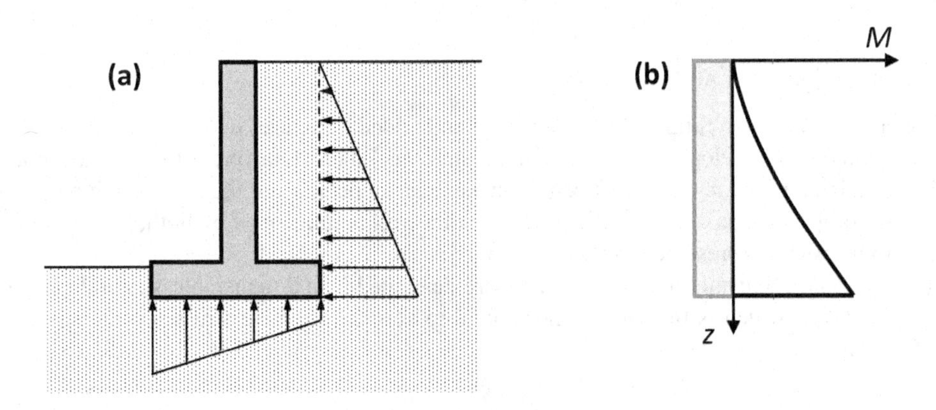

FIGURE 8.45 (a) Soil pressures and (b) bending moments for cantilever retaining walls.

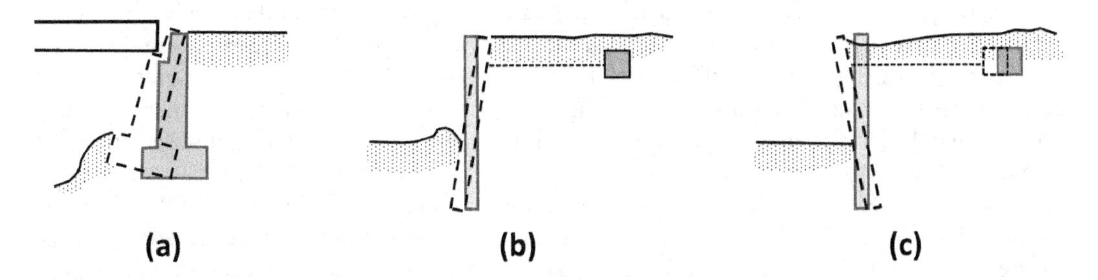

FIGURE 8.46 Potential modes of failure for braced walls: (a) rotation of bridge abutment about top; (b) rotation of anchored bulkhead due to lack of passive resistance ("kickout") at the toe; (c) lack of adequate anchor capacity.

Earthquake loading has the potential to produce any of the failure mechanisms described above. As described in Section 8.6.3, earthquakes change the earth pressures (demands) acting on embedded structures relative to those that exist under static conditions. These pressures can develop (1) during ground shaking for free-standing walls or other embedded structures, (2) from ground failure within backfill soil materials or below wall foundations (liquefaction, cyclic softening, etc.), or (3) from loads applied externally to the wall system from attached vibrating structures. These changes in demands can be exacerbated by reductions in capacity when the soil loses strength, for example from liquefaction of soils beneath wall foundations.

Observations of wall performance differ for the three broad categories of demand. In the case of demands from ground shaking, the observed performance of retaining structures and basement walls for structures is predominantly quite good (Lew et al., 1995, 2010). A notable exception is a series of wall failures in strongly shaken areas during the 1995 **M** 6.9 Kobe Japan earthquake, which involved both engineered and non-engineered gravity and cantilever walls, the failure of which was typically by rotation (Figure 8.44b) (Tatsuoka et al., 1996). Reinforced soil walls have also been observed to perform well, arguably even better than relatively stiff gravity or cantilever wall systems (Tatsuoka et al., 1996; Koseki and Shibuya, 2014), although some instances of poor performance have been observed (Ling et al., 2001; Koseki and Shibuya, 2014). In contrast, demands of the second type (from ground failure, especially soil liquefaction) have produced extensive failures of quay walls and wharf structures in translation, rotation, and global instabilities (e.g., Inagaki et al., 1996; Kerwin and Stone, 1997; Tanaka et al., 2001; Gazetas et al., 2004). The effects on walls of demands of the third type (from externally applied inertial forces) are relatively rarely documented for buildings but have been observed from pounding of bridge decks into abutment walls in several large events (**M** 7.6 1999 Chi Chi Taiwan and **M** 8.8 2010 Maule Chile; K Rollins and A Shamsabadi, personal communication).

8.6.2 Static Lateral Earth Pressures

The seismic response of retaining walls and other embedded structures depends on the total lateral earth pressures that develop during earthquake shaking. These total pressures include both the static gravitational pressures that exist before an earthquake occurs and the transient dynamic pressures induced by earthquakes. Since the response of a wall is influenced by both, a brief review of static earth pressures is presented in this section.

By convention, static lateral earth pressures are described by a dimensionless lateral earth pressure coefficient K, which is the ratio of horizontal-to-vertical effective stresses on a soil element (Figure 8.47a):

$$K = \frac{\sigma'_h}{\sigma'_v} \tag{8.55}$$

Static earth pressures on retaining structures are strongly influenced by wall and soil movements. When horizontal displacements (and lateral strains) are zero, lateral earth pressures are said to be *at-rest* and the earth pressure coefficient is denoted K_0 (Figure 8.47b). At-rest pressures are typically assumed for retaining walls restrained against horizontal movement, such as tieback walls, anchored bulkheads, basement walls, and integral bridge abutments (i.e., abutment walls integrated with the bridge deck).

As depicted in Figure 8.47b, when a wall is allowed to move away from the backfill soil, extensional lateral strain develops in the soil and lateral earth pressures drop (from at-rest) as shear stresses in the backfill increase to resist accompanying soil displacements. At the limit, the shear strength of the backfill soil is fully mobilized, at which point lateral earth pressures are referred to as *minimum active* and the lateral earth pressure coefficient is denoted K_A. The level of displacement required to mobilize minimum active earth pressures is typically on the order of 0.2% or less of the wall height for clean sandy backfills (Terzaghi, 1934, 1943). Because these displacements

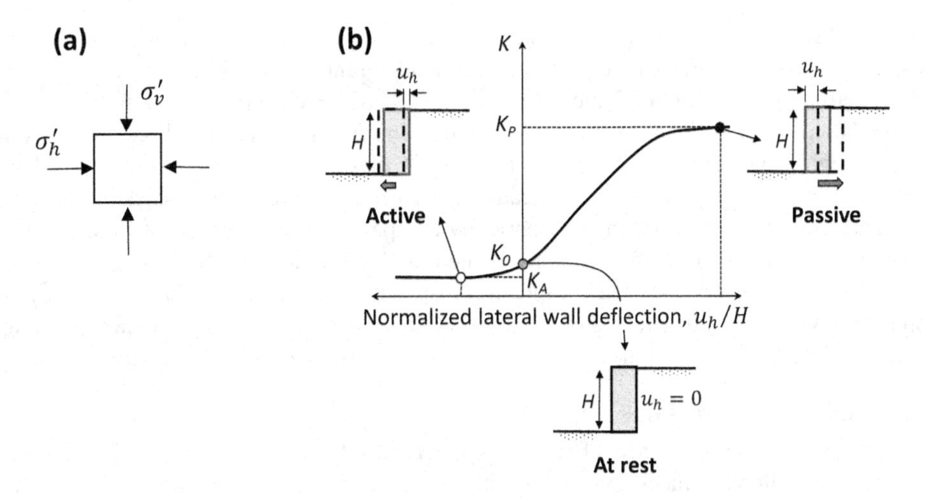

FIGURE 8.47 (a) Vertical and horizontal stresses on soil element (b) illustration of the relationship between lateral earth pressure coefficient K and normalized wall deflection. (Modified from Salgado, 2008; used with permission of R. Salgado.)

are very small, it is reasonable to assume that active earth pressures will develop for free-standing retaining walls and they are typically designed on this basis. When a retaining wall is displaced toward the backfill soil, compressive lateral strain occurs in the soil and the shear resistance of the soil is mobilized to resist this displacement. When the strength of the soil is fully mobilized, a state of *maximum passive* earth pressure acts against the wall with a corresponding lateral earth pressure coefficient denoted K_P. The level of displacement required to develop maximum passive earth pressures is much greater than for minimum active conditions, being on the order of 2%–5% of the wall height for clean sandy backfills (Tsagareli, 1965; Matsuo et al., 1978; Fang and Ishibashi, 1986; Lemnitzer et al., 2009).

The presence of water in the backfill behind a retaining wall influences the effective stresses and hence the lateral earth pressure that acts on the wall. For wall design, the hydrostatic pressure applied by the water must be added to the lateral earth pressure (since fluid pressures are equal in all directions, no lateral earth pressure coefficient is applied to the water pressure). Because the total lateral thrust on a wall retaining a saturated backfill is considerably greater than that on a wall retaining dry backfill, the provision of backfill drainage is an important part of retaining wall design.

The following subsections describe common simplified methods for prediction of at-rest, active, and passive earth pressures. These methods make certain simplifying assumptions (e.g., rigid walls and uniform soils) to enable the use of closed-form equations.

8.6.2.1 At-Rest Earth Pressures

Under normally consolidated conditions, at rest lateral earth pressures that develop in soil have been shown from empirical studies to be well described by (Jaky, 1948):

$$K_0 = 1 - \sin\phi' \tag{8.56}$$

where ϕ' is the effective stress friction angle of the soil. Historically, a strong effect of over consolidation on at-rest pressures has been used, which is typically described by (Mayne and Kulhawy, 1982):

$$K_0 = (1 - \sin\phi')(OCR)^{\sin\phi'} \tag{8.57}$$

Talesnick (2012) has challenged both of the above expressions based on testing performed with a non-displacing pressure cell installed within 'free-field' granular soil (i.e., away from walls) in a laboratory setting. He found that K_0 increases with soil density (contrary to the trend expected from Equation 8.56) and that K_0 does not appreciably change in vertical stress unload-reload cycles, which suggests no significant effect of over-consolidation. Of these two findings, the latter is of greater practical significance, because it can significantly influence K_0 estimates (Talesnick's measurements of K_0 ranged from 0.4 to 0.6, which is numerically comparable to predictions of Equation 8.56 for sands, whereas predictions of K_0 for overconsolidated sands from Equation 8.57 can be much higher). Further testing with sensors installed in rigid, non-displacing walls indicates that horizontal and vertical pressures were modestly affected by frictional effects along the walls, but broadly speaking similar trends for K_0 are found as in the free-field case (M. Talesnick, *personal communication*, 2016).

8.6.2.2 Active Earth Pressures

Solutions for active earth pressure coefficients have been developed by Rankine (1857) and Coulomb (1776) for planar failure surfaces and for logarithmic spiral surfaces (Rendulic, 1935; Taylor, 1937; Caquot and Kerisel, 1948). Each of the procedures make different assumptions about soil, wall, and/or failure surface geometry.

Rankine (1857) developed the simplest procedure for computing active earth pressures. By making assumptions about the stress conditions and strength envelope of the soil behind a retaining wall (the backfill soil), Rankine was able to render the lateral earth pressure problem determinate and directly computed the static pressures acting on retaining walls. Active pressures at a point on the back of a retaining wall were expressed by Rankine as:

$$p_A = K_A \sigma'_v - 2c'\sqrt{K_A} \tag{8.58}$$

where c' is the effective stress cohesion parameter. When the principal stress planes are vertical and horizontal (as in the case of a smooth vertical wall retaining a horizontal backfill), the coefficient of minimum active earth pressure is given by

$$K_A = \frac{1-\sin\phi'}{1+\sin\phi'} = \tan^2\left(45 - \frac{\phi'}{2}\right) \tag{8.59}$$

For the case of a cohesionless backfill inclined at an angle β with the horizontal, infinite slope solutions can be used (Terzaghi, 1943; Taylor, 1948) to compute K_A as

$$K_A = \cos\beta \frac{\cos\beta - \sqrt{\cos^2\beta - \cos^2\phi'}}{\cos\beta + \sqrt{\cos^2\beta - \cos^2\phi'}} \tag{8.60}$$

for $\beta \leq \phi'$. Equation (8.60) is equivalent to Equation (8.59) when $\beta = 0$. The pressure distribution on the back of the wall, as indicated by Equation (8.58), depends on the relative contributions of friction and cohesion to the backfill soil shear strength (Figure 8.48). Although the presence of cohesion indicates that tensile stresses will develop between the upper portion of the wall and the backfill, tensile stresses do not actually develop in the field. The creep, stress relaxation, and low-permeability characteristics of cohesive soils render them undesirable as backfill material for retaining structures, and their use in that capacity is avoided when possible. For dry homogeneous cohesionless backfill, Rankine theory predicts a triangular active pressure distribution oriented parallel to the backfill surface. The active earth pressure resultant (or thrust force), P_A, acts at a point located $H/3$ above the base of a wall of height H (Figure 8.48a) with magnitude

$$P_A = \frac{1}{2} K_A \gamma H^2 \tag{8.61}$$

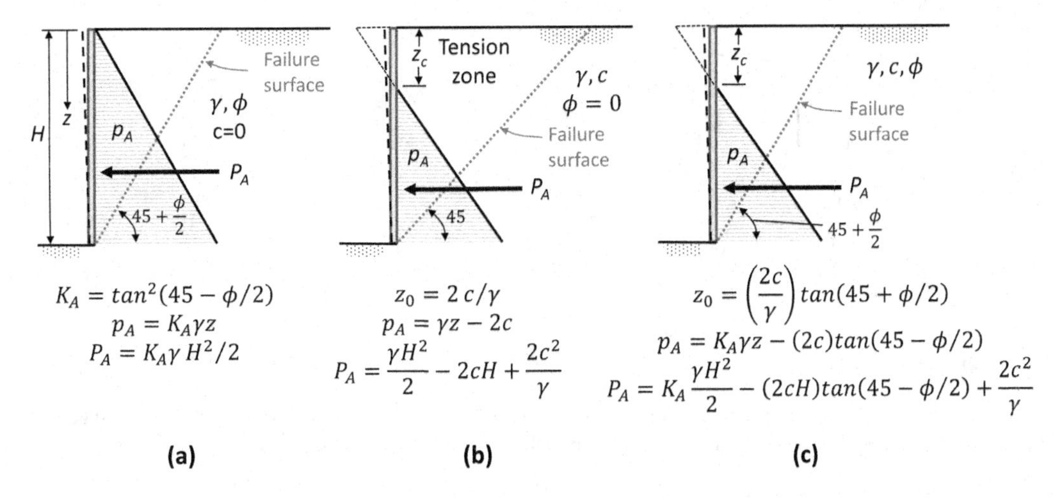

FIGURE 8.48 Rankine active earth pressure distributions for backfills with various combinations of frictional and cohesive strength: (a) friction resistance, no cohesion, (b) cohesive soil, no friction, (c) combined friction and cohesion. (After U.S. Navy, 1982.)

Example 8.7

A free-standing, unrestrained retaining wall at a Class *D* soil site has height H=6 m and retains soil with a unit weight of 18 kN/m^3, Poisson's ratio of 0.3, and active earth pressure coefficient of 0.25. The wall has good drainage, so that the water table will not build up behind the wall. What is the static earth pressure resultant force?

Solution:

Assuming a triangular distribution of static active earth pressure, the resultant force would be calculated using Equation (8.61) as: $P_A = \frac{1}{2} K_A \gamma H^2 = 0.5 \times 0.25 \times 18\ \text{kN/m}^3 \times (6\ \text{ft})^2 = 81\ \text{kN/m}$

Coulomb (1776) was the first to study the problem of lateral earth pressures on retaining structures. He assumed that the force acting on the back of a retaining wall is that required to maintain equilibrium of a wedge of soil above a planar failure surface (Figure 8.49). The active earth pressure resultant can be computed for a given a wall-soil interface friction angle δ, backfill angle, β, vertical angle of back side of wall, θ, and base angle of soil wedge, α_A. The problem is indeterminate in the sense that angle, α_A, is unknown and different wall reaction forces are derived for each α_A. The solution is obtained by using the surface that produces the greatest active thrust force. The thrust force is related to K_A using Equation (8.61) with K_A expressed as

$$K_A = \frac{\cos^2(\phi' - \theta)}{\cos^2\theta\cos(\delta + \theta)\left[1 + \sqrt{\frac{\sin(\delta + \phi')\sin(\phi' - \beta)}{\cos(\delta + \theta)\cos(\beta - \theta)}}\right]^2} \quad (\beta < \phi') \tag{8.62}$$

The critical failure surface is inclined at an angle

$$\alpha_A = \phi' + \tan^{-1}\left[\frac{\tan(\phi' - \beta) + C_1}{C_2}\right] \tag{8.63}$$

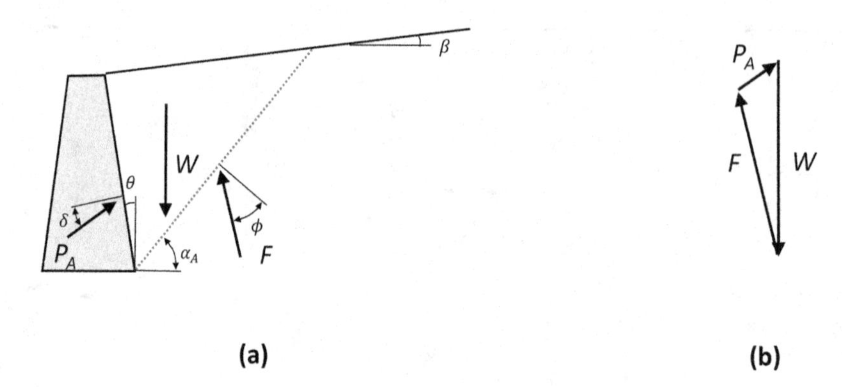

FIGURE 8.49 (a) Triangular active wedge bounded at base by planar sliding surface; (b) force polygon for active Coulomb wedge. The critical failure surface is that which maximizes P_A.

to the horizontal where

$$C_1 = \sqrt{\tan(\phi' - \beta)\left[\tan(\phi' - \beta) + \cot(\phi' - \theta)\right]\left[1 + \tan(\delta + \theta)\cot(\phi' - \theta)\right]} \tag{8.64a}$$

$$C_2 = 1 + \left\{\tan(\delta + \theta)\left[\tan(\phi' - \beta) + \cot(\phi' - \theta)\right]\right\} \tag{8.64b}$$

Coulomb theory does not explicitly predict the distribution of active pressure, but it can be shown to be triangular for linear backfill surfaces with no surface loads. In such cases, P_A acts at a point located $H/3$ above the base of a wall of height H and is inclined to a normal at the back of the wall at the interface friction angle, δ, which is typically taken as approximately 1/2–2/3 of ϕ' for soil-concrete interfaces and 1/3–1/2 of ϕ' for soil-steel interfaces.

The logarithmic spiral method for active earth pressures is illustrated in Figure 8.50. Log-spiral methods were introduced for limit equilibrium problems by Rendulic (1935), followed by Taylor (1937), and were first tabulated for use in the prediction of active and passive pressures on retaining walls by Caquot and Kerisel (1948). While the major principal stress may act nearly perpendicular to the backfill surface at some distance behind a rough ($\delta > 0$) wall, the presence of shear stresses on the wall-soil interface rotates principal stresses near the back of the wall. If the inclination of principal stresses varies within the backfill, the inclination of the failure surface must also vary. In other words, the failure surface must be curved, which can be described by a logarithmic spiral function. The critical failure surface consists of a curved portion near the back of the wall and a linear

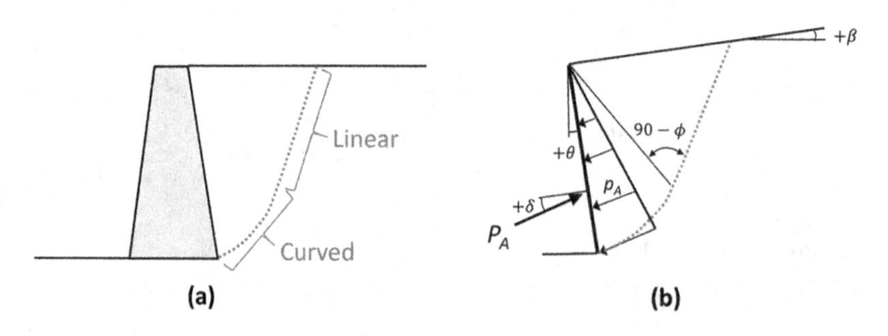

FIGURE 8.50 (a) Logarithmic spiral representation of the critical failure surface for active earth pressure conditions; (b) orientation of critical failure surface for nonvertical wall with inclined backfill surface.

portion that extends up to the ground surface (Figure 8.50a). The active earth pressure distribution is triangular (Figure 8.50b) for walls retaining cohesionless backfills. Thus the active soil thrust is given by Equation (8.61) using tabulated values of K_A from Table 8.5. The active earth pressure coefficients given by the log-spiral approach are generally considered to be slightly more accurate than those given by Rankine or Coulomb theory, but the difference is so small that the more convenient Coulomb approach is usually used.

8.6.2.3 Passive Earth Pressures

Solutions for passive earth pressure coefficients have been developed according to the same principles applied for active conditions by Rankine (1857), Coulomb (1776), and Terzaghi (1943)/Caquot and Kerisel (1948) (log-spiral). Rankine theory predicts wall pressures given by

$$p_P = K_P\sigma'_v + 2c'\sqrt{K_P} \tag{8.65}$$

where K_P is the coefficient of passive earth pressure. For smooth, vertical walls retaining horizontal backfills,

$$K_P = \frac{1+\sin\phi'}{1-\sin\phi'} = \tan^2\left(45+\frac{\phi'}{2}\right) \tag{8.66}$$

and

$$K_P = \cos\beta\frac{\cos\beta+\sqrt{\cos^2\beta-\cos^2\phi'}}{\cos\beta-\sqrt{\cos^2\beta-\cos^2\phi'}} \tag{8.67}$$

TABLE 8.5
Values of K_A and K_P for Log-Spiral Failure Surfaces

			ϕ											
			20°		25°		30°		35°		40°		45°	
δ	β	θ	K_A	K_P	K_A	K_P	K_A	K_P	K_A	K_P	K_A	K_P	K_A	K_P
0°	−15°	−10°	0.37	1.32	0.30	1.66	0.24	2.05	0.19	2.52	0.14	3.09	0.11	3.95
		0°	0.42	1.09	0.35	1.33	0.29	1.56	0.24	1.82	0.19	2.09	0.16	2.48
		10°	0.45	0.87	0.39	1.03	0.34	1.17	0.29	1.30	0.24	1.33	0.21	1.54
0°	0°	−10°	0.42	2.33	0.34	2.96	0.27	3.82	0.21	5.00	0.16	6.68	0.12	9.20
		0°	0.49	2.04	0.41	2.46	0.33	3.00	0.27	3.69	0.22	4.59	0.17	5.83
		10°	0.55	1.74	0.47	1.89	0.40	2.33	0.34	2.70	0.28	3.14	0.24	3.69
0°	15°	−10°	0.55	3.36	0.41	4.56	0.32	6.30	0.23	8.98	0.17	12.2	0.13	20.0
		0°	0.65	2.99	0.51	3.86	0.41	5.04	0.32	6.72	0.25	10.4	0.20	12.8
		10°	0.75	2.63	0.60	3.23	0.49	3.97	0.41	4.98	0.34	6.37	0.28	8.20
ϕ	−15°	−10°	0.31	1.95	0.26	2.90	0.21	4.39	0.17	6.97	0.14	11.8	0.11	22.7
		0°	0.37	1.62	0.31	2.31	0.26	3.35	0.23	5.04	0.19	7.99	0.17	14.3
		10°	0.41	1.29	0.36	1.79	0.31	2.50	0.27	3.58	0.25	5.09	0.23	8.86
ϕ	0°	−10°	0.37	3.45	0.30	5.17	0.24	8.17	0.19	13.8	0.15	25.5	0.12	52.9
		0°	0.44	3.01	0.37	4.29	0.30	6.42	0.26	10.2	0.22	17.5	0.19	33.5
		10°	0.50	2.57	0.43	3.50	0.38	4.98	0.33	7.47	0.30	12.0	0.26	21.2
ϕ	15°	−10°	0.50	4.95	0.37	7.95	0.29	13.5	0.22	24.8	0.17	50.4	0.14	115
		0°	0.61	4.42	0.48	6.72	0.37	10.8	0.32	18.6	0.25	39.6	0.21	73.6
		10°	0.72	3.88	0.58	5.62	0.46	8.51	0.42	13.8	0.35	24.3	0.31	46.9

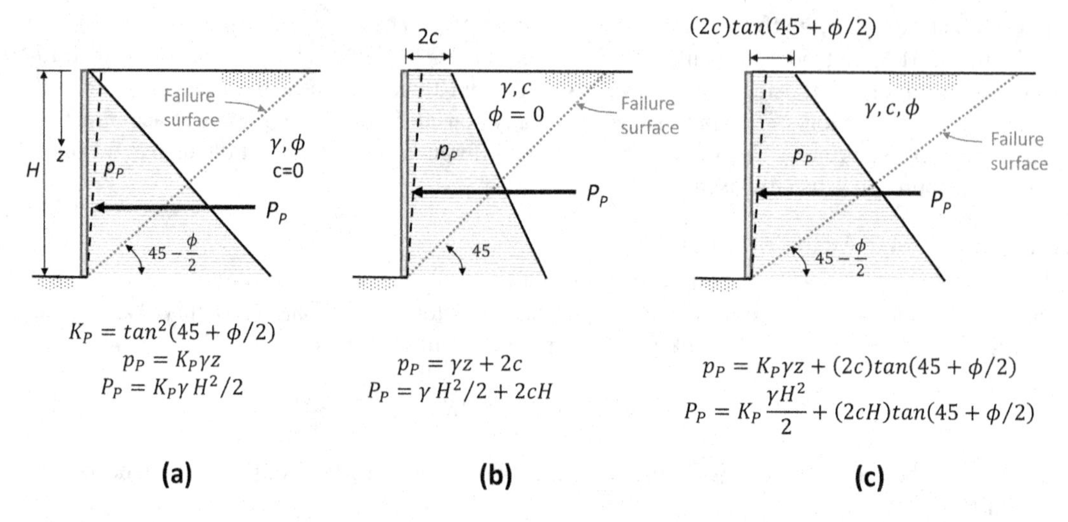

FIGURE 8.51 Rankine passive earth pressure distributions for backfills with various combinations of frictional and cohesive strength: (a) friction resistance, no cohesion, (b) cohesive soil, no friction, (c) combined friction and cohesion. (After U.S. Department of the Navy, 1982.)

for backfills inclined at angle β to the horizontal. Passive pressure distributions for various backfill strength characteristics are shown in Figure 8.51. For a dry homogeneous backfill, Rankine theory predicts a triangular passive pressure distribution oriented parallel to the backfill surface. The passive earth pressure resultant, or passive thrust, P_P, acts at a point located $H/3$ above the base of a wall of height H (Figure 8.51a) with magnitude

$$P_P = \frac{1}{2} K_P \gamma H^2 \tag{8.68}$$

The passive thrust force from Coulomb theory is also described by Equation (8.68), but with K_P computed as

$$K_P = \frac{\cos^2(\phi' + \theta)}{\cos^2\theta\cos(\delta - \theta)\left[1 - \sqrt{\dfrac{\sin(\delta + \phi')\sin(\phi' + \beta)}{\cos(\delta - \theta)\cos(\beta - \theta)}}\right]^2} \tag{8.69}$$

The angles used in the solution are defined in Figure 8.52 and mirror those for the active case, the only difference being the direction of the lateral earth pressure with respect to the back of the wall. The critical failure surface is again planar with an angle

$$\alpha_P = -\phi' + \tan^{-1}\left[\frac{\tan(\phi' + \beta) + C_3}{C_4}\right] \tag{8.70}$$

to the horizontal where

$$C_3 = \sqrt{\tan(\phi' + \beta)\left[\tan(\phi' + \beta) + \cot(\phi' + \theta)\right]\left[1 + \tan(\delta - \theta)\cot(\phi' + \theta)\right]} \tag{8.71a}$$

$$C_4 = 1 + \left\{\tan(\delta - \theta)\left[\tan(\phi' + \beta) + \cot(\phi' + \theta)\right]\right\} \tag{8.71b}$$

Whereas the positions of base of the sliding surface for the active case are similar for the Coulomb and log-spiral solutions, the differences can be much more significant in the case of passive earth pressures. As shown in Figure 8.53, the curved portion of the failure surface is much

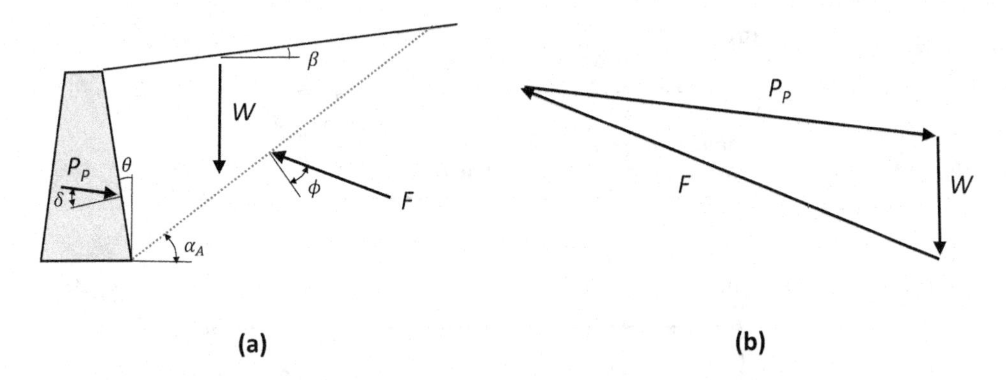

FIGURE 8.52 (a) Triangular passive wedge bounded at base by planar sliding surface; (b) force polygon for passive Coulomb wedge. The critical failure surface is that which minimizes P_P.

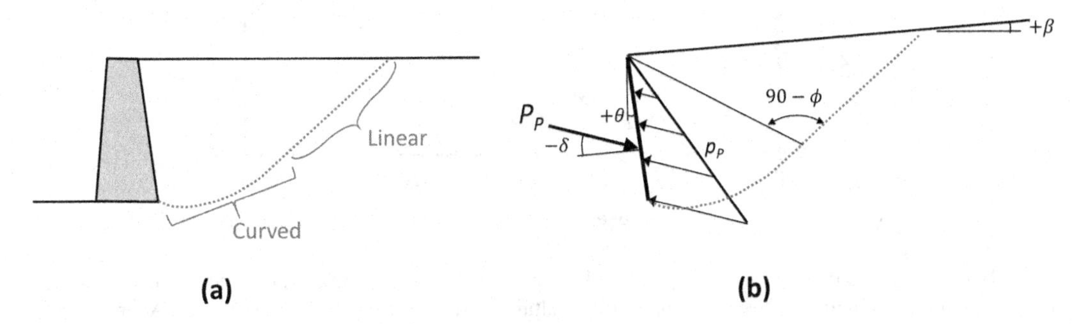

FIGURE 8.53 (a) Logarithmic spiral representation of the critical failure surface for passive earth pressure conditions; (b) orientation of critical failure surface for nonvertical wall with inclined backfill surface.

more pronounced than for the active condition and typically extends below the elevation of the wall heel – this makes for a larger failure surface but lower (and thus more critical) values of K_P. The passive thrust force from the log-spiral solution for granular backfills is given by Equation (8.68), with the K_P values tabulated in Table 8.5. Experimental results show that maximum passive earth pressure coefficients given by the log-spiral method are considerably more accurate than those given by Rankine or Coulomb theory, which tend to underpredict and overpredict, respectively (Mokwa and Duncan, 2001; Rollins and Cole, 2006; Rollins and Sparks, 2002; Lemnitzer et al., 2009). The maximum passive earth pressure resultants evaluated using log-spiral methods can be extended to load-deflection relationships using hyperbolic functions, details of which are presented by Shamsabadi et al. (2007) and Khalili-Tehrani et al. (2016).

For bridge abutment applications, it is often required to compute passive capacities for cases in which the inertial load applied to the abutment wall by a bridge deck is skewed relative to the alignment of the abutment. Figure 8.54a shows such a case, with skew angle, Θ, defined as the angle between the abutment wall alignment and the transverse direction of the bridge deck. As shown in Figure 8.54a, the bridge deck inertial load, P_L, is resisted within the abutment by a maximum passive earth pressure resultant, P_P, and a shear force, P_R. The value of P_P in this case is reduced relative to the values obtained without skew (i.e., Equations 8.68 with K_P taken from Table 8.5) by the factor R_{skew}, which depends on skew angle as shown in Figure 8.54b. The fit relationship for R_{skew} in Figure 8.54b is based on model tests and numerical simulations (Rollins and Jessee, 2013). For cohesionless backfills, shear resistance, P_R, is related to P_P as:

$$P_R = P_P \tan \delta \tag{8.72}$$

where δ, as before, is the wall-soil interface friction angle.

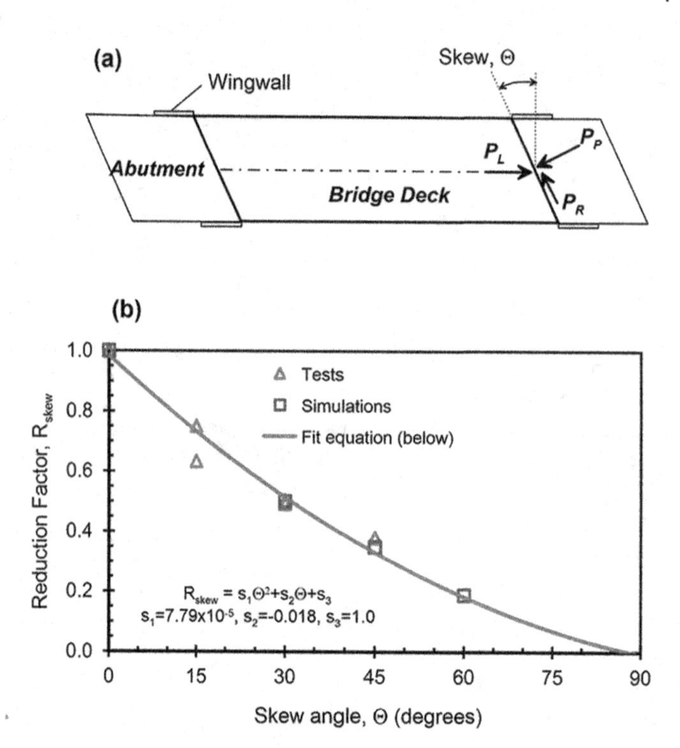

FIGURE 8.54 (a) Configuration of bridge with skewed abutment showing forces at deck-abutment interface; (b) variation of maximum passive earth pressure resultant reduction factor with skew angle. (Adapted from Rollins and Jessee, 2013.)

8.6.3 Seismic Lateral Earth Pressures

As described in Section 8.6.1, earthquake loading changes the earth pressures (demands) acting on embedded structures from those that exist under static conditions. Section 8.6.2 highlighted the profound effect of wall displacement on static lateral earth pressures. Seismic earth pressures can be understood in the same context, but with the additional complexity that both the wall and the free-field soil displace as a result of earthquake shaking. At any point during an earthquake, when the wall displaces away from the backfill, earth pressures can drop from the static condition (if not already in a minimum active state). Conversely, and more importantly for engineering application, at any point in time when the wall displaces toward the backfill, earth pressures will rise relative to the static condition. Since earthquake loading is transient, both increases and decreases in earth pressure are to be expected at different times over the duration of shaking; engineering procedures focus on the former for practical application.

Mechanisms that can produce relative wall-soil displacements include: (1) different levels of transient wall displacement relative to the free-field independent of the inertia of structures that may be connected to the wall; (2) free-field permanent ground displacements arising from soil strength loss (e.g., liquefaction, cyclic softening, etc.); and (3) forces applied to a wall/foundation system by attached vibrating structures, which in turn produce relative foundation/free-field displacements independent from those in (1). Mechanism (1) is a kinematic SSI problem having some similarity to the analysis of u_{FIM} and θ_{FIM} for embedded structures as described in Section 8.4.2. Mechanism (2) is a ground failure problem and is not discussed further here (analysis procedures for predicting the onset of ground failure are described in Chapter 9; the interaction of lateral spreads with foundations, which may include walls, is described in Section 8.5.2.2). Mechanism (3) is an inertial SSI problem that requires analysis of the response of a structure connected to a suitable foundation system using either direct (Figure 8.1) or substructure (Figures 8.2–8.4) methods of analysis.

The following subsections describe classical methods for the analysis of seismic earth pressure that are based on the acceleration of backfill soil, the kinematic wall response problem and available solutions, and considerations in the inertial wall-soil response problem.

8.6.3.1 Methods Based on Pseudo-Static Backfill Force

For many years, seismic earth pressures on retaining walls have been analyzed using a pseudo-static framework in which accelerations are applied to backfill materials, the wall is assumed to displace forming a soil wedge at a state of shear failure in the backfill (i.e., limit state), and resulting wall force demands are analyzed. Figure 8.55 illustrates the concept for the case of an active soil wedge with horizontal acceleration of $k_h g$ and vertical acceleration of $k_v g$ (where g is acceleration of gravity), which produce horizontal and vertical inertial forces in the backfill. The force polygon in Figure 8.55b illustrates how these inertial forces increase the wall soil interaction force by an amount, P_E, above the static force, P_A (assuming active conditions prior to the earthquake). By solving the equilibrium problem in Figure 8.55b for a variety of acceleration levels, the combined thrust P_A+P_E can be computed as

$$P_A + P_E = \frac{1}{2}\gamma H^2 K_{AE}(1-k_v) \tag{8.73}$$

where K_{AE} is a seismic earth pressure coefficient that combines the effects of gravity and the pseudo-static inertial forces in the wedge, and is given by (e.g., Koseki et al., 1998):

$$K_{AE} = \frac{\cos^2(\phi' - \theta - \psi)}{\cos\psi\cos^2\theta\cos(\delta-\theta-\psi)\left[1+\sqrt{\dfrac{\sin(\phi'+\delta)\sin(\phi'-\beta-\psi)}{\cos(\delta+\theta+\psi)\cos(\theta-\beta)}}\right]^2} \tag{8.74}$$

The angles ϕ', θ, β, and δ in Equation (8.74) are as defined in Figure 8.49, and ψ is related to the pseudo-static accelerations as:

$$\psi = \tan^{-1}\left(\frac{k_h}{1-k_v}\right) \tag{8.75}$$

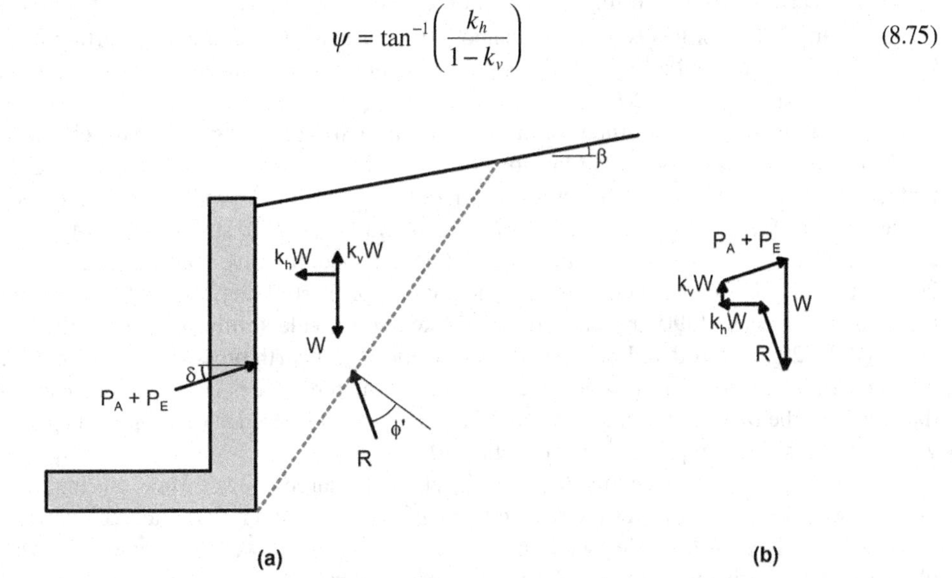

FIGURE 8.55 (a) Retaining wall with horizontal and vertical pseudo-accelerations applied to backfill wedge; (b) force polygon showing the effect of inertial forces from pseudo-acceleration on the wall-soil interaction force.

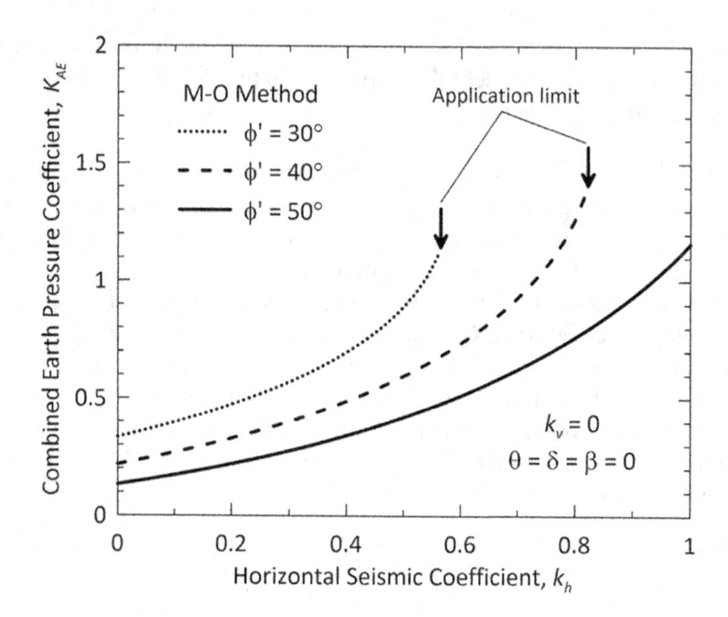

FIGURE 8.56 Active earth pressure coefficient K_{AE} as derived from M-O formulation for vertical frictionless wall with horizontal backfill and variable levels of horizontal acceleration. The seismic increment is the difference between K_{AE} for the case of finite horizontal acceleration and $k_h=0$. The angles defining the problem geometry are defined in Figure 8.49. (Adapted from results in Koseki et al., 1998; used with permission of Springer Nature BV.)

Figure 8.56 shows values of K_{AE} obtained using Equation (8.75) for a vertical ($\theta=0$) and smooth ($\delta=0$) wall with horizontal backfill ($\beta=0$) subjected to horizontal-only inertial loads ($k_v=0$). As expected, earth pressures increase with pseudo-acceleration k_h, but a notable feature of the solution is that earth pressures are undefined for large accelerations (indicated by the curves becoming nearly vertical), which is a result of soil failure in the backfill (Wood, 2023). Seed and Whitman (1970) observed that for levels of K_h up to about 0.4, the seismic increment of earth pressure could be reasonably approximated by $\Delta K_{AE}=0.75k_h$, which can then be used to estimate the combined earth pressure as $K_{AE} \approx K_A+\Delta K_{AE}$.

Pseudo-static limit state methods of this type for the active case originate in the classical work by Okabe (1926) and Mononobe and Matsuo (1929) [widely known as the "Mononobe-Okabe" (M-O) method]. Variants on the classical approach derived by means of kinematic limit analyses considering non-planar failure surfaces (Chen, 1975; Chen and Liu, 1990), stress fields (Mylonakis et al., 2007), backfill soils with cohesion and friction (National Cooperative Highway Research Program, 2008; Xu et al., 2015), and accounting for the phasing of inertial demands within the retained soil (Steedman and Zeng, 1990) are conceptually alike and provide similar results for the active case. Wood (1973, 2023) considered the case of pre-seismic at-rest earth pressures (as a consequence, the initial condition of the backfill is not at a limit state), which produces larger seismic earth pressures than those of the M-O method. Richart and Elms (1979) considered sliding block displacements of gravity walls (similar to procedures in Section 10.9.2).

A shortcoming of the conceptual framework behind this large body of work is its implicit assumption that seismic earth pressures are related to backfill acceleration rather than the relative displacement of the wall and soil. While pseudo-accelerations in the active wedge are associated with motion of the backfill, relative wall-soil motion is not considered because the formulation assumes the wall to be stationary during earthquake shaking. If both the backfill and wall move with comparable, in-phase displacement amplitudes, no appreciable seismic earth pressure will develop, despite the fact that the backfill is accelerating. Hence, there is a conceptual flaw in the classical methods.

Not surprisingly, when seismic earth pressures have been computed from direct analyses (Figure 8.1) or measured experimentally, they seldom conform to predictions from M-O-type procedures when PGA is used to estimate k_h. In some cases involving high-frequency ground motions, computed pressures exceed M-O predictions (e.g., Ostadan, 2005; Veletsos and Younan, 1994), while measured pressures in experiments (generally involving relatively low-frequency ground motions) typically fall below M-O predictions (e.g., Al Atik and Sitar, 2010; Hushmand et al., 2016; Wagner and Sitar, 2016; Candia et al., 2016). For these reasons, the continued use of pseudo-acceleration-based (e.g., M-O type) methods is problematic because of their inability to account for the problem physics.

8.6.3.2 SSI-Based Seismic Earth Pressures

The estimation of seismic earth pressures can be improved by recognizing that such pressures develop as a result of SSI involving the wall, supporting and retained soil, and any attached superstructures. As with other SSI problems, both kinematic and inertial components can be significant. Treating the problem that way allows consideration of important effects such as wall and soil stiffness and loading frequency to be taken into account, and produces pressures that are consistent with those obtained from direct analyses and model tests.

Kinematic Seismic Earth Pressures

Kinematic seismic earth pressures are produced by the combined seismic excitation of shallow layers of soil at a site and a relatively stiff wall system embedded within those soils. Figure 8.29 depicts a *U*-shaped building basement wall system. Figure 8.57 illustrates the problem for the case of a free-standing wall.

In the kinematic problem, the wall itself is assumed to have a certain flexural stiffness but no mass and to be connected to the soil but not to any structure. The ground motions producing the excitation occur over a wide frequency range, typically from about 0.2 to 10Hz. If these waves can be reasonably assumed as vertically propagating near the ground surface (such that the ground response is one-dimensional; Section 7.5), they produce waves of wavelength $\lambda = V_s/f$ with maximum amplitude at the ground surface (Figure 8.57 and Equation 8.41). Accordingly, at each frequency, and at a given point in time, the horizontal ground motion varies with depth. When this variation is small (i.e., for low frequencies with long wavelengths), the motion of the wall system nearly matches that of the free-field soil, relative wall-soil displacements are small, and seismic earth pressures are low (Brandenberg et al., 2015). Conversely, as shown in Figure 8.57, short wavelengths can produce

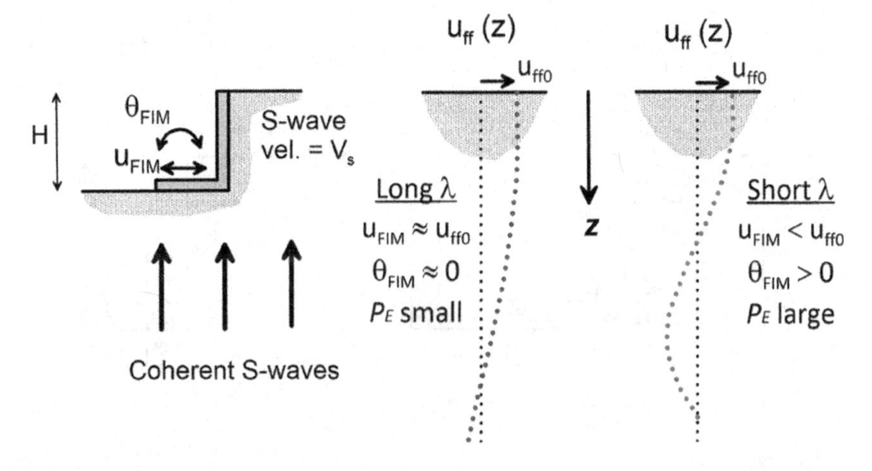

FIGURE 8.57 Schematic illustration of free-standing retaining wall subjected to seismic waves with different wavelengths. Displacement u_{FIM} applies at the foundation level of the wall, and θ_{FIM} represents the rotation of a rigid wall-footing system.

motions that decrease significantly with depth over the height of the wall. Under such conditions, the relatively stiff wall must displace differently from the free-field soil, and the differences in displacement produce potentially large seismic earth pressures. For these short wavelength conditions, the flexibility of the wall system, the distribution with depth of soil stiffness, and the ability of the wall to yield also affect seismic earth pressures.

The kinematic problem illustrated in Figure 8.57 can be solved using direct analysis in which the ground response and SSI are solved for simultaneously (Figure 8.1), and procedures of this sort are used for critical projects. Simplified methods are useful to help conceptualize the physics of the problem and for applications where more approximate solutions suffice. For the conditions represented in Figure 8.58, which include rigid wall elements and uniform, elastic backfill soils, an approximate solution for the amplitude of seismic earth pressure resultant P_E (normalized to remove dimensions) is given by Durante et al. (2022):

$$\frac{|P_E|}{k_y^i u_{ff0} H} = cos(kH) - \frac{\sin(kH)}{kH} \tag{8.76}$$

where k is the wavenumber (ω/V_S), u_{ff0} is the free-field ground surface displacement, and H is wall height (Figure 8.58). The normalized height of resultant h/H, where h is measured up to the resultant from the base of the foundation, is given for the same conditions as (Durante et al., 2022):

$$\frac{h}{H} = \frac{\cos(kH) - 1 + (kH)^2 \dfrac{\cos(kH)}{2}}{(kH)^2 \cos(kH) - (kH)\sin(kH)} \tag{8.77}$$

Equation (8.77) gives $h/H = 5/8$ for $\lambda/H \gtrsim 4$. The normal stress on the wall is described by the product of relative wall-soil displacement and stiffness intensity (introduced in Section 8.3.2.3) k_y^i, which is computed for application to walls as (Kloukinas et al., 2012):

$$k_y^i = \frac{\pi}{\sqrt{(1-\nu)(2-\nu)}} \frac{G}{H} \sqrt{1 - \left(\frac{4H}{\lambda}\right)^2} \tag{8.78}$$

where G and ν are the shear modulus and Poisson's ratio of the backfill, respectively.

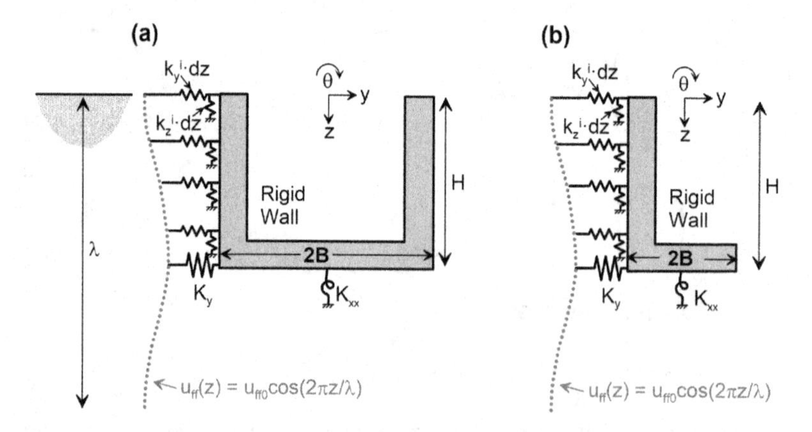

FIGURE 8.58 Schematic rigid wall systems with uniform backfill soils subjected to vertically propagating shear wave showing soil-structure stiffness terms for (a) *U*-shaped wall and (b) cantilever-type single wall.

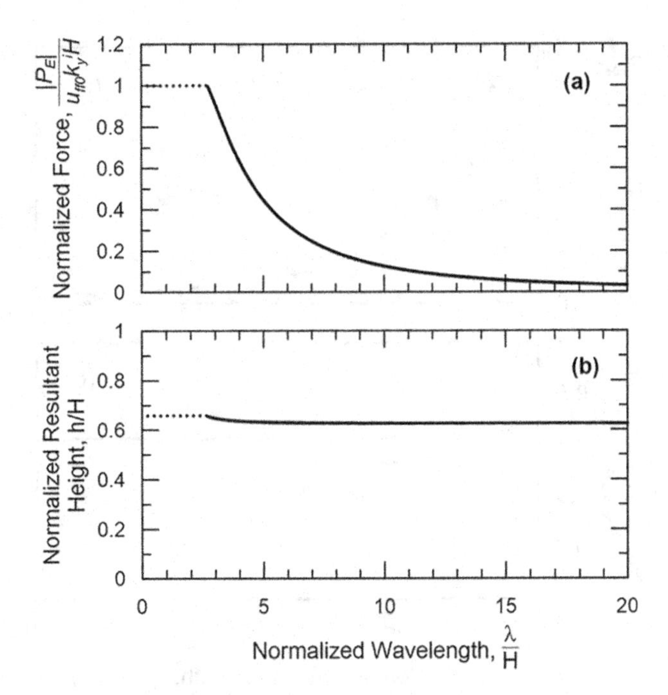

FIGURE 8.59 Variation of normalized amplitude of P_E with normalized wavelength λ/H for wall founded on rigid base (solution given in Equation 8.76). Dotted line at low λ/H is an approximation of the exact solution.

Equations (8.76–8.78) apply for the case of a stiff foundation (infinite base slab stiffnesses K_y and K_{xx}), rigid wall, and backfill of uniform stiffness. For those conditions, Figure 8.59 shows the variation of the normalized force amplitude $|P_E|/\left(k_y^i u_{ff0} H\right)$, and its normalized point of application h/H, with the ratio of wavelength to wall height, λ/H (note that this ratio increases with decreasing frequency). The portion of this curve for $\lambda/H \gtrsim 2.7$ typically contains the frequency range of engineering interest. Kinematic pressures are high near $\lambda/H=2.7$ due to large relative displacements of wall and soil. For smaller λ/H, relative displacements are large and normalized forces oscillate due to tensile and compressive stress changes acting on different portions of the wall height; this complex behavior is simplified to the dashed horizontal lines shown in Figure 8.59. As λ/H increases beyond 2.7, P_E decreases rapidly. In the limiting case where $\lambda/H \rightarrow \infty$, the deformed shape of the free-field soil profile becomes vertical, conforming to the shape of the rigid wall and producing zero kinematic interaction. For a given value of λ/H, the normalization of force accounts for the effects of soil stiffness and shaking amplitude on seismic earth pressures. The normalized resultant height has little variation with λ/H beyond 4.0.

In addition to wavelength, another critical factor that affects the development of seismic earth pressures is relative soil-to-wall flexibility. A rigid wall is less able to conform with free-field ground motions than is a flexible wall; the effect of increasing wall flexibility is to reduce both relative displacements and seismic earth pressures. Relative soil-to-wall flexibility is parameterized as (Novak, 1974):

$$\beta = \sqrt[4]{\frac{k_y^i}{4EI}} \tag{8.79}$$

where E is Young's modulus of the wall material and I is the moment of inertia for the wall section (the product represents the wall flexural stiffness). β has units of 1/length, and is multiplied by H to remove dimensions. A rigid wall has $\beta H=0$, but walls with $\beta H < 0.5$ respond in an effectively

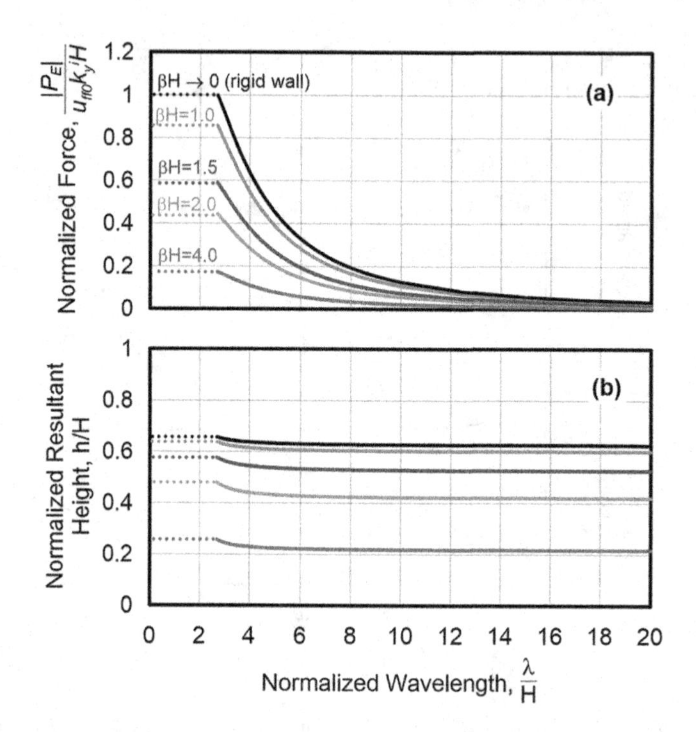

FIGURE 8.60 Effect of wall flexibility on (a) normalized wall resultant and (b) its point of application above the foundation. These results are for a rigid base and uniform backfill. (Modified from Durante et al., 2022; used with permission of SAGE Publications, Ltd.)

rigid manner. Rigid or nearly rigid walls are seldom encountered in practice; values of βH ranging from 2 to 4 are more common for modern building basement walls in seismically active regions. Figure 8.60 shows wall flexibility effects on the seismic earth pressure resultant and its height as derived from finite element analyses (Durante et al., 2022). As shown in Figure 8.60, P_E and its resultant height decrease substantially as wall stiffness decreases (i.e., with increasing βH).

Several additional factors affect seismic earth pressures, but to a lesser degree than wavelength and wall flexibility. One of these is non-uniform backfill conditions in which V_s increases with depth. The effects of velocity gradient have been investigated by holding the time-averaged shear wave velocity within the backfill soil as constant (Vrettos et al., 2016; Brandenberg et al., 2017). For rigid walls, backfill non-uniformity reduces $|P_E|$ and its resultant height; however, these effects are small for modest levels of wall flexibility. A second factor is nonlinearity in backfill soils, which both reduces stiffness intensity (Equation 8.78) and decreases wavelength due to modulus reduction. These effects largely offset and the net impact on seismic earth pressure resultant is typically small and can be neglected for simplified analyses (Durante et al., 2022). A third factor is non-rigid (i.e., compliant) foundation conditions, which strongly reduce $|P_E|$ for rigid walls (Brandenberg et al., 2015). However, these effects are more modest for flexible walls (Brandenberg et al., 2020), and may be neglected for simplified analyses (Durante et al., 2022).

The solutions presented in Figures 8.59 and 8.60 can be applied with a free-field ground motion by converting the time series to a Fourier series, computing earth pressure resultants (P_E) and moments (hP_E) for each frequency in the Fourier series, and computing the resultant force and moment time series using an inverse Fourier transformation. Routines for performing such calculations are presented by Brandenberg et al. (2020). A more approximate procedure that avoids the need for Fourier transforms simplifies the representation of ground motions by using the intensity measures of *PGV* (for amplitude) and mean period T_m (for frequency content; Section 3.3.2). This procedure is summarized as follows (Building Seismic Safety Council, 2020b; Durante et al., 2022),

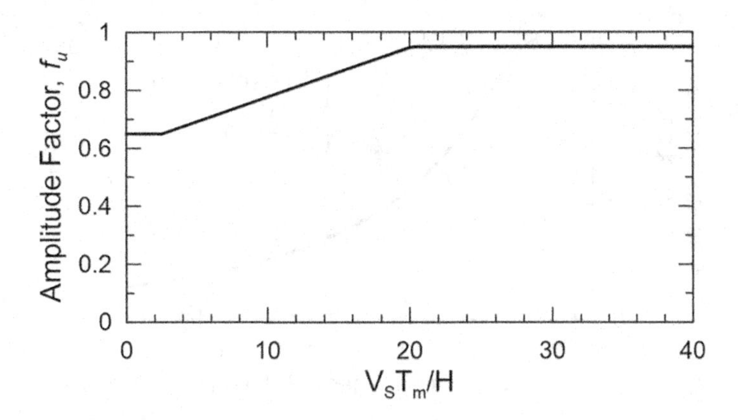

FIGURE 8.61 Ground motion amplitude adjustment factor for use with simplified method for evaluation of amplitude of seismic earth pressure resultant force, P_E. (Adapted from Durante et al., 2022.)

1. Perform a seismic hazard analysis (probabilistic or deterministic) to estimate *PGV* for the site.
2. If the estimation of *PGV* is based on probabilistic seismic hazard analysis, disaggregate the hazard (Section 4.4.3.5) at the return period of interest to obtain the controlling magnitudes and distances (either a single mean value or, as appropriate, multiple pairs for distinct contributing sources). Estimate the mean period, T_m, for this condition (Section 3.5.3.2). If *PGV* is derived using deterministic methods, use the selected magnitude and distance for the estimation of mean period. Compute the corresponding mean angular frequency, $\omega_m = 2\pi/T_m$.
3. Develop a shear wave velocity profile for the backfill soil. Compute the average shear wave velocity, $V_{s,av}$, as the ratio of wall height, *H*, to shear wave travel time through the backfill.
4. Estimate the ground surface displacement as $u_{g0} = f_u PGV/\omega_m$. The adjustment factor, f_u, depends on λ / H as shown in Figure 8.61. This factor has been calibrated to match the results of single-frequency analyses to more complete Fourier series analyses.
5. Estimate k_y^i as its static counterpart $\left(k_{y0}^i\right)$

$$k_{y0}^i = \frac{\pi}{\sqrt{(1-\nu)(2-\nu)}} \frac{\rho_s V_{s,av}^2}{H} \tag{8.80}$$

   This expression is modified from Equation (8.78) by removing the frequency-dependent term and by taking $G = \rho_s V_{s,av}^2$, where ρ_s is backfill mass density.
6. Estimate βH from the relative soil-to-wall stiffness (Equation 8.79) and wall height, *H*. In cases where βH is unknown because wall section sizes are undetermined, an initial estimate in the range of 1–2 can be applied and will often be conservative.
7. Compute the normalized force amplitude $|P_E|/\left(k_y^i u_{g0} H\right)$ and resultant height *h/H*. These quantities are obtained for rigid walls using Equations (8.76 and 8.77). The effects of wall flexibility can be incorporated using Figure 8.62 or the following expressions (which describe the change of resultant amplitude and its height with βH):

$$\frac{|P_E|}{|P_E|_{rigid\ wall}} = \begin{cases} 1-\exp\left(1-\dfrac{2.9}{\beta H}\right) & \beta H < 1 \\ \sin\left(-0.45+\dfrac{1.43}{\beta H}\right)+\cos\left(1.22+\dfrac{0.34}{\beta H}\right) & \beta H > 1 \end{cases} \tag{8.81}$$

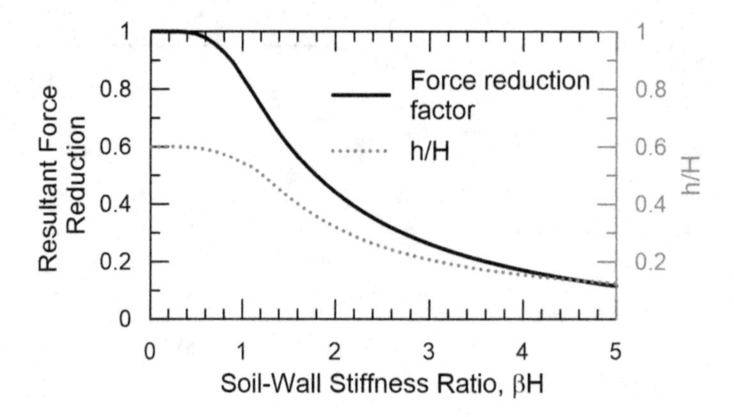

FIGURE 8.62 Effect of relative soil-to-wall stiffness (βH) on normalized resultant force and normalized resultant height. (Adapted from Durante et al., 2022.)

$$\frac{h}{H} = \begin{cases} 0.6 - \exp\left(-0.12 - \dfrac{2.8}{\beta H}\right) & \beta H < 1 \\ \sin\left(1.68 + \dfrac{1.5}{\beta H}\right) + \cos\left(2.87 - \dfrac{1.92}{\beta H}\right) & \beta H > 1 \end{cases} \tag{8.82}$$

The relations in Equations (8.81 and 8.82) and Figure 8.62 are curve-fits to finite element simulation results.

8. De-normalize to compute $|P_E|$. The moment at the base of the wall is computed as $|M| = h|P_E|$.

Simplifications associated with this procedure include the approximation of the ground motions with intensity measures, neglecting the frequency-dependence of wall-soil stiffness intensity, the treatment of the backfill as uniform and elastic, the treatment of the wall bending response as elastic, and neglecting wall inertia. Despite these assumptions, validation against centrifuge test data and the results of direct analyses (e.g., Al Atik and Sitar, 2010; Hushmand et al., 2016; Wagner and Sitar, 2016; Candia et al., 2016; Ostadan, 2005) show good (if slightly conservative) results, and markedly more reliable predictions than the M-O method (Durante et al., 2022).

Example 8.8

Consider again the wall from Example 8.7. Seismic hazard analyses are performed, and the *PGV* for the design return period is 50 cm/sec, with a controlling magnitude of 7.0 and source-to-site distance of 15 km. The average small-strain shear wave velocity of the sandy backfill materials over the 6.0 m wall height is 200 m/sec. The wall's relative soil/wall flexibility is $\beta = 0.33$/m. What is the resultant force from kinematic soil–structure interaction?

Solution:

To evaluate the seismic increment, the steps described in the text are applied as follows:

1. The seismic hazard analysis described in the problem statement provides the value of PGV=50 cm/sec.
2. The expected mean period of the ground motion is computed from the relations in Section 3.5.3.2 (Figure 3.44), the median T_m for the given magnitude and distance is 0.6 sec. The corresponding angular frequency is $\omega_m = 2\pi/T_m = 10.5$ rad/sec.
3. The shear wave velocity is 200 m/sec as given in the problem statement.

4. The ground surface displacement is computed as $u_{g0} = f_u PGV/\omega_m$ where the adjustment factor, f_u, depends on λ / H (Figure 8.61). Then, with $\lambda = V_s T_m = 120$ m, the value of $\lambda / H = 20$, so Figure 8.61 shows that $f_u = 0.95$ and u_{g0} can be calculated as

$$u_{g0} = (0.95)(200 \text{ m / sec}) / 10.5 \text{ sec}^{-1} = 0.045 \text{ m}$$

5. Recalling that $\nu = 0.3$ from Example 8.7, estimate k_{y0}^i using Equation (8.80) as

$$k_{y0}^i = \frac{\pi}{\sqrt{(1-0.3)(2-0.3)}} \frac{(18/9.81)200^2}{6} = 3.5 \times 10^4 \text{kN/m}^3$$

6. Given the relative soil/wall flexibility, $\beta H = 0.33 \times 6 = 2$
7. The normalized force amplitude for a rigid wall can then be read from Figure 8.59 using $\lambda / H = 20$, which gives a value of $|P_E| / (k_y^i u_{g0} H) = 0.033$. Since the wall is not rigid, the effects of its flexibility are accounted for by the reduction factor shown in Figure 8.62, which has a value of 0.5. The resulting normalized force amplitude is $0.033 \times 0.5 = 0.017$.
8. From de-normalization, the resultant in force units is $|P_E| = 0.017 (k_y^i u_{g0} H) = 157$ kN/m

From the above, the overall lateral earth pressure coefficient can be calculated using Equation (8.73) as $K_{AE} = (P_A + P_E) / \left(\frac{1}{2} \gamma H^2\right) = 0.7$. Given $K_A = 0.25$ (from Example 8.7), the seismic response can be seen to nearly triple the earth pressure from the static condition.

The static and seismic resultants will generally occur at different heights h above the base of the foundation; the static resultant is at $h / H = 1/3$ whereas the height of the seismic resultant is obtained from Figure 8.62. In this case, the seismic resultant is also at $h / H = 0.42$.

Inertial Seismic Earth Pressures

The kinematic solution presented above does not consider the effects of inertia. Actual retaining walls have mass, which will produce inertial forces when excited by earthquake shaking. That inertia will affect the wall response and therefore change seismic earth pressures relative to the kinematic case. Relative to the case where inertia is ignored, inertial effects can increase or decrease seismic earth pressures. The combined effects of kinematic and inertial interaction for free-standing walls or basement walls can be analyzed using frequency domain procedures (Brandenberg et al., 2020). Such analyses may well be justified when inertial effects may be important, such as when gravity walls are used.

Inertial seismic earth pressures are typically most significant when structures are connected to a foundation system containing walls and the lateral force resisting system below the ground level includes those walls (e.g., Figure 8.4b). Under these conditions, the base shear and moment generated by the vibrating structure cause the foundation system to displace horizontally and rotate, producing wall reaction stresses. Analysis of this problem does not require specialized procedures – direct or substructure approaches to the SSI problem can be employed as illustrated in Figures 8.1–8.4 and discussed in Section 8.2. Provided wall-soil interaction elements are included in the SSI model, the reaction stresses against the walls are a natural product of the analysis.

It is important to consider the degree to which subterranean walls participate in the below-ground lateral force resisting system in the analysis of inertial effects. Figure 8.63 shows a building in which lateral loads are resisted by a central core of shear walls that extend directly to the foundation mat. If the foundation for the core walls is not structurally connected to the foundations for the surrounding podium (i.e., a portion of the structure with a wider footprint, generally near and below ground line) and floor diaphragms within the podium are either not connected to the core walls or are relatively flexible, the podium's basement walls may see little of the inertial loading. In such cases, the loading of basement walls is likely to be controlled by the kinematic mechanisms described in the previous section.

Peak kinematic and inertial seismic demands on basement walls are unlikely to coincide in time. When both effects are expected, load combination rules such as SRSS can be applied.

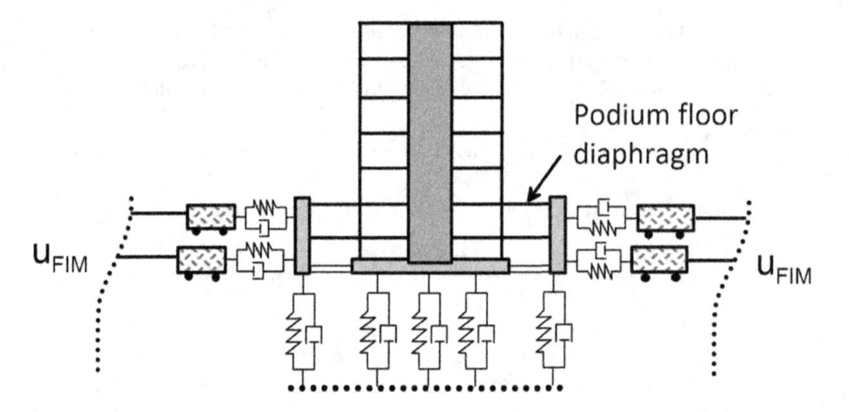

FIGURE 8.63 Structure with lateral loads resisted by a central core of shear walls supported on a mat foundation that is distinct from the foundations for podium basement walls. In such cases, the degree to which podium basement walls are affected by inertial interaction effects depends on the connectivity of floor diaphragms to the core walls and perimeter walls and their flexibility.

8.6.4 Underground Structures

Underground structures such as tunnels, culverts, and pipelines can comprise critical lifelines, the design and analysis of which is a SSI problem that often falls within the purview of geotechnical earthquake engineers. The distinction from retaining walls is that underground structures are buried (they do not extend to the ground surface), they are hollow, low-density inclusions, and their lengths can be much greater than their cross-sectional dimensions.

As with other earth-retaining structures, underground structures experience a combination of loading from static lateral (and vertical) earth pressures and additional demands imposed by earthquakes. Seismic demands in structural elements result from two principal sources: (1) spatially variable transient ground displacements that produce relative displacements between free-field and structure, which in turn produce force and moment demands in the lining of the underground structure; and (2) permanent displacements associated with ground failure mechanisms such as soil liquefaction, cyclic softening, and slope instability. The former is addressed in this section; the latter are ground failure problems that are the subject of Chapters 9 and 10.

Spatial variations of transient ground motions in both the vertical and horizontal directions can affect underground structures. Vertically variable demands are caused by one-dimensional ground response in a manner similar to that considered in kinematic analysis procedures for walls (Section 8.6.3.2). Horizontally variable demands are caused by wave passage and other spatial variability effects (Section 3.8). The manner by which these demands are considered for underground structures is described in the following subsections [more information can be found in Hashash et al. (2001) and Wang (1993)].

8.6.4.1 Demands from Vertically Propagating Shear Waves

As illustrated in Figure 8.64, vertical shear wave propagation can produce ovaling of circular tunnel sections and racking of rectangular sections (Owen and Scholl, 1981). Analysis procedures for the two cases consider the effects of the vertical variation of ground displacement, and the stiffness of the tunnel lining structure (relative to that of the soil), on the force and moment demands that develop in the lining. As with other SSI problems, these effects can be considered using direct SSI analysis procedures that simultaneously consider the site response and SSI (Section 8.2.1); the focus of this section is on simplified methods that illustrate some of the important physics of the problem.

Analysis of ovaling for circular sections begins with the estimation of peak soil shear strain (γ_{max}) over the depth range of the tunnel from site-specific ground response analysis. The ratio of

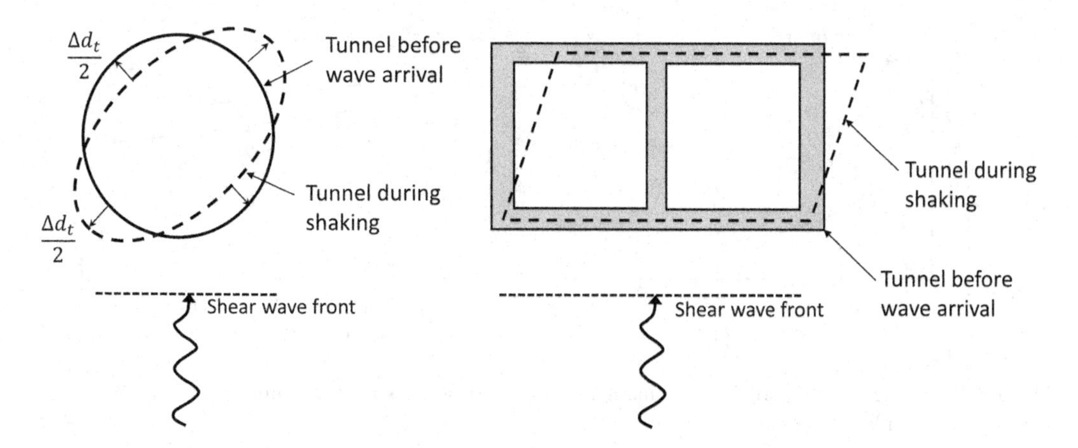

FIGURE 8.64 Deformation modes of tunnel or culvert cross-section from vertical shear wave propagation. (Adapted from Hashash et al., 2001 and Owen and Scholl, 1981; used with permission of Elsevier Science and Technology Journals.)

soil stiffness to tunnel lining stiffness in shear (Merritt et al., 1985) can be expressed in terms of the *flexibility ratio*

$$F = \frac{E_s\left(1-\nu_c^2\right)r_t^3}{6E_cI_c(1+\nu)} \tag{8.83}$$

where r_t is the tunnel radius, E_s is the soil Young's modulus (related to shear modulus, G, per Equation 8.29), ν is the soil Poisson's ratio, and E_c, I_c, and ν_c are the Young's modulus, moment of inertia, and Poisson's ratio of the tunnel lining material (generally concrete), respectively. Similarly, the ratio of soil medium to tunnel axial stiffness (Merritt et al., 1985) can be expressed by the *compressibility ratio*

$$C = \frac{E_s\left(1-\nu_c^2\right)r_t}{E_ct_c(1+\nu)(1-2\nu)} \tag{8.84}$$

where t_c is the thickness of the tunnel lining. The tunnel section diametric strain (computed as ratio of diameter change from ovaling, Δd_t, shown in Figure 8.64, to initial diameter) is then computed assuming full slip between soil and liner (Hashash et al., 2001) as

$$\frac{\Delta d_t}{2r_t} = \frac{1}{3}K_1F\gamma_{\max} \tag{8.85}$$

where K_1 is a response coefficient that depends on F and ν as shown in Figure 8.65a for the case where slip is allowed at the soil-tunnel interface. Demands within the tunnel liner, specifically the axial thrust force ($T_{\max}$) and moment ($M_{\max}$) per unit length (Hashash et al., 2001) are then computed as

$$T_{\max} = K_2\frac{E_s}{2(1+\nu)}r_t\gamma_{\max} \tag{8.86}$$

$$M_{\max} = \frac{1}{6}K_1\frac{E_s}{(1+\nu)}r_t^2\gamma_{\max} \tag{8.87}$$

where K_2 is a response coefficient that depends on F, C, and ν; results for ν=0.35 are shown in Figure 8.65b.

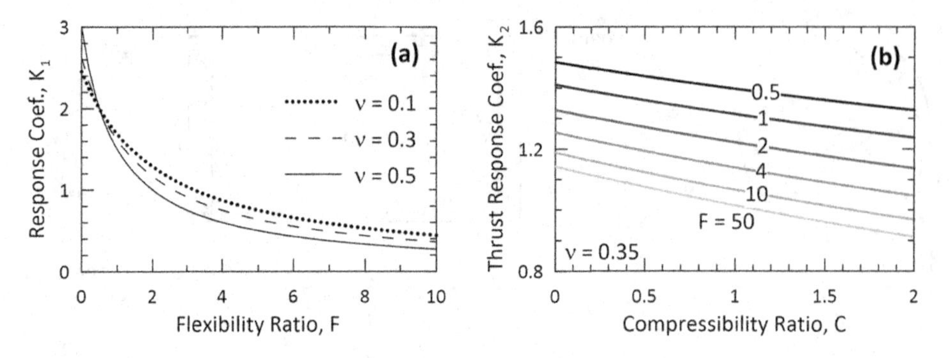

FIGURE 8.65 Response coefficients K_1 and K_2. Poisson's ratios shown in the figure are for soil (ν). (Plotted from equations in Wang, 1993.)

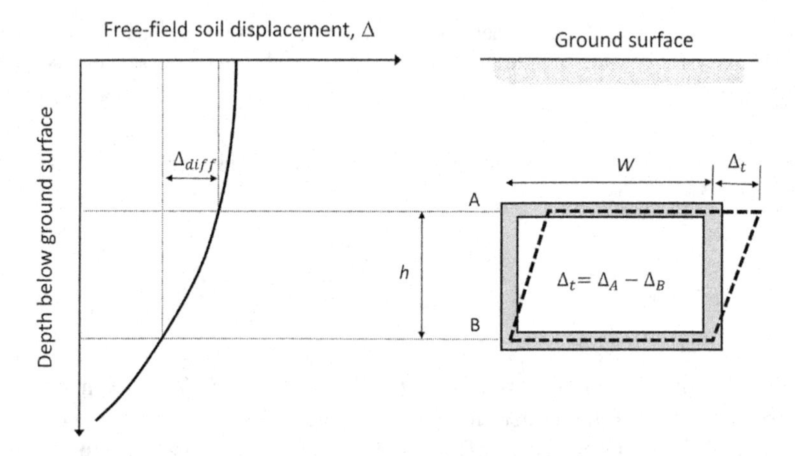

FIGURE 8.66 Free-field racking deformation imposed on a buried rectangular frame. (Adapted from Wang, 1993.)

Analysis of racking for rectangular sections begins with the evaluation of a ground displacement profile as shown in Figure 8.66 and the relative free-field displacement between the top and bottom of the tunnel section, Δ_{diff}. The racking stiffness of the tunnel section, K_t, is evaluated as the ratio of force (per unit length)/deflection to produce deformations of the type shown on the right side of Figure 8.66. The flexibility ratio, F_r, for this case is then computed as:

$$F_r = \frac{G}{K_t}\frac{w}{h} \tag{8.88}$$

where w and h are width and height of the tunnel, as shown in Figure 8.66. The ratio of tunnel/free-field displacement (Δ_t/Δ_{diff}) is then evaluated from F_r as shown in Figure 8.67. Internal member demands are evaluated from displacement Δ_t using standard structural analyses.

8.6.4.2 Demands from Horizontally Variable Ground Motions

Horizontally variable ground motions have the potential to produce axial and bending deformations of underground structures (Figure 8.68). As described in Section 3.8, these variations in ground shaking arise from horizontal wave passage (most predominantly in the direction from source-to-site) and more complex stochastic processes that produce spatial variations in amplitude and phase. Horizontally variable motions can also occur at boundaries between soils of different impedance, for example as might occur when a tunnel or pipeline passes from stiff soils beneath a hill into softer (and possibly even liquefiable) soils in an adjacent alluvial valley; sharp impedance contrasts at such material boundaries can impose high flexural demands on buried structures.

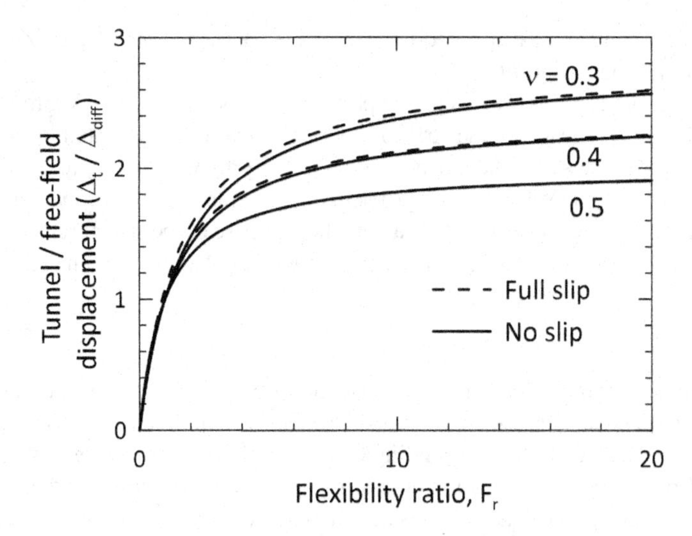

FIGURE 8.67 Dependence of tunnel-to-free-field displacement ratio with flexibility ratio F_r. Full slip results from Wang (1993) and Penzien (2000), no slip results from Penzien (2000). (Figure adapted from Power et al., 2006.)

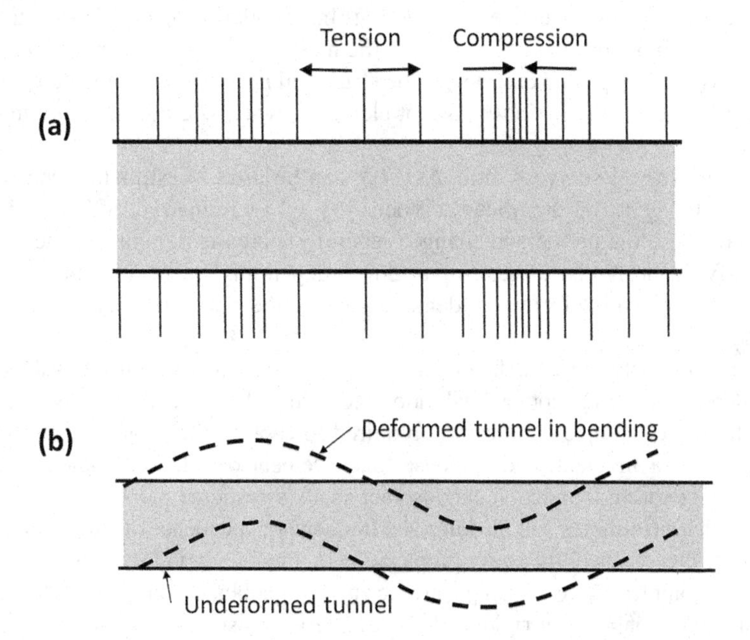

FIGURE 8.68 Deformation modes of underground structures due to traveling waves producing (a) tension/compression and (b) bending. (Adapted from Owen and Scholl, 1981.)

A first-order approximation of axial and bending strains induced in tunnel sections from horizontal ground motion variability is to assume they are equal to corresponding free-field ground strains (i.e., implicitly assuming that the tunnel is completely flexible). As described in Section 3.8.2, spatially variable ground strains from wave passage can cause a variety of shear and axial strains depending on wave type (p-waves, s-waves, surface waves) and direction of wave travel with respect to the longitudinal axis of the tunnel. The largest strains are most often associated with shear waves. However, as noted in Section 3.8.2, actual peak ground strains from analysis of dense array data are larger than can be attributed to shear wave passage alone, due to the other sources of horizontal ground motion variability.

When free-field ground strains are taken as approximations of strains in underground structures, the effects of SSI are neglected, causing the structural strains to be over-estimated. More realistic analyses use springs to model the structural stiffness of the underground structure and the stiffness of surrounding soil. As with other substructure SSI problems (e.g., Figures 8.2–8.4), spatially variable ground motions are applied at the ends of springs to evaluate the response of the structure. Additional details on analyses of this type for underground structures are described by Wang (1993) and Hashash et al. (2001).

8.7 SUMMARY

1. Seismic SSI analyses evaluate the response of three linked systems to seismic ground motion – a structure (such as a building or bridge), a foundation (shallow foundations or piles) and the geologic media surrounding the foundation. The purpose of these analyses is to provide a more realistic evaluation of seismic demands in structures and their foundations than is possible from fixed-base analysis in which the foundation support is assumed as rigid.
2. Seismic SSI analyses can follow a direct approach in which site response is modeled together with the response of the foundation and structure. Analyses of this sort require sophisticated treatment of spatially variable ground motions and accurate modeling of soil, foundation, structural, and interface elements. Alternatively, substructure methods separate the analysis of the effective excitation at the foundation level of a structure (termed the foundation input motion or FIM) from the modeling of foundation-soil flexibility and damping (using springs and dashpots). The structural response is computed using the FIM to excite the ends of the spring / dashpot elements, which are attached to a model of the structure.
3. The structure-to-soil stiffness ratio, $h/(V_sT_0)$, can be used to estimate when inertial SSI effects are likely to be significant. When $h/(V_sT_0)>0.1$, inertial SSI can significantly lengthen the building period and change (generally increase) damping in the system. This will modify the design base shear (up or down, depending on spectral shape) and the distribution of force and deformation demands within the structure, relative to a fixed-base analysis.
4. The interaction between foundation and soil is described by complex-valued and frequency-dependent springs for each vibration mode (translations and rotations in three directions). The use of complex numbers is required to capture the phase differences between demand (force or moment) and response (displacement or rotation), which are effects of damping. A vibrating foundation can also act as a wave source radiating body and surface waves into the surrounding soil medium; the frequency-dependence of stiffness and damping accounts for the effects of this wave propagation on the foundation response.
5. Models of complex-valued springs are termed impedance functions. The most basic impedance functions are for rigid shallow foundations resting on the surface of a uniform half-space. These can be extended for embedment effects, the impact of soil non-uniformity, and non-rigid structural foundation elements. Limiting capacities should be used with foundation springs to account for possible effects of soil yielding (e.g., bearing capacity).
6. Impedance models for pile foundations operate similarly to those for shallow foundations, representing the stiffness and damping of the foundation system at the pile head. The most significant pile impedances are for the vertical and lateral modes of vibration. Nonlinearities in pile-soil interaction, including soil-pile gapping effects, can be described with macro-element models distributed along the pile length.
7. Kinematic SSI causes FIMs to deviate from free-field ground motions as a result of spatially variable ground motions applied to stiff foundation systems. Base slab averaging results from spatial variability of ground motion in the horizontal direction within the perimeter of

the foundation system. Embedment and pile effects result from spatial variability of ground motions in the vertical direction. Models that have been validated against field recordings (especially for the cases of base slab averaging and embedment) are available for these effects. Each of these kinematic effects is most pronounced at high frequencies, generally reducing the amplitude of foundation motions relative to those in the free-field.

8. Practical applications of SSI vary by structure type. For buildings, consideration of SSI is generally optional in current guidelines documents and building codes for the design of new structures or retrofit of older structures. Consideration of SSI is more common for bridge structures, especially at the abutments. SSI is also relatively frequently considered in the analysis and design of nuclear structures. Seismic response is evaluated using force-based methods, nonlinear static pushover methods, and response history procedures. Both kinematic effects and inertial effects can be considered in each case.
9. Application of SSI principles is important for embedded structures including earth-retaining structures and underground structures such as tunnels. Under static conditions, lateral earth pressures depend strongly on the level of wall displacement, being minimized for the active case where the wall relaxes away from the backfill and maximized for the passive case in which the wall is advanced into the backfill. At-rest earth pressures are an intermediate case involving no horizontal displacement.
10. Transient ground shaking from earthquakes cyclically increases and decreases soil reactions against embedded structures due to relative displacements between the structure and free-field. In the absence of ground failure, these relative displacements can be caused by kinematic interaction and inertia applied to wall elements or underground structures from vibrations of attached superstructures. The primary factors affecting the development of kinematic seismic earth pressures are the amplitude of free-field ground shaking, the ratio of seismic wavelength to wall height, and the ratio of soil-to-wall stiffness. These same factors affect the demands applied to underground structures from vertically propagating shear waves. Additional demands for extended underground structures like tunnels are produced by spatial variations in ground motions and soil conditions in the horizontal direction.

REFERENCES

AASHTO. (2014). *AASHTO LRFD Bridge Design Specifications, Customary U.S. Units,* 7th edition, with 2015 and 2016 Interim Revisions, American Association of State Highway Transportation Officials, Washington, DC.

Abrahamson, N.A. (2005). "Effect of local site condition on spatial coherency," *Report No.RP2978-05*, Electric Power Research Institute, Palo Alto, CA.

Abrahamson, N.A., Schneider, J.F., and Stepp, J.C. (1991). "Empirical spatial coherency functions for application to soil-structure interaction analyses," *Earthquake Spectra*, Vol. 7, pp. 1–27. https://doi.org/10.1193/070913eqs198m

Al Atik, L. and Sitar, N. (2010). "Seismic earth pressures on cantilever retaining structures," *Journal of Geotechnical and Geoenvironmental Engineering*, Vol. 136, No. 10, pp. 1324–1333. https://doi.org/10.1061/(asce)gt.1943-5606.0000351

Allen, T.M. and Bathurst, R.J. (2018). "Application of the simplified stiffness method to design of reinforced soil walls," *Journal of Geotechnical and Geoenvironmental Engineering*, Vol. 144, No. 5, p. 13. https://doi.org/10.1061/(asce)gt.1943-5606.0001874

Allotey, N. and Naggar, M.H.E. (2003). "Analytical moment-rotation curves for rigid foundations based on a Winkler model," *Soil Dynamics and Earthquake Engineering*, Vol. 23, pp. 367–381. https://doi.org/10.1016/s0267-7261(03)00034-4

Allotey, N. and Naggar, M.H.E. (2007). "An investigation into the Winkler modeling of the cyclic response of rigid footings," *Soil Dynamics and Earthquake Engineering*, Vol. 28, pp. 44–57. https://doi.org/10.1016/j.soildyn.2007.04.003

American Petroleum Institute. (1993). "Recommended practice for planning, design, and constructing fixed offshore platforms," *API RP 2A-WSD*, 20th edition, API Publishing Services, Washington, DC.

Ancheta, T.D. and Stewart, J.P. (2015). "Conditional simulation of spatially variable motions on 2D grid," in T. Haukaas, ed., *Proceedings of the 12th International Conference on Applications of Statistics and Probability in Civil Engineering,* Vancouver, Canada, Paper No. 550 (electronic file). https://doi.org/10.14288/1.0076268

Ancheta, T.D., Stewart, J.P., and Abrahamson, N.A. (2011). "Engineering characterization of earthquake ground motion coherency and amplitude variability," *Proceedings, 4th International Symposium on Effects of Surface Geology on Seismic Motion, IASPEI/IAEE,* University of California Santa Barbara.

Anoyatis, G., Di Laora, R., Mandolini, A., and Mylonakis, G. (2013). "Kinematic response of single piles for different boundary conditions: Analytical solutions and normalization schemes," *Soil Dynamics & Earthquake Engineering*, Vol. 44, pp. 183–195. https://doi.org/10.1016/j.soildyn.2012.09.011

Apsel, R.J. and Luco, J.E. (1987). "Impedance functions for foundations embedded in a layered medium: An integral equation approach," *Earthquake Engineering and Structural Dynamics*, Vol. 15, No. 2, pp. 213–231. https://doi.org/10.1002/eqe.4290150205

Architectural Institute of Japan. (2001). *Recommendations for Design of Building Foundations*, AIJ, Tokyo (in Japanese).

ASCE. (2000). "Seismic analysis of safety-related nuclear structures and commentary," *ASCE 4-98,* American Society of Civil Engineers, Reston, VA. https://doi.org/10.1061/9780784404331

ASCE. (2005). "Seismic design criteria for structures, systems, and components in nuclear facilities," *ASCE/SEI 43-05,* American Society of Civil Engineers, Reston, VA. https://doi.org/10.1061/9780784407622

ASCE. (2016). "Seismic analysis of safety-related nuclear structures and commentary," *ASCE 4-16,* Reston, VA. https://doi.org/10.1061/9780784413937

ASCE. (2023). "Seismic evaluation and retrofit of existing buildings," *ASCE/SEI 41-23,* American Society of Civil Engineers, Reston, VA. https://doi.org/10.1061/9780784416112

Ashford, S.A., Boulanger, R.W., and Brandenberg, S.J. (2011). "Recommended design practice for pile foundations in laterally spreading ground," *Report No. PEER 2011/04,* Pacific Earthquake Engineering Research Center, Berkeley, CA.

ATC (1996). "Seismic evaluation and retrofit of concrete buildings," *ATC-40 Report*, Applied Technology Council, Redwood City, CA.

ATC (2010). "Modeling and acceptance criteria for seismic design and analysis of tall buildings," *ATC-72-1 Report*, Applied Technology Council, Redwood City, California.

Aviram, A., Mackie, K.R., and Stojadinović, B. (2008). "Guidelines for nonlinear analysis of bridge structures in California," *Report No. 2008/03*, Pacific Earthquake Engineering Research Center, Berkeley, CA.

Barghouthi, A.F. (1984). "Pile response to seismic waves," *Ph.D. Thesis*, University of Wisconsin, Madison, WI.

Berg, R.R., Christopher, B.R., and Samtani, N.C. (2009a). "Design of mechanically stabilized earth walls and reinforced soil slopes – Volume I," *Report No. FHWA-NHI-10-024*, National Highway Institute, Federal Highway Administration, Washington, DC, 332 pp.

Berg, R.R., Christopher, B.R., and Samtani, N.C. (2009b). "Design of mechanically stabilized earth walls and reinforced soil slopes – Volume II," *Report No. FHWA-NHI-10-024*, National Highway Institute, Federal Highway Administration, Washington, DC, 404 pp.

Bielak, J. (1975). "Dynamic behavior of structures with embedded foundations," *Earthquake Engineering and Structural Dynamics*, Vol. 3, pp. 259–274. https://doi.org/10.1002/eqe.4290030305

Bielak, J., Loukakis, K., Hisada, Y., and Yoshimura, C. (2003). "Domain reduction method for three-dimensional earthquake modeling in localized regions, Part I: Theory," *Bulletin of the Seismological Society of America*, Vol. 92, pp. 817–824. https://doi.org/10.1785/0120010251

Blaney, G.W., Kausel, E., and Roesset, J.M. (1975). "Dynamic stiffness of piles," *Proceedings, 2nd International Conference on Numerical Methods in Geomechanics,* Virginia Polytechnical Institute and State University, Blacksburg, VA, pp. 1010–1012.

Borja, W.I. and Wu, W.H. (1994). "Vibration of foundations on incompressible soils with no elastic region," *Journal of Geotechnical Engineering*, 120, pp. 1570–1592. https://doi.org/10.1061/(asce)0733-9410(1994)120:9(1570)

Boulanger, R.W., Chang, D., Brandenberg, S.J., Armstrong, R.J., and Kutter, B.L. (2007). "Seismic design of pile foundations for liquefaction effects," *4th International Conference on Earthquake Geotechnical Engineering,* pp. 277–302. https://doi.org/10.1007/978-1-4020-5893-6

Boulanger, R.W., Chang, D., Gulerce, U., Brandenberg, S.J., and Kutter, B.L. (2012). "Evaluating pile pinning effects on abutments over liquefied ground," in *Seismic Performance and Simulation of Pile Foundations in Liquefied and Laterally Spreading Ground, GSP 145*, ASCE, pp. 306–318. https://doi.org/10.1061/40822(184)25

Boulanger, R.W., Curras, C.J., Kutter, B.L., Wilson, D.W., and Abghari, A. (1999). "Seismic soil-pile-structure interaction experiments and analyses," *Journal of Geotechnical and Geoenvironmental Engineering*, Vol. 125, pp. 750–759. https://doi.org/10.1061/(asce)1090-0241(1999)125:9(750)

Brandenberg, S.J., Boulanger, R.W., Kutter, B.L., and Chang, D. (2007). "Static pushover analyses of pile groups in liquefied and laterally spreading ground in centrifuge tests," *Journal of Geotechnical and Geoenvironmental Engineering*, Vol. 133, pp. 1055–1066. https://doi.org/10.1061/(asce)1090-0241(2007)133:9(1055)

Brandenberg, S.J., Durante, M.G., Mylonakis, G., and Stewart, J.P. (2020). "Winkler solution for seismic earth pressures exerted on flexible walls by vertically inhomogeneous soil," *Journal of Geotechnical and Geoenvironmental Engineering*, Vol. 146, p. 04020127. https://doi.org/10.1061/(asce)gt.1943-5606.0002374

Brandenberg, S.J., Mylonakis, G., and Stewart, J.P. (2015). "Kinematic framework for evaluating seismic earth pressures on retaining walls," *Journal of Geotechnical and Geoenvironmental Engineering*, Vol. 141, pp. 04015031. https://doi.org/10.1061/(asce)gt.1943-5606.0001312

Brandenberg, S.J., Mylonakis, G., and Stewart, J.P. (2017). "Approximate solution for seismic earth pressures on rigid walls retaining inhomogeneous elastic soil," *Soil Dynamics and Earthquake Engineering*, Vol. 97, pp. 468–477. https://doi.org/10.1016/j.soildyn.2017.03.028

Building Seismic Safety Council, BSSC (2020a). *NEHRP Recommended Seismic Provisions for New Buildings and Other Structures. Volume I: Part 1 Provisions, Part 2 Commentary. Report No. FEMA P-2082-1*, Federal Emergency Management Agency, Washington DC.

Building Seismic Safety Council (2020b). "RP4: Seismic lateral earth pressures," in *NEHRP Recommended Seismic Provisions for New Buildings and Other Structures. Volume II: Part 3 Resource Papers. Report No. FEMA P-2082-2*, Federal Emergency Management Agency, Washington, DC, pp. 170–192.

Caltrans. (2013). *Caltrans Seismic Design Criteria Version 1.7*, California Department of Transportation, Sacramento, CA.

Caltrans. (2019). *Caltrans Seismic Design Criteria Version 2.0*, California Department of Transportation, Sacramento, CA.

Candia, G., Mikola, R.G. and Sitar, N. (2016). "Seismic response of retaining walls with cohesive backfill: Centrifuge model studies," *Soil Dynamics and Earthquake Engineering*, Vol. 90, 411–410. https://doi.org/10.1016/j.soildyn.2016.09.013

Caquot, A. and Kerisel, F. (1948). *Tables for the Calculation of Passive Earth Pressure, Active Earth Pressure and Bearing Capacity of Foundations*, Gauthier-Villars, Paris.

Chen, W.F. (1975). *Limit Analysis and Soil Plasticity: Developments in Geotechnical Engineering*, Elsevier, Amsterdam, the Netherlands. https://doi.org/10.1016/b978-0-444-41249-2.x5001-x

Chen, W.F. and Liu, X.L. (1990). *Limit Analysis in Soil Mechanics*, Elsevier, Amsterdam, the Netherlands.

Choi, J.I., Kim, M.M., and Brandenberg, S.J. (2015). "Cyclic *p-y* plasticity model applied to pile foundations in sand," *Journal of Geotechnical and Geoenvironmental Engineering*, Vol. 141, p. 04015013. https://doi.org/10.1061/(asce)gt.1943-5606.0001261

Chopra, A.K. and Goel, R.K. (1999). "Capacity-demand-diagram methods based on inelastic design spectrum," *Earthquake Spectra*, Vol. 15, pp. 637–656. https://doi.org/10.1193/1.1586065

Chopra, A.K. and Yim, S.C. (1985). "Simplified earthquake analysis of structures with foundation uplift," *Journal of Structural Engineering*, Vol. 111, pp. 906–930. https://doi.org/10.1061/(asce)0733-9445(1985)111:4(906)

Ciampoli, M. and Pinto, P.E. (1995). "Effects of soil-structure interaction on inelastic seismic response of bridge piers," *Journal of Structural Engineering*, Vol. 121, pp. 806–814. https://doi.org/10.1061/(asce)0733-9445(1995)121:5(806)

Clough, R.W. and Penzien, J. (1993). *Dynamics of Structures*, McGraw Hill, New York.

Coduto, D.P., Kitch, W.A., and Yeung, M.R. (2016). *Foundation Design, Principles and Practices*, 3rd edition, Pearson, Upper Saddle River, NJ.

Correia, A.A. and Pecker, A. (2021). "Nonlinear pile-head macro-element for the seismic analysis of structures on flexible piles," *Bulletin of Earthquake Engineering*, Vol. 18, pp. 1815–1849. https://doi.org/10.1007/s10518-020-01034-4

Coulomb, C.A. (1776). "Essai sur une application des regles des maximis et minimis a quelques problemes de statique relatifs a l'architecture," *Memoires de l'Academie Royale pres Divers Savants*, Vol. 7.

Cremer, C, Pecker, A., and Davenne, L. (2001). "Cyclic macro-element of soil structure interaction: Material and geometrical nonlinearities," *International Journal of Numerical and Analytical Methods in Geomechanics*, Vol. 25, pp. 1257–1284. https://doi.org/10.1002/nag.175

Day, S.M. (1978). Seismic response of embedded foundations," *Preprints of Conference Proceedings of ASCE Convention and Exposition,* Chicago, IL, Preprint No. 3450, American Society of Civil Engineers, New York.

Di Laora, R. and Rovithis, E. (2014). "Kinematic bending of fixed-head piles in nonhomogeneous soil," *Journal of Geotechnical and Geoenvironmental Engineering*, Vol. 141, No. 4, p. 04014126. https://doi.org/10.1061/(asce)gt.1943-5606.0001270

Dobry, R. and Gazetas, G. (1988). "Simple method for dynamic stiffness and damping of floating pile groups," *Géotechnique*, Vol. 38, pp. 557–574. https://doi.org/10.1680/geot.1988.38.4.557

Dobry, R., Vicente, E., O'Rourke, M.J., and Roesset, J.M. (1982). "Horizontal stiffness and damping of single piles," *Journal of the Geotechnical Engineering Division*, Vol. 108, pp. 439–459. https://doi.org/10.1785/BSSA0710062039

Durante, M.G., Stewart, J.P., Brandenberg, S.J., and Mylonakis, G. (2022). "Simplified solution for seismic earth pressures exerted on flexible walls," *Earthquake Spectra*, Vol. 38, pp. 1872–1892. https://doi.org/10.1177/87552930221083326

El-Gamal, M. and Siddharthan, R.V. (1998). "Nonlinear stiffness of abutments on spread footings for seismic design and retrofit," in P. Dakoulas, M. Yegian, and R.D. Holtz, eds., *Geotechnical Earthquake Engineering and Soil Dynamics-III*, ASCE Geotechnical Special Publication No. 75, pp. 1307–1318.

Ellison, K.C., Masroor, A.M., Almufti, I., Willford, M., and O'Riordan, N. (2015). "Structure-soil-structure interaction analysis in a highly seismic, dense urban regeneration zone," *Proceedings, 6th International Conference on Earthquake Geotechnical Engineering,* 1-4 November 2015, Christchurch, New Zealand.

El-Naggar, M.H. and Novak, M. (1994). "Non-linear model for dynamic axial pile response," *Journal of Geotechnical Engineering*, Vol. 120, pp. 308–329. https://doi.org/10.1061/(asce)0733-9410(1994)120:2(308)

El-Naggar, M.H. and Novak, M. (1996). "Nonlinear analysis for dynamic lateral pile response," *Soil Dynamics and Earthquake Engineering*, Vol. 15, pp. 233–244. https://doi.org/10.1016/0267-7261(95)00049-6

EPRI. (2007). "Hard-rock coherency functions based on the pinyon flat array data," *ADAMS Accession No. ML071980104*, Electric Power Research Institute, Palo Alto, CA.

Fadum, R.E. (1948). "Influence values for estimating stresses in elastic foundations, *Proceedings, 2nd International Conference on Soil Mechanics and Foundation Engineering,* Rotterdam, Vol. 3, pp. 77–84.

Fan, K., Gazetas, G., Kaynia, A., and Kausal, E. (1991). "Kinematic seismic response of single piles and pile groups," *Journal of Geotechnical Engineering*, Vol. 117, pp. 1860–1879. https://doi.org/10.1061/(asce)0733-9410(1991)117:12(1860)

Fang, Y. and Ishibashi, I. (1986). "Static earth pressures with various wall movements," *Journal of Geotechnical Engineering*, Vol. 112, No. 3, pp. 317–333. https://doi.org/10.1061/(asce)0733-9410(1986)112:3(317)

Faraji, S., Ting, J.M., Crovo, D.S., and Ernst, H. (2001). "Nonlinear analysis of integral bridges: Finite element model," *Journal of Geotechnical and Geoenvironmental Engineering*, Vol. 127, No. 5, pp. 454–461. https://doi.org/10.1061/(asce)1090-0241(2001)127:5(454)

FEMA. (2000). "Prestandard and commentary for the seismic rehabilitation of buildings," *FEMA 356*, prepared by the American Society of Civil Engineers for Federal Emergency Management Agency, Washington, DC.

FEMA. (2005). "Improvement of nonlinear static seismic analysis procedures," *FEMA 440*, prepared by the Applied Technology Council for Federal Emergency Management Agency, Washington, DC.

Fleming, W.G.K., Weltman, A.J., Randolph, M.F., and Elson, W.K. (2009). *Piling Engineering*, 3rd edition, Taylor and Francis, London and New York.

Flores-Berrones, R. and Whitman, R.V. (1982). "Seismic response of end-bearing piles," *Journal of the Geotechnical Engineering Division*, Vol. 108, pp. 554–569. https://doi.org/10.1061/AJGEB6.0001275

Gajan, S. and Kutter, B.L. (2008). "Capacity, settlement, and energy dissipation of shallow footings subjected to rocking," *Journal Geotechnical and Geoenvironmental Engineering*, Vol. 134, pp. 1129–1141. https://doi.org/10.1061/(ASCE)1090-0241(2008)134:8(1129)

Gajan, S. and Kutter, B.L. (2009). "A contact interface model for shallow foundations subjected to combined cyclic loading," *Journal Geotechnical and Geoenvironmental Engineering*, Vol. 135, pp. 407–419. https://doi.org/10.1061/(ASCE)1090-0241(2009)135:3(407)

Gajan, S., Hutchinson, T.C., Kutter, B.L., Raychowdhury, P., Ugalde, J.A., and Stewart, J.P. (2008). "Numerical models for analysis and performance-based design of shallow foundations subject to seismic loading," *Report No. PEER-2007/04*, Pacific Earthquake Engineering Research Center, University of California, Berkeley, CA.

Gajan, S., Raychowdhury, P., Hutchinson, T.C., Kutter, B.L., and Stewart, J.P. (2010). "Application and validation of practical tools for nonlinear soil-foundation interaction analysis," *Earthquake Spectra*, Vol. 26, pp. 111–129. https://doi.org/10.1193/1.3263242

Gazetas, G. (1991). "Foundation vibrations," in H.-Y. Fang, ed., *Foundation Engineering Handbook*, 2nd edition, Chapman and Hall, New York. https://doi.org/10.1007/978-1-4757-5271-7

Gazetas, G. (2006). "Seismic design of foundations and soil-structure interaction," *Proceedings, First European Conference on Earthquake Engineering and Seismology*, Geneva, Switzerland.

Gazetas, G. and Dobry, R. (1984a). "Horizontal response of piles in layered soil," *Journal of the Geotechnical Engineering Division*, Vol. 110, pp. 20–40. https://doi.org/10.1061/(ASCE)0733-9410(1984)110:1(20)

Gazetas, G. and Dobry, R. (1984b). "Simple radiation damping model for piles and footings," *Journal of the Geotechnical Engineering Division*, Vol. 110, pp. 937–956. https://doi.org/10.1061/(ASCE)0733-9399(1984)110:6(937)

Gazetas, G., and Roesset, J.M. (1976). "Forced vibrations of strip footings on layered soils," in W.E. Saul and A.H. Peyrol, eds., *Proceedings, National Structural Engineering Conference,* ASCE, Reston, VA, Vol. 1, pp. 115–131.

Gazetas, G., Psarropoulos, P.N., Anastasopoulos, I., and Gerolymos, N. (2004). "Seismic behaviour of flexible retaining systems subjected to short-duration moderately strong excitation," *Soil Dynamics and Earthquake Engineering*, Vol. 24, pp. 537–550. https://doi.org/10.1016/j.soildyn.2004.02.005

Givens, M.J., Mylonakis, G., and Stewart, J.P. (2016). "Modular analytical solutions for foundation damping in soil-structure interaction applications," *Earthquake Spectra*, Vol. 32, pp. 1749–1768. https://doi.org/10.1193/071115eqs112m

Guzina, B.B. and Pak, R.Y.S. (1998). "Vertical vibration of a circular footing on a linear-wave-velocity half space," *Geotechnique*, Vol. 48, pp. 159–168. https://doi.org/10.1680/geot.1998.48.2.159

Harden, C.W. and Hutchinson, T.C. (2009). "Beam-on-nonlinear-Winkler-foundation modeling of shallow, rocking-dominated footings," *Earthquake Spectra*, Vol. 25, pp. 277–300. https://doi.org/10.1193/1.3110482

Hardin, B.O. and Black, W.L. (1968). "Vibration modulus of normally consolidated clay," *Journal of the Soil Mechanics and Foundation Division*, Vol. 94, pp. 353–369. https://doi.org/10.1061/JSFEAQ.0001100

Hashash, Y.M.A., Hook, J.J., Schmidt, B., and Yao, J.I.-C. (2001). "Seismic design and analysis of underground structures," *Tunneling and Underground Space Technology*, Vol. 16, pp. 247–293. https://doi.org/10.1016/s0886-7798(01)00051-7

Hayashi, K. (1921): *Theorie des tr äagers auf elastischer unterlage und ihre anwendung auf den tiefbau,* Springer Verlag. Berlin

Hetenyi, M. (1946). *Beams on Elastic Foundation*, University of Michigan Press, Ann Arbor, MI.

Hushmand, A., Dashti, S., Davis, C., and Hushmand, B. (2016). "Seismic performance of underground reservoir structures: Insight from centrifuge modeling on the influence of structure stiffness," *Journal of Geotechnical and Geoenvironmental Engineering,* Vol. 142, No. 7. https://doi.org/10.1061/(ASCE)GT.1943-5606.0001477

Iguchi, M. and Luco, J.E. (1982). "Vibration of flexible plate on viscoelastic medium," *Journal of Engineering Mechanics*, Vol. 108, No. 6, pp. 1103–1120. https://doi.org/10.1061/JMCEA3.0002893

Inagaki, H., Iai, S., and Inatomi, T. (1996). "Performance of caisson type quay walls at Kobe Port," *Soils and Foundations*, Vol. 36, pp. 119–136. https://doi.org/10.3208/sandf.36.Special_119

Jakub, M. and Roesset, J.M. (1977). "Nonlinear stiffness of foundations," *Research Report R77-35*, Massachusetts Institute of Technology, Cambridge, MA.

Jaky, J. (1948). "Pressure in silos," *Proceedings, 2nd International Conference on Soil Mechanics and Foundation Engineering,* Balkema, Rotterdam, Vol. 1, pp. 103–107.

Jeremic, B., Jie, G.Z., Preisig, M., and Tafazzoli, N. (2009). "Time domain simulation of soil-foundation-structure interaction in non-uniform soils," *Journal of Earthquake Engineering and Structural Dynamics*, Vol. 38, pp. 699–718. https://doi.org/10.1002/eqe.896

Kagawa, T. and Kraft, L.M. (1980). "Seismic p-y response of flexible piles," *Journal of Geotechnical Engineering Division*, Vol. 106, No. 8, pp. 899–918. https://doi.org/10.1061/ajgeb6.0001018

Kagawa, T. and Kraft, L.M. (1981). "Lateral pile response during earthquakes," *Journal of Geotechnical Engineering*, Vol. 107, pp. 1713–1731. https://doi.org/10.1061/AJGEB6.0001222

Karatzia, X. and Mylonakis, G. (2017). "Horizontal stiffness and damping of piles in inhomogeneous soil," *Journal of Geotechnical and Geoenvironmental Engineering*, Vol. 143, No. 4, https://doi.org/10.1061/(ASCE)GT.1943-5606.0001621.

Kausel, E. (2010). "Early history of soil-structure interaction," *Soil Dynamics and Earthquake Engineering*, Vol. 30, No. 9, pp. 822–832. https://doi.org/10.1016/j.soildyn.2009.11.001

Kausel, E., Whitman, A., Murray, J., and Elsabee, F. (1978). "The spring method for embedded foundations," *Nuclear Engineering and Design*, Vol. 48, pp. 377–392. https://doi.org/10.1016/0029-5493(78)90085-7

Kaynia, A.M. and Kausel, E. (1982). "Dynamic stiffness and seismic response of pile groups," *Research Report R82-03*, Massachusetts Institute of Technology, Cambridge, MA.

Kaynia, A.M. and Novak, M. (1992). "Response of pile foundations to Rayleigh waves and to obliquely incident body waves," *Earthquake Engineering and Structural Dynamics*, Vol. 21, pp. 303–318. https://doi.org/10.1002/eqe.4290210403

Kerwin, S.T. and Stone, J.J. (1997). "Liquefaction failure and remediation: King Harbor Redondo Beach, California," *Journal of Geotechnical and Geoenvironmental Engineering*, Vol. 123, No. 8, pp. 760–769. https://doi.org/10.1061/(asce)1090-0241(1997)123:8(760)

Khalili-Tehrani, P., Shamsabadi, A., Stewart, J.P., and Taciroglu, E. (2016). "Physically parameterized backbone curves for passive lateral response of bridge abutments with homogeneous backfills," *Bulletin of Earthquake Engineering*, Vol. 14, pp. 3003–3023. https://doi.org/10.1007/s10518-016-9934-3

Khalili-Tehrani, P., Ahlberg, E.R., Rha, C., Lemnitzer, A., Stewart, J.P., Taciroglu, E., and Wallace, J.W. (2014). "Nonlinear load-deflection behavior of reinforced concrete drilled piles in stiff clay," *Journal of Geotechnical and Geoenvironmental Engineering*, Vol. 140, p. 04013022. https://doi.org/10.1061/(ASCE)GT.1943-5606.0000957

Kim, S. and Stewart, J.P. (2003). "Kinematic soil-structure interaction from strong motion recordings," *J. Geotechnical and Geoenvironmental Engineering*, Vol. 129, No. 4, pp. 323–335. https://doi.org/10.1061/(asce)1090-0241(2003)129:4(323)

Kloukinas, P., Langousis, M., and Mylonakis, G. (2012). "Simple wave solution for seismic earth pressures on gravity walls," *Journal of Geotechnical and Geoenvironmental Engineering*, Vol. 138, No. 12, pp. 1514–1519. https://doi.org/10.1061/(ASCE)GT.1943-5606.0000721

Koseki, J. and Shibuya, S. (2014). "Mitigation of disasters by earthquakes, tsunamis, and rains by means of geosynthetic reinforced retaining walls and embankments," *Transportation Infrastructure Geotechnology*, Vol. 1, No. 3, pp. 231–261. https://doi.org/10.1007/s40515-014-0009-0

Koseki, J., Tatsuoka, F., Munaf, Y., Tateyama, M., and Kojima, K. (1998). "A modified procedure to evaluate active earth pressure at high seismic loads," *Soils and Foundations*, Vol. 38, pp. 209–216. https://doi.org/10.3208/sandf.38.Special_209

Kutter, B.L., Moore, M., Hakhamaneshi, M., and Champion, C. (2016). "Rationale for shallow foundation rocking provisions in ASCE 41-13," *Earthquake Spectra*, Vol. 32, No. 2, pp. 1097–1119. https://doi.org/10.1193/121914eqs215m

Lemnitzer, A., Ahlberg, E.R., Nigbor, R.L., Shamsabadi, A., Wallace, J.W., and Stewart, J.P. (2009). "Lateral performance of full-scale bridge abutment wall with granular backfill," *Journal of Geotechnical and Geoenvironmental Engineering*, Vol. 135, pp. 506–514. https://doi.org/10.1061/(asce)1090-0241(2009)135:4(506)

Lew, M., Simantob, E., and Hudson, M.B. (1995). "Performance of shored earth retaining systems during the January 17, 1994, Northridge Earthquake," *Proceedings of the Third International Conference on Recent Advances in Geotechnical Earthquake Engineering and Soil Dynamics,* St Louis, MO, Vol. 3.

Lew, M., Sitar, N., Al-Atik, L., Pourzanjani, M., and Hudson, M.B. (2010). "Seismic earth pressures on deep building basements," *Structural Engineers Association of California, Proceedings of the Annual Convention.*

Li, Z., Kotronis, P., Escoffier, S., and Tamagnini, C. (2015). "Macroelement modeling for single vertical piles," *Proc. 6th International Conference on Earthquake Geotechnical Engineering,* Christchurch, New Zealand, Paper No. 302 (electronic file).

Ling, H.I., Leshchinsky, D., and Chou, N. (2001). "Post-earthquake investigation on several geosynthetic-reinforced soil retaining walls and slopes during the Chi-Chi earthquake in Taiwan," *Soil Dynamics and Earthquake Engineering*, Vol. 21, pp. 297–313. https://doi.org/10.1016/s0267-7261(01)00011-2

Liou, G.S. and Huang, P.H. (1994). "Effect of flexibility on impedance functions for circular foundations," *Journal of Engineering Mechanics*, Vol. 120, No. 7, pp. 1429–1446. https://doi.org/10.1061/(ASCE)0733-9399(1994)120:7(1429)

Luco, J.E. and Westmann, R.A. (1971). "Dynamic response of circular footings," *Journal of Engineering Mechanics*, Vol. 97, No. 5, pp. 1381–1395. https://doi.org/10.1061/JMCEA3.0001467

Lysmer, J., Ostadan, F., and Chen, C. (1999). *Computer Program SASSI2000*, Geotechnical Division, University of California, Berkeley, CA.

Mamoon, S.M. and Banerjee, P.K. (1990). "Response of piles and pile groups to traveling SH waves," *Earthquake Engineering and Structural Dynamics*, Vol. 19, pp. 597–610. https://doi.org/10.1002/eqe.4290190410

Maravas, A., Mylonakis, G., and Karabalis, D.L. (2014). "Simplified discrete systems for dynamic analysis of structures on footings and piles," *Soil Dynamics and Earthquake Engineering*, Vol. 61, pp. 29–39. https://doi.org/10.1016/j.soildyn.2014.01.016

Marcuson, W.F. and Wahls, H.E. (1972). "Time effects on dynamics shear modulus of clays," *Journal of Soil Mechanics and Foundations Division*, Vol. 98, pp. 1359–1373. https://doi.org/10.1061/JSFEAQ.0001819

Martin, G.R., March, M.L., Anderson, D.G., Mayes, R.L., and Power, M.S. (2002). "Recommended design approach for liquefaction induced lateral spreads," *Proceedings, 3rd National Seismic Conference and Workshop on Bridges and Highways*, Multidisciplinary Center for Earthquake Engineering Research, Document MCEER-02-SP04, Buffalo, NY.

Matlock, H. (1970). "Correlations for design of laterally loaded piles in soft clay," *Proceedings, 2nd Offshore Technology Conference*, Houston, TX, Paper No. 1204, Vol. 1, pp. 577–594.

Matsuo, M., Kenmochi, S., and Yagi, H. (1978). "Experimental study on earth pressure of retaining wall by field tests," *Soils and Foundations*, Vol. 18, No. 3, 27–41. https://doi.org/10.3208/sandf1972.18.3_27

Mayne, P.W. and Kulhawy, F.H. (1982). "K_0-OCR relationships in soil," *Journal Geotechnical Engineering Division*, Vol. 108, No. 6, pp. 851–872. https://doi.org/10.1061/AJGEB6.0001306

McClelland, B. and Focht, Jr, F.A. (1956). "Soil modulus for laterally loaded piles," *Journal of the Soil Mechanics and Foundations Division,* Vol. 82, No. 4, pp. 1–22. https://doi.org/10.1061/JSFEAQ.0000023

McGann, C.R., Arduino, P., and Mackenzie-Helnwein, P. (2011). "Applicability of conventional p-y relations to the analysis of piles in laterally spreading soil," *Journal of Geotechnical and Geoenvironmental Engineering*, Vol. 137, No. 6, pp. 557–567. https://doi.org/10.1061/(asce)gt.1943-5606.0000468

Merritt, J.L., Monsees, J.E., and Hendron, Jr, A.J. (1985). "Seismic design of underground structures," *Proceedings of the 1985 Rapid Excavation Tunneling Conference,* Vol. 1, pp. 104–131.

Michaelides, O., Gazetas, G., Bouckovalas, G., and Chrysikou, E. (1998). "Approximate nonlinear analysis of piles," *Geotechnique*, Vol. 48, pp. 33–54. https://doi.org/10.1680/geot.1998.48.1.33

Mikami, A., Stewart, J.P., and Kamiyama, M. (2008). "Effects of time series analysis protocols on transfer functions calculated from earthquake accelerograms," *Soil Dynamics and Earthquake Engineering*, Vol. 28, pp. 695–706. https://doi.org/10.1016/j.soildyn.2007.10.018

Mokwa, R.L. and Duncan, J.M. (2001). "Experimental evaluation of lateral-load resistance of pile caps," *Journal of Geotechnical and Geoenvironmental Engineering*, Vol. 127, No. 2, pp. 185–192. https://doi.org/10.1061/(asce)1090-0241(2001)127:2(185)

Mononobe, N. and Matsuo, M. (1929). "On the determination of earth pressures during earthquakes," *Proc. World Engineering Congress*, Vol. 9, pp. 179–187.

Mylonakis, G. (1995). "Contribution to static and seismic analysis of piles and pile-supported bridge piers," *Ph.D. Dissertation*, Department of Civil Engineering, University at Buffalo, State University of New York.

Mylonakis, G. and Crispin, J.J. (2021). "Simplified models for lateral static and dynamic analysis of pile foundations," In *Analysis of Pile Foundations Subjected to Static and Dynamic Loading,* CRC Press, 1st Edition. London. https://doi.org/10.1201/9780429354281

Mylonakis, G. and Gazetas, G. (1998). "Settlement and additional internal forces of grouped piles in layered soil," *Geotechnique*, Vol. 48, pp. 55–72. https://doi.org/10.1680/geot.1998.48.1.55

Mylonakis, G. and Gazetas, G. (1999). "Lateral vibration and internal forces of grouped piles in layered soil," *Journal Geotechnical and Geoenvironmental Engineering*, Vol. 125, pp. 16–25. https://doi.org/10.1061/(ASCE)1090-0241(1999)125:1(16)

Mylonakis, G. and Gazetas, G. (2000). "Seismic soil-structure interaction: Beneficial or detrimental," *Journal of Earthquake Engineering*, Vol. 4, pp. 377–401. https://doi.org/10.1080/13632460009350372

Mylonakis, G. and Roumbas, D. (2001). "Dynamic stiffness and damping of piles in inhomogeneous soil media," *Proceedings, 4th International Conference on Recent Advances in Geotechnical Earthquake Engineering and Soil Dynamics,* San Diego, CA, Paper No. 6.27.

Mylonakis, G., Kloukinas, P., and Papantonopoulos, C. (2007). "An alternative to the Mononobe-Okabe equations for seismic earth pressures," *Soil Dynamics and Earthquake Engineering*, Vol. 27, pp. 957–969. https://doi.org/10.1016/j.soildyn.2007.01.004

Mylonakis, G., Nikolaou, S., and Gazetas, G. (2006). "Footings under seismic loading: Analysis and design issues with emphasis on bridge foundations," *Soil Dynamics and Earthquake Engineering*, Vol. 26, pp. 824–853. https://doi.org/10.1016/j.soildyn.2005.12.005

National Cooperative Highway Research Program. (2008). "Seismic analysis and design of retaining walls, buried structures, slopes, and embankments," *Report 611*, D.G. Anderson, G.R. Martin, I.P. Lam, and J.N. Wang, eds., National Academies, Washington, DC. https://doi.org/10.17226/14189

Newmark, N.M. (1969). "Torsion of symmetrical buildings," *Proceedings, 4th World Conference on Earthquake Engineering,* Santiago, Chile.

Nikolaou, A., Mylonakis, G., Gazetas, G., and Tazoh, T. (2001). "Kinematic pile bending during earthquakes: Analysis and field measurements," *Geotechnique*, Vol. 51, pp. 425–440. https://doi.org/10.1680/geot.2001.51.5.425

NIST. (2012). "Soil-structure interaction for building structures," *Report No. NIST GCR 12-917-21*, National Institute of Standards and Technology, U.S. Department of Commerce, Washington, DC.

Nogami, T. (1983). "Dynamic group effect in axial response of grouped piles," *Journal of Geotechnical Engineering*, Vol. 109, pp. 228–243. https://doi.org/10.1061/(asce)0733-9410(1983)109:2(228)

Nogami, T. and Konagai, K. (1988). "Time-domain flexural response of dynamically loaded single piles," *Journal of Engineering Mechanics*, Vol. 114, pp. 1512–1525. https://doi.org/10.1061/(ASCE)0733-9399(1988)114:9(1512)

Nogami, T., Otani, J., Konagai, K., and Chen, H.-L. (1992). "Nonlinear soil-pile interaction model for dynamic lateral motion," *Journal of Geotechnical Engineering*, Vol. 118, pp. 89–106. https://doi.org/10.1061/(asce)0733-9410(1992)118:1(89)

Nova, R. and Montrasio, L. (1991). "Settlements of shallow foundations on sand," *Geotechnique*, Vol. 41, pp. 243–256. https://doi.org/10.1680/geot.1991.41.2.243

Novak, M. (1974). "Dynamic stiffness and damping of piles," *Canadian Geotechnical Journal*, Vol. 11, pp. 574–598. https://doi.org/10.1139/t74-059

Okabe, S. (1926). "General theory of earth pressure and seismic stability of retaining wall and dam," *Journal of the Japanese Society of Civil Engineering,* Vol. 12, No. 4, pp. 34–41.

Omrani, R., Mobasher, B., Liang, X., Günay, S., Mosalem, K.M., Zareian. F., and Taciroglu, E. (2015). "Guidelines for nonlinear seismic analysis of ordinary bridges: Version 2.0," *Caltrans Final Report No. 15-65A0454*, California Department of Transportation, Sacramento, CA.

OpenSees. (2011). *Open System for Earthquake Engineering Simulation: OpenSees*, University of California, Berkeley, https://opensees.berkeley.edu.

Ostadan, F. (2005). "Seismic soil pressure for building walls – an updated approach," *Soil Dynamics and Earthquake Engineering*, Vol. 25, pp. 785–793. https://doi.org/10.1016/j.soildyn.2004.11.035

Ostadan, F., Deng, N., and Kennedy, R. (2005). "Soil-structure interaction analysis including ground motion incoherency effects," *Proceedings, 18th International Conference on Structural Mechanics in Reactor Technology (SMiRT 18)*, Beijing, China.

Owen, G.N. and Scholl, R.E. (1981). "Earthquake engineering of large underground structures," *Report No. FHWA/RD-80/195*. Federal Highway Administration and National Science Foundation.

Pais, A. and Kausel, E. (1988). "Approximate formulas for dynamic stiffnesses of rigid foundations," *Soil Dynamics and Earthquake Engineering*, Vol. 7, pp. 213–227. https://doi.org/10.1016/s0267-7261(88)80005-8

Paulay, T., and Priestley, M.J.N. (1992). *Seismic Design of Reinforced Concrete and Masonry Buildings*, John Wiley & Sons, Inc., New York. https://doi.org/10.1002/9780470172841

Pecker, A. and Chatzigogos, C.T. (2010). Non-linear soil-structure interaction: Impact on the seismic response of structures," in M. Garevski and A. Ansal, eds., *Earthquake Engineering in Europe*, Springer, New York, pp. 79–103. https://doi.org/10.1007/978-90-481-9544-2_4

PEER. (2010). "Guidelines for performance-based seismic design of tall buildings," *Report No. 2010/05,* developed by the Pacific Earthquake Engineering Research Center as part of the Tall Buildings Initiative, University of California, Berkeley, CA.

PEER. (2017). "Guidelines for performance-based seismic design of tall buildings, Version 2," *Report No. 2017/06,* developed by the Pacific Earthquake Engineering Research Center as part of the Tall Buildings Initiative, University of California, Berkeley, CA.

Pender, M. (1993). "Aseismic pile foundation design analysis," *Bulletin of New Zealand National Society of Earthquake Engineering*, Vol. 26, pp. 49–160. https://doi.org/10.5459/bnzsee.26.1.49-160

Penzien, J. (2000). "Seismically-induced racking of tunnel linings," *Earthquake Engineering & Structural Dynamics*, Vol. 29, pp. 683–691. https://doi.org/10.1002/(SICI)1096-9845(200005)29:5<683::AID-EQE932>3.0.CO;2-1

Penzien, J., Scheffey, C.F., and Parmelee, R.A. (1964). "Seismic analysis of bridges on long pile," *Journal of the Engineering Mechanics Division*, Vol. 90, No. 3, pp. 223–254. https://doi.org/10.1061/JMCEA3.0000489

Perez-Rocha, L.E. and Aviles, J. (2003). "The evaluation of interaction effects in inflexible resistances," *Revista de Ingeniería Sìsmica*, No. 69 (in Spanish).

Poulos, H.G. and Davis, E.H. (1980). *Pile Foundation Analysis and Design*, Wiley, New York, NY.

Powell, G.H. (2006). "Static pushover methods – explanation, comparison and implementation," *Proceedings, 8th U.S. National Conference on Earthquake Engineering,* San Francisco, CA.

Power, M., Fishman, K., Makdisi, F., Musser, S., Richards, R., and Youd, T.L. (2006). "Seismic Retrofit Manual for Highway Structures: Part II - Retaining Structures, Slopes, Tunnels, Culverts and Roadways," *Report MCEER-06-SP11*, Multidisciplinary Center for Earthquake Engineering Research, Buffalo, NY.

Priestley, M.J.N. and Park, R. (1987). "Strength and ductility of concrete bridge columns under seismic loading," *ACI Structural Journal*, Vol. 84, pp. 61–76. https://doi.org/10.14359/2800

Randolph, M.F. (1981). "The response of flexible piles to lateral loading," *Géotechnique*, Vol. 31, pp. 247–259. https://doi.org/10.1680/geot.1981.31.2.247

Randolph, M.F. (2003). "Science and empiricism in pile foundation design," *Géotechnique*, Vol. 53, pp. 847–875. https://doi.org/10.1680/geot.53.10.847.37518

Randolph, M.F. and Wroth, C.P. (1978). "Analysis of deformation of vertically loaded piles," *Journal of Geotechnical Engineering*, Vol. 104, pp. 1465–1488. https://doi.org/10.1061/ajgeb6.0000729

Rankine, W. (1857). "On the stability of loose earth," *Philosophical Transactions of the Royal Society of London*, Vol. 147, pp. 185–187.

Raychowdhury, P. and Hutchinson, T.C. (2009). "Performance evaluation of a nonlinear Winkler-based shallow foundation model using centrifuge test results," *Earthquake Engineering and Structural Dynamics*, Vol. 38, pp. 679–698. https://doi.org/10.1002/eqe.902

Reese, L.C. and Welch, R. (1975). "Lateral loading of deep foundations in stiff clay," *Journal of the Geotechnical Engineering Division*, Vol. 101, pp. 633–649. https://doi.org/10.1061/AJGEB6.0000177

Reese, L.C., Isenhower, W.M., and Wang, S.-T. (2006). *Analysis and Design of Shallow and Deep Foundations*, Wiley, Hoboken, NJ. https://doi.org/10.1002/9780470172773

Rendulic, L. (1935). "Der hydrodynamische Spannungsangleich in zentral entwasserten Tonzylindren," *Wasserwirtsch. Tech*, Vol. 2, pp. 250–253 and pp. 269–273.

Richart, R. and Elms, D. (1979). "Seismic behavior of gravity retaining walls," *Journal of the Geotechnical Engineering Division*, Vol. 105, pp. 449–464. https://doi.org/10.1061/AJGEB6.0000783

Roesset, J.M. (1980). "Stiffness and damping coefficients of foundations, *Proceedings, ASCE Geotechnical Engineering Division National Convention,* pp. 1–30.

Rollins, K.M. and Cole, R.T. (2006). "Cyclic lateral load behavior of a pile cap and backfill," *Journal of Geotechnical and Geoenvironmental Engineering*, Vol. 132, No. 9, pp. 1143–1153. https://doi.org/10.1061/(asce)1090-0241(2006)132:9(1143)

Rollins, K.M. and Jessee, S.J. (2013). "Passive force-deflection curves for skewed abutments," *Journal of Bridge Engineering*, Vol. 18, No. 10, pp. 1086–1094. https://doi.org/10.1061/(ASCE)BE.1943-5592.0000439

Rollins, K.M. and Sparks, A.E. (2002). "Lateral load capacity of a full-scale fixed-head pile group," *Journal of Geotechnical and Geoenvironmental Engineering*, Vol. 128, No. 9, pp. 711–723. https://doi.org/10.1061/(ASCE)1090-0241(2002)128:9(711)

Rollins, K.M., Lane, J.D., and Gerber, T.M. (2005). "Measured and computed lateral response of a pile group in sand," *Journal of Geotechnical and Geoenvironmental Engineering*, Vol. 131, pp. 103–114. https://doi.org/10.1061/(asce)1090-0241(2005)131:1(103)

Rovithis, E.N., Pitilakis, K., and Mylonakis, G. (2009). "Seismic analysis of coupled soil-pile-structure systems leading to the definition of a pseudo-natural SSI frequency," *Soil Dynamics and Earthquake Engineering*, Vol. 29, pp. 1005–1015. https://doi.org/10.1016/j.soildyn.2008.11.005

Salgado, R. (2008). *The Engineering of Foundations*, McGraw-Hill, New York.

Scott, R. (1981). *Foundation Analysis*, Prentice Hall, Upper Saddle River, NJ.

Seed, H.B. and Whitman, R.V. (1970). "Design of earth retaining structures for dynamic loads," *Proceedings, ASCE Specialty Conference on Lateral Stresses in the Ground and Design of Earth Retaining Structures,* Cornell University, Ithaca, NY, Vol. 1, pp. 103–147.

Shamsabadi, A., Khalili-Tehrani, P., Stewart, J.P., and Taciroglu, E. (2010). "Validated simulation models for lateral response of bridge abutments with typical backfills," *Journal of Bridge Engineering*, Vol. 15, pp. 302–311. https://doi.org/10.1061/(ASCE)BE.1943-5592.0000058

Shamsabadi, A., Rollins, K.M., and Kapaskur, M. (2007). "Nonlinear soil abutment-bridge structure interaction for seismic performance-based design," *Journal of Geotechnical and Geoenvironmental Engineering*, Vol. 133(6), pp. 707–720. https://doi.org/10.1061/(asce)1090-0241(2007)133:6(707)

Sotiriadis, D., Klimis, N., Margaris, B., and Sextos, A. (2020). "Analytical expressions relating free-field and foundation ground motions in buildings with basement, considering soil-structure interaction," *Engineering Structures*, Vol. 216, p. 110757. https://doi.org/10.1016/j.engstruct.2020.110757

Soubra, A.-H. (1999). "Upper-bound solutions for bearing capacity of foundations," *Journal Geotechnical and Geoenvironmental Engineering*, Vol. 125, pp. 59–68. https://doi.org/10.1061/(ASCE)1090-0241(1999)125:1(59)

Steedman, R.S. and Zeng, X. (1990). "The influence of phase on the calculation of pseudo-static earth pressure on a retaining wall," *Geotechnique*, Vol. 40, No. 1, pp. 103–112. https://doi.org/10.1680/geot.1990.40.1.103

Stewart, J.P., Fenves, G.L., and Seed, R.B. (1999b). "Seismic soil-structure interaction in buildings II: Empirical findings," *Journal Geotechnical and Geoenvironmental Engineering*, Vol. 125, pp. 38–48. https://doi.org/10.1061/(ASCE)1090-0241(1999)125:1(38)

Stewart, J.P., Kim, S., Bielak, J., Dobry, R., and Power, M.S. (2003). "Revisions to soil structure interaction procedures in NEHRP design provisions," *Earthquake Spectra*, Vol. 19, pp. 677–696. https://doi.org/10.1193/1.1596213

Stewart, J.P., Seed, R.B., and Fenves, G.L. (1999a). "Seismic soil-structure interaction in buildings I: Analytical aspects," *Journal Geotechnical and Geoenvironmental Engineering*, Vol. 125, pp. 26–37. https://doi.org/10.1061/(ASCE)1090-0241(1999)125:1(26)

Stewart, J.P. and Tileylioglu, S. (2007). "Input ground motions for tall buildings with subterranean levels," *Structural Design of Tall and Special Buildings*, 16, pp. 543–557. https://doi.org/10.1002/tal.429

Syngros, K. (2004). *Contributions to the Static and Seismic Analysis of Piles and Pile Supported Bridge Piers Evaluated through Case Histories*, Ph.D. Dissertation, City University of New York, New York.

Taciroglu, E., Rha, C., and Wallace, J.W. (2006). "A robust macroelement model for soil-pile interaction under cyclic loads," *Journal of Geotechnical and Geoenvironmental Engineering*, Vol. 132, pp. 1304–1314. https://doi.org/10.1061/(asce)1090-0241(2006)132:10(1304)

Talesnick, M.L. (2012). "A different approach and result to the measurement of K_0 of granular soils," *Geotechnique*, Vol. 62, pp. 1041–1045. https://doi.org/10.1680/geot.11.p.009

Tanaka, Y., Yagiura, Y., Shimokawa, K., Higashi, S., Kishida, T., and Mizuwake, N. (2001). "Study of liquefaction damages of quay-walls and breakwaters during Kobe Earthquake," *International Conference on Recent Advances in Geotechnical Earthquake Engineering and Soil Dynamics*, Paper No. 33.

Tatsuoka, F., Tateyama, M., and Koseki, J. (1996). "Performance of soil retaining walls for railway embankments," *Soils and Foundations*, Vol. 36, No. Special, pp. 311–324. https://doi.org/10.3208/sandf.36.Special_311

Taylor, D.W. (1937). "The stability of earth slopes," *Journal of the Boston Society of Civil Engineers*, Vol. 24, No. 3, pp. 197–247.

Taylor, D.W. (1948). *Fundamentals of Soil Mechanics*, Wiley, New York, p. 700.

Terzaghi, K. (1934). "Large retaining wall tests," *Engineering News Record*, Feb. 1, Feb, 22, Mar. 8, Mar. 20, Apr. 19.

Terzaghi, K. (1943). *Theoretical Soil Mechanics*, Wiley, New York. https://doi.org/10.1002/9780470172766

Thomas, G.E. (1980). "Equivalent spring and damping coefficient for piles subjected to vertical dynamic loads," *M.S. Thesis*, Rensselar Polytechnic Institute, New York.

Thomas, J.M., Gajan, S., and Kutter, B.L. (2005). "Soil-foundation-structure interaction: Shallow foundations," *UCD/CGMDR-05/02*, Center for Geotechnical Modeling, University of California, Davis, CA.

Tileylioglu, S., Stewart, J.P., and Nigbor, R.L. (2011). "Dynamic stiffness and damping of a shallow foundation from forced vibration of a field test structure," *Journal of Geotechnical and Geoenvironmental Engineering*, Vol. 137, No. 4, pp. 344–353. https://doi.org/10.1061/(asce)gt.1943-5606.0000430

Tsagareli, Z.V. (1965). "Experimental investigation of the pressure of a loose medium on retaining walls with a vertical back face and horizontal backfill surface," *Journal of Soil Mechanics and Foundation Engineering*, Vol. 91, No. 4, pp. 197–200. https://doi.org/10.1007/BF01706095

Turner, B.J., Brandenberg, S.J., and Stewart, J.P. (2016). "Case study of parallel bridges affected by liquefaction and lateral spreading," *Journal of Geotechnical and Geoenvironmental Engineering*, Vol. 142, p. 05016001. https://doi.org/10.1061/(asce)gt.1943-5606.0001480

U.S. Department of the Navy. (1982). "Foundations and earth structures," *NAVFAC DM-7.2*, Naval Facilities Engineering Command, U.S. Government Printing Office, Washington, DC, 244 pp.

Veletsos, A.S. and Meek, J.W. (1974). "Dynamic behavior of building-foundation system," *Earthquake Engineering and Structural Dynamics*, Vol. 3, pp. 121–138. https://doi.org/10.1002/eqe.4290030203

Veletsos, A.S. and Nair, V.V. (1975). "Seismic interaction of structures on hysteretic foundations," *Journal of Structural Engineering*, Vol. 101, pp. 109–129. https://doi.org/10.1061/jsdeag.0003962

Veletsos, A.S. and Prasad, A.M. (1989). "Seismic interaction of structures and soils: Stochastic approach," *Journal of Structural Engineering*, Vol. 115, pp. 935–956. https://doi.org/10.1061/(asce)0733-9445(1989)115:4(935)

Veletsos, A.S., Prasad, A.M., and Wu, W.H. (1997). "Transfer functions for rigid rectangular foundations," *Earthquake Engineering and Structural Dynamics*, Vol. 26, pp. 5–17. https://doi.org/10.1002/(sici)1096-9845(199701)26:1<5::aid-eqe619>3.0.co;2-x

Veletsos, A.S. and Wei, Y.T. (1971). "Lateral and rocking vibrations of footings," *Journal of Soil Mechanics and Foundations Division*, Vol. 97, pp. 1227–1248. https://doi.org/10.1061/JSFEAQ.0001661

Veletsos, A.S. and Younan, A.H. (1994). "Dynamic soil pressures on rigid retaining walls," *Earthquake Engineering & Structural Dynamics*, Vol. 23, No. 3, pp. 275–301. https://doi.org/10.1002/eqe.4290230305

Vrettos, C. (1999). "Vertical and rocking impedances for rigid rectangular foundations on soils with bounded non-homogeneity," *Earthquake Engineering and Structural Dynamics*, Vol. 28, pp. 1525–1540. https://doi.org/10.1002/(sici)1096-9845(199912)28:12<1525::aid-eqe879>3.0.co;2-s

Vrettos, C., Beskos, D.E., and Triantafyllidis, T. (2016). "Seismic pressures on rigid cantilever walls retaining elastic continuously non-homogeneous soil: An exact solution," *Soil Dynamics and Earthquake Engineering*, Vol. 82, pp. 142–153. https://doi.org/10.1016/j.soildyn.2015.12.006

Wagner, N. and Sitar, N. (2016). "On seismic response of stiff and flexible retaining structures," *Soil Dynamics and Earthquake Engineering*, Vol. 91, pp. 284–293. https://doi.org/10.1016/j.soildyn.2016.09.025

Wang, J.-N. (1993). "Seismic design of tunnels: A state-of-the-art approach," *Monograph 7*, Parsons, Brinckerhoff, Quade and Douglas Inc., New York.

Watson, G.N. (1966). *A Treatise on the Theory of Bessel Functions*, 2nd edition, Cambridge University Press, United Kingdom.

Weaver, T.S., Ashford, S.A., and Rollins, K.M. (2005). "Response of liquefied sand to a 0.6-m CISS pile under lateral loading," *Journal of Geotechnical and Geoenvironmental Engineering*, Vol. 131, pp. 94–102. https://doi.org/10.1061/(ASCE)1090-0241(2005)131:1(94)

Wilson, D.W., Boulanger, R.W., and Kutter, B.L. (2000). "Observed seismic lateral resistance of liquefying sand," *Journal of Geotechnical and Geoenvironmental Engineering*, Vol. 126, pp. 898–906. https://doi.org/10.1061/(asce)1090-0241(2000)126:10(898)

Wilson, E.L., Der Kiureghian, A., and Bayo, E.P. (1981). "A replacement for the SRSS method in seismic analysis," *Earthquake Engineering and Structural Dynamics*, Vol. 9, pp. 187–192. https://doi.org/10.1002/eqe.4290090207

Winkler, E. (1867). *Die Lehre von der Elastizitat und Festigkeit*, Verlag H. Dominicus, Prague, Czechoslovakia.

Wolf, J.P. (1985). *Dynamic Soil-Structure Interaction*, Prentice-Hall, Upper Saddle River, NJ.

Wolf, J.P. and Song, C. (1996). "To radiate or not to radiate," *Earthquake Engineering and Structural Dynamics*, Vol. 25, pp. 1421–1432. https://doi.org/10.1002/(sici)1096-9845(199612)25:12>1421::>3.0.co;2-w

Wolf, J.P. and Von Arx, G.A. (1978). "Impedance function of a group of vertical piles," *Proceedings, Specialty Conference on Earthquake Engineering and Structural Dynamics*, Pasadena, CA, Vol. 2, pp. 1024–1041.

Wong, H.L. (1979). "Soil-structure interaction: A linear continuum mechanics approach (CLASSI users manual)," *Report No.* 79-04, Department of Civil Engineering, University of Southern California, Los Angeles, CA.

Wong, H.L. and Luco, J.E. (1985). "Tables of impedance functions for square foundations on layered media," *Soil Dynamics and Earthquake Engineering*, Vol. 5, pp. 149–158. https://doi.org/10.1016/0261-7277(85)90002-6

Wood, J.H. (1973). "Earthquake induced soil pressures on structures," *Report No. EERL 73-05*, California Institute of Technology, Pasadena, CA.

Wood, J.H. (2023). "Earthquake design loads for retaining walls," *Bulletin of the New Zealand Society of Earthquake Engineering*, Vol. 56, No. 4, 201-220.

Xu, S.Y., Shamsabadi, A., and Taciroglu, E. (2015). "Evaluation of active and passive seismic earth pressures considering internal friction and cohesion," *Soil Dynamics and Earthquake Engineering*, Vol. 70, pp. 30–47. https://doi.org/10.1016/j.soildyn.2014.11.004

Yamada, S., Hyodo, M., Orense, R.P., Dinesh, S.V., and Hyodo, T. (2008). "Strain-dependent dynamic properties of remolded sand-clay mixtures," *Journal Geotechnical and Geoenvironmental Engineering*, Vol. 134, pp. 972–981. https://doi.org/10.1061/(asce)1090-0241(2008)134:7(972)

Yoshimura, C., Bielak, J., Hisada, Y., and Fernandez, A. (2002). "Domain reduction method for three-dimensional earthquake modeling in localized regions, Part II: Verification and applications," *Bulletin of the Seismological Society of America*, Vol. 93, pp. 825–840. https://doi.org/10.1785/0120010252

9 Liquefaction and Cyclic Softening

9.1 INTRODUCTION

Liquefaction is one of the most important, interesting, complex, and historically controversial topics in geotechnical earthquake engineering. Its devastating effects sprang to the attention of geotechnical engineers in a three-month period in 1964 when the Good Friday earthquake (**M**=9.2) in Alaska was followed by the Niigata earthquake (M_S=7.5) in Japan. Both earthquakes produced spectacular liquefaction-induced damage, including slope failures, bridge and building foundation failures, and flotation of buried structures. In the decades since these earthquakes, liquefaction has been studied extensively by hundreds of researchers around the world. Much has been learned, but the road has not been smooth. Different terminologies, procedures, and methods of analysis have been proposed over the years as additional data has been obtained and knowledge gained, yet consensus has been elusive at times on the procedures that should be used to assess liquefaction hazards.

The term *liquefaction*, originally coined by Hazen (1920), has historically been used in conjunction with a variety of phenomena that involve soil deformations caused by monotonic, transient, or repeated disturbance of saturated, non-plastic, coarse-grained soils under undrained conditions. The generation of high excess pore pressure under undrained loading conditions is a hallmark of all liquefaction phenomena. *Cyclic softening* is a term used to describe a reduction of stiffness and strength typically displayed by saturated, plastic, fine-grained soils. Although the effects of cyclic softening may be similar to those of liquefaction for some soils, there are important differences between them, and they will be treated as different phenomena in this book. This chapter presents such terminology and a framework for conceptual understanding of liquefaction-related soil behavior and the various empirical and numerical procedures used to evaluate liquefaction- and cyclic softening-related hazards. Readers are encouraged to review Section 6.4 (particularly Sections 6.4.3 and 6.4.4) to prepare for the material presented in this chapter.

9.2 LIQUEFACTION-RELATED PHENOMENA

Over time, the term "liquefaction" has been used to describe different phenomena. More recently, observations of soil behavior in actual earthquakes have broadened the range of soil types that exhibit behavior commonly associated with that of liquefiable soils. These factors have led to a degree of confusion, or at least inconsistency, in the terminology used to describe liquefaction-related phenomena. Liquefaction and cyclic softening are both very important, and any evaluation of seismic hazards should carefully consider both. For the purposes of this textbook, phenomena associated with the softening/weakening of soils during earthquakes will be divided into three main groups: flow liquefaction, cyclic liquefaction, and cyclic softening.

9.2.1 Flow Liquefaction

Flow liquefaction produces the most dramatic consequences of all liquefaction-related phenomena – tremendous instabilities known as *flow failures*. Flow liquefaction can occur when the shear stress required for static equilibrium of a soil mass (i.e., the initial, static shear stress) is greater than the shear strength of the soil in its liquefied state. Once triggered, the large deformations produced by flow liquefaction are actually driven by static shear stresses. The cyclic stresses induced by earthquake

DOI: 10.1201/9781003512011-9

shaking simply bring the soil to an unstable state at which its strength drops sufficiently to allow the static stresses to drive the flow failure. It should be noted, however, that flow liquefaction can also be triggered by non-seismic loading (Terzaghi, 1950; Castro, 1969; Kramer and Seed, 1988). Flow liquefaction that occurs under static loading conditions is often referred to as *static liquefaction*.

Flow liquefaction failures are characterized by the sudden nature of their origin, the speed with which they develop, and the large distance over which the liquefied materials often move. The flow slide failures of the Lower San Fernando Dam (Figure 1.11) and Tapo Canyon Tailings Dam (Figure 1.12) are examples of flow liquefaction. The fluid nature of liquefied soil is illustrated in Figure 9.1.

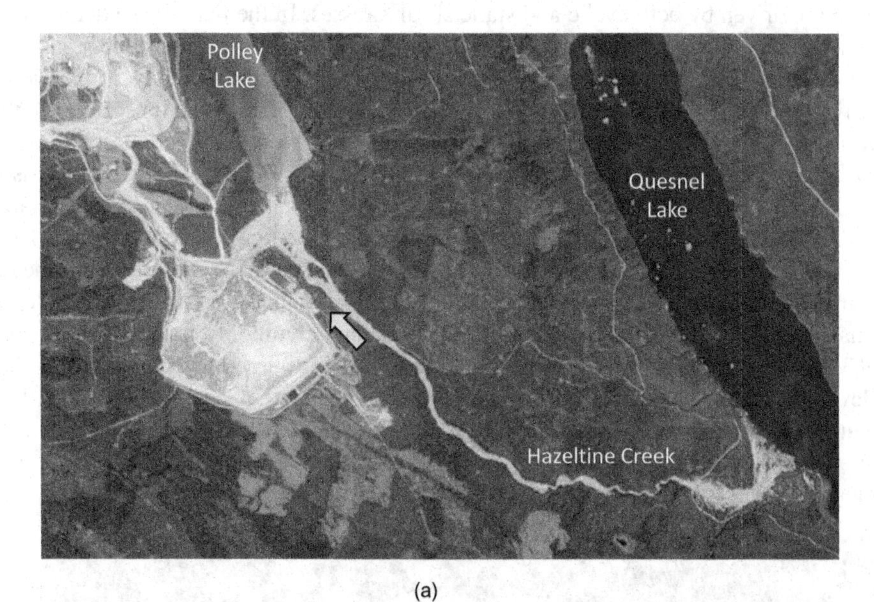

(a)

(b)

FIGURE 9.1 A flow slide from the 2014 failure of Mt. Polley tailings dam (left) near Likely, British Columbia: (a) satellite view (NASA Earth Observatory); arrow indicates orientation of photo below and (b) oblique aerial view showing left-to-right flow of liquefied tailings (with permission of The Canadian Press). The slide released 17 million m^3 of water and 8 million m^3 of tailings into Hazeltine Creek and Polley Lake.

9.2.2 Cyclic Liquefaction

Cyclic liquefaction is another phenomenon that can produce large, damaging permanent deformations during earthquake shaking. In contrast to flow liquefaction, cyclic liquefaction occurs when the static shear stress is less than the shear strength of the liquefied soil. Cyclic liquefaction is characterized by a substantial and often rapid reduction in soil stiffness and strength associated with the buildup of very high porewater pressure that occurs in granular soils. The deformations produced by cyclic liquefaction failures develop incrementally during earthquake shaking and largely cease when shaking has ended. In contrast to flow liquefaction, the deformations produced by cyclic liquefaction are driven by both cyclic and static shear stresses. In the field, these deformations can produce *lateral spreading* on very gently sloping ground or on virtually flat ground adjacent to bodies of water (Figure 9.2) and local settlement of structures supported on shallow foundations.

Cyclic liquefaction can occur under both level- and sloping-ground conditions. Because the static horizontal shear stresses that drive lateral deformations under sloping ground conditions do not exist, *level-ground liquefaction* can produce large, chaotic, transient movement known as *ground oscillation* during earthquake shaking, but typically produces relatively little permanent lateral soil movement. Level-ground liquefaction failures are caused by the upward flow of water that occurs when seismically induced excess pore pressures dissipate. Depending on the length of time required to reach hydraulic equilibrium, level-ground liquefaction failure may occur well after ground shaking has ceased. The ejection of subsurface sediments, commonly referred to as *sand boils* (Figure 9.3), excessive vertical settlement, and consequent flooding of low-lying land are characteristics of level-ground liquefaction failure. Damage to deep foundations, pipelines and the floating of buried structures can also be caused by level-ground liquefaction. When liquefaction is extensive, the volume of soil ejected from below the ground surface may be large (Figure 9.4) and may cause damaging additional settlement of the original ground surface.

FIGURE 9.2 Lateral spreading of flat ground toward the Motagua River following the 1976 Guatemala earthquake. Note orientation of ground surface cracks parallel to river bank (photo by G. Plafker, courtesy of USGS).

FIGURE 9.3 Sand boils near Niigata, Japan following the 1964 Niigata earthquake. Sand boils are often aligned along cracks in the ground. (Photo by K. Steinbrugge, courtesy of Earthquake engineering Research Center, University of California.)

FIGURE 9.4 Ejected sand covering roads and yards in a residential area of Christchurch, New Zealand following the 2011 Christchurch earthquake. (With permission of Stuff Limited.)

9.2.3 Cyclic Softening

As discussed in Section 6.4.4.3, fine-grained soils with significant plasticity can also suffer loss of stiffness and/or strength and develop significant deformations under repeated cyclic loading. While these consequences of *cyclic softening* are usually not as severe as those of liquefaction, they may still lead to damage and loss. The Fourth Avenue Slide in the 1964 Alaska earthquake, which caused displacements of up to 5 m (Figure 9.5) and involved all or parts of 14 city blocks in Anchorage, has been attributed to cyclic softening of the nearly normally consolidated Bootlegger Cove clay that underlays the area (Hansen, 1965; Idriss, 1985; Boulanger and Idriss, 2004). Procedures for evaluation of cyclic softening are presented in Section 9.8.

FIGURE 9.5 Fourth Avenue slide in Anchorage, Alaska following 1964 Good Friday earthquake. (Photo by U.S. Army, courtesy of USGS.)

9.3 EVALUATION OF LIQUEFACTION HAZARDS

Both flow liquefaction and cyclic liquefaction can produce damage at a particular site, and a complete evaluation of liquefaction hazards requires that the potential for each be addressed. When faced with such a problem, the geotechnical earthquake engineer can systematically evaluate potential liquefaction hazards by addressing the following questions:

1. Is the soil susceptible to liquefaction?
2. If the soil is susceptible, will liquefaction be triggered (i.e., initiated) by the anticipated earthquake shaking?
3. If liquefaction is triggered, will damage occur?

If the answer to the first question is no, the liquefaction hazard evaluation can be terminated with the conclusion that liquefaction hazards do not exist (although other hazards, such as cyclic softening, may exist). If the answer is yes, the next question must be addressed. If the answer to the second question is no (either because the soil is too dense or the anticipated shaking is too weak), the evaluation can be terminated. If the answers to all three are yes, a problem exists; if the anticipated level of damage is unacceptable, the site must be abandoned or improved (Chapter 11) or on-site structures strengthened. These questions pertain to the three most critical aspects of liquefaction hazard evaluation: *susceptibility*, *triggering*, and *consequences*. All three must be considered in a comprehensive evaluation of liquefaction hazards.

9.4 LIQUEFACTION SUSCEPTIBILITY

Not all soils are susceptible to liquefaction; consequently, the first step in a liquefaction hazard evaluation is usually the evaluation of liquefaction susceptibility. A liquefaction-susceptible soil is one that can exist in some state, i.e., some combination of density and effective stress (Section 6.4.3.6), in which liquefaction *can* be triggered if the soil is saturated and subjected to shaking of sufficient intensity and duration. Liquefaction-susceptible saturated soils may or may not liquefy in a particular earthquake depending on the level of shaking and the *in situ* state of the soil at the time of the earthquake.

In this book, susceptibility will be distinguished from triggering, i.e., susceptibility will not be taken to be influenced by the intensity or duration of ground motion or by the state of the soil. Those characteristics influence the potential for triggering of liquefaction (referred to hereafter as *liquefaction potential*), which is a separate issue. Susceptibility is taken as a function of soil composition, and it will be recognized that some soils for which the triggering of liquefaction may be very unlikely (e.g., an extremely dense clean sand in an area of low seismicity) are still susceptible to liquefaction. The fact that the soil is extremely dense means that liquefaction hazards may not exist, but they do not exist because liquefaction will not be triggered, not because that type of soil is not susceptible to liquefaction.

If the soil at a particular site is determined to be not susceptible, liquefaction hazards do not exist and the liquefaction hazard evaluation can be ended. The potential for significant cyclic softening, however, should still be evaluated. There are several indicators with which liquefaction susceptibility can be judged, and some are different for flow liquefaction and cyclic liquefaction. Historical and geological indicators can help identify sites where liquefaction-susceptible soils are likely to be found, and compositional indicators can be used to assess their actual susceptibility.

9.4.1 Historical Indicators

A great deal of information on liquefaction behavior has come from post-earthquake field investigations, which have shown that liquefaction often recurs at the same location (*recurrent liquefaction*) (Youd; 1984; Tohno and Shamoto, 1986; Yasuda and Tohno, 1988; Sims and Garvin, 1995; Obermeier, 1996; Wakamatsu, 2012; Quigley et al., 2013). Thus liquefaction case histories can be used to identify specific sites, or more general site conditions, with soils that may be susceptible to liquefaction in future earthquakes.

The occurrence of liquefaction is usually identified in the field by the presence of sand boils, ground cracking, and other surficial features. Such surficial evidence of liquefaction may disappear relatively quickly due to rainfall or clean-up operations, so it is important that it be observed and documented quickly following earthquakes. Also, liquefaction of deep and/or thin soil layers may not produce surface manifestation, so the absence of manifestation of liquefaction cannot be taken as evidence of the absence of liquefaction. Historical evidence may also be uncovered by *paleoliquefaction* investigations (Obermeier, 1996). Examination of natural exposures or subsurface investigations (e.g., trenching) may reveal buried evidence of past liquefaction in the form of dikes and sills of vented sand (Figure 9.6).

9.4.2 Geologic Indicators

Soil deposits that are susceptible to liquefaction are formed within a relatively narrow range of geological environments (Youd, 1991). The depositional environment, hydrological environment, and

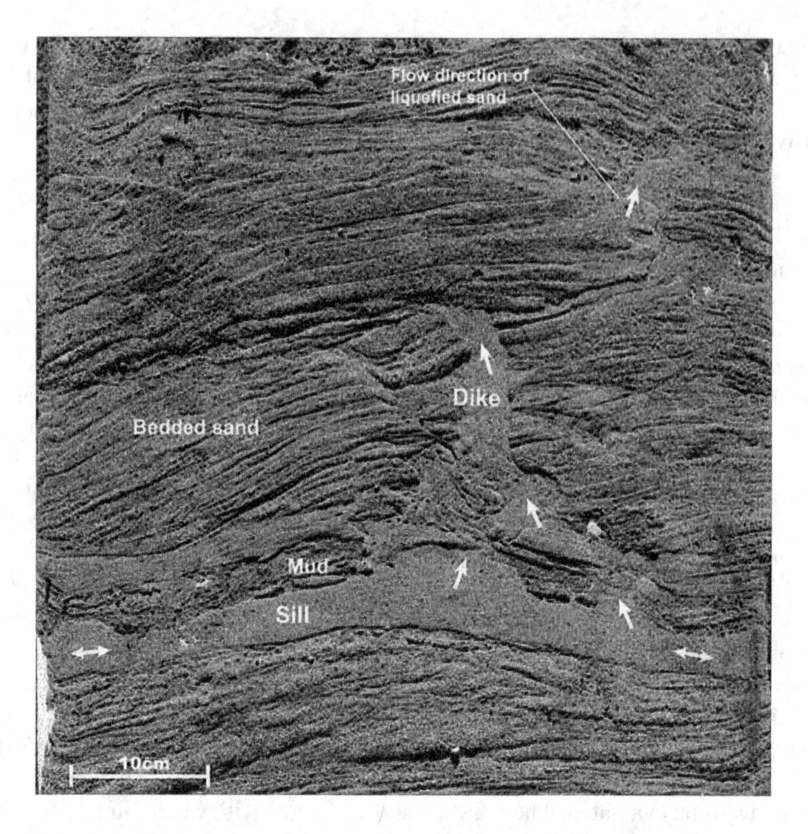

FIGURE 9.6 Intrusive sills and dikes within naturally bedded sand interpreted to mark paths of water expelled from liquefied sand beneath a mud-capped island along the Columbia River, Washington. (After Takada and Atwater, 2004 with permission of the Seismological Society of America.)

age of a soil deposit can all provide indications of its likely liquefaction susceptibility (Youd and Hoose, 1977) and potential consequences (Section 9.6.1).

Geologic processes that sort non-plastic soils into uniform grain size distributions and deposit them in loose states produce soil deposits with high liquefaction susceptibility. Consequently, fluvial deposits, colluvial, and aeolian deposits when saturated, are likely to be susceptible to liquefaction. Liquefaction has been observed to occur preferentially in old, buried river channels (e.g., Orense et al., 1991; Bennett and Tinsley, 1995; Wotherspoon et al., 2012) even when covered with more recent sediments or compacted fills. Liquefaction has also been observed in alluvial-fan, alluvial plain, beach, terrace, playa, and estuarine deposits, but not as consistently as in those listed previously. Older soil deposits, likely as a result of cementation between grains, are less likely to experience liquefaction than newer deposits. Soils of Holocene age are more likely to experience liquefaction than soils of Pleistocene age, and liquefaction of pre-Pleistocene deposits is extremely rare. However, these age effects are a consequence of increased resistance to triggering of liquefaction with age, whereas susceptibility is generally not affected by age.

Human-made soil deposits can also be susceptible to liquefaction. Loose fills, such as those dumped or otherwise placed without compaction, have liquefied in many past earthquakes. Hydraulically placed soils involve the sluicing or spraying of a mixture of soil particles suspended in water onto a site in an episodic series of placements (Figure 9.7). Following a given placement episode, the coarser particles settle out of suspension first and the finer particles later as the remaining water drains away. This process produces a layer with vertically variable particle size (and permeability). Hydraulic filling can be an inexpensive and efficient technique for moving and placing soil, and hydraulic fills can be found in some older dams, mine tailings piles, and reclaimed land.

FIGURE 9.7 Hydraulic filling during construction of Calaveras Dam. Note the localized nature of runoff. (California Department of Water Resources.)

Because the soil particles settle gently out of the fluid they are transported in, hydraulic fills tend to be very loose and saturated and therefore have high liquefaction potential. Furthermore, they tend to have a layered structure (Figure 9.8) that can exacerbate some consequences of liquefaction after it has been initiated.

9.4.3 Compositional Indicators

Since liquefaction requires the development of very high excess pore pressure, liquefaction susceptibility is influenced by the compositional characteristics that influence volume change behavior, specifically, those that affect the soil's tendency to reach a state of (nearly) zero effective stress. As discussed in Section 6.4.4.4, laboratory tests have shown that coarse-grained, non-plastic soils (e.g., clean sands and gravels), can generate such high pore pressure under undrained cyclic loading that effective stresses drop to nearly zero. Pore pressure generation in fine-grained, plastic soils, however, tends to be limited, i.e., laboratory tests show effective stresses dropping to a certain level and then remaining at that level during subsequent loading. While the limiting effective stress in such a soil may be a relatively small fraction of the initial effective stress and the resulting degree of cyclic softening may be significant, neither are as extreme as in the case of coarse-grained, non-plastic soils. The exception to this case is sensitive clay, which can suffer extreme strength loss due to cyclic softening and become unstable; the mechanisms by which clays develop sensitivity, and the behavior that sensitive clays exhibit, are different than those of liquefiable soil and, therefore, sensitive clays represent extreme cases of cyclic softening, not liquefaction. Procedures for evaluating seismic hazards associated with cyclic softening are described in Section 9.8.

Most questions regarding liquefaction susceptibility center on intermediate soils, i.e., silts, sandy silts, silty sands, and sand-silt-clay mixtures with low plasticity. As noted in Chapter 6, the use of a #200 sieve (0.074 mm) to distinguish between coarse-grained and fine-grained soils is more a matter of convenience than an indication of dramatic behavioral change at that specific particle size

FIGURE 9.8 Layering within hydraulic fill dam section exposed by excavation. (With permission of the National Information Service for Earthquake Engineering.)

level. Although coarse silts are classified as fine-grained soils, they are generally bulky-grained and non-plastic and therefore behave much like fine sands. As smaller silt particles are considered, the particle mass decreases faster than its surface area and the influence of electrical surface charges (Section 6.3) may increase, giving rise to plasticity. As plasticity increases, the ability of the silt to undergo a substantial reduction of stiffness and strength at a condition of low effective stress decreases.

Boulanger and Idriss (2006) developed a framework for evaluating soil response to cyclic loading that characterized soils as exhibiting "sand-like" or "clay-like" behavior and proposed separate, specific assessment procedures for sand-like and clay-like soils. The behavior of sand-like soils was considered best evaluated using field penetration test-based procedures and that of clay-like soils relying more on laboratory test-based procedures. The distinction between sand-like and clay-like soils is made on the basis of plasticity index with sand-like soils having $PI \leq 3$ and clay-like soils having $PI \geq 8$ (Figure 9.9). Intermediate soils, i.e. with $3 < PI < 8$, could potentially exhibit either sand-like or clay-like behavior, or an intermediate type of behavior best determined by laboratory testing. Idriss and Boulanger (2006) recommended a value of $PI=7$ as a relatively conservative upper bound for sand-like behavior. Using stress-strain behavior observed in laboratory tests, Bray and Sancio (2006) also found soil plasticity to be an important indicator of liquefaction susceptibility but considered soils with PI values up to 12 as being susceptible to liquefaction and soils with $PI \geq 18$ as being non-susceptible. Bray and Sancio also incorporated the ratio of water content to liquid limit, w_c/LL, into their characterization of liquefaction susceptibility (Figure 9.10). For a saturated inorganic soil, water content is essentially a measure of density, so the conclusion that soils of lower density (i.e., higher w_c/LL) are more likely to liquefy than soils of higher density (lower w_c/LL) is reasonable, but is, under the definition of susceptibility utilized here, an issue more appropriately used as a screening criterion for triggering (Section 9.5.3.2) rather than an indicator of susceptibility.

Subsequent laboratory testing (e.g., Beyzaei et al., 2018; Stuedlein et al., 2023) has shown that pore pressure and stress-strain behavior similar to that of clean sands extend to silty sands with plasticity indices that exceed the recommended boundary between sand-like and clay-like soils. The data for different soils is scattered, however, which, along with the stratigraphic complexity and mineralogical variability of transitional soils, suggests that susceptibility may not be accurately

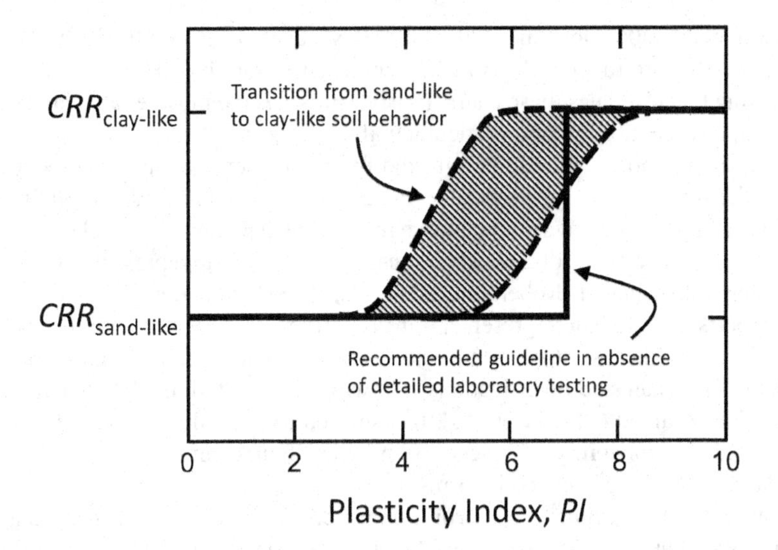

FIGURE 9.9 Range of plasticity indices corresponding to sand-like and clay-like behavior. (After Boulanger and Idriss, 2006 with permission of ASCE.)

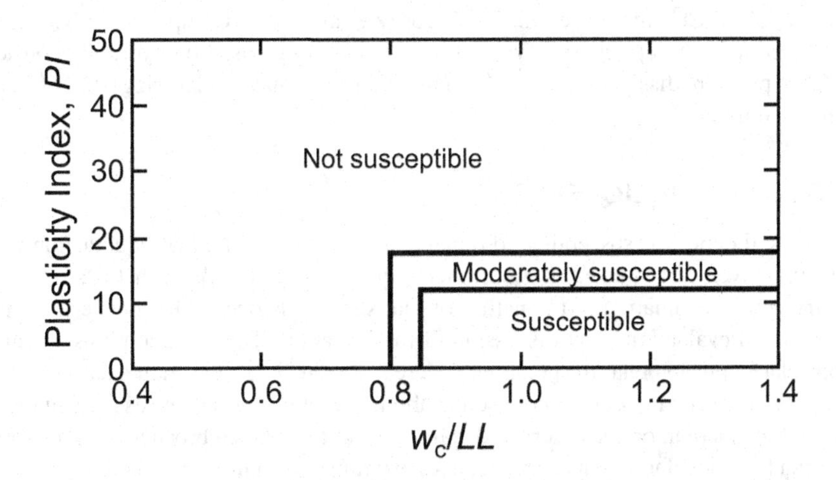

FIGURE 9.10 Ranges of plasticity index and w_c/LL ratio corresponding to different degrees of liquefaction susceptibility. (After Bray and Sancio, 2006 with permission of ASCE.)

indicated by a single deterministic quantity. As the emphasis here is on soil behavior rather than analytical tools, currently available data suggest that a *PI* of about 6–12 may represent a reasonable range, considering epistemic uncertainty, for the median value of a probabilistic susceptibility criterion. As it is an active area of current research, susceptibility characterization should be undertaken very carefully and is best evaluated from material-specific testing.

Silty soils with some plasticity are more readily sampled than non-plastic soils, and both Boulanger and Idriss (2006) and Bray and Sancio (2006) essentially agree that silty soils that can be sampled should be sampled and tested in the laboratory to evaluate susceptibility and for liquefaction/cyclic softening behavior. As described in Sections 6.4.3.3 and 6.4.3.4, clay-like (not susceptible) soils are likely to exhibit strength normalization in undrained monotonic shear, whereas sand-like (susceptible) are not. Similarly, as described in Section 6.4.4.5, the shapes of cyclic stress-strain loops (e.g., Bray and Sancio, 2006; Donahue et al., 2008; Beyzaei et al., 2018; Stuedlein et al., 2023), and the levels of pore pressure that develop in cyclic tests, differ for clay-like and sand-like materials. Laboratory testing that evaluates these characteristics provides the most reliable means by which to assess liquefaction susceptibility for a particular soil.

Liquefaction susceptibility can also be inferred from cone penetration tests (Section 6.5.3.3) by making use of the soil behavior type index, I_c (Equation 6.60). Sand-like soils usually exhibit relatively high tip resistances and low friction sleeve resistances (i.e., low I_c), and clay-like soils exhibit the opposite (high I_c). An I_c value of 2.60 has frequently been taken to represent a boundary between sand-like and clay-like behavior (Robertson and Wride, 1998) but the range of $2.50<I_c<2.70$ is now considered (Robertson, 2009b) to represent a transition from sand-like (essentially drained penetration at $I_c<2.50$) to clay-like (essentially undrained penetration at $I_c>2.70$). Examination of field and laboratory data from extensive investigations following the Canterbury Earthquake Sequence in Christchurch, New Zealand (Maurer et al., 2017) showed that I_c values ranging from 2.50 to 2.75 corresponded to 50% probability of susceptibility using four common CPT-based susceptibility relationships and suggested probabilistic forms of each.

At the other end of the grain size spectrum, liquefaction of gravelly soils has been observed in the laboratory (Wong et al., 1975; Evans and Seed, 1987; Kim et al., 2023) and in the field when the gravelly stratum is confined by low permeability soils (Coulter and Migliaccio, 1966; Chang, 1978; Wong, 1984; Youd et al., 1985; Stokoe et al., 1988; Yegian et al., 1994; Cao et al., 2010) or when the voids between gravel particles are filled with lower permeability soils (Cao et al., 2010; Nikolaou et al., 2014; Cubrinovski et al., 2017; Lopez et al., 2018). Compositionally, clean gravels should be treated as a susceptible soil type, although screening based on hydraulic considerations (e.g., rapid pore pressure dissipation due to high permeability) may render triggering of liquefaction to be extremely unlikely.

9.5 TRIGGERING OF LIQUEFACTION

The fact that a soil deposit is susceptible to liquefaction does not mean that liquefaction will necessarily occur in a given earthquake. Its occurrence requires ground shaking that is strong enough to initiate, or trigger, it. Evaluation of the nature of that shaking is one of the most critical parts of a liquefaction hazard evaluation. Any discussion of the initiation of liquefaction must specify which liquefaction-related phenomena are being considered. In the field, flow liquefaction occurs much less frequently than cyclic liquefaction or cyclic softening, but its consequences are usually far more severe. Cyclic liquefaction, on the other hand, can occur under a much broader range of soil and site conditions than flow liquefaction; its consequences can range from insignificant to highly damaging. In this book, the generic term "liquefaction" will be taken to include both flow liquefaction and cyclic liquefaction. Flow liquefaction and cyclic liquefaction will be identified individually when necessary. Many previous studies of liquefaction initiation have implicitly lumped flow liquefaction and cyclic liquefaction together, but since they are distinctly different phenomena, it is more appropriate to consider each separately. In the following sections, these conditions will be presented in a framework that allows the mechanics of both flow liquefaction and cyclic liquefaction to be clearly understood.

9.5.1 Flow Liquefaction

Although cyclic liquefaction is an earthquake-related phenomenon, flow liquefaction can be initiated in a variety of ways. Flow slides triggered by monotonic loading (static liquefaction) have been observed in natural soil deposits (Koppejan et al., 1948; Andresen and Bjerrum, 1968; Bjerrum, 1971; Kramer, 1988; Baille, 2014; Mason et al., 2021), man-made fills (Middlebrooks, 1942; Cornforth et al., 1975; Mitchell, 1984), and mine tailings piles (Kleiner, 1976; Jennings, 1979; Eckersley, 1985; AECOM, 2009; Dawson et al., 2011; Morgenstern et al., 2015, 2016; Jefferies et al., 2019; Robertson et al., 2019; Rana et al., 2021). Flow liquefaction has also been triggered by non-seismic sources of vibration, such as pile driving (Jakobsen, 1952; Broms and Bennermark, 1967), train traffic (Fellenius, 1953), geophysical exploration (Hryciw et al., 1990), and blasting (Conlon, 1966; Carter and Seed, 1988).

9.5.1.1 Flow Liquefaction Surface

The conditions at which flow liquefaction is initiated follow from the work of Arthur Casagrande and his students at Harvard University in the 1960s, and are most easily illustrated with the aid of the stress path (Section 6.2.2). The notion of the flow liquefaction surface follows from the type of laboratory testing described in Section 6.5.4.3. As discussed in the following sections, the effective stress conditions at which strain-softening behavior occurs in loose, saturated sands can be described very simply in stress path space (Hanzawa et al., 1979) using a three-dimensional surface that will be referred to hereafter as the *flow liquefaction surface* (FLS). While there are some practical difficulties in the measurement of the FLS for general stress paths, it provides (in conjunction with steady state concepts) a useful framework for conceptual understanding of the relationships between the various liquefaction phenomena.

Monotonic Loading

The conditions at the initiation of flow liquefaction can be seen most easily when the soil is subjected to monotonically increasing stresses. Consider, for example, the response of an isotropically consolidated specimen of very loose, saturated sand in undrained, stress-controlled triaxial compression (Figure 9.11). Immediately prior to undrained shearing (Point A), the specimen is in drained equilibrium under an initial effective confining stress, σ'_{3c}, with zero shear stress (Figure 9.11a, b) and zero excess pore pressure (Figure 9.11c). Since its initial state (Section 6.4.3.6) is well above the steady state line (SSL) (Figure 9.11d), the sand will exhibit contractive behavior when sheared. When undrained shearing begins, the specimen generates positive excess pore pressure as it mobilizes shearing resistance up to a peak value (Point B) that occurs at a relatively small strain. The excess pore pressure at Point B is also relatively low; the pore pressure ratio, $r_u = \Delta u/\sigma'_{3c}$ is well below 1.0. At Point B, however, the metastable structure of the sand specimen becomes unstable and collapses. As the specimen strains from Point B to Point C, the excess pore pressure increases dramatically. At and beyond Point C, the specimen is in a steady state of deformation and the effective confining stress is only a small fraction of the initial effective confining stress. This specimen has exhibited flow liquefaction behavior; the static shear stresses required for equilibrium (at Point B) were greater than the available shear strength (at Point C) of the liquefied soil. Flow liquefaction was initiated at the instant it became irreversibly unstable (i.e., at Point B). More formal criteria for the initiation of flow liquefaction (Lade, 1992, 1999;

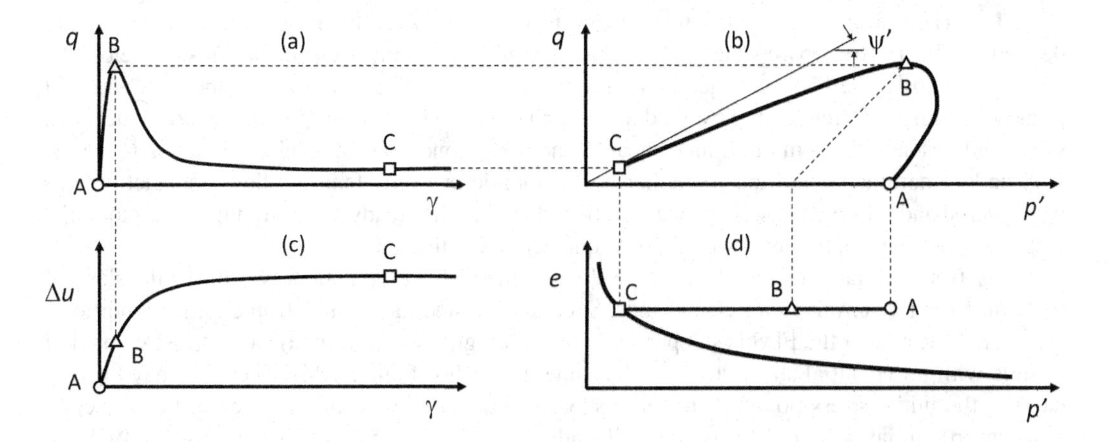

FIGURE 9.11 Response of isotropically consolidated-undrained triaxial specimen of loose, saturated sand: (a) stress-strain curve; (b) effective stress path; (c) excess pore pressure; (d) effective confining pressure.

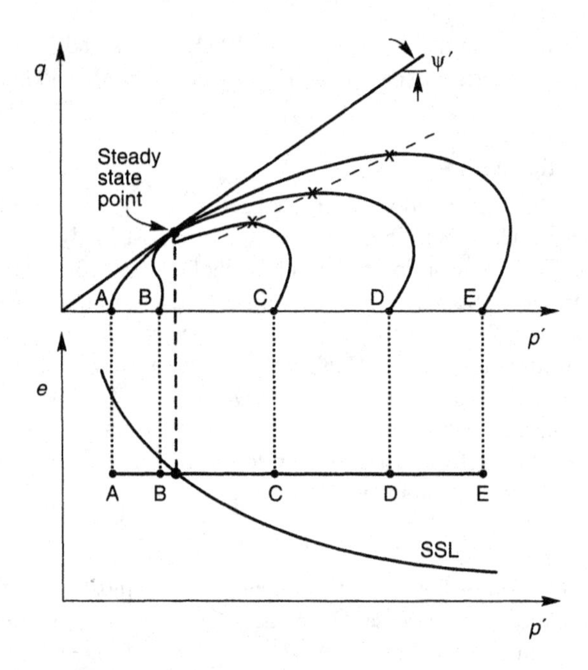

FIGURE 9.12 Response of five specimens isotropically consolidated to the same initial void ratio at different initial effective confining stresses. Flow liquefaction in specimens C, D, and E is initiated at the points marked with an X. The dotted line passing through these points is a line of constant principal effective stress ratio, K_L.

Borja, 2006; Andrade, 2009), expressed within the context of constitutive modeling, have also been proposed.

Now consider the response of a series of triaxial specimens initially consolidated isotropically to the same void ratio at different effective confining stresses. Since all of the specimens have the same void ratio, they will all reach the same effective stress conditions at the steady state, but they will get there by different stress paths. Figure 9.12 illustrates the response of each specimen under monotonic loading. The initial states of specimens A and B are below the SSL, so they exhibit dilative behavior upon shearing. Specimens C, D, and E all exhibit contractive behavior; each reaches a peak undrained strength after which they strain rapidly toward the steady state. For specimens C, D, and E, flow liquefaction is initiated at the peak of each stress path (at the points marked with an × symbol). The locus of points describing the effective stress conditions at the initiation of flow liquefaction is a straight line (the dotted line in Figure 9.12) that projects through the origin of the stress axes (Hanzawa et al., 1979; Vaid and Chern, 1983). Graphically, these points may be used to define the FLS in stress path space; since flow liquefaction cannot occur if the stress is below the steady state point, the FLS is truncated at that level (Figure 9.13). This form of the FLS was first proposed (with a different name) by Vaid and Chern (1985). The FLS marks the boundary between stable and unstable states in undrained shear. If the stress conditions in an element of soil reach the FLS under undrained conditions, whether by monotonic or cyclic loading, flow liquefaction will be triggered and the shearing resistance will be reduced to the steady state strength. Therefore, the FLS describes the conditions at which flow liquefaction is initiated.

For isotropic initial stress conditions, the slope of the FLS is often about two-thirds the slope of the drained failure envelope for clean sands. Specimens tested under anisotropic initial conditions, however, indicate that the FLS is steeper for soils with high initial (drained) shear stress compared to soils with lower initial shear stress at the same void ratio (Figure 9.14). The FLS may be very close to the initial stress point when initial shear stresses are large, in which case flow liquefaction may be initiated by only a very small undrained disturbance (Kramer and Seed, 1988) or increase in pore pressure. Case histories that have been attributed to "spontaneous liquefaction"

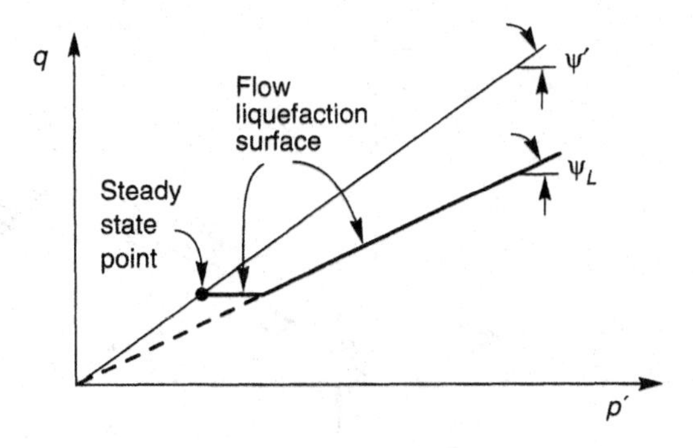

FIGURE 9.13 Orientation of the flow liquefaction surface in stress path space.

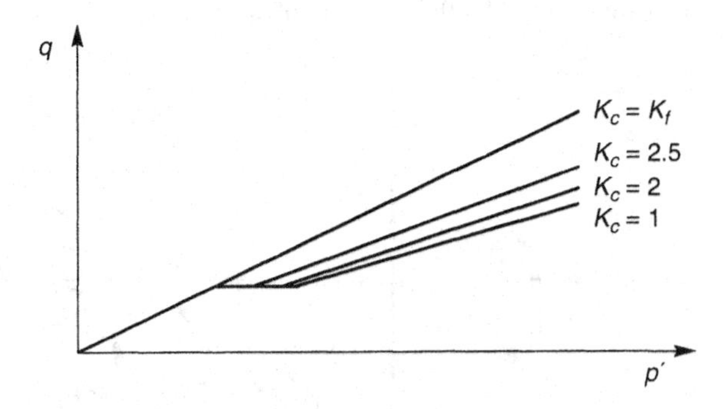

FIGURE 9.14 Variation of flow liquefaction surface inclination with initial principal effective stress ratio for constant void ratio.

(e.g., Terzaghi, 1956) probably involved initial shear stresses that were so high that the small stress changes required to trigger flow liquefaction were not recognized.

The behavior exhibited by specimen C (Figure 9.12), in which the shear stress drops after reaching a peak but then begins to increase again, is significant for cases in which the static shear stress increases (as in the case of monotonic loading described in this section). In such cases, the shearing resistance may drop to values at the point of phase transformation (Section 6.4.3.4), also known as *quasi-steady state*, that is lower than the steady state strength. This temporary drop in shearing resistance may produce shear strains of 5%–20% (Ishihara, 1993) and result in unacceptably large permanent deformations. Because the effects of initial stress conditions and soil fabric are not erased completely at these strain levels, they influence the quasi-steady state strength (Section 6.4.3.4). Procedures for estimation of quasi-steady state strength are given by Ishihara (1993).

Cyclic Loading

Vaid and Chern (1983) first showed that the FLS applied to both cyclic and monotonic loading, and a considerable amount of independent experimental evidence supports that observation. Other experimental evidence (e.g., Alarcon-Guzman et al., 1988) suggests that the effective stress path can move somewhat beyond the FLS before liquefaction is initiated by cyclic loading. Whether liquefaction is initiated precisely at the FLS under cyclic as well as monotonic loading is not currently known with certainty. Because the FLS is used as part of a conceptual model of liquefaction behavior in this book, and because it is slightly more conservative to do so, the FLS will be assumed to apply to both cyclic and monotonic loading.

Consider the responses of two identical, anisotropically consolidated, triaxial specimens of loose, saturated sand (Figure 9.15). Initially, the specimens are in drained equilibrium (Point A) under a static shear stress, τ_{static}, that is greater than the steady state strength, S_{su}. The first specimen

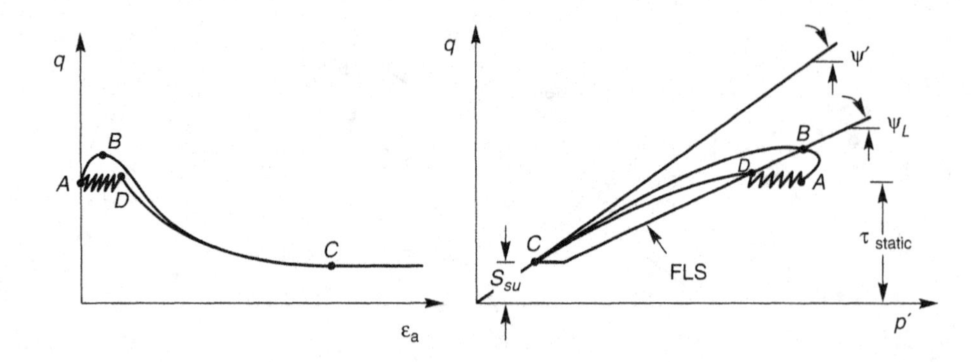

FIGURE 9.15 Initiation of flow liquefaction by cyclic and monotonic loading. Although the stress conditions at the initiation of liquefaction are different for the two types of loading (points B and D), both lie on the FLS.

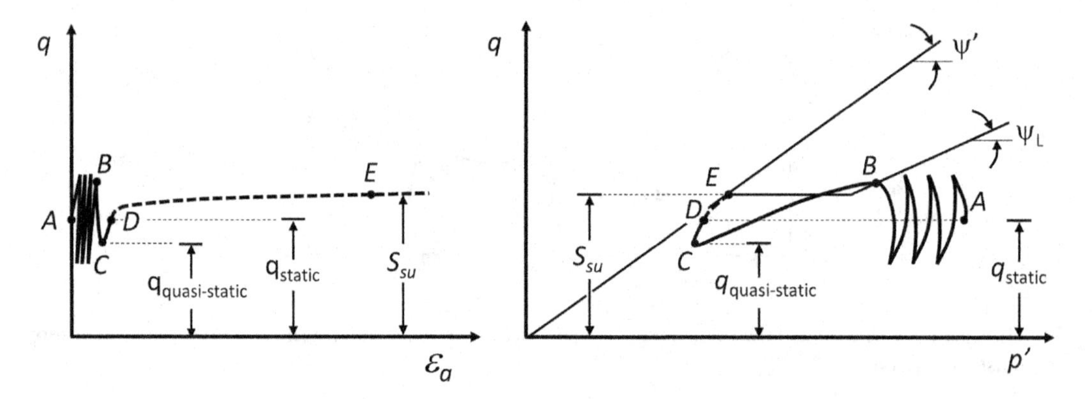

FIGURE 9.16 Schematic illustration of mobilization of quasi-static and steady state strengths under cyclic loading.

is loaded monotonically (under undrained conditions): the shearing resistance builds up to a peak value when the stress path reaches the FLS (Point B). At that point, the specimen becomes unstable and strains rapidly toward the steady state (Point C). The second specimen is loaded cyclically (also under undrained conditions); the effective stress path moves to the left as positive excess pore pressures develop and permanent strains accumulate. When the effective stress path reaches the FLS (at Point D), the specimen becomes unstable and strains toward the steady state of deformation (Point C). Although the effective stress conditions at the initiation of liquefaction (points B and D) were different, they fell in both cases on the FLS. The FLS, therefore, marks the onset of the instability that produces flow liquefaction. Lade (1992) provided a detailed description of this instability from a continuum mechanics standpoint.

A soil of intermediate density may have a quasi-static strength that is lower than the static shear stress but a steady state strength that is greater than the static stress; Figure 9.16 illustrates such a case. Cyclic loading takes an element of intermediate density from Point A to Point B at which it reaches the FLS. Since the quasi-static strength is lower than the static shear stress, the path moves from Point B to Point C, at which it begins to dilate producing a resistance that reaches the static shear stress at Point D. At that point straining would be expected to stop although inertial effects under stress-controlled conditions could lead to some additional strain. If the static stresses were increased after reaching Point D, the stress-strain and stress path curves would follow the dashed lines until the steady state strength is mobilized at Point E. The increment of strain from triggering (Point B) to cessation of straining (Point D) is much lower than the strain involved in flow (Point B to Point E).

9.5.1.2 Development of Flow Liquefaction

Flow liquefaction occurs in two stages. The first stage, which takes place at small strain levels, involves the generation of sufficient excess pore pressure to move the stress path from its initial position to the FLS. This excess pore pressure may be generated by undrained monotonic or cyclic loading or by changes in pore water pressure under constant total stress. When the effective stress path reaches the FLS, the soil becomes inherently unstable and the second stage begins. The second stage involves strain softening (and additional excess pore pressure generation) that is driven by the shear stresses required for static equilibrium. These shear stresses are the driving stresses (and must be distinguished from the locked-in stresses that develop during deposition and consolidation of the soil (Castro, 1991). Locked-in shear stresses, such as those that exist beneath level ground when $K_0 \neq 1.0$, cannot drive a flow liquefaction failure. Large strains develop in the second stage as the effective stress path moves from the FLS to the steady state. If the first stage takes the soil to the FLS under undrained, stress-controlled conditions, the second stage is inevitable.

The generation of excess pore pressure is the key to the initiation of flow liquefaction. Flow liquefaction can be initiated by cyclic loading only when the shear stress required for static equilibrium is greater than the steady state strength. In the field, these shear stresses are caused by gravity and remain essentially constant until large deformations develop. Therefore, initial states that plot in the shaded region of Figure 9.17 are vulnerable to flow liquefaction. The occurrence of flow liquefaction, however, requires an undrained disturbance strong enough to move the effective stress path from its initial point to the FLS.

If the initial stress conditions plot near the FLS, as they would in an element of soil subjected to large shear stresses under drained conditions, flow liquefaction can be triggered by small excess pore pressures (Kramer and Seed, 1988). The liquefaction resistance will be greater if the initial stress conditions are farther from the FLS. The FLS can be used to estimate the pore pressure ratio at the initiation of flow liquefaction; it decreases substantially with increasing initial stress ratio (Figure 9.18) for soils at a particular void ratio. At high initial stress ratios, flow liquefaction can be triggered by very small static or dynamic disturbances.

9.5.2 Cyclic Liquefaction

Situations in which soils loose enough to have a low steady state strength liquefy under initial shear stresses high enough to exceed that strength are relatively rare. As a result, flow slides resulting from flow liquefaction are relatively rare. Situations in which level or slightly sloping sites contain loose to medium-dense liquefiable soils are much more common. The static shear stresses in such situations are too low, or the steady state strength too high, to allow flow liquefaction, but cyclic

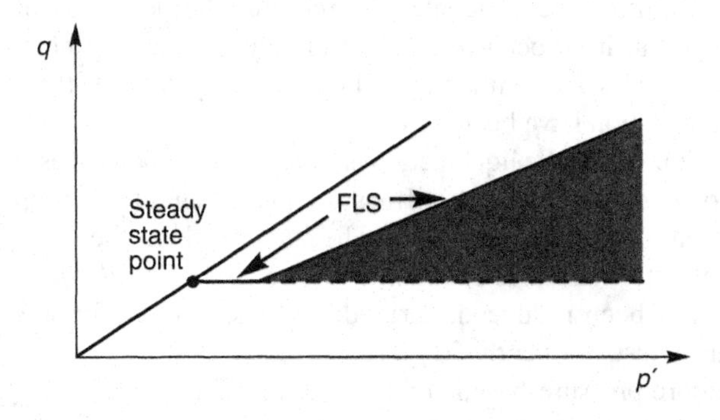

FIGURE 9.17 Zone of potential for flow liquefaction. If initial conditions fall within the shaded zone, flow liquefaction will occur if a disturbance brings the effective stress path from the point describing the initial conditions to the FLS.

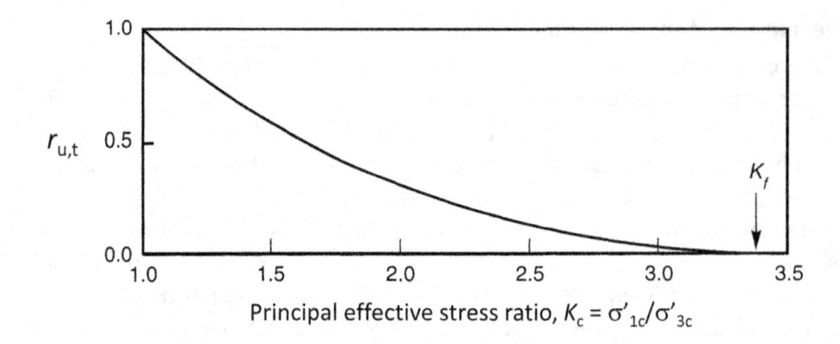

FIGURE 9.18 Variation of pore pressure ratio ($r_{u,t}=u_t/\sigma'_{3c}$) required to trigger flow liquefaction in triaxial specimens of Sacramento River Fine Sand (Kramer and Seed, 1988) with initial principal effective stress ratio.

liquefaction can still occur. Cyclic liquefaction occurs in the field far more frequently than flow liquefaction and is responsible for the great majority of liquefaction-related damage in earthquakes.

Understanding of liquefaction potential, i.e., the potential for triggering of liquefaction, has developed over the past 50 years. Both laboratory and field studies have led to the development of a variety of procedures for evaluation of liquefaction potential. Laboratory tests have proven very useful for illustrating the mechanics of liquefaction and identifying the factors that most strongly affect it. For reasons explained later in this section, procedures based on *in situ* testing and observations and interpretation of field case histories are generally used in lieu of laboratory-based triggering procedures in engineering practice. When major earthquakes occur, new liquefaction case histories are typically investigated and documented. These case histories often motivate the development of new and updating of existing, predictive models. Several such models have been proposed, debated, and implemented in practice, and new models will undoubtedly be developed in the future.

Given the rate at which new empirical models are developed, the approach taken in this chapter is to focus first on the fundamental soil behaviors associated with cyclic liquefaction. After covering these fundamentals, examples of current empirical and numerical models are presented. The examples described are those most commonly used in current U.S. practice for models in which soil state is characterized using cone penetration test (Boulanger and Idriss, 2015), standard penetration test (Boulanger and Idriss, 2012), and shear wave velocity (Kayen et al., 2013, 2015) data. By developing a solid, fundamental understanding of liquefiable soil behavior, geotechnical engineers should be able to evaluate the strengths, weaknesses, and applicability of both existing and new procedures for evaluation of liquefaction potential.

9.5.2.1 Cyclic Liquefaction Criteria

Prediction of the triggering of cyclic liquefaction requires the definition of exactly when cyclic liquefaction is considered to have occurred. In the literature regarding both laboratory- and field case history-based procedures for estimation of liquefaction potential, three primary criteria for triggering of cyclic liquefaction have been used.

As described in Section 9.2, cyclic liquefaction is a process in which excess pore pressure induced by cyclic loading becomes so high that the stiffness of an element of soil drops to a very low value, enabling the development of large shear strains. The state in which the pore pressures are so high that $r_u = 1.0$ (i.e., zero effective stress) is commonly referred to as *initial liquefaction* (Seed and Lee, 1966). This condition has been produced, verified, and taken as a criterion for the triggering of liquefaction in laboratory tests for nearly 50 years.

In some cases, the pore pressure-based definition of initial liquefaction is problematic. The low permeability of silty soil specimens, for example, may require testing at extremely low frequencies (much lower than those associated with earthquake shaking) to allow sufficient equilibration of pore pressures between the specimen and the pore pressure transducer for accurate determination of

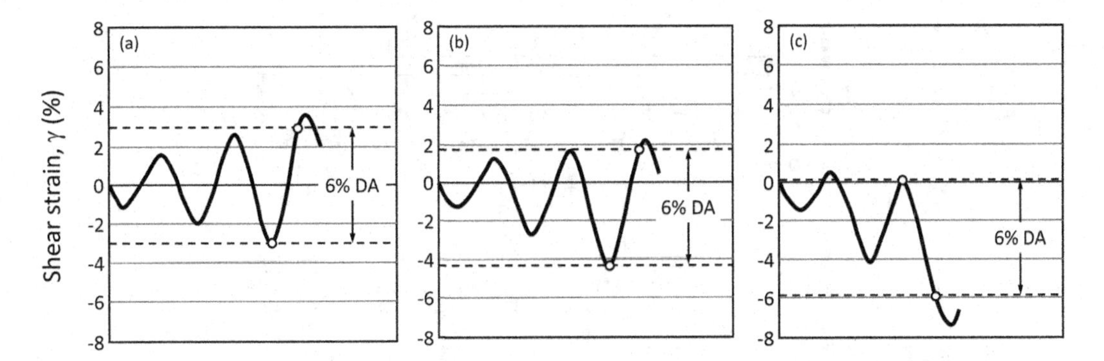

FIGURE 9.19 Triggering of cyclic liquefaction according to 6% double-amplitude (DA) criterion for: (a) symmetric response, (b) asymmetric response, and (c) highly asymmetric response. (After Wu et al., 2004 with permission of International Association of Earthquake Engineering.)

effective stresses. High pore pressure can lead to significant softening and the development of large cyclic and/or permanent strains in a loose, liquefiable soil even before a condition of initial liquefaction is reached. A silty liquefiable soil may, with sufficient numbers of loading cycles, achieve moderately large cyclic strains at pore pressure ratios well below 1.0. Dense sands under strong loading can also generate large strains at $r_u < 1.0$. In some laboratory investigations, "liquefaction" has been taken to have been initiated when the cyclic strain amplitude, γ_{max}, reaches some threshold value instead of when initial liquefaction is achieved. The threshold value is often expressed in terms of a "double-amplitude" strain defined as the maximum range of strain within a given loading cycle (Figure 9.19). Ishihara (1993) proposed a double-amplitude axial strain of 5% be used as a threshold for cyclic liquefaction in triaxial tests, and a single-amplitude strain of 3% in cyclic simple shear tests. Thus, both pore pressure and strain amplitude criteria have been used in the interpretation of laboratory test results. Wu et al. (2004) related pore pressure ratios to cyclic strain levels for different cyclic stress ratios and relative densities. For high cyclic stress ratios, double-acting cyclic shear strains of 6% could be reached at pore pressure ratios in the range of 0.80–0.85, i.e., well below the level associated with initial liquefaction.

Finally, the types of instrumentation that allow precise and consistent identification of the triggering of liquefaction in the laboratory do not exist in the field. The overwhelming majority of sites that have liquefied in past earthquakes have been uninstrumented, so evidence of liquefaction having occurred consists of surficial features such as sand boils, ground cracks, etc. that are observed following earthquake shaking. While these features are generally good indicators of liquefaction, it is possible for liquefaction to have occurred without leaving surficial evidence (or for the evidence to have disappeared before the site is examined), and it is possible for some forms of surface evidence to appear without initial liquefaction actually having occurred. When evaluating reports of liquefaction, whether in the laboratory or in the field, it is important to recognize the specific criterion for liquefaction being used, and to appreciate that not all definitions are equivalent or necessarily even consistent.

9.5.2.2 Historical Perspective

When liquefaction first came to the attention of geotechnical engineers in the mid-1960s, procedures for testing soils under cyclic loading conditions were primitive. Cyclic tests used for evaluating the suitability of pavement subgrade materials were available, but applied compression-only loading to triaxial specimens. Seed and Lee (1966) modified a pavement-testing apparatus to apply two-way "pulsating loads" to triaxial specimens. Although the applied loads were of a relatively crude, approximate square-wave form (Figure 9.20), the results of these first cyclic loading tests

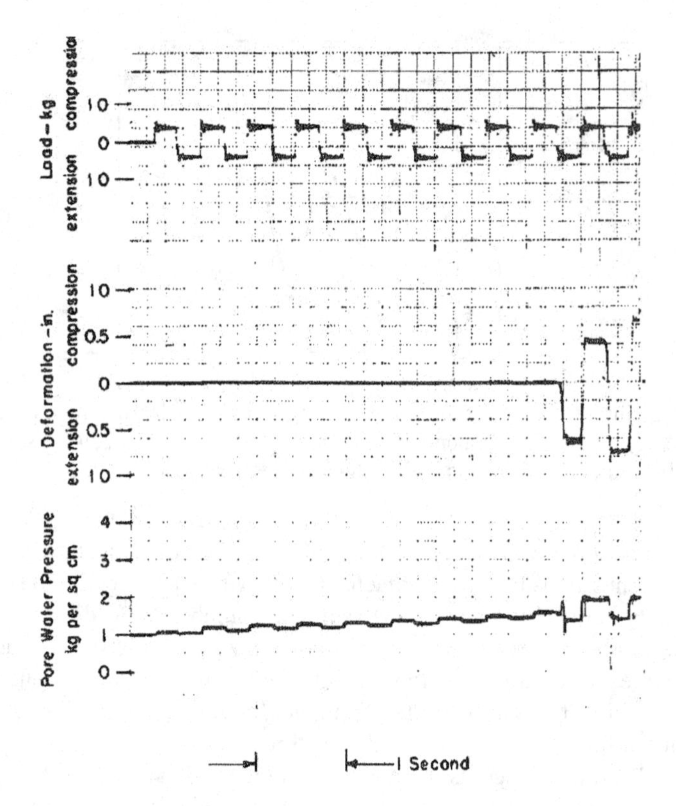

FIGURE 9.20 Pulsating load test. (Seed and Lee, 1966, with permission of ASCE.)

show many of the features of liquefiable soil behavior that are now well established – pore pressures build up incrementally from cycle to cycle, strains are initially small but rapidly become very large when initial liquefaction is reached.

Eventually, improved testing equipment was developed, and cyclic triaxial and cyclic simple shear tests were used to explore the effects of various liquefiable soil and site characteristics, such as those discussed in the following section, on pore pressure generation and stress-strain behavior.

Because sampling the loose, saturated, clean sands that are most susceptible to liquefaction is so difficult (samples are either densified or lost as soil falls out of the sampler during extraction; Section 6.5), laboratory tests are typically performed on reconstituted specimens. In the 1960s and 1970s, moist-tamped, dry-pluviated, and wet-pluviated specimens were tested by various investigators under the assumption that all would exhibit similar behavior if prepared to the same density under the same effective stress. However, subsequent work showed that cyclic stress-based measures of liquefaction resistance are influenced by factors other than the initial density and stress conditions. For example, liquefaction resistance is influenced by differences in the structure of the soil (or *soil fabric*) produced by different methods of specimen preparation (Ladd, 1974; Mulilis et al., 1975; Toki et al., 1986; Tatsuoka et al., 1986). Liquefaction resistance also tends to increase with increasing age of the soil (Youd and Hoose, 1977; Seed, 1979; Robertson et al., 2000; Andrus et al., 2009; Hayati and Andrus, 2009; Maurer et al., 2014a). Low-strain fabric and aging effects are easily destroyed by sampling disturbance and are very difficult to replicate in reconstituted specimens. Because of these factors, characterization of *in situ* liquefaction resistance by laboratory testing is extremely difficult and has been supplanted by methods based on field observations and *in situ* test results. Truly undisturbed sampling (e.g., by careful ground freezing and coring) is required for laboratory tests to be able to characterize liquefaction resistance reliably; because such sampling is extremely expensive, it is generally used only for large and complex projects for which the consequences of liquefaction are severe (Yoshimi et al., 1978; Davila et al., 1992; Hofmann et al., 2000).

9.5.2.3 Laboratory Testing of Liquefiable Soils

Laboratory tests can provide tremendous insight into the behavior of liquefiable soils. Cyclic triaxial and cyclic simple shear tests are most commonly used for liquefiable soils but, as described in Section 6.5.4.3, they impose different types of loading on soil specimens. The maximum shear stress in cyclic triaxial specimens is applied on planes inclined at 45° to horizontal whereas it is applied on nearly horizontal planes in the cyclic simple shear test. The loading applied to an element of soil in the field is typically expressed in terms of a dimensionless *cyclic stress ratio* as the ratio of horizontal cyclic shear stress amplitude to initial vertical effective stress, i.e.,

$$CSR = \frac{\tau_{cyc}}{\sigma'_{v0}} \tag{9.1}$$

This definition implicitly assumes level-ground initial conditions where the vertical effective stress, σ'_{v0}, is the initial major principal stress and τ_{cyc} is the (horizontal) shear stress acting on the major principal stress plane. It applies well to the cyclic simple shear test but not to the cyclic triaxial test in which no shear stress exists on horizontal planes. The *CSR* for the triaxial test is taken as the ratio of the maximum shear stress to the initial normal effective stress on the minor principal plane, i.e., as

$$CSR_{tx} = \frac{\sigma_{dc}}{2\sigma'_{3c}} \tag{9.2}$$

where σ_{dc} = cyclic deviator stress (twice the maximum shear stress) and σ'_{3c} = minor principal consolidation stress (equivalent to isotropic consolidation stress when the specimen is isotropically consolidated). The cyclic stress ratio required to initiate liquefaction is known as the *cyclic resistance ratio, CRR*. Cyclic resistance ratios measured in cyclic triaxial and cyclic simple shear tests are different, largely due to their different initial effective stress states and different principal stress axis rotation behavior. Cyclic triaxial test specimens are typically consolidated isotropically, so their initial lateral earth pressure coefficient, $K_0 = 1.0$. Cyclic simple shear tests are consolidated one-dimensionally with lateral restraint supplied by the apparatus (wire-reinforced membrane or stacked rings), so normally consolidated test specimens typically have K_0 values on the order of 0.45–0.5. Comparative tests (Ishihara, 1985) have shown that *CRR* varies approximately with the mean effective stress, so

$$CRR_{ss} = \frac{1+2(K_0)_{ss}}{3} CRR_{tx} \tag{9.3}$$

Comparison of cyclic simple shear and shaking table tests with one-directional and two-directional loading have shown that two-directional loading reduces the *CRR* by about 10%–15% (Pyke et al., 1975; Seed, 1979; Ishihara, 1996). Since ground motion in the field occurs in two horizontal directions, it is common to reduce uni-directional laboratory test *CRR*s by 10% when comparing them to field results. Therefore, comparisons of laboratory and field *CRR*s should be made (Idriss and Boulanger, 2008) assuming

$$CRR_{field} = 0.9\frac{1+2(K_0)_{field}}{3} CRR_{tx} = 0.9\frac{1+2(K_0)_{field}}{1+2(K_0)_{ss}} CRR_{ss} \tag{9.4}$$

If $K_0 = 0.5$ in both the field and the simple shear apparatus, $CRR_{field} = 0.6CRR_{tx} = 0.9CRR_{ss}$

9.5.2.4 Behavior of Liquefiable Soils

A number of important and fundamental aspects of the behavior of liquefaction-susceptible soils are described in Section 6.4. That section described the undrained response of sands and intermediate soils to both monotonic and cyclic loading and showed how the volume change tendencies of the soil affected effective stresses and stress-strain behavior both before and after triggering of cyclic liquefaction. Understanding those basic concepts is critical to understanding liquefaction, so that section should be reviewed before proceeding any further in this chapter. The behavior of saturated sands, as described in Section 6.4.4.2, is affected by a number of factors associated with both initial conditions and the characteristics of earthquake ground motions. The effects of a number of factors that can differ from site to site (and hence can affect cyclic liquefaction potential at any particular site) are described in the following sections. The effects are described and most easily visualized in terms of their influence on the results of cyclic simple shear tests, which simulate the loading on soils from vertically propagating shear waves, and can be used to investigate the effects of different factors under carefully controlled laboratory conditions.

The results of cyclic simple shear tests are commonly expressed in terms of stress-strain curves and stress path-like plots of horizontal shear stress vs. vertical effective stress. Figure 6.50 showed typical curves of these types for a single test on an element of soil that reached initial liquefaction in 22 cycles of loading. In that figure, the pore pressure can be seen to have increased (and the effective stress to have decreased) in a series of increments. Cyclic liquefaction, therefore, involves the incremental buildup of porewater pressure, and the consequent decrease in effective stress. Similar curves are used to illustrate the effects of loading amplitude, soil density, initial effective stress, initial static shear stress, and soil plasticity in the following sections.

Detailed examination of stress path and stress-strain behavior is required at this point. Figure 9.21 shows the response of an element of soil in its first cycle of loading (Points A–D) and a later cycle of loading (Points E–H). In the first cycle, the contractive nature of the soil causes the effective stress to drop a bit and the stiffness of the soil decreases by a small amount; the hysteresis loop has a conventional "football" shape. In the later cycle, enough pore pressure has built up that the effective stress path crosses the phase transformation (PT) line. At Point E, the stress path is on the PT line and the effective stress is at a local minimum. Looking at Point E on the stress-strain curve, the stiffness of the soil can be seen to have decreased substantially. As the shear stress increases from Point E to Point F, however, the soil dilates and the effective stress increases. As the effective stress increases, the stiffness of the soil increases. At Point F, the loading is reversed and the soil contracts until the stress path reaches the PT line at Point G. During this contractive phase, the effective stress drops and the stiffness decreases. At Point G, however, the soil becomes dilative and stiffens as the effective stress increases between Points G and H. This alternating sequence of contractive and dilative phases leads to "banana-shaped" stress-strain loops that are characteristic of cyclic mobility (Section 6.4.4.4). Note that the effective stress has not reached zero in this later cycle; additional cycles of loading may, as seen in subsequent illustrations, lead to a zero effective stress condition.

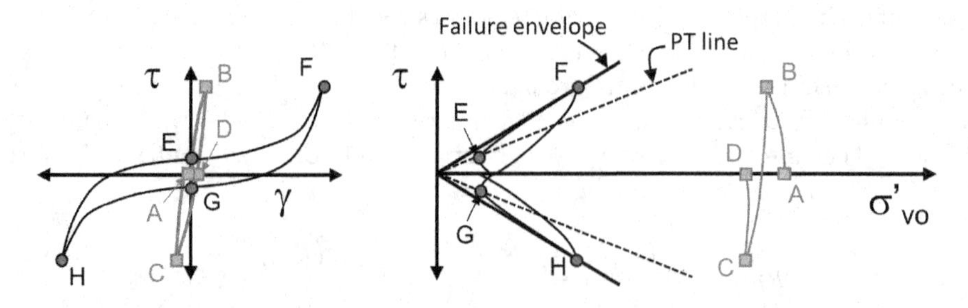

FIGURE 9.21 Schematic illustration of liquefiable soil response in initial (gray) and later (black) cycles of loading.

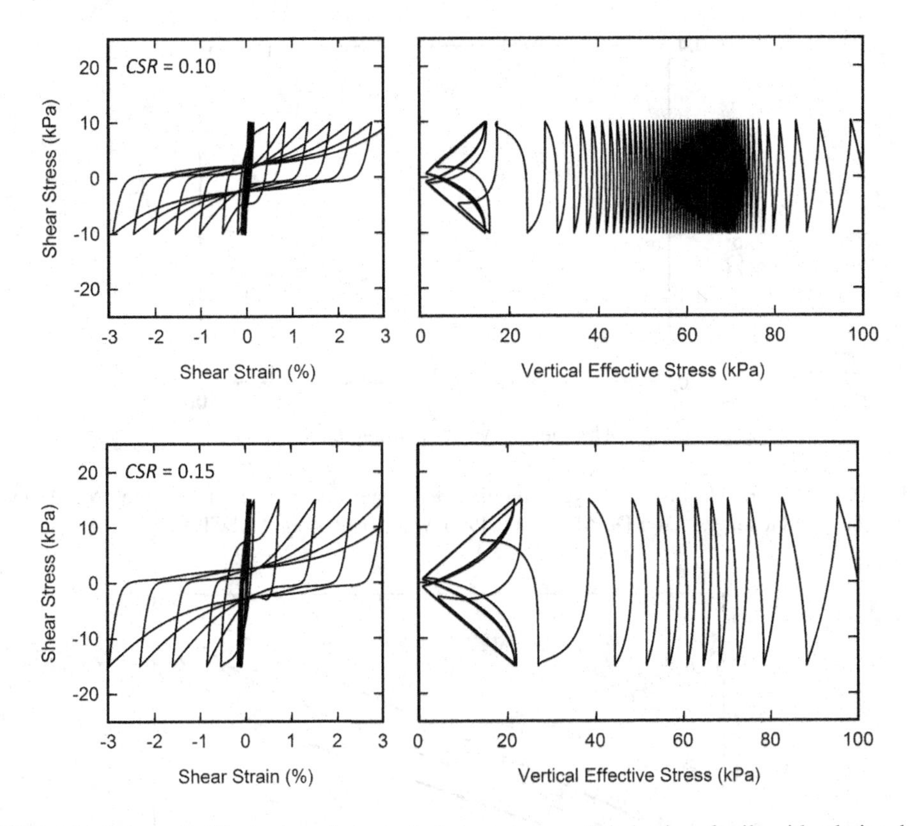

FIGURE 9.22 Schematic illustration of stress-strain and stress path behavior of soils with relative densities of 60% subjected to cyclic shear stresses of 15 kPa (top) and 20 kPa (bottom).

Effect of Load Amplitude

The excess pore pressure generated in a monotonic loading test on loose, saturated sand increases with increasing shear stress, so it is logical to assume that excess pore pressures under cyclic loading would increase at a faster rate as the amplitude of the cyclic loading increases. Figure 9.22 shows simulated stress-strain and stress path curves for two identical elements of soil at 60% relative density – one subjected to loading with a cyclic shear stress amplitude of 10 kPa and the other of 15 kPa. As the effective stress paths show, the pore pressure increases (and effective stress decreases) much faster for the element subjected to the higher load amplitude. The test with CSR=0.10 reached initial liquefaction in 63 cycles of loading while the test with CSR=0.15 liquefied in only 11 cycles. This pattern of decreasing number of cycles to liquefaction with increasing CSR is logical and is consistently observed for tests on individual soils at a particular density and initial effective stress, i.e., at the same initial state. Figure 9.23 shows the results of five cyclic simple shear tests on samples of Monterey #0 sand at a relative density of approximately 54%. The *cyclic strength curve* fit to the individual data points shows that liquefaction can be triggered by a small number of large-amplitude loading cycles (as may result from a small magnitude - hence, short duration - nearby earthquake) or by a large number of low-amplitude loading cycles (such as from a distant, but larger magnitude, earthquake).

The shapes of the stress paths in Figure 9.22 indicate that the pore pressure increases quickly in the first cycles of loading, then more slowly in the intermediate cycles, and then quickly again in the latter cycles that approach initial liquefaction. Cyclic tests with constant load amplitude have consistently shown this type of behavior (Figure 9.24). The transient shear stresses produced by actual earthquake ground motion, however, can be expected to produce more irregular rates of pore pressure generation due to variable stress amplitudes over the duration of shaking.

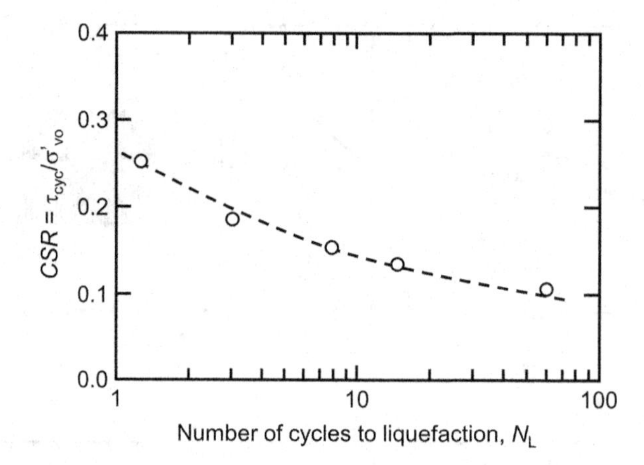

FIGURE 9.23 Variation of number of cycles required to trigger liquefaction of Monterey No. 0 sand with amplitude of applied loading. (After De Alba et al., 1976 with permission of ASCE.)

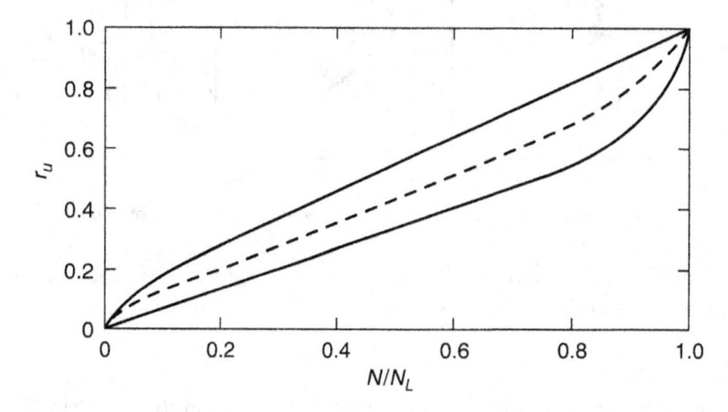

FIGURE 9.24 Rate of pore pressure generation vs. normalized number of cycles to initial liquefaction in constant amplitude cyclic simple shear tests. (After Seed, et al., 1975a with permission of ASCE.)

Effect of Soil Density

At a given effective stress level, increasing soil density corresponds to decreasing state parameter and therefore to decreasing levels of contractiveness. Hence, the rate of pore pressure generation can be expected to decrease with increasing soil density; this aspect of liquefiable soil behavior has been observed repeatedly in the laboratory and in the field. Figure 9.25 shows simulations of tests on specimens prepared to 55% and 65% relative densities and subjected to identical cyclic loading (CSR=0.18) under the same initial effective stress. The effective stress paths show that the rate of pore pressure generation decreases quickly with increasing soil density. The specimen prepared at 55% relative density reached initial liquefaction in seven cycles while the specimen at 65% relative density had not quite reached initial liquefaction after 25 cycles of loading. Figure 9.26 shows the results of tests on denser specimens of the same soil for which results were shown in Figure 9.23. The number of cycles required to reach initial liquefaction for a given CSR increases with increasing soil density. Put differently, the CSR required to initiate liquefaction in a given number of loading cycles increases with increasing soil density. While liquefaction can be initiated in only a few cycles in a loose specimen subjected to large cyclic shear stresses, hundreds of cycles may be required to initiate liquefaction in a dense specimen subjected to low cyclic shear stresses. It should also be noted that the sensitivity of the CSR required to trigger liquefaction (i.e., the slopes of the cyclic strength curves) increases with increasing soil density.

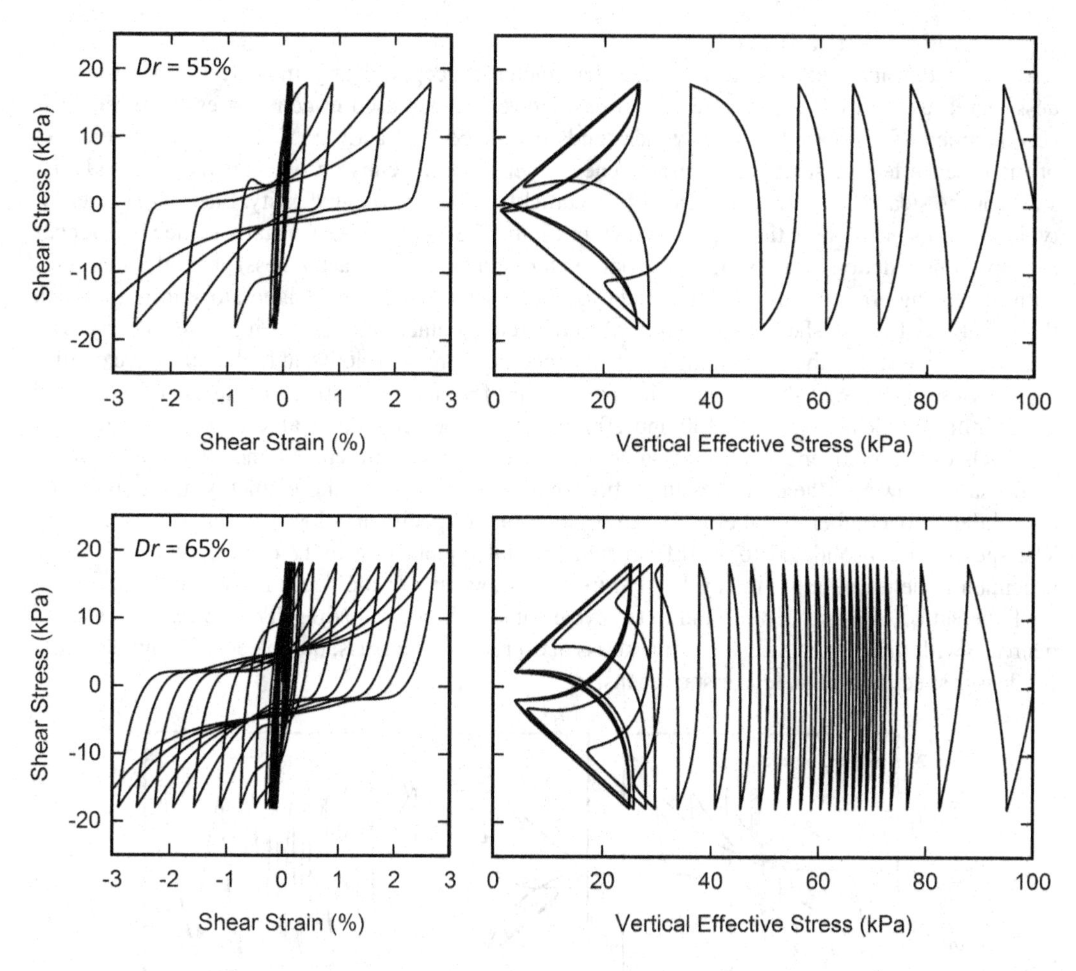

FIGURE 9.25 Schematic illustration of stress-strain and stress path behavior of soils with relative densities of 40% (top) and 50% (bottom) subjected to identical loading.

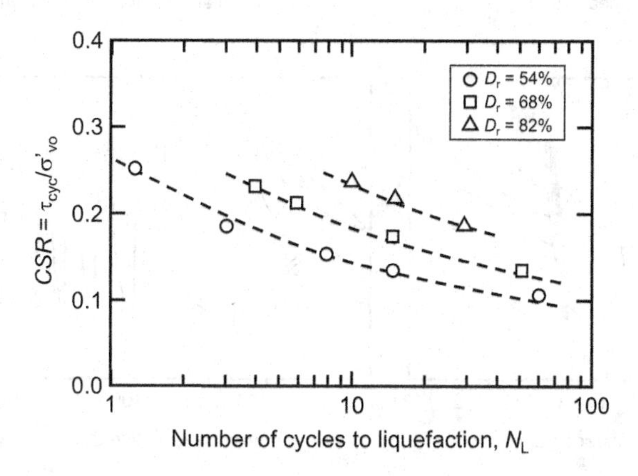

FIGURE 9.26 Variation of number of cycles required to trigger liquefaction of Monterey No. 0 sand with amplitude of applied loading for soils of different relative density. (After De Alba et al., 1976 with permission of ASCE.)

Effect of Initial Effective Stress

At a given density, steady state and state parameter concepts clearly indicate that contractiveness increases with increasing effective stress. However, increased effective stress in the field is accompanied by increased density, which tends to reduce contractiveness. The cumulative effect of increased effective stress on contractiveness, therefore, depends on whether the consolidation stress or the steady state stress increases more quickly with increasing density. Laboratory data for typical clean sands follow the pattern shown in Figure 6.38 – the SSL is usually somewhat steeper than the consolidation curve, so the state parameter (and hence contractiveness) generally increases with increasing effective stress. This effect has been seen repeatedly in laboratory tests in which the variation of cyclic shear stress required to trigger liquefaction does not increase linearly with initial effective stress, thus indicating that soils become more contractive at higher effective confining stresses. Figure 9.27 shows simulations of tests performed on identical specimens consolidated under initial vertical stresses of 100 and 200 kPa and subjected to the same cyclic stress ratio. If the levels of contractiveness were the same, the scaled stress-strain curves and stress paths would be the same; however, the pore pressure ratio can be seen to increase more quickly in the specimen consolidated to a higher initial effective stress than in the specimen at lower initial effective stress. The specimen consolidated to 100 kPa reaches initial liquefaction in 12 cycles, but the 200 kPa specimen liquefies in only nine cycles. Figure 9.28 shows the cyclic stress ratios required to initiate liquefaction of Fraser River Sand in ten cycles of simple shear loading. The *CSR* can be seen to decrease with increasing initial effective stress at all relative densities, but to decrease more rapidly for denser specimens than looser specimens.

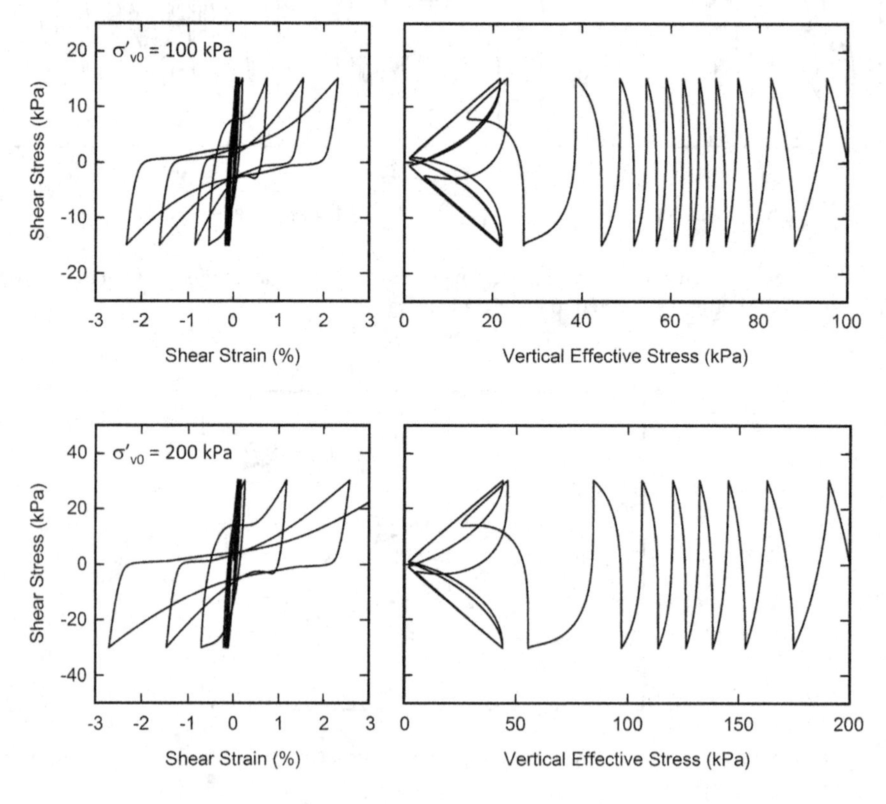

FIGURE 9.27 Schematic illustration of stress-strain and stress path behavior of soils with relative densities of 60% under initial vertical effective stresses of 100 kPa (top) and 200 kPa (bottom) subjected to identical cyclic stress ratios.

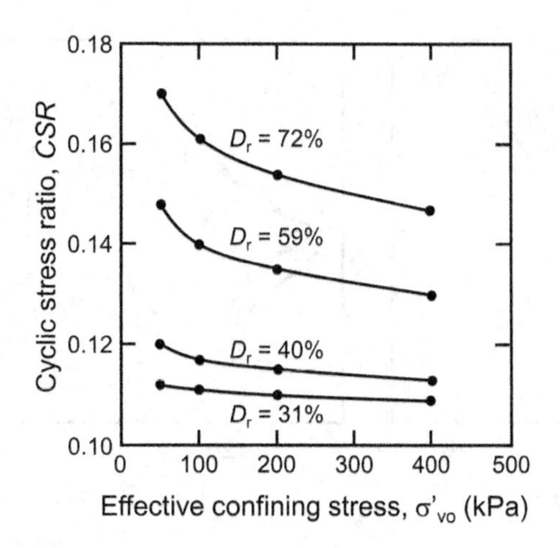

FIGURE 9.28 Effect of initial vertical effective stress on cyclic resistance ratio of Fraser Delta sand. (After Vaid and Sivathayalan, 1996 with permission of Canadian Science Publishing.)

Effect of Initial Shear Stress

For level-ground sites, no shear stress exists on planes parallel to the ground surface prior to earthquake shaking. At sites with sloping ground surfaces or upon which structural loads are imposed, an element of soil may be subjected to a significant initial, static shear stress on horizontal planes prior to earthquake shaking. The presence of such initial shear stresses can affect the rate at which pore pressure is generated in the soil. Experimental studies (Seed and Harder, 1990; Boulanger et al., 1991) have shown that liquefaction can occur more quickly in very loose soils in the presence of an initial shear stress. In medium-dense to dense soils, however, the presence of an initial shear stress tends to increase liquefaction resistance. The latter observation, which suggests the counter-intuitive conclusion that steeper slopes have lower liquefaction potential than flatter slopes, results from the lower degree of stress reversal that occurs when initial shear stresses exist. The generation of excess pore pressure is aided by soil particle reorientation which occurs more readily during the instants of time in which the soil is subjected to no shear stress, i.e., when the sign of the shear stress changes from positive to negative or vice versa. If the static shear stress is relatively low, the cyclic stresses will still cause the cyclic shear stress to change direction, and a state of initial liquefaction, with large reductions in stiffness, can still occur; under such conditions, the asymmetric stresses can induce large permanent strains in the soil. Figure 9.29 shows the response of two elements of soil with different static shear stresses when subjected to the same cyclic shear stress. The soil with the greater static shear stress builds up pore pressure slightly more slowly but accumulates shear strain more quickly than the soil with the lower static shear stress.

For high levels of initial shear stress, however, pulses of cyclic stress that are lower than the initial shear stress do not cause the shear stress to change sign, which tends to reduce their contribution to pore pressure development. If the initial shear stress exceeds the peak cyclic shear stress, the sign of the shear stress cannot change and the soil cannot reach a condition of zero effective stress. While the soil will soften and accumulate permanent strain, the reduction in stiffness is not nearly as extreme as in the case where initial liquefaction was reached.

Figure 9.30 shows the results of a series of cyclic simple shear tests on specimens of different densities subjected to cyclic loading with different levels of initial static shear stress. The cyclic stress ratios required to initiate liquefaction, normalized here by the required *CSR*s with no static shear stress, are seen to increase with increasing static shear stress for denser soils and to decrease

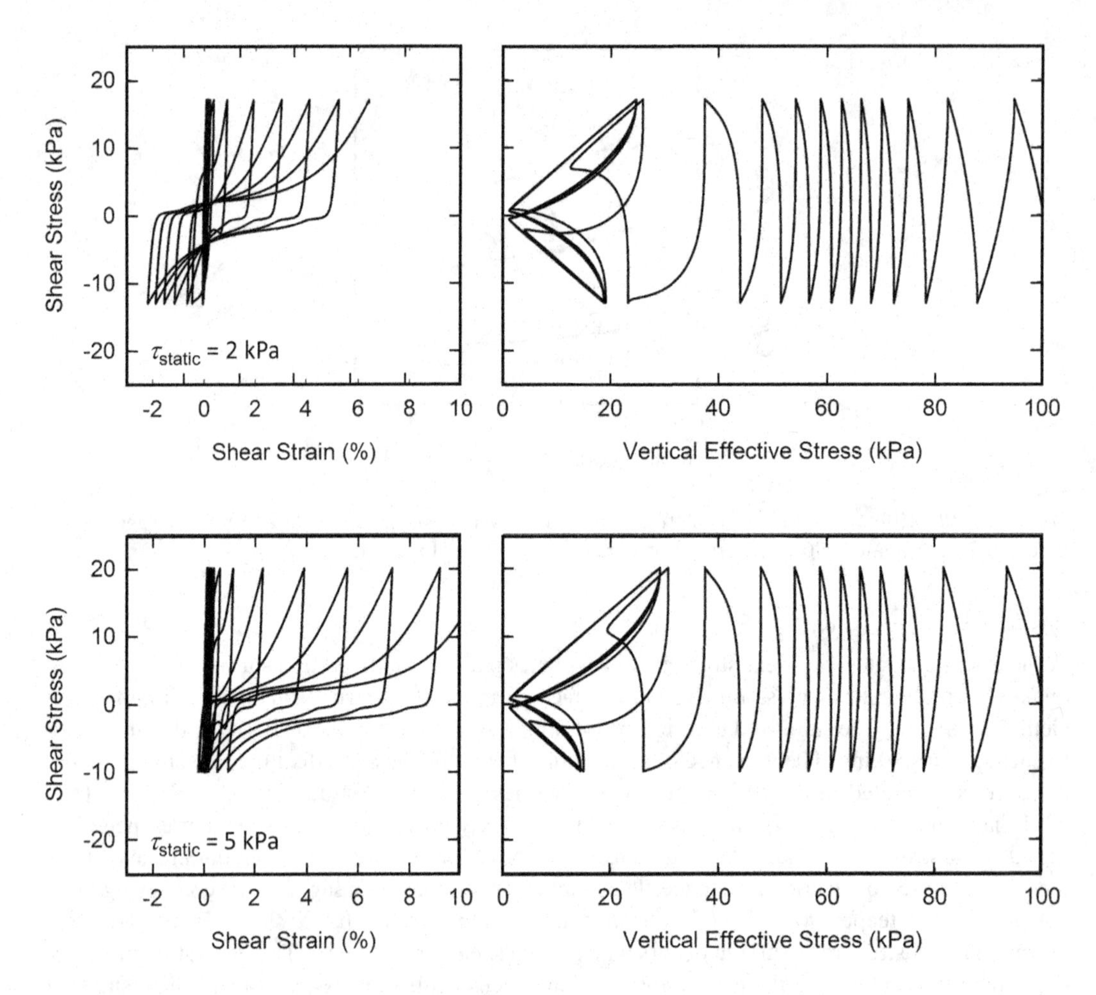

FIGURE 9.29 Schematic illustration of stress-strain and stress path behavior of soils with relative densities of 60% under initial static shear stresses of 5 kPa (top) and 8 kPa (bottom) subjected to identical cyclic stress ratios.

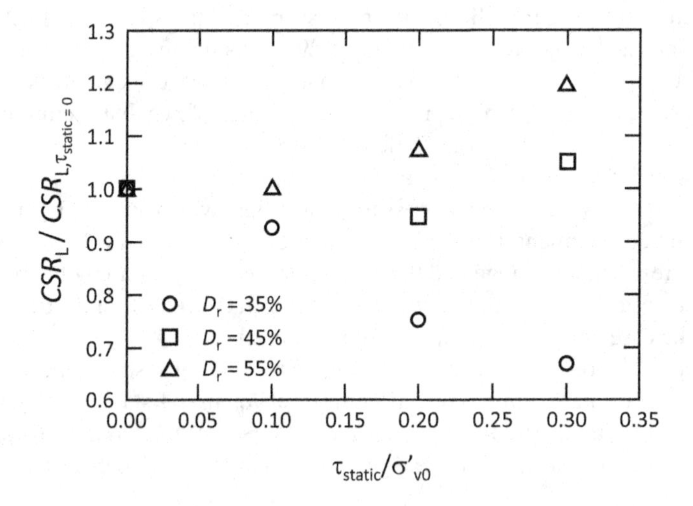

FIGURE 9.30 Effect of static shear stress on normalized cyclic stress ratio required to produce 3% shear strain in 10-30 cycles of simple shear loading. (After Boulanger et al., 1991.)

with increasing static shear stress for looser soils. Some studies (e.g., Ishihara and Yamazaki, 1980; Boulanger and Seed, 1995; Kammerer et al., 2004) have shown that three-dimensional effects can also be important, with pore pressure and strain in one horizontal direction being influenced by static initial shear stresses in the orthogonal horizontal direction.

Effect of Fines

The presence of fine-grained particles within a coarse-grained soil can have a strong, and complicated, effect on liquefaction hazard evaluation since they can influence both liquefaction resistance and the *in situ* test parameters to which liquefaction resistance is often correlated (Sections 6.5.3.1 and 6.5.3.3). Two factors are important in determining the influence of fines on liquefaction resistance – the amount of fines and their plasticity.

The amount of fines is typically expressed in terms of the fines content, FC, which is the fraction (by weight) of soil passing the No. 200 sieve. However, the influence of the fines depends on the relationship between the fines content and the limiting fines content, FC_L. As described in Section 6.3, the behavior of a silty sand will be controlled by the sand for $FC<FC_L$ and by the silt fraction for $FC>FC_L$. The liquefaction resistance of a mixture of sand and non-plastic silt has been shown to be essentially constant, and controlled by the relative density of the sand, for $FC<FC_L$, and to decrease significantly with increasing silt content at $FC>FC_L$ (Polito and Martin, 2001). Limiting fines contents vary from soil to soil but are generally within the range of about 20%–35% (Cubrinovski and Ishihara, 1999).

The extent to which a soil softens, and consequently the extent to which strain levels increase, is influenced by the minimum effective stress produced by cyclic loading and the extent to which the soil's stiffness is related to the effective stress level. As described in Section 9.4.3, the rate of pore pressure generation and the shapes of stress-strain loops are different in granular materials (gravels, sands, non-plastic silts) than in plastic soils. This difference is handled at the susceptibility stage (Section 9.4) in a liquefaction hazard assessment for a given site. Plastic fine-grained soils can build up pore pressure and soften, but the maximum pore pressure is lower and the minimum stiffness higher than that of non-plastic soils. As a result, the stress-strain loops are broader and more "football-shaped" than the "banana-shaped" loops of the non-plastic soil. Figure 9.31 shows stress-strain loops and effective stress paths from simulated cyclic simple shear tests on non-plastic and plastic silts. The shear stress intercept (at zero strain) of the more plastic soil is much higher and the rate at which stiffness increases upon dilation is much lower. The rate of energy dissipation, as evidenced by the broader stress-strain loops is also much higher in the plastic soil than the non-plastic soil. Such soils may exhibit a significant degree of cyclic softening (Section 9.8), with potential adverse effects, but they do not liquefy.

Effects of Other Factors

A number of other factors have been shown to affect liquefaction resistance. These factors may affect the contractiveness of the soil or the level of shear strain induced in the soil by cyclic loading. One of these factors is the history of prior seismic straining (*prestraining*) at levels below the volumetric threshold strain (Section 6.4.3.1), which laboratory studies have shown to increase liquefaction resistance relative to that of a specimen of the same density that has not been pre-sheared (Finn et al., 1970; Seed et al., 1975b; Suzuki and Toki, 1984). Such straining can be caused in the field by small or distant earthquakes that occur over time, or by low-level vibrations from other sources. Dobry et al. (2015) postulated that the apparently high liquefaction resistance (relative to that corresponding to measured penetration resistance) displayed by liquefiable sands in the Imperial Valley of California was due to preshaking by 60–70 earthquakes after their deposition in that area of high seismic activity. Also, liquefaction resistance increases with increasing overconsolidation ratio and lateral earth pressure coefficient (Seed and Peacock, 1971; Lee and Focht, 1975; Finn, 1981). Finally, the age, or length of time under sustained stress, has been shown (Ohsaki, 1969; Seed, 1979; Yoshimi et al., 1989; Robertson et al., 2000; Andrus et al., 2009; Hayati and Andrus, 2009; Maurer et al., 2014b) to increase liquefaction resistance. These additional parameters are all functions of the depositional

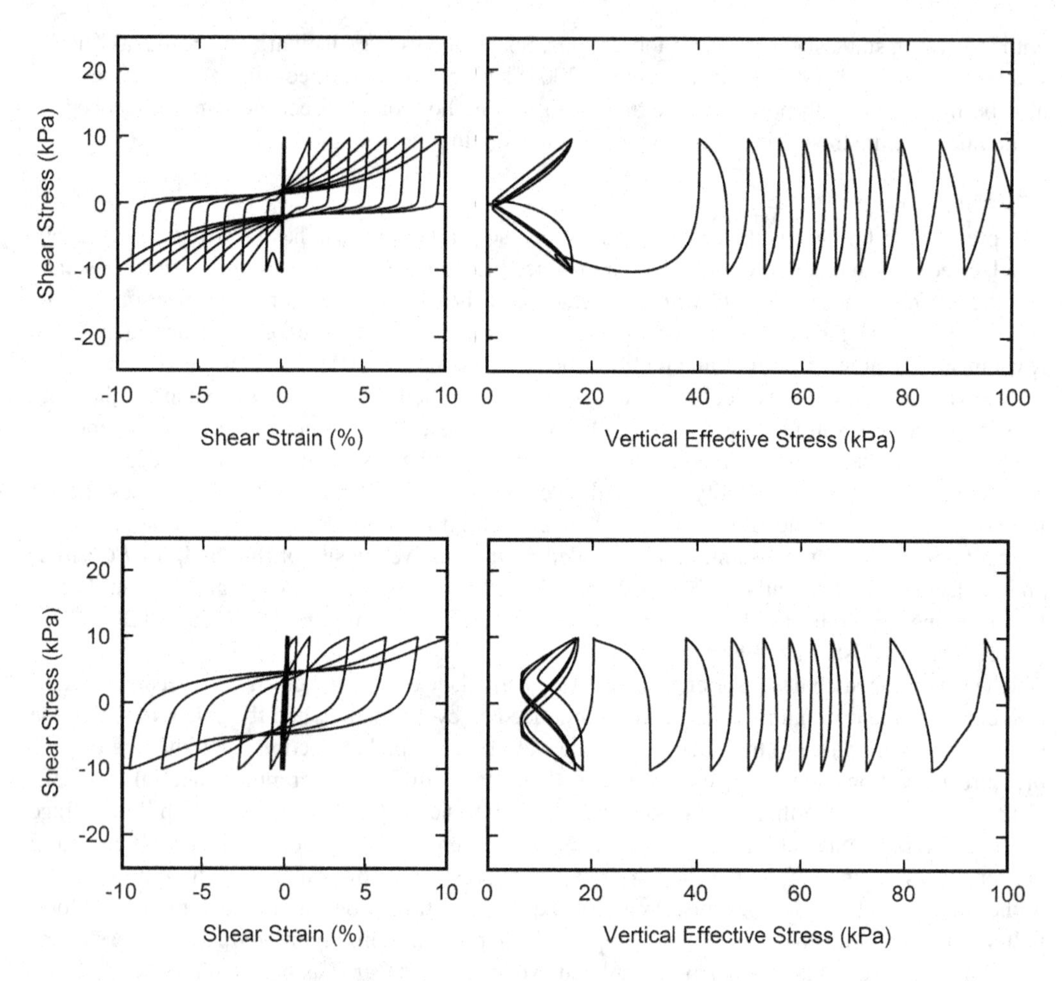

FIGURE 9.31 Schematic illustration of stress-strain and stress path behavior of soils with (a) sand-like, and (b) clay-like behavior. Note differences in shapes of stress-strain loops and minimum vertical effective stresses.

and historical environment of a soil deposit, and they tend to influence soil behavior primarily at the low strain levels associated with the triggering of liquefaction. Their influence on the larger strain behavior associated with the consequences of liquefaction is likely to be less pronounced.

Liquefaction resistance is influenced by gradation. Well-graded soils are generally less contractive than poorly graded soils; the filling of voids between larger particles by smaller particles in a well-graded soil results in lower volume change potential under drained conditions and, consequently, lower excess pore pressures under undrained conditions. Field evidence indicates that most liquefaction failures have involved uniformly graded soils. Particle shape can also influence liquefaction resistance. Soils with rounded particle shapes are known to densify more easily than soils with angular grains. Consequently, they are usually less resistant to liquefaction than angular-grained soils of the same particle size gradation.

9.5.3 Empirical Prediction of Cyclic Liquefaction

As discussed in Sections 6.4 and 9.5.2, liquefiable soils exhibit complicated, highly nonlinear responses to cyclic loading. Furthermore, those responses can be very sensitive to *in situ* conditions that are difficult to characterize in the field, replicate in the laboratory, or model analytically. Because of these factors, the most common approaches to the prediction of cyclic liquefaction potential are empirical, i.e., based on observations (or lack thereof) of surficial evidence (i.e., surface *manifestation*) of liquefaction in actual earthquakes. Empirical approaches are of two general types – purely empirical approaches that are based solely on readily available data (Section 9.5.8) and semi-empirical approaches that make use of subsurface data, soil mechanics principles, and laboratory test results as well as field data. Semi-empirical approaches have been much more commonly used in geotechnical earthquake engineering practice and are emphasized here.

The semi-empirical approach, first described by Whitman (1971), uses liquefaction case histories to characterize liquefaction resistance in terms of measured *in situ* test parameters. Previous case histories can be characterized by the combination of a loading parameter, $\mathcal{L}$, and a liquefaction resistance parameter, $\mathcal{R}$, which can be plotted with a symbol that indicates whether surface manifestation was or was not observed (Figure 9.32). A boundary can then be drawn between the $\mathcal{L}$ - $\mathcal{R}$ combinations that have and have not produced evidence of liquefaction in past earthquakes and taken to represent the liquefaction resistance of the soil. Ideally, the boundary would clearly separate all cases of liquefaction and no-liquefaction, but variability, uncertainty, and the inability of common measures of $\mathcal{L}$ and $\mathcal{R}$ to accurately represent the complex physics of triggering and its surface manifestation lead to some case histories falling on the wrong side of the boundary.

9.5.3.1 Characterization of Liquefaction Potential

All empirical procedures for characterization of liquefaction potential involve a comparison of some measure of the loading imposed on the soil by earthquake shaking with a consistent measure of the resistance of the soil to liquefaction. Different quantities – cyclic shear stress, cyclic shear strain, dissipated energy – have been used to describe loading and resistance, and each has advantages and limitations. Regardless of what quantities are used, the results of these empirical analyses are typically expressed in terms of a factor of safety against triggering of liquefaction, i.e.,

$$FS_L = \frac{\mathscr{R}}{\mathscr{L}} \tag{9.5}$$

where $\mathscr{R}$ is some measure of resistance and $\mathscr{L}$ is a consistent measure of loading such that FS_L is dimensionless. As shown in Figure 9.32, the case history data points tend to overlap each other and early procedures characterized $\mathscr{R}$ conservatively, i.e., by drawing a curve that fell below nearly all of the solid circles; subsequent statistical analyses of the data showed that these deterministic curves generally corresponded to approximately 15th percentile resistance values (or about mean – σ in ln $\mathscr{R}$). A factor of safety less than or equal to 1.0 indicates that the loading exceeds the resistance, so liquefaction is expected to occur. Factors of safety are usually calculated using some minimum allowable FS_L value (typically 1.2–1.5) to account for uncertainty and the consequences

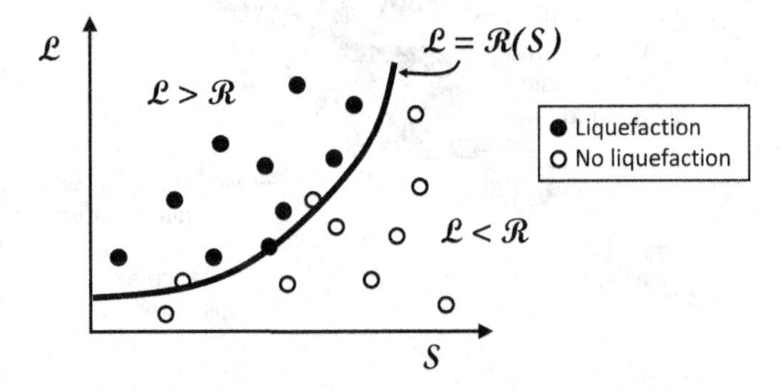

FIGURE 9.32 Typical plot showing combinations of loading parameter, $\mathscr{L}$, and soil parameter, S, for cases where evidence of liquefaction has been observed (solid circles) and not observed (open circles). Boundary indicates the value of resistance parameter, $\mathscr{R}$, i.e., loading parameter required to initiate liquefaction in a soil with a given soil parameter value.

of liquefaction. Alternatively, the potential for triggering of liquefaction for a given level of loading can be expressed as a probability of liquefaction

$$P_L = P[\mathcal{R} \le \mathcal{L}] = P[FS_L \le 1.0] \tag{9.6}$$

Probabilistic characterization of liquefaction hazards is particularly useful because there are significant uncertainties in estimates of both loading and resistance. The historical procedure of performing deterministic analyses with "conservative" input parameters and/or "conservative" interpretation of the results of the analyses can reduce the likelihood of liquefaction damage but generally does so in an inconsistent manner that produces designs with unknown probability of failure – in some cases, so high as to produce unacceptable risk and in other cases so low as to waste money or other resources.

9.5.3.2 Screening Procedures

For practical purposes, it can be helpful to screen cases where liquefaction is so unlikely to be initiated that a detailed, formal triggering analysis is not needed. The susceptibility evaluation, as described in Section 9.4, can be thought of as a first-level screening procedure as it eliminates soils whose compositions render them incapable of liquefying. It is also possible to screen susceptible soils, though, if they exist under conditions that make triggering extremely unlikely, e.g., if the loading they will be subjected to is very low and/or their resistance to triggering is very high.

Distance Screen

The level of loading imposed on an element of liquefaction-susceptible soil is a function of the level of ground shaking at the location of that element. As discussed in Section 3.4, ground shaking levels depend on a number of factors, the most influential of which are earthquake magnitude and source-to-site distance. For low magnitudes and/or large distances, ground shaking levels may be so low that even very loose soils will not liquefy. Figure 9.33 shows the greatest distances at which

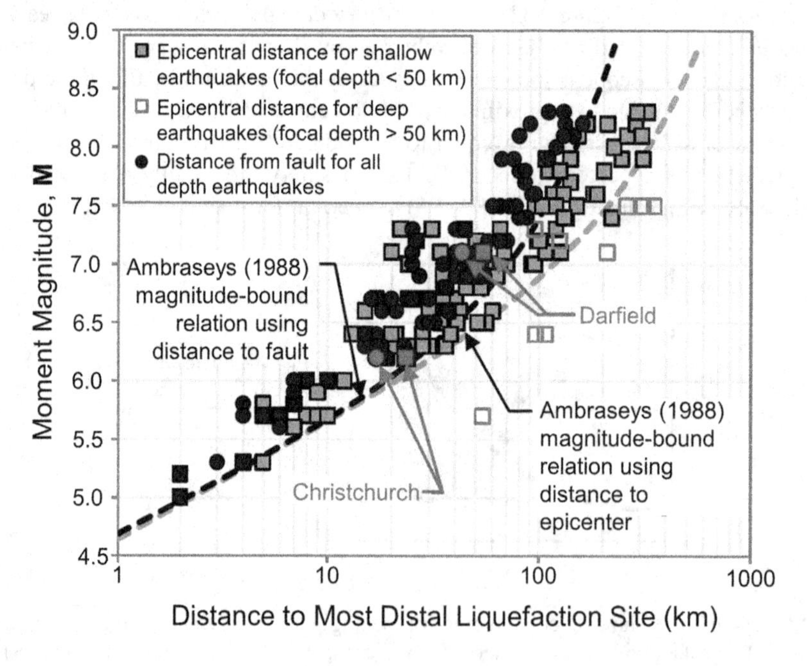

FIGURE 9.33 Distances within which liquefaction has occurred in historical earthquakes of different magnitude (after Maurer et al., 2014b with permission of Earthquake Engineering Research Institute).

liquefaction has been observed in past worldwide earthquakes of various magnitudes. The triggering of liquefaction at sites whose characteristics plot significantly below the bounding curve can be considered, on the basis of historical experience, to be unlikely. Aftershocks, however, can cause *recurrent liquefaction* of previously liquefied soils under conditions that plot well below the bounding curve in Figure 9.33.

Density Screen

Liquefaction-susceptible soils may also exist in a state that makes their likelihood of triggering so low that a formal evaluation of liquefaction potential is not necessary. As will be discussed in Section 9.5.4.4, liquefaction resistance is commonly related to various *in situ* test parameters that correlate to the density of the soil. If those parameters are sufficiently high, the soil can be considered to be so dense that triggering of liquefaction is extremely unlikely; under such conditions, the consequences of liquefaction, even if it were to be initiated, would likely be minimal. As part of a recommended susceptibility evaluation procedure, Bray and Sancio (2006) included a term (w_c/LL) that is related to the density of saturated soil. Since susceptibility is defined here as being based on composition (Section 9.4), the water content-related term of Bray and Sancio (2006) can be treated as a density screening criterion for soils with sufficient fines to allow Atterberg limits measurement. Soils with ratios of water content to liquid limit less than 0.80 are considered to have negligible liquefaction potential. Based on current empirical triggering models, screening parameters are shown in Table 9.1.

Shear Strain Screen

The concept of a volumetric threshold strain, γ_{tv} (Section 6.4.3.1), below which effectively no volume change occurs in a dry soil, can also be used as a screening criterion for cyclic liquefaction. If the volume of an element of soil does not change under drained loading conditions, it will not generate excess pore pressure at the same strain level under undrained conditions – and if the pore pressure does not change, cyclic liquefaction cannot occur. Screening of cyclic liquefaction with this criterion will generally require a site response analysis in order to determine the expected cyclic strain amplitudes; soils that are sufficiently stiff or subjected to weak shaking may develop peak strains that do not exceed the threshold shear strain.

Saturation Screen

Liquefaction generally occurs only in saturated, or very nearly saturated, soils, so the degree of saturation of a soil influences liquefaction potential. Liquefaction results from the contractive nature of soils, but that contraction does not translate to high pore pressure unless the pores of the soil are filled with a fluid of low compressibility. Shallow non-plastic soils in areas where groundwater levels are changing, or have changed, may be partially saturated and therefore not able to generate the high pore pressures required to initiate liquefaction.

TABLE 9.1
Practical Screening Thresholds for Non-Triggering of Liquefaction

Parameter	Screening Threshold
SPT resistance, $(N_1)_{60,cs}$	>40 blows/ft
CPT resistance, q_{c1Ncs}	>220
Shear wave velocity, V_{s1}	>300 m/sec
Water content/liquid limit, w/LL	<0.80
p-wave velocity, V_p	<300 m/sec

The effective stress-normalized parameters used here $\left((N_1)_{60,cs}, q_{c1Ncs}, V_{S1}\right)$ are defined in Section 9.5.4.3.

Experimental studies of porewater (and pore air) pressure generation in the laboratory (Tsukamoto et al., 2002; Yang, 2002, 2004; Kamata et al., 2007; Eseller-Bayat et al., 2013a, b; Tsukamoto et al., 2014; Baki et al., 2023) indicate that liquefaction can occur in partially saturated soils, but also that liquefaction resistance increases quickly as the degree of saturation of a soil drops below 100%. As discussed in Section 9.5.4.4, measured *p*-wave velocities can provide a good indication of partial saturation; since the bulk modulus of water is much higher than that of typical shallow soils, *p*-waves travel at their velocity in water (approximately 1,500 m/sec) in a saturated soil. Laboratory investigations have shown that liquefaction resistance becomes extremely high for soils with *p*-wave velocities less than 300 m/sec.

Drainage Screen

Drainage of pore fluids during earthquake shaking, which dissipates porewater pressure and suppresses liquefaction, is important in the case of gravels because the large particle (and void) size produces high permeability that enables rapid fluid flow. However, when layers of gravel are bounded by low permeability soils or are very thick, pore pressure dissipation cannot occur and the gravel remains undrained despite its high permeability. Also, gravel soils can have voids filled with sand, silt, and/or clay with lower permeabilities that prevent rapid pore pressure dissipation. Under such conditions, the gravel is likely to be vulnerable to liquefaction. Drainage may also limit pore pressures in portions of sand layers that are very close to drainage boundaries and may prevent triggering of liquefaction in thin, well-drained liquefiable layers.

9.5.3.3 Case Histories

As alluded to previously, the most common procedures for evaluation of liquefaction potential are based on field observations of the presence or absence of surficial manifestation of liquefaction (e.g., sand boils, ground cracking, settlement, etc.), as documented in case histories. This approach implicitly assumes that the presence or absence of surficial evidence accurately indicates the occurrence or non-occurrence of liquefaction, which, as discussed subsequently, is not always true. Because liquefaction sites are very rarely instrumented, most case history investigations simply note whether surficial evidence of liquefaction was or was not observed. These case histories, therefore, produce only binary, categorical data based on the presence or absence of surficial manifestation of liquefaction. It should be recognized, however, that surficial manifestation does not always definitively indicate the triggering of liquefaction (e.g., ground cracking can be caused by cyclic softening of a clay layer) and its absence does not definitively indicate the absence of triggering (triggering of a thin, deep layer may produce no sand boils). To evaluate existing and newly proposed empirical procedures for evaluation of liquefaction potential, it is important to have a good understanding of what constitutes good and useful case histories.

Most liquefaction case history investigations have two primary components: field reconnaissance and site characterization. The field reconnaissance effort generally takes place as soon as possible following a significant earthquake. Much of the evidence that indicates the occurrence of liquefaction, such as sand boils and ground cracking, is ephemeral and can be obscured or destroyed by rainfall, clean-up operations, or other activities that may take place shortly after an earthquake, so rapid mobilization is required. Geotechnical reconnaissance efforts have become much more accurate and sophisticated in recent years with the development of GPS, ground and satellite remote sensing, LiDAR, digital imaging, unmanned aerial vehicles (UAVs), and other technologies. The Geotechnical Extreme Events Reconnaissance (GEER) association has taken a leading role in organizing, conducting, and rapidly disseminating the results of geotechnical earthquake reconnaissance activities over the past 25 years. The site characterization phase involves the gathering of existing data on regional and local geology, ground water conditions, and subsurface data, usually followed by supplementary geophysical and subsurface explorations. In some cases, pre-earthquake subsurface data may be available but, more commonly, subsurface investigations take place after an earthquake has occurred. This situation complicates matters in an unpredictable manner; for example, post-liquefaction densification will tend to increase penetration resistance (usually in the lower part of a liquefied layer) but the upward flow of water and/or soil fabric degradation during liquefaction will tend to decrease it (more commonly in the upper part). As a result, post-earthquake measurements may not accurately reflect the conditions that existed at the time of the earthquake. Subsurface explorations typically involve *in situ* methods such as cone or standard penetration testing. When the CPT is used, some form of drilling and sampling is usually required to obtain samples for confirmation of soil type and measurement of grain size and plasticity characteristics, although CPT samplers are available for retrieval of disturbed samples. These samplers have outside diameters equal to the cone diameter and are pushed closed-ended with an internal rod to the desired depth at a location adjacent to a completed CPT sounding. The pushing rod is then retracted within the sampling tube, locked off, and the entire assembly is pushed downward to obtain the sample. After the sample is brought to the surface, a separate sampler can be pushed down the same hole to the next desired sampling depth. Geophysical explorations are typically oriented toward measurement of shear wave velocities (although p-wave velocity measurements can help identify or confirm groundwater depths).

Many commonly used triggering models are based on the concept of a "critical layer," i.e., the layer that liquefied (or the layer that would have liquefied first if the shaking had been stronger in the case of "no liquefaction" case histories) and was responsible for the observed consequences. The case history is then interpreted in terms of the loading and state (Section 6.4.3.6) of the critical layer. Identification of a critical layer is not always obvious – it can require considerable interpretation and judgment, and different investigators have often interpreted critical layers for the same case history differently. The depth of the critical layer is important because it influences both the level of ground shaking (loading) and the *in situ* test value (resistance) assigned to the case history. Available databases have many case histories with shallow critical layers – mean depths are on the order of 5–6 m and only about 10% of existing case histories have critical layers deeper than 8 m (National Research Council, 2016). Green and Olson (2015) proposed an approach to the identification of critical layers and characterization of the severity (e.g., none, marginal, moderate, or severe instead of the common binary yes/no classification) of surficial manifestation. In this approach, the severity of liquefaction helps guide identification of the critical layer – for example, a thin layer under level ground would not be expected to have been responsible for severe surficial manifestation even if it had a lower density than thicker liquefiable layers within the profile.

Some case histories are more valuable than others – a case history in which an extremely loose soil liquefies under very strong shaking does not reveal much new information about liquefaction resistance; neither does a case history of a dense soil that does not liquefy under weak shaking. The most useful case histories, from the standpoint of triggering of liquefaction, are those in which liquefaction just barely occurs or very nearly occurs – case histories such as these help define the boundary between the triggering and non-triggering of liquefaction (Figure 9.32).

An important aspect of a liquefaction case history is the loading, i.e., the level of ground shaking the site was subjected to. Different procedures are used to evaluate ground shaking demands depending on the availability of ground motion recordings (Stewart et al. 2016). If no recordings are available for the earthquake, ground motions can be estimated using a GMM, although this caries large uncertainty. If recordings are available, but not at the particular site of interest, event terms (Section 3.5.4.2) can be derived and added (as a natural logarithm) to the GMM median to estimate event-specific ground motions. Additional adjustments accounting for the spatial variability of ground motions for the event can be applied (e.g., Bradley and Hughes, 2012; Bradley, 2014; Kwak et al., 2016; Pretell et al. 2024). The adjustment of ground motions using event terms, and spatial adjustments, changes the ground motion estimate relative to the GMM median and reduces uncertainty. Recently, a new class of case history has been identified – that in which liquefaction occurred beneath a strong motion instrument (Kramer et al., 2013). At such sites, time-frequency analyses can be performed to identify the time at which liquefaction was initiated. That time can then be used to estimate the level of loading at the time of liquefaction, and that loading level has the smallest uncertainty among the options considered here. These case histories can be used to define liquefaction resistance in a manner that conventional binary case histories cannot (Kramer et al., 2016).

9.5.3.4 Desirable Characteristics of Empirical Procedures

The complexity of the liquefaction process and the scarcity of full-scale field data covering the full ranges of conditions for which geotechnical engineers need to evaluate liquefaction potential means that empirical predictive models need to interpolate and extrapolate over ranges of conditions that are not well constrained by actual data. Thus, the developers of empirical models are forced to make assumptions and approximations that affect their predictive capabilities. Different modelers have made different modeling decisions that lead to different results when applied to the same problem; provided that the models are all reasonable and error-free, those differences represent epistemic uncertainty in the prediction of liquefaction potential.

There are, however, certain desirable characteristics in an empirical model for prediction of liquefaction potential, and understanding how many of those characteristics are present in a particular model is helpful in evaluating its accuracy and applicability. These characteristics involve evaluation of loading and resistance from available data, and the expression of liquefaction potential.

Data

Empirical procedures are based on case history data and the amount and range of the data influence the form and applicability of each procedure. As liquefaction is observed in more and more earthquakes, case history data becomes increasingly available for the development and updating of triggering procedures. More recent procedures are generally based on more data that span a greater range of conditions of interest than older procedures. While early liquefaction triggering models were based on data collected and used by individual research groups, models in the last 20 years or so have been based on interpreted data made public by the model developers. More recently, the Next Generation Liquefaction (NGL) research project (Stewart et al., 2016) has established a public database (www.nextgenerationliquefaction.org; Brandenberg et al., 2020) of detailed, objective case history and laboratory test data that researchers and model developers can use and interpret as they wish.

As previously described, empirical procedures for triggering of liquefaction have historically been based on the implicit assumption that the presence of surface manifestation (e.g., sand boils or ground cracking) indicated that triggering had occurred, and its absence indicated that it had not. The development of surface manifestation, however, involves both mechanical and hydraulic properties of an entire soil profile that are not fully captured by the properties used in triggering models. As such, it is possible for liquefaction to be triggered without surface manifestation and also for evidence of surface manifestation to develop when liquefaction has not been triggered.

Loading

The loading, or demand, should be based on ground motion characteristics that the generation of excess pore pressure is sensitive to. Those characteristics are usually described by a ground motion intensity measure, *IM*, and it should be one that is both predictable and an efficient and sufficient predictor of pore pressure generation (Section 3.3). An *IM* that is an efficient predictor of liquefaction may not be optimal, however, if the uncertainty in its predicted value (i.e., $\sigma_{\ln IM}$), given some earthquake scenarios, is high (Kramer and Mitchell, 2006).

Estimation of the *IM* for an element of liquefaction-susceptible soil at some depth in a soil profile involves evaluation of site response. Ideally, sufficient data and resources to allow multi-dimensional, nonlinear site response analyses would be available. Such analyses would account for site-specific stratigraphy, topography, and groundwater conditions, for the nonlinear, inelastic behavior of the *in situ* soil, and would allow direct prediction of the ground motion intensity measure(s) of interest. For many projects, however, sufficient time and budget to allow such detailed, site-specific analyses are not available, so more simplified approaches for estimation of demand are used. Such approaches typically involve relationships based on regression of the results of many ground response analyses.

Those analyses should be performed on profiles that are representative of the range of site conditions likely to be encountered in liquefaction potential evaluations. The relationships regressed upon this data should be consistent with basic principles of site response, i.e., they should include variables known to affect the propagation of waves through soils.

Resistance

A desirable liquefaction resistance model should be based on a comprehensive database of well-documented case histories. The case histories should ideally span wide ranges of input parameter values that encompass all conditions of interest. In reality, however, available case histories do not cover the entire range of conditions that engineers are frequently interested in. As indicated in Section 9.5.3.3, case histories of liquefaction occurring at depths greater than about 8 m are relatively rare. Likewise, case histories involving denser soils (e.g., with $(N_1)_{60}>20$ blows/ft) are also rare. In practice, engineers are frequently called upon to predict the liquefaction behavior of soils at large depths and at intermediate and higher densities. Extending the range of the liquefaction resistance models to conditions that are not well represented in the case history database requires extrapolation, and extrapolated predictions depend strongly on the functional form of the predictive model. To extrapolate accurately, it is essential that the liquefaction resistance model be based on fundamental principles of soil mechanics and/or on high-quality laboratory tests that cover the range of conditions of interest. Some existing models do this better than others.

All empirical predictive models relate liquefaction resistance to some soil characteristics. Ideally, liquefaction resistance would be related to the state parameter, i.e., to density and effective stress, but direct measurement of the state parameter is not possible. Instead, procedures use various *in situ* test parameters in combination with effective stress as proxies for state parameter. Quantities such as CPT resistance, SPT resistance, and shear wave velocity can all be correlated to the density of coarse-grained soils and have all been used as predictors of liquefaction resistance. Because they are all affected by effective stress, they require some form of adjustment to account for the effective stress at the depth of the soil whose liquefaction resistance is being evaluated.

Liquefaction Potential

The development of predictive models for triggering of liquefaction requires the assignment of loading and resistance measures to all case histories in the database followed by regression to find a function that separates case histories in which liquefaction was observed from cases in which it was not observed. Such regression can be accomplished by categorical regression techniques such as logistic regression, or by maximum likelihood or Bayesian regression techniques. A complete predictive model should also provide an indication of uncertainty in the liquefaction potential. Moreover, the details of the reported uncertainty should be clearly explained – it is important that the user of a predictive model be able to properly account for both parametric uncertainty (i.e., uncertainty in the input parameters used by a model, such as penetration resistance) and model dispersion (a standard deviation term that represents the limited ability of the model to describe the physical problem under consideration; Section D.8). Model dispersion represents aleatory variability if it is not reducible by the development of better models, but would be at least partly epistemic otherwise.

9.5.4 Cyclic Stress Approach

In the 1960s and 1970s, many advances in the state of knowledge of cyclic liquefaction phenomena resulted from the research groups of H. B. Seed at the University of California at Berkeley and Kenji Ishihara at Tokyo University. This research was largely directed toward experimental evaluation of the cyclic loading conditions required to trigger liquefaction. The loading was described in terms of cyclic shear stresses, and liquefaction potential was evaluated based on the amplitude and number of cycles of earthquake-induced shear stress required to produce liquefaction. The general approach came to be known as the *cyclic stress approach.*

The cyclic stress approach is conceptually straightforward: excess pore pressure generation is assumed to be fundamentally related to the amplitude and number of cycles of shear stress applied to the soil. Therefore, the earthquake-induced loading, expressed in terms of cyclic shear stresses, is compared with the liquefaction resistance of the soil, also expressed in terms of cyclic shear stresses. Application of the cyclic stress approach, however, requires careful attention to the manner in which the loading conditions and liquefaction resistance are characterized. Historically, liquefaction resistance has typically been correlated to penetration resistance – initially SPT resistance but now more commonly CPT resistance. More recently, procedures that correlate liquefaction resistance to shear wave velocity have been proposed.

A number of empirical procedures for evaluation of liquefaction potential (Table 9.2) have been proposed, and more will undoubtedly be proposed as new approaches for model development are introduced and new earthquakes produce additional case history data. None of the procedures in Table 9.2 possess every desirable characteristic listed in Section 9.5.3.4. The intent of this section is to present the basic framework common to most procedures along with some guidance on model selection, given the reality that the models themselves will change in the years to come. This section presents selected penetration resistance- and shear wave velocity-based procedures for evaluation of liquefaction potential. The penetration test-based procedures are those proposed by Boulanger and Idriss (2012; 2015), which are consistent with fundamental principles and experimental data, and integrate CPT- and SPT-based procedures in a manner that produces a level of consistency not found in other available procedures. A shear wave velocity-based approach is also presented.

The cyclic stress ratio, or *CSR* (Equation 9.1), represents the loading parameter $\mathcal{L}$, in Figure 9.32, while the soil parameter, S (e.g., SPT- or CPT-based penetration resistance) is used to represent soil density. The resistance to liquefaction, $\mathcal{R}$, for a given S is the *cyclic resistance ratio*, or *CRR*. In a plot such as Figure 9.32, the vertical axis represents both the *CSR* (for the individual case history data points) and the *CRR* (for the curve that separates liquefaction from no-liquefaction

TABLE 9.2
Empirical Procedures for Evaluation of Liquefaction Potential

Soil Parameter	Reference	Type[a]	No. of Cases	Input Parameters
CPT	Robertson and Wride (1998); Robertson (2009)	Det.	227	PGA, σ'_{v0}, σ_{v0}, z, $\mathbf{M}$, q_c, f_s
CPT	Moss et al. (2006)	Prob.	188	PGA, σ'_{v0}, σ_{v0}, z, $\mathbf{M}$, q_c, f_s, PI, LL
CPT	B&I[b] (2015)	Prob.	253	PGA, σ'_{v0}, σ_{v0}, z, $\mathbf{M}$, q_c, f_s, FC
CPT	Saye et al. (2021)	Prob.	401	PGA, σ'_{v0}, σ_{v0}, z, $\mathbf{M}$, q_c, f_s
SPT	Youd et al. (2001)	Det.	126	PGA, σ'_{v0}, σ_{v0}, z, $\mathbf{M}$, $(N_1)_{60}$, FC, PI, LL
SPT	Cetin et al. (2018)	Prob.	210	PGA, σ'_{v0}, σ_{v0}, z, V_{s12}, $\mathbf{M}$, $(N_1)_{60}$, FC, PI
SPT	I&B, (2006; 2008; 2010) B&I[c] (2012)	Det., Prob	192	PGA, σ'_{v0}, σ_{v0}, z, $\mathbf{M}$, $(N_1)_{60}$, FC, PI
V_s	Andrus et al. (2004)	Det., Prob.	–180	PGA, σ'_{v0}, σ_{v0}, z, $\mathbf{M}$, V_{s1}, FC, K_0, age
V_s	Kayen et al. (2013)	Prob.	422	PGA, σ'_{v0}, σ_{v0}, z, $\mathbf{M}$, V_{s1}, FC, V_{s12}

[a] Det., Deterministic; Prob., Probabilistic.
[b] I&B, Idriss and Boulanger.
[c] B&I, Boulanger and Idriss.

observations). Liquefaction is then considered to occur when the loading exceeds the resistance, or when the factor of safety against triggering of liquefaction,

$$FS_L = \frac{CRR}{CSR} \tag{9.7}$$

is less than 1.0. It turns out that both the *CSR* and *CRR* are influenced by a number of factors that can vary from site to site and from point to point at a particular site. In order for the factor of safety to provide a reliable indication of liquefaction potential, it is critical that these factors be accounted for so that the *CSR* and *CRR* values used to compute it are consistent with each other.

Figure 9.34 illustrates the manner in which loading and resistance parameters can vary with depth within a soil profile. The loading, whether expressed in terms of cyclic shear stress or *CSR*, generally varies relatively smoothly with depth while the resistance (e.g., *CRR*) can vary irregularly due to stratigraphy and variations in soil density. The factor of safety drops below 1.0 when $CRR < CSR$; in this case indicating that liquefaction is expected to be triggered in a thin, shallow zone and a thick, deeper zone. As will be discussed in Section 9.6, the consequences of liquefaction are influenced by the depth, thickness, and density of layers that liquefy as well as the mechanical and hydraulic characteristics of non-liquefied layers (Cubrinovski et al., 2017).

For reasons discussed in the following sections, important factors affecting *CSR* and *CRR* include earthquake magnitude, initial vertical effective stress, and initial, static shear stress. Because these factors can vary so much from one case to another, it is common to express both *CSR* and *CRR* in terms of standardized reference values that can then be adapted to site-specific conditions by a series of adjustment factors. The factor of safety can then be computed using the site-specific or corresponding standardized values, i.e., as

$$FS_L = \frac{CRR_{M,\sigma'_{v0},\alpha}}{CSR_M} = \frac{CRR_{std}}{CSR_{std}} \tag{9.8}$$

where α is a factor (Equation 9.32) that represents the initial, static shear stress acting on the soil element of interest.

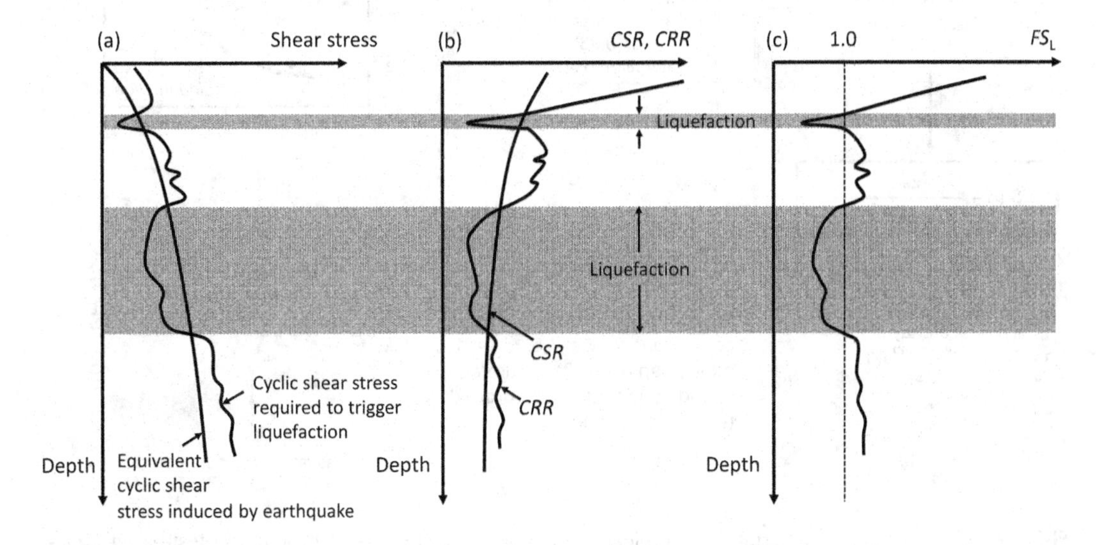

FIGURE 9.34 Schematic illustration of variation of (a) induced and resisting shear stresses, (b) cyclic stress and resistance ratios, and (c) factor of safety against triggering with depth. Shaded zone indicates depths at which liquefaction is expected to be triggered.

9.5.4.1 Basic Framework

As discussed in Section 6.4.3.6, the level of contractiveness of a soil depends on its state, i.e., the combination of density and effective stress it exists under. Penetration tests have been shown to measure quantities that correlate well to the relative density of coarse-grained soils. Descriptions of cone penetration and standard penetration tests are presented in Section 6.5.3. Both tests measure resistance to the advancement of a penetrometer into the soil. The penetrometers have sufficient volume that soil must be displaced in order for the penetrometer to move downward; this deformation involves relatively large strains beneath and adjacent to the tip of the penetrometer during which the soil may contract or dilate depending on its initial state. The measured penetration resistance is strongly affected by the degree to which the soil contracts or dilates as it is sheared. Soil characteristics that lead to increased liquefaction resistance typically also lead to increased penetration resistance, which helps explain why penetration resistance is useful for the evaluation of liquefaction potential.

Shear wave velocities are also correlated, though not as strongly, to relative density. Shear wave velocity has the advantage, however, of being measurable from the ground surface (e.g., by SASW or MASW). This feature allows shear wave velocity to be used in gravels or even coarser materials where conventional CPT or SPT testing is difficult. Shear wave velocity is related to shear modulus, which controls shear strain, and pore pressures are known to be closely related to cyclic shear strain amplitude (Section 9.5.5.2). These features have led to the use of shear wave velocity as an alternative soil parameter for characterization of liquefaction potential.

A number of adjustment factors are required to obtain site- and layer-specific values of *CSR* and *CRR*, and some are different for different *in situ* test parameters, which leads to a large number of possible combinations of factors and parameters. The basic framework of cyclic stress procedures is illustrated in Figure 9.35 and culminates in the calculation of a factor of safety, $FS_L(z)$ or, alternatively, a probability of liquefaction, $P_L(z)$. The framework separates the liquefaction potential

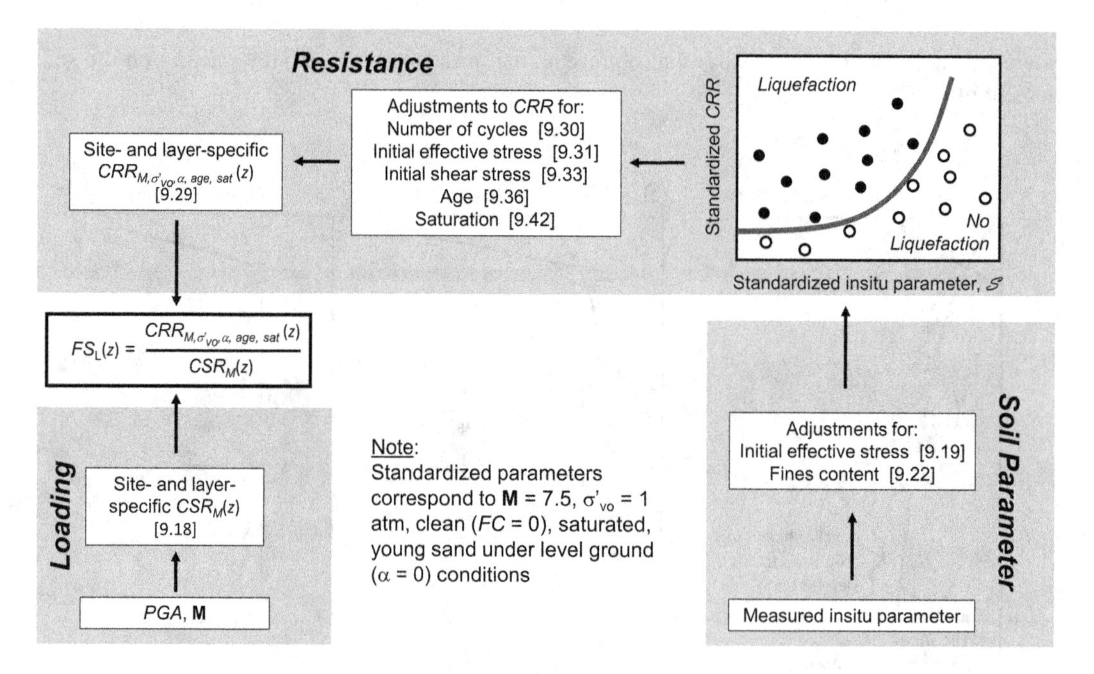

FIGURE 9.35 Schematic illustration of typical steps in a cyclic stress- and penetration test-based liquefaction potential evaluation. Evaluation of loading and resistance, with various adjustments for site-specific conditions, results in calculation of factor of safety against liquefaction. Numbers in square brackets refer to equation numbers for CPT-based evaluation described in the following sections.

evaluation into loading and resistance calculations and includes three sets of adjustments: (1) *CSR* adjustments, (2) soil parameter adjustments, and (3) *CRR* adjustments. The remainder of this section will be organized in a manner consistent with Figure 9.35. It will begin with characterization of loading, followed by characterization of *in situ* soil parameters, and then characterization of resistance, and will do so for liquefaction potential procedures based on CPT resistance, SPT resistance, and shear wave velocity.

It should be noted that numerous semi-empirical models for assessing the potential for triggering of liquefaction have been proposed and refined over the years. Each model has its own form with its own set of adjustment factors that are calibrated on its own case history database. Consequently, the adjustment factors are specific to the model; adjustment factors from different models *should not* be mixed and matched.

It should further be noted that the case history-based approach to evaluation of triggering resistance is common but not universal. Current practice in Japan largely revolves around correlation of triggering resistance to the results of laboratory tests (e.g., Tokimatsu and Yoshimi, 1983), often informed by testing of undisturbed samples obtained by ground freezing (Section 6.5.4.1). Upadhyaya et al. (2023) similarly propose a triggering model that distinguishes triggering from manifestation and is based solely on the results of cyclic laboratory tests.

9.5.4.2 Characterization of Earthquake Loading

Earthquake loading is characterized in terms of cyclic shear stress amplitude, which is related to the amplitude of the earthquake ground motion. The cyclic shear stress used in empirical procedures is the shear stress assumed to exist in the absence of pore pressure generation. While it may seem odd to evaluate liquefaction potential using stresses that do not actually exist in the soil being evaluated, it should be recognized that recorded ground motions (that would be influenced by pore pressure generation) are not available for the overwhelming majority of the case histories upon which empirical procedures are based. Therefore, the loading for case histories has historically been estimated from nearby recordings (at non-liquefied sites) or from GMMs that implicitly presume no pore pressure generation. As a result, and for consistency, the cyclic stresses in an empirical liquefaction potential evaluation should not be affected by pore pressure generation. The cyclic stresses can be predicted in two ways: by detailed, site-specific ground response analyses or with the use of a simplified approach.

Site Response Analysis

Ground response analyses (Sections 7.5–7.6) can be used to predict shear stress histories at various depths within a soil deposit. The analyses require input motions, which are typically selected and scaled or otherwise modified to be consistent with a scenario- or PSHA-derived response spectrum for the site condition (Section 4.5) of interest and with the distribution of magnitudes that contribute to that spectrum. The resulting shear stress histories should then have amplitudes and durations that are consistent with those expected for the return period of interest. Since the shear stresses used in empirical liquefaction analyses are those expected to develop in the absence of pore pressure generation, total stress site response analyses should be used. Such analyses produce shear stress histories with the transient, irregular characteristics of actual earthquake motions.

It should be noted that ground response analyses require proper site characterization, the development of representative input motions, and the use of a suitable computer program to perform the analysis. Obtaining this information requires some time and expense, but is part of a modern, high-quality liquefaction potential evaluation. As described in Chapter 7, site-specific ground response analyses can account for features of a particular site that cause it to systematically respond differently (more strongly or less strongly) than what is produced by standard (ergodic) models for site response, for example as contained in GMMs.

Simplified Method

In the early years of liquefaction potential evaluation, personal computers and graphical user interfaces were not available, so performing site-specific ground response analyses was much more difficult than it is at present. To eliminate the need for site-specific analyses, Seed and Idriss (1971) developed a "simplified method" that could be used without a site-specific ground response analysis. The basis for the simplified method is shown in Figure 9.36. In a rigid soil profile, the shear stress for a given level of peak ground acceleration would increase linearly with depth (in a soil of constant density). Since real soil profiles are not rigid, however, shear stresses will increase with depth at a different rate – one that is related to the wavelengths present in the ground motion (shorter wavelengths correspond to faster reduction with depth). In the simplified method, the compliance, or flexibility, of the soil profile is accounted for by a *depth reduction factor*, r_d, defined as shown in Figure 9.36.

The peak cyclic shear stress amplitude for the compliant soil can then be estimated (Seed and Idriss, 1971) as

$$\tau_{\max} = \frac{a_{\max}}{g}\sigma_{v0}r_d \tag{9.9}$$

where $a_{\max}$ is the peak ground surface acceleration, g is the acceleration of gravity, and σ_{v0} is the total vertical stress at the depth of interest. The cyclic stress ratio at that depth is then given by

$$CSR = 0.65\frac{a_{\max}}{g}\frac{\sigma_{v0}}{\sigma'_{v0}}r_d \tag{9.10}$$

Idriss and Boulanger (2008) used the results of analyses of multiple soil profiles subjected to multiple input motions (Golesorkhi, 1989; Idriss, 1999) to express the mean depth reduction coefficient as

$$r_d = \exp\left[\alpha(z) + \beta(z)\mathbf{M}\right] \tag{9.11}$$

where

$$\alpha(z) = -1.012 - 1.126\sin\left(\frac{z}{11.73} + 5.133\right)$$

$$\beta(z) = 0.106 + 0.118\sin\left(\frac{z}{11.28} + 5.142\right)$$

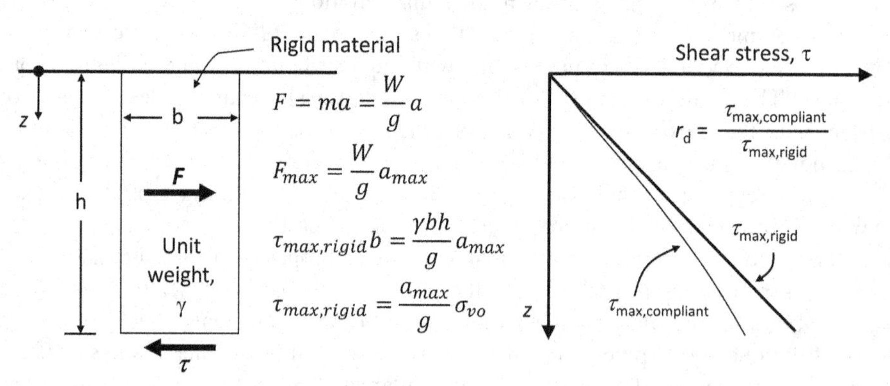

FIGURE 9.36 (a) Calculation of peak shear stress at the base of rigid slice of soil, and (b) definition of depth reduction factor, r_d.

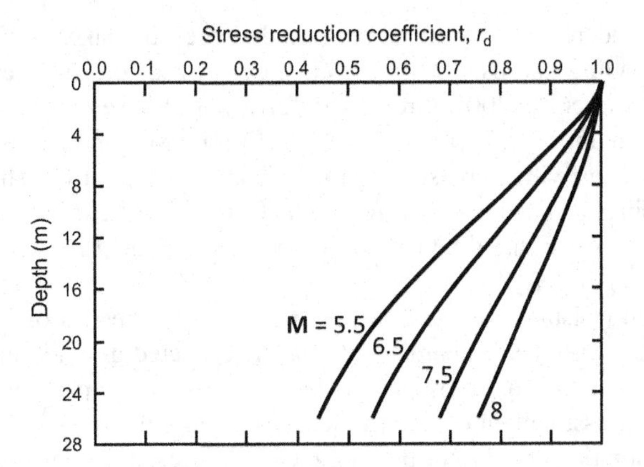

FIGURE 9.37 Variation of depth reduction factor, r_d, with depth. (After Idriss and Boulanger, 2010.)

and z is depth in meters. The resulting variation of r_d with depth is shown in Figure 9.37. The variation of r_d with magnitude accounts to some extent for the increased low-frequency (hence long wavelength) characteristics of large magnitude earthquakes; this relationship does not, however, account for shear wave velocity which also affects wavelength.

An alternative depth reduction factor model was developed from analyses of soil profiles consistent with soil profiles in the liquefaction case history database by Cetin et al. (2004) and used in the shear wave velocity model of Kayen et al. (2013),

$$r_d\left(z,\mathbf{M},a_{\max},V_{S12}\right)=\frac{\left(1+\dfrac{-23.013-2.949a_{\max}+0.999\mathbf{M}+0.0525V_{S12}}{16.258+0.201e^{0.341(-z+0.0785V_{S12}+7.586)}}\right)}{\left(1+\dfrac{-23.013-2.949a_{\max}+0.999\mathbf{M}+0.0525V_{S12}}{16.258+0.201e^{0.341(0.0785V_{S12}+7.586)}}\right)} \tag{9.12}$$

where V_{S12} is the time-averaged shear wave velocity in the upper 12 m of the site in m/sec. The uncertainty in r_d increases with depth, so the use of site-specific ground response analyses can be preferable for estimation of *CSR* at depths greater than 10–12 m. However, the soil profiles used to develop Equation (9.11) were somewhat stiffer and more uniform than those encountered in typical liquefiable soil profiles, which led to r_d values that are generally higher than those expected for typical liquefiable profiles (Cetin et al., 2004; Lasley et al., 2016). Because the liquefaction resistance models of Boulanger and Idriss (discussed subsequently) used Equation (9.11) to interpret the case histories, the effects of the bias are effectively canceled out when the same relationship is used to predict liquefaction. Some bias can be introduced, however, when site response analyses are used to compute *CSR* for profiles that differ from those used in model development

Magnitude Adjustment for Standardized CSR

As shown in Section 9.5.2.4, pore pressures in saturated soils build up incrementally with an increasing number of loading cycles in laboratory tests. The level of loading required to initiate liquefaction, therefore, is influenced by both the amplitude and duration of earthquake ground motion. In the era in which liquefaction triggering analyses were first being developed, the laboratory data from which liquefaction resistance was estimated were based on tests in which the cyclic shear stresses have uniform amplitudes. Therefore, comparison of earthquake-induced loading with laboratory-determined resistance required conversion of an irregular stress history to an equivalent series of uniform stress cycles. Seed et al. (1975a) applied a weighting procedure to a set of shear stress time histories from recorded strong ground motions to determine the number of equivalent, uniform stress cycles, N_{eq} (at an amplitude of 65% of the peak cyclic shear stress, i.e., $\tau_{cyc}=0.65\tau_{\max}$) that would produce an increase in pore pressure equivalent to that of the irregular time history.

Seed et al. (1975a) considered motions from magnitude 7.5 earthquakes to have an average of 15 equivalent cycles of loading and used $M=7.5$ as a standardized reference level for the effects of duration. Others (e.g., Liu et al., 2001; Green and Terri, 2005) have developed more refined procedures for evaluating numbers of equivalent cycles. In all cases, the number of equivalent uniform stress cycles increases with increasing earthquake magnitude (just as strong-motion duration increases with increasing earthquake magnitude). The factor of 0.65 used by Seed et al. (1975a) has been retained by custom since that time although other factors could just as easily be used (albeit with compensating changes in N_{eq}).

Because liquefaction resistance at the time was related to the results of laboratory tests, those results needed to be related to some characteristic of the expected ground motion that reflected its duration (or N_{eq}). Early liquefaction triggering analyses were typically performed for specific earthquake scenarios (i.e., the assumption of an earthquake of a particular magnitude occurring at a particular distance) and durations could not be predicted. As a result, the magnitude of the scenario earthquake was commonly used as a proxy for duration (or number of loading cycles). That approach has persisted, even though current ground motion hazards typically reflect contributions from many different magnitudes and distances that can be obtained from PSHA disaggregations (Section 4.4.3.5).

Using the concept of number of equivalent loading cycles, Seed et al. (1975a) introduced a *magnitude scaling factor* (*MSF*) that was intended to account for the effects of the duration of an actual, transient earthquake loading history on pore pressure generation. The *MSF*, which was formulated as a decreasing function of magnitude, allowed the cyclic stress, CSR_M, to be expressed as an equivalent cyclic stress ratio for a particular magnitude of interest. Therefore,

$$CSR_M = \frac{CSR}{MSF} = r_e \frac{a_{\max}}{g} \frac{\sigma_{v0}}{\sigma'_{v0}} \frac{r_d}{MSF} \tag{9.13}$$

where the equivalence ratio, r_e, is generally taken as 0.65.

Expressions for *MSF* have generally been based on the results of laboratory tests involving multiple cycles of uniform loading. Numbers of equivalent cycles are usually computed using the Palmgren-Miner cumulative damage hypothesis (Palmgren, 1924; Miner, 1945) from the results of laboratory tests performed on specimens subjected to different loading amplitudes. Assuming a cyclic strength curve can be expressed using a power law, e.g.,

$$CRR = aN^{-b} \tag{9.14}$$

the relative number of uniform loading cycles required to initiate liquefaction at two stress ratios, CSR_A and CSR_B, would be

$$\frac{N_A}{N_B} = \left(\frac{CSR_B}{CSR_A}\right)^{1/b} \tag{9.15}$$

This relationship allows the relative amounts of "damage" (i.e., pore pressure generation potential) from loading cycles of different amplitudes to be related to each other. Normalizing the amplitudes of all loading cycles by the amplitude of the strongest cycle, each cycle can be assigned a fraction of a loading cycle proportional to the damage it produces; summing the contributions of all loading cycles yields an equivalent number of loading cycles for the transient loading history. Selecting a magnitude of 7.5 provides a reference value of $N_{M=7.5}$ (generally 15) uniform loading cycles, the *MSF* can be defined as

$$MSF = \frac{CSR_M}{CSR_{M=7.5}} = \left(\frac{N_{M=7.5}}{N_M}\right)^b \tag{9.16}$$

The *MSF* relationship determined in this way is influenced by the parameter, b, which describes the slope of the cyclic strength curve (on a log-log plot of CSR vs N_L).

The *MSF* has been interpreted by various users as a loading parameter (i.e., one that increases *CSR* as **M** increases) or as a resistance parameter (that decreases *CRR* as **M** increases). Mathematically, both interpretations are equivalent with respect to their effect on the computed factor of safety (dividing the numerator in Equation (9.7) by *MSF* is the same as multiplying the denominator by the same number).

Another interpretation, however, is that magnitude is a descriptor of loading (used to account for the longer durations of ground motions caused by large magnitude earthquakes) that can also be related to liquefaction resistance (by correlation to the number of harmonic loading cycles in the laboratory test data that forms the basis for *MSF*). This alternative interpretation allows *MSF* to be defined by the laboratory-measured cyclic strength parameter slope, *b* (Equation 9.14). By basing *MSF* on *b* and considering the potential for *b* to depend on penetration resistance, the *MSF* relationship of Equation (9.16) becomes a mixed function of loading and penetration resistance. Boulanger and Idriss (2014) observed differences in the slopes of cyclic strength curves in laboratory tests on soils of different densities. This density dependence is ambiguous, however (Ulmer et al., 2018), with some testing programs showing dependence and others showing none. In order to distinguish between loading and resistance effects for conceptual purposes in this text, the Boulanger and Idriss (2014) *MSF* expression (Equation 9.16) will be written as the product of two terms — a ground motion duration-related loading term for a soil with a standardized reference penetration resistance and a resistance term that accounts for the influence of the number of harmonic loading cycles on the laboratory test-based liquefaction resistances of soils with penetration resistances that differ from the standard value. This form also allows use of laboratory-measured, soil-specific *b* parameters (Table 9.6) to be used in determination of *CRR*. The Boulanger-Idriss magnitude scaling factor, referred to here as MSF_{BI}, can then be written as

$$MSF_{BI} = MSF_L \left(\frac{MSF_{BI}}{MSF_L} \right) = MSF_L \cdot K_N \tag{9.17}$$

where the loading parameter, MSF_L, is the value of MSF_{BI} for a soil of a reference penetration resistance, taken here as $q_{c1Ncs} = 110$ (or $(N_1)_{60cs} = 15$ blows/ft), and the resistance parameter, K_N, describes the effects of number of loading cycles (via magnitude) on the liquefaction resistance of soils of different penetration resistance. With this approach, the magnitude-corrected cyclic stress ratio can be written as

$$CSR_M = 0.65 \frac{a_{max}}{g} \frac{\sigma_{v0}}{\sigma'_{v0}} \frac{r_d}{MSF_L} \tag{9.18}$$

and the K_N term can be used to adjust the cyclic resistance ratio. The values of MSF_L for CPT, SPT, and V_s-based liquefaction potential evaluation procedures are given in Table 9.3 and illustrated in Figure 9.38.

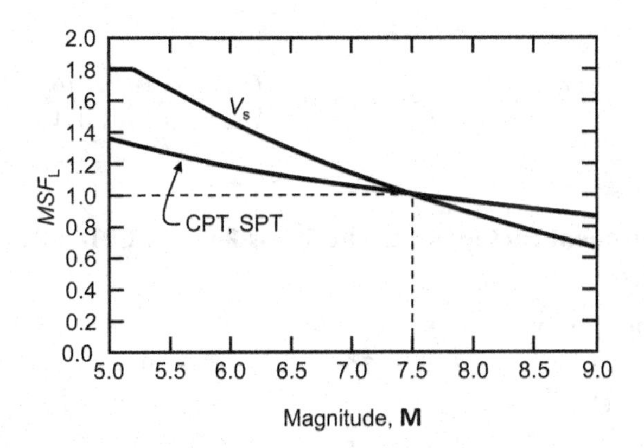

FIGURE 9.38 Magnitude scaling factor for loading.

The preceding discussion implicitly assumes free-field conditions, i.e., conditions far enough away from structures that the earthquake loading is not influenced by the presence of structures. It should be noted, however, that elements of soil beneath structures, and particularly near and beneath the corners and edges of structures, can be subjected to static shear stresses and considerably higher cyclic shear stresses induced by the inertial response of the structure itself as it moves laterally and rocks during earthquake shaking. Because of these high cyclic shear stresses and low static effective stresses (Travasarou et al., 2006), soils just outside of the footprint of a building can be subjected to very high *CSR*s; these soils may liquefy and lead to building settlement (Section 9.6.6.4) even when soils in the free-field do not.

Example 9.1

An investigation of a site located in California shows a liquefaction-susceptible layer of clean sand exists at 6–8 m depth. The liquefiable layer has a saturated unit weight of 20 kN/m³ and is overlain by 6 m of fill with a unit weight of 21.2 kN/m³. The groundwater table is at the fill/sand interface. A PSHA performed for the site indicates that a *PGA* of 0.58 g would have a return period of 2,475 years; disaggregation indicates a mean magnitude of 7.1 for that return period. Calculate the standardized *CSR* for the liquefiable sand layer, assuming it will be used with an SPT- or CPT-based triggering model.

Solution:

Representing the liquefiable layer by the conditions at its center, the total and effective stresses at 7 m depth are

$$\sigma_{v0} = (6\text{ m})(21.2\text{ kN/m}^3) + (1\text{ m})(20.0\text{ kN/m}^3) = 147.2\text{ kPa}$$

$$\sigma'_{v0} = (6\text{ m})(21.2\text{ kN/m}^3) + (1\text{ m})(20.0\text{ kN/m}^3 - 9.81\text{ kN/m}^3) = 137.4\text{ kPa}$$

Using the mean magnitude from the PSHA, the MSF is given by (Table 9.3)

$$MSF_L = 0.5803 + 2.7368\exp(-7.1/4) = 1.044$$

The depth reduction factor at 7 m depth is calculated using Equation (9.11) as

$$\alpha(z) = -1.012 - 1.126\sin\left(\frac{7.0}{11.73} + 5.133\right) = -0.421$$

$$\beta(z) = 0.106 + 0.118\sin\left(\frac{9}{11.28} + 5.142\right) = 0.047$$

$$r_d = \exp\left[\alpha(z) + \beta(z)M\right] = \exp\left[-0.421 + (0.047)(7.1)\right] = 0.916$$

Finally, from Equation (9.18)

$$CSR_{std} = 0.65\frac{a_{max}}{g}\frac{\sigma_{v0}}{\sigma'_{v0}}\frac{r_d}{MSF_L} = (0.65)(0.58)\left(\frac{147.2\text{ kPa}}{137.4\text{ kPa}}\right)\left(\frac{0.916}{1.044}\right) = 0.354$$

TABLE 9.3
Magnitude Adjustment Factors for Cyclic Stress Ratio in CPT-, SPT-, and V_s-Based Triggering Models

Model	Adjustment Parameter	Reference Value	Adjustment Factor
CPT	M	7.5	$MSF_L = 0.5784 + 2.7495\exp(-\mathbf{M}/4) \le 1.32$
SPT	M	7.5	$MSF_L = 0.5803 + 2.7368\exp(-\mathbf{M}/4) \le 1.32$
V_s	M	7.5	$MSF_L = 15\mathbf{M}^{-1.342}$

9.5.4.3 Characterization of *in situ* Soil Parameters

While liquefaction potential is known to be related to the density of the soil, it is extremely difficult to measure the *in situ* density of a liquefaction-susceptible soil without changing it by vibration or disturbance. As a result, procedures for evaluation of liquefaction potential commonly use more easily measured proxies for soil density; the most common of these are penetration resistances but shear wave velocity has also been used for this purpose. These parameters comprise the soil density parameter, $\boldsymbol{S}$, which is related to the resistance parameter, $\boldsymbol{\mathcal{R}}$, by the boundary curve in Figure 9.32.

In situ test parameters can be influenced by material characteristics and environmental factors. The material characteristics are most important for evaluation of liquefaction potential, so it is important to separate them from the environmental characteristics to the greatest extent possible. The penetration test parameters commonly used for evaluation of liquefaction potential are affected by effective stress level and fines content, so standardized values are obtained by applying effective stress and fines content adjustments. V_s is similarly adjusted for effective stress but is not adjusted for fines content in current procedures.

Effective Stress Adjustments for Soil Parameters

Penetration resistances and shear wave velocities are affected by both soil density and effective stress – for example, a loose sand at a high effective stress can have the same measured penetration resistance as a denser sand at a lower effective stress level. The different densities, however, can be distinguished by normalizing the soil parameters to a reference effective stress level, typically taken as 1 atm for liquefaction potential evaluations (Section 6.5.3.3). The effective stress-adjusted *in situ* parameters are given by

$$q_{c1N} = C_N q_{cN} = C_N \frac{q_c}{P_a} = C_N \frac{q_t - u_2(1-a_n)}{P_a} \tag{9.19}$$

$$(N_1)_{60} = C_N N_{60} \tag{9.20}$$

$$V_{S1} = C_{Vs} V_s \tag{9.21}$$

(note that CPT-based liquefaction triggering models are commonly expressed in terms of q_c rather than q_t; the two are related as indicated in Equation 6.53 and are virtually identical for sands). The effective stress adjustment factors have been developed with laboratory data and data from calibration chamber tests (Marcuson and Bieganousky, 1977a,b; Boulanger, 2003). Boulanger (2003) and Idriss and Boulanger (2003, 2008) proposed the CPT and SPT effective stress adjustment factors given in Table 9.4. Because the exponents used to compute q_{c1N} and $(N_1)_{60}$ depend on q_{c1N} and $(N_1)_{60}$, their values must be obtained iteratively.

TABLE 9.4
Effective Stress Adjustment Factors for CPT and SPT Resistances (Boulanger and Idriss, 2014) and for V_s (Kayen et al., 2013)

Procedure	Reference Value	Adjustment Factor
CPT	$\sigma'_{v0} = 1$ atm	$C_N = \left(\frac{p_a}{\sigma'_{v0}}\right)^m \leq 1.7$ where $m = 1.338 - 0.249 q_{c1Ncs}^{0.264}$
SPT	$\sigma'_{v0} = 1$ atm	$C_N = \left(\frac{p_a}{\sigma'_{v0}}\right)^m \leq 1.7$ where $m = 0.784 - 0.0768 (N_1)_{60cs}^{0.5}$
V_s	$\sigma'_{v0} = 1$ atm	$C_{Vs} = \left(\frac{P_a}{\sigma'_{v0}}\right)^{0.25}$

Fines Content Adjustments for Penetration Resistance

The presence of fines reduces the permeability and increases the compressibility of a sandy soil, both of which reduce penetration resistance when those soils are saturated. As a result, a silty sand will generally have a lower measured penetration resistance (or shear wave velocity) than a clean sand even if the sands are at the same relative density. Liquefaction resistance can also be influenced by the presence of fines. Laboratory tests have shown that the presence of fines actually decreases liquefaction resistance (Polito and Martin, 2001; Cubrinovsky et al., 2010) for soils with the same relative density. Therefore, fines content adjustments are made to account for the combined effects of fines on penetration resistance and liquefaction resistance. No corresponding adjustment is made to the normalized shear wave velocity, V_{s1}, used in shear wave velocity-based triggering models (although fines content is, as discussed subsequently, considered to affect V_s-based liquefaction resistance). Fines-adjusted penetration resistances, often referred to as the equivalent "clean sand" penetration resistances, are computed as

$$q_{c1Ncs} = q_{c1N} + \Delta q_{c1N} \tag{9.22}$$

$$(N_1)_{60cs} = (N_1)_{60} + \Delta(N_1)_{60} \tag{9.23}$$

where the penetration resistance adjustments, Δq_{c1N} and $\Delta(N_1)_{60}$, are given in Table 9.5 and illustrated in Figures 9.39 and 9.40. The "clean sand" penetration resistances given by Equations (9.22 and 9.23) represent the values that would produce the same *CRR* had the fines content of the soil been very low (<5%). Available data comes from tests on non-plastic fines and therefore apply to that condition; one would generally expect plastic fines to have greater effects than non-plastic fines, so the fines correction is generally applied to both cases with the recognition that it is likely somewhat conservative for plastic fines.

TABLE 9.5
Fines Adjustments for CPT- and SPT-Based Liquefaction Potential Evaluation Procedures

Procedure	Reference Condition	Adjustment Factor
CPT	$FC=0$	$\Delta q_{c1N} = \left(11.9 + \frac{q_{c1N}}{14.6}\right)\exp\left[1.63 - \frac{9.7}{FC+2} - \left(\frac{15.7}{FC+2}\right)^2\right]$
SPT	$FC=0$	$\Delta(N_1)_{60} = \exp\left[1.63 + \frac{9.7}{FC+0.01} - \left(\frac{15.7}{FC+0.01}\right)^2\right]$

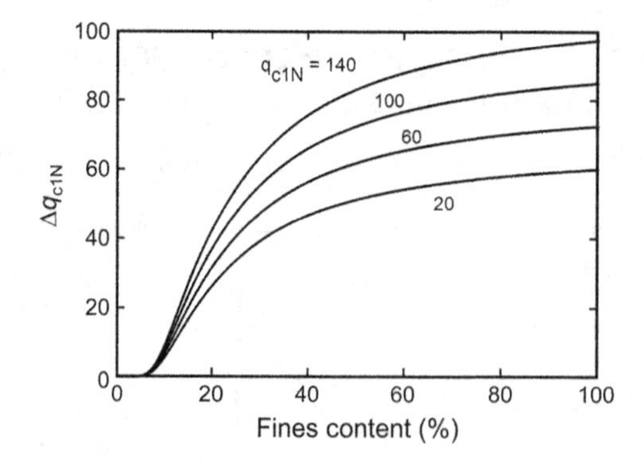

FIGURE 9.39 Fines content adjustment factor for CPT results. (After Boulanger and Idriss, 2014.)

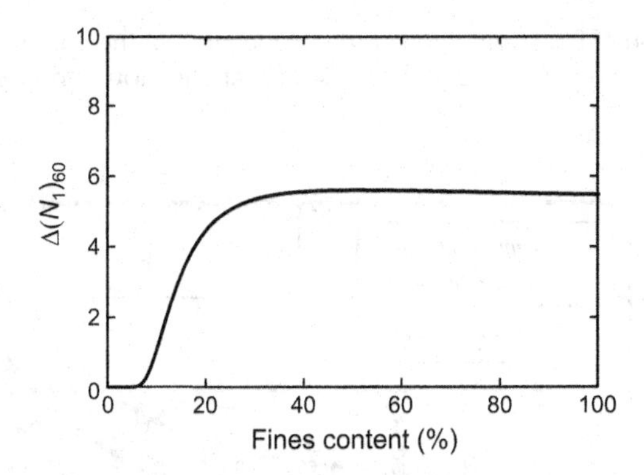

FIGURE 9.40 Fines content adjustment factor for SPT results. (After Boulanger and Idriss, 2014.)

Fines content should be measured directly whenever possible. When it cannot, it can be estimated, much more approximately, from the soil behavior type index, I_c (Section 6.5.3.3), measured in a cone penetration test. Boulanger and Idriss (2014) proposed that

$$I_c = \frac{(FC + 137)}{80} + \varepsilon \tag{9.24}$$

where ε is a normally distributed random variable with zero mean and a standard deviation of 0.29, and suggested that fines content could be estimated by inverting that equation, i.e.,

$$FC(\%) = 80(I_c + C_{FC}) - 137 \tag{9.25}$$

where C_{FC} is a fitting parameter that can be adjusted based on-site-specific data when available or taken as zero to match the general trend of available case history and laboratory test data (Figure 9.41). A value of C_{FC}=0.07, for example, approximates the relationship developed by Robinson et al. (2013) for liquefiable soils along the Avon River in Christchurch, New Zealand. The dashed lines in Figure 9.41 represent the mean ± one standard deviation values of I_c; examination of the data shows limited sensitivity of I_c to FC, which implies a high sensitivity of FC to I_c. The corresponding standard deviation of FC estimates from Equation (9.25) is 23%. An updated FC to I_c model based on a larger database in which the regression was targeted to estimate FC was presented by Hudson et al. (2024), although for consistency Eq. (9.25) should be used with the Boulanger and Idriss (2014) model.

Example 9.2

As indicated in Section 9.4.3, soils are considered to transition from sand-like to clay-like behavior at I_c values of approximately 2.5–2.7. Compute the range of fines contents that correspond to that range of I_c values for typical sandy soils.

Solution:

Using Equation (9.25) with C_{FC}=0.0,

$$FC(\%) = 80(I_c + C_{FC}) - 137 = \begin{cases} (80)(2.5) - 137 = 63\% & \text{for} \quad I_c = 2.5 \\ (80)(2.7) - 137 = 79\% & \text{for} \quad I_c = 2.7 \end{cases}$$

Thus, the apparent fines content range corresponding to the I_c-based transition range is 63%–79%. The actual soil behavior, however, depends on the nature (e.g., plasticity) of the fines so, while the

I_c-based fines correction of Equation (9.5) was developed within the context of the Boulanger and Idriss (2014) liquefaction triggering model (intended for application to non-plastic fines), its use is limited to that model.

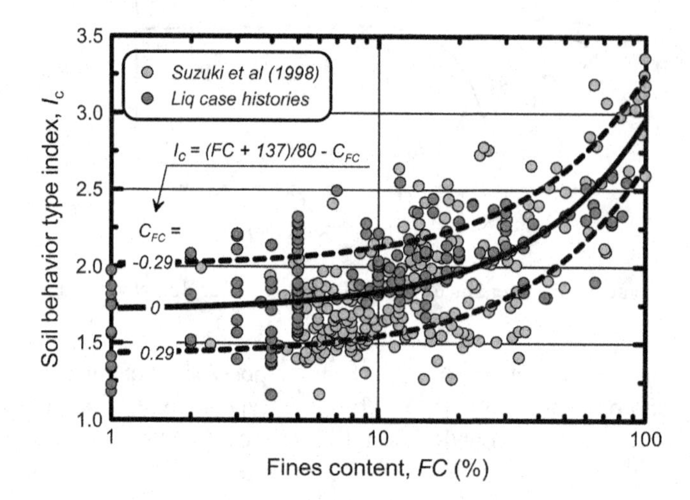

FIGURE 9.41 Data and suggested relationship for estimation of fines content from soil behavior type index. (After Boulanger and Idriss, 2014.)

9.5.4.4 Characterization of Liquefaction Resistance

In the cyclic stress method, liquefaction resistance is characterized by the *CRR*, which is correlated to the *in situ* state of the soil of interest. This correlation is affected by a number of factors that are considered in different ways by different triggering models. Liquefaction resistance is expressed in terms of a standardized *CRR* that applies to a set of reference conditions with adjustment factors that account for deviations of site-specific conditions from the reference conditions. The standardized *CRR* relationship is represented in Figure 9.32, whereby the *in situ* soil parameter, $\boldsymbol{S}$, is related to the standardized resistance parameter, $\mathcal{R}$.

Standardized Cyclic Resistance Ratio Relationships

Once the standardized penetration resistance or shear wave velocity has been determined, liquefaction resistance can be characterized in terms of a standardized *CRR*, which is equal to the *CSR* required to initiate liquefaction in a "standard" element of soil (i.e., a young, clean, saturated sand with $\sigma'_{v0}=1$ atm under level ground and subjected to shaking from a **M**7.5 earthquake). The *CRR* value for any combination of magnitude, vertical effective stress, and static shear stress that may exist at a particular site can then be related to the *CRR* value for "standard" conditions of a magnitude 7.5 earthquake, a vertical effective stress of 1 atm, and level-ground (zero initial shear stress) conditions. Standardized *CRR* values and the various adjustments required to obtain site-specific values are described for CPT-, SPT-, and V_S-based liquefaction potential procedures in the following sections.

CPT-Based Relationship

Boulanger and Idriss (2014) used both CPT and SPT data, principles of soil mechanics, and relevant experimental data within a relative state parameter index-based framework to develop deterministic standardized *CRR* relationships. The use of this framework ensures a level of consistency between CPT- and SPT-based evaluations of liquefaction potential. As the geotechnical engineering profession transitions from its historical use of SPT results for evaluation of liquefaction potential to CPT-based procedures, this approach helps protect against inconsistent

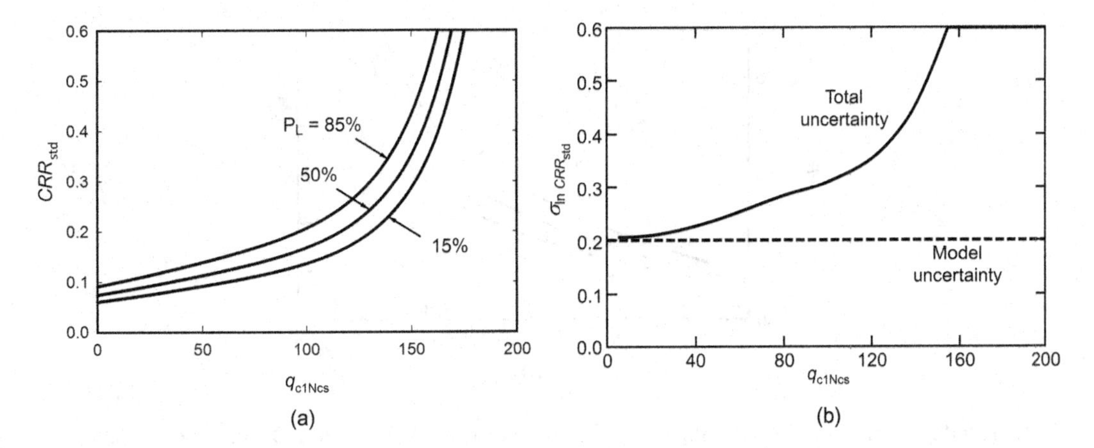

FIGURE 9.42 CPT-based liquefaction resistance (after Boulanger and Idriss (2016): (a) variation of *CRR* with corrected tip resistance for $M_w = 7.5$, $\sigma'_{v0} = 1$ atm, and $\tau_{static} = 0$, and (b) variation of model and total uncertainty in *CRR* with corrected tip resistance.

predictions of liquefaction potential at a particular site. The standardized *CRR* values based on standardized CPT resistance is given by

$$CRR_{std} = \exp\left[\frac{q_{c1Ncs}}{113} + \left(\frac{q_{c1Ncs}}{1,000}\right)^2 - \left(\frac{q_{c1Ncs}}{140}\right)^3 + \left(\frac{q_{c1Ncs}}{137}\right)^4 - 2.60 + \sigma_{\ln CRR}\Phi^{-1}(P_L)\right] \quad (9.26)$$

where P_L is the probability of liquefaction and $\sigma_{\ln CRR}$ is taken to have a minimum value of 0.2, representing model uncertainty, or larger values when total uncertainty is considered (Figure 9.42b). Boulanger and Idriss (2014) recommend that deterministic analyses be based on a cyclic resistance ratio one model uncertainty standard deviation below the mean, i.e., for $\Phi^{-1}(P_L) = -1.0$. The liquefaction resistance given by Equation (9.26) is shown graphically in Figure 9.42. The *CRR* can be seen to be sensitive to CPT resistance, particularly at q_{c1Ncs} values greater than about 125.

Example 9.3

A soil layer has an average q_{c1Ncs} value of 120. Compute the standardized *CRR* values that would have a 95% probability of exceedance considering (1) only model uncertainty, and (2) total uncertainty.

Solution:

From Equation (9.26), the median cyclic resistance ratio would be

$$CRR_{std} = \exp\left[\frac{q_{c1Ncs}}{113} + \left(\frac{q_{c1Ncs}}{1,000}\right)^2 - \left(\frac{q_{c1Ncs}}{140}\right)^3 + \left(\frac{q_{c1Ncs}}{137}\right)^4 - 2.60\right]$$

$$= \exp\left[\frac{120}{113} + \left(\frac{120}{1,000}\right)^2 - \left(\frac{120}{140}\right)^3 + \left(\frac{120}{137}\right)^4 - 2.60\right] = \exp[-1.565] = 0.209$$

The standard normal variate for a 95% probability of exceedance (or a 5% probability of non-exceedance) is $\Phi^{-1}(0.05) = -1.645$. From Figure 9.42b, the model and total standard deviations are 0.20 and 0.35, respectively. The standardized *CRR* values for the two uncertainty levels are then

$$CRR_{std} = \begin{cases} \exp[-1.565 + (-1.645)(0.20)] = 0.150 & \text{considering only model uncertainty} \\ \exp[-1.565 + (-1.645)(0.35)] = 0.118 & \text{considering total uncertainty} \end{cases}$$

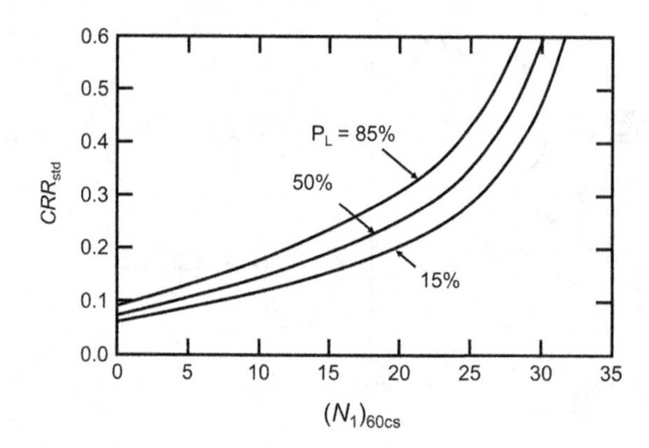

FIGURE 9.43 Variation of *CRR* with resistance parameter for M_w=7.5, σ'_{v0}=1 atm, and τ_0=0. (After Boulanger and Idriss, 2012.)

SPT-Based Procedure

The standardized, SPT-based cyclic resistance ratio of Idriss and Boulanger (2008) and Boulanger and Idriss (2014) is given by

$$CRR_{std} = \exp\left[\frac{(N_1)_{60cs}}{14.1} + \left(\frac{(N_1)_{60cs}}{126}\right)^2 - \left(\frac{(N_1)_{60cs}}{23.6}\right)^3 + \left(\frac{(N_1)_{60cs}}{25.4}\right)^4 - 2.60 + \sigma_{\ln CRR}\Phi^{-1}(P_L)\right] \quad (9.27)$$

and illustrated graphically in Figure 9.43. A value of $\sigma_{\ln CRR}$=0.20 was recommended by Boulanger and Idriss (2014) and used with $\Phi^{-1}(P_L)$=–1.0 for the deterministic *CRR* value shown in Figure 9.43.

V_s-Based

The potential for triggering of liquefaction has also been correlated to shear wave velocity. There are a number of attractive features to this approach – shear wave velocity is commonly measured, it can be measured from the ground surface using techniques such as SASW and MASW, it can be measured in materials, such as gravelly and cobbly soils, for which conventional penetration testing is difficult, and shear wave velocity is related to shear modulus, which relates ground motion amplitude to shear strain amplitude, which fundamentally controls pore pressure development. The drawbacks are that shear wave velocity describes the very low-strain characteristics of the soil, which, although correlated to higher-strain behavior, may not accurately represent the characteristics of the soil at the strain levels associated with triggering of liquefaction. Shear wave velocities are known to be relatively insensitive to relative density, a characteristic of critical importance in determining the contractiveness of a soil and also to be influenced by particle size in granular soils. Many methods of shear wave velocity measurement have significant uncertainty and provide limited spatial resolution since they average velocities over significant depth intervals and/or horizontal distance ranges. Shear wave velocity is also insensitive to the presence of fines, at least up to the limiting fines content, which is typically about 35% (Section 6.3).

An early V_s-based procedure in the cyclic stress-based approach to evaluation of liquefaction potential (Andrus and Stokoe, 2000; Andrus et al., 2004) took the same basic form as the penetration resistance-based procedures with a magnitude-adjusted *CSR* and an effective stress-adjusted, V_s-based *CRR* that is also adjusted for age. The procedure indicated an upper bound of V_{s1}=215 m/sec for clean sands ($FC \leq 5\%$) decreasing linearly to 200 m/sec for $FC \geq 35\%$. A subsequent model derived from an expanded database of 422 liquefaction case histories was developed (Kayen et al., 2013) by making shear wave velocity measurements at sites in the SPT (Cetin et al., 2004) and

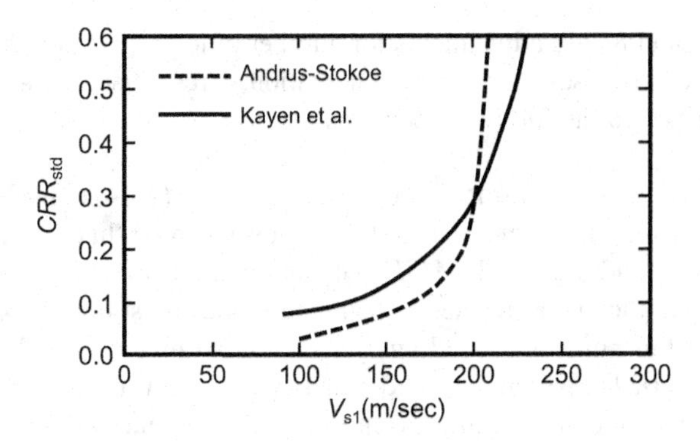

FIGURE 9.44 Median standardized cyclic resistance ratio according to Kayen et al. (2013) compared to M7.5 curve of Andrus and Stokoe (2000).

CPT (Moss et al., 2006) catalogs. In that model, *CSR* is defined as in Section 9.5.4.2 (with r_d from Equation 9.12), and the standardized *CRR* can be taken as,

$$CRR_{std} = \exp\left[\frac{-5.2726+\left(0.0073V_{s1}\right)^{2.8011}+0.0028FC-0.4809\Phi^{-1}\left(P_L\right)}{1.946}\right] \tag{9.28}$$

where V_{s1}=corrected shear wave velocity (Equation 9.21) and FC=fines content in percent. Unlike the Andrus and Stokoe relationship, the Kayen et al. (2013) *CRR* relationship (Figure 9.44) does not imply a specific upper bound V_{s1} value above which liquefaction does not occur.

Uncertainty in CRR

There are many sources of uncertainty in a liquefaction hazard evaluation, and in the calculation of liquefaction potential using the cyclic stress approach. The total uncertainty can be divided into parametric uncertainty, i.e., uncertainties in the values of the parameters that go into the calculations, and model dispersion, which results from the inability of the model to accurately and completely represent the physical system of interest (Section D.8). Parametric uncertainty can be evaluated by considering the uncertainties in parameters in all of the case histories in the database upon which the liquefaction procedure was developed (Cetin et al., 2002, 2004). The developers of the CPT- and SPT-based models described in the preceding sections did not feel that individual case history data was sufficient to characterize parametric uncertainty, so those models do not account explicitly for parametric uncertainty, e.g., uncertainty in CPT tip resistance, or any of the adjustment factors that affect *CSR* and *CRR*. Accordingly, the standard deviation term of $\sigma_{lnCRR} = 0.2$ given above represents model dispersion that includes contributions from both model uncertainty and irreducible model variability. For application to a particular site, the influences of both parametric uncertainty and model dispersion should be carefully considered to form a more complete picture of the total uncertainty (e.g., Figure 9.42b) in evaluation of the triggering of liquefaction.

Site-Specific Cyclic Resistance Ratio Adjustment Factors

The standardized *CRR* values (i.e., $CRR_{std} = CRR_{M=7.5,\sigma'_{v0}=1\text{ atm},\alpha=0,young,sat}$) presented in the preceding section apply to the specific conditions of $\mathbf{M}$=7.5, σ'_{v0}=1 atm, and τ_{static}=0 for young, saturated soils, which rarely exist in a given element of soil at a particular site. Deviations from standard conditions are accounted for using *CRR* adjustment factors so that a site- and layer-specific *CRR* can be expressed as

$$CRR_{M,\sigma'_{v0},\alpha,age,sat} = CRR_{std}\cdot K_N\cdot K_\sigma\cdot K_\alpha\cdot K_{DR}\cdot K_S \tag{9.29}$$

where K_N, K_σ, K_α, K_{DR}, and K_S are adjustments for number of loading cycles, effective stress, initial shear stress, diagenetic processes (e.g., age, cementation, stress history), and saturation, respectively, which are described in the following sections.

Number of Loading Cycles Adjustment Boulanger and Idriss (2014) considered soils of different density to be affected differently by number of loading cycles (note that the slopes of the curves in Figure 9.26 are not equal) and proposed a *MSF* that combined loading and resistance effects. Using the approach described in the preceding section on magnitude adjustment for cyclic stress ratio, the density-dependent effect of number of loading cycles on resistance can be described by an adjustment factor, $K_N = MSF_{BI}/MSF_L$, which can be defined as indicated in Table 9.6 and shown graphically in Figure 9.45. The value of K_N can be seen to reflect differences in the sensitivity of *CRR* to magnitude (as a proxy for number of loading cycles) for soils with CPT and SPT resistances other than 105 or 15, respectively. Table 9.6 also provides an expression for MSF_{max} when laboratory testing has established the value of the *b*-parameter for a specific soil. In the case of the V_S-based model of Kayen et al. (2013), and most other liquefaction triggering models in the literature, $K_N = 1.0$ (i.e., the magnitude adjustment is made in the demand parameter through the MSF_L term (Table 9.2).

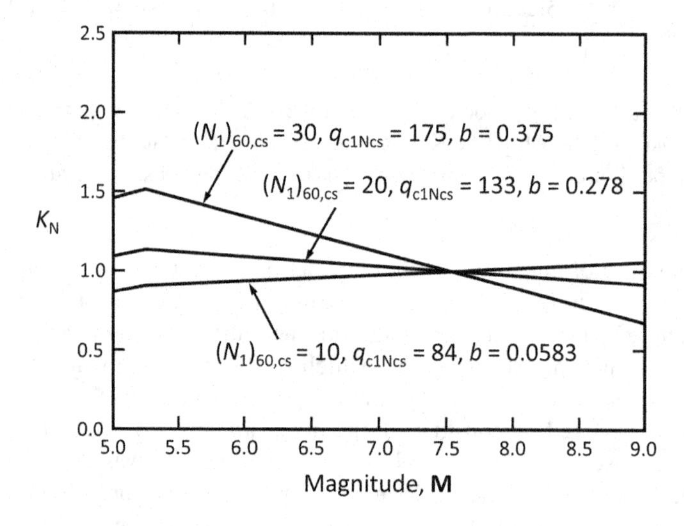

FIGURE 9.45 Magnitude adjustment factor for cyclic resistance ratio. Kinks in curves at low magnitude are due to limiting values in original Boulanger and Idriss relationship.

TABLE 9.6
Magnitude Adjustments to *CRR* for CPT- and SPT-Based Liquefaction Potential Evaluation Procedures

Procedure	Reference Condition	Adjustment Factor
CPT	$\mathbf{M} = 7.5$ $q_{c1Ncs} = 105$	$MSF_{max} = 1.09 + \left(\frac{q_{c1Ncs}}{180}\right)^3 \le 2.2$
SPT	$\mathbf{M} = 7.5$ $(N_1)_{60,cs} = 15$	$MSF_{max} = 1.09 + \left(\frac{(N_1)_{60cs}}{31.5}\right)^3 \le 2.2$
Lab	$\mathbf{M} = 7.5$ $b = 0.144$	$MSF_{max} = 0.65(20)^b \le 2.2$

$$K_N = \frac{\min\left\{1+(MSF_{\max}-1)\left[8.64\exp(-\mathbf{M}/4)-1.325\right],2.2\right\}}{0.5803+2.7368\exp(-\mathbf{M}/4)} \tag{9.30}$$

Effective Stress Adjustment As shown in Section 6.4.3.2, soils tend to become more contractive with increasing effective confining stress. Based on laboratory data and a relative state parameter index-based framework, and assuming a reference effective stress of 1 atm, Boulanger and Idriss (2014) proposed the effective stress adjustment factor for CPT- and SPT-based procedures given in Table 9.7. A similar factor is available for the V_s-based model from Kayen et al. (2013); that factor, however, was obtained from data regression and is close to 1.0, which implies much less sensitivity to effective stress than is generally observed in laboratory tests.

For the CPT- and SPT-based models, the upper bound on K_σ was a modeling decision made by Boulanger and Idriss (2014) motivated by a lack of data at low effective stress levels (below about 0.4 atm) but is inconsistent with expected soil behavior. The variation of K_σ with vertical effective stress is illustrated in Figure 9.46.

TABLE 9.7
Effective Stress Adjustments to *CRR* for CPT-, SPT-, and V_S-Based Liquefaction Potential Evaluation Procedures

Procedure	Reference Condition	Adjustment Factor	
CPT	$\sigma'_{v0}=1$ atm	$K_\sigma = 1-C_\sigma \ln\left(\sigma'_{v0}/P_a\right) \le 1.1$ where $C_\sigma = \frac{1}{37.3-8.27(q_{c1Ncs})^{0.264}} \le 0.3$	(9.31a)
SPT	$\sigma'_{v0}=1$ atm	$K_\sigma = 1-C_\sigma \ln\left(\sigma'_{v0}/P_a\right) \le 1.1$ where $C_\sigma = \frac{1}{18.9-2.55\sqrt{(N_1)_{60cs}}} \le 0.3$	(9.31b)
V_s	$\sigma'_{v0}=1$ atm	$K_\sigma = \left(\sigma'_{v0}/p_a\right)^{-0.0099}$	(9.31c)

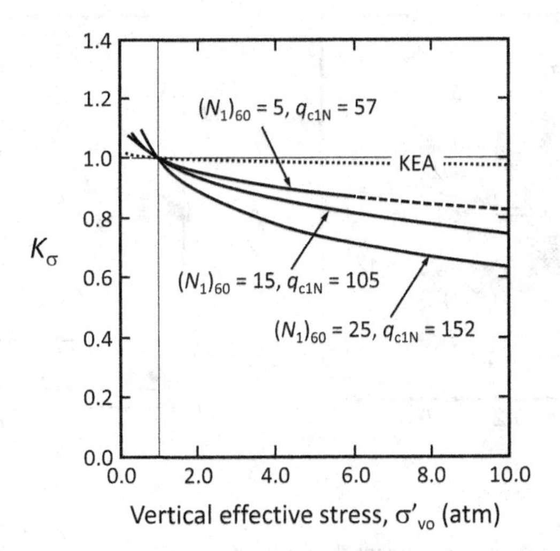

FIGURE 9.46 Effective stress adjustment for *CRR* (after Boulanger and Idriss, 2014). KEA13 indicates a relationship inferred from Kayen et al. (2013.)

Static Shear Stress Adjustment The presence of an initial, static shear stress can affect pore pressure generation in liquefaction-susceptible soils as discussed in Section 9.5.2.4. Laboratory studies show that pore pressures increase more rapidly when shear stress reversals occur, i.e., when the soil goes through an instant in which no shear stress exists. When a transient cyclic shear stress history is added to an existing static shear stress, fewer stress reversals may occur than would if the static shear stress did not exist; if the static shear stress is greater than the peak cyclic shear stress, no stress reversal will occur and initial liquefaction (zero effective stress) cannot occur. For very loose soils, however, increasing shear stress tends to bring the effective stress path closer to the FLS where pore pressures tend to increase more quickly. Thus increased static shear stress can tend to reduce the liquefaction resistance of very loose soils while it increases the liquefaction resistance of denser soils (Figure 9.47a).

The initial shear stress is usually characterized in terms of a static shear stress ratio,

$$\alpha = \frac{\tau_h}{\sigma'_{v0}} \tag{9.32}$$

Adjustments for the effects of initial static shear stresses depend on the level of static stress and the state of the soil. Using cyclic simple shear test data (Vaid and Finn, 1979; Boulanger et al., 1991). Boulanger (2003) developed an expression for the adjustment factor, K_α, as a function of the relative state parameter index (Section 6.4.3.6),

$$K_\alpha = a + b\exp\left(\frac{-\xi_R}{c}\right) \tag{9.33}$$

where

$$a = 1267 + 636\alpha^2 - 634\exp(\alpha) - 632\exp(-\alpha)$$

$$b = \exp\left(-1.11 + 12.3\alpha^2 + 1.31\ln(\alpha + 0.0001)\right)$$

$$c = 0.138 + 0.126\alpha + 2.52\alpha^3$$

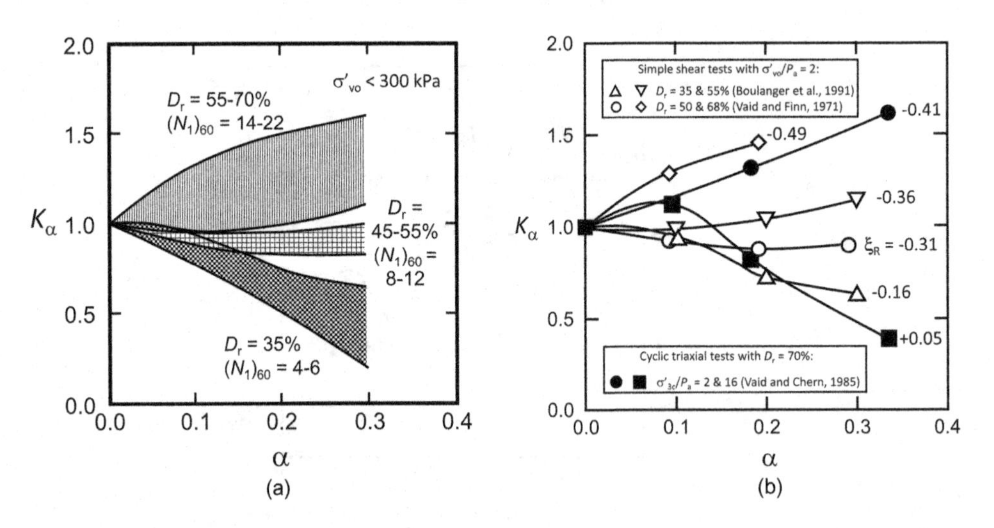

FIGURE 9.47 Static shear stress adjustment factor for *CRR* of sands expressed in terms of: (a) relative density (Harder and Boulanger, 2007 with permission of MCEER) and (b) relative state parameter index (Boulanger, 2003 with permission of ASCE).

which is illustrated in Figure 9.47b. The value of K_α can be computed from CPT and SPT resistances using the correlations

$$\xi_R = \frac{1}{Q - \ln\left(\dfrac{100(1+2K_0)\sigma'_{v0}}{3P_a}\right)} - \left(0.086\sqrt{Q_{tn}} - 0.334\right) \quad (9.34a)$$

and

$$\xi_R = \frac{1}{Q - \ln\left(\dfrac{100(1+2K_0)\sigma'_{v0}}{3P_a}\right)} - \sqrt{\frac{(N_1)_{60}}{46}} \quad (9.34b)$$

The adjustment factor for static shear stress is, compared to other adjustment factors, based on a limited amount of laboratory test data.

Age Adjustment Liquefaction has been observed to occur more frequently in younger soil deposits such as man-made fills and Holocene natural deposits than in older (e.g., Pleistocene and older) deposits (Youd and Hoose, 1977; Youd and Perkins, 1978; Obermeier et al., 1990; Lewis et al., 1999). Mechanisms such as cementation (chemical precipitation), particle reorientation under weak shaking in small earthquakes or other vibrations and increased lateral stresses may play roles in increased liquefaction resistance, but the actual mechanisms are not well understood (Mitchell, 1986, 2008; Schmertmann, 1991). Laboratory investigations of age effects are difficult because of the time involved and the difficulty of replicating aging mechanisms in the laboratory, but limited laboratory testing has shown that liquefaction resistance increases with age (Figure 9.48).

Using both field and laboratory test data, Hayati and Andrus (2009) proposed the use of an age-related deposit resistance adjustment factor, K_{DR}, intended to account for diagenetic processes (e.g., age, cementation, stress history) that affect low-strain behavior but are destroyed by larger strain deformations.

$$K_{DR} = 0.13\log t + 0.83 \quad (9.35)$$

where t is time since deposition (or last critical disturbance) in years. This relation gives a value of $K_{DR} = 1.0$ at a reference time of 23 years, which was considered appropriate since many

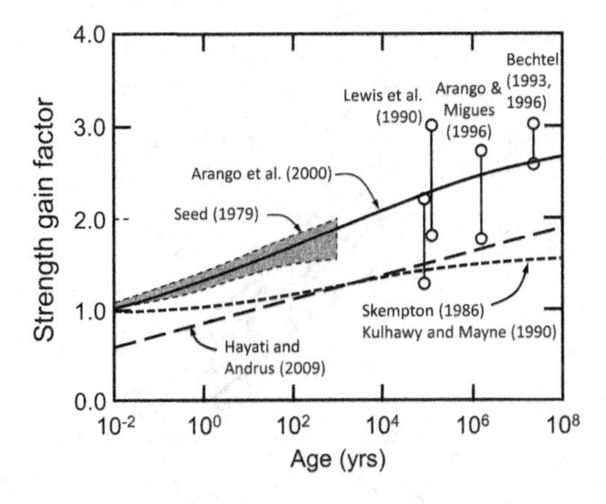

FIGURE 9.48 Relationship between strength gain and age. (Modified from Arango et al., 2000 with permission of Elsevier Science and Technology Journals). Strength gain factor is relative to strengths obtained from laboratory tests performed very shortly after specimen preparation except for that of Hayati and Andrus (2009) where the inclusion of laboratory data suggested a reference age of 23 years.

liquefaction case histories are associated with deposits with ages of 1–100 years at the time of shaking. The deposit resistance factor is then applied to the reference-age CRR as

$$CRR_{aged} = CRR_{ref} K_{DR} \tag{9.36}$$

Since the actual age of a soil deposit is usually unknown, and because it can be reset to zero after liquefaction caused by a previous earthquake, Hayati and Andrus (2009) proposed that the value of K_{DR} could be obtained from shear wave velocity measurements. Since the aging processes thought to increase liquefaction resistance also tend to increase shear wave velocity, and available shear wave velocity correlations are based primarily on "young" sands, the *measured to estimated velocity ratio, MEVR*, defined as

$$MEVR = \frac{V_{s1,measured}}{V_{s1,estimated}} \tag{9.37}$$

was taken as an indicator of age with $V_{s1,estimated}$ obtained by correlation to CPT or SPT resistance using Equation (6.77). The value of $MEVR$ can then be used to estimate K_{DR} as

$$K_{DR} = 1.08 \cdot MEVR - 0.08 \tag{9.38}$$

Schneider and Moss (2011) introduced the notion of a normalized rigidity index,

$$K_G = \frac{G_{\max}}{q_t} Q_{tn}^{0.75} \tag{9.39}$$

where $G_{\max}$ = maximum shear modulus (typically obtained from measured shear wave velocity), q_t is given in Equation (6.53) and has the same units as $G_{\max}$, and Q_{tn} is as defined in Equation (6.61). Soils that are very stiff at low strain levels relative to their penetration resistance (which is mobilized at high strain levels) tend to have high K_G values. Schneider and Moss (2011) showed laboratory and field data indicating that older soils exhibited higher K_G values than younger soils (Figure 9.49); Holocene sands tended to have K_G values in the range of 110–330 (with a median of 215) and aged, cemented, and calcareous sands had values of 330–1,100. Based on frozen samples of Holocene and Pleistocene soils (Roy, 2008), Schneider and Moss (2011) postulated that soils with higher K_G values (e.g., Pleistocene sands, which had a median K_G=408) had a higher liquefaction resistance than Holocene sands for q_{c1N} values less than about 150. Robertson (2016) proposed a modified version of the normalized rigidity index,

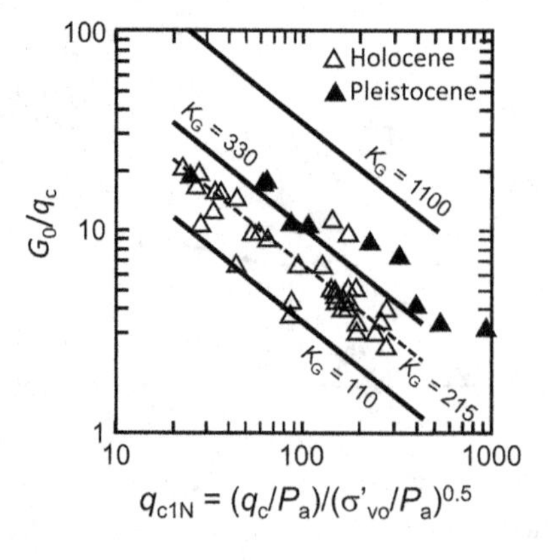

FIGURE 9.49 Correlations between normalized cone tip resistance and maximum shear modulus measured in seismic cone tests for various soil deposits. (Schneider and Moss, 2008 with permission of Emerald Publishing Limited.)

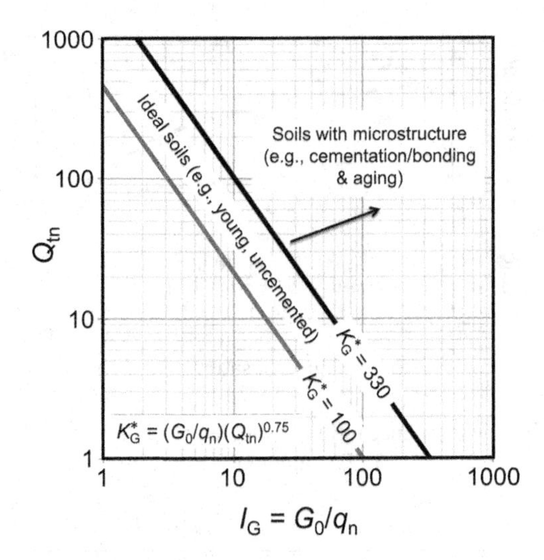

FIGURE 9.50 Proposed chart for identifying soils with microstructure. (Robertson, 2016 with permission of P.K. Robertson.)

$$K_G^* = \frac{G_{\max}}{q_n} Q_{tn}^{0.75} \tag{9.40}$$

where $q_n = q_t - \sigma_v$. Robertson used the term *microstructure* to describe post-depositional particle-scale features such as cementation, secondary compression, thixotropy, cold welding, and particle reorientation that tend to increase with the age of a soil deposit and found that soils with microstructure had K_G^* values greater than 330 and soils without had lower K_G^* values (Figure 9.50). Since they depend on essentially the same factors, *MEVR* and K_G^* should be related and Robertson suggested that

$$MEVR = \sqrt{\frac{K_G^*}{200}} \tag{9.41}$$

While laboratory studies and empirical evidence leave little question that liquefaction resistance increases with age, the data used to quantify that effect are still sufficiently sparse that the age correction is not commonly used in practice. It could be used in a logic tree framework, however, with a weighting factor interpreted as a subjective probability (i.e., a degree-of-belief) that will likely approach 1.0 as further evidence of age effects is brought to light.

Saturation Adjustment Seasonal or tidal fluctuations in groundwater level, or the generation of gases due to decaying organic material, can lead to zones of soil that are not completely saturated below the water table. As discussed in Section 6.2.4, the air bubbles that exist in the voids of partially saturated soils are so much more compressible than the water in the voids that the air/water mixture has a very high composite compressibility. The tendency for contraction of the soil skeleton, therefore, is accommodated by simple compression of the air bubbles, which occurs without producing significant pore pressure.

Cyclic triaxial tests on silty sands from Japan (Tsukamoto et al., 2014) showed that they generated pore pressure during the tests, but that their liquefaction resistances increased quickly when the degree of saturation dropped below 100% (Figure 9.51). Zhang et al. (2016) used laboratory data from tests on Toyoura sand and Nevada sand, along with a constitutive model, to investigate the effects of saturation on liquefaction resistance and developed a design chart for those soils

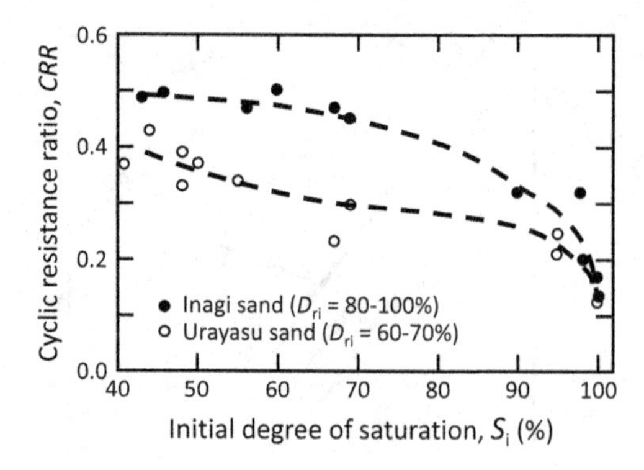

FIGURE 9.51 Variation of cyclic resistance ratio with degree of saturation for Inage sand and Urayasu sand. (Tsukamoto et al., 2014 with permission of Y. Tsukamoto.)

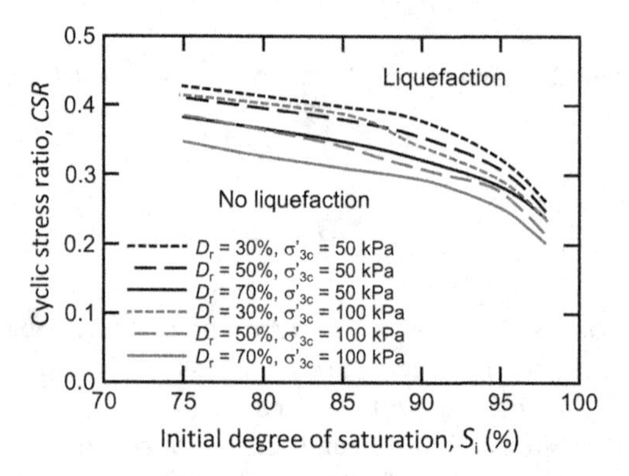

FIGURE 9.52 Proposed design chart for liquefaction resistance of unsaturated sand. (Zhang et al., 2016 with permission of ASCE.)

(Figure 9.52). These laboratory results give an indication of the sensitivity of liquefaction resistance to degree of saturation.

The propagation of p-waves through saturated sands occurs through the porewater rather than the softer soil skeleton (Section 6.5.2). As indicated in Section 6.5.2, the p-wave velocity of water (at 20°C) is 1,460 m/sec, so measured p-wave velocities lower than that value can indicate partial saturation. Laboratory tests on five liquefiable soils with mean grain sizes ranging from 0.04 to 1.25 mm and relative densities from 40% to 100% indicated (Hossain et al., 2013) that partial saturation increased liquefaction resistance by a saturation adjustment factor

$$K_S = \frac{CRR_{unsat}}{CRR_{sat}} = \begin{cases} \left\{0.95 + e^{\left[2-1.95\ln(V_p/100-1.3)\right]}\right\}\left(\dfrac{\mathbf{M}}{8}\right)^{(100-D_r)/\left[\left(0.03D_r^2+1.31D_r\right)\exp(2.85B)\right]} & \text{for } V_p < 1{,}400 \text{ m/sec} \\ 1.0 & \text{for } V_p \geq 1{,}400 \text{ m/sec} \end{cases} \tag{9.42}$$

where $B = \left[(V_p/V_s)^2 - 3.5\right]/\left[(V_p/V_s)^2 - 4/3\right]$ (Kokusho, 2000a). This adjustment, shown in Figure 9.53, is similar to that of Ishihara et al. (2001). Data from clean and silty sands from Christchurch (Baki et al., 2023) showed generally consistent behavior at p-wave velocities greater than 500 and 750 m/sec, respectively; the resistance of the silty sand increased very quickly at

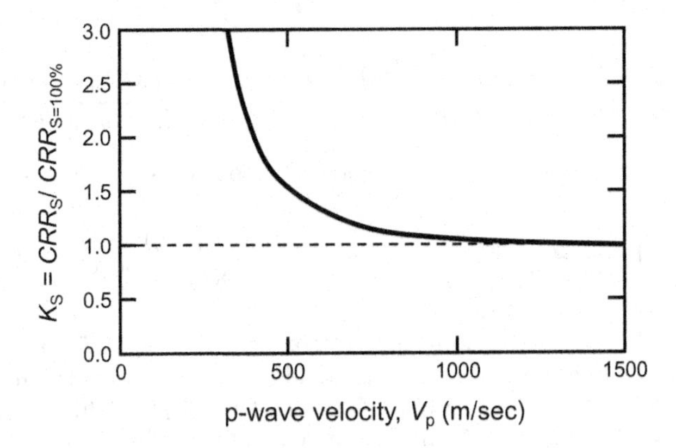

FIGURE 9.53 Variation of saturation adjustment factor, K_S, with p-wave velocity for **M**8 event.

p-wave velocities lower than 750 m/sec. As in the case of the age adjustment, the saturation adjustment is not explicitly used in typical current practice, as measurements of both V_p and V_s are not commonly available, and it is generally conservative to assume that all soils below the water table are saturated. In critical situations, or where the costs associated with the assumption of saturation are excessive, a saturation adjustment supported by field measurements can provide a more accurate indication of liquefaction potential.

Example 9.4

A 4-m-thick layer of dense clayey sand fill with a moist unit weight of 21.5 kN/m³ is underlain by a natural layer of silty sand at 4–5 m depth at a level-ground site. The silty sand has an average q_{c1Ncs} value of 120 and a dry unit weight of 17.5 kN/m³. Groundwater was encountered at the bottom of the fill. Historical records indicate that the silty sand layer was deposited during a flood 100 years ago; recent tests indicate that the p-wave velocity of the layer is 1,420 m/sec. Compute the expected cyclic resistance ratio of the silty sand layer when subjected to ground motion with a peak ground surface acceleration of 0.34g from a **M**6.8 earthquake.

Solution:

Calculating the vertical total and effective stress at the center of the silty sand layer, the dry unit weight of 17.5 kN/m³ corresponds to a saturated unit weight of 20.6 kN/m³ and a buoyant unit weight of 10.8 kN/m³ so the vertical stresses are

$$\sigma_{v0} = (4\text{ m})(21.5\text{ kN/m}^3) + (0.5\text{ m})(20.6\text{ kN/m}^3) = 96.3\text{ kPa}$$

$$\sigma'_{v0} = (4\text{ m})(21.5\text{ kN/m}^3) + (0.5\text{ m})(10.8\text{ kN/m}^3) = 91.4\text{ kPa}$$

From Equation (9.26), the median standardized cyclic resistance ratio is 0.209 (see calculation in Example 9.2). The *CRR* value for the actual conditions in the silty sand layer is calculated using the standardized values and the applicable adjustment factors. For this case, adjustments are needed for number of loading cycles, overburden stress, initial static shear stress, diagenetic (age-related) factors, and saturation.

The number of cycles correction is computed from Equation (9.30). The magnitude correction, using the decoupled version of MSF_{BI}, is expressed in terms of the number of cycles adjustment factor, K_N. From Table 9.6, it is calculated as,

$$MSF_{\max} = 1.09 + \left(\frac{q_{c1Ncs}}{180}\right)^3 = 1.09 + \left(\frac{120}{180}\right)^3 = 1.386$$

Then

$$K_N = \frac{1+(MSF_{\max}-1)[8.64\exp(-M/4)-1.325]}{0.5803+2.7368\exp(-M/4)} = \frac{1+(1.386-1)[8.64\exp(-6.8/4)-1.325]}{0.5803+2.7368\exp(-6.8/4)} = 1.016$$

The overburden adjustment factor (Table 9.7) can be computed as

$$K_\sigma = 1 - \frac{\ln(\sigma'_{vo}/P_a)}{37.3-8.27(q_{c1Ncs})^{0.264}} = 1 - \frac{\ln(91.4\text{ kPa}/101.3\text{ kPa})}{37.3-8.27(120)^{0.264}} = 1.013$$

The site is level, so the static shear stress adjustment factor, $K_\alpha = 1.0$. With no information other than the estimated age of the layer, the age-related adjustment factor (Equation 9.38) can be estimated as $K_D = 0.13\log t + 0.83 = 0.13\log(100) + 0.83 = 1.09$. Finally, the saturation adjustment factor (Equation 9.42), $K_S = 1.0$ since $V_p > 1{,}400$ m/sec. Therefore, the cyclic resistance ratio is

$$CRR_{M,\sigma'_{v0},\tau_{static},age,saturation} = CRR_{std} \cdot K_N \cdot K_\sigma \cdot K_\alpha \cdot K_{DR} \cdot K_S = (0.209)(1.016)(1.013)(1.0)(1.09)(1.0) = 0.234$$

9.5.4.5 Implications of Non-Triggering

As seen in laboratory and physical model tests, shear strains and deformations remain relatively small until pore pressures become very high and liquefaction is initiated. If liquefaction is not initiated, i.e., if $FS_L > 1.0$, some pore pressure may still exist at the end of shaking, and its dissipation may lead to some degree of settlement. The amount of excess pore pressure at level-ground sites can be estimated from laboratory tests as shown in Figure 9.54. Other procedures for estimation of pore pressure generation are based on cyclic stresses (Park and Ahn, 2013) cyclic strains (Dobry et al., 1982; Matasovic and Vucetic, 1993; Cetin and Bilge, 2012) and dissipated energy (Polito et al., 2008).

9.5.4.6 Limitations of the Simplified Method

The simplified method for evaluation of liquefaction potential has been widely and successfully used in engineering practice for over 50 years. It has been updated a number of times as large earthquakes have occurred and produced valuable case histories and as new research, tests, and

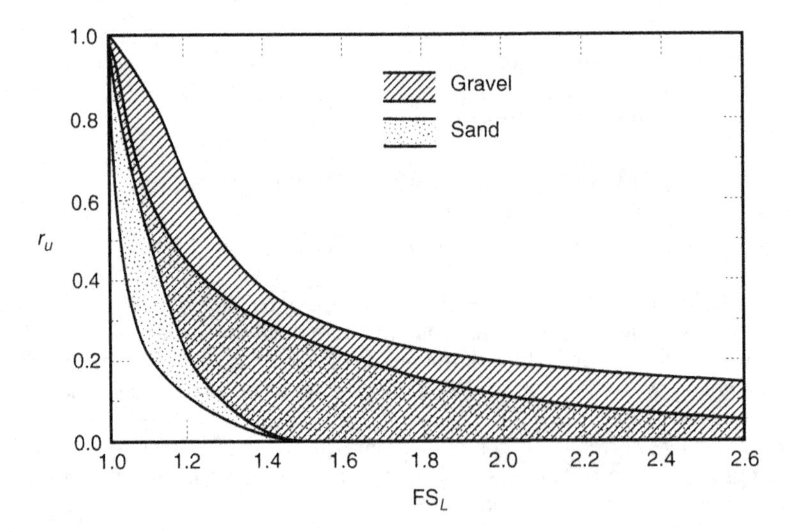

FIGURE 9.54 Relationship between residual excess pore pressure ratio and factor of safety against liquefaction for level-ground sites. (After Marcuson and Hynes, 1990.)

interpretive tools have advanced understanding of its basic mechanics. It remains limited, however, by a number of factors including: (a) the binary representation of triggering by the presence or absence of surface manifestation to represent the complex mechanical and hydraulic phenomena that occur beneath the surface during and after earthquake shaking, (b) the characterization of loading by a *PGA* value expected to have occurred in the absence of any pore pressure generation, a magnitude value that is loosely correlated to duration or number of loading cycles because path and site effects are not directly considered, and by the generic (i.e., ergodic) representation of the variation of shear stress with depth, (c) the historical characterization of case history sites by a single critical layer when multiple layers may contribute to the development, or prevention, of surface manifestation, and (d) the implicit assumption, in both case history interpretation and forward prediction, that all layers act independently of each other.

Most of these limitations are relics of the available data and measurement, interpretation, and computational capabilities that existed when the simplified method was first developed. Advances in strong motion instrumentation, field reconnaissance tools, ground motion modeling, site characterization, constitutive modeling, (nonlinear, effective stress) ground response analysis, and database management allow improved characterization and interpretation of liquefaction case histories and should pave the way for improved empirical triggering models in the near future.

9.5.5 Cyclic Strain Approach

As discussed in Section 9.5.2.2, stress-controlled cyclic tests showed that the *CRR* used to characterize liquefaction resistance in the cyclic stress approach was sensitive to a variety of *in situ* factors (age, soil fabric, prior shaking history, etc.) that could not be replicated in reconstituted laboratory test specimens. Subsequent strain-controlled testing showed that the volumetric strain of dry sands (Silver and Seed, 1971; Youd, 1972; Martin et al., 1975) and pore pressure generation in saturated sand (Dobry and Ladd, 1980; Dobry et al., 1982; Cetin and Bilge, 2012) were not significantly affected by those factors. These observations led to the now widespread recognition that pore pressures in liquefiable soils are more closely related to shear strain amplitude than shear stress amplitude.

In an effort to develop a more robust approach to the liquefaction problem, Dobry and Ladd (1980) and Dobry et al. (1982) described an approach that used cyclic strains rather than cyclic stresses to characterize earthquake-induced loading and liquefaction resistance. The cyclic strain approach has both advantages and disadvantages relative to the cyclic stress approach.

9.5.5.1 Characterization of Earthquake Loading

In the cyclic strain approach, earthquake-induced loading is expressed in terms of the amplitude of a series of equivalent constant strain loading cycles. As in the cyclic stress approach, a transient history, in this case of shear strain, must be converted to an equivalent series of uniform cycles. This leads to two primary difficulties in characterizing loading in the cyclic strain approach.

First, an equivalent cyclic shear strain amplitude must be estimated. This can be accomplished by performing a ground response analyses, but it should be recognized ground response analyses predict strains with much higher variability than they predict stresses, particularly for highly nonlinear conditions. The increased uncertainty in the shear strain induced in the soil by earthquake shaking reduces the benefits of decreased uncertainty in pore pressure generation given some level of cyclic shear strain. Dobry et al. (1982) proposed a simplified method for estimating the amplitude of the uniform cyclic strain from the amplitude of the uniform cyclic stress of Equation (9.9):

$$\gamma_{cyc} = 0.65 \frac{a_{\max}}{g} \frac{\sigma_v r_d}{G(\gamma_{cyc})} \tag{9.43}$$

where $G(\gamma_{cyc})$ is the secant shear modulus of the soil at $\gamma = \gamma_{cyc}$. Since γ_{cyc} appears on both sides of Equation (9.43), the value of $G(\gamma_{cyc})$ must be obtained iteratively from a measured G_{max} profile and appropriate modulus reduction curve (Section 6.6.3). At low strain levels, where the soil may exhibit relatively linear behavior, shear strains can be predicted more accurately when stiffnesses are obtained from measured shear wave velocities.

The second difficulty relates to the determination of the number of equivalent loading cycles. The conversion procedure is analogous to that used in the cyclic stress approach. The equivalent number of strain cycles, N_{eq}, depends on the earthquake magnitude. The implied consistency between number of stress cycles and number of strain cycles, however, may not exist (Carter and Seed, 1988; Green and Terri, 2005).

9.5.5.2 Characterization of Liquefaction Resistance

Despite its difficulties in characterizing loading, the cyclic strain approach has two significant advantages in characterizing liquefaction resistance. First, the existence of the volumetric threshold shear strain provides a screening level (Section 9.5.3.2) of response below which no pore pressure generation should be anticipated (if strains are too small to produce volumetric strain under drained conditions, they will be too small to produce excess pore pressure under undrained conditions). Second, laboratory tests have shown excess pore pressure to be closely related to strain amplitude, and to be quite insensitive to many of the factors known to significantly affect the shear stress amplitude required to initiate liquefaction. Figure 9.55a shows the pore pressure ratio produced by ten cycles of strain-controlled loading on two different sands prepared by three different methods at three different initial effective confining stresses. Figure 9.55b shows pore pressure ratios following stress-controlled loading cycles plotted against the peak shear strain produced by that loading; the different symbols represent different methods of specimen preparation, and the test conditions range from relative densities of 35%–100% and initial effective stresses of 40–400 kPa. The insensitivity of the generated pore pressure to factors other than cyclic strain amplitude illustrated in Figure 9.50 is a hallmark of the cyclic strain approach. The distribution of pore pressure ratio for a given maximum strain amplitude is relatively narrow, again indicating a close relationship between pore pressure and strain amplitude. Cyclic simple shear tests (e.g., Bhatia, 1980; Finn, 1981, Dobry et al., 1982; Stamatopoulos et al., 1999) and centrifuge tests (Sharp et al., 2000; Adalier and Elgamal, 2005) have shown, however, that the cyclic strain amplitude required to initiate liquefaction in a given number of cycles increases with increasing overconsolidation ratio. Preshaking or prestraining, either under drained conditions or under undrained conditions followed by pore pressure dissipation, can cause effects similar to those of overconsolidation and also increase liquefaction resistance (Finn et al., 1970; Seed, 1979; Bhatia, 1980; Finn, 1981; El-Sekelly et al., 2016).

In the cyclic strain approach, laboratory-based curves of the types shown in Figure 9.55 are used to characterize the liquefaction resistance of the soil. Note that initial liquefaction ($r_u = 1.0$) is reached at shear strains ranging from about 0.5% to as much as 50% in the tests shown in Figure 9.55.

Using a modulus reduction curve for sand, relationships between G_{max} and V_s, and the definitions of secant shear modulus and cyclic stress-based *CSR*, a relationship between *CRR*, V_{s1}, and cyclic shear strain, γ_c, can be established (Dobry and Abdoun, 2011; 2015). By comparing these relationships to the Andrus and Stokoe (2000) shear wave velocity-based *CRR* curve, the stress-based *CRR* curve was interpreted as a curve of approximately constant cyclic shear strain at $\gamma_c \approx 0.03\%$ for V_{s1} values less than about 160 m/sec. While a boundary curve drawn for that strain level was found to separate cases of liquefaction and non-liquefaction for case histories of uncompacted clean and non-plastic silty sand fills (*FC* up to about 34%) extracted from the Andrus and Stokoe (2000) database, the inferred triggering strain level is considerably lower than the strains corresponding to triggering in laboratory tests (e.g., in Figure 9.55). An investigation of natural silty sands with non-plastic fines from the Imperial Valley of California suggested that liquefaction triggered at shear strains of 0.1%–0.2%. On that basis, Dobry et al. (2015) suggested that a strain-based *CRR* of

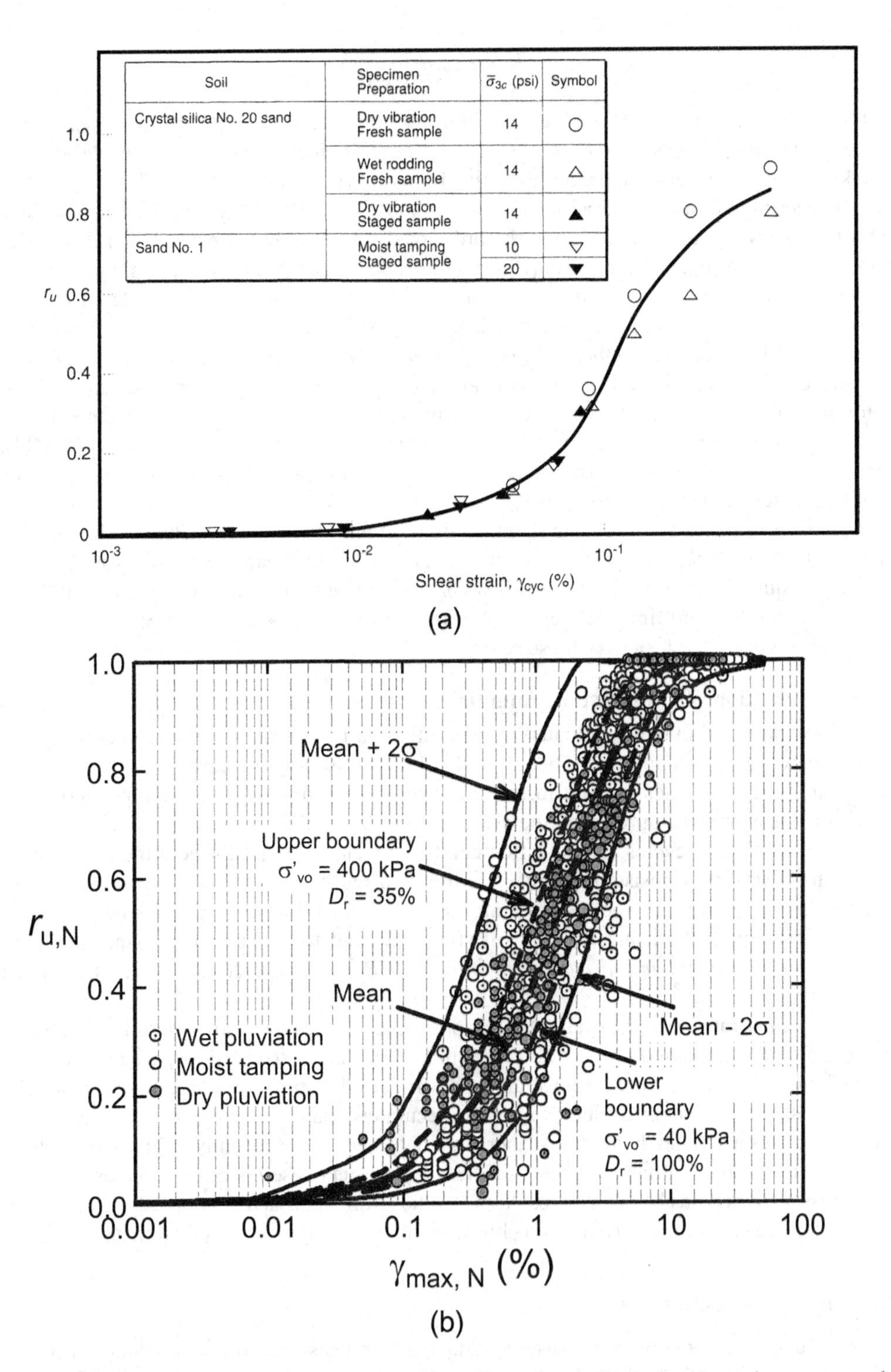

FIGURE 9.55 Variation of pore pressure ratio with cyclic strain amplitude for (a) specimens prepared by different methods and subjected to strain-controlled loading (after Dobry and Ladd, 1980), and (b) specimens prepared by different methods at different densities and effective stress levels and subjected to stress-controlled loading (after Cetin and Bilge, 2012 with permission of ASCE).

$$CRR = a \cdot MSF\left(\frac{V_{s1}}{100}\right)^2 \tag{9.44}$$

could be used at $V_{s1} < 200$ m/sec with $a = 0.033$ for uncompacted fills and 0.065 for Imperial Valley silty sands. The differences in resistance were indicated as being at least partially explained by the preshaking of the silty sands in the highly active Imperial Valley area.

The triggering strain corresponding to $CRR = 0.3$ (V_{S1} >~ 200 m/s) for the Andrus and Stokoe (2000) *CRR* curve is much higher – on the order of 0.27% to 0.56% - but still much lower than the values observed in laboratory tests (Dobry and Abdoun, 2015). These materials generally correspond to denser, overconsolidated, preshaken, or compacted sands. The large difference between the triggering strain levels measured in the laboratory and inferred from field case histories was attributed to differences in numbers of loading cycles (greater in the field than in the laboratory tests), two-directional (field) vs. one-directional (laboratory) shaking, and pore pressure redistribution (exists in field but not in laboratory tests). Corrections for the first two of these factors indicated that the triggering strain level for uncompacted fills would increase from 0.03% to 0.06%–0.12%. The upward flow of porewater during shaking, as measured in six centrifuge tests, was postulated to explain the remainder of the difference.

The apparent variability of the triggering strain level with V_{s1}, its inconsistency with laboratory data, and its dependence on hydraulic as well as ground motion parameters suggest that further research is required to better clarify its utility for the evaluation of liquefaction potential. It does, however, provide a useful link between penetration test-based and shear wave velocity-based triggering procedures, at least for very loose soils.

9.5.5.3 Evaluation of Liquefaction Potential

Liquefaction potential may be evaluated in the cyclic strain approach in a manner similar to that used in the cyclic stress approach. The cyclic loading imposed by the earthquake, characterized by the amplitude of a series of N uniform strain cycles, is compared with the liquefaction resistance, which is expressed in terms of the cyclic strain amplitude required to initiate liquefaction in the same number of cycles. Liquefaction can be expected at depths where the cyclic loading exceeds the liquefaction resistance. Since loading and resistance are characterized in terms of strains rather than stresses, the cyclic strain approach does not yield a stress-based factor of safety against liquefaction, but the ratio of strain demand to strain at triggering of liquefaction can be viewed as a *demand-capacity ratio*, or *DCR*, as often used in structural earthquake engineering.

The primary advantage of the cyclic strain approach derives from the strong relationship between pore pressure generation and cyclic strain amplitude. For a given soil, excess pore pressure can be predicted more accurately from cyclic strains than from cyclic stresses. As previously stated, however, cyclic strains are more difficult to predict accurately than cyclic stresses. The cyclic strain approach, at present, relies heavily on laboratory data and, more recently, high-quality physical (centrifuge) modeling, but has not been validated against field case histories to the extent that the cyclic stress approach has been. As a result, the cyclic strain approach is not used as commonly as the cyclic stress approach in current geotechnical earthquake engineering practice.

9.5.6 Energy Dissipation Approach

The use of dissipated energy as a measure of liquefaction resistance offers a number of potential advantages; it is related to both cyclic stresses and cyclic strains, it is a scalar quantity that reflects duration as well as amplitude, it can be related to fundamental earthquake parameters, and it can be related to inherently stochastic earthquake ground motions in a way that methods based on peak ground motion parameters alone cannot.

The densification of dry soil under cyclic loading involves rearrangement of grains under stress and hence the expenditure of energy. As a cyclically loaded dry soil densifies and approaches its minimum void ratio, the amount of energy required to further rearrange individual soil grains increases. If the soil is saturated, however, the tendency for densification causes the pore pressure to increase and the interparticle contact forces to decrease. As these contact forces decrease, the amount of energy needed to rearrange soil grains decreases. By combining these observations, Nemat-Nasser and Shokooh (1979) developed a simple, unified theory that related densification under drained conditions and pore pressure generation under undrained conditions to dissipated energy.

Davis and Berrill (1982) and Berrill and Davis (1985) built upon this idea, characterizing the energy density arriving at a particular site by means of the total energy released by the earthquake (Equation 2.4) adjusted for geometric spreading and crustal damping, and the rate of pore pressure generation as a function of SPT resistance. Calibration against field case history data yielded an expression for pore pressure ratio

$$r_u = \frac{120 A^{0.5} 10^{0.75\mathbf{M}}}{r\bar{N}^{1.5} \left(\sigma'_{vo}\right)^{0.75}} \tag{9.45}$$

where A is a normalized attenuation function that accounts for material damping, $\mathbf{M}$ is earthquake magnitude, r is the distance to the center of energy release, $\bar{N}$ is average corrected SPT resistance, and σ'_{v0} is initial vertical effective stress in kPa. Equation (9.45) combines loading and resistance to provide a direct prediction of pore pressure ratio rather than the factor of safety most commonly used to characterize liquefaction potential. This simple fundamental procedure has been supplanted by more detailed recent procedures.

9.5.6.1 Characterization of Earthquake Loading

Kokusho (2013) developed an energy-based procedure that explicitly characterized earthquake loading and liquefaction resistance. Two alternative procedures for estimating the energy density in a soil deposit were proposed. Both are based on the incoming energy, i.e., that associated with upward-traveling shear waves, in order to eliminate the effects of destructive interference between upward- and downward-traveling waves that can lead to the appearance of weak shaking at certain depths in a soil profile.

The first procedure, which can be used when ground motions at the base (taken as approximately 100 m depth) are not known, estimates the upward-traveling energy density in soil layer i using the Gutenberg-Richter relationship for total energy release (Equation 2.4), a spherical geometric spreading relationship, and the ratio of soil layer to base layer impedance, i.e.,

$$E_{u,i} = \left[\frac{(\rho V_s)_i}{(\rho V_s)_{base}}\right]^{0.7} \frac{10^{1.5\mathbf{M}+1.8}}{4\pi R^2} \tag{9.46}$$

where R is the distance to the center of energy release.

The second approach, which assumes that base motions are known or estimated and can be used to perform site response analyses, computes the upward-traveling energy density as

$$E_{u,z}(t) = (\rho V_s)_z \int_0^t \dot{u}_{u,z}^2 \, dt \tag{9.47}$$

where $(\rho V_s)_z$ = specific impedance of soil at depth, z, and $\dot{u}_{u,z}$ = particle velocity of upward-traveling wave at depth, z. The upward-traveling wave can be separated from the downward-traveling wave by means of equivalent linear analyses (as half the outcrop motion at a particular depth).

9.5.6.2 Characterization of Liquefaction Resistance

Soil specimens liquefied in cyclic triaxial tests show a consistent relationship between pore pressure ratio and normalized dissipated energy for a wide range of relative densities (Kokusho, 2013). Dissipated energy was computed by integration of stress-strain data up to the point of triggering (and thus includes the softening effects of pore pressure generation). Combining triaxial test data with SPT correlations, Kokusho (2013) described the normalized energy corresponding to 5% double-acting strain in 20 cycles of loading as

$$\frac{\Delta W}{\sigma_c'} = \frac{\int \sigma_d \, d\varepsilon_a}{\sigma_c'} = 0.032 - 0.48 R_{L20} + 2.40 R_{L20}^2 \tag{9.48}$$

where σ_c' is the isotropic effective stress during consolidation and R_{L20} is the cyclic resistance ratio at 20 cycles of loading, which is related to SPT resistance by

$$R_{L20} = \begin{cases} 0.0882\sqrt{N_1/1.7} & \text{for } N_1 < 14 \\ 0.0882\sqrt{N_1/1.7} + 1.6\times10^{-6}(N_1 - 14)^{4.5} & \text{for } N_1 \geq 14 \end{cases} \tag{9.49}$$

subject to a minimum value of R_{L20}=0.1, where N_1 is the effective stress-adjusted SPT resistance. Figure 9.56a illustrates the variation of normalized dissipated energy with SPT resistance for three strain levels.

Using the 253 case histories compiled by Boulanger and Idriss (2014) and a different procedure from Kokusho (2013) for defining ΔW and the triggering of liquefaction, Ulmer et al. (2023) developed a probabilistic CPT- and energy-based triggering model of the form

$$P_L = 1 - \Phi\left[\frac{1.224\times10^{-7}\cdot q_{c1Ncs}^{3.352} - 7.52 - \ln\left(\Delta W/\sigma_{vo}'\right)}{1.590}\right] \tag{9.50}$$

Dissipated energy was computed as the product of a total stress-interpreted, single-cycle work increment based on a representative shear stress amplitude and the number of equivalent cycles of loading (and consequently unaffected by pore pressure generation). The variation of normalized energy required to trigger liquefaction with CPT tip resistance is illustrated in Figure 9.56b.

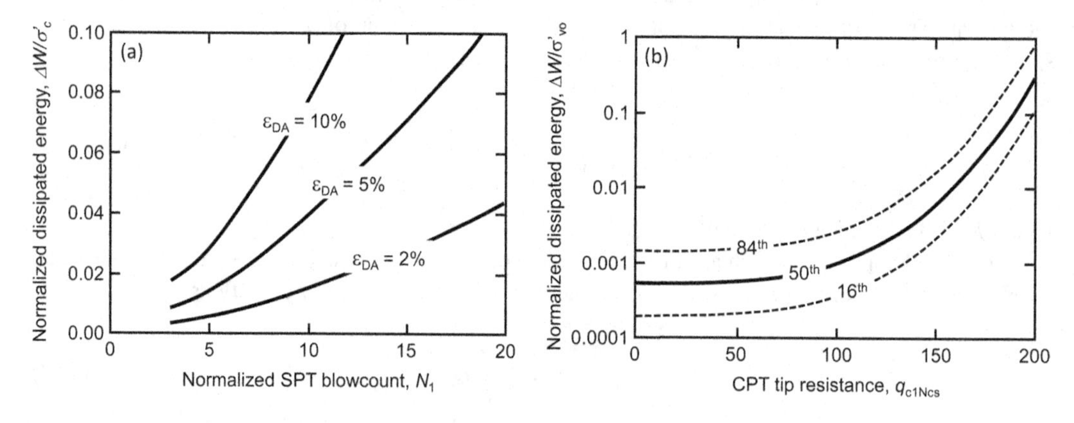

FIGURE 9.56 (a) Relationship between normalized dissipated energy to reach strain levels of 2%, 5%, and 10% and SPT resistance (after Kokusho, 2013 with permission of Canadian Science Publishing) and (b) percentiles of normalized energy required to trigger liquefaction by method of Ulmer et al. (2023). Note that the normalized energies for the two relationships shown here are defined differently.

TABLE 9.8
Expressions for energy density (J/m³) required to trigger liquefaction in various laboratory studies

Expression	Basis	Reference
$\log W = 2.002 + 0.00477\sigma'_{m0} + 0.0116D_r$	27 strain-controlled torsional shear tests on Reid Bedford sand	Figueroa et al. (1994)
$\log W = 2.4597 + 0.00448\sigma'_{m0} + 0.0115D_r$	30 centrifuge tests on Reid Bedford sand	Dief and Figueroa (2001)
$\log W = 2.1028 + 0.004566\sigma'_{m0} + 0.005685D_r + 0.001821FC - 0.02868C_u + 2.0214D_{50}$	284 cyclic triaxial, torsional, and simple shear tests on various sands	Baziar and Jafarian (2007)
$\log W = \frac{5}{4}\left(2\frac{\sigma'_{m0}}{300}\cdot\frac{D_r+40}{150}+\frac{D_r+40}{150}\cdot\frac{D_{50}}{0.5} \times\left(\frac{\sigma'_{m0}}{300}+\frac{D_{50}}{0.5}-\left(3\frac{\sigma'_{m0}}{300}-6\frac{FC+40}{150}+\frac{C_u}{1.5}\right)^2 - 1\right)+2\right)$	283 cyclic triaxial, torsional, and simple shear tests on various sands plus 18 centrifuge tests	Alavi and Gandomi (2012)

Source: After Alavi and Gandomi (2012).
C_u, coefficient of uniformity; D_{50}, mean grain size (mm); D_r, initial relative density (%); FC, fines content (%); σ'_{m0}, initial mean effective stress (kPa).

Several investigators have used laboratory test results to determine the dissipated energy density required to trigger liquefaction in the soils that were tested. Several representative relationships are listed in Table 9.8. These relationships characterize the soil in terms of mean effective stress and relative density, so an empirical correlation between relative density and some form of measured *in situ* parameter (e.g., $(N_1)_{60}$, q_{c1}, V_{s1}) would be required for practical application.

9.5.6.3 Evaluation of Liquefaction Potential

Liquefaction potential can be characterized in terms of an energy-based factor of safety defined as the ratio of energy capacity to energy demand. While interesting from the standpoint of the theoretical basis established by Nemat-Nasser and Shokooh (1979), the dissipated energy approach suffers from the difficulty in predicting the energy demands imposed on a particular element of soil at some depth below the ground surface, and on the lack of a direct empirical correlation between liquefaction resistance (in terms of energy capacity) and commonly measured *in situ* test parameters.

9.5.7 Effective Stress Response Analysis Approach

As discussed in Section 7.5.3.3, nonlinear effective stress site response models have the ability to predict pore pressure generation in potentially liquefiable layers within specific soil profiles subjected to specific input motions. In such analyses, the loading applied to the soil is computed on a site- and motion-specific basis rather than characterizing site response by a parameter (r_d) based on the average response of many different profiles, and the motion by relatively crude and simplistic parameters (a_{max} and **M**). Thus, for sites with strong impedance contrasts, velocity inversions, or other atypical features, or for motions with atypical features such as directivity pulses or very long (or short) durations, effective stress analyses can provide insight into liquefaction behavior that empirical procedures cannot. The accuracy of an effective stress analysis, however, depends on how accurately pore pressure generation, redistribution, and dissipation can be modeled.

Several approaches to the modeling of pore pressure generation have been proposed and implemented into nonlinear site response analyses. These range from relatively simple models in which pore pressures increase monotonically to sophisticated constitutive models that account for phenomena such as phase transformation (PT) behavior that cause pore pressures and shear moduli to fluctuate even within individual loading cycles.

9.5.7.1 Pore Pressure Models

Cyclic nonlinear stress-strain models (Section 6.4.5.2) use an empirical backbone curve and a series of unloading-reloading rules that govern cyclic behavior. Pore pressure prediction is accomplished by pore pressure models (e.g., Martin et al., 1975; Ishihara and Towhata, 1980; Finn and Bhatia, 1981; Dobry et al., 1985; Vucetic and Dobry, 1988) that can predict the generation of pore pressure under irregular cyclic loading conditions. In these models, the computed pore pressure is used to degrade, or soften, the backbone curve as the effective stress (and soil stiffness) decreases. Pore pressure models based on cyclic stresses, cyclic strains, and dissipated energy have been proposed.

Cyclic Stress-Based Pore Pressure Models

Based on cyclic simple shear data (De Alba et al., 1976), Seed et al. (1975) developed an expression for the pore pressure ratio produced by uniform amplitude harmonic loading

$$r_u = \frac{1}{2} + \frac{1}{\pi}\sin^{-1}\left[2\left(\frac{N}{N_L}\right)^{1/\alpha} - 1\right] \tag{9.51}$$

where N=number of loading cycles, N_L=number of cycles to liquefaction, and α=parameter related to soil properties and test conditions with an average value of 0.7. Polito et al. (2008) proposed that α be determined as a function of relative density, fines content, and cyclic stress ratio, i.e.,

$$\alpha = 0.5058 + 0.01166FC + 0.007397D_r + 0.01034CSR \tag{9.52}$$

where FC and D_r are in percent. This type of model suffers from the need to convert actual, transient loading histories to equivalent uniform loading histories, and the need to define the triggering of liquefaction in terms of a specific number of equivalent cycles. It predicts pore pressures that increase monotonically from the initial hydrostatic values.

Martin et al. (1975) proposed that increments of vertical strain from dry sands could be related to increments of pore pressure in saturated sands when subject to undrained loading. In a drained test, slip at the soil grain contacts produces an increment of contractive vertical strain, $\Delta\varepsilon_{vd}$ (Figure 6.111). Under undrained conditions, however, some of the vertical stress resisted by the soil skeleton is transferred to the more incompressible porewater, effectively unloading the soil skeleton and producing an increment of rebound volumetric strain, $\Delta\varepsilon_{vr}$. If the pore water is taken to be incompressible relative to the soil skeleton, the two volume change increments are equal in magnitude but opposite in sign,

$$\Delta\varepsilon_{vr} = -\Delta\varepsilon_{vd} \tag{9.53}$$

If the tangent constrained modulus in one-dimensional rebound is taken as $M_r = -d\sigma'/d\varepsilon_v$, and $du = -d\sigma'$, then $\Delta\varepsilon_v = \Delta u / M_r$. This allows the pore pressure increment under undrained conditions to be related to the vertical strain increment that would occur under drained conditions as:

$$\Delta u = -M_r\Delta\varepsilon_{vd} \tag{9.54}$$

This approach was used to develop the first simple models that could predict pore pressure generation under irregular cyclic loading.

Cyclic Strain-Based Pore Pressure Models

Dobry et al. (1982) used the results of cyclic torsional triaxial tests to predict the pore pressure ratio after N cycles of strain-controlled loading $r_{u,N}$. Vucetic and Dobry (1986) modified that work and proposed that

$$r_{u,N} = \frac{fpNF(\gamma_c - \gamma_{tv})^b}{1 + fpNF(\gamma_c - \gamma_{tv})^b} \tag{9.55}$$

where γ_c=cyclic strain amplitude, γ_{tv}=volumetric threshold shear strain (Section 6.4.3.1), f=1 for uni-directional shaking or 2 for bi-directional shaking, and p, F, and b are coefficients obtained by fitting to the results of laboratory tests. Cetin and Bilge (2012) used a database of 99 cyclic simple shear and cyclic triaxial tests to relate pore pressure ratio to peak shear strain, relative density, and effective stress

$$\ln(r_{u,N}) = \ln\left\{1 - \exp\left[-\frac{0.407\gamma_{\max,N}}{-0.486 + 0.025\ln\sigma'_{v0} - D_r/100}\cdot\left(\frac{0.620}{1+\gamma_{\max,N}}\right)\right]\right\} \tag{9.56}$$

where $\gamma_{\max,N}$=maximum shear strain in percent after N cycles, σ'_{v0}=initial vertical effective stress in kPa, and D_r=relative density in percent. Cetin and Bilge (2012) indicated that Equation (9.56) produced less biased and less uncertain pore pressure ratios than other stress-, strain-, and energy-based pore pressure models.

Dissipated Energy-Based Pore Pressure Models

Pore pressure generation can also be correlated to dissipated energy (Green et al., 2000). Laboratory tests have shown that pore pressure ratio increases linearly with the square root of normalized dissipated energy density, i.e.,

$$r_u = \sqrt{\frac{W_s}{PEC}} \le 1.0 \tag{9.57}$$

where W_s=dissipated energy per unit volume divided by initial effective confining stress and the pseudo energy capacity, PEC, is a calibration parameter obtained from the results of cyclic tests (cyclic triaxial or cyclic simple shear). The value of W_s can be computed as

$$W_s = \frac{1}{2\sigma'_{m0}}\int_0^\infty \sigma_d d\varepsilon_a = \frac{1}{2\sigma'_{m0}}\int_0^\infty \tau d\gamma \tag{9.58}$$

for cyclic triaxial tests (σ_d=deviator stress and ε_a=axial strain) and cyclic simple shear tests (τ =shear stress and γ=shear strain), respectively; for both, σ'_{m0}=initial mean effective stress. Green et al. (2000) found that laboratory data suggested that the pseudo energy capacity, PEC, could be estimated as

$$PEC = \frac{W_{s,r_u=0.65}}{0.4225} \tag{9.59}$$

where $W_{s,r_u=0.65}$ is the value of W_s required to produce r_u=0.65. Figure 9.57 shows a graphical interpretation of PEC as a measure of the W_s value corresponding to r_u=1.0. The linearity of the relationship between $\sqrt{W_s}$ and r_u is evident in the figure; although the experimental data saturates at higher r_u values, the linear portion of the data is projected to the r_u=1.0 (initial liquefaction) condition. Polito et al. (2008) related PEC to relative density for sand with fine contents lower than 35%

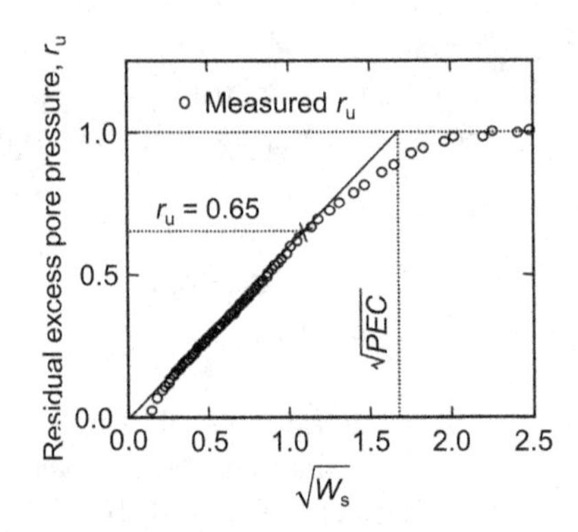

FIGURE 9.57 Relationship between pore pressure ratio and normalized dissipated energy with data from cyclic triaxial test on Yatesville clean sand. (Polito et al., 2008 with permission of ASCE.)

$$\ln PEC = \exp[0.0139 D_r] - 1.021 \tag{9.60}$$

Discussion

The type of pore pressure models described in the preceding paragraphs can capture first-order effects (i.e., soil softening due to pore pressure generation) of liquefaction, but the monotonically increasing nature of the pore pressures they predict does not allow consideration of well-known aspects of liquefiable soil behavior such as phase transformation and cyclic mobility. As discussed in Sections 6.4.3.4 and 9.5.2.4, a potentially liquefiable soil can dilate and stiffen above the PT line, even when the pore pressure ratio is high enough for initial liquefaction to have occurred. Thus, simple pore pressure models, and backbone curves that are simply softened (or degraded) in response to pore pressure generation can be inaccurate at levels of shaking that induce high pore pressures. When implemented into nonlinear, effective stress ground response programs, they can lead to very low stiffnesses in soils that generate high pore pressures; these low stiffnesses can essentially "base isolate" the overlying soils (Anderson et al., 2011) leading to severe underestimation of ground surface accelerations and severe overestimation of ground surface displacements.

9.5.7.2 Constitutive Models

Advanced constitutive models (Sections 6.4.5.3 and 7.5.3.3) provide a more rigorous approach to prediction of soil behavior under a wide variety of loading conditions. Such models describe the increments of volumetric and deviatoric strain produced by increments of volumetric and deviatoric stress. Advanced constitutive models can be incorporated into one-, two-, or three-dimensional nonlinear ground response and dynamic response analyses. As described in Section 7.5.3.3, advanced constitutive models balance the incremental volumetric strain of the soil skeleton with the incremental volumetric strain of the pore fluid through the bulk modulus of the pore fluid (Equation 7.67). When the soil is dry or partially saturated, the low bulk modulus (high compressibility) of the pore fluid (air or both water and air) allows incremental changes in stress to produce incremental changes in strain without resistance from the pore fluid. When the soil is saturated, however, the extremely low compressibility of water allows only an extremely small increment of volumetric strain that is accompanied by a large incremental increase in pore fluid pressure and corresponding incremental decrease in effective stress. With the inclusion of a dilatancy parameter (Equation 7.69), phase transformation behavior (Section 6.4.3.4) can be captured (Figure 9.58). When implemented into nonlinear, effective stress stress-deformation analyses, these models can predict the dilation-induced stiffening that produces dilation pulses such as those recorded after about 18-19 sec in Figure 9.59 and that restrain permanent deformations in phenomena like lateral spreading.

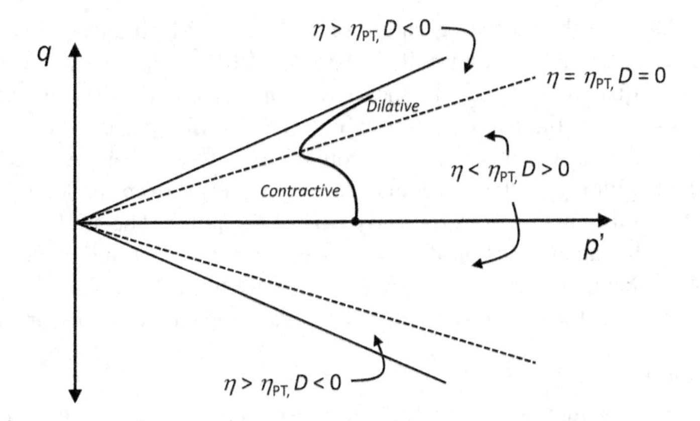

FIGURE 9.58 Illustration of phase transformation line and dilation function showing zones of contractive (σ'_v decreasing) and dilative (σ'_v increasing) behavior under undrained loading conditions.

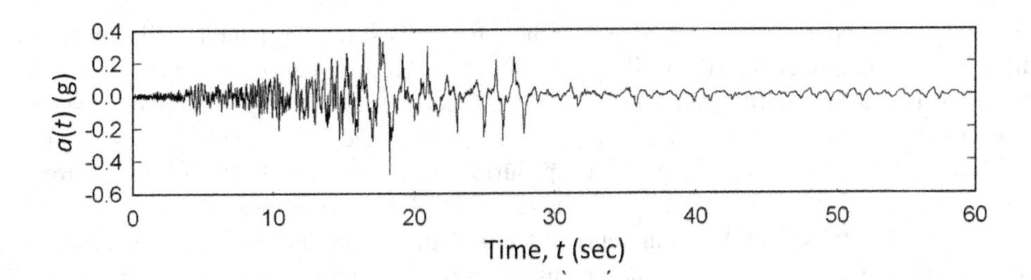

FIGURE 9.59 Kushiro Port record from Kushiro-Oki earthquake illustrating prominent dilation-induced acceleration spikes.

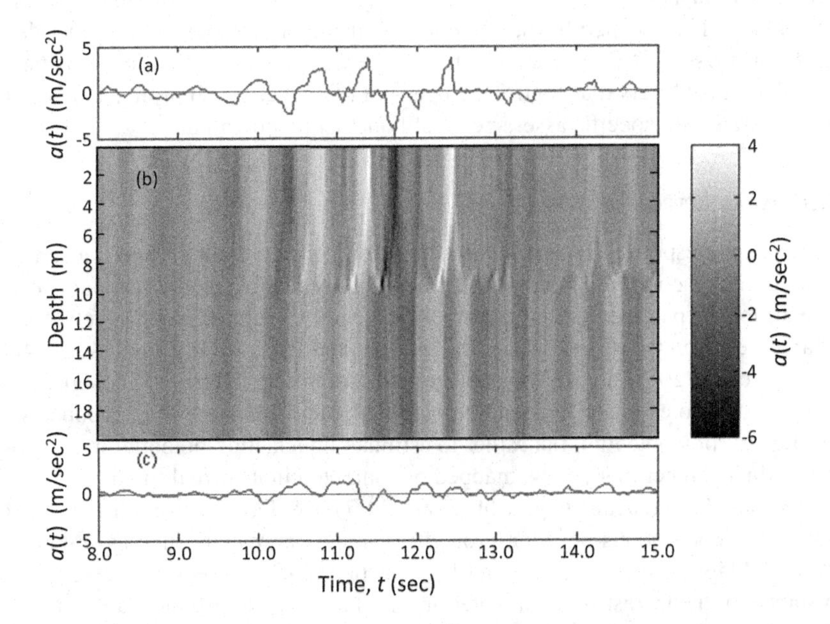

FIGURE 9.60 Development of dilation-induced acceleration pulses in one-dimensional, nonlinear, effective stress analysis of a 10-m-thick liquefiable soil layer underlain by 10 m of non-liquefiable soil: (a) computed surface motion, (b) grayscale indication of acceleration amplitude, and (c) input motion applied at 20 m depth.

Figure 9.60 shows how the stiffening nature of a dilating soil produces dilatancy pulses in a nonlinear, effective stress analysis of a profile consisting of 10 m of loose, saturated sand underlain by 10 m of stiff, non-liquefiable soil. As the shear stress increases at the base of a dilating layer, the shear modulus of the soil at the beginning of an upward-traveling pulse is, due to dilation, lower than it is at the end of the pulse. Thus, the stress from the end of the pulse travels upward faster than the stress from the beginning of the pulse, and the strain energy from the latter part of the pulse "catches up" to the strain energy from the early part of the pulse. The strain energy is therefore focused in time, which leads to increased response at that instant of time. Figure 9.60 shows how ground motion pulses become narrower in time as they move toward the surface, and how the characteristic dilation pulses at the ground surface coincide with the temporal focusing of strain energy.

9.5.7.3 Discussion

Nonlinear effective stress analyses have advanced considerably in recent years and computer programs that use well-calibrated constitutive models are becoming available. Such models have the potential to be used for prediction of triggering in response to specific earthquake input motions at the base of a soil profile (e.g., Ntritsos and Cubrinovski, 2020; Olson et al., 2020; Matasovic et al., 2024).

Relatively simple pore pressure models that monotonically soften liquefiable soils are capable of predicting pore pressure generation well up to pore pressure ratios of about 0.8 or so. Their inability to capture phase transformation behavior, however, affects their ability to predict pore pressures and site response accurately as liquefaction is approached and reached (Anderson et al., 2011). They can also lead to excessive softening that prevents upward-traveling waves from propagating through the oversoftened layer, thereby underpredicting the response of the overlying soils.

Advanced constitutive models can represent phase transformation behavior and the dilation pulses that can result from it. Data from which the post-triggering behavior of such models can be calibrated, however, are not plentiful at this time, and nonlinear effective stress models have not been through the type of independent verification and validation procedures (e.g., Stewart et al., 2008) that total stress nonlinear models have. Nevertheless, such models have the ability to account for the interaction of layers as pore pressures generate, redistribute, and dissipate during and after earthquake shaking. These capabilities, implemented using procedures such as that described by Ntritsos and Cubrinovski (2020), can account for the mechanical and hydraulic interactions between the "system" of multiple layers that control the behavior of a specific soil profile and offer the potential for more accurate, site-specific assessments of liquefaction potential.

9.5.8 Geospatial Approach

For purposes of rapid response, loss estimation, and emergency planning efforts, it can be useful to develop regional liquefaction mapping techniques that only require readily available data that can serve as proxies for the parameters that go into a site-specific evaluations of liquefaction potential. Geospatial analyses (Youd and Perkins, 1978; Knudsen and Bott, 2011; Zhu et al., 2014; Matsuoka et al., 2015; Zhu et al., 2017) utilize regional ground motion, topographic, geologic, geomorphic, and groundwater data to estimate liquefaction potential; their intent is not to produce site-specific estimates of liquefaction potential, but rather to estimate liquefaction potential over broad, or even global, regions, in a manner that can be mapped or computed in near-real time.

Modern geospatial approaches (e.g., Zhu et al., 2017) use digital elevation maps to characterize topography and correlations between ground slope and parameters such as soil density (Wald and Allen, 2007; Magistrale et al., 2012) and V_{S30} (Section 6.6.2.2). Precipitation records and maps showing distance to the nearest body of water are used to estimate groundwater depth. Combined with GMM-based estimates of ground motion characteristics, these data can be used to develop regional estimates of the spatial extent of liquefaction. The global probability of liquefaction (i.e., surface manifestation) is estimated (Zhu et al., 2017) as

$$P(X) = \begin{cases} \dfrac{1}{1+e^{-X}} & \text{if } PGV > 3 \text{ cm/sec and } V_{S30} < 620 \text{ m/sec} \\ 0 & \text{Otherwise} \end{cases} \tag{9.65}$$

where

$$X = 12.435 + 0.301 \ln PGV - 2.615 \ln V_{S30} + 5.556 \times 10^{-4} \cdot precip$$

$$-0.287(d_c)^{0.5} + 0.0666 d_r - 0.0369 d_r (d_c)^{0.5}$$

and PGV=peak ground velocity in cm/sec, V_{S30} is shear wave velocity in cm/sec, *precip* is mean annual precipitation in mm, d_c is distance to the coast in km, and d_r is distance to the nearest river in km. Rashidian and Baise (2020) found that field observations of the spatial extent of liquefaction could be improved by changing the bounds of the non-zero probability portion of Equation (9.62) to "PGV>3 cm/sec and PGA>0.1 g, and V_{S30}<620 m/sec." Calibration of modern geospatial approaches against data from individual earthquakes has shown that they can produce estimates that are about as accurate (in terms of minimizing false positive and false negative predictions) as current site-specific procedures (Geyin et al., 2020). However, application of geospatial models calibrated against data from one location has not performed well relative to site-specific analyses when applied to other locations. Nevertheless, the geospatial approach is useful for the types of regional evaluations for which it was developed.

9.5.9 Challenging Conditions

Engineers can encounter a number of conditions that complicate direct application of the triggering procedures described in the preceding sections (Bray et al., 2017). The previously described procedures are based almost exclusively on data from clean sands and non-plastic silty sands for which both the cyclic response and penetration resistance are relatively well understood. Some sites, however, contain significant amounts of soils whose response to cyclic loading is not well understood, and/or for which the results of penetration tests can be difficult to interpret. These include sites with large particles, thinly interbedded soils, tailings materials, and crushable soils, and sites subjected to unusual ground motions. Special care must be taken in the evaluation of liquefaction hazards in these cases.

9.5.9.1 Large Particles

Soils containing large particles, such as gravel- or cobble-sized particles, can be difficult to characterize for evaluation of liquefaction potential. Two situations are particularly important: (1) a soil unit consisting entirely of large particles and (2) a soil unit consisting of a mixture of sand and large particles where the large-particle content is less than a limiting fraction (Section 6.3) below which the large particles are "floating" in a matrix of the smaller particles.

In the first case, the high permeability of the large-particle soil may prevent the generation of high pore pressure since the pore pressure may dissipate nearly as quickly as it is generated. If the large-particle soil is bounded by less permeable soils or has voids filled with less permeable soils, however, it may be loaded under essentially undrained conditions during earthquake shaking, and high pore pressures may develop. In such cases, the state of the soil is usually inferred from Becker hammer penetration resistance (Section 6.5.3.2) using an empirical correlation to SPT resistance. The dynamic cone penetration test is an emerging alternative for characterization of gravelly soils and assessment of their liquefaction potential (Rollins et al., 2021). Alternatively, liquefaction resistance can be correlated to shear wave velocity instead of penetration resistance, although the relationship between shear wave velocity and liquefaction resistance is not as well established for large particles as it is for sands.

The second case is problematic because occasional large particles can influence both SPT and CPT resistances as well as liquefaction resistance of the soil itself. If the large particles are not plentiful and the sand they are in is of relatively uniform density, it may be relatively easy to identify spikes in measured CPT resistances as being due to the influence of individual large particles. Similarly, SPT resistance can be carefully measured in terms of blows per inch (instead of blows per foot) and the high resolution blowcount logs examined for evidence of poor sample recovery or large particle interference (anomalously high blowcounts and/or physical evidence of gravel in the sampler). For both the CPT and SPT, the use of a penetration resistance that approximates the lower bound of the measured resistance profile can provide a reasonable-to-conservative estimate of the penetration resistance of the sand, which will control liquefaction resistance. Obviously, anomalously high penetration resistances are easier to identify when the large particle content is relatively low; Ghafghazi et al. (2017) suggest that gravel influence is likely when more than approximately 20% gravel is found in split-spoon samplers. Cubrinovski et al. (2018) reported good results using 15 cm^2 CPT probes in well-graded fill containing 60%–70% gravel. Recalling the concept of limiting fines content, FC_L (Section 6.3) in which coarse particles remain in contact with each other for $FC < FC_L$ and are separated by fines for $FC > FC_L$, similar principles can be used to conceptualize the behavior of gravel. The key metric for the case of gravels is gravel content. At higher gravel contents, the gravel particles may be in contact with each other and form a load-resisting network of large particles. The smaller particles then fill the voids between the large particles, which control the actual response of the soil. Although it is influenced by the characteristics of the individual constituents, the transition from control by smaller particles to larger particles typically appears to take place when the large particle content exceeds about 35%. The presence of large particles can also complicate characterization of the behavior of the material, not only because of their effect on penetration resistance measurements but also because of their effect on maximum and minimum densities.

9.5.9.2 Thinly Interbedded Soils

Some depositional processes tend to produce stratified soil deposits. Tideflats can consist of alternating thin layers of coarse- and fine-grained soils (Figure 9.61). Overbank (floodplain) deposits can also consist of layers of coarse- and fine-grained soils. Constructed soil deposits such as hydraulic fills or tailings ponds can result in stratification as coarser particles settle out faster than finer particles following individual depositional episodes.

Interbedded deposits have several characteristics that are particularly important for liquefaction. The layering of coarser and finer particles leads to significant differences in permeability, which can affect pore pressure (and void) redistribution, phenomena that can strongly influence the consequences of liquefaction. Stratified soils, particularly when thinly bedded, are very difficult to characterize with the tools (e.g., CPT and SPT) commonly used to estimate liquefaction resistance. The penetration resistance of potentially liquefiable sand interlayers can be strongly influenced (generally reduced) by silt/clay interlayers above or below them. While thin-layer corrections for CPT resistance are available (Section 6.5.3.3), they primarily apply to the case of an isolated thin layer in a generally consistent material, not to a sequence of alternating interlayers whose thicknesses are small relative to the size of the penetrometer. Interface transitions tend to decrease CPT tip resistance in sandy layers less than about 1 m in thickness and increase it in clayey layers. In some cases, the transition in measured penetration resistance from a sand to a clay can lead to mistaken interpretation of a silt layer between the two. Thin, liquefiable layers may not be laterally continuous, a characteristic that is difficult to evaluate with typical large distances between borings or CPT soundings. These and other characteristics of interbedded deposits have often led to liquefaction hazards being overpredicted (Maurer et al., 2014a; Boulanger et al., 2016; Cubrinovski et al., 2019).

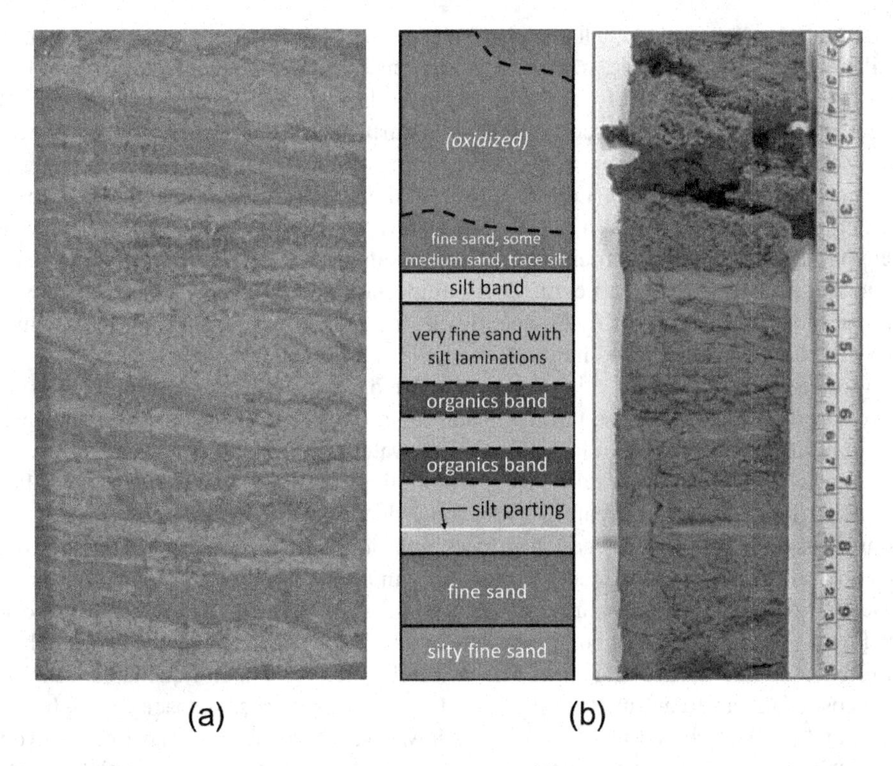

FIGURE 9.61 Interbedded soil deposits: (a) laminated sand (light color) and clay (dark color) from tidal deposits in Groningen area of The Netherlands (van der Linden, 2016), and (b) swamp deposits along Heathcote River in Christchurch, New Zealand. (After Beyzaei et al., 2017.)

9.5.9.3 Crushable Soils

Most cases of liquefaction occur in quartzitic soils with hard, strong, rounded to subrounded particles. Some soils, however, are comprised of particles that are so weak that they can fracture or crush under relatively low stresses. Calcareous soils, pumice, coral sands, fly ash, and other materials can display this type of behavior. The weakness of the particles can have a strong effect on both CPT and SPT resistance, which complicates their use as indicators of liquefaction resistance. It may also influence their actual liquefaction resistance and post-triggering behavior.

Although they can become cemented quickly (Coop and Airey, 2003), uncemented calcareous sands are up to 30 times more compressible than quartzitic sands (Morioka and Nicholson, 2000). Calcareous sands, which predominantly consist of the skeletal remains of marine organisms, tend to have more angular particle shapes than quartzitic sands, and these shapes allow formation of interlocking contacts that are generally more resistant to liquefaction for a given relative density and effective confining stress (Kaggwa et al., 1988; Morioka and Nicholson, 2000; Porcino and Marciano, 2010; Sandoval and Pando, 2012; Pando et al., 2012). On the other hand, calcareous sands are compressible and their CPT tip resistance is generally lower than that of quartzitic sands for the same relative density; Porcino and Marciano (2010) found that the combined effects of greater liquefaction resistance and lower penetration resistance allowed procedures developed for quartzitic sands to produce a slightly conservative estimate of the liquefaction resistance of a calcareous sand.

9.5.9.4 Mine Tailings

Mine tailings are materials that remain after separation of a mineral or other desired product from the raw ore extracted from a mine. In some cases, the separation takes place mechanically, in others it takes place chemically, and in many cases both mechanical and chemical processes are used.

As a result, tailings can include salts, acids, benzene, hydrocarbons, and other toxic chemicals. Mine tailings often have particle sizes and gradations similar to liquefaction-susceptible soils, and have been observed to liquefy and flow under both static and seismic loading (e.g., Figure 1.12). Some mine tailings can be used for beneficial purposes (e.g., as structural fills) but most are treated as waste materials and impounded behind tailings dams in tailings storage facilities (TSFs). Tailings dams are among the largest dams in the world and have failed at a higher rate than water supply reservoir dams (Morgenstern, 2018). The failure of tailings dams can lead to loss of life, severe environmental damage, and enormous cleanup and repair costs (Morgenstern et al., 2015; Morgenstern et al., 2016; Robertson et al., 2019). Important issues in TSF seismic stability include the susceptibility of tailings, their resistance to triggering, and their residual strength, which controls stability, runout distances, and inundation depths of potential flow slides.

Tailings dams differ from dams that retain water in a number of ways, one being that they are usually constructed sequentially as tailings materials are deposited. Tailings materials are often sluiced into place, i.e., mixed with a large quantity of water and pumped into the area behind a tailings dam, which is similar to hydraulic filling. The tailings particles then settle through the water, which is subsequently removed, creating a very loose deposit. While the water content is high, the tailings are loose and saturated, hence they have high liquefaction potential. There are three primary types of tailings dams – upstream, downstream, and centerline – that produce similar external geometries but differ significantly in terms of cost, tailings disposal capacity, and static/seismic stability. Figure 9.62 shows typical configurations for all three. The upstream method involves placing relatively small amounts of select fill, part or all of which rests upon liquefiable tailings materials. The cost of the upstream method is low and the tailings disposal capacity is high; however, both static and particularly seismic stability are low, and upstream dams have been involved in numerous stability failures. The downstream method involves constructing an initial embankment farther "upstream," and then raising the crest elevation by placing select fill on top of the existing select fill and on competent soil "downstream." As shown in Figure 9.62, the downstream method requires more select fill, which makes it more costly, and provides the least tailings storage capacity. However, it provides the highest level of stability under both static and seismic conditions. The centerline method is a hybrid that includes features of both the upstream and downstream methods and is intermediate to those methods in terms of cost, capacity, and stability.

The properties of tailings are influenced by the type of material that has been processed to produce them, and by the nature of the processes themselves. Coal can be extracted from ore at relatively large particle sizes (on the order of 0.5 mm) whereas mineral processing usually requires smaller particle sizes (near 0.075 mm, i.e., near the sand/silt boundary) to liberate the minerals of

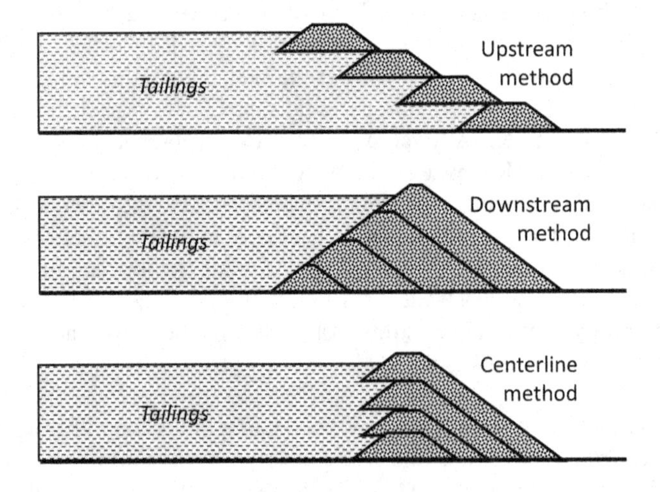

FIGURE 9.62 Types of tailings dams. (After Vick, 1983). For each type, construction begins with a small starter dam that is sequentially raised when additional capacity is needed.

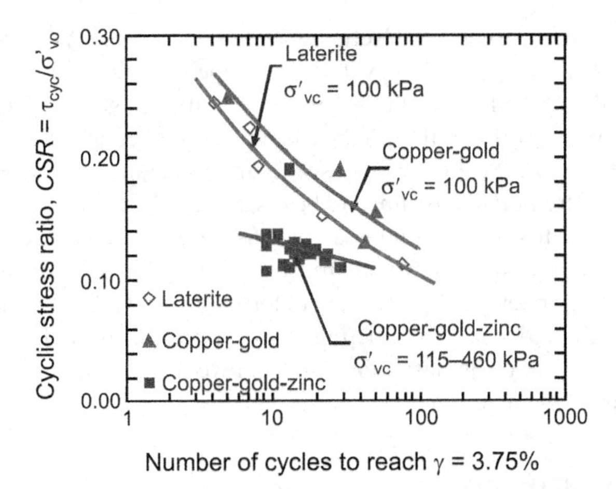

FIGURE 9.63 Laboratory liquefaction resistance of laterite, copper-gold, and copper-gold-zinc tailings prepared by identical methods (after Wijewickreme et al., 2005 with permission of Canadian Science Publishing).

interest. The fines of mineral tailings are generally non-plastic rock particles or "rock flour," which are themselves highly susceptible to liquefaction. As illustrated in Figure 9.63, tailings materials have been observed to display a wide range of liquefaction resistances in laboratory tests (Moriwaki et al., 1982; Ishihara et al., 1980, 1981; Poulos et al., 1985; Qiu and Sego, 2001; Wijewickreme et al., 2005, 2010; Al-Tarhouni et al., 2011; Geremew and Yanful, 2012). While many tailings materials behave very much like non-plastic, natural sands and silty sands, not all do so the applicability of conventional procedures for evaluation of liquefaction potential should be assessed for individual cases. Some types of tailings may have inherently viscous characteristics while others may have plasticity produced or altered by chemical processing.

9.5.9.5 Near-Fault Motions

As discussed in Section 3.7, ground motions in close proximity to generating faults produce motions with different time domain characteristics than at sites farther from the source. Forward directivity can produce ground motions with pulses of strong shaking and short duration that have been found to be more damaging to structures than motions without pulses; fling step and asperities can also produce ground motions with pulses. Directivity pulses are most pronounced in the fault normal direction.

Equivalent linear ground response analyses of a liquefiable soil profile with 27 pairs of motions – a fault normal motion with a pulse and corresponding fault-parallel motion without a pulse – showed that peak accelerations were generally higher for the fault normal (with pulse) than the fault parallel (without pulse) motions. The number of equivalent cycles, however, was generally found to be lower for fault normal than fault parallel motions (Green et al., 2008; Green and Lee, 2010; So et al., 2015). The observed differences in peak acceleration and MSF tended to compensate for each other.

In cases where pulse-like ground motions are expected, the temporal character of the motions should be considered, either by means of a cycle-counting-based MSF or by using pulse motions as inputs to nonlinear, effective stress response analyses.

9.6 CONSEQUENCES OF LIQUEFACTION

Liquefaction phenomena can affect buildings, bridges, dams, levees, buried pipelines, and other elements of infrastructure in many different ways. Liquefaction can strongly influence the nature of ground surface motions imparted to structures. Flow liquefaction can produce massive flow slides and contribute to the sinking or tilting of heavy structures, the floating of light buried structures, and to the failure of retaining structures. Cyclic liquefaction can cause movement of slopes, slumping of embankments, settlement of buildings, failure of pipelines, floating of buried structures, and failures of wharves and quay walls. Substantial ground oscillation, ground surface settlement, sand boils, and post-earthquake stability failures can develop at level-ground sites.

For many years, geotechnical engineers focused their attention on the triggering of liquefaction. Since its consequences were initially assumed to be disastrous, the approach to design and remediation focused on preventing triggering of liquefaction. As more data has become available, many of the consequences of liquefaction are now recognized as existing on a continuous range of response levels, typically expressed in terms of permanent deformations. The focus of modern geotechnical earthquake engineering practice is on predicting those deformations, and on designing structures and facilities to tolerate them or on designing soil improvement measures (Chapter 12) to reduce expected deformations to tolerable levels.

9.6.1 Geologic Considerations

The consequences of liquefaction are influenced by geologic factors in a manner that is important, but difficult to quantify. As discussed earlier, liquefaction occurs frequently in loosely deposited sandy soils, usually close enough to bodies of water that they are saturated at relatively shallow depths. Of particular importance is the vertical and lateral continuity of liquefiable layers, as case histories have shown that soil profiles with thick interconnected liquefiable layers tend to exhibit more severe consequences of liquefaction than profiles with thinner, isolated (vertically and horizontally) layers. In the latter case, liquefaction may be triggered in individual layers, but the profile may produce little surficial evidence of liquefaction having occurred. Thus, the spatial variability of the depositional environment influences the "system" response (Cubrinovski et al., 2019) of the soil profile.

Fluvial environments can be particularly complicated. Rivers are not stationary, and they rarely follow straight paths. River flow velocities vary with topography and are generally high in steep terrain (mountains and foothills) and slower in flatter terrain. The sizes of soil particles that can be suspended in flowing water depend on the velocity of flow, so large particles tend to drop out of suspension (i.e., to be deposited) in steep terrain and smaller particles in flatter terrain.

Soils deposited within a *river channel* (i.e., between *river banks*) form *bar* deposits. Bars tend to move as more soil is deposited, and their movement can change the position and shape of a river channel. Water velocities are higher on the outer portion of a river bend and slower on the inner portion. This causes erosion of the outer bank and deposition, in the form of *point bars* (Figure 9.64), on the inner bank; both processes tend to accentuate the curvature of the river over time. During

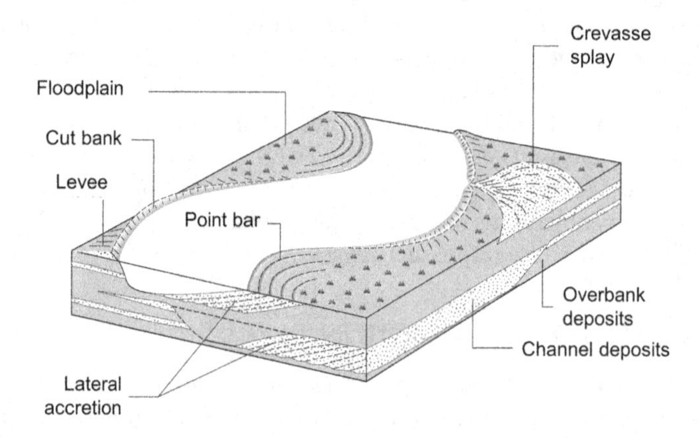

FIGURE 9.64 Deposition of point bars on inner bends of meandering river environment. (After Nichols, 2009 with permission of John Wiley & Sons.)

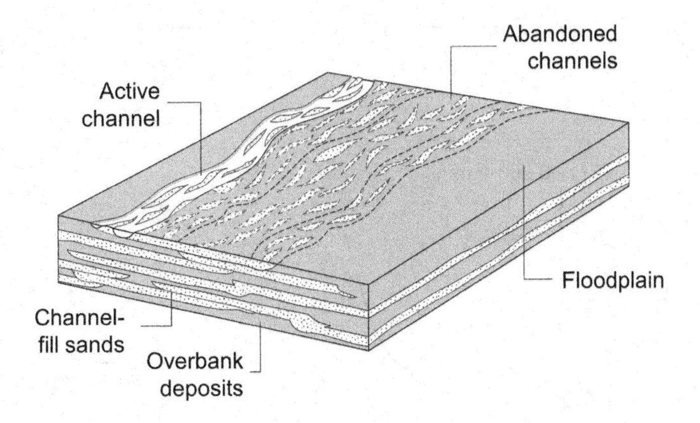

FIGURE 9.65 Deposition of channel (coarse-grained) and overbank (fine-grained) deposits in braided river environment. (After Nichols, 2009 with permission of John Wiley & Sons.)

flooding events, water may spill over the river banks onto the surrounding *floodplain*, where its velocity slows considerably, allowing deposition of fine-grained soils in *overbank* (flood) deposits. *Meandering* rivers influenced by this process tend to move over time across a floodplain. Old meanders within a floodplain can then be covered by overbank deposits and later re-incised by coarse-grained soils as the river meanders back across it. Such environments can lead to highly variable subsurface conditions with complex, continuous, and discontinuous layers of coarse- and fine-grained soils. Braided rivers contain multiple channels, separated by *channel bars*, within their banks. During flood events, the bars are submerged and migrate downstream as their upstream ends are eroded by the flowing water. When a channel moves laterally, a channel bar can emerge and be covered by overbank deposits in subsequent flood events (Figure 9.65). These processes show that fluvial depositional environments can contain irregularly distributed deposits of liquefiable soils, in many cases separated and/or covered by non-liquefiable, fine-grained soils. Evidence of liquefaction has been found to correlate well to old buried (and previously undetected) river and stream channel deposits in many earthquakes, thus reinforcing the need for good knowledge of the geologic history of a site and for detailed subsurface investigation of it.

9.6.2 Manifestation Severity Indicators

The consequences of liquefaction are influenced not only by the density of a liquefied soil, but also by its thickness and depth, among other factors. While the factor of safety against triggering of liquefaction at a particular point in a soil profile does not provide a direct indication of its consequences, knowledge of the liquefaction potential of the entire profile can be used to gain at least a rough idea of the potential damage that might occur. *Manifestation severity indicators* use triggering data and the characteristics of the entire profile to define a parameter that describes the severity of ejecta and ground cracking and is intended to correlate with potential damage.

Post-earthquake investigations of sites in Christchurch, New Zealand following the 2010–2011 Canterbury Earthquake Sequence (Cubrinovski et al., 2017) showed that the severity of surficial manifestation of liquefaction was related to the "system response" of a soil profile, and that profiles that contained relatively thick, hydraulically-connected liquefiable layers produced more severe manifestation than profiles with sequences of interbedded thinner layers of liquefiable and non-liquefiable soils, even if the total thicknesses and densities of the liquefiable soils were the same. Analyses suggested that sand boil formation involved not only pore pressure dissipation from the most critical layer in a soil profile but also from underlying, connected liquefiable soils that caused water to flow upward and aid in the formation of sand boils. The process is complex and involves mechanical and hydraulic interactions between various soil layers both during and after shaking.

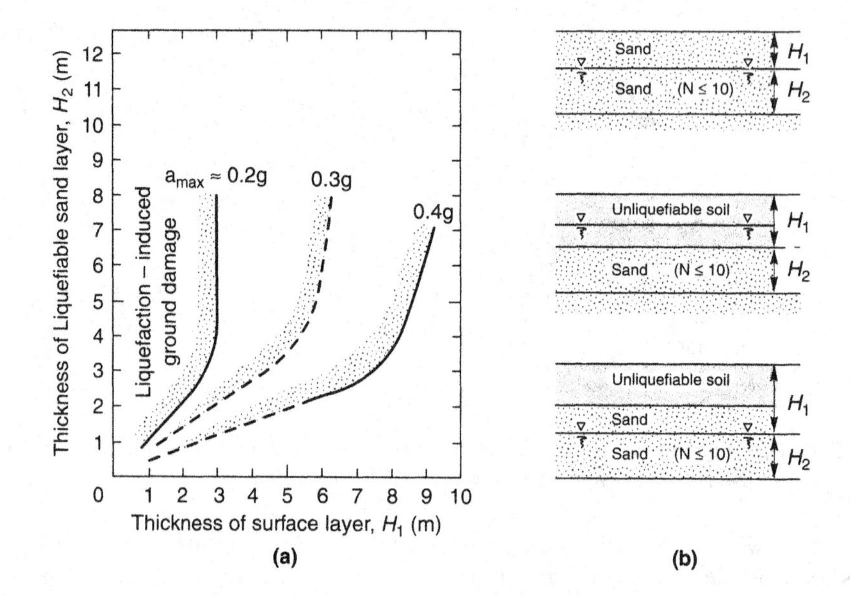

FIGURE 9.66 (a) Relationship between thickness of liquefiable layer and thickness of overlying layer at sites for which surface manifestation of level-ground liquefaction has been observed, and (b) guides to evaluation of respective layer thicknesses. (After Ishihara, 1985; with permission of Kluwer Academic Publishers.)

The original, and most basic, severity indicator simply indicates the potential for development of surficial evidence such as sand boils at level-ground sites. Ishihara (1985) examined the soil conditions associated with various liquefaction-related damage reports from the 1983 Nihonkai-Chubu earthquake (**M**=7.8) and 1976 Tangshan (**M**=7.6) earthquakes (Gao et al., 1983) and produced estimates of the thickness of the overlying layer required to prevent surface manifestation of liquefaction at level-ground sites (Figure 9.66). These estimates have been validated against larger databases and found to be ineffective for sites susceptible to ground oscillation or lateral spreading (Youd and Garris, 1995) and sites with multiple liquefied strata (Rateria and Maurer, 2022). Such limitations have led to the development of more generally applicable severity indicators; these indicators also provide continuous scales of severity that are more informative than the binary output of the original Ishihara (1985) indicator.

The *Liquefaction Potential Index* (Iwasaki et al., 1978, 1982) is based on a depth-weighted function of the factor of safety against liquefaction. It assumes that damage increases with a decreasing factor of safety since strains would be larger and liquefaction would likely be triggered earlier in a soil with a low factor of safety. The liquefaction potential index is defined as

$$LPI = \int_0^{20\text{ m}} w(z) \cdot F_1(z)\,dz \tag{9.66}$$

where the dimensionless weighting factor, $w(z) = 10 - 0.5z$ with z in meters and

$$F_1(z) = \begin{cases} 1 - FS_L & \text{for } FS_L \le 1.0 \\ 0 & \text{for } FS_L > 1.0 \end{cases}$$

While LPI as computed using Eq. (9.66) has units of meters, by convention it is considered to be dimensionless. The weighting factor, which decreases linearly from a value of 10 at the ground surface to zero at 20 m depth, indicates that a given layer of soil with $FS_L < 1.0$ at shallow depth has greater damage potential than it would with the same FS_L at a greater depth. Iwasaki et al. (1978)

TABLE 9.9
Liquefaction Risk Levels for Various Ranges of Liquefaction Potential Index

LPI	Liquefaction Risk
0	Very low
0–5	Low
5–15	High
>15	Very high

Source: After Iwasaki et al. (1978).

proposed that "liquefaction risk," interpreted as the development of significant surficial manifestation, was related to *LPI* as indicated in Table 9.9. Maurer et al. (2015) used data from the 2010–2011 Canterbury Earthquake Sequence to develop a version of *LPI* that used trends in the Ishihara (1985) curves (Figure 9.66) to include the effect of the relative thicknesses of liquefied and overlying non-liquefied layers. The resulting *Ishihara-inspired liquefaction potential index* is computed as

$$LPI_{ISH} = \int_{H_1}^{20\text{ m}} \frac{25.56}{z} \cdot F(FS_L)dz \tag{9.67}$$

where H_1 is the thickness of the non-liquefiable crust in meters, z is the depth in meters, $F(FS_L) = 1 - FS_L$ for $FS_L \leq 1.0$ and $H_1 \cdot m(FS_L) \leq 3$ m and otherwise is zero, and

$$m(FS_L) = \exp\left(\frac{5}{25.56(1 - FS_L)}\right) - 1 \tag{9.68}$$

Comparisons of *LPI* and LPI_{ISH} with observed liquefaction severities from 60 case histories from the 1989 Loma Prieta (California), 1994 Northridge (California), 1999 Kocaeli (Turkey), 1999 Chi Chi (Taiwan), 2010 Darfield (New Zealand), and 2011 Christchurch (New Zealand) earthquakes showed that LPI_{ISH} provided improved predictions of manifestation severity, particularly by reducing the number of false positive predictions (i.e., predictions of surface manifestations where none were observed).

The *Liquefaction Severity Number* (van Ballegooy et al., 2012, 2014) is a depth-weighted integral function of the expected volumetric strain caused by reconsolidation of liquefied soil, i.e.

$$LSN = 10 \int \frac{\varepsilon_v}{z} dz \tag{9.69}$$

where ε_v is computed free-field volumetric strain (in %). The volumetric strain can be estimated as a function of soil density and loading level using the technique of Zhang et al. (2002), which is described in Section 9.6.6.1. That technique accounts for the small volumetric strains that can occur when pore pressures are not high enough to initiate liquefaction, so it tends to increase more smoothly with decreasing FS_L than techniques that only consider volumetric strains of layers with $FS_L < 1.0$. Also, because volumetric strains tend to reach limiting values, a given soil profile will have a maximum *LSN* value. *LSN* can be related to surficial effects as indicated in Table 9.10.

In a manner analogous to that used to develop LPI_{ISH}, the Ishihara (1985) model was used to develop an Ishihara-inspired version of *LSN* (Ulmer et al., 2023).

TABLE 9.10
Predominant Performance Levels for Various Ranges of Liquefaction Severity Number (Tonkin and Taylor, 2013)

LSN Range	Predominant Performance
0–10	Little or no expression of liquefaction, minor effects
10–20	Minor expression of liquefaction, some sand boils
20–30	Moderate expression of liquefaction, with sand boils and some structural damage
30–40	Moderate to severe expression of liquefaction, settlement can cause structural damage
40–50	Major expression of liquefaction, undulations and damage to ground surface, severe total and differential settlement of structures
>50	Severe damage, extensive evidence of liquefaction at surface, severe total and differential settlements affecting structures, damage to services

$$LSN_{ISH} = \int_{H_1}^{20\text{ m}} F_{LSN_{ISH}}(\varepsilon_v) \cdot \frac{36.929}{z} dz \tag{9.70}$$

where ε_v is in percent and

$$F_{LSN_{ISH}} = \begin{cases} \dfrac{\varepsilon_v}{5.5} & \text{for } FS_L \leq 2 \text{ and } H_1 \cdot m(\varepsilon_v) \leq 3m \\ 0 & \text{otherwise} \end{cases}$$

with

$$m(\varepsilon_v) = \begin{cases} \exp\left(\dfrac{0.7447}{\varepsilon_v}\right) - 1 & \text{for } \varepsilon_v < 0.16 \\ 100 & \text{for } \varepsilon_v \geq 0.16 \end{cases}$$

The consistency of different manifestation severity indicators has been examined (Tonkin and Taylor, 2013) using a large database of damage observations from the 2010–2011 Canterbury Earthquake Sequence. The Ishihara (1985) criterion was found to provide inconsistent indications of manifestation severity, likely due to the difficulties in characterizing two simple layers (liquefied layer and non-liquefied crust layer) in the complex, interlayered geology of the Canterbury region (van Ballegooy et al., 2012). *LPI* was found to correlate well to liquefaction observations within individual events, but to be inconsistent from one event to another. *LSN* was found to have the strongest correlation to observed liquefaction and lateral spreading both within and between different earthquake events and was considered to be the most suitable parameter for predicting future performance in the study region. For insurance purposes, a value of $LSN=16$ was taken (Tonkin and Taylor, 2015) to indicate a site vulnerable to liquefaction-related damage after the Canterbury Earthquake Sequence. More recent studies (Upadhyaya et al., 2023), however, have shown that *LPI* and LPI_{ISH} produce fewer mispredictions of manifestation severity than *LSN* and LSN_{ISH} for case histories from the 2010–2011 Canterbury Earthquake Sequence.

Based on data from the Canterbury Earthquake Sequence and other global events, Geyin and Maurer (2020) developed fragility functions to allow estimation of the (lognormal) distributions of four different liquefaction qualitatively-described manifestation states (none, minor, moderate, and severe) as functions of the combinations of six triggering models and three manifestation severity indices.

Manifestation severity indices or manifestation states provide a relatively simple, quantitative indication of the potential severity of surficial effects of liquefaction at a particular free-field, level-ground site. Such indices can be mapped to illustrate the relative vulnerabilities of different areas to liquefaction damage (e.g., Sonmez, 2003; Baise et al., 2006; Holzer, 2008; Lenz and Baise, 2007; Cramer et al., 2008; Chung and Rogers, 2011; Holzer et al., 2011; Dixit et al., 2012), and to prioritize retrofitting, soil improvement, or vulnerability evaluations. It should be noted, however, that substantial ground movement and damage can be caused by shearing deformations of relatively thin liquefied layers below or near sloping ground surfaces, even though those thin layers would not contribute significantly to a manifestation severity index. Also, settlement of buildings supported on shallow foundations can be influenced by shear and normal stresses associated with the building itself, which are not accounted for in manifestation severity indices.

The behavior of a liquefiable soil profile during strong shaking is complex and not easily correlated to relatively simple characterizations of profile characteristics. At this time, there is no clear consensus on which, if any, manifestation indicator would be likely to provide a reliable indication of potential liquefaction-related damage at a particular site in a future earthquake. The use of multiple indicators in a logic tree format may provide the most useful indication of potential liquefaction severity. Manifestation severity indices should not be used as substitutes for detailed, site-specific evaluations of the effects of liquefaction.

9.6.3 Alteration of Ground Motion

The influence of the shear modulus and damping characteristics of soils on ground response is well established. Several examples of the effects of these characteristics presented in Chapter 7 showed that soft soil deposits respond differently than stiff soil deposits to the same motion, even under weak shaking. Under strong shaking, nonlinearity causes the stiffness of a non-liquefiable soil to drop at large strain levels, although the lack of pore pressure generation allows the stiffness to largely recover when strain levels decrease.

In a liquefiable soil, the development of positive excess pore pressures causes the soil stiffness to decrease due to reduced effective stress in addition to nonlinear behavior at large shear strain levels. A deposit of liquefiable soil that is relatively stiff at the beginning of the earthquake may be much softer by the end of the motion at both low and high strains. As a result, the amplitude and frequency content of the surface motion may change considerably (and suddenly) over the duration of a ground motion. In the most extreme case, the development of very high pore pressures can cause the stiffness (and strength) of a layer to be so low that the high-frequency components of a bedrock motion cannot be transmitted to the ground surface. An example of this effect is shown in Figure 9.67. It is not difficult to identify the point at which liquefaction-induced reduction of the stiffness of the underlying soil took place in this case – the acceleration amplitude and frequency content both changed dramatically about 7 sec after the motion began. For most cases, time-frequency analyses (Section A.4) are required to estimate the time at which liquefaction was triggered (Kramer et al., 2016). The fact that ground surface acceleration amplitudes decrease

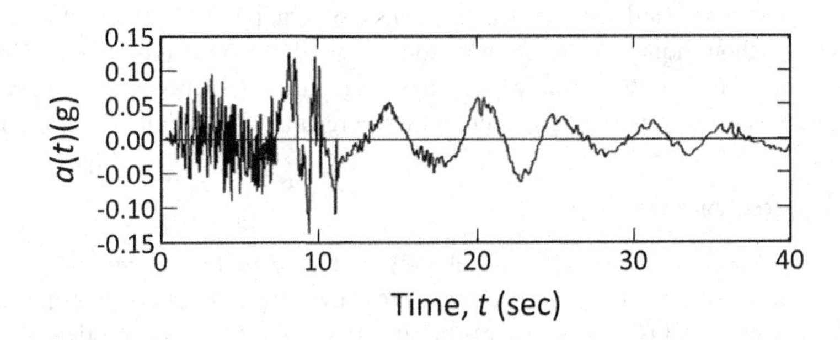

FIGURE 9.67 Recorded ground motion from site near apartment buildings resting on liquefiable soils (shown in Figure 1.9) in 1964 Niigata earthquake.

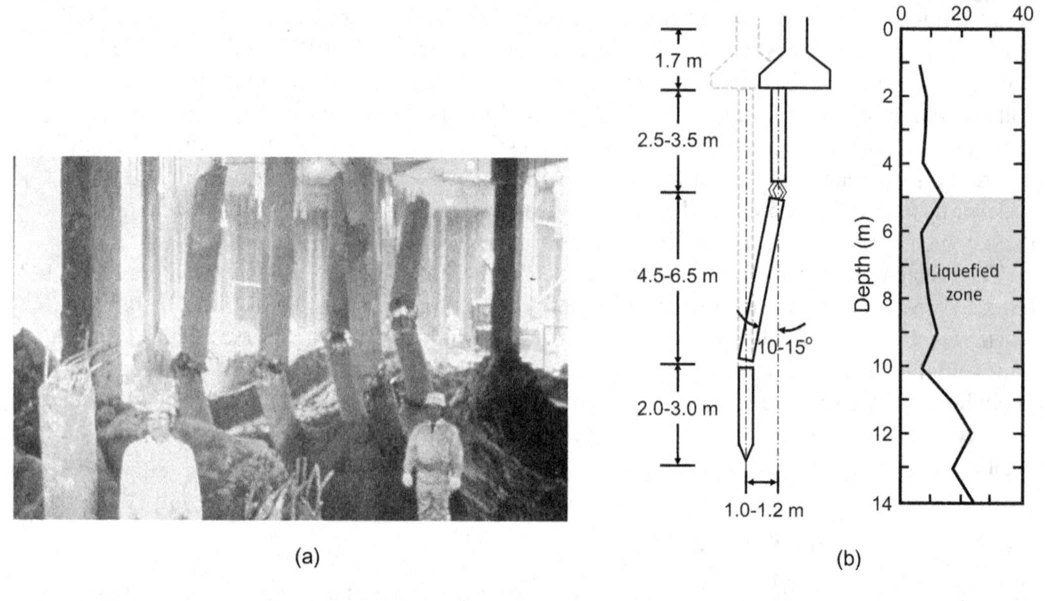

FIGURE 9.68 Pile damage due to liquefaction in 1964 Niigata earthquake: (a) fractured piles exposed beneath NHK Building during excavation for new building in 1985, and (b) post-earthquake pile configuration and SPT resistance. (After Kawamura et al., 1985 with permission of Nikkei Architecture.)

when pore pressures become large does not mean that damage potential is necessarily reduced because low acceleration amplitudes at low frequencies can still produce large displacements. These displacements may be of particular concern for buried structures, utilities, and structures supported on pile foundations that extend through liquefied soils (Figure 9.68).

The occurrence of liquefaction at depth beneath a level ground surface can effectively decouple the liquefied soils from the surficial soils and produce large, transient *ground oscillations*. The surficial soils are often broken into blocks separated by fissures that can open and close during the earthquake. Ground waves with amplitudes of up to several feet have been observed during ground oscillation, but permanent displacements are usually relatively small.

Most of the chaotic ground movements that fractured and buckled pavements in the Marina District of San Francisco during the 1989 Loma Prieta earthquake were attributed to ground oscillation (Youd, 1993). Prediction of the amplitude and pattern of ground oscillation at a particular site is very difficult; even detailed nonlinear ground response analyses can provide only crude estimates. Youd and Carter (2005) examined ground motion records from five instrumented sites in liquefiable soil and found a general reduction of short period spectral accelerations (spectral period < 0.7 sec) and amplification of long-period spectral accelerations (spectral period > 0.7 to 1.0 sec) at those sites compared to sites without liquefaction. Comparisons of hundreds of results from one-dimensional, nonlinear, total and effective stress analyses of hypothetical liquefaction sites (Hartvigsen, 2007; Kramer et al., 2011) showed general agreement with the results of Youd and Carter (2005).

9.6.4 Development of Sand Boils

Liquefaction is often accompanied by the development of sand boils. During and following earthquake shaking, seismically induced excess pore pressures are dissipated predominantly by the upward flow of porewater. This flow produces upward-acting forces on soil particles. If the hydraulic gradient driving the flow reaches a critical value, the vertical effective stress will drop to zero and the soil will be in a quick condition. In such cases, the water flow velocities may be sufficient

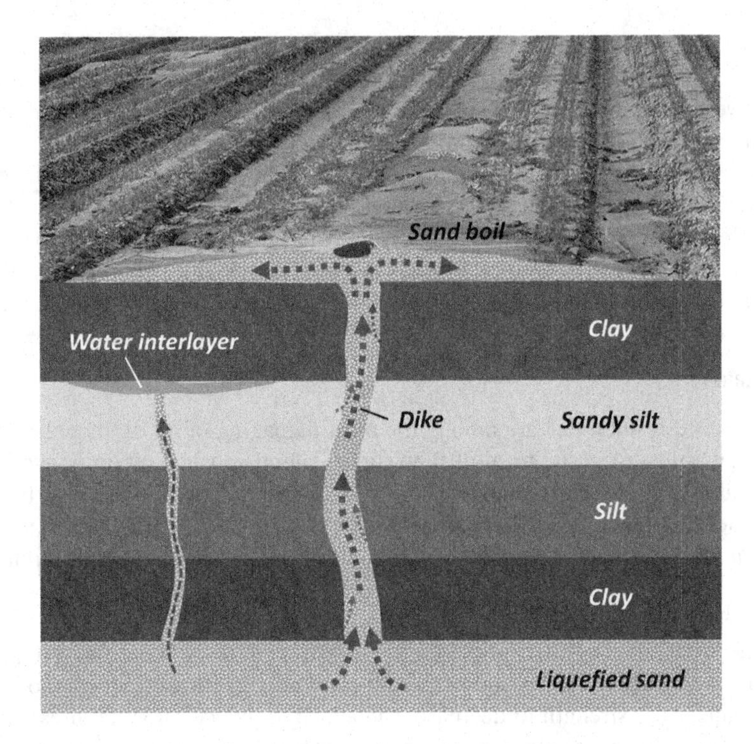

FIGURE 9.69 Upward flow of sand-entrained water through overlying crust layers to form sand boil. Note formation of water interlays at base of impermeable clay layer. (After Sims and Garvin, 1995 with permission of the Seismological Society of America.)

to carry soil particles to the surface (Figure 9.69). In the field, soil conditions are rarely uniform so the escaping porewater tends to flow at high velocity through localized cracks or channels. Soil particles can be carried through these channels and ejected at the ground surface to form sand boils. It should be recognized that sand boils often contain materials other than sand (e.g., silt) and may contain soil particles entrained by the upward flow from layers above the layers that actually liquefied. The development of sand boils is a complicated and somewhat random process; it depends on the magnitude of the excess pore pressure; the thickness, density, and depth of the zone of excess pore pressure; and the thickness, permeability, and intactness of any soil layers that overlie the zone of high excess pore pressure. At level-ground sites, the relationship between liquefiable layer thickness and overlying crust thickness (Figure 9.66) can indicate whether sand boils are likely to be observed.

There are pitfalls to reliance upon the presence of sand boils for evidence of liquefaction-related phenomena; liquefaction at great depths or in thin layers may not produce sand boils, but lower excess pore pressures (i.e., r_u close to but less than 1.0) in thick layers at shallow depths may produce flow velocities high enough to cause sand boils even though liquefaction was not triggered. Also, the low permeability of silty sand may prevent porewater from flowing quickly enough to produce sand boils, even if high excess pore pressures develop. The nature or source of the liquefied soil cannot be accurately judged from the nature of the materials that comprise a sand boil since the upward-flowing water may preferentially carry the smaller particles of a broadly distributed liquefied soil, or entrain particles from the overlying layers and also bring them to the ground surface.

Sand boils in undeveloped areas are generally of little engineering consequence by themselves, but they are useful indicators of high excess pore pressure generation. In developed areas, however, the extrusion of soil from beneath the original ground surface can have a number of damaging effects. The pure loss of volume of the extruded soil will result in settlement of the original ground surface (Section 9.6.6.2), which can affect structures and utilities supported above the source of the extruded soil. In many cases, the sources of ejecta are localized so differential settlements in their vicinity can be large. In other cases, however, liquefaction may be widespread with large volumes of soil being ejected over broad areas. In areas of Christchurch, New Zealand, liquefaction caused by four large earthquakes in a 15-month period brought over 500,000 tons of ejecta to the ground surface (Figure 9.4) covering large areas with 50–60 cm of soil. The ejecta, some of which was contaminated with raw sewage from broken sewer pipes, covered roads, impeded traffic and recovery efforts and presented health hazards. Ejecta can also come from beneath structures, leading to a component of settlement that is very difficult to predict. In areas of Christchurch, the great majority of the observed ground surface settlement was caused by extrusion of ejecta (van Ballegooy et al., 2014).

9.6.5 Instability

Liquefaction-induced instabilities are among the most damaging of all earthquake hazards. Their effects have been observed in the form of flow slides, lateral spreads, retaining wall failures, and foundation failures in many earthquakes throughout the world. Instabilities can be produced by different liquefaction phenomena and, because it is not always clear exactly which is most likely to occur at a given site, it is generally good practice to check all possible modes of failure.

9.6.5.1 Flow Failures

Slopes are stable when the shear strength of the soil is greater than the shear stress required to maintain equilibrium under static (gravity-induced) stresses. The triggering of liquefaction, however, can cause the available shear strength to decrease. Liquefaction-induced flow failures occur when the shear strength of the soil drops below the shear stress required to maintain static equilibrium. This situation can arise in several different ways; the National Research Council (1985) identified four conceptual mechanisms of flow failure.

Flow Liquefaction Failures (NRC Mechanism A)

Flow liquefaction represents an important flow failure mechanism. Flow liquefaction occurs under totally undrained conditions – no redistribution of pore water (or change in void ratio) is involved. As described in Section 9.5.1.2, flow liquefaction is initiated when sufficient pore pressure is generated to move the effective stress path of an element of soil from its initial position (Point A in Figure 9.70a) to the FLS (Point B). When that occurs, the soil skeleton becomes unstable and flow liquefaction failure begins. The soil fails and large strains are driven by the static shear stresses (corresponding to Point A), which exceed the available steady state strength (corresponding to Point C). The flow liquefaction failure, therefore, begins at the location(s) where liquefaction is first initiated by earthquake shaking. It should be noted that, although flow liquefaction can produce very large strains and deformations (Figure 9.70b), it is actually triggered at relatively low strain levels.

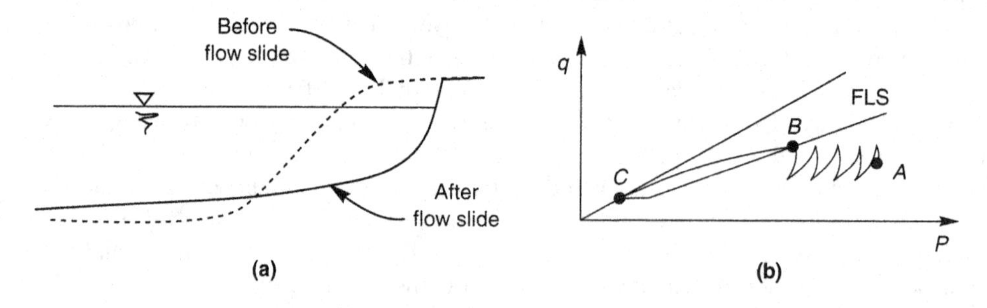

FIGURE 9.70 (a) Typical cross-section through a flow slide showing how liquefied materials can travel large distances and come to rest on very flat surfaces and (b) stress conditions at typical element of soil within failed mass. Prior to earthquake, element is in equilibrium at Point A under static shear stress greater than steady state (or residual) strength.

Under stress-controlled conditions, once the soil skeleton becomes unstable, a reduction in shearing resistance to the steady state strength is inevitable.

Flow liquefaction failures can develop progressively. When flow liquefaction is initiated at a particular location, the shearing resistance drops to the steady state strength. The static shear stresses that were previously resisted at that location must then be transferred to the surrounding soil, where they may initiate further flow liquefaction. As the redistribution of stresses proceeds, the zone of liquefaction grows. Eventually, through this process of *progressive failure*, a massive flow slide may develop. Thus, instability at a particular point can lead to a widespread failure even if the calculated factor of safety for the broader feature is greater than zero. Therefore, flow slide stability should be evaluated assuming that all of the liquefiable soil has liquefied and can only mobilize its liquefied strength.

Local Loosening Flow Failure (NRC Mechanism B)

Since the steady state strength is usually quite sensitive to soil density, a small amount of loosening can reduce the steady state strength substantially. In some cases, loosening may reduce the steady state strength to a value smaller than the shear stress required for equilibrium, thereby producing a flow failure.

If a sand layer is overlain by a less permeable material that does not permit drainage during the earthquake itself, the total volume of the sand will remain constant. If a condition of initial liquefaction (zero effective stress) is reached, however, the sand particles may rearrange under the action of gravity so that the lower part of the layer becomes denser and the upper part looser. This phenomenon is referred to as *void redistribution* (Fiegel and Kutter, 1994). If the upper part loosens sufficiently to reduce the steady state strength to a value smaller than the static shear stress, a local loosening flow failure can occur, as illustrated in Figure 9.71. In extreme cases, a *water interlayer* may form beneath the less permeable material (Liu and Qiao, 1984; Kokusho, 1999, 2000b; Malvick et al., 2008). Since the water interlayer would have zero shear strength, a flow failure could easily be produced.

Global Loosening Flow Failure (NRC Mechanism C)

High excess pore pressures generated at depth will cause porewater to flow toward drainage boundaries during and after an earthquake. As illustrated in Figure 9.72a, most of the flow is usually directed toward the ground surface and can result in a slightly different form of void redistribution. Shallow soils may rebound as the upward-flowing water causes the pore pressure to rise (and effective stress to decrease). The void ratio may increase to the extent that the steady state strength drops below the shear stress required to maintain equilibrium. In contrast with the local loosening case,

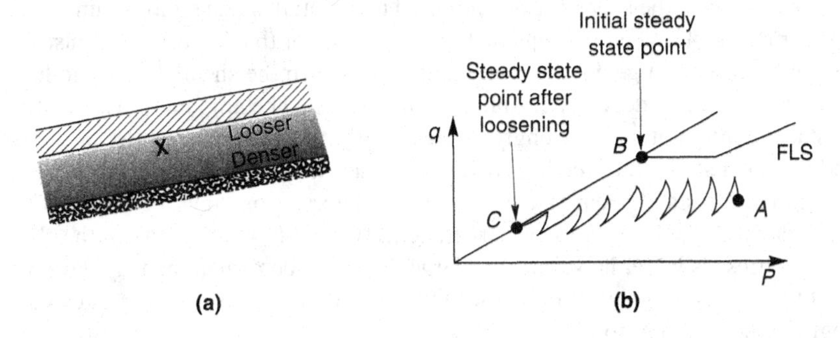

FIGURE 9.71 Flow failure due to loosening: (a) redistribution of grains in sand layer causes local volume change even though total volume remains constant; (b) stress conditions at point marked with x. Prior to earthquake, soil is in equilibrium at Point A with steady state strength (Point B) that is greater than static shear stress. As effective stresses are reduced during earthquake, loosening of soil reduces steady state strength to lower value (Point C), thereby allowing flow failure to occur.

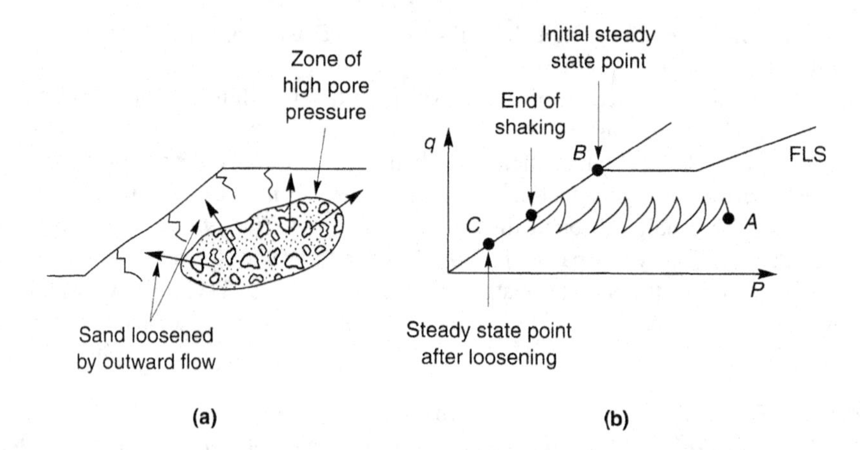

FIGURE 9.72 (a) Example of global loosening flow failure (after National Research Council, 1985) where earthquake-induced pore pressure at depth causes flow that loosens surficial soils; (b) possible effective stress path for element of surficial soil. Prior to earthquake, static shear stress (Point A) is less than steady state strength (Point B). Loosening due to outward flow of porewater reduces steady state strength (from Point B to Point C), allowing flow failure to occur.

this phenomenon does not require flow to be impeded by a lower permeability layer. The process is illustrated schematically in Figure 9.72b. Since the steady state strength is not reduced until water flows into the shallower soil, failure may not occur until well after earthquake shaking has ended. In some slopes and embankments, higher pore pressures may be generated in areas where static shear stresses are low than in areas where they are high. Dissipation of those pore pressures may cause water to flow into the areas of high static shear stress following earthquake shaking, where they can initiate flow sliding.

Interface Flow Failure (NRC Mechanism D)

Flow-type failures can also occur when the shear strength of the interface between a liquefiable soil and a structure becomes smaller than the shear stress required for equilibrium. Plunging failure of friction piles (De Alba, 1983) is an example of an interface flow failure. If the interface is smooth, as with steel or precast concrete piles, interface flow failure does not require volume change of the soil and therefore can occur in contractive or dilative sands.

Shear Strength of Liquefied Soil

Flow sliding occurs when the shear stresses required to maintain static equilibrium of a soil deposit (the static shear stresses) exceed the liquefied shear strength of that deposit. The unstable soil then accelerates until it reaches a geometric configuration in which the shear stresses no longer exceed the shear strength, and then decelerates until it comes to rest. The amount of deformation required to reach a stable configuration is strongly influenced by the difference between the static shear stresses and the shear strength of the liquefied soil. If the shear strength of the liquefied soil is only slightly lower than the static shear stress, only small deformations may be required to bring the shear stresses down to the level of the shear strength. If the difference between the shear strength and static shear stress is large, however, very large and rapid deformations may develop. Accurate evaluation of the effects of liquefaction-induced flow instability requires accurate estimation of the shear strength of the liquefied soil.

Conceptually, estimation of the strength of liquefied soil would involve measuring the ultimate steady state strength on undisturbed samples in the laboratory. Steady state concepts offer an excellent framework for understanding the effects of important variables on the undrained strength of soils that have reached the large strains associated with the steady state of deformation. Figure 6.27

illustrates the position of the SSL with respect to shear stress, effective stress, and void ratio, and how those quantities vary along the SSL. Figure 6.38 shows how the mean effective stress at the steady state, and hence the steady state shearing resistance, increases with increasing density (decreasing void ratio). As an element of soil is subjected to higher effective stresses during deposition, it is compressed to lower void ratios. The extent to which it becomes more or less contractive as it compresses, however, depends on the spatial relationship between its compression curve and its SSL; if the compression curve and SSL are parallel (in log p' space), the state parameter, ψ (Figure 6.42) will remain constant and the steady state strength, S_{us}, will be proportional to the effective stress. If the two curves diverge with increasing effective stress, the soil will become more contractive, i.e., the state parameter will increase.

As useful as the steady state framework may be for understanding large-strain shearing resistance, it is usually very difficult to obtain samples of liquefaction-susceptible soils for laboratory testing and steady state strengths are very sensitive to even small changes in density that are difficult to prevent during sampling. While rational laboratory test-based procedures for estimation of steady state strength have been developed (Poulos et al., 1985), their sensitivity to parameters that are difficult to quantify leads to high uncertainty in their results (Kramer, 1989). Also, the restrictive conditions associated with the steady state of deformation (Section 6.4.3.4) rarely exist during actual flow sliding events. As a flow failure develops in the field, drainage can occur leading to changes in effective stresses, volume, and density, strain rates are variable as the failing soil mass accelerates and then decelerates as it comes to rest, and different soils can be mixed together during flow. These factors also limit the direct applicability of laboratory-measured shear strengths to actual flow slides. As a result, the potential for flow sliding is generally evaluated using *residual strengths* that are back-calculated from flow slide case histories. It is also important to recognize that the back-calculated residual strength is influenced by a number of factors including boundary conditions, partial drainage, void redistribution, mixing (Naesgaard and Byrne, 2005), and hydroplaning (Mohrig et al., 1999; Harbitz et al., 2003; De Blasio et al., 2004) that vary from one case history site to another. The residual strength should therefore not be treated as a soil property, but rather as a system parameter that reflects the apparent shearing resistance mobilized in past case histories. While far from an ideal parameter, it is the best measure of expected resistance that is currently available and is widely used for evaluation of flow slide stability.

Three primary approaches to the estimation of residual strength of liquefied soils have been developed. Each is based on back-calculated strengths from flow slide case histories, and they differ predominantly in the manner in which the residual strength is assumed to vary with effective stress level. Because the flow slide case history database is relatively small at present, it is not possible to definitively determine which approach is most accurate. As such, it is generally prudent to estimate residual strengths using multiple approaches and to weigh the results of the approaches with consideration of their applicability to the site of interest and their consistency with basic principles of soil mechanics.

Direct Approach As an alternative to laboratory-based procedures, Seed (1987) developed a correlation to estimate the residual strength back-calculated from observed flow slides directly from corrected SPT resistance. Seed and Harder (1990) re-analyzed many of the flow slides and added new data to develop a graphical relationship between residual strength and an equivalent clean-sand SPT resistance (Figure 9.73). For sands with more than 10% fines, the equivalent clean-sand SPT resistance is obtained from

$$(N_1)_{60,cs} = (N_1)_{60} + N_{corr} \tag{9.71}$$

where N_{corr} is obtained from Table 9.11 and is only intended to correct the SPT resistance for the effects of fines.

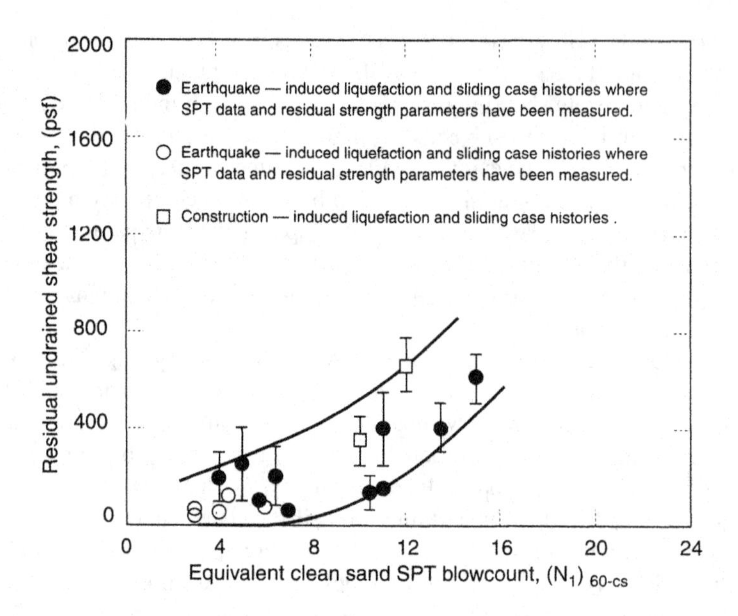

FIGURE 9.73 Estimation of residual strength from SPT resistance. (Seed and Harder, 1990.)

TABLE 9.11
Recommended Fines Content Correction for Seed and Harder (1990) Residual Strength Model

FC (%)	0	10	25	50	75
N_{corr}	0	1	2	4	5

The back-calculation procedures used to obtain the residual strength are not simple - many assumptions and approximations are involved. Combined with the variability of the SPT and the difficulty in selecting a single SPT value to represent each case history, these residual strengths must be considered approximate, as suggested by the wide band in Figure 9.66. The lower-third or lower quartile strength value in the Seed and Harder (1990) relationship has been commonly used in practice.

The direct approach is consistent with steady state concepts in that the residual strength is a function only of soil density, to the extent that the density of the soil is accurately represented by the corrected penetration resistance. The penetration resistance correction for effective stress, however, was not developed specifically for large-strain strength behavior, and it is not clear that it adequately captures the variation of density with depth in the field. More recent methods have used effective stress as an additional predictive variable.

Normalized Strength Approach: Residual Strength Ratio The concept of normalized strength is widely accepted in geotechnical engineering practice for cohesive soils (Section 6.4.3.3). Since the density of sand increases with increasing effective stress, the general concept has also been applied to the residual strength of liquefied soils. If the consolidation curve and steady state line of a liquefiable soil are parallel (on a standard semi-log plot of void ratio vs. effective stress), the steady state strength will be proportional to the consolidation stress (i.e., S_{su}/σ'_{v0}=constant). Application of this concept is complicated by the fact that sandy soils do not exhibit unique consolidation curves and because consolidation and steady state curves are typically not parallel for sands; hence, the ratio S_{su}/σ'_{v0} is not unique for a given soil.

The basic normalized strength approach requires site-specific measurement of the residual strength ratio. It is logical to expect that different soils would have different residual strength ratios; for example, the residual strength ratio of a well-graded angular sand should be greater than that of a uniform sand with rounded particles. Assuming that the factors that influence the residual strength ratio also influence penetration test results, it should be possible to relate the residual strength ratio to penetration resistance. Stark and Mesri (1992) used back-calculated and laboratory-measured residual strengths from field case histories to correlate values of residual strength ratio to SPT resistance. Olson and Stark (2002) expanded the case history database of Stark and Mesri (1992) and, performing more detailed analyses of many of them, did not find a systematic variation of normalized residual strength with fines content. Olson and Stark (2002) recommended that the residual strength ratio be estimated as

$$\frac{S_r}{\sigma'_{v0}} = 0.03 + 0.0075(N_1)_{60} \quad \text{for} \quad (N_1)_{60} \le 12 \tag{9.72}$$

$$\frac{S_r}{\sigma'_{v0}} = 0.03 + 0.0143 q_{c1} \quad \text{for} \quad q_{c1} \le 6.5 \text{ MPa} \tag{9.73}$$

with a listed error term for the ratio of ± 0.03 (but a specified $\sigma_{S_r/\sigma'_{v0}} = 0.025$). The Olson and Stark (2002) relationships are shown in Figure 9.74.

Idriss and Boulanger (2007) considered a subset of 18 flow slides from the Olson and Stark (2002) case history database. The 18 case histories were divided into three quality groups based on adequacy of *in situ* measurements and geometric information. Normalizing the average back-calculated strengths (from as many as available) computed by Seed (1987), Seed and Harder (1990), and Olson and Stark (2002) by initial vertical effective stress, Idriss and Boulanger (2007) produced a relationship for normalized residual strength ratio that can be expressed as

$$\frac{S_r}{\sigma'_{v0}} = \exp\left[\frac{(N_1)_{60cs-S_r}}{16} + \left(\frac{(N_1)_{60cs-S_r} - 16}{21.2}\right)^3 - 3.0\right] \times \left[1 + \beta \exp\left(\frac{(N_1)_{60cs-S_r}}{2.4} - 6.6\right)\right] \le \tan\phi' \tag{9.74}$$

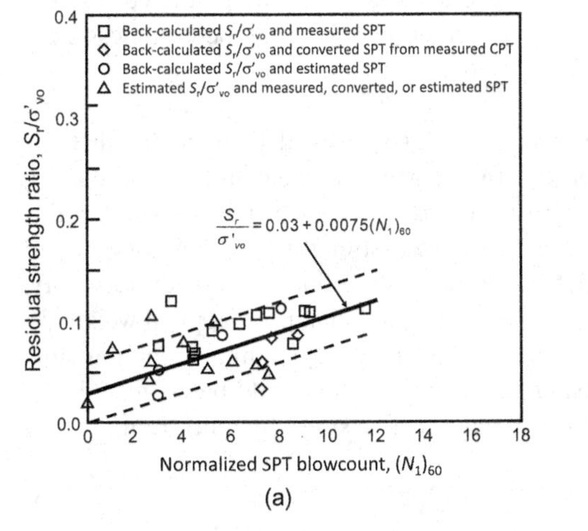

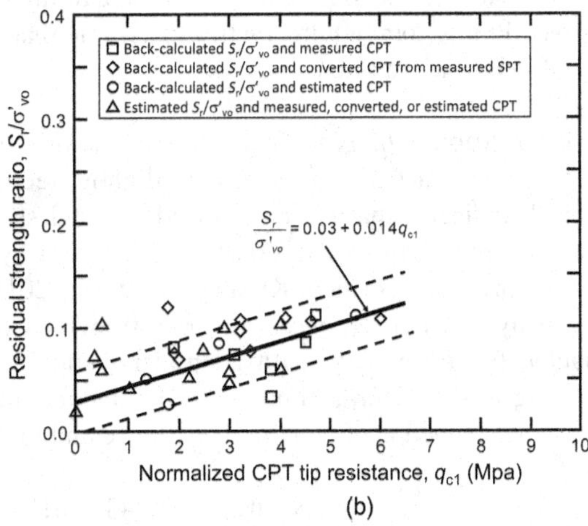

FIGURE 9.74 Residual strength ratio of liquefied soil as function of (a) corrected CPT resistance and (b) corrected SPT resistance. (After Olson and Stark, 2002 with permission of Canadian Science Publishing.)

TABLE 9.12
Fines Correction Factors for Idriss and Boulanger (2007) Residual Strength Relationship

Fines Content, *FC*	$(\Delta N_1)_{60cs\text{-}Sr}$	$\Delta q_{c1N\text{-}Sr}$
10	1	10
25	2	25
50	4	45
75	5	55

where $(N_1)_{60cs-S_r} = (N_1)_{60cs} + \Delta(N_1)_{60cs-S_r}$ and β=0 or 1 when void redistribution effects are significant or insignificant, respectively. Values of $\Delta(N_1)_{60cs-S_r}$ are given in Table 9.12. Using a conversion from corrected SPT resistance to corrected CPT resistance, Idriss and Boulanger (2007) produced the following equation for estimating residual strength ratio from CPT resistance

$$\frac{S_r}{\sigma'_{v0}} = \exp\left[\frac{q_{c1Ncs-S_r}}{24.5} - \left(\frac{q_{c1Ncs-S_r}}{61.7}\right)^2 + \left(\frac{q_{c1Ncs-S_r}}{106}\right)^3 - 4.42\right] \times \left[1 + \beta \exp\left(\frac{q_{c1Ncs-S_r}}{11.1} - 9.82\right)\right] \leq \tan\phi' \tag{9.75}$$

where $q_{c1Ncs-S_r} = q_{c1Ncs} + \Delta q_{c1N-S_r}$. Values of Δq_{c1N-S_r} are given in Table 9.12. The Idriss and Boulanger (2007) relationship is shown in Figure 9.75. The effects of void redistribution in Figure 9.75 are largely present beyond the range of the data, and hence those portions of the models in Equations (9.72 and 9.73) are mainly informed by judgment. The decision as to whether to consider void redistribution effects to be significant can be seen to have a large effect on the residual strength ratio for denser soils. The potential for void redistribution to have a significant effect on residual strength depends on a number of factors including the presence of significant permeability gradients, which can occur when clean sands are "capped" by a fine-grained layer or when non-uniformly graded soils are deposited hydraulically (by natural or artificial means), resulting in banded layers of coarser- and finer-grained materials. The thicknesses, densities, and stress conditions in the strata can also affect void redistribution potential. The fact that these factors are generally unknown indicates that there is considerable, and unquantified, epistemic uncertainty in the predicted residual strength; it is common in practice to assume that void redistribution effects can occur unless clear evidence exists to rule it out.

Mixed Approach: Non-Proportional Stress Dependence Laboratory data indicates that the steady state lines of many sands are slightly steeper than their corresponding consolidation curves, which indicates that the normalized residual strength should decrease with increasing effective stress rather than remain constant as implied by the normalized strength procedures described in the preceding section. Kramer and Wang (2015) relaxed the assumption of strength normality, re-analyzed a number of case histories from the Olson and Stark (2002) database, applied weighting factors for case history quality, considered conditions under which lateral spreading rather than flow sliding would occur, and proposed that median residual strength (in atm) could be estimated as a nonlinear function of both SPT resistance and initial vertical effective stress (in atm) from

$$\ln S_r = -8.444 + 0.109(N_1)_{60} + 5.379(\sigma'_{vo})^{0.1} \tag{9.76}$$

Median residual strengths are shown in Figure 9.76. The uncertainty in residual strength can be estimated as

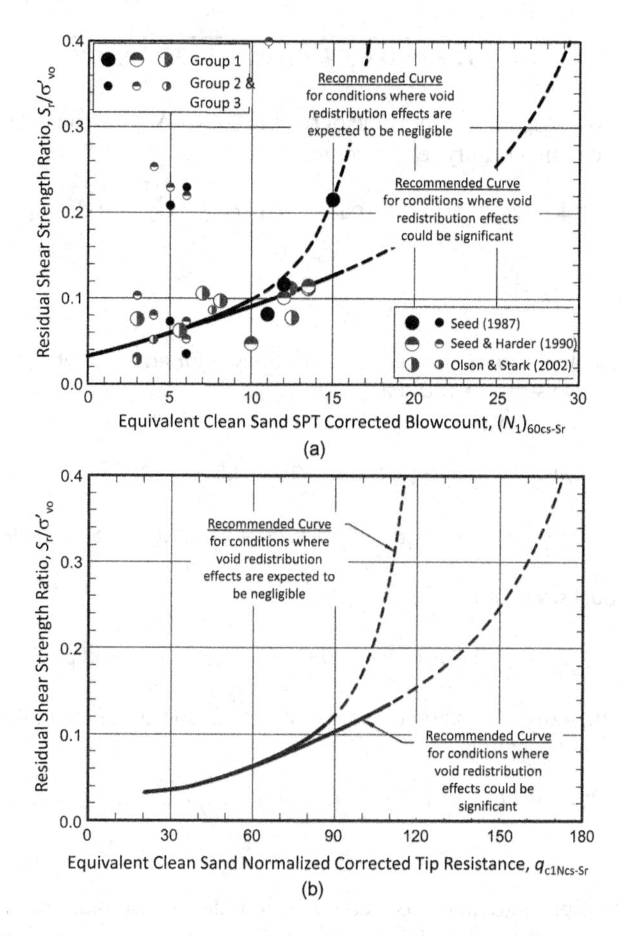

FIGURE 9.75 Residual strength ratio of liquefied soil as function of median values of (a) corrected SPT resistance and (b) corrected CPT resistance. (After Idriss and Boulanger, 2007 with permission of Elsevier Science and Technology Journals.)

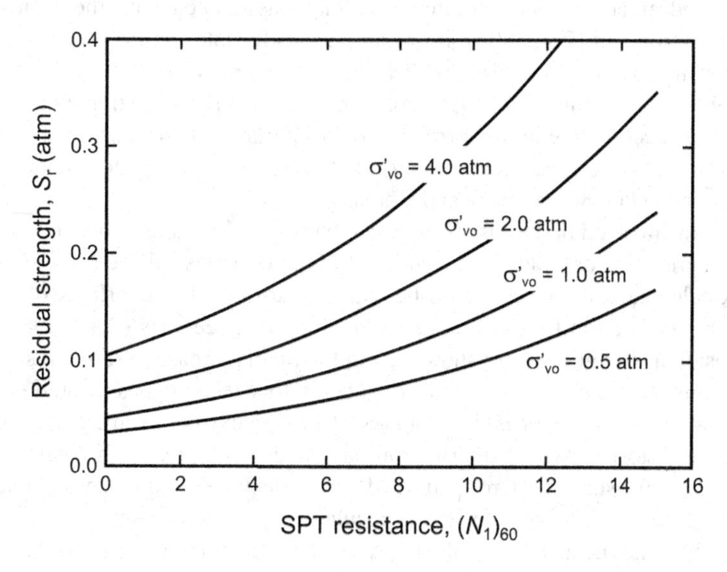

FIGURE 9.76 Median residual strength as function of SPT resistance in initial vertical effective stress. (After Kramer and Wang, 2015 with permission of ASCE.)

$$\sigma_{\ln Sr} = \sqrt{\sigma_m^2 + 0.00073N^2COV_N^2 + 4.935S^{0.2}COV_S^2} \tag{9.77}$$

where N=mean value of $(N_1)_{60}$, S=mean value of σ'_{v0} in atm, COV=coefficient of variation (Section D.6.2), and σ_m is a model uncertainty term computed as,

$$\sigma_m^2 = 1.627 + 0.00073N^2 + 0.0194N - 0.027NS^{0.1} - 3.099S^{0.1} + 1.621S^{0.2} \tag{9.78}$$

Example 9.5

Calculate the median and 10th percentile residual strength for an element of soil with $(N_1)_{60}=10$ under a vertical effective stress of 100 kPa.

Solution:

Using Equation (9.76) with $N=14$ and $S=100$ kPa/101.3 kPa=0.987

$$\ln S_r = -8.444 + 0.109(N_1)_{60} + 5.379(\sigma'_{v0})^{0.1} = -8.444 + (0.109)(10) + 5.379(0.987)^{0.1} = -1.982$$

so the median residual strength is

$$S_r = \exp[\ln S_r] = \exp[-1.982] = 0.138 \text{ atm} = 14.0 \text{ kPa}$$

Then, using $\sigma_{\ln Sr} = \sigma_m = 0.43$, the standard normal variate for the 10th percentile is z=−1.28 so the 10th percentile residual strength is

$$S_r = \exp\left[\ln S_r + \Phi^{-1}(P)\sigma_{\ln Sr}\right] = \exp\left[-1.982 + (-1.28)(0.43)\right] = 0.079 \text{ atm} = 8.1 \text{ kPa}$$

Void Redistribution

The importance of pore pressure and void redistribution, described under the local loosening flow failure (NRC Mechanism B) heading in Section 9.6.5.1, in the development of flow slides is now well recognized (Liu and Qiao, 1984; Whitman, 1985; Boulanger and Truman, 1996; Kokusho, 1999, 2000b, 2003; Kulasingam et al., 2004; Malvick et al., 2008; Kamai and Boulanger, 2010). Under the truly undrained (constant volume) conditions associated with the steady state, simple calculations (Robertson and Fear, 1995) show that the steady-state strength should increase rapidly due to dilation at higher relative densities. Undrained loading tests (e.g., Ishihara, 1993; Vaid and Sivathayalan, 1996; Yoshimine et al., 1999) confirm this behavior. In extreme cases of void redistribution, water may accumulate in the form of a water interlayer at the base of an impervious layer. Such interlayers would have zero shear strength, but are unlikely to be continuous or planar over long distances in the field (Montgomery and Boulanger, 2014).

Many of the soils involved in the case history database are silty sands deposited in water; as such, some degree of segregation or banding of coarser and finer particles existed prior to failure. In those cases, the back-calculated strengths would be expected to include the effects of void redistribution so that further correction for those effects would not be necessary (Seed, 1987). Indeed, case history-based residual strengths do not show the kind of rapid increase in residual strength at higher densities that steady state concepts and undrained laboratory tests suggest should occur. A reasonable interpretation of this behavior is that void redistribution exists to some degree in nearly *all* of the flow slide case histories. Void redistribution phenomena complicate the interpretation of case histories because the measured SPT resistances do not reflect the local loosening that produces the mobilized residual strength. Because the back-calculated strength corresponds to a looser condition than that inferred by the measurement of SPT resistance, the error can introduce a potential bias into the relationship between residual strength and SPT resistance. The extent to which subsurface density and permeability can be characterized sufficiently to determine whether void redistribution potential is or is not significant at a particular site is unclear at present. This situation provides additional motivation to consider more than one predictive procedure to estimate residual strength.

Discussion

Evaluation of the residual strength of liquefied sand remains one of the most difficult (and important) problems in contemporary geotechnical earthquake engineering practice. Steady state concepts are very useful for understanding the behavior of liquefiable soil, but steady state strengths correspond to conditions that seldom exist in actual flow slides. As a result, residual strengths are more commonly back-calculated from flow slide case histories. Because flow slides do not occur as frequently as lateral spreads, the flow slide case history database is much smaller and contains a number of older case histories that are poorly documented and require a higher level of interpretation than case histories for other aspects of liquefaction hazard evaluation. As a result, developers of residual strength models have back-calculated different residual strengths for the same case histories. They have also assigned different penetration resistances and different effective stress levels, and have made different assumptions about the effects of factors such as effective stress, fines content, and void redistribution on residual strength. Another challenge is that the case histories do not span the entire range of conditions for which residual strengths may need to be estimated. For example, the case history database does not include flow slides in materials with corrected SPT resistance greater than about 14; as a result, empirical strength models are extrapolated, in some cases based on judgment, to higher SPT resistances.

The inevitable result of this situation is a suite of residual strength models that are generally consistent but produce different estimates of residual strength for a given problem. The variability in estimated residual strengths is an example of model uncertainty, an important component of epistemic uncertainty. In the future, the availability of additional case histories that are better documented than those of the past should lead to a reduction in model uncertainty as predictive models are updated and new models are developed. In the meantime, model uncertainty should be recognized and accounted for using a logic tree with appropriate weighting factors.

Stability Analysis

Evaluation of stability with respect to flow sliding typically involves relatively conventional static, limit equilibrium slope stability analyses with residual strengths assigned to zones of soil in which liquefaction is expected to have been triggered and are discussed in detail in Section 10.8.1. Such analyses should also account for any expected cyclic degradation of the strengths of non-liquefied soils.

Because residual strengths are highly uncertain and the consequences of flow sliding are so severe, flow slide stability analyses are often performed and interpreted in a conservative, deterministic manner. In practice, this may take the form of (a) assigning conservative residual strengths to soil units with triggering factors of safety less than 1.0, (b) requiring a relatively high allowable factor of safety against sliding with residual strengths assigned to soils with triggering factors of safety below 1.0, or (c) using a low allowable factor of safety against sliding (FS = 1.0–1.2) computed after assigning residual strengths to soils with triggering factors of safety below relatively high threshold values (e.g., $FS_L < 1.2$–1.4). However, the use of conservative inputs to deterministic analyses does not provide any quantitative information on the probability of flow sliding or the damage and loss it can cause. Such probabilistic assessments are facilitated by aleatory variability distributions of residual strength, estimated using the Olson and Stark (2002) and Kramer and Wang (2015) models, in combination with model uncertainties accounted for using a logic tree.

9.6.5.2 Lateral Spreading

Not all liquefaction-related instabilities involve flow and very large displacements. At the element level, cyclic mobility in the presence of a static shear stress can cause significant permanent shear strain to accumulate (Figure 9.29), even if the shear stresses are lower than the shear strength of the soil. Integrating the incremental strains over the spatial extent of the liquefied zone in a soil profile

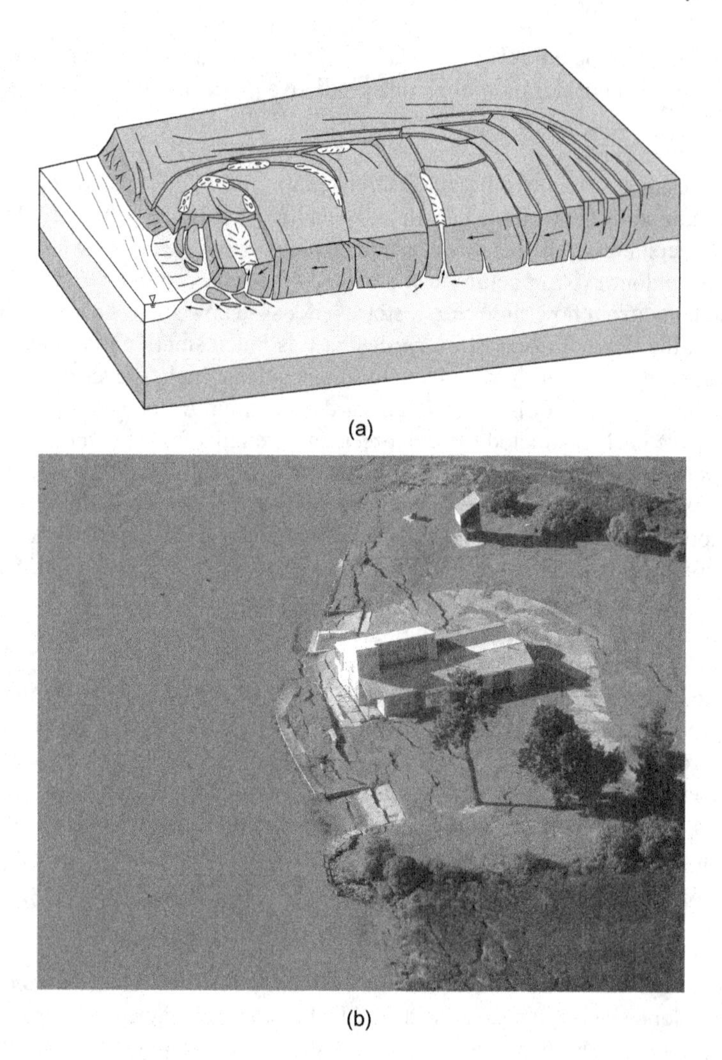

FIGURE 9.77 Lateral spreading: (a) typical deformation patterns in laterally spreading river bank soils (after Rauch, 1997; Varnes, 1978), and (b) ground cracking and damage to Pleasant Point Yacht Club along the Avon River during the February 2011 Christchurch earthquake. (Photo courtesy of M. Cubrinovski.)

produces incremental deformations that, by the end of an earthquake, may result in significant permanent ground deformation and extensive damage. *Lateral spreading* is an example of such a deformation failure. As illustrated in Figure 9.77, lateral spreading beneath a crust of unsaturated or non-liquefiable soil often causes the crust to break into blocks that progressively move downslope or toward a free face during earthquake shaking. The ground surface may develop fissures and scarps at the head of the lateral spread, shear zones along its lateral margins, and compression zones at the toe. Substantial amounts of soil may be ejected from these fissures, leading to further settlement of the original ground surface. The surficial blocks usually move irregularly in both horizontal and vertical directions; buildings and pipelines extending through the head of a lateral spread may be pulled apart, pipelines crossing the lateral margins may be sheared, and bridges or pipelines near the toe may be buckled. Deep foundations that extend through the laterally spreading soil may suffer bending failure (Figure 9.68). Lateral displacements usually range from a few centimeters to a meter or two, but may be larger if shaking is particularly strong or of long duration. Void redistribution may result in additional deformations that are of hydraulic/gravitational rather than inertial origin.

Lateral spreading is a complex phenomenon, not only due to the complexity of the stress-strain behavior of the soils that produce it, but also due to spatial variability, complex three-dimensional topography, and pore pressure and void redistribution which are sensitive to site characteristics that are difficult to quantify. A number of procedures have been proposed for estimation of deformations due to lateral spreading. These procedures generally fall into one of three categories: purely empirical, semi-empirical strain potential, and numerical.

Purely Empirical Procedures

Several investigators have used databases of lateral spreading case histories to develop predictive equations for lateral spreading displacement. Bartlett and Youd (1992) compiled a large database of lateral spreading case histories from Japan and the western United States. They identified a set of material, geometric, and loading parameters that correlated to lateral spreading displacement. Youd et al. (2002) used an expanded and corrected version of the 1992 database to develop a relationship to predict lateral spreading displacement from those parameters. The Youd et al. (2002) model requires that a slope be classified as either a *free-face case* (flat slope near a steep bank) or a *ground-slope case* (gently sloping ground). The lateral spreading displacement (in meters) can then be predicted from

$$\log D_H = b_0 + b_1\mathbf{M} + b_2 \log R^* + b_3 R + b_4 \log W + b_5 \log S + b_6 \log T_{15} + b_7 \log(100 - F_{15}) + b_8 \log(D50_{15} + 0.1\text{ mm}) \quad (9.79)$$

where $\mathbf{M}$=moment magnitude, $R^* = R + 10^{-0.89M-5.64}$, R=closest horizontal distance to the energy source, W=free-face ratio, S=ground slope inclination, T_{15}=cumulative thickness of soil layers with corrected SPT resistance, $(N_1)_{60}$, less than or equal to 15, in which liquefaction is expected to occur (i.e., $FS_L \le 1.0$), F_{15}=the average fines content (in percent), of the soil layers that contribute to T_{15}, and $D50_{15}$=mean grain size (in mm) of the soil layers that contribute to T_{15}. The values of W and S can be determined as indicated in Figure 9.78, and the values of the coefficients are presented in Table 9.13. Table 9.14 presents the recommended ranges of predictive variables for which the Youd et al. (2002) model is considered valid.

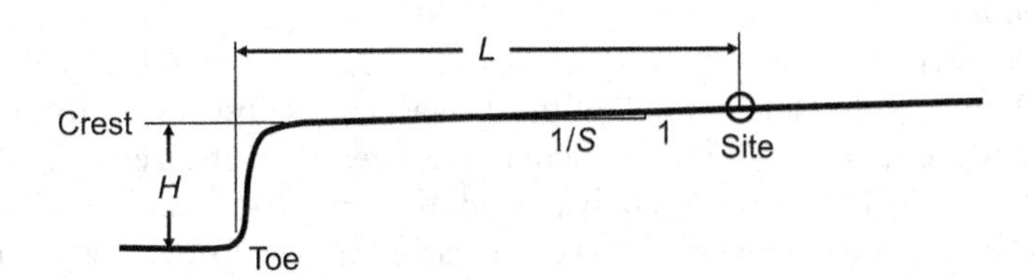

FIGURE 9.78 Slope Geometry Notation. (Youd et al., 2002.)

TABLE 9.13
Coefficients for Youd et al. (2002) Model

Case	b_0	b_1	b_2	b_3	b_4	b_5	b_6	b_7	b_8
Ground slope	−16.213	1.532	−1.406	−0.012	0	0.338	0.540	3.413	−0.795
Free face	−16.713	1.532	−1.406	−0.012	0.592	0	0.540	3.413	−0.795

TABLE 9.14
Recommended Range of Variable Values for the Youd et al. (2002) Predictive Equation

Variable	Description	Range
T_{15}	Equivalent thickness of saturated cohesionless soils (clay content ≤ 15%) in m	1–15 m
M	Moment magnitude of the earthquake	6.0–8.0
Z_T	Depth to top of shallowest layer contributing to T_{15}	1–15 m
W	Free face ratio	1%–20%
S	Ground slope	0.1%–6%
F_{15}, $D50_{15}$	See figure below	

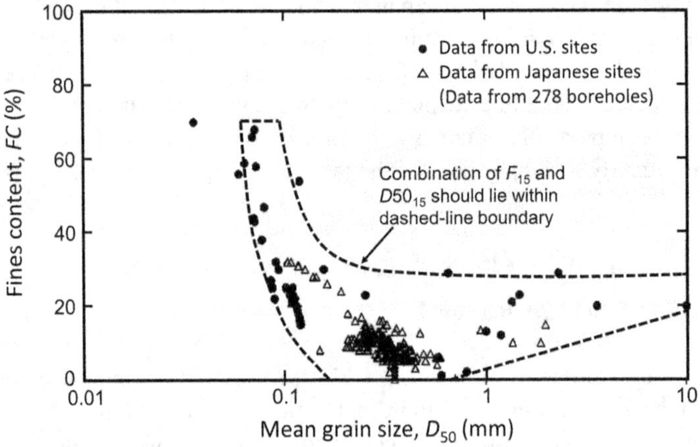

Youd et al. (2002) model implicitly assumes that sands with $(N_1)_{60} > 15$, although they can liquefy under sufficiently strong shaking, will be too dilative at the shallow depths at which lateral spreading generally occurs to contribute to lateral spreading displacement. It also implies, however, that lateral spreading displacement is insensitive to $(N_1)_{60}$ values below 15, which is inconsistent with behavior observed in laboratory tests.

Youd et al. (2002) reported that the great majority of lateral spreading displacement values given by Equation (9.79) were within a factor of 2 of the displacements observed at the case history sites. The variability in the Youd et al. (2002) predictions has been characterized by a logarithmic standard deviation, $\sigma_{\log D_H} = 0.197$, which is equivalent to $\sigma_{\ln D_H} = 0.464$ (Franke and Kramer, 2014).

While detailed information on fines content and mean grain size is often not reported in available case history data, descriptive information regarding the soils that comprise T_{15} is generally available. Gillins and Bartlett (2014) found that the T_{15} layer descriptions from the Youd et al. (2002) database fell into the five general soil index (SI) groups described in Table 9.15. Gillins and Bartlett (2014) then proposed that lateral spreading displacements could be predicted using the relative amounts of each SI group that constituted the T_{15} layer, i.e.

$$\log D_H = b_0 + b_1 M_w + b_2 \log R^* + b_3 R + b_4 \log W + b_5 \log S + b_6 \log T_{15} - 0.683 x_1 - 0.200 x_2 + 0.252 x_3 - 0.040 x_4 - 0.535 x_5 \quad (9.80)$$

where x_i is the fraction of T_{15} that is comprised of soils classified in SI group i and the common terms with Equation (9.79) are defined in the same way; the coefficients are as shown in Table 9.16. The coefficients on the x_i terms in Equation (9.80) indicate the relative contributions of each group to lateral spreading displacement; SI 3 soils can be seen to produce higher displacements than the other groups.

TABLE 9.15
Descriptions of Soil Index Groups Found in Youd et al. (2002) Database (Gillins and Bartlett, 2014)

SI	General USCS Symbol	Typical Soil Description
1	GM	Silty gravel with sand, silty gravel, fine gravel
2	GM-SP, SP	Very coarse sand, sand and gravel, gravelly sand, coarse sand, sand with some gravel
3	SP-SM	Sand, medium to fine sand, sand with some silt
4	SM, SM-ML	Fine sand, sand with silt, very fine sand, silty sand, dirty sand, silty/clayey sand
5	ML	Sandy silt, silt with sand

TABLE 9.16
Coefficients for Gillins and Bartlett (2014) Model

Model	b_0	b_1	b_2	b_3	b_4	b_5	b_6
Ground slope	−8.208	1.318	−1.073	−0.016	0	0.337	0.592
Free face	−8.208	1.308	−1.073	−0.016	0.445	0	0.592

Example 9.6

The gently sloping site shown below consists of a 2-m-thick layer of silty clay overlying a 4 m thick layer of loose, saturated silty sand. The sand has an average fines content of about 3% and an average $D50 = 0.22$ mm. Subsurface investigations indicate that the corrected SPT resistance of the silty sand is quite consistent with an average value of 11 blows/ft. Estimate the permanent displacement of the slope due to a **M**6.5 earthquake occurring at a (horizontal source-site) distance of 30 km.

Solution:

Using the Youd et al. (2002) approach (Equation 9.79), the relevant parameters are $\mathbf{M} = 6.5$, $S = \tan(2°) = 0.035 = 3.5\%$, $T_{15} = 4$ m, $F_{15} = 3$, and $D50_{15} = 0.22$ mm. Then, using the ground slope model coefficients from Table 9.13,

$$\begin{aligned}\log D_H &= b_0 + b_1\mathbf{M} + b_2 \log R^* + b_3 R + b_4 \log W + b_5 \log S + b_6 \log T_{15} + b_7 \log(100 - F_{15}) \\ &\quad + b_8 \log(D50_{15} + 0.1\text{ mm}) \\ &= -16.213 + (1.532)(6.5) + (-1.406)(\log 30) + (-0.012)(30) + (0.338)(\log 3.5) \\ &\quad + (0.540)(\log 4) + (3.413)(\log 97) + (-0.795)(\log(0.22 + 1)) = -1.471\end{aligned}$$

So $D_H = 10^{-1.471} = 0.034\text{ m} = 3.4\text{ cm}$

Semi-Empirical Strain Potential Procedures
Another approach to lateral spreading displacement estimation makes use of laboratory test data to characterize potential cyclic strain amplitude as a function of soil density and loading amplitude. The predicted cyclic strain is integrated over the thickness of a liquefiable layer to produce a displacement index that reflects the densities and thicknesses of the liquefied layers in a soil profile. That index is then used in a regression with site geometric parameters (e.g., ground slope) from field case histories of lateral spreading to produce a predictive relationship for lateral spreading displacement. This approach has an advantage over purely empirical procedures in that it also provides a stronger physical basis for estimating both surface displacements and subsurface displacement patterns.

Zhang et al. (2004) used the laboratory test-based relationship shown in Figure 9.79 to develop a cumulative shear strain model for prediction of lateral spreading displacement. The maximum cyclic shear strains in Figure 9.79 were defined by Ishihara and Yoshimine (1992) as the maximum shear strain (in any direction) under transient loading conditions. Figure 9.79 shows how cyclic shear strain amplitudes increase quickly for soils in which liquefaction is triggered (particularly at low relative densities), and how sensitive the strain amplitude is to the relative density of the soil. Zhang et al. (2004) capped the maximum cyclic shear strains by the limiting shear strains proposed by Seed (1979) and used empirical relationships between relative density and penetration resistance, i.e.,

$$\begin{aligned} D_r &= 14\sqrt{(N_1)_{60cs}} \qquad \text{for } (N_1)_{60cs} \leq 42 \\ D_r &= -85 + 76\log(q_{c1Ncs}) \ \text{ for } \ q_{c1Ncs} \leq 200 \end{aligned} \tag{9.81}$$

to allow computation of a *lateral displacement index* by integrating the expected maximum shear strain over the thickness of the liquefiable layer,

$$LDI = \int_0^{Z_{max}} \gamma_{max} dz \tag{9.82}$$

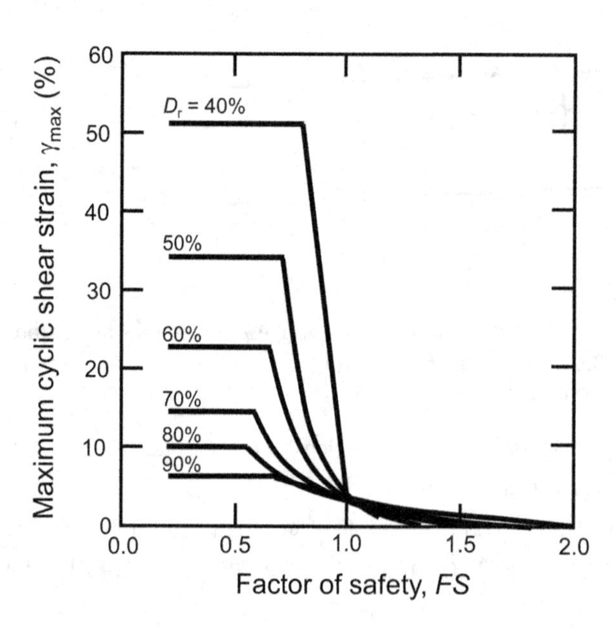

FIGURE 9.79 Variation of maximum cyclic shear strain with factor of safety and relative density. (After Zhang et al., 2004 with permission of ASCE.)

where Z_{max} is the depth below which all liquefaction-susceptible soils have $FS_L>2.0$ or 23 m (whichever is shallower).

The *LDI* value, therefore, accounts for the distribution of material properties (i.e., soil penetration resistance) at a site and their liquefaction response to ground motions. It does not, however, account for the initial shear stresses that drive actual displacements at ground slope and free-face sites. By calibrating the geometric parameters of Youd et al. (Figure 9.78) against empirical case history data, the expected lateral spreading displacement can be related to *LDI* as (Zhang et al., 2004)

$$D_H = \begin{cases} (S+0.2)\cdot LDI & \text{ground slope case} \\ 6W^{-0.8}\cdot LDI & \text{free-face case} \end{cases} \tag{9.83}$$

where S and W are defined as indicated in Figure 9.78 and subject to the respective limits $0.2\% < S < 3.5\%$ and $4 < W < 40$.

Care must be taken when using CPT data to estimate *LDI* in spatially variable soil deposits. Liquefiable soil deposits often have significant spatial variability, both in their texture and their density. While the vertical variability of liquefiable soils is readily seen in CPT tip resistance profiles, the soils are also variable in the horizontal direction (although the variability is not as rapid). Therefore, thin layers of loose, liquefiable soil may have limited lateral extents and exist as lenses bounded above, below, and to the sides by non-liquefiable soils. While such thin layers may liquefy, they may contribute little to lateral spreading deformations due to their isolated nature. The indiscriminate summation of all loose layers, no matter how thin, can lead to overestimation of *LDI* with consequent overestimation of lateral spreading displacement.

Example 9.7

The sand layer in Example 9.6 was explored with a CPT rig that produced the following data and triggering analysis results.

Depth	Mean q_{c1Ncs}	Mean I_c	Susceptible?	FS_L
2–3 m	140	2.3	Yes	0.85
3–4 m	120	2.4	Yes	0.78
4–5 m	95	2.9	No	n.a.
5–6 m	150	1.9	Yes	0.95

Estimate the lateral spreading displacement of the site using a semi-empirical procedure.

Solution:

The maximum cyclic shear strain for each sublayer can be determined from FS_L and the relative density of the soil using Figure 9.78. Relative density is estimated from mean q_{c1Ncs} using Equation (9.81). Since Layer 3 is non-susceptible, its contribution to *LDI* is zero. In tabular form,

Layer, i	Depth	q_{c1Ncs}	D_r (%)	FS_L	γ_{max} (%)	LDI_i	$D_{H,i}$
1	2–3 m	140	78.1	0.85	4.6	0.046	0.17
2	3–4 m	120	73.0	0.78	6.2	0.062	0.23
3	4–5 m	95	65.3	∞	0	0	0
4	5–6 m	150	80.4	0.95	3.8	0.038	0.14

Then the total predicted displacement is the sum of the layer displacements, 0.54 m, or 54 cm.

Numerical Procedures

Advances in understanding of liquefiable soil behavior, constitutive modeling of that behavior, and numerical analyses have made it possible to capture many of the most important aspects of the development of permanent deformations of soil profiles, including lateral spreading, in nonlinear effective stress analyses. That process, however, is complicated and can be sensitive to site characteristics that can be difficult to identify in advance. The ability of even the best of these procedures to consistently make accurate *a priori* predictions of lateral spreading displacements has not been demonstrated. They have advanced, however, to the point where they can show that a well-designed and constructed dam or embankment is unlikely to deform much, or that a poorly designed or constructed one will deform excessively. They can also be very useful for understanding the mechanism(s) and patterns of the expected deformations.

Any numerical analysis must solve the wave equation that governs the mechanical response of the model in one, two, or three dimensions. It is also desirable that the analysis solve the diffusion equation that governs the hydraulic response (i.e., pore pressure redistribution and dissipation) in the same number of dimensions. When the constitutive model is formulated in a critical state framework, this allows modeling of void redistribution effects. The analysis must also accurately model the constitutive behavior of the liquefiable soil in a manner that captures the primary behavior(s) of interest, and it is usually this aspect that distinguishes one numerical analysis from another.

A constitutive model used to estimate lateral spreading deformations should be able to model the pore pressure buildup that occurs prior to and at the triggering of liquefaction, and also capture cyclic mobility phenomena such as phase transformation behavior and fabric degradation that can occur near and after triggering. A critical aspect of the reliable use of numerical models is calibration of the constitutive model. Calibrated constitutive models should be checked at the element level to ensure that they produce rates of pore pressure generation, stress-strain behavior and strain amplitudes that are consistent with those observed in laboratory tests of representative soils of different densities subjected to different levels of static and cyclic loading under different effective confining stresses. For prediction of lateral spreading, it is particularly important to be able to model the effects of initial static shear stresses on pore pressure generation and post-triggering response. A limited cyclic test dataset (five tests on sands) indicated that the shear strain per post-triggering cycle decreased exponentially with increasing relative density (Tasiopoulou et al., 2020) as

$$\frac{\Delta\gamma}{\tau_{cyc}} = \exp(3.332 - 0.1D_r + 0.5\varepsilon) \tag{9.84}$$

where D_r is in percent and ε is a standard normal variable (zero mean and unit standard deviation). This relationship allows calibration of constitutive models to better predict cyclic and permanent deformations of liquefiable soils.

After implementing a calibrated constitutive model into a numerical analysis program, the program should be validated by application to selected case histories and/or physical model (e.g., centrifuge) test results. These types of analyses are typically performed by model developers – Ziotopoulou and Boulanger (2013) describe a thorough and methodical process of calibration/validation of the PM4sand constitutive model implemented into FLAC, Plaxis, and OpenSees. Other models capable of representing important aspects of liquefiable soil behavior in two-dimensional analyses include the Stress-Density Model (Cubrinovski and Ishihara 1998a,b) implemented in DIANA and OpenSees, UBCSAND (Byrne et al., 2004) implemented in FLAC, PDMY (Elgamal et al., 2002) and OpenSees, and multi-spring and cocktail glass models (Iai et al., 2011) implemented in FLIP.

Many liquefaction-susceptible soils are deposited in manners that lead to significant spatial variability that can influence deformations. The development of significant deformations requires shear straining within and across the zone of soil associated with a deformation mechanism and thus depends on the distribution of soil properties (e.g., penetration resistance) within that zone.

While vertical property variability can be readily measured at the specific locations of explorations such as borings and CPT soundings, lateral variability is much more difficult to characterize. The effects of both vertical and lateral variability can be assessed in two- and three-dimensional numerical analyses in which randomized, spatially correlated soil properties are assigned to individual elements (Section 5.8.1). Establishment of stable estimates of mean response (and aleatory variability when needed), however, requires multiple analyses with different random realizations of the properties. As a result, it is more common to represent the effects of spatial variability by using uniform representative soil properties that produce response consistent with the mean response of spatially variable analyses. Popescu et al. (1997) found that analyses with uniform 20th percentile properties produced pore pressures consistent with Monte Carlo analyses based on full property distributions. Popescu et al. (2005) found that maximum lateral displacements based on uniform 50th percentile properties were similar to those from Monte Carlo analyses. More recent analyses of lateral spreading of gently sloping ground suggest that representative properties ranging from 30th to 70th percentile values with the higher values corresponding to thinner crusts, thinner liquefiable layers, flatter slopes, and stronger levels of shaking (Montgomery and Boulanger, 2016) with 33rd percentile values being considered appropriate for deterministic analyses in practice. Application of numerical analyses with spatially randomized properties to case histories with liquefiable soils (Bassal and Boulanger, 2021; Paull et al., 2021, 2022) provides additional insight into the effects of spatial variability.

Discussion

Lateral spreading is an extremely complex phenomenon that can be sensitive to factors that are difficult to characterize in advance of an earthquake. Available empirical methods characterize loading, geometry, and material behavior with simplistic metrics, so it is not unexpected that factors not captured by those metrics should lead to significant uncertainty in predicted displacements. Recent case histories (e.g., Russell et al., 2017) suggest that lateral spreading hazards are also influenced by liquefiable layer continuity (vertical and horizontal), geomorphology, and sediment age, none of which are captured by existing empirical predictive models. Numerical analyses offer the ability to represent many of these factors and evaluate their effects on deformations, and their increasing use in practice is likely to continue.

Lateral spreading can have a profound effect on the performance of structures underlain by liquefiable soils. The ground surface movements associated with lateral spreading tend to be highly irregular and place severe demands on foundations and the structures they support. Structures supported by shallow foundations, particularly isolated foundation elements not tied together structurally, tend to be heavily damaged by lateral spreading (Figure 9.77b). The performance of such structures can be improved by various ground modification techniques described in Chapter 11. Structures known at the time of foundation design to be underlain by liquefiable soils, however, are often supported on deep foundations that extend through the liquefiable soil and penetrate sufficiently into denser and stiffer underlying soils to provide the required load capacity. Lateral spreading, however, imposes high kinematic demands on pile foundations that can cause excessive foundation movements (Figures 1.13 and 1.14) and damage to the piles themselves (Figure 9.68). Procedures for assessing the response of pile foundations to lateral spreading are presented in Section 8.5.2.2.

9.6.6 Settlement

The tendency of sands to densify when subjected to earthquake shaking is well documented and was discussed in Section 6.6.5. Subsurface densification is manifested at the ground surface in the form of settlement but lateral spreading can also have vertical components of ground movement. Earthquake-induced settlement frequently causes distress to structures supported on shallow foundations, damage to lifelines that are commonly buried at shallow depths, and damage to utilities that serve pile-supported structures.

Unsaturated sands tend to densify very quickly so their settlement is usually complete by the end of earthquake shaking. The settlement of a saturated sand deposit, however, requires more time since settlement occurs as earthquake-induced pore pressures dissipate. The time required for this settlement to occur depends on the permeability and compressibility of the liquefied soil, the permeability of the soils immediately above and below it, and on the length of the drainage path – it typically ranges from a few minutes up to a day or more.

Estimation of earthquake-induced settlement of sands is difficult. Errors of 25%–50% are common in static settlement predictions; even less accuracy should be expected for the more complicated case of seismic loading. Nevertheless, laboratory-based procedures have been shown to produce results that agree reasonably well with many cases of observed field behavior under free-field conditions.

9.6.6.1 Free-Field Reconsolidation Settlement

Free-field settlement, i.e., settlement of level ground away from the influence of structures, results from the tendency of soil to contract, or densify, when shaken. Procedures for estimating the volumetric strain of dry and saturated clean sands subjected to cyclic loading were described in Section 6.6.5.4. Under level-ground conditions, one-dimensional compression is assumed, so the vertical strain, ε_v, is equal to the volumetric strain, ε_{vol}. Reconsolidation settlement is then calculated by integrating vertical strain over the thickness of the soil profile, i.e.,

$$\Delta H_v = \int \varepsilon_v dz \tag{9.85}$$

It is generally more practical to divide a soil profile into a series of sublayers, compute the settlement of each sublayer as the product of its vertical strain and thickness, and sum the sublayer settlements,

$$\Delta H_v = \sum_{i=1}^{N} (\varepsilon_v)_i \Delta z_i \tag{9.86}$$

where $(\varepsilon_v)_i$ = vertical strain in i^{th} of n sublayers and Δz_i = thickness of ith sublayer. It should be noted that a one-dimensional integration of this form implicitly assumes infinite lateral homogeneity of soil layers, ignores arching effects that can exist in three-dimensional, spatially variable soil deposits, and assumes that volumetric strains at extremely large depths will contribute to settlement at the ground surface.

Zhang et al. (2002) used the laboratory-based vertical strain model of Ishihara and Yoshimine (1992) and an empirical correlation between relative density and CPT tip resistance to estimate volumetric strain as a function of q_{c1Ncs} and factor of safety against triggering of liquefaction (Figure 9.80). More recent volumetric strain relationships have been proposed by Yoshimine et al. (2006), Cetin et al. (2009), and Olaya and Bray (2022); the latter two of these include characterization of uncertainty in volumetric strain. Settlements computed by the Zhang et al. (2002) procedure were shown to agree well with observed settlements at sites in and near San Francisco in the 1989 Loma Prieta earthquake. However, settlements obtained by pure volumetric strain integration (Equation 9.85) were shown to underpredict smaller (less than about 6 cm) settlements and overpredict larger (greater than about 6 cm) settlements in the February 2011 Christchurch earthquake (Geyin and Maurer, 2019); this bias was greatly reduced by applying a depth weighting factor of the form $(1.0-0.1z)$ to the volumetric strain where z is depth in meters and volumetric strains at depths greater than 10 m are not considered to contribute to surface settlement.

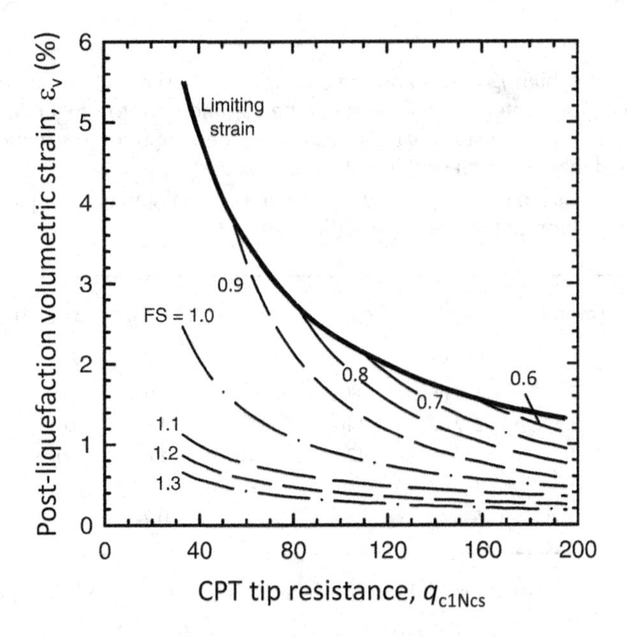

FIGURE 9.80 Relationship between post-liquefaction volumetric strain and clean sand CPT tip resistance for different factors of safety against triggering of liquefaction. (Zhang et al., 2002 with permission of Canadian Science Publishing.)

The volumetric strain relationship of Zhang et al. (2002) shown in Figure 9.80 can be approximated (Juang et al., 2013) by the relationship

$$\varepsilon_v(\%) = \begin{cases} 0 & \text{for} \quad FS_L \geq 2 \\ \min\left[\dfrac{a_0 + a_1 \ln q}{1/(2 - FS_L) - (a_2 + a_3 \ln q)}, b_0 + b_1 \ln q + b_2 (\ln q)^2\right] & \text{for} \quad 2 - \dfrac{1}{a_2 + a_3 \ln q} < FS_L < 2 \\ b_0 + b_1 \ln q + b_2 (\ln q)^2 & \text{for} \quad FS_L \leq 2 - \dfrac{1}{a_2 + a_3 \ln q} \end{cases} \tag{9.87}$$

where a_0=0.3773, a_1=–0.0337, a_2=1.5672, a_3=–0.1833, b_0=28.45, b_1=–9.3372, b_2=0.7975, and $q=q_{t1Ncs}$ in kg/cm^2.

Example 9.8

Consider the level ground soil profile shown below. Calculate the free-field settlement that would be expected within a few weeks of an earthquake that produces the indicated factors of safety against triggering.

Thickness	q_{c1Ncs}	I_c	FS_L
1 m	q_{c1Ncs} = 60	I_c = 3.2	FS_L = ∞
1 m	q_{c1Ncs} = 62	I_c = 3.2	FS_L = ∞
1 m	q_{c1Ncs} = 140	I_c = 2.2	FS_L = 0.85
1 m	q_{c1Ncs} = 120	I_c = 1.9	FS_L = 0.78
1 m	q_{c1Ncs} = 95	I_c = 2.9	FS_L = ∞
1 m	q_{c1Ncs} = 150	I_c = 2.1	FS_L = 0.95

Solution:

Layers 1, 2, and 5 have high I_c values and are therefore not susceptible to liquefaction; Layers 1 and 2 are also above the water table so they can be assumed not to trigger. While Layer 5 may generate some excess pore pressure that will result in settlement upon dissipation, it is likely sufficiently fine-grained that any such settlement would occur slowly.

The vertical strains of the coarse-grained layers can be estimated using Equation (9.87), which then allows the calculation of the free-field settlement.

Layer, i	Depth	q_{c1Ncs}	FS_L	ε_v (%)	ΔH_i (m)
1	0–1 m	60	∞	0	0
2	1–2 m	80	∞	0	0
3	2–3 m	140	0.85	1.05	0.011
4	3–4 m	120	0.78	1.64	0.016
5	4–5 m	95	∞	0	0
6	5–6 m	150	0.95	0.72	0.007

Then the total predicted settlement is the sum of the layer settlements, 0.034 m, or 3.4 cm.

The observed tendency for the Zhang et al. (2002) procedure to overestimate large observed settlements led to the introduction of an indicator variable (= 1 for triggering of liquefaction, 0 if there is no possibility of liquefaction) applied to volumetric strains; the mean and standard deviation of the indicator variable are P_L and $P_L(1-P_L)$, respectively (Juang et al., 2013). Then, for a soil profile of N layers, the ground surface settlement can be estimated by summing the probability-weighted contributions of each layer

$$\mu_{\Delta H_v} = \sum_{i=1}^{N} \varepsilon_{v,i} \Delta z_i P_{L,i} \tag{9.88a}$$

$$\sigma_{\Delta H_v} = \sqrt{\sum_{i=1}^{N} \varepsilon_{v,i}^2 \Delta z_i^2 P_{L,i} \left(1 - P_{L,i}\right)} \tag{9.88b}$$

where the i subscript represents the ith layer of the profile. The probability of triggering can be estimated using the procedures in Section 9.5.4 or, in cases where only FS_L is available, by a factor of safety mapping function (Ku et al., 2012) such as

$$P_L = 1 - \Phi\left[\frac{0.102 + \ln\left(FS_L\right)}{0.276}\right] \approx \frac{1}{1+\left(FS_L/0.9\right)^6} \tag{9.89}$$

which applies to factors of safety computed using the CPT procedures of Robertson and Wride (1998) or Robertson (2009a,b).

It should be recognized that the spatial variability of soils will influence the magnitude and pattern of post-liquefaction free-field settlement. When soil densities are measured or inferred from data measured in borings or CPT soundings, they represent the densities at those particular locations. The presence of looser or denser zones between such explorations can, depending on their size, depth, thickness, and nature, cause variability in total and differential settlements across a particular site.

9.6.6.2 Settlement Due to Ejecta

In a number of earthquakes, significant settlement has been caused by the ejection of soil from beneath the ground surface in the form of sand boils. When sand boils are formed, soils below the original ground surface end up above it, so the original ground surface settles. This often happens beneath or near the edges and corners of structures, where the resulting settlement can impose substantial demands on structures founded on shallow foundations.

The amount of settlement due to ejecta cannot be quantified at this time, but a number of factors have been observed to influence it. As discussed in Section 9.6.2, the presence of surficial evidence of liquefaction (i.e., ejecta) is related to the relative thicknesses of liquefiable soil layers and the overlying, non-liquefiable crust. The nature of the crust appears to be important – thickness, grain size and plasticity characteristics, number of penetrations (utility trenches, power poles, foundations, etc.) can all influence ejecta volumes. The depth of the water table and thickness/size of the liquefied zone can have an effect as can underlying layers that, while not fully liquefying, can produce excess pore pressures whose dissipation sustains the upward flow of shallower liquefied layers. Lateral spreading not only creates cracks but also appears to dissipate pore pressure through lateral extension. At a number of locations in Christchurch, NZ, sites that experienced lateral spreading produced less ejecta than level-ground sites that did not spread. At this time, the settlement due to ejecta, ΔH_e, is best estimated, albeit crudely, using manifestation severity indicators such as those described in Section 9.6.2 or by more recent quantitative procedures for estimating ejecta severity (Hutabarat and Bray, 2021a,b, 2022), and by comparison to relevant case histories (e.g., Bray and Sancio, 2009; Bray et al., 2014).

9.6.6.3 Settlement Due to Instability

Ground settlement can occur without significant volume change when shearing deformations occur via mechanisms that include coupled vertical and lateral movement. Embankments and levees that overlie liquefiable soils can settle when the soil beneath them moves laterally, in some cases being essentially squeezed out from beneath the higher ground (Figure 9.81). Simple, practical procedures for general cases do not exist at this time, but two- or three-dimensional, nonlinear, effective stress

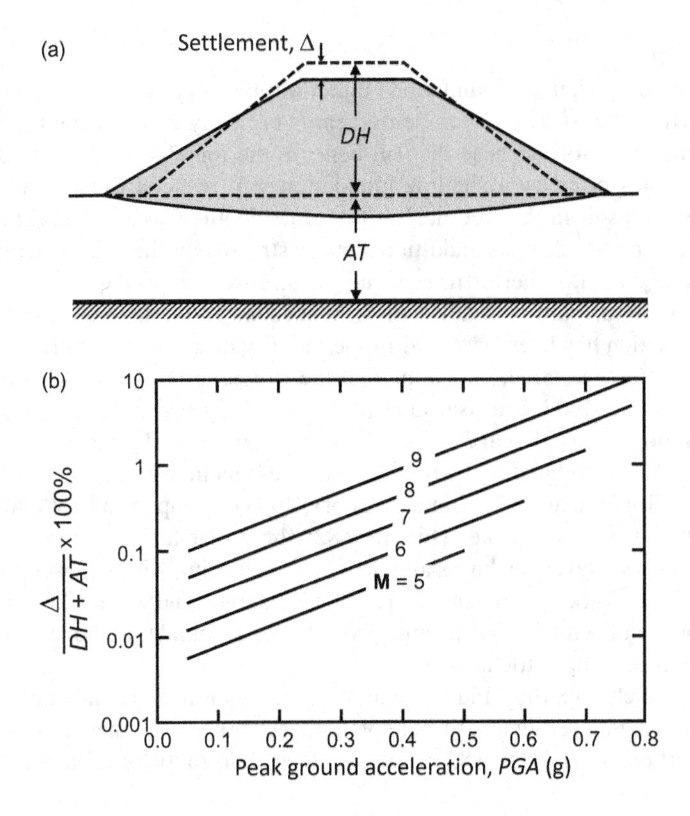

FIGURE 9.81 Chart for estimating crest settlement of dams. (After Swaisgood, 2003 with permission of J.R. Swaisgood.)

numerical analyses with a validated constitutive model can provide insight into likely deformation mechanisms and displacement levels.

Swaisgood (2003; 2014) compiled settlement data for 69 earth dam embankments and found that the average crest settlement, Δ, could be estimated as

$$\Delta = 0.01(DH + AT)\exp[6.07PGA + 0.57M_s + 8.0] \tag{9.90}$$

where DH=dam height, AT=alluvium thickness (Figure 9.81) and PGA is expressed as fraction of g. This settlement was interpreted as being caused by both volume change and spreading of the embankment and soils beneath it. Based on actual settlement data, this relationship was developed from a collection of dams with different characteristics at different sites in different geologic/tectonic environments and provides only a very rough indication of the expected settlement of any particular dam.

9.6.6.4 Foundation Settlement Effects

Structures located in areas underlain by potentially liquefiable sites are often supported by deep foundations. However, some older structures and/or relatively lightweight structures may be supported on shallow foundations that are underlain by liquefiable soils. Other structures may have been designed and constructed before liquefaction hazards at their sites were recognized. Such structures, and the utilities that connect to them, must be designed or retrofitted to accommodate settlement.

Deep foundations may also be adversely affected by liquefaction. The effects of lateral spreading on pile foundations is a complex soil-structure interaction problem that was discussed in Section 8.5.2.2. Pile foundations can also be affected by downdrag loads induced by post-earthquake settlement of liquefiable soils and the soils that overlie them.

Shallow Foundations

Structures supported by shallow foundations underlain by liquefiable soils have been observed to settle after earthquake shaking. Their settlement is typically greater than the free-field settlement of the surrounding soil because the soil beneath the foundation is not in a free-field situation. A structure supported by a shallow foundation will respond dynamically, and displace/rotate differently than soil in the free-field at the same depth. The horizontal displacement and rocking of the structure will impose additional shear stresses on the soil beneath it, particularly below its edges and corners. These stresses, and the strains they produce, can lead to additional pore pressure generation, additional fabric disturbance, and additional settlement. In some cases, evidence of liquefaction has been observed in the vicinity of a structure when it is not observed in the surrounding free-field area, indicating that the stresses from the building response caused liquefaction (Bray et al. 2004; Travasarou et al. 2006). Early investigations of shallow foundation settlement following earthquakes at sites generally with thick, clean sands noted that the amount of settlement was related to the thickness of the liquefied soil and the width of the structure (Yoshimi and Tokimatsu, 1977; Liu and Dobry, 1997). Compilations of data from the 1964 Niigata and 1990 Luzon earthquakes (Figure 9.82) show that foundation settlement increased with increasing liquefied layer thickness and decreasing building width. Observations from more recent earthquakes in Turkey, Taiwan, Chile, New Zealand, and Japan, however, have shown more complex behavior with loss of ground from beneath foundations due to ejecta playing a significant role in producing settlement.

More recent research (e.g., Bray and Macedo, 2017) has identified a number of mechanisms that can contribute to structure settlement through volumetric or shearing mechanisms. These include ejecta-induced settlement (Figure 9.83a), shear-related deformations (punching settlement and

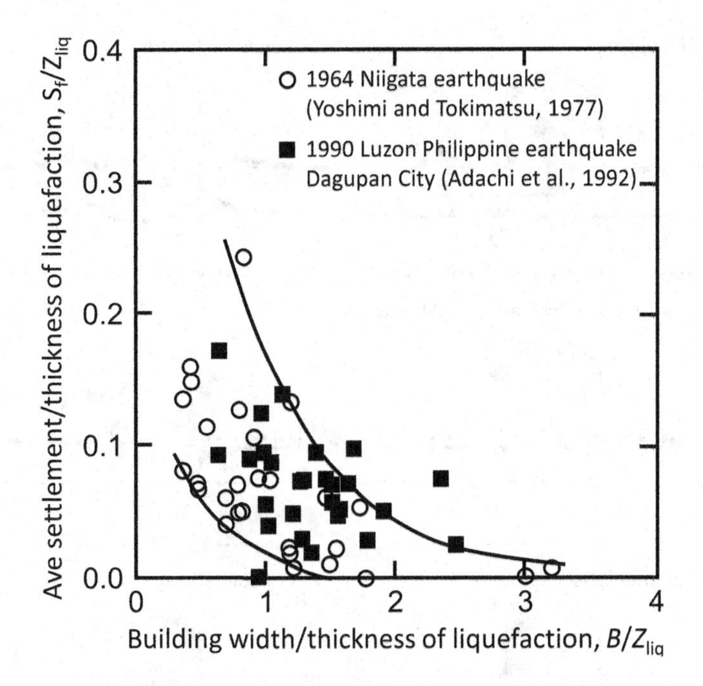

FIGURE 9.82 Variation of average foundation settlement at thick, clean sand sites with building width and thickness of liquefied soil in two earthquakes. (After Liu and Dobry, 1997 with permission of ASCE.)

tilting) due to partial bearing failure (Figure 9.83b), ratcheting deformations due to SSI-induced cyclic loading near the edges and corners of the foundation (Figure 9.83c), and volumetric-related deformations from sedimentation (Figure 9.83d) and consolidation-induced volumetric strains associated with three-dimensional pore pressure dissipation (Figure 9.83e) including partial drainage due to intense, transient hydraulic gradients that can develop during shaking. Centrifuge tests have shown that the interaction between a structure and underlying liquefiable soil is complex, and that it depends on the depth as well as thickness of the liquefiable soil; simple relationships such as that shown in Figure 9.82 do not apply when shearing mechanisms are more significant, such as cases where the structure is stiff or heavy or the liquefiable layer is thin or shallow. Shear-induced settlements increase with increasing liquefiable layer thickness up to a point beyond which they remain essentially constant, in contrast to the thickness proportionality implied in Figure 9.82.

An estimate of the liquefaction-induced settlement of buildings supported on shallow foundations can be made by predicting shear-induced settlement associated with a punching-type bearing capacity mechanism (Bray and Macedo, 2017). This component of settlement is added to the previously described components (reconsolidation and ejecta) to estimate the total settlement of the structure. Idealizing the soil profile as a two-layer system (Figure 9.84) consisting of a shallow non-liquefied crust underlain by a layer in which liquefaction is determined to have been triggered, the factor of safety against bearing capacity failure can be calculated as

$$FS_{BC} = \frac{q_{ult}}{q} \tag{9.91}$$

where q is the applied bearing pressure and the ultimate bearing capacity, q_{ult}, is computed (Meyerhof and Hanna, 1978) as

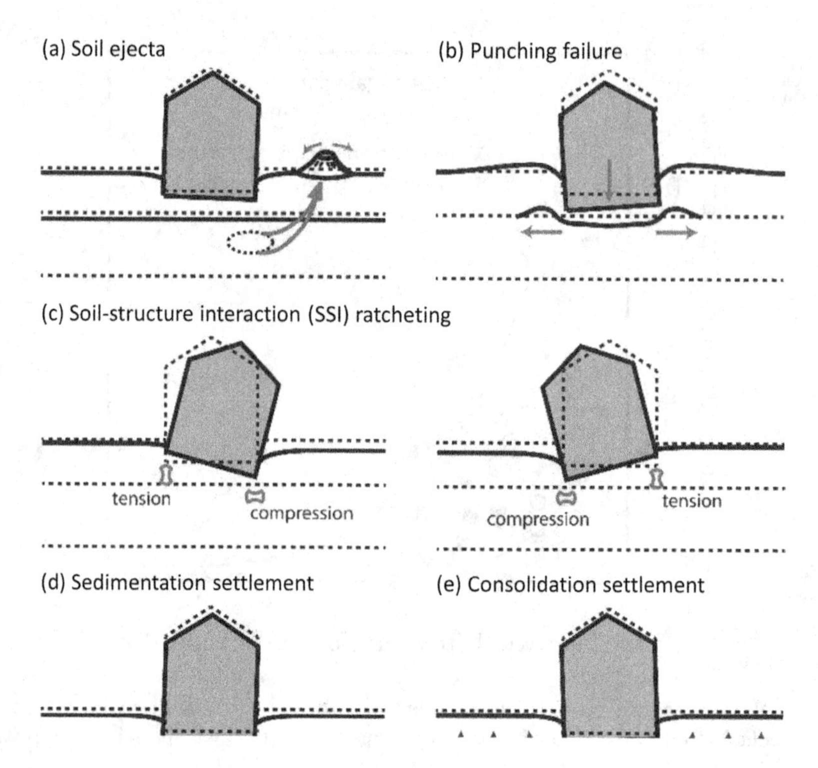

FIGURE 9.83 Liquefaction-induced building displacement mechanisms: (a) ground loss due to soil ejecta; shear-induced settlement from (b) punching failure, or (c) soil-structure-interaction (SSI) ratcheting; and volumetric-induced settlement from (d) sedimentation or (e) post-liquefaction reconsolidation. (Bray and Macedo 2017.)

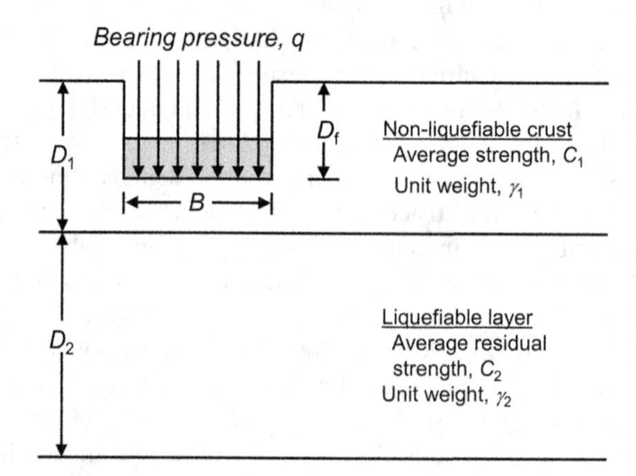

FIGURE 9.84 Illustration of shallow foundation underlain by liquefiable soil. Liquefiable layer assumed to be sufficiently thick that its thickness, D_2, does not affect foundation settlement. (After Bray and Macedo, 2017 with permission of Elsevier Science and Technology Journals.)

$$q_{ult} = 5.14C_2 + 2\frac{C_a D_1}{B} + \gamma_1 D_f \quad \leq 5.14C_1 + \gamma_1 D_f \tag{9.92}$$

with

$$C_a = C_1\left[0.612 + 0.96\left(\frac{C_2}{C_1}\right) - 0.58\left(\frac{C_2}{C_1}\right)^2\right] \tag{9.93}$$

Factor of safety values less than 1.0, indicating bearing capacity failure, have been associated with unacceptable performance in past earthquakes (Bray and Macedo, 2017).

If bearing capacity failure is not indicated, a liquefaction-induced building settlement index, *LBS*, can be computed as

$$LBS = \sum_{i=1}^{n} W_i \frac{\gamma_{\max,i}}{z_i} \Delta z_i \tag{9.94}$$

where the foundation-weighting factor W is 0.0 ($z < D_f$) or 1.0 ($z > D_f$), $\gamma_{\max}$ is obtained from Figure 9.79, and index i is for sublayers defined from the ground surface downward (n in total). The shear-induced building settlement (in mm) can then be estimated from

$$\ln \Delta H_s = c_1 + c_2 LBS + 0.58 \ln\left[\tan h\left(\frac{H_L}{6}\right)\right] + 4.59 \ln q - 0.42 (\ln q)^2 - 0.02B$$
$$+ 0.84 \ln CAV_{dp} + 0.41 \ln(S_a(1.0)) + \varepsilon \tag{9.95}$$

where $c_1 = -8.35$ and $c_2 = 0.072$ for $LBS \leq 16$ and $c_1 = -7.48$ and $c_2 = 0.014$ otherwise, H_L is the cumulative thickness of liquefiable layers in meters, q is the applied bearing (contact) pressure in kPa, B is the building width in meters, CAV_{dp} is a standardized version of cumulative absolute velocity (Campbell and Bozorgnia, 2011; Section 3.3.4.2) in g-sec, $S_a(1.0)$ is the 5%-damped spectral acceleration at $T = 1.0$ sec in g, and ε is a normally distributed random variable with zero mean and standard deviation of 0.50 in ln units.

Example 9.9

An office building with a 30 m×30 m stiff mat foundation is underlain by the soil profile shown below. The building imposes a bearing pressure of 100 kPa on the supporting soils. Estimate the median shear-induced settlement that would result from a ground motion with $S_a(1.0) = 0.30$ g and $CAV_{dp} = 0.58$ g-sec. Triggering analyses for free-field conditions produced the average FS_L values for each soil layer as provided in the figure.

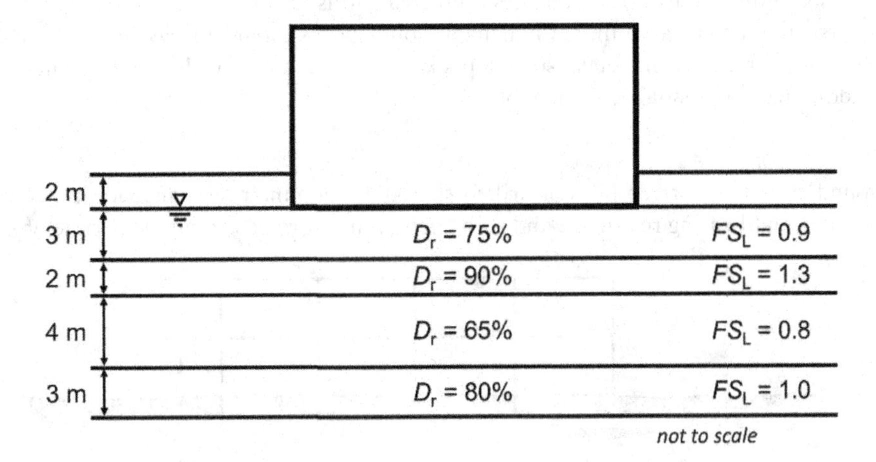

Solution:

Based on the FS_L values and relative densities, free-field maximum cyclic shear strains are estimated using Figure 9.79. The liquefaction-induced building settlement index layer components (*LBS*) are then computed using Equation (9.94).

Layer	Thickness (m)	Depth, z_i (m)	Strain, $\gamma_{max,i}$ (%)	*LBS*
1	2	1.0	n.a.	0
2	3	3.5	4.0	3.43
3	2	6.0	1.8	0.60
4	4	9.0	7.1	3.16
5	3	12.5	2.5	0.60

The layer components can be summed to produce $LBS = 7.79$. With that value established, the settlement can be computed using Equation (9.95)

$$\ln \Delta H_s = -8.35 + (0.072)(7.79) + (0.58)\ln\left[\tanh\left(\frac{10}{6}\right)\right] + (4.59)\ln 100 - (0.42)(\ln 100)^2 - (0.02)(30)$$
$$+0.84\ln 0.58 + 0.41\ln 0.30 = 2.848$$

The median settlement is then

$$\Delta H_s = \exp[\ln \Delta H_s] = \exp[2.848] = 17.3 \text{ mm}$$

The aforementioned procedure was derived to match foundation settlements observed in centrifuge tests and numerical analyses (Bray and Macedo, 2017) and has been shown to be reasonably consistent with those observed in regular (i.e., uniformly loaded) and short- to moderate-height (up to 24 m) buildings in a limited number of earthquakes. Because the procedure is empirical, it should be used with caution and engineering judgment for conditions that differ from those upon which the procedure has been calibrated.

For a given building, then, the total settlement can be estimated as the sum of the reconsolidation, ejecta, and shear-induced components, i.e.,

$$\Delta H_{tot} = \Delta H_v + \Delta H_e + \Delta H_s \tag{9.96}$$

Bullock et al. (2019) used physical model tests, numerical analyses, and case history observations to develop an alternative *CAV*-based procedure for estimation of building settlement that accounts for numerous soil profile, foundation, and structural characteristics.

Of course, the spatial variability of liquefiable soil deposits should always be taken into account. In some depositional environments, isolated pockets of liquefiable soil (Figure 9.85) may exist and lead to additional differential settlement of structures.

Deep Foundations

Deep foundations (e.g., driven piles or drilled shafts) develop their resistance to downward-acting loads through end-bearing resistance and skin friction that develops along the perimeter surface of

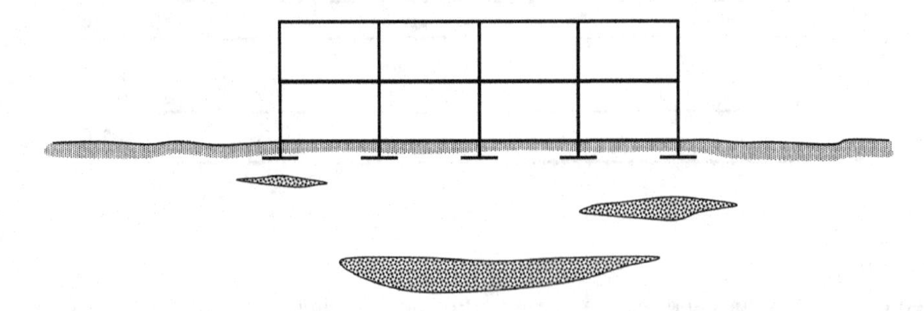

FIGURE 9.85 Pockets of liquefied soil that can lead to differential settlement. Small pockets at shallow depth or larger pockets at greater depth can cause different problems.

the foundation. The development of both components of resistance requires some level of downward movement of the foundation relative to the surrounding soil; mobilization of skin resistance requires much less relative movement (typically < 1 cm) than mobilization of end-bearing resistance (typically 10%–20% of pile or shaft diameter). As pore pressures build up during shaking, skin friction in liquefiable soil layers may decrease and lead to additional downward movement required to mobilize additional end-bearing resistance; this movement may be significant in large-diameter foundations such as drilled shafts in which an appreciable fraction of the capacity is derived from end bearing.

When deep foundations pass through compressible soils that undergo consolidation, the resulting settlement can cause the soil within and above the consolidating layer to move downward relative to the foundation element. When this occurs, the skin friction stresses act downward instead of upward; this *negative skin friction* imposes additional, downward-acting downdrag loads that must be resisted by the portions of the foundation below the consolidating layer. The pile or shaft itself must also have the structural capacity to resist the additional compressive force caused by negative skin friction. Two approaches to the estimation of the resulting *downdrag* loads have been used in practice.

In the "explicit" approach (Figure 9.86), the profile of settlement along the length of a pile (or shaft) is computed and the soil that experiences a settlement (relative to the pile or shaft) that exceeds some threshold value (10 mm is typical) is considered to cause negative skin friction (Fellenius and Siegel, 2008; AASHTO, 2017). Integration of that negative skin friction over the perimeter surface area on which it acts yields a downdrag force that is added to the load applied at the top of the pile for design.

In the neutral plane approach, two axial load profiles are plotted as functions of depth along the pile (Figure 9.87a). The first (loading) curve shows the axial load in the pile assuming that all soils develop negative skin friction; this curve begins with the applied dead load at the top of the pile and then increases with cumulative downdrag force at increasing depths. The second (resistance) curve begins at the tip of the pile with the mobilized end-bearing resistance, Q_{tm}, and assumes positive skin friction is developed in all soils; this curve increases with decreasing depth. The two curves intersect at the depth at which the applied dead load plus the cumulative downdrag force is equal to the mobilized base resistance plus cumulative positive skin resistance. This depth, at which there is

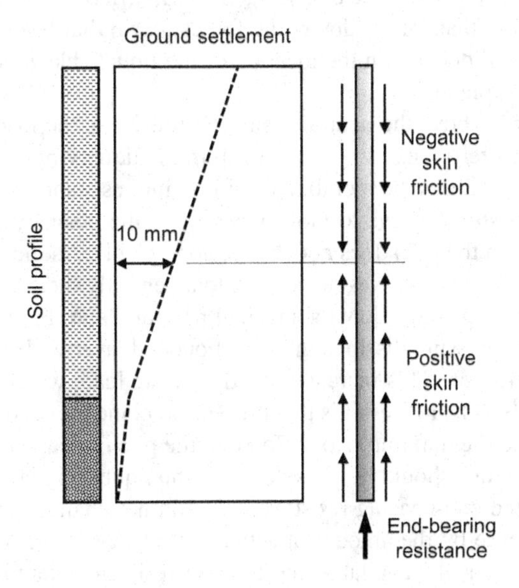

FIGURE 9.86 Illustration of positive and negative skin friction in explicit approach. All soil above depth of threshold ground settlement (10 mm in this case) is assumed to develop negative skin friction.

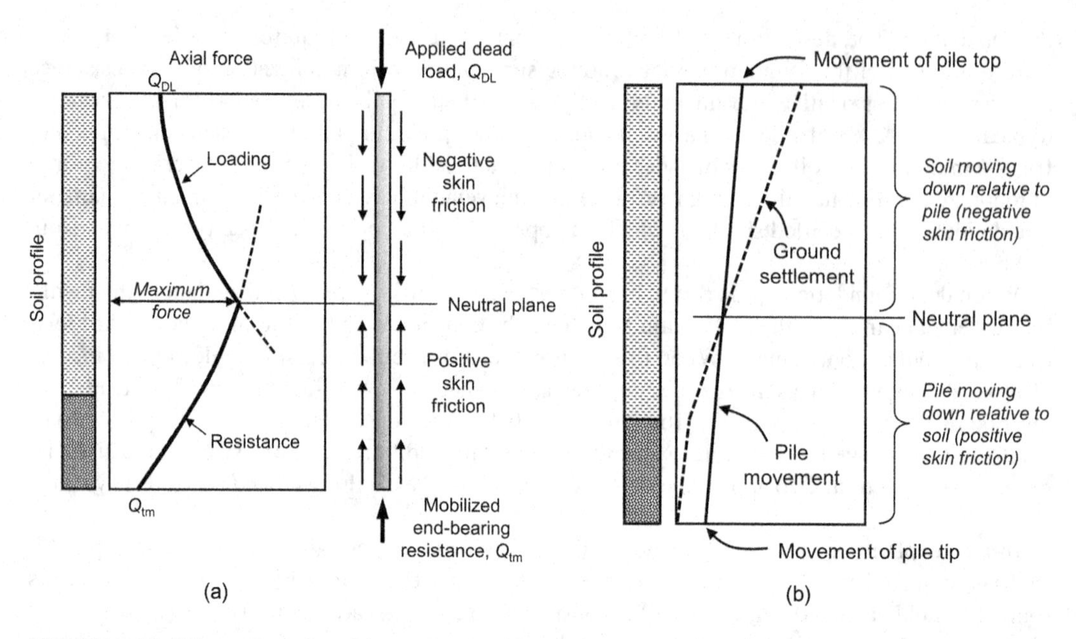

FIGURE 9.87 Illustration of positive and negative skin friction in the neutral plane approach: (a) loading and resistance curves showing axial load (bold curve) in pile, and (b) movements of ground and pile indicating position of neutral plane.

no relative movement between the pile and the soil, defines the *neutral plane* and is the location of the maximum axial force in the pile. Figure 9.87b illustrates the vertical movement of the soil and the pile (accounting for its elastic compression); negative skin friction is developed above the neutral plane and positive skin friction below it.

Post-earthquake settlement of liquefied soils can also induce significant downdrag loads in pile foundations, and the neutral plane approach can also be applied to that problem (Boulanger and Brandenberg, 2004; Rollins and Strand, 2006; Fellenius and Siegel, 2008; Muhunthan et al., 2017; Ziotopoulou et al. 2024). The occurrence of liquefaction will affect the skin friction in the liquefied layer and lead to settlement that causes downdrag forces above that level. The effects of liquefaction-induced settlement will depend on the location of the liquefiable zone relative to the pile and the pre-earthquake neutral plane.

If the liquefying layer is above the neutral plane (Figure 9.88), the loading curve will decrease as the skin friction goes to zero (i.e., as r_u goes to 1.0) in the liquefying layer. The reduced loading of the pile from downdrag reduces the mobilized end-bearing resistance, Q_{tm}. This unloading of the pile tip will cause the resistance curve to move upward, so the position of the neutral plane (and the maximum axial force in the pile) does not change appreciably. As the excess pore pressure dissipates, the loading and resistance curves move back to essentially their original positions.

If the liquefying layer is located below the original neutral plane (Figure 9.89), post-earthquake settlement caused by pore pressure dissipation in the liquefied layer will increase the length of pile subject to negative skin friction. This increases the downdrag load, which increases the mobilized end-bearing resistance, Q_{tm}. The net result is that the neutral plane moves downward, additional pile tip penetration occurs, and the maximum axial force in the pile increases.

Three additional conditions should be considered. If the liquefying layer lies entirely below the tip of the pile, the skin and end-bearing resistances should not change but the pile and soil above the liquefied layer will settle by the amount of settlement of that layer. Also, if the tip of the pile is within the liquefiable layer, the available end-bearing resistance may be significantly reduced; if the pile relies upon that resistance to support vertical loads, bearing capacity failure may occur. It should be noted that such a mechanism can occur in a deep layer of liquefiable soil, even if its

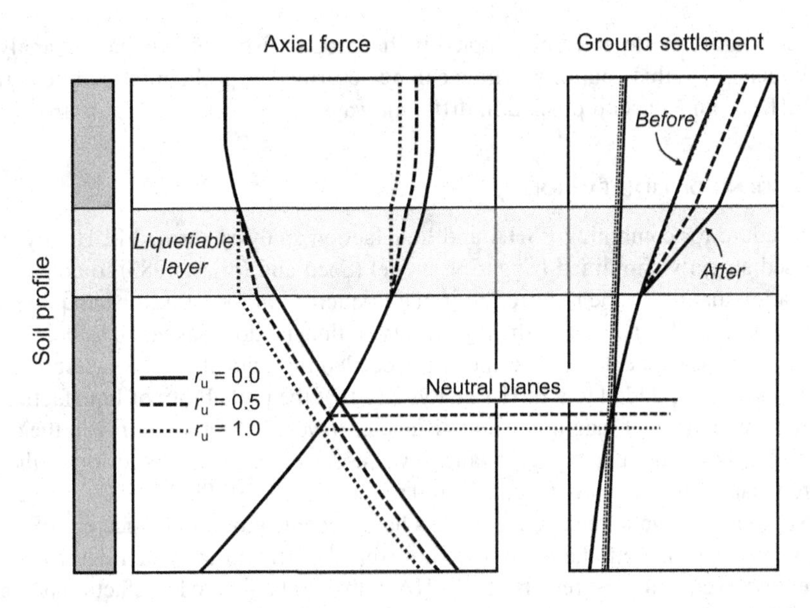

FIGURE 9.88 Loading and resistance curves, and settlement curves for downdrag with liquefiable layer above original neutral plane. (After Muhunthan et al., 2017.)

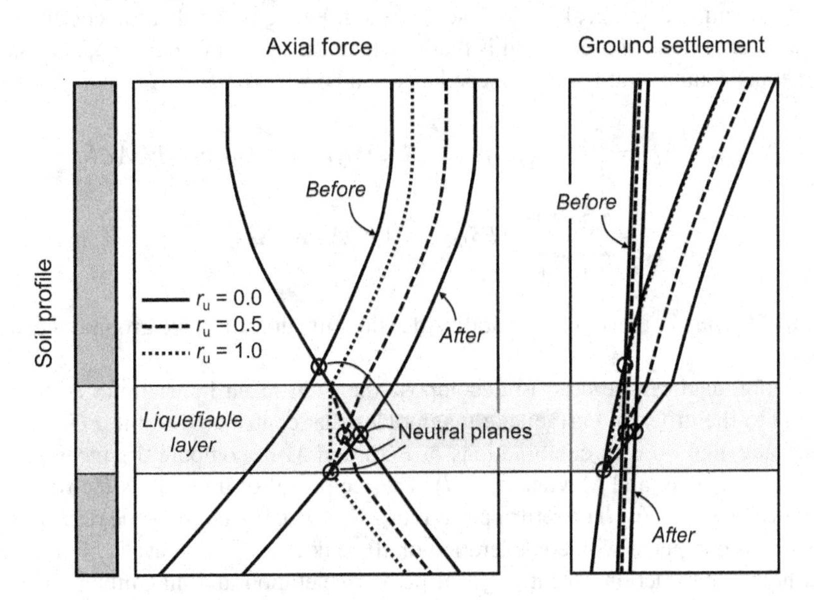

FIGURE 9.89 Loading, resistance, and settlement curves for downdrag with liquefiable layer below original neutral plane. (After Muhunthan et al., 2017.)

thickness is not sufficient to produce surficial evidence of liquefaction. Finally, liquefaction can cause porewater to flow upward along the perimeter of the pile and reduce the available skin friction both in the liquefied layer and in overlying layers (e.g., Ledezma et al., 2012).

9.7 PROBABILISTIC LIQUEFACTION HAZARD ANALYSIS

As discussed in Section 5.5, uncertainty in the prediction of response of some system to a given level of ground motion will affect the return period, and hence the mean annual probability, of the computed response. To predict the level of response that has a specific, desired return period, a probabilistic response model must be coupled with a probabilistic seismic hazard analysis. When applied to liquefaction, this coupling, which can be described as probabilistic liquefaction hazard analysis (PLHA), can be accomplished in different ways.

9.7.1 Triggering of Liquefaction

The first procedure for combining PSHA and liquefaction analysis, termed PLHA by Atkinson et al. (1984), used an early simplified triggering model (Seed and Idriss, 1983) to compute a critical peak acceleration that would be required to initiate liquefaction for a given earthquake magnitude and SPT resistance. In this approach, the probability of liquefaction was taken as being equal to the probability that the peak acceleration at the site exceeded the critical peak acceleration. The probability calculations were added to a PSHA analysis so that the probability of liquefaction was computed for all combinations of magnitude and distance. However, the uncertainty in the critical peak acceleration (i.e., in the liquefaction resistance) was not explicitly accounted for so the procedure falls short of what would currently be considered to represent a full PLHA.

As GMMs and earthquake recurrence models have become more sophisticated, PSHA software has become more specialized and complex. However, the general approach described in Section 5.5.2 can be used to combine the results of a PSHA with a probabilistic liquefaction potential model to achieve a full PLHA. Kramer and Mayfield (2007) modified that approach to account for the fact that the loading used for the evaluation of liquefaction potential depends on the joint distribution of earthquake magnitude and peak ground acceleration. Recognizing that the condition of interest in an evaluation of liquefaction potential is that in which the factor of safety, FS_L, is less than some value, fs_L, the mean annual rate of *non*-exceedance can be written as

$$\Lambda_{FS_L}\left(fs_L\right) = \int P[FS_L < fs_L] \mid PGA, \mathbf{M}] f_{PGA,M}(PGA, \mathbf{m})\, dPGA\, d\mathbf{M}$$
$$\approx \sum_{j=1}^{N_{\mathbf{M}}} \sum_{i=1}^{N_{a_{max}}} P[FS_L < fs_L \mid PGA_i, \mathbf{m}_j] \Delta\lambda_{PGA_i,\mathbf{m}_j} \tag{9.97}$$

where $f_{PGA,M}(PGA, m)$ is the joint probability density function of peak ground acceleration and magnitude.

Disaggregation data can be used to decompose the *PGA* hazard curve into a series of curves corresponding to the different earthquake magnitudes that contributed to the *PGA* hazard, which can then be integrated over all combinations of *PGA* and **M** to compute the mean annual rate of *non-exceedance* (Kramer and Mayfield, 2007). The reciprocal of that rate is the return period for non-exceedance of $FS_L = fs_L$. The return period for $FS_L \leq 1.0$, therefore, is the return period of liquefaction itself, as computed with consideration of all peak acceleration levels, all magnitudes, and the uncertainty in liquefaction potential given peak acceleration and magnitude. Factor of safety hazard curves, as illustrated in Figure 9.90, slope in the opposite direction to hazard curves for ground motion parameters since weak motions (which occur relatively frequently) produce high factors of safety and strong motions that occur more rarely produce low factors of safety. Considering a standard reference element in a reference soil profile, Kramer and Mayfield (2007) showed that conditions indicated as having identical liquefaction potential (i.e., the same FS_L values based on 475-year *PGA* and mean magnitudes in deterministic analyses) had actual return periods of liquefaction that varied by factors of up to 2 when applied at different locations across the continental United States. With the conventional approach, therefore, designs thought to provide the same level of liquefaction hazard can lead to very different levels of performance in different tectonic environments. These results show that actual liquefaction hazards, which can be characterized using a full PLHA (e.g., Equation 9.97), are influenced by the shape (particularly the slope) of the peak acceleration hazard curve and by the nature of the underlying magnitude distribution. Comparisons of the results

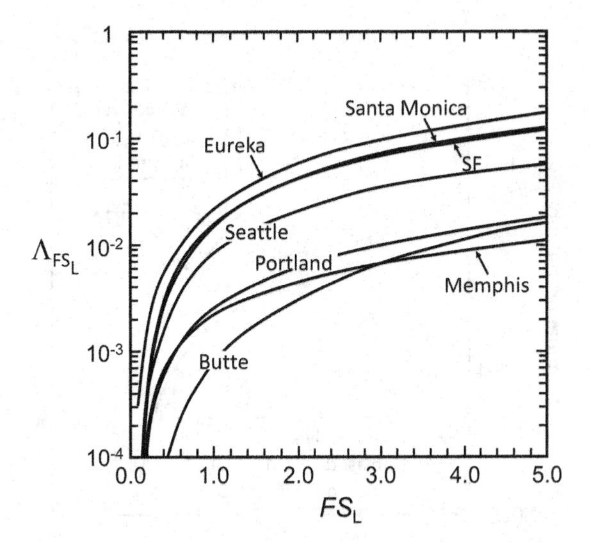

FIGURE 9.90 Results of PLHA-based hazard curves for triggering of liquefaction of reference element of soil (2 m depth, q_{c1Ncs}=80, V_{s30}=200 m/sec) at 10 U.S. locations (SF=San Francisco). (Calculations based on Idriss and Boulanger, 2015 liquefaction potential model, courtesy of A. Makdisi.)

of PLHA analyses using the probabilistic liquefaction triggering models of Cetin et al. (2004) and Boulanger and Idriss (2012) have shown that differences between the two exist but are smaller when used in PLHA analyses than when used in a conventional, deterministic manner (Franke and Wright, 2014).

Mayfield et al. (2010) showed that the voluminous calculations required to integrate liquefaction potential over the entire ranges of peak acceleration and magnitude could be encapsulated in a single parameter corresponding to a reference element in a reference soil profile. Site-specific values of that parameter (i.e., for different depths in different soil profiles) could then be accurately approximated by two simple adjustment factors. Mayfield et al. (2010) used FS_L and the SPT resistance required to resist liquefaction as liquefaction parameters and showed how values of those parameters corresponding to different return periods of liquefaction could be mapped across geographic regions. Ulmer and Franke (2015) performed the same type of analysis for *CSR* based on the Boulanger and Idriss (2014) liquefaction triggering model and showed that the full PLHA results could be approximated with good accuracy using a mapped *CSR* value with five adjustment factors. The development of liquefaction parameter maps of the types proposed by Mayfield et al. (2010), Ulmer and Franke (2015), and Makdisi and Kramer (2024) could allow geotechnical engineers to design for a specified return period of liquefaction itself, and to ensure equal liquefaction potential hazards at sites in different seismo-tectonic environments, using conventional liquefaction potential procedures.

9.7.2 Consequences of Liquefaction

With probabilistic response models, hazard curves can also be developed for effects of liquefaction such as manifestation severity, lateral spreading, and post-liquefaction settlement. Such curves show the mean annual rate of exceedance of various response parameters such as liquefaction potential index (Section 9.6.2), lateral spreading displacement (Section 9.6.5.2), and post-liquefaction settlement (Section 9.6.6). These types of analyses allow design based on an allowable level of response corresponding to a specified return period rather than a deterministic response corresponding to a ground motion with a specified return period.

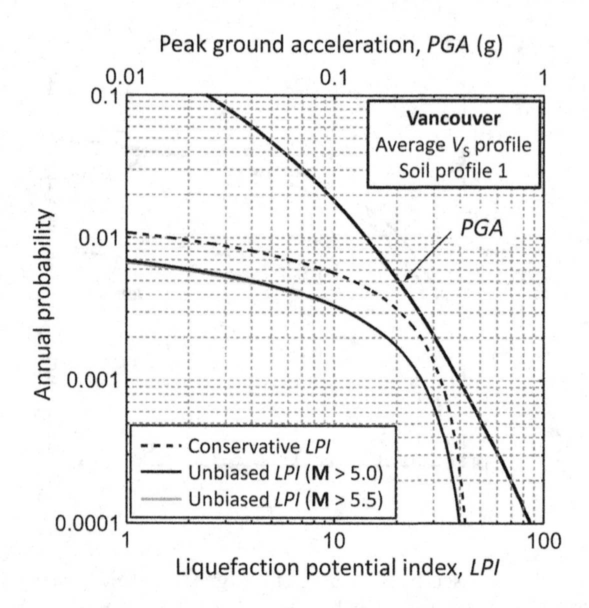

FIGURE 9.91 Liquefaction Potential Index (*LPI*) hazard curves for Vancouver, Canada. (After Goda et al., 2011 with permission of the Seismological Society of America).

Goda et al. (2011) extended the PLHA procedure of Atkinson et al. (1984) to include consideration of the uncertainty in liquefaction resistance and to compute values of liquefaction potential index (*LPI*). Using a shear wave velocity-based procedure (Andrus and Stokoe, 2000) with a probabilistic mapping procedure (Juang et al., 2005), a PSHA code was modified to compute probabilities of liquefaction within the hazard integral. Figure 9.91 shows *LPI* hazard curves for a representative soil profile located in Vancouver, British Columbia.

Geyin and Maurer (2019) used over 15,000 case histories from 24 earthquakes to develop fragility curves (Section 5.5.2.1) for different combinations of CPT-based triggering models, manifestation severity indices, and manifestation severities. Using a logic tree approach, weighted combinations of the triggering models and severity indices allowed direct calculation of hazard curves for minor, moderate, and severe manifestation. Such curves, or hazard maps based upon them, offer an objective and practical means of estimating potential infrastructure damage for planning and policy purposes.

Hazard curves for lateral spreading displacement can be computed in different ways depending on the nature of the displacement prediction model. Displacement prediction models of the general form of Youd et al. (2002) can be handled in a particularly efficient manner (Kramer et al., 2007). Franke and Kramer (2014) used such an approach to compute lateral spreading hazard curves (Figure 9.92) based on a probabilistic version of the Youd et al. (2002) lateral spreading model. After adjusting the penetration resistances of the soil profile shown in Figure 9.92 such that deterministic analyses based on 475-year PGAs and mean magnitudes produced 30 cm of displacement at ten sites across the United States, the actual return periods for 30 cm displacement at the sites were shown to vary by a factor of nearly 3. This result showed that complete and consistent predictions of lateral spreading displacement hazard using a predictive model of the form of Youd et al. (2002) required consideration of all combinations of magnitude and distance.

Settlement hazard curves for a hypothetical six-story reinforced concrete building are shown in Figure 9.93. The volumetric component of settlement (Section 9.6.6.1) was computed by the procedure of Juang et al. (2013) and the shear-induced component (Section 9.6.6.4) by that of Bray and Macedo (2017). The settlement can be seen to be dominated by volumetric strains at short return periods with shear strain-induced settlement becoming more important at return periods greater than about 3,000 years.

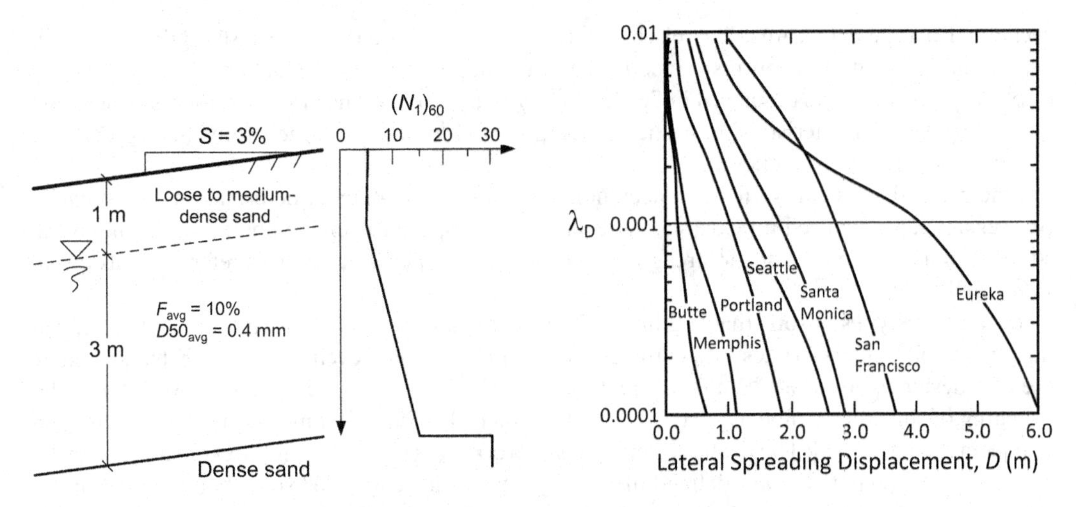

FIGURE 9.92 Lateral spread displacement hazard curves for gently sloping soil profile at ten locations across the United States. (After Franke and Kramer, 2014 with permission of ASCE.)

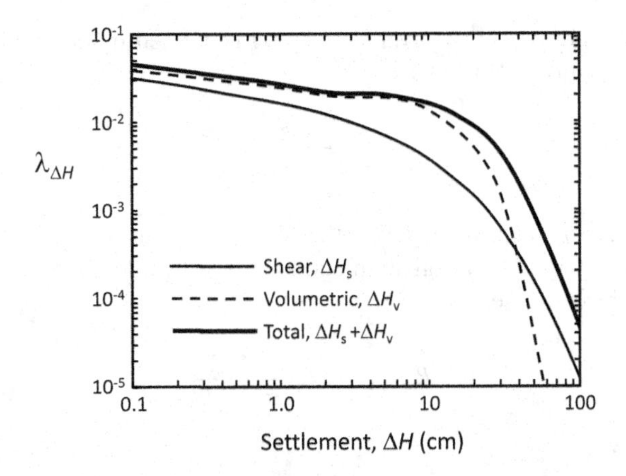

FIGURE 9.93 Post-liquefaction settlement hazard curves (Liu et al., 2021) for six-story building supported on 10 m by 20 m mat foundation embedded at a depth of 2 m below the ground surface.

9.7.3 Discussion

Though in their early stages of development, PLHA procedures that consider all levels of shaking and the uncertainty in response for each produce the most complete and accurate assessments of liquefaction hazards. The type of approaches developed by Mayfield et al. (2010) and Makdisi and Kramer (2024) offer engineers the ability to closely approximate the results of a full, site-specific PLHA for triggering with essentially the same level of effort required by conventional procedures. Extension beyond triggering to develop liquefaction consequence hazard curves such as those in Figures 9.91–9.93 represent response-level implementations of the performance-based concepts described in Section 5.6.3.

9.8 CYCLIC SOFTENING

As described in Sections 6.4.4.3, 6.4.4.5, and 9.3, saturated, plastic, fine-grained soils tend to generate less pore pressure than non-plastic fine- or coarse-grained soils when subjected to the same cyclic loading. However, they do generate excess pore pressure and they can develop significant cyclic and permanent strains. The effects of these phenomena on the shear strength of the soil, i.e., the large-strain shearing resistance required for slope stability, foundation capacity, and site response problems, were described in Sections 6.6.6.3 and 6.6.6.4. The effects on their stiffness and cyclic response characteristics, in particular the potential to develop large strains that can lead to strength loss, are described here.

Fine-grained soils can soften and accumulate significant permanent deformations when cyclic stresses are superimposed on existing static shear stresses, even if the combined static and cyclic stresses do not exceed the shear strength of the soil. This type of behavior is referred to as *cyclic softening* behavior.

Because the types of soils that exhibit cyclic softening are so much easier to sample than typical liquefiable soils, laboratory tests that provide useful information on cyclic softening behavior can be used for design applications. Boulanger and Idriss (2007) developed a useful framework for cyclic softening of soils they define as "clay-like" that parallels the simplified method used for evaluation of liquefaction potential. The cyclic softening framework makes use of cyclic stress and cyclic resistance ratios, although they are defined differently for plastic "clay-like" soils than the non-plastic "sand-like" soils that can liquefy, as discussed earlier in this chapter.

9.8.1 Characterization of Loading

The cyclic stress ratio used to evaluate cyclic softening is defined in a manner similar to that used for liquefiable soils (Equation 9.10), i.e., as

$$CSR = r_e \frac{\tau_{\max}}{\sigma'_{v0}} \tag{9.98}$$

where r_e is usually taken as 0.65. Interpreting the MSF as a measure of the number of equivalent loading cycles for an earthquake of a given magnitude, i.e., as a loading parameter, the reference cyclic stress ratio can be expressed as

$$CSR_{M=7.5} = 0.65 \frac{\tau_{\max}}{\sigma'_{v0} \cdot MSF} \tag{9.99}$$

where *MSF* is taken as

$$MSF = 1.12 \exp\left(-\frac{\mathbf{M}}{4}\right) + 0.828 \quad \leq 1.13 \tag{9.100}$$

Laboratory data show that clays generate pore pressure more slowly than sands, so the use of r_e=0.65 implies that 30 (rather than the 15 used for sand) equivalent loading cycles are consistent with a **M**=7.5 earthquake.

9.8.2 Characterization of Resistance

The cyclic resistance of a plastic, fine-grained soil is closely related to its undrained shear strength (Boulanger and Idriss, 2007). Figure 9.94 shows the number of cycles of loading required to produce 3% peak shear strain for rate-corrected cyclic stresses at different fractions of the soil's undrained strength. Despite having plasticity indices ranging from 13 to 73 and *OCR*s ranging from 1 to 4, the data fall within a relatively narrow band. The cyclic resistance ratio corresponding to a **M**=7.5 earthquake can then be expressed as

$$CRR_{M=7.5} = C_{2D} \left(\frac{\tau_{cyc}}{s_u}\right)_{M=7.5} \frac{s_u}{\sigma'_{v0}} K_\alpha \tag{9.101}$$

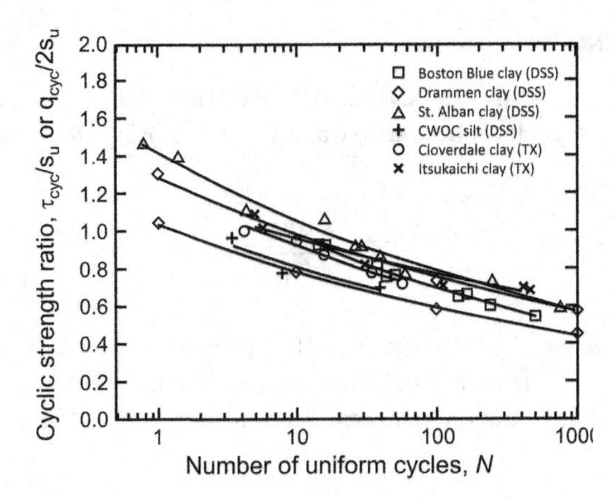

FIGURE 9.94 Cyclic strength ratios versus number of cycles required to cause 3% peak shear strain in six natural soils. (Boulanger and Idriss, 2007 with permission of ASCE.)

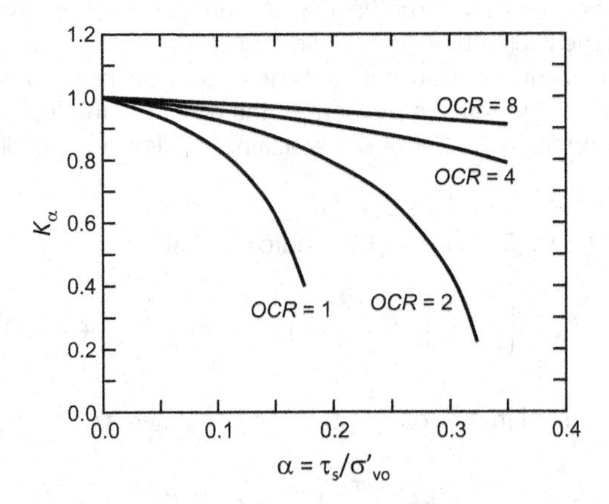

FIGURE 9.95 Variation of K_α with static shear stress ratio and overconsolidation ratio. (Boulanger and Idriss, 2007 with permission of ASCE.)

where C_{2D} is a correction for two-dimensional shaking estimated as 0.96 by Boulanger and Idriss (2004), $(\tau_{cyc}/s_u)_{M=7.5}$ is the ratio of cyclic shear stress to undrained strength for the number of cycles (30) representative of a **M** = 7.5 earthquake, and K_α is a static shear stress adjustment factor. Experimental results for several clays (Goulois et al., 1985; Andersen et al., 1988; Lefebvre and Pfendler, 1996) can be used to express K_α as a function of the measured undrained strength ratio, s_u/σ'_{v0}, and the static stress ratio, α (Equation 9.32)

$$K_\alpha = 1.344 - \frac{0.344}{\left(1 - \frac{\alpha \cdot \sigma'_{vo}}{s_u}\right)^{0.638}} \tag{9.102}$$

If the value of s_u for the on-site soils cannot be measured, it can be estimated using a common relationship for typical clays (Ladd, 1991)

$$s_u = 0.22\sigma'_{v0} \cdot OCR^{0.8} \tag{9.103}$$

Figure 9.95 shows how initial static shear stress ratios decrease K_α (and thus the cyclic resistance ratio) more strongly for normally consolidated soils than overconsolidated soils.

9.8.3 Cyclic Softening Potential

With the value of cyclic stress ratio and cyclic resistance ratio determined, the potential for cyclic softening (defined here as a peak cyclic shear strain exceeding 3%) can be expressed in terms of the factor of safety

$$FS = \frac{CRR_{M=7.5}}{CSR_{M=7.5}} \tag{9.104}$$

It should be noted that a factor of safety less than 1.0 does not necessarily imply catastrophic failure, but rather that cyclic shear strains are expected to exceed 3%; in sensitive soils, that level of disturbance could lead to fabric degradation that contributes to significant strength loss (Section 6.6.6.3).

Example 9.10

A slope in southern California is found to be underlain by a lightly overconsolidated (OCR=1.8) saturated clay. Laboratory tests have shown that the undrained strength of the clay is approximately 35% of the vertical effective stress, and site response analyses using ground motions appropriate for a **M**7.2 earthquake show cyclic shear stresses of approximately 40% of the monotonic undrained strength of the soil. An element of the clay near the toe of the slope has a static stress ratio, α =0.2. Compute the factor of safety against cyclic softening of that element of clay.

Solution:

For the **M**7.2 event of interest, Equation (9.100) indicates that

$$MSF = 1.12\exp\left(-\frac{\mathbf{M}}{4}\right) + 0.828 = 1.12\exp\left(-\frac{7.2}{4}\right) + 0.828 = 1.013$$

The cyclic stress ratio induced in the soil can then be computed as

$$CSR_{M=7.5} = 0.65\frac{\tau_{max}}{\sigma'_{v0} \cdot MSF} = 0.65\frac{0.40}{1.013} = 0.257$$

The static shear stress adjustment factor is computed as

$$K_\alpha = 1.344 - \frac{0.344}{\left(1 - \dfrac{\alpha \cdot \sigma'_{v0}}{s_u}\right)^{0.638}} = 1.344 - \frac{0.344}{\left(1 - \dfrac{0.2}{0.35}\right)^{0.638}} = 0.753$$

which allows the cyclic resistance ratio to be computed as

$$CRR_{M=7.5} = C_{2D}\left(\frac{\tau_{cyc}}{s_u}\right)_{M=7.5}\frac{s_u}{\sigma'_{vo}}K_\alpha = (0.96)(0.83)(0.35)(0.753) = 0.210$$

The resulting factor of safety against cyclic softening is

$$FS = \frac{CRR_{M=7.5}}{CSR_{M=7.5}} = \frac{0.210}{0.257} = 0.82$$

which indicates that shear strains greater than 3% are to be expected. The actual level of shear strain, and consequently the deformation of the slope, will depend on the sensitivity of the clay, the nature of the ground motion, and specific site conditions.

9.8.4 Discussion

The best approach to evaluating cyclic softening is to evaluate *CRR* in the laboratory by performing cyclic tests on high-quality samples from the site of interest. This approach, which combines cyclic testing (to obtain τ_{cyc}/S_u) with monotonic testing (to obtain S_u/σ'_{v0}), will provide the highest level of confidence in the *CRR* value and the lowest uncertainty in the cyclic softening prediction. The next best approach is to derive *CRR* using measured monotonic shear strengths derived from laboratory or field tests on soil from the site of interest and to take cyclic strength ratios (τ_{cyc}/S_u) from the results of prior studies (Figure 9.94). If no strength information is available, or if only a rough, preliminary estimate is desired, the cyclic resistance ratio can be estimated from Equation (9.106). Using simplifications described in Boulanger and Idriss (2007), the second and third approaches described above can produce cyclic resistance ratios of

$$CRR_{M=7.5} = 0.8\left(\frac{S_u}{\sigma'_{v0}}\right)_{NC} OCR^m \cdot K_\alpha \tag{9.105}$$

$$CRR_{M=7.5} = 0.18 \cdot OCR^{0.8} \cdot K_\alpha \tag{9.106}$$

CRR values estimated from Equation (9.105) should be recognized as being more uncertain than values obtained from laboratory testing, and values obtained from Equation (9.106) should be recognized as being more uncertain than values obtained from Equation (9.105).

9.9 SUMMARY

1. The term *liquefaction* refers to the loss of stiffness and strength of a loose, saturated, nonplastic soil. Liquefaction is most often triggered by transient loading such as that induced by earthquakes but can also be initiated by monotonic loading. The generation of high porewater pressure under undrained loading conditions is a hallmark of soil liquefaction.
2. Cyclic softening refers to the reduction of stiffness and strength of saturated, plastic, fine-grained soils that are not susceptible to liquefaction. These reductions can be significant but are generally not as severe as those associated with liquefaction.
3. Liquefaction and cyclic softening can occur over broad ranges of soil conditions. Both are capable of causing significant damage, and both should be evaluated as part of a complete seismic hazard assessment.
4. Flow liquefaction can occur when the shear strength of a liquefied soil, i.e., its residual strength, drops below the shear stresses required to maintain static equilibrium. Flow failures, or *flowslides*, can involve very large and rapid deformations, and may be triggered by transient or monotonic loading; although they can be tremendously damaging, flow slides are relatively rare.
5. Cyclic liquefaction results from the incremental generation of excess porewater pressure by transient loading while applied static shear stresses are lower than the shear strength of the soil. Cyclic liquefaction can produce small to large deformations of soil slopes and of soil–structure systems. Cyclic liquefaction of level-ground sites, where static shear stresses are zero, is generally manifested by ground oscillation, post-earthquake settlement, and the development of sand boils. Permanent lateral displacements due to level-ground liquefaction are usually small.
6. Evaluation of liquefaction hazards requires that liquefaction susceptibility, triggering, and effects be predicted. For a site to be considered free from significant liquefaction hazards, the soils must be non-susceptible to liquefaction, the anticipated loading must be insufficient to trigger liquefaction, or the effects of liquefaction must be tolerable.

7. Liquefaction susceptibility is a function of the composition of the soil itself, and is independent of the *state* (density and effective stress) or level of saturation that the soil exists in; state and saturation influence the resistance of triggering of liquefaction. Clean coarse-grained soils and non-plastic fine-grained soils are typically susceptible to liquefaction and highly plastic fine-grained soils are not susceptible. Liquefaction susceptibility is most commonly assessed in terms of the plasticity index of the soil. Historical and geologic indicators can be helpful in identifying locations of soils that are susceptible to liquefaction. Soils that are not susceptible to liquefaction are not exempt from potential problems, however, as they may undergo cyclic softening.
8. Under given loading conditions, any susceptible soil will reach a unique combination of effective stress, void ratio, and shear strength at large strains. The combination can be described graphically by a three-dimensional curve known as the *steady state line*. The position of the SSL is most strongly influenced by grain size and grain shape characteristics.
9. The resistance of a susceptible soil to triggering of liquefaction depends on its volume change characteristics, specifically its degree of contractiveness. The more contractive an element of soil is, the lower its liquefaction resistance will be. Volume change characteristics can be related to the *state parameter*, which is the difference between the void ratio and the void ratio at the steady state for the same effective stress. A soil with a high state parameter (generally greater than –0.05) will exhibit contractive behavior at large strain levels, and a soil with a lower state parameter will exhibit dilative behavior.
10. Steady state lines are generally somewhat steeper (in $e-\sigma'$ space) than consolidation curves, which means that the state parameter (hence, degree of contractiveness) increases with increasing effective confining stress.
11. Flow liquefaction can occur when the static shear stress acting on the soil is greater than its steady state shear strength (or, in the field, its residual strength). It is triggered when the principal effective stress ratio reaches a critical value under undrained, stress-controlled conditions. The stress state at the triggering of flow liquefaction can be described graphically in stress path space by the FLS. Once the effective stress path reaches the FLS, whether by static or cyclic loading, additional straining will induce additional excess pore pressure and the available shearing resistance will drop to the steady state strength. After triggering, deformations of the soil are driven by the difference between the static shear stress and the steady state shear strength. Under seismic conditions, the function of the earthquake is to bring the soil from its initial equilibrium state to the FLS; at that point, the soil skeleton collapses and deformations are driven by the static stresses.
12. Soils of intermediate density may experience a reduction in shearing resistance associated with a quasi-steady state beyond which dilation leads to increasing shearing resistance as the soil reaches the actual steady state at much larger strain amplitudes.
13. Cyclic liquefaction can be triggered in loose or dense susceptible soil since even dense soils tend to initially contract at low strain levels. The level of loading required to trigger liquefaction depends on their level of contractiveness (i.e., on their state parameter); because they are more contractive, looser soils at a particular effective stress will tend to liquefy at lower levels of cyclic shear stress (for a given number of loading cycles) than denser soils (and at a lower number of cycles for a given cyclic shear stress amplitude). The tendency of denser soils to dilate upon uni-directional (monotonic) straining can limit the severity of the effects of liquefaction even if it happens to be triggered. Although liquefaction can be triggered in very dense laboratory specimens, the level of loading required to do so may exceed that which can reasonably be expected from earthquake shaking; in such cases, screening criteria can be used to eliminate the need for a formal liquefaction potential evaluation.
14. The cyclic stress approach, which remains the most common procedure for the evaluation of liquefaction potential, characterizes both earthquake loading and soil liquefaction

resistance in terms of cyclic stresses. A transient earthquake motion is converted to an equivalent series of uniform cycles of shear stress. The number of equivalent cycles, a function of the duration of the motion, is correlated with the magnitude of the earthquake. Liquefaction resistance was originally obtained from laboratory (cyclic triaxial and cyclic simple shear) tests. The cyclic stress-based liquefaction resistance, however, is influenced by factors such as soil fabric, stress and strain history, and age that may be destroyed by sampling and are difficult to replicate in the laboratory.

15. *In situ* test-based procedures characterize liquefaction resistance in terms of *in situ* test parameters associated with soils at sites at which surficial manifestation of liquefaction has and has not been observed in past earthquakes; due to its repeatability and resolution, the CPT resistance is most commonly used but other *in situ* parameters, including SPT resistance (which has the benefit of providing soil samples) and shear wave velocity (which can often be measured economically from the ground surface), are also used. The cyclic stress approach allows the estimation of a factor of safety against liquefaction or a probability of liquefaction. Because of its basis on the presence or absence of surficial manifestation, common *in situ* test-based procedures can be interpreted as producing factors of safety against or probabilities of surficial manifestation.
16. Surficial manifestation of liquefaction usually takes the form of sand boils (ejecta) and/or ground cracking. Sand boil formation, however, can depend on the mechanical and hydraulic response of the entire soil profile; factors such as the depth, thickness, and void volume of the liquefied layer, the depth of the groundwater table, and the characteristics of overlying and underlying layers affect the hydraulic response and therefore surficial manifestation. Since liquefaction of a thin and/or silty layer at depth may not be expressed at the ground surface, the absence of sand boils does not necessarily indicate that level ground liquefaction has not occurred. Similarly, the large volume of water that can be expelled from a thick, shallow layer of loose soil in which liquefaction is not quite triggered can produce sand boils. Therefore, surficial manifestation as a criterion for liquefaction at depth can give rise to both false positive and false negative indications of liquefaction; this condition gives rise to some of the uncertainty in prediction of liquefaction potential.
17. Other approaches based on characteristics such as cyclic strain and dissipated energy have been proposed for the evaluation of liquefaction potential. Cyclic strain approaches are attractive in that pore pressure generation is inherently more closely related to cyclic strain amplitudes than cyclic stress amplitudes, but cyclic strain amplitudes are more difficult to predict than cyclic stress amplitudes. Dissipated energy demand is related to both stress and strain amplitudes and has been shown to correlate well to pore pressure generation in the laboratory but is also difficult to predict in the field and energy capacity is difficult to correlate to *in situ* parameters.
18. The generation of excess pore pressure and potential for liquefaction can also be investigated using nonlinear, effective stress ground response analyses. The development of improved numerical tools and constitutive models can allow evaluation and visualization of the coupled mechanical-hydraulic system response of an entire soil profile. The ability to "see inside" a profile as it responds to earthquake shaking can be very useful in understanding how and where pore pressures and deformations develop below as well as at the ground surface.
19. Rapid estimates of manifestation severity can be made using geospatial procedures that utilize regional ground motion, topographic, geologic, geomorphic, and groundwater data instead of site-specific geotechnical properties. Such estimates can be useful for planning and rapid response purposes.
20. The consequences of liquefaction are different for different liquefaction phenomena. Although flow liquefaction is capable of producing the most spectacular effects, cyclic liquefaction can also produce extensive damage.

21. Liquefaction can dramatically alter the amplitude and frequency content of ground surface motions. As the buildup of excess pore pressure causes a layer of liquefiable soil to soften, ground surface displacements may increase even when ground surface accelerations decrease. Dilation-induced stiffening can produce strong, high-frequency spikes in ground surface acceleration. Ground oscillation at level-ground sites may produce chaotic permanent movements of fractured blocks of surficial soil and significant amounts of sand can be ejected at the ground surface.
22. Ground surface settlement can develop during and/or after earthquakes due to the densification of dry or saturated sands. Settlement of dry sand occurs almost immediately, but settlement of saturated sands may not develop until well after earthquake shaking has ended. The magnitude of post-earthquake settlement depends on the density, thickness, and depth of the liquefiable soil, and on the amplitude and duration of shaking. Settlement can be caused by dissipation of excess pore pressure in the free-field or in the vicinity of structures, shear stresses associated with the response of structures supported on shallow foundations, and by loss of soil volume associated with ejecta.
23. The shear strength of a liquefied soil is sensitive to the void ratio, or density, of the soil. In the laboratory, where volume change can be eliminated, the shear strength mobilized at very large strain levels is often referred to as the steady state (or critical state) strength. In the field, where the requirements of the steady state of deformation are generally violated, the large-strain mobilized strength is referred to as the residual strength. Residual strengths are generally estimated as functions of penetration resistance and initial effective stress based on back-calculated strengths from flow slide case histories.
24. Excess pore pressures generated during earthquake shaking produce hydraulic gradients that cause porewater to flow during and after earthquake shaking. The resulting redistribution of pore pressure, and accompanying changes in soil void ratio, can have a strong effect on the consequences of liquefaction. This effect can be particularly important when flow is impeded by low-permeability soils (even thin lenses of such soils) that can cause void ratios to increase and steady state (or residual) strengths to decrease during or after shaking in the materials immediately below the lens. It is not uncommon for flow slides to occur minutes to hours after earthquake shaking has ended.
25. Deformation failures, such as lateral spreading, develop incrementally during, and in some cases partially after, the period of earthquake shaking. Dilation-induced stiffening can play an important role in limiting permanent deformations, and void redistribution can lead to increased deformations. For strong levels and/or long durations of shaking, deformation failures can produce large displacements and cause significant damage. Both empirical and numerical procedures are available to estimate displacements caused by deformation failures.
26. The triggering and consequences of liquefaction are influenced by ground motion characteristics and a complete evaluation of the performance of a soil profile containing liquefiable soils should consider all earthquake scenarios and the range of ground motions they can produce. Basing a liquefaction hazard analysis on one level of ground motion, i.e., that associated with a specific return period, ignores the fact that liquefaction can be triggered by weaker motions that occur more frequently and stronger motions that occur more rarely than those of the selected return period, and can result in highly inconsistent estimates of liquefaction hazard in different seismo-tectonic environments. Probabilistic liquefaction hazard analyses, or PLHAs, allow consideration of all earthquake scenarios and all associated levels of shaking to produce consistent and objective estimates of liquefaction hazards.
27. Saturated, plastic, fine-grained soils may not be susceptible to liquefaction but can generate excess pore pressure and soften under cyclic loading. The degree of softening is influenced by the undrained strength, static shear stress level, and cyclic shear stress level induced in the soil.

28. Plastic fine-grained soils can accumulate significant permanent deformations when cyclic stresses are superimposed on existing static shear stresses, even if the combined static and cyclic stresses do not exceed the shear strength of the soil.
29. Cyclic stress-based procedures that account for rate effects and estimate the potential for cyclic softening (characterized by shear strains in excess of 3%) are available. These soils characterize loading and resistance by means of cyclic stress and cyclic resistance ratios although both are defined differently than for sand-like soils. The cyclic resistance ratio is closely related to the static undrained shear strength ratio of the soil. The rate-dependence of clay-like soils should be considered in estimating the undrained strength that would be mobilized in the field (under rapid earthquake loading) from laboratory measurements (much slower loading).

REFERENCES

AASHTO. (2017). *AASHTO LRFD Bridge Design Specifications*, 8th edition, American Association of State Highway and Transportation Officials, Washington, DC, 1781 pp.

Adalier, K. and Elgamal, A. (2005). "Liquefaction of over-consolidated sand: A centrifuge investigation," *Journal of Earthquake Engineering*, Vol. 9, No. 1, pp. 127–150. https://doi.org/10.1080/13632460509350582

AECOM. (2009). "Root cause analysis of TVA Kingston dredge pond failure of December 22, 2008," *Report to Tennessee Valley Authority*, Knoxville, TN, 102 pp.

Alarcon-Guzman, A., Leonards, G.A., and Chameau, J.L. (1988). "Undrained monotonic and cyclic strength of sands," *Journal of Geotechnical Engineering*, Vol. 114, No. 10, pp. 1089–1108. https://doi.org/10.1061/(asce)0733-9410(1988)114:10(1089)

Alavi, A.H. and Gandomi, A.J. (2012). "Energy-based numerical models for assessment of soil liquefaction," *Geoscience Frontiers*, Vol. 3, No. 4, pp. 541–555. https://doi.org/10.1016/j.gsf.2011.12.008

Al-Tarhouni, M., Simms, P., and Sivathayalan, S. (2011). "Cyclic behavior of reconstituted and desiccated-rewet thickened gold tailings in simple shear," *Canadian Geotechnical Journal*, Vol. 48, pp. 1044–1060. https://doi.org/10.1139/t11-022

Andersen, K., Kleven, A., and Heien, D. (1988). "Cyclic soil data for design of gravity structures," *Journal of Geotechnical and Geoenvironmental Engineering*, Vol. 114, No. 5, pp. 517–539. https://doi.org/10.1061/(asce)0733-9410(1988)114:5(517)

Anderson, D.G., Shin, S., and Kramer, S.L. (2011). "Nonlinear, effective-stress ground motion response analyses following the AASHTO guide specifications for LRFD seismic bridge design," *Transportation Research Record*, Vol. 2251, pp. 144–154. https://doi.org/10.3141/2251-15

Andrade, J.E. (2009). "A predictive framework for liquefaction instability," *Geotechnique*, Vol. 59, No. 8, pp. 673–682. https://doi.org/10.1680/geot.7.00087

Andresen, A. and Bjerrum, L. (1968). "Slides in subaqueous slopes in loose sand and silt," *Norwegian Geotechnical Institute, Publication No. 81*, pp. 1–9.

Andrus, R.D., Hayati, H., and Mohanan, N.P. (2009). "Correcting liquefaction resistance for aged sands using measured to estimated velocity ratio," *Journal of Geotechnical and Geoenvironmental Engineering*, Vol. 135, No. 6, pp. 735–744. https://doi.org/10.1061/(asce)gt.1943-5606.0000025

Andrus, R.D. and Stokoe II, K.H. (2000). "Liquefaction resistance of soils from shear-wave velocity," *Journal of Geotechnical and Geoenvironmental Engineering*, Vol. 126, No. 11, pp. 1015–1025. https://doi.org/10.1061/(asce)1090-0241(2000)126:11(1015)

Andrus, R.D., Stokoe II, K.H., and Juang, C.H. (2004). "Guide for shear-wave based liquefaction potential evaluation," *Earthquake Spectra*, Vol. 20, No. 2, pp. 285–308. https://doi.org/10.1193/1.1715106

Arango, I., Lewis, M.R., and Kramer, C. (2000). "Updated liquefaction potential analysis eliminates foundation retrofitting of two critical structures," *Soil Dynamics and Earthquake Engineering*, Vol. 20, pp. 17–25. https://doi.org/10.1016/s0267-7261(00)00034-8

Atkinson, G.M., Finn, W.D.L., and Charlwood, R.G. (1984). "Simple computation of liquefaction probability for seismic hazard applications, *Earthquake Spectra*, Vol. 1, No. 1, pp. 107–123. https://doi.org/10.1193/1.1585259

Baille, T. (2014). "Breach of tailings pond results in 'Largest environmental disaster in modern Canadian history'," *Australasian Mine Safety Journal*, Vol. 8, 1 pp.

Baise, L.G., Higgins, R.B., and Brankman, C.M. (2006). "Liquefaction hazard mapping-statistical and spatial characterization of susceptible units," *Journal of Geotechnical and Geoenvironmental Engineering*, Vol. 132, No. 6, pp. 705–715. https://doi.org/10.1061/(asce)1090-0241(2006)132:6(705)

Baki, M.A.O., Cubrinovski, M., Stringer, M.E., van Ballegooy, S., and Ntritsos, N. (2023). "Effects of partial saturation on the liquefaction resistance of sand and silty sand from Christchurch," *Soils and Foundations*, Vol. 63, p. 101400. https://doi.org/10.1016/j.sandf.2023.101400

Bartlett, S.F. and Youd, T.L. (1992). "Empirical analysis of horizontal ground displacement generated by liquefaction-induced lateral spread," *Technical Report NCEER-92-0021*, National Center for Earthquake Engineering Research, Buffalo, NY.

Bassal, P. C., and Boulanger, R. W. (2023). "System response of an interlayered deposit with a localized graben deformation in the Northridge earthquake," *Soil Dynamics and Earthquake Engineering,* Vol. 165, 19 pp. 107668, 10.1016/j.soildyn.2022.107668.

Baziar, M.H. and Jafarian, Y. (2007). "Assessment of liquefaction triggering using strain energy concept and ANN model," *Soil Dynamics and Earthquake Engineering*, Vol. 27, pp. 1056–1072. https://doi.org/10.1016/j.soildyn.2007.03.007

Bennett, M.J. and Tinsley III, J.C. (1995). "Geotechnical data from surface and subsurface samples outside of and within liquefaction-related ground failures caused by the October 17, 1989, Loma Prieta earthquake, Santa Cruz and Monterey Counties, California," *U.S. Geological Survey Open-File Report 95-663*, 358 pp. https://doi.org/10.3133/ofr95663

Berrill, J.B. and Davis, R.O. (1985). "Energy dissipation and seismic liquefaction of sands: Revised model," *Soils and Foundations*, Vol. 25, No. 2, pp. 106–118. https://doi.org/10.3208/sandf1972.25.2_106

Beyzaei, C.Z., Bray, J.D., Cubrinovski, M., Reimer, M., and Stringer, M. (2018). "Laboratory-based characterization of shallow silty soils in southwest Christchurch," *Soil Dynamics and Earthquake Engineering*, Vol. 110, pp. 93–109. https://doi.org/10.1016/j.soildyn.2018.01.046

Beyzaei, C.Z., Bray, J.D., van Ballegooy, S., Cubrinovski, M., and Bastin, S. (2017). "Swamp depositional environment effects on liquefaction performance in Christchurch, New Zealand," *Proceedings, Third International Conference on Performance-Based Design, International Society for Soil Mechanics and Geotechnical Engineering*, Vancouver, BC, 9 pp.

Bhatia, S. (1980). "The verification of relationships for effective stress method to evaluate liquefaction potential of saturated sands," *Ph.D. Thesis*, Department of Civil Engineering, University of British Columbia, Vancouver, BC.

Bjerrum, L. (1971). "Subaqueous slope failures in Norwegian fjords," *Norwegian Geotechnical Institute, Publication No. 88.*

Borja, R.I. (2006). "Condition for liquefaction instability in fluid-saturated granular soils," *Acta Geotechnica*, Vol. 1, No. 4, pp. 211–224. https://doi.org/10.1007/s11440-006-0017-5

Boulanger, R.W. (2003). "Relating K_α to relative state parameter index," *Journal of Geotechnical and Geoenvironmental Engineering*, Vol. 129, No. 8, pp. 770–773. https://doi.org/10.1061/(asce)1090-0241(2003)129:8(770)

Boulanger, R.W. and Brandenberg, S.J. (2004). "Neutral plane solution for liquefaction induced down-drag on vertical piles," *Geotechnical Special Publication 126*, 9 pp. https://doi.org/10.1061/40744(154)32

Boulanger, R.W. and Idriss, I.M. (2004). "Evaluating the potential for liquefaction or cyclic failure of silts and clays," *Report No. UCD/CGM-04/01*, Center for Geotechnical Modeling, University of California, Davis, CA

Boulanger, R.W. and Idriss, I.M. (2006). "Liquefaction susceptibility criteria for silts and clays," *Journal of Geotechnical and Geoenvironmental Engineering*, Vol. 132, No. 11, pp. 1413–1426. https://doi.org/10.1061/(asce)1090-0241(2006)132:11(1413)

Boulanger, R.W. and Idriss, I.M. (2007). "Evaluation of cyclic softening in silts and clays," *Journal of Geotechnical and Geoenvironmental Engineering*, Vol. 133, No. 6, pp. 641–652. https://doi.org/10.1061/(asce)1090-0241(2007)133:6(641)

Boulanger, R.W. and Idriss, I.M. (2012). "Probabilistic SPT-based liquefaction triggering procedure," *Journal of Geotechnical and Geoenvironmental Engineering*, Vol. 138, No. 10, pp. 1185–1195. https://doi.org/10.1061/(asce)gt.1943-5606.0000700

Boulanger, R.W. and Idriss, I.M. (2014). "CPT and SPT based liquefaction triggering procedures," *Report No. UCD/CGM-14/01*, Center for Geotechnical Modeling, Department of Civil and Environmental Engineering, University of California, Davis, CA, 134 pp.

Boulanger, R. W., and Idriss, I. M. (2015). "CPT-based liquefaction triggering procedure," *Journal of Geotechnical and Geoenvironmental Engineering*, Vol. 142, No. 2, 11 pp.

Boulanger, R.W. and Idriss, I.M. (2016). "CPT-based liquefaction triggering procedure," *Journal of Geotechnical and Geoenvironmental Engineering*, Vol. 142, No. 2, p. 04015065. https://doi.org/10.1061/(asce)gt.1943-5606.0001388

Boulanger, R.W. and Seed, R.B. (1995). "Liquefaction of sand under bi-directional monotonic and cyclic loading," *Journal of Geotechnical and Geoenvironmental Engineering*, Vol. 121, No. 12, pp. 870–878. https://doi.org/10.1061/(asce)0733-9410(1995)121:12(870)

Boulanger, R.W. and Truman, S.P. (1996). "Void redistribution in sand under post-earthquake loading," *Canadian Geotechnical Journal*, Vol.33, pp. 829–833. https://doi.org/10.1139/t96-109-329

Boulanger, R.W., Moug, D.M., Munter, S.K., Price, A.B., and DeJong, J.T. (2016). "Evaluating liquefaction and lateral spreading in interbedded sand, silt and clay deposits using the cone penetrometer," *Australian Geomechanics*, Vol. 51, No. 4, pp. 109–128.

Boulanger, R.W., Seed, R.B., Chan, C.K., Seed, H.B., and Sousa, J. (1991). "Liquefaction behavior of saturated sands under uni-directional and bi-directional monotonic and cyclic simple shear loading," *Geotechnical Engineering Report No. UCB/GT/91-08*, University of California, Berkeley.

Bradley, B.A. (2014). "Site-specific and spatially-distributed ground-motion intensity estimation in the 2010-2011 Canterbury earthquakes," *Soil Dynamics and Earthquake Engineering*, Vol. 61–62, pp. 83–91. https://doi.org/10.1016/j.soildyn.2014.01.025

Bradley, B.A. and Hughes, M. (2012). "Conditional peak ground accelerations in the Canterbury earthquakes for conventional liquefaction assessment," *Technical Report prepared for the Department of Building and Housing*, 22 pp.

Brandenberg, S.J., Zimmaro, P., Stewart, J.P., Kwak, D.Y., Franke, K.W., Moss, R.E.S., Cetin, K.O., Can, G., Ilgac, M., Stamatakos, J., Weaver, T., and Kramer, S.L. (2020). "Next-generation liquefaction database," *Earthquake Spectra*, Vol. 36, pp. 939–959. https://doi.org/10.1177/8755293020902477

Bray, J., Cubrinovski, M., Zupan, J., and Taylor, M. (2014). "Liquefaction effects on buildings in the central business district of Christchurch," *Earthquake Spectra*, Vol. 30, pp. 85–109. https://doi.org/10.1193/022113eqs043m

Bray, J.D., Boulanger, R.W., Cubrinovski, M., Tokimatsu, K., Kramer, S.L., O'Rourke, T., Rathje, E., Green, R.A., Robertson, P.K., and Beyzaei, C.Z. (2017). "Liquefaction-induced ground movement effects," *Proceedings, U.S.-New Zealand_Japan International Workshop, PEER Report No. 2017/02*, Pacific Earthquake Engineering Research Center, University of California, Berkeley, 48 pp.

Bray, J.D. and Macedo, J. (2017). "6th Ishihara lecture: Simplified procedure for estimating liquefaction-induced building settlement," *Soil Dynamics and Earthquake Engineering*, Vol. 102, pp. 215–231. https://doi.org/10.1016/j.soildyn.2017.08.026

Bray, J.D. and Sancio, R.B. (2006). "Assessment of the liquefaction susceptibility of fine-grained soils," *Journal of Geotechnical and Geoenvironmental Engineering*, Vol. 132, No. 9, pp. 1165–1177. https://doi.org/10.1061/(asce)1090-0241(2006)132:9(1165)

Bray, J.D. and Sancio, R.B. (2009). "Performance of buildings in Adapazari during the 1999 Kocaeli, Turkey earthquake," in T. Kokusho, ed., *Earthquake Geotechnical Case Histories for Performance Based Design*, TC4 Committee, ISSMFE, CRC Press/Balkema, the Netherlands, pp. 325–340 & Data on CD-ROM. https://doi.org/10.1201/noe0415804844.ch20

Bray, J.D., Sancio, R.B., Durgunoglu, T., Onalp, A., Youd, T.L., Stewart, J.P., Seed, R.B., Cetin, O.K., Bol, E., Baturay, M.B., Christensen, C., and Karadayilar, T. (2004). "Subsurface characterization at ground failure sites in Adapazari, Turkey," *Journal of Geotechnical and Geoenvironmental Engineering*, Vol. 130, No. 7, pp. 673–685. https://doi.org/10.1061/(asce)1090-0241(2004)130:7(673)

Broms, B. and Bennermark, H. (1967). "Free discussion," *Proceedings, Geotechnical Conference*, Oslo, Norway, Vol. 2, pp. 118–120.

Bullock, Z., Dashti, S., Karimi, Z., Liel, A.B., Porter, K.A., and Franke, K. (2019). "Probabilistic models for residual and peak transient tilt of mat-founded structures on liquefiable soils," *Journal of Geotechnical and Geoenvironmental Engineering*, Vol. 145. https://doi.org/10.1061/(asce)gt.1943-5606.0002002

Byrne, P.M., Park, S.S., Beaty, M., Sharp, M.K., Gonzalez, L., and Abdoun, T. (2004). "Numerical modeling of liquefaction and comparison with centrifuge tests," *Canadian Geotechnical Journal*, Vol. 4, No. 2, pp. 193–211. https://doi.org/10.1139/t03-088

Campbell, K. W., and Bozorgnia, Y. (2011). "Predictive equations for the horizontal component of standardized cumulative absolute velocity as adapted for use in the shutdown of U.S. nuclear power plants," Nuclear Engineering Design, Vol. 241, pp. 2558–2569. https://doi.org/10.1016/j.nucengdes.2011.04.020.

Cao, Z.Z., Hou, L.Q., and Xu, H.M. (2010). "The distribution and characteristics of gravelly soils liquefaction in the Wenchuan Ms 8.0 earthquake," *Earthquake Engineering and Engineering Vibration*, Vol. 9, No. 2, pp. 167–175.

Carter, D.P. and Seed, H.B. (1988). "Liquefaction potential of sand deposits under low levels of excitation," *Report No. UCB/EERC-88/11*, Earthquake Engineering Research Center, University of California, Berkeley, 309 pp.

Castro, G. (1969). "Liquefaction of sands," *Harvard Soil Mechanics Series 87*, Harvard University, Cambridge, MA.

Castro, G. (1991). "On the behavior of soils during earthquakes – liquefaction," *Proceedings, NSF/EPRI Workshop on Dynamic Soil Properties and Site Characterization, EPRI NP-7337*, Vol. 2, Electric Power Research Institute, Palo Alto, CA, pp. 1–36.

Cetin, K.O. and Bilge, H.T. (2012). "Cyclic large strain and induced pore pressure models for saturated clean sands," *Journal of Geotechnical and Geoenvironmental Engineering*, Vol. 138, No. 3, pp. 309–323. https://doi.org/10.1061/(asce)gt.1943-5606.0000631

Cetin, K.O., Bilge, H.T., Wu, J., Kammerer, A.M., and Seed, R.B. (2009). "Probabilistic models for the assessment of cyclically induced reconsolidation (volumetric) settlements," *Journal of Geotechnical and Geoenvironmental Engineering*, Vol. 135, No. 3, pp. 387–398. https://doi.org/10.1061/(asce)1090-0241(2009)135:3(387)

Cetin, K.O., Der Kiureghian, A., and Seed, R.B. (2002) "Probabilistic models for the initiation of seismic soil liquefaction," *Structural Safety*, Vol. 24, pp. 67–82. https://doi.org/10.1016/s0167-4730(02)00036-x

Cetin, K.O., Seed, R.B., DerKiureghian, A., Tokimatsu, K., Harder, L.F., Jr., and Kayen, R.E. (2004). "SPT-Based probabilistic and deterministic assessment of seismic soil liquefaction potential," *Journal of Geotechnical and Geoenvironmental Engineering*, Vol. 130, No. 12, pp. 1314–1340. https://doi.org/10.1061/(asce)1090-0241(2004)130:12(1314)

Cetin, K.O., Seed, R.B., Kayen, R.E., Moss, R.E.S., Bilge, H.T., Ilgac, M., and Chowdhury, K. (2018). "Examination of differences between three SPT-based seismic soil liquefaction triggering relationships," *Soil Dynamics and Earthquake Engineering*, Vol. 113, pp. 75–86. https://doi.org/10.1016/j.soildyn.2018.03.013

Chang, K.T. (1978). "An analysis of damage of slope sliding by earthquake on the Paiho Main dam and its earthquake strengthening," *Tseng-hua Design Section Report*, Department of Earthquake-Resistant Design and Flood Control Command of Miyna Reservoir, Peoples Republic of China.

Chung, J.-W. and Rogers, J.D. (2011). "Simplified method for spatial evaluation of liquefaction potential in the St. Louis area," *Journal of Geotechnical and Geoenvironmental Engineering*, Vol. 137, pp. 505–515. https://doi.org/10.1061/(asce)gt.1943-5606.0000450

Conlon, R. (1966). "Landslide at Toulnustouc River, Quebec," *Canadian Geotechnical Journal*, Vol. 3, No. 3, pp. 113–144. https://doi.org/10.1139/t66-016

Coop, M.R. and Airey, D.W. (2003). "Carbonate sands," in T.S. Tan, K.K. Phoon, D.W. Hight, and S. Leroueil, eds., *Characterisation and Engineering Properties of Natural Soils*, A. A. Balkema, Publishers, pp. 1049–1086. https://doi.org/10.1201/noe0415426916

Cornforth, D.H., Worth, E.G., and Wright, W.L. (1975). "Observations and analysis of a flow slide in sand fill," *Field Instrumentation in Geotechnical Engineering*, John Wiley and Sons, New York, pp. 136–151.

Coulter, M. and Migliaccio, L. (1966). "Effects of the earthquake of March 27, 1964 at Valdez, Alaska," *Professional Paper No. 542-C*, U.S. Geological Survey, U.S. Department of the Interior, Washington, DC. https://doi.org/10.3133/pp542c

Cramer, C.H., Rix, G.J., and Tucker, K. (2008). "Probabilistic liquefaction hazard maps for Memphis, Tennessee," *Seismological Research Letters*, Vol. 79, No. 3, pp. 416–423. https://doi.org/10.1785/gssrl.79.3.416

Cubrinovski, M., Bray, J.D., de la Torre, C., Olsen, M., Bradley, B., Chiaro, G., Stocks, E., and Wotherspoon, L. (2017). "Liquefaction effects and associated damages observed at the Wellington CentrePort from the 2016 Kaikoura earthquake," *Bulletin of the New Zealand Society for Earthquake Engineering*, Vol. 50, No. 2, pp. 152–173. https://doi.org/10.5459/bnzsee.50.2.152-173

Cubrinovski, M., Bray, J.D, de la Torre, C., Olsen, M., Bradley, B., Chiaro, G., Stocks, E., Wotherspoon, L, and Krall, T. (2018). "Liquefaction-induced damage and CPT characterization of the reclamations CentrePort, Wellington," *Bulletin of the Seismological Society of America*, Vol. 108, No. 3B, pp. 1695–1708. https://doi.org/10.1785/0120170246

Cubrinovski, M. and Ishihara, K. (1998a). "Modelling of sand behaviour based on state concept," *Soils and Foundations*, Vol. 38, No. 3, pp. 115–127. https://doi.org/10.3208/sandf.38.3_115

Cubrinovski, M. and Ishihara, K. (1998b). "State concept and modified elastoplasticity for sand modelling," *Soils and Foundations*, Vol. 38, No. 4, pp. 213–225. https://doi.org/10.3208/sandf.38.4_213

Cubrinovski, M. and Ishihara, K. (1999). "Empirical correlation between SPT N-value and relative density for sandy soils," *Soils and Foundations*, Vol. 39, No. 5, pp. 61–71. https://doi.org/10.3208/sandf.39.5_61

Cubrinovsky, M., Rees, S., and Bowman, E. (2010). "Effects of non-plastic fines on liquefaction resistance of sandy soils," in M. Garevski and A. Ansal, eds., *Earthquake Engineering* in Europe, Geotechnical, Geological, and Earthquake Engineering, Vol. 17, pp. 125–144.

Cubrinovski, M., Rhodes, A., Ntritsos, N., and Van Ballegooy, S. (2019). "System response of liquefiable deposits," *Soil Dynamics and Earthquake Engineering*, Vol. 124, pp. 212–229. https://doi.org/10.1016/j.soildyn.2018.05.013

Davila, R.S., Sego, D.C., and Robertson, P.K. (1992). "Undisturbed sampling of sandy soils by freezing," *Proceedings, 45th Canadian Geotechnical Conference*, pp. 126–130.

Davis, R.O. and Berrill, J.B. (1982). "Energy dissipation and seismic liquefaction in sands," *Earthquake Engineering and Structural Dynamics*, Vol. 10, No. 1, pp. 59–68. https://doi.org/10.1002/eqe.4290100105

Dawson, R.F., Morgenstern, N.R., and Stokes, A.W. (2011). "Liquefaction flowslides in Rocky Mountain coal mine waste dumps," *Canadian Geotechnical Journal*, Vol. 35, No. 2, pp. 328–343. https://doi.org/10.1139/t98-009

De Alba, P. (1983). "Pile settlement in a liquefying soil deposit," *Journal of Geotechnical Engineering*, Vol. 109, No. 9, pp. 1165–1180. https://doi.org/10.1061/(asce)0733-9410(1983)109:9(1165)

De Alba, P., Seed, H.B., and Chan, C.K. (1976). "Sand liquefaction in large-scale simple shear tests," *Journal of the Geotechnical Engineering Division*, Vol. 102, No. GT9, pp. 909–927. https://doi.org/10.1061/ajgeb6.0000322

De Blasio, F.V., Engvik, L., Harbitz, C.B., and Elverhoi, A. (2004). "Hydroplaning and submarine debris flows," *Journal of Geophysical Research*, Vol. 109, p. C01002. https://doi.org/10.1029/2002jc001714

Dief, H.M. and Figueroa, J.L. (2001). "Liquefaction assessment by the energy method through centrifuge modeling," in X.W. Zeng, ed., *Proceedings of the NSF International Workshop on Earthquake Simulation in Geotechnical Engineering*, CWRU, Cleveland, OH.

Dixit, J., Dewaikar, D.M., and Jangrid, J.S. (2012). "Assessment of liquefaction potential index for Mumbai City," *Natural Hazards and Earth Science Systems*, Vol. 12, pp. 2759–2768. https://doi.org/10.5194/nhess-12-2759-2012

Dobry, R. and Abdoun, T. (2011). "An investigation into why liquefaction charts work: A necessary step toward integrating the states of art and practice," *Proceedings, 5th International Conference on Earthquake Geotechnical Engineering*, Santiago, Chile, pp. 13–44.

Dobry, R. and Abdoun, T. (2015). "Cyclic shear strain needed for liquefaction triggering and assessment of overburden pressure factor, K_σ," *Journal of Geotechnical and Geoenvironmental Engineering*, Vol 141, No. 11, p. 04015047. https://doi.org/10.1061/(asce)gt.1943-5606.0001342

Dobry, R., Abdoun, T., Stokoe, K.H., Moss, R.E.S., Hatton, M., and El Ganainy, H. (2015). "Liquefaction potential of recent fills versus natural sands located in high-seismicity regions using shear-wave velocity," *Journal of Geotechnical and Geoenvironmental Engineering*, Vol. 141, No. 3, p. 04014112. https://doi.org/10.1061/(asce)gt.1943-5606.0001239

Dobry, R. and Ladd, R. (1980). Discussion to "Soil Liquefaction and Cyclic Mobility Evaluation for Level Ground during Earthquakes," by H. B. Seed and "Liquefaction Potential: Science versus Practice," by R. B, Peck, *Journal of the Geotechnical Engineering Division*, Vol. 106, No. GT6, pp. 720–724. https://doi.org/10.1061/ajgeb6.0000984

Dobry, R., Ladd, R., Yokel, F., Chung, R., and Powell, D. (1982). "Prediction of pore water pressure buildup and liquefaction of sands during earthquakes by the cyclic strain method," *NBS Building Science Series 138*, National Bureau of Standards, U.S. Department of Commerce. https://doi.org/10.6028/nbs.bss.138

Dobry, R., Pierce, W.G., Dyvik, R., Thomas, G.E., and Ladd, R.S. (1985). "Pore pressure model for cyclic straining of sand," *Research Report*, Civil Engineering Department, Rensselaer Polytechnic Institute, Troy, NY, 56 pp.

Donahue, J.L., Bray, J.D., and Riemer, M.F. (2008). "Liquefaction susceptibility, resistance, and response of silty and clayey soils," *Report No. UCB/GT 2008-01*, University of California, Berkeley, 268 pp.

Eckersley, J.D. (1985). "Flowslides in stockpiled coal," *Engineering Geology*, Vol 22, No. 1, pp. 13–22. https://doi.org/10.1016/0013-7952(85)90034-1

Elgamal, A., Yang, Z., and Parra, E. (2002). "Computational modeling of cyclic mobility and post-liquefaction site response," *Soil Dynamics and Earthquake Engineering*, Vol. 22, No. 4, pp. 259–271. https://doi.org/10.1016/s0267-7261(02)00022-2

El-Sekelly, W., Abdoun, T., and Dobry, R. (2016). "Liquefaction resistance of a silty sand deposit subjected to preshaking followed by extensive liquefaction," *Journal of Geotechnical and Geoenvironmental Engineering*, Vol. 142, No. 4, p. 04015101. https://doi.org/10.1061/(asce)gt.1943-5606.0001444

Eseller-Bayat, E., Yegian, M. K., Alshawabkeh, A., and Gokyer, S. (2013a). "Liquefaction response of partially saturated sands. I: Experimental results," *Journal of Geotechnical and Geoenvironmental Engineering*, Vol. 139, No. 6, pp. 863–871. https://doi.org/10.1061/(asce)gt.1943-5606.0000815

Eseller-Bayat, E., Yegian, M. K., Alshawabkeh, A., and Gokyer, S. (2013b). "Liquefaction response of partially saturated sands. II: Empirical model," *Journal of Geotechnical and Geoenvironmental Engineering*, Vol. 139, No. 6, pp. 872–879. https://doi.org/10.1061/(asce)gt.1943-5606.0000816

Evans, M.D. and Seed, H.B. (1987). "Undrained cyclic triaxial testing of gravels – the effect of membrane compliance," *Report No. UCB/EERC-87/08*, Earthquake Engineering Research Center, University of California, Berkeley, CA.

Fellenius, B. (1953). "The landslide at Guntorp," *Geotechnique*, Vol. 5, No. 1, pp. 120–125. https://doi.org/10.1680/geot.1955.5.1.120

Fellenius, B.H. and Siegel, T.C. (2008). "Pile drag load and downdrag in a liquefaction event," *Journal of Geotechnical and Geoenvironmental Engineering*, Vol. 134, No. 9, pp. 1412–1416. https://doi.org/10.1061/(asce)1090-0241(2008)134:9(1412)

Fiegel, G.L. and Kutter, B.L. (1994). "Liquefaction mechanism for layered soils," *Journal of Geotechnical and Geoenvironmental Engineering*, Vol. 120, No. 4, pp. 737–755. https://doi.org/10.1061/(asce)0733-9410(1994)120:4(737)

Figueroa, J.L., Saada, A.S., Liang, L., and Dahisaria, M.N. (1994). "Evaluation of soil liquefaction by energy principles," *Journal of Geotechnical and Geoenvironmental Engineering*, Vol. 20, No. 9, pp. 1554–1569. https://doi.org/10.1061/(asce)0733-9410(1994)120:9(1554)

Finn, W.D.L. (1981). "Liquefaction potential: Developments since 1976," *Proceedings, International Conference on Recent Advances in Geotechnical Earthquake Engineering and Soil Dynamics*, University of Missouri, Rolla, MO, pp. 655–681.

Finn, W.D.L. and Bhatia, S.K. (1981). "Prediction of seismic pore-water pressures," *Proceedings, 10th International Conference on Soil Mechanics and Foundation Engineering, Rotterdam*, The Netherlands, Vol. 3, pp. 201–206.

Finn, W.D.L., Bransby, P.L., and Pickering, D.J. (1970). "Effect of strain history on liquefaction of sands," *Journal of the Soil Mechanics and Foundations Division*, Vol. 96, No. SM6, pp. 1917–1934. https://doi.org/10.1061/jsfeaq.0001478

Franke, K.W. and Kramer, S.L. (2014). "Procedure for the empirical evaluation of lateral spread displacement hazard curves," *Journal of Geotechnical and Geoenvironmental Engineering*, Vol 140, No. 1, pp. 110–120. https://doi.org/10.1061/(asce)gt.1943-5606.0000969

Franke, K. W., Wright, A. D., and Ekstrom, L. T. (2014). "Comparative study between two performance-based liquefaction triggering models for the standard penetration test." *Journal of Geotechnical and Geoenvironmental Engineering*, Vol, 140, No. 5, 4 pp.

Gao, Z., Ho, B., and Chang, D. (1983). "Some geological considerations for the damage during the Tangshan earthquake," *North China Earthquake Sciences*, Vol. 1, pp. 64–72 (in Chinese).

Geremew, A.M. and Yanful, E.K. (2012). "Laboratory investigation of the resistance of tailings and natural sediments to cyclic loading," *Geotechnical and Geological Engineering*, Vol. 30, pp. 431–437. https://doi.org/10.1007/s10706-011-9478-x

Geyin, M. and Maurer, B.W. (2019). "An analysis of liquefaction-induced free-field ground settlement using 1,000+ case-histories: Observations vs. state-of-practice predictions," *Geotechnical Special Publication 308*, pp. 489–498. https://doi.org/10.1061/9780784482100.049

Geyin, M. and Maurer, B.W. (2020). "Fragility functions for liquefaction-induced ground failure," *Journal of Geotechnical and Geoenvironmental Engineering*, Vol. 146, No. 12, p. 04020142. https://doi.org/10.1061/(asce)gt.1943-5606.0002416

Geyin, M., Baird, A.J., and Maurer, B.W. (2020). "Field assessment of liquefaction prediction models based on geotechnical versus geospatial data, with lessons for each," *Earthquake Spectra*, Vol. 36, No. 3, pp. 1386–1411. https://doi.org/10.1177/8755293019899951

Ghafghazi, M., DeJong, J.T., Sturm, A.P., and Temple, C.E. (2017). "Instrumented becker penetration test. II: iBPT-SPT correlation for characterization and liquefaction assessment of gravelly soils," *Journal of Geotechnical and Geoenvironmental Engineering*, Vol. 143, No. 9, p. 04017063. https://doi.org/10.1061/(asce)gt.1943-5606.0001718

Gillins, D.T. and Bartlett, S.F. (2014). "Multilinear regression equations for predicting lateral spread displacement from soil type and cone penetration test data," *Journal of Geotechnical and Geoenvironmental Engineering*, Vol. 140, No. 4, p. 04013047. https://doi.org/10.1061/(asce)gt.1943-5606.0001051

Goda, K., Atkinson, G.M., Hunter, J.A., Crow, H. and Motazedian, D. (2011). "Probabilistic liquefaction hazard analysis for four Canadian cities," *Bulletin of the Seismological Society of America*, Vol. 101, No. 1, pp. 190–201. https://doi.org/10.1785/0120100094

Golesorkhi, R. (1989). "Factors influencing the computational determination of earthquake-induced shear stresses in sandy soils," *Ph.D. Thesis*, University of California at Berkeley, 395 pp.

Goulois, A.M., Whitman, R.V., and Hoeg, K. (1985). "Effects of sustained shear stresses on the cyclic degradation of clay," in R.C. Chaney and K.R. Demars, eds., *Strength Testing of Marine Sediments: Laboratory and In Situ Strength Measurements, ASTM STP 883*, ASTM, Philadelphia, pp. 336–351. https://doi.org/10.1520/stp36344s

Green, R. and Terri, G. (2005). "Number of equivalent cycles concept for liquefaction evaluations-revisited," *Journal of Geotechnical and Geoenvironmental Engineering*, Vol. 131, No. 4, pp. 477–488. https://doi.org/10.1061/(asce)1090-0241(2005)131:4(477)

Green, R.A. and Lee, J. (2010). "The influence of near-fault rupture directivity on liquefaction," *Proceedings, 9th U.S. National and 10th Canadian Conference on Earthquake Engineering*, Toronto, Canada, Paper No. 1177, 11 pp.

Green, R.A., Lee, J., White, T.M., and Baker, J.W. (2008). "The significance of near-fault effects on liquefaction," *Proceedings, 14th World Conference on Earthquake Engineering*, Beijing, China, 8 pp.

Green, R.A. and Olson, S.M. (2015). "Interpretation of liquefaction field case histories for use in developing liquefaction triggering curves," Proceedings, 6th International Conference on Earthquake Geotechnical Engineering, Christchurch, New Zealand, 9 pp.

Green, R.A., Mitchell, J.K., and Polito, C.P. (2000). "An energy-based pore pressure generation model for cohesionless soils," in D.W. Smith and J.P. Carter, eds, *Proceedings, John Booker Memorial Symposium-Developments in Theoretical Geomechanics*, Balkema, Rotterdam, the Netherlands, pp. 383–390.

Hansen, W.R. (1965). "Effects of the earthquake of March 27, 1964, at Anchorage, Alaska," *U.S. Geological Survey Professional Paper 542-A*, p. A1-A68, 2 sheets, scale 1:480. https://doi.org/10.3133/pp542a

Hanzawa, H., Itoh, Y., and Suzuki, K. (1979). "Shear characteristics of a quick sand in the Arabian Gulf," *Soils and Foundations*, Vol. 19, No. 4, pp. 1–15. https://doi.org/10.3208/sandf1972.19.4_1

Harbitz, C.B., Parker, G., Elverhøi, A., Mohrig, D., and Harff, P. (2003). "Hydroplaning of subaqueous debris flows and glide blocks: Analytical solutions and discussions," *Journal of Geophysical Research*, Vol. 108, No. B7, pp. 23–49. https://doi.org/10.1029/2001jb001454

Harder, L. F., and Boulanger, R. W. (1997). "Application of Kσ and Kα correction factors." In T. L. Youd and I. M. Idriss, eds., *Proceedings, NCEER Workshop on Evaluation of Liquefaction Resistance of Soils*, Salt Lake City, UT, Technical Report NCEER-97-0022, National Center for Earthquake Engineering Research, SUNY, Buffalo, pp. 167–190.

Hartvigsen, A.J. (2007). "Influence of pore pressures in liquefiable soils on elastic response spectra," *Master's thesis*, University of Washington, Seattle, 150 pp.

Hayati, H. and Andrus, R.D. (2009). "Updated liquefaction resistance correction factors for aged sands," *Journal of Geotechnical and Geoenvironmental Engineering*, Vol. 135, No. 11, pp. 1683–1692. https://doi.org/10.1061/(asce)gt.1943-5606.0000118

Hazen, A. (1920). "Hydraulic fill dams," *Transactions of the American Society of Civil Engineers*, Vol. 83, pp. 1717–1745. https://doi.org/10.1061/taceat.0002944

Hofmann, B., Sego, D., and Robertson, P. (2000). "In situ ground freezing to obtain undisturbed samples of loose sand," *Journal of Geotechnical and Geoenvironmental Engineering*, Vol. 126, No. 11, pp. 979–989. https://doi.org/10.1061/(asce)1090-0241(2000)126:11(979)

Holzer, T.L. (2008). "Probabilistic liquefaction hazard mapping," *Proceedings, Geotechnical Earthquake Engineering and Soil Dynamics IV, American Society of Civil Engineers*, Sacramento, California, 32 pp. https://doi.org/10.1061/40975(318)30

Holzer, T. L., Noce, T. E., and Bennett, M. J. (2011). "Liquefaction probability curves for surficial geologic deposits," *Environmental & Engineering Geoscience*, Vol. 17, No. 1, pp. 1–21.

Hossain, A., Andrus, R., and Camp, W. (2013). "Correcting liquefaction resistance of unsaturated soil using wave velocity," *Journal of Geotechnical and Geoenvironmental Engineering*, Vol. 139, pp. 277–287. 10.1061/(ASCE)GT.1943-5606.0000770.

Hryciw, R.D., Vitton, S., and Thomann, T.G. (1990). "Liquefaction and flow failure during seismic exploration," *Journal of Geotechnical Engineering*, Vol. 116, No. 12, pp. 1881–1889. https://doi.org/10.1061/(asce)0733-9410(1990)116:12(1881)

Hudson, K.S., Ulmer, K.J., Zimmaro, P., Kramer, S.L., Stewart, J.P., and Brandenberg, S.J. (2024). "Relationship between fines content and soil behavior type index at liquefaction sites," *Journal of Geotechnical and Geoenvironmental Engineering*, Vol. 150 No. 5, 06024001. https://doi.org/10.1061/JGGEFK.GTENG-12146

Hutabarat, D. and Bray, J.D. (2021a). "Effective stress analysis of liquefiable sites to estimate the severity of sediment ejecta," *Journal of Geotechnical and Geoenvironmental Engineering*, Vol. 147, No. 5, p. 04021024. https://doi.org/10.1061/(asce)gt.1943-5606.0002503

Hutabarat, D. and Bray, J.D. (2021b). "Seismic response characteristics of liquefiable sites with and without sediment ejecta manifestation," *Journal of Geotechnical and Geoenvironmental Engineering*, Vol. 147, No. 6, p. 04021040. https://doi.org/10.1061/(asce)gt.1943-5606.0002506

Hutabarat, D. and Bray, J.D. (2022). "Estimating the severity of liquefaction ejecta using the cone penetration test," *Journal of Geotechnical and Geoenvironmental Engineering*, Vol. 148, No. 3, p. 04021195. https://doi.org/10.1061/(asce)gt.1943-5606.0002744

Iai, S., Tobita, T., Ozutsumi, O., and Ueda, K. (2011). "Dilatancy of granular materials in strain space multiple mechanism model," *International Journal of Numerical Methods in Geomechanics*, Vol. 35, pp. 360–392. https://doi.org/10.1002/nag.899

Idriss, I.M. (1985). "Evaluating seismic risk in engineering practice," *Proceedings, 11th International Conference on soil Mechanics and Foundation Engineering*, Vol. 1, Balkema, Rotterdam, the Netherlands, pp. 265–320.

Idriss, I.M. (1999). "Presentation notes: An update of the Seed-Idriss simplified procedure for evaluating liquefaction potential," *Proceedings, TRB Workshop on New Approaches to Liquefaction Anal., Publ. No. FHWARD-99-165*, Federal Highway Administration, Washington, D.C.

Idriss, I.M. and Boulanger, R.W. (2003). "Relating K_α and K_σ to SPT blow count and to CPT tip resistance for use in evaluating liquefaction potential," Proceedings, Dam Safety 2003, Association of State Dam Safety Officials, Lexington, KY.

Idriss, I.M. and Boulanger, R.W. (2006). "Semi-empirical procedures for evaluating liquefaction potential during earthquakes," *Journal of Soil Dynamics and Earthquake Engineering*, Vol. 26, pp. 115–130. https://doi.org/10.1016/j.soildyn.2004.11.023

Idriss, I.M. and Boulanger, R.W. (2007). "SPT- and CPT-based relationships for the residual shear strength of liquefied soils," in K.D. Pitilakis, ed., *Earthquake Geotechnical Engineering, 4th International Conference on Earthquake Geotechnical Engineering-Invited Lectures*, Springer, the Netherlands, pp. 1–22. https://doi.org/10.1007/978-1-4020-5893-6_1

Idriss, I.M. and Boulanger, R.W. (2008). *Soil Liquefaction during Earthquakes*, Earthquake Engineering Research Institute, Oakland, CA, 237 pp.

Idriss, I.M. and Boulanger, R.W. (2010). "SPT-based liquefaction triggering procedures," *Report No. UCD/CGM-10/02*. University of California, Davis, 136 pp.

Ishihara, K. (1985). "Stability of natural deposits during earthquakes," *Proceedings, Eleventh International Conference on Soil Mechanics and Foundation Engineering*, Vol. 1, pp. 321–376.

Ishihara, K. (1993). "Liquefaction and flow failure during earthquakes," *Geotechnique*, Vol. 43, No. 3, pp. 351–415. https://doi.org/10.1680/geot.1993.43.3.351

Ishihara, K. (1996). *Soil Behaviour in Earthquake Geotechnics*, Oxford Engineering Science Series 46, Oxford, UK, 350 pp. https://doi.org/10.1093/oso/9780198562245.001.0001

Ishihara, K. and Towhata, I. (1980). "One-dimensional soil response analysis during earthquakes based on effective stress model," *Journal of the Faculty of Engineering*, Vol. 35, No. 4, p. 656.

Ishihara, K., Troncoso, J., Kawase, Y., and Takahashi, Y. (1980). "Cyclic strength characteristics of tailings materials," *Soils and Foundations*, Vol. 20, No. 4, pp. 127–142. https://doi.org/10.3208/sandf1972.20.4_127

Ishihara, K., Tsuchiya, H., Huang, Y., and Kamada, K.,(2001). "Keynote Lecture: Recent Studies on Liquefaction Resistance of Sand-Effect of Saturation," *Proceedings, International Conferences on Recent Advances in Geotechnical Earthquake Engineering and Soil Dynamics*, Rolla, MO, pp. 1–7.

Ishihara, K. and Yamazaki, F. (1980). "Cyclic simple shear tests on saturated sand in multidirectional loading," *Soils and Foundations*, Vol. 20, No. 1, pp. 45–59. https://doi.org/10.3208/sandf1972.20.45

Ishihara, K., Yasuda, S., and Yokota, K. (1981). "Cyclic strength of undisturbed mine tailings," *Proceedings, International Conference on Recent Advances in Geotechnical Earthquake Engineering and Soil Dynamics*, University of Missouri-Rolla, St. Louis, MO, Vol. 1, pp. 53–58.

Ishihara, K. and Yoshimine, M. (1992). "Evaluation of settlements in sand deposits following liquefaction during earthquakes," *Soils and Foundations*, Vol. 32, No. 1, pp. 173–188. https://doi.org/10.3208/sandf1972.32.173

Iwasaki, T., Arakawa, T., and Tokida, K. (1982). "Simplified procedures for assessing soil liquefaction during earthquakes," *Proceedings, Conference on Soil Dynamics and Earthquake Engineering*, Southampton, pp. 925–939.

Iwasaki, T., Tatsuoka, F., Tokida, K.-I., and Yasuda, S. (1978). "A practical method for assessing soil liquefaction potential based on case studies at various sites in Japan," *Proceedings, 2nd International Conference on Microzonation*, San Francisco, pp. 885–896.

Jakobsen, B. (1952). "The landslide at Surte on the Gota River, September 29, 1952," *Proceedings No. 5*, Royal Swedish Geotechnical Institute, Stockholm.

Jefferies, M., Morgenstern, N.R., Van Zyl, D.V., and Wates, J., (2019). "Report on NTSF Embankment Failure. Cadia Valley Operations, for Ashurst Operations, South Orange, NSW, Australia," 119 pp.

Jennings, J.E., (1979). "The failure of a Slimes Dam at Bafokeng," *Civil Engineering in South Africa*, Vol. 6, pp. 135–140.

Juang, C.H., Ching, J., Wang, L., Khoshnevisan, S., and Ku, C.-S. (2013). "Simplified procedure for estimation of liquefaction-induced settlement and site-specific probabilistic settlement exceedance curve using cone penetration test (CPT)," *Canadian Geotechnical Journal*, Vol. 50, pp. 1055–1066. https://doi.org/10.1139/cgj-2012-0410

Juang, C.H., Yang, S.H., and Yuan, H. (2005). "Model uncertainty of shear wave velocity-based method for liquefaction potential evaluation," *Journal of Geotechnical and Geoenvironmental Engineering*, Vol. 131, pp. 1274–1282. https://doi.org/10.1061/(asce)1090-0241(2005)131:10(1274)

Kaggwa, W.S., Poulos, H.G., and Carter, J.P. (1988). "Response of carbonate sediments under cyclic triaxial test condition," *Proceedings, International Conference on Calcareous Sediments*, Balkema, Rotterdam, pp. 97–197.

Kamai, R. and Boulanger, R.W. (2010). "Characterizing localization processes during liquefaction using inverse analyses of instrumentation arrays," in Y.H. Hatzor, J. Sulem, and I. Vardoulakis, eds., *Meso-Scale Shear Physics in Earthquake and Landslide Mechanics*, CRC Press/Balkema, the Netherlands, pp. 219–238. https://doi.org/10.1201/b10826-23

Kamata, T., Tsukamoto, Y., Tatsuoka, F., and Ishihara, K. (2007). "Possibility of undrained flow in unsaturated sandy soils in triaxial tests," *Proceedings, 4th International Conference on Earthquake Geotechnical Engineering*, Thessaloniki, Greece, Paper No.1289.

Kammerer, A., Wu, J., Riemer, M., Pestana, J., and Seed, R. (2004). "Shear strain development in liquefiable soil under bi-directional loading conditions," *Proceedings, 13th World Conference on Earthquake Engineering*, Vancouver, BC, Canada, Paper No. 2081, 15 pp.

Kawamura, S., Nishizawa, T., and Wada, T. (1985). "Damage to piles due to liquefaction found by excavation twenty years after earthquake," in *Nikkei Architecture*, Vol. 27, pp. 130–134 (in Japanese).

Kayen, R., Moss, R.E.S., Thompson, E.M., Seed, R.B., Cetin, K.O., DerKiureghian, A., Tanaka, Y., and Tokimatsu, K. (2013). "Shear-wave velocity-based probabilistic and deterministic assessment of seismic soil liquefaction potential," *Journal of Geotechnical and Geoenvironmental Engineering*, Vol. 139, No. 3, pp. 407–419. https://doi.org/10.1061/(asce)gt.1943-5606.0000743

Kayen, R., Moss, R.E.S., Thompson, E.M., Seed, R.B., Cetin, K.O., DerKiureghian, A., Tanaka, Y., and Tokimatsu, K. (2015). "Erratum for 'Shear-wave velocity-based probabilistic and deterministic assessment of seismic soil liquefaction potential'," *Journal of Geotechnical and Geoenvironmental Engineering*, Vol. 141, No. 9, pp. 407–419. https://doi.org/10.1061/(asce)gt.1943-5606.0001390

Kim, J., Athanasopoulos-Zekkos, A., Cubrinovski, M. (2023). "Monotonic and cyclic simple shear response of well-graded sandy gravel soils from Wellington, New Zealand," *Journal of Geotechnical and Geoenvironmental Engineering*, Vol. 149, 04023046. https://doi.org/10.1061/JGGEFK.GTENG-10619

Kleiner, D.E. (1976). "Design and construction of an embankment dam to impound gypsum wastes," *Proceedings, 12th International Congress on Large Dams, International Committee on Large Dams*, Mexico City, pp. 235–249.

Knudsen, K. and Bott, J. (2011). "Geologic and geomorphic evaluation of liquefaction case histories for rapid hazard mapping," *Seismological Research Letters*, Vol. 82, p. 334.

Kokusho, T. (1999). "Formation of water film in liquefied sand and its effect on lateral spread," *Journal of Geotechnical and Geoenvironmental Engineering*, Vol. 125, No.10, pp. 817–826. https://doi.org/10.1061/(asce)1090-0241(1999)125:10(817)

Kokusho, T. (2000a). "Correlation of pore-pressure *B*-value with *P*-wave velocity and Poisson's ratio for imperfectly saturated sand or gravel," *Soils and Foundations*, Vol. 40, No. 4, pp. 95–102. https://doi.org/10.3208/sandf.40.4_95

Kokusho, T. (2000b). "Mechanism for water film generation and lateral flow in liquefied sand layer," *Soils and Foundations*, Vol. 40, No. 5, pp. 99–111. https://doi.org/10.3208/sandf.40.5_99

Kokusho, T. (2003). "Current state of research on flow failure considering void redistribution in liquefied deposits," *Soil Dynamics and Earthquake Engineering*, Vol. 23, No. 7, pp. 585–603. https://doi.org/10.1016/s0267-7261(03)00067-8

Kokusho, T. (2013). "Liquefaction potential evaluations: Energy-based method versus stress-based method," *Canadian Geotechnical Journal*, Vol. 50, pp. 1088–1099. https://doi.org/10.1139/cgj-2012-0456

Koppejan, A.W., Wamelan, B.M., and Weinberg, L.J. (1948). "Coastal flowslides in the Dutch Province of Zeeland," *Proceedings, 2nd International Conference on Soil Mechanics and Foundation Engineering*, Vol. 5, pp. 89–96.

Kramer, S.L. (1988). "Triggering of liquefaction flow slides in coastal soil deposits," *Engineering Geology*, Vol. 26, No. 1, pp. 17–31. https://doi.org/10.1016/0013-7952(88)90004-x

Kramer, S.L. (1989) "Uncertainty in steady state liquefaction evaluation procedures," *Journal of Geotechnical Engineering*, Vol. 115, No. 10, pp. 1402–1419. https://doi.org/10.1061/(asce)0733-9410(1989)115:10(1402)

Kramer, S.L., Astaneh, B., Ozener, P., and Sideras, S. (2013). "Effects of liquefaction on ground surface motions," in A. Ansal and M. Sakr, eds., *Perspectives on Earthquake Geotechnical Engineering, Geotechnical, Geological and Earthquake Engineering*, Springer International Publishing, Switzerland, Vol. 37, pp. 285–309. https://doi.org/10.1007/978-3-319-10786-8_11

Kramer, S.L. and Elgamal, A.-W. (2001). "Modeling soil liquefaction hazards for performance-based earthquake engineering," *Report PEER 2001/13*, Pacific Earthquake Engineering Research Center, University of California, 165 pp.

Kramer, S. L., Franke, K. W., Huang, Y.-M., and Baska, D. (2007). "Performance-based evaluation of lateral spreading displacement," *Proceedings, 4th International Conference on Earthquake Geotechnical Engineering, International Society for Soil Mechanics and Geotechnical Engineering*, London, Paper No. 1208.

Kramer, S.L., Hartvigsen, A.J., Sideras, S.S. and Ozener, P.T. (2011). " Site response modeling in liquefiable soils," *Proceedings, Effects of Surface Geology on Seismic Motion, Fourth IASPEI/IAEE International Symposium, Santa Barbara*, CA, 12 pp.

Kramer, S.L. and Mayfield, R.T. (2007). "The return period of liquefaction," *Journal of Geotechnical and Geoenvironmental Engineering*, Vol. 135, No. 7, pp. 802–813. https://doi.org/10.1061/(asce)1090-0241(2007)133:7(802)

Kramer, S.L. and Mitchell, R.A. (2006). "Ground motion intensity measures for liquefaction hazard evaluation," *Earthquake Spectra*, Vol. 22, No. 2, pp. 413–438. https://doi.org/10.1193/1.2194970

Kramer, S.L. and Seed, H.B. (1988). "Initiation of soil liquefaction under static loading condition," *Journal of Geotechnical Engineering*, Vol. 114, No. 4, pp. 412–430. https://doi.org/10.1061/(asce)0733-9410(1988)114:4(412)

Kramer, S.L., Sideras, S.S., and Greenfield, M.W. (2016). "The timing of liquefaction and its utility in liquefaction hazard evaluation," *Soil Dynamics and Earthquake Engineering*, Vol., 91, pp. 133–146. https://doi.org/10.1016/j.soildyn.2016.07.025

Kramer, S.L. and Wang, C.-H. (2015). "Empirical model for estimation of the residual strength of liquefied soil," *Journal of Geotechnical and Geoenvironmental Engineering*, Vol. 141, No. 9, p. 04015038. https://doi.org/10.1061/(asce)gt.1943-5606.0001317

Ku, C.-S., Juang, C.H., Chang, C.-W., and Ching, J. (2012). "Probabilistic version of the Robertson and Wride method for liquefaction evaluation: Development and application," *Canadian Geotechnical Journal*, Vol. 49, No. 1, pp. 27–44. https://doi.org/10.1139/t11-085

Kulasingam, R., Malvick, E.J., Boulanger, R.W., and Kutter, B.L. (2004). "Strength loss and localization at silt interlayers in slopes of liquefied sand," *Journal of Geotechnical and Geoenvironmental Engineering*, Vol. 130, No.11, pp.1192–1202. https://doi.org/10.1061/(asce)1090-0241(2004)130:11(1192)

Kwak, D.Y., Stewart, J.P., Brandenberg, S.J., and Mikami, A. (2016). "Seismic levee system fragility considering spatial correlation of demands and component fragilities," *Earthquake Spectra*, Vol. 32, No. 4, pp. 2207–2228. https://doi.org/10.1193/083115eqs132m

Ladd, C.C. (1991). "Stability evaluation during staged construction," *Journal of Geotechnical and Geoenvironmental Engineering*, Vol. 117, No. 4, pp. 540–615. https://doi.org/10.1061/(asce)0733-9410(1991)117:4(540)

Ladd, R.S. (1974). "Specimen preparation and liquefaction of sands," *Journal of the Geotechnical Engineering Division*, Vol. 100, No. GT10, pp. 1180–1184. https://doi.org/10.1061/ajgeb6.0000117

Lade, P.V. (1992). "Static instability and liquefaction of loose fine sandy slopes," *Journal of Geotechnical Engineering*, Vol. 118, No. 1, pp. 51–71. https://doi.org/10.1061/(asce)0733-9410(1992)118:1(51)

Lade, P.V. (1999). "Instability of granular materials," in P.V. Lade and J.A. Yamamuro, eds., *Physics and Mechanics of Soil Liquefaction*, Rotterdam, Balkema, pp. 3–16. https://doi.org/10.1201/9780203743317-1

Lasley, S.J., Green, R.A., and Rodriguez-Marek, A. (2016). "New stress reduction coefficient relationship for liquefaction triggering analyses," *Journal of Geotechnical and Geoenvironmental Engineering*, Vol. 142, No. 11, p. 06016013. https://doi.org/10.1061/(asce)gt.1943-5606.0001530

Ledezma, C., Hutchinson, T., Ashford, S.A., Moss, R., Arduino, P., Bray, J.D., Kayen, R., Olson, S.M., Hashash, Y.M.A., Frost, J.D., Verdugo, R., Sitar, N., and Rollins, K. (2012) "Effects of ground failure on bridges, roads, and railroads," *Earthquake Spectra*, Vol. 28, No. S1, pp. S119–S144. https://doi.org/10.1193/1.4000024

Lee, K.L. and Focht, J.A. (1975). "Liquefaction potential at Ekofisk Tank in North Sea," *Journal of the Geotechnical Engineering Division*, Vol. 101, No. GT1, pp. 1–18. https://doi.org/10.1061/ajgeb6.0000138

Lefebvre, G. and Pfendler, P. (1996). "Strain rate and preshear effects in cyclic resistance of soft clay," *Journal of Geotechnical and Geoenvironmental Engineering*, Vol. 122, No. 1, pp. 21–26. https://doi.org/10.1061/(asce)0733-9410(1996)122:1(21)

Lenz, A.J. and Baise, L.G. (2007). "Spatial variability of liquefaction potential in regional mapping using CPT and SPT data," *Soil Dynamics and Earthquake Engineering*, Vol. 27, No. 7, pp. 690–702. https://doi.org/10.1016/j.soildyn.2006.11.005

Lewis, M.R., Arango, I., Kimball, J.K., and Ross, T.E. (1999). "Liquefaction resistance of old sand deposits," *Proceedings, 11th Panamerican Conference on Soil Mechanics and Geotechnical Engineering*, Sao Paulo, Brazil, pp.821–829.

Liu, A., Stewart, J., Abrahamson, N., and Moriwaki, Y. (2001). "Equivalent number of uniform stress cycles for soil liquefaction analysis," *Journal of Geotechnical and Geoenvironmental Engineering*, Vol. 127, No.12, pp. 1017–1026. https://doi.org/10.1061/(asce)1090-0241(2001)127:12(1017)

Liu, C., Macedo, J., and Candia, G. (2021). "Performance-based probabilistic assessment of liquefaction-induced building settlements," *Soil Dynamics and Earthquake Engineering*, Vol. 151, p. 106955. https://doi.org/10.1016/j.soildyn.2021.106955

Liu, H. and Qiao, T. (1984). "Liquefaction potential of saturated sand deposits underlying foundation of structures," *Proceedings, 8th World Conference Earthquake Engineering*, San Francisco, Vol. 3, pp. 199–206.

Liu, L. and Dobry, R. (1997). "Seismic response of shallow foundation on liquefiable sand," *Journal of Geotechnical and Geoenvironmental Engineering*, Vol. 123, No. 6, pp.557–567. https://doi.org/10.1061/(asce)1090-0241(1997)123:6(557)

Lopez, J.S., Vera-Grunauer, X., Rollins, K., and Salvatierra, G. (2018). "Gravelly soil liquefaction after the 2016 Ecuador earthquake," *Proceedings, Geotechnical Earthquake Engineering and Soil Dynamics V, Geotechnical Special Publication 290*, ASCE, pp. 273–285. https://doi.org/10.1061/9780784481455.027

Magistrale, H., Rong, Y., Silva, W., and Thompson, E. (2012). "A site response map of the continental U.S.," *Proceedings, Fifteenth World Conference on Earthquake Engineering, International Association for Earthquake Engineering*, Lisbon, Portugal.

Makdisi, A.J., and Kramer, S.L. (2024). "Framework for mapping liquefaction hazard-targeted design ground motions," *Journal of Geotechnical and Geoenvironmental Engineering*. https:/doi.org/10.1061/JGGEFK/GTENG-12804

Malvick, E.J., Kutter, B.L., and Boulanger, R.W. (2008). "Postshaking shear strain localization in a centrifuge model of a saturated sand slope," *Journal of Geotechnical and Geoenvironmental Engineering*, Vol. 134, No. 2, pp. 164–174. https://doi.org/10.1061/(asce)1090-0241(2008)134:2(164)

Marcuson III, W.F. and Bieganousky, W.A. (1977a). "Laboratory standard penetration tests on fine sands," *Journal of the Geotechnical Engineering Division*, Vol 103, No. GT 6, pp. 565–588. https://doi.org/10.1061/ajgeb6.0000437

Marcuson III, W.F. and Bieganousky, W.A. (1977b). "SPT and relative density in coarse sands," *Journal of the Geotechnical Engineering Division*, Vol. 103, No. GT11, pp. 1295–1309. https://doi.org/10.1061/ajgeb6.0000521

Marcuson, W.F. and Hynes, M.E. (1990). "Stability of slopes and embankments during earthquakes," *Proceedings, ASCE/Pennsylvania Department of Transportation Geotechnical Seminar*, Hershey, PA.

Martin, G.R., Finn, W.D.L., and Seed, H.B. (1975). "Fundamentals of liquefaction under cyclic loading," *Journal of the Geotechnical Engineering Division*, Vol. 101, No. GT5, pp. 423–438.

Mason, H.B., Montgomery, J., Gallant, A.P., Hutabarat, D., Reed, A.N., Wartman, J., Irsyam, M., Simatupang, P.T., Alatas, I.M., Prakoso, W.A., Djarwadi, D., Hanifa, R., Rahardjo, P., Faizal, L., Harnanto, D.S., Kawanda, A., Himawan, A., and Yasin, W. (2021). "East Palu Valley flowslides induced by the 2018 MW7.5 Palu-Donggala earthquake," *Geomorphology*, No. 373, p. 107482. https://doi.org/10.1016/j.geomorph.2020.107482

Matasovic, N. and Vucetic, M. (1993). "Cyclic characterization of liquefiable sands," *Journal of Geotechnical and Geoenvironmental Engineering*, Vol. 119, No. 11, pp. 1805–1814. https://doi.org/10.1061/(asce)0733-9410(1993)119:11(1805)

Matasovic, N., Witthoeft, A., Borghei, A., and Elgamal, A.-W. (2024). "Seismic site response analysis with pore water pressure generation: Resources for evaluation," NCHRP Report 1092, National Cooperative Highway Research Program, Transportation Research Board, Washington, D.C., 105 pp. https://doi.org/10.17226/27536.

Matsuoka, M., Wakamatsu, K., Hashimoto, M., Senna, S., and Midorikawa, S. (2015). "Evaluation of liquefaction potential for large areas based on geomorphologic classification," *Earthquake Spectra*, Vol. 31, pp. 2375–2395. https://doi.org/10.1193/072313eqs211m

Maurer, B.W., Green, R.A., Cubrinovski, M., and Bradley, B.A. (2014b). "Evaluation of the liquefaction potential index for assessing liquefaction hazard in Christchurch, New Zealand," *Journal of Geotechnical and Geoenvironmental Engineering*, Vol. 140, No. 7, p. 04014032. https://doi.org/10.1061/(asce)gt.1943-5606.0001117

Maurer, B.W., Green, R.A., Cubrinovski, M., and Bradley, B.A. (2014a). "Assessment of aging correction factors for liquefaction resistance at sites of recurrent liquefaction: A study of the Canterbury (NZ) earthquake sequence," *Proceedings, 10th National Conference on Earthquake Engineering (10NCEE)*, Anchorage, AK, 21–25 July.

Maurer, B.W., Green, R.A., and Taylor, O.-D.S. (2015). "Moving towards an improved index for assessing liquefaction hazard: Lessons from historical data," *Soils and Foundations*, Vol. 55, No. 4, pp. 778–787. https://doi.org/10.1016/j.sandf.2015.06.010

Maurer, B.W., Green, R.A., van Ballegooy, S., and Wotherspoon, L. (2017). "Assessing liquefaction susceptibility using the CPT soil behavior type index," *Proceedings, Third International Conference on Performance-Based Design*, Vancouver, BC, 8 pp.

Mayfield, R.T., Kramer, S.L., and Huang, Y.-M. (2010). "Simplified approximation procedure for performance-based evaluation of liquefaction potential," *Journal of Geotechnical and Geoenvironmental Engineering*, Vol. 136, No. 1, pp. 140–150. https://doi.org/10.1061/(asce)gt.1943-5606.0000191

Meyerhof, G.G. and Hanna, A.M. (1978). "Ultimate bearing capacity of foundations on layered soils under inclined load," *Canadian Geotechnical Journal*, Vol. 15, No. 4, pp. 565–572. https://doi.org/10.1139/t78-060

Middlebrooks, T.A. (1942). "Fort peck slide," *Transactions*, Vol. 107, pp. 723–764. https://doi.org/10.1061/taceat.0005519

Miner, M.A. (1945). "Cumulative damage in fatigue," *Transactions*, Vol. 67, pp. A159–A164. https://doi.org/10.1115/1.4009458

Mitchell, D.E. (1984). "Liquefaction slides in hydraulically placed sands," *Proceedings, 4th International Symposium on Landslides*, Toronto.

Mitchell, J.K. (1986). "Practical problems from surprising soil behavior," *Journal of Geotechnical Engineering*, Vol. 112, No. 3, pp. 259–289. https://doi.org/10.1061/(asce)0733-9410(1986)112:3(255)

Mitchell, J.K. (2008). "Aging of sand -- A continuing enigma?" *Proceedings, 6th International Conference on Case Histories in Geotechnical Engineering*, Arlington, VA.

Mohrig, D., Elverøi, A., and Parker, G. (1999). "Experiments on the relative mobility of muddy subaqueous and subaerial debris flows, and their capacity to remobilize antecedent deposits," *Marine Geology*, Vol. 154, pp. 117–129. https://doi.org/10.1016/s0025-3227(98)00107-8

Montgomery, J. and Boulanger, R.W. (2014). "Influence of stratigraphic interfaces on residual strength of liquefied soil," *Proceedings, Dams and Extreme Events - Reducing Risk of Aging Infrastructure under Extreme Loading Conditions, 34th Annual United States Society on Dams Conference,* San Francisco, CA, pp. 101–111.

Montgomery, J. and Boulanger, R.W. (2016). "Effects of spatial variability on liquefaction-induced settlement and lateral spreading," *Journal of Geotechnical and Geoenvironmental Engineering*, Vol. 143, No. 1, p. 04016086. https://doi.org/10.1061/(asce)gt.1943-5606.0001584

Morgenstern, N.R. (2018). "Geotechnical risk, regulation, and public policy," *Soils and Rock*, Vol. 41, No. 2, pp. 107–129. https://doi.org/10.28927/sr.412107

Morgenstern, N.R., Vick, S.G., and Van Zyl, D. (2015). "Report on Mount Polley tailings storage facility breach," *Independent Expert Engineering Investigation and Review Panel*, Province of British Columbia, 156 pp.

Morgenstern, N.R., Vick, S.G., Viotti, C.B., and Watts, B.D. (2016). "Report on the immediate causes of the failure of the Fundao Dam," *Fundao Tailings Dam Review Panel*, 76 pp.

Morioka, B.T. and Nicholson, P.G. (2000). "Evaluation of the liquefaction potential of calcareous sand," Proceedings, International Offshore and Polar Engineering Conference, Seattle, WA, USA, pp. 494–500.

Moriwaki, Y., Akky, M.R., Ebeling, A.M., Idriss, I.M., and Ladd, R.S. (1982). "Cyclic strength and properties of tailings slimes," in *Dynamic Stability of Tailings Dams*, Preprint 82-539, American Society of Civil Engineers (ASCE), New York.

Moss, R.E., Seed, R.B., Kayen, R.E., Stewart, J.P., Der Kiureghian, A., and Cetin, K.O. (2006). "CPT-based probabilistic and deterministic assessment of in situ seismic soil liquefaction potential," *Journal of Geotechnical and Geoenvironmental Engineering*, Vol. 132, No. 8, pp. 1032–1051. https://doi.org/10.1061/(asce)1090-0241(2006)132:8(1032)

Muhunthan, B., Vijayatasan, N.V., and Abbasi, B. (2017). "Liquefaction-induced downdrag on drilled shafts," *Report WA-RD 865.1*, Washington State Department of Transportation, Olympia, WA, 163 pp.

Mulilis, J.P., Chan, C.K., and Seed, H.B. (1975). "The effects of method of sample preparation on the cyclic stress-strain behavior of sands," *Report No. EERC 75-18*, Earthquake Engineering Research Center, University of California, Berkeley.

Naesgaard, E. and Byrne, P.M. (2005). "Flow liquefaction due to mixing of layered deposits," *Proceedings, ICSMGE TC4 Satellite Conference on Recent Developments in Earthquake Engineering*, Osaka, Japan, pp. 103–108.

National Academies of Sciences, Engineering, and Medicine. 2021. State of the Art and Practice in the Assessment of Earthquake-Induced Soil Liquefaction and Its Consequences. Washington, DC: The National Academies Press. https://doi.org/10.17226/23474.

National Research Council. (1985). *Liquefaction of Soils during Earthquakes*, National Academy Press, Washington, DC, 240 pp. https://doi.org/10.17226/19275

National Research Council. (2016). *State of the Art and Practice in the Assessment of Earthquake-Induced Soil Liquefaction and Its Consequences*, National Research Council Report, 286 pp. https://doi.org/10.17226/23474

Nemat-Nasser, S. and Shokooh, A. (1979). "A unified approach to densification and liquefaction of cohesionless sand in cyclic shearing," *Canadian Geotechnical Journal*, Vol. 16, No. 4, pp. 649–678. https://doi.org/10.1139/t79-076

Nichols, G. (2009). *Sedimentology and Stratigraphy*, 2nd edition, Wiley-Blackwell, West Sussex, UK, 419 pp.

Nikolaou, S., Zekkos, D., Asimaki, D., and Gilsanz, R., eds. (2014). "GEER/EERI/ATC Reconnaissance: January 26th and February 2nd, 2014 Cephalonia, Greece Earthquakes," *Report GEER/EERI/ATC-034*, Version 1.

Ntritsos, N. and Cubrinovski, M. (2020). "A CPT-based effective stress analysis procedure for liquefaction assessment," *Soil Dynamics and Earthquake Engineering*, Vol. 131, p. 106063. https://doi.org/10.1016/j.soildyn.2020.106063

Obermeier, S.F. (1996). "Use of liquefaction-induced features for paleoseismic analysis - An overview of how seismic liquefaction features can be distinguished from other features and how their regional distribution and properties of source sediment can be used to infer the location and strength of Holocene paleo-earthquakes," *Engineering Geology*, Vol. 44, pp. 1–76. https://doi.org/10.1016/s0013-7952(96)00040-3

Obermeier, S.F., Jacobson, R.B., Smoot, J.P., Weems, R.E., Gohn, G.S., Monroe, J.E., and Powars, D.S. (1990). "Earthquake-induced liquefaction features in the coastal setting of South Carolina and in the fluvial setting of the New Madrid seismic zone," *U.S. Geological Survey Professional Paper 1504*, U.S. Government Printing Office, Washington, DC, 49 pp. https://doi.org/10.3133/pp1504

Ohsaki, Y. (1969). "The effects of local soil conditions upon earthquake damage," *Proceedings, Specialty Session 2, Seventh International Conference on Soil Mechanics and Foundation Engineering*, Mexico City, Mexico.

Olaya, F. and Bray, J.D. (2022). "Strain potential of liquefied soil," *Journal of Geotechnical and Geoenvironmental Engineering*, Vol. 148, No. 11, 16 pp. DOI: 10.1061/(ASCE)GT.1943-5606.0002896.

Olson, S.M., Mei, X., and Hashash, Y.M.A. (2020). "Nonlinear site response analysis with pore-water pressure generation for liquefaction triggering evaluation," *Journal of Geotechnical and Geoenvironmental Engineering*, Vol. 146, No. 2, p. 04019128. https://doi.org/10.1061/(asce)gt.1943-5606.0002191

Olson, S.M. and Stark, T.D. (2002). "Liquefied strength ratio from liquefaction flow failure case histories," *Canadian Geotechnical Journal*, Vol. 39, pp. 629–647. https://doi.org/10.1139/t02-001

Orense, R.P., Towhata, I., and Ishihara, K. (1991). "Soil liquefaction in Dagupan City during the 1990 Luzon, Philippines earthquake," *Proceedings, 26th Japan National Conference on Soil Mechanics and Foundation Engineering*, JSSMFE, pp. 871–874.

Palmgren, A. (1924). "Dielebensdauervonkugellageru," *ZVDI*, Vol. 68, No. 14, pp. 339–341.

Pando, M.A., Sandoval, E.A., and Catano, J. (2012). "Liquefaction susceptibility and dynamic properties of calcareous sands from Cabo Rojo, Puerto Rico," *Proceedings, 15th World Conference on Earthquake Engineering*, Lisbon, Portugal, 10 pp.

Park, D. and Ahn, J.-K. (2013) "Accumulated stress based model for prediction of residual pore pressure," *Proceedings, 18th International Conference on Soil Mechanics and Geotechnical Engineering*, Paris, pp. 1567–1570.

Paull, N.A., Boulanger, R.W., DeJong, J.T., and Friesen, S.J. (2021). "Using conditional random fields for a spatially variable liquefiable foundation layer in nonlinear dynamic analyses of embankments," *Journal of Geotechnical and Geoenvironmental Engineering*, Vol. 147, No. 11, p. 04021134. https://doi.org/10.1061/(asce)gt.1943-5606.0002610

Paull, N.A., Boulanger, R.W., DeJong, J.T., and Friesen, S.J. (2022). "Nonlinear dynamic analyses of Perris Dam using transition probability to model interbedded alluvial strata," *Journal of Geotechnical and Geoenvironmental Engineering*, Vol. 148, No. 1, p. 05021015. https://doi.org/10.1061/(asce)gt.1943-5606.0002663

Polito, C.P., Green, R.A., and Lee, J. (2008). "Pore pressure generation models for sands and silty soils subjected to cyclic loading," *Journal of Geotechnical and Geoenvironmental Engineering*, Vol. 134, No. 10, pp. 1490–1500. https://doi.org/10.1061/(asce)1090-0241(2008)134:10(1490)

Polito, C.P. and Martin, J.R. (2001). "The effects of non-plastic fines on the liquefaction resistance of sands," *Journal of Geotechnical and Geoenvironmental Engineering*, Vol. 127, No. 5, pp. 408–415. https://doi.org/10.1061/(asce)1090-0241(2001)127:5(408)

Popescu, R., Prevost, J.H., and Deodatis, G. (1997). "Effects of spatial variability on soil liquefaction: Some design recommendations," *Geotechnique*, Vol. 47, No. 5, pp. 1019–1036. https://doi.org/10.1680/geot.1997.47.5.1019

Popescu, R., Prevost, J.H., and Deodatis, G. (2005). "3D effects in seismic liquefaction of stochastically variable soil deposits," *Geotechnique*, Vol. 55, No. 1, pp. 21–31. https://doi.org/10.1680/ravige.34860.0008

Porcino, D. and Marciano, V. (2010). "Evaluating liquefaction resistance of a calcareous sand using the cone penetration test," *Proceedings, Fifth International Conference on Recent Advances in Geotechnical Earthquake Engineering and Soil Dynamics*, San Diego, CA, 10 pp.

Poulos, S.J., Robinsky, E.I., and Keller, T.O. (1985). "Liquefaction resistance of thickened tailings," *Journal of Geotechnical Engineering*, Vol. 111, No. 12, pp. 1380–1394. https://doi.org/10.1061/(asce)0733-9410(1985)111:12(1380)

Pretell, R., Brandenberg, S.J. Stewart, J.P. (2024). "Consistent framework for PGA estimation at liquefaction case history sites: Application to the 1989 M6.9 Loma Prieta earthquake," *Geo-Congress 2024: Geotechnics of Natural Hazards*, Vancouver, BC, February 2024, Geotechnical Special Publication No. 349, TM Evans, N Stark, and S Chang (Eds.), 161–170, ASCE Geo-Institute

Pyke, R., Seed, H.B., and Chan, C.K. (1975). "Settlement of sands under multi-directional loading," *Journal of the Geotechnical Engineering Division*, Vol. 101, No. GT4, pp. 379–398. https://doi.org/10.1061/ajgeb6.0000162

Qiu, Y. and Sego, D.C. (2001). "Laboratory properties of mine tailings," *Canadian Geotechnical Journal*, pp. 183–190. https://doi.org/10.1139/t00-082

Quigley, M.C., Bastin, S., and Bradley, B.A. (2013). "Recurrent liquefaction in Christchurch, New Zealand, during the Canterbury earthquake sequence," *Geology*, Vol. 41, pp. 419–422. https://doi.org/10.1130/g33944.1

Rana, N.M., Ghahramani, N., Evans, S.G., McDougall, S., Small, A., and Take, W.A. (2021). "Catastrophic mass flows resulting from tailings impoundment failures," *Engineering Geology*, Vol. 292, p. 106262. https://doi.org/10.1016/j.enggeo.2021.106262

Rashidian, V. and Baise, L.G. (2020). "Regional efficacy of a global geospatial liquefaction model," *Engineering Geology*, Vol. 272, p. 105644. https://doi.org/10.1016/j.enggeo.2020.105644

Rateria, G. and Maurer, B.W. (2022). "Evaluation and updating of Ishihara's (1985) model for liquefaction surface expression, with insights from machine and deep learning," *Soils and Foundations*, Vol. 62, pp. 1–17. https://doi.org/10.1016/j.sandf.2022.101131

Rauch, A.F. (1997). "EPOLLS: An empirical method for predicting surface displacements due to liquefaction-induced lateral spreading in earthquakes," *Ph.D. Thesis*, Virginia Tech, Blacksburg, VA, USA.

Robinson, K., Cubrinovski, M., and Bradley, B.A. (2013). "Comparison of actual and predicted measurements of liquefaction-induced lateral displacements from the 2010 Darfield and 2011Christchurch Earthquakes," *Proceedings, 2013 Conference of the New Zealand Society for Earthquake Engineering (NZSEE 2013)*, Wellington, New Zealand.

Robertson, P.K. (2009a). "Performance based earthquake design using the CPT," *Proceedings, IS-Tokyo2009: International Conference on Performance-Based Design in Earthquake Geotechnical Engineering - From Case History to Practice*, Tokyo, Japan, pp. 3–20.

Robertson, P.K. (2009b). "Interpretation of cone penetration tests - a unified approach," *Canadian Geotechnical Journal*, Vol. 46, No. 11, pp. 1337–1355. https://doi.org/10.1139/t09-065

Robertson, P.K. (2016). "Cone penetration test (CPT)-based soil behaviour type (SBT) classification system - an update," *Canadian Geotechnical Journal*, Vol. 53, No. 12, pp. 1910–1927. https://doi.org/10.1139/cgj-2016-0044

Robertson, P.K., de Melo, L., Williams, D.J., and Wilson, G.W. (2019). "Report of the expert panel on the technical causes of the failure of Feijao Dam I," *Expert Panel Report*, 80 pp.

Robertson, P.K. and Fear, C.E. (1995). "Liquefaction of sands and its evaluation," in K. Ishihara, ed., Proceedings, 1st International Conference on Earthquake Geotechnical Engineering, A. A. Balkema, Amsterdam.

Robertson, P.K. and Wride, C.E. (1998). "Evaluating cyclic liquefaction potential using the cone penetration test," *Canadian Geotechnical Journal*, Vol. 35, No. 3, pp. 442–459. https://doi.org/10.1139/t98-017

Robertson, P.K, Wride, C.E. (Fear), List, B.R., Atukorala, U., Biggar, K.W., Byrne, P.M., Campanella, R.G., Cathro, D.C., Chan, D.H., Czajewski, K., Finn, W.D.L., Gu, W.H., Hammamji, Y., Hofmann, B.A., Howie, J.A., Hughes, J., Imrie, A.S., Konrad, J.-M., Küpper, A., Law, T., Lord, E.R.F., Monahan, P.A., Morgenstern, N.R., Phillips, R., Piché, R., Plewes, H.D., Scott, D., Sego, D.C., Sobkowicz, J., Stewart, R.A., Watts, B.D., Woeller, D.J., Youd, T.L., and Zavodni, Z. (2000). "The CANLEX project: Summary and conclusions," *Canadian Geotechnical Journal*, Vol. 37, pp. 563–591. https://doi.org/10.1139/t00-046

Rollins, K.M., Roy, J., Athanasopoulos-Zekkos, A., Zekkos, D., Amoroso, S., and Cao, Z. (2021). "A new dynamic cone penetration test-based procedure for liquefaction triggering assessment of gravelly soils," *Journal of Geotechnical and Geoenvironmental Engineering*, Vol. 147, No. 12, p. 04021141. https://doi.org/10.1061/(asce)gt.1943-5606.0002686

Rollins, K.M. and Strand, S.R. (2006). "Downdrag forces due to liquefaction surrounding a pile," *Proceedings, 8th U.S. National Conference on Earthquake Engineering*, San Francisco, CA, April 18-22, Paper No. 1646.

Roten, D., Fah, D., and Bonilla, L.F. (2013). "High-frequency ground motion amplification during the 2011 Tohoku earthquake explained by soil dilatancy," *Geophysical Journal International*, Vol. 193, pp. 898–904. https://doi.org/10.1093/gji/ggt001

Roy, D. (2008). "Coupled use of cone tip resistance and small strain shear modulus to assess liquefaction potential," *Journal of Geotechnical and Geoenvironmental Engineering*, Vol. 134, No. 4, pp. 519–530. https://doi.org/10.1061/(asce)1090-0241(2008)134:4(519)

Russell, J., van Ballegooy, S., Ogden, M., Bastin, S., and Cubrinovski, M. (2017). "Influence of geometric, geologic, geomorphic and subsurface ground conditions on the accuracy of empirical models for prediction of lateral spreading," *Proceedings, 3rd International Conference on Performance-based Design in Earthquake Geotechnical Engineering*, International Society for Soil Mechanics and Geotechnical Engineering, Vancouver, BC, Canada.

Sandoval, E.A. and Pando, M.A. (2012). "Experimental assessment of the liquefaction resistance of calcareous biogenous sands," *Earth Sciences Research Journal*, Vol. 16, No. 1, pp. 55–63.

Saye, S.R., Olson, S.M., and Franke, K.E. (2021). "Common-origin approach to assess level-ground liquefaction susceptibility and triggering in CPT-compatible soils using Δ_Q," *Journal of Geotechnical and Geoenvironmental Engineering,* Vol. 147, No. 7, 14 pp.

Schmertmann, J.H. (1991). "The mechanical aging of soils," *Journal of Geotechnical Engineering*, Vol. 117, No. 9, pp. 1288–1330. https://doi.org/10.1061/(asce)0733-9410(1991)117:9(1288)

Schneider, J.A. and Moss, R.E.S. (2011). "Linking cyclic stress and cyclic strain based methods for assessment of cyclic liquefaction triggering in sands," *Géotechnique Letters*, Vol. 1 No. 1, pp. 31–36. https://doi.org/10.1680/geolett.11.00021

Seed, H.B. (1979). "Soil liquefaction and cyclic mobility evaluation for level ground during earthquakes," *Journal of the Geotechnical Engineering Division*, Vol. 105, No. GT2, pp. 201–255. https://doi.org/10.1061/ajgeb6.0000768

Seed, H.B. (1987). "Design problems in soil liquefaction," *Journal of the Geotechnical Engineering Division*, Vol. 113, No. 8, pp. 827–845. https://doi.org/10.1061/(asce)0733-9410(1987)113:8(827)

Seed, H.B. and Idriss, I.M. (1971). "Simplified procedure for evaluating soil liquefaction potential," *Journal of the Soil Mechanics and Foundations Division*, Vol. 107, No. SM9, pp. 1249–1274. https://doi.org/10.1061/jsfeaq.0001662

Seed, H.B., Idriss, I.M., and Arango, I. (1983). "Evaluation of liquefaction potential using field performance data," *Journal of the Geotechnical Engineering Division*, Vo. 109, No. 3, pp. 458–482.

Seed, H. B., Idriss, I. M., Makdisi, F., and Banerjee, N. (1975) "Representation of Irregular Stress Time Histories by Equivalent Uniform Stress Series in Liquefaction Analyses," *EERC 75-29*, Engineering and Vibration Research Centre University of California, Berkeley.

Seed, H.B. and Lee, K.L. (1966). "Liquefaction of sands during cyclic loading," *Journal of the Soil Mechanics and Foundations Division*, Vol. 92, No. SM6, pp. 105–134. https://doi.org/10.1061/jsfeaq.0000913

Seed, H.B., Martin, P.P., and Lysmer, J. (1975a). "Pore-water pressure changes during soil liquefaction," *Journal of the Geotechnical Engineering Division*, Vol. 102, No. 4, pp. 323–346. https://doi.org/10.1061/ajgeb6.0000258

Seed, H.B., Mori, K., and Chan, C.K. (1975b). "Influence of seismic history on the liquefaction characteristics of sands," *Report No. EERC 75-25*, Earthquake Engineering Research Center, University of California, Berkeley, 21 pp.

Seed, H.B. and Peacock, W.H. (1971). "Test procedures for measuring soil liquefaction characteristics," *Journal of the Soil Mechanics and Foundations Division*, Vol. 97, No. SM8, pp. 1099–1119. https://doi.org/10.1061/jsfeaq.0001649

Seed, R.B. and Harder, L.F. (1990). "SPT-based analysis of cyclic pore pressure generation and undrained residual strength," in J.M. Duncan, ed., *Proceedings, H. Bolton Seed Memorial Symposium*, University of California, Berkeley, Vol. 2, , pp. 351–376.

Sharp, M.K., Dobry, R., and Ledbetter, R. (2000). "Centrifuge research of liquefaction phenomena," *Proceedings, 12th World Conference on Earthquake Engineering*, Paper 2666, 7 pp.

Silver, M.L. and Seed, H.B. (1971). "Volume changes in sands during cyclic loading," *Journal of the Soil Mechanics and Foundations Division*, Vol. 97, No. SM9, pp. 1171–1189. https://doi.org/10.1061/jsfeaq.0001658

Sims, J.D. and Garvin, C.D. (1995). "Recurrent liquefaction induced by the 1989 Loma Prieta earthquake and 1990 and 1991 aftershocks: Implications for paleoseismicity studies," *Bulletin of the Seismological Society of America*, Vol. 85, No. 1, pp. 51–65.

So, M.M.L., Care, J.A., and Mote, T. (2015). "Near fault effects on near fault ground motion on soil amplification and liquefaction," *Proceedings, 10th Pacific Conference on Earthquake Engineering*, Sydney, Australia, 8 pp.

Sonmez, H. (2003). "Modification of the liquefaction potential index and liquefaction susceptibility mapping for a liquefaction-prone area (Inegol, Turkey)," *Environmental Geology*, Vol. 44, No. 7, pp. 862–871. https://doi.org/10.1007/s00254-003-0831-0

Stamatopoulos, C.A., Stamatopoulos, A.C., and Kotzias, P.C. (1999). "Decrease of liquefaction susceptibility by preloading measured in simple-shear tests," *Proceedings, Second International Conference on Earthquake Geotechnical Engineering*, Balkema, Lisbon, pp. 579–584.

Stark, T.D. and Mesri, G. (1992). "Undrained shear strength of sands for stability analysis," *Journal of Geotechnical Engineering*, Vol. 118, No. 11, pp. 1727–1747. https://doi.org/10.1061/(asce)0733-9410(1992)118:11(1727)

Stewart, J.P., Kramer, S.L., Kwak, D.Y., Greenfield, M.W. Kayen, R.E., Tokimatsu, K., Bray, J.D., Beyzaei, C.Z., Cubrinovski, M., Sekiguchi, T., Nakai, S., and Bozorgnia, Y. (2016). "PEER-NGL project: Open source global database and model development for the next-generation of liquefaction assessment procedures," *Soil Dynamics and Earthquake Engineering*, Vol. 91, pp. 317–328. https://doi.org/10.1016/j.soildyn.2016.07.009

Stewart, J.P., Kwok, A.O., Hashash, Y.M.A., Matasovic, N., Pyke, R., Wang, Z., and Yang, Z. (2008). "Benchmarking of nonlinear geotechnical ground response analysis procedures," *Report PEER 2008/04*, Pacific Earthquake Engineering Research Center, Berkeley, CA, 186 pp.

Stokoe II, K.H., Rix, G.J., Sanchez-Salinero, I., Andrus, R.D., and Nok, Y.J. (1988). "Liquefaction of gravelly soils during the 1983 Borah Peak, ID earthquake," *Proceedings, 9th World Conference in Earthquake Engineering*, Tokyo, Japan, pp. 183–188.

Stuedlein, A.W., Dadashiserej, A., Jana, A, and Evans, T.W. (2023). "Liquefaction susceptibility and cyclic response of intact nonplastic and plastic silts," *Journal of Geotechnical and Geoenvironmental Engineering*, Vol. 149, No. 1, p. 04022125. https://doi.org/10.1061/(asce)gt.1943-5606.0002935

Suzuki, T. and Toki, S. (1984). "Effects of preshearing on liquefaction characteristics of saturated sand subjected to cyclic loading," *Soils and Foundations*, Vol. 24, No. 2, pp. 16–28. https://doi.org/10.3208/sandf1972.24.2_16

Swaisgood, J.R. (2003). "Embankment dam deformations caused by earthquakes," *Proceedings, 2003 Pacific Conference on Earthquake Engineering*, Christchurch, New Zealand.

Swaisgood, J.R. (2014). "Behavior of embankment dams during earthquake," *Journal of Dam Safety*, American Society of Dam Safety Officials, Vol. 12, No. 2, pp. 35–44.

Takada, K. and Atwater, B.F. (2004). "Evidence for liquefaction identified in peeled slices of holocene deposits along the Lower Columbia River, Washington," *Bulletin of the Seismological Society of America*, Vol. 94, No. 2, pp. 550–575. https://doi.org/10.1785/0120020152

Tasiopoulou, P., Ziotopoulou, K., Humire, F., Giannakou, A., Chacko, J., and Travasarou, T. (2020). "Development and implementation of semiempirical Framework for modeling postliquefaction shear deformation accumulation in sands," *Journal of Geotechnical and Geoenvironmental Engineering*, Vol. 146, No. 1, 19 pp.

Tatsuoka, F., Ochi, K., Fujii, S., and Okamoto, M. (1986). "Cyclic undrained triaxial and torsional shear strength of sands for different sample preparation methods," *Soils and Foundations*, Vol. 26, No. 3, pp. 23–41. https://doi.org/10.3208/sandf1972.26.3_23

Terzaghi, K. (1950). "Mechanisms of landslides," *Engineering Geology (Berkey) Volume*, Geological Society of America. https://doi.org/10.1130/Berkey.1950.83

Terzaghi, K. (1956). "Varieties of submarine slope failures," *Proceedings, 8th Texas Conference on Soil Mechanics and Foundation Engineering*, University of Texas Bureau of Engineering Research, Special Publications, Vol. 29, pp. 1–41.

Tohno, I. and Shamoto, Y. (1986). "Liquefaction damage to the ground during the 1983 Nihonkai-Chubu earthquake in Aomori Prefecture," *Natural Disaster Science*, Vol. 8, No. 1, pp. 85–116. https://doi.org/10.1016/0148-9062(87)90536-5

Toki, S., Tatsuoka, F., Miura, S., Yoshimi, Y., Yasuda, S., and Makihara, Y. (1986). "Cyclic undrained triaxial strength of sand by a cooperative test program," *Soils and Foundations*, Vol. 26, No. 3, pp. 117–128. https://doi.org/10.3208/sandf1972.26.3_117

Tokimatsu, K and Yoshimi, Y. (1983). "Empirical correlation of soil liquefaction based on SPT N-value and fines content," *Soils and Foundations*, Vol. 23, No. 4, pp. 56–74. https://doi.org/10.3208/sandf1972.23.4_56

Tonkin & Taylor Ltd. (2013). "Liquefaction vulnerability study," *T&T Ref: 52020.0200/v.1.0*, Christchurch, New Zealand, 50 pp.

Tonkin & Taylor Ltd. (2015). "Canterbury earthquake sequence: Increased liquefaction vulnerability assessment methodology," *T&T Ref: 52010.140.v1.0*, 204 pp.

Travasarou, T., Bray, J.D., and Sancio, R.B. (2006). "Soil-structure interaction analyses of building responses during the 1999 Kocaeli earthquake," *Proceedings, 8th U.S. National Conference on Earthquake Engineering, the 1906 San Francisco Earthquake*, EERI, April, Paper 1877.

Tsukamoto, Y., Ishihara, K., Nakazawa, H., Kamada, K., and Huang, Y. (2002). "Resistance of partly saturated sand to liquefaction with reference to longitudinal and shear wave velocities," *Soils and Foundations*, Vol. 42, No. 6, pp. 93–104. https://doi.org/10.3208/sandf.42.6_93

Tsukamoto, Y., Kawabe, S., Matsumoto, J., and Hagiwara, S. (2014). "Cyclic resistance of two unsaturated silty sands against soil liquefaction," *Soils and Foundations*, Vol. 54, No. 6, pp. 1094–1103. https://doi.org/10.1016/j.sandf.2014.11.005

Ulmer, K. and Franke, K. (2015). "Modified performance-based liquefaction triggering procedure using liquefaction loading parameter maps," *Journal of Geotechnical and Geoenvironmental Engineering*, Vol. 142, No. 3, 11 pp. https://doi.org/10.1061/(asce)gt.1943-5606.0001421

Ulmer, K., Green, R.A., Rodriguez-Marek, A., and Mitchell, J.K. (2023). "Energy-based liquefaction triggering model," *Journal of Geotechnical and Geoenvironmental Engineering*, Vol. 149, No. 11, 14 pp. https://doi.org/10.1061/jggefk.gteng-11402

Ulmer, K., Upadhyaya, S., Green, R.A., Rodriguez-Marek, A., Stafford, P., Bommer, J., and van Elk, J.F. (2018). "A critique of b-values used for computing magnitude scaling actors," *Proceedings, Geotechnical Earthquake Engineering and Soil Dynamics V*, pp. 112–121. https://doi.org/10.1061/9780784481486.012

Upadhyaya, S., Green, R.A., Rodriguez-Marek, A., and Maurer, B.W. (2023). "True liquefaction triggering curve," *Journal of Geotechnical and Geoenvironmental Engineering*, Vol. 149, No. 3, p. 04023005. https://doi.org/10.1061/jggefk.gteng-11126

Vaid, Y.P. and Chern, J.C. (1983). "Effect of static shear on resistance of liquefaction," *Soils and Foundations*, Vol. 23, No. 1, pp. 47–60. https://doi.org/10.3208/sandf1972.23.47

Vaid, Y.P. and Chern, J.C. (1985). "Cyclic and monotonic undrained response of saturated sands," in V. Khosla, ed., *Advances in the Art of Testing Soils Under Cyclic* Conditions, ASCE, pp. 120–147.

Vaid, Y.P. and Finn, W.D.L. (1979) "Static shear and liquefaction potential," *Journal of Geotechnical Division*, Vol. 105, No. GT10, pp. 1233–1246. https://doi.org/10.1061/ajgeb6.0000868

Vaid, Y.P. and Sivathayalan, S. (1996). "Static and cyclic liquefaction potential of Fraser Delta sand in simple shear and triaxial tests," *Canadian Geotechnical Journal*, Vol. 33, pp. 281–289. https://doi.org/10.1139/t96-007

van Ballegooy, S., Malan, P., Lacrosse, V., Jacka, M.E., Cubrinovski, M., Bray, J.D., O'Rourke, T.D., Crawford, S.A., and Cowan, H. (2014). "Assessment of liquefaction-induced land damage for residential Christchurch," *Earthquake Spectra*, Vol. 30, No. 1, pp. 31–55. https://doi.org/10.1193/031813eqs070m

van Ballegooy, S., Malan, P.J., Jacka, M.E., Lacrosse, V.I.M.F., Leeves, J.R., and Lyth, J.E. (2012). "Methods for characterizing effects of liquefaction in terms of damage severity," *Proceedings, 15th World Conference on Earthquake Engineering*, Lisbon, Portugal, 10 pp.

van der Linden, T.I. (2016). "Influence of multiple thin soft layers on the cone resistance in intermediate soils," *M.Sc. Thesis*, Technical University, Delft.

Varnes, D.J. (1978). "Slope movement types and processes," *Landslides - Analysis: Special Report 176*, R.L. Schuster and R.J. Krizek, eds, Transportation Research Board, Washington DC.

Vick, S.G. (1983). *Planning, Design, and Analysis of Tailings Dams*, John Wiley & Sons, New York.

Vucetic, M. and Dobry, R. (1986). "Pore pressure build-up and liquefaction at level sandy sites during earthquakes," *Research Rep. CE-86-3*, Department of Civil Engineering, Rensselaer Polytechnic Institute, Troy, NY.

Vucetic, M. and Dobry, R. (1988). "Cyclic triaxial strain-controlled testing of liquefiable sands," in *Advanced Triaxial Testing of Soil and Rock*, ASTM STP 977, American Society of Testing and Materials, Philadelphia, pp. 475–485. https://doi.org/10.1520/stp29093s

Wakamatsu, K. (2012). "Recurrent liquefaction induced by the 2011 Great East Japan Earthquake compared with the 1987 earthquake," *Proceedings, International Symposium on Engineering Lessons Learned from the 2011 Great East Japan Earthquake*, pp. 675–686.

Wald, D.J. and Allen, T.I., (2007). "Topographic slope as a proxy for seismic site conditions and amplification," *Bulletin of the Seismological Society of America*, Vol. 97, pp. 1379–1395. https://doi.org/10.1785/0120060267

Whitman, R.V. (1971). "Resistance of soil to liquefaction and settlement," *Soils and Foundations*, Vol. 11, No. 4, pp. 59–68. https://doi.org/10.3208/sandf1960.11.4_59

Whitman, R.V. (1985). "On liquefaction," Proceedings, 11th International Conference on Soil Mechanics and Foundation Engineering, Balkema, San Francisco, pp. 1923–1926.

Wijewickreme, D., Khalili, A., and Wilson, G.W. (2010). "Mechanical response of highly gap-graded mixtures of waste rock and tailings. Part II: Undrained cyclic and post-cyclic shear response," *Canadian Geotechnical Journal*, pp. 566–582. https://doi.org/10.1139/t09-122

Wijewickreme, D., Sanin, M.V., and Greenaway, G.R. (2005). "Cyclic shear response of fine-grained mine tailings," *Canadian Geotechnical Journal*, Vol. 42, pp. 1408–1421. https://doi.org/10.1139/t05-058

Wong, W. (1984). "Earthquake damages to earth dams and levees in relation to soil liquefaction and weakness in soft clays," *Proceedings, International Conference on Case Histories in Geotechnical Engineering*, Vol. 1, pp. 511–521.

Wong, R.T., Seed, H.B., and Chan, C.K. (1975). "Liquefaction of gravelly soil under cyclic loading conditions," *Journal of the Geotechnical Engineering Division*, Vol. 101, No. 6, pp. 571–583. https://doi.org/10.1061/ajgeb6.0000174

Wotherspoon, L.M., Pender, M.J., and Orense, R.P. (2012). "Relationship between observed liquefaction at Kaiapoi following the 2010 Darfield earthquake and former channels of the Waimakariri River," *Engineering Geology*, Vol. 125, No. 1, pp. 45–55. https://doi.org/10.1016/j.enggeo.2011.11.001

Wu, J., Kammerer, A.M., Reimer, M.F., Seed, R.B., and Pestana, J.M. (2004). "Laboratory study of liquefaction triggering criteria," *Proceedings, 13th World Conference on Earthquake Engineering*, Vancouver, BC, Canada, Paper No. 2580, 14 pp.

Yang, J. (2002). "Liquefaction resistance of sand in relation to P-wave velocity," *Geotechnique*, Vol. 52, No. 4, pp. 295–298. https://doi.org/10.1680/geot.52.4.295.41021

Yang, J. (2004). "Evaluating liquefaction strength of partially saturated sand," *Journal of Geotechnical and Geoenvironmental Engineering*, Vol. 130, No. 9, pp. 975–979. https://doi.org/10.1061/(asce)1090-0241(2004)130:9(975)

Yasuda, S. and Tohno, I. (1988). "Sites of reliquefaction caused by the 1983 Nihonkai-Chubu earthquake," *Soils and Foundations*, Vol. 28, No. 2, pp. 61–72. https://doi.org/10.3208/sandf1972.28.2_61

Yegian, M.K., Gharaman, V.G., and Harutiunyan, R.N. (1994). "Liquefaction and embankment failure case histories, 1988 Armenian earthquake," *Journal of Geotechnical Engineering*, Vol. 120, No. 3, pp. 581–596. https://doi.org/10.1061/(asce)0733-9410(1994)120:3(581)

Yoshimi, Y., Hatanaka, H., and Oh-Oka, H. (1978). "Undisturbed sampling of saturated sands by freezing," *Soils and Foundations*, Vol. 18, No. 3, pp. 59–73. https://doi.org/10.3208/sandf1972.18.3_59

Yoshimi, Y. and Tokimatsu, K. (1977). "Settlement of buildings on saturated sand during earthquakes," *Soils and Foundations*, Vol. 17, No. 1, pp. 23–38. https://doi.org/10.3208/sandf1972.17.23

Yoshimi, Y., Tokimatsu, K., and Hasaka, Y. (1989). "Evaluation of liquefaction resistance of clean sands based on high-quality undisturbed samples," *Soils and Foundations*, Vol. 29, No. 1, pp. 93–104. https://doi.org/10.3208/sandf1972.29.93

Yoshimine, M., Nishizaki, H., Amano, K., and Hosono, Y. (2006). "Flow deformation of liquefied sands under constant shear load and its application to analysis of flow slide in infinite slope," *Soil Dynamics and Earthquake Engineering*, Vol. 26, pp. 253–264. https://doi.org/10.1016/j.soildyn.2005.02.016

Yoshimine, M., Robertson, P.K., and Wride (Fear), C.E. (1999). "Undrained shear strength of clean sands to trigger flow liquefaction," *Canadian Geotechnical Journal*, Vol. 36, No. 5, pp. 891–906. https://doi.org/10.1139/t99-047

Youd, T.L. (1972). "Compaction of sands by repeated shear straining," *Journal of the Soil Mechanics and Foundations Division*, Vol. 98, No. SM7, pp. 709–725. https://doi.org/10.1061/jsfeaq.0001762

Youd, T.L. (1984). "Recurrence of liquefaction at the same site," *Proceedings, 8th World Conference on Earthquake Engineering*, Vol. 3, pp. 231–238.

Youd, T.L. (1991). "Mapping of earthquake-induced liquefaction for seismic zonation," *Proceedings, Fourth International Conference on Seismic Zonation, Earthquake Engineering Research Institute*, Stanford University, Vol. 1, pp. 111–147.

Youd, T.L. (1993). "Liquefaction-induced lateral spread displacement," *Report TN-1862*, Naval Civil Engineering Laboratory, Port Hueneme, CA, 44 pp.

Youd, T.L. and Carter, B.L. (2005). "Influence of soil softening and liquefaction on spectral acceleration," *Journal of Geotechnical Engineering*, Vol. 131, pp. 811–825. https://doi.org/10.1061/(asce)1090-0241(2005)131:7(811)

Youd, T.L. and Garris, C.T. (1995). "Liquefaction-induced ground-surface disruption," *Journal of Geotechnical Engineering*, Vol. 121, No. 11, pp. 805–809. https://doi.org/10.1061/(asce)0733-9410(1995)121:11(805)

Youd, T.L., Hansen, C.M., and Bartlett, S.F. (2002) "Revised multilinear regression equations for prediction of lateral spread displacement," *Journal of Geotechnical and Geoenvironmental Engineering*, Vol. 128, No. 12, pp. 1007–1017. https://doi.org/10.1061/(asce)1090-0241(2002)128:12(1007)

Youd, T.L., Harp, E.L., Keefer, D.K., and Wilson, R.C. (1985). "The Borah Peak, Idaho earthquake of October 28, 1983--liquefaction," *Earthquake Spectra*, Vol. 2, No. 1, pp. 71–89. https://doi.org/10.1193/1.1585303

Youd, T.L. and Hoose, S.N. (1977). "Liquefaction susceptibility and geologic setting," *Proceedings, 6th World Conference on Earthquake Engineering*, New Delhi, Vol. 3, pp. 2189–2194.

Youd, T.L., Idriss, I.M., Andrus, R.D. Arango, I., Castro, G., Christian, J.T., Dobry, R., Finn, W.D.L., Harder, L.F., Hynes, M.E., Ishihara, K., Koester, J.P., Liao, S.S.C., Marcuson, W.F., Martin, G.R., Mitchell, J.K., Moriwaki, Y., Power, M.S., Robertson, P.K., Seed, R.B., and Stokoe, K.H. (2001). "Liquefaction resistance of soils: Summary report from the 1996 NCEER and 1998 NCEER/NSF workshops on evaluation of liquefaction resistance of soils," *Journal of Geotechnical and Geoenvironmental Engineering*, Vol. 127, No. 10, pp. 817–833. https://doi.org/10.1061/(asce)1090-0241(2001)127:10(817)

Youd, T.L. and Perkins, D.M. (1978). "Mapping liquefaction-induced ground failure potential," *Journal of the Geotechnical Engineering Division*, Vol. 104, No. 4, pp. 433–446. https://doi.org/10.1061/ajgeb6.0000612

Zeghal, M. and Elgamal, A.-W. (1994). "Analysis of site liquefaction using earthquake records," *Journal of Geotechnical Engineering*, Vol. 120, No. 6, pp. 996–1017. https://doi.org/10.1061/(asce)0733-9410(1994)120:6(996)

Zhang, B., Muraleetharan, K.K., and Liu, C. (2016). "Liquefaction of unsaturated sands," *International Journal of Mechanics*, Vol. 16, No. 6, p. D4015002. https://doi.org/10.1061/(asce)gm.1943-5622.0000605

Zhang, G., Robertson, P.K., and Brachman, R.W.I. (2002). "Estimating liquefaction-induced ground settlements from CPT for level ground," *Canadian Geotechnical Journal*, Vol. 39, pp. 1168–1180. https://doi.org/10.1139/t02-047

Zhang, G., Robertson, P.K., and Brachman, R.W.I. (2004). "Estimating liquefaction-induced lateral displacements using the standard penetration test or cone penetration test," *Journal of Geotechnical and Geoenvironmental Engineering*, Vol. 130, No. 8, pp. 861–871. https://doi.org/10.1061/(asce)1090-0241(2004)130:8(861)

Zhu, J., Daley, D., Baise, L.G., and Thompson, E.M. (2017). An updated geospatial liquefaction model for global application," *Bulletin of the Seismological Society of America*, Vol. 107, No. 3, pp. 1365–1385. https://doi.org/10.1785/0120160198

Zhu, J., Daley, D., Baise, L.G., Thompson, E.M., Wald, D.J., and Knudsen, K.L. (2014). "A geospatial liquefaction model for rapid response and loss estimation," *Earthquake Spectra*, Vol. 31, No. 3, pp. 1813–1837. https://doi.org/10.1193/121912eqs353m

Ziotopoulou, K. and Boulanger, R.W. (2013). "Calibration and implementation of a sand plasticity plane-strain model for earthquake engineering applications," *Soil Dynamics and Earthquake Engineering*, Vol. 53, pp. 268–280. https://doi.org/10.1016/j.soildyn.2013.07.009

Ziotopoulou, K., Sinha, S.K., Kutter, B.L. (2024). "Performance-based assessment of liquefaction-induced downdrag on piles," *Proceedings of the 8th International Conference on Earthquake Geotechnical Engineering*, Osaka, Japan, pp 41–53.

10 Ground Failure in Shear: Fault Movement and Seismic Slope Stability

10.1 INTRODUCTION

Geotechnical ground failure can produce earthquake damage by a number of mechanisms, several of which were discussed in Chapter 9. Liquefaction and cyclic softening can cause both vertical and horizontal ground movements due to both compression and shearing mechanisms. These movements result from the softening and weakening these soils undergo as high porewater pressures are generated during shaking; these movements develop incrementally and cause significant damage even when the strength of the soil is not exceeded.

Ground failure can also occur, however, in materials that do not generate significant porewater pressure during shaking. These failures occur when shear stresses on a potential failure surface exceed the available shear strength on that surface. This chapter describes ground movements that result from shearing on failure surfaces through two mechanisms – fault movement and slope instability – and presents procedures for predicting their likelihoods of occurrence and estimating the deformations they produce.

10.2 FAULT DISPLACEMENT

Perhaps the most direct form of seismic ground failure is that associated with fault rupture (Section 1.3.1) itself. As discussed in Section 2.4.2, fault rupture may or may not extend all the way to the ground surface. Ruptures that do extend to the ground surface can produce significant permanent displacements with vertical and/or horizontal offsets over short distances which can cause devastating damage to structures that lie above (e.g., buildings) or cross (e.g., dams, bridges or pipelines) the displaced zones. Ruptures that do not reach the ground surface can still produce significant warping, angular distortion, and extensional/compressive strain at and near the surface of sediments that overlie faulted rock. Fault rupture damage in the 1971 San Fernando earthquake led the California state legislature in 1972 to enact the Alquist-Priolo Earthquake Fault Zoning Act, which placed restrictions on the development of structures intended for human occupancy on properties near mapped active fault traces. This section describes fault rupture and ground surface displacement patterns and presents the basic components required for assessment of fault rupture hazards. More detailed treatments of such assessments can be found in Youngs et al. (2003), Petersen et al. (2011), Moss and Ross (2011), and Wells and Kulkarni (2014). Databases with information on field observations of surface rupture are presented by Sarmiento et al. (2021) and Nurminen et al. (2022), and models derived from these databases will be forthcoming.

10.2.1 Ground Surface Expression of Fault Rupture

Fault rupture damage is influenced by the distribution and nature of ground surface displacement, both of which are in turn influenced by local site conditions. At rock sites, fault rupture displacements are often sharp and readily visible. While the largest displacements may take place at depth on a specific primary fault surface, referred to as the principal fault, displacements near the surface may occur as rupture of the principal fault and as distributed ruptures across secondary

DOI: 10.1201/9781003512011-10

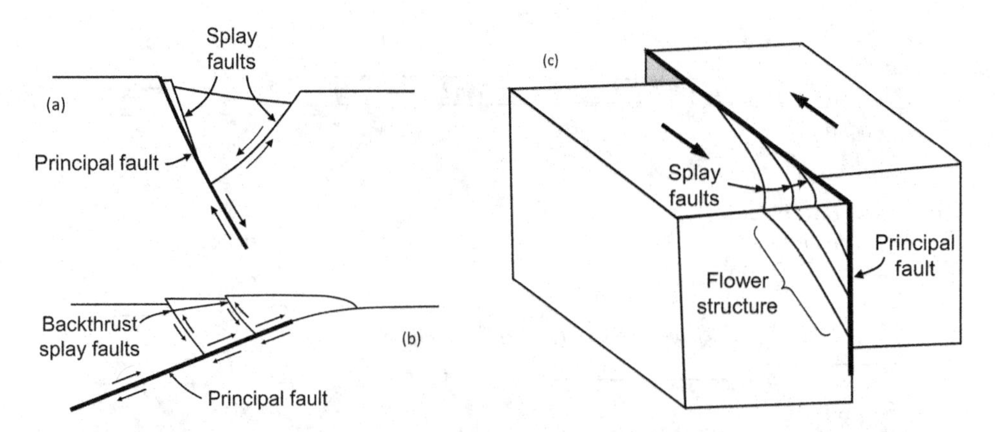

FIGURE 10.1 Schematic illustration of surface faulting from principal fault and distributed ruptures along splay faults at rock sites: (a) normal faulting with gravity graben, (b) reverse faulting, and (c) strike-slip faulting.

discontinuities such as *splay faults* that branch off from the principal fault at shallow depths and other faults and shears in the rock. The brittle nature of rock leads to geometries similar to those shown in Figure 10.1 (although subsequent weathering can smooth out their surface expression). Splay faults can exist in normal, reverse, and strike-slip environments and the discrete offsets they produce at rock sites can be damaging to structures and infrastructure even at significant distances from a main fault. The steep inclination of offshore backthrust splay faults in subduction zone environments, for example, means that more vertical movement of the seafloor, and therefore greater tsunami potential, occurs for a given amount of fault slip (Cummins and Kaneda, 2000; Felix et al., 2022).

When rock is covered by soil, the underlying fault movement may, depending on the thickness and ductility of the soil, reach the ground surface as discrete offsets on principal faults or as distributed shearing and rotation across a secondary deformation zone (Figure 10.2). Discrete offsets are obviously damaging to structures and buried infrastructure but distributed deformations can also cause structural distress or loss of functionality due to tilting of the ground surface.

It is important to remember that local fault movements are rarely purely strike-slip or dip-slip (normal or reverse). Faults are not perfectly planar and local variations in fault orientation lead to variations in stress conditions that influence local fault movement. For example, areas where predominantly strike-slip motion is accompanied by a component of compression (a condition referred to as *transpression*) can produce *en echelon* ridges often referred to as *mole tracks*. A component of extension along a strike-slip fault (*transtension*) can lead to localized subsidence, or trenches, along the fault. Such oblique local fault movement (Section 2.4.2.3) is both common and complex.

10.2.2 Prediction of Surface Rupture

As described in Section 4.2.1, several approaches are commonly used to identify earthquake sources and surface faults are commonly mapped in seismically active areas. When faults are mapped, geologists classify the accuracy of the mapping depending on the quality of the information used to locate the fault, with typical designations being strong, distinct, weak, and uncertain (Petersen et al., 2011; Scott et al., 2023). When earthquakes rupture on previously mapped faults, the mapped locations of rupture, when compared to the previously mapped fault, can be measured as strike-normal offset distances. Standard deviations of these offset distances decrease as fault confidence mapping increases, ranging from about 20 m for strong mapping to 90 m for uncertain mapping (Scott et al., 2023). Not all faults rupture all the way to the ground surface, either due to large hypocentral depths

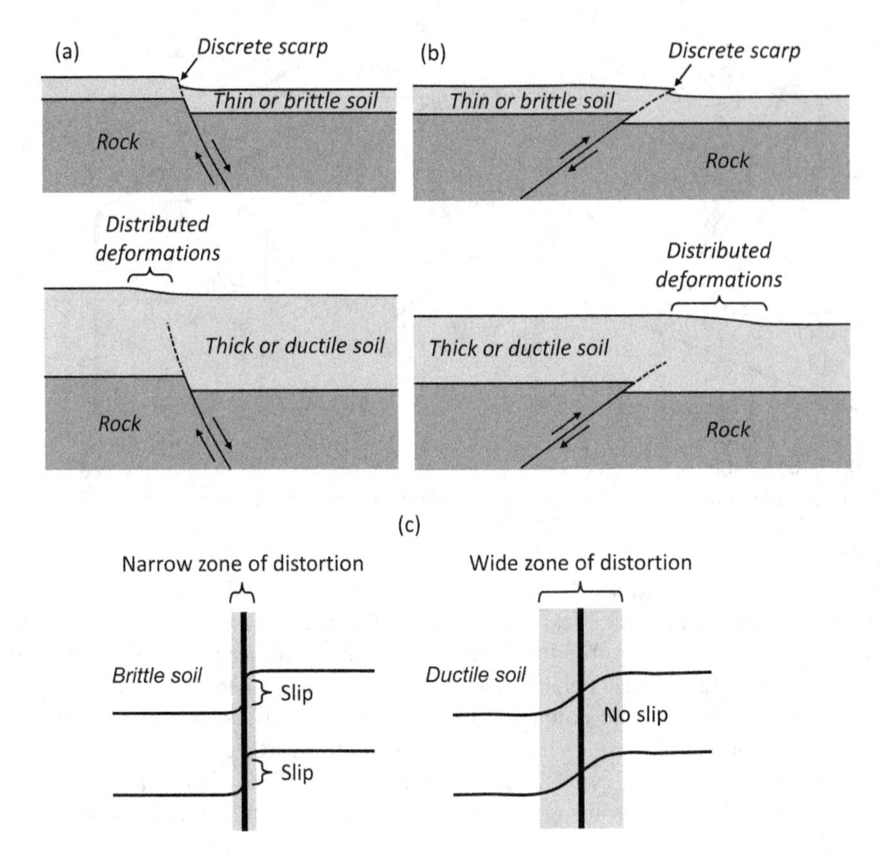

FIGURE 10.2 Illustration of effects of surficial materials on surface expression of faulting: (a) normal faulting, (b) reverse faulting, and (c) strike-slip faulting.

or, as illustrated in Figure 10.2, the presence of thick and/or ductile sediments that overlie the ruptured portion of the fault. The probability of surface rupture increases with increasing magnitude but, as illustrated in Figure 10.3, is different for different slip mechanisms.

The potential for distributed surface rupture off of the main fault must also be considered in a complete surface rupture hazard assessment. The identification and characterization of splay faults and other discontinuities that may experience distributed rupture can be difficult – in many cases, their locations are only revealed after rupture has occurred. It seems intuitive that the likelihood of observing distributed fault rupture at a given location should increase with increasing earthquake magnitude and decrease with increasing distance from the principal fault and empirical data bears this out. Figure 10.4 shows how the probability of secondary surface rupture on discontinuities varies with magnitude and distance from the principal fault for normal faulting; note the qualitative similarity to ground motion models.

10.2.3 Prediction of Fault Rupture Displacement

The potential for damage caused by surface rupture in a particular earthquake depends on the amount of displacement across the ruptured fault. In a manner similar to that used to characterize magnitude-rupture area relationships (Section 4.2.2.2), empirical relationships between magnitude and fault displacements have been developed (e.g., Wells and Coppersmith, 1994). Figure 10.5 illustrates the variation of average and maximum fault displacement with magnitude for different faulting mechanisms.

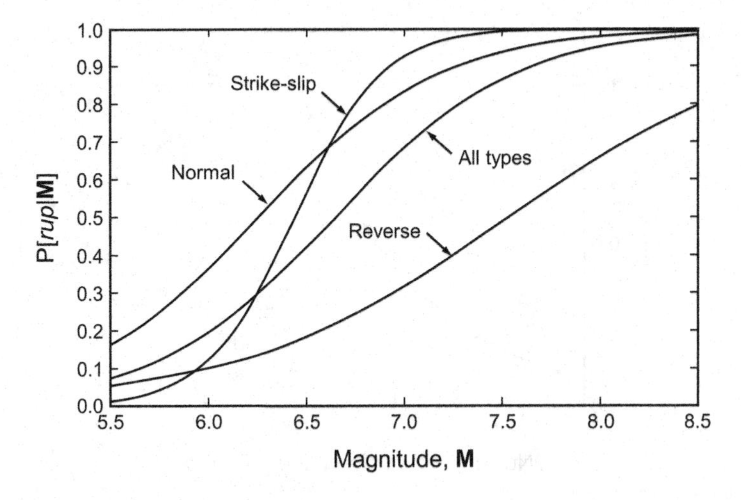

FIGURE 10.3 Probability of surface rupture (slip) on principal faults given magnitude for different slip types and all slip types. (Pizza et al. 2023.)

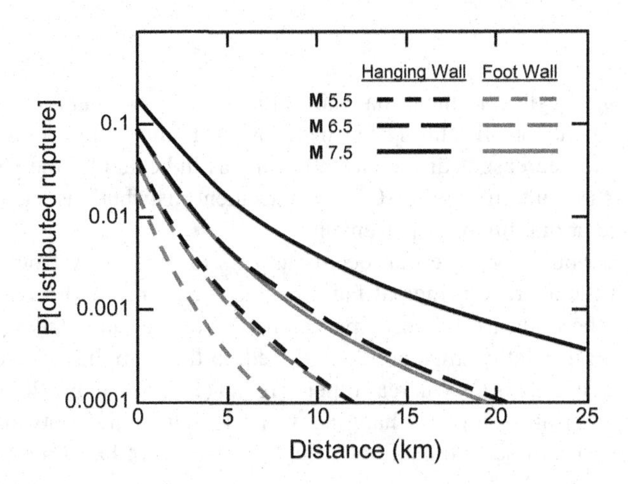

FIGURE 10.4 Variation of probability of distributed rupture across secondary discontinuities with magnitude and distance for normal faulting. (After Youngs et al., 2003.)

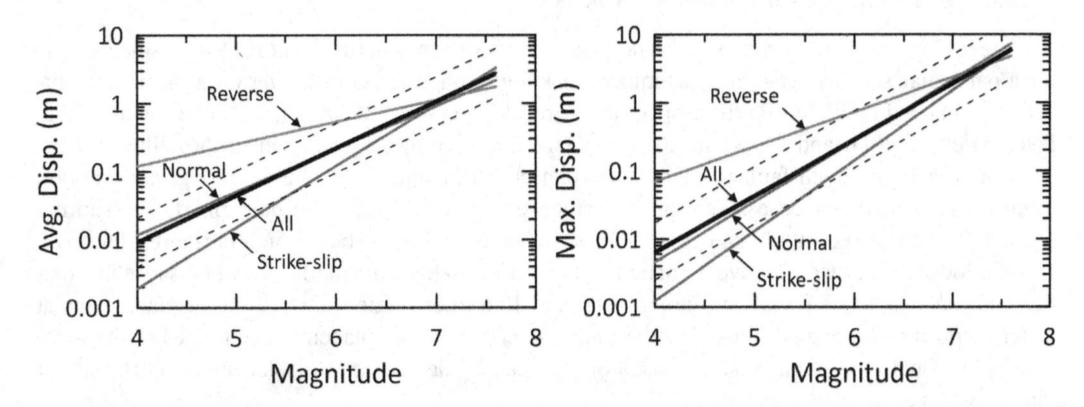

FIGURE 10.5 Variation of (a) average and (b) maximum principal surface rupture displacements with magnitude for all rupture types (Wells and Coppersmith, 1994), normal (Wells and Coppersmith, 1994), reverse (Moss and Ross, 2011), and strike-slip (Wells and Coppersmith, 1994; Petersen et al., 2011) faulting.

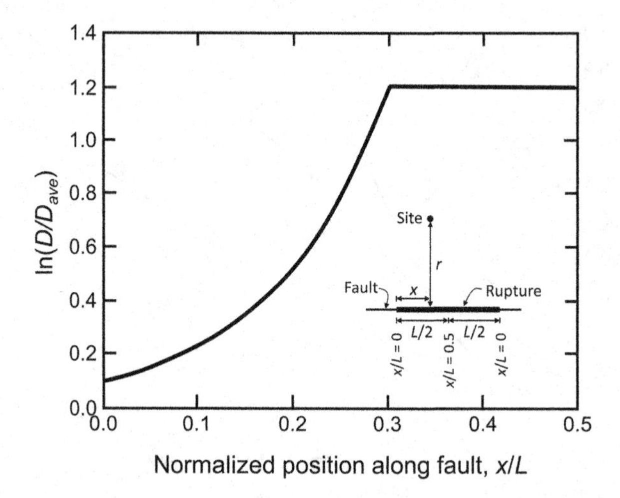

FIGURE 10.6 Variation of principal fault displacement with position along length of strike-slip fault (after Petersen et al., 2011). x-value is distance to closest end of rupture and therefore varies from zero at both ends to $L/2$ at center of rupture.

Mapping of fault rupture displacements has shown that the displacements do not occur uniformly along the length of a fault for all faulting mechanisms. Although data shows significant scatter, displacements appear to decrease near the ends of faults as indicated by the general relationships illustrated for strike-slip faults in Figure 10.6. Displacement distributions such as these are often normalized by average or maximum displacement.

The amount of distributed rupture that occurs at sites located away from the principal fault typically depends on the average principal fault displacement and the horizontal distance from the site of interest to the fault, as measured perpendicular to the fault strike (Figure 10.6 inset). Some distributed faulting relationships are conditioned on the mapping of a discontinuity at the site of interest (Youngs et al., 2003), whereas others are not (Petersen et al., 2011). For normal fault events, the displacement is greater on the hanging wall side than on the footwall (Figure 10.7). For strike-slip faults, the ratio of secondary to principal displacement has a mode of 0.03 (Petersen et al., 2011).

10.2.4 Fault Displacement Hazard Analysis

The preceding sections have provided a conceptual basis for predicting fault rupture displacements for a particular scenario (i.e., an earthquake of known magnitude occurring on a particular portion of a mapped fault). While these specific figures do not cover all cases (see Youngs et al., 2003; Petersen et al., 2011; and Moss and Ross, 2011 for more complete treatments), they illustrate the basic approach by which fault displacements can be estimated. As in the case of ground motion hazards, a deterministic assessment of fault rupture displacement can be performed by assuming a particular scenario earthquake occurs. The scenario event would occur on a particular fault at a specific location and would have a particular magnitude. That information can be used with relationships such as those shown in Figures 10.5 and 10.6 to estimate the resulting displacements at different positions along the length of the fault. Deterministic displacement estimates of the location and amount of slip may also be based on measured paleo-seismic displacements at the site of interest when available.

There is, however, a great deal of aleatory variability around the means predicted using these types of rupture likelihood and fault displacement relationships and, of course, there is uncertainty in the magnitude, rupture location, distance from the site to the principal fault trace, and

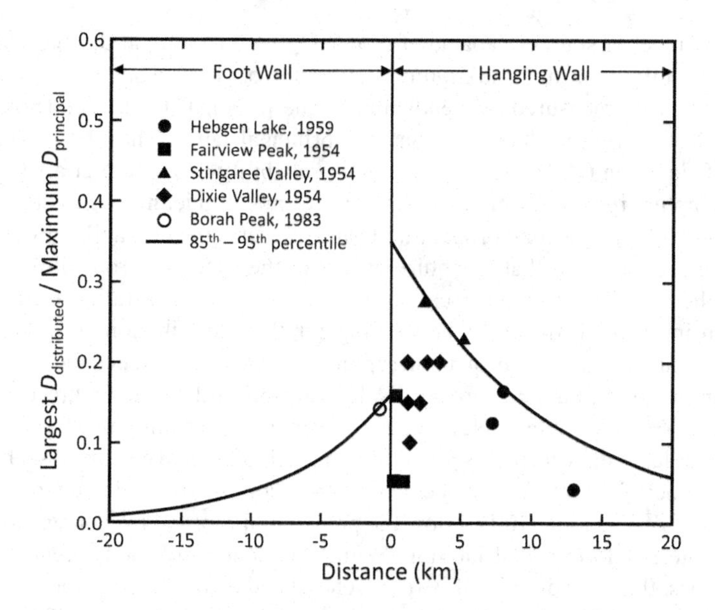

FIGURE 10.7 Variation of largest distributed displacements with distance from normal fault. Displacements are normalized by maximum displacement on principal fault. (After Youngs et al., 2003.)

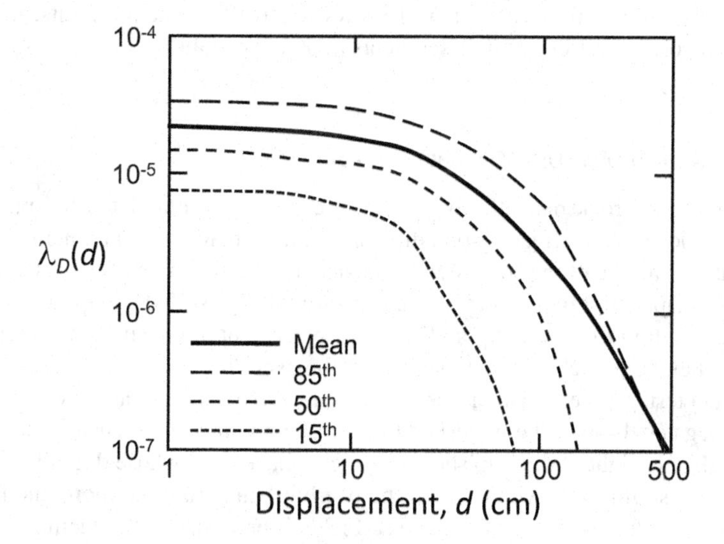

FIGURE 10.8 Fault displacement hazard curves for site on the Solitario Canyon fault at Yucca Mountain, Nevada. (After Youngs et al., 2003.)

rate of occurrence of earthquake scenarios that could occur. These factors lend themselves to the development and use of *probabilistic fault displacement hazard analyses* (PFDHAs). The most common form of a PFDHA (e.g., Youngs et al., 2003) follows the approach used for PSHAs (Section 4.4) and produces a hazard curve for surface or near-surface fault displacement calculated as

$$\lambda_D(d) = \sum_{i=1}^{N_s} \lambda_{\mathbf{M}_{\min,i}} \int_{\mathbf{M}_{\min,i}}^{\mathbf{M}_{\max,i}} f_{\mathbf{M},i}(\mathbf{m}) \left(\int_{r_{\min,i}}^{r_{\max,i}} f_i(r \mid \mathbf{m}) \cdot P\left[rup \mid \mathbf{m}, r\right] \cdot P\left[D > d \mid \mathbf{m}, r, rup\right] dr \right) d\mathbf{m} \quad (10.1)$$

where N_s is the number of sources capable of producing fault rupture at the site of interest, $\mathbf{M}_{\min,i}$ and $\mathbf{M}_{\max,i}$ are the minimum and maximum magnitudes of the ith source, and r is the distance to the site of interest as measured perpendicular to the principal fault trace (Figure 10.6; $r = 0$ for analyses of surface rupture hazards along the principal fault). The first probability term on the right side of Equation (10.1) is the conditional probability of rupture at a site located at distance, r, from a rupture resulting from an earthquake of magnitude, $\mathbf{m}$, occurring on the ith source (Figures 10.3 and 10.4). The second probability term represents the probability of exceeding a fault rupture displacement, d, given that a rupture occurs at the site; that probability is characterized by a joint distribution of magnitude, location and distance. By integrating over all magnitudes and all site-to-primary fault distances accounting for their distributions ($f_{\mathbf{M},i}(\mathbf{m})$ and $f_i(r \mid \mathbf{m})$, respectively), the mean annual rate of fault rupture displacement exceedance can be computed and expressed in the form of a fault rupture displacement hazard curve. Figure 10.8 shows hazard curves for a site at Yucca Mountain, Nevada. The epistemic uncertainty in exceedance rate can be seen to be large for large displacements, reflecting the lack of data to constrain such displacements (i.e., a range of models for rupture displacement distributions are possible due to limited data).

The calculations of Equation (10.1) follow the procedure proposed by Youngs et al. (2003), who were primarily interested in normal faulting. While components of the procedure are specific to different fault types, the procedure itself is fully general and can be applied for any type of surface faulting (Valentini et al., 2021). Model components oriented toward strike-slip (Petersen et al., 2011) and reverse (Moss and Ross, 2011) faulting are also available. Each of these references provides a more complete treatment of PFDHA than that presented here, including factors such as the accuracy of principal fault locations, the orientation of splays relative to principal faults, multiple rupture scenarios (e.g., involving different fault segments), spatial variability of slip displacements, and epistemic uncertainty.

10.2.5 Propagation of Fault Movement through Soils

Sharp offsets in faulted rock beneath a soil deposit can be accommodated by internal distortion within the soil, which has the effect of spreading out the deformations and reducing the size of individual offsets at the surface of the soil. A consequence of this spreading effect is that as the distinct rupture on the principal fault in rock propagates upward through soil, the amplitude of the principal rupture decreases within the overlying soils. The displacement at the surface is therefore affected by the rock offset and the thickness and ductility of the overlying soil.

Physical model tests, numerical analyses, and case history observations provide insight into how far rupture propagates through the materials that overlie a faulted rock surface. Figure 10.9 illustrates the normalized height of dip-slip shear rupture in clayey soil whose ductility is characterized by the axial failure strain in triaxial tests on the overlying material; the more ductile the soil, the lower the extent of rupture above the base offset. Field observations in the vicinity of such faulting have shown that ground surface deformations extend farther back from the fault on the hanging wall side than the foot wall (Bray, 2009).

10.2.6 Damage to Infrastructure

As with landslides, subsidence, and other forms of ground movement, engineered systems can be gravely damaged by ground deformations associated with fault displacement (Figure 1.2). Buried or embedded structures such as pipelines are typically in intimate contact with the ground on both sides of a fault and may experience very high compressive, tensile, shear, and bending demands as the surrounding ground shears and distorts (Figure 10.10a). The level of those demands depends on the relative stiffness of the pipeline and the surrounding ground (a soil-structure interaction problem similar to that of laterally loaded piles). Where possible, buried pipelines can be brought out of the ground and supported on bearings that allow lateral displacement; the combination of

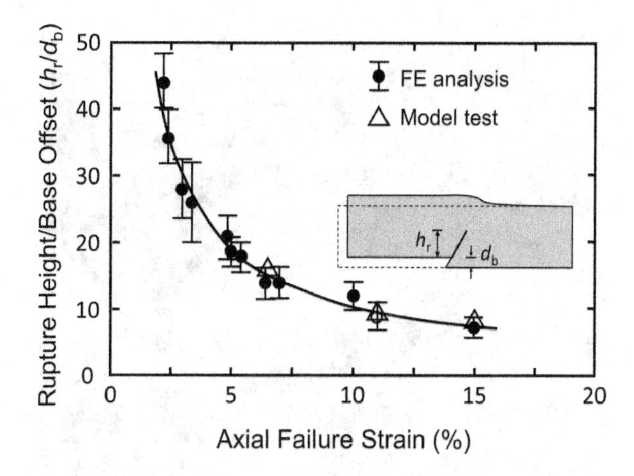

FIGURE 10.9 Variation of normalized height of shear rupture zone in clay overlying dip-slip faulting. (After Bray et al., 1994.)

Teflon-coated slider bearings, articulated joints, and a curved alignment over a 580-m-long corridor (Figure 10.10b) allowed the Trans-Alaska pipeline to survive fault displacements of 4.3 m horizontally and 80 cm vertically in the 2002 Denali earthquake (Cluff et al., 2003; Fuis et al., 2003) without any leakage.

While buildings that straddle faults are frequently destroyed by abrupt fault displacement, field observations have shown that buildings with very stiff and strong shallow foundation systems can survive and at least preserve life safety (Lettis et al., 2000; Anastasopoulos and Gazetas, 2007; Bray, 2009). The key to their survival has been their ability to move as intact bodies relative to the moving ground they are supported on. This relative movement may involve sliding on the base of the foundation and/or local failure of the surrounding soil as illustrated in Figure 10.11. Buildings supported on deep foundations, which tie the base of the structure tightly to the soil on each side of a fault, tend to be very heavily damaged by fault movement.

10.2.7 Mitigation of Fault Displacement Hazards

Most sites that geotechnical engineers deal with are far enough from active faults that fault displacement hazards are insignificant. However, some structures are inherently more likely to be located in such areas. For example, valleys associated with faulting may be attractive locations for dams or may need to be spanned by bridges that cross them. Many mountainous regions are bounded by faults, and infrastructure crossing the mountainous region such as tunnels and pipelines must cross the faults.

After identification of active faulting at a project site, mitigation of the displacement hazards associated with that faulting can be considered when the location of the fault is well established. Risk assessment and mitigation techniques have been developed for pipelines (Wham and Davis, 2019). Several techniques can be considered for mitigation of fault displacement hazards for structures at the ground surface such as buildings (Oettle and Bray, 2013); those techniques are briefly described here. Aside from avoidance, these mitigation measures are more effective when the predominant fault movement is horizontal (strike-slip earthquakes) than when it has a large vertical component (dip-slip and oblique earthquakes); the large vertical displacements from dip-slip events would be highly problematic for most structures.

10.2.7.1 Avoidance

The most common approach for reduction of fault displacement hazards is simply to avoid them by locating structures at some appropriate distance from active faults. The Alquist-Priolo Earthquake

(a)

(b)

FIGURE 10.10 Pipelines affected by fault rupture: (a) Pilarcitos pipeline extension failure at San Andreas crossing in the 1906 San Francisco earthquake (Bray and Kelson, 2006), and (b) defensive measures taken for the design of the Trans-Alaska pipeline crossing of the Denali Fault in Alaska. (Fuis et al., 2003.)

Fault Zoning Act of 1972, or A-P Act, passed following the 1971 San Fernando earthquake, specified that "no structure for human occupancy…be permitted to be placed across the trace of an active fault" and that "the area within fifty (50) feet of such active faults shall be presumed to be underlain by active branches of that fault unless proven otherwise…" (California Geological Survey, 2018). In many cases, however, it may not be possible to avoid an area subject to potential surface fault displacements.

10.2.7.2 Placement of Engineered Fill

The distribution of ground surface deformations can be modified by the excavation and replacement of near-surface soils with ductile engineered fill or placement of such fill on the surface of an

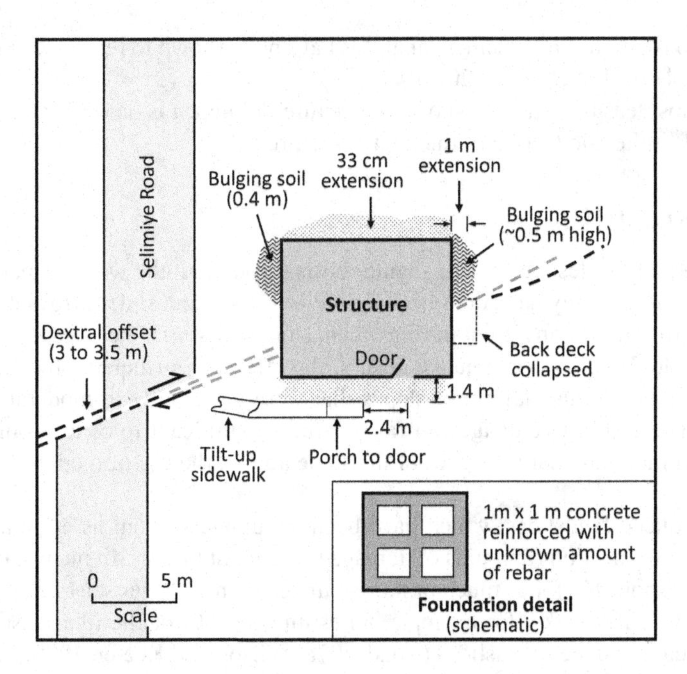

FIGURE 10.11 Two-story building with robust foundation undamaged by more than 3 m of nearby strike-slip fault movement in the 1999 Kocaeli earthquake. (Lettis et al., 2000.)

existing site. The engineered fill does not eliminate or even significantly reduce the amount of fault displacement but rather spreads it over a greater horizontal distance. The extent to which surface displacements will be smoothed by engineered fill depends on the thickness and ductility of the fill material. Oettle and Bray (2013) indicate that, as a rough rule of thumb, sharp offsets in rock will be spread over a horizontal distance approximately 1–2 times the thickness of a ductile engineered fill placed over the rock surface. Numerical analyses can be used to determine the thickness and fill characteristics that best modify potential ground surface movements for a particular site and structure. The ductility of engineered fill can be significantly enhanced with the use of reinforcement such as geosynthetics (e.g., Bray et al., 1993, Bray, 2001, Yang et al., 2020), which further distribute deformations, and in some cases can cause the fill to respond as an intact body that experiences limited damage in a similar manner to the structural foundation in Figure 10.11 (Olgun et al., 2001).

10.2.7.3 Diversion

In a number of earthquakes (Niccum et al., 1976; Berrill, 1983; Lettis et al., 2000; Ulusay et al., 2002; Faccioli et al., 2008), fault ruptures have moved around, rather than through, structures at soil sites. Structures with particularly strong foundations, such as thick, flexurally stiff mats, or with improved soil beneath them can provide sufficient resistance that a rupture surface or zone of concentrated deformation is diverted around them on its way to the surface, although rotation of the structure may occur (Bransby et al., 2008). The effectiveness of this approach depends on the stiffness and strength of the stiffened elements (i.e., mat foundation or improved ground) that are used, along with the type of faulting; diversion approaches are more effective in strike-slip ruptures than in dip-slip ruptures.

A different form of diversion can be obtained by introducing weak zones or interfaces near or beneath a structure. Structures can be isolated, at least to a degree, from differential horizontal ground movements by placing two layers of low-friction material (generally a geosynthetic) beneath them. The low-friction material, placed with bedding layers of sand above and beneath, provides a limiting shear stress on the interface between the layers (Kratzsch, 1983). Smooth, vertical, structural skirt walls surrounding shallow foundations can potentially divert fault rupture and reduce foundation rotation; finite element analyses have been shown to reproduce model foundation response reasonably well (Loli et al., 2018).

Diversion is most feasible when most of the structure of interest is on one side or the other of the fault trace so that limited deflection of the fault is required.

10.3 SLOPE STABILITY

Slope failures, i.e., landslides, occur on a regular basis throughout the world as part of the ongoing evolution of landscapes. Many landslides occur in natural slopes but slides also occur in constructed slopes from time to time. At any point in time, then, slopes exist in states ranging from very stable to marginally stable. When an earthquake occurs, the effect of earthquake-induced ground shaking is often sufficient to cause failure of slopes that were marginally to moderately stable before the earthquake. The resulting damage can range from insignificant to catastrophic depending on the geometric and material characteristics of the slope and on the use of the land on and below the landslide.

Earthquake-induced landslides, which have been documented from as early as 1789 B.C. (Li, 1990), have caused tremendous amounts of damage throughout history. In many earthquakes, landslides have been responsible for as much or more damage than all other seismic hazards combined. In the 1964 Alaska earthquake, for example, an estimated 56% of the total cost of damage was caused by earthquake-induced landslides (Youd, 1978; Wilson and Keefer, 1985). Kobayashi (1981) found that more than half of all deaths in large ($\mathbf{M}>6.9$) earthquakes in Japan between 1964 and 1980 were caused by landslides. The 1920 Haiyuan earthquake ($\mathbf{M}=8.5$) in the Ningxia Province of China produced hundreds of large landslides that caused more than 100,000 deaths (Close and McCormick, 1922). The 2008 Wenchuan earthquake in China, which killed 70,000 people and left 18,000 listed as missing even three years later (Dai et al., 2011), caused more than 43,000 landslides with a total area of 632 km^2 (Li et al., 2013), and were responsible for one-third of the total casualties (Wang et al., 2009). Evaluation of seismic slope stability is one of the most important activities of the geotechnical earthquake engineer.

This chapter describes different types of earthquake-induced landslides and the conditions under which they occur. It also reviews the basic principles of slope stability evaluation, including static stability analysis, and then presents several methods for seismic slope stability analysis.

10.4 TYPES OF EARTHQUAKE-INDUCED LANDSLIDES

Many factors, including geologic and hydrologic conditions, topography, climate, weathering, and land use, influence the stability of slopes and the characteristics of landslides. A number of procedures for classification of landslides have been proposed; that of Varnes (1978) is perhaps most widely used in the United States. Similar principles and terminology can be used to classify earthquake-induced landslides (Table 10.1) on the basis of material type (soil or rock), character of movement (disrupted or coherent), and other attributes, such as velocity, depth, and water content. Earthquake-induced landslides can be divided into three main categories: disrupted slides and falls, coherent slides, and lateral spreads and flows.

Disrupted slides and falls include rock falls, rock slides, rock avalanches, soil falls, disrupted soil slides, and soil avalanches. The earth materials involved in such failures are sheared, broken, and disturbed into a nearly random order. These types of failures, usually found in steep terrain, can produce extremely rapid movements and devastating damage; rock avalanches and rock falls have historically been among the leading causes of death from earthquake-induced landslides.

Coherent slides, such as rock and soil slumps, rock and soil block slides, and slow earth flows, generally consist of a few coherent blocks that translate or rotate on somewhat deeper failure surfaces in moderate to steeply sloping terrain. Most coherent slides occur at lower velocities than disrupted slides and falls.

TABLE 10.1
Types and Characteristics of Earthquake-Induced Landslides

Name	Type of Movement	Internal Disruption[a]	Water Content[b]				Velocity[c]	Depth[d]
			D	U	PS	S		
		Disrupted Slides and Falls						
Rock falls	Bounding, rolling, free fall	High or very high	X	X	X	X	Extremely rapid	Shallow
Rock slides	Translational sliding on basal shear surface	High	X	X	X	X	Rapid to extremely rapid	Shallow
Rock avalanches	Complex, involving sliding and/or flow, as stream of rock fragments	Very high	X	X	X	X	Extremely rapid	Deep
Soil falls	Bounding, rolling, free fall	High or very high	X	X	X	X	Extremely rapid	Shallow
Disrupted soil	Translational sliding on basal shear surface or zone of weakened, sensitive clay	High	X	X	X	X	Moderate to rapid	Shallow
Soil avalanches	Translational sliding with subsidiary flow	Very high	X	X	X	X	Very rapid to extremely rapid	Shallow
		Coherent Slides						
Rock slumps	Sliding on basal shear surface with component of headward rotation	Slight or moderate	X	X	X	X	Slow to rapid	Deep
Rock block slides	Translational sliding on basal shear surface	Slight or moderate	?	X	X	X	Slow to rapid	Deep
Soil slumps	Sliding on basal shear surface with component of headward rotation	Slight or moderate	?	X	X	X	Slow to rapid	Deep
Soil block slides	Translational sliding on basal shear surface	Slight or moderate	?	X	X	X	Slow to rapid	Deep
Slow earth flows	Translational sliding on basal shear surface with minor internal flow	Slight			X	X	Very slow to moderate with very rapid surges	Generally shallow, occasionally deep

(*Continued*)

TABLE 10.1 (*Continued*)
Types and Characteristics of Earthquake-Induced Landslides

Name	Type of Movement	Internal Disruption[a]	Water Content[b] D	U	PS	S	Velocity[c]	Depth[d]
		Lateral Spreads or Flows						
Soil lateral spreads	Translation on basal zone of liquefied sand, or silt or weakened, sensitive clay	Generally moderate, occasionally slight, occasionally high			X	X	Very rapid	Variable
Rapid soil flows	Flow	Very high	?	?	?	X	Very rapid to extremely rapid	Shallow
Subaqueous landslides	Complex, generally involving lateral spreading and/or flow; occasionally involving slumping and/or block sliding	Generally high or very high, occasionally moderate or slight			X	X	Generally rapid to extremely rapid, occasionally slow to moderate	Variable

Source: After Keefer, 1984.

[a] Internal disruption: "slight" signifies landslide consists of one or a few coherent blocks; "moderate" signifies several coherent locks; "high" signifies numerous small blocks and individual soil grains and rock fragments; "very high" signifies nearly complete disaggregation into individual soil grains or small rock fragments.

[b] Water content: D, dry; U, moist, but unsaturated; PS, partly saturated; S, saturated.

[c] Velocity:

extremely slow	0.6 m/yr	very slow	1.5 m/yr	slow	1.5 m/month	moderate	1.5 m/day	rapid	0.3 m/min	very rapid	3 m/sec	extremely rapid

[d] Depth: "shallow" signifies thickness generally < 3 m; "deep" generally > 3 m.

Lateral spreads and flows generally involve liquefiable soils, although sensitive clays can produce landslides with very similar characteristics. Due to the low residual strength of these materials, flow sliding can occur on remarkably flat slopes. Liquefaction-induced spreads and flow slides were discussed in detail in Chapter 9. The different types of earthquake-induced landslides occur with different frequencies. Rock falls, disrupted soil slides, and rock slides appear to be the most common types of landslides observed in historical earthquakes (Table 10.2). Subaqueous landslides, slow earth flows, rock block slides, and rock avalanches are least common, although the difficulty of observing subaqueous slides may contribute to their apparent rarity.

10.5 EARTHQUAKE-INDUCED LANDSLIDE ACTIVITY

For preliminary stability evaluations, knowledge of the conditions under which earthquake-induced landslides have occurred in past earthquakes is useful. It is logical to expect that the extent of earthquake-induced landslide activity should increase with increasing earthquake magnitude and that there could be a minimum magnitude below which earthquake-induced landsliding would rarely occur. It is equally logical to expect that the extent of earthquake-induced landslide activity should decrease with increasing source-to-site distance and that there could be a distance beyond which landslides would not be expected in earthquakes of a given size.

A study of 300 U.S. earthquakes between 1958 and 1977 (Keefer, 1984; 2002) showed that the smallest earthquakes noted to have produced landslides had local magnitudes of about 4.0. Minimum magnitudes for different types of landslides were estimated as shown in Table 10.3. Where magnitudes were not available, minimum Modified Mercalli Intensity (MMI) values of

TABLE 10.2
Relative Abundance of Earthquake-Induced Landslides from Study of 40 Historical Earthquakes Ranging from M_s=5.2 to M9.5

Abundance	Description
Very abundant (>100,000 in the 40 earthquakes)	Rock falls, disrupted soil slides, rock slides
Abundant (10,000 to 100,000 in the 40 earthquakes)	Soil lateral spreads, soil slumps, soil block slides, soil avalanches
Moderately common (1,000 to 10,000 in the 40 earthquakes)	Soil falls, rapid soil flows, rock slumps
Uncommon	Subaqueous landslides, slow earth flows, rock block slides, rock avalanches

Source: After Keefer, 1984, 2002.

TABLE 10.3
Estimates of the Smallest Earthquakes Likely to Cause Landslides

M_L	Description
4.0	Rock falls, rock slides, soil falls, disrupted soil slides
4.5	Soil slumps, soil block slides
5.0	Rock slumps, rock block slides, slow earth flows, soil lateral spreads, rapid soil flows, and subaqueous landslides
6.0	Rock avalanches
6.5	Soil avalanches

Source: After Keefer, 1984, 2002.

IV and V have been observed for disrupted slides or falls and other types of slides, respectively. Although these empirically based limits are useful, their approximate nature must be recognized; failure of slopes that are near the brink of failure under static conditions could be produced by quite weak earthquake shaking.

The maximum source-to-site distances at which landslides have been produced in historical earthquakes vary with magnitude and are different for different types of landslides. Disrupted slides or falls, for example, have rarely been found beyond epicentral distances of about 15 km for **M**=5 events but have been observed as far as about 200 km (124 mi.) in **M**=7 earthquakes. Figure 10.12 shows the maximum epicentral distances for different types of landslides from 40 earthquakes in active seismic areas, along with data from stable mid-continental regions that exceed the distances from the active seismic areas. Similarly, the area over which earthquake-induced landsliding can be expected also increases with increasing earthquake magnitude (Figure 10.13). Under conditions where ground motions attenuate relatively slowly with distance (leading to larger ground motion amplitudes at large distances), such as in central and eastern North America, the area of earthquake-induced landsliding can increase, as occurred in the 2011 Virginia earthquake (Figure 10.12).

Example 10.1

Estimate the maximum distances and areas over which rock slides and soil slumps would typically be observed in a **M**7.0 earthquake.

Solution:

Rock slides would and soil slumps would be classified as disrupted and coherent slides, respectively. From Figures 10.1 and 10.2, the typical maximum distances and areas would be

Type	Maximum Distance	Maximum Area
Rock slides	207 km	15,000–25,000 km^2
Soil slumps	159 km	15,000–25,000 km^2

10.6 CO-SEISMIC STABILITY HAZARDS

While earthquake-induced-triggered landslides can cause tremendous damage during earthquakes, their potential to cause problems does not end when the ground stops shaking. A number of secondary landslide-related hazards can and have caused extensive damage in the days, months, and even years after an earthquake. In some cases, these "secondary" hazards have resulted in even more damage than the original, seismically-triggered landslide.

The sliding of shallow colluvial soils in steep, mountainous terrain is often restrained by vegetation. The roots of shrubs and small trees provide a reinforcement effect and also draw moisture from the soil to sustain the plants, thereby increasing the strength of the soil. Seismically-triggered landslides can lead to secondary landsliding by removing vegetation – heavy rainfall following earthquakes, which is not uncommon in many parts of the world, can lead to extensive sliding and erosion of denuded slopes. In Taiwan, the 1999 Chi-Chi earthquake induced a significant number of landslides that denuded many slopes in the Choushui River watershed in central Taiwan. Typhoon Herb occurred 3 years before the Chi-Chi earthquake, and typhoon Toraji occurred 2 years after. Despite the fact that Herb brought more precipitation to that watershed than Toraji, Toraji produced 48.8 km^2 of landslides compared to only 9.8 km^2 caused by Herb (Lin et al., 2006). In the year after the Chi-Chi earthquake, the average rainfall thresholds for triggering of landslides in two other watersheds in central Taiwan dropped to less than 30% of their average values over the preceding 30–40 years (Shieh et al., 2009); by 2005–2006, the landslide triggering threshold had recovered to about 60% of the long-term average.

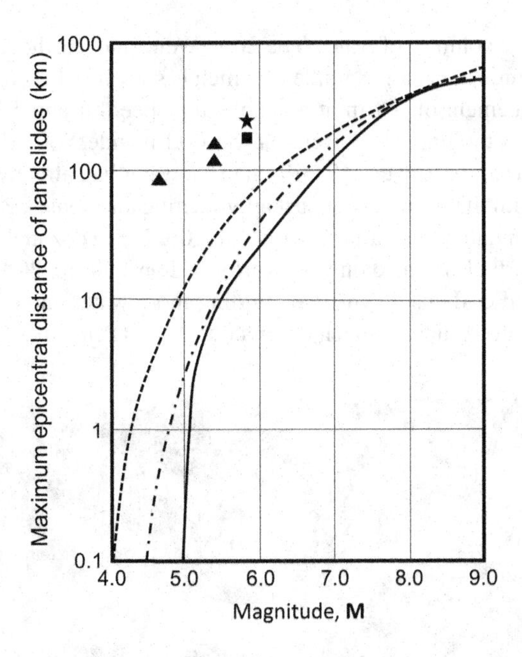

FIGURE 10.12 Maximum epicentral distance limit for different types of landslides from 40 historical earthquakes (after Keefer, 1984) with data for which maximum distances were exceeded: triangles are from Colorado Plateau (Keefer, 2002), square is 1988 Saguenay earthquake (Lefebvre et al., 1992), star is 2011 Virginia earthquake (Jibson and Harp, 2012).

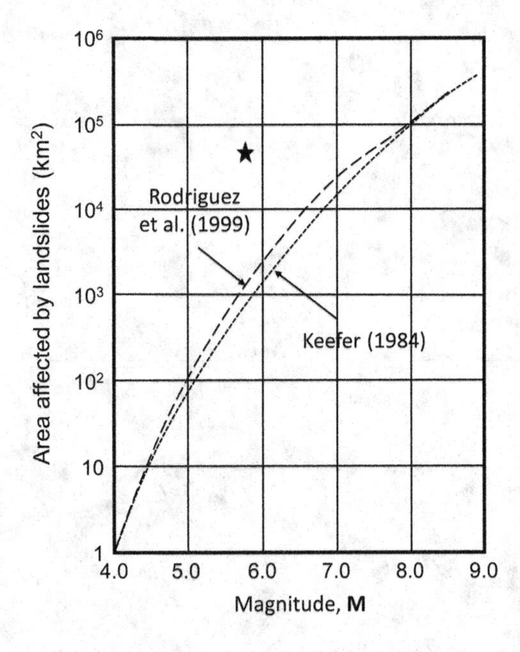

FIGURE 10.13 Upper bound of area affected by landslides (after Jibson and Harp, 2012). Star represents 2011 Virginia earthquake. (Jibson and Harp, 2012 with permission of the Seismological Society of America.)

In steep terrain, even unsaturated soils can slide over large distances. In a number of historical earthquakes, unstable materials have flowed or fallen into canyons or valleys where they have blocked rivers or streams. In some cases, the resulting landslide dams (Costa and Schuster, 1991) are small and quickly eroded away. In a number of cases, however, the dams have been quite large and have impounded very large volumes of water. The 2008 Wenchuan earthquake caused 34 landslides to dam rivers near the source zone, an example of which is shown in Figure 10.14. The Tangjiashan landslide involved displacement of 900 m at an estimated speed of about 30 m/sec. The landslide debris formed a dam across the Jianjiang River that was 600 m wide, 800 m long, and over 20 million m^3 in volume. In less than a month, over 240 million m^3 of water had already been impounded by the dam (Fan et al., 2019). With the rainy season just approaching and some 1.3 million people living in the area that would be inundated by a failure of the dam, discharge channels were excavated and the water level reduced. Nevertheless, the dam was overtopped on June 10, 2008 and the resulting flood wave overtopped two smaller dams downstream adding more water and causing tremendous damage in Beichuan, which had fortunately been evacuated earlier (Peng and Zhang, 2013a, b).

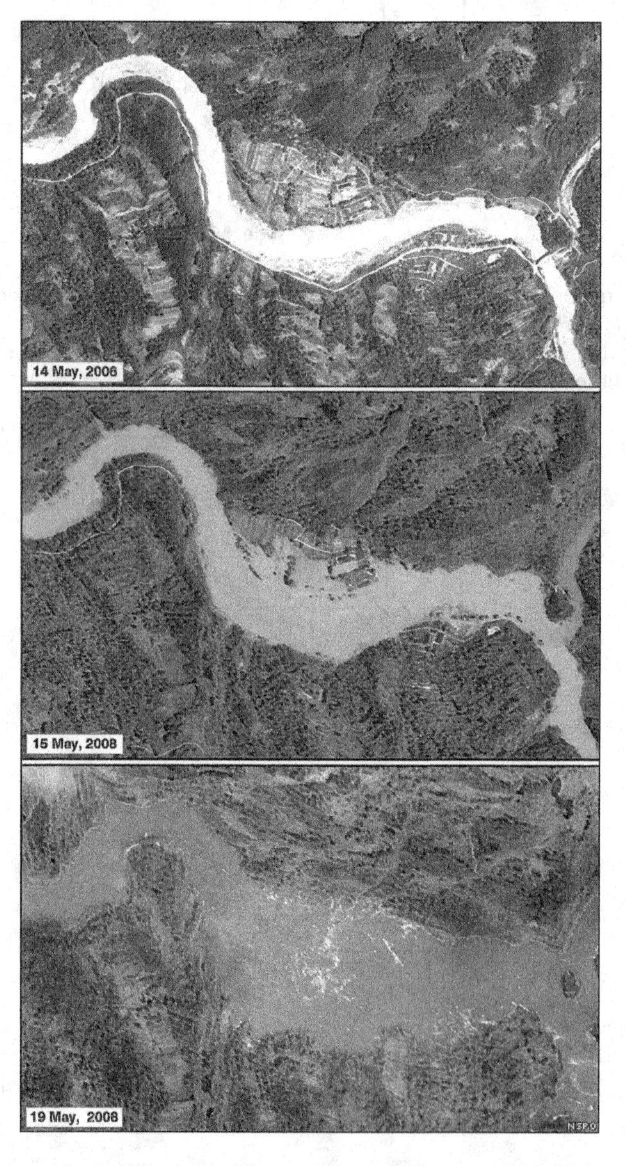

FIGURE 10.14 Satellite views of lake forming behind landslide dam in Beichuan County following the May 12, 2008 Wenchuan earthquake in China. Upper photo shows normal river conditions 2 years before the earthquake; lower photos show lake forming 3 and 7 days following the earthquake. (Formosat image © 2008 Dr. Cheng-Chien Liu, National Cheng-Kung University and Dr. An-Ming Wu, National Space Organization, Taiwan.)

10.7 EVALUATION OF SLOPE STABILITY

The stability of slopes is influenced by many factors, and a complete slope stability evaluation must consider the effects of each. Geological, hydrological, topographical, geometrical factors, as well as material characteristics and time all influence the stability of a particular slope. Information on these characteristics is needed to reliably perform and interpret the results of both static and seismic slope stability analyses. Review of available documents, field reconnaissance, field monitoring, subsurface investigation, and material testing can all be used to obtain this information.

For many sites, considerable useful information can be obtained from previously published documents such as geologic maps, soil survey and/or agricultural maps, topographic maps, natural hazard maps, and geologic and geotechnical engineering reports. Additional information may be obtained from aerial photographs (stereo-paired aerial photographs, Google Earth) and other forms of remote sensing.

Field reconnaissance involves careful observation and detailed mapping of a variety of site characteristics associated with past, existing, or potential slope instability. Features such as scarps; tension cracks; bulges; hummocky terrain; displaced ditches, channels, and fences; cracked foundations, walls, or pavements; and leaning trees or poles can be identified and mapped as evidence of instability. The locations of streams, springs, seeps, ponds, and moist areas, and differences in vegetative cover, can provide evidence of altered or disrupted water flow caused by slope instability. The potential for erosion by streams or wave action at the base of a slope can affect future slope stability.

If time permits, slope movement can be monitored. Historically, landslide monitoring usually involved repeated measurements at specific locations. Surface monuments can be installed at points on and near the slope and surveyed periodically to identify the magnitude and direction of surface movement. Photogrammetric methods can be used to determine relative movements from sets of stereo-paired aerial photographs taken at different times. In recent years, new technologies have produced tremendous advances in landslide monitoring (Rathje and Franke, 2016). Techniques such as Global Positioning System (GPS) technologies (Gili et al., 2000; Coe et al., 2003; Wang, 2011; Benoit et al., 2015), aerial and ground-based LIDAR (Glenn et al., 2006; Haneberg et al., 2009; Kayen et al., 2006), various synthetic aperture radar (SAR) techniques (Gabriel et al., 1989; Colesanti and Wasowski, 2006; Nobrega et al., 2013; Casagli et al., 2016), and Structure from Motion (SfM) optical techniques (Lucieer et al., 2014; Ruggles et al., 2016) have all led to advances in monitoring. Many of these technologies can now be implemented into unmanned aerial vehicles (UAVs), or "drones," that allow economical deployment with the ability to provide multiple perspectives on a landslide. Inclinometers are very useful for monitoring lateral deformation patterns below the ground surface. In many cases, crack gauges, tiltmeters, and extensometers can also be used to observe the effects of slope movement. When, as is commonly the case, pore pressures are important, piezometers and/or observation wells can provide important information on pore pressures and their variation with time.

Subsurface investigation can include excavation and mapping of test pits and trenches, boring and sampling, in situ testing, and geophysical testing. Such investigations can reveal the depth, thickness, density, strength, and deformation characteristics of subsurface units, and the depth and variation of the groundwater table. In situ and geophysical tests are particularly useful for determining the location of an existing failure surface.

Laboratory tests are often used to quantify the physical characteristics of the various subsurface materials for input into numerical slope stability analyses. Soil type, density, strength, and stress-strain behavior are of prime importance; other characteristics, such as grain size distribution, plasticity, permeability, and compressibility, are also useful.

Only after this information is estimated or obtained through testing can a stability analysis be performed. Although the remainder of this chapter focuses on methods of slope stability analysis, it is important to remember that the analysis itself is but a single part of a complete slope stability evaluation and that its accuracy will be reduced if careful attention is not given to the site characterization aspects of the evaluation.

10.8 STATIC SLOPE STABILITY ANALYSIS

Slopes become unstable when the shear stresses required to maintain equilibrium reach or exceed the available shearing resistance on some potential failure surface. For slopes in which the shear stresses required to maintain equilibrium under static gravitational loading are high, the additional dynamic stresses needed to produce instability may be low. Hence, the seismic stability of a slope is strongly influenced by its static stability. Because of this and the fact that the most commonly used methods of seismic stability analysis rely on static stability analyses, a brief review of static slope stability analysis is presented. The procedures for analysis of slope stability under static conditions are well established. An excellent, concise review of the state of the art for static analysis was presented by Duncan (1996). Detailed descriptions of specific methods of analysis can be found in standard texts (e.g., Abramson et al., 2001; Duncan et al., 2014; Cheng and Lau, 2014). Currently, the most commonly used methods of static slope stability analysis are limit equilibrium analyses and stress-deformation analyses.

10.8.1 Limit Equilibrium Analysis

Limit equilibrium analyses consider force and/or moment equilibrium of a mass of soil above a potential failure surface. The soil above the potential failure surface is assumed to be rigid. The same fraction of the available shear strength is assumed to be mobilized simultaneously at all points on the potential failure surface. Because the soil on the potential failure surface is assumed to be rigid-perfectly plastic (Figure 10.15), limit equilibrium analyses provide no information on slope deformations.

Slope stability is usually expressed in terms of an index, most commonly the factor of safety, which is usually defined as

$$FS = \frac{\text{Available shear strength}}{\text{Shear stress required to maintain equilibrium}} \tag{10.2}$$

Thus, the factor of safety is a ratio of capacity (the shear strength of the soil) to demand (the shear stress induced on the potential failure surface). The factor of safety can also be viewed as the factor by which the strength of the soil would have to be divided to bring the slope to the brink of

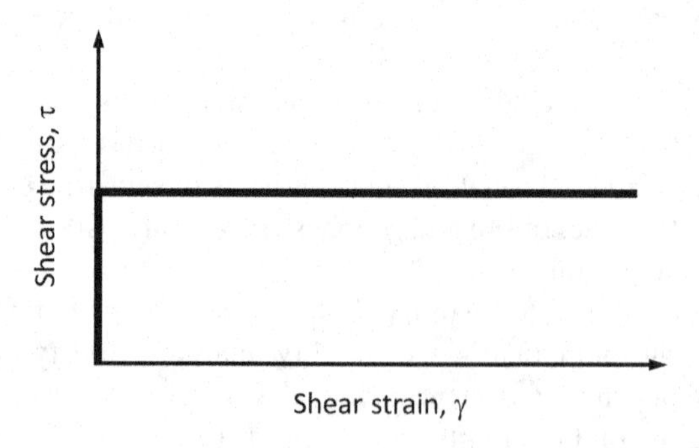

FIGURE 10.15 Stress-strain curve for a rigid-perfectly plastic material. No shear strain occurs until the strength of the material is reached, after which the material strains at constant shear stress.

instability. In contrast to the assumptions of limit equilibrium analysis, the strength of the soil in actual slopes is not reached at the same time at all points on the failure surface (i.e., the local factor of safety is not constant).

A variety of limit equilibrium procedures have been developed to analyze the static stability of slopes. In these procedures, stresses and strengths are typically integrated over the surfaces they are applied to with stability and then evaluated in terms of the resulting "driving" and "resisting" forces. Slopes that fail by translation on a planar failure surface (Figure 10.16a) such as a bedding plane, rock joint, or seam of weak material can be analyzed quite easily by the Culmann method (Taylor, 1948). Long slopes with shallow failure surfaces at constant depth are often idealized as infinite slopes and analyzed using the same principles as the Culmann method. Slopes in which failure is likely to occur on two or three planes (Figure 10.16b) can be analyzed by wedge methods (e.g., Perloff and Baron, 1976; Lambe and Whitman, 1969). In homogeneous slopes comprised of cohesive soil, the critical failure surface usually has a circular (Figure 10.16c) or log-spiral shape. Since the minimum factors of safety for circular and log-spiral failure surfaces are very close, homogeneous slopes are usually analyzed by methods such as the Ordinary Method of Slices (Fellenius, 1927) or Bishop's modified method (Bishop, 1955), which assumes circular failure surfaces. When subsurface conditions are not homogeneous (e.g., when layers with significantly different strength, highly anisotropic strength, or discontinuities exist), failure surfaces are likely to be noncircular (Figure 10.16d). In such cases, methods like those of Morgenstern and Price (1965), Spencer (1967), and generalized Janbu (1968) may be used. Nearly all limit equilibrium methods are susceptible to numerical problems under certain conditions. These conditions vary for different methods but are most commonly encountered where soils with high cohesive strength are present at the top of a slope or when failure surfaces emerge steeply at the base of slopes in soils with high frictional strength (Duncan, 1996).

In concept, any slope with a factor of safety above 1.0 should be stable. In practice, however, the level of stability is seldom considered acceptable unless the factor of safety is significantly greater than 1.0. Criteria for acceptable factors of safety recognize (1) uncertainty in the accuracy with which the input parameters (shear strength, groundwater conditions, slope geometry, etc.) are known, (2) uncertainty in the accuracy with which the slope stability analysis represents the actual mechanism of failure, (3) the likelihood and duration of exposure to various types of external loading, and (4) the potential consequences of slope failure. Typical minimum factors of safety used in slope design are about 1.5 for normal long-term loading conditions and about 1.3 for temporary slopes or end-of-construction conditions in permanent slopes (when dissipation of pore pressure can be expected to increase stability with time).

When the minimum factor of safety of a slope reaches a value of 1.0, the available shear strength of the soil is fully mobilized on some potential failure surface and the slope is at the point of incipient failure. Any additional loading will cause the slope to fail (i.e., to deform until it reaches a configuration in which the shear stresses required for equilibrium are less than or equal to the available

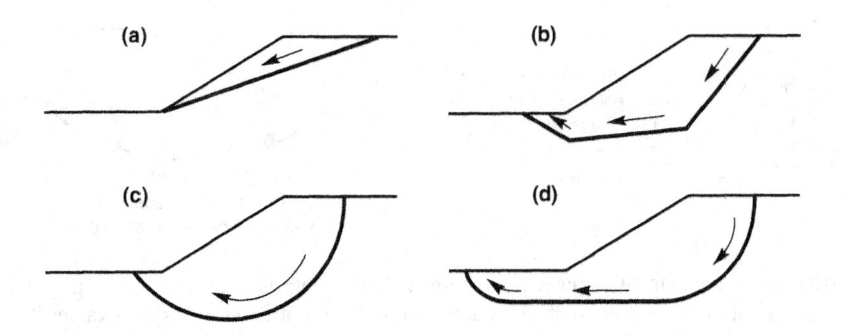

FIGURE 10.16 Common failure surface geometries: (a) planar; (b) multi-planar; (c) circular; (d) noncircular.

shear strength of the soil). The limit equilibrium assumption of rigid-perfectly plastic behavior suggests that the required deformation will occur in a ductile manner. Many soils, however, exhibit brittle, strain-softening stress-strain behavior. In such cases, the peak shear strength may not be mobilized simultaneously at all points on the failure surface.

When the peak strength of a strain-softening soil is reached, such as Point A in Figure 10.17a, the available shearing resistance will drop from the peak to the residual strength. As it does so, shear stresses related to the difference between the peak and residual strength of the soil at Point A can no longer be resisted at Point A and are transferred to the surrounding soil. These redistributed shear stresses may cause the peak strengths in the surrounding soil to be reached (Figure 10.17b) and exceeded, thereby reducing their available shearing resistances to residual values. As the stress redistribution process continues, the zone of failure may grow until the entire slope becomes unstable. This process is known as *progressive failure*, which is a frequent cause of slope instability in strain-softening soils, even when the limit equilibrium factor of safety (based on peak strength) is well above 1.0. Within the constraints of limit equilibrium analysis, the stability of slopes with strain-softening materials can be analyzed reliably only by using residual shear strengths. The factor of safety from such an analysis, however, may be extremely conservative.

Limit equilibrium analyses must be formulated with great care. Since the available shearing resistance of the soil depends on porewater drainage conditions, those conditions must be considered carefully in the selection of shear strengths (Section 6.6.6) and pore pressure conditions for the analysis. Duncan et al. (2014) provided guidelines for the selection of input parameters for limit equilibrium slope stability analyses.

10.8.2 Stress-Deformation Analyses

Stress-deformation analyses allow consideration of the stress-strain behavior of soil and rock and can be performed using finite element or finite difference methods. When applied to slopes, stress-deformation analyses can predict the magnitudes and patterns of stresses, movements, and pore pressures in slopes during and after construction/deposition. Nonlinear stress-strain behavior, complex

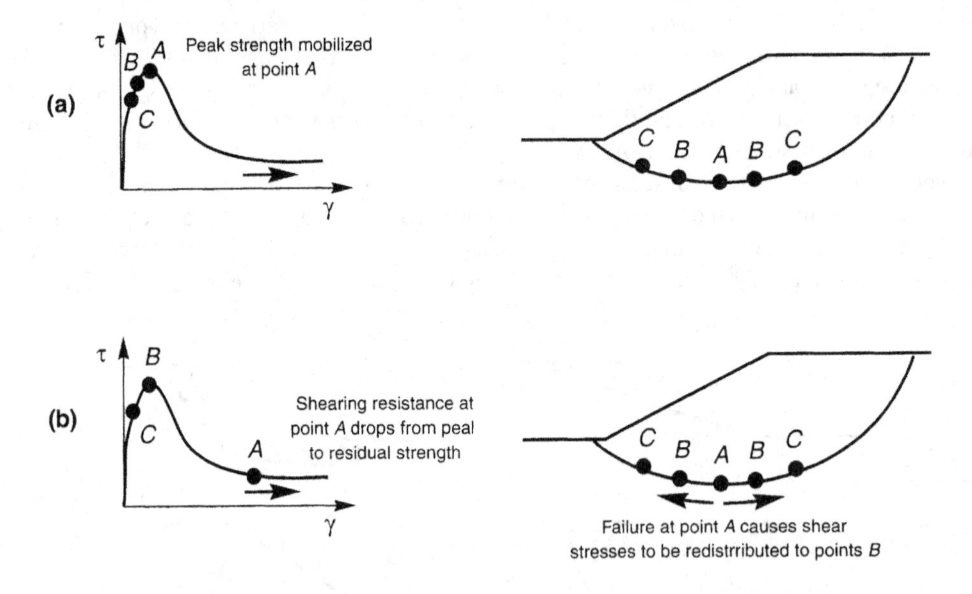

FIGURE 10.17 Development of progressive failure in slope comprised of strain-softening materials: (a) exceedance of peak strength at any point (A) reduces strength at that point to residual value; (b) redistribution of shear stresses from failure zone to surrounding area produces failure in surrounding zone (points B). Continued redistribution of stresses can eventually lead to failure of the entire slope (points C and beyond).

boundary conditions, groundwater, irregular geometries, and a variety of construction operations can all be considered in modern stress-deformation analyses. For static slope stability analysis, stress-deformation analyses offer the advantages of being able to identify the most likely mode of failure by predicting slope deformations up to (and in some cases beyond) the point of failure, of locating the most critically stressed zones within a slope, and of predicting the effects of slope failures. These advantages come at the cost of increased engineering time for problem formulation, characterization of material properties and interpretation of results, and increased computational effort.

The accuracy of stress-deformation analyses is strongly influenced by the accuracy with which the constitutive model represents actual material behavior. Many different constitutive models have been used for stress-deformation analysis of slopes; each has advantages and limitations. The accuracy of simple models is usually limited to certain ranges of strain and/or certain stress paths. Models that can be applied to more general stress and strain conditions are often quite complex and may require a large number of input parameters whose values can be difficult to determine. Historically, the hyperbolic model (Kondner, 1963; Duncan et al., 1980) has offered an appropriate compromise between simplicity and accuracy, although modern stress-deformation computer programs offer much more sophisticated models implemented in user-friendly computational platforms. As discussed in Section 10.9.3.2, mesh quality and representation of boundary conditions can also influence the accuracy of stress-deformation analyses.

Stress-deformation analyses are typically used for evaluation of static stability in one of two main ways. The first, as alluded to above, is based on the direct calculation of expected deformations using expected values of all soil parameters. The results of such analyses are interpreted in terms of the computed deformations relative to allowable deformations; allowable deformations are generally identified on a case-by-case basis with consideration of uncertainty in the computed deformations and the consequences of allowable deformation exceedance. The second approach, commonly referred to as the *strength reduction approach* (Zienkiewicz et al., 1975; Ugai and Leshchinsky, 1995; Dawson et al., 1999), involves performing stress-deformation analyses with all soil shear strengths divided by constant factors; the smallest value of the factor that leads to instability is taken as the factor of safety for the slope. Instability may be interpreted as an inability of the stress-deformation model to reach static equilibrium or as the exceedance of an allowable level of deformation.

10.9 SEISMIC SLOPE STABILITY ANALYSIS

The previously described procedures for static slope stability analysis have been used for many years and calibrated against many case histories of good and poor slope performance. The database against which seismic slope stability analyses can be calibrated is much smaller. Analysis of the seismic stability of slopes is further complicated by the need to consider the effects of (1) dynamic stresses induced by earthquake shaking, and (2) the effects of those stresses on the strength and stress-strain behavior of the slope materials. Seismic slope instabilities may be grouped into two categories based on which of these effects is predominant in a given slope. In *inertial instabilities,* the shear strength of the soil remains relatively constant, but slope deformations are produced by the accumulation of plastic strain and/or temporary exceedances of the strength by dynamic earthquake stresses. *Weakening instabilities* are those in which the earthquake serves to weaken the soil sufficiently that it cannot remain stable under earthquake-induced, or even static, shear stresses. Flow liquefaction and lateral spreading, both discussed in Chapter 9, and failures of sensitive clays are examples of weakening instability.

Earthquake motions can induce significant horizontal and vertical dynamic stresses in slopes. These stresses produce a complex pattern of dynamic normal and shear stresses along potential failure surfaces within a slope. When superimposed upon the previously existing static shear stresses, the dynamic shear stresses may exceed the available shear strength of the soil and produce inertial instability of the slope. Even if the combination of static and dynamic stresses does not reach the full shear strength of the soil, the static shear stress will cause the degree of nonlinearity of the soil to be greater in one direction than the other. As illustrated in Figure 6.42, this tendency will cause permanent strain to develop preferentially in one direction. Integrated over the volume of the slope, these permanent strains result in permanent deformations.

Several techniques for the analysis of inertial instability have been proposed. These techniques differ primarily in the accuracy with which the earthquake motion and the dynamic response of the slope are represented – the following sections describe several common approaches to the analysis of inertial instability. The first, pseudostatic analysis, produces a factor of safety against seismic slope failure in much the same way that static limit equilibrium analyses produce factors of safety against static slope failure. All the other approaches attempt to evaluate permanent slope displacements produced by earthquake shaking.

10.9.1 Pseudostatic Analysis

From as early as the 1920s, the seismic stability of earth structures has been analyzed by a pseudostatic approach in which the effects of an earthquake are represented by constant horizontal and/or vertical accelerations. The first explicit application of the pseudostatic approach to the analysis of seismic slope stability has been attributed to Terzaghi (1950).

In their most common form, pseudostatic analyses represent the effects of earthquake shaking by pseudostatic accelerations that produce horizontal and vertical inertial forces, F_h and F_v, which act through the centroid of the failure mass (Figure 10.18). The magnitudes of the horizontal and vertical pseudostatic forces are

$$F_h = \frac{a_h W}{g} = k_h W \tag{10.3a}$$

$$F_v = \frac{a_v W}{g} = k_v W \tag{10.3b}$$

where a_h and a_v are horizontal and vertical pseudostatic accelerations, k_h and k_v are dimensionless horizontal and vertical pseudostatic coefficients, and W is the weight of the failure mass. The magnitudes of the pseudostatic accelerations should be related to the severity of the anticipated ground motion; selection of pseudostatic accelerations for design, as discussed in the next section, is not a simple matter. Resolving the forces on the potential failure mass in a direction parallel to the failure surface shown in Figure 10.18, the pseudostatic factor of safety is computed as:

$$FS = \frac{\text{Resisting force}}{\text{Driving force}} = \frac{cl_{ab} + \left[(W - F_v)\cos\beta - F_h \sin\beta\right]\tan\phi}{(W - F_v)\sin\beta + F_h \cos\beta} \tag{10.4}$$

where c and ϕ are the Mohr-Coulomb strength parameters (Section 6.2.3) that describe the shear strength on the failure plane and l_{ab} is the length of the failure plane. The horizontal pseudostatic

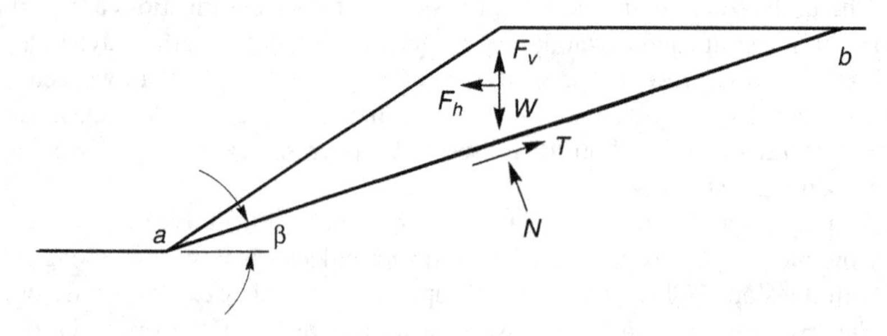

FIGURE 10.18 Forces acting on triangular wedge of soil above planar failure surface in pseudostatic slope stability analysis.

force clearly decreases the factor of safety – it reduces the resisting force (for $\phi>0$) and increases the driving force. The vertical pseudostatic force typically has less influence on the factor of safety since it reduces (or increases, depending on its direction) both the driving force and the resisting force – as a result, the effects of vertical accelerations are frequently neglected in pseudostatic analyses. The pseudostatic approach can be used to evaluate pseudostatic factors of safety for planar, circular, and noncircular failure surfaces. Many commercially available computer programs for limit equilibrium slope stability analysis have the option of performing pseudostatic analyses, including automated searching for critical (i.e., lowest FS) failure surfaces. It should be noted that such searches can be problematic in materials with uniform shear strength. Consider the case of a circular failure surface – the inertial driving force increases with the square of the radius while the resisting force increases in proportion to the radius, which drives the search toward very large radius values, and often failure surfaces so deep that the rigid block assumption is invalid.

Example 10.2

Assuming k_h=0.1 and k_v=0.0, compute the static and pseudostatic factors of safety for the 30-ft-high 2:1 (H:V) slope shown in Figure E10.2.

Solution:

Using a simple moment equilibrium analysis, the factor of safety can be defined as the ratio of the moment that resists the rotation of a potential failure mass about the center of a circular potential failure surface to the moment that is driving the rotation. The critical failure surface, defined as that which has the lowest factor of safety, is identified by analyzing a number of potential failure surfaces. Shown below are the factor of safety calculations for one potential failure surface.

Computation of the factor of safety requires evaluation of the overturning and resisting moments for both static and pseudostatic conditions. The overturning moment for static conditions results from the weight of the soil above the potential failure surface. The overturning moment for pseudostatic conditions is equal to the sum of the overturning moment for static conditions and the overturning moment produced by the pseudostatic forces. The horizontal pseudostatic forces are assumed to act in directions that produce positive (clockwise, in this case) driving moments. In the calculations shown in tabular form below, the soil above the potential failure surface is divided into two sections.

Overturning moment:

Section	Area (ft²)	γ (lb/ft³)	W (kips/ft)	Moment Arm (ft)	Static Moment (kip-ft/ft)	$k_h W$ (kips/ft)	Moment Arm (ft)	Pseudostatic Moment (kip-ft/ft)	Total Moment (kip-ft/ft)
A	1,360	110	149.6	30	4,488.1	15.0	38	570.0	5,058.0
B	2,300	125	287.5	5	1,437.5	28.8	62	1,785.6	3,223.1
					5,925.6				8,281.1

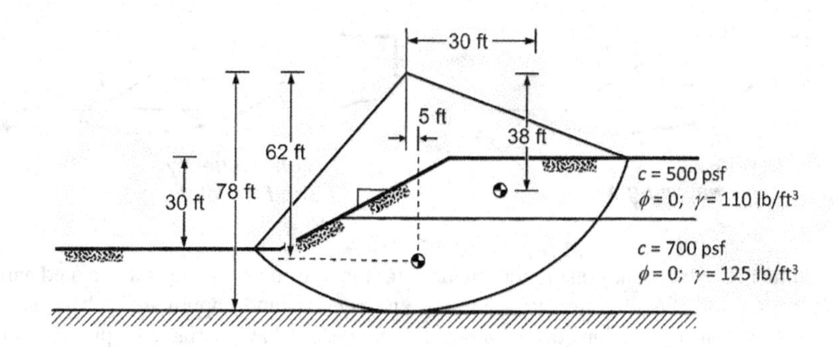

FIGURE E10.2

Resisting moment:

Section	Length (ft)	c (lb/ft²)	Force (kips/ft)	Moment Arm (ft)	Moment (kip-ft/ft)
A	11.5	500	5.75	78	448.5
B	129.3	700	90.5	78	7,059.8
					7,508.3

Factors of safety:

$$\text{Static FS} = \frac{\text{Resisting moment}}{\text{Static overturning moment}} = \frac{7{,}508.3 \text{ kip-ft/ft}}{5{,}925.5 \text{ kip-ft/ft}} = 1.27$$

$$\text{Pseudostatic FS} = \frac{\text{Resisting moment}}{\text{Static + pseudostatic overturning moment}} = \frac{7{,}508.3 \text{ kip-ft/ft}}{8{,}281.1 \text{ kip-ft/ft}} = 0.91$$

10.9.1.1 Nature of Inertial Forces

The notion of inertial forces, i.e., induced forces proportional to the acceleration of a body, was introduced in the derivations of the equations of motion for single-degree-of-freedom (Section B.3) and multiple-degree-of-freedom (Section B.10) systems. In both cases, the inertial forces acted on lumped masses that were assumed to be rigid. Similarly, the pseudostatic analysis described in the preceding section implicitly assumes that the soil above a failure surface is rigid. In a compliant material, however, accelerations are incoherent, i.e., they vary spatially. Reflections from the irregular ground surface, refractions at layer boundaries, and variable soil thicknesses produce complex wavefields that vary both vertically and horizontally throughout a slope. At any point in time, therefore, a complex pattern of variable accelerations may exist within a potentially unstable zone of soil.

For a potential failure mass that is shallow, comprised of stiff soils, and/or subjected to low-frequency motion, wavelengths will be long relative to its dimensions so accelerations throughout that mass will be nearly in phase, as shown in Figure 10.19a. Under such conditions, the rigid block assumption will be at least approximately satisfied. Accelerations in potential failure masses of slopes in deep and/or softer soils (and/or slopes subjected to higher frequency motion), however, have short wavelengths relative to failure mass dimensions so different locations within the failure mass will have different acceleration amplitudes and may be out of phase (Figure 10.19b). When this occurs, the inertial forces at different points within the potential failure mass will be different and may even act in opposite directions. Under these conditions, the resultant inertial force for the overall failure mass may be significantly smaller than that implied by the rigid block assumption. Accelerations within slopes with geometries such as those shown in Figure 10.19 will have additional incoherence (not illustrated in the figure) due to the differing thicknesses of the soil across the profile and inclined waves reflected from the face of the slope.

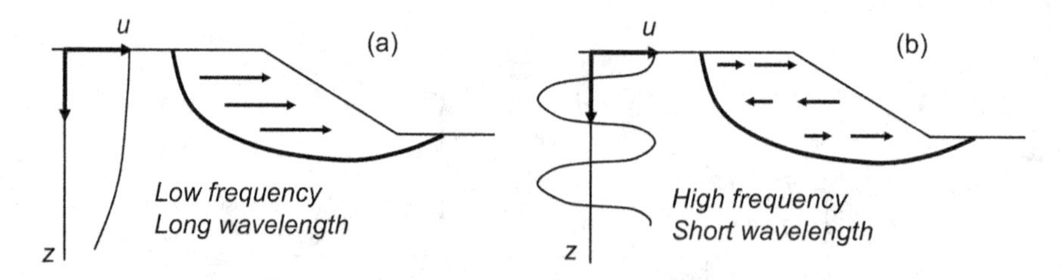

FIGURE 10.19 Influence of frequency on motions induced in slopes. Long wavelength associated with low-frequency motion (a) causes soil above failure surface to have nearly constant-amplitude, in-phase accelerations. For higher-frequency motion (b), portions of soil above failure surface have variable amplitude, out-of-phase accelerations. Note that amplitude and phase of accelerations within failure mass can also vary laterally.

10.9.1.2 Horizontal Equivalent Acceleration

The extent to which the rigid body assumption of the pseudostatic model is applicable can be examined using basic principles of wave propagation to provide insight into ground motion variations and the factors that affect the average acceleration. Consider the one-dimensional case of vertically propagating shear waves in a homogeneous, level soil deposit with shear wave velocity V_S and mass density ρ. For waves of displacement amplitude, A, at a circular frequency, ω, the displacement is given by

$$u(z,t) = 2A\cos kz \cdot e^{i\omega t} \tag{10.5}$$

where k is the wavenumber, i.e. $k = \omega/V_S$. The corresponding acceleration is

$$\ddot{u}(z,t) = -\omega^2 2A\cos kz \cdot e^{i\omega t} \tag{10.6}$$

and the shear stress is given by

$$\tau(z,t) = G\gamma(z,t) = G\frac{\partial u(z,t)}{\partial z} = -2GkA\sin kz \cdot e^{i\omega t} \tag{10.7}$$

Dividing the shear stress by the total vertical stress (assuming a constant density, ρ), the ratio of shear stress to total vertical stress at depth, z, is given by

$$\frac{\tau(z,t)}{\sigma_v} = \frac{2GkA\sin kz}{\rho g z}e^{i\omega t} = \frac{2V_s\omega A}{gz}\sin kz \cdot e^{i\omega t} \tag{10.8}$$

Now, the average acceleration of the soil above depth, z, can be computed as

$$\bar{a}(z,t) = \frac{1}{z}\int_0^z \ddot{u}(z,t)dz = -\frac{\omega^2 2Ae^{i\omega t}}{z}\int_0^z \cos kz = \frac{2\omega^2 A}{kz}\sin kz \cdot e^{i\omega t} \tag{10.9}$$

Recognizing that $k = \omega/V_s$, the right side of Equation (10.9) can be divided by the acceleration of gravity to express the average acceleration as a fraction of gravity,

$$\frac{\bar{a}(z,t)}{g} = \frac{2V_s\omega A}{gz}\sin kz \cdot e^{i\omega t} \tag{10.10}$$

Comparing Equations (10.10) and (10.8) shows that

$$\frac{\bar{a}(z,t)}{g} = \frac{\tau(z,t)}{\sigma_v} \tag{10.11}$$

i.e., the average acceleration of the soil above a particular depth, expressed as a fraction of gravity, is equal to the shear stress at that depth divided by the total vertical stress. This average acceleration is referred to as the *horizontal equivalent acceleration*, *HEA*(t) and can also be expressed in terms of a horizontal equivalent acceleration coefficient, $\bar{k}_h(t) = HEA(t)/g = \bar{a}(t)/g$. Note that the average acceleration above a particular depth differs from the actual acceleration at that depth due to the compliance of the overlying soil. The average acceleration accounts for the fact that the soil at different points above the depth of interest are moving by different amounts and out of phase with each other – some may even be moving to the left when others are moving to the right. For the simple case of harmonic loading, Equations (10.5), (10.7), and (10.10) can be manipulated to show that the ratio of *maximum horizontal equivalent acceleration*, *MHEA*, to peak ground surface acceleration, a_{max}, is given by

$$\frac{MHEA}{a_{max}} = \frac{|\sin(2\pi z/\lambda)|}{2\pi z/\lambda} \quad (10.12)$$

where λ is wavelength, i.e., $\lambda = 2\pi V_s/\omega$. Equation (10.12) indicates that the ratio is 1.0 at the ground surface, decreases with depth at a rate that increases with decreasing wavelength (i.e., increasing frequency and/or decreasing shear wave velocity), and approaches zero at very large depths (corresponding to many wavelengths). Figure 10.20 shows this variation as a function of depth/wavelength for a harmonic motion, with values equal to zero occurring between nodes of standing waves (Section 7.5.1.2). Because seismic waves include many frequencies, each of which has a different wavelength, the reduction in horizontal equivalent acceleration with actual depth will be more complicated than shown in Figure 10.20 for normalized depth but the trend of generally decreasing amplitude with depth illustrated in that figure still applies.

Equation (10.11) also shows that the average acceleration within a dynamically responding body can be obtained from the stresses acting on its boundaries. The same basic concepts used in the one-dimensional case here apply to the more realistic cases of two- and three-dimensional slopes. As previously described, waves can travel in many different directions in such cases leading to incoherent vertical and horizontal motions within a potentially unstable zone of soil. Approaches to handling the effects of such incoherence (Clough and Chopra, 1966; Seed and Martin, 1966) involve integrating the horizontal components of shear and normal stresses on a potential failure surface (Figure 10.21) and dividing the resulting horizontal force by the weight of the soil above the failure surface to obtain the horizontal equivalent acceleration coefficient, $\overline{k}_h(t) = HEA(t)/g$. The resultant inertial force acting on the unstable soil is the product of its mass and the horizontal equivalent acceleration (or of its weight and the horizontal equivalent acceleration coefficient).

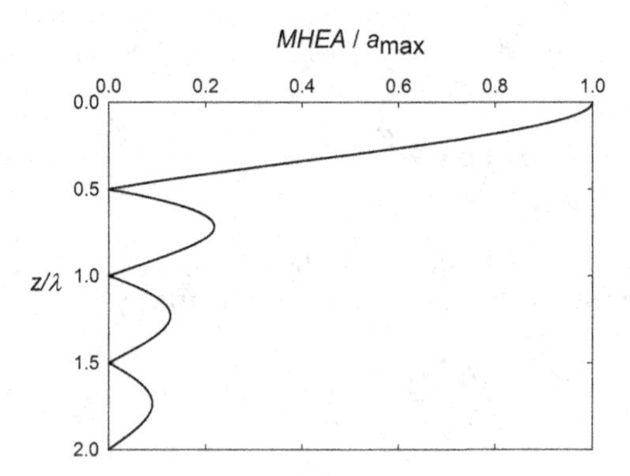

FIGURE 10.20 Reduction of horizontal equivalent acceleration amplitude with normalized depth for a single harmonic motion. Note that *MHEA* and a_{max} are defined as absolute values.

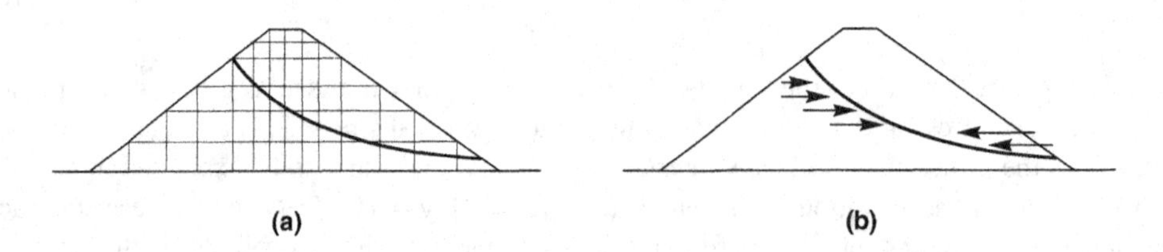

FIGURE 10.21 Evaluation of average acceleration for potential failure mass within an embankment. Two-dimensional numerical analysis predicts variations of shear and normal stresses on potential failure surface with time. Integration of horizontal components of stresses over failure surface gives resultant horizontal force acting within the potential failure mass. A time series of average acceleration is obtained by dividing resultant force by mass of potentially unstable soil.

10.9.1.3 Selection of Pseudostatic Coefficient

The results of pseudostatic analyses are critically dependent on the specified value of the seismic coefficient, k_h. Selection of an appropriate pseudostatic coefficient is the most important, and most difficult, aspect of a pseudostatic stability analysis. The seismic coefficient controls the pseudostatic force acting on the failure mass, so its value should be related to some measure of the amplitude of the inertial force induced in the potentially unstable material. In its original form, pseudostatic analyses were typically performed by applying the peak ground surface acceleration to a rigid failure mass. If the pseudostatic coefficient is based on the peak acceleration, i.e., $k_h = a_{max}/g$, a factor of safety of 1.0 would correspond to the point of incipient instability, and a factor of safety greater than 1.0 would imply no movement. A pseudostatic factor of safety less than 1.0 would imply failure with the unstable mass of soil accelerating in the downslope direction. However, the accelerations produced by earthquake shaking are transient rather than constant as assumed in the pseudostatic approach, and so the state of "failure" may not last for very long. As will be discussed in detail in Section 10.9.2, the result of a peak acceleration-based factor of safety temporarily falling below 1.0 will be a temporary acceleration of the unstable mass (relative to the material beneath it) and an increment of downslope movement.

As recognition grew that actual slopes are not rigid and that the peak acceleration exists for only an instant in time, the pseudostatic coefficients used in practice generally corresponded to acceleration values well below a_{max}. Terzaghi (1950) originally suggested the use of $k_h = 0.1$ for "severe" earthquakes (Rossi-Forel IX), $k_h = 0.2$ for "violent, destructive" earthquakes (Rossi-Forel X), and $k_h = 0.5$ for "catastrophic" earthquakes. Seed (1979) listed pseudostatic design criteria for 14 dams in ten seismically active countries; 12 required minimum factors of safety of 1.0–1.5 with pseudostatic coefficients of 0.10–0.12. Hynes-Griffin and Franklin (1984) applied the sliding block analysis method described in the following section to over 350 accelerograms and concluded that earth dams with pseudostatic factors of safety greater than 1.0 using $k_h = 0.5a_{max}$ would not develop "dangerously large" deformations, which they interpreted as being up to approximately 1 m.

As the preceding discussion indicates, there are no hard and fast rules for selection of a pseudostatic coefficient for design. In reality, the serviceability (or performance) of a slope is more closely related to the permanent deformations it undergoes than to the value of the factor of safety itself. A slope with a high pseudostatic factor of safety will generally displace less than a slope with a low pseudostatic factor of safety when subjected to the same earthquake motion. As will be discussed in Section 10.9.2.5, the relationship between pseudostatic factor of safety and permanent displacement can be used to identify a pseudostatic acceleration level that is consistent with a specified allowable level of permanent displacement. Such an approach represents the most appropriate use of the pseudostatic approach at this time.

10.9.1.4 Discussion

The pseudostatic approach has a number of attractive features. The analysis is relatively simple and straightforward; indeed, its similarity to the static limit equilibrium analyses routinely conducted by geotechnical engineers makes its computations easy to understand and perform. It produces a scalar index of stability (the factor of safety) that is analogous to that produced by static stability analyses.

However, representation of the complex, transient, dynamic effects of earthquake shaking by a single constant unidirectional pseudostatic acceleration is obviously quite crude. Even in its infancy, the limitations of the pseudostatic approach were clearly recognized. Terzaghi (1950) stated that "the concept it conveys of earthquake effects on slopes is very inaccurate, to say the least," and that a slope could be unstable even if the computed pseudostatic factor of safety was greater than 1. Indeed, pseudostatic analyses of a number of dams that failed in earthquakes (Table 10.4) show factors of safety well above 1.0. Detailed analyses of earthquake-induced landslides (e.g., Seed et al., 1969, 1975; Marcuson et al., 1979) have illustrated significant shortcomings of the pseudostatic

TABLE 10.4
Results of Pseudostatic Analyses of Earth Dams that Failed During Earthquakes

Dam	k_h	*FS*	Effect of Earthquake
Sheffield (CA)	0.10	1.2	Complete failure
Lower San Fernando (CA)	0.15	1.3	Upstream slope failure
Upper San Fernando (CA)	0.15	~2–2.5	Downstream shell, including crest, slipped about 6 ft downstream
Tailings (Japan)	0.20	~1.3	Failure of dam with release of tailings

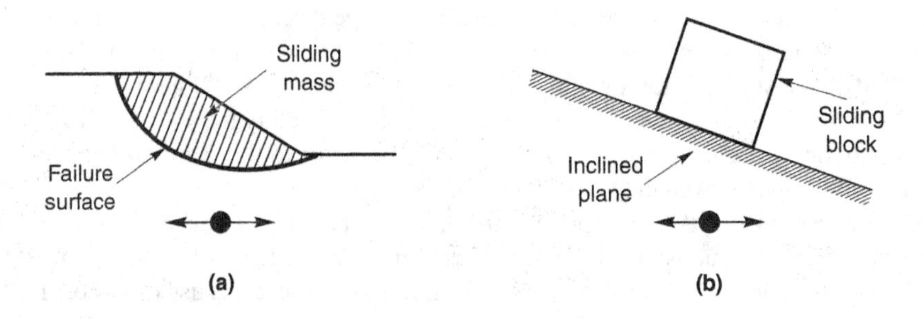

FIGURE 10.22 Analogy between (a) potential landslide and (b) block resting on inclined plane.

approach. Experience has clearly shown, for example, that pseudostatic analyses can be unreliable for soils that build up large pore pressures or show more than about 15% degradation of strength due to earthquake shaking.

Difficulty in the assignment of appropriate pseudostatic coefficients and in interpretation of pseudostatic factors of safety, coupled with the development of more realistic methods of analysis, have largely reduced the use of the pseudostatic approach for seismic slope stability analyses to screening applications. Methods based on evaluation of permanent slope deformation, such as those described in Section 10.9.2.5, are now commonly used for seismic slope stability analysis. Such deformation analyses can be used to identify pseudostatic coefficients that are consistent with specific levels of computed soil displacement. Using such coefficients in pseudostatic analyses provide factors of safety against exceeding desired allowable displacements. In this way, slopes that pass the screen are unlikely to exceed the allowable displacement.

10.9.2 Sliding Block Analysis

The pseudostatic method of analysis, like all limit equilibrium methods, provides an index of stability (the factor of safety) but no information on deformations associated with slope failure. Since the serviceability of a slope after an earthquake is controlled by deformations, analyses that predict slope displacements provide a more useful indication of seismic slope performance.

Earthquake-induced accelerations, and the inertial forces they produce, vary with time, so the factor of safety against slope failure will also vary over the duration of an earthquake. If the inertial forces acting on a potential failure mass become large enough that the total (static plus dynamic) driving forces exceed the available resisting forces, the factor of safety will drop below 1.0. When its factor of safety drops below 1.0, a potential failure mass is no longer in equilibrium; consequently, it will be accelerated by the unbalanced force. In 1963, Whitman laid out the principles by which a potentially unstable slope could be treated as a block resting on an inclined plane (Marcuson, 1994) subjected to horizontal shaking (Figure 10.22). Newmark (1965) further developed the sliding block model for prediction of the permanent displacement of a slope subjected to a specific ground motion.

Sliding block analyses allow calculation of permanent displacements for a very simple analog to a potentially unstable slope. The initial (pre-earthquake) stability of the slope is computed, the level of shaking required to initiate instability is determined, and permanent displacements are computed for specific ground motions. The required calculations are relatively simple, at least for the original, rigid block model. The limiting assumptions of the original, rigid block model have led to the development of more sophisticated sliding block models and procedures. The primary distinction between currently available sliding block models is related to the compliance, or flexibility, of the block that represents the soil above a potential failure surface. It should be noted that sliding block analyses predict deformations associated with shearing; additional deformations associated with volume change (Section 6.6.5.4) can occur and should be computed separately to determine the total deformation associated with earthquake shaking. Both components can be computed simultaneously in a numerical analysis with an appropriate constitutive model.

10.9.2.1 Rigid Block Analysis

The original sliding block model employed the same basic assumptions used in static and pseudo-static limit equilibrium analyses – the material above the failure surface was assumed to be rigid and the interface between the block and plane was assumed to exhibit rigid-perfectly plastic behavior.

Rigid Block Mechanics

Consider the rigid block in stable, static equilibrium on the inclined plane of Figure 10.23. Under static conditions, equilibrium of the block (in the direction parallel to the plane) requires that the static driving force, D_s, (Figure 10.23a) be less than the available static resisting force, R_s. Assuming that the block's resistance to sliding is purely frictional ($c=0$)

$$FS = \frac{\text{available resisting force}}{\text{static driving force}} = \frac{R_s}{D_s} = \frac{N_s \tan\phi}{W \sin\beta} = \frac{W \cos\beta \tan\phi}{W \sin\beta} = \frac{\tan\phi}{\tan\beta} \tag{10.13}$$

where W is the weight of the block, ϕ is the angle of friction between the block and the plane, and β is the inclination of the plane. Now consider the effect of inertial forces transmitted to the block by horizontal shaking of the inclined plane with acceleration, $a_h(t)=k_h(t)g$ (the effects of vertical accelerations will be neglected for simplicity). Because the block is rigid, the acceleration throughout it is constant. At a particular instant of time, horizontal acceleration of the block will induce a horizontal inertial force, k_hW (Figure 10.23b). When the inertial force acts in the downslope direction, resolving forces parallel to the inclined plane gives

$$FS_d(t) = \frac{\text{Available resisting force}}{\text{Pseudostatic driving force}} = \frac{R_d(t)}{D_d(t)} = \frac{[\cos\beta - k_h(t)\sin\beta]\tan\phi}{\sin\beta + k_h(t)\cos\beta} \tag{10.14}$$

where R_d and D_d are the dynamic resisting and driving forces, respectively.

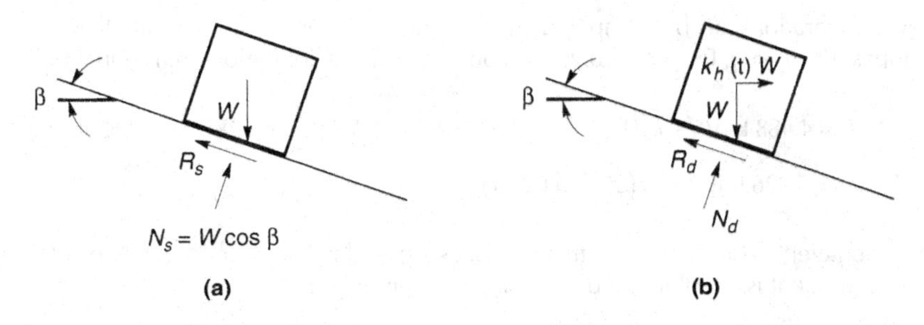

FIGURE 10.23 Forces acting on a block resting on an inclined plane: (a) static conditions; (b) dynamic conditions.

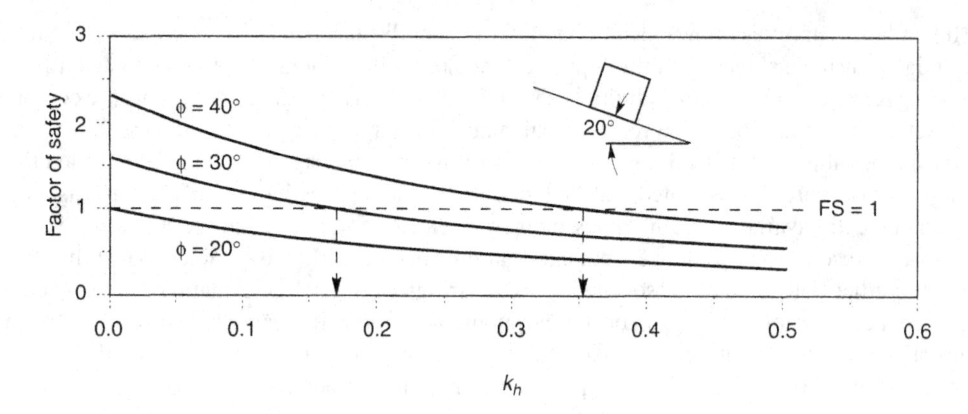

FIGURE 10.24 Variation of pseudostatic factor of safety with horizontal pseudostatic coefficient for block on plane inclined at 20°. For $\phi=20°$, block is at the point of failure ($FS=1$) under static conditions ($k_h=0$), so the yield coefficient is zero. For $\phi=30°$ and $\phi=40°$, yield coefficients are 0.17 and 0.36, respectively.

Obviously, the dynamic factor of safety decreases as k_h increases and there will be (for a statically stable block) some positive value of k_h that will produce a factor of safety of 1.0 (Figure 10.24). This coefficient, termed the *yield coefficient*, k_y, corresponds to the *yield acceleration*, $a_y=k_y g$. The yield acceleration is the minimum pseudostatic acceleration required to produce instability of the block. For the block of Figure 10.23, using Equation (10.14) with $FS_d=1.0$ gives the yield coefficient

$$k_y = \tan(\varphi - \beta) \tag{10.15}$$

for sliding in the downslope direction. For sliding in the uphill direction (which can occur when β and ϕ are small),

$$k_y = \frac{\tan\phi + \tan\beta}{1 + \tan\phi\tan\beta} \tag{10.16}$$

Example 10.3

Compute the yield acceleration for the slope described in Example 10.2.

Solution:

The yield acceleration can be computed by trial and error, or computed directly for relatively simple slopes. Reviewing Example 10.2, it is apparent that the total moment is equal to

$$\begin{aligned} M_t &= 4{,}488 \text{ k-ft/ft} + k_h(149.6 \text{ k/ft})(38 \text{ ft}) + 1{,}438 \text{ k-ft/ft} + k_h(287.5 \text{ k/ft})(62 \text{ ft}) \\ &= 5{,}926 \text{ k-ft/ft} + k_h(23{,}510 \text{ k-ft/ft}) \end{aligned}$$

The yield coefficient is the value of k_h that produces a pseudostatic factor of safety of 1.0. Because the resisting moment is equal to the overturning moment when $FS=1.0$,

$$5{,}926 \text{ k-ft/ft} + k_h(23{,}510 \text{ k-ft/ft}) = 10{,}624 \text{ k-ft/ft}$$

or

$$k_h = \frac{7{,}508 \text{ k-ft/ft} - 5{,}926 \text{ k-ft/ft}}{23{,}510 \text{ k-ft/ft}} = 0.067$$

Therefore, the yield acceleration is 0.067g.

When a block on an inclined plane is subjected to a pulse of acceleration that is lower than the yield acceleration, the block and plane move together with no relative displacement. If the acceleration of the plane exceeds the yield acceleration, the block cannot move as quickly as the plane and therefore moves relative to the plane – the plane is, in a sense, pulled out from under the block and the block moves to a lower position on the plane. To illustrate the procedure by which the resulting permanent displacements can be calculated, consider the case in which an inclined plane is subjected to a base acceleration history, $a_b(t)$, consisting of a single rectangular acceleration pulse of amplitude, A, and duration, Δt. If the yield acceleration, a_y, is less than A (Figure 10.25a), the acceleration of the block relative to the plane during the period from t_0 to $t_0 + \Delta t$ is

$$a_{rel}(t) = a_b(t) - a_y = A - a_y \qquad t_0 \le t \le t_0 + \Delta t \tag{10.17}$$

The relative movement of the block during this period can be obtained by integrating the relative acceleration twice, i.e.,

$$v_{rel}(t) = \int_{t_0}^{t} a_{rel}(t)\,dt = (A - a_y)(t - t_0) \qquad t_0 \le t \le t_0 + \Delta t \tag{10.18}$$

$$d_{rel}(t) = \int_{t_0}^{t} v_{rel}(t)\,dt = \frac{1}{2}(A - a_y)(t - t_0)^2 \qquad t_0 \le t \le t_0 + \Delta t \tag{10.19}$$

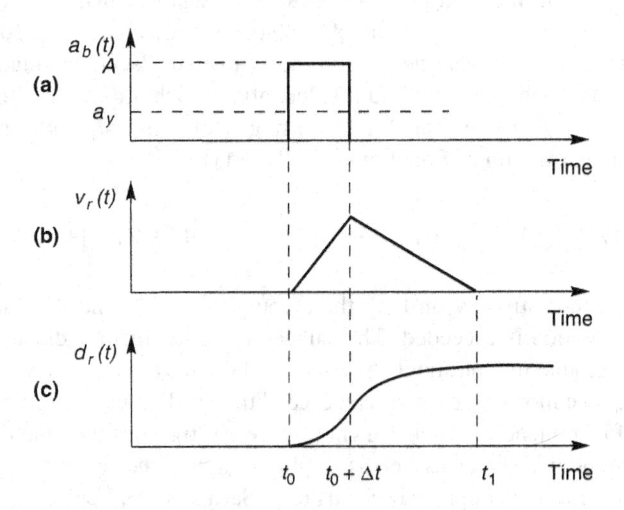

FIGURE 10.25 Variation of relative velocity and relative displacement between sliding block and plane due to rectangular pulse that exceeds yield acceleration between $t = t_0$ and $t = t_0 + \Delta t$.

At $t=t_o+\Delta t$, the relative velocity reaches its maximum value. At that time,

$$v_{rel}(t_0+\Delta t)=(A-a_y)\Delta t \tag{10.20}$$

$$d_{rel}(t_0+\Delta t)=\frac{1}{2}(A-a_y)\Delta t^2 \tag{10.21}$$

After the base acceleration, $a_b(t)$, drops below a_y (in this case to zero at $t=t_0+\Delta t$), the block does not stop sliding but rather the sliding block is decelerated by the friction force acting on its base. The block will continue to slide on the plane, but at a decreasing relative velocity that eventually reaches zero. The acceleration during this time is given by

$$a_{rel}(t)=a_b(t)-a_y=0-a_y=-a_y \qquad t_0+\Delta t\le t_1 \tag{10.22}$$

where t_1 is the time at which the relative velocity becomes zero (note that the block undergoes negative acceleration, or deceleration, during this period). Between $t_0+\Delta t$ and t_1, the relative velocity will decrease with time according to

$$v_{rel}(t)=v_{rel}(t+\Delta t)+\int_{t_0+\Delta t}^{t}a_{rel}(t)dt=A\Delta t-a_y(t-t_0) \qquad t_0+\Delta t\le t_1 \tag{10.23}$$

Setting the relative velocity equal to zero at $t=t_1$ gives

$$t_1=t_0+\frac{A}{a_y}\Delta t \tag{10.24}$$

Then, for $t_0+\Delta t\le t\le t_1$

$$d_{rel}(t)=\int_{t_0+\Delta t}^{t}v_{rel}(t)dt=A\Delta t(t-t_0-\Delta t)-\frac{1}{2}a_y\left[t^2-(t_0+\Delta t)^2\right]+a_yt_0(t-t_0-\Delta t) \tag{10.25}$$

After time t_1, the block and inclined plane move together again. During the total period of time between $t=t_0$ and $t=t_1$, the relative movement of the block is shown in Figure 10.25. Between t_0 and $t_0+\Delta t$, the relative velocity increases linearly and the relative displacement quadratically. At $t_0+\Delta t$, the relative velocity has reached its maximum value, after which it decreases linearly. The relative displacement continues to increase (but at a decreasing rate) until $t=t_1$. Note that the total relative displacement is given by the sum of Equations (10.19) and (10.25),

$$d_{rel}(t)=\frac{1}{2}(A-a_y)\Delta t^2+A\Delta t(t-t_0-\Delta t)-\frac{1}{2}a_y\left[t^2-(t_0+\Delta t)^2\right]+a_yt_0(t-t_0-\Delta t) \tag{10.26}$$

This displacement depends strongly on both the amount by which, and the length of time during which, the yield acceleration is exceeded. This suggests that the relative displacement caused by a single pulse of strong ground motion should be related to both the amplitude and frequency content of that pulse. An earthquake motion, however, can exceed the yield acceleration a number of times and produce a number of increments of displacement (Figure 10.26). Thus the total displacement will be influenced by strong-motion duration as well as amplitude and frequency content. Indeed, application of this approach to a variety of simple waveforms (e.g., Sarma, 1975; Yegian et al., 1991) has shown that the permanent displacement of a sliding block subjected to rectangular, sinusoidal, and triangular periodic base motions is proportional to the number of pulses and the square of their period.

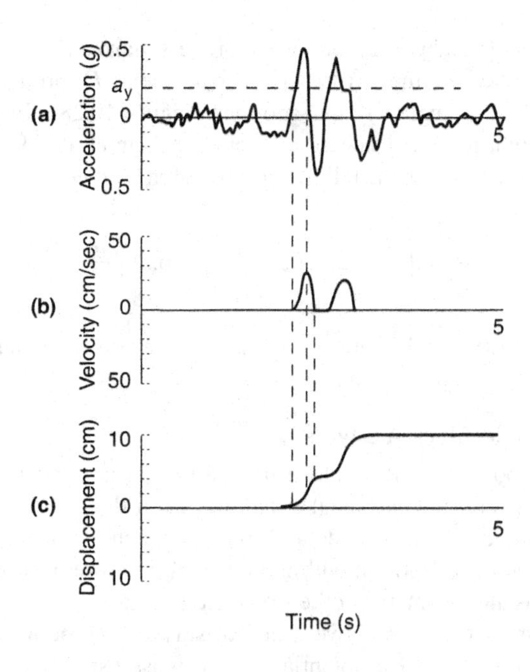

FIGURE 10.26 Development of permanent slope displacements for actual earthquake ground motion. (After Wilson and Keefer, 1985).

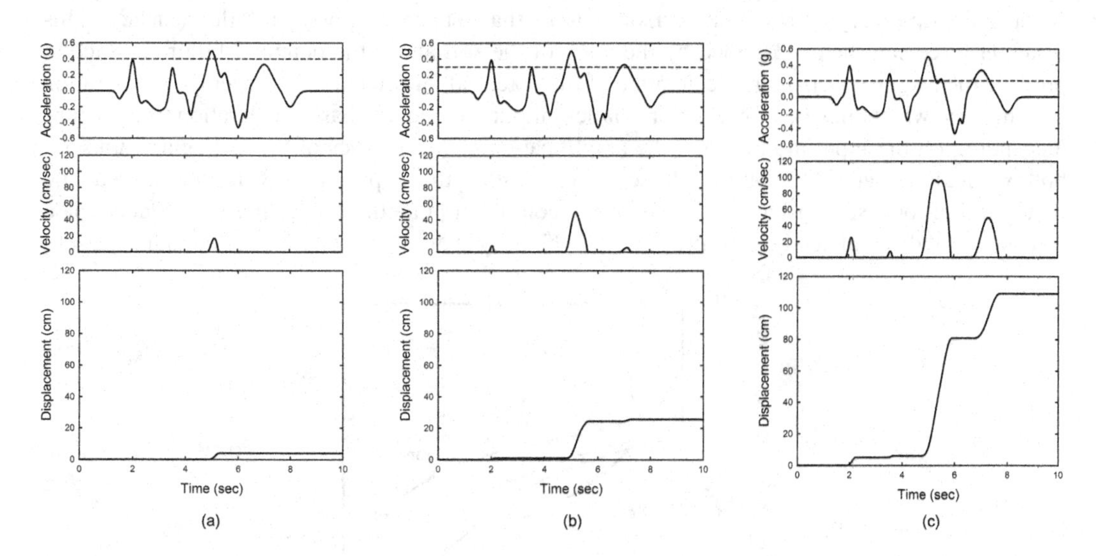

FIGURE 10.27 Illustration of influence of yield acceleration on sliding block displacements: (a) a_y=0.4g, (b) a_y=0.3g, (c) a_y=0.2g.

Rigid Block Behavior

Obviously, the sliding block model will predict zero permanent slope displacement if earthquake-induced accelerations never exceed the yield acceleration $\left(a_y/a_{\max}\right)$. Since the permanent displacement is obtained by double integration of the excess acceleration, the computed displacements for a slope with a relatively low yield acceleration (small $a_y/a_{\max}$) will be greater than that of a slope with a higher yield acceleration. Figure 10.27 illustrates how the computed displacements increase quickly with decreasing yield acceleration. The relationship between slope displacement and $a_y/a_{\max}$ has been investigated by a number of researchers. Sarma (1975) and Yegian et al. (1988) derived closed-form solutions for the permanent displacement, D, produced by simple periodic (triangular, sinusoidal, and rectangular) input motions (Figure 10.28). To allow measures of frequency content and duration to be considered explicitly, Yegian et al. (1991) used the database of Franklin and Chang (1977) to develop the following expression for median normalized displacement:

$$\log D_n = \log\left(\frac{D}{a_{\max} N_{eq} T^2}\right) = 0.22 - 10.12\frac{a_y}{a_{\max}} + 16.38\left(\frac{a_y}{a_{\max}}\right)^2 - 11.48\left(\frac{a_y}{a_{\max}}\right)^3 \quad (10.27)$$

with $\sigma_{\log D_n} = 0.45\left(\sigma_{\ln D_n} = 1.04\right)$. In Equation (10.27), N_{eq} is an equivalent number of cycles and T is the predominant period of the input motion.

10.9.2.2 Decoupled Rigid Block Analyses

Of course, the material above the failure plane in a soil slope is never truly rigid. Actual slopes are compliant – they deform during earthquake shaking. Their dynamic response depends on their geometry and stiffness and on the amplitude and frequency content of the motion of the underlying ground. The amplitude of the motion within a particular failure mass can be amplified or de-amplified by the materials and geometry of the slope itself.

The dynamic response of the soils above a failure surface can be accounted for and used to compute the average acceleration of a potential failure mass using dynamic stress-deformation analyses (Chopra, 1966). Using a dynamic analysis, the time-varying horizontal components of the dynamic stresses acting on a potential failure surface can be integrated over the failure surface to produce the time-varying resultant horizontal force that acts on the potential failure surface. This resultant force can then be divided by the mass of the soil above the potential failure surface to produce the average horizontal acceleration of the potential failure mass. The average acceleration time history, which may be of greater or smaller amplitude than the base acceleration time history (depending on the input motions and the amplification characteristics of the slope), accounts for both vertical and lateral variability of acceleration within the slope and provides a more realistic input motion for a sliding block analysis of the compliant potential failure mass. In a decoupled analysis (Makdisi and Seed, 1978; Bray et al., 1995; Bray et al., 1998, Rathje and Antonakos, 2011),

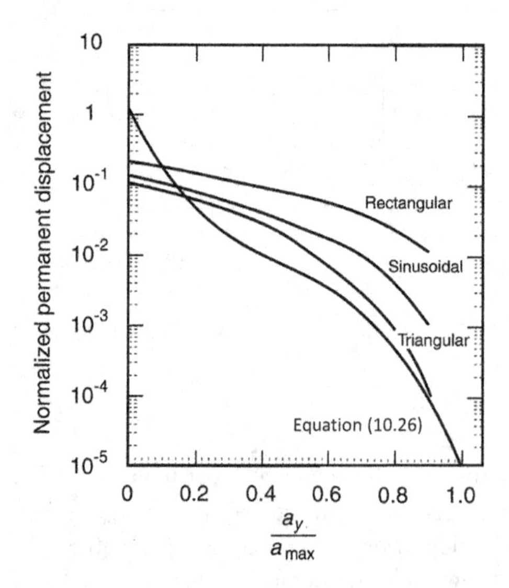

FIGURE 10.28 Variation of normalized permanent displacement with ratio of yield acceleration to maximum acceleration for simple waveforms. The normalized permanent displacement is defined in Equation (10.27). (After Yegian et al., 1991.)

the average acceleration time history computed in a dynamic analysis is used as the input motion in a rigid sliding block analysis.

The decoupled procedure can be viewed as a two-step process in which the sliding displacement analysis is decoupled from the dynamic response analysis. In the first step, a dynamic analysis of the compliant slope is performed, and its results are used to compute time histories of average acceleration corresponding to one or more potential failure surfaces passing through the slope. The slope is modeled as a continuum – no discrete failure surfaces are provided; in fact the analyses are usually performed using equivalent linear procedures so that no failure (or permanent deformation) is allowed to occur. These analyses typically idealize the slope as a soil column (one-dimensional) or as plane-strain (two-dimensional). It should be recognized that the average acceleration time histories for different potential failure surfaces will be different. In the second step, a conventional rigid sliding block analysis is performed with the average acceleration from the first step used as the input motion. Multiple potential failure surfaces should be examined in order to find the one that maximizes the slope displacement; the critical static surface will not necessarily produce the largest slope displacement under seismic loading conditions.

10.9.2.3 Coupled Compliant Block Analysis

The two-step decoupled procedure can account for the dynamic response of the soil above the failure surface but implicitly assumes that that response is unaffected by any slip that occurs on the failure surface. The validity of that assumption was investigated in the context of earth dam stability by Lin and Whitman (1986) and Gazetas and Uddin (1994). Lin and Whitman (1986) concluded that the decoupled procedure was somewhat conservative (i.e., predicted higher displacements) at frequencies above the fundamental frequency of the dam, and most conservative at and near the fundamental frequency. Gazetas and Uddin (1994) found generally good agreement between coupled and decoupled displacements, but found the decoupled procedure to produce conservative results for narrow-band motions that coincide with the natural frequency of the dam.

A compliant slope idealized as a one-dimensional lumped mass system (Kramer and Smith, 1997) to provide a first-order approximation of coupled sliding behavior showed that decoupled procedures overpredicted displacements of thin and/or stiff failure masses and underpredicted them for thick and/or soft ones. Normalizing the fundamental period of the failure mass, T_s, by the mean period of the input motion, T_m (Section 3.3.2.2), a more refined generalized SDOF model (Rathje and Bray, 1999) showed that the decoupled procedure was conservative (overpredicted displacements) at $a_y/a_{max}<0.6$ and slightly unconservative at higher a_y/a_{max} values for $T_s/T_m=1.0$. At $T_s/T_m=4.0$ (i.e., for relatively thick and/or soft failure masses), decoupled analyses were unconservative at all a_y/a_{max} ratios. Similarly, rigid block analyses were shown to underpredict permanent displacements at $T_s/T_m=1.0$ and overpredict them at $T_s/T_m=4.0$. The general behavior of coupled and decoupled models relative to rigid block behavior is illustrated in Figure 10.29.

Rathje and Bray (2000) used a one-dimensional nonlinear site response program to consider the effects of material nonlinearity on coupled sliding behavior and found results that were generally similar to those of Kramer and Smith (1997) and Rathje and Bray (1999); decoupled analyses were found to be potentially unconservative for systems with larger T_s/T_m ratios and a_y/a_{max} values greater than 0.4. As one-dimensional procedures, these analyses do not account for lateral incoherency of motions within the failure mass.

10.9.2.4 Sliding Block-Based Displacement Predictions

Sliding block analyses are not particularly difficult to perform in practice. Jibson et al. (2013) developed a program that can compute sliding block displacements for user-specified ground motions and yield accelerations. A number of researchers have performed and compiled the results of numerous sliding block analyses and used the results to develop predictive equations

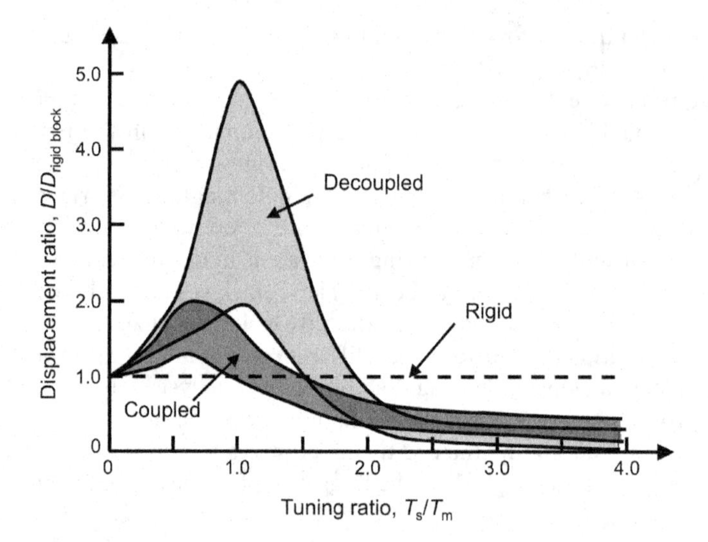

FIGURE 10.29 Comparison of approximate ranges of displacement computed by coupled and decoupled sliding block analyses to rigid sliding block displacements (after Wartman et al., 2003 with permission of ASCE).

for sliding block displacements. These equations relieve the user of having to perform the actual sliding block analyses, but produce results that are representative of the suite of ground motions used to develop the equations, which may or may not be representative of the ground motion hazards at a particular site of interest. The different predictive relationships make different assumptions about the nature of the potentially unstable material (e.g., rigid or compliant) and the manner in which the level of shaking is characterized (e.g., by various intensity measures and in terms of peak base or horizontal equivalent acceleration). The following subsections break these predictive relationships into rigid block and compliant block categories and provide examples of relationships based on scalar *IM*s, mixed scalar *IM*s, and vector *IM*s within each category.

Rigid Block Displacements

A number of investigators have used rigid block analyses to develop predictive relationships for sliding block displacements. For rigid blocks, horizontal equivalent accelerations are equivalent to base motions. These studies have been performed with increasingly broad ground motion databases and interpreted with increasingly sophisticated statistical analyses. The most current procedures are based on thousands of ground motions and allow estimation of probability distributions of predicted displacements. The procedures use different *IM*s, which differ in the extent to which they reflect amplitude, frequency content, and duration, to characterize the ground motions. Some models characterize ground motions with a single (scalar) *IM*, but others use multiple (vector) *IM*s to provide a more complete characterization of the motion. The models, therefore, differ in their complexity and in the dispersion of their predictions.

Scalar IMs The earliest scalar methods predicted permanent displacement as a function of peak acceleration, generally through the yield acceleration ratio, a_y/a_{max}. Ambraseys and Menu (1988) used recorded ground motions to show that plots of displacement vs. a_y/a_{max} showed shapes that were similar to those of the sinusoidal and triangular waves at a_y/a_{max} values greater than about 0.5. Ambraseys and Menu (1988) proposed that displacements could be predicted by an equation of the form

$$\log D = a + \log\left[\left(1 - \frac{a_y}{a_{max}}\right)^b \left(\frac{a_y}{a_{max}}\right)^c\right] \tag{10.28}$$

for $0.1 \le a_y/a_{max}$, $6.6 \le M_s \le 7.3$, with D in cm and a_y computed using residual soil strength. Using 50 motions from 11 earthquakes, Ambraseys and Menu (1988) found that median displacements were best predicted with $a=0.90$, $b=2.53$, and $c=-1.09$; the displacements were predicted with a variability of $\sigma_{\log D}=0.30$ ($\sigma_{\ln D}=0.69$). Using many more motions (2,270 recorded motions from 30 earthquakes) covering a broader range of earthquake magnitudes, Jibson (2007) found that Equation (10.28) matched the computed sliding block displacements best with $a=0.215$, $b=2.341$, and $c=-1.438$, giving $\sigma_{\log D}=0.510$ ($\sigma_{\ln D}=1.17$). Although the uncertainty in Jibson's calibration is higher, its median displacements are about one-third of those of Ambraseys and Menu (1988).

These $\sigma_{\ln D}$ values, which represent the fit of the predictive equation to the sliding block model displacements (not to actual observed slope displacements) are extremely high. The additional epistemic uncertainty of interest, i.e., the model uncertainty in actual slope displacement given a computed sliding block displacement, is not provided by these relationships.

Mixed Scalar IMs More recent investigations have found that improved predictions can be made by including the source parameter, moment magnitude, in addition to peak acceleration, which results in a "mixed" (ground motion *IM* and source parameter) measure of ground motion intensity. Jibson (2007) found that the dispersion of his scalar model could be reduced by including magnitude as

$$\log D = -2.710 + \log\left[\left(\left(1 - \frac{a_y}{a_{max}}\right)^{2.335}\left(\frac{a_y}{a_{max}}^{-1.478}\right)\right) + 0.424\mathbf{M}\right] \tag{10.29}$$

where D is in cm. For this relationship, $\sigma_{\log D}=0.454$ ($\sigma_{\ln D}=1.045$). The use of magnitude in the relationship introduces information on the frequency content and duration of the ground motion, both of which are relevant to sliding block displacements. As magnitude increases, seismic waves have longer periods and longer durations, both of which increase displacement for a given a_y/a_{max}.

Saygili and Rathje (2008) developed a model based on peak ground acceleration and earthquake magnitude with which permanent displacement could be computed as

$$\ln D = 4.89 - 4.85\left(\frac{a_y}{a_{max}}\right) - 19.64\left(\frac{a_y}{a_{max}}\right)^2 + 42.49\left(\frac{a_y}{a_{max}}\right)^3 - 29.06\left(\frac{a_y}{a_{max}}\right)^4 + 0.72\ln a_{max}/g + 0.89(\mathbf{M} - 6) \tag{10.30}$$

with

$$\sigma_{\ln D} = 0.732 + 0.789\left(\frac{a_y}{a_{max}}\right) - 0.530\left(\frac{a_y}{a_{max}}\right)^2$$

where D is in cm, and the yield acceleration, a_y, and peak acceleration, a_{max}, are in the same units. Figure 10.30 shows the variation of median displacement with yield acceleration ratio for different earthquake magnitudes. Noting the logarithmic nature of the displacement scale, the displacement can be seen to increase quickly as the peak acceleration exceeds the yield acceleration and to increase with increasing magnitude and, to a lesser degree, peak acceleration.

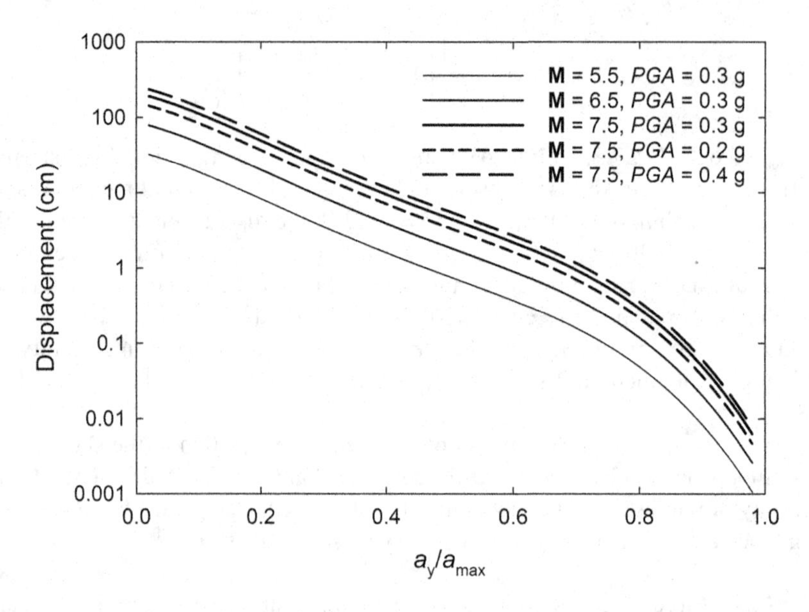

FIGURE 10.30 Variation of predicted sliding block displacement with yield acceleration ratio for different earthquake magnitudes and peak ground accelerations using Saygili and Rathje (2008) model.

Example 10.4

For the slope described in Example 10.3, estimate the 50th and 84th percentile permanent displacement when subjected to the Alhambra-Fremont School E-W record (see Figure 3.14) from the 1994 Northridge earthquake. Use the procedures of Jibson (Equation 10.29) and Saygili and Rathje (Equation 10.30).

Solution:

From Figure 3.14, the peak acceleration of the Alhambra-Fremont School motion was 0.10*g*. Therefore, the normalized yield acceleration, a_y/a_{max}=0.067*g*/0.1g=0.67. Using Equation (10.29), the Jibson (2007) 50th and 84th percentile displacements are

$$\log D_{50} = -2.710 + \log\left[(1-0.67)^{2.335}(0.67)^{-1.478}\right] + (0.424)(6.7)$$

$$= -0.7364 \rightarrow D_{50} = 10^{0.7364} = 0.18 \text{ cm}$$

$$\log D_{84} = \log D_{50} + \sigma_{\log D} = -0.7364 + 0.454 = -0.2824 \quad \rightarrow \quad D_{84} = 10^{-0.2824} = 0.52 \text{ cm}$$

Using Equation (10.30), the Saygili and Rathje (2008) permanent displacements are

$$\ln D_{50} = 4.89 - 4.85(0.67) - 19.64(0.67)^2 + 42.49(0.67)^3 - 29.06(0.67)^4$$

$$+ 0.72\ln(0.1) + 0.89(6.7-6) = -1.287 \rightarrow D_{50} = e^{-1.287} = 0.28 \text{ cm}$$

$$\sigma_{\ln D} = 0.732 + 0.789(0.67) - 0.530(0.67)^2 = 1.023$$

$$\ln D_{84} = \ln D_{50} + \sigma_{\ln D} = -1.287 + 1.023 = -0.264 \quad \rightarrow \quad D_{84} = e^{-0.264} = 0.77 \text{ cm}$$

Vector IMs Vector methods are based on the assumption that the use of additional ground motion information will lead to more accurate displacement predictions. Since sliding block displacements have been shown to be influenced by amplitude, frequency content, and duration, *IM*s that reflect those characteristics or, for example, reflect amplitudes over a range of different frequencies, should provide more accurate estimates of displacement.

Rathje and Saygili (2009) added peak ground velocity, *PGV*, to peak ground acceleration (and eliminating **M**) and produced a model of the form

$$\ln D = -1.56 - 4.58\left(\frac{a_y}{a_{max}}\right) - 20.84\left(\frac{a_y}{a_{max}}\right)^2 + 44.75\left(\frac{a_y}{a_{max}}\right)^3 - 30.5\left(\frac{a_y}{a_{max}}\right)^4 - 0.64\ln a_{max} + 1.55\ln PGV \quad (10.31)$$

$$\sigma_{\ln D} = 0.41 + 0.52\left(\frac{a_y}{a_{max}}\right)$$

where D is in cm, a_y and a_{max} are expressed as fractions of gravity, and *PGV* is in cm/sec. By including *PGV*, this model indirectly incorporates effects of frequency content, but does not incorporate duration. Median displacements from the *PGA-PGV* model are shown in Figure 10.31. The dispersion of the *PGA-PGV* model varies with a_y/a_{max} but is much lower than that of the *PGA*-**M** model, indicating a significantly greater efficiency in displacement prediction. As will be discussed in Section 10.12, the vector model leads to some complications in the prediction of ground motion hazards, but the improved efficiency can lead to substantial benefits in a performance-based slope displacement analysis.

Compliant Block Displacements

Compliant block models of both decoupled and coupled forms have also been used with large ground motion databases to compile predictive relationships for slope deformations.

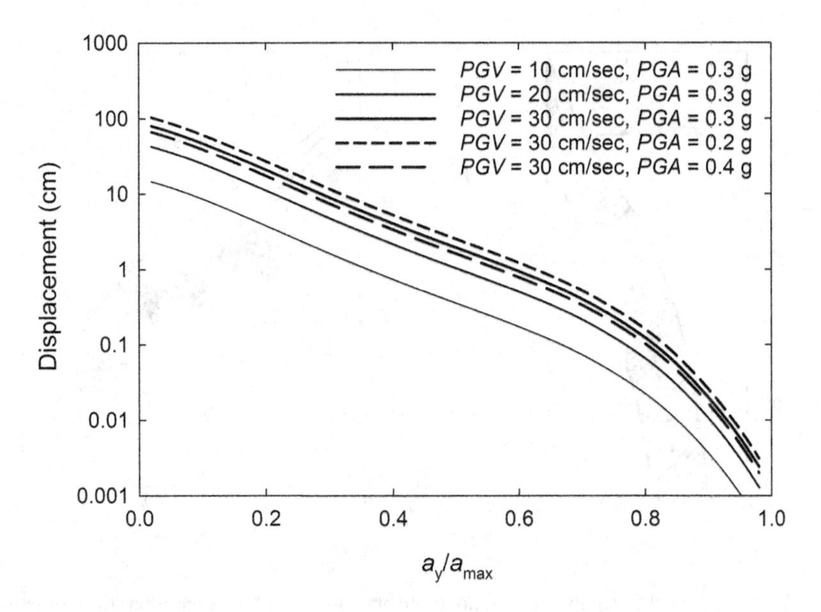

FIGURE 10.31 Variation of predicted sliding block displacement with yield acceleration ratio for different earthquake magnitudes and peak ground accelerations using Saygili and Rathje (2009) model.

Scalar IMs Makdisi and Seed (1978) used a decoupled procedure with average accelerations computed by the procedure of Chopra (1966) and sliding block analyses to compute earthquake-induced permanent deformations of earth dams and embankments. By using average accelerations instead of base accelerations, this approach directly accounts for the dynamic response of the compliant material within the failure mass.

By subjecting several real and hypothetical dams to several actual and synthetic ground motions scaled to represent different earthquake magnitudes, Makdisi and Seed computed the variation of permanent displacement with $a_y/\bar{a}_{\max}$, T_s, and magnitude. The use of magnitude accounts for the effects of duration and, albeit to a lesser degree, frequency content. Prediction of permanent displacements by the Makdisi-Seed procedure is accomplished with the charts shown in Figure 10.32. The bands shown in Figure 10.32b suggest a significant degree of variability in the predicted normalized displacements although the actual level of variability was not formally evaluated. The Makdisi-Seed procedure has been a cornerstone of practice in seismic slope stability evaluation since its introduction some 40 years ago but was based on what would now be considered an extremely small number of ground motions; newer predictive relationships are based on many more motions from many more earthquakes and provide much more robust predictions of slope displacement with quantified characterization of variability.

The effects of frequency content and duration can be accounted for more explicitly than by using magnitude as in Figure 10.32. Bray and Rathje (1998) accounted for both amplitude and frequency content through the use of a maximum equivalent horizontal acceleration coefficient, $k_{\max}=MHEA/g$, and duration through the significant duration, $D_{5\text{-}95}$ (Section 3.3.3).

$$\log\left(\frac{D}{k_{\max}D_{5-95}}\right)=1.87-3.477\frac{k_y}{k_{\max}} \tag{10.32}$$

where D=displacement in cm, k_y=yield coefficient, and the approximate (common log) standard deviation of the normalized displacement is 0.35 (Bray, 2007). The Bray and Rathje (1998) procedure (Figure 10.33) requires more effort than that of Makdisi and Seed (1978) but is based on many more ground motions and accounts for ground motions and dynamic response in a more comprehensive manner.

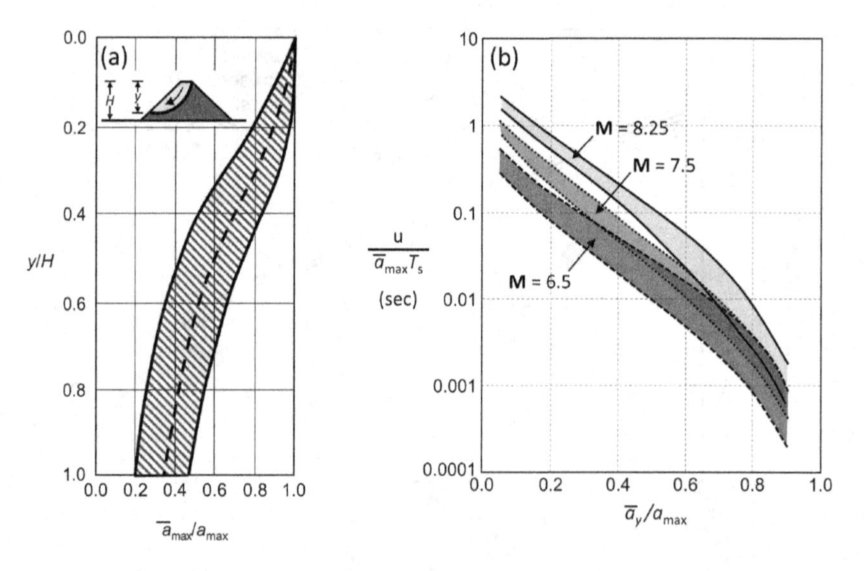

FIGURE 10.32 Variation of normalized permanent displacement with yield acceleration for earthquakes of different magnitudes: (a) summary for several earthquakes and dams/embankments; (b) average values. (After Makdisi and Seed, 1978 with permission of ASCE.)

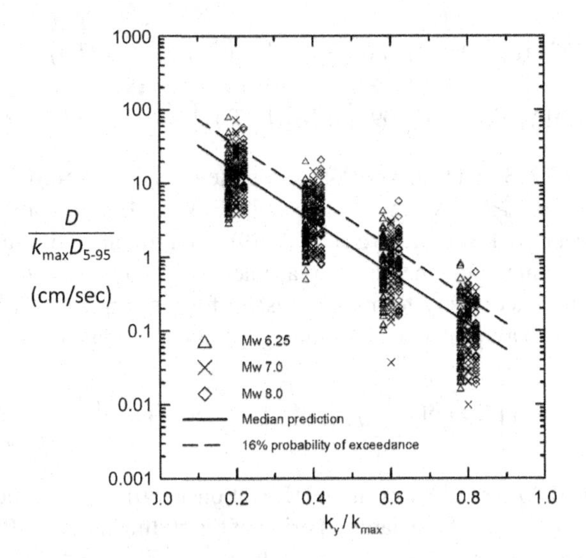

FIGURE 10.33 Variation of normalized displacement with k_y/k_{max}. (After Bray and Rathje, 1998 with permission of ASCE.)

Mixed Scalar IMs Bray and Travasarou (2007) found that compliant sliding block displacements for ground motions from shallow crustal earthquakes along active plate margins were better correlated to spectral accelerations at a lengthened period (that reflected soil stiffness degradation) than to peak acceleration. To account for softening due to soil nonlinearity, Bray and Travasarou (2007) identified a spectral acceleration at a multiple of the initial (low-strain) fundamental period of the slope, as an efficient intensity measure for sliding block displacement, and used it to develop a two-part model that treats the predicted displacement as a mixed (discrete-continuous) random variable with a probability density function

$$f_D(d) = \tilde{p}\delta(d - d_0) + (1 - \tilde{p})\bar{f}_D(d) \tag{10.33}$$

where $\tilde{p}$=discrete probability mass that the displacement is less than some value, d_0, considered to be negligible, $\delta(d-d_0)$=Dirac delta function (equal to 1.0 for $d \le d_0$ and zero elsewhere), and $\bar{f}_D(d)$=probability density function for $d>d_0$. Using a larger, updated crustal ground motion database, Bray and Macedo (2019) found that negligible predicted displacements (i.e., less than 0.5 cm – d_0<0.5 cm), the discrete probability mass can be computed as

$$\tilde{p} = \begin{cases} 1-\Phi\begin{bmatrix} -2.48 - 2.97\ln k_y - 0.12(\ln k_y)^2 - 0.72T_s \ln k_y \\ +1.70T_s + 2.78\ln[S_a(1.3T_s)] \end{bmatrix} & \text{for} \quad T_s \le 0.7\,\text{sec} \\ 1-\Phi\begin{bmatrix} -3.42 - 4.93\ln k_y - 0.30(\ln k_y)^2 - 0.35T_s \ln k_y \\ -0.62T_s + 2.86\ln[S_a(1.3T_s)] \end{bmatrix} & \text{for} \quad T_s > 0.7\,\text{sec} \end{cases} \tag{10.34}$$

where $S_a(1.3T_s)$=RotD50 5%-damped spectral acceleration at a period of $1.3T_s$ in units of g of design outcropping ground motion for site conditions below potential sliding mass [i.e., the value of $S_a(1.3T_s)$ for the earthquake ground motion at the elevation of the sliding surface if the potential sliding mass was removed]. The median displacement greater than 1 cm can then be computed (in cm) as

$$\ln D = a_1 - 2.482\ln k_y - 0.244\left(\ln k_y\right)^2 + 0.344\ln(k_y)\ln\left(S_a(1.3T_s)\right)$$
$$+ 2.649\ln\left(S_a(1.5T_s)\right) - 0.090\left[\ln\left(S_a(1.3T_s)\right)\right]^2 + a_2T_s + a_3\left(T_s\right)^2 + 0.603\mathbf{M} \pm \varepsilon_1 \quad (10.35)$$

where a_1=–5.981, a_2=3.223, and a_3=–0.945 for systems with $T_s \geq 0.10$ sec, and a_1=–4.684, a_2=–9.471, and a_3=0.0 for $T_s \geq 0.10$ sec. with $\sigma_{\ln D}$=0.72. Bray et al. (2018) presented a similar model for subduction zone events and Bray and Macedo (2019) also present models for rigid sliding blocks (i.e., T_s=0.0) and for maximum and median displacements caused by near-fault pulse motions.

With the median and uncertainty terms established for these models, the probability that the displacement exceeds some allowable displacement, d_{all}, can be computed as

$$P\left[D > d_{all}\right] = (1-\tilde{p})P\left[D > d_{all} \middle| D\rangle d_0\right] = (1-\tilde{p})\left[1 - \Phi\left(\frac{\ln d_{all} - \ln D}{\sigma_{\ln D}}\right)\right] \quad (10.36)$$

The Saygili and Rathje (2008) rigid block model (Equation 10.30) was modified to account for failure mass compliance in the form of the natural period of the sliding mass, T_s (Rathje and Antonakos, 2011). Using the average acceleration within the sliding mass, $\bar{a}_{\max}$, and magnitude to characterize the level of ground shaking, median displacements are predicted as

$$\ln D = 4.89 - 4.85\left(\frac{a_y}{\bar{a}_{\max}}\right) - 19.64\left(\frac{a_y}{\bar{a}_{\max}}\right)^2 + 42.49\left(\frac{a_y}{\bar{a}_{\max}}\right)^3 - 29.06\left(\frac{a_y}{\bar{a}_{\max}}\right)^4$$
$$+ 0.72\ln\left(a_{\max}/g\right) + 0.89(\mathbf{M}-6) + f_1\left(T_s\right) \quad (10.37)$$

with D in cm and

$$f_1\left(T_s\right) = \begin{cases} 3.69T_s - 1.22T_s^2 & T_s \leq 1.5s \\ 2.78 & T_s > 1.5s \end{cases}$$

$$\sigma_{\ln D} = 0.694 + 0.322\left(\frac{a_y}{\bar{a}_{\max}}\right)$$

The value of $\bar{a}_{\max}$ can be estimated from the relationship

$$\ln\left(\frac{\bar{a}_{\max}}{PGA}\right) = (0.459 - 0.702 \cdot PGA) \cdot \ln\left[\frac{T_s/T_m}{0.1}\right] + (-0.228 - 0.076 \cdot PGA) \cdot \left(\ln\left[\frac{T_s/T_m}{0.1}\right]\right)^2 \quad (10.38)$$

where PGA is the peak acceleration and T_m is the mean period (Section 3.3.2.2) of the base motion.

Example 10.5

A slope in southern California is determined to have a yield acceleration of 0.10g for a failure mass with a fundamental period, T_s=0.46 sec. If the slope was subjected to a repeat of the El Centro ground motion (EW component, Figure 3.9b) from the 1940 Imperial Valley (**M**6.9) earthquake, compute the median displacement and the probability that the displacement would exceed 30 cm.

Solution:

Using the procedure of Bray and Macedo (2019) for crustal ground motions, the elongated period of interest would be 1.3×0.46=0.60 sec. From Figure 3.9b, the spectral acceleration at the elongated period would be 610 cm/sec^2=0.622*g*. Then, from Equation (10.34), the probability of negligible (< 0.5 cm) displacement would be

$$\tilde{p} = 1 - \Phi(3.948) = 1 - 0.99996 = 0.00004$$

which indicates that there is essentially zero probability that the displacements will be negligible. Then, since T_s>0.1 sec, the median displacement is obtained from

$$\ln D = -5.981 - 2.482\ln 0.1 - 0.244(\ln 0.1)^2 + 0.344\ln(0.1)\ln 0.622 + 2.649\ln 0.622$$

$$-0.090[\ln 0.622]^2 + (3.223)(0.46) - 0.945(0.46)^2 + (0.603)(6.9) = 20.9\text{ cm}$$

The probability that the lateral displacement would exceed 30 cm is therefore

$$P[D > 30\text{ cm}] = 1 - \Phi\left[\frac{\ln 30 - \ln 20.9}{0.72}\right] = 0.308$$

Vector IMs The Rathje and Saygili (2009) rigid block vector model was also extended (Rathje and Antonakos, 2011) to account for failure mass compliance.

$$\ln D = -1.56 - 4.58\left(\frac{a_y}{\bar{a}_{\max}}\right) - 20.84\left(\frac{a_y}{\bar{a}_{\max}}\right)^2 + 44.75\left(\frac{a_y}{\bar{a}_{\max}}\right)^3 - 30.5\left(\frac{a_y}{\bar{a}_{\max}}\right)^4 \quad (10.39)$$

$$-0.64\ \ln\left(\bar{a}_{\max}/g\right) + 1.55\ \ln\left(k\text{-}vel_{\max}\right) + f_2(T_s)$$

With

$$f_2(T_s) = \begin{cases} 1.42T_s & T_s \le 0.5\text{ sec} \\ 0.71 & T_s > 0.5\text{ sec} \end{cases}$$

and

$$\sigma_{\ln D} = 0.40 + 0.284\left(\frac{a_y}{\bar{a}_{\max}}\right)$$

where the term $k\text{-}vel_{\max}$ is the peak value of a velocity-like parameter obtained by integrating the time history of normalized average acceleration, i.e.,

$$k\text{-}vel_{\max} = \max\left|\int_0^\infty \frac{\bar{a}(t)}{g}\,dt\right| \quad (10.40)$$

or, with knowledge of *PGV*, estimated from

$$\ln\left(\frac{k\text{-}vel_{\max}}{PGV}\right)=\begin{cases}0.240\cdot\ln\left[\frac{T_s/T_m}{0.2}\right]+(-0.091-0.171\cdot PGA)\cdot\left(\ln\left[\frac{T_s/T_m}{0.2}\right]\right)^2 & \text{for} \quad T_s/T_m\geq 0.2\\ 0 & \text{for} \quad T_s/T_m<0.2\end{cases} \tag{10.41}$$

Discussion

The preceding section presented a number of predictive relationships for permanent slope displacement based on sliding block models. The relationships differ in the physical characteristics of the sliding block (rigid or compliant) and the representation of ground shaking level (by scalar, mixed-scalar, and vector intensity measures). As such, they differ in the nature and amount of input data – scalar rigid block relationships require only a single ground motion intensity measure and a measure of the level of stability of the slope (e.g., yield acceleration) while vector compliant block relationships require multiple intensity measures computed from average accelerations that require some form of dynamic analysis, and hence distributions of soil density, stiffness, and failure surface geometry. The compliant block relationships are generally based on 1-D column representations of what are generally 2-D or 3-D profiles, hence they account for vertical, but not horizontal, spatial variability of ground motion. Methods based on horizontal equivalent accelerations are amenable to 2-D or 3-D analyses using the averaging approach illustrated in Figure 10.21.

The relationships also differ in the dispersions of their displacement predictions – the simpler relationships produce more dispersed displacement predictions and the more complex relationships have substantially less dispersion. Figure 10.34 shows the levels of dispersion in a number of the predictive relationships described in the preceding section. The dispersion values, as expected, tend to decrease as additional *IM*s are added to provide more complete descriptions of the ground motions. These reductions, which represent improvements in slope displacement prediction efficiency (Sections 3.3 and 5.6.2), can provide significant benefits in a fully probabilistic slope displacement hazard analysis (Section 10.12).

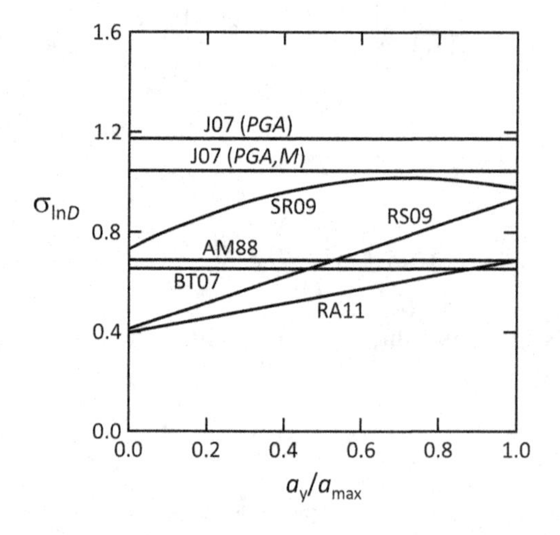

FIGURE 10.34 Dispersion associated with sliding block predictive relationships (J=Jibson, SR=Saygili and Rathje, RA=Rathje and Antonakos, AM=Ambraseys and Menu, BT=Bray and Travasarou; number indicates publication date). (Modified from Rathje and Antonakos (2011); used with permission of Elsevier Science and Technology Journals.)

It is important to recognize what these relationships represent and what they do not represent. They are models of sliding block displacements, and sliding block models are intended to represent potentially unstable slopes. The dispersions associated with these models, therefore, represent the effects of record-to-record variability, i.e., differences in the characteristics of the ground motions that are not represented by the *IM*s used in the predictive model. Uncertainties in the factors that control the inherent resistance of the slope to sliding, reflected in the yield acceleration (e.g., soil strength and density, slope geometry, etc.), are not considered. Furthermore, they represent uncertainty in the prediction of sliding block displacements and thus do not include uncertainty in the relationship between sliding block and actual slope displacements. Sliding block displacements have been shown to agree reasonably well with those observed in select individual case histories (e.g., Wilson and Keefer, 1983; Pradel et al., 2005) and to be positively correlated to regional landslide density (Jibson et al., 2000). While some efforts at investigating model uncertainty have been made (e.g., Bray, 2007), no general conclusions on model uncertainty for sliding block analyses have yet been developed. These factors provide a reminder that predicted sliding block displacements represent indices of actual seismic slope displacements and, under most conditions, should not be interpreted as accurate estimates of actual displacements.

Other Factors Influencing Slope Displacement

The standard sliding block analysis described in the preceding section is based on a variety of assumptions that are often violated by actual site conditions. Some of the conditions that can affect the applicability of the sliding block model are described in the following paragraphs.

Distributed Deformations Because of the rigid-perfectly plastic interface behavior they inherently assume, sliding block models only produce displacement when the driving shear stresses on a discrete failure surface exceed the available shearing resistance on that failure surface, and produce no displacement when they do not. However, soils exhibit nonlinear, inelastic behavior (Section 6.4.4) that can, in the presence of the static initial shear stresses that exist within slopes, produce permanent shear strains (Figure 6.42) even when the induced shear stresses do not reach the shear strength of the soil. Such shear strains are likely to be distributed throughout a potentially unstable slope and lead to a component of permanent deformation that is not associated with any discrete failure surface. Under strong shaking, a slope may deform through a combination of distributed permanent strains and localized straining on some discrete failure surface; sliding block models should be recognized as only being capable of representing the latter mechanism.

Variable Sliding Resistance The assumption of rigid-perfectly behavior on the interface is convenient in formulating sliding block analyses, but it may not be realistic for all interface materials. Laboratory tests on the types of geosynthetic materials used in landfill liner/cover construction, for example, often show a higher shearing resistance at low displacement levels than is mobilized at larger displacement levels (Wartman et al., 2003). A variable sliding resistance model that leads to a higher yield acceleration value at low displacements, a lower value at higher displacements, and a linear transition (Figure 10.35) between the two is illustrated in Figure 10.24. The displacements at which the transition begins and ends tend to increase with increasing normal stress. In the sliding block integration procedure, therefore, the higher yield acceleration controls the onset of permanent displacement in a particular loading pulse but the displacements are influenced by the transition and the lower, residual yield acceleration. Sliding block displacements computed with the variable sliding resistance are, as expected, higher than those computed using the peak resistance and lower than those computed using the residual displacement.

Directionality Sliding block analyses are typically performed using recorded earthquake ground motions, in some cases modified for compatibility with design criteria. Such recorded motions reflect ground motion along axes that may or may not correspond to the azimuthal direction a particular

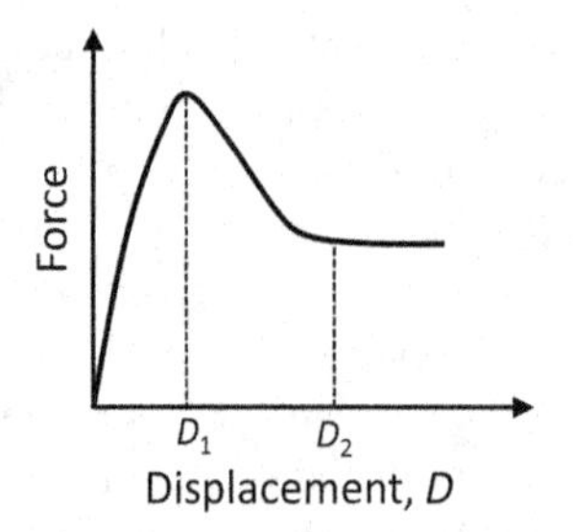

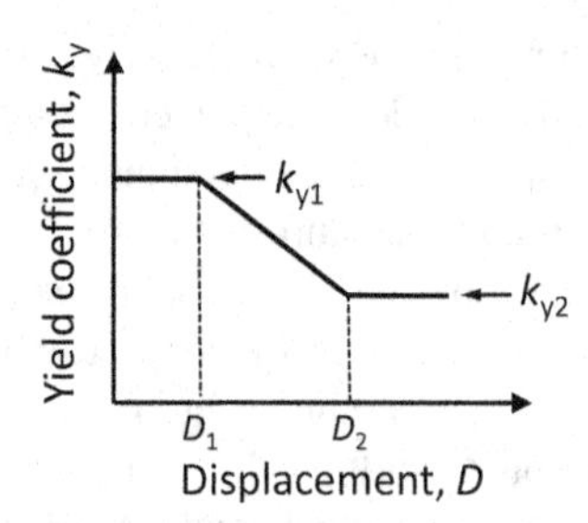

FIGURE 10.35 Schematic illustration of variable sliding resistance: (a) force-displacement behavior observed in laboratory tests, and (b) corresponding yield acceleration behavior. (After Matasovic et al., 1997; with permission of Emerald Publishing, Ltd.).

slope is facing. Because the intensity of shaking can be different in different directions (Figure 10.36a,b), the azimuthal orientation of a slope, relative to the orientation of the horizontal components of the recorded ground motion, can influence the computed slope displacement. Analyses of three-dimensional rigid blocks bounded by intersecting planes and subjected to all three components of 137 recorded ground motions (Kramer and Lindwall, 2004) showed strong variability in computed displacements depending on the azimuthal orientation of the line of intersection of the planes (Figure 10.36c,d). Permanent displacements were zero for some slope orientations, moderate for others, and relatively large for only a small range of slope orientations. In the near-fault region (Section 3.7), it may be possible to estimate directionally appropriate *IM*s for a particular slope with known dip direction, but *IM* amplitudes are essentially randomly oriented in the far field. Thus, the orientation of the slope relative to the ground motions can lead to another significant component of uncertainty in predicted slope displacements.

Directionality effects can be accounted for in slope deformation analysis by running ground motion hazard analysis for the case of arbitrary-component ground motion variability instead of the variability for combined horizontal components ($\sigma_{\ln IM}$ from Section 3.5.4.3):

$$\sigma_{\ln IM}^{arb} = \sqrt{\sigma_{\ln IM}^2 + \sigma_C^2} \tag{10.42}$$

where σ_C is approximately 0.2 (Watson-Lamprey and Boore, 2007) for peak accelerations and spectral accelerations over a range of period. With the hazard defined in this manner, seismic slope stability analyses can be run using the standard procedures presented earlier in this chapter.

Non-Planar Failure Surfaces The sliding block analysis assumes that sliding occurs on a planar surface of constant inclination, and is therefore best suited to situations where sliding occurs on joints, bedding planes, or shallow surfaces that can be idealized as infinite slopes. Actual failure surfaces, however, are frequently curved in such a way that (a) internal deformation (or, if the assumption of rigid-perfectly plastic behavior is maintained, development of internal failure surfaces) of the failure mass is required, and (b) the failure surface becomes flatter as displacements occur. Ambraseys and Srbulov (1995) proposed a multi-block (Figure 10.37a) stability analysis that accounts for internal failure surfaces. Michalowski (2007) used limit analysis to estimate yield accelerations for curved failure surfaces with multi-block mechanisms by assuming interface behavior to be governed by either associative or non-associative flow rules. Chained multiple-block (Figure 10.37b) models (e.g., Stamatopoulos, 1996) that account for variable failure surface slope without explicitly modeling internal failure mechanisms, have also been developed.

Rate Effects As discussed in Sections 6.6.2.3 and 6.6.6.4, some soils, particularly plastic clays and organic soils, can exhibit rate-dependent strength and stiffness. Since earthquake-induced shear

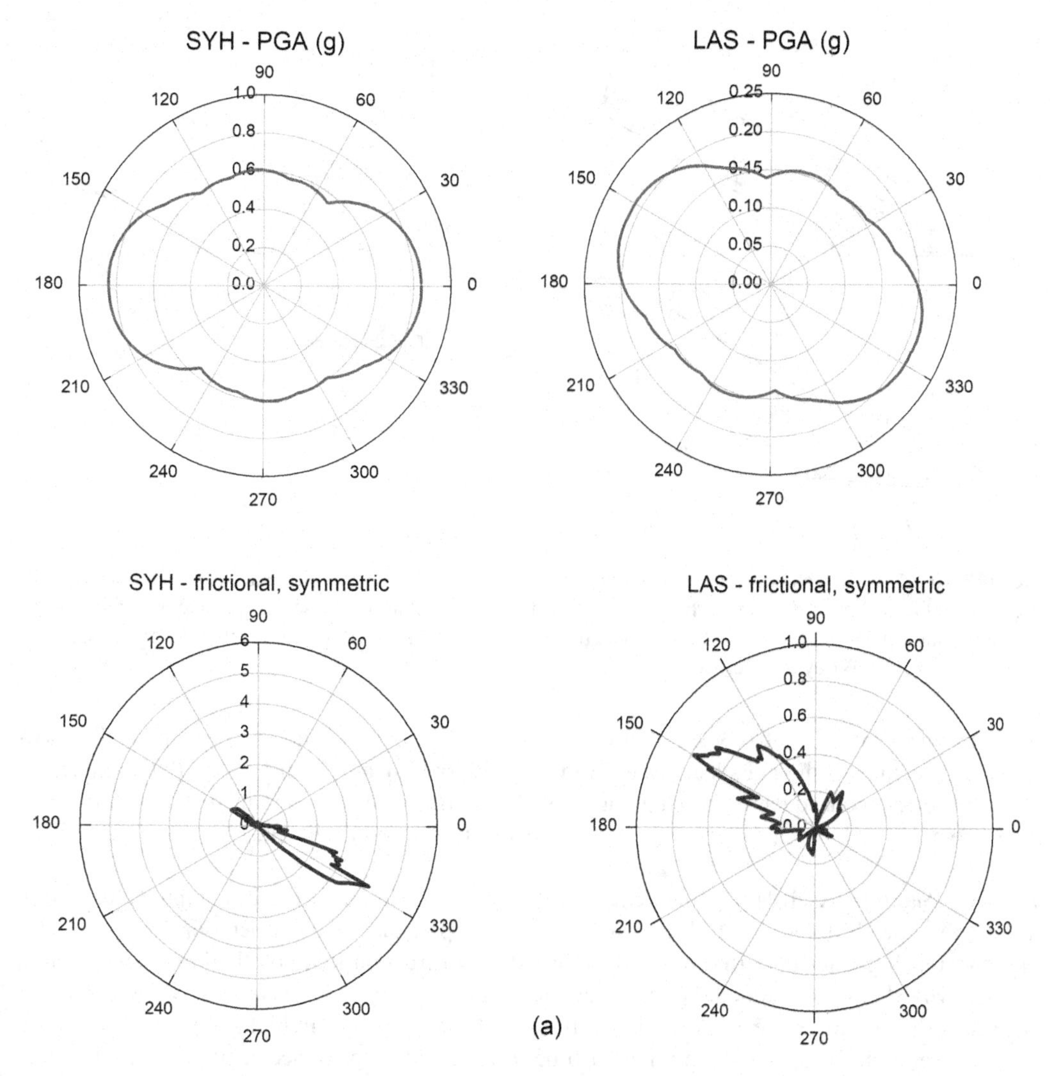

FIGURE 10.36 Azimuthal variability effects: (a) *PGA* at SYH station, (b) *PGA* at LAS station, (c) computed displacement (cm) for rigid block with $a_y/a_{max}=0.5$ at SYH, and (d) computed displacement (cm) for rigid block with $a_y/a_{max}=0.5$ at LAS (Kramer and Lindwall, 2004; used with permission of ASCE).

stresses are applied at different rates, the shear strength (and hence the yield acceleration) can vary with time over the duration of a ground motion (e.g., Hungr and Morgenstern, 1984; Lemos et al., 1985). Consideration of rate-dependent strength in a sliding block analysis should account for differences between strain rates in the field and in the laboratory tests used to measure the strength – this will generally result in an increase in the strengths used in analysis relative to those measured in the lab (at relatively slow rates). Lemos and Coelho (1991) and Tika-Vassilikos et al. (1993) suggested procedures for incorporating rate-dependent shearing resistances into numerical sliding block analyses.

Vertical Ground Motion As discussed in Section 10.9.1, vertical accelerations influence the driving and resisting forces acting on a mass of potentially unstable soils and do so in a manner that either increases or decreases both simultaneously. Because the numerator and denominator of the pseudostatic factor of safety (Equation 10.4) either both increase or both decrease, the vertical acceleration has relatively little effect on that factor of safety. Similarly, the vertical acceleration has

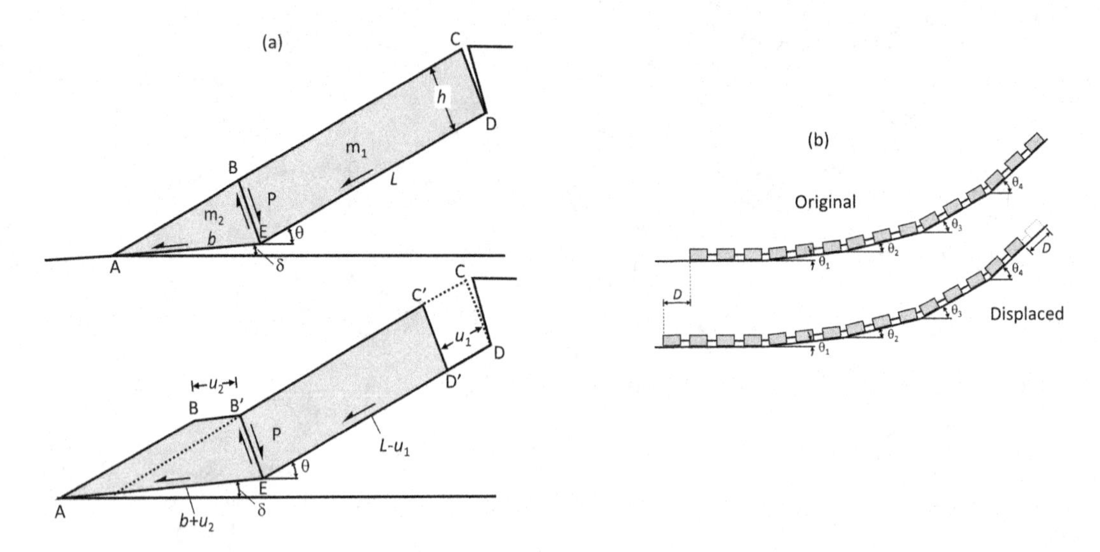

FIGURE 10.37 Multi-block models with non-planar failure surfaces: (a) bilinear base with internal failure surface (after Ambraseys and Srbulov, 1995; used with permission of Elsevier Science and Technology Journals), and (b) curved failure surface with chained multiple blocks. (After Stamatopoulos, 1996; used with permission of John Wiley and Sons.)

been shown to have a relatively small effect on the yield acceleration and on permanent displacements predicted by sliding block analyses (Yan et al., 1996; Ling and Leshchinsky, 1997; Matasovic et al., 1998) and, consequently, is usually ignored. Its significance, however, increases with increasing slope angle and increasing static factor of safety (Kramer and Lindwall, 2004).

Two-Way Sliding As discussed in Section 10.9.2.1, yield accelerations can be computed for sliding in both the downhill and uphill directions. For most slopes, the yield acceleration for uphill sliding is so much greater than that in the downhill sliding direction that uphill sliding is neglected, i.e., *one-way sliding* is assumed. For relatively flat slopes with low friction angles, which can occur when shear strengths are very low (on landfill liners, for example, or in highly softened clays), the yield acceleration for uphill sliding may also be lower than the peak acceleration. In such cases, *two-way sliding* can occur. Little difference between displacements computed for one-way and two-way sliding cases was found for a geosynthetic landfill cover configuration (Matasovic et al., 1998) when the yield accelerations for downhill and uphill sliding were quite different (0.14g and 0.82g, respectively). The application of 90 motions spanning 16 magnitude-distance bins to rigid blocks with weak interfaces on flat planes (Makdisi, 2016) showed that geometric mean displacements for slopes steeper than 2° were not significantly influenced by uphill sliding.

10.9.2.5 Displacement-Related Pseudostatic Analyses

As discussed in Section 10.9.1.3, pseudostatic analyses provide a rough indication of seismic slope stability but provide no information on the level of deformation associated with instability. Section 10.9.2 showed how a variety of sliding block models could be used to estimate permanent displacements associated with seismic instability. A number of the sliding block predictive relationships used the yield acceleration, a_y, or yield coefficient, $k_y = a_y/g$, to predict permanent displacements.

By definition of the yield acceleration, the pseudostatic factor of safety with $a_h = a_y$ is equal to 1.0 and no permanent displacement will occur. If a higher pseudostatic acceleration is used, the pseudostatic factor of safety will be less than 1.0 and some permanent displacement will occur. A sliding block model that predicts permanent displacement as a function of yield acceleration can therefore be inverted to compute an "apparent yield acceleration" that corresponds to a selected allowable (non-zero) level of permanent displacement. A pseudostatic analysis that produces $FS \geq 1.0$ using a pseudostatic acceleration equal to the apparent yield acceleration will correspond to a permanent displacement less than or equal to the allowable displacement.

Stewart et al. (2003) proposed a displacement-related pseudostatic analysis procedure based on the sliding block predictive model of Bray and Rathje (1998) as part of a screening procedure for seismic slope instability. The screening procedure allows pseudostatic analysis of factors of safety corresponding to tolerable slope displacements, $D = 5$ cm or $D = 15$ cm based on a pseudostatic coefficient

$$k = f_{eq} \cdot MHA \tag{10.43}$$

Values of f_{eq} for a tolerable slope displacement of 15 cm are shown in Figure 10.38. Bray and Travasarou (2009) used the Bray and Travasarou (2007) model to recommend a seismic coefficient compatible with an allowable displacement, D_a, as

$$k = \exp\left[\frac{-a + \sqrt{b}}{0.665}\right] \tag{10.44}$$

where

$$a = 2.83 - 0.566 \ln S_a(1.5T_s)$$

$$b = a^2 - 1.33\left\{\ln D_a + c - 3.04 \ln S_a(1.5T_s) + 0.244\left(\ln S_a(1.5T_s)\right)^2 - 1.5T_s - 0.278(M - 7) - \varepsilon\right\}$$

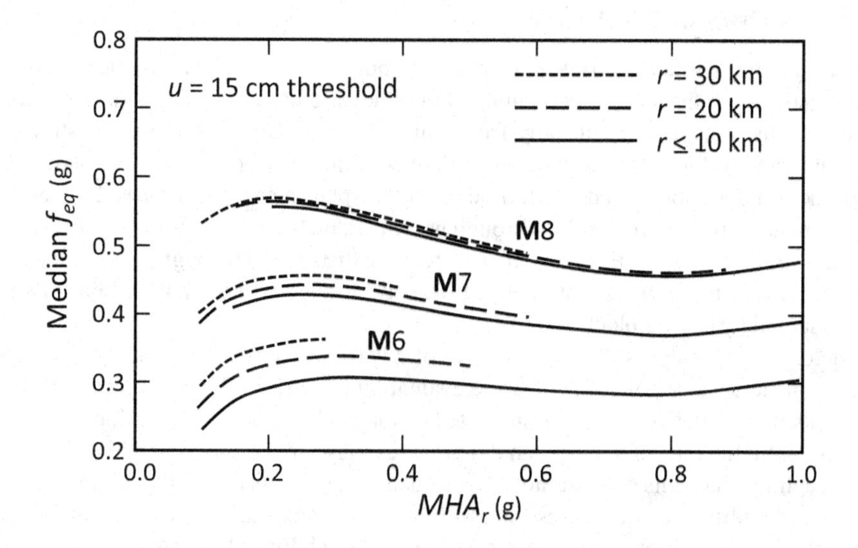

FIGURE 10.38 Variation of pseudostatic coefficient factor, f_{eq}, with mean horizontal acceleration for tolerable displacement of 15 cm. (After Stewart et al., 2003; used with permission of Earthquake Engineering Research Institute.)

$$c = \begin{cases} 1.10 & T_s \geq 0.05 \text{ sec} \\ 0.22 & T_s < 0.05 \text{ sec} \end{cases}$$

where T_s is the initial fundamental period of the sliding mass.

These procedures are based on a single ground motion level. However, by integrating over all ground motion levels, the displacement-compatible pseudostatic coefficient concept can be extended to a pseudostatic coefficient that is compatible with the displacement associated with a specified return period of exceedance (Macedo et al., 2018).

Example 10.6

Using the procedure of Stewart et al. (2003), compute the pseudostatic coefficient that is compatible with a tolerable displacement of 15 cm for a site at a distance of 20 km from a **M**7 earthquake. Calculate the MHA as the median PGA value given by Boore et al. (2014) as shown in Figure 3.34.

Solution:

From Figure 3.34, the median MHA for a M7.0 earthquake at a distance of 20 km is 0.15*g*. From Figure 10.38, the median value of f_{eq} for a M7 earthquake at 20 km is 0.425. The pseudostatic coefficient corresponding to a displacement of 15 cm, therefore, is

$$k = f_{eq} \cdot MHA = (0.425)(0.15) = 0.064$$

The displacement-related pseudostatic coefficient provides a rational basis for the performance of pseudostatic slope stability analyses that consider, at least in an approximate way, the potential effects of slope instability. The displacement-related pseudostatic coefficient for a specific slope could also be determined through a series of more rigorous deformation analyses, such as those described in the next section.

10.9.3 Stress-Deformation Analysis

Sliding block models, while both useful and popular, must be recognized as applicable to situations that are consistent with the assumptions upon which they are based. Sliding block models assume that permanent displacements result only from transient exceedances of the shear strength of the soil, and that they develop along discrete, well-defined failure surfaces. They also assume that the strength of the soil does not degrade with time or with displacement. As discussed in Section 6.4.4, however, permanent strains can develop through nonlinear, inelastic behavior at shear stresses lower than the shear strength of the soil. These strains are often distributed throughout an extensive region of the slope, leading to deformations that are broadly distributed rather than being localized as implicitly assumed by sliding block analyses.

Stress-deformation analyses, through the use of advanced constitutive models that capture the nonlinear, inelastic behavior of soil, allow the evaluation of seismic slope stability to be combined with the evaluation of the seismic response of the slope. Thus, the amplification behavior of the slope is considered, the effects of "slip" on dynamic response are considered, and deformations can develop through mechanisms that are not constrained to some a priori assumed geometry as they frequently are in sliding block analyses. Stress-deformation analyses, therefore, have the potential to predict actual slope deformations more accurately than sliding block analyses and to consider cases that are not amenable to sliding block analysis. Stress-deformation analyses can help identify deformation mechanisms, as well as displacement amplitudes, over wide ranges of cyclic and permanent soil deformations. They are, however, more complicated and time-consuming than sliding block analyses. They require more input data, some of which may be highly uncertain, and the interpretation of their results is more time-consuming. These complications represent the "cost" of obtaining a more accurate and informative indication of seismic slope stability. Stress-deformation analyses of seismic slope stability are usually performed using dynamic finite element or finite difference analyses. In such analyses, the seismically induced permanent strains in each element of a discretized mesh are integrated to obtain the permanent deformation of the slope.

10.9.3.1 One-Dimensional Analysis

Some one-dimensional nonlinear site response codes allow for the possibility of computing permanent deformations by imposing static initial stresses on each layer in the soil profile. The stresses are usually applied by specifying a ground surface slope, in which case the one-dimensional soil profile is essentially tilted to represent an infinite slope condition. Because it is a one-dimensional analysis, however, the ground motion is also tilted so that the waves travel in a direction perpendicular to the ground surface; such analyses should generally be limited to relatively gentle slopes when horizontal recorded ground motions are used as input motions.

Such analyses can take advantage of the constitutive model of the soil to compute permanent strains (and, hence, permanent displacements) that are distributed with depth, or they can be set up to constrain failure to an interface or layer at a particular depth, in which case the analysis is essentially an improved version of a coupled, compliant sliding block analysis. Figure 10.39 shows examples of displacement profiles for a gently sloping soft clay deposit obtained from two types of analysis. Figure 10.39b and c show the results of analyses in which all soil layers were modeled as nonlinear and inelastic, and Figure 10.39d and e show results of analyses in which all but one layer was modeled as linear elastic. The latter mixed linear-nonlinear analysis accounts for the compliance of the soil profile but simulates sliding block analyses by computing permanent deformations

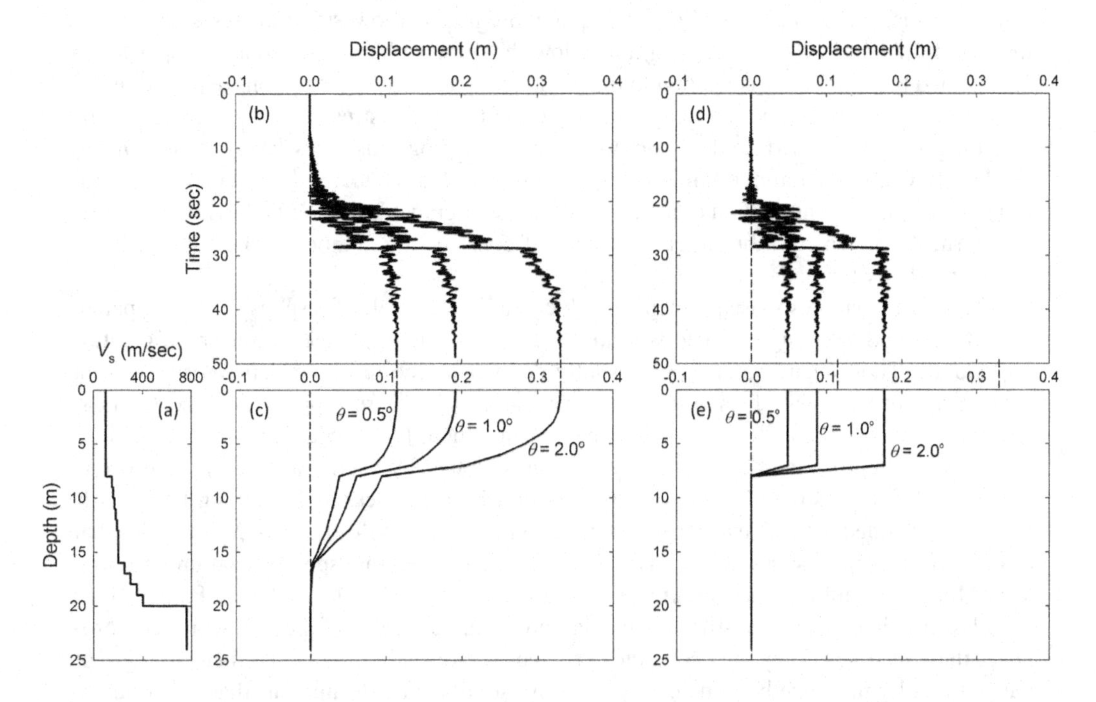

FIGURE 10.39 One-dimensional, nonlinear analyses of soil profile sloping at 0.5, 1.0, and 2.0 degrees: (a) shear wave velocity, (b) time histories of downslope ground surface displacement for nonlinear model, (c) final displacement profiles for nonlinear model, (d) time histories of downslope ground surface displacement for mixed linear-nonlinear model, and (e) final displacement profiles for mixed linear-nonlinear model.

only when the shear strength of the thin nonlinear layer (from 7 to 8 m depth) is reached. The fully nonlinear analysis, however, can also account for the development of permanent deformations in all layers and also those caused by shear stresses that do not fully mobilize the shear strength of the soil during earthquake shaking. Those distributed deformations can be seen to be nearly equal to the deformations resulting from full mobilization of the shear strength of the soil.

10.9.3.2 Multi-Dimensional Analysis

The same two- and three-dimensional dynamic stress-deformation analyses with nonlinear inelastic constitutive models used to predict site response can also be used to evaluate seismic slope stability. The analyses are subject to the same considerations described in Section 7.6.1 – indeed, the development of permanent deformations in a slope is part of the seismic response of the slope. The use of multi-dimensional stress-deformation analyses to estimate permanent deformations of slopes implies that relatively large stresses and strains are anticipated, and that condition brings additional considerations into play. Some of the most important of these are described below.

Discretization: A significant level of strain is required to mobilize the strength of an element of soil, so a significant level of deformation will frequently be involved in seismic stability analyses. The mesh used to model the slope should be discretized in a manner that allows stresses and strains to develop freely. The mesh should be refined with smaller elements in areas where stress and/or strain gradients are expected to be high and can be coarser in areas where they are expected to be low. Development and refinement of a mesh is one of the most important parts of a dynamic stress-deformation analysis, and adjustments of an initial mesh are often required as analyses proceed.

Initial Conditions: The stiffness, and hence dynamic response, of a soil profile can be strongly affected by its initial (pre-earthquake) stress conditions. Therefore, any dynamic stress-deformation analysis must be preceded by a static stress-deformation analysis to establish the initial stress state. Available computer programs allow static analyses to be performed in different ways. Some programs allow the mesh to be developed and properties to be assigned before gravity is applied to the model, and have different options (e.g., sudden or gradual) for the application of gravity – the nature of the initial stresses, however, may be different if gravity is applied suddenly or gradually. Other programs allow layers of elements to be placed (with gravity applied) in a sequential manner that simulates deposition of a natural soil deposit and/or construction of a man-made slope or embankment. By allowing the model to deform and reach equilibrium at each stage of construction, a more realistic initial stress state can usually be modeled.

Soil Model: For site response problems in which moderate levels of response are anticipated, the stiffness and damping characteristics of the soil at low to moderate strains are of greatest importance. These characteristics are usually defined by the results of laboratory resonant column and/or torsional shear tests (Section 6.5.4.2), which are typically limited to shear strains on the order of 0.5% or less. The manner in which the constitutive model transitions from a strain regime in which properties are determined by cyclic laboratory data to one controlled by shear strength (Section 6.6.3.3), for which stiffness and damping behavior is not as well defined, can have a significant effect on computed deformations. Many codes that can be used for nonlinear site response have material options that specify a failure criterion (e.g., Mohr-Coulomb) but assume linear elastic behavior at stress levels below failure. Such models will misrepresent the stiffness and damping characteristics of the soil, which compromises their accuracy. They only compute permanent deformations when the shear strength of the soil is fully mobilized; as such, they can miss significant deformations that occur due to nonlinear, inelastic behavior within the failure surface. On the other hand, exaggerated amplification in their linear ranges can lead, in some cases, to excessive estimates of displacement.

Mesh Locking: Numerical techniques such as finite element and finite difference analyses approximate the behavior of a continuum. Among the main approximations are those regarding the displacement patterns between nodes or grid points. These interpolation, or *shape function*, approximations are generally very reasonable under most conditions, but can be problematic under specific conditions. Problems can occur when the shape function cannot adequately describe the kinematics of deformation, for example when a constant strain element is used to model bending (where strains tend to vary linearly across the element). The deviation between actual and modeled strains can lead to the development of spurious shear strain, and to shear strain energy that causes the element to become excessively stiff in bending. Elements that behave this way produce very little deformation and are said to have locked. Numerical models can also have difficulties with incompressible materials – in soil mechanics, the behavior of saturated, fine-grained soils is referred to as undrained and, given the high bulk moduli of soil particles and water, are modeled as incompressible. As a material approaches incompressibility, Poisson's ratio, $\nu \rightarrow 0.5$, and the bulk modulus, $K \rightarrow \infty$. Depending on the formulation of the finite element itself, the elastic relationship between bulk modulus and shear modulus may not be preserved, resulting in excessively stiff element behavior also referred to as locking.

Large Deformations: In most stress-deformation analyses, the deformed geometry is defined with respect to the original configuration of the model and the stiffnesses of the individual elements are formulated with respect to their original shapes. As long as strains remain relatively small, element distortion is small and the element stiffness is well behaved. When deformations become large, however, elements may undergo large strain with highly distorted shapes, and the shape functions used to approximate the pattern of displacement between the nodal or grid points may become very inaccurate. Because the stiffness of the element depends on those functions, significant errors may result. Some stress-deformation codes use definitions of the deformed geometry that are updated as the deformations develop and can therefore handle large deformations in a reasonable manner. Empirical and numerical approaches to the estimation of flow slide runout deformations are described in Section 10.10.2.

One of the important advantages of stress-deformation analyses is their ability to identify deformation mechanisms, particularly when the geometry and/or material characteristics of a slope are complicated or highly variable. The shear strain contours from the finite difference analysis shown in Figure 10.40a illustrate the mechanisms by which settlement both upstream and downstream deformations are developing. The displacement vectors in Figure 10.40b show the distribution and

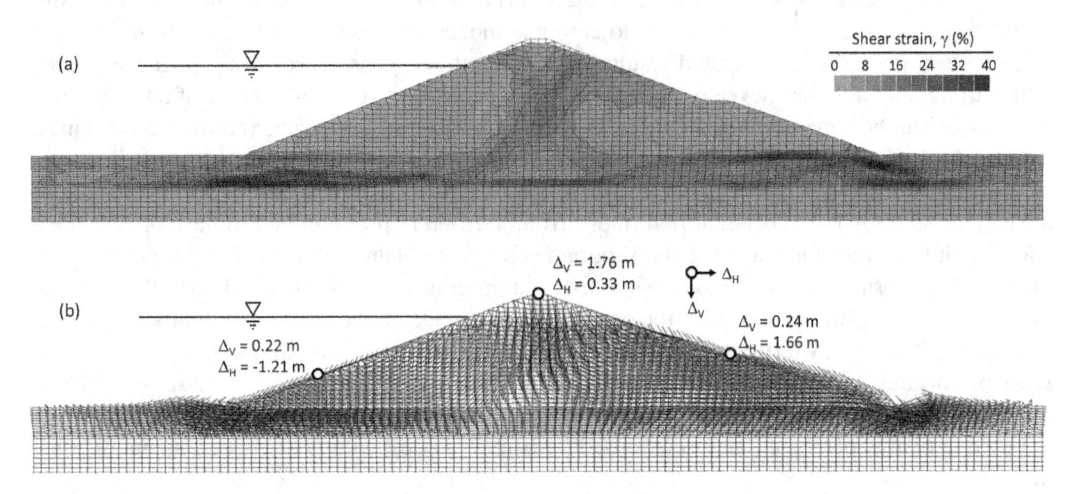

FIGURE 10.40 Illustration of deformation mechanisms of a hypothetical 45-m-tall dam with a clay core, sand shells, and a gravel berm on the downstream slope underlain by 12 m of liquefiable alluvium, the penetration resistance of which has been randomized: (a) shear strain contours and (b) displacement vectors. (Courtesy of R.W. Boulanger.)

directions of movement. Limit equilibrium programs can search for critical noncircular failure surfaces, but they do not consider the dynamic aspects of the response and do not indicate the type of internal deformations that may develop within an unstable mass of soil.

To provide additional confidence in the results of a two-dimensional nonlinear analysis, it is not uncommon to perform a two-dimensional equivalent linear analysis, extract the time-variations of average acceleration of the soil above a predetermined failure surface, and to use that acceleration history as the input to a sliding block analysis. These steps constitute a two-dimensional decoupled analysis (Section 10.9.2.2), but provide an approximate value of permanent displacement against which the displacements predicted directly by a two-dimensional nonlinear analysis can be compared for general reasonableness.

10.9.3.3 Considerations in Stress-Deformation Analyses

Soils exhibit complex nonlinear, inelastic behavior under cyclic loading even under controlled laboratory conditions in which they are subjected to simple, harmonic loading. That behavior becomes even more complex for soils that generate excess pore pressure under cyclic loading. In the field, where depositional processes cause spatial variability of grain size and other soil characteristics, and where actual earthquake loading is much more complex, soil behavior becomes even more complicated. This complex behavior places great demands on the numerical procedures used to perform stress-deformation analyses. They must represent the complex constitutive behavior of the soil, and accurately represent complex wave propagation phenomena for specific site geometries over broad ranges of frequencies. They must satisfy boundary conditions at the edges of the modeled domain, and also at boundaries with structural elements that exist on and/or beneath the ground surface. Finally, these analyses are typically performed with incomplete information, i.e., soil data from individual boring or soundings at specific, discrete locations between which soil properties must be inferred.

Despite these complications, nonlinear stress-deformation analyses have advanced considerably over the past 20 years. Such analyses are now commonly used on large or high-risk projects and are increasingly being used on more routine projects. Advances in both software and hardware have made such analyses more accessible and economical, and their use is expected to penetrate further into high-quality geotechnical earthquake engineering practice. At this time, there are a relatively small number of computer programs capable of performing nonlinear stress-deformation analyses. Most use finite element analyses, but finite difference programs are also available. These programs use different procedures to describe the spatial variation of strain within elements and use different numerical integration procedures to compute model response over time. These differences can lead to differences in computed results, even when the response is relatively linear. However, when strong nonlinearities develop, as in cases where permanent deformations are of interest, these differences can become more pronounced. Nonlinear analyses of all types (not just geotechnical earthquake engineering) often exhibit strong sensitivities to computational details as well as specific input parameters. It is therefore incumbent upon the analyst to recognize that the results from a single program may be different than those from a different program and to develop an understanding of how well suited a particular program is to the problem of interest. It is frequently good practice to perform parallel analyses using different programs to understand/confirm the behavior of the site, and to gain confidence in the decisions being made on the basis of the analyses.

Boulanger and Beaty (2016) provided an excellent discussion of issues that should be considered when performing and interpreting the results of nonlinear stress-deformation analyses of embankment dams; the basic principles, however, apply to many other problems where numerical analyses are used to predict permanent deformations. The primary issues, with brief comments on each, are described below.

Performance Objectives: Performance objectives should be discussed in advance of the analyses. Such objectives may be specified by regulation or code, or they may be agreed upon in collaboration and communication with owner/users. In current practice, performance is often defined in terms of permanent deformations, but performance can also be defined in terms of physical damage or loss (Section 5.3). Losses from earthquake-induced landslides can extend well off site, and in some cases do not develop until some period of time after the earthquake. While objectives can change depending on numerous circumstances, it is beneficial to begin the analyses with an understanding of the metrics used to judge performance.

Seismic Failure Modes and Important Phenomena: There are often multiple mechanisms by which permanent deformations can develop at a given site, and it is important to identify them before the analysis. Knowledge of the anticipated mechanism(s) can influence the manner in which a site model should be discretized, for example. They may also affect the choice of analytical method (simple vs. complex or Program *A* vs. Program *B*) based on their ability to model the anticipated phenomena of interest.

Verification and Validation Records for the Numerical Modeling Procedure: The extent of and conditions under which a particular numerical analysis code has been validated and verified (Section 3.6.3) should be reviewed before committing to apply it to a particular problem.

Site characterization Basis for Subsurface Geometry and Material Parameters: The spatial discretization of the modeled domain and material parameters assigned to different elements within the model are based on subsurface characterization of the site. To aid in the interpretation of results, the basis for the subsurface geometry and all material parameters should be documented and any implied uncertainties recognized and understood.

Calibration and Evaluation of the Constitutive Model: Different computer programs employ different constitutive models that have different input parameters. The material parameters, in conjunction with the constitutive model, define the behavior of the soil. Some parameters can be obtained from field measurements, some from laboratory measurements, and some from correlations to more readily measured soil parameters. Predictions of the constitutive model should be checked against laboratory test data – the ideal model should be able to predict modulus reduction and damping behavior, stress-strain behavior under monotonic and cyclic, drained and undrained loading conditions, and volume change (drained loading) and pore pressure (undrained loading) behavior under drained and undrained cyclic loading conditions with and without initial static shear stresses. Not all of these capabilities may be required for a particular problem, but the capabilities of the model should be checked against the phenomena of interest.

Numerical Modeling Procedures: The numerical analysis procedure should be documented and understood. Among the important features of a stress-deformation computer program are boundary conditions (absorbing boundaries, Section 7.6.1.2), damping (viscous, Rayleigh, hysteretic, numerical, Section 7.5.6.1), element size (relative to strain gradients, maximum frequency, Section 7.6.1.1), pore pressure considerations (permeability, fluid bulk modulus, drainage boundary conditions), large-strain modeling capabilities (remeshing, mesh locking, accommodation of residual strength), interface elements (shear and normal stiffness, shear strength, gapping behavior), and input motion application (rigid base or outcrop motion, acceleration or velocity time history).

Input Ground Motions: The input motions should be established using the procedures described in Section 4.5. The basis for, source(s) of, and any modification (scaling, spectral matching) of those motions should be understood and documented. The motions should be baseline-corrected.

Initial Static Stress Conditions: Seismic response, and particularly permanent deformations resulting from that response, are sensitive to initial stress conditions. Care should be taken to model the initial stress state accurately. As described in the preceding section, different programs have different capabilities in this regard. Lateral stresses, principal stress ratios, principal stress directions, and deformations should all be checked for reasonableness under static conditions prior to performing dynamic analyses.

Dynamic Response: The results of initial dynamic response analyses (for a single input motion) should be examined carefully to confirm that the program is performing properly and the results are generally consistent with expectations. It is often helpful to perform an initial analysis with an input motion scaled to a very low amplitude in order to minimize the effects of nonlinearity before proceeding with full-strength input motions. Acceleration histories, spectral amplification, and displacement and strain fields should be examined carefully. Deflections, shear, and bending responses of any structural elements should also be examined. Stress-strain and stress behavior should be examined for selected individual elements, particularly in areas likely to develop large shear strains. When possible, comparisons should be made with the results of equivalent linear analyses of the same site (comparison with one-dimensional analyses may be possible in portions of the site where multi-dimensional effects are not expected to be large). When the analyst is convinced that the weak-motion response is reasonable, the input motion can be scaled up to its full strength (if possible, in increments to examine the increasing effects of nonlinearity) and the response again carefully examined. When the response for the first input motion is understood and accepted, analyses with the remaining input motions can be performed.

Post-Shaking Deformations: Permanent deformations after shaking should be examined for reasonableness, and to understand the mechanism(s) leading to those deformations. The mechanism(s) should be compared with the results of limit equilibrium analyses. If there are any significant inconsistencies with limit equilibrium results, their causes should be determined. The potential for weakening of various elements of soil (e.g., due to cyclic softening and/or liquefaction) should be examined; a strength reduction approach can be used to evaluate the potential effects of such weakening (subject to the ability of the numerical analysis to represent the deformations that may result).

Sensitivity and Uncertainty: Some parameters will be known more accurately than others and some parameters will influence response more than others (Section D.9.1). Parameters related to shear strength and density can be expected to have a particularly strong influence on permanent deformations. Sensitivity analyses should be performed by varying the values of these parameters, for example by increasing and decreasing them by some percentage related to their uncertainty (e.g., by the amount of their standard deviation) and evaluating their effect on permanent deformations. Such analyses can help identify the most influential parameters for possible refinement by additional subsurface investigation or field/laboratory testing.

Reasonableness of Conclusions: The overall reasonableness of conclusions drawn from the results of the analyses should be evaluated. Comparison with case histories from other earthquakes, local experience, and informed engineering judgment are integral parts of such an evaluation.

10.9.4 Allowable Deformations

Both sliding block and stress-deformation analyses produce some measure of slope deformation. In the case of a sliding block analysis, the measure is a permanent displacement value that, due to the simplifying assumptions of the model, represents an index of potential slope movement. In the case of a stress-deformation analysis, the measure may be an entire displacement field, or deformed shape, of the slope and may have cyclic (during the earthquake) and permanent (after the earthquake) components.

Geotechnical earthquake engineers have historically evaluated seismic stability in terms of whether or not a slope has "failed." As discussed in Section 10.9.1, failure is generally taken as $FS < 1.0$ in a pseudostatic stability evaluation. The displacement computed in a dynamic analysis is usually compared with an allowable displacement; if the computed displacement is larger than the allowable displacement, failure is assumed to have occurred. The concept of "failure," therefore, should be interpreted as the failure to meet some desired performance objective specified in terms of an allowable level of deformation.

To characterize the performance of a slope in a more general manner, the computed deformations can be compared with one or more *limit state* levels. In this approach, each limit state can correspond to some expected response, damage, or loss level (Section 5.5); in current practice, limit states are usually defined in terms of deformations. Limit state displacements should be evaluated on a case-by-case basis – the level of displacement causing major damage to a slope traversed by a logging road is likely to be much larger than that required to cause major damage to a brittle, unreinforced masonry building near the crest of a different slope. Nevertheless, examples of limit state displacements can be found for different applications in the literature. As described in Section 10.9.1.3, displacements less than approximately 1 m have been described as not being "dangerously large" for well-constructed earth dams (Hynes-Griffin and Franklin, 1984). Considering areas developed with residential and commercial construction, Wieczorek et al. (1985) used 5 cm as a threshold for macroscopic ground cracking and failure in northern California, and Keefer and Wilson (1989) used 10 cm for coherent landslides in southern California. In southern California, Blake et al. (2002) suggested threshold displacements of 5 cm for failure surfaces intersecting relatively stiff improvements such as buildings and swimming pools, and 15 cm for failure surfaces in unimproved areas. Jibson and Michael (2009) defined landslide hazard levels for Anchorage, Alaska on the basis of sliding block displacements: Low (<1 cm), Moderate (1–5 cm), High (5–15 cm), and Very High (>15 cm). Seed and Bonaparte (1992) recommended limiting displacements of landfill liners to 15–30 cm. These recommendations indicate the problem-specific nature of allowable displacements for seismic slope stability evaluation. They also indicate a degree of subjectivity in the value of the allowable displacement; procedures for considering dispersion in "capacity" values in a probabilistic limit state evaluation were introduced in Section 5.6.3.

10.10 POST-SEISMIC STABILITY ANALYSIS

Slopes can experience flow failure during or after earthquake shaking has ended. Flow failures occur when the available shear strength of the soil drops from a level greater than the shear stress required for static equilibrium (a condition where $FS > 1.0$) to a level below that shear stress ($FS < 1.0$). The deformations, therefore, are driven by static shear stresses; the earthquake simply causes the reduction of shear strength that triggers the failure. Because they usually involve significant reduction in soil strength, flow failures usually produce large deformations and severe damage. Therefore, two problems need to be addressed - whether or not a flow failure will occur, and how far the materials will move if one does occur.

Instabilities have been observed to occur at some time after shaking has ended in a number of earthquakes. Many of these cases are associated with liquefaction where pore pressure and void redistribution can reduce the strength of highly stressed zones of soil within minutes to hours of an earthquake event, such as in the case of Lower San Fernando Dam (Figure 1.11). However, delayed slope failures can occur by other mechanisms. Following the 1980 Irpinia (Italy) earthquake, several large slides (with volumes up to 28 million m^3) began moving a few hours to a few days after the earthquake (Agnesi et al., 1982; Cotecchia, 1986) due to increasing spring flow associated with tectonic deformation in the area. Changes in regional groundwater conditions also contributed to delayed failures in the 1991 Racha (Republic of Georgia) earthquake (Jibson et al., 1994).

10.10.1 Analysis of Flow Slide Stability

Potential flow slide instability is most commonly evaluated by conventional static slope stability analyses using soil strengths based on end-of-earthquake conditions (Marcuson et al., 1990). Flow sliding is most commonly of concern for slopes that contain or are underlain by liquefiable soils. While liquefaction is generally triggered during or shortly after the strongest shaking, the full strength loss can be delayed by void redistribution effects that occur after earthquake shaking has ended. As a consequence, liquefaction-related flow slides typically occur after, rather than during, earthquakes. These factors support the decoupling of flow slide potential from dynamic analyses of slope response, i.e., of using static stability analyses with no dynamic or pseudostatic forces, in the analysis.

In a typical analysis, the factors of safety against triggering of liquefaction are computed at all points through which potential failure surfaces may pass. Residual strengths are then assigned to the zones of soil for which the factor of safety against liquefaction is less than 1.0 (although some engineers choose to assign residual strengths to zones with factors of safety slightly greater than 1.0 in an attempt to account for uncertainty in the prediction of triggering). At locations where the factor of safety against liquefaction is greater than 1.0, strength values are based on the effective stresses at the end of the earthquake (i.e., considering pore pressures generated during the earthquake) with consideration of cyclic softening or other means of strength reduction in non-liquefiable soils (Section 9.8). With these strengths, conventional limit equilibrium slope stability analyses are used to calculate an overall factor of safety against flow sliding. If the overall factor of safety is less than 1.0, flow sliding is expected. The possibility of progressive failure (Section 10.8.1) must be considered in stability evaluations of this type – the redistribution of stresses involved in progressive failure is not accounted for directly in limit equilibrium analyses.

Evaluation of the post-seismic stability of slopes in which excess pore pressure is generated can involve significant challenges. Higher excess pore pressure may be generated in a portion of a slope or embankment where static shear stresses are relatively low than in areas where those stresses are high. The spatially variable excess pore pressures produce hydraulic gradients that can cause porewater to flow from areas of low static shear stress to areas of high static shear stress following earthquake shaking (Figure 9.72). In the areas of high static shear stress, which are more critical from a slope stability standpoint, pore pressure may increase for a period of time following the end-of-earthquake shaking. As pore pressures increase, effective stresses decrease and, hence, strength and stiffness also decrease. The reduction in effective stress also causes rebound (i.e., expansion) of the soil skeleton to occur (Section 9.6.5.1). In a potentially liquefiable soil with a flat steady-state line, a relatively small amount of rebound can lead to a large reduction in steady state shear strength. If porewater flow is impeded by low-permeability soil layers, this process can be exacerbated with significant void redistribution (Section 9.6.5.1) occurring adjacent to the low-permeability layer. In the laboratory, such conditions have led to the development of water interlayers (National Research Council, 1985; Kokusho, 2000; Malvick et al., 2008) which have essentially no shear strength.

Analysis of the post-seismic stability of a slope in which void redistribution has occurred presents no significant challenges in limit equilibrium analyses. Conventional limit equilibrium analyses can be used to compute factors of safety based on the modeled slope geometry and assigned material properties. The difficulty lies in the assignment of strengths that reflect the effects of void redistribution. Void redistribution changes the density of the soil, to which the available strength is usually very sensitive, by an amount that depends on factors (e.g., spatial variability of permeability, presence, and continuity of fine-grained layers or lenses) that are virtually impossible to accurately characterize at present. As a result, judgments informed by case history behavior are typically required. These judgments introduce high levels of uncertainty into the residual strength; these uncertainties are typically accounted for in current practice by the use of very conservative strength parameters in post-seismic stability evaluations and/or by assigning residual strength to soils with factors of safety against triggering that fall below some threshold greater than 1.0.

10.10.2 Analysis of Flow Slide Deformations

The consequences of flow slides are often so devastating that the approach to design is to prevent their occurrence. In some cases, for example, evaluation of hazards at a site some distance below a slope susceptible to flow sliding, it is necessary to estimate how far the unstable materials will travel in the event that a flow slide occurs. Flow slides are generally triggered when the strength of the soil within a slope or in the vicinity of a structure is reduced significantly. The resulting deformations are driven by the unbalanced stresses that exist when the strength of the soil drops below the shear stress required to maintain static equilibrium; the deformations will continue until the soil reaches a new geometric configuration that is consistent with the available shear strength. If the shear strength is not significantly lower than the driving shear stress, only small deformations may be required to bring the shear stress down to the level of the shear strength. If the shear strength is well below the driving stress, however, large deformations may be required to reach a new condition of static equilibrium.

The reduction of strength and the initiation of flow sliding due to liquefaction was discussed in Sections 9.2.1 and 9.5.1. Strength loss can also occur in sensitive clays when disturbed by external loading or in soils subject to increased pore pressure due to changes in groundwater levels. Predicting the effects of flow sliding, particularly the distances over which the unstable materials are expected to flow, is an extremely difficult problem, as the slide material spreads, splits, and rejoins as it flows across complex three-dimensional terrain. Prediction of flow slide runout distance is fraught with uncertainty that must be considered in interpreting the results of runout analyses.

10.10.2.1 Empirical Procedures for Estimating Flow Slide Runout

Empirical procedures are typically based on statistical analysis of runout behavior observed in actual flow slides. One common empirical procedure is based on the angle of a line connecting the top of the landslide source zone to the most distant margin of the displaced material (Figure 10.41). Once the landslide source (location and volume) has been characterized this *angle-of-reach* can be used to estimate the runout length

$$L = \frac{H}{\tan\alpha} \tag{10.45}$$

The angle of reach has been found (e.g., Corominas, 1996; Devoli et al., 2009) to be a useful measure of landslide mobility that can be related to the volume of the landslide according to

$$\alpha = \tan^{-1}\left(10^{A+B\log V}\right) \tag{10.46}$$

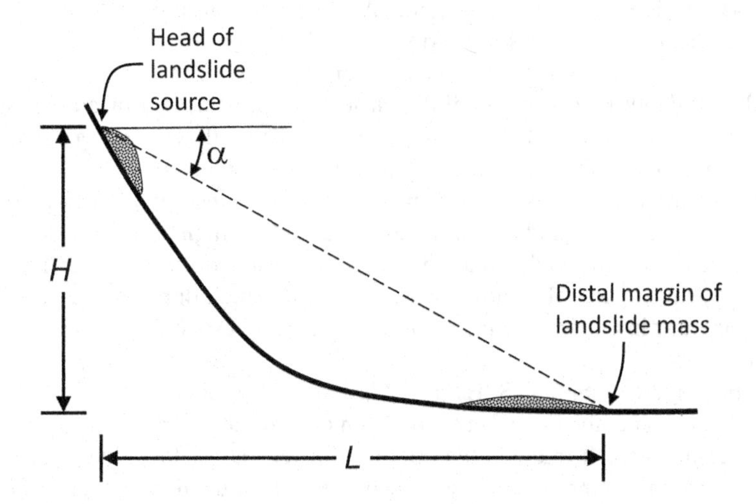

FIGURE 10.41 Illustration of angle of reach of landslide deposit.

TABLE 10.5
Coefficients and coefficient of determination, r^2, for estimation of angle of reach in Equation (10.46)

Landslide Type	Number of Cases	r^2	A	B
All	204	0.625	–0.047	–0.085
Rockfalls	47	0.759	0.210	–0.109
Translational slides	69	0.819	–0.159	–0.068
Debris flows	71	0.763	–0.012	–0.105
Earthflows and mudslides	17	0.648	–0.214	–0.070

Source: After Corominas, 1996.

where V is in units of m^3 and values of the coefficients A and B have been found to be different for different types of landslides, as shown for broad categories in Table 10.5. Values for subdivisions of each category are given in the reference. Empirical runout models have been developed for specific regions (e.g., Finlay et al., 1999; Miller and Burnett, 2008; Strîmbu, 2011; Tang et al., 2012) where similarity of materials and geologic processes can lead to more accurate runout predictions.

10.10.2.2 Numerical Procedures for Estimating Flow Slide Runout

Flow slides can also be analyzed using classes of numerical models capable of handling large strains and large deformations. Conventional continuum analyses, such as finite element and finite different analyses, are formulated in an Eulerian framework that can lead to mesh distortion at moderate to large strain levels. Lagrangian finite element and finite difference formulations can allow development of large strains, but require remeshing and remapping of variables (Li and Liu, 2002). Most current models for flow slides come from the computational fluid dynamics realm and treat the flowing soil as an equivalent fluid with relatively simple rheological properties that approximate the overall behavior of soil and fluid mixtures that constitute actual flow slides (McDougall and Hungr, 2004). Several computational approaches have been applied to the modeling of flow slides; Soga et al. (2016) provide a comprehensive review of many of them.

The computational burden of fully three-dimensional flow analyses can be reduced significantly by using depth-averaged models (Savage and Hutter, 1989; Hungr, 1995; Iverson, 1997). Depth-averaged models provide efficiency by integrating the equations of motion for open-channel flow over the depth of flow at each point. Although the variation of flow with depth is lost, depth-averaging effectively converts three-dimensional problems to two-dimensional problems. The equations of motion are based on shallow-water assumptions: (1) gradual flow depth variation, (2) small flow depth relative to areal dimensions of the slide, (3) incompressible material, and (4) stress-free flow surface. These assumptions imply that flow occurs generally parallel to the base of the flow so that shear stresses on planes perpendicular to the base are negligible and normal stresses on the base are essentially hydrostatic. The constitutive properties of the flowing soil are incorporated into the basal flow resistance. The Bingham model (Johnson, 1970) is often used to model the fluid and Coulomb friction is assumed on the base. This formulation allows the flow to leave its initial location (where its base is the failure surface) and flow across a three-dimensional terrain until it spreads, thins, and comes to rest either on flatter ground or upon reaching obstructions.

Smooth particle hydrodynamics (SPH) methods discretize a domain into particles with material properties that are smoothed by a kernel function over a spatial smoothing length (Monaghan, 1988). The procedure is mesh-free and thus can handle very large flow-type deformations. It has been applied to a number of problems in geomechanics (McDougall and Hungr, 2004; Bui et al., 2008; Augarde and Heaney, 2009; Dai and Huang, 2015, 2016). However, SPH can exhibit spatial instabilities and treatment of boundary conditions can be complicated. The need to search for neighboring particles renders the technique computationally demanding.

Another powerful approach that has become increasingly well-developed is the Material Point Method, or MPM (Sulsky et al., 1994; 1995). The MPM uses a hybrid Eulerian-Lagrangian approach in which material points representing soil move through a stationary background mesh. The material points have mass, density, volume, position, and velocity and carry stress, strain, and constitutive model variables with them as they move through the mesh. At each time step, the information carried by the material points is mapped, using shape functions, to the nodes of the background mesh. The equations of motion are then solved and new positions of the material points computed. The incremental strain is computed from the new material point positions and used with the constitutive model to compute the corresponding stress increments. The background mesh is then returned to its original position and the updated material point data mapped back to its nodal points (using the same shape functions) so the process can be repeated for the next time step. Because the geometry of the background mesh is re-established after each time step, it is never excessively deformed even though the material points may undergo large displacements. Figure 10.42 shows an illustration of a flow slide striking a square bridge column at different angles, and the forces the flow imparts upon the column, as modeled using the MPM.

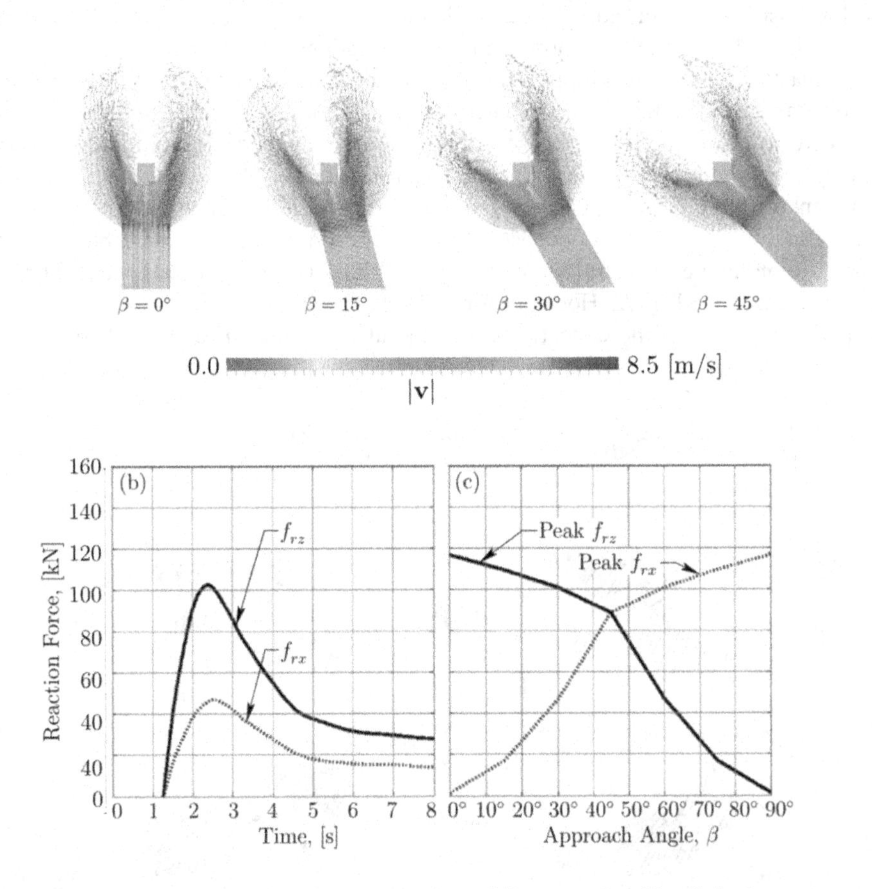

FIGURE 10.42 Flow slide striking a square column from different angles: (a) soil displacement pattern with particle velocity, (b) development of soil impact force on column faces, and (c) variation of peak impact force on column faces as function of approach angle. (After Mast et al., 2014; used with permission of Springer Nature BV).

10.11 ROCK SLOPE STABILITY

Rock slopes can also become unstable through a wide variety of mechanisms (Goodman and Kieffer, 2000) under both static and seismic loading. In contrast to soils, rock slope stability is often controlled by discontinuities such as joints, cracks, and fissures that bound individual blocks of rock. In some cases, however, instabilities can involve failure of intact rock, or by a combination of movement along discontinuities accompanied by failure of intact rock or fracture of intact blocks as they move.

The sizes of rock blocks depend on the spacing of the discontinuities that form them, which span across a spectrum of dimensions. In some cases, discontinuities are located far apart relative to the dimensions of the slope, in which case a stability evaluation may involve only a single block. In other cases, the discontinuities may be much more closely spaced (or the slope much taller) in which case a slope failure may involve so many blocks of rock that continuum methods, such as those used to evaluate the stability of soil slopes, may be appropriate. Rock slope stability is even more strongly influenced by the geometry of the discontinuities relative to the face of the slope. Joints or fissures that are oriented approximately parallel to the slope face can have a significantly adverse effect on stability while discontinuities oriented perpendicular to the face may have little effect.

The orientation and material characteristics of rock discontinuities and any material within them can determine the mode of failure, whether or not an individual block of rock can physically become unstable, and the level of seismic loading that would cause it to become unstable.

Rock slope instabilities can generally be divided into three modes, or categories – *slides*, *topples*, and *slumps* (Figure 10.43). Slides involve primarily translational motion, i.e., sliding on one or more relatively planar failure surfaces. Topples involve forward rotation of vertically or nearly vertically jointed blocks that tip and fall as their centers of gravity move beyond the edges of their bases. Slumps involve both downslope translation and backward rotation of single or multiple blocks often leaving some blocks acting as beams with edge-to-face contacts.

Detachment of rock blocks from a slope face can form rockfalls that have tremendously damaging, though usually localized, consequences. In order for a rockfall to occur, a block bounded by intersecting discontinuities must be kinematically capable of detaching from the face. Three criteria for detachment (Markland, 1972; Hoek and Bray, 1981) are: (1) potential joint planes that intersect (or "daylight" on) the face of the slope, (2) potential joint planes that dip at an angle that exceeds the

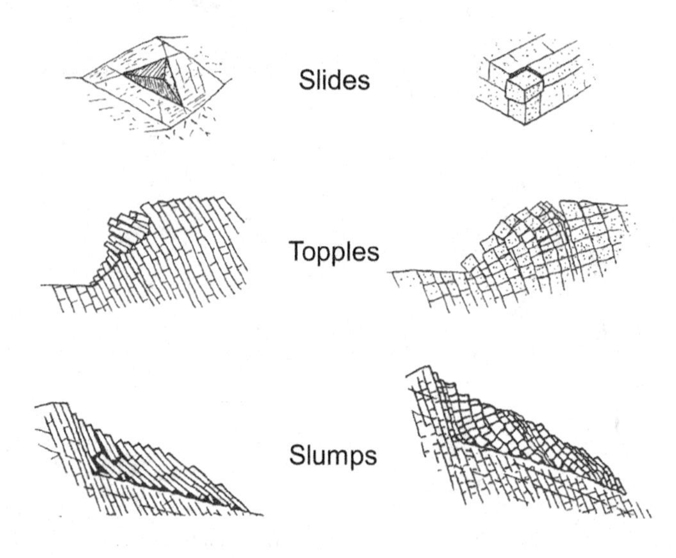

FIGURE 10.43 Illustration of common modes of rock slope failure. (After Goodman and Kieffer, 2000; used with permission of ASCE).

interface friction angle, and (3), dip directions of the joint sets and the face of the slope that differ by no more than 20°. If these conditions are not satisfied, the block may be kinematically incapable of detaching regardless of the strength of shaking. If they are, the block can potentially become detached and bounce/roll down the slope beneath it.

10.11.1 Empirical Stability Evaluation

A number of empirical procedures for qualitative evaluation of seismic rock slope stability hazards have been proposed. Keefer (1993) developed a decision tree framework based on geologic and topographic factors using observations from 60 slopes in 24 earthquakes. The evaluation depends on eight slope characteristics – height, inclination, degree of weathering, strength of induration, openness of fissures, spacing of fissures, presence of vegetation, and moisture conditions. As illustrated in Figure 10.44, the framework divides instability hazards into five categories – low, moderate, high, very high, and extremely high.

10.11.2 Numerical Stability Evaluation

While qualitative empirical methods are very useful for identification of potentially hazardous sites and prioritization of repair between multiple sites, quantification of rock slope instability hazards is often required. Such efforts generally rely upon numerical procedures for evaluation of stability. A number of sophisticated procedures are available; their results, however, should always be interpreted with consideration of their assumptions and the uncertainty in the input parameters required to use them.

10.11.2.1 Single Blocks

Single blocks, represented by a rectangular block resting on an inclined plane in Figure 10.45a, can slide and/or topple in response to seismic loading. If the strength of the intact block is much greater than the strength of the interface, the block can be assumed rigid and stability evaluated using common limit equilibrium techniques. Four states are possible with such a block: (1) stability, (2) sliding only, (3) toppling only, and (4) sliding and toppling. The state that a particular block will be in depends on its height/width, the angle of the slope, the friction angle of the interface, and the level

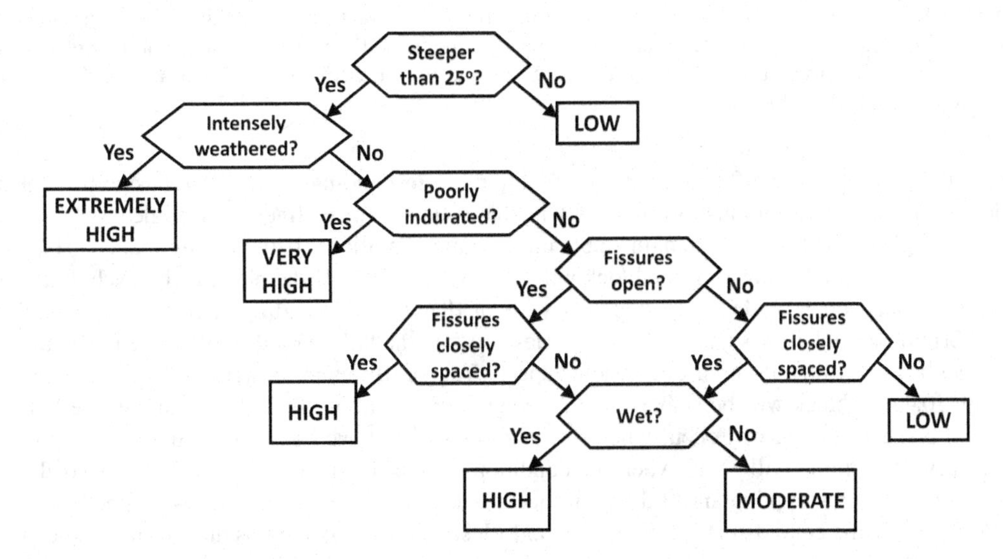

FIGURE 10.44 Decision tree for evaluation of rock slope stability hazard (after Keefer, 1993; used with permission of Elsevier Science and Technology Journals).

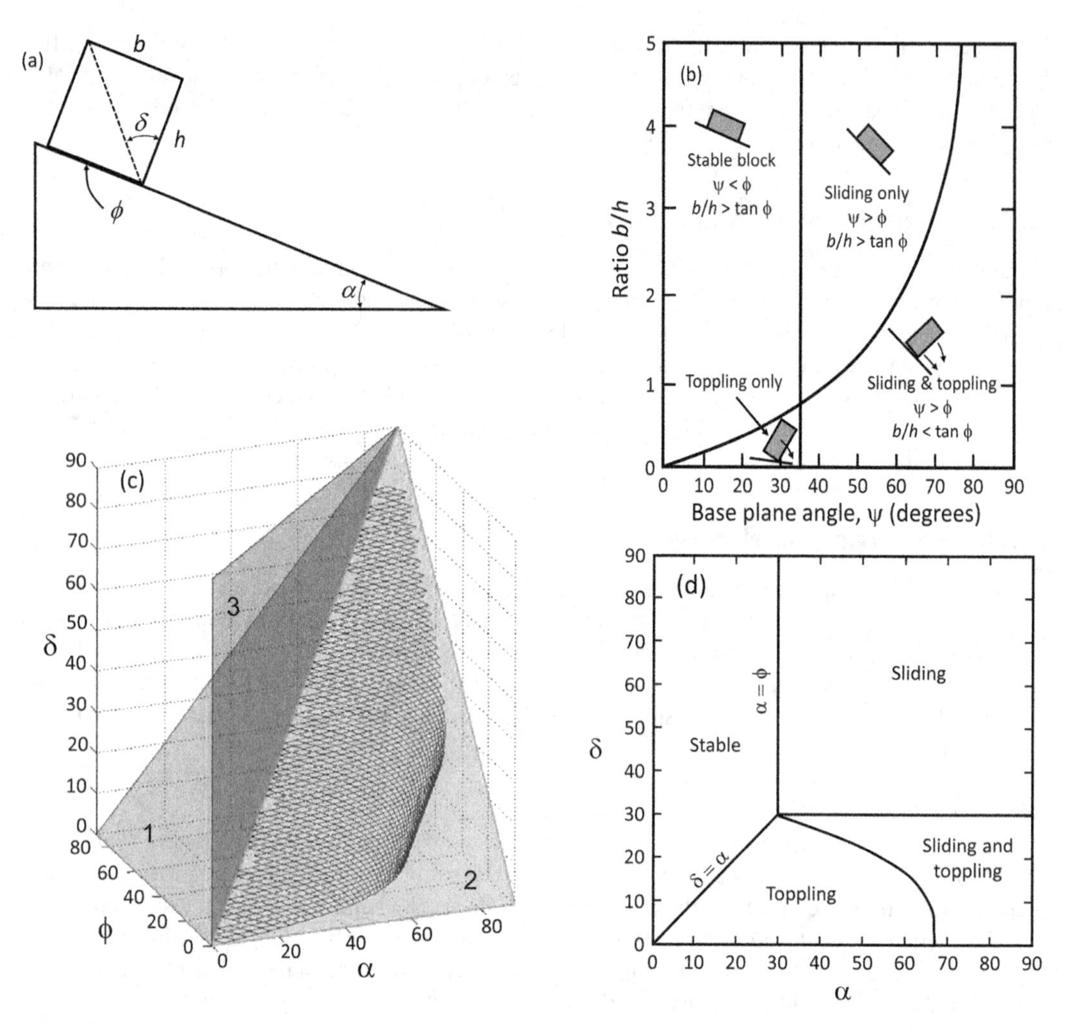

FIGURE 10.45 Single rectangular block resting on inclined plane: (a) notation, (b) kinematic failure modes for static conditions, (c) kinematic failure modes for dynamic conditions, and (d) kinematic failure modes for dynamic conditions with $\phi=30°$ (after Yagoda-Biran and Hatzor, 2013; used with permission of Elsevier Science and Technology Journals).

of inertial loading. Figure 10.45b shows the conditions corresponding to the different stability states under static loading conditions. A relatively wide block on a plane inclined at an angle less than the friction angle of the base will remain stable under static conditions. If the failure plane is steeper than the friction angle, however, the block will slide on the plane. If the block is relatively tall and narrow, it may topple with (steep slope) or without (flatter slope) sliding. Considering dynamic equilibrium, inertial effects cause the boundaries between the failure modes to change in a manner that can be represented in three-dimensional space (Yagoda-Biran and Hatzor, 2013). As shown in Figure 10.45c, a block will be stable when its conditions plot above boundary 1 and to the left of vertical plane 3. Sliding will occur when the conditions plot above boundary 2 and to the right of boundary 3. Toppling will occur when the conditions plot below inclined plane 1 and beyond the curved surface, and toppling and sliding will occur when the conditions are below inclined plane 3 and in front of the curved surface. Figure 10.45d illustrates the relative positions of the modes for and $\phi=30°$.

Example 10.7

A block of rock is three times as tall as it is wide (h=3b) and is in equilibrium on a 1:1 (H:V) slope with a friction angle of 30°. What are the most likely failure modes for static and dynamic conditions?

Solution:

The geometry of the rock block indicates that $\delta=\tan^{-1}(30°)=18.4°$ and the geometry of the block indicates $\alpha=45°$. Since $\delta<\alpha$ and $\phi<\alpha$, Figure 10.45 shows that sliding and toppling is the most likely static failure mechanism and toppling is most likely for dynamic conditions.

The condition of rectangular-shaped blocks can be relaxed to account for the response of blocks of parallelogram shape (Gibson et al., 2018). The geometry of these blocks can be defined by the base slope, β, and the backward and forward angles, α_1 and α_3, respectively, as illustrated in Figure 10.46a. Figure 10.46b illustrates the shapes of blocks defined by different combinations of backward and forward angle. Note the wide variety of block shapes that can be defined in this way; the condition of $\alpha_1=\alpha_3$ corresponds to a rectangular block.

This notation can also be used to define the conditions at which different modes of failure are expected. Figure 10.47a shows these conditions for cases in which the block is supported only on its base (i.e., where the back wall of the block is not in contact with the supporting medium) and where the block is supported by the base and back wall. The case of confined toppling (Figure 10.47b) occurs when the backwall overhangs the block. The transitions from one mode of failure to another occur at different combinations of block angles.

Yield accelerations can be computed using pseudostatic analyses for the different modes of failure illustrated in Figure 10.47. For the cases of ϕ=30° and 60°, Figure 10.48 shows pseudostatic yield coefficients for base- and backwall-supported blocks with base inclinations of 0° and 20°. Note that the shapes of the zones corresponding to different failure modes are similar to those for static failure (Figure 10.47b) but the sizes of the zones change with ϕ and β. The zone for sliding failure decreases with increasing interface friction angle and the yield accelerations decrease with increasing base inclination.

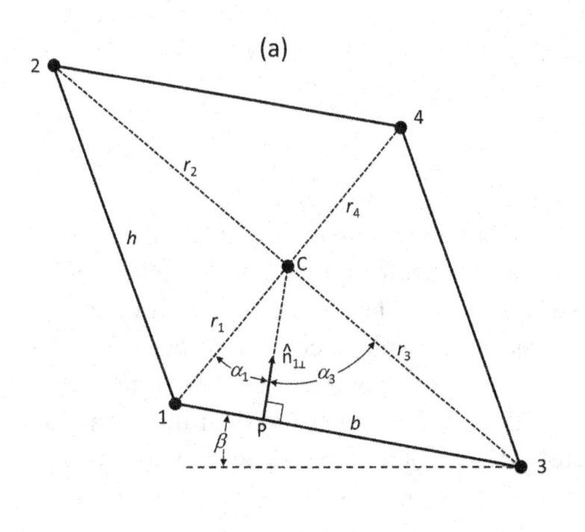

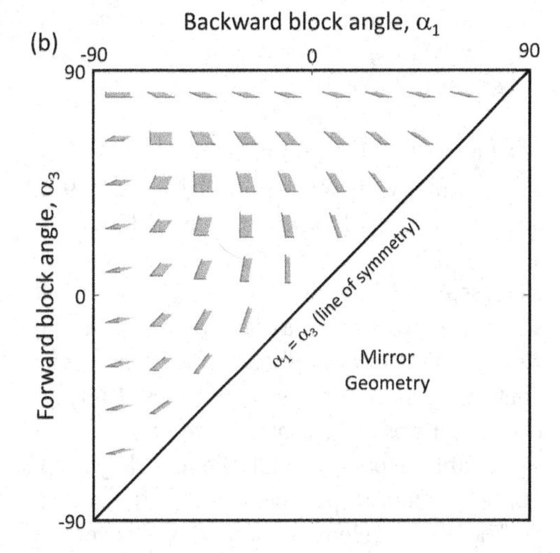

FIGURE 10.46 Notation for parallelogram-shaped blocks: (a) block geometry, and (b) examples of blocks defined by different backward and forward angles for β=0 (after Gibson et al., 2018; used with permission of Elsevier Science and Technology Journals).

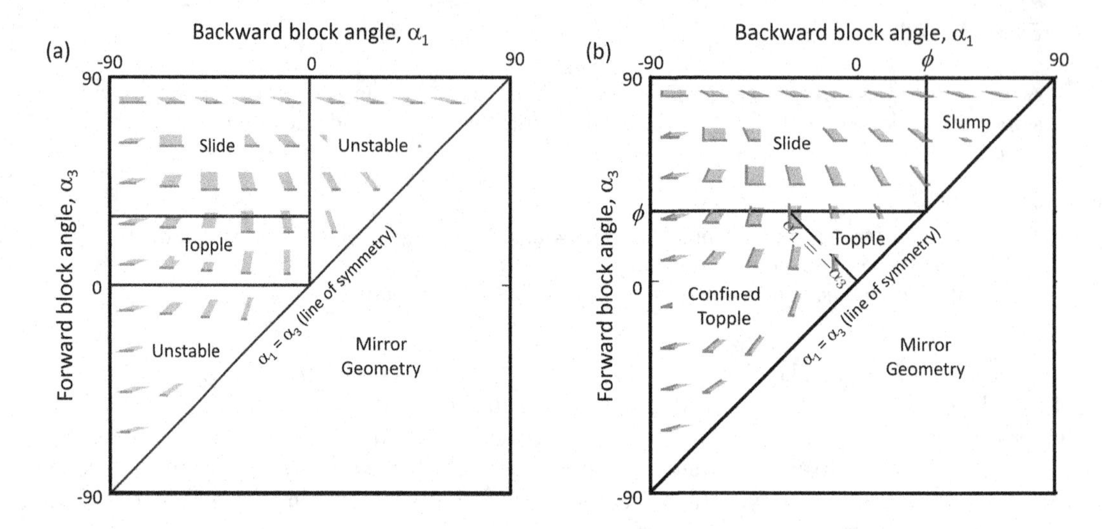

FIGURE 10.47 Modes of static failure for: (a) block supported only on its base, and (b) block supported on base and back wall (after Gibson et al., 2018; used with permission of Elsevier Science and Technology Journals).

10.11.2.2 Multiple Blocks

Most rock slopes contain multiple blocks bounded by multiple discontinuities (Figure 10.49). Analysis of the seismic stability of rock slopes with multiple blocks is more complex as the contacts and forces between closely spaced, interacting blocks must be accounted for. The potential for topographic amplification (Section 7.6.3) must also be accounted for. Multiple-block slopes can involve sliding, toppling, and/or slumping. The mechanisms may be relatively simple, as in Figure 10.45 or very complex, depending on the number, shapes, and arrangements of the blocks. Discrete element methods have been developed to handle problems such as these. The basic approach in discrete element methods is the formulation and solution of the dynamic equations of motion of all blocks within the system. The solution generally proceeds iteratively as blocks move and come into and out of contact with each other. Discrete element procedures must be capable of describing the behavior of all of the individual blocks themselves, and of all of the interfaces between the blocks – because there may be many thousands of each, discrete element analyses can be computationally demanding. Several variations of discrete element analyses are commonly used in practice (Eberhardt, 2003).

Distinct Element Analysis

The distinct element method (Cundall, 1971; Cundall and Strack, 1979) treats the rock mass as an assemblage of blocks connected by interface elements of user-defined stiffness. Early versions of this approach treated the blocks as rigid but later versions allowed them to be deformable. The assumption of rigidity may be reasonable at relatively low stress levels (i.e., in small blocks at shallow depths), but may not when stresses and/or distances between block contact points are large. The fact that the interface elements have some finite stiffness indicates that they represent "soft" contacts, which imply some degree of interpenetration of the blocks. While physical interpenetration is not possible, contacts are generally not completely rigid – otherwise, seismic waves would not be able to pass through them as they are known to do. Therefore, the soft contact assumption is not unreasonable, particularly in softer rock materials.

The distinct element method was developed to evaluate the two-dimensional response of jointed rock masses (Cundall, 1971), but has been extended to three-dimensional blocks. Distinct element codes track the behavior of blocks dynamically, i.e., by computing their movements and rotations in response to the net dynamic forces and moments acting upon them by forces at their contacts with

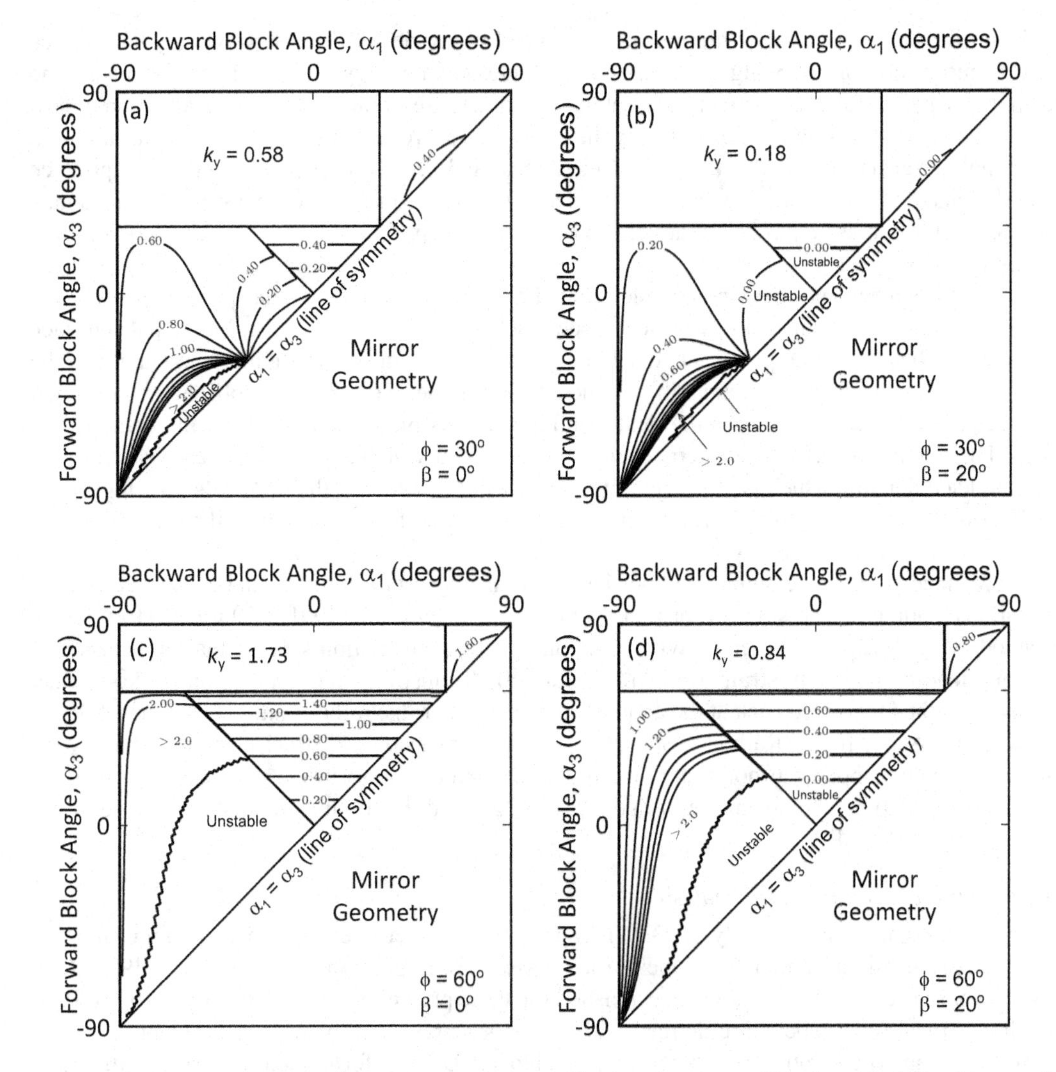

FIGURE 10.48 Contours of yield acceleration base- and backwall-supported blocks with: (a) level base and ϕ=30°, and (b) 20° base and ϕ=30°, (c) level base and ϕ=60°, and (d) 20° base and ϕ=60°. (After Gibson et al., 2018; used with permission of Elsevier Science and Technology Journals.)

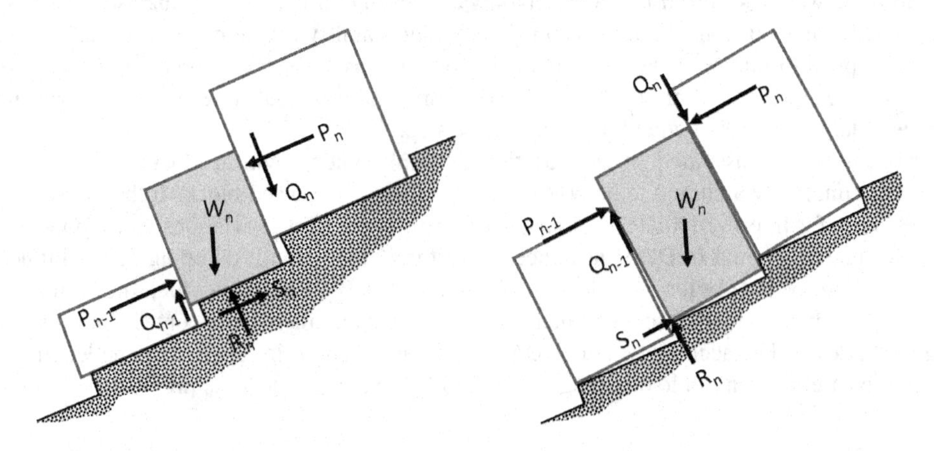

FIGURE 10.49 Multiple block potentially moving by: (a) sliding, and (b) toppling. (After Hoek and Bray, 1981; used with permission of ASCE.)

neighboring blocks. The block motions are computed in a series of time steps with dynamic force and moment equilibrium being satisfied and momentum and energy being conserved at each time step. At each time step, the contact forces at block contacts are evaluated (with consideration of new and lost contacts that might develop during the time step), net resultant forces and moments are used to compute translational and rotational accelerations, and accelerations are integrated to produce translational and rotational velocities and displacements at the end of the time step; iterations may be required to achieve equilibrium at the end of the time step. The process is then repeated for the next time step.

Distinct element analyses are complex and the computer programs that perform them must address a number of important technical issues including contact detection, block and interface constitutive behavior, damping, boundary conditions, visualization, and numerical stability. The last item is of considerable practical importance; the time integration is accomplished in an explicit manner (Section 7.5.3.2) so its numerical stability depends on the length of the integration time step. The maximum stable time step depends on the rate at which a disturbance can pass through the modeled domain, which is a function of the mass and stiffness of the materials. The higher the stiffnesses (of the blocks and/or the interfaces), the shorter the stable time step will be and the longer it will take to perform an analysis.

Nevertheless, distinct element methods have proven very useful for two- and even three-dimensional evaluations of the stability of rock slopes (e.g., Lorig et al., 1991; Nishimura et al., 2010; Rathod et al., 2012). The inherent dynamic nature of their calculation scheme makes inclusion of seismic loading relatively straightforward. Figure 10.50 illustrates the application of the distinct element method to jointed rock at the abutments of an arch bridge rail crossing of the Chenab River in Kashmir State, India (Rathod et al., 2012). Three-dimensional distinct element models were made based on mapped sub-horizontal foliation joints and two sub-vertical joint sets. The analyses included foundation loads from the bridge piers and showed that both static and seismic deformations were quite small.

Discontinuous Deformation Analysis

Discontinuous deformation analysis (DDA) is an implicit, discrete element procedure for evaluating static and dynamic movements of blocky rock masses (Shi, 1988; Shi and Goodman, 1989; Jing, 1998) whose accuracy has been well established for static problems (Doolin and Sitar, 2004). Two- or three-dimensional blocks are defined with contacts represented by stiff, bi-directional springs that prevent interpenetration or tension between blocks. DDA is formulated in terms of displacements with an implicit time integration scheme, so dynamic equilibrium equations are derived by minimizing the total potential energy of the system and solved simultaneously at each time step. This approach, which is similar to that commonly employed in finite element analyses, ensures that equilibrium is satisfied at all time steps and allows static stability problems to be solved in a single step. The implicit formulation also eliminates the potential instability that often requires very small time steps in explicit integration schemes for dynamic analyses, but it requires time-consuming solutions of large systems of equations at each time step.

The original DDA assumed linear variation of displacements within blocks, so stresses and strains were implicitly assumed to be constant. Later versions allowed blocks to be discretized by finite elements for improved analysis of cases with significant stress variations within blocks. The use of Newmark damping in DDA introduces algorithmic (numerical) damping that is influenced by the time step of integration, which in turn influences the block contact identification process. The numerical damping increases with increasing time step and increasing contact spring stiffness (Feng et al., 2017). The accuracy of the DDA process can then be influenced by the length of the time step so careful attention to damping is required for analysis of seismic problems.

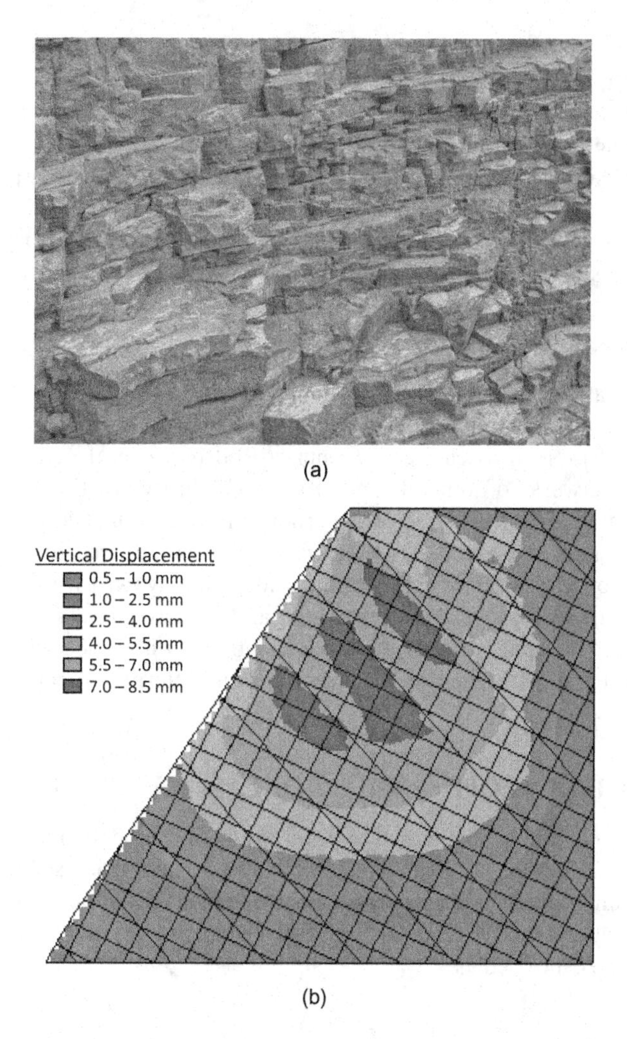

FIGURE 10.50 (a) Jointed dolomitic rock at left abutment of Chenab River crossing and (b) contours of vertical displacement from distinct element analysis of right abutment. (After Rathod et al., 2012; used with permission of ASCE.)

Bonded Particle Analysis

Rock behavior has been characterized as being similar to that of cemented assemblages of complex-shaped grains in which both the grains and cementing agents are deformable and capable of fracturing (Potyondy and Cundall, 2004). The bonded particle model (BPM) represents rock by dense packings of non-uniform circular (two-dimensional case) or spherical (three-dimensional case) particles bonded at their contact points. The particles and bonds are assumed to have finite stiffness and strength values, and fractures of the bonds under loading can coalesce into macroscopic cracks. Bonded particle models have been shown to capture many important aspects of the behavior of intact and fractured rock, including elasticity, fracturing, friction, nonlinearity, hysteresis, dilation, stress/strain history effects, and material anisotropy. By allowing bond fracture, BPM analyses can account for the fracturing of intact rock or intact rock blocks that may accompany (or kinematically permit) rock slope instability.

BPM analyses have been used for static rock slope stability (Wang et al., 2003; Camones et al., 2013; Stead and Wolter, 2015) and for wave propagation modeling (Olson et al., 1999; Resende et al., 2010; Marketos and O'Sullivan, C., 2013). Arnold et al. (2016) found that a BPM model with rigid particles was able to model the earthquake-induced collapse of two weathered and fractured cliffs in Christchurch, NZ that could not be predicted or explained by more traditional methods. The depth to which the collapse occurred was reasonably predicted as were the development of benched and overhanging post-failure geometries and the observed formation/dilation of cracks near the post-failure cliff face. Although providing good insight into these aspects of case history behavior, the BPM analyses severely underpredicted the vertical and horizontal deformations of the remaining slope. Nevertheless, the BPM appears to be a useful complementary numerical tool for the evaluation of rock slope stability.

10.11.3 Discussion

Assessment of the stability of rock slopes is complicated under static, let alone seismic, conditions. Characterization of variability and uncertainty in rock slope geometries and material behavior is important in such assessments (Jimenez-Rodriguez et al., 2006). Rock masses are often intersected by networks of joints that control their stability and the existence of removable blocks and there is always some degree of uncertainty in the geometric properties of joint networks, particularly below the surface. Also, the uncertain depth, degree, and extent of weathering affect the material properties of the joints and materials that may exist within them. The presence and effects of water in rock joints, in situ stresses, and joint roughness can also be significant sources of uncertainty in input parameters. When combined with model uncertainty, the results of stability assessments must be considered uncertain and interpreted carefully with respect to project performance criteria.

10.12 PROBABILISTIC SLOPE STABILITY HAZARD EVALUATION

The availability of probabilistic slope displacement prediction models means that the principles described in Section 5.5 can be applied to the problem of seismic slope stability evaluation. This type of evaluation can be described as a probabilistic slope stability hazard evaluation (PSSHA). With permanent displacement, D, taken as the *EDP*, Equation (5.16) for the mean annual rate of exceeding $D=d$ can be expressed as

$$\lambda_D(d) = \int_0^\infty P[D > d \mid IM] |d\lambda_{IM}| \tag{10.47}$$

A predictive relationship for rigid block models using *PGA* as a scalar intensity measure was presented in Equations (10.28). For that model, the conditional probability term can be computed as

$$P[D > d \mid IM] = 1 - \Phi\left[\frac{\ln d - \mu_{\ln D|IM}}{\sigma_{\ln D|IM}}\right] \tag{10.48}$$

where $\mu_{\ln D|IM}$ and $\sigma_{\ln D|IM}$ are the mean and standard deviation of ln D, respectively. More recent studies have shown that mixed scalar and vector predictive relationships show considerably lower variability in sliding block displacement predictions. For a mixed scalar relationship, such as that of Equation (10.30), the mean annual rate of displacement exceedance can be expressed as

$$\lambda_D(d) = \lambda_{\mathbf{M}_{\min}} \int_{\mathbf{M}_{\min}}^{\mathbf{M}_{\max}} \int_0^\infty P[D > d \mid pga, \mathbf{m}] f_{\mathbf{M},PGA}(pga, \mathbf{m}) d(pga) d\mathbf{m} \tag{10.49}$$

where $\lambda_{\mathbf{M}_{\min}}$ is the mean annual rate of earthquakes greater than the minimum magnitude and $f_{\mathbf{M},PGA}(pga, \mathbf{m})$ is the joint distribution of *PGA* and **M**. In discrete form, and using Equation (D.25) to break down the joint distribution of *PGA* and **M**, Equation (10.49) can be rewritten as

$$\lambda_D(d) = \lambda_{\mathbf{M}_{\min}} \sum_{i=1}^{N_{PGA}} \sum_{j=1}^{N_{\mathbf{M}}} P\left[D > d \mid PGA_i, \mathbf{m}_j\right] P\left[\mathbf{m}_j \mid PGA_i\right] P\left[PGA_i\right] \tag{10.50}$$

The conditional distribution of **M**|*PGA* can be extracted from PSHA disaggregation data. The probabilistic framework can be extended (e.g., Macedo et al., 2018) to include uncertainty in other parameters, such as yield acceleration or slope period, used in various permanent displacement prediction models.

Vector predictive relationships (e.g., Equation 10.31) have also been developed and shown to have significantly lower variability than scalar and mixed scalar relationships. In a manner similar to that expressed above for the mixed scalar approach, a vector response analysis (Section 5.7) can be used to compute a displacement hazard curve. For a *PGA-PGV*-based vector predictive relationship, the mean annual rate of exceeding specified levels of displacement can be expressed as

$$\lambda_D(d) = \lambda_{\mathbf{M}_{\min}} \int_0^{\infty}\int_0^{\infty} P\left[D > d \mid pga, pgv\right] f_{PGA,PGV}\left(pga, pgv\right) d\left(pga\right) d\left(pgv\right) \tag{10.51}$$

where $f_{PGA,PGV}\left(pga, pgv\right)$, the joint distribution of *PGA* and *PGV*, can be computed from a vector PSHA. If a vector PSHA is not available, Equation (10.51) can be rewritten in discrete form as

$$\lambda_D(d) = \lambda_{\mathbf{M}_{\min}} \sum_{i=1}^{N_{PGA}} \sum_{j=1}^{N_{PGV}} P\left[D > d \mid PGA_i, PGV_j\right] P\left[PGV_j \mid PGA_i\right] P\left[PGA_i\right] \tag{10.52}$$

where the conditional probability of *PGV*|*PGA* can be extracted from PSHA disaggregation data as

$$P\left[PGV_j \mid PGA_i\right] = \lambda_{\mathbf{M}_{\min}} \sum_{k=1}^{N_{\mathbf{M}}} \sum_{l=1}^{N_R} P\left[PGV_j \mid PGA_i, \mathbf{m}_k, R_l\right] P\left[\mathbf{m}_k, R_l \mid PGA_i\right] \tag{10.53}$$

which also requires GMPEs for *PGA* and *PGV* and knowledge of the correlation coefficient of *PGA* and *PGV*. Baker and Bradley (2017) indicate that the correlation coefficient for *PGA* and *PGV* in the NGA-West2 ground motion database is about 0.66.

Mixed scalar and vector predictive relationships have been used (Rathje et al., 2013) to develop permanent displacement hazard curves for a site in northern California. Site-specific PSHAs produced the *PGA* and *PGV* hazard curves shown in Figure 10.51a and b. A vector PSHA was performed to compute the joint distribution of *PGA* and *PGV*, and Equations (10.50) and (10.51) were integrated numerically to produce displacement hazard curves for a slope with a yield acceleration of 0.1*g*. The resulting displacement hazard curves, shown in Figure 10.51c, illustrate the significant effect of model variability on displacement hazard. For a return period of 1,000 years, the displacement using the mixed scalar predictive relationship is approximately 115 cm while the displacement using the vector predictive relationship is about 40 cm. The difference in these displacements is attributable to the reduced variability in the vector predictive relationship – using that relationship reduced the displacement for that hazard level by a factor of nearly 3. Looking at these results in a different manner, the curves of Figure 10.51c show that the mean annual rates of exceeding 1 m of permanent displacement are 0.0017 year^{-1} and 0.00021 year^{-1} (corresponding to return periods of 590 and 4,760 years, respectively) for the mixed scalar and vector predictive relationships, respectively. Using the Poisson assumption, the 100-year exceedance probabilities would be 15.6% and 2.1%, respectively – the lower variability in the vector predictive relationship has reduced the 100-year exceedance probability by a factor of 7.5. These results illustrate the tremendous benefits that

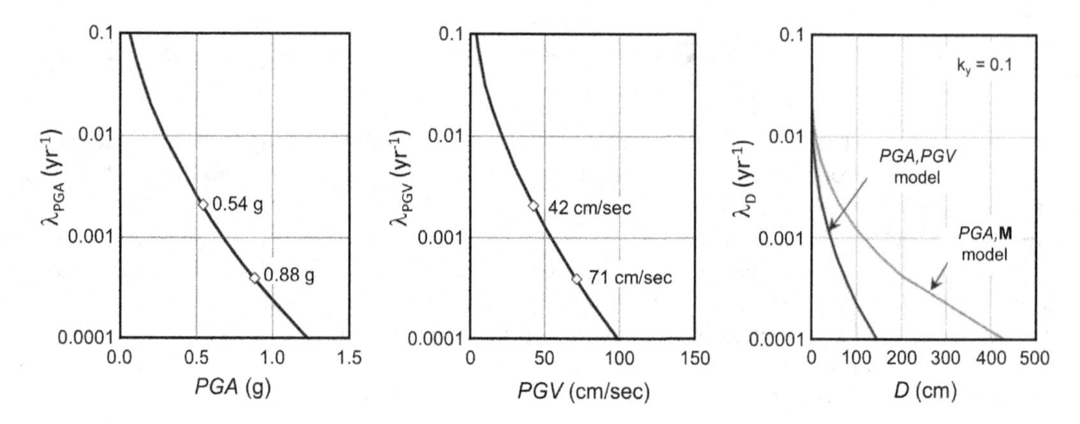

FIGURE 10.51 Results of performance-based slope stability hazard evaluation: (a) *PGA* hazard curve, (b) *PGV* hazard curve, and (c) permanent displacement hazard curves for mixed scalar (*PGA* and **M**) and vector (*PGA* and *PGV*) predictive relationships. (After Rathje et al., 2013; used with permission of Springer Nature BV.)

can be realized by developing improved probabilistic response analyses and coupling them with probabilistic seismic hazard analyses.

The PSSHA procedure has been extended (Rodriguez-Marek and Song, 2016) to include the effects of pulse-like near-fault forward directivity motions (Section 3.7.1.2), which can cause larger permanent slope displacements than ordinary motions. Pulse-like motions have been recorded in a number of events such as the 2008 Wenchuan (Wen et al., 2010), 2010 Darfield (Bradley, 2012), and 2011 Christchurch (Bradley and Cubrinovski, 2011) earthquakes, in which significant landsliding was observed.

10.13 SUMMARY

1. When the fault rupture that produces earthquakes reaches the ground surface, the resulting ground deformations can produce substantial damage to infrastructure crossing the principal faults or near the principal fault in a zone of distributed rupture.
2. The amount of displacement is typically largest on the principal fault trace. This displacement depends strongly on magnitude and to a lesser degree on location along the fault and focal mechanism (strike-slip, reverse, normal). Distributed ruptures on splay faults depend on the amount of rupture on the principal fault, distance to the principal fault, and focal mechanisms. These distributed deformations are often most significant in stepover zones for strike-slip faults and on the hanging wall of dip-slip faults.
3. Fault displacement hazard analysis, which follows similar procedures as seismic hazard analysis (Chapter 4), is used to predict time rates of exceedance for different fault displacement levels, which are expressed in the form of fault displacement hazard curves.
4. Historically, earthquake-induced landslides have been among the most damaging of all seismic hazards. Their characteristics are influenced by geologic, hydrologic, topographic, climatic, weathering, and land-use conditions. Slides can be classified on the basis of material type, type of movement, degree of internal disruption, water content, velocity, and depth. Earthquake-induced landslides are usually divided into three main categories: disrupted slides and falls, coherent slides, and lateral spreads and flows.
5. The seismic performance of a slope or embankment, expressed in terms of potential damage to infrastructure located above, on, below, or embedded within it, is most closely related to the permanent deformation it undergoes. Permanent deformations are much

more efficient predictors of physical damage than force/stress-related measures of response such as factor of safety.

6. A slope stability analysis is only one part of a comprehensive evaluation of slope stability. Prior to the analysis, detailed information on geologic, hydrologic, topographic, geometric, and material characteristics must be obtained. The accuracy of the analysis will be only as good as the accuracy of this information.
7. The dynamic shear stresses produced by earthquake shaking represent a source of loading and may also influence the strength and stress-strain behavior of the slope materials. Seismic slope instabilities may be grouped into two categories on the basis of which of these effects are predominant in a given slope. Inertial instabilities are those in which the shear strength of the soil remains essentially constant and slope deformations are caused by its temporary exceedance by dynamic earthquake stresses. Weakening instabilities occur when the earthquake serves to weaken the soil sufficiently that it cannot remain stable under earthquake-induced stresses. Issues associated with these instabilities, which are often associated with liquefaction or cyclic softening, are covered in Chapter 9.
8. Inertial instabilities are most commonly analyzed by pseudostatic, sliding block, or stress-deformation analyses. These analyses differ in complexity, accuracy, and the utility of their results with respect to performance assessment.
9. Pseudostatic analyses represent the effects of an earthquake simply and crudely by applying static horizontal and/or vertical accelerations to a potentially unstable mass of soil. The inertial forces induced by these pseudostatic accelerations increase the driving forces and may decrease the resisting forces acting on the soil. Pseudostatic analyses are not appropriate for soils that build up large pore pressures or show more than about 15% degradation of strength due to earthquake shaking. Stability is expressed in terms of a pseudostatic factor of safety calculated by limit equilibrium procedures; as such, pseudostatic analyses provide no information on slope movements. Selection of an appropriate pseudostatic acceleration requires great care; values considerably smaller than the peak acceleration of the sliding mass are usually used.
10. The pseudostatic acceleration required to bring a slope to the point of incipient failure is known as the yield acceleration. If earthquake-induced accelerations in a slope momentarily exceed the yield acceleration, the unstable soil will momentarily accelerate relative to the material beneath it. Sliding block analyses, which represent a mass of potentially unstable soil as a block resting on an inclined plane, can be used to calculate the amount of displacement that occurs. The block only moves relative to the plane when the plane's acceleration exceeds the yield acceleration. The total displacement depends on the amount by which the yield acceleration is exceeded (a function of the ground motion amplitude), the time over which the yield acceleration is exceeded (a function of the frequency content of the ground motion), and the number of times the yield acceleration is exceeded (a function of ground motion duration). Given the highly variable nature of ground motion characteristics, computed displacements can be highly variable.
11. Conventional sliding block analyses assume that the block representing the potentially unstable soil is rigid. This assumption is reasonable for shallow failure surfaces, very stiff soil, or low-frequency excitation (cases in which wavelengths are much greater than the depth of the unstable soil). For deeper failure surfaces, soft soils, or higher-frequency motions, however, different parts of the material above the failure surface may be responding out of phase, in which case the total inertial force acting on it will differ from that of the rigid block assumption. This behavior can be handled by decoupled or coupled analyses. Decoupled analyses are more amenable to problems in which the slope geometry is non-uniform (two-dimensional), whereas for one-dimensional problems (i.e., sliding mass can be accurately represented by a soil column) they may overpredict or underpredict the deformations given by more accurate coupled analyses. Sliding block analyses,

however, should be recognized as simplified approximations of actual soil systems and their results interpreted as indices of expected displacements rather than precise displacement predictions.

12. Compilations of the results of multiple sliding block analyses for one-dimensional systems have been expressed in the form of predictive relationships that allow rapid prediction of the distribution of estimated slope displacements. These are available for rigid and compliant blocks based on decoupled and coupled analyses with scalar and vector intensity measures. These relationships include descriptions of variability in their predicted displacements; this variability is with respect to the median predicted displacement, which can vary from the actual displacement.
13. Stress-deformation analyses are increasingly used to estimate permanent deformations caused by inertial instabilities. Stress-deformation analyses can account for deformation mechanisms, such as the accumulation of incremental plastic strains that develop at shear stresses lower than the shear strength of the soil. Although the computational effort is dramatically increased, permanent deformations can be analyzed more rigorously using nonlinear finite element and finite difference techniques. With advanced constitutive models being implemented in numerical platforms that include graphical interfaces, mesh generation, and parallel processing capabilities, stress-deformation analyses have become much more practical and offer insights into the mechanisms and patterns, as well as magnitudes, of slope deformations that are important to understand in geotechnical earthquake engineering practice.
14. Stress-deformation analyses should be carefully planned, conducted, and interpreted to produce the most accurate and useful information regarding slope performance. In advance of the actual analysis, geologic and geotechnical conditions should be understood and data/information gaps identified. Performance objectives should be determined and modes of deformation that could influence satisfaction of those objectives should be anticipated and considered, along with boundary conditions, in mesh development. Constitutive models should be calibrated and checked against available field and laboratory data for single elements, and static analyses should be performed with careful checking of the initial stress conditions. At that point, dynamic analyses with appropriate ground motions can be performed.
15. Flow failure instability is usually evaluated by limit equilibrium analysis. Residual strengths are applied to those portions of the failure surface that pass through liquefied soil. A factor of safety less than 1.0 suggests that flow failure is likely. While rough estimates of flow slide runout can be made with empirical procedures, new computational tools such as the Material Point Method offer the potential for site-specific analysis of flow slides.
16. The stability of rock slopes is governed largely by the orientation, spacing, and properties of discontinuities such as joints, cracks, and fissures that bound individual rock blocks; discontinuities oriented parallel to the slope face are particularly prone to instability. Multiple modes of failure, including sliding, toppling, and slumping can occur. Single blocks of rock can become dislodged and present a hazard by falling or rolling down a slope, or multiple blocks can fail together. Empirical procedures are available for qualitative assessment of rock slope stability. A number of computational methods that treat rock blocks and their interactions explicitly are available for assessment of rock slope stability.
17. Probabilistic slope stability hazard analyses involve the convolution of probabilistic slope displacement models with ground motion hazard curves as discussed in Chapter 5 to develop slope displacement hazard curves. Both scalar and vector methods have been developed. Implementation of the scalar method is straightforward but vector methods require a vector PSHA or disaggregation plus knowledge of the covariance matrix for the *IM*s being used.

REFERENCES

Abramson, L.W., Lee, T.S., Sharma, S., and Boyce, G.M. (2001). *Slope Stability and Stabilization Methods*, 2nd edition, John Wiley and Sons, New York, 736 pp.

Agnesi, V., Carrara, A., Macaluso, T., Monteleone, S., Pipitone, G., and Sorriso-Valvo, M. (1982). "Preliminary observations of slope instability phenomena induced by the earthquake of November 1980 on the upper valley of Sele river," *Geology and Applied Hydrology*, Vol. 17, pp. 79–93.

Ambraseys, N.N. and Menu, J.M. (1988). "Earthquake-induced ground displacements," *Earthquake Engineering and Structural Dynamics*, Vol. 16, pp. 985–1006. https://doi.org/10.1002/eqe.4290160704

Ambraseys, N.N. and Srbulov, M. (1995). "Earthquake induced displacements of slopes," *Soil Dynamics and Earthquake Engineering*, Vol. 14, pp. 59–71. https://doi.org/10.1016/0267-7261(94)00020-h

Anastasopoulos, I. and G. Gazetas (2007). "Foundation-structure systems over a rupturing normal fault: Part I. Observations after the Kocaeli 1999 earthquake," *Bulletin of Earthquake Engineering*, Vol. 5, pp. 253–275. https://doi.org/10.1007/s10518-007-9029-2

Arnold, L., Wartman, J., Massey, C., MacLaughlin, M., and Keefer, D. (2016). "Insights into the seismically-induced rock-slope failures in the Canterbury region using the discrete element method," *Proceedings, 6th International Conference on Earthquake Geotechnical Engineering*, Christchurch, New Zealand, 9 pp.

Augarde, C. and Heaney, C. (2009). "The use of meshless methods in geotechnics," *Proceedings, First International Symposium on Computational Geomechanics (ComGeo I)*, Juan-les-Pins, Cote d'Azur, France, pp. 311–320.

Baker, J.W. and Bradley, B.A. (2017). "Intensity measure correlations observed in the NGA-West2 database, and dependence of correlations on rupture and site parameters," *Earthquake Spectra*, Vol. 33, No. 1, pp. 145–156. https://doi.org/10.1193/060716eqs095m

Benoit, L., Briole, P., Martin, O., Thorn, C., Malet, J.-P., and Ulrich, P. (2015). "Monitoring landslide displacements with the Geocube wireless network of low-cost GPS," *Engineering Geology*, Vol. 195, pp. 111–121. https://doi.org/10.1016/j.enggeo.2015.05.020

Berrill, J.B. (1983). "Two-dimensional analysis of the effect of fault rupture on buildings with shallow foundations," *Soil Dynamics and Earthquake Engineering*, Vol. 2, No. 3, pp. 156–160. https://doi.org/10.1016/0261-7277(83)90012-8

Bishop, A.W. (1955). "The use of the slip circle in the stability analysis of slopes," *Geotechnique*, Vol. 5, No. 1, pp. 7–17. https://doi.org/10.1680/geot.1955.5.1.7

Blake, T.F., Hollingsworth, R.A., and Stewart, J.P. (2002). *Recommended Procedures for Implementation of DMG Special Publication 117-Guidelines for Analyzing and Mitigating Landslide Hazards in California*, Southern California Earthquake Center, Los Angeles, CA, 127 pp.

Boore, D.M., Stewart, J.P., Seyhan, E., and Atkinson, G.M. (2014)." NGA-West 2 Equations for predicting PGA, PGV, and 5%-damped PSA for shallow crustal earthquakes," *Earthquake Spectra*, Vol. 30, No. 3, pp. 1057–1085.

Boulanger, R.W. and Beaty, M.H. (2016). "Seismic deformation analyses of embankment dams: A reviewer's checklist." *Proceedings, Celebrating the Value of Dams and Levees - Yesterday, Today and Tomorrow, 36th USSD Annual Meeting and Conference,* United States Society on Dams, Denver, CO, pp. 535–546.

Bradley, B.A. (2012). "Strong ground motion characteristics observed in the 4 September 2010 Darfield, New Zealand earthquake," *Soil Dynamics and Earthquake Engineering*, Vol. 42, pp. 32–46. https://doi.org/10.1016/j.soildyn.2012.06.004

Bradley, B.A. and Cubrinovski, M., (2011). "Near-source strong ground motions observed in the 22 February 2011 Christchurch earthquake," *Seismological Research Letters*, Vol. 82, pp. 853–865. https://doi.org/10.1785/gssrl.82.6.853

Bransby, M.F., Davies, M.C.R., and El Nahas, A. (2008). "Centrifuge modelling of normal fault-foundation interaction," *Bulletin of Earthquake Engineering*, Vol. 6, pp. 585–605. https://doi.org/10.1007/s10518-008-9079-0

Bray, J.D., Seed, R.B., Cluff, L.S., & Seed, H.B. (1994). "Earthquake fault rupture propagation through soil," *Journal of Geotechnical Engineering*, Vol. 120, No. 3, pp. 543–561.

Bray, J.D. (2001). "Developing mitigation measures for the hazards associated with earthquake surface fault rupture," *Workshop on Seismic Fault-Induced Failures-Possible Remedies for Damage to Urban Facilities*, Japan Society for Promotion of Science, University of Tokyo Press, pp. 55–79.

Bray, J.D. (2007). "Chapter 14: Simplified seismic slope displacement procedures," in K.D. Pitilakis, ed., *Earthquake Geotechnical Engineering, 4th International Conference on Earthquake Geotechnical Engineering - Invited Lectures, in Geotechnical, Geological, and Earthquake Engineering Series,* Springer, Vol. 6, pp. 327–353. https://doi.org/10.1007/978-1-4020-5893-6_14

Bray, J.D. (2009). "Earthquake surface fault rupture design considerations," *Proceedings, Sixth International Conference on Urban Earthquake Engineering, Center for Urban Earthquake Engineering,* Tokyo Institute of Technology, Tokyo, Japan, pp. 37–45.

Bray, J.D., Ashmawy, A., Mukhopadhyay, G., and Gath, E.M. (1993). "Use of geosynthetics to mitigate earthquake fault rupture propagation through compacted fill," *Proceedings of the Geosynthetics '93 Conference,* ASCE, Vol. 1, pp. 379–392

Bray, J.D., Augello, A.J., Leonards, G.A., and Byrne, R.J. (1995). "Seismic stability procedures for solid-waste landfills," *Journal of Geotechnical Engineering,* Vol. 121, No. 2, pp. 139–151. https://doi.org/10.1061/(asce)0733-9410(1995)121:2(139)

Bray, J.D. and Kelson, K.I. (2006). "Observations of surface fault rupture from the 1906 earthquake in the context of current practice," *Earthquake Spectra,* Vol. 22, No. 2, pp. S69–S89. https://doi.org/10.1193/1.21814

Bray, J.D. and Macedo, J. (2019). "Procedure for estimating shear-induced seismic slope displacement for shallow crustal earthquakes," *Journal of Geotechnical and Geoenvironmental Engineering,* Vol. 145, No. 12, p. 04019106. https://doi.org/10.1061/(asce)gt.1943-5606.0002143

Bray, J.D., Macedo, J., and Travasarou, T. (2018). "Simplified procedure for estimating seismic slope displacements for subduction zone earthquakes," *Journal of Geotechnical and Geoenvironmental Engineering,* Vol. 144, No. 3, p. 04017124. https://doi.org/10.1061/(asce)gt.1943-5606.0001833

Bray, J.D. and Rathje, E.M. (1998). "Earthquake-induced displacements of solid-waste landfills," *Journal of Geotechnical and Geoenvironmental Engineering,* Vol. 124, No. 3, pp. 242–253. https://doi.org/10.1061/(asce)1090-0241(1998)124:3(242)

Bray, J.D., Rathje, E.M., Augello, A.J., and Merry, S.M. (1998). "Simplified seismic design procedures for geosynthetic-lined, solid waste landfills." *Geosynthetics International,* Vol. 5, No. 1–2, pp. 203–235. https://doi.org/10.1680/gein.5.0119

Bray, J.D. and Travasarou, T. (2007). "Simplified procedure for estimating earthquake-induced deviatoric slope displacements," *Journal of Geotechnical and Geoenvironmental Engineering,* Vol. 133, No. 4, pp. 381–392. https://doi.org/10.1061/(asce)1090-0241(2007)133:4(381)

Bray, J.D. and Travasarou, T, (2009), "Pseudostatic coefficient for use in simplified seismic slope stability evaluation," *Journal of Geotechnical and Geoenvironmental Engineering,* Vol. 135, No. 9, pp. 1336–1340. https://doi.org/10.1061/(asce)gt.1943-5606.0000012

Bui, H. H., Fukagawa, R., Sako, K., and Ohno, S. (2008). "Lagrangian meshfree particles method (SPH) for large deformation and failure flows of geomaterial using elastic-plastic soil constitutive model," *International Journal of Numerical and Analytical Methods in Geomechanics,* Vol. 32, No. 12, pp. 1537–1570. https://doi.org/10.1002/nag.688

California Geological Survey (2018). "Earthquake fault zones: A guide for government agencies, property owners/developers, and geoscience practitioners for assessing fault rupture hazards in California," *Publication 42,* California Geological Survey, 83 pp.

Camones, L.A.M., Vargas, E.D.A., de Figueiredo, R.P., and Velloso, R.Q. (2013). "Application of the discrete element method for modeling of rock crack propagation and coalescence in the step-path failure mechanism," *Engineering Geology,* Vol. 153, pp. 80–94. https://doi.org/10.1016/j.enggeo.2012.11.013

Casagli, N., Cigna, F., Bianchini, S., Holbling, D., Fureder, P., Righini, G., Sel Conte, S., Briedl, B., Schneiderbauer, S., Iasio, C., Vlcko, J., Greif, V., Proske, H., Granica, K., Falco, S., Lozzi, S., Mora, O., Arnaud, A., Novali, F., and Bianchi, M. (2016). "Landslide mapping and monitoring by using radar and optical remote sensing: Examples from the EC-FP7 project SAFER," *Remote Sensing Applications: Society and Environment,* Vol. 4, pp. 92–108. https://doi.org/10.1016/j.rsase.2016.07.001

Cheng, Y.M. and Lau, C.K. (2014). *Slope Stability Analysis and Stabilization - New Methods and Insight,* 2nd edition, CRC Press, Boca Raton, 426 pp.

Chopra, A.K. (1966). "Earthquake effects on dams," *Ph.D. dissertation,* University of California, Berkeley, CA.

Close, U. and McCormick, E. (1922). "Where the mountains walked," *National Geographic,* Vol. 41, No. 5, pp. 445–464.

Clough, R.W. and Chopra, A.K. (1966). "Earthquake stress analysis in earth dams," *Journal of the Engineering Mechanics Division,* Vol. 92, No. EM2, pp. 197–212. https://doi.org/10.1061/jmcea3.0000735

Cluff, L.S., Page, R.A., Slemmons, D.B., and Crouse, C.B. (2003). "Seismic hazard exposure for Trans-Alaska pipeline," *Proceedings, 6th U.S. Conference on Lifeline Earthquake Engineering,* Long Beach, CA, ASCE. https://doi.org/10.1061/40687(2003)55

Coe, J.A., Ellis, W.L., Godt, J.W., Savage, W.Z., Savage, J.E., Michael, J.A., Kibler, J.D., Powers, P.S., Lidke, D.J., and Debray, S. (2003). "Seasonal movement of the Slumgullion landslide determined from Global Positioning System surveys and field instrumentation, July 1998–March 2002," *Engineering Geology,* Vol. 68, pp. 67–101. https://doi.org/10.1016/s0013-7952(02)00199-0

Colesanti, C. and Wasowski, J. (2006). "Investigating landslides with space-borne Synthetic Aperture Radar (SAR) interferometry," *Engineering Geology*, Vol. 88, pp. 173–199. https://doi.org/10.1016/j.enggeo.2006.09.013

Corominas, J. (1996). "The angle of reach as mobility index for small and large landslides," *Canadian Geotechnical Journal*, Vol. 33, pp. 260–271. https://doi.org/10.1139/t96-005

Costa, J.E. and Schuster, R.L. (1991). "Documented historical landslide dams from around the world," *Open-File Report 91-239*, U.S. Geological Survey, 486 pp. https://doi.org/10.3133/ofr91239

Cotecchia, V. (1986). "Ground deformations and slope instability produced by the earthquake of 23 November, 1980 in Campania and Basilicata," *Geology and Applied Hydrology*, Vol. 21, No. 5, pp. 31–100.

Cummins, P.R., and Kaneda, Y. (2000). "Possible splay fault slip during the 1946 Nankai earthquake," *Geophysical Research Letters*, Vol. 27, No. 17, pp. 2725–2728. https://doi.org/10.1029/1999GL011139

Cundall, P.A. (1971). "A computer model for simulating progressive large-scale movements in blocky rock systems," *Proceedings, International Symposium on Rock Fracture*, Vol. 2, pp. 129–136.

Cundall, P.A. and Strack, O.D.L. (1979). "A discrete numerical model for granular assemblies." *Géotechnique*, Vol. 29, No. 1, pp. 47–65. https://doi.org/10.1680/geot.1979.29.1.47

Dai, F.C., Xu, C., and Yao, X. (2011). "Spatial distribution of landslides triggered by the 2008 Ms 8.0 Wenchuan earthquake, China," *Journal of Asian Earth Sciences*, Vol. 40, No. 4, pp. 883–895. https://doi.org/10.1016/j.jseaes.2010.04.010

Dai, Z. and Huang, Y. (2015). "The state of the art of SPH modeling for flow-slide propagation," in *Modern Technologies for Landslide Monitoring and Prediction,* Springer, Berlin, pp. 155–164. https://doi.org/10.1007/978-3-662-45931-7_8

Dai, Z. and Huang, Y. (2016). "A three-dimensional model for flow slides in municipal solid waste landfills using smoothed particle hydrodynamics," *Environmental Earth Sciences*, Vol. 75, p. 132. https://doi.org/10.1007/s12665-015-4923-4

Dawson, E.M., Roth, W.H., and Drescher, A. (1999). "Slope stability analysis by strength reduction," *Geotechnique*, Vol. 49, No. 6, pp. 835–840. https://doi.org/10.1680/geot.1999.49.6.835

Devoli, G., De Blasio, F.V., Elverhoi, A., and Hoeg, K. (2009). "Statistical analysis of landslide events in Central America and their run-out distance," *Geotechnical and Geological Engineering*, Vol. 27, pp. 23–42. https://doi.org/10.1007/s10706-008-9209-0

Doolin, D.M. and Sitar, N. (2004). "Time integration in discontinuous deformation analysis," *Journal of Engineering Mechanics*, Vol. 130, No. 3, pp. 249–258. https://doi.org/10.1061/(asce)0733-9399(2004)130:3(249)

Duncan, J.M. (1996). "State of the art: Limit equilibrium and finite-element analysis of slopes," *Journal of Geotechnical Engineering*, Vol. 122, No. 7, pp. 577–596. https://doi.org/10.1061/(asce)0733-9410(1996)122:7(577)

Duncan, J.M., Byrne, P.M., Wong, K.S., and Mabry, P. (1980). "Strength, stress-strain, and bulk modulus parameters for finite element analysis of stresses and movements in soil masses," *Report No. UCB/GT/80-01*, University of California, Berkeley, CA.

Duncan, J.M., Wright, S.G., and Brandon, T.L. (2014). *Soil Strength and Slope Stability*, John Wiley & Sons, New York, 336 pp.

Faccioli, E., Anastasopoulos, I., Gazetas, G., Callerio, A., and Paolucci, R. (2008). "Fault rupture-foundation interaction: Selected case histories," *Bulletin of Earthquake Engineering*, Vol. 6, pp. 557–583. https://doi.org/10.1007/s10518-008-9089-y

Eberhardt, E. (2003). "Rock slope stability analysis – utilization of advanced numerical techniques," *Report, Earth and Ocean Sciences at UBC, University of British Columbia*, Vancouver, 41 pp.

Fan, X., Scaringi, G., Korup, O., West, A.J., van Westen, C.J., Tanyas, H., Hovius, N., Hales, T.C., Jibson, R.W., Allstadt, K.E., Zhang, L., Evans, S.G., Xu, C., Li, G., Pei, X., Xu, Q., and Runqiu Huang, R. (2019). "Earthquake-induced chains of geologic hazards: Patterns, mechanisms, and impacts," *Reviews of Geophysics*, Vol. 57, No. 2, p. 421–503. https://doi.org/10.1029/2018rg000626

Felix, R.P., Hubbard, J.A., Moore, J.D.P., and Switzer, A.D. (2022). "The role of frontal thrusts in tsunami earthquake generation," *Bulletin of the Seismological Society of America*, Vol. 112, No. 2, pp. 680–694. https://doi.org/10.1785/0120210154

Fellenius, W. (1927). *Erdstatische berechnungen mit reibung und kohasion*, Ernst, Berlin.

Feng, X., Jiang, Q, Wu, A, Wei, W., and Zhang, X. (2017). "Study on the damping of the discontinuous deformation analysis based on two-block model," *Mathematical Problems in Engineering*, Vol. 2017, p. 7480937. https://doi.org/10.1155/2017/7480937

Finlay, P.J., Mostyn, G.R. and Fell, R. (1999). "Landslide risk assessment: Prediction of travel distance," *Canadian Geotechnical Journal*, Vol. 36, pp. 556–562. https://doi.org/10.1139/t99-012

Franklin, A.G. and Chang, F.K. (1977). "Permanent displacements of earth embankments by Newmark sliding block analysis," *Report 5, Miscellaneous Paper S-71-17*, U.S. Army Engineer Waterways Experiment Station, Vicksburg, MS.

Fuis, G.S., Wald, L.A., Hendley II, J.W., and Stauffer, P.H. (2003). "Rupture in south-central Alaska; the Denali Fault earthquake of 2002," *Fact Sheet 014-03*, U.S. Geological Survey, 4 pp.

Gabriel, A.K., Goldstein, R.M., and Zebker, H.A. (1989). "Mapping small elevation changes over large areas: Differential radar interferometry," *Journal of Geophysical Research*, Vol. 94, pp. 9183–9191. https://doi.org/10.1029/jb094ib07p09183

Gazetas, G. and Uddin, N. (1994). "Permanent deformation on pre-existing sliding surfaces in dams," *Journal of Geotechnical Engineering*, Vol. 120, No. 11, pp. 2041–2061. https://doi.org/10.1061/(asce)0733-9410(1994)120:11(2041)

Gibson, M.D., Wartman, J.P., MacLaughlin, M.M., and Keefer, D.K. (2018). "Pseudo-static failure modes and yield accelerations in rock slopes," *International Journal of Rock Mechanics and Mining Sciences*, Vol. 102, pp. 1–14. https://doi.org/10.1016/j.ijrmms.2017.11.001

Gili, J.A., Corominas, J., and Rius, J. (2000). "Using Global Positioning System techniques in landslide monitoring," *Engineering Geology*, Vol. 55, pp. 167–192. https://doi.org/10.1016/s0013-7952(99)00127-1

Glenn, N.F., Streutker, D.R., Chadwick, D.J., Thackray, G.D., and Dorsch, S.J. (2006). "Analysis of LiDAR-derived topographic information for characterizing and differentiating landslide morphology and activity," *Geomorphology*, Vol. 73, pp. 131–148. https://doi.org/10.1016/j.geomorph.2005.07.006

Goodman, R.E. and Kieffer, D.S. (2000). "Behavior of rock in slopes," *Journal of Geotechnical and Geoenvironmental Engineering*, Vol. 126, No. 8, pp. 675–684. https://doi.org/10.1061/(asce)1090-0241(2000)126:8(675)

Haneberg, W.C., Cole, W.F., and Kasali, G. (2009). "High-resolution lidar-based landslide hazard mapping and modeling," *Bulletin of Engineering Geology and the Environment*, Vol. 68, pp. 263–276. https://doi.org/10.1007/s10064-009-0204-3

Hoek, E.T. and Bray, J.W. (1981). *Rock Slope Engineering*, Institute of Mining and Metallurgy, CRC Press, 364 pp. https://doi.org/10.1201/9781482267099

Hungr, O. (1995). "A model for the runout analysis of rapid flow slides, debris flows, and avalanches," *Canadian Geotechnical Journal*, Vol. 32, No. 4, pp. 610–623. https://doi.org/10.1139/t95-063

Hungr, O. and Morgenstern, N.R. (1984). "High velocity ring shear tests on sand," *Geotechnique*, Vol. 34, No. 3, pp. 415–421. https://doi.org/10.1680/geot.1984.34.3.415

Hynes-Griffin, M.E. and Franklin, A.G. (1984). "Rationalizing the seismic coefficient method," *Miscellaneous Paper GL-84-13*, U.S. Army Corps of Engineers, Vicksburg, MS, 21 pp.

Iverson, R.M. (1997). "The physics of debris flows," *Reviews of Geophysics*, Vol. 35, No. 3, pp. 245–296. https://doi.org/10.1029/97rg00426

Janbu, N. (1968). "Slope stability computations," *Soil Mechanics and Foundation Engineering Report*, The Technical University of Norway, Trondheim, Norway.

Jibson, R.W. (2007). "Regression models for estimating coseismic landslide displacement," *Engineering Geology*, Vol. 91, pp. 209–218. https://doi.org/10.1016/j.enggeo.2007.01.013

Jibson, R.W. and Harp, E.L. (2012). "Extraordinary distance limits of landslides triggered by the 2011 Mineral, Virginia, earthquake," *Bulletin of the Seismological Society of America*, Vol. 102, No. 6, pp. 2368–2377. https://doi.org/10.1785/0120120055

Jibson, R.W., Harp, E.L., and Michael, J.A. (2000). "A method for producing digital probabilistic seismic landslide hazard maps," *Engineering Geology*, Vol. 58, pp. 271–289.

Jibson, R.W. and Michael, J.A. (2009). "Maps showing seismic landslide hazards in Anchorage, Alaska," *U.S. Geological Survey Scientific Investigations Map 3077*, 2 sheets (scale 1:25,000), 11 pp. https://doi.org/10.3133/sim3077

Jibson, R.W., Prentice, C.S., Borissoff, B.A., Rogozhin, E.A., and Langer, C.J. (1994). "Some observations of landslides triggered by the 29 April 1991 Racha earthquake, Republic of Georgia," *Bulletin of the Seismological Society of America*, Vol. 84, No. 4, pp. 963–973.

Jibson, R.W., Rathje, E.M., Jibson, M.W., and Lee, Y.W. (2013). "SLAMMER-seismic landslide movement modeled using earthquake records (ver.1.1, November 2014)," *U.S. Geological Survey Techniques and Methods, Book 12, Chapter B1*. https://doi.org/10.3133/tm12b1

Jimenez-Rodriguez, R., Sitar, N., and Chacon, J. (2006). "System reliability approach to rock slope stability," *International Journal of Rock Mechanics and Mining Sciences*, Vol. 43, pp. 847–859. https://doi.org/10.1016/j.ijrmms.2005.11.011

Jing L. (1998). "Formulation of discontinuous deformation analysis (DDA) - an implicit discrete element model for block systems," *Engineering Geology*, Vol. 49, No. 3–4, pp. 371–381. https://doi.org/10.1016/s0013-7952(97)00069-0

Johnson, A.M. (1970). *Physical Processes in Geology*. Freeman-Cooper, San Francisco, CA.

Kayen, R.E., Pack, R.T., Bay, J., Sugimoto, S., and Tanaka, H. (2006). "Terrestrial-LIDAR visualization of surface and structural deformations of the 2004 Niigata Ken Chuetsu, Japan earthquake," *Earthquake Spectra*, Vol. 22, No. S1, pp. S147–S162. https://doi.org/10.1193/1.2173020

Keefer, D.K. (1984). "Landslides caused by earthquakes," *Geological Society of America Bulletin*, Vol. 95, pp. 406–421. https://doi.org/10.1130/0016-7606(1984)95<406:lcbe>2.0.co;2

Keefer, D.K. (1993). "The susceptibility of rock slopes to earthquake-induced failure," *Bulletin of the Association of Engineering Geologists*, Vol. 30, No. 3, pp. 353–361. https://doi.org/10.1016/0148-9062(94)90727-7

Keefer, D.K. (2002). "Investigating landslides caused by earthquakes-A historical review," *Survey of Geophysics*, Vol. 23, No. 6, pp. 473–510. https://doi.org/10.1023/a:1021274710840

Keefer, D.K. and Wilson, R.C., (1989). "Predicting earthquake-induced landslides, with emphasis on arid and semi-arid environments," *Landslides in a Semi-arid Environment*, Inland Geological Society, Riverside, CA, Vol. 2, pp. 118–149.

Kobayashi, Y. (1981). "Causes of fatalities in recent earthquakes in Japan," *Journal of Disaster Science*, Vol. 3, pp. 15–22.

Kokusho, T. (1999). "Water film in liquefied sand and its effect on lateral spread." *Journal of Geotechnical and Geoenvironmental Engineering*, Vol. 125, No. 10, pp. 817–826.

Kondner, R.L. (1963). "Hyperbolic stress-strain response: Cohesive soils," *Journal of the Soil Mechanics and Foundations Division*, Vol 89, No. SM1, pp. 115–124. https://doi.org/10.1061/jsfeaq.0000479

Kramer, S.L and Lindwall, N. (2004). "Dimensionality and directionality effects in Newmark sliding block analyses." *Journal of Geotechnical and Geoenvironmental Engineering*, Vol. 130, No, 3, pp. 303–315. https://doi.org/10.1061/(asce)1090-0241(2004)130:3(303)

Kramer, S.L. and Smith, M.W. (1997). "Modified Newmark model for seismic displacements of compliant slopes," *Journal of Geotechnical and Geoenvironmental Engineering*, Vol. 123, No. 7, pp. 635–644. https://doi.org/10.1061/(asce)1090-0241(1997)123:7(635)

Kratzsch, H. (1983). *Mining Subsidence Engineering*, Springer, Berlin, Heidelberg. https://doi.org/10.1007/978-3-642-81923-0

Lambe, T.W. and Whitman, R.V. (1969). *Soil Mechanics*, John Wiley and Sons, New York.

Lefebvre, G., Leboeuf, D., and Hornych, P. (1992). "Slope failures associated with the Saguenay earthquake, Quebec, Canada," *Canadian Geotechnical Journal*, Vol. 29, No. 1, pp. 117–130. https://doi.org/10.1139/t92-013

Lemos, L.J.L. and Coelho, P.A.L.F. (1991). Displacements of slopes under earthquake loading, *Proceedings, 2nd International Conference on Recent Advances in Geotechnical Earthquake Engineering and Soil Dynamics*, St. Louis, MO, Vol. 2, pp. 1051–1056.

Lemos, L.J.L., Skempton, A.W., and Vaughan, P.R. (1985). "Earthquake loading of shear surfaces in slopes," *Proceedings, 11th International Conference on Soil Mechanics and Foundation Engineering*, San Francisco, Vol. 4, pp. 1955–1958.

Lettis, W., Bachhuber, J., Page, W., Witter, R., Barka, A., Bray, J., Page, W., and Swan, F. (2000). "Surface fault rupture," *Earthquake Spectra*, Vol. 16, No. Suppl. A, pp. 11–53. https://doi.org/10.1193/1.1586145

Li, S. and Liu, W. (2002). "Meshfree and particle methods and their applications," *Applied Mechanics Review*, Vol. 55, No. 1, pp. 1–34. https://doi.org/10.1115/1.1431547

Li, T. (1990). "Landslide management in mountain areas of China," *Occasional Paper 15*, International Centre for Integrated Mountain Development, Kathmandu, Nepal, 50 pp.

Li, W.-L., Huang, R.-Q., Chuan, T., Xu, Q., and van Westen, C. (2013). "Co-seismic landslide inventory and susceptibility mapping in the 2008 Wenchuan earthquake disaster area, China," *Journal of Mountain Science*, Vol. 10, No. 3, pp. 339–354. https://doi.org/10.1007/s11629-013-2471-5

Lin, C.-W., Liu, S.-H., Lee, S.-Y., and Liu, C.-C. (2006). "Impacts of the Chi-Chi earthquake on subsequent rainfall-induced landslides in central Taiwan," *Engineering Geology*, Vol. 86, pp. 87–101. https://doi.org/10.1016/j.enggeo.2006.02.010

Lin, J.S. and Whitman, R.V. (1986). "Earthquake induced displacements of sliding blocks," *Journal of Geotechnical Engineering*, Vol. 112, No. 1, pp. 44–59. https://doi.org/10.1061/(asce)0733-9410(1986)112:1(44)

Ling, H.I. and Leshchinsky, D. (1997). "Seismic stability and permanent displacement of landfill cover systems," *Journal of Geotechnical and Geoenvironmental Engineering*, Vol. 123, No. 2, pp. 113–122. https://doi.org/10.1061/(asce)1090-0241(1997)123:2(113)

Loli, M., Kourkoulis, R., and Gazetas, G. (2018). "Physical and numerical modeling of hybrid foundations to mitigate seismic fault rupture effects," *Journal of Geotechnical and Geoenvironmental Engineering,* Vol. 144, No. 11, 21 pp.

Lorig, L.J., Hart, R.D., and Cundall, P.A. (1991). "Slope stability analysis of jointed rock using distinct element method," *Transportation Research Record 1330*, Transportation Research Board, Washington, DC, 9 pp.

Lucieer, A., de Jong, S.M., and Turner, D. (2014). "Mapping landslide displacements using Structure from Motion (SfM) and image correlation of multi-temporal UAV photography," *Progress in Physical Geography*, Vol. 38, No. 1, pp. 97–116. https://doi.org/10.1177/0309133313515293

Macedo, J., Bray, J.D., Abrahamson, N., and Travasarou, T. (2018) "Performance-based probabilistic seismic slope displacement procedure," *Earthquake Spectra*, Vol. 34, No. 2, pp. 673–695. https://doi.org/10.1193/122516eqs251m

Makdisi, A.J. (2016). "The applicability of sliding block analyses for the prediction of lateral spreading displacements," *Masters Thesis*, University of Washington, Seattle, 431 pp.

Makdisi, F.I. and Seed, H.B. (1978). "Simplified procedure for estimating dam and embankment earthquake-induced deformations." *Journal of Geotechnical and Geoenvironmental Engineering Division*, Vol. 104, No. GT7, pp. 849–867. https://doi.org/10.1061/ajgeb6.0000668

Malvick, E. J., Kutter, B. L., and Boulanger, R. W. (2008). "Postshaking shear strain localization in a centrifuge model of a saturated sand slope." *Journal of Geotechnical and Geoenvironmental Engineering*, Vol. 134, No. 2, pp. 164–174.

Marcuson, W.F. (1994). "An example of professional modesty," *Proceedings, The Earth, Engineers, and Education, A Symposium in honor of Robert V.* Whitman, MIT, Cambridge, MA, pp. 200–202.

Marcuson III, W.F., Ballard, R.F., Jr., and Ledbetter, R.H. (1979). "Liquefaction failure of tailings dams resulting from the Near Izu Oshima earthquake, 14 and 15 January, 1978," *Proceedings, 6th Pan American Conference on Soil Mechanics and Foundation Engineering,* Lima, Peru.

Marcuson, W.F. III, Hynes, M.E., and Franklin, A.G. (1990). Evaluation and use of residual strength in seismic safety analysis of embankments, *Earthquake Spectra*, Vol. 6, No. 3, pp. 529–572.

Marketos, G. and O'Sullivan, C. (2013). "A micromechanics-based analytical method for wave propagation through a granular material," *Soil Dynamics and Earthquake Engineering*, Vol. 45, pp. 25–34. https://doi.org/10.1016/j.soildyn.2012.10.003

Markland, J. T. (1972). "A useful technique for estimating the stability of rock slopes when the rigid wedge slide type of failure is expected," *Imperial College Rock Mechanics Research Reprints*, Vol. 19, 10 pp.

Mast, C.M., Arduino, P., Miller, G.R., and Mackinzie-Hilnwein, P. (2014). "Avalanche and landslide simulation using the material point method: Flow dynamics and force interaction with structures," *Computational Geosciences*, Vol. 18, pp. 817–830. https://doi.org/10.1007/s10596-014-9428-9

Matasovic, N., Kavazanjian, E., and Giroud, J.P. (1998). "Newmark seismic deformation analysis for geosynthetic covers," *Geosynthetics International*, Vol. 5, No. 1–2, pp. 237–264. https://doi.org/10.1680/gein.5.0120

Matasovic, N., Kavazanjian, E., Jr. and Yan, L. (1997). "Newmark deformation analysis with degrading yield acceleration," *Proceedings, Geosynthetics '97,* IFAI, Long Beach, CA, Vol. 2, pp. 989–1000.

McDougall, S. and Hungr, O. (2004). "A model for the analysis of rapid landslide motion across three-dimensional terrain," *Canadian Geotechnical Journal*, Vol. 41, No. 6, pp. 1084–1097. https://doi.org/10.1139/t04-052

Michalowski, R.L. (2007). "Displacements of multiblock geotechnical structures subjected to seismic excitation," *Journal of Geotechnical and Geoenvironmental Engineering*, Vol. 133, No. 11, pp. 1432–1439. https://doi.org/10.1061/(asce)1090-0241(2007)133:11(1432)

Miller, D.J. and Burnett, K.M. (2008). "A probabilistic model of debris-flow delivery to stream channels, demonstrated for the Coast Range of Oregon, USA," *Geomorphology*, Vol. 9, No. 1–2, pp. 184–205. https://doi.org/10.1016/j.geomorph.2007.05.009

Monaghan, J. (1988). "An introduction to SPH," *Computer Physics Communications*, Vol. 48, pp. 89–96. https://doi.org/10.1016/0010-4655(88)90026-4

Morgenstern, N.R. and Price, V.E. (1965). "The analysis of the stability of general slip surfaces," *Geotechnique*, Vol. 15, No. 1, pp. 79–93. https://doi.org/10.1680/geot.1965.15.1.79

Moss, R.E.S. and Ross, Z.E. (2011). "Probabilistic fault displacement hazard analysis for reverse faults," *Bulletin of the Seismological Society of America*, Vol. 101, No. 4, pp. 1542–1553. https://doi.org/10.1785/0120100248

National Research Council (1985). *Liquefaction of soils during earthquakes*, National Academy Press, Washington, D.C., 240 pp.

Newmark, N. (1965). "Effects of earthquakes on dams and embankments," *Geotechnique*, Vol. 15, No. 2, pp. 139–160. https://doi.org/10.1680/geot.1965.15.2.139

Niccum, M.R., Cluff, L.S., Chamorro, F., and Wyllie, L. (1976). "Banco Central de Nicaragua: Acase history of a high-rise building that survived surface fault rupture." *Proceedings, Engineering Geology and Soils Engineering Symposium No. 4,* Idaho Transportation Department, Division of Highways, Boise, ID, pp. 133–144.

Nishimura, T., Jukuda, T., and Tsujino, K. (2010). "Distinct element analysis for progressive failure in rock slope," *Soils and Foundations*, Vol. 50, No. 4, pp. 505–513. https://doi.org/10.3208/sandf.50.505

Nobrega, R.A.A., Aanstoos, J., Gokaraju, B., Mahgooghy, M., Dabirru, L., and O'Hara, C.G. (2013). "Mapping weaknesses in the Mississippi river levee system using multi-temporal UAVSAR data," *Revista Brasileira de Cartografia*, Vol. 65, No. 4, pp. 681–694. https://doi.org/10.14393/rbcv65n4-43853

Nurminen, F., Baize, S., Boncio, P., Blumetti, A.M., Cinti, F.R., Civico, R. and Guerrieri, L. (2022). "SURE 2.0 – New release of the worldwide database of surface ruptures for fault displacement hazard analyses," *Scientific Data*, Vol. 9, p. 729, https://doi.org/10.1038/s41597-022-01835-z

Oettle, N.K. and Bray, J.D. (2013). "Geotechnical mitigation strategies for earthquake surface fault rupture," *Journal of Geotechnical and Geoenvironmental Engineering*, Vol. 139, No. 11, pp. 1864–1874. https://doi.org/10.1061/(ASCE)GT.1943-5606.0000933

Olgun, C.G., Martin II, J.R., Mitchell, J.K., Durgunoglu, H.T. (2001). "Improved ground performance during the 1999 Turkey earthquakes," *Proceedings, 15th International Conference on Soil Mechanics and Foundation Engineering (Istanbul).*

Olson, J., Narayanasamy, R., Holder, J., Rauch, A., and Comacho, B. (1999). "DEM study of wave propagation in weak sandstone," *Proceedings, Third International Conference on Discrete Element Methods, Geotechnical Special Publication 117,* ASCE, pp. 335–339.

Peng, M. and Zhang, L.M. (2013a). "Dynamic decision making for dam-break emergency management-Part 1: Theoretical framework," *Natural Hazards and Earth System Sciences*, Vol. 13, No. 2, pp. 425–437. https://doi.org/10.5194/nhess-13-425-2013

Peng, M. and Zhang, L.M. (2013b). Dynamic decision making for dam-break emergency management-Part 2: Application to Tangjiashan landslide dam failure," *Natural Hazards and Earth System Sciences*, Vol. 13, No. 2, pp. 439–454. https://doi.org/10.5194/nhess-13-439-2013

Perloff, W.H. and Baron, W. (1976). *Soil Mechanics*, The Ronald Press Company, New York.

Petersen, M.D., Dawson, T.E., Chen, R., Cao, T., Wills, C.J., Schwartz, D.P., and Frankel, A.D. (2011). "Fault displacement hazard for strike-slip faults," *Bulletin of the Seismological Society of America*, Vol. 101, No. 2, pp. 805–825. https://doi.org/10.1785/0120100035

Pizza, M., Ferrario, M.F., Thomas, F., Tringali, G., and Livio, F. (2023). "Likelihood of primary surface faulting: Updating of empirical regressions," *Bulletin of the Seismological Society of America*, Vol. 113, No. 5, pp. 2106–2118. https://doi.org/10.1785/0120230019

Potyondy, D.L. and Cundall, P.A. (2004). "A bonded-particle model for rock," *International Journal of Rock Mechanics and Mining Sciences*, Vol. 41, pp. 1329–1364. https://doi.org/10.1016/j.ijrmms.2004.09.011

Pradel, D.E., Smith, P.M., Stewart, J.P., and Raad, G. (2005). "Case history of landslide movement during the Northridge earthquake," *Journal of Geotechnical and Geoenvironmental Engineering*, Vol. 131, No. 11, pp. 1360–1369. https://doi.org/10.1061/(asce)1090-0241(2005)131:11(1360)

Rathje, E.M. and Antonakos, G. (2011). "A unified model for predicting earthquake-induced sliding displacements of rigid and flexible slopes," *Engineering Geology*, Vol. 22, pp. 51–60. https://doi.org/10.1016/j.enggeo.2010.12.004

Rathje, E.M. and Bray, J.D. (1999). "An examination of simplified earthquake induced displacement procedures for earth structures," *Canadian Geotechnical Journal*, Vol. 36, No. 1, pp. 72–87. https://doi.org/10.1139/t98-076

Rathje, E.M. and Bray, J.D. (2000). "Nonlinear coupled seismic sliding analysis of earth structures," *Journal of Geotechnical and Geoenvironmental Engineering*, Vol. 126, No. 11, pp. 1002–1014. https://doi.org/10.1061/(asce)1090-0241(2000)126:11(1002)

Rathje, E.M. and Franke, K. (2016). "Remote sensing for geotechnical earthquake reconnaissance," *Soil Dynamics and Earthquake Engineering*, Vol. 91, pp. 304–316. https://doi.org/10.1016/j.soildyn.2016.09.016

Rathje, E.M. and Saygili, G. (2009). "Probabilistic assessment of earthquake-induced sliding displacements of natural slopes," *Bulletin of the New Zealand Society for Earthquake Engineering*, Vol. 42, No. 1, pp. 18–27. https://doi.org/10.5459/bnzsee.42.1.18-27

Rathje, E.M., Wang, Y., Stafford, P.J., Antonakos, G., and Saygili, G. (2013). "Probabilistic assessment of the seismic performance of earth slopes," *Bulletin of Earthquake Engineering*, Vol. 12, No. 3, pp. 1071–1090. https://doi.org/10.1007/s10518-013-9485-9

Rathod, G.W., Varughese, Al, Shrivastava, A.K., and Rao, K.S. (2012). "3 dimensional stability assessment of jointed rock slopes using distinct element modelling," *Proceedings, GeoCongress 2012: State of the Art and Practice in Geotechnical Engineering,* ASCE, pp. 2382–2391. https://doi.org/10.1061/9780784412121.244

Resende, R., Lamas, L.N., Lemos, J.V., and Calcada, R. (2010). "Micromechanical modelling of stress waves in rock and rock fractures," *Rock Mechanics and Rock Engineering*, Vol. 43, pp. 741–761. https://doi.org/10.1007/s00603-010-0098-1

Rodriguez-Marek, A. and Song, J. (2016). "Displacement-based probabilistic seismic demand analyses of earth slopes in the near-fault region," *Earthquake Spectra*, Vol. 32, No. 2, pp. 1141–1163. https://doi.org/10.1193/042514eqs061m

Ruggles, S., Clark, J., Franke, K.W., Wolfe, D., Reimschiissel, B., Martin, R.A., Okeson, T.J., and Hedengren, J.D. (2016). "Comparison of SfM computer vision point clouds of a landslide derived from multiple small UAV platforms and sensors to a TLS-based model," *Journal of Unmanned Vehicle Systems*, Vol. 4, No. 4, pp. 1–13. https://doi.org/10.1139/juvs-2015-0043

Sarma, S.K. (1975). "Seismic stability of earth dams and embankments," *Geotechnique*, Vol. 25, pp. 743–761. https://doi.org/10.1680/geot.1975.25.4.743

Sarmiento, A., Madugo, D., Bozorgnia, Y., Shen, A., Mazzoni, S., Lavrentiadis, G., Dawson, T., Madugo, C., Kottke, A., Thompson, S., Baize, S., Milliner, C., Nurminen, F., Boncio, P., and Visini, F. (2021). "Fault displacement hazard initiative database," *Report No. 2021-08*, B. John Garrick Risk Institute, UCLA. https://doi.org/10.34948/N36P48

Savage, S.B. and Hutter, K. (1989). "The motion of a finite mass of granular material down a rough incline," *Journal of Fluid Mechanics*, Vol. 199, pp. 177–215. https://doi.org/10.1017/s0022112089000340

Saygili, G. and Rathje, E.M. (2008). "Empirical predictive models for earthquake-induced displacements of slopes," *Journal of Geotechnical and Geoenvironmental Engineering*, Vol. 134, No. 6, pp. 790–803. https://doi.org/10.1061/(asce)1090-0241(2008)134:6(790)

Scott, C., Adam, R., Arrowsmith, R., Madugo, C., Powell, J., Ford, J., Gray, B., Koehler, R., Thompson, S., Sarmiento, A., Dawson, T., Kottke, A., Young, E., Williams, A., Kozaci, O., Oskin, M., Burgette, R., Streig, A., Seitz, G., Page, W., Badin, C., Carnes, L., Giblin, J., McNeil, J., Graham, J., Chupik, D., and Ingersoll, S. (2023). "Evaluating how well active fault mapping predicts earthquake surface-rupture locations," *Geosphere*, Vol. 19, No. 4, pp. 1128–1156, https://doi.org/10.1130/GES02611.1.

Seed, H.B. (1979). "Considerations in the earthquake-resistant design of earth and rockfill dams," *Geotechnique*, Vol. 29, No. 3, pp. 215–263. https://doi.org/10.1680/geot.1979.29.3.215

Seed, H.B., Lee, K.L., Idriss, I.M., and Makdisi, F.I. (1975). "The slides in the San Fernando Dams during the earthquake of February 9, 1971," *Journal of the Geotechnical Engineering Division*, Vol. 101, No. GT7, pp. 651–688.

Seed, H.B. and Martin, G.R. (1966). "The seismic coefficient in earth dam design," *Journal of the Soil Mechanics and Foundations Division*, Vol. 92, No. SM3, pp. 25–58. https://doi.org/10.1061/jsfeaq.0000871

Seed, H.B., Lee, K.L., and Idriss, I.M. (1969). "An analysis of the Sheffield Dam failure," *Journal of the Soil Mechanics and Foundations Division*, Vol. 95, No. SM6, pp. 1453–1490. https://doi.org/10.1061/jsfeaq.0001352

Seed, R.B. and Bonaparte, R. (1992). "Seismic analysis and design of waste fills: Current practice," *Proceedings, Specialty Conference on Stability and Performance of Slopes and Embankments II,* ASCE Special Geotechnical Publication No. 31, pp. 1521–1545.

Shi, G.-H. (1988). "Discontinuous deformation analysis – A new numerical method for the statics and dynamics of block systems," *Ph.D. Dissertation*, University of California.

Shi, G.-H. and Goodman, R.E. (1989). "Generalization of two dimensional discontinuous deformation analysis for forward modeling," *International Journal of Numerical and Analytical Methods in Geomechanics*, Vol. 13, pp. 359–80. https://doi.org/10.1002/nag.1610130403

Shieh, C.L., Chen, Y.S., Tsai, Y.J., and Wu, J.H. (2009). "Variability in rainfall threshold for debris flow after the Chi-Chi earthquake in central Taiwan, China," *International Journal of Sediment Research*, Vol. 24, pp. 177–188. https://doi.org/10.1016/s1001-6279(09)60025-1

Soga, K., Alonso, E., Yerro, A., Kumar, K., and Bandara, S. (2016). "Trends in large-deformation analysis of landslide mass movements with particular emphasis on the material point method," *Geotechnique*, Vol. 66, No. 3, pp. 248–273. https://doi.org/10.1680/jgeot.15.lm.005

Spencer, E. (1967). "A method of analysis of the stability of embankments assuming parallel interslice forces," *Geotechnique*, Vol 17, No. 1, pp. 11–26. https://doi.org/10.1680/geot.1967.17.1.11

Stamatopoulos, C.A. (1996). "Sliding system predicting large permanent co-seismic movements of slopes," *Earthquake Engineering and Structural Dynamics*, Vol. 25, pp. 1075–1093. https://doi.org/10.1002/(sici)1096-9845(199610)25:10<1075::aid-eqe602>3.3.co;2–5

Stead, D. and Wolter, A. (2015). "A critical review of rock slope failure mechanisms: The importance of structural geology," *Journal of Structural Geology*, Vol. 74, pp. 1–23. https://doi.org/10.1016/j.jsg.2015.02.002

Stewart, J.P., Blake, T.F., and Hollingsworth, R.A. (2003). "A screen analysis procedure for seismic slope stability," *Earthquake Spectra*, Vol. 19, No. 3, pp. 697–712. https://doi.org/10.1193/1.1597877

Strîmbu, B. (2011). "Modeling the travel distances of debris flows and debris slides: quantifying hillside morphology," *Annals of Forestry Research*, Vol. 54, No. 1, pp. 119–134.

Sulsky, D., Chen, Z., and Schreyer, H.L. (1994), "A particle method for history-dependent materials," *Computer Methods in Applied Mechanics and Engineering*, Vol. 118, pp. 179–196. https://doi.org/10.1016/0045-7825(94)90112-0

Sulsky, D., Zhou, S.J., and Schreyer, H.L. (1995), "Application of a particle-in-cell method to solid mechanics," *Computer Physics Communications*, Vol. 87, pp. 236–252. https://doi.org/10.1016/0010-4655(94)00170-7

Tang, C., Zhu, J., Chang, M., Ding, J., and Qi, X. (2012). "An empirical-statistical model for predicting debris-flow runout zones in the Wenchuan earthquake area," *Quaternary International*, Vol. 250, pp. 63–73. https://doi.org/10.1016/j.quaint.2010.11.020

Taylor, D.W. (1948). *Fundamentals of Soil Mechanics*, John Wiley & Sons, New York, 700 pp.

Terzaghi, K. (1950). "Mechanisms of landslides," *Engineering Geology (Berkey) Volume*, Geological Society of America. https://doi.org/10.1130/berkey.1950.83

Tika-Vassilikos, T.E., Sarma, S.K., and Ambraseys, N. (1993). "Seismic displacements on shear surfaces in cohesive soils," *Earthquake Engineering and Structural Dynamics*, Vol. 22, pp. 709–721. https://doi.org/10.1002/eqe.4290220806

Ugai, K. and Leshchinsky, D. (1995). "Three-dimensional limit equilibrium and finite element analyses: a comparison of results," *Soils and Foundations*, Vol. 35, No. 4, pp. 1–7. https://doi.org/10.3208/sandf.35.4_1

Ulusay, R., Aydan, O., and Hamada, M. (2002). "The behavior of structures built on active fault zones: examples from the recent earthquakes of Turkey," *Structural Engineering and Earthquake Engineering*, Vol. 19, No. 2, pp.149–167. https://doi.org/10.2208/jsceseee.19.149s

Valentini, A, Fukushima, Y., Contri, P., Ono, M., Sakai, T., Thompson, S.C., Viallet, E., Annaka, T., Chen, R., Moss, R.E.S., Petersen, M.D., Visini, F., and Youngs, R.R. (2021). "Probabilistic fault displacement hazard assessment (PFDHA) for nuclear installations according to IAEA safety standards," *Bulletin of the Seismological Society of America*, Vol. 111, No. 5, pp. 2661–2672. https://doi.org/10.1785/0120210083

Varnes, D.J. (1978). "Slope movement types and processes," in R.L. Schuster and R.J. Krizek, eds., *Landslides – Analysis and Control, Transportation Research Board Special Report 176*, National Academy of Sciences, Washington, DC, pp. 12–33.

Wang, C., Tannant, D., and Lilly, P. (2003). "Numerical analysis of the stability of heavily jointed rock slopes using PFC2D," *International Journal of Rock Mechanics and Mining Sciences*, Vol. 40, pp. 415–424. https://doi.org/10.1016/s1365-1609(03)00004-2

Wang, F.W., Chen, Q.G., and Lynn, H. (2009). "Preliminary investigation of some large landslides triggered by the 2008 Wenchuan earthquake, Sichuan Province, China," *Landslides*, Vol. 6, No.1, pp. 47–54. https://doi.org/10.1007/s10346-009-0141-z

Wang, G. (2011). "GPS landslide monitoring: Single base vs network solutions – A case study based on the Purerto Rico and Virgin Islands permanent GPS network," *Journal of Geodetic Science*, Vol. 1, No. 3, pp. 191–203.

Wartman, J., Bray, J.D., and Seed, R.B. (2003). "Inclined plane studies of the Newmark sliding block procedure," *Journal of Geotechnical and Geoenvironmental Engineering*, Vol. 129, No. 8, pp. 673–684. https://doi.org/10.1061/(asce)1090-0241(2003)129:8(673)

Watson-Lamprey, J. A. and Boore, D.M. (2007). "Beyond SaGMRotI: Conversion to SN Sa, and MaxRot Sa," *Bulletin of the Seismological Society of America*, Vol. 97, pp. 1511–1524.

Wells, D.L. and Coppersmith, K.J. (1994) "New empirical relationships among magnitude, rupture length, rupture width, rupture area, and surface displacement," *Bulletin of the Seismological Society of America*, Vol. 84, No. 4, pp. 974–1002. https://doi.org/10.1785/BSSA0840040974

Wells, D.L. and Kulkarni, V.S. (2014). "Probabilistic fault displacement hazard analysis – Sensitivity analyses and recommended practices for developing design fault displacements," *Proceedings, 10th National Conference on Earthquake Engineering,* Earthquake Engineering Research Institute, Anchorage.

Wen, Z.P., Xie, J.J., Gao, M.T., Hu, Y.X., and Chau, K.T. (2010). "Near-source strong ground motion characteristics of the 2008 Wenchuan earthquake," *Bulletin of the Seismological Society of America*, Vol. 100, pp. 2425–2439. https://doi.org/10.1785/0120090266

Wham, B.P. and Davis, C.A. (2019). "Buried continuous and segmented pipelines subjected to longitudinal permanent ground deformation," *Journal of Pipeline Systems Engineering and Practice*, Vol. 10, No. 4, https://doi.org/10.1061/(ASCE)PS.1949-1204.0000400

Wieczorek, G.F., Wilson, R.C., and Harp, E.L. (1985). "Map showing slope stability during earthquakes in San Mateo County, California," *Miscellaneous Investigations Map I-1257-E, scale 1:62,500*, U.S. Geological Survey. https://doi.org/10.3133/i1257e

Wilson, R.C. and Keefer, D.K. (1983). "Dynamic analysis of a slope failure from the 6 August 1979 Coyote Lake, California, earthquake," *Bulletin of the Seismological Society of America*, Vol. 73, No. 3, pp. 863–877. https://doi.org/10.1785/bssa0730030863

Wilson, R.C. and Keefer, D.K. (1985) "Predicting areal limits of earthquake-induced landsliding," in J.I., Ziony, ed., *Evaluating Earthquake Hazards in the Los Angeles Region*, USGS Professional Paper 1360, pp. 317–345.

Yagoda-Biran, G. and Hatzor, Y.H. (2013). "A new failure mode chart for toppling and sliding with consideration of earthquake inertia force," *International Journal of Rock Mechanics and Mining Sciences*, Vol. 64, pp. 122–131. https://doi.org/10.1016/j.ijrmms.2013.08.035

Yan, L., Matasovic, N. and Kavazanjian, E., Jr. (1996). "Seismic response of rigid block on inclined plane to vertical and horizontal ground motions acting simultaneously," *Proceedings, Eleventh ASCE Engineering Mechanics Conference,* Fort Lauderdale, FL, Vol. 2, pp. 1110–1113.

Yang, K.S., Chiang, J., Lai, C.-W., Han, J., Lin, M.-L. (2020). "Performance of geosynthetic-reinforced soil foundations across a normal fault," *Geotextiles and Geomembranes*, Vol. 48, No. 3, pp. 357–373, https://doi.org/10.1016/j.geotexmem.2019.12.007.

Yegian, M.K., Marciano, E., and Ghahraman, V.G. (1991). "Earthquake-induced permanent deformations: Probabilistic approach," *Journal of Geotechnical Engineering*, Vol. 117, No. 1, pp. 35–50. https://doi.org/10.1061/(asce)0733-9410(1991)117:1(35)

Yegian, M.K., Marciano, E.A., and Gharaman, V.G. (1988). "Integrated seismic risk analysis for earth dams," *Report No. 88-15*, Northeastern University, Boston, MA.

Youd, T.L. (1978). "Major cause of earthquake damage is ground failure," *Civil Engineering*, Vol. 48, No. 4, pp. 47–51. https://doi.org/10.1016/0148-9062(78)91028-8

Youngs, R.R., Arabasz, W.J., Anderson, R.E., Ramelli, A.R., Ake, J.P., Slemmons, D.B., McCalpin, J.P., Doser, D.I., Fridrich, C.J., Swan III, F.H., Rogers, A.J., Yount, J.C., Anderson, L.W., Smith, K.D., Bruhn, R.L., Knuepfer, P.L.K., Smith, R.B., dePolo, C.M., O'Leary, D.W., Coppersmith, K.H., Pezzopane, S.K., Schwartz, D.P., Whitney, J.W., Olig, S.S., and Toro, G.R. (2003). "A methodology for probabilistic fault displacement hazard analysis (PFDHA)," *Earthquake Spectra*, Vol. 19, No. 1, pp. 191–219. https://doi.org/10.1193/1.1542891

Zienkiewicz, O.C., Humpheson, C. and Lewis, R.W. (1975). "Associated and non-associated visco-plasticity and plasticity in soil mechanics," *Geotechnique*, Vol. 25, No. 4, pp. 671–689. https://doi.org/10.1680/geot.1975.25.4.671

11 Soil Improvement for Mitigation of Seismic Hazards

11.1 INTRODUCTION

In many heavily populated and industrialized regions of the world, buildings, bridges, and other infrastructure are increasingly being constructed in geotechnically undesirable areas. Sites that may not have been considered suitable in the past are now being used because better sites have already been developed. The design and construction of earthquake-resistant structures on poor sites, however, frequently requires efforts to mitigate potential seismic hazards. If the effects of poor soil conditions cannot be easily or economically addressed by structural means, they can often be mitigated by improving the characteristics of the soils. These efforts are variously referred to as soil improvement, ground improvement, or ground modification; the first of those terms will be used here.

Soils have been modified to improve their engineering properties for hundreds, or even thousands, of years. Improved knowledge of soil behavior and geotechnical hazards, however, has led to the development and verification of many innovative soil improvement techniques. Increased recognition of seismic hazards and improved understanding of the factors that control them have allowed these techniques to be applied to the mitigation of seismic hazards in the past 50–60 years.

In both seismically active and inactive areas, soil improvement techniques are commonly used at sites where the existing soil conditions may lead to unsatisfactory performance. Unsatisfactory performance, whether expressed in measures of response, damage, or loss, usually involves unacceptably large cyclic and/or permanent soil movements. The movements may include horizontal and/or vertical components and may take place during or after earthquake shaking.

Because unacceptable movements usually result from insufficient soil strength and/or stiffness, most soil improvement techniques were developed to increase the soil strength and stiffness in order to control deformations during and after earthquake shaking. These techniques are described in detail in a number of useful references (e.g., Mitchell, 1981; Welsh, 1987; Van Impe, 1989; Hausmann, 1990; Broms,1991; Bell, 1993; Mitchell et al., 1995; Munfakh, 1997a,b; Schaefer et al., 1997; Terashi and Juran, 2000; Moseley and Kirsch, 2004; Indraratna and Chu, 2005; Towhata, 2008; Chu et al. 2009; Kirsch and Kirsch, 2010). Of particular importance is the ground modification manual published and periodically updated by the U.S. Federal Highway Administration (Schaefer et al., 2017a,b), which provides comprehensive and detailed information on the techniques described briefly in this chapter. Another excellent resource for detailed information on soil improvement is the web-based technology selection guidance system, *GeoTech Tools* (www.geotechtools.org). Soil improvement techniques that are not commonly used for the mitigation of seismic hazards are not discussed in this chapter.

11.2 SOIL IMPROVEMENT TECHNIQUES

During earthquakes, many factors can contribute to unacceptable performance. As previously mentioned, insufficient soil strength and/or stiffness can lead to large permanent movements under strong shaking. This is particularly true for non-plastic, coarse-grained soils in which the buildup of excess porewater pressure can lead to softening and weakening, and thus to very large deformations. Consequently, commonly used techniques for mitigation of seismic hazards often involve reducing the tendency of the soil to generate positive excess porewater pressure during earthquake shaking as well as increasing the strength and stiffness of the soil. The topic of soil improvement is

DOI: 10.1201/9781003512011-11

somewhat different from the topics presented in earlier chapters. Some techniques have well-established design procedures, some have proprietary design procedures, and others still await the development of formal design procedures. Advances in soil improvement technology have often resulted from the initiative and imagination of contractors. Research and explanatory "theories" have followed, rather than led, implementation; for some widely used techniques, proven theories have yet to be developed. In such cases, indirect or empirical evidence must be relied upon and the study of case histories is particularly important. This chapter does not attempt to describe all available soil improvement techniques in detail; instead, it presents an introduction to the soil improvement techniques that are most commonly used for the mitigation of seismic hazards. References to more complete descriptions of the techniques are presented. Because soil improvement technology changes rapidly as new techniques are developed and existing techniques are tested by actual earthquakes, the relevant geotechnical engineering literature should be reviewed on a regular basis. Methods for verification of the effectiveness of soil improvement techniques are also described.

Selection of a method of soil improvement at a particular site is influenced by a number of important factors including technical feasibility, soil type, performance criteria, constructability, cost, project schedule, proximity of existing structures and facilities, environmental and space considerations, local experience, and risk factors. All of the methods described in this chapter have conditions for which they perform well and conditions where their applicability is limited. Figure 11.1 illustrates the ranges of grain size characteristics that a number of soil improvement methods are best suited for.

On the basis of the mechanisms by which they improve the engineering properties of the soil, the most common methods of soil improvement can be divided into five major categories: densification techniques, reinforcement techniques, grouting/mixing techniques, drainage techniques, and the emerging area of biological techniques, recently termed *biogeotechnics*. However, some soil improvement techniques provide multiple benefits and do not fall neatly into a single category; in such cases, they will be described under the category in which their primary improvement mechanism lies.

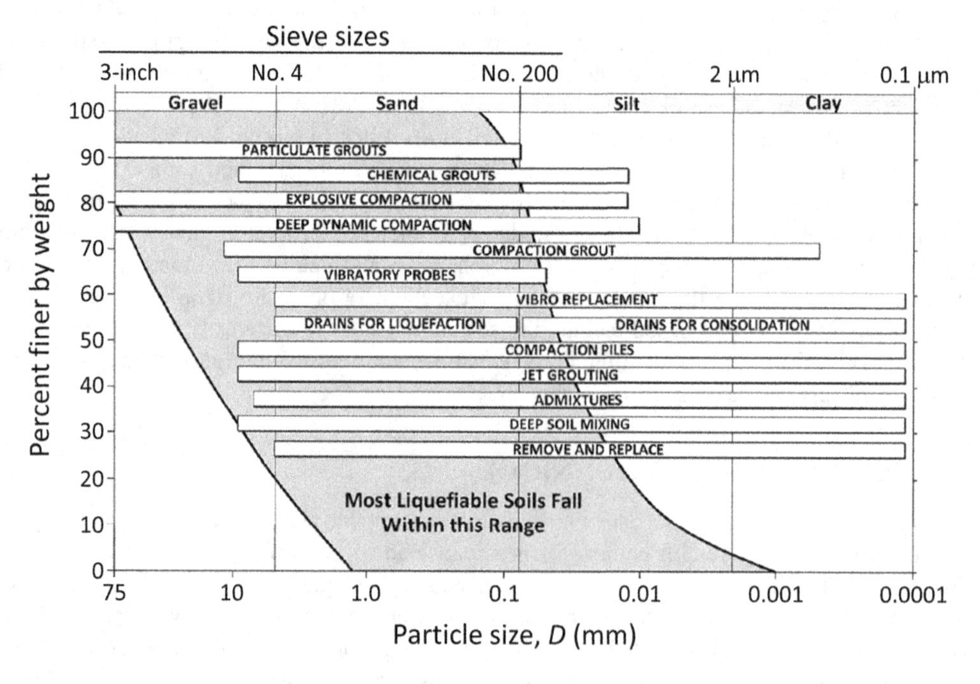

FIGURE 11.1 Applicability of different soil improvement techniques. (Mitchell, 2008; used with permission of ASCE.)

11.3 DENSIFICATION TECHNIQUES

As discussed in Chapter 6, the particles that comprise a particular soil can be arranged in many different ways, and the *state* of the soil, i.e., the density and effective stress under which it exists in the ground, will strongly influence its behavior under cyclic loading, particularly when saturated. Soils in which particles are packed in a dense configuration will have a negative state parameter, build pore pressure slowly under cyclic loading, and dilate quickly under monotonic loading – all characteristics that lead toward small deformations. Loose soils with positive state parameters, however, are contractive and can liquefy and produce the types of large lateral and vertical movements described in Chapter 9. Because the undesirable behavior of these soils is so closely related to their volume change behavior, densification is one of the most effective and commonly used means of improving soil characteristics for mitigation of seismic hazards.

Soil densification can be accomplished by traditional mechanical compaction in which high mechanical compressive stresses are used to push particles into a denser configuration. Densification of non-plastic, coarse-grained soils, however, can also be accomplished quite efficiently using vibratory techniques. Densification techniques can therefore be divided into categories based on vibration and compaction.

Densification produces permanent volume changes that often result in settlement of the ground surface. Different densification techniques produce different amounts of settlement; with some techniques, additional soil is placed at or beneath the ground surface during the process of densification in order to minimize the settlement. Despite such efforts, some densification techniques are limited to sites without existing structures or facilities that could be damaged by ground settlement.

11.3.1 Vibratory Techniques

Vibratory (or vibro) techniques use probes that are vibrated through a soil deposit in a grid pattern to densify the soil over a specified depth range. The vibration may be continuous or it may result from an impulsive disturbance. Vibro techniques are among the most commonly used techniques for mitigation of seismic hazards. They are usually the most economical techniques for the densification of liquefiable soils, providing that the fines content is less than about 20%. Soils with higher fines contents may be able to be densified by vibration or impact if closely spaced vertical drains are first installed (Dise et al., 1994; Luehring et al., 2001; Mitchell, 2008).

11.3.1.1 Vibro-Compaction

In vibro-compaction, a torpedo-like probe (the *vibroflot*) suspended by a crane penetrates the soil and is vibrated to densify a soil deposit. Vibroflots, usually 12–18 in. (30–46 cm) in diameter and about 10–16 ft (3.0–4.9 m) long, contain weights mounted eccentrically on a central shaft driven by electric or hydraulic power (Figure 11.2) to produce rapid horizontal vibrations in the surrounding soil.

The vibroflot is initially lowered to the bottom of the deposit by a combination of vibration and water or air jetting through ports in its pointed nose cone (Figure 11.3). The vibroflot is then incrementally withdrawn in 2–3 ft (60–90 cm) intervals at an overall rate of about 1 ft/min (30 cm/min) while still vibrating. Water may be jetted though ports in the upper part of the vibroflot to loosen the soil above the vibroflot temporarily and aid in its withdrawal. The vibrations produce a localized zone of temporary liquefaction that causes the soil surrounding the vibroflot to densify. A conical depression usually forms at the ground surface above the probe. This depression can be filled with granular material (such as clean sand or gravel) as the vibroflot is withdrawn. Alternatively, vibroflots with bottom-feed systems can introduce granular material through the tip of the vibroflot. As the vibroflot is removed, it leaves behind a column of densified soil. When gravel or crushed stone is introduced into the soil, the resulting *stone column* can provide benefits of reinforcement and drainage in addition to densification. The use of bottom-feed systems has increased rapidly in

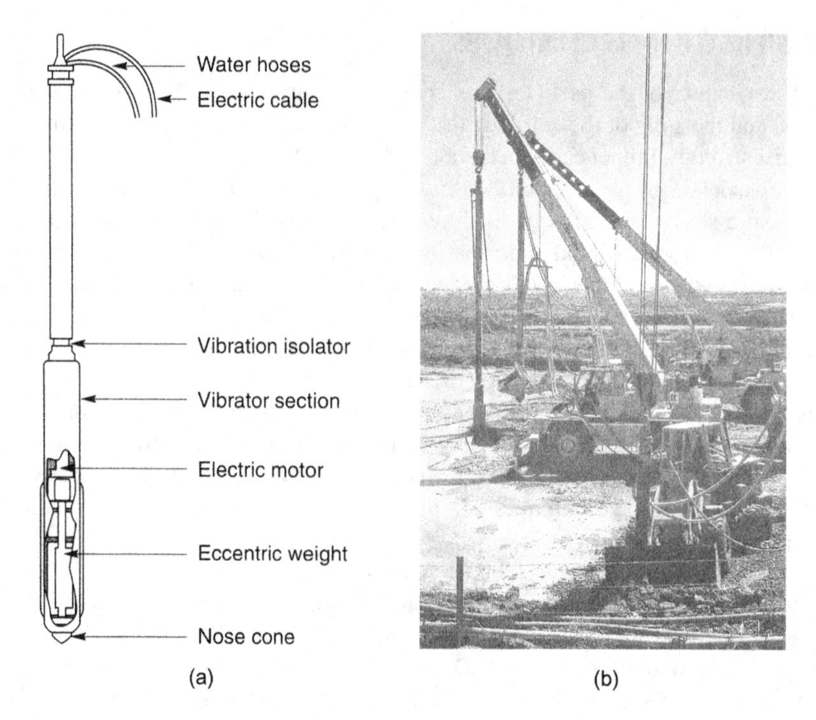

FIGURE 11.2 (a) Schematic illustration of a typical vibroflot (After Bell, 1993), and (b) vibroflots densifying liquefiable soils at a wastewater treatment facility in California (Photo courtesy of Hayward Baker).

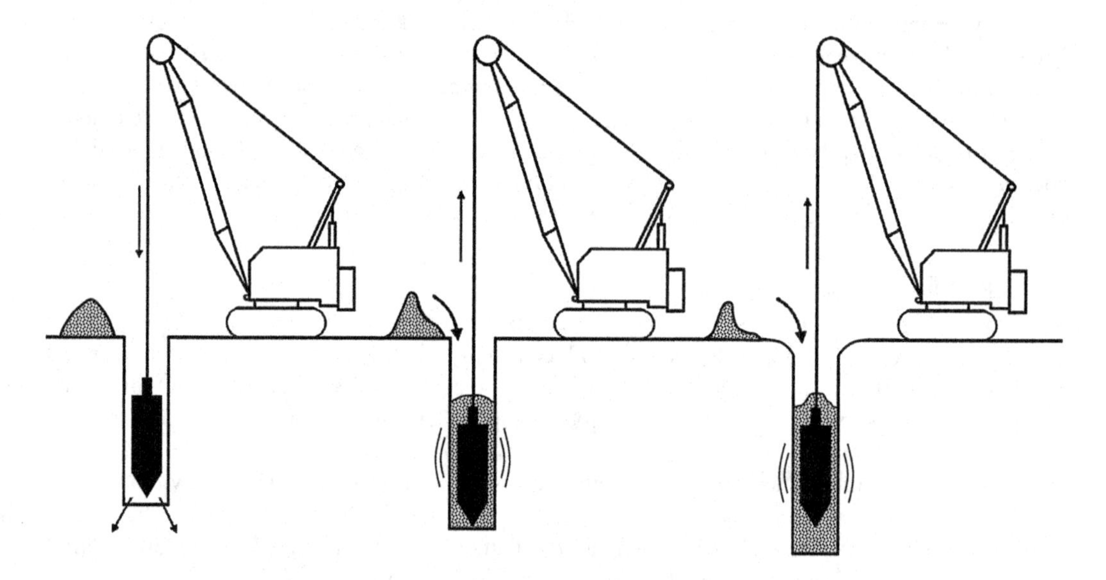

FIGURE 11.3 Schematic illustration of vibro-compaction. (After Schaefer et al., 2017a.)

recent years. Air delivery systems have become quite common and tend to be preferred over water delivery systems in environmentally sensitive areas.

Vibro-compaction is most effective in clean granular soils (Table 11.1) free from cobbles and boulders with fines contents less than 15% and clay contents below 2% (Schaefer et al., 2017a). In such soils, it typically produces high densities (relative densities of about 100%) within 12–18 in. (30–46 cm) of the vibroflot and lower densities (D_r = 70%–85%) at greater radial distances. Compared

TABLE 11.1
Suitability of Soil Types for Vibro-compaction

Soil Type	USCS	Comments on Suitability
Well-graded gravel	GW	Well suited for VC, potential penetration difficulties with less powerful machines
Poorly graded gravel	GP	If $D_{60}/D_{10} \leq 2$, compaction only marginal (trial compaction recommended)
Silty or clayey gravel	GM, GC	Compaction not possible if clay content >= 2$ and silt content > 10%
Well-graded sand	SW	Ideally suited
Poorly graded sand	SP	If $D_{60}/D_{10} \leq 2$, compaction only marginal (trial compaction recommended)
Silty sand	SM	Compaction inhibited if silt content > 8%
Clayey sand	SC	Compaction inhibited if clay content > 2%

Source: After Degen (1997); Kirsch and Kirsch (2010).

to other vibratory methods, the increase in density is relatively uniform with depth. To densify an entire site, vibro-compaction is performed in a grid pattern with a spacing that depends on the soil conditions and the power of the vibroflot; spacings of 6–10 ft (2–3 m) are common. Design guidance for spacings and target relative densities are provided in Schaefer et al. (2017a). Pre-drilling may be required in areas with hard crusts, and shallow treatment of the ground surface with vibratory rollers may be required after construction. Vibro-compaction has been used successfully to densify soils to depths of up to 165 ft (50 m). Case histories of vibro-compaction have been reported by Harder et al. (1984), Dobson (1987), Bo et al. (2005), Wehr (2005), Shao (2009), and Rollins et al. (2009).

Vibro-compaction is usually a cost-effective alternative to deep foundations or removal and replacement of problematic soils. By placing the energy (vibration) source at depth, it can be used to depths of about 100 ft (30 m), but has achieved densification to depths of approximately 165 ft (50 m). It is also effective both above and below the groundwater table. Its principal limitation results from the degradation of its performance in soils containing significant levels of fine-grained soils.

11.3.1.2 Vibro Rod

Vibro rod systems use a vibratory pile driving hammer to vibrate a long probe into the soil. The probe is then withdrawn while still being vibrated to densify the soil. To minimize densification-induced settlement, additional soil may be introduced at the ground surface or depth. Several types of probes have been used in vibro-compaction. In the Terraprobe system, a 30-in. (76-cm) open-ended steel pipe is vibrated into the ground; the vibrations densify the soil both inside and outside the pipe. The Vibro-Wing consists of a central rod with diametrically opposed 31-in. (80-cm) "wings" spaced 19 in. (50 cm) apart along the length of the rod (Figure 11.4). The Franki Y-probe consists of three 19-in. (50-cm)-wide steel plates welded to a central rod at 120° angles from each other. Horizontal cross-ribs may be welded to the faces of the steel plates to facilitate densification. By adjusting the frequency of vibration, these probes can be "tuned" to the resonant frequency of the soil-probe system to increase vibration amplitudes and densify the soil more efficiently (Massarch, 1991).

Vibro rod systems are most effective in soils similar to those for which vibro-compaction is most effective. Because vibro rods use vertical vibrations, their radius of influence is usually smaller than that observed for vibro-compaction. As a result, the grid spacing for soil improvement by vibro rods is generally smaller than for vibro-compaction. The effectiveness of vibro rods also appears to vary with depth (Janes, 1973). Case histories of vibro rod and similar systems have been reported by Massarch (1991), Neely and Leroy (1991), Senneset and Nestvold (1992), and Bo et al. (2005).

11.3.1.3 Blasting

Loose granular soils have also been compacted by blasting. Blasting densification involves the detonation of multiple explosive charges vertically spaced 10–20 ft (3–6 m) apart in drilled or jetted

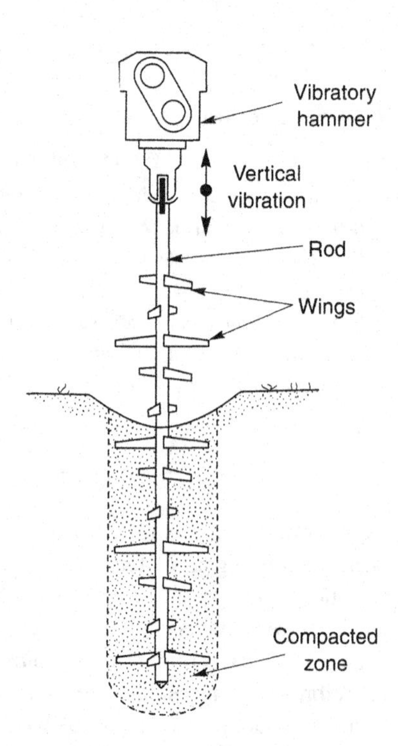

FIGURE 11.4 Vibro-wing system. Each pair of wings is oriented at a 120° angle to those located immediately above and below. (After Broms, 1991.)

boreholes; charges are usually placed about 2/3 of the thickness below the top of the layer to be densified. Charges can be placed at multiple depths in different boreholes (tiered charges) or at multiple depths within individual boreholes (decked charges). The boreholes are usually spaced between 25–50 ft (8–15 m) apart in undeveloped areas and 10–25 ft (3–8 m) in developed areas and backfilled prior to detonation. To increase the efficiency of the densification process for thicker (greater than about 30 ft, or 9.1 m) liquefiable layers, charges at different elevations may be detonated at small time delays, beginning with the lowest layers when tiered charges are used and the upper layers when decked charges are used. Immediately after detonation, the ground surface rises and gas and water are expelled from fractures (Figure 11.5) as the structure of the soil is broken down; supplementary vertical drains may also be used. The ground surface then settles as the excess gas and water pressure dissipates. Although the efficiency of densification decreases with each round of blasting, two or three rounds (with later rounds detonated at locations between those of the-earlier rounds) are often used to achieve the desired degree of densification.

Blasting can be effective to greater depths (> 100 ft, or 30 m) than many other densification methods. It is most effective in loose sands that contain less than 15% silt and less than 2% clay. Even small amounts of clay, or small clay seams, can substantially reduce the effectiveness of blasting. Blasting can be effective in dry soils, but the effects of capillary tension and gas bubbles in partially saturated soils virtually negate its effectiveness. As a result, blasting is most commonly used to densify completely saturated soils. In such soils, the shock wave produced by the charges produces localized, temporary liquefaction which allows the soil grains to move into a denser configuration. The use of instrumentation, such as Sondex tubes (corrugated plastic tubes with metal rings whose vertical spacings can be measured before and after blasting with a probe lowered through the tube), can provide rapid feedback on the effectiveness of the densification process.

Although blasting is quite economical, its use is limited by several practical considerations. It produces strong vibrations that may damage nearby structures or produce significant ground

FIGURE 11.5 Ground surface shortly after detonation of explosives during blast densification of loose soil beneath an abutment prior to construction of Coldwater Creek bridge near Mt. St. Helens in Washington state. (Photo by A.P. Kilian; used with permission.)

movements; as a result, it is good practice to monitor vibrations at selected locations along project boundaries. It requires the use of potentially hazardous explosives for which strict regulations on handling and storage usually apply. Improved knowledge of the blasting process has been gained from field studies (Gandhi et al., 1999; Gohl et al., 1998, 2000) and through theoretical/numerical analyses (Wu, 1995; Van Court and Mitchell, 1995; Gohl et al., 1998). Case histories of the use of blasting to mitigate seismic hazards have been described by Hachey et al. (1994), Gandhi et al. (1999), Dowding and Duplaine (2004), Shuwang et al. (2009), Narsillo et al (2009), and Finno et al. (2016).

11.3.2 Compaction Techniques

Compaction techniques use mechanical means to apply compressive and shearing stresses to the surface of a soil deposit, thereby pushing the soil particles into a denser configuration. Compaction of soil by rollers has been used for many years in the placement of fills with well-established principles and techniques (Shanklin et al., 2000; Holtz et al., 2011). The densification of soils by compaction differs from densification by consolidation (Section 6.4.2) in that it occurs by air, rather than water, being expelled from the voids of the soil. This section describes methods that can be used to compact soil deposits that are already in place (i.e., *in situ* soils) to improve their strength, stiffness, compressibility, and/or liquefaction resistance. Some of these methods, when they are applied very quickly to clean granular soils, may also induce vibrations that can help further densify the soils at some depth below the ground surface.

11.3.2.1 Dynamic Compaction

Dynamic compaction is performed by repeatedly dropping a heavy weight (a *tamper*) in a grid pattern on the ground surface (Figure 11.6). The tampers, usually constructed of solid steel plates or welded steel containers filled with concrete, generally range from 6 to 30 tons (53 to 267 kN), although weights of up to 170 tons (1,500 kN) have been used. Tampers are usually sized to produce static contact pressures of 5–10 psi; higher pressures can lead to excessive tamper penetration (with energy dissipated by local plastic deformation of the soil rather than compression/vibration) and lower pressures limit densification to very shallow depths. Lighter weights are used for thinner (~3 m) deposits and heavier weights for thicker (~10 m) deposits. The weights are usually flat-bottomed, although some studies (Feng et al., 2013; Arslan et al., 2007) indicate that improved densification has been achieved with conical-bottomed weights. Drop heights usually range from about 35 to 100 ft (10–30 m), although weights have been dropped from up to 130 ft (40 m). The weights are usually dropped three to eight times before moving to the next point on the grid. More detailed descriptions of dynamic compaction have been prepared by Lukas (1995) and Slocombe (2012).

At a particular site, dynamic compaction is generally performed in several stages, or passes. Empirical evidence suggests that the effective depth of influence (the depth to which significant improvement can be detected) increases with impact energy and that the greatest degree of improvement is usually observed at about half the effective depth of influence (Mayne et al., 1984). To avoid developing a shallow zone of dense soil that could inhibit the transmission of energy to greater

FIGURE 11.6 Aerial view of abandoned strip mining site undergoing soil improvement by dynamic compaction in Harrison County, Ohio. The grid pattern on which the dynamic compaction weight is dropped, and the need for subsequent regrading and surface compaction, are evident from this photograph. (Photo courtesy of Geotechnical Consultants, Inc.)

depths, the deepest soil is densified first with a series of high-energy (heavy weight and/or high drop height) drops on a widely spaced grid. After the craters produced by the first pass have been filled (preferably with well-graded granular soil), soils at intermediate depth are then compacted using a greater number of drops from a smaller height at closer spacing (often half the spacing of the original grid). Finally, the near-surface soils are compacted by dropping relatively light weights on a virtually continuous pattern to smooth or "iron" the ground surface.

The kinetic energy of the weight at impact produces stress waves that travel through the soil. The total energy delivered to the soil is a function of the weight, drop height, grid spacing, and number of drops per grid point. When the groundwater table is near the surface, placement of a gravel or sand blanket may be required prior to compaction. Although dynamic compaction has been used successfully for cohesive soils, its most common use for mitigation of seismic hazards is for potentially liquefiable soils. At each grid point, a series of drops causes the porewater pressure to increase so that the soil particles can more easily move into a denser configuration. Dissipation of the excess porewater pressure results in further densification within a short period (1–2 days for sands and gravels; 1–2 weeks for sandy silts) after treatment.

Dynamic compaction is generally effective to depths of about 30 ft (9 m), although extremely high impact energies may produce densification at greater depths. The depth of significant compaction (in meters) can be estimated as

$$D = n\sqrt{WH} \tag{11.1}$$

where n is a function of the equipment and soil type that ranges from 0.3 to 0.8 (Mayne et al., 1984) and is about 0.5 for most soils, and WH is the theoretical free-fall energy in tonne-m (1 metric tonne = 1.1 U.S. ton). Because the process is rather intrusive - it can produce considerable noise, dust, flying debris, and vibration - it is rarely used near occupied or vibration-sensitive structures. In addition to shallow groundwater, the effectiveness of dynamic compaction is inhibited by the presence of weak cohesive soils above or below the target layer. Case histories of dynamic compaction of potentially liquefiable soil have been described by Hussein and Ali (1987), Keller et al. (1987), Koutsoftas and Kiefer (1990), Mitchell and Wentz (1991), Mackiewicz and Camp (2007), Majdi et al. (2007), Lu et al. (2009) and others.

Dynamic compaction is most effective in dry granular materials but can also be effective in saturated soils of sufficient permeability to dissipate pore pressures between drops of the tamper. Densification of dry soils occurs immediately, and almost immediately in pervious soils below the water table. In saturated intermediate soils such as silty soils, dynamic compaction can be effective but care must be taken to provide time for pore pressure dissipation; compaction may need to take place in stages with time for dissipation (which can range from days to weeks) allowed between each. Pore pressure dissipation can be accelerated by the use of wick drains. Dynamic compaction can also densify partially saturated soils, but that densification occurs by conventional compaction mechanisms rather than vibration and thus does not extend to the same depths. Dynamic compaction can also be used to compact materials such as bouldery soils, construction debris, and solid waste that are otherwise not compactable.

The effectiveness of dynamic compaction can be monitored for each stage of treatment by measuring tamper penetration. In this sense, the tamper acts as a probe that can find particularly loose or soft zones at a site and can demonstrate densification and strengthening by the rate of penetration reduction as the compaction process proceeds. The process of lifting and dropping a weight is relatively simple and, unless very heavy weights or large drop heights are required, can often be accomplished by non-specialty contractors. The procedure can be disruptive, however, in terms of the vibrations it produces; the effects of such vibrations on people and infrastructure beyond the property lines of the site being treated may limit its use in urban areas. The procedure is also disruptive to the surface of the site being treated, and the craters typically left behind (Figure 11.6) must be filled and compacted after dynamic compaction has been completed. Surficial disruption is more severe with shallow groundwater levels, and dynamic compaction is not recommended for sites with groundwater at depths less than about 6 ft (1.8 m).

11.3.2.2 Rapid Impact Compaction

For sites requiring densification to depths of 15–20 ft (4.5–6 m), rapid impact compaction (Serridge and Synac, 2006) offers an efficient alternative to deep dynamic compaction. Rapid impact compaction uses equipment (Figure 11.7) mounted on an excavator that repeatedly raises and drops a 5–10 ton weight from a 3–4 ft (0.9–1.2 m) height onto a 5 ft (1.5 m) diameter tamper foot at a rate of 40–60 blows/minute. Rapid impact compaction is effective for densification of dry or free-draining saturated granular soils; for optimal results, groundwater levels should be at least 3 ft (1 m) below the ground surface. The vertical position of the tamper foot is monitored, and the equipment is moved to a new location when the vertical movement per blow falls below a user-selected threshold. Because the impact energy per individual blow is relatively low, vibrations attenuate quickly enough that rapid impact compaction can be used relatively close to structures.

Rapid impact compaction can achieve densification to depths of 10–20 ft (3–6 m), depending on compaction energy level, soil type and properties, and groundwater level. The technique is best

FIGURE 11.7 Rapid impact compaction: (a) schematic illustration, and (b) rapid impact compaction equipment. (Images courtesy of Cofra, Inc.)

suited to free-draining, non-plastic, granular soils with low fines content at sites where the groundwater is at least one meter deep.

11.3.2.3 Compaction Grouting

Soft or weak soils can be densified by injecting a very low slump [generally less than 1 in. (2.5 cm)] grout into the soil under high pressure, a process known as compaction grouting. Because the grout is highly viscous, it displaces the surrounding soil and forms an intact bulb or column that densifies the surrounding soil through a combination of shear and consolidation from confining pressure increase (Figure 11.8). Grout holes can be drilled vertically or at an angle to place grout beneath a structure or foundation. Grout mix design is critical in that excessively fluid grouts can lead to fracturing and lensing, in which case densification is reduced. Compaction grouting may be performed at a series of points in a grid or along a line. Grout point spacings ranging from 3 to 15 ft (1 to 4.6 m) have been used. Because higher overburden pressures allow the use of higher grout pressures, larger spacings are generally used when treating deeper soils. At shallow depths, little soil densification can be achieved, but compaction grouting may be used to lift settled slabs or structures; indeed, remediation of foundation settlement is probably the most common application of compaction grouting.

Compaction grouting may be performed from the top down (downstage grouting) or from the bottom up (upstage grouting). Upstage grouting is less expensive and more commonly used than downstage grouting. However, the downstage procedure is preferred (Stilley, 1982; Bell, 1993) for underpinning of structures or for sites where loose soils extend to the ground surface. By working from the top down, the placement of an upper grout bulb reduces the possibility of subsequent grout escaping at the surface and grout heave and also provides additional strength and confinement that allows the use of higher grouting pressures at greater depths.

Because it does not rely on vibration, compaction grouting can be used in all soil types. Since it requires drilling, it is not adversely affected by stiff or hard crusts. For seismic hazard mitigation, it is most commonly used in free-draining soils such as gravels, sands and non-plastic silts. Compaction grouting can be used to virtually any depth and can easily be used within a given range of depths. The size and shape of the grout bulb or column is influenced by the stiffness and strength of the soil and also by the rate and pressure at which the grout is injected. An important feature of compaction grouting is that its greatest effects occur where the soil is softest and weakest. Compaction grout masses with diameters greater than 3 ft (1 m) are not uncommon (Warner, 2003).

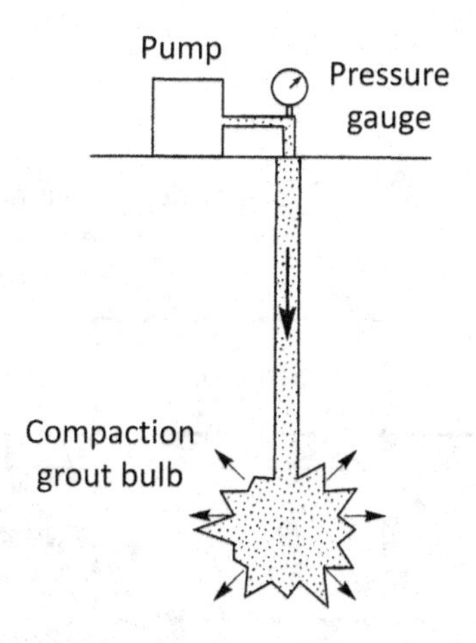

FIGURE 11.8 Compaction grouting. Low-slump ground is pumped under high pressure to form a bulb that displaces and densifies the surrounding soils. By raising the ground tube while pumping, a column of ground can be created in the soil. (After Hausmann, 1990.)

Compaction grouting has been used to depths of 100 ft (30 m). Case histories of compaction grouting have been described by Salley et al. (1987), Warner (2003), Graf (1992), Baez and Henry (1993), Boulanger and Hayden (1995), Miller and Roycroft (2004), and El-Kelesh et al. (2012).

11.3.3 Areal Extent of Densification

An important consideration in the densification of soils for the construction of individual structures and foundations is the areal extent of soil improvement required for satisfactory performance during earthquakes. The areal extent should be evaluated on a case-by-case basis since site-specific soil conditions, performance requirements, and failure consequences must be addressed.

The required areal extent of improvement depends on the mechanism of failure that the improvement is intended to eliminate. For potential stability failures, the areal extent of improvement will depend on the degree of improvement that can be achieved and on the extent of the potential failure surface(s). By estimating the residual strength of the soil after improvement, stability analyses can be used to estimate the extent of improvement that will produce an acceptable level of stability. To minimize post-earthquake settlement of a structure or foundation on loose, saturated sand, densification is usually performed within a zone defined by a 30°–45° line from the edge of the structure, as illustrated in Figure 11.9. Available research and field experience indicate that this approach is likely to produce a satisfactory extent of improvement (Iai et al., 1988), however, stress-deformation analyses can be used to confirm, or optimize, the required extent of improvement for an individual project.

11.4 REINFORCEMENT TECHNIQUES

In some cases, it is possible to improve the strength and stiffness of an existing soil deposit by installing discrete inclusions that reinforce the soil. These inclusions may consist of structural materials, such as steel, concrete, or timber, or geomaterials such as densified gravel. Reinforcement of new engineered fills using geosynthetic or metallic reinforcement is beyond the scope of this chapter.

Reinforcement techniques improve the soil by producing a stiffer and stronger "composite" material that consists of localized reinforced zones and the zones of soil between them. The reinforced zones are generally small and numerous enough that they are not considered individually but rather as components of a continuous mass of reinforced soil. While reinforcement techniques stiffen and strengthen the soil in which they are installed, they may also produce benefits by densifying surrounding soil during their installation and, for some methods, by promoting drainage through the reinforcing material.

11.4.1 Stone Columns

Soil deposits can be improved by the installation of dense columns of gravel known as stone columns. Stone columns may be used in both fine- and coarse-grained soils, and with both shallow

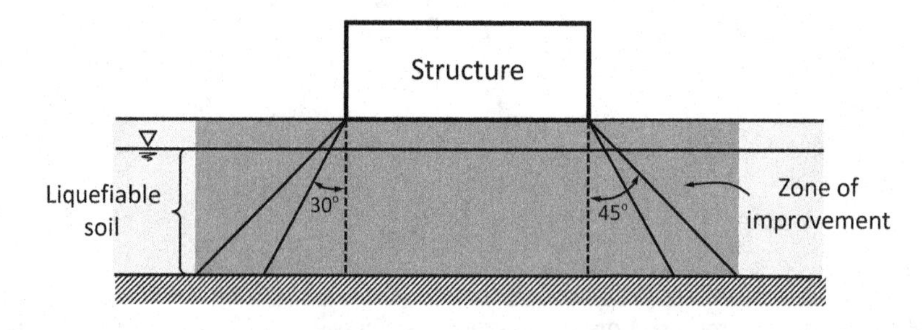

FIGURE 11.9 Typical areal extent of improvement for densification of potentially liquefiable soil beneath a structure.

and deep groundwater tables. In fine-grained soils, stone columns are usually used to increase shear strength beneath structures and embankments by accelerating consolidation (by allowing radial drainage) and introducing columns of stronger material. For mitigation of seismic hazards, they are commonly used for the improvement of liquefiable soil deposits.

Stone columns can be installed in a variety of ways. As discussed previously, stone columns may be constructed by introducing gravel during the process of vibro-compaction (Section 11.3.1.1). In one common approach, a probe is jetted to the desired depth of treatment, the uncased hole is flushed, and gravel is added in 0.3–1.2 m lifts by a vibrator at the bottom of the probe. Other approaches that do not use jetting and thus require a hole to stand open uncased are generally not applicable to the mitigation of liquefaction hazards.

Several other methods of installation are also available. In the Franki method, a steel casing initially closed at the bottom by a gravel plug is driven to the desired depth by an internal hammer (Figure 11.10). At that depth, part of the plug is driven beyond the bottom of the casing to form a bulb of gravel. Additional gravel is then added and compacted as the casing is withdrawn. The diameter of the resulting stone column depends on the stiffness and compressibility of the surrounding soil; in loose sand, a 19–28 in. (0.5–0.7 m) casing will typically produce a 0.8-m (31 in.) diameter column. Casings with trap doors at the bottom have also been used to install stone columns (Solymar and Reed, 1986). The trap door allows the casing to be driven as a closed-end pile but also allows gravel to be placed during the withdrawal of the casing. The gravel can be densified by pausing to redrive the casing at various intervals during the withdrawal process. Disruption of the ground surface may require shallow compaction with vibratory rollers following stone column installation.

Stone columns combine at least four different mechanisms for the improvement of liquefiable soil deposits. First, they improve the deposit by virtue of their own high density, strength, and stiffness – in this sense, they reinforce the soil deposit. Second, they provide closely spaced drainage boundaries that inhibit the development of high excess porewater pressures (Section 11.6) during earthquake shaking. Third, the processes by which they are installed densify the surrounding soil by the combined effects of vibration and displacement. Finally, the installation process increases the lateral stresses in the soil surrounding the stone columns. These multiple benefits have made the use of stone columns very popular. Case histories of seismic hazard mitigation by stone columns have been presented by Priebe (1991), Hayden and Welch (1991), Mitchell and Wentz (1991), Rollins et al. (2006), and Shao et al. (2013). Adalier and Elgamal (2004) summarize the literature on the application of stone columns for liquefaction mitigation.

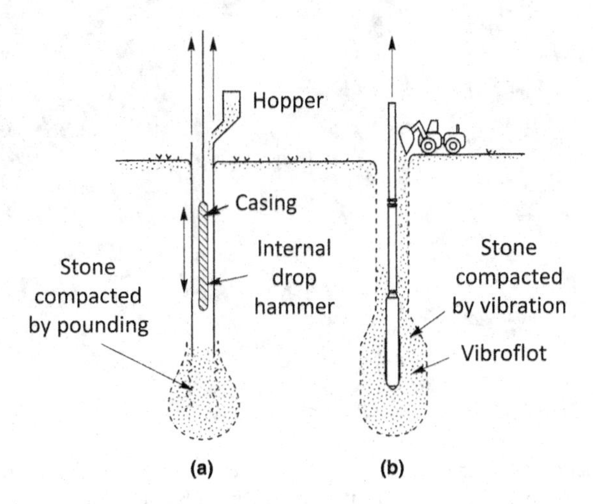

FIGURE 11.10 Methods of stone column installation: (a) Franki method and (b) by vibroflotation (After Broms, 1991.)

Stone column performance can be strongly influenced by construction procedures. The integrity of a stone column can be compromised by reductions in its diameter, which can result from excessive rates of casing withdrawal, squeezing of soft soils, or caving of unsupported walls of the hole. Stone columns can also be ineffective in profiles that contain peat or other organic materials, or soft clay layers that are thicker than the diameter of the column. Soil profiles with boulders, cobbles, or other obstructions may require pre-drilling in order for stone columns to be installed.

11.4.2 Compaction Piles

Granular soils can be improved by the installation of compaction piles. Compaction piles are displacement piles, usually prestressed concrete or timber, that are driven into a loose sand or gravel deposit in a grid pattern (Figure 11.11) and left there. Compaction piles improve the seismic performance of a soil deposit by three different mechanisms. First, the flexural strength of the piles themselves provides resistance to soil movement (reinforcement). Second, the vibrations and displacements produced by their installation cause densification. Finally, the installation process increases the lateral stresses in the soil surrounding the piles.

Compaction piles generally density the soil within a distance of 7–12 pile diameters (Robinsky and Morrison, 1964; Kishida, 1967), and consequently, are usually installed in a grid pattern. Between compaction piles, relative densities of up to 75%–80% can usually be achieved (Solymar and Reed, 1986). Improvement can be obtained with reasonable economy to depths of about 60 ft (18 m). Case histories describing the use of compaction piles have been presented by Lindqvist and Petaja (1981), Marcuson et al. (1991), Mitchell and Wentz (1991), Kramer and Holtz (1991), and Okamura et al. (2003).

FIGURE 11.11 Compaction piles driven into the upstream embankment of Sardis Dam to reduce liquefaction hazards. The contractor drove the piles to this level with a barge-mounted conventional hammer; the piles were later driven below the water surface with a different hammer. (Photo by T.D. Stark; used with permission.)

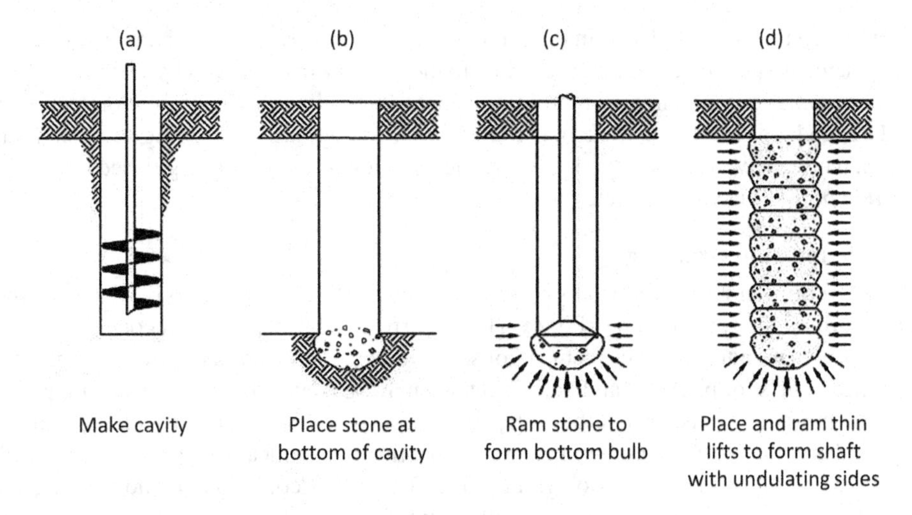

FIGURE 11.12 Steps in rammed aggregate pier construction: (a) augering hole, (b) placement of aggregate, (c) compaction of initial layer of aggregate, and (d) configuration after compacting multiple layers of aggregate. (After Wissmann et al., 2015.)

11.4.3 Rammed Aggregate Piers

A technique with some similarity to both stone columns and compaction piles is rammed aggregate pier construction (Lawton and Fox, 1994; Lawton et al., 1994). Rammed aggregate piers are constructed by drilling (up to 30 ft, or 9 m) or driving (typically up to about 45 ft, or 14 m) a 18–36-inch (46–92 cm) hollow mandrel to a desired depth of treatment. After driving, approximately 3 ft of coarse aggregate is dropped into the hold and rammed by repeated blows of a conical tamper foot (Figure 11.12). The tamper drives the aggregate both downward and outward into close contact with the surrounding soil. Additional lifts of aggregate are compacted as the casing is withdrawn. The soil is densified by both vibration and compression, and the lateral movement of the rammed soil increases the lateral stresses (and, hence, the stiffness) of the surrounding soils. A rammed aggregate pier can provide many of the same benefits as a stone column, but is significantly stiffer structurally. Case histories of rammed aggregate piers for liquefaction hazard mitigation have been described by Majchrzak et al. (2004) and Farrell et al. (2010).

11.5 GROUTING AND MIXING TECHNIQUES

The engineering characteristics of many soil deposits can be improved by injecting or mixing cementitious materials into the soil. These materials both strengthen the contacts between soil grains and fill the void space between the grains. Grouting techniques involve the injection of such materials into the voids of the soil or into fractures in the soil so that the particle structure of the majority of the soil remains intact. Mixing techniques introduce cementitious materials by physically mixing them with the soil, completely disturbing the particle structure of the soil. The mixing can be accomplished mechanically or hydraulically. Grouting and mixing techniques tend to be expensive but can often be accomplished with minimal settlement or vibration. As a result, grouting and mixing techniques can often be used in situations, such as under existing structures, where other soil improvement techniques cannot.

11.5.1 Grouting

The term grouting is used to describe a variety of processes by which cementitious material is introduced under pressure into the ground. Grouting techniques are often classified according to the method by which the grout is placed in the ground (Hausmann, 1990). In this chapter, however, soil improvement techniques are classified according to the primary mechanisms by which they produce improvement. As a result, compaction grouting is described with other densification techniques in Section 11.2, and jet grouting is treated as a mixing technique in the following section. With this convention, there are two primary types of grouting techniques that are distinguished by their levels of disruption of the existing soil skeleton.

11.5.1.1 Permeation Grouting

Permeation grouting involves the injection of low-viscosity liquid grout into the voids of the soil without disturbing the soil structure (Figure 11.13). Particulate grouts (i.e., aqueous suspensions of cement, fly ash, bentonite, microfine cement, or some combination thereof) or chemical grouts (e.g., silica and lignin gels, or phenolic and acrylic resins) may be used. Chemical grouts are pure solutions with no particles in suspension that can penetrate smaller voids than particulate grouts. Karol (2003) provides a thorough and comprehensive description of chemical grouting.

The suitability of different types of grouts for different soil conditions is most strongly influenced by the grain size (or, more accurately, the pore size) distribution of the soil. Virtually any type of grout, even relatively viscous regular Portland cement grouts, can be used in soil with large pores such as gravels and coarse sands. Medium to coarse sands can take microfine cement particulate grouts. Chemical grouts generally exhibit lower viscosity than particulate grouts (although the viscosity of microfine cement grouts may be as low as some chemical grouts) and can therefore be used in fine sands and some coarse silts (Semprich and Stadler, 2003). However, the presence of fines can significantly reduce the effectiveness of permeation grouting.

Grout pipes are typically installed in a grid pattern at spacings of 2.5–8 ft (0.8–2.5 m) with the spacing decreasing with decreasing soil permeability (Hayden, 1994). The grout may be injected in different ways. In stage grouting, a boring is advanced a short distance before grout is injected through the end of the drill rod. After the grout sets up, the boring is advanced another short distance and grouted again. This process continues until the grout has been placed to the desired depth. In the tube-a-manchette approach, a grout tube with injection ports every 12–24 in. (30–61 cm) along its length is installed in a borehole. Rubber sleeves (manchettes) that serve as one-way valves cover the injection ports on the outer surface of the grout tube and internal packer systems are used to control the depths at which grout is injected.

Colloidal silica is an aqueous solution of microscopic silica particles (7–22 nm) that has a viscosity similar to water at concentration of about 5% (by weight). Over a controllable period of time (typically up to a few months), the solution forms a gel that increases particle contact strength, thereby improving liquefaction and deformation resistance (Gallagher and Mitchell, 2002).

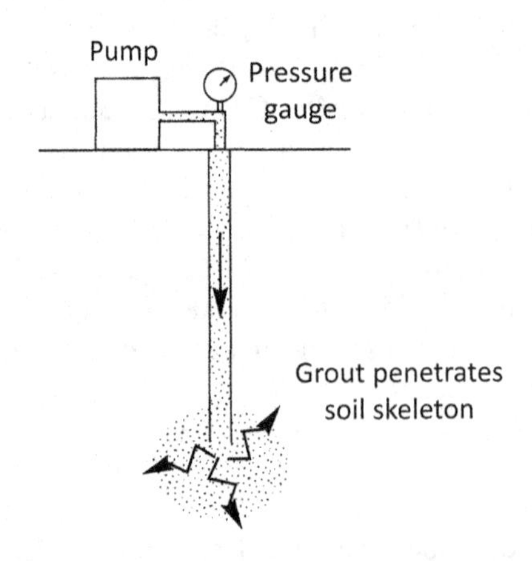

FIGURE 11.13 Permeation grouting. (After Hausmann, 1990.)

The slow elation time allows colloidal silica to be placed slowly and even transported across a site by flowing groundwater, which can allow treatment of areas (e.g., under structures) that are not otherwise accessible. Field testing (Gallagher et al., 2007) showed that high pore pressures were induced by blasting in an area treated with 7% colloidal silica solution but that settlements were significantly lower than in adjacent untreated areas.

Permeation grouting produces soil improvement by two primary mechanisms. First the grout tends to strengthen the contacts between individual soil grains, thereby producing a soil skeleton that is stronger and stiffer than that of the ungrouted soil. Second, the grout takes up space in the voids between soil particles, reducing the tendency for densification (or excess pore pressure generation) upon cyclic loading. Soils improved by permeation grouting can have shear strengths of 50–300 psi (345–2,070 kPa). Case histories in which permeation grouting was used to mitigate seismic hazards were described by Zacher and Graf (1979), Graf (1992), Bruce (1992), Karol (2003), and Powers et al. (2007).

11.5.1.2 Intrusion Grouting

In the process of intrusion grouting (Mitchell, 1970), fluid grout is injected under pressure to cause controlled fracturing of the soil (Figure 11.14). Because the grout is not intended to flow through the small voids between soil particles, relatively viscous (and strong) cement grouts can be used. In theory, the first fractures should be parallel to the minor principal stress planes, but observations show that they usually follow weak bedding planes. After allowing the initially placed grout to cure, repeated intrusion grouting fractures the soil along different planes. Eventually, a three-dimensional network of intersecting grout lenses can be formed. Some densification of the soil may occur, but the primary mechanism of improvement results from the increased stiffness and strength of the soil mass due to the hardened lenses of grout.

11.5.2 Mixing

Localized improvement of soil columns can be achieved by in situ mixing of the soil with cementitious material. Because the cementitious material is physically mixed with the soil, it need not have an extremely low viscosity - strong, cement slurries are commonly used. For mitigation of seismic hazards, this approach is most commonly accomplished by deep soil mixing and jet grouting. These techniques can be effective in soils that cannot be densified effectively or efficiently by vibratory techniques and for sites where vibratory methods are not feasible due to adjacent sensitive infrastructure.

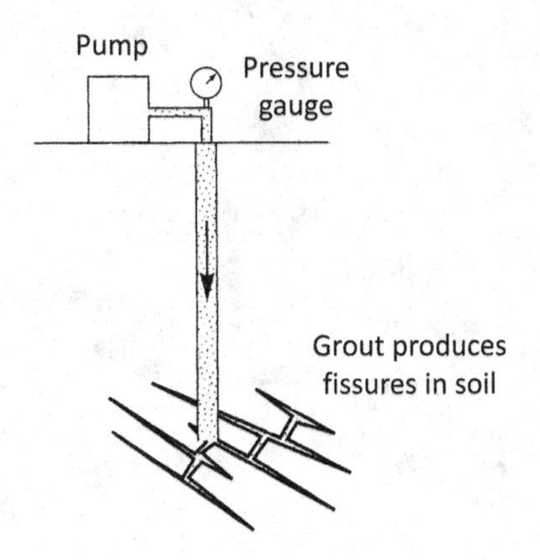

FIGURE 11.14 Intrusion grouting process. (After Hausmann, 1990.)

11.5.2.1 Deep Soil Mixing

The term *deep soil mixing* (DSM) describes a specific technique in which cementitious material is mechanically mixed into the soil using a hollow stem auger and mixing paddle arrangement (Figure 11.15). The cementitious material may be in powder form (dry DSM method) injected using compressed air or in the form of a slurry (wet DSM method). Typical soil mixing rigs may have single augers (1.6–13 ft, or 0.5–4 m) in diameter) or gangs of two to eight augers (usually about 3 ft, or 0.9 m) in diameter) that produce overlapping columns of mixed soil. Other types of mixing rigs use a chainsaw-like apparatus to make continuous panels of mixed soil. As the mixing augers are advanced into the soil, grout is pumped through their stems and injected into the soil under low pressure at their tips. The grout is thoroughly mixed with the soil by the auger flights and

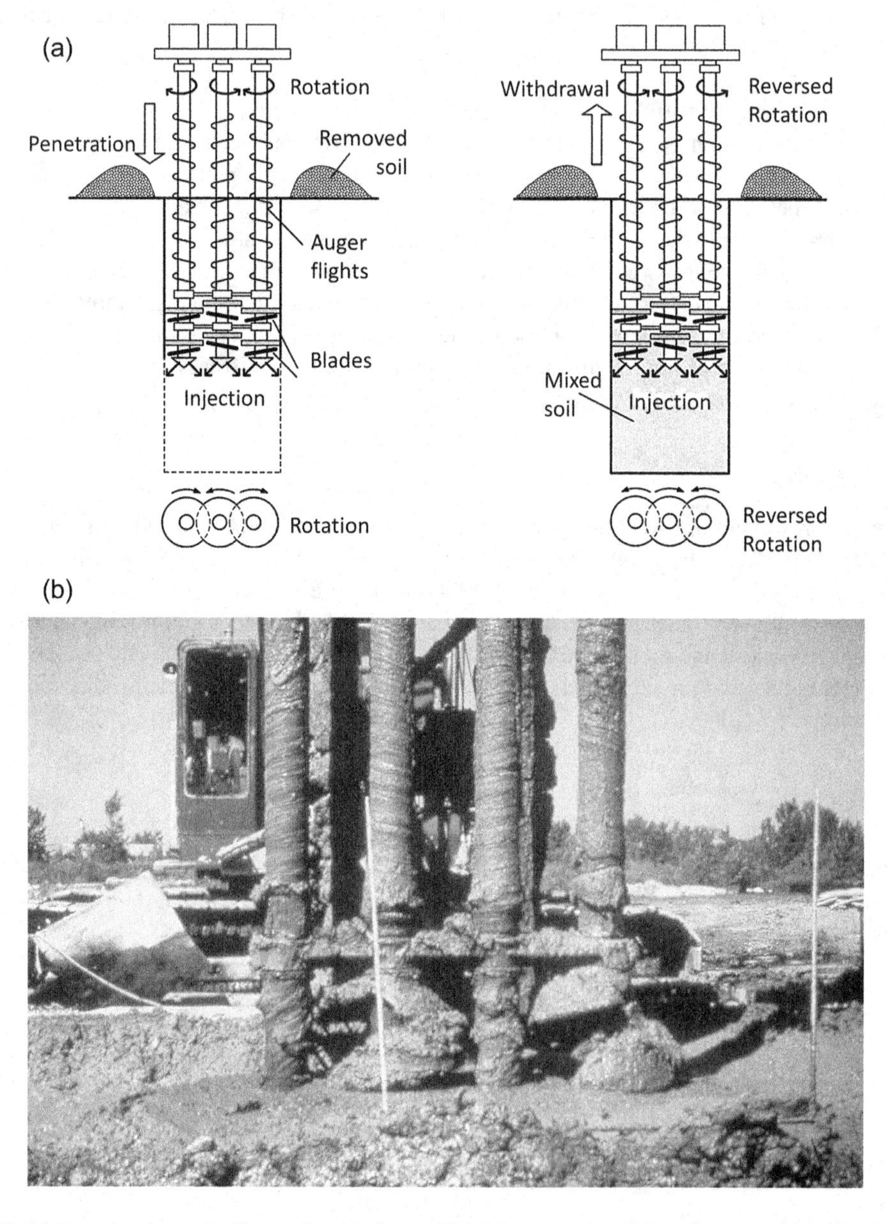

FIGURE 11.15 (a) schematic illustration of deep soil mixing process (After Chen et al., 2013; used with permission of ASCE), and (b) multi-auger wet-method mixing tool (Photo courtesy of The Collin Group).

mixing paddles. After the design depth has been reached, the augers are withdrawn while the mixing process continues. The soil mixing process leaves behind a uniform (constant width) column of mechanically mixed soil-cement. By overlapping the columns before the grout cures, walls and cellular structures that restrain movement of liquefiable soils can be constructed below the ground surface. Grids of soil-mixed panels constructed as overlapping columns have provided successful mitigation of liquefaction hazards in several earthquakes (Bruce, 2000; Bruce et al., 2013). Soil mixing equipment is large and heavy, which can complicate its use at soft sites; the site should also have sufficient space to accommodate grout batching operations.

Soil mixing can be used in virtually any type of inorganic soil that is free of boulders and obstructions. The dry method is typically used to depths of up to about 60 ft (18 m) and the wet method up to about 100 ft (30 m), although greater depths (up to 200 ft, or 60 m) have been used in Japan). The strength of the soil-cement mixture depends on the type of grout, type of soil, and degree of mixing; strengths of 200 psi (1,380 kPa) or more are commonly achieved. High plasticity soils can be difficult to mix thoroughly. Detailed descriptions of DSM design and construction processes are available in Bruce et al. (2013). Case histories involving the use of soil mixing for the mitigation of seismic hazards have been presented by Ryan and Jasperse (1989), Babasaki et al. (1991), Taki and Yang (1991), Yang et al. (2001), Porbaha et al. (2005), Yasui et al. (2005), and Hudson et al. (2014), among others.

11.5.2.2 Jet Grouting

In jet grouting, the soil is mixed with a cement slurry injected horizontally under high pressure in a previously drilled borehole (Figure 11.16). The injection nozzle is rotated to allow the grout to be placed in all horizontal directions. The grout slurry jet cuts and mixes with the *in situ* soil to form a soil-cement mixture sometimes referred to as soilcrete. Air or water may also be injected to aid in the mixing process. Jet grouting begins at the bottom of the borehole and proceeds to the top, leaving behind a relatively uniform column of mixed soil-cement. By overlapping the columns before the grout cures, walls and cellular structures can be constructed below the ground surface. Three jet grouting systems are commonly used. In the single-fluid system (Figure 11.16a), liquid grout is used as both the cutting fluid and the cementing medium. In the double-fluid system (Figure 11.16b), grout and air are ejected separately from nozzles on opposite sides of the rod; the air pressure helps disrupt the soil to improve cutting and can expand the diameter of the column by some 30% beyond that of the single-fluid system. The triple-fluid system (Figure 11.16c) uses air and water as cutting fluids in a zone just above the grout jet; the triple-fluid system is capable of producing the largest diameter jet-grouted columns.

The diameter of a jet-grouted column depends on the soil conditions and the manner in which the jet grouting is performed. Column diameters are generally greater in coarse-grained soils than in fine-grained soils. By varying the air, water, and grout pressures and the rates of rotation and lifting of the grout tubes, a jet grouting operator can control the effective dimensions of the column. Diameters ranging from 16–20 in. (40–50 cm) in clayey silt to 36–39 in. (90–100 cm) in sandy gravel can be expected using a single jet (grout only) system (Bell, 1993). Diameters of 30–40 in. (76–100 cm) in clayey silt and 6–8 ft (1.8–2.4 m) in sandy gravel can be expected with a triple-jet (air, water, and grout) system. Jet grouting is typically used at depths less than 100 ft (30 m), but has been used at depths greater than 165 ft (50 m). Jet grouting can be performed in any type of inorganic soil to depths limited only by the range of the drilling equipment. Case histories involving the use of jet grouting for mitigation of seismic hazards include Hayden (1994), Durgunoglu et al. (2003), Martin and Olgun (2006), and Wang et al. (2009).

11.5.2.3 Cellular Containment

The geometric flexibility of DSM and jet grouting allows for many options in the geometry of liquefiable soil treatment. In some cases, these techniques have been used to confine potentially liquefiable soils within grids of mixed or grouted walls. Such grids can improve the performance of a

FIGURE 11.16 Jet grouting configurations: (a) single-fluid system, (b) double-fluid system, and (c) triple-fluid system (a–c Courtesy of Keller), and (d) actual jet grouting rig with jets exposed (Image courtesy of railsystem.net).

treated area by (1) providing stiffness that reduces the shear strains induced in the contained soil, (2) providing resistance to lateral or vertical deformations if liquefaction is triggered in the contained soil, and (3) providing a hydraulic barrier against high pore pressures in unimproved soil outside the grid making their way into the grid (Nguyen et al., 2013).

The shallower foundation soils beneath Jackson Lake dam were improved with dynamic compaction, but deeper soils were reinforced with a honeycomb pattern of deep-soil-mixed walls (Ryan and Jasperse, 1989). Deep-soil-mixed grids were used to protect pile foundations supporting the 14-story Oriental Hotel located on a pier in Kobe, Japan, and were found to perform well (Tokimatsu et al., 1996; Hamada and Wakamatsu, 1997) in the1995 Kobe earthquake. A rectangular grid of DSM walls with interior columns was used to contain liquefiable soils beneath a reinforced concrete box culvert constructed for the expansion of the San Francisco Bay Area Rapid Transit (BART) system (Martin et al., 1999). Nguyen et al. (2013) found that strain compatibility-based methods produced low estimates of shear stress in the contained soil, and offered an improved method to estimate those stresses and tensile stresses in the grid elements.

11.5.2.4 Discussion

Quality control of the mixing process can be difficult in some cases. While the mechanical nature of DSM provides relatively uniform columns of consistent and predictable diameter, the hydraulic nature of jet grouting offers no such assurances. Excavated columns from jet grouting and DSM

FIGURE 11.17 Exhumation of jet-grouted (near) and soil mix (distant) columns at Tuttle Creek Dam (Courtesy of T.D. Stark). Three groups of nine jet-grouted columns were installed with double- and triple-fluid systems, different grout pressures, and different water-cement ratios. Inset shows surface exposed by diamond wire saw cut through three overlapping columns (sixth row from front) illustrating inclusions of untreated sand (light colored area in front of person) and clay (dark areas).

in a test section show the size and variability of the two types of columns at a particular site. The dimensions and uniformity of jet-grouted columns is much more likely to be influenced by soil type, consistency, density, and stress conditions than deep-soil-mixed columns. Figure 11.17 shows excavated deep-soil-mixed and jet-grouted columns from a test section at Tuttle Creek Dam near Manhattan, Kansas; the mixed columns had smaller but uniform diameters and were more thoroughly mixed and homogeneous than the jet-grouted columns, which had variable diameters and were plagued by voids and soil inclusions. If the in-situ soil is not completely mixed with the slurry, a jet-grouted column can be weakened by the presence of inclusions of soft, ungrouted soil; such inclusions are more likely to occur in fine-grained soil than coarse-grained soil, in large-diameter columns than small-diameter columns, and to be found near the bottom of the column than the top (Stark et al., 2009).

11.6 DRAINAGE TECHNIQUES

Groundwater plays an enormous role in the behavior of soils under both static and seismic conditions, so it is not surprising that soils can be improved by modifying porewater conditions. The initiation of liquefaction requires undrained conditions, so drainage techniques typically seek to control or retard the development of high pore pressures during earthquake shaking. Varying degrees of pore pressure control can be achieved by vertical drainage, lowering of the groundwater level, or modification of saturation.

11.6.1 Vertical Drains

The buildup of excess porewater pressure during earthquake shaking can be suppressed by providing vertical drains to which porewater can flow laterally during and after earthquake shaking. The installation of stone columns, for example, introduces columns of freely draining gravel into a liquefiable soil deposit in addition to densifying the soil. Earthquake-induced excess pore pressures may be rapidly dissipated by the horizontal flow of porewater into the drains. The rate of pore pressure dissipation depends on the diameter and spacing of the drains and on the permeability and compressibility of the surrounding soil. Several procedures (e.g., Seed and Booker, 1976, 1977; Pestana et al., 1997) for selecting the sizes and spacings of gravel drains (or stone columns) for mitigation of liquefaction hazards have been developed. The use of gravel drains for suppression of excess porewater pressure requires careful attention to installation procedures (since mixing of the gravel and the native soil during installation may reduce the permeability of the stone column), and to drain permeability and filtration behavior at the drain-soil boundary. Even though drainage techniques can mitigate liquefaction hazards by suppressing excess porewater pressure buildup, post-earthquake settlement may still occur. Case histories of the use of drainage techniques for mitigation of seismic hazards have been described by Ishihara et al. (1980), Aboshi et al. (1991), Iai et al. (1994), Rollins et al. (2006), and Rollins et al. (2009). However, drainage alone has not commonly been relied upon for the mitigation of liquefaction hazards due to the post-earthquake settlement that accompanies it and inconsistent performance.

A relatively recent innovation in liquefaction hazard mitigation by drainage is the use of relatively large-diameter, slotted, corrugated plastic drain pipes, or "earthquake drains." The pipes are wrapped tightly with a durable geotextile filter fabric and installed with a vibrating mandrel. The drains, typically 3–6 in. (7.5–15 cm) in diameter, are placed in a triangular pattern at spacings of 3–6 ft (0.9–1.8 m) depending on the permeability of the soil. Individual drains can handle very large flow volumes, and some degree of densification is achieved during installation. Rollins et al. (2004) tested full-scale earthquake drain test sections at sites in the San Francisco Bay Area and in Vancouver, B.C.; in both cases, liquefaction was induced by controlled blasting. Despite the fact that blasting produces a much more rapid increase in pore pressure than earthquake shaking, and hence less opportunity for drainage during shaking, settlements of treated areas ranged from 20% to 60% of the settlement that occurred in untreated areas. Shao et al. (2013) describe a case where both stone columns and earthquake drains were used to mitigate liquefaction and lateral spreading hazards at a site in Seattle.

11.6.2 Groundwater Level Lowering

Unacceptable movements of slopes, embankments, retaining structures, and foundations can frequently be eliminated by lowering the groundwater table prior to earthquake shaking. A number of dewatering techniques have been developed and proven useful in engineering practice. Procedures for the design of dewatering systems are well established and widely used (e.g., Cedergren, 1989; Powers, 1992). These standard techniques may be used to increase the stiffness and strength of a soil deposit for mitigation of seismic as well as non-seismic hazards.

11.6.3 Desaturation

Liquefaction is caused by the increase in porewater pressure that results from the tendency of a loose soil to contract, or densify, when subjected to cyclic loading. Equation (9.62) showed how the pore pressure associated with some increment of volumetric strain is related to the compressibility of the pore fluid. For a saturated soil, the pore fluid consists entirely of water, and the high bulk modulus of water means that a small increment of compressive volumetric strain will cause a large increment of positive porewater pressure. If the soil is not saturated, the bulk modulus of the pore fluid (water and air) will be greatly reduced (Santamarina et al., 2001) and the same increment of volumetric strain will cause only a small increase in porewater pressure.

The effect of incomplete saturation in retarding pore pressure generation under rapid loading has been recognized in laboratory tests for many years. Previous laboratory studies (e.g., Chaney, 1978; Yoshimi et al., 1989; Xia and Hu, 1991, Tsukamoto et al., 2002; Ishihara and Tsukamoto, 2004) have shown that liquefaction resistance increases quickly with a decreasing degree of saturation. Thus, desaturation, or induced partial saturation (Yegian et al., 2007) of a liquefiable soil, if it could be accomplished homogeneously, could help mitigate liquefaction hazards. Partially saturated soils can exist naturally as a result of biologic activity (Wheeler, 1988; Mitchell and Santamarina, 2005), by direct injection such as by air sparging (Camp et al., 2010), or it can be induced by construction activity. Numerous structures having soils that were partially saturated by the construction of pneumatic caisson foundations survived the 1995 Kobe and 2004 Niigata earthquakes without liquefaction damage. Okamura et al. (2006) measured degrees of saturation in samples taken several decades after sand compaction pile construction and found them to be as high as those measured immediately after construction. Nagao et al. (2007) investigated the use of micro-bubbles small enough (10–100 μm in diameter) to penetrate the voids between sand particles.

11.7 BIOLOGICAL TECHNIQUES

Biological processes can also be used to help modify the mechanical behavior of soils. Soils can be viewed as a combination of inert minerals and living entities (microbes of various forms) spread throughout. Some degree of biological activity is present in virtually every soil deposit – the top meter of a typical soil deposit contains more than 1 billion cells per gram and about 1 million cells per gram exist at 30 m depth (Whitman et al., 1998). The presence of soil micro-organisms can have a significant effect on soil behavior over time (Mitchell and Santamarina, 2005).

Bio-mediated soil improvement systems use biological activity to control the timing, rate, and spatial distribution of chemical reactions in the soil, and thereby control the beneficial by-products of those reactions (DeJong et al., 2010). Bio-mediated methods offer the potential for addressing challenging soil problems such as improvement beneath and adjacent to existing buildings and buried infrastructure (van Paassen, 2011; DeJong and Kavazanjian, 2019; San Pablo et al., 2020) with much smaller carbon footprints than many other types of soil improvement.

Biological treatment processes typically begin by introducing specific agents into the subsurface environment. The in-situ microbial environment is most commonly augmented by the injection of lab-cultured microbes (*bio-augmentation*) into the soil at high cell densities. In recent years, however, researchers have shown that the enrichment, or *bio-stimulation,* of existing micro-organisms with specific nutrients can achieve similar results with several important benefits (Fujita et al., 2000, 2008; Burbank et al., 2011; Gomez et al., 2016, 2018). When the activity level of the in-situ and/or injected microbes reaches a sufficient level, they initiate and sustain spatially distributed biogeochemical reactions whose by-products, such as calcite precipitation, bond particles at their contacts, roughen particle surfaces, and fill void space between the soil particles. This process of microbially induced calcite precipitation (MICP) has shown good results (improvements in strength, stiffness, and compressibility) in laboratory benchtop studies; optimal procedures for scaling the injection processes up to field-scale treatments and the permanence of its benefits are currently being investigated.

The natural heterogeneity of in-situ soils, compounded by complex depositional processes that produce interlayering of coarse- and fine-grained soils and other features that lead to spatially variable permeability, creates challenges with respect to the uniformity of field-scale biological treatment. However, spatial non-uniformity is a characteristic of other soil improvement techniques and both improvement of uniformity and determination of the fraction of soil that needs to be improved in order to obtain satisfactory performance of such techniques are active areas of research.

Other promising biological techniques include desaturation by the use of microbial dissimilatory nitrate reduction to generate biogas (van Paassen et al., 2017; O'Donnell et al., 2017) and production of biofilms (Kwon and Ajo-Franklin, 2013; Proto et al., 2016) that can increase resistance to particle rearrangement and enhance shear strength. Biological techniques are an emerging area of soil improvement that offers new mechanisms by which soils can be improved with significant environmental benefits.

11.8 VERIFICATION OF SOIL IMPROVEMENT

Applications of soil improvement should be checked to confirm that the desired modifications of soil properties have taken place. The most direct way of verifying the effectiveness of a particular soil improvement technique is to measure the soil characteristics that indicate deficient performance before improvement. For example, if the improvement was undertaken to increase the strength of the soil, measurement of the strength before and after improvement would provide the most direct verification of the effectiveness of the improvement process. However, it is not always feasible to measure the deficient characteristics directly. In such cases, verification is usually accomplished using related characteristics that are more easily measured.

Valuable information on the effectiveness of soil improvement can be obtained by carefully observing conditions during construction. Keeping records of quantities of injected or backfilled materials, ground movements, quantities of spoils, power/energy input, and consistency and viscosity of mixed materials can be helpful in inferring effectiveness. Instrumentation of the site can provide objective, quantitative information on the effectiveness of the improvement process, and on any off-site effects, during and after completion of treatment.

Following treatment, verification may be based on the results of laboratory or field tests. While laboratory tests were used for verification of soil improvement in the past, they have been replaced by the use of field testing techniques. Field testing techniques may be divided into in situ testing techniques and geophysical testing techniques. Common verification techniques were summarized by Ledbetter (1985).

11.8.1 Laboratory Testing Techniques

Laboratory testing techniques have a number of advantages over other methods for verification of soil improvement, but they also suffer from drawbacks that can significantly limit their usefulness for certain types of soil improvement. The requirement of obtaining a representative sample of the improved soil leads directly to many of the advantages of using laboratory testing techniques and also to many of the disadvantages. Obtaining a sample of improved soil allows visual inspection of the effects of improvement. For many improvement techniques (e.g., permeation grouting, soil mixing, etc.), the ability to inspect the treated soil provides direct and valuable evidence of the effectiveness of the treatment. Laboratory tests allow greater control and more accurate measurement of stress, strain, and environmental conditions than are possible in field tests. In some cases, this flexibility may allow more accurate characterization of the properties of the improved soil.

On the other hand, laboratory tests only provide verification at discrete points. When soil improvement is used to improve or eliminate localized zones or seams of weakness, verification by methods that require discrete sampling may be ineffective. Laboratory tests may also be influenced by the inevitable effects of sample disturbance, a problem that can be particularly significant in the improvement of liquefiable soils. The density changes produced by even thin-walled samplers (Marcuson et al., 1977; Seko and Tobe, 1977; Singh et al., 1979) can lead to considerable uncertainty in the evaluation of improvement effectiveness, particularly for liquefiable soils whose behavior is extremely sensitive to density.

11.8.2 In-Situ Testing Techniques

Many of the limitations of laboratory testing-based approaches to the verification of soil improvement effectiveness may be overcome by the use of in situ tests. Because many geotechnical seismic hazards are evaluated using in situ test parameters, those parameters can provide direct evidence of hazard mitigation. Indeed, soil improvement specifications may be written to require that a certain parameter value (e.g., a minimum CPT tip resistance) be achieved after improvement. The use of in situ tests for verification of soil improvement effectiveness has been addressed by several investigators (e.g., Mitchell, 1986; Welsh, 1986; Dove et al., 2000).

The effectiveness of some reinforcement techniques can be demonstrated with in-situ load tests. Rapidly performed plate load tests can be performed on individual stone columns, and zone load tests can be performed on groups of columns.

The SPT, CPT, and PMT (Section 6.5.3) can all be used for verification of soil improvement effectiveness. SPT and CPT are performed relatively quickly and inexpensively compared to sampling and laboratory testing. The CPT is particularly useful because it provides a continuous record with depth. However, CPT soundings can be difficult to interpret in the context of liquefaction hazard mitigation verification; for example, the fines contents inferred from tip and sleeve resistance measurements can appear to change (Siegel, 2013) as a result of improvement even when no physical change in gradation has taken place. The PMT is more expensive, but it also allows the measurement of lateral stresses and provides a more direct measurement of strength. For gravelly soils, the Becker hammer or iBPT penetration tests (Section 6.5.3.2) may be used for verification purposes.

Interpretation of soil improvement effectiveness from in situ test results must be performed carefully. The penetration resistance of granular soils, for example, is influenced not only by density and overburden stress but also by lateral stress. Soil improvement techniques that result in increased lateral stress may produce unconservative estimates of the density of the improved soil if the post-improvement stress state is not carefully considered in the interpretation of penetration test results. Because time-dependent changes in strength, stiffness, and penetration resistance are often observed after densification (Mitchell and Solymar, 1984; Mitchell, 1986; Schmertmann, 1986; Charlie et al., 1992), in situ tests performed immediately after densification may not reflect the actual degree of improvement of the soil. In most cases, penetration resistance increases following the completion of soil improvement as excess pore pressures dissipate, particles move to a new configuration, and lateral stresses increase. In some cases (e.g., Mejia and Boulanger, 1995), however, relaxation of lateral stresses can lead to reductions in penetration resistance with time. To reduce the impact of these time-dependent changes, verification testing is usually performed at least 72 hours after densification has taken place.

Many soil improvement techniques are applied at a grid of treatment points, and the degree of improvement usually decreases with distance from the treatment point. The relationship between the locations of in situ tests and the locations of treatment points should be considered in the interpretation of soil improvement effectiveness from in situ test results. In situ tests have limited effectiveness for the verification of grouting effectiveness (Welsh, 1986).

11.8.3 Geophysical Testing Techniques

Many soil improvement techniques increase the stiffness of the treated soil. The effectiveness of these techniques can be verified using seismic geophysical techniques (Section 6.5.2). In most cases, it is desirable to perform seismic tests both before and after improvement.

Cross-hole and downhole (including seismic cone) tests are most commonly used for verification of soil improvement. These techniques can measure p- or s-wave velocities over considerable distances, thereby providing spatially averaged stiffness measurements. However, each requires at least one borehole or CPT sounding. For sites where soil improvement has been performed over a large area, seismic reflection and seismic refraction tests may be useful for verification purposes. SASW tests provide similar information without the need for boreholes. At sites where stiffness changes irregularly in two or three dimensions or sites that contain inclusions, the results of SASW tests may be very difficult to interpret. Such tests must also be performed when background noise (including that produced by on-going soil improvement work) will not adversely affect their results. Tests that measure average wave propagation velocities may not accurately reflect the degree of improvement of thin, loose zones unless the distance over which velocities are averaged is quite small.

Other geophysical techniques may be used to verify certain characteristics of soil improvements. Liu et al. (2009) used electrical resistivity techniques to determine the uniformity of deep-soil-mixed columns. Shuwang et al. (2009) used ground-penetrating radar to evaluate densification by blasting.

11.9 OTHER CONSIDERATIONS

The application of soil improvement techniques to mitigate seismic hazards is relatively new compared with many aspects of geotechnical engineering and design. The theoretical underpinnings of many soil improvement techniques are poorly developed, and empirical observations of the performance of improved soil in actual earthquakes are few. Because of these factors, it is particularly important to review the relevant geotechnical engineering literature before attempting to mitigate seismic hazards by soil improvement.

Many soil improvement techniques are potentially damaging to nearby infrastructure or irritating to people in proximity to an improvement site. Noise, dust, vibrations, and other by-products of certain soil improvement techniques can limit their use in populated areas. A number of techniques use water or chemicals that produce environmentally damaging spoils that may need to be removed from a site for treatment.

The effectiveness of many soil improvement techniques can be difficult to predict in advance for a particular site. Furthermore, the equipment, procedures, experience, and skill of the soil improvement contractor can strongly influence soil improvement effectiveness. For these reasons, it is frequently beneficial to construct test sections before beginning production work or even before the final selection of a soil improvement technique. Test sections allow site- and procedure-specific evaluation of soil improvement effectiveness at a moderate cost; their use is advisable whenever possible.

Soil improvement strategies should consider the effect of the changed properties of the improved soil on the response of the soil and any structures affected by it. Stiffening and strengthening a soil deposit to reduce ground failure hazards can also change its dynamic response and, with it, the amplitude, frequency content, and even duration of its motion. A stiffened profile may transmit stronger high-frequency components of a ground motion to the surface, and induce greater demands on stiff structures and their contents.

11.10 SUMMARY

1. Unfavorable soil conditions can frequently be improved using soil improvement techniques. A variety of soil improvement techniques have been developed – some apply to long-term, static loading conditions and others also apply to seismic loading conditions.
2. The cost of different soil improvement techniques varies widely. Costs are influenced by the volume and extent of the soil to be treated, access to the site, site sensitivity to vibration and permanent ground movement, and other factors.
3. The presence of existing structures, pipelines, and other constructed facilities can eliminate many soil improvement techniques from consideration at a given site. The techniques that can be used at such sites tend to be among the more expensive.
4. Most soil improvement techniques are intended to increase the strength and stiffness of a soil deposit. Achieving those soil property changes typically requires increasing the density of a soil. Increased strength, stiffness, and density are generally desirable for both static and seismic loading conditions.
5. Current soil improvement techniques can be divided into five broad categories: densification, reinforcement, grouting/mixing, drainage, and biological techniques. Not all techniques fall entirely within a single category; for example, stone columns can improve a soil deposit by densification, reinforcement, and drainage functions.

6. Several soil improvement techniques that are commonly used to mitigate seismic hazards are intended to reduce the tendency of loose, saturated granular soils to generate excess porewater pressure during earthquake shaking. These techniques typically involve densification of the soil.
7. Densification is probably the most commonly used soil improvement technique for the mitigation of liquefaction hazards. Most densification techniques rely on the tendency of granular soils to densify when subjected to vibration. Densification can produce substantial settlement, although some procedures allow the introduction of new material to balance the volume change caused by densification.
8. Many densification techniques rely on vibrations that can be potentially damaging to structures, pipelines, and other constructed facilities. Such vibrations may also be too objectionable to people who live or work near sites that require improvement to allow their use.
9. Most vibratory techniques produce a temporary, localized zone of liquefaction in loose, saturated sand. Densification occurs as the sand particles are rearranged during reconsolidation. The presence of fines, particularly plastic fines, inhibits both the development of high pore pressures and the rearrangement of soil particles. As a result, vibratory techniques may have limited effectiveness in soils with significant fines contents.
10. Reinforcement techniques introduce discrete inclusions that stiffen and strengthen a soil deposit. The high stiffness and strength of the inclusions also tend to reduce the stresses imposed on the weaker material between the inclusions.
11. Cementitious materials may be injected or mixed into a soil deposit. The materials improve the soil by strengthening the contacts between individual grains and filling the space between the grains.
12. Grouting techniques involve the injection of cementitious materials into the voids of the soil or into fractures in the soil so that the particle structure of the majority of the soil remains intact. In permeation grouting, very low viscosity grouts are injected into the voids of the soil without disturbing the soil structure. In intrusion grouting, thicker and more viscous grouts are injected under pressure to cause controlled fracturing of the soil. Grouting techniques can often be used where other techniques cannot, such as under existing structures.
13. Mixing techniques introduce cementitious materials by physically mixing them with the soil, completely disturbing its particle structure. The mixing can be accomplished mechanically (soil mixing) or hydraulically (jet grouting). Both soil mixing and jet grouting leave behind relatively uniform columns of mixed soil-cement. By overlapping the columns, walls or cellular structures can be constructed below the ground surface.
14. Drainage techniques minimize the buildup of porewater pressure during earthquakes by allowing horizontal drainage along shortened drainage paths in a soil deposit. The installation of drains generally involves some degree of densification and the drains themselves may also provide some reinforcement.
15. Biological techniques supply nutrients to microbes that either exist in or are injected into the ground to augment existing microbes. The microbes can be stimulated to produce biogeochemical reactions that can, in various cases, bond particles together, fill voids, desaturate the soil by forming gas, or produce biofilms.
16. Verification of the effectiveness of soil improvement is an important part of seismic hazard mitigation. Direct or indirect measurement of stiffness, strength, or density characteristics both before and after improvement can allow reliable evaluation of soil improvement effectiveness. These characteristics may be measured by laboratory, in situ, or geophysical tests. The relative advantages and limitations of these types of tests, discussed in detail in Chapter 6, apply to their use in verification applications.

REFERENCES

Aboshi, H., Mizuno, Y., and Kuwabara, M. (1991). "Present state of sand compaction pile in Japan," in M.I. Esrig and R. C. Bachus, eds., *Deep Foundation Improvements: Design, Construction, and Testing*, ASTM STP 1089, ASTM, Philadelphia, pp. 32–46. https://doi.org/10.1520/stp25049s.

Adalier, K. and Elgamal, A.-W. (2004). "Mitigation of liquefaction and associated ground deformations by stone columns," *Engineering Geology*, Vol. 72, No. 3–4, pp. 275–291. https://doi.org/10.1016/j.enggeo.2003.11.001.

Arslan, H., Baykal, G., and Ertas, O. (2007). "Influence of tamper weight shape on dynamic compaction." *Ground Improvement*, Vol. 11, No. 2, pp. 61–66. https://doi.org/10.1680/grim.2007.11.2.61.

Babasaki, R., Suzuki, K., Saitoh, S., Suzuki, Y., and Tokitoh, K. (1991). "Construction and testing of deep foundation improvement using the deep cement mixing method," *Deep Foundation Improvements: Design Construction and Testing*, ASTM STP 1089, ASTM, pp. 224–233. https://doi.org/10.1520/stp25062s.

Baez, J.I. and Henry, J.F. (1993). "Reduction of liquefaction potential by compaction grouting at Pinoplis Dam, South Carolina," *Geotechnical Practice in Dam Rehabilitation*, Geotechnical Special Publication No. 35, ASCE, New York, pp. 430–466.

Bell, F.G. (1993). *Engineering Treatment of Soils*, E & FN Spon, London, 302 pp.

Bo, M.W., Chu, J. and Choa, V. 2005. "The Changi East reclamation project in Singapore," in B. Indraratna & J. Chu, eds., *Ground Improvement Case Histories*, Elsevier, Amsterdam, pp. 247–276.

Boulanger, R.W. and Hayden, R.F. (1995). "Aspects of compaction grouting of liquefiable soil." *Journal of Geotechnical Engineering*, Vol. 121, No. 12, pp. 844–855. https://doi.org/10.1061/(asce)0733-9410(1995)121:12(844).

Broms, B. (1991). "Deep compaction of granular soil," in H.Y. Fang, ed., *Foundation Engineering Handbook*, 2nd edition, Van Nostrand Reinhold, New York, pp. 814–832.

Bruce, D.A. (1992). "Progress and development in dam rehabilitation by grouting," *Grouting, Soil Improvement, and Geosynthetics*, Geotechnical Special Publication No. 30, ASCE, Vol. 1, pp. 601–613. https://doi.org/10.1016/0886-7798(92)90121-w.

Bruce, D.A. (2000). "An introduction to the deep soil mixing methods as used in geotechnical applications," *Report FHWA-RD-99-138*, Federal Highway Administration, U.S. Department of Transportation, 135 pp.

Bruce, M.E.C., Berg, R.R., Collin, J.G., Filz, G.M., Terashi, M., and Yang, D.S. (2013). "Deep mixing for embankment and foundation support," *Federal Highway Administration Design Manual FHWA-HRT-13-046*, Federal Highway Administration, Washington, DC, 244 pp.

Burbank, M.B., Weaver, T.J., Green, T.L., Williams, B.C., and Crawford, R.L. (2011). "Precipitation of calcite by indigenous microorganisms to strengthen liquefiable soils," *Geomicrobiology Journal*, Vol. 28, No. 4, pp. 301–312. https://doi.org/10.1080/01490451.2010.499929.

Camp, W.M., III, Camp, H.C., and Andrus, R.D. (2010). "Liquefaction mitigation using air injection," *Proceedings,* Fifth International Conference on Recent Advances in Geotechnical Earthquake Engineering and Soil Dynamics., San Diego, CA, Paper No. 4.39a, 6 pp.

Cedergren, H. (1989). *Seepage, Drainage, and Flow Nets*, 3rd edition, John Wiley & Sons, New York, 465 pp.

Chaney, R. (1978). "Saturation effects on the cyclic strength of sands," *Proceedings, ASCE Specialty Conference on Earthquake Engineering and Soil Dynamics*, New York, pp. 342–358.

Charlie, W.A., Rwebyogo, M.F.J., and Doehring, D.O. (1992). "Time-dependent cone penetration resistance due to blasting," *Journal of Geotechnical Engineering*, Vol. 118, No. 8, pp. 1200–1215. https://doi.org/10.1061/(asce)0733-9410(1992)118:8(1200).

Chen, J.-J., Zhang, L., Zhang, J.-F., Zhu, Y.-F., and Wang, J.-H. (2013). "Field tests, modification, and application of deep soil mixing method in soft clay," *Journal of Geotechnical and Geoenvironmental Engineering*, Vol. 139, No. 1, pp. 24–34. https://doi.org/10.1061/(asce)gt.1943-5606.0000746.

Chu, J., Varaksin, S., Klotz, U., and Menge, P. (2009). "Construction Processes, State of the Art Report." *Proceedings, 17th International Conference on Soil Mechanics and Geotechnical Engineering*, Alexandria, Egypt, 5–9 October 2009, 130 pp.

Degen, W. (1997). "56 m deep vibro-compaction and German lignite mining area," *Proceedings, 3rd International Conference on Ground Improvement Systems*, London, UK.

DeJong, J.T. and Kavazanjian, E. (2019). "Bio-mediated and Bio-inspired Geotechnics," in N. Lu and J.K. Mitchell, eds., *Geotechnical Fundamentals for Addressing New World Challenges*, Springer, Cambridge, pp. 193–207.

DeJong, J.T., Mortensen, B.M., Martinez, B.C., and Nelson, D.C. (2010). "Bio-mediated soil improvement," *Ecological Engineering*, Elsevier, Vol. 36, pp. 197–210. https://doi.org/10.1016/j.ecoleng.2008.12.029.

Dise, K., Stevens, M.G., and Von Thun, J.L. (1994). "Dynamic compaction to remediate liquefiable embankment foundation soils," Geotechnical Specialty Publication No. 45, ASCE, New York, pp. 1–25.

Dobson, T. (1987). "Case histories of the vibro systems to minimize the risk of liquefaction," *Soil Improvement - A Ten Year Update*, Geotechnical Special Publication No. 12, ASCE, New York, pp. 167–183.

Dove, J.E., Boxill, L.E.C., and Jarrett, J.B. (2000). "A CPT-based index for evaluating ground improvement," *Advances in Grouting and Ground Modification*, Geotechnical Special Publication No. 104, ASCE, pp. 296–310. https://doi.org/10.1061/40516(292)20.

Dowding, C.H. and Duplaine, H. (2004). "State of the art in blast densification or explosive compaction," *Proceedings, ASEP-GI 2004*, Vol. 2, Magnan (ed.), Presses de l'ENPC/LCPC, Paris, pp. 505–514.

Durgunoglu, H.T., Kulac, H.F., Oruc, K., Yildiz, R., Sickling, J., Boys, I.E., Altugu, T., and Emrem, C. (2003). "A case history of ground treatment with jet grouting against liquefaction, for a cigarette factory in Turkey," *Grouting and Ground Treatment*, Geotechnical Special Publication No. 120, ASCE, pp. 452–463. https://doi.org/10.1061/40663(2003)120.

El-Kelesh, A.M., Matsui, T., and Tokida, K.I. (2012) "Field investigation into effectiveness of compaction grouting." *Journal of Geotechnical and Geoenvironmental Engineering*, Vol. 138, No. 4, pp. 451–460. https://doi.org/10.1061/(asce)gt.1943-5606.0000540.

Farrell, T.M., Wallace, K., and Ho, J. (2010). "Liquefaction mitigation of three projects in California," *Proceedings, 5th International Conference on Recent Advances in Geotechnical Earthquake Engineering and Soil Dynamics and Symposium in Honor of Professor I.M. Idriss*, Paper No. 4.28a, San Diego, California, 8 pp.

Feng, S.-J., Tan, K., Shui, W.-H., and Zhang, Y. (2013). "Densification of desert sands by high energy dynamic compaction," *Engineering Geology*, Vol. 157, pp. 48–54. https://doi.org/10.1016/j.enggeo.2013.01.017.

Finno, R.J., Gallant, A.P., and Sabatini, P.J. (2016), "Evaluating ground improvement after blast densification, Performance at the Oakridge landfill," *Journal of Geotechnical and Geoenvironmental Engineering*, Vol. 142, No. 1, 13 pp. https://doi.org/10.1061/(asce)gt.1943-5606.0001365.

Fujita, Y., Ferris, F.G., Lawson, R.D., Colwell, F.S., and Smith, R.W. (2000). "Subscribed content calcium carbonate precipitation by ureolytic subsurface bacteria," *Geomicrobiology Journal*, Vol. 17, No. 4, pp. 305–318. https://doi.org/10.1080/782198884.

Fujita, Y., Taylor, J.L., Gresham, T.L., Delwiche, M.E., Colwell, F.S., McLing, T.L., … Smith, R. W. (2008). "Stimulation of microbial urea hydrolysis in groundwater to enhance calcite precipitation," *Environmental Science & Technology*, Vol. 42, No. 8, pp. 3025–3032. https://doi.org/10.1021/es702643g.

Gallagher, P.M., Conlee, C.T., and Rollins, K.M. (2007). "Full-scale field testing of colloidal silica grouting for mitigation of liquefaction risk," *Journal of Geotechnical and Geoenvironmental Engineering*, Vol. 133, No. 2, pp. 186–196. https://doi.org/10.1061/(asce)1090-0241(2007)133:2(186).

Gallagher, P.M. and Mitchell, J.K. (2002). "Influence of colloidal silica grout on liquefaction potential and cyclic undrained behavior of loose sand." *Soil Dynamics and Earthquake Engineering*, Vol. 22, No. 9–12, pp. 1017–1026. https://doi.org/10.1016/s0267-7261(02)00126-4.

Gandhi, S.R., Dey, A.K., and Selvam, S. (1999). "Densification of pond ash by blasting." *Journal of Geotechnical and Geoenvironmental Engineering*, Vol. 125, No. 10, pp. 889–899. https://doi.org/10.1061/(asce)1090-0241(1999)125:10(889).

Gohl, W.B., Jefferies, M.G., Howie, J.A., and Diggle, D. (2000). "Explosive compaction: design, implementation and effectiveness," *Geotechnique*, Vol. 50, No. 6, pp. 657–665. https://doi.org/10.1680/geot.2000.50.6.657.

Gohl, W.B., Tsujino, S., Wu, G., Yoshida, N., Howie, J.A., and Everard, J. (1998). "Field applications of explosive compaction in silty soils and numerical analysis," *Geotechnical Earthquake Engineering & Soil Dynamics III*, Geotechnical Special Publication 75, ASCE, New York, pp. 654–665.

Gomez, M.G., Anderson, C.M., DeJong, J.T., Nelson, D.C., Graddy, C.M.R., and Ginn, T.R. (2016). "Large-scale comparison of bioaugmentation and biostimulation approaches for bio-cementation of sands," *Journal of Geotechnical and Geoenvironmental Engineering*, https://doi.org/10.1061/(asce)gt.1943-5606.0001640.

Gomez, M.G., Graddy, C.M.R., DeJong, J.T., Nelson, D.C., and Tsesarsky, M. (2018). "Stimulation of native microorganisms for biocementation in samples recovered from field-scale treatment depths," *Journal of Geotechnical and Geoenvironmental Engineering*, https://doi.org/10.1061/(asce)gt.1943-5606.0001804.

Graf, E.D. (1992). "Earthquake support grouting in sands," *Grouting, Soil Improvement, and Geosynthetics*, Geotechnical Special Publication No. 30, ASCE, Vol. 2, pp. 879–888. https://doi.org/10.1016/0886-7798(92)90121-w.

Hachey, J.E., Plum, R.L., Byrne, R.J., Kilian, A.P., Jenkins, D.V. (1994). "Blast-Densification of thick, loose debris flow at Mt. St. Helen's, Washington." *Vertical and Horizontal Deformations of Foundations and Embankments, Geotechnical Specialty Publication 40*, ASCE, pp. 502–512.

Hamada, M., and Wakamatsu, K. (1997). "Liquefaction, ground deformation and their caused damage to structures," *The 1995 Hyogoken-Nambu Earthquake*, Japan Society of Civil Engineers, Tokyo, pp. 45–92.

Harder, L.F., Jr., Hammond, W.D., and Ross, P.S. (1984). "Vibroflotation compaction at Thermalito Afterbay," *Journal of Geotechnical Engineering*, Vol. 110, No. 1, pp. 57–72. https://doi.org/10.1061/(asce)0733-9410(1984)110:1(57).

Hausmann, M.R. (1990). *Engineering Principles of Ground Modification*, McGraw-Hill Publishing Company, New York, 632 pp.

Hayden, R.F. (1994). "Utilization of liquefaction countermeasures in North America," *Proceedings, 5th U.S. National Conference on Earthquake Engineering*, Chicago, Vol. 4, pp. 149–158.

Hayden, R.F. and Welch, J.P. (1991). "Design and installation of stone columns at naval air station," in M.I. Esrig and R.C. Bachus, eds., *Deep Foundation Improvement: Design, Construction, and Testing*, ASTM STP 1089, American Society for Testing and Materials, pp. 172–184. https://doi.org/10.1520/stp25058s.

Holtz, R.D., Kovacs, W.D., and Sheahan, T.C. (2011). *An Introduction to Geotechnical Engineering*, 2nd edition, Prentice-Hall, Englewood Cliffs, NJ.

Hudson, M.B., Shao, L., Murphy, M.A., and Lew, M. (2014). "Sustainable Foundation Support of Community Memorial Hospital against Liquefaction Hazards," *Geo-Characterization and Modeling for Sustainability*, ASCE, pp. 3836–3850. https://doi.org/10.1061/9780784413272.372.

Hussein, J.D. and Ali, S. (1987). "Soil improvement of the Trident submarine facility," *Soil Improvement - A Ten Year Update*, Geotechnical Special Publication No. 12, ASCE, New York, pp. 215–231.

Iai, S., Matsunaga, Y., Morita, T., Miyoda, M., Sakurai, H., Oishi, H., Ogura, H., Ando, Y., Tanaka, Y., and Kato, M. (1994). "Effects of remedial measures against liquefaction at 1993 Kushiro-Oki earthquake," *Proceedings, Fifth Japan-U.S. Workshop on Earthquake Resistant Design of Lifeline Facilities and Countermeasures for Soil Liquefaction*, Snowbird, Utah.

Iai, S., Noda, S., and Tsuchida, H. (1988). "Basic consideration for designing the area of the ground compaction as a remedial measure against liquefaction," *Proceedings, U.S.-Japan Joint Workshop on Remedial Measures for Liquefiable Soils*, Jackson, WY.

Indraratna, B. and Chu, J. (eds.) (2005). *Ground Improvement - Case Histories*, Elsevier, Amsterdam, 1115 pp.

Ishihara, K., Kawase, Y., and Nakajima, M. (1980). Liquefaction characteristics of sand deposits at an oil tank site during the 1978 Miyagiken-Oki earthquake, *Soils and Foundation*, Vol. 20, No. 2, pp. 97–111. https://doi.org/10.3208/sandf1972.20.2_97.

Ishihara, K. and Tsukamoto, Y. (2004). "Cyclic strength of imperfectly saturated sands and analysis of liquefaction." *Proceedings, Japan Academic Series B*, Vol. 80, No. 8, pp. 372–391.

Janes, H.W. (1973). "Densification of sand for drydock by Terra-probe," *Journal of the Soil Mechanics and Foundations Division*, ASCE, Vol. 99, No. SM6, pp. 451–470. https://doi.org/10.1061/jsfeaq.0001888.

Karol, R.H. (2003). *Chemical Grouting and Soil Stabilization*, Marcel Dekker, New York, 457 pp.

Keller, T.O., Castro, G., and Rogers, J.H. (1987). "Steel Creek Dam foundation densification," *Soil Improvement - A Ten Year Update*, Geotechnical Special Publication No. 12, ASCE, New York, pp. 136–166.

Kirsch, K. and Kirsch, F. (2010). *Ground Improvement by Deep Vibratory Methods*, Taylor and Francis, Boca Raton, FL, 189 pp.

Kishida, H. (1967). "Ultimate bearing capacity of piles driven into loose sand," *Soils and Foundations*, Vol. 7, No. 3, pp. 20–29. https://doi.org/10.3208/sandf1960.7.3_20.

Koutsoftas, D.C. and Kiefer, M.L. (1990). "Improvement of mine spoils in southern Illinois," *Geotechnics of Waste Fills*, Special Technical Publication 1070, ASTM, pp. 153–167. https://doi.org/10.1520/stp25305s.

Kramer, S.L. and Holtz, R.D. (1991). *Soil Improvement and Foundation Remediation with Emphasis on Seismic Hazards*, University of Washington, Seattle, Washington, DC, 106 pp.

Kwon, T.H. and Ajo-Franklin, J.B. (2013). "High-frequency seismic response during permeability reduction due to biopolymer clogging in unconsolidated porous media," *Geophysics*, Vol. 78, No. 6, pp. https://doi.org/10.1190/geo2012-0392.1.

Lawton, E.C. and Fox, N.S. (1994). "Settlement of structures supported on marginal or inadequate soils stiffened with short aggregate piers," in A.T. Yeung and G.Y. Felio, eds., *Vertical and Horizontal Deformations of Foundations and Embankments*, American Society of Civil Engineers, New York, Vol. 2, pp. 962–974.

Lawton, E.C., Fox, N.S., and Handy, R.L. (1994). "Control of settlement and uplift of structures using short aggregate piers." *Proceedings, In-Situ Deep Soil Improvement*, ASCE National Convention, Atlanta, Georgia, pp. 121–132.

Ledbetter, R.H. (1985). "Improvement of liquefiable foundation conditions beneath existing structures," *Technical Report REMR-GT-2*, U.S. Army Corps of Engineers, Waterways Experiment Station, Vicksburg, Mississippi, 51 pp.

Lindqvist, L. and Petaja, J. (1981). "Experience in the evaluation of the bearing capacity of tapered friction piles in postglacial sand and silt strata," *Proceedings, 10th International Conference on Soil Mechanics and Foundation Engineering*, Stockholm, Vol. 2, pp. 759–766.

Liu, S., Zhang, D., and Zhu, Z. (2009). "On the uniformity of deep mixed soil-cement columns with electrical resistivity method," in J. Han, G. Zheng, V.R. Schaefer, and M. Huang, eds., *Advances in Ground Improvement: Research to Practice in the United States and China*, Geotechnical Special Publication No. 188, ASCE, New York, pp. 140–149.

Lu, X., Filz, G., and Han, J. (2009). "Dynamic compaction of fill in a mountainous area," in J. Han, G. Zheng, V.R. Schaefer, and M. Huang, eds., *Advances in Ground Improvement: Research to Practice in the United States and China*, Geotechnical Special Publication No. 188, ASCE, New York, pp. 281–289.

Luehring, R., Snorteland, N., Mejia, L., and Stevens, M. (2001). "Liquefaction mitigation of a silty dam foundation using vibro-stone columns and drainage wicks: a case history at salmon lake dam," *Proceedings, 21st USSD Annual Meeting and Lecture*, Denver, CO.

Lukas, R.G. (1995). "Dynamic compaction," Geotechnical Engineering Circular No. 1, Publication No. FHWA-SA-95-037, Federal Highway Administration, Office of Engineering, Office of Technology Applications, Washington, D.C., 97 pp.

Mackiewicz, S.M. and Camp, W.M. (2007). "Ground modification: How much improvement?" *Soil Improvement*, Geotechnical Special Publication No. 172, ASCE, 9 pp. https://doi.org/10.1061/40916(235)14.

Majchrzak, M., Lew, M., Sorenson, K., and Farrell, T. (2004). "Settlement of shallow foundations constructed over reinforced soil: Design estimates vs. measurements," *Proceedings, 5th International Conference on Case Histories in Geotechnical Engineering*, Paper No. 1.64, New York, 10 pp.

Marcuson, W.F. III, Cooper, S.S., and Bieganousky, W.A. (1977). "Laboratory sampling study conducted on fine sands," *Proceedings of Specialty Session 2, 9th International Conference on Soil Mechanics and Foundation Engineering*, Tokyo, pp. 15–22.

Marcuson, W.F. III, Hadala, P.F., and Ledbetter, R.H. (1991). "Seismic rehabilitation of earth dams," *Geotechnical Practice in Dam Rehabilitation*, Geotechnical Special Publication No. 35, ASCE, pp. 430–466. https://doi.org/10.1061/(asce)0733-9410(1996)122:1(7).

Majdi, A., Soltani, A.S., and Litkouhi, S. (2007). "Mitigation of liquefaction hazard by dynamic compaction." *Ground Improvement*, Vol. 11, No. 3, pp. 137–143. https://doi.org/10.1680/grim.2007.11.3.137.

Martin, G.R., Arulmoli, K., Yang, L., Esrig, M.I., and Capelli, R.P. (1999). "Dry mix soil-cement walls: An application for mitigation of earthquake ground deformations in soft or liquefiable soils," *Dry Deep Mix Methods for Deep Soil Stabilization, Proceedings, International Conference on Dry Deep Mix Methods for Deep Soil Stabilization*, Stockholm, Sweden, pp. 27–44.

Martin, J.R. and Olgun, C.G. (2006) "Liquefaction mitigation using jet-grout columns - 1999 Kocaeli earthquake case history," *Ground Modification and Seismic Mitigation*, Geotechnical Special Publication 152, ASCE, pp. 349–358. https://doi.org/10.1061/40864(196)47.

Massarch, K.R. (1991). "Deep soil compaction using vibratory probes," in M.I. Esrig and R.C. Bachus, eds., *Deep Foundation Improvement: Design, Construction, and Testing*, ASTM STP 1089, American Society for Testing and Materials, pp. 297–319. https://doi.org/10.1520/stp25067s.

Mayne, P.W., Jones, J.S., and Dumas, J.C. (1984). "Ground Response to Dynamic Compaction," *Journal of Geotechnical Engineering*, Vol. 110, No. 6, pp. 757–774. https://doi.org/10.1061/(asce)0733-9410(1984)110:6(757).

Mejia, L.H. and Boulanger, R.W. (1995). "A long term test of compaction grouting for liquefaction mitigation," *Earthquake-Induced Movements and Seismic Remediation of Existing Foundations and Abutments*, Geotechnical Special Publication No. 55, ASCE, New York, pp. 94–109.

Miller, E.A. and Roycroft, G.A. (2004). "Compaction grouting test program for liquefaction control." *Journal of Geotechnical and Geoenvironmental Engineering*, Vol. 130, No.4, pp. 355–361. https://doi.org/10.1061/(asce)1090-0241(2004)130:4(355).

Mitchell, J.K. (1970). "In-place treatment of foundation soils," *Journal of Soil Mechanics and Foundations Division*, Vol. 97, No. SM1, pp. 73–110. https://doi.org/10.1061/jsfeaq.0001391.

Mitchell, J.K. (1981). "Soil improvement: State-of-the-art," *Proceedings, 10th International Conference on Soil Mechanics and Foundation Engineering*, Stockholm, Sweden, June, Vol. 4, pp. 509–565.

Mitchell, J.K. (1986). "Ground improvement evaluation by insitu tests," *Proceedings, Insitu '86*, Geotechnical Special Publication No. 6, ASCE, pp. 221–236.

Mitchell, J.K. (2008), "Mitigation of liquefaction potential of silty sands," *From Research to Practice in Geotechnical Engineering*, ASCE, New York, pp. 433–451.

Mitchell, J.K. and Santamarina, J.C. (2005). "Biological considerations in geotechnical engineering." *Journal of Geotechnical and Geoenvironmental Engineering*, Vol. 131, No. 10, pp. 1222–1233. https://doi.org/10.1061/(asce)1090-0241(2005)131:10(1222).

Mitchell, J.K. and Solymar, Z.V. (1984). "Time-dependent strength gain in freshly deposited or densified sand," *Journal of Geotechnical Engineering*, Vol. 104, No. GT7, pp. 995–1012. https://doi.org/10.1061/(asce)0733-9410(1984)110:11(1559).

Mitchell, J.K., Baxter, C.D.P., and Munson, T.C. (1995). "Performance of improved ground during earthquakes." *Soil Improvement for Earthquake Hazard Mitigation*, Geotechnical Special Publication No. 49, ASCE, New York, pp. 1–36.

Mitchell, J.K. and Wentz, F.L. (1991). "Performance of improved ground during the Loma Prieta earthquake," *Report No. UCB/EERC-91/12*, Earthquake Engineering Research Center, University of California, Berkeley.

Moseley, M.P. and Kirsch, K. (2004). *Ground Improvement*, 2nd edition, Spon Press, New York.

Munfakh, G.A. (1997a). "Ground improvement engineering-the state practice: Part 1. Methods." *Ground Improvement*, Vol. 1, No. 4, pp. 193–214. https://doi.org/10.1680/gi.1997.010402.

Munfakh, G.A. (1997b). "Ground improvement engineering-the practice: Part 2. Applications." *Ground Improvement*, Vol. 1, No. 4, pp. 215–222. https://doi.org/10.1680/gi.1997.010403.

Narsillo, G.A., Santamarina, J.C., Hebeler, T., and Bachus, R. (2009). "Blast densification: Multi-instrumented case history," *Journal of Geotechnical and Geoenvironmental Engineering*, Vol. 135, No. 6, pp. 723–734. https://doi.org/10.1061/(asce)gt.1943-5606.0000023.

Nagao, K., Azegami, Y., Yamada, S., Suemasa, N., and Katada, T. (2007). "A micro-bubble injection method for a countermeasure against liquefaction," *Proceedings, 4th International Conference on Earthquake Geotechnical Engineering*, Thessaloniki, Greece, Paper No. 1764, 12 pp.

Neely, W.J. and Leroy, D.A. (1991). "Densification of sand using a variable frequency vibratory probe," in M.I. Esrig and R.C. Bachus, eds., *Deep Foundation Improvement: Design, Construction, and Testing*, ASTM STP 1089, American Society for Testing and Materials, pp. 320–332. https://doi.org/10.1520/stp25068s.

Nguyen, T.V., Rayamajhi, D., Boulanger, R.W., Ashford, S.A., Lu, J., Elgamal, A., and Shao, L. (2013). "Design of DSM grids for liquefaction remediation," *Journal of Geotechnical and Geoenvironmental Engineering*, Vol. 139, No. 11, pp. 1923–1933. https://doi.org/10.1061/(asce)gt.1943-5606.0000921.

O'Donnell, S.T., Rittmann, B.E., and Kavazanjian Jr, E. (2017). "MIDP: Liquefaction mitigation via microbial denitrification as a two-stage process. I: Desaturation," *Journal of Geotechnical and Geoenvironmental Engineering*, Vol. 143, No. 12, 12 pp. https://doi.org/10.1061/(asce)gt.1943-5606.0001818.

Okamura, M., Ishihara, M., and Oshita, T. (2003). "Liquefaction resistance of sand deposit improved with sand compaction piles," *Soils and Foundations*, Vol. 43, No. 5, pp. 175–187. https://doi.org/10.3208/sandf.43.5_175.

Okamura, M., Ishihara, K., and Tamura, K. (2006). "Degree of saturation and liquefaction resistance of sand improved with sand compaction pile," *Journal of Geotechnical and Geoenvironmental Engineering*, Vol. 132, No. 2, pp. 258–264. https://doi.org/10.1061/(asce)1090-0241(2006)132:2(258).

Pestana, J.M., Hunt, C.E., and Goughnour, R.R. (1997). "FEQDrain: A finite element computer program for the analysis of the earthquake generation and dissipation of pore water pressure in layered sand deposits with vertical drains," *Report No. EERC 97-17*, Earthquake Engineering Res. Ctr., UC-Berkeley, CA.

Porbaha, A., Weatherby, D., Macnab, A., Lambrechts, J., Burke, G., Yang, D., and Puppala, A.J. (2005). "American practice of deep mixing technology," *Proceedings, International Conference on Deep Mixing-Best Practice and Recent Advances*, Stockholm.

Powers, J.P. (1992). *Construction Dewatering - New Methods and Applications*, John Wiley & Sons, Inc., New York, 492 pp.

Powers, P., Corwin, A.B., Herridge, C.J., Schmall, P.C., and Kaeck, W.E. (2007). *Construction Dewatering and Groundwater Control: New Methods and Applications*, 3rd edition, Wiley, Hoboken, NJ, 638 pp.

Priebe, H.J. (1991). "The prevention of liquefaction by vibro replacement," *Proceedings, International Conference on Earthquake Resistant Construction*, Berlin.

Proto, C.J., DeJong, J.T., and Nelson, D.C. (2016). "Biomediated permeability reduction of saturated sands," *Journal of Geotechnical and Geoenvironmental Engineering*, Vol. 142, No. 12, 11 pp. https://doi.org/10.1061/(asce)gt.1943-5606.0001558.

Robinsky, E.I. and Morrison, D.E. (1964). "Sand displacement and compaction around model friction piles," *Canadian Geotechnical Journal*, Vol. 1, No. 2, pp. 81–91. https://doi.org/10.1139/t64-002.

Rollins, K., Anderson, J.K.S., Goughnour, R.R., and McCain, A.K. (2004). "Liquefaction hazard mitigation using vertical composite drains," *Proceedings, 13th World Conference on Earthquake Engineering*, Vancouver, B.C., Paper No. 2880, 15 pp.

Rollins, K.M., Price, B.E., Dibb, E., and Higbee, J.B. (2006). "Liquefaction mitigation of silty sands in utah using stone columns with wick drains," *Ground Modification and Seismic Mitigation*, Geotechnical Special Publication 152, ASCE, pp. 343–348. https://doi.org/10.1061/40864(196)46.

Rollins, K.M., Quimby, M., Hohnson, S.R., and Price, B. (2009). "Effectiveness of stone columns for liquefaction mitigation of silty sands with and without wick drains," in J. Han, G. Zheng, V.R. Schaefer, and M. Huang, eds., *Advances in Ground Improvement: Research to Practice in the United States and China*, Geotechnical Special Publication No. 188, ASCE, pp. 160–169. https://doi.org/10.1061/41025(338)17.

Ryan, C.R. and Jasperse, B.H. (1989). "Deep soil mixing at the Jackson Lake dam," *Proceedings, Foundation Engineering: Current Principles and Practices, ASCE Specialty Conference*, Ithaca, New York, pp. 1–14.

Salley, J.R., Foreman, B., Baker, W.H., and Henry, J.F. (1987). "Compaction grouting test program Pinopolis West Dam," *Soil Improvement - A Ten Year Update*, Geotechnical Special Publication No. 12, ASCE, New York, pp. 245–269.

San Pablo, A.C.M, Lee, M., Graddy, C.M.R., Kolbus, C.M., Khan, M., Zamani, A., Martin, N., Acuff, C., DeJong, J.T., Gomez, M.G., and Nelson, D.C. (2020). "Meter-scale bio-cementation experiments to advance process control and reduce impacts: Examining spatial control, ammonium by-product removal, and chemical reductions," *Journal of Geotechnical and Geoenvironmental Engineering*, Vol. 146, No. 11, 14 pp. https://doi.org/10.1061/(asce)gt.1943-5606.0002377.

Santamarina, J.C., Kleinand, K.A., and Fam, M.A. (2001). *Soils and Waves*, John Wiley & Sons, New York, 488 pp.

Schaefer, V., Abramson, L.W., Hussin, J.D., and Sharp, K.D. (1997). *Ground Improvement, Ground Reinforcement, Ground Treatment: Developments 1987–1997*, Geotechnical Special Publication No. 69, ASCE, New York.

Schaefer, V.R., Berg, R.R., Collin, J.G., Christopher, B.R., DiMaggio, J.A., Filz, G.M., Bruce, D.A., and Ayala, D. (2017a). "Ground modification methods - Reference manual, Volume 1," Geotechnical Engineering Circular No. 13, *Report No. FHWA-NHI-16-027*, Federal Highway Administration, Washington, DC, 386 pp.

Schaefer, V.R., Berg, R.R., Collin, J.G., Christopher, B.R., DiMaggio, J.A., Filz, G.M., Bruce, D.A., and Ayala, D. (2017b). "Ground modification methods - Reference manual, Volume 2," Geotechnical Engineering Circular No. 13, *Report No. FHWA-NHI-16-027*, Federal Highway Administration, Washington, DC, 542 pp.

Schmertmann, J.H. (1986). "CPT/DMT QC of ground modification at a power plant," *Proceedings, In Situ '86, ASCE Specialty Conference*, Blacksburg, Virginia, pp. 985–1001. https://doi.org/10.1016/0148-9062(88)90325-7.

Seed, H.B. and Booker, J.R. (1976). "Stabilization of potentially liquefiable sand deposits using gravel drain systems," *Report No. EERC-76-10*, Earthquake Engineering Research Center, University of California, Berkeley,

Seed, H.B. and Booker, J.R. (1977). "Stabilization of potentially liquefiable sand deposits using gravel drains, *Journal of the Geotechnical Engineering Division*, Vol. 103, No. GT7, pp. 757–768. https://doi.org/10.1061/ajgeb6.0000453.

Seko, T. and Tobe, K. (1977). "An experimental investigation of sand sampling," *Proceedings of Specialty Session 2, 9th International Conference on Soil Mechanics and Foundation Engineering*, Tokyo, pp. 37–42.

Semprich, S. and Stadler, G. (2003). "Grouting in geotechnical engineering," in Smoltczyk, U., ed., *Geotechnical Engineering Handbook, 2: Procedures*, Ernst & Sohn, Berlin, pp. 57–90.

Senneset, K. and Nestvold, J. (1992). "Deep compaction by vibro wing technique and dynamic compaction," *Grouting, Soil Improvement, and Geosynthetics*, Geotechnical Special Publication No. 30, ASCE, New York, Vol. 2, pp. 889–901.

Serridge, C.J. and Synac, O. (2006). "Application of the Rapid Impact Compaction (RIC) technique for risk mitigation in problematic soil," *Proceedings, 10th IAEG International Congress*, Nottingham, United Kingdom.

Shanklin, D., Talbot, J., and Rademacher, K., eds. (2000). *Constructing and Controlling Compaction of Earth Fills*, Special Technical Publication 1384, ASTM International, Washington, DC, 341 pp.

Shao, L. (2009). "Soil improvement to support a warehouse building over difficult soils," in J. Han, G. Zheng, V.R. Schaefer, and M. Huang, eds., *Advances in Ground Improvement: Research to Practice in the United States and China*, Geotechnical Special Publication No. 188, ASCE, pp. 207–216. https://doi.org/10.1061/41025(338)22.

Shao, L., Taylor, D., and Koelling, M. (2013). "Stone columns and earthquake drain liquefaction mitigation for federal center south in Seattle, Washington," *Proceedings, Stability and Performance of Slopes and Embankments III*, Geotechnical Special Publication No. 231, pp. 864–878. https://doi.org/10.1061/9780784412787.089.

Shuwang, Y., Wei, D., and Juan, C. (2009). "Use of explosion in improving highway foundation," in J. Han, G. Zheng, V. R. Schaefer, and M. Huang, eds., *Advances in Ground Improvement: Research to Practice in the United States and China*, Geotechnical Special Publication No. 188, ASCE, pp. 290–297. https://doi.org/10.1061/41025(338)31.

Siegel, T.C. (2013). "Liquefaction mitigation synthesis report," *Journal of the Deep Foundations Institute*, Vol. 7, No. 1, pp. 13–31. https://doi.org/10.1179/dfi.2013.002.

Singh, S., Seed, H.B., and Chan, C.K. (1979). "Undisturbed sampling and cyclic load testing of sands," *Report No. UCB/EERC-79/33*, Earthquake Engineering Research Center, University of California, Berkeley, 131 pp.

Slocombe, B.C. (2012). "Dynamic compaction," *Ground Improvement*, 3rd edition, K. Kirsch and A. Bell, eds., Taylor & Francis Group, Boca Raton, FL, pp. 57–85.

Solymar, Z.V. and Reed, D.J. (1986). "A comparison of foundation compaction techniques," *Canadian Geotechnical Journal*, Vol. 23, No. 3, pp. 271–280. https://doi.org/10.1139/t86-040.

Stark, T.D., Axtell, P.J., Lewis, J.R., Dillon, J.C., Empson, W.B., Topi, J.E., and Walberg, R.C. (2009). "Soil inclusions in jet grout columns," *DFI Journal*, Deep Foundations Institute, Vol. 3, No. 1, pp. 33–44. https://doi.org/10.1179/dfi.2009.004.

Stilley, A.N. (1982). "Compaction grouting for foundation stabilization," *Proceedings, ASCE Specialty Conference on Grouting in Geotechnical Engineering*, New Orleans, pp. 923–937.

Taki, D. and Yang, D.S. (1991). "Soil-cement mixed wall technique," Geotechnical Special Publication No. 27, ASCE, Vol. 1, pp. 298–309.

Terashi, M. and Juran, I. (2000). "Ground improvement-State of the art," *Proceedings, GeoEng 2000*, 19–24 November 2000, Melbourne, Australia, Volume 1: Invited Papers, Technomic Publishing Company, Inc., Lancaster, PA, pp. 461–519.

Tokimatsu, K., Mizuno, H., and Kakurai, M. (1996). "Building damage associated with geotechnical problems." *Soils and Foundations*, Special Issue on Geotechnical Aspects of the January 17 1995 Hyogoken-Nanbu Earthquake, pp. 219–234. https://doi.org/10.3208/sandf.36.special_219.

Towhata, I. (2008). "Mitigation of liquefaction-induced damage," W. Wu and R. I. Borja, eds., *Geotechnical Earthquake Engineering*, Springer, Berlin, 684 pp. https://doi.org/10.1007/978-3-540-35783-4_26.

Tsukamoto, Y., Ishihara, K., Nakazawa, H., Kamada, K., and Huang, Y.N. (2002). "Resistance of partly saturated sand to liquefaction with reference to longitudinal and shear wave velocities." *Soils and Foundations*, Vol. 42, No. 6, pp. 93–104. https://doi.org/10.3208/sandf.42.6_93.

Van Court, N.W.A. and Mitchell, J.K. (1995). "New insights into compaction of loose, saturated, cohesionless soils," *Soil Improvement for Earthquake Hazard Mitigation*, ASCE, Geotechnical Special Publication No. 49, pp. 51–65.

Van Impe, W.F. (1989). *Soil Improvement Techniques and Their Evolution*, A.A. Balkema, Rotterdam, 125 pp.

van Paassen, L.A. (2011). "Bio-mediated ground improvement: from laboratory experiment to pilot applications," *Proceedings, Geo-Frontiers 2011: Advances in Geotechnical Engineering*, ASCE, pp. 4099–4108. https://doi.org/10.1061/41165(397)419.

van Paassen, L.A., Pham, V., Mahabadi, N., Hall, C.A., Stallings, E., and Kavazanjian, E. (2017). "Desaturation via biogenic gas formation as a ground improvement technique," *Proceedings, Pan-American Conference on Unsaturated Soil Mechanics*, Geotechnical Special Publication 300, ASCE, pp. 244–256. https://doi.org/10.1061/9780784481677.013.

Wang, J.L., Wang, D.F., and Wang, J.W. (2009). "State of jet grouting in Shanghai," in J. Han, G. Zheng, V.R. Schaefer, and M. Huang, eds., *Advances in Ground Improvement: Research to Practice in the United States and China*, Geotechnical Special Publication No. 188, ASCE, pp. 179–188. https://doi.org/10.1061/41025(338)19.

Warner, J.A. (2003). "Fifty years of low mobility grouting. Grouting and ground treatment," *Proceedings, Third International Conference*, L.F. Johnsen, D.A. Bruce, and M.J. Byle, eds., Geotechnical Special Publication No. 120, Geo-Institute of ASCE, New Orleans, LA, pp. 1–24. https://doi.org/10.1061/40663(2003)1.

Wehr, W.J. (2005). "Influence of the carbonate content of sand on vibrocompaction," *Proceedings, 6th International Conference on Ground Improvement Techniques*, Coimbra, Portugal, July.

Welsh, J.P. (1986). "Insitu testing for ground modification techniques," *Proceedings, Insitu '86*, Geotechnical Special Publication No. 6, ASCE, pp. 322–335.

Welsh, J.P. (1987). *Soil Improvement - A Ten Year Update*, Geotechnical Special Publication No. 12, ASCE, New York, 331 pp.

Wheeler, S.J. (1988). "A conceptual-model for soils containing large gas bubbles." *Geotechnique*, Vol. 38, No. 3, pp. 389–397. https://doi.org/10.1680/geot.1988.38.3.389.

Whitman, W.B., Coleman, D.C., and Wiebe, W.J., (1998). "Prokaryotes: the unseen majority," *Proceedings of the National Academy of Sciences*, Vol. 95, pp. 6578–6583. https://doi.org/10.1073/pnas.95.12.6578.

Wissmann, K.J., van Ballegooy, S., Metcalfe, B.C., Dismuke, J.N., and Anderson, C.K. (2015). "Rammed aggregate pier ground improvement as a liquefaction mitigation method in sandy and silty soils," *Proceedings, 6th International Conference on Earthquake Geotechnical Engineering*, Christchurch, NZ, 9 pp.

Wu, G. (1995). "A dynamic response analysis of saturated granular soils to blast loads using a single phase model," *Research Report*, Natural Sciences and Engineering Research Council (Canada), December, 1995.

Xia, H., and Hu, T. (1991). "Effects of saturation and back pressure on sand liquefaction." *Journal of Geotechnical Engineering*, Vol. 117, No. 9, pp. 1347–1362. https://doi.org/10.1061/(asce)0733-9410(1991)117:9(1347).

Yang, D.S., Scheibel, L.L., Lobedan, F. and Nagata, C. (2001). "Oakland airport roadway project," *Soil Mixing Specialty Seminar, 26th DFI Annual Conference*, St. Louis, MO, pp. 55–71.

Yasui, S., Yokozawa, K., Yasuoka, N., and Kondo, H. (2005). "Recent technical trends in dry mixing (DJM) in Japan," *Proceedings, International Conference on Deep Mixing-Best Practice and Recent Advances*, Stockholm.

Yegian, M.K., Eseller-Bayat, E., Alshawabkeh, A. and Ali, S. (2007). "Induced partial saturation for liquefaction mitigation: Experimental investigation," *Journal of Geotechnical and Geoenvironmental Engineering*, Vol. 133, No. 4, pp. 372–380. https://doi.org/10.1061/(asce)1090-0241(2007)133:4(372).

Yoshimi, Y., Tanaka, K., and Tokimatsu, K. (1989). "Liquefaction resistance of a partially saturated sand," *Soils and Foundations*, Vol. 29, No. 3, pp. 157–162. https://doi.org/10.3208/sandf1972.29.3_157.

Zacher, E.G. and Graf, E.D. (1979). "Chemical grouting of sand for earthquake resistance," *Civil Engineering*, ASCE, New York, Vol. 49, No. 1, pp. 67–69.

Appendix A
Vibratory Motion

A.1 INTRODUCTION

Many different types of dynamic loading can induce vibratory motion in soils and structures. To solve problems involving the dynamic response of soils and structures, it is necessary to be able to describe dynamic events. They can be described in different ways, and the geotechnical earthquake engineer must be familiar with each. This appendix provides a brief description of vibratory motion and introduces the nomenclature and mathematical forms by which it is usually described.

A.2 TYPES OF VIBRATORY MOTION

Vibratory motion can be divided into two broad categories: periodic motion and nonperiodic motion. Periodic motions are those that repeat themselves at regular intervals of time. Mathematically, a motion, $u(t)$, is periodic if there exists some period, T_f, for which $u(t + T_f) = u(t)$ for all t. The simplest form of periodic motion is simple harmonic motion in which displacement varies sinusoidally with time. Nonperiodic motions, which do not repeat themselves at constant intervals, can result from impulsive loads (e.g., explosions or falling weights) or from longer-duration transient loadings (e.g., earthquakes or traffic). Examples of periodic and nonperiodic motions are shown in Figure A.1.

Some forms of periodic motion (e.g., Figure A.1b) may appear to be much more complex than simple harmonic motion, but with the use of mathematical techniques described later in this appendix, they can be expressed as the sum of a series of simple harmonic motions. Even transient, nonperiodic motions such as those of Figure A.1c and d can be represented as periodic motions by assuming that they repeat themselves after some "quiet" zone during which no motion occurs (Figure A.2). Using this technique, even a transient motion can also be expressed as a periodic motion. This becomes a very powerful tool for the dynamic analysis of linear systems, where the

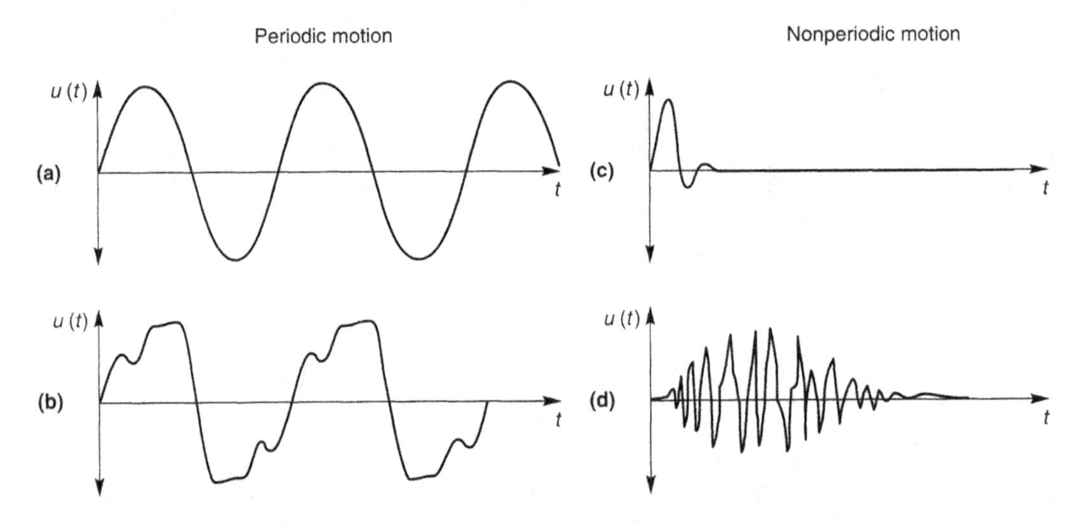

FIGURE A.1 Periodic and nonperiodic motions: (a) simple harmonic motion; (b) general periodic motion; (c) pulse-like motion (response to impact loading); and (d) earthquake ground motion.

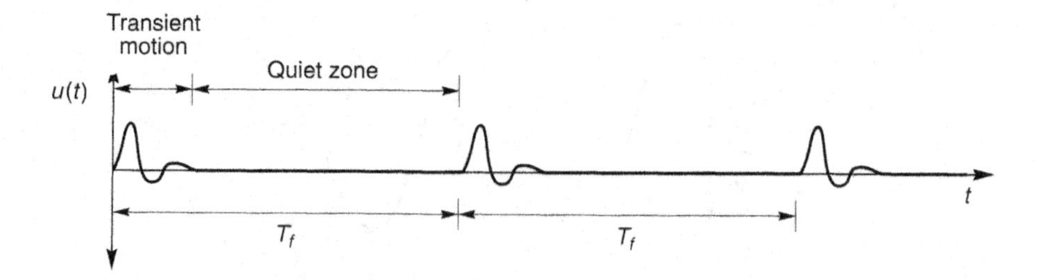

FIGURE A.2 Representation of a transient motion as a periodic motion using an artificial quiet zone. The motion repeats itself indefinitely at period T_f.

principle of superposition allows the response to transient loading to be expressed as the sum of the responses to a series of simple harmonic loads.

A.2.1 Simple Harmonic Motion

Simple harmonic motion can be characterized as a sinusoidal motion at constant frequency. Its most important features can be defined by three quantities: amplitude, frequency, and phase. Simple harmonic motion can be described in different ways, two of which will be presented in the following sections: using trigonometric notation or using complex notation. Both notations are equivalent, and both are commonly used in geotechnical earthquake engineering.

A.2.2 Trigonometric Notation for Simple Harmonic Motion

In its simplest form, simple harmonic motion can be expressed in terms of a displacement, $u(t)$, using trigonometric notation: for example,

$$u(t) = A\sin(\omega t + \phi) \tag{A.1}$$

where A represents the displacement amplitude, ω the circular frequency, and ϕ the phase angle. The displacement history of this simple harmonic motion is shown in Figure A.3. The amplitude, A, is occasionally referred to as the single amplitude to distinguish it from the double amplitude (which represents the peak-to-peak displacement) referred to in some of the older geotechnical earthquake engineering literature. The circular frequency describes the rate of oscillation in terms of radians per unit time, where 2π radians correspond to one cycle of motion. The phase angle describes the amount of time by which the peaks (and zero points) are shifted from those of a pure sine function (i.e., a sine function with zero phase shift), as illustrated in Figure A.4. The displacement will be

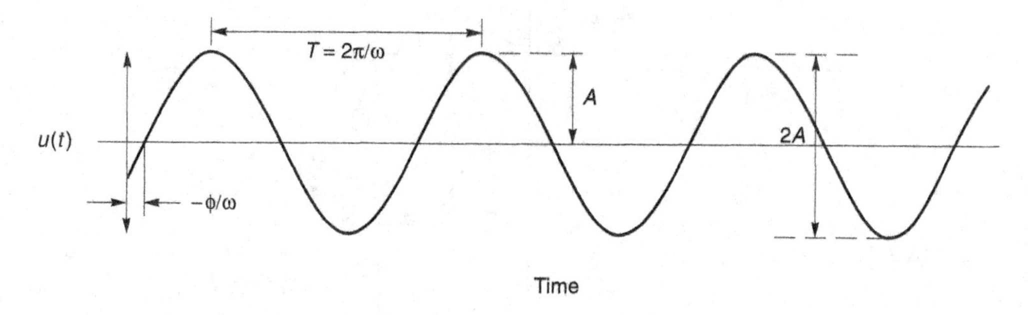

FIGURE A.3 Time history of simple harmonic displacement.

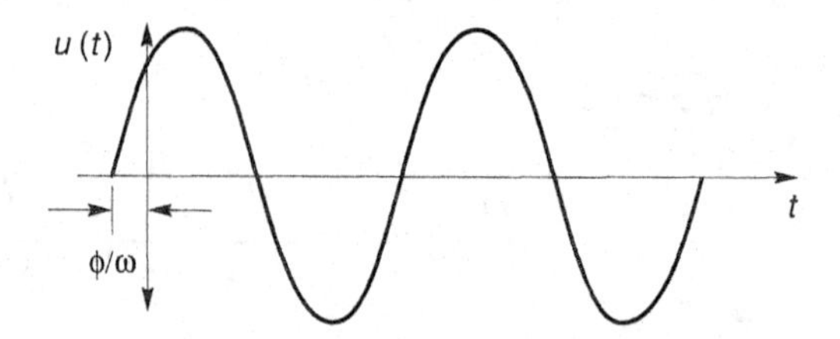

FIGURE A.4 Influence of phase angle on position of sinusoid.

zero when $\omega t + \phi = 0$ or, consequently, when $t = -\phi/\omega$. A positive phase angle indicates that the motion leads the pure sine function; it lags the pure sine function if the phase angle is negative.

The concept of circular frequency is more easily understood by considering the motion of the rotating vector of length A shown in Figure A.5. If the vector rotates counterclockwise about its origin at an angular speed, ω, from its initial horizontal position, the displacement, $u(t)$, is given by the vertical component of the vector

$$u(t) = A \sin \omega t \tag{A.2}$$

The vertical component increases to a maximum value at $\omega t = \pi/2$, then decreases through zero (at $\omega t = \pi$) and reaches its maximum negative value at $\omega t = 3\pi/2$. It continues back to its original position and then repeats the entire process.

The time required for the rotating vector to make one full revolution is the time required for one cycle of the motion. This time is referred to as the period of vibration, T, and is related to the circular frequency by

$$T = \frac{\text{angular change for one revolution}}{\text{angular speed}} = \frac{2\pi}{\omega} \tag{A.3}$$

Another common measure of the frequency of oscillation is expressed in terms of the number of cycles that occur in a particular period of time: Since the period of vibration represents the time per cycle, the number of cycles per unit time must be its reciprocal, i.e.,

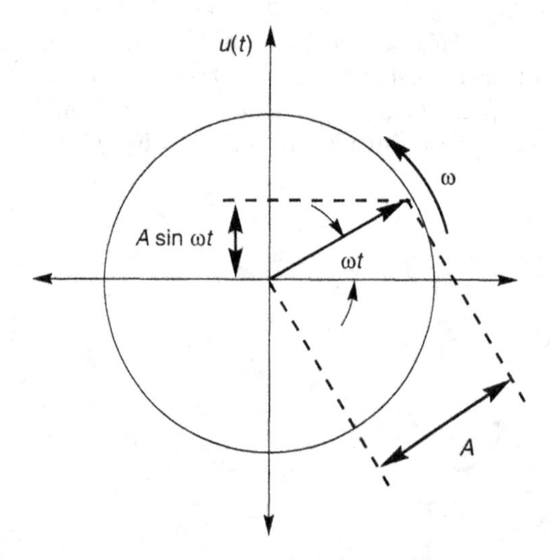

FIGURE A.5 Rotating vector representation of simple harmonic motion with zero phase angle.

$$f = \frac{1}{T} = \frac{\omega}{2\pi} \tag{A.4}$$

which is usually expressed in cycles per second or hertz (abbreviated Hz).

Simple harmonic motion can also be described as the sum of a sine function and a cosine function, i.e.,

$$u(t) = a\cos\omega t + b\sin\omega t \tag{A.5}$$

As shown in Figure A.6, the sum of the sine and cosine functions is also a sinusoid that oscillates at circular frequency, ω. However, its amplitude is not the simple sum of the amplitudes of the sine and cosine functions, and its peaks do not occur at the same times as those of the sine or cosine functions. The rotating vector representation of this function is illustrated in Figure A.7. Since $\cos\theta = \sin(\theta + 90°)$, the rotating vector of length, a, must be 90° ahead of the vector of length, b. The vertical components of vectors, a and b, are $a\cos\omega t$ and $b\sin\omega t$, respectively. As illustrated in Figure A.7a, the total value of $u(t)$ is given by Equation (A.5). The motion can be expressed in a different form by considering the resultant of vectors a and b, as shown in Figure A.7b. The length of the resultant will be $\sqrt{a^2 + b^2}$ and it will lead b by an angle $\phi = \tan^{-1}(a/b)$. Accordingly, the vertical component of the resultant is [Equation (A.1)]:

$$u(t) = A\sin(\omega t + \phi)$$

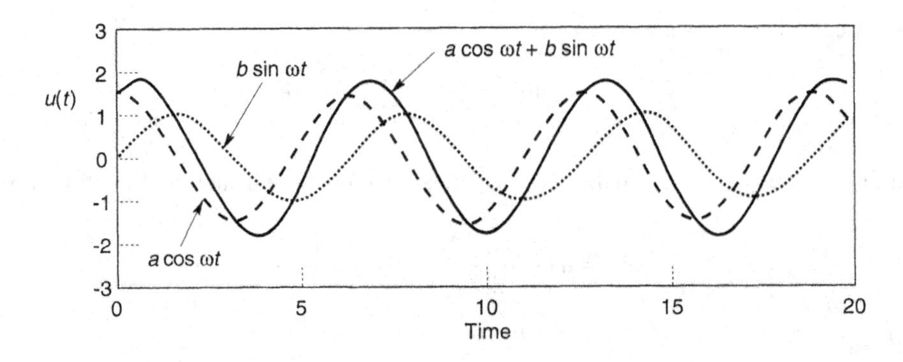

FIGURE A.6 Summation of sine and cosine functions of the dame frequency produces a sinusoid of the same frequency. Amplitude and phase of the sinusoid depends on the amplitudes of the sine and cosine functions.

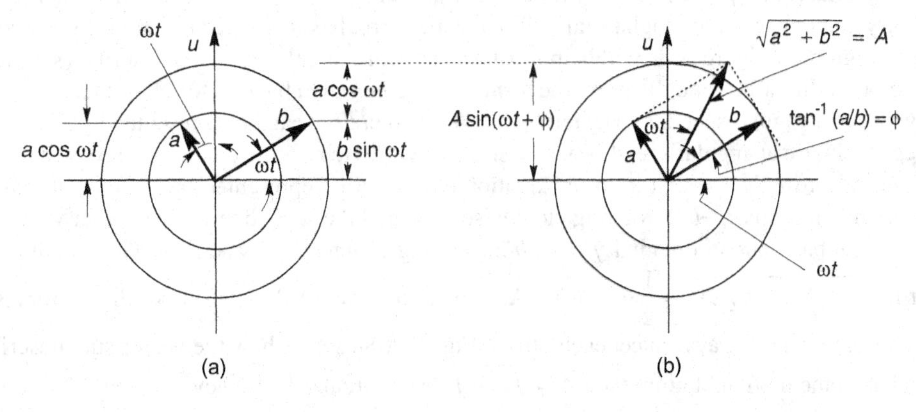

FIGURE A.7 Rotating vector representation of simple harmonic motion. Sum of vertical components of sine and cosine component in (a) is equal to vertical component of resultant of sine and cosine components in (b).

where $A = \sqrt{a^2 + b^2}$ is the amplitude and $\phi = \tan^{-1}(a/b)$ is the phase angle of the motion. Accordingly, Equations (A.1) and (A.5) are equivalent representation of sinusoidal motion.

A.2.3 Complex Notation for Simple Harmonic Motion

Trigonometric descriptions of simple harmonic motion use familiar functions that are easy to visualize. For many dynamic analyses, however, the use of trigonometric notation leads to very long and awkward equations. These analyses become much simpler when motions are described using complex notation (the word "complex" indicates that complex variables are used, not that the notation is particularly complicated). Complex notation can be derived directly from trigonometric notation using Euler's law:

$$e^{i\alpha} = \cos\alpha + i\sin\alpha \tag{A.6}$$

where i is the imaginary number $i = \sqrt{-1}$. The quantity $e^{i\alpha}$ is a complex number; it has two parts, a real part and an imaginary part, which can be written as

$$\mathrm{Re}\left(e^{i\alpha}\right) = \cos\alpha \tag{A.7a}$$

$$\mathrm{Im}\left(e^{i\alpha}\right) = \sin\alpha \tag{A.7b}$$

Euler's law can be used to show that

$$\cos\alpha = \frac{e^{i\alpha} + e^{-i\alpha}}{2} \qquad \sin\alpha = -i\frac{e^{i\alpha} - e^{-i\alpha}}{2} \tag{A.8}$$

Substituting these expressions into the general expression for harmonic motion (Equation A.5) gives

$$\begin{aligned} u(t) &= a\frac{e^{i\omega t} + e^{-i\omega t}}{2} - bi\frac{e^{i\omega t} - e^{-i\omega t}}{2} \\ &= \frac{a - ib}{2}e^{i\omega t} + \frac{a + ib}{2}e^{-i\omega t} \end{aligned} \tag{A.9}$$

This form of the displacement may be visualized as a pair of rotating vectors in an Argand diagram. An Argand diagram represents a complex number graphically as a vector with orthogonal real and imaginary components. Although usually drawn with the real axis oriented horizontally, the rotated Argand diagram of Figure A.8a will help illustrate how this complex notation describes simple harmonic motion. In the Argand diagram, the term $e^{i\omega t}$ is represented by a vector of unit length rotating clockwise at an angular speed, ω. The term $e^{-i\omega t} = e^{i(-\omega)t}$ therefore can be represented by a unit vector rotating clockwise at angular speed, $-\omega$, which is equivalent to rotating counterclockwise at angular speed, ω. Accordingly, the first term in Equation (A.9) can be represented by a vector of real part, $a/2$, and imaginary part, $-b/2$, rotating clockwise at ω, and the second term by another vector with the same real part, but an imaginary part, $b/2$, rotating counterclockwise at ω. The length of each vector is $\sqrt{(a/2)^2 + (b/2)^2} = \frac{1}{2}\sqrt{a^2 + b^2}$. As shown in Figure A.8a, the sum of the vectors is real (the imaginary parts always cancel each other). Figure A.8b shows how the vector sum describes a simple harmonic motion of amplitude $A = \sqrt{a^2 + b^2}$ and circular frequency ω.

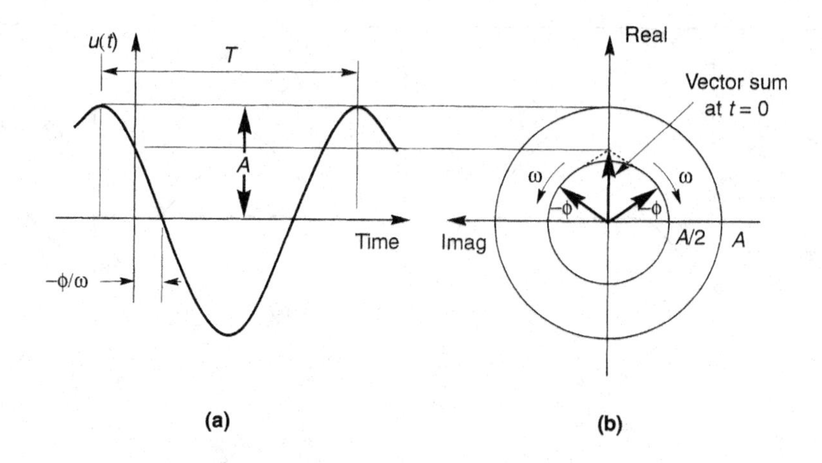

FIGURE A.8 (a and b) Illustration of how counterrotating vectors of length $A/2$ produce simple harmonic motion. Note that the phase angles are measured from the horizontal axis in the direction of vector rotation.

A.2.4 Other Measures of Motion

Displacement is not the only parameter that can be used to describe vibratory motion. In fact, other parameters are often of greater interest. If the variation of displacement with time is known, however, the other parameters of interest can be determined. Differentiating the expression for simple harmonic displacement produces expressions for velocity and acceleration:

$$u(t) = A\sin(\omega t + \phi) \quad \text{displacement} \tag{A.10a}$$

$$\dot{u}(t) = \frac{du}{dt} = \omega A\cos(\omega t + \phi) \quad \text{velocity} \tag{A.10b}$$

$$\ddot{u}(t) = \frac{d^2u}{dt^2} = -\omega^2 A\sin(\omega t + \phi) = -\omega^2 u \quad \text{acceleration} \tag{A.10c}$$

Note that when the displacement amplitude is A, the velocity amplitude is ωA, and the acceleration amplitude is $\omega^2 A$. Thus, frequency and the displacement, velocity, and acceleration amplitudes of a harmonic motion are related in such a way that knowledge of the frequency and any one amplitude, or knowledge of any two amplitudes, allows calculation of all other quantities. This important and useful property of harmonic motions allows the use of tripartite plots, in which a harmonic motion can be completely described in terms of frequency and displacement, velocity, and acceleration amplitudes by a single point. Since earthquake motions can be represented as the sum of a series of harmonic motions (Section A.3), *tripartite plots*, an example of which is shown in Figure A.9, can be used to describe earthquake ground motions.

Examination of Equation (A.10) reveals that in addition to having different amplitudes, the displacement, velocity, and acceleration are out of phase with each other (Figures A.10 and A.11). The velocity can be seen to lead the displacement by $\pi/2$ radians, or 90°, and the acceleration to lead the velocity by the same amount. The relationships between displacement, velocity, and acceleration for harmonic motions, in both trigonometric and complex notation, are (Figure A.11)

$$u(t) = A\sin\omega t \qquad u(t) = Ae^{i\omega t} \tag{A.11a}$$

$$\dot{u}(t) = \omega A\cos\omega t = \omega A\sin(\omega t + \pi/2) \qquad \dot{u}(t) = i\omega Ae^{i\omega t} \tag{A.11b}$$

$$\ddot{u}(t) = -\omega^2 A\sin\omega t = \omega^2 A\sin(\omega t + \pi) \qquad \ddot{u}(t) = i^2\omega^2 Ae^{i\omega t} = -\omega^2 Ae^{i\omega t} \tag{A.11c}$$

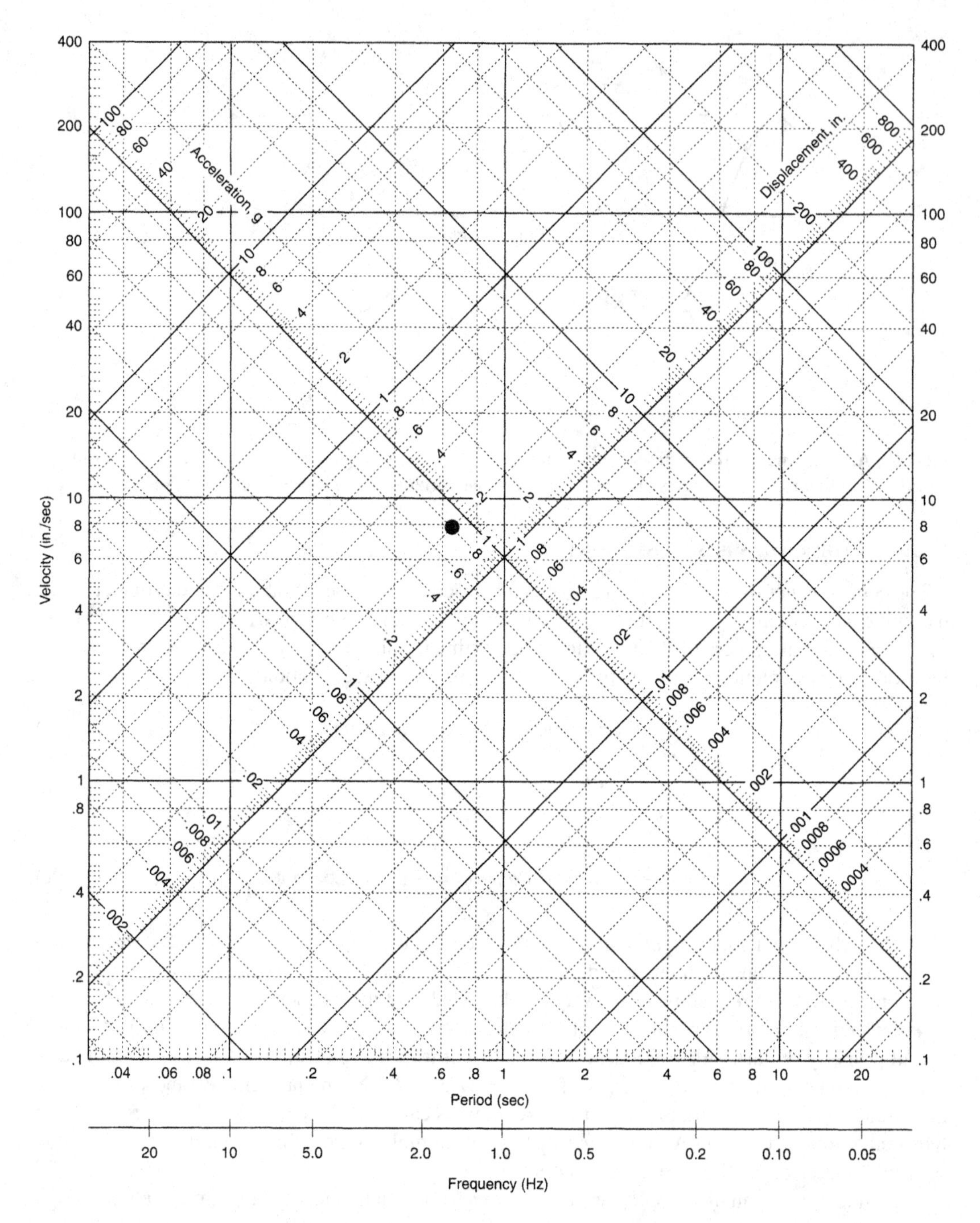

FIGURE A.9 Tripartite plot for harmonic motion. Point at center describes harmonic motion at a period of 0.65 sec with displacement amplitude of 0.8 in., velocity amplitude of 8.0 in./s, and acceleration amplitude of 0.20*g*. (After Richart et al., 1970).

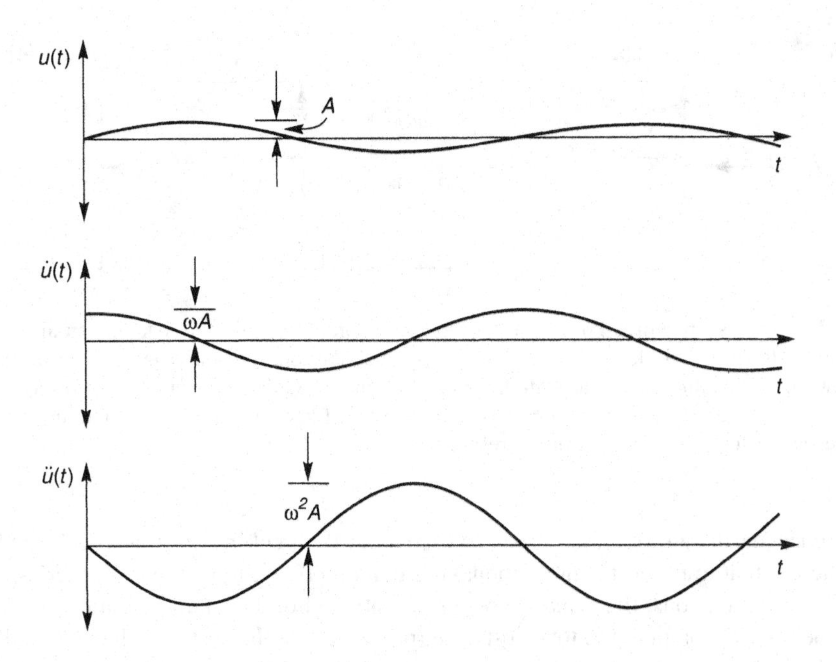

FIGURE A.10 Displacement, velocity, and acceleration histories. Note that acceleration leads velocity by one-quarter cycle and displacement by one-half cycle.

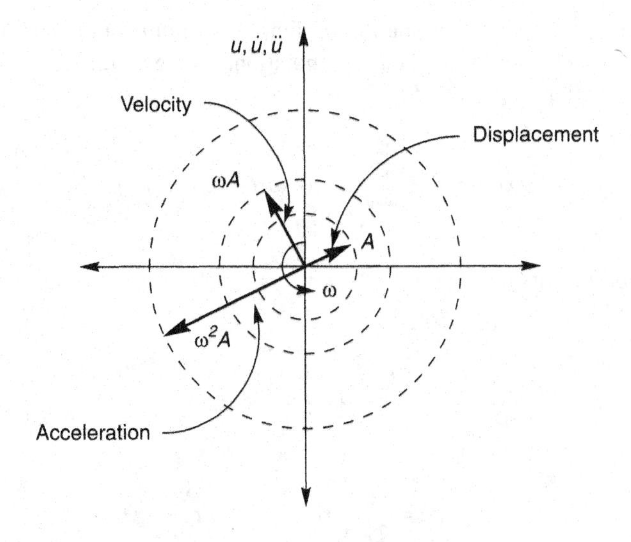

FIGURE A.11 Rotating vector representation of displacement, velocity, and acceleration Note how acceleration leads velocity by 90° and displacement by 180°.

A.3 FOURIER SERIES

While studying heat flow problems in the early nineteenth century, the French mathematician J.B.J. Fourier showed that any periodic function that meets certain conditions can be expressed as the sum of a series of sinusoids of different amplitude, frequency, and phase. Since the conditions for existence of a Fourier series are nearly always met for functions that accurately describe physical processes, it is an extraordinarily useful tool in many branches of science and engineering.

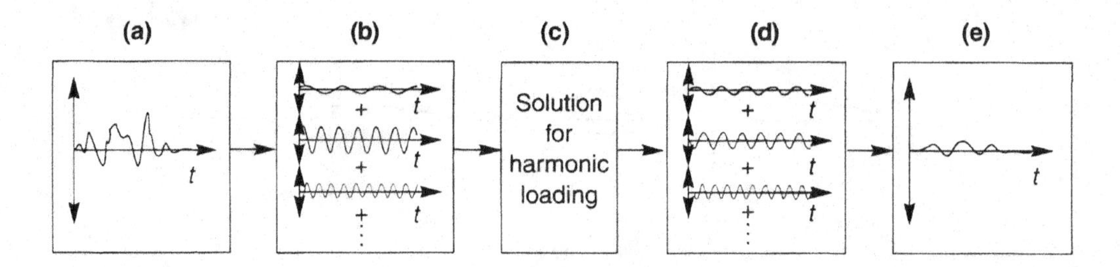

FIGURE A.12 Process by which Fourier series representation of complicated loading can allow relatively simple solutions for harmonic loading to be used to produce the total response: (a) time series of loading, (b) representation of loading history as sum of series of harmonic loads, (c) calculation of response for each harmonic load, (d) representation of response as sum of series of harmonic responses, and (e) summation of harmonic responses to produce time series of response.

Geotechnical earthquake engineering is no exception. By breaking down a complicated loading function such as that imposed by an earthquake ground motion into the sum of a series of simple harmonic loading functions, the principle of superposition allows available solutions for harmonic loading to be used to compute the total response (provided that the system is linear), as illustrated schematically in Figure A.12.

A.3.1 Trigonometric Form

Since a Fourier series is simply a summation of simple harmonic functions, it can be expressed using either trigonometric notation or complex notation. The general trigonometric form of the Fourier series for a function of period, T_f, is

$$x(t) = a_0 + \sum_{n=1}^{\infty}\left(a_n \cos \omega_n t + b_n \sin \omega_n t\right) \tag{A.12}$$

where the Fourier coefficients are

$$a_0 = \frac{1}{T_f}\int_0^{T_f} x(t)\,dt$$

$$a_n = \frac{2}{T_f}\int_0^{T_f} x(t)\cos \omega_n t\,dt$$

$$b_n = \frac{2}{T_f}\int_0^{T_f} x(t)\sin \omega_n t\,dt$$

and $\omega_n = 2\pi n / T_f$. The term a_0 represents the average value of $x(t)$ over the range $t = 0$ to $t = T_f$; its value is zero in many geotechnical earthquake engineering applications. Note that the frequencies, ω_n, are not arbitrary; rather, they are evenly spaced at a constant frequency increment, $\Delta\omega = 2\pi / T_f$.

Example A.1

The Fourier coefficients are not difficult to calculate for simple functions. Consider the square wave function shown in Figure EA.1. Over its period, T_f, the square wave is described by

$$x(t) = \begin{cases} +A & 0 < t \le \dfrac{T_f}{4} \\ -A & \dfrac{T_f}{4} < t \le \dfrac{3T_f}{4} \\ +A & \dfrac{3T_f}{4} < t \le T_f \end{cases}$$

Since the average value of $x(t)$ is easily seen to be zero, the coefficient $a_0 = 0$. The value of a_1 can be computed as

$$a_1 = \frac{2}{T_f}\int_0^{T_f} x(t)\cos\omega_1 t\,dt$$

$$= \frac{2}{T_f}\left[A\int_0^{T_f/4}\cos\omega_1 t\,dt - A\int_{T_f/4}^{3T_f/4}\cos\omega_1 t\,dt + A\int_{3T_f/4}^{T_f}\cos\omega_1 t\,dt\right]$$

$$= \frac{2A}{\omega_1 T_f}\left[\sin\frac{\omega_1 T_f}{4} - \left(\sin\frac{3\omega_1 T_f}{4} - \sin\frac{\omega_1 T_f}{4}\right) + \left(\sin\omega_1 T_f - \sin\frac{3\omega_1 T_f}{4}\right)\right]$$

Substituting $\omega_1 T_f = 2\pi$ yields

$$a_1 = \frac{A}{\pi}(1+2+1) = \frac{4A}{\pi}$$

Repeating for all n yields

$$a_n = \begin{cases} \dfrac{+4A}{n\pi} & n = 1,5,9,\ldots \\ \dfrac{-4A}{n\pi} & n = 3,7,11,\ldots \\ 0 & n = \text{even integers} \end{cases}$$

$$b_n = 0 \quad \text{all } n$$

$$b_n = 0 \qquad \text{all } n$$

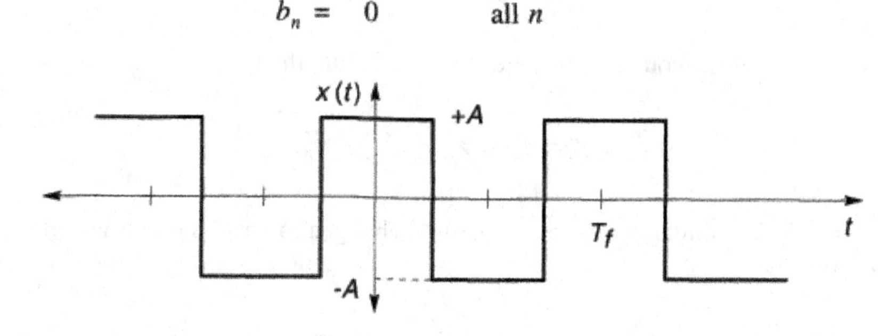

FIGURE EA.1 Square-Wave function.

so the Fourier series is

$$x(t) = \frac{4A}{\pi}\left(\cos\omega_1 t - \frac{1}{3}\cos 3\omega_1 t + \frac{1}{5}\cos 5\omega_1 t - \frac{1}{7}\cos 7\omega_1 t + \cdots\right)$$

where $\omega_1 = 2\pi/T_f$. The sine terms are all zero because the square wave, like the cosine function, is an even function [i.e., one for which $f(t) = f(-t)$]. For an odd function [$f(t) = -f(-t)$], the cosine terms are zero. For a function that is neither even nor odd, the Fourier series will contain both sine and cosine terms.

The Fourier series represents a function exactly only for $n = \infty$. If the series is truncated at some finite value of n, the Fourier series only approximates the function. For many functions, however, the approximation can be quite good even when n is relatively small. This characteristic is often used to great advantage in dynamic analyses of soils and structures.

From the equivalence of Equations (A.1) and (A.5), it is apparent that the Fourier series can also be expressed as

$$x(t) = c_0 + \sum_{n=1}^{\infty} c_n \sin(\omega_n t + \phi_n) \tag{A.13}$$

where $c_0 = a_0$, $c_n = \sqrt{a_n^2 + b_n^2}$, and $\phi_n = \tan^{-1}(a_n/b_n)$. In this form, c_n and ϕ_n are the amplitude and phase, respectively, of the nth harmonic. A plot of c_n versus ω_n is known as a *Fourier amplitude spectrum;* a plot of ϕ_n versus ω_n gives a *Fourier phase spectrum.* Fourier amplitude spectra are very useful in geotechnical earthquake engineering – as discussed in Chapter 3, they effectively describe the frequency content of an earthquake motion.

A.3.2 Exponential Form

The Fourier series can also be expressed in exponential form. Substituting Equations (A.8) into (A.12a) for all n gives

$$x(t) = a_0 + \sum_{n=1}^{\infty}\left(\frac{a_n - ib_n}{2}e^{i\omega_n t} + \frac{a_n + ib_n}{2}e^{-i\omega_n t}\right) \tag{A.14}$$

Defining new Fourier coefficients,

$$c_0^* = a_0$$

$$c_n^* = \frac{a_n - ib_n}{2}$$

$$c_{-n}^* = \frac{a_n + ib_n}{2}$$

where the * indicates the complex nature of the coefficient, the Fourier series can be rewritten as

$$x(t) = c_0^* + \sum_{n=1}^{\infty}\left(c_n^* e^{i\omega_n t} + c_{-n}^* e^{-i\omega_n t}\right) \tag{A.15}$$

Since $\omega_{-n} = -\omega_n$, the limits of summation can be changed to write the Fourier series in the more compact form

$$x(t) = c_0^* + \sum_{n=-\infty}^{\infty} c_n^* e^{i\omega_n t} \tag{A.16}$$

The complex Fourier coefficients, c_n^*, can be determined directly from $x(t)$ as

$$c_n^* = \frac{1}{T_f}\int_0^{T_f} x(t)e^{-i\omega_n t}\,dt \tag{A.17}$$

A.3.3 Discrete Fourier Transform

In many geotechnical earthquake engineering applications, loading or motion parameters are described by a finite number of data points rather than by an analytical function. In such cases, the Fourier coefficients are obtained by summation rather than integration. For a variable $x(t_k)$, $k = 1{:}N$, where $t_k = k\Delta t$, the discrete Fourier transform (DFT) is given by

$$X(\omega_n) = \Delta t \sum_{k=1}^{N} x(t_k)e^{-i\omega_n t_k} \tag{A.18a}$$

where $\omega_n = n\Delta\omega = 2\pi n/N\Delta t$. Using Euler's law, the DFT can also be written as

$$X(\omega_n) = \Delta t \sum_{k=1}^{N} \left[x(t_k)\cos\omega_n t_k - ix(t_k)\sin\omega_n t_k\right] \tag{A.18b}$$

Note that the Fourier coefficients of the DFT have units of the original variable multiplied by time. The DFT computed using Equation (A.18) produces amplitudes $X(\omega_n)$ that are dependent on record length $\Delta t \times N$; i.e., increasing the length of the record increases amplitudes. Fourier coefficients independent of record length can be computed by replacing Δt on the right side of Equation (A.18) with $\Delta t/\sqrt{\Delta t \times N}$. This adjustment changes the units of $X(\omega_n)$ to the units of the original variable multiplied by the square root of time.

The DFT can also be inverted; i.e., a set of data spaced at equal frequency intervals, $\Delta\omega$, can be expressed as a function of time, using the inverse discrete Fourier transform (IDFT):

$$x(t_k) = \frac{\Delta\omega}{2\pi} \sum_{n=1}^{N} X(\omega_n)e^{i\omega_n t_k} \tag{A.19a}$$

or

$$x(t_k) = \frac{\Delta\omega}{2\pi} \sum_{n=1}^{N} \left[X(\omega_n)\cos\omega_n t_k + iX(\omega_n)\sin\omega_n t_k\right] \tag{A.19b}$$

Either of these expressions can easily be programmed on a computer; since n takes on N different values, the summation operation will be performed N times. The time required for computation of a DFT (or IDFT), therefore, is proportional to N^2. If Fourier coefficients independent of record length are computed by dividing the right side of Eq. (18) by $\sqrt{\Delta t \times N}$, then the right side of Eq. (19) would need to be multiplied by $\sqrt{\Delta t \times N}$ to recover the same time series.

A.3.4 Fast Fourier Transform

The DFT was developed long before computers were available, and its use, for even modest values of N, was extremely labor-intensive. As early as 1805, the beginning of a more efficient approach to the DFT was described (Brigham, 1974). As digital computers were developed in the 1960s, Cooley and Tukey (1965) developed a computational algorithm for the case where N is a power of 2 that has become known as the fast Fourier transform (FFT). By performing repeated operations on groups that start with a single number and increase in size by a factor of 2 at each of j stages (where $N = 2^j$), the time required to complete the transform is proportional to $N \log 2N$. Consequently, the FFT is much more efficient than the DFT. For example, at $N = 2{,}048$, the FFT is more than 180 times faster than the DFT. The inverse fast Fourier transform (IFFT) operates with equal speed.

A.3.5 Power Spectrum

The Fourier amplitude spectrum illustrates how the amplitude of a quantity varies with frequency. This information can also be expressed in terms of power. The power of a signal, $x(t)$, that can be expressed in the form of Equations (A.12) or (A.13), is defined as

$$P(\omega_n) = \frac{1}{2}\left(a_n^2 + b_n^2\right) = \frac{1}{2}c_n^2 \tag{A.20}$$

Note that this definition of power can be applied to any signal (it is not related to mechanical processes in which force times velocity = power). Power can be plotted as a function of frequency to obtain a power spectrum. The total power of the signal is the same whether it is computed in the time domain or the frequency domain:

$$\text{Total power} = \sum_{n=1}^{\infty} P(\omega_n) = \int_0^{T_f} [x(t)]^2 dt = \frac{1}{2}\int_0^{\omega_n} c_n^2 d\omega \tag{A.21}$$

Power spectra can be used to describe earthquake-induced ground motions.

A.4 TIME-FREQUENCY REPRESENTATION OF VIBRATORY MOTION

A Fourier amplitude spectrum illustrates how the amplitude of a ground motion is distributed with respect to frequency, just as a ground motion time series illustrates the variation of amplitude with time. Fourier-based methods, however, operate on the entire time series and therefore represent the average frequency content over the entire motion. Earthquake ground motions, however, are generally non-stationary, i.e., their amplitudes and frequency contents vary with time. For example, first-arriving p-waves are generally richer in high frequencies than s-waves that arrive after p-waves. Surface waves, which usually arrive somewhat after s-waves, tend to be richer in low frequency motions. In some cases, it is useful to be able to track the evolving frequency content, as well as amplitude, of a ground motion. Time-frequency analyses (Cohen, 1995; Smith, 1999) allow both to be quantified.

A number of techniques for time-frequency decomposition are available but three have most commonly been used to characterize earthquake ground motions. By multiplying a recorded motion by a window (in time) function that is non-zero for only a short-time interval, a Fourier transform can be computed on the windowed motion, and the frequency content of the motion within the window can be determined. Plotting its amplitude vs. both frequency and the time corresponding to the center of the window produces a *short-time Fourier transform* (STFT) spectrogram, which generally provides very good resolution of frequency, but relatively poor resolution of time. A *wavelet transform* (Mallat, 1989; Gurley and Kareem, 1999) convolves a recorded motion with a wavelet function that is localized in time and centered about a particular frequency. The wavelet can be shifted in time and scaled in frequency by dilating (stretching) or compressing its width. The wavelet transform allows development of a wavelet scalogram that expresses wavelet amplitude as a function of time and central frequency. The scalogram generally provides very good time resolution, but relatively poor frequency resolution. A more recent, alternative form of time-frequency analysis is provided by the *Stockwell transform* (Stockwell et al., 1996). The Stockwell transform uses a moving window of frequency-dependent width to achieve a balance between the resolutions of the STFT and the wavelet transforms – it provides better time resolution than the STFT spectrogram and better frequency resolution than the wavelet scalogram. Figure A.13 shows all three time-frequency representations for an artificial signal with instantaneous changes in frequency. The width of the transitions, in both time and frequency, indicates the resolution of the different transforms.

Stockwell spectra for ground motions recorded by instruments underlain by liquefiable soil can show dramatic changes in frequency content over time as the softening of the soil by generated pore pressures affects the frequencies that are amplified and de-amplified by the soil profile (Ozener et al., 2020). To better illustrate the changing frequency content of a ground motion, variations in amplitude can be removed by normalizing the spectral amplitudes by the peak spectral amplitude at each time step. In some cases, such as the Kawagishi-cho record shown in Figure A.14a, the change

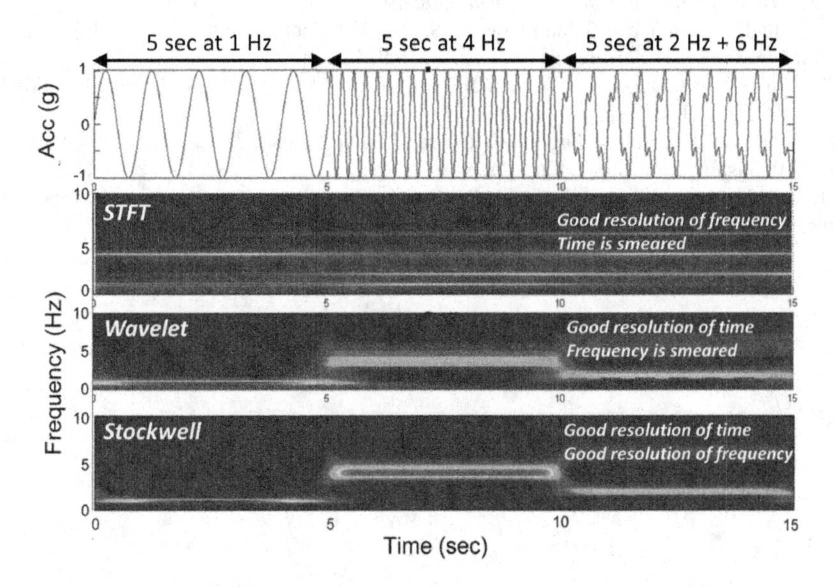

FIGURE A.13 Time-frequency representation of three-part function: (a) signal, (b) STFT spectrogram, (c) wavelet scalogram, and (d) Stockwell spectrum.

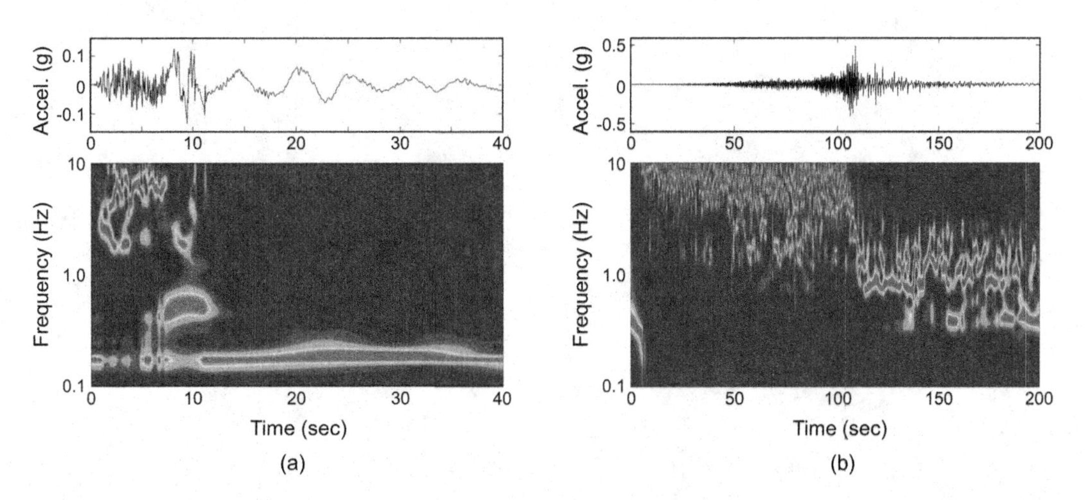

FIGURE A.14 Normalized Stockwell spectrograms for (a) Kawagishi-cho and (b) IBR014 records. Spectral values are normalized by peak Stockwell amplitude at each time step.

in frequency content is so drastic that it can be readily seen in the acceleration history. In other cases, such as the IBR014 record shown in Figure A.14b, the changes in frequency content are more subtle but readily distinguished using the Stockwell spectrum.

REFERENCES

Brigham, E.O. (1974). *The Fast Fourier Transform*, Prentice Hall, Englewood Cliffs, NJ, 252 pp.

Cohen, L. (1995). *Time-Frequency Analysis*, Prentice Hall, Englewood Cliffs, NJ, 299 pp.

Cooley, P.M. and Tukey, J.W. (1965). "An algorithm for the machine computation of complex Fourier series," *Mathematics of Computation*, Vol. 19, No. 4, pp. 297–301.

Gurley, K. and Kareem, A. (1999). "Applications of wavelet transforms in earthquake, wind and ocean engineering," *Engineering Structures*, Vol. 21, pp. 149–167.

Mallat, S. (1989). "A theory for multiresolution signal decomposition: the wavelet representation," *IEEE Transactions on Pattern Analysis and Machine Intelligence*, Vol. 11, pp. 674–693.

Ozener, P., Greenfield, M., Sideras, S., and Kramer, S. (2020). "Identification of time of liquefaction triggering," *Soil Dynamics and Earthquake Engineering*, Vol. 128, pp. 1–15.

Richart, F.E., Hall, J.R., and Woods, R.D. (1970). *Vibrations of Soils and Foundations*, Prentice-Hall, Inc., Englewood Cliffs, NJ, 401 pp.

Smith, S.W. (1999). *The Scientist & Engineer's Guide to Digital Signal Processing*, 2nd edition, California Technical Publishing, San Diego, CA, 628 pp.

Stockwell, R.G., Mansinha, L., and Lowe, R.P. (1996). "Localization of the complex spectrum: the S transform," *IEEE Transactions on Signal Processing*, Vol. 44, No. 4, pp. 998–1001.

Appendix B
Dynamics of Discrete Systems

B.1 INTRODUCTION

Many vibrating systems consist of discrete elements such as masses and springs, or can at least be idealized as such. For most practical problems of structural dynamics, the structure is idealized as a system of rigid masses connected by massless springs. Even continuous systems such as soil deposits have been idealized as assemblages of many discrete elements. Since the geotechnical earthquake engineer often provides input to the structural engineer, a firm understanding of the dynamic response of discrete systems is required. Also, many of the concepts and terminologies used in geotechnical earthquake engineering analyses are analogous to those of discrete system dynamics and are more easily introduced in that framework.

This appendix introduces the dynamics of discrete systems. It begins with very simple systems and then adds complicating factors such as damping, base motion, and nonlinearity. Analytical and numerical solutions in the time domain and frequency domain are presented. Finally, the response of multiple-degree-of-freedom systems is introduced. While many of the basic concepts of structural dynamics are presented, much more complete treatments may be found in a number of structural dynamics texts (e.g., Berg, 1989; Clough and Penzien, 1993; Paz and Kim, 2019; Chopra, 2022).

B.2 VIBRATING SYSTEMS

Vibrating systems can be divided into two broad categories: rigid systems and compliant systems. A rigid system is one in which no strains occur. All points within a rigid system move in phase with each other, and the description of rigid-body motion is a relatively simple matter of kinematics. In compliant systems, however, different points within the system may move differently (and out of phase) from each other. A given physical system may behave very nearly as a rigid system under certain conditions and as a compliant system under other conditions. Since neither soils nor structures are rigid, the dynamic response of compliant systems is central to the study of soil and structural dynamics and to earthquake engineering.

Compliant systems can be characterized by the distribution of their mass. Discrete systems are those whose mass can be considered to be concentrated at a finite number of locations, where the mass of a continuous system is distributed throughout the system. The number of independent variables required to describe the position of all the significant masses of a system is the number of dynamic degrees of freedom (DOF) of the system. Systems of interest in earthquake engineering may have anywhere from 1 to an infinite number of DOF. Figure B.1 illustrates several commonly encountered systems with varying numbers of DOF. Parts (a)–(c) in Figure B.1 depict discrete systems with finite numbers of DOF. Parts (d)–(e) depict continuous systems with infinite DOF. Certain types of analyses idealize continuous systems as discrete systems with large numbers of DOF, and other types represent discrete systems with many DOF as continuous systems.

B.3 SINGLE-DEGREE-OF-FREEDOM SYSTEMS

A discrete system whose position can be described completely by a single variable is known as a single-degree-of-freedom (SDOF) system. That single degree of freedom may represent translational

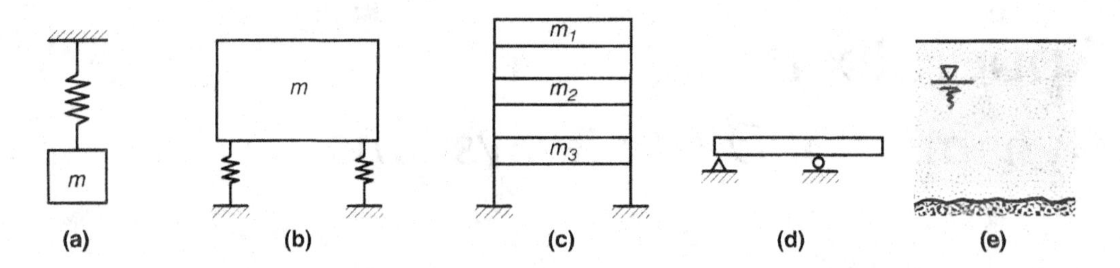

FIGURE B.1 Vibrating systems with various numbers of degrees of freedom: (a) one degree of freedom, vertical translation; (b) two DOF, vertical translation and rocking; (c) three DOF, horizontal translation; (d) infinite DOF; (e) infinite DOF.

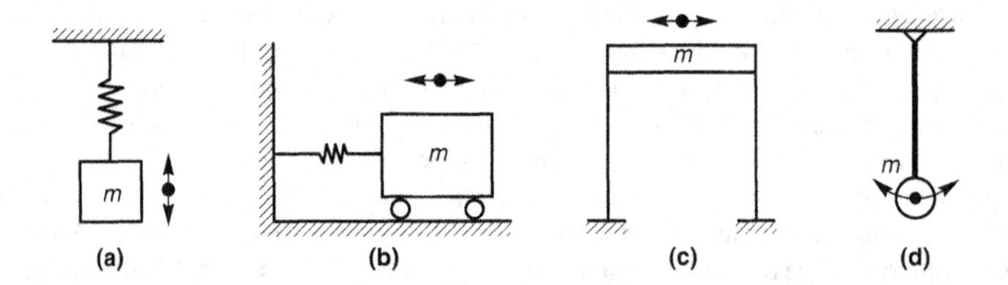

FIGURE B.2 Various SDOF systems. The degrees of freedom are (a) vertical translation, (b) and (c) horizontal translation, and (d) rotation.

displacement, as in the SDOF systems of Figure B.2a–c, or rotational displacement, as in the case of the pendulum of Figure B.2d.

A typical SDOF system is one in which a rigid mass, m, is connected in parallel to a spring of stiffness, k, and a dashpot of viscous damping coefficient, c, and subjected to some external load, $Q(t)$, as shown in Figure B.3. The spring and dashpot are assumed to be massless and the displacement origin to coincide with the static equilibrium position.

B.4 EQUATION OF MOTION FOR SDOF SYSTEM

Many SDOF systems are acted upon by externally applied loads. In earthquake engineering, dynamic loading often results from movement of the supports of the system. The dynamic response of an SDOF system such as that shown in Figure B.3 is governed by an equation of motion. The equation of motion can be derived in a number of ways; a simple, force equilibrium approach will be used here.

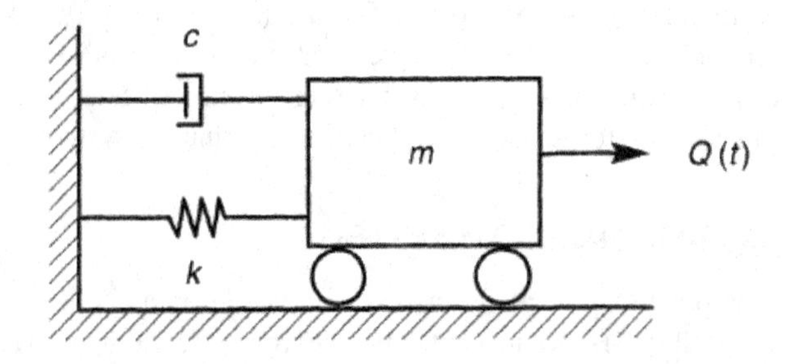

FIGURE B.3 Damped SDOF system subjected to external dynamic load, $Q(t)$.

B.4.1 Equation of Motion: External Loading

When a dynamic load is applied to the mass of an SDOF system (Figure B.3), the tendency for motion is resisted by the inertia of the mass and by forces that develop in the dashpot and spring. Thus the external load, $Q(t)$, acting in the positive x-direction is opposed by three forces (Figure B.4) that act in the negative x-direction: the *inertial force*, f_I, the viscous damping force, f_D, and the elastic spring force, f_S. The equation of motion can be expressed in terms of the dynamic equilibrium of these forces:

$$f_I(t) + f_D(t) + f_S(t) = Q(t) \tag{B.1}$$

These forces can also be expressed in terms of the motion of the mass. Newton's second law states that the inertial force acting on a mass is equal to its rate of change of momentum, which for a system of constant mass produces

$$f_I(t) = \frac{d}{dt}\left(m\frac{du(t)}{dt}\right) = m\frac{d^2u(t)}{dt^2} = m\ddot{u}(t) \tag{B.2a}$$

For a viscous dashpot, the damping force is proportional to the velocity of the mass:

$$f_D(t) = c\frac{du(t)}{dt} = c\dot{u}(t) \tag{B.2b}$$

and the force provided by the spring is simply the product of its stiffness and the amount by which it is displaced

$$f_S(t) = ku(t) \tag{B.2c}$$

The behavior of these forces is illustrated graphically in Figure B.5. The inertial force is proportional to the acceleration and the constant of proportionality is the mass. Similarly, the viscous damping force and the elastic spring force are proportional to the velocity and displacement with the damping and spring coefficients serving as the respective constants of proportionality.

Substituting Equations (B.2) into Equation (B.1), the equation of motion for the SDOF system can be written as

$$m\ddot{u}(t) + c\dot{u}(t) + ku(t) = Q(t) \tag{B.3}$$

This second-order differential equation is commonly used to describe the behavior of oscillating systems ranging from the mechanical systems considered in earthquake engineering problems to electrical circuits. The differential equation of motion is linear (i.e., all of its terms have constant coefficients). This linearity allows a closed-form analytical solution to be readily obtained and, importantly, it allows the principle of superposition to be used. When any of the coefficients are not

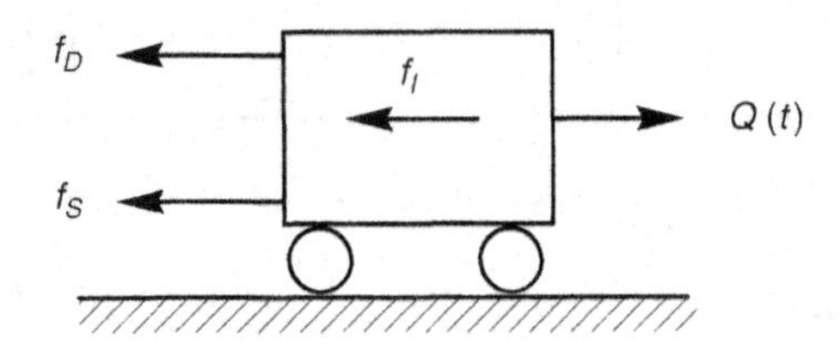

FIGURE B.4 Dynamic forces acting on mass from Figure B.3.

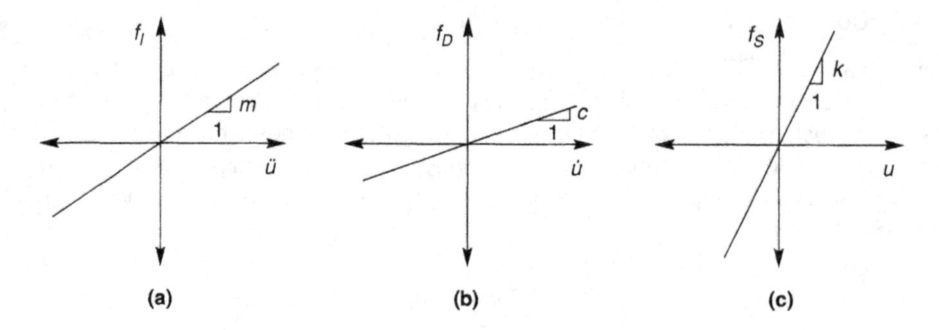

FIGURE B.5 Variation of (a) inertial, (b) viscous, and (c) elastic forces with acceleration, velocity, and displacement, respectively.

constant, the behavior is not linear and the solution becomes considerably more difficult. In most cases, the response of nonlinear systems must be evaluated numerically (Section B.9).

B.4.2 Equation of Motion: Vibration of Supports (Base Shaking)

For earthquake engineering problems, dynamic loading often results from vibration of the supports of a system rather than from dynamic external loads. To evaluate the response of such systems, it is necessary to develop an equation of motion for loading caused by base shaking. Consider the damped SDOF system shown in Figure B.6a. When subjected to dynamic base shaking, $u_b(t)$, it will deform into a configuration that might look like that shown in Figure B.6b at a particular time, t. The total displacement of the mass, $u_t(t)$, can be broken down as the sum of the base displacement, $u_b(t)$, and the displacement of the mass relative to the base, $u(t)$. The inertial force will depend on the total acceleration of the mass, while the viscous damping and elastic spring forces will depend on the relative velocity and displacement, respectively. Using the notation shown in Figure B.6b, the equation of motion can be written as

$$m\ddot{u}_t + c\dot{u} + ku = 0 \tag{B.4}$$

or substituting $\ddot{u}_t(t) = \ddot{u}_b(t) + \ddot{u}(t)$ and rearranging,

$$m\ddot{u} + c\dot{u} + ku = -m\ddot{u}_b \tag{B.5}$$

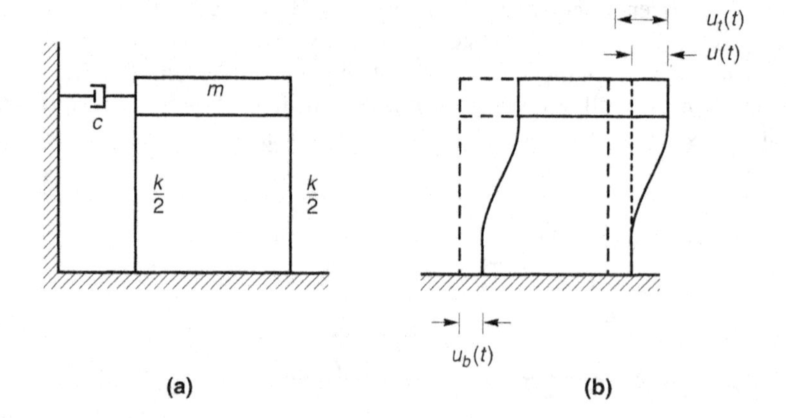

FIGURE B.6 Damped SDOF system subjected to base shaking.

In other words, the response of the system to base shaking is equivalent to the response that the system would have if its base was fixed and the mass was subjected to an external load $Q(t) = -m\ddot{u}_b(t)$. Thus any solutions for the response of an SDOF system subjected to external load can be used to evaluate the response of the system to base shaking.

B.5 RESPONSE OF LINEAR SDOF SYSTEMS

To evaluate the dynamic response of a linear SDOF system, the differential equation of motion must be solved. There are several types of conditions under which the dynamic response of SDOF systems are commonly calculated. Forced vibration occurs when the mass is subjected to some external loading, $Q(t)$. The loading may be periodic or nonperiodic and it may correspond to an actual physical force applied to the mass or to some known level of base shaking. Free vibration occurs in the absence of external loading or base shaking. It may result from the release of the mass from some initial displacement or may occur after some transient forced vibration has ended. The following sections will develop solutions to the equation of motion for cases in which damping is and is not present, and for cases in which external loading is and is not present. The resulting four permutations of these conditions are:

1. Undamped free vibrations: $c=0,\quad Q(t)=0$
2. Damped free vibrations: $c>0,\quad Q(t)=0$
3. Undamped forced vibrations: $c=0,\quad Q(t)\neq 0$
4. Damped forced vibrations: $c>0,\quad Q(t)\neq 0$

The solution of the equation of motion for each of these conditions will be presented in turn.

B.5.1 UNDAMPED FREE VIBRATIONS

An SDOF system undergoes free vibration when it oscillates without being acted upon by any external loads. When damping is not present ($c=0$) the equation of motion (for undamped free vibration) reduces to

$$m\ddot{u} + ku = 0 \tag{B.6}$$

or after dividing both sides by the mass,

$$\ddot{u} + \frac{k}{m}u = 0 \tag{B.7}$$

The solution to this simple differential equation can be found in any elementary text on differential equations as

$$u = C_1 \sin\sqrt{\frac{k}{m}}t + C_2 \cos\sqrt{\frac{k}{m}}t \tag{B.8}$$

where the values of the constants C_1 and C_2 depend on the initial conditions of the system. The quantity $\sqrt{k/m}$ is very important – it represents the *undamped natural circular frequency* of the system

$$\omega_0 = \sqrt{\frac{k}{m}} \tag{B.9}$$

The undamped natural frequency, f_0, and undamped natural period of vibration, T_0, can be written as

$$f_0 = \frac{\omega_0}{2\pi} = \frac{1}{2\pi}\sqrt{\frac{k}{m}} \tag{B.10}$$

$$T_0 = \frac{2\pi}{\omega_0} = 2\pi\sqrt{\frac{m}{k}} \tag{B.11}$$

The quantity ω_0 is usually expressed in radians/sec and f_0 is expressed in cycles/sec, or Hz (Hertz). Substituting Equation (B.9) into the solution for the equation of motion [Equation (B.8)] yields

$$u = C_1 \sin \omega_0 t + C_2 \cos \omega_0 t \tag{B.12}$$

which indicates that an undamped system in free vibration will oscillate harmonically at its undamped natural frequency. C_1 and C_2 can be evaluated by assuming the initial (t=0) conditions to be represented by an initial displacement, u_0, and initial velocity, $\dot{u}_0$. Then

$$u_0 = C_1 \sin(0) + C_2 \cos(0) = C_2$$

$$\dot{u}_0 = \omega_0 C_1 \cos(0) - \omega_0 C_2 \sin(0) = \omega_0 C_1$$

Therefore, $C_1 = \dot{u}_0/\omega_0$ and $C_2 = u_0$, so the complete solution to the undamped free vibration response of an SDOF system is given by

$$u = \frac{\dot{u}_0}{\omega_0} \sin \omega_0 t + u_0 \cos \omega_0 t \tag{B.13}$$

The response of such a system is shown in Figure B.7.

Referring back to Equation (A.1), the free vibration response can also be expressed as

$$u = A \sin(\omega_0 t + \phi) \tag{B.14}$$

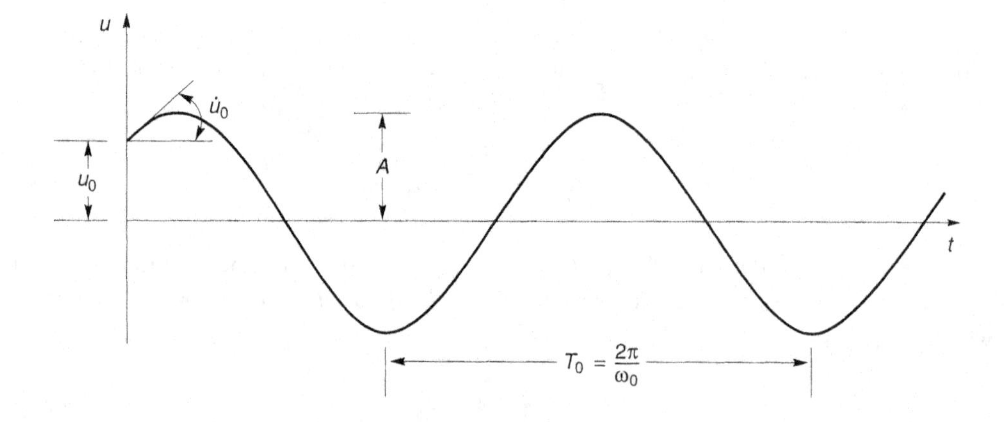

FIGURE B.7 Time history of undamped free vibration with initial displacement, u_0, and initial velocity, $\dot{u}_0$.

where the amplitude, A, and phase angle, ϕ, are given by

$$A = \sqrt{u_0^2 + \left(\frac{\dot{u}_0}{\omega_0}\right)^2}$$

$$\phi = \tan^{-1}\frac{u_0\omega_0}{\dot{u}_0}$$

The solution to the equation of motion of an undamped system indicates that the response of the system depends on its initial displacement and velocity. Note that the amplitude remains constant with time. Because no energy is lost in an undamped system, it will continue to oscillate forever. Obviously, truly undamped systems do not exist in the real world; however, some systems can have such low damping that their response over short periods of time may approximate that of an undamped system.

B.5.2 Damped Free Vibrations

In real systems, energy may be lost as a result of friction, heat generation, air resistance, or other physical mechanisms. Hence the free vibration response of a damped SDOF system will diminish with time. For damped free vibrations, the equation of motion is written as

$$m\ddot{u} + c\dot{u} + ku = 0 \tag{B.15}$$

or, dividing by m and substituting [from Equation (B.9)] $k = m\omega_0^2$, we have

$$\ddot{u} + 2\frac{c}{2\sqrt{km}}\omega_0\dot{u} + \omega_0^2 u = 0 \tag{B.16}$$

The quantity $2\sqrt{km}$, called the *critical damping coefficient*, c_c, allows the damping ratio, ξ, to be defined as the ratio of the damping coefficient to the critical damping coefficient, i.e.,

$$\xi = \frac{c}{c_c} = \frac{c}{2\sqrt{km}} = \frac{c}{2m\omega_0} = \frac{c\omega_0}{2k} \tag{B.17}$$

With this notation, the equation of motion can be expressed as

$$\ddot{u} + 2\xi\omega_0\dot{u} + \omega_0^2 u = 0 \tag{B.18}$$

The solution of this differential equation of motion depends on the value of the damping ratio. When $\xi < 100\%$ ($c < c_c$), the system is said to be *underdamped.* When $\xi = 100\%$ ($c = c_c$), the system is *critically damped*, and when $\xi > 100\%$ ($c > c_c$) the system is *overdamped.* Separate solutions must be obtained for each of the three cases, but structures of interest in earthquake engineering are virtually always underdamped.

For the case in which damping is less than critical, the solution to the equation of motion is of the form

$$u = e^{-\xi\omega_0 t}\left[C_1 \sin\left(\omega_0\sqrt{1-\xi^2}t\right) + C_2 \cos\left(\omega_0\sqrt{1-\xi^2}t\right)\right] \tag{B.19}$$

Note the exponential term by which the term in brackets is multiplied. This exponential term gets smaller with time and eventually approaches zero, indicating that the response of an underdamped system in free vibration decays exponentially with time. The rate of decay depends on the damping ratio — for small ξ the response decays slowly and for larger ξ the response decays more quickly. Defining the damped natural circular frequency of the system as $\omega_d = \omega_0\sqrt{1-\xi^2}$ the solution can be expressed as

$$u = e^{-\xi\omega_0 t}\left(C_1 \sin\omega_d t + C_2 \cos\omega_d t\right) \tag{B.20}$$

The natural frequency of a damped system is always lower than that of an undamped system, and it decreases with increasing damping ratio.

The coefficients C_1 and C_2 can be determined from the initial conditions in the same manner as for the undamped case. The initial displacement and velocity are

$$u_0 = e^{-\xi\omega_0(0)}\left[C_1 \sin(0) + C_2 \cos(0)\right] = C_2$$

$$\begin{aligned}\dot{u}_0 &= e^{-\xi\omega_0(0)}\left[\omega_d C_1 \cos\omega_d(0) - \omega_d C_2 \sin\omega_d(0)\right] - \xi\omega_0 e^{-\xi\omega_0(0)}\left[C_1 \sin\omega_d(0) + C_2 \cos\omega_d(0)\right]\\ &= \omega_d C_1 - \xi\omega_0 C_2\end{aligned}$$

Therefore, $C_1 = \left(\dot{u}_0 + \xi\omega_0 u_0\right)/\omega_d$ and $C_2 = u_0$, so the solution for damped free vibrations can be expressed as

$$u = e^{-\xi\omega_0 t}\left(\frac{\dot{u}_0 + \xi\omega_0 u_0}{\omega_d}\sin\omega_d t + u_0 \cos\omega_d t\right) \tag{B.21}$$

The free vibration response of an underdamped system is shown in Figure B.8. Note the exponential decay of displacement amplitude with time. The ratio of the amplitudes of any two successive peaks will be

$$\frac{u_n}{u_{n+1}} = \exp\left(2\pi\xi\frac{\omega_0}{\omega_d}\right) \tag{B.22}$$

Defining the logarithmic decrement as $\delta = \ln\,(u_n/u_n + 1)$; then

$$\delta = 2\pi\xi\frac{\omega_0}{\omega_d} = \frac{2\pi\xi}{\sqrt{1-\xi^2}} \tag{B.23}$$

Rearranging allows the damping ratio to be determined from the logarithmic decrement

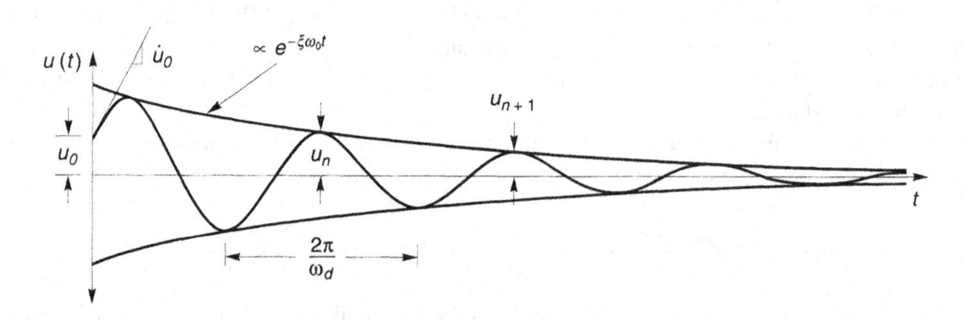

FIGURE B.8 Time history of damped free vibration with initial displacement, u_0, and initial velocity, $\dot{u}_0$.

$$\xi = \frac{\delta}{\sqrt{4\pi^2 + \delta^2}} \tag{B.24}$$

For small values of δ, $\xi \approx \delta / 2\pi$. Therefore, a simple way to estimate the damping ratio of an SDOF system is to perform a free vibration test, in which the logarithmic decrement is measured when a system is displaced by some initial displacement, u_0, and released with initial velocity $\dot{u}_0 = 0$.

B.5.3 Response of SDOF Systems to Harmonic Loading

An SDOF system is said to undergo forced vibration when acted upon by some external dynamic force, $Q(t)$. Dynamic loading may come from many different sources and may be periodic or non-periodic. For problems of soil and structural dynamics, the response to harmonic loading is very important. One form of simple harmonic loading can be expressed as $Q(t) = Q_0 \sin \bar{\omega} t$, where Q_0 is the amplitude of the harmonic load and $\bar{\omega}$ is the circular frequency at which the load is applied.

B.5.3.1 Undamped Forced Vibrations

The equation of motion for an undamped system subjected to such simple harmonic loading is

$$m\ddot{u} + ku = Q_0 \sin \bar{\omega} t \tag{B.25}$$

The general solution to this equation of motion is given by the sum of the complementary solution (for the homogeneous case in which the right side of the equation is zero) and the particular solution [which must satisfy the right side of Equation (B.25)]. The homogeneous equation is

$$m\ddot{u} + ku = 0$$

so the complementary solution is simply the solution to the undamped free vibration problem

$$u_c(t) = C_1 \sin \omega_0 t + C_2 \cos \omega_0 t \tag{B.26}$$

The portion of the response described by the complementary solution is that which results from the initial conditions of the system. It consists of a simple harmonic oscillation at the undamped natural frequency of the system.

The particular solution describes the portion of the response caused by the external loading. This portion of the response can be assumed to be of the same form and to be in phase with the harmonic loading (because of the condition of zero damping); thus

$$u_p(t) = U_0 \sin \bar{\omega} t \tag{B.27}$$

where U_0 is the amplitude of the harmonic response. Substituting Equation (B.27) into Equation (B.25) yields

$$-m\bar{\omega}^2 U_0 \sin \bar{\omega} t + kU_0 \sin \bar{\omega} t = Q_0 \sin \bar{\omega} t \tag{B.28}$$

Substituting $k/m = \omega_0^2$ and rearranging gives

$$U_0 = \frac{Q_0 / k}{1 - \bar{\omega}^2 / \omega_0^2} = \frac{Q_0 / k}{1 - \beta^2} \tag{B.29}$$

where $\beta = \bar{\omega} / \omega_0$ is referred to as the tuning ratio. Now the general solution of the equation of motion can be obtained by combining the complementary and particular solutions:

$$u(t) = u_c(t) + u_p(t) = C_1 \sin \omega_0 t + C_2 \cos \omega_0 t + \frac{Q_0 / k}{1 - \beta^2} \sin \bar{\omega} t \tag{B.30}$$

The general solution must satisfy the initial conditions. From Equation (B.30), the velocity can be written as

$$\dot{u}(t) = \frac{du}{dt} = \omega_0 C_1 \cos \omega_0 t - \omega_0 C_2 \sin \omega_0 t + \bar{\omega} \frac{Q_0 / k}{1 - \beta^2} \cos \bar{\omega} t \tag{B.31}$$

For a given initial displacement, u_0, and initial velocity, $\dot{u}_0$,

$$u_0 = C_1 \sin \omega_0 (0) + C_2 \cos \omega_0 (0) + \frac{Q_0 / k}{1 - \beta^2} \sin \bar{\omega}(0) = C_2 \tag{B.32}$$

and

$$\dot{u}_0 = \omega_0 C_1 \cos \omega_0 (0) - \omega_0 C_2 \sin \omega_0 (0) + \bar{\omega} \frac{Q_0 / k}{1 - \beta^2} \cos \bar{\omega}(0) = \omega_0 C_1 + \bar{\omega} \frac{Q_0 / k}{1 - \beta^2} \tag{B.33}$$

from which

$$C_1 = \frac{\dot{u}_0 - \bar{\omega}\left[(Q_0 / k) / \left(1 - \beta^2\right)\right]}{\omega_0} = \frac{\dot{u}_0}{\omega_0} - \frac{Q_0 \beta}{k\left(1 - \beta^2\right)} \tag{B.34}$$

Now the general response can finally be written as

$$u = \left[\frac{\dot{u}_0}{\omega_0} - \frac{Q_0 \beta}{k\left(1 - \beta^2\right)}\right] \sin \omega_0 t + u_0 \cos \omega_0 t + \frac{Q_0 / k}{1 - \beta^2} \sin \bar{\omega} t \tag{B.35}$$

It is interesting to consider the case in which the system is initially at rest in its equilibrium position, (i.e., $u_0 = \dot{u}_0 = 0$). For this case, the response is given by

$$u = \frac{Q_0}{k} \cdot \frac{1}{1 - \beta^2} \left(\sin \bar{\omega} t - \beta \sin \omega_0 t\right) \tag{B.36}$$

which indicates that the response has two components. One component occurs in response to the applied loading and occurs at the frequency of the applied loading. The other is a free vibration effect induced by the initial conditions; it occurs at the natural frequency of the system. It is useful to realize that the term Q_0/k in Equation (B.36) represents the displacement of the mass that would occur if the load Q_0 was applied statically. The term $1/(1-\beta^2)$ can then be thought of as a magnification factor that describes the amount by which the static displacement amplitude is modified by the harmonic load. The magnification factor varies with the tuning ratio, β, as shown in Figure B.9. Note that the displacement amplitude is greater than the static displacement for loading frequencies lower than $\sqrt{2}\omega_0$. At higher loading frequencies, the displacement amplitude is less than the static displacement and can become very small at high frequencies. However, the response of an undamped SDOF system becomes very large as $\bar{\omega}$ approaches ω_0. When harmonic loading is applied at the natural frequency of an undamped SDOF system, the response goes to infinity indicating resonance of the system. However, since truly undamped systems do not exist, true resonance is never

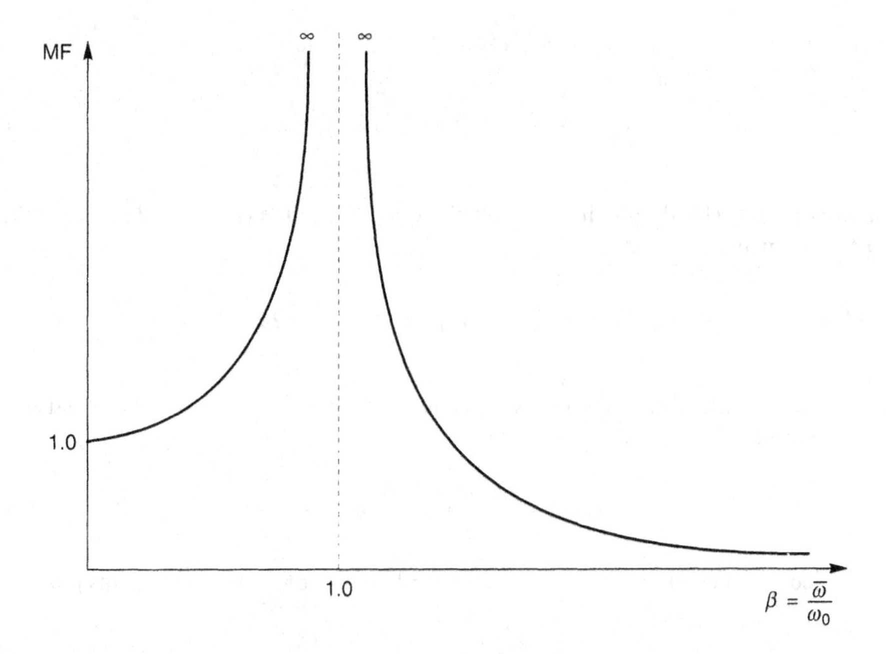

FIGURE B.9 Variation of magnification factor with tuning ratio for undamped SDOF system.

really achieved. The concept of the tuning ratio that relates the frequency of loading to the natural frequency of the system is an important one, as evidenced by its strong influence on the response.

B.5.3.2 Damped Forced Vibrations

The most general case is that of a damped system subjected to forced harmonic loading. Each of the three prior cases can be considered as a subset of this one since their equations of motion can be obtained by setting various terms of the equation of motion for damped forced vibrations shown below to zero. The equation of motion for a damped SDOF system subjected to simple harmonic loading of the form $Q(t) = Q_0 \sin \bar{\omega} t$ is

$$m\ddot{u} + c\dot{u} + ku = Q_0 \sin \bar{\omega} t \tag{B.37}$$

After dividing by m and using the relationships $\xi = c / 2m\omega_0$ and $\omega_0^2 = k / m$, Equation (B.37) can be rewritten as

$$\ddot{u} + 2\xi\omega_0\dot{u} + \omega_0^2 u = \frac{Q_0}{m} \sin \bar{\omega} t \tag{B.38}$$

The complementary solution represents the damped free vibration response, which was expressed for an underdamped system by Equation (B.20).

$$u_c(t) = e^{-\xi\omega_0 t}\left(C_1 \sin \omega_d t + C_2 \cos \omega_d t\right) \tag{B.39}$$

Since the response of a damped SDOF system is generally out of phase with the external loading, a harmonic particular solution of the form

$$u_p(t) = C_3 \sin \bar{\omega} t + C_4 \cos \bar{\omega} t \tag{B.40a}$$

can be assumed. The corresponding velocity and acceleration are

$$\dot{u}_p(t) = C_3\bar{\omega}\cos\bar{\omega}t - C_4\bar{\omega}\sin\bar{\omega}t \tag{B.40b}$$

$$\ddot{u}_p(t) = -\bar{\omega}^2 C_3\sin\bar{\omega}t - \bar{\omega}^2 C_4\cos\bar{\omega}t \tag{B.40c}$$

Substituting Equations (B.40) into the equation of motion [Equation (B.38)] and grouping the $\sin\bar{\omega}t$ and $\cos\bar{\omega}t$ terms gives

$$\left(C_3\omega_0^2 - C_3\bar{\omega}^2 - 2\xi\omega_0 C_4\bar{\omega}\right)\sin\bar{\omega}t + \left(C_4\omega_0^2 - C_4\bar{\omega}^2 + C_3\bar{\omega}2\xi\omega_0\right)\cos\bar{\omega}t = \frac{Q_0}{m}\sin\bar{\omega}t \tag{B.41}$$

Now, at the instances where $\bar{\omega}t = 0 + n\pi$ (where n is any positive integer), $\sin\bar{\omega}t = 0$ and $\cos\bar{\omega}t = 1$. Thus the relationship

$$C_4\omega_0^2 - C_4\bar{\omega}^2 + C_3\bar{\omega}2\xi\omega_0 = 0 \tag{B.42a}$$

must be satisfied. Further, at $\bar{\omega}t = \pi/2 + n\pi$, $\sin\bar{\omega}t = 1$ and $\cos\bar{\omega}t = 0$, which means that

$$C_3\omega_0^2 - C_3\bar{\omega}^2 - 2\xi\omega_0 C_4\bar{\omega} = \frac{Q_0}{m} \tag{B.42b}$$

Equations (B.42) represent two simultaneous equations with the two unknowns C_3 and C_4. Solving for the unknowns yields

$$C_3 = \frac{Q_0}{k}\frac{1-\beta^2}{\left(1-\beta^2\right)^2 + (2\xi\beta)^2} \tag{B.43a}$$

$$C_4 = \frac{Q_0}{k}\frac{-2\xi\beta}{\left(1-\beta^2\right)^2 + (2\xi\beta)^2} \tag{B.43b}$$

The general solution to the equation of motion for damped forced vibration can now be obtained by combining the complementary and particular solutions

$$u(t) = e^{-\xi\omega_0 t}\left(C_1\sin\omega_d t + C_2\cos\omega_d t\right) + \frac{Q_0}{k}\frac{1}{\left(1-\beta^2\right)^2 + (2\xi\beta)^2}\left[\left(1-\beta^2\right)\sin\bar{\omega}t - 2\xi\beta\cos\bar{\omega}t\right] \tag{B.44}$$

where the constants C_1 and C_2 depend on the initial conditions. There are several important characteristics of this solution. Note that the complementary solution (which represents the effects of the initial conditions) decays with time. The complementary solution therefore describes a transient response caused by the requirement of satisfying the initial conditions. After the transient response dies out, only the steady-state response described by the particular solution remains. The steady-state response occurs at the frequency of the applied harmonic loading but is out of phase with the loading.

The steady-state response could also be described by

$$u(t) = A\sin(\bar{\omega}t + \phi) \tag{B.45}$$

where

$$A = \frac{Q_0}{k}\frac{1}{\sqrt{\left(1-\beta^2\right)^2 + (2\xi\beta)^2}}$$

$$\phi = \tan^{-1}\left(-\frac{2\xi\beta}{1-\beta^2}\right)$$

The steady-state response can be visualized with the aid of rotating vectors, both for the deformation responses and for the forces induced in the system, as shown in Figure B.10. Note that the spring, dashpot, and inertial forces act opposite to the displacement, velocity and acceleration vectors, and that the displacement lags the applied loading vector by the negative phase angle, ϕ. For harmonic loading the phase angle varies with both damping ratio and tuning ratio, as shown in Figure B.11b.

The influence of the tuning ratio can be illustrated by the use of the magnification factor, again defined as the ratio of the amplitude to the static displacement:

$$M = \frac{A}{Q_0 / k} = \frac{1}{\sqrt{\left(1-\beta^2\right)^2 + (2\xi\beta)^2}} \tag{B.46}$$

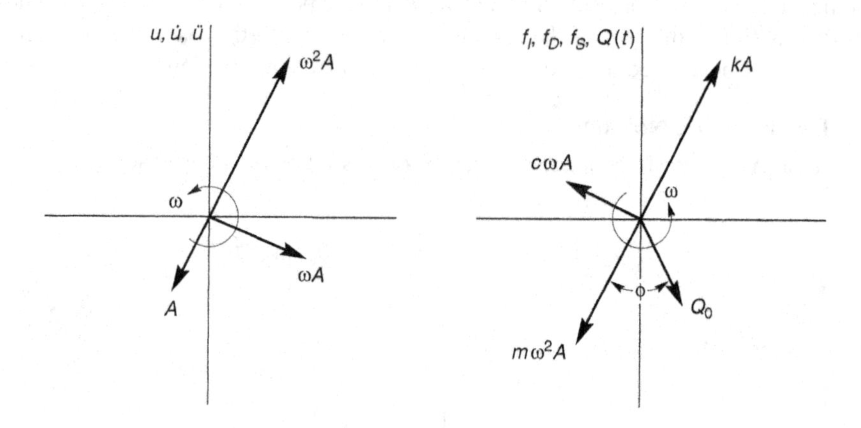

FIGURE B.10 Rotating vector representations of deformation response and forces in vibrating SDOF system.

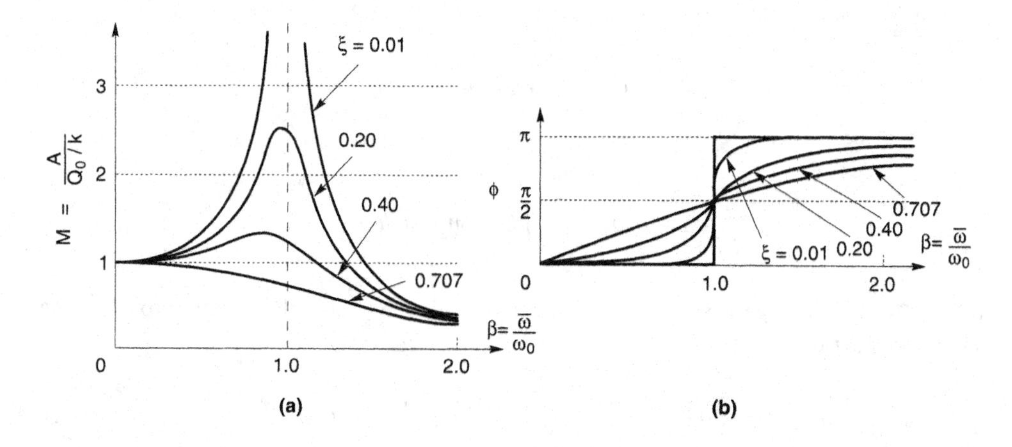

FIGURE B.11 Variation of (a) magnification factor, and (b) phase angle with damping ratio and tuning ratio.

The variation of the magnification factor with tuning ratio and damping ratio is shown in Figure B.11a. The damping ratio influences the peak magnification factor and also the variation of magnification factor with frequency. The magnification factor curves broaden with increasing damping ratio. Note that the magnification is unbounded (resonance) only for $\xi=0$ and $\beta=1$. For nonzero damping, there is some maximum magnification, M_{max},

$$M_{max} = \frac{1}{2\xi\sqrt{1-\xi^2}} \tag{B.47}$$

which occurs when the tuning ratio $\beta = \sqrt{1-2\xi^2}$. The shape of the magnification curve is obviously controlled by the damping ratio. Although a system with low damping may produce large magnification at a tuning ratio near 1, it will exhibit significant magnification over a smaller range of frequencies than a system with higher damping.

B.5.4 Response of SDOF Systems to Periodic Loading

The solutions for the response of an SDOF system to harmonic loading developed in the preceding section can be used to develop solutions for the more general case of periodic loading. As shown in Appendix A, periodic loading can be approximated by a Fourier series (i.e., as the sum of a series of harmonic loads). The response of an SDOF system to the periodic loading, using the principle of superposition, is simply the sum of the responses to each term in the loading series. The required calculations can be performed using trigonometric or exponential notation.

B.5.4.1 Trigonometric Notation

From Equation (A.12) a periodic load, $Q(t)$, can be expressed by the Fourier series

$$Q(t) = a_0 + \sum_{n=1}^{\infty} a_n \cos\omega_n t + b_n \sin\omega_n t \tag{B.48}$$

where the Fourier coefficients are

$$a_0 = \frac{1}{T_f}\int_0^{T_f} Q(t)\,dt$$

$$a_n = \frac{2}{T_f}\int_0^{T_f} Q(t)\cos\omega_n t\,dt$$

$$b_n = \frac{2}{T_f}\int_0^{T_f} Q(t)\sin\omega_n t\,dt$$

and $\omega_n = 2\pi n / T_f$. Using the steady-state portion of Equation (B.44), the response to each sine term in the Fourier series is

$$u_{n,\sin}(t) = \frac{b_n}{k}\frac{1}{\left(1-\beta_n^2\right)^2 + \left(2\xi\beta_n\right)^2}\left[\left(1-\beta_n^2\right)\sin\omega_n t - 2\xi\beta_n \cos\omega_n t\right] \tag{B.49a}$$

where $\beta_n = \omega_n T_f / 2\pi$. In the same way, the steady-state response to each cosine term can be shown to be

$$u_{n,\cos}(t) = \frac{a_n}{k} \frac{1}{\left(1-\beta_n^2\right)^2 + \left(2\xi\beta_n\right)^2} \left[\left(1-\beta_n^2\right)\cos\omega_n t + 2\xi\beta_n \sin\omega_n t\right] \quad \text{(B.49b)}$$

Since the steady-state response to the constant load term is the static displacement, $u_0 = a_0/k$, the total steady-state response is given by

$$u(t) = u_0 + \sum_{n=1}^{\infty} u_{n,\sin}(t) + u_{n,\cos}(t) \quad \text{(B.50)}$$

B.5.4.2 Exponential Notation

Periodic loading can also be described by the Fourier series in exponential form. Using Equation (A.16), a periodic load with a zero mean can be expressed as

$$Q(t) = \sum_{n=-\infty}^{\infty} \overset{*}{q}_n e^{i\omega_n t} \quad \text{(B.51)}$$

The complex Fourier coefficients, $\overset{*}{q}_n$, can be determined directly from $Q(t)$ as

$$\overset{*}{q}_n = \frac{1}{T_f} \int_0^{T_f} Q(t) e^{-i\omega_n t}\, dt \quad \text{(B.52)}$$

The response of an SDOF system loaded by the nth harmonic would be governed by the equation of motion

$$m\ddot{u}_n(t) + c\dot{u}_n(t) + ku_n(t) = \overset{*}{q}_n e^{i\omega_n t} \quad \text{(B.53)}$$

The response of the system can be related to the loading by

$$u_n(t) = H(\omega_n)\overset{*}{q}_n e^{i\omega_s t} \quad \text{(B.54)}$$

where $H(\omega_n)$ is a transfer function [i.e., a function that relates one parameter (in this case, the displacement of the oscillator) to another (the external load)]. Substituting Equation (B.54) into the equation of motion gives

$$-m\omega_n^2 H(\omega_n)\overset{*}{q}_n e^{i\omega_n t} + ic\omega_n H(\omega_n)\overset{*}{q}_n e^{i\omega_n t} + kH(\omega_n)\overset{*}{q}_n e^{i\omega_s t} = \overset{*}{q}_n e^{i\omega_n t} \quad \text{(B.55)}$$

which can be rearranged to find the transfer function

$$H(\omega_n) = \frac{1}{-m\omega_n^2 + ic\omega_n + k} = \frac{1}{k(-\beta_n^2 + 2i\beta_n\xi + 1)} \quad \text{(B.56a)}$$

Since $A^* = a + ib = Ae^{i\theta}$, where the modulus, $A = \sqrt{a^2 + b^2}$, and the phase, $\phi = \tan^{-1}(b/a)$, the transfer function can also be written as

$$H(\omega_n) = \frac{1/k}{\sqrt{\left(1-\beta_n^2\right)^2 + \left(2\xi\beta_n\right)^2}} \exp\left(i\tan^{-1}\frac{2\xi\beta_n}{\beta_n^2 - 1}\right) \quad \text{(B.56b)}$$

Note the close relationship between the modulus of the transfer function and the magnification factor of Equation (B.46). Because the transfer function can be used for any frequency in the series, the principle of superposition gives the total response as

$$u(t) = \sum_{n=-\infty}^{\infty} H(\omega_n) q_n^* e^{i\omega_n t} \tag{B.57}$$

Many different transfer functions can be developed. For example, a transfer function relating the acceleration of the SDOF system to the external load could have been developed just as easily. The advantages of the transfer function approach lie in its simplicity and in the ease with which it allows computation of the response to complicated loading patterns.

The transfer function may be viewed as a filter that acts upon some input signal to produce an output signal. In the case just considered, the input signal was the loading history, $Q(t)$, and the output was the displacement, $u(t)$. If the input signal has Fourier amplitude and phase spectra, $F_i(\omega_n)$ and $\phi_i(\omega_n)$, the Fourier amplitude and phase spectra of the output signal will be given by

$$F_0(\omega_n) = A\left[H(\omega_n)\right] \cdot F_i(\omega_n) \tag{B.58a}$$

$$\phi_0(\omega_n) = \phi\left[H(\omega_n)\right] + \phi_i(\omega_n) \tag{B.58b}$$

where the amplitude change and phase shift produced by $H(\omega_n)$ are given by $A[H(\omega_n)]$ and $\phi[H(\omega_n)]$, respectively. Thus, the procedure for Fourier analysis of SDOF system response can be summarized in the following steps:

1. Obtain the Fourier series for the applied loading (or base motion). In doing so, the loading (or base motion) is expressed as a function of frequency rather than a function of time.
2. Multiply the Fourier series coefficients by the appropriate value of the transfer function at each frequency. This will produce the Fourier series of the output motion.
3. Express the output motion in the time domain by obtaining the inverse Fourier transform of the output motion.

It is precisely this approach that forms the backbone of several of the most commonly used methods for the analysis of ground response and soil-structure interaction. These methods are presented in Chapters 7 and 8.

B.5.5 Response of SDOF Systems to General Loading

Not all loading is harmonic or even periodic. To determine the response of SDOF systems to general loading conditions, a more general solution of the equation of motion is required.

B.5.5.1 Response to Step Loading

Consider a damped SDOF system subjected to a step load of intensity, Q_0, which is applied instantaneously at $t=0$ and removed instantaneously at $t=t_1$ as shown in Figure B.12. For $t \le t_1$, the complementary solution to the equation of motion for this system [Equation (B.39)],

$$u_c(t) = e^{-\xi\omega_0 t}\left[C_1 \sin \omega_d t + C_2 \cos \omega_d t\right]$$

describes the transient response of the system. The equation of motion for the steady-state condition is given by

$$m\ddot{u}_p + c\dot{u}_p + ku_p = Q_0 \tag{B.59}$$

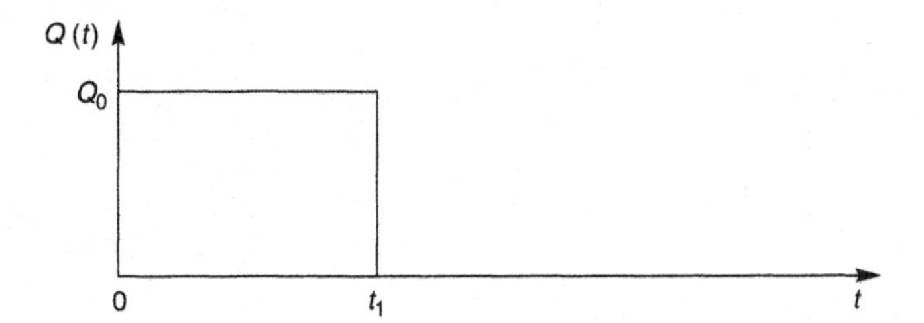

FIGURE B.12 Time history of step loading.

Since the applied load does not vary with time, the steady-state response will be a constant displacement,

$$u_p(t) = \frac{Q_0}{k} \tag{B.60}$$

The general solution to the step loading problem for $t \leq t_1$, can then be written as

$$u(t) = \frac{Q_0}{k} + e^{-\xi\omega_0 t}\left(C_1 \sin\omega_d t + C_2 \cos\omega_d t\right) \tag{B.61}$$

with free vibration occurring at $t > t_1$ (when no external load is applied). The constants are determined by the initial conditions, u_0 and $\dot{u}_0$. At $t=0$,

$$u_0 = \frac{Q_0}{k} + e^{-\xi\omega_0(0)}\left[C_1 \sin\omega_d(0) + C_2 \cos\omega_d(0)\right] = \frac{Q_0}{k} + C_2 \tag{B.62}$$

$$\dot{u}_0 = e^{-\xi\omega_0(0)}\left[\omega_d C_1 \cos\omega_d(0) - \omega_d C_2 \sin\omega_d(0)\right]$$

$$-\xi\omega_0 e^{-\xi\omega_0(0)}\left[C_1 \sin\omega_d(0) + C_2 \cos\omega_d(0)\right] = \omega_d C_1 - \xi\omega_0 C_2$$

from which it can be shown that $C_1 = \left[\dot{u}_0 + \xi\omega_0\left(u_0 - Q_0 / k\right)\right]/\omega_d$ and $C_2 = u_0 - \left(Q_0/k\right)$ so that

$$u(t) = \frac{Q_0}{k} + e^{-\xi\omega_0 t}\left[\frac{\dot{u}_0 + \xi\omega_0\left(u_0 - Q_0 / k\right)}{\omega_d}\sin\omega_d t + \left(u_0 - \frac{Q_0}{k}\right)\cos\omega_d t\right] \tag{B.63}$$

describes the response of the system up to the beginning of free vibration at $t=t_1$.

B.5.5.2 Dirac Pulse

A particular type of step loading can be described using a Dirac delta function. A Dirac delta function is one whose value is zero at all values of x except one at which it goes to infinity in such a way that the area under the function is unity. Mathematically, the Dirac delta function satisfies the conditions

$$\delta(x) = \begin{cases} 0 & \text{for } x \neq a \\ \infty & \text{for } x = a \end{cases} \tag{B.64a}$$

$$\int_{-\infty}^{\infty} \delta(x)\,dx = 1 \tag{B.64b}$$

FIGURE B.13 Dirac pulse loading.

Consider a Dirac pulse that consists of a constant force Q_0 applied over a duration t_1 that approaches zero as shown in Figure B.13. From impulse-momentum principles, $Q_0 t_1 = m\dot{u}_0(t_1)$. As t_1 approaches zero, the effect of the Dirac pulse is to cause an initial velocity, $\dot{u}_0 = Q_0 t_1 / m$, with no initial displacement. Thus, the steady-state response occurs only over an infinitesimal period of time, and the system is immediately set into free vibration. From Equation (B.21), if $t_1 = 0$ the response to a Dirac pulse disturbance at $t>0$ is given by

$$u(t) = e^{-\xi\omega_0 t}\left(\frac{Q_0 t_1}{m\omega_d}\sin(\omega_d t)\right) \tag{B.65}$$

B.5.5.3 Duhamel Integral

A general loading function such as that shown in Figure B.14 can be thought of as a series of load pulses, each of infinitesimal duration. Looking at one of these pulses, the pulse of duration $d\tau$ occurring at $t = \tau$ (Figure B.14), the response it causes at a later time, $t = \tilde{t}$, follows from Equation (B.66):

$$du(\tilde{t}) = e^{-\xi\omega_0(\tilde{t}-\tau)}\frac{Q_0(\tau)d\tau}{m\omega_d}\sin\left(\omega_d(\tilde{t}-\tau)\right) \tag{B.66}$$

The response induced by the entire train of load pulses can be obtained by summing the responses of all of the individual pulses up to the time $t = \tilde{t}$, i.e.,

$$u(\tilde{t}) = \frac{1}{m\omega_d}\sum_{i=1}^{n} Q(\tau_i)e^{-\xi\omega_0(\tilde{t}-\tau)}\sin\left(\omega_d\left(\tilde{t}-\tau_i\right)\right)d\tau \tag{B.67}$$

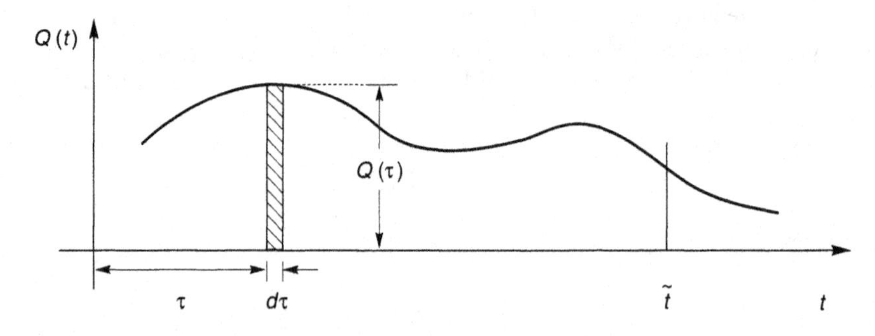

FIGURE B.14 Pulse of duration $d\tau$ occurring at $t=\tau$.

where n is the total number of pulses up to $t = \tilde{t}$. As $d\tau$ approaches zero, the summation becomes an integral with which the total response can be calculated as

$$u(\tilde{t}) = \frac{1}{m\omega_d} \int_0^{\tilde{t}} Q(\tau) e^{-\xi\omega_0(\tilde{t}-\tau)} \sin \omega_d(\tilde{t}-\tau)\, d\tau \tag{B.68}$$

This equation describing the response of a linear system is known as Duhamel's integral. It is usually very difficult to solve analytically, but can be integrated numerically by a variety of procedures. Its use, however, is constrained to linear systems.

B.6 DAMPING

Energy is dissipated in soils and structures by several mechanisms, including friction, heat generation, and plastic yielding. For specific soils and structures, however, the operative mechanisms are not understood sufficiently to allow them to be explicitly modeled. As a result, the effects of the various energy loss mechanisms are usually lumped together and represented by some convenient damping mechanism.

B.6.1 Viscous Damping

The most commonly used mechanism for representing energy dissipation is viscous damping. When a viscous damped SDOF system such as that shown in Figure B.3 is subjected to a harmonic displacement [Equation B.27]

$$u(t) = u_0 \sin \bar{\omega} t$$

the net force exerted on the mass by the spring and dashpot is

$$F(t) = ku(t) + c\dot{u}(t) = ku_0 \sin \bar{\omega} t + c\bar{\omega} u_0 \cos \bar{\omega} t \tag{B.69}$$

Evaluating these functions from time t_0 to time $t_0 + 2\pi/\bar{\omega}$ yields the force-displacement values for one cycle of a hysteresis loop. When the viscous damping coefficient, c, is zero, the force and displacement are in phase and proportional to each other, implying a linear elastic stress-strain relationship. For nonzero damping, however, the hysteresis loop is elliptical, as shown in Figure B.15. Note that when the displacement is zero, the spring force is zero and the net force comes entirely from the dashpot. Similarly, when the velocity is zero (at $\bar{\omega} t = \pi/2 + n\pi$), the dashpot force vanishes and the net force consists entirely of the spring force. The aspect ratio of the hysteresis loop decreases with increasing damping; the loop becomes a circle when $c = k/\bar{\omega}$.

Obviously, the shape of the hysteresis loop depends on the viscous damping coefficient and therefore on the damping ratio. Hence we should be able to determine the damping ratio from a known hysteresis loop. The energy dissipated in one cycle of oscillation is given by the area inside the hysteresis loop and can be obtained from

$$W_D = \int_{t_0}^{t_0+2\pi/\bar{\omega}} F \frac{du}{dt} dt = \pi c \bar{\omega} u_0^2 \tag{B.70}$$

At maximum displacement, the velocity is zero and the strain energy stored in the system is given by

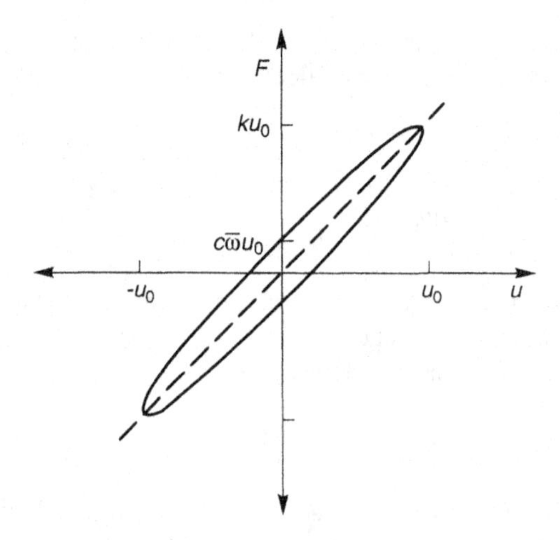

FIGURE B.15 Stress-strain behavior implied by viscous damping. Hysteresis loop is elliptical.

$$W_S = \frac{1}{2} k u_0^2 \tag{B.71}$$

Equations (B.70) and (B.71) show that $c = W_D / \left(\pi \omega u_0^2\right)$ and $k = 2W_S / u_0^2$. Substituting these into Equation (B.17) with $\bar{\omega} = \omega_0$ gives an expression

$$\xi = \frac{W_D}{4\pi W_S} \tag{B.72}$$

that is commonly used for graphical determination of the damping ratio from a measured hysteresis loop. Referring to Figure B.16, the damping ratio is taken as the ratio of the area of the hysteresis loop to the area of the shaded triangle, all divided by 4π. This graphical evaluation of the damping ratio is commonly used in the interpretation of many of the laboratory tests discussed in Chapter 6.

The damping characteristics of a linear system can also be evaluated from its frequency response characteristics. Setting the magnification factor expression [Equation (B.46)] equal to $M_{\max}/\sqrt{2}$, the half-power tuning ratios, shown in Figure B.17, can be approximated as

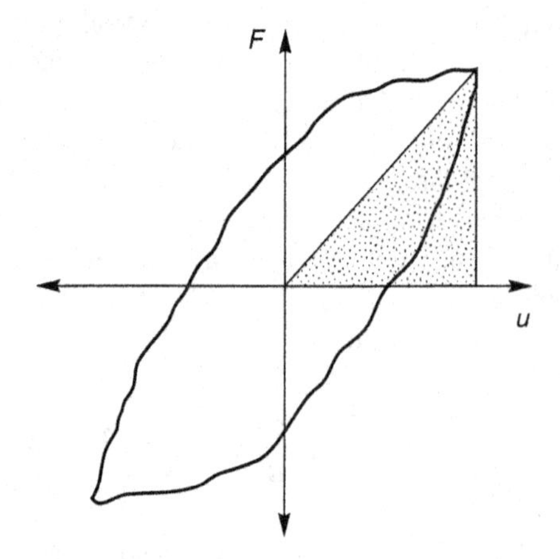

FIGURE B.16 Graphical evaluation of damping ration from measured hysteresis loop. The damping ration is proportional to the ratio of the area of the hysteresis loop to the area of the shaded triangle.

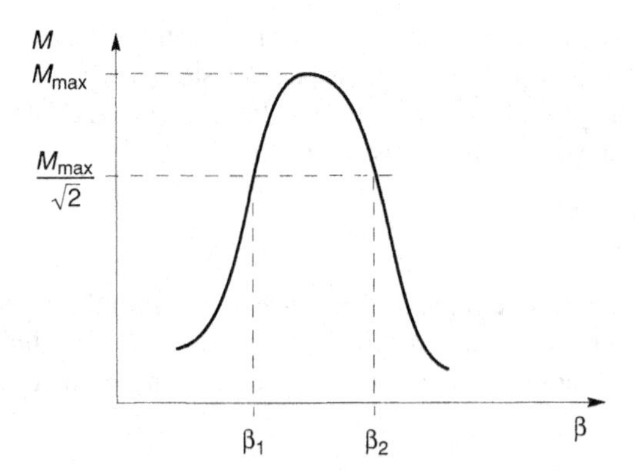

FIGURE B.17 Half-power tuning ratios for evaluation of damping ration from magnification curve.

$$\beta_1 \approx 1 - \xi - \xi^2 \tag{B.73a}$$

$$\beta_2 \approx 1 + \xi - \xi^2 \tag{B.73b}$$

Therefore, the damping ratio is given by half the difference between the half-power tuning ratios

$$\xi \approx \frac{\beta_2 - \beta_1}{2} \tag{B.74}$$

or, when the response is expressed in terms of frequency, where $\omega_1 = \beta_2\, \omega_0$ and $\omega_2 = \beta_2\, \omega_0$,

$$\xi \approx \frac{\omega_2 - \omega_1}{\omega_2 + \omega_1} \tag{B.75}$$

Thus the damping ratio of a system can be measured by exciting the system at different frequencies and determining the amplitude of the magnification factor at each frequency.

B.6.2 Other Measures of Energy Dissipation

In addition to the viscous damping ratio, ξ, a number of other parameters have been used to describe energy dissipation characteristics. Seismologists, for example, often work with the quality factor

$$Q = \frac{1}{2\xi} \tag{B.76}$$

In vibration analysis, the loss factor

$$\eta = 2\xi \tag{B.77}$$

and specific damping capacity

$$\psi = 2\pi\xi \tag{B.78}$$

are often used (Goodman, 1988).

It is important to remember that the damping ratio, and any of these other parameters, are simply parameters used to describe the effects of phenomena that are often poorly understood. They allow the effects of energy dissipation to be represented in a mathematically convenient manner. For most soils and structures, however, energy is dissipated hysteretically. In cases with large hysteretic damping caused by yielding or plastic straining of the material, the behavior is more accurately characterized by evaluating the nonlinear response of the system.

B.6.3 Complex Stiffness

A viscously damped system can be represented conveniently in a different but equivalent way for a class of techniques known as complex response analysis. Consider a damped SDOF system subjected to simple harmonic loading of amplitude, Q_0, and loading frequency, $\bar{\omega}$. The loading can be represented by

$$Q(t) = Q_0 e^{i\bar{\omega}t} \tag{B.79}$$

Assuming that $u(t)=U_0 e^{i\bar{\omega}t}$, the equation of motion is

$$m\ddot{u} + c\dot{u} + ku = Q_0 e^{i\bar{\omega}t} \tag{B.80}$$

and its steady-state solution is

$$u(t) = \frac{Q_0}{k - m\bar{\omega}^2 + ic\bar{\omega}} e^{i\bar{\omega}t} \tag{B.81}$$

Now consider the SDOF system of Figure B.18, which has no dashpot but which has a spring of complex stiffness $k^* = k_1 + ik_2$. The equation of motion for this system is

$$m\ddot{u} + k^* u = Q_0 e^{i\bar{\omega}t} \tag{B.82}$$

Again assuming that $u(t)=U_0 e^{i\bar{\omega}t}$, the steady-state solution can be expressed as

$$u(t) = \frac{Q_0}{k^* - \bar{\omega}^2 m} e^{i\bar{\omega}t} \tag{B.83}$$

Comparing Equations (B.81) and (B.83), it is apparent that

$$k^* = k + ic\bar{\omega} \tag{B.84}$$

By the appropriate choice of k^*, the displacement amplitude of Equation (B.83) can be made equal to that of Equation (B.81), (although a small phase difference between the two solutions will remain). To accomplish this, the complex stiffness is represented as

$$k^* = k\left(1 - 2\xi^2 + 2i\xi\sqrt{1-\xi^2}\right) \tag{B.85}$$

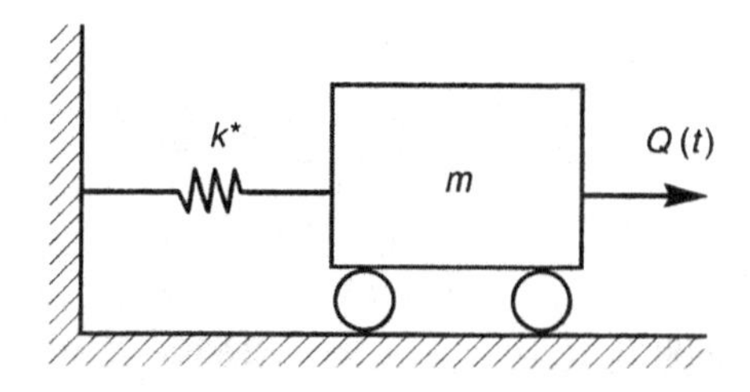

FIGURE B.18 SDOF system with spring of complex stiffness.

where $\xi \le 1$. For the usual small damping ratios considered in earthquake engineering problems, the ξ^2 terms can be neglected so that $k^* \approx k(1+2i\xi)$. Using this expression for k^*, the error in phase angle between the responses given by Equations (B.81) and (B.83) is $\Delta\theta = 2\xi/(1+\beta)$. As a result, a viscously damped system can be represented as an undamped system with complex stiffness. The use of this approach, however, is restricted to cases of harmonic motion. For problems in which loading is characterized as periodic (and therefore as the sum as a series of harmonic loads), the use of complex stiffness greatly simplifies the calculation of the response of damped systems.

For small damping ratios, the complex stiffness then consists of real and imaginary parts

$$\mathrm{Re}\left(k^*\right) = k$$

$$\mathrm{Im}\left(k^*\right) = 2k\xi \tag{B.86}$$

Consequently, the damping ratio can be expressed as

$$\xi = \frac{\mathrm{Im}\left(k^*\right)}{2\,\mathrm{Re}\left(k^*\right)} \tag{B.87}$$

which is useful to remember in the interpretation of quantities such as complex impedance functions (Section 8.3), which are usually expressed in terms of their real and imaginary parts.

B.7 RESPONSE SPECTRA

For earthquake-resistant design, the entire time series of response may not be required. For many applications, earthquake-resistant design may be based on the maximum (absolute) value of the response of a structure to a particular base motion. Obviously, the response will depend on the mass, stiffness, and damping characteristics of the structure and on the characteristics of the base motion.

The response spectrum describes the maximum response of a SDOF system to a particular input motion as a function of the natural frequency (or natural period) and damping ratio of the SDOF system (Figure B.19). The response may be expressed in terms of acceleration, velocity, or displacement. For a given ground motion, the maximum values of each of these parameters depend only on the natural frequency and damping ratio of the SDOF system. The maximum values of acceleration, velocity, and displacement are referred to as the spectral acceleration (S_a), spectral velocity (S_v), and spectral displacement (S_d), respectively. Note that an SDOF system of zero natural period (infinite natural frequency) would be rigid, and its spectral acceleration would be equal to the peak ground acceleration.

Application of the Duhamel integral to a linear elastic SDOF system produces expressions for the acceleration, velocity, and displacement time series that are proportional (by a factor of ω), except for a phase shift. Because the phase shift does not significantly influence the maximum response values, the spectral acceleration, velocity, and displacement can be approximately related to each other by the following simple expressions:

$$S_d = |u|_{\max} \tag{B.88a}$$

$$S_v = |\dot{u}|_{\max} \neq \omega_0 S_d = \mathrm{PSV} \tag{B.88b}$$

$$S_a = |\ddot{u}|_{\max} \approx \omega_0^2 S_d = \omega_0 \cdot \mathrm{PSV} = \mathrm{PSA} \tag{B.88c}$$

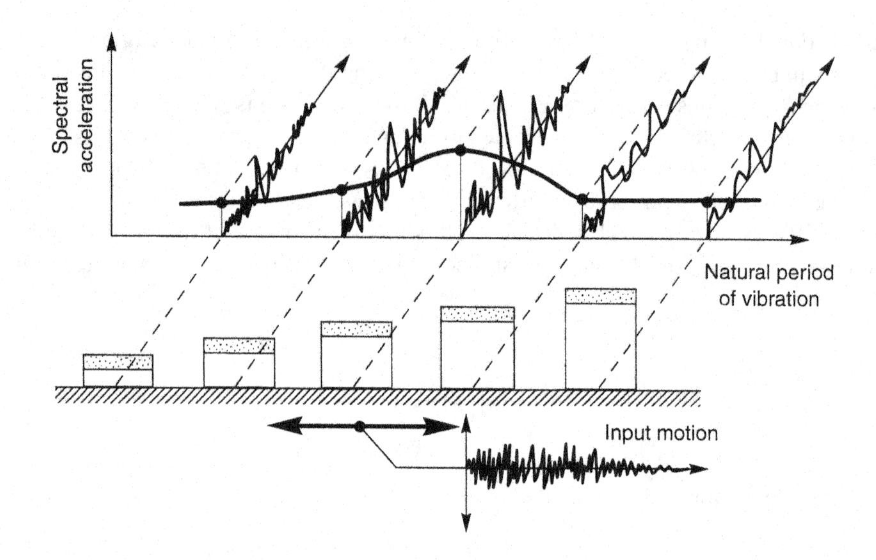

FIGURE B.19 Response spectrum. Spectral accelerations are the maximum absolute acceleration amplitudes of SDOF systems in response to the same input motion. The response spectrum is obtained by plotting the spectral accelerations against the periods of vibration of the SDOF systems.

where u and ω_0 are the displacement and natural frequency of the SDOF system, PSV is the pseudospectral velocity, and PSA is the pseudospectral acceleration. The PSV and PSA are not the true maximum values of velocity and acceleration, although in the case of PSA it is very close. In practice, the pseudospectral acceleration is generally assumed to be equal to the spectral acceleration and is generally referred to as S_a in this text. Note that there is no pseudospectral displacement, only spectral displacement. In most of the text, pseudospectral acceleration is written more concisely as "spectral acceleration."

B.8 DUCTILITY OF STRUCTURAL MATERIALS AND COMPONENTS

The ideal way to prevent damage due to a structure during earthquake shaking would be to ensure that it responds in a linear, elastic manner throughout an earthquake – if no yielding occurs, the structure (and all of its components) will return to its original position following shaking and there will be no structural damage. This objective may be easily achieved for low levels of shaking, but could require massive, expensive structural components for strong shaking.

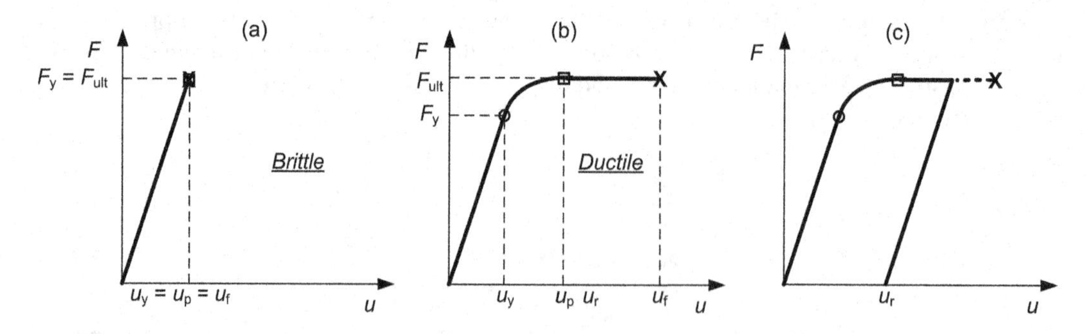

FIGURE B.20 Schematic illustration of different types of force-displacement behavior: (a) brittle, (b) ductile, and (c) ductile with removal of applied loading.

Consider an SDOF system with the force-displacement behavior shown in Figure B.20a. The system exhibits linear elastic behavior up to the yield force, F_y (and also the yield displacement, u_y), but the yield force is equal to the ultimate capacity, F_{ult}, so complete failure (with loss of subsequent load-resisting capability) occurs at that point. Such a system, in which load-carrying capacity is lost at, or very shortly after, yielding is described as being *brittle*. If a load smaller than the yield force is applied to this structure and then removed, the displacement will go back to zero and no damage will have occurred. A brittle system can perform well during earthquake shaking, as long as its strength exceeds all loads applied to it. If a brittle system fails, however, damage will usually be rapid and catastrophic.

Another SDOF system may have the force-displacement behavior shown in Figure B.20b. The ultimate capacity of this system is the same as that of the previous system, but it yields at a lower force, F_y (and yield displacement, u_y). After yielding, this system continues to pick up resistance (albeit with lower stiffness) with increasing displacement. At a larger displacement, u_p, the full plastic capacity of the system, F_{ult}, is reached and the system continues to deform, while maintaining its load-resisting capacity, until complete failure occurs at a larger failure displacement, u_f. This system, which can continue to resist load while deforming well past its yield point, is described as being *ductile*. Structural engineers generally quantify ductility as the ratio of some measure of deformation (such as displacement, curvature, etc.) to its value at yield, e.g., in the form of *ductility demand*, $\mu = u_{max}/u_y$. The ductility at failure, or *ductility capacity*, is defined as $\mu = u_f/u_y$. If the load applied to a ductile structure is released prior to failure (Figure B.20c), the displacement of the structure will decrease (usually nearly linearly) so that a residual displacement, u_r, occurs. The area under the force-displacement curve represents the energy dissipated by the plastic deformation of the structure. The structure would then be considered to have suffered some damage from the applied loading; the significance of this damage would depend on the amount of residual displacement and its effect on the utility of the specific structure. A ductile system can perform well during earthquake shaking even if the induced forces exceed its yield strength because it maintains load-carrying capacity and dissipates energy after yielding.

There are, therefore, two basic approaches that can be taken in the seismic design of a structure. One is to allow brittle behavior but make the structure very strong so that yielding (which would be followed shortly by failure) never occurs. The other is to make the structure very ductile (and allow controlled yielding) so that failure doesn't occur. The "strength vs. ductility" decision has a number of implications, both for design of new structures and retrofitting of existing structures. Designing a new structure to be stronger generally involves using more material – thicker columns, beams, girders, walls, etc. and/or more reinforcement – these increase the cost and mass of the structure and reduce the usable space within its boundaries. Designing a new structure to be ductile can generally be achieved using less material, and thereby provide a less expensive structure. However, it also means that some level of damage will be accepted and requires more detailed analysis of the structure (and its components and connections) to ensure that failure does not occur. The design of retrofitting measures for existing structures can be particularly challenging because many older structures, due to materials and/or form, are both brittle and weak.

Designing for ductility requires the ability to predict the seismic response of nonlinear systems. Nonlinear analyses are generally more complex and time-consuming, both from the standpoint of material characterization and computational demand, than linear analyses, as described in the following section.

B.9 RESPONSE OF NONLINEAR SDOF SYSTEMS TO GENERAL LOADING

Numerical integration of the Duhamel integral is very useful for the calculation of the response of linear systems to general loading. Many systems for which the seismic response is to be calculated, however, exhibit nonlinear behavior. In such systems, the mass is usually constant, but the damping coefficient and/or the stiffness may vary with time, deflection, or velocity. It will be useful to

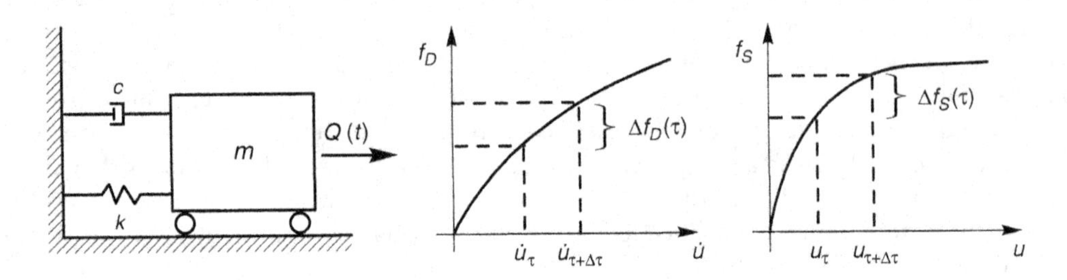

FIGURE B.21 SDOF system with nonlinear damping and spring forces.

develop methods for analysis of the response of nonlinear systems, recognizing that they will be appropriate for linear systems as well when damping and stiffness values are held constant.

The most common approach to nonlinear analysis is the direct integration of incremental equations of motion that govern the response of the system over small time increments. The response is calculated for each time increment after adjusting the stiffness and damping at the beginning of the increment. By using the conditions at the end of one time increment as the initial conditions for the next time increment, the nonlinear system is approximated as an incrementally changing linear system.

B.9.1 Incremental Equation of Motion

Consider the SDOF system shown in Figure B.21, which has a nonlinear spring and dashpot (i.e., the spring force is not proportional to displacement and the dashpot force is not proportional to velocity). Dynamic equilibrium at time τ requires that

$$f_I(\tau) + f_D(\tau) + f_S(\tau) = Q(\tau) \tag{B.89}$$

and that

$$f_I(\tau + \Delta\tau) + f_D(\tau + \Delta\tau) + f_S(\tau + \Delta\tau) = Q(\tau + \Delta\tau) \tag{B.90}$$

at time $\tau + \Delta\tau$. Defining

$$\Delta f_I(\tau) = f_I(\tau + \Delta\tau) - f_I(\tau) \tag{B.91a}$$

$$\Delta f_D(\tau) = f_D(\tau + \Delta\tau) - f_D(\tau) \tag{B.91b}$$

$$\Delta f_S(\tau) = f_S(\tau + \Delta\tau) - f_S(\tau) \tag{B.91c}$$

$$\Delta Q(\tau) = Q(\tau + \Delta\tau) - Q(\tau) \tag{B.91d}$$

and subtracting Equation (B.89) from Equation (B.90), the incremental equation of motion for the time interval from τ to $\tau + \Delta\tau$ is

$$\Delta f_I(\tau) + \Delta f_D(\tau) + \Delta f_S(\tau) = \Delta Q(\tau) \tag{B.92}$$

or expressing the incremental forces in terms of incremental displacements, velocities, and accelerations, as

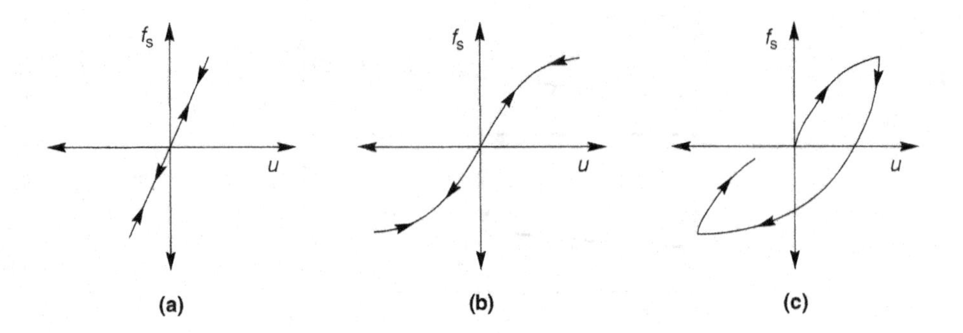

FIGURE B.22 Stress-strain behavior of (a) linear elastic, (b) nonlinear elastic, and (c) nonlinear, inelastic materials under cyclic loading conditions.

$$m\Delta\ddot{u}(\tau) + c(\tau)\Delta\dot{u}(\tau) + k(\tau)\Delta u(\tau) = \Delta Q(\tau) \tag{B.93}$$

By integrating this incremental equation of motion in a series of small time steps, the response of the nonlinear system can be obtained. It should be noted that this approach can be used to calculate the response of linear elastic, nonlinear elastic, or nonlinear inelastic materials with stress-strain behaviors shown in Figure B.22. The third of these is particularly important because it allows representation of the hysteretic damping displayed by cyclically loaded soils.

B.9.2 Numerical Integration

There are many ways to numerically integrate the incremental equation of motion. One of the simplest and most easily coded of these is the linear acceleration method. It is based on the assumption that the acceleration varies linearly within each time increment. If the acceleration in the time increment varies linearly, the velocity and displacement will vary quadratically and cubically, respectively.

$$\Delta\dot{u}(\tau) = \ddot{u}(\tau)\Delta t + \Delta\ddot{u}(\tau)\frac{\Delta t}{2} \tag{B.94}$$

$$\Delta u(\tau) = \dot{u}(\tau)\Delta t + \Delta\ddot{u}(\tau)\frac{\Delta t^2}{2} + \Delta\ddot{u}(\tau)\frac{\Delta t^2}{6} \tag{B.95}$$

Rearranging, the incremental acceleration and velocity can be expressed in terms of the incremental displacement

$$\Delta\ddot{u}(\tau) = \frac{6}{\Delta t^2}\Delta u(\tau) - \frac{6}{\Delta t}\dot{u}(\tau) - 3\ddot{u}(\tau) \tag{B.96a}$$

$$\Delta\dot{u}(\tau) = \frac{3}{\Delta t}\Delta u(\tau) - 3\dot{u}(\tau) - \frac{\Delta t}{2}\ddot{u}(\tau) \tag{B.96b}$$

Substituting Equations (B.97) into the incremental equation of motion [Equation (B.93)] provides

$$m\left[\frac{6}{\Delta t^2}\Delta u(\tau) - \frac{6}{\Delta t}\dot{u}(\tau) - 3\ddot{u}(\tau)\right] + c(\tau)\left[\frac{3}{\Delta t}\Delta u(\tau) - 3\dot{u}(\tau) - \frac{\Delta t}{2}\ddot{u}(\tau)\right] + k(\tau)\Delta u(\tau) = \Delta Q(\tau) \tag{B.97}$$

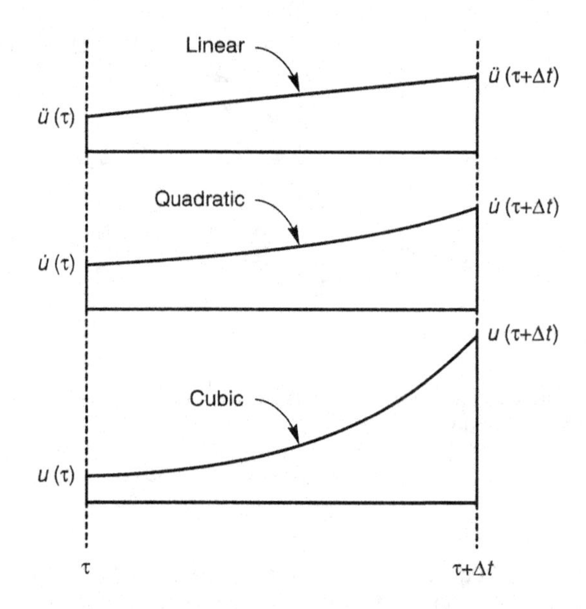

FIGURE B.23 Variation of acceleration, velocity, and displacement over time increment when the time-variation of acceleration is linear.

Rearranging terms and solving for incremental displacement provides

$$\Delta u(\tau) = \frac{\Delta Q(\tau) + m[(6/\Delta t)\dot{u}(\tau) + 3\ddot{u}(\tau)] + c(\tau)[3\dot{u}(\tau) + (\Delta t/2)\ddot{u}(\tau)]}{k(\tau) + \left(6/\Delta t^2\right)m + (3/\Delta t)c(\tau)} \tag{B.98}$$

Equation (B.98) shows that if the displacement, velocity, and acceleration at time τ are known, the incremental displacement during the succeeding time increment $\Delta\tau$ can be calculated based on the loading and the stiffness and damping during that time increment. From this incremental displacement, the incremental velocity and acceleration, and from these the displacement, velocity, and acceleration at the end of the time increment, can be determined. The conditions at the end of the time increment are then taken as the initial conditions for the next time increment and are used to calculate the appropriate stiffness and damping values for the next time increment. To prevent the accumulation of errors resulting from the assumptions of the linear acceleration method, the acceleration at the beginning of each time step should be calculated by subtracting the damping and spring forces from the total external load and dividing the result by the mass. This will ensure that total equilibrium is satisfied at each step of the analysis.

For numerical stability, it is necessary that the time steps be relatively small, typically less than about 55% of the shortest undamped natural period of the system. These small time steps can lead to considerable computational effort when the linear acceleration method is applied to multiple-degree-of-freedom systems. A number of other numerical integration techniques, including some that are unconditionally stable, are available; Berg (1989) describes the application of several to structural dynamics problems.

B.10 MULTIPLE-DEGREE-OF-FREEDOM SYSTEMS

In most physical systems, the motion of the significant masses cannot be described by a single variable; such systems must be treated as multiple-degree-of-freedom (MDOF) systems. With the exception of only the simplest cases, the types of buildings, bridges, and other structures that are of interest in earthquake engineering have multiple DOF. Some structures can be idealized with only a few DOF; others may require hundreds or even thousands.

In many respects, the response of MDOF systems is similar to the response of SDOF systems, and procedures for analysis are analogous to those described previously for SDOF systems. Although the additional DOF complicate the algebra, the procedures are conceptually quite similar. In fact, a very useful approach to the response of linear MDOF systems allows their response to be computed as the sum of the responses of a series of SDOF systems.

B.10.1 Equations of Motion

In evaluating the response of an MDOF system, the dynamic equilibrium of all masses must be ensured simultaneously. Consider the idealized two-story structure shown in Figure B.24. The structure has two DOF: horizontal translation of the upper mass and horizontal translation of the lower mass. For each mass the externally applied load must be balanced by the inertial, damping, and elastic forces that resist motion:

$$f_{I1} + f_{D1} + f_{S1} = q_1(t) \quad \text{(B.99a)}$$

$$f_{I2} + f_{D2} + f_{S2} = q_2(t) \quad \text{(B.99b)}$$

or, in matrix form,

$$\mathbf{f}_I + \mathbf{f}_D + \mathbf{f}_S = \mathbf{q}(t) \quad \text{(B.100)}$$

Each of the terms on the left side of Equation (B.100) are vectors that depend on the structural properties given in Figure B.24.

If the structure exhibits linear behavior, the principle of superposition is valid. Then the forces that resist motion at each level can be expressed in terms of coefficients by which the motion parameter at all levels are multiplied. For example, the elastic force resisting motion at level 1 can be expressed as

$$f_{S1} = k_{11}u_1 + k_{12}u_2 \quad \text{(B.101)}$$

where the stiffness coefficients k_{ij} represent the force induced at level i due to a unit displacement at level j (with the displacements at all levels except j held equal to zero). In matrix form

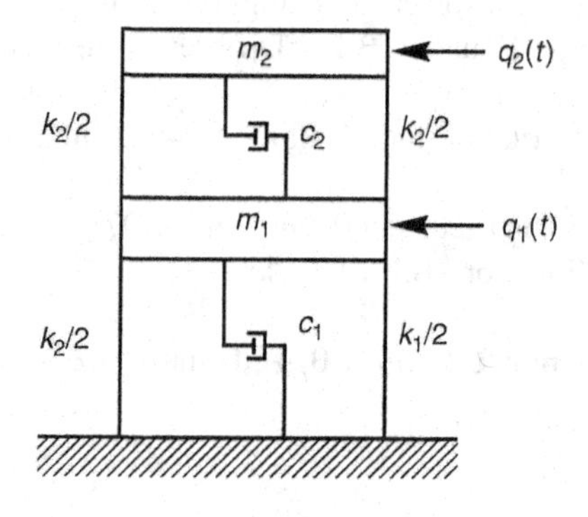

FIGURE B.24 Two-degree-of-freedom system. Displacements of masses 1 and 2 from equilibrium positions are u_1 and u_2, respectively.

$$\begin{Bmatrix} f_{S1} \\ f_{S2} \end{Bmatrix} = \begin{bmatrix} k_{11} & k_{12} \\ k_{21} & k_{22} \end{bmatrix} \begin{Bmatrix} u_1 \\ u_2 \end{Bmatrix} \tag{B.102}$$

or

$$\mathbf{f}_S = \mathbf{ku} \tag{B.103}$$

in which **k** is the stiffness matrix of the structure and **u** is a displacement vector.

Similarly, a damping matrix and a mass matrix can be developed in which the elements c_{ij} (or m_{ij}) represent the damping (or inertial) forces resisting motion at level i due to a unit velocity (or acceleration) of level j. Dynamic equilibrium of the MDOF system can then be described by a set of simultaneous equations of motion, which can be expressed in matrix form as

$$\mathbf{m\ddot{u}} + \mathbf{c\dot{u}} + \mathbf{ku} = \mathbf{q}(t) \tag{B.104}$$

MDOF systems also respond to base motions. The equation of motion for the case of base shaking is easily developed following the same procedure applied to the SDOF case in Section B.4.2. The resulting equation of motion is

$$\mathbf{m\ddot{u}} + \mathbf{c\dot{u}} + \mathbf{ku} = -\mathbf{m1}\ddot{u}_b(t) \tag{B.105}$$

where **1** is a column vector of ones. Equation (B.105) indicates that the response of an N-story structure to base motion is equal to the response to equivalent external loads, where $q_i = -m_i\ddot{u}_b(t)$ is the load applied to the ith floor.

B.10.2 Undamped Free Vibrations

For undamped free vibrations, all terms of the damping matrix are zero, so the equations of motion reduce to

$$\mathbf{m\ddot{u}} + \mathbf{ku} = \mathbf{0} \tag{B.106}$$

where **0** is a column vector of zeros. Assuming that the response of each mass (degree of freedom) is harmonic,

$$\mathbf{u}(t) = \mathbf{U}\sin(\omega t + \theta) \tag{B.107}$$

where **U** is a vector containing the displacement amplitudes, ω is an angular frequency of vibration, and θ is a phase angle (similar to Equation B.14). Differentiating Equation (B.107) twice gives

$$\mathbf{\ddot{u}}(t) = -\omega^2\mathbf{U}\sin(\omega t + \theta) = -\omega^2\mathbf{u}(t) \tag{B.108}$$

Substituting the expressions for displacement [Equation (B.107)] and acceleration [Equation (B.108)] into the equation of motion [Equation (B.106)] yields

$$-\mathbf{m}\omega^2\mathbf{U}\sin(\omega t + \theta) + \mathbf{kU}\sin(\omega t + \theta) = \mathbf{0} \tag{B.109a}$$

or

$$\left[\mathbf{k} - \omega^2\mathbf{m}\right]\mathbf{U} = \mathbf{0} \tag{B.109b}$$

which is a set of linear algebraic equations with unknown **U**. Applying methods from linear algebra, a nontrivial solution (one that gives values other than **U**=**0**) can be obtained only if

$$\det\left(\mathbf{k}-\omega^2\mathbf{m}\right)=\left|\mathbf{k}-\omega^2\mathbf{m}\right|=0 \tag{B.110}$$

Equation (B.110) is the frequency equation (or characteristic equation) of the system, which for a system of N DOF, will give a polynomial of Nth degree in ω^2. The N roots of the frequency equations $\{\omega_1^2,\omega_2^2,\omega_3^2,\ldots,\omega_N^2\}$ represent the frequencies at which the undamped system can oscillate in free vibration. These frequencies are called the *natural circular frequencies* of the system.

Each natural frequency is associated with a *mode of vibration* of the system. At the natural frequencies, the amplitude of the displacement vector, **U**, is indeterminate [scaling the displacements up or down by a constant factor will still satisfy Equation (B.110)]. The vector **U** does describe the shape of the vibrating system, which is different at each natural frequency. This shape is often made dimensionless by dividing the elements of **U** by the displacement of one (often the first, sometimes the largest) element. The resulting vector describes the *mode shape*; the mode shape for the nth mode of vibration would be

$$\boldsymbol{\phi}_n^T=\begin{bmatrix}\phi_{1n} & \phi_{2n} & \cdots & \phi_{Nn}\end{bmatrix}=\frac{1}{U_{Nn}}\begin{bmatrix}U_{1n} & U_{2n} & \cdots & 1\end{bmatrix} \tag{B.111}$$

All mode shapes (regardless of normalization) satisfy the relationship, $\left|\mathbf{k}-\omega_n^2\mathbf{m}\right|\boldsymbol{\phi}_n=\mathbf{0}$ for n=[1, N]. Example mode shapes for the structure in Figure B.24 are shown in Figure B.25. Note that in the first mode, both masses move in the same direction, whereas in the second mode, they move in opposite directions. Thus, a system of N degrees of freedom will have N natural frequencies corresponding to N modes of vibration. Each mode of vibration occurs at a particular natural frequency and causes the structure to deform with a particular mode shape. The mode corresponding to the lowest natural frequency is called the first mode or *fundamental mode*, the second lowest natural frequency is called the second mode, and so on. The mode shapes can be shown to be orthogonal, i.e., for $m \neq n$

$$\boldsymbol{\phi}_m^T\mathbf{m}\boldsymbol{\phi}_n=0 \tag{B.112a}$$

$$\boldsymbol{\phi}_m^T\mathbf{k}\boldsymbol{\phi}_n=0 \tag{B.112b}$$

B.10.3 Mode Superposition Method

For linear structures with certain types of damping, the response in each mode of vibration can be determined independently of the response in the other modes. The independent modal responses can then be combined to determine the total response. This is the basis of the *mode superposition method.*

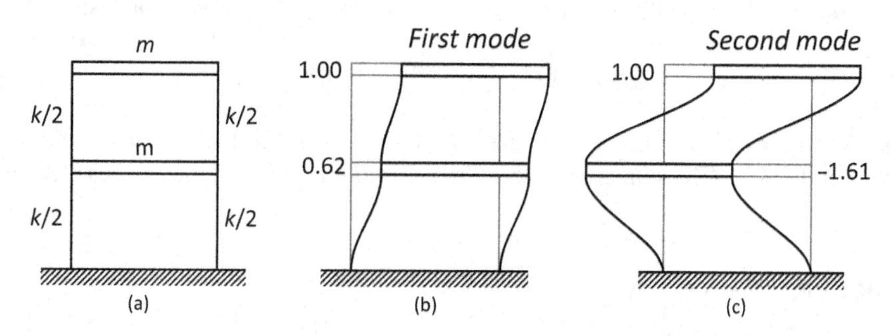

FIGURE B.25 Two-degree of freedom structure illustrating: (a) initial geometry and properties, (b) first mode shape with displacements normalized by top floor displacement, and (c) normalized second mode shape.

Recalling that the mode shape vector, $\boldsymbol{\phi}_n$, describes only the shape of the nth mode, the displacements can be expressed as the product of the mode shape and the modal amplitude, y_n:

$$\mathbf{U}_n(t) = \boldsymbol{\phi}_n y_n(t) \tag{B.113}$$

Then, by substituting Equation (B.113) into Equation (B.104) and pre-multiplying each term by ϕ_n^T, the equation of motion can be written for the nth mode of vibration as

$$M_n \ddot{y}_n + C_n \dot{y}_n + K_n y_n = Q_n(t) \tag{B.114}$$

where the *generalized* (or *modal*) *mass*, $M_n = \boldsymbol{\phi}_n^T \mathbf{m} \boldsymbol{\phi}_n$, *generalized damping coefficient*, $C_n = \boldsymbol{\phi}_n^T \mathbf{c} \boldsymbol{\phi}_n$, *generalized stiffness*, $K_n = \boldsymbol{\phi}_n^T \mathbf{k} \boldsymbol{\phi}_n$, and *generalized load*, $Q_n(t) = \boldsymbol{\phi}_n^T \mathbf{q}(t)$. This modal equation of motion is based on the assumption that the damping matrix is orthogonal (i.e., that $\boldsymbol{\phi}_m^T \mathbf{c} \boldsymbol{\phi}_n = 0$ for $m \neq n$. Rayleigh damping, in which the damping matrix can be broken into a component proportional to the mass matrix and a component proportional to the stiffness matrix, satisfies the orthogonality requirement. Other procedures are described in structural dynamics texts. Alternatively, the equation of motion can be written as

$$\ddot{y}_n + 2\xi_n \omega_n \dot{y}_n + \omega_n^2 y_n = \frac{Q_n(t)}{M_n} \tag{B.115}$$

where $\xi_n = C_n / (2M_n \omega_n)$. For the case of base shaking, the equation of motion can be expressed as

$$\ddot{y}_n + 2\xi_n \omega_n \dot{y}_n + \omega_n^2 y_n = -\frac{L_n}{M_n} \ddot{u}_b(t) \tag{B.116}$$

where $L_n = \sum_{j=1}^{N} m_j \phi_{jn}$. The quantity, L_n/M_n, often referred to as the *modal participation factor*, reflects the extent to which each mode contributes to the overall response.

By this process, the system of N simultaneous equations (the original equations of motion) is transformed into a system of N independent equations. Each of these independent equations can be solved for $y_n(t)$ using the SDOF procedures described earlier in this appendix. Then the total displacement is obtained by superposition of the modal contributions:

$$\mathbf{u}(t) = \boldsymbol{\phi}_1 y_1(t) + \boldsymbol{\phi}_2 y_2(t) + \cdots + \boldsymbol{\phi}_N y_N(t) \tag{B.117}$$

Once the displacements are known, they can be used to compute forces, stresses, and other parameters of interest. The displacements can also be used to compute a set of equivalent lateral forces, $f(t)$, which would produce the displacements $u(t)$ if they were applied as static loads:

$$\mathbf{f}(t) = \mathbf{k}\boldsymbol{\phi}_1 y_1(t) + \mathbf{k}\boldsymbol{\phi}_2 y_2(t) + \cdots + \mathbf{k}\boldsymbol{\phi}_N y_N(t) \tag{B.118}$$

Internal forces and moments can be computed by static analysis of the structure subjected to the equivalent lateral forces. These internal forces can be used for design of the various elements of the structure.

B.10.4 Response Spectrum Analysis

The mode superposition method produces the entire time series of structural response. For design purposes, however, the entire time series may not be needed; the maximum response values may be sufficient. Because each mode of vibration can be treated as an independent SDOF system, maximum values of modal responses can be obtained from the response spectrum. The modal maxima can then be combined to estimate the maximum total response.

B.10.4.1 Calculation of Modal Response Maxima

Let S_{dn}, S_{vn}, and S_{an} denote the spectral displacement, velocity, and acceleration associated with the nth mode of vibration, respectively (these values would be obtained from the response spectrum at a period, $T_n = 2\pi/\omega_n$). Then the maximum modal displacement is given by

$$(y_n)_{max} = \frac{L_n}{M_n} S_{dn} = \frac{L_n T_n^2}{4\pi^2 M_n} S_{an} \tag{B.119}$$

Using Equation (B.113), the maximum displacement of the jth floor from vibration in the nth mode would be

$$(U_{jn})_{max} = \frac{L_n}{M_n} S_{dn}\phi_{jn} = \frac{L_n T_n^2}{4\pi^2 M_n} S_{an}\phi_{jn} \tag{B.120}$$

The maximum value of the equivalent lateral force at the jth floor from vibration in the nth mode is

$$(f_{jn})_{max} = \frac{L_n}{M_n} m_j \phi_{jn} S_{an} \tag{B.121}$$

Maximum values of the internal forces and moments can then be computed by static analysis of the structure subjected to the maximum equivalent lateral forces.

B.10.4.2 Combination of Modal Response Maxima

Section B.10.4.1 showed how the response spectrum can be used to predict maximum values of various modal response parameters. The mode superposition method showed that time series of modal response can be combined by simple superposition to obtain the total time series of response. However, the combination of modal response maxima to obtain the maximum total response is not as straightforward.

The exact value of the maximum total response cannot be obtained directly from the modal maxima because the modal maxima occur at different times. Direct superposition of the modal maxima, which implies that the maxima do occur simultaneously, produces an upper bound to the maximum total response; for any response parameter $r(t)$,

$$r_{max} \leq \sum_{n=1}^{N} (r_n)_{max} \tag{B.122}$$

This upper bound value is usually too conservative and is rarely used for design. Instead, modal combination procedures based on random vibration theory are used. The simplest of these is the root-sum-square value

$$r_{max} = \sqrt{\sum_{n=1}^{N} (r_n)_{max}^2} \tag{B.123}$$

The root-sum-square procedure provides a good estimate of maximum total response when the natural periods are well separated (by a factor of about 1.5 or more for 5% damping). Procedures that account for correlation between modes are available (Newmark and Rosenblueth, 1971; Chopra, 1995) for cases of closely spaced modes.

B.10.5 Discussion

The mode superposition method and response spectrum analysis procedures both rely on the representation of an MDOF system by a set of SDOF systems. The characteristics of the set of SDOF systems are such that those corresponding to the lower natural frequencies generally contribute more to the total response than those corresponding to the higher natural frequencies. For practical purposes, the response of an MDOF system can be computed with reasonable accuracy by considering only the lower modes that contribute significantly to the total response of the structure. For some structures, only a small number of modes may need to be considered. It is common to exclude higher modes if their participation factors fall below a certain threshold of percent contribution to total response (often shear in the lowest story). All of the analyses described in this section apply to linear structures. Procedures for analysis of nonlinear MDOF structures are available but are beyond the scope of this appendix.

REFERENCES

Berg, G.V. (1989). *Elements of Structural Dynamics*, Prentice-Hall, Englewood Cliffs, NJ, 268 pp.

Chopra, A.K. (2022). *Dynamics of Structures; Theory and Applications to Earthquake Engineering*, 6th Edition, Pearson, London, UK.

Clough, R.W. and Penzien, J. (1993). *Dynamics of Structures*, McGraw Hill, New York.

Paz, M. and Kim, Y.H. (2019). *Structural Dynamics: Theory and Computation*, 6th ed., Springer, Berlin, Germany

Appendix C
Wave Propagation

C.1 INTRODUCTION

It is the continuous nature of geologic materials that causes soil dynamics and geotechnical earthquake engineering to diverge from their structural counterparts. While most structures can readily be idealized as assemblages of discrete masses with discrete sources of stiffness, geologic materials cannot. They must be treated as continua, and their response to dynamic disturbances must be described in the context of wave propagation.

Some basic concepts of wave propagation will be required to fully understand the material presented in Chapters 2–4; a more fundamental treatment of the basic concepts is presented in this appendix. The presentation follows a repeated pattern of simple-to-complex applications. The relatively simple problem of waves in unbounded media is followed by the more complicated problem of waves in bounded and layered media. Within each, the concepts are presented first for the simple case of one-dimensional wave propagation, and then for the more general three dimensional case. The careful reader will note that the basic techniques and principles used to solve the more complicated cases are generally the same as those used for the simple cases; the additional complexity simply results from the need to consider more dimensions.

C.2 WAVES IN UNBOUNDED MEDIA

The propagation of stress waves is most easily understood by first considering an unbounded, or "infinite," medium [i.e., one that extends infinitely in the direction(s) of wave propagation]. A simple, one-dimensional idealization of an unbounded medium is that of an infinitely long rod or bar. Using the basic requirements of equilibrium of forces and compatibility of displacements, and using strain-displacement and stress-strain relationships, a one-dimensional wave equation can be derived and solved. The process can be repeated, using the same requirements and relationships, for the more general case of wave propagation in a medium that extends infinitely in three orthogonal directions.

C.2.1 One-Dimensional Wave Propagation

Three different types of vibration can occur in a thin rod: longitudinal vibration during which the axis of the rod extends and contracts without lateral displacement; torsional vibration in which the rod rotates about its axis without lateral displacement of the axis; and flexural vibration during which the axis itself moves laterally. The flexural vibration problem has little application in soil dynamics and will not be considered further. For the first two cases, however, the operative wave equations are easily derived and solved.

C.2.1.1 Longitudinal Waves in an Infinitely Long Rod

Consider the free vibration of an infinitely long, linear elastic, constrained rod with cross-sectional area, A, Young's modulus, E, Poisson's ratio, ν, and mass density, ρ, as shown in Figure C.1. If the rod is constrained against radial straining, then particle displacements caused by a longitudinal wave must be parallel to the axis of the rod. Assume that cross-sectional planes will remain planar and that stresses will be distributed uniformly over each cross-section. As a stress wave travels along the rod and passes through the small element shown in Figure C.2, the axial stress at the left

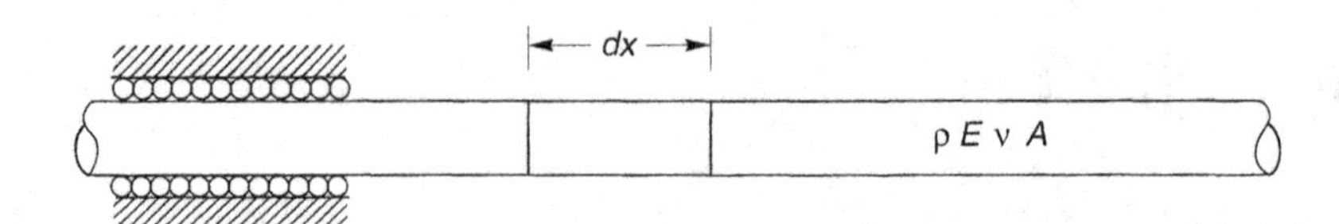

FIGURE C.1 Constrained, infinite rod for one-dimensional wave propagation. Constraint against radial straining schematically represented by rollers.

end of the element ($x=x_0$) is σ_x. At the right end ($x=x_0+dx$), the axial stress is $\sigma_x + (\partial\sigma_x / \partial x)dx$. Dynamic equilibrium of the element (Newton's second law) requires that

$$\left(\sigma_{x0} + \frac{\partial\sigma_x}{\partial x}dx\right)A - \sigma_{x0}A = \rho A dx \frac{\partial^2 u}{\partial t^2} \tag{C.1}$$

where u is the displacement in the x-direction. This simply states that the unbalanced external forces acting on the ends of the element [the left side of Equation (C.1)] must equal an inertial force induced by the acceleration of the mass of the element (the right side). Simplifying yields the one-dimensional equation of motion

$$\frac{\partial\sigma_x}{\partial x} = \rho\frac{\partial^2 u}{\partial t^2} \tag{C.2}$$

In this form, the equation of motion is valid for any stress-strain behavior but cannot be solved directly because it mixes stresses [on the left side of Equation (C.2)] with displacements (on the right side). To simplify the equation of motion, the left side can be expressed in terms of displacement by using the stress-strain relationship, $\sigma_x = M\varepsilon_x$, where the constrained modulus $M=(1-\nu)E/[(1+\nu)(1-2\nu)]$, and ε_x is obtained from the strain-displacement relationship, $\varepsilon_x = \partial u / \partial x$. These substitutions allow the one-dimensional equation of motion to be written in the familiar form of the one-dimensional longitudinal wave equation for a constrained rod:

$$\frac{\partial^2 u}{\partial t^2} = \frac{M}{\rho}\frac{\partial^2 u}{\partial x^2} \tag{C.3}$$

The one-dimensional wave equation can be written in the alternative form

$$\frac{\partial^2 u}{\partial t^2} = v_p^2\frac{\partial^2 u}{\partial x^2} \tag{C.4}$$

where v_P is the wave propagation velocity; for this case, the wave travels at $v_p = \sqrt{M/\rho}$. Note that the wave propagation velocity depends only on the properties of the rod material (its stiffness and density) and is independent of the amplitude of the stress wave or the nature of the loading. The wave propagation velocity increases with increasing stiffness and with decreasing density. The wave

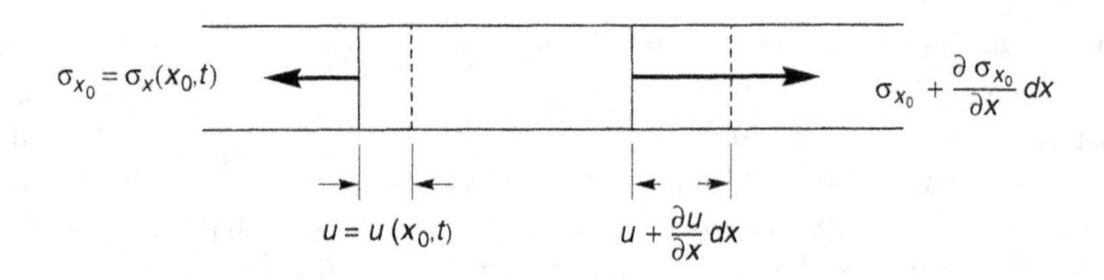

FIGURE C.2 Stresses and displacements at ends of element of length dx and cross sectional area, A.

propagation velocity is an extremely important material property that is relied upon heavily in soil dynamics and geotechnical earthquake engineering.

The wave propagation velocity is the velocity at which a stress wave would travel along the rod. It is not the same as the particle velocity, which is the velocity at which a single point within the rod would move as the wave passes through it. Knowing that $\partial u = \varepsilon_x \, \partial x$ (from the strain-displacement relationship), $\varepsilon_x = \sigma_x / M$ (from the stress-strain relationship), and $\partial x = v_p \, \partial t$ (from the definition of wave propagation velocity), the particle velocity u can be shown to be

$$\dot{u} = \frac{\partial u}{\partial t} = \frac{\varepsilon_x \, \partial x}{\partial t} = \frac{\sigma_x}{M}\frac{v_p \, \partial t}{\partial t} = \frac{\sigma_x}{M} v_p = \frac{\sigma_x}{\rho v_p^2} v_p = \frac{\sigma_x}{\rho v_p} \tag{C.5}$$

Equation (C.5) shows that the particle velocity is proportional to the axial stress in the rod. The coefficient of proportionality, ρv_p, is called the *specific impedance* of the material. The specific impedance is another important property that influences the behavior of waves at boundaries (Section C.4).

C.2.1.2 Torsional Waves in an Infinitely Long Rod

Torsional waves involve rotation of the rod about its own axis. In the case of the longitudinal wave, the direction of particle motion was parallel to the direction of wave propagation. For torsional waves, particle motion is constrained to planes perpendicular to the direction of wave propagation. The development of a wave equation for torsional vibrations, however, follows exactly the same steps as for longitudinal vibration. Consider the short segment of a cylindrical rod shown in Figure C.3 as a torsional wave of torque amplitude, T, travels along the rod. Dynamic torsional equilibrium requires that the unbalanced external torque [left side of Equation (C.6)] is equal to the inertial torque (right side):

$$\left(T_{x0} + \frac{\partial T}{\partial x} dx \right) - T_{x0} = \rho J dx \frac{\partial^2 \theta}{\partial t^2} \tag{C.6}$$

where J is the polar moment of inertia of the rod about its axis. This equilibrium equation can be simplified to produce the equation of motion

$$\frac{\partial T}{\partial x} = \rho J \frac{\partial^2 \theta}{\partial t^2} \tag{C.7}$$

Now, incorporating the torque-rotation relationship

$$T = GJ \frac{\partial \theta}{\partial x} \tag{C.8}$$

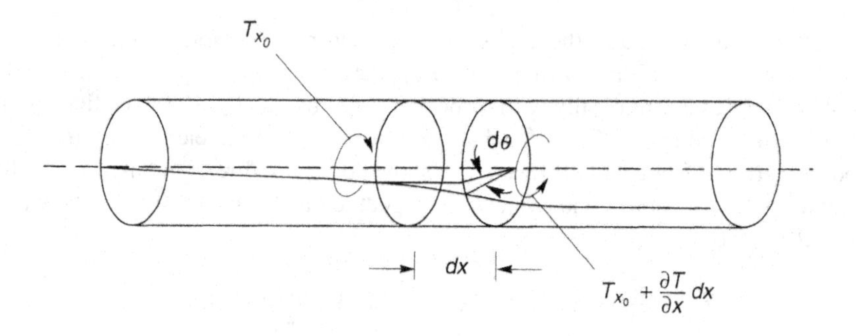

FIGURE C.3 Torque and rotation at ends of element of length dx and cross-sectional area, A.

where G is the shear modulus of the rod, the torsional wave equation can be written as

$$\frac{\partial^2 \theta}{\partial t^2} = \frac{G}{\rho}\frac{\partial^2 \theta}{\partial x^2} = v_s^2 \frac{\partial^2 \theta}{\partial x^2} \tag{C.9}$$

where $v_s = \sqrt{G/\rho}$ is the velocity of propagation of the torsional wave. Note that the form of the wave equation for torsional waves [Equation (C.9)] is identical to that for longitudinal waves [Equation (C.4)], but the wave propagation velocities are different. The wave propagation velocity depends both on the stiffness of the rod in the mode of deformation induced by the wave and on the material mass density but is independent of the amplitude of the stress wave.

C.2.1.3 Solution of the One-Dimensional Equation of Motion

The one-dimensional wave equation is a second-order partial differential equation of the form

$$\frac{\partial^2 u}{\partial t^2} = v^2 \frac{\partial^2 u}{\partial x^2} \tag{C.10}$$

where v represents the wave propagation velocity corresponding to the type of stress wave of interest. The solution of such an equation can be written in the form

$$u(x,t) = f(vt - x) + g(vt + x) \tag{C.11}$$

where f and g can be any arbitrary functions of $(vt-x)$ and $(vt+x)$ that satisfy Equation (C.10). Note that the argument of f remains constant when x increases with time (at velocity, v), and the argument of g remains constant when x decreases with time. Therefore, the solution of Equation (C.11) describes a displacement wave $f(vt-x)$ traveling at velocity v in the positive x-direction and another wave $g(vt+x)$ traveling at the same speed in the negative x-direction. It also implies that the shapes of the waves do not change with position or time.

If the rod is subjected to some steady-state harmonic stress $\sigma(t) = \sigma_0 \cos \bar{\omega} t$ where σ_0 is the stress wave amplitude and $\bar{\omega}$ is the circular frequency of the applied loading, the solution can be expressed using the wave number, $k = \bar{\omega}/v$, in the form

$$u(x,t) = A\cos(\bar{\omega}t - kx) + B\cos(\bar{\omega}t + kx) \tag{C.12}$$

Here the first and second terms describe harmonic waves propagating in the positive and negative x-directions, respectively. The wave number is related to the wavelength, λ, of the motion by

$$\lambda = v\bar{T} = \frac{v}{\bar{f}} = \frac{2\pi}{\bar{\omega}} v = \frac{2\pi}{k} \tag{C.13}$$

where $\bar{T} = 2\pi/\bar{\omega}$ is the period of the applied loading and $\bar{f} = 1/\bar{T}$. Note that at a given frequency, the wavelength increases with increasing wave propagation velocity. Equation (C.12) indicates that the displacement varies harmonically with respect to both time and position as illustrated in Figure C.4. Equation (C.13) and Figure C.4 show that wave number is to wavelength as circular frequency is to period of vibration. For a wave propagating in the positive x-direction only ($B=0$), differentiating $u(x, t)$ twice with respect to x and twice with respect to t and substituting into the wave equation [Equation (C.10)] gives

$$-\bar{\omega}^2 A\cos(\bar{\omega}t - kx) = -v^2 k^2 A\cos(\bar{\omega}t - kx) \tag{C.14}$$

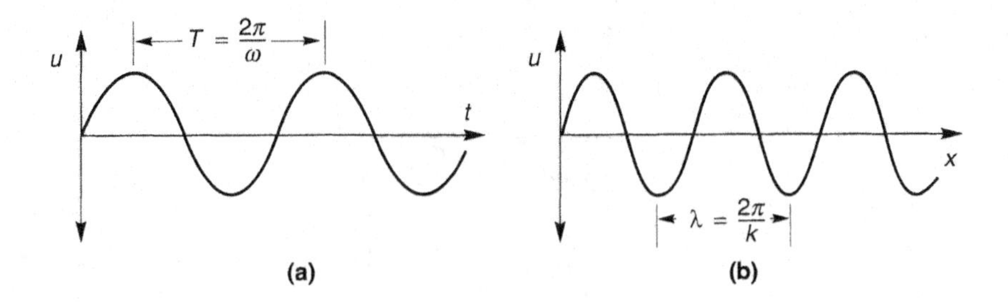

FIGURE C.4 Particle displacement (a) as function of time, and (b) as function of position along the rod.

which reduces to the identity $\bar{\omega} = kv$, thereby verifying Equation (C.12) as a solution to the wave equation.

Using complex notation (Appendix A), the equivalent form of the solution can be written as

$$u(x,t) = Ce^{i(\bar{\omega}t-kx)} + De^{i(\bar{\omega}t+kx)} \tag{C.15}$$

This form of the solution can be verified in the same way as the trigonometric form.

C.2.2 Three-Dimensional Wave Propagation

The preceding discussion of wave propagation in rods illustrates some of the basic principles of wave propagation, but an infinite rod is hardly an adequate model for describing the propagation of seismic waves through the Earth. Since the Earth is three-dimensional and sources of seismic energy are three-dimensional, seismic waves must be described in terms of three-dimensional wave propagation.

Derivations of three-dimensional equations of motion follow the same steps as those used for one-dimensional propagation; the equations of motion are formulated from equilibrium considerations, stress-strain relationships, and strain-displacement relationships. In the three-dimensional case, however, the various relationships are more complex and the derivation more cumbersome. Brief reviews of three-dimensional stress and strain notation and three-dimensional stress-strain behavior will precede the derivation of the equations of motion.

C.2.2.1 Review of Stress Notation

The stress at a point on some plane passing through a solid does not usually act normal to that plane but has both normal and shear components. Considering a small element with one corner at the center of an *x-y-z* Cartesian coordinate system (Figure C.5), a total of nine components of stress will act on its faces. These stresses are denoted by σ_{xx}, σ_{yy}, σ_{zz}, and so on, where the first and second letters in the subscript describe the direction of the stress itself and the axis perpendicular to the plane in which it acts, respectively. Thus σ_{xx}, σ_{yy}, σ_{zz} are normal stresses, while the other six components represent shear stresses. Moment equilibrium of the element requires that

$$\sigma_{xy} = \sigma_{yx} \quad \sigma_{xz} = \sigma_{zx} \quad \sigma_{yz} = \sigma_{zy} \tag{C.16}$$

which means that only six independent components of stress are required to define the state of stress of the element completely. In some references, the notation σ_x, σ_y, σ_z, τ_{xy}, τ_{yz}, and τ_{xz} is used to describe σ_{xx}, σ_{yy}, σ_{zz}, σ_{xy}, σ_{yz}, and σ_{xz}, respectively.

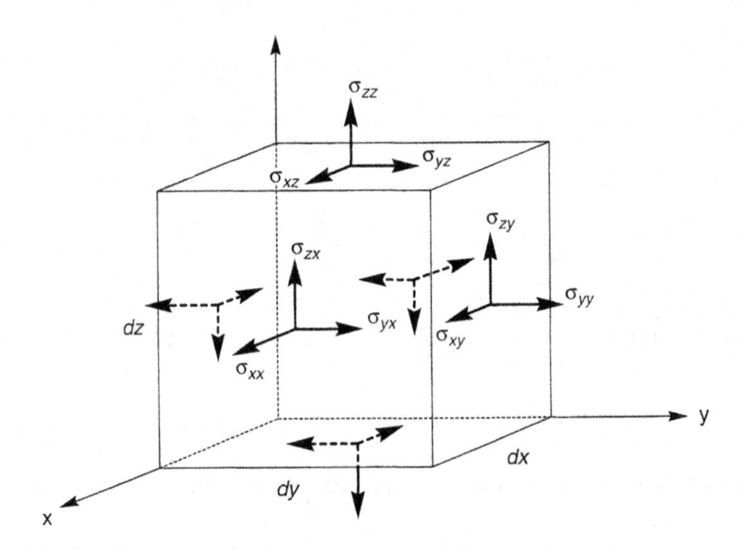

FIGURE C.5 Stress notation for element of dimensions dx by dy by dz.

C.2.2.2 Review of Strain Notation

Components of strain are easily visualized by considering the two-dimensional strain in the x-y plane shown in Figure C.6. The point P, at coordinates (x_0, y_0), is at one corner of the infinitesimal element $PQRS$ which has a square shape before deformation. After deformation, the infinitesimal element has been displaced, distorted, and rotated into the shape $P'Q'R'S'$. From Figure C.6, $\tan\alpha_1 = \partial v/\partial x$ and $\tan\alpha_2 = \partial u/\partial y$, where u and v represent displacements in the x- and y-directions, respectively. The shear strain in the x-y plane is given by $\varepsilon_{xy} = \alpha_1 + \alpha_2$. For small deformations, the angles may be taken equal to their tangents so that the relationship between the shear strain and the displacements is $\varepsilon_{xy} = \partial v/\partial x + \partial u/\partial y$. The rotation of the element about the z-axis is given by $\Omega = (\alpha_1 - \alpha_2)/2$.

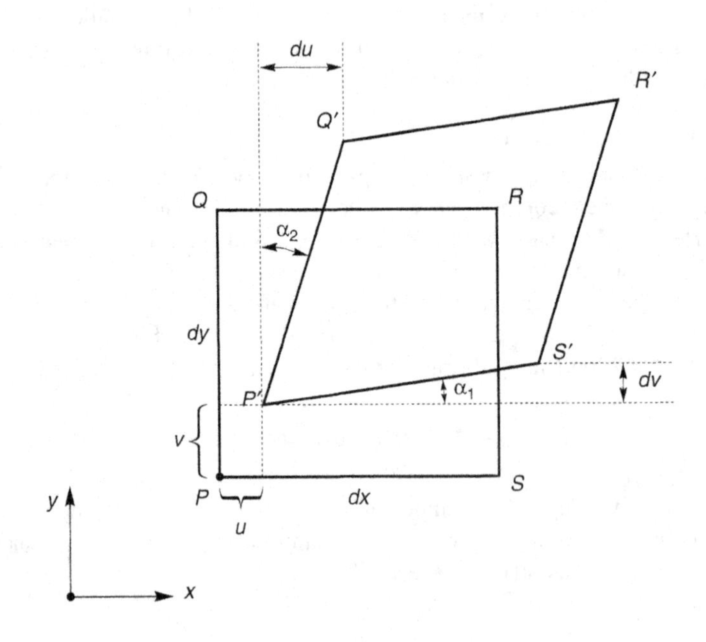

FIGURE C.6 Square element subjected to plane strain deformation.

Analogous definitions can be developed for the *x*-*z* and *y*-*z* planes. For the three-dimensional case, the strain-displacement relationships are defined by

$$\varepsilon_{xx} = \frac{du}{dx} \quad \varepsilon_{yy} = \frac{dv}{dy} \quad \varepsilon_{zz} = \frac{dw}{dz}$$

$$\varepsilon_{xy} = \frac{dv}{dx} + \frac{du}{dy} \quad \varepsilon_{yz} = \frac{dw}{dy} + \frac{dv}{dz} \quad \varepsilon_{zx} = \frac{du}{dz} + \frac{dw}{dx} \tag{C.17}$$

Rigid body rotation about the *x*-, *y*-, and *z*-axes are given by the rotation-displacement relationships

$$\Omega_x = \frac{1}{2}\left(\frac{dw}{dy} - \frac{dv}{dz}\right) \quad \Omega_y = \frac{1}{2}\left(\frac{du}{dz} - \frac{dw}{dx}\right) \quad \Omega_z = \frac{1}{2}\left(\frac{dv}{dx} - \frac{du}{dy}\right) \tag{C.18}$$

The first three quantities, ε_{xx}, ε_{yy}, and ε_{zz}, represent the extensional and compressional strains parallel to the *x*-, *y*-, and *z*-axes, and are called normal strains. The second three quantities, ε_{xy}, ε_{yz}, and ε_{zx}, represent the components of shear strain in the planes corresponding to their suffixes. These six quantities are the components of strain that correspond to the deformation at *P*. In some references, the notation ε_x, ε_y, ε_z, γ_{xy}, γ_{yz}, and γ_{zx} is used to describe ε_{xx}, ε_{yy}, ε_{zz}, ε_{xy}, ε_{yz}, and ε_{zx} respectively.

C.2.2.3 Review of Stress-Strain Relationships

Stresses and strains are proportional in a linear elastic body. The stress-strain relationship can be described by Hooke's law, which can be written in generalized form as

$$\sigma_{xx} = c_{11}\varepsilon_{xx} + c_{12}\varepsilon_{yy} + c_{13}\varepsilon_{zz} + c_{14}\varepsilon_{xy} + c_{15}\varepsilon_{yz} + c_{16}\varepsilon_{zx}$$

$$\sigma_{yy} = c_{21}\varepsilon_{xx} + c_{22}\varepsilon_{yy} + c_{23}\varepsilon_{zz} + c_{24}\varepsilon_{xy} + c_{25}\varepsilon_{yz} + c_{26}\varepsilon_{zx}$$

$$\sigma_{zz} = c_{31}\varepsilon_{xx} + c_{32}\varepsilon_{yy} + c_{33}\varepsilon_{zz} + c_{34}\varepsilon_{xy} + c_{35}\varepsilon_{yz} + c_{36}\varepsilon_{zx}$$

$$\sigma_{xy} = c_{41}\varepsilon_{xx} + c_{42}\varepsilon_{yy} + c_{43}\varepsilon_{zz} + c_{44}\varepsilon_{xy} + c_{45}\varepsilon_{yz} + c_{46}\varepsilon_{zx}$$

$$\sigma_{yz} = c_{51}\varepsilon_{xx} + c_{52}\varepsilon_{yy} + c_{53}\varepsilon_{zz} + c_{54}\varepsilon_{xy} + c_{55}\varepsilon_{yz} + c_{56}\varepsilon_{zx}$$

$$\sigma_{zx} = c_{61}\varepsilon_{xx} + c_{62}\varepsilon_{yy} + c_{63}\varepsilon_{zz} + c_{64}\varepsilon_{xy} + c_{65}\varepsilon_{yz} + c_{66}\varepsilon_{zx} \tag{C.19}$$

where the 36 coefficients represent the elastic constants of the material. The requirement that the elastic strain energy must be a unique function of the strain (which requires that $c_{ij}=c_{ji}$ for all *i* and *j*) reduces the number of independent coefficients to 21. If the material is isotropic, the coefficients must be independent of direction, so that

$$c_{12} = c_{21} = c_{13} = c_{31} = c_{23} = c_{32} = \lambda$$

$$c_{44} = c_{55} = c_{66} = \mu$$

$$c_{11} = c_{22} = c_{33} = \lambda + 2\mu \tag{C.20}$$

and all other constants are zero. Therefore, Hooke's law for an isotropic, linear, elastic material allows all components of stress and strain to be expressed in terms of the two *Lame constants*, λ and μ:

$$\sigma_{xx} = \lambda\bar{\varepsilon} + 2\mu\varepsilon_{xx} \quad \sigma_{xy} = \mu\varepsilon_{xy}$$

$$\sigma_{yy} = \lambda\bar{\varepsilon} + 2\mu\varepsilon_{yy} \quad \sigma_{yz} = \mu\varepsilon_{yz}$$

$$\sigma_{zz} = \lambda\bar{\varepsilon} + 2\mu\varepsilon_{zz} \quad \sigma_{zx} = \mu\varepsilon_{zx} \tag{C.21}$$

where the volumetric strain $\bar{\varepsilon} = \varepsilon_{xx} + \varepsilon_{yy} + \varepsilon_{zz}$. Note that the symbol λ is used universally for both Lame's constant and for wavelength; the context in which it is used should make its meaning obvious.

For convenience, several other parameters are often used to describe the stress-strain behavior of isotropic, linear, and elastic materials, each of which can be expressed in terms of Lame's constants. Some of the more common of these are

$$\text{Young's modulus: } E = \frac{\mu(3\lambda + 2\mu)}{\lambda + \mu} \tag{C.22a}$$

$$\text{Bulk modulus: } K = \lambda + \frac{2\mu}{3} \tag{C.22b}$$

$$\text{Shear modulus: } G = \mu \tag{C.22c}$$

$$\text{Constrained modulus: } M = \lambda + 2\mu \tag{C.22d}$$

$$\text{Poisson's ratio: } \nu = \frac{\lambda}{2(\lambda + \mu)} \tag{C.22e}$$

Hooke's law for an isotropic, linear, elastic material can be expressed using any combination of two of these parameters.

C.2.2.4 Equations of Motion for a Three-Dimensional Elastic Solid

The three-dimensional equations of motion for an elastic solid are obtained from equilibrium requirements in much the same way as for the one-dimensional rod, except that equilibrium must be ensured in three perpendicular directions. Consider the variation in stress across an infinitesimal cube aligned with its sides parallel to the *x-y-z* axes shown in Figure C.7. Assuming that the average stress on each face of the cube is represented by the stress shown at the center of the face, the resultant forces acting in the *x*-, *y*-, and *z*-directions can be evaluated; In the *x*-direction, the unbalanced external forces must be balanced by an inertial force in that direction, so that

$$\rho dxdydz\frac{\partial^2 u}{\partial t^2} = \left(\sigma_{xx} + \frac{\partial \sigma_{xx}}{\partial x}dx\right)dydz - \sigma_{xx}dydz + \left(\sigma_{xy} + \frac{\partial \sigma_{xy}}{\partial y}dy\right)dxdz - \sigma_{xy}dxdz$$
$$+\left(\sigma_{xz} + \frac{\partial \sigma_{xz}}{\partial z}dz\right)dxdy - \sigma_{xz}dxdy \tag{C.23}$$

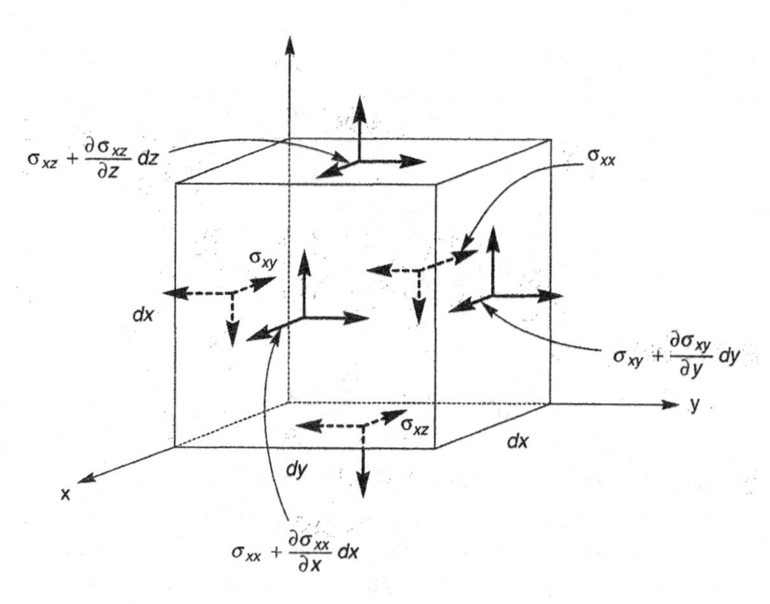

FIGURE C.7 Stresses in x-direction on infinitesimal cube.

which simplifies to

$$\rho \frac{\partial^2 u}{\partial t^2} = \frac{\partial \sigma_{xx}}{\partial x} + \frac{\partial \sigma_{xy}}{\partial y} + \frac{\partial \sigma_{xz}}{\partial z} \quad \text{(C.24a)}$$

Repeating this operation in the y- and z-directions

$$\rho \frac{\partial^2 v}{\partial t^2} = \frac{\partial \sigma_{yx}}{\partial x} + \frac{\partial \sigma_{yy}}{\partial y} + \frac{\partial \sigma_{yz}}{\partial z} \quad \text{(C.24b)}$$

$$\rho \frac{\partial^2 w}{\partial t^2} = \frac{\partial \sigma_{zx}}{\partial x} + \frac{\partial \sigma_{zy}}{\partial y} + \frac{\partial \sigma_{zz}}{\partial z} \quad \text{(C.24c)}$$

Equations (C.24) represent the three-dimensional equations of motion of an elastic solid. Note that these equations of motion were derived solely based on equilibrium considerations and thus apply to solids of any stress-strain behavior. To express these equations of motion in terms of displacements, it is again necessary to use a stress-strain relationship and a strain-displacement relationship. Using Hooke's law as developed in Section C.2.2.3, the first of the equations of motion [Equation (C.24a)] can be written in terms of strains as follows:

$$\rho \frac{\partial^2 u}{\partial t^2} = \frac{\partial}{\partial x}(\lambda \bar{\varepsilon} + 2\mu \varepsilon_{xx}) + \frac{\partial}{\partial y}(\mu \varepsilon_{xy}) + \frac{\partial}{\partial z}(\mu \varepsilon_{xz}) \quad \text{(C.25)}$$

Substituting the strain-displacement relationships

$$\varepsilon_{xx} = \frac{\partial u}{\partial x} \qquad \varepsilon_{xy} = \frac{\partial v}{\partial x} + \frac{\partial u}{\partial y} \qquad \varepsilon_{xz} = \frac{\partial w}{\partial x} + \frac{\partial u}{\partial z}$$

into Equation (C.25) produces the desired equation of motion in terms of displacements:

$$\rho \frac{\partial^2 u}{\partial t^2} = (\lambda + \mu) \frac{\partial \bar{\varepsilon}}{\partial x} + \mu \nabla^2 u \quad \text{(C.26a)}$$

where the Laplacian operator, ∇^2, represents

$$\nabla^2 = \frac{\partial^2}{\partial x^2} + \frac{\partial^2}{\partial y^2} + \frac{\partial^2}{\partial z^2}$$

Repeating this process in the y- and z-directions gives

$$\rho \frac{\partial^2 v}{\partial t^2} = (\lambda + \mu)\frac{\partial \bar{\varepsilon}}{\partial y} + \mu \nabla^2 v \tag{C.26b}$$

$$\rho \frac{\partial^2 w}{\partial t^2} = (\lambda + \mu)\frac{\partial \bar{\varepsilon}}{\partial z} + \mu \nabla^2 w \tag{C.26c}$$

C.2.2.5 Solutions of the Three-Dimensional Equations of Motion

Together, Equations (C.26) represent the three-dimensional equations of motion for an isotropic, linear, elastic solid. It turns out that these equations can be manipulated to produce two wave equations. Consequently, only two types of waves can travel through such an unbounded solid. The characteristics of each type of wave will be revealed by their respective wave equations.

The solution for the first type of wave can be obtained by differentiating each of Equations (C.26) with respect to x, y, and z and adding the results together to give

$$\rho\left(\frac{\partial^2 \varepsilon_{xx}}{\partial t^2} + \frac{\partial^2 \varepsilon_{yy}}{\partial t^2} + \frac{\partial^2 \varepsilon_{zz}}{\partial t^2}\right) = (\lambda + \mu)\left(\frac{\partial^2 \bar{\varepsilon}}{\partial x^2} + \frac{\partial^2 \bar{\varepsilon}}{\partial y^2} + \frac{\partial^2 \bar{\varepsilon}}{\partial z^2}\right) + \mu\left(\frac{\partial^2 \varepsilon_{xx}}{\partial x^2} + \frac{\partial^2 \varepsilon_{yy}}{\partial y^2} + \frac{\partial^2 \varepsilon_{zz}}{\partial z^2}\right) \tag{C.27a}$$

or

$$\rho \frac{\partial^2 \bar{\varepsilon}}{\partial t^2} = (\lambda + \mu)\nabla^2 \bar{\varepsilon} + \mu \nabla^2 \bar{\varepsilon} \tag{C.27b}$$

Rearranging yields the wave equation

$$\frac{\partial^2 \bar{\varepsilon}}{\partial t^2} = \frac{\lambda + 2\mu}{\rho}\nabla^2 \bar{\varepsilon} \tag{C.28}$$

Recalling that $\bar{\varepsilon}$ is the volumetric strain (which describes deformations that involve contraction/dilation but no shearing or rotation), this wave equation describes an irrotational, or dilatational, wave. It indicates that a dilatational wave will propagate through the body at a velocity

$$v_p = \sqrt{\frac{\lambda + 2\mu}{\rho}} \tag{C.29}$$

This type of wave is commonly known as a p-wave (or primary wave) and v_p is referred to as the *p-wave velocity* of the material. The general nature of p-wave motion is illustrated in Figure 2.1a. Note that particle displacements are parallel to the direction of wave propagation, just as they were in the constrained rod of Section C.2.1.1. The longitudinal wave in the constrained rod is actually a p-wave. Using Equations (C.22c) and (C.22e), v_p can be written in terms of the shear modulus and Poisson's ratio as

$$v_p = \sqrt{\frac{G(2 - 2\nu)}{\rho(1 - 2\nu)}} \tag{C.30}$$

As ν approaches 0.5 (at which point the body becomes incompressible, i.e., infinitely stiff with respect to dilatational deformations), v_p approaches infinity.

To obtain the solution for the second type of wave, $\bar{\varepsilon}$ is eliminated by differentiating Equation (C.26b) with respect to z and Equation (C.26c) with respect to y, and subtracting one from the other:

$$\rho\frac{\partial}{\partial t^2}\left(\frac{\partial w}{\partial y}-\frac{\partial v}{\partial z}\right)=\mu\nabla^2\left(\frac{\partial w}{\partial y}-\frac{\partial v}{\partial z}\right) \tag{C.31}$$

Recalling the definition of rotation [Equation (C.18)], Equation (C.31) can be written in the form of the wave equation

$$\frac{\partial^2\Omega_x}{\partial t^2}=\frac{\mu}{\rho}\nabla^2\Omega_x \tag{C.32}$$

which describes an equivoluminal, or distortional wave, of rotation about the x-axis. Similar expressions can be obtained by the same process for rotation about the y- and z-axes. Equation (C.32) shows that a distortional wave will propagate through the solid at a velocity

$$v_s=\sqrt{\frac{\mu}{\rho}}=\sqrt{\frac{G}{\rho}} \tag{C.33}$$

This type of wave is commonly known as an s-wave (or shear wave) and v_s is referred to as the *shear wave velocity* of the material. Note that the particle motion is constrained to a plane perpendicular to the direction of wave propagation, just as it was in the case of the torsional wave of Section C.2.1.2. Consequently, the torsional wave represented a form of an s-wave. The close relationship between s-wave velocity and shear modulus is used to advantage in many of the field and laboratory tests discussed in Chapter 6. The general nature of s-wave motion is illustrated in Figure 2.1b.

S-waves are often divided into two types, or resolved into two perpendicular components. SH-waves are s-waves in which particle motion occurs only in a horizontal plane. SV-waves are s-waves whose particle motion lies in a vertical plane. A given s-wave with arbitrary particle motion can be represented as the vector sum of its SH and SV components.

In summary, only two types of waves, known as body waves, can exist in an unbounded (infinite) elastic solid. P-waves involve no rotation of the material they pass through and travel at velocity, v_p. S-waves involve no volume change and travel at velocity, v_s. The velocities of p- and s-waves depend on the stiffnesses of the solid with respect to the types of deformation induced by each wave. Comparing the velocities [Equations (C.30) and (C.33)]

$$\frac{v_p}{v_s}=\sqrt{\frac{2-2\nu}{1-2\nu}} \tag{C.34}$$

the p-wave velocity can be seen to exceed the s-wave velocity by an amount that depends on the compressibility (as reflected in Poisson's ratio) of the body. For a typical Poisson's ratio of 0.3 for geologic materials, the ratio $v_p/v_s=1.87$. Thus, an observer at some distance from a disturbance that produces both would expect to feel p-waves arriving prior to s-waves.

C.3 WAVES IN A SEMI-INFINITE BODY

The Earth is obviously not an infinite body – it is a very large sphere with an outer surface on which stresses cannot exist. For near-surface earthquake engineering problems, the Earth is often idealized as a semi-infinite body with a planar free surface (the effects of the Earth's curvature are neglected). The boundary conditions associated with the free surface allow additional solutions to the equations of motion to be obtained. These solutions describe waves whose motion is concentrated in a shallow zone near the free surface (i.e., *surface waves*). Since earthquake engineering is concerned with the effects of earthquakes on humans and their environment, which are located on or very near the Earth's surface, and since they attenuate with distance more slowly than body waves, surface waves are very important.

Two types of surface waves are of primary importance in earthquake engineering. One, the *Rayleigh wave*, can be shown to exist in a homogeneous, elastic half-space. The other surface wave, the *Love wave*, requires a surficial layer of lower s-wave velocity than the underlying half-space. Other types of surface waves exist but are much less significant from an earthquake engineering standpoint.

C.3.1 Rayleigh Waves

Waves that exist near the surface of a homogeneous elastic half-space were first investigated by Rayleigh (1885) and are known to this date as Rayleigh waves. To describe Rayleigh waves, consider a plane wave (Figure C.8) that travels in the x-direction with zero particle displacement in the y-direction ($v=0$). The z-direction is taken as positive downward, so all particle motion occurs in the x-z plane. Two potential functions, Φ and Ψ, can be defined to describe the displacements in the x- and z-directions:

$$u = \frac{\partial \Phi}{\partial x} + \frac{\partial \Psi}{\partial z} \tag{C.35a}$$

$$w = \frac{\partial \Phi}{\partial z} - \frac{\partial \Psi}{\partial x} \tag{C.35b}$$

The volumetric strain, or dilatation, $\bar{\varepsilon}$, of the wave is given by $\bar{\varepsilon} = \varepsilon_{xx} + \varepsilon_{zz}$, or

$$\bar{\varepsilon} = \frac{\partial u}{\partial x} + \frac{\partial w}{\partial z} = \frac{\partial}{\partial x}\left(\frac{\partial \Phi}{\partial x} + \frac{\partial \Psi}{\partial z}\right) + \frac{\partial}{\partial z}\left(\frac{\partial \Phi}{\partial z} - \frac{\partial \Psi}{\partial x}\right) = \frac{\partial^2 \Phi}{\partial x^2} + \frac{\partial^2 \Phi}{\partial z^2} = \nabla^2 \Phi \tag{C.36}$$

The rotation in the x-z plane [Equation (5.18)] is given by

$$2\Omega_y = \frac{\partial u}{\partial z} - \frac{\partial w}{\partial x} = \frac{\partial}{\partial z}\left(\frac{\partial \Phi}{\partial x} + \frac{\partial \Psi}{\partial z}\right) - \frac{\partial}{\partial z}\left(\frac{\partial \Phi}{\partial z} - \frac{\partial \Psi}{\partial x}\right) = \frac{\partial^2 \Psi}{\partial z^2} + \frac{\partial^2 \Psi}{\partial x^2} = \nabla^2 \Psi \tag{C.37}$$

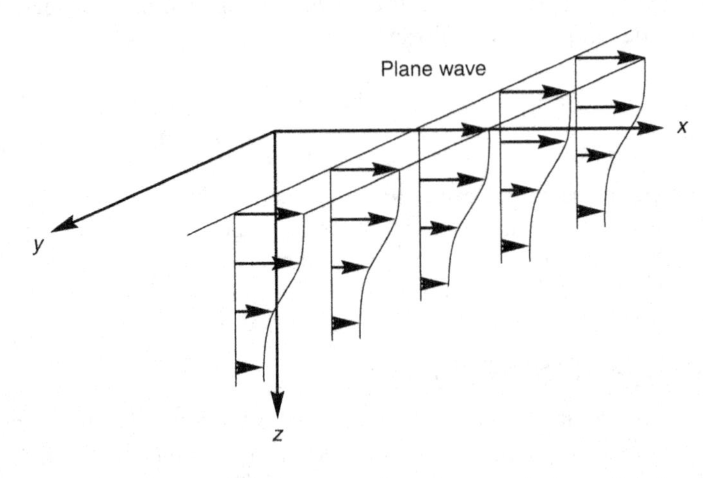

FIGURE C.8 Motion induced by a typical plane wave that propagates in the x-direction. Wave motion does not vary in the y-direction.

Use of the potential functions allows separation of the effects of dilatation and rotation [i.e., Equations (C.36) and (C.37) indicate that Φ and Ψ are associated with dilatation and rotation, respectively]. Therefore, Rayleigh waves can be thought of as combinations of p- and s-waves (SV waves for this case, since the x-z plane is vertical) that satisfy certain boundary conditions. Substitution of the expressions for u and w into the equations of motion as written in Equations (C.26a) and (C.26c) gives

$$\rho\frac{\partial}{\partial x}\left(\frac{\partial^2\Phi}{\partial t^2}\right)+\rho\frac{\partial}{\partial z}\left(\frac{\partial^2\Psi}{\partial t^2}\right)=(\lambda+2\mu)\frac{\partial}{\partial x}\left(\nabla^2\Phi\right)+\mu\frac{\partial}{\partial z}\left(\nabla^2\Psi\right) \tag{C.38a}$$

$$\rho\frac{\partial}{\partial z}\left(\frac{\partial^2\Phi}{\partial t^2}\right)-\rho\frac{\partial}{\partial x}\left(\frac{\partial^2\Psi}{\partial t^2}\right)=(\lambda+2\mu)\frac{\partial}{\partial z}\left(\nabla^2\Phi\right)-\mu\frac{\partial}{\partial x}\left(\nabla^2\Psi\right) \tag{C.38b}$$

Solving Equations (C.38) simultaneously for $\partial^2\Phi/\partial t^2$ and $\partial^2\Psi/\partial t^2$ shows

$$\frac{\partial^2\Phi}{\partial t^2}=\frac{\lambda+2\mu}{\rho}\nabla^2\Phi=v_p^2\nabla^2\Phi \tag{C.39a}$$

$$\frac{\partial^2\Psi}{\partial t^2}=\frac{\mu}{\rho}\nabla^2\Psi=v_s^2\nabla^2\Psi \tag{C.39b}$$

If the wave is harmonic with frequency ω and wave number k_R so that it propagates with Rayleigh wave velocity $v_R=\omega/k_R$, the potential functions can be expressed as

$$\Phi=F(z)e^{i(\omega t-k_R x)} \tag{C.40a}$$

$$\Psi=G(z)e^{i(\omega t-k_R x)} \tag{C.40b}$$

where F and G are functions that describe the manner in which the amplitude of the dilatational and rotational components of the Rayleigh wave vary with depth. Substituting these expressions for Φ and Ψ into Equations (C.39) gives

$$-\frac{\omega^2}{v_p^2}F(z)=-k_R^2F(z)+\frac{d^2F(z)}{dz^2} \tag{C.41a}$$

$$-\frac{\omega^2}{v_s^2}G(z)=-k_R^2G(z)+\frac{d^2G(z)}{dz^2} \tag{C.41b}$$

which can be rearranged to give the second-order differential equations

$$\frac{d^2F}{dz^2}-\left(k_R^2-\frac{\omega^2}{v_p^2}\right)F=0 \tag{C.42a}$$

$$\frac{d^2G}{dz^2}-\left(k_R^2-\frac{\omega^2}{v_s^2}\right)G=0 \tag{C.42b}$$

The general solution to these equations can be written as

$$F(z) = A_1 e^{-qz} + B_1 e^{qz} \tag{C.43a}$$

$$G(z) = A_2 e^{-sz} + B_2 e^{sz} \tag{C.43b}$$

where

$$q^2 = k_R^2 - \frac{\omega^2}{v_p^2}$$

$$s^2 = k_R^2 - \frac{\omega^2}{v_s^2}$$

The second terms in Equations (C.43a and C.43b) correspond to a disturbance whose displacement amplitude approaches infinity with increasing depth. Since this type of behavior is not realistic, B_1 and B_2 must be zero, and the potential functions can finally be written as

$$\Phi = A_1 e^{-qz+i(\omega t - k_R x)} \tag{C.44a}$$

$$\Psi = A_2 e^{-sz+i(\omega t - k_R x)} \tag{C.44b}$$

Since neither shear nor normal stresses can exist at the free surface of the half-space, $\sigma_{xz} = 0$ and $\sigma_{zz} = 0$ when $z = 0$. Therefore,

$$\sigma_{zz} = \lambda\bar{\varepsilon} + 2\mu\varepsilon_{zz} = \lambda\bar{\varepsilon} + 2\mu\frac{dw}{dz} = 0 \tag{C.45a}$$

$$\sigma_{xz} = \mu\varepsilon_{xz} = \mu\left(\frac{dw}{dx} + \frac{du}{dz}\right) = 0 \tag{C.45b}$$

Using the potential function definitions of u and w [Equations (C.35)] and the solution for the potential functions [Equation (C.44)], the free surface boundary conditions can be rewritten as

$$\sigma_{zz}(z=0) = A_1\left[(\lambda + 2\mu)q^2 - \lambda k_R^2\right] - 2iA_2\mu k_R s = 0 \tag{C.46a}$$

$$\sigma_{xz}(z=0) = 2iA_1 k_R q + A_2\left(s^2 + k_R^2\right) = 0 \tag{C.46b}$$

which can be rearranged to yield

$$\frac{A_1}{A_2}\frac{(\lambda + 2\mu)q^2 - \lambda k_R^2}{2i\mu k_R s} - 1 = 0 \tag{C.47a}$$

$$\frac{A_1}{A_2}\frac{2iqk_R}{s^2 + k_R^2} + 1 = 0 \tag{C.47b}$$

With these results, the velocities and displacement patterns of Rayleigh waves can be determined.

C.3.1.1 Rayleigh Wave Velocity

The velocity at which Rayleigh waves travel is of interest in geotechnical earthquake engineering. As discussed in Chapter 6, Rayleigh waves are often mechanically generated and their velocities are measured in the field to investigate the stiffness of surficial soils. Adding Equations (C.47) and cross-multiplying gives

$$4q\mu sk_R^2 = \left(s^2 + k_R^2\right)\left[(\lambda+2\mu)q^2 - \lambda k_R^2\right] \tag{C.48}$$

which, upon introducing the definitions of q and s and factoring out a $G^2k_R^8$ term, yields

$$16\left(1-\frac{\omega^2}{v_p^2k_R^2}\right)\left(1-\frac{\omega^2}{v_s^2k_R^2}\right) = \left(2-\frac{\lambda+2\mu}{\mu}\frac{\omega^2}{v_p^2k_R^2}\right)^2\left(2-\frac{\omega^2}{v_s^2k_R^2}\right)^2 \tag{C.49}$$

Defining K_{Rs} as the ratio of the Rayleigh wave velocity to the s-wave velocity

$$K_{Rs} = \frac{v_R}{v_s} = \frac{\omega}{v_sk_R} \tag{C.50a}$$

then

$$\frac{v_R}{v_p} = \frac{\omega}{v_pk_R} = \frac{\omega}{v_sk_R\sqrt{(\lambda+2\mu)/\mu}} = \alpha K_{Rs} \tag{C.50b}$$

where $\alpha = \sqrt{\mu/(\lambda+2\mu)} = \sqrt{(1-2\nu)/(2-2\nu)}$. Then Equation (C.49) can be rewritten as

$$16\left(1-\alpha^2K_{Rs}^2\right)\left(1-K_{Rs}^2\right) = \left(2-\frac{1}{\alpha^2}\alpha^2K_{Rs}^2\right)^2\left(2-K_{Rs}^2\right)^2 \tag{C.51a}$$

which can be expanded and rearranged into the equation

$$K_{Rs}^6 - 8K_{Rs}^4 + \left(24-16\alpha^2\right)K_{Rs}^2 + 16\left(\alpha^2-1\right) = 0 \tag{C.51b}$$

This equation is cubic in K_{Rs}^2, and real solutions for K_{Rs} can be found for various values of Poisson's ratio. These allow evaluation of the ratios of the Rayleigh wave velocity to both s- and p-wave velocities as functions of ν. The solution shown in Figure C.9 shows that Rayleigh waves travel slightly slower than s-waves for all values of Poisson's ratio except 0.5.

C.3.1.2 Rayleigh Wave Displacement Amplitude

Section C.3.1.1 showed how the velocity of a Rayleigh wave compares with that of p and s-waves. Some of the intermediate results of that section can be used to illustrate the nature of particle motion during the passage of Rayleigh waves. Substituting the solutions for the potential functions Φ and Ψ [Equations (C.44)] into the expressions for u and w [Equation (C.35)] and carrying out the necessary partial differentiations yields

$$u = -A_1ik_Re^{-qz+i(\omega t-k_Rx)} - A_2se^{-sz+i(\omega t-k_Rx)} \tag{C.52a}$$

$$w = \left(-A_1ik_Re^{-qz+i(\omega t-k_Rx)}\right) + A_2ik_Re^{-sz+i(\omega t-k_Rx)} \tag{C.52b}$$

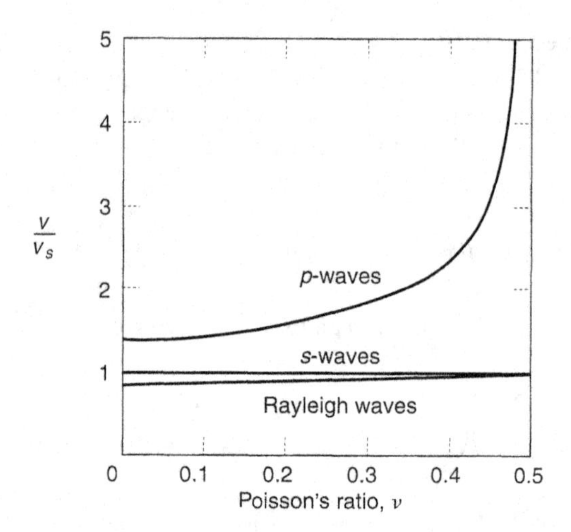

FIGURE C.9 Variation of Rayleigh wave and body wave propagation velocities with Poisson's ratio.

Rearranging Equation (C.47b) as $A_2 = -\frac{2qik_R}{s^2 + k_R^2} A_1$ and substituting into Equations (C.52), gives

$$u = A_1 \left(-ik_R e^{-qz} + \frac{2iqsk_R}{s^2 + k_R^2} e^{-sz} \right) e^{i(\omega t - k_R x)} \tag{C.53a}$$

$$w = A_1 \left(\frac{2qk_R^2}{s^2 + k_R^2} e^{-sz} - qe^{-qz} \right) e^{i(\omega t - k_R x)} \tag{C.53b}$$

where the terms in parentheses describe the variation of the amplitudes of u and w with depth. These horizontal and vertical displacement amplitudes are illustrated for several values of Poisson's ratio in Figure C.10. Examination of Equations (C.53) indicates that the horizontal and vertical displacements are out of phase by 90°. Hence the horizontal displacement will be zero when the vertical displacement reaches its maximum (or minimum), and vice versa. The motion of a particle near the surface of the half-space is in the form of a retrograde ellipse (as opposed to the prograde ellipse particle motion observed at the surface of water waves). The general nature of Rayleigh wave motion is illustrated in Figure 2.2a.

The Rayleigh waves produced by earthquakes were once thought to appear only at very large site-to-source distances (several hundred km). It is now recognized, however, that they can be significant at much shorter distances (a few tens of kilometers). The ratio of minimum epicentral distance, R, to focal depth, h, at which Rayleigh waves first appear in a homogeneous medium is given by

$$\frac{R}{h} = \frac{1}{\sqrt{(v_P / v_R)^2 - 1}} \tag{C.54}$$

where v_p and v_R are the wave propagation velocities of p-waves and Rayleigh waves, respectively (Ewing et al., 1957).

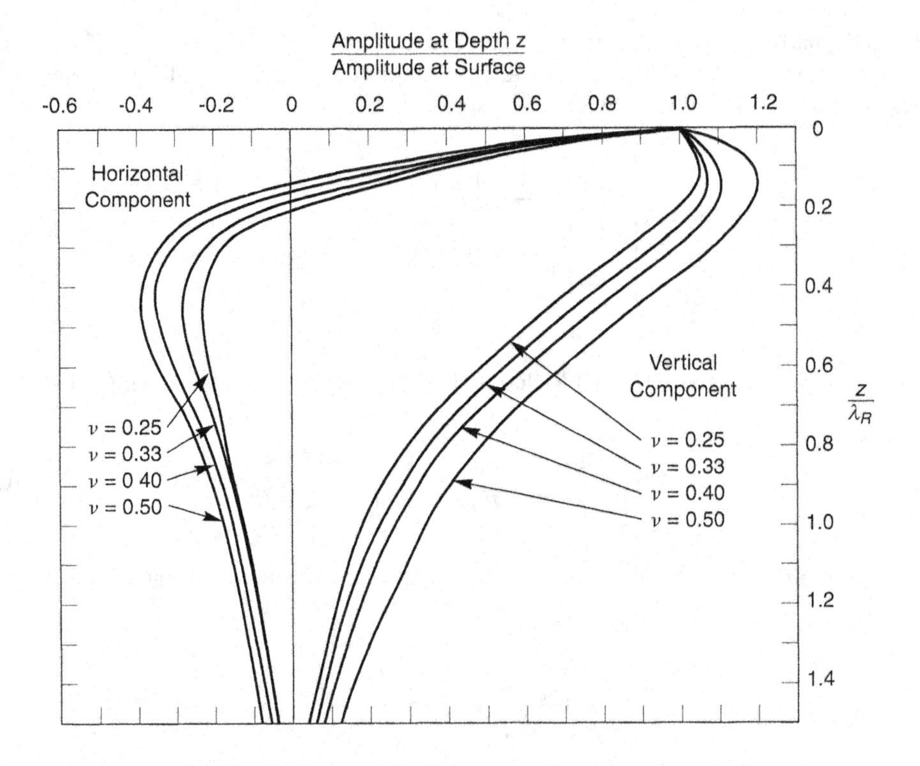

FIGURE C.10 Horizontal and vertical motion of Rayleigh waves. A negative amplitude ratio indicates that the displacement is in the opposite direction of the surface displacement. (After Richart et al., 1970.)

C.3.2 Love Waves

In a homogeneous elastic half-space, only p-waves, s-waves, and Rayleigh waves can exist. If the half-space is overlain by a layer of material with lower body wave velocity, however, Love waves can also develop (Love, 1927). Love waves essentially consist of SH-waves that are trapped by multiple reflections within the surficial layer. Consider the case of a homogeneous surficial layer of thickness, H, overlying a homogeneous half-space as shown in Figure C.11. A Love wave traveling in the +x-direction would involve particle displacements only in the y-direction (SH-wave motion), and could be described by the equation

$$v(x,z,t) = V(z)e^{i(k_L x - \omega t)} \tag{C.55}$$

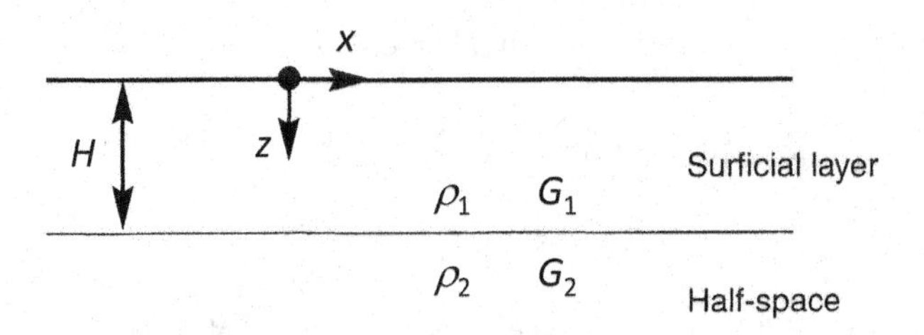

FIGURE C.11 Schematic illustration of softer surficial layer ($G_1/\rho_1 < G_2/\rho_2$) overlying elastic half-space, the simplest conditions for which Love waves can exist.

where v is the particle displacement in the y-direction, $V(z)$ describes the variation of v with depth, and k_L is the wave number of the Love wave. The Love wave must satisfy the wave equations for s-waves in both the surficial layer and the half-space

$$\frac{\partial^2 v}{\partial t^2} = \begin{cases} \dfrac{G_1}{\rho_1}\left(\dfrac{\partial^2 v}{\partial x^2} + \dfrac{\partial^2 v}{\partial z^2}\right) & \text{for} \quad 0 \le z \le H \\ \dfrac{G_2}{\rho_2}\left(\dfrac{\partial^2 v}{\partial x^2} + \dfrac{\partial^2 v}{\partial z^2}\right) & \text{for} \quad z \ge H \end{cases} \tag{C.56}$$

The amplitude can be shown (Aki and Richards, 1980) to vary with depth according to

$$V(z) = \begin{cases} A_1 e^{-\nu_1 z} + B_1 e^{\nu_1 z} & \text{for} \quad 0 \le z < H \\ A_2 e^{-\nu_2 z} + B_2 e^{\nu_2 z} & \text{for} \quad z \ge H \end{cases} \tag{C.57a}$$

where the A and B coefficients describe the amplitudes of downgoing and upgoing waves, respectively, and

$$\nu_1 = \sqrt{\frac{k_L^2 - \omega^2}{G_1 / \rho_1}} \qquad \nu_2 = \sqrt{\frac{k_L^2 - \omega^2}{G_2 / \rho_2}} \tag{C.57b}$$

Since the half-space extends to infinite depth, B_2 must be zero (no energy can be supplied or reflected at infinite depth to produce an upgoing wave). The requirement that all stresses vanish at the ground surface is satisfied if

$$\frac{\partial v}{\partial z} = \frac{\partial V(z)}{\partial z} e^{i(k_L x - \omega t)} = -A_1 \nu_1 e^{-\nu_1 z} + \nu_1 B_1 e^{\nu_1 z} = (A_1 - B_1)\nu_1 \left(e^{-\nu_1 z} + e^{\nu_1 z}\right) = 0 \tag{C.58}$$

in other words, if $A_1 = B_1$. The amplitudes can now be rewritten in terms of the two remaining unknown amplitudes as

$$V(z) = \begin{cases} A_1\left(e^{-\nu_1 z} + e^{\nu_1 z}\right) & \text{for} \quad 0 \le z < H \\ A_2 e^{-\nu_2 z} & \text{for} \quad z \ge H \end{cases} \tag{C.59}$$

At the $z = H$ interface, continuity of stresses requires that

$$2iG_1 \nu_1 A_1 \sin(i\nu_1 H) = G_2 \nu_2 A_2 e^{-\nu_2 H} \tag{C.60}$$

and compatibility of displacements requires that

$$2A_1 \cos(i\nu_1 H) = A_2 e^{-\nu_2 H} \tag{C.61}$$

Using Equations (C.60) and (C.61), A_2 can be expressed in terms of A_1 by

$$A_2 = \frac{2\cos(i\nu_1 H)}{e^{-\nu_2 H}} A_1 \tag{C.62}$$

Substituting Equations (C.59) and (C.60) into (C.55) gives

$$v(x,z,t) = \begin{cases} 2A_1 \cos\left[\omega\left(\frac{1}{v_{s1}^2} - \frac{1}{v_L^2}\right)^{1/2} z\right] e^{i(k_L x - \omega t)} & \text{for} \quad 0 \le z < H \\ 2A_1 \cos\left[\omega\left(\frac{1}{v_{s1}^2} - \frac{1}{v_L^2}\right)^{1/2} H\right] \exp\left[-\omega\left(\frac{1}{v_L^2} - \frac{1}{v_{s2}^2}\right)^{1/2} (z - H)\right] e^{i(k_L x - \omega t)} & \text{for} \quad z \ge H \end{cases} \tag{C.63}$$

where v_{s1} and v_{s2} are the shear wave velocities of materials 1 and 2, respectively, and v_L is the velocity of the Love wave. Equation (C.63) shows, as illustrated in Figure C.12, that the Love wave displacement amplitude varies with depth as a cosine function in the surficial layer and decays exponentially with depth in the underlying half-space. Because of this, Love waves are often described as SH-waves that are trapped in the surficial layer. The general nature of Love wave displacement is shown in Figure 2.2b.

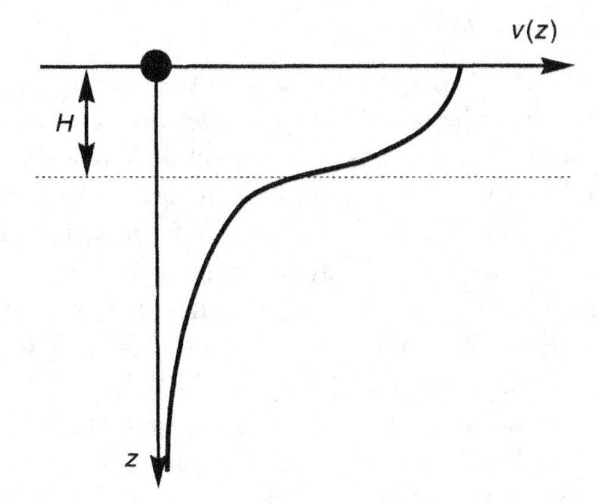

FIGURE C.12 Variation of particle displacement amplitude with depth for Love waves.

The Love wave velocity is given by the solution of

$$\tan \omega H\left(\frac{1}{v_{s1}^2} - \frac{1}{v_L^2}\right)^{1/2} = \frac{G_2}{G_1}\frac{\sqrt{1/v_L^2 - 1/v_{s2}^2}}{\sqrt{1/v_{s1}^2 - 1/v_L^2}} \tag{C.64}$$

which indicates, as illustrated in Figure C.13 that Love wave velocities range from the s-wave velocity of the half-space (at very low frequencies) to the s-wave velocity of the surficial layer (at very high frequencies). This frequency dependence indicates that Love waves are dispersive (Section C.3.4).

C.3.3 Higher-Mode Surface Waves

Any surface wave must (1) satisfy the equation of motion, (2) produce zero stress at the ground surface, and (3) produce zero displacement at infinite depth. Nontrivial solutions do not exist for arbitrary combinations of frequency and wave number; rather, a set of discrete and unique wave numbers exist for a given frequency. Each wave number describes a different displacement pattern, or mode, of the surface wave. The preceding derivations have been limited to the fundamental

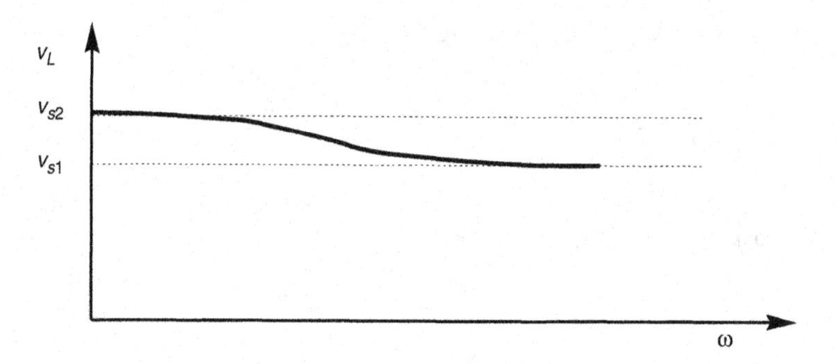

FIGURE C.13 Variation of Love wave velocity with frequency.

modes of Rayleigh and Love waves, which are the most important for earthquake engineering applications. Detailed treatment of higher-mode surface waves can be found in most advanced seismology texts.

C.3.4 Dispersion of Surface Waves

For a homogeneous half-space, the Rayleigh wave velocity was shown to be related to the body wave velocities by Poisson's ratio. Since the body wave velocities are constant with depth, the Rayleigh wave velocity in a homogeneous half-space is independent of frequency. The velocity of the Love wave, on the other hand, varies with frequency between an upper and a lower limit.

Dispersion is a phenomenon in which waves of different frequencies (and different wavelengths) propagate at different velocities. Hence Love waves are clearly dispersive, and Rayleigh waves in a homogeneous half-space are nondispersive. Near the Earth's surface, however, soil and rock stiffnesses usually increase with depth. Since the depth to which a Rayleigh wave causes significant displacement increases with increasing wavelength (Figure C.10), Rayleigh waves of long wavelength (low frequency) can propagate faster than Rayleigh waves of short wavelength (high frequency). Therefore, in the real world of heterogeneous materials, Rayleigh waves are also dispersive. The dispersion of Rayleigh waves can be used to evaluate subsurface stiffness profiles by field testing techniques described in Chapter 6.

Since the velocities of both Rayleigh waves and Love waves decrease with increasing frequency, the low-frequency components of surface waves produced by earthquakes can be expected to arrive at a particular site before their high-frequency counterparts. This tendency to spread the seismic energy over time is an important effect of dispersion.

C.3.5 Phase and Group Velocities

The solutions for Rayleigh wave velocity, v_R, and Love wave velocity, v_L, were based on the assumption of harmonic loading which produces an infinite wave train. These velocities describe the rate at which points of constant phase (e.g., peaks, troughs, or zero points) travel through the medium and are called *phase velocities*. A transient disturbance may produce a packet of waves with similar frequencies. This packet of waves travels at the group velocity, c_g, given by

$$c_g = c + k\frac{dc}{dk} \tag{C.65}$$

where c is the phase velocity (equal to v_R or v_L depending on which type of wave is being considered) and k is the wave number (equal to ω/v_R or ω/v_L). In a nondispersive material, $dc/dk=0$, so the group velocity is equal to the phase velocity. Since both v_R and v_L generally decrease with increasing frequency in geologic materials, dc/dk is less than zero and the group velocity is lower than the phase velocity. Consequently, a wave packet would appear to consist of a series of individual peaks that appear at the back end of the packet, move through the packet to the front, and then disappear. The opposite behavior can be observed by dropping a rock into a calm pond of water (for which $c<c_g$) and watching the resulting ripples carefully.

C.4 WAVES IN A LAYERED BODY

The model of a homogeneous elastic half-space is useful for explaining the existence of body waves and Rayleigh waves, and the addition of a softer surficial layer allows Love waves to be described. In the Earth, however, conditions are much more complicated with many different materials of variable thickness occurring in many areas. To analyze wave propagation under such conditions, and to understand the justification for idealizations of actual conditions when all features cannot be explicitly analyzed, the general problem of wave behavior at interfaces must be investigated.

C.4.1 One-Dimensional Case: Material Boundary in an Infinite Rod

Consider a harmonic stress wave traveling along a constrained rod in the $+x$ direction and approaching an interface between two different materials, as shown in Figure C.14. Since the wave is traveling toward the interface, it will be referred to as the incident wave. Since it is traveling in material 1, its wavelength will be $\lambda_1 = 2\pi/k_1$, and based on Eq. (C.12) it can be described by

$$\sigma_I(x,t) = \sigma_i e^{i(\omega t - k_1 x)} \tag{C.66a}$$

When the incident wave reaches the interface, part of its energy will be transmitted through the interface to continue traveling in the positive x-direction through material 2. This transmitted wave will have a wavelength, $\lambda_2 = 2\pi / k_2$. The remainder will be reflected at the interface and will travel back through material 1 in the negative x-direction as a reflected wave. The transmitted and reflected waves can be described by

$$\sigma_T(x,t) = \sigma_t e^{i(\omega t - k_2 x)} \tag{C.66b}$$

$$\sigma_R(x,t) = \sigma_r e^{i(\omega t + k_1 x)} \tag{C.66c}$$

Assuming that the displacements associated with each of these waves are of the same harmonic form as the stresses that cause them; i.e.,

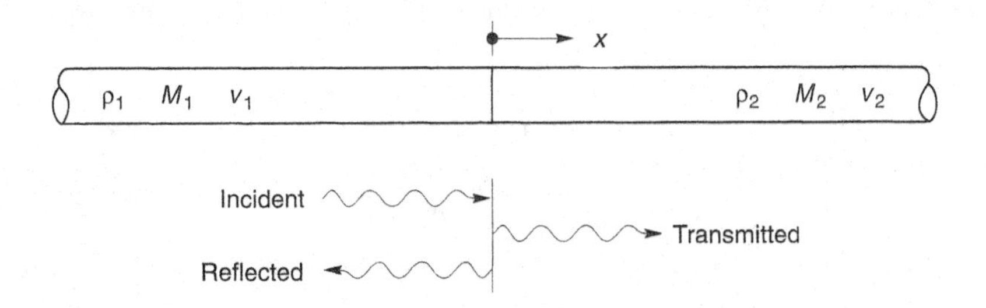

FIGURE C.14 One-dimensional wave propagation at material interface. Incident and reflected waves travel in opposite directions in material 1. The transmitted wave travels through material 2 in the same direction as the incident wave.

$$u_I(x,t) = A_i e^{i(\omega t - k_1 x)} \tag{C.67a}$$

$$u_R(x,t) = A_r e^{i(\omega t + k_1 x)} \tag{C.67b}$$

$$u_T(x,t) = A_t e^{i(\omega t - k_2 x)} \tag{C.67c}$$

Stress-strain and strain-displacement relationships can be used to relate the stress amplitudes to the displacement amplitudes:

$$\sigma_I(x,t) = M_1 \frac{\partial u_I(x,t)}{\partial x} = -ik_1 M_1 A_i e^{i(\omega t - k_1 x)} \tag{C.68a}$$

$$\sigma_R(x,t) = M_1 \frac{\partial u_R(x,t)}{\partial x} = +ik_1 M_1 A_r e^{i(\omega t + k_1 x)} \tag{C.68b}$$

$$\sigma_T(x,t) = M_2 \frac{\partial u_T(x,t)}{\partial x} = -ik_2 M_2 A_t e^{i(\omega t - k_2 x)} \tag{C.68c}$$

From these, the stress amplitudes are related to the displacement amplitudes by

$$\sigma_i = -ik_1 M_1 A_i \tag{C.69a}$$

$$\sigma_r = +ik_1 M_1 A_r \tag{C.69b}$$

$$\sigma_t = -ik_2 M_2 A_t \tag{C.69c}$$

At the interface ($x = 0$), both compatibility of displacements and continuity of stresses must be satisfied. The former requires that

$$u_I(0,t) + u_R(0,t) = u_T(0,t) \tag{C.70}$$

and the latter that

$$\sigma_I(0,t) + \sigma_R(0,t) = \sigma_T(0,t) \tag{C.71}$$

Substituting Equations (C.67) and (C.66) into Equations (C.70) and (C.71), respectively, indicates that

$$A_i + A_r = A_t \tag{C.72}$$

$$\sigma_i + \sigma_r = \sigma_t \tag{C.73}$$

at the interface. Substituting Equations (C.69) into Equation (C.73) and using the relationship $kM = \omega\rho v$, gives

$$-\rho_1 v_1 A_i + \rho_1 v_1 A_r = -\rho_2 v_2 A_t = -\rho_2 v_2 (A_i + A_r) \tag{C.74}$$

Equation (C.74) can be rearranged to relate the displacement amplitude of the reflected wave to that of the incident wave:

$$A_r = \frac{\rho_1 v_1 - \rho_2 v_2}{\rho_1 v_1 + \rho_2 v_2} A_i = \frac{1 - \rho_2 v_2 / \rho_1 v_1}{1 + \rho_2 v_2 / \rho_1 v_1} A_i \tag{C.75}$$

The displacement amplitude of the transmitted wave can be similarly derived as:

$$A_t = \frac{2\rho_1 v_1}{\rho_1 v_1 + \rho_2 v_2} A_i = \frac{2}{1 + \rho_2 v_2 / \rho_1 v_1} A_i \tag{C.76}$$

Remember that the product of the mass density and the wave propagation velocity is the specific impedance of the material. Equations (C.75) and (C.76) indicate that the partitioning of energy at the interface depends only on the ratio of the specific impedances of the materials on either side of the interface. Defining the impedance ratio as $\alpha_z = \rho_2 v_2 / \rho_1 v_1$, the displacement amplitudes of the reflected and transmitted waves are

$$A_r = \frac{1 - \alpha_z}{1 + \alpha_z} A_i \tag{C.77}$$

$$A_t = \frac{2}{1 + \alpha_z} A_i \tag{C.78}$$

After evaluating the effect of the interface on the displacement amplitudes of the reflected and transmitted waves, its effect on stress amplitudes can be investigated. From Equations (C.69)

$$A_i = -\frac{\sigma_i}{ik_1 M_1} \tag{C.79a}$$

$$A_r = \frac{\sigma_r}{ik_1 M_1} \tag{C.79b}$$

$$A_t = -\frac{\sigma_t}{ik_2 M_2} \tag{C.79c}$$

Substituting Equations (C.79) into Equations (C.77) and (C.78) and rearranging gives

$$\sigma_r = \frac{\alpha_z - 1}{1 + \alpha_z} \sigma_i \tag{C.80}$$

$$\sigma_t = \frac{2\alpha_z}{1 + \alpha_z} \sigma_i \tag{C.81}$$

The importance of the impedance ratio in determining the nature of reflection and transmission at interfaces can clearly be seen. Equations (C.77), (C.78), (C.80), and (C.81) indicate that fundamentally different types of behavior occur when the impedance ratio is less than or greater than 1. When the impedance ratio is less than 1, an incident wave can be thought of as approaching a "softer" material. For this case, the reflected wave will have a smaller stress amplitude than the incident wave and its sign will be reversed (an incident compression pulse will be reflected as a tensile pulse, and vice versa). If the impedance ratio is greater than 1, the incident wave is approaching a

TABLE C.1
Influence of Impedance Ratio on Displacement and Stress Amplitudes of Reflected and Transmitted Waves

	Displacement Amplitudes			Stress Amplitudes		
Impedance Ratio, α_z	**Incident**	**Reflected**	**Transmitted**	**Incident**	**Reflected**	**Transmitted**
0	A_i	A_i	$2A_i$	σ_i	$-\sigma_i$	0
¼	A_i	$3A_i/5$	$8A_i/5$	σ_i	$-3\sigma_i/5$	$2\sigma_i/5$
½	A_i	$A_i/3$	$4A_i/3$	σ_i	$-\sigma_i/3$	$2\sigma_i/3$
1	A_i	0	A_i	σ_i	0	σ_i
2	A_i	$-A_i/3$	$2A_i/3$	σ_i	$\sigma_i/3$	$4\sigma_i/3$
4	A_i	$-3A_i/5$	$2A_i/5$	σ_i	$3\sigma_i/5$	$8\sigma_i/5$
∞	A_i	$-A_i$	0	σ_i	σ_i	$2\sigma_i$

"stiffer" material in which the stress amplitude of the transmitted wave will be greater than that of the incident wave and the stress amplitude of the reflected wave will be less than, but of the same sign, as that of the incident wave. The displacement amplitudes are also affected by the impedance ratio. The displacement amplitude of a wave transmitted from a stiffer material into a softer material will be greater than that of the incident wave. The relative stress and displacement amplitudes of reflected and transmitted waves at boundaries with several different impedance ratios are illustrated in Table 5.1.

The cases of $\alpha_z = 0$ and $\alpha_z = \infty$ are of particular interest. An impedance ratio of zero implies that the incident wave is approaching a "free end" across which no stress can be transmitted ($\sigma_t = 0$). To satisfy this zero stress boundary condition, the displacement of the boundary (the transmitted displacement) must be twice the displacement amplitude of the incident wave ($A_t = 2A_i$). The reflected wave has the same amplitude as the incident wave but is of the opposite polarity ($\sigma_r = -\sigma_i$). In other words, a free end will reflect a compression wave as a tension wave of identical amplitude and shape and a tension wave as an identical compression wave. An infinite impedance ratio implies that the incident wave is approaching a "fixed end" at which no displacement can occur (u_t=0). In that case, the stress at the boundary is twice that of the incident wave ($\sigma_t = 2\sigma_i$) and the reflected wave has the same amplitude and polarity as the incident wave (A_r=$-A_i$).

The case of α_z=1, in which the impedances on each side of the boundary are equal, is also of interest. Equations (C.77), (C.78), (C.80), and (C.81) indicate that no reflected wave is produced and that the transmitted wave has, as expected, the same amplitude and polarity as the incident wave. In other words, all of the elastic energy of the wave crosses the boundary unchanged and travels away, never to return. Another way of looking at a boundary with an impedance ratio of unity is as a boundary between two identical, semi-infinite rods. A harmonic wave traveling in the positive x-direction (Figure C.15a) would impose an axial force [see Equation (C.5)] on the boundary:

$$F = \sigma_x A = \rho v_m A \dot{u} \tag{C.82}$$

This axial force is identical to that which would exist if the semi-infinite rod on the right side of the boundary were replaced by a dashpot (Figure C.15b) of coefficient c=$\rho v_m A$. In other words, the dashpot would absorb all the elastic energy of the incident wave, so the response of the rod on the left would be identical for both cases illustrated in Figure C.15. This result has important implications for ground response and soil-structure interaction analyses (Chapters 7 and 8), where the replacement of a semi-infinite domain by discrete elements such as dashpots can provide tremendous computational efficiencies.

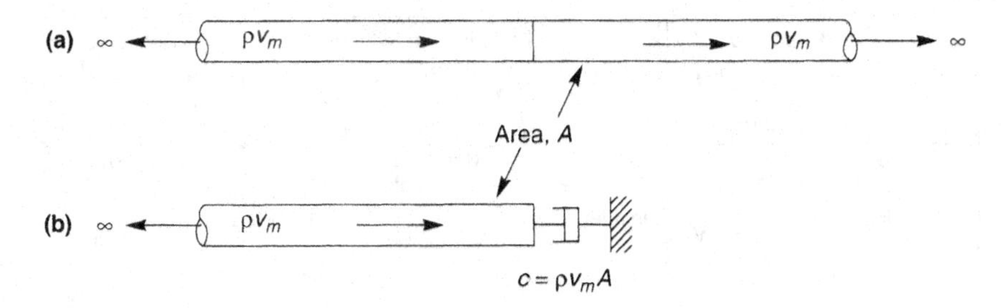

FIGURE C.15 (a) Harmonic wave traveling along two connected semi-infinite rods; (b) semi-infinite rod attached to dashpot. With proper selection of dashpot coefficient, response in semi-infinite rod on left will be identical for both cases.

C.4.2 Three-Dimensional Case: Inclined Waves

In general, waves will not approach interfaces at 90° angles as they did in Section C.4.1. The orientation of an inclined body wave can strongly influence the manner in which energy is reflected and transmitted across an interface. Fermat's principle defines the propagation time of a seismic pulse between two arbitrary points A and B as the minimum travel time along any continuous path that connects A and B. The path that produces the minimum travel time is called a *ray path*, and its direction is often represented by a vector called a *ray*. A *wavefront* is defined as a surface of equal travel time, consequently, a ray path must (in an isotropic material) be perpendicular to the wavefront as illustrated in Figure C.16. Snell considered the change of direction of ray paths at interfaces between materials with different wave propagation velocities. Using Fermat's principle, Snell showed that

$$\frac{\sin i}{v} = \text{constant} \tag{C.83}$$

where i is the angle between the ray path and the normal to the interface and v is the velocity of the wave (p- or s-wave) of interest. This relationship holds for both reflected and transmitted waves. It indicates that the transmitted wave will be refracted (except when $i=0$) when the wave propagation velocities are different on each side of the interface.

Consider the case of two half-spaces of different elastic materials in contact with each other. As for the previous case, the requirements of equilibrium and compatibility and the theory of elasticity can be used to determine the nature and distribution of energy among the reflected and transmitted waves for the cases of an incident p-wave, an incident SV wave, and an incident SH-wave.

The types of waves produced by incident p-, SV-, and SH-waves are shown in Figure C.17. Since incident p- and SV-waves involve particle motion perpendicular to the plane of the interface; they will each produce both reflected and refracted p- and SV-waves. An incident SH-wave does not

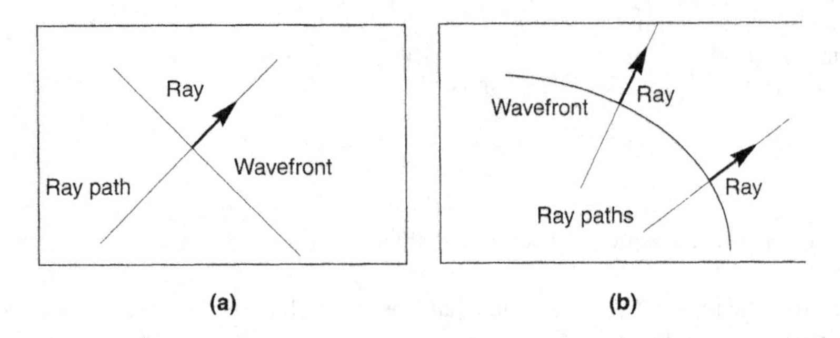

FIGURE C.16 Ray path, ray, and wavefront for (a) plane wave and (b) curved wavefront.

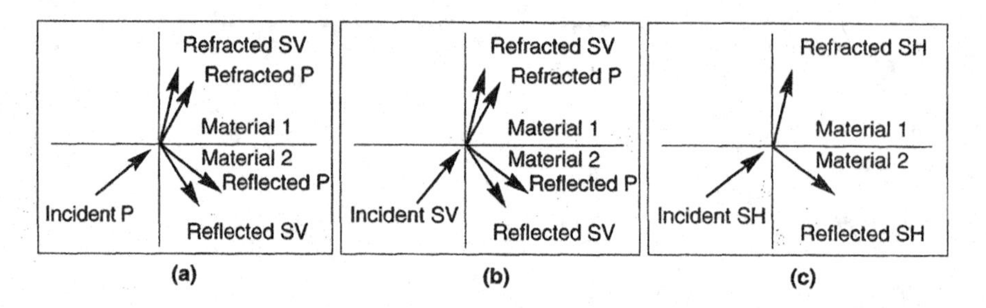

FIGURE C.17 Reflected and refracted rays resulting from incident (a) p-wave, (b) SV wave, and (c) SH-wave.

involve particle motion perpendicular to the interface; consequently, only SH-waves are reflected and refracted. The directions and relative amplitudes of the waves produced at the interface depend on both the direction and amplitude of the incident wave. Using Snell's law and the requirements of equilibrium and compatibility, these directions and amplitudes can be determined. Using the notation of Richter (1958):

Wave Type	Velocity	Amplitude	Angle with Normal
Incident p	U	A	a
Incident s	V	B	b
Reflected p	U	C	c
Reflected s	V	D	d
Refracted p	Y	E	e
Refracted s	Z	F	f

The directions of all waves are easily related to the direction of the incident wave using Snell's law:

$$\frac{\sin a}{U} = \frac{\sin b}{V} = \frac{\sin c}{U} = \frac{\sin d}{V} = \frac{\sin e}{Y} = \frac{\sin f}{Z} \tag{C.84}$$

Since incident and reflected waves travel through the same material, $a=c$ and $b=d$, which shows that the angle of incidence is equal to the angle of reflection for both p- and s-waves.

The angle of refraction is uniquely related to the angle of incidence by the ratio of the wave velocities of the materials on each side of the interface. Snell's law indicates that waves traveling from higher-velocity materials into lower-velocity materials will be refracted closer to the normal to the interfaces. In other words, waves propagating upward through horizontal layers of successively lower velocity (as is common near the Earth's surface) will be refracted closer and closer to a vertical path (Figure C.18). This phenomenon is relied upon heavily by many of the methods of ground response analysis presented in Chapter 7.

The critical angle of incidence, i_c, is defined as that which produces a refracted wave that travels parallel to the interface (e or $f=90°$). Therefore,

$$i_c = \sin^{-1}\frac{U}{Y} = \sin^{-1}\frac{V}{Z} \tag{C.85}$$

The concept of critical refraction is used in the interpretation of seismic refraction tests (Section 6.5.2.1).

Assuming that the incident wave is simple harmonic, satisfaction of the requirements of equilibrium and compatibility at the interface give rise to the following systems of simultaneous equations

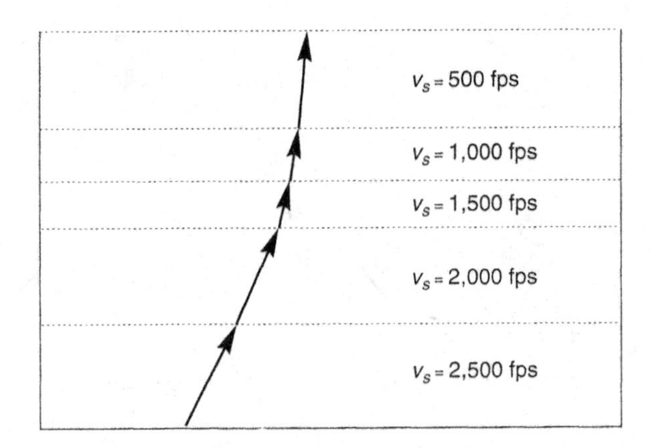

FIGURE C.18 Refraction of an SH-wave ray path through a series of successively softer (lower v,) layers. Note that orientation of ray path becomes closer to vertical as the ground surface is approached. Reflected rays are not shown.

(Richter, 1958), which allow the amplitudes of the reflected and refracted waves (*C*, *D*, *E*, and *F*) to be expressed in terms of the amplitude of the incident p-wave (*A*).

$$(A-C)\sin a + D\cos b - E\sin e + F\cos f = 0$$

$$(A+C)\cos a + D\sin b - E\cos e - F\sin f = 0$$

$$-(A+C)\sin 2a + D\frac{U}{V}\cos 2b + EK\left(\frac{Z}{V}\right)^2\frac{U}{Y}\sin 2e - FK\left(\frac{Z}{V}\right)^2\frac{U}{Z}\cos 2f = 0$$

$$-(A-C)\cos 2b + D\frac{V}{U}\sin 2b + EK\frac{Y}{U}\cos 2f - FK\frac{Z}{U}\sin(2f) = 0 \tag{C.86}$$

where $K=\rho_1/\rho_2$ (the subscripts 1 and 2 refer to materials 1 and 2, respectively). Note that the amplitudes are functions of the angle of incidence, the velocity ratio, and the density ratio. Figure C.19 shows the variation of amplitude with angle of p-wave incidence for the following conditions: $U=8.000$, $Y=2.003$, $K=0.606$, and $\nu=0.25$. The sensitivity of the reflected and refracted wave amplitudes to the angle of incidence is apparent. SV-waves are neither reflected nor refracted at angles of incidence of 0° and 90°, but can carry the majority of the wave energy away from the interface at intermediate angles.

For an incident SV-wave, both SV- and p-waves are reflected and refracted. The equilibrium/compatibility equations relating the relative amplitudes are

$$(B+D)\sin b + C\cos a - E\cos e - F\sin f = 0$$

$$(B-D)\cos b + C\sin a + E\sin e - F\cos f = 0$$

$$(B+D)\cos 2b - C\frac{V}{U}\sin 2a + EK\frac{Z^2}{VY}\sin 2e - FK\frac{Z}{V}\cos 2f = 0$$

$$-(B-D)\sin 2b + C\frac{U}{V}\cos 2b + EK\frac{Y}{V}\cos 2f + FK\frac{Z}{V}\sin 2f = 0 \tag{C.87}$$

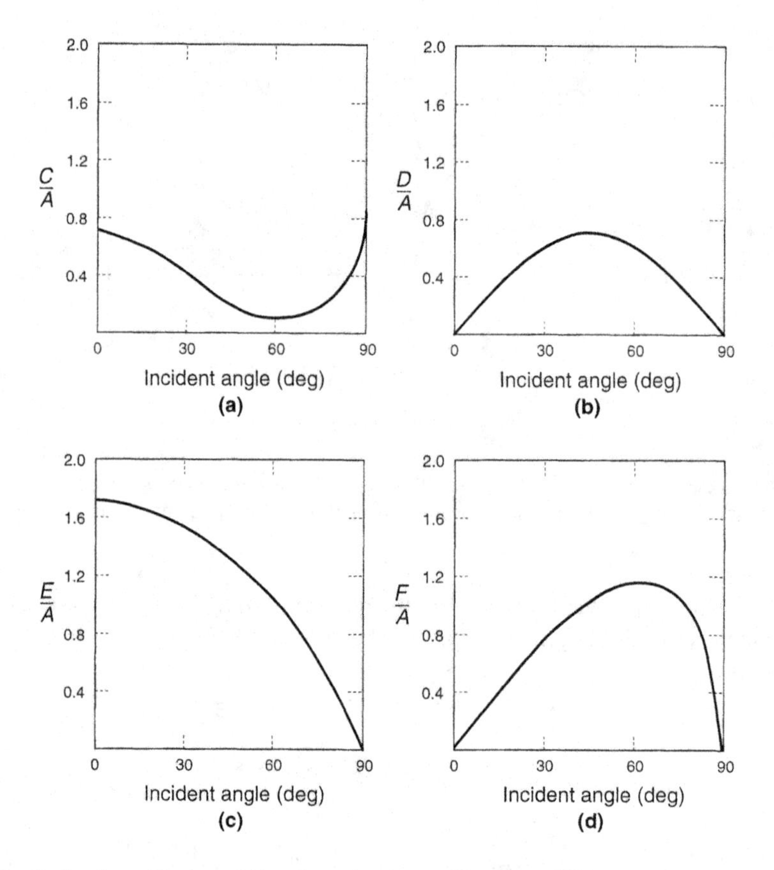

FIGURE C.19 Ratio of amplitudes of (a) reflected p-wave, (b) reflected SV-wave, (c) refracted p-wave, and (d) refracted SV-wave to amplitude of incident p-wave versus angle of incidence.

which produce the amplitude behavior shown in Figure C.20. For angles of incidence greater than $\sin^{-1}(V/U)$, about 36° in Figure C.20a, no p-wave can be reflected, so the incident wave energy must be carried away by the remaining waves. A more detailed discussion of this phenomenon can be found in McCamy et al. (1962).

An incident SH-wave involves no particle motion perpendicular to the interface; consequently, it cannot produce p-waves ($C=E=0$) or SV-waves. The equilibrium/compatibility equations are considerably simplified and easily solved as

$$D = \frac{1 - K\dfrac{Z}{V}\dfrac{\cos f}{\cos b}}{1 + K\dfrac{Z}{V}\dfrac{\cos f}{\cos b}} B$$

$$F = B\left(1 + \frac{1 - K\dfrac{Z}{V}\dfrac{\cos f}{\cos b}}{1 + K\dfrac{Z}{V}\dfrac{\cos f}{\cos b}}\right) \tag{C.88}$$

The preceding results show that the interaction of stress waves with boundaries can be quite complicated. As seismic waves travel away from the source of an earthquake, they invariably encounter heterogeneities and discontinuities in the Earth's crust. The creation of new waves and the reflection

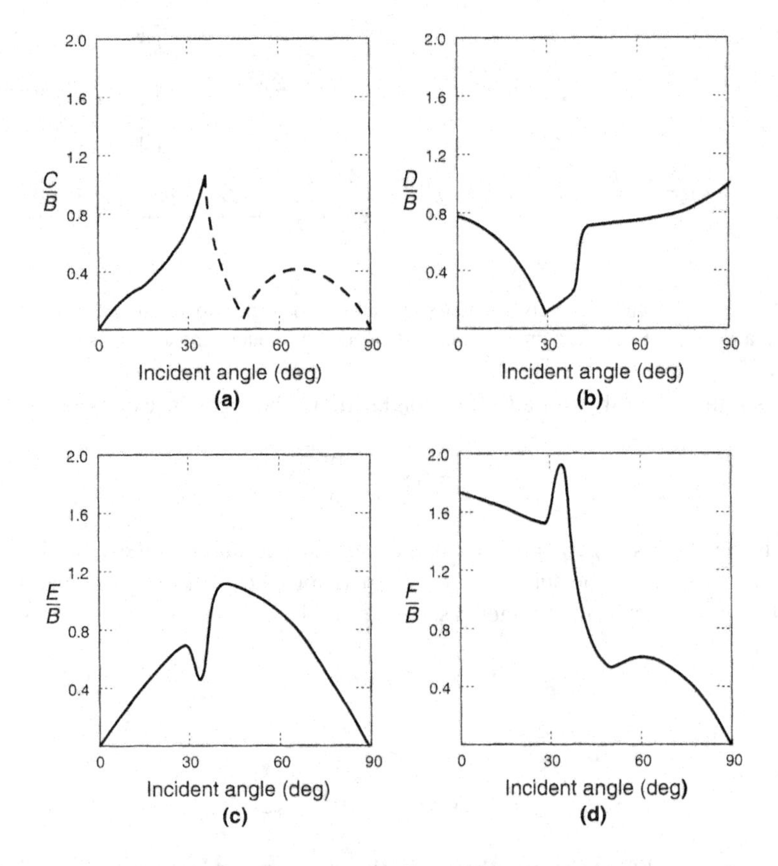

FIGURE C.20 Ratio of amplitudes of (a) reflected p-wave, (b) reflected SV-wave, (c) refracted p-wave, and (d) refracted SV-wave to amplitude of incident SV-wave versus angle of incidence.

and refraction of ray paths by these heterogeneities cause seismic waves to reach a site by many different paths. Since the paths have different lengths, the motion at the site is spread out in time by this wave-scattering effect. Scattering effects are discussed in more detail in Section C.5.3.

C.5 ATTENUATION OF STRESS WAVES

The preceding sections have considered only the propagation of waves in linear elastic materials. In a homogeneous linear elastic material, stress waves travel indefinitely without change in amplitude. This type of behavior cannot occur, however, in real materials. The amplitudes of stress waves in real materials, such as those that comprise the Earth, attenuate with distance. This attenuation can be attributed to two sources, one of which involves the materials through which the waves travel and the other the geometry of the wave propagation problem.

C.5.1 Material Damping

In real materials, part of the elastic energy of a traveling wave is converted to heat. The conversion is accompanied by a decrease in the amplitude of the wave. Viscous damping, by virtue of its mathematical convenience, is often used to represent this dissipation of elastic energy. For the purposes of viscoelastic wave propagation, soils are usually modeled as Kelvin-Voigt solids (i.e., materials whose resistance to shearing deformation is the sum of an elastic part and a viscous part). A thin element of a Kelvin-Voigt solid can be illustrated as in Figure C.21.

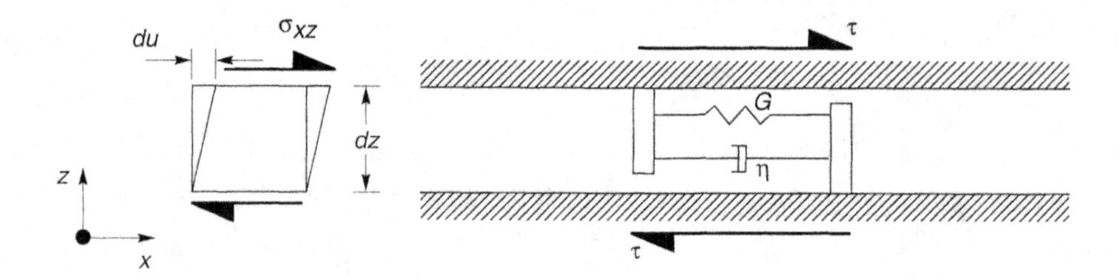

FIGURE C.21 Thin element of a Kelvin-Voigt solid subjected to horizontal shearing. Total resistance to shearing deformation is given by the sum of an elastic (spring) component and a viscous (dashpot) component.

The stress-strain relationship for a Kelvin-Voigt solid in shear can be expressed as

$$\tau = G\gamma + \eta \frac{\partial \gamma}{\partial t} \tag{C.89}$$

where $\tau = \sigma_{xz}$ is the shear stress, $\gamma = \partial u / \partial x$ is the shear strain, and η is the viscosity of the material. Thus the shear stress is the sum of an elastic part (proportional to strain) and a viscous part (proportional to strain rate). For a harmonic shear strain of the form

$$\gamma = \gamma_0 \sin \omega t \tag{C.90}$$

the shear stress will be

$$\tau = G\gamma_0 \sin \omega t + \omega \eta \gamma_0 \cos \omega t \tag{C.91}$$

Together, Equations (C.90) and (C.91) show that the stress-strain loop of a Kelvin-Voigt solid is elliptical. The elastic energy dissipated in a single cycle is given by the area of the ellipse, or

$$\Delta W = \int_{t_0}^{t_0 + 2\pi/\omega} \tau \frac{\partial \gamma}{\partial t} dt = \pi \eta \omega \gamma_0^2 \tag{C.92}$$

which indicates that the dissipated energy is proportional to the frequency of loading. Real soils, however, dissipate elastic energy hysteretically, by the slippage of grains with respect to each other. As a result, their energy dissipation characteristics are insensitive to frequency. For discrete Kelvin-Voigt systems (Section B.6.1), the damping ratio, ξ was shown to be related to the area within the force-displacement (or, equivalently, the stress-strain) loop as shown in Figure C.22. Since the peak energy stored in the cycle is

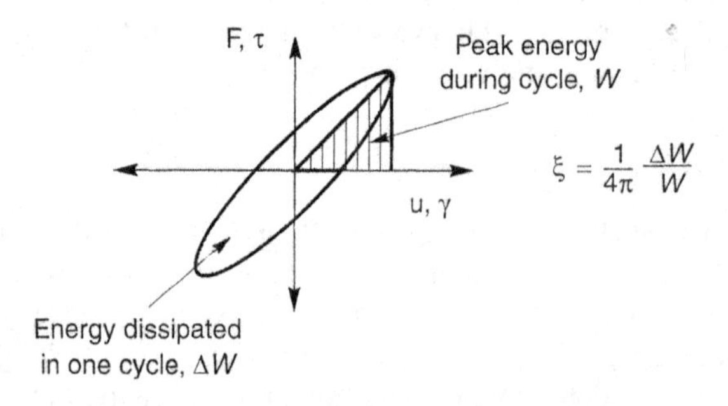

FIGURE C.22 Relationship between hysteresis loop and damping ratio.

$$W = \frac{1}{2} G \gamma_0^2 \tag{C.93}$$

Then

$$\xi = \frac{1}{4\pi} \frac{\pi \eta \omega \gamma_0^2}{\frac{1}{2} G \gamma_0^2} = \frac{\eta \omega}{2G} \tag{C.94}$$

To eliminate frequency dependence while maintaining the convenience of the viscoelastic formulation, Equation (C.94) is often rearranged to produce an equivalent viscosity that is inversely proportional to frequency. The use of this equivalent viscosity ensures that the damping ratio is independent of frequency:

$$\eta = \frac{2G}{\omega} \xi \tag{C.95}$$

A Kelvin-Voigt solid for vertically propagating SH-waves may be represented by a stack of infinitesimal elements of the type shown schematically in Figure C.21. The one dimensional equation of motion for vertically propagating SH-waves can be written as

$$\rho \frac{\partial^2 u}{dt^2} = \frac{\partial \sigma_{xz}}{\partial z} \tag{C.96}$$

Substituting Equation (C.89) into (C.96) with $\tau = \sigma_{xz}$ and $\gamma = \partial u / \partial z$, and differentiating the right side allows the wave equation to be expressed as

$$\rho \frac{\partial^2 u}{dt^2} = G \frac{\partial^2 u}{dz^2} + \eta \frac{\partial^3 u}{\partial z^2 \, \partial t} \tag{C.97}$$

For harmonic waves, the displacements can be written as

$$u(z,t) = U(z) e^{i\omega t} \tag{C.98}$$

where $U(z)$ is the depth-varying amplitude of a standing wave of frequency ω. Substituting this expression into the wave equation (C.97) yields the ordinary differential equation

$$(G + i\omega\eta) \frac{d^2 U}{dz^2} = -\rho \omega^2 U \tag{C.99}$$

$$G^* \frac{d^2 U}{dz^2} = -\rho \omega^2 U \tag{C.100}$$

where $G^* = G + i\omega\eta$ is the complex shear modulus. The complex shear modulus is analogous to the complex stiffness described in Section B.6.3. Using Equation (C.95) to eliminate frequency dependence, the complex shear modulus can also be expressed as $G^* = G(1 + 2i\xi)$. This equation of motion has the solution

$$u(z,t) = A e^{i(\omega t - k^* z)} + B e^{i(\omega t + k^* z)} \tag{C.101}$$

where A and B depend on the boundary conditions, $k^* = \omega / \sqrt{\rho / G^*}$ is the complex wave number, and time dependence has been added by multiplying the displacement terms by $e^{i\omega t}$. It can be shown (after Kolsky, 1963) that k^* is given by

$$k^* = k_1 + ik_2 \tag{C.102}$$

where

$$k_1^2 = \frac{\rho\omega^2}{2G\left(1+4\xi^2\right)}\left(\sqrt{1+4\xi^2}+1\right)$$

$$k_2^2 = \frac{\rho\omega^2}{2G\left(1+4\xi^2\right)}\left(\sqrt{1+4\xi^2}-1\right)$$

and only the positive root of k_1 and the negative root of k_2 have physical significance. Note that for the inviscid case ($\eta=0$), $k_2=0$ and $k_1=k$. For a wave propagating in the positive z-direction, the solution can be written as

$$u(z,t) = Ae^{k_2 z}e^{i(\omega t - k_1 z)} \tag{C.103}$$

which shows (since k_2 is negative) that material damping produces exponential attenuation of wave amplitude with distance.

Although the Kelvin-Voigt model is by far the most commonly used model for soils, it represents only one of an infinite number of rheological models. By rearranging and adding more springs and dashpots, many different types of behavior can be modeled, although the complexity of the wave equation solution increases dramatically as the number of springs and dashpots increases.

C.5.2 Radiation Damping

Since material damping absorbs some of the elastic energy of a stress wave, the specific energy (elastic energy per unit volume) decreases as the wave travels through a material. The reduction of specific energy causes the amplitude of the stress wave to decrease with distance. The specific energy can also be decreased by another common mechanism, which can be illustrated by the propagation of stress waves along an undamped conical rod.

Consider the unconstrained conical rod of small apex angle shown in Figure C.23 and assume that it is subjected to stress waves of wavelength considerably larger than the diameter of the rod in the area of interest. If the apex angle is sufficiently small, the normal stress will be uniform across each of two spherical surfaces that bound an element of width, dr, and will act in a direction virtually parallel to the axis of the rod. Letting u represent the displacement parallel to the axis of the rod, the equation of motion in that direction can be written, using exactly the same approach used in Section C.2.1.1, as

$$\rho r^2 \alpha dr \frac{\partial^2 u}{\partial t^2} = \left(\sigma + \frac{\partial \sigma}{\partial r}dr\right)(r+dr)^2\alpha - \sigma r^2 \alpha \tag{C.104}$$

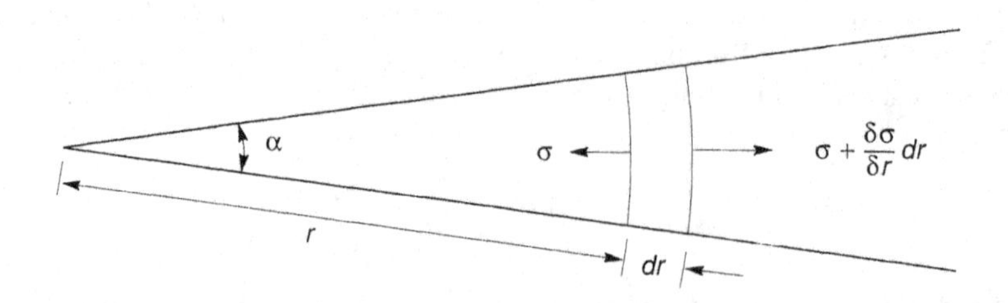

FIGURE C.23 Conical rod of apex angle, α.

which simplifies to

$$\rho r \frac{\partial^2 u}{\partial t^2} = r \frac{\partial \sigma}{\partial r} + 2\sigma \tag{C.105}$$

Substituting the stress-strain and strain-displacement relationships (assuming now that the ends of the element are planar) gives

$$\rho r \frac{\partial^2 u}{\partial t^2} = Er \frac{\partial^2 u}{\partial r^2} + 2E \frac{\partial u}{\partial r} \tag{C.106}$$

$$\frac{\partial^2 (ur)}{\partial t^2} = \frac{E}{\rho} \frac{\partial^2 (ur)}{\partial r^2} \tag{C.107}$$

$$u(r,t) = \frac{1}{r}[f(vt - r) + g(vt + r)] \tag{C.108}$$

where $v = \sqrt{G/\rho}$. Equation (C.108) indicates that the amplitude of the wave will decrease with distance (even though the total elastic energy remains the same). The reduction is of purely geometric origin, resulting from the decrease in specific energy that occurs as the area of the rod increases.

Even though elastic energy is conserved (no conversion to other forms of energy takes place), this reduction in amplitude due to spreading of the energy over a greater volume of material, which is often referred to as *radiation damping* (also as *geometric damping* and *geometric attenuation*). It should be distinguished from material damping in which elastic energy is actually dissipated by viscous, hysteretic, or other mechanisms.

When earthquake energy is released from a fault below the ground surface, body waves travel away from the source in all directions. If the rupture zone can be represented as a point source, the wavefronts will be spherical and the preceding analysis can easily be extended to show that geometric attenuation causes the amplitude to decrease at a rate of $1/r$. It can also be shown (Bullen, 1953) that geometric attenuation of surface waves causes their amplitudes to decrease at a rate of essentially $1/\sqrt{r}$; in other words, surface waves attenuate (geometrically) much more slowly than body waves. This explains the greater proportion of surface wave motion (relative to body wave motion) that is commonly observed at large epicentral distances. This partly explains the advantages of the surface wave magnitude, relative to body wave magnitude, for the characterization of distant earthquakes (although moment magnitude remains the preferred scale, when available).

For problems in which energy is released from a finite source, ranging from the large scale case of rupture along an earthquake fault to the smaller-scale case of a vibrating foundation, radiation damping can be extremely important. In such cases, the effects of radiation damping are often larger than those of material damping.

C.5.3 Scattering

Another mechanism, *wave scattering*, can lead to a reduction in the amplitude of a traveling wave without dissipation of elastic wave energy. Seismic wave scattering results from heterogeneities of density and/or stiffness in the geologic materials that waves travel through. Scattering in the Earth's crust contributes to path effects and scattering in near-surface materials contributes to site effects. While not affecting the total amount of energy it carries, scattering tends to disorganize a wave field by reflection and refraction at the boundaries of heterogeneities. The waves are spread out over a greater duration with their amplitudes, particularly at higher frequencies, being reduced. The factors that lead to this behavior are illustrated below using the concepts of Section C.4 to consider simple, individual inclusions in otherwise homogeneous materials; in reality, the three-dimensional spatial variability of soil and rock can be interpreted as multiple, geometrically-complex inclusions of stiffer and softer materials, which have much more complicated effects on wave propagation.

Consider a wave traveling along a constrained infinite elastic rod (Material 1) that contains an inclusion of length, L, whose properties (Material 2) are different than those of the rest of the rod. If an incident wave of displacement amplitude, A, travels from left to right (Figure C.24) along the rod, part of its energy will be reflected back to the left with amplitude, A_r, and part will be transmitted into the inclusion with amplitude, B, when it reaches the left side of the inclusion ($x=0$). However, when the transmitted wave reaches the right side of the inclusion ($x=L$), part of its energy will be reflected back with amplitude, B_r, and part will be transmitted with amplitude, C, into Material 1 on the right side of the inclusion. The effect of the inclusion can be seen by comparing the amplitude of the transmitted wave, C with that of the incident wave, A.

To make this comparison, the boundary conditions on the left and right sides of the inclusion must be considered. The various displacements of interest are described by

$$\begin{aligned} u_{i1}(x,t) &= Ae^{i(\omega t-k_1x)} \quad u_{i2}(x,t) = Be^{i(\omega t-k_2x)} \quad u_3(x,t) = Ce^{i(\omega t-k_1x)} \\ u_{r1}(x,t) &= A_re^{i(\omega t+k_1x)} \quad u_{r2}(x,t) = B_re^{i(\omega t+k_2x)} \end{aligned} \tag{C.109}$$

where k_1 and k_2 are the wave numbers in Materials 1 and 2, respectively. Using Equation (C.68), the corresponding stresses are given by

$$\begin{aligned} \sigma_{i1}(x,t) &= -ik_1M_1Ae^{i(\omega t-k_1x)} \quad \sigma_{i2}(x,t) = -ik_2M_2Be^{i(\omega t-k_2x)} \quad \sigma_3(x,t) = -k_1M_1Ce^{i(\omega t-k_1x)} \\ \sigma_{r1}(x,t) &= ik_1M_1A_re^{i(\omega t+k_1x)} \quad \sigma_{r2}(x,t) = k_2M_2B_re^{i(\omega t+k_2x)} \end{aligned} \tag{C.110}$$

where M_1 and M_2 are the respective constrained moduli of Materials 1 and 2. Continuity of displacements at the left side of the inclusion requires that $u_{i1}(x,t)+u_{r1}(x,t)=u_{i2}(x,t)+u_{r2}(x,t)$ which, since $x=0$ at that boundary, gives

$$A + A_r = B + B_r \tag{C.111}$$

At the right side of the inclusion, $u_{i2}(x,t)+u_{r2}(x,t)=u_3(x,t)$ so, since $x=L$ at that boundary,

$$Be^{-ik_2L} + B_re^{ik_2L} = Ce^{-ik_1L} \tag{C.112}$$

For equilibrium at the left side of the inclusion, $\sigma_{i1}(x,t)+\sigma_{r1}(x,t)=\sigma_{i2}(x,t)+\sigma_{r2}(x,t)$. Since $x=0$,

$$-k_1M_1A + k_1M_1A_r = -k_2M_2B + k_2M_2B_r \tag{C.113}$$

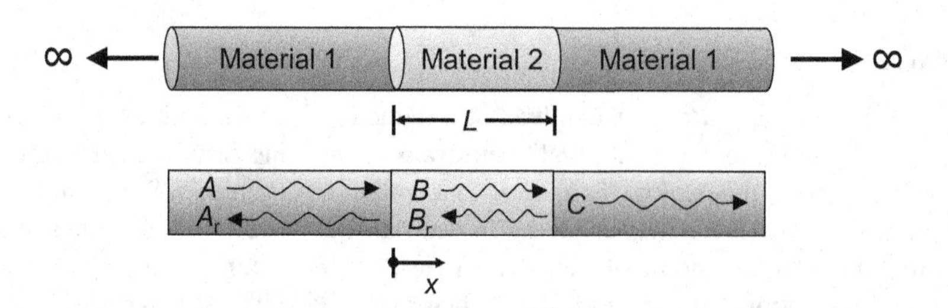

FIGURE C.24 Illustration of inclusion of length, L, in infinitely long rod indicating displacement amplitudes of left- and right-traveling waves.

Finally, equilibrium at the right side of the inclusion requires that $\sigma_{i2}(x,t)+\sigma_{r2}(x,t)=\sigma_3(x,t)$, or

$$-k_2M_2Be^{-ik_2L}+k_2M_2B_re^{-ik_2L}=-k_1M_1Ce^{-ik_1L} \tag{C.114}$$

Equations (C.111)–(C.114) are required to satisfy the boundary conditions at both ends of the inclusion. Dividing all four equations by the incident wave amplitude, A, and the last two by k_1M_1, and using the impedance ratio, $\alpha_z=\dfrac{k_2M_2}{k_1M_1}=\dfrac{\rho_2v_2}{\rho_1v_1}$ gives a set of four simultaneous equations

$$1+\frac{A_r}{A}-\frac{B}{A}-\frac{B_r}{A}=0 \tag{C.115a}$$

$$\frac{B}{A}e^{-ik_2L}+\frac{B_r}{A}e^{ik_2L}-\frac{C}{A}e^{-ik_1L}=0 \tag{C.115b}$$

$$-1+\frac{A_r}{A}+\alpha_z\frac{B}{A}-\alpha_z\frac{B_r}{A}=0 \tag{C.115c}$$

$$-\alpha_z\frac{B}{A}e^{-ik_2L}+\alpha_z\frac{B_r}{A}e^{ik_2L}+\frac{C}{A}e^{-ik_1L}=0 \tag{C.115d}$$

These equations involve four unknowns – A_r/A, B/A, A_r/A, and C/A. The amplitude of the wave that continues past the inclusion is of greatest interest and that amplitude, relative to that of the original incident wave, can be obtained from the solution of the equations (Semblat and Pecker, 2009) as

$$\left|\frac{C}{A}\right|=\frac{1}{\sqrt{\cos^2(2\pi L/\lambda_2)+\dfrac{1-\cos^2(2\pi L/\lambda_2)}{4}\left(\dfrac{1}{\alpha_z}+\alpha_z\right)^2}} \tag{C.116}$$

The displacement amplitude ratio can be seen to depend on the impedance ratio of the inclusion and the length of the inclusion relative to the wavelength of the motion. If the impedance ratio is 1.0, the amplitude ratio will be 1.0 (regardless of inclusion length or frequency) since waves would not be reflected at either end of the inclusion. Similarly, as the inclusion length and/or frequency go to zero, the impedance ratio goes to 1.0. If the impedance ratio is not equal to 1.0 (either smaller or larger), the amplitude ratio will be less than 1.0 (Figure C.25a). As the frequency increases, the wavelength shortens and the amplitude ratio decreases. Thus, the effect of the inclusion is to reduce the amplitude of the transmitted wave. The effect, however, is locally symmetric about $L/\lambda=0.25+0.50n$ (where n is a non-negative integer) and periodic at $L/\lambda=0.5$ (Figure C.25b).

For a given impedance ratio, the transmitted wave amplitude decreases with increasing L/λ ratio, i.e. increasing frequency for a given inclusion size or increasing inclusion size for a given frequency. Thus, earthquake motions will see their higher frequency (shorter wavelength) components more strongly affected by inclusions than their lower frequency (longer wavelength) components. If the wavelength of a wave is much longer than the dimension of an inclusion it encounters, the inclusion will simply translate almost as a rigid body as the wave passes through it. When waves encounter inclusions with dimensions similar to their wavelengths, the inclusions will respond dynamically, and the waves transmitted through them will be altered.

The situation becomes even more complicated when waves strike inclusions at angles other than 90°. Figure C.26 illustrates the refraction, reflection, and transmission of SH- and p-waves that strike an inclusion obliquely. In the case of an incident SH-wave, which has no component

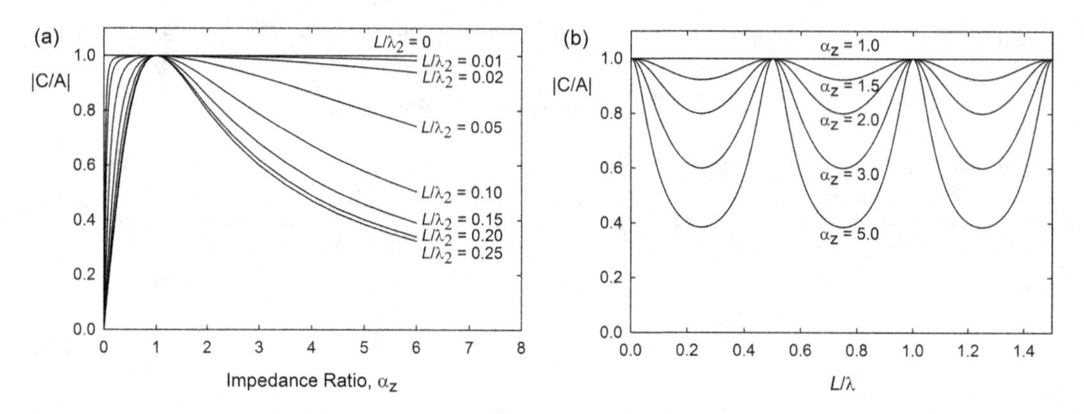

FIGURE C.25 Illustration of effects of inclusion on wave transmission: (a) as function of impedance ratio for $L/\lambda < 0.25$, and (b) as function of wavelength ratio for impedance ratios of 1.0–5.0.

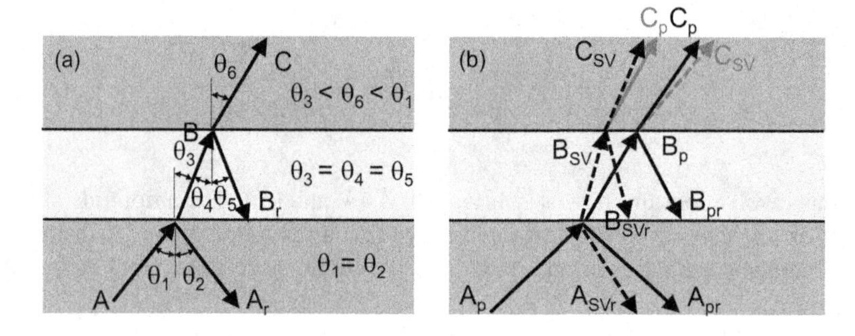

FIGURE C.26 Influence of inclusion on obliquely-incident waves: (a) incident SH-wave, and (b) incident p-wave.

of particle motion perpendicular to the boundaries, the transmitted wave is offset by refraction within the inclusion and propagates beyond the inclusion at a different orientation than the original incident wave. In the case of an incident p-wave, both p- and SV-waves are created at the inclusion boundaries, and all components travel in somewhat different directions. In the Earth's crust, naturally occurring heterogeneities act as inclusions, but without the simple, parallel-surface geometries shown in Figures C.25 and C.26. They therefore tend to scatter waves in very different directions so that the energy that reaches a particular site has traveled over a distribution of path lengths, thereby decreasing the amplitude and increasing the significant duration of the motion.

Scattering can cause even a simple, planar wave field to become quite complex, and can lead to substantial changes in the motions beyond the inclusion (Figure C.27). As in the one-dimensional case illustrated previously, the effects of the inclusion will depend on the ratio of its dimensions to the wavelengths of the incoming waves, with the result that high frequency (short wavelength) components will be scattered to a greater extent than low frequency (long wavelength) components. The reduction in wave amplitudes due to scattering, therefore, is greater at high frequency than low frequency. The high-frequency reduction in the Fourier amplitude spectra of recorded ground motions (Sections 3.4.2.2 and 3.6.1) is partially caused by this type of scattering phenomenon; the effects are similar to the effects that frequency-dependent damping would have.

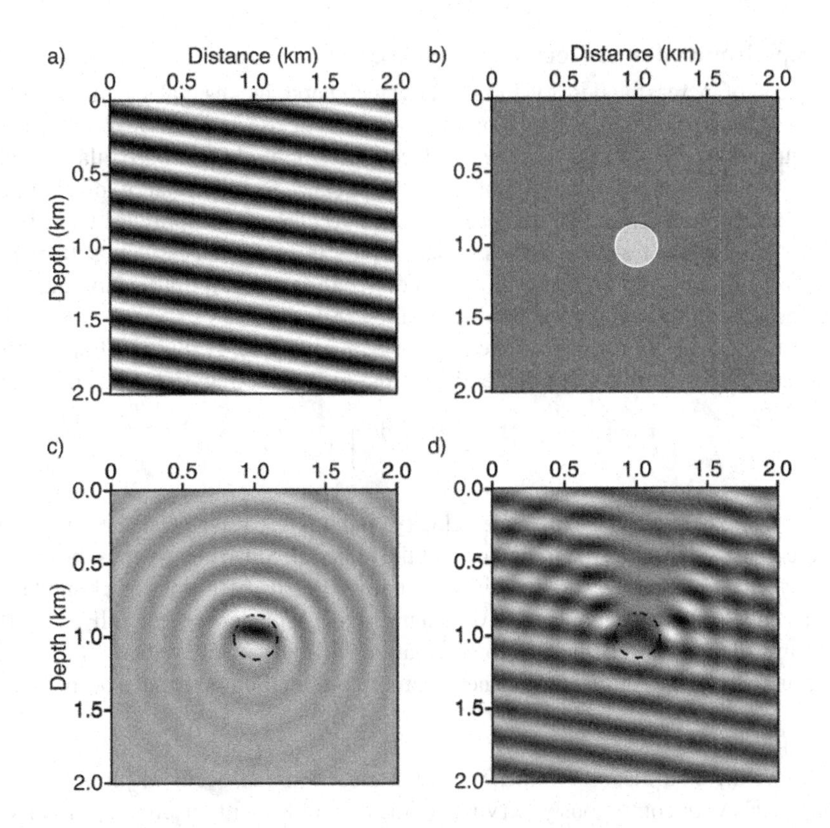

FIGURE C.27 Scattering of wavefield due to heterogeneity: (a) incident plane wave with no heterogeneity, (b) location of heterogeneity, (c) wavefield produced by oscillation of heterogeneity, and (d) total response [sum of (a) and (c)] showing alteration of plane wave by presence of heterogeneity (Pageot et al., 2013).

C.6 IMPLICATIONS FOR SITE RESPONSE

As discussed in Chapters 3 and 7, an important problem in geotechnical earthquake engineering is the evaluation of site response, i.e., the effect of local soil conditions on the nature of earthquake ground motions. The basic principles of wave propagation described in the earlier sections of this appendix can be used to gain insight into many aspects of the response of soil sites to the motions of the bedrock that underlies them. The following sections provide an introduction to the most important of these – they will help the reader understand the basis for site response models in ground motion models (GMMs) discussed in Section 3.5.2.3, and serve as an introduction to the more detailed concepts of site response analysis covered in Sections 7.4–7.7.

C.6.1 Ground Motion Amplification

The shallow geology at a particular site can have a strong effect on earthquake ground motions. Soil deposits overlying bedrock can amplify or de-amplify ground motion amplitudes and can do both at different frequencies. Upward-traveling seismic waves generally encounter progressively softer materials as they approach the ground surface, so they are refracted toward more and more vertical travel paths (Figure C.18). At shallow depths, seismic ground motions are often idealized as vertically propagating SH-saves, in which case all particle motion is horizontal. The primary mechanisms of ground motion amplification in such cases are associated with impedance gradients and resonance.

C.6.1.1 Amplification due to Decreasing Impedance

The general trend of increasing density and stiffness with depth means that specific impedance will also generally increase with depth. Based on the continuity of stresses and displacements at a layer boundary, Equation (C.78) showed that the displacement amplitude at a particular frequency will increase as a wave passes from a material of higher impedance to a material of lower impedance. If the displacement amplitude increases, the velocity and acceleration amplitudes will also increase. Thus, the fact that upward-traveling waves typically pass through successively softer materials as they approach the ground surface leads to a form of ground motion amplification manifested as a general increase in ground motion amplitude at shallower depths.

The same result can be obtained from energy considerations. Energy flux is defined as the amount of energy per unit time that flows through a given cross-section of material and can be defined as

$$E(t) = \frac{1}{2}\rho\left(\frac{\partial u}{\partial t}\right)^2 v \tag{C.117}$$

where ρ is the density, $\partial u/\partial t$ is the particle velocity, and v is the wave propagation velocity of the medium. Note that ρv is the specific impedance of the material. Since energy must be conserved at a boundary between two linear, elastic materials, the energy flux on both sides of a boundary between layers must be equal. Thus, when waves travel from a higher impedance (generally, stiffer) material into a lower impedance (softer) material, deformations (displacement, velocity, and acceleration) amplitudes must increase in order for the energy flux to be equal on both sides of the boundary.

C.6.1.2 Amplification due to Resonance

The type of amplification discussed in the preceding section can occur in profiles with discrete layers or in profiles with continuously varying stiffness. Profiles with discrete layers of substantial thickness can also display a tendency to strongly amplify motions at certain frequencies that are associated with the thicknesses and stiffnesses of the layers. The simplest case of this type would be a uniform soil layer overlying bedrock. Consider a layer of uniform elastic material (a simplistic idealization of "soil") underlain by a rigid base (a simplistic idealization of "rock"). If the rigid base moves horizontally, vertically propagating shear waves (SH-waves) will travel up through the soil, be reflected back downward at the ground surface (a free-end), and then be reflected back upward again by the rigid base (a fixed-end). The upward- and downward-traveling waves will interfere with each other, sometimes constructively and sometimes destructively, to produce the site response. The one-dimensional wave equation, along with the free- and fixed-end boundary conditions developed earlier in this appendix, allows the response of the soil layer to be computed for both undamped and damped soil conditions.

Undamped Soil

Consider a uniform layer of isotropic, linear elastic soil overlying rigid bedrock as shown in Figure C.28. Harmonic horizontal motion of the bedrock will produce vertically propagating shear

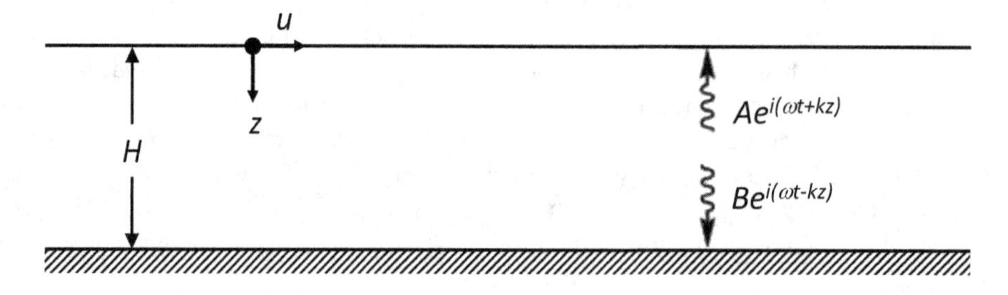

FIGURE C.28 Linear elastic soil deposit of thickness, H, underlain by rigid bedrock.

waves in the overlying soil. The resulting horizontal displacement can be expressed, using the results of Section C.2.1.3 (Equation C.15), as

$$u(z,t) = Ae^{i(\omega t+kz)} + Be^{i(\omega t-kz)} \tag{C.118}$$

where ω is the circular frequency of bedrock shaking, k the wave number ($= \omega/v_s$) and A and B the displacement amplitudes of waves traveling in the −z (upward) and +z (downward) directions, respectively. At the free surface ($z=0$), the shear stress, and consequently the shear strain, must vanish; i.e.,

$$\tau(0,t) = G\gamma(0,t) = G\frac{\partial u(0,t)}{\partial z} = 0 \tag{C.119}$$

Substituting (C.118) into (C.119) and differentiating yields

$$Gik(Ae^{ik(0)} - Be^{-ik(0)})e^{i\omega t} = Gik(A-B)e^{i\omega t} = 0 \tag{C.120}$$

which is satisfied (nontrivially) when $A=B$. The displacement can then be expressed as

$$u(z,t) = 2A\frac{e^{ikz}+e^{-ikz}}{2}e^{i\omega t} = 2A\cos(kz)e^{i\omega t} \tag{C.121}$$

which, describes a standing wave of amplitude $2A\cos(kz)$. Note that the amplitude of the wave is twice the amplitude of the upward-traveling wave, a characteristic known as the *free surface effect.* The standing wave is produced by the constructive interference of the upward and downward-traveling waves and has a fixed shape with respect to depth. Equation (C.121) can be used to define a transfer function that describes the ratio of displacement amplitudes at any two points in the soil layer. Choosing these two points to be the top and bottom of the soil layer gives the transfer function

$$F_1(\omega) = \frac{u(0,t)}{u(H,t)} = \frac{2Ae^{i\omega t}}{2A\cos(kH)e^{i\omega t}} = \frac{1}{\cos kH} = \frac{1}{\cos(\omega H/v_s)} \tag{C.122}$$

The modulus of the transfer function is the amplification function

$$|F_1(\omega)| = \sqrt{\{\text{Re}[F_1(\omega)]\}^2 + \{\text{Im}[F_1(\omega)]\}^2} = \frac{1}{|\cos(\omega H/v_s)|} \tag{C.123}$$

which indicates that the surface displacement is always at least as large as the bedrock displacement (since the denominator can never be greater than 1.0 and, at certain frequencies, is much larger). Thus $|F_1(\omega)|$ is the ratio of the free surface motion amplitude to the bedrock motion amplitude (or, since the bedrock is rigid in this case, the bedrock outcropping motion). As $\omega H/v_s$, approaches $\pi/2+n\pi$, the denominator of Equation (C.122) approaches zero, which implies that infinite amplification, or resonance, will occur (Figure C.29). Even this very simple model illustrates that the response of a soil deposit is highly dependent upon the frequency of the base motion, and that the frequencies at which strong amplification occurs depend on the geometry (thickness) and material properties (s-wave velocity) of the soil layer.

Damped Soil

Obviously, the type of unbounded amplification predicted by the previous analysis cannot physically occur. The previous analysis assumed no dissipation of energy, or damping, in the soil. Since damping is present in all materials, more realistic results can be obtained by repeating the analysis with damping. Assuming the soil to have the shearing characteristics of a Kelvin-Voigt solid, the wave equation can be written [Equation (C.97)] as

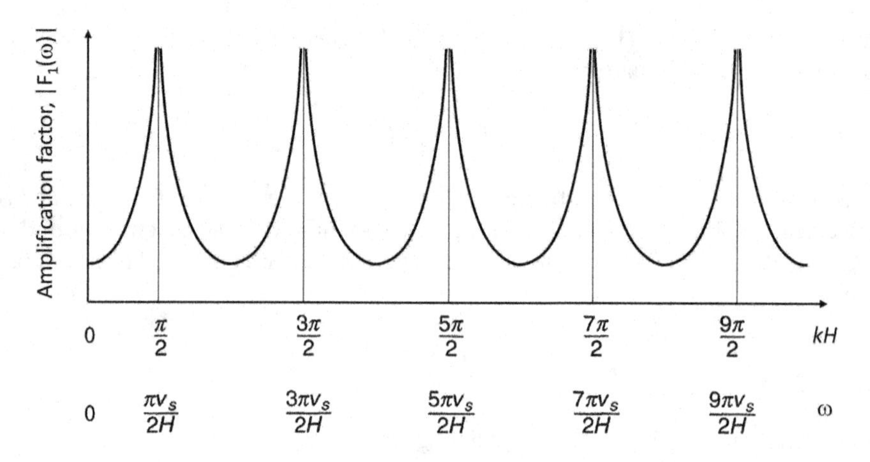

FIGURE C.29 Influence of frequency on the steady-state response of undamped linear elastic layer.

$$\rho \frac{\partial^2 u}{\partial t^2} = G \frac{\partial^2 u}{\partial z^2} + \eta \frac{\partial^3 u}{\partial z^2\, \partial t} \tag{C.124}$$

The solution to this equation can also be expressed in terms of a transfer function whose modulus can be expressed (see derivation in Section 7.5.1.2) as

$$|F_2(\omega)| = \frac{1}{\sqrt{\cos^2 kH + (\omega H / v_s)^2}} = \frac{1}{\sqrt{\cos^2(\omega H / v_s) + \left[\xi(\omega H / v_s)\right]^2}} \tag{C.125}$$

for small values of the damping ratio, ξ. This expression also indicates that amplification is frequency-dependent (Figure C.30) with local maxima (but not infinite in the case of an undamped material) at the *natural frequencies* of the layer.

The nth natural frequency of the soil deposit is given by

$$\omega_n \approx \frac{v_s}{H}\left(\frac{\pi}{2} + n\pi\right) \quad n = 0, 1, 2, \ldots, \infty \tag{C.126}$$

Since the peak amplification factor decreases with increasing natural frequency, the greatest amplification factor will occur approximately at the lowest natural frequency, i.e., the fundamental frequency, which can be expressed in radians/sec as

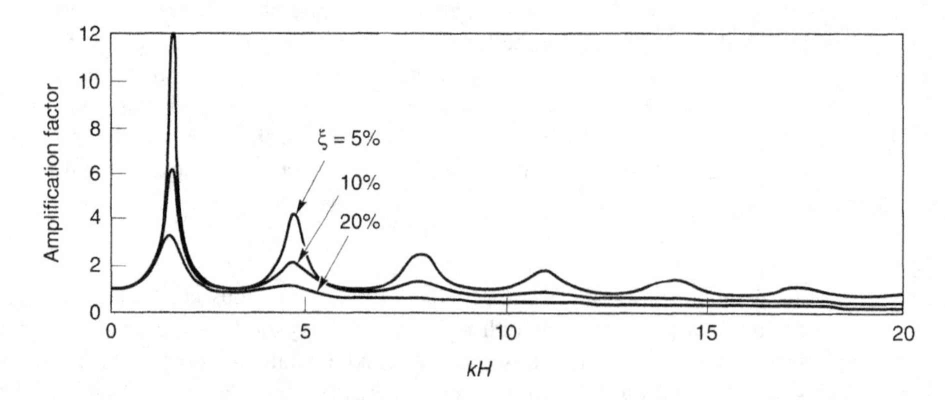

FIGURE C.30 Influence of frequency on the steady-state response of damped, linear elastic layer.

$$\omega_0 = \frac{\pi v_s}{2H} \tag{C.127}$$

or in Hz as

$$f_0 = \frac{v_s}{4H} \tag{C.128}$$

The fundamental period, or *characteristic site period,* then is

$$T_0 = \frac{4H}{v_s} \tag{C.129}$$

These concepts are used to help explain site effects terms in GMMs in Section 3.5.2.3 and are expanded upon in the development of more rigorous site response analysis procedures in Section 7.5.1.2.

C.6.2 Basin Effects

The refraction of seismic waves also plays an important role in the response of sites located in basins. Basins are localized zones of alluvium and sedimentary rock that are underlain by stiffer basement rock and can have dimensions ranging from kilometers to tens of kilometers. Figure C.31 schematically depicts two sedimentary basins – the narrow Basin 1 overlies the fault and Basin 2 is at some horizontal distance from it. The waves entering from beneath Basin 1 are propagating nearly vertically as they reach the bottom of the basin but are focused toward the center of the narrow basin by refraction at its sloping base. If Basin 1 had been broad with a relatively flat base, waves would have propagated vertically through the basin sediments and site response within the basin would be largely associated with the ground response effects described previously. Wave propagation in that case may be reasonably well represented by a one-dimensional wave propagation analyses (Section 7.5). On the other hand, because the seismic source is located outside the perimeter of Basin 2, waves can enter that basin from beneath (as with Basin 1) but also from the edge. Waves that enter the basin from the edge can be refracted in such a way that downward-traveling reflections strike the bottom of the basin at an incidence angle greater than the critical angle and are completely reflected back into the basin (total internal reflection). The waves then become trapped within the basin and generate Love waves that propagate across the basin.

Because basins are generally large (dimensions ranging from km to tens of km), they most strongly affect ground motions with long wavelengths, i.e., low-frequency ground motions. For this

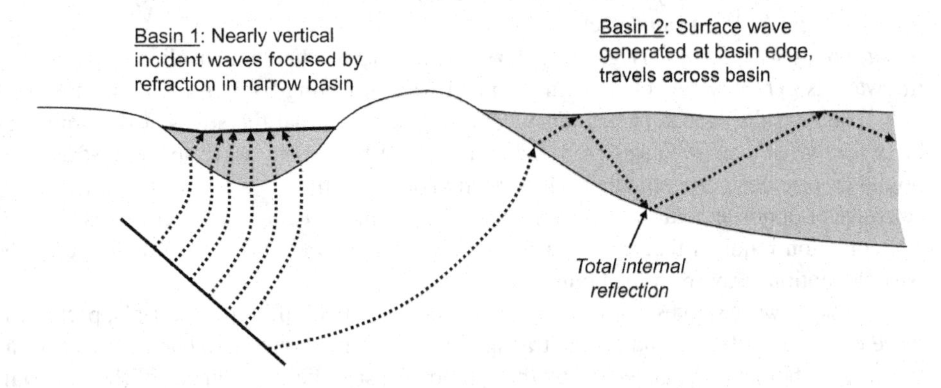

FIGURE C.31 Schematic showing seismic body waves entering basins from beneath and from the edge. The waves entering through the edge of Basin 2 undergo critical body wave reflections, which generate surface waves that travel across the basin. (Modified from Choi et al., 2005.)

reason, basin effects are typically only significant for periods greater than 1.0 sec. Because their specific effects are influenced by their three-dimensional geometries, simple engineering models that properly represent the physics of the problem do not presently exist. Instead, current models for basin-related amplification are a simple function of sediment depth, as discussed in Section 3.5.2.3. A more detailed description of basin effects is presented in Section 7.6.2.

C.7 SUMMARY

1. Only body waves can travel through an unbounded, homogeneous, solid. There are two types of body waves: p- and s-waves. P-waves are irrotational, or dilatational, waves-they induce volumetric but not shearing deformations in the materials they travel through. The direction of particle movement caused by p-waves is parallel to the direction in which the wave is traveling. S-waves, also known as shear waves, involve shearing but not volumetric deformations. The passage of an s-wave causes particle movement perpendicular to the direction of wave travel.
2. Body waves travel at velocities that depend on the stiffness and density of the material they travel through. Because geologic materials are stiffer in volumetric compression than in shear, p-waves travel faster through them than do s-waves.
3. The interaction of inclined body waves with the stress-free surface of the Earth produces surface waves. The motions produced by surface waves are concentrated in a shallow zone near the surface.
4. Rayleigh waves are the most important type of surface wave for earthquake engineering applications. In a homogeneous elastic half-space, Rayleigh waves would travel slightly more slowly than s-waves and would produce both vertical and horizontal particle motions that follow a retrograde elliptical pattern.
5. The depth to which Rayleigh waves induce significant motion is inversely proportional to the frequency of the wave. Low-frequency Rayleigh waves can produce particle motion at large depths, but the motions produced by high-frequency Rayleigh waves are confined to shallow depths.
6. When body wave velocities increase with depth, as they generally do in the Earth's crust, Rayleigh wave velocities are frequency-dependent. Low-frequency Rayleigh waves, which induce motion in deeper, stiffer materials, travel faster than high-frequency Rayleigh waves. Waves with frequency-dependent velocities are said to be dispersive.
7. Love waves are surface waves that can develop in the presence of a soft surficial layer. Love waves are dispersive – their velocities vary with frequency between the shear wave velocity of the surficial layer (at high frequencies) and the shear wave velocity of the underlying material (at low frequencies).
8. When a body wave strikes a rigid boundary oriented perpendicular to its direction of travel, the wave is perfectly reflected as an identical wave traveling back in the opposite direction. The zero-displacement boundary condition requires that the stress at the boundary be twice that of the wave away from the boundary. When a body wave strikes a stress-free boundary oriented perpendicular to its direction of travel, the wave is reflected as an identical wave of opposite polarity traveling back in the same direction. The zero-stress boundary condition requires that the particle motion at the boundary be twice as large as the particle motion away from the boundary.
9. When a body wave strikes a normal boundary between two different materials, part of the wave energy is reflected and part is transmitted across the boundary. The behavior of the wave at the boundary is governed by the ratio of the specific impedances of the materials on either side of the boundary. This impedance ratio determines the amplitudes and polarities of the reflected and transmitted waves.

10. When body waves strike boundaries between different materials at angles other than 90°, part of the wave energy is reflected and part is refracted as it crosses the boundary. If the direction of particle motion is parallel to the boundary, the reflected and refracted waves will be of the same form as the incident wave. If not, new types of waves can be created; for example, an inclined p-wave that strikes a horizontal boundary will produce reflected p- and SV-waves and also refracted p- and SV-waves.
11. When an inclined wave travels upward through horizontal layers that become successively softer, the portion of the wave that crosses each layer boundary will be refracted closer and closer to a vertical direction.
12. The amplitude of a stress wave decreases as the wave travels through the Earth's crust. There are two primary mechanisms that cause this attenuation of wave amplitude. The first, material damping, is due to the absorption of energy by the materials the wave is traveling through. The second, radiation damping, results from the spreading of wave energy over a greater volume of material as it travels away from its source.

REFERENCES

Aki, K. and Richards, P.G. (1980). *Quantitative Seismology: Theory and Methods*, Volumes 1 and 2, W.H. Freeman, San Francisco, California.

Bullen, K.E. (1953). *An Introduction to the Theory of Seismology*, Cambridge University Press, London, 296 pp.

Choi, Y., Stewart, J.P., Graves, R.W. (2005). "Empirical model for basin effects accounts for basin depth and source location," *Bulletin of the Seismological Society of America*, Vol. 95, No. 4, pp. 1412–1427.

Ewing, M., Jardesky, W. and Press, F. (1957). *Elastic waves in layered media*, McGraw-Hill Book Company, New York, 380 pp.

Kolsky, H. (1963). *Stress Waves in Solids*, Dover Publications, New York, 213 pp.

Love, A.E.H. (1927). *The Mathematical Theory of Elasticity*, 4th Edition, University Press, Cambridge.

McCamy, K., Meyer, R.P., and Smith, T.J. (1962). "Generally applicable solutions of Zoeppritz' amplitude equations," *Bulletin of the Seismological Society of America*, Vol. 52, No. 4, pp. 923–955.

Pageot, D., Operto, S., Vallee, M., Brossier, R., and Virieux, J. (2013). "A parametric analysis of two-dimensional elastic full waveform inversion of teleseismic data for lithospheric imaging," *Geophysical Journal International*, Vol. 193, pp. 1479–1505.

Richart, F.E., Hall, J.R., and Woods, R.D. (1970). *Vibrations of Soils and Foundations*, Prentice-Hall, Inc., Englewood Cliffs, NJ, 401 pp.

Richter, C.F. (1958). *Elementary Seismology*, W.H. Freeman and Company, San Francisco, California.

Semblat, J.F. and Pecker, A. (2009). *Waves and Vibrations in Soils*, IUSS Press, Pavia.

Appendix D
Probability Concepts

D.1 INTRODUCTION

Earthquake engineering problems are fraught with uncertainty, and performance-based earthquake engineering seeks to characterize and account for that uncertainty. As a result, the practice of performance-based earthquake engineering requires some level of familiarity with basic concepts of uncertainty and basic procedures of probabilistic analysis. Geotechnical earthquake engineering is particularly affected by uncertainty. At a particular site, bedrock motions depend on the size, location, and rupture behavior of the earthquake – none of which can be predicted with certainty. Because of the inherent variability of soils and the inevitable limits on exploration of subsurface conditions, the resistance of the soil to that loading is not known with certainty. When both loading and resistance are uncertain, the resulting effects are also uncertain. A number of geotechnical earthquake engineering analyses attempt to quantify the uncertainty in the various input parameters for a particular problem and compute the resulting uncertainty in the output.

This appendix provides a brief introduction to some of the basic concepts of probability and describes probability terms, distributions, and calculations that are used in the body of the book. It also discusses the propagation of uncertainty, i.e., how uncertainty in input parameters translates to uncertainty in output. More detailed information on these topics can be found in texts such as Benjamin and Cornell (2014), Baecher and Christian (2003), Ang and Tang (2007), and Fenton and Griffiths (2008).

D.2 SAMPLE SPACES AND EVENTS

Probability theory deals with the results, or outcomes, of processes that are usually described in a general sense as *experiments*. The set of all possible outcomes of an experiment is called the *sample space*, and each outcome of an experiment is called a *sample point*. The sample space therefore consists of all possible sample points. The sample space may be *continuous*, in which case the number of sample points is infinite, or it may be *discrete* as when the number of sample points is finite and countable.

An *event* is a subset of a sample space and therefore represents a set of sample points. A *single event* consists of a single sample point and a *compound event* consists of more than one sample point. If Ω represents a sample space and A an event, the *complementary event* $\bar{A}$ is the set of all sample points in Ω that are not in A. The interrelationships among sets can be conveniently illustrated by means of a *Venn diagram* of the type illustrated in Figure D.1. In Figure D.1, the sample space is represented by the rectangle Ω and event A by the circle. The complementary event $\bar{A}$ corresponds to the part of the rectangle that lies outside the circle. Two operations on events are of interest – the *union* of two events, A and B, consists of all sample points that are in either A or B (or both), and the *intersection* is the set of sample points that are in both A and B. Since no sample points are in both A and $\bar{A}$, the intersection of A and $\bar{A}$ is the null set $\varnothing$, i.e., $A \cap \bar{A} = \varnothing$. Similarly, the union of A and $\bar{A}$ is Ω, i.e., $A \cup \bar{A} = \Omega$. Two events, A and B, are said to be *mutually exclusive* if they share no common sample points, i.e. $A \cap B = \varnothing$. A set of events, $B_1, B_2, \ldots, B_n$ are *collectively exhaustive* if their union makes up the entire sample space, i.e., $B_1 \cup B_2 \cup \ldots \cup B_n = \Omega$.

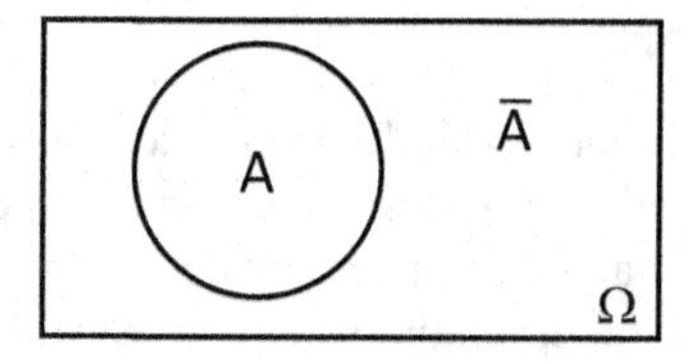

FIGURE D.1 Venn diagram illustrating event A in sample space Ω.

Example D.1

Consider the Venn diagram for the three events, *A*, *B*, and *C*, shown in Figure ED.1

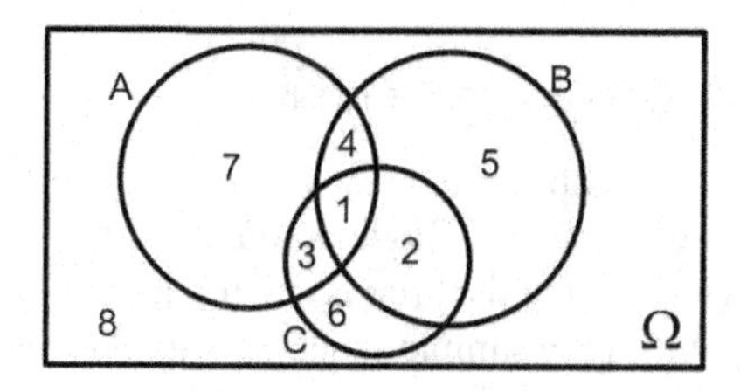

FIGURE ED.1 Venn diagram with three events considered in Example D.1.

$A \cap B$ = regions 1 and 4	$A \cap \bar{B}$ = regions 3 and 7
$B \cap C$ = regions 1, 2, 3, 4, 5, and 6	$(A \cup B) \cap C$ = regions 1, 2, and 3
$A \cap B \cap C$ = region 1	$\bar{A} \cap \bar{B} \cap \bar{C}$ = region 8

D.3 AXIOMS OF PROBABILITY

A probability measure, *P*, can be assigned to each sample point or set of sample points in a sample space. The probability of an event *A* is then denoted by the symbol *P*[*A*]. The entire theory of probability is based on the following three fundamental axioms:

Axiom 1. The probability of an event is represented by a number greater than or equal to zero but less than or equal to one.

$$0 \leq P[A] \leq 1 \tag{D.1a}$$

Axiom 2. The probability of an event equal to the entire sample space Ω is one

$$P[\Omega] = 1 \tag{D.1b}$$

Axiom 3. The probability of an event representing the union of two mutually exclusive events is equal to the sum of the probabilities of the events

$$P[A \cup B] = P[A] + P[B] \tag{D.1c}$$

These axioms can be used to develop the rules and theorems that comprise the mathematical theory of probability.

D.4 PROBABILITY OF EVENTS

Probabilities are often thought of in terms of relative frequencies of occurrence. If the existence of a water content greater than the optimum water content in a compacted fill is considered to be an event, the probability of that event can be estimated by determining the relative frequency of water content measurements that exceed the optimum water content. If the total number of water content measurements is small, the relative frequency may only approximate the actual probability, but as the number of measurements becomes large, the relative frequency will approach the actual probability. This frequentist point of view is not very helpful, however, for situations in which an experiment cannot be repeated. In such cases, probabilities can be viewed as relative likelihoods, as in the probability that the material at a certain depth in a boring is clay rather than sand or rock. The latter interpretation lends itself to the subjective evaluation of probability.

D.4.1 Probabilities of Unions and Intersections

Regardless of how probabilities are interpreted, the axioms of probability allow statements to be made about the probabilities of occurrence of single or multiple events. These can be visualized with the use of Venn diagrams drawn such that the area of the rectangle representing the sample space Ω is 1 and the areas of all events within the sample space are equal to their probabilities. Consider the non-exclusive events A and B in Figure D.2. The event $A \cap B$ is represented by the shaded region and $P[A \cap B]$ is given by the area of the shaded region in Figure D.2a. The event $A \cup B$ is represented by the shaded region in Figure D.2b; $P[A \cup B]$ is given by the area of that shaded region, or

$$P[A \cup B] = P[A] + P[B] - P[A \cap B] \tag{D.2}$$

If events A and B were mutually exclusive, their sets in a Venn diagram would not overlap, so $P[A \cap B] = 0$ and $P[A \cup B] = P[A] + P[B]$.

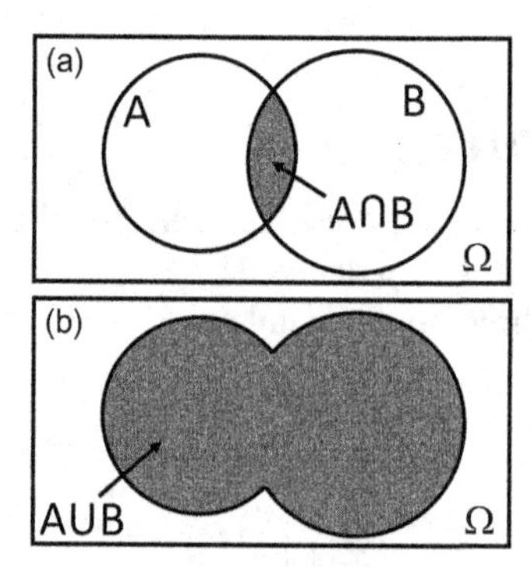

FIGURE D.2 (a) Intersecting events A and B in sample space Ω. If the area of the rectangle Ω is 1, the probability of $A \cap B$ is given by the shaded area. (b) Union of events A and B in sample space Ω. The probability of $A \cup B$ is given by the shaded area.

Example D.2

Consider the rolling of a single fair die as an experiment. Then the resulting sample space, $W = \{1, 2, 3, 4, 5, 6\}$, is the set of all possible outcomes of the experiment. Let the following three events be defined as

$A = \{1\}$ (a single roll produces a 1)
$B = \{1, 3, 5\}$ (a single roll produces an odd number)
$C = \{4, 5, 6\}$ (a single roll produces a number greater than 3)

Define the sets $A \cap B$, $A \cup B$, and $B \cup C$, and compute their probabilities.

Solution:

The set $A \cap B$ includes all outcomes that are in both A and B, so $A \cap B = \{1\}$. The set $A \cup B$ includes all outcomes that are in either A or B, so $A \cup B = \{1, 3, 5\}$. The set $B \cup C$ includes all outcomes that are in either B or C, so={1, 3, 4, 5, 6}. The probabilities of each set can be computed as

$$P[A \cap B] = P[A \mid B]P[B] = \left(\frac{1}{3}\right)\left(\frac{1}{2}\right) = \frac{1}{6}$$

$$P[A \cup B] = P[A] + P[B] - P[A \cap B] = \frac{1}{6} + \frac{1}{2} - \frac{1}{6} = \frac{1}{2}$$

$$P[B \cup C] = P[B] + P[C] - P[B \cap C] = \frac{1}{2} + \frac{1}{2} - \frac{1}{6} = \frac{5}{6}$$

D.4.2 Conditional Probability

In many instances, events are related so the probability of one event conditional upon the occurrence of another event is of interest. The *conditional probability* of event A given the occurrence of event B is denoted $P[A|B]$ and is defined (for $P[B>0]$) by

$$P[A \mid B] = \frac{P[A \cap B]}{P[B]} \tag{D.3a}$$

Similarly,

$$P[B \mid A] = \frac{P[A \cap B]}{P[A]} \tag{D.3b}$$

Example D.3

One hundred field compaction tests were performed in the early stages of the construction of an earth dam. The results of the tests are presented in terms of the numbers that satisfied specifications for minimum relative compaction and for compaction water content in the table below.

Water Content	Relative Compaction	
	Acceptable	**Not Acceptable**
Acceptable	80	10
Not Acceptable	6	4

Assume the contactor's performance in the future will be the same as in the first 100 tests and that the fill material does not change. Estimate the probability that the relative compaction specification will be satisfied in the next test if the water content specification is satisfied. Estimate that probability for the case in which the water content specification is not satisfied.

Solution:

Define two events, R and W, such that

R=relative compaction specification satisfied
W=water content specification satisfied

From the table the probability that both the relative compaction and water content specifications are satisfied can be estimated as $P[W \cap R] = 80/100$. Then the probability that the relative compaction specification will be satisfied in the next test if the water content specification is satisfied is the conditional probability $P[R|W]$, which can be computed as

$$P[R \mid W] = \frac{P[W \cap R]}{P[W]} = \frac{80/100}{80/100 + 10/100} = \frac{80}{90} = 0.889$$

The probability that the relative compaction specification is satisfied given that the water content specification is not satisfied can be estimated as $P[R \mid \bar{W}]$, or

$$P[R \mid \bar{W}] = \frac{P[\bar{W} \cap R]}{P[\bar{W}]} = \frac{6/100}{6/100 + 4/100} = \frac{6}{10} = 0.6$$

The conditional probability is easily visualized with the Venn diagram as the ratio of the area of $A \cap B$ to the area of B. Event A is *statistically independent* of event B if the occurrence does not affect the probability of occurrence of A, i.e., if

$$P[A \mid B] = P[A] \tag{D.4}$$

Rearranging Equation (D.3a), the probability of the intersection of the independent events A and B is given by

$$P[A \cap B] = P[A \mid B]\, P[B] \tag{D.5}$$

which, if A and B are statistically independent, becomes

$$P[A \cap B] = P[A]\; P[B] \tag{D.6}$$

This is known as the *multiplication rule* and can be extended to the multiple, mutually independent events $A, B, C, \ldots, N$ by

$$P[A \cap B \cap C \cap \ldots \cap N] = P[A] \cdot P[B] \cdot P[C] \cdot \ldots \cdot P[N] \tag{D.7}$$

The multiplication rule states that the probability of joint occurrence of independent events is equal to the product of their individual probabilities.

For a set of mutually exclusive, but collectively exhaustive, events $B_1, B_2, \ldots, B_N$, like that shown in the Venn diagram of Figure D.3, the probability of another event A can be expressed as

$$P[A] = P[A \cap B_1] + P[A \cap B_2] + \cdots + P[A \cap B_N] \tag{D.8}$$

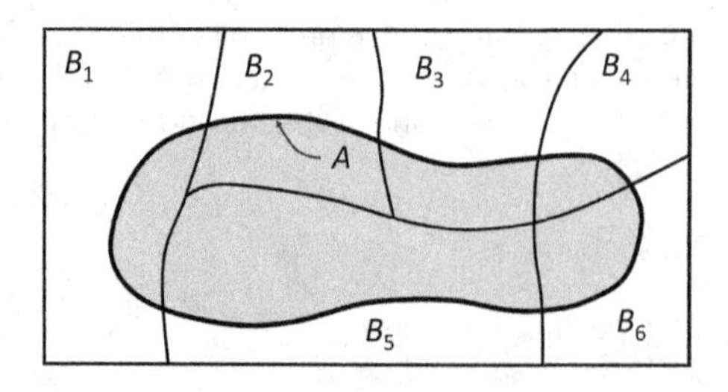

FIGURE D.3 Intersection of event A with collectively exhaustive events B_i.

Using Equation (D.4)

$$P[A] = P[A \mid B_1]P[B_1] + P[A \mid B_2]P[B2] + \cdots + P[A \mid B_N]P[B_N]$$
$$= \sum_{i=1}^{N} P[A \mid B_i]P[B_i] \qquad \text{(D.9)}$$

which is known as the *Total Probability Theorem.* The Total Probability Theorem allows the unknown probability of an event to be "built" from a series of known conditional probabilities and forms the backbone of the probability calculations required for probabilistic seismic hazard analyses (Chapter 4).

The Total Probability Theorem can also be extended to account for multiple dependencies. If, for example, Event C is conditionally dependent on Event B, which is in turn conditionally dependent on Event A, the probability of C can be computed as

$$P[C] = P[C \mid B] \cdot P[B \mid A] \cdot P[A] \qquad \text{(D.10)}$$

This type of "chain" of conditional probabilities can be extended to any number of events; such calculations are performed along individual branches of *logic trees* (Section 4.4.3.4) in seismic hazard analyses.

Example D.4

A structural engineer has determined that a structure will collapse in an earthquake that produces a peak acceleration of 0.3*g*. The probabilities that a given earthquake on fault *A*, *B*, or *C* would be strong enough to cause that level of ground shaking are 0.5, 0.2, and 0.1, respectively. The probabilities that such earthquakes will occur on faults *A*, *B*, and *C* during the life of the building are 0.01, 0.05, and 0.08, respectively. What is the probability that the structure will collapse in an earthquake?

Solution:

Define the following events as

A=the structure collapses in an earthquake
D_1=an earthquake capable of collapsing the structure occurs on fault *A*
D_2=an earthquake capable of collapsing the structure occurs on fault *B*
D_3=an earthquake capable of collapsing the structure occurs on fault *C*

Then the probability that the structure collapses in an earthquake is given by

$$P[A] = P[A \mid D_1]P[D_1] + P[A \mid D_2]P[D_2] + P[A \mid D_3]P[D_3]$$
$$= (0.5)(0.01) + (0.2)(0.05) + (0.1)(0.10)$$
$$= 0.025$$

D.4.3 Bayes' Theorem

Equations (D.3a) and (D.3b) can be solved for $P[A \cap B]$ and then set equal to obtain

$$P[B \mid A]P[A] = P[A \mid B]P[B] \qquad \text{(D.11)}$$

or

$$P[B \mid A] = \frac{P[A \mid B]P[B]}{P[A]} \qquad \text{(D.12)}$$

provided that $P[A] \neq 0$. Note that the order of the conditioning on the left and right sides of Equation (D.12) is reversed; Equation (D.12) allows calculation of $P[B \mid A]$ when only $P[A \mid B]$ is known.

If B consists of a number of mutually exclusive and collectively exhaustive events, B_1, B_2, … , B_N, (as in Figure D.3), the denominator in Equation (D.12) can be replaced by Equation (D.9), giving

$$P[B \mid A] = \frac{P[A \mid B]P[B]}{\sum_{i=1}^{N} P[A \mid B_i]P[B_i]} \tag{D.13}$$

Equation (D.13) is the most common form in which Bayes' Theorem is expressed. Bayes' Theorem provides another way of computing an unknown probability from a series of known conditional probabilities.

Example D.5

Suppose that you are investigating a site in an area where 10% of the previous borings have indicated the presence of a liquefiable layer. You are using a testing procedure, however, that is 80% accurate in detecting the presence of a liquefiable layer (when a liquefiable layer exists, the test will indicate its presence 80% of the time; however, the test also indicates the presence of a liquefiable layer 20% of the time when it does not actually exist). If the testing procedure indicates the presence of a liquefiable layer, what is the probability that a liquefiable layer actually exists?

Solution:

First, the events of interest can be defined:

Event A:	Liquefiable layer is detected by testing procedure
Event B_1:	Liquefiable layer actually exists
Event B_2:	Liquefiable layer does not exist

Note that events B_1 and B_2 are mutually exclusive (the liquefiable layer cannot both exist and not exist) and collectively exhaustive (the liquefiable layer either has to exist or not exist). From the problem description, then

$P[A \mid B_1] = 0.8$	(when liquefiable layer exists, test will indicate it 80% of time)
$P[A \mid B_2] = 0.2$	(when liquefiable layer absent, test will indicate it exists 20% of time)
$P[B_1] = 0.1$	(10% of previous borings have encountered liquefiable soil)
$P[B_2] = 0.9$	(90% of previous borings have encountered no liquefiable soil)

With this data, the probability that a liquefiable layer actually exists, in light of prior knowledge about subsurface conditions in the area and the observation that a particular testing procedure (of imperfect, but known, accuracy) has indicated its presence, can be computed as

$$P[B_1 \mid A] = \frac{P[A \mid B_1]P[B_1]}{P[A \mid B_1]P[B_1] + P[A \mid B_2]P[B_2]} = \frac{(0.8)(0.1)}{(0.8)(0.1) + (0.2)(0.9)} = 0.308$$

This probability is not intuitively obvious because it depends not only on the successful prediction (true-positive) rate of the testing procedure but also on the unsuccessful (false-positive) rate and the prior information about the existence of liquefiable soils in the region. Note that if the testing procedure gave 95% true-positive and 5% false-positive results, the probability that a liquefiable layer would exist given the test's indication would rise to 67.9%. If the testing procedure's accuracy is very low (say 20% true-positive and 80% false-positive), the probability of a liquefiable layer actually existing given its indication by the test would only be 2.7%.

D.5 RANDOM VARIABLES

All fields of science and engineering attempt to describe various quantities or phenomena with numerical values. In most cases, the precise numerical value cannot be predicted in advance of some process, or experiment, of interest. In such cases, the quantity or phenomenon can be described by a *random variable*. The random variable is used to describe an event in a sample space in quantitative terms.

D.5.1 Discrete Random Variables

A *discrete random variable* can take on only a finite or countable number of values. Each value has a probability, p_i, and the distribution of p_i values constitutes a *probability mass function*, or PMF, usually written as

$$p_X(x_i) = P[X = x_i] = p_i \tag{D.14}$$

All values of p_i must be non-negative and their collective sum must equal 1.0. Note that an upper-case letter is used to represent the random variable while lower case letters are used to describe specific values that the random variable can take on.

The cumulative distribution function, or CDF, of a discrete random variable, is defined as the probability that the discrete random variable is less than or equal to a particular value

$$F_X(x) = P[X < x] \tag{D.15}$$

which means that

$$F_X(x = a) = \sum_{\text{all } x_i \le a} P_X(x_i) \tag{D.16}$$

Examples of probability mass and cumulative distribution functions for a discrete random variable are shown in Figure D.4.

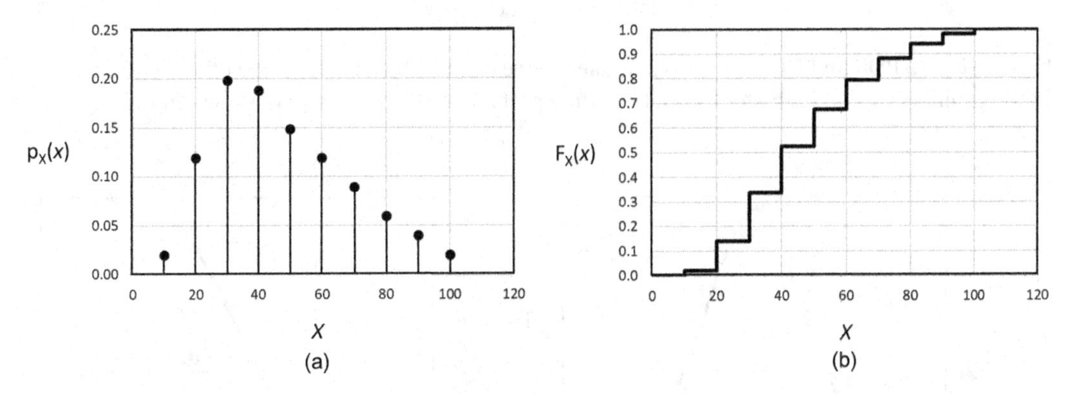

FIGURE D.4 (a) Probability mass function (PMF) and (b) Cumulative distribution function (CDF) for arbitrary discrete random variable.

D.5.2 Continuous Random Variables

A *continuous random variable* can take on any value within one or more intervals. Therefore, a continuous random variable can take on any of an infinite number of values and, as a consequence, the probability of it taking on any specific value is $1/\infty = 0$. The probability distribution of a continuous random variable can be described by its *probability density function* or PDF, $f_X(x)$, which must satisfy the conditions

$$f_X(x) \geq 0 \tag{D.17a}$$

$$\int_{-\infty}^{\infty} f_X(x)\,dx = 1 \tag{D.17b}$$

$$P[a \leq X \leq b] = \int_{a}^{b} f_X(x)\,dx \tag{D.17c}$$

According to these conditions, the area under the PDF between two values a and b represents the probability that the random variable will take on a value in the interval bounded by a and b. The probability distribution of a random variable can also be described by its CDF, which is given by

$$F_X(x) = P[X < x] = \int_{-\infty}^{x} f_X(x)\,dx \tag{D.18}$$

which, of course means that

$$f_X(x) = \frac{d}{dx} F_X(x) \tag{D.19}$$

Therefore, the probability that a random variable, X, falls between two values, a and b, is

$$P[a \leq X \leq b] = F_X(b) - F_X(a) \tag{D.20}$$

Obviously, the PDF and CDF are closely related – one can be obtained from the other by integration or differentiation. The PDF and CDF of a simple probability distribution are shown in Figure D.5.

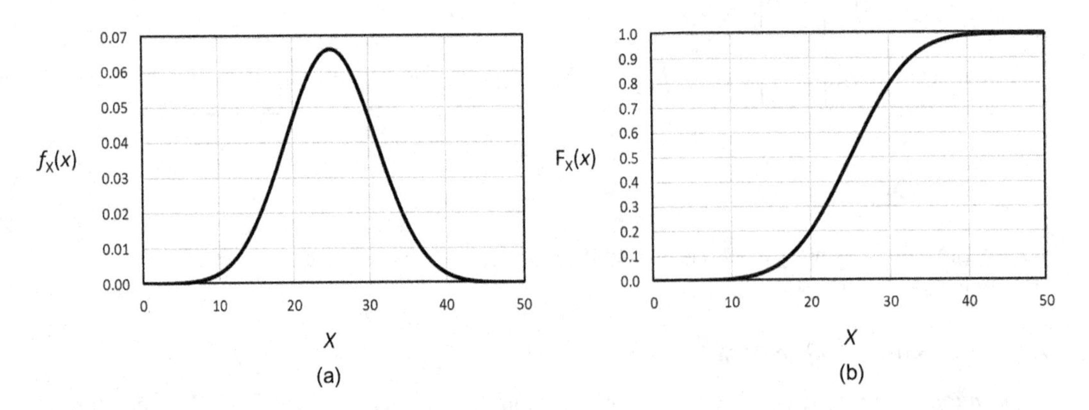

FIGURE D.5 (a) Probability density function (PDF) and (b) Cumulative distribution function (CDF) for an arbitrary continuous random variable.

Conditional probability concepts can be extended to continuous variables. For example, if Y depends on X and X can take on any possible value, the exceedance probability of Y, i.e., the probability that Y exceeds some value, y, can be computed as

$$P[Y > y] = \int_{-\infty}^{\infty} P[Y > y \mid X] f_X(x)\, dx \tag{D.21}$$

Thus, the desired probability is obtained by integrating over the entire distribution of X.

D.5.3 Jointly Distributed Random Variables

In some cases, the probability of more than one random variable occurring (or being exceeded) simultaneously is of interest. Such probabilities will depend not only on the distributions of the individual random variables but also on the relationships between the random variables. If two random variables, X and Y, are continuous, their joint distribution can be described by the joint PDF defined such that

$$f_{X,Y}(x,y)dxdy = P[x < X \le x + dx, y < Y \le Y + dy] \tag{D.22}$$

As in the case of a univariate PDF, a joint PDF must satisfy a number of conditions, namely that

$$f_{X,Y}(x,y) \ge 0 \tag{D.23a}$$

$$\int_{-\infty}^{\infty}\int_{-\infty}^{\infty} f_{X,Y}(x,y)\, dx\, dy = 1 \tag{D.23b}$$

$$P\left[a < X \le b, c < Y \le d\right] = \int_{c}^{d}\int_{a}^{b} f_{X,Y}(x,y)\, dx\, dy \tag{D.23c}$$

For continuous random variables, the extension of Equation (D.3) indicates *conditional distributions*

$$f_{X|Y}(x \mid y) = \frac{f_{X,Y}(x,y)}{f_Y(y)} \tag{D.24a}$$

and

$$f_{Y|X}(y \mid x) = \frac{f_{X,Y}(y,x)}{f_X(x)} \tag{D.24b}$$

from which the joint distribution can be expressed as

$$f_{X,Y}(x,y) = f_{X|Y}(x \mid y) f_Y(y) = f_{Y|X}(y \mid x) f_X(x) \tag{D.25}$$

In certain cases, it is convenient to determine an unknown joint distribution from known conditional and univariate distributions.

D.6 MOMENTS OF A PROBABILITY DISTRIBUTION

The characteristics of a random variable are completely described by its distribution, i.e., its PMF (for a discrete random variable) or its PDF or CDF (for a continuous random variable). The description of a complete distribution can be unwieldy, and in many cases, the complete distribution may not be known; in such cases, the important characteristics of a distribution can often be described by a small number of *moments* of the distribution.

The k-th moment of a probability distribution can be defined as

$$m_k = \sum_{i=1}^{n} p_i x_i^k \quad \text{if } X \text{ is discrete} \tag{D.26a}$$

$$m_k = \int_{-\infty}^{\infty} x^k f_X(x)\, dx \quad \text{if } X \text{ is continuous} \tag{D.26b}$$

D.6.1 Measures of Central Tendency

Since random variables can take on a wide range of possible values, some measure of the "central" value is of interest. The *mean*, or *expected value*, of a random variable, X, can be computed as the first moment of X

$$\mu_x = \sum_{i=1}^{n} p_i x_i \quad \text{if } X \text{ is discrete} \tag{D.27a}$$

$$\mu_x = \int_{-\infty}^{\infty} x\, f_X(x)\, dx \quad \text{if } X \text{ is continuous} \tag{D.27b}$$

The mean, therefore, is simply the weighted average of all X values, where the weighting factors are given by the PMF or PDF.

Other measures of central tendency include the *median*, $\hat{x}$, which is the value for which there is a 50% probability of being smaller and a 50% probability of being larger, i.e., the value of the variable for which $F_X(x)=0.5$. The *mode* of a random variable is the most probable single value, i.e., the value corresponding to the highest point on a PMF or PDF.

D.6.2 Measures of Dispersion

It is also useful to know how the values of a random variable are distributed with respect to its mean value. For that reason, the second and higher moments are usually taken about the mean, i.e., as

$$m_k = \sum_{i=1}^{n} p_i (x_i - \mu_x)^k \quad \text{if } X \text{ is discrete} \tag{D.28a}$$

$$m_k = \int_{-\infty}^{\infty} (x - \mu_x)^k f_X(x)\, dx \quad \text{if } X \text{ is continuous} \tag{D.28b}$$

The second moment about the mean is the *variance* of the random variable and the square root of the variance is the *standard deviation*.

$$Var(X) = \sigma_x^2 = \sum_{i=1}^{n} p_i (x_i - \mu_x)^2 \quad \text{if } X \text{ is discrete} \tag{D.29a}$$

$$Var(X) = \sigma_x^2 = \int_{-\infty}^{\infty} (x - \mu_x)^2 f_X(x)\,dx \quad \text{if } X \text{ is continuous} \tag{D.29b}$$

Since the standard deviation has the same units as the random variable itself, the dispersion can be conveniently normalized by the mean to produce the *coefficient of variation*,

$$COV = \sigma_x / \mu_x \tag{D.30}$$

D.6.3 Measures of Symmetry

The third moment can be used to define a *skewness coefficient*

$$\theta = \frac{\sum_{i=1}^{n} p_i (x_i - \mu_x)^3}{\sigma_x^3} \quad \text{if } X \text{ is discrete} \tag{D.31a}$$

$$\theta = \frac{\int_{-\infty}^{\infty} (x - \mu_x)^3 f_X(x)\,dx}{\sigma_x^3} \quad \text{if } X \text{ is continuous} \tag{D.31b}$$

A distribution for which $\theta=0$ is symmetric; when $\theta>0$, the distribution has a heavy positive tail and is said to be positively skewed.

D.6.4 Measures of Association

At times, more than one random variable at a time will be of interest, frequently as input parameters to some analysis or as outputs from an analysis. It is often important to know the level to which the variables are associated with each other. The *covariance* of two random variables is a measure of the degree to which their values are simultaneously above or below their respective means. The covariance between two random variables, X and Y, is defined by

$$Cov[X,Y] = E[(X - \mu_X)(Y - \mu_Y)]$$

$$= \sum_{i=1}^{nX} \sum_{j=1}^{nY} (x_i - \mu_X)(y_j - \mu_Y) P_{X,Y}(x_i, y_j) \quad \text{for discrete } X \text{ and } Y \tag{D.32a}$$

$$= \int_{-\infty}^{\infty} \int_{-\infty}^{\infty} (x - \mu_X)(y - \mu_Y) F_{X,Y}(x, y)\,dxdy \quad \text{for continuous } X \text{ and } Y \tag{D.32b}$$

Note that the covariance of a variable with itself is simply the variance of that variable. The covariance can also be computed as

$$Cov[X,Y] = E[XY] - \mu_X \mu_Y \tag{D.33}$$

It is important to distinguish between the covariance, *Cov* (usually written using lower-case letters), and the coefficient of variation, *COV* (usually written with all upper-case letters). It should be noted that the covariance between a random variable and itself is the variance of that random variable. The covariance can be normalized by the product of the standard deviations of the respective random variables, which produces the *correlation coefficient.*

$$\rho_{XY} = \frac{Cov[X,Y]}{\sigma_X \sigma_Y} \tag{D.34}$$

Figure D.6a shows an example of strong correlation between spectral accelerations (S_a) at closely spaced periods, while Figure D.6b shows an example of weaker correlation for S_a at more widely spaced periods. The correlation coefficient ranges from –1 (in which case X and Y are linearly related with $dY/dX<0$) to +1 (in which case X and Y are linearly related with $dY/dX>0$). Note that the covariance (and correlation coefficient) describe the degree of linear association between two random variables. Two random variables may be uniquely related by some nonlinear function and have a low correlation coefficient (e.g., $\rho_{XY}=0$ for $Y=\sin(X)$).

The level of association between multiple random variables may be expressed in terms of a *covariance matrix*, which for a series of random variables, $X_1, X_2, \ldots, X_n$ would be

$$\Sigma = \begin{bmatrix} Var(X_1) & Cov(X_1,X_2) & \cdots & Cov(X_1,X_n) \\ Cov(X_2,X_1) & Var(X_2) & \cdots & Cov(X_2,X_n) \\ \vdots & \vdots & \ddots & \vdots \\ Cov(X_n,X_1) & Cov(X_n,X_2) & \cdots & Var(X_n) \end{bmatrix} \tag{D.35}$$

The corresponding correlation matrix would be

$$\rho = \begin{bmatrix} 1 & \rho_{X_1,X_2} & \cdots & \rho_{X_1,X_n} \\ \rho_{X_2,X_1} & 1 & \cdots & \rho_{X_2,X_n} \\ \vdots & \vdots & \ddots & \vdots \\ \rho_{X_n,X_1} & \rho_{X_n,X_2} & \cdots & 1 \end{bmatrix} \tag{D.36}$$

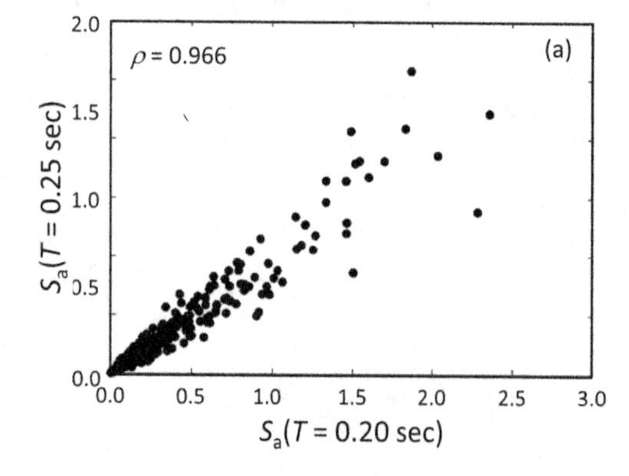

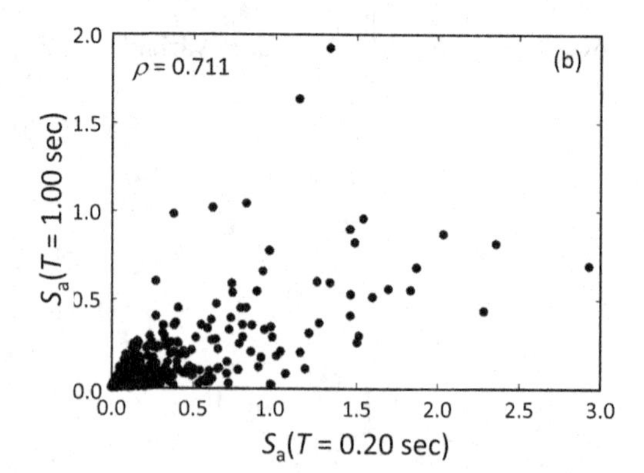

FIGURE D.6 Scatter plots of spectral accelerations at different periods showing different levels of correlation: (a) $T=0.25$ sec vs. $T=0.20$ sec, which are highly correlated ($\rho=0.97$) and (b) $T=1.00$ sec vs. $T=0.20$ sec, which are more weakly correlated ($\rho=0.71$). All data from NGA database for sites with $V_{S30}>500$ m/s.

D.6.5 Confidence Intervals

The moments of a set of data can be computed for very small samples of data from a large population – only a single data point is required, for example, to compute a mean. Of course, one would logically be concerned about how well that estimated sample mean would correspond to the population mean.

Consider a set of random samples from some large population, X, that is normally distributed with mean, μ_x, and standard deviation, σ_x. The sample values, $x_1, x_2, \ldots, x_n$ can be thought of as single realizations from a set of independent, identically distributed random variables, $X_1, X_2, \ldots, X_n$. Therefore, the *sample mean*

$$\bar{X} = \frac{1}{n}\sum_{i=1}^{n} x_i \tag{D.37}$$

is itself a random variable with an expected value

$$\mu_{\bar{X}} = E\left(\frac{1}{n}\sum_{i=1}^{n} x_i\right) = \frac{1}{n}(n\mu_x) = \mu_x \tag{D.38}$$

This indicates that the expected value of the sample mean is equal to the population mean, so $\bar{X}$ is an unbiased estimator of μ_x. The uncertainty in $\bar{X}$ may be of considerable interest; the variance of $\bar{X}$ is

$$Var(\bar{X}) = Var\left(\frac{1}{n}\sum_{i=1}^{n} x_i\right) = \frac{1}{n^2}Var\left(n\sum_{i=1}^{n} x_i\right) = \frac{1}{n^2}\left(n\sigma_x^2\right) = \frac{\sigma_x^2}{n} \tag{D.39}$$

Thus $\bar{X}$ is normally distributed with mean, μ, and standard deviation, $\sigma_x / \sqrt{n}$. This result clearly indicates that the estimate of the population mean improves with increasing sample size, n.

In most situations, the population variance, σ_x^2, is not known and must be estimated from the sample variance

$$s_x^2 = \frac{1}{n-1}\sum_{i=1}^{n}(x_i - \bar{X})^2 \tag{D.40}$$

When this is the case, the random variable, $\left(\bar{X} - \mu_x\right)/s_x\sqrt{n}$ will not be normally distributed, particularly if n is small. Instead, that variable will have the Student's t-distribution with $n-1$ degrees of freedom, which can be written as

$$f_T(t) = \frac{\Gamma\left(\dfrac{f+1}{2}\right)}{\sqrt{f\pi}\,\Gamma(f/2)}\left(1 + \frac{t^2}{f}\right)^{-(f+1)/2} \tag{D.41}$$

where f is the number of degrees of freedom and $\Gamma(\cdot)$ is the gamma function (values of the Student's t-distribution are generally looked up in published tables). The Student's t-distribution approaches the standard normal distribution as $f \to \infty$ and the two are very nearly the same for $f > 50$. For smaller numbers of degrees of freedom, the Student's t-distribution is bell-shaped but somewhat flatter/broader than the standard normal distribution. Then, the probability that the mean falls within certain bounds can be obtained from

$$P\left[t_{\alpha/2,n-1} < T \le t_{1-\alpha/2,n-1}\right] = 1 - \alpha \tag{D.42}$$

where $T = (\bar{X} - \mu_x)\sqrt{n}/s_x$ and $t_{\alpha/2,n-1}$ and $t_{1-\alpha/2,n-1}$ are the lower and upper critical values of the t-distribution with $n-1$ degrees of freedom at probabilities of $\alpha/2$ and $1-\alpha/2$. Alternatively, we can say that the confidence interval for μ_x is

$$\bar{x} \pm t_{1-\alpha/2,n-1}\frac{s_x}{\sqrt{n}} \tag{D.43}$$

Example D.6

A set of 10 SPT measurements were made at the same depth in a particular layer of sand. The sample mean and standard deviation were computed as 15 and 3.5, respectively. Over what interval would one have a 95% confidence of capturing the actual (population) mean?

Solution:

Using a Student's t-distribution table, the lower and upper critical values would be $t_{0.025,9}$=−2.2622 and $t_{0.975,9}$=+0.2622. Using the definition of T, the 95% confidence interval would be

$$\langle\mu\rangle_{0.95} = \left(15 - 2.2622\frac{3.5}{\sqrt{10}}; 15 + 2.2622\frac{3.5}{\sqrt{10}}\right) = (12.5;17.5)$$

If 20 measurements produced the same sample mean and standard deviation, the 95% confidence interval would be

$$\langle\mu\rangle_{0.95} = \left(15 - 2.0930\frac{3.5}{\sqrt{20}}; 15 + 2.0930\frac{3.5}{\sqrt{20}}\right) = (13.4;16.6)$$

The increased number of tests has not changed the estimated mean, but they have narrowed the confidence interval significantly. Note that these confidence intervals correspond to the mean – the individual data can extend well beyond these intervals.

D.7 COMMON PROBABILITY DISTRIBUTIONS

The results of statistical experiments often exhibit the same general type of behavior. As a result, the random variables associated with those experiments can be described by essentially the same PDF. Many PDFs exist, but only a few are required for the geotechnical earthquake engineering analyses described in this book.

D.7.1 Discrete Distributions

Earthquake engineers typically deal with quantities such as force, mass, displacement, and acceleration that are measured on continuous scales. Discrete random variables usually appear in counting processes, where the number of occurrences of some event or condition is noted, or when some continuous region is divided into a finite number of sub-regions.

D.7.1.1 Uniform Distribution

The simplest probability distribution is one in which all possible values of the random variable are equally likely. Such a random variable is described by a *uniform distribution.* The PMF for a discrete random variable, X, with n values uniformly distributed and equally spaced over the interval between two values a and b is

$$P_X(x) = \begin{cases} 0 & \text{for} \quad x < a \\ \dfrac{1}{n} = \dfrac{i}{b-a+1} & \text{for} \quad a \le x \le b \\ 0 & \text{for} \quad x > b \end{cases} \tag{D.44}$$

The first three moments of the PMF are

$$\mu_x = \frac{a+b}{2} \tag{D.45a}$$

$$\sigma_x^2 = \frac{(b-a+1)^2 - 1}{12} \tag{D.45b}$$

$$\theta = 0 \tag{D.45c}$$

D.7.1.2 Poisson Distribution

A Poisson distribution describes the probability of events that follow a Poisson process, i.e., one that yields values of a random variable describing the number of occurrences of a particular event in a specified time interval (or spatial region). Poisson processes have the following properties:

1. The number of occurrences in one time interval is independent of the number that occurs in any other time interval.
2. The probability of occurrence during a very short time interval is proportional to the length of the time interval.
3. The probability of more than one occurrence during a very short time interval is negligible.

If α is the average number of occurrences of the event of interest in a particular time interval, the number of occurrences in that interval will occur with probabilities

$$P[N=n] = \frac{\alpha^n e^{-\alpha}}{n!} \tag{D.46}$$

The first three moments of the PMF are

$$\mu_N = \alpha \tag{D.47a}$$

$$\sigma_N^2 = \alpha \tag{D.47b}$$

$$\theta = 1/\sqrt{\alpha} \tag{D.47c}$$

If the events are characterized as occurring at a rate, λ, such that $\lambda t = \alpha$, then the probability in a time interval, t, can be expressed as

$$P[N=n] = \frac{(\lambda t)^n e^{-\lambda t}}{n!} \tag{D.48}$$

D.7.2 Continuous Distributions

Most of the quantities dealt with by geotechnical earthquake engineers are measured on continuous scales. Many continuous distributions are available, but only a small number are commonly encountered in geotechnical earthquake engineering practice.

D.7.2.1 Uniform Distribution

The continuous uniform distribution is of similar form to its discrete counterpart with a PDF and CDF of the form

$$f_X(x) = \begin{cases} 0 & \text{for} \quad x < a \\ \dfrac{1}{b-a} & \text{for} \quad a \le x \le b \\ 0 & \text{for} \quad x > b \end{cases} \tag{D.49a}$$

$$F_X(x) = \begin{cases} 0 & \text{for} \quad x < a \\ \dfrac{x-a}{b-a} & \text{for} \quad a \le x \le b \\ 0 & \text{for} \quad x > b \end{cases} \tag{D.49b}$$

The first three moments of the uniform distribution are

$$\mu_x = \frac{a+b}{2} \tag{D.50a}$$

$$\sigma_x^2 = \frac{(b-a)^2}{12} \tag{D.50b}$$

$$\theta = 0 \tag{D.50c}$$

D.7.2.2 Normal Distribution

The most commonly used probability distribution in statistics is the *normal distribution* (or *Gaussian distribution*). Its PDF, which plots as the familiar bell-shaped curve of Figure D.5a, describes sets of data produced by a wide variety of physical processes. The normal distribution is completely defined by two parameters: the mean and standard deviation. Mathematically, the PDF of a normally distributed random variable X is given by

$$f_X(x) = \frac{1}{\sqrt{2\pi}\sigma_x} \exp\left[-\frac{1}{2}\left(\frac{x-\mu_x}{\sigma_x}\right)^2\right] \quad -\infty \le x \le \infty \tag{D.51}$$

The PDF and CDF for a normal distribution are illustrated in Figure D.5. Examples of normal PDFs for random variables with different means and standard deviations are shown in Figure D.7.

Integration of the PDF of the normal distribution does not produce a simple expression for the CDF, so values of the normal CDF are usually expressed in tabular form. The normal CDF is most

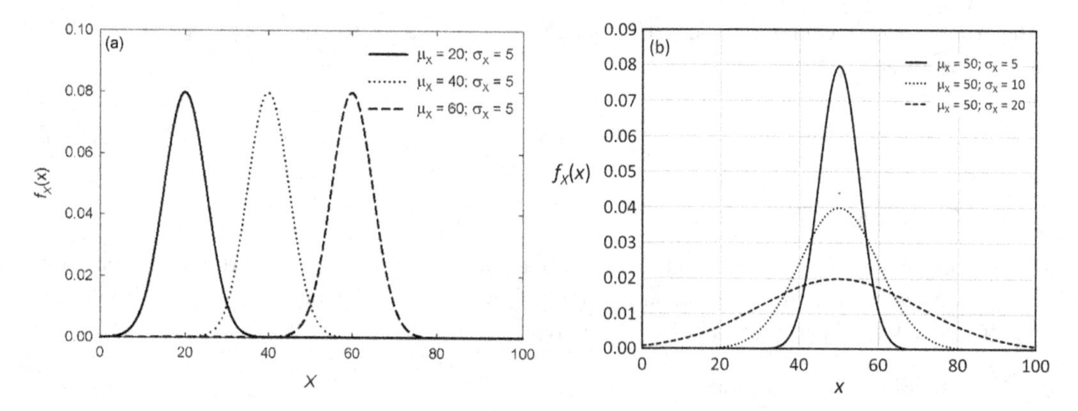

FIGURE D.7 Normal distribution: (a) different means and same standard deviation, (b) same mean and different standard deviations.

efficiently expressed in terms of the *standard normal variable,* which can be computed for any random variable. X, using the transformation

$$Z = \frac{X - \mu_x}{\sigma_x} \tag{D.52}$$

Whenever X has a value, x, the corresponding value of Z is $z=(x-\mu_x)/\sigma_x$. Thus, the mean value of Z is $\mu_z=0$ and the standard deviation is $\sigma_z=1$. Tabulated values of the standard normal CDF are presented in Table D.1.

TABLE D.1
Values of the CDF of the Standard Normal Distribution, $F_Z(z)=1-F_Z(-z)$

z	0.00	0.01	0.02	0.03	0.04	0.05	0.06	0.07	0.08	0.09
−3.4	0.0003	0.0003	0.0003	0.0003	0.0003	0.0003	0.0003	0.0003	0.0003	0.0002
−3.3	0.0005	0.0005	0.0005	0.0004	0.0004	0.0004	0.0004	0.0004	0.0004	0.0003
−3.2	0.0007	0.0007	0.0006	0.0006	0.0006	0.0006	0.0005	0.0005	0.0005	0.0005
−3.1	0.0010	0.0009	0.0009	0.0009	0.0008	0.0008	0.0008	0.0008	0.0007	0.0007
−3.0	0.0013	0.0013	0.0013	0.0012	0.0012	0.0011	0.0011	0.0011	0.0010	0.0010
−2.9	0.0019	0.0018	0.0017	0.0017	0.0016	0.0016	0.0015	0.0015	0.0014	0.0014
−2.8	0.0026	0.0025	0.0024	0.0023	0.0023	0.0022	0.0021	0.0021	0.0020	0.0019
−2.7	0.0035	0.0034	0.0033	0.0032	0.0031	0.0030	0.0029	0.0028	0.0027	0.0026
−2.6	0.0047	0.0045	0.0044	0.0043	0.0041	0.0040	0.0039	0.0038	0.0037	0.0036
−2.5	0.0062	0.0060	0,0059	0.0057	0.0055	0.0054	0.0052	0.0051	0.0049	0.0048
−2.4	0.0082	0.0080	0.0078	0.0075	0.0073	0.0071	0.0069	0.0068	0.0066	0.0064
−2.3	0.0107	0.0104	0.0102	0.0099	0.0096	0.0094	0.0091	0.0089	0.0087	0.0084
−2.2	0.0139	−0.0136	0.0132	0.0129	0.0125	0.0122	0.0119	0.0116	0.0113	0.0110
−2.1	0.0179	0.0174	0.0170	0.0166	0.0162	0.0158	0.0154	0.0150	0.0146	0.0143
−2.0	0.0228	0.0222	0.0217	0.0212	0.0207	0.0202	0.0197	0.0192	0.0188	0.0183
−1.9	0.0287	0.0281	0.0274	0.0268	0.0262	0.0256	0.0250	0.0244	0.0239	0.0233
−1.8	0.0359	0.0352	0.0344	0.0336	0.0329	0.0322	0.0314	00304	0.0301	0.0294
−1.7	0.0446	0.0436	0.0427	0.0418	0.0409	0.0401	0.0392	0.0384	0.0375	0.0367
−1.6	0.0548	0.0537	0.0526	0.0516	0.0505	0.0495	0.0485	0.0475	0.0465	0.0455
−1.5	0.0668	0.0655	0.0643	0.0630	0.0618	0.0606	0.0594	0.0582	0.0571	0.0559
−1.4	0.0808	0.0793	0.0778	0.0764	0.0749	0.0735	0.0722	0.0708	0.0694	0.0681
−1.3	0.0968	0.0951	0.0934	0.0918	0.0901	0.0885	0.0859	0.0853	0.0838	0.0823
−1.2	0.1151	0.1131	0.1112	0.1093	0.1075	0.1056	0.1038	0.1020	0.1003	0.0985
−1.1	0.1357	0.1335	0.1314	0.1292	0.1271	0.1251	0.1230	0.1210	0.1190	0.1170
−1.0	0.1587	0.1562	0.1539	0.1515	0.1492	0.1469	0.1446	0.1423	0.1401	0.1379
−0.9	0.1841	0.1814	0.1788	0.1762	0.1736	0.1711	0.1685	0.1660	0.1635	0.1611
−0.8	0.2119	0.2090	0.2001	0.2033	0.2005	0.1977	0.1949	0.1922	0.1894	0.1867
−0.7	0.2420	0.2389	0.2358	0.2327	0.2296	0.2266	0.2236	0.2206	0.2177	0.2148
−0.6	0.2743	0.2709	0.2676	0.2043	0.2611	02578	0.2546	0.2514	0.2483	0.2451
−0.5	0.3085	0.3050	0.3015	0.2981	0.2946	0.2912	0.2877	0.2843	0.2810	0.2776
−0.4	0.3446	0.3409	0.3372	0.3336	0.3300	0.3264	0.3228	0.3192	0.3156	0.3121
−0.3	0.3821	0.3783	0.3745	0.3707	0.3669	0.3032	0.3594	0.3557	0.3520	0.3483
−0.2	0.4207	0.4168	0.4129	0.4090	0.4052	0.4013	0.3974	0.3936	0.3897	0.3859
−0.1	0.4602	0.4562	0.4522	0.4483	0.4443	0.4404	0.4365	0.4325	0.4286	0.4247
−0.0	0.5000	0.4960	0.4920	0.4880	0.4840	0.4801	0.4761	0.4721	0.4681	0.4641

Table D.1 shows that 15.87% of a normally distributed random variable's values are more than one standard deviation above the mean, and an equal fraction are more than one standard deviation below the mean. This means that 68.26% of the values are within one standard deviation of the mean; similarly, 95.44% of the values would be within two standard deviations of the mean.

Example D.6

Given a normally distributed random variable, X, with $\mu_x=270$ and $\sigma_x=40$, compute the probabilities that (a) $X<300$, $X>350$, and (c) $200<X<240$.

Solution:

The required probabilities can be determined with the aid of Table D.1 after conversion to standard normal variables.

a. For $X=300$,

$$Z = \frac{X-\mu_X}{\sigma_X} = \frac{300-270}{40} = 0.75$$

Then $P[X<300]=P[Z<0.75]=1-F_Z(-0.75)=1-0.2266=0.7734$

b. For $X=350$

$$Z = \frac{X-\mu_X}{\sigma_X} = \frac{350-270}{40} = 2.0$$

Then $P[X>350]=P[Z>2.0]=1-F_Z(2.0)=F_Z(-2.0)=0.0228$

c. For $X=200$

$$Z = \frac{X-\mu_X}{\sigma_X} = \frac{200-270}{40} = -1.75$$

and for $X=240$

$$Z = \frac{X-\mu_X}{\sigma_X} = \frac{240-270}{40} = -0.75$$

Then $P[200<X<240]=P[-1.75<Z<-0.75=F_Z(-0.75)-F_Z(-1.75)=0.2266-0.0401$
$=0.1865$

D.7.2.3 Lognormal Distribution

Some problems, particularly those involving ground motion parameters (Chapter 3), are formulated in terms of the logarithm of a parameter rather than the parameter itself. If X is a random variable, then $Y=\ln X$ is also a random variable. If Y is normally distributed, then X is *lognormally distributed.* In other words, a random variable is lognormally distributed if its logarithm is normally distributed. The PDF of a lognormally distributed random variable X is given by

$$f_X(x) = \frac{1}{x\sqrt{2\pi}\sigma_{\ln x}} \exp\left[-\frac{1}{2}\left(\frac{\ln x-\mu_{\ln x}}{\sigma_{\ln x}}\right)^2\right] \quad 0 \le x \le \infty \tag{D.53a}$$

or, letting $\lambda_X = \mu_{\ln X}$ and $\zeta_x = \sigma_{\ln X}$,

$$f_X(x) = \frac{1}{x\sqrt{2\pi}\zeta_x} \exp\left[-\frac{1}{2}\left(\frac{\ln x - \lambda_X}{\zeta_x}\right)^2\right] \quad 0 \le x \le \infty \tag{D.53b}$$

The shape of the lognormal distribution is shown in Figure D.8. Note that the lognormal distribution assigns zero probability to negative values of the random variable. These characteristics can be very useful for some random variables [the normal distribution, for example, assigns nonzero probabilities for values ranging from $-\infty$ to $+\infty$; when applied to a random variable with a relatively high COV, it can assign some probability that the variable will have a negative value. For a parameter such as undrained shear strength, which may have a COV on the order of 35% (Phoon and Kulhawy, 1999), a normal distribution would imply a 0.2% probability (low, but potentially significant for design purposes where low probabilities of failure are of interest) that the undrained strength is negative, a result that is physically meaningless. It should also be noted that the lognormal distribution is skewed and that the amount of skew increases with increasing $\sigma_{\ln X}$.

Values of the CDF of the lognormal distribution are usually obtained from Table D.1, using the modified transformation

$$Z = \frac{\ln X - \mu_{\ln x}}{\sigma_{\ln x}} = \frac{\ln X - \lambda_X}{\zeta_X} \tag{D.54}$$

It therefore follows that the median and mean values of a lognormal distribution are given by

$$\hat{x} = \exp(\lambda_X) = \exp(\mu_{\ln X}) \tag{D.55a}$$

$$\mu_X = \exp\left(\lambda_X + \frac{1}{2}\zeta_X^2\right) \tag{D.55b}$$

Equation (D.55) indicates that the mean of a lognormally distributed random variable is always greater than the median. The values of X corresponding to common percentiles are given by

$$x_{84} = \exp(\lambda_X + \zeta_X) = \exp(\mu_{\ln X} + \sigma_{\ln X}) \tag{D.56a}$$

$$x_{16} = \exp(\lambda_X - \zeta_X) = \exp(\mu_{\ln X} - \sigma_{\ln X}) \tag{D.56b}$$

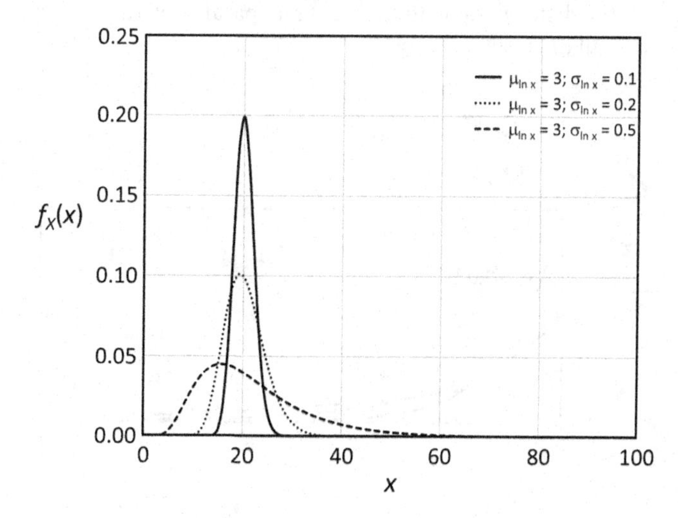

FIGURE D.8 Lognormal distribution. Note that degree of asymmetry increases with increasing $\sigma_{\ln X}$.

A lognormally distributed random variable with $\zeta = \sigma_{\ln X} = \alpha$ would have 68% of its values within a factor of e^{α} of its median value.

Example D.7

A random variable, *X*, is lognormally distributed with $\lambda = 5$ and $\zeta = 1.2$. Compute (a) the probability that $X < 100$, and (b) the value of *X* that has a 10% probability of being exceeded.

Solution:

As in the previous example, (a) for $X = 100$

$$Z = \frac{\ln X - \lambda}{\zeta} = \frac{\ln 100 - 5}{1.2} = -0.33$$

From Table D.1, $P[X < 100] = P[Z < -0.33] = F_Z(-0.33) = 0.3707$. For (b), Table D.1 indicates that the value of *Z* that would have a 10% probability of exceedance is 1.282, i.e., $F_Z(1.282) = 0.90$. Then, rearranging Equation (D.52) yields

$$\ln X = \lambda + Z\zeta = (1.282)(1.2) + 5 = 6.54$$

so

$$X = e^{6.54} = 691$$

D.7.2.4 Exponential Distribution

The exponential distribution is often used to model random variables in which small values occur more frequently than higher values. A common application in earthquake engineering is to model the distribution of earthquake magnitudes. It can also be used to model the time elapsed between events – in particular, it represents the probability distribution of the time between events in a Poisson process. A continuous random variable, *X*, is exponentially distributed with rate parameter, α, if its PDF is given by

$$f_X(x) = \begin{cases} \alpha e^{-\alpha x} & \text{for} \quad x > 0 \\ 0 & \text{for} \quad x \le 0 \end{cases} \tag{D.57}$$

The PDF of an exponential distribution with different rate parameter values is shown in Figure D.9. The CDF of an exponential distribution is given by

$$F_X(x) = \begin{cases} 1 - e^{-\alpha x} & \text{for} \quad x > 0 \\ 0 & \text{for} \quad x \le 0 \end{cases}$$

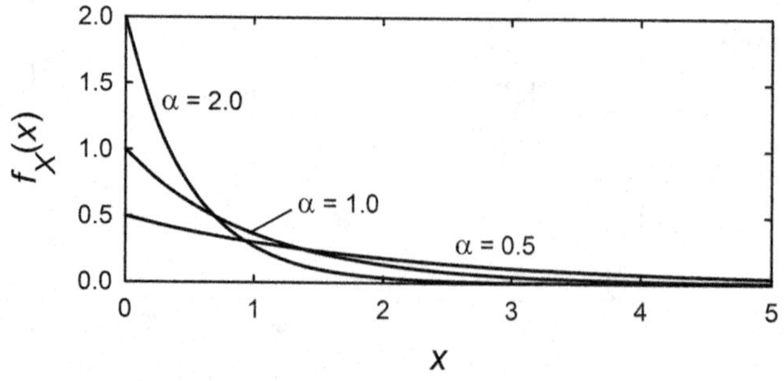

FIGURE D.9 PDFs for exponential distribution with rate parameters of 0.5, 1.0, and 2.0.

and the first three moments are

$$\mu_x = \frac{1}{\alpha} \tag{D.58a}$$

$$\sigma_x^2 = \frac{1}{\alpha^2} \tag{D.58b}$$

$$\theta = 2 \tag{D.58c}$$

D.7.2.5 Joint Normal Distribution

The normal distribution can be extended to multiple random variables. Two random variables, X and Y, that are joint normally distributed have a *joint probability density function*

$$f_{XY}(x,y) = \frac{1}{2\pi\sigma_X\sigma_Y\sqrt{1-\rho_{XY}^2}}\exp\left[-\frac{a}{2\left(1-\rho_{XY}^2\right)}\right] \tag{D.59}$$

where

$$a = \frac{(x-\mu_X)^2}{\sigma_X^2} - \frac{2\rho_{XY}(x-\mu_X)(y-\mu_Y)}{\sigma_X\sigma_Y} + \frac{(y-\mu_Y)^2}{\sigma_Y^2}$$

Note that the shape of the PDF is influenced by the correlation between X and Y. Figure D.10 shows joint normal distributions for two correlation coefficients. The conditional PDF of Y given $X=x$ is given by

$$f_{Y|X}(y \mid x) = \frac{1}{\sqrt{2\pi}\sigma_Y\sqrt{1-\rho_{XY}^2}}\exp\left[-\frac{1}{2}\left(\frac{y-\mu_Y-\rho_{XY}(\sigma_Y/\sigma_X)(x-\mu_X)}{\sigma_Y\sqrt{1-\rho_{XY}^2}}\right)\right] \tag{D.60}$$

which indicates that the conditional PDF is normal with

$$\mu_{Y|X=x} = \mu_Y - \rho_{XY}(\sigma_Y/\sigma_X)(x-\mu_X) \tag{D.61a}$$

and

$$\sigma_{Y|X=x} = \sigma_Y(1-\rho_{XY}^2) \tag{D.61b}$$

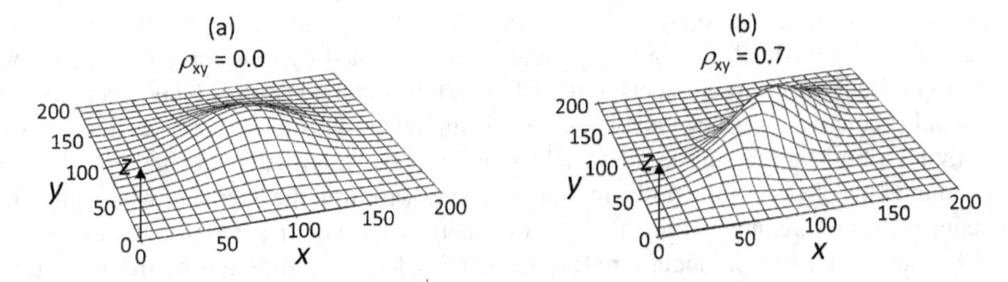

FIGURE D.10 Illustration of joint normal PDFs with (a) $\rho_{XY}=0$ and (b) $\rho_{XY}=0.7$. For both cases, $\mu_X=10$, $\sigma_X=3$, $\mu_Y=15$, $\sigma_Y=4$.

D.8 RANDOMNESS AND UNCERTAINTY

In hazard analysis and performance-based earthquake engineering, it is important to understand the various sources of uncertainty that can contribute to the uncertainty in performance predictions. The terms "randomness" and "uncertainty" are often used in a colloquial manner to cover a range of sources of variability or uncertainty, and it is sometimes necessary to break them down into specific categories.

Randomness, which is frequently described by the term *aleatory variability*, refers to the inherent or intrinsic variability of some quantity or phenomenon; as a result, it cannot be reduced by additional data or through more investigation. Randomness can manifest itself, for example, in the variability of response produced by different ground motions, even when scaled to the same level of intensity. This record-to-record variability, which results from the apparently random, unpredictable nature of earthquakes, is a very significant component of the overall uncertainty in a seismic performance evaluation. Uncertainty due to lack (or ignorance) of data or knowledge concerning a quantity or phenomenon is frequently referred to as *epistemic uncertainty*. Epistemic uncertainty differs from aleatory uncertainty in that it can be reduced by the acquisition of new information, e.g., by additional data, more extensive investigation, or by new research. Aleatory variability and epistemic uncertainty are handled differently in probabilistic performance-based earthquake engineering analyses and are discussed with respect to several topics within this text.

The distinction between aleatory and epistemic uncertainties can be difficult, ambiguous, and confusing. In practice, the distinction can depend on pragmatic as well as theoretical factors. While arguments can be made that all uncertainty is epistemic, practical considerations require that some be treated as aleatory; one could, for example, gain knowledge of the inherent variability of a natural soil deposit by drilling and sampling the entire site with boreholes on a 6-inch spacing – an action so obviously impractical (and destructive) that it illustrates why such variability is treated as aleatory. The assignment of aleatory vs. epistemic can also be situation-dependent. For example, suppose the shear wave velocity profile at a soil site has been measured using multiple defensible methods (e.g., downhole, suspension logging, surface wave inversions). Differences between these measured profiles are epistemic because the knowledge to identify which individual profile, or which combination of profiles, represents the "true" condition is lacking. Suppose one-dimensional ground response analyses are to be performed at this site. Such analyses assume laterally continuous layers with constant properties, which is never the case in natural soils. The inevitable heterogeneities that cause variations in the shear wave velocity profile, which cannot be accurately captured in one-dimensional analyses, are effectively irreducible and random and are generally considered to be aleatory.

The nature of the models used to predict performance will also affect the aleatory-epistemic distinction. All predictive models should be recognized as mathematical idealizations of reality – they are not perfect. Model uncertainty, i.e., errors in model predictions, has two primary components: (1) the effect of missing predictive variables, and (2) the effects of inaccurate model form. Missing variables may be those not recognized as being influential or those that cannot be measured or otherwise characterized. Inaccurate model form may result from practical consideration of computational complexity/effort or lack of understanding of the basic physics of the problem. For example, one-dimensional site response analyses are commonly used in engineering practice even though waves other than the vertically propagating shear waves assumed by those analyses are known to exist at nearly all sites. Both components of model uncertainty can potentially be reduced, by including additional predictive variables and/or the use of improved mathematical expressions, but there will usually be a limit to the number of variables that can be identified and/or measured or to the understanding of the physics of the problem of interest that will limit the degree to which uncertainty can be reduced. Therefore, model uncertainty will generally have both aleatory and epistemic components. The fact that different models are frequently of different forms and use different predictive variables means that they will predict different output values. The variability of mean (or median) predictions from different plausible models, therefore, represents another component of epistemic uncertainty. To properly account for epistemic uncertainty in response, multiple predictive models, where available, should also be used with their results combined using a logic tree where the weights assigned to the branches reflect judgments of the relative merits of the alternative models. It should be recognized, however, that the branches of a logic tree are those thought to be relevant by its developer and they may be incomplete if some unrecognized (hence ignored) but relevant physical mechanism is not included (Stafford, 2015); this form of uncertainty (due to "unknown unknowns") is known as *ontological uncertainty* (e.g., Marzocchi and Jordan, 2014).

Even when only the mean response is being used, however, it is still useful to consider which components of uncertainty can and cannot be reduced and to also consider the costs and benefits of doing so. Increasing uncertainty tends to drive the ground motions, response, damage, and losses for a given return period higher in a performance-based evaluation. The ability to show the benefits of increased investment, for example, in additional subsurface investigation or more sophisticated response modeling, represents a tremendous opportunity for geotechnical earthquake engineering practitioners.

More detailed treatments of randomness and uncertainty in hazard analysis and earthquake engineering can be found in Pate-Cornell (1996), Abrahamson and Bommer (2005), Faber (2005), Der Kiureghian and Ditlevsen (2009), and Stafford (2015), and Baker et al. (2021).

D.9 PROPAGATION OF VARIABILITY/UNCERTAINTY

Engineers are frequently interested in how the uncertainty in inputs affects uncertainty in the corresponding output of some analysis or process. Since the output may be required for design, or as input to further analyses, it is necessary to characterize the uncertainty in the output. In its simplest sense, the problem becomes one of computing the uncertainty in a function of one or more random variables given the uncertainty in the random variables themselves. In the discussion that follows, the propagation of uncertainty will be discussed in the framework of a response model (Figure D.11) into which one or more input variables are applied to produce a response variable. The intent of this section is not to present all methods by which uncertainties can be propagated but rather to provide the reader with some intuitive "feel" for how uncertainties work their way through problems likely to be encountered in earthquake engineering practice.

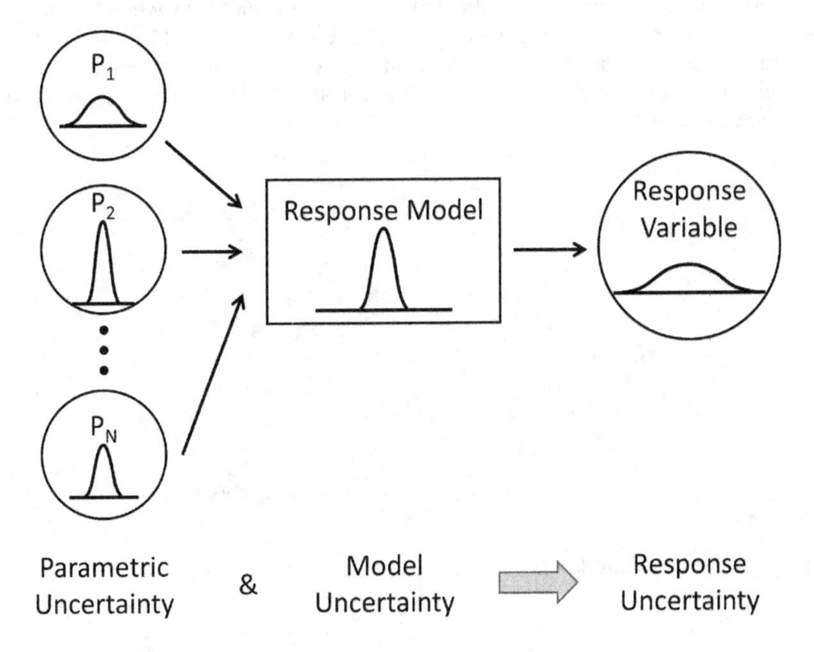

FIGURE D.11 Propagation of uncertainty.

D.9.1 Sensitivity Analysis and Tornado Diagrams

Before undertaking a detailed probabilistic analysis, it can be useful to investigate the sensitivity of the output of interest to the various input parameters used in a predictive model. Sensitivity analyses can be used to determine the relative importance of the inputs and to identify inputs for which errors or changes could significantly affect the model predictions and conclusions drawn from them.

The simplest form of a sensitivity analysis involves permuting each of the inputs individually by some common percentage (e.g., ±10%) of their expected value. The sensitivity is indicated by the range of the computed output values (the "swing") for each permuted input parameter.

A more informative analysis can be performed by tying the degree of the parameter permutation to the uncertainty/variability of the input parameter. In a tornado diagram analysis, the mean value of each input is individually permuted by the same fraction of its standard deviation (e.g., $\pm p \cdot \sigma$ where p would typically be 0.5 or 1.0). In this manner, the swing is influenced by both the sensitivity of the output to each input and the uncertainty in the input. The values of the output are then tabulated for each parameter and arranged graphically (Figure ED.8b) in order of their swing (highest on top) centered on the output value obtained with all inputs set to their mean values. Note that a high sensitivity to a parameter with low uncertainty, or a high uncertainty in a parameter to which the output is not sensitive, will produce a low swing. Input parameters with a high swing, particularly when produced by high uncertainty, may benefit from more detailed exploration or testing in order to reduce that uncertainty.

Example D.8

The ultimate settlement of a slightly overconsolidated soft clay site to be readied for development by placement of 5 ft of fill material is of interest. The site conditions, shown below, indicate the presence of a crust of desiccated clay with thickness, h_1, which is not expected to consolidate noticeably. The clay is underlain by a dense gravel, which will also not consolidate (Figure ED.8a).

A subsurface investigation revealed the properties shown in the table below. Many of the properties are uncertain, either due to scatter in the data from which they were developed, or due to the judgment of the engineers who were involved in the acquisition of the data. Nevertheless, the uncertainty associated with each property is indicated in the table and each is normally distributed. Construct a tornado diagram to show the sensitivity of settlement to variations in the tabulated input variables.

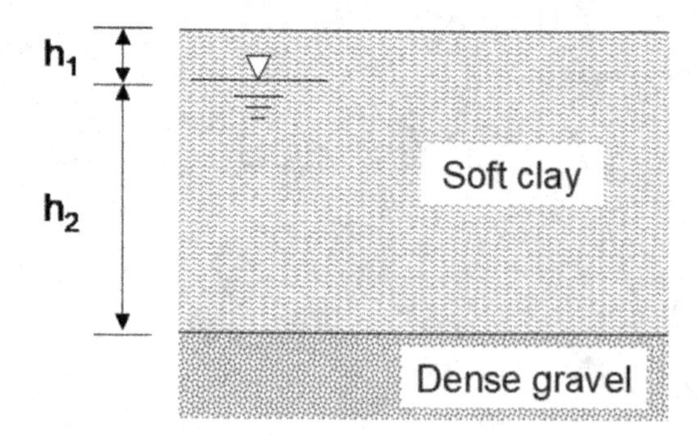

FIGURE ED.8A Site profile considered in Example D.8.

Parameter	Mean value, μ	COV (%)
h_1	3 ft	5
h_2	25 ft	5
C_c	0.75	20
e_0	1.54	7
C_r	0.05	20
$\sigma'_p - \sigma'_{v0}$	200 psf	50
γ_{fill}	130 pcf	7
h_{fill}	5 ft	2

Solution:

The tornado diagram is constructed by first calculating the ultimate settlement using the mean values of all input parameters – in this case, that settlement is 17.37 inches. The higher and lower values of each input parameter are then obtained by adding and subtracting one standard deviation from its mean value. The settlements are then calculated by changing the value of each parameter to its higher and then lower value while holding all other parameters at their mean values, as tabulated below.

Parameter	$\mu - \sigma$	$\mu + \sigma$	$\Delta H\mu_{-\sigma}$	$\Delta H\mu_{+\sigma}$	|Swing|
h_1	2.85	3.15	17.53	17.21	0.32
h_2	23.75	26.25	16.82	18.69	1.87
C_c	0.60	0.90	14.05	20.70	6.65
e_0	1.43	1.65	17.67	17.08	0.59
C_r	0.04	0.06	17.22	17.52	0.30
$\sigma'_p - \sigma'_{v0}$	100	300	22.14	13.23	8.91
γ_{fill}	120.9	139.1	16.00	18.69	2.69
h_{fill}	4.9	5.1	16.98	17.75	0.77

The computed settlements are then shown graphically, from top to bottom, in order of highest to lowest absolute swing as illustrated in Figure ED.8b. For this example, the amount by which the soft clay is overconsolidated is the most influential input parameter followed closely by the compression index, C_c.

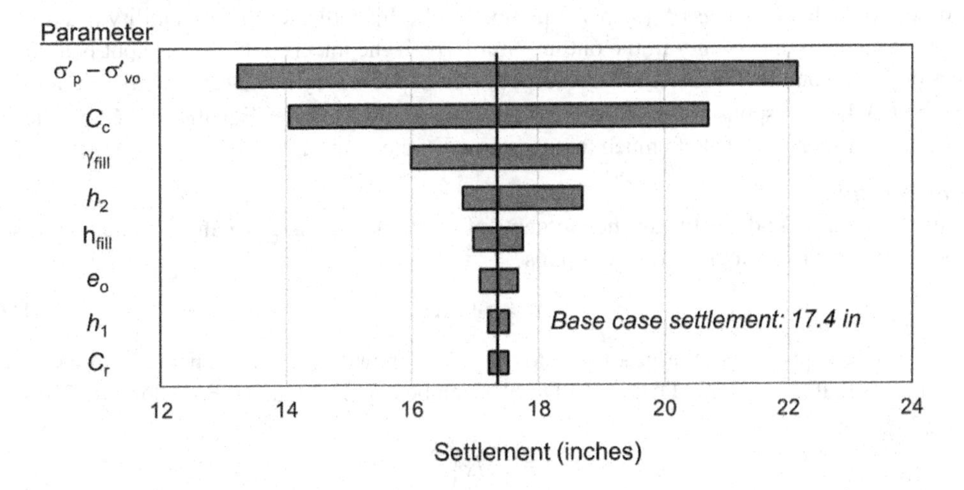

FIGURE ED.8B Tornado diagram indicating sensitivities of settlement to variations of input parameters.

These results indicate that additional consolidation testing to reduce uncertainty in the preconsolidation pressure, σ'_p, and compression index, C_c, may be the most efficient way to reduce uncertainty in the computed settlement.

D.9.2 Functions of Random Variables – Analytical Solutions

When a function of one or more random variables (i.e., a response model) is relatively simple, the uncertainty of the response variable can be expressed in terms of the uncertainties of the input variables in closed form. While nearly all practical problems are too complicated to allow an analytical solution, the analytical solution offers useful insight into general uncertainty propagation behavior.

D.9.2.1 Single Input Variable

In certain cases, a response parameter may be a function of a single input parameter, or a function of multiple parameters of which only one is uncertain. If the uncertain parameter is described by the random variable, X, and the response by the random variable, Y, the response model can be expressed as $Y = g(X)$. The inverse response function (which gives the input parameter that produces a particular level of response) can be expressed as $X = g^{-1}(Y)$. For a monotonically increasing function (i.e., increases in X produce increases in Y for all x), the probability that Y is less than or equal to some value, y, is equal to the probability that X is less than or equal to the corresponding value of x, i.e.,

$$P[Y \le y] = P[X \le g^{-1}(y)] \tag{D.62}$$

The probability that Y is between y and $y + dy$, then is equal to the probability that X is between the corresponding x and $x + dx$, which means that

$$f_Y(y) = f_X(x)\left|\frac{dx}{dy}\right| \tag{D.63}$$

where the absolute value is required to ensure positive values of the PDF when Y is a decreasing function of X. These relationships show that a response model that is simple enough to be analytically inverted and then differentiated can allow exact determination of the entire distribution of the response parameter if the distribution of the input parameter is known. More importantly, it succinctly illustrates an important characteristic of uncertainty propagation: uncertainty in the output depends on both the uncertainty in the input (represented by spread of the $f_X(x)$ term in Equation D.63) *and* the sensitivity of the output to the input (inverse of $|dx/dy|$ term). If the uncertainty in an input variable is small, its probability densities $f_X(x)$ will be large, which will produce large response probability densities ($f_Y(y)$ per Equation D.63); this indicates that variability in x does not contribute much to uncertainty in the output. Similarly, if the uncertainty in the input is high (low values of $f_X(x)$) but the sensitivity of the output to the input is very low (very low $|dy/dx|$ and very high $|dx/dy|$), large response probability densities will again occur per Equation (D.63), indicating that variable does not contribute much to uncertainty in the output.

Linear Function

When the response model is linear, the moments of the response variable can be related to those of the input variable in a simple manner. Suppose

$$Y = a + bX \tag{D.64}$$

where the intercept, a, and the gradient (slope), b are known constants. Then $g^{-1}(y) = (y - a)/b$ which means that $dx/dy = 1/b$. Then, the PDF of Y can be expressed, using Equation (D.63), as

$$f_Y(y) = \left|\frac{1}{b}\right| f_X\left(\frac{y-a}{b}\right) \tag{D.65}$$

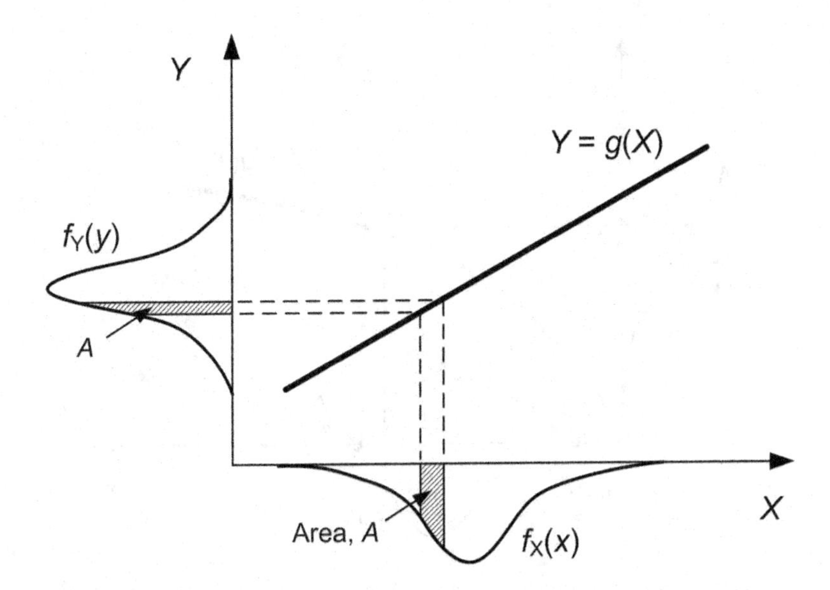

FIGURE D.12 Effect of a linear function on the propagation of uncertainty from X to Y.

It can further be shown for the linear model that

$$\mu_Y = a + b\mu_X \tag{D.66a}$$

$$\sigma_Y^2 = b^2\sigma_X^2 \tag{D.66b}$$

Figure D.12 graphically illustrates the relationship between $f_X(x)$ and $f_Y(y)$ when Y is a linear function of X. The shaded areas represent the probability that X is between the values at the left and right boundaries of the shaded area. That probability is equal to the probability that Y is between the corresponding values on the Y-axis, i.e. $f_X(x)dx = f_Y(y)dy$, which means that $f_Y(y) = f_X(x)dx/dy$. In Figure D.12, the probabilities are equal, so the shaded areas must be equal. Because the slope of the line is relatively flat, $dy < dx$, so the corresponding value of $f_Y(y)$ must be greater than that of $f_X(x)$. So for a linear model, the form of the distribution of Y will be the same as that of X, (e.g., if X is normally distributed, Y will also be normally distributed). The mean of Y is obtained by plugging the mean of X into the model, and the standard deviation of Y is equal to the standard deviation of X multiplied by the slope of the linear model, b. Note that σ_Y depends on σ_X and the sensitivity of Y to X, as captured by the gradient, b.

Nonlinear Function

When the response model is nonlinear, the gradient is not constant and, hence, the form of the Y distribution will be different than that of the X distribution. Figure D.13 illustrates the relationship graphically. Because Y increases monotonically with X, the probability that X is between two particular values of X is equal to the probability that y is between the corresponding values of Y, as in Equation (D.62). Hence, as in the linear case, the shaded areas, A_1, are equal as are the areas, A_2. The variable gradient, however, causes the shape of $f_X(x)$ to differ from that of $f_Y(y)$.

D.9.2.2 Multiple Input Variables

The propagation of uncertainty becomes more complicated when the response variable is a function of more than one random variable. Closed-form analytical solutions can be obtained only for a limited number of special cases.

Consider the general case where $Y = g(X_1, X_2)$ and X_1 and X_2 are jointly distributed with the pdf $f_{X_1,X_2}(x_1, x_2)$. If X_1 and X_2 are continuous, then

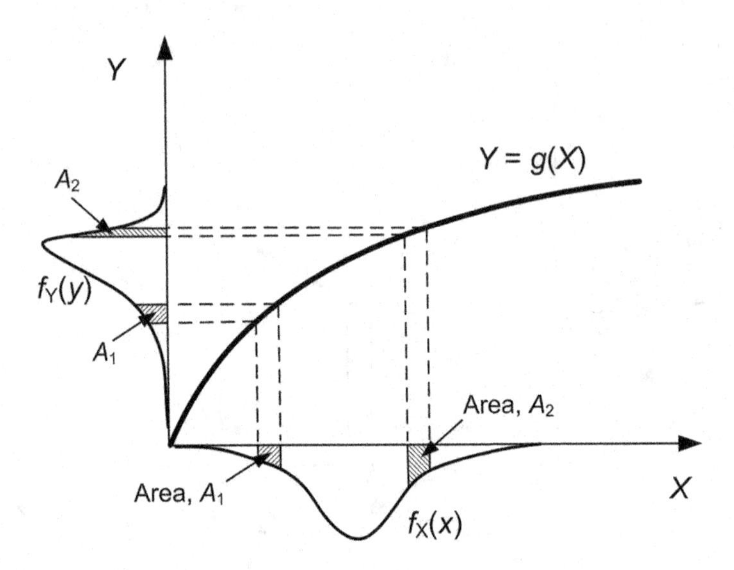

FIGURE D.13 Effect of a nonlinear function on the propagation of uncertainty from X to Y.

$$F_Y(y) = \iint\limits_{g(x_1,x_2)<y} f_{X_1,X_2}(x_1,x_2)\,dx_1\,dx_2 = \int_{-\infty}^{\infty}\int_{-\infty}^{g^{-1}(y,x_2)} f_{X_1,X_2}(x_1,x_2)\,dx_1\,dx_2 \tag{D.67}$$

where $x_1 = g^{-1}(y, x_2)$. Changing the variable of integration from x_1 to y, the CDF of Y can be expressed as

$$F_Y(y) = \int_{-\infty}^{\infty}\int_{-\infty}^{y} f_{X_1,X_2}\left(g^{-1}(y,x_2),x_2\right)\left|\frac{\partial g^{-1}(y,x_2)}{\partial y}\right| dx_2 \tag{D.68}$$

Taking the derivative with respect to y gives the PDF of Y

$$f_Y(y) = \int_{-\infty}^{\infty} f_{X_1,X_2}\left(g^{-1}(y,x_2),y\right)\left|\frac{\partial g^{-1}(y,x_2)}{\partial y}\right| dx_2 \tag{D.69}$$

Thus, the availability of a true closed-form solution requires the ability to analytically integrate the joint pdf in Equation (D.59), which is not possible for all cases. Two cases, however, occur frequently enough to warrant a brief description.

Sums of Random Variables

Many design problems involve response models whose output can be expressed as the sum of a series of random variables. If X_1 and X_2 are statistically independent and $Y = g(X_1, X_2) = X_1 + X_2$ then the joint PDF of X_1 and X_2 is equal to the product of the individual PDFs, so

$$F_Y(y) = \int_{-\infty}^{\infty}\int_{-\infty}^{y} f_{X_1}\left(g^{-1}(y,x_2)\right) f_{X_2}(x_2)\left|\frac{\partial g^{-1}(y,x_2)}{\partial y}\right| dy\,dx_2 \tag{D.70}$$

where $g^{-1}(y, x_2) = x_1 = y - x_2$. The corresponding PDF can then be calculated as

$$f_Y(y) = \int_{-\infty}^{\infty} f_{X_1}\left(g^{-1}(y,x_2)\right) f_{X_2}(x_2)\left|\frac{\partial g^{-1}(y,x_2)}{\partial y}\right| dx_2 \tag{D.71}$$

If X_1 and X_2 are normally distributed with means μ_1 and μ_2, and standard deviations, σ_1 and σ_2, the PDF can be integrated analytically to obtain

$$f_Y(y) = \frac{1}{\sqrt{2\pi}\sqrt{\sigma_{X_1}^2 + \sigma_{X_2}^2}} \exp\left[-\frac{1}{2}\left(\frac{y - (\mu_{X_1} + \mu_{X_2})}{\sqrt{\sigma_{X_1}^2 + \sigma_{X_2}^2}}\right)^2\right] \tag{D.72}$$

This result can be extended to show that the sum (or difference) of a set of n independent, normally distributed random variables is also a normally distributed random variable with mean,

$$\mu_Y = \mu_{X_1} + \mu_{X_2} + \cdots + \mu_{X_n} \tag{D.73a}$$

and standard deviation,

$$\sigma_Y = \sqrt{\sigma_{X_1}^2 + \sigma_{X_2}^2 + \cdots + \sigma_{X_n}^2} \tag{D.73b}$$

This result can be very useful in engineering practice. It means that the mean value of the output is the sum of the mean values of the input variables, and the variance of the output is equal to the sum of the variances of the inputs.

Products of Random Variables

Other response models can take the form of products of random variables, e.g., $Y = X_1 \cdot X_2 \cdot \ldots \cdot X_n$. In such cases, the fact that the logarithms of lognormally distributed random variables are normally distributed allows the results of the previous section to be applied, so the product can be rewritten as $\ln Y = \ln X_1 + \ln X_2 + \cdots + \ln X_n$. This fact indicates that the product (or ratio) of two independent, lognormally distributed random variables will also be lognormally distributed with logarithmic mean

$$\lambda_Y = \mu_{\ln Y} = \lambda_{X_1} + \lambda_{X_2} + \cdots + \lambda_{X_n} \tag{D.74a}$$

and logarithmic standard deviation

$$\zeta_Y = \sigma_{\ln Y} = \sqrt{\zeta_{X_1}^2 + \zeta_{X_2}^2 + \cdots + \zeta_{X_n}^2} \tag{D.74b}$$

D.9.3 Function of Random Variables – Approximate Solutions

Most problems in earthquake engineering are so nonlinear and/or otherwise complicated that closed-form solutions are not practical. In such cases, the propagation of uncertainty is usually handled numerically. A number of techniques, some with a variety of refinements, are available to evaluate the propagation of uncertainty. The following sections briefly introduce a few that require different levels of computational effort and provide different levels of information about the distribution of the response variable.

D.9.3.1 Direct Integration

The most accurate and informative approach would be to numerically integrate the response function over the ranges of all of the input variables. The required summation would be of the form

$$F_Y(y) = \sum_{i=1}^{N_i}\sum_{j=1}^{N_j}\cdots\sum_{n=1}^{N_n} f_{X_1,X_2,\ldots,X_n}(x_{1i}, x_{2j}, \ldots, x_{Nn})\Delta x_1 \Delta x_2 \ldots \Delta x_n \tag{D.75}$$

where N_1, N_2, … , N_n are the numbers of increments (of width Δx_1, Δx_2, … , Δx_n) that the range of each variable is divided into, x_i, x_j, … , x_n are the mid-values of each increment, and the summations are over ranges of variables that produce $Y<y$. For large values of Y, the number of function evaluations would be equal to $N_1 \cdot N_2 \cdot \ldots \cdot N_n$. This procedure would produce the entire distribution of Y with an accuracy that would increase with increasing numbers of input variable increments. With five variables divided into 20 increments each, 3.2 million response function evaluations would be required to obtain the complete distribution of Y. If the response was computed by an empirical model expressed as an algebraic equation requiring 1 ms to compute, direct integration would take about 53 minutes. However, if response was computed by a finite element model requiring 10 minutes for each analysis, about 61 years of computer time would be required. While direct integration can provide accurate and complete propagation of uncertainty, it is often prohibitively time-consuming.

D.9.3.2 First Order, Second Moment Method

Because direct integration is usually impractical for real design problems, a number of approximate procedures have been developed. The approximate procedures all make (or require) assumptions that reduce (often greatly) the number of response model evaluations required to characterize uncertainty in the response variable, but they also provide less complete information about the actual distribution of that variable. Nevertheless, the approximations are often reasonable and the results sufficiently accurate for many purposes.

In some cases, knowledge of the entire probability distribution may not be required. Instead, estimation of the moments (i.e., mean, variance, skewness, etc.) of the response variable distribution may be sufficient. If the moments are known and the form of the distribution (e.g., normal, lognormal, etc.) is known or can be assumed, an estimated response variable distribution can be computed.

The First Order Second Moment (FOSM) reliability method is based on a first-order Taylor series approximation of the response function linearized at the mean values of the random variables. FOSM uses only first and second moment statistics (means and covariances) of the input variables to compute the mean and variance of the response variable. FOSM is popular because it requires only a small number of response function evaluations and it reveals the relative contribution of each input variable to the computed uncertainty in the response variable.

The function $Y=g(\mathbf{X})=g(X_1,X_2,\ldots,X_n)$ can be expanded about the means of the random variables, μ_{X_i}, as a Taylor series

$$y = g(\mu_{X_1},\mu_{X_2},\ldots,\mu_{X_n}) + \frac{1}{1!}\sum_{i=1}^{n}(x_i-\mu_{X_i})\frac{\partial g}{\partial x_i} + \frac{1}{2!}\sum_{i=1}^{n}\sum_{j=1}^{n}(x_i-\mu_{Xi})(x_j-\mu_{Xj})\frac{\partial^2 g}{\partial x_i\,\partial x_j}$$
$$+\frac{1}{3!}\sum_{i=1}^{n}\sum_{j=1}^{n}\sum_{k=1}^{n}(x_i-\mu_{Xi})(x_j-\mu_{Xj})(x_k-\mu_{Xk})\frac{\partial^3 g}{\partial x_i\,\partial x_j\,\partial x_k}+\cdots \quad \text{(D.76)}$$

where all partial derivatives are taken at the mean values of the random variables. In the vicinity of the mean, the $(x_i-\mu_{X_i})$ terms will be small, so squares, cubes and higher powers of $(x_i-\mu_{X_i})$, will be much smaller and can, for many practical purposes, be neglected. Keeping only first-order terms, the truncated series provides the approximation (although it is exact when g is a linear function)

$$y \approx g(\mu_{X_1},\mu_{X_2},\ldots,\mu_{X_n}) + \sum_{i=1}^{n}(x_i-\mu_{X_i})\frac{\partial g}{\partial x_i} \quad \text{(D.77)}$$

The mean and variance of the approximated function can be computed to provide the approximate moments

$$\mu_y \approx g(\mu_{X_1},\mu_{X_2},\ldots,\mu_{X_n}) \quad \text{(D.78)}$$

and

$$\sigma_Y^2 \approx \sum_{i=1}^{n}\sum_{j=1}^{n} Cov(X_i, X_j)\frac{\partial g}{\partial x_i}\frac{\partial g}{\partial x_j} \tag{D.79}$$

Separating the variances (found on the diagonal of the covariance matrix) and using the correlation coefficient (Equation D.34)

$$\begin{aligned}\sigma_Y^2 &\approx \sum \sigma_{X_i}^2\left(\frac{\partial g}{\partial x_i}\right)^2 + \sum_{i=1}^{n}\sum_{j\neq i}^{n} Cov(X_i, X_j)\frac{\partial g}{\partial x_i}\frac{\partial g}{\partial x_j} \\ &= \sum_{i=1}^{n} \sigma_{X_i}^2\left(\frac{\partial g}{\partial x_i}\right)^2 + \sum_{i=1}^{n}\sum_{j\neq i}^{n} \rho_{X_i,X_j}\sigma_{X_i}\sigma_{X_j}\frac{\partial g}{\partial x_i}\frac{\partial g}{\partial x_j}\end{aligned} \tag{D.80}$$

where $\rho_{x_i x_j}$ is the correlation coefficient as described in Section D.6.4. For the case of independent variables, the off-diagonal terms of the covariance matrix are zero, so the expression for the variance simplifies to

$$\sigma_Y^2 \approx \sum_{i=1}^{n} \sigma_{X_i}^2\left(\frac{\partial g}{\partial x_i}\right)^2 \approx \sum_{i=1}^{n} \sigma_{X_i}^2\left(\frac{g(x_i+\Delta x_i)-g(x_i-\Delta x_i)}{2\Delta x_i}\right)^2 \tag{D.81}$$

The second part of Equation (D.81) uses a central difference approximation to the gradient where Δx_i is some small increment of x_i. With this approach, the FOSM method requires $2n+1$ response function evaluations to obtain the first two moments of Y. For the previous case of five input variables, a total of 11 (rather than 3 million for direct integration) function evaluations would be required. Note the similarity of Equation (D.81) to Equation (D.66b). In both cases, the uncertainty in the computed response depends on the uncertainty of the input variable and on the sensitivity of the response to the input variable. In this case, each input variable contributes to σ_y in relative amounts that are easily identified. The FOSM method can be used to identify which variables are more and less important from the standpoint of response model uncertainty.

Note that all partial derivatives are taken at the mean value of each input variable so the problem is "linearized" about the mean, which means that the FOSM approximation is most accurate for the "middle" of the distribution and is less accurate at the tails; the low probabilities of failure used in typical design scenarios, however, mean that the tails of the distributions are the regions of greatest interest. For design purposes, therefore, FOSM techniques have limited utility. Other approximate techniques, such as the first-order reliability method (FORM), which linearizes the problem in the region of interest, can be more useful for design purposes.

D.9.3.3 Monte Carlo Simulation

Another approach to the propagation of uncertainty is to use randomization techniques such as Monte Carlo simulation (MCS). Randomization involves the generation of multiple sets, or "realizations," of all random variables used as inputs to the problem of interest. The input variables are simulated in a series of "realizations" that match desired probability distributions (and, if necessary, desired covariances between correlated input variables). The problem is then solved deterministically for each realization. The resulting response variable values can be collected in a histogram which, as the number of simulations increases, eventually approximates the PDF of the response variable. The distribution of the response variable, therefore, reflects the distributions of the input variables and the sensitivities of the response to each of the input variables. The computed response values can be analyzed statistically, either by computation of their moments or by fitting a distribution to them.

A relatively small number of simulations may be required to obtain a reasonably accurate estimate of the mean or median response, but many more may be required to adequately characterize the tails of the distribution. Recognizing that hundreds, if not thousands, of simulations are typically performed in Monte Carlo analyses and that Student's t-distribution closely approximates the normal distribution for more than about 50 samples, Equation (D.43) can be rearranged and put in the form

$$n^* = \left(\frac{z_{1-\alpha/2} s_x}{(L-U)} \right)^2 \tag{D.82}$$

where z is the standard normal variate and L and U are the lower and upper values of the confidence interval, i.e., $L = \bar{x} - z_{1-\alpha/2} s_x$ and $U = \bar{x} + z_{1-\alpha/2} s_x$. On this basis, the minimum number of simulations required to provide $100(1-\alpha)\%$ confidence that μ_x is between L and U is the next integer greater than n^*.

Simulation of Single Variables

The process of simulating a single random variable is best illustrated graphically. In order to ensure that the simulated values follow the desired distribution, a set of random numbers, u_i, between 0 and 1 are generated. These values are then assigned as CDF values for the random variable being simulated, i.e., $F_X(x_i) = u_i$. Solving for $x_i = F_X^{-1}(u_i)$ produces simulated X values with the desired distribution. Figure D.14 illustrates this process for a generically distributed random variable. The reason why more simulations are required to accurately represent the tails of the distribution are apparent from the figure.

For example, normally distributed random variables can be generated easily in Excel using the statement

```
=NORMINV(RAND(), mean,stdev)
```

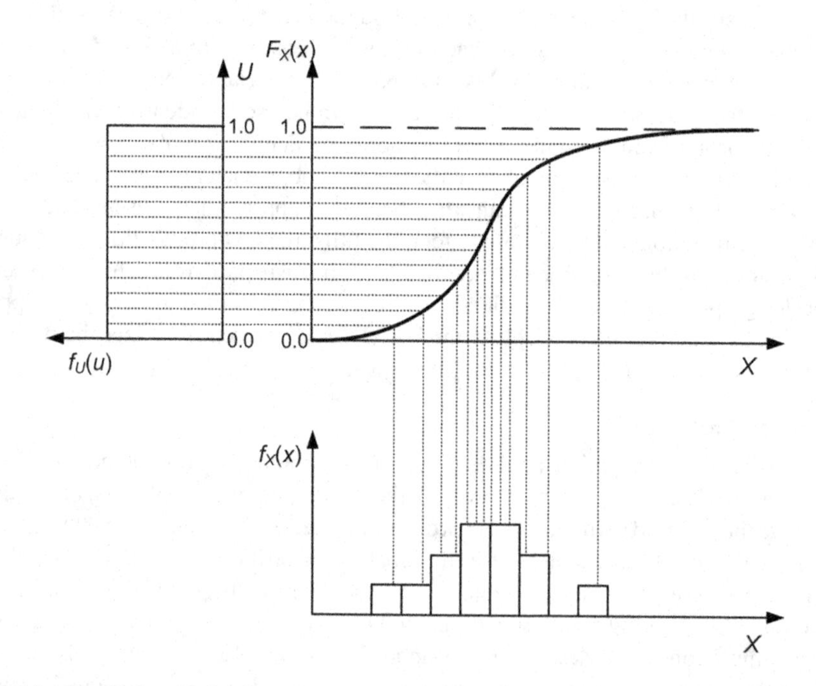

FIGURE D.14 Simulation of distributed random variable values using random numbers, U, and CDF of target distribution. As the number of random numbers increases, histogram of target values approaches PDF of target distribution.

where **mean** is the mean value of the variable and **stdev** is its standard deviation. A lognormally distributed random variable can be generated using

```
=LOGNORM.INV(RAND(),σln x,σln x)
```

where **σln x** and **σln x** are the mean and standard deviation of ln x.

Simulation of Multiple, Independent Random Variables

When the response is affected by more than one random variable, simulation of each random variable is required. If the random variables are independent, the simulation process can be performed individually as described in the preceding section. Before proceeding with the Monte Carlo simulations, checking the random variables for unintended correlation is advisable.

Simulation of Two Correlated Random Variables

It is often necessary to generate randomized pairs of variables with some desired level of correlation. This can be accomplished relatively easily. Assume it is desired to generate multiple realizations of two random variables, X and Y, which have means, μ_x and μ_y, standard deviations, σ_x and σ_y, and a correlation coefficient, ρ_{xy}. N pairs of properly correlated values of X and Y can be obtained from the following steps:

1. Generate N pairs of uncorrelated standard normal ($\mu=0$, $\sigma=1$) random variables, ε_1 and ε_2.
2. Define ε_3 as a linear combination of ε_1 and ε_2 using $\varepsilon_3 = \rho_{xy}\varepsilon_1 + \sqrt{1-\rho_{xy}^2}\,\varepsilon_2$.
3. Then compute

$$X = \mu_x + \sigma_x \varepsilon_1$$

$$Y = \mu_y + \sigma_y \varepsilon_3$$

X and Y will then approach their desired respective means, standard deviations, and correlation coefficient as N becomes large.

Discrete Representation of Continuous Distributions

In some cases, it is computationally convenient to represent continuous probability density functions by "equivalent" discrete PMFs. The equivalence of the two representations is generally evaluated in terms of differences in their statistical moments. Discrete approximations of continuous distributions are most commonly represented by a set of weights assigned to specific values of the variable of interest. Miller and Rice (1983) used Gaussian quadrature to solve for the weights and values of different numbers of discrete points representing common distributions, as shown in Table D.2.

TABLE D.2
Values and Associated Weighting Factors for Discrete Approximation of Continuous Distributions (from Miller and Rice, 1983)

Distribution	Number of Values	Value, x or z	Weight, w_x or w_z
Uniform $\{x\} = 1$ $0 \le x \le 1$	2	0.211325	0.500000
		0.788675	0.500000
	3	0.112702	0.277778
		0.500000	0.444444
		0.887298	0.277778
	4	0.069432	0.173927
		0.330009	0.326073
		0.669991	0.326073
		0.930568	0.173927
Standard Normal $\{z\} = \frac{1}{\sqrt{2\pi}} e^{-z^2/2}$ $-\infty \le z \le \infty$	2	−1.000000	0.500000
		1.000000	0.500000
	3	−1.732051	0.166667
		0.000000	0.666667
		−1.732051	0.166667
	4	−2.334414	0.045876
		−0.741964	0.454124
		0.741964	0.454124
		2.334414	0.045876
Exponential $\{x\} = e^{-x}$ $x \ge 0$	2	0.585768	0.853553
		3.414214	0.146447
	3	0.415775	0.711093
		2.294280	0.278518
		6.289945	0.010389
	4	0.322548	0.603154
		1.745761	0.357419
		4.536620	0.038888
		9.395071	0.000539

REFERENCES

Abrahamson, N.A. and Bommer, J.I. (2005). "Probability and uncertainty in seismic hazard analysis," *Earthquake Spectra*, Vol. 21, No. 2, pp. 603–617.

Ang, A.H.-S. and Tang, W.J. (2007). *Probability Concepts in Engineering: Emphasis on Applications in Civil & Environmental Engineering*, Wiley, New York, 406 pp.

Baecher, G.B. and Christian, J.C. (2003). *Reliability and Statistics in Geotechnical Engineering*, Wiley, New York, 616 pp.

Baker, J.W., Bradley, B.A., and Stafford, P.J. (2021). *Seismic Hazard and Risk Analysis*, Cambridge University Press, Cambridge, UK, 581 pp.

Benjamin, J.R. and Cornell, C.A. (2014). *Probability, Statistics, and Decision for Civil Engineers*, Dover Publications, Minneola, NY, 704 pp.

Der Kiureghian, A. and Ditlevsen, O. (2009). "Aleatory or epistemic? Does it matter?," *Structural Safety*, Vol. 31, pp. 105–112.

Faber, M.H. (2005). "On the treatment of uncertainties and probabilities in engineering decision analysis," *Journal of Offshore Mechanics and Arctic Engineering*, Vol. 127, No. 8, pp. 243–248.

Fenton, G.A. and Griffiths, D.V. (2008). *Risk Assessment in Geotechnical Engineering*, Wiley, New York, 480 pp.

Marzocchi, W. and Jordan, T.H. (2014). "Testing for ontological errors in probabilistic forecasts of natural systems," *Proceedings of the National Academy of Sciences*, Vol. 111, pp. 11973–11978.

Miller, A.C. and Rice, T.R. (1983). "Discrete approximations of probability distributions," *Management Science*, Vol. 29, No. 3, pp. 352–362.
Pate-Cornell, M.E. (1996). "Uncertainties in risk analysis: Six levels of treatment," *Reliability Engineering and System Safety*, Vol. 54, No. 2–3, pp. 95–111.
Phoon, K.K. and Kulhawy, F.H. (1999). "Characterization of geotechnical variability," *Canadian Geotechnical Journal*, Vol. 36, pp. 612–624.
Stafford, P.J. (2015). "Chapter 4: Variability and uncertainty in empirical ground-motion prediction for probabilistic hazard and risk analyses," in A. Ansal, ed., *Perspectives on European Earthquake Engineering and Seismology*, Springer Cham, Heidelberg, pp. 97–128.

Index

Note: **Bold** page numbers refer to tables and *italic* page numbers refer to figures.

Printed in the United States
by Baker & Taylor Publisher Services